DICTIONARY
OF
SCIENTIFIC BIOGRAPHY

PUBLISHED UNDER THE AUSPICES OF
THE AMERICAN COUNCIL OF LEARNED SOCIETIES

The American Council of Learned Societies, organized in 1919 for the purpose of advancing the study of the humanities and of the humanistic aspects of the social sciences, is a nonprofit federation comprising forty-five national scholarly groups. The Council represents the humanities in the United States in the International Union of Academies, provides fellowships and grants-in-aid, supports research-and-planning conferences and symposia, and sponsors special projects and scholarly publications.

MEMBER ORGANIZATIONS

AMERICAN PHILOSOPHICAL SOCIETY, 1743
AMERICAN ACADEMY OF ARTS AND SCIENCES, 1780
AMERICAN ANTIQUARIAN SOCIETY, 1812
AMERICAN ORIENTAL SOCIETY, 1842
AMERICAN NUMISMATIC SOCIETY, 1858
AMERICAN PHILOLOGICAL ASSOCIATION, 1869
ARCHAEOLOGICAL INSTITUTE OF AMERICA, 1879
SOCIETY OF BIBLICAL LITERATURE, 1880
MODERN LANGUAGE ASSOCIATION OF AMERICA, 1883
AMERICAN HISTORICAL ASSOCIATION, 1884
AMERICAN ECONOMIC ASSOCIATION, 1885
AMERICAN FOLKLORE SOCIETY, 1888
AMERICAN DIALECT SOCIETY, 1889
AMERICAN PSYCHOLOGICAL ASSOCIATION, 1892
ASSOCIATION OF AMERICAN LAW SCHOOLS, 1900
AMERICAN PHILOSOPHICAL ASSOCIATION, 1901
AMERICAN ANTHROPOLOGICAL ASSOCIATION, 1902
AMERICAN POLITICAL SCIENCE ASSOCIATION, 1903
BIBLIOGRAPHICAL SOCIETY OF AMERICA, 1904
ASSOCIATION OF AMERICAN GEOGRAPHERS, 1904
HISPANIC SOCIETY OF AMERICA, 1904
AMERICAN SOCIOLOGICAL ASSOCIATION, 1905
AMERICAN SOCIETY OF INTERNATIONAL LAW, 1906
ORGANIZATION OF AMERICAN HISTORIANS, 1907
AMERICAN ACADEMY OF RELIGION, 1909
COLLEGE ART ASSOCIATION OF AMERICA, 1912
HISTORY OF SCIENCE SOCIETY, 1924
LINGUISTIC SOCIETY OF AMERICA, 1924
MEDIAEVAL ACADEMY OF AMERICA, 1925
AMERICAN MUSICOLOGICAL SOCIETY, 1934
SOCIETY OF ARCHITECTURAL HISTORIANS, 1940
ECONOMIC HISTORY ASSOCIATION, 1940
ASSOCIATION FOR ASIAN STUDIES, 1941
AMERICAN SOCIETY FOR AESTHETICS, 1942
AMERICAN ASSOCIATION FOR THE ADVANCEMENT OF SLAVIC STUDIES, 1948
METAPHYSICAL SOCIETY OF AMERICA, 1950
AMERICAN STUDIES ASSOCIATION, 1950
RENAISSANCE SOCIETY OF AMERICA, 1954
SOCIETY FOR ETHNOMUSICOLOGY, 1955
AMERICAN SOCIETY FOR LEGAL HISTORY, 1956
AMERICAN SOCIETY FOR THEATRE RESEARCH, 1956
SOCIETY FOR THE HISTORY OF TECHNOLOGY, 1958
AMERICAN COMPARATIVE LITERATURE ASSOCIATION, 1960
AMERICAN SOCIETY FOR EIGHTEENTH-CENTURY STUDIES, 1969
ASSOCIATION FOR JEWISH STUDIES, 1969

DICTIONARY

OF

SCIENTIFIC BIOGRAPHY

CHARLES COULSTON GILLISPIE

Princeton University

EDITOR IN CHIEF

Volume 3

PIERRE CABANIS—HEINRICH VON DECHEN

CHARLES SCRIBNER'S SONS · NEW YORK

Copyright © 1970, 1971, 1972, 1973, 1974, 1975, 1976, 1978, 1980
American Council of Learned Societies.
First publication in an eight-volume edition 1981.

Library of Congress Cataloging in Publication Data

Main entry under title:

Dictionary of scientific biography.

"Published under the auspices of the American Council
of Learned Societies."
Includes bibliographies and index.
1. Scientists—Biography.   I. Gillispie, Charles
Coulston.   II. American Council of Learned Societies
Devoted to Humanistic Studies.
Q141.D5   1981   509'.2'2 [B]   80-27830
ISBN 0-684-16962-2 (set)

ISBN 0-684-16963-0 Vols. 1 & 2   ISBN 0-684-16967-3 Vols. 9 & 10
ISBN 0-684-16964-9 Vols. 3 & 4   ISBN 0-684-16968-1 Vols. 11 & 12
ISBN 0-684-16965-7 Vols. 5 & 6   ISBN 0-684-16969-X Vols. 13 & 14
ISBN 0-684-16966-5 Vols. 7 & 8   ISBN 0-684-16970-3 Vols. 15 & 16

Published simultaneously in Canada
by Collier Macmillan Canada, Inc.
Copyright under the Berne Convention.

All rights reserved. No part of this book
may be reproduced in any form without the
permission of Charles Scribner's Sons.

5 7 9 11 13 15 17 19   V/C   20 18 16 14 12 10 8 6 4

Printed in the United States of America

Editorial Board

Panel of Consultants

Contributors to Volume 3

The following are the contributors to Volume 3. Each author's name is followed by the institutional affiliation at the time of publication and the names of articles written for this volume. The symbol † indicates that an author is deceased.

HANS AARSLEFF
Princeton University
COMENIUS

GIORGIO ABETTI
Istituto Nazionale di Ottica
G. CALANDRELLI; I. CALANDRELLI

JOHN W. ABRAMS
University of Toronto
W.W. CAMPBELL

H. B. ACTON
University of Edinburgh
CARR

GARLAND E. ALLEN
Washington University, St. Louis
CASTLE; CONKLIN

WILBUR APPLEBAUM
University of Illinois
CRABTREE

A. ALBERT BAKER, JR.
Grand Valley State College
CLAISEN; CURTIUS

J. C. BEAGLEHOLE
Victoria University of Wellington
COOK

JOSEPH BEAUDE
CLERSELIER

SILVIO A. BEDINI
Smithsonian Institution
CAMPANI

JOHN BEER
University of Delaware
CARO

LUIGI BELLONI
University of Milan
CESTONI; COGROSSI

O. THEODOR BENFEY
Earlham College
COUPER

ALEX BERMAN
University of Cincinnati
CADET; CADET DE GASSICOURT; CADET
DE VAUX; CAVENTOU; J.B.A. CHEVALLIER

KURT-R. BIERMANN
German Academy of Sciences
CLAUSEN

ARTHUR BIREMBAUT
CARANGEOT; C. COMBES

JOSEPH L. BLAU
Columbia University
M. R. COHEN

HERMANN BOERNER
University of Giessen
CARATHÉODORY

UNO BOKLUND
*Royal Pharmaceutical Institute,
Stockholm*
CLEVE; CRONSTEDT

BRUNO A. BOLEY
Cornell University
CASTIGLIANO

FRANCK BOURDIER
École Pratique des Hautes Études
CHRISTOL; F. CUVIER; G. CUVIER;
DÉCHELETTE

E. J. BOWEN
University of Oxford
D. L. CHAPMAN

W. H. BROCK
University of Leicester
CROOKES

THEODORE M. BROWN
Princeton University
CHEYNE

K. E. BULLEN
University of Sydney
DAVID; DEBENHAM

VERN L. BULLOUGH
San Fernando Valley State College
CHAULIAC

IVOR BULMER-THOMAS
CONON OF SAMOS

WERNER BURAU
University of Hamburg
CLEBSCH

JOHN G. BURKE
University of California at Los Angeles
CLEAVELAND; CORDIER

JOHN C. BURNHAM
Ohio State University
CALKINS; CARROLL; CHILD

HAROLD L. BURSTYN
Carnegie-Mellon University
CROLL

H. L. L. BUSARD
University of Leiden
CARCAVI; CLAVIUS

JEROME BYLEBYL
University College, London
COLOMBO

RONALD S. CALINGER
University of Chicago
CASTILLON

GEORGES CANGUILHEM
University of Paris
CABANIS

ALBERT V. CAROZZI
University of Illinois
CAYEUX

ETTORE CARRUCCIO
Universities of Bologna and Turin
CASTELNUOVO; CATALDI; CAVALIERI

CARLOS CHAGAS
UNESCO, Brazilian Delegation
CHAGAS

JAMES CHALLEY
Princeton University
S. CARNOT

SEYMOUR L. CHAPIN
California State College at Los Angeles
CAMUS

GEORGES CHAUDRON
*Laboratoire de Recherches
Métallurgiques*
CHARPY; CHEVENARD

RICHARD J. CHORLEY
University of Cambridge
DAUBRÉE

THOMAS H. CLARK
McGill University
J. W. DAWSON

LORD COHEN OF BIRKENHEAD
University of Hull
E. DARWIN

ANDRÉ COPAUX
*École Supérieure de Physique et de
Chimie, Paris*
COPAUX

GEORGE W. CORNER
American Philosophical Society
CARREL

ALBERT B. COSTA
Duquesne University
CHANCEL; CHEVREUL; CIAMICIAN;
A. C. L. CLAUS; L. COHN; COURTOIS;
CRUM

PIERRE COSTABEL
École Pratique des Hautes Études
CHAZY; CORIOLIS; DEBEAUNE

MAURICE P. CROSLAND
University of Leeds
CHAPTAL

STANISLAW CZARNIECKI
Polish Academy of Sciences
CZEKANOWSKI; CZERSKI

KARL H. DANNENFELDT
Arizona State University
CALLINICOS OF HELIOPOLIS

EDWARD E. DAUB
University of Kansas
CLAUSIUS

HERBERT A. DAVIDSON
University of California at Los Angeles
CRESCAS

GAVIN DE BEER
CHARPENTIER; C. R. DARWIN

ALBERT DELAUNAY
Pasteur Institute
CALMETTE; CHAMBERLAND

ALFRED R. DESAUTELS, S.J.
Holy Cross College
CASTEL

D. R. DICKS
University of London
CLEOMEDES

SALLY H. DIEKE
The Johns Hopkins University
CURTISS

J. A. DIEUDONNÉ
University of Nice
CARTAN

JESSIE DOBSON
Hunterian Museum
CRUIKSHANK

CLAUDE E. DOLMAN
University of British Columbia
CREIGHTON

SIGALIA C. DOSTROVSKY
Worcester Polytechnic Institute
CHLADNI

AAGE G. DRACHMANN
CTESIBIUS

STILLMAN DRAKE
University of Toronto
CASTELLI; CESI

JOHN M. DUBBEY
University College, London
COTES

JOHN T. EDSALL
Harvard University
E. J. COHN

OLIN J. EGGEN
Australian National University
CHALLIS

CAROLYN EISELE
Hunter College
COUTURAT

JOSEPH EWAN
Tulane University
A. W. CHAPMAN; CLEMENTS;
DARLINGTON

EDUARD FARBER †
G. CLAUDE

A. S. FEDOROV
Soviet Academy of Sciences
CHERNOV

BERNARD S. FINN
Smithsonian Institution
J. L. CLARK; CUMMING

MENSO FOLKERTS
Technical University of Berlin
CENSORINUS

JAROSLAV FOLTA
Czechoslovak Academy of Sciences
ČECH; CESÁRO

ERIC FORBES
University of Edinburgh
CARRINGTON

GEORGE S. FORBES
Harvard University
COOKE

ATTILIO FRAJESE
University of Rome
CABEO

PIETRO FRANCESCHINI
CASTALDI; CHIARUGI

HANS FREUDENTHAL
University of Utrecht
CAUCHY

GEORGE F. FRICK
University of Delaware
CATESBY; COLLINSON

JOSEPH S. FRUTON
Yale University
CORI

PAUL GANIÈRE
CORVISART

GERALD L. GEISON
University of Minnesota
F. COHN; CUSHNY

WILMA GEORGE
University of Oxford
F. DARWIN

CHARLES COULSTON GILLISPIE
Princeton University
L. CARNOT; CONDILLAC

C. STEWART GILLMOR
Wesleyan University
COULOMB; D'ARCY

OWEN GINGERICH
Harvard University
A. J. CANNON

MARIO GLIOZZI
University of Turin
CARDANO

EDWARD D. GOLDBERG
Scripps Institution of Oceanography
F. W. CLARKE

JERRY B. GOUGH
Washington State University
CHARLES

GILLES GRANGER
University of Aix-Marseilles
CONDORCET; COURNOT

FRANK GREENAWAY
Science Museum, London
H. C. H. CARPENTER

SAMUEL L. GREITZER
Rutgers University
CREMONA

ASHOT T. GRIGORIAN
Soviet Academy of Sciences
CHAPLYGIN; DAVIDOV

JEAN-CLAUDE GUÉDON
York University, Toronto
CORNETTE

FRANCISCO GUERRA
*Wellcome Historical
Medical Library*
CASAL JULIAN; CERVANTES

PIERRE G. HAMAMDJIAN
University of Paris
DALIBARD

OWEN HANNAWAY
The Johns Hopkins University
DAVISON

RICHARD S. HARTENBERG
Northwestern University
CULMANN

THOMAS HAWKINS
Swarthmore College
CALLANDREAU

JOHN L. HEILBRON
University of California at Berkeley
CANTON; CAVALLO

STERLING B. HENDRICKS
*United States Department
of Agriculture*
COTTRELL

JOHN W. HERIVEL
Queen's University, Belfast
CORNU

RICHARD G. HEWLETT
*United States Atomic Energy
Commission*
K. T. COMPTON

JULIAN W. HILL
The Crystal Trust
CAROTHERS

BROOKE HINDLE
New York University
COLDEN

ERICH HINTZSCHE
University of Bern
CALDANI; A. CORTI

HO PENG YOKE
University of Malaya
CHIN CHIU-SHAO; CHU SHIH-CHIEH

JOSEPH E. HOFMANN
University of Tübingen
M. B. CANTOR; CURTZE; CUSA

MICHAEL A. HOSKIN
University of Cambridge
CURTIS; DAWES

PIERRE HUARD
École Pratique des Hautes Études
CRUVEILHIER

WŁODZIMIERZ HUBICKI
Marie Curie-Skłodowska University
CENTNERSZWER

KARL HUFBAUER
University of California at Irvine
CRELL

JEAN ITARD
CHUQUET; CLAIRAUT

PHILLIP S. JONES
University of Michigan
CRAMER

SHELDON JUDSON
Princeton University
DAVIS

ROBERT H. KARGON
The Johns Hopkins University
CHARLETON

GEORGE B. KAUFFMAN
California State College at Fresno
C. E. CLAUS

MORRIS KAUFMAN
*Rubber and Plastics Processing Industry
Training Board*
CHARDONNET

MARTHA B. KENDALL
Indiana University
C. DAWSON

MILTON KERKER
Clarkson College of Technology
CLAPEYRON

JOHN S. KIEFFER
St. John's College, Annapolis
CALLIPPUS

GEORGE KISH
University of Michigan
CORONELLI

DAVID M. KNIGHT
University of Durham
H. DAVY

MANFRED KOCH
Bergbau Bücherei, Essen
CANCRIN; DECHEN

KENKICHIRO KOIZUMI
University of Pennsylvania
DAVISSON

ZDENĚK KOPAL
University of Manchester
G. DARWIN

ELAINE KOPPELMAN
Goucher College
CHASLES

A. N. KOST
University of Moscow
CHICHIBABIN

VLADISLAV KRUTA
Purkinje University, Brno
CZERMAK

FRIDOLF KUDLIEN
University of Kiel
CELSUS

G. D. KUROCHKIN
Soviet Academy of Sciences
CHERNYSHEV

YVES LAISSUS
Museum National d'Histoire Naturelle
CELS; COMMERSON

BENGT-OLOF LANDIN
University of Lund
CLERCK

LAURENS LAUDAN
University of Pittsburgh
COMTE

WILLIAM LE FANU
Royal College of Surgeons of England
CLIFT

HENRY M. LEICESTER
University of the Pacific
W. M. CLARK; CANNIZZARO

JACQUES R. LÉVY
Paris Observatory
F. A. CLAUDE; COSSERAT

G. A. LINDEBOOM
Free University, Amsterdam
CAMPER

STEN H. LINDROTH
University of Uppsala
CELSIUS

ROBERT BRUCE LINDSAY
Brown University
C. G. DARWIN

ESMOND R. LONG
University of Pennsylvania
COUNCILMAN

J. M. LÓPEZ DE AZCONA
*Comisión Nacionale de Geologia,
Madrid*
CORTÉS DE ALBACAR

D. J. LOVELL
Massachusetts College of Optometry
COBLENZ

RUSSELL McCORMMACH
University of Pennsylvania
CAVENDISH

ERIC McDONALD
D'ARCET

ROBERT M. McKEON
Tufts University
COURTIVRON

DUNCAN McKIE
University of Cambridge
E. D. CLARKE

VICTOR A. McKUSICK
The Johns Hopkins University
CHAUVEAU

PATRICK J. McLAUGHLIN
St. Patrick's College, Maynooth
CALLAN

MICHAEL McVAUGH
*University of North Carolina at
Chapel Hill*
CONSTANTINE THE AFRICAN

BRIAN G. MARSDEN
Smithsonian Astrophysical Observatory
CHANDLER; CROMMELIN

KIRTLEY F. MATHER
Harvard University
CHAMBERLIN; DALY

HERBERT MESCHKOWSKI
Free University, Berlin
G. CANTOR

ELLEN J. MOORE
Natural History Museum, San Diego
CONRAD

EDGAR W. MORSE
University of California at Davis
S. H. CHRISTIE

ANN MOZLEY
University of Chicago Press
W. B. CLARKE

GIULIO MURATORI
University of Ferrara
CANANO

JOHN NICHOLAS
University of Pittsburgh
N. R. CAMPBELL

ROBERT LAING NOBLE
University of British Columbia
COLLIP

JOHN D. NORTH
University of Oxford
CAYLEY; T. CHEVALLIER; CHILDREY;
W. H. M. CHRISTIE; CLIFFORD

LUBOŠ NOVÝ
Czechoslovak Academy of Sciences
ČECH; CESÁRO

HERBERT OETTEL
G. CEVA, T. CEVA

ROBERT OLBY
University of Leeds
D. CAMPBELL; CORRENS

CHARLES D. O'MALLEY †
CAIUS

JACQUES PAYEN
École Pratique des Hautes Études
CAGNIARD DE LA TOUR; CAHOURS;
CAILLETET; CLÉMENT

LEONARD M. PAYNE
Royal College of Physicians of England
CROONE

ENRIQUE PÉREZ-ARBELÁEZ
*Jardín Botánico "Jósé Celestino Mutis,"
Bogotá*
CALDAS

VINCENTE R. PILAPIL
California State College at Los Angeles
COMAS SOLÁ

P. E. PILET
University of Lausanne
A. CANDOLLE; A.-P. CANDOLLE;
CHODAT; R. COMBES

DAVID PINGREE
University of Chicago
DÁSABALA

LUCIEN PLANTEFOL
University of Paris
COSTANTIN

LORIS PREMUDA
University of Padua
CASSERI; B. CORTI

HANS PRESCHER
*Staatliche Museum für Mineralogie und
Geologie, Dresden*
COTTA

DEREK J. DE SOLLA PRICE
Yale University
CHAUCER

JAN A. PRINS
Technological University of Delft
CLAY; COSTER

MANUEL PUIGCERVER
University of Barcelona
CABRERA

CARROLL PURSELL
University of California at Santa Barbara
DANFORTH

VARADARAJA V. RAMAN
Rochester Institute of Technology
CATALÁN

P. RAMDOHR
University of Heidelberg
CARNALL; CREDNER

P. M. RATTANSI
University of Cambridge
CUDWORTH

NATHAN REINGOLD
Smithsonian Institution
CATTELL

MARIA LUISA RIGHINI-BONELLI
*Istituto e Museo di Storia della Scienza,
Florence*
DANTI

GLORIA ROBINSON
Yale University, School of Medicine
CHAMBERLAIN; DE BARY

JOEL M. RODNEY
Elmira College
S. CLARKE

FRANCESCO RODOLICO
University of Florence
COCCHI; DAINELLI

COLIN A. RONAN
COMMON

PAUL G. ROOFE
University of Kansas
COGHILL

B. VAN ROOTSELAAR
*State Agricultural University,
Wageningen*
CHWISTEK

EDWARD ROSEN
City College, City University of New York
COMMANDINO; COPERNICUS

MARTIN J. RUDWICK
University of Cambridge
CONYBEARE

DONALD HARRY SADLER
Royal Greenwich Observatory
COMRIE; COWELL

WILLIAM L. SCHAAF
*Brooklyn College, City University of
New York*
DECHALES

ROBERT SCHLAPP
University of Edinburgh
CHRYSTAL

CHARLES B. SCHMITT
University of Leeds
DALÉCHAMPS

RUDOLF SCHMITZ
University of Marburg
E. CORDUS; V. CORDUS

GERALD SCHRÖDER
*Technische Hochschule Carolo
Willhelmina*
CROLLIUS

DOROTHY M. SCHULLIAN
Cornell University Library
COITER; COTUGNO

ERNEST LEONARD SCOTT
T. CLARK; COOPER

JOSEPH FREDERICK SCOTT
*St. Mary's College of Education,
Middlesex*
CRAIG

CHRISTOPH J. SCRIBA
Technical University, Berlin
CRELLE

EDITH SELOW
CASSIRER

CONTRIBUTORS TO VOLUME 3

ROBERT S. SHANKLAND
Case Western Reserve University
A. H. COMPTON

ELIZABETH NOBLE SHOR
COUES; DALL; DAVENPORT; DAY; DEAN

ROBERT SIEGFRIED
University of Wisconsin
J. DAVY

W. A. SMEATON
University College, London
CHARDENON; COLLET-DESCOTILS

CHARLES P. SMYTH
Princeton University
DEBYE

H. A. M. SNELDERS
University of Utrecht
CALLENDAR; CHENEVIX; E. J. COHEN

Y. I. SOLOVIEV
Soviet Academy of Sciences
CHUGAEV

ROBERT SPENCE
University of Kent
COCKROFT

PIERRE SPEZIALI
University of Geneva
CROUSAZ

WILLIAM H. STAHL †
CALCIDIUS

WILLIAM STANTON
University of Pittsburgh
DANA

DIRK J. STRUIK
Massachusetts Institute of Technology
CEULEN; CHRISTOFFEL; CODAZZI;
COOLIDGE; DANDELIN; DARBOUX

CHARLES SÜSSKIND
University of California at Berkeley
DE FOREST

FERENC SZABADVÁRY
Technical University of Budapest
DEBRAY

GIORGIO TABARRONI
University of Bologna
CAPRA

RENÉ TATON
École Pratique des Hautes Études
CASSINIS (I, II, III, IV); CHÂTELET;
CLOUET; DALENCÉ

KENNETH L. TAYLOR
University of Oklahoma
COTTE

ANDRÉE TETRY
École Pratique des Hautes Études
CAULLERY; CHABRY; J.-B. CHARCOT;
J.-M. CHARCOT; CUÉNOT

ARNOLD THACKRAY
University of Pennsylvania
CARLISLE; DALTON; DANIELL; DAUBENY

JEAN THÉODORIDÈS
Centre National de la Recherche Scientifique
DAVAINE

K. BRYN THOMAS
Royal Berkshire Hospital, Reading
W. B. CARPENTER

PHILLIP DRENNON THOMAS
University of Wichita
CASSIODORUS

ELIZABETH H. THOMSON
Yale University, School of Medicine
CUSHING

VICTOR E. THOREN
Indiana University
CASSEGRAIN

HEINZ TOBIEN
University of Mainz
CLOOS

RUTH TODD
Smithsonian Institution
CUSHMAN

GERALD JAMES TOOMER
Brown University
CAMPANUS OF NOVARA

D. N. TRIFONOV
Soviet Academy of Sciences
CHERNYAEV

GERALD R. VAN HECKE
Harvey Mudd College
COLLIE

J. J. VERDONK
CHRISTMANN; DASYPODIUS

JUAN VERNET
University of Barcelona
CARAMUEL; CAVANILLES; AL-DAMĪRĪ

HUBERT BRADFORD VICKERY
Connecticut Agricultural Experimental Station
CHITTENDEN

MAURICE B. VISSCHER
University of Minnesota
CARLSON

WILLIAM A. WALLACE, O.P.
Catholic University of America
CELAYA; CIRUELO; CORONEL

DEBORAH JEAN WARNER
Smithsonian Institution
A. CLARK; COMSTOCK

ADRIENNE R. WEILL
M. CURIE

DEREK T. WHITESIDE
Whipple Science Museum
COLLINS

W. P. D. WIGHTMAN
King's College, Aberdeen
CULLEN

MARY E. WILLIAMS
Skidmore College
COLE

WESLEY C. WILLIAMS
Case Western Reserve University
CHAMBERS

CURTIS WILSON
University of California at San Diego
CYSAT

FRANK H. WINTER
University College, London
CONGREVE

HARRY WOOLF
The Johns Hopkins University
CHAPPE D'AUTEROCHE

JEAN WYART
University of Paris
P. CURIE

HATTEN S. YODER, JR.
Carnegie Institution of Washington
CROSS

A. P. YOUSCHKEVITCH
Soviet Academy of Sciences
CHEBOTARYOV; CHEBYSHEV

DICTIONARY
OF
SCIENTIFIC BIOGRAPHY

DICTIONARY OF
SCIENTIFIC BIOGRAPHY

CABANIS—DECHEN

CABANIS, PIERRE-JEAN-GEORGES (*b*. Cosnac, Corrèze, France, 5 June 1757; *d*. Rueil, near Paris, France, 5 May 1808), *philosophy, medicine, history and sociology of medicine.*

Cabanis's father was a landed proprietor who was interested in agricultural innovations and experiments. He was also a friend of Turgot and it was through the latter that the young Pierre Cabanis was introduced, in 1771, into Parisian society, after studying in the local church-run schools. From 1773 to 1775 Cabanis lived in Poland as secretary to Prince Massalsky, bishop of Vilna. From 1777 to 1783 he studied medicine at Saint-Germain-en-Laye, under the guidance of a noted doctor, Léon Dubreuil. On 22 September 1784 Cabanis became a doctor of medicine at Rheims. From 1785 to 1789 he lived in the immediate neighborhood of Mme. Helvétius at Auteuil and often attended her salon, where he became friendly with Volney and Dominique Garat. It was they who, after the taking of the Bastille by the people of Paris, introduced him to Mirabeau, whose doctor he became.

On the strength of his *Observations sur les hôpitaux* (1790) Cabanis was named a member of the Commission de Réforme des Hopitaux (1791–1793). In 1792 Condorcet moved to Auteuil, and he and Cabanis became very close friends. Cabanis helped Condorcet escape the pursuit of the Convention, although ultimately Condorcet was unsuccessful. (Arrested on 27 March 1794, he poisoned himself two days later.) In 1796 Cabanis married Charlotte de Grouchy, the sister of Mme. Condorcet and Emmanuel de Grouchy.

In 1794 the Convention had organized the Écoles Centrales, created by a decree of 1793, and Cabanis was named professor of hygiene. In 1795 he was elected a member of the Institut de France, in the class of moral sciences. Following the creation of the Écoles de Santé, which replaced the Facultés de Médecine in Paris, Montpellier, and Strasbourg, Cabanis held successively, in Paris, the positions of assistant professor at the École de Perfectionnement, of assistant

to Corvisart in the chair of internal medicine, and of titular professor in the chair of the history of medicine and of legal medicine.

In 1797 Cabanis was elected to the Conseil des Cinq-Cents. He approved of Bonaparte's coup d'etat of 18 Brumaire and was named senator. But his relations with Bonaparte, as first consul and then as emperor of the French, deteriorated as a result of distrust and mutual hostility. Cabanis refrained from attending the sessions of the Senate. On 22 April 1807 Cabanis suffered his first attack of apoplexy. He died on 5 May 1808, at the age of fifty years and eleven months.

Cabanis applied medicine to philosophy and philosophy to medicine from a purely theoretical point of view, even when he acted as a reformer.

As a philosopher, Cabanis sought in medicine an instrument for the analysis of ideas, that is to say, for the reconstruction of their genesis. His fundamental philosophical work, *Rapports du physique et du moral de l'homme,* is presented as "simple physiological researches."[1] It is composed of twelve *Mémoires* (the first six of which were first read in the sessions of the Institute) collected in one volume in 1802. In this work Cabanis sets forth a psychology and an ethical system based on the necessary effects of an animal's organization upon its relationships with its environment. Even the unlimited perfectibility of the human species, which renders it "capable of all things," derives from the fact that "man is undoubtedly the most subject to the influence of exterior causes."[2]

Even more than his friends the Idéologues—Antoine Destutt de Tracy, Joseph Garat, Marie-Joseph Degérando, Pierre Laromiguière—Cabanis deemed as too abstract and limited Condillac's method of analysis, which regarded all psychic functions as transformations of sensations.[3] Sensation, he contended, cannot be studied in isolation from organic needs and from sensibility (in the physiological sense of the term) in its relations to motor irritability.

As a physician, Cabanis considered, in the seventh memoir of the *Rapports,* the influence of illnesses on the formation of ideas and values. The text is a summary of his physiological and medical conceptions. It is without originality, especially in regard to the theory of fevers. Still, it helps us to understand the importance he attributed, from a moral and social point of view, to perfecting the art of medicine, "the basis of all the moral sciences." Borrowing the word from the German philosophers, Cabanis termed the science of man *anthropologie,* the methodical joining of the physical history and the moral history of man.[4]

In the epoch of the "Lumières," all philosophy in France merged with politics. Cabanis's medical philosophy was no exception. In seeking the most rational means of making men more reasonable by improvement of public health, Cabanis simultaneously sought to render physicians more knowledgeable and more effective by the reform of medical instruction. The reorganization of the hospitals seemed to meet this twofold requirement. This explains Cabanis's interest in the question, which concerns both public health and medical pedagogy.

In 1790 Cabanis published his *Observations sur les hôpitaux,* in which he advocated the establishment of small hospital units outside the large cities because, according to him, large hospitals preclude individual care, are conducive to the spread of contagious diseases, and ultimately make impossible "the fulfillment of the purpose for which they were founded." From this time on, Cabanis desired that there be annexed to the hospitals practical medical schools, modeled on the teaching clinic founded by Gerard Van Swieten in Vienna, where the lessons were given in the hospital and "it is the different illnesses that serve as the textbook."

Cabanis's *Du degré de certitude de la médecine* (1798) contains a defense of medical empiricism enriched by a history of medical practice through the centuries; in this account the Hippocratic concept of nature is once more paramount, and the clearest conclusion consists in the rejection of theoretical systems. This rejection is based on the philosophical conviction that the human mind, incapable of discovering causes, should content itself with organizing, without preconceived ideas, relations of facts.

Despite its title, *Coup d'oeil sur les révolutions et sur la réforme de la médecine* (written in 1795, published in 1804), the principal work of Cabanis the physician, remains a purely speculative treatise. A history of medicine, retraced by Cabanis in the beginning of the book (ch. 2), allows him to affirm that the succession of nosological systems and the erroneous application of other sciences (physics, chemistry, mathematics) to medicine have discredited the art of healing. The revolutions in medicine, he says, have been only revolutions in ideas and have done nothing but engender the skepticism of the public and the arrogance of the charlatans. An effective reform is now indispensable. The new medical doctrine will be constituted by the relations of order and of logical sequence established between methodically gathered tables of observations and experiments. These relations will be extracted by philosophical analysis, combining the two procedures of decomposition and recomposition. Likewise, medical instruction ought to be given according to the method of analysis. Cabanis states that he attaches the greatest importance to making "complete collections of observations on all the human infirmities,"[5] and to their comparison, in the clinical schools attached to the hospitals.

Although Cabanis perceived, after or along with a great many others—Vicq d'Azyr and Jacques Tenon, for example—that the hospital was the place where the reform of medicine must occur, he did not understand that this reform was not only one of observation. In making possible the permanent consultation of numerous cases of illnesses identified by the cross-checking of clinical examination and anatomico-pathological autopsies, the hospital dethroned the centuries-old practice of observing individual sick people.

Although he was a friend of Condorcet, Cabanis did not understand the meaning and the interest of the latter's researches in the application of the mathematics of probability to the analysis of social facts. The statistical method applied, over a great number of cases, to the relationship of symptom and lesion, or even to the effects of a certain treatment, would show itself, in the near future, to be more pertinent and more effective as an instrument of analysis than the genetic analysis of ideas inherited from Condillac. At the very moment when the France of the Revolution, of the Consulate, and of the Empire, under the guidance of Pinel and Corvisart, was successfully experimenting with new practices in its hospitals, Cabanis, appearing to be a reformer of public health and of medical pedagogy, remained a theoretician of a barely reformed classical medicine.

NOTES

1. Preface to the 1st ed. (1802).
2. Memoir 8, § 3.
3. Memoir 10, sec. II, § 11.
4. *Coup d'oeil sur les révolutions et sur la réforme de la médecine,* ch. 1, § 2.
5. *Ibid.,* ch. 4, § 4.

BIBLIOGRAPHY

I. ORIGINAL WORKS. Cabanis's writings include *Observations sur les hôpitaux* (Paris, 1790); *Journal de la maladie et de la mort d'H.-G.-V. Riquetti Mirabeau* (Paris, 1791); *Du degré de certitude de la médecine* (Paris, an VI [1798]); *Rapports du physique et du moral de l'homme*, 2 vols. (Paris, an X [1802]; 2nd ed., an XIII [1805]); *Coup d'oeil sur les révolutions et sur la réforme de la médecine* (Paris, an XII [1804]); and *Lettre (posthume et inédite) de Cabanis à Mr. F. sur les causes premières* (Paris, 1824).

These works have been collected, along with many articles, discourses, reports, *éloges,* and notices, in *Oeuvres complètes de Cabanis,* François Thurot, ed., 5 vols. (Paris, 1823–1825); and *Oeuvres philosophiques de Cabanis,* C. Lehec and J. Cazeneuve, eds., 2 vols. (Paris, 1956), which is Corpus Général des Philosophes Français, XLIV, 1.

II. SECONDARY LITERATURE. Cabanis or his work is discussed in E. H. Ackerknecht, *Medicine at the Paris Hospital 1794–1848* (Baltimore, 1967); F. Colonna d'Istria, "Cabanis et les origines de la vie psychologique," in *Revue de métaphysique et de morale* (1911), 177 ff.; "Les formes de la vie psychologique et leurs conditions organiques selon Cabanis," *ibid.* (1912), 25 ff.; "L'influence du moral sur le physique d'après Cabanis et Maine de Biran," *ibid.* (1913), 451 ff.; "La logique de la médecine d'après Cabanis," *ibid.* (1917), 59 ff.; J. M. Guardia, *Histoire de la médecine d'Hippocrate à Broussais et à ses successeurs* (Paris, 1884), pp. 218–227, 442–453; A. Guillois, *Le salon de Mme. Helvétius. Cabanis et les Idéologues* (Paris, 1894); Pierre-Louis Ginguené, "Cabanis," in Michaud, ed., *Biographie universelle,* VI (Paris, 1812) 426–433; P. Janet, "Schopenhauer et la physiologie française, Cabanis et Bichat," in *Revue des deux-mondes* (1 May 1880), 35 ff.; M. Laignel-Lavastine, "La médecine française sous la Révolution," in *Progrès medical* (1935), no. 3, 115 ff.; and C. Lehec and J. Cazeneuve, introduction to their ed. of *Oeuvres philosophiques de Cabanis,* vol. I (Paris, 1956).

GEORGES CANGUILHEM

CABEO, NICCOLO (*b.* Ferrara, Italy, 26 February 1586; *d.* Genoa, Italy, 30 June 1650), *meteorology, magnetism, mathematics.*

A Jesuit, Cabeo taught moral theology and mathematics in Parma, then was a preacher in various Italian cities until he settled in Genoa, where he taught mathematics. He published two major works, *Philosophia magnetica* and *In quatuor libros meteorologicorum Aristotelis commentaria.*

Cabeo is remembered mainly because in Genoa he became acquainted with Giovanni Battista Baliani, who at the fortress of Savona had experimented with falling weights, which, although of different heaviness, took almost the same length of time to reach the ground. Cabeo interpreted these experiments perhaps too broadly and was therefore the indirect cause of

other experiments conducted by Vincenzo Renieri, who refers to them in a letter of 13 March 1641 to Galileo. These experiments, however, showed considerable differences in time of descent because of air resistance. Renieri wrote that he had undertaken them "because a certain Jesuit writes that [two different weights] fall in the same length of time." Galileo wrote in the *Discorsi e dimostrazioni matematiche intorno a due nuove scienze* (*Edizione Nazionale delle Opere,* VIII, 128) that to conduct such experiments "involves some difficulties" and referred to descent along an inclined plane and to the oscillations of a pendulum. Thus Vincenzo Viviani's account of the results of Galileo's experiments that involved dropping different weights from the top of the bell tower in Pisa seems to be completely unfounded.

BIBLIOGRAPHY

I. ORIGINAL WORKS. Cabeo's books are *Philosophia magnetica in qua magnetis natura penitus explicatur, et omnium quae hoc lapide cernuntur, causae propriae afferuntur . . .* (Ferrara, 1629); and *In quatuor libros meteorologicorum Aristotelis commentaria . . .* (Rome, 1646).

II. SECONDARY LITERATURE. Galileo Galilei, *Discorsi e dimostrazioni matematiche intorno a due nuove scienze,* Adriano Carugo and Ludovico Geymonat, eds. (Turin, 1958), p. 689, note; and see Carlos Sommervogel, *Bibliothèque de la Compagnie de Jésus,* II, pt. 1 (Brussels–Paris, 1891).

ATTILIO FRAJESE

CABRERA, BLAS (*b.* Lanzarote, Canary Islands, Spain, 20 May 1878; *d.* Mexico City, Mexico, 1 August 1945), *physics.*

Cabrera obtained his *licenciatura* in sciences in 1898 and his Ph.D. with honors in 1901 at the University of Madrid. He became an assistant professor there and in 1905 was appointed professor of electricity and magnetism.

Early in his career Cabrera published several papers on the properties of electrolytes. He also began working on the magnetic properties of matter, which later became his major interest. As a result of his papers, in 1910 Cabrera was appointed director of the newly established Spanish Physical Research Institute. From 1910 to 1912 he worked with Pierre Weiss, then at the Polytechnical School of Zurich. This period was crucial in Cabrera's scientific career: most of his research thereafter dealt with the study of weakly magnetic substances.

From 1912 on, he carried out intensive research in magnetochemistry. He tried to provide an experimental check on Weiss's magneton theory, which he mentions in almost every paper. A result of this work was a Rockefeller Foundation grant to build a new National Institute of Physics and Chemistry. Appointed its director in 1932, Cabrera encouraged teamwork between physicists and chemists, and work in magnetochemistry proceeded vigorously.

Cabrera published over 110 papers between 1912 and 1934. He contributed to establishment of Hund and Van Vleck's molecular field, established the variation of the atomic magnetic moment versus atomic number, modified the Curie-Weiss law for the rare earths, derived an equation for the atomic magnetic moment including the temperature effect and improved many experimental devices he worked with. Some of his magnetic susceptibility measurements have not been improved upon.

In 1936 Cabrera, with Cotton, started a laboratory for magnetic research in Paris. From 1941 to 1945 he was a professor at the University of Mexico. At his death, he had published over 150 papers.

BIBLIOGRAPHY

I. ORIGINAL WORKS. Cabrera's most important papers include "Sobre la relación que liga la susceptibilidad con la permeabilidad magnética," in *Anales de la Sociedad española de física y química,* **3** (1905), 34–35; "Sobre la relación del magnetismo permanente con la temperatura," *ibid.,* **5** (1907), 152–168, 214–222; "La teoría de los magnetones y la magnetoquímica de los compuestos férricos," *ibid.,* **9** (1911), 316–344, 394–430, written with E. Moles; "Magnetoquímica de los compuestos de hierro," *ibid.,* **11** (1913), 398–419, written with E. Moles; "Instalación para la medida de la susceptibilidad de los cuerpos fuertemente paramagnéticos," *ibid.,* **12** (1914), 512–525; "La magnetoquímica de las sales de cobre y la teoría del magnetón," *ibid.,* 373, written with E. Moles; "La magnetoquímica de los compuestos de níquel y la teoría del magnetón," *ibid.,* 131–142, written with E. Moles and J. Guzman; "La magnetoquímica de las sales de cobalto y la teoría de los magnetones," *ibid.,* **14** (1916), 357–373, written with E. Jimeno and M. Marquina; "Magnetoquímica de los compuestos del cromo," *ibid.,* **15** (1917), 199–209, written with M. Marquina; "Magnétochimie," in *Journal de chimie physique,* **16** (1918), 442–460; "El paramagnetismo de las sales sólidas y la teoría del magnetón," in *Anales de la Sociedad española de física y química,* **16** (1918), 436–449; "La magnétochimie des sels de manganèse," in *Journal de chimie physique,* **16** (1918), 11–20, written with E. Moles and M. Marquina; "La magnetoquímica de las sales cromosas y oxicrómicas," in *Anales de la Sociedad española de física y química,* **17** (1919), 149–167, written with S. Piña; "Variación de la constante magnética del catión oxicrómico por la acción del ácido sulfúrico," *ibid.,* **20** (1922), 175–181, written with S. Piña; "Los magnetones de Weiss y de Bohr y la constitución del átomo," *ibid.,* **21** (1923), 505–526; "La constante Δ de la ley de Curie modificada," *ibid.,* **22** (1924), 463–475; "Les terres rares et la question du magnéton," in *Comptes rendus de l'Académie des sciences,* **180** (1925), 668–680; "Variación del paramagnetismo con la temperatura," in *Anales de la Sociedad española de física y química,* **24** (1926), 297–317, written with J. Palacios; "Sur le paramagnétisme des familles du palladium et du platine," in *Comptes rendus de l'Académie des sciences,* **185** (1927), 414, written with A. Duperier; "Sobre la teoría general de la propiedades magnéticas de la materia," in *Anales de la Sociedad española de física y química,* **26** (1928), 50–71; "Valor del magnetón de Weiss deducido de los cuerpos paramagnéticos," *ibid.,* **28** (1930), 431–447; "Diamagnétisme et température," in *Comptes rendus de l'Académie des sciences,* **197** (1933), 379, written with H. Fahlenbrach; "Diamagnetismus von Wasser bei verschiedenen Temperaturen. I," in *Zeitschrift für Physik,* **82** (1933), 759, written with H. Fahlenbrach; "Über den Diamagnetismus des flüssingen und festen schweren Wassers und seinen Temperaturverlauf," in *Die Naturwissenschaften,* **22** (1934), 417–424, written with H. Fahlenbrach; "Über den Diamagnetismus organischer Verbindungen im Hinblick auf den Einfluss von Temperatur und Konstitution. II," in *Zeitschrift für Physik,* **89** (1934), 682–694, written with H. Fahlenbrach; "Magnetische Untersuchung der gegenseitigen Beeinflussung von Kaliumjodid und Wasser in der Lösung," *ibid.,* 166–175, written with H. Fahlenbrach; "Further Results on the Magnetism of Chlorides of the Palladium and Platinum Triads of Elements," in *Proceedings of the Physical Society* (London), **51** (1939), 845, written with A. Duperier; "Sur le paramagnétisme des terres rares," in *Journal de chimie physique,* **36** (1939), 273; "Diamagnétisme et structure moléculaire," *ibid.,* **38** (1941), 1; and "Les susceptibilités diamagnétiques des alcools butyliques," in *Comptes rendus de l'Académie des sciences,* **213** (1941), 108, written with H. Colson.

II. SECONDARY LITERATURE. See C. E. Hodgman, ed., *Handbook of Physics and Chemistry,* 37th ed. (Cleveland, Ohio, 1955), p. 2392; and A. H. Morrish, *The Physical Principles of Magnetism* (New York, 1965), p. 69.

MANUEL PUIGCERVER

CADET (or **CADET DE GASSICOURT** or **CADET-GASSICOURT), LOUIS-CLAUDE** (*b.* Paris, France, 24 July 1731; *d.* Paris, 17 October 1799), *chemistry.*

Cadet was the son of Claude Cadet, a surgeon at the Hôtel-Dieu in Paris who died in 1745, leaving an impoverished widow and thirteen children who were adopted by friends in various localities. Later in life members of the Cadet family appropriated the names of communities in which they had been raised. Such was the case with Louis-Claude, who had been sent to the village of Gassicourt, near Mantes-la-Jolie. Upon completion of his studies at the Collège des Quatre-Nations, Cadet served an apprenticeship in

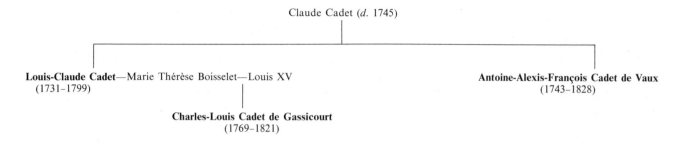

Claude Cadet (*d.* 1745)

Louis-Claude Cadet—Marie Thérèse Boisselet—Louis XV
(1731–1799)

Antoine-Alexis-François Cadet de Vaux
(1743–1828)

Charles-Louis Cadet de Gassicourt
(1769–1821)

The Cadet Family. Names in boldface are discussed in articles.

pharmacy and chemistry at the establishment of Claude Humbert Piarron de Chamousset. He then found employment in the prestigious apothecary shop owned by Claude-Joseph Geoffroy and his son, Claude-François, both members of the Academy of Sciences.

In 1753 Cadet received a six-year appointment as *apothicaire-major* at the Hôtel Royal des Invalides. At the conclusion of his term at the Invalides he purchased an apothecary shop on the rue St. Honoré that achieved an excellent reputation and provided him with a good income. Cadet also served with the military outside of France, in 1761 reorganizing the pharmaceutical services of the French armies stationed in Germany. He collaborated with Berthollet and Lavoisier at the Paris mint and served as royal commissioner at the Sèvres porcelain works. Cadet was regarded by his contemporaries as a chemist of repute, as evidenced by his election to the Royal Academy of Sciences in 1766 as *adjoint chimiste,* with promotions to *associé chimiste* in 1770 and *pensionnaire chimiste* in 1777.

Cadet's earliest research, dating from 1755 to 1757, concerned the analysis of mineral water and was carried on initially with his teacher, Guillaume-François Rouelle. In 1757 he discovered a "fuming liquor" resulting from the distillation of arsenous oxide with potassium acetate. His discovery of this impure cacodyl substance was presented to the Academy of Sciences that same year, reported on favorably by L. C. Bourdelin and Lassone in January 1758, and published in the Academy's *Mémoires de mathématique et de physique* in 1760. Using "Cadet's fuming liquor," Bunsen in 1837 began his important investigation of cacodyl compounds, which led to his isolation and elucidation of the cacodyl radical.

Cadet's attempt in 1759 to show the chemical nature of borax was unsuccessful. He did succeed, however, in developing more efficient methods for producing potassium acetate and ether. In 1774 his claim that mercuric oxide was reduced by heat to mercury was challenged by Baumé but supported by Sage,

Mathurin Brisson, and Lavoisier in a report to the Academy. Particularly noteworthy among Cadet's many investigations made jointly with his fellow academicians were those performed in 1772 to test the effect of heat on diamonds, in which he collaborated with Macquer and Lavoisier.

Influenced by the teachings of the Geoffroys and the Rouelles, Louis-Claude Cadet worked in the mainstream of pharmaceutical, mineralogical, and analytical chemistry long cultivated on the Continent. He was comfortably settled in this tradition and, unlike his colleagues Sage and Baumé, did not attack the New Chemistry, preferring to remain silent.

BIBLIOGRAPHY

I. ORIGINAL WORKS. The bulk of Cadet's work was published alone or in collaboration with others in *Mémoires de mathématique et de physique, présentés à l'Académie royale des sciences par divers sçavans et lus dans ses assemblées; Histoire de l'Académie royale des sciences;* and in Abbé Rozier's *Observations sur la physique, sur l'histoire naturelle et sur les arts.* A detailed bibliography of Cadet's publications, comprising some fifty-two items, is given in Paul Dorveaux, "Apothicaires membres de l'Académie royale des sciences. X. Louis-Claude Cadet, dit Cadet de Gassicourt, alias Cadet-Gassicourt," in *Revue d'histoire de la pharmacie,* **4,** no. 88 (Dec. 1934), 385–397; **5,** no. 89 (Mar. 1935), 1–13; bibliography, pp. 10–13.

II. SECONDARY LITERATURE. In addition to the Dorveaux article cited above, see also Pierre F. G. Boullay, *Notice historique sur la vie et les travaux de L.-Cl. Cadet-Gassicourt* (Paris, 1805); Eusèbe Salverte, *Notice sur la vie et les ouvrages de Louis-Claude Cadet-Gassicourt . . . lue à la rentrée du Lycée républicain le 1er Frimaire an VIII . . .* (Paris, 1799); L. G. Toraude, *Étude scientifique, critique et anecdotique sur les Cadet, 1695–1900* (Paris, n.d.), repr., rev., and enl. from *Bulletin des sciences pharmacologiques,* **6** (1902); Alex Berman, "The Cadet Circle: Representatives of an Era in French Pharmacy," in *Bulletin of the History of Medicine,* **40,** no. 2 (Mar.–Apr. 1966), 101–111; and J. R. Partington, *A History of Chemistry,* III (London–New York, 1962), 96. For Cadet's role in the diamond experiments, see Henry Guerlac, *Lavoisier—The Crucial Year* (Ithaca, N.Y., 1961).

ALEX BERMAN

CADET DE GASSICOURT (or **CADET**), **CHARLES-LOUIS** (*b.* Paris, France, 23 January 1769; *d.* Paris, 21 November 1821), *chemistry, public health.*

The natural son of Louis XV and Marie-Thérèse Boisselet, wife of Louis-Claude Cadet, Charles-Louis received his preliminary education at the Collège Mazarin. As a youth he was introduced to many distinguished scientists who visited the home of the elder Cadet, such as Joseph Lalande, d'Alembert, Condorcet, Fourcroy, Bailly, and Vicq d'Azyr. Although Cadet studied law and subsequently engaged in its practice, his main interests appear to have been literary, political, and scientific. When, in 1787, Fourcroy began his lectures at the Lycée on the rue de Valois, Cadet soon became one of his most ardent students and an enthusiastic partisan of the New Chemistry. Cadet took a prominent part in the insurrection of 13 Vendémiaire *an* IV (5 October 1795) against the Convention and for a time was forced into hiding. Upon the death of the elder Cadet in 1799, Charles-Louis decided to abandon the profession of law for that of pharmacy; and in 1800 he qualified as a pharmacist. Partly because of the distinction of the Cadet name but mainly because of his own exceptional talents, he quickly emerged as one of the most important leaders in pharmaceutical circles and was one of the founders, in 1809, of the influential *Bulletin de pharmacie.* Appointed pharmacist to Napoleon, he was present at the battle of Wagram in 1809 and shortly after was awarded the order of Chevalier of the Empire. In 1821 Cadet was admitted to the Academy of Medicine.

As a scientist Cadet is noteworthy for his part in the diffusion and popularization of the New Chemistry rather than for any specific discovery. His most important work, the four-volume *Dictionnaire de chimie,* published in 1803 and dedicated to Fourcroy, replaced the older chemical dictionary of Macquer. Cadet's *Dictionnaire* clearly elucidated the revolutionary changes that had occurred in chemistry and in chemical nomenclature. Although Cadet was a prolific writer, his scientific publications constituted only a part of his total published work, much of which dealt with literature, politics, pharmacy, and other subjects. Virtually all of his scientific writings, including some fifty articles that revealed him to be an able analytical chemist, were published after 1800.

Restless by temperament and intellectually innovative, Cadet was never at a loss for new ideas. Typi-

cal were his published suggestions for the application of science to national defense, as well as his ideas regarding the creation of an Institut Nomade that would travel around France to promote scientific, technological, and industrial development. None of his suggestions had any noticeable influence. On the other hand, it was at his suggestion that a health council (*conseil de salubrité*) was organized for Paris in 1802 and subsequently dealt with a great number of health problems: disinfection, resuscitation, industrial hygiene, medical statistics, food adulteration, sewage, epidemics, and a host of other health concerns. Cadet was an active member of this council, which became a model for similar health councils established in other French cities.

BIBLIOGRAPHY

I. ORIGINAL WORKS. Cadet's scientific papers are listed in the Royal Society of London, *Catalogue of Scientific Papers (1800–1863),* I (London, 1867), 752–753. In addition to his *Dictionnaire de chimie* (Paris, 1803), several other scientific publications are worth noting. Cadet made a bold but inconclusive attempt to study the interrelations of science in a doctoral dissertation: *De l'étude simultanée des sciences, ou dissertation sur cette proposition: Pour perfectionner une seule des sciences physiques et naturelles, il est nécessaire de connaître la philosophie de toutes les autres* (Paris, 1812). See also his *Sur les moyens de destruction et de résistance que les sciences physiques pourraient offrir dans une guerre nationale . . .* (n.p., n.d.); and *Projet d'un institut nomade* (Paris, 1820).

II. SECONDARY LITERATURE. On Cadet's life and work see J. J. Virey, "Notice sur la vie et les travaux de Charles-Louis Cadet de Gassicourt," in *Journal de pharmacie et des sciences accessoires,* **8** (1822), 1–15; Eusèbe Salverte, *Notice sur la vie et les ouvrages de Charles-Louis Cadet Gassicourt . . .* (Paris, 1822); E. Pariset, "Éloge de C.-L. Cadet de Gassicourt," in *Histoire des membres de l'Académie royale de médecine,* II (Paris, 1850), 130–163; L. G. Toraude, *Étude scientifique, critique et anecdotique sur les Cadet, 1695–1900* (Paris, n.d.), repr., rev., and enl. from *Bulletin des sciences pharmacologiques,* **6** (1902); and Alex Berman, "The Cadet Circle: Representatives of an Era in French Pharmacy," in *Bulletin of the History of Medicine,* **40,** no. 2 (Mar.–Apr. 1966), 101–111.

ALEX BERMAN

CADET DE VAUX (or **CADET-DEVAUX** or **CADET LE JEUNE**), **ANTOINE-ALEXIS-FRANÇOIS** (*b.* Paris, France, 11 January 1743; *d.* Nogent-les-Vierges, France, 29 June 1828), *chemistry, agriculture, nutrition, public health.*

Following the example of his older brother, Louis-Claude Cadet, Cadet de Vaux was first apprenticed to the philanthropist Piarron de Chamousset and in 1759 replaced the elder Cadet as *apothicaire-major* at the Hôtel Royal des Invalides. From 1769 to 1781 he practiced pharmacy on the rue St. Antoine. In 1777 he became one of the cofounders of the first daily newspaper in Paris, *Le journal de Paris,* which threw its support during the Revolution to the Club des Feuillants and its leaders: Barnave, Lafayette, Bailly, André Chénier, and Mirabeau. This resulted in the sacking of the *Journal* offices in 1792 by Jacobin sympathizers.

Many of Cadet de Vaux's activities before the Revolution were concerned with the disinfection of cesspools and wells, the reform of sanitary conditions in prisons, industrial hygiene, and the removal of cemeteries from the center of Paris, particularly the Cimetière des Innocents. In 1787 Cadet de Vaux and his brother Louis-Claude were elected to membership in the American Philosophical Society. Cadet de Vaux's letters to Benjamin Franklin, now in the possession of the American Philosophical Society, reflect not only a friendship with Franklin, who then lived in Passy, but also some of Cadet de Vaux's major interests at that time, such as Franklin's stove, publication of correspondence by Franklin in the *Journal de Paris,* Montgolfier's balloon, Indian corn, experiments on the preservation of wheat and flour, breadmaking, and the École de Boulangerie, which Cadet de Vaux and Parmentier had been instrumental in founding in 1780. Cadet de Vaux was active in local politics for a short time, serving from 1791 to 1792 as president of the department of Seine-et-Oise. His close friendship with Parmentier, who had succeeded him in 1766 at the Invalides, led to a fruitful collaboration over the years. In 1820 he was elected a member of the Academy of Medicine in Paris.

From the beginning of his career Cadet de Vaux had a strong interest in chemistry and science, which he sought to apply to such fields as agriculture, nutrition, and public health. In 1771 and 1772 he taught chemistry at the Royal Veterinary School in Alfort, and in 1770 he produced an annotated French translation of Jacob Reinbold Spielmann's *Institutiones chemiae.* It was also primarily as a chemist that Cadet de Vaux, along with his colleagues Laborie and Parmentier, was invited by the French government to recommend safe methods for cleaning out cesspools, a hazardous occupation frequently resulting in workers' being overcome, sometimes fatally, by noxious gases. Their findings, in which they recommended, among other things, the use of quicklime, a ventilator, and furnaces, were presented by Cadet de Vaux in 1788 before the Royal Academy of Sciences and reported on favorably the same year by Lavoisier,

Milly, and Fougeroux de Bondaroy. In 1783 Cadet de Vaux and his two colleagues gave similar advice in connection with large-scale disinterments in the northern port city of Dunkerque.

Chiefly because of his chemical expertise, Cadet de Vaux was admitted to membership in the Society of Agriculture in 1785; there, in 1789, he and Fourcroy jointly issued an enthusiastic report on Lavoisier's *Traité élémentaire de chimie.* At the École de Boulangerie, where Cadet de Vaux and Parmentier were professors, their lectures dealt with such subjects as the analysis of wheat and flour, methods of preservation, and the technology of baking.

In 1788 Cadet de Vaux purchased an estate in Fraconville, not far from Paris, where he spent most of the remaining forty years of his life. His multifarious projects during these four decades included agriculture (methods for preserving crops, prevention of mole infestation, cultivation of fruit and tobacco, extraction of sugar from sugar beets, and forest conservation) and home economics and nutrition (paints, steam laundries, disinfection of walls, winemaking, potato bread, coffee, gelatin and bouillon, and soup kitchens for the poor).

A product of the Enlightenment, utilitarian in his scientific outlook, Cadet de Vaux numbered among his friends Benjamin Franklin, Condorcet, and La Rochefoucauld-Liancourt, with all of whom he shared many interests.

BIBLIOGRAPHY

I. Original Works. Of the numerous publications by Cadet de Vaux, the following are representative: *Instituts de chymie de M. Jacques-Reinbold Spielmann . . . traduits du latin, sur la 2e édition, par M. Cadet le jeune, . . .* 2 vols. (Paris, 1770); *Observations sur les fosses d'aisance, . . . par MM. Laborie, Cadet le jeune et Parmentier, . . .* (Paris, 1778); *Discours prononcés à l'ouverture de l'École gratuite de boulangerie, le 8 juin 1780, par MM. Parmentier et Cadet de Vaux . . .* (Paris, 1780); *Mémoire historique et physique sur le cimetière des Innocents . . . lu à l'Académie royale des sciences en 1781 . . .* (n.p., n.d.); "Rapport de MM. Laborie, Parmentier et Cadet de Vaux, relatif à l'exhumation des cadavres d'une partie de l'Église paroissiale de Saint Éloy de Dunkerque," in *Recueil des pièces concernant les exhumations faites dans l'enceinte de l'Église de Saint Éloy de la ville de Dunkerque* (Paris, 1783); *Recueil de rapports, de mémoires et d'expériences sur les soupes économiques et les fourneaux à la Rumford . . . par les citoyens Cadet-Devaux, Decandolle, Delessert, Money et Parmentier* (Paris, 1801); *Instruction populaire sur le blanchissage domestique à la vapeur . . .* (Paris, 1805); *Dissertation sur le café . . .* (Paris, 1806); *Traité de la culture du tabac et de la préparation de sa feuille . . .* (Paris, 1810); *Aperçu économique et chimique sur l'extraction des betteraves* . . . (Paris, 1811); *Moyens de prévenir le retour des disettes* . . . (Paris, 1812); *Des bases alimentaires et de la pomme de terre* . . . (Paris, 1813); *De la gélatine des os et de son bouillon* . . . (Paris, 1818); and *L'art oenologique réduit à la simplicité de la nature par la science et l'expérience* . . . (Paris, 1823).

For the enthusiastic report by Cadet de Vaux and Fourcroy on the chemical theories and nomenclature of Lavoisier, see "Extrait des registres de la Société royale d'agriculture du 5 février 1789," in A. L. Lavoisier, *Traité élémentaire de chimie,* II (Paris, 1789), 650–653. The letters of Cadet de Vaux to Benjamin Franklin are cited in *Calendar of Papers of Benjamin Franklin in the Library of the American Philosophical Society,* I. Minis Hays, ed., 5 vols. (Philadelphia, 1908). More extensive listings of Cadet de Vaux's publications will be found in *Nouvelle biographie générale,* VII (Paris, 1843), 69–70; and *Catalogue général des livres imprimés de la Bibliothèque nationale,* XXII (Paris, 1928), 194–203.

II. SECONDARY LITERATURE. A comprehensive account of Cadet's life and work is André Vaquier, "Un philanthrope méconnu Cadet de Vaux (1743–1828)," in *Paris et Île-de-France,* vol. IX in the series Mémoires de la Fédération des Sociétés Historiques et Archéologiques de Paris et de l'Île-de-France (Paris, 1958) pp. 365–467. See also L. G. Toraude, *Étude scientifique, critique et anecdotique sur les Cadet, 1695–1900* (Paris, n.d.), repr., rev., and enl. from *Bulletin des sciences pharmacologiques,* **6** (1902); Alex Berman, "The Cadet Circle: Representatives of an Era in French Pharmacy," in *Bulletin of the History of Medicine,* **40,** no. 2 (Mar.–Apr. 1966), 101–111. For an excellent study of the important contributions made by Cadet de Vaux and Parmentier to the École de Boulangerie, see Arthur Birembaut, "L'École gratuite de boulangerie," in René Taton, ed., *Enseignement et diffusion des sciences en France au XVIIIe siècle* (Paris, 1964), pp. 493–509.

ALEX BERMAN

CAGNIARD DE LA TOUR, CHARLES (*b.* Paris, France, 31 March 1777; *d.* Paris, 5 July 1859), *physics.*

Cagniard studied at the École Polytechnique and the École du Génie Géographe. He was later *auditeur* to the Council of State, director of special projects for the city of Paris, and a member of the board of directors of the Société d'Encouragement. His honors included membership in the Legion of Honor and knighthood of the Order of St. Michel.

Because of the great diversity of the subjects he dealt with, it is slightly difficult to present a complete picture of Cagniard's scientific career. He first worked in mechanics, beginning, in 1809, with a heat engine. Between 1809 and 1815 he produced a new hydraulic engine, a new air pump, a waterwheel mounted horizontally and turned by the current of a river, a portable military mill, and a heat-driven winch. Until 1819 these machines were constantly being improved; after 1820 a curved-cylinder pump was added to the list.

Between 1820 and 1823 Cagniard began his research in physics, starting with the discovery of the existence of a critical state in the vaporization of liquids; at the same time he became construction chief of the Crouzoles aqueduct in the Puy-de-Dôme. Between 1824 and 1827, after the invention of his siren, Cagniard began research in acoustics and the mechanism of voice production and devoted much effort to this field from then on. Between 1828 and 1831 new interests appeared: studies on the crystallization and the effect of acids on carbon; studies on phosphorus; and studies on silica and its crystallization and the hardening of mortar. Between 1832 and 1835 Cagniard worked on adapting the principle of the Archimedean screw to the function of an air pump and then began research on alcoholic fermentation; this work reached its culmination between 1836 and 1838. Toward the end of his career, while still pursuing all his interests, Cagniard contributed to mechanics a dynamometric device giving the average dynamic effect of a machine operating during an interval bounded by two successive observations.

The most original aspects of Cagniard's work include the heat engine, the critical state of vaporization, the siren, the Archimedean screw, and alcoholic fermentation.

One can best characterize the principle of the heat engine proposed by Cagniard by quoting from Lazare Carnot's report on the subject delivered to the Académie des Sciences on 8 May 1809: "A mass of cold air introduced into the bottom of a tank filled with hot water expands, and in its attempt to rise to the surface, acts as a weight, but from bottom to top." The cold air was conducted down to the bottom of the tank by means of a partially immersed Archimedean screw (this is the cagniardelle discussed below); heated by the water bath, it acted from below on the inverted cups of a paddle wheel immersed in hot water. The energy supplied by the shaft of the paddle wheel was five times that exerted on the shaft of the screw. As Carnot remarked, "The effect of the heat is therefore to increase fivefold the natural effect of the motive force," but he did not determine the quantity of heat absorbed. Both the machine and the remarks it occasioned are characteristic of this very early period of thermodynamics.

Gay-Lussac had demonstrated that vapors subjected to pressure much lower than their saturation pressure behave like ideal gases. About 1822, while attempting to vaporize liquids in a sealed vessel, maintaining a specific ratio between the volume of the liquid and that of the vessel, Cagniard proved that, above a certain temperature, a liquid contained in a hermetically sealed vessel could be completely vaporized. He further determined the temperatures and pressures corresponding to that critical state for a certain number of substances. For sulfuric ether he found, for example, 175° C. and thirty-eight atmospheres; for alcohol, 248° C. and 119 atmospheres. This research was taken up again by Andrews in 1867, but with more effective laboratory equipment and more accurate methods of measurement.

In his lifetime Cagniard's reputation was made by the acoustical siren. Besides demonstrating the nature of sound, this device, equipped with a speedometer for measuring the rate of revolution, allowed ready determination of the frequency of vibration of any sonorous body: it had only to be put in harmony with the body being studied. According to Cagniard's original design, there was no auxiliary motor; the apertures were arranged obliquely so that the perforated disk would automatically rotate like a turbine when air under pressure was applied. It had the disadvantage that the pressure had to be increased in order to provide a sharper sound, so that the siren was soft for low notes and shrill for high ones. Furthermore, it was difficult to keep the device at a given pitch, since air pressure was difficult to regulate with such accuracy. The siren was gradually perfected and took final shape in the hands of Helmholtz, who added independent feed (several rings of holes were arranged concentrically on a single disk, thereby allowing variation in the pitch in a known ratio, without changing speed).

The cagniardelle consists of an Archimedean screw partially immersed in a liquid; when rotated, it creates a forced draft or air blast. The first one was set up as early as 1827 by Koechlin et Cie., of Mulhouse; however, Cagniard had been using the principle since 1809. This prototype—it was three meters long, two and a half meters in diameter, had four turns, and made six revolutions per minute—produced thirty-five cubic meters of blast per minute at a pressure of twenty-seven millimeters of mercury. This was sufficient to supply twenty-two forges and two cupola furnaces. Later cagniardelles could attain pressures of up to sixty millimeters of mercury. This machine had an excellent yield, but it was cumbersome and ceased to be used when production of compressed air at higher pressure became widespread.

Cagniard's studies on alcoholic fermentation have unquestionably remained the most valuable of his works. Begun as early as 1835, they led him, toward the end of 1836, to see that there was certainly a living substance in brewer's yeast. Schwann came to the same conclusion at the same time, but Liebig's violent attacks forced this point of view into the background for twenty years. (It was in 1857 that Pasteur shed light on the question.) Reporting on research done with the finest microscopes available, Cagniard wrote, "Ferments . . . are composed of very simple organized microscopic bodies . . . brewer's yeast is a mass of small globulous bodies capable of reproducing themselves . . . it is very probably through some effect of their growth that they release carbon dioxide and . . . convert [a sugary solution] into a spirituous liquor."

BIBLIOGRAPHY

I. Original Works. Cagniard himself compiled the list in the *Notice sur les travaux scientifiques . . .*, which he had published anonymously and used to support his candidacy for the Académie des Sciences in 1851 (see below). It has by no means been proved that this list is exhaustive, for Cagniard's works are numerous and each was published several times. For those who wish to do research within this labyrinth, the information below will be of help.

Cagniard regularly presented the results of his studies to the Académie des Sciences or the Société Philomatique, or to both at the same time. No fewer than 116 references concerning him are in the publications of the Académie des Sciences. One might begin with the following: for 1808–1835, the *Procès-verbaux des séances de l'Académie* (4–10), 6 vols. (Hendaye, 1913–1922), table of proper names in each volume. For the period after 1835, see *Table générale des Comptes rendus des séances de l'Académie des sciences, tomes Ier à XXXI, 3 août 1835 à 30 décembre 1850* (Paris, 1853) and *Table générale . . . tomes XXXII à LXI, 6 janvier 1851 à 30 décembre 1865* (Paris, 1870).

Cagniard's correspondence with the Société Philomatique was published in the form of extracts from the society's minutes, which appeared weekly in *L'Institut. Journal général des sociétés et travaux scientifiques de la France et de l'étranger. Première section: Sciences mathématiques, physiques et naturelles*. The principal ones are (*a*) works on fermentation: **4,** no. 158 (18 May 1836), 157; no. 159 (25 May 1836), 165; no. 164 (29 June 1836), 209; no. 165 (6 July 1836), 215; no. 166 (13 July 1836), 224–225; no. 167 (20 July 1836), 236–237; no. 185 (25 Nov. 1836), 389–390; **5,** no. 199 (1 Mar. 1837), 73; (*b*) works on acoustics, the formation of human speech, the artificial glottis, and the *anche à torsion* (twisting reed): **6,** no. 229 (17 May 1838), 162–163; **10,** no. 453 (1 Sept. 1842), 311; **11,** no. 482 (23 Mar. 1843), 93; no. 485 (13 Apr. 1843), 122; no. 490 (18 May 1843), 165; no. 498 (13 July 1843), 233; **12,** no. 536 (3 Apr. 1844), 116–117; **13,** no. 577 (15 Jan. 1845), 24; **14,** no. 639 (1 Apr. 1846), 106–107; **18,** no. 845 (13 Mar. 1850), 84–85; no. 883 (4 Dec. 1850), 390; (*c*) works on the carbonization of plants in a sealed vessel: **6,** no. 229 (17 May 1838), 163; **18,** no. 861 (3 July 1850), 214–215; and no. 866 (7 Aug. 1850), 253–254.

Other publications are "Sur la sirène, nouvelle machine

d'acoustique destinée à mesurer les vibrations de l'air qui constituent le son," in *Annales de chimie,* **12** (1819), 167–171; "Exposé de quelques résultats obtenus par l'action combinée de la chaleur et de la compression sur certains liquides, tels que l'eau, l'alcool, l'éther sulfurique et l'essence de pétrole rectifiée," *ibid.,* **21** (1822), 127–132, 178–182; "Expériences à une haute pression avec quelques substances," *ibid.,* **23** (1823), 267–269; "Nouvelle note . . . sur les effets qu'on obtient par l'application simultanée de la chaleur et de la compression à certains liquides," *ibid.,* **22** (1823), 410–415; "Considérations diverses sur la vibration sonore des liquides," *ibid.,* **56** (1834), 280–294; *Exposé relatif à la vis soufflante connue . . . sous le nom de Cagniardelle, lu à l'Académie des sciences le 26 mai 1834* (Paris, 1834); "Mémoire sur la fermentation vineuse présenté à l'Académie des sciences le 12 juin 1837," in *Annales de chimie,* **68** (1838), 206–222; and *Expériences sur la cristallisation du charbon présentées à l'Académie des sciences le 12 juillet 1847* (Paris, 1847).

II. SECONDARY LITERATURE. Works on Cagniard are A. C. Becquerel, *Funérailles de M. le baron Cagniard de Latour. Discours . . . prononcé . . . le jeudi 7 juillet 1859* (Paris, n.d. [1859]); "Cagniard-Latour," and "Note de Mr Cagniard-Latour," 10 pps. in autograph, dated 5 Nov. 1827; and *À Messieurs les membres de l'Académie des sciences* (Paris, 1851); Marcel Florkin, *Naissance et déviation de la théorie cellulaire dans l'oeuvre de Théodore Schwann* (Paris, 1960), pp. 51, 53, 54, 91; Jacob, *Biographie de Cagniard de Latour, membre de l'Institut,* extract from *La science* (Paris, n.d. [after 1851]); Thomas S. Kuhn, "Sadi Carnot and the Cagniard Engine," in *Isis,* **52** (1961), 567; *Notice des travaux du Bᵒⁿ Cagniard de Latour* (Paris, n.d. [after 1822]); *Notice sur les travaux scientifiques de M. Cagniard-Latour* (Paris, 1851); *Notice sur M. le baron Cagniard de Latour, membre de l'Institut,* extract from Vol. III of *Études critiques et biographiques* (Paris, 1856); and René Taton, *Histoire générale des sciences. Tome III: La science contemporaine,* I (Paris, 1961), 198, 261, 271, 444, 456.

JACQUES PAYEN

CAHOURS, AUGUSTE ANDRÉ THOMAS (*b.* Paris, France, 2 October 1813; *d.* Paris, 17 March 1891), *chemistry.*

Cahours's father was a tailor on the rue de Provence in Paris. Cahours studied at the École Polytechnique from 1833 to 1835, graduating as a staff officer. He had already decided to study chemistry, however, and resigned his commission in 1836 in order to enter Chevreul's laboratory at the Muséum d'Histoire Naturelle as a *préparateur.* Cahours became *docteur-ès-sciences* at the Faculty of Sciences in 1845 and was professor of chemistry at the École Polytechnique and the École Centrale.

At the Muséum, Cahours devoted himself, in 1839–1840, to the thorough study of potato oil, a substance analogous in composition to ethyl alcohol,

which had been discovered by Scheele and analyzed by Dumas. Cahours suspected that it might well behave like a true alcohol because it was isomorphic with ethyl alcohol. Starting with only a liter of the substance, he succeeded in producing a great number of derivatives, all of which were quite analogous to the corresponding derivatives of ethyl alcohol. In this work he followed the method indicated by Dumas and Eugène Péligot's research on methyl alcohol (1839). He next used the research of Regnault and Faustino Malaguti on the action of chlorine on the ethers in order to study derivatives formed by substituting chlorine for hydrogen in the series, which he had named amyl.

Another important group of Cahours's researches concerned abnormal vapor densities. He sought to elucidate the principle that the stability of a given density in a certain temperature interval characterized the molecular groupings. In 1845 he began to study the variation in the density of acetic acid vapor between 124° and 336°C., then that of phosphorus pentachloride; in 1863 he conducted the same type of study on the substitution derivatives of acetic acid, and in 1866 he made a more complete examination of acetic acid vapor. He established that it behaves as a perfect gas from 240° to 440°C. and then decomposes into methane and carbon dioxide. "These classic works," wrote Armand Gautier in 1891, "have become one of the solid foundations on which we today base the fundamental proposition that as a perfect gas the [gram] molecular weight of most substances occupies the same volume in the vapor state."

Among Cahours's other accomplishments the most important are the discovery (1834) of toluene, identified among the products of the dry distillation of benzoin, and the study (1844) of the oil of *Gaultheria procumbens,* known to have the same composition and properties as methyl salicylate; Cahours demonstrated by synthesis that it was the same substance. Cahours also discovered anisic acid, anisole, and the polysulfides of alcohol; achieved the etherification of the phenols; and studied tetravalent sulfur. Finally, while preparing acid chlorides by using phosphorus pentachloride, he paved the way for Gerhardt's discovery of the acid anhydrides.

Cahours became assayer at the Monnaie in 1853, replacing Auguste Laurent. In 1851 he had become Jean-Baptiste Dumas's *suppléant* at the Sorbonne; and when Dumas became perpetual secretary of the Académie des Sciences in 1868, Cahours succeeded to his seat there.

The second half of Cahours's life was marked by a series of sorrows: between 1866 and 1871 he lost his brother, his wife, and his two sons, who were hardly more than twenty years old. Although greatly affected by these ordeals, he did not give up his research, which then took on a more fragmentary and episodic character. A second marriage, late in life, brought some serenity to Cahours's last years.

Among his honors were the Prix Jecker of the Institut de France in 1860 and 1867, corresponding membership in the Berlin Academy, the rank of commander in the Legion of Honor, and membership in the Académie des Sciences in 1868.

BIBLIOGRAPHY

I. ORIGINAL WORKS. Most of Cahours's research results were published in the *Comptes rendus hebdomadaires des séances de l'Académie des sciences;* lists can be found in *Tables générales des Comptes rendus de l'Académie des sciences . . . tomes Ier à XXXI . . .* (Paris, 1853), p. 109; *Tables générales . . . tomes XXXII à LXI . . .* (Paris, 1870), pp. 85–86; *Tables générales . . . tomes LXII à XCI . . .* (Paris, 1888), pp. 98–99; and *Tables générales . . . tomes XCII à CXXI . . .* (Paris, 1900). Among his most important works are "Mémoire sur l'huile volatile de pommes de terre et ses combinaisons," in *Annales de chimie,* 70 (1839), 81–104, and 75 (1840), 193–204; "Recherches chimiques sur le salicylate de méthylène et l'éther salicylique," *ibid.,* 10 (1844), 327–369; "Recherches sur les acides volatils à six atomes d'oxygène," *ibid.,* 13 (1845), 87–115, and 14 (1845), 485–507; *Recherches sur les huiles essentielles et sur une classification de ces produits en familles naturelles, fondée sur l'expérience* (Paris, 1845), his doctoral thesis; *Leçons de chimie générale élémentaire professées à l'École centrale des arts et manufactures,* 2 vols. (Paris, 1855–1856), subsequent eds. entitled *Traité de chimie générale élémentaire* (2nd. ed., 3 vols., 1860; 3rd ed., 3 vols., 1874–1875; 4th ed., 3 vols., 1879); "Recherches sur les radicaux organométalliques," in *Annales de chimie,* 58 (1860), 5–82, and 62 (1861), 257–350; and "Recherches sur les densités de vapeur anomales," in *Comptes rendus hebdomadaires des séances de l'Académie des sciences,* 56 (1863), 900–912, and 63 (1866), 14–21.

II. SECONDARY LITERATURE. Cahours's work is discussed in Maurice Delacre, *Histoire de la chimie* (Paris, 1920), pp. 441, 443, 497; Armand Gautier, "L'oeuvre de M. A. Cahours," in *Revue scientifique,* 48 (Jan.–July 1891), 385–387; Edouard Grimaux, "L'oeuvre scientifique d'Auguste Cahours," *ibid.,* 49 (Jan.–July 1892), 97–101; Raoul Jagnaux, *Histoire de la chimie,* 2 vols. (Paris, 1891), *passim;* Albert Ladenburg, *Histoire du développement de la chimie depuis Lavoisier jusqu'à nos jours,* trans. from the 4th German ed. by A. Corvigny, 2nd French ed. with a supplement by A. Colson (Paris, 1919), *passim;* and *Notice sur les travaux scientifiques de M. Auguste Cahours* (Paris, 1868).

JACQUES PAYEN

CAILLETET, LOUIS PAUL (*b.* Châtillon-sur-Seine, Côte-d'Or, France, 21 September 1832; *d.* Paris, France, 5 January 1913), *physics, technology.*

Cailletet, the son of a metallurgist, began his studies at the *collège* of his native city, completed them at the Lycée Henri IV in Paris, and then became a day student at the École des Mines. Upon his return to Châtillon he managed the forges and rolling mills of his father's business, paying special attention to the operation of the blast furnaces; from this interest resulted his first series of investigations in metallurgy.

The permeability of metals to gases, notably that of iron to hydrogen, enabled Cailletet to explain some of the accidents that occurred in the tempering of incompletely forged pieces of iron, which, when retaining gas in dissolution, can be highly unstable in certain cases.

Some chemists, including J. J. Ebelmen, had concerned themselves with the analysis of gases from blast furnaces. However, they drew off the gases under conditions that produced gradual cooling, so that the dissociated elements were able to recombine. Cailletet undertook new investigations in which the gases were suddenly cooled at the very moment at which they were collected. The gases obtained under these conditions had a composition very different from that established by Ebelmen: little exhaust gas, little carbon dioxide, some hydrogen, some oxygen and carbon monoxide, and a large proportion of very finely divided carbon. The investigations of Henri Sainte-Claire Deville on dissociation had given Cailletet the idea of correcting an experimental result invalidated by the very conditions of the experiment.

Cailletet is most famous for his investigations on the compression and liquefaction of gases. At the time there were still six gases that were considered permanent: oxygen, nitrogen, hydrogen, nitrogen dioxide, carbon monoxide, and acetylene. Liquefaction had not been achieved despite the use of what were considered enormous pressures. At the end of 1877 and the beginning of 1878 Cailletet liquefied all these gases shortly before Raoul Pictet, who employed a completely different procedure.

Cailletet had, following Andrews, recognized the importance of the critical temperature, above which liquefaction of a gas does not take place. In order to produce the necessary cooling, he had recourse to expansion, sometimes employing several expansions in a staged process. It was in this manner that he succeeded in liquefying oxygen by beginning with liquid ethylene and passing through the stage of liquid methane.

These results removed the last obstacle to a unified conception of the role of heat in changes in the physical state, thus realizing the prophetic views that opened Lavoisier's *Traité élémentaire de chimie.* For these investigations Cailletet received the Prix Lacaze of the Académie des Sciences in 1883.

Other accomplishments of Cailletet include the installation of a 300-meter manometer on the Eiffel Tower; a study on the same structure of air resistance to the fall of bodies; the construction of devices, such as automatic cameras and air-sample collectors, for the study of the upper atmosphere by sounding balloons; and the study of a liquid-oxygen respiratory apparatus designed for high-altitude ascents. These investigations led to Cailletet's being chosen president of the Aéro Club de France.

Cailletet was elected *correspondant* of the Académie des Sciences on 17 December 1877 and became *académicien libre* on 27 December 1884.

BIBLIOGRAPHY

I. ORIGINAL WORKS. Cailletet presented all the reports of his investigations to the Académie des Sciences in the form of notes published in the *Comptes rendus hebdomadaires des séances de l'Académie des sciences.* See the following *Tables générales des Comptes rendus . . .:* for vols. **32–61** (Paris, 1870), p. 86; for vols. **62–91** (Paris, 1888), p. 99; for vols. **92–121** (Paris, 1900), pp. 106–107; for vols. **122–151** (Paris, 1927), p. 114; and for vols. **152–181** (Paris, 1931), p. 124. His most important works are "Sur la condensation des gaz réputés incoercibles," in *Comptes rendus . . .,* **85** (31 Dec. 1877), 1270–1272; "De la condensation de l'oxygène et de l'oxyde de carbone," *ibid.* (24 Dec. 1877), 1213–1214, see also 1212, 1214–1219; "Sur la liquéfaction de l'acétylène," *ibid.* (5 Nov. 1877), 851–852; "Liquéfaction du bioxyde d'azote," *ibid.* (26 Nov. 1877), 1016–1017; "Recherches sur la liquéfaction des gaz," in *Annales de chimie,* 5th ser., **15** (1878), 132–144, illustrated; "Sur l'emploi des gaz liquéfiés et en particulier de l'éthylène pour la production des basses températures," in *Comptes rendus . . .,* **94** (1 May 1882), 1224–1226; and "Nouvel appareil pour la liquéfaction des gaz. Emploi des gaz liquéfiés pour la production des basses températures," in *Annales de chimie,* 5th ser., **29** (1883), 153–164.

II. SECONDARY LITERATURE. On Cailletet or his work, see *Notices sur les travaux scientifiques de M. L. Cailletet* (Paris, 1883), edited by Cailletet himself to support his candidacy for membership in the Académie des Sciences, and E. Colardeau, "Louis Cailletet," in *Nature* (Paris), **41,** no. 1 (25 Jan. 1913), 143–144.

JACQUES PAYEN

CAIUS (pronounced and sometimes written **KEYS**), **JOHN** (*b.* Norwich, England, 6 October 1510; *d.* London, England, 29 July 1573), *medicine.*

After preparatory studies in Norwich, John, son of Robert and Alice Caius, entered Gonville Hall, Cambridge, in 1529. He was graduated in 1533 and received the M.A. in 1535. Thereafter he studied medicine and in 1539 transferred to the University of Padua, where he received the M.D. degree on 13 May 1541.

Despite his anatomical training under the iconoclastic Vesalius, with whom he lived for eight months in Padua, Caius firmly believed that once Galen's writings had been properly reconstructed, they would make medical research unnecessary. "Except for certain trivial matters nothing was overlooked by him, and all those things that recent authors consider important could have been learned solely from Galen" (*De libris suis,* f. 10r). An excellent Greek scholar, Caius sought to contribute to the corpus Galenicum, and in the summer of 1542 began an extensive trip through Italy, studying the Galenic manuscripts in the principal libraries. He returned to England in 1545 by way of Switzerland (where he began a lifelong friendship with Conrad Gesner) and then Germany and Belgium. The first results of his investigations were two books published in Basel in 1544: (1) *Libri aliquot Graeci,* a collection of emendated Greek texts of Galen's writings, notably the hitherto unpublished first book of the *Concordance of Plato and Hippocrates,* the *Anatomical Procedures,* and the *Movement of Muscles;* (2) *Methodus medendi,* a general work on medical treatment based on the doctrines of Galen and Giambattista da Monte, Caius's professor of clinical medicine at Padua. A number of emendated Greek texts and Latin translations of Galenic and Hippocratic writings, products of Caius's further studies after his return to England, were published under the titles *Galeni de tuenda valetudine* (Basel, 1549), *Opera aliquot et versiones* (Louvain, 1556), and *Galeni Pergameni libri* (Basel, 1557).

On 22 December 1547 Caius was admitted to the College of Physicians of London, of which he soon became a fellow and, in 1550, an elect. In 1555 he was chosen president, an office to which he was reelected for the ninth and final time in 1571. He was a strict disciplinarian who sought not only to strengthen the power of the college in its control of medical licensing in London but also to extend that control over all England. Although he was not always successful, nevertheless he did gain a greater respect for the profession of medicine in England; and in 1569 the college was able to force the powerful Lord Burghley to agree to banishment of a quack he had been shielding and to declare that he held no animosity against the college and "had the highest opinion of all the Fellows." As part of his well-intended but frequently strongly opposed efforts to raise the level of medical education in England, Caius sought to prevent the universities of Oxford and Cambridge from granting medical degrees to those of dubious ability. It was also through the urgings of Caius that in 1565 the College of Physicians, like the United Company of Barber-Surgeons in 1540, was annually awarded by the crown the bodies of four executed criminals for anatomical demonstration; his old college at Cambridge received two.

Meanwhile, in 1546, Caius had been appointed anatomical demonstrator to the Company of Barber-Surgeons, a position that he held for seventeen years, during which time he made notable contributions to the development of this basic science in England. About the beginning of 1548 he began the practice of medicine in London and was appointed physician, successively, to Edward VI, Mary, and Elizabeth. His services also were frequently demanded outside London by the nobility and gentry, and it was on the occasion of such a visit to Shrewsbury in 1551 that he observed the ravages of the "sweating sickness," possibly a form of influenza, in its fifth outbreak in England. The result was his *Boke or counseill against the disease called the sweate* (1552), a minor classic of medical literature and the first original description of a disease to be written in England and in English. Caius studied the history of "the sweat," established a diagnosis and was able to prove the disease quite unlike that of any earlier epidemic, described its course, and provided primitive statistics on the mortality rates. Despite his dislike of the use of the vernacular, in this instance he believed that the seriousness of the pestilence required him to reach as wide a public as possible. Later he wrote in Latin on the same subject for the medical profession: *De ephemera Britannica,* published in his *Opera aliquot et versiones* (1556).

The remainder of Caius's published works were either composed or printed toward the end of his life. *De rariorum animalium atque stirpium historia* (1570), a description of fauna and flora that came to his attention in and around London, was originally composed for inclusion in Conrad Gesner's *Historia animalium* but omitted by reason of the latter's death. It appeared with *De canibus Britannicis,* likewise originally intended for Gesner's work, and *De libris suis,* Caius's literary autobiography. The English rendering by Abraham Fleming of the second of these works, *Of Englishe dogges* (1576), is a far from exact translation of the original Latin text. Three further works, which have no immediate relation to Caius's scientific activities, were *De antiquitate Cantabrigiensis Academiae* (1568), *Historia Cantabrigiensis Academiae* (1574), and *De pronunciatione Graecae* (1574). Caius

also emendated or translated into Latin still other Greek medical texts that, however, remained unpublished and are now known only through his references to them in *De libris suis*. He also refers there to an unpublished work on the baths of England, *De thermis Britannicis,* the earliest treatise of its kind. His record of the College of Physicians of London from 1518 to 1572, *Annalium Collegii medicorum Londini liber,* was first published in 1912.

In 1557 Caius was empowered by letters patent to refound his old college at Cambridge as Gonville and Caius College. He accepted its mastership in 1559 and provided large benefactions for rebuilding. Nevertheless his position became untenable because he had remained faithful to Catholicism, and he resigned in June 1573. Upon his death the following month he was interred in the college chapel with the simple inscription *Fui Caius*.

BIBLIOGRAPHY

John Venn, "John Caius," in *The Works of John Caius, M.D.,* E. S. Roberts, ed. (Cambridge, 1912), pp. 1–78, includes a bibliography of Caius's writings and an appendix of documents. The *Works* contains all of Caius's published writings, as well as the hitherto unpublished *Annales* of the College of Physicians of London, but not the emendated texts and translations of classical medicine. There is a modern facsimile edition of *The Sweate,* Archibald Malloch, ed. (New York, 1937).

C. D. O'Malley emphasizes the medical aspects of Caius's life in *English Medical Humanists* (Lawrence, Kan., 1965), pp. 26–46; contributions to zoology are dealt with in Edward C. Ash, *Dogs; Their History and Development* (London, 1927), I, 68–70, 74–84; II, 656–658; J. W. Barber-Lomax, "De canibus Britannicis," in *Journal of Small Animals,* **1** (1960), 24–31, 109–114; and Charles E. Raven, *English Naturalists From Neckam to Ray* (Cambridge, 1947), pp. 138–147.

C. D. O'MALLEY

CALANDRELLI, GIUSEPPE (*b.* Zagarolo, near Rome, Italy, 22 May 1749; *d.* Rome, 24 December 1827), *astronomy*.

Calandrelli was professor of mathematics at the Gregorian University of the Jesuit-run Collegio Romano in Rome and director of the observatory that he had built there in 1787. The thin square tower that he used as an observatory (which was replaced by the observatory built by Angelo Secchi) may still be seen. Calandrelli confined himself exclusively to positional astronomy and, in collaboration with his codirectors, Andrea Conti and Giacomo Ricchebach, he published the series Opuscoli Astronomici (1803–1824). Those of 1806, dedicated to Pope Pius VII, include *Osservazioni e riflessioni sulla parallasse annua dell'alfa della Lira* and *Soluzione esatta del problema delle altezze corrispondenti.*

In 1816, again in collaboration with Conti and Ricchebach, Calandrelli published the *Tavola delle parallassi di altezza di longitudine e di latitudine,* also dedicated to Pius VII, who had become so keenly interested in the work that he provided the observatory with "perfect machines," among them the transit of Reichenbach. Calandrelli also wrote historical articles on the Gregorian calendar and on Roman astronomy.

BIBLIOGRAPHY

A list of Calandrelli's writings is in Poggendorff, I, 361.
The *Bollettino di bibliografia e di storia del Boncompagni* (1868–1887) states that Calandrelli maintained correspondence with d'Alembert. See also G. Abetti, *Storia dell'astronomia* (Florence, 1963), p. 389; and Tipaldo, *Biografia degli Italiani illustri,* III (1836), 243.

GIORGIO ABETTI

CALANDRELLI, IGNAZIO (*b.* Rome, Italy, 27 October 1792; *d.* Rome, 12 February 1866), *astronomy, mathematics.*

A nephew of Giuseppe Calandrelli, Ignazio was professor of astronomy and director of the observatory of the University of Bologna from 1845 to 1848. Before and after this he held a similar position at the Pontifical University and at the Observatory of the Campidoglio in Rome. The latter was founded by Pope Leo XII (through the bull *Quod divina sapientia*), which provided for the establishment of a good astronomical observatory for the compilation of a calendar computed for Rome. The observatory, which the pontiff conceived of as constituting the "first true Roman observatory," was completed in 1827, on the eastern tower of the Senatorial Palace of the Campidoglio.

Calandrelli's scientific work was confined almost exclusively to positional astronomy. In 1853 he provided the observatory with Ertel's meridian circle, with which he made observations on the determination of latitude; prepared a catalog of stars; and carried out studies on refraction. He also performed numerous calculations of the orbits of small planets and comets.

In 1858 Calandrelli published a memoir on the proper motion of Sirius and his observations on the solar eclipse of 15 March of that year. He observed the occultation of Saturn on 8 May 1859. Calandrelli

was a member of the Accademia dei Nuovi Lincei in Rome.

BIBLIOGRAPHY

Calandrelli's works are listed in Poggendorff, III, 226. Among them is *Lezioni elementari di astronomia teorico-pratica ad uso dei giovani studenti nelle due Università dello Stato Pontificio* (Bologna, 1848).
Further information may be found in G. Abetti, *Storia dell'astronomia* (Florence, 1963), pp. 381, 388, 389; and *Osservatori astrofisici e astronomici italiani* (Rome, 1956).

GIORGIO ABETTI

CALCIDIUS (fourth, possibly fifth, century A.D.), *Platonist commentary.*

Calcidius' Latin translation of the first two-thirds of Plato's *Timaeus* was the only extensive text of Plato known to western Europe for 800 years. Latin cosmology, throughout the early Middle Ages, was based upon the *Timaeus;* and Calcidius' version, together with his commentary upon it, provided scholars with their best contact with the master.

Nothing is known of Calcidius' life. Hosius, or Osius, the close friend to whom he dedicated his work, has generally been identified with the bishop of Córdoba who was prominent at the First Council of Nicaea (A.D. 325). Waszink, the latest editor of Calcidius, prefers to identify Osius with a Milanese patrician and official at the end of the fourth century. Waszink also regards Calcidius as certainly Christian.

Calcidius' commentary is eclectic in character. His Platonic concepts appear to have been derived from Porphyry and from writers of the Middle Platonist school, his Aristotelianism from Adrastus, and his Christian doctrines from Origen. An extended section on astronomy closely follows and at times translates the second half of Theon of Smyrna's commentary on Plato, but it is agreed that both Calcidius and Theon were using Adrastus here.

Calcidius' commentary is six times as long as his translation of the *Timaeus* and deals almost exclusively with passages in the middle third of Plato's treatise. The opening chapters are devoted to explicating enigmatic passages about the creation of the universe (31c–32c) and the origin and constitution of the World Soul (34c–35b). Chapters 3 and 4 deal with the numerical ratios of the harmonic intervals in the musical scale used in the fabrication of the World Soul (36a–36c). Chapter 5, ostensibly commenting upon two passages in the *Timaeus* about intelligibles and sensibles (37a–37c), turns out to be a conventional handbook treatment of astronomy. The highlight of

Calcidius' discussion comes in the following chapter, when he explains the epicyclic motions of Venus (presumably of Mercury too) and attributes the system to Heraclides Ponticus. Although mistaken in assuming that Heraclides was using a geometrical demonstration instead of hypothesizing actual orbits of those planets about the sun, Calcidius was the most influential authority in keeping alive geoheliocentric views in the Middle Ages and thus laying the foundations for Copernicanism. His Latin version of Plato's account of Atlantis (24D–25D) was also vital in preserving that myth of a lost continent. Calcidius' theory of matter is a conflation of Platonic and Aristotelian concepts. *Silva* to Calcidius has the character of both "space in which" (Plato) and of "matter out of which" (Aristotle). Retaining at times the Platonic *in qua*, Calcidius also struggles to avoid the Aristotelian "merely possible" concept of matter.

Manuscripts of Calcidius are abundant. Few were the medieval libraries of any standing that did not have a copy of his work. His part in transmitting classical cosmology to the Latin West culminated with the Scholastics of Chartres in the twelfth century.

BIBLIOGRAPHY

The definitive ed., *Timaeus a Calcidio translatus commentarioque instructus,* J. H. Waszink and P. J. Jensen, eds. (London–Leiden, 1962), includes exhaustive discussions of MSS (about 150 are filiated) and of Calcidius' sources. J. H. Waszink, *Studien zum Timaioskommentar des Calcidius I* (Leiden, 1964), deals mostly with the sources of the first half of the commentary. Waszink believes that Calcidius drew from both Adrastus and Porphyry. J. C. M. van Winden, *Calcidius on Matter; His Doctrines and Sources* (Leiden, 1959), prefaces his exhaustive discussion of matter with a running account of the contents of the commentary. W. H. Stahl, *Roman Science; Origins, Development, and Influence to the Later Middle Ages* (Madison, Wis., 1962), places Calcidius in the traditions of Latin cosmology and digests his account of mathematics and astronomy. T. L. Heath, *Aristarchus of Samos* (Oxford, 1913), discusses Calcidius' imputations of epicyclic concepts to Heraclides. Pierre Duhem, *Le système du monde,* III (Paris, 1954), devotes 119 pages to Calcidius, Macrobius, and Martianus Capella as transmitters of the "Heraclidean" system to the medieval world.

WILLIAM H. STAHL

CALDANI, LEOPOLDO MARCANTONIO (*b.* Bologna, Italy, 21 November 1725; *d.* Padua, Italy, 30 December 1813), *anatomy, physiology.*

It was mainly in consequence of the work of Caldani that the initial, very powerful resistance to Haller's doctrine of differential sensibility and irritability of animal tissues and organs was overcome in Italian physiological circles. (Despite Haller's great reputation, that opposition was at first very powerful, in Italy as elsewhere.) Caldani completed his education in his native Bologna, receiving the Ph.D. and M.D. on 12 October 1750. After several years of further education and practical experience he became a professor of medicine in Bologna.

At this time Caldani was concerned mainly with his own animal experiments, which were designed to verify and amplify Haller's findings. In connection with these experiments he first used an electric current to stimulate muscle tissue. His first publication, a report on seventy-three new experiments, was read at the end of 1756 to the Istituto delle Scienze in Bologna; it met with Haller's approval but encountered stiff opposition from some of Caldani's colleagues. Discouraged by the disputes that resulted, Caldani thought of leaving Bologna. He decided to remain, however, partly in consequence of Haller's warning that to move abroad might entail his conversion—a prospect from which, as a devout Catholic, Caldani shrank.

To improve his knowledge of anatomy, Caldani spent the first months of 1758 in Padua with Morgagni, whom he hoped to succeed. It was somewhat unwillingly, though, that he returned to Bologna and resumed his advocacy of Haller's teachings. Grateful for this support, Haller arranged Caldani's election to membership in the Gesellschaft der Wissenschaften in Göttingen in 1759 and later sponsored him for membership in the Royal Society of London. In the fall of 1759 Caldani, following the schedule of rotation, took over the anatomical demonstrations in Bologna; these were followed in the early part of 1760 by public anatomical disputations.

In these disputations Caldani certainly had the scientific capability to defend himself against the attacks of Paolo Balbi and Tommaso Laghi, severe though they were. Nevertheless, he resigned from his teaching position and, although his means were slender, left his home for Venice. He had to wait until the summer of 1764 to be appointed at Padua as professor of theoretical medicine, specifically of Boerhaave's theory. He alternately lectured on pathology and physiology; and from the beginning of 1773 he lectured on anatomy as well. Caldani was confirmed as Morgagni's successor in the chair of anatomy at Padua by the Venetian senate on 11 November 1773. He prepared his lectures for the three books called *Institutiones*—the *Institutiones physiologiae* was often reprinted—which were written in the style of Haller; they were widely read because several universities introduced them as textbooks.

Most of Caldani's later writings are more medical than biological. Toward the end of his life he published, with his nephew Floriano Caldani, a collection of anatomical drawings, some of which were made from their own preparations and some of which were drawings by others. He did not relinquish his professorial chair in Padua until 1805, when advanced age made it necessary to do so.

BIBLIOGRAPHY

I. ORIGINAL WORKS. Caldani's writings include *Sull'insensitività e irritabilità di alcune parti degli animali. Lettera scritta al chiarissimo sig. Alberto Haller* (Bologna, 1757), trans. as "Lettre de Mr. Marc Antoine Caldani à Mr. Albert de Haller sur l'insensibilité et l'irritabilité de quelques parties des animaux," in A. Haller, *Mémoires sur les parties sensibles et irritables du corps animal,* III (Lausanne, 1760), 1–156; "Sur l'insensibilité et l'irritabilité de Mr. Haller. Seconde lettre de Mr. Marc-Antoine Caldani," *ibid.,* 343–485; *Lettera terza del sig. dott. Leopoldo Marc'Antonio Caldani sopra l'irritabilità e sensitività halleriana* (Bologna, 1759); *Riflessioni fisiologiche sopra due dissertazioni del sig. Claudio Nicola Le Cat* (Venice, 1767); *Esame del capitolo settimo contenuto nella XII parte dell'ultima opera del chiarissimo sig. Antonio De Haen* (Padua, 1770); *Institutiones physiologiae* (Padua, 1773); *Icones anatomicae, quotquot sunt celebriores, ex optimis neotericorum operibus summa diligentia depromptae et collectae opera et studio . . .,* 4 vols. (Venice, 1801–1813), written with his nephew Floriano Caldani; and *Iconum anatomicarum explicatio,* 5 vols. (Venice, 1802–1814).

II. SECONDARY LITERATURE. Works on Caldani are Floriano Caldani, *Memorie intorno alla vita e alle opere di Leopoldo Marco Antonio Caldani* (Modena, 1822); "Caldani," in Dezeimeris *et al., Dictionnaire historique de la médecine ancienne et moderne,* I, pt. 2 (Paris, 1831), 595–600, with a complete list of publications; "Leopoldo Caldani," in Pietro Capparoni, *Profili bio-bibliografici di medici e naturalisti celebri Italiani del sec. XV° al sec. XVII°,* II (Rome, 1928), 92–96; A. von Haller, *Epistolarum ab eruditis viris ad Albertum Hallerum scriptarum Pars I Latine,* IV–VI (Bern, 1774–1775); Erich Hintzsche, ed., *Albrecht von Haller-Marc Antonio Caldani: Briefwechsel 1756–1776* (Bern-Stuttgart, 1966); and Giuseppe Ongaro, "Leopoldo Marc' Antonio Caldani e Albrecht von Haller," in *Atti del XXIII Congresso nazionale di storia della medicina (Modena 22–24 settembre 1967),* also reprinted separately.

ERICH HINTZSCHE

CALDAS, FRANCISCO JOSÉ DE (*b.* Popayán, New Granada [now Colombia], 1768; *d.* Santa Fe [now Bogotá], New Granada, 29 October 1816), *botany, astronomy, geography.*

Caldas taught himself mathematics and astronomy. His father sent him to Santa Fe, capital of the vice-royalty of New Granada, to study law, but family circumstances forced him to go into the transportation business. At Quito in 1802 he met Alexander von Humboldt and Aimé Bonpland, who had become familiar with his work in Popayán and considered him a genius. Caldas spent six months with the travelers and learned much of what he needed to become an astronomer, geodesist, volcanologist, geographer, and botanist. He hoped to accompany Humboldt on the rest of his journey, and it is not known why he did not do so. At this time José Mutis, who had sent Caldas money to finance his trip with Humboldt, made him a member of a botanical expedition and commissioned him to collect plants, mainly cinchonas, from the southern part of New Granada. Later he named Caldas director of a newly built astronomical observatory (the first in South America), geographer of the viceroyalty, and his successor.

In December 1805 Caldas arrived in Bogotá, where he took over directorship of the observatory and began work on an improved map of the viceroyalty. In January 1808 he began publication of *Semanario del nuevo reino,* which continued until 1811 and contained studies that are still important.

When the province of Bogotá proclaimed its independence in 1810, Caldas was among the most active rebels. He published the *Diario político,* enlisted in the army of liberation as an engineer, directed the army training school, and organized the arsenal for manufacturing rifles, gunpowder, and ammunition. When the rebellion was suppressed in 1816, Caldas sought refuge on his ranch, Paispamba, near Popayán. He was seized and taken to Bogotá, where he was tried and shot.

Caldas' greatest scientific achievements were his discovery of the method for measuring altitude by the boiling point of pure water, his discovery of many species and varieties of cinchona, and his collaboration with Humboldt on the study of the distribution of plants according to altitude.

BIBLIOGRAPHY

See *Semanario de la Nueva Granada (corregido y aumentado con opúsculos diversos por Joaquín Acosta)* (Paris, 1849).

ENRIQUE PÉREZ ARBELÁEZ

CALKINS, GARY NATHAN (*b.* Valparaiso, Indiana, 18 January 1869; *d.* Scarsdale, New York, 4 January 1943), *zoology.*

Calkins came of old New England stock and was the son of John Wesley Calkins and Emma Frisbie

Smith. In 1894 he married Anne Marshall Smith, and in 1909 Helen Richards Colton. Calkins took his B.S. in biology from the Massachusetts Institute of Technology in 1890. He was appointed microscopist and assistant biologist at the Massachusetts Board of Health, lecturing in biology at M.I.T. at the same time. In 1893 he began graduate work at Columbia University, from which he received the Ph.D. in 1897. While a student there, he started teaching in 1894, as tutor in biology. He rose rapidly through the ranks to a professorship of zoology in 1904. Two years later his title was changed to professor of protozoology. He remained at Columbia for the rest of his life, retiring as professor emeritus in 1939.

Beginning in 1893 Calkins worked for many years at the Marine Biological Laboratory at Woods Hole and was a pillar of that institution. In 1896–1897 he had charge of two expeditions to the Northwest and Alaska. During his early years at Columbia he was also interested in statistics and held office in the American Statistical Association. He had a lifelong interest in cancer research and served as consulting biologist to the New York State Department of Health Cancer Laboratory from 1902 to 1908.

Calkins's interest in the entire field of biology and experimental medicine resulted in a general textbook of biology (1914), but he was best known as a student of protozoan life. He was the author of *The Protozoa* (1901), one of the two earliest modern works on the subject and the first in English. Over the years he produced several extremely influential textbooks of protozoology, useful not only as teaching aids but as important synthetic statements in the science. His views reached maturity in the widely cited *The Biology of the Protozoa* (1926). Many honors came to him during his lifetime, and he died at the age of seventy-three in Scarsdale, New York, where he had made his home for many years.

Although fully aware of the pathogenic importance of one-celled organisms—indeed, it was the focus of some of his earliest work—Calkins was concerned during most of his career with the general biological and purely scientific aspects of unicellular animals. Each of Calkins's general treatises included important revisions of and improvements in the taxonomy of the protozoa. He suggested, for example, redefining the protozoans to exclude chlorophyll-bearing flagellates, and his basic concern was to maintain a clearly zoological portion of the protista kingdom.

Calkins regarded vital processes as more individual and inexplicable than did many of his colleagues; and although he was therefore close to the neovitalism that flourished early in the twentieth century, he explicitly stated his expectation that physical and chemical explanations for life processes would be found. He himself, from the time of his Ph.D. dissertation, did the largest part of his research on the reproduction and regeneration of protozoa. For decades he was engaged in a major controversy over whether or not ciliates can continue indefinitely to maintain themselves by division without conjugation. Calkins consistently provided evidence and argument to suggest that the generational rhythms of the organisms will prove fatal without conjugation and that conjugation stimulates physiological processes.

Calkins distinguished between ontogenetic and phylogenetic regeneration, and he believed that the former will decrease in vitality without conjugation. Although the question of indefinite maintenance of these organisms without conjugation had not been completely settled even after many years, during his lifetime Calkins's views were considered by his colleagues to represent a one-sided approach to the problems of both vital processes and the processes of reproduction and regeneration. Calkins was a major figure and an influential and prolific writer in his own era. At his death the electron microscope and other innovations were already transforming the entire science of the protista that he had pioneered.

BIBLIOGRAPHY

In addition to standard directories, information is available in a biographical article in the *National Cyclopaedia of American Biography*, XXXIII (1947), 50–51. A complete bibliography has not been collected, and one must rely upon standard sources such as *Biological Abstracts*. H. S. Jennings, *Genetics of the Protozoa* (The Hague, 1929), provides an extensive review of some of Calkins's most important work.

J. C. BURNHAM

CALLAN, NICHOLAS (*b.* near Dundalk, Ireland, 20 December 1799; *d.* Maynooth, Ireland, 14 January 1864), *electromagnetics.*

Son of "Wee" Denis Callan and the former Margaret Smith, Nicholas Callan was one of the Callans of Dromiskin, a widely spread County Louth Catholic family of means. Callan had his final schooling at the Dundalk Presbyterian Academy before entering St. Patrick's College, Maynooth, where he was ordained priest in 1823. After postgraduate studies in Maynooth and Rome, in 1826 he was appointed professor of natural philosophy at Maynooth. Here he spent the rest of his life, a life dedicated to the formation of priests and the teaching of science. Small in stature, he was a dynamo of effort. As a young professor he was vigorous and aggressively active. As a priest he was scrupulous and zealous and became known for his "large benevolences" during the famine. Pope's verse was adapted to describe him in his later years:

> Of manners gentle, of affections mild
> In wit a man: simplicity a child
> With priestly virtue tempering scientific rage
> He helped the poor, electrified the age.

Callan is a reminder that science lost as well as gained when it fell into the hands of the professionals. Until recently he has been a neglected figure in the history of science. He was a pioneer in the development of electromagnetism as a source of power. He built large batteries and constructed huge electromagnets, invented the induction coil in 1836, and in 1838 discovered the principle of self-excitation in dynamo-electric machines. However, as Heathcote points out: "Credit for discoveries and inventions properly due to Callan has been given to others, to Ruhmkorff for the invention in 1851 of the induction coil and to Werner Siemens for the discovery in 1866 of the principle of the self-induced dynamo" (p. 145).

Callan's inventions, especially that of the induction coil, became widely known in his lifetime and greatly influenced the growth of electricity as a power source. A contemporary wrote, in an obituary notice that appeared in the *Dundalk Democrat* (16 Jan. 1864): "Perhaps no man, after Faraday and Wheatstone, contributed more to the progress of electricity or deserves a higher tribute in its annals."

It was through private means rather than assistance from the college authorities that Callan was able to conduct his researches. His students lacked elementary knowledge; and except for the talented few, they did not appreciate him. Most colleagues saw him as a furious experimenter obsessed with wires and magnets, a Baconian rather than a Cartesian, a visionary who somewhat foolishly spoke of a day when electricity would be of benefit to man by reducing human drudgery. Regarded as a "character," an amiable eccentric, he was the subject of countless stories. Under pressure from colleagues and the need to combat rampant proselytism, this first-class scientist passed laborious years in making available in English the simple devotions of St. Alphonsus Liguori. He returned to his scientific pursuits in his declining years, when he was acclaimed by the Royal Irish Academy and, in 1857, was honored at the Dublin meeting of the British Association for the Advancement of Science.

BIBLIOGRAPHY

I. ORIGINAL WORKS. Rare extant copies of Callan's writings, preserved in Maynooth, include *Electricity and Galvanism* (Dublin, 1832) and "Manuscript on Physics." An account of the second is given by P. J. McLaughlin in "The Prelections of Nicholas Callan," in *Irish Astronomical Journal*, **6** (1964), 249–252. Fairly complete lists of Callan's scientific papers are to be found in the secondary literature.

II. SECONDARY LITERATURE. The following studies do much to bring out the historical significance of Callan's work: Niels H. de V. Heathcote, "N. J. Callan, Inventor of the Induction Coil," in *Annals of Science*, **21** (1965), 145–167; and P. J. McLaughlin, "Some Irish Contemporaries of Faraday and Henry," in *Proceedings of the Royal Irish Academy*, **64A** (1964), 17–35; and *Nicholas Callan: Priest-Scientist* (London, 1965).

P. J. McLaughlin

CALLANDREAU, PIERRE JEAN OCTAVE (*b.* Angoulême, France, 18 September 1852; *d.* Paris, France, 13 February 1904), *astronomy*.

After completing his studies at the École Polytechnique in 1874, Callandreau was induced by Le Verrier to become an *aide astronome* at the Paris Observatory. While there, Callandreau attended Victor Puiseux's lectures on celestial mechanics at the Sorbonne. They influenced the young astronomer to devote his energies to perfecting the theories of celestial mechanics. Confident that the germs for the solutions of most problems in mathematical astronomy are to be found in Laplace's *Traité de mécanique céleste*, Callandreau contributed more than 100 papers on various problems in this field.

Callandreau's work was strongly influenced by two of the leading mathematical astronomers of the second half of the nineteenth century, François Tisserand and Hugo Gyldén. In 1879 Callandreau translated, from the Swedish, Gyldén's memoir on perturbation theory; and during the next few years he published several papers, including his doctoral thesis, in which he developed and applied Gyldén's methods. Callandreau also made significant contributions related to the capture hypothesis concerning Jupiter's comets. Tisserand, who had published the first general theory based on this hypothesis in 1889, encouraged Callandreau to develop the theory further and to resolve apparent objections to it. For his successful efforts Callandreau received the Prix Damoiseau from the Académie des Sciences in 1891. Callandreau also applied the capture theory ideas to the theory of shooting stars.

Several of Callandreau's contributions to mathematical astronomy were incorporated by Tisserand into his *Traité de mécanique céleste* (1888–1896). In 1892 Callandreau became a member of the Académie des Sciences, and in 1893 he was made professor of astronomy at the École Polytechnique. He was an editor of the *Bulletin astronomique* from 1884 until his death, and it was under his impulsion as president of the Société Astronomique de France from 1899 to 1900 that the first systematic observation of shooting stars was undertaken in France.

BIBLIOGRAPHY

Callandreau's works have not been published in collected form. See, however, his *Notice sur les titres scientifiques de M. O. Callandreau* (Paris, 1884, 1892). Another listing of his works is Poggendorff, IV, 214.

The following notices, occasioned by Callandreau's death, represent the extent of the literature on him: *Astronomische Nachrichten,* **164** (1904), 387; *Bulletin astronomique,* **21** (1904), 129–138; *Revue générale des sciences pures et appliquées,* **15** (1904), 281–282; *Vierteljahrsschrift der Astronomischen Gesellschaft,* **39** (1904), 3–6.

THOMAS HAWKINS

CALLENDAR, HUGH LONGBOURNE (*b.* Hatherop, Gloucestershire, England, 18 April 1863; *d.* Ealing, England, 21 January 1930), *physics, engineering.*

Callendar received his early education at Marlborough. After classical and mathematical studies at Trinity College, Cambridge, from which he graduated both in classics (1884) and in mathematics (1885), Callendar became a fellow there in 1886. After serving as professor of physics at the Royal Holloway College, Egham (1888–1893), Callendar accepted an appointment as professor at McGill University, Montreal (1893). In 1898 he returned to England as Quain professor of physics at University College, London; in 1902 he became professor of physics at the Royal College of Science (in 1907 incorporated in the Imperial College of Science and Technology), a post he held until his death. In 1894 Callendar was elected a fellow of the Royal Society. He received the Watt Medal of the Institution of Civil Engineers, for his work with Nicolson on the laws of condensation of steam, in 1898; the Rumford Medal of the Royal Society in 1906; and the Hawksley Gold Medal of the Institution of Mechanical Engineers, for his investigations of the flow of steam through nozzles and throttles, in 1915.

Callendar's scientific work was mainly in the field of experimental science, particularly heat and thermodynamics, in which he devised and carried out accurate methods of measurement and designed new apparatus. Especially important were his introduction

of the platinum resistance thermometer as a new standard of accuracy for physical and engineering measurements and his investigations on the thermal properties of water and steam.

Callendar's first publication, which was communicated to the Royal Society in 1886, dealt with platinum resistance thermometry. Sir Humphry Davy had discovered the dependence of the electrical resistance of metals on temperature (1821), and the German engineer Ernst Werner von Siemens had used this phenomenon in the construction of a platinum resistance thermometer (1861). Callendar made elaborate experiments on this subject at the Cavendish Laboratory in Cambridge, in which he compared the platinum resistance thermometer with Regnault's normal air thermometer and from which he deduced that the resistance of a properly made platinum wire can be related to the reading of the air thermometer by a parabolic formula that was accurate within 1 percent. In a later paper (1891), Callendar gave practical suggestions for the construction of stable platinum thermometers with mica insulation to avoid strains in the platinum wire. The use of the platinum thermometer as a laboratory instrument instead of the more complicated air thermometer was especially important for the research worker as an easy and accurate method of measuring temperature. It also gave the engineer a convenient and practicable method of heat regulation in industrial operations. Callendar also added an automatic recording bridge to the platinum thermometer (1898).

In 1899 the Committee on Electrical Standards, which was headed by Lord Rayleigh, accepted Callendar's proposals for a standard scale of temperature based on the platinum thermometer. The platinum thermometer is now the recognized international means of interpolation between the boiling point of liquid oxygen ($-182.97°$C.) and the melting point of antimony ($630.5°$C.). Above the latter temperature a thermocouple is used.

Callendar used the platinum thermometer in a number of investigations. In a joint study with Howard Turner Barnes he determined the variations, with temperature and with strength of solution, of the electromotive force of different forms of the Clark standard cell (1898). With John Thomas Nicolson he determined the temperature of steam expansion behind a piston (1897). In the latter investigation clear evidence was obtained for the existence of steam in a supersaturated state, a state already known at that time but considered of negligible importance in the field of steam engineering.

During his stay in Montreal, Callendar developed his method of continuous electrical calorimetry, the first experimental application being the measurement by his assistant Howard Turner Barnes of the specific heat of water between its freezing point and boiling point, together with a determination of the mechanical equivalent of heat (1902). With this continuous-flow method a new way was opened for measuring specific heats of liquids, while the large corrections for heat capacity of the apparatus that were necessary in normal calorimetry were reduced to a minimum.

At McGill University, Callendar also studied engineering problems connected with steam turbines and internal combustion engines. With John Thomas Nicolson he made many valuable experiments on the heat of transmission and on leakage losses from steam-engine cylinders. Callendar did important work on the properties of vapors. With his continuous-flow method he determined the specific heat of steam and its variation with the pressure. He also measured the total heat of steam and the specific heat of water at any pressure and temperature. In 1900 he studied the thermodynamic properties of gases and vapors, as deduced from a modified form of the Joule-Kelvin equation, with special reference to the properties of steam; in his report he first put forward his equation for an imperfect gas, which has been very useful in representing the properties of steam. This paper formed the basis of Callendar's subsequent work on steam, which led him to the formulation of his steam equation and the publication of *Callendar Steam Tables* (1915 ff.), which give the properties of steam up to and beyond the critical pressure.

Besides his work on thermometry and vapors, Callendar wrote a number of papers on such subjects as the osmotic pressures of solutions (1908 ff.), the absolute expansion of mercury (with Herbert Moss, 1911 ff.), and the determination of the boiling point of sulfur (with Ernst Howard Griffiths, 1890, and Herbert Moss, 1912). In conjunction with the staff of the Air Ministry Laboratory, Callendar published papers on dopes and detonation (1925, 1926). In 1926 he published on the cause of knock in gasoline engines and the effect of antiknock compounds on engine knock.

Callendar was the inventor of a system of shorthand which was in fairly general use in some parts of the former British colonies. In 1889–1890 he published *A Manual of Cursive Shorthand* and *A System of Phonetic Spelling Adopted to English.*

BIBLIOGRAPHY

I. ORIGINAL WORKS. Callendar's scientific papers include "On the Practical Measurements of Temperature: Experiments Made at the Cavendish Laboratory, Cambridge," in *Philosophical Transactions of the Royal Society,* **178A** (1887), 161–230; "On the Construction of Platinum Thermometers," in *Philosophical Magazine,* **32** (1891), 104–113; "On a New Method of Determining the Specific Heat of a Liquid in Terms of the International Electrical Units," in *British Association Reports* (1897), pp. 552–553, written with H. T. Barnes; "Experiments on the Condensation of Steam," *ibid.,* pp. 418–424, written with J. T. Nicolson; "On the Bridge Method of Comparing Low Resistances," in *The Electrician* (London), **41** (1898), 354; "An Alternating Cycle-Curve Recorder," *ibid.,* 582–586; "On the Variation of the Electromotive Force of Different Forms of the Clark Standard Cell With Temperature and With Strength of Solution," in *Proceedings of the Royal Society,* **62** (1898), 117–152, written with H. T. Barnes; "On the Law of Condensation of Steam Deduced From Measurements of Temperature-Cycles of the Walls and Steam in the Cylinder of a Steam-Engine," in *Minutes of Proceedings of the Institution of Civil Engineers* (London), **131** (1898), 147–206, written with J. T. Nicolson; "Proposals for a Standard Scale of Temperature on the Platinum Resistance Thermometer," in *British Association Reports* (1899), pp. 242–243; "Notes on Platinum Thermometry," in *Philosophical Magazine,* **47** (1899), 191–222; "On a Practical Thermometric Standard," *ibid.,* **48** (1899), 519–547; "Note on the Variation of the Specific Heat of Water Between 0° and 100°C.," in *The Physical Review,* **10** (1900), 202–214, written with H. T. Barnes; "On the Thermodynamical Properties of Gases and Vapours as Deduced From a Modified Form of the Joule-Kelvin Equation, With Special Reference to the Properties of Steam," in *Proceedings of the Royal Society,* **67** (1900), 266–286; "Continuous Electrical Calorimetry," in *Philosophical Transactions of the Royal Society,* **199A** (1902), 55–148; and "On the Variation of the Specific Heat of Water, With Experiments by a New Method," *ibid.,* **212A** (1912), 1–32.

II. SECONDARY LITERATURE. For obituary notes, see *Engineering,* **129** (1930), 115–117; *Nature,* **125** (1930), 173–174; and *Proceedings of the Royal Society,* **134A** (1932), xviii–xxvi. Biographies of Callendar are in *Chambers' Dictionary of Scientists,* A. V. Howard, ed. (London, 1950), col. 84; and *Dictionary of National Biography,* supp. 4 (1930), 152–154. Bibliographies are given in the Royal Society of London's *Catalogue of Scientific Papers,* XIV (1915), 18–19; and Poggendorff, IV (1904), 214–215, 1703; V (1926), 197; VI (1936), 389.

H. A. M. SNELDERS

CALLINICOS OF HELIOPOLIS (*fl. ca.* A.D. 673), *chemistry.*

Callinicos ("handsome winner"), an architect, is credited with having invented or perfected "Greek fire." According to most authorities, he was a native of Heliopolis (Baalbek) in Syria. Georgius Cedrenus, a Greek monk of the eleventh century, reported in his Σύνοψις ἱστοριῶν (*Compendium historiarum*) that

Callinicos came from Heliopolis in Egypt, and this view is also taken by Gibbon (ch. 52). It is probable that he was a Jewish refugee who was forced to flee to Constantinople.

Greek fire, a terrifying new weapon, apparently was first used successfully at the battle of Cyzicus (*ca.* A.D. 673), in which the Byzantines under Emperor Constantine IV defeated the attacking Saracens. The most detailed and authoritative statement, although brief, on the life and activities of Callinicos is that of Theophanes, a Greek monk who wrote about A.D. 815. In his Χρονογραφία (*Chronographia*) he states: "At this time the architect Callinicos of Heliopolis in Syria fled to the Romans and invented the sea fire, which set on fire the boats of the Arabs and burned them completely. And in this way the Romans turned them back in victory, and this the sea fire procured."

A number of incendiary materials had been used on a limited scale in warfare before the seventh century, and the invention of Callinicos was probably a secret ingredient added to the chemical mixtures already available. This liquid or semiliquid was propelled from the siphons, or flame projectors, with which the Greek ships were fitted even before the use of Greek fire. The actual ingredients probably consisted of sulfur, pitch, petroleum, and some unknown substances. Purified saltpeter was not known in the West until the early thirteenth century and most likely was not an ingredient of the Greek fire of Callinicos.

BIBLIOGRAPHY

Brief Byzantine references to Callinicos are found in Theophanes, *Chronographia,* I. Classen, ed., 2 vols., in Corpus Scriptorum Historiae Byzantinae, G. B. Niebuhr, ed. (Bonn, 1839–1841), I, 540–542; II, 178; and Cedrenus, *Compendium historiarum,* I. Bekker, ed., 2 vols., in Corpus Scriptorum Historiae Byzantinae (Bonn, 1838–1839), I, 765. These and other references are fully discussed in J. R. Partington, *A History of Greek Fire and Gunpowder* (New York, 1960), pp. 12–14, 30–32; Maurice Mercier, *Le feu grégeois. Les feux de guerre depuis l'antiquité. La poudre à canon* (Paris, 1952), pp. 11–15; and Ludovic Lalanne, *Recherches sur le feu grégeois et sur l'introduction de la poudre à canon en Europe,* 2nd ed. (Paris, 1845), pp. 15–17. See also George Sarton, *Introduction to the History of Science,* I (Baltimore, 1927), 494–495; and Charles Oman, *A History of the Art of War in the Middle Ages,* II (New York, 1924), 46–47.

KARL H. DANNENFELDT

CALLIPPUS (*b.* Cyzicus, Turkey, *ca.* 370 B.C.), *mathematics, astronomy.*

One of the great astronomers of ancient Greece,

Callippus belonged to the distinguished line of mathematicians and astronomers who, with Eudoxus at their head, were associated with the Academy and the Lyceum. References to him, although rare, clearly establish the magnitude of his achievements. He continued the work of Eudoxus on the motion of the planets, made accurate determinations of the lengths of the seasons, and constructed a seventy-six-year cycle to harmonize the solar and lunar years. This "Callippic period" was used by all later astronomers to record and date observations of heavenly phenomena.

The main biographical information on Callippus is found in Simplicius' commentary on Aristotle's *De caelo*. It states: "Callippus of Cyzicus, having studied with Polemarchus, Eudoxus' pupil, following him to Athens dwelt with Aristotle, correcting and completing, with Aristotle's help, the discoveries of Eudoxus" (Heiberg, ed., p. 49). It seems Callippus was in Athens in the decade before the death of Alexander in 323 B.C. Ptolemy says in the *Almagest* (Heiberg, ed., I, p. 206) that the year 50 of the first Callippic period was forty-four years after the death of Alexander, thus placing the beginning of the period in 330/329 B.C.

The discoveries of Eudoxus that Callippus corrected and completed are described by Aristotle in *Metaphysics*, Λ. In order to "save the appearances," as Plato had proposed, Eudoxus fixed each planet on a sphere that rotated on poles attached inside another sphere rotating in a different direction at a different rate, and this sphere in another until enough concentric spheres were so arranged and moving as to account by their combined uniform motions for the observed irregularities of the planet's motion. Aristotle tells us that Callippus found it necessary to add two spheres each for the sun and moon, and one for each of the other planets except Jupiter and Saturn. These changes in Eudoxus' system are perhaps testimony to Callippus' careful observations.

A papyrus, the so-called *Ars Eudoxi*, states that Callippus had determined the lengths of the seasons more accurately than Meton, giving them as ninety-four, ninety-two, eighty-nine, and ninety days, respectively, from the vernal equinox. The error is much less than that in Meton's determinations, made a century earlier.

From such calculations Callippus was led to see that Meton's nineteen-year cycle was a trifle too long. He therefore combined four nineteen-year periods into one cycle of seventy-six years and dropped one day from the period. Thus he brought the measure of the year closer to its true value, and his period became the standard for later astronomers. Many of the observations cited by Ptolemy are given in reference to the Callippic period. Hipparchus, in Ptolemy's references, seems to have used that period regularly.

Although the system of concentric spheres gave way to epicycles and eccentrics, Callippus' period became the standard for correlating observations accurately over many centuries, and thus contributed to the accuracy of later astronomical theories.

BIBLIOGRAPHY

Ancient sources are Aristotle, *Metaphysics,* Sir David Ross, ed. (Oxford, 1924), ref. in text to 1073b32–38; Simplicius, *Commentary on Aristotle's De caelo,* J. L. Heiberg, ed. (Berlin, 1894), vol. VII of *Commentaria in Aristotelem Graeca,* which gives details from Eudemus' history of the work of Eudoxus and Callippus; Theon of Smyrna, *Expositio rerum mathematicarum ad legendum Platonem utilium,* E. Hiller, ed. (Leipzig, 1878), which covers much the same ground as Simplicius in his section on the planets; Geminus, *Elementa astronomiae,* or *Isagoge,* C. Manitius, ed. (Leipzig, 1898), pp. 120–122, an account of the period of Callippus, with number of months, including intercalary, and days in the cycle; *Ars Eudoxi,* Blass, ed. (Kiel, 1887), which gives lengths of seasons as determined by Callippus; and Ptolemy, *Almagest,* J. L. Heiberg, ed. (Leipzig, 1898), and Taliaferro, tr., Great Books of the Western World, vol. XVI, which frequently mentions the Callippic period.

Modern works are Sir Thomas Heath, *Aristarchus of Samos* (Oxford, 1913), which contains the best English-language appreciation of Callippus and refers to the relevant modern literature on him; and Pauly-Wissowa, X, pt. 2, cols. 1662 f., which gives an exhaustive account of the seventy-six-year cycle, with full references to ancient sources and modern discussions.

JOHN S. KIEFFER

CALMETTE, ALBERT (*b.* Nice, France, 12 July 1863; *d.* Paris, France, 29 October 1933), *bacteriology.*

Calmette entered the Naval Medical Corps at the age of twenty and made several voyages. In 1889 he attended one of the first courses in microbiology given by Émile Roux at the Pasteur Institute in Paris and shortly afterward accepted, at Pasteur's request, the directorship of a research laboratory in Saigon. In that post he immediately revealed his great talent, particularly through his study of snake poisons.

Calmette returned to France in 1895 to become director of the Pasteur Institute that had just been founded in Lille; he held the post for nearly twenty-five years. During this period he continued his research on venoms and showed—just after Behring and Roux had perfected serotherapy—that with these substances one could give animals an immunity similar to that produced by an injection of microbic toxins.

In this regard Calmette appears to have been one of the very first to prepare antivenin serums. Equally interested in problems of public health, he studied the purification of sewage, entered the battle against ancylostomiasis, and made several improvements in fermentation techniques. While still in Lille, Calmette and his associate Camille Guérin made an intensive study of tuberculosis that led to the preparation of the antituberculosis vaccine BCG (Bacillus Calmette-Guérin). In 1917 he was appointed assistant director of the Pasteur Institute in Paris but was unable to leave northern France, which was occupied by the Germans (Mme. Calmette had been deported). At the Pasteur Institute in Paris he founded the tuberculosis department. His studies and those of his many students—among whom were Nègre and Boquet—drew the attention of the entire world.

The perfection of BCG remains Calmette's greatest contribution. This vaccine contains a strain of Koch's bovine bacillus that has been subjected to repeated culture on a nutritive medium mixed with bile. It is effective for both animals and man because tubercle bacilli of the bovine and human types are sufficiently related that they can produce cross-immunity. In addition, BCG is a safe vaccine because the live bacilli it contains remain at low virulence.

The first clinical trials were made in 1922. In 1930 the death of many children at Lübeck after a BCG vaccination caused a stir; however, it was quickly established that the sole cause—as the parties responsible admitted—was a serious error made during the preparation of the vaccine. Originally BCG was administered orally shortly after birth, but today, following Nègre and Bretey, it is administered by injection. This may cause local lesions, but they are always minor; and the insignificant annoyance they represent is largely compensated for by the high degree of protection.

Calmette died suddenly in the midst of his scientific work. The wish that he had expressed in his youth had been thus realized—to be able to devote himself to his work until his death.

BIBLIOGRAPHY

I. ORIGINAL WORKS. Calmette's writings were published, for the most part, in the *Annales de l'Institut Pasteur* (1892–1933). Two of his books are *Les venins, les animaux venimeux et la sérothérapie anti-venimeuse* (Paris, 1907); and *L'infection bacillaire et la tuberculose chez l'homme et chez les animaux* (Paris, 1920; 4th ed., rev. by A. Boquet and L. Nègre, Paris, 1936).

II. SECONDARY LITERATURE. On Calmette and his work, see Noel Bernard, *La vie et l'oeuvre d'Albert Calmette* (Paris,

1961); and Roger Kervran, *Albert Calmette et le BCG* (Paris, 1962).

ALBERT DELAUNAY

CAMERARIUS, RUDOLF JAKOB (*b*. Tübingen, Germany, 12 February 1665; *d*. Tübingen, 11 September 1721), *botany*.

For a detailed study of his life and work, see Supplement.

CAMPANELLA, TOMMASO (*b*. Stilo, Calabria, Italy, 5 September 1568; *d*. Paris, France, 21 May 1639), *philosophy, geology*.

For a detailed study of his life and work, see Supplement.

CAMPANI, GIUSEPPE (*b*. Castel San Felice, near Spoleto, Italy, 1635; *d*. Rome, Italy, 28 July 1715), *astronomy, microscopy, horology*.

Campani was a member of a peasant family and left his native village as a youth to obtain an education in Rome. While learning the new profession of lens grinding, he worked with his two brothers, Matteo Campani degli Alimeni, pastor of the church of San Tommaso in Via Parione, and Pier Tommaso, a clockmaker in the Vatican palaces, in the invention of a silent night clock. Presented to Pope Alexander VII in 1656, the clock brought Giuseppe into prominence; and he went on to produce lenses and telescopes whose superior workmanship earned him recognition from such patrons as Archduke Ferdinand II of Tuscany, Cardinal Francesco Barberini, and Giovanni Domenico Cassini at the Royal Observatory at Paris.

In 1663–1664 Campani invented the composite lens eyepiece and constructed a telescope with four lenses, consisting of a triple ocular and an objective. In 1664 he developed a lens-grinding machine lathe that could grind and polish lenses without first casting blanks in molds. With it he produced telescopic instruments of great focal lengths that were widely used.

Using his own instruments, Campani made significant astronomical observations of the satellites of Jupiter and of the rings of Saturn in 1664–1665 and published the results. Also interested in the microscope, he developed a screw-barrel type of instrument that could be made of metal or wood and permitted greater precision of adjustment than had previously been possible.

Campani was active in the production of lenses and optical instruments for more than fifty years. After his death the contents of his workshop in Rome were purchased by Pope Benedict XIV for the Istituto delle Scienze at Bologna. A substantial number of his clocks, telescopes, microscopes, and lenses are preserved in major collections, including the Istituto e Museo di Storia della Scienza in Florence, the Conservatoire National des Arts et Métiers and the Observatoire National in Paris, and the Medical Museum of the Armed Forces Institute of Pathology in Washington, D.C.

BIBLIOGRAPHY

I. ORIGINAL WORKS. Campani's writings are *Discorso di Giuseppe Campani Intorno a Suoi' muti Oriuoli, alle nuove Sfere Archimedee, e ad un'altra rarissima & Utilissima inventione di Personaggio cospicuo* (Rome, 1660); *Ragguaglio di Due Nuove Osservazioni Una Celeste in Ordine alla Stella di Saturno: e Terrestre l'altra in Ordine agli Strumenti Medesimi, co'quali si e Fatta l'una e l'altra Osservazione, dato al Serenissimo Principe Mattia di Toscana* (Rome, 1664); and *Lettera di Giuseppe Campani Intorno all'Ombre delle Stelle Medicee nel volto di Giove, ed altri nuovi Fenomeni Celesti scoperti co'suoi Occhiali, al Signor Gio: Domenico Cassini Primario Astronomo nell'inclito Studio di Bologna* (Rome, 1665).

II. SECONDARY LITERATURE. On Campani or his work, see S. A. Bedini, "Die Todesuhr," in *Uhrmacher und Goldschmied*, no. 12 (1956); and "The Optical Workshop of Giuseppe Campani," in *Journal of the History of Medicine and Allied Sciences*, **16**, no. 1 (1961), 18–38.

S. A. BEDINI

CAMPANUS OF NOVARA (*b*. [probably] Novara, Italy, first quarter of thirteenth century; *d*. Viterbo, Italy, 1296), *mathematics, astronomy*.

Our scanty information on the life of Campanus is derived from a few references in contemporary documents and writers supplemented by inferences from his own works. His full style is Magister Campanus Nouariensis (there is no authority for the forename Iohannes occasionally applied to him from the sixteenth century on). He refers to himself as Campanus Nouariensis.[1] This presumably indicates that Novara was his birthplace. Contemporary documents usually refer to him as Magister Campanus. The title Magister would in this period usually mean that the holder was a member of a faculty at a university; but we have no other evidence connecting Campanus with any university. His birthdate can be only approximately inferred from the fact that he was holding ecclesiastical office and writing major works in the late 1250's and early 1260's.

The earliest piece of biographical evidence is contained in one manuscript of Campanus' edition of Euclid, which seems to connect the work with Jacques

Pantaléon, patriarch of Jerusalem. This would date it between 1255 and 1261,[2] the years of Pantaléon's tenure of that position. The connection is a likely one, since when Pantaléon was elected to the papacy in 1261, becoming Urban IV, he took Campanus as one of his chaplains, as we learn from the latter's preface to his *Theorica planetarum,* which he dedicated to Urban. The letters of Urban reveal that he conferred other benefices on Campanus, including the rectorship of the Church of Savines in the diocese of Arles (reconferred 1263) and a canonicate in the cathedral of Toledo (1264). Urban died in 1264, but Campanus had another powerful patron in the person of Ottobono Fieschi, cardinal deacon of St. Adrian's and papal legate to England (later Pope Adrian V). Campanus was Ottobono's chaplain in 1263–1264, and it was probably through the influence of the cardinal that he became parson of Felmersham in Bedfordshire, England (he is attested as holding this benefice by a document of 1270).

Campanus' scientific reputation was already great enough by 1267 for Roger Bacon to name him as one of the four best contemporary mathematicians (although not one of the two "perfect").[3] Benjamin has suggested that Campanus may have accompanied Cardinal Ottobono when the latter was in England from 1265 to 1268, and there made Bacon's acquaintance. This hypothesis, although attractive, remains unproven, and indeed Campanus may never have left Italy, since the holding of benefices in absence was a common practice.

Later documents show that Campanus held a canonicate of Paris and was chaplain to popes Nicholas IV (1288–1292) and Boniface VIII (1294–1303). It is probable that he spent his later years at the convent of the Augustinian Friars at Viterbo. A letter of Boniface VIII, dated 17 September 1296, informs us that Campanus had just died at Viterbo; and in his will Campanus gave instructions for the construction of a chapel to St. Anne in the Church of the Holy Trinity there. The general impression one gets from these scattered pieces of information is of a life spent tranquilly cultivating the mathematical sciences under the patronage of powerful ecclesiastical figures, assisted by the income from a plurality of benefices that had made him a comparatively wealthy man at his death.

Of Campanus' numerous surviving works only one (the *Theorica planetarum*) exists in a modern critical edition. The unsatisfactory printed editions of others belong to the fifteenth and early sixteenth centuries; some exist only in manuscript. Much scholarly work remains to be done on these. Therefore the present survey, which is based mostly on arbitrarily selected manuscripts of the works in question, can be no more than an interim report on Campanus' output. It is not yet possible to set up even a relative system of dating of his writings as a preliminary to tracing his mathematical development. Certain individual works, however, can be quite closely dated. We have already seen that the Euclid can probably be dated to 1255–1259 and that the *Theorica* certainly belongs to the period of Urban IV's papacy (1261–1264). The planetary tables predate 1261, since they are cited in that year by Petrus Peregrinus of Maricourt. The *Computus maior* is dated to 1268 from a computation occurring in it. The *Sphere* is later than 1268, since it cites the *Computus.* An introductory letter to Simon of Genoa's *Clavis sanationis* belongs to the papacy of Nicholas IV (1288–1292), and the letter to Raner of Todi is datable to about the same period on internal evidence. But I am unable to draw any significant conclusions from the above dates, and therefore restrict myself to a description of the content of the surviving works.

The work by which Campanus is best known is his (Latin) edition of Euclid's *Elements* (in fifteen books, including the non-Euclidean books XIV and XV). This is the text of the *editio princeps* of Euclid, and it was reprinted at least thirteen more times in the fifteenth and sixteenth centuries. There are also very many manuscripts. It was undoubtedly the version in which Euclid was usually studied in the later Middle Ages. Yet surprisingly little is known of its general characteristics, and many false or misleading statements about it can be found in reference works. Thus it is frequently referred to as his "translation" of Euclid, and sometimes as his "commentary" on Euclid. Neither is accurate. We can state with confidence that Campanus did not possess the linguistic competence to translate from either Arabic or Greek: in none of his other works does he display any knowledge of either language (despite the fact that as chaplain of Urban IV he almost certainly came into close contact with the most competent Grecist of the time, William of Moerbeke, who was attached to the papal court as *poenitentiarius minor*). Campanus' edition is in fact a free reworking of an earlier translation or translations, at least one of which was made from the Arabic, as is evidenced by his retention of such terms as (h)elmuhaym (= Arabic al-mu'ayyan) for "rhombus" and (h)elmu(r)arife (= Arabic al-munḥarif) for "trapezium." He may also have used existing translations from the Greek. Determination of the exact relationship of Campanus' edition to its antecedents must await publication and examination of the numerous versions of Euclid that were current in the Middle Ages. But it already seems certain that he was heavily dependent on one or more of the versions attributed to Adelard of Bath (early twelfth century).

If one compares Campanus' edition with the standard edition of the original Greek by Heiberg, one finds that the content (although not the wording) is much the same, except for some rearrangement of material and numerous additions, both small and large, in the Campanus text. Some of these additions can be found in other Greek or Arabic versions, but a number appear to stem from Campanus himself, and indeed in some manuscripts some are explicitly headed *commentum Campani*. They are, however, a commentary only in the medieval sense of "proof of an enunciated theorem," since they consist of alternative proofs, corollaries, and additional theorems. These additions are not necessarily original contributions of Campanus.

Clagett has pointed out that two demonstrations added to book I, prop. 1, are taken from the commentary of al-Nairīzī (Anaritius, translated from the Arabic by Gerard of Cremona),[4] and has printed *in extenso* a theorem on the trisection of an angle, added to book IV in some manuscripts, that seems to be related to a solution of the problem in the *De triangulis* of Jordanus de Nemore.[5] Murdoch has noted Campanus' dependence on the same author's *Arithmetica*, particularly in book VII.[6] The whole question of Campanus' originality in the *Elements* has still to be answered. But at the very least he produced from existing materials a textbook of elementary geometry and arithmetic that was written in a readily comprehensible form and language (unlike many versions of Euclid then current). Its popularity for the next 300 years is attested by the large number of manuscripts and printed copies still extant.

No other work of pure mathematics can be definitely attributed to Campanus. Most of his other writings are concerned with astronomy. Of these the most influential was the *Theorica planetarum*. It is a description of the structure and dimensions of the universe according to the Ptolemaic theory, together with instructions for the construction of an instrument for finding the positions of the heavenly bodies at any given moment (such an instrument was later known as an equatorium). Campanus' main source was the *Almagest* of Ptolemy, which he must have studied closely (probably in Gerard of Cremona's translation from the Arabic). He reproduces accurately the geometrical models evolved by Ptolemy for the explanation of the apparent motions of the heavenly bodies but refers the reader to Ptolemy's own writings for their justification (i.e., the groundwork of theory and observation).

In the *Almagest* Ptolemy had given no absolute dimensions, except for the distances of sun and moon: for the other five planets he had given only relative parameters, i.e., the size of epicycle and eccentricity in terms of a standard deferent circle with a radius of sixty units. Campanus, however, gives in addition the absolute dimensions of the whole system (in both earth radii and "miles," where one "mile" = 4,000 cubits). The mathematical basis for this lies in the assumptions that the order of the planetary spheres is known and that there is no space wasted in the universe. Then the farthest distance from the central earth reached by a planet is equal to the nearest distance to the earth reached by the planet next in order above it. Given an absolute distance of the lowest body, the moon (which Campanus takes from the *Almagest*), one can then compute all the other absolute distances. The author of this ingenious idea is also Ptolemy, in his *Planetary Hypotheses*. Campanus derived it from the work of the ninth-century Arabic author al-Farghānī (Alfraganus), which he knew in the Latin translation of John of Seville. From the latter he also took other data, such as his figure for the size of the earth and the relative sizes of the bodies of the planets. But the computations are all his own and are carried out with meticulous (indeed absurd) accuracy. Campanus obviously delighted in long arithmetical calculations, and went so far as to determine the area of the sphere of the fixed stars in square "miles."

Among the most interesting features of the work are the parts describing the construction of an equatorium. This is the first such description known in Latin Europe. It seems improbable, however, that Campanus conceived the idea independently. Descriptions of equatoria were written nearly 200 years earlier in Islamic Spain by Ibn al-Samḥ and al-Zarqāl.[7] It seems likely that Campanus got the idea, if not the particular form, of his equatorium from some (as yet unknown) Latin translation of an Arabic work. The instrument he describes is the simplest possible, being merely a scale model of the Ptolemaic system for each planet (motion in latitude is neglected, and hence a two-dimensional model suffices). Thus in Figure 1, which depicts the model for an outer planet, the outer graduated circle represents the ecliptic, the next smaller the equant, and the smallest the planet's epicycle. The double and dotted lines represent parts of the instrument that can be freely rotated. The user arranges the instrument to imitate exactly the positions of the various parts of the planet's mechanism at the given moment according to Ptolemaic theory, and then stretches a thread attached to *D*, which represents the observer (earth), through the point representing the planet and reads off its position on the outer circle.

The disadvantage of such an instrument is that it would be extremely laborious to construct, bulky to

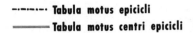

**INSTRVMENTVM EQVATIONIS
SATVRNI IOVIS MARTIS ET VENERIS**

FIGURE 1

transport, and clumsy to use. The earliest treatise on the equatorium after Campanus, that of John of Lignières (early fourteenth century), is an explicit attempt to improve on Campanus' instrument.[8] This and later works on the subject describe much more compact and sophisticated instruments, and preserved examples of equatoria from the Middle Ages are far superior to Campanus' crude model.[9] But we cannot doubt that Campanus' description was a major (although not the sole) influence toward the subsequent development of the equatorium.

The *Theorica*, although it omits Ptolemy's proofs, is nevertheless a highly technical work; and its influence seems to have been confined mostly to professional astronomers. Campanus later produced a more popular work on the *Sphere* (*Tractatus de spera*). This is a description of the universe in language intelligible to the layman, requiring no great geometrical skill, and with none of the precise details of the *Theorica*. It is similar in plan to the earlier works of the same name by Johannes de Sacrobosco and Robert Grosseteste, and shows no originality.[10]

Although the equatorium of the *Theorica* is explicitly intended for those who find operating with conventional astronomical tables too wearisome or diffi-

cult,[11] Campanus also describes in detail in that work how to use such tables. It is obvious that his examples are drawn from the Toledan Tables, an incongruous hodgepodge of tables carelessly extracted from the works of al-Khwārizmī, al-Battānī, and al-Zarqāl, probably translated by Gerard of Cremona and widely used in western Europe from the twelfth to the fourteenth centuries. This use of the Toledan Tables is not surprising, since Campanus had already produced an adaptation of them to Julian years, the Christian era, and the meridian of Novara (the originals are constructed for Arabic years, era of the Hegira, and the meridian of Toledo). The result is, naturally, no more satisfactory than the original, but the conversion is carried out fairly accurately. As we should also conclude from reading the relevant passages of the *Theorica*, Campanus was not merely able to use astronomical tables but understood the underlying structure of most of them, an uncommon accomplishment in his time.

Campanus' third major work besides the Euclid and the *Theorica* is his *Computus maior*. The "computus" or "compotus" is a form of literature of which literally hundreds of examples were composed in the Middle Ages. Its origins lie in the difficulties of early Christians in computing the date of Easter. Works were written explaining the rather complicated rules governing such computation (on which different doctrines were held at different times by various branches of the Church). Since the computation involved both a lunar and a solar calendar, it was natural for dissertations on the calendar to be included in such works and for a sketch of the astronomical basis to be added. Bede's *De temporum ratione* (written in 725) gave the computus a form that was to be widely imitated and is essentially that adopted by Campanus.

After giving a definition of "time," Campanus discusses the various subdivisions of time: day, hour, week, month, and year. This naturally involves a good deal of astronomical discussion, and it is this portion that distinguishes Campanus' treatise from other such works, since he is able to introduce the most "modern" astronomical doctrine. The lengthy chapter 10 ("On the Solar Year") is especially notable. It contains an extended (and slightly erroneous) description of the theory of trepidation (oscillation of the equinoxes with respect to the fixed stars) of Thābit ibn Qurra, and criticisms of Ptolemy and Robert Grosseteste. From chapter 13 on, the treatise is more strictly calendrical, leading up to the computation of Easter by various methods, including the use of tables. The work concludes with a calendar giving, besides the main feast days, the tables necessary for the calculations described in the text. This part, although providing little scope for originality, is written in Campanus' usual clear and orderly way, and the work as a whole is one of the most successful examples of its genre. Since Campanus himself refers to it as "my greater computus,"[12] it is probable that he is the author of the shortened version that is found in some manuscripts.

The interest in astronomical instruments that is apparent in the *Theorica* led Campanus to compose two other short works, on the quadrant and on the astrolabe. The first is an instrument for measuring angles at a distance, and its principal use was to measure the elevation of the sun, although it could be applied in many other ways, e.g., to find the angular elevation of the top of a tower in order to compute its height. The second is a schematic representation of the celestial sphere on a plane by means of stereographic projection, and was used for the solution of problems involving the rising, setting, and transits of stars or parts of the celestial sphere that would otherwise have required spherical trigonometry. Campanus' *De quadrante* does little more than explain how to solve certain types of astronomical and mensurational problems by means of the quadrant, and is very similar to other medieval works on the subject, such as that attributed to Robert the Englishman (dated 1276). The slight work on the astrolabe, however, is unusual in that it is neither a description of the construction of the astrolabe nor of its use, but rather a series of theoretical geometrical problems connected with the astrolabe as an example of stereographic projection. I am not convinced that its attribution to Campanus is correct.

In later times Campanus had a considerable reputation as an astrologer.[13] Although the details of his prowess may be legendary, there is no doubt that like every highly educated man of his time he was well versed in astrological doctrine (this is obvious from, e.g., the *Computus* or the *Quadrant*). It seems that Campanus also wrote specifically astrological works, for a method of his for dividing the heavens into the twelve "houses" is mentioned by Regiomontanus and others, but no such work survives that can definitely be assigned to him. A long anonymous work on astrology that is found in three manuscripts is attributed to Campanus in the margin of one, but its authorship still awaits investigation.

A number of other mathematical and astronomical treatises and commentaries are occasionally attributed to Campanus in manuscripts or early printed editions. Of these the only one that is certainly his is a letter addressed to the Dominican friar Raner of Todi in response to a query of the latter on a point in Cam-

panus' *Computus*. Others are certainly not his, e.g., a work on the instrument known as the solid sphere, and the so-called "Quadrature of the Circle."[14] A verdict on the others must await detailed examination of their contents, vocabulary, and style.

Whatever the result of such an examination, it is unlikely to have much effect on our judgment on Campanus as a man of science. He was a highly competent mathematician for his time and was thoroughly acquainted with the most up-to-date works on the subject, including the available translations from the Arabic. Thus he had read and understood Ptolemy's *Almagest,* and was able in his *Theorica* not only to summarize its conclusions but even on occasion to apply one of its techniques in another context.[15] Similarly, he was able to produce from the often obscure earlier translations of and commentaries on Euclid a version embodying their mathematical content in a more acceptable form. He had a gift for clear and plain exposition. But although he had a good understanding of his material and made few errors, he can hardly be called an original or creative scientist. His philosophical position was an unreflective Aristotelianism; his mathematics and cosmology were equally conventional for his time. His talent was for presenting the work of others in a generally intelligible form.

As such, Campanus was a writer of considerable influence. His Euclid was almost the canonical version until the sixteenth century, when it was gradually superseded by translations made directly from the Greek.[16] The continuing popularity of his *Computus* and *Sphere* is attested by their being printed several times in the sixteenth century. The *Theorica* was never printed, probably because, unlike the others, it was not a popular work but a technical one that would appeal only to those with a professional interest. It was nonetheless influential: this is shown both by the large number of surviving manuscripts and by the references to it in astronomical works of the fourteenth and fifteenth centuries. By the sixteenth century, however, it seems to have been little read (one may suspect that its role as a summary of the *Almagest* had been taken over by the Peurbach-Regiomontanus epitome of Ptolemy's work). Perhaps its greatest single contribution was the popularization of the idea of the planetary equatorium (the introduction of this is also Campanus' strongest claim to originality, but, as stated above, I believe that he owes the idea to some hitherto undiscovered Arabo-Latin source). The history of that instrument after Campanus is a good illustration of the technical ingenuity of the astronomy of the later Middle Ages and early Renaissance.

NOTES

1. E.g., *Theorica planetarum*, §1.
2. The earliest known manuscript of the work is dated 11 May 1259 (Murdoch, *Revue de synthèse*, 3rd ser., **89** [1968], 73, n. 18).
3. Bacon, *Opus tertium*, XI, 35.
4. *Isis*, **44** (1953), 29, n. 31 (4).
5. Clagett, *Archimedes*, I, 678–681.
6. *Revue de synthèse*, **89** (1968), 80, n. 41; 82, n. 53; 89, n. 84; 92, n. 100.
7. See Price, *Equatorie*, pp. 120–123.
8. *Ibid.*, p. 188.
9. See, e.g., the photograph of the equatorium at Merton College, Oxford, *ibid.*, frontispiece.
10. For a detailed outline of the contents, see Thorndike, *Sphere*, pp. 26–28.
11. *Theorica*, §§59–61.
12. "In compoto nostro maiori," MS Bibl. Naz., Conv. Soppr. J X 40, f. 47v.
13. See especially the work of Symon de Phares (late fifteenth century) on famous astrologers, Wickersheimer, ed., pp. 167–168.
14. A text of the "Quadratura circuli" has been published by Clagett, in *Archimedes*, I, 581–609, with an introduction in which the editor gives his reasons for doubting the authorship of Campanus.
15. See my note on §1280 of the *Theorica*.
16. The first version from the Greek to be printed was that of Zambertus (Venice, 1505); but Campanus' version continued to be printed for another fifty years.

BIBLIOGRAPHY

On Campanus' life, the unsatisfactory biographies in standard reference works are superseded by Benjamin's account in the Benjamin and Toomer ed. of the *Theorica planetarum*, intro., ch. 1. This contains references to all known sources. My summary is drawn from it. Roger Bacon mentions Campanus in *Fr. Rogeri Bacon opera quaedam hactenus inedita*, J. S. Brewer, ed., I (*Opus tertium*), in the "Rolls Series" (London, 1859), cap. 11, 35. The muddled account of Campanus by Symon de Phares is published in *Recueil des plus célèbres astrologues et quelques hommes faict par Symon de Phares du temps de Charles VIII*, E. Wickersheimer, ed. (Paris, 1929), pp. 167–168.

A full bibliography of works by or attributed to Campanus is given by Benjamin in the Benjamin and Toomer ed. of the *Theorica*, intro., ch. 2, with details of all printed eds. of such works in ch. 2, app. 1 (the evidence for the dates of writing is in ch. 1). The version of Euclid's *Elements* was first printed at Venice in 1482 by Erhardus Ratdolt. For later eds. see, besides Benjamin, Charles Thomas-Stanford, *Early Editions of Euclid's Elements* (London, 1926). The standard ed. of the Greek text is by J. L. Heiberg, *Euclidis opera omnia*, I–V (Leipzig, 1883–1888). No adequate study has been made of the relationship of Campanus' version to earlier medieval translations, but John E. Murdoch, "The Medieval Euclid: Salient Aspects of the Translations of the ELEMENTS by Adelard of Bath and Campanus of Novara," in *Revue de synthèse*, 3rd ser., **89**

(1968), 67–94, is valuable, particularly for his demonstration of the didactic intent of many of the changes and additions made by Campanus. Much relevant material will also be found in Marshall Clagett, "The Medieval Latin Translations from the Arabic of the Elements of Euclid, with Special Emphasis on the Versions of Adelard of Bath," in *Isis*, **44** (1953), 16–41, and some in the same author's *Archimedes in the Middle Ages*, I, *The Arabo-Latin Tradition* (Madison, Wis., 1964); see general index under "Campanus of Novara."

The *Theorica planetarum* has been critically edited on the basis of almost all known manuscripts, with translation, commentary, and introduction, by Francis S. Benjamin, Jr., and G. J. Toomer (Madison, Wis., in press). For an account of the Ptolemaic system, see the intro. to that ed., ch. 4. The Greek text of the *Almagest* was edited by J. L. Heiberg, *Claudii Ptolemaei opera quae exstant omnia*, I, *Syntaxis mathematica*, 2 vols. (Leipzig, 1898–1903). Gerard of Cremona's translation from the Arabic was printed at Venice by Petrus Liechtenstein in 1515 as *Almagestum Cl. Ptolemaei Pheludiensis Alexandrini . . . opus ingens ac nobile*. Ptolemy's *Planetary Hypotheses* are printed in J. L. Heiberg, ed., *Claudii Ptolemaei opera quae exstant omnia*, II, *Opera astronomica minora* (Leipzig, 1907), 70–145; but the most important passage in connection with Campanus is found only in Bernard R. Goldstein, "The Arabic Version of Ptolemy's Planetary Hypotheses," in *Transactions of the American Philosophical Society*, n.s. **57**, pt. 4 (1967). The translation of al-Farghānī used by Campanus was published in multigraph by Francis J. Carmody, *Alfragani differentie in quibusdam collectis scientie astrorum* (Berkeley, 1943). The origin and history of the system described above for determining the absolute dimensions of the universe are studied by Noel Swerdlow, *Ptolemy's Theory of the Distances and Sizes of the Planets*, Ph.D. diss. (Yale, 1968). On the history of the equatorium see D. J. Price, *The Equatorie of the Planetis* (Cambridge, 1955), esp. pp. 119–133; for the later Middle Ages and the Renaissance, Emmanuel Poulle, "L'équatoire de la Renaissance," in *Le soleil à la Renaissance, science et mythes, colloque de Bruxelles, avril 1963* (Brussels, 1965), pp. 129–148; and, for further bibliography, Poulle's "L'équatoire de Guillaume Gilliszoon de Wissekerke," in *Physis*, **3** (1961), 223–251, and "Un équatoire de Franciscus Sarzosius," *ibid.*, **5** (1963), 43–64, written with Francis Maddison.

Campanus' *Sphere* was printed by L. A. de Giunta at Venice in 1518, and three more times up to 1557 (see *Theorica*, intro., ch. 2, app. 1). Similar works by Sacrobosco and others can be found in the modern ed. of Lynn Thorndike, *The Sphere of Sacrobosco and Its Commentators* (Chicago, 1949). Grosseteste's *Sphere* is printed in *Die philosophischen Werke des Robert Grosseteste, Bischofs von Lincoln*, Ludwig Baur, ed., *Beiträge zur Geschichte der Philosophie des Mittelalters*, IX (Münster, 1912), 10–32. There is no printed ed. of Campanus' astronomical tables, nor of their parent Toledan Tables. An extensive analysis of the latter is given by G. J. Toomer in *Osiris*, **15** (1968), 5–174. The *Computus maior* was printed twice at Venice in 1518, by L. A. de Giunta and by Octavianus Scotus. On the technical content of the medieval computus, the best treatment is W. E. van Wijk, *Le nombre d'or* (The Hague, 1936). Bede's *De temporum ratione,* edited by Charles W. Jones as *Bedae opera de temporibus* (Cambridge, Mass., 1943), contains on pp. 3 ff. a useful account of the historical development of the computus. Campanus' works on the quadrant and astrolabe have never been printed. For the former, compare "Le traité du quadrant de Maître Robert Anglès . . . texte latin et ancienne traduction grecque publié par M. Paul Tannéry," in *Notices et extraits des manuscrits de la Bibliothèque nationale,* **35** (1897), 561–640, reprinted in Tannéry's *Mémoires scientifiques* (Paris, 1922), pp. 118–197. The work on astrology is attributed to Campanus only in MS Vat. Pal. 1363, where it is found on ff. 66r–88r. The letter to Raner of Todi is in only one MS: Florence, Biblioteca Nazionale Centrale J X 40, ff. 46v–56r. For other works that may or may not be by Campanus, see Benjamin in the Benjamin and Toomer ed. of the *Theorica,* intro., ch. 2.

No detailed study of the influence of Campanus' work exists. Some information will be found in the Benjamin and Toomer ed. of the *Theorica,* intro., ch. 8. The Peurbach-Regiomontanus epitome of the *Almagest* was printed at Venice in 1496 (colophon "Explicit Magne Compositionis Astronomicon Epitoma Johannis de Regio monte"), and several times subsequently.

G. J. Toomer

CAMPBELL, DOUGLAS HOUGHTON (*b.* Detroit, Michigan, 16 December 1859; *d.* Palo Alto, California, 24 February 1953), *botany.*

Campbell was the fifth of the six children of James V. Campbell, one of the first three justices of the Michigan supreme court. The first three decades of his life were spent in Michigan, first at Detroit High School (until 1878), then at the University of Michigan, Ann Arbor (1878–1882), where he received his master's degree. He then returned to Detroit High School to teach botany (1882–1886). He received his doctorate from Michigan in 1886 on the basis of studies carried out on the structure and fertilization of fern gametophytes. His working habits at this time set the pattern for the rest of his life. He taught in the mornings, researched in the afternoons and evenings, and by living at home accumulated enough money to support himself for two years of research in Germany (1886 to 1888). After working in Berlin, Bonn, and Tübingen, he returned to the United States and was appointed professor of botany at the University of Indiana (1888–1891). His second and final appointment was to the chair of botany (1891–1925) at the newly founded Stanford University. In the remaining twenty-seven years he lived in his own house on the campus as an emeritus professor.

Campbell's first paper appeared in 1881 when he was in his early twenties, his last in 1947, when he was nearly ninety. He was an editor of the *American Naturalist,* a member of the American Academy of Sciences (from 1910), and president of both the Botanical Society of America (1913) and the Pacific section of the American Association for the Advancement of Science in 1930. Campbell never married.

Botany in the 1880's was dominated by the German school of cytologists led by E. A. Strasburger. The cell theory of M. J. Schleiden and T. Schwann had been corrected and the role of the nucleus and the structure of the nuclear threads (chromosomes) were being explored, chiefly by W. Flemming. The relevance of the cell theory to plant embryology and phylogeny had been brilliantly explored by W. Hofmeister. By detailed and far-ranging studies of the life histories of the lower plants he was able to dispose of the erroneous analogies which had been made by men like Schleiden between the spores of fungi, algae, liverworts, mosses, and ferns and the pollen grains of flowering plants. Instead Hofmeister identified the true analogies between lower and higher plants (cryptogams and phanerogams) with the sexual organs of the gametophyte generation. Campbell read Hofmeister's *The Higher Cryptogamia* (1862) in his student days at Michigan. It fired him with enthusiasm for the study of the Cryptogamia.

Filling in the details of Hofmeister's scheme and constructing a phylogenetic tree for the plant kingdom occupied the attention of a growing number of botanists and cytologists during the latter part of the nineteenth century and the first two decades of the twentieth. To this program of research Campbell contributed both during his two years in Germany and afterward. Along with F. O. Bower, the British expert on plant phylogeny, Campbell contributed to the theoretical discussions on the subject of the origins of a land flora. His book *The Structure and Development of the Mosses and Ferns* (1895) was a landmark in the history of the subject and supplied the first systematic review since Hofmeister's classic.

Campbell's textbook was used in virtually all university botany departments so that before the turn of the century its author was universally known. His reputation was further enhanced by his *Lectures on the Evolution of Plants* (1899). These textbooks combined original findings with summaries of published work in a form suitable for the student. The exposition is clear and is all directed to one purpose—the clarification of the phylogenetic relationships of the lower plants. Campbell's true love of natural history gave to these books a welcome touch, and his deep knowledge of ecological factors enabled him to see evolutionary significance in many obscure structures.

As a schoolboy, Campbell had been greatly impressed by A. R. Wallace's *Malay Archipelago* (1869), and during his frequent journeys to the tropics, especially to Hawaii and to the mountain laboratory of the botanical garden at Buitenzorg, Java, he gathered valuable data on the growth and structure of tropical members of the Cryptogamia, especially of the *Anthocerotaceae*. From his time in Strasburger's laboratory he was made aware of the exciting discovery of the details of cell division, and in Pfeffer's laboratory he developed the use of vital stains. In later publications he drew the attention of American botanists to the ease with which the details of cell division can be demonstrated, but he failed at this time to apply his knowledge of chromosome mechanics to the explanation of alternation of generation. His style of writing was inferior to that of his friend Bower, which may partly explain why to all but a few Campbell is a forgotten name, a man who lingered on while the fashion of the science he knew and loved changed almost beyond recognition. In part this is because he did not broaden his work in the direction of palaeobotany to anything like the extent that Bower did. Whereas Bower wrote for the advanced student, Campbell wrote for the beginning student. Some of his popular writings—*An Outline of Plant Geography,* for instance—were judged very superficial.

Campbell's original papers, however, were anything but superficial. In a highly competitive field, to which a great number of botanists contributed, Campbell supplied numerous details gained from very careful study of material. While working in German laboratories, Campbell was introduced to the techniques of microtome sectioning of tissue embedded in paraffin wax. He quickly wrote up an account of this technique as applied to botanical material and published it in the *Botanical Gazette* for the benefit of American botanists. In his own hands it yielded him a most important find—the discovery of the precise manner of formation of the archegonia and antheridia in the eusporangiate ferns. On the factual side this led him to make many contributions to the knowledge of the details of gametophyte structure and habit from practical advice about the germination of spores to details of the division behavior of the apical cell. His view that the sporophyte of eusporangiate ferns is derived from an adventitious bud arising endogenously within the primary root is not now accepted, and his denial of the application of the stelar theory to the vascular organization of the Ophioglossaceae is questioned.

Campbell's most important contribution was to link the ferns with the liverworts by way of *Anthoceros.* When his book *The Structure and Development of Mosses and Ferns* appeared, popular opinion, which included Bower's, had it that the leptosporangiate ferns were the more primitive and were descended directly from the algae and that the mosses and liverworts constitute a parallel evolution. Campbell was the first to suggest that the eusporangiate pteridophytes are the more primitive and have originated from the strange hepatic group Anthocerotales and to give as evidence the embedded condition of the antheridia and archegonia in *Anthoceros* and in eusporangiate ferns.

His analogy between the germ-cell portion of the eusporangiate antheridium and the whole of the anthocerotean antheridium, and between the eusporangiate jacket layer and the anthocerotean antheridial chamber is now the accepted view. Campbell went on to show that the sporophyte of *Anthoceros* can sustain itself for several months when separated from the gametophyte, that its chlorophyll-containing tissue is capable of photosynthesis to a limited extent, and that given a small supply of carbohydrates it can grow and sporulate (1917). In 1924 he described nine-month-old, sixteen-centimeter sporophytes of *Anthoceros,* showing differentiation and growing on withered gametophytes in the San Jose Canyon. These very important discoveries led Campbell to give increasing support over the years to the antithetic theory of the origin of the sporophyte generation. This theory, first outlined by L. Celavosky in 1874 and developed by Bower in 1890, regards the sporophyte of plants higher than mosses and liverworts as a new structure intercalated into the life cycle. In the course of evolution, it was suggested, the fusion nucleus formed from the antherozoid and oogonium, instead of rapidly producing a sporogonium filled with spores, as in mosses and liverworts, has progressively delayed the spore-forming process and has rendered more and more of the sporogonial tissue nonspore-forming.

According to the rival homologous theory, the sporophyte is a direct modification of the gametophyte, and there is no special significance in that the spore-bearing generation always results from the product of sexual reproduction—the zygote. This view of the origin of the sporophyte was favored by J. M. Coulter but opposed by Campbell (1903). Today the antithetic theory is firmly established on the basis of the chromosome cycle.

Campbell's advocacy of the antithetic theory went hand-in-hand with his belief in the anthocerotean origin of the eusporangiate pteridophytes. Little credence was given to the latter suggestion until R. Kidston and W. H. Lang discovered well-preserved fossil representatives of the psilophytons of the early Devonian in 1917. Before that time the gap between a rootless, leafless sporophyte of *Anthoceros* and the root- and leaf-bearing, free-living sporophyte of the pteridophytes seemed too large. In 1895 Campbell had referred to the strange fossil psilophyton which J. W. Dawson had discovered as early as 1859, the structure of which suggested a leafless and rootless sporophyte organization, but the incomplete nature of the specimen and the doubts thrown on it by H. Solms-Laubach, which Campbell appeared to accept, prevented the idea of a link between *Anthoceros* and the pteridophytes being established until better specimens became available (after 1917).

By the early 1920's, it had become clear that plants lacking roots and leaves did exist as independent sporophytes in the early Devonian period. Campbell's suggestion of the anthocerotean origin of the eusporangiate pteridophytes is now generally accepted, and it is for this idea and for his associated belief in the primitive character of the Eusporangiatae that he is remembered.

BIBLIOGRAPHY

I. ORIGINAL WORKS. Campbell published more than 150 papers and reviews, seven books, one monograph, and a few pamphlets. Most of these are listed in Gilbert Smith's biography (see below). Campbell's books are *Elements of Structural and Systematic Botany for High Schools and Elementary College Courses* (Boston, 1890); *The Structure and Development of Mosses and Ferns (Archegoniatae)* (New York–London, 1895; 2nd. ed., 1905; 3rd. ed., 1918); *Lectures on the Evolution of Plants* (New York–London, 1899); *A University Textbook of Botany* (New York–London, 1902); *An Outline of Plant Geography* (New York, 1926); and *The Evolution of Land Plants (Embryophyta)* (Stanford–London, 1940).

The development of Campbell's views on the origin of the ferns can be traced in "On the Affinities of the Filicineae," in *Botanical Gazette,* **15** (1890), 1–7; "Notes on the Archegonium of Ferns," in *Bulletin of the Torrey Botanical Club,* **18** (1891), 16; "On the Relationships of the Archegoniata," in *Botanical Gazette,* **16** (1891), 323–333; and "The Origin of the Sexual Organs of the Pteridophyta," in *Botanical Gazette,* **20** (1895), 76–78. His work on the eusporangiate ferns is beautifully presented in "The Eusporangiatae. The Comparative Morphology of the Ophioglossaceae and Marattiaceae," in *Carnegie Institution of Washington Publications,* no. 140 (1911). His important studies of *Anthoceros* can be found in "Studies on Some Javanese Anthocerotaceae, I," in *Annals of Botany,* **21** (1907), 467–486; pt. II, *ibid.,* **22** (1908), 91–102; "Growth of Isolated Sporophytes of *Anthoceros,*" in *Proceedings of the National Academy of Sciences,* **3** (1917), 494–496; and "A Remarkable Development of the Sporophyte in *Anthoceros Fusiformis Ausr.,*" in *Annals of Botany,* **38** (1924), 473–481.

II. SECONDARY LITERATURE. Neither the general scientific periodicals nor the specialist botanical journals appear to have carried obituary notices of Campbell, but biographical information can be found in the essay by G. M. Smith, "Douglas Houghton Campbell," in *Biographical Memoirs, National Academy of Sciences,* **29** (1956), 45–63, with portrait and bibliography.

For an authoritative discussion of Campbell's contributions, see Gilbert M. Smith, *Cryptogamic Botany,* vol. II of *Bryophytes and Pteridophytes* (New York–London, 1938).

ROBERT OLBY

CAMPBELL, NORMAN ROBERT (*b.* Colgrain, Dumbarton, Scotland, 1880; *d.* Nottingham, England, 18 May 1949), *physics, philosophy of science.*

Campbell was educated at Eton and became a scholar of Trinity College, Cambridge. In 1904 he became a fellow of Trinity College, where he worked mainly as a student of J. J. Thomson on the ionization of gases in closed vessels. In addition to performing successful experimental research in this field, he established, in collaboration with A. Wood, the radioactivity of potassium.

Campbell was then appointed to the Cavendish research fellowship at Leeds where, in 1913, he became honorary fellow for research in physics. While at Leeds he continued research on the ionization of gases by charged particles and on secondary radiation. After the start of World War I, he joined the research staff of the National Physical Laboratories, working on problems concerning the mechanism of spark discharge in plugs for internal combustion engines. Reports of this work were submitted to the Advisory Committee for Aeronautics.

In 1919 Campbell joined the research laboratories of the General Electric Company, Ltd. As well as continuing his work on electrical discharge in gases, he worked on photoelectric photometry and color matching, the standardization and theory of photoelectric cells, statistical problems, the adjustment of observations, and the production of "noise" in thermionic valves and circuits. He published nine books and over eighty papers. In 1912 he married Edith Sowerbutts; the couple adopted two children. The last fifteen years of his life were spent in retirement—in ill health.

Although Campbell distinguished himself as an experimental physicist, he devoted himself to careful study of both the theoretical and philosophical aspects of his science. Profoundly influenced by J. J. Thomson and the ideas of Faraday and Maxwell, he was basically a proponent of a mechanical view of physics. His reasons for this inclination were sophisticated and based not only on the "intrinsic interest" of mechanical explanations, natural enough for an associate of Thomson, but also on profound deliberations on the nature of scientific knowledge. His major work was done at a time when physical science was changing in a very radical way and his continuing examination of this process can be seen in the progressive editions of his *Modern Electrical Theory*.

Although Campbell saw himself as an experimental physicist, it is as a philosopher of science that he is best known. Insisting that a basic understanding of the nature of science was essential—perhaps most of all to an experimentalist—he turned to the writings of Mach, Duhem, Kelvin, Tait, Helmholtz, Stallo, and Poincaré, most of whom were physicists who concerned themselves with the broader nature of science as well as its particularities. He felt that only practitioners of science should undertake such an analysis and remarked of Mill, a nonscientist, that he "never knew a law when he saw one" (1) and that Mill's views were often suggestive if only "because they are erroneous" (2).

The most notable of Campbell's theses for the theory of science was his strongly urged distinction of laws and theories. He saw this as pivotal for even a beginning understanding of the nature and status of scientific propositions.

Laws, he asserted, are propositions that can be established by experiment and observation, which does not mean that there are, in the main, overly simple relations between these laws and the fundamental propositions concerning our naïve observations. Rather, scientific laws depend, for both their justification and their "significance," on other sets of laws as well as complex collections of such simple and immediate judgments of sensations. Thus, if we use a variety of different means to determine the extension in a Hooke's law experiment—for example, optical lever, interference apparatus, micrometer screw gauge, and so forth—we depend on the assumption that certain laws hold, and in particular, that all the methods yield the same result. Indeed Campbell went further than claiming that the proof of such a law depends on the truth of other laws; the meanings of the terms involved—"extension," "force"— require that certain of those laws hold. If we say anything about electrical resistance, Campbell insisted, we assume that Ohm's law is true. Were it not true, "resistance" would be "without any meaning." Terms that depend in this way for their meaning on the truth of laws are termed "concepts." Campbell was at pains to emphasize that almost all the laws of physics state relations between such concepts, and not between judgments of simple sensations. He was not unaware that there are fundamental laws, but he

was reluctant to specify exactly how they are related to fundamental judgments, for it is at this level that, almost paradoxically, the concepts in question are too familiar for ease of analysis.

With the basis of his account of the nature of scientific laws, Campbell elaborated his major thesis of the structure of theories and their distinction from laws. A theory is a connected set of propositions which fall into two different categories. The basis of the theory, and that which basically establishes its identity, is the set of propositions, termed "the hypothesis," which concern some collection of "ideas" characteristic of that theory. In isolation, this hypothesis is incapable of either proof or disproof—it is, in a sense, merely arbitrary. The second group of the propositions that constitute a theory Campbell termed "the dictionary." These propositions assert the relation of terms of the hypothesis ("hypothetical ideas") with the terms of scientific laws ("concepts") whose truth or falsity is determinate. Campbell considered a fabricated, trivial example of such a combination of hypothesis and dictionary. The hypothesis states that a and b are constants for all values of the independent variables u, v, w and that $c = d$, where c and d are dependent variables. The dictionary asserts for this meager hypothesis that $(c^2 + d^2)a = R$, where R is the resistance of a particular piece of metal. Further, the dictionary states that $cd/b = T$, where T is the temperature of the same piece. It is an immediate consequence of the hypothesis that $(c^2 + d^2)a/(cd/b) = 2ab$, which is a constant. From this conclusion and the dictionary it may be concluded that the resistance of the metal is directly proportional to the temperature. This statement asserts the relation of observational "concepts" in the sense referred to before and is the law which the theory explains—at least in the provisional sense that it has the law as a consequence.

A paradigm example of a highly significant physical theory is the dynamical theory of gases by means of which Boyle's law and Gay-Lussac's law may be explained. With his thesis of the nature of theories, Campbell reconstructed in almost perfect accord this theory and the way in which it related to the physical laws mentioned. He emphasized that this theory does not exhibit the artificiality of his explicative example and offered, in contrast with it, a genuine mode of scientific explanation. The difference, for Campbell, rested on his conclusion that the propositions of the hypothesis of the dynamical theory of gases display an analogy that the corresponding propositions of the other theory do not display. This analogy is the third essential constituent of any physical theory. Although Campbell equivocated about the specific nature of the analogy, he insisted that in the case of the dynamical theory of gases statements of the hypothesis take such a form that, examining a system of particles in a box, etc., we would find that such particles obey physical laws analogous to those principles. Provided that we associate the appropriate measurable physical concepts with the various symbols in the statements of the hypothesis, we would be able observationally to establish the truth of those propositions as laws.

Campbell emphasized the role that such an analogy has in theory construction by indicating that the propositions of the dictionary are suggested by it. One particular term in the hypothesis is identified with the pressure since in what would be our lawlike (in Campbell's strict sense of "law") analogue, it would represent the average pressure on the walls of the observed box. He noted that, for the most part, philosophers of science have misunderstood the role of analogy with respect to hypothesis and represented it as an "aid," in a plainly heuristic sense, to the construction of theories serving only a "suggestive" function (3). Campbell, however, insisted that rather than being such a merely heuristic and basically dispensable aid it is utterly essential to theories.

Indeed, theories would be without any value if such analogies were absent. Physical science is not purely logical, and physicists cannot rest content with "a set of propositions all true and logically connected but characterized by no other feature" (4). Campbell insisted that frequently the analogy is the "greatest hindrance" to the establishment of theories rather than being an aid in that process. It must be recognized that theories such as that in his explicative example are trivial not only in the sense alluded to but also in the ease with which they may be constructed. To construct, on the other hand, a theory that exhibits a significant analogy is a creative act of major importance.

Having developed his analysis of the nature of theories and having shown its applicability to at least a paradigm example based on mechanical analogy, Campbell faced the criticism of those theoreticians in the tradition of Stallo and Mach who militated against not only the essential role of analogies in theories but even against their desirability. The major problem for Campbell's reconstruction was to give an account of a paradigm example of those "mathematical theories" which the opposing school of thinkers cited as exemplars. Fourier's theory of heat conduction supplied the basis for Campbell's reply.

Campbell noted two important differences between the theory of gases and Fourier's theory. Although it was clear that both theories exhibited the hypotheses and dictionaries so basic to Campbell's the-

sis, they diverged in the following respects. In the first place, it seemed that every idea in the hypothesis of Fourier's theory was related directly by means of the dictionary to a corresponding concept; while in the theory of gases only functions of those ideas occurred in the dictionary. In the second place, it seemed to be the case that any analogy of the sort pertinent to the theory of gases was absent. Campbell's views on this second difference were equivocal. He vacillated between countenancing the total absence of any analogy and maintaining that there was analogy but of a rather different kind from that exhibited in the theory of gases. It is clear that Campbell wished to find some basis for the claim that analogy had some essential role in theories such as Fourier's. If it were to be conceded that there was no analogy at all, no distinction could be made—at least in Campbellian terms of reference—between Fourier's theory and Campbell's trivial explicative example; and, consequently, no grounds could be given for claiming that Fourier's theory provided a significant scientific explanation.

Campbell offered a number of views on what form the analogy pertinent to Fourier's theory might take. First, if there is an analogy, he claimed, it is between the propositions of the hypothesis and the laws which the theory is to explain. For the theory of gases, on the other hand, the analogy is between the hypothetical propositions and a set of laws which are found to be true, but which are distinct from those to be explained. Second, regarding theories of the type of Fourier's as "generalizations" of certain experimental laws, he asserted that it is not the case that any generalization will be adequate. The basic constraint which prohibits such license is the requirement that only the simplest generalizations be acceptable. Thus it is that "just as it is the analogy which gives its value" to the theories like the theory of gases "so it is the simplicity which gives its value" to the theories of the mathematical type. Despite the fact that only a few pages earlier he insisted that "*some* analogy is essential" to Fourier's theory, Campbell eventually concluded that "it may be true" that mathematical theories "are not characterized by the analogy and do not derive their value from it." It is, he continued, "characterized by a feature which is as personal and arbitrary as the analogy." It is from this element, a certain "intellectual simplicity . . . and ease with which laws may be brought within the same generalization," that it derives its value.

"Value" and "explanatory power" were almost interchangeable for Campbell. This was so precisely because the primary object of theories is the explanation of laws. Explanation is, in Campbellian terms, the substitution of more satisfactory for less satisfactory ideas. The intrinsic unsatisfactoriness of ideas may be due to confusion or complexity, or, alternatively, to a lack of familiarity. One kind of explanation of laws is by laws. The concepts involved are more satisfactory in the sense of being simpler, which in turn is a function of the generality of the explaining laws. The explanation of laws by laws, although a significant scientific process in the ordering of phenomena, is not as basic as the explanation of laws by theories. It is really only with this latter kind of explanation, Campbell professed, that genuine intellectual satisfaction is achieved.

Mathematical theories like that of Fourier do indeed give intellectual satisfaction in a way that extends beyond that of explanation by laws. It should be noted, however, that the almost indefinable counterpart of analogy in mathematical theories is just that elegant synthesis of simplicity and generality which Campbell identified with the explanatory power of mere laws. It is not without reason that the role of mathematical theories "is hardly different from that played by laws." Theories similar to the theory of gases, exhibiting analogy in the stronger sense which Campbell gave, have a value which is a function of the familiarity of the analogy. This yields a kind of explanation of laws which is distinct from the already mentioned type. Laws could not provide explanations comparable in kind.

The explanatory power of a theory is dependent on the "intrinsic interest" which it may have for a particular scientist or school of scientists. Campbell confessed that in admitting this he relativized much of the debate between the protagonists of mathematical theories and those who insist on the strong sense of analogy to considerations of a subjective kind, although not totally. Campbell viewed as mistaken the argument that mathematical theories are less likely to lead the scientist into error, on the grounds that they are, as Mach put it, "purely phenomenal," since considerations of simplicity of the kind which Campbell identified as characteristic of mathematical theories cannot be considered as determined solely by "phenomena." Indeed, in an important sense, mechanical theories are more "purely phenomenal" since the propositions of their hypotheses are analogous to true observational laws.

Despite Campbell's own inclination toward mechanical theories, he was at pains to point out that one can hold theories which exhibit the strong analogy—which he found so important—without their being strictly mechanical. It is likely that the general appeal of such analogies stems from the close relation and relevance of the laws of mechanics to our voluntary actions. Although this is the case, such appeal could be displaced to other laws, say, electrical ones, for it is not the case that all changes in the world with which we may be familiar are changes of matter and motion. The important requirement, though, is still maintained for this class of theory; that is, there should be an analogy between propositions of the hypothesis and some true observational laws which have "intrinsic interest."

One of the major stimuli for Campbell's careful scrutiny of the distinction of theories and laws was his concern for the semantic status of theoretical propositions. Although he, once again, equivocated in his views on exactly how those propositions derived their meaning, he felt it important to distinguish between the meaning given the theories by virtue of the dictionary, which was closer to "meaning" in the sense of empirical testability, from that "significance" which the analogy provided. His basic concern was the meaning of those terms of the theoretical propositions which the dictionary only obliquely or partially made either meaningful or testable. This was clearly a most pressing problem for the kind of theory, such as the theory of gases, in which the dictionary provided relations with functions of ideas and not the ideas themselves.

Although Campbell pressed for the role of analogy in these paradigm theories, he did not believe that they provided theoretical propositions with naïve common-sensical meanings. He emphasized that "the-velocity-of-the-electron" should not be thought of as meaning the same kind of thing as "the velocity of this billiard ball," for to think so would be to be deluded by the grammatical form. He was careful to hyphenate such phrases in order to accentuate the logical indivisibility of the notion.

As is characteristic of such an experimentalist, Campbell's inclinations were toward a quantitative view of his science. Indeed, for him physics was the "science of measurement," and he devoted much space in his texts to considerations of the nature of physical measurement and its relation to purer mathematics. Once again his firm commitment to the role of theories appeared in his insistence that "no new measurable quantity has ever been introduced into physics except as a result of the suggestions of some theory."

BIBLIOGRAPHY

1. *Modern Electrical Theory* (Cambridge, 1907, 1913, 1923).

2. *Principles of Electricity* (London–Edinburgh, 1912).

3. *Physics, the Elements* (Cambridge, 1920), repr. as *The Foundations of Science* (New York, 1957).

4. *What is Science?* (London, 1921).

5. *An Account of the Principles of Measurement and Calculation* (London–New York, 1928).

6. *Photoelectric Cells* (London, 1929, 1934), written with Dorothy Ritchie.

JOHN NICHOLAS

CAMPBELL, WILLIAM WALLACE (*b.* Hancock County, Ohio, 11 April 1862; *d.* San Francisco, California, 14 June 1938), *astronomy, education.*

Campbell, who was called Wallace, was the sixth of seven children born to Robert Wilson Campbell and Harriet Welsh. Robert Campbell died in 1866, and Wallace was brought up on the family farm by his mother. He attended local schools, where his ability was sufficiently recognized that he was urged to attend a major university. After a short period of teaching school he was admitted to the University of Michigan in civil engineering. During his junior year he came across Simon Newcomb's *Popular Astronomy,* which so captivated him that he decided to make astronomy his career. Under the tutelage of J. M. Schaeberle he became a skilled observer and in his senior year served as an assistant in the university observatory. He also calculated comet orbits after having read Watson's *Theoretical Astronomy.* He graduated with the B.S. in 1886 and from 1886 to 1888 was professor of mathematics at the University of Colorado. In 1888 he returned to the University of Michigan as instructor to fill the vacancy left by Schaeberle, who had joined the newly opened Lick Observatory of the University of California, at Mount Hamilton, California. Campbell served as volunteer assistant at Lick in the summer of 1890 and joined that observatory in May 1891.

Campbell was married in 1892 to Elizabeth Ballard Thompson. There were three sons of this marriage: Wallace, Douglas, and Kenneth. At the Lick Observatory, Campbell was a very active observer during his ten years (1891–1901) as staff astronomer. His contribution to his major field, spectroscopic observation, was fostered in 1893 when D. O. Mills donated funds for the construction of an adequate spectrograph for the thirty-six-inch Lick refractor, to be built to Campbell's specifications. The Mills spectrograph was a design classic and played an important role in Campbell's career.

In 1898 Campbell went to India on the first of seven eclipse expeditions in which he actively participated: India (1898), Thomaston, Georgia (1900), Spain (1905), Flint Island (near Tahiti) (1908), Kiev (1914), Goldendale, Washington (1918), and Western Aus-

tralia (1922). On the death in 1900 of James E. Keeler, director of the Lick Observatory and earlier Campbell's mentor in spectroscopic observations, Campbell was made acting director of the Lick Observatory, an appointment that was confirmed as of 1 January 1901. It is a measure of Campbell's stature in his profession that he was the nominee recommended by all of the twelve leading astronomers whose advice had been sought by the president of the university, Benjamin Ide Wheeler.

As director of the Lick Observatory, Campbell not only maintained its prominent position but in 1910, with the financial aid initially of Mills and later of others, established a southern station in Chile in order to obtain radial velocity observations of stars in the southern sky, to be used for the determination of the solar motion, a major area of Campbell's research.

In 1923 Campbell entered another phase of his career by accepting appointment as president of the University of California. Although he retained the nominal directorship of the Lick Observatory, the nature of his publications changed in that he began to discuss problems of scientific organization and some problems of more popular scientific interest, along with his spectrographic work (done principally in collaboration with Joseph H. Moore) and his eclipse observations (done principally with Robert J. Trumpler); both collaborators were astronomers at the Lick Observatory. His publications also demonstrated an interest in the history of astronomy.

In failing health at the time of his retirement in 1930 and having recently lost the sight of one eye, Campbell returned with his wife to Mount Hamilton, planning to resume his scientific work. He was, however, invited to accept nomination for the presidency of the National Academy of Sciences, which position he assumed on 1 July 1931. During his term of office, 1931–1935, Campbell attempted to restore the preeminence of the National Academy of Sciences as an advisory agency to the government. His efforts were rewarded by the creation within the academy of the Government Relations and Science Advisory Committee. He must thus be viewed as a strong and successful exponent of the use of scientific advice by government at a time when this was not generally accepted practice.

Campbell and his wife returned from Washington to Mount Hamilton in 1935 but soon moved to San Francisco. In retirement he continued his interest in astronomical events, delivered an address in tribute to Simon Newcomb at the Hall of Fame at New York University in 1936, and published an obituary of Ambrose Swasey (who had made the mounting for the thirty-six-inch refractor at the Lick Observatory) in 1937. In deteriorating health and fearing a total loss of vision, he did not wish to become a burden upon others and took his life on 14 June 1938.

Campbell's early scientific career was in orbit computation. Using his own observations along with those of others, he gained that dedication to precision of observation and refinement of technique which is the mark of a great observational astronomer. Campbell did not make any notable advances in this field, although in 1891 he published "Corrections to Watson's *Theoretical Astronomy,*" a list of errata for this standard work. That year he also published *A Handbook of Practical Astronomy for University Students and Engineers*. This was later revised and enlarged, reappearing as *The Elements of Practical Astronomy* (1899). It ran through numerous editions and has remained a standard text for many years.

At Lick Observatory, Campbell initially was introduced to spectroscopic observation through association with Keeler while a volunteer assistant in 1890. On joining the observatory's permanent staff, he continued to work in this area; and although his equipment was not ideal and did not become truly satisfactory until the Mills spectrograph became fully operational in 1896, he made notable contributions in the observation of Nova Aurigae (1892), noting the changes in the spectrum from continuous to bright-line. He also observed the characteristic bright-band emission spectra of Wolf-Rayet stars and made the first observations of the variation in spectral intensity of the F line of hydrogen and of the green nebular lines. At the opposition of Mars in 1894 he observed its spectrum and concluded that the atmosphere of Mars was deficient in oxygen and water vapor and unable to support life. His observational results were not in accord with general belief, but Campbell vigorously defended them. In 1909 he led an expedition to observe Mars from Mount Whitney, so as to minimize the absorption spectrum of the earth's atmosphere. In 1910 he again observed Mars from Mount Hamilton, this time at quadrature, when the relative radial velocity could best be used to separate lines of Martian origin from those arising in the terrestrial atmosphere. His earlier findings were confirmed.

Campbell was a most careful observer and a designer of techniques of observation and of reduction, and he put forth his findings with confidence, irrespective of their accord with accepted theory. He did not, however, hesitate to repeat his observations if he felt this necessary. An example is his eclipse observations to detect the sun's deflection of light from the stars, which had been predicted by Einstein. An attempt at the Russian eclipse of 1914 was thwarted by weather.

Although the eclipse equipment was kept in Russia because of the war, a further attempt was made in 1918 at Goldendale, Washington. The results were negative. When Eddington reported a confirmation of Einstein's predictions in 1919, Campbell returned to the task and attempted with Trumpler to observe the predicted deflection at the 1922 eclipse in Western Australia. This time the results fully confirmed the predictions of the general theory.

Campbell is probably best known today for his inauguration of a systematic program of radial velocity observation to result in a catalog, primarily to provide data for the determination of the sun's path among the stars. This was begun at Lick Observatory in 1896. The results of this program with its numerous extensions were not only to provide data for the initial task envisaged but also to lead to greatly improved observational techniques, discovery of numerous spectroscopic binaries, and the compilation of the basic data that were used later in the more detailed analysis of stellar motion, including galactic rotation.

The successful observation of radial velocities at Lick Observatory was a model that was emulated in many other observatories. Lick Observatory, under Campbell, continued as one of the major training centers for astronomers. As graduates moved to other observatories, many continued an interest in radial velocity and spectrographic observation, for which Campbell and his colleagues were exemplars. His influence thus pervaded astronomy and has continued.

Campbell was William Ellery Hale Lecturer for the National Academy of Sciences in 1914 and Halley Lecturer at Oxford in 1925. After he was Silliman Lecturer at Yale in 1909–1910, his lectures were published as *Stellar Motions; With Special Reference to Motions Determined by Means of the Spectrograph* (1913).

Campbell received honorary degrees from six American universities, the University of Western Australia, and Cambridge. In addition to his term as president of the National Academy of Sciences, he was president of the International Astronomical Union (1922–1925), the American Association for the Advancement of Science (1915), the American Astronomical Society (1922–1925), and the Astronomical Society of the Pacific (1895, 1910). Campbell was a foreign member of the Royal Society of London and was awarded honorary membership in many American and foreign organizations. He received the Lalande Medal (1903) and the Janssen Medal (1910) of the Paris Academy of Sciences, the Annual Medal of the Royal Astronomical Society of London (1906), the Draper Medal of the National Academy of Sciences (1906), and the Bruce Medal of the Astronomical Society of the Pacific (1915).

BIBLIOGRAPHY

I. ORIGINAL WORKS. Campbell's writings include "Corrections to Watson's *Theoretical Astronomy*," in *Publications of the Astronomical Society of the Pacific*, **3** (1891), 87–91; *A Handbook of Practical Astronomy for University Students and Engineers* (Ann Arbor, Mich., 1891), rev. and enl. as *The Elements of Practical Astronomy* (New York, 1899); "The Reduction of Spectroscopic Observations of Motions in the Line of Sight," in *Astronomy and Astrophysics,* **2** (1892), 319–325, repr. in J. Scheiner, *A Treatise on Astronomical Spectroscopy*, trans. E. B. Frost (Boston, 1894), pp. 338–344; *Stellar Motions; With Special Reference to Motions Determined by Means of the Spectrograph* (New Haven–London, 1913); "A Brief History of Astronomy in California," in Z. S. Eldredge, *History of California,* V (New York, 1915), 231–271; "International Relations in Science," in University of California, *Semi-Centennial Publications* (Berkeley, 1918), pp. 390–413; and *Newton's Influence on the History of Astrophysics,* spec. pub. no. 1, History of Science Society (Baltimore, 1928).

II. SECONDARY LITERATURE. A bibliography of some 330 articles, besides those listed above, is in *Biographical Memoirs, National Academy of Sciences,* **25** (1949), 58–74; it is preceded by a biography by W. H. Wright, pp. 35–58. See also University of California, *In Memoriam* (Berkeley, 1938), pp. 3–10. Obituaries are Robert G. Aitken, in *Science,* **88**, no. 2271 (1938), 25–28, repr. in *Publications of the Astronomical Society of the Pacific,* **50** (1938), 204–209; Henry Crozier Plummer, in *Nature,* **142**, no. 3585 (16 July 1938), 102–103; Frank Schlesinger, in *Popular Astronomy,* **47** (1939), 2–5; and Robert J. Trumpler, in *Sky,* **3**, no. 2 (Dec. 1938), 18.

JOHN W. ABRAMS

CAMPER, PETER (PETRUS) (*b.* Leiden, Netherlands, 11 May 1722; *d.* The Hague, Netherlands, 7 April 1789), *anatomy, surgery, obstetrics, anthropology, ophthalmology.*

Camper's father, Florentius Camper, a minister, married Sara Geertruida Ketting at Batavia, Java (now Djakarta, Indonesia), while serving there. For many years he lived in well-to-do retirement at Leiden, often receiving men of science and fame, among them Hermann Boerhaave. He took care that his gifted son was taught carpentry, as well as the arts of design and painting, at an early age. When he was twelve Camper was accepted at Leiden University, where he studied classics, natural sciences, and medicine for twelve years. Among his teachers were W.

J. 'sGravesande, Pieter van Musschenbroek, and, in the medical faculty, Albinus the Younger, Herman Oosterdijk Schacht, and H. D. Gaub. Camper practiced midwifery under the guidance of C. Trioen, the teacher of the city's midwives. On 14 October 1746 he took his degrees in science and in medicine with two theses, *De visu* and *De quibusdam oculi partibus,* both published in Leiden in the same year.

After having been a physician, Camper left Leiden in 1748 for a long journey through England, France, Switzerland, and Germany; during the trip he met many foreign scientists, contacts that he later cultivated. In England he attended William Smellie's course in midwifery and was accepted as a member of the Painters' Academy. In September 1749, while traveling from Paris to Geneva, the twenty-seven-year-old scientist was notified that he had been appointed professor of philosophy at Franeker University, in the province of Friesland. Soon he was also awarded the chairs of medicine and surgery. His assumption of the chairs was delayed by illness until 28 April 1751; in his inaugural address (*De mundo optimo*) he dealt with the best of all possible worlds.

Four years later, in 1755, Camper was appointed professor of anatomy and surgery at the Athenaeum Illustre in Amsterdam. His inaugural address (*De anatomes in omnibus scientiis usu*) concerned the use of anatomy in all sciences. In Amsterdam he performed many dissections for the surgeons' guild and was portrayed with the governors of that guild by Tibout Regters. In 1758 he was appointed professor of medicine and delivered an oration on certainty in medicine (*De certo in medicina*). However, in 1761, at the urging of his wife, Johanna Bourboom, he resigned his professorship (retaining the title *professor honorarius*) to settle at her country house in Klein Lankum, near Franeker. There he conducted research in comparative anatomy and completed his two-volume *Demonstrationum anatomico-pathologicarum* (1760–1762) as well as minor papers.

After two years the restless man accepted an appointment in theoretical medicine, anatomy, surgery, and botany at Groningen University. His broad outlook on nature is apparent from the subject of his inaugural oration (*De analogia inter animalia et stirpes*), in which he discussed the analogy between animals and plants. For some ten years Camper devoted himself to his extensive teaching duties (he also gave a course in forensic medicine) and his medical consultations. In the academic year 1765–1766 he acted as vice-chancellor.

In 1773 Camper again retired to return to Klein Lankum. The death of his wife following an operation for breast cancer (1776) left him depressed. He attempted to improve his frame of mind by traveling in Germany and Belgium.

As a landowner Camper was concerned about the flooding of a great part of Friesland in 1776; he wrote a booklet critical of the building of dikes that provoked a polemical answer. Gradually he became more involved in public affairs; he became burgomaster of Workum and member of the States of Friesland, which sent him as a representative to the States-General in The Hague. There, as a faithful supporter of the house of Orange, he was nominated a member of the State Council. However, the man who had received so many honors in the scientific world and was not at all devoid of vanity—as well as having a rather quick temper—now had to suffer vehement political attacks that soured his last years. He died in 1789 at The Hague and was buried in the Pieterskerk, Leiden.

During his lifetime Camper was one of the most famous scientists of western Europe. On his numerous journeys (he visited England in 1748, 1752, and 1785) he made the acquaintance of many outstanding scientists. Camper had an encyclopedic mind and contributed to many fields.

For some thirty years Camper practiced midwifery and performed several experimental symphysiotomies on pigs, an operation that, however, he never performed on a living human. In anatomy he described structures that are still associated with him: the *processus vaginalis* (*peritonei* and *testis*), Camper's fascia, and Camper's *chiasma tendinum digitorum manus*. In comparative anatomy he discovered the air spaces in the bones of birds and studied the hearing of fishes and the croaking of frogs. He performed careful dissections of the elephant, the rhinoceros, and the orangutan.

Goethe sent his treatise on the *os inter maxillare* to Camper. His measurement of the facial angle and the introduction of Camper's line contributed to the foundation of anthropology. In surgery Camper recommended a procedure of cutting for bladder stone in two operations. He was a supporter of inoculation, and his thesis on vision was a good piece of work.

Camper had a great talent for illustrating, as is shown in many of his publications. Moreover, he provided the illustrations for Smellie's well-known book on midwifery. He discussed the relations between anatomy and the arts of drawing, painting, and sculpting in a two-volume treatise that was translated by Cogan five years after his death.

Camper was highly esteemed in the scientific world, as is shown by his membership in most learned societies of western Europe. His comprehensive knowledge, his inquiring mind, his industry in research and in writing, and his skill in graphic arts procured him a well-merited fame attained by few other contemporary scientists.

BIBLIOGRAPHY

I. ORIGINAL WORKS. Many of Camper's works were written in Dutch; the most important academic lectures are mentioned above. There is no complete collection of his works, although there are some collections of his minor pieces available in German and French. Among his writings are *Demonstrationum anatomico-pathologicarum libri duo* (Amsterdam, 1760–1762); *Sämmtliche kleine Schriften, die Arzney-, Wundarzneykunst und Naturgeschichte betreffend,* trans. and with supp. by J. F. M. Herbell, 3 vols. (Leipzig, 1784–1790); *The Works of the Late Professor Camper, on the Connexion between the Science of Anatomy and the Arts of Drawing, Painting, Statuary,* T. Cogan, trans. (London, 1794); *Dissertationes decem,* 2 vols. (Lingen, 1798–1800); *Oeuvres de P. Camper, qui ont pour objet l'histoire naturelle, la physiologie et l'anatomie comparée,* H. J. Jansch, trans., 3 vols. (Paris, 1883), with atlas; *De oculorum fabrica et morbis,* with trans. into German, no. 2 in the series Opuscula selecta Neerlandicorum de arte medica (Amsterdam, 1913); *Petri Camperi itinera in Angliam. 1748–1785,* no. 15, *ibid.* (Amsterdam, 1939), which contains his lecture notes on Smellie's course on midwifery and his diaries, with English translation, introduction in Dutch and English by B. W. T. Nuyens; and *Optical Dissertation on Vision* (1746), G. ten Doesschate, trans., no. 3 in the series Dutch Classics of the History of Science (Nieuwkoop, 1962), with facsimile and introduction.

II. SECONDARY LITERATURE. There is no full biography of Camper in English. The introductions to the editions of his optical thesis and diaries cited above offer many biographical data, however. See also the biographical sketch of Camper in *Edinburgh Medical and Surgical Journal,* **3** (1803), 257–262; A. G. Camper, *Levensschets van P. Camper* (Leeuwarden, 1791); C. E. Daniëls, *Het leven en de verdiensten van Petrus Camper* (Utrecht, 1880); C. J. Doets, *De heelkunde van Petrus Camper 1722–1789,* thesis (Leiden, 1948); and E. B. Kaplan, *Peter Camper 1722–1789,* in *Bulletin of the Hospital for Joint Diseases,* **17** (1956), 371–385, which contains an extensive bibliography. *Nederlands tijdschrift voor geneeskunde,* **83** (6 May 1939), 2039–2149, contains twelve articles on various aspects of Camper's work written by Dutch historians of medicine to commemorate the sesquicentennial of Camper's death.

G. A. LINDEBOOM

CAMUS, CHARLES-ÉTIENNE-LOUIS (*b.* Crécy-en-Brie, France, 25 August 1699; *d.* Paris, France, 4 May 1768), *mathematics, mechanics, astronomy.*

Camus was the son of Marguerite Maillard and

Étienne Camus, a surgeon. He early evidenced mathematical and mechanical abilities that induced his parents to send him to the Collège de Navarre. He subsequently continued to study mathematics (with Varignon) and also undertook work in civil and military architecture, mechanics, and astronomy.

In 1727 Camus entered the Academy of Sciences' prize competition for the best manner of masting vessels. His memoir on this subject won half the prize money and was published by the Academy; more important, it was mainly responsible for bringing him election to that body as an assistant mechanician on 13 August 1727.

During the next forty years Camus served the Academy as administrator (he was its director in 1750 and 1761), as frequent commissioner for diverse examinations, and as active scientist. In the last capacity he presented some purely mathematical memoirs, although the greatest number of his contributions dealt with problems of mechanics. These included treatments of toothed wheels and their use in clocks, studies of the raising of water from wells by buckets and pumps, an evaluation of an alleged solution to the problem of perpetual motion, and works on devices and standards of measurement. His most important scientific service was with Maupertuis, Clairaut, and Lemonnier on the Academy's 1736 expedition to Lapland to determine the shape of the earth. He subsequently served with the same people to determine the amplitude of the arc of Picard's earlier measure and, several years later, with Bouguer, Pingré, and Cassini de Thury in closely related operations. He was also involved in Cassini de Thury's famous cartographical venture, which produced the *Carte de la France* published by the Academy in 1744–1787.

In 1730 Camus was named to the Academy of Architecture and became its secretary shortly thereafter. There he gave public lessons to aspiring architects as the Academy's professor of geometry. These lessons later served as the basis of a *Cours de mathématiques* that he drew up for the use of engineering students, a task he assumed in 1748 in conjunction with the creation of the École du Génie at Mézières. A standard examination procedure was also established, and Camus was named the examiner of engineering students.

According to his instructions, Camus's course was to consist of four parts—arithmetic, geometry, mechanics, and hydraulics; the *Cours,* published in three parts from 1749–1751, covered all but the last. (Among the large number of manuscripts left at his death were a work on hydraulics, apparently intended to complete the *Cours,* and a treatise on practical

geometry differing from what he had published.) In 1755, when Camus was also named the examiner for artillery schools, this *Cours* became the standard work for artillery students. Its great success was, therefore, due more to Camus's monopoly on examinations than to its intrinsic merit. In point of fact, the *Cours* came under increasing attack in the 1760's as inappropriate for artillery students and too elementary for those at Mézières.

BIBLIOGRAPHY

I. ORIGINAL WORKS. Although published separately in 1728, Camus's prize-winning "De la mâture des vaisseaux" is most conveniently found in *Pièces qui ont remporté les prix de l'Académie royale des sciences,* II (Paris, 1732). His subsequent contributions to the *Mémoires de l'Académie royale des sciences* include the mathematical notices "Solution d'un problème de géométrie, proposé par M. Cramer, Professeur de mathématiques à Genève" (1732), 446–451; and "Sur les tangentes des points communs à plusieurs branches d'une même courbe" (1747), 272–286; the notices on mechanics "Sur la figure des dents des roües, et des ailes des pignons, pour rendre les horloges plus parfaites" (1733), 117–140; "Sur l'action d'une balle de mousquet, qui perce une pièce de bois d'une épaisseur considerable sans lui communiquer de vitesse sensible" (1738), 147–158; "De la meilleure manière d'employer les séaux pour élever de l'eau" (1739), 157–188; "Sur les meilleures proportions des pompes, et des parties qui les composent" (1739), 297–332; "Sur un problème de statique, qui a rapport au mouvement perpétuel" (1740), 201–209; "Sur un instrument propre à jauger les tonneaux et les autres vaisseaux qui servent à contenir des liqueurs" (1741), 382–402; and, with Hellot, "Sur l'étalon de l'aune du Bureau des Marchands Mercier de la ville de Paris" (1746), 607–617; and the geodetic report, with P. Bouguer, C. F. Cassini de Thury, and A. G. Pingré, "Opérations faites par l'ordre de l'Académie pour mesurer l'intervalle entre les centres des pyramides de Villejuive et de Juvisy, en conclurre la distance de la tour de Montlhéri au clocher de Brie-Comte-Robert, et distinguer entre les différentes déterminations que nous avons du degré du méridien aux environs de Paris, celle qui doit être préférée" (1754), 172–186.

The *Cours de mathématiques* appeared as *Élémens d'arithmétique* (Paris, 1749), *Élémens de géométrie théorique et pratique* (Paris, 1750), and *Élémens de méchanique statique,* 2 vols. (Paris, 1750–1751).

II. SECONDARY LITERATURE. The standard biographical sources on Camus are the *éloge* by Grandjean de Fouchy in *Histoire de l'Académie royale des sciences* (Paris, 1768), pp. 144–154; and Théophile Lhuillier, "Essai biographique sur le mathématicien Camus, né à Crécy-en-Brie," in *Almanach historique de Seine-et-Marne pour 1863* (Meaux, 1863). A brief evaluation of his work, but with several errors, is available in Niels Nielsen, *Géomètres français du dix-huitième siècle* (Paris, 1935), pp. 81–83. For more important considerations of his *Cours* and his role as examiner, see Roger Hahn, "L'enseignement scientifique aux écoles militaires et d'artillerie," in *Enseignement et diffusion des sciences en France au XVIIIe siècle*, René Taton, ed. (Paris, 1964), pp. 513–545; and René Taton, "L'École royale du Génie de Mézières," *ibid.*, pp. 559–615.

SEYMOUR L. CHAPIN

CANANO, GIOVAN BATTISTA (*b.* Ferrara, Italy, 1515; *d.* Ferrara, January 1579), *anatomy.*

Canano's father, Ludovico, was a notary; his mother was Lucrezia Brancaleoni. His grandfather, for whom he was named, was lecturer in medicine at Ferrara and physician at the court of Matthias Corvinus, king of Hungary and Bohemia.

In 1534 Canano matriculated at the University of Ferrara. He attended the lectures in humanism of G. B. Giraldi and in practical medicine of Antonio Musa Brasavola and of his relative Antonio Maria Canano. About 1540, while he was still a student, Canano performed private anatomical dissections in his home (3 Turco Square) with Antonio Canano. These were attended by other Ferrarese physicians and the students Bartolomeo Nigrisoli, G. B. Susio, Jacobo Antonio Buoni, Arcangelo Piccolomini (who later taught anatomy in Rome), Ippolito Boschi, and Franciscus Vesalius, brother of Andreas. In his home Canano received several visits from Andreas Vesalius and also from Gabriele Falloppio (lecturer in medical botany at Ferrara in 1548) and John Caius; the latter described Canano's library as one of the richest private collections he had seen in Italy.

In 1541 or 1543, Canano published his book *Musculorum humani corporis picturata dissectio* (a bibliographical rarity), which concerns the muscles of the arm and is illustrated with twenty-seven copperplates by the Ferrarese painter Girolamo da Carpi.

On 18 April 1543, Canano, sponsored by Brasavola, graduated in arts and medicine at Ferrara. In 1541, although still a student, he appears in the records of the University of Ferrara as an unsalaried lecturer in logic. He is not listed among the faculty of the university in 1542 and 1543, but he is found again as a lecturer in practical medicine, or more frequently in surgery, from 1544 until 1552; Canano had charge of the public anatomical demonstrations, which were held in a theater at the Convent of St. Dominic and were attended by the duke of Ferrara and his court. His salary was reduced in 1544 because in that spring he went to the imperial encampments in France as physician to Francesco d'Este. There he observed the siege of St.-Dizier and again met Andreas Vesalius, whom he told about his observation of the valves of deep veins. In 1552 Canano relinquished his teaching post at the University of Ferrara and went to Rome as physician to Pope Julius III, from whom he received benefices and canonicates.

In 1555, at the death of the pope, Canano returned to Ferrara and was made chief physician of the Este principality. He died at Ferrara in 1579 and was buried on 29 January in the Convent of St. Dominic, in a tomb that was built during his lifetime and may still be seen.

Canano flourished in a period when the study of natural phenomena aroused great interest in Italy and when, particularly in Ferrara, there was a great effort to oppose the traditional dogmatic method. The work of Leonardo certainly remained unknown to the medical school of Ferrara. Canano extended to anatomy the criticism and experimental control advocated by Leoniceno (1492) in medical botany and improved upon the work of Berengario da Carpi (1521), who introduced iconography and started independent anatomical observation at Bologna. Probably the medical school of Ferrara influenced Vesalius' conception and completion of his *Fabrica*.

The *Picturata dissectio* is a small book of only forty pages, but of outstanding importance for its originality. It is the first part of a treatise on myology, of which the remainder never appeared, probably because at almost the same time Vesalius published the *Fabrica* (1543), in which the muscles are depicted in detail.

The *Picturata dissectio* is based exclusively on direct observation of the structures of the human body ("in mortuorum dissectione omnes hominis partes brevi tempore cognoscimus") and of living animals ("animantium . . . dissectione") and not, like the work of Galen, on dissection of the ape ("dissecandarum . . . scimiae partium"). The text consists of a dedication, an introduction, and the explanations of the plates. Copperplates, used for the first time, allowed presentation of finer details than did the woodcuts of Berengario and Vesalius. The *Picturata dissectio* contains the first anatomical drawings of the lumbricales (pl. 18) and of the interossei (pl. 21) of the hand, and the first description and drawing of the "m. palmaris brevis" (short palmar muscle) (pl. 19E) and of the oblique head of the adductor pollicis (pl. 20), which Vesalius did not observe and was unknown to Galen ("de quo non meminit Galenus").

When depicting the muscles, Canano followed the same order as Galen ("Galenum sequentes muscolos") and often used his words, but was quick to point out Galen's omissions, errors, or contradictions.

Another important contribution by Canano was the

observation of the valves of the deep veins (the azygos, the renal veins, and the sacral veins) and the assertion that they serve to prevent the reflux of the blood, as Vesalius referred to it ("hasque sanguinis refluxui obstare asseruit"). Like Ingrassia when he discovered the stapes, Canano made these observations public only orally, explaining them to such leading anatomists as Falloppio and Vesalius, who give us unequivocal testimony of it. Falloppio, Eustachi, and Vesalius were not able to confirm Canano's observations, probably because the valves are infrequent in the deep veins; and Amatus Lusitanus interpreted their function erroneously. Thus, little attention was paid to this research of Canano until Girolamo Fabrizio described the valves in the superficial veins of the limbs, attributing the discovery to himself. Nevertheless, in the eighteenth century such authorities as Morgagni and Haller correctly credited Canano with this discovery, which was to open the way for the further investigations that led to greater knowledge of the blood circulation.

BIBLIOGRAPHY

Canano's only published work is *Musculorum humani corporis picturata dissectio* (Ferrara, 1541 [or 1543]). According to Cushing and Streeter, only 11 copies are extant. A facsimile ed., with notes by H. Cushing and E. C. Streeter, is part of the series Monumenta Medica, E. H. Sigerist, ed. (Florence, 1925). Another facsimile ed., with Italian and English trans., historical intro., and references from Galen and Vesalius, was edited by Giulio Muratori (Florence, 1962); it also appeared in *Archivio italiano di anatomia,* **67** (1962), 1–109.

On Canano or his work, see J. Caius, *De libris suis* (London, 1570); V. Ducceschi, "Giambattista Canano," in *Gli scienziati italiani,* I, pt. 2 (1923), 285–292; G. Falloppio, *Observationes anatomicae* (Venice, 1561); A. von Haller, *Elementa physiologiae corporis humani,* I (Venice, 1768); J. B. Morgagni, *Valsalvae opera. Epistolarum anatomicarum duodeviginti* (Venice, 1741), II, epist. xv; G. Muratori, "The Academic and Anatomical Teaching of G. B. Canani and the Anatomical Theatres of the University of Arts and Medicine in Ferrara," in *Acta anatomica* (in press); G. Muratori and D. Bighi, "A. Vesalio, G. B. Canano e la rivoluzione rinascimentale dell'anatomia e della medicina," in *Acta medicae historiae patavina,* **10** (1964), 51–95; G. Muratori and A. Franceschini, "Nuovi documenti riguardanti l'attività dell'anatomico ferrarese G. B. Canani," in *Atti e memorie Deputazione provinciale ferrarese di storia patria,* III (1966), 89–132; A. Vesalius, *Anatomicarum G. Falloppii observationum examen* (Venice, 1564); and N. Zaffarini, *Scoperte anatomiche di G. B. Canani* (Ferrara, 1809).

GIULIO MURATORI

CANCRIN (CANCRINUS), FRANZ LUDWIG VON (*b.* Breitenbach, Hesse, Germany, 21 February 1738; *d.* St. Petersburg, Russia, 1812 or 1816), *mining, metallurgy.*

Cancrin was descended from a mining family; when he was very young, his father moved the family to Bieber, Hesse, where he had become director of a mine. Cancrin received his primary education in the schools of Bieber, while his father trained him in mathematical and scientific subjects. Despite family tradition, Cancrin wished to study jurisprudence. Prince William of Hesse ordered the elder Cancrin to train one of his sons in the science of mining, however, and Franz Ludwig was chosen. He spent several years in these more advanced studies with his father and then from 1759 to 1762 attended the University of Jena, where he continued his mathematical studies and began his legal ones.

In 1764 Cancrin became a clerk in Hanau and in 1767 was made assessor in the revenue office there. At this time he also tutored Prince Frederick in mathematics and engineering and published his first book, *Abhandlung von der Zubereitung und Zugutemachung der Kupfererze* (1766), a treatise on copper smelting. He then moved on to become professor of mathematics at the Hanau Military Academy (from which he withdrew on the plea of ill health) and held sundry official positions, including director of the mint in 1774.

Despite the press of his civil offices, Cancrin found time (1773–1791) to write an encyclopedic work in twenty-one volumes that covered all aspects of the mining of metals and salt—including mineralogy, assaying, mathematics, and mechanics. This work brought him a European reputation; he was made a member of scientific societies in Giessen and Berlin, and in 1782 accepted an appointment as *Regierungsdirektor* of Altenkirchen from the margrave of Ansbach.

In 1783 Cancrin entered the service of the Russian empress Catherine II; his duties included management of the Staraya Russa saltworks. He spent the rest of his life in Russia. From 1796 he was assisted at Staraya Russa by his son (who became the famous Russian minister Count Georg von Cancrin). He was finally called to St. Petersburg, where he served as a financial officer and died almost unnoticed.

BIBLIOGRAPHY

I. ORIGINAL WORKS. Cancrin's most important works include *Abhandlung von der Zubereitung und Zugutemachung der Kupfererze* (Frankfurt, 1766); *Beschreibung der vorzüglichsten Bergwerke in Hessen, in dem Waldeckischen, an dem Harz, in dem Mannsfeldischen, in Chursachsen und in dem Saalfeldischen* (Frankfurt, 1767); *Erste Gründe der Berg- und Salzwerkskunde,* 21 vols. (Frankfurt, 1773–1791); *Beschreibung eines Cupolofens, als Anhang zur Schmelzkunst* (Frankfurt, 1784; 2nd ed., Marburg, 1791); *Gründliche Anleitung zur Schmelzkunst und Metallurgie* (Frankfurt, 1784); *Geschichte und systematische Beschreibung der in der Grafschaft Hanau-Münzenberg etc. gelegenen Bergwerke* (Leipzig, 1787); *Abhandlung von der Natur und Einrichtung einer Bergbelehnung* (Giessen, 1788); *Abhandlung von der Zubereitung des Roh-Eisens in Schmiede-Eisen, auch des Stahleisens in Stahl, beides in einem Hammer mit Flammenfeuer* (Giessen, 1788); *Kleine technologische Werke,* 7 vols. (Giessen, 1788–1811); *Abhandlung vom Torfe, dessen Ursprung, Nachwuchs, Aufbereitung, Gebrauch und Rechten* (Giessen, 1789); *Abhandlung von dem Wasserrechte,* 4 vols. (Halle, 1789–1800); *Abhandlung von einer feuersparenden Fruchtdarre* (Halle, 1790), 2nd ed. entitled *Abhandlung von einer feuerfesten, am Brand sparenden Fruchtdarre oder Fruchtriege, mit einem Anhange, wie diese Darre zum Trocknen des Heues und Klees gebraucht werden könne* (Leipzig, 1799); *Abhandlung von der Natur und Gebrauch des Gyps- und Lederkalks* (Giessen, 1790); *Abhandlung von dem Bau einer neu eingerichteten Pottaschensieberey* (Frankfurt, 1791); *Erste Gründe der Berg-, Cameral- und Bergpolizeywissenschaft* (Frankfurt, 1791); *Von der Anlage und dem Bau unserer Städte etc.* (Frankfurt, 1792); *Von der Anlage, Bau und Unterhaltung der Röhrenbrunnen* (Frankfurt, 1792); *Abhandlung vom Recht, Anlage, Bau und Verwaltung der Ziegelhütten* (Marburg, 1795); *Praktische Abhandlung von dem Bau der Oelmühlen* (Frankfurt-Leipzig, 1798); *Abhandlung vom Bau nicht rauchender feuerfester Schornsteine* (Marburg, 1799); *Abbildung und Beschreibung eines neuen Spleiss- und Treib-Ofens* (Giessen–Halle, 1800); *Abhandlung von der vorteilhaften Grabung, der guten Fassung und dem rechten Gebrauch der süssen Brunnen, um reines und gesundes Wasser zu bekommen* (Marburg, 1800; 2nd ed., Giessen, 1804); *Wie man das beste Eisen erhalten kann* (Giessen–Halle, 1800); and *Abhandlung von sehr vorteilhaften Brand sparenden Oefen und Kochherden* (Marburg, 1808).

II. Secondary Literature. Biographical articles are "Franz Ludwig von Cancrin," in H. E. Scriba, *Biographisch-literarisches Lexikon der Schriftsteller des Grossherzogtums Hessen im 19. Jahrhundert,* pt. 2 (Darmstadt, 1843), p. 112; F. W. Strieder, "Franz Ludwig von Cancrin," in *Grundlage zu einer hessischen Gelehrten- und Schriftstellergeschichte,* II (Kassel-Marburg, 1781), p. 108; Gümbel, "Franz Ludwig von Cancrin," in *Allgemeine Deutsche Biographie,* III (Leipzig, 1876), pp. 740–742; and Walter Serlo, "Franz Ludwig von Cancrin," in *Männer des Bergbaus* (Berlin, 1937), p. 33.

M. Koch

CANDOLLE, ALPHONSE DE (*b.* Paris, France, 17 October 1806; *d.* Geneva, Switzerland, 4 April 1893), *botany, phytogeography.*

Alphonse de Candolle was the son of the Genevan botanist Augustin-Pyramus de Candolle and Fanny Torros. He spent the first years of his life at Paris, then at Montpellier, where his father taught at the university. In 1816 he went to Geneva, where he followed a classical curriculum and, in 1825, received a bachelor's degree in science. In spite of his devotion to botany, he turned to the study of jurisprudence, earning his doctorate in law in 1829. Under the guidance of his father, Candolle continued his research in taxonomy and became honorary professor at the Geneva Academy in 1831. In 1835 he succeeded his father in the chair of botany at the university and in the directorship of the botanical gardens. In the same year Candolle published his standard botanical text, *Introduction à l'étude de la botanique,* which was translated into German and Russian. By 1850 he had given up teaching to concentrate on research.

Candolle did not forget his training in law, and took part in the public life of the city. He was responsible for the introduction of postage stamps into Switzerland. In 1866, however, he withdrew completely from public affairs.

Besides his interest in politics, Candolle was passionately devoted to the history of science and in 1873 published a remarkable book, *Histoire des sciences et des savants depuis deux siècles.* The book displays both the naturalist's objectivity and the jurist's clarity. Darwin had just published his own works when Candolle wrote the *Histoire;* and Candolle was enthusiastic over the thesis of natural selection, which he applied with keen intelligence to the moral and intellectual characteristics of man and of human societies.

Until his death Candolle maintained his interest in the learned societies of his country, especially the Société Helvétique des Sciences Naturelles, the oldest of the itinerant associations in Europe. His personal library and herbarium, in the family residence in the Cour St.-Pierre, was a "sanctuary" of botany, where numerous researchers from all over the world came to submit their work and to discuss their ideas with the Genevan master. Candolle had the satisfaction of seeing his son, Casimir, follow in his footsteps and continue the line of the Candolle botanists.

Candolle's most outstanding work is his *Géographie botanique raisonnée* (1855), which is still the key work of phytogeography. He ushered in new methods of investigation and analyzed with exactitude the causes of the distribution of plant life over the surface of the globe. Besides his research he carried on the publication of the *Prodromus,* a vast treatise on phytotaxonomy begun by his father in 1824. Mention should also be made of *Origine des plantes cultivées* (1882).

Most of Candolle's numerous publications are

monographs dealing with important plant families. Among them are those on the Campanulaceae (1830), the Myrsinaceae (1834), the Apocynaceae (1843), the sandalwoods (1857), and the Begoniaceae (1860). Other reports concern the laws of nomenclature, and still others are on geobotany and the origin of species and of cultivated plants. Candolle also wrote books and articles on plant biology, showing an interest in raphides (1825), hothouses (1829), seed germination, the age of trees (1831), diseases of the grapevine, the dormancy of plants (1851), the culture of fruit trees, and the movement of plant life.

Candolle's name fails to appear in any of the important discoveries that underline the evolution of a science, yet most of his works show great lucidity as well as a critical and objective spirit and remain classics in the field.

BIBLIOGRAPHY

I. ORIGINAL WORKS. Among Candolle's works are *Prodromus systematis naturalis regni vegetabilis,* VII–XVII (Paris, 1824–1873), the completion of his father's work; *Géographie botanique raisonnée,* 2 vols. (Paris–Geneva, 1855); *Lois de la nomenclature botanique* (Paris, 1867); *La phytogéographie ou l'art de décrire les végétaux considérés sous différents points de vue* (Paris, 1880); *Origine des plantes cultivées* (Paris, 1882, 1883, 1886), trans. into Italian (1883) and into English and German (1884); and *Nouvelles remarques sur la nomenclature botanique* (Geneva, 1883).

II. SECONDARY LITERATURE. Works on Candolle are E. Blanchard, "A. de Candolle," in *Journal des savants* (1894), 353; G. Bonnier, "Alphonse de Candolle," in *Revue scientifique,* **51** (1893); J. Briquet, "Biographies des botanistes à Genève (1500–1931)," in *Bulletin de la Société botanique suisse,* **50a** (1940), 130–147; H. Christ, "Notice biographique sur A. de Candolle," in *Bulletin herbier Boissier,* **1** (1893), 203; O. Drude, "A. de Candolle," in *Léopoldina,* **31,** no. 1 (1895); G. L. Goodale, "A Sketch of A. de Candolle," in *American Journal of Science,* **46** (1893), 236; and M. Micheli, "Alphonse de Candolle et son oeuvre scientifique," in *Archives des sciences physiques et naturelles,* **30** (1893).

P. E. PILET

CANDOLLE, AUGUSTIN-PYRAMUS DE (*b.* Republic of Geneva, 4 February 1778; *d.* Geneva, 9 September 1841), *botany, agronomy.*

The son of Augustin de Candolle, a magistrate of the Republic of Geneva, and of the former Louise Eléonore Brière, Candolle developed a love for the "science aimable" while still quite young. His family moved from Geneva to Grandson, in the Vaud district, on the shore of the Lake of Neuchâtel. At the age of fourteen Candolle undertook solitary, ambitious botanizing expeditions and in his plant-study notebooks described, with remarkable attention to detail, the flora of the Swiss Plateau and of the Jura.

In 1794 the Candolle family was again in Geneva, and Augustin was attending the Collège de Calvin. He had decided to become a botanist but obeyed his father's wish that he first study medicine. For two years Candolle followed the courses at the Academy of Geneva. He was nevertheless able to continue his botanical excursions into the countryside around Geneva and in nearby Savoy, gathering numerous specimens and making observations that were later useful in preparing the monographs that brought him fame.

At about this time Candolle became acquainted with the botanist Jean Pierre Vaucher, pastor of the church of St. Gervais, who later wrote a remarkable work on freshwater algae. Vaucher had great influence on Candolle, with whom he shared his observations on the fertilization of the *Confervae;* these observations gave rise to Candolle's classic memoir, *Histoire des conferves d'eau douce* (1803). Vaucher also showed Candolle part of the manuscript of a work, not published until late in his life, that gave orientation to Candolle's biological research. This work of Vaucher's, *Histoire physiologique des plantes d'Europe,* was indisputably the inspiration for the *Prodromus,* the first sections of which Candolle published in 1824.

Another acquaintance of Candolle's was Jean Sénebier, also a pastor who was interested in plant physiology. Through Sénebier, Candolle became aware of the importance of the life processes of plants, and through his contact with this remarkable biologist—whose major contribution was his significant observations on photosynthesis—Candolle was to become not merely a distinguished taxonomist but also a botanist of far wider accomplishment.

A third influential friend in this period was Horace-Bénédict de Saussure, one of the most illustrious Genevan scholars of the eighteenth century. Through Saussure, Candolle was initiated into the study of mathematics, philosophy, and physics; he became interested in geology and fossils and went on to study the flora of the past as well as of the present.

In 1796 Candolle went to Paris to study both the natural sciences and medicine; there, as in Geneva, he sought contact with the greatest scholars in the city. He was a frequent visitor to the Muséum d'Histoire Naturelle and had exciting discussions with Georges Cuvier on the origin of species and on the new science of paleontology. He also became friendly with Lamarck. At this point Candolle decided definitely to abandon the study of medicine for the natural sciences, particularly botany.

In 1797 Candolle began to think seriously of publishing the results of his research. His early memoirs dealt more with experimental biology than with descriptive taxonomy. He made a careful study of the germination of legumes and a detailed report on the absorption of water by seeds under various conditions. He was interested in lichens and tried, by rudimentary means, to analyze their nutrition. He gathered important information on medicinal plants. Finally, in 1798 Candolle decided to publish his first botanical paper; it was on *Reticularia rosea,* a plant that he had discovered in the Jura. During the years 1799–1802 Candolle brought out his first important work (in twenty sections), *Plantarum historia succulentarum,* to which he added eight final sections in 1803.

Meanwhile, the political situation in Geneva was growing worse, and Candolle decided to settle in Paris, where he remained until 1808. Besides his botanical work, which he pursued enthusiastically, he occupied himself with the welfare of the poor and published several essays on philanthropy and political economy. In 1808 Candolle was called to Montpellier, to the chair of botany at the École de Médecine and the Faculté des Sciences; he lived there until 1816. The peaceful life of this provincial town allowed him ample opportunity to organize the countless research papers that he had accumulated during his ten years in Paris. Eventually, however, he found such a life monotonous. The Academy of Geneva had made Candolle an honorary professor in 1800; in 1802 the directors had offered him a chair in zoology, which he declined on the ground that his work was more oriented toward botany. In 1816, however, when a chair was established for him in natural history, he left Montpellier for Geneva.

With the support of the Republic and Canton of Geneva (which had become part of Switzerland in 1815), as well as of the people of Geneva, Candolle completely reorganized the botanical gardens created by the Société de Physique in 1791; these gardens opened on 19 November 1817 and served as a model of the genre for many years. Candolle played an active role in the creation of a museum of natural history that, through the support of Henri Boissier, rector of the Academy, had begun as early as 1811 to receive the physics and natural history collections of many Genevan scholars. He was also responsible for the founding of the Conservatoire Botanique. The excellent collections of this center of taxonomic research permitted it to make important contributions to the development of botany in the nineteenth century. During his twenty-five years in Geneva, Candolle was active in many fields besides botany. He was associated with the public library, the Société des Arts,

and the Musée des Beaux-Arts. He was concerned with the politics of his canton and from 1816 to 1841 was a member of its representative body. He was rector of the Academy from 1831 to 1832.

When Candolle retired from the Academy in 1835, his chair was divided: his son Alphonse succeeded to the chair of botany and his best student, Jean François Pictet, to that of zoology. The six years remaining to Candolle were clouded by illness. His death was greatly mourned by his colleagues and fellow citizens.

The writings of Candolle are considerable and touch on many areas of plant biology. The world's botanists have shown their admiration of and gratitude to the Genevan naturalist by dedicating more than 300 plants to his memory: one family (Candolleaceae) and two genera (*Candollea* and *Candollina*) have been named for him. Candolle signed nearly 180 memoirs and other works, and at his death he left some forty unfinished manuscripts. Two works were published posthumously: several sections of the *Prodromus* (Volumes VIII–XI) and the *Mémoire sur la famille des Myrtacées* (1842).

Among the works published by Candolle are his complete revision of Lamarck's *Flore française* (1805, 1815); his *Théorie élémentaire de la botanique,* which first appeared in 1813 and was frequently republished; and his *Cours de botanique* (1827), which was republished several times. His early and well-deserved reputation, however, was due principally to his excellent monographs on extremely diverse families of plants, including those on the Leguminosae (1800), the Crassulaceae (1801), the Compositae (1810), the Cruciferae (1821), the Cucurbitaceae (1825), the Portulacaceae (1828), and the Cactaceae (1829). Candolle also published several superbly illustrated large works that are still considered classics, such as his four-volume work on the Liliaceae (1802–1808) with 240 plates by the fashionable painter Pierre Redouté, who had taught Marie Antoinette, and his book on the Leguminosae (1825).

Candolle's most significant work was the *Prodromus systematis naturalis regni vegetabilis,* which was published in several sections from 1824 to 1839 and was continued by various collaborators after his death. The *Prodromus* is a huge botanical treatise that differs from those published previously because of Candolle's far-reaching interests. In this work there are no dry descriptions of plants, such as were customary; instead the author, whose knowledge of botany was remarkable, treats all aspects of the science. The *Prodromus* contains allusions to problems of evolution as well as discussions, well in advance of the times, of ecology, phytogeography, biometry, and agronomy.

Candolle was fascinated by everything touching the world of plants, and not merely by taxonomy. Intrigued by questions of phytochemistry, he published papers including *Observations sur une espèce de gomme qui sort des bûches du hêtre* (1799), *Examen d'un sel recueilli sur le Reaumuria* (1804), and *Notice sur la matière organique qui a coloré en rouge les eaux du lac de Morat* (1826). He was interested in plant pathology, particularly in parasitic mushrooms, on which he wrote noteworthy memoirs in 1807 and 1817. Candolle was the first to analyze fossil mushrooms; in 1817 he published a valuable memoir on the genera *Asteroma, Polystigma,* and *Stilbospora.* Teratology was also one of his interests, but it was not until 1841 that he published his *Monstruosités végétales.* Doubtless due to the influence of Jean Sénebier, Candolle did extensive research in physiology; his most important works in this area are *Premier essai sur la nutrition des Lichens* (1797), *Expériences relatives à l'influence de la lumière sur quelques végétaux* (1800), *Note sur la cause de la direction des tiges vers la lumière* (1809), and *De l'influence de la température atmosphérique sur le développement des arbres au printemps* (1831).

Candolle was a pioneer in the science of agronomy, and some of his publications on this subject are still considered classics. His best works in the field include *Mémoire sur la fertilisation des dunes* (1803), *Avis aux propriétaires de vignobles* (1816), *Premier rapport sur les pommes de terre* (1822), *Instruction sur l'emploi des engrais liquides* (1825), *Notice sur la culture de l'olivier* (1825), *Les pépinières du canton de Genève* (1828), *Considérations sur les forêts de la France* (1830), and *Essai sur la théorie des assolements* (1831).

His early studies led Candolle to medical botany. He wrote numerous papers on pharmacology, of which the best-known are *Recherches botanico-médicales sur les différentes espèces d'Ipécacuanha* (1802), *Notice sur la racine de Caïnca, nouveau médicament du Brésil* (1829), and *Note sur l'huile de Ramtilla* (1833). He also summarized the pharmacological knowledge of the day in an outstanding treatise, *Essai sur les propriétés médicales des plantes, comparées avec leurs formes extérieures et classification naturelle* (1804).

Candolle undoubtedly made his most original contribution to what was later known as phytogeography; he figures incontestably as a forerunner in this totally new science. Among his most noteworthy publications are *Mémoire sur la géographie des plantes de France, considérée dans ses rapports avec la hauteur absolue* (1817), *Essai élémentaire de géographie botanique* (1820), and *Projet d'une flore physico-géographique de la vallée du Léman* (1821). A worthy pupil of Lamarck, Candolle accurately describes the relationships between plants and soil, which affect the distribution of vegetation. He later had the opportunity to verify some of his "physico-geographical" theories while studying the flora of Brazil (1827), eastern India (1829), and northern China (1834).

Also an excellent biographer, Candolle left many notices of great value in the history of science; one of the best-known is *La vie et les écrits de Fr. Huber* (1832), which was translated into English by Silliman in 1833. He wrote also on political economy and statistics; a notable example is his classic memoir *Sur la statistique du royaume des Pays-Bas* (1830). In addition he was involved in education, administration, and public welfare.

BIBLIOGRAPHY

I. ORIGINAL WORKS. Among Candolle's major writings are *Plantarum historia succulentarum,* 28 secs. (1–20, Paris, 1799–1802; 21–28, Paris, 1803); *Astragalogia* (Paris, 1802); *Les liliacées,* 4 vols. (Paris, 1802–1808); *Essai sur les propriétés médicales des plantes, comparées avec leurs formes extérieures et classification naturelle* (Paris, 1804, 1816), trans. into German by K. J. Perleb (1818); *Synopsis plantarum in flora gallica descriptarum* (Paris, 1806); *Icones plantarum Galliae rariorum* (Paris, 1808); *Théorie élémentaire de la botanique* (Montpellier, 1813); *Regni vegetabilis systema naturale,* 2 vols. (Paris, 1817–1821); *Prodromus systematis naturalis regni vegetabilis,* 7 vols. (Paris, 1824–1839); and *Cours de botanique: Organographie* (Paris, 1827), trans. into German by M. Meissner (Tübingen, 1827) and into English by M. Kingdon (London, 1839; New York, 1840).

II. SECONDARY LITERATURE. On Candolle or his work, see J. Briquet, "Bibliographies des botanistes à Genève (1500–1931)," in *Bericht der Schweizerischen botanischen Gesellschaft,* **50,** sec. A (1940), 114–130; Alphonse de Candolle, *Mémoires et souvenirs de A. P. de Candolle* (Geneva, 1862); P. Gervais, *Discours prononcé à l'inauguration du buste de M. de Candolle dans le jardin botanique* (Montpellier, 1854); and A. de la Rive, *Notice sur la vie et les écrits de A. P. de Candolle* (Geneva, 1851).

P. E. PILET

CANNIZZARO, STANISLAO (*b.* Palermo, Sicily, 13 July 1826; *d.* Rome, Italy, 10 May 1910), *chemistry.*

Cannizzaro was the youngest of the ten children of Mariano Cannizzaro, a magistrate and minister of police in Palermo, and Anna di Benedetto, who came from a family of Sicilian noblemen. Sicily was under the rule of the Bourbon kings of Naples, and the Cannizzaro family supported the regime. One of Stanislao's sisters became a lady-in-waiting to the queen. On his mother's side, however, there were a number of political liberals. Three of Cannizzaro's maternal uncles were later killed in the campaigns of Garibaldi, and he himself became a strong antimonarchist.

Cannizzaro's early education in the schools of Palermo was essentially classical, although it included some mathematics. In 1841 he entered the University of Palermo as a medical student. Here he met the physiologist Michele Foderà, who introduced him to biological research. With Foderà he attempted to work out a distinction between centrifugal and centripetal nerves. In the course of this work Cannizzaro realized his need for more understanding of chemistry, which was very poorly taught at the university.

In 1845, at the Congress of Italian Scientists in Naples, Cannizzaro reported the results of his physiological studies and met the physicist Macedonio Melloni, in whose laboratory he worked for a short time. He confided his lack of chemical training to Melloni and as a result was introduced to Raffaele Piria, professor of chemistry at the University of Pisa and the leading Italian chemist of the day. He took Cannizzaro as his laboratory assistant, not only teaching him chemistry but also allowing him to take part in investigations of natural substances. It was at Pisa, between 1845 and 1847, that Cannizzaro decided to devote himself to chemistry. Here also he became a close friend of Cesare Bertagnini, a very promising pupil of Piria's. Although Bertagnini died at thirty, he and Cannizzaro, along with Piria, were influential in founding an Italian school of chemistry during the early 1850's.

In the summer of 1847 Cannizzaro returned to Palermo, intending to resume his studies at Pisa in the autumn. He soon found that a revolution against the Bourbons was in preparation; and in spite of the conservatism of his family, he joined the revolutionaries. In January 1848 the Bourbons were driven from Naples and the kingdom of Sicily was established. Young Cannizzaro became an artillery officer and a representative in the Chamber of Commons and took an active part in the fighting. When the rebellion finally failed in April 1849, he was forced to flee to Marseilles.

From Marseilles he made his way to Paris, where, through the influence of Piria, he met Cahours, who introduced him into Chevreul's laboratory in the Jardin des Plantes. Here he resumed his chemical studies, working with Stanislaus Cloëz on cyanamide and its derivatives.

In 1851 Cannizzaro was able to return to Italy as professor of physics, chemistry, and mechanics at the Collegio Nazionale in Alessandria. Although the facilities were poor, Piria urged him to accept the position because it could—and indeed did—lead to better

appointments. Cannizzaro built up the research laboratory and carried out some of his best work in organic chemistry there.

As a result of his work at Alessandria, Cannizzaro was appointed professor of chemistry at the University of Genoa in 1855. There was no laboratory at the university; and Cannizzaro, an excellent teacher, was able for a time to devote much thought to his course in theoretical chemistry. It was from Genoa that in 1858 he sent the letter describing the course on which his fame chiefly rests. In September 1860 he attended the Karlsruhe Congress, at which he made known his ideas to the chemical world. In 1856 or 1857 Cannizzaro married, in Florence, Henrietta Withers, the daughter of an English pastor. They had one daughter and one son, who became an architect.

Political events again changed the course of Cannizzaro's career. Garibaldi's Sicilian revolt in 1860 was successful, and Cannizzaro returned to his native Palermo to take part in the new government. This time he did not participate in the actual fighting, but he became a member of the Extraordinary Council of the state of Sicily. In 1861 he was appointed professor of inorganic and organic chemistry at the University of Palermo. Once more he had to organize and build a laboratory, since the only facility for chemical research was the same small room that had been available in his student days. Cannizzaro was so successful in his efforts that Palermo became the center of chemical education in Italy. Such men as Wilhelm Körner, who devised the method of locating the position of substituents in the benzene ring, and Adolf Lieben, later a noted organic chemist in Vienna, were among his students. At the same time he was active in establishing schools of various types in Palermo, and during an epidemic of cholera he served as commissioner of public health.

With the unification of Italy, Cannizzaro made his last move, to the University of Rome in 1871. As before, he found that laboratory facilities had been neglected. He therefore founded the Italian Institute of Chemistry in the old Convent of San Lorenzo. In the functioning laboratory that he established he was able to continue the work on the constitution of natural substances that he had begun with Piria. His efforts during the latter part of his life were devoted to determining the structure of santonin, which he showed to be one of the few natural compounds derived from naphthalene. With his move to Rome, Cannizzaro was made a senator of the kingdom. As in Palermo, he spent much time on public and civic duties.

Cannizzaro continued to give his lectures with great enthusiasm and success until nearly the end of his life, discontinuing them only the year before his death at eighty-three. During the latter part of his life he was honored by most of the important scientific societies of Italy and the rest of Europe. On the centenary of his birth in 1926, during the Second National Italian Congress of Pure and Applied Chemistry, his body was transferred to the Pantheon at Palermo.

Cannizzaro carried out all of his experimental work in the field of organic chemistry. Whenever he had a laboratory available, he continued the work on natural substances that he had begun at Pisa. He also devoted much time to the study of aromatic alcohols, a class of compounds little known before his work. In 1853, while studying the behavior of benzaldehyde, he discovered its reaction with potassium hydroxide, in which an oxidation-reduction produces both benzoic acid and benzyl alcohol. This is still known to organic chemists as the "Cannizzaro reaction." He was also the first to propose the name "hydroxyl" for the OH radical.

Cannizzaro's lasting fame depends, however, on the letter that he wrote in 1858 to his friend Sebastiano de Luca, who had succeeded Bertagnini in Piria's chair at Pisa. This was the famous "Sunto di un corso di filosofia chimica fatto nella Reale Università di Genova," published in the journal *Nuovo cimento*, established at Pisa by Piria, in the same year and reprinted as a pamphlet in 1859. It has frequently been republished and translated.

The complicated condition of chemistry that led Cannizzaro to compose his letter stemmed from events and personalities going back as far as fifty years before the "Sunto" appeared. When Dalton published the first volume of the book explaining his atomic theory in 1808, he considered but rejected the idea that equal volumes of gases under the same conditions contained equal numbers of particles. Only a few years later, in 1811, Amedeo Avogadro took up this idea. By making a clear distinction between atoms (which he called "elementary molecules") and molecules ("integral molecules"), he was able to draw a number of important conclusions. Three years later Ampère proposed a similar idea. If the conclusions deduced from this hypothesis had been accepted at the time they were suggested, chemists would have been spared half a century of confusion. However, the papers were not well understood; and the chemical facts known were not sufficient to provide all the evidence needed to confirm the hypothesis. Even more important, the authorities who dominated chemical thinking during the first half of the nineteenth century, Berzelius and Dumas, did not accept the idea.

Berzelius did not distinguish atoms from molecules, speaking indifferently of an atom of hydrogen or an atom of alcohol. His electrochemical (dualistic) theory, to which he tried to make all facts conform, required that chemical compounds be held together by opposite electrical charges. Thus, there could not be combination of electrically similar atoms, and hydrogen and oxygen could not be diatomic. Berzelius' analytical determinations of atomic weights were based on Gay-Lussac's law of combining volumes of gases and were in most cases quite accurate; however, he was unable to apply this law consistently to solid compounds, and so a number of his values for atomic weights were incorrect.

Dumas recognized that vapor density determinations could be used for determining atomic weights; but since he too confused atoms and molecules, he wrote of water as composed of "an atom of hydrogen" and "half an atom of oxygen." (To Berzelius the concept of half an atom was ridiculous.) Dumas determined the vapor densities of mercury, phosphorus, arsenic, and sulfur and found "atomic" weights that he believed were impossibly high. He therefore discarded Avogadro's hypothesis. In 1843 Berzelius accepted Dumas's experimental results and definitely rejected the Avogadro concept. The influence of these two men was so strong that the hypothesis of atomic weights had little chance of being accepted.

In the meantime, in 1813, Wollaston had proposed the use of equivalent weights as the fundamental units of chemistry. Equivalent weights appealed to many chemists because they seemed to be experimentally determinable without recourse to any theory. Confusion was increased because there was no standardization of meaning for many formulas employed to represent chemical compounds. Symbols involving barred or double atoms came to mean different things to different chemists. When Laurent and Gerhardt tried in the 1840's to return to Avogadro's principle, they went too far and introduced new confusion into chemistry. A few men, such as M. A. A. Gaudin, a calculator in the Bureau des Longitudes in France, appreciated the Avogadro hypothesis and published work depending on it; but they were outside official circles and had no influence.

Thus, when Cannizzaro wrote the "Sunto," there was no agreement among chemists as to what values should be adopted for atomic, molecular, or equivalent weights; no possibility of systematizing the relationship of the various elements; and no unanimity as to how organic compounds should be formulated.

The deficiency in laboratory facilities in the various universities in which he had taught and his own enthusiasm for teaching had combined to cause Cannizzaro to devote much thought to the courses he gave. He well recognized the difficulty his students encountered in learning chemistry when they found that even masters of the science could not agree as to what constituted the fundamental structure of chemical compounds. Believing that he understood how this confusion had arisen, he set himself to explain as simply and clearly as he could what the true basis of chemistry should be. His being an Italian perhaps permitted Cannizzaro to see more clearly than foreign chemists what his countryman Avogadro had suggested almost fifty years earlier. In his theoretical course he now proposed to clear up the difficulties that had arisen. His letter to Luca outlined the development of his pedagogical ideas.

Cannizzaro was well-read in the history of chemistry and was therefore able to develop his course historically. He not only gave credit to the work of well-known figures but also devoted time to such little-known authors as Gaudin. His first four lectures were purely historical, to give his students the background for understanding the current situation of chemistry.

Cannizzaro began by stressing the distinction between atoms and molecules made by Avogadro and Ampère. He then explained the theories of Berzelius and how they had misled the master analyst. He also showed how Dumas had felt forced to conclude that there were different rules governing inorganic and organic chemistry. He reviewed the contributions of many chemists closer to his own time, showing how often they had approached the truth without realizing it completely. Throughout this historical review he repeatedly insisted that application of Avogadro's hypothesis explained the inconsistencies noted by others and that no facts contradicting it were known.

He was then ready, in his fifth lecture, to show how Avogadro's hypothesis could be used. Most of what he pointed out had been stated, or at least implied, by Avogadro; but Cannizzaro brought it out much more clearly and was able to supply a wealth of examples from cases that had not been known earlier. He stressed that since all atomic weights are relative, one standard weight had to be chosen with which all other values could be compared. He chose hydrogen as this standard, but since he knew it to be diatomic, he used "half a molecule of hydrogen" as unity. In using this term he avoided Dumas's error, the "half atom of hydrogen" that had so disturbed Berzelius.

Cannizzaro next told his students, "Compare the various quantities of the same element contained in the molecule of free substance and in those of all its different compounds, and you will not be able to escape the following law: The different quantities of the same element contained in different molecules are all multiples of one and the same quantity, which, always

being entire, has the right to be called an atom." This he called the law of atoms, and Partington says that it deserves to be called the Cannizzaro principle. He gave numerous examples of the application of this law, especially to metals, the atomic weights of which were in a particular state of confusion.

The method of determining molecular weights by the use of vapor densities depended on the existence of volatile compounds. When such compounds were not known for a given element, Cannizzaro used analogies or depended on the relation between atomic weight and specific heat discovered by Dulong and Petit. In the case where both methods could be used, he showed that they gave the same result. This strengthened his argument. In his discussion of organic radicals Cannizzaro stressed their similarity in combining power to atoms of various elements. This approach came very close to a statement of the theory of valence, which had not yet been clearly enunciated. He pointed out that radicals like methyl are monatomic, like hydrogen, while radicals like ethylene resemble mercuric or cupric compounds. "The analogy between mercuric salts and those of ethylene or propylene has not been noted, so far as I know, by any other chemist."

Thus, in his "Sunto" Cannizzaro not only called attention once more to Avogadro's hypothesis, made the distinction between atoms and molecules fully clear, and showed how vapor densities could be used to determine molecular weights (and atomic weights), but he laid to rest completely the idea that inorganic and organic chemistry functioned by different rules. As Tilden summed up his work in the Cannizzaro Memorial Lecture to the Chemical Society, "There is, in fact, but one science of chemistry and one set of atomic weights."

When the "Sunto" was first published, it attracted little attention, possibly because of the place and language of its publication. Chemists grew more and more frustrated in their attempts to systematize their science. This was particularly true of the younger workers, who were most active in research and who most felt the need for a sound theoretical background for their studies. A leading spirit in this search for a background was August Kekulé, who had just published his epochal paper on the linking of carbon chains and the tetratomicity of carbon. In the spring of 1860 he proposed to his friend Carl Weltzien, professor of chemistry at the Technische Hochschule in Karlsruhe, that an international congress of chemists be called to establish, among other things, more precise definitions of the concepts of "atom, molecule, equivalent, atomicity, alkalinity, etc." In association with Charles Wurtz of Paris, Kekulé and Weltzien organized the first international chemical congress, which met at Karlsruhe for three days, beginning on 3 September 1860. Most of the men attending were the younger chemists, active in research and therefore anxious to clarify the basis of their studies. Many of the well-established older men, such as Liebig and Wöhler, more sure of their theoretical ideas, did not come. Dumas was the most important of the older workers who did attend, but he spent much of his time reiterating the idea of the difference between inorganic and organic chemistry.

On the first day of the meeting, discussion centered on the distinction between physical molecules, which meant particles of a gas, liquid, or solid; chemical molecules, the smallest part of a body taking part in a reaction but capable of being divided; and atoms, which could not be divided. Although Kekulé supported this distinction, Cannizzaro stated that he could see no difference between physical and chemical molecules. On the second day questions of nomenclature were discussed, and on the third day there was a lively consideration of whether the principles of Berzelius should be adopted for purposes of nomenclature. Cannizzaro delivered a lengthy refutation of this proposal in which he summarized the arguments he had used in the "Sunto." He strongly defended Avogadro's hypothesis and pointed out that anomalous vapor pressures of some substances could be explained by the phenomenon of dissociation at higher temperatures, which had recently been discovered by Deville. In the discussion that followed, the prevailing opinion was that no vote could be taken on scientific questions and that each scientist should be allowed full freedom to use the system he preferred.

Cannizzaro left at the end of the meeting, probably feeling that his efforts had been futile. However, his friend Angelo Pavesi, professor of chemistry at the University of Pavia, remained behind and distributed copies of the "Sunto" which Cannizzaro had brought with him. This was the decisive step, for it brought Cannizzaro's clear and logical arguments to the attention of the chief chemists of the day. Since these arguments had been prepared to introduce students to chemistry, they omitted no step in the reasoning or deductions and thus were ideally suited to convince even practicing chemists whose preconceptions might have prevented them from following a more condensed version.

One of the first to see the significance of the paper was Lothar Meyer, who read the pamphlet on his way back to Breslau. As he expressed it, the scales fell from his eyes and he was convinced. His book *Die modernen Theorien der Chemie,* published in 1864, utilized

Cannizzaro's ideas throughout and exerted a strong influence on the chemical world. Mendeleev also attended the congress and later wrote of the defense that Cannizzaro had presented for Avogadro's hypothesis. It was the recognition of true atomic weights that permitted Meyer and Mendeleev to formulate the periodic law at the end of the 1860's.

In organic chemistry the confusion of formulas that had originated in the disagreement over whether to use atomic or equivalent weights of carbon and oxygen also disappeared. The way was opened for the full development of the structural theory developed by Butlerov and others in the decade following the Karlsruhe Congress. In 1860 the chemical world was ready for the revival of Avogadro's hypothesis, but it was the great logic and clarity of Cannizzaro's presentation that made its acceptance easy.

BIBLIOGRAPHY

I. ORIGINAL WORKS. There is a bibliography of Cannizzaro's papers on experimental chemistry in *Bulletin. Société chimique de France,* 4th ser., 7 (1910), VII–XIII. The Cannizzaro reaction is described by Cannizzaro himself in "Ueber den der Benzoësäure entsprechenden Alkohol," in *Justus Liebig's Annalen der Chemie,* 88 (1853), 129–130; 90 (1854), 252–254. "Sunto di un corso di filosofia chimica fatto nella Reale Università di Genova" appeared in *Nuovo cimento,* 7 (1858), 321–366, and was republished as a pamphlet (Pisa, 1859). An English translation is Alembic Club Reprints, no. 18 (Edinburgh, 1910); and a German translation is Ostwald's Klassiker der Exacten Wissenschaften, no. 30 (Leipzig, 1891).

II. SECONDARY LITERATURE. Extensive biographical material is in W. A. Tilden, "Cannizzaro Memorial Lecture," in *Journal of the Chemical Society,* 101 (1912), 1677–1693; and Domenico Marotta, "Stanislao Cannizzaro," in *Gazetta chimica italiana,* 69 (1939), 689–717. A shorter biography is A. Gautier, "Stanislas Cannizzaro, " in *Bulletin. Société chimique de France,* 4th ser., 7 (1910), I–VI. Cannizzaro's part in the Karlsruhe Congress is described by Clara de Milt, "Carl Weltzien and the Congress at Karlsruhe," in *Chymia,* 1 (1948), 153–169.

HENRY M. LEICESTER

CANNON, ANNIE JUMP (*b.* Dover, Delaware, 11 December 1863; *d.* Cambridge, Massachusetts, 13 April 1941), *astronomy.*

Miss Cannon's father, Wilson Lee Cannon, was a man of wide influence who became state senator in Delaware; her mother was Mary Elizabeth Jump Cannon. When Miss Cannon entered Wellesley College in the class of 1884, she became one of the first girls of her native state to go away to college. In 1894 she returned to Wellesley for graduate study in mathematics, physics, and astronomy. The following year she enrolled as a special student in astronomy at Radcliffe College, probably at the suggestion of Edward C. Pickering, who had already employed several talented women astronomers at Harvard College Observatory. Miss Cannon joined the staff at Harvard in 1896 and worked there for the rest of her life.

Miss Cannon quickly recognized the immense opportunity offered by the study of astronomical photographs, an endeavor in which Harvard Observatory had taken a foremost position under Pickering's aggressive leadership. Much of her early work dealt with variable stars, but her greatest contributions were in the field of stellar spectral classification. On the photographic plates she discovered more than 300 variable stars, and a large number of these were detected from their spectral characteristics.

In the early work at Harvard, the spectra of stars had been sorted into various groups designated by the letters A, B, C, and so on. Miss Cannon developed the definitive Harvard system of spectral classification by rearranging these groups, omitting some letters, adding a few, and further subdividing the others. Her work proved that the vast majority of stars are representatives of but a few species; and she also demonstrated that these few spectral types, with rather rare exceptions, could be arranged in a continuous series.

In 1901, after five years of research, Miss Cannon published a description of the spectra of 1,122 of the brighter stars, a volume that proved to be the cornerstone on which her larger catalogs were based. *The Henry Draper Catalogue,* published as volumes **91-99** of the *Annals of Harvard College Observatory,* is Miss Cannon's outstanding contribution to astronomy; it contains spectral classifications of virtually all stars brighter than ninth or tenth magnitude—a colossal enterprise embracing 225,300 stars.

Although Miss Cannon began classifying spectra in her first year at the observatory, the classifications for *The Henry Draper Catalogue* were made in the relatively brief interval from 1911 to 1915; but because checking and arranging the material for publication required several additional years, the final volume was not issued until 1924. Unsated, she continued her work, publishing about 47,000 additional classifications in the *Henry Draper Extension (Annals,* **100** [1925–1936]) and several thousand more in the *Yale Zone Catalogue* and *Cape Zone Catalogue;* another 86,000 were published posthumously in the *Annals* (**112** [1949]).

Miss Cannon's ability to classify stellar spectra from low-dispersion objective prism plates was quite phe-

nomenal: her rate of more than three stars a minute and her ability to duplicate the classifications later also attest to her unusual skill and determination. While she came to recognize at a glance the characteristics that placed a star in the general sequence, she rarely failed to note the peculiarities in the spectrum. Cecilia Payne-Gaposchkin has remarked that "Miss Cannon was not given to theorizing; it is probable that she never published a controversial word or a speculative thought. That was the strength of her scientific work—her classification was dispassionate and unbiased."

From 1911 Miss Cannon was curator of astronomical photographs in charge of the ever-growing collection of Harvard plates. In 1938 she became the William Cranch Bond Astronomer, one of the first women to receive an appointment from the Harvard Corporation. Her personal charm and cheerful excitement conveyed a spirit of enthusiasm and evoked the admiration of all who knew her, and she became universally recognized as the dean of women astronomers. Throughout her career she was almost completely deaf unless assisted by a hearing aid, a handicap that must have contributed to her immense powers of concentration.

Among her numerous honorary degrees was the first honorary doctorate awarded to a woman by Oxford University. She was for a while the only woman member of the Royal Astronomical Society—an honorary member, because women were not then admitted to regular membership. She was one of the few women ever elected to the American Philosophical Society. In 1931 the National Academy of Sciences awarded her the Draper Gold Medal, and in 1932 she received the Ellen Richards Research Prize, which—with characteristic generosity—she turned over to the American Astronomical Society to establish the Annie Jump Cannon Prize for women astronomers.

BIBLIOGRAPHY

For the most part Miss Cannon's publications appeared in the *Annals of Harvard College Observatory* (note references in text). Her daily work is recorded in 201 record books preserved at Harvard College Observatory.

On Miss Cannon or her work, see Leon Campbell, "Annie Jump Cannon," in *Popular Astronomy,* **49** (1941), 345–347; Owen Gingerich, "Laboratory Exercises in Astronomy—Spectral Classification," in *Sky and Telescope,* **28** (1964), 80–82; and Cecilia H. Payne-Gaposchkin, "Miss Cannon and Stellar Spectroscopy," in *The Telescope,* **8** (1941), 62–63.

OWEN GINGERICH

CANNON, WALTER BRADFORD (*b.* Prairie du Chien, Wisconsin, 19 October 1871; *d.* Franklin, New Hampshire, 1 October 1945), *physiology.*

For a detailed study of his life and work, see Supplement.

CANTON, JOHN (*b.* Stroud, England, 31 July 1718; *d.* London, England, 22 March 1772), *physics.*

Canton, who was to earn his living as a schoolmaster, received little formal education himself, for his father removed him from school just as he reached the rudiments of algebra and astronomy in order to set him to the family trade of broadcloth weaving. Young John, however, refused to abandon his studies and continued to labor over his books at night, despite his father's refusal to allow him a candle for the purpose. His learning and mechanical talent, as expressed in an accurate sundial that the elder Canton proudly set up in front of the house, brought John to the attention of several neighboring gentlemen, including Dr. Henry Miles, who decisively affected his career. Miles, a native of Stroud, occupied a Dissenting pulpit in Tooting, Surrey, on the outskirts of London; and he persuaded the father to allow John to reside with him while he arranged employment more suitable than weaving. Accordingly, in March 1737, Canton proceeded to Tooting, where he remained until the following May, when he articled himself to one Samuel Watkins, master of a school in Spital Square, London. In 1745 he succeeded Watkins and kept the school himself until his death in 1772.

Canton's first contributions to science were routine calculations of the times of lunar eclipses, published in *Ladies' Diary* for 1739 and 1740. Thereafter, probably through the intervention of Miles, who became a fellow of the Royal Society in 1743, Canton met several of London's best young "experimental philosophers," men like the apothecary William Watson and the clockmaker John Ellicott, who occupied social positions not much different from his own. By 1747 or 1748 Canton also enjoyed a reputation as an experimentalist, largely for his invention of a new method of making strong artificial magnets. Several French and English philosophers were then occupied with the subject, inspired by the success of Gowin Knight (1746), who refused to disclose his profitable technique. Similarly, Canton, who hoped to derive some income from his invention, kept it secret until 1751, about a year after the publication of John Michell's *A Treatise of Artificial Magnets* (1750). It then appeared that Canton's procedure paralleled Michell's very closely. Michell cried plagiarism, which did not prevent the Royal Society, which had made Canton a fellow in 1749, from awarding him the Copley Medal for 1751. Doubtlessly Canton had a method before Michell's book appeared; the question is how closely it coincided with the one he later published as his own. All that is known of his character testifies to his innocence of Michell's charge.

Canton's entry into his special field, electricity, was also ill-omened. His first extended appearance in print—two paragraphs signed "A.B." in the *Gentleman's Magazine* for 1747—posed two electrical puzzles. Benjamin Wilson and his friend John Smeaton immediately supplied an answer, a patronizing sneer, and the intelligence that the queries contained nothing new. Canton withdrew to his magnets and to meteorological measurements until, in the early summer of 1752, he learned of the French experiments confirming Franklin's conjectures about lightning. He was the first in England to repeat the experiments successfully; and in the process he discovered, independently of Lemonnier, Beccaria, and Franklin, that clouds came electrified both positively (as theory suggested) and negatively (which remained a puzzle). The problem of determining the sign of a cloud's charge apparently led Canton to design the well-known experiments on electrostatic induction that have earned him a permanent place in the history of electricity. In a typical arrangement, an insulated tin tube receives a small positive charge, which causes two cork balls suspended from one end of the tube to repel one another; an electrified glass rod is then brought under the balls, which collapse as the rod approaches and diverge as it recedes. Canton accounted for these phenomena by modifying the old doctrine of atmospheres in much the way Beccaria was to adopt, i.e., by distinguishing what the Italian called "aerial" and "proper" electricity. Franklin later (1755) redesigned Canton's experiments and so modified the atmospheres as to render them all but superfluous. Their total abolition, an essential step toward modern theories of electrostatics, was first advocated by F. U. T. Aepinus and Johann Karl Wilcke in 1759. Although Canton did not subscribe to their program, his experiments figured importantly in its development.

Canton also enriched the study of electricity with the notable discovery that glass does not always charge positively by friction: the sign of the electricity developed depends upon the nature of the substance rubbed over it and the condition of the surface of the glass. (He found, e.g., that rough glass became negative when rubbed with flannel and positive when rubbed with silk.) Among his many smaller contributions to the subject were a portable pith-ball electroscope (1754); a method for electrifying the air by communication (1754); a careful account of some of the electrical regularities of that bewildering stone, the tourmaline (1759); and a valuable improvement in the electrical machine, coating its cushion with an amalgam of mercury and tin (1762). Throughout, Canton was guided by Franklin's theories, which he zealously defended against attacks emanating from Wilson's circle.

Like most gifted amateur physicists of his time, Canton attended to a great many subjects seldom cultivated together today. He examined the luminosity of seawater and identified its cause, the putrefaction of organic material; he invented a strongly phosphorescent compound, "Canton's phosphor," made of sulfur and calcined oyster shells (CaS); he kept a meteorological journal; he recorded the diurnal variation of the compass; he carefully observed transits, occultations, and eclipses; and he demonstrated the compressibility of water. This last investigation, a notable achievement in eighteenth-century physics, depended upon measurements so minute that several philosophers challenged Canton's revolutionary interpretation of them. But his results, which compare favorably with modern determinations, stood the scrutiny of a special committee of the Royal Society and earned him a second Copley Medal in 1765.

Judging from the clarity, brevity, and precision of his scientific papers, Canton was an admirable pedagogue. In leisure hours he enjoyed good conversation and a good glass, particularly at the Club of Honest Whigs in the company of Franklin, fellow teachers, and Dissenting ministers like Joseph Priestley, whose *History and Present State of Electricity* owed much to his patient assistance. Liberal in politics, latitudinarian in religion, devoted to his profession, schoolmaster Canton was one of the most distinguished of the group of self-made, self-educated men who were the best representatives of English physics in the mid-eighteenth century.

BIBLIOGRAPHY

I. ORIGINAL WORKS. Canton's most important papers are "A Method of Making Artificial Magnets Without the Use of Natural Ones," in *Philosophical Transactions of the Royal Society,* **47** (1751-1752), 31-38; "Electrical Experiments, With an Attempt to Account for Their Several Phaenomena, Together With Observations on Thunder Clouds," *ibid.,* **48:1** (1753), 350-358; "Some New Electrical Experiments," *ibid.,* **48:2** (1754), 780-784; [On the Tourmaline], in *Gentleman's Magazine,* **29** (1759), 424-425; "Experiments to Prove Water Is Not Incompressible," in *Philosophical Transactions of the Royal Society,* **52** (1762), 640-643; "Experiments and Observations on the Compressibility of Water and Some Other Fluids," *ibid.,* **54** (1764), 261-262; "An Easy Method of Making a Phospho-

rus That Will Imbibe and Emit Light, Like the Bolognian Stone, With Experiments and Observations," *ibid.,* **58** (1768), 337-344. Poggendorff lists all Canton's papers printed in the *Philosophical Transactions.* A complete bibliography would also include many brief notes in *Gentleman's Magazine, Ladies' Diary,* and *Gazeteer;* occasional contributions to books; and scattered astronomical, meteorological, and electrical data in papers published by Henry Miles, William Watson, James Short, and Joseph Priestley in the *Philosophical Transactions.* The Royal Society of London has preserved Canton's correspondence, some of which has been published: e.g., by Augustus De Morgan, in *The Athenaeum* (1849), pp. 5-7, 162-164, 375; by R. E. Schofield, in *A Scientific Autobiography of Joseph Priestley* (Cambridge, Mass., 1966); and by W. C. Walker, "The Detection and Estimation of Electric Charges in the Eighteenth Century," in *Annals of Science,* **1** (1936), 66-100.

II. SECONDARY LITERATURE. The fullest biography of Canton is the notice prepared by his son William for the 2nd ed. of *Biographia Britannica,* A. Kippis, ed., III, 215-222. See also vol. I of the Canton papers at the Royal Society; *The Papers of Benjamin Franklin,* L. W. Labaree *et al.,* eds., (New Haven, Conn., 1961-1968), vols. IV-XII; the articles by De Morgan cited above; and V. W. Crane, "The Club of Honest Whigs: Friends of Science and Liberty," in *William and Mary Quarterly,* **23** (1966), 210-233. For Canton's scientific work consult P. Rivoire, ed. and trans., *Traités sur les aimans artificiels* (Paris, 1753), preface; J. Priestley, *The History and Present State of Electricity,* 3rd ed., 2 vols. (London, 1775), I, *passim;* I. B. Cohen, *Franklin and Newton* (Philadelphia, 1956), pp. 516-543; E. N. Harvey, *A History of Luminescence* (Philadelphia, 1957), *passim;* and A. Wolf, *A History of Science, Technology and Philosophy in the Eighteenth Century,* 2nd ed. (New York, 1952), pp. 251, 272-273.

JOHN L. HEILBRON

CANTOR, GEORG (*b.* St. Petersburg, Russia, 3 March 1845; *d.* Halle, Germany, 6 January 1918), *mathematics, set theory.*

Cantor's father, Georg Waldemar Cantor, was a successful and cosmopolitan merchant. His extant letters to his son attest to a cheerfulness of spirit and deep appreciation of art and religion. His mother, Marie Böhm, was from a family of musicians. Her forebears included renowned violin virtuosi; and Cantor described himself also as "rather artistically inclined," occasionally voicing regrets that his father had not let him become a violinist.

Like his father, Cantor was a Protestant; his mother was Catholic. The link with Catholicism may have made it easier for him to seek, later on, support for his philosophical ideas among Catholic thinkers.

Cantor attended the Gymnasium in Wiesbaden, and later the Grossherzoglich-Hessische Realschule in Darmstadt. It was there that he first became interested in mathematics. In 1862 he began his university studies in Zurich, resuming them in Berlin in 1863, after the sudden death of his father. At that time Karl Weierstrass, famed as a teacher and as a researcher, was attracting many talented students to the University of Berlin. His lectures gave analysis a firm and precise foundation, and later many of his pupils proudly proclaimed themselves members of the "Berlin school" and built on the ideas of their teacher.

Cantor's own early research on series and real numbers attests to Weierstrass' influence, although in Berlin he also learned much from Kummer and Kronecker. His dissertation, *De aequationibus secundi gradus indeterminatis,* dealt with a problem in number theory and was presented to the department by Kummer. In those days it was still the custom for a doctoral candidate to have to defend his scholarly theses against some of his fellow students. Worthy of note is Cantor's third thesis, presented on receiving his doctor's degree in 1867: *In re mathematica ars proponendi pluris facienda est quam solvendi.* And indeed his later achievements did not always consist in *solving* problems. His unique contribution to mathematics was that his special way of asking questions opened up vast new areas of inquiry, in which the problems were solved partly by him and partly by successors.

In Berlin, Cantor was a member (and from 1864 to 1865 president) of the Mathematical Society, which sought to bring mathematicians together and to further their scientific work. In his later years he actively worked for an international union of mathematicians, and there can have been few other scholars who did as much as he to generate and promote the exchange of ideas among scientists. He conceived a plan to establish an Association of German Mathematicians and succeeded in overcoming the resistance to it. In 1890 the association was founded, and Cantor served as its first president until 1893. He also pressed for international congresses of mathematicians and was responsible for bringing about the first ever held, in Zurich in 1897.

Thus Cantor was no hermit living within his own narrow science. When, later, he did sever ties with many of his early friends—as with H. A. Schwarz in the 1880's—the reasons lay in the nature of his work rather than in his character. During the Berlin years a special friendship had grown up between him and Schwarz, who was two years his senior. Both revered their teacher, Weierstrass; and both were concerned to gain the good opinion of Kronecker, who frequently criticized the deductions of Weierstrass and his pupils as unsound. These first years of Cantor's early research were probably the happiest of his life. His letters from that period radiate a contentment

seldom granted him in later times, when he was struggling to gain acceptance for his theory of sets.

In 1869 Cantor qualified for a teaching position at the University of Halle, soon becoming associate professor and, in 1879, full professor. He carried on his work there until his death. His marriage in 1874 to Vally Guttmann was born of deep affection, and the sunny personality of the artistically inclined "Frau Vally" was a happy counter to the serious, often melancholy, temperament of the great scholar. They had five children, and an inheritance from his father enabled Cantor to build his family a house. In those days a professor at Halle was so poorly paid that without other income he would have been in financial straits. It was Cantor's hope to obtain a better-endowed, more prestigious professorship in Berlin, but in Berlin the almost omnipotent Kronecker blocked the way. Completely disagreeing with Cantor's views on "transfinite numbers," he thwarted Cantor's every attempt to improve his standing through an appointment to the capital.

Recognition from abroad came early, however. Cantor's friend Mittag-Leffler accepted his writings for publication in his then new *Acta mathematica.* He became an honorary member of the London Mathematical Society (1901) and of other scientific societies, receiving honorary doctor's degrees from Christiania (1902) and St. Andrews (1911).

The closing decades of Cantor's life were spent in the shadow of mental illness. Since 1884 he had suffered sporadically from deep depression and was often in a sanatorium. He died in 1918 in Halle University's psychiatric clinic. Schoenfliess was of the opinion that his health was adversely affected by his exhausting efforts to solve various problems, particularly the continuum problem, and by the rejection of his pioneering work by other eminent mathematicians.

Cantor has gone down in history as the founder of set theory, but the science of mathematics is equally indebted to him for important contributions in classical analysis. We mention here his work on real numbers and on representation through number systems. In his treatise on trigonometric series, which appeared in 1872,[1] he introduced real numbers with the aid of "fundamental series." (Today we call them fundamental sequences or Cauchy sequences.) They are sequences of rational numbers $\{a_n\}$ for which, given an arbitrarily assumed (positive, rational) ϵ, we have an integer n_1 such that

$$|a_{n+m} - a_n| < \epsilon$$

if $n \geqq n_1$ and m is an arbitrary positive integer. To series having this property Cantor assigned a "limit" *b.* If $\{a_n'\}$ is a second sequence of the same kind and if $|a_n - a_n'| < \epsilon$ for sufficiently large n, then the same limit b is assigned to this second sequence. We say today that the real number b is defined as an equivalence class of fundamental sequences.

Cantor further showed[2] that any positive real number r can be represented through a series of the type

$$r = c_1 + \frac{c_2}{2!} + \frac{c_3}{3!} + \frac{c_4}{4!} + \cdots \qquad (1)$$

with coefficients c_γ that satisfy the inequality

$$0 \leqq c_\gamma \leqq \gamma - 1.$$

Series such as (1) are known today as Cantor series. The work also contains a generalization of representation (1) and representations of real numbers in terms of infinite products. With these writings (and with several remarkable studies on the theory of Fourier series) Cantor established himself as a gifted pupil of the Weierstrass school. His results extended the work of Weierstrass and others by "conventional" means.

In November 1873, in an exchange of letters with his colleague Dedekind in Brunswick, a question arose that would channel all of Cantor's subsequent scientific labor in a new direction. He knew that it was possible to "count" the set of rational numbers, i.e., to put them into a one-to-one correspondence with the set of natural numbers, but he wondered whether such one-to-one mapping were not also possible for the set of real numbers. He believed that it was not, but he "could not come up with any reason." A short time later, on 2 December, he confessed that he "had never seriously concerned himself with the problem, since it seemed to have no practical value," adding, "I fully agree with you when you say that it is therefore not worth very much trouble."[3] Nevertheless, Cantor did further busy himself with the mapping of sets, and by 7 December 1873 he was able to write Dedekind that he had succeeded in proving that the "aggregate" of real numbers was uncountable. That date can probably be regarded as the day on which set theory was born.[4] Dedekind congratulated Cantor on his success. The significance of the proof had meanwhile become clear, for in the interim Cantor (and probably Dedekind, independently) had succeeded in proving that the set of real algebraic numbers is countable. Here, then, was a new proof of Liouville's theorem that transcendental numbers do exist. The first published writing on set theory is found in *Crelle's Journal* (1874).[5] That work, "Über eine Eigenschaft des Inbegriffes aller reellen algebraischen Zahlen," contained more than the title

indicated, including not only the theorem on algebraic numbers but also the one on real numbers, in Dedekind's simplified version, which differs from the present version in that today we use the "diagonal process," then unknown.[6]

Following his initial successes, Cantor tackled new and bolder problems. In a letter to Dedekind dated 5 January 1874, he posed the following question:

> Can a surface (say, a square that includes the boundaries) be uniquely referred to a line (say, a straight-line segment that includes the end points) so that for every point of the surface there is a corresponding point of the line and, conversely, for every point of the line there is a corresponding point of the surface? Methinks that answering this question would be no easy job, despite the fact that the answer seems so clearly to be "no" that proof appears almost unnecessary.[7]

The proof that Cantor had in mind was obviously a precise justification for answering "no" to the question, yet he considered that proof "almost unnecessary." Not until three years later, on 20 June 1877, do we find in his correspondence with Dedekind another allusion to his question of January 1874. This time, though, he gives his friend reasons for answering "yes." He confesses that although for years he had believed the opposite, he now presents Dedekind a line of argument proving that the answer to the following (more general) question was indeed "yes":

> We let $x_1, x_2, \cdots, x_\rho$ be ρ independent variable real quantities, each capable of assuming all values $\geqq 0$ and $\leqq 1$. We let y be a $\sqrt{(\rho + 1)}$ st variable real quantity with the same free range $\left(y \begin{smallmatrix} \geqq 0 \\ \leqq 1 \end{smallmatrix}\right)$. Does it then become possible to map the ρ quantities $x_1, x_2, \cdots, x_\rho$ onto the one y so that for every defined value system $(x_1, x_2, \cdots, x_\rho)$ there is a corresponding defined value y and, conversely, for every defined value y one and only one defined value system $(x_1, x_2, \ldots, x_\rho)$?[8]

Thus for $\rho = 2$ we again have the old problem: Can the set of points of a square (having, say, the coordinates x_1 and x_2, $0 \leqq x_\gamma \leqq 1$) be mapped in a one-to-one correspondence onto the points of a line segment (having, say, the coordinates y, $0 \leqq y \leqq 1$)?

Today we are in a position to answer this question affirmatively with a very brief proof.[9] Cantor's original deduction[10] was still somewhat complicated, but it was correct; and with it he had arrived at a result bound to seem paradoxical to the mathematicians of his day. Indeed, it looked as if his mapping had rendered the concept of dimension meaningless. But Dedekind recognized immediately that Cantor's map of a square onto a line segment was discontinuous, suspecting that a continuous one-to-one correspondence between sets of points of different dimensions was not possible. Cantor attempted to prove this, but his deduction[11] did not stand up. It was Brouwer, in 1910, who finally furnished a complete proof of Dedekind's supposition.

Cantor's next works, dealing with the theory of point sets, contain numerous definitions, theorems, and examples that are cited again and again in modern textbooks on topology. The basic work on the subject by Kuratowski[12] contains in its footnotes many historical references, and it is interesting to note how many of the now generally standard basic concepts in topology can be traced to Cantor. We mention only the "derivation of a point set," the idea of "closure," and the concepts "dense" and "dense in itself." A set that was closed and dense in itself Cantor called "perfect," and he gave a remarkable example of a perfect discontinuous set. This "Cantor set" can be defined as the set of all points of the interval $]0; 1[$ "which are contained in the formula

$$z = \frac{c_1}{3} + \frac{c_2}{3^2} + \frac{c_3}{3^3} + \cdots$$

where the coefficients c_γ must arbitrarily assume the values 0 or 2, and the series may consist of both a finite and an infinite number of terms."[13] It was Cantor, too, who provided the first satisfactory definition of the term "continuum," which had appeared as early as the writings of the Scholastics. A continuous perfect set he called a continuum, thereby turning that concept, until then very vague, into a useful mathematical tool. It should be noted that today a continuum is usually introduced as a compact continuous set, a definition no longer matching Cantor's. The point is, though, that it was he who provided the first definition that was at all usable.

With his first fundamental work, in 1874,[14] Cantor showed that, with the aid of the one-to-one correspondence, it becomes possible to distinguish differences in the infinite: There are *countable* sets and there are sets having the power of a *continuum*. Of root importance for the development of general set theory was the realization that for every set there is a set of higher power. Cantor substantiated this initially through his theory of ordinal numbers. It can be seen much more simply, though, through his subset theorem, which appears in his published writings in only one place,[15] and there for only one special case. But in a letter to Dedekind dated 31 August 1899,[16] we find a remark to the effect that the so-called "diagonal process," which Cantor had been using, could be

applied to prove the general subset theorem. The essence of his proof was the observation that there can be no one-to-one correspondence between a set L and the set M of its subsets. To substantiate this, Cantor introduced functions $f(x)$ that assign to the elements x of a set L the image values 0 or 1. Each such function defines a subset of L—the set, say, for which $f(x) = 0$—and to each subset $L' \subset L$ there belongs a function $f(x)$, which becomes 0 precisely when x belongs to L'.

Now, if there were a one-to-one correspondence between L and M, the set of functions in question could be written in the form $\varphi(x,z)$

> ... so that through each specialization of z an element $f(x) = \varphi(x,z)$ is obtained and, conversely, each element $f(x)$ of M is obtained from $\varphi(x,z)$ through a particular specialization of z. But this leads to a contradiction, because, if we take $g(x)$ as the single-valued function of x which assumes only the values 0 or 1 and is different from $\varphi(x,x)$ for each value of x, then $g(x)$ is an element of M on the one hand, while, on the other, $g(x)$ cannot be obtained through any specialization $z = z_0$, because $\varphi(z_0,z_0)$ is different from $g(z_0)$.[17]

According to the subset theorem, for each set there is a set of higher power: the set of subsets or the power set $P(M)$. The question of a general "set theory" thus became acute. Cantor regarded his theory as an expansion of classical number theory. He introduced "transfinite" numbers (cardinal numbers, ordinal numbers) and developed an arithmetic for them. With these numbers, explained in terms of transfinite sets, he had, as Gutzmer remarked on the occasion of Cantor's seventieth birthday in 1915, opened up "a new province" for mathematics.

Understandably, the first probing steps taken in this new territory were shaky. Hence, the definition of the basic concept has undergone some noteworthy modifications. In Cantor's great synoptic work in the *Mathematische Annalen* of 1895 we read:

> We call a power or cardinal number that general concept which with the aid of our active intelligence is obtained from the set M by abstracting from the nature of its different elements m and from the order in which they are given. Since every individual element m, if we disregard its nature, becomes a 1, the cardinal number M itself is a definite set made up merely of ones.[18]

Modern mathematics long ago dropped this definition, and with good reason. Today two sets are called "equal" if they contain the same elements, however often they are named in the description of the set. So if between the braces customarily used in defining these sets we place several 1's, we then have the set with the single element 1: e.g., $\{1,1,1,1,1\} = \{1\}$.

Cantor himself may have sensed the inadequacy of his first definition. In discussing a book in 1884, and later in 1899 in a letter to Dedekind, he called a power "that general concept that befits it and all its equivalent sets." Today we would say, more simply, that "a cardinal number is a set of equivalent sets." But this second definition turns out to be inadequate, too. We know, of course, that the concept of a "set of all sets" leads to contradictions. From this, though, it follows that the concept of a "set of all sets equivalent to a given set M" is also inconsistent. We let M, for example, be a given infinite set and

$$M^* = M \cup \{\mathfrak{M}\},$$

where $\mathfrak{M}$ encompasses the set of all sets. The set $\{\mathfrak{M}\}$ then naturally has the power 1, and the sets M^* (which contain only one element more than M) are all equivalent to M. Accordingly, the system of the sets M is a genuine subset of the "set of all sets equivalent to M." But since we can map this system into the elements of the "set of all sets," we thus have a concept that must lead to antinomies.

In short, in all of Cantor's works we find no usable definition of the concept of the cardinal number. The same is true of the ordinal number.

But the story does not end there. A third Cantor definition of a cardinal number appears in a report by Gerhard Kowalewski, included in his biography, *Bestand und Wandel,* of meetings that he had had with Georg Cantor. The eighty-year-old Kowalewski wrote the book shortly before his death, around 1950. With graphic vividness he tells of events and meetings that had occurred a half century before. Around 1900, Kowalewski was a *Privatdozent* in Leipzig. In those days the mathematicians from Halle and Leipzig used to meet about twice a month, first in one city, then in the other. On these occasions they would discuss their work. Although by then Cantor was no longer publishing, he often, according to the report of a young colleague, impressively held forth at the meetings on his theory of manifolds. This included his studies on the "number classes," the set of ordinal numbers that belong to equivalent sets. The numbers of the second number class were, for example, the ordinal numbers of the countable sets. Kowalewski then discusses powers (which were called "alephs"):

This "power" can also be represented, as was Cantor's wont, by the smallest or initial number of that number class, and the alephs can be identified with these initial numbers so that such and such would be the case if we wished to use those designations for the initial terms of the second and third number class from the Schoenfliess report on set theory.[19]

In a modern book on set theory we find the following definition of a cardinal number: "A cardinal number is an ordinal number which is not similar to any smaller ordinal number."[20] So modern mathematics has adopted Cantor's belated definition, which is not to be found anywhere in his published writings. It is unlikely that Stoll, author of the above book, ever read Kowalewski's biography. The modern view of the cardinal-number concept lay dormant for a time, to be embraced by younger scholars. One should mention, though, that even Cantor himself finally did arrive at the definition of a cardinal number considered "valid" today. True, this modern version of the concept does presuppose the availability of an ordinal-number concept. Here again we cannot accept Cantor's classical definition, for roughly the same reasons that prevent our accepting the early versions of the cardinal-number concept. Today, according to John von Neumann, an ordinal number is described as a well-ordered set w in which every element $v \in w$ is equal to the segment generated by v:

$$v = A_v.$$

To sum up, we are not indebted to Cantor only for his initiative in developing a theory of transfinite sets. He proved the most important theorems of the new theory himself and laid the groundwork for the present-day definitions of the concept. It would be silly to hold against him the fact that his initial formulations did not fully meet modern precision requirements. One who breaks new ground in mathematics needs a creative imagination, and his initial primitive definitions cannot be expected to stand up indefinitely. When Leibniz and Newton founded infinitesimal calculus, their definitions were crude indeed compared with the elegant versions of centuries later, refined by Weierstrass and his pupils. The same is true of the beginnings of set theory, and it is worthy of note that Cantor himself was inching ever closer to the definitions that the present generation accepts as "valid."

By his bold advance into the realm of the infinite, Cantor ignited twentieth-century research on the fundamentals. Hilbert refused to be driven out of the "paradise" that Cantor had created. But Cantor himself was not an axiomaticist. His way of thinking belonged more to the classical epoch. In the annotations to his *Grundlagen einer allgemeinen Mannigfaltigkeitslehre* (1883) he expressly acknowledged his adherence to the "principles of the Platonic system," although he also drew upon Spinoza, Leibniz, and Thomas Aquinas.

For Cantor the theory of sets was not only a mathematical discipline. He also integrated it into metaphysics, which he respected as a science. He sought, too, to tie it in with theology, which used metaphysics as its "scientific tool."[21] Cantor was convinced that the actually infinite really existed "both concretely and abstractly." Concerning it he wrote: "This view, which I consider to be the sole correct one, is held by only a few. While possibly I am the very first in history to take this position so explicitly, with all of its logical consequences, I know for sure that I shall not be the last!"[22]

Mathematicians and philosophers oriented toward Platonic thought accepted actual infinity abstractly, but not concretely. In a remarkable letter to Mittag-Leffler, Cantor wrote that he believed the atoms of the universe to be countable, and that the atoms of the "universal ether" could serve as an example of a set having the power of a continuum. Present-day physicists are not likely to be much interested in these quaint opinions. Today his philosophical views also appear antiquated. When we now ask what is left of Cantor's work, we can answer very simply: Everything formalizable is left. His statements in the realm of pure mathematics have been confirmed and extended by subsequent generations, but his ideas and conceptions in that of physics would not be acceptable to most of the present generation.

To the end Cantor believed that the basis of mathematics was metaphysical, even in those years when Hilbert's formalism was beginning to take hold. Found among his papers after his death was a shakily written penciled note (probably from 1913) in which he reaffirms his view that "without some grain of metaphysics" mathematics is unexplainable. By metaphysics he meant "the theory of being." There are several important theorems in set theory that were first stated by Cantor but were proved by others. Among these is the Cantor-Bernstein equivalence theorem: "If a set A is equivalent to a subset $B' \subset B$ and B to a subset $A' \subset A$, then A and B are equivalent." A simple proof of this, first demonstrated by Cantor's pupil Bernstein, is found in a letter from Dedekind to Cantor.[23] That every set can be well ordered was first proved by Zermelo with the aid of the axiom of choice. This deduction provoked many disagreements because a number of constructivists objected to pure "existence theorems" and were critical of the paradoxical consequences of the axiom of choice.

More fundamental, though, were the discussions about the antinomies in set theory. According to a theorem proved by Cantor, for every set of ordinal numbers there is one ordinal number larger than all the ordinal numbers of the set. One encounters a contradiction when one considers the set of all ordinal numbers. Cantor mentioned this antinomy in a letter to Hilbert as early as 1895. A much greater stir was caused later by the Russell antinomy, involved in the "set of all sets which do not contain themselves as an element." It was chiefly Hilbert who was looking for a way out of the impasse, and he proposed a strict "formalization" of set theory and of all mathematics. He hoped thereby to save the "paradise" that Cantor had created. We do not have time or space here to dwell upon the arguments that raged around Hilbert's formalism. Suffice it to say that today's generally recognized structural edifice of mathematics is formalistic in the Hilbertian sense. The concept of the set is preeminent throughout.

When we pick up a modern book on probability theory or on algebra or geometry, we always read something about "sets." The author may start with a chapter on formal logic, usually followed by a section on set theory. And this specialized discipline is described as the theory of certain classes of sets. An algebraic structure, say, is a set in which certain relations and connections are defined. Other sets, defined by axioms concerning "neighborhoods," are called spaces. In probability theory we are concerned with sets of events and such.

In Klaua's *Allgemeine Mengenlehre* there is a simple definition of mathematics: "Mathematics is set theory." Actually, we can regard all mathematical disciplines as the theory of special classes of sets. True, a high price (in Cantor's eyes) was paid for this development. Modern mathematics deals with formal systems; and Cantor, probably the last great Platonist among mathematicians, never cottoned to the then nascent formalism. For him the problem of the continuum was a question in metaphysics. He spent many years attempting to show that there can be no power between that of the countable sets and the continuum. In recent time it has been shown (by Gödel and Cohen) that the continuum hypothesis is independent of the fundamental axioms of the Zermelo-Fraenkel system. That solution of the problem would not have been at all to Cantor's taste. Yet had he not defended, against his antagonist Kronecker, the thesis that the essence of mathematics consists in its freedom?[24] Does this not include the freedom for a theory created by Cantor to be interpreted in a way not in conformity with his original ideas? The fact that his set theory has influenced the thinking of the twentieth century in a manner not in harmony with his own outlook is but another proof of the objective significance of his work.

NOTES

1. *Gesammelte Abhandlungen,* pp. 92 ff.
2. *Ibid.,* pp. 35 ff.
3. E. Noether and J. Cavaillès, *Briefwechsel Cantor–Dedekind,* pp. 115–118.
4. The first version of Cantor's proof (in his letter to Dedekind) was published in E. Noether and J. Cavaillès, *Briefwechsel Cantor–Dedekind,* pp. 29–34, and by Meschkowski, in *Probleme des Unendlichen,* pp. 30 ff.
5. *Gesammelte Abhandlungen,* pp. 115 ff.
6. See, for instance, Meschkowski, *Wandlungen des mathematischen Denkens,* pp. 31 ff.
7. E. Noether and J. Cavaillès, *Briefwechsel Cantor–Dedekind,* pp. 20–21.
8. *Ibid.,* pp. 25–26.
9. Meschkowski, *Wandlungen des mathematischen Denkens,* pp. 32 ff.
10. *Crelle's Journal,* **84** (1878), 242–258; *Gesammelte Abhandlungen,* pp. 119 ff.
11. *Gesammelte Abhandlungen,* pp. 134 ff.
12. K. Kuratowski, *Topologie I, II* (Warsaw, 1952).
13. *Gesammelte Abhandlungen,* p. 193.
14. "Über eine Eigenschaft des Inbegriffes aller reellen algebraischen Zahlen," in *Crelle's Journal,* **77** (1874), 258–262; also in *Gesammelte Abhandlungen,* pp. 115 ff.
15. *Gesammelte Abhandlungen,* pp. 278 ff.; *Jahresbericht der Deutschen Mathematikervereinigung,* **1** (1890–1891), 75–78.
16. *Gesammelte Abhandlungen,* p. 448.
17. *Ibid.,* p. 280.
18. *Mathematische Annalen,* **46** (1895), 481
19. G. Kowalewski, *Bestand und Wandel,* p. 202.
20. R. Stoll, *Introduction to Set Theory and Logic* (San Francisco, 1961), p. 317.
21. See Meschkowski, *Probleme des Unendlichen,* ch. 8.
22. *Gesammelte Abhandlungen,* p. 371.
23. *Ibid.,* p. 449.
24. *Ibid.,* p. 182.

BIBLIOGRAPHY

I. Original Works. Cantor's important mathematical and philosophical works are contained in *Gesammelte Abhandlungen mathematischen und philosophischen Inhalts,* E. Zermelo, ed. (Berlin, 1930; repr. Hildesheim, 1962), with a biography of Cantor by A. Fraenkel.

There are several other publications on mathematical and theological questions and on the Shakespeare-Bacon problem, which greatly interested Cantor for a time: "Vérification jusqu'à 1000 du théorème empirique de Goldbach," in Association Française pour l'Avancement des Sciences, *Comptes rendus de la 23ᵐᵉ session (Caen 1894),* pt. 2 (1895), 117–134; "Sui numeri transfiniti. Estratto d'una lettera di Georg Cantor a G. Vivanti, 13 dicembre 1893," in *Rivista di matematica,* **5** (1895), 104–108 (in German); "Lettera di Georg Cantor a G. Peano," *ibid.,* 108–109 (in German); "Brief von Carl Weierstrass über das Dreikörperproblem," in *Rendiconti del Circolo matematico di Palermo,* **19** (1905), 305–308; *Resurrectio Divi Quirini Francisci Baconi Baronis de Verulam Vicecomitis Sancti Albani CCLXX annis post obitum eius IX die aprilis anni MDCXXVI* (Halle, 1896), with an English preface by Cantor; *Confessio fidei Francisci Baconi Baronis de Verulam . . . cum versione Latina a. G. Rawley . . . nunc denuo typis excusa cura et impensis G. C.*
(Halle, 1896), with a preface in Latin by Cantor; *Ein Zeugnis zugunsten der Bacon-Shakespeare-Theorie mit einem Vorwort herausgegeben von Georg Cantor* (Halle, 1897); *Ex oriente lux* (Halle, 1905), conversations of a teacher with his pupil on important points of documented Christianity; and *Contributions to the Founding of the Theory of Transfinite Numbers* (Chicago–London, 1915).

Cantor's letters have been published in a number of works listed below. A fairly large number of letters and a complete list of all the published letters appear in Meschkowski's *Probleme des Unendlichen* (see below).

II. Secondary Literature. On Cantor or his work see J. Bendiek, "Ein Brief Georg Cantors an P. Ignatius Jeiler OFM," in *Franziskanische Studien,* **47** (1965), 65–73; A. Fraenkel, "Georg Cantor," in *Jahresberichte der Deutschen Mathematikervereinigung,* **39** (1930), 189–266; H. Gericke, "Aus der Chronik der Deutschen Mathematikervereinigung," *ibid.,* **68** (1966), 46–70; G. Kowalewski, *Bestand und Wandel* (Munich, 1950), 106–109, 198–203, 207–210; *Lied eines Lebens: 1875–1954* (n.p., n.d.), pp. 1–10, an anonymous, privately printed biography of Else Cantor; W. Lorey, "Der 70. Geburtstag des Mathematikers Georg Cantor," in *Zeitschrift für mathematischen und naturwissenschaftlichen Unterricht,* **46** (1915), 269–274; H. Meschkowski, *Denkweisen grosser Mathematiker* (Brunswick, 1961), pp. 80–91; "Aus den Briefbüchern Georg Cantor," in *Archive for the History of Exact Sciences,* **6** (1965), 503–519; and *Probleme des Unendlichen. Werk und Leben Georg Cantors* (Brunswick, 1968); E. Noether and J. Cavaillès, *Briefwechsel Cantor–Dedekind* (Paris, 1937); B. Russell, *Portraits From Memory and Other Essays* (London, 1956), pp. 24–25; and *The Autobiography of Bertrand Russell 1872–1914* (London, 1967), pp. 217–220; A. Schoenfliess, "Zur Erinnerung an Georg Cantor," in *Jahresberichte der Deutschen Mathematikervereinigung,* **31** (1922), 97–106; and "Die Krisis in Cantors mathematischen Schaffen," in *Acta mathematica,* **50** (1927), 1–23; J. Ternus, "Zur Philosophie der Mathematik," in *Philosophisches Jahrbuch der Görres-Gesellschaft,* **39** (1926), 217–231; and "Ein Brief Georg Cantors an P. Joseph Hontheim S.J.," in *Scholastik,* **4** (1929), 561–571; A. Wangerin, "Georg Cantor," in *Leopoldina,* **54** (1918), 10–13, 32; and W. H. Young, "The Progress of Mathematical Analysis in the Twentieth Century," in *Proceedings of the London Mathematical Society,* **24** (1926), 412–426.

H. Meschkowski

CANTOR, MORITZ BENEDIKT (*b.* Mannheim, Germany, 23 August 1829; *d.* Heidelberg, Germany, 9 April 1920), *mathematics.*

Cantor's father, Isaac Benedikt Cantor, was from Amsterdam; his mother, Nelly Schnapper, was the daughter of a money changer. Cantor married Tilly Gerothwohl, from Frankfurt am Main.

Cantor was first taught by private tutors and completed his secondary education at the Gymnasium in Mannheim. In 1848 he began studying at Heidelberg under Franz Schweins and Arthur Arneth, and from 1849 to 1851 he worked at Göttingen under Gauss and Moritz Stern. He took his degree at Heidelberg in 1851 with the thesis *Ein wenig gebräuchliches Coordinatensystem.* During the summer semester of 1852 he studied in Berlin under Dirichlet and Jakob Steiner and qualified for inauguration at Heidelberg in 1853 with *Grundzüge einer Elementar-Arithmetik.*

Cantor was greatly influenced by Arneth's *Geschichte der reinen Mathematik in ihrer Beziehung zur Entwicklung des menschlichen Geistes* and was encouraged in his work by Stern and the cultural philosopher E. M. Roeth. During a stay in Paris he became a close friend of Chasles and of Joseph Bertrand. From 1860 he lectured on the history of mathematics. In 1863, as a result of his *Mathematischen Beiträge zum Culturleben der Völker,* Cantor was appointed extraordinary professor. The *Römischen Agrimensoren und ihre Stellung in der Geschichte der Feldmesskunst* (1875) led to his appointment as honorary professor; in 1908 he became full professor, and in 1913 he became emeritus.

From 1856 to 1860 Cantor was coeditor of *Kritischen Zeitschrift für Chemie, Physik und Mathematik,* and from 1860 of *Zeitschrift für Mathematik und Physik;* from 1877 to 1899 he edited *Abhandlungen zur Geschichte der Mathematik.* He published many short papers and reviews in periodicals devoted to pure mathematics and the history of science and, from 1875, wrote most of the biographies of mathematicians in *Allgemeine Deutsche Biographie.*

Together with Curtze and Günther, Cantor was one of the leading historians of mathematics in Germany at the turn of the century. He is best known for the once highly praised *Vorlesungen über Geschichte der Mathematik,* which, despite many contemporary emendations (such as those of Braunmühl, Gustaf Eneström, and Wieleitner), has not been equaled in content and extent. Although the *Vorlesungen* is now dated, it gave a definite impetus to the development of the history of mathematics as a scholarly discipline.

BIBLIOGRAPHY

I. ORIGINAL WORKS. Cantor's best-known work is *Vorlesungen über Geschichte der Mathematik,* 4 vols. (Leipzig, 1880–1908). His other writings include *Ein wenig gebräuchliches Coordinatensystem* (Frankfurt, 1851); *Grundzüge einer Elementar-Arithmetik* (Heidelberg, 1855); *Mathematischen Beiträge zum Culturleben der Völker* (Halle, 1863); and *Römischen Agrimensoren und ihre Stellung in der Geschichte der Feldmesskunst* (Leipzig, 1875).

II. SECONDARY LITERATURE. Writings on Cantor or his work are Karl Bopp, "M. Cantor †, Gedächtnisrede, gehalten im Mathematischen Verein zu Heidelberg am 19.VI.1920," in *Sitzungsberichte der Heidelberger Akademie der Wissenschaften,* Math.-nat. Kl., Abt. A (1920); and "Cantor, M.," in *Deutsches biographisches Jahrbuch,* II (Stuttgart-Berlin-Leipzig, 1928), 509–513, with complete bibliography; M. Curtze, "Verzeichnis der mathematischen Werke, Abhandlungen und Recensionen von M. Cantor," in *Zeitschrift für Mathematik und Physik,* **44,** supp. (1899), 625–650, with portrait; and "Zum 70. Geburtstage M. Cantors," in *Bibliotheca mathematica,* 3rd ser., **1** (1900), 227–231; and J. E. Hofmann, "Cantor, M. B.," in *Neue Deutsche Biographie,* III (Berlin, 1957), 129.

J. E. HOFMANN

CAPRA, BALDASSAR or **BALDESAR** (*b.* Milan, Italy, *ca.* 1580; *d.* Milan, 8 May 1626), *astronomy.*

Capra was the son of Aurelio Capra and Ippolita Della Croce. Corte states that he was a good philosopher, doctor, and astronomer; Argelati adds that he received his degree in medicine. Both these authors, besides giving a list of his works in Latin, emphasize the nobility of his family, which included Galeazzo Capra (or Capella), who was secretary and historian to Duke Francesco II, the last of the Sforzas of Milan. But Capra is known today only because he was one of the first opponents of Galileo, whom he attacked unjustly in 1605 regarding his observations on the new star which had appeared the year before, and whom he gravely offended two years later by plagiarizing the first book Galileo submitted for publication. He published as his own a brief Latin treatise on the proportional compass, which proved to be scarcely more than a translation of Galileo's *Le operazioni del compasso geometrico e militare,* a translation not without defects and, in addition, sprinkled with malevolent insinuations against the real author.

When he was a little more than ten years old, Capra was taken to Padua by his father, who augmented his income by teaching fencing and surgery. Galileo himself, according to letters and the *Difesa* that he wrote in 1607 "against the calumnies and fraud of the Milanese Baldessar Capra," had procured for Aurelio Capra a student and a profitable consultation. Yet Capra's father entrusted his son to Simon Mayr, a teacher from abroad, who Latinized his surname to Marius and, coming from Prague, boasted of having studied under Tycho Brahe and Johannes Kepler (even though he had stayed in that city for less than a year). In Padua, where he arrived at the end of 1601, Mayr attended the course in medicine and, later, Galileo's classes; and although he had enemies who called him an "unskilled and bad astrologer," he must have distinguished himself and gained importance, for he was councillor of the German nation in 1604 and in 1605, the year in which he returned to his native land, where he became the court mathematician to the margrave of Brandenburg, who had financed his studies. Mayr's influence on Capra and on his works is evident not only in the work published in the year of his departure but also in the subsequent ones, which certainly had been begun before he left Italy.

Capra's first work, published in Padua in 1605 (and reprinted later in the National Edition of Galileo's *Works*, vol. II), is entitled *Consideratione astronomica circa la nova et portentosa stella che nell'anno 1604 a dì 10 ottobre apparse, con un breve giudicio delli suoi significati, di Baldesar Capra, gentil'homo milanese, studioso di astronomia e medicina*. Although it is not among the works listed by Corte and Argelati, it is perhaps the one of greatest interest and value. It is written in Italian, for which he apologizes in the dedication to his maternal uncle, Giovanni Antonio Della Croce, stating, "If one considers the material which is dealt with, it should have, for every reason, been described [written entirely] in Latin, which is more excellent and worthy." This is probably an excuse, for he finds justification in his having to refute the nonsense written, also in Italian, by a Peripatetic (probably Antonio Lorenzini of Montepulciano, in his *Discorso intorno alla nuova stella*). Galileo had polemized against Lorenzini in his classes and in a popular dialogue, which appeared at the same time under the pseudonym of Cecco de' Ronchitti.

Capra also maintains, against the Aristotelian thesis, that the new star belongs to the sphere of fixed objects and gives ample and precise information on its position and the variations of its color and its size, demonstrating that he made careful observations. At the end there is an attempt at an astrological interpretation, which should not be surprising, since astrology was still popular—in fact, astrology is also present in the similar booklet by Kepler. In very bad taste, on the other hand, are the frequent taunts at Galileo, on secondary questions and often with an air of pretext, almost as if the author were looking for controversy. For the time being, however, Galileo did not feel it necessary to answer, nor did he wish anyone else to do so.

In 1606, in Padua, Capra published two booklets in Latin, both dedicated to Cardinal Federigo Borromeo and each consisting of fourteen pages. Modest also in their content, they were entitled *Tyrocinia astronomica*, in which Capra prudently appeals as much to the authority of Tycho Brahe as to that of Ptolemy, and *Disputationes duae, una de logica et eius partibus, altera de enthymemate* (the enthymeme, the well-known abbreviated syllogism, implies one of the premises).

The work to which he devoted the greatest effort and the most diligence, *Usus et fabrica circini cuiusdam proportionis, per quem omnia fere tum Euclidis, tum mathematicorum omnium problemata facili negotio resolvuntur*, consisting of fifty-six pages with numerous woodcuts and one copperplate on the title page, was published at the beginning of 1607 but was probably the result of several years of work, even though it shows very little originality. It explains the use and construction of a proportional compass by means of which, with a few simple operations, one can solve almost all the problems of both geometry and mathematics; and the clear engravings on the title page illustrate the use of the proportional compass with the aid of a standard compass. The work is dedicated to Joachim Ernest of Brandenburg, to whose court Simon Mayr had returned after completing his studies. Because of the close analogies with Galileo's *Il compasso*, which had appeared a few months before, this book was clamorously condemned to sequestration and destruction, but for the same reason it was fully analyzed by Galileo's students and included in all editions of his works, beginning with the one published at Bologna in 1655.

To all that has been said on this subject, it is possible to add that probably Capra, while his teacher was still in Padua, had set out to give greater authority (by putting them in Latin) to the handwritten instructions on use of the compass that had been publicly dictated by Galileo but generally were circulated in the "adespota" form (anonymously), which meant they were both without author and without owner. This at least explains the imprudent and foolish attempt by Capra, who, if he had only cited his source and renounced the absurd pretext of priority, would have had both fame and advantage, including economic benefit, from that work. That unfortunate attempt was instead his ruin. Having left Padua quite precipitously, he apparently returned to Milan, where he kept in contact with the anti-Galileans, including Horky; but at the end of 1620, when he asked to register at the medical college of Milan, Lodovico Settala energetically opposed him because of his behavior toward Galileo. There is no further news of him until his death, which came shortly after his forty-fifth birthday.

BIBLIOGRAPHY

On Capra or his work, see the following (listed chronologically): Bartolomeo Corte, *Notizie istoriche intorno a' medici scrittori milanesi . . .* (Milan, 1718); Filippo Argelati, *Bibliotheca scriptorum Mediolanensium* (Milan, 1745); Pietro Riccardi, *Biblioteca matematica italiana* (Modena, 1870–1880); Antonio Favaro, *Galileo Galilei e lo Studio di Padova*, I (Florence, 1883), 182–192; *Scampoli galileiani*,

CVI, *Intorno alla stampa della Difesa contro il Capra* (Padua, 1906); *Amici e corrispondenti di Galileo,* XVII (Venice, 1906); and *Per la storia del compasso di proporzione* (Venice, 1908); and the Edizione Nazionale of Galileo's *Opere,* XX (Florence, 1909), biographical index.

GIORGIO TABARRONI

CARAMUEL Y LOBKOWITZ, JUAN (*b.* Madrid, Spain, 23 May 1606; *d.* Milan, Italy, 7 September 1682), *mathematics.*

The son of Lorenzo Caramuel y Lobkowitz, a Bohemian engineer in Spain, and Catalina de Frisia, Caramuel became a member of the Cistercian Order and studied at Alcalá and Salamanca, the principal Spanish universities of the time. He received the doctorate in theology at Louvain. He held various important posts within the Cistercian Order and spent most of his life in Flanders, Bohemia, and Italy. Caramuel was in the service of Emperor Ferdinand III and then of Pope Alexander VII, who appointed him bishop of Campagna (near Amalfi). He died while serving in that post.

Caramuel's some seventy works treat many subjects. One of the more important is *Mathesis biceps: Vetus et nova* (Campagna, 1670), which, although it contains no sensational discovery, presents some original contributions to the field of mathematics. In it he expounded the general principle of the numbering systems of base n (illustrated by the values 2, 3, . . ., 10, 12, and 60), pointing out that some of these might be of greater use than the decimal. He also proposed a new method of approximation (although he did not say so) for trisecting an angle. Caramuel developed a system of logarithms of which the base is 10^9, the logarithm of 10^{10} is 0, and the logarithm of 1 is 10. Thus, his logarithms are the complements of the Briggsian logarithms to the base 10 and therefore do not have to use negative characteristics in trigonometric calculations. In these particulars Caramuel's logarithms prefigure cologarithms, but he was not understood by his contemporaries; some, such as P. Zaragoza, raised strenuous objections.

A man of encyclopedic knowledge, Caramuel tried to apply a mechanical formulation to astronomy, relegating astrology to the domain of superstition and criticizing some of the statements of Tycho Brahe. In addition, he made meteorological observations, investigated the globe's physical properties, and theorized about the possibility of aerial navigation.

BIBLIOGRAPHY

Writings on Caramuel are enumerated in Ramón Ceñal, S. J., "Juan Caramuel. Su epistolario con Atanasio Kircher S. I.," in *Revista de filosofía,* **12,** no. 44 (1953), 101–147;

David Fernández Diéguez, "Juan Caramuel," in *Revista matemática hispano-americana,* **1** (1919), 121–127, 178–189, 203–212; and Patricio Peñalver y Bachiller, *Bosquejo de la matemática española en los siglos de la decadencia* (Seville, 1930), pp. 29–33.

JUAN VERNET

CARANGEOT, ARNOULD (*b.* Rheims, France, 12 March 1742; *d.* Meaux, France, 18 November 1806), *crystallography, entomology.*

Carangeot was the son of Ponce Carangeot, a weaver, and his wife, Liesse Tellier. He was educated in an unimportant seminary; but at the age of twenty, he went to Paris, where his integrity and orderliness won him several patrons who made him their business manager. The fashionable bourgeois whom Carangeot served were deeply interested in the natural sciences; thus, the young administrator dedicated a great deal of his leisure time to studying botany, mineralogy, and entomology.

In 1772, Carangeot's teacher, Romé de L'Isle, published *Essai de cristallographie ou Description des figures géométriques propres à différens corps du règne minéral connus vulgairement sous le nom de cristaux,* in which he emphasized those characteristics that define mineralogy as an individual discipline. The success of this work encouraged Romé de L'Isle to prepare a much enlarged second edition, which included 438 illustrations of minerals from his extensive collection. Further, he commissioned the engraver Swebach-Desfontaines to reproduce these minerals in terra-cotta, which reproductions were to be sent as premiums to subscribers.

Swebach asked for clay models from which to work, so Romé de L'Isle asked Carangeot and Claude Lermina to fabricate them. In so doing, Carangeot experienced some difficulty in measuring the dihedral angles of a quartz sample; he was thus inspired to construct a goniometer—a device combining the qualities of the proportional compass and the protractor. Carangeot's prototype measured seventy-eight millimeters in exterior diameter. Carangeot then asked Nicolas Vinçard, an engineer who specialized in mathematical instruments, to build two models of this new gauge (one was of silver, and cost thirty-six *livres,* while the other, of copper, cost half as much).

On 11 April 1782, Carangeot presented the goniometer to a group of scientists and artists assembled by Pahin de La Blancherie. Through the use of this instrument, he was the first to observe the invariability of dihedral angles in mineral species. When Vinçard died in 1788, his assistant, Jean-Baptiste-Pierre François Férat, continued to manufacture Carangeot's goniometer, which was then duplicated by Richer and others.

Carangeot is also known as an entomologist. When the financier Gigot d'Orcy published the *Papillons d'Europe peints d'après nature,* Jacques-Louis-Florentin Engramelle wrote the text for the first three volumes (1779–1784) and Carangeot wrote the text of volumes IV through VIII (1785–1793).

Although Carangeot remained in Paris throughout the Revolution, he avoided all and any participation in public life. On 10 August 1785 he was elected an associate member of the Académie de La Rochelle; on 13 April 1798 he was named curator of the museum at Meaux, where he founded and became permanent secretary of the Société d'Agriculture, Sciences et Arts of Seine-et-Marne.

BIBLIOGRAPHY

I. ORIGINAL WORKS. Carangeot's writings include "Un goniomètre ou mesure-angle," in *Nouvelles de la République des lettres et des arts,* no. 14 (17 April 1782), 111; "Goniomètre, ou mesure-angle," in *Observations sur la physique, sur l'histoire naturelle et sur les arts,* **22** (Mar. 1783), 193–197; "Lettre . . . à M. de La Mètherie sur le goniomètre," *ibid.,* **29** (Sept. 1786), 226–227; "Lettre à M. Kaestner . . . sur de prétendues erreurs dans la description du goniomètre," *ibid.,* **31** (Sept. 1787), 204–206; *Papillons d'Europe peints d'après nature,* IV–VIII (Paris, 1785–1793); "Lettre au rédacteur," in *Journal des mines,* no. 25 (*an* V), 78–80; *Réflexions sur les musées et leur organisation* (Meaux, *an* VIII; 2nd ed., Meaux, 1901); and *Principes de lecture* (Meaux, 1810).

II. SECONDARY LITERATURE. For Carangeot's life and work, see also A. Birembaut, "Les frères Engramelle," in *Actes du VIIIe Congrès international d'histoire des sciences* (Florence), I (Paris, 1958), 149–155; and R. Homberg, "Some Unknown Plates in Ernst and Engramelle's '*Papillons d'Europe peints d'après nature,*' 1779–1793," in *The Journal of the Society for the Bibliography of Natural History,* **3** (1953), 28–53.

The 2nd ed. of Carangeot's *Réflexions sur les musées . . .* (Meaux, 1901) contains introductory biographical material by A. Le Blondel (pp. 3–7) and S.-N. Guyardin (pp. 9–14).

ARTHUR BIREMBAUT

CARATHÉODORY, CONSTANTIN (*b.* Berlin, Germany, 13 September 1873; *d.* Munich, Germany, 2 February 1950), *mathematics.*

Carathéodory was descended from an old Greek family that had lived for several generations in Adrianople (now Edirne), Turkey. His grandfather had been a professor at the Academy of Medicine in Constantinople and attending physician to two Turkish sultans. His father, who had been a diplomat in the Turkish embassies in St. Petersburg and Berlin, was the Turkish ambassador in Brussels from 1875 on.

Carathéodory won prizes in mathematics while still in secondary school. From 1891 to 1895 he attended the École Militaire de Belgique. After completing his education, he went to Egypt in the employ of the British government as assistant engineer at the Asyut dam. In 1900, however, Carathéodory suddenly decided to go to Berlin and study mathematics. There he was stimulated by the students of Hermann Amandus Schwarz. In 1902 he followed his friend Erhard Schmidt to Göttingen, where, under Hermann Minkowski, he received the doctorate in 1904. Encouraged by Klein and Hilbert, who had recognized his genius, he qualified as a university lecturer a year later.

After having taught in Bonn, Hannover, Breslau, Göttingen, and Berlin, Carathéodory was called to Smyrna by the Greek government in 1920, to direct the completion of its university. In 1922, when Smyrna was burned by the Turks, Carathéodory was able to rescue the university library and take it to Athens. He then taught for two years at Athens University, and in 1924 accepted an invitation to the University of Munich as the successor of C. L. F. Lindemann. He remained there for the rest of his life.

For many years, Carathéodory was the editor of *Mathematische Annalen.* He was a member of scientific societies and academies in many countries; his membership in the Papal Academy was particularly noteworthy.

Carathéodory married a distant cousin, Euphrosyne Carathéodory. The marriage produced one son and one daughter.

Carathéodory was the most notable Greek mathematician of recent times, and the only one who does not suffer by comparison with the famous names of Greek antiquity. He made significant contributions to several branches of mathematics, and during the period of his activity he worked in all of them more or less simultaneously.

The first field was the calculus of variations, the theory of maxima and minima in curves. In his dissertation and in his habilitation thesis, Carathéodory drew up a comprehensive theory of discontinuous solutions (curves with corners). Previously, only the so-called Erdmann corner condition was known, but Carathéodory showed that all of the theory known for smooth curves can also be applied to curves with corners. He was also concerned with the fields of solution curves, which play a central part in the theory. Thoroughly familiar with the history of the subject, he drew upon many ideas of such mathematicians as Christiaan Huygens and Johann I Bernoulli. Inspired by these ideas, he restudied the relationship between the calculus of variations and first-order partial differential equations. The result of this was

Variationsrechnung und partielle Differentialgleichungen erster Ordnung, which includes a quite surprising new "entry" to the theory.

By means of this method, Carathéodory was able to make significant progress in solving the so-called problem of Lagrange, i.e., variation problems with differential side conditions. He also wrote fundamental papers on the variation problems of *m*-dimensional surfaces in an *n*-dimensional space. Except for the case $m = 1$ of the curves, far-reaching results had until then been available only for the case $m = n - 1$. Carathéodory was the first to tackle the general case successfully. Here again, the consideration of fields played a decisive part.

Carathéodory was particularly interested in the application of the calculus of variations to geometrical optics. This interest is best shown in his "Elementare Theorie des Spiegelteleskops von B. Schmidt," for which he carried out complete numerical calculations.

In the second main field, the theory of functions, Carathéodory's achievements are in many areas: the problems concerning Picard's theorem, coefficient problems in expansions in a power series, and problems arising from Schwarz's lemma; he also significantly advanced the theory of the functions of several variables. His most important contributions, however, are in the field of conformal representation. The so-called main theorem of conformal representation of simply connected regions on the circle of unit radius had been proved rigorously for the first time shortly before World War I, and Carathéodory was able to simplify the proof greatly. His main achievement in this area was his theory of boundary correspondence, in which he investigated the geometrical-set theoretic properties of these boundaries in a completely new way.

The third main field consists of the so-called theory of real functions and the theory of the measure of point sets and of the integral. Carathéodory's book on this subject, *Vorlesungen über reelle Funktionen* (1918), represents both a completion of the development begun around 1900 by Borel and Lebesgue and the beginning of the modern axiomatization of this field. He returned to it in the last decade of his life, when he carried the axiomatization or, as he called it, the algebraization, of the concepts one step further.

In applied mathematics Carathéodory produced papers on thermodynamics and on Einstein's special theory of relativity.

BIBLIOGRAPHY

Carathéodory's articles were collected in *Gesammelte mathematische Schriften,* 5 vols. (Munich, 1954–1957); vol. V has autobiographical notes up to 1908 (pp. 387–408) and an obituary by Erhard Schmidt (pp. 409–419). Carathéodory's books are *Vorlesungen über reelle Funktionen* (Leipzig-Berlin, 1918; 2nd ed., 1928); *Conformal Representation,* Cambridge Tracts in Mathematics and Mathematical Physics, no. 28 (Cambridge, 1932); *Variationsrechnung und partielle Differentialgleichungen erster Ordnung* (Leipzig-Berlin, 1935; 2nd. ed., 1956); *Geometrische Optik,* IV, pt. 5 of the series Ergebnisse der Mathematik und ihrer Grenzgebiete (Berlin, 1937); *Reelle Funktionen,* I (Leipzig-Berlin, 1939); *Funktionentheorie,* 2 vols. (Basel, 1950); and *Mass und Integral und ihre Algebraisierung* (Basel, 1956).

An article on Carathéodory is Oskar Perron, "Constantin Carathéodory," in *Jahresbericht der Deutschen Mathematikervereinigung,* **55** (1952), 39–51.

H. BOERNER

CARCAVI, PIERRE DE (*b.* Lyons, France, *ca.* 1600; *d.* Paris, France, April 1684), *mathematics.*

The son of a banker named Trapezita, he was made a member of the *parlement* of Toulouse on 20 July 1632, and in 1636 left for Paris after having bought the office of member of the *grand conseil* there. In 1645 he entered the renewed dispute over the quadrature of the circle. John Pell was involved in a controversy with the Danish astronomer Longomontanus, who claimed to have effected the quadrature of the circle; Carcavi sent Pell a refutation in which he claimed that this was impossible. About 1648 he was forced to sell his office as member of the *grand conseil,* in order to be able to pay his father's debts, and entered the service of the Duke of Liancourt. A protégé of the duke, Amable de Bourzeis, presented Carcavi to Colbert, who in 1663 charged him with the classification of his library and made him custodian of the Royal Library (later the site of the meetings of the French Academy of Sciences). At the Academy's first official meeting there, 22 December 1666, Carcavi announced the king's decision to protect the new institution. On 30 May 1668 Colbert charged Huygens, Roberval, Carcavi, Auzout, Picard, and Gallois with research on the method of determining longitude at sea that had been proposed by R. de Neystt. On 6 June 1668 the commission rejected the method. After the death of Colbert in 1683 Carcavi was replaced by Gallois at the Academy and the Royal Library.

Carcavi rendered great services to science. His polite and engaging manner brought him many friends, including Huygens, Fermat, and Pascal. He carried on an extensive correspondence and thus was a medium for the communication of scientific intelligence. His friendship with Fermat dates from his time at Toulouse, when both were members of the *parlement.* He was probably the first to recognize Fermat's extraordinary scientific abilities. After Carcavi went to

Paris, Fermat corresponded with him and sent him many treatises; for instance, in the autumn of 1637 Carcavi received the text of Fermat's *Isagoge ad locos planos et solidos,* a short introduction into analytical geometry written in 1636, a year before Descartes published his *Géométrie.*

After the death of Mersenne in 1648, Carcavi offered Mersenne's correspondence to Descartes. In his letter of 11 June 1649 the philosopher thanked Carcavi and asked him about the experiment of Pascal, who had had a barometer carried to the top of the Puy de Dôme. This experiment showed that the greater the altitude, the lower the air pressure. Descartes claimed that he had given Pascal the idea two years before. In his answer of 9 July 1649, Carcavi said that the report of the experiment had been printed some months previously. At the same time he informed Descartes of Roberval's objections to his *Géométrie.* On 17 August 1649 Descartes replied with a refutation of Roberval's assertions. After Carcavi's answer of 24 September 1649, in which he defended Roberval, Descartes broke with him.

In the spring of 1650 Fermat sent Carcavi a treatise entitled *Novus secundarum et ulterioris ordinis radicum in analyticis usus,* in which he corrected the process given by Viète in his *De aequationum recognitione et emendatione* (written 1591) by treating the method of elimination of one or more unknowns in several equations. This is the first known method of elimination. When Fermat began to fear that his discoveries might be lost, he tried to find collaborators and asked Pascal and Carcavi to publish his papers. This attempt failed (letter of 27 October 1654 from Pascal to Fermat), as did the second attempt by Carcavi, who on 22 June 1656 proposed to Huygens that the papers be published by Elsevier in Amsterdam. Carcavi made a new attempt in his letter of 14 August 1659 to Huygens. He informed him that he had a collection of Fermat's papers, corrected by the author himself, ready for publication. He also enclosed a paper by Fermat on his discoveries concerning the theory of numbers. In his answer of 4 September 1659 Huygens promised that he would deal with the publisher. In the years 1659–1662 Carcavi sent Huygens more treatises by Fermat that mentioned Huygens' new results; Huygens was not pleased with them. This may be one reason why the papers were not published. It was not until 1679 that Samuel de Fermat succeeded in publishing the *Varia opera mathematica,* which did not contain all Fermat's discoveries.

Carcavi was also a friend of Pascal, who gave him his calculating machine. When in 1658 Pascal sent all mathematicians a challenge offering prizes for the first two solutions of some problems concerning the cycloid, he lodged the prizes and his own solutions with Carcavi, who, with Roberval, was to act as a judge.

BIBLIOGRAPHY

Carcavi's letters can be found in collections of the correspondence of Galileo, Mersenne, Torricelli, Descartes, Fermat, Pascal, and Huygens.

Some information on Carcavi's life and work is in Charles Henry, "Pierre de Carcavy, intermédiaire de Fermat, de Pascal, et de Huygens," in *Bollettino di bibliografia e storia delle scienze matematiche e fisiche,* **17** (1884), 317–391.

H. L. L. BUSARD

CARDANO, GIROLAMO (*b.* Pavia, Italy, 24 September 1501; *d.* Rome, Italy, 21 September 1576), *medicine, mathematics, physics, philosophy.*

Cardano was the illegitimate son of Fazio Cardano and Chiara Micheri, a widow of uncertain age who was both ignorant and irascible. The early years of his life were characterized by illness and mistreatment. Encouraged to study the classics, mathematics, and astrology by his father, a jurist of encyclopedic learning and a friend of Leonardo da Vinci, Cardano began his university studies in 1520 at Pavia and completed them at Padua in 1526 with the doctorate in medicine. Almost immediately he began to practice his profession in Saccolongo, a small town near Padua, where he spent nearly six years; he later recalled this period as the happiest of his life. Having been cured of impotence, which had afflicted him throughout his youth, he married Lucia Bandareni in 1531; they had two sons and a daughter.

In 1534, sponsored by noblemen who were friends of his father, Cardano became a teacher of mathematics in the "piattine" schools of Milan. (These schools, founded by a bequest of Tommaso Piatti [*d.* 1502], taught Greek, dialectics, astronomy, and mathematics.) He simultaneously practiced medicine, achieving such success that his colleagues became envious. His first work, *De malo recentiorum medicorum usu libellus* (Venice, 1536), was directed against them. Within a few years Cardano became the most famous physician in Milan, and among the doctors of Europe he was second only to Vesalius. Among his famous patients was John Hamilton, archbishop of Edinburgh, who suffered from asthma. Cardano remained in Scotland for most of 1552 in order to treat the archbishop and a number of other English noblemen.

In 1539, while awaiting the publication of *Practica arithmetice,* his first book on mathematics, Cardano learned that Nicolò Tartaglia knew the procedure for solving third-degree equations. He succeeded in obtaining this information by promising, possibly under oath, not to reveal it. After having kept the promise for six years, he considered himself released from it when he learned that the credit for the discovery actually belonged to Scipione dal Ferro. He therefore published the method in his *Artis magnae sive de regulis algebraicis liber unus* (1545), commonly called *Ars magna,* his greatest work in mathematics. Its publication angered Tartaglia, who in his *Quesiti et inventioni diverse* (1546) accused Cardano of perjury and wrote of him in offensive terms that he repeated in *General trattato di numeri et misure* (1556–1560). The latter work was well known among mathematicians and thus contributed greatly to posterity's low opinion of Cardano.

In 1543 Cardano accepted the chair of medicine at the University of Pavia, where he taught until 1560, with an interruption from 1552 to 1559 (when the stipend was not paid). In 1560 his elder son, his favorite, was executed for having poisoned his wife. Shaken by this blow, still suffering public condemnation aroused by the hatred of his many enemies, and embittered by the dissolute life of his younger son, Cardano sought and obtained the chair of medicine at the University of Bologna, to which he went in 1562.

In 1570 Cardano was imprisoned by the Inquisition. He was accused of heresy, particularly for having cast the horoscope of Christ and having attributed the events of His life to the influence of the stars. After a few months in prison, having been forced to recant and to abandon teaching, Cardano went in 1571 to Rome, where he succeeded in obtaining the favor of Pope Pius V, who gave him a lifetime annuity. In Rome, in the last year of his life, he wrote *De propria vita,* an autobiography—or better, an *apologia pro vita sua*—that did not shrink from the most shameful revelations. The *De propria vita* and the *De libris propriis* are the principal sources for his biography.

Cardano wrote more than 200 works on medicine, mathematics, physics, philosophy, religion, and music. Although he was insensitive to the plastic arts, his was the universal mentality to which no branch of learning was inaccessible. Even his earliest works show the characteristics of his highly unstable personality: encyclopedic learning, powerful intellect combined with childlike credulity, unconquerable fears, and delusions of grandeur.

Cardano's fame rests on his contributions to mathematics. As early as the *Practica arithmetice,* which is devoted to numerical calculation, he revealed uncommon mathematical ability in the exposition of many original methods of mnemonic calculation and in the confidence with which he transformed algebraic expressions and equations. One must remember that he could not use modern notation because the contemporary algebra was still verbal. His mastery of calculation also enabled him to solve equations above the second degree, which contemporary algebra was unable to do. For example, taking the equation that in modern notation is written $6x^3 - 4x^2 = 34x + 24$, he added $6x^3 + 20x^2$ to each member and obtained, after other transformations,

$$4x^2(3x + 4) = (2x^2 + 4x + 6)(3x + 4),$$

divided both members by $3x + 4$, and from the resulting second-degree equation obtained the solution $x = 3$.

His major work, though, was the *Ars magna,* in which many new ideas in algebra were systematically presented. Among them are the rule, today called "Cardano's rule," for solving reduced third-degree equations (i.e., they lack the second-degree term); the linear transformations that eliminate the second-degree term in a complete cubic equation (which Tartaglia did not know how to solve); the observation that an equation of a degree higher than the first admits more than a single root; the lowering of the degree of an equation when one of its roots is known; and the solution, applied to many problems, of the quartic equation, attributed by Cardano to his disciple and son-in-law, Ludovico Ferrari. Notable also was Cardano's research into approximate solutions of a numerical equation by the method of proportional parts and the observation that, with repeated operations, one could obtain roots always closer to the true ones. Before Cardano, only the solution of an equation was sought. Cardano, however, also observed the relations between the roots and the coefficients of the equation and between the succession of the signs of the terms and the signs of the roots; thus he is justly considered the originator of the theory of algebraic equations. Although in some cases he used imaginary numbers, overcoming the reluctance of contemporary mathematicians to use them, it was only in 1570, in a new edition of the *Ars magna,* that he added a section entitled "De aliza regula" (the meaning of *aliza* is unknown; some say it means "difficult"), devoted to the "irreducible case" of the cubic equation, in which Cardano's rule is extended to imaginary numbers. This was a recondite work that did not give solutions to the irreducible case, but it was still important for the algebraic transformations which it employed and for the presentation of the solutions of at least three important problems.

His passion for games (dice, chess, cards) inspired

Cardano to write the *Liber de ludo aleae,* which he completed in his old age, perhaps during his stay at Bologna; it was published posthumously in the *Opera omnia.* The book represents the first attempt at a theory of probability based on the philosophical premise that, beyond mere luck, laws and rules govern any given case. The concept of probability was introduced, expressed as the ratio of favorable to possible cases; the law of large numbers was enunciated; the so-called "power law" (if p is the probability of an event, the probability that the event will be repeated n times is p^n) was presented; and numerous problems relating to games of dice, cards, and knucklebones were solved. The book was published, however, subsequent to the first research into the theory of games developed in the correspondence between Fermat and Pascal in 1654; it had no influence on the later development of the field.

Cardano published two encyclopedias of natural science: *De subtilitate libri XXI* (1550) and *De rerum varietate* (1557), a supplement to *De subtilitate.* The two works, written in an elliptical and often obscure Latin, contain a little of everything: from cosmology to the construction of machines; from the usefulness of natural sciences to the evil influence of demons; from the laws of mechanics to cryptology. It is a mine of facts, both real and imaginary; of notes on the state of the sciences; of superstition, technology, alchemy, and various branches of the occult. The similarities between the scientific opinions expressed by Cardano in these two works and those of Leonardo da Vinci, at that time unpublished, has led some historians, particularly Pierre Duhem, to suppose that Cardano has used Leonardo's manuscript notes; others insist that the similarity is entirely coincidental. Be that as it may, Cardano must always be credited with having introduced new ideas that inspired new investigations. In the sixteenth century there were five editions of *De rerum varietate* and eight of *De subtilitate,* as well as seven editions of the French translation of the latter.

Cardano reduced the elements to three (air, earth, water), eliminating fire, which he considered a mode of existence of matter; and he reduced the four qualities to two (hot and moist). His magic was, above all, an attempt to interpret natural phenomena in terms of sympathy and antipathy.

In mechanics, Cardano was a fervent admirer of Archimedes. He studied the lever and the inclined plane in new ways and described many mechanical devices, among them "Cardano's suspension," known in classical antiquity, which he attributed to a certain Jannello Turriano of Cremona. Cardano followed a middle road between the partisans of the theory of impetus and the supporters of the Aristotelian theory, who attributed the movement of projectiles to pushing by the air: he favored the idea that at the beginning of its trajectory the projectile was moved by the impetus of the firing mechanism but subsequently was accelerated by the movement of the air. Notable is his observation that the trajectory described by a projectile is not rectilinear at the center, but is a line "which imitates the form of a parabola." Cardano's chief claim to fame, however, was his affirmation of the impossibility of perpetual motion, except in heavenly bodies.

Cardano's contributions to hydrodynamics are important: counter to contemporary belief, he observed that in a conduit of running water, the water does not rise to the level from which it started, but to a lower level that becomes lower as the length of the conduit increases. He also refuted the Aristotelian "abhorrence of a vacuum," holding that the phenomena attributed to this abhorrence can be explained by the force of rarefaction. Cardano investigated the measurement of the capacity of streams and stated that the capacity is proportional to the area of the cross section and the velocity. He observed that a stream presses against its banks and, counter to contemporary opinion, he held that the upper levels of moving water move faster than the lower levels.

In his *Opus novum de proportionibus,* Cardano turned to problems of mechanics, with the principal aim of applying quantitative methods to the study of physics. His use of the concept of moment of a force in his study of the conditions of equilibrium in balance and his attempt to determine experimentally the relation between the densities of air and water are noteworthy. The value that he obtained, $1:50$, is rough; but it is the first deduction to be based on the experimental method and on the hypothesis that the ratio of the distances traveled by bullets shot from the same ballistic instrument, through air and through water, is the inverse of the ratio between the densities of air and water.

Geology is indebted to Cardano for several theories: that the formation of mountains is often due to erosion by running water; that rise of the ocean floor is indicated by the presence of marine fossils in land that was once submerged; and the idea—then novel —that streams originate from rainwater, which runs back to the sea and evaporates from it, to fall back to earth as rain, in a perpetual cycle.

The many editions of Cardano's works and the citations of them by writers of the second half of the sixteenth century demonstrate their influence, especially as a stimulus to the study of the particular and the concrete.

BIBLIOGRAPHY

I. ORIGINAL WORKS. Nearly all of Cardano's writings are collected in the *Opera omnia,* Charles Sponi, ed., 10 vols. (Leiden, 1663). The published works to which scholars most often refer are *Practica arithmetice et mensurandi singularis* (Milan, 1539); *Artis magnae, sive de regulis algebraicis liber unus* (Nuremberg, 1545); *De subtilitate liber XXI* (Nuremberg, 1550; 6th ed., 1560), trans. by Richard Le Blanc as *De la subtilité . . .* (Paris, 1556; 9th ed., 1611), and bk. 1, trans., with intro. and notes, by Myrtle Marguerite Cass (Williamsport, Pa., 1934); *Liber de libris propriis* (Leiden, 1557); *De rerum varietate libri XVII* (Basel, 1557; 5th ed., 1581); *De subtilitate . . . cum additionibus. Addita insuper Apologia adversus calumniatorem* (Basel, 1560; 4th ed., 1611); and *Opus novum de proportionibus numerorum, motuum, ponderum, sonorum, aliarumque rerum mensurandarum. . . . Item de aliza regula liber* (Basel, 1570). The autobiography was published by Gabriel Naudé as *De propria vita liber . . .* (Paris, 1643; 2nd ed., Amsterdam, 1654); it was translated into Italian (Milan, 1821, 1922; Turin, 1945); German (Jena, 1914); and English (New York, 1930). The French translation by Jean Dayre (Paris, 1936) includes the Latin text with the variants of a 17th-century MS preserved in the Biblioteca Ambrosiana in Milan. The *Liber de ludo aleae* was first published in the *Opera omnia* and translated into English by Sidney Henry Gould as *The Book on Games of Chance* (New York, 1961).

II. SECONDARY LITERATURE. On Cardano himself, the following works contain many bibliographic references: Angelo Bellini, *Girolamo Cardano e il suo tempo* (Milan, 1947); and Henry Morley, *The Life of Girolamo Cardano of Milan, Physician,* 2 vols. (London, 1854). His mathematical work is analyzed in Ettore Bortolotti, *I contributi del Tartaglia, del Cardano, del Ferrari e della scuola matematica bolognese alla teoria algebrica delle equazioni cubiche,* no. 9 in the series Studi e Memorie per la Storia dell'Università di Bologna (Bologna, 1926), pp. 55–108, and *I cartelli di matematica disfida,* no. 12 in the series Studi e Memorie per la Storia dell'Università di Bologna (Bologna, 1935), pp. 3–79; Moritz Cantor, *Vorlesungen über Geschichte der Mathematik,* 2nd ed. (Leipzig, 1899), II, 484–510, 532–541; and Pietro Cossali, *Origine e trasporto in Italia dell'algebra,* II (Parma, 1797), 159–166, 337–384. The most profound study of Cardano's contribution to the theory of games is Oystein Ore, *Cardano the Gambling Scholar* (Princeton, 1953), which concludes with Gould's translation of the *Liber de ludo aleae.*

Cardano's physics is presented in Raffaello Caverni, *Storia del metodo sperimentale in Italia,* I (Florence, 1891), 47–50, and IV (Florence, 1895), 94–95, 197–198, 385–386 (entire work repr. Bologna, 1969). The hypothesis of his intellectual debt to Leonardo is defended by Pierre Duhem in *Les origines de la statique,* I (Paris, 1895), 237–238, 242; and *Études sur Léonard de Vinci,* I (Paris, 1906), 223–245. On Cardano's work in magic, alchemy, and the arts of divination, see Lynn Thorndike, *A History of Magic and Experimental Science,* V (New York, 1951), 563–579; on his contributions to cryptology, see David Kahn, *The Codebreakers* (London, 1967).

MARIO GLIOZZI

CARLISLE, ANTHONY (*b.* Stillington, Durham, England, 15 February 1768; *d.* London, England, 2 November 1840), *medicine, galvanism.*

The third son of Thomas Carlisle and Elizabeth Hubback, who died in childbirth, Anthony was brought up by his stepmother, Susannah Skottowe. Little is known of his early circumstances, and nothing of his schooling, but family connections were sufficient to place him in York under Dr. Anthony Hubback, his maternal uncle. After Hubback's death (*ca.* 1783), Carlisle went to study under William Green, founder of the Durham Dispensary. Transferring to London to complete his medical education, he attended the lectures of George Fordyce and William Cruikshank, and worked under John Hunter at the Windmill Street School of Anatomy. Carlisle was an apt pupil who quickly won Hunter's admiration and support. After spending some time under Henry Watson, the resident surgeon at the Westminster Hospital, Carlisle succeeded to his position in 1792. Thus began a fruitful and lifelong association with the hospital.

Carlisle had great skill as a surgeon. He devised effective modifications to several instruments, such as the amputating knife. The great bulk of his almost fifty published papers deal with anatomical and surgical questions, but contain no fundamental insights. His extraordinary success was due, rather, to his great ability to win and keep friends. Elected to the Royal College of Surgeons in 1800 (the year he married Martha Symmons), his social connections ensured his election to the professorship of anatomy at the Royal Academy in 1808. These same connections led to his appointment as surgeon to the duke of Gloucester, and later as surgeon extraordinary to the Prince Regent, on whose accession to the throne Carlisle was knighted. Carlisle served on the Court of Assistance (Council) of the Royal College of Surgeons from 1815 until his death. He twice gave the Hunterian Oration (1820, 1826), was vice-president of the College four times, and president in 1828 and 1837. He was also a member of the Linnean Society (1793), the Royal Society (1804; Croonian Lecturer, 1805, 1806), the Horticultural Society (1812), and the Geological Society (1820).

Carlisle's medical work was competent but pedestrian; and he owes his enduring scientific reputation not to this, but to a famous experiment in the then new and sensationally developing science of galvanism. Volta's discovery of the chemical means of generating a steady electric current was communicated to the Royal Society in 1800, in a two-part letter. Publication of the whole was delayed until the second part arrived. Meanwhile, Sir Joseph Banks showed the first part to scientific friends, including Carlisle. A flurry of experiments resulted. The first to be reported, and the most startling, was that in which Carlisle and William Nicholson together electrolyzed water into its constituents. Using "a pile consisting of 17 half crowns [of silver], with a like number of pieces of zinc, and of pasteboard, soaked in salt water . . . [the whole being] arranged in the order of silver, zinc, card, & c.," . . . "Mr. Carlisle observed a disengagement of gas" where "a drop of water upon the upper plate" was used to complete the circuit. Rearrangement of the apparatus led to Nicholson's collection of hydrogen and oxygen in roughly the proportions that combine to give water. An early triumph of electrochemistry, the experiment was seized on and dramatically improved by Humphry Davy.

BIBLIOGRAPHY

The classic paper is Nicholson's "Account of the New Electrical or Galvanic Apparatus of Sig. Alex. Volta, and Experiments Performed with the Same," in his *Journal of Natural Philosophy, Chemistry and the Arts* (quarto ser.), **4** (1800), 179–187. An informative obituary is in *Gentleman's Magazine,* n.s. **14** (1840), 660. A contemporary notice of Carlisle is in T. J. Pettigrew, *Biographical Memoirs of the Most Celebrated Physicians,* I (London, 1838); and J. F. Clarke, *Autobiographical Recollections of the Medical Profession* (London, 1874), pp. 283–294, contains a lively insight into his declining years. R. J. Cole, "Sir Anthony Carlisle, F.R.S.," in *Annals of Science,* **8** (1952), 255–270, contains much detail and a comprehensive bibliography of Carlisle's publications.

ARNOLD THACKRAY

CARLSON, ANTON JULIUS (*b.* Bohuslan, Sweden, 29 January 1875; *d.* Chicago, Illinois, 2 September 1956), *physiology.*

Besides being a creative scientist, Carlson was a gifted teacher, a philosopher, a civil libertarian, an academic statesman, and a public educator of unusual talent and influence.

Born into a farm family, Carlson came to the United States alone in 1891, knowing scarcely a word of English, to join an older brother, Albin, who had immigrated earlier to become a carpenter in Chicago and had sent him money for his passage. He spent the next two years as a carpenter's helper, earning $1.25 a day for ten hours of work and, more important, learning enough English to be able to enroll in Augustana Academy, a Swedish Lutheran school associated with Augustana College and Theological Seminary in Rock Island, Illinois. Carlson was an unusually serious and ambitious student and completed the work of the academy and college in five and a half years. His original intention was to become a minister in the Lutheran Church, and after graduation he became a substitute pastor in the Swedish Lutheran church of Anaconda, Montana, where he gave sermons (in Swedish) and also taught classes in science and philosophy. Although dubious about the acceptability of the rigid dogma of the church even while he was in college, Carlson had nevertheless been elected valedictorian by his classmates; however, the election was voided by the faculty because he had already given cause for concern as to his doctrinal positions. Within a year after assuming his ministerial post he had decided that he could not maintain his intellectual integrity while pretending adherence to dogma to which he no longer subscribed. He therefore announced to his congregation that he was leaving the ministry to study neurophysiology, stating his reasons quite candidly.

Carlson borrowed money from a friend to go to Stanford University, where he began his graduate studies in either 1899 or 1900. His first research was on conduction velocity in nerves, and he demonstrated that in the motor nerves of certain invertebrates, such as those in the pseudopod of the slug, the conduction time was increased with elongation of the organ. In 1902 he earned a Ph.D. at Stanford with a dissertation based on his conduction velocity studies. Carlson became a research associate of the Carnegie Institution for two years, spending part of the time at the Woods Hole Marine Biological Station. There, in 1904, he began his studies on the heart of the horseshoe crab *Limulus,* which established the neurogenic mechanism of automaticity and of conduction of excitation in the heart of *Limulus.* This work led to his being offered a position as associate in the department of physiology of the University of Chicago, where he remained for the rest of his working life. He was promoted through the ranks to professor in 1914, chairman of the department in 1916, and emeritus professor in 1940.

Carlson continued his work on the comparative physiology of automaticity in the heart and on mechanisms of conduction for several years, publishing thirty-seven papers on the comparative physiology of the heart in invertebrates, including a definitive review in 1909 in the *Ergebnisse der Physiologie.* After 1909 his publications dealt mainly with other problems, including studies on lymph formation, on several endocrine organs, and on the physiology of the stomach. Long before the discovery of insulin he made the interesting and important observation that during the latter part of pregnancy it was not possible to produce diabetes by pancreatectomy in the gravid dog but that symptoms occurred promptly after delivery. These observations contributed to the proof that the antidiabetic substance of the pancreas is carried by the bloodstream and is capable of passing the placental barrier.

Carlson's first publications on the physiology of the stomach, which appeared in 1912, began a long series of studies of gastric physiology with a number of colleagues, especially Arno B. Luckhardt, who worked with him at the University of Chicago for the remainder of his career. Carlson carried on extensive studies on the nervous control of the hunger mechanism and established the role of the vagal innervation of the stomach in its motor activity. In 1915 he published a sixty-four-page summary in the Harvey Society Lecture Series, summarizing the work of others and himself on all aspects of the hunger mechanism.

During World War I, Carlson enlisted as an officer in the Sanitary Corps of the U.S. Army and became concerned with the nutritional problems of the military. After the war he was assigned to work with the Hoover Commission, which was involved in feeding the undernourished in Europe. He played an important part in this relief work and came into intimate personal contact with the consequences of the devastation of towns and villages all over central Europe. These experiences probably accounted for his later antiwar views, which resulted in his strong early opposition to American participation in World War II.

Besides his numerous contributions to science, Carlson played an important part in many public service enterprises. He testified for the Food and Drug Administration concerning lead residue from application of lead arsenate as an insecticide and thus became an early champion of antipollution legislation. He was also the leader in American scientific circles in the fight against antivivisectionists. His scientific work during the years after 1915 was carried on largely with dogs and cats which he obtained from among the unclaimed animals at the Chicago city pound. He was responsible for the passage of ordinances by the city council which authorized such practice but had to carry on a running battle to keep these sources of supply open. His experiences with the antivivisectionists encouraged Carlson to organize the National Society for Medical Research, of which he was president for twelve years. With a former pupil, Andrew C. Ivy, he organized virtually all of the biological science societies into the Council of the NSMR and obtained funds to carry on a vigorous educational campaign to alert the American people to the hazards to the public welfare inherent in legislation which would cripple research in medicine and other biological fields.

Carlson, an ardent civil libertarian, served a term as president of the American Association of University Professors. He was active in the Association for the Protection of the Foreign Born and was a sponsor of the Humanist Manifesto of the American Humanist Association. He received many honors, including the Distinguished Service Award of the American Medical Association in 1946. Carlson was elected a member of the National Academy of Sciences in 1920 and was a member of the American Philosophical Society. He served as president of the American Association for the Advancement of Science, the American Physiological Society, and the Federation of American Societies for Experimental Biology.

His colorful personality made Carlson a stimulating teacher and a sought-after lecturer for both professional and academic audiences. In discussing scientific problems, his first question in connection with any theory or conclusion was, "What is the evidence?" He died of cancer of the prostate after a long illness.

One of Carlson's former students and colleagues, Lester R. Dragstedt, wrote of him:

> In the death of Dr. Carlson the world has lost a vigorous voice for human freedom, science has lost a biologist of great critical judgment and intuition, and the United States a great citizen. . . . His gift for keen analysis, his ready wit and pungent criticism, so often displayed at scientific meetings, gave him an acknowledged place of leadership in biological and medical societies. It is probable that no man in America not engaged in clinical practice had so great an effect on medicine.

BIBLIOGRAPHY

I. ORIGINAL WORKS. Carlson's works include "The Rate of Nervous Impulse in Certain Molluscs," in *American Journal of Physiology,* **8** (1903), 251–268, written with O. P. Jenkins; "The Rhythm Produced in the Resting Heart of Molluscs by the Stimulation of the Cardioaccelerator Nerves," *ibid.,* **12** (1904), 55–66; "The Nervous Origin of the Heartbeat in Limulus and the Nervous Nature of Coordination or Conduction in the Heart," *ibid.,* 67–74; "Further Evidence of the Fluidity of the Conducting Substance in Nerve," *ibid.,* **13** (1905), 351–357; "On the Mechanism of Co-ordination and Conduction in the Heart with Special Reference to the Heart of Limulus," *ibid.,* **15** (1906), 99–120; "Comparative Physiology of the Invertebrate

Heart. VI. The Excitability of the Heart During the Different Phases of the Heart-beat," *ibid.,* **16** (1906), 67–84; "Vergleichende Physiologie der Herznerven und der Herzganglien bei den Wirbellosen," in *Ergebnisse der Physiologie (biologischen Chemie und experimentellen Pharmakologie),* **8** (1909), 371–462; "The Effects of Stretching the Nerve on the Rate of Conduction of the Nervous Impulse," in *American Journal of Physiology,* **27** (1910), 323–330; "Contributions to the Physiology of the Stomach. II. The Relation Between the Contractions of the Empty Stomach and the Sensation of Hunger," *ibid.,* **31** (1912), 175–192; "The Secretion of Gastric Juice in Man," *ibid.,* **37** (1915), 50–73; *On the Nervous Control of the Hunger Mechanism,* Harvey Lecture, (Philadelphia–London, 1917), 37–100; "The Endocrine Function of the Pancreas and Its Relation to the Sex Life of Women," in *Surgery, Gynecology, and Obstetrics,* **25** (1917), 283–293; "A Few Observations of Certain Conditions in Europe After the War," in *Journal of the Missouri Medical Association,* **17** (1920), 229; "Experimentation and Medicine; Man's Debt to the Animal World," in *Hygeia,* **13** (1935), 126–128; "Fundamental Sciences—Their Role in Medical Progress (Elias Potter Lyon Lecture)," in *Journal of the American Medical College,* **15** (1940), 351–358; "Science and the Supernatural (William Vaughan Moody Lecture)," in *Scientific Monthly,* **59** (1944), 85–95; and "Science of Biology and Future of Man," in *Science,* **100** (1944), 437–439.

II. SECONDARY LITERATURE. The major source on Carlson and his work is Lester R. Dragstedt, "Anton Julius Carlson," in *Biographical Memoirs. National Academy of Sciences,* **35** (1961).

MAURICE B. VISSCHER

CARNALL, RUDOLF VON (*b.* Glatz, Germany [now Klodzko, Poland], 9 February 1804; *d.* Breslau, Germany [now Wroclaw, Poland], 17 November 1874), *geology.*

Carnall was trained to be a Prussian civil mining engineer. After working underground for one year in various mines, he studied in Berlin; he then was appointed *Bergassessor* to Upper Silesia, where he lectured on mining science at the Bergschule at Tarnowitz. In 1845 he went to Bonn as *Oberbergrat,* and soon afterward returned to Berlin, where with Leopold von Buch und Gustav Rose, he initiated the Deutsche Geologische Gesellschaft. As an administrative officer of the mining department he was in charge of the Prussian Knappschaftsordnung, which regulations still govern much of the German mining industry. He simultaneously lectured at the University of Berlin until 1855, when he took the Ph.D. there. In 1854 he became a *Referent* in the Prussian ministry of commerce and in 1855 was promoted to *Berghauptmann.*

His scientific work was diverse. He founded the official German mining journal, which (until 1945)

published important papers on administrative, technical, mining, and geological problems. His scientific publications include *Der bergmännische Strebbau auf der Bleierzgrube Friedrich bei Tarnowitz;* "Geognostische Beschreibung von einem Teil des Nieder- und Oberschlesischen Gebirges," a geognostic description of a part of the mountains of Lower and Upper Silesia, which was published in *Karstens Archiv für Mineralogie; Geognostischer Vergleich zwischen Nieder- und Oberschlesischen Gebirgsformationen,* a geognostical comparison between the Lower and Upper Silesian formations; and *Sprünge im Steinkohlengebirge* (1835), which dealt with faults in the Carboniferous era. Heinrich Rose named the mineral carnallite after him.

P. RAMDOHR

CARNOT, LAZARE-NICOLAS-MARGUERITE (*b.* Nolay, Côte-d'Or, France, 13 May 1753; *d.* Magdeburg, Germany, 2 August 1823), *mechanics, mathematics, engineering.*

Known to French history as the "Organizer of Victory" in the wars of the Revolution and to engineering mechanics for the principle of continuity in the transmission of power, Carnot remains one of the very few men of science and of politics whose career in each domain deserves serious attention on its own merits. His father, Claude, lawyer and notary, was among the considerable bourgeois of the small Burgundian town of Nolay in the vicinity of Beaune though on the wrong side of the ridge for the vineyards. The family still owns the ancestral house. His most notable descendants have been his elder son Sadi, famous in thermodynamics, and a grandson, the latter's nephew, also called Sadi, president of the French Republic from 1887 until his assassination in 1894.

Carnot had his early education in the Oratorian *collège* at Autun. Thereafter, his father enrolled him in a tutoring school in Paris which specialized in preparing candidates for the entrance examinations to the service schools that trained cadets for the navy, the artillery, and the Royal Corps of Engineers. Strong in technique and weak in prestige, the Corps of Engineers was the only branch of military service in which a commoner might still hold a commission. Carnot graduated from its school at Mézières after the normal course of two years. Gaspard Monge was then at the height of his influence over the cadets as professor of mathematics and physics, but although Carnot's handling of problems always bore the mark of the engineer, it does not appear that he was one of Monge's favorite pupils. Carnot's approach to mathematics and mechanics was in a curious way more concerned with both fundamentals and operations than that of Monge. Its actual mathematical yield was consequently much less while its significance for the physics of work and energy and for the evolution of engineering mechanics was much greater.

Promotion was slower for engineers than for line officers, and Carnot's outlook on the world, like that of many able and industrious men in the last years of the old regime, was compounded of frustrated talent and civic affirmation, some of which he expressed in tolerable verse. After routine assignments at Calais, Cherbourg, and Béthune, he was posted to Arras, where society was livelier. There in 1787 he became acquainted with Maximilien de Robespierre, a fellow member of the literary and philosophic society of the Rosati. During these years of garrison duty, Carnot sought reputation by writing of mechanics, mathematics, and military strategy in essays prepared for competitions of the kind regularly set by learned societies in the eighteenth century. In 1777 the Académie des Sciences in Paris proposed the "theory of simple machines with regard to friction and the stiffness of cordage" for a prize. Carnot entered a memoir in that contest and revised it in 1780 for resubmission in a second round opened by the Academy when none of the entries in the first proved worthy of an award. It then received an honorable mention following a memoir on friction by Coulomb which won the prize. Carnot developed its theoretical portion into his first publication, *Essai sur les machines en général,* which appeared in 1783. In retrospect it is evident that that modest work inaugurated the peculiarly French literature of engineering mechanics, but it commanded little attention until after its author had become great and famous in the politics and military affairs of the Revolution.

The fate of an early mathematics memoir was similar. In 1784 the Academy of Berlin invited entries in a competition for a "clear and precise" justification of the infinitesimal calculus. The essay that Carnot submitted forms the basis of the *Réflexions sur la métaphysique du calcul infinitésimal* that he published many years later, in 1797. Once again Carnot's entry received an honorable mention but no other notice.

It was, indeed, only through his writings on military strategy that Carnot won a certain minor recognition prior to the Revolution. In 1784 the Academy of Dijon set as subject for its annual prize the career of the founder of the Royal Corps of Engineers, the Maréchal de Vauban (like Carnot a Burgundian), whose theory and practice of fortification and siege-craft had guided the strategy of Louis XIV and become standard doctrine in the limited warfare of the Enlightenment. Carnot's *Éloge de Vauban* carried off

that award. Its publication brought him into the crossfire of a skirmish of the books that had broken out among members of the French armed forces wherein the political interests and social prestige of the several arms and services were entangled with opposing theories of warfare. Writing out of the specialist tradition of the engineering corps, Carnot upheld the purportedly humane view of warfare that made its purpose the defense of civilization in the wise employment of prepared positions rather than the barbarous destruction of an enemy in conflict to the death. His argument criticized the emphasis of the combat arms on gallantry, movement, and command under fire, aspects recommended anew to elements among the line officers by the recent example of Frederick the Great and the resurgence of aristocratic pretensions in French society. It is ironic, therefore, that the latter was the type of warfare that Carnot mastered and waged when charged with direction of the Revolutionary war effort. Not that he had any choice—in the emergency of 1793–1794 Carnot disposed not of the disciplined, professional, and cautious forces presupposed by conservative strategy, but of untrained levies, some under arms because of patriotism and some because of conscription. Patriotism produced dash and conscription mass, and in both respects the armies of the Republic were quite unlike those of the eighteenth century.

Carnot entered politics in 1791 when he was elected a deputy to the Legislative Assembly from the Pas-de-Calais. No flaming radical, his idea of social justice was the career open to talent. As the monarchy proved its untrustworthiness in 1791 and 1792, he became a republican out of a kind of civic commitment natural to his class and family background. Following the outbreak of war in April 1792, the services of a patriotic deputy competent in military matters were at a premium. There was an integrity in Carnot that made itself felt and trusted in dangerous times. He combined it with the engineer's ability to improvise arrangements and organize procedures. His mission to the Army of the Rhine immediately after the overthrow of the monarchy in August 1792 imposed the sovereign authority of the Republic upon the officers and local agents whose allegiance had been to the Crown. Amid the military disasters in Belgium in the spring of 1793, it was Carnot who, as representative of the people incarnating its revolutionary will, had to override the demoralized generals and organize first the defense and then the attack to his own presciption. Never an ideologist, never really a democrat, he had the reputation of a tough and reliable patriot when, in August 1793, he was called by the more politically minded men already constituting the emergency Committee of Public Safety to serve among its membership of twelve men who, ruling France throughout the Jacobin Terror, converted anarchy into authority and defeat into victory.

His main responsibility was for the war, and he alone of his erstwhile colleagues continued in office after the fall of Robespierre in July 1794 through the ensuing reaction and on into the regime of the Directory that followed in 1795. For two more years Carnot was the leading member of that body, in which office he, together with four incompatible colleagues, exercised the executive power of the French Republic. In 1797 the leftist coup d'état of 18 Fructidor (4 September) displaced Carnot from government. He took refuge in Switzerland and Germany, returning in 1800 soon after Napoleon's seizure of power.

Carnot had given Napoleon command of the Army of Italy in 1797, and now Napoleon named his sometime patron minister of war. The Bonapartist dispensation proved uncongenial to an independent spirit, however, and after a few months Carnot resigned.

Thereupon he devoted his older years to the technical and scientific interests of his younger days, having qualified for membership in the Institut de France in 1796 by virtue of his prominence if not of any wide comprehension of his youthful work in mechanics, his only scientific work then in print. Throughout the Napoleonic period he served on numerous commissions appointed by the Institute to examine the merits of many of the mechanical inventions that testify to the fertility of French technical imagination in those years of conquest and warfare. He was never too old for patriotism, however. Amid the crumbling of the Napoleonic system, he offered his services when the retreat from Moscow reached the Rhine. In those desperate circumstances Napoleon appointed him governor of Antwerp. Carnot commanded the defense. He rallied to the emperor again during the Hundred Days and served as his last minister of the interior. That act bespoke the consistency of the old revolutionary yet more decisively than Carnot's having voted death to Louis XVI some twenty-two years before. He was not forgiven by a monarchy that had had to be restored twice, and he fled into exile once again, leaving in France his elder son, Sadi, an 1814 graduate of the École Polytechnique, and taking the younger, Hippolyte, to bring up in Magdeburg, in which tranquil city he lived out his days corresponding with old associates and publishing occasional verse.

Scientists often become public figures, but public figures seldom produce science, and a glance at the Bibliography below will suggest that the range and originality of his published work in mechanics, ge-

ometry, and the foundations of the calculus in the ten years following his fall from power in 1797 must be unique in the annals of statesmen in political eclipse. The phenomenon is the more interesting in that its beginnings go back to his days as a young engineering officer who had failed to win a hearing for the approach that began to be appreciated only in the days of his later prominence. The *Essai sur les machines en général* of 1783 contains all the elements of his engineering mechanics, in which subject it was the first truly theoretical treatise. It attracted no detectable attention prior to the revision he published in 1803 under the title *Principes fondamentaux de l'équilibre et du mouvement*. Even the analysis given the problem of machine motion in the latter work began to affect the actual treatment of problems only in the 1820's through the theoretical practice of his son Sadi and his contemporary polytechnicians, notably Navier, Coriolis, and Poncelet. The lag is not to be explained on the grounds either of some mathematical sophistication in the analysis or of the unexpectedness of some signal discoveries. On the contrary, the style and argument seem naïve and literal compared to the mathematical mode of treatment that Lagrange set as the standard of elegance for rational mechanics in *Mécanique analytique* (1788), a work to which Carnot never aspired to compare his own. The unfamiliarity of Carnot's approach derived from his purpose, not his content, from what he thought to do with the science of mechanics rather than what he added to it.

His purpose was to specify in a completely general way the optimal conditions for the operation of machines of every sort. Instead of adapting the laws of statics and dynamics to the properties of the standard classes of simple machine—i.e., the lever, the wedge, the pulley, the screw, and so forth—as did conventional manuals of applied mechanics, he began with a completely abstract definition. A machine is an intermediary body serving to transmit motion between two or more primary bodies that do not act directly on one another. The problem of a truly general mechanics of machines could then be stated:

> Given the virtual motion of any system of bodies (i.e., that which each of the bodies would describe if it were free), find the real motion that will ensue in the next instant in consequence of the mutual interactions of the bodies considered as they exist in nature, i.e., endowed with the inertia that is common to all the parts of matter.

In developing the reasoning, Carnot gave important impetus to what was to become the physics of energy, not by building out from the mathematical frontier of analytical mechanics, but by starting behind those front lines with the elementary principles of mechanics itself. From among them he selected the conservation of live force or *vis viva*, the quantity MV^2 or the product of the mass of a body multiplied by the square of its velocity, as the basis from which propositions adaptable to the problem of machine motion might most naturally be derived. The principal explicit finding of the *Essai sur les machines en général* was that it is a condition of maximum efficiency in the operation of machines that power be transmitted without percussion or turbulence (in the case of hydraulic machines). That principle of continuity was usually known as Carnot's until after the middle of the nineteenth century, when its parentage became obscured in the generality of conservation of energy, of which felt but unstated law it was one of many early partial instances.

In the course of deriving his own principle of continuity from that of live force, Carnot recognized its equivalent, the product of the dimensions of force multiplied by distance—MGH in the gravitational case, for he usually preferred to reason on the example of weights serving as loads or motive forces—to be the quantity that mechanics might most conveniently employ to estimate the efficacy of machines. In 1829 Coriolis proposed to designate by the word "work" any quantity thus involving force times distance. Carnot had called it "moment of activity." Although he did not explicitly state the measure of power to be the quantity of live force (energy) expended or moment of activity (work) produced in a given time, that usage is implicit in the argument, of which the central thrust in effect transformed the discussion of force and motion into an analysis of the transmission of power. Dimensionally, of course, there was nothing new about it: the equivalence of $MV^2/2$ to MGH derived from Galileo's law of fall as the means of equating the velocity that a body would generate in falling a certain height with the force required to carry it back on rebound.

It was the engineer in Carnot that saw the advantage of winning from that trivial dimensional equivalence an application of the science of mechanics to analysis of the operation of machines. In the ideal case, live force reappears as moment of activity; or, in much later words, input equals output. It was simply a condition of perfect conversion that nothing be lost in impacts and that motion be communicated smoothly—hence his principle. His approach carried over the employment of live-force conservation from hydrodynamics into engineering mechanics generally. Following publication in 1738 of Daniel Bernoulli's *Hydrodynamica*, that subject had been the only one that regularly invoked the principle of live force in

the solution of engineering problems; and except for certain areas of celestial mechanics, it was the only sector of the science in which it had survived the discredit of the metaphysical disputes between partisans of the Newtonian and Leibnizian definitions of force early in the century. The evidence runs through all of Carnot's writings on mechanics and into those of his son Sadi that the hydraulic application of principles and findings, although ostensibly a special case, actually occupied a major if not a primary place in their thinking.

Carnot did not reach his conclusions so easily as this, however, and certain features of his reasoning point much further than its mere results, namely in the direction of dimensional and vector analysis and also toward the concept of reversible processes. Wishing to attribute to machines no properties except those common to all parts of matter, Carnot envisaged them as intrinsically nothing more than systems composed of corpuscles. He began his analysis with the action of one corpuscle upon another in machine motions, and obtained an equation stating that in such a system of "hard" (inelastic) bodies, the net effect of mutual interaction among the corpuscles constituting the system is zero:

$$\int mVU \cos Z = 0,$$

where the integral sign means summation; m is the mass of each corpuscle; V its actual velocity; U the velocity that it "loses", i.e., the resultant of V and its virtual velocity; and Z the angle between the directions of V and U. The notion of balancing the forces and motions "lost" and "gained" from the constraints of the system was central to his analysis of the manner in which he imagined forces transmitted by shafts, cords, and pulleys constraining and moving points within systems composed of rigid members. In his way of seeing the problem, he had necessarily to find constructs that would incorporate the direction as well as the intensity of forces in expressions of their quantity; and in combining and resolving such quantities, he habitually adopted simple trigonometric relations in a kind of proto-vector analysis that permitted him to represent the projection of the quantity of a force or velocity upon a direction other than its own. In the *Principes fondamentaux de l'équilibre et du mouvement*, he proposed that the projection $\overline{Aa'}$ of one force or velocity $\overline{Aa}$ upon an intersecting straight line $\overline{AB}$ be represented by the notation

$$\overline{Aa'} = \overline{Aa} \, \cos \overline{Aa} \overset{\wedge}{\,} \overline{AB},$$

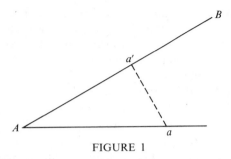

FIGURE 1

where the last term represents the angle, a convention that would have been obvious although cumbersome had it been much adopted (Fig. 1). Not that Carnot had given his reader the assistance of such a diagram in the *Essai sur les machines,* but adapting it to the equation above makes it easier to see at a glance what was then in his mind, and also to appreciate his strategy: since

$$U \cos Z = V,$$

the relation reduces to

$$\int M V^2 = 0,$$

which is to say the principle of live-force conservation.

Given the generality of his statement of the problem of machine motion, Carnot could not simply proceed to a direct application of that principle, for along with the legacy of eighteenth-century matter theory he inherited that of seventeenth-century collision theory. According to the former, solid matter consists in impenetrable, indeformable corpuscles connected by rods and shafts: rigid ones in "hard" or inelastic bodies and springs in elastic bodies. Micromachines Carnot imagined the former to be, and took them for the term of comparison to which real machines ideally reduced in nature. As for the other states of matter, liquids were fluids congruent with hardness in mechanical properties since they were incompressible although deformable (a circumstance reinforcing the primacy of hydrodynamics in his thought) and gases were fluids mechanically congruent with elasticity. His difficulty was that in classical collision theory live force was conserved only in the interaction of elastic bodies; in the supposedly more fundamental case of "hard" body, live force was conserved only when motion was communicated smoothly—by "insensible degrees" in his favorite phrase, to which processes was restricted the application of the fundamental equation stated above.

He had, therefore, to convert that equation into an expression applicable to all interactions in which motion was communicated, whatever the nature of the body or the contact. To that end Carnot introduced

the notion that he always regarded as his most significant contribution to mathematics and to mechanics: the idea of geometric motions, which he defined as displacements depending for their possibility only on the geometry of a system quite independently of the rules of dynamics. The concept was that which in later mechanics was called virtual displacement. In the *Principes fondamentaux de l'équilibre et du mouvement* of 1803, Carnot simplified his definition so that it amounted to specifying geometric motions as those that involve no work done on or by the system. But historically his first elucidation is the more interesting for it exhibits that what suggested to Sadi the idea of a reversible process must almost certainly have been his father's concept of geometric motion.

Imagine, Carnot charged the reader of the *Essai sur les machines en général,* that any system of hard bodies be struck by an impact, and further that its actual motion be stilled at just this instant, and it be made instead to describe two successive movements, arbitrary in character, but subject to the condition that they be equal in velocity and opposite in direction. Such an effect could be accomplished in infinitely many ways and (this was the essential matter) by purely geometric operations. An important, though not exhaustive class of such motions would be those that involved the constituent bodies of a system in no displacement relative to one another. In such a generalized system, the velocities of the neighboring corpuscles relative to each other would be zero, and Carnot could derive the further fundamental equation

$$\int m u U \cos z = 0,$$

which differs from the former in that the actual physical velocity V has been replaced by an idealized, geometric velocity u. It would apply, therefore, to interactions of elastic or of inelastic bodies whether gradual or sudden since by definition such motions were independent of the dynamical considerations that excluded inelastic collision. In later terms, what Carnot had done was to derive conservation of moment of momentum or torque from conservation of energy or work by considering the ideal system within which no energy or work was lost. What he himself claimed to have done was to derive a generalized indeterminate solution to the problem of machines from which could be deduced such established though partial principles as d'Alembert's and Maupertuis's least action. In actual cases, among all the motions of which a system was geometrically capable, that which would physically occur would be the geometric motion for which sum of the products of each of the masses by the square of the velocity "lost" was a minimum, i.e., for which

$$d\int mU^2 = 0.$$

His conception of the science of machines as a subject, however, rather than his somewhat jejune solution to its generalized problem, was what made Carnot an important influence upon the science of mechanics. Through that and similar influence it became in fact and not just in precept the basis of the profession of mechanical engineering. His analysis, which balanced accounts between "moment of activity produced" (work done on) and "moment of activity consumed" (work done by) the system, was the kind that the physics of work and energy have found useful ever since those topics became explicit in the 1820's, 1830's, and 1840's. Its most recognizable offspring was the heat cycle of Sadi Carnot, which considered a system in view of what had been done to it or by it in shifting from an initial to a final state. The family resemblance was marked in the abstractedness of the systems imagined; in the discussion of force in terms of what it can do, taken usually over distance when it was a question of its measure, and over time when it was a question of its realization in mechanical processes; in the notion that process consisted in the transition between successive "states" of a system; in the requirement that for purposes of theory this transition occur in infinitesimal and reversible changes (which for Lazare Carnot was to say that all motions be geometric); in the indifference (given these conditions) to the details of rate, route, or order of displacements; in the restrictive mode of reasoning by which, perpetual motion being excluded as physically unthinkable, the maximum possibilities in operations were thereby determined in the ideal case; in the relevance of this extreme schematization to the actuality of tools, engines, and machinery operating discontinuously and irreversibly in physical fact; and, finally, in thus making theoretical physics out of the engineering practice and industrial reality of the age.

The operationalism of the engineer distinguishes Carnot's mathematical writings in a similar manner. They may be summarized more briefly because, although more voluminous and evidently more important to him in his middle years, his work in mathematics did not enter into the texture of the subject as it did in mechanics. Not that it lacked for a public: his *Réflexions sur la métaphysique du calcul infinitésimal,* first published in 1797 while he was still in political office as director, was quickly translated into Portuguese, German, English, Italian, and Russian.

Carnot revised and enlarged a second edition for publication in 1813; and that version was republished from time to time in Paris, most recently in 1921. His book frankly acknowledged the difficulties that infinitesimal analysis raises for common sense, and although it was reserved to the reforms initiated by Cauchy, Bolzano, and Gauss to put the calculus on a rigorous footing, Carnot's justification evidently answered for well over a century to the needs of a public that wished to understand its own use of the calculus.

The genius of the infinitesimal calculus, in Carnot's account, lay in its capacity to compensate in its own procedures for errors that it deliberately admitted into the process of computation for the purpose of facilitating a solution. The sort of error in question is that which arises from supposing that a curve can be considered equivalent to a polygon composed of a very large number of very short sides, but the compensation of error that justified the calculus meant neither the approximation of rectilinear elements to a curve by a method of exhaustion nor the cancellation of error through some balancing of excess against defect. Such procedures would merely reduce error to the tolerable or insignificant, whereas what Carnot meant by compensation actually eliminated error and made the procedures of analysis as rigorous as those of synthetic demonstration. What the calculus really involved, properly understood, was the auxiliary use of infinitesimal quantities in order to find relations between given quantities, and its results contained no errors at all, not even infinitesimal errors.

In explaining what he meant, Carnot proposed substituting for the conventional division of quantities into determinate and indeterminate a tripartite classification into quantities (1) that were invariant and given by the conditions of the problem, (2) that although variable by nature acquired determinate values because of the conditions of the problem, and (3) that were always indeterminate. To the last class belonged all infinitely great or small quantities, and also all those involving the addition of a finite and an infinitesimal quantity. Quantities of the first two classes Carnot called designated and those of the third nondesignated. What characterized these last was not that they were minute or negligible, but that they could be made arbitrarily small at the will of the calculator. Yet at the same time they were not merely arbitrary. On the contrary, they were related to quantities of the second class by one system of equations just as the latter were to quantities of the first class by another, not unrelated, system of equations. The systems containing only designated quantities Carnot called complete or perfect. Those containing terms in nondesignated quantities he called incomplete or imperfect. The art of the infinitesimal calculus, therefore, consisted in transforming insoluble or difficult complete equations into manageable incomplete equations and then managing the calculation so as to eliminate all nondesignated quantities from the result. Their absence proved its correctness.

The congruence between Carnot's point of view in calculus and in mechanics appears to excellent advantage in his development of these comparisons. Consider, he asked the reader, any general system of quantities, some constant and some variable, and suppose the problem is to find the relations between them. Let any specified state of the general system be taken for datum. Its quantities and the variables depending exclusively on them would be the designated ones. If now the system were to be considered in some different state invoked for the purpose of determining more readily the relations between the designated quantities of the fixed system, this latter state would serve an auxiliary role and its quantities would be auxiliary quantities. If further the auxiliary state be approached to the fixed state so that all the auxiliary quantities approach more and more closely to their designated analogues, and if it is in our power to reduce these remaining differences as far as we please, then the differences would be what is meant by infinitesimal quantities. Since they were merely auxiliary, these arbitrary quantities might not figure in the solution to the problem, which was to determine the relation between the designated quantities. Reciprocally, that no arbitrary quantities occur in the result was a proof of its validity. Thus had the error willfully introduced in order to solve the problem been eliminated. If it did persist, it could only be infinitesimal. But that was impossible since the result contains no infinitesimals. Hence the procedures of the calculus had eliminated it, and that was the secret of their success.

In effect, as Carnot made clear in a historical excursion, his doctrine of compensation of errors was an attempt to combine the analytical advantages of the Leibnizian method of infinitesimals with the rigor of the Newtonian method of limits or first and last ratios of vanishing quantities.

Carnot's main geometric writings were also motivated largely by the attempt to make reasonable the employment of unreasonable quantity in analysis, although with the focus on negative rather than infinitesimal quantity. Indeed, it is clear from the manuscript remains of his mature years that he regarded the unthinking manipulation of quantities that could have no literal physical meaning as an impropriety, not to say a scandal, in mathematics that made the obscurity attending infinitesimal quantity relatively venial. There was no doubt that infinitesimals existed. The problem was only how to understand and manage them. On the other hand the difficulty with negative and imaginary numbers concerned finite analysis, went deeper, and occurred in almost every algebraic operation. In Carnot's own view his resolution of that anomaly in *De la corrélation des figures de géométrie* of 1801 and its extension, the *Géométrie de position* of 1803, constituted his most significant clarification of the procedures of mathematics.

Carnot found absurd the notion that a quantity itself could be less than zero. "Every quantity is a real object such that the mind can be seized of it, or at least its representation in calculation, in an absolute manner," he held robustly in *Géométrie de position* (p. 7), and insisted in *Corrélation des figures* on distinguishing between a quantity properly speaking and the algebraic value of a function. It was equally unacceptable to interpret the minus sign as meaning simply that a quantity was to be taken in a direction opposite to a positive one. A secant to the arc of a circle in the third quadrant was indistinguishable from one in the first in magnitude and direction. According to this latter interpretation, it should be positive in sign. In fact it is negative. Neither could this view explain why it is impossible to take the square root of a negative quantity: There is no difficulty in taking the square root of a left ordinate, after all. In short, none of the usual conventions could stand scrutiny.

There were only two senses, in Carnot's view, in which negative quantity could be rightly understood. One was the trivial sense in which it was the magnitude of a value governed by a minus sign, and this usage could be correct only when it was preceded by a positive value of greater magnitude. The deeper and more revealing sense was that a negative quantity was a magnitude governed by a sign that was wrong. Consider, for example, the formula

$$\cos (a + b) = \cos a \cdot \cos b - \sin a \cdot \sin b. \quad (1)$$

Now, in the first or literal sense, the last term here is negative. In the second sense it is positive, however, so long as a, b, and $(a + b)$ are each less than 90°, because since the equation is exact in the first quadrant, the sign is correct. But if $(a + b)$ were greater than 90°, the equation would be incorrect. For then it would turn out that

$$\cos (a + b) = \sin a \cdot \sin b - \cos a \cdot \cos b. \quad (2)$$

If formula (1) were to apply to this case, the first term, $\cos (a + b)$, would have to be regarded as governed by a sign contrary to what it ought to have been.

For changing the sign would give

$$-\cos(a+b) = \cos a \cdot \cos b - \sin a \cdot \sin b, \quad (3)$$

which reduces to equation (2). In effect, therefore, applying equation (1) to the case in which $(a+b)$ was greater than 90° subjected the term, $\cos(a+b)$, to the wrong sign and made it a negative quantity in the second sense.

For quantities of this sort Carnot proposed the term "inverse" in contradistinction to "direct quantities" that bear their proper sign. How, then, did these inversions get into the process of calculation? Carnot said by error, but not quite the kind of error that introduced infinitesimals into the calculus. The occasion of error in finite analysis was rather a mistaken assumption about the basic conditions of the problem. A credit might be mistaken for a debt in actuarial work. The algebraic expression of what was to be paid or received would then be reversed, and we would recognize the error in the solution through knowledge of whether money was owed or owing. But the question was not merely one of correcting trivial misapprehensions. Often it was necessary to introduce quantities governed by a false sign into a calculation in order to formulate a problem at all. The ordinate of a curve might be required without the geometer's knowing in which quadrant it lay. Making an assumption would permit him to get an absolute value, and if he had been wrong about the sign, the error would show up as an absurdity in the result, which he could correct without redoing the computation. Indeed, what distinguished analysis from synthesis in Carnot's view was uniquely its capacity to employ negative and imaginary forms and to eliminate the unreal entities in the course of calculation after they had served their purpose of auxiliaries. In the end whatever was unintelligible was made to disappear, and there remained a result that in principle could have been discovered synthetically. But it had been obtained more easily and directly, almost mechanically. The difference between infinitesimal and finite analysis is that errors in the calculus are eliminated through the very process of computation, whereas in algebra they remain in the solution, where we recognize them by comparison with a rational or metrical reality, and that is where geometry comes in.

The service that geometry might render to analysis lay mainly through the study of the correlation of figures, and that in turn exhibits a certain congruence with his point of view in the mechanics of machine processes. Its subject matter is the comparison of geometric systems in various states that can evolve into each other. The state taken for term of comparison he called the primitive system and any other state

a transformed system. An elementary example from the 1801 treatise *Corrélation des figures* will exhibit the nature of the problems and their relevance to resolving the anomaly of inverse quantity. In the triangle ABC (Fig. 2), the foot of the perpendicular AD falls between B and C in the primitive system.

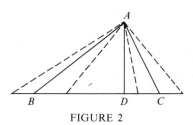

FIGURE 2

It is possible to transform the system by moving C to the left, however. In effect, the figure constitutes a system in which $\overline{CD}$ is a variable and $\overline{BD}$ a constant. In both primitive and transformed states, $\overline{BC} > \overline{BD}$ until C has passed to the left of D. Their difference, $\overline{CD}$, is a direct quantity, and $\overline{BC}$ and $\overline{BD}$ are in "direct order" in the two systems of which, therefore, the correlation was direct. Let the point C pass to the left of D, however, and then in the transformed system $\overline{BD} > \overline{BC}$, their difference $\overline{CD}$ is inverse, and the quantities $\overline{BC}$ and $\overline{BD}$ are in inverse order in the two systems, the correlation of which thereby becomes indirect.

By correlative systems Carnot meant all those that could be considered as different states of a single variable system undergoing gradual transformation. It was not necessary that all correlative systems should actually have been evolved out of the primitive system. It sufficed that they might be assimilated to it by changes involving no discontinuous mutations. The whole topic may be taken as the geometric operation of Carnot's favorite reasoning device—a comparison of systems between which the nexus of change is a continuum. When the correlation was direct, any train of reasoning that was valid for the primitive system would hold for the correlative system. When it was indirect, the formulas of the primitive system were applicable to the other by changing the signs of all the variables that had become inverted. Reciprocally, the procedure might be used in solving problems. Suppose, in the example it was given that the three sides were in the proportion

$$\overline{AC} = \overline{BC} = \frac{2}{3}\overline{AB},$$

and the segment $\overline{CD}$ was to be found. Trying the hypothesis that D was between B and C, we would have

$$\overline{AB}^2 - \overline{AC}^2 = \overline{BD}^2 - \overline{CD}^2,$$

which works out to give a negative value

$$\overline{CD} = -\frac{1}{12}\overline{AB}.$$

The minus sign signals that the hypothesis had been wrong, and that in fact C fell between B and D, for which assumption the problem yields a positive value in the solution.

In *Géométrie de position* Carnot developed what he had at first intended as a somewhat fuller edition of the *Corrélation des figures* into a vastly more extensive exploration of the problem-solving reaches of geometry. The method bore a marked resemblance to the porisms of Euclid, an affinity that Carnot recognized, for he was well versed in the history of geometry. He frequently expressed appreciation for eighteenth-century British geometry, and particularly the work of Colin Maclaurin, Robert Simson, John Landen, and Matthew Stewart—the last three names now largely forgotten in the history of mathematics. He no longer limited the scope of his work to correlations of particular geometric systems, but proposed to associate in a single treatment relations of magnitude as studied by the ancients with relations of position as studied more characteristically by the moderns, and thus to compare and unify the two main types of geometric relation. The *Géométrie de position* wears the appearance of a sort of engineering handbook of geometric systems that, were it ever to be completed, would permit resolving problems by considering unknown systems as correlatives of the set of primitive systems of which the properties were known. The formulas were to contain only real and intelligible expressions—no imaginary and no inverse quantities.

The entire subject Carnot imagined as preliminary to that which was closest to his heart and on which he promised a further book: the science of geometric motion. He redeemed the promise that same year by publishing the *Principes fondamentaux de l'équilibre et du mouvement,* which treats the science of machines in terms of geometric motions. For it is extremely interesting that it was in this that he thought to unify his geometric and mechanical interests. In *Géométrie de position,* he defines such motions as he had in the *Essai sur les machines en général,* motions depending on the geometry and not the dynamics of a system such that "the contrary motion is always possible." Such a science would be intermediary between geometry and mechanics and would rid mechanics and hydrodynamics of their analytical difficulties because it would then be possible to base both sciences entirely on the most general principle in the communication of motion, the equality and opposition of action and reaction.

BIBLIOGRAPHY

I. ORIGINAL WORKS. Carnot's scientific works are as follows:

(1) *Essai sur les machines en général, par un officier du Corps Royal du Génie* (Dijon, 1783). Although absent from the first edition, the name of the author did appear on the second, otherwise identical printing (Dijon–Paris, 1786).

(2) *Réflexions sur la métaphysique du calcul infinitésimal* (Paris, 1797). The identical text was also published, together with (1) above, in *Oeuvres mathématiques du citoyen Carnot* (Basel, 1797). In 1813 Carnot published a revised and enlarged edition, which has been reprinted without change in all later French editions.

(3) *Lettre du citoyen Carnot au citoyen Bossut, contenant quelques vues nouvelles sur la trigonométrie,* in Charles Bossut, *Cours de mathématiques,* new ed., vol. II (Paris, 1800).

(4) *De la corrélation des figures de géométrie* (Paris, 1801).

(5) *Géométrie de position* (Paris, 1803).

(6) *Principes fondamentaux de l'équilibre et du mouvement* (Paris, 1803).

(7) *Mémoire sur la relation qui existe entre les distances respectives de cinq points quelconques pris dans l'espace, suivi d'un essai sur la théorie des transversales* (Paris, 1806), to which is appended a summary of the theory of negative quantity, "Digression sur la nature des quantités dites négatives."

Between 1800 and 1815 Carnot served on many commissions of the Institute to which numerous mechanical inventions and mathematical writings were submitted. His reports on these subjects will be found in Institut de France, Académie des Sciences, *Procès-verbaux des séances de l'Académie tenues depuis la fondation de l'Institut jusqu'au mois d'août 1835,* vols. I–V (Paris, 1910–1914). Each volume contains a full index of its contents. Carnot's most important reports were those on the Niepce and the Cagniard engines, III, 465–467, and IV, 200–202. The archives of the Academy in Paris contain the text of an unpublished "Lettre sur les aerostats" that Carnot submitted to the Academy on 17 January 1784 following the flight engineered by the brothers Montgolfier on 5 June 1783. The problem was that of propulsion.

II. SECONDARY LITERATURE. There is a recent and largely reliable political biography, Marcel Reinhard, *Le grand Carnot,* 2 vols. (Paris, 1950–1952).

The memoir by Carnot's younger son, Hippolyte Carnot, *Mémoires sur Carnot,* 2 vols. (Paris, 1861–1863), is a work of anecdote animated by family piety but remains indispensable.

Carnot's scientific work forms the subject of a monograph, *Lazare Carnot savant,* by the undersigned (in press—scheduled for publication by Princeton University Press in 1971). The appendix of this work reprints in facsimile the theoretical portions of the two manuscripts on mechanics that Carnot submitted to the *Académie des sciences* in 1778 and 1780, and the entire text of the "Dissertation" on the infinitesimal calculus that he submitted to the Académie Royale des Sciences, Arts et Belles-Lettres de Berlin in 1785. The originals are contained in the archives of the respective academies. The forthcoming work also contains a commentary on this latter manuscript by A. P. Youschkevitch, who published an extensive introduction and commentary to the Russian translation of the *Réflexions sur la métaphysique du calcul infinitésimal: Lazar Karno, Razmyshlenia o metafizike ischislenia beskonechnomalykh,* N. M. Solovev, trans., with a critical introduction by A. P. Youschkevitch and a biographical sketch by M. E. Podgorny (Moscow, 1933).

Persons who consult accounts of Carnot's work in mechanics in the secondary literature—e.g., René Dugas, *Histoire de la mécanique* (Neuchâtel, 1950), or Émile Jouguet, *Lectures de mécanique,* 2 vols. (Paris, 1908–1909), should be warned that their authors have sometimes confused item I, 6, with item I, 1, above, and attributed passages from the later, more widely read work to the earlier one. There is a brief notice of Carnot's geometric work in Michel Chasles, *Aperçu historique sur l'origine et le développement des méthodes en géométrie* (2nd ed., Paris, 1875), and Felix Klein, *Vorlesungen über die Entwicklung der Mathematik im 19. Jahrhundert,* 2 vols. (Berlin, 1926) I, 79–80.

CHARLES C. GILLISPIE

CARNOT, NICOLAS LÉONARD SADI (*b.* Paris, France, 1 June 1796; *d.* Paris, 24 August 1832), *thermodynamics.*

The eldest son of Lazare Carnot, Sadi was born in the Palais du Petit-Luxembourg, where his father lived as a member of the Directory. The powerful and often turbulent worlds of French politics and science were an integral part of the environment in which Sadi and his younger brother Hippolyte spent their youth. Withdrawing from public life in 1807, Lazare Carnot concentrated on science and the education of his sons. Through his studies Sadi acquired not only his taste and aptitude for mathematics but also a solid training in physics, the natural sciences, languages, and music.

Because of his rapid progress it was decided that Sadi should attend the elite École Polytechnique upon attaining the age of sixteen, the minimum for admission. Following a few months' preparation at the Lycée Charlemagne in Paris, he passed the entrance examination and was admitted to the Polytechnique. His studies there from 1812 to 1814 stressed analysis, mechanics, descriptive geometry, and chemistry, taught by a distinguished faculty including Poisson, Gay-Lussac, Ampère, and Arago.

In 1813 Sadi addressed a letter to Napoleon on behalf of his fellow students, asking permission to join the fight against the invading Allies, and in March 1814 he was among the students who fought bravely, though in vain, at Vincennes. Ranking sixth in his class, he finished his studies at the Polytechnique in October 1814 and was immediately sent to the École du Génie at Metz as a student second lieutenant. During the two-year course in military engineering Sadi wrote several scientific papers, now lost but which his brother said were well received. During the Hundred Days, Lazare Carnot was Napoleon's minister of the interior, and Sadi became an object of special attention from his superiors. This ended in October 1815, when Lazare was exiled by the Restoration.

In late 1816 Sadi finished his studies and began serving as a second lieutenant in the Metz engineering regiment. For the next two years he was shifted about from garrison to garrison, inspecting fortifications and drawing up plans and reports doomed to bureaucratic oblivion.[1] In spite of some connections with high officials, his father's name and reputation became a burden to him in the first years of the Restoration, and his intellectual development was frustrated by the tedium of military garrisons. In 1819 he seized an opportunity to escape by passing a competitive examination for appointment to the army general staff corps in Paris. He immediately obtained a permanent leave of absence and took up residence in his father's former Paris apartment.

Relieved of the constraints of military life, Carnot began the wide range of study and research that continued, despite numerous interruptions, until his death. In addition to private study he followed courses at the Sorbonne, the Collège de France, the École des Mines, and the Conservatoire des Arts et Métiers. At the latter he became a friend of Nicolas Clément, who taught the course in applied chemistry and was then doing important research on steam engines and the theory of gases. One of Carnot's particular interests was industrial development, which he studied in all of its ramifications. He made frequent visits to factories and workshops, studied the latest theories of political economy, and left in his notes detailed proposals on such current problems as tax reform. Beyond this, his activity and ability embraced mathematics and the fine arts.

In 1821 Carnot interrupted his studies to spend a few weeks with his exiled father and brother in Magdeburg. It was apparently after this visit that, once again in Paris, he began to concentrate on the problems of the steam engine. After Lazare's death in August 1823, Hippolyte returned to Paris to find his brother at work on the manuscript of the *Réflexions.* In an attempt to make his work comprehensible to a wide audience, Sadi forced Hippolyte to read and criticize portions of the manuscript. On 12 June 1824 the *Réflexions sur la puissance motrice du feu et sur les machines propres à développer cette puissance* was published by Bachelier, the leading scientific publisher in France. By all reasonable standards the book was well received. On 14 June it was formally presented to the Académie des Sciences, and on 26 July P.-S. Girard read a lengthy and very favorable review to the Academy. This review, printed in the August issue of the *Revue encyclopédique,* emphasized the book's conclusions and its applications for steam-engine construction. Although the major theorems were cited, there was no discussion of the highly original reasoning that Carnot had employed.

Following the publication of his book, Carnot continued his research, fragments of which are preserved in his manuscript notes. A reorganization of the general staff corps, however, forced Carnot to return to active service in 1827 with the rank of captain. After less than a year of routine duty as a military engineer in Lyons and Auxonne, Carnot resigned permanently and returned to Paris. He again focused his attention on the problems of engine design and the theory of heat. In 1828 a contemporary referred to Carnot as a "builder of steam engines,"[2] although there is no record of his formal connection with any firm. Except for his informal contact with Clément, Carnot always worked independently and rarely discussed his research.

Although sensitive and perceptive, he appeared extremely introverted, even aloof, to all but a few close friends, most of whom, like Claude Robelin and the geometer Michel Chasles, were his classmates at the Polytechnique.

True to his father's republican principles, Carnot welcomed the July Revolution, but was soon disappointed with the new government. Nonetheless, he was highly regarded in some political circles, for shortly after the Revolution he was mentioned as a possible member of the Chambre des Pairs. He objected to the hereditary nature of this position, however, and refused to be nominated.

In 1831 Carnot began to investigate the physical properties of gases and vapors, especially the relationship between temperature and pressure. In June 1832, however, he contracted scarlet fever. This was followed by "brain fever," which so undermined his fragile health that on 24 August 1832 he fell victim to a cholera epidemic and died within the day, at the age of thirty-six. In accordance with the custom, his personal effects, including nearly all of his papers, were burned. Although for eight years his work had been almost completely ignored, he was not forgotten. The news of his death received a front-page article

in the 27 August *Moniteur,* and a note describing his book as "remarkable for its original views" appeared in the February 1833 issue of the *Annales de chimie et de physique.* The only full obituary was published in the August 1832 issue of the *Revue encyclopédique,* of which Hippolyte Carnot was then the editor.

The extant scientific work of Sadi Carnot includes three major items: the "Recherche" manuscript, the *Réflexions,* and the manuscript notes. Fragments of mathematics, his translations of two of Watt's papers, and lecture notes from various mathematics and physics courses have also survived.

The earliest of the major works is a twenty-one-page manuscript written probably in 1823 and entitled "Recherche d'une formule propre à représenter la puissance motrice de la vapeur d'eau."[3] As the title indicates, the paper was an attempt to find a mathematical expression for the motive power (work, in modern terms) produced by one kilogram of steam. Explicitly seeking a general solution covering all types of steam engines, Carnot reduced their operation to three basic stages: an isothermal expansion as the steam is introduced into the cylinder, an adiabatic expansion, and an isothermal compression in the condenser. Although Carnot claimed the cycle was complete, this was true only with respect to the motion of the piston and not to the working substance, which was not returned to the temperature of the boiler. Employing Clément's law for saturated vapors and devising an approximation function for Dalton's table of vapor pressures, Carnot neatly derived the motive power as a function of the initial and final temperatures and pressures of the steam. Since Dalton's table relates pressure to temperature, Carnot in effect expressed the motive power of a unit quantity of steam as a function of temperature alone.

The essay, both in methods and in objectives, is similar to the many papers published between 1818 and 1824 by such "power engineers" as Hachette, Navier, Petit, and Combes. Carnot's work, however, is distinguished for his careful, clear analysis of the units and concepts employed and for his use of both an adiabatic working stage and an isothermal stage in which work is consumed. The polished nature of the paper, in contrast with his rough notes, makes it appear intended for publication, although it remained unknown, in manuscript, until 1966.

The *Réflexions,* the only work that Carnot published in his lifetime, appeared in 1824 as a modestly priced essay of only 118 pages. After a concise review of the industrial, political, and economic importance of the steam engine, Carnot raised two problems that he felt prevented further development of both the utility and the theory of steam engines. Does there exist an assignable limit to the motive power of heat, and hence to the improvement of steam engines? Are there agents preferable to steam in producing this motive power? As Carnot conceived it, the *Réflexions* was nothing more, nor less, than a "deliberate examination" of these questions. Both were timely problems and, although French engineers had investigated them for a decade, no generally accepted solutions had been reached. In the absence of a clear concept of efficiency, proposed steam-engine designs were judged largely on practicality, safety, and fuel economy.[4] Air, carbonic acid, and alcohol had all been advocated by some engineers as a better working substance than steam. The usual approach to these problems was either an empirical study of the fuel input and the work output of individual engines or the application of the mathematical theory of gases to the abstract operations of a specific type of engine. In his choice of problems Carnot was firmly in this engineering tradition; his method of attacking them, however, was radically new and is the essence of his contribution to the science of heat.

Previous work on steam engines, as Carnot saw it, had failed for want of a sufficiently general theory, applicable to all imaginable heat engines and based on established principles. As the foundations for his study Carnot carefully set out three premises. The first was the impossibility of perpetual motion, a principle that had long been assumed in mechanics and had recently played an important role in the work of Lazare Carnot. As his second premise Carnot used the caloric theory of heat, which, in spite of some opposition, was the most accepted and most developed theory of heat available. In the *Réflexions,* heat (*calorique*) was always treated as a weightless fluid that could neither be created nor be destroyed in any process.[5] As an element in Carnot's demonstrations this assumption asserted that the quantity of heat absorbed or released by a body in any process depends only on the initial and final states of the body. The final premise was that motive power can be produced whenever there exists a temperature difference. The production of motive power was due "not to an actual consumption of caloric, but to its transportation from a warm body to a cold body."[6] Making the analogy with a waterwheel, Carnot observed that this motive power must depend on both the amount of caloric employed and the size of the temperature interval through which it falls. In his concept of reversibility Carnot also implicitly assumed the converse of this premise, that the expenditure of motive power will return caloric from the cold body to the warm body.

The analysis of heat engines began where the "Re-

cherche" ended, with an abstract, three-stage steam-engine cycle. The incompleteness of this cycle proved troublesome, and Carnot pushed the abstraction one step further, producing the ideal heat engine and the cycle that now bear his name. The "Carnot engine" consisted simply of a cylinder and piston, a working substance that he assumed to be a perfect gas, and two heat reservoirs maintained at different temperatures. The new cycle incorporated the isothermal and adiabatic expansions and the isothermal compression of the steam engine, but Carnot added a final adiabatic compression in which motive power was consumed to heat the gas to its original, boiler temperature. In describing the engine's properties, Carnot introduced two fundamental thermodynamic concepts, completeness and reversibility. At the end of each cycle the engine and the working substance returned to their original conditions. This complete cycle not only provided an unambiguous definition of the input and output of the engine, but also rendered superfluous the detailed examination of each stage of the cycle. With each cycle the engine transferred a certain quantity of caloric from the high-temperature reservoir to the low-temperature reservoir and thereby produced a certain amount of motive power. Since each stage of the cycle could be reversed, the entire engine was reversible. Running backward, the engine consumed as much motive power as it produced running forward and returned an equal amount of caloric to the high-temperature reservoir. Joined together but operating in opposite directions, two engines would therefore produce no net effect.

Carnot then postulated the existence of an engine that, by virtue of design or working substance, would produce more motive power than a "Carnot engine" operating over the same temperature interval and with the same amount of caloric. A reversed "Carnot engine" would be able to return to the boiler all of the caloric transported to the condenser by the hypothetical engine. Yet the reversed "Carnot engine" would consume only a portion of the motive power produced by the hypothetical engine, leaving the remainder available for external work. Together these two engines would form a larger engine whose only net effect was the production of motive power in unlimited quantities. Since such a perpetual motion machine violated his first premise, Carnot concluded that no engine whatsoever produced more motive power than a "Carnot engine." Formulating the result now known as "Carnot's theorem," he stated that "the motive power of heat is independent of the agents employed to realize it; its quantity is fixed solely by the temperatures of the bodies between which is effected, finally, the transfer of caloric."[7]

To elucidate further the motive power of heat, Carnot turned his attention to the physical properties of gases, a subject in which there had been considerable activity for over a decade. By 1823 a sizable body of experimental data on adiabatic and isothermal processes and on specific heats had been assimilated into the caloric theory of heat and mathematized by Laplace and Poisson. Combining the results of this activity with the concepts involved in his fundamental theorem, Carnot derived a series of seven theorems. With the exception of a long footnote in which he attempted to cast his results in algebraic form, Carnot developed his theorems in a synthetic, geometric manner that, although clear and logically rigorous, was in sharp contrast with the mathematical analysis dominant in the scientific community. Nonetheless, at least three of the theorems represented major advances.[8] The first, that the quantity of heat absorbed or released in isothermal changes is the same for all gases, was experimentally established by Dulong in 1828, but without any reference to Carnot. In a very subtle verbal argument, Carnot also demonstrated that "the difference between specific heat under constant pressure and specific heat under constant volume is the same for all gases."[9] The final theorem proved that the fall of caloric produces more motive power when the temperature interval is located lower rather than higher on the temperature scale. Although aware of the uncertainties introduced by some assumptions and experimental data for specific heat changes, Carnot was able to calculate motive power values and to verify the theorem. The works of Clapeyron and William Thomson were in part motivated by the desire to derive an algebraic expression ("Carnot's function") for this theorem and to verify it with more accurate data.

In the final section of the *Réflexions,* Carnot returned to his original questions on steam engines. With experimental data taken from the current literature he verified that all gases produce the same amount of motive power and was able to estimate the ideal limit for its production. In a review of the most common types of steam engines, Carnot sought to apply his findings to the practical questions of steam-engine design and operation. His contributions, however, fell short of his original goal. His conclusions— that steam ought to be used expansively (adiabatically), over a large temperature interval, and without conduction losses—were already widely recognized by engineers of his time. Because of difficulties in engine construction even the problem of the best working substance was not conclusively answered.

Although the *Réflexions* was regarded by contemporaries as primarily an essay on steam engines,

Carnot's most important innovations lay in a new approach to the study of heat. While he accepted, and in some theorems furthered, the theory of heat developed by Laplace and Poisson, Carnot also shifted the emphasis from the microscopic to the macroscopic. Rather than build upon the notion of gas particles surrounded by atmospheres of caloric, he began with the directly measurable entities of volume, pressure, temperature, and work.

Of his concepts of an ideal engine, completeness, and reversibility, there were some vague anticipations. The notion of an abstract heat engine was approximated in the work of Hachette and was more clearly present in the studies by Cagniard de La Tour and Clément of the motive power produced by a bubble of gas rising adiabatically in water. Jacob Perkins' steam engine, widely discussed in 1823, represented an attempt to design a closed, complete system, and engineers were aware that certain types of hydraulic engines were reversible. The most striking parallels, however, are found in two ideas stressed by Lazare Carnot: the concept of geometrical (reversible) motions in mechanical machines and the necessity of computing the work done by a machine only after a complete cycle of operation. In addition, the style of Sadi Carnot's work, the synthetic, abstract generality that made his work distinctively different from that of both engineers and physicists, was in large measure due to the example set by the scientific work of Carnot *père.* Thus, in applying the concepts of ideal engines, completeness, and reversibility to the study of heat, Sadi Carnot gave them an unprecedented precision and generality and placed them in a highly original and fruitful combination.

Although the exact reasons are impossible to determine, the *Réflexions* had almost no influence on contemporary science. The original edition had not sold out by 1835; by 1845 booksellers had forgotten it completely. Aside from the reviews in 1824 and the references in obituaries, Carnot's work was mentioned only twice between 1824 and 1834. Clément recommended the book in his 1824–1825 lectures and Poncelet, writing sometime before 1830, cited it in his *Introduction à la mécanique industrielle* (Paris, 1839). In Carnot's obituary Robelin attributed the neglect of the *Réflexions* to its difficulty, an explanation that would have applied only to engineers and craftsmen unfamiliar with contemporary physics and mathematics. Another explanation points to the failure of the *Réflexions* to reach conclusions of real value to steam engineers. The silence on the part of physicists like Dulong, who later retraced portions of Carnot's work, is more difficult to explain. One probable factor, however, was Carnot's use of the caloric theory and of experimental results such as Clément's law of saturated vapors. His work was thus especially vulnerable, as he realized himself, when Clément's law was disproved in 1827 and when problems of radiant heat initiated a period of "agnosticism" concerning the nature of heat.

In 1834 Clapeyron, with whom Carnot may have been acquainted in 1832, published an analytical reformulation of the *Réflexions*. While Clapeyron preserved the premises, the theorems, and some of the specific arguments, the emphasis and style were considerably altered. He related the Carnot cycle to the pressure-volume indicator diagram and, emphasizing Carnot's function, translated Carnot's synthetic work from the world of heat engines to the realm of the mathematical theory of gases. Carnot's work attracted no further attention until C. H. A. Holtzman in 1845 and William Thomson in 1848 began working on special aspects of Clapeyron's paper. Between 1848 and 1850 Thomson, working directly from the *Réflexions*, published a series of papers that both extended and confirmed Carnot's results. These papers constituted a strong defense for Carnot's work, including his use of the caloric theory, at a time when Joule, Julius Mayer, and Helmholtz were establishing the convertibility of heat and work and the principle of energy conservation. In 1850 Clausius showed that Carnot's theorem was correct as stated but that Carnot's proof, which assumed no heat was lost, needed modification. Clausius added the statements that in the Carnot engine a certain quantity of heat is destroyed, another quantity is transferred to the colder body, and both quantities stand in a definite relation to the work done. With these additions, which Thomson also adopted in 1851, Carnot's theorem became the second law of thermodynamics.

The third major item of Carnot's scientific work is a bundle of twenty-three loose sheets of manuscript notes that escaped destruction after his death. Rough and disjointed, they contain notes from journal articles, outlines of experiments, and drafts of research results. Most concern one of three issues: adiabatic processes, the source of heat generated by friction, and the nature of radiant heat. From the development of these themes and some internal evidence the notes have been chronologically ordered, and many appear to have been written between 1824 and 1826. Several passages in the *Réflexions* indicate that Carnot had some serious reservations about the validity of the caloric theory. In the notes he sharpened these doubts and gathered relevant evidence. The difficulty of explaining certain adiabatic phenomena, Fresnel's vibrational theory of light, and the widespread speculation on the similarity of light and radiant heat all appear in the early notes. Carnot accepted Fresnel's theory and employed it against the caloric theory with the argument that motion (radiant heat) could never produce matter (caloric). Shortly after this he conducted an extensive literature search in which the works of Rumford and Davy figure prominently. Also among these notes are plans for experiments to test the temperature effects of friction in liquids and gases, several of which are nearly identical with those Joule performed almost twenty years later.

Finally convinced that "heat is nothing else than motive power, or rather motion which has changed its form,"[10] Carnot began to work out the details of a kinetic theory. He was aware that this new theory of heat negated the arguments of the *Réflexions,* but the notes contain no specific attempts to reformulate his earlier work. Although his procedure is missing from the notes, he calculated a conversion coefficient for heat and work and went on to assert that the total quantity of motive power in the universe was constant. These notes anticipated nearly all of the groundwork for the first law of thermodynamics. They remained undiscovered and unpublished until 1878, however.

NOTES

1. Two of his later military papers have survived in MS: "Route de Coulommiers à Couilly" (13 pp., dated 25 Oct. 1824) and "Essai sur l'artillerie de campagne" (82 pp., dated Mar. 1828).

2. A. Fourcy, *Histoire de l'École polytechnique* (Paris, 1828), p. 445.

3. Gabbey and Herivel give the period 1816–1824 as the date of composition. The lower bound is set by Carnot's citation of Biot's *Traité de physique* (Paris, 1816), and the incomplete cycle and the lack of any explicit comparison between heat employed and work produced would suggest a date prior to the *Réflexions*. A reference to Clément as a professor at the Conservatoire des Arts et Métiers, however, indicates the "Recherche" was written sometime after 1819. Carnot's use of "dynames" as the unit of motive power (one kilogram raised one meter) suggests 1823 as the most probable date of composition, since Dupin coined that term in a report to the Academy in Apr. 1823.

4. Since heat and work were regarded as separate entities, there was no way to measure engine output in terms of input and, hence, no indication of what 100 percent efficiency would be. Carnot's search for a limit to the motive power of heat was essentially an attempt to establish such a criterion.

5. Carnot's use of the two words "chaleur" and "calorique" has led to several misinterpretations of his work. In 1910 H. L. Callendar, in the article "Heat" for the 11th and subsequent eds. of the *Encyclopaedia Britannica*, denied that Carnot regarded heat as a substance conserved in all processes. A more extreme position was advanced by V. K. La Mer (*American Journal of Physics*, **22** [1954], 20–27), in which he argued that by "calorique" Carnot meant "entropy" and by "chaleur," "heat." This interpretation, however, has no historical foundation. Several theorems in the *Réflexions* clearly used "calorique" as a material fluid that is conserved, and Carnot explicitly stated that "quantité de chaleur" and "quantité de calorique" were interchangeable expressions. Cf. T. S. Kuhn, "Carnot's Version of 'Carnot's Cycle,'" in *American Journal of Physics*, **23** (1955), 91–95.

6. *Reflections,* E. Mendoza, ed., p. 7.

7. *Ibid.,* p. 20.

8. In their original form all but one of Carnot's theorems are regarded as valid today. The incorrect theorem, in modern notation, asserts that $c = c_0 + k \log v/v_0$ where c is the specific heat of a gas at constant volume. This increase in specific heat with expansion was universally accepted on the basis of experiments performed in 1812 by Bérard and Delaroche.

9. *Reflections,* E. Mendoza, ed., p. 24.

10. *Sadi Carnot, biographie et manuscrit,* p. 81.

BIBLIOGRAPHY

I. Original Works. The 1st ed. of *Réflexions sur la puissance motrice du feu et sur les machines propres à développer cette puissance* (Paris, 1824) is very rare. There is, however, a photographic reprint (Paris, 1953). An earlier reprint (Paris, 1878) also contains Hippolyte Carnot's biography of his brother and an abridged version of the MS notes. The standard English translation is R. H. Thurston, *Reflections on the Motive Power of Heat* (New York, 1897), which also contains major portions of Hippolyte's biography and the MS notes. This translation has been reprinted many times, but the most useful is that appearing in the collection edited by E. Mendoza, *Reflections on the Motive Power of Fire by Sadi Carnot and Other Papers on the Second Law of Thermodynamics by E. Clapeyron and R. Clausius* (New York, 1960). The "Recherche" MS has been published by W. A. Gabbey and J. W. Herivel as "Un manuscrit inédit de Sadi Carnot," in *Revue d'histoire des sciences,* **19** (1966), 151–166. The only complete ed. of the notes is in *Sadi Carnot, biographie et manuscrit* (Paris, 1927), with an introduction by Émile Picard and an ordering of the notes by C. Raveau.

II. Secondary Literature. The main biographical source is Hippolyte Carnot, "Notice biographique sur Sadi Carnot," first published in 1878 (see above). This is an expanded version of a letter written by Hippolyte's son, also named Sadi, and published by the Conte di St. Robert in "Notice biographique sur Sadi Carnot," in *Atti della R. Accademia delle scienze di Torino,* **4** (1868), 151–170. Other sources are Hippolyte Carnot, *Mémoires sur Carnot par son fils* (Paris, 1863); a note by Michel Chasles in *Comptes rendus de l'Académie des sciences,* **68** (1869), 115–117; and the obituary by Robelin, "Notice sur Sadi Carnot," in *Revue encyclopédique,* **55** (1832), 528–530.

Background to Carnot's work can be found in Robert Fox, "The Background to the Discovery of Dulong and Petit's Law," in *The British Journal for the History of Science,* **4** (1968), 1–22; Charles C. Gillispie, *The Edge of Objectivity* (Princeton, N. J., 1960), ch. 9; and T. S. Kuhn, "Energy Conservation as an Example of Simultaneous Discovery," in Marshall Clagett, ed., *Critical Problems in the History of Science* (Madison, Wis., 1959), pp. 321–356.

Two useful commentaries on the *Réflexions* are Martin Barnett, "Sadi Carnot and the Second Law of Thermodynamics," in *Osiris,* **13** (1958), 327–357; and F. O. Koenig, "On the History of Science and the Second Law of Thermodynamics," in Herbert Evans, ed., *Men and Moments in History of Science* (Seattle, Wash., 1959), pp. 57–111. For a contemporary review of the *Réflexions,* see Pierre-Simon Girard, *Revue encyclopédique,* **23** (1824), 411–414.

On various technical aspects of Carnot's work see D. S. L. Cardwell, "Power Technology and the Advance of Science, 1700–1825," in *Technology and Culture,* **4** (1965), 188–207; Milton Kerker, "Sadi Carnot and the Steam Engine Engineers," in *Isis,* **51** (1960), 257–270; T. S. Kuhn, "The Caloric Theory of Adiabatic Compression," *ibid.,* **49** (1958), 132–140; "Engineering Precedent for the Work of Sadi Carnot," in *Archives internationales d'histoire des sciences,* **13** (1960), 251–255; and "Sadi Carnot and the Cagniard Engine," in *Isis,* **52** (1961), 567–574; and E. Mendoza, "Contributions to the Study of Sadi Carnot and His Work," in *Archives internationales d'histoire des sciences,* **12** (1959), 377–396.

Also of interest is Robert Fox, "Watt's Expansive Principle in the Work of Sadi Carnot and Nicholas Clément," in *Notes and Records of the Royal Society,* **24** (Apr. 1970), 233–243, which, however, appeared too late to be utilized in writing this biography.

JAMES F. CHALLEY

CARO, HEINRICH (*b.* Poznan, Poland, 13 [17?] February 1834; *d.* Dresden, Germany, 11 October 1910), *dye chemistry.*

Caro's career mirrors the dynamic expansion of the synthetic dye industry and illustrates the process by which science entered into close and permanent cooperation with industry. The son of a successful Jewish grain dealer who moved to Berlin in 1842, Caro was given a vocational education, proceeding in 1852 from the Gymnasium to the Gewerbeinstitut, where he was trained as a dyer. Simultaneously he attended chemistry lectures at the University of Berlin. In 1855 he was hired as a colorist by a calico printing firm in Mülheim an der Ruhr, where natural dyes and secret recipes were still in use. After demonstrating the power of scientific solutions to certain production problems, Caro, like many a technician of his generation, was sent to England in 1857 (the year following William Perkin's discovery of mauve from aniline) to learn the most up-to-date dyeing techniques.

Two years later he was back in England, having resigned his Mülheim job, eager to plunge into the exciting new field of synthetic dyes. As analytical chemist for the Manchester chemical firm Roberts, Dale & Co., he soon earned admission into partnership by discovering a more efficient synthesis of Perkin's mauve. Through engineering experience gained in commercializing this process and the simultaneous study of A. W. Hofmann's researches on the synthesis and constitution of the new aniline dyes, Caro rounded out his training as an industrial organic chemist. While maintaining close touch with other budding dye chemists—all of them at one time Hofmann's students or assistants, and including among them several Germans destined for future leadership in the dye industry—Caro now launched on several lines of research that ultimately led to his major scientific contributions. Following up Peter Griess's pioneering studies on diazo compounds, he discovered induline, a useful aniline dye; and in 1864–1865, together with C. A. Martius, found and developed a commercial synthesis for Bismarck brown and Martius yellow. Meanwhile, Caro also began a decade of intermittent studies that contributed significantly to elucidating the triphenylmethane structure of rosaniline dyes.

Personal success and marriage to a British subject, Edith Eaton, were not enough to hold Caro in England. Like most of his German chemical colleagues in England, he saw greater opportunity at home. In 1866 he settled in Heidelberg to pursue fundamental research in Bunsen's laboratory. Two years later he accepted the directorship of what was probably the first true industrial research organization, the Badische Anilin- & Soda-Fabrik in Ludwigshafen.

Continuing his previous practice of working closely with fundamental researchers in aromatic chemistry at universities and elsewhere, Caro and his organization hurried from one commercial triumph to the next. When, in 1869, Graebe and Liebermann showed him their new laboratory synthesis of alizarin, he found a much cheaper process to make this coveted dye commercially. In 1874 he brominated Baeyer's fluorescein to produce eosin, a fluorescent, red dyestuff. Further collaboration with Baeyer in azotizing tertiary amines (e.g., dimethylaniline) yielded methylene blue and related dyes. Substantial mastery over azo reactions achieved during the mid-1870's by Griess, Otto Witt, and others yielded a spate of valuable azo dyes. Of these Caro codiscovered and brought into production chrysoidine, orange, and fast red. Besides numerous lesser discoveries he also made fruitful suggestions to colleagues and associates. An intensive collaboration, begun in 1880 with Baeyer and the Farbwerke Hoechst to effect a commercial synthesis of indigo, led Caro to a technically but not commercially feasible solution in which indigo was to be made from cinnamic acid, which in turn had been derived from benzal chloride. After 1889, when Caro resigned as director of research, the indigo project was continued by BASF chemists under the leadership of his successor, Heinrich Brunck, until success came at last in 1897. Among Caro's last major finds were naphthol yellow (1879) and persulfuric acid, H_2SO_5 (1898). The latter, also known as Caro's acid, is a combination of sulfuric acid and hydrogen peroxide and is a strong oxidizing agent.

By 1883 Caro had become a leading spokesman for the German chemical industry. He made outstanding contributions to the formulation of effective patent laws and practices for the protection of chemical inventions. In several articles and personal tributes he also recorded for history the spectacular growth of the dye industry during his own lifetime.

BIBLIOGRAPHY

I. ORIGINAL WORKS. All of Caro's technical publications are in the form of patents and articles. His published speeches were gathered into a book by his daughter Amalie Caro, *Gesammelte Reden und Vorträge von Heinrich Caro* (Leipzig, 1913). The Deutsches Museum in Munich holds much of his correspondence with Adolf von Baeyer, Graebe, Liebermann, and other chemists. Three of his historical articles are of great value: his obituary of Peter Griess, in *Berichte der Deutschen chemischen Gesellschaft,* **24** (1891), i–xxxviii; "Über die Entwicklung der chemischen Industrie von Mannheim-Ludwigshafen a. Rh.," in *Zeitschrift für angewandte Chemie,* **17** (1904), 1343–1362; and "Über die Entwicklung der Teerfarben-Industrie," in *Berichte der Deutschen chemischen Gesellschaft,* **25,** no. 3 (1892), 955–1105.

II. SECONDARY LITERATURE. Obituaries and tributes are the main biographical sources regarding Caro's life. The most important and lengthy of these is the obituary by A. Bernthsen, in *Berichte der Deutschen chemischen Gesellschaft,* **45,** no. 2 (1912), 1987–2042. In addition see A. Bernthsen, "Zum 70. Geburtstage von H. Caro," in *Zeitschrift für angewandte Chemie,* **24** (1911), 1059–1064; and Carl Duisberg, "Heinrich Caro," *ibid.,* 1057–1058. Also useful is E. Darmstaedter, "Heinrich Caro (1834–1910)," in Gunther Bugge, ed., *Das Buch der grossen Chemiker,* 2nd ed., II (Weinheim, 1955), 298–309, which draws on new sources and provides a good bibliography.

A brief description of Caro's correspondence at the Deutsches Museum is to be found in Kurt Schuster's biographical essay "Heinrich Caro," in Kurt Oberdorffer, ed., *Ludwigshafener Chemiker,* II (Düsseldorf, 1960), 45–83. Caro's contribution to German patent legislation and practices is expertly reviewed in Paul A. Zimmermann, *Patentwesen in der Chemie* (Ludwigshafen, 1965). For background information and additional bibliography see John J. Beer, *The Emergence of the German Dye Industry* (Urbana, Ill., 1959).

JOHN J. BEER

CAROTHERS, WALLACE HUME (*b.* Burlington, Iowa, 27 April 1896; *d.* Philadelphia, Pennsylvania, 29 April 1937), *chemistry*.

Wallace Carothers was the eldest of the four children of Ira Hume and Mary Evelina McMullin

Carothers. His early education, in Des Moines, Iowa, included a final year at Capital City Commercial College, where his father taught and later served as vice-president. In 1915 he entered Tarkio College in Tarkio, Missouri, supporting himself as a student by teaching accounting, English, and his major interest, chemistry. He received his bachelor's degree at Tarkio in 1920 and his master's degree at the University of Illinois in Urbana in 1921. After teaching for a year at the University of South Dakota, he returned to Urbana and in 1924 earned his doctorate, with a major in organic chemistry and a dissertation on hydrogenations with modified platinum-oxide-platinum-black catalysts, under the direction of Roger Adams. He remained there for two years as instructor in organic chemistry and in 1926 accepted a like position at Harvard University. In the period 1923–1927 he published his first four independent papers, of which the most notable—on the electronic nature of the double bond—confirmed his early promise of brilliance and originality.

A turning point in Carothers' career came in 1928, when he joined E. I. du Pont de Nemours and Company to lead the organic group in a program of fundamental research at its central laboratories in Wilmington, Delaware. At the time this was a radical departure in industrial research, and the opportunity thus offered to undertake basic research with a group of postdoctoral associates and maximum physical support overcame Carothers' reluctance to leave the academic world. The ensuing nine years until his premature death were extraordinarily productive, resulting not only in major contributions to theoretical organic chemistry but also in the founding of two industries, the production of synthetic rubber and of wholly synthetic fibers for textile and industrial purposes. In 1936 Carothers was elected to the National Academy of Sciences, the first organic chemist associated with industry to be so recognized.

Although he was modest to the point of shyness, Carothers' personal warmth, generosity of spirit, and sense of humor inspired deep affection in his friends. He was married 21 February 1936 to Helen Everett Sweetman of Wilmington, Delaware, and a daughter, Jane, was born 27 November 1937. His death, by his own hand, followed a long history of intensifying mental depressions, despite the best medical advice and care.

The first area that Carothers chose for intensive investigation at Du Pont was the synthesis of polymers of high molecular weight by means of simple, well-understood reactions, such as esterification and amide formation. Using as reactants compounds that can be generalized as xAx, yBy, and xCy, where x represents hydroxyl (—OH) or amino (—NH$_2$) groups and y the carboxyl (—COOH) group, his team prepared and studied in detail a great number and variety of linear polymers of general formula

$$xAzBz(AzBz)_nAzBy \text{ or } xC(Cz)_nCy.$$

A, B, and C represent chains of carbon atoms, and z represents an ester (—OCO—) or amide (—NHCO—) group formed by reaction of an hydroxyl or amino group with carboxyl. When the factor n attained some minimum high value, the polymers proved capable of being formed into fibers that, after being stretched to orient the chain molecules into parallel configuration, were strong, flexible, and tough. The outcome of these studies was a great advance in the understanding of the chemistry and properties of both natural and synthetic polymers and the opening of new avenues of experimentation in polymerization. A massive development program on one of these new fiber-forming polymers, based on hexamethylene diamine and adipic acid, led to the first nylon, the manufacture of which the Du Pont Company started in 1939 at Seaford, Delaware.

Carothers' second major area of investigation was the chemistry of vinylacetylene and divinylacetylene, low polymers of acetylene that were available from the discoveries of Nieuwland and work in other Du Pont laboratories. These studies led to a wide range of derivatives, the most important of which was chloroprene (2-chlorobutadiene), made by adding hydrogen chloride to vinylacetylene. This compound, the chloro analogue of isoprene, was shown to convert readily to a polymer having elastic properties like those of rubber. Du Pont started manufacture of this product, called neoprene, in 1931. Alone and with collaborators, Carothers published sixty-two scientific papers and was granted sixty-nine U.S. patents, many of which were equivalent in content to scientific papers.

BIBLIOGRAPHY

Carothers' papers were brought together in *Collected Papers of Wallace H. Carothers on Polymerization*, H. Mark Whitby and G. S. Whitby, eds. (New York, 1940). Roger Adams, in *Biographical Memoirs. National Academy of Sciences*, **20** (1939), 293–309, contains a complete list of Carothers' scientific papers and most of his U.S. patents. See also *Dictionary of American Biography*, supp. 2 (New York, 1958), 96–97.

JULIAN W. HILL

CARPENTER, HENRY CORT HAROLD (*b.* Bristol, England, 6 February 1875; *d.* Swansea, South Wales, 13 September 1940), *metallurgy.*

Carpenter was the second son of William Lant Carpenter and Annie Viret, the grandson of the naturalist William Benjamin Carpenter, and the great-great-grandson of Henry Cort, who invented the puddling process for iron. His intellectual development was influenced by his uncle, the theologian J. Estlin Carpenter. He was educated at St. Paul's School and at Eastbourne College and took a degree in science at Merton College, Oxford, in 1896. In the company of Frederick Soddy he took a Ph.D. at Leipzig in organic chemistry, a subject he then pursued with W. H. Perkin at Owens College, Manchester, until 1901, when he was appointed head of the chemistry and metallurgy departments of the National Physical Laboratory. From 1905, the year of his marriage to Ethel Mary Lomas, his interests were confined to metallurgy.

His subsequent career was distinguished by original contributions to the application of physicochemical principles to the metallurgy of iron and steel and of nonferrous metal alloys, as well as by his promotion of corporate professional action among metallurgists. A recurrent theme in his work is the changes that take place in metals under repeated or cyclic temperature changes. He perceived this first as a technical problem but later isolated it as fundamental in metal structure. In 1906 he elucidated the process of embrittlement of nickel wire used as the heating element in electric furnaces and first observed in this the crystalline growth that was later to form the basis of one of his most important studies. Also in 1906 he was appointed to a new chair of metallurgy at the Victoria University of Manchester. There he solved with Rugan the problem of the "growth" of cast iron on repeated heating and cooling, a matter sensationally brought to notice by the San Francisco fire, in which buildings incorporating cast iron supports failed to recover their shape on cooling. Carpenter and Rugan showed that this was mainly due to oxidation within the body of the casting brought about by oxygen that had penetrated via the disseminated graphite. Another field of technology that benefited from Carpenter's study was hydraulic mechanics. The erratic behavior of copper-aluminum alloys, sometimes able to resist high pressures and sometimes—unpredictably—unable to, proved to be traceable to the history of the individual casting.

In 1913 Carpenter was appointed to a chair at the Royal School of Mines in London. Among his work there was that (with L. Taverner) on cold-rolled aluminum sheet, a possible forerunner of his work, begun in 1920 with Miss Elam, on the production of aluminum and the study of its remarkable properties; this was the beginning of a new era in the study not only of this metal but of metallography in general.

Almost as important as his experimental work was his activity in professional and learned bodies. The Iron and Steel Institute had existed since 1869, but the study of nonferrous metals was not as well organized. Carpenter was a leader in the founding of the Institute of Metals, serving it in office for many years and as president from 1918 to 1920. He was elected a fellow of the Royal Society in 1918 and was knighted in 1929. He was president of the Institution of Mining and Metallurgy in 1934 and of the Iron and Steel Institute (1935–1936). He was a member of many committees, notably the Advisory Council of the Department of Scientific and Industrial Research, when it was establishing its influential role in British scientific life; he served as chairman of a treasury committee to inquire into the employment conditions of professional men in public services. He contributed to the history of metallurgy with his Royal Society of Arts Cantor Lectures in 1917 on progress in the metallurgy of copper. At the beginning of World War II his department was evacuated to Swansea, where he died suddenly at the height of his intellectual powers.

BIBLIOGRAPHY

His major work is *Metals,* 2 vols. (London, 1939), written with J. MacIntyre Robertson. Numerous papers were published in the metallurgical periodicals.

Biographical notices appeared in the *Times* (16 September 1940); *Obituary Notices of Fellows of the Royal Society,* no. 10 (December 1941), with portrait; *Journal of the Iron and Steel Institute,* **142** (1940); and *Metallurgia,* **22** (October 1940).

FRANK GREENAWAY

CARPENTER, WILLIAM BENJAMIN (*b.* Exeter, England, 29 October 1813; *d.* London, England, 19 November 1885), *medicine, natural history.*

Carpenter was the son of Dr. Lant Carpenter, a Unitarian minister of strong principles. His sister Mary was founder of the ragged school movement, and his younger brother, Philip Pearsall, was noted as a conchologist. His son Philip Herbert, a master at Eton, was a zoologist who assisted his father and wrote extensively on fossils.

Carpenter attended his father's school at Bristol and

was apprenticed to John Bishop Estlin, a medical practitioner of that city. After traveling to the West Indies as companion to a patient, he enrolled at University College, London, and became a member of the Royal College of Surgeons in 1835 and M.D. at Edinburgh in 1839. From 1840 to 1844 Carpenter worked in Bristol as a medical practitioner; in the latter year he moved to London, gave up medicine, and devoted the remainder of his life to research. He married Louisa Powell in 1840 and had five sons.

In 1845 Carpenter became Fullerian professor of physiology at the Royal Institution, professor of forensic medicine at University College, and lecturer in physiology at the London Hospital. He was made fellow of the Royal Society in 1844 and was Royal Medalist in 1861. From 1856 to 1879 he was registrar of London University: upon retirement from this post he became a member of the Senate of that university and was created Companion of the Order of the Bath. He staunchly maintained his Unitarian views and wrote many articles on topics in which science touches religion, including one on Charles Darwin, and on teetotalism.

From 1839, when he qualified in medicine, Carpenter's output of writing on physiology, and later on zoology, was prodigious. Among early works was his graduation thesis, *The Physiological Inferences to Be Deduced From the Structure of the Nervous System in the Invertebrate Classes of Animals* (1839), noteworthy for its new ideas on the function of the ventral cord ganglia of the Arthropoda and for its translation by the physiologist Johannes Müller in 1840.

Also in 1839 Carpenter published his *Principles of General and Comparative Physiology,* which went to four editions by 1854. From this work he developed further penetrating works on the physiology of man and animals. In particular, his *Principles of Mental Physiology* (1874) introduced new ideas on the working of nervous mechanisms and launched its author into the controversy that arose over his concept of "unconscious cerebration," by which he meant that "thought and feeling could be regarded as an expression of Brain-change," a physiological phenomenon: "[I] . . . have especially applied myself to the elucidation of the share which the Mind has not only in the *interpretation* of sense-impressions, but in the *production* of sensorial states not less real to the Ego who experiences them than are those called forth by external objects . . ." (*Principles of Mental Physiology* [1874], pp. viii–ix).

Priority in this matter was, however, claimed by Thomas Laycock, whose concept of reflex function of the brain (1844) is similar to that of Carpenter although their principles appear different, the latter

believing in a physiological stimulus arising outside the brain, while Laycock postulated a reflex action of the brain itself. Their dispute was responsible for further advances in the field.

It is evident that Carpenter had his own reservations about a purely physiological expression of mental activity, for he stated: "I cannot regard myself either Intellectually or Morally, as a mere puppet . . . any more than I can *dis*regard that vast body of Physiological evidence, which proves the direct and immediate relation between Mental and Corporeal agency" (*ibid.,* p. x). He claimed that his theories on "Brain-change" were acceptable to men of such diverse views as John Stuart Mill, who saw them in a physiological sense, and Sir William Hamilton, the metaphysician.

Carpenter's compromise in the face of the main scientific dilemma of his day is found in his ambivalent acceptance of natural selection, which he saw as modifying an ordained creative process. His acceptance and criticism of Darwinism are well shown in an essay review on the *Origin of Species* in the *British and Foreign Medico-Chirurgical Review,* of which he was editor. His realization of the brilliance of Darwin's work was unequivocal; but although it extended to agreement with the detail of selection, he displayed an inability to reconcile the full implications of descent with his basic religious beliefs. He wrote in his review of *Origin of Species:* "supposing that we concede to Mr. Darwin that all Birds have descended from one common stock—and we cannot see that there is any essential improbability in such an idea, so small are the divergencies from a common type presented by any members of that group—yet it by no means thence follows that Birds and Reptiles or Birds and Mammals should have had a common ancestry. . . ." and "there seems to us so much in the psychical capacity of Man, to separate him from the nearest of the Mammalian class, that we can far more easily believe him to have originated by a distinct creation, than suppose him to have had a common ancestry with the Chimpanzee, and to have been separated from it by a series of progressive modifications" (p. 404).

That Charles Darwin knew of this view and appreciated Carpenter's support—incomplete though it was—is shown in a letter to Carpenter. With typical modesty he wrote, "It is a great thing to have got a great physiologist on our side, I look at it as immaterial whether we go to quite the same length" (*Life and Letters of Charles Darwin,* F. Darwin, ed., II [1888], 223). Thus Carpenter's weight and influence were on the side of Darwin's theory.

Again, as registrar of the University of London,

Carpenter was in a unique position to propound and execute important developments in the teaching of science; and his ideas were welcomed and expanded by educationalists such as Herbert Spencer. In addition, Carpenter fostered the idea of popular scientific education by his support of the Gilchrist lectures, which took science to the working people. His ideas on the "training" of animals and young children by bearing upon them with strong enough "motives" have a distinctly Pavlovian feeling. Although not acceptable to his contemporaries, such arguments have today largely been crystallized in the work of Pavlov himself and that of the behaviorist school.

In the 1850's Carpenter moved into microscopy and zoology; his book *The Microscope and Its Revelations* reached its eighth edition in 1901. His chief work, however, lay in his encouragement of and scientific contributions to marine zoology. Of particular note are his descriptions and classification of the Foraminifera, both fossil and recent, as exemplified by four papers in the *Philosophical Transactions* and the splendidly illustrated monograph produced by the Ray Society in 1862. Later investigators have modified his classification of these interesting organisms, but his original work on their morphology is still quoted. Although the study of the Foraminifera has become more specialized and regionalized, owing largely to their importance to the petroleum industry, it is to the pioneering efforts of scientists like Carpenter that much of such modern industrial application is due.

The interest in marine study led to a most productive association with Charles Wyville Thomson, professor of natural history at the University of Edinburgh. Between 1868 and 1871 they took part in dredging cruises off Scotland and Ireland, which led to publications on the Crinoidea; but it was the voyage of H.M.S. *Challenger* (1873–1876), suggested by the Royal Society and organized by the British government, that threw much new light on so many aspects of oceanography. Advancing years probably prevented Carpenter from active participation, although he was greatly concerned in the preparations for the three-year circumnavigation. Thomson was appointed to organize the report on the voyage, which appeared in fifty volumes (1880–1895). Carpenter did not himself contribute. The article on the Foraminifera (1884) came from Henry Bowman Brady, an earlier collaborator of Carpenter's, while the latter's son, Philip Herbert, wrote (1888) the paper on the *Comatulae* (feather stars), a subject upon which the father had previously published a paper.

Carpenter was a founder member of the Marine Biological Association and was closely associated with the establishment of its important marine research laboratory on Plymouth Sound. His deep-sea investigations led also to an interest in marine physics, and he developed a pioneer doctrine of general oceanic circulation in a paper to the Royal Society. Among other papers read to the society was an important series on the animal nature of *Eozoon canadense*, although his conclusions have since been shown to be incorrect.

Carpenter thus represents the "complete naturalist," perhaps almost the last of the type. His original work in fields as far apart as mental physiology and marine biology, coupled with an enormous output, renders him an important figure in nineteenth-century science; and his knowledge of literature served to make him a lucid writer and lecturer. The detail of his work may now be largely forgotten, but it was of a substance sufficient to provide firm bases for much later development.

BIBLIOGRAPHY

I. ORIGINAL WORKS. Carpenter's papers are listed in the Royal Society's *Catalogue of Scientific Papers*, I (1867), VII (1877), IX (1891), XII (1902), and XIV (1915). Among his works are *On the Physiological Inferences to Be Deduced From the Structure of the Nervous System in the Invertebrate Classes of Animals* (Edinburgh, 1839); *Principles of General and Comparative Physiology* (London, 1839); "Asphyxia," in *Library of Medicine*, A. Tweedie, ed., III (London 1840), 212–249; *The Microscope and Its Revelations* (London, 1856); *Zoology. A Systematic Account of the General Structure, Habits, Instincts and Uses of the Principal Families of the Animal Kingdom . . .*, 2 vols., in Bohn's Scientific Library (London, 1857); his review of Darwin, Wallace, and Baden-Powell, in *British and Foreign Medico-Chirurgical Review*, **25** (1860), 367–404; and *Introduction to the Study of the Foraminifera* (London, 1862), written with W. K. Parker and T. R. Jones.

II. SECONDARY LITERATURE. For more on Carpenter or his work, see E. R. Lancaster, "William Benjamin Carpenter," in *Nature*, **33** (1885–1886), 83–85; Thomas Laycock, *Mind and Brain*, 2 vols. (Edinburgh, 1860); and C. W. Thomson, *The Depths of the Sea, an Account of the Dredging Cruises . . . Under the Scientific Direction of Dr. Carpenter* (London, 1873). Obituaries are in *British Medical Journal* (1885), **2**, 139; and *Lancet* (1885), **2**, 928.

K. BRYN THOMAS

CARPI, BERENGARIO DA. See **Berengario da Carpi, Jacopo.**

CARR, HERBERT WILDON (*b.* London[?], England, 16 January 1857; *d.* Los Angeles, California, 8 July 1931), *philosophy.*

The eldest son of Benjamin William Carr, he was

educated privately and at King's College, University of London. He became a successful and wealthy businessman and during this period belonged to the Aristotelian Society for the Systematic Study of Philosophy, founded in London in 1880. As secretary, he knew scientists such as George John Romanes and philosophers such as Alexander Bain, Bernard Bosanquet, and Bertrand Russell. He was president of the society from 1916 to 1918, professor of philosophy at King's College, London, from 1918, and visiting professor at the University of Southern California in 1925, where he lived during his last years. He married Margaret Geraldine Spooner and had three daughters.

Wildon Carr, as he was invariably called, developed his philosophical views in discussions at the then rather unacademic but scientifically oriented Aristotelian Society. He accepted the philosophical idealism that was influential during the 1890's and early twentieth century; but whereas such leading idealists as Bosanquet and Francis Herbert Bradley argued for a suprapersonal Absolute in which persons were absorbed, Carr held that the ultimate realities are active persons. His inspiration here was Leibniz. Carr believed that the history of philosophy is inseparable from the history of science, and his teaching and writing emphasized this. Belief in absolute space and time, he held, provided a foundation for the materialist view that mind is a determined and evanescent feature of an objective world of atoms in space. He believed that Leibniz, in his correspondence with Clarke on Newton's *Principia* and in other writings, had given conclusive philosophical reasons for a relative theory of space and time and for the metaphysical primacy of active perceiving monads. He welcomed the general theory of relativity for providing, as he thought, scientific reasons for the monadism that he had inherited from Leibniz, whom he regarded as the founder of idealism. He stated this thesis in "The Principle of Relativity and its Importance for Philosophy" and expanded it in *The General Principle of Relativity* (1920). Carr did not say that the principle of relativity in itself proved idealism but rather that it gave support to it in showing that there is no conception of an objective physical world apart from a point of reference and system of measurement. The idea of matter as a space occupancy, he wrote in *Cogitans Cogitata* (1930), has been replaced by "the idea of fields of force arising from the deformable nature of a space-time continuum." Carr also criticized mechanistic theories in biology, arguing that both memory and foresight imply a spontaneity that material particles could not have. Here he was influenced

by Bergson, whose *Energie Spirituelle* he translated into English. *The Unique Status of Man* (1928), based on his lectures in California, and *Cogitans Cogitata* contain the most systematic expositions of his philosophy.

BIBLIOGRAPHY

I. ORIGINAL WORKS. Carr's writings are *Henri Bergson. The Philosophy of Change* (London, 1911; 2nd ed., 1919); *The Philosophy of Change* (London, 1914); *The Philosophy of Benedetto Croce* (London, 1918); *The General Principle of Relativity in its Philosophical and Historical Aspect* (London, 1920; 2nd ed., 1922); *A Theory of Monads. Outlines of the Philosophy of the Principle of Relativity* (London, 1922); *The Scientific Approach to Philosophy* (London, 1924); *Changing Backgrounds in Religion and Ethics* (London, 1927); *The Unique Status of Man* (London, 1928); *The Freewill Problem* (London, 1928); *Leibniz* (London, 1929); *Cogitans Cogitata* (London, 1930); and *The Monadology of Leibniz* (London, 1931). He translated into English Henri Bergson, *Mind Energy* (London, 1920), and Giovanni Gentile, *The Theory of Mind as Pure Act* (London, 1921).

Alfred North Whitehead includes Carr's paper "The Problem of Simultaneity" in *Relativity, Logic and Mysticism*, Aristotelian Society supp. III (London, 1923), 34–41.

II. SECONDARY LITERATURE. On Carr and his work see "In Memoriam. Herbert Wildon Carr," in *Mind*, n. s. **40** (1931), 535–536.

H. B. ACTON

CARREL, ALEXIS (*b.* Lyons, France, 28 June 1873; *d.* Paris, France, 5 November 1944), *surgery, experimental biology.*

He was the eldest child of Alexis Carrel-Billiard, a textile manufacturer, and his wife, Anne-Marie Ricard, both from bourgeois Roman Catholic families. When Alexis was five years old, his father died, and the three children were brought up by their devout and solicitous mother. Alexis was sent to a Jesuit day school and college near his home in Lyons. As a schoolboy he showed an interest in biology by dissecting birds. Encouraged by an uncle, he conducted experiments in chemistry. After taking his baccalaureate he entered the University of Lyons in 1890 as a student of medicine. He was attached to hospitals at Lyons from 1893 to 1900, except for a year as surgeon in the French army's Chasseurs Alpins. His talent for anatomy and operative surgery became apparent when in 1898 he was attached to the laboratory of the celebrated anatomist J.-L. Testut. In 1900 he received his formal medical degree from the University of Lyons.

Carrel became interested in surgery of the blood

vessels about 1894, inspired, it is said, by the death of President Carnot from an assassin's bullet, which cut a major artery. Such wounds could not at that time be successfully repaired. He developed extraordinary skill in using the finest needles and devised a method of turning back the ends of cut vessels like cuffs, so that he could unite them end-to-end without exposing the circulating blood to any other tissue than the smooth lining of the vessel. By this device and by coating his instruments, needles, and thread with paraffin jelly, he avoided blood clotting that might obstruct flow through the sutured artery or vein. He avoided bacterial infection by a most exacting aseptic technique. His first successes in suturing blood vessels were announced in 1902.

This brilliant achievement did not spare Carrel from difficulties, brought on partly by his critical attitude toward what he considered the antiquated traditions and political atmosphere of the Lyons medical faculty. Finding a university career blocked by local opposition, he left Lyons and after a year of advanced medical studies in Paris went in 1904 to the United States. At the University of Chicago, where he was given an assistantship in physiology, he resumed his experiments in blood-vessel surgery, applying his methods to such difficult feats as kidney transplants in animals. His growing reputation for surgical skill, bold experimentation, and technical originality won him in 1906 appointment as Member of the Rockefeller Institute for Medical Research (now Rockefeller University) in New York. There he resumed his surgical experimentation. Subsequent progress in surgery of the heart and blood vessels and in transplantation of organs has rested upon the foundation he laid down between 1904 and 1908.

Carrel's pioneer successes with organ transplants led him to dream of cultivating human tissues and even whole organs as substitutes for diseased or damaged parts. He seized at once upon the work of Ross G. Harrison of Yale University, who announced in 1908 the cultivation of frog's nerve cells *in vitro*. Bringing to this kind of research his own dexterity, inventiveness, and command of asepsis, Carrel succeeded in cultivating the cells of warm-blooded animals outside the body. To prove his results in the face of skepticism, he began his famous undertaking to keep such a culture alive and growing indefinitely, using a bit of tissue from the heart of an embryo chick. He kept this strain of connective-tissue cells alive for many years; in the care of one of his assistants it outlived Carrel himself. Although he did not add greatly, by his largely methodological achievement, to the understanding of cellular physiology, in other

hands tissue culture has contributed greatly not only to scientific theory but to practice as well—for example, the growing of virus cultures in animal cells and the preparation of vaccines. In 1912 he received the Nobel Prize for his surgical and cell-culture experiments.

Carrel was married in 1913 to Anne de la Motte de Meyrie, a devout Roman Catholic widow with one son. The couple had no children of their own. Recalled in 1914 to service in the French army during World War I, Carrel conducted a hospital and research center near the front lines, where Mme. Carrel assisted him as a surgical nurse. With the aid of a chemist, Henry B. Dakin, he developed a method of treating severely infected wounds, which although often effective was too complicated for general use and has been supplanted by the use of antibiotics.

About 1930 Carrel undertook another far-reaching experimental program, aimed at the cultivation of whole organs. In this work he was aided by the celebrated aviator Charles A. Lindbergh, who devised a sterilizable glass pump for circulating culture fluid through an excised organ. Carrel was thus enabled to keep such organs as the thyroid gland and kidney alive and, to a certain extent, functioning for days or weeks. This was a pioneer step in the development of apparatus now used in surgery of the heart and great vessels.

Carrel's naturally religious, even mystical, temperament led him to speculate on the great problems of human destiny. In a widely read book, *Man the Unknown* (1935), he expressed the hope that scientific enlightenment might confer upon mankind the boons of freedom from disease, long life, and spiritual advancement, under the leadership of an intellectual elite.

He retired from the Rockefeller Institute in 1938 and, after the outbreak of World War II, returned to Paris, hoping to serve his native country by a grandiose program to safeguard and improve the population by scientific nutrition, public hygiene, and eugenics. During the German occupation he remained in Paris at the head of a self-created Institute for the Study of Human Problems. His acceptance of support from Vichy and his negotiations with the German command on behalf of his institute led to exaggerated charges of collaborationism. His death from heart failure aggravated by the hardships of life in wartime Paris spared him the indignity of arrest. Although not a fully orthodox churchman, he received the last rites of the Roman Catholic Church and was interred in a chapel at his home on the island of St. Gildas, off the coast of Brittany.

BIBLIOGRAPHY

I. ORIGINAL WORKS. Carrel's writings include *Man the Unknown* (New York, 1938); *The Culture of Organs* (New York, 1938), written with Charles A. Lindbergh; *La Prière* (Paris, 1944), English trans., *Prayer* (New York, 1948); *Le Voyage à Lourdes* (Paris, 1949), English trans., with an intro. by Charles A. Lindbergh, *Voyage to Lourdes* (New York, 1950); and *Réflexions sur la vie* (Paris, 1952), English trans., *Reflections on Life* (London, 1952). A bibliography of his numerous scientific and popular articles is included in Soupault's biography, cited below.

II. SECONDARY LITERATURE. The definitive biography is Robert Soupault, *Alexis Carrel, 1873-1944* (Paris, 1952), in French, with portraits and a full bibliography. See also Mme. Carrel's preface to *Reflections on Life;* George W. Corner, *History of the Rockefeller Institute* (New York, 1965); Henriette Delaye-Didier-Delorme, *Alexis Carrel, Humaniste Chrétien* (Paris, 1964); Joseph T. Durkin, *Hope for Our Time: Alexis Carrel on Man and Society* (New York, 1965); and Alfonso M. Moreno, *Triunfo y ruina de una vida: Alexis Carrel* (Madrid, 1961).

Carrel's correspondence and scientific records are at the library of Georgetown University, Washington, D.C.

GEORGE W. CORNER

CARRINGTON, RICHARD CHRISTOPHER (*b.* London, England, 26 May 1826; *d.* Churt, Surrey, England, 27 November 1875), *astronomy.*

Carrington had the good fortune to belong to a wealthy family and to receive his basic education at a private school, located at Hedley and run by a Mr. Faithful. His father, the proprietor of a large brewery at Brentford, was desirous that his son should be prepared for the respectable profession of the Church and therefore arranged that he spend some time in the house of a clergyman named Blogard before beginning his studies in theology at Trinity College, Cambridge, in 1844. Even before this time, however, Carrington had come to realize that his natural aptitudes lay rather in the pursuit of the physical sciences, particularly in those aspects involving observation and mechanical ingenuity.

Between 1842 and the completion of his B.A. degree in 1848, Carrington submitted no fewer than six quite substantial contributions to collections of problems in various branches of mathematics, physics, and astronomy. The final impulse to his resolution to be a practical astronomer came when he attended the lectures of the Plumian professor, James Challis, on that subject; and when his father raised no objection to this choice of career, he accepted the post of observer at the University of Durham and began work there in October 1849 under the direction of the Rev. Temple Chevalier. Carrington's early contributions to the

Monthly Notices of the Royal Astronomical Society and to the *Astronomische Nachrichten,* containing the preliminary results of observations of minor planets and comets, ensured his prompt admittance to membership in the Royal Astronomical Society when his application was considered on 14 March 1851.

By this time Carrington was becoming discontent with his duties in Durham and saw little prospect of better instruments being purchased to enable him to extend the scope of his activities. He wished in particular to complete the work of Bessel and Argelander by systematically surveying the spherical zone of the heavens within 9° of the north celestial pole and preparing a catalog of circumpolar stars brighter than the eleventh magnitude—a deficiency of which his observations of minor planets had made him well aware. He therefore resolved to use his substantial means to set up his own observatory and buy his own instruments, and thus he resigned his appointment in Durham in March 1852.

The site that Carrington selected for his private observatory was Redhill, near Reigate, Surrey. The building was completed and his instruments, a 5.5-foot-focus transit circle and a 4.3-foot-focus equatorial (both made by William Simms), were installed and adjusted by July 1853, although what Carrington terms "the real work of observing" did not start until the beginning of the following year. The transit circle, a scale model of the large meridian instrument then at Greenwich, was for finding the absolute positions of the brightest stars; the equatorial enabled comparisons to be made of the positions of the fainter stars relative to those of the brighter ones. Assisted by George Harvey Simmonds, who had been with him for some months at Durham, Carrington worked steadily with "talent and zeal, untiring devotion and industry, and an unsparing but prudent application of private resources" during the course of the next three years to produce his *Catalogue of 3735 Circumpolar Stars* (1857), which was printed at public expense by the Lords of the Admiralty and received the Gold Medal of the Royal Astronomical Society in 1859. Its excellence of design and execution and the unquestioned reliability of the results were largely responsible for its author's election as a fellow of the Royal Society on 7 June 1860.

Carrington is better remembered by posterity, however, for his pioneering investigations of the motions of sunspots. While his observatory at Redhill was in the course of construction, he spent some of his time examining drawings and records of sunspots possessed by the Royal Astronomical Society and was struck by the scarcity of systematic solar observations. Since he was well aware that no public observatory was then occupied with work of this nature, he resolved to remedy the observational deficiency by "close and methodical research" during the daylight hours whenever the time required to reduce his stellar observations should permit. The method that he devised and employed throughout the seven-and-one-half-year period to which his sunspot observations refer was one that required no micrometer and no clockwork. He projected a solar image about twelve inches in diameter and the perpendicular cross hairs of a micrometer inclined at an angle of 45° to the direction of the daily motion onto a sheet of paper, fixing the telescope so that the image passed across the field of view. By noting the times of contact of the wires with the limbs of the sun and the boundaries of each spot, the heliocentric positions of the spots were generally obtained with an accuracy of a few minutes of arc.

His father's death in July 1858 made it necessary for Carrington to take over the management of the brewery and prevented his further personal participation in this program; the continual pressure of the business subsequently caused him to abandon the project altogether. His *Observations of the Spots on the Sun From Nov.ᵗ 9, 1853, to March 24, 1861, Made at Redhill,* a ponderous quarto volume, was published in 1863 with the aid of Royal Society funds. In this book he determined the position of the sun's axis with unprecedented accuracy and established the important empirical laws of sunspot distribution and the variation in solar rotation as functions of the heliocentric latitude, which served to revolutionize ideas on solar physics just as effectively as the results of spectrum analysis.

Despite his devotion to the above-mentioned tasks, Carrington took time to travel, although his motives for doing so were inevitably linked to his astronomical interests. As a young man of twenty-five he went from Durham to the small town of Lilla Edet in Sweden to observe a total solar eclipse; he described this experience in an Admiralty pamphlet printed and circulated in May 1858 to those who might be in South America to witness a similar eclipse later that year. After a second visit to the Continent in 1856, he drew up a valuable report on the condition of a number of German observatories (*Monthly Notices of the Royal Astronomical Society,* **17**). He made a number of other contributions to the *Monthly Notices,* including three articles on the motions and distribution of sunspots and another on the distribution of cometary orbits with regard to the direction of solar motion. He also described a solar flare that he had observed for a few minutes on 1 September 1859, a phenomenon which at that time was quite unknown.

In a letter to the vice-chancellor of Cambridge, dated 13 April 1861, Carrington claims to have visited and carefully inspected, and considered the construction and adaptation of, many of the leading Continental observatories; all of those in England, Scotland, and Ireland; and very many private establishments at home and abroad. Moreover, he knew nearly all the opticians and their works and nearly all the observers, and was unusually familiar with materials and construction of both buildings and instruments. He also remarks: "I have devoted nearly the whole of thirteen years since I left the University to establishing . . . a reputation for success in the labours of practical Astronomy, specially with the object of obtaining such a position as is about to become vacant at Cambridge. . . ." The failure of his bid for the directorship of the Cambridge Observatory on the occasion of Challis' resignation, to which he is here referring, came as a great blow to him; indeed, judging from a comment in a letter to John Herschel written five weeks earlier, it would appear to have been directly responsible for his decision to sell the Redhill observatory about this time.

Nevertheless, Carrington was unable to suppress his love for astronomy and was soon contemplating the prospect of going to Chile to observe the stars in the neighborhood of the south celestial pole; however, a severe illness in 1865 left his health permanently impaired and this plan never materialized. Instead, he disposed of his brewery and retired from business to Churt, Surrey, where he established a new observatory on top of an isolated conical hill in a lonely and picturesque spot known as the Middle Devil's Jump. No records exist of observations made with the large altazimuth instrument that he installed there, and his ambition to erect a "double altazimuth"—the subject of the last of his many contributions to the *Monthly Notices of the Royal Astronomical Society*—was never fulfilled. He died at his home on 27 November 1875, apparently as the result of a brain hemorrhage, only ten days after his wife Rosa had been found dead in bed after taking an overdose of chloral hydrate. In his will Carrington left £2,000 to the Royal Astronomical Society, with which he had been actively associated for nearly half his lifetime and which he had served so conscientiously as secretary from 1857 to 1862. The society also possesses his manuscript books of sunspot observations and reductions, with a folio volume of drawings.

BIBLIOGRAPHY

A fairly comprehensive list of Carrington's contributions to astronomy is in Poggendorff, III, 240. Lists of his papers also appear in the Royal Society's *Catalogue of Scientific Papers,* I and VII. Carrington's major works are his *Catalogue of 3735 Circumpolar Stars* (London, 1857) and *Observations of the Spots on the Sun from Nov.ʳ 9, 1853, to March 24, 1861, Made at Redhill* (London, 1863). An account of his observatory at Redhill is given by C. André and G. Rayet in *L'astronomie pratique et les observatoires en Europe et en Amérique depuis le milieu du XVIIᵉ siècle jusqu'à nos jours* (Paris, 1874), pp. 108–114. Further biographical information is contained in the *Dictionary of National Biography* and in an obituary notice in the *Monthly Notices of the Royal Astronomical Society,* **36** (Feb. 1876), 137–142. This last periodical is the one to which Carrington most frequently contributed: it contains no fewer than twenty articles or notes by him between the years 1850 and 1874 inclusive.

In addition to these printed sources mentioned, there are twenty letters from Carrington to John Herschel (11 Dec. 1856–17 Dec. 1866) and two of Herschel's replies, plus three letters from Carrington to Edward Sabine, preserved in the archives of the Royal Society of London. Nine letters from Carrington to George Airy and seven of Airy's replies are contained in the latter's miscellaneous correspondence at the Royal Greenwich Observatory, Herstmonceux Castle, Sussex.

ERIC G. FORBES

CARROLL, JAMES (*b.* Woolwich, England, 5 June 1854; *d.* Washington, D.C., 16 September 1907), *bacteriology.*

Carroll was the son of James and Harriet Chiverton Carroll. He attended Albion House Academy, Woolwich, in preparation for an engineering career in the navy. At the age of fifteen, however, he imigrated to Canada and in 1874, at the age of twenty, he enlisted in the U.S. army. He remained in the army until his death. In 1883 Carroll became a hospital steward, a position he held officially until 1898, when, during the Spanish-American War, he was appointed acting assistant surgeon. In 1886–1887 he attended medical school at the University of the City of New York (now New York University), and in 1889 he continued at the University of Maryland, from which he received the M.D. in 1891.

From 1891 to 1893 Carroll attended the new postgraduate classes in bacteriology and pathology at Johns Hopkins University, coming into contact with several brilliant Americans who later became famous in medicine. In 1893, at the Army Medical School in Washington, Carroll met Walter Reed and in 1895 was assigned to assist Reed, then curator of the Army Medical Museum. While stationed in Washington, Carroll taught at Columbian (now George Washington) University. After Reed's death in 1902, Carroll succeeded him at the Army Medical Museum and

at George Washington, reaching the full professorship of pathology and bacteriology in 1905.

In 1899 Reed and Carroll disproved Sanarelli's theory that *Bacillus icteroides* is the specific agent in yellow fever. In 1900 Carroll was made second in command to Reed of the now-famous commission sent to Cuba to study yellow fever. Reed planned a series of experiments to determine how the disease is spread and what causes it. Since Reed had to be in Washington a large part of the time, Carroll was responsible for much of the actual work carried out. On 27 August 1900 Carroll, who did not put much stock in the theory of Carlos Finlay that a mosquito acts as the vector in yellow fever, caused an infected mosquito to bite his arm. Four days later he came down with the first experimental case of yellow fever. He nearly died then, and he acquired a heart disease from which he did die a few years later, one of the genuine martyrs of science. Much bitterness was engendered because Congress refused adequate compensation to his wife, Jennie M. George Lucas Carroll, and their seven children. Before his death Carroll did receive promotions to the rank of major, and the universities of Maryland and Nebraska awarded him honorary degrees.

Carroll's most important published contributions to science consist of works written jointly with Walter Reed and others defining the method of the propagation of yellow fever, based upon experiments that were executed largely by Carroll, and a series of papers on the etiological agent in that disease. His most important personal contribution was the demonstration, at Reed's suggestion, that the bacteriological agent of yellow fever is a filterable virus. Prior to that time, ultramicroscopic phenomena had been identified clearly as pathogens only in tobacco mosaic disease and cattle foot-and-mouth disease. Carroll's report was among the first to establish a virus etiology of a human disease.

BIBLIOGRAPHY

A list of Carroll's publications is in John C. Hemmeter, *Master Minds in Medicine* (New York, 1927), pp. 319–320. This list should be supplemented by standard indexes of the day.

For biographical material, see James M. Phalen's article in the *Dictionary of American Biography,* III, 525–526; the article by Caroline W. Latimer in Howard Atwood Kelly and Walter Lincoln Burrage, eds., *American Medical Biography* (New York, 1920); and Hemmeter, *op. cit.,* pp. 297–336, which contains documents and letters as well as some interview material. Good obituaries include that in *Bulletin of the Johns Hopkins Hospital,* **19** (1908),

1–12; and Howard A. Kelly, in *Proceedings of the Washington Academy of Sciences,* **10** (1908), 204–207.

JOHN C. BURNHAM

CARTAN, ÉLIE (*b.* Dolomieu, France, 9 April 1869; *d.* Paris, France, 6 May 1951), *mathematics.*

Cartan was one of the most profound mathematicians of the last hundred years, and his influence is still one of the most decisive in the development of modern mathematics. He was born in a village in the French Alps. His father was a blacksmith, and at that time children of poor families had almost no opportunity to reach the university. Fortunately, while he was still in elementary school, his intelligence impressed the young politician Antonin Dubost, who was then an inspector of primary schools (and was later president of the French Senate); Dubost secured for Cartan a state stipend enabling him to attend the *lycée* in Lyons and later to enter the École Normale Supérieure in Paris. After graduation he started his research with his now famous thesis on Lie groups, a topic then still in its very early stages. He held teaching positions at the universities of Montpellier, Lyons, Nancy, and finally Paris, where he became a professor in 1912 and taught until his retirement in 1940. In 1931 he was elected a member of the French Academy of Sciences, and in his later years he received many honorary degrees and was elected a foreign member of several scientific societies.

Cartan's mathematical work can be described as the development of analysis on differentiable manifolds, which many now consider the central and most vital part of modern mathematics and which he was foremost in shaping and advancing. This field centers on Lie groups, partial differential systems, and differential geometry; these, chiefly through Cartan's contributions, are now closely interwoven and constitute a unified and powerful tool.

Cartan was practically alone in the field of Lie groups for the thirty years after his dissertation. Lie had considered these groups chiefly as systems of analytic transformations of an analytic manifold, depending analytically on a finite number of parameters. A very fruitful approach to the study of these groups was opened in 1888 when Wilhelm Killing systematically started to study the group in itself, independent of its possible actions on other manifolds. At that time (and until 1920) only local properties were considered, so the main object of study for Killing was the Lie algebra of the group, which exactly reflects the local properties in purely algebraic terms. Killing's great achievement was the determination of all simple complex Lie algebras; his proofs, however, were often defective, and Cartan's thesis was devoted mainly to

giving a rigorous foundation to the "local" theory and to proving the existence of the "exceptional" Lie algebras belonging to each of the types of simple complex Lie algebras Killing had shown to be possible. Later Cartan completed the "local" theory by explicitly solving two fundamental problems, for which he had to develop entirely new methods: the classification of simple real Lie algebras and the determination of all irreducible linear representations of simple Lie algebras, by means of the notion of weight of a representation, which he introduced for that purpose. It was in the process of determining the linear representations of the orthogonal groups that Cartan discovered in 1913 the spinors, which later played such an important role in quantum mechanics.

After 1925 Cartan grew more and more interested in topological questions. Spurred by Weyl's brilliant results on compact groups, he developed new methods for the study of global properties of Lie groups; in particular he showed that topologically a connected Lie group is a product of a Euclidean space and a compact group, and for compact Lie groups he discovered that the possible fundamental groups of the underlying manifold can be read from the structure of the Lie algebra of the group. Finally, he outlined a method of determining the Betti numbers of compact Lie groups, again reducing the problem to an algebraic question on their Lie algebras, which has since been completely solved.

Cartan's methods in the theory of differential systems are perhaps his most profound achievement. Breaking with tradition, he sought from the start to formulate and solve the problems in a completely invariant fashion, independent of any particular choice of variables and unknown functions. He thus was able for the first time to give a precise definition of what is a "general" solution of an arbitrary differential system. His next step was to try to determine all "singular" solutions as well, by a method of "prolongation" that consists in adjoining new unknowns and new equations to the given system in such a way that any singular solution of the original system becomes a general solution of the new system. Although Cartan showed that in every example which he treated his method led to the complete determination of all singular solutions, he did not succeed in proving in general that this would always be the case for an arbitrary system; such a proof was obtained in 1955 by Kuranishi.

Cartan's chief tool was the calculus of exterior differential forms, which he helped to create and develop in the ten years following his thesis, and then proceeded to apply with extraordinary virtuosity to the most varied problems in differential geometry, Lie groups, analytical dynamics, and general relativity. He discussed a large number of examples, treating them in an extremely elliptic style that was made possible only by his uncanny algebraic and geometric insight and that has baffled two generations of mathematicians. Even now, some twenty years after his death, students of his results find that a sizable number of them are still in need of clarification; chief among these are his theory of "equivalence" of differential systems and his results on "infinite Lie groups" (which are not groups in the usual sense of the word).

Cartan's contributions to differential geometry are no less impressive, and it may be said that he revitalized the whole subject, for the initial work of Riemann and Darboux was being lost in dreary computations and minor results, much as had happened to elementary geometry and invariant theory a generation earlier. His guiding principle was a considerable extension of the method of "moving frames" of Darboux and Ribaucour, to which he gave a tremendous flexibility and power, far beyond anything that had been done in classical differential geometry. In modern terms, the method consists in associating to a fiber bundle E the principal fiber bundle having the same base and having at each point of the base a fiber equal to the group that acts on the fiber of E at the same point. If E is the tangent bundle over the base (which since Lie was essentially known as the manifold of "contact elements"), the corresponding group is the general linear group (or the orthogonal group in classical Euclidean or Riemannian geometry). Cartan's ability to handle many other types of fibers and groups allows one to credit him with the first general idea of a fiber bundle, although he never defined it explicitly. This concept has become one of the most important in all fields of modern mathematics, chiefly in global differential geometry and in algebraic and differential topology. Cartan used it to formulate his definition of a connection, which is now used universally and has superseded previous attempts by several geometers, made after 1917, to find a type of "geometry" more general than the Riemannian model and perhaps better adapted to a description of the universe along the lines of general relativity.

Cartan showed how to use his concept of connection to obtain a much more elegant and simple presentation of Riemannian geometry. His chief contribution to the latter, however, was the discovery and study of the symmetric Riemann spaces, one of the few instances in which the initiator of a mathematical theory was also the one who brought it to its completion. Symmetric Riemann spaces may be defined in various ways, the simplest of which postulates the existence around each point of the space of a "symmetry" that is involutive, leaves the point fixed, and preserves distances. The unexpected fact discovered by Cartan is that it is possible to give a complete description of these spaces by means of the classification of the simple Lie groups; it should therefore not be surprising that in various areas of mathematics, such as automorphic functions and analytic number theory (apparently far removed from differential geometry), these spaces are playing a part that is becoming increasingly important.

Cartan's recognition as a first-rate mathematician came to him only in his old age; before 1930 Poincaré and Weyl were probably the only prominent mathematicians who correctly assessed his uncommon powers and depth. This was due partly to his extreme modesty and partly to the fact that in France the main trend of mathematical research after 1900 was in the field of function theory, but chiefly to his extraordinary originality. It was only after 1930 that a younger generation started to explore the rich treasure of ideas and results that lay buried in his papers. Since then his influence has been steadily increasing, and with the exception of Poincaré and Hilbert, probably no one else has done so much to give the mathematics of our day its present shape and viewpoints.

BIBLIOGRAPHY

I. ORIGINAL WORKS. Cartan's papers have been collected in his *Oeuvres complètes,* 6 vols. (Paris, 1952–1955). He published the following books: *Leçons sur les invariants intégraux* (Paris, 1922); *La géométrie des espaces de Riemann,* fasc. 9 of Mémorial des Sciences Mathématiques (Paris, 1925); *Leçons sur la géométrie des espaces de Riemann* (Paris, 1928, 1946); *Leçons sur la géométrie projective complexe* (Paris, 1931); *Les espaces métriques fondés sur la notion d'aire,* no. 1 of Exposés de Géométrie (Paris, 1933); *Les espaces de Finsler,* no. 2 of Exposés de Géométrie (Paris, 1934); *La théorie des groupes finis et continus et la géométrie différentielle* (Paris, 1937); *Leçons sur la théorie des spineurs,* 2 vols., no. 11 of Exposés de Géométrie (Paris, 1938); and *Les systèmes différentiels extérieurs et leurs applications géométriques,* no. 994 of Actualités Scientifiques et Industrielles (Paris, 1945).

II. SECONDARY LITERATURE. Two excellent obituary notices are S. S. Chern and C. Chevalley, in *Bulletin of the American Mathematical Society,* **58** (1952); and J. H. C. Whitehead, in *Obituary Notices of the Royal Society* (1952).

JEAN DIEUDONNÉ

CARTESIUS, RENATUS. See **Descartes, René.**

CARUS, PAUL (*b.* Ilsenburg, Germany, 18 July 1852; *d.* La Salle, Illinois, 11 February 1919), *philosophy.*

For a detailed study of his life and work, see Supplement.

CASAL JULIAN, GASPAR ROQUE FRANCISCO NARCISO (*b.* Gerona, Spain, 31 December 1680; *d.* Madrid, Spain, 10 August 1759), *medicine, natural history.*

Casal was the son of Federico Casal y Dajón and Magdalena Julian. He spent his childhood in Utrillas, in the province of Soria, his mother's birthplace. There is no record of Casal's having studied medicine at a university, although he stated that he had practiced medicine in several villages of Guadalajara province between 1707 and 1712; evidence indicates that he learned medicine in Atienza, Guadalajara, by apprenticeship to Juan Manuel Rodríguez de Lima, formerly apothecary to Pope Innocent XI and a distinguished pupil of Doncelli.

In 1713 Casal received the bachelor of arts degree from the nearby University of Sigüenza. He then left the province of Guadalajara and practiced medicine in Madrid. For reasons of health Casal left Madrid in 1717 and established his medical practice in Oviedo, Asturias, first as the city's official doctor and later as physician to the local hospitals. During his residence in Oviedo, Casal formed a close friendship with two enlightened Benedictine writers, Fr. Benito Feyjóo y Montenegro and Fr. Martín Sarmiento, who greatly stimulated his scientific studies. In 1751 he returned to Madrid and was appointed royal physician, and in 1752 he became a member of the Board of the Protomedicate and the Royal Academy of Medicine. Casal's first wife, María Ruiz, bore him two sons, Andrés Simón and Pablo; after her death he married María Álvarez Rodríguez Arango and had two more children, Ventura Benito and Magdalena.

Casal wrote a natural and medical history of Asturias, which was published posthumously in 1762. The first part, "Historia physico-médica de el principado de Asturias," describes the geography, climate, plants, animals, and diseases most frequently observed in the province; the diseases include intestinal parasites, fevers, endemic goiter, scabies, leprosy, and the "malady of the rose," or pellagra. Casal also discussed hygiene and preventive measures. The second part, a discussion of Hippocratic doctrines, confirms Casal's excellent powers of observation. The third part deals with several epidemics that occurred in Asturias between 1719 and 1750, and Casal describes cases observed. The fourth part, "Historia affectionum quarundam regionis hujus familiarum," describes scabies, identifying the *Sarcoptes* and recommending treatment with sulfur ointment. Leprosy, sometimes confused in the text with severe dermatological afflic-

tions, and the "mal de la rosa," described in every clinical manifestation, are illustrated by an engraved plate.

BIBLIOGRAPHY

Casal's *Historia natural y médica del principado de Asturias. Obra posthuma,* Juan José García Sevillano, ed. (Madrid, 1762), was repr. with an excellent biographical study by Fermín Canella, preface by Ángel Pulido, and notes by A. Buylla and R. Sarandeses (Oviedo, 1900). For discussions of Casal or his work, see Anastasio Chinchilla, *Historia de la medicina española,* III (Valencia, 1846), 309–347; Antonio Hernández Morejón, *Historia bibliográfica de la medicina española,* VII (Madrid, 1852), 252–259; Juan Catalina García, *Biblioteca de escritores de la provincia de Guadalajara* (Madrid, 1899), p. 433; Fermín Canella, "Noticias biográficas de Don Gaspar Casal," in *Casal* (Oviedo, 1900), the best biographical study, which has been the basis of subsequent works; M. López Sendón, "Gaspar Casal, breve estudio de su vida y de su obra," in *Trabajos. Catedra de historia crítica de la medicina, Universidad de Madrid,* **1** (1933), 313–324; Jaime Peyrí Rocamora, *Mal de la rosa, su historia, causa, casos, curación* (Masnou, 1936), preface, which provides the record of his birth and studies, as well as a clinical appraisal of Casal's contributions to dermatology; and Rafael Sancho de San Roman, "Vida y obra de Gaspar Casal," in *Publicaciones del Seminario de historia de la medicina, Universidad de Salamanca,* **2**, no. 3 (1959), 153–183, a survey of previous literature with an objective critical analysis of Casal as physician and scientist.

FRANCISCO GUERRA

CASSEGRAIN (*fl.* Chartres, France, 1672), *physics.*

Virtually all that is known about Cassegrain is that he conceived the arrangement of telescope mirrors that bears his name. So obscure is his life that even his identity itself is in question. One tradition credits him with a professorship of physics at the Collège de Chartres; another identifies him with a man who served Louis XIV as a sculptor and founder.

Cassegrain's appearance on the scientific scene was occasioned by a memoir on the megaphone published in February 1672. Having his own thoughts on the subject, Cassegrain quickly submitted them to the Paris Academy of Sciences. In the covering letter, his intermediary, one Henri de Bercé of Chartres, mentioned that Cassegrain had also conceived a refracting telescope, different from the one recently described by Newton in the *Philosophical Transactions.* The design consisted of a convex secondary mirror placed so as to intercept the rays from the objective before they arrive at the focus and to reflect them straight back to the eyepiece through a hole in the objective.

It was set forth briefly in the proceedings for 25 April 1672. The accompanying implications of priority and claims of superiority were easily dispatched by Newton, who pointed out that the idea was only a minor modification of one published by James Gregory in 1663. The real virtue of the design—partial cancellation of the spherical aberrations introduced by the two mirrors—was established by Ramsden a century later. Since that time, the Cassegrain focus has been very popular in the construction of large refractors.

BIBLIOGRAPHY

I. ORIGINAL WORKS. His writings are "Extrait d'une lettre de M. de Bercé écrite de Chartres à l'auteur de ces mémoires, touchant la trompette à parler de loin, dont on a donné l'explication dans le second mémoire; & touchant la nouvelle lunette de M. Newton, dont il a esté parlé dans le troisième," in *Journal des savants,* **3** (1672–1674), 121–123; and "Extrait d'une lettre écrite de Chartres par Monsieur Cassegrain; Sur les proportions des trompettes à parler de loin," *ibid.,* 131–137.

II. SECONDARY LITERATURE. See Isaac Newton, "Some Considerations Upon Part of a Letter of M. de Bercé, Concerning His Reflecting Telescope Pretended To Be Improv'd by M. Cassegrain," in *Philosophical Transactions of the Royal Society of London,* **7** (1672), 4056–4059; and "N. Cassegrain," in Hoefer, *Nouvelle biographie générale* (Paris, 1855). See also "Guillaume Cassegrain," in *La grande encyclopédie,* IX, 696; since this Cassegrain appears in the state records from 1666 to 1684, he at least constitutes a candidate who is known to have actually existed. His identification as the writer of the material in question, however, appears to be either entirely circumstantial or heavily dependent on several references to *fondeurs de cloches* in the discussion of the megaphone.

VICTOR E. THOREN

CASSERI (or **CASSERIO**), **GIULIO** (*b.* Piacenza, Italy, ca. 1552; *d.* Padua, Italy, 8 March 1616), *anatomy, surgery.*

Casseri was born into a humble and quite poor family. His father, Luca, died at an early age; his mother, Margherita, survived her son. While still a youngster Casseri moved from Piacenza to Padua. It is likely that he married. He also was employed by several well-to-do students. In Padua he had the opportunity to serve the renowned Girolamo Fabrizio, public lecturer in anatomy and surgery at the University of Padua from 1565.

Fabrizio had the habit of performing preparatory dissections of corpses before presenting them at the public lectures. He was assisted in this procedure by young Casseri, who undoubtedly had revealed ability and readiness. Fabrizio therefore encouraged Casseri's talents. He quickly proved himself, studying literature, especially the classics. He then enrolled in the Facoltà Artista and received his doctorate in medicine and philosophy at Padua around 1580. Besides Fabrizio, he also studied with Mercuriale. Around 1585 he was an established surgeon in full professional practice in Padua.

Shortly before 1590 Casseri's financial situation improved, and he brought his mother and brother Teodoro to Padua. Their residence, in a palace in the Santa Sofia district, was rented from the counts Canale for 100 ducats per year, a considerable sum for those times.

Casseri continued to dedicate himself to anatomical research and teaching. Sometimes he was called upon to substitute for the ailing Fabrizio in the anatomy classes, and he also gave private lessons that were well attended and well received. This, however, did not please Fabrizio, who did not always look kindly upon his pupil's successes. The situation regarding the official teaching of anatomy and surgery—both conducted by Fabrizio—became insupportable around 1608. Fabrizio was no longer able to meet his teaching obligations.

In decrees dated 25 August 1609, Fabrizio was given only the post of special lecturer in anatomy, and the title of public lecturer in surgery was conferred on Casseri. We find among the latter's syllabi the subjects "De ulceribus" and "De vulneribus." Casseri also continued to give private lessons in anatomy, a subject to which he felt especially drawn and in which he would certainly have succeeded Fabrizio, had he not died before him.

Shortly after Casseri's appointment as lecturer in surgery, Caspar Bartholin, who maintained friendship and admiration for Casseri, came to Padua. He also had the esteem and the veneration of the German students, whom he had defended when an attempt was made to deny them, as Protestants, the privilege of receiving their medical degrees through private instruction.

On 3 January 1614, Fabrizio's poor health caused him to be unable to deliver the anatomy lectures on corpses and Casseri took his place. He did not, however, want to hold the lectures in the famous anatomy theater, founded by Fabrizio in 1595, because he declared that there he would have taught only as an ordinary lecturer. Therefore, Casseri's demonstrations were held in a theater, a good part of whose construction costs he had paid out of his own pocket, in the Palazzo del Capitanio.

In January 1616, at the insistence of the Riformatori dello Studio and of the Capitanio, Casseri consented to hold his lectures in the theater in the Palazzo Centrale of the university, which still exists. The course lasted three weeks and covered problems of angiology, neurology, myology, and osteology and related areas of pathology. The lectures were warmly received by the large audiences.

Suddenly, a serious feverish illness struck Casseri. Five days later he died, on the evening of 8 March 1616. He was buried in the Church of the Hermits, almost opposite the door of the Mantegna Chapel.

Casseri achieved such fame as an anatomist during his lifetime that the universities of Parma and Turin offered him the chairs of anatomy. He always refused because he was convinced he would succeed Fabrizio at Padua, in the chair that had been held by Vesalius, Colombo, and Falloppio. He was named Cavaliere of San Marco by the Republic of Venice.

Casseri's scientific achievements are collected in three anatomical works: *De vocis auditusque organis historia anatomica* (Ferrara, 1600–1601; Venice, 1607), *Pentaestheseion, hoc est De quinque sensibus liber* (Venice, 1609; reprinted in Frankfurt and Venice), and *Tabulae anatomicae LXXIIX, omnes novae nec antehac visae,* published posthumously in Venice in 1627 under the editorship of Dr. Daniel Rindofleisch, better known by the pseudonym of Danieles Bucretius. Published with it was a treatise by Adriaan van den Spieghel, "De humani corporis fabrica," with illustrations by Casseri. It was reprinted, without illustrations, in two Latin editions and two editions with German translation.

De vocis contains two treatises—one on the anatomy of the larynx and the other on the anatomy of the ear—and thirty-four plates. The first treatise, based on comparative anatomy, consists of 192 pages and is divided into three books. The first book, containing twenty chapters, concerns the anatomy of the larynx. Human vocal organs are studied in relation to those of other mammals, birds, amphibians, and even insects. The research is extended to the superficial and deep muscles. For the first time a precise description of the two cricoid-thyroid muscles is given. The description of the superior and inferior laryngeal nerves is accurate, as are his assumption that they originate from cranial nerves and his statement of the function of the laryngeal nerves. In chapter 20 laryngotomy is illustrated, and its importance is specified in acute forms of glottal occlusion.

The second book deals with phonation: the nature of sound, the history of concepts regarding the nature of the voice, and a comparative examination of the mechanisms of phonation.

The third book concerns the importance of the larynx in general and the reasons for its shape, posi-

tion, and structure. Casseri mistakenly holds that the function of the thyroid is to moisten the larynx.

The second treatise, which concerns the anatomy of the ear, is also divided into three books. The first presents the comparative and descriptive anatomy of the ear. The description of the tympanic membrane is thorough. The illustration of the semicircular canals is brief, and their number is correctly determined. The treatise also deals with vascularization and the innervation of the middle and inner ear.

The second book deals with the auditory function and problems of acoustics.

The third book concerns the physiology of hearing. It generally reflects the knowledge of the times and is often well-founded. In chapter 12 the relation between the shape of the earlobe and the tendency to criminal behavior, a correlation accepted by many criminal anthropologists, is also indicated.

The *Pentaestheseion*, a 346-page treatise on esthesiology, includes thirty-three plates. It is divided into five books on the organs of touch, taste, smell, hearing, and vision. For over half of the seventeenth century the work was considered the best that science could offer on the unquestionably fascinating subject of the sense organs, both because of the easy-flowing style of the text and because of the liveliness of the plates.

The *Tabulae anatomicae* was to have formed a complete atlas of human anatomy, and Casseri undoubtedly had thought of its publication since 1593. His illustrations in general, but especially in the *Tabulae,* made a vigorous and concrete contribution to the development of anatomical illustration. They include plates that reproduce with unusual accuracy the muscles of the back, the overall view of the abdominal viscera, the distribution of the portal vein in the liver, and the formation of the superior hepatic veins. In one plate, besides the urachus and the lateral umbilical ligaments, the inguinal fossae and the peritoneum, with the lower peritoneal tissue detached from the abdominal wall, are illustrated for the first time. Casseri's illustrations represent the last word, solemn and authoritative, uttered by the Paduan anatomical school at the twilight of the golden century of its existence.

In Haller's words, Casseri was *Felix chirurgus, insignis anatomicus.*

BIBLIOGRAPHY

On Casseri or his work, see D. Bertelli, "Giulio Casseri da servo a professore universitario," in *Settimo centenario della Università di Padova* (Padua, 1922), pp. 9–10; P. Capparoni, "Giulio Casserio," in *Profili bio-bibliografici di medici e naturalisti celebri italiani del secolo XV al secolo*

XVIII, II (Rome, 1928), 49–52; G. Ghilini, *Teatro d'huomini letterati,* I (Venice, 1647), 130; A. von Haller, *Biblioteca anatomica,* I (Leiden, 1774), 289–290; N. C. Papadopoli, *Historia gymnasii Patavini,* I (Venice, 1726), bk. 3, p. 346; A. Portal, *Histoire de l'anatomie et de la chirurgie,* II (Paris, 1770), 229; L. Premuda, *Storia dell'iconografia anatomica* (Milan, 1957), pp. 161–163; Charles Singer, *A Short History of Anatomy and Physiology From the Greeks to Harvey* (New York, 1957), pp. 161–163, *passim;* I. P. Tomasini, *Gymnasium Patavinum* (Udine, 1654), p. 336; and R. von Töply, "Geschichte der Anatomie," in Th. Puschmann, *Handbuch der Geschichte der Medizin,* II (Jena, 1903), 236.

LORIS PREMUDA

CASSINI I—Gian Domenico Cassini, 1625–1712
CASSINI II—Jacques Cassini, 1677–1756
CASSINI III—César-François Cassini de Thury, 1714–1784
CASSINI IV—Jean-Dominique Cassini, 1748–1845

The Cassinis

CASSINI, GIAN DOMENICO (JEAN-DOMINIQUE) (CASSINI I) (*b.* Perinaldo, Imperia, Italy, 8 June 1625; *d.* Paris, France, 14 September 1712), *astronomy, geodesy.*

The first of a family of astronomers who settled in France and were prominent in directing the activities of the French school of astronomy until the Revolution, Cassini was the son of Jacopo Cassini, a Tuscan, and Julia Crovesi. Raised by a maternal uncle, he studied at Vallebone and then at the Jesuit college in Genoa and at the abbey of San Fructuoso. He showed great intellectual curiosity and was especially interested in poetry, mathematics, and astronomy. He was attracted at first by astrological speculations, but reading Pico della Mirandola's pamphlet *Disputationes Joannis Pici Mirandolae adversus astrologiam divinatricem* persuaded him of the frivolity of that pseudoscience. Yet, paradoxically, the beginning of his scientific career benefited from the reputation he acquired for his knowledge of astrology. The Marquis Cornelio Malvasia, a rich amateur astronomer and senator of Bologna who calculated ephemerides for astrological purposes, invited him to come to work in his observatory at Panzano, near Bologna.

In accepting this position Cassini initiated the first part of his career, which lasted until his departure for France in February 1669. Thanks to the marquis's aid, he thus made use, from 1648, of several instruments that allowed him to begin his first researches. He was also able to complete his education under the tutelage of two excellent scientists, the Bolognese Jesuits Giovan Battista Riccioli—who was then finishing his great treatise, the *Almagestum novum* (1651)—and

Francesco Maria Grimaldi, who later became famous for his discovery of the phenomenon of diffraction, published in his posthumous work *De lumine* (1665). Although one cannot exactly determine their influence on the young Cassini, it appears that they convinced him of the importance of precise and systematic observation and of the necessity of a parallel improvement in instruments and methods. They probably likewise contributed, less happily, to making him wary of the new theories—especially of Copernicus' system—and to reinforcing in him the conservative tendencies that he displayed throughout his life.

With his first works Cassini won the esteem of his fellow citizens to such an extent that in 1650 the senate of Bologna, on the recommendation of its patron, designated him to occupy the principal chair of astronomy at the university, which had been vacant since Bonaventura Cavalieri's death in 1647. Cassini was actively interested in planetary astronomy and in 1653 wrote to Pierre Gassendi requesting precise observations concerning the superior planets. In 1652–1653 the passage of a comet attracted his attention. In the account of his observations he accepted that the earth is at the center of the universe, that the moon possesses an atmosphere, and that the comets, which are situated beyond Saturn, are formed as a result of emanations originating from the earth and the planets. But he affirmed later that comparison with other observations soon led him to reject the latter theory, of Aristotelian inspiration, and to adopt that of Apollonius of Myndos; thus he now considered the comets as heavenly bodies analogous to the planets but describing trajectories of very great eccentricity.

A happy circumstance permitted him to reveal his practical abilities. Since the determination of certain essential astronomical data is tied to the movement of the sun (solstices, obliquity of the ecliptic, and so forth) and thus requires the daily observation of the height of that body at the time of its passage to the meridian, astronomers for a long time had tried to increase the precision of these observations by employing high structures—churches in particular—as supports for large sundials, called meridians. Such was the case at the church of San Petronio of Bologna, where an important meridian had been constructed in 1575 by a predecessor of Cassini in the chair of astronomy at the university, Egnatio Danti. Unfortunately, structural modifications necessitated by the enlargement of the church had recently rendered this meridian unusable by blocking the orifice through which the solar rays entered. In 1653, Cassini, wishing to employ such an instrument, sketched a plan for a new and larger meridian but one that would be difficult to build. His calculations were precise; the construction succeeded perfectly; and its success made Cassini a brilliant reputation.

During the following years Cassini made with this meridian numerous observations on the obliquity of the ecliptic, on the exact position of the solstices and the equinoxes, on the speed of the sun's apparent motion and the variation of its diameter, and even on atmospheric refraction; for all these phenomena he provided increasingly more precise measurements. His principal observations, published in *Specimen observationum Bononiensium. . .* (1656), are dedicated to Queen Christina of Sweden, then in exile in Italy. In later publications he drew upon other of the measurements he made by means of the meridian of San Petronio.

Activities of a more technical nature, however, were to oblige Cassini to abandon astronomical research to some extent. As an official expert delegated by the Bolognese authorities, he participated in 1657 in the settlement, directed by Pope Alexander VII, of a dispute between the cities of Bologna and Ferrara concerning the course of the Reno River. On this occasion he composed several memoirs on the flooding of the Po River and on the means of avoiding it; moreover, he also carried out experiments in applied hydraulics. In the course of the following years he was charged with various further missions and important technical functions. In 1663 he was named superintendent of fortifications and in 1665 inspector for Perugia.

In 1663 Cassini defended the views of the papal authorities before the grand duke of Tuscany at the time of the controversies regarding the regularization of the waters of the Chiana River. He returned to Tuscany in 1665 for the same purpose, with the title of superintendent of the waters of the ecclesiastical states. Requested by the pope to take holy orders, he declined to do so and endeavored to reconcile the exercise of his functions at the papal court with his teaching at the University of Bologna. He was resolved not to give up his purely scientific activity, and he accordingly took advantage of his numerous trips to participate in certain meetings of the Accademia del Cimento in Florence, to make observations on insects, and to carry out experiments on blood transfusion at Bologna.

Astronomy, however, remained his preoccupation. In 1659 he presented a model of the planetary system that was in accord with the hypothesis of Tycho Brahe; in 1661 he developed a method, inspired by Kepler's work, of mapping successive phases of solar eclipses; and in 1662 he published new tables of the sun, based on his observations at San Petronio. He also elaborated the first major theory of atmospheric refraction founded on the sine law. Although his

model of the atmosphere was incorrect, the tables that he made in 1662 were later successfully employed in the construction of the ephemerides, before being corrected in accordance with the observations made by Jean Richer in Cayenne in 1672. In 1664 Cassini published an observation of a solar eclipse made at Ferrara. The study of comets, however, continued to hold his special interest. In 1664–1665 he observed one of them in the presence of Queen Christina and formulated on this occasion a new theory (in agreement with the Tychonian system) in which the orbit of the comet is a great circle whose center is situated in the direction of Sirius and whose perigee is beyond the orbit of Saturn.

A new and fertile direction now opened up for Cassini's observations. Through his friendship with the famous Roman lensmakers Giuseppe Campani and Eustachio Divini, Cassini, beginning in 1664, was able to obtain from them powerful celestial telescopes of great focal length. He used these instruments—very delicate and extremely accurate for the time—with great skill, and made within several years a remarkable series of observations on the planetary surfaces, which led him to important discoveries. In July 1664 he detected the shadow of certain satellites on Jupiter's surface and was thus able to study the revolution of the satellites and to demonstrate that of the planet; the period that he attributed to the latter, $9^h 56^m$, is close to the presently accepted value. At the same time, he described the whole group of the planet's bands, as well as its spots, and observed its flattening. This discovery involved him in polemics which, far from diminishing his activity, incited him to pursue his research and his observations. At the beginning of 1666 he observed the spots on Mars and investigated the rotation of that planet, whose period he calculated at $24^h 40^m$ (three minutes less than the value presently accepted). He made the same observations regarding Venus in 1667, but in a less precise form.

Cassini likewise worked on establishing tables of movements of the satellites of Jupiter, a task that Galileo had undertaken primarily in order to obtain a solution to the problem of the determination of longitudes. While Galileo was unable fully to develop these tables owing to a lack of sufficiently precise and complete observations, and while his direct successor, Vincenzo Renieri, similarly failed, Cassini succeeded in this enterprise and published in 1668 his *Ephemerides Bononienses mediceorem siderum*. These ephemerides were employed for several decades by astronomers and navigators, until they were replaced by the more precise tables that Cassini published in Paris in 1693; in particular, they were used by Olaus Römer in his demonstration, in 1675, that light has a finite speed.

The fame that these tables, as well as his important discoveries concerning the planets, brought to Cassini was to change his destiny and to open up for him a new and brilliant career at Paris, at the recently founded Académie Royale des Sciences. Desiring to enhance the prestige of the Academy, Colbert endeavored to attract to France several famous foreign scientists. Thus, after having recruited Christiaan Huygens before the Academy actually opened, in 1667, he offered Cassini membership as a regular correspondent.

Cassini accepted, and in 1668 Colbert proposed that he come to Paris for a limited period, under attractive financial conditions, to help set up the observatory, the construction of which had just begun. Several persons took part in this negotiation, including the astronomer Adrien Auzout; the terms settled upon were an annual pension of 9,000 livres (Huygens himself received only 6,000 livres), free lodging, and a travel allowance of 1,000 écus. After a second, diplomatic discussion, the senate of Bologna and the pope authorized Cassini at the end of 1668 to accept the invitation, while maintaining both the various titles he had acquired in Italy and their corresponding emoluments. In fact, his departure from Bologna on 25 February 1669 marked not the beginning of a long foreign mission but the end of his Italian career.

Cassini arrived in Paris on 4 April and was very cordially received by the king five days later. He immediately began to participate in the activities of the Academy, taking an active role in the enterprises already under way. Since he had a particular interest in the construction of the observatory, he strove in vain to modify the plans, which had been conceived by Claude Perrault and approved by the Academy. Cassini thought to remain in France for only the brief time arranged and then to resume his previous duties and way of life; therefore he at first made little effort to accustom himself to French life. Moreover, he spoke French only haltingly; and his rather authoritarian character and privileged situation, due to the favor of the Crown, provoked considerable hostility from the moment of his arrival.

He gradually got used to speaking French, however. He was delighted by the living and working conditions provided for him. The ambition to organize and direct the important research program of the Academy fortified his resolve. With all this, Cassini succeeded in overcoming much of the opposition he had encountered and in winning essential collaboration. In September 1671 he moved into the apartment prepared especially for him in the new observatory, where work was now beginning. Although this establishment had in theory been placed under the collective responsibility of the astronomers of the Academy, Cassini assumed the effective direction of it. He then decided to settle in France, and on 14 July 1673 he obtained the benefits of French citizenship. In 1674 he married Geneviève de Laistre, the daughter of the lieutenant general of the *comté* of Clermont, whose dowry of valuable landholdings included the château of Thury in the Oise, which became the family's summer residence. From this marriage Cassini had two sons; the younger, Jacques, succeeded him as astronomer and geodesist under the name of Cassini II.

The important work that Cassini accomplished in France encompassed quite diverse aspects. Some were related to the continuation of his Italian projects and to the exploitation of the new paths that he had opened; others pointed toward new directions brought to light by discussions among the Academicians and by the possibilities offered by the new observatory.

While remaining faithful to certain traditional methods (he had a gnomon constructed in the great hall of the observatory), Cassini strove to follow the rapid progress of technology and to utilize recent inventions and improvements: lenses of high focal length, the micrometer, and the attachment of eyepieces to measuring instruments. A large official subsidy allowed the purchase of new instruments which were thus employed for observations regularly made at the observatory as well as in preparing the ephemerides, in improving the celestial map, and in various researches: they were further used in the course of the numerous geographic, geodesic, and astronomical expeditions carried out under the patronage of the observatory. These instruments included quadrants, octants, equatorials, telescopes, and azimuth compasses and such original contrivances as a main mast and a wooden tower 120 feet high erected on top of the observatory to permit the use of the most powerful lenses.

Cassini continued the observational work begun in Italy using a lens made by Campani with a focal length of seventeen feet that he had brought from Italy, as well as others even more powerful (up to a focal length of 136 feet), commissioned from either Campani or Divini, or from French lensmakers. In September 1671 he discovered a second satellite of Saturn, Iapetus (VIII), and explained that the variations in its brightness were due to its always turning the same face toward Saturn. In 1672 he observed a third satellite, Rhea (V), and on 21 March 1684, two others, Tethys (III) and Dione (V). Moreover, his remarkable abilities as an observer allowed him to discern a band on the surface of the planet and to discover, in 1675, that its ring is subdivided into two parts, separated by a narrow band (Cassini's division). He suggested that the two parts are constituted by the aggregation of a very great number of corpuscles, each of which is invisible and behaves like a tiny satellite; this hypothesis has been verified by spectroscopy. Between 1671 and 1679 he observed the features of the lunar surface and sketched an atlas that enabled him to draw a large map of the moon, which he presented to the Academy in 1679. In 1683 he observed, following Kepler, the zodiacal light and had the merit of considering this phenomenon as being of a cosmic, not a meteorological, order. It is true, however, that he linked it in part to a completely false theory of the solar structure.

In 1680 the appearance of a particularly spectacular comet led Cassini back to one of his favorite subjects. Yet while Newton drew decisive arguments from this occasion for his gravitational theory, Cassini saw in it the confirmation of the cogency of his method of studying cometary trajectories and of his theory limiting these trajectories to a band of the celestial vault, the cometary zodiac.

The tables of the eclipses of the satellites of Jupiter that Cassini had published in 1666 were utilized for the determination of longitudes in the course of numerous worldwide expeditions undertaken by French astronomers (in Denmark, the coast of France, Cayenne, Egypt, the Cape Verde Islands, and the Antilles, among other places). As initiator of the new method, Cassini made the observations at Paris to serve as controls and coordinated the results on a large planisphere. Beyond its geographical implications, Richer's expedition to Cayenne in 1672–1673 had several astronomical objectives, of which the most important was the determination of the parallax of Mars during its opposition of 1672; it was accomplished through the simultaneous observations made by Richer at Cayenne and by Cassini and Jean Picard in Paris. The result obtained, 25″, enabled them to fix the parallax of the sun at 9.5″ (instead of 8.8″) and to calculate for the first time with a reasonable approximation the mean earth–sun distance and the dimensions of the planetary orbits. The members of this expedition were also able to study atmospheric refraction near the equator and to correct the tables previously published by Cassini. Finally, Richer observed that the length of a pendulum with a frequency of once a second is less at Cayenne than at Paris, an unexpected fact whose interpretation provoked ardent polemics for two thirds of a century. Whereas Richer thought this phenomenon could be explained by the flattening of the earth and while Huygens—quickly followed by Newton but through a different approach —arrived at this same conclusion, Cassini believed in the sphericity of the earth and attempted to explain the phenomenon by temperature differences. Settlement of the debate required better measurements of

arcs of meridian than those taken by Picard between Paris and Amiens from 1668 to 1670. In 1683 Cassini obtained an agreement from Colbert and the king to extend the earlier measurement (an arc of approximately 1°21′) to an arc of 8°30′ between the northern and southern frontiers of France. Assisted by several collaborators, he immediately undertook to extend the meridian of Paris toward the south, while Philippe de la Hire carried out the same operation toward the north. But in 1684 the death of Colbert and the difficult situation of the public treasury interrupted these activities at a time when Cassini had reached only the vicinity of Bourges. It was not until 1700 that the king decided to resume the project. With the aid of several collaborators, including his son Jacques and his nephew Giacomo Filippo Maraldi, Cassini measured the arc of meridian from Paris to Perpignan and, in addition, conducted various associated geodesic and astronomical operations, which he reported on to the Academy. The result of this last great expedition directed by Cassini led him to adopt the hypothesis of the lengthening of the terrestrial spheroid, which was viewed favorably by the Cartesians. His direct successors, moreover, were to defend this hypothesis with a certain obstinacy.

The traditionalist character shown by Cassini's position in this controversy is characteristic of the majority of his theoretical conceptions. While it seems that in 1675 he narrowly preceded Römer in formulating the hypothesis of the finite speed of light to explain certain irregularities in the apparent movements of Jupiter's satellites, he soon rejected this explanation and, as a resolute Cartesian, combated Römer's theory, which had the support of Huygens. Likewise Cassini was a determined opponent of the theory of universal gravitation. Moreover, while he seems to have renounced Tycho Brahe's planetary system, his Copernicanism remained very limited, especially as he proposed to replace the Keplerian ellipses by curves of the fourth degree (ovals of Cassini), a locus of points of which the product of the distances to two fixed points is constant.

At the beginning of the eighteenth century, Cassini's activities declined rapidly, and his son Jacques gradually replaced him in his various functions. His last two years were saddened by the total loss of his sight.

Judgments on Cassini's work vary greatly. While many historians, following Jean-Baptiste Delambre, accuse him of having found his best ideas in the writings of his predecessors and of having oriented French astronomy in an authoritarian and retrograde direction, others insist on the importance of his work as observer and organizer of the research at the Observatory. Although Cassini's control did restrict the

Observatory's studies and although he did fight against most of the new theories, his behavior does not seem as uniformly tyrannical and baleful as Delambre described it. He was not a theoretician; he was, however, a gifted observer and his indisputable discoveries are sufficient to win him a high position among the astronomers of the pre-Newtonian generation.

BIBLIOGRAPHY

I. ORIGINAL WORKS. Most of Cassini's publications and memoirs are listed in the *Catalogue général des livres imprimés de la Bibliothèque Nationale,* XXIV (Paris, 1905), cols. 678–682, or in the *Table générale des matières contenues dans l'Histoire et dans les Mémoires de l'Académie Royale des Sciences,* I–III (Paris, 1729–1734). Almost complete lists are given in A. Fabroni, *Vitae Italorum doctrina excellentium,* IV (Pisa, 1779), 313–335, and V. Riccardi, *Biblioteca matematica italiana,* I (Bologna, 1887), cols. 275–285; the latter, which has been repr. in facsimile (Milan, 1952), does not cite the articles in the *Journal des Savants* or in the *Philosophical Transactions.*

A large part of Cassini's publications subsequent to his arrival in France are collected in *Recueil d'observations faites en plusieurs voyages par ordre de S. M. pour perfectionner l'astronomie et la géographie avec divers traités astronomiques par Messieurs de l'Académie Royale des Sciences* (Paris, 1693), and in *Mémoires de l'Académie Royale des Sciences depuis 1666 jusqu'en 1699* (Paris, 1730), vol. VIII ("Oeuvres diverses"). Many MSS by Cassini or initialed by him are preserved in the Archives de l'Observatoire de Paris and at the Bibliothèque de l'Institut.

II. SECONDARY LITERATURE. On Cassini or his work, see F. Arago, *Notices biographiques,* III (Paris, 1855), 315–318; F. S. Bailly, *Histoire de l'astronomie moderne,* II–III (Paris, 1779); J. B. Biot, in *Biographie universelle,* VII (Paris, 1813), 297–301, and in new ed., VII (Paris, 1844), 133–136; J. D. Cassini IV, *Mémoires pour servir à l'histoire des sciences et à celle de l'Observatoire de Paris . . .* (Paris, 1810); J. de Lalande, *Astronomie,* 2nd ed., I (Paris, 1771), 217–220, and *Bibliographie astronomique* (Paris, 1802); J. B. J. Delambre, in *Histoire de l'astronomie moderne,* II (Paris, 1821), 686–804, and table, I, LXVII–LXIX; A. Fabroni, in *Vitae Italorum doctrina excellentium,* IV (Pisa, 1779), 197–325, B. Fontenelle, "Éloge de J. D. Cassini," in *Histoire de l'Académie royale des Sciences* [pour] *1712* (Paris, 1714), and *ibid.,* 84–106; F. Hoefer, in *Nouvelle biographie générale,* IX (Paris, 1835), cols. 38–51; C. G. Jöcher, in *Allgemeines gelehrten Lexicon,* III (Leipzig, 1750), cols. 1732–1733; J. F. Montucla, *Histoire des mathématiques,* II (Paris, an VII [1798–1799]), 559–567; and J. P. Nicéron, in *Mémoires pour servir à l'histoire des hommes illustres . . .,* VII (Paris, 1729), 287–322.

RENÉ TATON

CASSINI, JACQUES (CASSINI II) (*b.* Paris, France, 18 February 1677; *d.* Thury, near Clermont, Oise, France, 15 April 1756), *astronomy, geodesy.*

The son of Gian Domenico Cassini (Cassini I) and Geneviève de Laistre, Cassini began his studies at the family's home in the Paris observatory and then entered the Collège Mazarin, where, in August 1691, he defended a thesis in optics under the direction of Varignon. Oriented very early toward astronomy, he was admitted as a student to the Académie des Sciences in 1694 (he became an associate at the time of the reorganization of 1699 and succeeded his father as a *pensionnaire* on 29 November 1712). In 1695 Cassini accompanied his father on a journey through Italy, where he made numerous scientific observations and carried out several geodesic operations, collaborating in particular on the restoration of the meridian of the church of San Petronio in Bologna. He traveled next to Flanders, the Netherlands, and England, making various geodesic and astronomical measurements, which he published on his return; while visiting these countries he became acquainted with Newton, Halley, and Flamsteed and was admitted to the Royal Society.

In 1700–1701 Cassini participated in his father's project of extending the meridian of Paris as far as the southern border of France. After having criticized the measurement of the arc of meridian made by Snell in 1617 and having presented a new method for the determination of longitudes (by means of the eclipses of the stars and the planets by the moon), in 1713 he took a position clearly supporting the hypothesis of the elongation of the terrestrial ellipsoid. He based his view on the results of earlier measurements of arcs of meridian—especially on that of the meridian of Paris in which he had participated—which seemed to show "that the degrees of a terrestrial meridian grow smaller from the equator toward the pole." Until his retirement Cassini expended great effort in defending this point of view, thus taking part in the battle led by the last Cartesians against the expansion of Newtonianism.

Important administrative positions were soon added to Cassini's purely scientific work. Although at this time the Paris observatory had no official director, Cassini succeeded his father as its manager from the time that the latter's state of health reduced his activity (before 1710) and abandoned this position only in the last years of his life, in favor of his son César-François. Moreover, despite his quite modest legal background, in 1706 Cassini was named *maître ordinaire* of the *chambre des comptes.* In this office he acquired a reputation for seriousness and honesty, but also for indecision. Designated magistrate of the *cour de justice* in 1716, he necessarily obtained the title of *avocat;* and that of *conseiller d'état* was awarded him in 1722.

In 1710 Cassini married Suzanne-Françoise Charpentier de Charmois, by whom he had three sons (Dominique-Jean, later *maître ordinaire* of the *chambre des comptes;* César-François, who succeeded him as astronomer and geodesist and was known as Cassini III or Cassini de Thury; and Dominique-Joseph, a brigadier general) and two daughters (Suzanne-Françoise and Elisabeth-Germaine).

In 1718 Cassini managed to complete the determination of the meridian of Paris, extending as far as Dunkirk the Paris–Amiens axis measured by Picard. Relying on the results of this operation, which were presented in the *Histoire de l'Académie des sciences* for 1718, in 1722 he published the important work *De la grandeur et de la figure de la terre,* in which he confirmed his support for the hypothesis of the elongation of the terrestrial ellipsoid and opposed that of its flattening, which was defended by the supporters of Newton and Huygens. While Mairan sought to justify this apparent disagreement between theory and observation, the Newtonians criticized Cassini's position: Desaguliers in 1725, Maupertuis in a veiled form in 1732, and Giovanni Poleni in 1733. In order to reply to them Cassini, with the aid of his sons and of other collaborators, undertook in 1733–1734 the determination of the perpendicular to the meridian of Paris, from Saint-Malo to Strasbourg. Although the results of this operation seemed to confirm his point of view, the young Newtonians of the Academy arranged for the sending of geodesic expeditions to Peru and to Lapland in order to obtain measurements of arcs of meridian of latitudes sufficiently different to permit a settlement of the dispute. After having proposed a few more improvements in geodesic methods, Cassini left the pursuit of the polemic to his son César-François, who had participated in the last operations and had improved certain methods of measurement; Cassini limited himself to replying in 1738 to a direct attack by Celsius.

In astronomy proper Cassini's work is vast. Besides working patiently as an observer and directing frequently effective work while head of the Paris observatory he published a great number of memoirs in the *Histoire de l'Académie* and two books on astronomy (1740): a collection of tables and a manual. Cassini's principal areas of interest were the study of the planets and their satellites—particularly the inclination of the orbits of the satellites and the structure of Saturn's ring—the observation and the theory of the comets, and the tides.

Certainly these fields yielded valuable observations,

particularly, in 1738, the revelation of the proper motions of the stars; the presentation of improved instruments and of several new methods; and some original hypotheses of limited scope. Their theoretical value, however, is considerably lessened by Cassini's biases. A timid Copernican but a convinced Cartesian and a fervent disciple of his father, Cassini fought unceasingly to defend the work of his father and to reconcile the facts of observation with the theory of vortices; he also never admitted the value of the theory of gravitation. Furthermore, as Delambre notes, the use he made of graphical methods often did not allow one to compensate for the insufficiency of his calculations.

Beginning in 1740, perhaps realizing the futility of his opposition to the triumphant Newtonianism, Cassini progressively abandoned his scientific activity, leaving to his son the task of pursuing the family work in a less outmoded perspective. He restricted himself to some astronomical and physical observations and to collaborating on the map of France, an undertaking directed by his son.

BIBLIOGRAPHY

I. Original Works. A list of the memoirs presented by Cassini to the Academy of Sciences is given in *Table générale des matières contenues dans l'Histoire et dans les Mémoires de l'Académie royale des sciences,* I–VI (Paris, 1729–1758). His books are *Theses mathematicae de optica . . .* (Paris, 1691); *De la grandeur et de la figure de la terre. Suite des Mémoires de l'Académie royale des sciences. Année 1718* (Paris, 1720 [actually 1722]); *Réponse à la dissertation de M. Celsius sur les observations faites pour déterminer la figure de la terre* (Paris, 1738); *Éléments d'astronomie . . .* (Paris, 1740); and *Tables astronomiques du soleil, de la lune, des planètes, des étoiles fixes et des satellites de Jupiter et de Saturne . . .* (Paris, 1740).

II. Secondary Literature. On Cassini or his work see the following, listed chronologically: J. P. Grandjean de Fouchy, "Éloge de J. Cassini," in *Histoire de l'Académie royale des sciences. Année 1756* (Paris, 1762), "Histoire," 134–147; J. S. Bailly, in *Histoire de l'astronomie moderne,* II and III (Paris, 1779–1782), see III, index; C. M. Pillet, in Michaud's *Biographie universelle,* VII (Paris, 1813), 301–302, also in new ed., VII (Paris, 1854), 136; J. B. J. Delambre, *Histoire de l'astronomie au XVIIIe siècle* (Paris, 1827), pp. 250–275; R. Grant, *History of Physical Astronomy* (London, 1852), pp. 234, 543, 545, 554; F. Hoefer, in *Nouvelle biographie générale,* IX (Paris, 1855), cols. 51–52; Poggendorff, I, cols. 390–391; J. C. Houzeau and A. Lancaster, *Bibliographie générale de l'astronomie,* 3 vols. (Brussels, 1882–1889; repr. London, 1964), see index; P. Ravel, *La chambre de justice de 1716* (Paris, 1928); P. Brunet, *Maupertuis,* II (Paris, 1929), chs. 2, 3; and M. Prévost, in

Dictionnaire de biographie française, VII (1956), cols. 1329–1339.

RENÉ TATON

CASSINI, JEAN-DOMINIQUE (CASSINI IV) (*b.* Paris, France, 30 June 1748; *d.* Thury, near Clermont, Oise, France, 18 October 1845), *astronomy, geodesy.*

The son of César-François Cassini (Cassini III) and Charlotte Drouin de Vandeuil, Cassini was born at the Paris observatory and received his secondary education at the Collège du Plessis in Paris and at the Oratorian *collège* at Juilly. He then studied under the physicist Nollet, the mathematician Antoine Mauduit, and the astronomers Giovanni Maraldi and J. B. Chappe d'Auteroche. In 1768, on a cruise in the Atlantic, he was in charge of continuing the attempts to test a marine chronometer of Pierre Le Roy. Elected *adjoint* by the Académie des Sciences on 23 July 1770 (he became *associé* in 1785), he was put in charge of the publication of Chappe's *Voyage en Californie.* Assured from 1771 of succession to the directorship of the Paris observatory, a post created for his father, he gradually assumed its responsibilities before being officially appointed in 1784, upon the death of Cassini III.

In 1773 Cassini married Claude-Marie-Louise de la Myre-Mory, who died in 1791, leaving him with five young children: Cécile, Angélique, Aline, Alexis, and Alexandre Henri Gabriel, later a jurist and botanist, with whom the French branch of the Cassini family died out.

In 1784 Cassini persuaded Louis XVI to agree to the restoration and reorganization of the observatory, a project, however, that he was able to realize only partially. He occupied himself with the completion of the great map of France undertaken by his father and in 1787, with A.-M. Legendre and Méchain, participated in the geodesic operations joining the Paris and Greenwich meridians.

Cassini accepted some political duties at the beginning of the Revolution, and directed the execution of a portion of the new administrative maps and participated for several months, in 1791, in the work of the commission of the Academy responsible for preparing a new metrological system. He was firmly attached to the monarchy, however, and, little by little, adopted an attitude of hostile reserve toward the Revolution. From March 1793 he opposed the reforms that the authorities wanted to introduce at the observatory, and he attempted to maintain his former authority. However, after bitter polemics he gave up his duties on 6 September 1793 and a few weeks later left the Paris observatory, which for 120 years had practically been the property of his family.

Denounced by the revolutionary committee of Beauvais, Cassini was arrested in Paris on 14 February 1794 and imprisoned. On 5 August 1794, he retired to the family château of Thury, in the Oise. He participated in local affairs as a member of the board of examiners of the primary schools and of the École Centrale de l'Oise but declined his nomination to the Bureau des Longitudes at the end of 1795 and to the astronomy section of the new Institut National in January 1796.

A few years later Cassini attempted to resume his career. He accepted election as *associé* of the experimental physics section on 24 April 1798 and then as member of the astronomy section of the Institute on 24 July 1799, but he strove in vain to secure renomination to the Bureau des Longitudes. Renouncing further pursuit of his scientific work, he assumed the presidency of the Conseil Général de l'Oise from 1800 to 1818. Pensioned and decorated by Napoleon and Louis XVIII, Cassini devoted himself to local politics and to polemical writings aimed at combating liberal ideas, defending his family's scientific prestige, and justifying his attitude.

Besides his efforts to modernize and reorganize the Paris observatory during the last years of the *ancien régime* and to complete the map of France, Cassini's personal works are basically limited to accounts of astronomical and geodesic expeditions and to reports of observations.

BIBLIOGRAPHY

I. ORIGINAL WORKS. An important part of Cassini's scientific work is the many memoirs and communications presented to the Académie des Sciences. Their titles may be found in *Table générale des matières contenues dans l'Histoire et dans les Mémoires de l'Académie des sciences,* VIII–X (Paris, 1774–1809). His principal works are *Voyage fait par ordre du roi en 1768 pour éprouver les montres marines inventées par M. Le Roy* . . . (Paris, 1770); *Exposé des opérations faites en France, en 1787, pour la jonction des observatoires de Paris et de Greenwich* (Paris, n.d. [1792]), written with Méchain and A.-M. Legendre; and *Mémoires pour servir à l'histoire des sciences et à celle de l'Observatoire de Paris, suivis de la vie de J. D. Cassini* (Paris, 1810). He was editor of Chappe d'Auteroche's *Voyage en Californie pour l'observation du passage de Vénus sur le disque du soleil, le 3 juin 1769* . . . (Paris, 1772); *Extrait des observations astronomiques et physiques, faites à l'Observatoire royal,* 5 vols. for 1785, 1787–1791 (Paris, 1786–1791), in collaboration with J. Perny de Villeneuve, N.-A. Nouet, and A. Ruelle; and *Carte géométrique et topographique de la France* . . . *publiée sous les auspices de l'Académie des sciences* . . . (Paris, 1784–1793). In addition, Cassini collaborated on the *Dictionnaire de physique* of the *Encyclopédie méthodique,* 4 vols. (Paris, 1793–1822). Many of his MSS are in the library of the Paris Observatory, the library of the Institut de France, and the library of Clermont, Oise.

II. SECONDARY LITERATURE. On Cassini or his work see the following, listed chronologically: J. F. S. Devic, *Histoire de la vie et des travaux scientifiques et littéraires de J. D. Cassini IV* (Clermont, 1851); C. Delacour, "Le dernier des Cassini," in *Mémoires de la Société académique . . . de l'Oise,* **2** (1853), 67–92; F. Hoefer, in *Nouvelle biographie générale,* IX (Paris, 1854), cols. 52–54; Poggendorff, I, cols. 392–393; J. Houzeau and A. Lancaster, *Bibliographie générale de l'astronomie,* 3 vols. (Brussels, 1882–1889; repr. London, 1964), see index; J. Guillaume, *Procès-verbaux du Comité d'instruction publique de la Convention nationale,* 7 vols. (Paris, 1891–1957), see VII, index, 232; C. Wolf, *Histoire de l'Observatoire de Paris, de sa fondation à 1793* (Paris, 1902); G. Boquet, *Histoire de l'astronomie* (Paris, 1924), pp. 471–473; N. Nielsen, *Géomètres français sous la Révolution* (Copenhagen, 1929), pp. 54–57; F. Marguet, *Histoire générale de la navigation du XVe au XVIIIe siècle* (Paris, 1931), *passim;* and M. Prévost, in *Dictionnaire de biographie française,* VII (Paris, 1956), cols. 1331–1332.

RENÉ TATON

CASSINI DE THURY, CÉSAR-FRANÇOIS (CASSINI III) (*b.* Thury, near Clermont, Oise, France, 17 June 1714; *d.* Paris, France, 4 September 1784), *geodesy, astronomy.*

Cassini de Thury, the son of Jacques Cassini (Cassini II) and Suzanne-Françoise Charpentier de Charmois, studied at the family home in the Paris observatory under the guidance of his granduncle G. F. Maraldi, who developed his very precocious gift for astronomy. His career began when the debate in France between the Cartesians and the Newtonians reached its peak. In particular the quarrel over the form of the terrestrial spheroid—essentially the question of whether it was elongated or flattened along the line of the poles—had taken on a symbolic value at the Académie des Sciences, the Cartesians defending the first point of view and the Newtonians the second. From 1683 to 1718 Cassini I and Cassini II had directed major geodesic operations designed to determine the meridian of Paris, an 8°30′ arc of the great circle stretching between the northern and southern frontiers of France. The brilliant confirmation of the Cartesian view that Cassini II (in 1718 and 1722) thought he had derived from the results of these measurements appeared to be a victory for the Cartesians.

After some hesitation the Newtonians counterattacked, contesting the accuracy and the conclusiveness of these results. In an article published in the *Journal historique de la république des lettres* (January–February 1733), Giovanni Poleni showed that since the possible errors in the measurements were of the same order as the differences established between the lengths of the different degrees of the meridian, more precise techniques of measurement might lead to the opposite conclusion. He therefore suggested resorting to complementary verifications, such as the measurement of an arc of the parallel between two well-established points of longitude. Feeling the necessity of presenting new arguments in favor of the theory of elongation, Cassini II decided to undertake the determination of the arc of the great circle perpendicular to the meridian of Paris, an operation also of great interest for the establishment of the map of France.

With the aid of several collaborators, including his son César-François, Cassini II determined the western portion of this perpendicular (between Paris and Saint-Malo) in 1733, and in 1734 the eastern portion (between Paris and Strasbourg). These operations familiarized César-François with the theory and practice of geodesic operations and as early as 1733 he presented an argument for the importance of such operations before the Academy of Sciences. The Academy named him supernumerary assistant on 12 July 1735. He became a regular assistant on 22 January 1741, an associate in the section of mechanics on 22 February 1741, and a full or "pensioned" member of the section of astronomy on 25 December 1745.

Although these new measurements seemed to confirm the hypothesis of the elongation of the terrestrial spheroid, the Académie des Sciences decided to organize major geodesic expeditions to Peru (1735–1744) and to Lapland (1736–1737) in order to settle the debate by measuring arcs of meridians at very different latitudes. In 1735–1736 Cassini III directed new geodesic operations designed both to bring new elements into this dispute and to complete the guidelines of a rough draft of a new map of France; assisted by his father's principal collaborators, he determined the two demiperpendiculars to the meridian of Paris, which were run to the west at a distance of 60,000 *toises* north and south of the Paris Observatory. (In the old French measures there were 12 *lignes* in the *pouce* [inch], 12 *pouces* in the *pied* [foot], and 6 *pieds* in the *toise* [fathom]; although not quite equal to the English foot, the *pied du roi* was of the same order of magnitude.)

But the return of the Lapland expedition at the end of August 1737 and the presentation, a few weeks later, of its first results, which supported the Newtonian theory of flattening, did not settle the argument. While Cassini II did not renounce the theory of elongation that he had always defended, Cassini III admitted that the value of the earlier geodesic operations, on which the Cartesians had based their arguments, was now in doubt. As a result, attempts to verify the meridian of Paris were undertaken in 1739 and 1740. In these the Abbé N. L. de Lacaille seems to have played the essential role. Meanwhile, Cassini III had become opposed to the theory previously held by his family and took responsibility for conclusions resulting from this position (*La méridienne de l'Observatoire royal de Paris vérifiée dans toute l'étendue du royaume,* 1744). Having succeeded in this difficult conversion, he gave up all argument on the subject and devoted himself to astronomical observation and to the establishment of a new map of France, at the same time leading a rather brilliant social life. In 1747 Cassini III married Charlotte Drouin, daughter of Louis-François, *seigneur* of Vaudeuil; they had a son, Jean-Dominique, who succeeded his father as astronomer and geodesist under the name Cassini IV, and a daughter, Françoise-Elisabeth. In 1748 he was named *maître ordinaire* at the *chambre des comptes* and also obtained the office of *conseiller du roi.* A foreign member of the Royal Society and of the Berlin Academy, he later renewed his ties with Italy and with his family's native city, Bologna.

Cassini III's astronomical work is not very remarkable: observations of lunar and solar eclipses and of occultations of stars and planets, determinations of solstices, the study of the trajectories of comets, and improvements in the details of the construction and use of instruments. The financial aid that he requested to restore the observatory building, to purchase observational instruments, and to improve the functioning of that establishment were not granted until 1785, as part of a complete reorganization of its regime. Cassini IV had just succeeded his father as director of the observatory, a post created by the king in 1771. He had assisted his father during his last years, which Cassini III spent in semiretirement.

Cassini III's essential work, however, was in cartography. The various geodesic operations undertaken in France from 1733 to 1740 by the teams from the observatory, as well as other triangulation projects conducted along the coasts and land borders, had already provided the essential elements for a new map of France. The 400 principal triangles constructed on eighteen bases, mentioned by Cassini III in 1740, were completed during the following years, permitting him to produce the new map in eighteen sheets on the scale 1:870,000. (Announced to the Academy in November 1745, it was not published until 1746 or 1747.)

This map, however, was actually only a preliminary work. In 1746–1747 Cassini carried out important cartographical operations in Flanders and in the Netherlands which were highly valued by the king,

who delegated to him the responsibility of establishing a new map of France on this model. Adopting the scale of 1:86,400 (one *ligne* for 100 *toises*) and relying on triangulations already made, Cassini III drew up the plan of a map in 182 sheets; he predicted that its cost of execution would be more than 700,000 *livres* and that it would require the work of several groups of specialists for about twenty years. The project having been officially adopted, Cassini III ordered the necessary material and undertook the training of various specialists. The surveying began in 1750, but the discontinuance of financial aid from the state interrupted the work in 1756. Then, under the patronage of the Académie des Sciences, Cassini III organized a private financial society which permitted the resumption of activity. In spite of numerous difficulties the work progressed, thanks especially to the help of devoted collaborators: geometers, engineers, and engravers, such as Pierre-Charles Capitaine (*d.* 1778) and his son Louis (1749–1797). At the time of the death of Cassini III in 1784, only the map of Brittany remained to be done. Cassini IV continued this great work, which was almost finished when the Revolution began.

The magnitude and the quality of this work, the first modern map of France, overshadow Cassini III's other accomplishments, such as the projects for the extension of the French network of triangulations toward Vienna (1761, 1775) and toward England (accomplished after his death) and his participation in measurements of the speed of sound (1738).

While he was a good geodesist and a talented cartographer, Cassini III was only a second-rate astronomer; and the name of this third representative of the Cassini dynasty at the Paris Observatory will remain associated with the first map of France produced according to modern principles.

BIBLIOGRAPHY

I. Original Works. Nearly complete lists of Cassini's publications can be found in *Table générale des matières contenues dans l'Histoire et dans les Mémoires de l'Académie royale des sciences,* V–X (Paris, 1747–1809) and in *Catalogue général des livres imprimés de la Bibliothèque nationale, auteurs,* XXIV (Paris, 1905), cols. 686–689. His major works are *Carte des triangles de la France* (Paris, 1744), written with D. Maraldi; *La méridienne de l'Observatoire royal de Paris, vérifiée dans toute l'étendue du royaume . . .* (Paris, 1744); *Avertissement ou introduction à la carte générale et particulière de la France* (Paris, 1755); *Addition aux tables astronomiques de M. Cassini* (Paris, 1756); *Relations de deux voyages faits en Allemagne . . .* (Paris, 1763; 2nd ed., 1775); *Opérations faites . . . pour la vérification du degré du méridien compris entre Paris et Amiens* (Paris, 1757), written with P. Bouguer *et al.; Almanach perpetuel pour trouver l'heure par tous les degrés de hauteur du soleil* (Paris, 1770); *Opuscules divers* (Paris, 1774); *Description géométrique de la terre* (Paris, 1775); and *Description géométrique de la France* (Paris, 1783).

Various MSS can be found in the library of the Paris observatory, the archives of the Académie des Sciences, the library of the Institut de France, the Bibliothèque Nationale in Paris, and the library of Clermont de l'Oise.

II. SECONDARY LITERATURE. See J. S. Bailly, in *Histoire de l'astronomie moderne,* III (Paris, 1782), *passim;* Henri Marie Auguste Berthaut, *La carte de France (1750–1898),* I (Paris, 1898), 16–65; A. Beuchot, in Michaud, ed., *Biographie universelle,* VII (1813), 302–304, new ed., VII (1854), 136–137; G. Bigourdan, *La querelle Cassini-Lalande* (Paris, 1927); M. J. Caritat, marquis de Condorcet, "Éloge de C. F. Cassini," in *Histoire de l'Académie des sciences* for 1784 (1787), "Histoire," 54–63; F. de Dainville, "La carte de Cassini et son intérêt géographique," in *Bulletin de l'Association des géographes français,* no. 521 (1955), pp. 138–147; J. B. J. Delambre, *Histoire de l'astronomie au XVIIIe siècle* (Paris, 1827), pp. 275–309; L. Drapeyron, "La vie et les travaux géographiques de Cassini de Thury," in *Revue de géographie,* 20 (1896), 241–254; "Projet de jonction géodésique entre la France et l'Italie," in *Bulletin de la Société de géographie de l'Ain,* 18 (1899), 38–45, also in *Bulletin de l'Union géographique du Nord de la France,* 19 (1898), 289–295; and *Cassini de Thury, le Capitaine et l'enquête sur la grande carte topographique de France* (Paris, 1899), taken from *Revue de géographie;* L. Gallois, "L'Académie des sciences et les origines de la carte des Cassini," in *Annales de géographie,* 18 (1909), 193–204, 289–310; F. Hoefer, in *Nouvelle biographie générale,* IX (1854), col. 52; J. C. Houzeau and A. Lancaster, *Bibliographie générale de l'astronomie,* vols. I and II of 3 (Brussels, 1882–1889; repr. London, 1964), *passim;* N. Nielsen, *Géomètres français du XVIIIe siècle* (Copenhagen–Paris, 1935), pp. 113–121; J. C. Poggendorff, I, cols. 391–392; M. Prévost, in *Dictionnaire de biographie française,* VII (1956), cols. 1328–1329; and Sueur-Merlin, "Mémoire sur les travaux géographiques de la famille Cassini," in *Journal des voyages, découvertes et navigations modernes . . .,* 15 (1812), 174–191.

RENÉ TATON

CASSIODORUS SENATOR, FLAVIUS MAGNUS AURELIUS (*b.* Scylacium [now Squillace], Italy, ca. 480; *d.* Vivarium, Italy, ca. 575), *preservation and dissemination of knowledge.*

Cassiodorus' contributions to the development of science reside in the encouragement he gave to the study and preservation of the works of classical authors and to his acknowledgment of the need for a cleric to study secular subjects.

A member of a distinguished southern Italian family from the town of Scylacium, Cassiodorus, like his father, served the Ostrogoth kings. In return for this service, which lasted from 503 to 537, he was granted both rank and honor. In 507 he became quaestor, in 514 consul, in 523 master of the offices, and from 533 to 537 praetorian prefect. A record of his official career is preserved in his *Variae,* a collection of 468 official letters and documents that he issued while in office. With the reconquest of Italy by the forces of Justinian his public career came to an end. Shortly after 537 Cassiodorus withdrew from politics and devoted himself to the establishment of a double monastery (which contained one section for the solitary hermit and one for cenobite communal monks) near his home town of Scylacium in the south Italian province of Bruttium.

Although Cassiodorus probably did not become a monk, he was nevertheless the driving force behind the establishment of the monastic community of Vivarium, named after the artificial fishponds, vivaria, that he constructed for the monks. The rule of the order seems to have been rather informal and based upon the recommendations of Cassian. The primary obligation of the monks at Vivarium was to serve God by studying and copying the scriptures, the works of the Church Fathers, and the classics of antiquity. Although the copying of manuscripts had been practiced in other monasteries, Cassiodorus established the labors of the scriptorium as a regular part of monastic life and equated them in dignity to manual work in the fields. Secular as well as ecclesiastical works were copied, for Cassiodorus contended that profane knowledge frequently illuminated the meaning of the Scriptures.

Cassiodorus himself joined in the labors of Vivarium. He prepared a lengthy commentary on the Psalms; collated, commented upon, and revised other works; and, at the age of ninety-three, composed a work on orthography for the monks who were engaged in transcribing manuscripts. His greatest work, the *Institutiones divinarum et humanarum litterarum,* was a manual of instruction for the members of the order. This treatise is divided into two books. Book I describes a course of study for understanding the nuances of the Scriptures, emphasizing the need for monks to study the seven liberal arts in their quest for spiritual truth; Book II surveys the subjects of the trivium and quadrivium. In the trivium Cassiodorus devoted most of his efforts to describing rhetoric and dialectic. Indeed, these subjects compose almost half of Book II. Arithmetic and music receive most of his attention in his description of the subjects of the quadrivium. The remaining liberal arts are examined in a cursory fashion. Cassiodorus continues the encyclopedic tradition of the Roman world in the *Institutiones,* and it is as an encyclopedist maintaining the importance of the seven liberal arts that he had his greatest impact upon medieval intellectual life. Through his encouragement of the copying of manuscripts in monasteries and by his support of the study of secular subjects as a prelude to scriptural investigation, Cassiodorus established a tradition of scholarship that was continued in later centuries.

BIBLIOGRAPHY

I. ORIGINAL WORKS. Cassiodorus was the author of a number of works on diverse subjects. Unfortunately, some of them are no longer extant and others exist only in fragments. His *Gothic History* is lost and is known only through Jordanes' *Origins and Deeds of the Goths.* Collected editions of his writings were made by G. Fornerius (Paris, 1579) and Johannes Garetius (Rouen, 1679; Venice, 1729). The latter edition was reprinted by J. P. Migne as *Magni Aureli Cassiodori opera omnia,* vols. LXIX and LXX of Patrologiae Cursus Completus, Series Latina (Paris, 1865). Theodor Mommsen edited the *Variae* as vol. XII of Monumenta Germaniae Historica, Auctores Antiquissimi (Berlin, 1894). Thomas Hodgkin translated and paraphrased the *Variae* as *The Letters of Cassiodorus* (London, 1886). Cassiodorus' work on orthography was edited by H. Keil in *Grammatici latini* (Leipzig, 1880). R. A. B. Mynors edited Cassiodorus' most important work as *Cassiodori Senatoris Institutiones Edited From the Manuscripts* (Oxford, 1937). This excellent edition of the *Institutiones* provided the basis for the very delightful translation of this work by L. W. Jones, *An Introduction to Divine and Human Readings* (New York, 1946).

II. SECONDARY LITERATURE. Surveys of Cassiodorus' career and writings can be found in A. M. Franz, *Aurelius Cassiodorus Senator* (Breslau, 1872); E. K. Rand, *Founders of the Middle Ages* (Cambridge, 1928); R. W. Church, "Cassiodorus," in his *Miscellaneous Essays* (London, 1888); Dom M. Cappuyns, "Cassiodore Senator," in *Dictionnaire d'histoire et de géographie ecclésiastiques,* I (Paris, 1949); and L. W. Jones's introduction to his translation of the *Institutiones,* which contains an important survey of Cassiodorus' activities.

His literary endeavors are discussed in Max Manitius, *Geschichte der lateinischen Literatur des Mittelalters,* I (Munich, 1911). William H. Stahl presents an interesting appraisal of Cassiodorus' relationship to earlier scientific traditions in his *Roman Science: Origins, Development and Influence to the Later Middle Ages* (Madison, Wis., 1962). E. K. Rand, "The New Cassiodorus," in *Speculum,* **13** (1938), 433–447; and A. Van de Vyver, "Cassiodore et son oeuvre," *ibid.,* **6** (1931), 244–292; and "Les *Institutiones* de Cassiodore et sa fondation à Vivarium," in *Revue bénédictine,* **63** (1941), 59–88, contain useful descriptions of his influence.

PHILLIP DRENNON THOMAS

CASSIRER, ERNST ALFRED (*b.* Breslau, Germany, 28 July 1874; *d.* New York, N.Y., 13 April 1945), *philosophy.*

Cassirer was the son of Eduard Cassirer, a merchant, and Jenny Cassirer. He married his cousin Tony Bondy; they had two sons and one daughter. He was awarded an honorary doctor of law degree by the University of Glasgow and won the gold medal of the Kuno Fischer Institute of the Heidelberg Academy of Sciences in 1914.

Cassirer was the most important of the younger circle of the so-called Marburg school, which advocated the logical-transcendental tendency of neo-Kantianism. He came from a well-to-do Jewish merchant family that included many children. His cousin Bruno Cassirer was a well-known Berlin publisher. After graduating from the Johannes Gymnasium in Breslau, Cassirer began to study law in Berlin but soon changed to philosophy and literature. He studied history, art history, and natural science at the universities of Leipzig, Heidelberg, and Munich. Through Georg Simmel he became aware of Hermann Cohen and of neo-Kantianism, which had originated as a reaction against the reigning Hegelianism and its decline into materialism. After intensive work on the writings of Kant, Cassirer went to Marburg and there, through Hermann Cohen, the founder and head of the Marburg school, was exposed to important philosophical influences. He received his doctorate in 1899 with a dissertation on Descartes, which became the starting point for an outstanding work on Leibniz. The latter was awarded a prize by the Berlin Academy of Sciences. From 1906 Cassirer was a *Privatdozent* in Berlin. In 1919 he moved to Hamburg, where he carried out an effective program of pedagogical and scholarly work, becoming rector in 1930. While a professor at Hamburg he also strengthened the collaboration with the important library founded by Aby Warburg, which later became the Warburg Institute in London.

Political developments in Germany caused Cassirer to emigrate in 1933. He went first to England (All Souls College, Oxford) then in 1935 to Göteborg, Sweden. In 1941 he became visiting professor at Yale University and in 1944 went to Columbia University. He died in New York City in 1945.

Cassirer's work, which is based on a profound knowledge of the history of philosophy and on intensive mathematical and scientific study, bears the mark of neo-Kantianism. Cassirer began his scholarly career as an historian of philosophy. For him the history of philosophy did not consist in the collecting and stringing together of facts and ideas; it ought, rather, to make clear the principal sense and thrust of the set of ideas and show their meaning systematically. In this way the historian attains the method through which the fundamental problems and concepts and the formative powers of cognition can be crystallized.

In his major systematic work, *Das Erkenntnisproblem in der Philosophie und Wissenschaft der neueren Zeit,* Cassirer presents an analysis of the formation of philosophic concepts and shows the connection between the theory of knowledge and the general intellectual culture. In tracing the development from objective to functional-rational thoughts, which has culminated in modern logic and in the mathematical and scientific disciplines, he also provides a history of human knowledge in general.

At the origin of every cognition is the formation of a concept. To the Aristotelian concept, which, formed by ever greater abstraction, becomes ever emptier, Cassirer opposes a new functional concept. It is characterized by the appearance and recognition of the particularities of a group of attributes and of their necessary connection. Cassirer saw the function of the scientific concept as the presentation of a rule by means of which the concrete details of a group of attributes are chosen. He examined the concepts of mathematics and of the natural sciences because the fundamental principles of cognition can be known most clearly through them. He considered chemistry in this light for the first time; according to him, it had developed, through the conception of energy, from a descriptive into an exact science. In his philosophic-historical investigations Cassirer also considered the ideas and conclusions of Kepler, Galileo, Huygens, and Newton, as well as other scientists.

On the fundamental problems of mathematics and natural science, such as number, space, time, and causality, Cassirer took a definite position. Mathematics—for Cassirer a theory of symbols, not of things—presented, as a theory of combinations, the possible modes of combination and of mutual dependence. He held, in opposition to Bertrand Russell, among others, an "ordinal" theory of numbers. As ordinal numbers, the integers designate positions in an ordered sequence, and their meaning consists in their reciprocal relations. The concept of number is a direct result of the laws of thought themselves and presupposes only the ability of the human mind to relate one thing to another thing.[1] Mathematical concepts arise through genetic definition—that is, there must exist a definite mode of production, and the desired property must be shown to result therefrom by means of a strict deductive proof (real definition). This applies to all of mathematics.

Cassirer held that with their basic concepts the natural sciences can express the empirically given in relationships. Mass, force, atom, ether, absolute space, and absolute time become instruments of thought, with the aid of which appearances are ordered and reduced to an organized and measurable whole that can be comprehended.[2] Physics does not seek a representation of reality; rather, it considers the structure of all events from the point of view of measurability and strives to reduce the structure to a numerical order. The true objects of cognition are relationships; the concepts of things are only a means of establishing relationships.

The book *Determinismus und Indeterminismus in der modernen Physik* was designed primarily to provide logical clarification of the newly emerging problems of modern physics. Cassirer was one of those who cautioned against precipitately burdening the indeterminism of quantum theory with metaphysical speculations. He did not wish to see the concept of cause replaced by that of purpose. Just as every perceptual experience can be seen from various points of view (mathematical, aesthetic, mythical), so physics must use new explanatory and representational schemata in order to describe microphysical facts.

Cassirer made thorough studies of the problems of space and time in connection with Einstein's theory of relativity. He thought he could show that Einstein's theories did not conflict with those of Kant, even though with the theory of relativity "a step beyond Kant" had been taken.[3] In depriving space and time of the last remnants of physical objectivity, the standpoint of critical idealism obtains "the most definite application and accomplishment within the empirical science itself."[4] The spatiotemporal order is never given directly and sentiently; rather, it is the product of an intellectual construction. For Cassirer, Euclidean space is one possibility among many. We establish reality and experience by means of pure possibilities, and we achieve from various mathematical systems—for example, non-Euclidean geometries—the knowledge that we are dealing with pure possibilities, which cannot be derived from sense perception.[5]

Even though the problem of knowledge is Cassirer's chief concern, his investigations in the history of philosophy—of Descartes, Leibniz, Kant, and the Renaissance—as well as his works on the history of literature and culture are marked by a comprehensive knowledge of all the subjects treated and by a feeling for the essential and a profound insight into the relationships of cultural history.

After World War I, Cassirer began to free himself from the narrow conception of neo-Kantianism. For over a decade he labored on a work that he regarded as "Prolegomena to any future philosophy of culture"; in the *Philosophie der symbolischen Formen,* summarized in the later work *Was ist der Mensch?,* he developed the theory that human reason alone does not provide access to reality. The mind in its totality, with all its functions, feeling, willing, thinking, is responsible for the union of subject with object. Man grasps reality with the help of "symbolic forms," such as language, myth, art, science, and religion, which place themselves between him and the universe. These symbols must form the basic principles for a theory of man—that is, for an adequate philosophical anthropology.

All of Cassirer's writings display a basically humanistic attitude, which also expressed itself in his life and actions. His work also testifies to creative imagination, constant reflection on and reorganization of an immense amount of material, and an unfailing knowledge and tolerance.

Political circumstances in Germany prevented wide dissemination of Cassirer's philosophical work. Moreover, after World War II phenomenology, *Lebensphilosophie,* and existentialism largely dominated the field. Nevertheless, Cassirer enriched the history of philosophy and the classic problem of the theory of knowledge; in addition, he promoted basic research and provided valuable starting points for a philosophical anthropology and history of culture.

NOTES

1. *Das Erkenntnisproblem,* IV, 74.
2. *Substanzbegriff und Funktionsbegriff,* p. 220.
3. *Zur Kritik der Einsteinschen Relativitätstheorie,* p. 82.
4. *Ibid.,* p. 79.
5. *Die Philosophie der symbolischen Formen,* III, 489.

BIBLIOGRAPHY

I. ORIGINAL WORKS. Cassirer's writings are *Descartes' Kritik der mathematischen und naturwissenschaftlichen Erkenntnis* (Marburg, 1899), his inaugural diss., repr. as an intro. to *Leibniz' System in seinen wissenschaftlichen Grundlagen* (Marburg, 1902; 2nd ed., Darmstadt-Hildesheim, 1962); *Substanzbegriff und Funktionsbegriff* (Berlin, 1910); *Das Erkenntnisproblem in der Philosophie und Wissenschaft der neueren Zeit,* 3 vols., I (Berlin, 1906; 3rd ed., 1922); II (Berlin, 1907; 3rd ed., 1922); III (Berlin, 1920; 2nd ed., 1923)—the MS of Cassirer's proposed vol. IV appeared first, in English, as *The Problem of Knowledge: Philosophy, Science, and History Since Hegel* (New Haven-London, 1950), trans. by W. H. Woglom and C. W. Hendel, with forward by Hendel, and later, in German, as vol. IV of *Erkenntnisproblem,* with the title *Von Hegels Tod bis zur Gegenwart 1832–1932* (Stuttgart, 1957); *Zur Kritik der Einsteinschen Relativitätstheorie. Erkenntnistheoretische Betrachtungen* (Berlin, 1921; 2nd ed., 1925); *Die Philosophie der symbolischen Formen,* 3 vols. and index (Berlin, 1923–1931); *Kants Leben und Lehre* (Berlin, 1918; 4th ed., 1924), published as vol. XI of *Immanuel Kants Werke,* ed. by Cassirer and Hermann Cohen (Berlin, 1912); *Determinismus und Indeterminismus in der modernen Physik. Historische und systematische Studien zum Kausalproblem* (Göteborg, 1937); *Zur Logik der Kulturwissenschaften* (Göteborg, 1942); *Was ist der Mensch? Versuch einer Philosophie der menschlichen Kultur* (Stuttgart, 1960), originally publ. as *An Essay on Man* (New Haven, 1944).

II. SECONDARY LITERATURE. A bibliography to 1949 is given in *Philosophy of Ernst Cassirer,* P. A. Schilpp, ed. (Evanston, Ill., 1949); the German ed. (Stuttgart–Berlin–Cologne–Mainz, 1966) contains a bibliography to 1957. See also F. Ueberweg, *Grundriss der Geschichte der Philosophie,* IV (13th ed., Tübingen, 1951), 443–444.

EDITH SELOW

CASTALDI, LUIGI (*b.* Pistoia, Italy, 14 February 1890; *d.* Florence, Italy, 12 June 1945), *anatomy.*

Castaldi was the son of Vittorio Castaldi, an army officer, and of Vincenza Giovacchini-Rosati. He attended school in Pistoia and then studied medicine in Florence; after graduating in 1914, he became a physician at the Santa Maria Nuova Hospital in Florence before serving as an army medical officer in the war against Austria. In 1919 he was appointed assistant at the Institute of Human Anatomy of Florence, where Giulio Chiarugi was director and where he had worked while a student. In 1922 he began to teach human anatomy and in 1923 was named professor of human anatomy at the University of Perugia. He was appointed professor of anatomy at the University of Cagliari in 1926 and was kept there, against his wishes, until 1943 because he was at odds with the Fascist government. After Mussolini's fall he obtained a transfer to the University of Genoa, but the political and military situation obliged him to remain in Florence, where he died after a long illness.

Castaldi began his scientific activity as a histologist with a systematic study of the connective tissue of the liver, "Il connettivo nel fegato dei Vertebrati" (1920), a work based on the microscopical examination of the liver of forty-one species (including man) and developed through the critical revision of an extensive bibliography (300 works). This book was used as a text for some years because of its illustrations of the capillaries of the liver.

Even in neurology, with his work on the mesencephalon (1922–1928), Castaldi paid homage to more traditional descriptive morphology. Influenced by a concern for so-called postencephalitic parkinsonism, then very common and characterized by interference with voluntary movement, Castaldi was induced to reexamine the anatomical knowledge of the mesencephalon. For this purpose he decided to compare the descriptions by different authors with its development in a single species (*Cavia cobaya*). Castaldi's was the most precise and documented work on the structure of the mesencephalon then known. He clarified the structure of the corpora and brachia quadrigemina, the nuclei of the oculomotor nerve, the lateral and medial lemnisci, the red nucleus, the substantia nigra, and the cerebral peduncle. In addition, he confirmed the validity of Beccari's observations on the so-called tegmental centers: they also are very well developed in the mesencephalon. Castaldi illustrated their importance for extrapyramidal motility in his exhaustive 1937 report to the Italian Society of Anatomy on the extrapyramidal pathways of the central nervous system.

Castaldi also carried out remarkable experimental work on the influence of the endocrine gland on morphogenesis. Working from the demonstrated role of iodine deficiency in cretinism and goiter, from 1920 to 1928 Castaldi studied the histology of human thyroid glands from areas where goiter occurred and from those where it did not. Thus he confirmed the strict relationship between the iodine content of the thyroid and its activity and established its influence on the development and height of man. Castaldi also studied the physiology of the adrenal cortex. From 1924 to 1926 he analyzed the effects on very young animals (*Cavia cobaya*) of dried, salted adrenal cortex taken from an ox and administered with the food. He thereby confirmed that the adrenal cortex can stimulate growth of the muscular and skeletal systems.

Using the biostatistical methods of Adolphe Quetelet and Francis Galton, Castaldi in 1923 began his work on biometrical evaluation in man with the study of the weights of thymus glands in relation to age, sex, body weight, and height. In 1924 he repeated this research on ovaries and in 1927 (with D. Vannucci) extended the same study to the principal organs (heart, lungs, liver, spleen, kidneys, encephalon, thyroid, thymus, adrenals, hypophysis, testes, ovaries) of 300 corpses of both sexes and all ages. Their average values were related, in twenty-nine tables and twenty-five graphs, to average values of the external anthropometrical measurements.

Convinced that the traditional method of teaching anatomy by dissecting cadavers was insufficient, Castaldi also taught the morphology of the living man, seeking to show the morphology of the different constitutional types of the two sexes at various ages. Thus he could accurately trace the human growth curve and connect variations in early postnatal growth patterns with the different constitutional types of human adults established by Achille De Giovanni.

BIBLIOGRAPHY

Castaldi's early histological paper is "Il connettivo nel fegato dei Vertebrati," in *Archivio Italiano di Anatomia e di Embriologia,* **17** (1920), 373–506 L, with 5 colored plates.

Among his books are *Influenza della ghiandola tiroide sull'accrescimento corporeo* (Milan, 1923); *Accrescimento corporeo e costituzioni dell'uomo* (Florence, 1928); *Compendio pratico di anatomia umana,* 3 vols. (Naples, 1931–1941); *Trent'anni di vita della Società Italiana di Storia delle Scienze Mediche e Naturali* (Siena, 1938); *L'anatomia in Italia* (Milan, 1939), a special number of *Acta Medica Italica; Atlante cromomicrofotografico di splancnologia dell' uomo e di altri mammiferi* (Florence, 1939).

Papers on the thyroid are in *Lo Sperimentale,* **74** (1920), **76** (1922), **78** (1924); *Monitore Zoologico Italiano,* **32** (1921); *Archivio Italiano di Anatomia e di Embriologia,* supp. **18** (1922); Scritti Biologici, no. 3 (Siena, 1928); and *Rassegna Internazionale di Clinica e Terapia,* **20** (1939).

The adrenal cortex is the subject of *Archivio di Fisiologia,* **20** (1922); *Rendiconti Accademia dei Lincei,* cl. sc. fis. mat. nat., 5th ser., **33** (1924); *Archivio Italiano di Anatomia e di Embriologia,* **22** (1925); *Revista Sudamericana de Endocrinologia, Immunologia y Quimioterapia,* **9** (1926); *Giornale del medico pratico,* **11** (1928); *Folia Clinica e Biológica* (São Paulo), **2** (1930); and *Bollettino della Accademia Medica Pistoiese,* **5** (1932).

On the hypophysis, see *Lo Sperimentale,* **77** (1923).

The thymus is discussed in *Monitore Zoologico Italiano,* **34** (1923).

On the ovary, see *Lo Sperimentale,* **78** (1924).

The mesencephalon is treated in *Lo Sperimentale,* **76** (1922); *Bollettino di Oculistica,* **1** (1922); *Archivio Italiano di Anatomia e di Embriologia,* **20** (1923), **21** (1924), **23** (1926), **25** (1928); *Rivista di Patologia Nervosa e Mentale,* **29** (1924); and Scritti Biologici, no. 5 (Siena, 1930). His most important work in this area is "Il sistema nervoso motore dei centri e delle vie extrapiramidali," in *Monitore Zoologico Italiano,* supp. **48** (1938), 11–58.

For his papers on biometrics and constitutional anatomy, see *Monitore Zoologico Italiano,* **32** (1921), **34** (1923); *Rivista Critica di Clinica Medica,* **25** (1924); *Lo Sperimentale,* **78** (1924); Scritti Biologici, no. 2 (Siena, 1927).

In the history of medicine, see "Una centuria di rivendicazioni di priorità ad italiani in contributi scientifici nella medicina e chirurgia," in *La Riforma Medica* (1929). See also "Filippo Pacini," in *Rivista di Storia delle Scienze Mediche e Naturali,* **15** (1923); *Bernardino Genga uno dei rivendicatori di Colombo e di Cesalpino per la scoperta della circolazione del sangue* (Florence, 1941); *I microscopi construiti da Galileo Galilei 1610–1630* (Florence, 1942); and *Francesco Boi 1767–1860 primo cattedratico di anatomia* *umana in Cagliari e le cere anatomiche fiorentine di Clemente Susini* (Florence, 1947).

Other writings include *Concezione moderna dell'anatomia umana* (Perugia, 1923); *L'eredità* (Florence, 1925); *Wilhelm Roux 1850–1924* (Florence, 1925); *La figura umana in Leonardo da Vinci* (Cagliari, 1926), see also Scritti Biologici, no. 2 (Siena, 1927), and *Rivista di Storia delle Scienze Mediche e Naturali,* **18** (1927); *L'Uomo Sardo* (Cagliari, 1932); *Il problema estetico della maternità* (Pistoia, 1935), see also Scritti Biologici, no. 10 (Siena, 1935); "Nel bicentenario della nascita di Luigi Galvani," in *La Riforma Medica* (1937); *In memoria di Dino Vannucci 1895–1937* (Cagliari, 1937), also in Scritti Biologici no. 13 (Siena, 1938); and *L'Italia culla dell'anatomia classica* (Cagliari, 1942), also in *Rassegna Internazionale di Clinica e Terapia,* **24** (1943).

PIETRO FRANCESCHINI

CASTEL, LOUIS-BERTRAND (*b.* Montpellier, France, 15 November 1688; *d.* Paris, France, 11 January 1757), *physics, mathematics.*

Castel was probably the most vociferous opponent of Newtonian science during the second quarter of the eighteenth century in France. He failed to block the gradual acceptance of Newton's ideas because the Cartesian rationalism that he tried to establish found diminishing favor with French scientists, more and more influenced by the merits of the experimental approach.

The second son of Guillaume Castel, a physician, Louis-Bertrand received his early education at the Jesuit school of Toulouse and entered the Jesuit order at the age of fifteen. His obituary in the Jesuit periodical *Journal de Trévoux* states that Castel's early writings came to the attention of Fontenelle, the eminent Cartesian philosopher and scientist, who is credited with influencing Castel to leave Toulouse for the more intellectual climate of Paris in 1720. His being immediately chosen an associate editor of the *Journal de Trévoux* clearly indicates that Castel had already shown promise as a scholar. While working on the monthly, Castel was associated with the faculty of the Jesuit school in the rue Saint-Jacques, the present Lycée Louis-le-Grand, where he taught physics, mathematics, specialized courses in infinitesimal calculus, and mechanics. Once installed at Louis-le-Grand, he never left Paris except for one trip to southern France toward the end of his life. The political philosopher Montesquieu honored him with his friendship, did not hesitate to submit his manuscripts to him before publication, and even chose him to be his son's tutor for a time, a post that made Castel inordinately proud.

Upon his arrival in Paris, Castel's first article was published in the *Mercure de France.* The "Lettre à M. de ***" stressed that truth was one, and that therefore astronomy and religion could never come into conflict because both are true. In 1724 his *Traité de la pesanteur* attracted a great deal of attention, particularly because it was hostile to Newton. In 1730, through his friendship with the English oculist J. T. Woolhouse, Castel was elected to the Royal Society of London. He entered the Bordeaux Academy in 1746, and in 1748 he was elected to the academies of Rouen and Lyons.

Although Castel published a creditable anti-Newtonian scientific theory that succeeded in delaying the acceptance of Newton's ideas in France, he is remembered as the spokesman of French scientists who saw in Newton a threat to the prestige of their national hero, Descartes, and a threat to their religious faith. While Descartes's metaphysical system had generally been abandoned by the thinkers of the Enlightenment, Newton's growing prestige brought about a gradual rally to the physics and astronomy of Descartes. Even as late as 1738 most French scholars still supported Descartes; with the exception of Maupertuis and Clairaut, the Academy of Sciences was composed entirely of Cartesians. Even though Castel felt competent to refute Cartesian science, he never abandoned Descartes's a priori, rationalistic approach to science—hence his impatience with a science based on experimentation rather than on a logical process. Pascal's fundamental objection to Cartesian physics, almost a century before Castel's system, was that Descartes had reasoned a priori in physics instead of observing and experimenting. It was the latter approach to science that so many physicists and astronomers of the eighteenth century, with Castel at their head, found repugnant. As a consequence of this attitude, Cartesian physicists had rendered the French scientists indolent; they preferred an attractively reasoned system, with daring ideas based on the logical process, to seeking scientific truth painfully and laboriously. The net result of the Cartesian approach was the relative stagnation of research in France.

This leads one to appreciate Castel's reaction to Newton: he complained about the numerous experiments that formed the basis of Newton's theories because they were not within the reach of the common man, and he reproached Newton with wanting to reduce man to "using only his eyes." Physics, for Castel, must be based on reason instead of observations. Hence his contempt for the "complicated laboratories" of the disciples of Newton. Castel's second brief against Newton was that his system of the world was suspect to the religious man because it smacked of materialism. Castel's accusation was clearly expressed in *Journal de Trévoux* (July 1721, pp. 1233, 1236): influenced by Democritus and Epicurus, Newton sought to give a philosophical basis to materialism by substituting the void for divine intelligence. On the other hand, Voltaire was genuinely persuaded that Newton's discoveries of nature's secrets conclusively proved the existence of God.

There is little point in presenting an outline of the system Castel proposed to replace Newton's. It was an attempt to harmonize philosophy, scientific curiosity, and religious dogma by means of rationalism. Newton gradually secured a foothold in France, and Voltaire was not the last of the propagandists on his behalf.

Castel's ocular harpsichord helped to spread his fame much more than his scientific reputation did. The best sources available for an explanation of his invention are two articles in the *Journal de Trévoux* (1735): "Nouvelles expériences d'optique et d'acoustique" and "L'optique des couleurs fondée sur les simples observations." It was a scheme for making colors and musical tones correspond. By 1742 the fame of Castel and of his invention had reached as far as St. Petersburg and had been brought to the attention of the empress. The instrument was completed in July 1754, and on 21 December of the same year Castel gave a private demonstration of it before fifty guests. The spectators were enthusiastic and applauded several times (*Mercure de France* [July 1755], p. 145). The idea of the color organ did not die with him, since several varieties of it have appeared in Europe and in the United States at various times.

BIBLIOGRAPHY

Castel's writings include *Lettre philosophique pour rassurer l'univers contre les bruits populaires d'un dérangement dans le cours du soleil* (Paris, 1736); *Traité de physique sur la pesanteur universelle des corps,* 2 vols. (Paris, 1724); his 1735 art. in *Journal de Trévoux* cited above repr. as *L'optique des couleurs, fondée sur les simples observations, et tournée surtout à la pratique de la peinture* (Paris, 1740); *Dissertation philosophique et littéraire où, par les vrais principes de la physique et de la géométrie, on recherche si les règles des arts sont fixes ou arbitraires* (Amsterdam, 1741); and *Le vrai système de physique générale de M. Isaac Newton exposé et analysé en parallèle avec celui de Descartes, à la portée des physiciens* (Paris, 1743).

Castel and his work are discussed in Jean Ehrard, *L'idée de nature en France dans la première moitié du XVIIIe siècle* (Paris, 1963), pp. 117–121; 155–156.

A. R. DESAUTELS

CASTELLI, BENEDETTO (*b.* Brescia, Italy, 1578; *d.* Rome, Italy, 19[?] April 1643), *hydraulics, astronomy, optics.*

Antonio, eldest son of Annibale and Alda Castelli,

took the name Benedetto upon entering the Benedictine order at Brescia in September 1595. At some time before 1604 he moved to the monastery of Santa Giustina, at Padua, where he studied under Galileo. In 1607 he was at Cava, and in 1610 had returned to Brescia. Shortly after receiving from Galileo a copy of the *Sidereus Nuncius,* he applied for transfer in order to work with Galileo. Late in 1610 he wrote to Galileo, expressing Copernican convictions and suggesting telescopic observation of Venus to detect its phases, observations which were then being made by Galileo.

In April 1611 Castelli was moved to Florence. In 1612 he suggested to Galileo the method of recording sunspot observations by drawing them on a paper screen parallel to the eyepiece of a telescope. Galileo's controversy with philosophers over bodies in water and the phenomena of floating resulted in the publication of several polemics in 1612; it was to Castelli that Galileo entrusted the publication of replies (largely written by Galileo himself) in 1615.

Late in 1613 Castelli was present at the table of the grand duke of Tuscany at Pisa when Galileo's Copernican views were attacked; Castelli defended them and notified Galileo, who replied at length concerning his views of the relation between science and religion. This letter to Castelli was later important in Galileo's dealings with the Roman Inquisition, and in an expanded form it became the famed *Letter to the Grand Duchess Christina,* circulated in 1615 and eventually published at Strasbourg in 1636.

At Galileo's recommendation Castelli became professor of mathematics at the University of Pisa in 1613, a chair to which he was confirmed for life in 1624. In 1626, however, he resigned the post. At Pisa, for want of a Benedictine monastery, he lived in one belonging to the Jesuate (not Jesuit) order. There he met the young Milanese student Bonaventura Cavalieri, whom he introduced both to the study of mathematics and to the personal acquaintance of Galileo. Cavalieri's "geometry by indivisibles" was an important step toward the infinitesimal calculus. Later, at Rome, Castelli was the teacher of Evangelista Torricelli and of Giovanni Alfonso Borelli; he was also the instructor of Galileo's son, Vincenzio.

While Castelli was still at Pisa, he became interested in the study of water in motion. Early in 1626 he sent to Florence, for Galileo's comment, two treatises on the motion of rivers, in one of which he corrected the classic work of Frontinus, *De aqueductibus urbis Romae.* About this time he was called to Rome by Pope Urban VIII as papal consultant on hydraulics, tutor to Taddeo Barberini, and professor of mathematics at the University of Rome. In 1628 he published the book *Della misura dell'acque correnti,* considered to be the beginning of modern hydraulics. Its fundamental propositions related the areas of cross sections of a river to the volumes of water passing in a given time. He also discussed the relation of velocity and head in flow through an orifice. A posthumous edition included the proposition that where a stream was dammed, the velocity of flow over the top was in direct proportion to the depth of water so flowing. Castelli's defective "proof" of this was deleted in most copies as a result of his dissatisfaction with it, expressed in some of his letters. Castelli's pioneer work in hydraulics was carried on much further and with greater accuracy by his pupil Torricelli.

Some writers have declared that Castelli owed his knowledge of hydraulics to the manuscripts of Leonardo da Vinci and in particular to the compilation of Leonardo's writings on that subject by Luigi Maria Arconati, now in the Barberini archives. In fact, however, that compilation was dated 1643, and the manuscripts of Leonardo did not pass to the Biblioteca Ambrosiana, where Arconati consulted them, until 1637. Castelli's correspondence shows quite clearly that his studies of hydraulics were chiefly of an experimental character. It is of interest in this connection that he obtained from Galileo the length of an approximate seconds pendulum for use in his experiments and devised a cylindrical rain gauge.

In 1634, as a result of discussions with friends about the mutual illumination of the earth and the moon, Castelli arrived at the conclusion that the illumination given by any two lights of different intensity and surface area is directly proportional to those two factors and inversely proportional to the squares of the distances of the lights from the illuminated body. (This photometric law, in a less detailed form, had been given by Johannes Kepler in 1604 but seems not to have been noticed elsewhere until 1638, when it was published by Ismael Boulliau at Paris.) Proceeding with his lunar observations, Castelli wrote to Galileo in 1637 that he was convinced of the existence of large land bodies in the South Seas, a conclusion approved by Galileo in his reply and confirmed later by the vast extent of the Australian continent.

Castelli's optical investigations were continued in a treatise sent to Giovanni Ciampoli in 1639 and published posthumously in 1669. Included are many observations and conclusions with respect to the persistence of optical images, by which Castelli explained the perception of motion, the illusion of forked tongues in serpents, and other phenomena. In the same treatise he recommended the use of diaphragms in telescopes to impede transverse rays, anticipating Hevelius. His discussions of the camera obscura, the inversion of images on the retina, and

of cataract (from which Galileo had recently lost his sight), although less novel, are not without interest.

More celebrated is Castelli's discussion of heat in a series of letters to Galileo (1637–1638) and particularly his experiments with the absorption of radiant and transmitted heat by black and white objects. Two of these letters, in which the pursuit of experimental science is even more clearly described than in Galileo's work on bodies in water, were published in 1669. Castelli was also interested in algebraic researches, particularly on the use of negative quantities in the solution of problems in the theory of numbers.

Castelli's importance to science lay not only in his extension and dissemination of Galileo's work and methods, but also in his long and faithful service to Galileo during the two periods of crisis with the Inquisition. It was to Castelli that Galileo addressed his first discussion of religion and science, and it was on Castelli's advice in 1630 that Galileo transferred the printing of the 1632 *Dialogue* from Rome to Florence, a maneuver without which that important work might never have issued from the press.

BIBLIOGRAPHY

I. ORIGINAL WORKS. With Galileo, Castelli wrote *Risposta alle opposizioni del S. Lodovico delle Colombe e del S. Vincenzio di Grazia . . .* (Florence, 1615). *Della misura dell'acque correnti* (Rome, 1628) includes, as the second part, *Demonstrazioni geometriche della misura dell'acque correnti;* to the 2nd ed. (Rome, 1639) are added two appendixes and a letter to Galileo relating to hydrology; to the 3rd ed. (Bologna, 1660), several hydrological treatises. An English trans. of Castelli's works on hydraulics and hydrology, based on the 3rd ed., was published by Thomas Salusbury in *Mathematical Collections and Translations, the First Tome* (London, 1661; repr. 1967). A French trans. of the work on hydraulics is contained in *Traité de la mesure des eaux courantes de B. Castelli par Saporta . . .* (Barcouda, 1664). The Italian text has been frequently reprinted. *Alunci opuscoli filosofici . . .* (Bologna, 1669) contains *Discorso sopra la vista, Del modo di conservare i grani,* and *Due lettere . . . sopra'l differente riscaldimento . . . .* Castelli's short commentary on Galileo's *La bilancetta* was first published in *Opere di Galileo Galilei* (Florence, 1718), III, 309–311. Castelli's lectures on geometrical pavement and on questions of algebra were published in an appendix to A. Favaro, "Amici e corrispondenti di Galileo Galilei, XXI.,—Benedetto Castelli," in *Atti del Reale Istituto veneto di scienze, lettere ed arti,* **67,** pt. 2 (1907–1908), 1–130. Most of Castelli's correspondence was published in *Opere di Galileo Galilei,* Edizione Nazionale, A. Favaro, ed. (Florence, 1890–1910; repr. 1929–1939, 1964–1966), which also contains his commentary on Giorgio Coresio's polemic against Galileo over floating bodies (V, 245–285).

II. SECONDARY LITERATURE. The principal source is

A. Favaro, "Amici . . .," cited above; see also Favaro, "Intorno al trattato di Leonardo da Vinci sul moto e misura dell'acqua," *Reale Accademia dei Lincei. Rendiconti*, **27** (1918), 17 Nov., and "Galileo Galilei, Benedetto Castelli e la scoperta delle fasi di Venere," in *Archeion*, **1** (1920), 283–296. See also G. L. Masetti Zannini, *La vita di Benedetto Castelli* (Brescia, 1961), which includes an extensive bibliography of works by and about Castelli.

STILLMAN DRAKE

CASTELNUOVO, GUIDO (*b.* Venice, Italy, 14 August 1865; *d.* Rome, Italy, 27 April 1952), *mathematics.*

Castelnuovo studied mathematics at the University of Padua under Veronese and was graduated in 1886. He then spent a year in Rome on a postgraduate scholarship and subsequently became assistant to E. D'Ovidio at the University of Turin, where he had important exchanges of ideas with Segre. From 1891 until 1935 he taught projective and analytical geometry at the University of Rome. After the death of Cremona, Castelnuovo also taught advanced geometry and, later, courses in the calculus of probability. When Jewish students were barred from state universities, Castelnuovo organized courses for them.

After the liberation Castelnuovo was special commissioner of the Consiglio Nazionale delle Ricerche and then president of the Accademia Nazionale dei Lincei until his death. He was also a member of many other academies. On 5 December 1949 he was named senator of the Italian republic for life.

Castelnuovo's scientific activity was principally in algebraic geometry and was important in the Italian school of geometry, which included Cremona, Segre, Enriques, and Severi. His mathematical results particularly concerned algebraic surfaces and the theories constituting their background.

In connection with these results is the theorem of Kronecker-Castelnuovo: "If the sections of an irreducible algebraic surface with a doubly infinite system of planes turn out to be reducible curves, then the above surface is either ruled or the Roman surface of Steiner" (*Memorie scelte*, pp. 223–227). The origin of this theorem was explained by Castelnuovo as follows: Kronecker, during a meeting of the Accademia dei Lincei in 1886, had communicated verbally one of his theorems on irreducible algebraic surfaces having infinite plane sections that are split into two curves. But the written note was not sent by Kronecker to the Accademia dei Lincei, nor was it published elsewhere. Further, Castelnuovo held that perhaps Kronecker had not finished the final draft. On the basis of information furnished to him by Cremona and Cerruti, Castelnuovo reconstructed the demonstration of the theorem. He also developed the theo-

rem that every unruled irreducible algebraic surface whose plane sections are elliptical curves (or are of genus 1) is rational (*Memorie scelte,* pp. 229–232).

Besides writing on algebraic geometry, the calculus of probability, and the theory of relativity, Castelnuovo also delved into the history of mathematics, producing *Le origini del calcolo infinitesimale nell'era moderna* (1938), which contains a quick and effective summary of the evolution of infinitesimal methods from the Renaissance to Newton and Leibniz.

BIBLIOGRAPHY

I. ORIGINAL WORKS. Castelnuovo's writings include *Lezioni di geometria analitica* (Milan-Rome-Naples, 1903; 7th ed., 1931); *Calcolo della probabilità* (Rome, 1919; 3rd ed., Bologna, 1948); *Spazio e tempo secondo le vedute di A. Einstein* (Bologna, 1923); *Memorie scelte* (Bologna, 1937), with a list of publications on pp. 581–584, to which should be added those in the obituary by A. Terracini (see below); *La probabilité dans les différentes branches de la science* (Paris, 1937); and *Le origini del calcolo infinitesimale nell'era moderna* (Bologna, 1938; 2nd ed., Milan, 1962). A picture of Castelnuovo's results in the geometry of surfaces may be obtained from the two articles he wrote in collaboration with Federigo Enriques: "Grundeigenschaften der algebraischen Flächen" and "Die algebraischen Flächen vom Gesichtspunkte der birationalen Transformationen aus," both in *Encycklopädie der mathematischen Wissenschaften* (Leipzig, 1903–1915), III, pt. 2.

II. SECONDARY LITERATURE. Obituaries include F. Conforto, in *Rendiconti di matematica e delle sue applicazioni,* 5th ser., **11** (1952), i–iii; and A. Terracini, "Guido Castelnuovo," in *Atti della Accademia delle scienze* (Turin), **86** (1951–1952), 366–377. Castelnuovo's numerous results in algebraic geometry are discussed in F. Conforto, *Le superficie razionali* (Bologna, 1939), index; and F. Enriques and O. Chisini, *Lezioni sulla teoria geometrica delle equazioni e delle funzioni algebriche,* 4 vols. (Bologna, 1915–1934), see indexes of the various volumes under "Castelnuovo."

ETTORE CARRUCCIO

CASTIGLIANO, (CARLO) ALBERTO (*b.* Asti, Italy, 9 November 1847; *d.* Milan, Italy, 25 October 1884), *structural engineering.*

During the second half of the nineteenth century there flourished in Italy a large group of structural engineers and elasticians in great measure responsible for establishing and popularizing the various methods of structural analysis based on the concepts of energy and work. This group included men of a variety of callings, and the following list of the principal names is remarkable as much for the versatility of the individuals it contains as for the evidence it presents of the intellectual and scientific vigor of the times: Alessandro Dorna (1825–1866, engineer and astronomer), Luigi Menabrea (1809–1896, general and statesman), Emilio Francesco Sabbia (1838–1914, general), Angelo Genocchi (1817–1889, mathematician), Enrico Betti (1823–1892, mathematician and engineer), Vincenzo Cerruti (1850–1909, engineer and mathematician), Francesco Crotti (1839–1896, engineer), Luigi Donati (1846–1932, physicist), and, of course, Castigliano. Beyond this point the list would not be useful in illustrating the background of Castigliano's world, partly because of his untimely death and partly because of the larger variety of problems which engaged the attention of the later Italian elasticians.

Castigliano left his native Piedmont at the age of nineteen to teach mechanics and machine design at the Technical Institute of Terni in Umbria for four years, returning only in 1870 to the Polytechnic of Turin. As a student there he began his work on the theory of structures, leading to his first publication, that of his celebrated dissertation, in 1873. He then found employment with the Northern Italian Railroads and soon became chief of the office responsible for artwork, maintenance, and service. He maintained that position until his death, all the while continuing to study and write.

The principal contributions of Castigliano are the two theorems known by his name. The first of these, contained in his dissertation, *Intorno ai sistemi elastici,* states that the partial derivative of the strain energy (*lavoro molecolare*), considered as a function of the applied forces (or moments) acting on a linearly elastic structure, with respect to one of these forces (or moments), is equal to the displacement (or rotation) in the direction of the force (or moment) of its point of application. He included the case of external reactions, not prescribed, noting that when the support corresponding to these reactions is unyielding, the partial derivative is zero and that his theorem then reduces to Menabrea's "principle of least work." In 1875 he published his second theorem, in which the strain energy is considered a function of the unprescribed displacements of discrete boundary points; its derivative with respect to one of these displacements gives the corresponding force acting there.

Earlier milestones in the history of energy principles of this type were the proof by Clapeyron, in 1827, of the principle of the conservation of work, equating the work performed by the applied external forces with the internal work performed by the stresses; Menabrea's development of his principle of least work; and Cotterill's independent proof (unknown to Castigliano) of Castigliano's theorems. It is clear that Menabrea's principle may be considered to be included in Castigliano's theorems; furthermore, Me-

nabrea's proofs were not satisfactory and were in fact repeatedly modified by him as a result of considerable criticism. A new demonstration of Menabrea's principle was given by him in 1875 on the basis of some of the newly published results of Castigliano, who, however, was referred to only in a footnote. Castigliano strongly objected to this lack of sufficient recognition in a letter full of youthful indignation to the president of the Accademia dei Lincei. Menabrea replied in the reasoned and somewhat condescending tones befitting an elder statesman, pointing out the priority of his work. The mathematician and engineer Luigi Cremona, acting as chairman of a meeting of the Academy, gave a solomonic judgment on the controversy, stating that

> he believes that Mr. Castigliano's complaint is not sufficiently well founded: the theorem in question precedes the work of both authors, and the proofs do not seem free of every objection. It is thus his opinion that there is no matter for dispute, and concludes: Mr. Castigliano can have the honor of having done a good piece of work: no one will be able to take away from Member Menabrea the merit of having made popular and of common use a general principle, which is certainly destined to receive ever more extensive application.

The entire correspondence and reply are found in *Atti della Reale Accademia Nazionale dei Lincei* (2nd ser., **2**[1874/75], 59). The controversy, while not subsiding completely, then lost most of its virulence. The principal continuations of Castigliano's work were Crotti's extension to nonlinear elastic systems (principle of complementary energy) and Donati's work on the mathematical and conceptual basis of energy methods.

Among Castigliano's minor contributions, one might mention an engineer's handbook; works on the theory of leaf and torsion springs (published in a book in Vienna, 1884), on masonry arches, and on water hammers; and the invention of an extensometer.

Castigliano's principal work, while not free of conceptual shortcomings, represented a definite advance over that of his predecessors. To assess the importance of his contribution, however, it is important to note that, although there is some validity in Cremona's attribution of the popularization of energy methods to Menabrea, it is precisely in this respect that Castigliano excels. He solved an amazing number of important structural problems by his methods, pointing out by comparisons with previously known *ad hoc* solutions both their superiority and their correctness and establishing once and for all their convenience and versatility. As he states in the preface to his *Théorie de l'équilibre des systèmes élastiques, et ses applications,* this was indeed one of his explicit goals, and the success with which he achieved it is remarkable both because of his short career (he died at thirty-seven) and his lack of formal academic or other strong "establishment" ties.

BIBLIOGRAPHY

I. ORIGINAL WORKS. Castigliano's principal works are his diss., *Intorno ai sistemi elastici* (Turin, 1873), and the book *Théorie de l'équilibre des systèmes élastiques, et ses applications* (Turin, 1879). The diss. and the main part of the book were repr. on the fiftieth anniversary of his death as *Alberto Castigliano, Selecta,* G. Colonnetti, ed. The book itself was published in English trans. by E. S. Andrews as *Elastic Stresses in Structures* (London 1919). Mention should also be made of three papers by Castigliano, all in the *Atti della Reale Accademia della scienze* (Turin), 2nd ser.: "Intorno all'equilibrio dei sistemi elastici," in **10** (1875), 10; "Nuova teoria intorno all'equilibrio dei sistemi elastici," in **11** (1875), 127; and "Intorno ad una proprietà dei sistemi elastici," in **17** (1882), 705.

II. SECONDARY LITERATURE. Derivations and applications of Castigliano's theorems may be found in any standard text on strength of materials or theory of structures. It should be noted in this connection that the references to the first and second theorems given in the text correspond to the most usual modern practice but are opposite to those employed by Castigliano. Futhermore, Menabrea's principle of least work is occasionally referred to as Castigliano's second theorem. A comprehensive historical review of the *Historical Development of Energetical Principles in Elastomechanics* was given by G. Ae. Orawas and L. McLean in *Applied Mechanics Reviews,* **19**, no. 8 (Aug. 1966), 647–658, and no. 11 (Nov. 1966), 919–933. Clapeyron's proof of the principle of the conservation of work is in his "Mémoire sur le travail des forces élastiques dans un corps solide élastique déformé par l'action des forces extérieures," in *Comptes rendus de l'Académie des sciences,* **46** (1827), 208. A eulogy of Castigliano was given by F. Crotti, "Commemorazione di Alberto Castigliano," in *Politecnico,* **32**, nos. 11/12 (Nov./Dec. 1884), 597.

BRUNO A. BOLEY

CASTILLON, JOHANN (Giovanni Francesco Melchiore Salvemini) (*b.* Castiglione, Tuscany, Italy, 15 January 1704; *d.* Berlin, Germany, 11 October 1791), *mathematics.*

The son of Giuseppe Salvemini and Maria Maddalena Lucia Braccesi, Castillon studied mathematics and law at the University of Pisa, where he received a doctorate in jurisprudence in 1729. Shortly afterward he went to Switzerland, where for some unknown reason he changed his name to Johann Castillon, after his birthplace. In 1737 he became the director of a humanistic school in Vevey, and in 1745 he took a teaching position in Lausanne. In that year he married Elisabeth du Frèsne, who died in 1757. He married Madeleine Ravène in 1759. From 1749 to 1751 Castillon taught in both Lausanne and Bern. In the summer of 1751 he received offers of positions in St. Petersburg and Utrecht; in December 1751, after much thought, he accepted the invitation of the prince of Orange to lecture on mathematics and astronomy at the University of Utrecht, where he acquired a doctorate in philosophy in 1754 and rose to professor of philosophy in 1755 and rector in 1758. In 1764 he traveled to Berlin to accept a position in the Mathematics Section of the Academy of Sciences there. In the following year he became the royal astronomer at the Berlin Observatory.

During his lifetime Castillon was known as an able geometer and a general philosopher. His work in mathematics, however, did not go far beyond elementary considerations. His first two mathematical papers dealt with the cardioid curve, which he named. He also studied conic sections, cubic equations, and artillery problems. After publishing the letters of Leibniz and Johann I Bernoulli in 1745, he edited Euler's *Introductio in analysin infinitorum* in 1748. In 1761 he published his useful commentary on Newton's *Arithmetica universalis.* Throughout his mathematical work there is a preference for synthetic, as opposed to analytic, geometry, which is perhaps a reflection of his preoccupation with Newton's mathematics. In addition to this mathematical research, Castillon delved into the study of philosophy. In general he opposed Rousseau and his supporters and leaned toward the thinkers of the English Enlightenment. He translated Locke's *Elements of Natural Philosophy* into French.

Castillon became a member of the Royal Society of London and the Göttingen Academy in 1753 and a foreign member of the Berlin Academy of Sciences in 1755; he was elected to full membership in the Berlin Academy in 1764, upon the personal recommendation of Frederick the Great. In 1787 he succeeded Lagrange as director of the Mathematics Section of the Berlin Academy, a post he held until his death.

BIBLIOGRAPHY

Castillon wrote *Discours sur l'origine de l'inégalité parmi les hommes . . .* (Amsterdam, 1756); and *Observations sur le livre intitulé Système de la nature* (Berlin, 1771). He edited the following: *Isaaci Newtoni, equitis aurati, opuscula mathematica, philosophica & philologica,* 3 vols. (Lausanne, 1744); *Virorum celeberr. Got. Gul. Leibnitii & Johan. Bernoullii commercium philosophicum & mathematicum*

(Lausanne, 1745); and Euler's *Introductio in analysin infinitorum* (Lausanne, 1748). He translated Locke's *Elements of Natural Philosophy* as *Abrégé de physique* (Amsterdam, 1761); George Campbell's *A Dissertation on Miracles: Containing an Examination of the Principles Advanced by David Hume, Esq; in an Essay on Miracles* as *Dissertation sur les miracles, contenant l'examen des principes posés par M. David Hume dans son Essai sur les miracles* (Utrecht, 1765); and Francesco Algarotti's *Opere varie del Conte Francesco Algarotti* as *Mémoires concernant la vie & les écrits du comte François Algarotti* (Berlin, 1772).

Castillon's articles include "De curva cardiode," in *Philosophical Transactions of the Royal Society*, **41**, no. 461 (1741), 778–781; "De polynomia," *ibid.*, **42**, no. 464 (1742), 91–98; "Deux descriptions de cette espèce d'hommes, qu'on appelle negres-blans," in *Histoire de l'Académie royale des sciences et des belles lettres de Berlin* (*avec mémoires*), **18** (1762), 99–105; "Descartes et Locke conciliés," *ibid.* (1770), 277–282; "Mémoire sur les équations résolues par M. de Moivre, avec quelques réflexions sur ces équations et sur les cas irréductibles," *ibid.* (1771), 254–272; "Sur une nouvelle propriété des sections coniques," *ibid.* (1776), 284–311; "Sur un globe mouvant qui représente les mouvements de la Terre," *ibid.* (1779), 301–306; "Sur la division des instruments de géométrie et d'astronomie," *ibid.* (1780), 310–348; "Rapport sur une lettre italienne de M. le Professeur Moscati, concernant une végétation électrique nouvellement découverte," *ibid.* (1781), 22–23; "Mémoire sur la règle de Cardan, et sur les équations cubiques, avec quelques remarques sur les équations en général," *ibid.* (1783), 244–265; "Premier mémoire sur les parallèles d'Euclide," *ibid.*, **43** (1786–1787) and "Second mémoire . . .," **44** (1788–1789), 171–203; "Recherches sur la liberté de l'homme," *ibid.*, **43** (1786–1787), 517–533; "Examen philosophique de quelques principes de l'algèbre" (two memoirs), *ibid.*, **45** (1790–1791), 331–363; and "Essai d'une théorie métaphysico-mathématique de l'expérience," *ibid.*, 364–390.

An obituary is Friedrich von Castillon, "Éloge de M. de Castillon, père," in *Histoire de l'Académie royale des sciences et des belles lettres de Berlin* (*avec mémoires*), **46** (1792–1793), 38–60.

RONALD S. CALINGER

CASTLE, WILLIAM ERNEST (*b.* Alexandria, Ohio, 25 October 1867; *d.* Berkeley, California, 3 June 1962), *biology.*

Castle was the fourth of six children born to William Augustus and Sarah Fassett Castle. Of modest means, Castle's father had been a schoolteacher in Johnstown, Ohio, before turning to farming. Castle received a B.A. degree from Denison University in 1889, and after teaching for three years entered Harvard University, where he received a second B.A. in 1893, an M.A. in 1894 (under C. B. Davenport), and a Ph.D. in 1895 (under E. L. Mark). After several years of teaching biology at the University of Wisconsin (1895–1896) and at Knox College, Galesburg, Illinois (1896–1897), he was appointed instructor in biology at Harvard (1897–1903), assistant professor (1903–1908), and professor of biology (1908–1936). From his retirement in 1936 until his death, Castle was research associate in mammalian genetics at the University of California, Berkeley. Castle was married in 1896 to Clara Sears Bosworth, who remained his constant companion until her death in 1940. The vitality that Castle brought to his work was somewhat masked by an external appearance of reserve and formality, yet he was strongly liked by his students and respected by his colleagues.

Castle's interest in natural history began early. As a farm boy, he collected wild flowers and learned to graft trees and to reconstruct skeletons of animals. At Denison, Clarence J. Herrick, an enthusiastic teacher of geology, zoology, and botany, introduced Castle to Darwin's theory of natural selection, a subject that fired the young man's interest in biological ideas. Academic emphasis at Denison was on classics, however, and Castle majored in Latin. After graduation he taught classics at Ottawa University, Ottawa, Kansas, from 1889 to 1892. In the fall of that year he entered Harvard University to pursue further his interests in natural history. He received his Ph.D. with a thesis focusing on a cell-lineage study of the ascidian *Ciona intestinalis*. This work was directed toward the elucidation of certain disputed questions concerning the mode of origin of the primary germ layers of tunicates. Carefully and methodically tracing the embryonic state of every cell from first cleavage through late gastrula, Castle took issue with the prevailing view of the origin of the chordate mesoderm. He concluded correctly that the mesoderm in *Ciona* and other primitive chordates originates from pouches in the infolded endoderm of the gastrula, in a manner similar to that in the echinoderms. *Ciona* is a hermaphrodite in which individuals are self-sterile (i.e., sperm of an individual will not fertilize eggs from the same individual). In the course of his studies Castle also discovered that self-fertilization is prevented not, as had been supposed, by ripening of sperm and eggs at different times but by physiological incompatibility between the gametes. This phenomenon had already been observed in certain flowering plants but never before in animals; Castle speculated that such a block to self-fertilization was probably chemical in nature.

Before 1900 Castle described himself as an "experimental evolutionist." While he continued work on some developmental problems in invertebrates, he also began to study the role of heredity in determining the sex ratio in mice and guinea pigs. As a result of breeding experiments and his reading of Darwin,

Weismann, De Vries, and Bateson, Castle was prepared to grasp the significance of Mendel's work when it was rediscovered in 1900. From that time on, the direction of Castle's research was clear; he became an enthusiastic Mendelian and devoted the remainder of his career to investigating either aspects of Mendel's laws or the relationship between Mendelian inheritance and evolution. For example, within a year or two after becoming familiar with Mendel's work, he attempted to apply the concept of dominance and recessiveness to the inheritance of sex (1903). Somewhat complex, his interpretation involved a subsidiary assumption of selective fertilization (i.e., only certain sperm will fertilize certain eggs) and did not throw much light on the problem of sex inheritance. At the same time, however, Castle was carrying out a long series of experiments (1901–1906) on inbreeding and outbreeding in the fruit fly *Drosophila*. Castle was the first to use this organism for any extensive laboratory breeding experiments, and it was through his work that T. H. Morgan's attention was initially called to *Drosophila*.

In 1951 Castle noted that four general questions had motivated his work between 1900 and 1920: (1) How generally applicable are Mendel's laws? (2) Is Mendel's assumption of the purity of gametes true? (3) Are the germ cells and somatic cells basically different, as Weismann asserted? (4) Can the fundamental nature of genes be modified by selection? The first question was attacked in a series of breeding experiments with guinea pigs between 1903 and 1907. Castle showed that the inheritance of coat colors followed strictly Mendelian lines, a conclusion that was verified by the work of Bateson, Punnett, A. D. Darbishire, Doncaster, C. C. Hurst, and Cuenot on a variety of other organisms. This evidence convinced Castle that Mendel's conclusions could be generalized throughout the animal and plant worlds.

Unlike William Bateson, however, Castle was by no means a complete convert to all the Mendelian assumptions. For example, Castle tended to answer the second question posed above in the negative. He believed that it was possible for two contrasting genes in a heterozygote to affect or contaminate each other in such a way that neither gene expressed itself in the same manner in subsequent generations. Thus, as a corollary, Castle maintained that it was possible to produce permanent blends of contrasting characters, a conclusion that was somewhat substantiated by his studies of inheritance of mammalian coat colors. Castle's belief in the modifiability of gametes was also the basis of his view that permanent genetic change can be effected through selection.

In 1909 Castle, with the expert surgical help of his co-worker J. C. Phillips, showed dramatically the validity of Weismann's distinction between germ and somatic tissues. Castle and Phillips transplanted the ovaries from a pure black guinea pig into a pure white guinea pig whose own ovaries had been removed. The transplant took, and in three subsequent breedings the female produced only black offspring. Thus, the genetic composition of the ovaries was unaffected by their being in the body of a different genetic type. This experiment represents one of Castle's most significant and unequivocal contributions.

The question that dominated much of Castle's life between 1900 and 1920 was the modifiability of Mendelian factors by selection. In June 1906 Hansford MacCurdy, along with Castle, completed a study of the inheritance of color patterns in rats. MacCurdy had shown that the darkly pigmented, uniform coloration pattern characteristic of wild rats was dominant over the piebald or "hooded" pattern. Hooded rats are white with a colored area (darkly pigmented) on the head and a narrow line down the back. A hooded rat crossed with a wild rat produced all wild F_1. In the F_2 the ratio of wild to hooded was 3:1. The hooded individuals, however, showed a much greater variability as a result of this cross. This suggested to MacCurdy and Castle that the recessive hooded gene was modified by existing in the hybrid along with genes for normal, wild pigmentation. Using this variability, Castle, with MacCurdy and J. C. Phillips, carried out a series of selection experiments on hooded rats between 1907 and 1914. In one line (called "plus") Castle selected for an increase in the extent of the hooded pattern, and in another line (called "minus") for a decrease in the hooded pattern. Selection was effective and ultimately yielded individuals far beyond the limits of the variability of the original parent series. After thirteen generations, selection in the "minus" strain produced almost solid white rats, while selection in the "plus" strain produced almost totally pigmented individuals. Castle wondered next whether such changes were permanent—i.e., whether they would disappear rapidly as soon as selection was relaxed. After an equivalent number of generations of backward selection (using sixth-generation "minus" strain rats) Castle found that the individuals produced were still much less highly pigmented than the parents from which he had originally started in the forward selection. In an important paper of 1914, "Piebald Rats and Selection," he pointed out that selection permanently modified genes by bringing them to reside in germ cells with other genes, which contaminated them. Castle's colleague at the Bussey Institution (the Harvard graduate school of biology at Jamaica Plain, Massachusetts), E. M.

East, suggested that the concept of "multiple factor" might also account for these selection results. However, Castle rejected this notion on the ground that he had already obtained a nearly all-black race of rats without having witnessed any significant reduction in variability. He reasoned that if selection simply increased or decreased the number of modifying factors bearing on the extent of hoodedness, then he should expect to see the limit of variability decrease as selection in any one direction proceeded.

Castle's conclusion, striking as it did at a fundamental assumption of Mendelian genetics (i.e., purity of the gametes), drew sharp criticism from T. H. Morgan's group at Columbia, principally from A. H. Sturtevant and H. J. Muller. As a result of these criticisms Castle repeated his selection experiments between 1914 and 1919. They suggested to Castle that the hypothesis of modifying factors was the most consistent with all of his various results. In a paper of 1919 ("Piebald Rats and Selection: A Correction") Castle conceded gracefully to the criticisms of the Morgan school. Characteristically, as L. C. Dunn remarks, Castle made the correction in a seminar attended by his students and colleagues, some of whom had also disagreed with him on this matter.

Castle's concession put to final rest the assertion of certain workers in the late nineteenth and early twentieth centuries that selection itself created new and specific variability in the germ plasm. The Danish botanist Wilhelm Johannsen had challenged this idea in 1903 and 1909 with a series of selection experiments with plants. However, until the work of Castle and his colleagues the idea had not been sufficiently investigated in animals. Aside from confirming Johannsen's results with animals, Castle's experiments with hooded rats also gave strong support to the increasingly important multiple-factor hypothesis in genetics.

Also in 1919 Castle posed another challenge. Questioning the foundation of the chromosome theory—linear arrangements of genes and the process of genetic mapping—Castle argued (in *Proceedings of the National Academy of Sciences*) that a nonlinear, branching model was more in agreement with some of the breeding data. Sturtevant, Bridges, and Morgan replied heatedly that Castle had failed to understand the key to the linear theory: the phenomenon of double crossing-over. Ultimately he was forced to concede on this point as well. Nevertheless, Castle's challenge forced the *Drosophila* school to reexamine its fundamental assumptions in light of the evidence. While Castle did indeed, at the time, seem to misunderstand the basis of chromosome mapping, others shared his confusion and were enlightened by the explanations necessary to resolve the controversy.

Castle's work in the last forty years of his life was devoted to three problems: (1) the inheritance of quantitative characters, such as size, in mammals; (2) the construction of genetic maps for small mammals, such as the rat and the rabbit; and (3) the inheritance of coat colors in large mammals, such as the horse. Once he had understood and accepted the idea of linearity in chromosome structure, Castle carried out a number of breeding experiments to determine the genetic maps of several common mammals. His work and that of his student L. C. Dunn contributed significantly to the understanding of mammalian genetics. Questions about the inheritance of quantitatively varying characters, such as size, had arisen considerably earlier in Castle's career, especially in relation to his advocacy of blending inheritance. In work carried out between 1929 and 1934 Castle, in collaboration with P. W. Gregory, showed that difference in size between different races of rabbits was due not to permanent blends between parental factors for size but, rather, to different rates of development of the fertilized egg. This rate was clearly determined by a genetic factor (or factors) in the sperm and was inherited in a Mendelian fashion.

After his retirement Castle became research associate in mammalian genetics at the University of California, Berkeley. In these years, partly as a relief from the arduous breeding experiments, he became interested in the inheritance of coat colors in various breeds of horses. Data for his analyses could be obtained from studbooks, and his work on this subject contributed to a more sound knowledge of breeding lines in horses. Castle's last published paper (1961) concerned the inheritance of coat pattern and gene interaction in the palomino breed. Significantly, this study brought to full circle Castle's interest in mammalian coat color inheritance, a subject that he had begun investigating sixty years before.

Outside of his strict interest in experimental genetics, Castle maintained throughout his career a strong concern for eugenics, the application of breeding data to human inheritance. His first publication on this subject was *Genetics and Eugenics* (1916). In the 1920's and 1930's, as racial and immigration matters drew heavily on eugenic findings, Castle spoke out frequently. He argued strongly for objective evaluation of the roles of heredity and environment in drawing conclusions about human physical, psychological, or intellectual characteristics. Castle felt that there was an important place in biology for sound eugenic ideas but, like many biologists at the time, he believed that doctrines of racial superiority and segregation, based on supposed genetic facts, were unfounded and detrimental to social progress.

Through his lecture courses at Harvard, Castle introduced many undergraduates to the problems of genetics and evolution. His primary influence, however, was on graduate students; in twenty-eight years at the Bussey Institution he trained over twenty doctoral candidates. At the Bussey, Castle was responsible for mammalian genetics and E. M. East for plant genetics. However, as L. C. Dunn remarks, there was little formal division between the fields. Students of East's could, and frequently did, confer with Castle and vice versa. Among Castle's most important students were J. H. Detlefsen, C. C. Little, Sewall Wright, L. C. Dunn, Gregory Pincus, and P. W. Gregory. At the same time East trained such students as R. A. Emerson, Edgar Anderson, Paul Mangelsdorf, and E. R. Sears. Between them Castle and East had a significant and very widespread influence on American genetics.

Castle always enjoyed the individuality of research and opposed the strong organization of science. Similarly, he disliked bureaucracy and avoided it as much as possible. Warm and generous, Castle was interested in the individual in science, not the organization. His door was always open to his graduate students, but he spent little of his active career in any organizational capacity. Nevertheless, Castle took on administrative duties when necessary. For example, he helped to found the American Breeders Association (1903) and to reorganize it as the American Genetics Association (1913), and to found the joint section on genetics of the American Society of Zoologists and the Botanical Society of America (1922) and to reorganize it as the Genetics Society of America (1932). He also was vice-president of the American Genetics Association, chairman of the joint section on genetics (1924), vice-president of the eastern branch of the American Society of Zoologists (1905–1906), and president of the American Society of Naturalists (1919). In addition, Castle was a member of the editorial board of the *Journal of Experimental Zoology* from its founding (1904) until his death. He also helped to found the *Journal of Heredity* (1913) as a new form of the *American Breeders' Magazine,* and with ten colleagues he founded the journal *Genetics* in 1916.

Castle received a number of academic honors. He was elected a member of Phi Beta Kappa at Harvard (1893), the American Academy of Arts and Sciences (1900), the American Philosophical Society (1910), and the National Academy of Sciences (1915). In addition, he was awarded an Sc.D. by the University of Wisconsin and an LL.D. by Denison University in 1921, and the Kimber Genetics Award of the National Academy of Sciences, the first time this award was given (1955).

Castle was important to genetics in the twentieth century in three ways: (1) His strong support of Mendel's work between 1900 and 1910 provided (in part) in America the same service that Bateson provided in England. While later Mendelians, such as T. H. Morgan, ultimately made more of the theory than Castle was able to do, his interest and work carried the young science through the early days of skepticism and doubt. (2) Castle helped to extend to mammals the conclusions that other Mendelians had reached with insects or plants. (3) Castle immediately saw, after 1900, the relationship between Mendel's work and the Darwinian theory of selection. Although for many years he argued for what came to be a discredited view, his foresight did much to turn the attention of biologists to the importance of understanding natural selection in genetic terms.

In many respects, however, Castle's work was out of the mainstream of genetics after 1915. He approached problems of heredity essentially through breeding experiments and never entered into the study of the physical basis of heredity, which led Morgan and his school to establish the chromosome theory. While this may have been to some extent a result of Castle's long-standing association with the practical breeding tradition of the Bussey Institution, it was also a function of his personality. He had an aversion to complex experiments and abstractions. Breeding results were definite, while the theoretical constructions of the chromosome theory, brilliant as they were, seemed to him symbolic. He was frequently unable to see to the heart of an abstract matter and as a result often argued on the losing side of a controversy. Nevertheless, his practical common sense, enormous energy, and open-mindedness were important in establishing the Mendelian theory on a firm basis for a number of kinds of organisms.

BIBLIOGRAPHY

Works by Castle include "On the Cell Lineage of the Ascidian Egg. A Preliminary Notice," in *Proceedings of the American Academy of Arts and Sciences,* **30** (1894), 200–216; *Heredity of Coat Characters in Guinea-Pigs and Rabbits,* Carnegie Institution of Washington Publication no. 23 (Washington, D.C., 1905); "Color Varieties of the Rabbit and of Other Rodents: Their Origin and Inheritance," in *Science,* **26** (1907), 287–291; *Selection and Cross-Breeding in Relation to the Inheritance of Coat Pigments and Coat Patterns in Rats and Guinea-Pigs,* Carnegie Institution of Washington Publication no. 70 (Washington, D.C., 1907), written with H. MacCurdy; "A Successful Ovarian Transplantation in the Guinea-Pig, and Its Bearing on Problems of Genetics," in *Science,* **30** (1909), 312–313, written with J. C. Phillips; "The Effect of Selection Upon Mendelian

Characters Manifested in One Sex Only," in *Journal of Experimental Zoology,* **8** (1910), 185–192; *Piebald Rats and Selection: An Experimental Test of the Effectiveness of Selection and of the Theory of Gametic Purity in Mendelian Crosses,* Carnegie Institution of Washington Publication no. 195 (Washington, D.C., 1914), written with J. C. Phillips; *Genetics and Eugenics* (Cambridge, Mass., 1916); "Is the Arrangement of the Genes in the Chromosomes Linear?," in *Proceedings of the National Academy of Sciences of the United States of America,* **5** (1919), 25–32; "Piebald Rats and Selection, A Correction," in *American Naturalist,* **53** (1919), 265–268; "The Embryological Basis of Size Inheritance in the Rabbit," in *Journal of Morphology,* **48** (1929), 81–93, written with P. W. Gregory; "Race Mixture and Physical Disharmonies," in *Science,* **71** (1930), 603–606; "Further Studies on the Embryological Basis of Size Inheritance in the Rabbit," in *Journal of Experimental Zoology,* **59** (1931), 199–210, written with P. W. Gregory; "The Genetics of Coat Color in Horses," in *Journal of Heredity,* **31** (1940), 127–128; and "The Palomino Horse," in *Genetics,* **46** (1961), 1143–1150, written with W. R. Singleton.

A more detailed biography is L. C. Dunn, "William Ernest Castle, October 25, 1867–June 3, 1962," in *Biographical Memoirs. National Academy of Sciences,* **38** (1965), 31–80.

GARLAND E. ALLEN

CATALÁN, MIGUEL ANTONIO (*b.* Zaragoza, Spain, 9 October 1894; *d.* Madrid, Spain, 11 November 1957), *experimental spectroscopy.*

Catalán entered the University of Zaragoza in 1908, at the age of fourteen; from there he went to the University of Madrid, where he completed his studies with a D.Sc. under A. Del Campo. He then joined the Laboratorio de Investigaciones Físicas in Madrid, where he worked mainly on atomic spectroscopy, a field that had attracted him from his student days and to which he devoted the major part of his creative efforts for the next forty years.

A Spanish government fellowship enabled Catalán to spend the year 1920/1921 in Alfred Fowler's laboratory at Imperial College, London, then a leading center of atomic spectroscopy outside of Germany and the Netherlands. This field, which for many years had looked upon spectra as no more than the language of atoms, was slowly moving into the very center of physical research, spurred by the Bohr atomic model, which provided an interpretation of that language.

On the experimental side, spectroscopic studies up to 1922 had revealed the following important features:

1. The spectra of most elements, especially of the first three groups of the periodic table, are made up of distinct lines that could be classified into at least four different series, which are known as the sharp (*s*), principal (*p*), diffuse (*d*), and fundamental (*f*) series.

2. Each line was known to be made up of one, two, or three closely spaced lines. These were referred to as singlets, doublets and triplets, respectively. The physical origin of these was still obscure.

3. In general the alkali metals, such as sodium and potassium, gave rise to doublets, and the frequency difference was almost the same in all the sharp and diffuse series. In the second group of the periodic table, which consists of the alkaline earth metals, many of the lines are actually triplets, although singlets are also present. These characteristics refer to arc spectra, or spectra from neutral atoms. In the case of spark spectra, or spectra of ionized atoms, the elements behaved as if they belonged to the next group, in accordance with the so-called displacement law.[1]

From these facts one could tabulate the nature of complex spectra, as follows:[2]

Group:	Arc Sp.
VIII,0	complex and trip.
I	doub.
II	trip.
III	doub.
IV	trip.(?)
V	doub.(?)
VI	trip.(?)
VII	?

Manganese belongs to group VII in the periodic table, and from the pattern observed, its spectrum should be made up of doublets. Triplets had, however, been detected in the manganese spectrum.

It must be realized that the spectra of heavier elements are very complex and had not been fully resolved into their components at that time. But since at least some of these lines could be arranged into series, it was generally surmised that all complex spectra could eventually be resolved into doublets and triplet-singlets.

Catalán felt that the investigation of a spectrum characterized by multitudinous lines could reveal more general regularities that in turn would probably account for the recalcitrant lines in the first two columns of the periodic table as well as for the complex spectra in the center. He therefore undertook a detailed study of the manganese spectrum in Fowler's laboratory.[3]

Catalán's extensive studies of the manganese spectrum revealed in it a system of four triplets that were analogous to the ones found in the alkaline earth spectra. He also found a system of three narrower triplet series that were of the sharp, diffuse, and principal types and corresponded to the singlet system of the alkaline earths. And finally there was yet another, still narrower system of triplets that was parallel to the preceding ones.

Each diffuse triplet in the manganese spectrum was actually analyzed into nine lines. Catalán explained this situation, which he described as multiplets, by assuming the d terms to be fivefold, thus departing from the accepted views about the structure of atomic spectra.

While he was back in Spain, Catalán made similar studies of chromium, selenium, molybdenum, and other elements, and showed that the phenomenon observed in manganese was not restricted to that element but, rather, was a general regularity.

Arnold Sommerfeld, a leading theorist in the field, met Catalán during a visit to Spain in April 1922. Impressed by Catalán's discovery, he took back to Germany the manuscript of Catalán's epochal paper, which appeared the following year in the *Philosophical Transactions*.[4] He then showed that his "inner" (total angular momentum) quantum numbers could be assigned to Catalán's multiplet terms.[5] Very soon thereafter both theoreticians and experimentalists began to investigate multiplets with great enthusiasm and success, for they now had the key for unraveling the spectra of the elements in the center of the periodic table. The intense interest in this field during the period 1923–1925 was due especially to the lack of an adequate explanation of the physical origin of multiplets, indeed even of doublets and triplets. It was widely anticipated that the investigation of this problem would lead to the uncovering of the quantum mechanics of the atom. These researches did contribute to establishing the structure of the atom, but the actual explanation of the multiplets are found to lie in the spin of the electron, and quantum mechanics emerged from a different line of research.

Sponsored by Sommerfeld, Catalán received an International Education Board (Rockefeller Foundation) fellowship to work with Sommerfeld at Munich during the year 1924/1925. Here, along with Karl Bechert, Catalán analyzed several other spectra; and when he returned to Madrid in the fall of 1925, Bechert followed him with an I.E.B. fellowship. Thus Catalán maintained a close cooperation with the Munich school of theoretical spectroscopy.

In 1930 Catalán became the chief of the section on atomic spectroscopy of the Rockefeller Institute in Madrid. He won the admiration and affection of the many students under his direction, for besides his abilities as a scientist, Catalán was an extremely cordial person and a gifted teacher—indeed, he often considered himself a teacher first. Realizing the importance of maintaining bridges between the scientist's world and that of the layman, he often gave popular lectures on science to nontechnical audiences.

Catalán's numerous scientific papers firmly established his reputation on the international scene. After World War II he visited the United States a number of times, lecturing and working in various centers. In 1950 he was made head of the department of spectroscopy at the Institute of Optics in Madrid. Two years later he became a member of the Joint Commission for Spectroscopy of the International Council of Scientific Unions. In 1955 he was elected to the Royal Academy of Sciences in Spain. He also made a number of lecture tours in Latin America.

Catalán was active until his last year. His death, not long after his sixty-third birthday, was sudden.

NOTES

1. Arnold Sommerfeld, *Atombau und Spektrallinien*, 3rd ed. (Brunswick, 1922), pp. 266–471.
2. F. D. Foote and F. L. Mohler, *The Origin of Spectra* (New York, 1922), p. 42.
3. The earliest studies on the manganese spectrum had been made by Sir William Huggins in 1864.
4. Catalán, "Series and Other Regularities in the Spectrum of Manganese," in *Philosophical Transactions of the Royal Society*, **223** (1923), 127–173.
5. Arnold Sommerfeld, "Uber die Deutung verwickelter Spektren," in *Annalen der Physik*, **70** (1922), 32–62.

BIBLIOGRAPHY

A bibliography of Catalán's works is in Poggendorff, VI, pt. 1 (1936), 413; and VIIb, pt. 2 (1967), 742–743. Letters relating to Catalán's work are listed in T. S. Kuhn *et al.*, *Sources for History of Quantum Physics. An Inventory and Report* (Philadelphia, 1967), 28b, 120b, 130c, 135a.

There is an article on Catalán in *Enciclopedia universal ilustrado europeo americana* (Bilbao, 1931), app., pp. 1266–1267, and ann. supp. for 1957–1958 (1961), pp. 199–200. Obituaries are *Journal of the Optical Society of America*, **48** (1958), 138; R. Velasco *et al.*, in *Anales de la R. Sociedad española de física y química*, ser. A, **53** (1957), 217, repr. in ser. B (1957), 773; and R. Velasco, in *Nature*, **181** (1958), 234.

VARADARAJA V. RAMAN

CATALDI, PIETRO ANTONIO (*b.* Bologna, Italy, 15 April 1552; *d.* Bologna, 11 February 1626), *mathematics*.

Cataldi was the son of Paolo Cataldi, also of Bologna. Little else is known of his life. His first teaching position (1569–1570) was at the Florentine Academy of Design. He then went to Perugia to teach mathematics at the university; his first lecture was given on 12 May 1572. He also taught at the Academy of Design in Perugia and was a lecturer in mathematics at the Studio di Bologna from 1584 until his death. Cataldi showed his benevolence by giving the superiors of various Franciscan monasteries the task of distributing free copies of his *Pratica aritmetica* to monasteries, seminaries, and poor children.

In the history of mathematics Cataldi is particularly remembered for the *Trattato del modo brevissimo di trovar la radice quadra delli numeri,* finished in 1597 and published in 1613. In this work the square root of a number is found through the use of infinite series and unlimited continued fractions. It represents a notable contribution to the development of infinite algorithms.

In the orientation of mathematical thinking of the late Renaissance and the seventeenth century toward infinitesimal questions, along with geometric methods (such as Cavalieri's principle of indivisibles), in the field of arithmetic, the passage from the finite to the infinite appears in processes of iteration in calculus. In this area Cataldi started with the practical rules furnished by ancient treatises on arithmetic for finding the square root of any natural number N that is not a perfect square. These treatises gave first the basic rule for finding the natural number a such that

$$a^2 < N < (a + 1)^2.$$

Then rules were given for finding approximate rational values of $\sqrt{N}$, expressed respectively by the formulas

$$a + \frac{N - a^2}{2a},$$

$$a + \frac{N - a^2}{2a + 1}.$$

The first of these formulas, which goes back at least as far as Hero of Alexandria (and probably to the Babylonians), coincides with the arithmetic mean $1/2\,(a + N/a)$ of the two values a and N/a (which have as their geometric mean $\sqrt{N}$, which is being sought). The second formula is obtained by a process of linear interpolation:

$$x_1 = a^2 \qquad x_2 = (a + 1)^2 \qquad x = a^2 + r$$
$$y_1 = a \qquad y_2 = a + 1.$$

The y corresponding to x is determined by means of the equation

$$\frac{x - x_1}{x_2 - x_1} = \frac{y - y_1}{y_2 - y_1},$$

from which one obtains

$$y = a + \frac{r}{2a + 1}.$$

That is, setting $x = N = a^2 + r$, the result is

$$y = a + \frac{N - a^2}{2a + 1}.$$

Returning to the rounded maximum value (already considered), we write

$$a_1 = a + \frac{N - a^2}{2a}.$$

Starting with a_1, by an analogous procedure we obtain a new value, a_2, a closer approximation than a_1:

$$a_2 = a_1 - \frac{a_1^2 - N}{2a_1} = a + \frac{N - a^2}{2a} - \frac{a_1^2 - N}{2a_1}.$$

Setting $r_1 = a_1^2 - N \cdots r_{n-1} = a_{n-1}^2 - N$, one obtains

$$a_2 = a + \frac{r}{2a} - \frac{r_1}{2a_1}.$$

By iteration of the indicated procedure we obtain

$$a_n = a + \frac{r}{2a} - \frac{r_1}{2a_1} - \frac{r_2}{2a_2} - \cdots - \frac{r_{n-1}}{2a_{n-1}}.$$

As Cataldi established, r_n may be rendered as small as one wishes, provided that n is large enough. The formula

$$a_1 = \frac{1}{2a}\,(a^2 + N)$$

results in

$$r_1 = a_1^2 - N = \frac{1}{4a^2}[(a^2 + N)^2 - 4a^2\,N]$$
$$= \frac{(a^2 - N)^2}{4a^2} = \left(\frac{r}{2a}\right)^2.$$

Analogously, in general

$$r_n = \left(\frac{r_{n-1}}{2a_{n-1}}\right)^2,$$

from which one obtains

$$\frac{r_n}{r_{n-1}} = \frac{r_{n-1}}{4a_{n-1}^2}.$$

Given $r_{n-1} < a_{n-1}^2$, one will have

$$\frac{r_n}{r_{n-1}} < \frac{1}{4},$$

which guarantees the rapid convergence of the series representing $\sqrt{N}$:

$$a + \frac{r}{2a} - \frac{r_1}{2a_1} - \frac{r_2}{2a_2} - \cdots - \frac{r_{n-1}}{2a_{n-1}} - \cdots.$$

Cataldi presents analogous considerations in relation to the other series, which is obtained from the rounded minimum values of $\sqrt{N}$:

$$a + \frac{r}{2a + 1} + \frac{r_1}{2a_1 + 1} + \frac{r_2}{2a_2 + 1} + \cdots$$

$$+ \frac{r_{n-1}}{2a_{n-1} + 1} + \cdots.$$

We shall now see how, starting with the rounded maximum value (already considered) of $\sqrt{N}$,

$$a_1 = a + \frac{r}{2a},$$

Cataldi arrives at continued fractions. With the aim of obtaining a better approximation than that reached with a_1, Cataldi adds to the denominator a value x, and then considers the expression

$$a + \frac{r}{2a + x}.$$

He observes that this expression has a rounded minimum value of $\sqrt{N}$ when (and only when) $a + x$ has a rounded maximum value of $\sqrt{N}$. That is (as can be verified by simple calculation),

$$\left(a + \frac{r}{2a + x}\right)^2 < a^2 + r$$

when and only when $(a + x)^2 > a^2 + r$. Therefore, given

$$\left(a + \frac{r}{2a}\right)^2 > a^2 + r,$$

one will have

$$\left(a + \frac{r}{2a + \dfrac{r}{2a}}\right)^2 < a^2 + r.$$

That is,

$$a_2 = a + \frac{r}{2a + \dfrac{r}{2a}}$$

has a rounded minimum value of $\sqrt{N}$, while, for reasons analogous to those already given, the rounded maximum value of the same root will be

$$a_3 = a + \frac{r}{2a + r/(2a + r/2a)},$$

and so on, indefinitely. If we write

$$a_1 = \frac{p_1}{q_1}, \; a_2 = \frac{p_2}{q_2} \cdots a_n = \frac{p_n}{q_n},$$

we will see that the result p_n/q_n for order n can be expressed in terms of the result for order $n - 1$ by

means of the formula

$$\frac{p_n}{q_n} = \frac{r}{2a + (p_{n-1}/q_{n-1})}.$$

From it are derived the recurrent formulas habitually used by Cataldi:

$$p_n = rq_{n-1}, \; q_n = 2aq_{n-1} + p_{n-1}.$$

Moreover, Cataldi finds the fundamental relation,

$$p_n q_{n-1} - p_{n-1} q_n = (-1)^n r^n,$$

from which can be obtained the expression of the difference between two consecutive results:

$$\frac{p_n}{q_n} - \frac{p_{n-1}}{q_{n-1}} = (-1)^n \frac{r^n}{q_n q_{n-1}}.$$

In sum, Cataldi compares the results of the series just studied with the results of the continued fraction, considering the same $\sqrt{N}$, and establishes that the results of the series obtained by starting with the rounded minimum values reproduce the results of even orders of the continued fraction (a_2, a_4, $\cdots$); the results of the series obtained by starting with the maximum values reproduce the results of order

$$(2^n - 1) \; (n = 1, 2, 3, \cdots).$$

Having examined Cataldi's contribution to the theory of continued fractions, clearly explained in various writings by E. Bortolotti, it remains for us to consider the place of these contributions in the history of mathematical thought. The question is complex and has given rise to many discussions and polemics. Given the great number of questions that can lead to consideration of continued fractions, hints of the theory of continued fractions are presented many times and in presumably independent ways in the course of mathematical history. It is not our task to reconstruct the history of continued fractions; we shall limit ourselves to indicating some elements of it in order to clarify Cataldi's position. We are led to consideration of this question when we come to Euclid's procedures for determining the greatest common divisor of two natural numbers (bk. VII, prop. 2), which lead to consideration of a limited continued fraction, while the Euclidean criterion for establishing whether two homogeneous magnitudes are commensurable or incommensurable leads in the latter case to an unlimited continued fraction (*Elements,* bk. X, prop. 2); in neither case is the algorithm presented explicitly.

The successive reductions of the continued fraction that is expressed by $\sqrt{2}$ appear in the work of the Neoplatonic philosopher Theon of Smyrna (second century A.D.), *Expositio rerum mathematicarum ad legendum Platonem utilium.* An examination of this text, however, leads one to conclude that such values were calculated with an aim different from that indicated above. For the sake of brevity we shall omit mention of other appearances of the continued fractions, which can be found in Greek, Indian, and Arabic writings.

In the Renaissance, Rafael Bombelli gave a procedure for the extraction of the square root of a number that is not a square, such that the successive steps in the procedure lead to the calculation of the successive results of a continued fraction. However, Bombelli, who refers to numerical cases, performs the calculations that have been considered, in such a way that the final result retains no traces of the algorithm implicitly defined by the procedure of iteration that is applied (see Bombelli's *L'algebra,* E. Bortolotti, ed. [Bologna, 1929], pp. 26–27).

The use of limited continued fractions in the expression of relationships between large numbers is found in the *Geometria practica* of Daniel Schwenter (1627), published soon after Cataldi's death.

The term "continued fraction" was introduced by John Wallis, who gave a systematic treatment of it in his *Arithmetica universalis* (1655). It contains an example of the development of a transcendental number, under the form that originated with William Brouncker:

$$\frac{4}{\pi} = 1 + \cfrac{1^2}{2 + \cfrac{3^2}{2 + \cfrac{5^2}{2 + \cfrac{7^2}{2 + \cdots}}}}$$

A theory of continued fractions was devised by Euler, and Lagrange formulated the theorem concerning the periodic character of continued fractions that represent square roots.

Continued fractions have been of great use to mathematicians in the investigation of the nature of numbers—for instance, Liouville's work on the existence of transcendental numbers (1844) was based on the use of the algorithm considered above.

Cataldi also has a place in the history of the criticism of Euclid's fifth postulate, which led to construction of a non-Euclidean geometry. In his *Operetta delle linee rette equidistanti et non equidistanti,* he attempted to demonstrate the fifth postulate on the basis of remainders. The defect of his argument is found in his definition of equidistant straight lines: "A given straight line is said to be equidistant from another straight line in the same plane when the two shortest lines that are drawn from any two different points on the first line to the second line are equal." Cataldi did not realize that two conditions are imposed on the first, given line, which are not stated as compatible: (1) that it is a locus of points at a constant distance from the second line and (2) that it is a straight line. The admission of such a compatibility constitutes a postulate equivalent to Euclid's fifth postulate.

Cataldi's other works, which are mainly didactic, are concerned with theoretical and practical arithmetic, algebra, geometry, and astronomy, and furnish good documentation of the mathematical knowledge of his time.

BIBLIOGRAPHY

I. ORIGINAL WORKS. Cataldi's writings are *Prima lettione fatta pubblicamente nello studio di Perugia il 12 maggio 1572* (Bologna, 1572); *Due lettioni fatte nell'Accademia del disegno di Perugia* (Bologna, 1577); *Pratica aritmetica,* 4 pts. (Bologna, 1602–1617); *Operetta delle linee rette equidistanti et non equidistanti* (Bologna, 1603); *Trattato dei numeri perfetti* (Bologna, 1603); *Aggiunta all'operetta delle linee rette equidistanti et non equidistanti* (Bologna, 1604); *Trattato dell'algebra proportionale* (Bologna, 1610); *Trasformatione geometrica* (Bologna, 1611); *Trattato della quadratura del cerchio* (Bologna, 1612); *Due lettioni di Pietro Antonio Cataldi date nella Accademia erigenda; dove si mostra come si trovi la grandezza delle superficie rettilinee* (Bologna, 1613); *I primi sei libri de gl'Elementi d'Euclide ridotti alla prattica* (Bologna, 1613); *Tavola del levar del sole et mezo di per la città di Bologna* (Bologna, 1613); *Trattato del modo brevissimo di trovar la radice quadra delli numeri* (Bologna, 1613); *Aritmetica universale* (Bologna, 1617); *Algebra discorsiva numerale et lineale* (Bologna, 1618); *Operetta di ordinanze quadre di terreno et di gente* (Bologna, 1618); *Regola della quantità, o cosa di cosa* (Bologna, 1618); *Nuova algebra proportionale* (Bologna, 1619); *Diffesa d'Archimede, trattato del misurare o trovare la grandezza del cerchio* (Bologna, 1620); *Elementi delle quantità irrazionali, o inesplicabili necessarij alle operazioni geometriche et algebraiche* (Bologna, 1620); *I tre libri settimo ottavo et nono de gli Elementi aritmetici d'Euclide ridotti alla pratica* (Bologna, 1620); *Trattato geometrico . . . dove si esamina il modo di formare il pentagono sopra una linea retta, descritto da Alberto Durero. Et si mostra come si formino molte figure equilatere, et equiangole sopra ad una proposta linea retta* (Bologna, 1620); *Algebra applicata, dove si mostra la utilissima applicazione d'essa alla inventione delle cose recondite nelle diverse scienze* (Bologna, 1622); *Decimo libro degli Elementi d'Euclide ridotto alla pratica* (Bologna, 1625); and *Difesa d'Euclide* (Bologna, 1626).

II. SECONDARY LITERATURE. On Cataldi or his work see the following by E. Bortolotti: "Le antiche regole empiriche del calcolo approssimato dei radicali quadratici e le prime serie infinite," in *Bollettino della mathesis*, **11** (1919), 14–29; "La scoperta delle frazioni continue," *ibid.*, 101–123; "La storia dei presunti scopritori delle frazioni continue," *ibid.*, 157–188; "Ancora su la storia delle frazioni continue," *ibid.*, **12** (1920), 152–162; "La scoperta dell'irrazionale e le frazioni continue," in *Periodico di matematiche*, 4th ser., **11**, no. 3 (1931); "Cataldi, P. A.," in *Enciclopedia Italiana Treccani*, IX (1931), 403; "Frazione," *ibid.*, XVI (1932), 45–47; and "I primi algoritmi infiniti nelle opere dei matematici italiani del secolo XVI," in *Bollettino dell'Unione matematica italiana*, 2nd ser., **1**, no. 4 (1939), 22.

Works by other authors are G. Fantuzzi, *Notizie degli scrittori Bolognesi*, III (Bologna, 1733), 152–157; A. Favaro, "Notizie storiche su le frazioni continue," in *Bullettino Boncompagni*, **7** (1874), 451–502, 533–589; S. Gunther, "Storia dello sviluppo della teoria delle frazioni continue fino all'Euler," *ibid.*, 213–254; G. Libri, *Histoire des sciences mathématiques en Italie* (Paris, 1838–1841), IV, 87; and P. Riccardi, *Biblioteca matematica italiana*, I (Modena, 1893), 302–310.

ETTORE CARRUCCIO

CATESBY, MARK (*b.* Castle Hedingham, Essex, England, 3 April 1683; *d.* London, England, 23 December 1749), *natural history.*

His father, John, was a lawyer and landowner and four-term mayor of Sudbury, Suffolk; his mother, Anne Jekyll, came from a family of Essex antiquaries and lawyers. Catesby may have attended Sudbury Grammar School or, more likely, received some schooling in Castle Hedingham. His correspondence demonstrates a high degree of English literacy and a knowledge of Latin. Through his uncle, Nicholas Jekyll, Catesby as a young man came to know the premier English naturalist of his time, John Ray, and Ray's friend and collaborator, the apothecary Samuel Dale of Braintree, Essex.

The real beginning of Catesby's career as a naturalist came in 1712 when he joined the family of his sister Elizabeth in Virginia. Through her husband, the physician William Cocke, Catesby met the botanizing and gardening gentlemen of the province who would later supply him with materials and information. During this stay in the New World, he explored Virginia as far west as the Blue Ridge and also visited Bermuda and Jamaica. While Catesby later remembered these years as being barren, the plants he sent back to Samuel Dale and to the gardener Thomas Fairchild brought him to the attention of the English natural history circle. When Catesby returned to England in the fall of 1719, Dale began to enlist the support of William Sherard—whose will established the Oxford chair in botany which bears his name and who was then at work on an attempt to revise the *Pinax* of Gaspard Bauhin, for which he wanted addi-

tional collections. Sherard secured subscribers, notably Sir Hans Sloane, to send Catesby to South Carolina in 1722 on the journey which would result in the *Natural History of Carolina, Florida, and the Bahama Islands.* Catesby spent the next three years collecting and sketching in the Carolina low country and Piedmont (the "Florida" of the title would soon be called Georgia) and then spent parts of 1725 and 1726 in the Bahamas.

Catesby's career as a field naturalist was ended; he would spend most of the rest of his life in the production of his *Natural History.* From Joseph Goupy, the French-born watercolorist and etcher, Catesby learned to etch his own plates and produced all save two of the 220 illustrations in the work, in addition to which he wrote the accompanying text and oversaw the coloring of all the plates. Ultimately he was able to earn a meager living from the sale of the book partly because Peter Collinson, the Quaker patron of American science, lent him money to keep the *Natural History* out of the hands of booksellers. Finally, his widow, Elizabeth Rowland, whom he married on 2 October 1747, and his children, Mark and Anne, were able to live on the proceeds of the sale of the plates and drawings until her death in 1753. Catesby himself seems to have maintained good health until late in life. In July 1749 he seemed near death from dropsy, although he survived for another five months and then died after a fall in the street.

Although Catesby was primarily a botanist, his major contribution was in ornithology and in bird illustration. Compared with his contemporaries, his drawings excelled the stiff profiles of Eleazer Albin and, although they lacked the precision and decorative quality of George Edward's figures, Catesby's botanical backgrounds, which were often ecologically correct, were superior to Edward's stylized settings. Catesby's work was saved from the neglect accorded many pre-Linnaean studies by the fact that Linnaeus made use of it as the source for many designations of American birds. Fewer of Catesby's plants and animals other than birds were the bases of binomials, although designations based on the *Natural History* were made well into the nineteenth century. As a descriptive and theoretical naturalist, Catesby was traditional in using deluvial explanations, in advocating climatic uniformity by latitude, and in his use of three of the four elements to organize his "Account of Carolina." He was nonetheless critical of tall tales, as in his "Of Birds of Passage," which he read to the Royal Society, to which he was admitted a fellow in 1732. There he posited cold and lack of food as causes of migration and denied that swallows hibernated in ponds.

BIBLIOGRAPHY

I. ORIGINAL WORKS. The *Natural History* (London, 1731–1743, 1729–1747) went through new eds. in London in 1754 and 1771. European copies of the work are discussed in Frick and Stearns, p. 110. "Of Birds of Passage" was printed in *Philosophical Transactions,* **44** (1747), 435–444, and was extracted in *Gentleman's Magazine,* **17** (1748), 447–448. The posthumous *Hortus Britanno-Americanus* (London, 1763) was reissued as *Hortus Europae Americanus* (London, 1767). Its figures are derived largely from the *Natural History.*

II. SECONDARY LITERATURE. George F. Frick and Raymond P. Stearns, *Mark Catesby* (Urbana, Illinois, 1961)—which should also be consulted for additional bibliography—and Elsa G. Allen, "History of American Ornithology Before Audubon," in *Transactions of the American Philosophical Society,* **41** (Philadelphia, 1951), deal with Catesby's life and science. J. E. Dandy, *The Sloane Herbarium* (London, 1958), pp. 110–113, adds botanical material.

Catesby's family is treated in Anthony R. Wagner, *English Genealogy* (Oxford, 1960); and Paul H. Hulton and David B. Quinn, *The American Drawings of John White* (London, 1964), cover Catesby's borrowings from White.

GEORGE F. FRICK

CATTELL, JAMES McKEEN (*b.* Easton, Pennsylvania, 25 May 1860; *d.* Lancaster, Pennsylvania, 20 January 1944), *psychology, scientific publishing.*

Cattell was the first person in the world to have the title "professor of psychology" (University of Pennsylvania, 1888). After being asked to leave Johns Hopkins University in 1883 (his colleagues regarded him as a nuisance), he went to Leipzig, where he studied with Wilhelm Wundt, whose first laboratory assistant he was. He received his doctorate in 1886. Part of the next two years was spent studying at Galton's anthropometric laboratory in London. In 1891 Cattell left Pennsylvania for Columbia University, where he remained until 1917, when his pacifist views resulted in his expulsion—one of the more spectacular academic freedom cases of the period.

Cattell's psychological work shows the influence of his two masters. Like Wundt he was interested in experimental studies of human behavior; like Galton he was interested in statistical studies, of scientists in Cattell's case. Cattell and his students (e.g., E. L. Thorndike, R. S. Woodworth) greatly influenced American psychology in its bias toward experimentation and quantification.

Cattell's principal importance, however, is as a great scientific publisher and editor and as a powerful force in the development of the scientific and intellectual community. From 1895 until his death Cattell owned and edited *Science.* With J. M. Baldwin he founded

Psychological Review in 1894 and continued as co-editor until 1903. Cattell took over *Popular Science Monthly* in 1900, relinquishing his editorship in 1915; he was not able to arrest the periodical's decline, however. To supplant *Popular Science Monthly,* in 1915 he founded *Scientific Monthly,* which he edited until his death. In the same year he founded another journal, *School and Society,* which remained under his aegis for the rest of his life. With the initial support of the Carnegie Institution of Washington, Cattell produced the first edition of *American Men of Science* in 1906 and edited all editions through the sixth (1938). Two editions of a related directory, *Leaders in Education,* appeared in 1932 and 1940. To handle all these publications he founded the Science Press of Lancaster, Pennsylvania, in 1923. At a time when founding and maintaining scientific journals was quite risky, Cattell's successful entrepreneurship earned him much gratitude from the scientific community.

Related to Cattell's editorial work was his role as a would-be reformer of the American intellectual scene. In both his writings and his conduct Cattell upheld the rights of professors against deans and university presidents; he was instrumental in the founding of the American Association of University Professors in 1915. Selecting the American Association for the Advancement of Science as his chosen instrument for the reorganization of the scientific community, Cattell made *Science* its official organ in 1900 and later arranged for title to the periodical to pass to the Association on his death. In his scheme of things, the Association was to be the "House of Commons" of science, with representatives of each professional society and other organizations. The National Academy was to be an increasingly powerless "House of Lords." His statistical analyses in *American Men of Science* and his attempt to "star" the leading scientists in the directory were used to buttress his position.

These compilations of statistical data on scientists are often all that later scholars have to fall back on. His attempts to determine scholarly productivity by having panels of scientists vote on comparative merits of investigators in particular fields are summed up by Stephen Visher in *Scientists Starred* (Baltimore, 1947). In general Cattell was somewhat more cautious than Visher in using the starred scientists, but both display a fair degree of gullibility. Cattell's findings (and those of Visher) generated a good many myths that required correction in the development of national science policy after World War II—for example, small colleges are more productive of scientists than universities.

Cattell was a conscious opponent of George Ellery Hale's program for revitalizing the National Academy of Sciences and making it the leader of the scientific community. After World War I, when Hale's views prevailed, Cattell's influence waned outside the A.A.A.S. In 1921 he founded the Psychological Corporation and Science Service, the former to give scientists an opportunity to reap the benefits of their discoveries, the latter to communicate news of science to the lay public.

BIBLIOGRAPHY

Cattell's personal papers are in the Library of Congress. The collection is discussed in Grover C. Batts, "The James McKeen Cattell Papers," in Library of Congress, *Quarterly Journal of Current Acquisitions,* **17** (1960), 170–174, and is described in greater detail in the published register of the collection, *James McKeen Cattell, a Register of His Papers in the Library of Congress* (Washington, D.C., 1962).

No good bibliography of Cattell's writings exists, but the most useful is the one appended to Pillsbury's memoir (see below). Although admittedly selective, it is reasonably good for his psychological research but rather skimpy for Cattell as a man of scientific affairs. After Cattell's death, A. T. Poffenberger edited a splendid collection of Cattelliana, *James McKeen Cattell, Man of Science, 1860–1944,* 2 vols. (Lancaster, Pa., 1947), which contains much useful bibliographic information. The Cattell papers in the Library of Congress have files of his papers and speeches, including unpublished items. Simply by browsing through *Science* and other publications associated with Cattell one can turn up other writings, some quite consequential.

The best brief biographical account is W. B. Pillsbury, "Biographical Memoir of James McKeen Cattell, 1860–1944," in *Biographical Memoirs. National Academy of Sciences,* **25** (1949), 1–16. More ambitious but less reliable on some details is R. S. Woodworth, "James McKeen Cattell, 1860–1944," in *Psychological Review,* **51** (1944), 201–209, repr. in Poffenberger, *op. cit.,* I, 1–12. A memorial issue of *Science* contains interesting reminiscences and appreciations: **99** (1944), 151–165. For contemporary estimates of Cattell as a psychologist, see *The Psychological Researches of James McKeen Cattell: A Review by Some of His Students,* no. 30 of Archives of Psychology (New York, 1914); and E. G. Boring, *A History of Experimental Psychology* (New York, 1929), pp. 519–528. A discussion of the Cattell-Hale relationship is in Nathan Reingold, "National Aspirations and Local Purposes," in *Transactions of the Kansas Academy of Science,* **71** (1969), 235–246.

NATHAN REINGOLD

CAUCHY, AUGUSTIN-LOUIS (*b.* Paris, France, 21 August 1789; *d.* Sceaux [near Paris], France, 22 May 1857), *mathematics, mathematical physics, celestial mechanics.*

Life. Cauchy's father, Louis-François Cauchy, was born in Rouen in 1760. A brilliant student of classics at Paris University, after graduating he established himself as a barrister at the *parlement* of Normandy. At the age of twenty-three he became secretary general to Thiroux de Crosnes, the *intendant* of Haute Normandie. Two years later he followed Thiroux to Paris, where the latter had been appointed to the high office of *lieutenant de police.*

Louis-François gradually advanced to high administrative positions, such as that of first secretary to the Senate. He died in 1848. In 1787 he married Marie-Madeleine Desestre, who bore him four sons and two daughters. She died in 1839. Of their daughters, Thérèse died young and Adèle married her cousin G. de Neuburg. She died in 1863. The youngest son, Amédée, died in 1831, at the age of twenty-five; Alexandre (1792–1857) held high judicial posts; and Eugène (1802–1877) held administrative posts and became known as a scholar in the history of law. Augustin was the eldest child.

Cauchy enjoyed an excellent education; his father was his first teacher. During the Terror the family escaped to the village of Arcueil, where they were neighbors of Laplace and Berthollet, the founders of the celebrated Société d'Arcueil. Thus, as a young boy Augustin became acquainted with famous scientists. Lagrange is said to have forecast his scientific genius while warning his father against showing him a mathematical text before the age of seventeen.

After having completed his elementary education at home, Augustin attended the École Centrale du Panthéon. At the age of fifteen he completed his classical studies with distinction. After eight to ten months of preparation he was admitted in 1805 to the École Polytechnique (at the age of sixteen). In 1807 he entered the École des Ponts et Chaussées, which he left (1809?) to become an engineer, first at the works of the Ourcq Canal, then the Saint-Cloud bridge, and finally, in 1810, at the harbor of Cherbourg, where Napoleon had started building a naval base for his intended operations against England. When he departed for Cherbourg, his biographer says, Cauchy carried in his baggage Laplace's *Mécanique céleste,* Lagrange's *Traité des fonctions analytiques,* Vergil, and Thomas à Kempis' *Imitatio.* Cauchy returned to Paris in 1813, allegedly for reasons of health, although nothing is known about any illness he suffered during his life.

Cauchy had started his mathematical career in 1811 by solving a problem set to him by Lagrange: whether the angles of a convex polyhedron are determined by its faces. His solution, which surprised his contemporaries, is still considered a clever and beautiful piece of work and a classic of mathematics. In 1812 he solved Fermat's classic problem on polygonal numbers: whether any number is a sum of *n* *n*gonal numbers. He also proved a theorem in what later was called Galois theory, generalizing a theorem of Ruffini's. In 1814 he submitted to the French Academy the treatise on definite integrals that was to become the basis of the theory of complex functions. In 1816 he won a prize contest of the French Academy on the propagation of waves at the surface of a liquid; his results are now classics in hydrodynamics. He invented the method of characteristics, which is crucial for the theory of partial differential equations, in 1819; and in 1822 he accomplished what to the heterodox opinion of the author is his greatest achievement and would suffice to assure him a place among the greatest scientists: the founding of elasticity theory.

When in 1816 the republican and Bonapartist Gaspard Monge and the "regicide" Lazare Carnot were expelled from the Académie des Sciences, Cauchy was appointed (not elected) a member. (Even his main biographer feels uneasy about his hero's agreeing to succeed the highly esteemed and harshly treated Monge.) Meanwhile Cauchy had been appointed *répétiteur,* adjoint professor (1815), and full professor (1816) at the École Polytechnique [11];[1] at some time before 1830 he must also have been appointed to chairs at the Faculté des Sciences and at the Collège de France.[2] His famous textbooks, which date from this period, display an exactness unheard of until then and contain his fundamental work in analysis, which has become a classic. These works have been translated several times.

In 1818 Cauchy married Aloïse de Bure, daughter (or granddaughter) of a publisher who was to publish most of Cauchy's work. She bore him two daughters, one of whom married the viscount de l'Escalopier and the other the count of Saint-Pol. The Cauchys lived on the rue Serpente in Paris and in the nearby town of Sceaux.

Cauchy's quiet life was suddenly changed by the July Revolution of 1830, which replaced the Bourbon king, Charles X, with the Orléans king of the bourgeoisie, Louis-Philippe. Cauchy refused to take the oath of allegiance, which meant that he would lose his chairs. But this was not enough: Cauchy exiled himself. It is not clear why he did so: whether he feared a new Terror and new religious persecutions, whether he meant it as a demonstration of his feelings against the new authority, or whether he simply thought he could not live honestly under a usurper.

Leaving his family, Cauchy went first to Fribourg, where he lived with the Jesuits. They recommended him to the king of Sardinia, who offered him a chair at the University of Turin.[3] Cauchy accepted. In 1833,

however, he was called to Prague, where Charles X had settled, to assist in the education of the crown prince (later the duke of Chambord). Cauchy accepted the offer with the aim of emulating Bossuet and Fénelon as princely educators. In due time it pleased the ex-king to make him a baron. In 1834 Mme. Cauchy joined her husband in Prague—the biographer does not tell us for how long.

The life at court and journeys with the court took much of Cauchy's time, and the steady flow of his publications slowed a bit. In 1838 his work in Prague was finished, and he went back to Paris. He resumed his activity at the Academy, which meant attending the Monday meeting and presenting one or more communications to be printed in the weekly *Comptes rendus;* it is said that soon the Academy had to put a restriction on the size of such publications. In the course of less than twenty years the *Comptes rendus* published 589 notes by Cauchy—and many more were submitted but not printed. As an academician Cauchy was exempted from the oath of allegiance. An effort to procure him a chair at the Collège de France foundered on his intransigence, however. In 1839 a vacancy opened at the Bureau des Longitudes which legally completed itself by cooption. Cauchy was unanimously elected a member, but the government tied the confirmation to conditions that Cauchy again refused to accept. Biot [7] tells us that two subsequent ministers of education vainly tried to build golden bridges for Cauchy. Bertrand [11] more specifically says that the only thing they asked of him was to keep silent about the fact that he had not been administered the oath. But, according to Biot [7, p. 152], "even such an appearance terrified Cauchy, and he tried to make it impossible by all diplomatic finesses he could imagine, finesses which were those of a child."

When the February Revolution of 1848 established the Second Republic, one of the first measures of the government was the repeal of the act requiring the oath of allegiance. Cauchy resumed his chair at the Sorbonne (the only one that was vacant). He retained this chair even when Napoleon III reestablished the oath in 1852, for Napoleon generously exempted the republican Arago and the royalist Cauchy.

A steady stream of mathematical papers traces Cauchy's life. His last communication to the Academy closes with the words "C'est ce que j'expliquerai plus au long dans un prochain mémoire." Eighteen days later he was dead. He also produced French and Latin poetry, which, however, is better forgotten. More than a third of his biography deals with Cauchy as a devout Catholic who took a leading part in such charities as that of François Régis for unwed mothers, aid for starving Ireland, rescue work for criminals, aid to the Petit Savoyards, and important activity in the Society of Saint Vincent de Paul. Cauchy was one of the founders of the Institut Catholique, an institution of higher education; he served on a committee to promote the observance of the sabbath; and he supported works to benefit schools in the Levant. Biot [7] tells us that he served as a social worker in the town of Sceaux and that he spent his entire salary for the poor of that town, about which behavior he reassured the mayor: "Do not worry, it is only my salary; it is not my money, it is the emperor's."

Cauchy's life has been reported mainly according to Valson's work [8], which, to tell the truth, is more hagiography than biography. It is too often too vague about facts which at that time could easily have been ascertained; it is a huge collection of commonplaces; and it tends to present its hero as a saint with all virtues and no vices. The facts reported are probably true, but the many gaps in the story will arouse the suspicion of the attentive reader. The style reminds one of certain saccharine pictures of saints. Contrary to his intention Valson describes a bigoted, selfish, narrow-minded fanatic. This impression seems to be confirmed by a few contemporary anecdotal accounts. N. H. Abel [10] called Cauchy mad, infinitely Catholic, and bigoted. Posthumous accounts may be less trustworthy, perhaps owing their origin or form to a reaction to Valson's book.

A story that is hardly believable is told by Bertrand [9] in a review of Valson. Bertrand, who deeply admired Cauchy's scientific genius, recalled that in 1849, when Cauchy resumed his chair of celestial mechanics at the Sorbonne,

> . . . his first lessons completely deceived the expectation of a selected audience, which was surprised rather than charmed by the somewhat confused variety of subjects dealt with. The third lesson, I remember, was almost wholly dedicated to the extraction of the square root, where the number 17 was taken as an example and the computations were carried out up to the tenth decimal by methods familiar to all auditors but believed new by Cauchy, perhaps because he had hit upon them the night before. I did not return; but this was a mistake, since the next lectures would have introduced me ten years earlier to some of the most brilliant discoveries of the famous master.

This story was a vehement reaction to Valson's statement that Cauchy was an excellent teacher who ". . . never left a subject until he had completely exhausted and elucidated it so he could satisfy the demands of the most exacting spirits" [8, I, 64]. Clearly, this is no more than another of Valson's many cliché epithets dutifully conferred upon his subject. In fact Cauchy's

manner of working was just the contrary of what is here described, as will be seen.

According to Valson, when Mme. Cauchy joined her husband in Prague [8, I, 90], he complained that he was still separated from his father and mother and did not mention his daughters. Yet according to Biot [7] his wife brought their daughters with her and the family stayed in Prague and left there together. It is characteristic of Valson that after the report of their birth he never mentions Cauchy's daughters, except for the report that one of them was at his bedside when he died. His wife is not mentioned much more. The result of such neglect is that despite the many works of charity one feels a disturbing lack of human relationships in Cauchy's life. Possibly this was less Cauchy's fault than that of his biographer. But even Bertrand [11], who was more competent than Valson, more broad-minded, and a master of the *éloge*, felt uneasy when he had to speak about Cauchy's human qualities. Biot [7] was more successful. It is reassuring to learn from him that Cauchy befriended democrats, nonbelievers, and odd fellows such as Laurent. And it is refreshing to hear Biot call Cauchy's odd behavior "childish."

A prime example of his odd behavior is his exile. One can understand his refusal to take the oath, but sharing exile with the depraved king is another matter. It could have been a heroic feat; unfortunately it was not. The lone faithful paladin who followed his king into exile while all France was gratified at the smooth solution of a dangerous crisis looks rather like the Knight of the Rueful Countenance. Yet his quixotic behavior is so unbelievable that one is readily inclined to judge him as being badly melodramatic. Stendhal did so as early as 1826, when he said in *New Monthly Magazine* (see [13], p. 192) about a meeting of the Académie des Sciences:

> After the lecture of a naturalist, Cauchy rose and protested the applause. "Even if these things would be as true as I think they are wrong"—he said—"it would not be convenient to disclose them to the public . . . this can only prejudice our holy religion." People burst out laughing at this talk of Cauchy, who . . . seems to seek the role of a martyr to contempt.

Probably Stendhal was wrong. Biot knew better: Cauchy was a child who was as naïve as he looked. Among his writings one finds two pieces in defense of the Jesuits [8, pp. 108–121] that center on the thesis that Jesuits are hated and persecuted because of their virtue. It would not be plausible that the man who wrote this was really so naïve if the author were not Cauchy.

Another story about Cauchy that is well confirmed

comes from the diary of the king of Sardinia, 16 January 1831 [13, p. 160]. In an audience that the king granted Cauchy, five times Cauchy answered a question by saying, "I expected Your Majesty would ask me this, so I have prepared to answer it." And then he took a memoir out of his pocket and read it.

Cauchy's habit of reading memoirs is confirmed by General d'Hautpoul [12; 13, p. 172], whose memoirs of Prague shed an unfavorable light on Cauchy as an educator and a courtier.

Sometimes, in the steady flow of his Academy publications, Cauchy suddenly turned or returned to a different subject; after a few weeks or months it would become clear why he did so. He would then submit to the Academy a report on a paper of a *savant étranger* (i.e., a nonmember of the Academy), which had been sent to him for examination. Meanwhile he had proved anew the author's results, broadened, deepened, and generalized them. And in the report he never failed to recall all his previous investigations related to the subject of the paper under consideration. This looks like extremely unfair behavior, and in any other case it would be—but not with Cauchy. Cauchy did not master mathematics; he was mastered by it. If he hit on an idea—and this happened often—he could not wait a moment to publish it. Before the weekly *Comptes rendus* came into being this was not easy, so in 1826 he founded a private journal, *Exercices de mathématiques*. Its twelve issues a year were filled by Cauchy himself with the most improbable choice of subjects in the most improbable order. Five volumes of the *Exercices* appeared before he left Paris. In Turin he renewed this undertaking—and even published in the local newspaper—and continued it in Prague and again in Paris, finally reaching a total of ten volumes. He published in other journals, too; and there are at least eighteen memoirs by him published separately, in no periodical or collection, as well as many textbooks. Sometimes his activity seems explosive even by his own standards. At the meetings of the Academy of 14, 21, and 28 August 1848 he submitted five notes and five memoirs—probably to cover the holiday he would take until 9 October. Then in nine meetings, until 18 December, he submitted nineteen notes and ten memoirs. He always presented many more memoirs than the Academy could publish.

On 1 March 1847 Lamé presented to the Academy a proof of Fermat's last theorem. Liouville pointed out that the proof rested on unproved assumptions in the arithmetic of circle division fields. Cauchy immediately returned to this problem, which he had considered earlier. For many weeks he informed the Academy of all his abortive attempts to solve the problem (which is still unsolved) by proving Lamé's assumption. On 24 May, Liouville read a letter from Ernst Kummer, who had disproved Lamé's assumption. Even such an incident would not silence Cauchy, however, and a fortnight later he presented investigations generalizing those of Kummer.

The story that Abel's Paris memoir went astray [10] through Cauchy's neglect rests on gossip. It has been refuted by D. E. Smith, who discovered Legendre and Cauchy's 1829 report on Abel's work [10a; 10b]. In general it would be wrong to think that Cauchy did not recognize the merits of others. When he had examined a paper, he honestly reported its merits, even if it overlapped his own work. Of all the mathematicians of his period he is the most careful in quoting others. His reports on his own discoveries have a remarkably naïve freshness because he never forgot to sum up what he owed to others. If Cauchy were found in error, he candidly admitted his mistake.

Most of his work is hastily, but not sloppily, written. He was unlike Gauss, who published *pauca sed matura* only—that is, much less than he was able to, and many things never. His works still charm by their freshness, whereas Gauss's works were and still are turgid. Cauchy's work stimulated new investigations much earlier than Gauss's did and in range of subject matter competes with Gauss's. His publishing methods earned for Gauss an image of an almost demonic intelligence who knew all the secrets better and more deeply than ordinary men. There is no such mystery around Cauchy, who published lavishly—although nothing that in maturity could be compared with Gauss's publications. He sometimes published the same thing twice, and sometimes it is evident that he was unfamiliar with something he had brought out earlier. He published at least seven books and more than 800 papers.

More concepts and theorems have been named for Cauchy than for any other mathematician (in elasticity alone [35] there are sixteen concepts and theorems named for Cauchy). All of them are absolutely simple and fundamental. This, however, is an objective assessment and does not consider the subjective value they had for Cauchy. In the form that Cauchy discovered and understood them, they often were not so simple; and from the way that he used or did not use them, it often appears that he did not know that they were fundamental. In nearly all cases he left the final form of his discoveries to the next generation. In all that Cauchy achieved there is an unusual lack of profundity. He was one of the greatest mathematicians—and surely the most universal—and also contributed greatly to mathematical physics. Yet he was the most superficial of the great mathematicians, the one who had a sure feeling for what was simple and fundamental without realizing it.

Writings. Cauchy's writings appeared in the publications of the Academy, in a few scientific journals, separately as books, or in such collections as the *Exercices*. Some of his courses were published by others [3; 4]. There were, according to Valson, eighteen *mémoires détachés*. A Father Jullien (a Jesuit), under the guidance of Cauchy, once catalogued his work. The catalog has not been published. Valson's list was based on it, but he was not sure whether it was complete;[4] and it is not clear whether Valson ever personally saw all of the *mémoires détachés* or whether all of them really existed. Some of them, which have been lithographed, are rare.[5] Cauchy must have left an enormous quantity of anecdota, but nothing is known of what happened to them. What the Academy possesses seems to be insignificant.

In 1882 the Academy began a complete edition of Cauchy's work [1]. Volume XV of the second series is still lacking. In the second of the second series, which appeared as late as 1957, the commissioners of the Academy declared their decision to cease publication after still another volume. They did not say whether the edition would be stopped because it is finished or whether it will remain unfinished. The missing volume seems to have been reserved for the *mémoires détachés,* among which are some of Cauchy's most important papers. Fortunately, one of them has been reprinted separately [2; 2a; 2b].

The Academy edition of Cauchy's works contains no anecdota. Cauchy's papers have been arranged according to the place of original publication. The first series contains the Academy publications; the second, the remainder. This makes use of the edition highly inconvenient. Everything has been published without comment: there is no account of how the text was established and no statement whether printing errors and evident mistakes were corrected (sometimes it seems that they were not). Sometimes works have been printed twice (such as [1, 1st ser., V, 180–198], which is a textual extract of [1, 2nd ser., XI, 331–353]). In other cases such duplications have been avoided, but such avoidances of duplication have not regularly been accounted for. Since Cauchy and his contemporaries quote the *Exercices* according to numbers of issues it is troublesome that this subdivision has not been indicated. This criticism, however, is not to belittle the tremendous value of the Academy edition.

Important bibliographic work on Cauchy was done by B. Boncompagni [5].

Since Valson's biography [8] and the two biographic sketches by Biot and Bertrand [7; 11], no independent biographical research on Cauchy has been undertaken except for that by Terracini [13]. (It would be troublesome but certainly worthwhile to establish a faithful picture of Cauchy from contemporary sources. He was one of the best-known people of his time and must have been often mentioned in newspapers, letters, and memoirs.[6]) Valson's analysis of Cauchy's work is unsatisfactory because sometimes he did not understand Cauchy's mathematics; for instance, he mistook his definition of residue. Lamentably no total appreciation of Cauchy's work has been undertaken since. There are, however, a few historical investigations of mathematical fields that devote some space to Cauchy. See Casorati [19] on complex functions (not accessible to the present reporter); Verdet [20] on optics; Studnicka [21] on determinants (not accessible); Todhunter [21a] on elasticity; Brill and Noether [22] on complex functions (excellent); Stäckel and Jourdain [23, 24, 25] on complex functions; the *Encyclopädie der mathematischen Wissenschaften* [26] on mathematical physics and astronomy; Burkhardt [27] on several topics (chaotic, with unconnected textual quotations, but useful as a source); Miller [28] on group theory; Jourdain [29] on calculus (not accessible); Love [30] on elasticity (fair); Lamb [31] on hydrodynamics (excellent); Whittaker [32] on optics (excellent); Carruccio [33] on complex functions; Courant and Hilbert [34] on differential equations (fair); Truesdell and Toupin [35] on elasticity (excellent).

Because of the great variety of fields in which Cauchy worked it is extremely difficult to analyze his work and properly evaluate it unless one is equally experienced in all the fields. One may overlook important work of Cauchy and commit serious errors of evaluation. The present author is not equally experienced in all the fields: in number theory less than in analysis, in mathematical physics less than in mathematics, and entirely inexperienced in celestial mechanics.

Calculus. The classic French *Cours d'analyse* (1821), descended from Cauchy's books on calculus [1, 2nd ser., III; IV; IX, 9–184], forcefully impressed his contemporaries. N. H. Abel [17] called the work [1, 2nd ser., III] "an excellent work which should be read by every analyst who loves mathematical rigor." In the introduction Cauchy himself said, "As to the methods, I tried to fill them with all the rigor one requires in geometry, and never to revert to arguments taken from the generality of algebra." Cauchy needed no metaphysics of calculus. The "generality of algebra," which he rejected, assumed that what is true for real numbers is true for complex numbers; that what is true for finite magnitudes is true for infinitesimals;

that what is true for convergent series is true for divergent ones. Such a remark that looks trivial today was a new, if not revolutionary, idea at the time.

Cauchy refused to speak about the sum of an infinite series unless it was convergent, and he first defined convergence and absolute convergence of series, and limits of sequences and functions [1, 2nd ser., III, 17–19]. He discovered and formulated *convergence criteria*: the Cauchy principle of $s_{m+n} - s_n$ becoming small[7] [1, 2nd ser., VII, 269], the root criterion using the lowest upper limit of $\sqrt[n]{|a_n|}$ [1, 2nd ser., III, 121], the quotient criterion using that of $|a_{n+1}|/|a_n|$ [1, 2nd ser., III, 123], their relation, the integral criterion [1, 2nd ser., VII, 267–269]. He defined upper and lower limits [1, 2nd ser., III, 121], was first to prove the convergence of $(1 + 1/n)^n$, and was the first to use the limit sign [1, 2nd ser., IV, 13 f.]. Cauchy studied convergence of series under such operations as addition and multiplication [1, 2nd ser., III, 127–130] and under rearrangement [1, 1st ser., X, 69; 1st ser., IX, 5–32]. To avoid pitfalls he defined convergence of double series too cautiously [1, 2nd ser., III, 441; X, 66]. Explicit estimations of convergence radii of power series are not rare in his work. By his famous example $\exp(-x^{-2})$ he warned against rashness in the use of Taylor's series [1, 2nd ser., II, 276–282]. He proved Lagrange's and his own remainder theorem, first by integral calculus [1, 2nd ser., IV, 214] and later by means of his own generalized mean-value theorem [1, 2nd ser., IV, 243, 364; VI, 38–42], which made it possible to sidestep integral calculus. In the first proof he used the integral form of the remainder that is closely connected to his famous formula [1, 2nd ser., IV, 208–213],

$$(1) \quad \int^t \cdots_n \int^t f(t)\, dt = \frac{1}{(n-1)!} \int^t (t - \tau)^{n-1} f(\tau)\, d\tau.$$

An important method in power series arising from multiplication, inversion, substitution, and solving differential equations was Cauchy's celebrated *calcul des limites* (1831–1832), which in a standard way reduces the convergence questions to those of geometrical series [1, 2nd ser., II, 158–172; XI, 331–353; XII, 48–112].

Cauchy invented our notion of continuity[8] and proved that a continuous function has a zero between arguments where its signs are different [1, 2nd ser., III, 43, 378], a theorem also proved by Bolzano. He also did away with multivalued functions. Against Lagrange he again and again stressed the limit origin of the differential quotient. He gave the first adequate definition of the definite integral as a limit of sums [1, 2nd ser., IV, 122–127] and defined improper integrals [1, 2nd ser., IV, 140–144], the well-known

Cauchy principal value of an integral with a singular integrand [1, 1st ser., I, 288–303, 402–406], and closely connected, singular integrals (i.e., integrals of infinitely large functions over infinitely small paths [δ functions]) [1, 1st ser., I, 135, 288–303, 402–406; 2nd ser., I, 335–339; IV, 145–150; XII, 409–469]. Cauchy made much use of discontinuous factors [1, 1st ser., XII, 79–94] and of the Fourier transform (see under Differential Equations). Cauchy also invented what is now called the Jacobian, although his definition was restricted to two and three dimensions [1, 1st ser., I, 12].

In addition, Cauchy gave the proof of the fundamental theorem of algebra that uses the device of lowering the absolute value of an analytic function as long as it does not vanish [1, 2nd ser., I, 258–263; III, 274–301; IV, 264; IX, 121–126]. His investigations (1813, published in 1815) on the number of real roots [1, 2nd ser., I, 170–257] were surpassed by Sturm's (1829). In 1831 he expressed the number of complex roots of $f(z)$ in a domain by the logarithmic residue formula, noticing that the same expression gives the number of times $\mathrm{Re}\, f(z)/\mathrm{Im}\, f(z)$ changes from $-\infty$ to ∞ along a closed curve—in other words how often the f-image of the curve turns around 0—which led to a new proof of the fundamental theorem that was akin to Gauss's first, third, and fourth (reconstructed from [8, II, 85–88]—see also [1, 1st ser., IV, 81–83], since the 1831 *mémoire détaché* has not yet been republished). In [1, 2nd ser., I, 416–466] the proof has been fashioned in such a way that it applies to mappings of the plane into itself by pairs of functions.

With unsurpassed skill and staggering productivity he calculated and transformed integrals and series developments.

In mathematics Cauchy was no dogmatist. Despite his insistence on the limit origin of the differential quotient, he never rejected the formal approach, which he called symbolic [1, 2nd ser., VII, 198–254; VIII, 28–38] and often justified by Fourier transformation. On a large scale he used the formal approach in differential and difference equations. Cauchy admitted semiconvergent series, called "limited" [1, 1st ser., VIII, 18–25; XI, 387–406], and was the first to state their meaning and use clearly. By means of semiconvergent series in 1842 he computed all the classic integrals such as $\int_v^\infty \cos 1/2\pi v^2\, dv$ [1, 1st ser., VII, 149–157] and, in 1829 [1, 1st ser., II, 29–58], asymptotics of integrals of the form $\int u^n v\, dx$, particularly those such as $\int (1 - x)^m x^n f(x)\, dx$, where beta functions are involved if $f(x)$ is duly developed in a series [1, 1st ser., IX, 75–121; II, 29–58]; he used rearrangements of conditionally convergent series [1, 1st ser., IX, 5–14] in the same way.

In a more profound sense Cauchy was rather more flexible than dogmatic, for more often than not he sinned against his own precepts. He operated on series, Fourier transforms, and improper and multiple integrals as if the problems of rigor that he had raised did not exist, although certainly he knew about them and would have been able to solve them. Although he had been first to define continuity, it seems that Cauchy never proved the continuity of any particular function. For instance, it is well known that he asserted the continuity of the sum of a convergent series of continuous functions [1, 2nd ser., III, 120]; Abel gave a counterexample, and it is clear that Cauchy himself knew scores of them. It is less known that later Cauchy correctly formulated and applied the uniform convergence that is needed here [1, 1st ser., XII, 33]. He proved

$$\lim \left(1 + \frac{1}{n}\right)^n = \sum \frac{1}{n!}$$

by a popular but unjustified interchange of limit processes [1, 2nd ser., III, 147], although he was well acquainted with such pitfalls; it is less well-known that he also gave a correct proof [1, 2nd ser., XIV, 269–273]. Terms like "infinitesimally small" prevail in Cauchy's limit arguments and epsilontics still looks far away, but there is one exception. His proof [1, 2nd ser., III, 54–55] of the well-known theorem

$$\text{If } \lim_{x \to \infty} (f(x + 1) - f(x)) = \alpha,$$
$$\text{then } \lim_{x \to \infty} x^{-1} f(x) = \alpha$$

is a paragon, and the first example, of epsilontics—the character ε even occurs there. It is quite probable that this was the beginning of a method that, after Cauchy, found general acceptance. It is the weakest point in Cauchy's reform of calculus that he never grasped the importance of uniform continuity.

Complex Functions. The discoveries with which Cauchy's name is most firmly associated in the minds of both pure and applied mathematicians are without doubt his fundamental theorems on complex functions.

Particular complex functions had been studied by Euler, if not earlier. In hydrodynamics d'Alembert had developed what are now called the Cauchy-Riemann differential equations and had solved them by complex functions. Yet even at the beginning of the nineteenth century complex numbers were not yet unanimously accepted; functions like the multivalued logarithm aroused long-winded discussions. The geometrical interpretation of complex numbers, although familiar to quite a few people, was made explicit by Gauss as late as 1830 and became popular under his name. It is, however, quite silly to doubt whether, earlier, people who interpreted complex functions as pairs of real functions knew the geometric interpretation of complex numbers. Gauss's proofs of the fundamental theorem of algebra, although reinterpreted in the real domain, implicitly presupposed some facts from complex function theory. The most courageous ventures in complex functions up to that time were the rash ideas of Euler and Laplace of shifting real integration paths in the complex domain (for instance, that of e^{-x^2} from $-\infty$ to ∞) to get new formulas for definite integrals [24], then an entirely unjustified procedure. People sometimes ask why Newton or Leibniz or the Bernoullis did not discover Cauchy's integral theorem and integral formula. Historically, however, such a discovery should depend first on some geometrical idea on complex numbers and second on some more sophisticated ideas on definite integrals. As long as these conditions were not fulfilled, it was hardly possible to imagine integration along complex paths and theorems about such kinds of integrals. Even Cauchy moved slowly from his initial hostility toward complex integration to the apprehension of the theorems that now bear his name. It should be mentioned that Gauss knew most of the fundamental facts on complex functions, although he never published anything on them [22, pp. 155–160].

The first comprehensive theory of complex numbers is found in Cauchy's *Cours d'analyse* of 1821 [1, 2nd ser., III, 153–256]. There he justified the algebraic and limit operations on complex numbers, considered absolute values, and defined continuity for complex functions. He did not teach complex integration, although in a sense it had been the subject of his *mémoire* submitted to the French Academy in 1814 and published in 1825 [1, 1st ser., I, 329–506]. It is clear from its introduction that this *mémoire* was written in order to justify such rash but fruitful procedures as those of Euler and Laplace mentioned above. But Cauchy still felt uneasy in the complex domain. He interpreted complex functions as pairs of real functions of two variables to which the Cauchy-Riemann differential equations apply. This meant bypassing rather than justifying the complex method. Thanks to Legendre's criticism Cauchy restored the complex view in footnotes added to the 1825 publication, although he did not go so far as to admit complex integration paths. The problem Cauchy actually dealt with in this *mémoire* seems strange today. He considered a differentiable function $f = u + iv$ of the complex variable $z = x + iy$ and, using one of the Cauchy-Riemann differential equations, formed the double integral

$$(2) \qquad \iint u_x \, dx \, dy = \iint v_y \, dx \, dy$$

over a rectangle $x_0 \leqslant x \leqslant x_1, y_0 \leqslant y \leqslant y_1$. Performing the integrations, he obtained the fundamental equality

$$(3) \qquad \int_{y_0}^{y_1} (u(x_1,y) - u(x_0,y))\, dy =$$

$$\int_{x_0}^{x_1} (v(x,y_1) - v(x,y_0))\, dx.$$

Using the other Cauchy-Riemann differential equation, he obtained a second equality; and together they yielded

$$(4) \qquad i\int_{y_0}^{y_1} (f(x_1,y) - f(x_0,y))\, dy =$$

$$\int_{x_0}^{x_1} (f(x,y_1) - f(x,y_0))\, dx,$$

the Cauchy integral theorem for a rectangular circuit, as soon as one puts the i between the d and the y.

Of course regularity is supposed in this proof. Cauchy had noticed, however, that (3) and (4) may cease to hold as soon as there is a singularity within the rectangle; this observation had even been his point of departure. He argued that when drawing conclusions from (2), one had interchanged integrations; and he decided that this was not generally allowed. He tried to compute the difference between the two members of (4), but his exposition is quite confused and what he means is elucidated elsewhere [1, 2nd ser., VI, 113–123].

Let $a + ib$ be the (simple polar) singularity. Then the integrals in (3) and (4) have to be understood as their principal values, e.g.,

$$\int_{y_0}^{y_1} = \lim_{\varepsilon=0}\left(\int_{y_0}^{b-\varepsilon} + \int_{b+\varepsilon}^{y_1}\right).$$

This means that the first member of (3) is the limit of the sum of the double integrals over the rectangles

$$x_0 \leqslant x \leqslant x_1, y_0 \leqslant y \leqslant b - \varepsilon;$$
$$x_0 \leqslant x \leqslant x_1, b + \varepsilon \leqslant y \leqslant y_1;$$

and the difference between both members of (3) and (4) is the limit of the integral over

$$x_0 \leqslant x \leqslant x_1, b - \varepsilon \leqslant y \leqslant b + \varepsilon.$$

In other words,

$$\lim_{\varepsilon=0}\int_{x_0}^{x_1} (f(x, b + \varepsilon) - f(x, b - \varepsilon))\, dx,$$

where x_0, x_1 may still be replaced by arbitrary abscissae around a. This expression is just what Cauchy calls a singular integral. In his 1814 paper he allows the singularity to lie on the boundary of the rectangle,

and even in a corner. (To make the last step conclusive, one should define principal values in a more sophisticated way.)

Of course if there is one singularity $a + ib$ within the rectangle, then according to the residue theorem the difference of both members of (4) should be $2\pi i$ times the coefficient of $(z - (a + ib))^{-1}$ in the Laurent series of $f(z)$. This knowledge is still lacking in Cauchy's 1814 paper. He deals with simple polar singularities only, taking $f(z)$ as a fraction $g(z)/h(z)$ and proving that the difference equals

$$(5) \qquad 2\pi i\,\frac{g(a + ib)}{h'(a + ib)}.$$

In the 1825 footnotes he adds the expression

$$(6) \qquad 2\pi i \lim_{\varepsilon=0} \varepsilon f((a + ib) + \varepsilon),$$

which had already appeared in 1823 [1, 2nd ser., I, 337].

Cauchy's most important general result here is the computation of

$$\int_{-\infty}^{\infty} f(z)\, dz$$

(over the real axis) as a sum of expressions (5) from the upper half-plane; singularities on the real axis are half accounted for in such sums. The conditions under which he believes one is entitled to pass from the rectangle as an integration path to the real axis are not clearly formulated. It seems that he requires vanishing of $f(z)$ at infinity, which of course is too much; in any case, he applies the result to functions with an infinity of poles, where this requirement does not hold. In 1826 he stated more sophisticated but still too rigid conditions [1, 2nd ser., VI, 124–145]; strangely enough, at the end of this paper he returned to the useless older ones. In 1827 [1, 2nd ser., VII, 291–323] he discovered the "good conditions": $zf(z)$ staying bounded on an appropriate sequence of circles with fixed centers and with radii tending to infinity.

Even in the crude form of the 1814 mémoire, Cauchy's integral theorem proved to be a powerful instrument; a host of old and new definite integrals could be verified by this method. The approach by double integrals looks strange, but at that time it must have been quite natural; in fact, in his third proof of the fundamental theorem of algebra (1816), Gauss used the same kind of double integrals to deal with singularities [22, pp. 155–160].

Genuine complex integration is still lacking in the 1814 mémoire, and even in 1823 Poisson's reflections on complex integration [23] were bluntly rejected by Cauchy [1, 2nd ser., I, 354]. But they were a thorn in his side; and while Poisson did not work out this idea, Cauchy soon did. In a *memoire détaché* of 1825 [2], he took a long step toward what is now called Cauchy's integral theorem. He defined integrals over arbitrary paths in the complex domain; and through the Cauchy-Riemann differential equations he derived, by variation calculus, the fact that in a domain of regularity of $f(z)$ such an integral depended on the end points of the path only. Curiously enough he did not introduce closed paths. Further, he allowed the changing path to cross a simple polar singularity γ, in which case the integral had to be interpreted by its principal value. Of course, the variation then would differ from zero; its value, equal toward both sides, would be

$$\lim_{\varepsilon=0} \varepsilon f(\gamma + \varepsilon)\pi i.$$

In the case of an *m*tuple polar singularity γ the integrals over paths on both sides of γ would differ by

$$(6) \qquad \frac{1}{(m-1)!} \lim_{\varepsilon=0} \frac{d^{m-1}\varepsilon^m f(\gamma + \varepsilon)}{d\varepsilon^{m-1}} 2\pi i,$$

a formula that goes back at least as far as 1823 [1, 2nd ser., I, 337 n.]. (Notice that at this stage Cauchy did not know about power series development for analytic functions.) The foregoing yields the residue theorem with respect to poles; it was extended to general isolated singularities by P. A. Laurent in 1843 [22].

(It is a bewildering historical fact that by allowing for simple singularities upon the integration path, Cauchy handled his residue theorem as a much more powerful tool than the one provided by modern textbooks, with their overly narrow formulation.)

The important 1825 *memoire* was neither used nor quoted until 1851 [1, 1st ser., XI, 328], a circumstance utterly strange and hard to explain. Did Cauchy not trust the variational method of proof? Was he bothered by the (unnecessary) condition he had imposed on the paths, staying within a fixed rectangle? Did he not notice that the statement could be transformed into the one about closed paths that he most needed? Or had he simply forgotten about that *memoire détaché*? In any case, for more than twenty-five years he restricted himself to rectangular paths or circular-annular ones (derived from the rectangular kind by mapping), thus relying on the outdated 1814 *memoire* rather than on that of 1825.

The circle as an integration path and Cauchy's integral formula for this special case had in a sense already been used in 1822 and 1823 [1, 2nd ser., II, 293–294; I, 338, 343, 348], perhaps even as early as 1819 [1, 2nd ser., II, 293 n.]. The well-known integral expression for the *n*th derivative also appeared, although of course in the form

$$(7) \qquad f^{(n)}(0) = \frac{n!}{2\pi} \int_0^{2\pi} (re^{i\phi})^{-n} f(re^{i\phi}) \, d\phi,$$

since complex integration paths were still avoided. In 1840 (perhaps as early as 1831) such an expression would be called an average (over the unit circle) and, indeed, constructed as the limit of averages over regular polygons [1, 2nd ser., XI, 337].

Indirect applications of Cauchy's definition of integrals were manifold in the next few years. In the *Exercices* of 1826–1827 [1, 2nd ser., V–VI] many papers were devoted to a rather strange formal calculus of residues. The residue of $f(z)$ at γ is defined as the coefficient of πi in (6); the residue in a certain domain, as the sum of those at the different points of the domain. A great many theorems on residues are proved without recurring to the integral expressions, and it often seems that Cauchy had forgotten about that formula.

By means of residues Cauchy arrived at the partial fractions development of a function $f(z)$ with simple poles,

$$f(z) = \sum \frac{\alpha_\nu}{z - \gamma_\nu}.$$

The trouble with this series is the same as that with the residue theorem. Originally the asymptotic assumption under which this would hold, reads: vanishing at infinity. This is much too strong and surely is not what Cauchy meant when he applied the partial fractions development under much broader conditions. The condition in [1, 2nd ser., VII, 324–362] is still too strong. It is strange that in this case Cauchy did not arrive at the "good condition"; and it is stranger still that in 1843 he again required continuity at infinity, which is much too strong [1, 1st ser., VIII, 55–64].

From the partial fractions development of meromorphic functions it was a small step to the product representation of integral functions; it was taken by Cauchy in 1829–1830 [1, 2nd ser., IX, 210–253]. In special cases Cauchy also noticed the exponential factor, needed in addition to the product of linear factors; the general problem, however, was not solved until Weierstrass. Poles and roots in such investigations used to be simple; Cauchy tried multiple ones as well [1, 2nd ser., IX, 223], but this work does not testify for a clear view.

Cauchy skillfully used residues for many purposes. He expressed the number of roots of a function in a domain by logarithmic residues [1, 2nd ser., VII, 345–362] and, more generally, established a formula for sums over the roots z_i of $F(z)$,

(8) $\Sigma\phi(z_i)$ = sum of residues of $\dfrac{\phi(z)F'(z)}{F(z)}$,

which had many applications. He was well aware of the part played by arrangement in such infinite sums. In 1827 he derived the Fourier inversion formula in this context [1, 2nd ser., VI, 144; VII, 146–159, 177–209].

In 1827 Cauchy devised a method to check the convergence of a special power series for implicit functions, the so-called Lagrange series of celestial mechanics [1, 1st ser., II, 29–66]. It is the method that in the general case leads to the power series development: a function in the complex domain with a continuous derivative can be developed into a power series converging in a circle that on its boundary contains the next singularity. It seems to have been proved in the Turin *mémoires détachés* of 1831–1832; a summary of these papers was published in 1837 as *Comptes rendus* notes [1, 2nd ser., IV, 48–80] and the papers themselves, or a substantial part of them, were republished in 1840–1841 [1, 2nd ser., XI, 331–353; XII, 48–112; see also XI, 43–50]. Here Cauchy first derives in a remarkable way his integral formula from his integral theorem by means of

$$\oint \frac{f(\zeta) - f(z)}{\zeta - z} d\zeta = 0,$$

which is formulated for circular paths only, although it also applies to arbitrary circuits. The development of the integrand of

$$\frac{1}{2\pi i} \oint \frac{f(\zeta)}{\zeta - z} d\zeta = f(z)$$

according to powers of z yields the power series development of $f(z)$. Cauchy also finds an integral expression of the remainder if the development is terminated and the power series coefficients theorem

$$|a_n| \leqslant \max_{|z|=r} |f(z)| \cdot r^{-n}$$

(see also [1, 1st ser., VIII, 287–292]), which was to become the cornerstone of the powerful *calcul des limites.*

The results were applied to implicitly given functions. Using (8) a simple zero w of $F(z,w) = 0$ or a sum of simple roots or a sum $\Sigma\phi(w_i)$ over simple roots, w_i is developed into a power series according to z. Cauchy also noticed that the power series for a simple root will converge up to the first branching point, which is obtained by $\partial F(z,w)/\partial w = 0$—of course it should be one of the same sheet, but this was not clear at the time. In one of his 1837 notes [1, 1st ser., IV, 55–56] Cauchy had gone so far as to state that in all points developments according to fractional powers

were available; in 1840–1841, however, he did not come back to this point.

The foregoing summarizes some of Cauchy's tremendous production in this one area of his work. It is awe-inspiring and yet, in a sense, disappointing. One feels that Cauchy had no clear overall view on his own work. Proofs are usually unnecessarily involved and older papers, superseded by newer results, are repeatedly used and quoted. Often he seems to be blindfolded; for example, he did not notice such a consequence of his work as that a bounded regular function must be constant [1, 1st ser., VIII, 366–385] until Liouville discovered this theorem in the special case of doubly periodic functions—this is why it is now (incorrectly) called Liouville's theorem. One can imagine that Cauchy felt ashamed and confused, so confused, indeed, that he missed the point to which he should have connected Liouville's theorem. Instead of using the power series coefficients theorem he handled it with partial fractions development, which does not work properly because of the asymptotic conditions.

Cauchy also failed to discover Laurent's theorem and the simple theorem about a function with an accumulating set of roots in a regularity domain, which he knew only in crude forms [1, 1st ser., VIII, 5–10]. He would have missed much more if others had cared about matters so general and so simple as those which occupied Cauchy. Most disappointing of all is, of course, the fact that he still did not grasp the fundamental importance of his 1825 *mémoire.* He confined himself to rectangular and circular integration paths and to a special case of his integral formula.

A sequence of *Comptes rendus* notes of 1846 [1, 1st ser., X, 70–74, 133–196] marks long-overdue progress. Cauchy finally introduced arbitrary closed integration paths, although not as an immediate consequence of his 1825 *mémoire,* which he did not remember until 1851 [1, 1st ser., XI, 328]; instead, he proved his integral theorem anew by means of what is now called Green's formula—a formula dating from 1828 but possibly rediscovered by Cauchy. A still more important step was his understanding of multivalued analytic functions. The history of this notion is paradigmatic of what often happens in mathematics: an intuitive notion that is fruitful but does not match the requirements of mathematical rigor is first used in a naïve uncritical fashion; in the next phase it is ignored, and the results to which it led are, if needed, derived by cumbersome circumvention; finally, it is reinterpreted to save both the intuitive appeal and the mathematical rigor. In multivalued functions Cauchy embodied the critical phase. From 1821 he treated multivalued functions with a kill-or-cure remedy: if branched at the origin, they would be admitted in the upper half-plane only [1, 2nd ser., III, 267]. Fortunately, he more often than not forgot this gross prescription, which if followed would lead him into great trouble, as happened in 1844 [1, 1st ser., VIII, 264] —strangely enough, he wrote this confused paper just after he had taken the first step away from this dogmatism. Indeed [1, 1st ser., VIII, 156–160], he had already allowed for a plane slit by the positive axis as the definition domain of functions branched at the origin; and he had even undertaken integrations over paths pieced together from $|z| = r$ in the positive sense, $r \leqslant z \leqslant R$ in the positive sense, $|z| = R$ in the negative sense, $r \leqslant z \leqslant R$ in the negative sense, where the two rectilinear pieces are combined into one over the jump function. Such paths had long since been obtained in a natural way by a mapping of rectangular paths.

The progress Cauchy achieved in 1846 consisted in restoring the intuitive concept of a multivalued function. Such a function may now freely be followed along rather arbitrary integration paths, which are considered closed only if both the argument and the function return to the values with which they started (of course this was not yet fully correct). Integration over such closed paths produces the *indices de périodicité* that are no longer due to residues.

This is a revival of the old idea of the multivalued function, with all its difficulties. In 1851—the year of Riemann's celebrated thesis—after Puiseux's investigations on branchings, which again depended on Cauchy's work, Cauchy came back with some refinements [1, 1st ser., XI, 292–314]. He slit the plane by rectilinear *lignes d'arrêt* joining singularities and, as in the 1844 paper, proposed to compute the *indices de périodicité* by means of the integrals of the jump functions along such slits. This is too crude, and it gave Cauchy wrong ideas about the number of linearly independent periods. The correct reinterpretation of multivalued functions is by means of Riemann surfaces, with their *Querschnitte;* Cauchy's *lignes d'arrêt* are drawn in the plane, which means that they may be too numerous.

Nevertheless, the progress made in Cauchy's 1846 notes was momentous. The periodicity of elliptic and hyperelliptic functions had previously been understood as an algebraic miracle rather than by topological reasons. Cauchy's crude approach was just fine enough for elliptic and hyperelliptic integrals, and his notes shed a clear light of understanding upon those functions. Notwithstanding Riemann's work, this seemed sufficient for the near future. Thus, Briot and Boucquet [18], when preparing the second edition of their classic work, saw no advantage in using Riemann

surfaces and still presented Cauchy's theory in its old form.

Cauchy's work on complex functions has to be pieced together from numerous papers; he could have written a synthetic book on this subject but never did. The first to undertake such a project were Briot and Boucquet [18]. Nevertheless, complex function theory up to Riemann surfaces, with the sole exceptions of Laurent's theorem and the theorem on accumulating zeros, had been Cauchy's work. Of course he also did less fundamental work in complex function theory, such as generalizing Abel's theorem [1, 1st ser., VI, 149–175, 187–201], investigating "geometrical factorials" [1, 1st ser., VIII, 42–115] and so on.

Error Theory. Cauchy also made three studies of error theory, which he presented as logically connected; this, however, is misleading since to understand them one has to consider them as not connected at all.

The first seems to date from 1814, although it was not published until 1824 and 1831 [1, 2nd ser., II, 312–324; I, 358–402]. Laplace [14, II, 147–180] had tried to fit a set of n observational data $\ulcorner x_i, y_i \urcorner$ to a linear relation $y = ax + b$. Before Laplace, calculators proceeded by first shifting the average

$$\ulcorner \frac{1}{n} \Sigma x_i, \frac{1}{n} \Sigma y_i \urcorner$$

to the origin to make the problem homogeneous, and then estimating a by

$$\Sigma \delta_i y_i / \Sigma \delta_i x_i, \text{ where } \delta_i = x_i / |x_i|.$$

Laplace proposed a choice of a and b that would make the maximal error $|y_i - ax_i - b|$ (or, alternatively, the sum of the absolute errors minimal). To do so Laplace developed a beautiful method, the first specimen of linear programming. Cauchy, following a suggestion of Laplace, extended his method to fitting triples of observational data $\ulcorner x_i, y_i, z_i \urcorner$ to a relation $z = ax + by + c$; where Laplace had reasoned by pure analysis, Cauchy presented his results in a geometrical frame, which shows him to be, as often, motivated by considerations of geometry.

At the time when Cauchy took up Laplace's problem, fitting by least squares had superseded such methods. Nevertheless, in 1837 [1, 1st ser., II, 5–17] Cauchy attempted to advocate the pre-Laplacian method. He postulated the maximal error (among the $|y_i - ax_i|$) to be "minimal under the worst conditions." It does not become clear what this means, although it is a principle vague enough to justify the older methods. Actually Cauchy now dealt with a somewhat different problem: fitting systems of observational data to polynomials (algebraic, Fourier, or some other kind)

$$u = ax + by + cz + \cdots$$

where the number of terms should depend on the goodness of fit, reached during the course of the computation. What Cauchy prescribes is no more than a systematic elimination of $a, b, c, \cdots$. In 1853, when Cauchy again drew attention to this method [1, 1st ser., XII, 36–46], he was attacked by Bienaymé [16], a supporter of "least squares." Cauchy [1, 1st ser., XII, 63–124] stressed the advantage of the indeterminate number of terms in his own method, obviously not noticing that "least squares" could easily be adapted to yield the same advantage; it is, however, possible that it was the first time he had heard of "least squares."

In this discussion with Bienaymé, Cauchy took a strange turn. What looks like an argument in favor of his second method is actually a third attempt in no way related to the first and second. Cauchy assumes the errors

$$\varepsilon_i = u_i - ax_i - by_i - cz_i - \cdots$$

to have a probability frequency f. The coefficients k_i by which a has to be eliminated from $\Sigma k_i \varepsilon_i = 0$ should be chosen to maximize the probability of $\Sigma k_i \varepsilon_i$ falling within a given interval $(-\eta, \eta)$. This is an unhealthy postulate, since generally the resulting k_i will depend on the choice of η. Cauchy's remedy is to postulate that f should be so well adapted that k_i would not depend on η. This is quite a strange assumption, since f is not an instrument of the observer but of nature; but it does produce a nice result: the only f that obey these requirements are those with a Fourier transform ϕ such that

$$\phi(\xi) = \exp(-\alpha \xi^N),$$

where α and N are constants. For $N = 1$ these are the celebrated Cauchy stochastics with the probability frequency

$$f(\xi) = \frac{\gamma}{\pi} \frac{1}{1 + \gamma^2 \xi^2}.$$

Their paradoxical behavior of not being improved by averaging was noticed by Bienaymé and forged into an argument against Cauchy. In the course of these investigations Cauchy proved the central limit theorem by means of Fourier transforms in a much more general setting than Laplace had done. The present author adheres to the heterodox view that Cauchy's proof was rigorous, even by modern standards.

This was a muddy chapter of Cauchy's work, which shows him coining gold out of the mud.

Algebra. Cauchy published (1812) the first comprehensive treatise on determinants [1, 2nd ser., I, 64–169]; it contains the product theorem, simultaneously discovered by J. Binet; the inverse of a matrix; and theorems on determinants formed by subdeterminants. He knew "Jacobians" of dimension 3 [1, 1st ser., I, 12]; generally defined "Vandermonde determinants"; and in 1829, simultaneously with Jacobi, published the orthogonal transformation of a quadratic form onto principal axes [1, 2nd ser., IX, 172–195], although he must have discovered it much earlier in his work on elasticity. Through his treatise the term "determinant" became popular, and it is strange that he himself later switched to "resultant." A more abstract approach to determinants, like that of Grassmann's algebra, is found in [1, 2nd ser., XIV, 417–466].

Cauchy gave the first systematic theory of complex numbers [1, 2nd ser., III, 153–301]. Later he confronted the "geometric" approach with the abstract algebraic one of polynomials in x mod $x^2 + 1$ [1, 2nd ser., XIV, 93–120, 175–202].

One of Cauchy's first papers [1, 2nd ser., I, 64–169] generalized a theorem of Ruffini's; he proved that if under permutations of its n variables a polynomial assumes more than two values, it assumes at least p values, where p is the largest prime in n—in other words, that there are no subgroups of the symmetric group of n permutands with an index i such that $2 < i < p$. Bertrand here replaced the p with n itself for $n > 4$, although to prove it he had to rely on a hypothetical theorem of number theory (Bertrand's postulate) that was later verified by P. L. Chebyshev. Cauchy afterward proved Bertrand's result without this assumption [1, 1st ser., IX, 408–417]. His method in [1, 2nd ser., I, 64–169], further developed in [1, 2nd ser., XIII, 171–182; 1st ser., IX, 277–505; X, 1–68], was the *calcul des substitutions,* the method of permutation groups. Fundamentals of group theory, such as the order of an element, the notion of subgroup, and conjugateness are found in these papers. They also contain "Cauchy's theorem" for finite groups: For any prime p dividing the order there is an element of order p. This theorem has been notably reinforced by L. Sylow.

In 1812 Cauchy attacked the Fermat theorem on polygonal numbers, stating that every positive integer should be a sum of n $n$gonal numbers. At that time proofs for $n = 3,4$ were known. Cauchy proved it generally, with the addendum that all but four of the summands may be taken as 0 or 1 [1, 2nd ser., VI,

320–353]. Cauchy's proof is based on an investigation into the simultaneous solutions of

$$\Sigma x_i{}^2 = n, \ \Sigma x_i = m.$$

Cauchy contributed many details to number theory and attempted to prove Fermat's last theorem. A large treatise on number theory is found in [1, 1st ser., III].

Geometry. Cauchy's most important contribution to geometry is his proof of the statement that up to congruency a convex polyhedron is determined by its faces [1, 1st ser., II, 7–38]. His elementary differential geometry of 1826–1827 [1, 2nd ser., V] strongly influenced higher instruction in mathematics. Of course his elasticity theory contains much differential geometry of mappings and of vector and tensor fields, and the notions of grad, div, rot, and their orthogonal invariance.

Differential Equations. What is fundamentally new in Cauchy's approach to differential equations can be expressed in two ideas: (1) that the existence of solutions is not self-evident but has to be proved even if they cannot be made available in an algorithmic form and (2) that uniqueness has to be enforced by specifying initial (or boundary) data rather than by unimportant integration constants. The latter has become famous as the Cauchy problem in partial differential equations. It may have occurred to Cauchy in his first great investigation (1815), on waves in liquids [1, 1st ser., I, 5–318]. Indeed, the difficulty of this problem—and the reason why it had not been solved earlier—was that to be meaningful it had to be framed into a differential equation with initial and boundary data.

To solve ordinary differential equations Cauchy very early knew the so-called Cauchy-Lipschitz method of approximation by difference equations, although its proof was not published until 1840 [1, 2nd ser., XI, 399–404]. Several instances show that he was also acquainted with the principle of iteration [1, 1st ser., V, 236–260; 2nd ser., XI, 300–415 f.; 3, II, 702]. With analytic data the celebrated *calcul des limites* led to analytic solutions of ordinary differential equations [1, 1st ser., VI, 461–470; VII, 5–17; 3, II, 747].

Cauchy discovered (1819), simultaneously with J. F. Pfaff, the characteristics method for first-order partial differential equations [1, 2nd ser., II, 238–252; see also XII, 272–309; 1st ser., VI, 423–461]. His method was superior to Pfaff's and simpler, but it still appears artificial. The geometrical language in which it is taught today stems from Lie. Of course Cauchy also applied the *calcul des limites* to partial differential equations [1, 1st ser., VII, 17–68]. It is not quite clear

which class of equations Cauchy had in mind. In 1875 Sonja (Sophia) Kowalewska precisely formulated and solved the problem by an existence theorem that usually bears the names of Cauchy and Kowalewska.

Another way to solve a system

$$\frac{dx_i}{dt} = X_i(x_1, \cdots, x_n)$$

was by means of exp tZ, with

$$Z = \Sigma X_i(x) \frac{\partial}{\partial x_i},$$

and by an analogous expression if X_i depended on t as well [1, 2nd ser., XI, 399–465; 1st ser., V, 236–250, 391–409]; the convergence of such series was again obtained by *calcul des limites*.

The greater part of Cauchy's work in differential equations was concerned with linear partial equations with constant coefficients, which he encountered in hydrodynamics, elasticity, and optics. The outstanding device of this research was the Fourier transform. It occurs in Cauchy's work as early as 1815, in his work on waves in liquids [1, 1st ser., I, 5–318], as well as in 1817 [1, 2nd ser., I, 223–232] and 1818 [1, 2nd ser., II, 278–279]. Fourier's discovery, while dating from 1807 and 1811, was published as late as 1824–1826 [27], so Cauchy's claim that he found the inversion formula independently is quite acceptable. It is remarkable that he nevertheless recognized Fourier's priority by calling the inversion formula Fourier's formula. Cauchy put the Fourier transform to greater use and used it with greater skill than anybody at that time and for long after—Fourier and Poisson included; and he was the first to formulate the inversion theorem correctly. He also stressed the importance of principal values, of convergence-producing factors with limit 1, and of singular factors (δ functions) under the integral sign [1, 2nd ser., I, 275–355]. His use of the Fourier transform was essentially sound—bold but not rash—but to imitate it in the pre-epsilontic age one had to be another Cauchy. After Weierstrass, Fourier transforms moved into limbo, perhaps because other methods conquered differential equations. Fourier transforms did not become popular until recently, when the fundamentals of Fourier integrals were proved with all desirable rigor; but so much time had elapsed that Cauchy's pioneering work had been forgotten.

From 1821 on, Cauchy considered linear partial differential equations in the operational form

$$(9) \qquad F\left(\frac{\partial}{\partial x_1}, \cdots, \frac{\partial}{\partial x_n}, \frac{\partial}{\partial t}\right)\omega = 0,$$

with F as a polynomial function in $u_1, \cdots, u_n, s$. Such a differential equation has the exponential solutions

$$\exp(\Sigma u_i x_i + st),$$

which are functions of x_i, t for every system of u_i, s fulfilling

(10) $\qquad F(u_1, \cdots, u_n, s) = 0.$

The Fourier transform method aims at obtaining the general solution by continuously superposing such exponential solutions, with imaginary $u_1, \cdots, u_n, s$. For wave equations this means wave solutions by superposition of plane harmonic waves. In the 1821 and 1823 papers [1, 2nd ser., II, 253–275; I, 275–333] a kind of interpolation procedure served to satisfy the initial conditions for $t = 0$. Another approach would be solutions arising from local disturbances (spherical waves under special conditions); they may be obtained from plane waves by superposition and in turn may give rise to general solutions by superposition. This idea is present in the 1815 papers on waves; it is neglected but not absent in the 1821 and 1823 papers.

In 1826 the residue calculus is introduced as a new device, first for solving linear ordinary differential equations with constant coefficients [1, 2nd ser., VI, 252–255; VII, 40–54, 255–266]. The general solution

$$F\left(\frac{d}{dt}\right)\omega = 0$$

is obtained as the integral

$$\frac{1}{2\pi i} \oint \frac{\phi(\zeta)e^{\zeta t}}{F(\zeta)} d\zeta$$

performed around the roots of F, with an arbitrary polynomial $\phi(\zeta)$. Several times Cauchy stressed that this formula avoids casuistic distinctions with respect to multiple roots of F [1, 1st ser., IV, 370].

In 1830, when Cauchy went into optics, this formula was applied to partial differential equations, which meant (10) and (9) explicitly and elegantly solved with respect to s and t, whereas with respect to $x_1, \cdots, x_n$ the Fourier transform method prevailed. Again due care was bestowed on the initial conditions at $t = 0$. The formula, obtained in polar coordinates, is involved and not quite clear; its proof is not available because the *memoire* of which the 1830 paper is a brief extract seems never to have been published and possibly is lost. The construction of wave fronts rested upon intuitive arguments, in fact upon Huygens' principle, although he did not say so and never proved it. According to the same principle Cauchy constructed ray solutions as a superposition of planar disturbances in planes that should be slightly inclined toward each other, as Cauchy says.

From June 1839 to March 1842 Cauchy, again drawing on optics, tried new approaches to linear partial differential equations with constant coefficients. This work was instigated by P. H. Blanchet's intervention (see [1: 1st ser. IV, 369–426; V, 5–20; VI, 202–277, 288–341, 375–401, 404–420; 2nd ser., XI, 75–133, 277–264; XII, 113–124]). It now starts with a system of first-order linear ordinary differential equations, in modern notation

(11) $\qquad \dfrac{dx}{dt} = Ax,$

where x is an n-vector and A a linear mapping. One considers

$$S(s) = \det(A - s)$$

and defines the *fonction principale* Θ as the solution of

$$S\left(\frac{d}{dt}\right)\Theta = 0, \ \Theta^{(i)}(0) = 0 \text{ for } i \leqslant n - 2,$$

$$\Theta^{(n-1)}(0) = 1,$$

which is obtained as an integral

$$\frac{1}{2\pi i} \oint \frac{\phi(s)e^{st}}{S(s)} ds$$

around the roots of S. From Θ the solution of (11) with the initial vector $\ulcorner \alpha_1, \cdots, \alpha_n \urcorner$ is elegantly obtained by using

$$Q(s) = \det(A - s) \cdot (A - s)^{-1}$$

and applying $Q(d/dt)$ to the vector

$$(-1)^{n-1} \ulcorner \alpha_1 \Theta, \cdots, \alpha_n \Theta \urcorner.$$

This method is extended to partial differential equations, where it again suffices to solve (9) under the initial conditions

$$\left(\frac{d}{dt}\right)^i \omega = 0 \text{ for } 0 \leqslant i \leqslant n - 2, \ \left(\frac{d}{dt}\right)^{n-1} \omega.$$

The formula obtained is much simpler than that of 1830, particularly if F is homogeneous. (See the quite readable presentation in [1, 1st ser., VI, 244–420].) If, moreover, the initial value of $(d/dt)^{n-1}\omega$ prescribed is a function Π of

$$\sigma = \Sigma \, u_i x_i,$$

one obtains, by first assuming Π to be linear and then using the homogeneity of F,

(12) $\quad \omega(x,t) = \left(\dfrac{d}{dt}\right)^{1-n} \dfrac{1}{2\pi i} \oint \dfrac{s^{n-1}}{F(u,s)} \Pi(\sigma + st) \, ds,$

where $(d/dt)^{-1}$ means the integration over t from 0. An analogous formula is obtained if the initial datum is assumed as a function of a quadratic function of x, say of $r = (\Sigma x_i^2)^{1/2}$.

The direct method of 1830 is again applied to the study of local disturbances and wave fronts, with Huygens' principle on the background. Cauchy apparently believed that disturbances of infinitesimal width stay infinitesimal, although this belief is disproved by $(d/dt)^{1-n}$ in (12) and by the physical argument of two or even three forward wave fronts in elasticity. It was only after Blanchet's intervention that Cauchy admitted his error.

It is doubtful whether Cauchy's investigations on this subject exerted strong immediate influence; perhaps the generations of Kirchhoff, Volterra, and Orazio Tedone were terrified by his use of Fourier transforms. Solutions arising from local disturbances, substituted into Green's formula, seemed more trustworthy. But even after Hadamard's masterwork in this field Cauchy's attempts should not be forgotten. The only modern book in which they are mentioned and used, although in an unsatisfactory and somewhat misleading fashion, is Courant and Hilbert [34].

It should be mentioned that Cauchy also grasped the notions of adjointedness of differential operators in special cases and that he attacked simple boundary problems by means of Green's function [1, 1st ser., VII, 283–325; 2nd. ser., XII, 378–408].

Mechanics. Cauchy can be credited with some minor contributions to mechanics of rigid bodies, such as the momental ellipsoid and its principal axes [1, 2nd ser., VII, 124–136]; the surfaces of the momentaneous axes of rigid motion [1, 2nd ser., VII, 119], discovered simultaneously with Poinsot; and the first rigorous proof that an infinitesimal motion is a screw motion [1, 2nd ser., VII, 116]. His proper domain, however, was elasticity. He created the fundamental mathematical apparatus of elasticity theory.

The present investigations have been suggested by a paper of M. Navier, 14 August 1820. To establish the equation of equilibrium of the elastic plane, the author had considered two kinds of forces, the ones produced by dilatation or contraction, the others by flection of that plane. Further, in his computations he had supposed both perpendicular to the lines or faces upon which they act. It came into my mind that these two kinds could be reduced to one, still called tension or pressure, of the same nature as the hydrodynamic pressure exerted by a fluid against the surface of a solid. Yet the new pressure should not be perpendicular upon the faces which undergo it, nor be the same in all directions at a given point. . . . Further, the pressure or tension exerted against an arbitrary plane is easily derived as to magnitude and direction from the pressures or ten-

sions exerted against three rectangular planes. I had reached that point when M. Fresnel happened to speak to me about his work on light, which he had only partially presented to the Academy, and told me that concerning the laws according to which elasticity varies in the different directions through one point, he had obtained a theorem like mine. However, this theorem was far from sufficient for the purpose I had in mind at that time, that is, to form the general equation of equilibrium and internal motion of a body, and it is only recently that I have succeeded in establishing new principles, suited to lead to this goal and the object of my communication. . . .

These lines, written by Cauchy in the fall of 1822 [1, 2nd ser., II, 300–304], announced the birth of modern elasticity theory. Rarely has a broad mathematical theory been as fully explained in as few words with as striking a lack of mathematical symbols. Never had Cauchy given the world a work as mature from the outset as this.

From Hooke's law in 1660 up to 1821, elasticity theory was essentially one-dimensional. Euler's theory of the vibrating membrane was one of the few exceptions. Another was the physical idea of internal shear stress, which welled up and died twice (Parent, 1713; Coulomb, 1773) with no impact upon the mathematical theory. Even in this one-dimensional setting, elasticity was a marvelous proving ground for Euler's analysis of partial derivatives and partial differential equations (see [36]). In 1821 Navier's paper on equilibrium and vibration of elastic solids was read to the Academy (published in 1827) [30, 32]. Navier's approach constituted analytic mechanics as applied to an isotropic molecular medium that should obey Hooke's law in molecular dimensions: any change in distance between two molecules causes a proportional force between them, the proportionality factor rapidly decreasing with increasing distance. Cauchy was one of the examiners of Navier's paper. It was not, however, this paper of Navier's to which Cauchy alluded in the above quotation. Cauchy's first approach was independent of Navier's; it was nonmolecular but rather geometrically axiomatic.

Still another advance had taken place in 1821. Thomas Young's investigations on interference in 1801 had made it clear that light should be an undulation of a hypothetical gaseous fluid, the ether. Consequently light waves were thought of as longitudinal, like those of sound in air, although the phenomena of polarization pointed to transverse vibrations. In 1821 Fresnel took the bold step of imagining an ether with resistance to distortion, like a solid rather than a fluid; and marvelously enough he found transmission by transverse waves (although longitudinal ones would subsist as well [32]). Fresnel's results encouraged Cauchy to pursue his investigations.

The short communication from which the above extract was taken was followed by detailed treatises in 1827–1829 [1, 2nd ser., VII, 60–93, 141–145; VIII, 158–179, 195–226, 228–287; IX, 342–369], but nearly all fundamental notions of the mechanics of continuous media were already clear in the 1822 note: the stress tensor (and the concept of tensor at all), the strain tensor, the symmetry of both tensors, their principal axes, the principle of obtaining equilibrium and motion equations by cutting out and freezing an infinitesimal piece of the medium, and the striking idea of requiring Hooke's law for the principal stresses and strains. For homogeneous media this led to Navier's equations, with one elastic constant, but independently of Navier's molecular substructure. Soon Cauchy introduced the second elastic constant, which arose from an independent relation between volume stress and volume strain. It led to the now generally accepted elasticity theory of isotropic media. For anisotropic media Cauchy was induced by Poisson's intervention to admit a general linear dependence between stress and strain, involving thirty-six parameters. The only fundamental notion then lacking was the elastic potential, which allows one to reduce the number of parameters to twenty-one; it is due to G. Green (1837) [30; 32].

Meanwhile, in 1828–1829 Cauchy had pursued Navier's molecular ideas and had arrived at a fifteen-parameter theory for anisotropic media [1, 2nd ser., VIII, 227–277; IX, 162–173].[9] The nineteenth-century discussions are long-since closed in favor of the axiomatic "multi-constant" theory and against the molecular "rari-constant" theory, if not against any molecular theory of elasticity at all.

Cauchy applied the general theory to several special problems: to lamellae [1, 2nd ser., VIII, 288–380], to the rectangular beam [1, 2nd ser., IX, 61–86] (definitively dealt with by Saint Venant), and to plane plates [1, 2nd ser., VIII, 381–426; IX, 9–60], in which Kirchhoff finally succeeded. The application on which Cauchy bestowed more pains than on any other subject was elastic light theory. The mathematical context of this theory was partial differential equations with constant coefficients. In the history of physics it was one of the great pre-Maxwellian efforts necessary before physicists became convinced of the impossibility of any elastic light theory.

Cauchy developed three different theories of reflection and refraction (1830, 1837, 1839) [1, 1st ser., II, 91–110; 2nd ser., II, 118–133; 1st ser., IV, 11–38; V, 20–39]. The problems were to explain double refraction (in which he succeeded fairly well), to adjust the elastic constants to the observational data on the velocity of light under different conditions in order to obtain Fresnel's sine and tangent laws of polarization by suitable boundary conditions, and to eliminate the spurious longitudinal vibrations. Whether he assumed the transverse vibrations to be parallel or orthogonal to the polarization plane (as he did in his first and second theories, respectively), he obtained strange relations between the elastic constants and was forced to admit unmotivated and improbable boundary conditions. His third theory, apparently influenced by Green's work, was based upon the curious assumption of an ether with negative compressibility—later called labile by Lord Kelvin—which does away with longitudinal waves. In 1835 Cauchy also attempted dispersion [1, 2nd ser., X, 195–464]; the problem was to explain the dependence of the velocity of light upon the wavelength by a more refined evaluation of the molecular substructure.

Celestial Mechanics. One not acquainted with the computational methods of astronomers before the advent of electronic apparatus can hardly evaluate Cauchy's numerous and lengthy contributions to celestial mechanics. In handbooks of astronomy he is most often quoted because of his general contributions to mathematics. Indeed, it must have been a relief for astronomers to know that the infinite series they used in computations could be proved by Cauchy to converge. But he also did much detailed work on series, particularly for the solution of the Kepler equation [1, 1st ser., VI, 16–48] and developments of the perturbative function [1, 1st ser., V, 288–321; VII, 86–126]; textbooks still mention the Cauchy coefficients. Cauchy's best-known contribution to astronomy (1845) is his checking of Leverrier's cumbersome computation of the large inequality in the mean motion of Pallas by a much simpler method [1, 1st ser., IX, 74–220; see also XI, 385–403]. His tools consisted of formulas for the transition from the eccentric to the mean anomaly [1, 1st ser., VI, 21]; the Cauchy "mixed method" [1, 1st ser., VIII, 168–188, 348–359], combining numerical and rational integrations when computing negative powers of the perturbative function; and asymptotic estimations of distant terms in the development of the perturbative function according to multiples of the mean anomaly—such asymptotics had interested Cauchy as early as 1827 [1, 1st ser., II, 32–58; see also IX, 5–19, 54–74; XI, 134–143].

NOTES

1. All references in brackets are to numbered works in the bibliography.
2. The data on his career in [7; 8; 9; 11] are incomplete and contradictory, even self-contradictory; but it still would not be

too difficult to check them. According to Cauchy's own account [1, 2nd ser., II, 283] he taught in 1817 as Biot's *suppléant* at the Collège de France. However, on the title pages of his books published until 1830 he never mentions a chair at the Collège de France.

3. Turin was the capital of Piedmont; the dukes of Piedmont and Savoy had become kings of Sardinia.

4. In the library of the University of Utrecht I came across an unknown print of a work by Cauchy that must have appeared in a periodical. I noticed several quotations from papers missing from Valson's list and the Academy edition (e.g. [1, 2nd ser., II, 293, note]).

5. I know only two of them.

6. [1, 1st ser., IX, 186–190] shows him quarreling with the press.

7. Of course he did not prove it; he simply applied it. It had been discovered by Bolzano in 1817.

8. His own account of this invention [1, 1st ser., VIII, 145], although often repeated, is incorrect. With Euler and Lagrange, he said, "continuous" meant "defined by one single law." Actually, there was no serious definition of continuity before Cauchy.

9. I do not understand on what grounds Müller and Timpe [26, IV, 23] and Love [30] claim that Cauchy mistook his fifteen-parameter theory for the twenty-one-parameter theory. The common source of this criticism seems to be Clausius (1849). Although Cauchy was sometimes less outspoken on this point, the charge is at least refuted by [1, 2nd ser., IX, 348].

BIBLIOGRAPHY

1. Cauchy, *Oeuvres complètes,* 1st ser., 12 vols., 2nd ser., 14 vols. (Paris, 1882–   ). The final volume, 2nd ser., XV, is due to appear in 1970.

2. Cauchy, *Mémoire sur les intégrales définies prises entre des limites imaginaires* (Paris, 1825). To be republished in [1, 2nd ser., XV].

2a. Reprint of [2] in *Bulletin des sciences mathématiques,* 7 (1874), 265–304; 8 (1875), 43–55, 148–159.

2b. Cauchy, *Abhandlung über bestimmte Integrale zwischen imaginären Grenzen,* trans. of [2] by P. Stäckel, no. 112 in Ostwald's Klassiker der exacten Wissenschaften (Leipzig, 1900).

3. F. N. M. Moigno and M. Lindelöf, eds., *Leçons du calcul différentiel et de calcul intégral, rédigées d'après les méthodes et les oeuvres publiés ou inédits d'A.-L. Cauchy,* 4 vols. (Paris, 1840–1861).

4. F. N. M. Moigno, *Leçons de mécanique analytique, rédigées principalement d'après les méthodes d'Augustin Cauchy. . . . Statique* (Paris, 1868).

5. B. Boncompagni, "La vie et les travaux du baron Cauchy," in *Bollettino di bibliografia e di storia delle scienze matematiche e fisiche,* 2 (1869), 1–102. Review of [8].

6. K. Rychlik, "Un manuscrit de Cauchy aux Archives de l'Académie tchécoslovaque des sciences," in *Czechoslovak Mathematical Journal,* 7 [82] (1957), 479–481.

7. J. B. Biot, *Mélanges scientifiques et littéraires,* III (Paris, 1858), 143–160.

8. C. A. Valson, *La vie et les travaux du baron Cauchy,* 2 vols. (Paris, 1868).

9. J. Bertrand, "La vie et les travaux du baron Cauchy par C. A. Valson," in *Journal des savants* (1869), 205–215, and *Bulletin des sciences mathématiques,* 1 (1870), 105–116. Review of [8].

10. C. A. Bjerknes, *Niels-Henrik Abel, tableau de sa vie et de son action scientifique* (Paris, 1885), pp. 268–322, 342–347.

10a. D. E. Smith, "Among My Autographs, 29. Legendre and Cauchy Sponsor Abel," in *American Mathematical Monthly,* 29 (1922), 394–395.

10b. F. Lange-Nielsen, "Zur Geschichte des Abelschen Theorems. Das Schicksal der Pariser Abhandlung," in *Norsk matematisk tidsskrift,* 9 (1927), 55–73.

11. J. Bertrand, *Éloges académiques,* new ed. (Paris, 1902), pp. 101–120.

12. A. d'Hautpoul, *Quatre mois à la cour de Prague* (Paris, 1912). Not accessible to the present reporter.

13. A. Terracini, "Cauchy a Torino," in *Rendiconti del Seminario matematico* (Turin), 16 (1956–1957), 159–205.

14. P. S. Laplace, *Traité de la mécanique céleste,* II (Paris, 1804).

15. C. Sturm, "Analyse d'un mémoire sur la résolution des équations numériques," in *Bulletin universel des sciences et de l'industrie,* 11 (1829), 419–422.

16. J. Bienaymé, "Considerations . . ., Remarques . . .," in *Comptes rendus de l'Académie des sciences,* 37 (1853), 5–13, 68–69, 309–324.

17. N. H. Abel, "Untersuchungen über die Reihe . . .," in *Journal für reine und angewandte Mathematik,* 1 (1826), 311–359, also published as no. 71 in Ostwald's Klassiker der exacten Wissenschaften (1895).

18. S.-H. Briot and J. C. Boucquet, *Théorie des fonctions doublement périodiques et, en particulier, des fonctions elliptiques* (Paris, 1859); 2nd ed., *Théorie des fonctions elliptiques* (Paris, 1873–1875).

19. F. Casorati, *Teorica delle funzioni di variabili complesse* (Pavia, 1868). Not accessible to the present reporter.

20. E. Verdet, *Leçons d'optique physique,* vols. V–VI of Verdet's *Oeuvres* (Paris, 1869–1872).

21. F. Studnicka, *A. L. Cauchy als formaler Begründer der Determinantentheorie* (Prague, 1876). Not accessible to the present reporter.

21a. I. Todhunter, *A History of the Theory of Elasticity and of the Strength of Materials from Galilei to the Present Time,* vol. I (Cambridge, 1886).

22. A. Brill and M. Noether, "Die Entwicklung der Theorie der algebraischen Funktionen in älterer und neuerer Zeit," in *Jahresbericht der Deutschen Mathematiker-Vereinigung,* 3 (1894), esp. 155–197.

23. P. Stäckel, "Integration durch das imaginäre Gebiet," in *Bibliotheca mathematica,* 3rd ser., 1 (1900), 109–128.

24. P. Stäckel, "Beitrag zur Geschichte der Funktionentheorie im 18. Jahrhundert," in *Bibliotheca mathematica,* 3rd ser., 2 (1901), 111–121.

25. P. Jourdain, "The Theory of Functions With Cauchy and Gauss," in *Bibliotheca mathematica,* 3rd ser., 6 (1905), 190–207.

26. The following articles in the *Encyklopädie der mathematischen Wissenschaften:* C. H. Müller and A. Timpe, "Grundgleichungen der mathematischen Elastizitätstheorie," IV (1907), 1; A. Wangerin, "Optik, ältere Theorie," V (1909), 21; K. F. Sundman, "Theorie der

Planeten," VI, pt. 2 (1912), 15; H. von Zeipel, "Entwicklung der Störungsfunktion," VI, pt. 2 (1912), 557–665.

27. H. Burkhardt, "Entwicklungen nach oscillierenden Funktionen und Integration der Differentialgleichungen der mathematischen Physik," in *Jahresbericht der Deutschen Mathematiker-Vereinigung,* **10** (1904–1908).

28. E. A. Miller, "Historical Sketch of the Development of the Theory of Groups of Finite Order," in *Bibliotheca mathematica,* 3rd ser., **10** (1909), 317–329.

29. P. Jourdain, "The Origin of Cauchy's Conception of a Definite Integral and of the Continuity of a Function," in *Isis,* **1** (1913–1914), 661–703. Not accessible to the author.

30. A. E. H. Love, *A Treatise on the Mathematical Theory of Elasticity* (Oxford, 1927), esp. the introduction.

31. H. Lamb, *Hydrodynamics* (Cambridge, 1932), pp. 384 f.

32. E. T. Whittaker, *A History of the Theories of Aether and Electricity* (Edinburgh, 1951), esp. pp. 128–169.

33. E. Carruccio, "I fondamenti delle analisi matematica nel pensiero di Agostino Cauchy," in *Rendiconti del Seminario matematico* (Turin), **16** (1956–1957), 205–216.

34. R. Courant and D. Hilbert, *Methods of Mathematical Physics,* II (1962), esp. pp. 210–221.

35. C. Truesdell-R. Toupin, "The Classical Field Theories," in *Handbuch der Physik,* III (Berlin-Göttingen-Heidelberg, 1960), pp. 226–793, esp. 259–261, 270–271, 306–308, 353–356, 536–556.

36. C. Truesdell, "The Rational Mechanics of Flexible and Elastic Bodies, 1638–1788," in *Leonhardi Euleri Opera Omnia,* 2nd ser., XI, sec. 2 (Zurich, 1960), 7–435.

HANS FREUDENTHAL

CAULLERY, MAURICE (*b.* Bergues, France, 5 September 1868; *d.* Paris, France, 13 July 1958), *zoology, biology.*

The son of an army captain, Caullery belonged to an old family of northern France. He was a brilliant student at the *lycée* of Douai and was accepted by both the École Polytechnique and the École Normale Supérieure (1887). He chose the latter and rapidly obtained *licences* in mathematics (1888), physics (1889), and natural sciences (1890). He passed the *agrégation* in natural sciences in 1891 and received the *doctorat ès sciences naturelles* with a thesis on the compound Ascidiacea in 1895. He then became *préparateur agrégé* (zoology) at the École Normale; assistant in physics, chemistry, and natural sciences; lecturer at Lyons from 1896 to 1900; full professor at Marseilles from 1900 to 1903; and lecturer at Paris from 1903 to 1909, when he became professor at the Laboratoire d'Évolution des Êtres Organisés of the Faculté des Sciences, Paris, until his retirement in 1939. In this chair of evolution he succeeded Alfred Giard, whose teaching had inspired him as a student and had oriented his career toward biology. He also was director of the Wimereux marine laboratory.

As a zoologist Caullery left many works, written either alone or in collaboration with his friend Félix Mesnil. He published numerous notices and reports, nearly all of them on marine species. He described new species, but he gave the closest study to animals whose morphology, mode of reproduction, and ecology were of special interest or posed problems from the evolutionary point of view. His early research on the Tunicata enabled him to specify the different origins of the organs of the oozooid and the blastozooid. In the fixed polychaete Annelida he analyzed the important transformations connected with the epigamic metamorphosis that manifest themselves during sexual activity. He explained the evolutionary cycle of the Orthonectida, discovered by Giard in 1877. While studying the Annelida gathered by the Dutch ship *Siboga* from the bottom of the Malay Archipelago, he described a strange organism that he named *Siboglinum weberi;* it was later recognized as the representative of a new branch created in 1944, the Pogonofora.

Parasites greatly interested Caullery, and his important works on parasitic Protozoa (Gregarinida and Actinomyxidia) have precisely defined their sexual cycles. In the Orthonectida, the Turbellaria, and the epicarid Crustacea he observed the transformations brought about by parasites, polymorphism, morphological deterioration, anatomical regression, and hermaphroditism.

A zoologist and researcher of wide scope, Caullery was also an excellent teacher, a brilliant lecturer, and a lucid and rigorous scientific writer. His lectures were models of didactic exposition and served as the basis for numerous richly documented works that focused on the development of biology. He also was attracted to the history of science, particularly to the history of biology and the evolution of ideas.

The many problems set forth in publications also held Caullery's attention. With authority he concerned himself with the *Bulletin biologique de la France et de la Belgique,* successor to the *Bulletin scientifique du département du Nord,* founded by Giard. He enlivened the series of works in general biology issued by Doin and was one of the founders of the Presses Universitaires de France.

BIBLIOGRAPHY

Caullery's papers include "Contribution à l'étude des Ascidies composées," in *Bulletin scientifique de la France et de la Belgique,* **27** (1895), 1–157; "Les formes épitoques et l'évolution des Cirratuliens," in *Annales de l'Université de Lyon,* **39** (1898); "Recherches sur les Actinomyxidies," in *Archiv für Protistenkunde,* **6** (1905), 272–308; "Recherches sur le cycle évolutif des Orthonectides," in *Bulletin scientifique de la France et de la Belgique,* **46** (1912), 139–171; "Histoire des sciences biologiques," in *Histoire de la nation française,* **15** (1925), 3–326; and "Le genre *Siboglinum,*" in *Traité de zoologie,* **11** (1948), 395–399.

Among his books are *Les universités et la vie scientifique aux États-Unis* (Paris, 1917); *Le parasitisme et la symbiose* (Paris, 1922, 1950), also trans. into English (London, 1952); *Le problème de l'évolution* (Paris, 1931); *La science française depuis le XVIIème siècle* (Paris, 1933, 1946), also trans. into English (New York, 1933); *Les conceptions modernes de l'hérédité* (Paris, 1935, 1950); *Les progrès récents de l'embryologie expérimentale* (Paris, 1939); *Biologie des abeilles* (Paris, 1941); *Les étapes de la biologie* (Paris, 1941); *L'embryologie* (Paris, 1942); *Organisme et sexualité* (Paris, 1942, 1951); *Génétique et hérédité* (Paris, 1943); and *Biologie des jumeaux* (Paris, 1945).

A notice on the life and works of Caullery is E. Fauré-Fremiet, in *Académie des sciences* (Paris, 1960), which includes a bibliography, information on his scientific jubilee in 1939, and references to obituaries.

ANDRÉE TÉTRY

CAVALIERI, BONAVENTURA (*b.* Milan, Italy, probably 1598; *d.* Bologna, Italy, 30 November 1647), *mathematics.*

Cavalieri's date of birth is uncertain; the date given above is the one cited by Urbano d'Aviso, a disciple and biographer of Cavalieri. The name Bonaventura was not his baptismal name but rather that of his father. It is the name the mathematician adopted when, as a boy, he entered the Jesuati religious order, adherents of the rule of St. Augustine. Cavalieri was received in the minor orders in Milan in 1615 and in 1616 transferred to the Jesuati monastery in Pisa, where he had the good fortune of meeting the Benedictine monk Benedetto Castelli, who had studied with Galileo at Padua and was at the time a lecturer in mathematics at Pisa. Through him Cavalieri was initiated into the study of geometry. He quickly absorbed the classical works of Euclid, Archimedes, Apollonius, and Pappus, demonstrating such exceptional aptitude that he sometimes substituted for his teacher at the University of Pisa. He was introduced by Castelli to Galileo, whose disciple he always considered himself. He wrote Galileo at least 112 letters, which are included in the national edition of the *Opere di Galileo;* only two of Galileo's letters to Cavalieri have come down to us, however.

In 1620 Cavalieri returned to Rome under orders of his superiors, and in 1621 he was ordained a deacon to Cardinal Federigo Borromeo, who held Fra' Bonaventura in great esteem and gladly discussed mathematics with him; the cardinal subsequently wrote a letter commending him to Galileo. Cavalieri was hardly twenty-one when he taught theology at the monastery of San Girolamo in Milan, attracting attention by his profound knowledge of the subject.

During his Milan period (1620–1623) Cavalieri developed his first ideas on the method of indivisibles, his major contribution to mathematics. From 1623 to 1626 he was the prior of St. Peter's at Lodi. Later he was a guest in Rome of Monsignor Ciampoli, to whom he later dedicated his *Geometria.* From 1626 to 1629 he was the prior of the monastery of the Jesuati in Parma, hoping in vain to be appointed lecturer in mathematics at the university there. In the autumn of 1626, during a trip from Parma to Milan, he fell ill with the gout, from which he had suffered since childhood and which was to plague him to the end of his life. This illness kept him at Milan for a number of months. On 16 December 1627 he announced to Galileo and Cardinal Borromeo that he had completed his *Geometria.* In 1628, learning that a post of lecturer at Bologna had become vacant through the death of the astronomer G. A. Magini, he wrote Galileo for assistance in securing the appointment. Galileo, in 1629, wrote to Cesare Marsili, a gentleman of Bologna and member of the Accademia dei Lincei, who had been commissioned to find a new lecturer in mathematics. In his letter, Galileo said of Cavalieri, "few, if any, since Archimedes, have delved as far and as deep into the science of geometry." In support of his application to the Bologna position, Cavalieri sent Marsili his geometry manuscript and a small treatise on conic sections and their applications in optics. Galileo's testimonial, as Marsili wrote him, induced the "Gentlemen of the Regiment" to entrust the first chair in mathematics to Cavalieri, who held it continuously from 1629 to his death.

At the same time he was appointed prior of a convent of his own order in Bologna, specifically, at the Church of Santa Maria della Mascarella, enabling him to pursue without any impediment both his work in mathematics and his university teaching. During the period in which Cavalieri taught at Bologna, he published eleven books in that city, including the *Geometria* (1635).

Cavalieri's theory, as developed in this work and in others subsequently published, relates to an inquiry in infinitesimals, stemming from revived interest in Archimedes' works, which during the Renaissance were translated from Greek into Latin, with commentaries. The translations of Tartaglia, Maurolico, and Commandino are cited since they served as a point of departure for new mathematical developments.

The only writings of Archimedes known to seventeenth-century mathematicians were those based upon the strict method of exhaustion, by which the ancient mathematicians dealt with questions of an infinitesimal character without recourse to the infinite or to the actual infinitesimal. Nevertheless, the great mathematicians of the seventeeth century were so thoroughly pervaded with the spirit of Archimedes as to appreciate that in addition to the "method of exhaustion" the ancient geometricians must have known a more manageable and effective method for research. On this point Torricelli wrote:

> I should not dare affirm that this geometry of indivisibles is actually a new discovery. I should rather believe that the ancient geometricians availed themselves of this method in order to discover the more difficult theorems, although in their demonstration they may have preferred another way, either to conceal the secret of their art or to afford no occasion for criticism by invidious detractors. Whatever it was, it is certain that this geometry represents a marvelous economy of labor in the demonstrations and establishes innumerable, almost inscrutable, theorems by means of brief, direct, and affirmative demonstrations, which the doctrine of the ancients was incapable of. The geometry of indivisibles was indeed, in the mathematical briar bush, the so-called royal road, and one that Cavalieri first opened and laid out for the public as a device of marvelous invention [*Opere,* I, pt. I, 139–140].

In 1906 J. L. Heiberg found, in a palimpsest belonging to a Constantinople library, a small work by Archimedes in the form of a letter to Eratosthenes, which explained a method by which areas, volumes, and centers of gravity could be determined. This method, which in turn was related to the procedures of Democritus of Abdera, considered a plane surface as made up of chords parallel to a given straight line, and solids as made up of plane sections parallel to one another. In addition, according to Archimedes, principles of statics were applied, whereby the figures, thought of as heavy bodies, were weighed in an ideal scale. "I do believe," said Archimedes, "that men of my time and of the future, and through this method, might find still other theorems which have not yet come to my mind" (Rufini, *Il "Metodo" di Archimede e le origini del calcolo infinitesimale nell'antichità* [Milan, 1961], p. 103). The challenge that Archimedes extended was not taken up, as we know, by his contemporaries and fell into oblivion for many centuries.

The concept of indivisibles does sometimes show up fleetingly in the history of human thought: for example, in a passage by the eleventh-century Hebrew philosopher and mathematician Abraham bar Ḥiyya (Savasorda); in occasional speculations—more philosophical than mathematical—by the medieval Scholastics; in a passage by Leonardo da Vinci; in Kep-

ler's *Nova stereometria doliorum* (Linz, 1615). By a conception differing from Cavalieri's, indivisibles are treated by Galileo in his *Discorsi e dimostrazioni matematiche intorno a due nuove scienze.*

In Cavalieri we come to a rational systematization of the method of indivisibles, a method that not only is deemed useful in the search for new results but also, contrary to what Archimedes assumed, is regarded as valid, when appropriately modified, for purposes of demonstrating theorems.

At this point a primary question arises: What significance did Cavalieri attribute to his indivisibles? This mathematician, while perfectly familiar with the subtle philosophical questions connected with the problem of the possibility of constituting continuous magnitudes by indivisibles, seeks to establish a method independent of the subject's hypotheses, which would be valid whatever the concept formed in this regard. While Galileo asserted, "The highest and the ultimate, although primary components of the continuous, are infinite indivisibles" (*Opere,* VII, 745–750), Cavalieri did not dare to assert that the continuous is composed of indivisible elements, about which he did not give an explicit definition, nor did he clarify whether they were actual or potential infinitesimals. It is also probable that Cavalieri's conception of his indivisibles underwent a change and that these were born as actual infinitesimals (like those of Galileo) and grew to become potential infinitesimals (see G. Cellini). It must be further pointed out, according to L. Lombardo Radice, that the Cavalieri view of the indivisibles has given us a deeper conception of the sets: it is not necessary that the elements of the set be assigned or assignable; rather it suffices that a precise criterion exist for determining whether or not an element belongs to the set.

Quite aside from any philosophical considerations on the nature of indivisibles, the determinations of area and volumes made by Cavalieri are based on the principle bearing his name, which may be formulated as follows:

> If two plane figures cut by a set of parallel straight lines intersect, on each of these straight lines, equal chords, the two figures are equivalent; if the chords pertaining to a single straight line of the set have a constant ratio, the same ratio obtains between the two figures.
>
> Similarly, in space: if the sections of two solids obtained by means of planes that are parallel to each other are equivalent two by two, the two solids are equivalent; if the two sections obtained with a given plane have a constant ratio when the plane is varied, the two solids have a ratio that is equal to that of two of their sections obtained with one same plane.

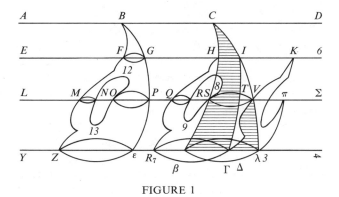

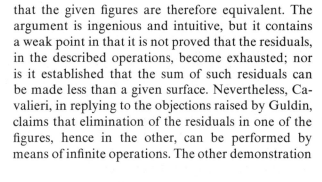

FIGURE 1

that the given figures are therefore equivalent. The argument is ingenious and intuitive, but it contains a weak point in that it is not proved that the residuals, in the described operations, become exhausted; nor is it established that the sum of such residuals can be made less than a given surface. Nevertheless, Cavalieri, in replying to the objections raised by Guldin, claims that elimination of the residuals in one of the figures, hence in the other, can be performed by means of infinite operations. The other demonstration

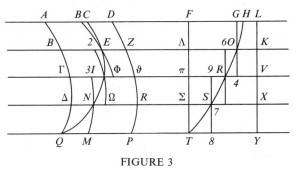

FIGURE 3

From the viewpoint of modern infinitesimal analysis, the Cavalieri principle affirms in substance that two integrals are equal if the integrands are equal and the integration limits are also equal. Furthermore, a constant that appears as a multiplier in the integrand may be carried out of the sign of integration without causing the value of the integral to vary.

However, the concept of the integral, according to the definition of A. Cauchy, was not precisely in the mathematical thought of Cavalieri, but rather was looked into by P. Mengoli, his disciple and successor in the chair at Bologna. Cavalieri pursued many paths to demonstrate his principle, and they are to be found in Book VII of his *Geometry*.

Let us consider the case in plane geometry, where, on the hypotheses of the stated principle, the corresponding chords of the given figures are equal in pairs (see Fig. 1). Cavalieri then, through a translation in the direction of the parallel straight lines in question, superimposes two equal chords. The parts of the figure which thus are superimposed are therefore equivalent or, rather, equal, because they are congruent. The remaining parts, or residuals, which are not superimposed, will still satisfy the conditions relative to the chords that were satisfied in the original figure. In this way, one can proceed with successive superpositions by translation, and it is impossible at a given point in the successive operations that one figure be exhausted unless the other is also. Cavalieri concludes

of the Cavalieri principle is made by the ancients' method of exhaustion and is a rigorous one for the figures that satisfy certain conditions: that is, the demonstration is valid for figures which, in addition to satisfying the hypothesis of the principle, fall into one of the following classes:

(1) Generalized parallelograms, namely, figures included between straight parallel lines p and l which intersect chords of constant length on straight lines running in the same direction as p and l (see Fig. 2).

(2) The *figurae in alteram partem deficientes* ("figures deficient in another part") are included between two parallel lines p and l and, in addition, the chords intercepted by a transverse line parallel to p diminish as the distance of the transverse from straight line p increases (see Fig. 3).

(3) Figures capable of being broken down into a finite number of parts belonging to either of the aforementioned two classes (see Fig. 4).

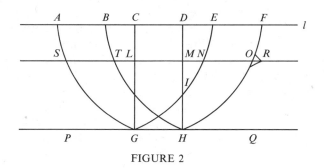

FIGURE 2

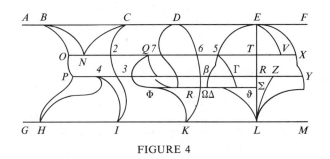

FIGURE 4

Notwithstanding the demonstrations mentioned and the success of the method of indivisibles, contemporary mathematicians, who were more attached to the traditions of classical mathematics, entered into a polemic with Cavalieri, unaware that Archimedes himself had already used methods similar to those that they were opposing. Such is the case of Guldin, who had an interesting discussion with Cavalieri that is summed up in exercise III of the *Exercitationes geometricae sex.*

Many results that were laboriously obtained by the method of exhaustion were obtained simply and rapidly through the Cavalieri principle: for example, the area of an ellipse and the volume of a sphere. Through his methods, Cavalieri had found the result which in today's symbols would be expressed as:

$$\int_0^a x^n \, dx = \frac{a^{n+1}}{n+1}$$

for any natural number $n (n = 1, 2, 3, \cdots)$. Cavalieri was unaware that this result, which appears in the *Centuria di varii problemi* (1639), had already been found as early as 1636 by Fermat and Roberval, who had arrived at it by other means.

By means of the method of indivisibles and based upon a lemma established by his pupil G. A. Rocca, Cavalieri proved the Guldin theorem on the area of a surface and the volume of rotating solids. This theorem, which also appears in certain editions of Pappus' works, although held to be an interpolation, was enunciated in the *Centrobaryca* of Guldin, who proved its correctness in certain particular cases, without, however, providing the general proof.

The most significant progress in the field of infinitesimal analysis along the lines set forth by Cavalieri was made by Evangelista Torricelli. In his *Arithmetica infinitorum* (1655), John Wallis also makes use of indivisibles.

Especially interesting is the opinion of the Cavalieri method expressed by Pascal in his *Lettres de Dettonville* (1658): "Everything that is demonstrated by the true rules of indivisibles will also and necessarily be demonstrated in the manner of the ancients. For which reason, in what follows, I shall not hesitate to use the very language of indivisibles." Although in the following years in the field of analysis of the infinitesimal, new ideas replaced the old on the indivisibles, the methods of Cavalieri and Torricelli exerted a profound influence, as Leibniz acknowledged in a letter to G. Manfredi: ". . . in the sublimest of geometry, the initiators and promotors who performed a yeoman's task in that field were Cavalieri and Torricelli; later, others progressed even further

by availing themselves of the work of Cavalieri and Torricelli." Moreover, Newton, while assuming in his *Principia* a critical attitude in the matter of indivisibles, did nevertheless in his *Tractatus de quadratura curvarum,* use the term *fluens* to indicate a variable magnitude—a term previously used by Cavalieri in his *Exercitationes geometricae sex.*

In proposition I of Book I of the *Geometria,* we find in geometric form the theorem of mean value, also known as the Cavalieri theorem. The theorem is presented as the solution of the following problem: Given a plane curve, provided with a tangent at every point and passing through two points A and B, to find a straight line parallel to AB and tangent to the curve at some point on the curve between A and B. Analytically we have: If the real function $f(x)$ of the real variable x is continuous in the interval (a, b) and at every point within this interval it is differentiable, at least one point ζ exists such that $a < \zeta < b$, so that

$$\frac{f(b) - f(a)}{b - a} = f'(\zeta).$$

Logarithms were introduced into mathematics in the work of Napier in 1614. In Italy such valuable auxiliaries to numerical calculation were introduced by Cavalieri, together with noteworthy developments in trigonometry and applications to astronomy. In this connection we might mention *Directorium generale uranometricum* (1632), *Compendio delle regole dei triangoli* (1638), *Centuria di varii problemi* (1639), *Nuova pratica astrologica* (1639), and *Trigonometria plana, et sphaerica, linearis et logarithmica* (1643). The *Directorium,* the *Pratica,* and the *Trigonometria* contain, moreover, excellent logarithmic-trigonometric tables.

In the *Centuria,* Cavalieri dealt with such topics as the general definition of cylindrical and conical surfaces, formulas to determine the volume of a barrel and the capacity of a vault with pointed arches, and the means of obtaining from the logarithms of two numbers the logarithm of the sum or of the difference, a problem that was subsequently taken up by various mathematicians, Gauss among others. *Lo specchio ustorio* ("The Burning Glass") contains some interesting historical data on the origin of the theory of the conics among the Greeks; according to Cavalieri, the origins are to be found in the gnomonic requirements. In this work, we find a theory of conics with applications to optics and acoustics. Among the former, we note the idea of the reflecting telescope, of which—according to Piola and Favaro—Cavalieri was the first inventor, preceding Gregory and Newton; determination of the focal length of a lens of uneven sphericity; and explications of the burning glass of Archimedes. In the field of acoustics, Cavalieri attempted the archaeological reconstruction of the resonant vases mentioned by Vitruvius and used in theaters for amplifying sound.

In this work, various pointwise constructions of conics appear. More interesting still are the constructions given in the *Geometria* and in the *Exercitationes,* obtained by means of projective pencils which antedated the work of Steiner.

A delicate question relates to the astrological activities that Cavalieri engaged in by virtue of his office, but, as pointed out by D'Aviso, he was opposed to predictions based upon the position of the stars and states so at the end of his *Pratica astrologica.*

BIBLIOGRAPHY

I. Original Works. Cavalieri's works include *Directorium generale uranometricum* (Bologna, 1632); *Geometria indivisibilibus continuorum nova quadam ratione promota* (Bologna, 1635; 2nd ed., 1653). Translated into Russian by S. J. Lure (Moscow-Leningrad, 1940). Translated into Italian, by Lucio Lombardo-Radice, as *Geometria degli indivisibili di Bonaventura Cavalieri,* with introduction and notes (Turin, 1966). *Compendio delle regole dei triangoli con le loro dimostrationi* (Bologna, 1638); *Centuria di varii problemi* (Bologna, 1639); *Nuova pratica astrologica* (Bologna, 1639); *Tavola prima logaritmica. Tavola seconda logaritmica. Annotationi nell'opera, e correttioni de gli errori più notabili* (Bologna, n. d.); *Appendice della nuova pratica astrologica* (Bologna, 1640); *Trigonometria plana, et sphaerica, linearis et logarithmica* (Bologna, 1643); *Trattato della ruota planetaria perpetua* (Bologna, 1646); *Exercitationes geometricae sex* (Bologna, 1647).

II. Secondary Literature. See U. D'Aviso, "Vita del P. Buonaventura [sic] Cavalieri," in *Trattato della Sfera* (Rome 1682); G. Piola, *Elogio di Bonaventura Cavalieri* (Milan, 1844); A. Favaro, *Bonaventura Cavalieri nello studio di Bologna* (Bologna, 1885); E. Bortolotti, "I progressi del metodo infinitesimale nell'opera geometrica di Torricelli," in *Periodico di matematiche,* 4th ser., **8** (1928), 19–59; "La Scoperta e le successive generalizzazioni di un teorema fondamentale di calcolo integrale," in *Archivio di Storia della scienza* (1924), pp. 205–227; F. Conforto, "L'opera scientifica di Bonaventura Cavalieri e di Evangelista Torricelli," in *Atti del Convegno di Pisa* (23–27 Sept. 1948), pp. 35–56; A. Masotti, "Commemorazione di Bonaventura Cavalieri," in *Rendiconti dell'Istituto Lombardo di scienze e lettere, parte generale e atti ufficiali,* **81** (1948), 43–86; G. Castelnuovo, *Le origini del calcolo infinitesimale nell'era moderna* (Milan, 1962), pp. 43–53; G. Cellini, "Gli indivisibili nel pensiero matematico e filosofico di Bonaventura Cavalieri," in *Periodico di matematiche,* 4th ser., **44** (1966), 1–21; "Le dimostrazioni di Cavalieri del suo principio," *ibid.,* pp. 85–105.

Ettore Carruccio

CAVALLO, TIBERIUS (*b.* Naples, Italy, 30 March 1749; *d.* London, England, 21 December 1809), *physics.*

Cavallo, the son of a Neapolitan physician, did his scientific work in England, where he settled in 1771 intending to acquire some experience in business. He apparently put aside his plans on becoming acquainted with some amateur English physicists, particularly William Henley, who introduced him to experimental philosophy and encouraged him in its pursuit. He became a fellow of the Royal Society in 1779.

Cavallo's first studies (1775–1776) concerned atmospheric electricity, which he explored with Franklin kites and with improved detectors of his own invention, fashioned after Canton's pith-ball electroscope. Although little came of his investigations (beyond the intelligence that rain often carries a negative charge), they required a course of self-instruction that culminated in Cavallo's most important work, *A Complete Treatise on Electricity in Theory and Practice* (1777). A second edition, with revisions, appeared in 1782 and a third, in three volumes, between 1786 and 1795. An excellent compendium, the *Treatise* served the needs of both the neophyte and the initiate, who found in its appendixes valuable details about medical electricity; about Beccaria's obscure theories; and about Cavallo's forte, the design and operation of electrostatic instruments.

With the publication of the *Complete Treatise* Cavallo switched his attention to the physics of the atmosphere and to the constitution of "permanently elastic fluids." Again his strengths appeared in instrumentation—an improved air pump and a modified eudiometer—and in smoothing the way for others. This time his course of self-instruction, *A Treatise on the Nature and Properties of Air and Other Permanently Elastic Fluids* (1781), was a judicious examination of contemporary work, particularly Priestley's, presented from a nondogmatic phlogistic point of view. He always retained a lively interest in pneumatic physics and chemistry, whose applications to ballooning and to medicine became subjects of two later books.

Cavallo also gave some attention to magnetism, on which he delivered a Bakerian lecture to the Royal Society in 1786 and published a treatise the following year; and he had a sustained interest in music—he was an amateur violinist—which led him to an investigation of the temperament of fretted instruments (1788). The inevitable text on the physics of music appeared in Cavallo's last, most ambitious, and least successful exposition: his wordy, overly elementary *Elements of Natural Philosophy* (1803).

BIBLIOGRAPHY

I. ORIGINAL WORKS. Cavallo's most important works are "Some New Electrical Experiments," in *Philosophical Transactions of the Royal Society,* **67:1** (1777), 48–55; "New Electrical Experiments and Observations, With an Improvement of Mr. Canton's Electrometer," *ibid.,* **67:2** (1777), 388–400; *A Complete Treatise on Electricity in Theory and Practice With Original Experiments* (London, 1777; 2nd ed., rev., 1782; 3rd ed., 3 vols., 1786–1795); *An Essay on the Theory and Practice of Medical Electricity* (London, 1780; 2nd ed., 1781); *A Treatise on the Nature and Properties of Air and Other Permanently Elastic Fluids.* . . . (London, 1781); "Description of an Air Pump," in *Philosophical Transactions of the Royal Society,* **73** (1783), 435–452; *The History and Practice of Aerostation* [ballooning] (London, 1785); *A Treatise on Magnetism in Theory and Practice* (London, 1787; 2nd ed., 1800); "Of the Methods of Manifesting the Presence, and Ascertaining the Quality, of Small Quantities of Natural or Artificial Electricity," in *Philosophical Transactions of the Royal Society,* **78** (1788), 1–22, the Bakerian lecture; "Description of a New Electrical Instrument Capable of Collecting Together a Diffused or Little Condensed Quantity of Electricity," *ibid.,* 255–260; "Of the Temperament of Those Musical Instruments, in Which the Tones, Keys or Frets, Are Fixed . . .," *ibid.,* 238–254; *An Essay on the Medical Properties of Factitious Air.* . . . (London, 1798); and *Elements of Natural Philosophy* (London, 1803).

No satisfactory bibliography for Cavallo exists; Poggendorff omits several contributions even to the *Philosophical Transactions.* There is a collection of Cavallo's correspondence (115 letters, 1782–1809) at the British Museum (Add MSS 22,897 and 22,898).

II. SECONDARY LITERATURE. Biographical information about Cavallo must be culled from his correspondence at the British Museum; *obiter dicta* in his papers; the Canton MSS II, 103, 107–108, at the Royal Society; *Dictionary of National Biography,* III, 1246–1247; and *Nouvelle biographie générale,* IX, 285. For his most significant scientific work see W. C. Walker, "The Detection and Estimation of Electric Charges in the Eighteenth Century," in *Annals of Science,* **1** (1936), 66–100. For his contributions to chemistry and animal electricity see, respectively, J. R. Partington, *A History of Chemistry,* III (London, 1962), 89, 300, 324; and P. C. Ritterbush, *Overtures to Biology. The Speculations of Eighteenth-Century Naturalists* (New Haven–London, 1964), *passim.*

JOHN L. HEILBRON

CAVANILLES, ANTONIO JOSÉ (*b.* Valencia, Spain, 16 January 1745; *d.* Madrid, Spain, 10 May 1804), *botany.*

In his youth Cavanilles specialized in the study of mathematics and physics and obtained a doctorate in theology. Because he failed to secure the professorship to which he aspired, he accepted the post of guardian to the sons of the duke of Infantado and accompanied the duke to Paris in 1777. In 1780 he became fascinated with botany and took courses given by the renowned naturalists A. Laurent de Jussieu and Lamarck. He published, in 1785, the first of ten monographs that constitute his *Monadelphiae.*

His return to Spain in 1789 marked the beginning of a rivalry with the director of the Madrid Botanical Gardens, Casimiro Gómez Ortega, and the botanist Hipólito Ruiz. In 1791 he was ordered to travel throughout the peninsula to study its botanical wealth; and, starting in 1799, he collaborated on the newly created *Anales de historia natural.* In 1801 he was given a professorship and the directorship of the Madrid Botanical Gardens; and the Spanish government ordered that any expeditions exploring Spanish America at that time send the Botanical Gardens examples of any herbs, seeds, and other plant forms that might be collected.

His disciples were Mariano Lagasca (1776–1839) and Simon de Rojas Clemente (1777–1827).

BIBLIOGRAPHY

I. ORIGINAL WORKS. A complete listing is in the study made by E. Alvarez López. The following are the most important: the series of ten monographs entitled *Monadelphiae classis dissertationes* (Paris, 1785; Madrid, 1790); *Icones et descriptiones plantarum quae aut sponte in Hispania crescunt aut in hortis hospitantur,* 6 vols. (Madrid, 1791–1801), in which 712 plants were listed according to the Linnaean classification and which gives data on some American plants; *Observaciones sobre la historia natural, geografía, agricultura, población y frutos del reyno de Valencia,* 2 vols. (Madrid, 1795–1797; new Zaragoza edition, 1958), a work in which he attempts to fix the natural wealth of the region studied; and *Descripción de las plantas que Don A. J. Cavanilles demostró en las lecciones públicas del año 1801* (Madrid, 1803, 1804).

II. SECONDARY LITERATURE. Works concerning Cavanilles are Luis Valdés Cavanilles, *Archivo del ilustre botánico D. Antonio Joseph Cavanilles* (Madrid, 1946); E. Alvarez López, "Antonio José Cavanilles," in *Anales del Jardín botánico de Madrid,* **6**, pt. 1 (1946), 1–64, an important study; Eduardo Reyes Prósper, *Dos noticias históricas del inmortal botánico y sacerdote hispano-valentino Don Antonio José Cavanilles, por Don Antonio Cavanilles y*

Centí y Don Mariano Lagasca, con anotaciones y los estudios biobibliográficos de Cavanilles y Centí y de Lagasca (Madrid, 1917).

JUAN VERNET

CAVENDISH, HENRY (*b.* Nice, France, 10 October 1731; *d.* London, England, 24 February 1810), *natural philosophy.*

In an age when leading British scientists were largely middle-class, Henry Cavendish stood out for his high aristocratic lineage. Although without title (he was, however, often addressed by the courtesy title "Honourable"), he was descended from dukes on both sides. His father, Lord Charles Cavendish, was the fifth son of the second duke of Devonshire. His mother, formerly Lady Anne Grey, was the fourth daughter of the duke of Kent. His mother's health was poor, for which reason she went to Nice, where Henry was born. She died two years later, shortly after giving birth to her second son, Frederick.

At eleven Cavendish was sent to Dr. Newcome's Academy at Hackney, a school attended mainly by children of the upper classes. He proceeded to St. Peter's College, Cambridge, in 1749, entering as a Fellow Commoner. He remained at Cambridge until 1753, leaving without a degree, a practice frequent among Fellow Commoners. It has been suggested that Cavendish objected to the religious tests at Cambridge, but in fact nothing is known about his religious convictions or lack of them. After leaving Cambridge he lived with his father in Great Marlborough Street, London, where he fitted out a laboratory and workshop. When his father died in 1783, Cavendish transferred his main residence and laboratory to Clapham Common. He never married.

Cavendish had independent means all of his life and never had to prepare for a profession; at some point he became immensely wealthy through bequests from relatives. At no time did he show an interest in entering the nonscientific world open to one with his assets of wealth and class. He shunned conventional society, which, by all contemporary accounts, he found difficult. Instead he devoted himself almost exclusively to scientific pursuits. His father, a distinguished experimentalist and prominent figure in the counsels of the Royal Society, encouraged his scientific bent. He put his instruments at his son's disposal and, most important, introduced him into London's scientific circles. In 1758 he took Henry to meetings of the Royal Society and to dinners of the Royal Society Club. Henry was elected to membership in these organizations in 1760, and he rarely missed a meeting.

Like his father, Cavendish was heavily involved in the work of the Council and committees of the Royal Society. He was a member of the Royal Society of Arts (1760) and a fellow of the Society of Antiquaries (1773). He was a trustee of the British Museum (1773) and a manager of the Royal Institution (1800). His career in general was distinguished by a wide and usually active participation in the organized scientific and intellectual life of London. Toward the latter part of his career he was esteemed at home and abroad (he was elected foreign associate of the Institut de France) as the most distinguished British man of science.

Henry Cavendish had fitful habits of publication that did not at all reveal the universal scope of his natural philosophy. He wrote no books and fewer than twenty articles in a career of nearly fifty years. Only one major paper was theoretical, a study of electricity in 1771; the remainder of his major papers were carefully delimited experimental inquiries, the most important of which were those on pneumatic chemistry in 1766 and 1783–1788, on freezing temperatures in 1783–1788, and on the density of the earth in 1798. The voluminous manuscripts uncovered after his death show that he carried on experimental, observational, and mathematical researches in literally all of the physical sciences of his day. They correct the impression derived from his few published writings that his interests were predominantly experimental and chemical.

Many of his interests—pure mathematics, mechanics, optics, magnetism, geology, and industrial science—that are strongly represented in his private papers are barely reflected in his published works. Cavendish left unpublished whatever did not fully satisfy him, and that included the great majority of his researches. The profundity of his private studies has exercised an immense fascination on subsequent workers in the fields that Cavendish explored. Fragments of his unpublished work were gradually revealed throughout the nineteenth century, culminating in James Clerk Maxwell's great edition of Cavendish's electrical researches in 1879. Far less successful was the attempt in 1921 by a group of scientific specialists to select for publication certain of Cavendish's nonelectrical manuscripts to complement Maxwell's edition. The totality of Cavendish's researches was too vast for that design.

The unifying ideas underlying Cavendish's numerous and varied basic researches relate to the Newtonian framework in which he chose to work. While he drew immediate stimulus from his contemporaries, the ultimate source of his inspiration was Newton. In the preface to the *Principia,* after explaining how he had derived the law of gravitation from astronomical phenomena and how he had deduced from it the motions of the planets, comets, and the seas, Newton expressed his wish that the rest of nature could be derived from the attracting and repelling forces of particles and the results cast in the deductive mode of the *Principia.* It was the conception of natural philosophy as the search for the forces of particles that guided Cavendish's scientific explorations. (His one important difference with Newton was his preference for the point–particles of John Michell and Bošković over Newton's extended corpuscles.) The *Principia* was forever his model of exact science; when this fact is appreciated, his various and seemingly disconnected researches are seen to form a rational, coherent whole.

Little is known about Cavendish's scientific activities between his leaving Cambridge and his first publication in 1766. His extant manuscripts suggest that he devoted much early effort to dynamics. The most important dynamical study, "Remarks Relating to the Theory of Motion," contains a full statement of his theory of heat. He subscribed to Newton's view that heat is the vibration of particles but went beyond Newton in rendering the vibration theory precise: Heat, Cavendish said, is the "mechanical momentum," or *vis viva,* of vibrating particles. He proved that the time average of the mechanical momentum of a collection of particles remains sensibly constant, provided the forces have certain symmetry properties. He related this theorem to another conservation law. It was well known that when two bodies are placed in thermal contact, the heat lost by one equals that gained by the other. Cavendish interpreted this to mean that the mechanical momentum lost by the particles of one body equals that gained by the particles of the second. But he was not satisfied. He had observed phenomena—such as fermentation, dissolution, and combustion—that involve quantities of heat which are inexplicable, even when the "additional" mechanical momentum of elastic compression is taken into account. Cavendish turned to heat experiments, which indicated a way around the theoretical impasse.

Cavendish drafted in fair copy, but did not publish, a long manuscript entitled "Experiments on Heat," based on laboratory work done in and possibly before 1765. Although he knew something of the work of Joseph Black and his circle, he essentially rediscovered the basic facts of specific heats (a term he later privately endorsed) and latent heats (a term he also privately endorsed but only after divorcing it from its connotation of a material theory of heat). The

difference in specific heats of mixtures or compounds and their component parts helped him explain the anomalous heats in the reactions violating his Newtonian heat theory. He thought that the difference in the specific heats accounted entirely for the addition or subtraction of sensible heat in reactions. Cavendish broke off his accounts of both specific and latent heats with inconclusive experiments on airs. In the one case he tried to find the specific heat of air by passing it through a worm tube encased in hot water, measuring the increase in the heat of the air. In the other he measured the cold produced by dissolving alkaline substances in acids, releasing fixed air, a phenomenon that he viewed as similar to evaporation.

In 1766 Cavendish published his first paper, for which he received the Royal Society's Copley Medal. It was on "factitious" airs, that is, airs that are contained inelastically in other bodies but are capable of being freed and made elastic. Cavendish's careful gravimetric discrimination of several factitious airs, together with the work of Black on fixed air, put forward strong evidence against the notion of a single, universal air. Cavendish produced fixed air by dissolving alkaline substances in acids, and by dissolving metals in acids he released inflammable air. He collected the airs that animal and vegetable substances yield on putrefaction and fermentation. (These agents—metals, alkalies, animal and vegetable matter—and their associated airs were the ones that Cavendish treated in the context of his last heat experiments.) He collected airs by inverting a bottle filled with water (or mercury for water-soluble fixed air) in a trough of water (or mercury); a tube led from the mouth of the inverted bottle to another, in which the reactants were placed. After collecting the airs he observed their combustibility, water solubility, and specific gravity. He found that fixed air is 1.57 times heavier than common air and that inflammable air is about eleven times lighter than common air. He showed that fermented organic substances give off a mixture of airs which includes a heavier inflammable air. From the fact that the same weight of a metal (zinc, iron, tin) produced the same volume of inflammable air regardless of the acid used (diluted sulfuric or hydrochloric acid), Cavendish concluded that the inflammable air came from the metal, not the acid. He suggested that the inflammable air of metals is pure phlogiston. In 1767 he published a related study of the composition of water from a certain pump, proving that the calcareous earth in the water is held in solution by fixed air.

In 1771, guided by his knowledge of elastic airs, Cavendish published a mathematical, single-fluid theory of electricity. In a preliminary draft he introduced the term "compression" in speaking of the state of tension of the electric fluid. Although he omitted the expression from his published theory, he retained the notion that the electric fluid within a body resembles an air compressed in a container. This was the central idea of his theory, providing an intensity measure in addition to a quantity measure of the electric fluid. There were essential differences, too, between the elastic fluids of electricity and air; and Cavendish stressed these as well as their resemblances. He proved that the particles of the electric fluid did not follow Newton's inverse first-power law of force of air particles. Just as his experimental discrimination among factitious airs helped discredit the notion that there is only one true, permanent air, so his electrical investigations indicated that, contrary to the common belief, there are elastic fluids in nature which must be represented by different laws of force. Cavendish was able to mathematize fully only one elastic fluid, that of electricity; elastic airs proved too complex.

Joseph Priestley, having stated the inverse-square law for the electric force in 1767, may have provided the occasion for Cavendish to elaborate his ideas on electricity. From the beginning Cavendish was partial to the inverse-square law, although in 1771 he had not yet performed his now famous hollow-globe experiments to settle the question of the exact numerical power. He postulated instead an elastic, electric matter of electricity, the particles of which repel one another and attract the particles of all other matter with a force varying inversely as some power of the distance less than the cube; in a symmetric manner the particles of all other matter repel each other and attract those of the electric fluid according to the same law. Cavendish's object was to exhibit the consequences of a variety of long-range electric forces and then to select the actual law from all possible laws by comparing their consequences with experience. His electrical researches are a direct expression, and partial vindication, of Newton's vision of the future of natural philosophy. From certain phenomena Cavendish deduced, but did not publish, the exact law of electric attraction and repulsion between particles; and from that law he derived a rich store of new, quantitative electrical phenomena. His greatest predictive achievement lay buried in manuscript: it was the calculation and experimental confirmation of the precise quantities of electric fluid that bodies of different geometrical form and size can contain at any electrical tension. His confirmatory experiments, together with an extended theoretical development, constituted the design of an unfinished, unpublished treatise, a work that was intended to stand as the electrical sequel to the gravitational "System of the World" of the *Principia*.

Cavendish examined minutely the facts of specific inductive capacity (not his term), an electrical corollary of chemical differences. His efforts at understanding this empirical phenomenon, which seemed at first to contradict his theory, diverted him from completing his original design. So did the fact that he came to seek a dynamics as well as a statics of the electric fluid. He attempted without success to find the relation between force, resistance, and velocity in the passage of the electric fluid through various substances. His researches trailed off into largely inconclusive experiments on conductivities. He revealed certain of his dynamical findings in a second electrical publication, a study in 1776 of the properties of a model of an electric fish, the torpedo. His electrical researches, the most sustained and organized effort of his career, came to an end in 1781.

Priestley's account in 1781 of his and John Warltire's experiments prompted Cavendish to return to the subject of elastic airs. The first of his new publications on the subject was a study of the principles of eudiometry in 1783. The most important fruit of his renewed interest was his celebrated publication in 1784 on the synthesis of water from two airs. Warltire had electrically fired mixtures of common and inflammable airs in a closed vessel, recording a weight loss that he attributed to the escape of ponderable heat. He and Priestley observed a deposit of dew inside the vessel.

Cavendish repeated the experiments and found dew but no loss in weight. He then undertook experiments to discover the cause of the diminution of common air when it is fired with inflammable air and when it is phlogisticated by any other means. He found that when inflammable and common air are exploded, all of the inflammable air and about four-fifths of the common air are converted into dew and that this dew is pure water. What Cavendish was basically interested in was the constitution of the airs; he concluded that inflammable air is phlogiston united to water and that dephlogisticated air is water deprived of phlogiston. In several papers through 1788 he pursued investigations stemming from those of 1784, concluding that phlogisticated air is nitrous acid united to phlogiston. Cavendish's publications on pneumatic chemistry in 1783–1788 involved the agency of electricity, the transition between elastic and inelastic states of matter, and the generation of heat; they drew, therefore, on the basic themes of his research for the previous quarter century.

Concurrently with his work on airs Cavendish published several papers on the freezing points of mercury, vitriolic acid, nitrous acid, and other liquids. This work was an extension of his published study of the Royal Society's meteorological instruments in 1776, and it drew heavily upon his early knowledge of latent heats. The most important of his conclusions was that the extraordinarily low readings that had been recorded on mercury thermometers were due merely to the shrinkage of solidifying mercury.

Cavendish published five papers between 1784 and 1809 relating to his astronomical interests. With one exception they were comparatively minor productions, concerned with the height of the aurora, a reconstruction of the Hindu civil year, a calculation in nautical astronomy, and a method of marking divisions on circular astronomical instruments. The exception was his determination of the density of the earth (or weighing of the world) in 1798, by means of John Michell's torsion balance. The apparatus consisted of two lead balls on either end of a suspended beam; these movable balls were attracted by a pair of stationary lead balls. Cavendish calculated the force of attraction between the balls from the observed period of oscillation of the balance and deduced the density of the earth from the force. He found it to be 5.48 times that of water. Cavendish was the first to observe gravitational motions induced by comparatively minute portions of ordinary matter. The attractions that he measured were unprecedentedly small, being only 1/500,000,000 times as great as the weight of the bodies. By weighing the world he rendered the law of gravitation complete. The law was no longer a proportionality statement but a quantitatively exact one; this was the most important addition to the science of gravitation since Newton.

Cavendish's career marked the culmination and the end of the original British tradition in mathematical physics. By the 1780's, British natural philosophy had moved away from any central concern with mathematical interparticulate forces. It had become concerned with the ethereal mode of communication of forces and with imponderable fluids and the question of their separateness or unity. These directions were antithetical to Cavendish's thought. Likewise, chemistry tended to follow Lavoisier's direction, about which Cavendish had strong reservations. Cavendish was intellectually isolated long before the end of his career. He was not a teacher; he formed or inspired no school. Rather his place in British natural philosophy is as the first after Newton to possess mathematical and experimental talents at all comparable to Newton's. In intellectual stature Cavendish was without peer in eighteenth-century British natural philosophy.

BIBLIOGRAPHY

I. ORIGINAL WORKS. Cavendish's two electrical papers from the *Philosophical Transactions* and the bulk of his electrical MSS are published in *The Electrical Researches of the Honourable Henry Cavendish*, J. Clerk Maxwell, ed. (Cambridge, 1879), also in Cass Library of Science Classics (London, 1967). A rev. ed. of this work is *The Scientific Papers of the Honourable Henry Cavendish, F.R.S., 1: The Electrical Researches*, J. Clerk Maxwell, ed., rev. with notes by Sir Joseph Larmor (Cambridge, 1921). There is also a companion volume containing the rest of Cavendish's papers from the *Philosophical Transactions*, together with a selection of his MSS on chemistry, heat, meteorology, optics, mathematics, dynamics, geology, astronomy, and magnetism: *The Scientific Researches of the Honourable Henry Cavendish, F.R.S., 2: Chemical and Dynamical*, ed. with notes by Sir Edward Thorpe (Cambridge, 1921), with contributions by Charles Chree, Sir Frank Watson Dyson, Sir Archibald Geikie, and Sir Joseph Larmor. Some additional Cavendish MSS relevant to the water controversy are printed in the Rev. W. Vernon Harcourt's address in the *British Association Report* (1839), pp. 3-68 plus 60 pp. of lithographed facsimiles. The correspondence between Cavendish and Joseph Priestley is published in *A Scientific Autobiography of Joseph Priestley, 1733–1804*, ed. with commentary by Robert E. Schofield (Cambridge, Mass., 1966). The vast bulk of Cavendish's MS papers and correspondence has not been published; it is deposited in Chatsworth, in the possession of the duke of Devonshire.

II. SECONDARY LITERATURE. On Cavendish or his work, see A. J. Berry, *Henry Cavendish: His Life and Scientific Work* (London, 1960); J. B. Biot, "Cavendish (Henri)," in *Biographie universelle*, 2nd ed., VII, 272–273; Henry Brougham, "Cavendish," in *Lives of Men of Letters and Science Who Flourished in the Time of George III* (Philadelphia, 1845), pp. 250–259; James Gerald Crowther, *Scientists of the Industrial Revolution: Joseph Black, James Watt, Joseph Priestley, Henry Cavendish* (London, 1962); Georges Cuvier, "Henry Cavendish," trans. D. S. Faber, in *Great Chemists*, E. Faber, ed. (New York, 1961), pp. 229–238; Humphry Davy's estimate of Cavendish, in John Davy, *Memoirs of the Life of Sir Humphry Davy, Bart.*, I (London, 1836), 220–222; Russell McCormmach, "The Electrical Researches of Henry Cavendish," unpub. diss. (Case Institute of Technology, 1967); "John Michell and Henry Cavendish: Weighing the Stars," in *British Journal for the History of Science*, IV (1968), 126–155; "Henry Cavendish: A Study of Rational Empiricism in Eighteenth-Century Natural Philosophy," in *Isis* (1970); J. R. Partington, "Cavendish," in *A History of Chemistry*, III (London, 1962), 302–362; Robert E. Schofield, *Mechanism and Materialism: British Natural Philosophy in an Age of Reason* (Princeton, 1970), ch. 10; Thomas Thomson's estimate of Cavendish, in his *History of Chemistry*, I (London, 1830),

336–349; George Wilson, *The Life of the Honourable Henry Cavendish* (London, 1851); and Thomas Young, "Life of Cavendish," in *Scientific Papers of the Honourable Henry Cavendish,* I, 435–447.

RUSSELL MCCORMMACH

CAVENTOU, JOSEPH-BIENAIMÉ (*b.* Saint-Omer, France, 30 June 1795; *d.* Paris, France, 5 May 1877), *chemistry, toxicology.*

Son of Pierre-Vincent Caventou, military pharmacist and chief pharmacist of the civil hospital of Saint-Omer, Joseph-Bienaimé Caventou decided early in life to follow his father's profession. After some preliminary training with his father he left for Paris, where he obtained an apprenticeship in a pharmacy and began course work at the School of Pharmacy and the Faculty of Sciences. In 1815 he competed successfully for an internship in hospital pharmacy, but the news of Napoleon's return from Elba aroused his patriotic feelings to such an extent that he resigned his appointment to enlist as a military pharmacist. Caventou's military service was of short duration. The small garrison where he was stationed in Holland surrendered soon after the French defeat at Waterloo, and before the end of 1815 he was back in Paris to resume his studies.

Caventou had by this time developed a keen interest in chemistry and, in order to supplement the meager allowance from his father, conceived the idea of writing a book on chemical nomenclature according to the classification adopted by Thenard. The work, *Nouvelle nomenclature chimique,* appeared in 1816 as a practical handbook designed especially for beginners in chemistry and for those who were unfamiliar with the newest chemical terminology. In the meantime Caventou again competed successfully for an internship in hospital pharmacy and in 1816 received his appointment at the Saint-Antoine Hospital, where laboratory facilities to carry on his research were available. He published a chemical analysis of the daffodil by the end of 1816, of laburnum in 1817, and a treatise on pharmacy in 1819. These were followed in 1821 by an annotated French translation, made jointly with J. B. Kapeler, physician at the Saint-Antoine Hospital, of a German work by Johann Christoph Ebermaier on drug adulteration.

In 1826 Caventou became a member of the teaching staff of the École Supérieure de Pharmacie, in 1830 associate professor of chemistry, and in 1834 full professor of toxicology, a post he held until his retirement at the end of 1859. Despite the demands of teaching and research, he found time to direct a pharmacy on the rue Gaillon. In 1821 Caventou was admitted to the Academy of Medicine. In 1827 he and Pierre-Joseph Pelletier shared the Montyon Prize of 10,000 francs, awarded by the Academy of Sciences, for their discovery of quinine.

It was in 1817 that Caventou published his first joint paper with Pelletier, a twenty-nine-year-old owner of a pharmacy on the rue Jacob, who had already attracted favorable attention by his chemical analyses of plant substances. The young men had been drawn together by their mutual scientific interests, and until Pelletier's death in 1842 their frequent collaboration resulted in a number of important discoveries in alkaloid chemistry. It is idle to speculate on Caventou's development as a scientist had he not collaborated with Pelletier; but his most impressive scientific accomplishments came from this association, particularly during the years from 1817 to 1821. By the age of twenty-six, the achievements which would bring him most fame were already behind him. During this period both scientists had embarked on the investigation of natural products: the description of a new acid formed by the action of nitric acid on the nacreous material of human biliary calculi (1817); a study of the green pigment in leaves, which they named chlorophyll (1817); the separation of crotonic acid from croton oil (1818); the examination of carmine, the coloring matter in cochineal (1818); and the isolation of ambrein from ambergris (1820).

Far more significant, however, was their extraction of alkaline nitrogenous substances (alkaloids) from plants. When Pelletier and Caventou began this phase of their work, the stage had already been set for dramatic developments in alkaloid chemistry by the pioneer work on opium by such scientists as Derosne, Armand Seguin, and especially Sertürner, who was the first to recognize the alkaline nature of morphine and whose findings, published from 1805 to 1817, established him as its discoverer. In rapid succession Pelletier and Caventou isolated strychnine in 1818, brucine and veratrine (independently of Karl Meissner) in 1819, and cinchonine and quinine in 1820. They discovered caffeine in 1821, independently of Robiquet and Runge.

The discovery of quinine was by far the most dramatic result of their collaboration, and soon there was worldwide demand for quinine as a therapeutic agent. In a letter written to the Academy of Sciences in 1827 Pelletier and Caventou pointed out that by 1826 a burgeoning French industry was annually producing approximately 90,000 ounces of quinine sulfate from cinchona bark, enough to treat more than a million individuals. The basic and salifiable nature of these new alkaloids, as well as their physical characteristics, were elucidated by Pelletier and Caventou, who demonstrated that they contained oxygen, hydrogen, and carbon, but who failed, initially, to find nitrogen. Antoine Bussy's meticulous analysis of morphine in 1822 and further joint investigations by Jean Dumas and Pelletier proved conclusively the presence of nitrogen in alkaloids.

Caventou's lifelong interest in phytochemistry is reflected in numerous papers that he published in this field. In 1830 he collaborated with Pelletier and François to isolate the bitter crystalline principle, cahinca acid, from cahinca root (*Chiococca racemosa* L. *Rubiaceae*). As early as 1825, and for several years thereafter, he attempted to develop chemical tests for pathological conditions, especially tuberculosis. Although his results were inconclusive, his descriptions of his experiments, published much later in the *Annales de chimie et de physique* (1843), are of historical interest. His expertise in toxicology was recognized by his colleagues in the Academy of Medicine when he was called upon in 1839 and 1841 to report on cases dealing with arsenical poisoning. But Caventou's major scientific output clearly belonged to his early years, when he firmly established a reputation as a gifted and original investigator in alkaloid chemistry.

BIBLIOGRAPHY

I. ORIGINAL WORKS. In addition to his *Nouvelle nomenclature chimique* (Paris, 1816), mentioned in the text, Caventou's earliest publications were "Recherches chimiques sur le narcisse des prés, *Narcissus pseudo-narcissus*, Linn.," in *Journal de pharmacie et des sciences accessoires*, **2** (1816), 540–549; "Examen chimique des fleurs du cytise des Alpes, *Cytisus laburnum*, Linn.," *ibid.*, **3** (1817), 306–309; and "Observations chimiques faites dans l'analyse d'un calcul cystique," *ibid.*, 369–373. Among Caventou's most important publications were the following, published jointly with Pelletier: "Sur l'action qu'exerce l'acide nitrique sur la matière nacrée des calculs biliaires humains, et sur le nouvel acide qui en résulte," *ibid.*, 292–305; "Essai analytique sur la graine du médicinier cathartique, *Jatropha curcas*," *ibid.*, **4** (1818), 289–297; "Note sur un nouvel alcali," in *Annales de chimie et de physique*, 2nd ser., **8** (1818), 323–324; "Sur la matière verte des feuilles," *ibid.*, **9** (1818), 194–196; "Mémoire sur un nouvel alcali végétal (la strychnine) trouvé dans la fève de Saint-Ignace, la noix vomique, etc.," *ibid.*, **10** (1819), 142–177; "Mémoire sur une nouvelle base salifiable organique trouvée dans la fausse angusture, *Brucaea antidysenterica*," *ibid.*, **12** (1819), 113–148; "Examen chimique de plusieurs végétaux de la famille des colchicées, et du principe actif qu'ils renferment . . .," *ibid.*, **14** (1820), 69–83; "Recherches chimiques sur les quinquinas," *ibid.*, **15** (1820), 289–318, 337–365; *Analyse chimique des quinquina* (Paris, 1821); "Lettre de MM. Pelletier et Caventou à MM. les membres de l'Académie royale des sciences, sur la fabrication du sulfate de quinine," in *Annales de chimie et de physique*, 2nd ser., **34** (1827), 331–335; "Recherches chimiques sur quelques matières animales saines et morbides," *ibid.*, 3rd ser., **8** (1843), 321–346; and, with Pelletier and François, "Nouveau principe amer acide, cristallisé contenu dans l'écorce de la racine de kahinça," *ibid.*, 2nd ser., **44** (1830), 291–296. Caventou's view that all pharmacy students should receive a rigorous training in chemistry is elucidated in his *Traité élémentaire de pharmacie théorique d'après l'état actuel de la chimie . . .* (Paris, 1819). His work in toxicology is reflected in two reports: *Rapport sur un empoisonnement par l'acide arsénieux, fait à l'Académie royale de médecine . . .* (Paris, 1839); and *Rapport sur les moyens de constater la présence de l'arsenic dans les empoisonnements par ce toxique, au nom de l'Académie royale de médecine . . .* (Paris, 1841). For a fuller listing of Caventou's papers, see Royal Society of London, *Catalogue of Scientific Papers (1800–1863)*, I (London, 1867), 847–848.

II. SECONDARY LITERATURE. See Étienne Jules Bergeron, *Éloge de Caventou* (Paris, 1897); and Marcel Delépine, "Joseph Pelletier and Joseph Caventou," in *Journal of Chemical Education*, **28** (1951), 454–461.

ALEX BERMAN

CAYEUX, LUCIEN (*b.* Semousies, Nord, France, 26 March 1864; *d.* Mauves-sur-Loire, Loire-Inférieure, France, 1 November 1944), *sedimentary petrography, stratigraphy.*

Cayeux's youth was spent in the Avesnois, the country surrounding Semousies, where, after completing his secondary education, he taught for a short period at the primary school of Avesnes-sur-Helpe.

Désiré Piérart, an enthusiastic collector of fossils from the neighboring village of Dourlers, had a strong influence on the farmer Xavier Cayeux and his son Lucien. He awoke the son's interest in geology and prompted the father to send him to Lille to study under Jules Gosselet and Charles Barrois. While working for his master's degree, Cayeux was appointed their assistant, or more correctly their *préparateur*, in 1887. This is a peculiar position in the French academic system that combines the preparation of materials for classroom and laboratory demonstrations with curatorial duties, participation in administrative work, and personal research. Such an assistant is supposed to spend equal time on professional services and on his own research, a combination that develops a strong interest in original investigations and a deep sense of academic duty. It was the ideal start for an energetic and industrious man like Cayeux.

In 1891 Marcel Bertrand offered Cayeux the position of *préparateur* at the National School of Mines in Paris. Torn for a while between Gosselet's exhortation to stay at Lille and Barrois's advice to take advantage of the vast scientific resources of the capital, Cayeux eventually went to Paris and never regretted his decision. While at the School of Mines he completed his doctoral dissertation and defended it in Lille in 1897. Appointed in 1898 to the National Agronomical Institute, Cayeux occupied several positions at that institution and in 1901 became professor of agricultural geology. His teaching in this new field of applied geology met with great success.

In 1904, when Marcel Bertrand had to resign from the School of Mines for reasons of health, Cayeux replaced him as assistant professor; and after Bertrand's death in 1907 he became, contrary to a well-established tradition, professor-in-charge. Five years later he was called to the chair of geology of the Collège de France, to take the place of Auguste Michel-Lévy. In this final position of his career Cayeux spent twenty-four years developing the teaching of the petrography of sedimentary rocks, which closely reflected his personal research.

In 1936 new administrative regulations forced Cayeux's retirement, depriving him of three years of teaching during which he was planning to complete his carefully organized academic program. This situation greatly embittered Cayeux, who then dedicated himself entirely to research in his own laboratory at the School of Mines. At the outbreak of the war in 1939 he secluded himself on his property at Mauves-sur-Loire, where in previous years he had spent long periods. During this last portion of his life Cayeux combined the delicate and demanding task of mayor of a town under German occupation with the completion of some of his scientific works. He wrote the synthesis of his long and unique experience, a remarkable booklet entitled *Causes anciennes et causes actuelles en géologie,* a subject he had planned to discuss during the last year of his teaching at the Collège de France.

Cayeux's fundamental contribution pertains to the petrography of sedimentary rocks, a field of which he is considered one of the founders. His doctoral dissertation was the first of an impressive series of memoirs dealing with such rocks. It consists of a study of the chalk of the Paris basin and of certain siliceous rocks (*gaizes, meules, tuffeaux*) peculiar to the Mesozoic and Cenozoic formations of northern France and Belgium. When Cayeux began his investigation of the sedimentary rocks, they had been largely neglected in comparison with metamorphic and igneous rocks, on which rapid progress had been accomplished under the leadership of such scientists as Zirkel, F. Fouqué, Michel-Lévy, Rosenbusch, and Lacroix. Only H. C. Sorby and Sir John Murray in Great Britain had understood the potential importance of the petro-

graphic study of the sedimentary rocks. In order to succeed in such a task, one had to be a petrographer, a paleontologist, a stratigrapher, and a chemist; Cayeux met all these prerequisites.

In his doctoral dissertation Cayeux outlined his superb analytical method of investigation and stated his aims, which were to remain the same in his later works. He first describes the components of a sedimentary rock: detrital minerals, authigenic or secondary minerals generated within the rock itself, organic remains, and cement or matrix holding these components together. Adding the results of chemical analyses to the spectrum of microscopic observations, his synthesis attempts to determine the original characters of the sediment and the provenance of its detrital components. Finally, he follows through geological time the diagenetic modifications undergone by the sediment after its deposition.

For Cayeux the science of sedimentary rocks was essentially the complete natural history of ancient and recent sediments. It was therefore natural that he relied heavily upon the wealth of data collected by the *Challenger* expedition, using them for comparative purposes in order to understand the genesis of ancient sediments on the basis of recent ones. Hence he created a new approach: paleo-oceanography.

After his investigation of the chalk, in which he demonstrated that in spite of apparent analogies with the recent deep-sea Globigerina ooze, it was a shallow-water sediment, Cayeux undertook a series of comprehensive and well-illustrated monographs, concerned particularly with the sedimentary rocks of France and its colonies. He successively applied his unusual analytical gifts to the Tertiary sandstones of the Paris basin (1906), the Paleozoic and Mesozoic oolitic iron ores (1909, 1922), the siliceous rocks or cherts (1929), the phosphates (1939, 1941, 1950 [posthumous]), and the carbonate rocks: limestones and dolomites (1935). Although Cayeux's interpretation of the oolitic iron ores did not gain general acceptance, his synthetic views on the genesis of siliceous rocks, phosphates, and carbonates show an amazingly modern character. Particularly in carbonate rocks he anticipated a great number of modern trends concerning early diagenetic changes, such as induration, recrystallization, and dolomitization.

As a result of his years of teaching at the Collège de France, Cayeux produced his *Introduction à l'étude pétrographique des roches sédimentaires,* published in 1916, out of print in 1927, and reprinted in 1931. In this beautifully illustrated monograph that obviously filled a need among geologists, he gave an exhaustive account of all the methods of analyzing sedimentary rocks: physical, microchemical, and chromatic. He then proceeded to distinguish all the constitutive elements of these deposits, describing each individual mineral species as well as the various organisms that contributed to their genesis. Of particular interest is a long chapter on the microstructure of the skeletal remains of living and fossil organisms. Sedimentary petrographers are still waiting for a modern equivalent of this masterpiece.

These outstanding monographs, usually preceded by the publication of numerous short papers—which total more than 270—should not, however, lead one to ignore another interesting aspect of Cayeux's long career. Between 1900 and 1904 he served as geological adviser to the archaeologists working in Greece and in the archipelago, sponsored by the French School of Athens. In that capacity Cayeux undertook a detailed study of the island of Delos, determining, among other things, that the mean sea level on the shores of the eastern Mediterranean had not changed appreciably since remotest antiquity. He also undertook an interesting petrographic study of the building and decorative materials that the ancient Greeks used in construction and adornment of the famous temple of Apollo and the determination of their geographic and geological origin. Cayeux's survey of Crete and of the Peloponnesus remained unachieved. He did, however, demonstrate the occurrence of overthrusts in the mountains of the southern part of the country.

Cayeux's philosophical approach to geology underwent a remarkable evolution throughout his life. In his first work on the chalk of the Paris basin, he attributed a great importance to the time factor in diagenesis, assuming that the chalk would undergo, after deposition, many changes toward a stable condition, through a long span of geological time and under the action of the everlasting causes postulated by the concept of uniformitarianism. His subsequent studies on iron ores, cherts, phosphates, and carbonates gradually revealed to him that many fundamental features of these deposits had been generated contemporaneously or penecontemporaneously with deposition by submarine processes of reworking, transportation, induration, and recrystallization, and that such deposits underwent no other major changes after emergence. Furthermore, Cayeux stressed that most of these processes, so clearly expressed in ancient sediments, were not active in present-day oceans. Therefore, he concluded that serious consideration should be given to the dual concept of past and present causes in geology. This fundamental challenge to uniformitarianism ended Cayeux's scientific contribution with an issue that remains to be resolved by geologists.

BIBLIOGRAPHY

I. ORIGINAL WORKS. Cayeux's writings include "Contribution à l'étude micrographique des terrains sédimentaires. I. Étude de quelques dépôts siliceux, secondaires et tertiaires du bassin de Paris. II. Craie du bassin de Paris," *Mémoires de la Société géologique du Nord,* **4,** no. 2 (1897); *Structure et origine des grès du Tertiaire parisien,* in the series Études des gîtes minéraux de la France (Paris, 1906); *Les minerais de fer oolithique de France. Fascicule I: Minerais de fer primaires,* in the series Études des gîtes minéraux de la France (Paris, 1909); *Description physique de l'île de Délos,* vol. IV of *Exploration archéologique de Délos, . . .* (Paris, 1911); *Introduction à l'étude pétrographique des roches sédimentaires,* in the series Mémoires explicatifs de la Carte géologique détaillée de la France, 2 vols. (Paris, 1916; repr. 1931); *Les minerais de fer oolithique de France. Fascicule II: Minerais de fer secondaires,* in the series Études des gîtes minéraux de la France, (Paris, 1922); *Les roches sédimentaires de France. Roches siliceuses,* in the series Mémoires explicatifs de la Carte géologique détaillée de la France (Paris, 1929); *Les roches sédimentaires de France. Roches carbonatées (calcaires et dolomies)* (Paris, 1935), annotated English trans. by Albert V. Carozzi (New York, 1970); *Les phosphates de chaux sédimentaires de France (France métropolitaine et d'outre-mer),* I (Paris, 1939); *Causes anciennes et causes actuelles en géologie* (Paris, 1941), annotated English trans. by Albert V. Carozzi (New York, 1970); *Les phosphates de chaux sédimentaires de France (France métropolitaine et d'outre-mer),* II, *Égypte, Tunisie, Algérie* (Paris, 1941), pp. 351–659, pls. XVI–XXXIII; and *Les phosphates de chaux sédimentaires de France (France métropolitaine et d'outre-mer),* III, *Maroc et conclusions générales* (Paris, 1950), pp. 661–970, pls. XXXIV–LIV.

II. SECONDARY LITERATURE. For further information on Cayeux and his work, see L. Bertrand, "Notice sur l'oeuvre de Lucien Cayeux (1864–1944)," in *Notices et discours. Académie des sciences,* II, *1937–1948* (Paris, 1949), 607–632; M. Leriche, "La vie et l'oeuvre de Lucien Cayeux (1864–1944)," in *Bulletin de la Société belge de géologie,* **55B** (1946), 259–314, with complete list of publications; and "Lucien Cayeux (1864–1944)," in *Bulletin de la Société géologique de France,* 5th ser., **17** (1947), 349–377, with complete list of publications; and E. Margerie, "Memorial to Lucien Cayeux," in *Proceedings of the Geological Society of America* for 1947 (1948), 131–133.

ALBERT V. CAROZZI

CAYLEY, ARTHUR (*b.* Richmond, Surrey, England, 16 August 1821; *d.* Cambridge, England, 26 January 1895), *mathematics, astronomy.*

Cayley was the second son of Henry Cayley, a merchant living in St. Petersburg, and Maria Antonia Doughty. He was born during a short visit by his parents to England, and most of his first eight years were spent in Russia. From a small private school in London he moved, at fourteen, to King's College School there. At seventeen he entered Trinity College, Cambridge, as a pensioner, becoming a scholar in 1840. In 1842 Cayley graduated as senior wrangler and took the first Smith's prize. In October 1842 he was elected a fellow of his college at the earliest age of any man of that century. He was tutor there for three years, spending most of his time in research. Rather than wait for his fellowship to expire (1852) unless he entered holy orders or took a vacant teaching post, he entered the law, studying at Lincoln's Inn. He was called to the bar in 1849.

During the fourteen years Cayley was at the bar he wrote something approaching 300 mathematical papers, incorporating some of his best and most original work. It was during this period that he first met the mathematician J. J. Sylvester, who from 1846 read for the bar and, like Cayley, divided his time between law and mathematics. In 1852 Sylvester said of Cayley that he "habitually discourses pearls and rubies," and after 1851 each often expressed gratitude to the other in print for a point made in conversation. That the two men profited greatly by their acquaintance is only too obvious when one considers the algebraic theory of invariants, of which they may not unreasonably be considered joint founders. They drifted apart professionally when Cayley left London to take up the Sadlerian professorship but drew together again when, in 1881–1882, Cayley accepted Sylvester's invitation to lecture at Johns Hopkins University.

In 1863 Cayley was elected to the new Sadlerian chair of pure mathematics at Cambridge, which he held until his death. In September 1863 he married Susan Moline, of Greenwich; he was survived by his wife, son, and daughter. During his life he was given an unusually large number of academic honors, including the Royal Medal (1859) and the Copley Medal (1881) of the Royal Society. As professor at Cambridge his legal knowledge and administrative ability were in great demand in such matters as the drafting of college and university statutes.

For most of his life Cayley worked incessantly at mathematics, theoretical dynamics, and mathematical astronomy. He published only one full-length book, *Treatise on Elliptic Functions* (1876); but his output of papers and memoirs was prodigious, numbering nearly a thousand, the bulk of them since republished in thirteen large quarto volumes. His work was greatly appreciated from the time of its publication, and he did not have to wait for mathematical fame. Hermite compared him with Cauchy because of his immense capacity for work and the clarity and elegance of his analysis. Bertrand, Darboux, and Glaisher all com-

pared him with Euler for his range, his analytical power, and the great extent of his writings.

Cayley was the sort of courteous and unassuming person about whom few personal anecdotes are told; but he was not so narrow in outlook as his prolific mathematical output might suggest. He was a good linguist; was very widely read in the more romantic literature of his century; traveled extensively, especially on walking tours; mountaineered; painted in watercolors throughout his life; and took a great interest in architecture and architectural drawing.

Characteristically, as explained in the bibliography of his writings, Cayley frequently gave abundant assistance to other authors (F. Galton, C. Taylor, R. G. Tait, G. Salmon, and others), even writing whole chapters for them—always without ostentation. Salmon, who corresponded with him for many years, gave *Esse quam videri* as Cayley's motto. Although Cayley disagreed strongly with Tait over quaternions (see below), their relations were always amicable; and the sixth chapter of the third edition of Tait's *Quaternions* was contributed by Cayley, much of it coming verbatim from letters to Tait. Cayley was above all a pure mathematician, taking little if any inspiration from the physical sciences when at his most original. "Whose soul too large for vulgar space, in *n* dimensions flourished," wrote Clerk Maxwell of Cayley. So far as can be seen, this was a more astute characterization than that of Tait, by whom Cayley was seen in a more pragmatic light, "forging the weapons for future generations of physicists." However true Tait's remark, it was not an indication of Cayley's attitude toward his own work.

A photograph of Cayley is prefixed to the eleventh volume of the *Collected Papers*. A portrait by Lowes Dickenson (1874, Volume VI) and a bust by Henry Wiles (1888) are in the possession of Trinity College, Cambridge. A pencil sketch by Lowes Dickenson (1893) is to be found in Volume VII.

Cayley's mathematical style was terse and even severe, in contrast with that of most of his contemporaries. He was rarely obscure, and yet in the absence of peripheral explanation it is often impossible to deduce his original path of discovery. His habit was to write out his findings and publish without delay and consequently without the advantage of second thoughts or minor revision. There were very few occasions on which he had cause to regret his haste. (References below to the *Collected Mathematical Papers*, abbreviated *C.M.P.*, contain the volume number, followed by the number of the paper, the year of original publication, and the page numbers of the reprint.)

Cayley is remembered above all else for his contributions to invariant theory. Following Meyer (1890–1891), the theory may be taken to begin with a paper by Boole, published in 1841, hints of the central idea being found earlier in Lagrange's investigation of binary quadratic forms (1773) and Gauss's similar considerations of binary and ternary forms (1801). Lagrange and Gauss were aware of special cases in which a linear homogeneous transformation turned a (homogeneous) quadratic into a second quadratic whose discriminant is equal to that of the original quadratic multiplied by a factor which was a function only of the coefficients of the transformation. Cauchy, Jacobi, and Eisenstein all have a claim to be mentioned in a general history of the concept of invariance, but in none of their writings is the idea explicit. Boole, on the other hand, found that the property of invariance belonged to all discriminants, and he also provided rules for finding functions of "covariants" of both the coefficients and the variables with the property of invariance under linear transformation.

In 1843 Cayley was moved by Boole's paper to calculate the invariants of nth-order forms. Later he published a revised version of two papers he had written. The first, with the title "On the Theory of Linear Transformations" (*C.M.P.,* I, no. 13 [1845], 80–94), dealt only with invariants; the second, "On Linear Transformations" (*C.M.P.,* I, no. 14 [1846], 95–112), introduced the idea of covariance. In this second paper Cayley set out "to find all the derivatives of any number of functions, which have the property of preserving their form unaltered after any linear transformations of the variables." He added that by "derivative" he meant a function "deduced in any manner whatever from the given functions." He also attempted to discover the relations between independent invariants—or "hyperdeterminants," as he called them at first, looking upon algebraic invariance as a generalized form of the multiplication of determinants. When writing the notes to his *Collected Papers,* he remarked that what he had done in this paper was to be distinguished from Gordan's "Ueberschiebung," or derivational theory. Cayley may be regarded as the first mathematician to have stated the problem of algebraic invariance in general terms.

Cayley's work soon drew the attention of many mathematicians, particularly Boole, Salmon, and Sylvester in England and Aronhold, Clebsch, and, later, Gordan in Germany. (Jordan and Hermite followed in France; and Brioschi in Italy was to carry the new ideas into the realm of differential invariants, in the study of which his compatriots later excelled.) Salmon's many excellent textbooks (in particular, see

his *Modern Higher Algebra,* 1859, dedicated to Cayley and Sylvester), which were translated into several languages, diffused Cayley's results, to which Cayley himself constantly added. Sylvester was, among other things, largely responsible for the theory's luxuriant vocabulary; and in due course Aronhold related the theory to Hesse's applications of determinants to analytical geometry. The vocabulary of the subject is today one of the greatest obstacles to a discussion of invariant theory, since following Gordan's theorem of 1868 and Hilbert's generalizations of it, the tendency has been away from developing techniques for generating and manipulating a multiplicity of special invariants, each with its own name. Notice, however, that Cayley's "quantic" is synonymous with the "form" of later algebraists. As a typical source of terminological confusion we may take the contravariant (or the curve represented by the contravariant equation), called by Cayley the "Pippian" and known elsewhere (following Cremona) as the "Cayleyan."

Beginning with an introductory memoir in 1854, Cayley composed a series of ten "Memoirs on Quantics," the last published in 1878, which for mathematicians at large constituted a brilliant and influential account of the theory as he and others were developing it. The results Cayley was obtaining impressed mathematicians by their unexpectedness and elegance. To take three simple examples, he found that every invariant vanishes, for a binary p-ic which has a linear factor raised to the rth power, if $2r > p$; that a binary p-ic has a single or no p-ic covariant of the second degree in the coefficients according as p is or is not a multiple of 4; and that all the invariants of a binary p-ic are functions of the discriminant and $p-3$ anharmonic ratios, each formed from three of the roots together with one of the remaining $p-3$ roots. A more renowned theorem concerned the number of linearly independent seminvariants (or invariants) of degree i and weight w of a binary p-ic. Cayley found an expression giving a number which he proved could not be less than that required; and for a long time he treated this as the required number although admitting his inability to prove as much. Sylvester eventually gave the required proof.

An irreducible invariant (covariant) is one that cannot be expressed rationally and integrally in terms of invariants (covariants and invariants) of degree lower in the coefficients than its own, all invariants belonging to the same quantic or quantics. At an early stage Cayley appreciated that there are many cases in which the number of irreducible invariants and covariants is limited. Thus in his "Second Memoir on Quantics" (*C.M.P.,* II, no. 141 [1856], 250–275) he determined the number (with their degrees) of "asyzygetic" invariants for binary forms of orders 2 to 6, and he gave similar results for asyzygetic systems of irreducible covariants. Cayley made the mistake, however, of thinking that with invariants of forms of order higher than 4, the fundamental system is infinite. The error (which arose from his wrongly taking certain syzygies to be independent, thus increasing the number of invariants and covariants allowed for) stood for thirteen years, until Gordan (*Crelle's Journal,* **69** [1869], 323–354) proved that the complete system for a binary quantic of any order has a finite number of members. Hilbert, in 1888 and later, simplified and greatly generalized Gordan's findings.

Perhaps the best known of Cayley's "Memoirs on Quantics" was the sixth (*C.M.P.,* II, no. 158 [1859], 561–592; see also the note on 604–606, where he compares his work with that of Klein, which followed), in which Cayley gave a new meaning to the metrical properties of figures. Hitherto, affine and projective geometry had been regarded as special cases of metric geometry. Cayley showed how it was possible to interpret all as special cases of projective geometry. We recall some of the more important results of earlier geometrical studies. Poncelet (*ca.* 1822) had evolved the idea of the absolute involution determined by the orthogonal lines of a pencil on the line at infinity and having the "circular points" (so called because they are common to all circles in the plane) as double points. Beginning with the idea that perpendicularity could be expressed in terms of the formation of a harmonic range with the circular points, Laguerre (*ca.* 1853) showed that the numerical value of the angle of two lines of the Euclidean plane expressed in radian measure is $1/2i$ times the natural logarithm of the cross ratio which they form with the lines of their pencil through the circular points. Cayley now found that if P and Q are two points, and A and B are two further points in which the line PQ cuts a conic, then (if A and B are a real point pair; otherwise, where they are conjugate imaginaries we multiply by i) their separation could be expressed as a rather involved arc cosine function involving the coordinates, which space does not permit to be detailed here (see *C.M.P.,* II, no. 158 [1859], 589). A clear idea of the importance of his paper is obtained if we consider Klein's substitution of a logarithmic function for the arc cosine (which Cayley later admitted to be preferable), in which case

$$2c \log \left(\frac{AP \cdot BQ}{AQ \cdot BP} \right),$$

where c is a constant for all lines, may be taken as the generalized distance (which we may here call

$\delta[P,Q]$) between P and Q, in the sense that the following fundamental requirements are met by the function: $\delta(P,Q) = 0$ if and only if P and Q are identical; $\delta(P,Q) = \delta(Q,P)$; $\delta(P,Q) + \delta(Q,R) \geq \delta(P,R)$, the equality holding when P, Q, and R are collinear. Cayley referred to the arbitrarily assumed conic as the "Absolute."

In his definition of distance Cayley has frequently been accused of circularity (recently, for example, by Max Jammer, in *Concepts of Space* [Cambridge, Mass., 1954], p. 156). Cayley anticipated such criticism, however, explaining in his note to the *Collected Papers* that he looked upon the coordinates of points as quantities defining only the ordering of points, without regard to distance. (This note shows that Klein drew his attention to Staudt's work in the same vein, of which he was ignorant when writing the sixth memoir.) Thus if x_a and x_b are coordinates belonging respectively to the points A and B, the corresponding coordinate of P may be written $\lambda_1 x_a + \lambda_2 x_b$, and similarly for the remaining points and coordinates. The function $\delta(P,Q)$ then reduces to one in which no trace of the ordinary (Euclidean) metric distance remains.

The full significance of Cayley's ideas was not appreciated until 1871, when Klein (*Mathematische Annalen,* **4** [1871], 573–625) showed how it was possible to identify Cayley's generalized theory of metrical geometry with the non-Euclidean geometries of Lobachevski, Bolyai, and Riemann. When Cayley's Absolute is real, his distance function is that of the "hyperbolic" geometry; when imaginary, the formulas reduce to those of Riemann's "elliptic" geometry. (The designations "hyperbolic" and "elliptic" are Klein's.) A degenerate conic gives rise to the familiar Euclidean geometry. Whereas during the first half of the century geometry had seemed to be becoming increasingly fragmented, Cayley and Klein, through the medium of these ideas, apparently succeeded for a time in providing geometers with a unified view of their subject. Thus, although the so-called Cayley-Klein metric is now seldom taught, to their contemporaries it was of great importance.

Cayley is responsible for another branch of algebra over and above invariant theory, the algebra of matrices. The use of determinants in the theory of equations had by his time become a part of established practice, although the familiar square notation was Cayley's (*C.M.P.,* I, no. 1 [1841], 1–4) and although their use in geometry, such as was provided by Cayley from the first, was then uncommon. (They later suggested to him the analytical geometry of n dimensions.) Determinants suggested the matrix notation; and yet to those concerned with the history of the "theory of multiple quantity" this notational innovation, even with its derived rules, takes second place to the algebra of rotations and extensions in space (such as was initiated by Gauss, Hamilton, and Grassmann), for which determinant theory provided no more than a convenient language.

Cayley's originality consisted in his creation of a theory of matrices that did not require repeated reference to the equations from which their elements were taken. In his first systematic memoir on the subject (*C.M.P.*, II, no. 152 [1858], 475–496), he established the associative and distributive laws, the special conditions under which a commutative law holds, and the principles for forming general algebraic functions of matrices. He later derived many important theorems of matrix theory. Thus, for example, he derived many theorems of varying generality in the theory of those linear transformations that leave invariant a quadratic or bilinear form. Notice that since it may be proved that there are $n(n + 1)/2$ relations between them, Cayley expressed the n^2 coefficients of the *n*ary orthogonal transformation in terms of $n(n - 1)/2$ parameters. His formulas, however, do not include all orthogonal transformations except as limiting cases (see E. Pascal's *Die Determinanten* [1919], paras. 47 ff.).

The theory of matrices was developed in two quite different ways: the one of abstract algebraic structure, favored by Cayley and Sylvester; the other, in the geometrical tradition of Hamilton and Grassmann. Benjamin Peirce (whose study of linear associative algebras, published in 1881 but evolved by him much earlier, was a strong influence on Cayley) and Cayley himself were notable for their ability to produce original work in both traditions. (It is on the strength of his work on linear associative algebras that Peirce is often regarded as cofounder of the theory of matrices.) In his many informal comments on the relation between matrices and quaternions (see, for example, his long report to the British Association, reprinted in *C.M.P.*, IV, no. 298 [1862], 513–593; and excerpts from his controversial correspondence with his friend P. G. Tait, printed in C. G. Knott's *Life and Scientific Work of P. G. Tait* [Cambridge, 1911], pp. 149–166) Cayley showed a clearer grasp of their respective merits than most of his contemporaries, but like most of them he found it necessary to favor one side rather than the other (coordinates rather than quaternions in his case) in a heated controversy in which practical expediency was the only generally accepted criterion. He had no significant part in the controversy between Tait and J. W. Gibbs, author of the much simpler vector analysis. In passing, we notice Cayley's statement of the origins of his matrices (Knott, *op. cit.*,

p. 164, written 1894): "I certainly did not get the notion of a matrix in any way through quaternions: it was either directly from that of a determinant; or as a convenient mode of expression of the equations [of linear transformation]. . . ."

That Cayley found geometrical analogy of great assistance in his algebraic and analytical work—and conversely—is evident throughout his writings; and this, together with his studied avoidance of the highly physical interpretation of geometry more typical of his day, resulted in his developing the idea of a geometry of *n* dimensions. It is not difficult to find instances of the suggested addition of a fourth dimension to the usual trio of spatial dimensions in the work of earlier writers—Lagrange, d'Alembert, and Moebius are perhaps best known. (But only Moebius made his fourth dimension spatial, as opposed to temporal.) Grassmann's theory of extended magnitude, as explained in *Ausdehnungslehre* (1844), may be interpreted in terms of *n*-dimensional geometry; and yet by 1843 Cayley had considered the properties of determinants formed around coordinates in *n*-space. His "Chapter in the Analytical Geometry of (*n*) Dimensions" (*C.M.P.*, I, no. 11 [1843], 55–62) might have been considered at the time to have a misleading title, for it contained little that would then have been construed as geometry. It concerns the nonzero solutions of homogeneous linear equations in any number of variables.

By 1846 Cayley had made use of four dimensions in the enunciation of specifically synthetic geometrical theorems, suggesting methods later developed by Veronese (*C.M.P.*, I, no. 50 [1846], 317–328). Long afterward Cayley laid down in general terms, without symbolism, the elements of the subject of "hyperspace" (*cf.* his use of the terms "hyperelliptic theta functions," "hyperdeterminant," and so on) in his "Memoir on Abstract Geometry" (*C.M.P.*, VI, no. 413 [1870], 456–469), showing that he was conscious of the metaphysical issues raised by his ideas in the minds of his followers but that as a mathematician he was no more their slave then than when remarking in his paper of 1846 (published in French): "We may in effect argue as follows, *without having recourse to any metaphysical idea as to the possibility of space of four dimensions* (all this may be translated into purely analytic language). . . ."

As an example of Cayley's hypergeometry, we might take the result that a point of $(m - n)$-space given by a set of linear equations is conjugate, with respect to a hyperquadric, to every point whose coordinates satisfy the equations formed by equating to zero a certain simple set of determinants (involving the partial differential coefficients of the hyperquadric function). Cayley and Sylvester subsequently developed these ideas.

In 1860 Cayley devised the system of six homogeneous coordinates of a line, now usually known as Plücker's line coordinates. Plücker, who published his ideas in 1865 (*Philosophical Transactions of the Royal Society,* **155** [1865], 725–791), was working quite independently of Cayley (*C.M.P.,* IV, no. 284 [1860], 446–455, and no. 294 [1862], 490–494), who neglected to elaborate upon his own work. Influenced not by Cayley but by Plücker, Klein (Plücker's assistant at the time of the latter's death in 1868) exploited the subject most fully.

Cayley wrote copiously on analytical geometry, touching on almost every topic then under discussion. Although, as explained elsewhere, he never wrote a textbook on the subject, substantial parts of Salmon's *Higher Plane Curves* are due to him; and without his work many texts of the period, such as those by Clebsch and Frost, would have been considerably reduced in size. One of Cayley's earliest papers contains evidence of his great talent for the analytical geometry of curves and surfaces, in the form of what was often known as Cayley's intersection theorem (*C.M.P.,* I, no. 5 [1843], 25–27). There Cayley gave an almost complete proof (to be supplemented by Bacharach, in *Mathematische Annalen,* **26** [1886], 275–299) that when a plane curve of degree r is drawn through the mn points common to two curves of degrees m and n (both less than r), these do not count for mn conditions in the determination of the curve but for mn reduced by

$$(m + n - r - 1)(m + n - r - 2).$$

(The Cayley-Bacharach theorem was subsequently generalized by Noether. See Severi and Löffler, *Vorlesungen über algebraische Geometrie,* ch. 5.) He found a number of important theorems "on the higher singularities of a plane curve" (the title of an influential memoir; *C.M.P.,* V, no. 374 [1866], 520–528), in which they were analyzed in terms of simple singularities (node, cusp, double tangent, inflectional tangent); yet the methods used here did not find permanent favor with mathematicians. A chapter of geometry which he closed, rather than opened, concerns the two classifications of cubic curves: that due to Newton, Stirling, and Cramer and that due to Plücker. Cayley systematically showed the relations between the two schemes (*C.M.P.,* V, no. 350 [1866], 354–400).

It is possible only to hint at that set of interrelated theorems in algebraic geometry which Cayley did so much to clarify, including those on the twenty-eight bitangents of a nonsingular quartic plane curve and the theorem (first announced in 1849) on the twenty-seven lines that lie on a cubic surface in three dimensions (*C.M.P.,* I, no. 76 [1849], 445–456). (Strictly speaking, Cayley established the existence of the lines and Salmon, in a correspondence prior to the paper, established their number. See the last page of the memoir and G. Salmon, *The Geometry of Three Dimensions,* 2nd ed. [Dublin, 1865], p. 422.) Although no longer in vogue this branch of geometry, in association with Galois theory, invariant algebra, group theory, and hyperelliptic functions, reached a degree of intrinsic difficulty and beauty rarely equaled in the history of mathematics. The Cayley-Salmon theorem is reminiscent of Pascal's mystic hexagram, and indeed Cremona subsequently found a relation between the two (see B. Segre, *The Nonsingular Cubic Surface* [Oxford, 1942] for a survey of the whole subject). Cayley's twenty-seven lines were the basis of Schläfli's division of cubic surfaces into species; and in his lengthy "Memoir on Cubic Surfaces" Cayley discussed the complete classification with masterly clarity, adding further investigations of his own (*C.M.P.,* VI, no. 412 [1869], 359–455).

As might have been expected from his contributions to the theory of invariants, Cayley made an important contribution to the theory of rational transformation and general rational correspondence. The fundamental theorem of the theory of correspondence is difficult to assign to a particular author, for it was used in special cases by several writers; but Chasles (*Comptes rendus,* **58** [1864], 175) presented the theorem that a rational correspondence $F(x,y) = 0$ of degree m in x and n in y (x and y being, if necessary, parameters of the coordinates of two points) between spaces or loci in spaces gives in the general case $m + n$ correspondences. (For a history of the subject see C. Segre, "Intorno alla storia del principio di corrispondenza," in *Bibliotheca mathematica,* 2nd ser., **6** [1892], 33–47; Brill and Noether, "Bericht über die Entwicklung der Theorie der algebraischen Funktionen in älterer und neuerer Zeit," in *Jahresbericht der Deutschen Mathematiker-Vereinigung,* **3** [1894], secs. 6, 10.) Soon after this, Cayley generalized Chasles's theorem to curves of any genus (*C.M.P.,* V, no. 377 [1866], 542–545), but his proof was not rigorous and was subsequently amended by A. Brill. The Chasles-Cayley-Brill theorem states that an (m,n) correspondence on a curve of genus p will have $m + n + 2p\gamma$ coincidences, where γ is known as the "value of the correspondence." (The points corresponding to a point P, together with P taken γ times, is to be a group of a so-called linear point system.)

Cayley's many additions to the subject of rational correspondences have for the most part passed into anonymity, although the name "Cayley-Plücker equations" is a reminder to geometers of how early appreciated were the connections between the order, the rank, the number of chords through an arbitrary point, the number of points in a plane through which two tangents pass, and the number of cusps of a curve in space and corresponding quantities (class, rank, and so on) of its osculating developable. These equations are all due to Cayley but were deduced from Plücker's equations connecting the ordinary singularities of plane curves.

Cayley devoted a great deal of his time to the projective characteristics of curves and surfaces. Apart from his intricate treatment of the theory of scrolls (where many of his methods and his vocabulary still survive), the Cayley-Zeuthen equations are still a conspicuous reminder of the permanent value of his work. Given an irreducible surface in three-dimensional space, with normal singularities and known elementary projective characters, many other important characteristics may be deduced from these equations, which were first found empirically by Salmon and later proved by Cayley and Zeuthen. For further details of Cayley's very extensive work in algebraic geometry, an ordered if unintentional history of his thought is to be found almost as a supporting framework for Salmon's *Treatise on the Analytic Geometry of Three Dimensions* (of the several editions the third, of 1882, with its preface, is historically the most illuminating). (For a more general history of algebraic geometry see "Selected Topics in Algebraic Geometry," which constitutes *Bulletin of the National Research Council* [Washington, D.C.], **63** [1929] and supp. **96** (1934), written by committees of six and three, respectively.)

Cayley's wide mathematical range made it almost inevitable that he should write on the theory of groups. Galois's use of substitution groups to decide the algebraic solvability of equations, and the continuation of his work by Abel and Cauchy, had provided a strong incentive to many other mathematicians to develop group theory further. (Thus Cayley wrote "Note on the Theory of Permutations," *C.M.P.*, I, no. 72 [1849], 423–424.) Cayley's second paper on the theory (1854), in which he applied it to quaternions, contained a number of invaluable insights and provided mathematicians with what is now the accepted procedure for defining a group. In the abstract theory of groups, where nothing is said of the nature of the elements, the group is completely specified if all possible products are known or determinable. In

Cayley's words: "A set of symbols, 1, α, β, . . . all of them different, and such that the product of any two of them (no matter in what order), or the product of any one of them into itself, belongs to the set, is said to be a *group*." From the first Cayley suggested listing the elements in the form of a multiplication table ("On the Theory of Groups, as Depending on the Symbolic Equation $\theta^n = 1$," *C.M.P.*, II, no. 125 [1854], 123–130; second and third parts followed, for which see *C.M.P.*, II, no. 126 [1854], 131–132, and IV, no. 243 [1859], 88–91). This formulation differed from those of earlier writers to the extent that he spoke only of symbols and multiplication without further defining either. He is sometimes said to have failed to appreciate the step he had taken, but this seems unlikely when we consider his footnote to the effect that "The idea of a group *as applied to* permutations or substitutions is due to Galois . . ." (italics added). He went on to give what has since been taken as the first set of axioms for a group, somewhat tacitly postulating associativity, a unit element, and closure with respect to multiplication. The axioms are sufficient for finite, but not infinite, groups.

There is some doubt as to whether Cayley ever intended his statements in the 1854 paper to constitute a definition, for he not only failed to use them subsequently as axioms but later used a different and unsatisfactory definition. (See, for instance, an article for the *English Cyclopaedia*, in *C.M.P.*, IV, no. 299 [1860], 594–608; *cf.* the first two of a series of four papers in *C.M.P.*, X, no. 694 [1878], 401–403.) In a number of historical articles G. A. Miller (see volume I of his *Collected Works* [Urbana, Ill., 1935]) has drawn attention to the unsatisfactory form of a later definition and indeed has criticized other mathematicians for accepting it; but there are few signs that mathematicians were prepared for the postulational definition until well into the present century. In 1870 Kronecker explicitly gave sets of postulates applied to an abstract finite Abelian group; but even Lie and Klein did most of their work oblivious to the desirability of such sets of axioms, as a result occasionally using the term "group" in what would now be reckoned inadmissible cases.

In addition to his part in founding the theory of abstract groups, Cayley has a number of important theorems to his credit: perhaps the best known is that every finite group whatsoever is isomorphic with a suitable group of permutations (see the first paper of 1854). This is often reckoned to be one of the three most important theorems of the subject, the others being the theorems of Lagrange and Sylow. But perhaps still more significant was his early appreciation of the way in which the theory of groups was capable of drawing together many different domains of mathematics: his own illustrations, for instance, were drawn from the theories of elliptic functions, matrices, quantics, quaternions, homographic transformations, and the theory of equations. If Cayley failed to pursue his abstract approach, this fact is perhaps best explained in terms of the enormous progress he was making in these subjects taken individually.

In 1845 Cayley published his "Mémoire sur les fonctions doublement périodiques," treating Abel's doubly infinite products (*C.M.P.*, I, no. 25 [1845], 156–182; see his note on p. 586 of the same volume). Weierstrass subsequently (1876, 1886) simplified the initial form and in doing so made much of Cayley's work unnecessary (see Cayley's later note, *loc. cit.*). His work on elliptic functions, pursued at length and recurred to at intervals throughout his life, nevertheless contains ample evidence of Cayley's ability to simplify the work of others, an early instance being his establishment of some results concerning theta functions obtained by Jacobi in his *Fundamenta nova theoriae functionum ellipticarum* of 1829 (*C.M.P.*, I, no. 45 [1847], 290–300). Cayley's only full-length book was on elliptic functions, and he made a notable application of the subject to geometry when he investigated analytically the property of two conics such that polygons may be inscribed by one and circumscribed about the other. The property was appreciated by Poncelet and was discussed analytically by Jacobi (using elliptic functions) when the conics were circles. Using his first paper of 1853 and gradually generalizing his own findings, by 1871 Cayley was discussing the problem of the number of polygons which are such that their vertices lie on a given curve or curves of any order and that their sides touch another given curve or curves of any class. That he was able to give a complete solution even where the polygons were only triangles is an indication of his great analytical skill.

Cayley wrote little on topology, although he wrote on the combinatorial aspect, renewed the discussion of the four-color-map problem, and corresponded with Tait on the topological problems associated with knots. He wrote briefly on a number of topics for which alone a lesser mathematician might have been remembered. He has to his credit an extremely useful system of coordinates in plane geometry which he labeled "circular coordinates" (*C.M.P.*, VI, no. 414 [1868], 498) and which later writers refer to as "minimal coordinates." There is also his generalization of Euler's theorem relating to the numbers of faces, vertices, and edges of the non-Platonic solids. He wrote to great effect on the theory of the numbers of partitions, originated by Euler. (His interest in this

arose from his need to apply it to invariant theory and is first evident in his second memoir on quantics, *C.M.P.*, II, no. 141 [1856], 250–281.) His short paper "On the Theory of the Singular Solutions of Differential Equations of the First Order" (*C.M.P.*, VIII, no. 545 [1873], 529–534) advanced the subject considerably and was part of the foundation on which G. Chrystal's first satisfactory treatment of the *p*-discriminant was based (*Transactions of the Royal Society of Edinburgh*, **138** [1896], 803 ff.).

Cayley long exploited the theory of linear differential operators (previously used by Boole to generate invariants and covariants), as when he factored the differential equation $(D^2 + pD + q)y = 0$ as $(D + \alpha[x])(D + \beta[x])y' = 0$, with $\alpha + \beta = p$ and $\alpha\beta + \beta' = q$ both being theoretically soluble (*C.M.P.*, XII, no. 851 [1886], 403). This technique is linked to that of characterizing invariants and covariants of binary quantics as the polynomial solutions of linear partial differential equations. (The differential operators were in this context known as annihilators, following Sylvester.) He wrote occasionally on dynamics, but his writings suggest that he looked upon it as a source of problems in pure mathematics rather than as a practical subject. Thus in five articles he considered that favorite problem of the time, the attraction of ellipsoids; and in a paper of 1875 he extended a certain problem in potential theory to hyperspace (*C.M.P.*, IX, no. 607 [1875], 318–423). That he kept himself informed of the work of others in dynamics is evident from two long reports on recent progress in the subject which he wrote for the British Association (*C.M.P.*, III, no. 195 [1857], 156–204; IV, no. 298 [1862], 513–593).

Cayley wrote extensively on physical astronomy, especially on the disturbing function in lunar and planetary theory; but the impact of what he wrote on the subject was not great, and Simon Newcomb, who spoke of Cayley's mathematical talents with extraordinary deference, did not allude to them in his *Reminiscences of an Astronomer* (London–New York, 1903, p. 280). (It is interesting to note that when he met Cayley at an Astronomical Society Club dinner, Newcomb mistook Cayley's garb for that of an attendant.) Cayley nevertheless performed a great service to his countryman John Couch Adams, who in 1853, taking into account the varying eccentricity of the earth's orbit, had obtained a new value for the secular acceleration of the moon's mean motion. Adams' figure, differing from Laplace's, was contested by several French astronomers, including Pontécoulant. Cayley looked into the matter independently, found a new and simpler method for introducing the variation of the eccentricity, and confirmed the value

Adams had previously found (*C.M.P.,* III, no. 221 [1862], 522–561). Here was yet another instance of the truth of the remark made about Cayley by Sylvester: ". . . whether the matter he takes in hand be great or small, 'nihil tetigit quod non ornavit'" (*Philosophical Transactions,* 17 [1864], 605). And yet Cayley deserves to be remembered above all not for those parts of mathematics which he embellished, but for those which he created.

BIBLIOGRAPHY

I. ORIGINAL WORKS. The great majority of Cayley's mathematical writings (966 papers in all, with some short notes subsequently written about them) are in *The Collected Mathematical Papers of Arthur Cayley,* 13 vols. indexed in a 14th (Cambridge, 1889–1898). The printing of the first seven vols. and part of the eighth was supervised by Cayley himself. The editorial task was assumed by A. R. Forsyth when Cayley died. His excellent biography of Cayley is in vol. VIII, which also contains a complete list of the lectures Cayley gave in Cambridge as Sadlerian professor. The list of writings in vol. XIV includes the titles of several articles which Cayley contributed to the *Encyclopaedia Britannica.* See, e.g., in the 11th ed., "Curve" (in part), "Determinant," "Equation," "Gauss," "Monge," "Numbers, Partition of," and "Surface" (in part). A work in which Cayley's part was not negligible is G. Salmon, *A Treatise on the Higher Plane Curves,* 2nd and 3rd (1879) eds. Upward of twenty sections and the whole of ch. 1 were written by Cayley for the 2nd ed., and further additions were made in the 3rd ed. See Salmon's prefaces for further details. Cayley frequently gave advice and assistance to other authors. Thus he contributed ch. 6 of P. G. Tait's *An Elementary Treatise on Quaternions* (Cambridge, 1890), as well as making improvements. There is no systematic record as such of Cayley's less conspicuous work. He composed a six-penny booklet, *The Principles of Book-Keeping by Double Entry* (Cambridge, 1894). His *An Elementary Treatise on Elliptic Functions* (London, 1876) was issued in a 2nd ed. which, owing to his death, was only partly revised.

II. SECONDARY LITERATURE. There are few works dealing historically with Cayley's mathematics alone. General histories of mathematics are not listed here, nor are mathematical works in which historical asides are made. The best biographical notice is by A. R. Forsyth, reprinted with minor alterations in *The Collected Mathematical Papers of Arthur Cayley,* VIII (1895), ix–xliv, from the "Obituary Notices" in *Proceedings of the Royal Society,* 58 (1895), 1–43. Forsyth also wrote the article in the *Dictionary of National Biography,* XXII (supp.), 401–402. Another admirable and long obituary notice is by M. Noether, in *Mathematische Annalen,* 46 (1895), 462–480. Written during Cayley's lifetime was G. Salmon's "Science Worthies no. xxii.—Arthur Cayley," in *Nature,* 28 (1883), 481–485. Of general value are Franz Meyer, "Bericht über den gegenwärtigen Stand der Invariantentheorie," in *Jahresbericht der Deutschen Mathematiker-Vereinigung,* 1 (1890–1891), 79–288; and A. Brill and M. Noether, "Bericht über die Entwicklung der Theorie der algebraischen Functionen in älterer and neuerer Zeit," *ibid.,* 3 (1894), 107–566. The best specifically historical studies of aspects of Cayley's mathematics are Luboš Nový, "Arthur Cayley et sa définition des groupes abstraits-finis," in *Acta historiae rerum naturalium necnon technicarum* (Czechoslovak Studies in the History of Science, Prague), spec. issue no. 2 (1966), 105–151; and "Anglická algebraická školá," in *Dějiny věd a techniky,* 1, no. 2 (1968), 88–105.

J. D. NORTH

ČECH, EDUARD (*b.* Stračov, Bohemia, 29 June 1893; *d.* Prague, Czechoslovakia, 15 March 1960), *mathematics.*

Eduard Čech was the fourth child of Čeněk Čech, a policeman, and Anna Kleplová. After studying at the Gymnasium in Hradec Kralove, he attended lectures on mathematics at the Charles University of Prague from 1912 to 1914. In 1920 Čech took his degree in mathematics at the University of Prague. Even his first works showed his mathematical talent. He began to study differential projective properties of geometrical figures and became interested in the work of Guido Fubini. He obtained a scholarship for the school year 1921–1922 that enabled him to study with Fubini in Turin. Later they wrote *Geometria proiettiva differenziale* (1926–1927) and *Introduction à la géométrie projective différentielle des surfaces* (1931). In 1922 Čech was appointed associate professor of mathematics at the University of Prague; on this occasion he presented a study on differential geometry. From 1923 he was professor of mathematics at the Faculty of Natural Sciences of the University of Brno, lecturing on mathematical analysis and algebra.

From 1928 on, Čech was interested in topology, inspired by the works of mathematicians who contributed to the Polish journal *Fundamenta mathematicae.* His work from 1932 on, devoted to the general theory of homology in arbitrary spaces, the general theory of varieties, and theorems of duality, showed him to be one of the foremost experts in combinatorial topology. In September 1935 he was invited to lecture at the Institute for Advanced Study at Princeton. Čech returned to Brno in 1936 and founded a topology seminar among the young mathematicians there. During the three years the seminar was in existence, the works of P. S. Alexandrov and Pavel Uryson were studied and twenty-six papers were written. The group disbanded at the closing of Czech universities following the German occupation in 1939.

In his paper "On Bicompact Spaces" Čech stated

precisely the possibilities of utilizing a new type of topological space (defined by Tichonow in 1930), which later came to be known as Čech's bicompact envelope (βS of a completely regular space S) or as Stone and Čech's compact envelope. Čech's interpretation became a very important tool of general topology and also of some branches of functional analysis.

Čech was also concerned with the improvement of the teaching of mathematics in secondary schools. He organized courses for secondary school teachers in Brno in 1938–1939; the results are shown in a series of mathematics textbooks for secondary schools that were written under his guidance after World War II.

In 1945 Čech went to the faculty of Natural Sciences of Charles University in Prague. There he was instrumental in founding two research centers: the Mathematical Institute of the Czechoslovak Academy of Science (1952) and the Mathematical Institute of Charles University.

In topology, in addition to the theory of topological spaces, Čech worked on the theories of dimension and continuous spaces. In combinatory topology he was concerned primarily with the theory of homology and general varieties. He was most active in differential geometry from 1921 to 1930, when he became one of the founders of systematic projective differential geometry; he dedicated himself chiefly to problems of the connection of varieties, to the study of correspondences, and to systematic utilization of duality in projective spaces. After 1945 he returned to problems of differential geometry and developed a systematic theory of correspondences between projective spaces. His attention was then drawn to the problems of congruences of straight lines that play a significant role in the theory of correspondences. Somewhat different is his work on the relations between the differential classes of points of a curve and the object attached to it. A number of his ideas were elaborated in the works of his students. They can also be found in some of his manuscripts that have been preserved, published in part in 1968.

BIBLIOGRAPHY

I. ORIGINAL WORKS. The bibliography of Čech's scientific works compiled by Katětov, Novák, and Švec lacks only the rev. and enl. ed. of *Topological Spaces* and a number of articles on education. His papers include "O křivkovém a plošném elementu třetího řádu" ("On the Curve and Surface Element of the Third Order"), his thesis at the University of Prague, in *Časopis pro pěstování matematiky a fysiky,* **50** (1921), 219–249, 305–306; and "On Bicompact Spaces," in *Annals of Mathematics,* **38** (1937),

823–844. With Guido Fubini he wrote *Geometria proiettiva differenziale,* 2 vols. (Bologna, 1926–1927); and *Introduction à la géométrie projective différentielle des surfaces* (Paris, 1931). His *Topological Spaces* (Prague, 1959) was revised by Zdeněk Frolík and Miroslav Katětov (Prague, 1966).

II. SECONDARY LITERATURE. M. Katětov, J. Novák, and A. Švec, "Akademik Eduard Čech," in *Časopis pro pěstování matematiky,* **85** (1966), 477–491, includes an almost complete bibliography on 488–491; it also appears in Russian in *Chekhoslovatsky matematichesky zhurnal* ("Czechoslovak Mathematical Journal"), **10** (1960), 614–630, with bibliography on 627–630. Two articles by K. Koutský discuss Čech's work: "Čechuv topologický seminář v Brně z let 1936–1939" ("Čech's Topological Seminar at Brno in 1936–1939"), in *Pokroky matematiky, fyziky a astronomie* (1964), 307–316; and "O Čechových snahách ve středoškolské matematice" ("Čech's Endeavors for the Reform of Secondary School Mathematics"), in *Sborník pro dějiny přírodních věd a techniky,* **11** (1967), 217–230. See also P. S. Aleksandrov, in *Uspekhi matematicheskikh nauk,* **15,** no. 2 (1960), 25–95; and J. Kelley, *General Topology* (New York–Toronto–London, 1955), p. 298.

LUBOŠ NOVÝ
JAROSLAV FOLTA

CELAYA, JUAN DE (*b.* Valencia, Spain, *ca.* 1490; *d.* Turia, Spain, 6 December 1558), *logic, natural philosophy.*

Celaya probably began his education at the University of Valencia, then transferred to Paris in the early years of the sixteenth century, apparently enrolling in the Collège de Montaigu and finishing the arts course there about 1509. Among his professors at Montaigu were Gasper Lax and John Dullaert of Ghent, both in turn students of the Scot John Maior, whose influence is detectable in Celaya's thought. One of Celaya's fellow students in arts, the Aragonian Juan Dolz del Castellar, composed three works on logic by 1513, the last of which was attacked by Celaya in his *Summulae logicales* (Paris, 1515); Dolz replied to Celaya in an extended rebuttal prefaced to his *Cunabula omnium . . . difficultatum in proportionibus et proportionalibus* ("The Origin of All . . . Difficulties in Ratios and Proportions"), printed at Montauban in 1518. Celaya meanwhile had continued studies in theology, lecturing on the Scriptures from 1515 to 1517 and on the *Sentences* during the academic year 1517/1518; he received the licentiate in theology on 24 March 1522 and the doctorate on 21 June 1522.

As a master of arts Celaya taught in the Collège de Coqueret at Paris from about 1510 to 1515, where his associates were the Portuguese Alvaro Thomaz and the Scot Robert Caubraith; the former, especially, impressed Celaya with his "calculatory" techniques in treating physical problems. In 1515 Celaya passed to the College of Santa Barbara, where he remained until about 1524; among his students there were the Segovian Francisco de Soto (later to become a Dominican friar and change his name to Domingo) and the Portuguese Juan Ribeyro, both of whom became faithful disciples. During his entire stay at Paris, Celaya was a prolific writer, turning out a large number of works on logic and expositions, with questions, of Aristotle's *Physics* (Paris, 1517), *De caelo* (Paris, 1517), and *De generatione* (Paris, 1518). The *Physics* commentary, in particular, is important for its influence on the development of modern science; its treatment of motion in the third book, which spans seventy-one folios of the 201 that make up the volume, is extensive, summarizing the main contributions of the English Mertonians, the Paris terminists, and the Paduan "calculatores," as well as the teaching current at Paris in the nominalist (i.e., Ockhamist), realist (i.e., Scotist), and Thomist schools.

Following a dispute with the German Gervase Wain, Celaya returned to Spain about 1524 and in 1525 took up the post of rector *in perpetuo* and professor of theology at the University of Valencia. In his later years he seems to have lost interest in nominalist teachings and to have devoted himself instead to the Aristotelian-Thomistic tradition, which soon came to dominate in the Spanish universities with the rise of the "second Scholasticism."

BIBLIOGRAPHY

I. ORIGINAL WORKS. Celaya's writings are listed in Ricardo G. Villoslada, S. J., *La universidad de Paris durante los estudios de Francisco de Vitoria,* (1507–1522), Analecta Gregoriana XIV (Rome, 1938), 180–215, esp. 207; this is the best study on Celaya available. None of Celaya's works is translated from the Latin, and copies of the originals are quite rare; the University of Chicago has acquired a copy of his commentaries on the *Physics, De caelo,* and *De generatione.*

II. SECONDARY LITERATURE. Pierre Duhem, *Études sur Léonard de Vinci,* III (Paris, 1913), *passim;* Hubert Élie, "Quelques maîtres de l'université de Paris vers l'an 1500," in *Archives d'histoire doctrinale et littéraire du moyen âge,* **18** (1950–1951), 193–243; and William A. Wallace, O.P., "The Concept of Motion in the Sixteenth Century," in *Proceedings of the American Catholic Philosophical Association,* **41** (1967), 184–195.

WILLIAM A. WALLACE, O.P.

CELS, JACQUES-PHILIPPE-MARTIN (*b.* Versailles, France, 15 June 1740; *d.* Montrouge, Hauts-de-Seine, France, 15 May 1806), *botany, horticulture.*

The son of an employee of the *bâtiments du roi,* Cels

entered the office of the Ferme Générale in Paris in 1759. In 1761 he became collector of revenues at the Little Saint Bernard Pass and a short while later that of the *octroi* at Saint-Jacques gateway. With the latter position came the use of a house with about three-quarters of an acre of land, on which he laid out a rich botanical garden. Cels had early shown a methodical temperament and a pronounced taste for botany. He studied under Bernard de Jussieu and Louis-Guillaume Le Monnier, and he was sympathetic to the ideas of Rousseau.

The Revolution abolished his office in 1790, forcing Cels to find a new way of making a living. He retired to Montrouge and devoted himself to the cultivation and sale of plants. He encountered serious difficulties, however. Made responsible by the Convention for a considerable sum stolen from his strongbox in 1789, upon the looting of the *octroi barrière,* he had to sell his large library in order to acquit himself; moreover, on several occasions he had to transfer his cultivation to other locations. At the beginning of the nineteenth century Cels's holdings were in the plain near Montrouge, about one kilometer from Paris, near "the Jansenist windmill"; they finally totaled about eighteen acres, covered by greenhouses, beds, borders, and especially by a tree nursery and small ornamental and fruit trees. In 1800–1802 Étienne Ventenat published *Description des plantes nouvelles et peu connues cultivées dans le jardin de J. M. Cels,* enriched with 100 beautiful color plates by P. J. Redouté, and, in 1803–1808, *Choix de plantes dont la plupart sont cultivées dans le jardin de Cels,* illustrated by the same artist.

In this garden, which became one of the most beautiful in Europe, were drawn and described several of the species published in the *Stirpes novae* (1784–1791) of Charles-Louis L'Héritier, in the *Histoire naturelle des plantes grasses* (1799) and the *Astragologia* (1802) of Augustin-Pyramus de Candolle, and in the *Liliacées* (1802–1816) of P. J. Redouté. Cels applied himself particularly to propagating the rare plants introduced into France by the naturalist voyagers of his time (several of whom were his personal friends): Joseph Dombey, André Michaux, Louis Bosc, Guillaume-Antoine Olivier, Jean Bruguières, Pierre Auguste Broussonet, Jacques de La Billardière, Hippolyte Nectoux, and Anselme Riedlé. He carried on an important correspondence with the principal botanists of Europe, who provided him with new species, and he generously placed his abundant plant collection at their disposal. He also sent plants to the botanical garden at Kew.

A member of the Commission of Agriculture and then of the Council of Agriculture, Cels prevented the destruction of parks and châteaux during the Revolution; he also contributed to the formation and the direction of the agricultural institutions at Le Raincy, Sceaux, and Versailles. He published opinions and instructions on various branches of agriculture, took an important part in the project of the *code rural,* and collaborated on several dictionaries and other works. The majority of his writings appeared anonymously. He was elected to the First Class of the Institut National on 22 Frimaire *an IV* (13 December 1795), and he became a member of the Société d'Agriculture de la Seine in 1799.

His son, François, was his collaborator and continued his work.

BIBLIOGRAPHY

I. ORIGINAL WORKS. Cels's writings include *Coup d'oeil éclairé d'une bibliothèque à l'usage de tout possesseur de livres* (Paris, 1773), written with A. M. Lottin; *Observations sur les jardins de luxe . . .* (Paris, n.d. [1794?]); *Instruction sur les effets des inondations et débordemens des rivières, relativement aux prairies, aux récoltes de foin et à la nourriture des animaux . . .* (Paris, *an IV*), written with F.-H. Gilbert; *Avis sur les récoltes des grains . . .* (Paris, *an VII*); *Notes du citoyen Cels sur ses cultures* (n.p., n.d.); and *Mémoire sur quelques inconvéniens de la taille des arbres à fruits, et nouvelle méthode de les conduire pour assurer la fructification . . .* (Paris, 1806), by Cadet de Vaux, followed by a report by Cels and Thouin.

Cels collaborated on *Annuaire du cultivateur* (Paris, 1794). He also presented several memoirs and reports to the First Class of the Institut de France; those published include "Rapport sur le projet précédent [John Sinclair, 'Projet d'un plan pour établir des fermes expérimentales et pour fixer les principes des progrès de l'agriculture . . .'] . . .," in *Mémoires présentés à l'Institut . . . par divers savants,* Sciences mathématiques et physiques, **1** (1806), 17–32; and "Notice historique sur la plante nommée *Robinia viscosa* (*Robinia visqueux*) . . .," in *Mémoires de l'Institut national . . ., Sciences mathématiques et physiques,* **5** (*an XII*), 110–113.

Cels is mentioned quite frequently in the *Procès-verbaux des séances de l'Académie des sciences tenues depuis la fondation de l'Institut jusqu'au mois d'août 1835,* 3 vols. (Hendaye, 1910–1913). MSS by Cels or concerning him include the following: in the library of the Institut de France, an autograph note on the life of L'Héritier de Brutelle (MS 3180) and several pieces used by Cuvier in his *éloge* of Cels (MS 3184)—see Henri Dehérain, *Catalogue des manuscrits du fonds Cuvier,* fasc. 1 (Paris, 1908), no. 180, p. 30, and no. 184, p. 31; in the archives of the Académie des Sciences, "Mémoire sur le froid du 21 vendémiaire *an XIV* [13 October 1805]," in the file for the meeting of 31 March 1806; in the central library of the Muséum National d'Histoire Naturelle, two letters: one of 27 November 1794 to Abbé Grégoire, on the botanical garden of Montpellier (MS 1972, no. 246); one of 20 February 1806 to Lezermes, director of the nursery at Roule, on the latter nursery (MS 1972, no. 247); and at the British Museum, four letters to Sir Joseph Banks dated between 1791 and 1801 (Add. MSS 8097–8099)—see Warren R. Dawson, *The Banks Letters: A Calendar of the Manuscript Correspondence of Sir Joseph Banks . . .* (London, 1958), p. 209.

II. SECONDARY LITERATURE. On Cels's life, see Georges Cuvier, "Éloge historique de Jacques Martin Cels . . .," in *Mémoires de la classe des sciences mathématiques et physiques de l'Institut,* for 1806, 2nd semester (1807), 139–158, repr. in *Magasin encyclopédique,* VI (1806), 64–84. See also Le Texnier, *Notices sur les jardiniers célèbres et les amateurs de jardins. Une famille de jardiniers, les Cels* (Paris, 1909).

YVES LAISSUS

CELSIUS, ANDERS (*b.* Uppsala, Sweden, 27 November 1701; *d.* Uppsala, 25 April 1744), *astronomy.*

Celsius' father was professor of astronomy at the University of Uppsala, and his son early followed in his footsteps. He studied astronomy, mathematics, and experimental physics; and in 1725 he became secretary of the Uppsala Scientific Society. After teaching at the university for several years as professor of mathematics, in April 1730 Celsius was appointed professor of astronomy. From 1732 to 1736 he traveled extensively in other countries to broaden his knowledge. He visited astronomers and observatories in Berlin and Nuremburg; in the latter city he published a collection of observations of the aurora borealis (1733). He went on to Italy, and then to Paris; there he made the acquaintance of Maupertuis, who was preparing an expedition to measure a meridian in the north in hopes of verifying the Newtonian theory that the earth is flattened at the poles and disproving the contrary Cartesian view. Celsius joined the Maupertuis expedition, and in 1735 he went to London to secure needed instruments. The next year he followed the French expedition to Torneå, in northern Sweden (now Tornio, Finland). During 1736–1737, in his capacity as astronomer, he helped with the planned meridian measurement; and Newton's theory was confirmed. He was active in the controversy that later developed over what Maupertuis had done and fired a literary broadside, *De observationibus pro figura telluris determinanda* (1738), against Jacques Cassini.

On his subsequent return to Uppsala, Celsius breathed new life into the teaching of astronomy at the university. In 1742 he moved into the newly completed astronomical observatory, which had been under construction for several years and was the first modern installation of its kind in Sweden.

Although he died young, Celsius lived long enough to make important contributions in several fields. As an astronomer he was primarily an observer. Using a purely photometric method (filtering light through glass plates), he attempted to determine the magnitude of the stars in Aries (*De constellatione Arietis*, 1740). During the lively debate over the falling level of the Baltic, he wrote a paper on the subject based on exact experiments, "Anmärkning om vatnets förminskande" (1743). Today Celsius is best known in connection with a thermometer scale. Although a 100-degree scale had been in use earlier, it was Celsius' famous observations concerning the two "constant degrees" on a thermometer, "Observationer om twänne beständiga grader på en thermometer" (1742), that led to its general acceptance. As the "constant degrees," or fixed points, he chose the freezing and boiling points of water, calling the boiling point zero and the freezing point 100. The present system, with the scale reversed, introduced in 1747 at the Uppsala observatory, was long known as the "Swedish thermometer." Not until around 1800 did people start referring to it as the Celsius thermometer.

BIBLIOGRAPHY

Celsius' most important writings are *De observationibus pro figura telluris determinanda* (Uppsala, 1738); *De constellatione Arietis* (Stockholm, 1740); "Observationer om twänne beständiga grader pa en thermometer," in *Kungliga Svenska vetenskapsakademiens handlingar* (1742), 121–180; and "Anmärkning om vatnets förminskande," *ibid.* (1743), 33–50. "Observationer . . ." may be found in German as no. 57 in Ostwald's Klassiker der exakten Wissenschaften (Leipzig, 1894). Many of his minor writings were published as academic treatises or appeared in *Kungliga Svenska vetenskapsakademiens handlingar, Philosophical Transactions of the Royal Society,* and other journals. His personal papers, including letters from Maupertuis, J. N. Delisle, and Le Monnier, are at the Uppsala University library.

There is a comprehensive biography by N. V. E. Nordenmark, *Anders Celsius* (Uppsala, 1936), and a shorter version by the same author in S. Lindroth, ed., *Swedish Men of Science, 1650–1950* (Stockholm, 1952).

STEN LINDROTH

CELSUS, AULUS CORNELIUS (*fl.* Rome, *ca.* A.D. 25), *collection of knowledge.*

Celsus lived in Rome in the time of Emperor Tiberius. His presence there is attested to by the records for the year A.D. 25/26.[1] Since he belonged to the *gens* Cornelia, he was undoubtedly a member of Rome's leading circles. He compiled an encyclo-

pedia entitled *Artes* and ranks, along with his predecessor Varro, as Rome's most important master of this literary form. Columella, Quintilian, and Pliny the Elder, all of whom lived in the first century, cite Celsus with considerable praise.[2] Nothing more of Celsus' life or personality is known.

We have no clear idea of the contents or arrangement of the *Artes*. It is certain that besides eight extant books on medicine, there were five books on agriculture and also sections of unknown length on military science and on rhetoric. Whether there were also books on philosophy and jurisprudence remains as uncertain as the division and arrangement of the material.[3]

Concerning the character of *De medicina octo libri* (book I, historical review and general dietetics; book II, pathology and general therapy; books III and IV, special therapy; books V and VI, pharmacology; book VII, surgery; book VIII, bone diseases), an intense, almost disproportionately great, scholarly quarrel developed.[4] Was Celsus himself a doctor? This question has from the first generally been answered (correctly, in my opinion) in the negative, although many details of the text could be interpreted to support the opposite position.[5] It has been pointed out that in the time of Celsus, medical knowledge was regarded as an important part of general education. In ancient times, as in the Middle Ages, it certainly was not unusual for someone who was not a doctor to write about medicine. Moreover, if Celsus had been a doctor, then Pliny the Elder surely would have referred to him as *medicus* and would not have placed him among the *auctores*.

Now, however, another question arises. Celsus' *De medicina* is undoubtedly written in good, very clear, and at times even brilliant Latin; the ancient authorities generally attested that Celsus had this ability,[6] and modern philologists have confirmed that this is especially true in the medical writings.[7] Yet, in content and presentation *De medicina* also ranks high and has been called "brilliant," even a "masterpiece" (Wellmann). Could a purely literary man be capable of independently producing a medical work so excellent in both style and content, as Celsus did?

On this point Wellmann, more than anyone else, has been an energetic advocate of the thesis that Celsus merely translated a certain Greek text into Latin. His argument is, to be sure, open to serious criticism on many points. None of the men (e.g., Cassius Dionysius and Tiberius Claudius Menecrates) that Wellmann has proposed as author of the presumed Greek text is convincing; we know hardly anything about any of them, and certainly nothing that would justify the idea of an important medical work. Friedrich Marx, likewise a supporter of the one-source theory, has gotten no further with similar proposals.

More likely is the theory that Celsus compiled his book from several sources, an idea first advocated by Wellmann[8] and later primarily by J. Ilberg. It is plausible because in other, nonmedical parts of the *Artes,* Celsus sometimes obviously employed several sources.[9] Nevertheless, another question remains: Assuming Celsus employed several sources, was he a compiler who simply translated the relevant excerpts? Or did he possess judgment, critical appreciation, and possibly even his own medical point of view?

Meinecke in particular has stressed that in Celsus' style "not the slightest vestige of a translation" can be discovered.[10] We should accept this and finally abandon the idea that Celsus simply translated from the Greek. The concept of "compilation" then ought to lose its pejorative overtones, thus leaving more room for the intellectual achievement of Celsus the "compiler." I agree completely with Temkin that this intellectual achievement, which lies in the working up of the sources according to Celsus' own point of view, should be considered very important and that one can go so far as to speak of Celsus' "intrinsic originality."[11]

Finally, if one asks what, precisely, was the medical outlook of the nonphysician Celsus, it seems inadvisable to pin him too firmly to any one medical position (empiricism or methodology).[12] Whether Celsus really was, as Temkin holds,[13] especially committed to methodism, will not be examined; but I do agree with Temkin that in any case Celsus drew upon many different sources. To this extent he was a genuine eclectic—an attitude that in the first century was typical in medicine as well as in other sciences. Moreover, Celsus' strong Hippocratism is striking.[14] However, this was typical of medicine in the imperial period; long before Galen, the pneumatic physicians (with whom Celsus was not closely connected, since he did not mention them at all) in particular had inaugurated a kind of second Hippocratic renaissance (after the Hellenistic).

Celsus, therefore, even as a nonphysician—whether one classifies him, following Luigi Castiglioni, as a "scholar" or, following Meinecke, as an "artifex medicinae"—fits very well into the medical world of imperial Rome: an aristocratic, Greek-educated Roman who, within the framework of a general education, was very strongly interested in medicine and who held a completely individual point of view that was nevertheless typical of the time; one might call him a Hippocratic eclectic.

NOTES

1. C. Cichorius, *Römische Studien* (Berlin, 1922), p. 412.
2. B. Meinecke, "Aulus Cornelius Celsus . . .," pp. 291 f.
3. M. Wellmann, in Pauly-Wissowa, IV, pt. 1 (1900), 1274–1276; G. Baader, "Überlieferungsprobleme des A. Cornelius Celsus," p. 215, with further literature.
4. The essential earlier literature has been compiled by O. Temkin in "Celsus 'On Medicine' . . . ."
5. See W. G. Spencer, ed., *Celsus De medicina,* I, xi f.
6. See Meinecke, *loc. cit.*
7. See Baader, pp. 217 f.
8. Wellmann, p. 1274.
9. *Ibid.*
10. Meinecke, p. 288.
11. Temkin, pp. 255 f.
12. Baader, pp. 215.
13. Temkin, pp. 255 f.
14. See the list of his Hippocratic citations in Spencer's ed., III, 624–627.

BIBLIOGRAPHY

I. ORIGINAL WORKS. Editions of *De medicina* are F. Marx, *A. Cornelii Celsi quae supersunt,* vol. I of Corpus Medicorum Latinorum (Leipzig–Berlin, 1915); and W. G. Spencer, *Celsus De medicina,* with an English trans., 3 vols., in the Loeb Classical Library (London–Cambridge, Mass., 1960–1961).

II. SECONDARY LITERATURE. On Celsus or his work, see G. Baader, "Überlieferungsprobleme des A. Cornelius Celsus," in *Forschungen und Fortschritte,* **34** (1960), 215–218; B. Meinecke, "Aulus Cornelius Celsus—Plagiarist or Artifex Medicinae?," in *Bulletin of the History of Medicine,* **10** (1941), 288–298; and O. Temkin, "Celsus 'On Medicine' and the Ancient Medical Sects," *ibid.,* **3** (1935), 249–264.

FRIDOLF KUDLIEN

CENSORINUS (*fl.* Rome[?] first half of the third century A.D.), *grammar, collection of knowledge.*

Censorinus wrote two works. The one entitled *De accentibus,* which dealt with grammatical questions and was praised by Flavius Magnus Cassiodorus Senator and Priscian, has been lost. The second, *De die natali,* which is extant, is dedicated as a birthday present to a little-known patron, Quintus Caerellius; according to Censorinus' own statements (ch. 18, sec. 12; ch. 21, sec. 6), it was written in 238. The content of this comprehensive work can be divided into three parts: a general introduction (chs. 1–3), one part on the life of man (chs. 4–15), and one on time and its divisions (chs. 16–24). The first part, which is based on the Roman savant Varro, deals chiefly with human procreation and pregnancy, with excursuses on the influence of the stars and on music. In the second part, which is based on a lost work by Varro, Censorinus treats the different divisions of time (age, year, month, day, etc.). As in the first part, he mentions the doctrines of Greek philosophers.

Not all of *De die natali* has survived; the manuscript of the conclusion has been lost, and thus the beginning of the next treatise is unknown, as are its author and title. This work, now entitled *Fragmentum Censorini*, is more important for the history of science than is *De die natali*. It contains a series of short tractates from an encyclopedic work on astronomy, geometry, music, and metrics. The chapter on geometry, which deals with the definitions, postulates, and axioms of book I of Euclid's *Elements*, differs greatly from the other known translations of Euclid. The chapters on metrics are very detailed. This part contains the oldest known information on Roman metrics and may be based on a work by Varro. Thus *De die natali* and *Fragmentum Censorini* enrich our view of Greek and Roman science in some respects and increase our knowledge of treatises by Varro.

BIBLIOGRAPHY

I. ORIGINAL WORKS. The source of the various editions of *De die natali* and *Fragmentum Censorini* is a Cologne MS of the seventh century now in Cologne's Dombibliothek, MS 166. All other codices and MS Vaticanus 4929, which was treated by Hultsch as an independent source, are based on it. L. Carrion was the first to use it (Paris, 1583). A much more complete edition is that of O. Jahn (Berlin, 1845). Some details were corrected and supplemented in the edition by F. Hultsch (Leipzig, 1867).

II. SECONDARY LITERATURE. Censorinus or his work is discussed by M. Schanz and C. Hosius, in *Geschichte der römischen Literatur*, 3rd ed., III (Munich, 1922), 219–222; W. Teuffel, in *Geschichte der römischen Literatur*, 6th ed., III (Leipzig, 1913), 150–152; L. Urlichs, "Zur Kritik des Censorinus," in *Rheinisches Museum für Philologie*, **22** (1867), 465–476; and G. Wissowa, in Pauly-Wissowa, III (1899), cols. 1908–1910.

MENSO FOLKERTS

CENTNERSZWER, MIECZYSŁAW (*b.* Warsaw, Poland, 10 July 1871; *d.* Warsaw, 27 March 1944), *chemistry.*

Centnerszwer was the son of Gabriel Centnerszwer, a well-known bookseller and dealer in old books, and Rebeka Silberfeld. He graduated from secondary school in 1891, then studied chemistry at the University of Leipzig, where in 1896–1898 he did advanced work under Wilhelm Ostwald and obtained the Ph.D. (1898). Next Centnerszwer became assistant to Paul Walden at the Polytechnic Institute in Riga. He worked with Walden on Arrhenius' theory of electrolytic dissociation as applied to nonaqueous solvents, especially to liquid sulfur dioxide.

In 1902 he married Franciszka Anna Beck; they had a daughter, Jadwiga. In 1904 Centnerszwer obtained the M.S. at the University of St. Petersburg, and in 1905 he began to lecture on inorganic chemistry at the Polytechnic of Riga, where he was appointed associate professor in 1917 and full professor in 1919.

In 1929 he became doctor *honoris causa* of the Polytechnic of Riga and moved to Warsaw, where he was given the chair of physical chemistry at the University of Warsaw. In 1930 he was elected to the Krakow Academy of Sciences. During the German occupation of Poland he went into hiding because he was Jewish. On 27 March 1944 he was killed under mysterious circumstances.

Centnerszwer's scientific writing includes about 120 papers and books, mostly on experimental subjects, in various fields of physical chemistry. He was concerned mainly with chemical kinetics and investigated the thermal dissociation of salt, publishing a number of papers on the corrosion of metals and putting forward a hypothesis concerning the kinetics of the solution of metals in acids. He also conducted ebulliometric investigation of concentrated solutions. His most important publications concern experiments on the solubility and dissociation of substances in waterless solutions of liquid sulfur dioxide, hydrocyanic acid, and cyanogen.

Centnerszwer was the founder of an important and large school of chemists, called the Baltic school. He was the author of several books on inorganic chemistry and physical chemistry, which were published in, or translated into, Polish, Russian, German, French, Spanish, Finnish, and Latvian.

BIBLIOGRAPHY

I. ORIGINAL WORKS. Centnerszwer's writings include *Szkic z historii chemii* (Warsaw, 1909), trans. into Russian (Odessa, 1912; Leningrad, 1927); *Teoria jonów* (Warsaw, 1909); *Podręcznik do ćwiczeń z chemii fizycznej, termochemii i elektrochemii* (Warsaw, 1912), written with W. Świętosławski, trans. into Russian (Riga, 1912), French (Paris, 1914), and Spanish (Barcelona, 1922); *Das Radium und die Radioaktivität* (Lipsk, Poland, 1913, 1921), trans. into Russian (Leningrad, 1925) and Finnish (Helsinki, 1915); *Die chemische Verwandschaft und ihre Bedeutung für die Technik* (Riga, 1914), trans. into Russian (Petrograd, 1915); *Praktikum po chimii* (Riga, 1919); *Chemia fizyczna*, vol. I, *Przemiany materii* (Warsaw, 1922); and *Lekcja po nieorganiczeskoj chimii*, 2 vols. (Riga, 1923–1924), trans. into Latvian by J. Krustinson (Riga, 1922–1924).

II. SECONDARY LITERATURE. See S. Łoza, *Czy wiesz kto to jest?* (Warsaw, 1938), p. 95; M. Łaźniewski, "Mieczyslaw Centnerszwer (1871–1944)," in *Przemysł chemiczny*, **37** (1958), 246–251; W. Lampe, *Zarys historii chemii w Polsce* (Krakow, 1948), p. 31; Poggendorff, IV, 233; and a short biography in Gutenberg, *Ilustrowana encyklopedia powszechna*, II (Krakow, 1930), 17.

WŁODZIMIERZ HUBICKI

CERVANTES, VICENTE (*b.* Zafra, Badajoz, Spain, 1755; *d.* Mexico City, Mexico, 26 July 1829), *pharmacy, botany*.

Of humble origin, Cervantes began the study of Latin and botany while still an apprentice in pharmacy; after being licensed as an apothecary, he was patronized by Casimiro Gómez Ortega and became a member of the Royal College of Pharmacy. Cervantes had four children, Mariana, Julian (who was also a botanist), Vicente, and Antonio.

In 1786, while pharmacist to the General Hospital in Madrid, Cervantes was made a member of the royal botanical expedition to New Spain led by Martín Sessé. Cervantes was a founder of the botanical garden in Mexico City in 1788 and in that year became professor of botany at the University, a position he held until his death. In 1802 he became director of the botanical garden, member of the Real Tribunal del Protomedicato, inspector of pharmacies, and pharmacist to the Hospital de San Andrés, where he was also director for eighteen years. Cervantes considerably expanded the knowledge of Mexican botany and in his *Hortus mexicanus* described 1,400 botanical species. He started the training of Mexican botanists in the Linnaean system; for his communications on medicinal plants to the Royal Academy of Medicine, Madrid, he was elected a member of that body.

BIBLIOGRAPHY

I. ORIGINAL WORKS. Cervantes' best-known work is the *Ensavo a la materia médica vegetal de México*, written in 1791 but published posthumously (1889). The *Gazeta de México* published in 1788 the first *Ejercicios públicos de botánica*, in which Cervantes reviewed the history of botany up to Linnaeus and the work of the botanical expedition in Mexico. Several other papers appeared in the same journal and in the *Gazeta de literatura* (Mexico, 1788–1792), where Cervantes and his pupils presented their studies in the botanical garden. The description of the rubber tree *Castilloa elastica* was published in 1794, and many more monographs were published before the Mexican War of Independence.

II. SECONDARY LITERATURE. For biographical and bibliographical data on Cervantes, see Silvio Ibarra Cabrera, *Dr. D. Vincente Cervantes. Boceto biográfico* (Mexico, 1936) and Rafael Roldan Guerrero, *Diccionario biográfico y bibliográfico de autores españoles* (Madrid, 1958). An annotated survey of Cervantes' works is included in Ida K.

Langman, *A Selected Guide to the Literature on Flowering Plants of Mexico* (Philadelphia, 1964).

FRANCISCO GUERRA

CESALPINO, ANDREA (*b.* Arezzo, Italy, 6 June 1519; *d.* Rome, Italy, 23 February 1603), *botany, medicine, physiology*.

For a detailed study of his life and work, see Supplement.

CESÀRO, ERNESTO (*b.* Naples, Italy, 12 March 1859; *d.* Torre Annunziata, Italy, 12 September 1906), *mathematics*.

Cesàro was the son of Luigi Cesàro and Fortunata Nunziante, his second wife. The elder Cesàro owned a farm and shop in Torre Annunziata; he was one of the first farmers in Italy to introduce agricultural machinery, a supporter of Italian unification, and a backer of Garibaldi's revolution of 1860—all of which led him into financial difficulties.

Ernesto Cesàro completed the first class of the Gymnasium in Naples, studied in the seminary of Nola for two years, then returned to Naples to finish the fourth class of the Gymnasium in 1872. In 1873 his father sent him to Liège to join his older brother Giuseppe, who had gone there in 1867. Cesàro stayed for a year with his brother, who in the meantime had become lecturer in mineralogy and crystallography at the École des Mines, and then entered the École himself on a scholarship. He matriculated there, then applied unsuccessfully for admission to an Italian university, following which he was forced to enter the École des Mines of Liège. He studied mathematics with Eugène Catalan, who noticed Cesàro's talent and helped him publish his first mathematical paper in *Nouvelle correspondance de mathématiques,* of which Catalan was editor.

In 1879 Cesàro's father died, the family's financial troubles increased, and Cesàro was forced to return for some time to Torre Annunziata. Nevertheless, he finished the fourth form at Liège in 1881, and at the same time prepared his first major mathematical work, "Sur diverses questions d'arithmétique," which was published in 1883 in *Mémoires de l'Académie de Liège*. This paper brought Cesàro to the attention of the mathematical public.

In 1882 Cesàro returned once more to Belgium, having won another scholarship to continue his studies at Liège. Shortly thereafter he returned to Italy, where he married his stepbrother's daughter Angelina. At this period he also accompanied a friend, the son of the Prince de Soissons, to Paris, where he spent several months. He attended the lectures of Hermite,

Darboux, Serret, Briot, Bouquet, and Chasles at the Sorbonne; he attracted the attention of Hermite in particular. In 1883 the latter cited Cesàro's results. Darboux's lectures led Cesàro to formulate his "intrinsic geometry."

Cesàro did not finish his studies at Liège, perhaps because of a personal quarrel with a professor Deschamps. He returned to Torre Annunziata and once again sought to continue his work in Italy. His mathematical works and the recommendations of Cremona, Battaglini, and Dino secured him a scholarship to the University of Rome, which he entered in 1884 in the fourth form in pure mathematics. Here, in addition to attending a great many lectures, he wrote some eighty works—on infinite arithmetics, isobaric problems, holomorphic functions, theory of probability, and, particularly, intrinsic geometry—in the two years 1884–1886. Despite his intensive activity, he did not earn an advanced degree at this time. (The University of Rome gave him a doctorate, with honors, in February 1887.)

In 1886 Cesàro won a competition for the position of professor of mathematics at the Lycée Terenzio Mamiani in Rome; in similar competitions at the universities of Messina and Naples he placed first and second, respectively. On Cremona's advice, Cesàro left the Lycée Mamiani after a month to fill the vacant chair of higher algebra at the University of Palermo. He stayed at Palermo until 1891, when he accepted the chair of mathematical analysis at Naples. He held this chair until his death, never realizing his intention of going over to the chair of theoretical mechanics.

Cesàro's bibliography is extensive; indeed, the author of the most complete bibliography available, A. Perna, mentions 259 works and expresses doubt whether his list is complete. Cesàro's topics are varied. In 1878, when he was nineteen, he attempted to master certain topological problems in a non-traditional way in his *Forme poliedrichi regolari e semi-regolari in tutti gli spazii* (published in Lisbon in 1888). The most prominent of his early works, however, deal with the sums of divergent series, for Cesàro, Borel, Fejér, and Voronoj were together creating the techniques for the elaboration of such problems. One of Cesàro's first published works, *Sur diverses questions d'arithmétique* (Liège, 1882), and, more importantly, his series of nine articles published in *Annali di mathematica pura ed applicata* (**13** [1885], 235–351) are related to the theory of numbers. He was here concerned with such problems as the determination of the number of common divisors of two numerals, determination of the values of the sum totals of their squares, the probability of incommensurability of three arbitrary numbers, and so on; to these he attempted to apply obtained results

in the theory of Fourier series. Later he occupied himself with prime numbers of a certain type and tried to make Chebyshev's formulas more precise ("Sulla distributione dei numeri primi," in *Rendiconti dell'Accademia delle scienze fisiche e matemetiche* (Naples), **2** [1896], 297–304).

Despite the generally sophisticated level of his mathematics, Cesàro reverted to such elementary problems as, for example, his work on constructions using limited geometrical means (1899) which repeats results already known. His textbooks, on the other hand, are rather exacting. They were successful and influential in their time; *Corso di analisi algebrica con introduzione al calculo infinitesimale* was published in Turin in 1894 and *Elementa di calcolo infinitesimale* appeared in Naples in 1899. Both texts were the outgrowth of Cesàro's lectures in Palermo and Naples and both were distinguished by the pertinent and novel exercises that they contained.

The two textbooks also reveal Cesàro's interest in the problems of mathematical physics. In addition, his textbook *Introduzione alla teoria matematica della elasticità* (Turin, 1894) had dealt with the theory of elasticity in an elementary way; there is no doubt that he planned to investigate mathematical physics in more detail, since he prepared two works on this subject, "Teoria matematica dels calore" and "Lezione sull'idrodinamica"; he died before he could publish these, however, and they remain unpublished to this day.

Cesàro's most important contribution remains his intrinsic geometry. It has been noted that he began to develop it while he was in Paris in 1883; it occupied him, with interruptions, from that time on. His earlier work on the subject is summed up in his monograph *Lezione di geometria intrinseca* (Naples, 1896), in which he proceeds from a utilization of Darboux's method of a mobile coordinate trihedral—formed by the tangent, the principal normal, and the binormal at a variable point of a curve—and used it to simplify the analytic expression and make it independent of extrinsic coordinate systems. By this means, Cesàro stressed the intrinsic qualities of the objects examined. This method proved fertile for him, and he systematically elaborated and propagated it, while at the same time pointing out further applications. The *Lezione* also describes the curves that bear Cesàro's name; Cesàro later expanded his method to the curves devised by H. von Koch, which are continuous but so constructed as to have no tangent at any point.

The last part of *Lezione* deals with the theory of surfaces and multidimensional spaces in general. Cesàro returned to this subject in the last years of his life and emphasized the independence of his geometry from the axiom of parallels. The special selection of the square of the linear element enabled him to extend the results to multidimensional spaces with constant curvature. He further established other bases on which to build non-Euclidean geometry, which he described in *Rendiconti della R. Accademia dei Lincei* ("Sui fondamenti della interseca non-euclidea," **13** [1904], 438–446) and more especially in "Fondamento intrinseco della pangeometria," in *Memorie della R. Accademia dei Lincei* (**5** [1904], 155–183).

Throughout his life, the variety of Cesàro's interests was always remarkable, ranging from elementary geometrical problems to the application of mathematical analysis; from the theory of numbers to symbolic algebra; from the theory of probability to differential geometry. Moreover, his admiration for Maxwell, whose faithful interpreter in theoretical physics he became, is worthy of note.

In recognition of his work he was named an honorary member of many learned and scientific societies.

Cesàro died of injuries sustained while coming to the aid of his seventeen-year-old son who was drowning in the rough sea near Torre Annunziata.

BIBLIOGRAPHY

I. ORIGINAL WORKS. In addition to the individual works cited in the text, extensive lists of Cesàro's writings may be found in A. Perna and P. del Pezzo, below, and in Poggendorff.

II. SECONDARY LITERATURE. Works on Cesàro include C. Alasia, "Ernesto Cesàro, 1859–1906," published simultaneously in French and Italian in *Rivista di fisica, matematica e scienze naturali,* **15** (1907), 23–46, and *Enseignement mathématique,* **9** (1907), 76–82; V. Cerruti, "Ernesto Cesàro, Commemorazione," in *Rendiconti della R. Accademia dei Lincei,* **16** (1907), 76–82; A. Perna, "Ernesto Cesàro," in *Giornale di matematiche di Battaglini,* **45** (1907), 299–319, which includes a bibliography of 259 items as well as a presentation of some of Cesàro's problems and a review of his solutions; and P. del Pezzo, "Ernesto Cesàro," in *Rendiconti dell'Accademia delle Scienze fisiche e matematiche,* 3rd ser., **12** (1906), 358–375, which includes a list of 254 of Cesàro's works.

LUBOŠ NOVÝ
JAROSLAV FOLTA

CESI, FEDERICO (*b.* Rome, Italy, 13 March 1585; *d.* Acquasparta, Italy, 1 August 1630); *botany, scientific organization.*

The historical place of Federico Cesi as a precursor of modern scientific botany has been eclipsed by his stature as the founder of the first truly modern scientific academy, the Accademia dei Lincei. He was nevertheless a pioneer in systematic botanical classification as well as in the microscopic study of plant structure.

Cesi, hereditary marquis of Monticelli and duke of Acquasparta, acquired those titles in 1628 and 1630. He was made prince of San Angelo and San Polo in 1613 and assumed administration of all family estates in 1618. His noble status, together with the moral and financial support of his mother, Olimpia Orsini, assisted him in carrying out his scientific program against his father's strong opposition and, in later years, against growing antagonism toward science in theological and academic circles. Educated at Rome by private tutors, Cesi founded the Lincean Academy there in 1603, with four members. In 1610 Giovanni Battista Porta was enrolled, and after the election of Galileo in 1611 the Academy grew rapidly. Among its eventually thirty-two members were such foreign scientists as Mark Welser, Theophilus Müller, and Johannes Faber. Cesi was its principal administrator and sole financial supporter until his death, which was soon followed by the condemnation of Galileo and the collapse of the Academy.

The *Praescriptiones Lynceae,* drafted by Cesi in 1605 and published in 1624, fixed the objectives of the Academy as the study of science and mathematics, the pursuit of new knowledge, and the publication of scientific discoveries. The Academy's first important publication was Galileo's book on sunspots (1613). That work had in turn been inspired by Welser's publication at Augsburg in 1612 of Christopher Scheiner's letters on sunspots, sent by Welser to Galileo, whom he knew through Cesi. In 1616, when the Holy Office forbade the Copernican doctrine, Cesi supported freedom of opinion within the Academy against the Lincean mathematician Luca Valerio, who moved to dissociate the membership from the forbidden views. The Academy published Galileo's *Saggiatore* in 1623 at Cesi's expense and would have published the later *Dialogue* also, if Cesi had not died before the imprimatur was obtained.

Either personally or through his Academy, Cesi sponsored publication of scientific works by Porta, Johannes Eck, and Francesco Stelluti. His wide correspondence circulated scientific information in early seventeenth-century Italy and Germany, as did Mersenne's shortly afterward in France and England. Cesi also informed his colleagues of currents favorable or adverse to science at Rome and represented their interests in tactful discussions with authorities there. He tried (without success) to establish branches of the Academy at Naples and in other cities, both in Italy and abroad, envisioning an international scientific society. He labored for many years with three fellow academicians on a *Theatrum totius naturae*—a projected "Cosmos" in the Humboldtian sense. All that was eventually published of this was a folio broadside, the *Apiarium,* dedicated in 1625 to Pope Urban VIII and containing the first anatomical drawing made with the microscope, and the *Nova plantarum et mineralium mexicanorum historia,* edited from manuscripts of Francisco Hernandez. To the latter work was appended Cesi's *Phytosophicae tabulae;* although printed in 1630, the work was not published until 1651, funds for its completion and assembly having been cut off by Cesi's death.

Cesi's phytosophic tables anticipated by more than a century the work of Linnaeus in formulating a rational system for the classification and nomenclature of plants. Not only did Cesi conceive a natural system based on morphology and physiology, but he is reported to have discovered the spores of cryptogams and described the sexuality of plants in the course of his studies of microscopic plant anatomy. Except for the tables, however, his botanical work remains in manuscript.

Cesi was married in 1614 to Artemisia Colonna, who died late in 1615. From his marriage to Isabella Salviati in 1617, he left two surviving daughters. His own health had long been precarious, and in 1630 he died after a brief high fever. His library, scientific instruments, and manuscripts, dispersed after his death, have now been largely reacquired by the present Accademia dei Lincei at Rome, which was reconstituted as a national academy in 1875.

BIBLIOGRAPHY

I. ORIGINAL WORKS. Cesi's major works are *Praescriptiones Lynceae* (Rome, 1624; repr. Rome, 1745); and *Phytosophicae tabulae,* in *Nova plantarum et mineralium mexicanorum historia a Francisco Hernandez . . . compilata* . . . (Rome, 1651; Cesi's tables extracted and repr. Rome, 1904). *Il carteggio Linceo,* Giuseppe Gabrieli, ed. (Rome, 1938-1944), includes all of Cesi's surviving correspondence, much of which was also included by A. Favaro in *Le opere di Galileo Galilei* (Florence, 1890-1910; repr. Florence, 1929-1939; a 2nd repr., completed in Florence in 1965, was almost entirely destroyed by flood before issuance).

II. SECONDARY LITERATURE. Biographical sources on Cesi are Baldassare Odescalchi, *Memorie istorico-critiche dell'Accademia dei Lincei* (Rome, 1806); Domenico Carutti, *Breve storia dell'Accademia dei Lincei* (Rome, 1883); R. Pirotta and E. Chiovenda, *Flora Romana* (Rome, 1900-1901), *parte storica,* pp. 146-150; and Giuseppe Gabrieli, "Federico Cesi Linceo," in *Nuova antologia,* **350** (July-Aug. 1930), 352-369. See also the prefatory sections to the various parts of *Il carteggio Linceo,* above, and bibliographies therein.

STILLMAN DRAKE

CESTONI, GIACINTO (*b.* Montegiorgio, Italy, 10 or 13 May 1637; *d.* Leghorn, Italy, 29 January 1718), *natural history.*

Cestoni was born of poor parents, who did, however, send him to school until 1648. In that year he was apprenticed to a pharmacist, thus embarking on the profession he was to pursue the rest of his life. In 1650 he entered the service of a Roman pharmacist.

He left Rome in 1656 and set out to travel, using his modest savings. So he reached Leghorn, taking up again the practice of his profession. In 1666 he set out again, traveling via Marseilles and Lyons to Geneva, where he practiced pharmacy for about four months. Returning to Leghorn, Cestoni resumed his old job. He married in 1668 and spent the remainder of his life in Leghorn.

The grand ducal court summered in Leghorn, which facilitated his encounters with Francesco Redi, the chief physician to the grand duke. In 1680 Cestoni and Redi began a lively correspondence, which is known chiefly through Redi's letters.

When Redi died in 1697, Cestoni began to write weekly letters to Antonio Vallisnieri. Most of Cestoni's observations contained in the correspondence seem to have been made prior to 1697, however. A real friendship developed between the two men, and Vallisnieri visited Cestoni in Leghorn in 1705.

Vallisnieri published only part of Cestoni's observations, inserting them into his own works or into such journals as the *Galleria di Minerva* and the *Giornale de' letterati d'Italia.* Among the few observations published abroad were those on the metamorphic cycle of the flea which appeared in the *Philosophical Transactions of the Royal Society.*

Cestoni's principal subject of study was the generation of insects, and in this field Redi and Vallisnieri found him a valuable ally in working to disprove the theory of spontaneous generation. Connected with his studies on the generation of insects was the discovery of the acarid etiology of mange in collaboration with G. C. Bonomo. The latter published their discovery in the form of a letter to Redi, *Osservazioni intorno a' pellicelli del corpo umano* (Florence, 1687).

The Cestoni-Vallisnieri correspondence (for the most part published in 1940) merits systematic critical analysis in order to assign each of Cestoni's numerous observations its precise place in the general development of science. His observations on the parthenogenesis and the viviparity of the aphides seem especially relevant.

BIBLIOGRAPHY

The major source of information on Cestoni is *Epistolario ad Antonio Vallisnieri,* 2 vols. (Rome, 1940-1941).

An extract of a letter written by Cestoni to Martin Lister (24 Nov. 1698), concerning Redi's manuscripts and the generation of fleas, was published as "A New Discovery of the Origin of Fleas, Made by Signior D'iacinto Cestoni of Leghorn," in *Philosophical Transactions of the Royal Society,* **21** (1699), 42–43.

LUIGI BELLONI

CEULEN, LUDOLPH VAN (*b.* Hildesheim, Germany, 28 January 1540; *d.* Leiden, Netherlands, 31 December 1610), *mathematics.*

Van Ceulen's name may originally have been Ackerman, latinized as Colonus and gradually modified to Van Ceulen. He was the son of a merchant; and after traveling widely, he settled in Holland, first, perhaps, in Breda and Amsterdam. In 1580 he was in Delft, where he became a fencing master and mathematics instructor. During 1589 he spent some time in Arnhem, and in 1594 Van Ceulen received permission to open a fencing school in Leiden. In 1600 he was appointed teacher of arithmetic, surveying, and fortification at the engineering school founded in Leiden by Prince Maurice of Nassau. He held this position until his death. His second wife, Adriana Symons, whom he married in 1590, brought out Latin versions of two of his works posthumously, with the aid of Van Ceulen's pupil Willebrord Snell.

Van Ceulen was an indefatigable computer and concentrated on the computation of π, sometimes called Ludolph's number. This brought him into controversy with the master reckoner Simon Van der Eycke, who had published an incorrect quadrature of the circle (1584–1586). Then he became acquainted with Archimedes' *The Measurement of the Circle,* which his friend Jan Cornets de Groot, a mayor of Delft and father of Hugo Grotius, translated for him from the Greek. Now Van Ceulen began to work in the spirit of Archimedes, computing the sides of more regular polygons inscribed within and circumscribed about a circle than Archimedes had and inventing a special short division for such computation. In his principal work, *Van den Circkel* (1596), he published π to twenty decimal places by computing the sides of a regular polygon of 15×2^{31} sides. He continued to work on this subject; and in his *Arithmetische en geometrische fondamenten* (1615), published by his widow, he reached thirty-three decimal places, always enclosing π between an upper and a lower limit. Finally, Willebrord Snell, in his *Cyclometricus* (1621), published Van Ceulen's final triumph: π to thirty-five decimal places. This was inscribed on his tombstone in the Pieterskerk in Leiden.

The *Van den Circkel* consists of four sections. The first contains the computation of π. The second shows how to compute the sides of regular polygons of any number of sides, which in modern terms amounts to the expression of sin nA in terms of sin A (n is an integer). The third section contains tables of sines up to a radius of 10^7 (not an original achievement), and the fourth has tables of interest.

The first and second sections are the most original; they contain not only the best approximation of π reached at that time but also show Van Ceulen to be as expert in trigonometry as his contemporary Viète. In 1595 the two men competed in the solution of a forty-fifth degree equation proposed by Van Roomen in his *Ideae mathematicae* (1593) and recognized its relation to the expression of sin $45A$ in terms of sin A.

Van Ceulen's tables of interest were not the first to be published. He was anticipated by others, including his friend Simon Stevin (1583). Van Ceulen probably had computed his tables before he knew of Stevin's work.

BIBLIOGRAPHY

I. ORIGINAL WORKS. Two of Van Ceulen's books are *Van den Circkel* (Delft, 1596); and *Arithmetische en geometrische fondamenten* (Leiden, 1615). For the titles of Van Ceulen's early polemical writings and the full titles of his books, see Bierens de Haan and Bosmans (below). Snell's Latin versions are *Fundamenta arithmetica et geometrica* (Leiden, 1615; Amsterdam, 1619), a translation; and *L. à Ceulen De circulo et adscriptis liber* (Leiden, 1619), a modified version of the original. Van Ceulen also wrote a manuscript entitled "Algebra," which seems to have been lost.

II. SECONDARY LITERATURE. On Van Ceulen or his work, see D. Bierens de Haan, *Bouwstoffen voor de Geschiedenis der Wis-en Natuurkundige Wetenschappen in de Nederlanden,* 2 vols. (Amsterdam, 1876–1878), nos. 8, 9, 17 (repr. from *Verslagen en mededeelingen der Koninklijke Akademie van Wetenschappen Amsterdam*); H. Bosmans, "Un émule de Viète," in *Annales de la Société scientifique de Bruxelles,* **34**, pt. 2 (1910), 88–139, with an analysis of *Van den Circkel;* and "Ludolphe van Ceulen," in *Mathésis. Recueil mathématique à l'usage des écoles spéciales* (Ghent), **39** (1925), 352–360, with a portrait; and C. de Waard, "Ceulen," in *Nieuw Nederlandsch Biographisch Woordenboek,* VII (1927), cols. 291–295. See also P. Beydals, in *Nieuwe Rotterdamsche courant* (1 Oct. 1936). On his tables of interest, see Simon C. Stevin, *Tafelen van Interest* (Antwerp, 1583), repr. with English translation in *The Principal Works of S. Stevin,* IIa, Dirk J. Struik, ed. (Amsterdam, 1958), 13–24; and C. Waller Zeeper, *De oudste interesttafels* (Amsterdam, 1927). The fate of the tombstone is discussed by C. de Jong and W. Hope-Jones in *Mathematical Gazette,* **22** (1938), 281–282.

DIRK J. STRUIK

CEVA, GIOVANNI (*b.* Milan, Italy, 1647 or 1648; *d.* Mantua, Italy, 1734), *mathematics.*

Ceva's dates must be inferred from incomplete information; he died sometime in 1734 at the age of eighty-six years and six months and therefore must have been born in 1647 or 1648. In his correspondence with Antonio Magliabecchi, the librarian of the grand duke of Tuscany, Ceva states that he studied in Pisa, and in the preface to one of his books he gives particular praise to Donato Rossetti, who was a professor of logic there until he moved to Turin in 1674. He further mentions Alessandro Marchetti (1633–1714) and his son Angiolo Marchetti (1674–1753), both professors of mathematics at Pisa. Ceva was married and a daughter was born to him on 28 October 1685. His father was still living with him in Mantua in 1686. His brother was the poet, philosopher, and mathematician Tommaso Ceva. The bishop of Tortona, Carlo Francesco Ceva, was his cousin.

Ceva described his youth as saddened by "many kinds of misfortune" and his later work as distracted by "serious cares and affairs of his friends and family." At the time of his death his name was carried in a register of the salaried employees of the royal court as "Commissario dell'arciducale Camera et Matematico cesareo." He was buried in the church of St. Teresa de' Carmelitani Scalzi.

Ceva's efforts concerning the problem of diverting the river Reno into the Po deserve special attention; his opposition to this plan of the Bolognese led to the abandonment of the project.

Ceva's most important mathematical work is *De lineis rectis* (Milan, 1678), dedicated to Ferdinando Carlo. Chasles mentions it with praise in his *Aperçu historique*. In this work Ceva used the properties of the center of gravity of a system of points to obtain the relation of the segments which are produced by straight lines drawn through their intersections. He further utilized these properties in many theorems of the theory of transverse lines—for example, in placing at the points of intersection of the straight lines weights that are inversely proportional to the segments. From the relations of the weights, which are determined by the law of the lever, the relation of the segments is then derived. Ceva first applied his method, which is a combination of geometry and mechanics, to five basic figures, which he called "elements." He then used these in special problems, in which given relations are used to calculate others. The theorem of Menelaus concerning the segments produced by a transverse line of a triangle is proved, as is the transversal theorem concerning the concurrency of the transverse lines through the vertices of a triangle, which is named after Ceva. This theorem was established again by Johann I Bernoulli. Ceva worked with proportions and proved their expansion; he calculated many examples in detail and for all possible cases. (Occasionally he treated examples in a purely geometrical manner to demonstrate the advantage of his method.)

In the second book of the *De lineis,* Ceva went on to more complex examples and applied his method to cylindrical sections, ellipses in the triangle, and conic sections and their tangents.

In a geometrical supplement, not related to either of the first two books, Ceva dealt with classical geometrical theorems. He solved problems on plane figures bounded by arcs of circles and then calculated the volumes and centers of gravity of solid bodies, such as the paraboloid and the two hyperboloids of rotation. Cavalieri's indivisibles are used successfully in this case.

Ceva's mastery of all the other geometrical problems of his time is shown by other works which are dedicated to the mathematician Cardinal Ricci. Among these, the *Opuscula mathematica* (Milan, 1682) met with particular acclaim.

The *Opuscula* is in four parts. In the first part Ceva investigated forces and formed the parallelogram of forces and the resultants of many different forces. Geometrical proofs accompany the mechanical considerations. "Geometrical proofs can themselves provide a verification for that which we have determined mechanically," Ceva wrote in the scholium to the sixth proposition. He then considered levers at greater length and obtained proportions for the quadrangle by means of lever laws. He further discussed centers of gravity for surfaces and bodies. In all the problems he showed how geometry can be used profitably in statics.

In the second part of the book, Ceva investigated pendulum laws—here he refers to Galileo. In the third proposition of this section, Ceva arrived at the erroneous conclusion (which he later corrected) that the periods of oscillation of two pendulums are in the ratio of their lengths.

Solid bodies and perforated vessels are observed in flowing water in the last part of the work. Flow velocity is measured by means of the motion of pendulums suspended in the flow. Finally the amount of water is measured by means of flow cross sections. In the last pages, Ceva again added a geometrical appendix. He examined a ring with a semicircular cross section and proved that certain sections cannot be elliptical. He then calculated the center of gravity of the surface of a hemisphere by using Cavalieri's indivisibles and indicated that it is not necessary to work with parallel sections in this method. Central sections, "small cones," could also be used in the calculation. Ceva's infinitesimal method is sketched in this section with clear and detailed demonstrations.

A third work, *Geometria motus* (Bologna, 1692), is

also of great interest. Here Ceva attempted to determine the nature of motions geometrically, stating that he has always been "interested without restraint" in such studies. His further prefatory remark—that geometry brings unadulterated truth—again indicated his interest in pure geometry. Ceva worked with coordinate systems

$$s = f(v); \; s = f(t), \text{ etc.}$$

The areas defined by the curves are determined by the Cavalieri method. Although he preferred the geometrical method, he did not hesitate to use indivisibles and he considered the points of curves to be quantities "smaller than any of those specified." After he had examined individual motions, he went on to a comparison of motions. This brought him to parabolas and hyperbolas, and he made particular reference to the work of Stephano de Angelis (*Miscellaneum hyperbolicum et parabolam,* Venice, 1659). Ceva assumed that, over an "infinitely short distance," motion can be considered uniform. He gave no more precise substantiation of this view, however.

Ceva treated composite motions in the second book of the *Geometria motus.* Here he also discussed the laws of pendulums and (in the scholium on the fifteenth proposition, theorem XI) corrected his error in the *Opuscula mathematica.* In considering motions of points along curves, Ceva was led to a comparison between parabolas and spirals with equal arcs. He considered the lines as "flows of points." He also investigated bodies formed by the rotation of certain figures and considered the falling of bodies along inclined planes, the subject of a great deal of his previous work. The final part of the *Geometria* consists of studies on the stretching and motion of ropes in which weights suspended by ropes are experimentally raised and dropped.

Although Ceva used archaic and complicated formulations, the *Geometria* anticipates or at least suggests elements of infinitesimal calculus.

Ceva's interest in a variety of problems led him to produce *De re numeraria* (Mantua, 1711), a work praised by Cinelli for its great accuracy.

Ceva frequently became involved in controversies on physical problems. In particular he criticized Paster Vanni's erroneous conception of the distribution of forces on an inclined plane.

BIBLIOGRAPHY

I. ORIGINAL WORKS. Ceva's extant writings include *De lineis rectis se invicem secantibus statica constructio* (Milan, 1678); *Opuscula mathematica* (Milan, 1682); *Geometria motus* (Bologna, 1692); *Tria problemata geometrice proposita* (Mantua, 1710); *De re numeraria, quod fieri potuit, geometrice tractata* (Mantua, 1711); *De mundi fabrica, una gravitatis principio innixo* (Mantua, 1715); his polemical works concerning the diversion of the Reno into the Po, *Le conseguenze del Reno, se con l'aderire al progretto de' Signori Bolognesi si permetesse in Po grande* (Mantua, 1716; Bologna, 1716), *Replica de Giovanni Ceva indifesa delle sue dimostrazioni, e ragioni, per quali non debassi introdurre Reno in Po, contro la riposta data dal Sig. Eustachio Manfredi* (Mantua, 1721), and *Riposta de Giovanni Ceva alle osservazioni dal Signor dottor Eustachio Manfredi contro la di lui replica in proposito dell'immissione de Reno in Po grande pretesa da' Signori Bolognesi* (Mantua, 1721); *Hydrostatica* (Mantua, 1728); and two letters, one addressed to Vincenzo Viviani and the other to Antonio Magliabecchi, in the Royal Library in Florence.

II. SECONDARY LITERATURE. See Gino Loria, "Per la biografia de Giovanni Ceva," in *Rendiconti dell'istituto lombardo di scienze e lettere,* **48** (1915), 450–452.

HERBERT OETTEL

CEVA, TOMASSO (*b.* Milan, Italy, 20 December 1648; *d.* Milan, 3 February 1737), *mathematics.*

Tomasso Ceva came from a rich and famous Italian family; he was the brother of Giovanni Ceva. In 1663 he entered the Society of Jesus and at an early age became professor of mathematics at Brera College in Milan.

Ceva's first scientific work, *De natura gravium* (1669), deals with physical subjects—such as gravity, the attraction of masses for each other, free fall, and the pendulum—in a philosophical and even theological way. (For example, several pages are devoted to the concept of the *spatium imaginarum.*) Ceva wrote the treatise in two months of steady work; in his "Conclusion," he asks his readers for emendations.

Ceva's only truly mathematical work is the *Opuscula mathematica* (1699; parts were published separately in the same year as *De ratione aequilibri, De sectione geometrico-harmonia et arithmetica,* and *De cycloide; de lineis phantasticis; de flexibilibus*). The book is discussed in *Acta eruditorum* (1707); its particular importance is that it is the summation of all of Ceva's mathematical work. It is concerned with gravity, arithmetic, geometric-harmonic means, the cycloid, division of angles, and higher-order conic sections and curves. It also contains a report on an instrument designed to divide a right angle into a specified number of equal parts; this same instrument was described in 1704 by L'Hospital—who makes no mention, however, of Ceva.

Higher-order curves are also the primary subject of an extensive correspondence between Ceva and Guido Grandi. Ceva proposed the problem; Grandi reported that such curves had well-defined properties. Grandi replied to Ceva's questions not only in letters, but also in a work on the logarithmic curve published in 1701 with an appended letter by Ceva.

Ceva's contribution to mathematics was modest; he is perhaps better remembered as a poet. Although some of his verse is mathematical and philosophical, he is best known for his religious poem *Jesus Puer,* which went through many printings and was translated into several languages. The German poet Lessing called Ceva a great mathematician as well as a great poet, while Schubart, writing in 1781, considered him the greatest Jesuit poet-genius.

BIBLIOGRAPHY

Ceva's mathematical and scientific works are *De natura gravium libri duo Thomae Cevae* (Milan, 1669); *Instrumentum pro sectione cujuscunque anguli rectilinei in partes quotcunque aequales* (Milan, 1695; repr. in *Acta eruditorum* [1695], p. 290); and *Opuscula mathematica Thomae Cevae e Soc. Jesu* (Milan, 1699), discussed in *Acta eruditorum* (1707), pp. 149–153.

Other works are *Jesus Puer, Poema* (Milan, 1690, 1699, 1704, 1718, 1732, 1733), translated into German (Augsburg, 1844), French, and Italian; *Sylvae. Carmina Thomae Cevae* (Milan, 1699, 1704, 1733); and *Carmina videlicet philosophia novo-antiqua* (Milan, 1704; Venice, 1732).

Ceva's correspondence with Grandi is in the Braidense Library (eight letters) and the Domus Galilaeana, Pisa (485 letters).

An important secondary source is Guido Grandi, *Geometrica demonstratio theorematum Hugenianorum circa logisticam, seu logarithmicam lineam, addita epistola geometrica ad P. Thomam Cevam* (Florence, 1701).

HERBERT OETTEL

CHABRY, LAURENT (*b.* Roanne, France, 19 February 1855; *d.* Paris, France, 23 December 1893), *physiology, embryology.*

The son of a master tinsmith, Chabry showed remarkable aptitude at a very early age and quickly finished his secondary studies. In order to continue his education he went, despite meager resources, to St. Petersburg in 1876.

Upon returning home Chabry became an ardent participant in the socialist movement and collaborated on the weekly publication *L'égalité;* he was also involved in the events stemming from the International Socialist Congress of 1878.

His passion for politics did not prevent Chabry from pursuing his medical studies. Drawn toward mechanical physiology as a result of attending classes given at the Collège de France by the great physiologist Marey, he presented and defended a doctoral thesis in medicine (1881) on the movement of the ribs and sternum. Chabry worked at the Sorbonne under the guidance of Lacaze-Duthiers and then with G. Pouchet, professor of zoology at the Muséum d'Histoire Naturelle. Later he was at the maritime laboratory of Concarneau. In 1887 he defended his thesis, on the embryology of Ascidiacea, for the doctorate in science at the Sorbonne. He was appointed lecturer in zoology and embryology at Lyons in 1888. Two years later he left science to practice medicine. He was soon disappointed and, returning to science, studied bacteriology at the Pasteur Institute.

Chabry's work on ascidians, more or less summarized in his doctoral thesis, assured him an eminent place in the history of embryology. Examining the spontaneous monstrosities of ascidians, he showed that each of the primary cells of the embryo has a predetermined fate independent of any subsequent circumstances of development. Furthermore, he succeeded in experimentally producing half-embryos by destroying, through *piqûre,* one of the two primary cells. In order to perform this very delicate operation, he constructed a simple and precise micromanipulator.

E. G. Conklin noted in 1905 the importance and originality of this work with *piqûre:* Chabry had shown himself to be both a master and a creator.

With G. Pouchet, Chabry published some interesting experimental research on the embryology of sea urchins. Between 1881 and 1886 he collaborated with Pouchet and Charles Robin in microscopic anatomical research on the teeth and the embryology of cetaceans. He also published a series of articles on animal mechanics, jumping, balance, the swimming of fish, and insect wings, and contributed articles on these subjects to Robin's *Dictionnaire de médecine.*

In bacteriology Chabry's later research, which remained incomplete, led him—as he himself stated—"to discoveries that created hope."

Despite his short career, Chabry left work that clearly makes him one of the founders of experimental embryology; he was the first to perform experimental operations on such a small fertilized egg (the ascidian egg measures .10–.20 mm.).

BIBLIOGRAPHY

I. ORIGINAL WORKS. Chabry's writings include *Contribution à l'étude du mouvement des côtes et du sternum* (Paris, 1881), his M.D. thesis; "Contribution à l'embryologie normale et tératologique des Ascidies simples," in *Journal de l'anatomie et de la physiologie normales et pathologiques de l'homme et des animaux,* **23** (1887), 167–316; "L'eau de mer artificielle comme agent tératologique," *ibid.,* **25** (1889), 298–307, written with Pouchet; and "De la production des larves monstrueuses d'oursin par privation de chaux," in *Comptes rendus hebdomadaires des séances de l'Académie des sciences,* Paris, **108** (1889), 196–198, written with Pouchet.

II. SECONDARY LITERATURE. A biographical sketch with a list of publications was published by Pouchet in *Journal de l'anatomie et de la physiologie normales et pathologiques de l'homme et des animaux,* **29** (1893), 735–739. E. G. Conklin gave a flattering commentary on Chabry's work in "Mosaic Development in Ascidian Eggs," in *Journal of Experimental Zoology,* **2** (1905), 145–223 (see 197–198). A biographical article by Maurice Caullery is in *Revue scientifique,* **78** (1940), 230–232.

ANDRÉE TÉTRY

CHAGAS, CARLOS RIBEIRO JUSTINIANO (*b.* Oliveira, Minas Gerais, Brazil, 9 July 1879; *d.* Rio de Janeiro, Brazil, 8 November 1934), *medicine.*

Chagas was the first son of a coffee planter whose family had arrived in Brazil from Portugal around the middle of the sixteenth century. His paternal granduncle, João das Chagas Andrade, was a medical doctor, as was his mother's brother. The latter, Carlos Ribeiro de Castro, graduated from the Medical School of Rio de Janeiro in 1888 and in the early 1890's established, in Oliveira, a hospital in which the methods of Lister were used for the first time in Brazil. This young surgeon was to exert a great influence on his nephew.

In 1914 Chagas enrolled in the School of Mining Engineering in the old capital of the state of Minas Gerais. Having been taken ill in 1895 (with a typical case of B-deficiency avitaminosis), he spent some time in Oliveira, where, under the influence of his uncle, he decided to take up medicine. He started medical training at the Medical School of Rio de Janeiro in 1896 and finished the course there in 1902. He earned the M.D. in 1903 with the completion of a thesis, "Estudos hematológicos do impaludismo," at the Instituto Oswaldo Cruz.

As a student, Chagas' main concern was malaria and yellow fever, which were endemic at the time in Rio de Janeiro. He became familiar with the laboratory methods for diagnosing malaria—these methods having been introduced into Brazil by his first teacher, Francisco Fajardo.

In 1907 Chagas became a full-time staff member of the Instituto Oswaldo Cruz. His friendship with that institution's founder, which was to end only with Cruz's death in 1917, was a major influence in his career. For the first years of his professional practice, Chagas established himself as a general practitioner in Rio de Janeiro.

In 1905 financial difficulties forced Chagas to accept a mission to Santos in order to fight malaria, endemic to the workers employed for the construction of the port. In Itatinga, a small village nearby, Chagas undertook the first successful Brazilian campaign for malaria control. He did not use the already well-established technique of destroying larvae, but instead used pyrethrum to disinfect households and thus achieved a surprising success. He published his technique in *Prophylaxia do impaludismo,* a work which was revived forty years later as a guide to the use of new synthetic insecticides.

In 1909 Cruz asked Chagas to undertake a new antimalaria campaign in the village of Lassance, about 375 miles from Rio, which was at that time the terminal of the projected Central Railways System. Chagas stayed in Lassance for two years (1909–1910), living in a railroad car. During this time he discovered and described the disease that bears his name.

A few days after his arrival in Lassance, Chagas' attention was drawn to the large number of bugs that infested the walls of the huts there. Upon examining them he found that they frequently harbored a trypanosome. Chagas identified this trypanosome as a new genus and species and proposed to name it *Schizotrypanum cruzi* (known generally as *Trypanosoma cruzi*), the species name being in honor of his friend and teacher. He then verified the pathogenicity of this trypanosome in monkeys, dogs, and cats. The positive infectiousness of the trypanosome in these animals led him to consider that such an infection in humans would explicate the pattern of disease that he found in the region. Chagas then looked for the trypanosome in human blood; he found it in an infant patient who presented the symptoms of an acute infection.

Chagas described and studied the manifestations of the disease as well as its pathogenic mechanism. He depicted the disease as consisting of an acute phase, marked by fever and a generalized hard edema, which might be followed by one of three chronic forms. Chagas classified these later stages by their symptomatology, the most frequent form being characterized by cardiac disturbances, while the other two presented gastrointestinal and neural syndromes, respectively. While in Lassance, Chagas performed more than 100 postmortem examinations on patients who exhibited the chronic form of the disease, although he was able to observe and describe only twenty-two cases of the acute form.

Chagas further established a basis for epidemiological studies related to the new disease. He described its vectors, studied the various grades of infectiousness of *Triatoma* in general, and observed the armadillo to be the most frequent reservoir of the microorganism. Chagas' work may therefore be pointed out as a unique feat in the history of medicine—it is the only instance in which a single investigator has described the infection, its agent, its vector, its manifestations, its epidemiology, and some of the hosts of the pathogenic genus.

After his stay in Lassance, Chagas made a ten-month trip to the Amazon Basin to study tropical diseases further. In his later years, he was responsible for the reorganization of the Department of Health in Brazil, and was very active in the establishment of international centers for preventive medicine and in the reformulation of Brazilian medical education.

Chagas was the first Brazilian to be awarded an honorary degree by Harvard University and by the University of Paris. Among his many other honors was an award given him at the Pasteur centennial exhibition in Strasbourg, in 1922.

BIBLIOGRAPHY

A complete bibliography of Chagas' work and of writings on Chagas' disease up to 1959 is in *Annals of the International Congress of Chagas' Disease,* 5 vols. (Rio de Janeiro, 1960).

Chagas' most important writings are *Prophylaxia do impaludismo* (Rio de Janeiro, 1905); "Beitrag zur Malaria-prophylaxis," in *Zeitschrift für Hygiene und Infektions-krankheiten,* **60** (1908), 321–334; "Nouvelle espèce de Trypanosomiase humaine," in *Bulletin de la Société de pathologie exotique,* **2** (1909), 304–307; "Nova trypanozo-miaze humana. Estudos sobre a morfologia e o ciclo evolu-tivo do Schizotrypanum cruzi n. gen., n. sp., agente etio-lógico de nova entidade mórbida do homem," in *Memórias do Instituto Oswaldo Cruz,* **1** (1909), 159–218, text in Portu-guese and German; "Ueber eine neue Trypanosomiasis des Menschen. Zweite vorläufige Mitteilung," in *Archiv für Schiffs- und Tropenhygiene,* **13** (1909), 351–353; "Les formes nerveuses d'une nouvelle Trypanosomiase (Trypanosoma Cruzi inoculé par Triatoma megista) (Maladie de Chagas)," in *Nouvelle iconographie de la Salpêtrière, clinique des maladies du système nerveux,* **26** (1913), 1–9; *Revision of the Life Cycle of "Trypanosoma cruzi." Supplementary Note* (Rio de Janeiro, 1913); "Processos patogénicos da tripa-nosomiase americana," in *Memórias do Instituto Oswaldo Cruz,* **8** (1915), 5–36; *Cardiac Form of American Trypa-nosomiasis* (1920); "American Trypanosomiasis. Study of the Parasite and of the Transmitting Insect," in *Proceedings of the Institute of Medicine of Chicago,* **3** (1921), 220–242; *Amerikanische Trypanosomiases (Chagasche Krankheit). Kurse aetiologische und klinische Betrachtungen* (1925); "Ueber die amerikanische Trypanosomiasis (Chagaskrank-heit)," in *Münchener medizinische Wochenschrift,* **72** (1925), 2039–2040; "Quelques aspects evolutifs du Trypanosoma cruzi dans l'insect transmetteur," in *Comptes rendus de la Société biologique* (Paris), **97** (1927), 824–832; "Sur les al-térations du coeur dans la trypanosomiase américaine (Maladie de Chagas)," in *Archives des maladies du coeur, des vaisseux et du sang,* **21** (1928), 641–655; and "Ameri-kanische Trypanosomenkrankheit, Chagaskrankheit," in

Carl Mense, *Handbuch der Tropenkrankheiten,* 3rd ed., V, pt. 1 (Leipzig, 1929), 673–728, written with Enrico Villela and H. da Rocha Lima.

CARLOS CHAGAS

CHALCIDIUS. See **Calcidius.**

CHALLIS, JAMES (*b.* Braintree, Essex, England, 12 December 1803; *d.* Cambridge, England, 3 December 1882), *astronomy.*

The fourth son of John Challis, James received his early training in a private school at Braintree run by the Reverend Daniel Copsey, who recognized his potential and secured him a place at Mill Hill School, near London. He entered Trinity College, Cambridge, in 1821 as a sizar and graduated as senior wrangler and first Smith prizeman in 1825. Elected a fellow of Trinity in 1826, he resided there until ordained in 1830, when he took up the college living of Papworth Everard. He vacated his fellowship in 1831 by marry-ing the widow of Daniel Copsey, daughter-in-law of his early patron. He was reelected a fellow of Trinity in 1870 and retained his fellowship until his death.

When George Biddell Airy was elected astronomer royal in 1835, he was succeeded as Plumian profes-sor and director of the Cambridge Observatory by Challis, who took up residence in 1836. Challis was succeeded at the observatory by John Couch Adams in 1861 but retained the Plumian chair until his death.

Challis was a spectacular failure as a scientist, and ironically, this failure has immortalized him. His most distinguished contemporaries reacted to the peculiar-ity of his scientific views, and he was saved from intellectual ostracism only by his amiability. Although he possessed very little physical sense, he nevertheless claimed to have proved a far wider generalization than that shown earlier by Isaac Newton. In a work published privately in 1869, entitled *Notes on the Principles of Pure and Applied Calculation; and Appli-cation of Mathematical Principles to Theories of the Physical Forces,* he attempted to show that all physical phenomena are mathematically deducible from a few simple laws. At a later time, or under less amiable circumstances, he would have been branded a char-latan. He would now be as forgotten as his peculiar ideas had not the events surrounding the discovery of Neptune in 1845 given him a genuine opportunity for scientific immortality. But he fumbled it.

John Couch Adams had approached Challis in 1844 to help him obtain from Airy the errors in the motion of the planet Uranus, which had been under observa-tion for some time at the Greenwich Observatory. He needed these to predict the orbit of an unknown planet that might be perturbing the orbit of Uranus. After computing the orbit of the new planet, Adams presented Challis in September 1845 with some predictions as to where it might be found. By July 1846 Airy was pressing Challis very hard to begin a search for the planet because the Northumberland equatorial telescope at Cambridge was the best instrument for the task and because Continental astronomers were pressing the search, using the predictions of the French mathematician J. Leverrier, which were similar to those of Adams. He offered to supply Challis with an assistant for the purpose but Challis apparently misunderstood the offer and rebuffed him with the words "I understand your proposal [of supplying an assistant] to be made on the supposition that I decline undertaking the search myself" (Challis to Airy, 18 July 1846).

The subsequent history of the discovery is detailed more fully elsewhere in the *Dictionary* in the articles on John Couch Adams, Leverrier, and Airy, but Challis' contribution and his failure are well summarized in his letter to Airy of 12 October 1846 (the italics are mine):

> I had heard of the discovery of the planet on Oct. 1. I have been greatly mortified to find that my observations would have shewn me the planet in the early part of August if I had only discussed them. I commenced observing on July 29, attacking first of all, as it was prudent to do, the position which Mr. Adams' calculations assigned as the most probable place of the Planet. On July 30 . . . I took all the stars to the 11th magnitude in a zone 9′ in breadth. . . . On account of moonlight I did not observe again till Aug. 12. On that day I went over again the zone of 9′ breadth which I examined on July 30 . . . on comparing the observations of those two days I found that the zone of July 30 contained every star in the corresponding position of the zone of Aug. 12 except one star of the 8th magnitude. This . . . must have been a Planet. By this statement you will see that after four days of observing the Planet was in my grasp if only I had examined or mapped the observations. I delayed doing this partly because I thought the probability of discovery was very small till a much larger portion of the heavens was scrutinized, *but chiefly because I was making a grand effort to reduce the vast numbers of comet observations which I have accumulated and this occupied the whole of my time.*

The judgment on Challis by history and by his more objective contemporaries is expressed by H. C. Schumacher, Danish astronomer and founder-editor of the *Astronomische Nachrichten,* in a letter to Airy dated 24 October 1846:

> I scarcely know what I shall say about Mr. Challis. He sees the great probability that the predicted planet

must exist by the near coincidence of two totally independent predictions. You request sweeps with the Northumberland equatorial from him. He makes such sweeps on July 30 and Aug. 12, but lays them aside without looking at them. Now such observations are only made to see if a star has changed its relative place to the others. First when he hears that the new planet has been observed Sept. 23 at Berlin he examined his observations made nearly two months before, and lo! the planet is observed Aug. 12 and so would have asserted the honour of one of the most brilliant discoveries to his Nation, to his University and to his Observatory.

OLIN J. EGGEN

CHAMBERLAIN, CHARLES JOSEPH (*b.* Sullivan, Ohio, 23 February 1863; *d.* Chicago, Illinois, 5 January 1943), *botany.*

A contributor in the field of morphology of the angiosperms and gymnosperms and to methods for the study of plant cells, Chamberlain was drawn early in his career to the study of the cycads and centered his researches for over forty years on these plants, whose origin had been little known. Resembling the palms but related to the ferns and still retaining the fern leaf, they were numerous and widely distributed during the Triassic and Jurassic periods. Today they are native only to certain tropical and subtropical regions and are represented by the Cycadaceae, a family of the class Gymnospermae. Because they have survived virtually unchanged, they have often been called "living fossils."

The son of Esdell W. and Mary Spencer Chamberlain, he attended Oberlin College, where he studied botany under the geologist A. A. Wright, who also taught zoology. Chamberlain was perhaps influenced then toward his later interest in the application of histological methods to botanical study and in the origin of plants whose line was to be traced from remote ancestors in the Paleozoic era. He was graduated from Oberlin in 1888 and that year was married to Martha E. Life; they had one daughter. He then taught in the public schools of Ohio and Minnesota and for several years was principal of the high school in Crookston, Minnesota. During the summers, meanwhile, he continued his botanical studies toward the master's degree, which he received from Oberlin in 1894. In 1893 Chamberlin matriculated at the University of Chicago, where in 1897 he received the first doctorate conferred in the department of botany. At the University of Chicago he was in charge of the botanical laboratories from their establishment and was a member of the faculty for over forty-five years. In 1901–1902 he engaged in research at Bonn, in the laboratory of the renowned botanist Eduard Strasburger. At Chicago, Chamberlain became professor of morphology and cytology in 1915 and professor emeritus in 1929. In 1931 his wife died, and in 1938 he married Martha Stanley Lathrop.

Chamberlain contributed to botanical publications and in 1902 became American editor for cytology of the *Botanisches Zentralblatt*. Over the years he received numerous honors, including the honorary Sc.D. from Oberlin in 1923. A member of international botanical societies, he was vice-president and chairman of the botanical section of the American Association for the Advancement of Science in 1923 and president of the Botanical Society of America in 1931–1932.

In tracing the evolution of the cycads, Chamberlain studied the evidence of paleobotany and examined the structural relationships of past and living representatives, following the stages of their life histories. He saw the once-flourishing cycads as now restricted and struggling for survival. In order to see them as they grew in their natural surroundings, Chamberlain made a number of expeditions. He visited Mexico in 1904, 1906, 1908, and 1910; and in 1911–1912 he traveled around the world, to Australia, New Zealand, and South Africa, observing all of the oriental cycads during this journey. Later he made two such trips to Cuba. *The Living Cycad* is an account of these field expeditions to study the cycads. While collecting plant material and photographs and meeting other botanists interested in these plants, Chamberlain recorded a range of observations on the zoological features and the customs of countries he visited. He sent back to Chicago plant material for microscopic as well as macroscopic study in the laboratory, and specimens for the botanical garden.

Enriched by his contributions over the years, the collection of living cycads in the botanical garden at the University of Chicago became foremost in the world; and at Chamberlain's death it contained all of the nine genera which now survive and half of the known species. His lifework was to have led to a monograph, then near completion, on the morphology and phylogeny of the cycads.

BIBLIOGRAPHY

I. ORIGINAL WORKS. In collaboration with John Merle Coulter, Chamberlain wrote *Morphology of Spermatophytes* (New York, 1901); *Morphology of Angiosperms* (New York, 1903); and *Morphology of Gymnosperms* (Chicago, 1910, 1917). Interested in the microscopic examination of plant tissue as an aid to botanical study, he published *Methods in Plant Histology* (Chicago, 1901, 1905, 1915, 1924, 1932). *The Living Cycad* (Chicago, 1919) recounts his botanical expeditions. Chamberlain also wrote *Elements of Plant Science* (New York, 1930) and *Gymnosperms: Structure and Evolution* (Chicago, 1935; repr. New York, 1957). Among his articles are "Spermatogenesis in *Dioon edule*," in *Botanical Gazette*, **47** (1909), 215–236; "The Living Cycads and the Phylogeny of Seed Plants," in *American Journal of Botany*, **7** (1920), 146–153; and "Hybrids in Cycads," in *Botanical Gazette*, **81** (1926), 401–418.

II. SECONDARY LITERATURE. See J. T. Bucholz, "Charles Joseph Chamberlain," in *Botanical Gazette*, **104** (1943), 369–370; and Arthur W. Haupt, "Charles Joseph Chamberlain," in *Chronica botanica VII*, **8** (1943), 438–440.

GLORIA ROBINSON

CHAMBERLAND, CHARLES EDOUARD

(*b*. Chilly le Vignoble, Jura, France, 12 March 1851; *d*. Paris, France, 2 May 1908), *bacteriology*.

One of Pasteur's most famous associates, Chamberland was later to become an expert himself, enriching the techniques of bacteriology with important apparatus as well as setting down useful rules for public health.

After studying the classics at the *lycée* in Lons-le-Saunier and a period at the Collège Rollin in Paris, Chamberland was admitted in 1871 to both the École Polytechnique and the École Normale Supérieure. He chose the latter and was appointed professor at the *lycée* of Nîmes in 1874. A year later he returned to the École Normale and remained there until 1888, first as an assistant in Pasteur's laboratory and then as one of the laboratory's assistant directors. In 1885 he was elected a deputy from the Jura.

As soon as Chamberland returned to Paris in 1875, the controversy between Pasteur and Bastian concerning spontaneous generation erupted. Pasteur asked his new assistant to investigate the causes of error in Bastian's experiments. Soon afterward Chamberland was able to explain why acid organic liquids heated to 100°C. could be preserved without changing, even though they were full of microbes, when enough sterile potassium was added to make them alkaline. He also showed that in order to kill certain spores it is first necessary to heat the liquid to a temperature of 115°C. for twenty minutes. This important observation led him to perfect both rules and new methods for the sterilization of culture media. His work resulted in the autoclave, which soon became an indispensable tool in bacteriology departments, hospitals, and disinfection stations.

Next Chamberland showed how porous walls are capable of retaining fine particles in suspension and substituted for the filtration process then in use the slightly warmed porcelain filter. In doing so he set forth an excellent sterilization procedure for liquids that could be changed by heat. The filter, which was of immediate use in laboratories, a few years later facilitated the discovery of microbic exotoxins and the first viruses. Because it made possible the purification of drinking water, it was of great value to public health.

While at the École Normale, Chamberland participated in Pasteur's studies: the attenuation of viruses and preventive inoculations, the etiology and prophylaxis of anthrax, and vaccination against hog cholera and rabies. In 1888, when the Pasteur Institute was opened in Paris, he became director of one of the six departments created at that time: that of microbiology applied to hygiene; one of its main functions was to prepare, on a large scale, various Pasteur vaccines. Chamberland headed the department until his death (from 1904 he was also assistant director of the Pasteur Institute and a member of the Academy of Medicine).

On the rue Dutot he studied the possibilities of disinfecting places and objects with compounds containing chlorine (with E. Fernbach) and the antiseptic properties of *essences* and hydrogen peroxide. His studies with Jouan on microbes of the *Pasteurella* type have remained classic. Chamberland thought that the atmosphere did not play the main role in transmitting infectious germs. He hoped that there would be more concern with soiled objects, clothing, and the hands of doctors and their assistants.

Chamberland's common sense and creative mind have often been emphasized. Calmette said of him: "He was a master of his trade and a friend full of charming good humor, with extraordinary kindness and exceptionally penetrating intelligence. His premature death—he was only fifty-seven years old—was a grievous loss to the Institute." He was tall and slender, with handsome features. He was married and had one child.

BIBLIOGRAPHY

Various papers by Chamberland appeared in the *Bulletin de l'Académie de médecine,* the *Comptes rendus hebdomadaires des séances de l'Académie des sciences,* and the *Annales de l'Institut Pasteur* between 1878 and 1908. Among them are "La théorie des germes et ses applications à la médecine et à la chirurgie," in *Bulletin de l'Académie de médecine,* 2nd ser., **7** (1878), 432, written with Pasteur and Joubert; and "Recherches sur l'origine et le développement des organismes microscopiques," *Annales scientifiques de l'École normale supérieure,* 2nd ser., **7** (1879), supp., his doctoral thesis.

A notice on Chamberland's life and work is in *Annales de l'Institut Pasteur,* **22** (1908), 369.

ALBERT DELAUNAY

CHAMBERLIN, THOMAS CHROWDER (*b.* Mattoon, Illinois, 25 September 1843; *d.* Chicago, Illinois, 15 November 1928), *geology, cosmology.*

Chamberlin's father, John Chamberlin, left North Carolina because of his strong feelings against slavery and went west. In Illinois he married Cecilia Gill, from Kentucky. They soon settled on a farm near Mattoon and there, on the crest of the Shelbyville terminal moraine, "T. C." was born. When he was about three years old, the family moved to a farm overlooking the rolling prairie near Beloit, Wisconsin.

In his later years Chamberlin sometimes referred to the geological setting of his birthplace as prophetic of his career. The selection of that career seems, however, to have been influenced by the presence of a limestone quarry near his boyhood home and the intellectual atmosphere in that home. Helping his older brothers quarry stone for a new farmhouse, young Chamberlin was intrigued by the fossils they found. Many were the discussions as to how the "snails" and "snakes" could have gotten inside the rock. The father was by this time a Methodist circuit rider who earned his living on the farm but devoted his Sundays to preaching. Although without much academic education, he was an avid reader and an incisive thinker, strict in his views and vigorous in any argument. He also insisted that his sons think for themselves.

On graduating from Beloit College with an A.B. degree in 1866, Chamberlin spent the next two years as principal of the Delavan, Wisconsin, high school, where he introduced elementary geology and astronomy into the curriculum. In 1867 he married Alma Isabel Wilson of Beloit; they had one son, Rollin Thomas, also a geologist and teacher.

Dissatisfied with his intellectual equipment for teaching the natural sciences and foreseeing the coming wave of scientific development, Chamberlin spent the academic year 1868–1869 in graduate study at the University of Michigan, where he worked in geology under Alexander Winchell. Returning to Wisconsin, Chamberlin was professor of natural science at the state normal school at Whitewater from 1869 to 1873.

In 1873 the Wisconsin Geological Survey was established by the state legislature, and Chamberlin was appointed assistant geologist in charge of the southeast section of the state. That region of nearly flat, early Paleozoic sedimentary rocks is covered deeply by glacial drift; thus Chamberlin was practically forced to become a student of glaciation, an enterprise that eventually led him into the fundamental problems of geology and cosmology. From 1876 to the completion of the survey in 1882, he was its chief geologist. The four volumes of the survey's report,

Geology of Wisconsin (1877–1883), are models of pioneer geological research in which the chief geologist's contributions deal especially with the glacial deposits, the lead and zinc ores, and the Silurian coral reefs. (Chamberlin apparently was the first geologist to identify certain structures in the Silurian limestones of Wisconsin as ancient coral reefs.) In the midst of these labors he served also as part-time professor of geology at Beloit College from 1873 to 1882 and delivered lectures on the relations between science and religion to audiences that filled the Second Congregational Church of Beloit.

Even before the completion of Chamberlin's work with the Wisconsin Geological Survey he was appointed, on 17 June 1881, to the United States Geological Survey as geologist in charge of its newly organized glacial division, a post that he held until 1904. The series of notable memoirs by Chamberlin and his associates that appeared in the *Report of the United States Geological Survey* between 1882 and 1888 are among the classics of geological literature. Chamberlin became president of the University of Wisconsin in 1887; in the next five years he transformed that institution from a college into a true university, both in organization and in spirit. Nevertheless, in 1891, when William Rainey Harper offered him the chairmanship of the department of geology to be organized at the new University of Chicago, he accepted. From 1892 to 1918 Chamberlin devoted his energy unstintingly to that institution of higher learning. There he established the *Journal of Geology,* and under his supervision there developed one of the most distinguished geology departments in any university, American or foreign.

In June 1918, shortly before his seventy-fifth birthday, Chamberlin retired from his university duties, continuing only his service as senior editor of the *Journal of Geology.* But retirement meant merely greater freedom for continuing his research, and the last ten years of his life were among the most fruitful. One of the finest of all his publications was *The Two Solar Families,* which came from the press on his eighty-fifth birthday, only a few weeks before his death.

Chamberlin was tall and broad-shouldered, a rugged and vigorous figure throughout most of his life. His characteristically serious mien, softened by a kindly and serene expression, gave him an appearance of great dignity, enhanced in his later years by snow-white hair and full beard. He was not without physical afflictions, however. In 1909 he and his son were the scientists in a four-man team commissioned by John D. Rockefeller to reconnoiter the vast and little-known Chinese empire as a potential site for the humani-

tarian projects of the nascent Rockefeller Foundation. His health broken by months of arduous travel and poor food, the first symptoms of serious stomach trouble appeared while the Chamberlins were en route from Mukden (now Shen-yang) to Moscow. His condition worsened, and at several times during the next three or four years it caused alarm. Fortunately, skillful treatment prevailed, and a rigid regimen of diet and exercise, worked out largely by Chamberlin himself, slowly restored his health, except for his steadily failing eyesight, during the last fifteen years of his life.

Chamberlin was active in many scientific organizations and was the recipient of many honors. He was president of the Wisconsin Academy of Science, Arts, and Letters in 1885–1886, of the Geological Society of America in 1895, of the Chicago Academy of Sciences from 1897 to 1915, of the Illinois Academy of Sciences in 1907, and of the American Association for the Advancement of Science in 1908–1909. He was a fellow of the National Academy of Sciences, the American Academy of Arts and Sciences, and the American Philosophical Society, as well as a corresponding member of the British Association for the Advancement of Science and of the geological societies of Edinburgh, London, Sweden, and Belgium. Honorary doctoral degrees were conferred on him by the University of Michigan, Beloit College, Columbia University, the University of Wisconsin, the University of Toronto, and the University of Illinois. He received the medal for geological publications at the Paris expositions in 1878 and 1893, the Culver Medal of the Geographical Society of Chicago in 1910, the Hayden Medal of the Philadelphia Academy of Natural Sciences in 1920, the Penrose Medal of the Society of Economic Geologists in 1924, and the Penrose Medal of the Geological Society of America in 1927.

The titles in the selected bibliography of Chamberlin's published works indicate something of his breadth of interest and the scope of his mind. He was one of the first glaciologists to see that the Ice Age was a time of multiple glaciation. This led him to focus attention upon the causes of glacial climates and the problem of the changing climates of the geological past. He stressed the importance of variations in the composition of the earth's atmosphere, with special reference to its CO_2 content. His 1897 paper on this subject is still a stimulus for all concerned with this problem, which has not yet been fully resolved. Most significantly, this consideration of the condition of the atmosphere early in the earth's history forced Chamberlin to doubt the validity of the then generally approved Laplacian hypothesis concerning the origin and early history of the earth. It was one of the first steps along the path that brought him to the heights in cosmology and fundamental geological philosophy.

About this time an incident occurred that indicated Chamberlin's extraordinary prophetic vision. In a famous address delivered in 1897, Lord Kelvin announced that the sun could not have supplied heat sufficient for life on the earth for more than 20,000,000–30,000,000 years and chided geologists for their extravagant postulates concerning time. Replying in 1899, Chamberlin criticized the "unqualified assumptions" on which Kelvin had based his calculations and went on to say that the internal constitution of the atoms was still an open question and that it was not improbable that they were complex organizations possessed of enormous energies.

To test critically the Laplacian hypothesis and to develop meticulously the planetesimal hypothesis as its replacement, Chamberlin needed the assistance of someone who could supplement his own naturalistic approach by bringing mathematical methods of analysis to bear upon the problems. This he found in his colleague at the University of Chicago, F. R. Moulton, whose mastery of celestial mechanics and profound mathematical insight enabled him to contribute brilliantly to the success of their joint enterprise. By 1900 they had completely abandoned the Laplacian hypothesis, and by 1905 they had molded the planetesimal hypothesis into the form in which it appeared in *Yearbook of the Carnegie Institution of Washington* (no. 3). The concept of planetary growth by accretion of planetoidal particles and bodies (planetesimals) seems to be standing the test of time.

This latter concept has far-reaching implications in geological theory, and Chamberlin proceeded to follow them to their logical conclusions in two series of publications: one, with the general title Diastrophism and the Formative Process, encompassed fifteen papers that appeared in the *Journal of Geology* between 1913 and 1920; the other, under the rubric Study of Fundamental Problems of Geology, was published in the Yearbooks of the Carnegie Institution of Washington between 1904 and 1928. Throughout these and other contributions to geological lore he consistently used the "method of multiple working hypotheses," setting an example in scientific methodology that has been followed ever since by researchers in all scientific disciplines.

BIBLIOGRAPHY

I. ORIGINAL WORKS. Among Chamberlin's writings are "Preliminary Paper on the Terminal Moraine of the Second Glacial Epoch," in *Report of the United States Geological Survey* (1882), 291–402; "The Requisite and Qualifying Conditions of Artesian Wells," in *Report of the United States Geological Survey* (1884), 125–173; "Preliminary Paper on the Driftless Area of the Upper Mississippi Valley," in *Report of the United States Geological Survey* (1885), pp. 205–322, written with R. D. Salisbury; "Glacial Phenomena of North America," in James Geikie, *The Great Ice Age*, 3rd ed. (New York, 1895), pp. 724–774; "A Group of Hypotheses Bearing on Climatic Changes," in *Journal of Geology*, **5** (1897), 653–683; "The Method of Multiple Working Hypotheses," *ibid.*, 837–848, repr. *ibid.*, **39** (1931), 155–165; "On Lord Kelvin's Address on the Age of the Earth as an Abode Fitted for Life," in *Science*, **9** (1899), 889–901, and **10** (1899), 11–18, repr. in *Report of the Board of Regents of the Smithsonian Institution* for 1899 (1900), pp. 223–246; "On the Habitat of the Early Vertebrates," in *Journal of Geology*, **8** (1900), 400–412; *Geology*, 3 vols. (New York, 1904–1906), written with R. D. Salisbury; "Fundamental Problems of Geology," in *Yearbook of the Carnegie Institution of Washington*, no. 3 (Washington, D.C., 1905), pp. 195–238; "Early Terrestrial Conditions That May Have Favored Organic Synthesis," in *Science*, **28** (1908), 897–911, written with R. T. Chamberlin; *College Geology* (New York, 1909), written with R. D. Salisbury; "Diastrophism as the Ultimate Basis of Correlation," in *Journal of Geology*, **17** (1909), 685–690; "The Development of the Planetesimal Hypothesis," in *Science*, **30** (1909), 643–645, written with F. R. Moulton; "The Future Habitability of the Earth," in *Report of the Board of Regents of the Smithsonian Institution* for 1910 (1911), pp. 371–389; *The Origin of the Earth* (Chicago, 1916); "Certain Phases of Megatectonic Geology," in *Journal of Geology*, **34** (1926), 1–28; and *The Two Solar Families: The Sun's Children* (Chicago, 1928).

II. SECONDARY LITERATURE. Biographies of Chamberlin include Rollin T. Chamberlin, in *Biographical Memoirs. National Academy of Sciences*, **15** (1934), 307–407, with a bibliography of 251 titles; Kirtley F. Mather, in *Proceedings of the American Academy of Arts and Sciences*, **70** (1936), 505–508; and Bailey Willis, in *Bulletin of the Geological Society of America*, **40** (1929), 23–44.

KIRTLEY F. MATHER

CHAMBERS, ROBERT (*b.* Peebles, Scotland, 10 July 1802; *d.* St. Andrews, Scotland, 17 March 1871), *biology, geology.*

The son of a Scottish cotton manufacturer, Chambers was largely self-educated. After operating his own bookstore for a period, he joined his brother William in 1832 to form the well-known Edinburgh publishing firm bearing their names. He wrote many items of antiquarian interest and concerning local, that is, Scottish, history and literature. From the mid-1840's through the 1850's Chambers wrote a number of papers dealing with Scottish geology, particularly glacial action and erosion phenomena. These papers, competent although not highly original, earned him a reputation as a scientist; and he was given credit for helping to gain acceptance for the glacial theory.

The scientific work for which Chambers is now best known, *Vestiges of the Natural History of Creation* (1844), was published anonymously. By 1860 it had sold over 20,000 copies in eleven British editions plus editions in the United States, Germany, and the Netherlands. Chambers had planned to write no more on the subject; but late in 1845, largely in response to Adam Sedgwick's review of the *Vestiges* in the *Edinburgh Review,* he wrote *Explanations: A Sequel to "Vestiges of the Natural History of Creation."* The *Explanations* was appended to later editions of the *Vestiges,* several of which underwent substantial revisions.

The main thesis of the *Vestiges* is that the organic world is controlled by the law of development, just as the inorganic is controlled by gravitation. Within each realm the respective law is the factor unifying all the phenomena. Chambers suggests, although not explicitly, that a higher generalization will someday be found that will unify the phenomena of both the organic and the inorganic realms. He begins with a discussion of the nebular hypothesis, which by the mid-1840's had lost many of its more scientific supporters but still had many believers among the general public. Chambers emphasized the developmental aspects of that hypothesis, arguing that the solar system had developed from "a universal Fire Mist" to its present configuration.

Chambers then devoted about a quarter of the volume to geology and paleontology, considering each era or formation in turn, beginning with the oldest. This is probably the strongest section of the book, geology being the one area of science in which he had had any firsthand experience. While discussing the formation of the strata, he also discussed the fossil fauna and flora contained therein. Chambers demonstrated that the fossils show a general progression from lower to higher types, with extinctions and new appearances taking place until the "superficial formations" and the appearance of the present species. The appearance of man is a very recent event; and within the human species there has been a development that finally produced the highest race, the Caucasian. To support his doctrine of development Chambers pointed to analogies between three sets of organic phenomena: the order of geological succession of forms, the general taxonomic arrangement of these forms, and the stages through which each embryo passes during development.

Much of the remainder of the volume is devoted to a wide range of biological phenomena, about which Chambers often demonstrates his lack of firsthand knowledge and his naïveté. It must be remembered that essentially all of Chambers' research for the *Vestiges* was done in a library and that he had little or no experience with which to evaluate his sources. The topics that he considers range from phrenology and the spontaneous generation of life by means of an electric charge to geographical distribution and taxonomy. For example, he accepts as demonstrated that oats can and do transform into rye, and he argues that birds gave rise to the duck-billed platypus, from which the other mammals arose. On a sounder basis, he relies on Augustin de Candolle's work on the geographical distribution of vegetation and recognizes the peculiarity of the fauna of Australia. In the first edition Chambers tended strongly toward William Macleay's circular system of taxonomy. However, he shifted in later editions to a linear branching scheme for representing the relationships of different groups and thus was able to indicate cases of parallel development.

The system that Chambers created was contrary to the contemporary theology. He thought the idea of having God create each species individually at the time at which it appeared in the geological record was belittling to God and unduly anthropomorphized Him. To Chambers it was far more noble to envisage the Creator as working through natural laws and having the organic world develop from humble beginnings. Such a system also provided the basis for an original unity of the entire organic world. Chambers argued that the operation of the natural laws is inherently good because they are God's laws. However, exceptions to the usual operation of the laws of nature arise as a result of localized conditions; and these exceptions are interpreted as apparent evils. Established religion did not take long to react to Chambers' views of the Deity and the cause of evil.

Often naïve, often gullible, Chambers still managed in the *Vestiges* to bring together a large variety of data from geology and the life sciences that bore on the problem of the origin of species. He was writing the *Vestiges* at the same time that Darwin was writing his "Sketch of 1842" and his "Essay of 1844." Darwin would not publish until he had accumulated a great deal more supporting evidence. Chambers, with far less experience in science, did not feel such a concern. The *Vestiges* played a significant role in mid-nineteenth-century biology. By presenting an evolutionary view of nature, it received the first wave of reaction and thus eased the way for Darwin's *On the Origin of Species* fifteen years later.

BIBLIOGRAPHY

Chambers' most important scientific work was *Vestiges of the Natural History of Creation* (London, 1844). Ten editions appeared by 1853, and Chambers' last revision was for the eleventh edition (1860), after Darwin had published *On the Origin of Species.* Chambers appended *Explanations: A Sequel to "Vestiges . . ."* (London, 1845) to the later editions. A facsimile of the first edition of *Vestiges,* with an introduction by Sir Gavin de Beer, is also available (Leicester, 1969). Chamber's only volume devoted wholly to geology was *Ancient Sea-Margins, as Memorials of Changes in the Relative Level of Sea and Land* (Edinburgh, 1848). His geological papers are listed in Millhauser.

A very thorough study of Chambers' life, the background to his work, and his impact in Victorian England is Milton Millhauser, *Just Before Darwin. Robert Chambers and "Vestiges"* (Middletown, Conn., 1959). This contains a bibliography of Chambers' principal works.

WESLEY C. WILLIAMS

CHAMISSO, ADELBERT VON, also known as **Louis Charles Adélaïde Chamisso** (*b.* Château de Boncourt, Champagne, France, 30 January 1781; *d.* Berlin, Germany, 21 August 1838), *botany.*

For a detailed study of his life and work, see Supplement.

CHANCEL, GUSTAV CHARLES BONAVENTURE (*b.* Loriol, France, 18 January 1822; *d.* Montpellier, France, 5 August 1890), *chemistry.*

After studying at the École Centrale in Paris, Chancel worked in the laboratory of Jules Pelouze. In 1846 he was an assistant in chemistry at the École des Mines, and when Charles Gerhardt left Montpellier, Chancel replaced him as professor of chemistry (1851). He became dean of the faculty of sciences in 1865 and rector of the Montpellier Academy in 1879. He was a corresponding member of the Académie des Sciences from 1880 and received its Jecker Prize in 1884 for his work in organic chemistry.

Chancel's researches were primarily in analytical and organic chemistry. His analytical studies dealt with the separation and analysis of metals in solution. In 1858 he introduced a new method of precipitation. Instead of adding the precipitating agent to a solution, Chancel slowly generated it in the solution itself. The principal advantage lay in the formation of a dense crystalline precipitate of high purity. By means of this method he precipitated aluminum as the hydroxide by means of sodium thiosulfate in the presence of iron.

Chancel investigated ketones in his first organic chemical researches. In 1844 he obtained butyrone from butyric acid and proved that it was analogous to acetone; he also discovered butyral and butyramide. The following year he isolated valerone and valeral from valeric acid. By 1845 several unsuccessful attempts had been made to clarify the nature of ketones and their relation to other organic classes. Chancel set the stage for the establishment of their constitution by proposing that they were combinations of an aldehyde and a hydrocarbon. Gerhardt adopted this idea in 1853, proposing four types of inorganic molecules from which all organic compounds could be derived by substitution of radicals for hydrogen atoms. One such type was the hydrogen type $\left.{H \atop H}\right\}$. It included all hydrocarbons:

$$\left.{CH_3 \atop H}\right\} \qquad \left.{C_2H_5 \atop H}\right\}$$
$$\text{methane} \qquad\quad \text{ethane}$$

It also included aldehydes and ketones:

$$\left.{C_2H_3O \atop H}\right\} \qquad \left.{C_2H_3O \atop CH_3}\right\} \qquad C_2H_3O =$$
$$\text{acetaldehyde} \qquad \text{acetone} \qquad \text{acetyl radical}$$

Chancel collaborated often with Gerhardt. In 1852 they proposed that the sulfonyl group could replace the carbonyl group and announced the preparation of benzenesulfonyl chloride from benzoyl chloride. Chancel helped to confirm Gerhardt's homologous series for alcohols. After noting that the alcohol series had missing links between ethyl and amyl alcohols, he succeeded in preparing propyl alcohol (1853).

Chancel also collaborated with Auguste Laurent. They discovered butyronitrile in 1847 and phenylurea and diphenylurea in 1849. Chancel confirmed Laurent's proposal of alcohol and ether as members of the water type (1850). Most chemists followed Liebig in regarding ether and alcohol as ethyl oxide and hydrated ethyl oxide, respectively. Laurent in 1846 proposed the water type:

$$\left.{H \atop H}\right\}O \qquad \left.{C_2H_5 \atop H}\right\}O \qquad \left.{C_2H_5 \atop C_2H_5}\right\}O$$

The experiments of Chancel, and independently those of Alexander Williamson, decided the issue in favor of Laurent. Chancel prepared ether from potassium ethylate and ethyl sulfate. He also prepared methylethyl ether from potassium methylate and ethyl sulfate. In the latter case Liebig's theory predicted a mixture of dimethyl and diethyl ethers, whereas Laurent's theory predicted methylethyl ether. Chancel claimed no priority of discovery but simply stated that he had reached the same results as Williamson and that these discoveries confirmed the ideas of Gerhardt and Laurent on the constitution of alcohol and ether.

BIBLIOGRAPHY

I. ORIGINAL WORKS. Chancel wrote a guide for his students: *Cours élémentaire d'analyse chimique* (Montpellier, 1851). He collaborated with Gerhardt on *Précis d'analyse chimique qualitative* (Paris, 1855; 3rd ed., 1874) and *Précis d'analyse chimique quantitative* (Paris, 1859; 3rd ed., 1875). His important papers include "Mémoire sur la butyrone," in *Comptes rendus hebdomadaires des séances de l'Académie des sciences,* **18** (1844), 1023–1028; "Théorie de la formation et de la constitution des produits pyrogénés," *ibid.,* **20** (1845), 1580–1587; "Recherches sur l'acide valérique," *ibid.,* **21** (1845), 905–911; "Sur l'étherification et sur une nouvelle classe d'éthers," *ibid.,* **31** (1850), 521–523, and **32** (1851), 587–589; "Recherches sur les combinaisons de l'acide sulfurique avec les matières organiques," *ibid.,* **35** (1852), 690–694, written with Gerhardt; "Recherches sur l'alcool propionique," *ibid.,* **37** (1853), 410–412; and "De l'emploi des hyposulfites dans l'analyse," *ibid.,* **46** (1858), 987–990.

II. SECONDARY LITERATURE. There is a bibliography of Chancel's papers with an account of his life and work by M. R. Forcrand: "Notice sur la vie et les travaux de G. Chancel," in *Bulletin de la Société chimique de Paris,* **5** (1891), i–xx. See also the notice by P. Hamon, in *Dictionnaire de biographie française,* VIII (Paris, 1959), 363–364.

ALBERT B. COSTA

CHANCOURTOIS, A. E. BÉGUYER DE. See Béguyer de Chancourtois, A. E.

CHANDLER, SETH CARLO (*b.* Boston, Massachusetts, 17 September 1846; *d.* Wellesley Hills, Massachusetts, 31 December 1913), *astronomy.*

Chandler was educated at the English High School in Boston. After graduation (1861) he became private assistant to Benjamin Apthorp Gould and involved with the U.S. Coast Survey, which he joined in 1864. Rather than accompany Gould to Argentina, Chandler worked for many years as an actuary and did not resume his scientific activities until 1881.

He then became associated with the Harvard College observatory, evolving with J. Ritchie the Science Observer Code (1881), a system for transmitting by telegraph information about newly discovered comets. Chandler published many papers on comets and variable stars and compiled several useful catalogs of the latter objects.

His most important contribution to science was the discovery of the variation of latitude. Soon after arriving at Harvard, Chandler devised the almucantar, an instrument by means of which one relates positions of stars not to the meridian but to a small circle centered at the zenith. In his discussion (1891) of observations made with this instrument during 1884–1885 he concluded that the latitude varied with amplitude 0″.3 in a period of fourteen months. Observations made about the same time in Berlin by Küstner had shown a similar variation. From an exhaustive rediscussion of observations made as far back as Bradley's time, Chandler was able to verify the fourteen-month period and to show that there was in addition a variation having a period of twelve months. His announcements met with considerable opposition, initially because Euler had shown that any variation would have a period of ten months; but as Newcomb pointed out, since the earth is not completely rigid, the period would be longer.

Chandler was editor of the *Astronomical Journal* from 1896 to 1909 and subsequently an associate editor. He received an LL.D. degree from DePauw University (1891), the Watson Medal of the National Academy of Sciences (1895), and the Gold Medal of the Royal Astronomical Society (1896).

BIBLIOGRAPHY

Chandler's "On the Variation of Latitude" is in *Astronomical Journal,* **11** (1891), 59–61, 65–70, 75–79, 83–86; **12** (1892), 17–22, 57–62, 65–72, 97–101; **13** (1893), 159–162. Chandler's many papers in the *Astronomical Journal* are listed in the General Index to the *Journal,* **1–50** (1948), 15–17.

Chandler is discussed in H. H. Turner, *Astronomical Discovery* (London, 1904), pp. 177–217. See also Turner's obituary notice of Chandler in *Monthly Notices of the Royal Astronomical Society,* **75** (1915), 251–256.

BRIAN G. MARSDEN

CHAPLYGIN, SERGEI ALEKSEEVICH (*b.* Ranenburg [now Chaplygin], Russia, 5 April 1869; *d.* Moscow, U.S.S.R., 8 October 1942), *mechanics, engineering, mathematics.*

Chaplygin was born into the family of a shop assistant. His father, Aleksei Timofeevich Chaplygin, died suddenly of cholera in 1871, when his son was two. In 1886 Chaplygin graduated from the Voronezh Gymnasium and immediately enrolled at Moscow University, from which he graduated with a brilliant record in 1890; at the request of N. E. Zhukovsky he was retained there to prepare for a teaching career. He became an assistant professor at Moscow University in 1894. From 1896 until 1906 he taught mechanics at Moscow Technical College and from 1901 was professor of mechanics at Moscow Women's College, which he headed from 1905 until 1918.

Chaplygin's first scientific papers, which were written under Zhukovsky's influence, were devoted to hydromechanics. In 1893 he wrote a long article, "O nekotorykh sluchayakh dvizhenia tverdogo tela v zhidkosti" ("On Certain Cases of Movement of a Solid Body in a Liquid"), which was awarded the Brashman Prize. In 1897 there appeared a second article with the same title; this was his master's dissertation. In these papers Chaplygin gave a geometric interpretation of those cases of the movement of a body in a liquid that had earlier been studied from a purely analytic standpoint by the German scientists Clebsch and Kirchhoff, as well as by the Russian scientist Steklov. In this regard Zhukovsky has written that Chaplygin "demonstrated in his two excellent papers what strength the cleverly conceived geometrical methods of investigation can possess."

Even at the beginning of his scientific career Chaplygin devoted much attention to the development of the general methods of classical mechanics. A whole series of his papers, which appeared at the turn of the century, has among its topics the problem of a body's motion in the presence of nonintegrable relationships and the motion of a solid body around a fixed point. In the article of 1897, "O dvizhenii tverdogo tela vrashchenia na gorizontalnoy ploskosti" ("On the Motion of a Solid Body of Revolution in a Horizontal Plane"), general equations for the motion of nonholonomic systems were first obtained; these equations are a generalization of the Lagrangian equation. The Petersburg Academy of Sciences awarded Chaplygin a gold medal in 1899 for his investigations of the movement of a solid body.

Among Chaplygin's papers a special place is occupied by his investigation of the mechanics of liquids and gases. Even in the 1890's he had shown a great interest in the study of jet streams. At that time jet flow theory was the basis for study of the laws of motion of bodies in a fluid. In 1899 Chaplygin, using Zhukovsky's investigations as a base, solved somewhat differently the problem of a stream of incompressible fluid passing around a plate ("K voprosu o struyakh v neszhimaemoy zhidkosti" ["To the Problem of Currents in an Incompressible Fluid"]). The problem of gas passing around bodies was especially interesting to him.

In the nineteenth century Russian scientists as well as others had published a number of papers on the theory of a high-speed stream of gas. For example, in 1839 St. Venant had investigated the phenomenon of the escape of gas through an opening at a great rate of flow. In 1858 N. V. Maievsky established the influence of the compressibility of air on the resistance to the motion of a shell for a flight velocity close to the speed of sound.

In 1902 Chaplygin published his famous paper "O gazovykh struyakh" ("On Gas Streams"), in which he developed a method permitting the solution, in many cases, of the problem of the noncontinuous flow of a compressible gas. With this paper he opened the field of high-velocity aeromechanics. The method devised by Chaplygin made it possible to solve the problem of the flow of a gas stream if, under the limiting conditions, the solution to the corresponding problem of an incompressible liquid is known. The equations derived by Chaplygin for the motion of a compressible fluid are valid for the case in which the velocity of the current never exceeds the speed of sound. He applied this theory to the solution of two problems concerning the stream flow of a compressible fluid: escape from a vessel and flow around a plate that is perpendicular to the direction of flow at infinity.

Chaplygin found precise solutions to the problems he examined; they are still the only instances of precise solutions to problems in gas dynamics. He compared the results of his theoretical investigations on the escape of a gas and on the flow around a plate with experimental data and obtained qualitative confirmation of his theory.

Chaplygin also developed a method of approximation for the solution of problems in gas dynamics that was noteworthy for its simplicity; however, it is possible to apply this method only when the velocity of the gas flow does not exceed approximately half the speed of sound.

"O gazovykh struyakh" was Chaplygin's doctoral dissertation. At the time it did not receive wide recognition, partly because at the velocities then obtaining in aviation there was no need to consider the influence of the compressibility of air; on the other hand, in artillery great interest was centered on investigations at velocities greater than the speed of sound.

The significance of this paper for solving problems in aviation came to light at the beginning of the 1930's, when it became necessary to create a new science about the motion of bodies at velocities equal to and greater than the speed of sound, and for the flow patterns past them. The bases of this new science, gas dynamics, had been laid down by Chaplygin, who thus was more than thirty years ahead of the necessary technology.

In 1910 Chaplygin began his important investigations into the theory of the wing. That February he reported to the Moscow Mathematical Society on the aerodynamic forces acting on an airplane wing. He stated the results of these investigations in his paper "O davlenii plosko-parallelnogo potoka na pregrazhdayushchie tela (k teorii aeroplana)" ("On the Pressure Exerted by a Plane-parallel Stream on an Impeding Body [Toward a Theory of the Airplane]"), which was published that same year. The postulate concerning the determination of the rate of circulation around a wing was first precisely stated in this paper. This postulate—the so-called Chaplygin-Zhukovsky postulate—gives a complete solution to the problem of the forces exerted by a stream on a body passing through it. This article includes the fundamentals of plane aerodynamics, particularly Chaplygin's celebrated formulas for calculating the pressures exerted by the stream of a fluid on an impeding body. These formulas were applied by Chaplygin to the calculation of the stream pressure on various wing profiles for which he gives the construction.

In "O davlenii plosko-parallelnogo . . ." Chaplygin obtained a number of other remarkable results. He was the first to study thoroughly the question of the longitudinal moment acting on a wing, considering this question an essential element of the theory of the wing. On the basis of a study of the general formula for the moment of the lifting force he established a simple relationship between the longitudinal moment and the angle of attack; this relationship was not obtained experimentally until several years later and subsequently proved to be one of the fundamental aerodynamic characteristics of a wing.

After the October Revolution, Chaplygin immediately sided with the Soviet government and, with Zhukovsky, actively participated in the organization (1918) of the Central Aerohydrodynamic Institute; after the death of his friend and teacher, Chaplygin became the director of this prominent scientific center. In 1924 he was elected a corresponding member, and in 1929 a full member, of the Academy of Sciences of the U.S.S.R. In 1929 the title Honored Scientist of the R.S.F.S.R. was conferred on him. His scientific, technological, and organizational services were recognized in 1941, when he was awarded the title Hero of Socialist Labor.

Chaplygin's subsequent scientific papers were devoted to the development of aerohydrodynamics. His fundamental investigations of wing cross sections, wing profiles, a wing's irregular motion, and the theory of structural framework had great significance for the development of aerodynamics throughout the world.

Chaplygin's works also enriched mathematics: his studies of methods of approximation for solving differential equations are achievements of mathematical thought.

BIBLIOGRAPHY

Chaplygin's works have been brought together in *Sobranie sochineny* ("Collected Works"), 4 vols. (Moscow-Leningrad, 1948–1950).

On Chaplygin or his work, see V. V. Golubev, *Sergei Alekseevich Chaplygin (1869–1942)* (Moscow, 1951), in Russian, which includes a bibliography of Chaplygin's published works and critico-biographical literature about him; A. T. Grigorian, *Die Entwicklung der Hydrodynamik und Aerodynamik in der Arbeiten von N. E. Shukowski und S. A. Tschaplygin,* no. 5 in the series Naturwissenschaften, Technik und Medizin (Leipzig, 1965), pp. 39–62; and A. A. Kosmodemyansky, "Sergei Alekseevich Chaplygin," in *Lyudi russkoy nauki* ("People of Russian Science"), I (Moscow, 1961), 294–302.

A. T. GRIGORIAN

CHAPMAN, ALVAN WENTWORTH (*b.* Southampton, Massachusetts, 23 September 1809; *d.* Apalachicola, Florida, 6 April 1899), *botany.*

Chapman's *Flora of the Southern United States* (1860) faithfully carried forward Torrey and Gray's plan for a comprehensive flora of the nation. Revisions served until John Kunkel Small's *Flora* (1903).

The youngest of five children born to Paul Chapman, a tanner, and Ruth Pomeroy Chapman, Alvan Chapman graduated B.A. from Amherst College in 1830. He arrived in Savannah, Georgia, in May 1831 and became a tutor on nearby Whitemarsh Island. While president of an academy at Washington, Georgia, he studied medicine under Albert Reese from 1833 until 1835, when he moved to Quincy, Florida; in 1847 he moved to Apalachicola. The place and date of his M.D. degree are obscure, but the University of Louisville awarded him an honorary M.D. in 1846, and the University of North Carolina an LL.D. in 1886. In November 1839 he married Mary Ann Simmons Hancock, who died in 1879. Their daughter, Ruth, died in infancy.

Chapman botanized on the Apalachicola River in 1837 with Hardy Bryan Croom, discoverer of the endemic gymnosperm *Torreya.* The manuscripts left unfinished at Croom's accidental death led Chapman to redouble his botanical activities. His first botanical publication (1845) was a list of plants growing in the vicinity of Quincy.

During the Civil War, Chapman was a "Union man"; his wife was a secessionist. She lived in Marianna, Florida, while he extended aid to runaway slaves and refugees. During guerrilla raids he hid in Trinity Church. He later said of the war years: "I would not have given a sixpence for my life during those four years." Being the only surgeon in town undoubtedly saved him.

Chapman's *Flora* appeared without his knowledge,

Torrey having read proof. It was 1878 before he could publish species he had overlooked. He also assembled three herbaria.

In 1875 Asa Gray spent a week with this "excellent, loyal man all through." Handsome and blue-eyed, Chapman stood over six feet tall and was "even a bit of a dandy." Although increasingly deaf in later years, he wrote with a clear, strong hand to the end.

BIBLIOGRAPHY

I. ORIGINAL WORKS. The *Flora* was first published in New York in 1860. The 2nd ed. (1883) had a seventy-page supplement and was reissued (1892) with a second supplement (pp. 655–703); this printing is scarce. The 3rd ed., completely reset, appeared in 1897. The largest collections of Chapman's letters are preserved in the Torrey correspondence at the New York Botanical Garden, the Gray letters at Harvard University, and the Engelmann letters at the Missouri Botanical Garden. His diary has not been located. He ran a continuing "Professional Notices" in the Apalachicola *Commercial Advertiser* at least during 1847. His "A List of the Plants Growing Spontaneously in the Vicinity of Quincy, Florida," in *Western Journal of Medicine and Surgery,* **3** (1845), 461–483, was also privately printed and repaged. Although Chapman's first name often appears as "Alvin," his own report to Amherst College alumni of Dec. 17, 1872, reads "Alvan."

II. SECONDARY LITERATURE. The various eds. of the *Flora* are discussed by E. D. Merrill, "Unlisted Binomials in Chapman's *Flora of the Southern United States,*" in *Castanea,* **13** (1948), 61–70. Three portraits accompany an obituary by John G. Ruge, who wrote from close friendship, in *Gulf Fauna and Flora Bulletin,* **1,** no. 1 (1899), 1–5. See also William Trelease, "Alvin [*sic*] Wentworth Chapman," in *American Naturalist,* **33** (1899), 643–646, with portrait; and Donald Culross Peattie, "Alvan Wentworth Chapman," in *Dictionary of American Biography,* IV (1930), 16–17. The most perceptive account is Winifred Kimball, "Reminiscences of Alvan Wentworth Chapman," in *Journal of the New York Botanical Garden,* **22** (1921), 1–11. Lloyd H. Shinners provides a critique of Chapman's role in botany in "Evolution of the Gray's and Small's Manual Ranges," in *Sida,* **1** (Nov. 1962), 9–10.

JOSEPH EWAN

CHAPMAN, DAVID LEONARD (*b.* Wells, Norfolk, England, 6 December 1869; *d.* Oxford, England, 17 January 1958), *chemistry.*

Chapman studied at Manchester Grammar School and Christ Church, Oxford, graduating in 1893 with first-class honors in chemistry. For ten years he was on the staff at Owens College, Manchester, where he worked out the "Chapman equation," interpreting theoretically the gaseous explosion rates being measured there by H. B. Dixon.

In 1907 Chapman was elected fellow at Jesus College, Oxford, where he taught, conducted research, and superintended the college laboratory until his retirement in 1944. He opened a new field by being the first to carry out rate measurements on a homogeneous gas reaction, the thermal decomposition of ozone. From this he turned to a lengthy study of the photochemical reaction between hydrogen and chlorine, which had given many workers trouble because of its variable behavior. With his students Chapman showed that the mysterious "induction periods" observed were caused by minute traces of nitrogenous materials in the water used in the apparatus and was able to eliminate them.

Next Chapman examined the kinetics and the inhibiting effect of oxygen. For hydrogen-chlorine mixtures of high purity contained in large vessels, he demonstrated that the reaction chain ended in bimolecular recombination of the carriers (chlorine atoms), by showing that rates were proportional to the square root of the light intensity. With hydrogen-bromine mixtures, which also follow the square-root law, he experimented with light interrupted by a variable-speed sector and was the first to work out the mathematical theory of the rate changes, which made it possible to estimate the lifetime of the bromine or chlorine atoms which acted as chain carriers.

Later, Chapman was interested in catalysis of gas reactions by metals; and during World War II some work for the "Tube Alloy" program (the British program to further development of an atom bomb) was carried out in his laboratory.

He was elected fellow of the Royal Society in 1913.

BIBLIOGRAPHY

A fuller account of Chapman's work and a complete list of his publications (43 papers) is given in *Biographical Memoirs of Fellows of the Royal Society,* **4** (1958), 35.

E. J. BOWEN

CHAPPE D'AUTEROCHE, JEAN-BAPTISTE (*b.* Mauriac, Cantal, France, 23 March 1728; *d.* San José del Cabo, Cape Lucas, Baja California, 1 August 1769), *astronomy.*

An outstanding record at the Collège Louis-le-Grand in Paris brought Chappe to the attention of Jacques Cassini at the Paris observatory. Under the latter's direction he published one of the newly augmented editions of Halley's astronomical tables that had begun to appear in the 1750's.[1]

Chappe's fame rests essentially on his role in the observation of the transits of Venus of 1761 and 1769, but his first important scientific communication was connected with the antecedent though not unrelated transit of Mercury of 1753.[2] Between this event and his election to the Académie des Sciences as adjoint astronomer in 1759, Chappe undertook surveys in Lorraine that involved latitude determinations derived from measurement of the meridian altitudes of selected stars or the sun's limb, and longitude determinations from lunar eclipses and the occultations of stars.[3] Following his appointment to the Academy, he joined the Paris observatory staff, rapidly acquiring a reputation as a skilled observer in close association with J. D. Maraldi, Cassini de Thury, and Lalande.

The twin transits of Venus were the capstone of eighteenth-century observational activity, and Chappe shared in these great events, in the former through his participation in a Siberian winter expedition in 1761 and in the latter through his participation in an expedition to southern California to observe the transit of Venus in 1769. The voyage to Siberia led to studies beyond astronomy and controversies with Empress Catherine II and her spokesmen over the ecological and biological determinants in Russian life and character. Ancillary activity in science was part of the California expedition too, although beyond some minor oceanographic studies (partly inspired by Lavoisier) and an associated natural history of the region around Mexico City, little resulted from Chappe's final effort. However, his astronomical observations entered into the calculations that J. D. Cassini[4] employed in 1772 to arrive at a solar parallax of 8.5″.

Between the years of the two transits, Chappe was also involved in the sea tests of Ferdinand Berthoud's chronometer. Chappe's report on this never appeared in either the *Histoire* or *Mémoires* of the Academy, but was appended afterwards to Berthoud's own work.[5] Observations of the meridian transit of Mercury in 1764, of the eclipses of the satellites of Jupiter in 1760–1764, of the solar eclipses of 1765 and 1766 and the lunar eclipse of 1768, and of the comet of 1766 also occupied Chappe's time between the two great voyages that are the high points of his career. A middling miscellany of other activities in science also marks his achievements, but these are best gleaned from his own works or from the secondary studies listed below.

He died of an unknown epidemic disease that killed all but one of the group sent to California.

NOTES

1. *Tables astronomiques de M. Halley* (Paris, 1752; 2nd ed., Paris, 1754), consisting of the portion of Halley's tables dealing with the sun and moon with full, explanatory notes by Chappe. The

French ed. of the tables dealing with comets and planets came out in 1759, ed. by Lalande.
2. Chappe observed the event from the inner terrace of the Paris observatory. See *Procès verbaux de l'Académie royale des sciences* (1753), fols. 173–179.
3. These constituted his first contributions to the *Mémoires de l'Académie royale des sciences*. See *Mémoires* (1760), pp. 158 ff.
4. *Histoire de l'Académie royale des sciences* (1769), p. 172.
5. F. Berthoud, *Traité des horloges marines* (Paris, 1773), pp. 539 ff.

BIBLIOGRAPHY

I. ORIGINAL WORKS. Further observations are in "Addition au mémoire précédent, sur les remarques qui ont rapport à l'anneau lumineux, et sur le diamètre de Vénus, observé à Tobolsk le 6 Juin 1761," in *Mémoires de l'Académie royale des sciences* (1761), 373–377. His travels to Siberia and California are covered in "Extrait du voyage fait en Sibérie, pour l'observation de Vénus sur le disque du Soleil, faite à Tobolsk le 6 Juin 1761," *ibid.*, 337–372, freely trans. as "Extract from a Journey to Siberia, for Observing the Transit of Venus over the Sun," in *Gentleman's Magazine,* **33** (Nov. 1763), 547–552; "The Same Transit at Tobolsk in Siberia," in *Philosophical Transactions of the Royal Society,* **70** (1761), 254 ff.; *Mémoire du passage de Vénus sur le Soleil; contenant aussi quelques autres observations sur l'astronomie, et la déclinaison de la boussole faites à Tobolsk en Sibérie l'année 1761* (St. Petersburg, 1762); *Voyage en Californie pour l'observation du passage de Vénus sur le disque du Soleil, le 3 Juin 1769; contenant les observations de ce phénomène, et la description historique de la route de l'auteur à travers le Mexique* (Paris, 1772); *A Journey into Siberia Made by Order of the King of France* (London, 1774); *Voyage en Sibérie, fait par ordre du Roi en 1761; contenant les moeurs, les usages des Russes, et l'état actuel de cette puissance; la description géographique et le nivellement de la route de Paris à Tobolsk* (Paris, 1778); and *A Voyage to California to Observe the Transit of Venus* (London, 1778).

II. SECONDARY LITERATURE. On Chappe d'Auteroche or his work, see Angus Armitage, "Chappe d'Auteroche: A Pathfinder for Astronomy," in *Annals of Science,* **10** (1954), 277–293; J. P. G. de Fouchy, "Éloge de M. l'Abbé Chappe," in *Histoire de l'Académie royale des sciences* (1769), 163–172; "Éloge de M. l'Abbé Chappe," in *Nécrologe des hommes célèbres de France par une société de gens de lettres* (Paris, 1771), VI, 133–157; and Harry Woolf, *The Transits of Venus. A Study of Eighteenth-Century Science* (Princeton, 1959).

HARRY WOOLF

CHAPTAL, JEAN ANTOINE (*b.* Nojaret, Lozère, France, 5 June 1756; *d.* Paris, France, 30 July 1832), *applied chemistry.*

Chaptal's parents, Antoine Chaptal and Françoise Brunel, were small landowners. It was, however, his uncle, Claude Chaptal, a wealthy and successful physician at Montpellier, who was to exercise the decisive influence on Chaptal's education.

Chaptal was educated at the *collèges* of Mende and Rodez. In 1774 he enrolled in the Faculty of Medicine at Montpellier; on 5 November 1776 he submitted a thesis for the degree of bachelor of medicine and three months later was received as a doctor of medicine. He persuaded his uncle, who wanted him to go into general practice, to allow him to go to Paris to obtain wider experience. In Paris, Chaptal not only attended courses in medicine but also followed courses in chemistry given by Bucquet, Mitouard, and Sage. His enthusiasm for chemistry led to his appointment to a specially created chair at Montpellier (1780).

Some idea of Chaptal's religious position may be obtained from a political tract published by him in 1790, *Catechisme à l'usage des bons patriotes,* in which he inferred the brotherhood of man from the fatherhood of God. Later Chaptal showed his respect for the work of the church in France. As minister of the interior he effected the return of nuns to hospital work. In 1802 he presented a silver monstrance to the cathedral at Mende, and in 1827 he remembered the parish church where he had been baptized with another generous gift.

In 1781 Chaptal married Anne-Marie Lajard, the daughter of a merchant dealing in cotton. Apart from a substantial dowry, Chaptal received 120,000 francs from his uncle. This gave him considerable independence and, equally important, it gave him capital to invest in the chemical industry that he was to found in the next few years. After a temporary setback at the time of the Revolution, Chaptal soon regained his former fortune. As Bonaparte's minister of the interior he was a rich man; and when he left this post in 1804, he became a member of the Senate, again at a large salary. In 1802 Chaptal bought a magnificent château at Chanteloup in the Loire valley. He spent a considerable sum on the restoration of this property, and in 1808 he claimed that it was fit to receive a king. On the estate he raised sheep, distilled brandy, and experimented with the cultivation of sugar beets. Chaptal spent a considerable part of each year at his country house in the period 1804–1810 and again after Napoleon's final exile in 1815. Later he was obliged to sell his property to pay the large debts incurred by his son in the chemical industry.

As evidence of Chaptal's early interest in medicine we have his membership in the Société Royale de Médecine of Paris (3 October 1781) and one or two later medical memoirs. More important was his membership in the Société Royale des Sciences de Montpellier (from 24 April 1777), where he began to present a succession of memoirs. Most of this early work was of little value and was never published. In 1787, however, a memoir by Chaptal on fermentation was considered by the Montpellier society to be worthy of a prize of 300 livres. By then, though, Chaptal was looking for national recognition, and his work began to appear in the *Mémoires* of the Paris Académie Royale des Sciences. Chaptal was elected a nonresident associate in the chemistry section of the First Class of the Institute on 28 February 1796. On his removal to Paris he was elected as a resident member to a vacancy in the chemistry section on 24 May 1798. He was president of the First Class in 1803–1804. As minister of the interior at this time Chaptal drew up a project for the reform of the Institute, by which the post of permanent secretary was reestablished and each class of the Institute was given greater autonomy. Chaptal was again elected president in 1825 of what was once more known as the Académie des Sciences. He joined the Société Philomatique in 1798 but was never a very active member. Chaptal was a close friend of Berthollet, whose house at Arcueil he often visited. It was probably only in 1810, however, that he joined the famous Society of Arcueil. There was only one society to which Chaptal devoted himself wholeheartedly. This was the Société d'Encouragement pour l'Industrie Nationale, of which he was president from its founding in 1801.

At the beginning of the Revolution, Chaptal showed himself to be a liberal in politics; but in 1793, when extremists gained the upper hand, he was imprisoned for a short time. In December 1799 a consulate was established in France. One of the consuls, Jean Cambacérès, was a close friend of Chaptal and appreciated his abilities, so Chaptal was appointed a councillor of state. On 6 November 1800 Chaptal was made acting minister of the interior, an appointment that was confirmed on 21 January 1801. For four crucial years in the reconstruction of post-revolutionary France Chaptal held the key post in the government. As minister of the interior he was responsible, through the system of prefects, for the general administration of the whole of France. In particular he was responsible for education, religion, public works, customs and excise, theaters, state factories, palaces and museums, hospitals, and prisons. Chaptal exercised patronage generally in the field of industry and commerce.

Chaptal's administration was notable for its advocacy of an expansion of technical education. He had plans for the establishment of a school of dyeing at Lyons, a school for pottery at Sèvres, and a school at Montpellier concerned with distillation and the preparation of acids. He was not able to realize all these projects, but he did bring about the removal of the

École des Mines from Paris to a mining locality in 1802 so that students should have a coordinated program of lectures and practical work. He was able to extend the Conservatoire des Arts et Métiers in Paris into an institution for technical education, including technical drawing, spinning, and weaving. It was Chaptal who was responsible for the first major exhibition of French industry, which was held in Paris in 1801. Even after he resigned from the post of minister of the interior (8 July 1804), Chaptal continued to be available to the French government for technical advice—for example, as a member of the Council for Commerce and Industry. In the final months of Napoleon's rule Chaptal was given special powers to organize conscription. Napoleon's confidence in Chaptal's loyalty and ability was again shown during the Hundred Days, when he was appointed minister of agriculture, commerce, and industry. Under the Restoration, Chaptal at first withdrew completely from public life. On 7 December 1815 William Lee, as United States consul, wrote to the French chemist, inviting him to immigrate to the United States, where the government fully appreciated his talents. Chaptal declined this invitation, as he had an earlier one after the Revolution. In 1818 Chaptal was nominated to the Chambre des Pairs.

In 1780 Chaptal took up his appointment to the new chair of chemistry at Montpellier. In an inaugural lecture to the States of Languedoc, he pointed to the danger of science's becoming a language intelligible only to a small circle of initiates. In the spirit of Diderot, he claimed the need for someone to act as a link between the academies and the people and to explain in simple language the value of science to society. There was a great potential application for chemistry in the south of France. The mines, salt deposits, large grape harvest, and other natural resources were waiting to be exploited. These ideas were repeated in his first book, *Mémoires de chimie,* which, dating from November 1781, was too early a work to constitute a significant contribution to chemistry. In this book he understandably showed his concern with the chemical industry—for example, in a memoir on a means of reducing the consumption of soda in glassworks—but there were also many contributions to pure chemistry.

Chaptal held two further teaching appointments. In 1795 he was *professeur de chimie appliquée aux arts* for a few months at the newly founded École Polytechnique, where he shared a course with Berthollet. He then returned to Montpellier to take up the post of professor of chemistry at the reorganized medical school there.

Chaptal made few original contributions to pure

chemistry, but he was one of the greatest chemical manufacturers of his age. He was always ready to apply the lessons of the chemistry laboratory to the factory. For example, he improved his process for the manufacture of sulfuric acid by applying Lavoisier's new chemical theory based on oxygen. He distinguished three degrees of oxidation in the acid corresponding to sulfurous, sulfuric, and fuming sulfuric acids, respectively. Nevertheless, he compares unfavorably as a man of science with someone like his friend Berthollet. Whereas Berthollet was concerned with establishing general principles and applying them to particular cases, Chaptal wrote as an industrialist with great practical experience, whose concern with the fundamental understanding of nature was subordinate to his interest in controlling chemical reactions. Nevertheless, Chaptal's voice was an important and influential one in advocating the introduction of science into the old craft procedures. On the practical importance of chemistry he wrote:

Chemistry bears the same relation to most of the arts, as the mathematics have to the several parts of the science which depend on their principles. It is possible, no doubt, that works of mechanism may be executed by one who is no mathematician; and so likewise it is possible to dye a beautiful scarlet without being a chemist; but the operations of the mechanic and of the dyer are not the less founded upon invariable principles, the knowledge of which would be of infinite utility to the artist.

We continually hear in manufactories of the caprices and uncertainty of operations; but it appears to me that this vague expression owes its birth to the ignorance of the workman with regard to the true principles of their art. For nature itself does not act with determination and discernment, but obeys invariable laws; and the inanimate substance which we make use of in our manufactures, exhibits necessary effects, in which the will has no part and consequently in which caprices cannot take place. Render yourselves better acquainted with the materials you work upon, we might say to the artists; study more intimately the principles of your art; and you will be able to foresee, to predict, and to calculate every effect. . . .[1]

Chaptal was also an economist who showed a detailed concern for questions of cost, transport, use of chemicals, and the sources of labor to be used. He was also deeply concerned with agriculture, which he described as "the basis of public welfare." He insisted on the utility of chemical knowledge in improving the yield of crops. As minister of the interior he tried by imports to raise the standard of French cattle and sheep. In the production of wine he proposed that in years in which there had been insufficient sun, and fermentation of the grape juice took place only with difficulty, an improvement could be made by adding sugar. This simple but effective method of improving the yield of wine became known as "chaptalization." Chaptal showed that in fermentation it was not necessary to leave the vats open to the air, and by enclosing them he prevented the evaporation of the alcohol. He proposed improvements in the apparatus for distilling alcohol in which heat was retained and fuel consumption was drastically reduced.

The manufacture of acids was a particularly important part of Chaptal's contribution to industry. He sold his first nitric acid and sulfuric acid in 1785, and in 1786 he added spirit of salt. By carefully dissolving hydrogen chloride in water he was able to produce hydrochloric acid successfully and comparatively cheaply, so that, according to Fourcroy, he was able to supply not only towns in France, including Lyons and Paris, but was able to export it to England and Germany.

In January 1786 Chaptal discovered that oil of vitriol could form hexahedral crystals, given suitable conditions of concentration and cold. This memoir was first presented to the Academy of Sciences at Montpellier, which considered it of sufficient merit to submit it to the Paris Academy for publication. In this way it became the first memoir of Chaptal's to be published. In 1787 Chaptal reported on his efforts to manufacture acids of great purity at minimum cost. In 1788 he asked the States of Languedoc for a subsidy to cover further research on the production of alum and mineral acids.

Apart from the mineral acids, Chaptal manufactured large quantities of oxalic acid; sal ammoniac; blue, green, and white vitriol; white lead; and, especially, alum. Alum was a particularly important product, being used not only as a mordant in dyeing but also in tanning and in the preparation of paper and cloth. In 1784 Chaptal discovered a source of alum in Languedoc; and by suitable treatment with air and water, followed by calcination, extraction, and crystallization, he was able for the first time to produce alum in reasonable quantities on French soil. Manufacturers continued to prefer alum of Tolfa, imported from Italy, but Chaptal and others were later able to show by chemical analysis that alum produced in France need not be inferior.

Although Lavoisier had suggested in 1780 that alum was not a simple double salt, this possibility was overlooked. Chaptal, for example, writing in 1790, considered alum as the sulfate of alumina but said that alkali had to be added to neutralize the excess of acid which would otherwise prevent the formation of crystals. Berthollet, in his *Élémens de l'art de la teinture* (1791), reported the opinion of François

Descroizilles that the sulfate of potash and the sulfate of alumina enter into the composition of alum in some way. It was left, however, to Vauquelin and Chaptal independently in 1797 to conclude that alum was a triple salt. In his textbook of 1790 Chaptal was concerned with the composition of several salts. He correctly stated that fuming liquor of Libavius (stannic chloride) is a "muriate" in which the acid is in the state of "oxymuriatic acid." He was concerned with the impurity of many salts sold in France, notably Epsom salts and antimony compounds. Because of the dangers involved in the medicinal use of the latter he advocated governmental inspection.

Although Chaptal was not in Paris at the time that Lavoisier collaborated with Guyton de Morveau, Berthollet, and Fourcroy in the reform of chemical nomenclature (1787), he was able to suggest a simple improvement in the new nomenclature by the logical substitution of *nitrogène* (1790) for the *azote* of the Paris chemists. About this time Chaptal became convinced of the superiority of the new oxygen theory, and in 1791 Lavoisier wrote to him to express his satisfaction at having won him over from the old phlogiston theory. Chaptal was happy to testify to the improvement the oxygen theory made not only in theoretical chemistry but also in the practical chemistry of which he was such an influential exponent: "It is this doctrine alone which has led me to simplify most of the processes, to bring some of them to perfection, and to rectify all my ideas."[2]

Chaptal made few contributions to animal chemistry, although it should be remembered that he had first embarked upon a study of chemistry through medicine. Because of his medical qualifications it is all the more interesting to examine his ideas on vitalism, a subject to which he devoted several pages in one of his technical works. He summed up his views as follows:

> Chemistry in its application to living bodies may therefore be considered as a science which furnishes new means of observation, and permits us to verify the results of vitality by the analysis of its products. But let us beware of intermeddling in the peculiar province of vitality. Chemical affinity is there blended with the vital laws which defy the power of art.[3]

Chaptal carried out work on the extraction of saltpeter, being one of many French chemists called upon in 1793–1794 to apply their knowledge to national defense. In the southwest of France he organized the extraction of saltpeter from suitable natural sources of decaying organic matter. He also insisted on the construction of artificial niter beds to safeguard future supplies. Equally useful to the national economy was

Chaptal's work on bleaching. The use of "oxymuriatic acid" (chlorine) to bleach cotton had been discovered by Berthollet. In a memoir published in 1787 Chaptal extended this process to the treatment of paper. He proposed the use of chlorine for treating old paper and rags so that they could be used in the manufacture of white writing paper, previously made exclusively from unprinted linen. He suggested that chlorine could also be used in the restoration of faded books and prints and in the removal of ink stains. Chaptal later favored the use of chlorine as a disinfectant—for instance, in prisons—and suggested that it was superior to the hydrogen chloride favored by his colleague Guyton de Morveau. He also introduced the method of bleaching by soaking the material in alkali and then treating it with steam.

Chlorine was prepared from hydrochloric acid, of which Chaptal was a major manufacturer. After his resignation as minister of the interior in 1804 he was again able to take a close interest in his chemical factories. The one at Ternes, near Paris, of which he made Darcet manager, produced oxalic acid, tartaric acid, corrosive sublimate, potassium and sodium arsenates, copper sulfate, tin and lead salts, alcohol, ether, and ammonia. Soda was refined there and sulfuric acid was concentrated in platinum vessels. According to Pigeire, under the Empire he had at his factory at La Folie, near Nanterre, 150 workmen engaged in the annual production of about 600 tons of hydrochloric acid, 400 tons of sulfuric acid, sixty tons of nitric acid, 1,200 tons of crude soda, 400 tons of potassium carbonate, sixty tons of potassium sulfate, and 600 tons of alum. Chaptal's third factory was near Martigues, on the Mediterranean coast of France, and here soda was one of the main products.

Chaptal, a constant advocate of France's self-sufficiency, was an early enthusiast of the possibility of replacing cane sugar with beet sugar. In 1811 he was a member of a committee appointed by the First Class of the Institute to examine the possible production of beet sugar. It was not until 1815, when the end of the war permitted the resumption of trade with the West Indies and threatened the ruin of the sugar beet industry, that Chaptal presented a memoir on the subject to the Institute. He was anxious to show that the industry, if efficiently run, could justify itself economically. It was largely due to his efforts, with the later support of Thenard, that this industry continued to function in France.

In his book *Chimie appliquée aux arts* (1807), Chaptal warned the industrialist of the danger, on the one hand, of ignoring the potential help of science. On the other hand, Chaptal warned him not to found a new factory just on the basis of a laboratory experiment—suitable large-scale trials were necessary. The

chemist could be no more than a consultant. It was the manufacturer who knew the market and who must decide to what extent and in what ways science was to be applied. A manufacturer who found his goods undercut by a competitor should not ask for government regulations to protect himself. Nor should the manufacturer be protected by tariffs on imported goods, "for the manufacturer does not strive to improve, unless he has before him articles of a better or more economical manufacture than his own." Chaptal also discussed the location of industry and the supply of labor.

NOTES

1. *Elements of Chemistry,* W. Nicholson, trans., 2nd ed. (London, 1795), I, xliii.
2. *Ibid.,* p. iv.
3. *Chemistry Applied to Arts and Manufactures,* W. Nicholson, trans. (London, 1807), I, 50.

BIBLIOGRAPHY

1. ORIGINAL WORKS. Chaptal's articles include "Observations sur la crystallisation de l'huile de vitriol," in *Mémoires de l'Académie royale des sciences* (Paris) (1784), 622–630; "Observations sur l'acide muriatique oxigéné," *ibid.* (1787), 611–616; "Observations sur la manière de former l'alun par la combinaison directe de ses principes constituans," *ibid.* (1788), 768–777; "Sur les moyens de fabriquer de la bonne poterie à Montpellier, et sur un vernis qu'on peut employer pour les enduire," in *Annales de chimie,* **2** (1789), 73–85; "Sur quelques phénomènes que nous présente la combustion du soufre," *ibid.,* 86–91; "Sur les caves et le fromage de Roquefort," *ibid.,* **4** (1790), 31–61; "Instruction sur un nouveau procédé pour le raffinage du salpêtre," in *Journal de physique,* **45** (1794), 146–152, written with Champy and Bonjour; "Observations sur le savon de laine et sur ses usages dans les arts," in *Mémoires de l'Institut,* **1** (1796), 93–101; "Analyse comparée des quatre principales sortes d'alun connues dans le commerce; et observations sur leur nature et leur usage," in *Journal des mines,* **5** (1796–1797), 445–456; "Sur la production artificielle du froid," in *Annales de chimie,* **22** (1797), 280–296; "Vues générales sur la formation du salpêtre," *ibid.,* **20** (1797), 308–355; "Considérations chimiques sur l'effet des mordans dans la teinture en rouge du coton," *ibid.,* **26** (1798), 251–258; "Considérations chimiques sur l'usage des oxides de fer dans la teinture du coton," *ibid.,* 266–277; "Sur la nécessité et les moyens de cultiver la barille en France," *ibid.,* 178–187; "Observations chimiques sur la couleur jaune qu'on extrait des végétaux," in *Mémoires de l'Institut,* **2** (1798), 507–516; "Sur la manière dont on fertilise les montagnes dans les Cevennes," in *Annales de chimie,* **31** (1799), 41–47; "Essai sur le perfectionnement des arts chimiques en France," in *Journal de physique,* **50** (1800), 217–233; "Mémoire sur le vin," *ibid.,* **51** (1800),

133–149; "Notice sur une nouvelle méthode de blanchir le coton," *ibid.,* 305–309; "Rapport des expériences sur le sucre contenu dans la betterave," *ibid.,* 371–389; "Sur les vins," in *Annales de chimie,* **35** (1800), 240–299; **36** (1800), 3–50, 113–142, 225–258; **37** (1800), 3–37; "Notice sur un nouveau moyen de blanchìr le linge dans nos ménages," *ibid.,* **38** (1801), 291–296; "Vues générales sur l'action des terres dans la végétation," in *Mémoires de la Société d'agriculture de la Seine,* **4** (1802), 5–14; "Rapport sur la question de savoir si les manufactures qui exhalent une odeur désagréable peuvent être nuisibles à la santé!," in *Annales de chimie,* **54** (1805), 86–103, written with Guyton de Morveau; "Rapport sur deux mémoires de M. Gratien Lepère, relatifs aux pouzzolanes naturelles et artificielles," *ibid.,* **64** (1807), 273–285; "Notice sur quelques couleurs trouvées à Pompeïa," in *Mémoires de l'Institut* (1808), 229–235; "Observations sur la distillation des vins," *ibid.,* 170–194; "Mémoire sur le sucre de betterave," in *Annales de chimie,* **95** (1815), 233–293; n.s. **7** (1817), 191–193; and "Recherches sur la peinture encaustique des anciens," *ibid.,* **93** (1815), 298–313.

Chaptal's books are *Mémoires de chimie* (Montpellier, 1781); *Éléments de chimie,* 3 vols. (Montpellier, 1790; Paris, 1794, 1796, 1803), also trans. into English by W. Nicholson as *Elements of Chemistry,* 3 vols. (London, 1791, 1795, 1800, 1803; Philadelphia, 1796; Boston, 1806); *L'art de faire, gouverner et perfectionner le vin* (Paris, 1801; repr. 1807, 1811; 2nd ed., 1819; 3rd ed., 1839), also trans. into Italian (1812–1813); *Art de la teinture du coton en rouge* (Paris, 1807); *Chimie appliquée aux arts,* 4 vols. (Paris, 1807; 2nd ed., Brussels, 1830), trans. as *Chemistry Applied to Arts and Manufactures,* 4 vols. (London, 1807) and as *Die Chemie in ihrer Anwendung auf Künste,* 2 vols. (Berlin, 1808); *Principes chimiques sur l'art du teinturier-dégraisseur* (Paris, 1808); *De l'industrie française,* 2 vols. (Paris, 1819); and *Chimie appliquée à l'agriculture,* 2 vols. (Paris, 1823, 1829), trans. by W. P. Page as *Chymistry Applied to Agriculture* (Boston, 1835, 1838; New York, 1840).

II. SECONDARY LITERATURE. On Chaptal or his work see J. A. Chaptal, *La vie et l'oeuvre de Chaptal. Mémoires personnels rédigés par lui-même de 1756 à 1804. Continués, d'après ses notes, par son arrière-petit-fils jusqu'en 1832* (Paris, 1893); M. P. Crosland, *The Society of Arcueil. A View of French Science at the Time of Napoleon I* (Cambridge, Mass., 1967), *passim;* P. Flourens, "Éloge historique de Jean Antoine Chaptal" (read at public meeting of 28 Dec. 1835), in *Mémoires de l'Académie des sciences de l'Institut,* **15** (1838), 1–39; J. M. de Gerando, "Notice sur Chaptal" (read at general meeting of the Société d'encouragement, 22 Aug. 1832); J. Pigeire, *La vie et l'oeuvre de Chaptal (1756–1832)* (Paris, 1932); and R. Tresse, "J. A. Chaptal et l'enseignement technique de 1800 à 1819," in *Revue d'histoire des sciences,* **10** (1957), 167–174.

M. P. CROSLAND

CHARCOT, JEAN-BAPTISTE (*b.* Neuilly-sur-Seine, France, 15 July 1867; *d.* at sea, 16 September 1936), *medicine, navigation, geography, oceanography, exploration.*

The son of the renowned doctor and professor at the Faculté de Médecine, J.-M. Charcot, he chose a medical career. A brilliant student at the Faculté de Médecine, he was an *interne* at the Hôpitaux de Paris, received the M.D. degree at the Faculté de Médecine in 1895, was chief of its neurological clinic, and was associated with the Institut Pasteur.

Between 1887 and 1901 Charcot published works devoted to neurology, on such topics as various forms of epilepsy, tuberculosis of the paracentral region, motor agraphia, and Benedikt's syndrome. In the textbook *Manuel de médecine* he wrote articles on aphasia and lead poisoning.

Yet Charcot abandoned his medical career to devote himself to the sea. He admitted that his medical knowledge was subsequently very useful to him and that he was able to pursue his maritime explorations and researches ". . . thanks to the education and the training in the scientific method that I had the good fortune to undergo at the harsh and unyielding school of my father, Professor Charcot, and at the Institut Pasteur, from my teachers and masters, Dr. Roux and Professor Metchnikoff" (*Notice sur les titres* [1921], p. 3).

Accustomed to handling a boat since his childhood, Charcot trained himself in preparation for the explorations that he undertook beginning in 1903. He had already sailed on a small schooner in the latitudes of the Faeroe Islands in 1901; and on a sailing ship, he set out in 1902 for the Faeroes and went as far as the polar island of Jan Mayen. He brought back valuable information on the whale fisheries, on the hydrography of the region, and on the physical conditions of the water in the vicinity of the ice floes.

Voyages of exploration and scientific expeditions followed these preparatory missions, and Charcot proved his ability as an organizer and leader. He had constructed, according to his directions and plans, the *Français* and the *Pourquoi Pas?,* the first polar vessels to be built in France. The *Pourquoi Pas?* was both the model for ships built for scientific polar exploration and a research ship for all latitudes. From 1911 the *Pourquoi Pas?,* with Charcot as its director, was considered the floating marine research laboratory of the École Pratique des Hautes Études and was often used by the Muséum National d'Histoire Naturelle. The *Pourquoi Pas?* was also used as a school for candidates for the certificate of competence as captain of a merchant vessel.

The cruises, always under Charcot's command, included physicists, oceanographers, biologists, and geologists, both French and foreign. Among them were the first French Antarctic expedition, aboard the *Français* (1903–1905); the second French Antarctic expedition, aboard the *Pourquoi Pas?* (1908–1910);

and missions in the Atlantic, the Arctic polar regions, and the Gulf of Gascony (1912, 1913, 1914, 1920); the North Atlantic (1921, 1922); the western Mediterranean and the Gulf of Gabès (1923); the English Channel, the Irish Sea, and the western British archipelago (1924); the waters around Jan Mayen Island and Jameson Land in Greenland, the Scoresby Sound (1925, 1926); the North Sea and the Baltic (1927, 1930); and the waters around Greenland (1928, 1929, 1932). In 1933 he explored the coast of Blosseville Kyst, and in 1934 he led back to Angmassalik the Victor mission (which had to winter in Greenland). In 1936 he returned to Greenland to find the Victor mission again to bring it back to France. It was his last voyage. On the morning of 16 September 1936 the *Pourquoi Pas?*, buffeted for twelve hours by a storm, foundered on the Borgarfjord reefs off Iceland. The sole survivor, the master helmsman, wrote an account of the shipwreck which appeared in several newspapers (*Le Temps,* 18 Sept. 1936; *Paris Midi,* 17 Sept. 1936; *Le Petit Parisien,* 18 Sept. 1936).

Funeral services were held for the twenty-two scientists and sailors lost at sea at St. Malo on 11 October 1936 and at Paris on the following day. Homage was paid to Charcot and his companions at the great amphitheater of the Sorbonne on 25 November 1936.

During the expeditions work was done in hydrography, meteorology, atmospheric electricity, gravitation and terrestrial magnetism, actinometry, the chemistry of the atmosphere, the tides, zoology, botany, geology and mineralogy, glaciology, and bacteriology. New lands were discovered; the problem of the South American Antarctic was resolved; and the collections and documents brought back were both numerous and original. Two new sciences were created: submarine geology, with the establishment of marine geological maps, and geological oceanography. Charcot's organization of systematic and practical oceanographic studies, a new field for France, had been inspired by the work of Albert I of Monaco.

After World War I, Charcot was greatly concerned with the application of oceanography to commercial fishing. He contributed to the creation and development of the Office Scientifique et Technique des Pêches, and in 1923 he published the first French fishing map of the North Sea.

In 1932 Charcot led a "Polar Year" expedition to Greenland that remained there for a year, the first French polar expedition to use winter quarters. Charcot was also instrumental in French participation in the exploration of Antarctica.

All those who sailed on the *Pourquoi Pas?* were impressed with Charcot's energy, courage, refined and profound sensibility, total unselfishness, great kindness, youthful spirit, and simplicity. He was a member of the Académie des Sciences (1926), the Académie de Médecine, and the Académie de Marine. He was a grand officer of the Legion of Honor and an officer of several foreign orders. In addition, he received the gold medals given by the geographical societies of Paris, London, New York, Brussels, Antwerp, and St. Petersburg. The Académie des Sciences awarded him the first biennial prize of the Prince of Monaco.

BIBLIOGRAPHY

I. ORIGINAL WORKS. Charcot's articles include "Tuberculose de la région paracentrale. Fréquence et raisons anatomiques de cette localisation," in *Bulletin de la Société anatomique* (8 May 1891), p. 274; "Coup de feu dans l'oreille. Paralysie faciale. Hémiplégie. Obstruction de la carotide interne," *ibid.* (18 Dec. 1891), p. 679; "Sur un appareil destiné à évoquer les images motrices graphiques chez les sujets atteints de cécité verbale. Application à la démonstration d'un centre moteur graphique fonctionnellement distinct," in *Progrès médical,* **1** (1892), 478; and *Comptes rendus. Société biologique* (11 June 1892), p. 235; "Sur un cas d'agraphie motrice, suivi d'autopsie," in *Comptes rendus. Société biologique* (1 July 1893), p. 129; "Trois cas d'arthropathie tabétique bilatérale et symétrique," in *Nouvelle iconographie de la Salpêtrière* (1894), p. 221; "Contribution à l'étude de l'atrophie musculaire progressive (type Duchenne-Aran)," in *Archives de médecine expérimentale et d'anatomie pathologique* (1 July 1895), p. 441; "Quelques observations du trouble de la marche. Dysbasies d'origine nerveuse," in *Archives de neurologie* (Feb. 1895), p. 81; "De l'hémarthrose tabétique et de deux symptômes rares au cours du tabes dorsualis," in *Nouvelle iconographie de la Salpêtrière* (1896), p. 265; "Une cause nouvelle d'intoxication saturnine," in *Comptes rendus. Société biologique* (20 June 1896), p. 639; "Rapport du Dr. J. B. Charcot sur le voyage de Punta Arenas à l'île Déception," in *Géographie,* **19** (1909), 279–281; "The Second French Antarctic Expedition," in *Geographical Journal* (Mar. 1911); and *Scottish Geographical Magazine,* **27** (Mar. 1911); "Quelques considérations sur le 2ème expédition antarctique française," in *Revue scientifique* (10 June 1911); "L'expédition antarctique française 1908–1910," in *Géographie,* **23** (1911), 5–16; "Le laboratoire des recherches maritimes scientifique du *Pourquoi Pas?,*" in *Comptes rendus hebdomadaires des séances de l'Académie des sciences* (20 Nov. 1911); "Au sujet de l'île Jan-Mayen," *ibid.* (14 Mar. 1921); "Les missions du *Pourquoi Pas?,*" in *Bulletin de la Société d'océanographie de France,* nos. 2, 3, ff. (1921); "Sur un appareil peu coûteux destiné à permettre aux pêcheurs de prendre la température de l'eau de mer entre la surface et 100 mètres" in *VI° Congrès national des pêches maritimes à Tunis,* I (1921), 23; "L'îlot de Rockall," in *Nature* (Paris), no. 2478 (1 Oct. 1921); "Sur l'étude géologique du fond de la Manche," in *Comptes rendus hebdomadaires des séances de l'Académie des sciences*

(13 Nov. 1922); and "Rapport preliminaire sur la campagne du *Pourquoi Pas?* en 1922," in *Annales hydrographiques*, no. 1880 (1922).

His books include *Le "Francais" au Pôle Sud, 1903–1905* (Paris, 1906); *Pourquoi faut-il aller dans l'Antarctique* (Paris, 1907); *Le "Pourquoi Pas?" dans l'Antarctique, 1908–1910* (Paris, 1910); *Autour du Pôle Sud,* 2 vols. (Paris, 1912); *Notice sur les titres et travaux scientifiques du Dr. J. B. Charcot* (Paris, 1921, 1923, 1926, 1929); *Christophe Colomb vu par un marin* (Paris, 1928); and *La mer du Groendland* (Paris, 1929).

Publications resulting from various cruises include *Expédition antarctique française 1903–1905. Sciences naturelles, documents scientifiques,* 17 fasc. (Paris, n.d.); *Expédition antarctique française 1903–1905. Hydrographie et physique du globe* (Paris, n.d.); *Rapports préliminaires sur les travaux exécutés dans l'Antarctique par la mission commandée par le Dr. J. B. Charcot de 1908 à 1910* (Paris, 1910); and *2ème expédition antarctique française 1908–1910. Sciences naturelles et sciences physiques. Documents scientifiques,* 25 fasc. (Paris, n.d.). Many scientific notes appeared in specialized periodicals.

II. SECONDARY LITERATURE. Among the many tributes to Charcot are "Hommages au Commandant Charcot et aux victimes du naufrage du *Pourquoi Pas?*," in *La géographie,* **66** (Nov. 1936), 201–226, with an account of the funeral services; "Hommage national à J. Charcot et à ses compagnons," in *Bulletin du Muséum d'histoire naturelle,* 2nd ser., **8** (1936), 449–472; *Cérémonie commemorative en hommage à J. B. Charcot à la Sorbonne le 2 oct. 1937,* Institut de France, L'Académie des sciences, fasc. 23 bis (Paris, 1937); L. Dangeard, "Charcot et son oeuvre: Les croisières du *Pourquoi Pas?*," in *Geologie der Meere und Binnengewässer,* I (1937), 361–373; "30ème anniversaire de la mort de Charcot," in *Comptes rendus hebdomadaires des séances de l'Académie des sciences* (21 Nov. 1966), p. 250; and "Le centenaire de la naissance du Commandant Charcot célébré à la Sorbonne le 22 novembre 1967," in *Annales de l'Université de Paris,* **38** (1968), summary account on p. 250.

ANDRÉE TÉTRY

CHARCOT, JEAN-MARTIN (*b.* Paris, France, 29 November 1825; *d.* Lake Settons, Nièvre, France, 16 August 1893), *medicine, neurology.*

Charcot was the son of a wheelwright. After studying at the Faculty of Medicine of Paris, he interned in the Hôpitaux in 1848. He was awarded the M.D. in 1853 upon presentation of an outstanding doctoral thesis on gout and chronic rheumatism. In 1860 he was *agrégé;* in 1862 he became resident doctor at the Salpêtrière, where he created a major neurological department. Charcot was appointed professor of anatomical pathology at the Paris Faculty of Medicine in 1872, and in 1882 he accepted the chair for the study of nervous disorders at the Salpêtrière clinic, with Joseph Babinski as his director.

During the next twenty years Charcot published a series of memoirs that attracted wide attention among neurologists. At this time, he also practiced extensively the clinical anatomical method that correlated the symptoms observed in the sick patient and the lesions discovered at the time of autopsy. Through these he proved that the cells of the dorsal horn of the spinal cord possess certain trophic properties and then analyzed the lesions found in these cells as a result of infantile paralysis. In collaboration with C. J. Bouchard he studied the secondary degeneration of the spinal cord.

Charcot then described the neurogenic arthropathies popularly known as Charcot's disease; in later studies he delineated the difference between patchy sclerosis and the effects of Parkinson's disease. He noted that the pseudohypertrophic paralysis of Duchenne is quite different from myelopathic muscular atrophy; he recorded the history of lateral amyotrophic sclerosis (also known as Charcot's disease); and he discovered a progressive neuropathic muscular atrophy, later named Charcot-Marie (after Pierre Marie) amyotrophy.

Charcot vigorously supported and defended the theory of cerebral localizations in man; several of his outstanding courses dealt with this theory and its application to Jacksonian epilepsy, aphasia, and Beard's neurasthenia. In 1872 he initiated work on hysteria and hysterical hemianesthesia, and on the link between traumatism and local hysteria. At the time of his sudden death he was engaged in work on hysteria, convulsive attacks in hysterical patients, and hypnotic therapy. Indeed, Charcot must be considered one of the first to demonstrate the clear and fruitful relationship between psychology and physiology (his work on hysteria stimulated Freud's investigations) as well as, with Duchenne, one of the fathers of modern neurology.

BIBLIOGRAPHY

Charcot's most influential works include *Leçons sur les maladies du système nerveux,* 5 vols. (Paris, 1872–1893) and *Leçons du mardi à la Salpêtrière,* 2 vols. (Paris, 1889–1890). His works are collected in *Oeuvres complètes,* 9 vols. (Paris, 1886–1890).

A detailed biography, with a complete bibliography, is G. Guillain, *J.-M. Charcot (1825–1893) La vie, son oeuvre* (Paris, 1955). See also H. Meige, "Charcot artiste. Ceremonie du centenaire de la naissance de J. Charcot à la Sorbonne," *Notice de l'Académie des sciences,* no. 14 (Paris, 1925).

ANDRÉE TÉTRY

CHARDENON, JEAN PIERRE (*b.* Dijon, France, *ca.* 22 July 1714 [date of baptism]; *d.* Dijon, 16 March 1769), *chemistry, medicine.*

The son of Guillaume Chardenon and Marguerite Canquoin, Chardenon became a surgeon in Paris, but poor health led him to adopt the less arduous profession of physician. He practiced in Dijon, and in 1744, four years after its foundation, he was elected to the Académie des Sciences, Arts et Belles-Lettres de Dijon, in which he served as scientific secretary from 1752 to about 1762. He read papers on medical and surgical topics and was interested in the chemical aspects of medicine.

While discussing the way in which fat was formed in animals, Chardenon considered the nature of oil. Some early chemists had thought that oil was a chemical element or principle, since it was found in all three kingdoms of nature, but Chardenon refuted this by showing that it could be decomposed by heating into acid, water, phlogiston (some of which was in the charcoal that remained in the residue after combustion), and earth. These four substances could not be recombined by affinity (*rapport*), and they must therefore have originally been combined by an agency (*action*), which could only have operated in the vegetable and animal kingdoms—an indication that Chardenon believed in some kind of vitalism. Mineral oil, he said, must have been formed by the decay of vegetable matter below the surface of the earth.

In 1761 Chardenon discussed the way in which mercury compounds acted as poisons and medicaments. He concluded that they affected bodily humors because of the high density of mercury, for it had been demonstrated in physics that the action of one body on another was related to its mass.

This Newtonian approach to certain phenomena is also apparent in Chardenon's discussion of the cause of the weight increase of metals on calcination, deposited with the Dijon Academy in a sealed note on 6 August 1762 and read at several meetings in 1763 and 1764. He criticized two earlier theories: one, credited to Boyle, that the increase was caused by absorption of fire particles; the other, advanced by Laurent Béraut in 1747, that "salts" from the air were deposited on the metal. Chardenon believed that the only change during calcination was the loss of phlogiston, and he explained the gain in weight by assuming that phlogiston was specifically lighter than air and that a metal weighed in air would therefore become heavier when it lost its phlogiston, just as a fishing net suspended by cork, which is specifically lighter than water, would become heavier if the cork were detached. This naïve theory took no account of the volume of the cork or of the relative volumes of

the metal and the calx, and it was further complicated by Chardenon's belief that gravity did not act uniformly on all kinds of matter.

The theory implied that the gain in weight of a metal on calcination would be proportional to its phlogiston content, and Chardenon intended to test this by measuring the amount of phlogiston, in the form of charcoal, required to prepare each metal from its calx. He died before he could do this, but his work was continued by Guyton de Morveau, his colleague in the Dijon Academy. Chardenon's theory was published but received little attention; Guyton's revised version, however, was more widely read and criticized.

BIBLIOGRAPHY

I. ORIGINAL WORKS. Chardenon's only works to be published in his lifetime were an abstract of his memoir on calcination, "Extrait de la séance publique de l'Académie de Dijon, le 9 décembre 1764," in *Mercure de France* (July 1765), pt. 2, pp. 127–134; and a further criticism of Béraut's theory, "Lettre de M. Chardenon sur l'augmentation de poids des matières calcinées," in *Journal des Sçavans* (1768), pp. 648–658. The complete memoir was published posthumously as "Mémoire sur l'augmentation de poids des métaux calcinés," in *Mémoires de l'Académie de Dijon,* **1** (1769), 303–320. A memorial lecture and abstracts of three other papers by Chardenon were also published: "Éloge de Mr. [Jean-Baptiste] Fromageot [1724–1753]," *ibid.,* cxiii–cxxix; "Usage des énervations des muscles droits du bas-ventre," *ibid.,* lxxxiv–lxxxvi; "Sur les huiles," *ibid.,* **2** (1774), ix–xiv; and "Sur les noyés," *ibid.,* lv–lvi.

The archives of the Dijon Academy, now in the Archives départementales de la Côte-d'Or, Dijon, contain in dossier no. 123 correspondence concerning Chardenon's theory of calcination, as well as the manuscripts of the complete text of "Sur les huiles" and an unpublished "Analise d'une résidence trouvée dans un vase qui contenoit de l'eau de neige évaporée à l'air libre." Abstracts of two unpublished papers are included in the minute books: "Réflexions sur les poisons et particulièrement sur le sublimé corrosif," in Registre II (26 June 1761), ff. 18r–18v; and "Essai sur les embarras du canal intestinal, et sur les moyens de les détruire ou d'en éloigner les funestes effets," Registre II (10 August 1764), ff. 156v–157v. Other papers are recorded in the minute books only by their titles, with no details of their contents.

II. SECONDARY LITERATURE. There are several references to Chardenon in H. Maret, "Histoire de l'Académie des sciences, arts et belles lettres de Dijon," in *Mémoires de l'Académie de Dijon,* **1** (1769), i–xli; his early career is mentioned in Richard de Ruffey, *Histoire secrète de l'Académie de Dijon (de 1741 à 1770),* M. Lange, ed. (Paris, 1909), pp. 106–107. Chardenon's theory of calcination is discussed in detail by J. R. Partington and D. McKie in "Historical Studies on the Phlogiston Theory. Part I," in *Annals of*

Science, 2 (1937), 373–379. They give the dates of his baptism and death in a note at the end of "Historical Studies on the Phlogiston Theory. Part II," in *Annals of Science,* 3 (1938), 58; on the death certificate he was described as a doctor of medicine of Montpellier, but the University of Montpellier has no record of the dates of his attendance or graduation.

W. A. SMEATON

CHARDONNET, LOUIS-MARIE-HILAIRE BERNIGAUD, COMTE DE (*b.* Besançon, France, 1 May 1839; *d.* Paris, France, 12 March 1924), *cellulose technology.*

Although Chardonnet was not the first to think of the possibility of making "artificial silk," he was the pioneer of the technology necessary to establish the industry. The influences that prepared him for this, his most important contribution to science and technology, were his training as an engineer at the École Polytechnique and the work of his illustrious fellow townsman Louis Pasteur. In particular, Pasteur's preoccupation at one time with a disease of the silkworm led Chardonnet, as he later said, to think of "imitating as closely as possible the work of the silkworm." As a result he devised a process in which a solution of cellulose nitrate was extruded through very fine glass capillaries to form continuous filaments. He applied for his first patent and submitted his memoir entitled "Une matière textile artificielle ressemblant à la soie" to the Academy of Sciences in 1884. After another five years, during which he was much concerned to reduce the flammability of his material, he felt his process and product were sufficiently developed for public display. In 1889 he was awarded the grand prize at the Paris Exposition.

In the same year he established with the support of capitalists of his hometown, the Société de la Soie de Chardonnet at Besançon and another factory in Satvar, Hungary, in 1904. At the turn of the century the success of his company stimulated investigation of alternative materials and methods for the manufacture of "artificial silk" (later known as rayon), and these subsequently led to the replacement of Chardonnet silk. Chardonnet made some minor contributions in other scientific fields, including studies of the absorption of ultraviolet light, telephony, and the behavior of the eyes of birds. He published no major work but presented all his researches in the *Comptes rendus* of the Academy of Sciences.

BIBLIOGRAPHY

Chardonnet's address given when he received the Perkin Medal of the Society of Dyers and Colourists is in *Journal of the Society of Dyers and Colourists,* 30 (1914), 176–179.

Secondary literature includes the *Dictionnaire de biographie française,* VI, col. 120; and the obituary of Chardonnet by the president of the Academy of Sciences, in *Comptes rendus hebdomadaires des séances de l'Académie des sciences,* 178 (1924), 977–978.

M. KAUFMAN

CHARLES, JACQUES-ALEXANDRE-CÉSAR (*b.* Beaugency, France, 12 November 1746; *d.* Paris, France, 7 April 1823), *experimental physics.*

Almost nothing is known of Charles's family or of his upbringing, except that he received a liberal, nonscientific education. As a young man he came to Paris, where he was employed as a petty functionary in the bureau of finances. In a period of governmental austerity, Charles was discharged from this position, and owing to the pervasive influence of Franklin (who was visiting France in 1779), he set about learning the elements of nonmathematical, experimental physics. In 1781, after only eighteen months of study, Charles began giving a public course of lectures which, because of the eloquence of his discourse and the variety and precision of his experimental demonstrations, soon attracted a wide audience of notable patrons.

Charles was named a resident member of the Académie des Sciences on 20 November 1795. He was professor of experimental physics at the Conservatoire des Arts et Métiers, librarian of the Institute, and, from 1816, president of the Class of Experimental Physics at the Academy.

In 1804, Charles married Julie-Françoise Bouchard des Hérettes, an attractive young lady who achieved notoriety through her intimate friendship with the poet Lamartine. She died in 1817 after a long illness.

Charles was known to his contemporaries primarily through his contributions to the science of aerostation (ballooning). Shortly after the famous balloon experiments of the Montgolfier brothers, Charles conceived the idea of using hydrogen ("inflammable air") instead of hot air as a medium of displacement. With the aid of a pair of clever Parisian artisans, the brothers Robert, Charles developed nearly all the essentials of modern balloon design. He invented the valve line (to enable the aeronaut to release gas at will for a descent), the appendix (an open tube through which expanded gas could freely escape, thus preventing rupture of the balloon sack), and the nacelle (a wicker basket suspended by a network of ropes covering the balloon and held in place by a wooden hoop). To prevent the subtle hydrogen gas from escaping through the balloon, Charles covered the taffeta sack with an impervious mixture of rubber dissolved in turpentine.

On 1 December 1783 Charles and the elder Robert ascended from Paris in their newly constructed *Charlière*. (An unmanned trial balloon had been successfully launched in August.) The two aeronauts landed in a small village twenty-seven miles from Paris, and Charles continued the voyage on his own three miles farther. The king, who was at first opposed to this dangerous experiment, afterward granted Charles lodgings in the Louvre.

Charles published almost nothing of significance. The law which bears his name was discovered by him in about 1787, but it was first made public by Gay-Lussac (and at about the same time by Dalton). In his article on the expansion of gases by heat, Gay-Lussac described, criticized, and considerably improved upon Charles's experimental procedure.

Apart from his experiments on gaseous expansion and his contributions to aerostation, Charles's achievements in science were relatively minor. He is usually attributed with the invention of the megascope, a device to magnify large objects. He also made an improved hydrometer and invented a goniometer for measuring the angles of crystals.

Assertions to the contrary notwithstanding, there is no evidence that Charles knew anything but the rudiments of mathematics. Through an unfortunate confusion of names, biographers and bibliographers have completely confounded J.-A.-C. Charles with another contemporary known only as Charles le Géomètre. This obscure mathematician—whose first names, date, and place of birth are unknown—was elected *associé géomètre* of the Academy in 1785; most biographers of J.-A.-C. Charles have falsely asserted that he entered the Academy in this year. Charles le Géomètre was royal teacher of hydrodynamics, author of numerous articles on mathematical subjects, and one of the editors of the mathematical section of the *Encyclopédie méthodique*. Since Charles le Géomètre died in 1791, a year for which (because of the Revolution) the Academy published no memoirs, the usual *éloge* by the *secrétaire perpétuel* does not exist for him.

BIBLIOGRAPHY

I. Original Works. Charles published no works of major significance. The articles that Poggendorff attributes to him—with the exception of one on electricity in the *Journal de physique*—were actually written by Charles le Géomètre.

II. Secondary Literature. An article on Charles appears in Charles Brainne, *Les hommes illustres de l'Orlé-anais*, 2 vols. (Orleans, 1852); see also J.-J. Fourier, *Éloge*, in *Mémoires de l'Académie Royale des Sciences de l'Institut de France*, **8** (1829), pp. lxxiii–lxxxvii; all the important facts concerning Charles and his wife as well as a few letters have been published by Anatole France in his *L'Élvire de Lamartine, Notes sur M. & Mme. Charles* (Paris, 1893).

For Charles's experiments on gaseous expansion, see *Annales de chimie*, **43** (1802), 157 ff.; on the distinction between J.-A.-C. Charles and Charles le Géomètre, we have only the information in *Index biographique des membres et correspondants de l'Académie des Sciences* (Paris, 1954).

J. B. Gough

CHARLETON, WALTER (*b.* Shepton Mallet, Somerset, England, 13 February 1620; *d.* London, England, 6 May 1707), *natural philosophy, medicine.*

Charleton's father, Reverend Walter Charleton, was rector of the church at Shepton Mallet. It was he who assumed the responsibility for the future physician's early education and prepared him carefully for the university. In 1635 the young man was sent to Oxford, where he enrolled at Magdalen Hall (later Hertford College). At Oxford, Charleton made the acquaintance of the famous John Wilkins, later bishop of Chester, who was well versed in the new philosophy. Under Wilkins' tutelage, Charleton demonstrated a talent for philosophy and logic, although it is said he distinguished himself rather more by his diligence than by his originality.

For a career Charleton chose medicine and was awarded the degree of doctor of physick in January 1643. Shortly afterward he was made physician-in-ordinary to the king. At the time Charleton was considered an extraordinary genius by many and, owing to his precocity, became the object of envy and resentment, which (Anthony à Wood reports) prevented his election to the College of Physicians until 1676.

In the 1650's Charleton turned his talents to writing, mostly on medicine, natural philosophy, and related topics, although he became famous for his works on Stonehenge and on Epicurean ethics. His first efforts included translation and amplification (a genre he found particularly congenial) of works by the chemist J. B. van Helmont. An early effort was *A Ternary of Paradoxes* (1650), which discussed magnetic cures; in the same year he produced his own *Spiritus gorgonicus,* in which he ascribed the formation of stones in the human body to a stone-forming spirit.

Soon after the appearance of these works Charleton, perhaps under the influence of his friend Hobbes, turned from Helmont to Gassendi, Descartes, and other "new philosophers." The most important of these was the atomist Gassendi, and Charleton became intrigued with atomic explanation in natural philosophy and its theological implications. The products of these interests include *The Darknes of Atheism Dispelled by the Light of Nature: A Physico-theologicall Treatise* (1652) and the *Physiologia Epicuro-Gassendo-Charltoniana* (1654), a translation and amplification of the physical part of Gassendi's previously published *Animadversions on the Tenth Book of Diogenes Laertius* (1649). The *Physiologia* became a book of minor reputation but was read by such important natural philosophers as Boyle and Newton. It was an important part of Gassendi's program to purify and render acceptable to Christians the atomic philosophy.

The period following his "conversion" to Epicureanism was, in terms of writings, a most prolific one for Charleton, although his medical career sagged. In addition to the two works cited above, Charleton published *The Ephesian and Cimmerian Matrons* (1668), *Epicurus's Morals* (1656), and *The Immortality of the Human Soul Demonstrated by the Light of Nature* (1657), the last of which contains a long section lauding the College of Physicians as a worthy example of Solomon's House. In 1659 he published a major work on physiology, *Natural History of Nutrition, Life and Voluntary Motion,* one of the first English textbooks on physiology.

During the interregnum Charleton privately and publicly remained faithful to the Crown. For his steadfastness he was rewarded in 1660 by a marked upturn in his fortunes. He remained physician to Charles II, about whom he wrote his *Imperfect Pourtraicture of His Sacred Majesty Charles the II* (1661). In this eulogy Charleton portrayed the flamboyant Charles as possessing the qualities of piety, courage, and justice "in an excellent, harmonious perfect mixture."

The *Chorea gigantum* (1663) was Charleton's most famous work. It concerns the origins of Stonehenge, about which there was a great deal of discussion in the seventeenth century. Charleton wrote against the theory of Inigo Jones, who claimed that the rocks were the remains of a Roman temple. Charleton argued to the effect that Stonehenge was not a Roman temple but, rather, the ruined meeting place of ancient Danish chieftains.

In the Restoration period Charleton enjoyed his greatest reputation and prosperity. He continued his prolific publishing, became an active original member of the Royal Society, was elected to the Royal College of Physicians, and eventually served as president of the Royal College (1689–1691). His fame reached Europe, and he was reported to have received from the University of Padua an offer of a professorship, which he declined.

After his tenure as president of the Royal College of Physicians, Charleton's fortunes declined markedly. His practice dwindled and he was forced, owing to his straitened circumstances, to retire to Jersey, from which he returned to London only in his last years. He died destitute in May 1707.

Charleton's importance as a natural philosopher (his medical works aside) rests primarily on his role as expositor of the atomic philosophy in a period during which its reception as a viable doctrine was in doubt. Charleton's *Physiologia* was widely read in the early 1650's as a convenient substitute for Gassendi's scarce works. Charleton enabled atomism to reach a wider audience and helped prepare that audience for the reception of the atomic doctrine into Christian natural philosophy.

In his early, precocious years Charleton very well displayed the spirit of the new intellectual age, however little he added to it. In his works—most of them derivative—he mirrored all the controversy, enthusiasm, and ferment of the turbulent 1650's. With respect to atomism, Charleton was a disseminator rather than a creator. His forte was the exposition and elucidation of the ideas of others, and he did not pretend otherwise.

BIBLIOGRAPHY

I. Original Works. Among Charleton's most important nonmedical works are *The Darknes of Atheism Dispelled by the Light of Nature: A Physico-theologicall Treatise* (London, 1652); *Physiologia Epicuro-Gassendo-Charltoniana* (London, 1654); *The Immortality of the Human Soul* (London, 1657); and *Chorea gigantum* (London, 1663). For a fuller list see Rolleston (below).

His medical works include *Oeconomia animalis* (London, 1659); *Exercitationes pathologicae* (London, 1661); and *Enquiries Into Human Nature* (London, 1680).

II. Secondary Literature. The fullest biographical account is still in the *Biographia Britannica* (London, 1747–1766), II, 1286–1292. Anthony à Wood's short piece on Charleton, in P. Bliss, ed., *Athenae Oxonienses,* 4 vols. (London, 1812–1820), IV, 152, was thought to be detrimental to him. John Aubrey, *Brief Lives,* Andrew Clark, ed., 2 vols. (Oxford, 1898), I, 16; and Thomas Hearne, *Remarks and Collections,* 11 vols. (Oxford, 1884–1918), contain biographical fragments. See also H. Rolleston, "Walter Charleton, D.M., F.R.C.P., F.R.S.," in *Bulletin of the History of Medicine,* **8** (1940), 403–416. Charleton as a physicotheologian is discussed by R. Westfall, in *Science and Religion in Seventeenth Century England* (New Haven, 1958), pp. 118–120. Charleton's atomism is treated in R. Kargon, "Walter Charleton, Robert Boyle and the Acceptance of Epicurean Atomism in England," in *Isis,* **55** (1964),

184–192; and *Atomism in England: From Hariot to Newton* (Oxford, 1966), pp. 77–92. C. Webster, "The College of Physicians: 'Solomon's House' in Commonwealth England," in *Bulletin of the History of Medicine,* **41** (1967), 393–412, discusses Charleton and his account of the College of Physicians. For Charleton's physiology, see T. M. Brown's as yet unpublished dissertation, "The Mechanical Philosophy and the 'Animal Oeconomy'" (Princeton, 1968).

ROBERT KARGON

CHARPENTIER, JOHANN (JEAN) DE (*b.* Freiberg, Germany, 7 December 1786; *d.* Bex, Switzerland, 12 September 1855), *mining engineering, glaciology.*

Charpentier's father, Johann Friedrich Wilhelm de Charpentier, was a professor at the Mining Academy of Freiberg; his mother was Luise Dorothea von Zobel. He followed the family profession of geology and entered the Mining Academy, which was directed by the famous neptunist Abraham Gottlob Werner, who had also taught Humboldt and Buch.

Graduating with distinction, Charpentier began his career in the coal mines at Waldenburg, Lower Silesia (of which his brother Toussaint was director), where he solved an engineering problem that had baffled Humboldt. From 1808 to 1813 Charpentier traveled in the Pyrenees, where he directed a copper mine at Baigorry. His researches on Pyrenees geology were published in 1823 in a work that received the prize of the Paris Academy of Sciences. In 1813 his friend Charles Lardy, also a student of Werner's, secured Charpentier the post of director of the salt mines of Bex, in the canton of Vaud.

These salt mines held a special position in the Swiss economy, for they were the only ones in the Confederation and importation of salt was not only expensive but also subject to political conditions, especially if imported from France. The efficiency of the Bex mines was therefore a matter of national concern, and Charpentier was successful in raising it substantially. Instead of continuing the practice of searching for salt springs, he cut galleries into the rock and extracted the rock salt, which was taken to basins where water was added to form a saturated solution; it was then pumped into other containers and evaporated.

In 1818 the disaster of the Glacier de Giétroz in the Val de Bagnes (it had dammed a lake which burst through the ice with great loss of life) turned Charpentier's attention to glaciers. He knew the views of Jean-Pierre Perraudin and of his friend Venetz on the formerly great extent of glaciers and rejected them as impossible; when he investigated them, however, he found to his astonishment that all the evidence supported such views. Immediately behind Charpentier's house at Les Dévens, near Bex, two huge blocks

of rock lay unconformably on the slope of a hill. He was forced to the conclusion that only one means of transport could have brought these blocks from the mountains behind Bex to their present position: a glacier. (It was not then known on the Continent that this view had been published by James Hutton in 1795, John Playfair in 1802, and Sir John Leslie in 1804.) Charpentier then devoted his time to a study of the Rhone valley and basin and the blocks of rock scattered in them all the way from the Alps to the Jura.

Erratic blocks had been recognized in Switzerland by Lang in 1708 and in 1715 by Laurens Roberg in Scandinavia, where Swedenborg and Linnaeus also observed them. The problem was to discover where they came from and how they had come. Some Swiss, such as Louis Bourguet and Moritz Anton Cappeler, thought that they were meteorites; this, however, would fail to explain why they were mineralogically identical with some peaks in the Alps, whence, as John Strange recognized, they had come.

In a masterpiece of absurdity, Jean-André Deluc imagined that there had been caverns in the earth which, when their roofs collapsed, catapulted the blocks—without evidence of the existence of such caverns or calculation of the energy required to hurl 100,000 cubic feet of granite 100 miles. In any case it would not explain why the blocks were spread not in all directions, but in groups opposite the mouths of the valleys containing mountains of similar composition.

This arrangement led many to conclude that the blocks had been brought to their positions by floodwaters; this was the view of Saussure, Buch, H. C. Escher von der Linth, and Élie de Beaumont, all geologists of repute who failed to consider where such vast quantities of water had come from; where the water had gone; how blocks weighing hundreds of tons could have been moved a hundred miles and, in the Jura, uphill; why such waters had done no other recognizable damage; and why some of the blocks had sharp edges, unworn by rolling.

Another hypothesis ascribed the transport of blocks to icebergs or ice rafts on rivers, lakes, or an arm of the sea. Subscribers to this view included Daniel Tilas (1738), Jens Esmark, Sir James Hall, Charles Lyell, Roderick Murchison, and Charles Darwin. All of them failed to see that the surface of the lake, river, or sea would have had to maintain a constant level for a considerable time; that erratic blocks should be found exclusively at levels at which they were not; and that if some icebergs had floated blocks from the Alps to the Jura, then others must have floated Jura blocks to the Alps—where there are none.

Charpentier analyzed all of these hypotheses and

refuted them with evidence of his own finding. The debacle of the Glacier de Giétroz showed him that masses of water do not move big blocks. To ensure Venetz's priority in the idea of the previous extent of glaciers, he had Venetz's paper of 1821, which had been forgotten, printed by the Swiss Natural Science Society in 1833 and himself read a paper before that organization in 1834. It was met with incredulity and scorn.

Undismayed, Charpentier continued his observations and invited the incredulous to visit him and see the evidence for themselves. Among his visitors was Louis Agassiz, who was soon carried away with such enthusiasm for the theory of the Ice Age that he visited a number of glaciers and blocks and rushed into print, ahead of Charpentier, with his *Études sur les glaciers* (1840). Agassiz asserted that Europe had been covered with ice before the uplift of the Alps; that when the Alps were elevated, they lifted the plate of ice and pierced it in spots, so that blocks slid down the slope from the Alps to the Jura (the slope is 1°8′); and that erratic blocks on the Jura were different in composition from those carried by glaciers (he had not observed that some erratic blocks may have been water-worn before they were transported by a glacier). In other words, he denied that glaciers extended from the Alps to the Jura. He also claimed that the stratification of the snowfields was carried throughout the glacier (he mistook the veined structure resulting from pressure for stratification) and that the direction of crevasses at the sides of glaciers, pointing outward and downhill, meant that the sides of glaciers traveled faster than their centers (the opposite of the truth). Agassiz afterward retracted these errors, but such is the power of publicity and enthusiasm that the origin of the theory of the Ice Age is today commonly attributed to Agassiz and not to its rightful scientific parent, Charpentier.

Charpentier received Agassiz's book on 28 October 1840, three days before he finished his own *Essai sur les glaciers,* which was published in February 1841. The scrupulous care with which he weighed the evidence and described the phenomenon of erratic blocks and the function of glaciers in transporting them makes this book a classic.

There are, however, two aspects of Charpentier's glacier studies in which he failed. The first is a matter that he rightly saw as of prime importance: the meteorological conditions requisite for the former great extent of glaciers. He saw that glaciers were formed from increased precipitation after the uplift of the Alps; and he sought the cause of this increase in the percolation of water into fissures in the earth's crust following the uplift until terrestrial heat converted this water into steam and vapor that subsequently fell on the Alps as snow. The problem still awaits solution, even after Robert Falcon Scott's suggestion that heating of the Arctic Ocean makes its water available for evaporation by winds and deposition as snow.

The other aspect was the mechanism of glacier motion, which Charpentier ascribed to the expansion of water as it falls onto glaciers and is converted into ice. James Forbes of Edinburgh showed that glaciers move even in winter, when all water is frozen; and that when water freezes, the latent heat of ice melts the surrounding ice in a glacier (he found that the glacier stays at 0°C.). Forbes called on Charpentier in 1844 at Les Dévens; and Charpentier, always open-minded toward new evidence, was prepared to admit the correctness of Forbes's demonstrations.

Charpentier was a man of great personality and charm, always helpful and kind. His home became a place of pilgrimage for men of science to discuss glaciology and to see his great collections of minerals, shells, and plants, which are now in the Lausanne Museum. When his fame had become established, Charpentier received numerous offers to lucrative posts elsewhere but refused them all, preferring to stay in "his" mountains. He had married Thérèse Louise von Gablenz of Dresden, but she died young, during childbirth.

BIBLIOGRAPHY

I. Original Works. Charpentier's writings include *Mémoire sur la nature et le gisement du Pyroxene en roche, connu sous le nom d'Herzolithe* (Paris, 1812); *Essai sur la constitution géognostique des Pyrénées* (Paris–Strasbourg, 1823); "Notice sur la cause probable du transport des blocs erratiques de la Suisse," in *Annales des mines,* **8** (1835), 219; and *Essai sur les glaciers et sur le terrain erratique du bassin du Rhône* (Lausanne, 1841).

II. Secondary Literature. On Charpentier or his work see Jules Marcou, *Life, Letters, and Works of Louis Agassiz* (New York, 1896); E. Payot, "Souvenirs d'hommes utiles au pays. Jean de Charpentier," in *Revue historique vaudoise* (Lausanne), **21** (1913), 161; and J. C. Shairp, P. G. Tait, and A. Adams-Reilly, *Life and Letters of James David Forbes* (London, 1873).

Gavin de Beer

CHARPY, AUGUSTIN GEORGES ALBERT (*b.* Oullins, Rhône, France, 1 September 1865; *d.* Paris, France, 25 November 1945), *metallurgy, chemistry.*

After graduating from the École Polytechnique in 1887, Charpy, in the school's chemistry laboratory, prepared a doctoral thesis on the variations in volume and density of solutions as a function of temperature and concentration. In 1892 he became an engineer at the Laboratoire d'Artillerie de la Marine and directed his research toward the study of metals and their alloys. He joined the Compagnie de Chatillon Commenty in 1898 and quickly became its technical director. In 1920 he was named director general of the Compagnie des Acieries de la Marine et Homécourt. He was elected to the Académie des Sciences in 1918, in the newly created section of industrial applications of science.

Along with Osmond and Henry Le Chatelier, Charpy was one of the three founders of the science of alloys in France. At the Laboratoire de la Marine he studied the quenching of steel and elucidated the influence of the quenching temperature by correlating the mechanical properties and the temperature values of the critical points. He recognized the lamellar form of eutectic alloys and deduced from it an explanation of the structure of these steels. He gave experimental results for several regions of the solubility diagram of carbon in iron.

Charpy was the first to introduce pyrometry into industrial practice in France, and he constructed the first electrical resistance furnace with platinum elements.

After studying the production of steels, Charpy turned his attention to measuring their brittleness; to this end he devised the pendulum drop hammer that bears his name.

In 1902 Charpy was asked to examine silicon steel sheets. He showed that the magnetic hysteresis of a steel decreases as the size of the crystalline particle increases. He also studied the alloys of iron with silicon, aluminum, and antimony, and proposed a composition of 2 or 3 percent silicon, which was universally adopted for the low-leakage magnetic circuits fundamental in the manufacture of modern electrical equipment.

Charpy's brilliant industrial career did not prevent him from dedicating part of his time to teaching; he was a very highly regarded professor at the École Supérieure des Mines in Paris and at the École Polytechnique.

BIBLIOGRAPHY

I. ORIGINAL WORKS. Among Charpy's articles, all of them published in *Comptes rendus hebdomadaires des séances de l'Académie des sciences,* are the following: "Sur la détermination des équilibres chimiques dans les systèmes dissous," **114** (1892), 665; "Influence de la température de recuit sur les propriétés mécaniques et la structure du laiton," **116** (1893), 1131; "Sur la transformation produite dans le fer par une déformation permanente à froid," **117** (1893), 850; "Sur la transformation allotropique du fer sous l'influence de la chaleur," **118** (1894), 418, 868, 1258; **119** (1894), 735; "Sur l'acier au bore," **120** (1895), 130; "Sur

la constitution des alliages ternaires," **126** (1898), 1645; "Sur l'équilibre chimique des systèmes fer-carbone," **134** (1902), 103, written with L. Grenet; "Sur la dilatation des aciers aux températures élevées," *ibid.,* 540, written with L. Grenet; "Dilatométrie des aciers," *ibid.,* 598, written with L. Grenet; "Sur la dilatation des aciers trempés," **136** (1903), 92, written with L. Grenet; "Sur le diagramme d'équilibre des alliages fer-carbone," **141** (1905), 948; "Sur la solubilité de graphite dans le fer," **145** (1907), 1277; "Sur l'action de l'oxyde de carbone sur le chrome, le nickel, leurs alliages et leurs oxydes," **148** (1909), 560; "Sur la cémentation du fer par le carbone solide," **150** (1910), 389, written with S. Bonnerot; "Sur les gaz des aciers," **152** (1911), 1247; "Sur un dilatomètre non différentiel," **212** (1941), 112; and "Sur les traitements thermiques des aciers," **213** (1941), 421.

His books include *Leçons de chimie* (Paris, 1892), written with H. Gautier; and *Notions élémentaires de sidérurgie* (Paris, 1946), written with P. Pingault.

II. SECONDARY LITERATURE. Two notices are R. Barthélemy, "Notice sur la vie et l'oeuvre de Georges Charpy," in *Notices et discours. Académie des sciences* (June 1947); and M. Caullery, "Annonce du décès de M. G. Charpy," in *Comptes rendus hebdomadaires des séances de l'Académie des sciences,* **221** (1945), 677.

GEORGES CHAUDRON

CHASLES, MICHEL (*b.* Épernon, France, 15 November 1793; *d.* Paris, France, 18 December 1880), *synthetic geometry, history of mathematics.*

Chasles was born into an upper-middle-class Catholic family, settled in the region of Chartres. He was given the name Floréal, but it was changed to Michel by court order, 22 November 1809. His father, Charles-Henri, was a lumber merchant and contractor who became president of the *chambre de commerce* of Chartres. Chasles received his early education at the Lycée Impérial and entered the École Polytechnique in 1812. In 1814 he was mobilized and took part in the defense of Paris. After the war he returned to the École Polytechnique and was accepted into the engineering corps, but he gave up the appointment in favor of a poor fellow student. After spending some time at home, he obeyed his father's wishes and entered a stock brokerage firm in Paris. However, he was not successful and retired to his native region, where he devoted himself to historical and mathematical studies. His first major work, the *Aperçu historique,* published in 1837, established his reputation both as a geometer and a historian of mathematics. In 1841 he accepted a position at the École Polytechnique, where he taught geodesy, astronomy, and applied mechanics until 1851. In 1846 a chair of higher geometry was created for him at the Sorbonne, and he remained there until his death.

Chasles was elected a corresponding member of the

Academy of Sciences in 1839 and a full member in 1851. His international reputation is attested to by the following partial list of his affiliations: member of the Royal Society of London; honorary member of the Royal Academy of Ireland; foreign associate of the royal academies of Brussels, Copenhagen, Naples, and Stockholm; correspondent of the Imperial Academy of Sciences at St. Petersburg; and foreign associate of the National Academy of the United States. In 1865 Chasles was awarded the Copley Medal by the Royal Society of London for his original researches in pure geometry.

Chasles published highly original work until his very last years. He never married, and his few interests outside of his research, teaching, and the Academy, which he served on many commissions, seem to have been in charitable organizations.

Chasles's work was marked by its unity of purpose and method. The purpose was to show not only that geometry, by which he meant synthetic geometry, had methods as powerful and fertile for the discovery and demonstration of mathematical truths as those of algebraic analysis, but that these methods had an important advantage, in that they showed more clearly the origin and connections of these truths. The methods were those introduced by Lazare Carnot, Gaspard Monge, and Victor Poncelet and included a systematic use of sensed magnitudes, imaginary elements, the principle of duality, and transformations of figures.

The *Aperçu historique* was inspired by the question posed by the Royal Academy of Brussels in 1829: a philosophical examination of the different methods in modern geometry, particularly the method of reciprocal polars. Chasles submitted a memoir on the principles of duality and homography. He argued that the principle of duality, like that of homography, is based on the general theory of transformations of figures, particularly transformations in which the cross ratio is preserved, of which the reciprocal polar transformation is an example. The work was crowned in 1830, and the Academy ordered it published. Chasles requested permission to expand the historical introduction and to add a series of mathematical and historical notes, giving the result of recent researches. His books and almost all of his many memoirs are elaborations of points originally discussed in these notes. One of his weaknesses—that he did not know German—is apparent here too, and as will be seen below, many of the results claimed as new had been wholly or partly anticipated. It was this expanded work which was published.

Chasles wrote two textbooks for his course at the Sorbonne. The first of these, the *Traité de géométrie*

supérieure (1852), is based on the elementary theories of the cross ratio, homographic ranges and pencils, and involution, all of which were originally defined and discussed in the *Aperçu historique;* the cross ratio in note 9, involution in note 10. In the case of the cross ratio, which Chasles called the anharmonic ratio, he was anticipated by August Moebius, in his *Barycentrische Calcul* (1827). However, it was Chasles who developed the theory and showed its scope and power. This book, Chasles felt, showed that the use of sensed magnitudes and imaginary elements gives to geometry the freedom and power of analysis.

The second text, the *Traité sur des sections coniques* (1865), applied these methods to the study of the conic sections. This was a subject in which Chasles was interested throughout his life, and he incorporated many results of his own into the book. For example, he discussed the consequences of the projective characterization of a conic as the locus of points of intersection of corresponding lines in two homographic pencils with no invariant line, or dually as the envelope of lines joining corresponding points of two homographic ranges with no invariant point. Chasles had originally given this in notes 15 and 16 of the *Aperçu historique,* but here also he must share credit with a mathematician writing in German, the Swiss geometer Jacob Steiner.

The book also contains many of Chasles's results in what came to be called enumerative geometry. This subject concerns itself with the problem of determining how many figures of a certain type satisfy certain algebraic or geometric conditions. Chasles considered first the question of systems of conics satisfying four conditions and five conditions (1864). He developed the theory of characteristics and of geometric substitution. The characteristics of a system of conics were defined as the number of conics passing through an arbitrary point and as the number of conics tangent to a given line. Chasles expressed many properties of his system in formulas involving these two numbers and then generalized his results by substituting polynomials in the characteristics for the original values. There are many difficulties in this type of approach, and although Chasles generalized his results to more general curves and to surfaces, and the subject was developed by Hermann Schubert and Hieronymus Zeuthen, it is considered as lacking in any sound foundation.

Chasles did noteworthy work in analysis as well. In particular, his work on the attraction of ellipsoids led him to the introduction and use of level surfaces of partial differential equations in three variables (1837). He also studied the general theory of attraction (1845), and though many of the results in this paper had been anticipated by George Green and Carl Gauss, it remains worthy of study.

Chasles wrote two historical works elaborating points in the *Aperçu historique* (notes 12 and 3 respectively) which had given rise to controversy. The *Histoire d'arithmétique* (1843) argued for a Pythagorean rather than a Hindu origin for our numeral system. Chasles based his claim on the description of a certain type of abacus, which he found in the writings of Boethius and Gerbert. The second work was a reconstruction of the lost book of *Porisms* of Euclid (1860). Chasles felt that the porisms were essentially the equations of curves and that many of the results utilized the concept of the cross ratio. Neither of these works is accepted by contemporary scholars.

In 1867 Chasles was requested by the minister of public education to prepare a *Rapport sur le progrès de la géométrie* (1870). Although the work of foreign geometers is treated in less detail than that of the French, the *Rapport* is still a very valuable source for the history of geometry from 1800 to 1866 and for Chasles's own work in particular.

Chasles was a collector of autographs and manuscripts, and this interest allied with his credulity to cause him serious embarrassment. From 1861 to 1869 he was the victim of one of the most clever and prolific of literary forgers, Denis Vrain-Lucas. Chasles bought thousands of manuscripts, including a correspondence between Isaac Newton, Blaise Pascal, and Robert Boyle which established that Pascal had anticipated Newton in the discovery of the law of universal gravitation. Chasles presented these letters to the Academy in 1867 and took an active part in the furor that ensued (1867–1869), vigorously defending the genuineness of the letters. In 1869 Vrain-Lucas was brought to trial and convicted. Chasles was forced to testify and had to admit to having purchased letters allegedly written by Galileo, Cleopatra, and Lazarus, all in French.

But this misadventure should not be allowed to obscure his many positive contributions. He saw clearly the basic concepts and their ramifications in what is now known as projective geometry, and his texts were influential in the teaching of that subject in Germany and Great Britain as well as in France. Finally, with all its faults, the *Aperçu historique* remains a classic example of a good history of mathematics written by a mathematician.

BIBLIOGRAPHY

I. ORIGINAL WORKS. The first edition of the *Aperçu historique sur l'origine et le développement des méthodes en géométrie, particulièrement de celles qui se rapportent à la géométrie moderne, suivi d'un mémoire de géométrie sur deux principes généraux de la science, la dualité et l'homographie* appeared in *Mémoires couronnés par l'Académie de Bruxelles,* vol. **11** (1837) and is very rare; but it was reprinted, without change, in Paris in 1875 and again in 1888. A German translation, *Geschichte der Geometrie, hauptsächlich mit Bezug auf die neueren Methoden,* L. Sohncke, trans. (Halle, 1839), omits the last section.

The *Traité de géométrie supérieure* (Paris, 1852), reprinted in 1880, was freely translated into German under the title *Die Grundlehren der neueren Geometrie. Erster Theil: Die Theorie des anharmonischen Verhältnisses, der homographischen Theilung und der Involution, und deren Anwendung auf die geradlinigen und Kreis-Figuren,* C. H. Schnuse, trans. (Brunswick, 1856). Principal parts of the work also appear in Benjamin Witzchel, *Grundlinien der neueren Geometrie mit besonderer Berücksichtigung der metrischen Verhältnisse an Systemen von Punkten in einer Geraden und einer Ebene* (Leipzig, 1858), and in Richard Townsend, *Chapters on the Modern Geometry,* vol. II (Dublin, 1865). The second part of the *Traité des sections coniques faisant suite au traité de géométrie supérieure* (first part, Paris, 1865) never appeared.

The major results in enumerative geometry are summarized in "Considérations sur la méthode générale exposée dans la séance du 15 février. Différences entre cette méthode et la méthode analytique. Procédés généraux de démonstration," in *Comptes rendus hebdomadaires des séances de l'Académie des sciences,* **58** (1864), 1167–1176. The most important papers in analysis are "Mémoire sur l'attraction des ellipsoïdes," in *Journal de l'École polytechnique,* **25** (1837), 244–265; "Sur l'attraction d'une couche ellipsoïdale infiniment mince; Des rapports qui ont lieu entre cette attraction et les lois de la chaleur, dans un corps en équilibre de température," *ibid.,* pp. 266–316; and "Théorèmes généraux sur l'attraction des corps," in *Connaissance des temps ou des mouvements célestes pour l'année* (*1845*), pp. 18–33.

The historical works are *L'histoire d'arithmétique* (Paris, 1843); *Les trois livres de Porismes d'Euclide, rétablis pour la première fois, d'après la notice et les lemmes de Pappus, et conformément au sentiment de R. Simson sur la forme des énoncés de ces propositions* (Paris, 1860); and *Le rapport sur le progrès de la géométrie* (Paris, 1870). Chasles's contributions to the Pascal-Newton controversy are scattered throughout the *Comptes rendus:* **65** (1867); **66** (1868); **67** (1868); **68** (1869); and **69** (1869).

The *Rendiconti delle sessioni dell'Accademia delle scienze dell'Istituto di Bologna* (1881), pp. 51–70; the *Catalogue of Scientific Papers,* Royal Society of London, I (1867), 880–884; VII (1877), 375–377; IX (1891), 495–496; and Poggendorff, I (1863), 423, and III (1898), 261–264 all contain extensive lists of Chasles's works. The first named is the most complete, although it omits all works that appeared in the *Nouvelles annales de mathématique.*

II. SECONDARY LITERATURE. There are many biographical sketches of Chasles. Among the more valuable are Joseph Bertrand, *Éloges académiques* (Paris, 1902), pp. 27–58; Eduarde Merlieux, *Nouvelle biographie générale* (Paris, 1863), VIII, 38; and Claude Pichois, *Dictionnaire de biographie française* (Paris, 1959), X, 694, which contains a short bibliography.

Chasles's works are discussed in "Commemorazione di Michele Chasles" by Pietro Riccardi, in *Rendiconti* (1881), 37–51; Edward Sabine, "Presidential Address," in *Proceedings of the Royal Society*, London, **14** (1865), 493–496; and Gino Loria, "Michel Chasles e la teoria delle sezioni coniche," in *Osiris*, **1** (1936), 421–441.

Details of the Vrain-Lucas affair can be found in J. A. Farrar, *Literary Forgeries* (London, 1907), pp. 202–214.

ELAINE KOPPELMAN

CHÂTELET, GABRIELLE-ÉMILIE LE TONNELIER DE BRETEUIL, MARQUISE DU (*b.* Paris, France, 17 December 1706; *d.* Lunéville, Meurthe-et-Moselle, France, 10 September 1749), *scientific commentary.*

The daughter of Louis-Nicolas Le Tonnelier de Breteuil, baron of Preuilly, chief of protocol at the royal court, and of Gabrielle Anne de Froulay, who came from a family of the nobility of the sword, Gabrielle-Émilie received a literary, musical, and scientific education. On 22 June 1725 she married Florent-Claude, marquis du Châtelet and count of Lomont, who, after spending several years with her when he was governor of the city of Semur-en-Auxois, pursued a military career and visited her only briefly. From this union three children were born: Gabrielle-Pauline (1726), who married the duke of Montenero in 1743; Louis-Marie-Florent (1727), future duc du Châtelet and an ambassador of France, who was guillotined in 1793; and Victor-Esprit (1733), who lived only a few months.

After returning to Paris in 1730, Émilie du Châtelet led a glittering existence and had several affairs before becoming intimate, in 1733, with Voltaire, who had just completed his *Lettres philosophiques.* An ardent Newtonian, Voltaire devoted several of these *lettres* to Newton's philosophy. The manuscript of those *lettres* which dealt with Newton had been given for review to Maupertuis, the author of the first French work devoted to the Newtonian world system, the *Discours sur la figure des astres* (1732). Mme. du Châtelet in her turn struck up a very cordial friendship with Maupertuis and with another ardent Newtonian, Alexis-Claude Clairaut. The mathematics lessons that she received from Maupertuis at the beginning of 1734 awakened her scientific inclinations.

In June 1734 Voltaire, threatened with arrest, withdrew to one of Mme. du Châtelet's properties, the château at Cirey in Champagne, whose restoration he undertook. The marquise spent a few months there at the end of 1734 and then made several prolonged stays. Devoting their time variously to their literary endeavors; metaphysical, philosophical, and scientific discussions; and a very refined worldly existence, she and Voltaire made the château at Cirey one of the most brilliant centers of French literary and philosophical life.

The stay at Cirey, at the end of 1735, of Francesco Algarotti, who was preparing a popularization of Newtonian optics, *Il newtoniasmo per le dame,* which appeared in 1737, incited Voltaire and Mme. du Châtelet to undertake a work propagandizing Newtonianism and science. Her knowledge of Latin, Italian, and English had enabled Mme. du Châtelet to write several literary and philosophical works: a translation of Bernard de Mandeville's *Fable of the Bees;* and the composition of a *Grammaire raisonnée* and an *Examen de la Genèse.* At that time she began a systematic study of Newton's work, writing an *Essai sur l'optique,* of which a fragment is preserved, and participating in the elaboration of the *Éléments de la philosophie de Newton,* published by Voltaire in 1738 but composed in large part as early as the end of 1736. It is to this book that she devoted her "Lettre sur les élémens de la philosophie de Newton" (1738), a report on and defense of that part of the work which discusses Newtonian attraction.

At the end of August 1737 Mme. du Châtelet finished an important memoir on fire (*Dissertation sur la nature et la propagation du feu*), written for a prize competition organized by the Académie des Sciences. Voltaire entered the same contest, creating for this purpose a small chemistry laboratory at Cirey, but Mme. du Châtelet succeeded in preparing her memoir and sending it to the Academy without his knowledge. The results of the competition were announced on 16 April 1738: the prize was divided among Euler and two authors of second rank; only their memoirs were to be published. However, Voltaire arranged for his memoir and that of Mme. du Châtelet to be included with the winning memoirs; the first edition—identical to the definitive edition of 1752—appeared in April 1739. In 1744 Mme. du Châtelet secured the publication of a slightly modified edition of her memoir.

During this period Mme. du Châtelet was writing a book of Newtonian inspiration on the principles of physics and mechanics, designed for the instruction of her son. However, her conversion to Leibniz's doctrine of *forces vives* in 1738, under the influence of Johann I Bernoulli and Maupertuis, obliged her to interrupt her work in order to give more prominence in it to the Leibnizian epistemology. She likewise attempted, but in vain, to modify a note in her memoir on fire, the publication of which was controlled by the Académie des Sciences. In order to become more familiar with Leibniz's philosophy, she obtained the aid of Samuel König, a disciple of Christian Wolf. König worked with her from April to November 1739, revising the manuscript of her book and providing her with the desired documentation. After she had broken with König over a question of money and had solicited Johann I Bernoulli's help in vain, she completed her work, which she published anonymously at the end of 1740 as *Institutions de physique* (a revised edition appeared in Amsterdam in 1742). The chapter of this book dealing with the problem of *forces vives* vigorously defends Leibniz's point of view and severely criticizes a memoir by Dortous de Mairan (1728) that condemned this principle. Mairan, who had become perpetual secretary of the Académie des Sciences, published a harsh reply in February 1741. Mme. du Châtelet answered with a very direct attack, but—quite curiously—Voltaire publicly defended Mairan. In correspondence with the most noted physicists of the time—Euler, Maupertuis, Clairaut, Musschenbroek, s'Gravesande, Jurin, Cramer, and others—Mme. du Châtelet discussed this problem of *forces vives,* trying to obtain their support in a quarrel that she had helped to sharpen but that surpassed her competence.

Beginning in 1745, however, she dedicated all of her scientific activity to perfecting a French translation of Newton's *Principia.* It was to be enriched by a commentary on the work inspired by the one accompanying the Latin edition of T. Le Sueur and F. Jacquier and by theoretical supplements drawn essentially from the most recent works of Clairaut. Whether at Cirey, Paris, Brussels, or Lunéville, Mme. du Châtelet almost always remained by Voltaire's side, lavishing on him her valuable advice concerning his writings as well as his defense against attacks of all kinds. As early as 1746 she obtained Clairaut's collaboration as adviser, as reviser of her translation and her commentaries, and as author of theoretical supplements to her work. In the spring of 1747 the definitive plan was settled upon, the translation completed, and the printing begun. But Clairaut then found himself involved in a major discussion on the modifications eventually to be made in the law of universal gravitation in order to explain an anomaly observed in the movement of the moon's apogee. He was a partisan of modification from November 1747 until February 1749, which made the writing of commentaries on the *Principia* inadvisable. Moreover, the marquise's long visit at the court of Stanislas I, former king of Poland, at Lunéville and her affair with a young officer, the marquis J. F. de Saint-Lambert, prevented her from doing much work. In February 1749 she came to Paris to finish her book in collabo-

ration with Clairaut. The revelation of an unexpected and late pregnancy increased her desire to complete the project before the confinement that she dreaded. At the end of June, fleeing indiscreet stares, she left for Lunéville, where she died of childbed fever. Before her death she entrusted the manuscript of her annotated translation of the *Principia* to the librarian of the Bibliothèque du Roi in Paris. This work, which appeared in 1759—after partial publication in 1756—remains the sole French translation of the *Principia.*

Known throughout intellectual Europe as Émilie, the name popularized by Voltaire, Mme. du Châtelet—beyond the influence that she had for some fifteen years on the orientation of Voltaire's work and on his public activity—contributed to the vitality of French scientific life and to the parallel diffusion of Newtonianism and Leibnizian epistemology. Her affairs entertained the fashionable world of her period, yet her last moments revealed the sincerity of her scientific vocation. Although she limited her efforts to commentary and synthesis, her work contributed to the great progress made by Newtonian science in the middle of the eighteenth century.

BIBLIOGRAPHY

I. ORIGINAL WORKS. Mme. du Châtelet's published writings are "Lettre sur les élémens de la philosophie de Newton," in *Journal des sçavans* (Sept. 1738), 534–541; *Dissertation sur la nature et la propagation du feu* (Paris, 1739, 1744), repr. in *Recueil de pièces qui ont remporté le prix de l'Académie royale des sciences . . .,* IV (Paris, 1752), 87–170, 220–221; *Institutions de physique* (Paris, 1740; London, 1741; rev. ed., Amsterdam, 1742), also trans. into Italian (Venice, 1743); *Réponse de Madame *** à la lettre que M. de Mairan . . . lui a écrite . . . sur la question des forces vives* (Brussels, 1741; new ed., with author's name, Brussels, 1741; repub. 1744); "Mémoire touchant les forces vives adressé en forme de lettre à M. Jurin . . .," in C. Giuliani, ed., *Memorie sopra la fisica & istoria naturale . . .,* III (Lucca, 1747), 75–84; *Principes mathématiques de la philosophie naturelle* (partial ed., Paris, 1756; full ed., Paris, 1759; repr. Paris, 1966), the French trans. of Newton's *Principia,* followed by Mme. du Châtelet's commentaries and supplements by Clairaut; "Réflexions sur le bonheur," in J.-B. J. Suard and J. Bourlet de Vauxcelles, eds., *Opuscules philosophiques et littéraires* (Paris, 1796), pp. 1–40, also a critical ed. by R. Manzi (Paris, 1961); and "Réponse à une lettre diffamatoire de l'abbé Desfontaines," in A. J. Q. Beuchot, ed., *Mémoires anecdotiques sur Voltaire* (Paris, 1838), pp. 423–431. An analysis of *Examen de la Genèse,* part of her translation *La fable des abeilles,* ch. 4 ("De la formation des couleurs") of *Essai sur l'optique,* and chs. 6–8 of *Grammaire raisonnée* can be found in I. O. Wade, ed., *Studies on Voltaire with Some Unpublished*

Papers of Mme du Châtelet (Princeton, 1947), pp. 48–107, 131–187, 188–208, and 209–241, respectively.

Mme. du Châtelet's correspondence went through several partial editions, which have been superseded by E. Asse, ed., *Lettres de la M^{ise} du Châtelet . . .* (Paris, 1878), which contains 246 letters, some of them incomplete; and especially by T. Besterman, ed., *Les lettres de la marquise du Châtelet,* 2 vols. (Geneva, 1958), 486 letters. Even the latter work must be completed by some letters addressed to Mme. du Châtelet that have been included in vols. 3 and 17 of Besterman's 107-vol. edition of Voltaire's correspondence (Geneva, 1953–1965).

The Bibliothèque Nationale, Paris, has MSS of the *Institutions de physique* and the French translation of the *Principia* (fonds fr. 12265–12268) and of rough drafts of letters to Maupertuis (fonds fr. 12269); the public library of Leningrad has MSS of most of the works published by I. O. Wade; and the municipal library of Troyes has a nonautograph MS of *Examen de la Genèse.*

II. SECONDARY LITERATURE. Mme. du Châtelet or her work is discussed in the following (listed chronologically): Voltaire, "Éloge historique de Mme. la marquise du Châtelet . . .," in *Bibliothèque impartiale* (Leiden) (Jan.–Feb. 1752), 136–146; J. F. Montucla, *Histoire des mathématiques,* new ed., III (Paris, 1802), 629–643; Hochet, "Notice historique sur Madame du Châtelet," in *Lettres inédites de Madame la marquise du Chastelet à M. le comte d'Argental . . .* (Paris, 1806); Mme. de Graffigny, *La vie privée de Voltaire et de Mme. du Châtelet* (Paris, 1820); G. Desnoiresterres, *Voltaire et la société au XVIIIe siècle. Voltaire à Cirey* (Paris, 1871), passim; E. Asse, "Notice sur la marquise du Châtelet," in *Lettres de la M^{ise} du Châtelet . . .* (Paris, 1878), pp. i–xliv; F. Hamel, *An Eighteenth Century Marquise: A Study of Emilie du Châtelet and Her Times* (London, 1910); E. Jovy, *Le P. François Jacquier et ses correspondants . . .* (Vitry-le-François, 1922), pp. 22–29; A. Maurel, *La marquise du Châtelet, amie de Voltaire* (Paris, 1930); N. Nielsen, *Géomètres français du XVIIIe siècle* (Copenhagen–Paris, 1935), pp. 125–126; M. S. Libby, *The Attitude of Voltaire to Magic and the Sciences* (New York, 1935); I. O. Wade, *Voltaire and Mme. du Châtelet; an Essay on the Intellectual Activity at Cirey* (Princeton, 1941); and *Studies on Voltaire With Some Unpublished Papers of Mme. du Châtelet* (Princeton, 1947); R. L. Walters, *Voltaire and the Newtonian Universe. A Study of the "Éléments de la philosophie de Newton"* (Princeton, 1954), unpub. diss.; M. L. Dufrénoy, "Maupertuis et le progrès scientifique," in *Studies on Voltaire . . .,* XXV (Geneva, 1963), 519–587, esp. 531–548; H. Frémont, "Gabrielle-Émilie Le Tonnelier de Breteuil" (Du Châtelet, 16), in *Dictionnaire de biographie française,* XI (Paris, 1966), cols. 1191–1197; R. L. Walters, "Chemistry at Cirey," in *Studies on Voltaire . . .,* LVIII (Geneva, 1967), 1807–1827; W. T. Barber, "Mme du Châtelet and Leibnizianism. The genesis of the *Institutions de physique,*" in *The Age of the Enlightenment. Studies Presented to T. Besterman* (Edinburgh-London, 1967), pp. 200–222; C. Kiernan, *Science and the Enlightenment in Eighteenth Century France,* vol. LIX of *Studies on Voltaire . . .* (Geneva, 1968); I. O. Wade, *The Intellectual Develop-*ment of Voltaire (Princeton, N. J., 1969), pp. 253–570; I. B. Cohen, "The French Translation of Isaac Newton's *Philosophia Naturalis Principia Mathematica* (1756, 1759, 1966)," in *Archives internationales d'histoire des sciences,* **21** (1968), 261–290; and René Taton, "Madame du Châtelet, traductrice de Newton," in *Archives internationales d'histoire des sciences,* **22** (1969).

RENÉ TATON

CHAUCER, GEOFFREY (*b.* London, England, *ca.* 1343–1344; *d.* London, 25 October 1400), *literature, astronomy.*

His father, John, and his grandfather and stepgrandfather were vintners and wine merchants with some wealth and standing at court and some experience in public office. For his mother, probably Agnes de Copton, this was the second of three marriages. The earliest records of Geoffrey Chaucer show him in 1357 in the household, probably as a page, of Elizabeth, countess of Ulster, and her husband, Prince Lionel. Thereafter he was with the English army in France.

He was married by 1366 to Philippa Roet, sister of a wife of John of Gaunt. After holding many public offices, Chaucer became controller of customs for wools and hides in the port of London, then clerk of the kings' works; he held several other high offices and sinecures. In 1399 he leased for fifty-three years a house in the garden of Westminster Abbey, but died there the next year.

Chaucer's popular masterpiece, the *Canterbury Tales,* is of course the chief reason for his fame, so much so that his translation of Boethius' *De consolatione philosophiae* and his other books are often forgotten. In particular, his two astronomical works, *A Treatise on the Astrolabe* (1391) and *Equatorie of the Planetis* (1392), both present considerable problems. They are both securely dated to his later years, after the bulk of the poetry, and they are remarkable as being quite self-consciously written in the vernacular instead of the usual Latin of such scholarship. Both books are clearly pedagogic in intent, and both display clear technical mastery of the special line of instrumentation associated at this time principally with Merton College, Oxford, which had specialized in instruments and in Alfonsine astronomy for two generations before. The *Astrolabe,* modeled on a treatise by Messahalla, describes the construction and use of this popular instrument for computing the position of any bright star as seen from any particular place at any time of any day of the year. It is an analogue method of computation that did for the medieval astronomer what a slide rule does, in our time, for the engineer. The *Equatorie* supplies similar details for a companion instrument that allows analogue computation for the places of the planets and the moon in the ecliptic following the Ptolemaic theory and using a set of standard astronomical tables. Unfortunately, details of Chaucer's scientific education are completely unknown; he is not recorded at Oxford, and if educated in London, perhaps by a Merton man, the details are quite lost and cannot even be guessed.

BIBLIOGRAPHY

I. ORIGINAL WORKS. Chaucer's astronomical works are Derek J. Price, ed., *The Equatorie of the Planetis* (Cambridge, 1955); and Walter W. Skeat, ed., *A Treatise on the Astrolabe* (London, 1872). His literary works are collected in F. N. Robinson, ed., *The Works of Geoffrey Chaucer* (Boston, 1957), with a full bibliography on pp. 641–645.

II. SECONDARY LITERATURE. Chaucer's astronomical works are discussed in Walter Clyde Curry, *Chaucer and the Mediaeval Sciences* (New York, 1960); P. Pintelon, *Chaucer's Treatise on the Astrolabe* (Antwerp, 1940).

DEREK J. DE SOLLA PRICE

CHAULIAC, GUY DE (*b.* Chauliac, Auvergne, France, *ca.* 1290; *d.* in or near Lyons, France, *ca.* 1367–1370), *medicine.*

Guy's family were of the peasant class, and he was aided in his studies by the lords of Mercoeur. He studied medicine first at Toulouse, then at Montpellier, and finally in Bologna. It was at Bologna that he perfected his knowledge of anatomy under the guidance of his master Nicolaus Bertrucius (Bertruccio), and his description of Bertrucius' teaching methods has often been quoted. After leaving Bologna, Guy traveled to Paris before taking up residence in Lyons, where he was appointed canon of St. Just. He was later appointed a canon of Rheims and of Mende. At this period the popes resided in Avignon, and Guy became private physician to Clement VI (1342–1352), Innocent VI (1352–1362), and Urban V (1362–1370). His service to the popes was valued enough to earn him an appointment as a papal clerk (*capellanus*). While serving at Avignon, Guy made the acquaintance of Petrarch.

In his works Guy calls himself "cyrurgicus magister in medicine," and he received his master's degree in medicine (equivalent to the M.D. of Bologna) from Montpellier. He was one of the most influential surgeons of the fourteenth century; in fact, some nineteenth-century medical historians went so far as to rank him second only to Hippocrates in his influence on surgery. His chief work was the *Inventorium sive collectorium in parte chirurgiciali medicine,* which is usually referred to by its shorter title of *Chirurgia,* or sometimes *Chirurgia magna.* Guy completed it in 1363 and dedicated it to his colleagues at Montpellier, Bologna, Paris, and Avignon, places where he had either practiced medicine or been a student. The seven parts or books that make up the *Chirurgia* passed through numerous editions and served at least until the seventeenth century as the standard work on the subject. The book or parts of it were translated early from Latin into Provençal, French, English, Italian, Dutch, and Hebrew. The prologue ("Capitulum singulare") is an invaluable essay on general facts that Guy thought every surgeon should know about liberal arts, diet, surgical instruments, and the manner of conducting an operation. It also included a brief history of medicine and surgery in the form of notes on earlier physicians and surgeons and is the source of much information about Guy himself.

Guy regarded his book as a collection of the best medical ideas of his time, and he modestly stated that only a few things were original with him. E. Nicaise, the editor of the standard scholarly edition of the work, found that of some 3,300 quotations made by Guy, no fewer than 1,400 were from Arab writers and 1,100 from ancient authors. All told, some 100 different writers were cited or quoted. Galen led the list with 890 different citations, but frequent references were also made to Hippocrates, Aristotle, Al-Rāzī (Rhazes), Abul Kasim (Albucasis), Ibn Sīnā (Avicenna), Ibn Rushd (Averroës), and many other Arab or classical writers. Chauliac also mentioned and assessed his contemporaries. He thought highly of William of Saliceto but little of Lanfranchi, and he ridiculed John of Gaddesden.

Chauliac urged surgeons to study anatomy and went so far as to say that surgeons who were ignorant of anatomy carved the human body in the same way that a blind man carved wood. In spite of his emphasis on anatomy, his section (Tractatus I) on the subject is the weakest part of his book; Guy shows little real understanding, even though he had undoubtedly assisted at dissections and carried out postmortem examinations. His work may be said to reflect more the teacher than the scientist and is more didactic than scientific; he probably accurately represents the state of medical knowledge at his time.

His work is a mine of information on all kinds of things. In his section on carbuncles, abscesses, tumors, and so forth, Guy also included buboes in the armpits. He then went on to describe the plagues of 1348 and 1360 at Avignon. He pointed out the prevalence of plague in Asia and Europe, indicated the differences between the pneumonic and bubonic types, and then revealed himself to be a man of his time by blaming the disease on the Jews—who, he said, wished to poison the world—or on certain conjunctions of the planets. He did, however, recognize the contagious nature of the plague and recommended purification of the air, as well as venesection and a good diet for the afflicted.

Much of the historical controversy on the place of Guy de Chauliac in the history of medicine has raged over the question of his views on infection. In many standard histories of medicine, Guy is accused of believing that pus (laudable pus) was a necessary part of the healing process. He is also accused of practicing meddlesome medicine by prescribing all sorts of salves, plasters, and so forth instead of depending upon the healing powers of nature. This labeling of Guy as a medical reactionary has in part been discounted by historians such as Jordan Haller, who are extremely critical of the standard treatments of Guy's medical attitudes.

BIBLIOGRAPHY

I. ORIGINAL WORKS. Chauliac's original MS has long been lost, and so has the archetype of the translation that was made into French. There are, however, several versions dating from the last quarter of the fifteenth century, as well as fragments, commentaries, and abridgments. The first published ed. was a French one, *De la pratique de cyrurgie,* collated with the Latin text by Nicolas Panis (Lyons, 1478). The first Latin ed. was published at Venice in 1498. An Italian trans. appeared at Venice in 1480; a Dutch trans. at Antwerp in 1500; and Catalan and Spanish trans. at Barcelona in 1492. Most of these went through several eds. An English ed. published in 1541 has not survived, although Robert Coplan's trans. of the *Chirurgia parva,* dating from 1541–1542, is still extant. This *Chirurgia parva,* although often ascribed to Chauliac, is simply a rather poor compendium of some parts of the *Chirurgia magna.* Guy also wrote a treatise on astrology, *Practica astrolabii* (*De astronomia*), dedicated to Clement VI, and two other works that are now lost, *De ruptura,* a treatise on hernia, and *De subtilianti diaeta,* a treatise on cataract with a regimen for the patient. The best ed. is that of E. Nicaise, which is a trans. into modern French under the title *La grande chirurgie* (Paris, 1890). There is an English trans. of Tractatus III, *On Wounds and Fractures,* by W. A. Brennan (Chicago, 1923). See also *The Middle English Translation of Guy de Chauliac's Anatomy,* Björn Wallner, ed. (Lund, Sweden, 1964).

II. SECONDARY LITERATURE. George Sarton, *Introduction to the History of Science,* III (Baltimore, 1948), 1690–1694, summarizes much of the information about Guy and includes a discussion of various eds. of his work. Also valuable is E. Gurlt, *Geschichte der Chirurgie,* 3 vols. (Berlin, 1898), II, 77–107. See also Leo Zimmerman and Ilza Veith,

Great Ideas in the History of Surgery (Baltimore, 1961), pp. 149–157; and Vern L. Bullough, *The Development of Medicine as a Profession* (Basel, 1966), pp. 64–65, 93–95. A good summary article on whether Guy was or was not a medical reactionary is Jordan Haller, "Guy de Chauliac and His *Chirurgia magna,*" in *Surgery,* **55** (1964), 337–343.

VERN L. BULLOUGH

CHAUVEAU, JEAN-BAPTISTE AUGUSTE (*b.* Yonne, France 21 November 1827; *d.* Paris, France, 4 January 1917), *physiology, veterinary medicine.*

Chauveau was the son of a blacksmith. After preparation in public schools, he entered Alford, the well-known veterinary school near Paris in 1844. He graduated first in his class in 1848 and was then appointed *chef de travaux* in anatomy and physiology at the veterinary school in Lyons.

The first part of his scientific career was devoted to the comparative anatomical studies of animals. These studies culminated in his highly regarded text on the anatomy of domestic animals, which went through several editions. At this time he also made experimental observations. Operating on a limited budget, he used decrepit horses for physiological studies in the morning and for dissection in the afternoon. He did cardiovascular experiments, including cardiac catheterization, to make observations on intracardiac pressures and the correlation of the heart sounds with cardiac (especially valvular) events. Like most good experimentalists he was clever at the design of equipment, and had a special talent for designing apparatus for graphic recording.

In 1863 Chauveau succeeded Lecoq in the chair of anatomy and physiology at Lyons. In 1876, when the department was split into two, Chauveau opted for the physiology chair; his friend and protégé, Arloing, became professor of anatomy. Meanwhile Chauveau's interest in bacteriology had been stimulated by the pioneering work of Pasteur.

In 1886 Chauveau left Lyons for Paris to succeed Henri Bouley as inspector general of the veterinary schools. He also succeeded Bouley in the Academy of Sciences and at the Museum of Natural History in the chair of comparative pathology. The remainder of his scientific career was devoted to a study of the sources and transformations of energy during the work of muscles.

Chauveau was president of the Academy of Sciences in 1907 and of the Academy of Medicine in 1913. The degree of doctor of medicine *honoris causa* had been conferred on him at Lyons in 1877.

Chauveau was a large and muscular man (he was an ardent mountain climber). Tall and majestic in bearing, with a massive head set on a strongly built body, he was a striking figure at public events. Chauveau was noted for the catholicity of his outlook, which is reflected in the breadth of his investigations, and for his encouragement of young scientists.

Chauveau's scientific career can be divided into three broad phases, which encompass his cardiovascular, microbiologic, and bioenergetic studies. During the first, or cardiovascular, phase he performed some of the earliest cardiac catheterizations on horses, rendering the subject motionless by transection of the upper spinal cord. He showed that the heart beat palpated on the surface of the chest occurs with ventricular systole, not diastole as previously thought. In experiments in which he inserted his finger into the beating heart, he demonstrated the relationship between closure of the atrioventricular valve and the first heart sound. He also directed his attention to the origin of heart murmurs and the vesicular breath sounds (respiratory murmur). Much of this work was in collaboration with Joseph Faivre and later with Marey.

Chauveau infected cattle with the human tuberculosis organism, and on the basis of this implemented the inspection of slaughterhouses. The guiding concept in Chauveau's work on infectious diseases was the possibility of a vaccine against each, either a naturally attenuated microorganism after Jenner's method or an artificially attenuated one after Pasteur's.

Chauveau's work in the field of energy metabolism in relation to muscle effort was productive of several important discoveries. He showed that there is an arteriovenous glucose difference across muscle, indicating that muscle metabolizes glucose and not protein. Across the lung, he found no arteriovenous glucose difference. These findings were a refutation of Claude Bernard's view that glucose was metabolized in the lung but not in muscle. In his study of muscle contraction, he measured heat production during muscle contraction and showed an increase in the metabolism of glucose. He showed that the metabolism of protein requires energy. In dogs fed a rich ration of meat he demonstrated heat production and referred to it as the specific dynamic action of protein.

BIBLIOGRAPHY

On Chauveau and his work, see E. Gley, "A. Chauveau (1827–1917)," in *Journal de physiologie et de pathologie générale,* **17** (1917), 1–2; F. Maignon, *Éloge de Jean-Baptiste Auguste Chauveau* (Paris, 1927); and the unsigned obituary in *Lancet* (1917), **1**, 121–122.

VICTOR A. McKUSICK

CHAZY, JEAN FRANÇOIS (*b.* Villefranche-sur-Saône, France, 15 August 1882; *d.* Paris, France, 9 March 1955), *celestial mechanics.*

The author of a thesis on differential equations of the third and higher orders, which he defended at the Sorbonne on 22 December 1910, Chazy seemed destined for very specialized work in the field of analysis. His services in sound ranging during World War I revealed his gifts as a calculator; and his *citation à l'ordre de l'Armée,* which praised his remarkable accuracy in determining the position of the Big Bertha cannons that fired on Paris in May-June 1918, assured him the respect of his colleagues and students. Among those who took Chazy's course in classical mechanics between 1930 and 1940, very few suspected that they had a teacher distinguished by a rather uncommon quality, the ability to change his field. A mathematician familiar with the problems of analysis and also gifted with a very fine awareness of numerical problems, he did great work in celestial mechanics.

Chazy was born into a family of small provincial manufacturers. After brilliant secondary studies at the *collège* of Mâcon and then at the *lycée* in Dijon, he entered the École Normale Supérieure in 1902. The *agrégation* in mathematical sciences in 1905 directed him immediately toward research in the field of differential equations; Paul Painlevé had recently obtained beautiful results in this area, and Chazy hoped to extend them. A lecturer first at Grenoble, then at Lille after his doctoral thesis in 1910, he returned to the Faculty of Sciences in Lille as a professor at the end of the war. Appointed to the Sorbonne in 1925, he was given the course in mechanics in 1927 and assumed in succession the chair of rational mechanics and then, until his retirement in 1953, those of analytical mechanics and celestial mechanics. He was elected a member of the astronomy section of the Académie des Sciences on 8 February 1937, and became a titular member of the Bureau des Longitudes in 1952.

Chazy's scientific work had been associated with the Academy since 1907, and the long list of his communications in its *Comptes rendus* reveals both his development and the importance of his works in his field of specialization. In 1912 he shared the Grand Prix des Sciences Mathématiques with Pierre Boutroux and René Garnier for work on perfecting the theory of second- and third-order algebraic differential equations whose general integral is uniform. In 1922 Chazy received the Prix Benjamin Valz for his work in celestial mechanics, particularly for his memoir "Sur l'allure du mouvement dans le problème des trois corps quand le temps croît indéfiniment."

Beginning in 1919, Chazy applied to the famous

three-body problem of Newtonian mechanics the mastery he had acquired in the study of the singularities of solutions to differential equations when the initial conditions vary. He was able to determine the region of the twelve-dimensional space defined by the positions and velocities of two of the bodies relative to the third within which the bounded trajectories can only exist—and thus to achieve a representation of planetary motions.

In 1921 Chazy became interested in the theory of relativity, to which he devoted two important works. He became famous through his critique of Newcomb's calculations for the advance of the perihelion of Mercury, considered in the framework of Newtonian theory, and furnished for this secular advance a value that was later confirmed as very nearly correct by the powerful mechanical calculation developed by Gerald Clemence in 1943–1947.

He foresaw the necessity of attributing a speed of propagation to attraction, and when this new conception was affirmed as a consequence of relativity, he devoted a substantial amount of work to it, utilizing with good results the notion of isothermal coordinates, first perceived by Einstein in 1916 and determined exactly by Georges Darmois in 1925–1927.

While penetrating in an original and profound way the field of research opened by the relativity revolution, Chazy nevertheless remained a classically trained mathematician. With solid good sense, he held a modest opinion of himself. The reporters of his election to the Academy were, however, correct to stress how much, in a period of crisis, celestial mechanics needed men like him, who were capable of pushing to its extreme limits the model of mathematical astronomy that originated with Newton. Thus, beyond the lasting insights that Chazy brought to various aspects of the new theories, his example remains particularly interesting for the philosophy of science.

BIBLIOGRAPHY

Chazy's writings include "Sur les équations différentielles du 3ème ordre et d'ordre supérieur dont l'intégrale générale a ses points critiques fixes," his doctoral thesis, in *Acta mathematica*, **34** (1911), 317–385; "Sur l'allure du mouvement dans le problème des trois corps quand le temps croît indéfiniment," in *Annales scientifiques de l'École normale supérieure*, 3rd ser., **39** (1922), 29–130; *La théorie de la relativité et la mécanique céleste*, 2 vols. (Paris, 1928–1930); *Cours de mécanique rationnelle*, 2 vols. (Paris, 1933; I, 4th ed., 1952; II, 3rd ed., 1953); and *Mécanique céleste: Équations canoniques et variation des constantes* (Paris, 1953).

Among the articles, contributions, and communications by Chazy one may cite "Sur les vérifications astronomiques de la relativité," in *Comptes rendus hebdomadaires des séances des séances de l'Académie des sciences*, **174** (1922), 1157–1160; "Sur l'arrivée dans le système solaire d'un astre étranger," in *Proceedings of the International Mathematical Congress*, II (Toronto, 1924); "Sur les satellites artificiels de la terre," in *Comptes rendus hebdomadaires des séances de l'Académie des sciences*, **225** (1947), 469–470; and "Au sujet des tirs sur la lune," *ibid.*, **228** (1949), 447–450.

A secondary source is Georges Darmois, *Notice sur la vie et les travaux de Jean Chazy (1882–1955)* (Paris, 1957).

PIERRE COSTABEL

CHEBOTARYOV, NIKOLAI GRIGORIEVICH (*b.* Kamenets-Podolsk [now Ukrainian S.S.R.], 15 June 1894; *d.* Moscow, U.S.S.R., 2 July 1947), *mathematics.*

Chebotaryov became fascinated by mathematics while still in the lower grades of the Gymnasium. In 1912 he entered the department of physics and mathematics at Kiev University. Beginning in his second year at Kiev, Chebotaryov participated in a seminar given by D. A. Grave which included O. Y. Schmidt, B. N. Delaunay, A. M. Ostrowski and others. Chebotaryov's scientific interests took definite shape in this group. After graduating from the university in 1916, he taught and did research.

From 1921 to 1927 Chebotaryov taught at Odessa, where he prepared a paper on Frobenius' problem; he defended this paper as a doctoral dissertation in Kiev in 1927. In that year he was appointed a professor at Kazan University. In January 1928, he assumed his post at the university, where he spent the rest of his life and where he founded his own school of algebra. In 1929 Chebotaryov was elected a corresponding member of the Academy of Sciences of the Union of Soviet Socialist Republics, and in 1943 the title Honored Scientist of the Russian Soviet Federated Socialist Republic was conferred upon him. For his work on the theory of resolvents he was posthumously awarded the State Prize in 1948.

Chebotaryov's main works deal with the algebra of polynomials and fields (Galois's theory); the problem of resolvents (first raised by Felix Klein and David Hilbert)—that is, the problem of the transformation of a given algebraic equation with variable coefficients to an equation whose coefficients depend on the least possible number of parameters (1931 and later); the distribution of the roots of an equation on the plane (1923 and later); and the theory of algebraic numbers. In 1923 he published a complete solution to Frobenius' problem concerning the existence of an infinite set of prime numbers belonging to a given class of substitutions of Galois's group of a given normal algebraic field. This problem generalized Dirichlet's famous theorem concerning primes among natural numbers in arithmetic progressions. The method applied was utilized by E. Artin in 1927 in proving his generalized law of reciprocity. In 1934 Chebotaryov, applying the methods of Galois's theory and p-adic series, made significant advances toward a solution of the question—first posed by the ancient Greeks—of the possible number of lunes that are bounded by two circular arcs so chosen that the ratio of their angular measures is a rational number and that can be squared using only a compass and a straightedge. One of Chebotaryov's disciples, A. V. Dorodnov, completed the investigation of this famous problem in 1947. Chebotaryov also did work on the theory of Lie groups, in geometry (translation surfaces), and in the history of mathematics.

BIBLIOGRAPHY

I. ORIGINAL WORKS. Chebotaryov's works may be found in the three-volume *Sobranie sochineny* ("Collected Works"; Moscow and Leningrad, 1949–1950).

II. SECONDARY LITERATURE. More on Chebotaryov and his work is in *Nauka v SSSR za pyatnadtsat let. Matematika* ("Fifteen Years of Science in the U.S.S.R.: Mathematics"; Moscow–Leningrad, 1932), see index; and *Matematika v SSSR za tridtsat let* ("Thirty Years of Mathematics in the U.S.S.R."; Moscow–Leningrad, 1948), see index. *Matematika v SSSR za sorok let* ("Forty Years of Mathematics in the U.S.S.R."; Moscow–Leningrad, 1959), II, 747–750, contains a bibliography of Chebotaryov's works. An obituary of Chebotaryov is in *Uspekhi matematicheskikh nauk*, **2**, no. 6 (1947), 68–71. See also V. V. Morozov, A. P. Norden, and B. M. Gagaev, "Kazanskaya matematicheskaya shkola za 30 let" ("Thirty Years of the Kazan School of Mathematics"), *ibid.*, pp. 3–20; B. L. Laptev, "Matematika v Kazanskom universitete za 40 let (1917–1957)" ("Forty Years of Mathematics at Kazan University"), in *Istoriko-matematicheskie issledovania,* **12** (1959), 11–58; and *Istoria otechestvennoi matematiki* ("History of Native [Russian] Mathematics"), vols. II–III (Kiev, 1967–1968), see index.

A. P. YOUSCHKEVITCH

CHEBYSHEV, PAFNUTY LVOVICH (*b.* Okatovo, Kaluga region, Russia, 16 May 1821; *d.* St. Petersburg, Russia, 8 December 1894), *mathematics.*

Chebyshev's family belonged to the gentry. He was born on a small estate of his parents, Lev Pavlovich Chebyshev, a retired army officer who had participated in the war against Napoleon, and Agrafena Ivanovna Pozniakova Chebysheva. There were nine children, of whom, besides Pafnuty, his younger brother, Vladimir Lvovich, a general and professor at the Petersburg Artillery Academy, was also well known. Vladimir paid part of the cost of publishing the first collection of Pafnuty's works (1, 1a).

In 1832 the Chebyshevs moved to Moscow, where Pafnuty completed his secondary education at home. He was taught mathematics by P. N. Pogorelski, one of the best tutors in Moscow and author of popular textbooks in elementary mathematics.

In 1837 Chebyshev enrolled in the department of physics and mathematics (then the second section of the department of philosophy) of Moscow University. Mathematical disciplines were then taught brilliantly by N. D. Brashman and N. E. Zernov. Brashman, who always directed his pupils toward the most essential problems of science and technology (such as the theory of integration of algebraic functions or the calculus of probability, as well as recent inventions in mechanical engineering and hydraulics), was especially important to Chebyshev's scientific development. Chebyshev always expressed great respect for and gratitude to him. In a letter to Brashman, discussing expansion of functions into a series by means of continued fractions, Chebyshev said: "What I said illustrates quite sufficiently how interesting is the topic toward which you directed me in your lectures and your always precious personal talks with me" (2, II, 415). The letter was read publicly at a meeting of the Moscow Mathematical Society on 30 September 1865 and printed in the first issue of *Matematichesky sbornik* ("Mathematical Collection"), published by the society in 1866. Chebyshev was one of the first members of the society (of which Brashman was the principal founder and the first president).

As a student Chebyshev wrote a paper, "Vychislenie korney uravneny" ("Calculation of the Roots of Equations"), in which he suggested an original iteration method for the approximate calculation of real roots of equations $y = f(x) = 0$ founded on the expansion into a series of an inverse function $x = F(y)$. The first terms of Chebyshev's general formula are

$$x = \alpha - \frac{f(\alpha)}{f'(\alpha)} - \left[\frac{f(\alpha)}{f'(\alpha)}\right]^2 \frac{f''(\alpha)}{2f'(\alpha)} - \cdots,$$

where α is an approximate value of the root x of the equation $f(x) = 0$ differing from the exact value by sufficiently little. Choosing a certain number of terms of the formula and successively calculating from the chosen value α further approximations $\alpha_1, \alpha_2, \cdots$, it is possible to obtain iterations of different orders. Iteration of the first order is congruent with the widely known Newton-Raphson method: $\alpha_{n+1} = \alpha_n - f(\alpha_n)/f'(\alpha_n)$. Chebyshev gives an estimation of error for his formula. This paper, written by Chebyshev for a competition on the subject announced by the department of physics and mathematics for the year 1840–1841, was awarded a silver medal, although it undoubtedly deserved a gold one. It was published only recently (2, V).

In the spring of 1841 Chebyshev graduated from Moscow University with a candidate (bachelor) of mathematics degree. Proceeding with his scientific work under Brashman's supervision, he passed his master's examinations in 1843, simultaneously publishing an article on the theory of multiple integrals in Liouville's *Journal des mathématiques pures et appliquées* and in 1844 an article on the convergence of Taylor series in Crelle's *Journal für die reine und angewandte Mathematik* (see 1*a*, I; 2, II). Shortly afterward he submitted as his master's thesis, *Opyt elementarnogo analiza teorii veroyatnostey* ("An Essay on an Elementary Analysis of the Theory of Probability"; (see 2, V). The thesis was defended in the summer of 1846 and was accompanied by "Démonstration élémentaire d'une proposition générale de la théorie des probabilités" (*Journal für die reine und angewandte Mathematik,* 1846; see 1*a*, I; 2, II), which was devoted to Poisson's law of large numbers. These works aimed at a strict but elementary deduction of the principal propositions of the theory of probability; of the wealth of mathematical analysis Chebyshev used only the expansion of $ln(1 + x)$ into a power series. In the article on Poisson's theorem we find an estimation of the number of tests by which it is possible to guarantee a definite proximity to unit probability of the assumption that the frequency of an event differs from the arithmetic mean of its probabilities solely within the given limits. Thus, even in Chebyshev's earliest publications one of the peculiar aspects of his work is manifest: he aspires to establish by the simplest means the most precise numerical evaluations of the limits within which the examined value lies.

It was almost impossible to find an appropriate teaching job in Moscow, so Chebyshev willingly accepted the offer of an assistant professorship at Petersburg University. As a thesis *pro venia legendi* he submitted "Ob integrirovanii pomoshchyu logarifmov" ("On Integration by Means of Logarithms"), written, at least in the first draft, as early as the end of 1843. The thesis, defended in the spring of 1847, incidentally solved a problem of integration of algebraic irrational functions in the final form that had been posed shortly before by Ostrogradski. The thesis was published posthumously, as late as 1930 (see 2, V), but Chebyshev included its principal results in his first publication on the subject in 1853.

In September 1847, at Petersburg University, Chebyshev began lecturing on higher algebra and the theory of numbers. Later he lectured on numerous other subjects, including integral calculus, elliptic functions, and calculus of finite differences; but he taught the theory of numbers as long as he was at the university (until 1882). From 1860 he regularly lectured on the theory of probability, which had previously been taught for a long time by V. Y. Bunyakovski. A. M. Lyapunov, who attended Chebyshev's lectures in the late 1870's, thus characterized them:

> His courses were not voluminous, and he did not consider the quantity of knowledge delivered; rather, he aspired to elucidate some of the most important aspects of the problems he spoke on. These were lively, absorbing lectures; curious remarks on the significance and importance of certain problems and scientific methods were always abundant. Sometimes he made a remark in passing, in connection with some concrete case they had considered, but those who attended always kept it in mind. Consequently, his lectures were highly stimulating; students received something new and essential at each lecture; he taught broader views and unusual standpoints [4, p. 18].

Soon after Chebyshev moved to St. Petersburg, he was hired by Bunyakovski to work on the new edition of Euler's works on the theory of numbers that had been undertaken by the Academy of Sciences. This edition (*L. Euleri Commentationes arithmeticae collectae,* 2 vols. [St. Petersburg, 1849]) comprised not only all of Euler's previously published papers on the subject but also numerous manuscripts from the Academy's archives; in addition Bunyakovski and Chebyshev contributed a valuable systematic review of Euler's arithmetical works. Probably this work partly inspired Chebyshev's own studies on the theory of numbers; these studies and the investigations of Chebyshev's disciples advanced the theory of numbers in Russia to a level as high as that reached a century before by Euler. Some problems of the theory of numbers had been challenged by Chebyshev earlier, however, in his thesis *pro venia legendi*. He devoted to the theory of numbers his monograph *Teoria sravneny* ("Theory of Congruences"; 7), which he submitted for a doctorate in mathematics. He defended it at Petersburg University on 27 May 1849 and a few days later was awarded a prize for it by the Academy of Sciences. Chebyshev's systematic analysis of the subject was quite independent and contained his own discoveries; it was long used as a textbook in Russian universities. It also contained the first of his two memoirs on the problem of distribution of prime numbers and other relevant problems; the second memoir, submitted to the Academy of Sciences in 1850, appeared in 1852. Through these two works, classics in their field, Chebyshev's name became widely known in the scientific world. Later Chebyshev returned only seldom to the theory of numbers.

In 1850 Chebyshev was elected extraordinary professor of mathematics at Petersburg University; in 1860 he became a full professor. This was a decade of very intensive work by Chebyshev in various fields. First of all, during this period he began his remarkable studies on the theory of mechanisms, which resulted in the theory of the best approximation of functions. From his early years Chebyshev showed a bent for construction of mechanisms; and his studies at Moscow University stimulated his interest in technology, especially mechanical engineering. In 1849–1851 he undertook a course of lectures on practical (applied) mechanics in the department of practical knowledge of Petersburg University (this quasi-engineering department existed for only a few years); he gave a similar course in 1852–1856 at the Alexander Lyceum in Tsarskoe Selo (now Pushkin), near St. Petersburg. Chebyshev's mission abroad, from July to November 1852, was another stimulus to his technological and mathematical work. In the evenings he talked with the best mathematicians of Paris, London, and Berlin or proceeded with his scientific work; morning hours were devoted to the survey of factories, workshops, and museums of technology. He paid special attention to steam engines and hinge-lever driving gears. He began to elaborate a general theory of mechanisms and in doing so met, according to his own words, certain problems of analysis that were scarcely known before (2, V, 249). These were problems of the theory of the best approximation of functions, which proved to be his outstanding contribution; in this theory his technological and mathematical inclinations were synthesized.

Back in St. Petersburg, Chebyshev soon submitted to the Academy of Sciences his first work on the problem of the best approximation of functions, prepared mainly during his journey and published in 1854. This was followed by another work on the subject, submitted in 1857 and published in 1859. These two papers marked the beginning of a great cycle of work in which Chebyshev was engaged for forty years. While in Europe, Chebyshev continued his studies on the integration of algebraic functions. His first published work on the problem, far surpassing the results at which he had arrived in the thesis *pro venia legendi,* appeared in 1853. Chebyshev published papers on this type of problem up to 1867, the object of them being to determine conditions for integration in the final form of different classes of irrational functions. Here, as in other cases, research was associated with university teaching; Chebyshev lectured on elliptic functions for ten years, until 1860.

In 1853 Chebyshev was voted an adjunct (i.e., junior academician) of the Petersburg Academy of Sciences with the chair of applied mathematics. Speaking

for his nomination, Bunyakovski, Jacobi, Struve, and the permanent secretary of the Academy, P. N. Fuss, emphasized that Chebyshev's merits were not restricted to mathematics; he had also done notable work in practical mechanics. In 1856 Chebyshev was elected an extraordinary academician and in 1859 ordinary academician (the highest academic rank), again with the chair of applied mathematics.

From 1856 Chebyshev was a member of the Artillery Committee, which was charged with the task of introducing artillery innovations into the Russian army. In close cooperation with the most eminent Russian specialists in ballistics, such as N. V. Maievski, Chebyshev elaborated mathematical devices for solving artillery problems. He suggested (1867) a formula for the range of spherical missiles with initial velocities within a certain limit; this formula was in close agreement with experiments. Some of Chebyshev's works on the theory of interpolation were the result of the calculation of a table of fire effect based on experimental data. Generally, he contributed significantly to ballistics.

Simultaneously Chebyshev began his work with the Scientific Committee of the Ministry of Education. Like Lobachevski, Ostrogradski, and a number of other Russian scientists, Chebyshev was active in working for the improvement of the teaching of mathematics, physics, and astronomy in secondary schools. For seventeen years, up to 1873, he participated in the elaboration of syllabi for secondary schools. His concise but solid reviews were of great value to the authors of textbooks that the Scientific Committee was supposed, as one of its principal functions, to approve or reject.

From the middle of the 1850's, the theory of the best approximation of functions and the construction of mechanisms became dominant in Chebyshev's work. Studies on the theory of functions embraced a very great diversity of relevant problems: the theory of orthogonal polynomials, the doctrine of limiting values of integrals, the theory of moments, interpolation, methods of approximating quadratures, etc. In these studies the apparatus of continued fractions, brilliantly employed by Chebyshev in many studies, was further improved.

From 1861 to 1888 Chebyshev devoted over a dozen articles to his technological inventions, mostly in the field of hinge-lever gears. Examples of these devices are preserved in the Mathematical Institute of the Soviet Academy of Sciences in Moscow and in the Conservatoire des Arts et Métiers in Paris.

In the 1860's Chebyshev returned to the theory of probability. One of the reasons for this new interest was, perhaps, his course of lectures on the subject started in 1860. He devoted only two articles to the theory of probability, but they are of great value and designate the beginning of a new period in the development of this field. In the article of 1866 Chebyshev suggested a very wide generalization of the law of large numbers. In 1887 he published (without extensive demonstration) a corresponding generalization of the central limit theorem of Moivre and Laplace.

Besides the above-mentioned mathematical and technological fields, which were of primary importance in Chebyshev's life and work, he showed lively interest in other problems of pure and applied mathematics. (His studies in cartography will be mentioned later.) His paper of 1878, "Sur la coupe des vêtements" (1a, II; 2, V), provided the basis for a new branch of the theory of surfaces. Chebyshev investigated a problem of binding a surface with cloth that is formed in the initial flat position with two systems of nonextensible rectilinear threads normal to one another. When the surface is bound with cloth the "Chebyshev net," whose two systems of lines form curvilinear quadrangles with equal opposite sides, appears. Wrapping a surface in cloth is a more general geometrical transformation than is deformation of a surface, which preserves the lengths of all the curved lines; distances between the points of the wrapped cloth that are situated on different threads are, generally speaking, changed in wrapping. In recent decades Chebyshev's theory of nets has become the object of numerous studies.

Theoretical mechanics also drew Chebyshev's attention. Thus, in 1884 he told Lyapunov of his studies on the problem of the ring-shaped form of equilibrium of a rotating liquid mass the particles of which are mutually attracted according to Newton's law. It is hard to know how far Chebyshev advanced in this field, for he published nothing on the subject. Still, the very problem of the form of equilibrium of a rotating liquid mass, which he proposed to Lyapunov, was profoundly investigated by the latter, who, along with Markov, was Chebyshev's most prominent disciple.

Among Chebyshev's technological inventions was a calculating machine built in the late 1870's. In 1882 he gave a brief description of his machine in the article "Une machine arithmétique à mouvement continu" (Ia, II; 2, IV). The first model (*ca.* 1876) was intended for addition and subtraction; he supplemented it with an apparatus enabling one to multiply and divide as well. Examples of the machine are preserved in the Museum of History in Moscow and in the Conservatoire des Arts et Métiers in Paris.

During this period Chebyshev was active in the work of various scientific societies and congresses.

Between 1868 and 1880 he read twelve reports at the congresses of Russian naturalists and physicians, and sixteen at the sessions of the Association Française pour l'Avancement des Sciences between 1873 and 1882; it was at these sessions that he read "Sur la coupe des vêtements" and reported on the calculating machine. He gave numerous demonstrations of his technological inventions both at home and abroad. Chebyshev was in contact with the Moscow and St. Petersburg mathematical societies and with the Moscow Technological College (now Bauman Higher Technological College).

In the summer of 1882, after thirty-five years of teaching at Petersburg University, Chebyshev retired from his professorship, although he continued his work at the Academy of Sciences. Nonetheless, he was constantly in touch with his disciples and young scientists. He held open house once a week, and K. A. Posse states that "hardly anybody left these meetings without new ideas and encouragement for further endeavor" (2, V, 210); it was sufficient only that the problem be relevant to the fields in which Chebyshev had been interested. When Chebyshev was over sixty, he could not work at his former pace; nevertheless, he published some fifteen scientific papers, including a fundamental article on the central limit theorem. He submitted his last work to the Academy of Sciences only a few months before his death at the age of seventy-three.

Besides being a member of the Petersburg Academy of Sciences, Chebyshev was elected a corresponding (1860) and—the first Russian scientist to be so honored—a foreign member (1874) of the Academy of Sciences of the Institut de France, a corresponding member of the Berlin Academy of Sciences (1871), a member of the Bologna Academy (1873), and a foreign member of the Royal Society of London (1877), of the Italian Royal Academy (1880), and of the Swedish Academy of Sciences (1893). He was also an honorary member of all Russian universities and of the Petersburg Artillery Academy. He was awarded numerous Russian orders and the French Legion of Honor.

Petersburg Mathematical School. Chebyshev's importance in the history of science consists not only in his discoveries but also in his founding of a great scientific school. It is sometimes called the Chebyshev school, but more frequently the Petersburg school because its best-known representatives were almost all educated at Petersburg University and worked either there or at the Academy of Sciences. The Petersburg mathematical school owes its existence partly to the activity of Chebyshev's elder contemporaries, such as Bunyakovski and Ostrogradski; nevertheless, it was

Chebyshev who founded the school, directed and inspired it for many years, and influenced the trend of mathematics teaching at Petersburg University. For over half a century mathematical chairs there were occupied by Chebyshev's best disciples or their own disciples. Thus the mathematics department of Petersburg University achieved a very high academic level. Some disciples of Chebyshev took his ideas to other Russian universities.

Chebyshev was highly endowed with the ability to attract beginners to creative work, setting them tasks demanding profound theoretical investigation to solve and promising brilliant results. The Petersburg school included A. N. Korkin, Y. V. Sohotski, E. I. Zolotarev, A. A. Markov, A. M. Lyapunov, K. A. Posse, D. A. Grave, G. F. Voronoi, A. V. Vassiliev, V. A. Steklov, and A. N. Krylov. Chebyshev, however, was such a singular individual that he also influenced scientists who did not belong to his school, both in Russia (e.g., N. Y. Sonin) and abroad. During the latter half of the nineteenth and the beginning of the twentieth centuries the Petersburg mathematical school was one of the most prominent schools in the world and the dominant one in Russia. Its ideas and methods formed an essential component of many divisions of pure and applied mathematics and still influence their progress.

Although a great variety of scientific trends were represented in the Petersburg mathematical school, the work of Chebyshev and his followers bore some important common characteristics. Half seriously, Chebyshev shortly before his death said to A. V. Vassiliev that previously mathematics knew two periods: during the first the problems were set by gods (the Delos problem of the duplication of a cube) and during the second by demigods, such as Fermat and Pascal. "We now entered the third period, when the problems were set by necessity" (16, p. 59). Chebyshev thought that the more difficult a problem set by scientific or technological practice, the more fruitful the methods suggested to solve it would be and the more profound a theory arising in the process of solution might be expected to be.

The unity of theory and practice was, in Chebyshev's view, the moving force of mathematical progress. He said in a speech entitled "Cherchenie geograficheskikh kart" ("Drawing Geographical Maps"), delivered at a ceremonial meeting of Petersburg University in 1856:

Mathematical sciences have attracted especial attention since the greatest antiquity; they are attracting still more interest at present because of their influence on industry and arts. The agreement of theory and practice brings most beneficial results; and it is not exclusively the practical side that gains; the sciences are advancing

under its influence as it discovers new objects of study for them, new aspects to exploit in subjects long familiar. In spite of the great advance of the mathematical sciences due to works of outstanding geometers of the last three centuries, practice clearly reveals their imperfection in many respects; it suggests problems essentially new for science and thus challenges one to seek quite new methods. And if theory gains much when new applications or new developments of old methods occur, the gain is still greater when new methods are discovered; and here science finds a reliable guide in practice [2, V, 150].

Among scientific methods important for practical activity Chebyshev especially valued those necessary to solve the same general problem:

How shall we employ the means we possess to achieve the maximum possible advantage? Solutions of problems of this kind form the subject of the so-called theory of the greatest and least values. These problems, which are purely practical, also prove especially important for theory: all the laws governing the movement of ponderable and imponderable matter are solutions of this kind of problem. It is impossible to ignore their special influence upon the advance of mathematical sciences [ibid.].

These statements (reminding one somewhat of Euler's ideas on the universal meaning of the principle of least action) are illustrated by Chebyshev with examples from the history of mathematics and from his own works on the theory of mechanisms and the theory of functions. The speech quoted above is specially devoted to solution of the problem of searching for such conformal projections upon a plane of a given portion of the earth's surface under which change in the scale of image (different in various parts of a map) is the least, so that the image as a whole is the most advantageous. The projection sought bears a characteristic particularity: on the border of the image the scale preserves the same value. Demonstration of this theorem of Chebyshev's was first published by Grave in 1894.

Chebyshev's general approach to mathematics quite naturally resulted in his aspiration toward the effective solution of problems and the discovery of algorithms giving either an exact numerical answer or, if this proved impossible, an approximation ready for scientific and practical applications. He interpreted the strictness of the theory in the event of approximate evaluations as a possibility of precise definition of limits not trespassed by the error of approximation. Chebyshev was a notable representative of the "mathematics of inequalities" of the latter half of the nineteenth century. His successors held similar views.

Lastly, not the least characteristic feature of

Chebyshev's works from the early period on was his inclination toward possibly elementary mathematical apparatus, particularly his almost exclusive use of functions in the real domain. He was especially adept at using continued fractions. Many other scholars of his school made wider use of contemporary analysis, especially of the theory of functions of complex variables.

The Petersburg school was concerned with quite a number of subjects. It dealt primarily with the domains of pure and applied mathematics investigated by Chebyshev himself and developed by his disciples. However, Chebyshev's followers worked in other areas of mathematical sciences that were far from the center of Chebyshev's own interests. Growing in numbers and ability, the Petersburg school gradually became an aggregate of scientific schools brought together by similar principles of study and closely connected both on the level of ideas and on the personal level; they differed mainly in their predominant mathematical subjects: the theory of numbers, the theory of the best approximation of functions, the theory of probability, the theory of differential equations, and mathematical physics.

Although the work of Chebyshev and his school was independent in the formulation of numerous problems and in the elaboration of methods to solve them and discovered large new domains of mathematical study, it was closely related to mathematics of the eighteenth and the first half of the nineteenth centuries. In Russia the school developed the tradition leading back to Euler, whose works were thoroughly studied and highly valued. Much as they differed as individuals (the difference in mathematics of their respective epoches was no less great), Chebyshev and Euler had much in common. Both were interested in a great variety of problems, from the theory of numbers to mechanical engineering. Both were aware of the profound connection of mathematical theory with its applications and tended to set themselves concrete problems as a source of theoretical conclusions that were later generalized; both, on the other hand, understood the vital necessity of developing mathematics in its entirety, including problems the solution of which did not promise any immediate practical gain. Finally, both were always seeking most effective solutions that approached computing algorithms.

It is important to note that at the beginning of the present century younger representatives of the Chebyshev school started to bring about its contact with other trends in mathematics, which soon led to great progress.

Theory of Numbers. It had been proved in ancient Greece that there exist infinitely many prime numbers

2, 3, 5, 7, $\cdots$. This principal result in the doctrine of the distribution of prime numbers remained an isolated result until the end of the eighteenth century, when the first step in investigation of the frequency of prime numbers in the natural number series 1, 2, 3, 4, $\cdots$ was made: Legendre suggested in 1798–1808 the approximate formula

$$\pi(x) \approx \frac{x}{\ln x - 1.08366}$$

to express the number of prime numbers not exceeding a given number x, e.g., for the function designated $\pi(x)$. This formula accorded well with the table of prime numbers from 10,000 to 1,000,000. In his article "Sur la fonction qui détermine la totalité des nombres premiers inférieurs à une limite donnée" (1849; see 1a, I; 2, I) Chebyshev, making use of the properties of Euler's zeta function in the real domain, proved the principal inaccuracy of Legendre's approximate formula and made a considerable advance in the study of the properties of the function $\pi(x)$. According to Legendre's formula, the difference $x/\pi(x) - \ln x$ for $x \to \infty$ has the limit -1.08366. Chebyshev demonstrated that this difference cannot have a limit differing from -1. With sufficiently great x the integral $\int_2^x dx/\ln x$ gives better approximations to $\pi(x)$ than Legendre's and similar formulas; besides, the difference

$$\frac{x}{\ln x - 1.08366} - \int_2^x \frac{dx}{\ln x},$$

inconsiderable within the limits of the tables used by Legendre, reaches the minimum for $x \approx 1,247,689$ and then increases without limit with the increase of x.

It also followed from Chebyshev's theorems that the ratio of the function $\pi(x)$ to $\int_2^x dx/\ln x$ or to $x/\ln x$ cannot, for $x \to \infty$, have a limit differing from unity.

Chebyshev later continued the study of the properties of $\pi(x)$. In his "Mémoire sur les nombres premiers" (1850, pub. 1852; see 1a, I; 2, II) he demonstrated that $\pi(x)$ can differ from $x/\ln x$ by no more than approximately 10 percent—more exactly, that

$$0.92129x/\ln x < \pi(x) < 1.10555x/\ln x$$

(although he stated this result in a slightly different form).

Other remarkable discoveries were described in these two articles. The second article demonstrates Bertrand's conjecture (1845) that for $n > 3$, between n and $2n - 2$ there is always at least one prime number. In it Chebyshev also proved some theorems on convergence and on the approximate calculation of the sums of infinite series the members of which are functions of successive prime numbers (the first series of this kind were studied by Euler). In a letter to P. N. Fuss published in 1853 (1a, I; 2, II) Chebyshev set a problem of estimating the number of prime numbers in arithmetical progressions and gave some results concerning progressions with general members of the form $4n + 1$ and $4n + 3$.

Chebyshev's studies on the distribution of prime numbers were developed by numerous scientists in Russia and abroad. Important advances were made in 1896 by Hadamard and Vallée-Poussin, who, employing analytical functions of a complex variable, proved independently the asymptotic law of distribution of prime numbers:

$$\lim_{x \to \infty} \frac{\pi(x)}{x/\ln x} = 1.$$

Studies in this field of the theory of numbers are being intensively conducted.

Among Chebyshev's other works on the theory of numbers worthy of special attention is his article "Ob odnom arifmeticheskom voprose" ("On One Arithmetical Problem"; 1866; 1a, I; 2, II), which served as a point of departure for a series of studies devoted to a linear heterogeneous problem of the theory of diophantine approximations. The problem, in which Hermite, Minkowski, Remak, and others were later interested, was completely solved in 1935 by A. Y. Khintchine.

Integration of Algebraic Functions. Chebyshev's studies on the integration of algebraic functions were closely connected with the work of Abel, Liouville, and, in part, Ostrogradski. In the article "Sur l'integration des différentielles irrationelles" (1853; see 1a, I; 2, II) Chebyshev succeeded in giving a complete solution of the problem of defining the logarithmic part of the integral

$$\int \frac{f(x)\,dx}{F(x)\sqrt[m]{\theta(x)}}$$

for the case when it is expressed in final form—here the functions $f(x)$, $F(x)$, $\theta(x)$ are integral and rational, and m is any positive integer. But the article is known mostly for the final solution of the problem of the integration of the binomial differential $x^m(a + bx^n)^p\,dx$ it contains; here m, n, p are rational numbers. Generalizing Newton's result, Goldbach and Euler showed that this type of integral is expressible in elementary functions in any of the three cases when p is an integer; $(m + 1)/n$ is an integer; and $(m + 1)/n + p$ is an integer. Chebyshev demonstrated that the three cases are the only cases when the integral mentioned is taken in the final form.

This theorem is included in all textbooks on integral calculus.

In the theory of elliptic integrals Chebyshev substantially supplemented Abel's results. Integration of any elliptic differential in the final form is reduced to integration of a fraction that has a linear function $x + A$ in the numerator and a square root of a polynomial of the fourth degree in the denominator. Chebyshev considered the problem in the article "Sur l'intégration de la différentielle

$$\frac{(x + A)\,dx}{\sqrt{x^4 + \alpha x^3 + \beta x^2 + \gamma x + \delta}}\text{''}$$

(1861; see 1a, I; 2, II). It is supposed that the polynomial in the denominator has no multiple roots (in the case of multiple roots the differential is integrated quite easily). The elliptic differential in question either is not integrable in elementary functions or is integrable for one definite value of the constant A. Abel had shown that the latter case occurred if a continued fraction formed by expansion of

$$\sqrt{x^4 + \alpha x^3 + \beta x^2 + \gamma x + \delta}$$

is periodic. However, Abel could offer no final criterion enabling one to ascertain nonperiodicity of such an expansion. Chebyshev gave a complete and efficient solution of the problem with rational numbers $\alpha, \beta, \gamma, \delta$: he found a method enabling one to ascertain nonperiodicity of such an expansion by means of the finite number of operations, which in turn depends on the finite number of integer solutions of a system of two equations with three unknowns; he also determined the limit of the number of operations necessary in case of integrability. He did not publish any complete demonstration of his algorithm; this was done by Zolotarev in 1872. Soon Zolotarev suggested a solution of the same problem for any real coefficients α, β, γ, δ, for which purpose he devised his own variant of the theory of ideals.

Chebyshev also studied the problem of integrability in the final form of some differentials containing a cube root of a polynomial. In Russia this direction of study was followed by I. L. Ptashitski and I. P. Dolbnia, among others.

Theory of the Best Approximation of Functions. It was said that Chebyshev had approached the theory of best approximation of functions from the problems of the theory of hinge mechanisms, particularly from the study of the so-called Watt parallelogram employed in steam engines and other machines for the transformation of rotating movement into rectilinear movement. In fact, it is impossible to obtain strictly rectilinear movement by this means, which produces a destructive effect. Attempting somehow to reduce the deviation of the resultant movement from the rectilinear, engineers searched empirically for suitable correlation between the parts of mechanisms. Chebyshev set the task of elaborating a sound theory of the problem, which was lacking; he also devised a number of curious mechanisms that, although they could not strictly secure rectilinear movement, could compete with "strict" mechanisms because the deviation was very small. Thus, the virtually exact seven-part rectifying device suggested by Charles Peaucellier and independently by L. I. Lipkin is, in view of the complexity of construction, less useful in practice than Chebyshev's four-part device described in the article "Ob odnom mekhanizme" ("On One Mechanism"; 1868; see 1a, II; 2, IV).

Chebyshev made a profound investigation of the elements of a hinge mechanism, setting out to achieve the smallest deviation possible of the trajectory of any points from the straight line for the whole interval studied. A corresponding mathematical problem demanded that one choose, from among the functions of the given class taken for approximation of the given function, that function with which the greatest modulo error is the smallest under all considered values of the argument. Some special problems of this kind had previously been solved by Laplace, Fourier, and Poncelet. Chebyshev laid foundations for a general theory proceeding from the approximation of functions by means of polynomials. The problem of approximation of the given function by means of polynomials might be formulated differently. Thus, in an expansion of the given function $f(x)$ into a Taylor series of powers of the difference $x - a$, the sum of the first $n + 1$ members of the series is a polynomial of nth degree that in the neighborhood of the value $x = a$ gives the best approximation among all polynomials of the same degree. Chebyshev set the task of achieving not a local best approximation but the uniform best approximation throughout the interval; his object was to find among all polynomials $P_n(x)$ of nth degree such a polynomial that the maximum $|f(x) - P_n(x)|$ for this interval is the smallest.

In the memoir "Théorie des mécanismes connus sous le nom de parallélogrammes" (1854; see 1a, I; 2, II), which was the first in a series of works in this area, Chebyshev considered the problem of the best approximation of the function $f(x) = x^n$ by means of polynomials of degree $n - 1$; that is, he considered the problem of the determination of the polynomial of the nth degree, $x^n + p_1 x^{n-1} + \cdots + p_n$ with the leading coefficient equal to unity, deviating least from zero. This formulation of the problem engendered the frequently used term "theory of polynomials deviating least from zero" (the term was used by Chebyshev himself). In the case of the interval $(-1, +1)$ this polynomial is

$$\frac{1}{2^{n-1}} T_n(x) = \frac{1}{2^{n-1}} \cos (n \arccos x)$$

$$= x^n - \frac{n}{1!} \frac{x^{n-2}}{2^2}$$

$$+ \frac{n(n-3)}{2!} \frac{x^{n-4}}{2^4} - \frac{n(n-4)(n-5)}{3!} \frac{x^{n-6}}{2^6} + \cdots,$$

and its maximum deviation from zero is $\frac{1}{2}^{n-1}$. Polynomials $T_n(x)$, named for Chebyshev, form an orthogonal system with respect to a weight function

$$\frac{1}{\sqrt{1 - x^2}}.$$

In his next long memoir, "Sur les questions de minima qui se rattachent à la représentation approximative des fonctions" (1859; see 1a, I; 2, II), Chebyshev extended the problem to all kind of functions $F(x, p_1, p_2, \cdots, p_n)$ depending on n parameters, expressed general views concerning the method of solution, and gave a complete analysis of two cases when F is a rational function; in this work some curious theorems on the limits of real roots of algebraic equations were obtained. Varying restrictions might be imposed upon the function of the best approximation that is to be determined; Chebyshev also solved several problems of this type.

Gradually considering various problems either directly relevant to the theory of the best approximation of functions or connected with it, Chebyshev obtained important results in numerous areas.

(a) The theory of interpolation and, especially, interpolation on the method of least squares (1855–1875).

(b) The theory of orthogonal polynomials. Besides polynomials $T_n(x)$ deviating least from zero, Chebyshev in 1859 proceeded from consideration of different problems to the study of other orthogonal systems, such as Hermite and Laguerre polynomials. However, he did not take up the determination of polynomials under the condition of orthogonality in the given interval with respect to the given weight function; he introduced polynomials by means of an expansion in continued fractions of certain integrals of the

$$\int_a^b \frac{p(x)\, dx}{z - x}$$

type, where $p(x)$ is the weight function.

(c) The theory of moments was first treated by Chebyshev in his article "Sur les valeurs limites des intégrales" (1874; see 1a, I; 2, III). Here the following problem is considered: given the values of moments of different orders of an unknown function $f(x)$, e.g., of the integrals

$$\int_A^B f(x)\, dx = C_0, \quad \int_A^B xf(x)\, dx = C_1, \cdots,$$

$$\int_A^B x^m f(x)\, dx = C_m$$

in an interval (A,B) where $f(x) > 0$, one is required to find the limits within which the value of the integral

$$\int_a^b f(x)\, dx$$

lies $(A < a < b < B)$. Chebyshev once more connected the investigation of the problem with expansion of the integral

$$\int_A^B \frac{f(x)\, dx}{z - x}$$

in continued fraction; in conclusion he gave a detailed solution of the problem for a special case $m = 2$ formulated in mechanical interpretation: "Given length, weight, site of the center of gravity, and the moment of inertia of a material straight line with an unknown density that changes from one point to another, one is required to find the closest limits with respect to the weight of a certain segment of this straight line" (2, III, 65). Chebyshev's work on estimations of integrals received important application in his studies on the theory of probability.

(*d*) Approximate calculus of definite integrals. In his article "Sur les quadratures" (1874; 1a, II; 2, III) Chebyshev, proceeding from work by Hermite, suggested new general formulas of approximate quadratures in which all the values of the integrand are introduced under the same coefficient or, at least, under coefficients differing only in sign. The aforementioned characteristic of the coefficients renders Chebyshev's quadrature formulas very suitable for the calculus in many cases; A. N. Krylov used them in his studies on the theory of ships. One of the formulas is

$$\int_{-1}^1 f(x)\, dx \approx k[f(x_1) + f(x_2) + \cdots + f(x_n)];$$

the values $x_1, x_2, x_3, \cdots$ are obtained from an equation of the nth degree: $k = \frac{2}{n}$. Chebyshev himself calculated these values for $n = 2, 3, 4, 5, 6, 7$; for $n = 8$ the equation has no real roots and the formula is unusable; for $n = 9$ the roots are real again but, as S. N. Bernstein demonstrated in 1937, beginning at $n = 10$ the formula is again unusable.

This large cycle of Chebyshev's works was carried further by Zolotarev, Markov, Sonin, Posse, Steklov, Stieltjes, Riesz, and many others; work in these and new fields is being continued. It is worth special mention that at the beginning of the twentieth century the theory of the best approximation acquired essentially new features through the connections between Chebyshev's ideas and methods, on the one hand, and those developed in western Europe, on the other. Works of S. N. Bernstein and his disciples were of primary importance here.

Theory of Probability. In his article "O srednikh velichinakh" ("On Mean Values"; 1866; see 1a, I; 2, II) Chebyshev, using an inequality previously deduced by J. Bienaymé, gave a precise and very simple demonstration of the generalized law of large numbers that might be thus expressed in modern terms: If $x_1, x_2, x_3, \cdots$ are mutually independent in pairs of random quantities, with expectation values $a_1, a_2, a_3, \cdots$ and dispersions $b_1, b_2, b_3, \cdots$, the latter being uniformly limited—e.g., all $b_n \leq C$—then for any $\varepsilon > 0$ the probability P of an inequality

$$\left| \frac{x_1 + x_2 + \cdots + x_n}{n} - \frac{a_1 + a_2 + \cdots + a_n}{n} \right| < \varepsilon$$

is $\geqslant 1 - (C/n\varepsilon^2)$. From this it follows immediately that

$$\lim_{n \to \infty} P\left(\left| \frac{x_1 + x_2 + \cdots + x_n}{n} - \frac{a_1 + a_2 + \cdots + a_n}{n} \right| < \varepsilon \right)$$
$$= 1.$$

The theorems of Poisson and Jakob Bernoulli are only special cases of Chebyshev's law of large numbers for sequences of random quantities.

Developing his method of moments and of estimation of the limit values of integrals, Chebyshev also managed to extend to sequences of independent random quantities the central limit theorem of Moivre and Laplace: within the framework of former suppositions, supplemented with the condition that there exist expectation values (moments) of any order and that they all are uniformly limited,

$$\lim_{n \to \infty} P\left(t_1 < \frac{x_1 + x_2 + \cdots + x_n - a_1 - a_2 - \cdots - a_n}{\sqrt{2(b_1 + b_2 + \cdots + b_n)}} < t_2 \right)$$
$$= \frac{1}{\sqrt{\pi}} \int_{t_1}^{t_2} e^{-t^2}\, dt.$$

This theorem and the draft of its demonstration were published by Chebyshev in the article "O dvukh teoremakh otnositelno veroyatnostey" ("On Two Theorems Concerning Probability"; 1887; see 1a, II; 2, III); the first theorem mentioned in the title is the law of large numbers. Chebyshev's second theorem enabled one to apply, on a larger scale, the theory of probability to mathematical statistics and natural sciences; both regard the phenomenon under study as resulting from common action of a great number of random factors, each factor displaying considerably smaller influence independently in comparison with their influence as a set. According to this theorem, such common action closely follows the normal distribution law. Chebyshev's demonstration was supplemented a decade later by Markov.

Chebyshev's studies on limit theorems were successfully developed by his disciples and successors, first by Markov, Lyapunov, and Bernstein and later by numerous scientists. A. N. Kolmogorov, a foremost authority in the field, says:

> From the standpoint of methodology, the principal meaning of the radical change brought about by Chebyshev is not exclusively that he was the first mathematician to insist on absolute accuracy in demonstration of limit theorems (the proofs of Moivre, Laplace, and Poisson were not wholly consistent on the formal logical grounds in which they differ from those of Bernoulli, who demonstrated his limit theorem with exhaustive arithmetical accuracy). The principal meaning of Chebyshev's work is that through it he always aspired to estimate exactly in the form of inequalities absolutely valid under any number of tests the possible deviations from limit regularities. Further, Chebyshev was the first to estimate clearly and make use of the value of such notions as "random quantity" and its "expectation (mean) value." These notions were known before him; they are derived from fundamental notions of the "event" and "probability." But random quantities and their expectation values are subject to a much more suitable and flexible algorithm [21, p. 56].

BIBLIOGRAPHY

I. Original Works. During his lifetime Chebyshev's works appeared mainly in publications of the Petersburg Academy of Sciences—*Mémoires* or *Bulletin de l'Académie des sciences de St. Pétersbourg*—in Liouville's *Journal de mathématiques pures et appliquées,* in *Matematichesky sbornik,* and others; some are in Russian and some in French. A complete bibliography of Chebyshev's works, except some minor articles published in (8), is in (2), V, 467–471.

Chebyshev's works are the following:

(1) *Sochinenia,* A. A. Markov and N. Y. Sonin, eds., 2 vols. (St. Petersburg, 1899–1907). This Russian ed. is nearly complete but does not contain his master's thesis (6) and doctor's thesis (7) and some articles included in (2).

(1a) *Oeuvres de P. L. Tchebychef,* A. A. Markov and N. Y. Sonin, eds., 2 vols. (St. Petersburg, 1899–1907), a French version of (1). Both (1a) and (1) contain a biographical note based entirely on (9).

(2) *Polnoe sobranie sochineny,* 5 vols. (Moscow-Leningrad, 1944–1951): I, *Teoria chisel;* II–III, *Matematichesky analiz;* IV, *Teoria mekhanizmov;* V, *Prochie sochinenia. Biograficheskie materïaly.* This ed. contains very valuable scientific commentaries that are largely completed and developed in (3).

(3) *Nauchnoe nasledie P. L. Chebysheva,* 2 vols. (Moscow-Leningrad, 1945): I, *Matematika;* II, *Teoria mekhanizmov.* This contains important articles by N. I. Akhiezer, S. N. Bernstein, I. M. Vinogradov and B. N. Delaunay, V. V. Golubev, V. L. Goncharov, I. I. Artobolevsky and N. I. Levitsky, Z. S. Blokh, and V. V. Dobrovolsky.

(4) *Izbrannye matematicheskie trudy* (Moscow-Leningrad, 1946).

(5) *Izbrannye trudy* (Moscow, 1955).

(6) *Opyt elementarnogo analiza teorii veroyatnostey* ("An Essay on an Elementary Analysis of the Theory of Probability"; Moscow, 1845; also [2], V), his master's thesis.

(7) *Teoria sravneny* ("Theory of Congruences"; · St. Petersburg, 1849, 1879, 1901; also [2], I), his doctoral thesis. Trans. into German as *Theorie der Congruenzen* (Berlin, 1888) and into Italian as *Teoria delle congruenze* (Rome, 1895).

(8) V. E. Prudnikov, "O statyakh P. L. Chebysheva, M. V. Ostrogradskogo, V. Y. Bunyakovskogo i I. I. Somova v Entsiklopedicheskom slovare, sostavlennom russkimi uchenymi i literatorami,'" in *Istoriko-matematicheskie issledovania,* no. 6 (1953).

II. Secondary Literature. On Chebyshev or his work, see the following:

(9) K. A. Posse, "P. L. Chebyshev," in S. A. Vengerov, *Kritiko-bibliografichesky slovar russkikh pisateley i uchenykh,* VI (St. Petersburg, 1897–1904; see also [1], II; [1a], II; [2], I).

(10) A. M. Lyapunov, *Pafnuty Lvovich Chebyshev* (Kharkov, 1895; see also [4], pp. 9–21).

(11) A. Wassiliev, "P. Tchébychef et son oeuvre scientifique," in *Bollettino di bibliografia e storia delle scienze matematiche,* I (Turin, 1898). Trans. into German as *P. L. Tchebychef und seine wissenschaftliche Leistungen* (Leipzig, 1900).

(12) N. Delaunay, *Die Tschebyschefschen Arbeiten in der Theorie der Gelenkmechanizmen* (Leipzig, 1900).

(13) V. G. Bool, *Pribory i mashiny dlya mekhanicheskogo proizvodstva matematicheskikh deystvy* (Moscow, 1896).

(14) L. E. Dickson, *History of the Theory of Numbers,* 3 vols. (Washington, D.C., 1919–1927).

(15) V. A. Steklov, *Teoria i praktika v issledovaniakh P. L. Chebysheva* (Petrograd, 1921), also in *Uspekhi matematicheskikh nauk,* **1,** no. 2 (1946), 4–11.

(16) A. V. Vassiliev, *Matematika,* pt. 1 (Petrograd, 1921), 43–61.

(17) L. Bianchi, *Lezzioni di geometria differenziale,* I, pt. 1 (Bologna, 1922), 153–192.

(18) A. N. Krylov, *Pafnuty Lvovich Chebyshev. Biografichesky ocherk* (Moscow–Leningrad, 1944).

(19) N. I. Akhiezer, "Kratky obzor matematicheskikh trudov P. L. Chebysheva," in (4), pp. 171–188.

(20) S. N. Bernstein, "Chebyshev, yego vlianie na razvitie matematiki," in *Uchenye zapiski Moskovskogo gosudarstvennogo universiteta,* no. 91 (1947), 35–45.

(21) A. N. Kolmogorov, "Rol russkoy nauki v razvitii teorii veroyatnostey," *ibid.,* 53–64.

(22) B. N. Delone, *Peterburgskaya shkola teorii chisel* (Moscow–Leningrad, 1947), pp. 5–42.

(23) B. V. Gnedenko, *Razvitie teorii veroyatnostey v Rossii.* Trudy Instituta istorii yestestvoznania, II (Moscow–Leningrad, 1948), 394–400.

(24) Y. L. Geronimus, *Teoria ortogonalnykh mnogochlenov. Obzor dostizheny otechestvennoy matematiki* (Moscow–Leningrad, 1950), *passim.*

(25) L. Y. Sadovsky, "Iz istorii razvitia mashinnoy matematiki v Rossii," in *Uspekhi matematicheskikh nauk,* **5,** no. 2 (1950), 57–71.

(26) M. G. Kreyn, "Idei P. L. Chebysheva i A. A. Markova v teorii predelnykh velichin integralov i ikh dalneyshee razvitie," *ibid.,* **6,** no. 4 (1951), 3–24.

(27) S. A. Yanovskaya, *Dva dokumenta iz istorii Moskovskogo universiteta,* Vestnik Moskovskogo Universiteta, no. 8 (1952).

(28) N. I. Akhiezer, "P. L. Chebyshev i yego nauchnoe nasledie," in (5), pp. 843–887.

(29) K. R. Biermann, *Vorschläge zur Wahl von Mathematikern in die Berliner Akademie* (Berlin, 1960), pp. 41–43.

(30) I. Y. Depman, "S.-Peterburgskoe matematicheskoe obshchestvo," in *Istoriko-matematicheskie issledovania,* no. 13 (1960), 11–106.

(31) *Istoria yestestvoznania v Rossii,* N. A. Figurovsky, ed. II (Moscow, 1961).

(32) A. A. Gusak, "Predystoria i nachalo razvitia teorii priblizhenia funktsy," in *Istoriko-matematicheskie issledovania,* no. 14 (1961), 289–348.

(33) L. Y. Maystrov, "Pervy arifmometr v Rossii," *ibid.,* 349–354.

(34) A. A. Kiselev and E. P. Ozhigova, "P. L. Chebyshev na siezdakh russkikh yestestvoispytateley i vrachey," *ibid.,* no. 15 (1963), 291–317.

(35) *Istoria Akademii nauk SSSR,* II (Moscow–Leningrad, 1964).

(36) V. E. Prudnikov, *P. L. Chebyshev, ucheny i pedagog,* 2nd ed. (Moscow, 1964).

(37) A. N. Bogolyubov, *Istoria mekhaniki mashin* (Kiev, 1964).

(38) V. P. Bychkov, "O razvitii geometricheskikh idey P. L. Chebysheva," in *Istoriko-matematicheskie issledovania,* no. 17 (1966), 353–359.

(39) A. P. Youschkevitch, "P. L. Chebyshev i Frantsuzskaya Akademia nauk," in *Voprosy istorii yestestvoznania i tekhniki,* no. 18 (1965), 107–108.

(40) *Istoria otechestvennoy matematiki,* I. Z. Shtokalo, ed. II (Kiev, 1967).

(41) A. P. Youschkevitch, *Istoria matematiki v Rossii do 1917 g.* (Moscow, 1968).

A. P. YOUSCHKEVITCH

CHENEVIX, RICHARD (*b.* Ballycommon, near Dublin, Ireland, 1774; *d.* Paris, France, 5 April 1830), *chemistry, mineralogy.*

Chenevix, of Huguenot ancestry, was educated at the University of Glasgow. Soon thereafter he revealed his abilities as an analytical chemist in his analyses of a new variety of lead ore (1801); of arsenates of copper (1801); and of sapphire, ruby, and corundum (1802). The publication of *Remarks Upon Chemical Nomenclature, According to the Principles of the French Neologists* in 1802 won for him the reputation of a pioneer in that field.

At the same time, Chenevix was acquiring notoriety for his heated attacks upon his scientific colleagues, especially those in Germany. He was particularly critical of the German school of *Naturphilosophie:* He took to task Oersted's *Materialien zu einer Chemie des neunzehnten Jahrhunderts* (1803), in which the dualistic system of the Hungarian chemist J. J. Winterl was defended, and opposed the dynamical theory of crystallization of the mineralogist Christian Samuel Weiss (1804). His pugnacity was to lead him into the most disastrous enterprise of his career.

In 1803 an anonymous handbill was circulated among British scientists. It announced the isolation of a new chemical element, palladium or "new silver," and offered the metal for sale. Chenevix, believing the announcement a fraud, purchased the complete stock. He set about analyzing it with the preconceived notion that it was an alloy of platinum and mercury. After a series of laborious experiments, he concluded that palladium was in fact an amalgam of platinum made in some peculiar way. His report to the Royal Society caused a sensation. Not long after, Wollaston read to the society a paper in which he declared himself the author of the handbill and the discoverer of two new elements in crude platinum ore—namely, palladium and rhodium. About 1804, with his scientific reputation badly damaged, Chenevix left England and went to France, where he lived for the remainder of his life.

In these years he published a significant paper, "Sur quelques méthodes minéralogiques" (1808), in which he disputed Abraham Gottlob Werner's classification of minerals by their chemical composition. He himself adopted R. J. Haüy's criterion of classification, the physical characteristics of minerals. In 1809 he turned again to chemistry, developing a method for the preparation of acetone by the distillation of acetates. Increasingly, however, Chenevix turned to literary work and wrote several novels, plays, and poems.

BIBLIOGRAPHY

I. ORIGINAL WORKS. Chenevix' articles about Oersted and Weiss are in *Annales de chimie,* **50** (1804), 173–199, and **52** (1804), 307–339. His major works are *Remarks Upon Chemical Nomenclature, According to the Principles of the French Neologists* (London, 1802); and "Sur quelques méthodes minéralogiques," in *Annales de chimie,* **65** (1808), 5–43, 113–160, 225–277.

II. SECONDARY LITERATURE. On the palladium incident, see A. M. White and H. B. Friedman, "On the Discovery of Palladium," in *Journal of Chemical Education,* **9** (1932), 236–245; and D. Reilley, "Richard Chenevix (1774–1830) and the Discovery of Palladium," *ibid.,* **32** (1955), 37–39.

H. A. M. SNELDERS

CHERNOV, DMITRI KONSTANTINOVICH (*b.* St. Petersburg, Russia, 1 November 1839; *d.* Yalta, U.S.S.R., 2 January 1921), *metallurgy.*

The son of a physician, Chernov graduated from the St. Petersburg Practical Technological Institute in 1858 and remained there until 1866 as a teacher of mathematics and assistant to the head of the scientific and technical library. From 1866 to 1880 he worked in the Obukhovsky steel casting plant in St. Petersburg, first as an engineer in the forging shop and then as metallurgical assistant to the head of the plant. Chernov prospected for deposits of rock salt in the Donets Basin from 1880 to 1884, returning in the latter year to St. Petersburg. From 1889 he was professor of metallurgy at the Mikhailovsky Artillery Academy in St. Petersburg.

As a result of his study of the reasons for waste in the manufacture of gun forgings, and also of his penetrating analysis of the work of P. P. Anosov, P. M. Obukhov, A. S. Lavrov, and N. V. Kalakutsky (who were concerned with the smelting, casting, and forging of steel ingots), in 1866–1868 Chernov proposed that the structure and properties of steel depend on its treatment by mechanical means (pressure) and heat. He discovered the critical temperatures at which phase transformations in steel occur; these transformations alter the essential structure and properties of the metal as the result of its being heated or cooled in the solid state. Introduced onto a thermometric scale, these temperatures formed a series of points called "Chernov points," the most important being the critical points *a* and *b,* associated with the polymorphic transformation of iron. Point *a* was defined by Chernov as the temperature below which steel does not harden independently of the speed of cooling.

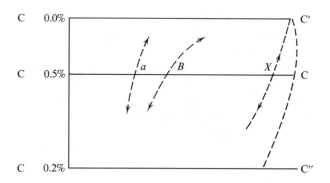

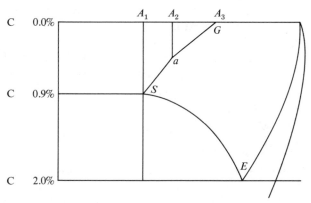

A comparison of Chernov's critical points with the "iron-carbon" diagram later constructed by Roberts-Austin.

This point (now designated A_1) corresponds to the temperature of the eutechtoid transformation of steel. The next critical point, *b* (now designated A_3), is defined as the temperature at which, in the opinion of a majority of scientists, the dissolution of ferrite in austenite ceases during the heating of steel. Chernov stated that to obtain the most desirable fine-grained structure, steel should be heated to point *b* or higher and then cooled. Chernov's critical points depend on the chemical composition of the steel, especially on its carbon content. He graphically represented the influence of carbon on the position of the critical points by making the first rough sketch of the most important lines of the "'iron-carbon' state" diagram. At the end of the nineteenth and the beginning of the twentieth centuries many metallurgists in various countries were working on the development and improvement of this classic diagram.

Chernov presented the results of his remarkable research in April and May 1868 at the sessions of the Russian Technical Society and published them in the same year. In 1876 this work was published in England, and a year later in France. It formed the scientific basis for contemporary metallurgy and the processes for heat treatment of metals. The French metallurgist Floris Osmond, in developing Chernov's work, discovered the precise significance of the critical points (1886) by using the thermoelectric pyrometer, invented in the same year by Henry Le Chatelier. Osmond introduced the designation of the critical points by the capital letter *A* with its corresponding indices, and the description of the character of the microstructural changes that occur in the transition through the critical points. At the beginning of the 1920's the transformations produced in steel by the action of heating and cooling also received experimental confirmation from radiographic methods.

In his next scientific work, "Issledovania, otnosyashchiesya do struktury litykh stalnykh bolvanok" ("Investigations Relating to the Structure of Cast Steel Ingots"), Chernov presented a coherent theory of the crystallization of steel ingots. He investigated in detail the origin and growth of crystals, constructed a diagram of the structural zones of ingots, developed a theory of successive crystallization, thoroughly studied the defects of cast steel, and suggested effective measures to overcome them. This research greatly aided the transformation of metallurgy from a practical art into a theoretically based scientific discipline.

Of great importance for the progress of steel metallurgy was Chernov's work on expanding the use of metallurgical processes and on improvements in the techniques of production. He confirmed the importance of the complete deoxidation of steel during smelting and the advisability of using complex deoxidizers, and he recommended a system of measures for obtaining dense forms of the metal free of air bubbles. He advanced the idea of stirring the metal during crystallization, proposing a rotating crucible for this purpose. Chernov did much to improve the converter method of producing cast steel. In 1872 he proposed warming liquid low-silicon iron, previously considered unprofitable for the Bessemer process, in a cupola furnace before it was aerated in the converter. Later this method was widely used in both Russian and foreign factories. Chernov was the first to use the spectroscope to determine when to stop the Bessemer process. He warmly supported Akerman's proposal to use oxygen-enriched air for aerating liquid iron in the converter. He also worked on the problem of obtaining steel directly from ore, bypassing the blast furnace. Chernov was responsible for much important research in artillery production: the obtaining of high-quality barrels on weapons; steel-armored projectiles; and research on the corrosion of the bores of weapons, resulting from the action of gases released when the weapon is fired. He also wrote a number of works on mathematics and aviation.

Chernov was the founder of an important school of metallurgists and metallurgical scientists. He had many pupils and followers—including A. Baykov, N. Kurnakov, A. Rzheshotarsky, and N. Gudtsov—who developed his ideas and made their own contributions to metallurgy. Among Chernov's foreign followers were Henry M. Howe, E. Hein, F. Osmond, and A. Portevin. Chernov was elected honorary president of the Russian Metallurgical Society, honorary vice-president of the English Institute of Iron and Steel, honorary member of the American Institute of Mining Engineers, honorary corresponding member of the Royal Society of Arts (London), and member of many other scientific organizations.

BIBLIOGRAPHY

I. ORIGINAL WORKS. Chernov's basic works are "Kritichesky obzor statey Lavrova i Kalakutskogo o stali i stalnykh orudiakh i sobstvennye D. K. Chernova issledovania po etomu zhe predmetu" ("A Critical Survey of the Articles of Lavrov and Kalakutsky on Steel and Steel Weapons and D. K. Chernov's Own Research on the Same Subject"), in *Zapiski Russkogo tekhnicheskogo obshchestva* ("Notes of the Russian Technical Society") (1868), no. 7, 339–440; V. Anderson, trans., in *Engineering* (1876), issues of 7 and 14 July, and M. Tchernoff, trans., *Note sur la constitution et le travail de l'acier* (Paris, 1877); "Issledovania, otnosyashchiesya do struktury litykh stalnykh bolvanok" ("Investigations Relating to the Structure of Cast Steel Ingots"), *ibid.* (1879), no. 1, 1–24; *Staleliteynoe delo* ("The Cast Steel Business"; St. Petersburg, 1898); and "Materialy dlya izuchenia bessemerovania" ("Material for the Study of the Bessemer Process"), *ibid.* (1915), no. 1, 55–90. These and a number of other works by Chernov are included in the collection *D. K. Chernov i nauka o metallakh* ("D. K. Chernov and the Science of Metals"; Leningrad-Moscow, 1950).

II. SECONDARY LITERATURE. On Chernov or his work see I. P. Bardin, "Osnovopolozhnik sovremennogo metallovedenia, vydayushchysya russky ucheny-metallurg D. K. Chernov" ("The Founder of Contemporary Metal Science, the Distinguished Russian Scientist and Metallurgist D. K. Chernov"), in *Izvestiya Akademii nauk SSSR. Otdelenie tekhnicheskikh nauk* (1951), no. 6, 900–906; A. A. Baykov, "Veliky russky metallurg D. K. Chernov" ("The Great Russian Metallurgist D. K. Chernov"), in *Stal* (1939), no. 10–11, ix–xiii; A. A. Bochvar, "Raboty D. K. Chernova v oblasti metallovedenia i ikh znachenie v sovremennoy nauke" ("The Works of D. K. Chernov in the Field of Metallurgy and Their Significance in Contemporary Science"), in *Izvestiya Akademii nauk SSSR. Metally* (1969), no. 1, pp. 3–13; A. S. Fedorov, "D. K. Chernov. K stoletiyu so dnya rozhdenia" ("D. K. Chernov. On the Centenary of His Birth"), in *Izvestiya Akademii nauk SSSR. Otdelenie tekhnicheskikh nauk* (1939), no. 9, 95–108; and "Yarkaya stranitsa v nauke o metalle. K 100-letiyu otkrytia D. K. Chernovym fazovykh prevrashcheny v stali" ("A Bright

Page in the Science of Metal. On the 100th Anniversary of the Discovery by D. K. Chernov of Phase Transformation in Steel"), in *Priroda* (1968), no. 4, 109–115; A. F. Golovin, *D. K. Chernov—osnovopolozhnik nauki o metallakh* ("D. K. Chernov—Founder of the Science of Metals"), no. 5 in the series Trudy po Istorii Tekhniki (Moscow, 1954), 39–57; and "K stoletiyu otkrytia D. K. Chernovym polifmorfnykh prevrashcheny v stali. 1868–1968" ("On the 100th Anniversary of the Discovery by D. K. Chernov of Polymorphic Transformation in Steel. 1868–1968"), in *Metallovedenie i termicheskaya obrabotka metallov* (1968), no. 5, 2–8; G. Z. Nesselshtraus, "Zhizn i deyatelnost D. K. Chernova—osnovatelya metallografii" ("The Life and Work of D. K. Chernov, Founder of Metallography"), in *D. K. Chernov i nauka o metallakh* (see above), 7–59; and A. M. Samarin, "D. K. Chernov i sovremennoe stalevarenie" ("D. K. Chernov and Contemporary Steel Foundry"), in *Izvestiya Akademii nauk SSSR. Metally* (1969), no. 1, pp. 77–82.

A. S. FEDOROV

CHERNYAEV, ILYA ILYICH (*b.* Spasskoye, Vologda district, Russia, 21 January 1893; *d.* Moscow, U.S.S.R., 30 September 1966), *chemistry.*

Chernyaev, a pupil of L. A. Chugaev, graduated in 1915 from St. Petersburg University and was retained in the department of inorganic chemistry by Chugaev, so that he could prepare for a teaching career. During the same year he published his first scientific work, which dealt with the chemistry of complex platinum compounds.

Chernyaev was one of the Russian scientists who took an active part in the organization and development of science from the beginning of the existence of the Soviet state. From 1917 on, he worked in the platinum department of the Commission for the Study of the Productive Capacity of Russia; from 1918 to 1933 he worked at the Institute for Platinum and Other Precious Metals; and from 1923 to 1925 and again from 1937 to 1938 he did important work in the field of refining platinum metals. Chernyaev's crowning achievement was the development of methods to obtain platinum, palladium, and rhodium in a spectroscopically pure state. During the period from 1934 to 1941 he was in charge of the department of complex compounds of the Institute of General and Inorganic Chemistry of the Academy of Sciences of the U.S.S.R., and from 1941 to 1964 he was director of the Institute. In 1943 he became a full member of the Academy of Sciences of the U.S.S.R.

Chernyaev's scientific activities were closely connected with his work as a teacher. From 1925 to 1935 he lectured on inorganic chemistry at the University of Leningrad and also gave special courses on the chemistry of platinum metals and structural theory of complex compounds. In 1932 he was named professor of inorganic chemistry at Leningrad University. Chernyaev also taught at the Petroleum Institute of the University of Moscow.

Chernyaev's main contribution to the development of science lay in the chemistry of complex compounds, especially of platinum metals. In 1926 he discovered (while working with nitroso compounds of bivalent platinum) the transeffect principles, which played a large part in the development of the chemistry of complex compounds. On the basis of these principles Chernyaev, with the help of his students and co-workers, achieved a number of directional syntheses of new complex compounds. Future development of the stereochemistry of complex compounds was based largely on the principle of transeffects and their consequences.

Chernyaev completed an important work cycle in the study of the optical activity of complex platinum compounds. He divided the tetravalent ethylenediamine compounds $(PtEnNH_3NO_2Cl_2)Cl$ (En = ethylenediamine) into optical antipodes and proved the correctness of the geometric configuration theory concerning the dependence of the optical activity of certain compounds on their steric structure. Chernyaev divided about ten different complex compounds of tetravalent platinum into optically active antipodes. Furthermore, he discovered the phenomenon of a change in the rotation symbol of the polarization plane in optically active amino compounds of tetravalent platinum at the time of their transmutation into amido (or imido) compounds.

Chernyaev studied a number of oxidation-reduction processes in the presence of complex platinum compounds. These investigations contributed to the further interpretation of the structure of similar compounds and of the reaction mechanism in which they are involved. Especially, it was shown that in nitro compounds of platinum the latter is combined with an NO_2 group over nitrogen.

Of special importance was Chernyaev's work in the analytical chemistry of platinum metals: methods of identifying and purifying platinum metals and a method for the quantitative identification of nitrogen included in the composition of intraspherical nitro groups.

During the last decade of his life Chernyaev studied the complex compounds of thorium and uranium. These studies contributed greatly to the establishment of the nuclear fuel industry in the U.S.S.R. Chernyaev was one of the founders and leaders in the U.S.S.R. of an important school of chemists specializing in the study of chemical complexes.

BIBLIOGRAPHY

I. ORIGINAL WORKS. Chernyaev's writings include "Voprosy khimii kompleksnykh soedineny" ("Problems in the Chemistry of Complex Compounds"), in *Uspekhi khimii,* **5** (1936), 1169; "O geometricheskoy izomerii soedineny chetyrekhvalentnoy platiny" ("On the Geometric Isomerism of Tetravalent Platinum Compounds"), *ibid.,* **16** (1947), 385; and also "Eksperimentalnoe obosnovanie zakonomernosti transvliania" ("Experimental Basis of the Principles of Transeffects"), in *Izvestia sektora platiny i drugikh blagorodnykh metallov,* no. 28 (1954), p. 14. He also was editor of *Kompleksnye soedinenia urana* ("Complex Uranium Compounds"; Moscow, 1964).

II. SECONDARY LITERATURE. On Chernyaev or his work see A. V. Babaeva, "Vydayushchysya sovetsky ucheny I. I. Chernyaev (k 60-letiyu so dnya rozhdenia)" ("The Outstanding Soviet Scientist I. I. Chernyaev [on the Occasion of His Sixtieth Birthday]"), in *Zhurnal obshchey khimii,* **23** (1953), 5; *Ilya Ilyich Chernyaev* (Moscow–Leningrad, 1948), no. 9 in the series A.N.S.S.R. Materialy k Bibliografii Uchenykh S.S.S.R., Seria Khimicheskikh Nauk (Soviet Academy of Sciences, Materials for the Bibliographies of the Scientists of the U.S.S.R., Chemical Sciences Series); V. M. Vdovenko, "Radiokhimia" ("Radiochemistry"), in *Razvitie obshchey, neorganicheskoy i analiticheskoy khimii v SSSR* ("The Development of General, Inorganic, and Analytical Chemistry in the U.S.S.R."; Moscow, 1967), p. 196; and O. E. Zvyagintsev, et al., "Khimia kompleksnykh soedineny" ("The Chemistry of Complex Compounds"), *ibid.,* p. 137.

D. N. TRIFONOV

CHERNYSHEV, FEODOSY NIKOLAEVICH (*b.* Kiev, Russia, 12 September 1856; *d.* St. Petersburg, Russia, 2 January 1914), *geology, stratigraphy, paleontology.*

Chernyshev's parents were teachers in secondary schools in Kiev. He attended the First Kiev Gymnasium and in 1872 entered the Cadet Corps of the Naval College in St. Petersburg. Shortly before he was due to graduate from the Cadet Corps, Chernyshev left the Naval College for the St. Petersburg Mining Institute, from which he graduated in 1880.

In March 1882 Chernyshev was selected as junior geologist on the recently organized Geological Committee of Russia. He was assigned to make a geological survey of the territory on the western slope of the southern Urals. At that time material and financial support for field geologists was modest, and the conditions of work in unpopulated forest and mountain areas were very difficult. Chernyshev traveled for weeks with one worker on one horse, and a scanty supply of food. He conducted research on the Urals and central Russia. His fieldwork was accompanied by laboratory study of a wide range of geological and paleontological materials. Before Chernyshev's research the upper Silurian deposits of the Urals were known (through Murchison and others) and noted on the geological map of the range. Chernyshev's investigations showed that these deposits included layers with fauna characteristic of the lower and middle Devonian periods. He restudied collections of Devonian deposits of European Russia, Siberia, western Europe, and America. The results of his research were of great significance for the clarification of the physical and geographical conditions of the Devonian period over a large part of the earth's surface, from Western Europe to America. As the basis for the organization of these materials, Chernyshev worked out a new scheme of the stratigraphy of the Urals, showing the sequence of beds and deposits and the composition of the rock of which the Urals and its foothills are composed.

In 1885, for his scientific work in investigating the geological structure of the Urals and other territory of Russia, Chernyshev was selected as senior geologist of the Geological Committee, and his monographs were awarded prizes by the Academy of Sciences and the Mineralogical Society.

His discovery of the Devonian strata of the Urals led to the discovery of analogous deposits in the mountains of central Asia, the Altai, eastern Siberia, and a number of other areas of Russia.

In 1889–1890 Chernyshev led a geological expedition to the Timan ridge, under the most difficult conditions, through almost impassable swamps and unpopulated areas. The field investigation was carried out by traveling in small boats and on foot through swarms of midges and mosquitoes. A large amount of geological and paleontological material was collected, the treatment of which allowed Chernyshev to publish one of the most important monographs on the Urals and the Timan (1902).

In 1892 Chernyshev was commissioned to head a scientific group to study the geological structure of the Donets coal basin. Using the stratigraphical-paleontological method of research, Chernyshev and the group of geologists that he organized (L. I. Lutugin and others) produced a highly detailed analysis of the coal deposits of the Donbas and made clear the possibility of their graphical representation on a one-verst map of the basin. Chernyshev spent much time and energy in compiling a geological map of European Russia on the scale of 60 versts (40 miles) to 1 inch.

At the 1897 session of the International Geological Congress, which took place in St. Petersburg, Chernyshev was chosen secretary-general of the Congress. At the 1906 session in Mexico his monograph *Verkhnekamennougolnye brakhiopody Urala i Timana*

("The Upper Coal Layers of Brachiopods of the Urals and the Timan"), published in 1902, won a prize.

Chernyshev contributed much to the science of the physical-geographical conditions of the Devonian period, which played an important role in the evolutionary development of life on the earth.

For his productive scientific activity in the field of research on the geological structure of major areas of Russia and the territories of other countries, Chernyshev was made adjunct of the Academy of Sciences in St. Petersburg (Russia) in 1897, extraordinary academician in 1899, and academician in 1909.

In 1899 Chernyshev was selected as one of the scientific leaders of an expedition sent by the Russian and Swedish academies of science to measure the latitude of Spitsbergen. He conducted research on the geological structure of Spitsbergen and vividly described its geology and glacial landscape.

In 1903 Chernyshev was made director of the Geological Committee of Russia. At this time he was also chosen director of the Mineralogical Museum of the Academy of Sciences. Chernyshev's scientific works were not limited to stratigraphy, paleontology, and physical geology. He also wrote on mineralogy, petrography, ore deposits, and other areas of geological prospecting. Chernyshev was made honorary doctor by a number of foreign universities and was elected a member of many scientific societies in his native country and abroad. A mountain ridge in the northern Urals and a mountain range in the Amur region are named for him.

BIBLIOGRAPHY

I. ORIGINAL WORKS. Chernyshev's works include "Materialy k izucheniyu devonskikh otlozheny Rossii" ("Materials for the Study of the Devonian Deposits of Russia"), in *Trudy Geologicheskago komiteta* (St. Petersburg), **1,** no. 3 (1884), 1–82; "Fauna nizhnego devona zapadnogo sklona Urala" ("Lower Devonian Fauna of the Western Slope of the Urals"), *ibid.,* **3** (1885), 1–107; "Fauna srednego i verkhnego devona zapadnogo sklona Urala" ("Middle and Upper Devonian Fauna of the Western Slope of the Urals"), *ibid.,* **3,** no. 3 (1887), 1–208; "Obshchaya geologicheskaya karta Rossii, list 139. Opisanie tsentralnoy chasti Urala i zapadnogo yego sklona" ("A General Geological Map of Russia, sheet 139. A description of the Central Part of the Urals and the Western Slope"), *ibid.,* **3,** no. 4 (1889), 1–393; "Fauna devona vostochnogo sklona Urala" ("Lower Devonian Fauna of the Eastern Slope of the Urals"), *ibid.,* **4,** no. 3 (1893), 1–221; "Geologicheskie raboty, proizvedennye v Donetskom basseyne v 1893 g." ("Geological Work Done in the Donets Basin in 1893"), in *Izvestiya Geologicheskago komiteta,* **13,** no. 4–5 (1894), 117–127; *Geologicheskaya karta Timanskogo kryazha, na*
trekh listakh ("Geological Map of the Timan Mountain Ridge, in Three Sheets"; St. Petersburg, 1900); and "Verkhnekamennougolnye brakhiopody Urala i Timana" ("The Upper Coal Brachiopods of the Urals and the Timan"), in *Trudy Geologicheskago komiteta,* **16,** no. 2 (1902), 1–63.

II. SECONDARY LITERATURE. On Chernyshev or his work, see K. I. Bogdanovich, "Pamyati Feodosia Nikolaevicha Chernyshova" ("Recollections of Feodosy Nikolaevich Chernyshev"), in *Izvestiya Geologicheskago komiteta,* **33,** no. 1 (1914), 1–12; V. A. Feyder, *Feodosy Nikolaevich Chernyshev. Bibliografichesky ukazatel i materialy k biografii* ("Feodosy Nikolaevich Chernyshev. Bibliography and Materials for a Biography"; Leningrad, 1961); A. P. Karpinsky, "Feodosy Nikolaevich Chernyshev," in *Izvestiya Akademii nauk,* no. 3 (1914), 167–184; D. V. Nalivkin, "Feodosy Nikolaevich Chernyshev," in *Lyudi russkoy nauki* ("People of Russian Science"), I (Moscow-Leningrad, 1948), 108–114; and the article in *Bolshaya Sovetskaya Entsiklopedia* ("Great Soviet Encyclopedia"), 2nd ed., XLVII (1957), 202–203.

G. D. KUROCHKIN

CHERWELL, LORD. See **Lindemann, F. A.**

CHEVALLIER, JEAN-BAPTISTE-ALPHONSE (*b.* Langres, France, 19 July 1793; *d.* Paris, France, 29 November 1879), *chemistry, public health.*

Chevallier's claim to fame rests on his successful application of chemical expertise to a variety of public health problems, such as food and drug adulteration, industrial hygiene, toxicology, and disinfection.

Coming to Paris as a youth, he worked as a laboratory assistant to Vauquelin and Laugier at the Museum of Natural History, then as an intern in hospital pharmacy, and later became the owner of his own pharmacy. In 1835 Chevallier decided to leave the practice of pharmacy to accept a teaching position at the École de Pharmacie and to establish a private analytical laboratory, where he also gave instruction. By this time his numerous publications and his active role in public health had brought him wide recognition. This was reflected in his election to the Academy of Medicine in 1824 and his appointment in 1831 to the Council on Hygiene and Health of the Seine department.

The volume of Chevallier's publications was prodigious. In addition to numerous books his articles, numbering in the hundreds, appeared in many journals, such as *Journal de chimie médicale, de pharmacie et de toxicologie; Annales d'hygiène publique et de médecine légale;* and *Journal de pharmacie et de chimie.* Chevallier was frequently in demand as an expert chemist in judicial proceedings; he collaborated with persons prominent in toxicology and forensic medicine, such as Orfila, Barruel, Lassaigne, Bayard, Flandin, Lesueur, and Charles-Prosper Ollivier. As for his contributions to public health, these were embodied in a multitude of memoirs, reports, and monographs dealing with such diverse subjects as safeguarding workers and consumers against toxic substances, disinfection with chlorine compounds, chemical detection of adulterated foods and drugs, and many other related problems. Chevallier's *Dictionnaire des altérations et falsifications des substances alimentaires, médicamenteuses et commerciales* (1850–1852) was the most important work of its kind to appear in France.

BIBLIOGRAPHY

I. ORIGINAL WORKS. Probably the most influential of all Chevallier's publications was his *Dictionnaire des altérations et falsifications des substances alimentaires, médicamenteuses et commerciales,* 2 vols. (Paris, 1850–1852, and many subsequent eds.). Of his longer works, the following are of particular interest: *Traité élémentaire des réactifs* . . . (Paris, 1822; 2nd ed., 1825; 3rd ed., 1829), written with A. Payen; *L'art de préparer les chlorures de chaux, de soude et de potasse* . . . (Paris, 1829); *Dictionnaire des drogues simples et composées* . . ., 5 vols. (Paris, 1827–1829), written with A. Richard and J. A. Guillemin; and *Traité de toxicologie et de chimie judiciaire* (Paris, 1868). For a comprehensive listing of Chevallier's publications, see A. Goris, *Centenaire de l'internat en pharmacie des hôpitaux et hospices civils de Paris* (Paris, 1920), pp. 300–316, and the secondary literature cited below.

II. SECONDARY LITERATURE. G. Sicard, *Notice sur la vie et les travaux de M. Jean-Baptiste-Alphonse Chevallier* . . . (Paris, 1880), contains a bibliography of Chevallier's publications compiled by his son, A. Chevallier *fils,* for the period 1822–1862 and a supplementary list by Sicard for the period 1863–1879. See also T. Gallard, "Alphonse Chevallier," in *Annales d'hygiène publique et de médecine légale,* 3rd ser., **3** (1880), 181–187.

ALEX BERMAN

CHEVALLIER, TEMPLE (*b.* Bury St. Edmunds (?), England, 19 October 1794; *d.* Durham, England, 4 November 1873), *astronomy, education.*

After attending grammar schools at Bury St. Edmunds and Ipswich, Temple Chevallier proceeded to Pembroke College, Cambridge, where he was successively Bell scholar (1814), second wrangler and second Smith's prizeman (1817), and fellow of Pembroke College and Catharine Hall. He was ordained in 1818, remaining in Cambridge until 1835 when he moved to Durham, where he held various offices in the cathedral and university. He was made professor of mathematics (1835), professor of astronomy (1841),

and reader in Hebrew (1841). His wide learning enabled him to publish translations from authors such as Clement of Rome, Polycarp, Ignatius, Justin Martyr, and Tertullian, besides many sermons and religious works, and his Halsean Lectures (1827) are said to have suggested to Whewell the idea of his Bridgewater treatise.

Chevallier wrote about thirty astronomical papers. The observatory at Durham University was largely planned by him. He made use of its facilities, including the Fraunhofer equatorial telescope, in his work on the sun's diameter, solar eclipses, the planets, and meteorological phenomena.

Chevallier seems to have enjoyed calculating the elements of the orbits of objects within the solar system, and some of his most useful writings, although very brief, concern simple graphical and other approximative computing methods. He showed little research interest in physics, although he offered a paper "On an Analogy Between Heat and Electricity" to the British Association (1855). His *Easy Lessons on Mechanics* (written for the Society for the Propagation of Christian Knowledge) is perhaps a better guide to Chevallier's merits than his more strictly scientific writings, for he was fervent in wishing to introduce science into education. He founded a class at Durham for mining and civil engineering in 1838, and later a department of physics. In the year of his retirement (1871), he helped to found the College of Science at Newcastle (Armstrong College), which was long associated with the University of Durham. The college has since become part of the University of Newcastle.

BIBLIOGRAPHY

I. ORIGINAL WORKS. Apart from works referred to in the text, eighteen of Chevallier's papers on astronomy may be found in the *Journal of the Royal Astronomical Society.* Others appear in *Astronomische Nachschrift,* some written jointly with G. Rümker. His firm beliefs in the desirability of the study of mathematics are set down in *The Study of Mathematics as Conducive to the Development of the Intellectual Powers* (Durham, 1836).

II. SECONDARY LITERATURE. For additional details of Chevallier's career see the entry in the *Dictionary of National Biography* by Robert Hunt, whose sources included private information.

J. D. NORTH

CHEVENARD, PIERRE ANTOINE JEAN SYLVESTRE (*b.* Thizy, Rhône, France, 31 December 1888; *d.* Fontenay-aux-Roses, France, 15 August 1960), *metallurgy.*

Chevenard studied at the École des Mines at

Saint-Étienne, from which he graduated in 1910. He then joined the Société de Commentry Fourchambault et Decazeville, where he remained throughout his career, becoming its scientific director in 1935. In 1946 he was elected to the Académie des Sciences, in the section of industrial applications of science.

From his first assignment for the Société de Commentry Fourchambault, at its works at Imphy, Chevenard pursued the investigations begun by Charles Edouard Guillaume, discoverer of the alloys Invar and Elinvar. He explored the physical properties of nickel and was subsequently able to provide the clockmaking industry with a remarkable series of alloys.

In 1917 Chevenard brought out several alloys that showed good resistance to creep at high temperatures and especially high resistance to corrosion. In particular the alloy A.T.G., with 60 percent nickel, 10 percent chromium, and 4 percent tungsten, was of great value in the chemistry of high pressures and temperatures—for example, in the synthesis of ammonia under very high pressure, developed by Georges Claude. Other applications of these alloys have included the fabrication of valves for internal combustion engines and blades for steam engines.

The first ferronickel alloys, however, were found to be susceptible to cracking. Chevenard showed that the heterogeneity of the solid solution was the main problem. He also perfected equipment to measure the degree of heterogeneity in iron-nickel-chromium-carbon alloys and determined its quantitative influence on cracking.

Chevenard was also responsible for steam-resistant alloys (A.T.V.), solid solutions rich in chromium and very low in carbon. Certain additions render such alloys capable of sustaining structural hardening without damage to their resistance to chemicals.

The studies he carried out in this area enabled Chevenard to suggest new research. It is to this group of studies, in which the techniques of measurement played a primary role, that the name "precision metallurgy" has been given. In order to gain the precision necessary for such research on new alloys and for their preparation, Chevenard created a series of self-recording devices, almost all of which were based on the optical tripod.

Chevenard's micromachines, constructed for test samples 1,000 times smaller than the classic ones, permitted the examination of heterogeneous pieces with an exactness never attained before. Chevenard's instruments facilitated the use of certain physical methods to such a degree that his name has passed from the device to the method itself: this is the case with the dilatometric and thermoponderal methods,

through his differential dilatometer and his thermobalance.

In his dilatometric investigations Baros, which possesses a regular and reversible dilatation, was replaced by another standard alloy, Pyros, which is more rigid when heated and has an even more regular dilatation.

From 1923 to 1929 Chevenard, in collaboration with Albert Portevin, often transposed certain of his techniques and results to the study of alloys other than those of steel and nickel. It was then that he investigated by dilatometry the structural hardening of light alloys of aluminum.

BIBLIOGRAPHY

Among Chevenard's papers, the following were published in the *Comptes rendus hebdomadaires des séances de l'Académie des sciences:* "Dilatation des aciers au nickel dans un grand intervalle de température," **158** (1914), 175; "Dilatomètre différentiel enregistreur," **164** (1917), 916; "Changement thermique des propriétés élastiques des aciers au nickel," **170** (1920), 1499; "Influence de l'écrouissage et de la trempe sur les propriétés élastiques de divers métaux et alliages," **181** (1925), 716, written with A. Portevin; "Anomalie du frottement interne des ferronickels réversibles," **184** (1927), 378; "Causes de la variation de volume accompagnant le durcissement des alliages légers aluminium-cuivre," **186** (1928), 144, written with A. Portevin; "Influence du revenu sur la dilatation et la dureté des alliages aluminium silicium trempés," **191** (1930), 252, written with A. Portevin; "Micromachine à enregistrement photographique pour l'essai mécanique des métaux," **200** (1935), 212; "Très petite machine de traction à enregistrement photographique et son application à l'étude des fibres textiles," **203** (1936), 841; "Corrosion intercristalline des ferronickels chromés carburés, écrouis après hypertrempe," **204** (1937), 1167, written with X. Waché; "Nouveaux alliages du type élinvar susceptibles de durcissement structural," *ibid.,* 1231, written with L. Huguenin, X. Waché, and A. Villachon; "Amplificateur mécanique à grandissement supérieur à 1000. Application à l'enregistrement de la déformation visqueuse des métaux aux températures élevées," **205** (1937), 107, written with E. Joumier; "Thermoélasticimètre enregistreur," **211** (1940), 548, written with E. Joumier; and "Influence de la vitesse sur la forme des cycles couple-torsion d'un métal étudié à l'état visqueux. Hystérésigraphe de torsion à enregistrement photographique," **214** (1942), 415, written with C. Crussard.

Papers published in the *Revue de métallurgie* include "Contribution des aciers au nickel," **11** (1914), 841; "Dilatomètre différentiel enregistreur," **14** (1917), 610; "Remarques et observations concernant les phénomènes de trempe des aciers," **18** (1921), 428, written with A. Portevin; "Nouvelles applications du pyromètre à dilatation à l'analyse thermique des alliages," **19** (1922), 546; "Dilatomètre différentiel à enregistrement mécanique," **23**

(1926), 92; "Henry Le Chatelier et l'organisation scientifique des usines," **34** (1937), 87; and "Les propriétés mécaniques secondaires. Frottement interne. Relaxation visqueuse. Réactivité. Coefficient de Poisson," **40** (1943), 289.

Papers that appeared elsewhere are "Applications des alliages spéciaux à la pyrométrie," in *Chaleur et industrie,* spec. no. (July 1923); "A Dilatometric Study of the Transformations and Thermal Treatment of Light Alloys of Aluminium," in *Journal of the Institute of Metals,* **30** (1923), 329, written with A. Portevin; "A Dilatometric Study of Some Univariant Two-Phase Reactions," *ibid.,* **42** (1929), 337, written with A. Portevin and X. Waché; "L'analyse dilatométrique des matériaux et ses recents progrès," in *Revue des aciers spéciaux,* **3**; "Nouvelles recherches sur la corrosion intercristalline des ferronickels chromés," in *Métaux et corrosion,* **12** (Feb. 1937), 23; "L'oeuvre métallurgique de Ch. Ed. Guillaume: L'étude et les applications des ferronickels," in *Brochure sur la vie et l'oeuvre de Charles Edouard Guillaume,* p. 25; "Nouveaux alliages du type élinvar pour spiraux de chronomètres," in *Annales de chronométrie,* **7** (1937), 259, written with X. Waché and A. Villachon; "Étude de la corrosion sèche des métaux au moyen d'une thermobalance," in *Bulletin. Société chimique de France,* 5th ser., **10** (1943), 41, written with X. Waché and R. de la Tullaye; and "Nouvelles études dilatométriques des minéraux," in *Bulletin de la Société française de minéralogie,* **66** (1943), 131, written with A. Portevin.

GEORGES CHAUDRON

CHEVREUL, MICHEL EUGÈNE (*b.* Angers, France, 31 August 1786; *d.* Paris, France, 9 April 1889), *chemistry.*

After his early education in Angers, where his father was the director of the medical school, Chevreul went to Paris in 1803 to study chemistry under Nicolas Vauquelin at the Muséum d'Histoire Naturelle, thereby beginning a career of nearly ninety years at that institution. Chevreul rose rapidly through the scientific ranks during the 1820's and 1830's, his two most important positions being those of professor of chemistry at the Museum from 1830 and director of dyeing at the Manufactures Royales des Gobelins, the national tapestry workshop, where he succeeded Claude Berthollet in 1824. For nearly sixty years he taught courses in chemistry at these two institutions. He was director of the Museum from 1864 to 1879. Chevreul became a member of the Académie des Sciences in 1826 and was elected to the presidency in 1839 and again in 1871.

His countrymen admired the centenarian for his abilities in many different fields. Chevreul displayed his versatility in his many books and papers in history, philosophy, and psychology. He contributed many papers to the *Journal des savants* on a variety of subjects, of which the most important were his studies on alchemy, the early history of chemistry, and the history of medicine. He wrote two books on these subjects: *Histoire des connaissances chimiques* (1866) and *Résumé d'une histoire de la matière* (1878).

Chevreul's interest in psychology culminated in 1853, when the Académie des Sciences appointed him chairman of a committee to investigate séances and other psychic phenomena, as well as the use of divining rods and exploring pendulums in locating water or mineral deposits. He published an exposé of psychic phenomena and of these devices in his *De la baguette divinatoire* (1854). Chevreul also produced many papers on philosophy and on scientific method. His ideas on scientific method found their most complete expression in his *De la méthode a posteriori expérimentale* (1870).

Chevreul was very regular and methodical throughout his long life. In his later years he continued to work at the Museum and to attend meetings of the Academy, presenting his last communication to this body in 1888, when he was 102. He had married Sophie Davallet, the daughter of a tax official, in 1818. She died in 1862 and Chevreul forsook almost all social life, living his remaining years in the Museum and going out only to scientific meetings. Because of his achievements and great age, his colleagues accorded him great respect and regarded him as one of the most distinguished scientists of the century. The French nation celebrated Chevreul's centenary on 31 August 1886. It was a great national event conducted in the presence of the president of the Republic and of scientific delegations from all over the world. The Museum was the locale for the main ceremonies, which featured the unveiling of Chevreul's statue and many discourses by dignitaries extolling his achievements. The festivities concluded with a torchlight parade through the streets of Paris and special performances at Paris theaters. His funeral some two years later at the Cathedral of Notre Dame attracted thousands of people.

Chevreul's scientific researches on dyes, color theory, and the chemistry of natural fats all stemmed from his association with Vauquelin at the Museum. Vauquelin introduced Chevreul to the study of organic substances in 1807 by having his student investigate plant dyes. This was Chevreul's first important series of researches. He isolated and examined several natural dyes. On the completion of these studies in 1811 Chevreul turned to the chemistry of natural fats as his next project. This study occupied him until 1823, and he returned to the study of dyes upon his appointment the following year as director of dyeing at Gobelins.

Chevreul's immediate task at Gobelins was to work

240

on the improvement of color intensity and fastness in wools. He had been selected for this position because he was an outstanding chemist; and his initial studies were on the chemical aspects of dyes and dyeing, attempting to place the art of dyeing on a more rational basis than the complicated and empirical procedures then employed. He embarked upon a thorough study of the properties of natural dyes and between 1838 and 1864 published several important papers in the *Mémoires de l'Académie des Sciences,* reviewing the chemistry and technology of dyeing. These papers and his two-volume *Leçons de chimie appliquée à la teinture* (1829–1830) rendered an important service to the dye industry during the years prior to the advent of synthetic dyestuffs.

In addition to the chemical aspects of dyeing Chevreul made an intensive study of the principles governing the contrast of colors, which resulted in his monumental *De la loi du contraste simultané des couleurs* (1839), the most influential of his many books. This book was the outcome of his discovery that the apparent intensity and vigor of colors depended less on the pigmentation of the material used than on the hue of the neighboring fabric. After many experiments on color contrast Chevreul formulated for the first time the general principles and effects of simultaneous contrast, the modification in hue and tone that occurs when juxtaposed colors are seen simultaneously. According to Chevreul, "where the eye sees at the same time two contiguous colours, they will appear as dissimilar as possible, both in their optical composition and in the height of their tone. We have then, *at the same time,* simultaneous contrast of colour properly so called, and contrast of tone" (*The Principles of Harmony and Contrast of Colours,* p. 11).

Under simultaneous contrast Chevreul included all the modifications that differently colored objects appear to undergo in their composition and tone. Furthermore, he diagramed these variations on a chromatic circle on which, out of the three primary colors of red, yellow, and blue, he defined almost 15,000 tones by first placing these three colors on equidistant radii of the circle and interpolating twenty-three color mixtures in each of the sectors. He thereby obtained a chromatic circle of seventy-two colors representing the entire visible spectrum. He then prepared additional circles by toning down colors with known proportions of black. To compare tones for each color, he mixed the normal colors with known proportions of black in one direction and white in the other, producing complete scales of colors, with pure white and pure black forming the extremes and the pure color the middle of the scale. In this way

Chevreul believed he had met the need for precise standards in the definition and use of colors and a way of faithfully reproducing any tone of color. His circles and scales were valuable to the painter and dyer because they represented every possible color modification.

Indeed, Chevreul designed his *De la loi du contraste simultané* less for scientists than for painters, designers, and decorators. He devoted much of the book to the applications of the principles of contrast to the various problems that the artist and designer encounter in the use of color and to the harmonizing of colors and their use as agents of pictorial harmony.

De la loi du contraste simultané was one of the most influential treatises on color written during the nineteenth century, even though it was never revised by its author and by the 1860's was already antiquated as a scientific treatise. Physicists had studied color on the basis of rigorous measurements with optical instruments—in contrast to Chevreul's reliance on empirical observation and experiments on color behavior as he observed it in nature, paintings, and tapestries. Nevertheless, Chevreul continued to be acclaimed for the discovery of several important principles governing color behavior.

Finally, one important aspect of Chevreul's color studies must be mentioned: their influence in the fine arts. The neo-impressionist painters derived their methods of painting from Chevreul's principles, applying separate touches of pure colors to the canvas and allowing the eye of the observer to combine them. Several artists hoped that the instinct of impressionists like Manet and Monet, who had dissociated tones on the canvas to enhance their brilliance, might be scientifically verified by the study of the physics and physiology of color perception. This "scientific" impressionism emerged in the 1880's. It was based on the resolution of the colors of nature into the colors of the spectrum and their representation on the canvas by dots of unmixed pigments, these dots at a distance giving the effect and brilliance of natural light and not of pigments. The leaders of neo-impressionism were Georges Seurat and Paul Signac. They found the scientific basis for the division of tones in Chevreul's principles of simultaneous contrast. Chevreul's arguments for painting into pictures the effects of contrast based on nature's behavior, his proofs, and his principles were all adopted by these artists. They limited their palettes to Chevreul's circle of fundamental colors and intermediate tones and applied colors scientifically to their canvases as opposed spots. In May 1886, to make more explicit their unified aims as scientific impressionists, Seurat, Signac, Pissarro, and others hung their work together and created a new artistic language with canvases appearing for the first time painted solely with pure, separate hues in equilibrium, mixing optically according to a rational method.

While his color studies made him one of the most influential scientists of the nineteenth century, Chevreul's work on the chemical nature of the natural fats established him also as one of the major figures in the early development of organic chemistry. When he began these studies in 1811, research in organic chemistry was in a very rudimentary state. Chemists had isolated many materials from animal and plant sources and had investigated their chemical properties. These immediate principles were of a more complex nature than inorganic bodies and supposedly were the products of special forces in the organism, forces that were different from those in the inanimate world. The whole subject of animal and vegetable chemistry was in a state of confusion. Among the prerequisites to a removal of this confusion was a means of determining the elementary composition of organic materials, which were just beginning to succumb to elementary analysis when Chevreul embarked upon his investigation of animal fats. This was the first area of organic chemistry to receive a thorough examination, and Chevreul's researches between 1811 and 1823 resulted in the natural fats' becoming the first class of naturally occurring organic substances whose fundamental character was understood.

While animal fats and vegetable oils had been utilized since antiquity, especially in the making of soaps, chemists had accomplished little in clarifying the nature of the saponification process or in understanding the chemical nature of the ingredients and products. Chevreul began his exploration of this subject when he analyzed a potassium soap made from pig fat. Upon acid treatment this soap yielded a crystalline material with acidic properties. It was the first of the several fatty acids he was to discover, and it led him into a systematic investigation of animal fats. His studies appeared in a series of articles in the *Annales de chimie* and were finally collected, augmented, and synthesized into an account of the chemistry of natural fats in his *Recherches chimiques sur les corps gras* (1823).

Chevreul obtained a whole series of new organic acids by decomposing the soaps derived from a variety of animal fats. He isolated, studied, and named many of the members of the fatty acid series from butyric to stearic acid. By 1816 he had established that all animal fats yielded both fatty acids and glycerol on saponification with alkali, the glycerol having

earlier been observed by Scheele as a conversion product of several fats and oils.

After his discovery of the fatty acids Chevreul attacked the saponification process itself, and by a masterly interpretation of reactions, he unraveled its nature. All fats were resolvable into fatty acids and glycerol. Each of these saponification products was isolated and submitted to elementary analysis. Chevreul found that the sum of the weights of the saponification products surpassed that of the original fat by 4 to 6 percent. Furthermore, his analyses revealed that these products contained the same quantity of carbon but more hydrogen and oxygen, in the proportion of water, than the fat. Chevreul concluded that saponification was essentially the chemical fixation of water. During this chemical fixation the alkali replaced the glycerol in its combination with the fatty acid. Thus, soaps were combinations of a fatty acid with an inorganic base and were therefore true salts.

These conclusions about saponification and the nature of fats and soaps were the outcome of a number of original techniques that Chevreul introduced into organic chemistry. He successfully isolated, purified, and analyzed fats, soaps, fatty acids, and glycerol, and achieved an understanding of the chemical nature of these different types of organic substances. To accomplish these difficult tasks, Chevreul had to develop his own methods of separating the constituents of natural fats and of identifying closely related organic acids in mixtures of considerable complexity. By the careful use of solvents, he developed methods of fractional solution to separate the immediate principles in fats. He separated the mixtures of fatty acids on the basis of their differing solubility in a given solvent, purified them by repeated crystallization, and determined their purity by the constancy of their melting point. Chevreul thus was chiefly responsible for introducing the melting point as a useful means of establishing the identity and purity of organic substances.

His *Recherches chimiques sur les corps gras* is mainly experimental and descriptive. Chevreul was reluctant to go beyond his observations and the immediate inferences from them. It was only at the end of the book that he proposed his conjectures relative to the arrangement of the elements in natural fats. In addition to the conclusions concerning saponification, Chevreul inferred that fats were comparable to esters. They were compounds of a fatty acid and the alcohol glycerol. Like esters, they were neutral substances, and under the influence of alkali they yielded an alcohol and an acid. Exactly what kind of alcohol was glycerol and what kind of esters were fats remained an open question until Marcellin Berthelot in the 1850's synthesized the immediate principles of fats by combining glycerol with different fatty acids, showing that one equivalent of glycerol can unite with three equivalents of fatty acid and that it has, therefore, a triple alcoholic function.

Chevreul's *Recherches* is a full account of the chemistry of fats. He had revealed the nature of a large and important class of organic substances and had shown that they were composed of a few chemical species which were amenable to analysis and obeyed the same laws of chemical combination followed by the simpler substances in the inorganic realm. The book is a model of complete, exhaustive research in organic chemistry. Few areas of chemistry at this time had been so thoroughly explored. Although his studies were followed by those of many chemists, who extended the number of fatty acids isolatable from fats and oils, there was no one comparable to Chevreul in this field until Thomas Percy Hilditch in Great Britain, a century later.

Following the completion of his investigation of natural fats, Chevreul wrote a work of a more general nature, *Considérations générales sur l'analyse organique* (1824), which is a reflection on the many years of research with organic materials. Here Chevreul considered the methods of research in organic chemistry, methods that he himself used with such success in his investigations. The primary analytical problems were how to determine whether one had a pure substance or a mixture and how to resolve the frequently complex animal or plant material into its immediate principles. It was Chevreul who placed this immediate analysis on a rigorous basis. He considered the methods used in handling natural products, in isolating pure substances from them in unaltered form, and in recognizing their purity. He gave precise criteria for what constituted a pure organic compound and presented for the first time a clear and accurate account of the methods of immediate analysis that must necessarily precede elementary analysis. Indeed, the foremost chemists of his time—Berzelius, Liebig, Dumas, Wurtz, Berthelot—all testified to Chevreul's achievement, recognizing that he introduced into organic chemistry the fundamental concept of the immediate principle endowed with constant and definite properties and composition, from which no other material could be separated without altering the principle.

Prior to Chevreul's work there was no method of submitting the products of analysis to a system of tests to determine whether the products were pure principles. Chevreul applied such a system of tests in his study of fats, which thereby became the first exact model of analytical research in organic chemistry. In the opinion of his contemporaries, Chevreul's studies constituted the best work done in organic chemistry, wherein quantitative and atomic relations had been established, constitutions discovered, relationships discerned, and the way shown toward the clarification of the difficulties of organic chemistry.

BIBLIOGRAPHY

I. ORIGINAL WORKS. An extensive bibliography of Chevreul's published writings was issued on the occasion of his centenary by Godefroy Malloizel, *Oeuvres scientifiques de Michel-Eugène Chevreul, 1806–1886* (Paris, 1886). There is also a collection of unpublished material available: E. Chevreul, *Quelques notes et lettres de M. E. Chevreul* (Dijon, 1907). Of Chevreul's many books and articles the most significant are his two books on organic chemistry and the one on color theory: *Recherches chimiques sur les corps gras d'origine animale* (Paris, 1823); *Considérations générales sur l'analyse organique et sur les applications* (Paris, 1824); *De la loi du contraste simultané des couleurs et de l'assortiment des objets colorés* (Paris, 1839). There are two English versions of the latter book: *The Principles of Harmony and Contrast of Colours*, Charles Martel, trans. (London, 1854; 3rd ed., 1872), and *The Laws of Contrast of Colour*, John Spanton, trans. (London, 1858).

Mention should be made of Chevreul's historical, philosophical, and psychological studies: *De la baguette divinatoire, du pendule dit explorateur et des tables tournantes, au point de vue de l'histoire, de la critique et de la méthode expérimentale* (Paris, 1854); *Histoire des connaissances chimiques* (Paris, 1866); *De la méthode a posteriori expérimentale et de la généralité de ses applications* (Paris, 1870); *Résumé d'une histoire de la matière depuis les philosophes grecs jusqu'à Lavoisier inclusivement* (Paris, 1878).

II. SECONDARY LITERATURE. There are many essays on the life and work of Chevreul. An excellent biography is Georges Bouchard's *Chevreul (1786–1889), le doyen des savants qui vit au cours d'un siècle quatre rois, deux empereurs, trois républiques, quatre révolutions* (Paris, 1932). Chevreul's granddaughter, Mme. de Champ, collected a number of personal remembrances in her *Michel-Eugène Chevreul, vie intime* (Paris, 1930). Marcellin Berthelot analyzed the principal scientific contributions of Chevreul in the official biography of the Académie des Sciences: "Notice historique sur la vie et les travaux de M. Chevreul," in *Mémoires de l'Académie des sciences,* **47** (1904), 388–433. A modern study that emphasizes Chevreul's importance and contribution to organic chemistry is Albert B. Costa's *Michel Eugène Chevreul: Pioneer of Organic Chemistry* (Madison, Wis., 1962).

For Chevreul's historical writings see George Sarton, "Hoefer and Chevreul," in *Bulletin of the History of Medicine,* **8** (1940), 419–445; and Hélène Metzger, "Eugène Chevreul, historien de la chimie," in *Archeion,* **14** (1932),

6–11. Joseph Jastrow examined Chevreul's study of psychic phenomena in "Chevreul as Psychologist," in *Scientific Monthly,* **44** (1937), 487–496. An excellent study of scientific impressionism which provides an analysis of Chevreul's color experiments and theory is William Innes Homer, *Seurat and the Science of Painting* (Cambridge, Mass., 1964), pp. 20–29.

ALBERT B. COSTA

CHEYNE, GEORGE (*b.* Aberdeenshire, Scotland, 1671; *d.* Bath, England, 12 April 1743), *medicine, mathematics, theology.*

At first educated for the ministry, Cheyne was influenced by the Scottish iatromechanist Archibald Pitcairn to take up medicine instead. He studied with Pitcairn in Edinburgh and then, in 1702, moved to London, where he joined the Royal Society and established a medical practice. Cheyne was soon an at least peripheral member of a prominent circle of medical and scientific writers that included the astronomers David Gregory and Edmund Halley and the physicians Richard Mead and John Arbuthnot. He spent several active years in London, winning a major reputation also as a wit and drinking companion in the tavern and coffeehouse set. Some years later, probably by 1720, he renounced his earlier life and moved permanently to Bath as a sober and dedicated medical practitioner. Cheyne spent the major part of his last decades advising his patients and correspondents (the novelist Samuel Richardson, for one) to lives of sober and pious moderation, while conveying his general precepts to the public in a series of popular medical tracts. Through these later works he became one of England's most widely read medical writers.

Cheyne's intellectual career was divided into two phases. During the first, which coincided with his association with Pitcairn in Scotland and his early years in London, he was a principal representative of British "Newtonianism" in its many cultural facets. His first book, *A New Theory of Fevers* (1702), was an elaborate, quasi-mathematical explication of febrile phenomena in terms of Pitcairn's supposedly "mathematical" and "Newtonian" variety of iatromechanism. Cheyne followed Pitcairn in positing a theory of the "animal oeconomy" based on a view of the body as a system of pipes and fluids, and, in fact, he called for the composition of a *Principia medicinae theoreticae mathematica,* which would treat such topics as the hydraulics of circulation and the elastic behavior of vascular walls with the same mathematical rigor that Newton applied to celestial mechanics.

In 1703 Cheyne followed his call for medical mathematicization with a treatise of his own on Newtonstyle mathematics, the *Fluxionum methodus inversa.* A work on the calculus of dubious mathematical validity (David Gregory counted 429 errors), the *Fluxionum* brought Cheyne more anguish than positive reputation. Abraham de Moivre responded with a thorough refutation, and the great Newton himself—so Gregory claimed—was sufficiently provoked to publish his work on "quadratures" in the 1704 edition of the *Opticks.*

Cheyne pressed ahead nevertheless, in 1705 turning his attention to the theological significance of Newtonian science. In *Philosophical Principles of Natural Religion,* along with several other arguments for the existence and continued superintendence of the Deity, he claimed that the observed phenomena of attraction in the universe argued for a Supreme Being. Since attraction was not a property essential to the mere being of brute and passive matter, its very occurrence, whether in planetary gravitation or in the simple cohesion of terrestrial materials, therefore gave immediate testimony to the hand of God in designing and maintaining the universe. Cheyne's argument proved very popular with his contemporaries, perhaps somewhat impressing even Newton. Sir Isaac included a discussion of the phenomena of attraction in the new and lengthy twenty-third "Query" to the 1706 edition of his *Opticks* (famous as the thirty-first "Query" of later editions) that seems to reflect some of Cheyne's examples and vocabulary. Cheyne at least thought so, for according to an entry in Gregory's *Memoranda,* "Dr. Cheyne uses to say among his Chronys that all the additions (made by Sr Isaac to his book of Light and Colours in the latin version) were stolen from him."

In the second phase of his intellectual career, which coincided with his residence at Bath, Cheyne repudiated his youthful mathematical brashness and excessive Newtonian enthusiasm. Although he never gave up his intense interest in philosophical and theological speculations or even in Newtonian science, in the works composed while practicing at Bath, Cheyne turned his attention largely to medical subjects. In 1720 he published *An Essay on the Gout,* in 1724 *An Essay of Health and Long Life,* in 1740 *An Essay on Regimen,* and in 1742 *The Natural Method of Cureing the Diseases of the Body and the Disorders of the Mind Depending on the Body.* All these treatises were essentially practical guides that placed considerable emphasis on the medical wisdom of moderation in diet and drink. But Cheyne also devoted some space in each of these books to philosophical and theological issues. In medical theory, for example, he was much committed to directing attention from the body's fluids to its fibrous solids, his uncited guide in this matter almost certainly being the influential Leiden professor Hermann Boerhaave.

Cheyne's most elaborate development of his views on the bodily fibers was contained in the treatise *De natura fibrae* (1725). He was simultaneously concerned with the relationship between the immaterial, musician-like soul and the material, instrument-like body. Although opinions on this subject can be found in all his later writings, the most extensive account of his views was contained in *The English Malady* (1733). Through his later medical works generally, and especially through these last two, Cheyne seems to have aroused much interest in Britain in further investigation of the bodily fibers and in exploration of the metaphysical relationship of mind and body.

BIBLIOGRAPHY

Cheyne's principal writings have been mentioned above; no complete edition of his works exists. For useful biographical and bibliographical summaries that emphasize the later medical writings, see "George Cheyne," in *Dictionary of National Biography* and Charles F. Mullet, Introduction to *The Letters of Dr. George Cheyne to the Countess of Huntingdon* (San Marino, Calif., 1940). The best source for Cheyne's early activities as a "Newtonian" is W. G. Hiscock's edition of Gregory's *Memoranda*, published as *David Gregory, Isaac Newton and Their Circle* (Oxford, 1937). For an interesting summary of Cheyne's theological views, see Hélène Metzger, *Attraction universelle et religion naturelle chez quelques commentateurs anglais de Newton* (Paris, 1938). General summaries of some of Cheyne's medical theories can be found in Albrecht von Haller, *Bibliotheca medicinae practicae*, IV (Basel, 1788), 435–438; Kurt Sprengel, *Histoire de la médecine*, V (Paris, 1815), 167–170; and Charles Daremberg, *Histoire des sciences médicales*, II (Paris, 1870), 1207–1214.

THEODORE M. BROWN

CHIARUGI, GIULIO (*b.* Siena, Italy, 28 January 1859; *d.* Florence, Italy, 17 March 1944), *anatomy*.

Chiarugi, one of the most eminent embryologists of the last hundred years, not only helped shape present-day anatomy; he also influenced thousands of medical students as a great teacher.

The son of Pietro Chiarugi, a small trader, and of Elisa Del Puglia, Chiarugi studied medicine at the University of Siena and graduated in 1882 from the University of Turin. In 1884 he was appointed an assistant at the Institute of Anatomy in Siena, and in 1886 he was elected professor of anatomy at the University of Siena. Chiarugi was named director of the Institute of Human Anatomy in Florence and professor of descriptive, microscopical, and topo-graphical anatomy in the Medical School of Florence in 1890; from 1900 he was a famous teacher of anatomy for artists; and in 1908 he began to teach embryology as well. He retired from teaching in 1934 but continued his work almost until his death. In 1902 he had founded the *Archivio italiano di anatomia e di embriologia,* the publication of which he always supervised.

Chiarugi began his anatomical work with studies of the human encephalon: *Osservazioni sulle circonvoluzioni frontali* (1885) and *Studio critico sulla genesi delle circonvoluzioni cerebrali* (1886). The conclusions of these first studies confirmed the assumptions found in works by Theodor Bischoff, Paul Broca, and Nicolaus Rüdinger, that the formation of the cerebral convolutions (the gyri of the cerebral cortex), once acquired by the species, is maintained and passed on by heredity.

In embryology, Chiarugi was the first to demonstrate (1887) that the activity of the heart begins in the very first stages of ontogeny, when the nervous connections are not organized, and even before the morphological differentiation of the primitive myocardium: in chick embryos of nine–ten somites on the second day of incubation, the pulsation of the heart begins before the appearance of myofibrils. Chiarugi's observation thus proved the theory of cardiac automatism.

Chiarugi also sought in embryology the explanation of the significance of cerebral nerves and studied the early development of the vagus, accessory, and hypoglossus nerves. Through this research (1888–1894) he showed the dual origin—neural and branchial—of the vagus nerve, and also that its branchial component originates from at least two somites.

Chiarugi understood that experimental embryology might help to explain the action of some environmental factors on the early stages of the fertilized egg. From 1895 to 1901 he made a series of observations on the effect of such physical agents as light and temperature on the segmentation of the fertilized egg of *Salamandrina*. He also studied the pathology of the human embryo. For this embryological work Chiarugi was nominated a member of the International Institute of Embryology at Utrecht. In 1926 he published a book on twins, and in 1929 the first volume of his monumental *Trattato di embriologia*, which was completed in 1944.

BIBLIOGRAPHY

I. ORIGINAL WORKS. Bibliographies of Chiarugi's writings are in *Archivio italiano di anatomia e di embriologia,* **18,** supp. (1922), vii–xxviii; and **39** (1938), i–xi. Among his works are *Istituzioni di anatomia dell'uomo,* 3 vols. (Milan, 1904–1917; 2nd ed., 4 vols., 1921–1926; 5th ed., 1938; 9th ed., 1959; 10th ed., 1968), the first modern Italian treatise on anatomy; *Atlante di anatomia dell'uomo ad uso degli artisti* (Florence, 1908), with 20 life-size plates; *I gemelli* (Turin, 1926); *Trattato di embriologia,* 4 vols. (Milan, 1929–1944); and "Il peso dell'encefalo e delle sue principali parti negli Italiani," in *Atti della R. Accademia d'Italia,* Memorie della Classe di Scienze Fisiche, Matematiche e Naturali, **14** (1943), 227–391.

II. SECONDARY LITERATURE. For reliable discussions of Chiarugi's life and work, see N. Beccari, "La vita e l'opera di Giulio Chiarugi," in *Archivio italiano di anatomia e di embriologia,* **50** (1945), 232–257; Pietro Franceschini, "L'automatismo del cuore nelle osservazioni di Giulio Chiarugi e di Giulio Fano," in *Physis,* **2** (1960), 163–183; and "Giulio Chiarugi," in *Scientia medica italica,* **8** (1960), 269–283; and G. Levi, "Giulio Chiarugi ricercatore e maestro," in *Rendiconti dell'Accademia nazionale dei Lincei,* Classe di Scienze Fisiche, Matematiche e Naturali, 7th ser., **27** (1959), fasc. 5.

PIETRO FRANCESCHINI

CHICHIBABIN, ALEXEI YEVGENIEVICH (*b.* Kuzemino, Poltava gubernia [now Ukrainian SSR], Russia, 29 March 1871; *d.* Paris, France, 15 August 1945), *chemistry.*

Chichibabin, the son of a lower-echelon government employee, graduated from Moscow University in 1892. From his first year there he studied under Markovnikov, investigating the reduction of propylbenzene with hydrogen iodide. After graduation he investigated privately, under the direction of M. I. Konovalov, the nitration of alkylpyridines and from then on concentrated his research on pyridine. In 1905–1906, he was assistant professor at the University of Warsaw, and from 1909 to 1930 full professor of chemistry at Moscow Higher Technical School. From 1918 to 1923 he was also professor at the University of Moscow. In 1930, after the death of his only daughter in a chemical accident, he moved to Paris, where he worked in the organic chemistry laboratory at the Collège de France. His fundamental textbook, *Osnovnye nachala organicheskoy khimii,* has gone through seven editions since 1924 and has been translated into English, French, Spanish, Czech, Slovak, Hungarian, and Chinese.

After the beginning of World War I, Chichibabin organized the production of pharmaceuticals in Russia; he was responsible for the construction of the first Russian alkaloid plant and the production of opium, morphine, codeine, atropine, cocaine, caffeine, aspirin, phenyl salicylate, and phenacetin.

In 1926 Chichibabin was the first Soviet chemist to receive the Lenin Prize, for his work with pyridines and in pharmaceutical chemistry.

Chichibabin studied naphthenic acids in Caucasian oil (he was the first to find aliphatic acids in this oil). He also worked on phenol alkylation; established the structure of musk ketone and found a new synthesis of amber musk; discovered the synthesis of thiodiglycol by the action of hydrogen sulfide on ethylene oxide; synthesized pilocarpine and santonin. In other areas of chemistry Chichibabin showed that under certain conditions organo-magnesium compounds can react with ortho esters, acetals, and even ethers. He was also concerned with triphenylmethane compounds and the theoretical problems connected with them. To explain the special properties of triarylmethyls, Chichibabin developed the concept of trivalent carbon and even attempted to explain olefins by means of it.

In pyridine chemistry, the field in which he did his most notable work, Chichibabin investigated the rearrangement of 1-benzylpyridine salts and found that they can regroup—with the benzyl radical occupying not only positions 2 and 4, as previously believed, but also position 3. His studies of the reaction of aldehyde and acetylene with ammonia led to new methods of synthesizing pyridine or alkylpyridine with the alkyl group in any position. Most important is the synthesis from acetaldehyde and ammonia of 2-methyl-5-ethyl-pyridine, used for the preparation of 5-methyl-5-vinylpyridine (for copolymerization with styrene). Important methods of synthesizing furan, pyrrole, and thiophene from acetylene were also devised.

In 1913 Chichibabin and Seide discovered that a hydrogen atom in position 2 or 4 of the pyridine nucleus can be directly substituted for an amino group. Later Chichibabin showed that the same was true of a hydroxyl radical. The direct amination, now known as the Chichibabin reaction, is applied to industrial preparation of 2-aminopyridine from pyridine. A study of the Chichibabin amination was extended to different pyridines, natural nicotine, and compounds of the quinoline and isoquinoline series.

Chichibabin investigated the laws governing the introduction of a second substituent into aminopyridines. A new type of amino-imino tautomerism was discovered in aminopyridines and it was shown that, depending on conditions, alkylation can attack the amino group or the hetero-nitrogen atom. In its reactions with difunctional compounds 2-aminopyridine reacts as an amidine, forming condensed bicyclical structures.

BIBLIOGRAPHY

I. ORIGINAL WORKS. Chichibabin's books include *Issledovania po voprosu o trekhvalentnom uglerode i stroenii prosteyshikh okrashennykh proizvodnykh trifenilmetana* ("Study on Trivalent Carbon and on the Structure of the Simplest Colored Triphenylmethane Derivatives"; Moscow, 1912) and *Osnovnye nachala organicheskoy khimii* ("Basic Principles of Organic Chemistry"; Moscow, 1925). A few of his more than 350 papers are "Novy obshchy metod poluchenia aldegidov" ("A New General Method for Preparing Aldehydes"), in *Zhurnal Russkago fiziko-khimicheskago obshchestva,* **35** (1903), 1284–1286; "Sintezy piridinovykh osnovany iz aldegidov predelnogo ryada i ammiaka" ("Synthesis of Pyridine Bases from Aldehydes of the Saturated Series and Ammonia"), *ibid.,* **37** (1905), 1229–1253; "Novaya reaktsia soedineny, soderzhashchikh piridinovoe yadro" ("New Reactions of Compounds Containing a Pyridine Nucleus"), *ibid.,* **46** (1914), 1216–1236; "Tautomeria α-aminopiridina" ("Tautomerism of α-amino-pyridine"), in *Berichte der Deutschen chemische Gesellschaft,* **57** (1924), 1168–1172, and in *Zhurnal Russkago fiziko-khimicheskago obshchestva,* **57** (1925), 399–425; "Sintez pilopovykh kislot i stroenie pilokarpina" ("Synthesis of Pilopic Acids and Structure of Pilocarpine"), in *Doklady Akademii nauk SSSR,* **2** (1930), 25–32; *Berichte der Deutschen chemische Gesellschaft,* **63** (1930), 460–470; and "Hétérocycles hexatomiques avec un atome d'azote. Groupes de la pyridine et de la pipéridine," in V. Grignard, ed., *Traité de chimie organique,* Vol. XX (Paris, 1953), 33–375.

II. SECONDARY LITERATURE. On Chichibabin or his work, see M. Delépine, "Hommage de la Société chimique de France à Alexis Tchitchibabine (1871–1945)," in *Bulletin de la Société chimique de France* (1958), p. 407; Y. S. Musabekov, *Istoria organicheskogo sinteza v Rossii* ("History of Organic Synthesis in Russia"; Moscow, 1958), which contains material on Chichibabin's scientific activities; and P. M. Yevteeva, in *Trudy Instituta istorii estestvoznaniya i tekhniki* ("Transactions of the Institute of Natural Sciences and Technological History"), **18** (1958), 296–356, a biography and review of Chichibabin's scientific work, with a list of 346 writings.

A. N. KOST

CHILD, CHARLES MANNING (*b.* Ypsilanti, Michigan, 2 February 1869; *d.* Palo Alto, California, 19 December 1954), *zoology.*

Child was the only surviving child of Charles Chauncey Child and Mary Elizabeth Manning Child, both of substantial and prosperous old New England families. Keenly interested in natural history in all its aspects, he grew up in the family home at Higganum, Connecticut, and attended nearby Wesleyan University as a commuter student. There he finally chose the field of biology rather than chemistry and received the Ph.B. in 1890 and the M.S. in biology in 1892. He was an outstanding student and held a graduate assistantship from 1890 to 1892. After the death of his parents, Child went to Leipzig, where he worked for a time in the psychology laboratory of Wilhelm Wundt before taking his Ph.D. in 1894, in zoology, with Rudolf Leuckart.

In 1895, after some research at the famous Naples Zoological Station, Child joined the faculty of the University of Chicago, where he rose through the ranks to a professorship in 1916. He carried on research at Woods Hole, at Naples in 1902–1903, and at marine stations on the Pacific Coast. He married Lydia Van Meter in 1899. At Chicago, Child was made department head and was retained after reaching retirement age in 1934. In 1928 he founded the journal *Physiological Zoology.* He retired to California in 1937 but continued research and writing until his death. A prodigious worker, traveler, and hiker, he had close friends but was generally reserved in his personal relationships.

Child's work in zoology involved two themes: sensitivity and reactivity of animal organisms, and the problems of reproduction and development. His doctoral dissertation was a morphological study of insect sense organs, and by the late 1890's he was working on reproduction and embryology in simple organisms. His main contribution, the gradient theory, combined the two interests. He became intrigued by the classic controversy of preformation versus epigenesis, and he turned his attention to the development of the form of various organisms. In his search for the mechanism by which the structure of the organism develops from the germinal material, Child came to emphasize physiological factors, positing a dynamic process by which given elements in the cells respond to the environment in such a way as to produce molar development. Child believed the communication between protoplasmic units to be primarily excitatory and only secondarily chemical.

The gradient theory took its name from Child's observation that regeneration of the organism occurs in graded physiological stages along the axis (originally observed in planaria), with each gradient physiological process seemingly connected to those immediately adjacent to it. The controversial concept was important in the science of morphogenesis after about 1911. Many of Child's colleagues felt that he generalized too much from experiments with simple animals, but Child had nevertheless identified a basic problem in organic development. He spent most of the rest of his life exploring his idea in numerous publications.

Child gained substantial reputation and recognition for his work and received many honors in his lifetime. Although he lectured ably to undergraduates, his forte was research and he trained a number of graduate students in his laboratory. Not the least of his contributions was as head of the department of zoology at Chicago. Although Child's dominating major interest was narrow, his colleagues were nevertheless a vigorous and diverse group.

BIBLIOGRAPHY

Biographical information and a full bibliography of Child's writings may be found in Libbie H. Hyman, "Charles Manning Child, 1869–1954," in *Biographical Memoirs of the National Academy of Sciences,* **30** (1957), 73–103.

J. C. BURNHAM

CHILDREY, JOSHUA (*b.* Rochester, England, 1623; *d.* Upwey, Dorsetshire, England, 26 August 1670), *meteorology, natural history.*

Joshua Childrey was educated at Rochester Grammar School and Magdalen College, Oxford, where his studies were interrupted by the outbreak of the civil war in 1642. He returned to take his B.A. in 1646, but was expelled by the Parliamentary Visitors in 1648. He then became a schoolmaster in Faversham, Kent. After becoming a doctor of divinity in 1661, he was appointed chaplain to Lord Herbert and held ecclesiastical appointments in the West Country. In 1664 he was made rector of Upwey, a position he held until his death. He married in 1665.

Childrey's first work was an astrological tract of sixteen pages, *Indago astrologica; or A Brief and Modest Enqiry into Some Principal Points of Astrology* (1652). He followed it with an ephemeris, *Syzygiasticon instauratum* (1653). The purpose of both was to prove that heliocentric astrology is more effective than geocentric. In the latter, which is not altogether untypical of the almanacs of the time, this effectiveness was claimed especially as regards the meteorological consequences of planetary aspects. Childrey apologized profusely for having omitted in his former work to calculate planetary latitudes as well as longitudes, and now rectified his fault. His ambition was to extend the Baconian method to astrology. One result of his "experimental astrology" was to confirm an old idea that there is a thirty-five-year cycle in the weather. He was assisted in this by his friend Richard Fitz-Smith, who did many of the calculations.

Better known in his own day was his *Britannia Baconia: or, The Natural Rarities of England, Scotland and Wales, According as They Are to be Found in Every Shire. Historically Related, According to the Precepts of Lord Bacon* (1660). This is largely a compilation of previous writers, but has some original observations. As its title hints, it is drawn from notebooks kept with the classification used by Francis Bacon for his "Histories" at the end of *Novum organum.* In character, although much shorter than either, it resembles Plot's *Natural History of Oxfordshire,* which is said by Anthony Wood to have been inspired by it.

Childrey made numerous observations on the weather and tides when living in Weymouth. The manuscripts are lost, but his controversy with John Wallis on the matter of tides resulted in a letter to Seth Ward, at one time bishop of his diocese, which contains many interesting observations, historical and otherwise (*Philosophical Transactions of the Royal Society,* no. 64, pp. 2061–2068). Wallis's original paper and reply, in the same volume, show a much keener analytical approach to the difficult problem, but Childrey presented information of whose significance neither Wallis nor any of their contemporaries was aware, such as the meteorological dependence of tides whose periods are multiples or submultiples of a tropical year; the tidal inequality due to the inclination of the moon's orbit and its varying distance, and so on. The discussion was bedeviled by the fact that the examples cited were all narrow channels, where certain atypical tidal patterns prevail.

Childrey was not a man of great scientific originality—he was not even abreast of the scientific movement of his century—but he stimulated discussion of meteorological and related topics which were in his day often neglected.

BIBLIOGRAPHY

I. ORIGINAL WORKS. Childrey's chief writings are mentioned in the text. The *Britannia Baconia* was also published in a French translation (Paris, 1662, 1667). A number of letters to and from Childrey are in A. R. and M. B. Hall, *The Correspondence of Henry Oldenburg,* vol. V (Madison, Wis., 1968); see esp. pp. 384–386 and 454–456.

II. SECONDARY LITERATURE. There is no biography of Childrey. Details of his career may be found in the article in the *Dictionary of National Biography,* which is based on Anthony Wood's *Athenae Oxonienses,* 2nd enl. ed. (London, 1721), p. 467. There is some supplementary information in the Oldenburg letters cited (see, for instance, an editor's note in vol. I, on Childrey as a maker and improver of telescopes).

J. D. NORTH

CH'IN CHIU-SHAO (*b.* Szechuan, China, *ca.* 1202; *d.* Kwangtung, China, *ca.* 1261), *mathematics.*

Ch'in Chiu-shao (literary name Tao-ku) has been described by George Sarton as "one of the greatest mathematicians of his race, of his time, and indeed of all times." A genius in mathematics and accomplished in poetry, archery, fencing, riding, music, and architecture, Ch'in has often been judged an intriguing and unprincipled character, reminding one of the sixteenth-century mathematician Girolamo Cardano. In love affairs he had a reputation similar to Ibn Sīnā's. Liu K'e-chuang, in a petition to the emperor, described him as a person "as violent as a tiger or a wolf, and as poisonous as a viper or a scorpion." He was also described as an ill-disciplined youth. During a banquet given by his father, a commotion was created when a stone suddenly landed among the guests; investigation disclosed that the missile had come from the direction of Ch'in, who was showing a *fille de joie* how to use a bow as a sling to hurl projectiles. Chou Mi, in his supplementary volume to the *Kuei-yu tsa-chih,* tells us how Ch'in deceived his friend Wu Ch'ien in order to acquire a plot of his land, how he punished a female member of his household by confinement and starvation, and how he became notorious for being inclined to poison those he found disagreeable. We are also told that on his dismissal from the governorship of Ch'iung-chou in 1258 he returned home with immense wealth—after having been in office for just over a hundred days.

According to a recent study by Ch'ien Pao-tsung and others Ch'in was born in the city of P'u-chou (now An-yüeh) in Szechuan province. Ch'in called himself a native of Lu-chün, in Shantung province; but he was simply referring to the place his ancestors came from rather than to his place of birth. His father, Ch'in Chi-yu (literary name Hung-fu), was a civil servant. In 1219 Ch'in joined the army as the head of a unit of territorial volunteers and participated in curbing a rebellion staged by Chang Fu and Mo Chien. In 1224–1225 he followed his father when the latter was transferred to the Sung capital, Chung-tu (now Hang-chow). There he had the opportunity to study astronomy at the astronomical bureau under the guidance of the official astronomers. Shortly afterward, however, his father was sent to the prefecture of T'ung-ch'üan (now San-t'ai in Szechuan province), and Ch'in had to leave the capital. About 1233 he served as a sheriff in one of the subprefectures in Szechuan.

The Mongols invaded Szechuan in 1236, and Ch'in fled to the east, where he first became a vice-administrator (*t'ung-p'an*) in Ch'i-chou prefecture (now Ch'i-ch'un in Hupeh province) and then governor of Ho-chou (now Ho-hsien in Anhwei province). In the latter part of 1244 Ch'in was appointed one of the vice-administrators of the superior prefecture of Chien-k'ang-fu (now Nanking), but some three months later he relinquished this post because of his mother's death. He returned to Hu-chou (now Wu-hsing in Chekiang province), and it was probably there that he wrote his celebrated mathematical treatise *Shu-shu chiu-chang* ("Mathematical Treatise in Nine Sections"), which appeared in 1247. In the preface of this book Ch'in mentions that he learned mathematics from a certain recluse scholar, but he does not give his identity.

In 1254 Ch'in returned to Chung-tu to reenter civil service, but for some unknown reason he soon resigned and went back to his native home. He paid a visit to Chia Shih-tao, an influential minister at that time, and was appointed governor of Ch'iung-chou (in modern Hainan) in 1258. A few months later, however, Ch'in was dismissed because of charges of bribery and corruption. Nevertheless, he managed to find another job as a civil aide to an intimate friend of his, Wu Ch'ien (literary name Li-chai), who was then in charge of marine affairs in the district of Yin (near modern Ningpo in Chekiang province). Wu Ch'ien eventually became a minister, but in 1260 he lost favor and was given a lesser assignment in south China. Ch'in followed his friend to Kwangtung province and received an appointment in Mei-chou (now Mei-hsien), where he died shortly afterward. The year of Ch'in's death has been estimated as 1261, for there was an edict in the following year banning Wu Ch'ien and his associates from the civil service.

The title of Ch'in's *Shu-shu chiu-chang* has given rise to some confusion. According to Ch'en Chen-sun, a contemporary of Ch'in Chiu-shao's and owner of a copy of the work, the original title was *Shu shu* ("Mathematical Treatise"). However, in his *Chih chai shu lu chieh ti* he gives the title as *Shu-shu ta lüeh* ("Outline of Mathematical Methods"). During the thirteenth century this treatise was referred to as the *Shu-shu ta-lüeh* or the *Shu-hsüeh ta-lüeh* ("Outline of Mathematics"), while during the Ming period (1368–1644) it was known under the names *Shu-shu chiu-chang* and *Shu-hsüeh chiu-chang*. This has led Sarton to conclude that the *Shu-shu chiu-chang* and the *Shu-hsüeh ta-lüeh* were separate works. The treatise is now popularly known as the *Shu-shu chiu-chang*.

It appears that the *Shu-shu chiu-chang* existed only in manuscript form for several centuries. It was copied and included in the great early fifteenth-century encyclopedia *Yung-lo ta-tien* under the title *Shu-hsüeh chiu-chang*. This version was revised and included in the seventeenth-century imperial collection *Ssu-k'u ch'üan-shu,* and later a commentary was added to it by the Ch'ing mathematician Li Jui. There was also a handwritten copy of the *Shu-shu chiu-chang* during the early seventeenth century. A copy from the text belonging to the Wen-yüan k'o library was first made by Wang Ying-lin. And in 1616 Chao Ch'i-mei wrote that he had made a copy of the text that he borrowed from Wang Ying-lin and had added a new table of contents to it. Toward the beginning of the nineteenth century this handwritten copy came into the possession of the mathematician Chang Tun-jen and attracted much attention during the time when interest in traditional Chinese mathematics was revived. Many copies were reproduced from the text owned by Chang Tun-jen. It seems that blocks were also made for the printing of the book, but it is not certain whether it was actually printed. Also in the early nineteenth century Shen Ch'in-p'ei began to make a textual collation of the *Shu-shu chiu-chang,* but he died before his work was finished. One of his disciples, Sung Ching-ch'ang, completed it; and the result appeared in the *Shu-shu chiu-chang cha-chi* ("Notes on the Mathematical Treatise in Nine Sections"). In 1842 both the *Shu-shu chiu-chang* and the *Shu-shu chiu-chang cha-chi* were published and included in the *I-chia-t'ang ts'ung shu* collection. Later editions of both these books, such as those included in the *Ku-chin suan-hsüeh ts'ung shu,* the *Kuo-hsüeh chi-pen ts'ung shu,* and the *Ts'ung shu chi-ch'eng* collections, are based on the version in the *I-chia-t'ang ts'ung-shu* collection.

Each of the nine sections in the *Shu-shu chiu-chang* includes two chapters made up of nine problems. These sections do not correspond in any way to the nine sections of the *Chiu-chang suan-shu* of Liu Hui. They consist of (1) *ta yen ch'iu i shu,* or indeterminate analysis; (2) *t'ien shih,* which involves astronomical, calendrical, and meteorological calculations; (3) *t'ien yü,* or land measurement; (4) *ts'e wang,* referring to surveying by the method of triangulation; (5) *fu i,* or land tax and state service; (6) *ch'ien ku,* or money and grains; (7) *ying chien,* or structural works; (8) *chün lü,* or military matters; and (9) *shih wu,* dealing with barter and purchase. The complete text has not yet been translated or investigated in full, although some individual problems have been studied.

With Ch'in's *Shu-shu chiu-chang* the study of indeterminate analysis in China reached its height. It had first appeared in Chinese mathematical texts about the fourth century in a problem in the *Sun-tzu suan ching:*

> There is an unknown number of things. When counted in threes, they leave a remainder of two; when counted

by fives, they leave a remainder of three; and when counted by sevens, they leave a remainder of two. Find the number of things.

The problem can be expressed in the modern form

$$N \equiv 2 \ (\mathrm{mod}\ 3) \equiv 3 \ (\mathrm{mod}\ 5) \equiv 2 \ (\mathrm{mod}\ 7),$$

where the least integer for N is required. The *Sun-tzu suan ching* gives the following solution:

$$N = 2 \times 70 + 3 \times 21 + 2 \times 15 - 2 \times 105 = 23.$$

There is no explanation of the mathematical method in general, but the algorithmical procedure is given as follows:

> If you count by threes and have the remainder 2, put 140.
>
> If you count by fives and have the remainder 3, put 63.
>
> If you count by sevens and have the remainder 2, put 30.
>
> Add these numbers, and you get 233.
>
> From this subtract 210, and you have the result.

A brief explanation of the procedure follows:

> For each 1 as a remainder, when counting by threes, put 70.
>
> For each 1 as a remainder, when counting by fives, put 21.
>
> For each 1 as a remainder, when counting by sevens, put 15.
>
> If the sum is 106 or more, subtract 105 from it, and you have the result.

(The number 105 to be subtracted is, of course, derived from the product of 3, 5, and 7.)

The next example of indeterminate problem appeared in that of the "hundred fowls," found in the fifth-century mathematical manual *Chang Ch'iu-chien suan ching.* It gives three different possible answers but no general solution. Since it is in the form of two simultaneous linear equations of three unknowns, i.e.,

$$ax + by + cz = 100$$
$$a'x + b'y + c'z = 100,$$

this problem is not of the same nature as that given in the *Sun-tzu suan ching.*

As early as the middle of the third century, Chinese calendar experts had taken as their starting point a certain date and time in the past known as the Grand Cycle (*Shang yuan*), which was the last time that the winter solstice fell exactly at midnight on the first day of the eleventh month, which also happened to be the first day (*chia-tzu,* cyclical day) of a sixty-day cycle. If *a* denotes the tropical year, R_1 the cyclical-day number of the winter solstice (i.e., the number of days in the sixty-day cycle between winter solstice and the last *chia-tzu* preceding it), b the synodic month, and R_2 the number of days between the first day of the eleventh month and the winter solstice, then N, the number of years since the Grand Cycle, can be found from the expression

$$aN \equiv R_1 \ (\text{mod } 60) \equiv R_2 \ (\text{mod } b).$$

For several hundred years, calendar experts in China had been working out the Grand Cycle from new astronomical data as it became available. None, however, has passed on the method of computation. The earliest elucidation of the method available to us comes from Ch'in Chiu-shao. Problem 12 in his *Shu-shu chiu-chang* deals exactly with the above and may be stated in the modern form

$$6,172,608 \ N \equiv 193,440 \ (\text{mod } 60 \times 16,900)$$
$$\equiv 16,377 \ (\text{mod } 499,067),$$

taking $a = 365 \dfrac{4,108}{16,900}$ days

$$b = 29 \dfrac{8,967}{16,900} \text{ days.}$$

Ch'in's method of solving indeterminate analysis may be explained in the modern form as follows:

Given $N \equiv R_1 \ (\text{mod } a_1) \equiv R_2 \ (\text{mod } a_2) \equiv R_3 \ (\text{mod } a_3) \equiv \cdots \equiv R_n \ (\text{mod } a_n)$ where $a_1, a_2, a_3, \cdots, a_n$ have no common factors.

If $k_1, k_2, k_3, \cdots k_n$ are factors such that

$$k_1 a_2 a_3 \cdots a_n \equiv 1 \ (\text{mod } a_1)$$
$$k_2 a_3 \cdots a_n a_1 \equiv 1 \ (\text{mod } a_2)$$
$$k_3 a_1 a_2 a_4 \cdots a_n \equiv 1 \ (\text{mod } a_3), \text{ and}$$
$$k_n a_1 a_2 a_3 \cdots a_{n-1} \equiv 1 \ (\text{mod } a_n),$$

then $N \equiv (R_1 k_1 a_2 a_3 \cdots a_n) + (R_2 k_2 a_3 \cdots a_n a_1)$
$$+ (R_3 k_3 a_1 a_2 a_4 \cdots a_n) + \cdots$$
$$+ (R_n k_n a_1 a_2 a_3 \cdots a_{n-1})(\text{mod } a_1 a_2 a_3 \cdots a_n),$$

or, putting it more generally,

$$N \equiv \sum_1^n R_i k_i \frac{M}{a_i} - pM$$

where $M = a_1 a_2 a_3 \cdots a_i$, i.e., the least common multiple and p is the integer that yields the lowest value for N.

The rule is given in a German manuscript from Göttingen (*ca.* 1550), but it was not rediscovered in Europe before Lebesque (1859) and Stieltjes (1890). The identity of the Chinese rule with Gauss's formula has also been pointed out by Matthiessen in the last century, after Ch'in's study of indeterminate analysis was first brought to the attention of the West by Alexander Wylie.

The *Shu-shu chiu-chang* also deals with numbers which have common factors among them, in other words the more general form

$$N \equiv R_i \ (\text{mod } A_i)$$

where $i = 1, 2, 3, \cdots, n$ and where A_i has common factors with A_j, A_k, and so on.

The method involves choosing $A_1, A_2, \cdots, A_n$, which are relative primes in pairs such that each A_i divides the corresponding a_i and that further the *LCM* of $A_1, A_2, \cdots, A_n$ equals that of $a_1 a_2 \cdots a_n$. Then every solution of

$$N \equiv R_1 \ (\text{mod } A_1) \equiv R_2 \ (\text{mod } A_2)$$
$$\equiv \cdots \equiv R_n \ (\text{mod } A_n)$$

also satisfies

$$N \equiv R_1 \ (\text{mod } a_1) \equiv R_2 \ (\text{mod } a_2)$$
$$\equiv \cdots \equiv R_n \ (\text{mod } a_n).$$

The above is valid only under the condition that each difference $R_i - R_j$ is divisible by d, the *GCD* of the corresponding moduli Ai and Aj, i.e., $R_i - R_j = 0 \ (\text{mod } d)$. The Chinese text does not mention this condition, but it is fulfilled in all the examples given by Ch'in Chiu-shao. Ch'in would go about searching for the least integral value of a multiple k_i such that

$$k_i \frac{M}{a_i} \equiv 1 \ (\text{mod } a_i).$$

This is an important intermediate stage in the process of solving problems of indeterminate analysis; hence the Chinese term *ch'iu i shu* ("method of searching for unity") for indeterminate analysis. Over time the process became known as the *ta yen ch'iu i shu* ("the Great Extension method of searching for unity"). The term *ta yen* ("Great Extension") came from an obscure statement in the *Book of Changes*. In an ancient method of divination, fifty yarrow stalks were taken, and one was set aside before the remaining forty-nine were divided into two random heaps. The *Book of Changes* then says:

> The numbers of the Great Extension [multiplied together] make fifty, of which [only] forty-nine are used [in divination]. [The stalks representing these] are divided into two heaps to represent the two [emblematic lines, or heaven and earth] and placed [between the little finger and ring finger of the left hand], that there may thus be symbolized the three [powers of heaven, earth, and man]. [The heaps on both sides] are manipulated by fours to represent the four seasons. . . .

Ch'in sought to explain the term *ta yen ch'iu i shu* in the first problem of his book by introducing the so-called "Great Extension number" 50 and the number 49 and showing how they could be arrived at from the numbers 1, 2, 3, and 4—as mentioned above.

Since Ch'in also introduced many technical terms used in conjunction with indeterminate analysis, it will be worthwhile to follow, step by step, the actual process he used in working out the problem that may be expressed as $N \equiv 1 \pmod 1 \equiv 1 \pmod 2 \equiv 1 \pmod 3 \equiv 1 \pmod 4$.

Ch'in first arranged the given numbers 1, 2, 3, and 4, known as *yuan-shu* ("original number"), in a vertical column. He placed the number 1 to the left of each of these numbers, as in Fig. 1.

1	1	24	1
1	2	12	2
1	3	8	3
1	4	6	4
Celestial Monads	Original Numbers	Operation Numbers	Original Numbers

FIGURE 1   FIGURE 2

This he called the *t'ien-yuan* ("celestial monad" or "celestial element"). Next, he cross-multiplied each celestial monad by the original numbers not pertaining to it, thus obtaining the *yen-shu* ("operation numbers"), which were then placed to the left of the corresponding original numbers, as in Fig. 2.

He next removed all the common factors in the original numbers, retaining only one of each. Thus the original numbers became prime to one another and were known by the term *ting-mu* ("definite base numbers"), as in Fig. 3. Each celestial monad was cross-multiplied by the definite base numbers not pertaining to it, giving another set of operation numbers, which were then placed to the left of the corresponding definite base numbers as in Fig. 4.

1	1	12	1
1	1	12	1
1	3	4	3
1	4	3	4
Celestial Monads	Definite Base Numbers	Operation Numbers	Definite Base Numbers

FIGURE 3   FIGURE 4

Ch'in then took the definite base numbers as moduli and formed the congruences with their respective operation numbers:

$$12 \equiv 1 \pmod 1$$
$$12 \equiv 1 \pmod 1$$
$$4 \equiv 1 \pmod 3$$
$$3 \equiv 3 \pmod 4.$$

For the residues (*ch'i shu*) that were unity, the corresponding multipliers (*ch'eng lü*) were taken as unity. A residue that was not unity was placed in the upper-right space on the counting board, with the corresponding definite base number below it. To the

left of this residue Ch'in placed unity as the celestial monad, as in Fig. 5.

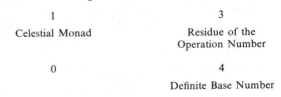

1	3
Celestial Monad	Residue of the Operation Number
0	4
	Definite Base Number

FIGURE 5

Dividing the definite base number by the residue yielded unity in this case. Ch'in next multiplied the celestial monad by unity and placed the result (in this case, also unity) at the bottom left. This he called the *kuei-shu* ("reduced number"). The space for the definite base number was then filled by the residue of the base number, as shown in Fig. 6.

1	3
Celestial Monad	Residue of the Operation Number
1	1
Reduced Number	Residue of the Base Number

FIGURE 6

Next the residue of the operation number in the upper right-hand corner was divided by the residue of the base number so that a quotient could be found to give a remainder of unity. If a quotient could not be found, then the process had to be repeated, taking the number in the lower right-hand corner and that in the upper right-hand corner alternately until such a quotient was found. In this case, however, a quotient of 2 would give a remainder of unity, as in Fig. 7.

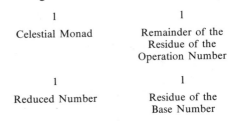

1	1
Celestial Monad	Remainder of the Residue of the Operation Number
1	1
Reduced Number	Residue of the Base Number

FIGURE 7

Multiplying the quotient of 2 by the reduced number and adding the result to the celestial monad gave 3, the corresponding multiplier, as in Fig. 8.

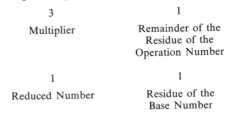

3	1
Multiplier	Remainder of the Residue of the Operation Number
1	1
Reduced Number	Residue of the Base Number

FIGURE 8

Having found all the multipliers, Ch'in then arranged them side by side with their operation numbers and definite base numbers, as in Fig. 9.

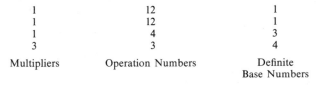

Multipliers	Operation Numbers	Definite Base Numbers
1	12	1
1	12	1
1	4	3
3	3	4

FIGURE 9

Then he multiplied the operation numbers by their corresponding multipliers and found the so-called "reduced use numbers" (*fan-yung-shu*). These were placed to the left of the corresponding definite base numbers, as in Fig. 10.

Reduced Use Numbers	Definite Base Numbers
12	1
12	1
4	3
9	4

FIGURE 10

The operation modulus (*yen-mu*), or the least common multiple, was obtained by multiplying all the definite base numbers together. If common factors had been removed, they had to be restored at this stage. The products of these factors and the corresponding definite base numbers and the reduced use numbers were restored to their original numbers, and the reduced use numbers became the definite use numbers (*ting-yung-shu*), as shown in Fig. 11.

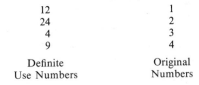

Definite Use Numbers	Original Numbers
12	1
24	2
4	3
9	4

FIGURE 11

In this particular problem Ch'in tried to explain that the sum of the operation numbers amounted to 50, while that of the definite use numbers came to 49. The former is the *ta yen* number, and the latter is the number that was put into use as stated in the *Book of Changes*. To obtain N, the definite use numbers were multiplied by the respective remainders given in the problem. Their sum, after it had been diminished repeatedly by the least common factor, would ultimately give the required answer.

A list of the technical terms used by Ch'in for indeterminate analysis is given in Fig. 12.

No.	Chinese Term	Translation		Modern Equivalent
		After Mikami	After Needham	
1.	*wen shu* *yuan-shu*	original number		A_i
2.	*ting-mu*	definite base number	fixed denominator	a_i
3.	*yen-mu*	operation modulus	multiple denominator	M
4.	*yen-shu*	operation number	multiple number	$\dfrac{M}{a_i}$
5.	*ch'eng lü*	multiplier	multiplying term	k_i
6.	*yu shu*			R_i
7.	*fan-yung-shu*	reduced use number		$\dfrac{k_i M}{a_i}$
8.	*ting-yung-shu*	definite use number		$\dfrac{k_i M}{a_i} \dfrac{A_i}{a_i}$
9.	*yung shu*		use number	either (7) or (8), depending on context
10.	*tsung teng*			$\dfrac{k_i M}{a_i} \cdot R_i$
11.	*tsung shu*			$\Sigma k_i \dfrac{M}{a_i} R_i$
12.	*ch'i shu*			$\dfrac{M}{a_i} - q_i a_i$
13.	*kuei shu*	reduced number		
14.	*t'ien-yuan*	celestial monad		

FIGURE 12

It should be pointed out that Ch'in's use of the celestial monad or celestial element (*t'ien-yuan*) differs from the method of celestial element (*t'ien-yuan-shu*) that was employed by his contemporary Li Chih and that later was known in Japan as the *tengen jutsu*. In the former, the celestial element denotes a known number, while in the latter it represents an unknown algebraic quantity.

Ch'in represented algebraic equations by placing calculating rods on the countingboard so that the absolute term appeared on the top in a vertical column; immediately below it was the unknown quantity, followed by increasing powers of the unknown quantity. Originally, Ch'in used red and black counting rods to denote positive and negative quantities, respectively; but in the text, negative quantities are denoted by an extra rod placed obliquely over the first figure of the number concerned. The *Shu-shu chiu-chang* also is the oldest extant Chinese mathematical text to contain the zero symbol. For example, the equation $-x^4 + 763,200x^2 - 40,642,560,000 = 0$ is represented by calculating rods placed on a countingboard as in Fig. 13, and can be expressed in Arabic numerals as in Fig. 14.

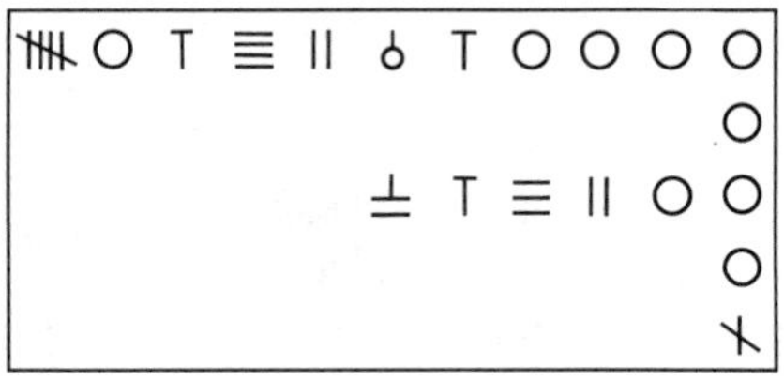

FIGURE 13

FIGURE 14

More than twenty problems in the *Shu-shu chiu-chang* involve the setting up of numerical equations. Some examples are given below.

$$4.608x^3 - 3,000,000,000 \times 30 \times 800 = 0$$
$$-x^4 + 15,245x^2 - 6,262,506.25 = 0$$
$$-x^4 + 1,534,464x^2 - 526,727,577,600 = 0$$
$$400x^4 - 2,930,000 = 0$$
$$x^{10} + 15x^8 + 72x^6 - 864x^4 - 11,664x^2 - 34,992 = 0$$

Ch'in always arranged his equations so that the absolute term was negative. Sarton has pointed out that this is equivalent to Thomas Harriot's practice (1631) of writing algebraic equations so that the ab-

solute term would stand alone in one member. Ch'in used a method called the *ling lung k'ai fang,* generally known by the translation "harmoniously alternating evolution," by which he could solve numerical equations of any degree. The method is identical to that rediscovered by Paolo Ruffini about 1805 and by William George Horner in 1819. It is doubtful that Ch'in was the originator of this method of solving numerical equations of higher degrees, since his contemporary Li Chih was also capable of solving similar equations, and some two decades later Yang Hui also described a similar method without mentioning Ch'in or Li, referring instead to several Chinese mathematicians of the twelfth century. Wang Ling and Joseph Needham have indicated that if the text of the *Chiu-chang suan-ching* (first century) is very carefully followed, the essentials of the method are there.

In the *Shu-shu chiu-chang* various values for π are used. In one place we find the old value $\pi = 3$, in another we come across what Ch'in called the "accurate value" $\pi = 22/7$, and in yet another instance the value $\pi = \sqrt{10}$ is given. This last value was first mentioned by Chang Heng in the second century.

Formulas giving the areas of various types of geometrical figures are also mentioned in the *Shu-shu chiu-chang,* although some of them are not very accurate. The area, A, of a scalene triangle with sides a, b, and c is obtained from the expression

$$A^4 - \frac{1}{4}\left\{a^2c^2 - \left(\frac{c^2 + a^2 - b^2}{2}\right)^2\right\} = 0.$$

That is,

$$A = \sqrt[4]{\frac{1}{4}\left\{a^2c^2 - \left(\frac{c^2 + a^2 - b^2}{2}\right)^2\right\}}$$
$$= \sqrt{s(s - a)(s - b)(s - c)},$$

where $s = \frac{1}{2}(a + b + c).$

The area, A, of a quadrangle with two pairs of equal sides, a and b, with c the diagonal dividing the figure into two isosceles triangles (see Fig. 15) is given by

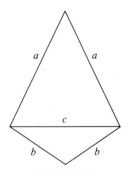

FIGURE 15

the expression

$$-A^4 + 2(X + Y)A^2 - (Y - X)^2 = 0,$$

where

$$X = \left\{ b^2 - \left(\frac{c}{2}\right)^2 \right\} \left(\frac{c}{2}\right)^2$$

and

$$Y = \left\{ a^2 - \left(\frac{c}{2}\right)^2 \right\} \left(\frac{c}{2}\right)^2.$$

The area, A, of a so-called "banana-leaf-shaped" farm (see Fig. 16) formed by two equal circular arcs with

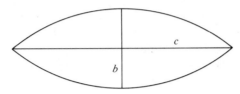

FIGURE 16

a common chord, c, and a common sagitta, b, is incorrectly given by the expression

$$(2A)^2 + 2A\left\{ \left(\frac{c}{2}\right)^2 - \left(\frac{b}{2}\right)^2 \right\} - 10(c + b)^3 = 0.$$

This formula is put in another form by Mikami but has been wrongly represented by Li Yen. An earlier expression for the area of a segment (Fig. 17) given

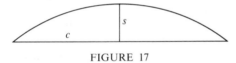

FIGURE 17

the chord, c, and the arc sagitta, s, is given in the *Chiu-chang suan-shu* in the form

$$A = \frac{1}{2}s(s + c).$$

This has been in use in China for 1,000 years. Ch'in's formula is a departure from the above, but it is not known how he arrived at it. In 1261 another formula was given by Yang Hui in the form

$$-(2A)^2 + 4Ab^2 + 4db^3 - 5b^4 = 0,$$

where d is the diameter of the circle.

Sometimes Ch'in made his process unusually complicated. For example, a problem in chapter 8 says:

Given a circular walled city of unknown diameter with four gates, one at each of the four cardinal points. A tree lies three *li* north of the northern gate. If one turns and walks eastward for nine *li* immediately leaving the southern gate, the tree becomes just visible. Find the circumference and the diameter of the city wall.

If x is the diameter of the circular wall, c the distance of the tree from the northern gate, and b the

distance to be traveled eastward from the southern gate before the tree becomes visible (as shown in Fig.

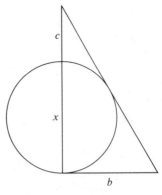

FIGURE 18

18), then Ch'in obtains the diameter from the following equation of the tenth degree:

$$y^{10} + 5cy^8 + 8c^2y^6 - 4c(b^2 - c^2)y^4 - 16c^2b^2y^2 - 16c^3b^3 = 0,$$

where $x = y^2$.

It is interesting to compare this with an equivalent but simpler expression given by Ch'in's contemporary Li Chih in the form

$$x^3 + cx^2 - 4cb^2 = 0.$$

In one of the problems in chapter 4, Ch'in intends to find the height of rainwater that would be collected on the level ground when the rain gauge is a basin with a larger diameter at the opening than at the base. The diameters of the opening and the base, a and b; the height, h, of rainwater collected in the basin; and the height, H, of the basin are given. The height of rain collected on level ground, h', is given by the formula $h' =$

$$\frac{h\{Hb[Hb + (a - b)h] + H^2b^2 + [Hb + (a - b)h]^2\}}{3(aH)^2}.$$

The *Shu-shu chiu-chang* also concerns itself with series, such as

$$\sum_1^n (a + \overline{n - 1}\, b) = na + \frac{n(n + 1)}{2} b$$

and

$$\sum_1^n n = \frac{(2m - 1)2m}{2},$$

where

$$m = \frac{n + 1}{2}.$$

One of the problems in chapter 13 deals with finite difference, a subject that had attracted considerable attention from Chinese mathematicians and calendar makers.

Linear simultaneous equations are also discussed in Ch'in's book. The numbers are set up in vertical columns. For example, the simultaneous equations

$$(1) \qquad 140x + 88y + 15z = 58{,}800$$
$$(2) \qquad 792x + 568y + 815z = 392{,}000$$
$$(3) \qquad 64x + 30y + 75z = 29{,}400$$

are set up on the countingboard as follows:

29,400	392,000	58,800
64	792	140
30	568	88
75	815	15

These equations are solved by first eliminating z in equations (1) and (3), and then in (2) and (3). From these x and y are obtained. Finally z is found by substituting the values of x and y in equation (3).

Ch'ien Pao-tsung has pointed out that from the *Shu-shu chiu-chang* one can gather much information on the sociological problems in thirteenth-century China, from finance and commerce to the levy of taxes. The book gives information not only on the merchandise imported from overseas but also on the use of the rain gauge by the Sung government, which was greatly concerned with rainfall because of the importance of agriculture.

BIBLIOGRAPHY

Further information on Ch'in Chiu-shao and his work may be found in Ch'ien Pao-tsung, *Ku-suan kao-yüan* ("Origin of Ancient Chinese Mathematics") (Shanghai, 1935) pp. 7, 43–66; and *Chung-kuo shu-hsüeh-shih* ("History of Chinese Mathematics") (Peking, 1964) pp. 157–167, 206–209; Ch'ien Pao-tsung *et al., Sung Yuan shu-hsüeh-shih lun-wên-chi* ("Collected Essays on Sung and Yuan Chinese Mathematics") (Peking, 1966) pp. 60–103; Hsü Shun-fang, *Chung-suan-chia ti tai-shu-hsüeh yen-chiu* ("Study of Algebra by Chinese Mathematicians") (Peking, 1951) pp. 16–21; Li Yen, *Chung-kuo shu-hsüeh ta-kang* ("Outline of Chinese Mathematics") I (Shanghai, 1931) 117–136, and *Chung-kuo suan-hsüeh-shih* ("History of Chinese Mathematics") (Shanghai, 1937; repr. 1955) pp. 98–99, 128–132; Li Yen and Tu Shih-jan, *Chung-kuo ku-tai shu-hsüeh chien-shih* ("A Short History of Ancient Chinese Mathematics") I (Peking, 1963) II (Peking, 1964) 49–151, 210–216; Yoshio Mikami, *The Development of Mathematics in China and Japan* (Leipzig, 1913; repr. New York) pp. 63–78; Joseph Needham, *Science and Civilisation in China,* III (Cambridge, 1959) 40–45, 119–122, 472, 577–578; George Sarton, *Introduction to the History of Science,* III (Baltimore, 1947) 626–627; D. E. Smith and Yoshio Mikami, *A History of Japanese Mathematics* (Chicago, 1914) pp. 48–50, 63; and Alexander Wylie, *Chinese Researches* (Shanghai, 1897; repr. Peking, 1936; Taipei, 1966) pp. 175–180.

HO PENG-YOKE

CHITTENDEN, RUSSELL HENRY (*b.* New Haven, Connecticut, 18 February 1856; *d.* New Haven, 26 December 1943), *physiological chemistry.*

Chittenden was the only child of Horace Horatio Chittenden, the superintendent of a small factory in New Haven, and Emily Eliza Doane of Westbrook, Connecticut. He attended private and public schools in New Haven and entered the Sheffield Scientific School of Yale University in 1872, planning a career in medicine.

His undergraduate thesis problem was assigned to him by Professor Samuel W. Johnson, who asked him to find out why scallops, when served warmed over from a previous meal, always seemed sweeter than when freshly cooked. Chittenden discovered that the scallop muscle contains glycogen and glycine and was thus the first to observe the occurrence of free glycine (then called glycocoll) in nature. At Johnson's suggestion a German translation of the paper was submitted to Liebig's *Annalen* in 1875; in 1878 W. Kühne, who had read the paper, invited Chittenden to be his student and lecture assistant at the University of Heidelberg, where Chittenden had gone for a year of advanced training.

In 1880, a year after his return from Germany, Chittenden was awarded the Ph.D. and two years later he was appointed professor of physiological chemistry in the Sheffield Scientific School. Shortly thereafter he became a member of the governing board of the school and its permanent secretary.

Chittenden attained wide recognition as a teacher and investigator and his students carried his principles and methods to almost every other medical school in the country. In a monograph of 1930, *The Development of Physiological Chemistry in the United States,* Chittenden argued that this science, as represented in the current faculties in the medical schools of the country, had grown almost entirely from seeds planted in New Haven.

Chittenden's administrative talents were as great as his pedagogical ones. In 1898 he succeeded Brush—whom he had assisted as an undergraduate—as director of the Sheffield Scientific School and as its treasurer. He thus assumed responsibility for the entire administration of the school—the development of policies, all appointments, the enlargement of facilities, the management of funds, and especially the complex relationship between the financially independent school and Yale College. After his retirement in 1922 he retained his membership of the Board of Trustees of the school and continued his service as treasurer until 1930.

Chittenden's scientific work falls into two major categories. Under the influence of Kühne, he became interested in the action of the enzymes which render food materials soluble and available for absorption by the organism; and for a number of years he engaged in a collaboration with Kühne to discover the action of proteolytic enzymes. Kühne and Chittenden made parallel experiments in Heidelberg and in New Haven in an effort to discover what happens when insoluble protein in the digestive tract is rendered soluble and is absorbed by the body. They exchanged data and published their findings jointly. Such collaboration between an American and a European scientist was almost unique at that period. Although later methods of analysis do not substantiate their hypotheses concerning the nature of proteins, Chittenden's and Kühne's basic premise—that digestion is a gradual process—was sound and laid a firm technical foundation for the study of enzymes.

Chittenden's second area of investigation was concerned with toxicology. He was further concerned with the scientific legal evidence in a number of cases of poisoning and was involved in research upon the effects of heavy metals and of many drugs. His long-continued investigations on the effects of alcohol on man received international recognition as did his work from 1908 to 1915 with the Referee Board of the Secretary of Agriculture on sodium benzoate and other commercially used additives to human food.

Chittenden's most important contribution to physiological chemistry is probably his study of the protein requirement of man, a study that had an amusing origin when in 1902 Chittenden learned of a rich American, Mr. Horace Fletcher, who maintained that his extraordinary physical development and good health were a consequence of long-continued mastication of every mouthful of food he took ("Fletcherism"). Mr. Fletcher was a guest in Chittenden's home for several months; Chittenden shrewdly noted that his consumption of protein food was remarkably low. At a period when the necessary intake of protein was assumed to be 118 grams per day, or the so-called Voit standard, Fletcher consumed about forty grams of protein a day, yet could compete to advantage with young students in the gymnasium at Yale. Chittenden made experiments on himself and a group of young subjects (a detachment of volunteers from the Hospital Corps of the army) and concluded that men could be maintained in good health on a diet that contained about fifty grams of protein and yielded 2,500 to 2,600 calories per day. (He ignored the excessive mastication advocated by Fletcher.) He further hypothesized that various ailments might be the result of high-protein diets due to toxic products of the decomposition of protein in the colon. During World War I Chittenden was one of the American committee especially concerned with the supply of protein food to the Allies; the results of his experiment established that a standard diet supplying 3,300 calories per day was adequate for men doing average work.

Chittenden served for many years as a member of the editorial boards of several journals, in particular the *American Journal of Physiology,* the *Journal of Experimental Medicine,* and, later, the *Journal of Biological Chemistry.*

Chittenden maintained throughout his life that physiological chemistry is a pure biological science, on a par with zoology and botany. Although it is a part of the science of physiology it should not be restricted to the needs of any applied science (such as medicine). Were physiological chemistry allowed to develop on a broad basis, useful applications would surely be found—but specialized development would be undesirable. The present position of what is today known as biochemistry is clear justification of Chittenden's views.

BIBLIOGRAPHY

Between 1875 and 1917, Chittenden published some hundred papers in scientific journals, at first mainly in the *American Journal of Science* and the *American Chemical Journal.* For a number of years after 1884 many papers appeared in the *Transactions of the Connecticut Academy of Arts and Sciences.* Reprints of these, together with reprints from other journals, were bound for private circulation as *Studies from the Laboratory of Physiological Chemistry, Yale University* (3 vols., 1885–1888) and are to be found in many university libraries.

Chittenden's books include *Digestive Proteolysis* (New Haven, 1895); *Physiological Economy in Nutrition with Special Reference to the Minimal Proteid Requirement of the Healthy Man, An Experimental Study* (New York, 1904); *The Nutrition of Man* (New York, 1907); *The Influence of Sodium Benzoate on the Nutrition and Health of Man,* U.S. Department of Agriculture Report No. 88 (Washington, D.C., 1909); pp. 13–292; *Influence of Vegetables Greened with Copper Salts on the Nutrition and Health of Man,* U.S. Department of Agriculture Report No. 97 (Washington, D.C., 1913); *History of the Sheffield Scientific School of Yale University 1846–1922,* 2 vols. (New Haven, 1928); *The Development of Physiological Chemistry in the United States,* American Chemical Society Monograph No. 54 (New York, 1930); and *The First Twenty-five Years of the American Society of Biological Chemists* (New Haven, 1945). In addition Chittenden wrote a number of biographical memoirs of deceased colleagues prepared for the National Academy of Sciences.

A complete bibliography of Chittenden's papers may be found in the memoir by the present writer in *Biographical Memoirs of the National Academy of Sciences,* XXIV (Washington, D.C., 1945), 95–104. His manuscript autobiography is deposited in the Historical Manuscripts Room of the Yale University Library.

HUBERT BRADFORD VICKERY

CHLADNI, ERNST FLORENZ FRIEDRICH (*b.* Wittenberg, Germany, 30 November 1756; *d.* Breslau, Germany, 3 April 1827), *physics.*

Except for a few publications on meteorites, in which he proposed their extraterrestrial origin, Chladni devoted his research to the study of acoustics and vibration. His most important work was in providing demonstrations of the vibrations of surfaces, using the sand pattern technique which he devised.

The son of Ernst Martin Chladni, a jurist whose family had come from Hungary in the seventeenth century, and Johanna Sophia Chladni, he himself was obliged, against his will, to study jurisprudence. He received his degree at Leipzig in 1782; but after the death of his father he turned to science, choosing acoustics as his particular area of experimental investigation. His choice was almost certainly influenced by his interest in music. An amateur musician, Chladni designed and constructed two keyboard instruments, the euphonium and the clavicylinder, both being variations of the glass harmonica. The results of his first project, a study of the vibrations of plates, were published in 1787. For the remainder of his life he alternated between periods at his home in Wittenberg, where he continued his experimental studies and wrote several treatises on acoustics, and travel in Europe, demonstrating the vibrations of plates by his sand figure method (see below) and performing on his musical instruments. His travels also brought him into personal contact with Goethe, Lichtenberg, Olbers, Laplace, and others.

During the eighteenth century, scientists had undertaken experimental and theoretical work on the vibration of strings and had made attempts to study the vibration of rods and membranes. The vibrations of solid plates had not been treated, and Chladni, in his first report, emphasized that although the vibration of strings was understood, the production of sound by solid plates was not.

By spreading sand over plates and running a violin bow over their edges, Chladni was able to observe the structure of the resulting vibrations, because the sand collected along the nodal curves where there was no motion. Patterns formed in this way were symmetrical and often spectacular, the lines of sand forming circles, stars, and other geometric patterns. Chladni first used circular and rectangular plates of glass and copper, three to six inches in diameter. Later he extended his observations to ellipses, semicircles, triangles, and

six-sided polygons. He generally fixed the plates at one internal point, which became a node, and left the sides free. In a few cases a stationary point or line occurred on the edge of the plate.

Chladni analyzed the sand patterns by classifying them according to geometrical shape and noting for each the corresponding pitch. Thus he was able to emphasize that the patterns and sounds of a vibrating plate are analogous to the shapes and tones of the modes in the harmonic series of a string.

In addition to his analysis of surface vibrations, Chladni studied the vibrations of cylindrical and prismatic rods. For the latter, he again used the sand figure method. He deduced the velocity of sound in solids from the pitch that a long rod of a given material produces when made to vibrate longitudinally. He measured the velocity of sound in gases other than air in a similar way: he compared the pitch of a wind instrument filled with the gas being studied with the normal pitch of the instrument under standard atmospheric conditions.

The visible demonstration of surface vibration received much attention in the 1820's and 1830's. In Germany, Friedrich Strehlke continued experimental investigations; Wilhelm and Ernst Weber dedicated their treatise on wave motion to Chladni, since they believed that the acoustical figures had stimulated the contemporary interest in the subject; and in the middle of the century Gustav Kirchoff worked on the mathematical theory of vibration. Thomas Young in England had been interested in the experiments as early as 1800, and Charles Wheatstone later analyzed them by a geometrical superposition principle. Michael Faraday knew about the phenomenon and experimented with the figures, considering especially the surrounding air currents.

French scientists showed even more interest in Chladni's work. Félix Savart, who was truly Chladni's professional successor, carried on experimental investigations. Jean Baptiste Biot and Siméon Denis Poisson collaborated on a study of the sound vibrations of gases and gave much attention to Chladni's relevant experiments. The most organized activity was in the study of the theory of the vibration of surfaces. Chladni had visited the Paris Academy in 1808 and had demonstrated the vibration patterns before an audience that included not only the leading French scientists but Napoleon himself. The physicists at the Academy responded by announcing a prize competition for the best mathematical study of elastic vibrations. They noted a similarity between Chladni's demonstration of the nodal curves in vibrating surfaces and Joseph Sauveur's demonstrations of the nodal points in vibrating strings a century earlier.

Since Sauveur's efforts had led to fruitful mathematical work, it was expected that the sand figures would have as beneficial an effect on theory. Sophie Germain was an early contributor to the large body of work on the mathematical theory of vibration, which, until the middle of the nineteenth century, is associated principally with Louis Navier, Poisson, and Augustin Cauchy.

BIBLIOGRAPHY

I. ORIGINAL WORKS. Chladni first described his early experiments using the sand figures in *Entdeckungen über die Theorie des Klanges* (Leipzig, 1787). They were presented with additional observations in *Die Akustik* (Leipzig, 1802), which is also a general acoustics text containing very complete historical material. This appeared in French translation in 1809 (*Traité d'acoustique*). There was a second edition, the *Neue Beiträge zur Akustik* (Leipzig, 1817). The work on the instruments was described in *Beiträge zur praktischen Akustik und zur Lehre vom Instrumentbau* (Leipzig, 1821). Chladni described his ideas about meteorites in *Über den Ursprung der von Pallas gefundenen und anderen ähnlichen Eisenmassen* (Leipzig, 1794). For a complete listing of Chladni's works, see Poggendorff.

II. SECONDARY LITERATURE. Chladni gives an autobiographical summary in the introduction to his *Traité d'acoustique*. There are biographical essays by H. Schimanck, in *Sudhoffs Archiv für Geschichte der Medizin und der Naturwissenschaften,* **37** (1953) and by Kurt Loewenfeld, in *Naturwissenschaftlicher Verein. Abhandlungen* (Hamburg, 1929). There is also a biography by the acoustician F. Melde, *Chladni's Leben und Wirken* (Marburg, 1888). An interesting account of recent Chladni figure experiments, with comparisons to other nineteenth-century results and with photographs and analyses of them, will be found in Mary D. Waller, *Chladni Figures: A Study in Symmetry* (London, 1961).

SIGALIA C. DOSTROVSKY

CHODAT, ROBERT (*b*. Moutier-Grandval, Switzerland, 6 April 1865; *d*. Geneva, Switzerland, 28 April 1934), *botany, algology, plant physiology.*

Chodat received the doctorate in systematic botany at Geneva in 1887. He became *Privatdozent* at the University of Geneva in 1888; was appointed professor of medical and pharmaceutical botany in 1889; and succeeded Jean Marc Thury as director of the Institut de Botanique in 1900. His early research culminated in the voluminous *Monographia Polygalacearum* (1893), which attracted worldwide attention. Besides floristics and ecology—not only of France and Switzerland, but also of Paraguay, Argentina, Spain, and the Balearic Islands—Chodat was interested in plant anatomy, cytology, teratology, and pathology.

It was in algology that Chodat showed his remarkable abilities. As early as 1893 he began studying the structure and polymorphism of lacustrine algae and was one of the first to obtain a pure culture of isolated algae, thereby opening a new chapter in the study of the metabolism and genetics of algae. He was interested in symbiosis and made a culture of the algae linked with the fungus mycelia in lichens. Chodat became interested in the selection of yeasts, and consequently in fermentation in general. He was also a superb chemist, as is shown in his research on the catalyst peroxidase.

Chodat was a brilliant plant anatomist and was considered one of the finest flowering plant experts of his day. Deeply involved with applied botany, he discovered the cause of the vine disease known as *court noué* and showed that its cause was an acarid; from then on, the disease was called acariasis. In cytology he dealt with cellular division, plasmolysis, and accumulated reserves in the cytoplasm. He was one of the first to apply biometry to genetic problems. Chodat also conducted practical research in horticulture and grain culture. He traveled extensively and was always interested in the flora of the countries he visited. He collaborated on the monumental *Handbuch der biologischen Arbeitsmethoden* from its very inception.

BIBLIOGRAPHY

I. ORIGINAL WORKS. Chodat's first major work was *Monographia Polygalacearum* (Geneva, 1893). His *Principes de botanique* appeared in 3 eds. (Geneva–Paris, 1907, 1911, 1920). A complete bibliography of 462 items is in the article by Lendner (below).

II. SECONDARY LITERATURE. On Chodat or his work, see E. Fischer, "L'oeuvre scientifique de R. Chodat," in *Bulletin de la Société botanique de Genève,* **25** (1932), 23–33; and A. Lendner, "Prof. Dr. R. Chodat," in *Actes de la Société helvétique des sciences naturelles* (1934), 259–550.

P. E. PILET

CHRISTIE, SAMUEL HUNTER (*b*. London, England, 22 March 1784; *d*. Twickenham, London, 24 January 1865), *magnetism.*

Christie was the only son of James Christie, founder of the well-known auction galleries, and his second wife, formerly Mrs. Urquhart. Samuel was educated at Walworth School, Surrey, and Trinity College, Cambridge, which he entered as a sizar in October 1800. He was active in athletics and was a brother officer with Lord Palmerston in the grenadier company of University Volunteers. In 1805 he took his bachelor's degree as second wrangler and shared the

Smith's prize with Thomas Turton. Appointed third mathematical assistant in Woolwich Military Academy in July 1806, Christie became professor of mathematics there in June 1838. He made major revisions in the curriculum, raising it to a high standard.

Christie was elected a fellow of the Royal Society on 12 January 1826, frequently served on the Society's council, and was its secretary from 1837 to 1854. He married twice and by his second wife, Margaret Malcom, was the father of the future Sir William H. M. Christie, astronomer royal from 1881 to 1922. Samuel H. Christie was a vice-president of the Royal Astronomical Society and one of the visitors of the Royal Observatory, Greenwich. Owing to ill health, he retired from his professorship in 1854 and moved to Lausanne.

Almost all of Christie's investigations were related to terrestrial magnetism. In June 1821, while studying the influence of an unmagnetized iron plate on a compass, he discovered "that the simple rotation of the iron had a considerable influence on its magnetic properties."[1] Although he delayed making a detailed announcement of his results until June 1825, Christie did refer to this discovery in June 1824.[2] His work was independent of, and prior to, Arago's report of the magnetic influence of rotating metals. From his experiments he concluded that since "the direction of magnetic polarity, which iron acquires by rotation about an axis . . . has always reference to the direction of the terrestrial magnetic forces, . . . this magnetism is communicated from the earth."[3] He then went on to speculate that the earth in turn receives its magnetism from the sun.

In other papers Christie reported on a method for separating the effects of temperature from observations of the diurnal variation of the earth's magnetic field. In addition he speculated that this variation is caused by thermoelectric currents in the earth produced by the sun's heating.[4] Christie also observed a direct influence of the aurora on the dip and horizontal intensity of the earth's magnetic field.[5] As a recognized authority, Christie prepared a "Report on the State of Our Knowledge Respecting the Magnetism of the Earth" for the 1833 meeting of the British Association, reported on the magnetic observations made by naval officers during various polar voyages, and was the senior reporter on Alexander von Humboldt's proposal that cooperating magnetic observatories be founded in British possessions.[6]

Christie's paper "Experimental Determination of the Laws of Magneto-electric Induction . . ." was the Bakerian lecture for 1833. In it Christie showed that "the conducting power, varies as the squares of [the wires'] diameters directly, and as their lengths in-versely." He also concluded that voltaelectricity, thermoelectricity, and magnetoelectricity are all conducted according to the same law, which lent further support to the theory that all these electricities are identical.[7] In this paper Christie also gave the first description of the instrument that came to be known as the Wheatstone bridge.

NOTES

1. *Edinburgh Journal of Science,* **5** (July 1826), 12–13.
2. *Philosophical Transactions of the Royal Society,* **115** (1825), 347–415; *ibid.,* 58n.
3. *Ibid.,* 411.
4. *Philosophical Transactions,* **113** (1823), 342–392; *ibid.,* **115** (1825), 1–65; *ibid.,* **117** (1827), 308–354.
5. *Journal of the Royal Institution of Great Britain,* **2** (Nov. 1831), 271–280.
6. *Report of the British Association for the Advancement of Science* (1833), pp. 105–130; *Philosophical Transactions,* **116** (1826), pt. 4, 200–207; *ibid.,* **126** (1836), 377–416; *Proceedings of the Royal Society,* **3** (1837), 418–428.
7. *Philosophical Transactions,* **123** (1833), 130, 132.

BIBLIOGRAPHY

I. ORIGINAL WORKS. These are listed in the Royal Society's *Catalogue of Scientific Papers.* He also published *An Elementary Course of Mathematics for the Use of the Royal Military Academy and for Students in General,* 2 vols. (London, 1845–1847). The library of the Royal Society has some of his manuscript letters and reports.

The following are Christie's most important papers, and information in this article is based upon them: "On the Laws According to Which Masses of Iron Influence Magnetic Needles," in *Transactions of the Cambridge Philosophical Society,* **1** (1822), 147–173; "On the Diurnal Deviations of the Horizontal Needle When Under the Influence of Magnets," in *Philosophical Transactions of the Royal Society,* **113** (1823), 342–392; "On the Effects of Temperature on the Intensity of Magnetic Forces; and on the Diurnal Variation of the Terrestrial Magnetic Intensity," *ibid.,* **115** (1825), 1–65; "On the Magnetism of Iron Arising From Its Rotation," *ibid.,* 347–417; "On Magnetic Influence in the Solar Rays," *ibid.,* **116** (1826), 219–239; "Experimental Determination of the Laws of Magneto-electric Induction in Different Masses of the Same Metal, and of Its Intensity in Different Metals," *ibid.,* **123** (1833), 95–142; "Report on the State of Our Knowledge Respecting the Magnetism of the Earth," in *Report of the British Association for the Advancement of Science* (1833), 105–130; with G. B. Airy, "Report Upon a Letter Addressed by M. le Baron de Humboldt to His Royal Highness the PRS, and Communicated by HRH to the Council," in *Proceedings of the Royal Society,* **3** (1837), 418–428.

II. SECONDARY LITERATURE. Very little secondary literature exists on the history of magnetism, and Christie is infrequently mentioned in that. Biographical information can be found in the *Dictionary of National Biography* and in the sources listed there, as well as in William Roberts, *Memorials of Christie's*, 2 vols. (London, 1897), and H. D. Buchan-Dunlop, ed., *Records of the Royal Military Academy, 1741–1892* (Woolwich, 1895). The most useful, although very old, account of the history of magnetism up to about 1831 is [Sir David Brewster] "Magnetism," in *Encyclopaedia Britannica,* 7th and 8th eds.

EDGAR W. MORSE

CHRISTIE, WILLIAM HENRY MAHONEY (*b.* Woolwich, England, 1 October 1845; *d.* at sea, 22 January 1922), *astronomy.*

Christie was perhaps more important in the administration of astronomy than in advancing the theoretical foundations of the subject. Under his care, the observatory at Greenwich prospered more materially than it had done at any time since its foundation in 1675.

W. H. M. Christie was the eldest son of S. H. Christie, a professor of mathematics at the Royal Military Academy, Woolwich, and secretary to the Royal Society (1837–1854). The famous London firm of auctioneers was founded by Christie's grandfather. William was educated at King's College, London, and Trinity College, Cambridge, where he was fourth wrangler in 1868. He was elected a fellow of his college in 1869, but a year later he left Cambridge to become chief assistant at the Royal Observatory at Greenwich. There he was engaged primarily in positional astronomy, and one of his early minor achievements was to improve Airy's transit circle in several respects.

Despite Airy's influence over him, Christie was anxious that Greenwich should undertake physical as well as positional observations, and with the help of E. W. Maunder he undertook daily sunspot observations (correlating them in due course with terrestrial magnetic observations) and also made many attempts, although not highly successful ones, to measure the radial velocities of stars. He made and reported many planetary observations, as, for example, that of the Mercury halo during the transit of 6 May 1878, and made many eclipse expeditions, often bringing back very fine photographic records. Nevertheless, Christie made no remarkable advances outside the realm of "fundamental" astronomy.

Work on this subject, which had been the prime justification for the foundation of the observatory, necessarily continued as before, after Christie was made astronomer royal in 1881, but it is to his credit, and that of the Admiralty, that the scope of observations during this period went far beyond utilitarian needs. Christie equipped the observatory with many new instruments, notably a twenty-eight-inch visual refracting telescope (completed 1894), and a twenty-six-inch photographic refractor with a twelve-and-three-fourth-inch guiding telescope. This had a nine-inch photoheliograph on the same mounting (the gift of the surgeon Sir Henry Thompson in 1894) and enabled Greenwich to participate in the international photographic survey. He added several new buildings to the existing range, and moved the magnetic pavilion to an isolated building in Greenwich Park. With the twenty-eight-inch refractor, Christie initiated a large program for the micrometric observation of double stars.

Christie was elected a fellow of the Royal Society in 1881 and was president of the Royal Astronomical Society from 1890 to 1892. Among his other honors he was created C.B. in 1897 and K.C.B. in 1904. In 1910 he retired to Downe, in Kent. His wife (Violet Mary Hickman) died in 1888, after only seven years of marriage, and one son survived him. Christie himself died at sea on a voyage to Mogador, Morocco, in 1922.

BIBLIOGRAPHY

I. ORIGINAL WORKS. More than a dozen volumes of astronomical and other observations were published by Christie, or under his direction, under the auspices of the Greenwich Observatory. His most substantial work was the six-volume *Astrographic Catalogue for 1900.0* (Greenwich section, 1904–1932), completed after his death. Other works of interest are *Telegraphic Determinations of Longitude Made in the Years 1888–1902 Under the Direction of Sir W. H. M. Christie* (Edinburgh, 1906); and *Temperature of the Air as Determined From the Observations and Records of the Fifty Years 1841 to 1890, Made at the Royal Observatory, Greenwich, Under the Direction of W. H. M. Christie* (London, 1895), with a second volume for 1891–1905 (London, 1906). Other volumes concern photographic records of the barometer and thermometer, photoheliographic results, observations of the minor planet Eros, and a reduction of Groombridge's circumpolar catalogue for 1810.0.

II. SECONDARY LITERATURE. Apart from an entry by F. W. Dyson in the *Dictionary of National Biography* (vol. for 1922–1930), the main source of information remains the numerous obituary notices of Christie. For information concerning his innovations at Greenwich, see the numerous editions of the descriptive handbooks issued by the observatory (published by Her Majesty's Stationery Office) and E. W. Maunder's *The Royal Observatory Greenwich* (London, 1900), with contemporary photographs of many of the buildings.

JOHN NORTH

CHRISTMANN, JACOB (*b.* Johannesberg, Rheingau, Germany, November 1554; *d.* Heidelberg, Germany, 16 June 1613), *oriental studies, mathematics, chronology, astronomy.*

Christmann studied oriental subjects at Heidelberg and became a teacher there in 1580. Shortly thereafter, however, he had to leave that university because he, as a Calvinist, could not subscribe to the concordat-formulary set down by the Lutheran Elector Ludwig VI. Christmann traveled for some time, then settled down to teach in a Reformed school in Neustadt, Pfalz. He was a teacher there in 1582. The death of Ludwig (12 October 1583) enabled him to return to Heidelberg, where he was appointed professor of Hebrew on 18 June 1584. From 1591 on he taught Aristotelian logic. He was made rector of the university in 1602.

In 1608 Frederick IV appointed Christmann professor of Arabic. Christmann thus became the second teacher of that subject in Europe, the first having been Guillaume Postel at Paris in 1538. This appointment must have given great satisfaction to its recipient, since in 1590, in the preface of his *Alfragani chronologica et astronomica elementa,* Christmann had advocated the establishment of a chair of Arabic "to open possibilities for teaching philosophy and medicine from the [original] sources." Indeed, Christmann had demonstrated his scholarly interest in the Arabic language as early as 1582, with the publication of his *Alphabeticum Arabicum,* a small book of rules for reading and writing Arabic. Besides Arabic, he is said to have known Syrian, Chaldaic, Greek, Latin, French, Italian, and Spanish. He was an extremely modest man despite his learning, with a passion for work that may well have hastened his death of jaundice.

On the death of Valentine Otho, Christmann inherited the entire library of G. J. Rhäticus, which had been in Otho's keeping. This collection contained trigonometric tables more extensive than those that Rhäticus had published in the *Opus Palatinum* of 1596 (adapted by B. Pitiscus as the basis for his *Thesaurus mathematicus* of 1613) as well as the original manuscript of Copernicus' *De revolutionibus orbium coelestium.* The inclusion of instruments in the bequest stimulated Christmann to begin making astronomical observations. In 1604 he proposed to Kepler that they should exchange the results of their researches. Christmann was the first to use the telescope in conjunction with such instruments as the sextant or Jacob's staff (1611), with the results reported in his *Theoria lunae* and *Nodus gordius.* These last works also show him to be a competent astronomical theorist. He gave a good treatment of prosthaphaeresis, the best method of calculating trigonometric tables to be developed before the invention of logarithms, which he based on such formulas as

$$2 \sin \alpha \sin \beta = \cos (\alpha - \beta) - \cos (\alpha + \beta);$$

he then went on to prove that this method had been devised by Johann Werner.

In his *Tractatio geometrica de quadratura circuli,* Christmann defended against J. J. Scaliger the thesis that the quadrature of the circle could be solved only approximately. In his books on chronology—a topic of great concern at a time of radical calendar reform—he disputed the work of not only Scaliger but also J. J. Lipsius. He further criticized Copernicus, Tycho Brahe, and Clavius—some such criticisms may be found in some detail in manuscript annotations of his own copy of *Alfragani chronologica et astronomica elementa,* which is now in the library of the University of Utrecht.

BIBLIOGRAPHY

I. ORIGINAL WORKS. Christmann's works are *Alphabetum Arabicum cum isagoge scribendi legendique Arabice* (Neustadt, 1582); *Epistola chronologica ad Iustum Lipsium, qua constans annorum Hebraeorum connexio demonstratur* (Heidelberg, 1591; Frankfurt, 1593); *Disputatio de anno, mense, et die passionis Dominicae* (Frankfurt, 1593, combined with the 2nd ed. of *Epistola*); *Tractatio geometrica, de quadratura circuli* (Frankfurt, 1595); *Observationum solarium libri tres* (Basel, 1601); *Theoriae lunae ex novis hypothesibus et observationibus demonstrata* (Heidelberg, 1611); and *Nodus gordius ex doctrina sinuum explicatus, accedit appendix observationum* (Heidelberg, 1612).

He translated from a Hebrew translation and commented on *Muhamedis Alfragani Arabis chronologica et astronomica elementa, additus est commentarius, qui rationem calendarii Romani . . . explicat* (Frankfurt, 1590, 1618) and translated and commented on *Uri ben Simeon, calendarium Palaestinorum* (Frankfurt, 1594). He edited, with translation and comments, *Is. Argyri computus Graecorum de solennitate Paschatis celebranda* (Heidelberg, *ca.* 1612).

II. SECONDARY LITERATURE. On Christmann's life, see Melchior Adam, *Vitae Germanorum philosophorum* (Heidelberg, 1615), pp. 518–522; on his Arabic studies, Johann Fück, *Die arabischen Studien in Europa* (Leipzig, 1955), pp. 44–46; on his instruments, H. Ludendorff, "Über die erste Verbindung des Fernrohres mit astronomischen Messinstrumenten," in *Astronomische Nachrichten,* **213** (1921), cols. 385–390; on the prosthaphaeresis, A. von Braunmühl, *Vorlesungen über Geschichte der Trigonometrie,* I (Leipzig, 1900), see index. See also Daniël Miverius, *Apologia pro Philippo Lansbergio ad Jacobum Christmannum* (Middelburg, 1602); and J. Kepler, *Gesammelte Werke* (Munich, 1949–1954), esp. II (1939), 14–16—XV (1951), 41 f., gives a letter from Christmann to Kepler, dated 11 April 1604 (old style).

J. J. VERDONK

CHRISTOFFEL, ELWIN BRUNO (*b.* Montjoie [now Monschau], near Aachen, Germany, 10 November 1829; *d.* Strasbourg, France [then Germany], 15 March 1900), *mathematics.*

Christoffel studied at the University of Berlin, where he received his doctorate in 1856 with a dissertation on the motion of electricity in homogeneous bodies. He continued his studies in Montjoie. In 1859 he became lecturer at the University of Berlin, in 1862 professor at the Polytechnicum in Zurich, and in 1869 professor at the Gewerbsakademie in Berlin. In 1872 he accepted the position of professor at the University of Strasbourg, newly founded after its acquisition by the Germans. Here he lectured until 1892, when his health began to deteriorate.

Christoffel has been praised not only as a very conscientious mathematician but also as a conscientious teacher. Politically he represented the traditional Prussian academician loyal to emperor and army. This may have contributed to his choice of Strasbourg and his endeavor to create a great German university in that city.

Scientifically, Christoffel was primarily a follower of Dirichlet, his teacher, and of Riemann, especially of the latter. Their ideas inspired his early publications (1867, 1870) on the conformal mapping of a simply connected area bounded by polygons on the area of a circle, as well as the paper of 1880 in which he showed algebraically that the number of linearly independent integrals of the first kind on a Riemann surface is equal to the genus p. The posthumous "Vollständige Theorie der Riemannschen θ-Function" also shows how, rethinking Riemann's work, Christoffel came to an independent approach characteristic of his own way of thinking. Also in the spirit of Riemann is Christoffel's paper of 1877 on the propagation of plane waves in media with a surface of discontinuity, an early contribution to the theory of shock waves.

Another interest of Christoffel's was the theory of invariants. After a first attempt in 1868, he succeeded in 1882 in giving necessary and sufficient conditions for two algebraic forms of order p in n variables to be equivalent. Christoffel transferred these investigations to the problem of the equivalence of two quadratic differential forms, again entering the Riemannian orbit. In what well may be his best-known paper, "Über die Transformation der homogenen Differentialausdrücke zweiten Grades," he introduced the three index symbols

$$\begin{bmatrix} g\ h \\ k \end{bmatrix} \text{ and } \begin{Bmatrix} g\ h \\ k \end{Bmatrix}, k, g, h = 1, 2, \cdots, n,$$

now called Christoffel symbols of the first and second order, and a series of symbols of more than three indices, of which the four index symbols, already introduced by Riemann, are now known as the Riemann-Christoffel symbols, or coordinates of the Riemann-Christoffel curvature tensor. The symbols of an order higher than four are obtained from those of a lower order by a process now known as covariant differentiation. Christoffel's reduction theorem states (in modern terminology) that the differential invariants of order $m \geq 2$ of a quadratic differential form

$$\Sigma a_{ij}(x)\, dx^i\, dx^j$$

are the projective invariants of the tensors a_{ij}, its Riemann-Christoffel tensor, and its covariant derivatives up to order $(m - 2)$. The results of this paper, together with two papers by R. Lipschitz, were later incorporated into the tensor calculus by G. Ricci and T. Levi-Cività.

Christoffel also contributed to the differential geometry of surfaces. In his "Allgemeine Theorie der geodätischen Dreiecke" he presented a trigonometry of triangles formed by geodesics on an arbitrary surface, using the concept of reduced length of a geodesic arc. When the linear element of the surface is $ds^2 = dr^2 + m^2\, dx^2$, m is the reduced length of arc r. In this paper Christoffel already uses the symbols $\begin{Bmatrix} g\ h \\ k \end{Bmatrix}$, but only for the case $n = 2$.

BIBLIOGRAPHY

I. ORIGINAL WORKS. Christoffel's writings were brought together as *Gesammelte mathematische Abhandlungen,* L. Maurer, ed., 2 vols. (Leipzig–Berlin, 1910). Among his papers are "Ueber einige allgemeine Eigenschaften der Minimumsflächen," in *Journal für die reine und angewandte Mathematik,* **67** (1867), also in *Gesammelte Abhandlungen,* I, 259–269; "Allgemeine Theorie der geodätischen Dreiecke," in *Abhandlungen der Königlichen Akademie der Wissenschaften zu Berlin* (1868), 119–176, also in *Gesammelte Abhandlungen,* I, 297–346; "Über die Transformation der homogenen Differentialausdrücke zweiten Grades," in *Journal für die reine und angewandte Mathematik,* **70** (1869), 46–70, 241–245, also in *Gesammelte Abhandlungen,* I, 352–377, 378–382; "Ueber die Abbildung einer einblättrigen, einfach zusammenhängenden, ebenen Fläche auf einem Kreise," in *Nachrichten der Königlichen Gesellschaft der Wissenschaften zu Göttingen* (1870), 283–298, see also 359–369, also in *Gesammelte Abhandlungen,* II, 9–18, see also 19–25; "Ueber die Fortpflanzung von Stössen durch elastische feste Körper," in *Annali di matematica,* 2nd ser., **8** (1877), 193–243, also in *Gesammelte Abhandlungen,* II, 81–126; "Algebraischer Beweis des Satzes von der Anzahl der linearunabhangigen Integrale erster Gattung," *ibid.,* **10** (1883), also in *Gesammelte Abhandlungen,* II, 185–203; and "Vollständige Theorie der Riemannschen θ-Function," in *Mathematische Annalen,* **54** (1901), 347–399, also *Gesammelte Abhandlungen,* II, 271–324.

II. SECONDARY LITERATURE. Christoffel and his work are discussed in C. F. Geiser and L. Maurer, "E. B. Christoffel," in *Mathematische Annalen,* **54** (1901), 328–341; and W. Windelband, *ibid.,* 341–344; there is a bibliography on 344–346. These articles are excerpted in *Gesammelte Abhandlungen,* I, v–xv. The papers by Lipschitz that were incorporated into the tensor calculus are "Untersuchungen in Betreff der ganzen homogenen Functionen von *n* Differentialen," in *Journal für die reine und angewandte Mathematik,* **70** (1869), 71–102; **72** (1870), 1–56; and "Entwickelung einiger Eigenschaften der quadratischen Formen von *n* Differentialen," *ibid.,* **71** (1870), 274–287, 288–295. Beltrami's comment on Christoffel's "Allgemeine Theorie der geodätischen Dreiecke" is in his *Opere matematiche,* 4 vols. (Milan, 1904), II, 63–73.

Mlle. L. Greiner of the Bibliothèque Nationale et Universitaire, Strasbourg, informs me that the *Handschriftlicher Nachlass* mentioned in the *Verzeichniss der hinterlassenen Büchersammlung des Herrn Dr. E. B. Christoffel* (n.p., 1900) is not in this library, contrary to what might be expected from a statement in the Geiser-Maurer article.

D. J. STRUIK

CHRISTOL (CRISTOL), JULES DE (*b.* Montpellier, France, 25 August 1802; *d.* Montpellier, 25 June 1861), *paleontology.*

A collector of fossils, Christol presented a late and, by his own admission, hasty thesis for his *doctorat ès-sciences* at Montpellier in 1834. He was made professor of geology in 1837 at the newly created Faculté des Sciences in Dijon and soon after was also asked to take over the duties of secretary of the Faculty; he became dean in 1853. His administrative duties greatly hampered his scientific activity.

Christol confined his studies to the south of France. A rather mediocre geologist, sometimes placing fossils of different ages and origins in the same faunal group, Christol was, however, a competent osteologist and corrected certain of Cuvier's errors. A disciple of Marcel de Serres and of William Buckland, he specialized in the study of the Tertiary and Quaternary mammals of the south of France, particularly those from caves and osseous breccia. He was greatly interested in the rhinoceros, the hyena of the Quaternary, and in particular the Equidae, in which family he identified the important genus *Hipparion.*

Christol has often been cited as one of the founders of the science of prehistory, for in 1829 he had dared to state in *Notice sur les ossements . . . fossiles . . .,* a pamphlet printed at his own expense, that the human bones found in the cave of Pondres (Hérault) were contemporaneous with "extinct races" of lions, hyenas, and bears. Although based on observations found to be stratigraphically incorrect, this was a courageous statement, for it risked incurring the op-

position of Cuvier, who denied the existence of fossil man. It must be remembered, however, that the existence of this fossil man had been upheld previously—particularly by Christol's friend Buckland and, especially with respect to the caves around Montpellier, by L. A. d'Hombres-Firmas.

BIBLIOGRAPHY

Christol's pamphlet on human bone fossils is *Notice sur les ossements humains fossiles . . . du Gard, présentée à l'Académie des sciences . . .* (Montpellier, 1829).

An MS list of the titles and works of Christol (Bibliothèque du Muséum, Paris, MS 2358-3) seems to have been used by Paul Gervais for his "Discours prononcé aux funérailles de M. de Christol," in *Mémoires de l'Académie des sciences de Montpellier,* **5** (1861), 75–79. This list is completed by that in the Royal Society of London, *Catalogue of Scientific Papers,* vol. I. Both spellings of the name should be consulted in any source.

The existence of fossil man is supported by L. A. d'Hombres-Firmas in Bibliothèque Universelle de Genève, XVII (Geneva, 1821), 33–41; and in *Journal de physique,* **92** (1821), 227–253.

FRANCK BOURDIER

CHRYSTAL, GEORGE (*b.* Mill of Kingoodie, near Old Meldrum, Aberdeenshire, Scotland, 8 March 1851; *d.* Edinburgh, Scotland, 3 November 1911), *education, mathematics, physics.*

Chrystal's father, William, a self-made man, was successively a grain merchant, a farmer, and a landed proprietor. His mother was the daughter of James Burr of Mains of Glack, Aberdeenshire. Chrystal attended the local parish school and later Aberdeen Grammar School, from which he won a scholarship to Aberdeen University in 1867. By the time he graduated in 1871, he had won all the available mathematical distinctions and an open scholarship to Peterhouse, Cambridge. Entering Peterhouse in 1872, Chrystal came under the influence of Clerk Maxwell, and when the Cavendish Laboratory was opened in 1874 he carried out experimental work there. In the mathematical tripos examination of 1875 he was bracketed (with William Burnside) second wrangler and Smith's prizeman and was immediately elected to a fellowship at Corpus Christi College, Cambridge. In 1877 he was appointed to the Regius chair of mathematics in the University of St. Andrews and in 1879 to the chair of mathematics at Edinburgh University; in the same year he married a childhood friend, Margaret Ann Balfour.

Chrystal's thirty-two-year tenure of the Edinburgh chair saw a progressive and substantial rise in the standard of the mathematical syllabus and teaching at the university, especially after the institution of specialized honors degrees by the Universities (Scotland) Act of 1899. The main burden of formulating policies and drafting regulations under the act fell on Chrystal as dean of the Faculty of Arts, an office he held from 1890 until his death. He was an outstanding administrator, with an exceptionally quick grasp of detail, tactful, fair-minded, and forward-looking. He also contributed much to preuniversity education throughout Scotland, acting as inspector of secondary schools, initiating a scheme for a standard school-leaving-certificate examination, and negotiating the transfer of the teacher-training colleges from control by the Presbyterian churches to a new provincial committee, of which he was the first chairman.

Notwithstanding his administrative and teaching burdens, Chrystal found time for scientific work. His wide-ranging textbook on algebra, with its clear, rigorous, and original treatment of such topics as inequalities, limits, convergence, and the use of the complex variable, profoundly influenced mathematical education throughout Great Britain and beyond its borders. He published some seventy articles, about equally divided between scientific biography, mathematics, and physics. Many of the biographies, written for the *Encyclopaedia Britannica,* are still of considerable value. In the mathematical papers his strength lay particularly in lucid exposition and consolidation. Of the physics papers the most important are two long survey articles, "Electricity" and "Magnetism," in the ninth edition of the encyclopaedia, and his later hydrodynamic and experimental investigations of the free oscillations (known as seiches) in lakes, particularly the Scottish lochs, using the results of a recent bathymetric survey.

Chrystal held honorary doctorates from Aberdeen and Glasgow and was awarded a Royal Medal of the Royal Society of London just before his death. He was buried at Foveran, Aberdeenshire, and was survived by four sons and two daughters.

BIBLIOGRAPHY

I. ORIGINAL WORKS. A full list of Chrystal's publications is appended to the obituary notice in *Proceedings of the Royal Society of Edinburgh,* **32** (1911–1912), 477. The following items are important: *Algebra. An Elementary Textbook for the Higher Classes of Secondary Schools and for Colleges,* I (Edinburgh, 1886; 5th ed., London, 1904), II (Edinburgh, 1889; 2nd ed., London, 1900); *Introduction to Algebra for the Use of Secondary Schools and Technical Colleges* (London, 1898; 4th ed., London, 1920); the contributions (mentioned in the text) to the ninth edition of the *Encyclopaedia Britannica;* and the papers on seiches, in *Proceedings of the Royal Society of Edinburgh,* **25** (1904–1905), and *Transactions of the Royal Society of Edinburgh,* **41** (1905), **45** (1906), and **46** (1909).

II. SECONDARY LITERATURE. Besides the obituary notice referred to above, there is an excellent notice in *The Scotsman* (4 Nov. 1911), p. 9. See also A. Logan Turner, ed., *History of the University of Edinburgh 1883–1933* (Edinburgh, 1933), *passim.*

ROBERT SCHLAPP

CHU SHIH-CHIEH (*fl.* China, 1280–1303), *mathematics.*

Chu Shih-chieh (literary name, Han-ch'ing; appellation, Sung-t'ing) lived in Yen-shan (near modern Peking). George Sarton describes him, along with Ch'in Chiu-shao, as "one of the greatest mathematicians of his race, of his time, and indeed of all times." However, except for the preface of his mathematical work, the *Ssu-yüan yü-chien* ("Precious Mirror of the Four Elements"), there is no record of his personal life. The preface says that for over twenty years he traveled extensively in China as a renowned mathematician; thereafter he also visited Kuang-ling, where pupils flocked to study under him. We can deduce from this that Chu Shih-chieh flourished as a mathematician and teacher of mathematics during the last two decades of the thirteenth century, a situation possible only after the reunification of China through the Mongol conquest of the Sung dynasty in 1279.

Chu Shih-chieh wrote the *Suan-hsüeh ch'i-meng* ("Introduction to Mathematical Studies") in 1299 and the *Ssu-yüan yü-chien* in 1303. The former was meant essentially as a textbook for beginners, and the latter contained the so-called "method of the four elements" invented by Chu. In the *Ssu-yüan yü-chien,* Chinese algebra reached its peak of development, but this work also marked the end of the golden age of Chinese mathematics, which began with the works of Liu I, Chia Hsien, and others in the eleventh and the twelfth centuries, and continued in the following century with the writings of Ch'in Chiu-shao, Li Chih, Yang Hui, and Chu Shih-chieh himself.

It appears that the *Suan-hsüeh ch'i-meng* was lost for some time in China. However, it and the works of Yang Hui were adopted as textbooks in Korea during the fifteenth century. An edition now preserved in Tokyo is believed to have been printed in 1433 in Korea, during the reign of King Sejo. In Japan a punctuated edition of the book (Chinese texts were then not punctuated), under the title *Sangaku keimo kunten,* appeared in 1658; and an edition annotated by Sanenori Hoshino, entitled *Sangaku keimo chūkai,* was printed in 1672. In 1690 there was an extensive commentary by Katahiro Takebe, entitled *Sangaku keimō genkai*, that ran to seven volumes. Several abridged versions of Takebe's commentary also appeared. The *Suan-hsüeh ch'i-meng* reappeared in China in the nineteenth century, when Lo Shih-lin discovered a 1660 Korean edition of the text in Peking. The book was reprinted in 1839 at Yangchow with a preface by Juan Yuan and a colophon by Lo Shih-lin. Other editions appeared in 1882 and in 1895. It was also included in the *ts'e-hai-shan-fang chung-hsi suan-hsüeh ts'ung-shu* collection. Wang Chien wrote a commentary entitled *Suan-hsüeh ch'i-meng shu i* in 1884 and Hsu Feng-k'ao produced another, *Suan-hsüeh ch'i-meng t'ung-shih,* in 1887.

The *Ssu-yüan yü-chien* also disappeared from China for some time, probably during the later part of the eighteenth century. It was last quoted by Mei Ku-ch'eng in 1761, but it did not appear in the vast imperial library collection, the *Ssu-k'u ch'üan shu,* of 1772; and it was not found by Juan Yuan when he compiled the *Ch'ou-jen chuan* in 1799. In the early part of the nineteenth century, however, Juan Yuan found a copy of the text in Chekiang province and was instrumental in having the book made part of the *Ssu-k'u ch'üan-shu.* He sent a handwritten copy to Li Jui for editing, but Li Jui died before the task was completed. This handwritten copy was subsequently printed by Ho Yüan-shih. The rediscovery of the *Ssu-yüan yü-chien* attracted the attention of many Chinese mathematicians besides Li Jui, Hsü Yu-jen, Lo Shih-lin, and Tai Hsü. A preface to the *Ssu-yüan yü-chien* was written by Shen Ch'in-p'ei in 1829. In his work entitled *Ssu-yüan yü-chien hsi ts'ao* (1834), Lo Shih-lin included the methods of solving the problems after making many changes. Shen Ch'in-p'ei also wrote a so-called *hsi ts'ao* ("detailed workings") for this text, but his work has not been printed and is not as well known as that by Lo Shih-lin. Ting Ch'ü-chung included Lo's *Ssu-yüan yü-chien hsi ts'ao* in his *Pai-fu-t'ang suan hsüeh ts'ung shu* (1876). According to Tu Shih-jan, Li Yen had a complete handwritten copy of Shen's version, which in many respects is far superior to Lo's.

Following the publication of Lo Shih-lin's *Ssu-yüan yü-chien hsi-ts'ao,* the "method of the four elements" began to receive much attention from Chinese mathematicians. I Chih-han wrote the *K'ai-fang shih-li* ("Illustrations of the Method of Root Extraction"), which has since been appended to Lo's work. Li Shan-lan wrote the *Ssu-yüan chieh* ("Explanation of the Four Elements") and included it in his anthology of mathematical texts, the *Tse-ku-shih-chai suan-hsüeh,* first published in Peking in 1867. Wu Chia-shan wrote the

Ssu-yüan ming-shih shih-li ("Examples Illustrating the Terms and Forms in the Four Elements Method"), the *Ssu-yüan ts'ao* ("Workings in the Four Elements Method"), and the *Ssu-yüan ch'ien-shih* ("Simplified Explanations of the Four Elements Method"), and incorporated them in his *Pai-fu-t'ang suan-hsüeh ch'u chi* (1862). In his *Hsüeh-suan pi-t'an* ("Jottings in the Study of Mathematics"), Hua Heng-fang also discussed the "method of the four elements" in great detail.

A French translation of the *Ssu-yüan yü-chien* was made by L. van Hée. Both George Sarton and Joseph Needham refer to an English translation of the text by Ch'en Tsai-hsin. Tu Shih-jan reported in 1966 that the manuscript of this work was still in the Institute of the History of the Natural Sciences, Academia Sinica, Peking.

In the *Ssu-yüan yü-chien* the "method of the celestial element" (*t'ien-yuan shu*) was extended for the first time to express four unknown quantities in the same algebraic equation. Thus used, the method became known as the "method of the four elements" (*ssu-yüan shu*)—these four elements were *t'ien* (heaven), *ti* (earth), *jen* (man), and *wu* (things or matter). An epilogue written by Tsu I says that the "method of the celestial element" was first mentioned in Chiang Chou's *I-ku-chi,* Li Wen-i's *Chao-tan,* Shih Hsin-tao's *Ch'ien-ching,* and Liu Yu-chieh's *Ju-chi shih-so,* and that a detailed explanation of the solutions was given by Yuan Hao-wen. Tsu I goes on to say that the "earth element" was first used by Li Te-tsai in his *Liang-i ch'un-ying chi-chen,* while the "man element" was introduced by Liu Ta-chien (literary name, Liu Jun-fu), the author of the *Ch'ien-k'un kua-nang;* it was his friend Chu Shih-chieh, however, who invented the "method of the four elements." Except for Chu Shih-chieh and Yüan Hao-wen, a close friend of Li Chih, we know nothing else about Tsu I and all the mathematicians he lists. None of the books he mentions has survived. It is also significant that none of the three great Chinese mathematicians of the thirteenth century—Ch'in Chiu-shao, Li Chih, and Yang Hui—is mentioned in Chu Shih-chieh's works. It is thought that the "method of the celestial element" was known in China before their time and that Li Chih's *I-ku yen-tuan* was a later but expanded version of Chiang Chou's *I-ku-chi.*

Tsu I also explains the "method of the four elements," as does Mo Jo in his preface to the *Ssu-yüan yü-chien.* Each of the "four elements" represents an unknown quantity—u, v, w, and x, respectively. Heaven (u) is placed below the constant, which is denoted by *t'ai,* so that the power of u increases as it moves downward; earth (v) is placed to the left of the constant so that the power of v increases as it moves toward the left; man (w) is placed to the right of the constant so that the power of w increases as it moves toward the right; and matter (x) is placed above the constant so that the power of x increases as it moves upward. For example, $u + v + w + x = 0$ is represented in Fig. 1.

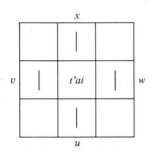

FIGURE 1

Chu Shih-chieh could also represent the products of any two of these unknowns by using the space (on the countingboard) between them rather as it is used in Cartesian geometry. For example, the square of

$$(u + v + w + x) = 0,$$

i.e.,

$$u^2 + v^2 + w^2 + x^2 + 2ux + 2vw + 2vx + 2wx = 0,$$

can be represented as shown in Fig. 2 (below). Obviously, this was as far as Chu Shih-chieh could go, for he was limited by the two-dimensional space of the countingboard. The method cannot be used to represent more than four unknowns or the cross product of more than two unknowns.

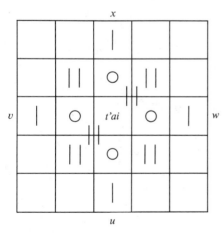

FIGURE 2

Numerical equations of higher degree, even up to the power fourteen, are dealt with in the *Suan-hsüeh ch'i-meng* as well as the *Ssu-yüan yü-chien*. Sometimes a transformation method (*fan fa*) is employed. Al-

though there is no description of this transformation method, Chu Shih-chieh could arrive at the transformation only after having used a method similar to that independently rediscovered in the early nineteenth century by Horner and Ruffini for the solution of cubic equations. Using his method of *fan fa,* Chu Shih-chieh changed the quartic equation

$$x^4 - 1496x^2 - x + 558236 = 0$$

to the form

$$y^4 + 80y^3 + 904y^2 - 27841y - 119816 = 0.$$

Employing Horner's method in finding the first approximate figure, 20, for the root, one can derive the coefficients of the second equation as follows:

1	0	−1496	−1	+558236	20
	20	400	−21920	−438420	
	20	−1096	−21921	119816	
	20	800	−5920		
	40	−296	−27841		
	20	1200			
	60	904			
	20				
	80				

Either Chu Shih-chieh was not very particular about the signs for the coefficients shown in the above example, or there are printer's errors. This can be seen in another example, where the equation $x^2 - 17x - 3120 = 0$ became $y^2 + 103y + 540 = 0$ by the *fan fa* method. In other cases, however, all the signs in the second equations are correct. For example,

$$109x^2 - 2288x - 348432 = 0$$

gives rise to

$$109y^2 + 10792y - 93312 = 0$$

and

$$9x^4 - 2736x^2 - 48x + 207936 = 0$$

gives rise to

$$9y^4 + 360y^3 + 2664y^2 - 18768y + 23856 = 0.$$

Where the root of an equation was not a whole number, Chu Shih-chieh sometimes found the next approximation by using the coefficients obtained after applying Horner's method to find the root. For example, for the equation $x^2 + 252x - 5292 = 0$, the approximate value $x_1 = 19$ was obtained; and, by the method of *fan fa,* the equation $y^2 + 290y - 143 = 0$. Chu Shih-chieh then gave the root as $x = 19(143/$

$1 + 290$). In the case of the cubic equation $x^3 - 574 = 0$, the equation obtained by the *fan fa* method after finding the first approximate root, $x_1 = 8$, becomes $y^3 + 24y^2 + 192y - 62 = 0$. In this case the root is given as $x = 8(62/1 + 24 + 192) = 8\ 2/7$. The above was not the only method adopted by Chu Shih-chieh in cases where exact roots were not found. Sometimes he would find the next decimal place for the root by continuing the process of root extraction. For example, the answer $x = 19.2$ was obtained in this fashion in the case of the equation

$$135x^2 + 4608x - 138240 = 0.$$

For finding square roots, there are the following examples in the *Ssu-yüan yü-chien:*

$$\sqrt{265} = 16\frac{9}{2 \times 16 + 1} = 16\frac{9}{33}$$

$$\sqrt{74} = 8.6\frac{4}{2 \times 8.6 + 1} = 8.6\frac{4}{17.3}$$

Like Ch'in Chiu-shao, Chu Shih-chieh also employed a method of substitution to give the next approximate number. For example, in solving the equation $-8x^2 + 578x - 3419 = 0$, he let $x = y/8$. Through substitution, the equation became $-y^2 + 578y - 3419 \times 8 = 0$. Hence, $y = 526$ and $x = 526/8 = 65\text{-}3/4$. In another example, $24649x^2 - 1562500 = 0$, letting $x = y/157$, leads to $y^2 - 1562500 = 0$, from which $y = 1250$ and $x = 1250/157 = 7\ 151/157$. Sometimes there is a combination of two of the above-mentioned methods. For example, in the equation $63x^2 - 740x - 432000 = 0$, the root to the nearest whole number, 88, is found by using Horner's method. The equation $63y^2 + 10348y - 9248 = 0$ results when the *fan fa* method is applied. Then, using the substitution method, $y = z/63$ and the equation becomes $z^2 + 10348z - 582624 = 0$, giving $z = 56$ and $y = 56/63 = 7/8$. Hence, $x = 88\ 7/8$.

The *Ssu-yüan yü-chien* begins with a diagram showing the so-called Pascal triangle (shown in modern form in Fig. 3), in which

$$(x + 1)^4 = x^4 + 4x^3 + 6x^2 + 4x + 1.$$

Although the Pascal triangle was used by Yang Hui in the thirteenth century and by Chia Hsien in the twelfth, the diagram drawn by Chu Shih-chieh differs from those of his predecessors by having parallel oblique lines drawn across the numbers. On top of the triangle are the words *pen chi* ("the absolute term"). Along the left side of the triangle are the values of the absolute terms for $(x + 1)^n$ from $n = 1$ to $n = 8$, while along the right side of the triangle are the values of the coefficient of the highest power of x. To the left, away from the top of the triangle, is the explanation that the numbers in the triangle should be used horizontally when $(x + 1)$ is to be raised to the power n. Opposite this is an explanation that the numbers inside the triangle give the *lien,* i.e., all coefficients of x from x^2 to x^{n-1}. Below the triangle are the technical terms of all the coefficients in the polynomial. It is interesting that Chu Shih-chieh refers to this diagram as the *ku-fa* ("old method").

The interest of Chinese mathematicians in problems involving series and progressions is indicated in the earliest Chinese mathematical texts extant, the *Chou-pei suan-ching* (*ca.* fourth century B.C.) and Liu Hui's commentary on the *Chiu-chang suan-shu.* Although arithmetical and geometrical series were subsequently handled by a number of Chinese mathematicians, it was not until the time of Chu Shih-chieh that the study of higher series was raised to a more advanced level. In his *Ssu-yüan yü-chien* Chu Shih-chieh dealt with bundles of arrows of various cross sections, such as circular or square, and with piles of balls arranged so that they formed a triangle, a pyramid, a cone, and so on. Although no theoretical proofs are given, among the series found in the *Ssu-yüan yü-chien* are the following:

(1) $1 + 2 + 3 + 4 + \cdots + n = \dfrac{n(n + 1)}{2!}$

(2) $1 + 3 + 6 + 10 + \cdots + \dfrac{n(n + 1)}{2!}$

$\qquad = \dfrac{n(n + 1)(n + 2)}{3!}$

(3) $1 + 4 + 10 + 20 + \cdots + \dfrac{n(n + 1)(n + 2)}{3!}$

$\qquad = \dfrac{n(n + 1)(n + 2)(n + 3)}{4!}$

(4) $1 + 5 + 15 + 35 + \cdots + \dfrac{n(n + 1)(n + 2)(n + 3)}{4!}$

$\qquad = \dfrac{n(n + 1)(n + 2)(n + 3)(n + 4)}{5!}$

(5) $1 + 6 + 21 + 56 + \cdots$

$\qquad + \dfrac{n(n + 1)(n + 2)(n + 3)(n + 4)}{5!}$

$\qquad = \dfrac{n(n + 1)(n + 2)(n + 3)(n + 4)(n + 5)}{6!}$

```
              1
            1   1
          1   2   1
        1   3   3   1
      1   4   6   4   1
    1   5  10  10   5   1
```

FIGURE 3

(6) $\quad 1^2 + 2^2 + 3^2 + \cdots + n^2 = \dfrac{n(n+1)}{2!} \cdot \dfrac{(2n+1)}{3}$

(7) $\quad 1 + 5 + 14 + \cdots + \dfrac{n(n+1)}{2!} \cdot \dfrac{(2n+1)}{3}$

$$= \dfrac{n(n+1)(n+2)}{3!} \cdot \dfrac{(2n+2)}{4}$$

(8) $\quad 1 + 6 + 18 + 40 + \cdots + \dfrac{n(n+1)}{2!} \cdot \dfrac{(3n+0)}{3}$

$$= \dfrac{n(n+1)(n+2)}{3!} \cdot \dfrac{(3n+1)}{4}$$

(9) $\quad 1 + 8 + 30 + 80 + \cdots$

$$+ \dfrac{n(n+1)(n+2)}{3!} \cdot \dfrac{(4n+0)}{4}$$

$$= \dfrac{n(n+1)(n+2)(n+3)}{4!} \cdot \dfrac{(4n+1)}{5}$$

After Chu Shih-chieh, Chinese mathematicians made almost no progress in the study of higher series. It was only after the arrival of the Jesuits that interest in his work was revived. Wang Lai, for example, showed in his *Heng-chai suan-hsüeh* that the first five series above can be represented in the generalized form

$$1 + (r+1) + \dfrac{(r+1)(r+2)}{2!} + \cdots$$

$$+ \dfrac{n(n+1)(n+2)\cdots(n+\overline{r-2})(n+\overline{r-1})}{r!}$$

$$= \dfrac{n(n+1)(n+2)\cdots(n+\overline{r-1})(n+r)}{(r+1)!}$$

where r is a positive integer.

Further contributions to the study of finite integral series were made during the nineteenth century by such Chinese mathematicians as Tung Yu-ch'eng, Li Shan-lan, and Lo Shih-lin. They attempted to express Chu Shih-chieh's series in more generalized and modern forms. Tu Shih-jan has recently stated that the following relationships, often erroneously attributed to Chu Shih-chieh, can be traced only as far as the work of Li Shan-lan.

If $U_r = \dfrac{r^{|p|}}{1^{|p|}}$, where r and p are positive integers, then

(a) $\quad \displaystyle\sum_{r=1}^{n} \dfrac{r^{|p|}}{1^{|p|}} (n+1-r) = \sum_{r=1}^{n} \dfrac{r^{|p+1|}}{1^{|p+1|}},$

with the examples

$$\sum_{r=1}^{n} r(n+1-r) = \sum_{r=1}^{n} \dfrac{r(r+1)}{2},$$

$$\sum_{r=1}^{n} \dfrac{r(r+1)(r+2)}{1 \cdot 2 \cdot 3}$$

$$= \sum_{r=1}^{n} \dfrac{(n+1-r)(n+2-r)(n+3-r)}{1 \cdot 2 \cdot 3},$$

and

(b) $\quad \displaystyle\sum_{r=1}^{n} \dfrac{r^{|p|}}{1^{|p|}} \cdot \dfrac{(n+1-r)^{|q|}}{1^{|q|}} = \sum_{r=1}^{n} \dfrac{r^{|p+q|}}{1^{|p+q|}},$

where q is any other positive integer.

Another significant contribution by Chu Shih-chieh is his study of the method of *chao ch'a* ("finite differences"). Quadratic expressions had been used by Chinese astronomers in the process of finding arbitrary constants in formulas for celestial motions. We know that this method was used by Li Shun-feng when he computed the Lin Te calendar in A.D. 665. It is believed that Liu Ch'uo invented the *chao ch'a* method when he made the Huang Chi calendar in A.D. 604, for he established the earliest terms used to denote the differences in the expression

$$S = U_1 + U_2 + U_3 + \cdots + U_n,$$

calling $\Delta = U_1$ *shang ch'a* ("upper difference"),

$\quad \Delta^2 = U_2 - U_1$ *erh ch'a* ("second difference"),

$\quad \Delta^3 = U_3 - (2\Delta^2 + \Delta)$ *san ch'a* ("third difference"),

$\quad \Delta^4 = U_4 - [3(\Delta^3 + \Delta^2) + \Delta]$ *hsia ch'a* ("lower difference").

Chu Shih-chieh illustrated how the method of finite differences could be applied in the last five problems on the subject in chapter 2 of *Ssu-yüan yü-chien*:

If the cube law is applied to [the rate of] recruiting soldiers, [it is found that on the first day] the *ch'u chao* [Δ] is equal to the number given by a cube with a side of three feet and the *tz'u chao* [$U_2 - U_1$] is a cube with a side one foot longer, such that on each succeeding day the difference is given by a cube with a side one foot longer than that of the preceding day. Find the total recruitment after fifteen days.

Writing down Δ, Δ^2, Δ^3, and Δ^4 for the given numbers, we have what is shown in Fig. 4. Employing the conventions of Liu Ch'uo, Chu Shih-chieh gave *shang ch'a* (Δ) = 27; *erh ch'a* (Δ^2) = 37; *san ch'a* (Δ^3) = 24;

Number of days	Total recruitment	Δ	Δ^2	Δ^3	Δ^4
1st day	27	$3^3 = 27$			
			37		
2nd day	91	$4^3 = 64$		24	
			61		6
3rd day	216	$5^3 = 125$		30	
			91		6
4th day	432	$6^3 = 216$		36	
			127		—
5th day	775	$7^3 = 343$		—	
———	—	———	—		

FIGURE 4

and *hsia ch'a* (Δ^4) = 6. He then proceeded to find the number of recruits on the *n*th day, as follows:

Take the number of days [n] as the *shang chi*. Subtracting unity from the *shang chi* [$n - 1$], one gets the last term of a *chiao ts'ao to* [a pile of balls of triangular cross section, or $S = 1 + 2 + 3 + \cdots + (n - 1)$]. The sum [of the series] is taken as the *erh chi*. Subtracting two from the *shang chi* [$n - 2$], one gets the last term of a *san chiao to* [a pile of balls of pyramidal cross section, or $S = 1 + 3 + 6 + \cdots + n(n - 1)/2$]. The sum [of this series] is taken as the *san chi*. Subtracting three from the *shang chi* [$n - 3$], one gets the last term of a *san chio lo i to* series

$$\left[S = 1 + 4 + 10 + \cdots + \frac{n(n - 1)(n - 2)}{3!} \right].$$

The sum [of this series] is taken as the *hsia chi*. By multiplying the differences [*ch'a*] by their respective sums [*chi*] and adding the four results, the total recruitment is obtained.

From the above we have:

Shang chi = n

$erh\ chi = 1 + 2 + 3 + \cdots + (n - 1) = \frac{1}{2!}n(n - 1)$

$san\ chi = 1 + 3 + 6 + \cdots + \frac{1}{2}n(n - 1)$

$$= \frac{1}{3!}n(n - 1)(n - 2)$$

$hsia\ chi = 1 + 4 + 10 + \cdots + \frac{n}{3!}(n - 1)(n - 2)$

$$= \frac{1}{4!}n(n - 1)(n - 2)(n - 3).$$

Multiplying these by the *shang ch'a, erh ch'a, san ch'a,* and *hsia ch'a,* respectively, and adding the four terms, we get

$$f(n) = n\Delta + \frac{1}{2!}n(n - 1)\Delta^2 + \frac{1}{3!}n(n - 1)(n - 2)\Delta^3$$

$$+ \frac{1}{4!}n(n - 1)(n - 2)(n - 3)\Delta^4.$$

The following results are given in the same section of the *Ssu-yüan yü-chien:*

$$\sum_{n=1}^{n} f(r) = \frac{1}{2!}n(n + 1)\Delta + \frac{1}{3!}(n - 1)n(n + 1)\Delta^2$$

$$+ \frac{1}{4!}(n - 2)(n - 1)n(n + 1)\Delta^3$$

$$+ \frac{1}{5!}(n - 3)(n - 2)(n - 1)n(n + 1)\Delta^4$$

$$\sum \frac{1}{3!}r(r + 1)(2r + 1) = \frac{1}{12}n(n + 1)(n + 1)(n + 2)$$

$$\sum \frac{1}{2!}r(r + 1) \cdot r = \frac{1}{4!}n(n + 1)(n + 2)(3n + 1)$$

$$\sum \frac{1}{3!}r(r + 1)(r + 2) \cdot r$$

$$= \frac{1}{5!}n(n + 1)(n + 2)(n + 3)(4n + 1)$$

$$\sum r\{a + (r - 1)b\} = \frac{1}{3!}n(n + 1)\{2bn + (3a - 2b)\}$$

$$\sum r(a + \overline{n - r} \cdot b) = \frac{1}{3!}n(n + 1)\{bn + (3a - b)\}$$

$$\sum \frac{1}{2!}r(r + 1) \cdot (a + \overline{r - 1} \cdot b)$$

$$= \frac{1}{4!}n(n + 1)(n + 2)\{3bn + (4a - 3b)\}$$

$$\sum r^2 \cdot (a + \overline{n - r} \cdot b)$$

$$= \frac{1}{3!}n(n + 1)(2n + 1)a + \frac{2n}{4!}(n - 1)n(n + 1)b$$

The *chao ch'a* method was also employed by Chu's contemporary, the great Yuan astronomer, mathematician, and hydraulic engineer Kuo Shou-ching, for the summation of power progressions. After them the *chao ch'a* method was not taken up seriously again in China until the eighteenth century, when Mei Wen-ting fully expounded the theory. Known as *shōsa* in Japan, the study of finite differences also received considerable attention from Japanese mathematicians, such as Seki Takakazu (or Seki Kōwa) in the seventeenth century.

BIBLIOGRAPHY

For further information on Chu Shih-chieh and his work, consult Ch'ien Pao-tsung, *Ku-suan k'ao-yüan* ("Origin of Ancient Chinese Mathematics") (Shanghai, 1935), pp. 67–80; and *Chung-kuo shu-hsüeh-shih* ("History of Chinese Mathematics") (Peking, 1964), 179–205; Ch'ien Pao-tsung *et al., Sung Yuan shu-hsüeh-shih lun-wen-chi* ("Collected Essays on Sung and Yuan Chinese Mathematics") (Peking, 1966), pp. 166–209; L. van Hée, "Le précieux miroir des quatre éléments," in *Asia Major,* **7** (1932), 242; Hsü Shun-fang, *Chung-suan-chia ti tai-shu-hsüeh yen-chiu* ("Study of Algebra by Chinese Mathematicians") (Peking, 1952), pp. 34–55; E. L. Konantz, "The Precious Mirror of the Four Elements," in *China Journal of Science and Arts,* **2** (1924), 304; Li Yen, *Chung-kuo shu-hsüeh ta-kang* ("Outline of Chinese Mathematics"), I (Shanghai, 1931), 184–211; "*Chiu-chang suan-shu pu-chu*" *Chung-suan-shih lun-ts'ung* (German trans.), in *Gesammelte Abhandlungen über die Geschichte der chinesischen Mathematik,* III (Shanghai, 1935), 1–9; *Chung-kuo suan-hsüeh-shih* ("History of Chinese Mathematics") (Shanghai, 1937; repr. 1955), pp. 105–109, 121–128, 132–133; and *Chung Suan-chia ti nei-ch'a fa yen-chiu* ("Investigation of the Interpolation Formulas in Chinese Mathematics") (Peking, 1957), of which an English trans. and abridgement is "The Interpolation Formulas of Early Chinese Mathematicians," in *Proceedings of the Eighth International Congress of the History of Science* (Florence, 1956), pp. 70–72; Li Yen and Tu Shih-jan, *Chung-kuo ku-tai shu-hsüeh chien-shih* ("A Short History of Ancient Chinese Mathematics"), II (Peking, 1964), 183–193, 203–216; Lo Shih-lin, *Supplement to the Ch'ou-jen chuan* (1840, repr. Shanghai, 1935), pp. 614–620; Yoshio Mikami, *The Development of Mathematics in China and Japan* (Leipzig, 1913; repr. New York), 89–98; Joseph Needham, *Science and Civilisation in China,* III (Cambridge, 1959), 41, 46–47, 125, 129–133, 134–139; George Sarton, *Introduction to the History of Science,* III (Baltimore, 1947), 701–703; and Alexander Wylie, *Chinese Researches* (Shanghai, 1897; repr. Peking, 1936; Taipei, 1966), pp. 186–188.

HO PENG-YOKE

CHUGAEV, LEV ALEKSANDROVICH (*b.* Moscow, Russia, 17 October 1873; *d.* Gryazovets, Vologoskaya oblast, U.S.S.R., 23 September 1922), *chemistry.*

Chugaev, who studied under Zelinski, graduated from Moscow University in 1895. From 1896 to 1904 he was chairman of the chemistry section of the Bacteriology Institute in Moscow; from 1904 to 1908 professor of chemistry at the Moscow Technical College; and from 1908 until 1922 professor of inorganic chemistry at St. Petersburg (later Leningrad) University, while simultaneously professor of organic chemistry at the St. Petersburg Institute of Technology. At Chugaev's initiative, the Institute for the Study of Platinum and Precious Metals was created in 1918; he served as its director and carried out the organizational details of the development of the platinum industry in the Soviet Union until 1922. He was one of the founders of the Russian Scientific and Technical Institute of Nutrition (1918) and of the Institute of Applied Chemistry (1919). Among Chugaev's students were I. I. Chernyaev, A. A. Grinberg, V. G. Khlopin, V. V. Lebedinsky, and F. V. Tserevitinov.

Chugaev's research covered problems in biochemistry and in organic, inorganic, and physical chemistry. In his first series of experiments in biochemistry and bacteriology special attention was given to the action of poisons on microorganisms, to the physiology of phosphorescent bacteria, and to triboluminescence.

In 1899 Chugaev began a series of experiments on the chemistry of terpenes; this work was generalized in his master's dissertation, "Issledovania v oblasti terpenov i kamfory" ("Research in the Field of Terpenes and Camphor," 1903). Using the fundamental result of this research, the development of the xanthogen method, Chugaev first synthesized a number of terpene hydrocarbons. His method is one of dehydration of complex unsaturated alcohols without concomitant isomerization. For example, the action of CS_2 on the sodium salt of the alcohol $C_{10}H_{19}ONa$ gives the xanthate $C_{10}H_{19}OC\overset{S}{{-}}SNa$, which can be converted into the ether with methyl iodide and then, by careful heating, made to yield the corresponding hydrocarbon, $C_{10}H_{18}$. Chugaev's method was a classic of organic chemistry and has been applied successfully for decades.

A study of the optical properties of organic compounds allowed Chugaev to formulate in 1908 the "rule of distance," according to which the optical activity of a compound increases as a function of the decrease of the distance of the inactive part of the molecule from its asymmetric center. In 1911 he discovered a new type of rotary dispersion that is conditioned by the internal superimposition of the opposing optical effects of the individual asymmetric hydrocarbon atoms in a molecule of the compound.

The work of Chugaev and his students was in two areas: (1) the synthesis and stereochemistry of the inner complex salts of copper, nickel, silver, cobalt, and the platinum metals and (2) a study of the specific reactions of the platinum metals that can be used for their separation and purification. His studies of the inner complex metallic salts notably included experiments with the organic imide complexes of copper, nickel, and silver and the stable α-dioxime complexes of nickel, copper, iron, cobalt, platinum, and palladium. In 1906 Chugaev demonstrated that the most stable organic inner complex salts contain five- or six-member cycles; he concluded in 1908 that the formation of cyclically structured complex compounds depends on the stereochemical configuration of their components.

Chugaev established the structure of the isomers of divalent platinum with organic sulfides and isonitriles, and he investigated isomerism in the complex compounds of platinum with nitriles, hydrazine, and hydroxylamine. In 1920 he synthesized the tetravalent platinum pentamine chloride $[Pt(NH_3)_5Cl]_3$, which had been predicted on Alfred Werner's coordination theory.

In 1905 Chugaev introduced dimethylglyoxime as a new and sensitive analytical reagent for divalent nickel. This initiated the use of organic reagents into analytical chemistry. Chugaev also discovered specific sensitive organic reagents for iridium (malachite green), platinum (carbylamine), and osmium (thiocarbamide). The last reaction is highly specific and one in which the metals of the platinum group do not interfere.

Chugaev's research indicated the valuable qualities of chelates, compounds that are formed by the interaction of organic reagents and metal ions. These compounds often possess properties that are important from an analytical point of view: sharp color, insignificant solubility in water, good solubility in organic solvents, and so forth. Because of these properties they are widely used in modern analytic chemistry.

BIBLIOGRAPHY

I. Original Works. A collection of Chugaev's works is *Izbrannye trudy* ("Selected Works"), 3 vols. (Moscow, 1954–1962). Vol. II includes a full bibliography of Chugaev's works (429 titles).

II. Secondary Literature. On Chugaev and his work see I. I. Chernyaev, "Lev Aleksandrovich Chugaev (1873–1922)," in *Lyudi russkoy nauki* ("People of Russian Science"), vol. I (Moscow, 1948); *L. A. Chugaev. Sbornik rechey i doladov, posvyashchennykh yego pamyati* ("L. A. Chugaev. A Collection of Speeches and Papers Dedicated to His Memory," Leningrad, 1924); G. B. Kauffman, "Terpenes to Platinum. The Chemical Career of Lev Aleksandrovich Chugaev," in *Journal of Chemical Education,* **40,** no. 12 (1963), 656–664; and O. Y. Zvyagintsev, Y. I. Soloviev, and P. I. Staroselsky, *Lev Aleksandrovich Chugaev* (Moscow, 1965).

Y. I. Soloviev

CHUQUET, NICOLAS (*b.* Paris [?], France; *fl.* second half fifteenth century), *algebra.*

Nicolas Chuquet is known only through his book, which as an entity has remained in manuscript; one part, the "Triparty," on the science of numbers, was published by Aristide Marre in 1880. The following year Marre published the statement of, and replies to, a set of 156 problems that follow the "Triparty" in the manuscript. The analysis of these problems remains unpublished, as do an application to practical geometry and a treatise on commercial arithmetic.

The conclusion of the "Triparty" indicates that it was composed by one Nicolas Chuquet of Paris, holder of the baccalaureate in medicine, at Lyons, in 1484. (Marre believes that he can date the work precisely: one problem in the treatise on business arithmetic permits fixing the work no later than May 1484.)

Only one copy of the book (the work of a firm of copyists) is known to exist; although it remained in manuscript until 1880, several passages from it were copied slavishly by "Master Étienne de la Roche, also called Villefranche, native of Lyons on the Rhone," in his arithmetic text of 1520, of which there was a second edition in 1538.

We owe confirmation of the existence and importance of Nicolas Chuquet to this unscrupulous Master Étienne, who appears to have been well established in Lyons, where his name was included on the tax rolls of 1493. He cites Chuquet and his "Triparty" at the beginning of his text—without, however, adding that he is plagiarizing outrageously.

Chuquet called himself a Parisian. He spent his youth in that city, where he was probably born and where the name is yet known. There he pursued his extensive studies, up to the baccalaureate in medicine (which implies a master of arts as well). It is difficult to say more about his life. He was living in Lyons in 1484, perhaps practicing medicine but more probably teaching arithmetic there as "master of algorithms." The significant place given to questions of simple and compound interest, the repayment of debts, and such in his work leads one to suppose this. However, he used these questions only as pretexts for exercises in algebra.

Chuquet's mathematical learning was solid. He cites by name Boethius—whom everyone knew at that time—Euclid, and Campanus of Novara. He knew the propositions of Archimedes, Ptolemy, and Eutocius, which he stated without indicating his sources (referring to Archimedes only as "a certain wise man"). In geometry his language seems to be that of a translator, transposing terms taken from Greek or Latin into French. By contrast, in the parts devoted solely to arithmetic or algebra there is no borrowing of learned terminology. Everything is written in simple, direct language, with certain French neologisms that have not been preserved elsewhere. The only exception is the ponderous nomenclature for the various proportions encumbering the first pages of the "Triparty," for Chuquet was respecting a style that goes back to Nicomachus and his translator Boethius and that still infested the teaching of mathematics in the seventeenth century.

On the whole Chuquet wrote a beautiful French that is still quite readable. His simple, very mathematical style does not lack elegance, although occasional affectation led to the use of three or four synonyms in order to avoid monotonous repetition. Marre purports to find many Italianisms in Chuquet's French. He attributes this peculiarity to the close relations between Lyons and the cities of northern Italy and to its large and prosperous Italian colony. Upon examination, many of these so-called Italianisms appear to be nothing more than Latinisms, however, and the French used by Chuquet seems as pure as that of his contemporaries.

As for his mathematical work, in order to judge it fairly, we must compare it with the work of such contemporaries as Regiomontanus, whose *De triangulis omnimodis* was written twenty years earlier, in 1464 (although it was not printed until 1533), and most notably with that of Luca Pacioli, whose *Summa de arithmetica, geometria proportioni et proportionalita* was published ten years later, in 1494.

Chuquet made few claims of priority. The only thing that he prided himself on as his personal discovery was his "règle des nombres moyens." On one further occasion he seemed also to be claiming for himself the "règle des premiers." ("What I call first numbers the ancients called 'things.'") But he says no more on this point—where, as far as we know, his originality is obvious.

Chuquet engaged in very little controversy, contenting himself with twice criticizing a certain Master Berthelemy de Romans, also cited by a contemporary French arithmetician, Jehan Adam (whose arithmetic manuscript is dated 1475). "Master Berthelemy de Romans, formerly of the Order of Preaching Friars [Dominicans] at Valence and Doctor of Theology" may well have been one of the mathematics professors of Chuquet and Adam. Nothing definite is known of this matter, however.

The "Triparty" is a treatise on algebra, although the word appears nowhere in the manuscript. This algebra deals only with numbers, but in a very broad sense of the term. The first part concerns rational numbers. Chuquet's originality in his rules for decimal numeration, both spoken and written, is immediately obvious. He introduced the practice of division into groups of six figures and used, besides the already familiar million, the words billion (10^{12}), trillion (10^{18}), quadrillion (10^{24}), etc. Here is an example of his notation:

$$745324^3804300^2700023^1654321.$$

In this example Chuquet's exponents are simply commas, but he points out that one can use ^{1} in place of the first comma, ^{2} in place of the second, ^{3} in place of the third, and so on.

Fractions, which Chuquet called "nombres routz" (*sic*), or broken numbers, were studied clearly and without complicated rules. Like all his contemporaries, however, he always used a numerator smaller than the denominator (and hence mixed numbers instead of improper fractions), a practice that led to unnecessary complications in his calculations.

Chuquet's study of the rules of three and of simple and double false position, clear but commonplace, served as pretext for a collection of remarkable linear problems, expounded in a chapter entitled "Seconde partie d'une position." Here he did not reveal his methods but reserved their exposition for a later part of the work, where he then said that after having solved a problem by the usual methods—double position or algebra (his "règle des premiers")—one must vary the known numerical quantities and carefully analyze the sequence of computations in order to extract a canon (formula). This analysis generally led him to a correct formula, although at times he was mistaken and gave methods applicable only for particular values.

Another original concept occurred in this group of problems. In a problem with five unknowns, Chuquet concluded: "I find 30, 20, 10, 0, and minus 10, which are the five numbers I wished to have." He then pointed out that zero added to or subtracted from a number does not change the number and reviewed the rules for addition and subtraction of negative numbers. In the thirteenth century Leonardo Fibonacci had made a similar statement but had not carried it as far as Chuquet in the remainder of his work.

The first part of the "Triparty" ends with the "règle des nombres moyens," the only discovery to which Chuquet laid claim. According to him—and he was right—this rule allows the solution of many problems that are unapproachable by the classic rule of three or the rules of simple or double false position. It consisted of establishing that between any two given fractions a third can always be interpolated that has for numerator the sum of the numerators of the other two fractions, and for denominator the sum of their denominators. It has been demonstrated in modern times that by repeating this procedure it is possible to arrive at all the rational numbers included between the two given fractions. It is obvious, therefore, that this rule, together with a lot of patience, makes it possible to solve any problem allowing of a rational solution. Further on, Chuquet utilizes it in order to approach indeterminately the square roots, cube roots, and so on, of numbers that do not have exact roots. But here he used it to solve the equation

$$x^2 + x = 39\tfrac{13}{81}.$$

Successively interpolating five fractions between 5 and 6, he found the exact root, $5\frac{7}{9}$. Since he was, moreover, little concerned about rapid methods of approximation, Chuquet throughout his work used nothing but this one rule and only rarely at that.

The second part of the "Triparty" deals with roots and "compound numbers." There is no trace of Euclidean nomenclature, of "binomials," or of "apostomes." The language has become simpler: there is no question of square roots or cube roots, but of second, third, fourth roots, "and so on, continuing endlessly." The number itself is its own first root. Moreover, everything is called a "number"—whole numbers, rational numbers, roots, sums, and differences of roots—which, in the fifteenth century, was audacious indeed. The notation itself was original. Here are several examples:

Chuquet's Notation	Modern Notation
R^1 12	12
$3 \, \bar{p} \, R^2 \, 5$	$3 + \sqrt{5}$
$R^2 \, 2 \, \bar{p} \, R^2 \, 5$	$\sqrt{2} + \sqrt{5}$
$R^2 \, \underline{14 \, \bar{p} \, R^2 \, 180}$	$\sqrt{(14 + \sqrt{180})}$
$R^2 \, \underline{14 \, \bar{m} \, R^2 \, 180}$	$\sqrt{(14 - \sqrt{180})}$

Unfortunately, Chuquet, a man of his times after all, occasionally became involved in computations that to us might seem inextricable. For example, he wrote the product of $\sqrt[6]{7} \times 2$ as R^6 448, whereas after Descartes it would be written $2\sqrt[6]{7}$.

The third part is by far the most original. It deals with the "règle des premiers," a "truly excellent" rule that "does everything that other rules do and, in addition, solves a great many more difficult problems." It is "the gateway and the threshold to the mysteries that are in the science of numbers." Such were the enthusiastic terms in which Chuquet announced the algebraic method. First, he explained his notation and his computational rules. The unknown, called the "first number" (*nombre premier*), is written as 1^1. Therefore, where Chuquet wrote 4^0, we should read 4; if he wrote 5^1, we should read $5x$; and if he wrote 7^3, we should read $7x^3$.

By a daring use of negative numbers, he wrote our $-12x$ as $\bar{m}12^1$, our $-\sqrt[3]{12} \cdot x^3$ as $\bar{m} \, R^3 \, 12^3$. For $12x^{-1}$ or $12/x$, he wrote $12^{1\bar{m}}$; for $-12x^{-1}$, he wrote $\bar{m}12^{1\bar{m}}$. In order to justify his rules of algebraic computation, and particularly those touching the product of the powers of a variable, he called upon analogy. He considered the sequence of the powers of 2 and showed, for example, that $2^2 \times 2^3 = 2^5$. He wished only to make clear, by an example that he considered commonplace and that goes back almost to Archi-

medes, the algebraic rule that if squares are multiplied by cubes the result is the fifth power.

For division he announced that the quotient of

36^3	by 6^1	is 6^2	$36x^3 \div 6x$	$= 6x^2$
72^0	by 8^3	is $9^{3\bar{m}}$	$72 \quad \div 8x^3$	$= 9x^{-3}$
$84^{2\bar{m}}$	by $7^{3\bar{m}}$	is 12^1	$84x^{-2} \div 7x^{-3}$	$= 12x.$

Further on—but only once in the entire work—he wrote down a rational function:

$$\frac{30 \, m \, 1^1}{1^2 \, p \, 1^1} \, , \text{ i.e., } \left(\frac{30 - x}{x^2 + x} \right).$$

In accordance with the custom of Chuquet's time, all these rules of computation were simply set forth, illustrated by a few examples, and at times justified by analogy with more elementary arithmetic, but never "demonstrated" in the modern sense of the term. Having set down these preliminaries, Chuquet dealt with the theory of equations, which he called the "method of equaling."

In setting up an equation, he specified that 1^1 should be taken as the unknown and that this "premier" should be operated on "as required in 'la raison,'" that is, the problem under study.

One can, moreover, take 2^1, 3^1, etc., as the unknowns in place of 1^1. One should end with two expressions equal or similar to each other. These are the two "parts" of the equation. This is followed by the classic rules for solving binomial equations of the type $ax^m = b$.

These rules include procedures for reducing equations to the binomial form. Chuquet emphasized the importance of recognizing "equipollent numbers," i.e., expressions of the same power as $\sqrt{x^2}$ and x or $\sqrt[5]{x^{10}}$ and x^2. He notes that when the two parts of the equation are "similar" it may either have infinitely many solutions ($ax^m = ax^m$) or be impossible of solution ($ax^m = bx^m; a \neq b$). Several of the many numerical problems that follow—certain of them having several unknowns—come to one of these results. When the two parts are not similar, the equation has for Chuquet at most one solution. He is, however, of two minds regarding negative solutions; sometimes he accepts them, sometimes not.

Subsequently Chuquet solved problems that led to equations of the form

$$ax^{2m} + bx^m + c = 0.$$

Like all his contemporaries he distinguished three cases (a, b, c positive):

$$c + bx^m = ax^{2m}$$
$$bx^m + ax^{2m} = c$$
$$c + ax^{2m} = bx^m.$$

While he knew and stated that in the last case the equation might be insoluble, and admitted two solutions when it was soluble, he found only one solution in the other two cases. In this regard he did not show any progress over his predecessors.

On the subject of division into mean and extreme ratio, he showed, in fact, a profound ignorance of the theory of numbers.

Indeed, having to solve $144 + x^2 = 36x$, Chuquet obtained the answer $x = 18 - \sqrt{180}$, and, in effect, declared: "Campanus, in the ninth book of Euclid, at the end of Proposition 16, affirms that such a problem is impossible of solution in numbers, while in fact, as we have just seen, it is perfectly capable of solution."

Now in fact Campanus demonstrated, very elegantly, that the solution cannot be rational. Chuquet, who had a very broad concept of numbers, did not grasp the subtlety. In several other cases he also showed a lack of understanding, particularly regarding the problems that he included in his "rule of apposition and separation." These problems derive from Diophantine analysis of the first order, the exact theory of which had to await Bachet de Méziriac and the seventeenth century.

The "Triparty" ends with an admission of inadequacy: there remains the task of studying equations of a more general type, wrote the author, but he would leave them for those who might wish to go further with the subject. This attitude was common among the majority of the algebraists of the fifteenth century, who would have liked to go beyond the second degree but were not sure how to go about it.

Summarizing the "Triparty," we may say that it is a very abstract treatise on algebra, without any concrete applications, in which one can see a great extension of the concept of number—zero, negative numbers, and roots and combinations of roots all being included. There appear in it excellent notations that prefigure, in particular, those of Bombelli. But weak points remain. In linear algebra the cases of indetermination and insolubility were poorly set forth. In the theory of equations the importance of degree was not realized, and for equations of the second degree the old errors persisted. Moreover, Chuquet's taste for needlessly complicated computations—and beyond that the way in which the necessity for these computations arose in the shortcomings of numerical algebra—is painfully evident. At the end of a problem one may, for example, find

$$\mathbf{R}^3 \; 13\frac{865}{1728} \, p \; \mathbf{R}^2 \; 182\frac{699841}{2985984},$$

"which, if abbreviated by extraction of the second

and third roots, comes to 3." Such an answer required a certain courage in the person making the calculation!

The "Triparty" has become well known since its publication in 1880. The same is not true of the sequel to the 1484 manuscript. It is true that Marre gave, in 1881, the statement of the 156 problems that follow the "Triparty," accompanied by answers and some remarks that Chuquet added to them. But this incomplete publication let several important points go by.

By contrast with the problems of the "Triparty," the greater part of those treated in this appendix appear to be concrete. But this is merely an appearance, and one can imagine the author coldly cutting an heir into several pieces if a problem concerning a will fails to come out even. Moreover, the questions dealt with were not all original; many were part of a long tradition going back at least as far as Metrodorus' anthology.

Marre indicated those problems that were taken over—almost word for word—by the plagiarist Étienne de la Roche and those that may be found in an almost contemporary treatise written in the Languedoc dialect of the region of Pamiers. One can find many other such duplications. Chuquet's originality lay not in his choice of topics but in his way of treating them. Indeed, he very often made use of his "règle des premiers," that is, of algebra.

A very important fact seems to have gone unnoticed by historians, no doubt because of the incomplete publication of this part of the work. It is known that, in his *Summa* of 1494, Luca Pacioli made use in certain cases of not one unknown but two: the *cosa* and the *quantita*. Now, what has passed unnoticed is that Chuquet employed the same device in 1484, and on at least five occasions. The first time was in the following problem:

> Three men have some coins. If the first man took 12 coins from the two others, he would have twice the amount remaining to them, plus 6 coins. If the second man took 13 coins from the first and the third, he would have four times as many as remained to them, plus 2 coins. If the third man took 11 coins from the two others, he would have three times what remained to them, plus 3 coins. How many coins does each man have?

Chuquet solved the problem by a mixed method, a combination of the rule of two false positions and his "règle des premiers." Someone (it must be Étienne de la Roche) has written in the margin "by the rule of two positions and by the rule of the 'thing' together." This annotator, perhaps under the influence of Pacioli, used the language of the Italian algebraists.

Chuquet posited 6 as the holdings of the first man,

and with the first datum of the problem deduced that the three men have 24 coins among them. Assigning to the third the value 1^1 and basing himself on the third condition, he found, using the "règle des premiers," that the third man has $7\frac{3}{4}$ coins and therefore the second has $10\frac{1}{4}$.

By following the then classic method of false position, he established that the use of 6 for the first man's holdings gives, for the second condition, an excess of $18\frac{1}{4}$. He started again with 12, and by applying the classic rule, obtained the answer: the holdings are 8, 9, and 10.

But, a bit further on, he announced "another way of solving it, using only the 'règle des premiers.'" The annotator has written in the margin, "This rule is called the rule of quantity." Here it is plain that the annotator was familiar with Pacioli's *Summa*.

As we have already said, Chuquet liked to vary the numerical data of his problems. This was a necessity for him, and it mitigated to a degree the absence of literal computation. Thus he could better analyze the sequence of computations and draw from them the rules to follow for solving automatically problems of the same kind. Thus, he presented the following problem:

If the first of three men took 7 coins from the others, he would have five times what remained to them, plus 1. If the second took 9 coins from the others, he would have six times what remained to them, plus 2, and if the third took 11 coins from the first two, he would have seven times as many as remained to them, plus 3.

Chuquet then posited 1^1—i.e., x—for the holdings of the first, and for the first datum found that they have, in all,

$$1^1\frac{1}{5} \ \bar{p} \ 8\frac{1}{5} \ , \ \text{i.e.,} \ \left(\frac{6}{5}x + \frac{41}{5}\right).$$

"To find the portion belonging to the second man, I assign to him 1^2." This symbol no longer represents the square, x^2, of the unknown x but another unknown, y. Then, making use of the second condition of the problem, he found that

$$y = 1^1\frac{1}{35} \ \bar{m} \ 1\frac{24}{35} \ , \ \text{i.e.,} \ \left(\frac{36}{35}x - \frac{59}{35}\right).$$

Finally, to find the third man's portion, he used the third condition and took 1^2 for the third unknown. This unknown is not the y of the preceding computation but a new unknown, z. He found

$$z = 1^1\frac{1}{20} \ \bar{m} \ 3\frac{9}{20} \ , \ \text{i.e.,} \ \left(\frac{21}{20}x - \frac{69}{20}\right).$$

Writing

$$x + y + z = 1^1\frac{1}{5} \ \bar{p} \ 8\frac{1}{5},$$

he arrived at the final equation:

$$3^1\frac{11}{140} \ \bar{m} \ 5\frac{19}{140} \ \text{equals} \ 1^1\frac{1}{5} \ \bar{p} \ 8\frac{1}{5},$$

which remains only to be solved.

Several problems on geometric progressions and compound interest were not completely solved until the introduction of logarithms. Chuquet was aware of the difficulty without finding a way to solve it.

A vessel with an open spout loses one-tenth of its contents each day. In how many days will the vessel be half-empty?

After having answered that it will be $6\frac{31441}{531441}$ days, he adds, "Many people are satisfied with this answer. However, it seems that between six days and seven, one should search for a certain proportional number that, for the present, is unknown to us."

Among the numerous problems studied is a question of inheritance that is found in Bachet de Méziriac (1612) and again in Euler's *Algebra* (1769) and that, by its nature, demands a whole number as the answer.

The oldest son takes 1 coin and one-tenth of the rest, the second takes 2 coins and one-tenth of the new remainder, and so on. Each of the children receives the same sum. How many heirs are there?

After having varied the numerical data, Chuquet gave a general rule for the solution of the problem, then treated eleven other numerical cases—in which the answers include fractions of an heir. One detects here not a number theorist, but a pure algebraist.

Many of Bachet de Méziriac's *Problèmes plaisans et délectables* can be found in Chuquet, especially in the chapter entitled "Jeux et esbatemens qui par la science des nombres se font." But they formed part of a thousand-year-old tradition, and a manuscript by Luca Pacioli, *De viribus quantitatis*, also dealt with them.

The geometric part of Chuquet's work is entitled "Comment la science des nombres se peut appliquer aux mesures de géométrie." It formed part of a tradition that goes back to the Babylonians and that developed over the centuries, in which figured notably Hero of Alexandria and, in the thirteenth century, Leonardo Fibonacci. It comprised, first of all, the measurement of length, area, and volume. Then it became more scientific and, in certain respects, a veritable application of algebra to geometry.

Rectilinear measure was either direct or effected by means of the quadrant, which was represented in practice by a figure on the back of the astrolabe that utilized a direct and inverse projection or shadow. No recourse was made to trigonometry, nor did any measurement of angles appear. In this we are removed from the astronomic tradition so brilliantly represented by Regiomontanus. The quadrant was used for measuring horizontal distances, depths, and heights. Other elementary techniques were also indicated, such as those of the vertical rod and the horizontal mirror.

As for curves, "the circular line is measured in a way that will be described in connection with the measuring of circular surfaces. Other curved lines are reduced as much as possible to a straight line or a circular line." For the circle the approximations $3\frac{1}{7}$ for π and $\frac{11}{14}$ for $\pi/4$ were the only ones indicated.

Chuquet separated triangles into equilateral and nonequilateral triangles. Each triangle consisted of one base and two hypotenuses. The "cathète" descends perpendicular to the base. This terminology was quite unusual and certainly different from classical terminology.

Chuquet knew Euclid's *Elements* very well, however. For him the following proposition was the fundamental one: "If the square of the perpendicular is subtracted from the square of the hypotenuse, the square root of the remainder will be the length of the part of the base that corresponds to the hypotenuse." He gave no demonstration. The first part of this practical geometry was completed by a few notes on the "measurement of hilly surfaces" and the volumes of spheres, pyramids, and cubes. These are scarcely developed. This work does not have the amplitude of Pacioli's treatment of the same topic in the *Summa* and is scarcely more than a brief summary of what can be found in the pages of Leonardo Fibonacci. But "the application of the aforesaid rules" brought forth a new spirit, and here the algebraist reappeared. For example, Chuquet applied Hero's rule on the expression of the area of a triangle as a function of its three sides to the triangle 11, 13, 24. He found the area to be zero, and from this concluded the nonexistence of the triangle.

Again, two vertical lines of lengths 4 and 5 are horizontally distant by 12. Chuquet sought a point on the horizontal that was equidistant from the two apexes. This led him to the equation

$$169m \; 24^1 \; \bar{p} \; 1^2 \text{ equals } 1^2 \; \bar{p} \; 16 \; \cdots .$$

There are still other, similar algebraic exercises. Two of them are important.

In the first it is proposed to find the diameter of the circle circumscribed about a triangle whose base is 14 and whose sides are 13 and 15. To do this it was first necessary to compute the projections x and y of the sides on the base, using the relation $x^2 - y^2 = 15^2 - 13^2$. The perpendicular then follows directly. Now, taking as the unknown the distance from the center to the base and expressing the center as equidistant from the vertices, one is led to the equation

$$148m \; 24^1 \; \bar{p} \; 1^2 \text{ equals } 49 \; \bar{p} \; 1^2.$$

In the second exercise Chuquet proposed to compute the diameter of a circle circumscribed about a regular pentagon with sides of length 4. He first took as the unknown the diagonal of the pentagon. Ptolemy's theorem on inscribed quadrilaterals then leads to the equation

$$4^1 \; \bar{p} \; 16 \text{ equals } 1^2,$$

which permitted him to make his computations. By this method he concluded that Euclid's proposition is correct: the square of the side of the hexagon added to that of the decagon gives that of the pentagon inscribed in the same circle.

Two analogous applications of algebra to geometry are in the *De triangulis* of Regiomontanus. Others are to be found in the geometrical part of Pacioli's *Summa.* Thus, it appears that this tradition was strongly implanted in the algebraists of the fifteenth century.

Without any vain display of erudition Chuquet showed his extensive learning in borrowings from Eutocius for the graphic extraction of cube roots. He also gave constructions with straight-edge and compasses and devoted a rather weak chapter to research on the squaring of the circle. Ramón Lull's quadrature was recalled here. Chuquet showed that it was equivalent to taking the value $2 + \sqrt{2}$ for π. As for the approximation 22/7, it is given by "a wise man. . . . But this is a thing that cannot be proved by any demonstration."

If Étienne de la Roche had been more insensitive—especially if he had plagiarized the "règle des premiers" and its applications—mathematics would perhaps be grateful to him for his larcenies. He was unfortunately too timid, and in his arithmetic text he returned to the classical errors of his time, and thus the most original part of Chuquet's work remained unknown. One cannot, however, assert that the innovations that Chuquet introduced were entirely lost. His notation can be found again, for example, in Bombelli. Was it rediscovered by the Italian algebraist? Was there a connection between the two, direct or indirect? Or did both derive from a more ancient source? These are the questions that remain open.

BIBLIOGRAPHY

The MS of Chuquet's book is in the Bibliothèque Nationale, Fonds Français, no. 1346. Aristide Marre's ed. of the "Triparty" is "Le Triparty en la science des nombres par Maistre Nicolas Chuquet, parisien, d'après le manuscrit fonds français, n°. 1346 de la Bibliothèque nationale de Paris," in *Bullettino di bibliografia e di storia delle scienze matematiche e fisiche,* **13** (1880), 593–659, 693–814, preceded by Marre's "Notice sur Nicolas Chuquet et son Triparty en la science des nombres," pp. 555–592; the appendix to the "Triparty" is *ibid.,* **14** (1881), 413–416, with extracts from the MS on pp. 417–460. Numerous passages from Chuquet are plagiarized in *Larismethique novellement composée par Maistre Estienne de la Roche dict Villefranche natif de Lyon sur le Rosne . . .* (Lyons, 1520, 1538).

Works comparable in certain ways to that of Chuquet are Bibliothèque Nationale, Fonds Français, nouvelles acquisitions no. 4140, a fifteenth-century arithmetic treatise in the Romance dialect of the Foix region; and Bibliothèque Ste. Geneviève, Paris, MS fr. no. 3143. On the latter, see Lynn Thorndike, "The Arithmetic of Jehan Adam, 1475 A.D.," in *The American Mathematical Monthly,* **33** (1926), 24–28; and *Isis,* **9,** no. 29 (Feb. 1927), 155. Chuquet's notations are discussed in Florian Cajori, *A History of Mathematical Notations,* 2 vols. (Chicago, 1928–1929). Studies on Chuquet are Moritz Cantor, *Vorlesungen über Geschichte der Mathematik,* II (Leipzig, 1892), pt. 1, ch. 58; and Charles Lambo, "Une algèbre française de 1484, Nicolas Chuquet," in *Revue des questions scientifiques,* **2** (1902), 442–472.

Jean Itard

CHWISTEK, LEON

CHWISTEK, LEON (*b.* Zakopane, Poland, 13 January 1884; *d.* Berwisza, near Moscow, U.S.S.R., 20 August 1944), *philosophy, logic, aesthetics.*

An acute thinker who was strongly opposed to metaphysics and idealistic philosophy, Chwistek was professor of logic at the University of Lvov from 1930 to 1940, when he took refuge in the Soviet Union. In 1921 he published his theory of the plurality of realities. Rejecting the idea of one reality, he distinguished four main concepts of reality: natural reality, physical reality, reality of sensation, and reality of images. These concepts should not be confused, and each has its proper sphere of application. He used this theory to classify movements and styles in art. He maintained that aesthetic evaluation should be based not on reality but on form. From 1919 to 1920 he was coeditor of the periodical *Formiści.*

Under the influence of Poincaré, Chwistek developed a strictly nominalistic attitude toward science, particularly logic and mathematics. In 1921 he observed that for the removal of Russell's paradox in the theory of classes, the simplified theory of types suffices. Dissatisfied with Russell's foundation of mathematics, in which he rejected such idealistic elements as the axiom of reducibility in the theory of types, Chwistek proposed his theory of constructive types in 1924. His main contribution was the foundation of logic and mathematics on his system of rational semantics.

Rational semantics is a system of expressions constructed from the symbols $*$ and c according to two rules: (1) c is an expression and (2) if E and F are expressions, then $*EF$ is an expression. The role of everyday language is restricted to the use of (*a*) E is an expression, (*b*) E is a theorem, (*c*) If X, then Y, in cases where E denotes an expression and X and Y denote propositions of form (*a*), (*b*), or (*c*).

Integers $.0L$, $.1L$, $.2L$, $\cdots$ of type L (an expression) occur as expressions $*LL$, $**LL*L$, $***LL*LL**LL*LL$, $\cdots$. The expression $*cc$ is abbreviated as 0. With the help of the expressions $*.0L.1L$ and $*.1L.0L$, abbreviated as $.IL$ and $.IIL$, respectively, the fundamental pattern of Sheffer's stroke $/EF$ is introduced as $*.IE.IIF$. Essential is the fundamental pattern of substitution $(EFGH)[L]$, an expression of the form $****.IL.0E.0F.0G.0H$. It reads: H is the result of substituting G for F in E. The type is indicated by L.

From the expressions theorems are singled out by certain rules, e.g., by $(0*EF\,G\,0)\,[c]$ is a theorem and $(EEFF)\,[c]$ is a theorem. Variables, quantification, and the axioms of logic are introduced by patterns. Quantification is always over constructed expressions. There results a collection of systems $[MN]$ described by $= [MN]E$, short for $**.IM.0E.IIN$, where M is an integer greater than N and E an expression. For instance, the fact that $.01$ is an expression of type 1 in system $[20]$ is reflected by the definite expression $= [20]$ Expr $[1]$ $.01$.

Chwistek succeeded in constructing a theory of classes based on types. The members of a class are of a higher type than the class, and there is no highest type. Chwistek and Hetper reconstructed the arithmetic of natural and rational numbers and considered the possibility of developing analysis. For a detailed and overall picture of Chwistek's scientific activity see his posthumous *Limits of Science* (1948). His ideas have been taken up and carried further by J. Myhill.

BIBLIOGRAPHY

I. Original Works. Chwistek's writings are "Antynomej logiki formalnej" ("Antinomies of Formal Logic"), in *Przegląd filozoficzny,* **24** (1921), 164–171; *Wielość rceczywistości* ("The Plurality of Realities"; Cracow, 1921); "Über die Antinomien der Prinzipien der Mathematik," in *Mathematische Zeitschrift,* **14** (1922), 236–243; "Zastowanie metody konstrukcyjnej do teorji poznania" ("The Application of the Constructive Method in the Theory of Knowledge"), in *Przegląd filozoficzny,* **26** (1923), 175–187, and **27** (1927), 296–298; "The Theory of Constructive Types. Principles of Logic and Mathematics," in *Annales de la Société polonaise de mathématique,* **2** (1924), 9–48, and **3** (1925), 92–141; "Pluralité des réalités," in *Atti del V Congresso internazionale di filosofia* (Naples, 1925), pp. 19–24; "Neue Grundlagen der Logik und Mathematik," in *Mathematische Zeitschrift,* **30** (1929), 704–724; "Zweite Mitteilung," *ibid.,* **34** (1932), 527–534; "Die nominalistische Grundlegung der Mathematik," in *Erkenntnis,* **3** (1932/1933), 367–388; "Fondements de la métamathématique rationnelle," in *Bulletin de l'Académie polonaise des sciences et des lettres,* Classe des sciences mathématiques et naturelles, ser. A (1933), 253–264, written with W. Hetper and J. Herzberg; *Granice nauki* ("The Limits of Science"; Lvov–Warsaw, 1935); "New Foundations of Formal Metamathematics," in *Journal of Symbolic Logic,* **3** (1938), 1–36, written with W. Hetper; "A Formal Proof of Gödel's Theorem," *ibid.,* **4** (1939), 61–68; and *The Limits of Science,* Helen C. Brodie, ed. (London, 1948).

II. Secondary Literature. On Chwistek or his work, see A. A. Fraenkel, *Abstract Set Theory* (Amsterdam, 1953); A. A. Fraenkel and Y. Bar-Hillel, *Foundations of Set Theory* (Amsterdam, 1958); and J. Myhill, review of *The Limits of Science,* in *Journal of Symbolic Logic,* **14** (1949), 119–125; "Report on Some Investigations Concerning the Consistency of the Axiom of Reducibility," *ibid.,* **16** (1951), 35–42; and "Towards a Consistent Set-Theory," *ibid.,* 130–136.

B. van Rootselaar

CIAMICIAN, GIACOMO LUIGI (*b.* Trieste, Italy, 27 August 1857; *d.* Bologna, Italy, 2 January 1922), *chemistry.*

Ciamician was of Armenian descent and studied at Vienna and Giessen, where he received the Ph.D. in 1880. He became Cannizzaro's assistant at the University of Rome (1880) and professor of general chemistry at Padua (1887) and then at Bologna (1889). The excellence and importance of his work was such that Emil Fischer several times proposed him for the Nobel Prize.

Ciamician's earliest researches (1877–1880), conducted while he was still a student in Vienna, were in spectroscopy. His first organic chemical studies were on the components of natural resins. In 1880 he began a lengthy investigation of the chemistry of pyrrole and related compounds that resulted in eighty papers and lasted until 1905. He established the nature of pyrrole as a secondary amine and clarified its ring structure by synthesizing it from succinimide. Ciamician prepared many derivatives of pyrrole, pyrroline, and pyrrolidine, including the synthesis of iodol (tetraiodopyrrole), which proved to have therapeutic use as a substitute for iodoform.

Ciamician was one of the founders of photochemistry, making the first systematic study of the behavior of organic compounds toward light. Between 1900 and 1915 Ciamician and his assistant Paolo Silber published fifty photochemical papers. Among the many photochemical reactions that he discovered were the reciprocal oxidation-reduction of alcohols and carbonyl compounds, the hydrolysis of cyclic ketones into fatty acids and unsaturated aldehydes, the condensation of hydrocyanic acid with carbonyl compounds, the polymerization of unsaturated compounds, and numerous photochemical syntheses. Ciamician was also interested in the applications of photochemistry and discussed the potentialities of the utilization of solar energy in desert regions and in the large-scale photochemical syntheses of valuable plant substances.

Another important area of Ciamician's investigation was plant chemistry. From 1888 to 1899 he determined the constitution of several essential oils: eugenol, safrole, and apiole from the oils of cloves, sassafras, and parsley and celery, respectively. During his photochemical studies he became convinced that the future of organic chemistry lay in its application to biology. From 1908 to 1922 Ciamician, in collaboration with Ciro Ravenna, professor of agricultural chemistry at Pisa, published twenty-one papers on the origin and function of organic substances in plants. They achieved the synthesis of glycosides in plants by inoculating them with the proper raw materials. Injection of plants with amino acids stimulated the production of alkaloids. The two chemists made many studies of the influence of organic compounds on the development of plants. They found that inoculation of plants with alkaloids such as caffeine and theobromine increased the activity of chlorophyll and led to the overproduction of starch. Ciamician and Ravenna concluded that alkaloids were not excretory products of plants but had a function similar to that of hormones in animals.

BIBLIOGRAPHY

I. Original Works. Ciamician's published writings include two books, *I problemi chimici del nuovo secolo* (Bologna, 1903; 2nd ed., Bologna, 1905) and *La chimica organica negli organismi* (Bologna, 1908). He collected the results of his investigations on the chemistry of pyrrole in "Il pirrolo e i suoi derivate," in *Atti della Reale Accademia dei Lincei. Memorie, Classe di scienze fisiche, mathematiche e naturali,* **4** (1887), 274–377. A later summary, "Über die Entwicklung der Chemie des Pyrrols im letzten Vierteljahrhundert," is in *Berichte der Deutschen chemischen Gesellschaft,* **37** (1904), 4200–4255. Most of the photochemical papers appeared in *Berichte . . .,* under the title

"Chemische Lichtwirkungen," between 1901 and 1915. There are two convenient reviews of this work: "Sur les actions chimiques de la lumière," in *Bulletin de la Société chimique de France,* 4th ser., **3-4** (1908), i–xxvii, and "Actions chimiques de la lumière," in *Annales de chimie et de physique,* 8th ser., **16** (1909), 474–520, written with Paolo Silber. An address delivered in New York on the potential uses of photochemistry, "La photochimica dell'avenire," appears in both Italian and English in *Transactions. Eighth International Congress of Applied Chemistry* (Washington, D.C.), **28** (1912), 135-168. For his work on essential oils see "Studi sui principii aromatici dell' essenza di sedano," in *Gazzetta chimica italiana,* **28**, pt. 1 (1898), 438–481, written with Paolo Silber. Many of the papers on the syntheses and functions of organic substances in plants are in *Gazzetta chimica italiana* from 1917 to 1922. A summary of the work with alkaloids, "La genèse de alcaloïdes dans les plantes," is in *Annales de chimie et de physique,* 8th ser., **25** (1912), 404–421, written with Ciro Ravenna.

II. SECONDARY LITERATURE. Among the many Italian studies on the life and work of Ciamician are Giuseppi Bruni, "Giacomo Ciamician," in *Rendiconti delle sessioni della Reale Accademia della scienze dell'Istituto di Bologna,* **27** (1922–1923), supp.; Luigi Mascarelli, *Giacomo Ciamician* (Turin, 1922); Giuseppe Plancher, "Giacomo Ciamician," in *Gazzetta chimica italiana,* **54** (1924), 3–22; and Guido Timeus, *In memoria di Giacomo Ciamician* (Trieste, 1925). There are several informative essays in other languages, including René Fabré, "Notice sur Giacomo Ciamician," in *Bulletin de la Société chimique de France,* 4th ser., **41** (1927), 1562–1566, trans. by Eduard Farber in Eduard Farber, ed., *Great Chemists* (New York, 1961), pp. 1085–1092; Paul Jacobson, "Giacomo Ciamician," in *Berichte der Deutschen chemischen Gesellschaft,* **55A** (1922), 19–20; William McPherson, "Giacomo Ciamician 1857-1922," in *Journal of the American Chemical Society,* **44** (1922), 101–106; Raffaelo Nasini, "Giacomo Luigi Ciamician," in *Journal of the Chemical Society* (London), **129** (1926), 996–1004; and T. E. Thorpe, "Prof. Giacomo Ciamician," in *Nature,* **109** (1922), 245–246.

ALBERT B. COSTA

CIRUELO, PEDRO, *also known as* **Pedro Sánchez Ciruelo** (*b.* Daroca, Spain, 1470; *d.* Salamanca [?], Spain, 1554), *mathematics, logic, natural philosophy.*

Ciruelo learned logic and arts at Salamanca during the latter part of the fifteenth century, then proceeded to the University of Paris to complete his education. He studied theology there from 1492 to 1502, supporting himself by "the profession of the mathematical arts." In 1495 he published at Paris a treatise on practical arithmetic, *Tractatus arithmeticae practicae,* that went through many subsequent printings; in the same year he also published revised and corrected editions of Bradwardine's *Arithmetica speculativa* and *Geometria speculativa* that enjoyed a similar success. Dating from this same period are Ciruelo's

editions of the *Sphere* of Sacrobosco, including the questions of Pierre d'Ailly (Élie cites editions of 1494, 1498, and 1515; Villoslada mentions others of 1499, 1505, and 1508 under a modified title). By 1502 Ciruelo was teaching mathematics at Paris; presumably he continued this career until about 1515, when he returned to Spain, attracted to the newly founded University of Alcalá.

At Alcalá, Ciruelo taught the theology of Aquinas; among his students was the young Domingo de Soto. Ciruelo maintained an interest in mathematics and philosophy, however, and in 1516 published his *Cursus quatuor mathematicarum artium liberalium;* this included a paraphrase of Boethius' *Arithmetica,* "more clearly and carefully edited than that of Thomas Bradwardine"; a brief compendium of Bradwardine's geometry, "with some additions"; another short compendium of John Peckham's *Perspectiva communis,* "to which also have been added a few glosses"; a treatise on music; and two short pieces on squaring the circle, both of which he recognized as defective. The *Cursus* appeared at both Paris and Alcalá, being reprinted at Alcalá in 1518 and in both places in 1523, 1526, and 1528. At Alcalá, Ciruelo also published his edition of the *Apotelesmata astrologiae christianae* in 1521 and a new edition of the *Sphere* in 1526. He transferred to the University of Salamanca some time after this; his first work to be published there was a *Summulae* (1537) that is a more mature treatment of logic than his *Prima pars logices* (Alcalá, 1519). Also published at Salamanca was his *Paradoxae quaestiones* (1538), two questions on logic, three on physics, and five on metaphysics and theology; these contain his somewhat singular views on gravity and impetus and his criticisms of the Jewish cabala, particularly as evaluated by Giovanni Pico della Mirandola.

BIBLIOGRAPHY

Ciruelo's works are listed and appraised in fragmentary fashion in the following: Vicente Muñoz Delgado, O.M., "La lógica en Salamanca durante la primera mitad del siglo XVI," in *Salmanticensis,* **14** (1967), 171–207, esp. 196–198 (logic); Hubert Élie, "Quelques maîtres de l'université de Paris vers l'an 1500," in *Archives d'histoire doctrinale et littéraire du moyen âge,* **18** (1950–1951). 193–243, esp. 236–237 (general); J. Rey Pastor, *Los matemáticos españoles del siglo XVI* (Toledo, 1926), pp. 54–61 (mathematics); David E. Smith, *Rara arithmetica* (Boston–London, 1908), see index; Lynn Thorndike, *The Sphere of Sacrobosco and Its Commentators* (Chicago, 1949), see index; and R. G. Villoslada, S. J., *La universidad de Paris durante los estudios de Francisco de Vitoria (1507–1522),* Analecta Gregoriana, no. 14 (Rome, 1938), pp. 402–404, *passim* (general).

WILLIAM A. WALLACE, O.P.

CLAIRAUT, ALEXIS-CLAUDE (*b.* Paris, France, 7 May 1713; *d.* Paris, 17 May 1765), *mathematics, mechanics, celestial mechanics, geodesy, optics.*

Clairaut's father, Jean-Baptiste Clairaut, was a mathematics teacher in Paris and a corresponding member of the Berlin Academy. His mother, Catherine Petit, bore some twenty children, few of whom survived. Of those who did, two boys were educated entirely within the confines of the family and showed themselves to be notably precocious children. The younger died around 1732 at the age of sixteen, however.

Alexis-Claude would probably have learned the alphabet from the figures in Euclid's *Elements.* When he was nine years old his father had him study Guisnée's *Application de l'algèbre à la géométrie.* Guisnée, who had studied under Varignon, taught mathematics to several important people, notably Pierre-Rémond de Montmort, Réaumur, and Maupertuis. His work, which was subsidized by Montmort, is a good introduction to the pioneering mathematics of that era: analytical geometry and infinitesimal calculus.

At the age of ten Clairaut delved into L'Hospital's posthumous *Traité analytique des sections coniques* and his *Analyse des infiniment petits,* which was based on Johann Bernoulli's lessons. Clairaut was barely twelve when he read before the Académie des Sciences "Quatre problèmes sur de nouvelles courbes," later published in the *Miscellanea Berolinensia.*

Around 1726 the young Clairaut, together with Jean Paul de Gua de Malves, who was barely fourteen years old; Jean Paul Grandjean de Fouchy, nineteen; Charles Marie de la Condamine, twenty-five; Jean-Antoine Nollet, twenty-six; and others founded the Société des Arts. Even though this learned society survived for only a few years, it was nonetheless a training ground for future members of the Académie des Sciences.

Around this time also Clairaut began his research on gauche curves. This work culminated in 1729 in a treatise (published in 1731) that led to his election to the Academy. The Academy proposed his election to the Crown on 4 September 1729, but it was not confirmed by the king until 11 July 1731, at which time he was still only eighteen.

In the Academy, Clairaut became interested in geodesy through Cassini's work on the measurement of the meridian. He allied himself with Maupertuis and the small but youthful and pugnacious group supporting Newton. It is difficult, however, to specify the moment at which he turned toward this new area of physics, for which his mathematical studies had so well prepared him.

He became a close friend of Maupertuis and was much in the company of the marquise du Châtelet and Voltaire. During the fall of 1734 Maupertuis and Clairaut spent a few months with Johann Bernoulli in Basel, and in 1735 they retired to Mont Valérien, near Paris, to concentrate on their studies in a calm atmosphere.

On 20 April 1736 Clairaut left Paris for Lapland, where he was to measure a meridian arc of one degree inside the arctic circle. Maupertuis was director of the expedition, which included Le Monnier, Camus, the Abbé Outhier, and Celsius. This enthusiastic group accomplished its mission quickly and precisely, in an atmosphere of youthful gaiety for which some reproached them. On 20 August 1737 Clairaut was back in Paris.

His work turned increasingly toward celestial mechanics, and he published several studies annually in *Mémoires de l'Académie des sciences.* From 1734 until his death, he also contributed to the *Journal des sçavans.* Clairaut guided the marquise du Châtelet in her studies, especially in her translation of Newton's *Principia* and in preparing the accompanying explanations. Even though the work was so sufficiently advanced in 1745 that a grant could be applied for and awarded, it did not appear until 1756, seven years after the marquise's death.

This translation of the *Principia* is elegant and, on the whole, very faithful to the original. It is, furthermore, most valuable because of its second volume, in which an abridged explanation of the system of the world is found. It illustrates, summarizes, and completes, on certain points, the results found in the *Principia,* all in about 100 pages. There is also an "analytical solution of the principal problems concerning the system of the world." Clairaut's contribution to this volume is a fundamental one, even greater than that to the translation. Many of his works are used in it. For example, in the section devoted to an explanation of light refraction there is a condensed version of his "Sur les explications cartésienne et newtonienne de la réfraction de la lumière," and the approximately sixty pages devoted to the shape of the earth constitute a very clear résumé of the *Figure de la terre.*

Clairaut's contribution to science, however, lies mainly in his own works: *Théorie de la figure de la terre* (1743), *Théorie de la lune* (1752), *Tables de la lune* (1754), *Théorie du mouvement des comètes* (1760 [?]), and *Recherches sur la comète* (1762). When we add to these works on celestial mechanics the two didactic works *Élémens de géométrie* (1741) and *Élémens d'algèbre* (1746) and the controversies that arose around these works, we have an idea of the intensity of Clairaut's effort.

Vivacious by nature, attractive, of average height but well built, Clairaut was successful in society and, it appears, with women. He remained unmarried. He was a member of the Royal Society and the academies of Berlin, St. Petersburg, Bologna, and Uppsala. Clairaut maintained an almost continuous correspondence with Gabriel Cramer and Euler and with six corresponding members of the Académie des Sciences: Samuel Koenig, whom he had met in Basel during his visit to Johann Bernoulli; François Jacquier and Thomas Le Seur, publishers of and commentators on the *Principia;* Samuel Klingenstierna, whom he had met during his trip to Lapland; Robert-Xavier Ansart du Petit Vendin; and the astronomer Augustin d'Arquier de Pellepoix.

In 1765 Clairaut wrote, concerning his theory of the moon: "A few years ago I began the tedious job of redoing all the calculations with more uniformity and rigor, and perhaps some day I shall have the courage to complete it. . . ." He was lacking not in courage, but in time; he died that year, at fifty-two, following a brief illness.

Although the expression "double-curvature curve" (gauche curve) is attributed to Henri Pitot (1724), Clairaut's treatise on this type of figure is no less original, representing the first serious analytical study of it. The curve is determined by two equations among the three orthogonal coordinates of its "point courant" (locus). Assimilation of the infinitesimal arc to a segment of a straight line permits determination of the tangent and the perpendiculars. It also permits rectification of the curve. This work also includes quadratures and the generation of some special gauche curves. "Sur les courbes que l'on forme en coupant une surface courbe quelconque par un plan donné de position" (1731) deals with plane sections of surfaces. The essential implement in this is a change in the reference point of the coordinates. Clairaut used this device mainly for explaining Newton's enumeration of curves of the third degree and the generation of all plane cubics through the central projection of one of the five divergent parabolas.

Clairaut's "Sur quelques questions de maximis et minimis" (1733) was noteworthy in the history of the calculus of variations. It was written, as were all his similar works, in the style of the Bernoulli brothers, by likening the infinitesimal arc of the curve of three elementary rectilinear segments. In the same vein was the memoir "Détermination géométrique . . . à la méridienne . . ." (1733), in which Clairaut made an elegant study of the geodesics of quadrics of rotation. It includes the property already pointed out by Johann Bernoulli: the osculating plane of the geodesic is normal to the surface.

What we call "Clairaut's differential equations" first appeared in the *Mémoires de l'Académie des sciences* for 1734. They are solved by differentiation and, in addition to the general integral, also allow for a singular solution. Brook Taylor had made this discovery in 1715.

In his research on integral calculus (1739–1740) Clairaut showed that for partial derivatives of the mixed second order, the order of differentiation is unimportant; and he established the existence of an integrating factor for linear differential equations of the first order. This factor had appeared in 1687, in the work of Fatio de Duillier. A lesson given by Johann Bernoulli to L'Hospital on this subject was summarized by Reyneau, in 1708. Euler also dealt with this subject at the same time as Clairaut.

As a result of the experiments of Dortous de Mairan, Clairaut proved in 1735 that slight pendulum oscillations remain isochronous, even if they do not occur in the same plane. His most significant paper on mechanics, however, is probably "Sur quelques principes . . . de dynamique" (1742), on the relative movement and the dynamics of a body in motion. Even though he did not clarify Coriolis's concept of acceleration, he did at least indicate a method of attacking the problem.

In 1743 the *Théorie de la figure de la terre* appeared. It was to some degree the theoretical epilogue to the Lapland expedition and to the series of polemics on the earth's shape—an oblate ellipsoid, according to Newton and Huygens, and a prolate ellipsoid, according to the Cassinis. Newton had merely outlined a proof; and Maclaurin, in his work on tides, which was awarded a prize by the Académie des Sciences in 1740, had set forth a general principle of hydrostatics that allowed this proof to be improved upon. Clairaut gave this principle an aspect that he felt was more general but that d'Alembert later showed to be the same as Maclaurin's. He drew from it an analytical theory that since George Green (1828) has been called the theory of potential. It allowed him to study the shape of the stars in more general cases than those examined by his predecessors. Clairaut no longer assumed that the fluid composing a star was homogeneous, and he considered various possible laws of gravitation. It was also in this work, considered a scientific classic, that the formula named for Clairaut, expressing the earth's gravity as a function of latitude, is found.

It was through the various works submitted for the 1740 competition on the theory of tides, and this work of Clairaut's, that Newton's theory of gravitation finally won acceptance in French scientific circles.

In 1743 Clairaut read before the Academy a paper

entitled "L'orbite de la lune dans le système de M. Newton." Newton was not fully aware of the movement of the moon's apogee, and therefore the problem had to be reexamined in greater detail. However, Clairaut—and d'Alembert, and Euler, who were also working on this question—found only half of the observed movement in their calculations.

It was then that Clairaut suggested completing Newton's law of attraction by adding a term inversely proportional to the fourth power of the distance. This correction of the law elicited a spirited reaction from Buffon, who opposed this modification with metaphysical considerations on the simplicity of the laws of nature. Clairaut, more positive and more a pure mathematician, wanted to stick to calculations and observations. The controversy that arose between these two academicians appears in the *Mémoires* of the Academy for 1745 (published long afterward). Nevertheless, the minimal value of the term added soon made Clairaut think that the correction—all things considered—could apply to the calculations but not to the law. While the latter-day Cartesians were delighted to see Newtonianism at bay, Clairaut found toward the end of 1748, through a consideration that was difficult to hold in suspicion, given the state of mathematical analysis at the time, that in Newton's theory the apogee of the moon moved over a time period very close to that called for by observations. This is what he declared to the Academy on 17 May 1749.

He wrote to Cramer around that time:

> Messrs. d'Alembert and Euler had no inkling of the strategem that led me to my new results. The latter twice wrote to tell me that he had made fruitless efforts to find the same thing as I, and that he begged me to tell him how I arrived at them. I told him, more or less, what it was all about [Speziali, in *Revue d'histoire des science et de leurs applications,* p. 227, letter of 26 July 1749].

This first approximate resolution of the three-body problem in celestial mechanics culminated in the publication of the *Théorie de la lune* in 1752 and the *Tables de la lune* in 1754.

The controversies arose anew when Clairaut, strengthened by his first success, turned to the movement of comets.

In 1705 Edmund Halley had announced that the comet observed in 1552, 1607, and 1682 would reappear in 1758 or 1759. He attributed the disparity in its period of appearance to perturbations caused by Jupiter and Saturn. The attention of astronomers was therefore drawn toward this new passage, and Clairaut found in it a new field of activity. The conditions favorable, in the case of the moon, to solving the three-body problem did not exist here. The analysis had to be much more precise. It was essentially a question of calculating the perturbations brought about by the attractions of Jupiter and Saturn, in the movement of the comet. Clairaut thus began a race against time, attempting to calculate as accurately as possible the comet's passage to perihelion before it occurred. For purely astronomical calculations such as those of the positions of the two main planets, he enlisted the services of Lalande, who was assisted by a remarkable woman, Nicole-Reine Étable de Labrière Lepaute.

These feverishly completed calculations resulted in Clairaut's announcement, at the opening session of the Academy in November 1758, that passage to perihelion would occur about 15 April 1759. "You can see," he said, "the caution with which I make such an announcement, since so many small quantities that must be neglected in methods of approximation can change the time by a month."

The actual passage to perihelion took place on 13 March. By reducing his approximations even more, Clairaut calculated the date as 4 April. Then, in a paper awarded a prize by the Academy of St. Petersburg in 1762, and through use of a different method, he arrived at the last day of March as the date of passage. It was difficult to do any better. Nevertheless, there were arguments, particularly from d'Alembert.

Several anonymous articles appeared in 1759 in the *Observateur littéraire,* the *Mercure,* and the *Journal encyclopédique.* The latter, which Clairaut attributed to d'Alembert, was actually by Le Monnier. Clairaut offered a general reply in the November issue of the *Journal des sçavans.* "If," he stated, "one wishes to establish that Messrs. d'Alembert et al. can solve the three-body problem in the case of comets as well as I did, I would be delighted to let him do so. But this is a problem that has not been solved before, in either theory or practice."

In the fourteenth memoir of his *Opuscules mathématiques* (1761), d'Alembert attacked Clairaut ruthlessly. In December of that year Clairaut replied in the *Journal des sçavans:* "Since M. d'Alembert does not have the patience to calculate accurate tables, he should let alone those who have undertaken to do so." D'Alembert replied in the February 1762 issue of the *Journal encyclopédique,* and Clairaut put an end to the debate in the June issue of the *Journal des sçavans* by congratulating his adversary for his research on the comet "since its return."

In 1727 James Bradley made public his discovery of the aberration of light, and the Lapland expedition scrupulously took this phenomenon into consideration in the determination of latitudes. Bradley had not given any theoretical proof, however; and in a paper written in 1737, immediately following his return from the expedition, Clairaut offered the neglected proofs. He dealt with this question several times again; and in 1746 he indicated how to make corrections for the aberration of planets, comets, and satellites.

In 1739 Clairaut became interested in the problem of refraction in Newton's corpuscular theory of light. The theory had just been presented in France by Voltaire (probably inspired by Clairaut) in his *Élémens de philosophie de Newton*. It was the first time that Newton's theory of light had been presented in its entirety by a member of the Académie des Sciences.

The unequal refrangibility of light rays made the astronomical (refracting) telescope imperfect, and Newton had therefore preferred the reflecting telescope. In 1747 Euler suggested making up two-component glass object lenses with water enclosed between them. On the strength of Newton's experiments the English optician John Dollond rejected that solution, and Clairaut supported him. But in 1755 Klingenstierna sent a report to Dollond that led the latter to redo the Newtonian experiments and to go back to his original opinion. Thus, he used object lenses made up of two glasses having different optical qualities. His technique, however, still remained mysterious. Clairaut, thinking that it would be expedient to provide a complete theory on the question, undertook precise experiments in which he was assisted by the optician George and his colleague Tournière. In April 1761 and June 1762 Clairaut presented to the Académie des Sciences three mémoires under the title "Sur les moyens de perfectionner les lunettes d'approche par l'usage d'objectifs composés de plusieurs matières différemment réfringentes."

Clairaut's intention to publish a technical work on this subject for craftsmen was interrupted by his death. Shortly afterward d'Alembert published three extensive studies on the achromatic telescope. The Academy of St. Petersburg had made this question the subject of a competition in 1762, and the prize was won by Klingenstierna. Clairaut translated Klingenstierna's paper into French and had it published in the *Journal des sçavans* in October and November of that year.

Clairaut's mathematical education, conforming so little to university traditions, influenced his ideas on pedagogy. The parallel with Blaise Pascal is obvious. However, we know Pascal's ideas only through a few pages preserved by Leibniz and through what survives in the Port Royal *Géométrie*. As for Clairaut, he left us two remarkable didactic works.

In geometry Euclid's authority had undergone its first serious attack in France in the sixteenth century by Petrus Ramus. The influence of this reformer, however, was felt most in the Rhineland universities. Criticism of Euclid's geometric concepts reappeared in the following century in the Port Royal *Logique* and gave rise to the Port Royal *Géométrie*. At that point one could observe the appearance in mathematical teaching of a more liberal and intuitionist school (alongside a dogmatic school and in contrast with it) represented especially by the teachings of the Oratorians.

Clairaut's *Élémens de géométrie* is linked to the more liberal school. He wanted geometry to be rediscovered by beginners and therefore put together, for himself and for them, a very optimistic history of mathematical thought that reflected the concepts of his contemporaries Diderot and Rousseau.

> I intended to go back to what might have given rise to geometry; and I attempted to develop its principles by a method natural enough so that one might assume it to be the same as that of geometry's first inventors, attempting only to avoid any false steps that they might have had to take [*Élémens de géométrie,* preface].

He found that the measurement of land would have been the most suitable area to give rise to the first propositions. Through analogy, he went from there to more advanced and more abstract research, all the while resolving problems without setting forth theorems and admitting as obvious all truths that showed themselves to be self-evident.

This work is pleasant and has definite pedagogical value, but it remains on a very elementary level. The subject matter corresponds to books 1–4, 6, 11, and 12 of Euclid.

The *Élémens d'algèbre* was conceived in the same spirit. Clairaut again endeavored to follow a pseudo-historical path, or at least a "natural" one. First he deals with elementary problems and appeals to common sense alone. Then, gradually increasing the difficulty, he brings forth the necessity of an algebraic symbolism and technique. Much more scholarly than his geometry, his algebra extends all the way to the solution of fourth-degree equations.

The *Élémens d'algèbre* influenced the instructional technique of the *écoles centrales* of the Revolution. S. F. Lacroix, who republished and commented upon it, considered Clairaut to be the "first who, in blazing a philosophical path, shed a bright light on the principles of algebra."

However, the algebra of Bézout, which represented the dogmatic tendency in this field, frequently was victorious over Clairaut's. Euler's *Algebra* (1769) was

in no way inferior to Clairaut's from the pedagogical point of view and greatly surpassed it from the scientific point of view. For example, it is very rich in material concerning the theory of numbers, an aspect of mathematics completely foreign to Clairaut's research and mathematical concepts.

As he says in the preface to the *Élémens d'algèbre*, Clairaut wanted to complete his didactic works with an application of algebra to geometry in which, among other things, he would have been able to study conic sections. Nothing more of this work seems to exist, and it may not have been undertaken. However, there is a manuscript entitled "Premières notions sur les mathématiques à l'usage des enfants," sent by Diderot to Catherine II of Russia as a work by Clairaut.

BIBLIOGRAPHY

I. ORIGINAL WORKS. Clairaut's books are *Recherches sur les courbes à double courbure* (Paris, 1731); *Élémens de géométrie* (Paris, 1741; 5th ed., 1830; 11th ed., 2 vols., 1920), trans. into Swedish (Stockholm, 1744–1760), German (Hamburg, 1790), and English (London, 1881); *Théorie de la figure de la terre tirée des principes de l'hydrostatique* (Paris, 1743, 1808), trans. into German as Ostwald's Klassiker no. 189 (Leipzig, 1903), Italian (Bologna, 1928), and Russian (Leningrad, 1947); *Élémens d'algèbre* (Paris, 1746; 4th ed., 1768; 2 vols., 1797, 1801), trans. into German (Berlin, 1752) and Dutch (Amsterdam, 1760); *Théorie de la lune déduite du seul principe de l'attraction réciproquement proportionnelle aux quarrés des distances* (St. Petersburg, 1752), which won the prize offered by the Royal Academy of St. Petersburg in 1750, 2nd ed. (Paris, 1765) includes "Tables de la lune construites sur une nouvelle révision de toutes les espèces de calculs dont leurs équations dépendent"; *Tables de la lune, calculées suivant la théorie de la gravitation universelle* (Paris, 1754); *Réponse de Mr Clairaut à quelques pièces, la plupart anonymes, dans lesquelles on a attaqué le Mémoire sur la comète de 1682, lu le 14 novembre 1758* (Paris, 1759); *Théorie du mouvement des comètes, dans laquelle on a égard aux altérations que leurs orbites éprouvent par l'action des planètes . . .* (Paris, 1760 [?]); and *Recherches sur la comète des années 1531, 1607, 1682 et 1759, pour servir de supplément à la théorie, par laquelle on avait annoncé en 1758 le temps de retour de cette comète . . .* (St. Petersburg, 1762).

An exhaustive account of Clairaut's papers in the publications of learned societies is in the bibliography added by René Taton to Pierre Brunet's study on Clairaut (see below). This bibliography was taken up again and completed by Taton in *Revue d'histoire des sciences et des techniques*, **6**, no. 2 (April–June 1953), 161–168. Among Clairaut's papers are "Sur les courbes que l'on forme en coupant une surface courbe quelconque par un plan donné de position," in *Mémoires de l'Académie des sciences* (1731), 483–493; "Détermination géométrique de la perpendicu-laire à la méridienne, tracée par M. Cassini, avec plusieurs méthodes d'en tirer la grandeur et la figure de la terre," *ibid.* (1733), 406–416; "Sur quelques questions de maximis et minimis," *ibid.*, 186–194; "Quatre problèmes sur de nouvelles courbes," in *Miscellanea Berolinensia*, **4** (1734), 143–152; "Solution de plusieurs problèmes où il s'agit de trouver des courbes dont la propriété consiste dans une certaine relation entre leurs branches, exprimée par une équation donnée," in *Mémoires de l'Académie des sciences* (1734), 196–215; "Examen des différentes oscillations qu'un corps suspendu par un fil peut faire lorsqu'on lui donne une impulsion quelconque," *ibid.* (1735), 281–298; "De l'aberration apparente des étoiles causée par le mouvement progressif de la lumière," *ibid.* (1737), 205–227; "Investigationes aliquot, ex quibus probetur terrae figuram secundam leges attractionibus in ratione inversa quadrati distanciarum maxime ad ellipsin accedere debere," in *Philosophical Transactions of the Royal Society* (1737), 19–25; "Recherches sur la figure des sphéroïdes qui tournent autour de leur axe, comme la Terre, Jupiter et le Soleil, en supposant que leur densité varie du centre à la périphérie," *ibid.* (1738); "Sur les explications cartésienne et newtonienne de la réfraction de la lumière," in *Mémoires de l'Académie des sciences* (1739), 258–275; "Sur la manière la plus simple d'examiner si les étoiles ont une parallaxe et de la déterminer exactement," *ibid.*, 358–369; "Recherches générales sur le calcul intégral," *ibid.*, 425–436; "Sur l'intégration ou la construction des équations différentielles du premier ordre," *ibid.* (1740), 293–323; "Sur quelques principes qui donnent la solution d'un grand nombre de problèmes de dynamique," *ibid.* (1742), 1–52; "De l'orbite de la lune dans le système de M. Newton," *ibid.* (1743), 17–32; "De l'aberration de la lumière des planètes, des comètes et des satellites," *ibid.* (1746), 539–568; "De l'orbite de la lune en ne négligeant pas les quarrés des quantités de même ordre que les perturbations," *ibid.* (1748); "Construction des tables du mouvement horaire de la lune," *ibid.* (1752), 593–622; "Construction des tables de la parallaxe horizontale de la lune," *ibid.*, 142–160; "Sur l'orbite apparente du soleil autour de la terre, en ayant égard aux perturbations produites par les actions de la lune et des planètes principales," *ibid.* (1754), 521–564; "Sur les moyens de perfectionner les lunettes d'approche, par l'usage d'objectifs composés de plusieurs matières différemment réfringentes," *ibid.* (1756), 380–437; (1757), 524–550; (1762), 578–631; and "Mémoire sur la comète de 1759, dans lequel on donne les périodes qu'il est le plus à propos d'employer, en faisant usage des observations faites sur cette comète dans les quatre premières apparitions," *ibid.* (1759), 115–120.

Also of note are *Principes mathématiques de la philosophie naturelle*, 2 vols. (Paris, 1756; new printing, 1966), which he helped the marquise du Châtelet to translate and explain; and "Premières notions sur les mathématiques à l'usage des enfants," Paris, Bibliothèque de l'Institut, MS 2102, a copy of the MS sent to Catherine II by Diderot.

Some of Clairaut's correspondence has been published. That with Lesage is in P. Prevost, *Notice sur la vie et les écrits de G. L. Lesage* (Geneva, 1805), pp. 362–372; with

d'Alembert, in Charles Henry, "Correspondance inédite de d'Alembert avec Cramer, Lesage, Clairaut, Turgot, Castillon, Beguelin," in *Bollettino di bibliografia e di storia delle scienze matematiche e fisiche,* **18** (1886), 3–112; with Cramer, in Pierre Speziali, "Une correspondance inédite entre Clairaut et Cramer," in *Revue d'histoire des sciences et de leurs applications,* **8,** no. 3 (1955), 193–237. His correspondence with Euler is included in Euler's complete works (in press).

II. SECONDARY LITERATURE. On Clairaut or his work, see Jean Sylvestre Bailly, *Histoire de l'astronomie moderne,* III (Paris, 1785); Joseph Bertrand, "Clairaut, sa vie et ses travaux," in *Éloges académiques,* 2nd ser. (1902), 231–261; Charles Bossut, *Histoire générale des mathématiques depuis leur origine jusqu'à l'année 1808* (Paris, 1810), II, 346–350, 383–386, 418–433; Pierre Brunet, *La vie et l'oeuvre de Clairaut* (Paris, 1952); Moritz Cantor, *Vorlesungen über Geschichte der Mathematik,* III (Leipzig, 1901), 778–786; J. L. Coolidge, *A History of Geometrical Methods* (Oxford, 1940), pp. 135, 172, 321; René Dugas, *Histoire de la mécanique* (Paris–Neuchâtel, 1950), pp. 268–273, 354–357; "Éloge de Clairaut," in *Journal des sçavans* (1765); Jean Grandjean de Fouchy, "Éloge de Clairaut," in *Histoire de l'Académie des sciences* (1765), pp. 144–159; Lesley Hanks, *Buffon avant l' "Histoire naturelle"* (Paris, 1966), p. 296 and *passim;* J. E. Montucla, *Histoire des mathématiques,* 2nd ed. (Paris, 1802; photographically repro., 1960), III, IV; Niels Nielsen, *Géomètres français du XVIIIème siecle* (Copenhagen–Paris, 1935), pp. 132–153; J. C. Poggendorff, *Histoire de la physique* (Paris, 1883), pp. 464–482; René Taton, *L'oeuvre scientifique de Monge* (Paris, 1951), pp. 106–108 and *passim;* Félix Tisserand, *Traité de mécanique céleste,* II (Paris, 1891); and Isaac Todhunter, *History of the Theories of Attraction and the Figure of the Earth,* I (London, 1873).

Works by Clairaut's father are listed in Taton's article in *Revue d'histoire des sciences et des techniques,* **6,** no. 2 (1953), 167. Works by his younger brother are "Méthode de former tant de triangles qu'on voudra, de sorte que la somme des quarrés des deux côtés soit double, triple, etc. . . . du quarré de la base, d'où suivent des quadratures de quelques espèces de lunules" (read before the Academy in 1730), in *Journal des sçavans* (1730), 273f. and (1733), 279; and *Diverses quadratures circulaires, élliptiques et hyperboliques par M. Clairaut le cadet* (Paris, 1731).

JEAN ITARD

CLAISEN, LUDWIG (*b.* Cologne, Germany, 14 January 1851; *d.* Godesberg-am-Rhein, Germany, 5 January 1930), *chemistry.*

The son of Heinrich Wilhelm Claisen, a notary, and of the former Emilia Theresa Berghaus, Claisen had two brothers and a sister. After graduating in 1869 from the Gymnasium of the Apostles Church in Cologne, he enrolled in the University of Bonn, where he studied chemistry under Kekulé and physics under Clausius. His studies were interrupted by the Franco-

Prussian War, and Claisen enlisted as a medical corpsman. At Göttingen he attended Tollens' lectures on organic chemistry and worked in Wöhler's laboratory. He returned to Bonn in 1872 and received the doctorate there in 1874.

Claisen taught at Bonn from 1875 to 1882 and at Owens College in Manchester, England, from 1882 to 1885. During the next four years he worked in Baeyer's laboratory at Munich, leaving to teach at the Aachen Technische Hochschule from 1890 to 1897. He was associate professor at the University of Kiel from 1897 to 1904 and worked in Emil Fischer's laboratory in Berlin from 1904 to 1907. In 1907 Claisen established his own laboratory at Godesberg, continuing work in organic synthesis until 1926.

Claisen's early interest in condensation reactions led him to develop the so-called Claisen condensation for synthesizing keto esters and 1,3-diketones. He used sodium hydroxide, hydrogen chloride, sodium ethoxide, and sodamide as condensing agents and found that nitriles, as well as carbonyls, may be condensed. Claisen contributed greatly to the understanding of tautomerism when, in 1893, he reported the isolation of dibenzoyl acetone in two solid modifications, identifying the acidic form as *enol* and the neutral as *keto* and extending these terms to apply to all tautomers. In 1912 Claisen began work on the rearrangement of allyl aryl ethers into phenols. The mechanism of this rearrangement was the subject of the last (1926) of his 125 published papers.

Eulogizing Claisen, Richard Anschütz said that he was endowed with first-rate talent and that his discoveries placed him among the most renowned researchers in organic chemistry.

BIBLIOGRAPHY

For a complete list of Claisen's publications, see the Anschütz article (below), pp. 165–169.

For a carefully written biography, a discussion of Claisen's contributions to chemistry, and a list of his publications and patents, see Richard Anschütz, "Ludwig Claisen, ein Gedenkblatt," in *Berichte der Deutschen chemischen Gesellschaft,* **69** (1936), 97–170. See also Ernest E. Gorsline, *A Study of the Claisen Condensation* (Easton, Pa., 1908); and Ernest H. Huntress, "Centennials and Sesquicentennials During 1951 With Interest for Chemists and Physicists," in *Proceedings of the American Academy of Arts and Sciences,* **79** (1951), 10–11.

A. ALBERT BAKER, JR.

CLAPEYRON, BENOIT-PIERRE-ÉMILE (*b.* Paris, France, 26 February 1799; *d.* Paris, 28 January 1864), *civil engineering.*

Clapeyron graduated from the École Polytechnique in 1818 and then attended the École des Mines. In 1820 he and his friend and classmate Gabriel Lamé went to Russia, where they taught pure and applied science at the École des Travaux Publics in St. Petersburg and did construction work (1). While in Russia they published a number of papers in the *Journal des voies de communication de Saint-Pétersbourg,* the *Journal du génie civil,* and the *Bulletin Ferussac,* as well as various works that came out in France (2). They left, following the July Revolution of 1830, when their position became somewhat difficult because of their well-known liberal tendencies.

Upon returning to France, Clapeyron engaged in railroad engineering (3), specializing in the design and construction of steam locomotives. In 1836 he traveled to England to order some locomotives that would negotiate a particularly long continuous grade along the St.-Germain line. When the illustrious Robert Stephenson declined to undertake the commission because of its difficulty, the machines were built in the shops of Sharp and Roberts, according to Clapeyron's designs (4). He extended his activities to include the design of metal bridges, making notable contributions in this area (5).

Clapeyron was elected to the Academy of Sciences in 1848, replacing Cauchy, and served on numerous committees of the Academy, including that which awarded the prize in mechanics and those which investigated the project for piercing the Isthmus of Suez and the application of steam to naval uses.

Clapeyron had a continuing interest in steam-engine design and theory throughout his career. His most important research paper (6) dealt with regulation of the valves in a steam engine. From 1844 Clapeyron was a professor at the École des Ponts et Chaussées, where he taught the course on the steam engine.

Clapeyron is best known today for the relationship, which bears his name, between the temperature coefficient of the equilibrium vapor pressure over a liquid or solid and the heat of vaporization. This was an application of Sadi Carnot's principle, as developed by Carnot in his memoir *Réflexions sur la puissance motrice du feu* (1824). Carnot's work found hardly an echo among his contemporaries until 1834, when Clapeyron published a paper (7) that is a detailed exposition of the *Réflexions.* In it he transformed Carnot's verbal analysis into the symbolism of the calculus and represented the Carnot cycle graphically by means of the Watt indicator diagram, familiar to engineers. The paper also appeared in translation in England and Germany, so that despite the rarity of the original, Carnot's work was generally

available and associated with the name of Clapeyron. However, not only was Clapeyron's original paper ignored by the other engineers, but he himself made only one passing reference to it until the work of Kelvin and Clausius made its true significance generally known as the basis for the second law of thermodynamics (8).

NOTES AND BIBLIOGRAPHY

1. Toward the end of 1809 Emperor Alexander I created a corps of engineers that was to deal with highways and bridges as well as military engineering. He requested some engineers from the French government to provide a nucleus for this corps and to engage in instruction. This program was reinstituted after the Restoration. See article on Clapeyron in A. Fourcy, *Histoire de l'École polytechnique* (Paris, 1828).

2. G. Lamé and B. P. E. Clapeyron, *Mémoire sur la stabilité des voûtes avec un rapport de M. de Prony sur ce mémoire* (Paris, 1823); *Mémoire sur les chemins de fer considérés sous le point de vue de la défense des territoires* (Paris, 1832); and *Vues politiques et pratiques sur les travaux publics* (Paris, 1832), written with E. Flachat.

3. Clapeyron and Lamé entered the railroad business at an early stage. The period 1823–1833 saw several concessions to private companies; but except for some short lines radiating from St-Étienne, these were failures. In 1833, under pressure from a group of brilliant young followers of Saint-Simon, 500,000 francs were authorized for an engineering study of the problem, including the sending of engineers to England and the United States for study and observation. The engineering talent for this and the earlier assistance to the "concessionaires" was provided by the Corps des Mines and included Clapeyron and Lamé. Clapeyron conceived the idea of a railroad from Paris to St.-Germain; but while waiting for financing he went to St.-Étienne as professor at the École des Mineurs, where he taught the course in construction. In 1835, upon authorization of a line from Paris to St.-Germain, Clapeyron and Lamé were charged with direction of the work. Lamé left shortly thereafter to accept the chair of physics at the École Polytechnique.

4. B. P. E. Clapeyron, "Expériences faites sur le chemin de fer de Saint-Germain avec une nouvelle locomotive," in *Bulletin de la Société d'encouragement de l'industrie nationale,* **45** (1846), 413–414, and "Note sur une expérience faite le 17 Juin, 1846 au chemin de fer de Saint-Germain," in *Comptes rendus de l'Académie des sciences,* **22** (1846), 1058–1061.

5. B. P. E. Clapeyron, "Calcul d'une poutre élastique reposant librement sur des appuis inégalement espacés," in *Comptes rendus de l'Académie des sciences,* **45** (1857), 1076–1080, and "Mémoire sur le travail des forces élastiques dans un corps solide déformé par l'action de forces extérieures," *ibid.,* **46** (1858), 208–212.

6. "Mémoire sur le règlement des tiroirs dans les machines à vapeur," in *Comptes rendus de l'Académie des sciences,* **14** (1842), 632–633.

7. B. P. E. Clapeyron, "Mémoire sur la puissance motrice de la chaleur," in *Journal de l'École polytechnique,* **14** (1834), 153–190. The English trans. is by Richard Taylor, in *Scientific Memoirs, Selected From the Transactions of Foreign Academies of Science and Learned Societies and From Foreign Journals,* **1** (1837), 347–376. The German trans. by Poggendorff appeared in *Annalen der Physik und Chemie,* **59** (1843), 446–451.

8. M. Kerker, "Sadi Carnot and the Steam Engine Engineers," in *Isis,* **51** (1960), 257–270.

Milton Kerker

CLARK, ALVAN (*b.* Ashfield, Massachusetts, 8 March 1804; *d.* Cambridgeport, Massachusetts, 19 August 1887); **CLARK, ALVAN GRAHAM** (*b.* Fall River, Massachusetts, 10 July 1832; *d.* Cambridgeport, Massachusetts, 9 June 1897); **CLARK, GEORGE BASSETT** (*b.* Lowell, Massachusetts, 14 February 1827; *d.* Cambridgeport, Massachusetts, 20 December 1891), *astronomical instrumentation.*

Three instrument makers—Alvan Clark and his sons, George Bassett and Alvan Graham—figured importantly in the great expansion of astronomical facilities that occurred during the second half of the nineteenth century. Almost every American observatory built during this period, and some observatories abroad, housed an equatorial refracting telescope, and often auxiliary apparatus as well, made by the Clarks. Five times the Clarks made the objectives for the largest refracting telescope in the world; and the fifth of their efforts, the forty-inch lens at the Yerkes Observatory, has never been surpassed. Their optical work, which was recognized as unexcelled, was the first significant American contribution to astronomical instrument making. American telescopes had been made before, but none compared with those of European manufacture; by the end of the nineteenth century, however, partly because of the example set by Alvan Clark & Sons, several other Americans were making fine astronomical instruments that did indeed compare.

In their factory at Cambridgeport, Massachusetts, Alvan and Alvan Graham specialized in optical work, while George supervised the mechanical constructions. Most of the telescopes produced were visual achromats, with objectives patterned after Fraunhofer's lenses; the Clarks did not join in the contemporary mathematical search for more perfect lens configurations. The preliminary lens grinding was done by workmen using simple horizontal turntables. Then Alvan or Alvan Graham would locate the errors, usually by examining the image of an artificial star at the focus of the lens, and remove them by "local correction"—by manually retouching the offending areas.

Before the development of the Hartmann test the standard way of describing the perfection of a lens was in terms of its actual defining power. The Clarks tested their lenses by searching for new and difficult double stars. In 1862, while testing the 18½-inch objective later installed in the Dearborn Observatory in Chicago, Alvan Graham discovered the predicted but hitherto unseen companion of Sirius; for this find he won the Lalande Prize of the Paris Academy of Sciences.

BIBLIOGRAPHY

For further information see Deborah Jean Warner, *Alvan Clark & Sons—Artists in Optics* (Washington, D.C., 1968).

DEBORAH JEAN WARNER

CLARK, JOSIAH LATIMER (*b.* Great Marlow, Buckinghamshire, England, 10 March 1822; *d.* London, England, 30 October 1898), *technology.*

With early experience as a chemist, Clark joined his brother Edwin first as a civil engineer and then, in 1850, as an assistant engineer in a telegraph company. From that time on, his major interest was electricity, especially submarine telegraphy. He exhibited an extreme bent for practicality, perhaps best illustrated by a letter to the editor of *Engineer* in 1885 that argued that a separate set of easily remembered symbols should be established for working electricians, using the initial letters of volt, ampere, ohm for voltage, current, resistance, and so on. He was instrumental in the founding of the Society of Telegraphic Engineers and Electricians (later the Institute of Electrical Engineers) and was its fourth president.

Clark's interest in practical telegraphic matters led him into investigations of some scientific importance. In 1863 he demonstrated that the speed of a current pulse was independent of the voltage applied. A decade earlier he had shown that the retardation effects in telegraph cables were caused by induction; this information was then expanded upon by Faraday (*Experimental Researches,* III, 508–517). Clark developed a zinc-mercury standard cell that was widely used for both technological and scientific purposes. In 1861 he and Charles Bright presented a paper to the British Association suggesting the establishment of standard electrical units; their reasoning, again, was highly practical. This led to the formation of the important Committee on Standards, on which Clark served until it was temporarily disbanded in 1870.

Clark had a strong avocational interest in astronomy; in 1857 he helped Airy, the astronomer royal, develop a telegraphic system for reporting Greenwich mean time throughout the country.

BIBLIOGRAPHY

I. ORIGINAL WORKS. Clark's articles include the paper presented to the British Association, "Measurement of Electrical Quantities and Resistance," in *A Journal of Telegraphy,* **1** (1861), 3–4, written with Charles Bright; and "On a Standard Voltaic Battery," in *Philosophical Transactions of the Royal Society,* **164** (1874), 1–14. Other papers are listed in the *Royal Society Catalogue of Scientific Papers.*

His books include *An Elementary Treatise on Electrical Measurement* (London, 1868); and *Dictionary of Metric and Other Useful Measures* (London, 1891). Other books are cited in the *Dictionary of National Biography* entry mentioned below.

II. SECONDARY LITERATURE. Among the obituary notices the best is in *Journal of the Institute of Electrical Engineers,* **28** (1899), 667–672; that in *Electrician,* **42** (1899), 33, is accompanied by a photograph. There is an article on Clark in *Dictionary of National Biography,* Supp. 1, and another in J. T. Humphreys, *Celebrities of the Day* (London, 1881).

See also the introduction to William D. Weaver, *Catalogue of the Wheeler Gift of Books, Pamphlets, and Periodicals to the Library of the American Institute of Electrical Engineers* (New York, 1909), **1**, 15–46; the collection was originally assembled by Clark.

BERNARD S. FINN

CLARK, THOMAS (*b.* Ayr, Scotland, 31 March 1801; *d.* Glasgow, Scotland, 27 November 1867), *chemistry.*

Clark is best known for his method of softening water ("Clark's process"); in pure chemistry his discovery of sodium pyrophosphate had far-reaching implications.

Clark was educated at Ayr Academy; although at first apparently considered dull, he later showed himself to be gifted in mathematics. In 1816 he began ten years of employment in firms founded by two leading Glasgow industrialists—first with Charles Macintosh, who invented waterproof cloth in 1823 (after Clark had been in his employ) and then, after finding chemistry more to his taste than accountancy, with Charles Tennant, who patented the preparation of bleaching powder.

In 1826 Clark was appointed lecturer in chemistry at the Glasgow Mechanics' Institution, a post for which Thomas Graham was also a candidate. It is not known how or where Clark learned his chemistry; but judging from an account left by one of his students at the institution,[1] his teaching showed a degree of sophistication that has puzzled historians of chemistry. Griffin reproduced a set of tables, prepared by Clark for his courses of 1827–1828, of elements and compounds, with symbols and formulas, and atomic weights or "combining proportions," and what we would now call molecular weights (in the case of compounds). They impress one immediately by the liberal use and modernity of symbols. The symbols differ somewhat from those proposed by Berzelius in 1814, which were little used in Britain until about 1833. The most striking feature of Clark's formulas, however, is their suggestion of structure—an inchoate concept at this time.[2] By using an uncommon set of atomic weights ($H = 1/2$, $O = 8$, $C = 6$), Clark arrived at the correct formulas of a number of compounds, which he expressed as shown:

Alcohol	HO; CH, H; CH, H, H
Acetic acid	CH, H, H; CO, O; H

Neither what he had in mind nor the significance of the "punctuation marks" is known. They appear to be used the way we now use full stops and colons in structural formulas to denote single and double bonds, but we do not know what Clark actually intended them to mean. His views on the constitution of acids and salts also seem more advanced than those of most of his contemporaries.

Clark became a medical student at Glasgow University in 1827, with a view solely to teaching chemistry in a medical school. He resigned his post at the Mechanics' Institution in 1829 (Graham taking his place) and became apothecary to Glasgow Infirmary. He obtained his doctorate in 1831 and in 1833 was appointed professor of chemistry at Marischal College in Aberdeen, a post which he held nominally until his college was joined with King's College to form the University of Aberdeen.

Clark had by this time published the paper describing how he had obtained a new compound, which he called sodium pyrophosphate, from ordinary sodium phosphate at red heat, by the loss of water:

$$2Na_2HPO_4 = Na_4P_2O_7 + H_2O.$$

This paved the way for Graham's study of the phosphoric acids,[3] which led to the concept of polybasic acids and ultimately to a clearer understanding of the constitution of acids in general.

Clark patented his water-softening process (by the addition of a calculated quantity of milk of lime) in 1841. The hardness of water was tested by finding the quantity of a standard solution (which Clark called the "soap-test") which would produce a lather lasting for five minutes when added to a specified quantity of the water. He also devised a scale of water hardness (one degree of hardness was that produced in one gallon of water by one grain of chalk or magnesia, held in solution as a bicarbonate). Although highly praised by Graham and others, it was some time before the softening process was much used by water companies.

What seems to have been a promising career was effectively terminated sometime in 1842 or 1843 by the onset of a brain disease from which Clark never fully recovered. He did not resign but appointed a series of assistants to carry out his teaching duties. During periods of partial recovery he interested himself in a variety of subjects—the philology of English, spelling reform, and textual criticism. In 1849 Clark married Mary M'Ewen; their only child died while still a boy.

NOTES

1. See J. J. Griffin, *The Radical Theory in Chemistry* (London, 1858), pp. 5–17. Griffin reproduced the tables in support of his contention that Clark had anticipated, and Griffin had developed, some of the theories that Gerhardt had advanced as his own. Unfortunately, we have only Griffin's word for what Clark taught, but it seems to be generally agreed that the tables are genuine.
2. For a recent comment on this point, in connection with Clark's work, see W. V. Farrar, "Dalton and Structural Chemistry," in D. S. L. Cardwell, ed., *John Dalton and the Progress of Science* (Manchester–New York, 1968), pp. 294–295.
3. Thomas Graham, "Researches on the Arseniates, Phosphates and Modifications of Phosphoric Acid," in *Philosophical Transactions of the Royal Society,* **123** (1833), 253–284.

BIBLIOGRAPHY

I. ORIGINAL WORKS. Clark's more important papers are listed in The Royal Society's *Catalogue of Scientific Papers,* I (1867), 932–933. Those particularly relevant to the text are "On the Pyrophosphate of Soda, One of a New Class of Salts Produced by the Action of Heat on the Phosphates," in *Edinburgh Journal of Science,* **7** (1827), 298–309; and "On the Examination of Water for Towns, for Its Hardness, and for the Incrustation It Deposits on Boiling," in *Chemical Gazette,* **5** (1847), 100–106. He also published a pamphlet on his process, *A New Process for Purifying the Waters Supplied to the Metropolis by the Existing Water Companies . . .* (London, 1841).

II. SECONDARY LITERATURE. The main source is A. Bain, "Biographical Memoir of Dr. Thomas Clark," in *Transactions of the Aberdeen Philosophical Society,* **1** (1884), 101–115. The paper was written in 1879; Bain, then professor of logic at Aberdeen, had been a pupil of Clark's. Other accounts, partially derivative, are A. Findlay, "The Teaching of Chemistry in the University of Aberdeen," in *Aberdeen University Studies,* **112** (1935), 18–36; and J. H. S. Green, "Thomas Clark (1801–1867), A Biographical Study," in *Annals of Science,* **13** (1957), 164–179.

E. L. SCOTT

CLARK, WILLIAM MANSFIELD (*b.* Meyersville, New York, 17 August 1884; *d.* Baltimore, Maryland, 19 January 1964), *chemistry.*

Clark, descended on both sides of his family from clergymen, became interested in chemistry while at Williams College. In his senior year he taught chemistry in the local high school. After graduation in 1907, he took a master's degree at Williams and then went to Johns Hopkins University, where he received a

Ph.D. in physical chemistry in 1910. During his summer vacations he worked at Woods Hole under Carl Alsberg and, later, D. D. Van Slyke. These men aroused his interest in biochemistry.

After receiving the Ph.D., Clark became chemist in the Dairy Division of the U.S. Department of Agriculture. His work on the bacteriology of milk products led him to study problems involving acidity, and he became adept in the use of the hydrogen electrode. Struck by the variability of titration indicators then available, he developed, with Herbert Lubs, a set of thirteen indicators that covered almost the entire pH range used in titrations. These studies led him to write *The Determination of Hydrogen Ions* (1920), which became a classic in its field.

In 1920 Clark became chief of the Division of Chemistry of the Hygiene Laboratory, Public Health Service, the forerunner of the National Institutes of Health. Here he began studies on oxidation-reduction potentials of dyes, a field that, together with studies of metalloporphyrins, occupied him for the rest of his life. In 1927 he became professor of physiological chemistry in the Johns Hopkins Medical School.

Clark was an excellent teacher, stressing especially the importance of physical chemistry for medical students. He was active in many government bureaus, particularly during World War II. After his retirement in 1952 he devoted himself to writing, especially to his monograph on oxidation-reduction systems (1960).

BIBLIOGRAPHY

There is a bibliography of Clark's works in *Biographical Memoirs. National Academy of Sciences,* **39** (1967), 27–36. His chief works are *The Determination of Hydrogen Ions* (Baltimore, 1920; 2nd ed., 1923; 3rd ed., 1928); and *Oxidation-Reduction Potentials of Organic Systems* (Baltimore, 1960). An autobiographical sketch by Clark is "Notes on a Half-Century of Research, Teaching and Administration," in *Annual Review of Biochemistry,* **31** (1962), 1–24.

More detailed information may be found in Hubert Bradford Vickery, "William Mansfield Clark," in *Biographical Memoirs. National Academy of Sciences,* **39** (1967), 1–26.

HENRY M. LEICESTER

CLARKE, EDWARD DANIEL (*b.* Willingdon, Sussex, England, 5 June 1769; *d.* London, England, 9 March 1822), *mineralogy, geology.*

His father, Edward, was the son of "mild" William Clarke, fellow of St. John's College, Cambridge, prebendary of Chichester Cathedral, and a distinguished antiquary. His mother, Anne, was the daughter of

Thomas Grenfield of Guildford, Surrey. Edward, Sr., also a fellow of St. John's, published a statistical account of Spain in 1763 and, after service as chaplain in Madrid and Minorca, settled down to a quiet literary life in his Sussex parish.

Edward Daniel, the second of a family of three sons and one daughter, was at Tonbridge School, Kent, from 1779 until he entered Jesus College, Cambridge, in 1786. Immediately after taking an undistinguished B.A. in 1790 he began the first of several tutorships to sons of the nobility, some of which enabled him to travel extensively. In 1791 the nucleus of his mineralogical collection was formed in the course of a tour of southern England, Wales, and Ireland. A prolonged stay in Naples in 1792–1793 enabled him to make a thorough study, in company with the pioneer volcanologist Sir William Hamilton, of Vesuvius in eruption. In 1794 he was on the Continent again collecting vigorously, and in 1797 he made an extended tour of Scotland and the Western Isles. On his return he became a fellow and bursar of Jesus College.

To Clarke's restless spirit college life was irksome, and in May 1799 he set off on a tour of Scandinavia, Russia, and the Middle East that was to take three years. He was accompanied throughout by a pupil who paid the expenses of the journey and in the earlier stages by two fellows of Jesus, William Otter and the economist Robert Malthus. Clarke collected avidly everywhere: 800 mineral specimens from Siberia alone and plants, seeds, antiquities, and manuscripts. His frequent letters to friends in Cambridge excited considerable attention, and on his return he was awarded an LL.D. Some of his marbles were presented to the University of Cambridge; others were purchased by the British Museum. Later his medieval manuscripts were sold to the Bodleian and his coins to a private collector. His mineral collection was purchased posthumously by the University of Cambridge for £1,500.

In 1805 he took holy orders. On 25 March 1806 he married Angelica, fifth daughter of Sir William Rush, baronet of Wimbledon, by whom he had five sons and two daughters. Mrs. Clarke was greatly admired by Cambridge society; but she had one foible, extravagance; and, although her husband's income in his later years was substantial, the family was never well-off.

In March 1807 Clarke began the first of his spectacularly successful annual courses of lectures on mineralogy. Although he had then made no notable contribution to mineralogy, he presented his material in modern terms and evinced some depth of understanding; he was, moreover, a stylish lecturer, his boundless enthusiasm simply infectious. The professorship of mineralogy was created for him in December 1808.

Clarke's scientific fieldwork ended with his return to Cambridge in 1802, and he devoted much of the remainder of his life to preparing his topographical observations on geology, botany, and other subjects for publication. The first volume of the *Travels* appeared in 1810 and subsequent volumes in 1812, 1814, 1816, 1819, and 1823. Four editions were published in England and one in America; volume I, which deals with Russia, had two Scottish editions (Edinburgh, 1839; Aberdeen, 1848) and was translated into German (Weimar, 1817) and French (Paris, 1812).

In 1816 Clarke began to study the response of a wide range of refractory substances (*The Gas Blowpipe* [1819] lists experiments on ninety-six minerals) to high temperatures with the oxyhydrogen blowpipe. Volcanological analogies were drawn and some important improvements were made to the blowpipe, but this work made little lasting impact.

In 1817 Clarke was appointed university librarian, an office he held in plurality with his professorship and his two benefices, Harlton and Yeldham. The following year his health began to fail; nevertheless he continued to lecture and preach, to perform chemical experiments, and to write. He died at the London house of his father-in-law on 9 March 1822 at the age of fifty-two and was buried in the chapel of Jesus College, where a memorial by Chantrey was erected to him. The college possesses a particularly fine portrait by John Opie, R.A.

Clarke never held high office in any scientific society. The Geological Society of London elected him to honorary membership at its second meeting on 4 December 1807. In 1819 he played a leading part in the foundation of the Cambridge Philosophical Society. His correspondents and friends in Europe and England were many; Pallas, Haüy, Faujas de St. Fond, Gadolin, Davy, Wollaston, Pennant, Hamilton, Kidd, Jameson, and Thomas Thomson, to name but a few.

Clarke is essentially a satellite figure in science in that he made no notable discovery, but he did make a significant contemporary impact by his teaching of mineralogy in terms of crystallography and the new chemistry, by the topographical geology and volcanological observations in his *Travels,* and by his enthusiasm which fired the interest of contemporaries and undergraduates in minerals and in scientific fieldwork generally. His historical position is difficult to assess in that the importance of his influence so far outweighs that of his actual scientific achievement.

BIBLIOGRAPHY

I. ORIGINAL WORKS. Clarke's scientific writings are encompassed in his *A Tour Through the South of England, Wales, and Part of Ireland, Made During the Summer of 1791* (London, 1793); *A Methodical Distribution of the Mineral Kingdom Into Classes, Orders, Genera, Species, and Varieties* (Lewes, 1806); *A Syllabus of Lectures in Mineralogy* (Cambridge, 1807; 2nd ed., London, 1818; 3rd ed., 1820); *Travels in Various Countries of Europe, Asia, and Africa,* 6 vols. (London, 1810–1823); *The Gas Blow-pipe, or Art of Fusion by Burning the Gaseous Constituents of Water* (London, 1819).

A list of papers in various journals is given in Otter (see below).

II. SECONDARY LITERATURE. A very full but uncritical biography was written by William Otter shortly after Clarke's death, *The Life and Remains of Edward Daniel Clarke* (London, 1823; 2nd ed., 2 vols., 1825).

DUNCAN MCKIE

CLARKE, FRANK WIGGLESWORTH (*b.* Boston, Massachusetts, 19 March 1847; *d.* Chevy Chase, Maryland, 23 May 1931), *geochemistry.*

Clarke was perhaps the most distinguished of the early American geochemists. His compilations of rock, mineral, and natural water analyses provided a basis to explain the chemical processes occurring at the earth's surface. Many of his concepts concerning the chemical evolution of geological systems are still valid.

He was the son of Henry W. Clarke, a Boston hardware merchant and later a dealer in iron-working machinery, and Abby Fisher Clarke, who died when he was about ten days old.

Clarke received his primary and secondary schooling in the Boston area. In March 1865 he entered the Lawrence Scientific School of Harvard University where he studied chemistry under Wolcott Gibbs. He received the B.S. in 1867. He remained at Harvard for an additional year at which time his first scientific paper, describing some new techniques in mineral analysis, was published. During the subsequent year, 1869, he acted as an assistant to J. M. Crafts at Cornell University. For the next four years he lectured on chemistry in the Boston Dental College and supplemented his meager earnings with newspaper and magazine articles. He reported Tyndall's Lowell Institute lectures and the proceedings of meetings of the American Association for the Advancement of Science for the *Boston Advertiser.* He wrote popular scientific reviews, many for periodicals catering to young readers. Foreshadowing the later direction of his career, he proposed in 1873, in *Popular Science Monthly,* a scheme of evolution of the heavy elements from the light. In spite of these varied and time-consuming activities, he maintained an involvement in scientific researches. He initiated a series of articles under the general title of "Constants of Nature" for the Smithsonian Institution with the contribution "A Table of Specific Gravities, Boiling Points and Melting Points of Solids and Liquids."

Clarke accepted an appointment as professor of chemistry and physics at Howard University in Washington, D.C., in 1873. Following his marriage to Mary P. Olmsted of Cambridge, Massachusetts, in 1874, he went to the University of Cincinnati as professor of chemistry and physics, a position he held for nine years. In 1883 he was appointed chief chemist to the U. S. Geological Survey in Washington, D.C., and honorary curator of minerals of the United States National Museum. He retired from these offices on 31 December 1924.

Clarke received honorary degrees from Columbian University in 1891, from Victoria in 1903, from Aberdeen in 1906, and from Cincinnati in 1914. He was honored with the *chevalier* of the Legion of Honor in 1900.

The forty-one years Clarke spent at the Geological Survey witnessed the development of the United States as one of the foremost centers of geochemical research. As chief chemist he was responsible for the analyses of thousands of samples of crustal rock, water, and air. Such investigations were carried out with great accuracy using the best available techniques. The laboratories gained international distinction and provided a training site for many young aspiring chemists both from the United States and abroad. The associates of Clarke in these enterprises involved the elite of analytical and geological chemistry, including W. F. Hillebrand, F. A. Gooch (designer of the Gooch crucible), Eugene Sullivan (the inventor of Pyrex glass), W. T. Schaller, H. N. Stokes, T. M. Chatard, and George Steiger. Both major and trace elemental concentrations were sought in the materials under study. Clarke emphasized the importance of all available data stating, "The rarest substances, however, whether elementary or compound, supply data for the solution of chemical problems."

It is interesting to note that Clarke himself had little proficiency in the laboratory; yet his interest in mineralogy dates to his initial researches. His first published paper in 1868 bore the title "On a New Process in Mineral Analysis."

The studies provided the base upon which Clarke put forth his interpretation of natural phenomena. An annual report of the work done in the Washington laboratory of the chemistry and physics division of the U. S. Geological Survey contained the results from the examination of minerals, rocks, and waters. The vehicles for the transmission to the scientific community of such work and its significance in understanding geological phenomena at the earth's surface were the volumes *The Data of Geochemistry,* first published as U. S. Geological Survey Bulletin no. 331 (1908; 5th ed., Bulletin no. 770, 1924).

Clarke's impact upon environmental science derived from his thesis that the chemistry and mineralogy of a rock is a record of past chemical reactions, which in part serves to describe geological events connected with the formation or alteration of the solid phase. The earth's crust is the site of such reactions and is composed of three domains: the atmosphere, the hydrosphere, and the lithosphere. Since the compositions of the first two zones are relatively homogenous and well-defined, the solid earth was the focus of his studies. The analysis and synthesis of minerals and rocks were the techniques used to define the chemical reactions.

Clarke emphasized the importance of equilibrium considerations in the prediction and description of chemical reactions in nature. In the fifth edition of *The Data of Geochemistry* (p. 10) he states:

> The reactions which took place during the formation of the rock were strivings toward chemical equilibrium, and a maximum of stability under the existing conditions was the necessary result. . . . To determine what changes are possible, how and when they occur, to observe the phenomena which attend them, and to note their final results are the functions of the geochemist.

Such advice has been used by subsequent generations of geochemists who have successfully applied equilibrium thermodynamics to the formation and alteration of igneous, sedimentary, and metamorphic rocks.

To provide a baseline for their investigations, Clarke and his associate H. S. Washington sought the average composition of the earth's crust to a depth of ten miles. They assumed that such an average could be approximated by the average composition of igneous rocks since contributions from metamorphic and sedimentary rocks appeared to be trivial. Their estimate for the composition of the crust was 95% igneous rock, 4% shale, 0.75% sandstone, and 0.25% limestone.

From 8,600 elemental analyses of rocks, Washington selected 5,197 superior assays on the basis of criteria he had devised. The overall average, recalculated to 100% with the elimination of water and minor constituents, turned out to be:

Species	SiO_2	Al_2O_3	Fe_2O_3	FeO	MgO	CaO	Na_2O	K_2O	TiO_2	P_2O_5
Wt. %	60.18	15.61	3.14	3.88	3.56	5.17	3.91	3.19	1.06	0.30

Petrographical analyses upon 700 igneous rocks gave an average mineralogical composition: quartz, 12.0%; feldspars, 59.5%; pyroxene and hornblende, 16.8%; biotite, 3.8%; titanium minerals, 1.5%; apatite, 0.6%; other accessory minerals, 5.8%.

This composition is intermediate to that of a granite and to that of a basalt, the two principal rock types of the earth's crust. Although many criticisms have been voiced against their computation, usually on the grounds that the sampling of the rocks was neither random nor representative, these results, with only minor modifications, are still valid.

Since silicon is the most important metallic element in crustal materials, it was not unexpected that Clarke's attention would be directed to the problems of silicates and their compositions and structures. His first paper in this area, "Researches on Lithia Micas" appeared in 1886. Although the laboratory syntheses, alteration products, and pseudomorphs of minerals provided Clarke with what he called the natural history of a mineral, he extended his knowledge with the preparation of artificial substances, especially the ammonia analogues of sodium, potassium, and calcium minerals. This new field of investigation provided additional information on the constitution of minerals.

Such work essentially closed the era of silicate mineralogy based upon chemical composition and reactions. The subsequent development of x-ray diffraction and electron microscopy was to provide details on atomic architecture which would give a new foundation to such studies.

Clarke also calculated the average composition of sedimentary rocks, using such data to compute the total amount of sedimentation occurring during geologic time. The argument given is that the total amount of igneous rock eroded is equal to the total amount of sedimentary rock produced plus the dissolved salts of the sea. He calculated that ninety-three million cubic miles of sediments had been produced over geological time.

He extended his interest in inorganic sediments to those precipitated biologically with investigations upon the inorganic constituents of marine invertebrates. He was aware that many important rocks were composed of solids precipitated by the animals and plants of the sea, furnishing, through their shells, silicon dioxide, calcium carbonate, and calcium phosphates. He was able to provide some of the first estimates on the relative contributions of such materials to the geological column.

Clarke was closely involved in the formation of the American Chemical Society. Before 1873 chemistry was not accorded much importance in the A.A.A.S.

In that year, at a meeting in Portland, Maine, Clarke, with C. E. Munroe, W. McMurtrie, and H. W. Wiley, proposed the establishment of a subsection of chemistry. This suggestion was accepted and the section first met the following year in Hartford, Connecticut.

In 1876 the New York chemists formed a local society with the name of the American Chemical Society; eight years later the Washington, D.C., chemists created a similar organization. In 1886 Clarke wrote to Munroe, who was serving as chairman of the A.A.A.S. section of chemistry, that it would be most reasonable to form a national chemical society. Three years later a proposal of the New York group was accepted—their name and charter would form the basis of a national society to include both the New York and Washington groups as local sections. Clarke became president of the American Chemical Society in 1901.

BIBLIOGRAPHY

A complete compilation of the publications of Frank Wigglesworth Clarke, prepared by Lucia K. Williams, is in *Biographical Memoirs. National Academy of Sciences,* XV (1932), 146–165. Of Clarke's many works the following are of particular interest: "Evolution and the Spectroscope," in *Popular Science Monthly,* 2, no. 9 (Jan. 1873), 320–326; a letter to the editor of *Popular Science Monthly* advocating a subsection of chemistry in the *Proceedings of the American Association for the Advancement of Science, ibid.,* 7, no. 39 (July 1875), 365; "A Preliminary Notice of the Revision of the Atomic Weights," a paper presented before the chemistry subsection of the American Association for the Advancement of Science and reviewed in *American Chemical Journal,* 1, no. 4 (Oct. 1879), 295–296; "The Constants of Nature, pt. 5—Recalculation of the Atomic Weights," Smithsonian miscellaneous publications no. 441; *Elements of Chemistry* (New York, 1884); "The Minerals of Litchfield, Maine," in *American Journal of Science,* 3rd ser., 31 (1886), 262–272; and "The Relative Abundance of the Chemical Elements," in *Bulletin of the U. S. Geological Survey,* 78 (1891), 34–42. See also "Report of the Committee on Determination of Atomic Weights," in *Journal of the American Chemical Society,* 16 (1894), 179–193; "The Constitution of the Silicates," in *Bulletin of the U. S. Geological Survey,* 125 (1895), 1–125; "The New Law of Thermochemistry," a paper read before the Philosophical Society of Washington (see *Proceedings,* 14 [1902], 399); "The Data of Geochemistry," in *Bulletin of the U. S. Geological Survey,* 330 (1908), 716 pp.; and *Chemical Abstracts,* 2, no. 9 (10 May 1908), 1252–1253; "Some Geophysical Statistics," in *Proceedings of the American Philosophical Society,* 51 (July 1912), 214–234; "The Evolution and Disintegration of Matter," U. S. Geological Survey paper 132-D (9 April 1924); and "The Internal Heat of the Earth," in *Scientific American,* 134 (June 1926), 370–371.

With Henry S. Washington he wrote "The Average Composition of Igneous Rocks," paper no. 462 from the Geophysical Laboratory, Carnegie Institute of Washington, and "The Composition of the Earth's Crust," U. S. Geological Survey paper no. 127.

EDWARD D. GOLDBERG

CLARKE, SAMUEL (*b.* Norwich, England, 11 October 1675; *d.* London, England, 17 May 1729), *metaphysics, mathematics.*

The son of Edward Clarke, an alderman of Norwich, and Hannah Clarke, Samuel Clarke attended the Norwich Free School and entered Gonville and Caius College, Cambridge, in 1690. He became a scholar of his college in 1691 and received the B.A. in 1695. His major interests were physics and theology, and he mastered the contents of Sir Isaac Newton's *Principia* while at Gonville and Caius. He was elected a fellow of the college in 1696 and retained the office until 1700.

The standard physics text used at Cambridge during this period was Théophile Bonnet's clumsy Latin translation of Jacques Rohault's *Physics.* Clarke's tutor, Sir John Ellis, urged him to prepare a more elegant version of the work. Making use of his familiarity with Newtonian physics, Clarke appended to his edition a series of notes that had the novel effect of turning a Cartesian treatise into a vehicle for disseminating the ideas of Newton. His translation remained the major text at Cambridge for over forty years and was translated into English at late as 1723, by Clarke's brother, John.

A chance meeting (in a Norwich coffeehouse) with the mathematician and Arian theologian William Whiston, in 1697, led to Clarke's introduction to Dr. John Moore, bishop of Norwich and Whiston's patron. In 1698 Clarke succeeded Whiston as Moore's chaplain and received his M.A. from Cambridge. His interests now turned to the primitive church, a subject of pivotal importance in the internecine quarrel between orthodox Anglicans, who accepted the Athanasian Creed, and Arian churchmen, who rejected it. Clarke's Arian bent did not hinder his rise in the church at this time. Bishop Moore appointed him rector of Drayton and granted him a parish within Norwich itself.

At Norwich, Clarke gained a reputation as a preacher of clear, learned sermons; this led him to be chosen to present the Boyle lectures for 1704.[1] He made such a favorable impression that he was asked to deliver an additional series the next year. The sixteen sermons were subsequently published as *The*

Being and Attributes of God.[2] He opposed Descartes's idea that "motion is essential to all things" and his insistence on "an infinite *plenum.*" Clarke cited Newton's concept of the void to show that matter was no more "necessary" than motion, and it was clear that "The Self-existent and original Cause of All Things is Not a Necessary Agent, but a Being indued with Liberty and Choice." Liberty stood at the very roots of the universe, for without it there could be no "Cause, Mover, Principle, or Beginning of Motion anywhere."[3]

Thus the evidence of the physical world confuted the ideas of atheists. Clarke's sermons of 1705 were aimed at the deists; his vindication of natural religion demonstrated not merely God's existence but His presence in the world.

In 1706 Sir Isaac Newton commissioned Clarke to translate his *Opticks* into Latin for a broader audience. He was so satisfied with Clarke's "pure and intelligible Latin" that he paid him £500. In the same year Clarke was appointed to the rectory of St. Benet's in London, was introduced to Queen Anne, and was made one of her chaplains-in-ordinary.

Clarke's move to a London parish and his contact with the court involved him in two bitter controversies in which he defended Newton's ideas as well as his own. The first began with the High Church apologist Henry Dodwell, whose *Epistolary Discourse . . . that the Soul is . . . Naturally Mortal* was a high point of the Tory revival that marked Anne's reign.[4] Dodwell considered the soul as a material entity and appealed to the authority of Athanasius, Descartes, and Spinoza in support of his position. Had he been a man of lesser repute, and of worse connections, his work might have been ignored as a ridiculous attempt to cast doubt on both natural and revealed religion. His stature forced those who disagreed with him to reply, and "Mr. Clarke was thought the most proper person for the work."[5]

Clarke denied that the soul could be mortal, since it could not possibly be material. Matter was particulate; the soul was not, and thus was totally independent of matter. At the end of his treatise he cautioned Dodwell to reconsider the implication of his ideas, for Dodwell had furnished "A Weapon for the hands of skeptical men . . . to make profane men rejoice."[6]

The advice was sound but was given far too late. The materialist Anthony Collins now claimed to be Dodwell's defender. In his "Remarks on a (pretended) demonstration of the immateriality and natural immortality of the soul," he maintained that men were "meer machines," incapable of exercising free will.[7] Dodwell faded from the scene, and Collins emerged

as Clarke's chief antagonist. In his initial response to Collins, Clarke sought to annihilate his adversary's basic argument by denying that matter could possibly think.[8] Collins replied that if soul existed, it needed extension; if it were extended, then it had to be corporeal.[9]

As the correspondence continued, Collins cited Descartes in support of his idea that consciousness was merely a "mode of motion," and Clarke turned to Newton's work in rebuttal. He accused Collins of having failed to perceive that "the present operations of nature, depending on Gravitation, cannot be Mechanical Effects of Matter in Constant Motion perpetually striking one Part against each other."[10] His final reply to Collins also revealed a debt to Newton in his assertion that the spirit makes use of space as its medium. Clarke's attack on Collins had succeeded in defending Newton's ideas in a metaphysical sense. Collins had had enough, and the dispute was over by 1710.[11]

During the course of this exchange, the queen had named Clarke rector of St. James's, Westminster, a most prestigious position. He was awarded the D.D. by Cambridge at the same time (1709) and ably defended the thesis that "no article of faith is opposed to right reason."

Clarke's Arianism became obvious with the publication of his *Scripture Doctrine of the Trinity* in 1712. This unitarian work led to a long pamphlet war with such orthodox divines as Daniel Waterland[12] and to a complaint being made about it by the Lower House of Convocation in 1714. Clarke succeeded in defending himself before the Upper House and was not censured.[13] He did, however, agree to write and preach nothing further on this topic, a course of action of which the more outspoken Arians did not wholly approve.

George I's accession to the English throne upon the death of Anne in 1714 effectively destroyed the Tory party for fifty years. The Tory leader, Bolingbroke, who had attacked Clarke, was powerless, and the High Church faction was subdued. For Clarke, the most significant aspect of the Hanoverian succession lay not in these developments, however welcome, but in his meeting and subsequent friendship with the princess of Wales (later queen), Caroline. A close friend of Leibniz, the princess originally sought to have Clarke translate his *Theodicy.* Clarke refused, since the work contained an attack on Newton's concept of gravity. Impressed by Clarke's persuasive powers, Caroline sent copies of his works to Leibniz and asked for his opinion of them. His reply, shown by the princess to Clarke, sparked an exchange of letters that lasted until Leibniz's death in November 1716.

In the course of his five letters, Leibniz accused the Newtonians of having made "Natural religion itself . . . to decay [in England]," of turning God into a watchmaker who "wants to wind up his watch from time to time," of making gravity either an occult quality or a "perpetual miracle," and of having failed to understand the principle of "sufficient reason" by their belief in the vacuum, in atoms, and in the reality of space and time. Clarke undertook to answer these charges, drawing on the writings of Newton and occasionally obtaining his advice.[14] Thus, in his fourth paper Clarke cited not only Newton but the work of earlier experimenters to prove his contentions that the vacuum did exist, as did hard, impenetrable particles ("physical atoms"), or there could be "no matter at all in the Universe." Clarke also maintained that time and space were real entities and not merely the order of successive and coexistent phenomena, respectively, as suggested by his opponent.[15]

Clarke saw the conflict with Leibniz as involving not merely a differing interpretation of the physical universe and its phenomena but as a far more basic one implying a struggle between freedom and necessity. Leibniz's insistence on philosophical necessity always signified "absolute necessity" to Clarke, for Leibniz, no less than Collins, would have reduced men to "meer machines."[16]

Clarke's most direct contribution to physics during the course of this correspondence came in a footnote to his fifth paper, in which he considered the problem of computing the force of a moving body. He discussed the Newtonian position that such force could be expressed as the product of mass and velocity (mv) and that of the Leibnizians, who expressed it as the product of mass and the square of the velocity (mv^2). He saw the issue as one between the concepts of momentum and kinetic energy. He developed these ideas in his last published paper, a letter to Benjamin Hoadly, "Concerning the proportion of force to velocity in bodies in motion," in which he strongly advocated the Newtonian position.[17]

Despite Caroline's continuing favor, Clarke's heterodoxy had halted his rise in the church, and a bishopric, or an even higher position, escaped him.[18] He accepted only one further office: the mastership of Wigston Hospital, in Leicestershire, in 1718. He had powerful friends among the more latitudinarian clergy. During the 1720's George Berkeley, Benjamin Hoadly, and William Sherlocke met with him frequently at court. He retained his popularity as a preacher and was much in demand to deliver sermons away from St. James's until his death in May 1729.

Clarke used his talents as a translator to prepare editions of Caesar's *Commentaries* (dedicated to the duke of Marlborough) in 1712 and, by royal command, of the first twelve books of the *Iliad* in 1729. His political convictions were strongly Whiggish, for the Whigs were supporters of political liberty.[19]

Clarke married Katherine Lockwood and had five children who survived him. His son, Samuel, became a fellow of the Royal Society.

For twenty-five years Clarke had held to his position and had vindicated that of Newton. His contemporaries ranked him almost with Newton in the force of his intellect; certainly, of Newton's circle, Clarke was best fitted for the role of defender and publicist.

NOTES

1. Founded by the Honorable Robert Boyle "to assert and vindicate the great fundamentals of Natural Religion," the lectures consisted of eight sermons a year preached in a London church. The tone of the lectures was set by the first incumbent, Richard Bentley, who demonstrated how Newton's *Principia* gave proof of the existence of God and of free will.
2. "A Demonstration of the Attributes of God" and "A Discourse Concerning Natural Religion," in *A Defense of Natural and Revealed Religion: Being a Collection of Sermons at the Boyle Lectures,* vol. II.
3. *Ibid.,* pp. 12–32, *passim.*
4. By 1706 this movement was well under way. A long war abroad, the linking of the Whigs with the cause of dissent and broad church policies, and the queen's undoubted preference for the rituals of High Anglican tradition led to a realliance between the Tories and the extreme Episcopalians.
5. B. Hoadly, "Life of Clarke," prefixed to *The Works of Samuel Clarke,* I, v.
6. S. Clarke, "A Letter to Mr. Dodwell," *Works,* III, 749; *cf.* British Museum, Add. Mss 4370, fols. 1, 2.
7. A. Collins, "A Letter to the Learned Henry Dodwell," *Works,* III, 752 ff.
8. S. Clarke, "Defense of An Argument . . .," *Works,* III, 761.
9. A. Collins, "Reply to Clarke's Defense," *Works,* III, 775.
10. S. Clarke, "Third Defense," *Works,* III, 848.
11. Collins did reappear after the death of Leibniz with his *Philosophical Enquiry Concerning Human Liberty* (1717), and again Clarke answered him.
12. Clarke was aided by his brother, John, and by John Jackson; his opponents included Waterland, Edward Wells, and Joseph Clarke. The controversy is referred to briefly in H. G. Alexander, *The Leibniz-Clarke Correspondence,* pp. xli, xlii; and in detail in J. Rodney, *God, Freedom, and the Cosmos* (to be published).
13. The case is to be found in Clarke, *Works,* IV; *cf.* British Museum, Add. Mss 4370.
14. Cited in Alexander, *op. cit., passim.*
15. *Ibid.,* pp. xiii, 100.
16. Clarke, "Fifth Reply," in Alexander, *op. cit.,* p. 110.
17. *Ibid.,* pp. 121–125; *Works,* IV, 737–740; reprinted in *Philosophical Transactions of the Royal Society* (1728).
18. Voltaire (*Oeuvres,* LV, 96) maintains that Caroline was prevented from appointing Clarke archbishop of Canterbury by Bishop Gibson's telling her that he had but one defect: "He was not a Christian."
19. The triumph of the Whigs in 1688 and in 1714 implies the acceptance politically of the philosophical tenet of human liberty. Clarke's political zeal may be seen in his willingness to dedicate his work to Marlborough at a time when the duke was out of favor; he remained a friend of Sarah, Duchess of Marlborough.

BIBLIOGRAPHY

I. ORIGINAL WORKS. Clarke's translations of scientific works are *Jacobi Rohaulti, Physica: Latine vertit Samuel Clarke, recensuit et uberioribus, ex illustrissimi Isaac Newtoni philosophia maximam partem hausti, amplificavit et ornavit* (London, 1697); and *Optice: Sive de reflexionibus, refractionibus, inflexionibus & coloribus . . . authore Isaaco Newton, equite aurato, vertit Samuel Clarke* (London, 1706).

His writings on theological and philosophical topics, including the full texts of the correspondence with Dodwell, Collins, and Leibniz, as well as his letter to Hoadly, are brought together in *The Works of Samuel Clarke,* John Clarke, ed., 4 vols. (London, 1738).

II. SECONDARY LITERATURE. The best contemporary biographical sketch is Benjamin Hoadly's "Some Account of the Life, Writings, and Character of [Samuel Clarke]," prefixed to the 1738 ed. of Clarke's *Works.* William Whiston's *Memoirs of the Life of Dr. Samuel Clarke* (London, 1730) is longer but far less reliable.

There is as yet no full-length modern study of Clarke. H. G. Alexander's *Clarke-Leibniz Correspondence* (Manchester, 1956) contains a brief sketch of Clarke's life and an analysis of the correspondence.

Modern articles on Clarke are equally scarce: John Gay's "Matter and Freedom in the Thought of Samuel Clarke," in *Journal of the History of Ideas* (1963), and J. Rodney's "Newton Revisited: Foes, Friends, and Thoughts," in *Research Studies, Washington State University,* **36** (1968), 351–360, are the most recent.

JOEL M. RODNEY

CLARKE, WILLIAM BRANWHITE (*b.* East Bergholt, England, 2 June 1798; *d.* Sydney, Australia, 16 June 1878), *geology.*

Clarke was a Church of England clergyman who emigrated to New South Wales in 1839 and laid the foundations of Australian geology. The son of William Clarke, the parish schoolmaster of East Bergholt, and the former Sarah Branwhite, Clarke was educated at Dedham Grammar School and Jesus College, Cambridge, from which he received the B.A. in 1821 and the M.A. in 1824. While at Cambridge he attended the lectures of Edward Clarke, professor of mineralogy, and of Adam Sedgwick, newly appointed Woodwardian professor of geology. In 1826 he was elected a fellow of the Geological Society of London, and for the next thirteen years he contributed papers on English and European geology and meteorology to British scientific journals. He was also active in the debate on geology and natural theology and, in reviews in theological journals in the 1830's, argued for a clear distinction between the claims of science and Scripture.

Clarke reached Australia with his family in May 1839. Very little was known of the geology of the continent at that time, and he at once began to collect rocks and fossils and explore in a widening arc from Sydney. He examined the colony's coal deposits and in 1841 discovered evidence of gold. He publicly predicted that the country would be found "wonderfully rich in metals," a prediction that was verified in the rush to the goldfields in 1851. From 1851 to 1853 Clarke was engaged by the government to conduct a gold survey of New South Wales and, on foot and horseback, traversed 60,000 square miles of country, reporting on the metalliferous districts and offering observations on the stratigraphy and structure of the country in twenty-eight reports. He published *Plain Statements and Practical Hints Respecting the Discovery and Working of Gold in Australia* in 1851 and *Researches in the Southern Goldfields of New South Wales* in 1860.

Clarke's major contribution to Australian geology centered on his work on the age of the great carboniferous basins of New South Wales. From extensive fieldwork he postulated that there was a perfect conformity between the upper (botanical) beds and underlying (marine) beds in the coal formations and that both types of beds were lower in the Paleozoic sequence than the coal deposits of Europe and India. This age assignment brought him into sharp conflict with Frederick McCoy, professor of natural history at Melbourne University, who argued, on the paleontological evidence, that the coal deposits were from the Mesozoic (Jurassic) age. The dispute, hammered out in British and colonial journals, continued for thirty years. Clarke's stratigraphic determinations were at length upheld by the European paleontologists Laurent de Koninck (1876–1877) and Ottaker Feistmantel (1878–1879) in two monographs based on Clarke's collections of fossils. In 1877 Clarke was awarded the Murchison Medal of the Geological Society of London for his coal researches.

Clarke published his findings on Australian stratigraphy in his *Remarks on the Sedimentary Formations of New South Wales* (4th ed., 1878). In addition he published some eighty monographs and papers on the geology and mineralogy of his adopted country and found time to contribute to the study of Australian meteorology, particularly on the laws of winds and storms.

Much of the impact of Clarke's scientific labor stemmed from his continuing contact with science abroad. He corresponded regularly with Sedgwick, J. B. Jukes, and Roderick Murchison and with leaders of French geology. His friendship with James Dana, whom he met during the visit of the United States Exploring Expedition to Sydney in 1839–1840, brought him fruitful communication with American science. He was also one of the first Australian scientists to give an open-minded reception to *The Origin of Species* and corresponded with both Charles Darwin and Richard Owen. Clarke was a founder of the Royal Society of New South Wales in 1867 and its active vice-president until 1876. The society's Clarke Medal, struck in 1878 as Australia's first scientific honor, commemorates his work. He was elected a fellow of the Royal Society of London in 1876.

BIBLIOGRAPHY

I. ORIGINAL WORKS. Clarke's publications are listed in the Royal Society's *Catalogue of Scientific Papers,* I, 937–938; VII, 396; IX, 528; and in R. L. Jack, *Catalogue of Works . . . on the Geology . . . etc. of the Australian Continent and Tasmania* (Sydney, 1882).

Among his works are *Plain Statements and Practical Hints Respecting the Discovery and Working of Gold in Australia* (Sydney, 1851); *Researches in the Southern Goldfields of New South Wales* (Sydney, 1860); and *Remarks on the Sedimentary Formations of New South Wales* (Sydney, 1867; 4th ed., 1878).

Clarke's numerous private papers are deposited in the Mitchell Library, Sydney.

II. SECONDARY LITERATURE. A detailed description of Clarke's papers can be found in Ann Mozley, *A Guide to the Manuscript Records of Australian Science* (Canberra, 1966). Articles by the same author are "James Dwight Dana in New South Wales, 1839–1840," in *Journal and Proceedings, Royal Society of New South Wales,* **97** (1964), 185–191; "The Foundations of the Geological Survey of New South Wales," *ibid.,* **98** (1965), 91–100; "Evolution and the Climate of Opinion in Australia, 1840–76," in *Victorian Studies,* **10**, no. 4 (1967), 411–430; and the article on Clarke in *Australian Dictionary of Biography,* III (1969), 420–422. James Jarvis published a brief but uncritical biography, *Rev. W. B. Clarke, M.A., F.R.S., F.G.S., F.R.G.S. The Father of Australian Geology* (Sydney, 1944).

ANN MOZLEY

CLAUDE, FRANÇOIS AUGUSTE (*b.* Strasbourg, France, 30 December 1858; *d.* Paris, France, 15 July 1938), *astronomy.*

Although Alsace had been annexed by Germany in 1871, Claude chose French citizenship as soon as he came of age; he immigrated to France and there served in the army. After being released from his military duties, he left his profession of industrial designer, at which he had worked for six years, and went to the Bureau des Longitudes on his own initiative.

His entire career was spent at the observatory of the Bureau des Longitudes in the Parc de Montsouris in Paris: he became assistant calculator in 1884, chief calculator in 1898, assistant director in 1910, and director in 1929, which post he held until his death. Four times he was awarded a prize by the Académie des Sciences; nevertheless, since he held no university degrees, he was only an associate member of the Bureau des Longitudes (1906).

He was a gracious and reserved man, who devoted himself in a generous way to improving the techniques used in operations for which the Bureau des Longitudes had responsibility: telephonic and radiotelegraphic transmissions of signals and observations of the passage and altitude of stars. His main contribution was the invention of the prismatic astrolabe (1899).

This instrument is made up of a horizontal telescope with an equilateral prism mounted in front of it. The light rays given off by a star reach the prism, in part directly and in part indirectly by reflection from a basin of mercury, thus providing two images to the telescope. These images coincide at the moment when the altitude of the star reaches 60°. Observation of the passage of several stars allows one to determine the geographical coordinates of a certain place.

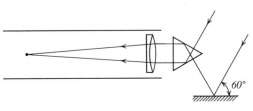

FIGURE 1

The first results, published in 1900, led the hydrographer L. Driencourt to collaborate with Claude in the perfection of this instrument and to adapt it to the needs of geodetic engineers. In 1938 A. Danjon built a high-precision instrument called the impersonal astrolabe on the same principle, and it has become one of today's most important instruments of fundamental astronomy, facilitating the study of the earth's movement around its center of gravity and the determination of the coordinates of fundamental stars.

BIBLIOGRAPHY

I. ORIGINAL WORKS. Claude's most important publications concerning the astrolabe are "Sur l'emploi d'un prisme à réflexion dans les lunettes," in *Bulletin astronomique,* **17** (1900), 19–29; "La méthode des hauteurs égales en astronomie de position," in *Revue générale des sciences pures et appliquées,* **16** (1905), 972–983, written with L. Driencourt; "L'instrument des hauteurs égales ou astrolabe à prisme," *ibid.,* pp. 1071–1083, with L. Driencourt; and

Description et usage de l'astrolabe à prisme (Paris, 1910), with L. Driencourt.

His principal contributions to the *Comptes rendus hebdomadaires des séances de l'Académie des sciences* are "Description d'un niveau autocollimateur," **143** (1906), 394–397, with L. Driencourt; "Comparaisons téléphoniques et radiotélégraphiques de signaux . . . ," **150** (1910), 306–309; **151** (1910), 935–938; **152** (1911), 1152–1155, written with L. Driencourt and G. Ferrié; "L'orthostathmescope, ou instrument pour observer le passage par le zénith de deux étoiles," **155** (1912), 574–577, with L. Driencourt; "L'astrolabe à prisme type SOM," **171** (1920), 847–849, with L. Driencourt.

II. SECONDARY LITERATURE. A historical study of instruments of the astrolabe type was made by E. Chandon and A. Gougenheim, *Les instruments pour l'observation des hauteurs égales en astronomie* (Paris, 1935). An interesting letter from L. Driencourt to A. Danjon is reproduced by the latter in *Bulletin astronomique,* **18** (1955), 254–255.

JACQUES LÉVY

CLAUDE, GEORGES (*b.* Paris, France, 24 September 1870; *d.* St. Cloud, France, 23 May 1960), *technology.*

His father rose from humble beginnings to assistant director at the Manufactures des Glaces de Saint-Gobin. Georges graduated from the municipal school of physics and chemistry in 1889 and then worked at the Usines Municipales d'Électricité des Halles. He married in 1893 and had three children.

After a near-fatal accident with a high-tension wire, he devised better safety measures, which he presented in 1893 to his former teacher d'Arsonval. From 1896 to 1902 Claude was with the Compagnie Française Houston-Thomson, where he developed the method of handling acetylene by dissolving it in acetone and began his work on the manufacture of liquid air. He used thermodynamic principles to improve methods that used the energy of the expanding gas for producing electricity. In the later stages of this work, from 1917 to 1920, he concentrated on the separation and utilization of the rare gases in the atmosphere. During World War I, he used liquid oxygen in explosives and initiated a new approach to ammonia synthesis. Thermodynamic calculations showed him the advantages of going from the then usual 200 atmospheres to 1,000; engineering skill overcame the great practical difficulties. He separated hydrogen from illuminating gas in several stages of cooling, using liquid carbon monoxide as the coolant in the last stage. He used the compressed hydrogen for driving a motor lubricated by a small proportion of injected nitrogen, which not only remained efficient down to $-211°$ C., where all other lubricants failed, but in addition did not contaminate the gas for this particular

use. He solved another great problem by transferring the heat of reaction to the incoming mixture of hydrogen and nitrogen to bring them to the 500° C. required for the synthesis. Thermodynamic considerations also led him into one project that, in 1933, after eight years of effort, resulted in dismal and costly failure: the generation of energy from an apparently inexhaustible source, the ocean, by transferring the heat of warm surface water to the lower temperature of the depths.

In 1903, Claude was honored by the John Scott Medal of the Franklin Institute. Many other honors followed; they corresponded with Claude's inclinations toward public affairs. He presented the results of his work in books, articles, and talks; he founded industrial companies; he became directly involved in politics, when he joined the Action Française in 1919 and stood for election to public office in 1928, without success. Unfortunately, he conducted what he described as "Conférences sous l'occupation allemande: Mes imprudences et mes malheurs" in the years 1940–1945. After the war, he was accused of collaboration with the enemy. He pleaded guilty, was sentenced, and his name was removed from the roster of the Legion of Honor. Through the efforts of friends he was released, after four and a half years, under police surveillance. Little is known about the inventions he devised afterward in retirement.

BIBLIOGRAPHY

His first book, *L'air liquide* (Paris, 1903), with a preface by d'Arsonval, trans. by Henry E. P. Cottrell (Philadelphia, 1913), had several eds. Articles by Claude appear in *Moniteur des produits chimiques,* **9,** no. 91 (1926), 9–12; *ibid.,* no. 92 (1926), 6–11; *Zeitschrift für angewandte Chemie und Zentralblatt für technische Chemie,* **43** (1930), 417–423; and *Chimie et Industrie* (Special No.; March, 1932), pp. 449–452. Among his autobiographical books are *Souvenirs et enseignements d'une campagne électorale* (Paris, 1932); and *Ma vie et mes inventions* (Paris, 1957).

EDUARD FARBER

CLAUS, ADOLF CARL LUDWIG (*b.* Kassel, Germany, 6 June 1838; *d.* Gut Horheim, near Freiburg im Breisgau, Germany, 4 May 1900), *chemistry.*

Claus was one of the first organic chemists to appreciate the importance of understanding the structural relationships of organic compounds, and in his books and papers he helped to make the Kekulé-Couper structural ideas known to all chemists. Claus entered the University of Marburg in 1858, intending to become a physician, but the lectures of Hermann Kolbe turned his interest to chemistry. He then studied under Wöhler at Göttingen. In 1862 he accepted an assistantship at the University of Freiburg im Breisgau, remaining there for the rest of his life. He became the director of the chemical institute at Freiburg in 1883.

With his many students Claus published almost 400 papers in organic chemistry. His experimental researches included the study of alkaloids; the preparation and systematic investigation of quinoline compounds, especially the halogen and sulfonic acid derivatives; the preparation and study of mixed aliphatic-aromatic ketones; and syntheses in the naphthalene, anthracene, and phenanthrene series. His best-known experimental discovery was the synthesis of phenazine in 1872.

Claus was greatly interested in the theoretical problems of organic chemistry and contributed many papers on structural theory. He was a critic of van't Hoff's stereochemical views, maintaining that formulas depicting the arrangement of atoms in space were too hypothetical. During the 1890's he wrote several papers against Hantzsch's application of stereochemical ideas to the isomeric oximes. Claus argued that the oximes of benzil were not structurally identical and that there was no need to apply geometrical formulas to them. Of the two benzil monoximes, only one was truly an oxime,

$$C_6H_5—C=NOH$$
$$C_6H_5—C=O,$$

the other being the isomeric nitroso alcohol,

$$C_6H_5—C—NO$$
$$C_6H_5—C—OH.$$

Until the end of his life Claus fought for his structural formula of benzene against the rival representations of Kekulé, Ladenburg, Dewar, Baeyer, and others. In 1865 Kekulé proposed his hexagonal ring structure for benzene. In attempting to solve the problem of the disposition of the fourth valence of each carbon atom in the benzene ring, Kekulé suggested an arrangement of alternate single and double bonds:

The presence of the double bond was questionable, since benzene did not display the typical addition reactions of double-bonded carbon compounds. Alternatives to Kekulé's structure appeared almost as soon as it was published, Claus advancing his diagonal formula in 1866. He presented two alternatives:

and

In both of these alternatives, of which he preferred the first, each carbon atom was bound to three others, thereby removing the need for double bonds.

Claus's structure was a strong rival to Kekulé's. Kekulé himself declared that Claus's formula was an illuminating one, and Baeyer acknowledged that it agreed best with the facts, including those which Kekulé's formula did not explain. In the 1890's Claus attempted to reconcile his ideas to those of Kekulé by suggesting that the inner chemical bonds of benzene were labile and readily formed double or diagonal bonds, depending on the particular chemical change involved.

BIBLIOGRAPHY

I. ORIGINAL WORKS. Claus wrote two books on theoretical organic chemistry: *Theoretische Betrachtungen und deren Anwendung zur Systematik der organischen Chemie* (Freiburg, 1866) and *Grundzüge der modernen Theorien in der organischen Chemie* (Freiburg, 1871). His criticism of stereochemical explanations is found in "Zur Kenntnis der Oxime und der sogenannten Stereochemie," in *Journal für praktische Chemie,* **44** (1891), 312–335, and **45** (1892), 1–20; and in "Zur Charakteristik der sogenannten Stereochemie des Stickstoffs," *ibid.,* **46** (1892), 546–559. Among his many papers on the structure of benzene are "Über die Constitution des Benzols und des Naphthalins," in *Berichte der Deutschen chemischen Gesellschaft,* **15** (1882), 1405–1411; and "Über die Constitution des Benzols," in *Journal für praktische Chemie,* **42** (1890), 260–267; **43** (1891), 321–344; **48** (1893), 576–595; and **49** (1894), 505–524.

II. SECONDARY LITERATURE. Two informative notices on Claus are Georg W. A. Kahlbaum, in *Biographisches Jahr-*

buch und deutscher Nekrolog, **9** (1906), 348–349; and G. N. Vis, "Adolf Claus," in *Journal für praktische Chemie,* **62** (1900), 127–133.

ALBERT B. COSTA

CLAUS, CARL ERNST (*b.* Dorpat, Russia [now Tartu, Estonian S. S. R.], 23 January 1796; *d.* Dorpat, 24 March 1864), *chemistry, pharmacy.*

Although of German origin, Claus was born and died in Russia and carried out all his work there, so the Russian form of his name (Karl Karlovich Klaus) might be considered more appropriate. Orphaned at an early age, he was forced to earn his own living at fourteen, becoming an apprentice to a pharmacist in St. Petersburg. In 1815 he went back to Dorpat, where he passed the pharmacy examination at the university. He then returned to St. Petersburg. An interest in the botanical aspects of pharmacy caused him to move in 1817 to Saratov, where he spent ten years as a pharmacist's assistant and devoted his leisure to studying the flora and fauna of the Volga steppes. In 1821 he married. He opened his own pharmacy in Kazan in 1826 and in a few years was regarded as an authority on the botany and ecology of the steppes. In 1831 he became an assistant in the chemistry department at the University of Dorpat; there he began to study chemistry, in which he soon received his bachelor's and master's degrees. He applied for the vacant chair of pharmacy at the University of Kazan but instead was called to the new chair of chemistry there, an event that seems to have turned him from a pharmacist and botanist into a chemist. Under his direction, the chemistry laboratory, which was opened in 1838, soon acquired a national reputation. On receiving his doctorate in pharmacy in 1839, Claus was made *professor extraordinarius* of chemistry. In 1844, the year of his discovery of ruthenium, he was made *professor ordinarius.* In 1852 he moved back to Dorpat to occupy a newly created chair of pharmacy at the university, where he continued his work on the platinum metals. In 1861 he became a corresponding member of the Russian Academy of Sciences, and in 1863 the Russian government sent him to Western Europe to visit laboratories and platinum refineries. In 1864, on returning home from a scientific meeting in St. Petersburg, he caught a chill, fell ill, and died.

Beginning in 1840 Claus started work on the insoluble waste residues from the St. Petersburg platinum refineries, previously investigated (1828) by J. J. Berzelius and G. W. Osann, formerly professor of chemistry and physics at the University of Dorpat. In these residues, Berzelius had found only rhodium, palladium, osmium, and iridium, but Osann claimed the presence of three new metals—pluran, ruthen,

and polin. Claus resolved the issue by two years of intensive work (1842–1844); from two pounds of residue, he was able to extract six grams of the last platinum metal to be discovered. He named it ruthenium from the Latin word for Russia, thus honoring his homeland and Osann (1).

Claus also carried out extensive research on iridium, rhodium, and osmium, and in 1854 all his papers on this subject were collected and published as "Beiträge zur Chemie der Platinmetalle" (2) in a jubilee volume in honor of the fiftieth anniversary of the founding of the University of Kazan. His monograph on the history, chemistry, and applications of the platinum metals, left unfinished at his death, was eventually published posthumously in 1883 as *Fragment einer Monographie des Platins und der Platinmetalle* (3). Despite his isolation in a frontier university of eastern Russia, Claus achieved a worldwide reputation for his research on the platinum metals.

Claus's second best-known contribution to chemistry resulted from his research on platinum ammines. In 1856 he proposed a theory that attempted to explain the formation of such compounds and that recognized the analogy between metal ammines and salt hydrates (4). He believed that on combination with metallic oxides, ammonia does not affect the saturation capacity of the metal and that it loses its basicity and becomes "passive." Claus's theory encountered vigorous opposition, but his views were later vindicated when they appeared in only slightly modified form in Alfred Werner's coordination theory (1893) (5).

BIBLIOGRAPHY

I. ORIGINAL WORKS. The works referred to in the text are:

(1) *Memoirs of the Imperial University of Kazan* (Kazan, 1844), pt. III, 15–200 (in Russian); *Journal de Pharmacie et de Chimie,* **8** (1845), 381–385 (in French); *Annalen der Chemie und Pharmacie,* **56** (1845), 257 (in German); *Philosophical Magazine,* **27** (1845), 230 (in English).

(2) *Festschrift zur Jubelfeier des 50 jährigen Bestehens der Universität Kasan.* Copies of this book, which furnished the foundation for all later work on the platinum metals, later became all but unobtainable. A reprint edition was published by the Chemische Fabrik Braunschweig G. M. B. H. in 1926.

(3) Commissionäre der Kaiserlichen Akademie der Wissenschaften (St. Petersburg, 1883). Only three chapters of Claus's projected book were found on his death. In 1864 these were sent to the Academy of Sciences, and printing was begun but, for some unknown reason, stopped. In 1883 Butlerov, Claus's former pupil and successor at Kazan, completed the book. James Lewis Howe used the biblio-

graphical section in the compilation of his *Bibliography of the Metals of the Platinum Group.* Claus's monograph is extremely rare; presumably only 300 copies were printed, and the only copy still in existence in 1946 outside the Soviet Union was Howe's copy in the library of the chemistry department at Washington and Lee University, Lexington, Virginia.

(4) "Ueber die Ammoniummoleküle der Metalle," in *Annalen der Chemie und Pharmacie,* **98** (1856), 317–333. An English translation appears in G. B. Kauffman, *Classics in Coordination Chemistry, Part II: Selected Papers (1798–1935)* (New York, in press).

(5) A. Werner, "Beitrag zur Konstitution anorganischer Verbindungen," in *Zeitschrift für anorganische Chemie,* **3** (1893), 267–330. An English translation appears in G. B. Kauffman, *Classics in Coordination Chemistry, Part I: The Selected Papers of Alfred Werner* (New York, 1968), pp. 9–88.

II. Secondary Literature. The source of most biographical data on Claus is B. N. Menshutkin, in *Izvestiya Instituta po izucheniyu platiny i drugikh blagorodnykh metallov,* **6** (1928), 1–11. Biographical data, with emphasis on the discovery of ruthenium, are found in M. E. Weeks, *Journal of Chemical Education,* **9** (1932), 1017–1034; *Discovery of the Elements,* 6th ed. (Easton, Pa., 1956), pp. 440–447; and D. McDonald, in *Platinum Metals Review,* **8** (1964), 67–69.

GEORGE B. KAUFFMAN

CLAUSEN, THOMAS (*b.* Snogbaek, Denmark, 16 January 1801; *d.* Dorpat, Estonia [now Tartu, Estonian S.S.R.], 23 May 1885), *mathematics, astronomy.*

Thomas Clausen was the eldest of the eight children of Claus Clausen and Cecilia Rasmussen Clausen. The elder Clausen was a poor peasant farmer in northern Schleswig. A local pastor, who was interested in the natural sciences and for whom young Clausen herded cattle, became interested in the young man and instructed him in Latin, Greek, mathematics, astronomy, and natural science over a seven-year period. On his own, Clausen studied French, English, and Italian. Upon the pastor's recommendation, H. C. Schumacher made Clausen his assistant at his Altona observatory in 1824. Clausen, a very individualistic man, had a falling out with his superior; and at the end of 1828 he moved to Munich as the appointed successor to Fraunhofer at the Joseph von Utzschneider Optical Institute. He held this position in name only, however; Utzschneider allowed Clausen to devote himself completely to his mathematical and astronomical calculations and publications, which soon gained him the attention and recognition of such authorities as Olbers, Gauss, Bessel, Steinheil, Hansen, Crelle, A. von Humboldt, Arago, and W. Struve.

In the middle of 1840, after a severe bout with mental illness, Clausen returned to Altona, where he spent two years in seclusion and reached the zenith of his scientific creativity. He also engaged in a mathematical argument with C. G. J. Jacobi. In 1842 he was appointed observer at the Dorpat observatory, and in 1844 he took his doctorate *honoris causa* under Bessel. On 1 January 1866 he was made director of the Dorpat observatory and professor of astronomy at Dorpat University. He went into retirement at the end of 1872. Clausen never married. In 1854 he received through Gauss a corresponding membership in the Göttingen Academy. In 1856 he received the same class of membership from the St. Petersburg Academy.

Like many astronomers of the first half of the nineteenth century, Clausen was a self-made man. He differed from most of his professional colleagues in that he was in a position to make a major contribution to those mathematical problems with which the leading intellectuals of his time were preoccupied. He possessed an enormous facility for calculation, a critical eye, and perseverance and inventiveness in his methodology. As a theoretician he was less inclined toward astronomy. Gauss soon recognized the "outstanding talents" of Clausen. The Copenhagen Academy awarded him a prize for his work "Determination of the Path of the 1770 Comet."[1] Bessel commented: "What a magnificent, or rather, masterful work! It is an achievement of our time which our descendants will not fail to credit him with."[2] Clausen's approximately 150 published works are devoted to a multitude of subjects, from pure and applied mathematics to astronomy, physics, and geophysics. He repeatedly solved problems that were posed to him publicly by other mathematicians or proved theories that had been published without proof, as was the custom at that time, and corrected mistakes and errors in others. Special mention should be made in this connection of the calculation of fourteen paths of comets, as well as of the theorem, named for Staudt and Clausen, dealing with Bernoullian numbers.[3] Fermat had hypothesized that all numbers of the form $F(n) = 2^{2^n} + 1$ were prime numbers. Euler disproved this hypothesis in 1732 by factoring Fermat's number $F(5) = 2^{2^5} + 1$. In 1854 Clausen factored $F(6) = 2^{2^6} + 1$ and thus proved that this number also is not a prime number. Clausen used as a basis his own new method—still unpublished—of factoring numbers into their prime factors.[4] There is substantial reason for believing that Clausen gained a deeper insight into the field of number theory than the material published by him would indicate.

NOTES

1. *Astronomische Nachrichten,* **19** (1842), 121–168.
2. *Ibid.,* pp. 335–336.
3. *Ibid.,* **17** (1840), 351–352.
4. *Journal für die reine und angewandte Mathematik,* **216** (1964), 185.

BIBLIOGRAPHY

I. ORIGINAL WORKS. In addition to the works cited in the text, see *Astronomische Nachrichten, Journal für die reine und angewandte Mathematik, Archiv der Mathematik und Physik,* and *Publications de l'Académie impériale des sciences de St. Pétersbourg,* in which almost all of Clausen's works appeared. See also the bibliographical aids in the Biermann work cited below.

II. SECONDARY LITERATURE. For a study of Clausen, see Kurt-R. Biermann's "Thomas Clausen, Mathematiker und Astronom," in *Journal für die reine und angewandte Mathematik,* **216** (1964), 159–198. This biography contains a virtually complete list (pp. 193–195) of the literature dealing with Clausen and a list of his MSS and unedited letters (pp. 191–193). Moreover, reference should be made to J. Gaiduk, "Thomas Clausen ja tema matemaatika-alane looming," in *Matemaatika ja kaasaeg,* **12** (1967), 116–122; and U. Lumiste, "Täiendusi Th. Clauseni biograafiale," *ibid.,* pp. 123–124.

KURT-R. BIERMANN

CLAUSIUS, RUDOLF (*b.* Köslin, Prussia [now Koszalin, Poland], 2 January 1822; *d.* Bonn, Germany, 24 August 1888), *physics.*

After receiving his early education at a small private school that his father had established and was serving as pastor and principal, Clausius continued his studies at the Stettin Gymnasium before entering the University of Berlin in 1840. Although strongly attracted by Leopold von Ranke's lectures in history, he settled on a career in mathematics and physics, completing his doctor of philosophy degree at Halle in 1847.

His famous paper on the theory of heat (1850) led to his first major teaching position at the Royal Artillery and Engineering School in Berlin; his ensuing publications resulted in an invitation to be professor of mathematical physics at the new Polytechnicum in Zurich in 1855. In Zurich he joined a faculty which would soon number among its younger members Richard Dedekind in mathematics, Gustav Zeuner in mechanics, and Franz Reuleaux in machine design; Albert Mousson taught experimental physics and became Clausius' lifelong friend. During Clausius' years at Zurich, he enjoyed frequent visits with John

Tyndall, whom he had first met in 1851 as a fellow member of the circle of students gathered about Gustav Magnus in Berlin.

In 1867 Clausius accepted a professorship at the University of Würzburg. (He expressed regrets at leaving Zurich but longed to return to Germany.) He moved on to Bonn two years later, remaining there for the rest of his life and serving as rector of the university during his final years. Clausius received many scientific honors; he was elected an honorary member of numerous scientific societies and received many awards, of which the most notable was the Copley Medal of the Royal Society in 1879.

Clausius' major contributions to physics seem to have come predominantly during the years prior to his going to Bonn, and two events may have hampered his later life as a scholar. He was wounded in 1870 while leading a student ambulance corps in the Franco-Prussian War and suffered continuing pain from that injury. More tragically, he had to assume sole responsibility for the care of his family in 1875, when his wife died while giving birth to their sixth child; he did not remarry until 1886, shortly before the close of his career.

The significant beginning of that career, of course, dates from 1850, when Clausius established the foundations for modern thermodynamics in his first great paper on the theory of heat, "Ueber die bewegende Kraft der Wärme,"[1] but several earlier papers are of interest as a revealing prelude to that major work. In one paper, dealing mainly with the problem of reflected light in the sky, Clausius' distinctive approach to many physical questions was already apparent: an excellent grasp of the fundamental facts and equations relevant to the phenomena, a microscopic model to account for them, and an attempt to correlate the two with mathematics. For example, Clausius imagined that the blue color of the sky arose from the preferential reflection of blue light from thin films of water. He postulated, therefore, the existence of water bubbles in the earth's atmosphere and investigated mathematically the requisite conditions of number, size, and thickness. Expressing a similar type of approach in another early paper, Clausius proposed that changes in molecular arrangement might explain certain anomalies encountered in the experimental study of elastic solids. In fact, that paper may mark the beginning of the line of thought which led to his rejection, in 1850, of the caloric theory in favor of the new principle of the equivalence of work and heat, for in his consideration of the possible microscopic explanations of these anomalies he criticized several proposals that were based on the concepts of free and bound molecular heats, concepts fundamental to the prevailing view.

Two ideas were central to the caloric theory: (1) the total heat in the universe is conserved and (2) the heat present in any substance is a function of the state of that substance. In thermodynamics the state of a simple homogeneous substance is defined by any two of the three variables, pressure, volume, and temperature; and any property that is similarly determined, and thus called a function of state, is subject to a certain well-developed mathematical treatment. It was therefore the second assumption—that the heat present in a substance is a function of the state of that substance—which transformed the caloric theory from a vague supposition about an imponderable fluid to a sophisticated mathematical system in which permanently valid relations could be derived. For example, Poisson was able to establish the correct pressure and volume relation for the adiabatic expansion of gases, and Clapeyron, the variation of vapor pressure with temperature.

The caloric theory further provided a conceptual framework for explaining the behavior of gases and vapors in terms of a distinction between free and latent states of heat. Free heat could be sensed and measured by a thermometer; latent heat, however, because it was intimately bound to the molecules, could not. Therefore, the temperature of a gas rose when it was compressed because part of the bound latent heat apparently was squeezed free. Clausius not only denied the fundamental assumptions of the caloric theory but also provided a new mechanical explanation for the traditional concepts of free and latent heat.

The denial was based, of course, on what has become the first law of thermodynamics—that whenever work is produced by heat, a quantity of heat equivalent to that amount of work is consumed—a premise that Clausius believed had been firmly established by Joule's experiments. The total heat in the universe, therefore, could not be conserved; and the usual concept of heat in a substance, representing the total heat added to that substance, could no longer be considered a function of state. In Clausius' reinterpretation, the only kind of heat that could have any real existence in a substance was the free heat; and free heat was understood as the *vis viva* (kinetic energy) of the fundamental particles of matter and the determiner of temperature. Latent heat, in contrast, was heat that no longer existed, having been destroyed by conversion into work—internal work against intermolecular forces and external work against the surrounding pressure. Clausius made an important distinction between these two forms of

work: The internal work, being determined by molecular configuration, is a state function, subject solely to the initial and final conditions of change; the external work, however, depends on the conditions under which the change occurs.

In 1850 Clausius did not give mathematical expression to these interpretive ideas of heat in a body and internal work but rather simply illustrated them with an explanation of the vaporization of water; they did, however, form the conceptual framework for his theory of heat. Even in the final edition of *Die mechanische Wärmetheorie,* which appeared in 1887, he still introduced the first law of thermodynamics as

$$dQ = dH + dJ + dW,$$

where the increment of heat dQ added to a body is equal to the sum of the changes in the heat in the body dH, the internal work dJ, and the external work dW. Only after confessing ignorance of the exact expression for the internal work dJ did Clausius introduce the classic thermodynamic expression

$$dQ = dU + dW,$$

in which U was simply the energy in the body, without any attempt to differentiate that energy into molecular forms. The function U played a very important role in the new mechanical theory of heat, and it was Clausius who introduced this new state property into thermodynamic thought, another major contribution of his 1850 paper.

Clausius' manner of establishing the function U as a state property of a substance reveals another distinguishing characteristic of his thought. In his approach to the theory of heat, general concepts should not be dependent upon particular molecular models because those very concepts form the structure within which models must operate. Hence, although Clausius could have introduced the function U simply as the sum of H and J (as he did later in his developed theory), he chose to assure complete generality for his original derivation in 1850 by employing an extremely tedious analysis of an infinitesimal Carnot cycle. It was just such an approach, independent of molecular assumptions, that became normative in thermodynamic thought.

For example, when Kelvin developed his dynamical theory of heat in 1851, he did not explain the function U in terms of molecular energy states. He simply argued that since Q and W must be equal whenever a substance undergoes a full cycle of changes, a consequence of the theorem of the equivalence of work and heat, and since every function that is characteristic of a substance shows no net change in a cycle, $Q - W$ must represent some new function of the substance. Thus, he approached the question purely on the macroscopic level and coined the name "intrinsic energy" for U, since it represented the total mechanical work that might be theoretically obtained from the substance.

Clausius, of course, had discovered the function U a year earlier than Kelvin, but he had never given it a name. The reason is clear: The heat in a body H and the internal work J were his fundamental concepts. For example, in order to simplify the function U for gases and to derive Poisson's equation for the adiabatic behavior of gases, he argued that intermolecular forces are negligible in gases because of the relative uniformity of their pressure, volume, and temperature relations. He was able, therefore, to reduce the function U to the heat absorbed at constant volume and demonstrated that this heat could only depend on the temperature. Such a simplification of the function U normally requires the use of the second law of thermodynamics, but Clausius originally proposed the idea solely on the basis of the first law and his molecular model for heat and internal work. That model for heat perhaps also provided the rationale for his brilliant revision of the traditional Carnot argument in thermodynamics, a revision that made it possible to incorporate the most significant results of the caloric theory within the new theoretical structure.

It was the idea of a universal function of temperature, introduced by Carnot but fully developed only by Clapeyron, that had proved the genius of the caloric theory. The idea depended on Carnot's theorem that all ideal engines must produce equal amounts of work with equal amounts of heat when operating between the same temperatures, whatever the substance being employed. The proof was an indirect one—that is, denying the premise would lead to a contradiction of the traditional mechanical principle that work cannot be created from nothing. In the mechanical theory of heat, however, there would no longer be a contradiction of that principle, since any work produced must be accompanied by the disappearance of an equal quantity of heat. Some other principle would have to be contradicted. Clausius rearranged the traditional argument so that any denial of the premise would require that heat be transferred from a colder to a warmer body, a conclusion which Clausius stated would be an obvious deviation from the normal behavior of heat. Clausius was therefore able to derive Clapeyron's equation for the universal temperature function C and to continue Clapeyron's fruitful handling of vapor–liquid equilibrium. Using his assumption that internal forces are absent in gases, he established the relation

$$C = A(a + t),$$

where A is the mechanical equivalent of heat and a the coefficient of expansion for gases. Clausius did not introduce the symbol T for this universal function of temperature $(a + t)$ until his second important paper, "Ueber eine veränderte Form des zweiten Hauptsatzes der mechanischen Wärmetheorie,"[2] in which he developed the concept of entropy.

Clausius did not propose the name "entropy" at this time, calling the new theorem simply the principle of the equivalence of transformations.[3] This accorded well with the unique engine cycle that he had conceived to establish the theorem, a cycle portraying two important types of transformation: a conversion of heat into work at one temperature and a transfer of heat from a higher to a lower temperature. The two transformations were called equivalent because they could replace one another. For example, suppose that the transfer of heat had occurred. By operating the cycle in reverse, Clausius argued, the heat could be restored to its original temperature and there would be a conversion of work into heat. Clausius, by assigning positive transformation values to these processes and equal negative values to their opposites, established by his modified Carnot argument that the transformation values could only be universal functions of heat and temperature, $Qf(t)$ and $Q_1F(t_1,t_2)$.

Clausius still had to establish the nature of the function $F(t_1,t_2)$. By combining several cycles, he demonstrated that the transformation value for a flow of heat could be reduced to the same form as a conversion of heat into work, namely,

$$Q_1F(t_1,t_2) = -Q_1f(t_1) + Q_1f(t_2),$$

so that every exchange of heat could be treated identically. The sum of transformation values for his cycle, therefore, was simply the sum of $Qf(t)$ at all temperatures; and, since the cycle consisted of a transformation plus the inverse of its equivalent,

$$\sum Qf(t) = 0.$$

Extending the relation to any reversible cycle, Clausius created a new function of state,

$$\int dQ/T = 0.$$

The complexity of this derivation should appear strange to a modern mind, for we see no essential difference between the heats occurring at t, t_1, and t_2 in Clausius' original cycle and wonder why he differentiated them. The occurrence of Q_1 at both t_1 and t_2 should not necessitate treating those heats as any more functionally related than the heat at t. Undoubtedly, Clausius was still under the influence of the caloric theory, in which the work accomplished by an engine cycle was supposedly caused by the fall of heat. His derivation of the new function of state, $\int dQ/T$, put this remaining idea at an end.

Actually, his proposed new function was not unique with Clausius, for Rankine had introduced a similar function; and suggestions of such a relation also appeared in Kelvin's thought. Clausius made his distinct contribution by considering the case of irreversible processes. He concluded that since a negative transformation value would correspond to a flow of heat from a lower temperature to a higher temperature, contrary to the normal behavior of heat, the sum of transformation values in any cycle could only be zero (reversible) or positive (irreversible). He later capsuled this idea in his famous couplet for the two laws of thermodynamics: "Die Energie der Welt ist constant; die Entropie strebt einen Maximum zu."[4]

Clausius often called upon this theorem of transformation values when challenged to defend the principle on which it was based, that heat cannot of itself (*von selbst*) go from a cold body to a hot body. For example, when Gustav Hirn proposed an intriguing thought experiment in which a gas would be heated beyond 100°C. by using a source of heat at 100°, Clausius showed that the net transformation value for the process would nonetheless be positive and therefore would not contradict his fundamental principle. The most rabid and persistent critic of that principle was the controversial Peter Guthrie Tait of Edinburgh.

Their clash was perhaps inevitable. Both were chauvinistic; both were involved in the Tait-Tyndall dispute over the relative priority claims of Joule and Julius Mayer to the discovery of equivalence of work and heat. Clausius, having innocently become involved by sending Tyndall, prior to the controversy, the complete set of Mayer's publications (at Tyndall's request), did not remain neutral. He affirmed Mayer's priority and claimed that Mayer had thus secured for Germany a national priority as well.[5]

This was not the first occasion on which Clausius championed German achievements against apparent British infringements. For example, in 1856 he criticized William Thomson for quoting an earlier letter from Joule containing Joule's suggestion that Carnot's function *C* might be the absolute gas temperature.[6] Clausius reminded the English that a German, Karl Holtzmann, had been the first to establish that relationship in 1845 and then proceeded to recount his own more recent contributions as well. In 1872, however, when Clausius began his bitter controversy with Tait by claiming that the British seemed intent on claiming more than their rightful share of the theory

of heat,[7] he was arguing for recognition of his rights alone.

Their first exchanges appeared in the *Philosophical Magazine,* but the continuing argument entered the prefaces and appendices of the various editions of their books. Clausius appears to have rebutted every denial of his axiom that Tait could muster. For example, Tait finally argued, perhaps in some desperation, that Maxwell's demon could contradict Clausius' principle by separating the faster molecules from the slower molecules. The German replied that his principle concerned what heat *von selbst* might do—and not with the help of demons.

The immediate occasion for Clausius' comments in 1872 on the British approach to the theory of heat was the appearance of Maxwell's *Theory of Heat,* in which the word "entropy," following Tait, was associated with the available energy in a system. This was directly contrary to Clausius' own interpretation of his concept, and Maxwell responded by making certain revisions in his treatment. Maxwell had been more than fair to Clausius some years earlier, when, acknowledging his debt for certain fundamental ideas in the kinetic theory of gases, he hailed Clausius as the founder of a new science.

Clausius' first venture into the kinetic theory was "Ueber die Art der Bewegung, welche wir Wärme nennen,"[8] and his image of molecular motion went far beyond the "billiard ball" model of such previous writers as Krönig. He ascribed rotatory and vibratory, as well as translational, motion to the molecules, a complexity that led to an important conclusion. The conservation of translational kinetic energy in collision could no longer be assumed, because collisions might cause transformations of one form of motion into another. Quite obviously, Clausius argued, the idea of a constant equal velocity for all molecules must be untenable. By supposing that translational velocities would vary among the molecules, Clausius offered an explanation for the evaporation of a liquid. Since only molecules with higher than average velocities possess sufficient kinetic energy to escape the attractive forces of the liquid, there is a loss of energy and a drop in liquid temperature.

In his discussion of the complexity of molecular motions, Clausius did not merely suggest that motions other than translational were a possibility; he demonstrated that such motions must exist by showing that translational motion alone could not account for all of the heat present in a gas. He began, of course, by deriving the fundamental equation in the kinetic theory of gases,

$$3/2 \cdot pv = nmu^2/2,$$

which related the pressure p, the volume v, the number of molecules n, their mass m, and the average velocity u. The total translational kinetic energy K therefore was $3/2 \cdot pv$. The total heat H, however, corresponding to the heat added at constant volume to an ideal gas (intermolecular forces being negligible), was $[c/(c'-c)]pv$, where c was the heat capacity at constant volume and c' the heat capacity at constant pressure. The ratio of translational energy to total heat was therefore

$$K/H = 3/2(c'/c - 1);$$

and since the specific heat ratio was known to be about 1.42 for simple gases, the translational energy K could be only a fraction of the heat. Other motions, Clausius maintained, must therefore exist in the molecular realm. Thus he established the first significant tie between thermodynamics and the kinetic theory of gases.

This 1857 paper also marked another important beginning in physical theory, for it presented the first physical argument in support of Avogadro's hypothesis that equal volumes of gases at the same temperature and pressure contain equal numbers of molecules. Clausius argued that if it were assumed that all types of molecules possess the same translational energy at equal temperatures, then, since all gases have the same relationship between pressure, volume, and temperature, they would necessarily contain equal numbers of molecules in equal volumes at the same temperature and pressure. Avogadro's hypothesis, therefore, found support in the mechanical theory of heat, independently of the usual chemical arguments.

When, however, Clausius began to treat the chemical evidence in order to argue the case for diatomic molecules, it is surprising to learn that he apparently thought he was the first to make that suggestion. For example, he italicized his statement of Avogadro's hypothesis as though it were completely new and, in a note later appended to the paper, admitted that it was only upon reading comments by Verdet and Marignac that he learned of similar views advanced earlier by Dumas, Laurent, and Gerhardt. Still later he wished to claim, in a note added in 1866 to his paper on ozone, that he was the first to state unequivocally that the oxygen molecule was diatomic, arguing that Gerhardt sometimes wrote his formula O^3O and sometimes OO. Fortunately, no such priority concerns ever colored his second important paper on kinetic theory, "Uber die mittlere Länge der Wege,"[9] for there he developed the idea of the mean length of path of a moving molecule, an idea no one else could claim.

The occasion for Clausius' investigation of the progress of molecular collisions was the objection raised against the kinetic theory of gases by Buys Ballot: that if molecules actually possessed the velocities ascribed to them, the diffusion of gases should occur practically instantaneously. Clausius replied that the apparent discrepancy could be explained by assuming that collisions occur so frequently among the molecules that their forward progress is continually interrupted.

In order to analyze the process, Clausius adopted a simplified model for his admittedly complicated molecule. He assumed that whatever the actual patterns of intermolecular forces, one could suppose that there is some average distance between the centers of molecules which would represent a general boundary between attractive and repulsive forces. If two molecules were to approach each other within that boundary, repulsion would generally occur. Thus the very complex problem of intermolecular action was reduced to a "billiard ball" model.

In order to derive an expression for the average path of a molecule between collisions, Clausius imagined only one moving molecule, with the remainder fixed in an essentially lattice framework, and set the probability of a collision proportional to the fraction of area cut off by the repulsive spheres of action. He found that for this hypothetical case, the mean length of path would be λ^3/ρ^2, where λ is the average distance between the stationary molecules in the lattice framework and ρ is the radius of the repulsive collision sphere about each molecule. It was only when Clausius considered the true case of all molecules in motion that he came upon an apparently significant relationship. Thus, after stating (without giving the detailed proof) that the true mean length of path would be reduced by a factor of 3/4 because the relative velocity is 4/3 the actual average velocity, Clausius found the surprising consequence that

$$\frac{l}{\rho} = \frac{\lambda^3}{4/3\pi\rho^3},$$

where l is the mean free path. It appeared, therefore, that a neat relationship existed, the ratio between the mean length of path and the radius of the collision sphere being equal to the ratio of the average space between molecules and the volume of the collision sphere for each molecule. Perhaps Clausius found this unexpected result to be a guarantee of the validity of his approach, for he rather abruptly challenged Maxwell for proposing a change in 1860 and completely failed to grasp the significance of the new direction in Maxwell's reasoning.

The controversy concerned Maxwell's first brilliant

paper on the kinetic theory of gases, in which he proposed his famous law for the distribution of velocities among colliding spherical molecules. By using this new distribution function, Maxwell was able to establish that the mean relative velocity between molecules would be larger than their average velocity by a factor of $\sqrt{2}$. Clausius, apparently considering this an affront to his own claims, sent a very curt note to the *Philosophical Magazine*[10] in which he detailed how he had derived 4/3, thus ignoring completely that Maxwell had adopted a wholly new approach. Maxwell never replied publicly; but he did note somewhat ironically to Tait, in recounting the history of the question, "Clausius supposing Maxwell's knowledge of the integral calculus is imperfect writes to Phil Mag showing how to do the integration on the assumption $v = $ const."[11] For some strange reason, Clausius never did adopt Maxwell's distribution function and continued to operate with a uniform velocity.

Clausius later gave more careful scrutiny to Maxwell's arguments and found an error that Maxwell was to admit as far more serious.[12] In his initial approach to the conduction of heat in gases, Maxwell drew a brilliant analogy between diffusion (a transfer of mass) and conduction (a transfer of kinetic energy), thereby making it possible to use the form of his diffusion equation to represent conduction, simply replacing the mass of a molecule with its kinetic energy. Clausius criticized this adoption of the diffusion equation, because, given the assumptions, mass transfer would accompany the heat conduction and the process would not be one of energy transfer alone. He then offered a revised theory of conduction that was more painstaking than brilliant, its only net result being replacement of a factor of 1/2 with 5/12. No further significant contributions to kinetic theory came from Clausius' pen, as he turned his thought to understanding the second law of thermodynamics.

Clausius began that search for understanding in 1862 by introducing the concept of disgregation, a concept that, he said, was based on an idea he had long held: that the force of heat for performing mechanical work (both internal and external together) was proportional to the absolute temperature.[13] Clausius had never stated this idea explicitly before, although he had argued in 1853, by adopting an analogy between a reversible steam engine and a thermocouple, that the potential difference at a thermocouple junction should be proportional to the absolute temperature. In any event, he now wished to assert that the work which can be done by heat in any change of the arrangement of a body is proportional to the absolute temperature multiplied by a function of molecular arrangement, the disgregation Z. Given this

assumption and his postulate that the heat in a body H was only a function of temperature, he was able (1) to prove his theorem of the equivalence of transformations and (2) to separate the equivalence function (entropy) into a temperature-dependent term and a configurational-dependent term,

$$dQ/T = dH/T + dZ.$$

The mathematical expression itself was not new, Rankine having established an identical relation several years before. Clausius, however, gave a more concrete meaning to the terms involved by relating Z directly to the configuration of the particles. He seemed to be trying to handle entropy in the same way he had conceptualized the function U, breaking it into temperature-dependent and configurational-dependent terms. Some years were to pass, however, before he was able to derive an equation in pure mechanics that bore some correspondence to his idea.

His first venture led to the virial equation (1870).[14] The derivation followed simply from a well-known equation of classical mechanics,

$$m/2(dx/dt)^2 = 1/2Xx + m/4\, d^2(x^2)/dt^2,$$

where the force X and the motions of particles are only in the x direction. Clausius eliminated the second term by taking the time average over an extended interval,

$$m/2t \int_0^t (dx/dt)^2\, dt = -1/2t \int_0^t Xx\, dt$$
$$+ m/4t[dx^2/dt_{t=t} - dx^2/dt_{t=0}],$$

and observing that in the theory of heat, neither position nor velocity could ever increase indefinitely, so that the last term becomes negligible as the time interval approaches infinity. The virial for a large number of points in three-dimensional space became

$$\sum m/2\overline{v^2} = -\frac{1}{2}\sum \overline{(Xx + Yy + Zz)},$$

an equation which since Van der Waals has become fundamental for deriving equations of state for real gases. Clausius, however, never extended his equation in any direction; to him the virial was but one possible expression of his idea that the force of heat—that is, the translational kinetic energy of the molecules—was proportional to the absolute temperature. He soon abandoned this suggestion in favor of another approach that provided a more promising mechanical analogue for disgregation.[15]

Adopting a model in which moving mass points traverse certain periodic closed paths under internal and external forces defined by a function U (the potential energy function), Clausius analyzed the case in which an increment of kinetic energy is given to a particle and found that the variation in path would be governed by the relationship

$$\delta \bar{U} = h \delta \log (h i^2),$$

in which δ represents the amount of variation in the quantities indicated, h is the average kinetic energy of the particle (absolute temperature), and i the period of its cyclical motion. Since the change in potential energy U traditionally represented work, the change in disgregation, by analogy, should be related to $\delta \log (h i^2)$. If, Clausius suggested, one further assumed that the particle moved with constant speed, then the variation in disgregation would be proportional to the variation in the logarithm of the path length, an intimation, therefore, of some correlation with molecular configuration. No sooner did Clausius present this successful interpretation of disgregation than a young Austrian physicist, Ludwig Boltzmann, disclosed that he had published essentially the same reduction of the second law in 1866.

Clausius, however, challenged Boltzmann's claim to priority, arguing that he had taken account of a complication which, he felt, Boltzmann had ignored: that the potential energy function U was itself subject to change during the variation, and that the variation of the potential energy function U could therefore be equated with the work done only if that additional factor were proved to be negligible. In traditional mechanics this problem did not arise because the external forces were always assumed to be constant and fixed in space. In the theory of heat, however, either pressure or volume is bound to vary and thus to change either the intensity or the location of the external forces. Clausius claimed that he had taken this complication into account but that Boltzmann had not.[16]

Clausius now devoted several years to the elaboration of what he thought represented a new and unique contribution to theoretical mechanics, his idea of a variation in the force function itself. He ignored the new directions in Boltzmann's thought and, surprisingly, never once sought to find a mechanical explanation for the irreversible increase of entropy. In fact, in his final attempt, he even adopted a model in which he reduced the admittedly disordered collisions of molecules to a case of noncolliding mass points in ordered motion.[17] Clausius made no further attempt to probe for a molecular interpretation of the second law, and his last significant contribution to the theory was a proposal for an amended version of Van der Waal's equation of state. His arguments, however, showed little relation to any molecular model; and his final equation represented the outcome of a rather methodical search for an improved empirical correlation of the data for carbon dioxide.[18]

Clausius' major effort in mathematical physics after 1875 involved his quest for an adequate electrodynamic theory. He spelled out the fundamental tenets to his approach in 1875: (1) Weber's law was incorrect for the case where only one kind of electricity is assumed to move, since the equation entails that a current exert a force on a charge at rest; (2) a revision would be possible if one assumed that the electrodynamic action occurred via an intervening medium, for then electric particles that are not moving relative to each other (moving at equal velocities) could still exert forces on one another by virtue of their absolute motion in the medium; (3) forces should not be restricted to the line joining two charges.[19] In 1876 Clausius simplified the equation he had previously proposed by subjecting it to the condition of the conservation of energy.[20] In doing so, however, he ignored the possibility that energy changes might occur in the intervening medium.

His most elaborate treatment appeared in 1877,[21] and he there included Riemann's equation within the indictment he had earlier aimed at Weber's. Actually, Clausius framed an equation that showed marked similarities to Riemann's, except that his noncentral forces were parallel and proportional to the absolute velocities and accelerations, rather than the relative values.[22] Clausius quite obviously had not tailored this theory to fit any explicit model for a single type of electricity in motion. In fact, he indicated in 1877 that his equation was not restricted to the case of a single mobile electricity but would be valid also for two electricities, whether equal or unequal in strength. His thought on electrodynamics was ruled by the earlier tradition plus two principles, the conservation of energy and simplicity, and simplicity played an important role. Thus, in his 1877 derivation, after obtaining a very general form for the equation by applying the conservation of energy as a condition, he introduced a simplifying constant and then set that constant equal to zero to obtain the desired form. He justified the choice by saying that the final equation was "in einfachster und daher wahrscheinlichster Form" ("in the simplest and therefore most probable form").

Not many scientists followed Clausius' lead, and a number of penetrating criticisms arose. One argument, ironically enough, posed the difficulty that Clausius' equation would also entail a force on a charge at rest on the earth, since the earth is moving in space. Clausius replied that the absolute velocity in his equation was relative to the surrounding medium and not to space, so that if the medium were to move with the earth, then the earth itself would be the referent for absolute motion and no such supposed force should occur.[23] In one of those strange quirks of history, H. A. Lorentz, for whom the ether was absolutely at rest, adopted Clausius' electrodynamic equation for deriving the force on an electron moving in that supposedly immovable ether.[24] Perhaps Lorentz was aware that Emil Budde had answered the objection to Clausius' equation without supposing any convection of the ether. Budde had argued that the movement of the earth in the ether would cause separations within the electricity and that those separations would cause electrostatic forces equal and opposite to the induced electrodynamic ones, thus eliminating the supposed force due to the motion of the earth.[25]

Wiedemann brought Budde's argument to Clausius' attention in 1880, at the very time Clausius was penning his own reply. Clausius granted in a footnote that Budde's answer would suffice but nonetheless chose to offer his conjecture of ether convection. Clausius' ignoring of a valid quantitative answer in favor of his own speculative generality is significant, for we find a similar indifference to ideas being expressed in thermodynamics. Clausius' great legacy to physics is undoubtedly his idea of the irreversible increase in entropy, and yet we find no indication of interest in Josiah Gibbs's work on chemical equilibrium or Boltzmann's views on thermodynamics and probability, both of which were utterly dependent on his idea. It is strange that he himself showed no inclination to seek a molecular understanding of irreversible entropy or to find further applications of the idea; it is stranger yet, and even tragic, that he expressed no concern for the work of his contemporaries who were accomplishing those very tasks.

NOTES

1. "Ueber die bewegende Kraft der Wärme und die Gesetze welche sich daraus für die Wärmelehre selbst ableiten lassen," in *Annalen der Physik,* **79** (1850), 368–397, 500–524.
2. Ibid., **93** (1854), 481–506.
3. "Ueber verschiedene für die Anwendung bequeme Formen der Hauptgleichungen der mechanischen Wärmetheorie," *ibid.,* **125** (1865), 353–400. Clausius chose the name "entropy" from the Greek word for transformation.
4. Ibid., p. 400.
5. "Rescension der Mayer'schen Schriften," in *Literarisches Zentralblatt für Deutschland* (Leipzig, 1868), pp. 832–834.
6. "On the Discovery of the True Form of Carnot's Function," in *Philosophical Magazine,* **11** (1856), 388–390.
7. "A Contribution to the History of the Mechanical Theory of Heat," *ibid.,* **43** (1872), 106–115. See also the preface to the 2nd ed. of Tait's *Sketch of Thermodynamics* (1877) and the section entitled "Tendenz des Buches *Sketch of Thermodynamics* von Tait," in Clausius' *Die mechanische Wärmetheorie,* 2nd ed., II (Brunswick, 1879), 324–330.
8. *Annalen der Physik,* **100** (1857), 497–507.

9. "Ueber die mittlere Länge der Wege, welche bei der Molecularbewegung gasförmiger Körper von den einzelnen Molecülen zurückgelegt werden," *ibid.,* **105** (1858), 239–258.
10. "On the Dynamical Theory of Gases," **19** (1860), 434–436.
11. Maxwell-Tait correspondence, Cambridge University Library.
12. "Ueber die Wärmeleitung gasförmiger Körper," in *Annalen der Physik,* **115** (1862), 1–56.
13. "Ueber die Anwendung des Satzes von der Aequivalenz der Verwandlungen auf die innere Arbeit," *ibid.,* **116** (1862), 73–112.
14. "Ueber einen auf die Wärme anwendbaren mechanischen Satz," *ibid.,* **141** (1870), 124–130.
15. "Ueber die Zurückführung des zweiten Hauptsatzes der mechanischen Wärmetheories auf allgemeine mechanische Prinzipien," *ibid.,* **142** (1871), 433–461.
16. "Bemerkungen zu der Prioritätsreclamation des Hrn. Boltzmann," *ibid.,* **144** (1872), 265–274. This was in reply to Boltzmann's article, "Zur Priorität der Beziehung zwischen dem zweiten Hauptsatze und dem Prinzipe der kleinsten Wirkung," *ibid.,* **143** (1871), 211–230.
17. "Ueber den Satz vom mittleren Ergal und seine Anwendung auf die Molecularbewegungen der Gase," *ibid.,* supp. **7** (1876), 215–280.
18. "Ueber das Verhalten der Kohlensäure in Bezug auf Druck, Volumen, und Temperatur," in *Annalen der Physik,* **169** (1880), 337–357.
19. "Ueber ein neues Grundgesetz der Elektrodynamik," *ibid.,* **156** (1875), 657–660.
20. "Ueber das Verhalten des electrodynamischen Grundgesetzes zum Prinzip von der Erhaltung der Energie und über eine noch weitere Vereinfachung des ersteren," *ibid.,* **157** (1876), 489–494.
21. "Ueber die Ableitung eines neuen elektrodynamischen Grundgesetzes," in *Journal für die reine und angewandte Mathematik,* **82** (1877), 85–130.
22. For an excellent comparison of the theories of Gauss, Weber, Riemann, and Clausius, see J. J. Thomson's article, "Report on Electrical Theories," in *Report of the British Association for the Advancement of Science* (1886), pp. 97–155, esp. 107–111.
23. "Ueber die Vergleichung der elektrodynamischen Grundsetze mit der Erfahrung," in *Annalen der Physik,* **170** (1880), 608–618.
24. See Mary Hesse, *Forces and Fields* (London, 1965), p. 219.
25. Emil Budde, "Das Clausius'sche Gesetz und die Bewegung der Erde in Raume," in *Annalen der Physik,* **170** (1880), 553–560.

BIBLIOGRAPHY

The best original source for studying the development of Clausius' early ideas is the 1st edition of his *Die mechanische Wärmetheorie,* 2 vols. (Brunswick, 1865–1867), a compilation of his original papers in thermodynamics, kinetic theory, and electricity, with additional dated comments. The Royal Society's *Catalogue of Scientific Papers* gives an exhaustive listing of his published papers and cites numerous obituaries, I, 945–947; VII, 400–401; IX, 533–534; and XIV, 259–260. Poggendorff also lists his major papers and books, I, cols. 454–455; III, 281–282; and IV, 258.

There is no biography of Clausius other than the short sketch in *Allgemeine deutsche Biographie,* LV, 720–729, but a recent article by Grete Ronge, "Die Züricher Jahre des Physikers Rudolf Clausius," in *Gesnerus,* **12** (1955), 73–108, includes some new information about his personal life. J. W. Gibbs's evaluation of Clausius' scientific work shows particular excellence—"Rudolf Julius Emmanuel Clausius," in *Proceedings of the American Academy of Arts and*

Sciences, n.s. **16** (1889), 458–465—and François Folie's comments on Clausius and his family—"R. Clausius. Sa vie, ses travaux et leur portée métaphysique," in *Revue des questions scientifiques,* **27** (1890), 419–487—are of great value, since we know so little about him apart from his scientific writings.

There is no extensive secondary literature on Clausius himself, but helpful treatments of his ideas appear in most general histories of physics and especially in those on the kinetic theory and thermodynamics. Ferdinand Rosenberger devoted considerable space to Clausius in his *Die Geschichte der Physik,* vol. III (Brunswick, 1890), and Charles Brunold reviewed Clausius' concept of entropy in his *L'entropie* (Paris, 1930), pp. 58–105. G. H. Bryan offered a very positive appraisal of Clausius' contribution to mechanical interpretations of the second law of thermodynamics in his article, "Researches Relating to the Connection of the Second Law with Dynamical Principles," in *Report of the British Association for the Advancement of Science* [Cardiff, 1891] (London, 1892).

More recently, Stephen Brush has evaluated Clausius' work in the kinetic theory of gases in two articles: "The Development of the Kinetic Theory of Gases. III. Clausius," in *Annals of Science,* **14** (1958), 185–196; and "Foundations of Statistical Mechanics, 1845–1915," in *Archive for History of Exact Sciences,* **4** (1967), 145–183. Martin J. Klein has given a perceptive and sensitive appraisal of Gibbs's views on Clausius in "Gibbs on Clausius," in *Historical Studies in the Physical Sciences,* **1** (1969), 127–149. For my views on Rankine and Clausius, see "Atomism and Thermodynamics," in *Isis,* **58** (1967), 293–303; and, for a further discussion on Boltzmann and Clausius, "Probability and Thermodynamics," *ibid.,* **60** (1969), 318–330. In my forthcoming article, "Entropy and Dissipation," due to appear in *Historical Studies in the Physical Sciences,* **2,** I discuss at length the relations between Clausius and the British thermodynamic tradition of Thomson, Maxwell, and Tait.

EDWARD E. DAUB

CLAVASIO. See **Dominic de Clavasio.**

CLAVIUS, CHRISTOPH (*b.* Bamberg, Germany, 1537; *d.* Rome, Italy, 6 February 1612), *mathematics, astronomy.*

Clavius entered the Jesuit order at Rome in 1555 and later studied for a time at the University of Coimbra (Portugal), where he observed the eclipse of the sun on 21 August 1560. He began teaching mathematics at the Collegio Romano in Rome in 1565, while still a student in his third year of theology; and for all but two of the next forty-seven years he was a member of the faculty as professor of mathematics or as scriptor. From October 1595 until the end of 1596, he was stationed in Naples.

In 1574 Clavius published his main work, *The*

Elements of Euclid. (With the help of native scholars, Matteo Ricci, between 1603 and 1607, translated into Chinese the first six books of Clavius' *Elements.*) His contemporaries called Clavius "the Euclid of the sixteenth century." The *Elements,* which is not a translation, contains a vast quantity of notes collected from previous commentators and editors, as well as some good criticisms and elucidations of his own. Among other things, Clavius made a new attempt at proving the "postulate of parallels." In his *Elements* of 1557, the French geometer Peletier held that the "angle of contact" was not an angle at all. Clavius was of a different opinion; but Viète, in his *Variorum de rebus mathematicis responsorum* of 1593, ranged himself on the side of Peletier. In a scholion to the twelfth proposition of the ninth book of Euclid, Clavius objects to Cardanus' claim to originality in employing a method that derives a proposition by assuming the contradictory of the proposition to be proved. According to Clavius, Cardanus was anticipated in this method by Euclid and by Theodosius of Bithynia in the twelfth proposition of the first book of his *Sphaericorum.*

As an astronomer, Clavius was a supporter of the Ptolemaic system and an opponent of Copernicus. In his *In Sphaeram Ioannis de Sacro Bosco commentarius* (Rome, 1581) he was apparently the first to accuse Copernicus not only of having presented a physically absurd doctrine but also of having contradicted numerous scriptural passages. The friendship between Clavius and Galileo, according to their correspondence, began when Galileo was twenty-three and remained unimpaired throughout Clavius' life. In a report of April 1611 to Cardinal Bellarmine of the Holy Office, Clavius and his colleagues confirmed Galileo's discoveries, published in the *Sidereus nuncius* (1610), but they did not confirm Galileo's theory.

In his *Epitome arithmeticae practicae* (Rome, 1583), Clavius gave a distinct notation for "fractions of fractional numbers," but he did not use it in the ordinary multiplication of fractions. His $\frac{3}{5} \cdot \frac{4}{7} \cdot$ means $\frac{3}{5}$ of $\frac{4}{7}$. The distinctive feature of this notation is the omission of the fractional line after the first fraction. The dot cannot be considered as the symbol of multiplication. He offered an explanation for finding the lowest common multiple, which before him only Leonardo Fibonacci in his *Liber abaci* (1202) and Tartaglia in his *General trattato di numeri et misure* (1556) had done. In his *Astrolabium* (Rome, 1593) Clavius gives a "tabula sinuum," in which the proportional parts are separated from the integers by dots. However, his real grasp of that notation is open to doubt, and the more so because in his *Algebra* (Rome, 1608) he wrote all decimal fractions in the form of common fractions. Apart from that, his *Algebra* marks the appearance in Italy of the German plus (+) and minus (−) signs and of algebraic symbols used by Stifel. He was one of the very first to use parentheses to express aggregation of terms. As symbol of the unknown quantity, he used the German radix (𝔄). For additional unknowns he used 1*A*, 1*B*, etc.; for example, he wrote 3𝔄 + 4*A*, 4*B* − 3*A* for 3*x* + 4*y*, 4*z* − 3*y*. In his *Algebra*, Clavius did not take notice of negative roots, but he recognized that the quadratic $x^2 + c = bx$ may be satisfied by two values of *x*. His geometrical proof for this statement was one of the best and most complete. The appendix of his commentary on the *Sphaericorum* of Theodosius (Rome, 1586)—containing a treatise on the sine, the tangent, and the secant—and the rules for the solutions of both plane and spherical triangles in the *Astrolabium*, the *Geometria practica* (Rome, 1604), and the *Triangula sphaerica* (Mainz, 1611) comprehend nearly all the contemporary knowledge of trigonometry; in the *Astrolabium*, for example, is his treatment of the so-called prosthaphaeresis method, by which addition and subtraction were substituted for multiplication, as in

$$\sin a \sin b = \frac{1}{2}[\cos(a - b) - \cos(a + b)].$$

In this he also gives a graphic solution of spherical triangles based on the stereographic projection of the sphere.

Mention must also be made of Clavius' improvement of the Julian calendar. Pope Gregory XIII brought together a large number of mathematicians, astronomers, and prelates, who decided upon the adoption of the calendar proposed by Clavius, which was based on Reinhold's *Prussian Tables*. To rectify the errors of the Julian calendar it was agreed to write in the new calendar 15 October immediately after 4 October of the year 1582. The Gregorian calendar met with a great deal of opposition from scientists such as Viète and Scaliger and from the Protestants.

BIBLIOGRAPHY

Clavius' collected works, *Opera mathematica*, 5 vols. (Mainz, 1611/1612), contain, in addition to his arithmetic and algebra, his commentaries on Euclid, Theodosius, and Sacrobosco; his contributions to trigonometry and astronomy; and his work on the calendar.

The best account of Clavius' works and their several editions can be found in C. Sommervogel, *Bibliothèque de la Compagnie de Jésus*, II (Brussels–Paris, 1891). Some information on his life and work can be found in B. Boncompagni, "Lettera di Francesco Barozzi al P. Christoforo Clavio," in *Bollettino di bibliografia e storia delle scienze matematiche e fisiche*, **17** (1884), 831–837; A. von Braunmühl, *Vorlesungen über Geschichte der Trigonometrie*, I (1900), 189–191; F. Cajori, "Early 'Proofs' of the Impossibility of a Fourth Dimension of Space," in *Archivio di storia della scienza*, **7** (1926), 25–28; J. Ginsburg, "On the Early History of the Decimal Point," in *American Mathematical Monthly*, **35** (1928), 347–349; O. Meyer, "Christoph Clavius Bambergensis," in *Fränkisches Land*, **9** (1962), 1–8; J. E. Montucla, *Histoire des mathématiques*, 2nd ed., I (Paris, 1799), 682–687; E. C. Philips, "The Correspondence of Father Christopher Clavius S. I.," in *Archivum historicum Societatis Iesu*, **8** (1939), 193–222; and J. Tropfke, "Zur Geschichte der quadratischen Gleichungen über dreieinhalb Jahrtausend," in *Jahresbericht der Deutschen Mathematikervereinigung*, **44** (1934), 117–119.

H. L. L. BUSARD

CLAY (CLAIJ), JACOB (*b*. Berkhout, Netherlands, 18 January 1882; *d*. Bilthoven, Netherlands, 31 May 1955), *physics.*

Clay studied physics at Leiden University and was assistant to Kamerlingh Onnes from 1903 to 1907. He received his doctorate in 1908 with the thesis "De galvanische weerstand van metalen en legeeringen bij lage temperaturen" ("The Galvanic Resistance of Metals and Alloys at Low Temperatures"). He taught in a secondary school at Leiden in 1906 and at Delft from 1907 to 1920. At the Delft Technological University he was *privaat-docent* in natural philosophy from 1913. In 1920 he became professor of physics at the Bandung Technological University (now in Indonesia) and in 1929 took the same post at the University of Amsterdam.

With interest in general physics and its teaching Clay combined a predilection for philosophy, starting from Hegel, and on the experimental side, for atmospheric electricity. In Bandung, assisted by his physicist wife and his children, he investigated the then rather new subject of cosmic radiation. On voyages from Indonesia to the Netherlands he discovered the latitude effect, a diminution in the intensity of cosmic radiation in the equatorial regions that is caused by the earth's magnetic field, thus establishing the presence of charged particles in primary cosmic radiation. Against doubts of other investigators he firmly established the latitude effect and, with the aid of pupils, made further investigations after moving to Amsterdam University. In this connection he worked for the improvement of electric measurements of ionization in general.

Clay's straightforward nature and honest diplomacy made him a good executive as director of scientific institutions. Most of his scientific work is published in *Physica* (The Hague).

BIBLIOGRAPHY

I. Original Works. Among Clay's books are *Rayons cosmiques* (Paris, 1938), written with P. M. S. Blackett and G. Lemaître; *De ontwikkeling van het denken* (Utrecht, 1950); *Atmosferische electriciteit* (The Hague, 1951); and *Wetenschap en maatschappij* (Amsterdam, 1952), as well as books on philosophy and measurement of radioactivity.

II. Secondary Literature. Clay is discussed in *Gedenkboek Athenaeum-Universiteit Amsterdam* (1932); *Nederlands tijdschrift voor natuurkunde,* **18** (1952), 241, and **21** (1955), 149; and *Jaarboek van de K. Nederlandse Akademie van wetenschappen, gevestigd te Amsterdam* (1955–1956), p. 209.

J. A. Prins

CLEAVELAND, PARKER (b. Rowley [Byfield Parish], Massachusetts, 15 January 1780; d. Brunswick, Maine, 15 October 1858), *mineralogy.*

Cleaveland was the author of *An Elementary Treatise on Mineralogy and Geology,* the first important mineralogical text published in the United States.

A descendant of the early Puritan settlers of northern Massachusetts, Cleaveland remained a religious conservative. He was graduated from Harvard College in 1799 and taught in secondary schools at Haverhill, Massachusetts, and York, Maine, until 1803. During this period he vacillated between entering the ministry or the legal profession. In 1803 he became a tutor in mathematics and natural philosophy at Harvard and finally elected to pursue a scientific career in 1805, when he accepted an appointment as professor of mathematics and natural philosophy at the recently established Bowdoin College at Brunswick, Maine. Four early publications in the *Memoirs of the American Academy of Arts and Sciences* described meteorological phenomena and fossil shells.

The exploitation of local mineral deposits stimulated Cleaveland's interest in mineralogy, and he began to teach courses in mineralogy and chemistry in 1808. The growing interest in the United States in the study of both practical and theoretical mineralogy, as evidenced by the popularity of public lectures and the formation of mineral cabinets, convinced Cleaveland that the publication of an elementary work on American mineralogy was desirable. In 1812 he commenced a wide correspondence to collect information concerning the locations and types of mineral deposits in the various states and territories. Almost every contemporary American naturalist responded and aided Cleaveland in the project.

The theoretical part of Cleaveland's work was compiled from the writings of European scientists. Thus, it fully described Haüy's crystallographic theory and method and closely followed Brongniart's systematic mineralogy. It also displayed a neptunist bias, in that Cleaveland not only assumed an aqueous origin for basalt but also classified rocks according to the Wernerian chronological-stratigraphic system. However, because it was the first important American mineralogical text and contained much valuable information concerning the minerals of the United States, it was favorably received and praised on both sides of the Atlantic. A greatly enlarged, two-volume, second edition that included a reprint of Maclure's map was published in 1822. Cleaveland continued to plan a third edition until 1842, but increased academic responsibilities prevented him from completing the necessary revisions.

Cleaveland was interested more in the orderly arrangement of scientific data than in scientific investigation or the development of theory. As a teacher he was a stern disciplinarian and rigidly conservative. Thus, he avoided teaching geological theory when he perceived that it was becoming a threat to the biblical account of creation. Because of the recognition given to his *Treatise,* however, he received teaching offers from almost every major American college; but he continued to teach at Bowdoin until the day of his death. He is honored in the eponymous Cleavelandite, a member of the feldspar family so named by H. L. Alling in 1936.

BIBLIOGRAPHY

I. Original Works. Cleaveland's *magnum opus* was *An Elementary Treatise on Mineralogy and Geology* (Boston, 1816; 2nd ed., 2 vols., 1822). See also *Results of Meteorological Observations Made at Brunswick, Maine, Between 1807 and 1859 . . .* (Washington, D.C., 1867). There is an extensive collection of Cleaveland's MS material in the archives of the library of Bowdoin College, Brunswick, Maine.

II. Secondary Literature. On Cleaveland or his work, see Nehemiah Cleaveland, *Address Made at the Opening of the Cleaveland Cabinet, July 10, 1873* (Boston, 1873); Thomas A. Riley, "Goethe and Parker Cleaveland," in *Publications of the Modern Language Association of America,* **67** (1952), 350–374; and Leonard Woods, *The Life and Character of Parker Cleaveland* (Brunswick, Me., 1860).

John G. Burke

CLEBSCH, RUDOLF FRIEDRICH ALFRED (b. Königsberg, Germany [now Kaliningrad, U.S.S.R.], 19 January 1833; d. Göttingen, Germany, 7 November 1872), *mathematics.*

In 1850 Clebsch entered the University of Königsberg, where the school of mathematics founded by Jacobi was then flourishing. His teachers included the mathematical physicist Franz Neumann and the mathematicians Friedrich Richelot and Ludwig Otto Hesse, both pupils of Jacobi. After graduation, in 1854, he went to Berlin, where he was taught under the direction of Karl Schellbach at various schools. Clebsch's academic career began in 1858, when he became *Privatdozent* at the University of Berlin. Soon afterward he moved to Karlsruhe, where he was a professor at the Technische Hochschule from 1858 to 1863. From 1863 to 1868 he was professor at the University of Giessen, collaborating with Paul Gordan. From 1868 until his death, he was professor at the University of Göttingen and in the forefront of contemporary German mathematics. In 1868 he and his friend Carl Neumann, son of Franz Neumann, founded the *Mathematische Annalen.*

Clebsch's doctoral dissertation at the University of Königsberg concerned a problem of hydrodynamics, and the main problems considered in the first period of his scientific career were in mathematical physics, especially hydrodynamics and the theory of elasticity. His book on elasticity (1862) may be regarded as marking the end of this period. In it he treated and extended problems of elastic vibrations of rods and plates. His interests concerned more the mathematical than the experimental side of the physical problems. He soon moved on to pure mathematics, where he achieved a dominant place.

Clebsch's first researches in pure mathematics were suggested by Jacobi's papers concerning problems in the theory of variation and of partial differential equations. He had not known Jacobi personally but collaborated in the edition of his *Gesammelte Werke.* For general problems in the calculus of variations, Clebsch calculated the second variation and promoted the integration theory of Pfaffian systems, surpassing results that Jacobi had obtained in these fields.

Although in these analytical papers Clebsch already proved himself to be highly skilled in calculus, his fame as a leader of contemporary scientists was first gained through his contributions to the theory of projective invariants and algebraic geometry. In the nineteenth century these fields were called modern geometry and modern algebra. The term "modern" or "new" geometry was applied to the projective geometry developed in synthetic form by Poncelet, Steiner, and Staudt, and in analytical form by Plücker and Hesse. Clebsch wrote a biographical sketch of Plücker, giving evidence of the author's deep insight into mathematical currents of the nineteenth century. In numerical geometry we still speak of the Plücker-Clebsch principle of the resolubility of several algebraic equations.

The name "modern algebra" was applied to the algebra of invariants, founded by the English mathematicians Cayley, Salmon, and Sylvester. One of the first German contributors to this discipline was Aronhold. It was especially the papers of Aronhold that incited Clebsch to his own researches in the theory of invariants, or "algebra of quantics," as it was called by the English. The results in the theory of invariants are to be interpreted by geometric properties of algebraic curves, surfaces, and so on. This connection between algebra and geometry attracted Clebsch in a special way. He was soon a master of the difficult calculations with forms and determinants occurring in the theory of invariants. In this he surpassed his teacher Hesse, whose ability and elegance in analytical geometry were praised at the time.

Clebsch completed the symbolic calculus for forms and invariants created by Aronhold, and henceforth one spoke of the Clebsch-Aronhold symbolic notation. Clebsch's own contributions in this field of algebraic geometry include the following. With the help of suitable eliminations he determined a surface of order $11n - 24$ intersecting a given surface of order n in points where there is a tangent that touches the surface at more than three coinciding points. For a given cubic surface he calculated the tenth-degree equation on the resolution of which the determination of the Sylvester pentahedron of that surface depends. For a plane quartic curve Clebsch found a remarkable invariant that, when it vanishes, makes it possible to write the curve equation as a sum of five fourth-degree powers. At the end of his life Clebsch inaugurated the notion of a "connex," a geometrical object in the plane obtained by setting a form containing both point and line coordinates equal to zero.

The general interest in the theory of invariants began to abate somewhat in 1890, when Hilbert succeeded in proving that the system of invariants for a given set of forms has a finite basis. In 1868 Gordan had already proved the precursory theorem on the finiteness of binary invariants. The theory of binary invariants thus being more complete, Clebsch published a book on this part of the theory (1872), giving a summary of the results obtained.

In the last year of his life Clebsch planned the publication of his lectures on geometry, perhaps to include those on n dimensions, results by no means as self-evident then as now. After Clebsch's death his pupil Karl Lindemann published two volumes of these lectures (1876–1891), completed with his additions but confined to plane and three-dimensional geometry. Between 1906 and 1932 a second edition of volume I, part 1 of this work appeared under the name of both Clebsch and Lindemann. The first volume contained almost all of the known material on plane algebraic curves and on Abelian integrals and invariants connected with them.

In 1863 Clebsch began his very productive collaboration with Gordan by inviting him to Göttingen. In 1866 they published a book on the theory of Abelian functions. Papers by Clebsch alone on the geometry of rational elliptic curves and the application of Abelian functions in geometry may be regarded as ancillary to the book. All these works are based on Riemann's fundamental paper (1857) on Abelian functions. In this celebrated work Riemann based the algebraic functions on the Abelian integrals defined on the corresponding Riemann surface, making essential use of topological ideas and of the so-called Dirichlet principle taken from potential theory.

Riemann's ideas were difficult for most contemporary mathematicians, Clebsch included. In the following years mathematicians sought gradually to eliminate the transcendental elements from the Riemann theory of algebraic functions and to establish the theory on pure algebraic geometry. Clebsch's papers were an essential step in this direction. The application of Abelian functions to geometry in his principal paper (1865) is to be understood as the resolution of contact problems by means of Abel's theorem, i.e., the determination of systems of curves or surfaces touching a curve in a plane or in space in given orders, such as the double tangents of a plane quartic curve.

Clebsch and Gordan's book had the following special features: (a) use of homogeneous coordinates for the points of an algebraic curve and for the Abelian integrals defined on it and (b) definition of the genus p for a plane algebraic curve of order n possessing as singularities only d double points and s cusps. They were the first to define the genus p by the expression

$$p = \frac{(n-1)(n-2)}{2} - d - s,$$

whereas Riemann had defined this expression as the topological genus of the corresponding Riemann surface. Also, as the title indicates, the transcendental point of view is prevalent in the book. It treats the Jacobian problem of inversion, introduces the theta functions, and so on. On the whole, to a modern reader a century later, the book may seem old-fashioned; but it must be remembered that it appeared long before Weierstrass' more elegant lectures on the same object.

As successors to Clebsch there arose the German school of algebraic geometry, led by Brill and M. Noether, both regarded as his pupils. At the end of the nineteenth century, algebraic geometry moved to Italy, where particular attention was paid to the difficult theory of algebraic surfaces. But the beginnings of the theory of algebraic surfaces go back to Cayley, Clebsch, and Noether, for Clebsch described the plane representations of various rational surfaces, especially that of the general cubic surface. Clebsch must also be credited with the first birational invariant of an algebraic surface, the geometric genus that he introduced as the maximal number of double integrals of the first kind existing on it.

BIBLIOGRAPHY

Among Clebsch's works are *Theorie der Elastizität fester Körper* (Leipzig, 1862); "Über die Anwendung der Abelschen Funktionen auf die Geometrie," in *Journal für die reine und angewandte Mathematik,* **63** (1865), 189–243; *Theorie der Abelschen Funktionen* (Leipzig, 1866), written with Gordan; "Zum Gedächtnis an Julius Plücker," in *Abhandlungen der Königliche Gesellschaft der Wissenschaften zu Göttingen* (1871); *Theorie der binären algebraischen Formen* (Leipzig, 1872); and *Vorlesungen über Geometrie,* Karl Lindemann, ed., 2 vols. (Leipzig, 1876–1891). Karl Lindemann brought out a 2nd ed. of vol. I, pt. 1 in 3 sections (1906, 1910, 1932).

On Clebsch, see "Rudolf Friedrich Alfred Clebsch, Versuch einer Darstellung und Würdigung seiner wissenschaftlichen Leistungen, von einigen seiner Freunde," in *Mathematische Annalen,* **7** (1874), 1–40.

WERNER BURAU

CLÉMENT, NICHOLAS (*b.* Dijon, France, end of 1778 or beginning of 1779; *d.* Paris, France, 21 November 1841), *chemistry.*

When Clément had finished his primary education in Dijon, he went to Paris as clerk to an uncle who was a notary. Since childhood he had devoted all his leisure time to chemistry, and in Paris he was able to attend the courses at the Jardin des Plantes. Chance altered the course of his career; he won the lottery and from then on was able to devote himself entirely to chemistry. He developed quite extensive connections within the first group to be promoted at the École Polytechnique and then throughout the scientific world. He later married the daughter of Charles-Bernard Desormes, an assistant in the laboratory of Guyton de Morveau at the École Polytechnique. A lasting scientific collaboration was established between the two men, and Clément often used the name Clément-Desormes. Hence, some accounts report that Clément and Desormes were the same person.

The scientific work of Clément and Desormes was numerous and varied; here we must be content to enumerate only the most important. It began in

1801–1802 with the exact determination of the composition of carbon monoxide and of carbon disulfide. A controversy took place on this subject with Berthollet, who harbored a lasting grudge against Clément. Berthollet held that carbon could not exist without containing hydrogen and that carbon disulfide was identical to hydrogen sulfide, and he presupposed "water combined in gases" to take account of the composition of certain barium salts, for which Clément and Desormes gave the correct analysis. In 1806 the analysis of ultramarine was achieved; by establishing definitively that the blue color was not due to iron, Clément prepared the way for the synthesis that J. B. Guimet carried out twenty-one years later.

Research on alum in the same year established that the preference given to Italian alums was based solely on prejudice for certain purely exterior characteristics: for example, the color resulting from inert impurities was considered essential; consequently, alum that was too pure was not accepted. This finding permitted the development of a French alum industry. Also in this period Clément achieved one of his most remarkable successes by giving an exact account of the chemical reactions produced in the manufacture of sulfuric acid; the quality of its production was improved, which aided the entire chemical industry.

Another part of the work of Clément and Desormes concerned the study of heat (*ca.* 1812–1819). Their investigations led them to give a number of a satisfactory order of magnitude for absolute zero; their attempts to estimate the quantities of heat existing outside of a material support seem to have been severely criticized by Gay-Lussac, even though the law he formulated was the basis for their calculations. In addition, they published papers on the mechanical power of fire. They hoped to discover its basis in the law, named for Watt, according to which the quantity of heat in a given weight of water vapor is independent of its temperature and volume—a law that the limits of contemporary practice allowed one to consider as exact. Using this relationship and Gay-Lussac's law, Clément and Desormes correctly deduced that it is advantageous to operate the machine at high pressures and to allow maximum expansion to take place. From the lecture notes of a student at the Conservatoire des Arts et Métiers, it is known that Sadi Carnot attended Clément's lectures; it is also known that the latter lent Carnot the manuscript of an important memoir of 1819, unpublished and now lost, in which the detailed results of his work on steam engines were presented (the 1819 memoir is cited many times in Carnot's *Réflexions sur la puissance motrice du feu,* which Clément recommended that his students read when it appeared in 1824). It is certain that Carnot owed a great deal to Clément, beginning perhaps with his choice of subject matter and title.

The year 1819, moreover, marked the peak of Clément's scientific career; he became a professor at the Conservatoire des Arts et Métiers, holding one of the three chairs in higher technical education established that year. Clément was all the more suited because he was an experienced business manager; he had to his credit, among other things, a beet-sugar refinery, a distillery that manufactured alcohol from potatoes, and an alum refinery. During his later years Clément acquired the habit of traveling during the summer to study the principal business enterprises in France and abroad. As a result of his competence, he was sought as consulting engineer by, among others, the Salines de l'Est and the Compagnie de Saint-Gobain. It was with the latter that he achieved a fatal success: he had made a contract on a percentage basis, which obliged the company to pay him considerable sums, and it therefore terminated its connection with him as soon as the processes he introduced were sufficiently developed for his services to be unnecessary. This disappointment seems to have shortened Clément's life, for soon afterward, in 1839, he had a stroke, following which he was deprived of his faculties. He died suddenly on 21 November 1841.

BIBLIOGRAPHY

I. ORIGINAL WORKS. Almost all of Clément's works were presented to the Institute and are enumerated from day to day in the *Procès-verbaux des séances de l'Académie tenues depuis la fondation de l'Institut jusqu'au mois d'août 1835,* 10 vols. (Hendaye, 1910–1922).

Clément's other publications include the following, without collaborators: "Remarques sur l'évaporation de l'eau par l'air chaud. Observations sur un procédé économique de M. Curaudau inséré dans les *Annales des arts et manufactures,* no. 118 (avril 1811)," in *Annales de chimie,* **79** (1811), 84–89; "Sur la quantité de matière ligneuse existante dans quelques racines et dans quelques fruits. Note lue le 3 février 1816 à la Société Philomatique," in *Annales de chimie et de physique,* **1** (1816), 173–176; "Sur la fermentation alcoolique," *ibid.,* **5** (25 Aug. 1817), 422–423; *Appréciation du procédé d'éclairage par le gaz hydrogène du charbon de terre par M. Clément-Desormes, manufacturier* (Paris, 1819); "Sur un procédé pour découvrir la magnésie," in *Annales de chimie et de physique,* **20** (1822), 333; "Sur la découverte d'une pierre propre à la fabrication de ciment romain," *ibid.,* **24** (10 Oct. 1823), 104–106; "Cascade chimique," in *Annales des mines,* **9** (1824), 194–196; "Note sur des lingots de cuivre obtenus par la voie humide," in *Annales de chimie et de physique,* **27** (1824), 440–442; *Théorie générale de la puissance mécanique de la vapeur d'eau* (n.p., n.d. [Paris, 1826]), of which the only copy is in Musée Carnavalet, Paris, série topographique, grands cartons XIV; *Programme du cours de chimie appliquée* ([Paris], 1829); and *Influence du bas prix du sel sur sa consommation* (n.p., 1834). Another published work is *Manuel de chimie appliquée* (n.p., n.d.); a work available in MS form is "Travaux de M. Clément," with an autobiography written when he was a candidate for the Institute, undoubtedly 27 January 1823; an original autograph, presented to the Conservatoire des Arts et Métiers on 20 September 1854 by Anselme Payen is in the Conservatory's archives—ser. 10, liasse 494, presently in the Conservatory's library. There is also "Chimie industrielle, Journal des cours de 1825 à 1830," with introductory material, table of contents, and remarks by J. M. Baudot, 3 MS vols. at the library of the Conservatory: cote 8° Fa 40.

In collaboration with Charles-Bernard Desormes he wrote "Sur la réduction des métaux par le charbon. Anomalie qu'elle présente. Découverte d'un gaz nouveau," *ibid.,* **38** (*an* IX), 285–290, on carbon monoxide; "Mémoire sur la réduction de l'oxyde blanc de zinc par le charbon et sur le gaz oxyde qui s'en dégage," in *Annales de chimie,* **39** (*an* IX), 26–64; "Expériences sur le charbon," *ibid.,* **42** (*an* X), 121–152, containing an announcement of carbon disulfide; "Expériences sur l'eau contenue dans les gaz et sur quelques sels barytiques," *ibid.,* **43** (*an* X), 284–305; "Mémoire sur l'outremer lu à l'Institut le 27 janvier 1806," *ibid.,* **57** (1806), 317–326; "Théorie sur la fabrication de l'acide sulphurique. Mémoire lu à l'Institut le 20 janvier 1806," *ibid.,* **59** (1806), 329–339; "Fabrication du blanc de plomb (procédé de Montgolfier)," *ibid.,* **80** (1811), 326–329; "Sur le nouveau procédé de congélation de M. Leslie, et sur les applications de ce procédé considéré comme un moyen d'évaporation, *ibid.,* **78** (1811), 183–202; "De l'épuration des corps par la cristallisation," *ibid.,* **92** (1814), 248–253; "Détermination expérimentale du zéro absolu de la chaleur et du calorique spécifique des gaz," in *Journal de physique,* **89** (1819), 321–346, 428–455, reprinted in shorter form in *Bibliothèque universelle des sciences,* année 5, **13** (1820), 95–111; and "Mémoire sur la théorie des machines a feu. Extrait," in *Bulletin des sciences par la Société philomatique de Paris* (1819), 115–118.

Another work done in collaboration, this time with L. de Freycinet, is "Mémoire sur la distillation de l'eau de mer et sur les avantages qui en résultent pour la navigation," in *Annales de chimie et de physique,* **4** (1817), 225–241.

II. SECONDARY LITERATURE. On Clément or his work, see Jean Chaptal and N.-L. Vauquelin, "Rapport du mémoire sur l'alun de MM. Desormes et Clément, fait à l'Institut le 27 janvier 1806," in *Annales de chimie,* **57** (1806), 327–333; Pierre Costabel, "Le 'Calorique du vide' de Clément et Desormes (1812–1819)," in *Archives internationales d'histoire des sciences,* **21** (1968), 3–14; Charles Dunoyer, "Clément-Desormes," in *Journal des economistes,* **1** (1842), 208–213; Louis-Bernard Guyton de Morveau and Marie Riche de Prony, "Rapport sur un appareil établi à la Monnaie pour faire consumer la fumée des machines (Institut, 16 janvier 1809)," in *Annales de chimie,* **69** (1809), 189–203; Anselme Payen, "Notice sur Clément-Desormes," in *Bulletin de la Société d'encouragement,* **41** (1842), 377–380; Jacques Payen, "Une source de la pensée de Sadi

Carnot," in *Archives internationales d'histoire des sciences,* **21** (1968), 15–37; Paul Thénard, *Traité de Chimie,* 5th ed., I (Paris, 1827), 77–81, an edition of a fragment of the memoir of 1819 on *machines à feu*—see also the commentary added by Thénard, pp. 81–82.

JACQUES PAYEN

CLEMENTS, FREDERIC EDWARD (*b.* Lincoln, Nebraska, 16 September 1874; *d.* Santa Barbara, California, 26 July 1945), *botany.*

Clements was the most influential ecologist of the first half of the twentieth century; protagonist of "plant succession" and its component concepts of formation and climax; glossarist (he introduced "sere," "ecad," etc.); and adviser to the U.S. government on policies of range management, forestry, and Dust Bowl rehabilitation. He was wholly a product of the Great Plains by birth, education, and outlook. A protégé of C. E. Bessey and a classmate of Roscoe Pound at the University of Nebraska, Clements gave direction in the United States to emergent plant ecology launched by Warming and others abroad.

Clements' father, Ephraim George Clements, was a photographer who had served in the Civil War with the New York Infantry; his mother was Mary Angeline Scoggin. At nineteen Clements published his first paper, which described new species of fungi. The following year he received his B.S. degree at the University of Nebraska and was appointed an assistant in botany. He participated in the phytogeographic survey of Nebraska, the findings of which were published in 1898 in collaboration with Roscoe Pound. Clements quickly rose in the academic ranks—instructor in 1897 and adjunct professor in 1899 following the award of his Ph.D. in 1898—to become professor of plant physiology in 1906. Edith Gertrude Schwartz, who took a Ph.D. in botany at Nebraska as Clements' student, wrote in 1961 that they had dreamed of an "alpine laboratory" in the summer of 1899 on their honeymoon visit to Colorado (they had been married on 30 May in Lincoln). The purchase of the cabin site in Engelmann Canyon on Pikes Peak, made with income from the sale to museums of Colorado *exsiccatae* collected during the following three summers, led to the establishment of the laboratory there.

In 1907 Clements became head of the botany department at the University of Minnesota, remaining there ten years. He left the university to become a full-time research associate of the Carnegie Institution of Washington, spending winters at experimental gardens in Tucson, Arizona, and then Santa Barbara, California, and summers at Alpine Laboratory, where he directed a staff of nine permanent and seasonal (usually student) assistants. Scientists came to observe what he liked to distinguish as "dynamic ecology." The Clementses spent the summer of 1911 in Europe examining Gaston Bonnier's transplant program and the alpine flora. In 1913 they played a prominent part in the International Phytogeographic Excursion.

Clements' *Plant Succession* (1916) is generally considered by ecologists to be his greatest work. It was based on his *Development and Structure of Vegetation* (1904). Shantz (1945) considered Clements botanically to be "essentially a philosopher" and his "analysis and synthesis" to be his greatest contribution. From his earliest fieldwork Clements espoused a conservative, inclusive interpretation of plant species and sought to demonstrate experimentally that some species might be transformed under the impact of environment. Using grasses and native caespitose perennials, he divided and transplanted them to contrasting habitats at different elevations, from the plains at 5,500 feet to the summit of Pikes Peak, 14,110 feet. He insisted that he had "converted" alpine timothy into a lowland timothy. His neo-Lamarckian views, minimizing the role of chromosomes, challenged cytogeneticists; and his findings, not substantiated today in the way he promulgated, stimulated the "new systematics."

Clements' *Genera of Fungi* (1909, rev. with C. L. Shear in 1931) was a landmark in mycology. Popular identification manuals on Rocky Mountain and Sierra wild flowers, illustrated with original paintings by his wife, were first published in the *National Geographic Magazine* and were reprinted by demand. These works disseminated Bessey's phylogenetic views.

It was characteristic of Clements to outfit himself as nattily for the field as for addressing a banquet of the Soil Conservation Service. He spoke precisely, affecting classic Latin pronunciation for plant genera. Although he relished debate, he was reasonable in argument. He was kind and considerate; esteem came more easily to him than comradeship. Success through the years gave him a confidence that bordered on arrogance. Above all there persisted his sharp intellect and capacity for unremitting work.

Inseparably associated with her husband (the story of her typing reports as he drove the automobile while dictating his observations is not apocryphal), Edith Clements organized his *reliquiae* and published summaries of their work after his death.

BIBLIOGRAPHY

I. ORIGINAL WORKS. Among his publications are *Development and Structure of Vegetation* (Lincoln, Nebr., 1904);

Research Methods in Ecology (Lincoln, Nebr., 1905); *Minnesota Mushrooms* (Minneapolis, Minn., 1910); *Plant Succession* (Washington, D.C., 1916); *Plant Indicators* (Washington, D.C., 1920); and *Climatic Cycles and Human Populations on the Great Plains* (Washington, D.C., 1938).

Collaborations include *Experimental Pollination* (Washington, D.C., 1923), written with Frances L. Long; *Phylogenetic Method in Taxonomy. The North American Species of Artemisia, Chrysothamnus, and Atriplex* (Washington, D.C., 1923), written with H. M. Hall; *Flower Families and Ancestors* (New York, 1928), written with Edith Clements; *Plant Ecology* (New York, 1929), written with J. E. Weaver; and *Bio-ecology* (New York, 1939), written with Victor Shelford. Additional titles are listed in the *National Cyclopedia* article (below).

The Clements papers, photographs, field books, and memorabilia are preserved in the Department of Archives, University of Wyoming, in 154 document boxes.

II. SECONDARY LITERATURE. An encomium by Edith Clements appeared in *National Cyclopedia of American Biography*, **34** (1948), 266–267; she also wrote an anecdotal biography, in *Adventures in Ecology* (New York, 1960), *passim;* and an article on Alpine Laboratory, in *Nebraska Alumnus*, **57**, no. 6 (1961), 12–16. Biographical sketches by others include Joseph Ewan, in LeRoy Hafen, *Colorado and Its People*, II (New York, 1948), 24; Joseph Ewan, *Rocky Mountain Naturalists* (Denver, 1950), pp. 183–184; H. L. Shantz, in *Ecology*, **26** (1945), 317–319; and A. G. Tansley, in *Journal of Ecology*, **34** (1947), 194–196, excellent and balanced.

JOSEPH EWAN

CLEOMEDES, *astronomy.*

Cleomedes wrote an elementary two-part handbook of astronomy entitled Κυκλικὴ θεωρία μετεώρων ("Circular Theory of the Heavens"). Nothing is known of his personal circumstances, and even his dates are uncertain. Since the latest authority he quotes is Posidonius (*ca.* 135–50 B.C.) and since he nowhere mentions Ptolemy (fl. A.D. 127–141), Cleomedes must have lived not earlier than the first century B.C. and not later than the early second century A.D. Attempts have been made to determine his dates astronomically from his statement (I, xi, 59) that at diametrically opposite places on the horizon Aldebaran could be observed setting as Antares rose, each at 15° of its sign (Taurus and Scorpio, respectively); Ptolemy in his star catalog (*ca.* A.D. 138) puts these stars at 12 2/3° of their signs (*Almagest,* VII, 5, 88, Heiberg, ed.; VIII, 1, 110). Therefore, it is suggested, the difference of 2 1/3° must correspond to the difference in date between Cleomedes and Ptolemy because, by the phenomenon known as the precession of the equinoxes, stellar longitudes increase with time by an amount that

Ptolemy estimated as 1° in 100 years; hence, 2 1/3° corresponds to 233 years, giving a date for Cleomedes of *ca.* A.D. 371 (or A.D. 306 by the true figure for precession, 1° in seventy-two years).

Unfortunately, the phenomena described could not possibly have been observed. Since the two stars have different latitudes, they can never be seen at the same time at diametrically opposite points of the horizon; and in any case refraction effects, which near the horizon cause celestial objects to appear over 1/2° higher than their true positions, would preclude their being 180° apart in longitude. Aside from this, it is highly improbable that Cleomedes' "fifteenth degree" of the signs means other than simply the "middle" of them, a loose designation that may include anything from 12° to 18°. He was not writing a scientific treatise but a popular handbook; he almost certainly made no observations himself; and other numerical data he gave are often far from accurate—for example, he described (I, viii, 42–43) the head of Draco (γ) as being "in the zenith" at Lysimachia (lat. 41°N.), whereas its true declination in the period A.D. 1–500 was between +52.8° and +52.3°, an error of over 20 percent; similarly, he stated that from Lysimachia to Syene (now Aswan) (lat. 24°N.), assumed to be on the same meridian, is 1/15 the whole circumference, an error of over 25 percent.

Thus, to base conclusions concerning his date on an apparent discrepancy of 2 1/3° (or 17 percent) is unrealistic. Other internal evidence—the extent and virulence of his diatribes against the Epicureans (in the longest chapter of his work he compared them unfavorably with rats, reptiles, and worms and poured scorn on their scientific naïveté [II, i, 86 ff., 91 f.]) and his omission of any reference to Ptolemy (hardly conceivable for a Greek writer on astronomy after the second century)—argues strongly against a date as late as the fourth century but is consistent with one in the first century.

Cleomedes' work belongs to the class of handbook written to popularize the main ideas in the purely technical treatises of the scientists (particularly astronomers); such books were common in Alexandrian and later Greek literature and exerted considerable influence on Roman and medieval writers (*cf.* W. H. Stahl, *Roman Science* [Madison, Wis., 1962], p. 32f.). The treatment of much of the material (usually derived not from the scientific works at first hand but through intermediary sources; e.g. Cleomedes evidently knew Hipparchus' work only at second hand [II, i, 83]) became standardized, and there are many correspondences in both style and subject matter between extant examples of the handbook tradition, as in the astronomical works of Geminus (first century

B.C.), Cleomedes, Theon of Smyrna (early second century A.D.), and Achilles (third century A.D.).

Cleomedes' chief authority is Posidonius, and it is unlikely that he himself added anything original (two sentences affixed to the end of the manuscripts expressly state this), but he also used other sources that sometimes disagreed with Posidonius' views (see I, vi, 32–33). His own astronomical knowledge was that of the well-educated Stoic writer of his time, and its limitations are sometimes apparent—in I, vi, 28, after giving a highly inaccurate arithmetical scheme for calculating the length of the day, he asserted that the zodiac intersects the equator "nearly at right angles"; in II, iv, 105 he rejected altogether the possibility of annular solar eclipses; and in II, vi, 123 that of the "paradoxical" case, when the eclipsed moon rises while the sun appears to be still above the horizon. Where he understood his sources, however, he gave a clear and useful account of basic astronomical phenomena.

In Book I, Cleomedes described the Stoic view of the spherical cosmos permeated by "pneuma" and surrounded by the limitless void (for which he argues, against Aristotle), with the spherical earth stationary at the center; he explained the main circles of the celestial sphere (equator, tropics, "arctic" and "antarctic" circles, the latter being the limits of the stars always visible or always invisible at a particular latitude), the corresponding zones of the terrestrial globe, their different climates, and changes in the length of day and night and in the positions of the circles relative to the horizon at different latitudes. He gave approximately correct values for the astronomical seasons and knew that the sun's orbit is eccentric to the earth, that the sidereal day is shorter than the solar day (first discovered by Hipparchus), and that the sun rises four hours earlier for the Persians than for the Spaniards (I, vi, 29, 30; I, viii, 41). He (or rather his source) showed understanding of the effects of refraction at the horizon (II, i, 66 f.; see II, vi, 124, which he might have used to account for the "paradoxical" lunar eclipse), explained the moon's phases correctly, intelligently criticized current notions about lunar phenomena, and understood the mechanism of eclipses (I, iv, v, vi). Planetary phenomena were only briefly touched on; there is no mention of epicycles, but lunar and planetary deviations in latitude were recognized (II, vii); approximately correct figures for the zodiacal periods (except for Mars) were given (I, iii—known to the Greek astronomers since the end of the fifth century B.C.); and remarkably accurate values for the synodic periods (Mercury, 116 days; Venus, 584; Mars, 780; Jupiter, 398; Saturn, 378—II, vii) are presented. Finally, Cleomedes was the only

Greek writer whose extant work gives details of the methods used by Eratosthenes and Posidonius for estimating the circumference of the earth (I, x).

BIBLIOGRAPHY

The text of the handbook, with Latin trans. and commentary by J. Bake, was ed. by R. Balfour (Lyons, 1820); and, with Latin trans., was ed. by H. Ziegler (Leipzig, 1891).

See also A. Rehm, "Kleomedes," in Pauly-Wissowa, XI (1922), cols. 679–694.

D. R. DICKS

CLERCK, CARL ALEXANDER (*b*. Stockholm, Sweden, 1709; *d*. Stockholm, 22 July 1765), *entomology*.

Surprisingly little is known about Clerck's life, and not even his birth date can be ascertained accurately. He studied at Uppsala University in 1726; but it is believed that during his short time there he had no contact with his contemporary Linnaeus, who later became his friend and supporter. Financial problems caused him to abandon his studies at Uppsala in 1727 and find work in the capital, where he subsequently became a tax agent. From early youth he showed an interest in natural history, but it was not until the age of thirty that he began the serious study of insects and spiders as a sideline to his daily work.

All during his life Clerck was in need of money, mainly because he himself had to pay for the color illustrations of his books. At his death he was so deeply in debt that his collections of animals, plants, and minerals had to be sold to satisfy his creditors. The insect collection was purchased by Petter Jonas Bergius, who bequeathed it to the Royal Swedish Academy of Sciences. (At present it is in the Naturhistoriska Riksmuseet in Stockholm.)

Clerck was a practical man, not only observing the life of spiders and insects in the field but also inventing apparatus for collecting and preserving them. Among his inventions are "butterfly tongs" (a basic tong structure with a pair of flat, mesh "paddles" at the ends) and an ingenious box for collecting spiders.

Clerck's fame is based mainly on two works, *Aranei Suecici* (1757) and *Icones insectorum rariorum* (1759–1764). The former became a standard work for spider nomenclature and is the only exception to the accepted fact that modern scientific nomenclature began in 1758 with Linnaeus' *Systema naturae. Icones*, one of the greatest rarities among entomological books, deals with Swedish and tropical butterflies and moths; the illustrations were made from specimens in the Queen Lovisa Ulrika collection. The butterfly plates, which Clerck designed with great skill, were hand-colored by C. M. Rising, Erik Borg, and J. A. Aleander; the artistic quality is uneven, however, and the reproductions frequently vary from copy to copy.

BIBLIOGRAPHY

Clerck's two main works are *Aranei Suecici—Svenska spindlar uti sina Hufvud-Slägter indelte* (Stockholm, 1757), with 6 hand-colored plates; and *Icones insectorum rariorum cum nominibus eorum trivialibus, locisque e C. Linnaei syst. nat. allegatis*, 2 vols. (Stockholm, 1759–1764). Vol. I has 16 hand-colored plates and vol. II has 39; both have indexes. Vol. III, unfinished, has 7 plates, 3 of which are colored; there is no text. There are several variants of the *Icones,* differing mainly in the title page and the separate plates; uncolored as well as partly colored copies exist.

An article describing the "butterfly tongs" is "Några Anmärkningar, angående Insecter. 1. Beskrifning på en Phalaena. 2. Beskrifning på en Tång, at fånga Fjärillar och Insecter. 3. Om Kårk-Bottnars nytta i Insect-Cabinetter," in *Kungliga Svenska vetenskapsakademiens handlingar,* **16** (1755), 214–216, tab. 1; and one on the box for collecting spiders is "Om Spindlars fångande och födande," in *Kungliga Svenska vetenskapsakademiens handlingar,* **22** (1761), 243–245, tab. 1.

Important information on the *Icones* is in P. C. Zeller, "Caroli Clerck *Icones insectorum rariorum* 1759, kritisch bestimmt," in *Stettiner entomologische Zeitung,* **14** (1853), 199–214, 239–254, 271–294.

BENGT-OLOF LANDIN

CLERSELIER, CLAUDE (*b*. Paris, France, 21 March 1614; *d*. Paris, 13 April 1684), *publication of scientific works*.

Clerselier's fame rests solely on his unswerving admiration of and boundless devotion to Descartes. The son of Claude Clerselier, adviser and secretary to the king, and of Marguerite l'Empereur, Clerselier was a counsel to the Parlement of Paris.

His sister Marguerite was married to Pierre Chanut, the French ambassador to Sweden, who brought Descartes to Queen Christina's court. Through love of Cartesianism, Clerselier permitted the marriage of his daughter Geneviève to Jacques Rohault (whose *Oeuvres postumes* he published in 1682) even though the Clerseliers saw this marriage as a misalliance, Rohault being of a much lower social class. Clerselier's fortune was very large—on 5 November 1630 he had married Anne de Virlorieux, who had brought him a considerable dowry. He was not at all miserly in publishing Descartes's works and was even less sparing of his time and efforts.

Clerselier was responsible for the first edition of the French translation of the *Méditations* (1647); he himself translated the "Objections" and the "Réponses."

He completely revised and corrected the second edition (1661). After Descartes's death he published three volumes of *Lettres* (1657–1667). In 1659 he brought out in the same volume *L'homme* and the *Traité de la formation du foetus*. In 1677 he published a second edition, to which he added *Le monde ou Traité de la lumière,* based on the original manuscript, which he had in his possession (the first edition of *Le monde* had been based on a copy).

In his zeal to defend Cartesianism, Clerselier was sometimes lacking in critical judgment; but without him a portion of Descartes's work would be unknown to us. Descartes himself said of Clerselier, in regard to his quarrels with Gassendi, that Clerselier had been "at once his translator, his apologist, and his mediator."

BIBLIOGRAPHY

Legal documents clarifying the kinship of Clerselier and Chanut are to be found at the Bibliothèque Nationale, Paris, Département des manuscrits, *pièces originales,* 786. See also "Extrait d'une lettre écrite à l'auteur de ces nouvelles," in *Nouvelles de la république des lettres* (June 1684), 431–433; and "Observations de Monsieur Clerselier, touchant l'action de l'âme sur le corps. (Lettre à Monsieur de La Forge du 4 décembre 1660)," in *Lettres de Mr Descartes,* III (Paris, 1667), 640–646.

See also Charles Adam, "Clerselier éditeur des lettres de Descartes 1657–1659–1667," in *Séances et travaux de l'Académie des sciences morales et politiques,* n.s. 45 (1896), p. 722; Pierre Bayle, "Dissertation où on défend contre les Peripatéticiens les raisons par lesquelles quelques Cartésiens ont prouvé que l'essence des corps consiste dans l'étenduë," in *Oeuvres diverses de Mr Pierre Bayle,* 2nd ed., IV (The Hague, 1731), 109–132; and René Descartes, *Oeuvres,* Charles Adam and Paul Tannery, eds. (Paris, 1896–1913; new ed., rev., 1964–　　), esp. *Correspondance,* passim.

JOSEPH BEAUDE

CLEVE, PER TEODOR (*b.* Stockholm, Sweden, 10 February 1840; *d.* Uppsala, Sweden, 18 June 1905), *mineralogy, chemistry, oceanography.*

Cleve began his studies of chemistry and botany in 1858 in Uppsala, having learned the basic principles of mineralogy in Stockholm from Mosander, the discoverer of lanthanum, didymium, erbium, and terbium. In his dissertation Cleve discussed mineral analysis; he was awarded the Ph.D in 1863. Through his works in widely separated areas of natural science, Cleve assumed a leading role in Swedish research in the natural sciences during the last decades of the nineteenth century and surrounded himself with an ever increasing number of disciples.

After only five years of research Cleve was appointed assistant professor in chemistry at Uppsala University. He also taught chemistry at the Technological Institute in Stockholm until 1874, when he became professor in general and agricultural chemistry in Uppsala. He was the president of the Royal Swedish Academy of Science's Nobel Prize committee for chemistry from 1900 to 1905 and was a member of several foreign learned societies.

His first work, "Några ammoniakaliska chromföreningar" ("Some Compounds," 1861), was soon followed by four other papers on complex metal compounds, and in still others, he described syntheses of a multiple of new complex compounds, until in 1872 he ended this series of analyses with a detailed epitome in English, "On Ammoniacal Platinum Bases."

Cleve then began a series of analyses of the rare earth metals, in particular ytterbium, erbium, lanthanum, and didymium. He prepared numerous new compounds of these metals and could, as a consequence, confirm Mendeleev's prediction that they would prove to be trivalent. He also expressed the suspicion that didymium was not an element, which was confirmed eleven years later, in 1885, when Welsbach divided it into neodymium and praseodymium. Of the new element scandium, which Nilsson had discovered in 1879, Cleve isolated, in the same year, a quantity big enough to determine reliably its atomic weight; this made it possible for him to identify the element with Mendeleev's ekabor, the existence of which had been predicted eighteen years earlier. Cleve's exhaustive researches on the chemistry of the rare earth metals was crowned in 1879 with his discovery of two more new elements, holmium and thulium, and with the publication of a monograph on samarium, discovered by Boisbaudran in the same year.

Cleve was active in organic chemistry as well, and several of his papers testify to his interest in the chemistry of naphthalene, which he enriched with, among other things, his discoveries of six of the ten possible dichlorine naphthalenes. He also discovered those aminosulfon acids that were known for some time as "Cleve's acids."

Cleve devoted the last fifteen years of his life almost exclusively to completing the biological works that he had started in his youth. His earliest studies were of the Swedish freshwater algae, to which he had devoted two monographs. Little by little he began to specialize in the plankton that create diatoms; his intensive researches soon brought him to the position of being the greatest authority of his time in this area. His method of determining the age and order of deposits in late glacial and postglacial stratifications, based on the diatomaceous flora in mud, proved to be scientifically useful. His idea that diatoms make good index fossils was further stated in the hypothesis that the streams in the oceans could be characterized by the plankton they transport and, conversely, that through the existence of one type of plankton one can determine the origin of the stream. His main work on this subject, *The Seasonal Distribution of Atlantic Plankton Organisms,* is a basic text of oceanography.

BIBLIOGRAPHY

Cleve's works include "Mineral-analytiska undersökningar" (Ph.D. diss., Uppsala, 1862); "Bidrag till kännedomen om Sveriges sötvattensalger af familjen Desmidieae," in *Öfversigt af Kongliga vetenskapsakademiens förhandlingar,* **20** (1863), 481–497; "Förelöpande underrättelser om några brom-och jodhaltiga ammoniakaliska platinaföreningar," *ibid.,* **22** (1865), 487–500; "Svenska och norska diatomacéer," *ibid.,* **25** (1868), 213–239; "Om några isomera platinabaser. Med anmärkningar av C. W. Blomstrand," *ibid.,* **27** (1870), 777–796; "On Ammoniacal Platinum Bases," in *Kungliga vetenskapsakademiens handlingar,* **10,** no. 9 (1872); "Bidrag till jordmetallernas kemi" (diss. for professorship, Uppsala, 1874); "Om tvänne nya modifikationer af diklornaftalin," in *Öfversigt af Kongliga vetenskapsakademiens förhandlingar,* **32** (1875), 35–37; "Om några lantan-och didymföreningar," *ibid.,* no. 5 (1878), 9–25; "Cerium, Lanthan, Didym, Yttrium, Erbium, Beryllium," in Gmelin-Kraut's *Handbuch der Chemie,* vol. II, pt. 1 (Heidelberg, 1878), written with K. Kraut; "Om skandium," in *Öfversigt af Kongliga vetenskapsakademiens fördhandlingar* **36,** no. 7 (1879), 3–10; "Om tillvaron af tvänne nya grundämnen i erbinjorden," *ibid.,* **36,** no. 7 (1879), 11–14; "Om samariums föreningar," *ibid.,* **42,** no. 1 (1885), 15–20; "Nya undersökningar öfver didyms föreningar," *ibid.,* pp. 21–27; "Karaktäristik af Atlantiska oceanens vatten å grund af dess mikroorganismer," *ibid.,* **54** (1897), 95–102; and *The Seasonal Distribution of Atlantic Plankton Organisms* (Göteborg, 1900).

Uno Boklund

CLIFFORD, WILLIAM KINGDON (*b.* Exeter, England, 4 May 1845; *d.* Madeira, 3 March 1879), *mathematics.*

Clifford is perhaps most widely remembered as a popular writer on mathematics and physics, his work being colored by highly personal philosophical overtones. He played an important part, nevertheless, in introducing the ideas of G. F. B. Riemann and other writers on non-Euclidean geometry to English mathematicians. Clifford added a number of his own ideas to the subject, and these were highly regarded at the time, as were his papers on biquaternions, the classi-fication of loci, and the topology of Riemann surfaces.

The son of William Clifford, Clifford was educated at a small private school in Exeter until, at the age of fifteen, he was sent to King's College, London. In October 1863 he took up a minor scholarship to Trinity College, Cambridge, where he read mathematics. Clifford was second wrangler in the mathematical tripos, and second Smith's prizeman, in 1867. A year later he was elected professor of applied mathematics at University College, London, and in 1874 he became a fellow of the Royal Society.

Clifford was a first-class gymnast, whose repertory apparently included hanging by his toes from the crossbar of a weathercock on a church tower, a feat befitting a High Churchman, as he then was. His health began to fail, however, when Clifford was barely thirty, and the lectures for which he had earned some celebrity gave way to a series of popular review articles, concerned especially with the interrelation of metaphysics, epistemology, and science. The *jeux d'esprit* and (according to Stephen) "strong sense of the ridiculous" characterizing his lectures—often given before distinguished audiences—are evident in these writings, as well as in the published versions of the lectures themselves. Philosophically speaking, he was something of a Spinozist and argued that the mind is the one ultimate reality. An early liking for Aquinas was dispelled by his reading Darwin and Spencer. He was always hostile to those who put ecclesiastical system and sect above humanity.

In 1875 Clifford married Lucy Lane, of Barbados. His health deteriorated rapidly, and despite long visits to Spain, Algeria, the Mediterranean, and finally Madeira, he died of pulmonary tuberculosis before he was thirty-four. His wife and two daughters survived him, and Mrs. Clifford subsequently became well-known as a novelist and dramatist.

In mathematics, Clifford was first and foremost a geometer; as an undergraduate at Cambridge, and as a member of a club known as the Apostles, he had inveighed against the current Cambridge bias towards analysis. At a later date he was atypical in arguing—under the influence of Riemann and Lobachevski—that geometrical truth is a product of experience. It is significant that Clifford, through a translation published in *Nature* (1873), should have drawn attention in England to Riemann's famous *Über die Hypothesen welche der Geometrie zugrunde liegen* (1854). (This had been delivered before a nonmathematical audience, and hence was shorn of the underlying analysis.) Riemann had broadly indicated a way in which matter might be regarded as an efficient cause of spatial structure, and Clifford went further in making matter (and its motion), electrical phenomena, and so forth

322

a manifestation of the varying curvature of space. (See *The Common Sense of the Exact Sciences,* chap. 4; "On the Space-Theory of Matter," in *Proceedings of the Cambridge Philosophical Society,* **2** [1876], 157–158.)

Clifford's writings in geometry were largely on projective geometry; but in non-Euclidean geometry he did some of his best work, investigating the consequences of adjusting the definitions of parallelism (especially by abandoning the condition of coplanarity). Thus he found that parallels not in the same plane can exist (within current non-Euclidean terms) only in a Riemannian (elliptic) space, and that they do exist. (See "Preliminary Sketch of Bi-quaternions" [1873], in his *Mathematical Papers,* pp. 181–200.) He showed how a certain three parallels define a ruled second-order surface that has a number of interesting properties. The properties of such "Clifford's surfaces," as they were later known, were not investigated very deeply by Clifford himself, but Bianchi and Klein made much of them, considering especially an interpretation under which the geometry of the surface was Euclidean.

Elsewhere in his geometrical writings Clifford left memorable results, as in his investigations of the geometrical consequences of extending a method of Cayley's for forming a product of determinants, in his research into quaternion representations of the most general rigid motion in space, and in his application of the techniques of higher-dimensional geometry to a problem in probability. Simultaneously with Max Noether he proved (1870) that a Cremona transformation may be regarded as a compound of quadratic transformations, and toward the end of his life (1877) he established some important topological equivalences for Riemann surfaces. In all this, Clifford justifies the commonly expressed belief of contemporaries that his early death deprived the world of one of the best mathematicians of his generation.

BIBLIOGRAPHY

I. ORIGINAL WORKS. Most of Clifford's writings were published posthumously. *The Common Sense of the Exact Sciences* (London, 1885) was edited and completed by Karl Pearson. *Elements of Dynamics* (London, 1879; 1887), *Lectures and Essays* (London, 1879), and *Science and Thinking* (London, 1879) were all popular treatments. *Lectures and Essays* has an introduction by Sir Frederick Pollock, with a brief biography. The *Mathematical Papers,* edited by R. Tucker (London, 1882), has an introduction by H. J. S. Smith, and a good bibliography.

II. SECONDARY LITERATURE. See also Sir Leslie Stephen's biography in *Dictionary of National Biography* for personal detail—it apparently was written, however, more for Clifford's widow than for posterity. See also A. Macfarlane, *Lectures on Ten British Mathematicians of the Nineteenth Century* (New York, 1916).

JOHN D. NORTH

CLIFT, WILLIAM (*b.* Burcombe, Cornwall, England, 14 February 1775; *d.* London, England, 20 June 1849), *comparative anatomy, paleontology.*

Clift was the youngest child of Robert Clift, a miller, who died in 1784, leaving his wife (Joanna Coutts, a carpenter's daughter) and seven children in poverty. William showed marked artistic promise and manual skill. Through local patronage he was apprenticed to the great London surgeon and biologist John Hunter, working as dissection assistant and recorder in lieu of fee, but Hunter's sudden death in 1793 deprived him of surgical training. Hunter's will directed that his scientific collections be offered for sale to the government as a unit. During negotiations (1793–1799) his executors, Matthew Baillie and Everard Home, retained Clift as curator of the collections. He maintained and developed the Hunterian Museum for fifty years and perpetuated Hunter's method of medical education through research in comparative anatomy.

The government bought the collections and placed them in trust with the Royal College of Surgeons in London; Clift moved them from Hunter's former home in 1806, settled at the college, and became administrator of the Hunterian Museum, as the collections were known. He equipped a new museum in 1813 (rebuilt in 1834–1837); formed a scientific library; acquired specimens by gift and purchase; dissected, mounted, and described them; provided anatomical and pathological material for the college's lecturers; and explained the museum to visitors. Clift had educated himself by studying Hunter's preparations and manuscript records, of which he made calligraphic copies, and was employed by Baillie and Home to illustrate their scientific writings. Home used Hunter's manuscripts for his own voluminous contributions to the *Philosophical Transactions* and destroyed most of the originals in 1823, greatly to the disadvantage of Clift's descriptive cataloguing of the museum.

Hunter's purpose in forming the museum had been to display the processes of life through examples drawn from the whole animal kingdom, both extant and fossil, arranged according to the functional systems of the body—skeletal, muscular, nervous, digestive, and reproductive—in both normal and pathologic conditions. Clift's achievement was the organization of this educative display, applying by his own manual skill the best technical methods of the day and adding to Hunter's collection in order to embrace advancing knowledge, without overloading it or altering Hunter's scheme. Under Clift's charge the museum attracted worldwide interest and redirected the method and purpose of museum display. Clift became an acknowledged authority in comparative anatomy, especially in the identification of fossil bones, and helped to formulate the scientific basis of paleontology. He was active in the anatomical, geological, and zoological societies, and especially in the Animal Chemistry Society (1809–1825). He was elected a fellow of the Royal Society in 1823 and served on its council in 1833–1834.

Clift's lifework and memorial was the Hunterian Museum. He published a few papers, mainly descriptive, but left a mass of unpublished records of his curatorship. The unsigned catalogs of the museum, printed between 1830 and 1840, were planned and partly written by him but were completed by Richard Owen. Clift's help was acknowledged by many prominent scientists—including Banks, Brodie, Cuvier, Davy, Lyell, and Mantell—in addition to those whose writings he illustrated with accomplished draftsmanship.

In 1801 Clift married Caroline Pope; their son William Home Clift, who died at twenty-nine, was trained to succeed him. Their daughter Caroline married his assistant and successor Richard Owen, who became the greatest British comparative anatomist of the century.

Clift was not self-seeking, but extremely generous of his time and knowledge. Accustomed from his youth to working for others, he never displayed his tenacity and independence until he denounced Home's destruction of Hunter's manuscripts before the Parliamentary Committee on Medical Education in 1834 and showed that Home had hoped to destroy evidence of his plagiarism. Clift's artistic abilities were notable. Hunter taught him to dissect and mount specimens and provided him with professional lessons in drawing and calligraphy; he also became a competent watercolorist. He was a keen amateur musician, playing several string and wind instruments. A small man, rather broad of face, he bore a fortuitous resemblance to Hunter.

BIBLIOGRAPHY

I. ORIGINAL WORKS. Clift's published articles are "Experiments to Ascertain the Influence of the Spinal Marrow on the Motion of the Heart in Fishes," in *Philosophical Transactions of the Royal Society*, **105** (1815), 91–96; "A Description of the Bones," in Joseph Whidbey, "On Some Fossil Bones Found in the Caverns in the Limestone Quarries of Oreston," *ibid.*, **113** (1823), 78–90 (Clift's "Description" 81–90), plates 8–12 by Clift; "On the Fossil Remains of the Two New Species of Mastodon and of Other Vertebrated Animals Found on the Left Bank of the Irawadi," in *Transactions of the Geological Society*, **2** (1829), 369–376, plates 36–44 (36, 39, and 40 are signed by Clift); and "Some Account of the Remains of the Megatherium Sent to England from Buenos Aires by Woodbine Parish," *ibid.*, **3** (1835), 437–450, plates 43–46 (45–46 signed by Clift).

Published drawings (listed chronologically) are in Matthew Baillie, *A Series of Engravings to Illustrate Morbid Anatomy* (London, 1799–1803), seventy-three plates; Patrick Russell, *A Continuation of an Account of Indian Serpents* (London, 1801), plates 11, 12, 13, and 16 in color; Everard Home, *Lectures on Comparative Anatomy*, vols. I–II (1814), 132 plates, vols. III–IV (1823), 171 plates; Nathaniel Highmore, *Case of a Foetus Found in the Abdomen of a Young Man* (London, 1815), two plates; William Buckland, "Account of Fossil Teeth and Bones Discovered in a Cave at Kirkdale in Yorkshire," in *Philosophical Transactions*, **112** (1822), parts of plates 18, 20, and 21 signed by Clift; William Maiden, *An Account of a Case of Recovery after the Shaft of a Chaise Had Been Forced Through the Thorax . . . the Health of the Sufferer until his Decease* (London, 1824), three plates (the *Recovery* was first published in 1812 without these illustrations); and George James Guthrie, *On Some Points Connected with the Anatomy and Surgery of Inguinal and Femoral Herniae* (London, 1833), 3 plates. Many of the 320 plates in Home's 100 papers in the *Philosophical Transactions* were drawn by Clift, from "The Kangaroo," **85** (1795), plates 18–21, to "The Membrana Tympani of the Elephant," **113** (1823), plates 3–5. After their quarrel Home employed Franz Bauer as artist.

Clift's diaries (1811–1842), a working record of daily duties, attendance at meetings, etc., with occasional comments, are at the Royal College of Surgeons; letters, bills, and leaflets are loosely inserted. Clift's voluminous memoranda, notebooks, etc. are listed in V. G. Plarr, *Catalogue of the Manuscripts in the Library of the Royal College of Surgeons of England* (London, 1928), pp. 12–21. There are many autograph letters by Clift in the Owen papers at the British Museum (Natural History).

II. SECONDARY LITERATURE. Notices by contemporaries are Sir William Lawrence, *Hunterian Oration* (London, 1846), pp. 59–64; unsigned memoir in *Proceedings of the Royal Society* (London), **5** (1849), 876–880 (Owen's draft is at the Royal College of Surgeons); and F. C. Skey, *Hunterian Oration* (London, 1850), pp. 5–10. Sir Benjamin Brodie discussed Clift in his "Autobiography," in *The Works of . . . Brodie*, C. Hawkins, ed., I (London, 1865), 65–67.

Modern studies are N. G. Coley, "The Animal Chemistry Society," in *Notes and Records, Royal Society of London*, **22** (1967), 173–185; Jessie Dobson, *William Clift* (London, 1954), the only biography based on the original sources, with references and reproduction of portraits; Sir Arthur Keith, "The Vicary Lecture on the Life and Times of William Clift," in *British Medical Journal* (1923), **2**, 1127–1132; and Jane M. Oppenheimer, *New Aspects of John and William Hunter* (New York, 1946), pt. 1, pp. 3–105

("Everard Home and the Destruction of the John Hunter Manuscripts"), which gives an adverse view of Clift.

Two pencil drawings of Clift by his son and a pencil drawing (1831) by Sir Francis Chantrey are in the British Museum; an oil painting by Henry Schmidt is at the Royal Society; and a plaster bust (1845) is at the Royal College of Surgeons.

WILLIAM LE FANU

CLOOS, HANS (*b.* Magdeburg, Germany, 8 November 1885; *d.* Bonn, Germany, 26 September 1951), *geology.*

Cloos's father, an architect, died young, leaving the upbringing of the three children to their mother, a highly gifted woman who contributed greatly to her son's intellectual development. Upon completing his secondary studies at Saarbrücken and Cologne, Cloos began to study architecture at the Technische Hochschule in Aachen but soon became interested in geology. After a short period of study in Bonn and Jena he went to Freiburg im Breisgau, where he received his doctorate under Wilhelm Deecke. Between 1909 and 1913 he did work in applied geology in South West Africa and Java. In 1914 Cloos qualified as a lecturer at the University of Marburg, and in 1919 he was appointed professor of geology and paleontology at the University of Breslau (now Wrocław, Poland). He was called to the University of Bonn in 1926 to succeed Gustav Steinmann. He married Elli Grüters, the daughter of an orchestra conductor, who bore him four children.

Cloos was a corresponding member of the Academy of Sciences of Berlin and of Göttingen and an honorary member of the geological societies of Germany, Finland, London, Sweden, the United States, and Peru and of the Natural Science Society of Schaffhausen, Switzerland.

In his dissertation, in which he examined the tectonic relationships between the folded mountains and plateaus of the Jura south of Basel, Cloos treated subjects that appeared in his later work: tectonics, structural geology, and internal dynamics. His stay in South West Africa presented Cloos with problems involving the mechanics of magmatic intrusion when he investigated the granite massifs in the Erongo Mountains. As a petroleum geologist in Java he was able to study active volcanoes and their structure.

During his years in Breslau, Cloos developed the field that became known as granite tectonics, the reconstruction of the dynamics of emplacement of a mobilized pluton from its internal structure (1921, 1922, 1925). Its basis was provided by the great granite massifs of Silesia. Cloos discovered that the granite, considered until then as massive and essentially struc-tureless, bears clearly oriented features gained during or directly after its intrusion. These include the linear or laminar flow textures of the solidifying magma, the regular relationship of the joint and vein systems to the flow textures, and the cleavage of the granite (well known to quarry men). Cloos applied granite tectonics to the Bavarian forest near Passau, and in Norway and North America (1928). The results obtained during the years at Breslau are among Cloos's most significant achievements.

At the same time that Cloos developed granite and magmatic tectonics, still another procedure for investigating similar problems was discovered by Bruno Sander—structural petrology, which examines, among other things, adjustment and behavior of mineral granules in a solidifying, magmatic liquid solution under tectonic influence. For a time there was sharp opposition between the two lines of inquiry. This resulted in scientific polemics between Cloos and Sander, which are now settled.

When Cloos moved from Breslau to Bonn in 1926, he turned his attention to other tectonic problems: jointing and cleavage as typical of the deformation of solid rocks, especially in the Paleozoic formations of the Rheinische Schlefergebirge (the area around Bonn) folded and faulted in the Hercynian orogeny, as well as the reproduction of tectonic phenomena and processes in the laboratory. He may have been the first to make extensive use of wet clay. His experiments on rift valley formation using wet clay (1931) have found a place in the literature and are particularly impressive. Many trips, including those to Africa and North America, widened his knowledge and experience and enlarged his view. The results can be seen in his outstanding textbook on internal dynamics (1936). Moreover, the years in Bonn witnessed more new knowledge: the division of the old continental masses into polygonal fields, the existence of *Grundschollen* (ground blocks) and *Erdnähten* (geofractures) which put Cloos in opposition to Alfred Wegener (1947)—the significance of the buckling processes in the earth's outer crust, the mechanics of volcanic processes in the so-called "embryonic volcanoes" in Württemberg (1939), and many lesser studies on the mechanics of folding and faulting.

Cloos always placed tectonic considerations foremost, regarding the earth's crust as an architectonic edifice. In his view the significance of the structural form and its analysis were primary; the investigation of the historical development of the mechanical processes that led to this structural form were of lesser interest. In many respects Cloos had an artistic temperament, filled with passion and enthusiasm for his science; and he knew how to convey this feeling to his students. His popular book *Gespräch mit der Erde* (1951) is a further expression of this ability. Through his tectonic researches, especially those concerning the relationship between tectonics and magma, Cloos provided geology with new knowledge and methods.

BIBLIOGRAPHY

I. ORIGINAL WORKS. Cloos's writings are "Der Mechanismus tiefvulkanischer Vorgänge," in *Sammlung Vieweg* (Brunswick), **57** (1921), 1–95; "Tektonik und Magma. Untersuchungen zur Geologie der Tiefen; Band I Einleitung: Über Ausbau und Anwendung der granittektonischen Methode," in *Abhandlungen der Preussischen geologischen Landesanstalt,* n.s. **89** (1922), 1–18; *Einführung in die tektonische Behandlung magmatischer Erscheinungen (Granittektonik), I, spezieller Teil: Das Riesengebirge in Schlesien* (Berlin-Bonn, 1925); "Bau und Bewegung der Gebirge in Nordamerika, Skandanavien und Mitteleuropa," in *Fortschritte der Geologie und Palaeontologie,* **7,** no. 21 (1928), 241–327; "Zur experimentellen Tektonik II. Brüche und Falten," in *Naturwissenschaften,* **19** (1931), 242–247; *Einführung in die Geologie: Ein Lehrbuch der inneren Dynamik* (Berlin, 1936; repr. 1963); "Hebung—Spaltung—Vulkanismus; Elemente einer geometrischen Analyse irdischer Grossformen," in *Geologische Rundschau,* **30,** special no. 4A (1939), 405–527; "Grundschollen und Erdnähte. Entwurf eines konservativen Erdbildes," *ibid.,* **35** (1947), 133–154; and *Gespräch mit der Erde,* 3rd ed. (Munich, 1951), trans. by E. B. Garside as *Conversation with the Earth,* ed. and slightly abr. by Ernst Cloos and Curt Dietz (New York, 1953).

II. SECONDARY LITERATURE. See also R. Balk, "Memorial to Hans Cloos (1886–1951)," in *Proceedings. Geological Society of America* (1963), 87–94, with portrait; Erich Bederke, "Hans Cloos," in *Zeitschrift der Deutschen geologischen Gesellschaft,* **104** (1953), 553–557; S. von Budnoff, "Requiem auf Hans Cloos," in *Geologische Rundschau,* **41** (1953), 1–10, with complete bibliography and portrait; and E. Hennig, "Cloos, Hans," in *Neue deutsche Biographie,* III (Berlin, 1957), 294.

HEINZ TOBIEN

CLOUET, JEAN-FRANÇOIS (*b.* Singly, Ardennes, France, 11 November 1751; *d.* Cayenne, French Guiana, 4 July 1801), *chemistry, metallurgy.*

Clouet was the son of Norbert Clouet, a farmer, and the former Marie-Jeanne Tayant. After studying classics at the *collège* of Charleville, he took the elementary technical courses at the École du Génie at Mézières. After several journeys devoted to study he established a pottery works at Singly and experimented with the manufacture of enamels. Although his enterprise had some success, Clouet abandoned it in 1783 in order to become a drawing teacher and a *préparateur* in physics and chemistry at the École du Génie. He assisted Gaspard Monge, a professor at the school, in experiments on the composition of water, the liquefaction of sulfur dioxide, and the flight of balloons. At the end of 1784 he succeeded Monge as professor of physics and chemistry, and from that time on, he directed his research toward metallurgical problems, such as the composition of siderite, the study of arsenious iron, the manufacture of Damascus blades, and the preparation of cast steel.

During the Revolution, Clouet was politically active. In January 1793 he was put in charge of reorganizing several metallurgical establishments. Called to Paris in 1795 to expand his attempts to manufacture cast steel, he worked in various laboratories, including those at the École Polytechnique and the Conservatoire des Arts et Métiers. On 10 April 1798 he presented his results to the Institut de France and was honored at the exposition of September 1798.

Associate of the mechanical arts section of the Institut National (28 February 1796), member of the Bureau Consultatif des Arts et Manufactures, and esteemed by the leading chemists of Paris, Clouet nevertheless rebelled against social constraints and joined the circle of *idéologues* that gathered around Claude Henri de Saint-Simon. In November 1799 he left for Guiana in order to escape from civilized life and to undertake researches on the conditions of life in tropical regions. He died at Cayenne on 4 June 1801.

Although unable to complete his work, Clouet helped to orient French chemical research toward concrete problems, particularly the modernization of the metallurgical industry.

BIBLIOGRAPHY

I. ORIGINAL WORKS. Clouet's writings include "Résultat des expériences et observations de MM. Ch . . . et Cl . . . sur l'acier fondu," in *Journal de physique,* **38** (1788), 46–47; "Mémoires sur la composition colorante du bleu de Prusse," in *Annales de chimie,* **11** (1791), 30–35; "Résultats d'expériences sur les différents états du fer," in *Journal des mines,* **9** (1798), 3–12; "Recherches sur la composition des émaux," in *Annales de chimie,* **34** (1800), 200–224; and "Instruction sur la fabrication des lames figurées ou des lames dites de Damas," in *Journal des mines,* **15** (1804), 421–435. A list of his works is in the Royal Society's *Catalogue of Scientific Papers,* I (1867), 959.

II. SECONDARY LITERATURE. It should be noted that many secondary works confuse certain aspects of the life and work of Clouet with those of others with the same name, particularly two contemporaries who also wrote on chemistry: the Abbé Pierre-Romain Clouet (1748–1810), who was librarian of the École des Mines, and Jean-Baptiste-Paul-Antoine Clouet (1739–1816), who under the *ancien* régime was *régisseur* of powders and saltpeter and worked with Lavoisier. Several authors incorrectly give Jean-François the Christian name Louis; others have propagated incorrect information.

On Clouet or his work, see J. B. Biot, in *Biographie universelle*, IX (1813), 123–131; and in *Nouvelle biographie universelle*, VIII (1854), 479–481, which contains many exaggerations; the Abbé Bouillot, in *Biographie ardennaise*, I (Paris, 1830), 254–264; J. N. P. Hachette, in *Annales de chimie*, **46** (1807), 97–104; A. Hannedouche, in *Les illustrations ardennaises* (Sedan, 1880), pp. 46–52; and René Taton, "Jean-François Clouet, chimiste ardennais. Sa vie, son oeuvre," in *Présence ardennaise*, no. 10 (1952), 5–30; and "Quelques précisions sur le chimiste Clouet et deux de ses homonymes," in *Revue d'histoire des sciences et leurs applications*, **5** (1952), 309–367.

RENÉ TATON

CLUSIUS, CAROLUS. See **L'Écluse, Charles de.**

COBLENTZ, WILLIAM WEBER (*b.* North Lima, Ohio, 20 November 1873; *d.* Washington, D. C., 15 September 1962), *infrared spectroscopy.*

Coblentz, the older son of David Coblentz and Catherine Good, was raised on a farm under relatively primitive conditions. He was graduated from the Case School of Applied Science (now Case Western Reserve University) in 1900. He then entered Cornell University, receiving an M.S. in 1901 and a Ph.D. in 1903. Upon graduation Coblentz accepted an appointment as research associate of the Carnegie Institution of Washington, to work two years as an honorary fellow at Cornell. He began an extension of his thesis topic by measuring the spectral absorptions of thousands of molecular substances.

In 1904 Coblentz joined the staff of the National Bureau of Standards, where he continued his work on the infrared spectroscopy of various molecules. Moreover, he made numerous important contributions to other areas of radiation measurement. He was the first to determine accurately the radiation constants of a blackbody and thus to verify Planck's law. In 1914 he used the Crossley reflector at the Lick Observatory to measure the heat radiated from 110 stars, as well as from planets and nebulae. Later he extended these measurements, using the facilities at Lowell Observatory and at Mount Wilson Observatory. The thermopiles he constructed at the National Bureau of Standards, sought after for their exceptional sensitivity, were used in studies significant to botany, physiology, and psychology, as well as to physics. However, Coblentz is known more for the studies he made with these instruments than for the instruments themselves.

Coblentz was chief of the radiometry section of the National Bureau of Standards from 1905 until 1945. He was widely honored for his pioneering contributions, which extended from infrared radiometry and spectroscopy through astrophysics to medical problems and ultraviolet therapy. The thoroughness and accuracy of his work established great confidence in him among physicists, so that it has been said that he was responsible for the adoption of radiometric standards, from the extreme ultraviolet to the far infrared.

BIBLIOGRAPHY

I. ORIGINAL WORKS. Among Coblentz's papers are "Some Optical Properties of Iodine," in *Physical Review,* **16** (1903), 35–50, 72–93; **17** (1903), 51–59, his doctoral thesis; "Methods of Measuring Radiant Energy," *ibid.,* **17** (1903), 267–276, written with E. L. Nichols; *Investigations of Infrared Spectra. Part I. Absorption Spectra; Part II. Emission Spectra,* Carnegie Institution of Washington, publication no. 35 (Washington, D.C., 1905); *Investigations of Infrared Spectra. Part III. Transmission Spectra; Part IV. Reflection Spectra,* Carnegie Institution of Washington, publication no. 65 (Washington, D.C., 1906); *Investigations of Infrared Spectra. Part V. Reflection Spectra; Part VI. Transmission Spectra; Part VII. Emission Spectra,* Carnegie Institution of Washington, publication no. 97 (Washington, D.C., 1908); "Instruments and Methods Used in Radiometry. I," in *Bulletin of the Bureau of Standards,* **4** (1908), 391–480; "The Blanket Effect of Clouds," in *Monthly Weather Review. U.S. Department of Agriculture,* **37** (1909), 65–66; *A Physical Study of the Firefly,* Carnegie Institution of Washington, publication no. 164 (Washington, D.C., 1911); "The Constants of Spectral Radiation of a Uniformly Heated Enclosure or So-called Black Body. I," in *Bulletin of the Bureau of Standards,* **10** (1913), 1–77; "Note on the Radiation from Stars," in *American Astronomical Society Publications,* **3** (1914), 76–78; "A Comparison of Stellar Radiometers and Radiometric Measurements of 110 Stars," in *Bulletin of the Bureau of Standards,* **11** (1915), 613–656; "Relative Sensibility of the Average Eye to Light of Different Colors and Some Practical Applications to Radiation Problems," *ibid.,* **14** (1917), 167–236, written with W. B. Emerson; "A Radiometric Investigation of the Germicidal Action of Ultraviolet Radiation," in *Scientific Papers of the United States Bureau of Standards,* **19** (1924), 641–680, written with H. R. Fulton; "Is There Life on Other Planets?" in *The Forum,* **74** (1925), 688–696; "Spectral Energy Distribution of the Light Emitted by Plants and Animals," in *Scientific Papers of the United States Bureau of Standards,* **21** (1926), 521–534, written with C. W. Hughes; "Instruments for Measuring Ultraviolet Radiation and the Unit of Dosage in Ultraviolet Therapy," in *Medical Journal and Record,* **130** (1929), 691–695; "The Biologically Active Component of Ultraviolet in Sunlight and Daylight," in *Transactions of the Illuminating Engineering Society,* **26** (1931), 572–578; "The Emergence of the Cicada," in *Sci-*

entific Monthly, **43** (1936), 239–243; and "Circulation of Ozone in the Upper Atmosphere," in *Bulletin. American Meteorological Society,* **20** (1939), 92–95.

His autobiography, *From the Life of a Researcher* (New York, 1951), provides a keen insight into his character.

II. SECONDARY LITERATURE. Many additional facts regarding Coblentz's career are in *Biographical Memoirs. National Academy of Sciences,* **49** (1967), 55–102, which also gives a complete bibliography of his scientific writings. The November 1963 issue of *Applied Optics* was a commemorative volume to him and contains several assessments of his work.

D. J. LOVELL

COCCHI, IGINO (*b.* Terrarossa, Massa e Carrara, Italy, 27 October 1827; *d.* Leghorn, Italy, 18 August 1913), *geology.*

Cocchi was the son of Giuseppe Cocchi and Anna Vighi. After receiving his degree at Pisa in 1848 and further study in Paris and London, he spent several years at Pisa, working under Paolo Savi and Giuseppe Meneghini; from 1859 to 1873 he taught geology at the Institute of Higher Studies (now the University) of Florence. He belongs to that group of courageous and eager scholars who, after the proclamation of the kingdom of Italy (1861), renewed and organized the geological study and mapping of Italy through two institutions that are still active: the Italian Geological Committee (now Service), of which Cocchi was the first president and, for many years, an active collaborator; and the Italian Geological Society, of which he was a founding member and twice president.

Cocchi was responsible for the compilation of a geological map, on a scale of 1:600,000, of almost all of Italy. Exhibited at the Paris International Exposition of 1867, it was a decisive improvement on the first attempt at such a map, undertaken in 1846 by Giacinto Provana di Collegno. Tuscany, however, was always his favorite territory for geological research. During his stay in Paris, Cocchi prepared a general paper on the magmatic and sedimentary rocks of Tuscany (published in France in 1855); later he studied in depth the geology of four areas, each of which presents a continuum from the oldest to the most recent deposits of the region: Elba, where he thoroughly investigated the mineral deposits; Monte Argentario and the surrounding territory; the Magra valley and the Apuan Alps, where he was the first to recognize the presence of glacial traces; and the basin of the Upper Valdarno and the environs of Arezzo.

Cocchi also studied paleontology (from the systematic study of a family of fish, the Pharyngodopilidae, to that of two Tuscan monkeys), but he was perhaps more attracted to the then nascent study of paleoethnology. After having explored caves in the Apuan

Alps and on the Leghorn coast, he had a fortunate opportunity to examine a fossil human cranium discovered in 1862 at Olmo, near Arezzo. Wishing to make an exhaustive study of this important relic, which had provoked heated discussion, Cocchi united geological and paleoethnological investigation for the first time in the closest way: this represents, without a doubt, his most original contribution to scientific progress.

Cocchi was endowed with a vast and varied humanistic knowledge. In his last years, following a trip to Finland, he became enthusiastically interested in the traditions and customs of that nation and learned its language well enough to produce the first Italian translation of the *Kalevala,* the Finnish national epic. It was published in 1913, the year of his death.

BIBLIOGRAPHY

Cocchi's complete bibliography, preceded by biographical notices, is in *Bollettino del reale Comitato geologico d'Italia,* **5** (1913–1914), 1–9; and *Bollettino della Società geologica italiana,* **32** (1913), xcix–ci.

FRANCESCO RODOLICO

COCHON DE LAPPARENT. See **Lapparent, Albert.**

COCHRANE, ARCHIBALD. See **Dundonald, Archibald Cochrane, Earl of.**

COCKCROFT, JOHN DOUGLAS (*b.* Todmorden, Yorkshire, England, 27 May 1897; *d.* Cambridge, England, 18 September 1967), *physics.*

Cockcrofts have lived in the Calder Valley of the West Riding of Yorkshire for generations. During the nineteenth century they owned a cotton mill in Todmorden, but the business declined and in 1899 was transferred to Birks Mill, situated a few miles away at Walsden. Cockcroft's father, J. A. Cockcroft devoted himself to rebuilding the prosperity of the firm, with the wholehearted support of his wife, daughter of a millowner. John, eldest of their five sons, was sent first to the Walsden Church of England School and then, after attending the Todmorden Elementary School, went to the Todmorden Secondary School in 1909. Perhaps because physics and mathematics were well taught there, he read more widely in the physical sciences, and accounts of the work of J. J. Thomson, Rutherford, and others gave him an ambition to do research. Cockcroft entered the University of Manchester in the autumn of 1914 with a county major scholarship, chiefly to study mathematics under

Horace Lamb, and was fortunate enough to be able to attend some first-year lectures given by Rutherford.

In the summer of 1915, after only one year at the university, Cockcroft volunteered for war service with the Y.M.C.A. Later in the year he was called up for military service and became a signaler in the Royal Field Artillery. Cockcroft took part in several of the major battles on the western front, was commissioned in the spring of 1918, and was released from the army when peace came in the autumn. He did not return to the science faculty of the university but entered the college of technology to study electrical engineering, possibly because as a signaler he had developed a professional interest in this subject. In 1920 he became a college apprentice at the Metropolitan Vickers Electrical Company, where he carried out some work in the research department under the direction of Miles Walker. He was awarded the degree of M.Sc.Tech. in 1922. On Walker's advice he went on to Cambridge to study mathematics at St. John's College and distinguished himself in part two of mathematical tripos, obtaining a B.A. degree in 1924. Cockcroft thus had an exceptionally long period of training—seven years excluding the war years—before joining Rutherford's team at the Cavendish Laboratory.

He married Eunice Elizabeth Crabtree at Todmorden on 26 August 1925. She was the daughter of Herbert Crabtree of Stansfield Hall, a cotton manufacturer, and John had known her since childhood. They had four daughters and one son. Their happy, close-knit family life was a source of great strength to Cockcroft.

His work with Miles Walker on the harmonic analysis of voltage and current wave forms at commercial power frequencies was published in 1925 and was followed by two further papers describing some detailed studies of boundary effects in electrical conductors. By this time Cockcroft had acquired a deep insight into certain aspects of electrical engineering and commanded some powerful theoretical techniques. He helped the Russian physicist Peter Kapitza, who was in Cambridge, with his work on very high magnetic fields by designing extremely efficient magnet coils in which the stresses were minimized. Later, he designed an electromagnet for α-ray spectroscopy for Rutherford and a permanent magnet for β-ray spectroscopy. Cockcroft also carried out an elegant investigation into the properties of molecular beams, following the classical work of Otto Stern and Immanuel Estermann. It was shown that the Frenkel theory of surface condensation applied, and attention was drawn to the role of adsorbed gaseous impurities. This work, published in 1928, afforded valuable ex-

perience in vacuum technology that was later to be crucial.

In the meantime, Thomas Allibone and E. T. S. Walton had been experimenting with different methods of accelerating electrons, and after Cockcroft had calculated from George Gamow's theory of α emission that protons of a few hundred kilovolts' energy should have an appreciable probability of penetrating the energy barriers of light nuclei, he became interested in the possibility of designing an accelerator for protons. Walton decided to join him, and with Rutherford's support they eventually constructed a voltage multiplier of a type originally proposed by Greinacher in 1920 and connected this to an accelerating tube provided with a proton source designed by M. L. E. Oliphant. Four stages of voltage multiplication yielded a steady potential of 710,000 volts at the accelerating tube. The maintenance of the necessary degree of vacuum in the system was difficult, and the success achieved was, in part, attributable to current improvements in vacuum techniques. When a lithium target was exposed to the proton beam and the voltage was raised progressively, scintillations due to α particles appeared on the zinc sulfide detector screen at 100,000 volts, increasing rapidly in number at higher voltages. The identification of α particles was confirmed by several methods, and it was also shown that two particles are ejected simultaneously in opposite directions, each with 8.6 MEV energy, in accordance with the equation

$$^{7}_{3}\text{Li} + ^{1}_{1}\text{H} \rightarrow ^{4}_{2}\text{He} + ^{4}_{2}\text{He} + 17.2 \text{ MEV}.$$

This was the first nuclear transformation to be brought about by artificial means. Similar effects were observed when other elements, such as boron and fluorine, were bombarded.

At the same time E. O. Lawrence and M. S. Livingston were constructing another device for accelerating protons, later known as the cyclotron, at the University of California, and they were soon able to confirm Cockcroft and Walton's results. By adding further stages to the voltage multiplier, the beam energy of the Cockcroft-Walton machine could be increased to about 3 MEV, whereas the cyclotron was capable of reaching much higher energies.

The year 1932 became an outstanding one for the Cavendish Laboratory with the discovery of the neutron by James Chadwick. In 1933 Rutherford received, through Cockcroft, some of the first heavy water from G. N. Lewis at Berkeley, and Cockcroft and Walton bombarded lithium, boron, and carbon with deuterons from their accelerating apparatus. After the announcement of the artificial production of radioactivity by Irène Joliot-Curie and Frédéric

Joliot in 1934, Cockcroft and Walton showed that radioactive nuclei were produced from boron and carbon exposed to proton and deuteron beams from the machine. Finally, a more detailed study of the disintegration of boron by protons and deuterons was carried out in collaboration with W. B. Lewis, using highly purified deuterium and with magnetic separation of unwanted ions from the beam.

From 1935 until the outbreak of war, Cockcroft was made personally responsible by Rutherford for a major reconstruction and reequipping of the Cavendish Laboratory, including the building of a cyclotron. Although habitually economical in the use of words, he was always genial and approachable yet made firm, impartial, and prompt decisions that were accepted almost without question. These qualities proved to be extremely valuable when Cockcroft became involved in the development of radar during the war. After building radar stations for detection of submarines and low-flying aircraft at various remote sites, he went to the United States in August 1940 as a member of the famous Tizard mission to negotiate scientific and technical exchanges of military importance. On his return he was appointed chief superintendent of the Air Defence Research Development Establishment at Christchurch, Hampshire, and during the same year he became a member of the committee which had been formed to investigate the possible applications of nuclear fission. After three more grueling years on radar development, Cockcroft was sent as director to the Anglo-Canadian atomic energy research laboratory in Montreal in April 1944. There he was faced not only with technical and scientific problems but also with a delicate diplomatic situation involving Canadian, British, American, and French interests. The NRX reactor, built under his direction at Chalk River, Ontario, was for a long time a unique instrument for nuclear research as well as for technology.

Cockcroft returned to Britain in 1946 to become director of the new Atomic Energy Research Establishment at Harwell, where he remained until 1959. During this period, in the latter part of which he was member for research of the Atomic Energy Authority, he guided and stimulated nuclear developments of all kinds, from basic research to power stations. He was especially concerned with the succession of particle accelerators constructed at Harwell and was largely responsible for obtaining approval to build the 7 GEV proton synchrotron and the Rutherford Laboratory at nearby Chilton. Cockcroft also did a great deal to promote science through international organizations such as CERN, the European laboratory for high-energy physics, and in many other ways. Nor did these activities cease when he became master of Churchill College, Cambridge, in 1959. Indeed, he was elected president of the well-known Pugwash Conferences on Science and World Affairs just before his death on 18 September 1967.

Cockcroft shared the 1951 Nobel Prize for physics with E. T. S. Walton and received many other awards, including a knighthood and the O.M., in recognition of his work on radar and more especially for his outstanding role in promoting the peaceful uses of atomic energy.

BIBLIOGRAPHY

Among Cockcroft's many papers are "An Electric Harmonic Analyser," in *Journal of the Institution of Electrical Engineers,* **63** (1925), 69–113, written with R. T. Coe, J. A. Lyacke, and Miles Walker; "The Design of Coils for the Production of Strong Magnetic Fields," in *Philosophical Transactions of the Royal Society,* **A227** (1928), 317–343; "The Effect of Curved Boundaries on the Distribution of Electrical Stress Round Conductors," in *Journal of the Institution of Electrical Engineers,* **66** (1928), 385–409; "On Phenomena Occurring in the Condensation of Molecular Streams on Surfaces," in *Proceedings of the Royal Society,* **A119** (1928), 293–312; "Skin Effect in Rectangular Conductors at High Frequencies," *ibid.,* **A122** (1929), 533–542; "Experiments With High Velocity Positive Ions," *ibid.,* **A129** (1930), 477–489, written with E. T. S. Walton; "Experiments With High Velocity Positive Ions. I. Further Developments in the Method of Obtaining High Velocity Positive Ions," *ibid.,* **A136** (1932), 619–630, written with Walton; "Experiments With High Velocity Positive Ions. II. The Disintegration of Elements by High Velocity Protons," *ibid.,* **A137** (1932), 229–242, written with Walton; "A Permanent Magnet for β-Ray Spectroscopy," *ibid.,* **A135** (1932), 628–636, written with C. D. Ellis and H. Kershaw; "Disintegration of Light Elements by Fast Neutrons," in *Nature,* **131** (1933), 23, written with Walton; "A Magnet for α-Ray Spectroscopy," in *Journal of Scientific Instruments,* **10** (1933), 71–75; "Experiments With High Velocity Positive Ions. III. The Disintegration of Lithium, Boron and Carbon by Heavy Hydrogen Ions," in *Proceedings of the Royal Society,* **A144** (1934). 704–720, written with Walton; "Experiments With High Velocity Positive Ions. IV. The Production of Induced Radioactivity by High Velocity Protons and Diplons," *ibid.,* **A148** (1935), 225–240, written with C. W. Gilbert and Walton; "Experiments With High Velocity Positive Ions. V. Further Experiments on the Disintegration of Boron," *ibid.,* **A154** (1936), 246–261, written with W. B. Lewis; "Experiments With High Velocity Positive Ions. VI. The Disintegration of Carbon, Nitrogen, and Oxygen by Deuterons," *ibid.,* 261–279, written with Lewis; "High Velocity Positive Ions. Their Application to the Transmutation of Atomic Nuclei and the Production of Artificial Radioactivity," in *British Journal of Radiology,* **10** (1937), 159–170; "The Cyclotron and Betatron," in *Journal of Scientific Instruments,* **21** (1944), 189–193; "Rutherford: Life and

Work After the Year 1919, With Personal Reminiscences of the Second Cambridge Period," in *Proceedings of the Physical Society,* **58** (1946), 625–633, the second Rutherford memorial lecture; "The Development of Linear Accelerators and Synchrotrons for Radiotherapy, and for Research in Physics," in *Proceedings of the Institution of Electrical Engineers,* **96,** pt. 1 (1949), 296–303; "Modern Concepts of the Structure of Matter," *ibid.,* **98,** pt. I (1951), 301–308, the forty-second Kelvin lecture; "Experiments on the Interaction of High Speed Nucleons With Atomic Nuclei," in *Les prix Nobel en 1951* (Stockholm, 1952), pp. 101–118; "The Scientific Work of the Atomic Energy Research Establishment," in *Proceedings of the Royal Society,* **A211** (1952), 155–168; and "High Energy Particle Accelerators," in *Endeavour,* **24** (1955), 61–70, written with T. G. Pickavance.

There is a notice on Cockcroft by M. L. E. Oliphant and W. G. Penney in *Biographical Memoirs of Fellows of the Royal Society,* **14** (1968), 139–188.

ROBERT SPENCE

CODAZZI, DELFINO (*b.* Lodi, Italy, 7 March 1824; *d.* Pavia, Italy, 21 July 1873), *mathematics.*

Codazzi first taught at the Ginnasio Liceale of Lodi; then at the *liceo* of Pavia; and, from 1865 to his death, at the University of Pavia as professor of complementary algebra and analytic geometry and, for a while, of theoretical geodesics. His best-known paper, on the applicability of surfaces, was written as an entry in a prize competition sponsored in 1859 by the French Academy. All three entries in the contest—those of Edmond Bour, Ossian Bonnet, and Codazzi—have been very valuable. Bour's and Bonnet's papers were published long before that of Codazzi, which appeared in 1883; however, its main ideas were incorporated in a paper in *Annali di matematica pura ed applicata* (1867–1872). Here we find the formulas that Bonnet (1863) called "les formules de M. Codazzi." They were not new, for Codazzi's colleague at Pavia, Gaspare Mainardi, had derived them in his paper "Su la teoria generale delle superficie" (1856); however, Codazzi's formulation was simpler and his applications were wider. Bonnet (1867) used Codazzi's formulas to prove the existence theorem in the theory of surfaces.

Codazzi also published on isometric lines, geodesic triangles, equiareal mapping, and the stability of floating bodies.

BIBLIOGRAPHY

Codazzi's papers are in *Annali di scienze matematiche e fisiche,* **7** (1856) and **8** (1857), continued in *Annali di matematica pura ed applicata,* **1** (1858) and 2nd ser., **1–5** (1867–1872). His best-known paper is "Mémoire relatif à l'appli-

cation des surfaces les unes sur les autres," in *Mémoires présentés par divers savants à l'Académie des sciences de l'Institut de France,* **27,** no. 6 (1883), 29–45; its main ideas are incorporated in the paper "Sulle coordinate curvilinie d'una superficie e dello spazio." in *Annali di matematica pura ed applicata,* 2nd ser., **1** (1867–1868), 293–316, **2** (1868–1869), 101–119; **3** (1869–1870), 269–287; **4** (1870–1871), 10–24; **5** (1871–1872), 206–222.

The only biographical material on Codazzi seems to be U. Amaldi, in *Enciclopedia italiana,* app. I (Rome, 1938), 438; and E. Beltrami, a biographical notice, in *Memorie per la storia dell'Università di Pavia,* ser. Ia (1878), 459. On the history of Codazzi's formulas see G. Darboux, *Leçons sur la théorie générale des surfaces,* II (Paris, 1889), 369, n. 1. Codazzi's formulas were derived in Gaspare Mainardi, "Su la teoria generale delle superficie," in *Giornale del R. Istituto lombardo,* **9** (1856), 385–398. Bonnet's relevant papers are "Note sur la théorie de la déformation des surfaces," in *Comptes rendus hebdomadaires des séances de l'Académie des sciences,* **57** (1863), 805–813; and "Mémoire sur la théorie des surfaces applicables sur une surface donnée," in *Journal de l'École polytechnique,* **42** (1867), 1–151.

Some of the data in this article were supplied by professors A. Pensa and S. Cinquini of the University of Pavia.

DIRK J. STRUIK

COGHILL, GEORGE ELLETT (*b.* Beaucoup, Illinois, 17 March 1872; *d.* Gainesville, Florida, 12 July 1941), *embryology, anatomy, psychobiology.*

Coghill was the fifth of seven children born to John Waller Coghill and the former Elizabeth Tucker. Since he had a rural Baptist background, he found his attempt to understand the nature of man's mind a formidable task. After receiving an A.B. from Brown University in 1896, he became an instructor at the University of New Mexico, where he was greatly influenced by its president, Clarence L. Herrick, under whom he received a master's degree in biology. In order to decide his future course in life, Coghill spent three months in the Mesa Verde area, thinking things over. He decided that to understand the mind he must fathom its mechanisms, especially their origin and development. He received the Ph.D. from Brown University in 1902.

After teaching at three universities, Coghill became an associate professor of anatomy at the University of Kansas (Lawrence) Medical School in 1913, attaining the department chairmanship and becoming secretary to the faculty in 1918. In 1925 he was elected professor of comparative anatomy at the Wistar Institute of Anatomy and Biology, Philadelphia, a post that he held until 1935. The University of Pittsburgh conferred an honorary Sc.D. on him in 1931, as did Brown University in 1935, the year he was elected to the U. S. National Academy of Sciences.

Coghill's analysis of the developing nervous system

of *Ambystoma* permitted him to formulate a universal law which states that the total pattern of development of the central nervous system and behavior (integrative action) dominates the partial patterns (reflex or analytical). This single idea was best formulated in his three lectures on anatomy and the problem of behavior, delivered at University College, London, in 1928 and published as *Anatomy and the Problem of Behavior* in 1929. This concept was enthusiastically accepted by psychologists and psychiatrists and formed the basis for much of C. Judson Herrick's thinking on the nature and origins of human mentation. The American school of psychobiology was founded by Clarence L. Herrick, C. Judson Herrick, and Coghill.

BIBLIOGRAPHY

All Coghill's works are listed and commented upon by C. Judson Herrick in his biography of Coghill (below). A similar but abbreviated list of publications appears in Herrick's biographical memoir of Coghill (below).

C. Judson Herrick's biography of Coghill is *George Ellett Coghill, Naturalist and Philosopher* (Chicago, 1949). Herrick also wrote the biographical memoir of Coghill in *Biographical Memoirs. National Academy of Sciences,* **22** (1942), memoir 12.

PAUL G. ROOFE

COGROSSI, CARLO FRANCESCO (*b.* Caravaggio, Italy, 5 July 1682; *d.* Crema, Italy, 13 January 1769), *medicine.*

Cogrossi studied at the University of Padua, where he benefited especially from the teaching of Domenico Guglielmini, an iatrophysicist and student of Malpighi. Receiving his degree in philosophy and in medicine in 1701, Cogrossi practiced medicine, first in Padua, where he also dedicated himself "to the anatomical dissection of sickly corpses of that hospital" and attended the courses of B. Ramazzini and A. Vallisnieri (to whom Cogrossi dedicated his major work). He then practiced in Venice and Crema. He was called to a chair of medicine at the University of Padua and began his teaching on 19 January 1721 with the oration "Pro medicorum virtute adversus fortunam medicam." In 1733 he went into retirement for reasons of health; he withdrew to Crema, where he maintained a modest private practice.

Cogrossi's most important work is the *Nuova idea del male contagioso de' buoi,* which appeared in Milan in 1714 and constitutes a clear formulation of the "contagium vivum seu animatum," based on some fundamental achievements of the seventeenth century, such as the introduction of the microscope, the

negation of the theory of spontaneous generation of insects, the studies of parasitism, the discovery of the microbiological world, and the acarid etiology of mange. The starting point of the *Nuova idea* was furnished to Cogrossi by an outbreak of cattle plague that wreaked havoc in Italy from the summer of 1711 to 1714.

The acarid etiology of mange brought parasitism to the limits of visibility of the naked eye, explaining not only the cause and the cure but also the mechanism of contagion. Cogrossi conjectured that parasitism might occur at the microscopic level; he thought that beings invisible to the naked eye—Leeuwenhoek's microorganisms—could be the cause of highly contagious diseases, thereby accounting for the ease with which diseases are transmitted from man to man until they acquire the actual physiognomy of epidemics.

The living nature of the infection permits an explanation of some fundamental characteristics of contagious diseases: the receptivity of one animal species and the immunity of another to a specific infection; the susceptibility of certain individuals of the receptive species and the immunity of other individuals of the same species; and the susceptibility of a people or of a country and the immunity of another people or country. Concerning this immunity, the author hypothesizes that the life of the contaminating agent is dependent on the climatic conditions of the country and the hygienic habits of the people.

To rid a nation of a contagious disease, it is necessary to isolate and cure those infected, to isolate the suspect ones, and to disinfect the personal belongings of both, in order to exterminate the causative agent and its eggs. An epidemic that has been put down can start anew if the pathogenic germ is reintroduced to an area by an infected individual or a carrier. Renewal is much more likely if the weather conditions favor the life and the multiplication of the germ. Suggestive is the analogy with the periodicity of the invasions of grasshoppers and other insects, which are to be ascribed to their monstrous multiplication. Only the unchecked reproduction of microorganisms can explain the rapid diffusion of infection throughout vast territories, even after the importation of only one infected animal.

Cogrossi also attempts to explain how such a minuscule being can inflict illness and death on an animal the size of a steer. He rightly invokes the alterations provoked by the microorganism in the liquids and solids of the infected animal. He also speculates on the entrance routes of the infection and its transmission through the secretions and excretions of the infected animal.

BIBLIOGRAPHY

I. ORIGINAL WORKS. Cogrossi's major work is the *Nuova idea del male contagioso de' buoi* (Milan, 1714), repr. in facs., with Eng. trans. by Dorothy M. Schullian and forward by Luigi Belloni, *New Theory of the Contagious Disease Among Oxen* (Milan, 1953). For a list of his minor works, see Belloni and Schullian below.

II. SECONDARY LITERATURE. Works on Cogrossi are Luigi Belloni and Dorothy M. Schullian, "Una autobiografia (1735) di Carlo-Francesco Cogrossi (1682–1769) nel suo epistolario con G. M. Mazzuchelli," in *Rivista di storia delle scienze mediche e naturali*, **45** (1954), 105–113; Luigi Belloni, "Le 'contagium vivum' avant Pasteur," in *Les conférences du Palais de la découverte* (Paris, 1961), pp. 12–14; and "I secoli italiani della dottrina del contagio vivo," in *Simposi clinici*, **4** (1967), liv.

LUIGI BELLONI

COHEN, ERNST JULIUS (*b.* Amsterdam, Netherlands, 7 March 1869; *d.* Auschwitz [Oświęcim], Poland, *ca.* 5 March 1944), *physical chemistry, history of chemistry.*

Cohen was a son of the German chemist Jacques Cohen, a pupil of Liebig and Bunsen, who later became a Dutch citizen. After leaving secondary school Cohen studied Latin and Greek and in 1888 passed the state university entrance examination. He then studied chemistry at the University of Amsterdam. In 1890 he passed the "candidate" examination and went, on the advice of his tutor, J. H. van't Hoff, to Paris, where he worked in the laboratory of Henri Moissan. In 1892 Cohen passed the doctoral examination and in 1893 defended his dissertation for the doctorate in chemistry, which concerned the electrical method of determining transition points and the electromotive force of chemical reactions. Cohen was a devoted pupil and disciple of van't Hoff, who sponsored him.

Cohen was appointed assistant in van't Hoff's laboratory. In 1896 he became lecturer, in 1901 extraordinary professor of chemistry in Amsterdam, and a year later ordinary professor of physical chemistry and director of the chemical laboratory at the University of Utrecht, which position he held until his retirement in 1939. As a Jew, Cohen was a victim of the Nazi regime. In 1944 he was arrested, sent to Westerbork, and then transported to Auschwitz, where he died in a gas chamber on or about 5 March. Cohen was well-known inside and outside his native country, especially through his numerous speaking engagements. He published his impressions of the United States, formed during a visit in 1926, in the book *Impressions of the Land of Benjamin Franklin* (1928).

Cohen was a prolific author, publishing more than 400 papers. His earliest investigations were concerned with photography, in which he was very interested: he studied the action of hydrogen on silver bromide gelatin plates, the solubility of silver halides in various solutions, and the supposed influence of gelatin on the double decomposition of salts (1897). With Georg Bredig and van't Hoff he worked on transition elements. In 1897 Cohen published a report on his detailed research on the dissociation of dissolved compounds in mixtures of alcohol and water. However, Cohen was known primarily for his famous series of investigations on physical isomerism.

Throughout his life Cohen studied the allotropy of tin. It was known that white tin undergoes a partial transformation into a gray powder, which causes the crumbling of organ pipes and other tin objects ("tin pest" or "tin disease"). Cohen and C. van Eyk showed that the transformation occurred because tin has two allotropes, white tin and gray tin, with very different properties. White tin is stable above 13.2°C., gray tin below that transition point (Cohen first found the value 18°C.). The difference in specific volume of white and gray tin is about 25 percent, with the greater value for the gray tin. Thus even a small transformation is easily seen by the formation of "warts" on the white tin. Cohen and van Eyk studied the reversible reaction between white and gray tin by measuring the electromotive force of the galvanic combination of a solution of a tin salt and two electrodes, one of white tin and one of gray tin. The electromotive force changes its sign on passing the transition point. Another way to determine the transition point is to use a dilatometer at a fixed temperature: a positive change of volume means the formation of gray tin from white tin, and vice versa. Cohen and van Eyk found that under ordinary conditions the change does not occur, the white tin being then metastable. The change may, however, be greatly accelerated by lowering the temperature, by the presence of certain solvents, by the "inoculation" of gray tin, and so on.

In a long series of investigations Cohen studied what he called cases of physical isomerism: enantiotropy, with a transition point, as in the case of tin, and monotropy, in which one modification is always stable and the other always metastable (e.g., antimony). The substances investigated by Cohen and his students included phosphorus, tellurium, cadmium, bismuth, zinc, copper, lead, silver, potassium, sodium, antimony iodide, cadmium iodide, silver iodide, thallous picrate, and ammonium nitrate. Attention was paid especially to the existence of metastable phases and to their usual formation under conditions where thermodynamically the stable phase would be expected. Cohen came to the conclusion that enantiotropy and monotropy were exceedingly common among both elements and compounds and that most metals in common use were, if not entirely metastable, at least mixtures of stable and metastable phases.

Cohen also studied intensively the electrochemistry of the galvanic cells, such as the standard Clark and Weston cells, the Daniell cell, and the calomel cell. He investigated the reactions occurring in such cells, the condition of stability or metastability of all phases present, and the values of the cells as standard galvanic elements. He also investigated thermodynamic relations, measurement of affinity in chemical reactions, and the use of transition cells for determining the temperature at which a transition occurs—for example, between different hydrates of the same salt. Cohen and his co-workers in Utrecht also did work in piezochemistry. Their investigations concerned the effect of pressures up to 1,500 atmospheres on the reactions and equilibria occurring in condensed (liquid and solid) systems. They studied the effect of high pressure on the velocity of chemical reactions, on the transition temperatures of solid phases, on the electromotive force of galvanic cells, on the solubilities of solids, on Faraday's first law, on the velocity of diffusion in liquid systems, on the viscosity of liquids, and on the electrical conductivity of solutions. In 1919 a survey of this work was published by Cohen and W. Schut as *Piezochemie kondensierter Systeme* and augmented in his *Physico-chemical Metamorphosis and Some Problems in Piezochemistry* (1928), which contains Cohen's Baker Lectures, given at Cornell University in 1925–1926.

Besides his numerous investigations in physical chemistry, which clearly showed him to be an experimenter first and foremost, Cohen was deeply interested in the history of chemistry. He published a great many articles in the field, most of them biographical. He was not interested so much in the historical development of chemical theories as in the lives of chemists and in the discoverers of chemical operations and experiments. In 1907 Cohen published a book on the history of laughing gas, and in 1912 his still-important biography of van't Hoff appeared. Cohen published on such chemists as Gerardus Mulder, professor of chemistry at Utrecht; the German physiological chemist Hartog Jakob Hamburger; the chemist Cornelis Adriaan Lobry de Bruyn; Hermann Boerhaave; and Daniel Fahrenheit. Most of Cohen's historical articles were originally published in Dutch and later translated into German.

Cohen also published an enlarged edition of van't Hoff's famous book *Études de dynamique chimique* under the title *Studien zur chemischen Dynamik* (1896). In Abegg's *Handbuch der anorganischen Chemie* he published an article on tin (1909), and he also wrote two textbooks for medical students.

BIBLIOGRAPHY

I. ORIGINAL WORKS. A complete survey of Cohen's publications is given in *Chemisch Weekblad,* **15** (1918), 1452–1470; **24** (1927), 489–492; **36** (1939), 519–522; and **41** (1945), 128–129. Cohen's dissertation was *Het bepalen van overgangspunten langs electrischen weg en de electromotorische kracht bij scheikundige omzetting* (Amsterdam, 1893). An excerpt appeared as "Die Bestimmung von Umwandlungspunkten auf elektrischem Wege und die elektromotorische Kraft bei chemischer Zersetzung," in *Zeitschrift für physikalische Chemie,* **14** (1894), 53–92.

His scientific works include "Das Umwandlungselement und eine neue Art seiner Anwendung," in *Zeitschrift für physikalische Chemie,* **14** (1894), 535–547, written with G. Bredig; "Zur Theorie des Umwandlungselements ohne metastabile Phase," *ibid.,* **16** (1895), 453–457, written with J. H. van't Hoff and G. Bredig; *Studien zur chemischen Dynamik. Nach J. H. van't Hoff's Études de dynamique chimique* (Amsterdam, 1896); *Experimentaluntersuchung über die Dissociation gelöster Körper in Alkohol-Wassergemischen* (Rotterdam, 1897); "Physikalisch-chemische Studien am Zinn," in *Zeitschrift für physikalische Chemie,* **30** (1899), 601–622, written with C. van Eyk; **33** (1900), 57–62; **35** (1900), 588–597; **36** (1901), 513–523; **48** (1904), 243–245; **50** (1904), 225–237, written with E. Goldschmidt; **63** (1908), 625–634; **68** (1909), 214–231; **127** (1927), 178–182, written with K. Douwes Dekker; and **173A** (1935), 32–34, written with A. K. W. A. van Lieshout; "Zinn," in Richard Abegg, *Handbuch der anorganischen Chemie,* III, pt. 2 (Leipzig, 1909), 531–610; *Piezochemie kondensierter Systeme* (Leipzig, 1919), written with W. Schut; "The Influence of Pressure on Chemical Transformations," in *Contemporary Developments in Chemistry* (New York, 1927); and *Physico-chemical Metamorphosis and Some Problems in Piezochemistry* (New York, 1928).

Among Cohen's historical works are *Das Lachgas. Eine chemisch-kulturhistorische Studie* (Leipzig, 1907); *Jacobus Henricus van't Hoff. Sein Leben und Wirken* (Leipzig, 1912); *Herman Boerhaave en zijne beteekenis voor de chemie* (n.p., 1918); "Herman Boerhaave und seine Bedeutung für die Chemie," in *Janus,* **23** (1918), 223–290; and "Wat leeren ons de archieven omtrent Gerrit Jan Mulder," in *Verhandelingen der Koninklijke akademie van wetenschappen, afdeeling natuurkunde,* **19,** no. 2 (1948), 1–73.

II. SECONDARY LITERATURE. On Cohen or his work, see the following, listed chronologically: H. R. Kruyt, "In Memoriam Ernst Cohen," in *Chemisch Weekblad,* **41** (1945), 126–129; F. G. Donnan, "Ernst Julius Cohen (1869–1944)," in *Obituary Notices of Fellows of the Royal Society of London,* **5** (1948), 667–687; C. A. Browne, "Dr. Ernst Cohen as Historian of Science," in *Journal of Chemical Education,* **25** (1948), 302–307; and A. L. T. Moesveld, "The Scientific Work of Ernst Cohen," *ibid.,* 308–314.

H. A. M. SNELDERS

COHEN, MORRIS RAPHAEL (*b*. Minsk, Russia, 25[?] June 1880; *d*. Washington, D.C., 25 January 1947), *philosophy of science*.

Cohen came to New York City from his native Russia in 1892. He attended the public schools and was graduated with highest honors from the College of the City of New York in 1900. In his early youth he was influenced by Thomas Davidson at the Educational Alliance on New York's Lower East Side. This influence and his natural bent led Cohen to the graduate study of philosophy at Harvard under Hugo Münsterberg, Josiah Royce, and William James. He received his Ph.D. in 1906 and taught mathematics at City College for several years. In 1912 he was finally appointed to teach philosophy and rapidly became a prominent figure in the emerging naturalistic movement, a first-rate teacher, and a developer of latent philosophic talents. In 1938, after retiring from City College, Cohen served as professor of philosophy at the University of Chicago until 1941. His health, never robust, then broke down completely. He spent the rest of his life completing, with the aid of his children, the books that summed up his major intellectual interests in science, law, and social ethics.

Cohen's philosophy was a scientifically grounded naturalism. Unlike other philosophic naturalists of his time, he turned from an empiricist interpretation of scientific method to an older conception of rationalism as the ally of science in combating the forces of superstition, supernaturalism, and authoritarianism. He did not deny the reality of the concrete data of experience; he argued, rather, that only when the principles suggested by these data were universalized by rational logical or mathematical methods did they have any value in formulating a world view. His differences with his fellow naturalists were a matter of emphasis and a wholesome corrective to exaggerated stress upon the data of sensation.

Cohen went beyond this methodological principle to assert an ontological principle of rationality. He believed that man does not merely impose the universal laws of science upon the external world of nature, but discovers them there. These laws express the invariant relations underlying all particulars. Cohen recognized the perspectival and contingent elements in the relations among particulars but rejected the romantic tendency to elevate contingency and change into basic ontological principles. Instead, he retained from his study of Hegel what he called the Principle of Polarity, that is, the principle that opposites involve each other. He applied this principle in explaining why he could not accept change as fundamental by maintaining that change can be attributed only with reference to some constant.

Aware as he was of the inexorable particularity of gross experience, Cohen was careful to point out that it is only the intelligibility of things, not their existence, that is dependent upon invariant scientific laws. Knowledge is of universals alone, but these universals are only as real as experienced particulars. Ultimately, Cohen insisted, we must recognize as one particular characteristic of the universe that it supports the quest for intelligibility. Thus, in the end, Cohen reconciled the polar elements of particularity and universality without importing explanatory principles from outside his system.

BIBLIOGRAPHY

I. ORIGINAL WORKS. Cohen's major books are *Reason and Nature* (New York–London, 1931); *Law and the Social Order* (New York, 1933); *An Introduction to Logic and Scientific Method* (New York–London, 1934), written with Ernest Nagel; *Preface to Logic* (New York, 1944); *The Meaning of Human History* (Lasalle, Ill., 1947), the Carus Lectures; and *A Dreamer's Journey* (Boston, 1949), his autobiography.

II. SECONDARY LITERATURE. Works on Cohen are Salo W. Baron, Ernest Nagel, and Koppel S. Pinson, eds., *Freedom and Reason: Studies in Philosophy and Jewish Culture in Memory of Morris Raphael Cohen* (Glencoe, Ill., 1951), esp. the articles in pt. 1; Joseph L. Blau, *Men and Movements in American Philosophy* (New York, 1952), ch. 9, sec. 3: "Rationalistic Naturalism: Morris R. Cohen"; Yervant Krikorian, "Morris Raphael Cohen," in *Encyclopedia of Philosophy,* Paul Edwards, ed., II, 128–129; and Leonora Cohen Rosenfield, *Portrait of a Philosopher: Morris R. Cohen in Life and Letters* (New York, 1962).

JOSEPH L. BLAU

COHN, EDWIN JOSEPH (*b*. New York, N.Y., 17 December 1892; *d*. Boston, Massachusetts, 1 October 1953), *biochemistry*.

Cohn was a notable leader in the study of proteins. After taking his Ph.D. (1917) with L. J. Henderson of Harvard and F. R. Lillie of the University of Chicago, he worked with T. B. Osborne in New Haven and S. P. L. Sørensen in Copenhagen. Returning to Harvard Medical School in 1920, he joined the new department of physical chemistry, then headed by Henderson (and later by Cohn himself). He initiated active research on the physical chemistry of proteins, especially on their solubilities in different media and their acidic and basic properties. From 1926 to 1932 he also studied the factor in liver that G. R. Minot had shown to be effective against pernicious anemia. Cohn obtained a highly active preparation of great clinical use but failed to isolate the pure active principle (vitamin B-12), which was obtained by others in 1948.

About 1930 Cohn turned his attention to the amino acids and peptides, the smaller building blocks of which proteins are composed. These are extraordinarily polar molecules, containing widely separated centers of positive and negative charge. They also contain various polar and nonpolar side chains. Over a period of ten years Cohn and his associates, G. Scatchard, J. G. Kirkwood, T. L. McMeekin, J. P. Greenstein, J. Wyman, J. T. Edsall, J. L. Oncley, and others, established many systematic relations between the structures of these molecules and their physical properties—dipole moments, solubilities, apparent molal volumes, ionization constants, and infrared and Raman spectra. They showed how to describe influences of electric charge and dipole moment, and of various polar and nonpolar side chains, in quantitative terms, thus laying a foundation for the further study of proteins.

This work provided the background essential for the next major phase of Cohn's activities, which coincided with the outbreak of World War II. Supported by the Office of Scientific Research and Development, he initiated and directed a major program, involving biochemists, clinicians, and many others, for the large-scale fractionation of human blood plasma. This yielded purified serum albumin for treatment of shock, gamma globulin for passive immunization against measles and hepatitis, fibrinogen and fibrin for neurosurgery, and numerous other protein fractions of blood plasma. The methods developed in the laboratory and its pilot plant were rapidly applied on an industrial scale, and the products were distributed for large-scale use by the armed forces and later in civilian medicine. Apart from its practical results, this program led to a great advance in knowledge of the chemistry and physiology of the multifarious components of blood plasma.

After the war the fame of this work attracted young investigators from all over the world to Cohn's laboratory, and he continued to contribute actively to the advancement of protein chemistry until his death.

BIBLIOGRAPHY

I. ORIGINAL WORKS. Cohn's writings include "The Physical Chemistry of Proteins," in *Physiological Reviews,* **5** (1925), 349–437; "Die physikalische Chemie der Eiweisskörper," in *Ergebnisse der Physiologie* (*biologischen Chemie und experimentellen Pharmakologie*), **33** (1931), 781–882; "The Chemistry of the Proteins and Amino Acids," in *Annual Review of Biochemistry,* **4** (1935), 93–148; *Proteins, Amino Acids and Peptides as Ions and Dipolar Ions,* American Chemical Society Monograph no. 90 (New York, 1943), written with J. T. Edsall; and "The History of Plasma Fractionation," in *Advances in Military Medicine,* I (Boston, 1948), 364–443. Cohn published numerous papers in the *Journal of Biological Chemistry,* the *Journal of the American Chemical Society,* and elsewhere, between 1920 and 1953.

II. SECONDARY LITERATURE. There are two biographical articles on Cohn by J. T. Edsall: in *Ergebnisse der Physiologie* (*biologischen Chemie und experimentellen Pharmakologie*), **48** (1955), 23–48; and in *Biographical Memoirs. National Academy of Sciences,* **35** (New York, 1961), 47–84. Both contain essentially complete bibliographies of Cohn's publications. The former also lists a number of publications from his laboratory to which his name was not attached; the latter article provides more extensive details about his life and work.

JOHN T. EDSALL

COHN, FERDINAND JULIUS (*b.* Breslau, Lower Silesia [now Wroclaw, Poland], 24 January 1828; *d.* Breslau, 25 June 1898), *botany, bacteriology.*

Cohn was born to impoverished young parents in Breslau's Jewish ghetto. His father, Issak, soon achieved success as a merchant and was able to nourish and encourage Ferdinand's precocious talents. It is said that Cohn could read at age two and that he was familiar with the basic doctrines of natural history by age three. He first attended school at age four; and after entering the Breslau Gymnasium in 1835, he advanced rapidly until, at age ten or eleven, a hearing defect began to slow his incredible pace. A shy, studious, and sensitive child, Cohn suffered from an acute sense of physical and emotional retardation that he did not begin to overcome until his last year in the Gymnasium.

In 1842 he entered the philosophical faculty of the University of Breslau, uncertain as to his career goals although he was inclined to the professions. In time, botany became his chief interest, mainly because of the influence of the Breslau professors Heinrich Goeppert and Christian Nees von Esenbeck. Although Cohn grew up in a period of partial liberalization of earlier restrictions on Jews, he was nevertheless barred as a Jew from the degree examinations at Breslau. After his petition for removal of the restriction was denied by the government, Cohn went to the University of Berlin in October 1846; there he received his doctorate in botany on 13 November 1847, at the age of nineteen. At Berlin he was stimulated by the teaching of Eilhard Mitscherlich, Karl Kunth, Johannes Müller, and especially Christian Ehrenberg, who introduced him to the study of microscopic animals as well as plants. When revolution rocked Berlin in March 1848, Cohn was in passionate sympathy with the revolutionaries. Although he did not himself assume an active role, Cohn's academic career may well have suffered because of his political opinions, as well as because he was a Jew. In 1849 he returned to Breslau, which remained his home for the rest of his life. In 1850 Cohn was recognized as a *Privatdozent* at the University of Breslau. He was appointed extraordinary professor of botany there in 1859 and ordinary professor in 1872. He married Pauline Reichenbach in 1867.

Cohn began his career in the midst of an intellectual revolution in botany produced by Matthias Schleiden's cell theory and Hugo von Mohl's description of protoplasm in the plant cell. In these circumstances Cohn's decision to focus his interests on the lowest plants, especially unicellular algae, probably resulted in part from his conviction of the value of cellular studies and his belief that the best way to gain insight into the cellular processes of higher organisms was to begin by carefully studying the cellular processes of the simplest organisms. In 1848 Cohn's former teacher Goeppert asked Cohn to devote himself especially to algae, in the hope that he would then contribute to a projected flora of the cryptogamous plants of Silesia. By the time this projected flora began to be published in 1876, it was Cohn, not Goeppert, who directed the work and edited the first two volumes.

In the meantime Cohn had gained early fame for his work of 1850 on the unicellular alga *Protococcus pluvialis,* especially for his novel suggestion that the protoplasm in plants and the "sarcode" in animals were "if not identical, at least highly analogous formations." The motile forms of *P. pluvialis,* with their long protoplasmic flagella, reminded Cohn of such flagellated infusoria as *Euglena* and the *Monades.* From this base, he argued against attempts to distinguish animals from plants on the basis that the former possessed differentiated organ systems or a contractile substance peculiar to themselves. Dujardin and Siebold had already shown that infusoria and rhizopods—although classified as animals—did not contain differentiated organ systems; and Cohn now suggested that the optical, chemical, and physical properties which Dujardin had ascribed to animal "sarcode" were also possessed by plant protoplasm.

Several others were working toward this same conclusion, particularly Alexander Braun, who had implied the identity of plant and animal protoplasm in a work sent to Cohn four months before the latter's first communication on *P. pluvialis.*[1] But Cohn, who was more familiar with the zoological literature, was the first to draw explicit attention to the identity between the contractile contents of plant and animal cells. This represented an important step toward the belief that the basic attributes of all life were to be sought in a single substance called protoplasm. This

protoplasmic theory of life received its classic expression in a paper published in 1861 by the German histologist Max Schultze,[2] and only near the end of the century did this conception of protoplasm as the unitary substance of life break down, to be replaced by the notion that protoplasm is a dynamic emulsion of several distinct substances and therefore has only a morphological significance.

Cohn's work on *P. pluvialis* also marked an important advance toward a redefinition of the cell and toward the recognition that Schleiden had placed too much emphasis on the cellulose cell wall. By suggesting, like Braun, that there existed primordial plant cells (motile swarm cells) devoid of cellulose walls, Cohn confirmed and expanded the suggestions of Nägeli, Mohl, Braun, and others that the essential constituent of the cell was its protoplasmic contents.

In 1854 Cohn incorporated much of his earlier work into a large treatise on the developmental history of microscopic algae and fungi. His main conclusion—that algae and fungi should be united into one class—was soon discredited, but the treatise contained much of lasting value. Of particular interest was the section on bacteria—then called Vibrionia. Up to this time, no one had questioned the animal nature of the Vibrionia, but Cohn argued that they were plants by virtue of their close relationship to known algae. The Vibrionia had been considered animals primarily because of their active, apparently voluntary, movement; but Cohn pointed out that the ciliated swarm cells of algae and fungi performed similar movements. He also suggested that bacteria followed the same developmental course as algae, demonstrating this by comparing the development of *Bacterium termo* Dujardin with the alga *Palmella.* Cohn did not claim definitive observations for the other genera of Vibrionia, but suggested that the larger bacteria seemed also to belong to the plant kingdom and seemed to display an especially close relationship to the Oscillaria.

In his article of 1855 on the unicellular alga *Sphaeroplea annulina,* Cohn contributed importantly to developing notions about the sexuality of the algae. Following the demonstration of sexuality in the brown marine alga *Fucus* by Thuret (1855) and in the freshwater alga *Vaucheria* by Pringsheim (1855), Cohn extended these conclusions to *Sphaeroplea,* another freshwater alga. He observed in *Sphaeroplea* the formation of spermatozoa and followed their progress all the way to the egg, although he was unable to see them actually penetrate it. In 1856 Cohn demonstrated another case of sexuality in algae—this time in *Volvox globator,* a motile form; previously sexuality had been demonstrated only in filamentous algae.

In the decade from 1856 to 1866, Cohn published an important paper on the contractile tissues of plants (1860), reemphasizing his conviction that contractility was not confined to animal tissue, and a model experimental study on phototropism in microscopic organisms (1865). But even though he remained an active investigator during this decade, Cohn's contributions to the botanical literature were less original and less important than his earlier work. He devoted much of his time to consolidating his earlier work and to nonresearch matters. At the Schlesische Gesellschaft für Vaterländische Kultur, of which he had been a member since 1852, Cohn accepted the chairmanship of the botanical section in 1856 and remained active in this post for the next thirty years.

His most important activity from 1856 to 1866 was his agitation, eventually successful, for an institute of plant physiology. This had been a matter of top priority for Cohn since 1847, when he had defended at Berlin the thesis that Germany needed institutes of plant physiology. In 1866 the Breslau authorities finally acceded to Cohn's long-standing request and acquired a nearby building that had once been a prison. In these inauspicious surroundings Cohn founded the first institute for plant physiology in the world, and soon launched the second great creative period of his career.

One of the earliest apparatuses installed in Cohn's institute was a small, simple marine aquarium that yielded material for much of his later work. In 1866 he reported on some new infusoria that he had found in this aquarium and revealed his method of cultivating marine plants. In 1867 Cohn suggested that since the red algae of the Oscillaria family could survive in primitive environments fatal to other plants, they must have been the first inhabitants of earth and the plants from which the rest of the plant world evolved. This led him to attempt a classification of previously neglected lower plants. Although not entirely successful, it was a pioneering attempt to base a classificatory system on Darwinian evolutionary principles.

About 1870 Cohn turned his attention primarily to bacteria, and it is for his researches in this area that he is best known. In 1870 he founded a journal, *Beiträge zur Biologie der Pflanzen*, designed primarily to publish the work that came out of his institute. In this journal appeared the founding papers of modern bacteriology. Cohn initiated the movement with his classic work of 1872, of which William Bulloch wrote that "its perusal makes one feel like passing from ancient history to modern times." In this treatise Cohn tried to bring order out of the chaos caused by the use (especially by Pasteur) of vague and arbitrary names for bacteria and by the frequent introduction of new terms. He defined bacteria as "chlorophyll-free cells of spherical, oblong, or cylindrical form, sometimes twisted or bent, which multiply exclusively by transverse division and occur either isolated or in cell families" and distinguished four groups of bacteria on the basis of their constancy of external form: (1) Sphaerobacteria (round), (2) Microbacteria (short rods or cylinders), (3) Desmobacteria (longer rods or threads), and (4) Spirobacteria (screw or spiral). Under these four groups Cohn recognized six genera of bacteria, with at least one genus belonging to each group. He repeated his conclusion of 1854 that bacteria belong to the plant kingdom by virtue of their affinity with well-known algae and suggested—in opposition to Hallier—that there was no genetic relationship between bacteria and the fungi with which they often appeared.

Cohn also showed, by researches on the nutrition of *Bacterium termo,* that bacteria were like green plants in that they obtained their nitrogen from simple ammonia compounds but unlike green plants in that they were unable to take their carbon from carbonic acid, requiring instead carbohydrates and their derivatives. Arguing that putrefaction was a chemical process excited by the growth of *Bacterium termo,* Cohn also maintained that there was a clear distinction between these bacteria of putrefaction and pathogenic bacteria.

Then, in a long series of experiments, Cohn proved that a temperature of 80°C. effectively destroyed the life of virtually all bacteria and prevented their development in an organic infusion; he admitted, however, that some doubt remained regarding *Bacillus subtilis,* the bacteria of butyric fermentation, which were more resistant to heat than *B. termo.* His experiments on the effect of low temperatures showed that although bacteria were rendered torpid by long exposure to freezing temperatures, they were not killed and regained their former vitality with the return of higher temperatures. He suggested, finally, that Pasteur's difficulty in dealing with the French supporters of equivocal generation resulted primarily from the fact that certain conditions relating to bacteria remained poorly understood.

Although later regarded as a classic work of science, Cohn's treatise of 1872 did not immediately convince everyone. In particular, its conclusions were disputed by those, such as Theodor Billroth, who believed that the various external shapes of bacteria did not really correspond to distinct genera and species but were merely various stages in the developmental history of a single plant form. In his work on *Coccobacteria septica* (1874) Billroth argued that all bacteria belonged to a single plant species whose various shapes appeared mainly in response to altered environmental circumstances. The various forms were, he argued, ultimately convertible one into the other.

In 1875, in the second of his "Researches on Bacteria," Cohn defended his earlier work and rejected Billroth's conclusions. He pointed out that he had based his classificatory scheme on external form primarily because he had found that certain characteristic physiological phenomena—especially specific fermentative activities—were associated with specific and apparently constant forms of bacteria. He did not insist that distinctions based on the external form and fermentative activity of bacteria were necessarily natural divisions; but he strongly defended his method of focusing on the lower and simpler life conditions of the fermentative organisms, since the question was one of ascertaining the general biological relationships of bacteria.

In one long and interesting section of this treatise of 1875, Cohn discussed Bastian's controversial experiments on turnip–cheese infusions. Bastian had found that bacteria could appear in such infusions even after ten minutes of boiling in a sealed flask. Assuming that the boiling destroyed all organisms previously living in the flask, Bastian concluded that living organisms could originate through abiogenesis (a sophisticated form of spontaneous generation).

When Cohn repeated Bastian's experiments, he obtained the same results but did not accept Bastian's conclusion. He argued that there might be a special developmental stage or germ that survived the boiling. He showed that the bacteria which appeared after boiling in cheese infusions were not the common putrefactive bacteria (B. termo) but, rather, bacillus rods or threads, which he identified as *Bacillus subtilis* (Pasteur's butyric ferment), whose resistance to temperature he had already mentioned in his work of 1872. Close observation of the bacilli in boiled cheese infusions revealed that after a short time many of them swelled at one end and became filled with "oval or roundish, strongly refractive little bodies" which multiplied continuously. Cohn asserted his conviction that these little bodies represented a stage in the life cycle of the bacilli and proposed as highly probable that they were "real spores, from which new Bacilli may develop." Since it was known that other spores were thermoresistant, it seemed likely that it was the spores in *Bacillus subtilis* which survived the boiling and germinated to form bacteria in Bastian's boiled and sealed turnip–cheese infusions.

In 1876 Cohn discussed in greater detail the implications of his discovery of thermoresistant endospores in *Bacillus subtilis* for the controversy over sponta-

neous generation. He believed that his discovery could at last explain the well-known anomalies presented by boiled infusions of hay and turnip–cheese. In all such infusions, Cohn demonstrated, it was the thermoresistant spores of *Bacillus subtilis* which had caused the difficulties. He showed that boiled hay infusions, like boiled infusions of turnip and cheese, contained bacillus spores, which were capable of surviving strong heat and then germinating to form new bacilli.

Cohn also showed that the complete growth and development of bacilli, and especially the formation of their spores, depended on the presence of air. When air was excluded, the activities of the bacilli induced butyric fermentation. Since normal putrefaction took place only in the presence of *B. termo,* while butyric fermentation took place only in the presence of *Bacillus subtilis,* Cohn felt that this work gave new support to his thesis that there existed distinct, independent, and incontrovertible genera of bacteria with different courses of development, different biological properties, and different fermentative activities.

In another section of this treatise of 1876, Cohn described the results of a series of experiments designed to reveal the effects of temperatures of less than 100°C. on the development of bacilli in hay infusions. He reported that bacilli, unlike any other bacteria or fungi in hay infusions, were capable of normal activity at temperatures up to 50°C. Between 50° and 55°, all multiplication and development of mature bacillary threads ceased, as did spore formation; but any spores already formed survived and retained their ability to germinate. Hay infusions were generally sterilized after twenty-four hours of heating at 60°C., but individual bacillary spores retained the ability to germinate even after three to four days of heating at 70° to 80°C.

Cohn promised a more exact determination of the temperature limits within which the development of bacilli could take place, but this project was forestalled by John Tyndall's important researches on the sterilization of hay infusions by discontinuous heating. Utilizing a superior experimental design, Tyndall carried the argument against spontaneous generation to a new level of completeness; but it should be remembered that in the crucial case of the hay bacillus (*Bacillus subtilis*) he had been forearmed with the results of Cohn's work of 1876.[3] Cohn and Tyndall together contributed as much as Pasteur to the final overthrow of the old doctrine of spontaneous generation.

At the end of his paper of 1876, Cohn referred to the pathogenic significance of the bacilli that appeared in the blood of animals and men afflicted with anthrax; he then introduced Robert Koch's classic paper on *Bacillus anthracis,* which followed immediately. Cohn stated his conviction that there could be no uncertainty about Koch's results. After Koch became famous for his contributions to bacteriology, a myth developed, especially in Breslau, to the effect that Koch had been a student of Cohn's and that his ideas owed much to Cohn's influence. In 1890 Cohn himself clarified this situation in an accurate and characteristically generous manner by pointing out, as he had when he introduced Koch's paper in 1876, that Koch had come to Cohn's institute at Breslau only to demonstrate results which he had already reached on his own and to ask Cohn and his colleagues for their judgment of his work. Cohn's role was essentially that of stimulating and encouraging Koch's work and of providing a place for its publication.[4]

Cohn's paper of 1876 was his last important contribution to bacteriology, except insofar as his direct and indirect influence is revealed in the subsequent "Researches on Bacteria" that appeared in his journal.[5] He had made four contributions of fundamental importance to bacteriology: (1) his system of classification (1872), (2) his discovery of spores (1875), (3) his discussion of the implication of his discovery of spores for the question of spontaneous generation (1876), and (4) his *Beiträge zur Biologie der Pflanzen,* in which the founding papers of modern bacteriology appeared.

Besides his scientific monographs and treatises, Cohn published many popular lectures and the widely read book *Die Pflanze* (1882), which was graced with history, biographical notes, and Goethe-inspired poetry and was credited with making innumerable converts to botany. In 1887 the University of Breslau provided him with a new institute of plant physiology in the Breslau botanical gardens. Cohn held an honorary doctorate from the faculty of medicine at the University of Tübingen and was named a corresponding member of the Accademia dei Lincei in Rome, the Institut de France in Paris, and the Royal Society of London. In 1885 he was awarded the Leeuwenhoek Gold Medal and in 1895 the Gold Medal of the Linnean Society.

NOTES

1. Braun implied the identity in his *Betrachtungen über die Verjüngung in der Natur, besonders in dem Leben und Entwicklung der Pflanzen* (Leipzig, 1851 [preface dated 1850]). Braun sent a copy of this work to Cohn on 10 May 1850, and although Cohn was already at work on *Protococcus pluvialis,* he did not present his first communication on it until September of that year (see *Ferdinand Cohn, Blätter der Erinnerung,* pp. 83–87).

2. Max Schultze, "Ueber Muskelkörperchen und dass was man eine Zelle zu nennen habe," in *Archiv für Anatomie, Physiologie und wissenschaftliche Medicin* (1861), 1–27.

3. See John Tyndall, "Further Researches on the Deportment and Vital Resistance of Putrefactive and Infective Organisms, From a Physical Point of View," in *Philosophical Transactions of the Royal Society,* **167** (1877), 149–206. On p. 152 Tyndall reports that in the autumn of 1876 Cohn "placed in my hands" his treatise of 1876. See also Tyndall's "On Heat as a Germicide When Discontinuously Applied," in *Proceedings of the Royal Society* (London), **25** (1877), 569–570.

4. Cohn clarified his relationship with Koch in the newspaper *Breslauer Zeitung* (17 Dec. 1890).

5. Between 1872 and 1885 a dozen "Researches on Bacteria" were published in Cohn's *Beiträge.* Three of these he wrote himself, one he coauthored, and most of the rest were contributed by his students.

BIBLIOGRAPHY

I. ORIGINAL WORKS. Cohn's writings discussed in the text are "Nachträge zur Naturgeschichte des *Protococcus pluvialis Kützing*" [1850], in *Nova acta physico-medica Academiae Caesareae Leopoldino Carolinae germanicae naturae curiosorum,* **22** (for 1847), 605–764; "Untersuchungen über die Entwicklungsgeschichte der mikroskopischen Algen und Pilze," *ibid.,* **24** (1854), 103–256; "Ueber die Fortpflanzung von *Sphaeroplea annulina,*" in *Bericht über die zur Bekanntmachung geeigneten Verhandlungen der K. Preussischen Akademie der Wissenschaften zu Berlin* (1855), 335–351; "Beobachtungen über den Bau und die Fortpflanzung von *Volvox globator,*" in *Übersicht der Arbeiten und Veränderungen der Schlesischen Gesellschaft für vaterländische Kultur* (Breslau) (1856), 77–83; "Contractile Gewebe im Pflanzenreiche" [1860], in *Jahresbericht des Akademischen naturwissenschaftlichen Vereins zu Breslau (Abhandlungen)* (1861), 1–48; "Ueber die Gesetze der Bewegung der mikroscopischen Pflanzen und Thiere unter Einfluss des Lichtes," in *Bericht über die Versammlung der Deutschen Naturforscher und Aerzte,* **40** (1865), 219–222; "Neue Infusorien im See-aquarium" [1865], in *Zeitschrift für wissenschaftliche Zoologie,* **16** (1866), 253–302; "Beiträge zur Physiologie der Phycochromaceen und Florideen," in *Archiv für mikroskopische Anatomie und Entwicklungsmechanik,* **3** (1867), 1–60; "Untersuchungen über Bacterien," in *Beiträge zur Biologie der Pflanzen,* **1,** no. 2 (1872), 127–224; "Untersuchungen über Bacterien, II.," *ibid.,* no. 3 (1875), 141–207; "Untersuchungen über Bacterien, IV. Beiträge zur Biologie der Bacillen," *ibid.,* **2,** no. 2 (1876), 249–276; and *Die Pflanze: Vorträge aus dem Gebiete der Botanik* (Breslau, 1882, 1897).

At least one paper not discussed in the text deserves special mention: "*Empusa muscae* und die Krankheit der Stubenfliegen: Ein Beitrag zur Lehre von den durch parasitische Pilze charakterisirten Epidemieen," in *Nova acta physico-medica Academiae Caesareae Leopoldino Carolinae germanicae naturae curiosorum,* **25** (1855), 299–360, which is noteworthy as the first careful study of a disease in animals (the housefly) caused by a fungus. Cohn published nearly 200 papers and books; almost all of them can be located by judicious use of these three bibliographies: (1)

the bibliography in *Ferdinand Cohn, Blätter der Errinerung,* pp. 261–266 (see below); (2) *Royal Society Catalogue of Scientific Papers,* II, 8–10; VII, 413–414; IX, 547–548; XIV, 294; (3) *British Museum General Catalogue of Printed Books,* XLI (1966), 322–323.

II. SECONDARY LITERATURE. The basic source for Cohn's life and work is *Ferdinand Cohn, Blätter der Errinerung* (Breslau, 1901), collected by his wife Pauline Cohn, with contributions by Felix Rosen. This truly remarkable book is filled with intimate insights into Cohn, his work, and his times and is based upon his boyhood diary (discontinued in 1852), his family letters, scientific correspondence, congratulatory notes, and descriptions of his travels throughout Europe. In his last few months Cohn began an autobiography, but he had completed only a few pages at the time of his death.

Fairly detailed accounts of Cohn's life and work are also given by Felix Rosen, in *Berichte der Deutschen botanischen Gesellschaft,* **17** (1899), 172–201; and also by C. Mez, in *Biographisches Jahrbuch und Deutscher Nekrolog,* III (1900), 284–296. For still other biographical sketches, see *Neue Deutsche Biographie,* III (1957), 313–314; *Münchener medicinische Wochenschrift,* **45,** pt. 2 (1898), 1005–1007; and the references given in the *Royal Society Catalogue of Scientific Papers,* XIV, 294.

For Cohn's place in the history of botany, see Julius von Sachs, *Geschichte der Botanik vom 16. Jahrhundert bis 1860* (Münich, 1875), pp. 225, 228, 478; and John R. Baker, "The Cell-Theory: A Restatement, History, and Critique, Part II," in *Quarterly Journal of Microscopical Science,* **90** (1949), 87–108 (94–96). For his place in the history of bacteriology, see William Bulloch, *The History of Bacteriology* (London, 1938), pp. 106, 113, 116–119, 150, 174–177, 187–188, 192–195, 198, 200–203, 207–210, 213–214, 217–218, 296, 319, 328–330, 358.

A brief sketch of Cohn's life and an assessment of his bacteriological researches can be found in Morris C. Leikind's introduction to Cohn's "Bacteria, the Smallest of Living Organisms," [1872], Charles S. Dolley, trans. [1881], in *Bulletin of the History of Medicine,* **7** (1939), 49–92, to which is appended a reprint of the bibliography of Cohn's works found in *Blätter der Errinerung.*

GERALD L. GEISON

COHN, LASSAR (*b.* Hamburg, Germany, 6 September 1858; *d.* Königsberg, Germany [now Kaliningrad, U.S.S.R.], 9 October 1922), *chemistry.*

Cohn, who published under the name of Lassar-Cohn, studied chemistry at Heidelberg, Bonn, and Königsberg. He was *Privatdozent* (1888) and then professor of chemistry (1894–1897) at Königsberg, at Munich (1897–1898), and again at Königsberg (1902–1909). Cohn was associated with various industrial firms throughout his career; he conducted researches in organic and physiological chemistry and in chemical technology.

The most elaborate of Cohn's investigations was the isolation of the acids in ox and human bile by means of saponification, acidification, and solvent extraction (1892–1898). By oxidizing cholic and dehydrocholic acids Cohn prepared bilianic and isobilianic acids in 1899. From the latter he obtained another oxidation product called cilianic acid. The understanding of the structures of the bile acids and their oxidation products came with the work of Heinrich Wieland in 1912.

Cohn also studied the electrolysis of organic potassium salts (1889) and prepared several halogenated derivatives of salicylic acid (1905). He developed an improved nitrometer for the determination of nitrogen by the Dumas method (1901) and invented a new saccharimeter (1922). He also published papers dealing with the utilization and disposal of chemical waste materials from industrial processes. Cohn made a careful study of the sulfite waste liquor from cellulose factories, proposing that the waste matter, instead of polluting rivers, could be made beneficial to crops. After chemical treatment to reduce acidity, it could be pumped into canals and thence to irrigated fields.

Cohn was a successful writer of popular books on chemistry. Two books were especially widely read: *Die Chemie im täglichen Leben* (1896), which appeared in twelve German editions and was translated into many languages, and *Einführung in die Chemie in leichtfasslicher Form* (1899), which appeared in seven German editions. Both books were written for the general public, the former stressing the practical applications of chemistry, the latter the theoretical principles. His *Arbeitsmethoden für organisch-chemische Laboratorien* (1891) was a valuable compilation of all the methods for particular laboratory operations in organic chemistry: drying, distillation, extraction, filtration, molecular weight determination, sulfonation, halogenation, nitration, and so forth. The book was widely used in Germany and was also successful in its English translations.

BIBLIOGRAPHY

I. ORIGINAL WORKS. Cohn's *Arbeitsmethoden für organisch-chemische Laboratorien* (Hamburg, 1891; 5th ed., Leipzig, 1923) appeared in two English versions: *A Laboratory Manual of Organic Chemistry,* translated by Alexander Smith (London, 1895) and *Organic Laboratory Methods,* translated by Ralph E. Oesper and edited by Roger Adams and Hans T. Clarke (Baltimore, 1928). *Die Chemie im täglichen Leben* (Hamburg-Leipzig, 1896; 12th ed., Leipzig, 1930) was translated by M. M. Pattison Muir as *Chemistry in Daily Life* (London, 1896; 5th ed., 1917). Cohn's *Praxis der Harnanalyse* (Hamburg, 1897) had nine German editions and an English translation by H. W. F. Lorenz, *Praxis of Urinary Analysis* (New York, 1903). The work on bile acids was compiled in *Die Säuren der Rinder-galle und der Menschengalle* (Hamburg, 1898). *Einführung in die Chemie in leichtfasslicher Form* (Hamburg–Leipzig, 1899; 7th ed., Leipzig, 1927) was translated by M. M. Pattison Muir as *An Introduction to Modern Scientific Chemistry* (London, 1901).

II. SECONDARY LITERATURE. There is a brief notice on Cohn by Friedrich Klemm in *Neue deutsche Biographie,* III (Berlin, 1956), 316–317.

ALBERT B. COSTA

COITER, VOLCHER (*b.* Groningen, Netherlands, 1534; *d.* Champagne, France, 2 June 1576), *anatomy, physiology, ornithology, embryology, medicine.*

Son of a jurist, Coiter was favored with an excellent education in his native city at St. Martin's School, where the learned Regnerus Praedinius was master; there he first became acquainted with Galen and dissection. His ability was such that in 1555 the city fathers awarded him a stipend for five years of study at foreign universities. During this period, although dates and movements are not always clear, we know that his good fortune in enjoying eminent teachers continued. He probably studied with Leonhard Fuchs at Tübingen. In 1556 he was briefly at Montpellier; he mentions Guillaume Rondelet and Laurent Joubert, and he also knew Felix Platter. Gabriele Falloppio taught him at Padua, and Bartolomeo Eustachi at Rome. By 1560, and possibly even by 1559, he was at Bologna, where he received the doctorate in medicine on 24 March 1562 and where his researches were guided by Ulisse Aldrovandi and Giulio Cesare Aranzio. At Bologna he lectured on logic and surgery; and at Bologna his first two publications, tables on human anatomy, were issued. For a time he also taught at Perugia.

A brilliant career seemed assured, but letters written in 1566 by his friend Joachim Camerarius the Younger tell of Coiter's arrest and imprisonment, first in Rome and then in Bologna. It is generally assumed that his Protestantism was responsible and that he offended the Inquisition. By the fall of 1566 he was back in Germany, where Camerarius smoothed the path for him. He served Pfalzgraf Ludwig VI at Amberg and taught there until 1569, when he became physician to the city of Nuremberg. Documents in Nuremberg and Erlangen, passages in his works, and inscriptions in copies of these works that he presented to friends attest his anatomical and medical activities in Nuremberg and provide details concerning the publication of the treatises which appeared in that city. Bodies of criminals furnished opportunities for dissection, and his daily medical practice fostered an attention to pathology. Among physicians he associated not only with Camerarius but also with Georg Palm, Heinrich Wolff, Melchior Ayrer, Franz Renner, and Thomas Erastus of Heidelberg and Basel. On more than one occasion he had sharp brushes with barber surgeons and quacks. The eminent families Imhof and Kress were served by him. He often traveled outside Nuremberg to treat magisterial, noble, and ecclesiastical patients.

Coiter's longest excursion from Nuremberg took place from the fall of 1575 to the spring of 1576, when he attended Pfalzgraf Johann Casimir on his expedition to France in support of the Huguenot cause. Coiter did not return from this journey; his death was the result not of military action but of illness, possibly typhus, after peace had been declared and the army was returning to Germany. He left a widow, Helena, who was a foreigner and who was granted permission to remain in Nuremberg two years longer without citizenship.

In a short life of forty-two years Coiter effected significant advances in biological knowledge. His early works on human anatomy were expressly intended for students, as he himself states and as their tabular form favors. In them he still adheres to Galenic doctrine, and they have no essential significance for the history of anatomy. The treatises published at Nuremberg, however, display a greater maturity; a realization that dissection is the most important part of anatomy (both normal and pathological); an appreciation of the work of Vesalius (of whose contributions to anatomy he said, "Incomparabili industria stupendoque ingenio, hanc artem omnibus quasi suis numeris absoluit" ["With incomparable industry and amazing genius he perfected this art in almost every respect."]), Fallopio (notes on whose lectures he published in 1575), and Eustachi; and original thought. The treatise on the skeleton of the fetus and of a child six months old points up the differences between these and adult skeletons and shows where ossification begins.

With a solid grounding in human (especially skeletal) anatomy, Coiter was well prepared for exploration in comparative anatomy. He covered almost the entire vertebrate series—amphibians, reptiles, birds, and mammals—and was the first to raise this field to independent status in biology, although he emphasized points of difference from human anatomy rather than points of similarity. He recorded what he saw in the living hearts of cats, reptiles, frogs, and several fishes. He called attention to the orbicular muscle of the hedgehog and described the poison gland of the adder. Coiter's investigation of avian anatomy is particularly significant: he depicted the skeletons of the crane, the starling, the cormorant, and the parrot; provided a general classification, in tabular form, of the birds known to him; and discovered the tongue and hypobranchial apparatus of the woodpecker.

Coiter knew the value of good drawings and himself signed most of the finely drawn copper engravings that illustrate his anatomical work. Case histories are provided in the interesting miscellany of anatomical and surgical observations, and in these for the first time Coiter described the spinal ganglia and the musculus corrugator supercilii. An interest in anatomical nomenclature and etymology and in balneology is often apparent in his works.

Unillustrated but nonetheless epochal are Coiter's studies on the development of the chick, begun in Bologna with the encouragement of Aldrovandi and published in Nuremberg; based on observations made on twenty successive days, they presented the first systematic statement since the three-period description (after three days of incubation, after ten, after twenty) provided by Aristotle two millennia before. Coiter worked without a lens but also without Scholastic influence: "There is no futile quarreling over trifles, no pompous arguments, no theoretical bias, and, except for a few references to Aristotle, Lactantius, Columella and Albertus no appeal to authority, but simply a calm, dispassionate, impartial and concise record of his observations with implicit confidence that the truth will prevail" (Adelmann; p. 333). Twenty-eight years later (1600) Aldrovandi published his account; Girolamo Fabrizio followed in his posthumous work of 1621, Nathaniel Highmore and William Harvey in 1651, and Marcello Malpighi in 1673 and 1675.

BIBLIOGRAPHY

I. ORIGINAL WORKS. *Tabulae externarum partium humani corporis* (Bologna, 1564) is included, with some changes, in the *Externarum et internarum principalium humani corporis partium tabulae,* as is, in revised form, *De ossibus, et cartilaginibus corporis humani tabulae* (Bologna, 1566). *Externarum et internarum principalium humani corporis partium tabulae* (Nuremberg, 1572, 1573) contains "Introductio in anatomiam," "Externarum humani corporis partium tabulae," "Internarum humani corporis partium tabulae," "De ovorum gallinaceorum generationis primo exordio progressuque, et pulli gallinacei creationis ordine" (trans. and ed. with notes and intro. by Howard B. Adelmann, in *Annals of Medical History,* n.s. **5** [1933], 327–341, 444–457; trans. and analyzed seriatim at many points, listed in the index, in Adelmann's *Marcello Malpighi and the Evolution of Embryology,* 5 vols. [Ithaca, N.Y., 1966]), "Tabulae ossium humani corporis," "Ossium tum humani foetus adhuc in utero existentis, vel imperfecti abortus, tum infantis dimidium annum nati brevis historia atque explicatio" (also in Hendrik Eysson, *Tractatus anatomicus &*

medicus, De ossibus infantis, cognoscendis, conservandis, & curandis [Groningen, 1659], pp. 169–201; and, after Eysson, in Daniel Le Clerc and Jean Jacques Manget, *Bibliotheca anatomica* [Geneva], **2** [1699], 509–512), "Analogia ossium humanorum, simiae, et verae, et caudatae, quae cynocephali similis est, atque vulpis," "Tabulae oculorum humanorum," "De auditus instrumento," and "Observationum anatomicarum chirurgicarumque miscellanea varia, ex humanis corporibus . . . deprompta." *Lectiones Gabrielis Fallopii de partibus similaribus, ex diversis exemplaribus a Volchero Coiter . . . collectae. His accessere diversorum animalium sceletorum explicationes, iconibus . . . illustratae . . . autore eodem Volchero Coiter* (Nuremberg, 1575) has appended to it the treatise "De avium sceletis et praecipuis musculis." Extensive selections from the published works, with English translation, are furnished in *Volcher Coiter,* B. W. Th. Nuyens and A. Schierbeek, eds., no. 18 of Opuscula Selecta Neerlandicorum de Arte Medica (Amsterdam, 1955). Herrlinger (1952) published, from MSS in Nuremberg, Coiter's "Gutachten zur Sondersiechenschau" and "Ein ordenttlich Regiementt wie man sich im Wildt Badt haltten sol." His "Vom rechten Gebrauch dess Carls Padt bei Ellenbogen" appeared in the *Anzeiger* of the Germanisches Nationalmuseum (1891), 10–18.

II. SECONDARY LITERATURE. On Coiter or his work, see B. W. Th. Nuyens, "Doctor Volcher Coiter, 1534–1576?" in *Nederlands tijdschrift voor geneeskunde,* **77** (1933), 5383–5401; and "De laatste tien Jaren van Volcher Coiter's Leven," *ibid.,* **79** (1935), 2653–2659; Walton B. McDaniel, II, "Notes on the 'Tractatus de ossibus foetus' of Volcher Coiter," in *Annals of Medical History,* n.s. **10** (1938), 189–190; F. J. Cole, *A History of Comparative Anatomy* (London, 1949), pp. 73–83; Dorothy M. Schullian, "New Documents on Volcher Coiter," in *Journal of the History of Medicine and Allied Sciences,* **6** (1951), 176–194; Elsa Guerdrum Allen, *The History of American Ornithology Before Audubon* (Philadelphia, 1951), pp. 405–410 (this is *Transactions of the American Philosophical Society,* n.s. **41,** pt. 3); and Robert Herrlinger, *Volcher Coiter, 1534–1576* (Nuremberg, 1952); and "News on Coiter," in *Journal of the History of Medicine and Allied Sciences,* **12** (1957), 79–80.

Coiter's portrait (1575) in oils, attributed to Nicolas Neufchatel and representing him demonstrating the muscles of the arm, with the *écorché* he had constructed on his left and a shelf of medical classics behind him, is preserved in the Germanisches Nationalmuseum at Nuremberg; there are later portraits at Weimar and Amsterdam, and there is a copper engraving (1669) by J. F. Leonhart, copies of which are in Nuremberg, Coburg, Copenhagen, and Ithaca, New York.

DOROTHY M. SCHULLIAN

COLDEN, CADWALLADER (*b.* Ireland, 7 February 1688; *d.* Flushing, New York, 20 September 1776), *botany, medicine, physics.*

Although born in Ireland while his mother was on a visit, Cadwallader Colden was raised in Berwickshire, Scotland, where his father, the Reverend Alexander Colden of the Church of Scotland, had the church in Duns. Colden graduated from the University of Edinburgh in 1705, studied medicine in London, and, feeling that he lacked funds to pursue a career in England, removed to Philadelphia in 1710. There he practiced medicine and engaged in business for a time; but, with the encouragement of Governor Robert Hunter, a fellow Scot, he moved to New York in 1718 and speedily rose to prominence in its politics and government. He was appointed surveyor general in 1720 and a member of the governor's council in 1721. In 1761 he became lieutenant governor and several times thereafter served as acting governor. About 1727 he began to develop Coldengham, a country seat west of Newburgh; after 1739 he spent most of his time there until 1762, when he acquired the Spring Hill estate in Flushing. He remained a major force in government, aroused patriot wrath at the time of the Stamp Act, and died loyal to the crown.

Colden returned once to Scotland and there married Alice Christy in 1715;· they had eight children who lived to maturity. Of these, Jane followed her father's tutoring and made botanical contributions of her own, while David sought to extend some of his father's scientific ideas, publishing most usefully in electricity.

Colden consistently aspired to scientific achievement. In Philadelphia he enjoyed the acquaintance of James Logan; and before removing to New York, he began a correspondence with British members of an international circle engaged in the study of natural history. He soon became a correspondent of Peter Collinson in England, J. F. Gronovius in Holland, and Carl Linnaeus in Sweden. With a group of American naturalists he participated in making known to the European members of the circle new species and genera of plants.

Having studied some botany under Charles Preston at Edinburgh, Colden pursued the improvement of his knowledge in America. He profited from the study of Gronovius's *Flora Virginica* and even more from the writings of Linnaeus, whose system of classification he mastered. He did not, however, accept the Linnaean system easily, nor was he ever satisfied with it, feeling continuing need for a natural system of classification. Despite these feelings he sent Linnaeus a carefully cataloged and described collection of plants, which Linnaeus published under the title "Plantae Coldenghamiae" in the *Acta Upsaliensis*. Moreover, in *Flora Zeylanica* (1747), he conferred the accolade of naming a plant the Coldenia. Colden's work was carefully done and was admired by taxonomists of his own day and later.

Colden interested himself in a great variety of sciences. A significant portion of his correspondence related to medicine; and he wrote several medical essays, some of which were published after his death. Probably his most influential publication was *An Abstract From Dr. Berkley's* [sic] *Treatise on Tar-Water* (1745), to which he added important reflections of his own. Among other contemporary publications were articles on the medical virtues of pokeweed and the New York diphtheria epidemic, which were published in London journals. Other of his ideas appeared as newspaper articles. Because Colden early gave up medical practice, he was sometimes speculative and incorrect in his observations—as in his manuscript essay on yellow fever.

Colden maintained an alertness to astronomical events of importance and corresponded on the application of astronomy to cartography and surveying. He wrote essays on light, on optics, and on waterspouts and published a piece on fluxions. His knowledge of mathematics was good, and he displayed significant insights in several fields. His *History of the Five Indian Nations,* although much of it was based on French accounts, was considered as reliable on both sides of the Atlantic.

Colden's major scientific effort was the most ambitious ever attempted in the colonies. Conscious that Newton had admitted his inability to understand the cause of gravity, Colden sought to supply this deficiency by devising his own explanation. He began publishing on this subject with *An Explication of the First Causes of Action in Matter,* which appeared in 1746. The same year, a pirated edition was brought out in London; a German translation came out in 1748, and a French translation in 1751. In that year Colden issued a second, expanded edition in London under the title *The Principles of Action in Matter.* Several magazines published extracts and synopses; but when he produced a third, corrected and further expanded, edition in 1755, no publisher would touch it. He had to be content with a couple of magazine articles drawn from it and with the permanence anticipated from depositing the manuscript in the library of the University of Edinburgh.

Although specific reactions appeared slowly, Colden's effort ultimately provoked much comment of widely varying character. None then realized that his method had been to elaborate upon uncoordinated texts drawn primarily from the "Queries" in Newton's *Opticks,* where the great author had sought to suggest developmental possibilities (such phenomena as fermentation, putrefaction, and fertilization) residing in the corpuscular philosophy. Colden never realized that his work was in direct conflict with the laws of motion of the *Principia,* which he did not fully comprehend. He therefore resented being called an opponent of Newton and sought further to apply his method to explanations of heat, fermentation, putrefaction, colors, and fertilization. At the end, several qualified commentators dismissed the enterprise; Leonhard Euler pronounced it incompetent.

On the basis of his constructive achievements in botany and other fields, as well as his wide correspondence and European reputation, Colden was able to play a significant role in the American scientific community. He was an original member of the 1743 American Philosophical Society. He had useful relationships with his fellow Scottish émigrés James Alexander, Alexander Garden, and William Douglass; and he knew well not only Logan but also John Bartram, Benjamin Franklin, and most other Americans active in science.

BIBLIOGRAPHY

I. ORIGINAL WORKS. Colden's published botanical contributions were made in "Plantae Coldenghamiae," in *Acta Societatis regiae scientiarum Upsaliensis,* **4** (1743), 81–136; **5** (1744–1750), 47–82. Among his published medical writings, the heavily annotated *An Abstract From Dr. Berkley's* [sic] *Treatise on Tar-Water* (New York, 1745) was published separately as well as in a newspaper. In periodicals, he published "Extract of a Letter From Cadwallader Colden, Esq. to Dr. Fothergill Concerning the Throat Distemper, Oct. 1, 1753," in *Medical Observations and Inquiries,* **1** (1757), 211–229; "The Cure of Cancers," in *Gentlemen's Magazine,* **21** (1751), 305–308; and "Farther Account of Phytolacca, or Poke-Weed," *ibid.,* **22** (1752), 302. Other medical essays appeared in newspapers and after his death. His *History of the Five Indian Nations* (New York, 1727; 2nd ed., London, 1747) was of ethnological as well as political significance; a continuation was published long after his death. The work he regarded as his magnum opus was *An Explication of the First Causes of Action in Matter* (New York, 1745 [actually 1746]). This was translated into German by A. G. Kastner (Hamburg, 1748) and also appeared in French (Paris, 1751). Colden issued an expanded edition under the title *The Principles of Action in Matter* (London, 1751) and published articles extending and explaining it in English periodicals.

The great bulk of Colden MSS are among the Colden Papers of the New-York Historical Society; most of these have been published: *The Colden Letter Books,* vols. 9 and 10 of New-York Historical Society, *Collections* (New York, 1876–1877) and *The Letters and Papers of Cadwallader Colden,* vols. 50–56, 67–68 of New-York Historical Society, *Collections* (New York, 1917–1923, 1934–1935); some materials, especially scientific papers, remain unpublished. The third, MS ed. of *The Principles of Action in Matter* and some other items are in the library of the University of Edinburgh. Additional MSS are held by the Newberry Library, Chicago, and the Huntington Library, San Marino, Calif.

Colden's MS copybook (1737–1753) is in the Rosenbach Foundation, Philadelphia.

II. SECONDARY WRITINGS. The only biography is Alice M. Keys, *Cadwallader Colden* (New York, 1906), which does little with his scientific career. Brooke Hindle, *The Pursuit of Science in Revolutionary America* (Chapel Hill, N.C., 1956), pp. 38–48, describes his general role; one aspect of his science is examined in Brooke Hindle, "Cadwallader Colden's Extension of the Newtonian System," in *William and Mary Quarterly,* 3rd ser., **15** (1956), 459–475. Saul Jarcho has written on aspects of Colden's medical career: "Cadwallader Colden as a Student of Infectious Disease," in *Bulletin of the History of Medicine,* **29** (1955), 99–115; "The Correspondence of Cadwallader Colden and Hugh Graham on Infectious Fevers," *ibid.,* **30** (1956), 195–211; "Obstacles to the Progress of Medicine in Colonial New York," *ibid.,* **36** (1962), 450–461; and "The Therapeutic Use of Resin and of Tar Water by Bishop George Berkeley and Cadwallader Colden," in *New York State Journal of Medicine,* **55** (1955), 834–840.

BROOKE HINDLE

COLE, FRANK NELSON (*b.* Ashland, Massachusetts, 20 September 1861; *d.* New York, N.Y., 26 May 1926), *mathematics.*

Cole was the third son of Otis and Frances Maria Pond Cole. He was graduated from Harvard College with an A.B. in 1882, second in a class of 189. Awarded a traveling fellowship, he spent two years at Leipzig studying under Felix Klein. In 1885 Cole returned to Harvard as a lecturer in the theory of functions. The next year he received his Ph.D. from Harvard; his dissertation was entitled "A Contribution to the Theory of the General Equation of the Sixth Degree." Cole's mathematical enthusiasm, according to W. F. Osgood, who, with M. Bocher, was among Cole's students, was contagious enough to inaugurate a new era in graduate instruction at Harvard.

In the fall of 1888 Cole went to the University of Michigan; he remained there until 1895, when he went to Columbia University. He was to have retired from Columbia in October 1926 but died the preceding May, survived by his wife and three children. He had married Martha Marie Streiff of Göttingen in 1888, but he had largely isolated himself from his family since 1908. At the time of his death Cole lived in a rooming house under the name of Edward Mitchell and claimed to be a bookkeeper.

Cole's most productive years were those at Ann Arbor. His research dealt mainly with prime numbers, number theory, and group theory. He was a leader in the organization of the American Mathematical Society and active in its affairs until his death. He was its secretary from 1896 to 1920 and a member of the editorial staff of its *Bulletin* from 1897 to 1920. His

appreciation of scholarship and his literary skill exerted a great influence on the *Bulletin,* which in turn did much to establish the American Mathematical Society as an important scientific organization. In 1920, when Cole resigned as secretary and as editor, he was given a purse commemorating his long service to the society. He contributed this money to help establish the Frank Nelson Cole prize in algebra. A second prize in theory of numbers was established in Cole's name by the society in 1929. The *Bulletin* for 1921 was dedicated to Cole, and his portrait was the frontispiece.

BIBLIOGRAPHY

I. Original Works. Cole revised and translated, with the author's permission, E. Netto's *The Theory of Substitutions and Its Applications to Algebra* (Ann Arbor, Mich., 1892). Twenty of his articles are listed in R. C. Archibald, *A Semicentennial History of the American Mathematical Society, 1888–1938* (New York, 1938), p. 103. See also his reports as secretary of the American Mathematical Society, in *Bulletin of the American Mathematical Society,* **3–27** (1896–1920).

II. Secondary Literature. Cole and his work are discussed in R. C. Archibald, *A Semicentennial History of the American Mathematical Society, 1888–1938* (New York, 1938), pp. 100–103 and *passim;* T. S. Fiske, in *Bulletin of the American Mathematical Society,* **33** (1927), 773–777; and D. E. Smith, in *National Cyclopaedia of American Biography* (1933), p. 290. See also *American Men of Science,* 3rd ed. (1921), p. 137; "Class of 1882," in *Harvard College Alumni Report* (1883–1926); *New York Times* (27 May 1926), p. 25; (28 May 1926), p. 1; (29 May 1926), p. 14; (31 May 1926), p. 14; (3 June 1926), p. 24; (7 June 1926), p. 18; and *Who's Who in America* (1920–1921).

Mary E. Williams

COLLET-DESCOTILS, HIPPOLYTE-VICTOR (*b.* Caen, France, 21 November 1773; *d.* Paris, France, 6 December 1815), *chemistry.*

The son of Jean Collet-Descotils, a lawyer and high official in the provincial administration, and Marie-Marguerite Le Cocq, Collet-Descotils (often called only Descotils) was educated at the Collège du Bois of the University of Caen. About 1789 he was taken by his father to Paris, where he studied physics under J. A. C. Charles and chemistry under Vauquelin. In 1792 he was drafted into the navy and served at Cherbourg; he obtained his discharge in 1794 after passing the entrance examination for the École des Mines in Paris. The Société Philomathique accepted Collet-Descotils as a member in 1796, and he was invited to join the scientific group, led by Berthollet and Monge, that accompanied Napoleon's expedition to Egypt in 1798. He stayed in Egypt until 1801 as a member of the Institut d'Égypte.

On returning to France, Collet-Descotils succeeded Vauquelin as professor at the École des Mines; he gave only one lecture course, however, for teaching ceased there when two mining schools were founded in the provinces. Nevertheless, he remained as director of the laboratory; and in 1809 he became one of the engineers in chief in the Department of Mines. He was appointed acting director of the École des Mines when it reopened (on a limited basis) in Paris in 1814, only a year before his death. He left a widow and two sons.

Collet-Descotils joined the editorial board of *Annales de chimie* in 1804 and later became its secretary; and from 1807 he was a member of the Société d'Arcueil, which met at Berthollet's country house. He was respected as a very competent chemist; but although he was in touch with leading French scientists, he did no outstanding research.

In the course of his official duties Collet-Descotils visited mines in various parts of France and Italy, investigated metallurgical problems, and analyzed numerous ores. In 1805 he examined a brown lead ore from Mexico that A. M. del Rio believed to contain a new metal. He concluded that it was chromium, and other chemists agreed; thus the true identity of vanadium remained unknown until its rediscovery in 1830 by N. G. Sefström.

Collet-Descotils was more fortunate in his research on crude platinum, which contained a black constituent that normally seemed insoluble in aqua regia. In 1803 he showed that it dissolved slowly, especially in the presence of excess nitric acid, to form a solution that gave a red precipitate with ammonium chloride. Pure platinum gave a yellow precipitate, so a new metal must be present in the red compound. Fourcroy and Vauquelin independently reached the same conclusion, and a more detailed study by Smithson Tennant revealed that there were in fact two metals in the less soluble part of crude platinum—iridium (which gave the red compound) and osmium. In 1807 Collet-Descotils found that iridium could be precipitated from a platinum solution by making it alkaline and oxidizing it by exposure to air; it was only in the twentieth century, however, that this procedure, in an improved form, was generally adopted by platinum refiners.

BIBLIOGRAPHY

I. Original Works. Collet-Descotils's twenty-six contributions to journals are listed in *The Royal Society Catalogue of Scientific Papers,* II (1868), 20–21. He also pub-

lished "Observations sur les propriétés tinctoriales du henné," in *Mémoires sur l'Égypte, publiés pendant les campagnes du Général Bonaparte,* I (Paris, 1800), 280–283, written with C. L. Berthollet; and "Description de l'art de fabriquer le sel ammoniaque," in *Description de l'Égypte, ou recueil des observations . . . publié par les ordres de Sa Majesté l'Empereur Napoléon le Grand, État Moderne,* I (Paris, 1809), 413–426.

II. SECONDARY LITERATURE. L.-J. Gay-Lussac's "Notice sur Hippolyte-Victor Collet-Descotils," in *Annales de chimie et de physique,* **4** (1817), 213–220, also appeared in English, in *Annals of Philosophy,* **9** (1817), 417–421. Gay-Lussac did not give the names of Collet-Descotils's parents, which have been obtained from the baptismal register in the municipal archives of Caen. Further biographical information is given in M. P. Crosland, *The Society of Arcueil* (Cambridge, Mass., 1967); and the particular aspects of Collet-Descotils's work mentioned above are discussed, with references, in D. McDonald, *A History of Platinum* (London, 1960), pp. 101–103, 138; and M. E. Weeks and H. M. Leicester, *The Discovery of the Elements,* 7th ed. (Easton, Pa., 1968), pp. 351–355, 414–415.

W. A. SMEATON

COLLIE, JOHN NORMAN (*b.* Alderley Edge, Cheshire, England, 10 September 1859; *d.* Sligachan, Isle of Skye, Scotland, 1 November 1942), *chemistry.*

The second son of John and Selina Mary Collie, John Norman learned early in life to love the outdoors so much that mountaineering vied with chemistry as his chief interest in life. In 1877 he entered University College, Bristol, where he studied under E. A. Letts. Collie continued his studies as assistant to Letts when the latter moved to Queen's College, Belfast, in 1879, working on reactions of tetrabenzyl phosphonium salts. In 1883 Collie went to Würzburg to obtain his Ph.D. under Wislicenus, studying thermal decomposition products of phosphonium and ammonium salts. Here Collie also began the study of β-aminocrotonic ester that led him ultimately into his major works on polyketides and pyrones. After completing his degree in 1885, Collie became science lecturer at Ladies' College, Cheltenham, and remained there until 1887, when he went to University College, London, as assistant to William Ramsay. Collie was named professor of chemistry at the Pharmaceutical College, Bloomsbury Square, and fellow of the Royal Society in 1896. In 1902 he returned to University College, London, as the first professor of organic chemistry, a position he held until his retirement in 1928 as emeritus professor and honorary fellow of the college.

When Collie finished his studies on phosphonium salts, the lutidone derivative he had obtained earlier from heating β-aminocrotonic ester still piqued his interest. He managed to obtain this lutidone from dehydracetic acid and ammonia as well. This observation could not be reconciled with the then accepted structure for dehydracetic acid and Collie began to study this acid to understand its structure and chemistry. He proposed as dehydracetic acid 2-hydroxy, 6-acetonyl, γ-pyrone, or its δ-lactone tautomer, to explain its reactions. The acetone and carbon dioxide observed during the production of dehydracetic acid from tetraacetic acid were explained by the decomposition of unstable diacetic acid. This suggestion led to Collie's attempt to isolate triacetic acid or its lactone. He successfully obtained 4-hydroxy, 6-methyl, α-pyrone, the tautomeric δ-lactone of triacetic acid, by hydrolysis of dehydracetic acid with sulfuric acid. With the finding of the postulated triacetic acid, Collie suggested the appropriate conditions for the condensation of tetra-, tri-, di-, and acetic acids into lactones.

Dehydracetic acid with hydrochloric acid yielded 2,6-dimethyl, γ-pyrone rather than a lactone. This finding, along with the observation that this pyrone ring could be opened to form diacetylacetone, suggested further studies. From diacetylacetone Collie succeeded in obtaining orcinol, symmetrical dimethyl lutidone, and various naphthalene and isoquinoline derivatives.

Collie culminated his work on dehydracetic acid, 2,6-dimethylpyrone, and diacetylacetone in his generalization of the multiple ketene group. What Collie called the ketene group ($-CH_2CO-$) and its enol tautomer ($-CH=COH-$) can, in one comprehensive scheme, relate pyrone, coumarin, benzopyrone, pyridine, isoquinoline, and naphthalene to the polyacetic acids. He also speculated on the formation of sugars and fats, with the multiple ketene group as the fundamental building block for these biological materials. Pentose sugars would be formed from pyrones and fats or acetogenins from acetic acid, depending on the hydrolysis of the multiple ketene group. The breadth of this early suggestion is just now being appreciated in biochemistry.

BIBLIOGRAPHY

I. ORIGINAL WORKS. No collected volume of Collie's works exists. Nearly all of his scientific papers were published in *Journal of the Chemical Society* from 1885 to 1920. Two key works are "Dehydracetic Acid," in *Journal of the Chemical Society,* **71,** pt. 2 (1900), 971–977; and "Derivatives of the Multiple Ketene Group," *ibid.,* **91** (1907), 1806–1813. Numerous articles appeared in various mountaineering journals, mainly in *Alpine Journal.* He also wrote two books, *Climbing on the Himalaya and other Mountain Ranges* (Edinburgh, 1902) and *Climbs and Explorations in the Canadian Rockies* (London, 1903).

II. Secondary Literature. An excellent biographical sketch, including a complete bibliography, appears in *Obituary Notices of Fellows of the Royal Society of London,* **4,** no. 12 (Nov., 1943), 329–356. A notice on Collie appears in *Dictionary of National Biography,* supp. 6 (1941–1950), 167–168. A fine sketch highlighting Collie's mountaineering appears in *Alpine Journal,* **54,** no. 266 (May, 1943), 59–65.

                                          Gerald R. Van Hecke

COLLINS, JOHN (*b.* Wood Eaton, near Oxford, England, 5 March 1625; *d.* London, England, 10 November 1683), *algebra.*

Removed from local grammar school after the death of his father, a "poore Minister," orphaned him, Collins was briefly apprenticed (at thirteen) to an Oxford bookseller "who failing I lived three yeares at Court [as kitchen clerk] and in this space forgot the Latin I had." From 1642, on the outbreak of the Civil War, he spent seven years in the Mediterranean as a seaman "in the Venetian service against ye Turke." On his return to London he set himself up as "Accountant philomath" (mathematics teacher). After the Restoration in 1660 Collins held a variety of minor government posts, notably "in keeping of Accompts" in the Excise Office, and for fifteen years managed the Farthing Office, but after its closure became once more a lowly accountant with the Fishery Company. He thought long about becoming a stationer but lacked the necessary capital; yet in a private capacity he did much to revive the London book trade after the disastrous 1666 fire, using to the full his own limited resources and the foreign contacts his employment gave him. Although he had "no Universitie education," he was deservedly elected fellow of the Royal Society in 1667.

On his own assessment Collins' mathematical attainments were "meane, yet I have an ardent love to these studies . . . endeavouring to raise a Catalogue of Math^ll Bookes and to procure scarce ones for the use of the Royall Society and my owne delight." His published works—*Merchants Accompts, Decimal Arithmetick, Geometricall Dyalling,* and *Mariners Plain Scale,* among others—are essentially derivative but reveal his competence in business arithmetic, navigational trigonometry, sundial construction, and other applications of elementary mathematics; his papers on theory of equations and his critique of Descartes's *Géométrie* are uninspiring.

Collins' scientific importance lies rather in his untiring effort, by correspondence and word of mouth, to be an efficient "intelligencer" of current mathematical news and to promote scientific learning: with justice, Isaac Barrow dubbed him "Mersennus Anglus." Between 1662, when he first met Barrow, and 1677, when the deaths of Barrow and Oldenburg (following on that of James Gregory in 1675), coupled with a growing reluctance on the part of Newton and Wallis to continue a letter exchange and the worries of his own straitened financial circumstances, effectively terminated it, Collins carried on an extensive correspondence with some of the finest exact scientists of his day, not only with his compatriots but with Bertet, Borelli, and (through Oldenburg) with Huygens, Sluse, Leibniz, and Tschirnhausen. Further, deploying his specialized knowledge of the book trade to advantage, Collins saw through press in London such substantial works as Thomas Salusbury's *Mathematical Collections,* Barrow's *Lectiones* and *Archimedes,* Wallis' *Mechanics* and *Algebra,* Horrocks' *Opera posthuma,* and Sherburne's *Manilius.* He sought likewise, but in vain, to have several of Newton's early mathematical works published. For the modern historian of science, Collins' still-intact collection of some 2,000 books and uncounted original manuscripts of such men as Newton, Barrow, and Halley is a major primary source.

BIBLIOGRAPHY

I. Original Works. "The first thing I published was about a quire of Paper concerning Merchants Accompts [a rare folio broadsheet (London, 1652), repr. in G. de Malyne, *Consuetudo: vel, Lex Mercatoria* (London, 1656)] which upon later thoughts I found myself unable to amend and was reprinted in May last" (to Wallis, 1666). This reappeared in augmented form as *An Introduction to Merchants-Accompts* (London, 1674).

Next Collins wrote "a despicable treatise of quadrants [*The Sector on a Quadrant . . . Accomodated* [sic] *for Dyalling: For the Resolving of All Proportions Instrumentally* (London, 1659), a revision of his *Description and Use of a General Quadrant* (London, 1658)]. . . . And among these Luxuriances I met with a Dyalling Scheame of Mr Fosters and commented upon y^t [*Geometricall Dyalling* (London, 1659)] which it is too late to wish undone." Also in 1659 he issued his *Mariners Plain Scale New Plain'd* and *Navigation by the Mariners Plain Scale.*

Posthumously there appeared his *Decimal Arithmetick, Simple Interest, &c* (London, 1685), an augmentation of his rare 1669 single-sheet equivalent, "Compendium for a Letter Case." Of his other nonscientific publications his monograph *Salt and Fishery* (London, 1682) deserves mention for its passages on salt refining and fish curing. In addition, the *Philosophical Transactions* has several unsigned reviews by Collins and four short articles—**2,** no. 30 (Dec. 1667), 568–575; **4,** no. 46 (Apr. 1669), 929–934; **6,** no. 69 (Mar. 1675), 2093–2096; **14,** no. 159 (May 1684), 575–582—dealing with topics in arithmetic and algebra.

His correspondence is preserved in private possession (Shirburn 101.H.1–3), except for a group of letters relating to Newton's invention of fluxions deposited from it, at Newton's request, in the Royal Society's archives in 1712 (now, with some losses, MS LXXXI [Collins' Descartes critique is no. 39]) and smaller collections in the British Museum, the University of St. Andrews, and Cambridge University library.

II. SECONDARY LITERATURE. Edward Sherburne, in his *Sphere of Manilius Made an English Poem* (London, 1675), app., pp. 116–118 (also found as a separate broadsheet), gives a creditable contemporary impression of Collins' work up to 1675. Brief sketches by Agnes M. Clerke, in *Dictionary of National Biography*, XI (1887), 368–369; and H. W. Turnbull, in *James Gregory Memorial Volume* (London, 1939), pp. 16–18, must serve in place of a standard biography. Collins' letters are printed in Newton's *Commercium Epistolicum* (London, 1712); S. P. Rigaud's *Correspondence of Scientific Men of the Seventeenth Century* (Oxford, 1841); and in modern editions of the letters of Newton, James Gregory, and Oldenburg.

D. T. WHITESIDE

COLLINSON, PETER (*b.* London, England, 28 January 1693/94; *d.* Mill Hill, Middlesex, England, 11 August 1768), *natural history, dissemination of science.*

His father, also called Peter, was a haberdasher and citizen of London; his mother, Elizabeth Hall, was the daughter of a Southwark mealman. Collinson was sent at the age of two to live with his grandmother at Peckham, in Surrey, where he "received the first liking to gardens and plants." While he was largely self-taught, such education as he had was received from Quakers, possibly in the school which Richard Scoryer established in Wandsworth, Surrey, and which later moved to Southwark.

In 1724 Collinson married Mary Russell, the daughter of a prosperous weaver and landowner. Two of their children, Michael and Mary, survived them, while two sons, both named for their father, died as infants. At about the time of his marriage, Collinson and his brother, James, took charge of the family business, which their mother had continued after her husband's death in 1711. They kept shop on Gracechurch Street as mercers and haberdashers until James's death in 1762, and then Peter continued it alone for four more years. Collinson was not a man of great wealth, although the business brought him a comfortable living until it declined in the 1760's.

Both brothers performed duties which were expected of citizens of London. Although they were Quakers, they served as churchwardens and vestrymen in the parish of St. Benet, Gracechurch Street, which they seem to have treated as local government in avoiding purely religious functions. Peter also held offices in the Bridge Within Ward as constable and bencher and, late in life, was overseer of the poor.

While James was more involved in the affairs of the Society of Friends than he, Peter was active in the work of the Society, serving from 1753 until his death as a correspondent of the London Meeting for Sufferings, the meeting most concerned with the relations of Quakers with the state. Here his contacts with political figures made him an effective lobbyist.

Collinson's health was good, except for an occasional attack of gout. His death came as the result of a strangury which he suffered while on a visit to Thorndon Hall, the seat of Lord Petre, near Brentwood, Essex. He was buried in the Quaker burying ground on Long Lane in Bermondsey.

Collinson was, above all, a gardener who had considerable success in domesticating foreign plants. His gardens at Peckham and, after 1749, at Ridgeway House in the village of Mill Hill north of London, were noted for their exotics. One biographer credits him with the introduction or reintroduction of 180 new species into England, of which more than fifty were American.

Although eventually Collinson would have a worldwide network of "philosophical" correspondents, he depended largely on American merchants, with whom he did an extensive business, to send him seeds as well as orders for dry goods. Occasionally he did find a more dependable source, such as Mark Catesby, whom Collinson helped secure patrons while the future author of the *Natural History of Carolina* was in America between 1722 and 1726 and whom he aided further by lending money without interest while that work was in the process of completion.

Possibly through his aid to Catesby and certainly because of his growing reputation as a collector of curiosities, Collinson moved into the circle of one of Catesby's benefactors, Sir Hans Sloane, the greatest collector of that time and president of both the College of Physicians and the Royal Society, whose sponsorship secured his election as a fellow in 1728. For the remainder of his life, Collinson was an active participant in the affairs of the Royal Society. He rarely missed meetings from October through May, when his business kept him in London, and frequently brought "curious" visitors with him, many of whom he sponsored for membership. Unlike many of the eighteenth-century fellows, he participated often in the proceedings of the Society: making comments, reading letters from his scientific correspondents in England and abroad, and, less often, giving his own contributions. Most of the latter were insignificant, even at a time when much trivia found its way into the *Philosophical Transactions*.

His most important paper, "A Letter to the Honorable J. Th. Klein . . . Concerning the Migration of Swallows" (*Philosophical Transactions,* **51** [1760], 449–464), was a well-reasoned argument based on correspondence and on travelers' accounts of migration against the commonly held belief that the birds hibernated under water. This argument, which he also sent to Linnaeus, he had to urge on the great Swedish taxonomist for years afterward. Collinson's importance to the Society, though, lay largely in promoting its aims and activities, and to this end he served fourteen years on its council, a considerable expenditure of time for a busy mercer.

The most significant result of Collinson's American business connections was his contact with Philadelphia's first scientific institution, the Library Company of Philadelphia, which he served as unpaid London agent beginning in 1732. By his choice of books and instruments and by his reports of what interested his friends in the Royal Society, he was able to give direction to the scientific interests of the city which, during these years, took a central place in American cultural development. Through the company's first secretary, Joseph Breintnall, Collinson initiated a correspondence with John Bartram of Kingsessing, Pennsylvania, which would produce one of the more fruitful botanical partnerships of the eighteenth century. Collinson secured customers for the seeds and plants which Bartram collected: first of all, Robert James, eighth Lord Petre, and, later, the dukes of Richmond, Norfolk, Bedford, and Argyll. Ultimately the list would include most of the planting lords and gentlemen of England at a time when many of them were remodeling the English landscape along naturalistic lines. This gave Collinson entree to the powerful men of his time; but he acted in all of this only for the New World plants which he received from Bartram.

Through the Library Company also, Collinson came to correspond with Benjamin Franklin, who was its first mover and most active member. While it is hardly true that Collinson was responsible for making Franklin's reputation to the same degree that he was the creator of John Bartram as a botanical collector, nonetheless Collinson played a significant role in Franklin's life. Collinson sent the company, probably in 1745, when he distributed similar materials to other correspondents, a glass tube and some accounts of German electrical experiments, which started Franklin and his associates in the Library Company on their study of electricity. Franklin transmitted his accounts of their activities to Collinson, who read some of them to the Royal Society and eventually saw to the publication of *Experiments and Observations on Electricity* (London, 1751–1754).

While Collinson was never again so spectacularly

effective as he was with Franklin, his part in the Philadelphia experiments was typical of his place in the eighteenth-century scientific community. He was at once gadfly, middleman, and entrepreneur. In the period of his greatest activity, from the 1730's to near the time of his death, Collinson was at the center of a network of scientific intelligence which reached from Peking to Philadelphia, but which was concerned principally with the communications of Americans with European and English savants. He informed his correspondents about those things which interested his friends in the Royal Society, transmitted their collections to English and European naturalists, and either read their letters and formal papers to the Royal Society or secured their publication in the *Gentleman's Magazine* or occasionally in Continental journals. While he was engaged in this, Collinson was also active in the Society of Antiquaries, to which he was elected in 1732 and on whose council he served eight terms.

To the Antiquaries, who were closely interrelated with the Royal Society, he transmitted reports and artifacts sent him by those of his friends who were interested in archaeology and read a speculative paper, "Observations on the Round Towers of Ireland," which was printed posthumously in the first volume of *Archaeologia* (pp. 329–331). He did not join Henry Baker and a number of his close associates from the Royal Society in that group's more practical equivalent, the Royal Society of Arts, possibly because he lacked sufficient time and resources. Nonetheless, he again shared with them both his advice and his correspondents, including the Connecticut agricultural experimenter Jared Eliot, who received their gold medal for his work in metallurgy.

Virtually the only American naturalist of any significance who did not benefit from Collinson's efforts in the period of his greatest activity was Alexander Garden of Charleston, South Carolina, who, although he corresponded with Collinson, transacted his business largely through Collinson's friend, John Ellis, or dealt directly with Linnaeus. Collinson was involved, however, along with Mark Catesby, in the collaboration of John Clayton of Virginia with Johann Friedrich Gronovius which produced the *Flora Virginica* (Leiden, 1739–1743) and also contributed to the Leiden revision of 1762, even though that edition put an end to his efforts to publish Clayton's own expanded version in London. Collinson performed a similar service for another Virginia naturalist, John Mitchell, the future cartographer, whose botanical writings he sent to Christian J. Trew in Nuremberg, where they were published in the *Acta physicomedica.*

To the northward, Collinson served as the principal

connection with London of James Logan of Philadelphia. Logan's accounts of his experiments with plant hybridization passed through Collinson's hands to publication in London and Leiden. In the case of Cadwallader Colden of New York, Collinson acted as scientific correspondent, literary agent, and, as was so often true of his American friends, as a source of influence for the acquisition of office. Collinson encouraged Colden's botanical studies, which elicited the admiration of Linnaeus, and was responsible for the London publication of Colden's *History of the Five Indian Nations*. He tried, without success, to secure favorable English comment on Colden's most ambitious effort, his *Principles of Action in Matter*. For Colden, as for many lesser Americans with scientific pretensions, Collinson was the major tie to a world of learning beyond the Atlantic.

Certainly Collinson, aided by John Ellis and others of his circle, helped to move both English and American botany in a Linnaean direction. Collinson, himself, was notably unsystematic, and had to depend upon Gronovius in Leiden, J. F. Dillenius in Oxford, and, later, upon Linnaeus' pupil, Daniel C. Solander (whom he helped to bring to England), to name the specimens which Bartram sent him. He complained to Linnaeus of the complexity of his classification, but nevertheless helped to circulate the works of the great taxonomist in both England and America, as he also supplied him with specimens and information.

Collinson's generous promotion of the works of others was the real source of the admiration which men of his time held for him and accounts for his election to the Uppsala and Berlin academies. In his own interests, he was very much the eighteenth-century amateur. At his death, he left cabinets of curiosities and a wide range of notes on scientific and antiquarian subjects. His library contained one of the major collections of Americana in England, and he amassed a large amount of material on the American Indian. Collinson's only published treatment of the Indians, however, was a typically Quaker proposal for a mild regulation of trade and settlement (*Gentleman's Magazine,* **33** [1763], 419–420). As was the case with a history of American pines, which his friends expected of him, his learning was put to utilitarian purposes. In the latter (*Ibid.,* **25** [1753], 503–504, 550–551), the practical gardener again demonstrated his concern with the adaptability of plants to the English soil and climate.

BIBLIOGRAPHY

I. ORIGINAL WORKS. Collinson's major writings are summarized in Norman G. Brett-James, *The Life of Peter Collinson* (London, 1925).

II. SECONDARY LITERATURE. Brett-James's biography is marred with inaccuracy and should be supplemented with Earl G. Swem, "Brothers of the Spade . . .," in *American Antiquarian Society Proceedings,* **68** (1948), 17–190, which also has a good treatment of secondary works on Collinson. Raymond P. Stearns, "Colonial Fellows of the Royal Society of London, 1661–1788," in *Notes and Records of the Royal Society of London,* **8** (1951), 178–246, and Brooke Hindle, *The Pursuit of Science in Revolutionary America, 1735–1789* (Chapel Hill, N.C., 1956), contain useful materials.

GEORGE F. FRICK

COLLIP, JAMES BERTRAM (*b.* Thurold Township, near Belleville, Ontario, 20 November 1892; *d.* London, Ontario, 19 June 1965), *endocrinology.*

Collip's father, James Dennis Collip, operated a vegetable and flower shop in Belleville; his mother, Mahala Vance, was a former schoolteacher. At the age of fifteen he began his education in science at Trinity College, University of Toronto. In 1912 he received his B.A., graduating at the head of his class, then went on to obtain the M.A. in 1913 and a Ph.D. in biochemistry in 1916. His graduate training was under the direction of A. B. Macallum, whom he succeeded in 1928 as professor of biochemistry at McGill University.

In 1915 Collip received his first academic appointment, as lecturer in biochemistry at the University of Alberta. Here he had a heavy teaching load in a department depleted of staff by the war, but he was nevertheless able to continue his research. In 1916 he published the first of his many contributions to medical literature, a paper entitled "Internal Secretions," which he had presented to the Alberta Medical Association. (An earlier abstract of the paper with A. B. Macallum given at the British Association for the Advancement of Science was published at London in 1914; in their reports of 1913, p. 673, Collip's name appears as "Collop.") Although in the nature of a review, it reveals Collip's thinking in this field, which was to be his major interest. His early papers, however, were concerned mainly with the comparative blood chemistry of vertebrates and invertebrates. He recorded some new findings on the alkali reserve of plasma, acid-base exchange, and osmotic pressure of blood serum. In 1920 he wrote *On the Formation of Hydrochloric Acid in the Gastric Tubercules of the Vertebrate Stomach,* which included many of his own observations made while working toward his Ph.D.

In 1921 Collip was awarded a Rockefeller Traveling Fellowship to visit laboratories in North America and England. This proved to be a decisive turning point toward a career in endocrine research. His first visit was to Toronto, where there was intense interest in the development of a pancreatic extract to combat diabetes. This possibility so caught Collip's imagination that he gave up the fellowship to become an assistant professor in the department of pathological chemistry at Toronto. He was subsequently asked by J. J. R. Macleod, head of the department of physiology, to join the group working with Banting and Best on the new hormone "insulin."

Historically there was strong evidence to suggest that the pancreas had a major role in the control of sugar metabolism. It was known that surgical removal of the pancreas in the dog led to a diabetic condition as a result of which the animal rapidly died. Many unsuccessful attempts had been made to extract the pancreas and obtain a fraction which would effectively maintain diabetic dogs. Unfortunately the pancreas contains—in addition to cells producing insulin—cells that produce powerful digestive enzymes and these presumably destroyed the hormone in extraction processes. In the autumn of 1921 extracts had been prepared from dogs whose pancreatic ducts had been ligated (such a procedure is followed by a degeneration of the enzyme-producing tissue but not of the insulin-producing cells) and from fetal pancreas obtained from slaughterhouses. Those showed activity in depancreatized dogs but toxic impurities and difficulty of preparation prevented their practical application to the treatment of patients with diabetes. Collip rapidly developed a method for the preparation of insulin from cattle or hog pancreas, employing alcohol in varying concentrations to obtain a differential precipitation of impurities. The resulting extract was sufficiently pure to allow the clinical group to test its action in humans. The first clinical results with insulin were published in March 1922 by Banting, Best, Collip, Campbell, and Fletcher, under the title "Pancreatic Extracts in the Treatment of Diabetes Mellitus." Over the short interval of two years the Toronto group published fifteen fundamental papers and ten communication-abstracts that constituted an extensive contribution to knowledge of insulin and carbohydrate metabolism. In 1923 Banting and Macleod were the first Canadians to receive the Nobel Prize in physiology and medicine; in recognition of the parts played by their collaborators, they shared the monetary gifts with Best and Collip.

With the breakup of the group of collaborators in Toronto, Collip returned to Edmonton in 1922 to continue research as professor of biochemistry; he earned the D.Sc. in 1924 and the M.D. in 1926. During five years in Edmonton, he made some of his most important contributions to medicine by relating the hormonal control of calcium and phosphorus metabolism to an active principle in the parathyroid gland.

This original work was published in 1926 as "The Extraction of a Parathyroid Hormone Which Will Prevent or Control Parathyroid Tetany and Which Regulates the Level of Blood Calcium." The extensive discoveries made in this field at Edmonton are presented in Collip's Harvey Lectures of 1925–1926.

In 1927, at the age of thirty-five, Collip became chairman and head of the department of biochemistry at McGill University. The next eleven years were the most productive of his career and, with a large group of distinguished collaborators that included David L. Thomson and Hans Selye, he published more than 200 papers. They contributed to nearly every facet of endocrinology, particularly to pituitary function. Most of the projects centered on Collip, who had the remarkable ability to handle large concentrates of glands, purify them to manageable proportions, and separate out various hormone fractions. His restless and inquisitive nature led him to explore one area rapidly and then pass on to another. Many other laboratories benefited, therefore, from his original observations as they proceeded to develop them. Throughout his life Collip explored areas of research in which it seemed directly possible to discover new treatments for human diseases.

The confusing field of female sex hormones and "ovary-stimulating substances" of the placenta attracted the attention of the McGill group; and as a result of their studies one of the group, J. S. L. Browne, isolated the female sex hormone estriol and prepared the first orally active estrogenic preparation that could be used clinically.

Collip's interest in placental gonadotrophic hormones led him to the complex problems of the production by the anterior lobe of the pituitary of gonadotrophic and other trophic hormones. In the early 1930's little was known of the nature or even the number of different hormones this part of the pituitary produced, and new animal models (chiefly the hypophysectomized rat) had to be developed to allow their assay. By 1933 Collip had separated pituitary growth hormone essentially free from both adrenocorticotrophic hormone (ACTH) and thyrotrophic hormone (TSH), which tended to run closely together in pituitary gland extracts. The preparation of separated "pure" hormones allowed extensive investigations of their often complicated actions. ACTH was prepared in sufficiently pure form to be administered to patients. Unfortunately, it was not used to treat any of the diseases against which it is now known to be effective.

The preparations of ACTH, although not pure, were tolerated by patients and the activity was such that corticoid hormones were stimulated in sufficient amounts to be active against inflammatory reactions, rheumatism, arthritis, and various collagen diseases—conditions which were later shown by Hench and Kendall at the Mayo Clinic to respond to corticoid treatment. During many experiments with fractions containing TSH it was noticed that after repeated injections the initial stimulating response was lost and that the blood serum of the animal would inhibit the action of the hormone in other animals. Collip envisioned the development of "anti-hormones" as a normal physiological means of maintaining hormone homeostasis. This theory created great interest; and although it was later shown that closely bound nonspecific proteins and not the pure hormones probably led to this reaction, it focused attention on the possible influence of immune reactions in endocrine therapy.

In 1938 Collip abandoned research to devote most of his time to the organization of medical research in Canada, first as a member of the newly created Associate Committee for Medical Research of the National Research Council and later as chairman, succeeding Sir Frederick Banting. He was also medical liaison officer to the United States. He was decorated for his work by both the Canadian and United States governments. After the war Collip continued his administrative responsibilities as director and chairman of the Medical Advisory Committee of the National Research Council until his retirement at the age of sixty-five.

In 1941 Collip resigned his position in the department of biochemistry to become the Gilman Cheney professor of endocrinology and director of the Institute of Endocrinology at McGill. Although not participating personally in the research in his new laboratory, he closely followed its work. During these years the staff of the laboratory published thirty-eight papers containing significant contributions to such widely separated fields as experimental traumatic shock, motion sickness, audiometry, and blood preservation.

In 1947 Collip resigned from McGill to become dean of medicine at the University of Western Ontario and director of the department of medical research at the new Collip Medical Research Laboratory. Here he formed an active research group, and many graduate students benefited from training in his department. By 1965 more than 125 publications had originated from this laboratory. Collip retired as dean of medicine in 1961 but continued as director of the laboratory until his sudden death in 1965.

Many honors were bestowed on Collip for his pioneer investigations in endocrinology. His election in 1925 to the Royal Society of Canada and in 1933 to the Royal Society of London recognized his early achievements. He received honorary degrees from nine Canadian universities and from Harvard (1936), Oxford (1946), and the University of London (1948). Among Collip's awards are the Canadian Medical Association's F. N. G. Starr Award (1936), the Flavell Medal of the Royal Society of Canada, the Cameron Prize at Edinburgh, and the Banting Medal (1960). He was a fellow of several medical colleges and a member or honorary member of many Canadian and foreign scientific societies.

Collip never aspired to be a public figure; and his modesty and inherent shyness made him reluctant to lecture or give papers, even to scientific audiences. At first meeting he gave the impression of abruptness, but one soon realized that behind this manner were shyness and great personal charm. Collip had strong family ties and made many lasting friendships. He found little time for his hobbies of bridge, billiards, badminton, and golf.

BIBLIOGRAPHY

I. ORIGINAL WORKS. Collip's writings, alone or with collaborators, include "Internal Secretions," in *Canadian Medical Association Journal* (Dec. 1916), 1063–1069; *On the Formation of Hydrochloric Acid in the Gastric Tubercules of the Vertebrate Stomach,* University of Toronto Studies, Physiological Series, no. 35 (Toronto, 1920), 1–46; "Pancreatic Extracts in the Treatment of Diabetes Mellitus," in *Canadian Medical Association Journal,* **12** (Mar. 1922), 141–146, written with F. G. Banting, C. H. Best, W. R. Campbell, and A. A. Fletcher; "The Original Method as Used for the Isolation of Insulin in Semipure Form for the Treatment of the First Clinical Cases," in *Journal of Biological Chemistry,* **55** (Feb. 1923), 40–41; "The Extraction of a Parathyroid Hormone Which Will Prevent or Control Parathyroid Tetany and Which Regulates the Level of Blood Calcium," *ibid.,* **63** (Mar. 1925), 395–438; *The Parathyroid Glands* (New York, 1925–1926), the Harvey Lectures; "The Ovary-stimulating Hormone of the Placenta," in *Nature* (March 22, 1930), 1–2; "Placental Hormones," in *Proceedings of the California Academy of Medicine* (1930), 38–73; "Further Clinical Studies on the Anterior Pituitary-like Hormone of the Human Placenta," in *Canadian Medical Association Journal,* **25** (1931), 9–19, written with A. D. Campbell; "The Adrenotrophic Hormone of the Anterior Pituitary Lobe," in *Lancet,* **12** (1933), 347–350, written with E. M. Anderson and D. L. Thomson; "Preparation of a Purified and Highly Potent Extract of Growth Hormone of Anterior Pituitary Lobe," in *Proceedings of the Society for Experimental Biology and Medicine of New York,* **30** (1933), 544–546, written with H. Selye and D. L. Thomson; "Thyrotrophic Hormone of Anterior Pituitary," *ibid.,* 680–683, written with E. M. Anderson; "Preparation and Properties of an Antithyrotropic Substance," in *Lancet,* **30** (1934), 784–790, written with E. M. Anderson;

"The Production of Serum Inhibitory to the Thyrotropic Hormone," *ibid.,* 76–79, written with E. M. Anderson; "Diabetogenic, Thyrotropic, Adrenotropic and Parathyrotropic Factors of the Pituitary," in *Journal of the American Medical Association,* **104** (1935), 827, 916; "John James Rickard Macleod (1876–1935)," in *Biochemical Journal,* **29** (1935), 1253–1256; "Recent Studies on Anti-hormones," in *Annals of Internal Medicine,* **9** (1935), 150–161; "Endocrine Organs," in Hawk and Bergeim, eds., *Practical Physiological Chemistry,* 11th ed. (Philadelphia, 1937), ch. 26, written with D. L. Thomson; "Results of Recent Studies on Anterior Pituitary Hormones," in *Edinburgh Medical Journal,* **45** (1938), 782–804, the Cameron Lecture; "The Anti-hormones," in *Biological Review,* **15** (1940), 1–34, written with D. L. Thomson and H. Selye; "Adrenal and Other Factors Affecting Experimental Traumatic Shock in the Rat," in *Quarterly Journal of Experimental Physiology,* **31** (1942), 201–210, written with R. L. Noble; "Recollections of Sir Frederick Banting," in *Canadian Medical Association Journal,* **47** (1942), 401–405; "Science and War," *ibid.,* **49** (1943), 206–209; "Alexander Thomas Cameron 1882–1947," in *Biochemical Journal,* **43** (1948), 1, written with F. D. White; and "Professor E. G. D. Murray—An Appreciation," in *Canadian Medical Association Journal,* **92** (1965), 95–97.

II. SECONDARY LITERATURE. A biographical notice is R. L. Noble, "Memories of James Bertram Collip," in *Canadian Medical Association Journal,* **93** (1965), 1356–1364. Papers by members of his laboratory staff include J. S. L. Browne, "The Chemical and Physiological Properties of Crystalline Oestrogenic Hormones," in *Canadian Journal of Research,* **8** (1933), 180–197; E. G. Burr and H. Mortimer, "Improvements in Audiometry at the Montreal General Hospital," in *Canadian Medical Association Journal,* **40** (1939), 22–27; O. F. Denstedt, Dorothy E. Osborne, Mary N. Roche, and H. Stansfield, "Problems in the Preservation of Blood," *ibid.,* **44** (1941), 448–462; and R. L. Noble, "Treatment of Experimental Motion Sickness in Humans," in *Canadian Journal of Research and Experiment,* **24** (1946), 10–22.

R. L. NOBLE

COLOMBO, REALDO (*b.* Cremona, Italy, *ca.* 1510; *d.* Rome, Italy, 1559), *anatomy, physiology.*

Relatively little is known of Colombo's life; he is frequently given the first name Matteo, but this seems to be an error. He was the son of Antonio Colombo, an apothecary, and received his undergraduate education at Milan. For a short time he seems to have pursued his father's trade but then to have become an apprentice to Giovanni Antonio Lonigo, a leading Venetian surgeon, with whom he remained for seven years. By 1538 he had gone on to study at the University of Padua, whose archives for that year refer to him as "an outstanding student of surgery." While still a medical student he occupied a chair of sophistics at Padua for the academic year 1540/1541. He probably received his degree in 1541, and by 1542 he had returned to Venice to assist Lonigo.

Late in 1542 Andreas Vesalius, the professor of surgery and anatomy at Padua, went to Switzerland to oversee the printing of his *Fabrica* (1543); and when he did not return in time for the annual anatomical demonstrations early in 1543, Colombo was appointed as his temporary replacement. Vesalius subsequently relinquished his chair at Padua, and it was given to Colombo on a regular basis in 1544. At the invitation of Cosimo I de' Medici, Colombo left Padua in 1545 to teach anatomy at Pisa. In 1548 he made an extended visit to Rome, where he engaged in anatomical studies with Michelangelo. Their intention was to publish an illustrated anatomy that would rival the *Fabrica,* but the artist's advanced age prevented them from fulfilling this plan. Colombo returned to Pisa for a time; but later in 1548 he settled permanently in Rome, where he taught at the Sapienza. He gained favor at the papal court and performed autopsies on a number of leading ecclesiastics, including Cardinal Cibo and Ignatius of Loyola. He remained in Rome for the rest of his life.

Colombo is commonly said to have been a disciple of Vesalius; but in fact the details of his training in anatomy and his early relationship with Vesalius are unclear, largely because both men commented on this subject only after they had become bitter enemies. Renewed interest in human dissection at Venice and Padua preceded Vesalius' arrival, and Colombo's involvement in anatomy probably stemmed from this movement. He himself regarded Lonigo as his most important teacher, apparently in anatomy as well as in surgery, and Lonigo's conducting a month-long course of human dissections at Padua in 1536 would seem to lend some credence to this view. During the same period an anatomical revival took place at Paris along more strongly Galenic lines. In the autumn of 1537 Vesalius, a student of the Parisian anatomists, came to Padua to complete his medical education, bringing with him a deep knowledge of Galen's anatomical teachings and a flair for demonstrating them by dissections and drawings. After only a few months he received his degree and was immediately appointed professor of surgery, and in that capacity he began giving his own anatomical demonstrations.

Over the next few years Colombo attended Vesalius' course of dissections several times, but he seems also to have carried on private dissections of his own. In a part of the *Fabrica* probably written in 1541, Vesalius refers to some observations that had been made "by my very good friend Realdo Colombo, now professor of sophistics at Padua and most studious of anatomy." To judge from this statement, the relationship between the two men was that of friends and colleagues (Colombo was probably the older of the two) rather than that of master and disciple. Colombo undoubtedly learned a great deal from Vesalius; but it is probably not true that he owed all that he knew to him, as Vesalius later claimed. Nor can we assume that the influence was entirely one-sided, since the most important developments in Vesalius' thought took place some time after he came to Padua.

Later in 1541 Colombo made an unsuccessful bid to obtain one of the two chairs of surgery held by Vesalius; this may have marked the beginning of friction between the two men, although the main falling out occurred in 1543. In his public demonstrations of that year Colombo pointed out some errors in Vesalius' teaching, most notably his attribution of certain features of the cow's eye to that of the human. Late in 1543 Vesalius visited Padua, and on learning of these criticisms he became quite incensed. He publicly ridiculed Colombo; and in his *China Root Letter* (1546) he denounced him as an ignoramus and a scoundrel, asserting that he himself had taught Colombo what little he knew of anatomy. This and a similar statement in Vesalius' *Examen* (1564) have led to the belief that Colombo was a "disciple" of Vesalius or had even been his assistant, but this view seems inconsistent with Vesalius' specific emphasis on the meagerness of Colombo's formal training in anatomy.

Thus Colombo was the first anatomist to criticize Vesalius, not for his rejection of Galen's authority but for his own anatomical errors. In his public lectures at Padua, Pisa, and Rome, Colombo presented numerous additional corrections and discoveries. As mentioned, the aim of his work with Michelangelo was to produce a new, more correct anatomy text that would supersede the *Fabrica*. In 1556 Colombo's friend and former student Juan de Valverde published a Spanish anatomy text, *Historia de la composición del cuerpo humano,* which was avowedly based on the *Fabrica* but also incorporated many of Colombo's corrections and new discoveries. In 1559 Colombo published his own unillustrated text, *De re anatomica,* consisting of fifteen books. Of these he seems to have written the first four during the early 1550's as a separate treatise on bones, cartilages, and ligaments. The next nine books, dealing with the remaining parts of the body, seem to have been added rather hastily in 1558, perhaps because Colombo anticipated his impending death. The last two books are devoted to vivisection and pathological observations, respectively. Colombo evidently died just as the book was being published, since in most copies his two sons

replaced his dedicatory letter with one of their own mentioning his recent demise.

Colombo seems to have eschewed the deep Galenic learning shared by other leading contemporary anatomists, but to judge from the *De re anatomica* he more than compensated for this by his rich experience in dissection, vivisection, autopsy, and the practice of surgery. Quite naturally the *Fabrica* provided the main framework for his studies, and he made numerous improvements in Vesalius' descriptions besides reporting a number of new discoveries of his own. The many pathological and anomalous observations he described likewise reflect his wide experience and attention to detail. He also had a strong interest in physiology and seems to have been unsurpassed among his contemporaries in his skill at vivisection. Colombo was not at all inclined to underestimate his own achievements and was sometimes careless in what he wrote. Nevertheless, he succeeded in giving a good account of human anatomy that was both brief and clear; these qualities probably explain the considerable popularity of the *De re anatomica* during the later sixteenth century.

One man who was not pleased by the book, however, was Gabriele Falloppio, professor of anatomy at Padua, who found that Colombo's new observations overlapped with his own on a number of points. In 1561 he published his *Observationes anatomicae,* at least in part to regain priority for himself. Falloppio said that he had written his own work four years prior to its publication, but this is belied by its numerous thinly veiled references to the *De re anatomica.* Falloppio insinuated that Colombo had plagiarized discoveries made by himself and other anatomists, including the *levator palpebrae superioris* muscle, which Falloppio claimed for himself, and the stapes, which he claimed for Ingrassia. In 1574 Falloppio's student and friend G. B. Carcano explicitly charged Colombo with plagiarism; however, it is difficult to judge the validity of the charge because independent discovery is especially common in anatomy, and Colombo certainly had the capacity to make these discoveries on his own. The net result of Colombo's unfavorable relations with Falloppio and Vesalius, as well as his lack of classical learning, was that he was held in low esteem by the Italian anatomists of the later sixteenth century, although his work was quite well thought of outside of Italy, especially in Germany.

Colombo is best known for his discovery of the pulmonary circuit, that is, the passage of blood from the right cardiac ventricle to the left through the lungs. This idea was presented by Valverde in his *Historia* as well as by Colombo in the *De re anatomica.* Not until the late seventeenth century was it found that

Michael Servetus had described the pulmonary circuit in his *Christianismi restitutio,* a theological work printed in 1553 but almost totally destroyed by the censors prior to publication. In the early twentieth century A. D. Tatawi discovered that both men had been anticipated by Ibn al-Nafis, an Arab of the thirteenth century. The resulting priority controversies have generated a voluminous literature, but it has yet to be shown that Ibn al-Nafis' account was available in Europe during the Renaissance or that Colombo had knowledge of Servetus' work. Moreover, there is good evidence in the accounts of Colombo and Valverde that Colombo actually made the discovery on his own through vivisectional observations. It is also clear that as a result of such observations Colombo's more general understanding of the operations of the heart, lungs, and arteries was superior to that of the other two men.

In the Galenic physiology of the sixteenth century, the right cardiac ventricle was thought to receive blood from the vena cava and to send it into the pulmonary artery to nourish the lungs. The left ventricle was supposed to ventilate the innate heat of the heart by breathing in and out through the pulmonary vein. In addition, the left ventricle was thought to receive blood from the right ventricle through minute pores in the intervening septum. From this blood and from some of the air received from the lungs, the left ventricle generated vital spirits and arterial blood, which it distributed to the entire body through the arteries to preserve life.

In his treatise *On the Uses of the Parts of the Body of Man,* Galen had described anastomoses between the pulmonary artery and vein, and the passage of blood from the former to the latter, although he apparently did not conceive of this as part of a pathway from the right cardiac ventricle to the left. Vesalius, however, questioned Galen's teaching that blood can pass directly from the right ventricle to the left through minute pores in the cardiac septum, and some historians have suggested that Vesalius' doubts plus Galen's description of the pulmonary anastomoses formed the main bases for the idea that blood passes from the right ventricle to the left through the lungs. This may have been true for Servetus, but it does not seem to have been so for Colombo. There is some evidence that Colombo, unlike Servetus, was not directly familiar with the relevant passages in the *Uses of Parts.* Moreover, Colombo did not posit direct connections between the two pulmonary vessels (as he might have done had he been influenced by Galen's statements) but thought that blood oozes through the flesh of the lungs in passing from one vessel to the other. Finally, it appears that a desire to find an alternative to the Galenic septal pores did not provide the original motivation for Colombo's investigations. Instead, he seems to have been troubled by the serious conflict between the commonly held view that the left ventricle inhales and exhales through the pulmonary vein and the fact that this vessel contains blood. He found that in live animals the pulmonary vein is completely filled with blood, which seemed to disprove the idea that the left ventricle breathes through it. He probably went on to try to determine the source of this blood and concluded that it comes from the lungs, and ultimately from the right ventricle, through the pulmonary artery.

Colombo realized that his discovery had eliminated the need for the Galenic septal pores, but it was also clear to him that the pulmonary circuit is an important phenomenon in its own right. He particularly emphasized that it is in the lungs, rather than in the heart, that the venous blood is mixed with air and converted to arterial blood. The arterial blood was thought to preserve the life of all parts of the body, and the unique ability to generate this important substance had been one of the traditional attributes of the heart. By transferring this power to the lungs, Colombo was quite consciously diminishing the status of the heart, whose main task was now to distribute the arterial blood rather than to generate it.

Through his studies in vivisection Colombo also made considerable progress in understanding the heartbeat. His predecessors had generally thought that the heart functions like a bellows whose main action is a strenuous dilation, by which it draws materials into its two ventricles; contraction was considered a less vigorous expulsion. With few exceptions what they took to be cardiac dilation is actually contraction, and vice versa. Thus, they thought that the heart and arteries dilate and contract at the same time and that, like the heart, the arteries pulsate actively. Colombo's observations convinced him that the traditional designation of the phases of the heartbeat should be reversed and that contraction, by which the heart expels materials, is more strenuous than dilation, by which it receives them. Thus the arteries dilate when the heart contracts; and Colombo may even have thought that the arterial pulse is actually caused by the impulsion of materials from the heart, although he was not entirely clear on this point.

Colombo maintained the traditional view that nutritive blood flows outward from the liver through the venous system, but otherwise his work represents a significant advance in understanding the operations of the heart, lungs, and arteries. The idea of the pulmonary circuit was moderately well received prior to the publication of Harvey's *De motu cordis* (1628).

Over twenty favorable reactions to the discovery of the pulmonary circuit were published during this period, although it was also opposed by some important authorities. Less attention was paid to Colombo's observations on the heartbeat; but it appears that they formed the actual starting point for Harvey's vivisectional studies on the heart, which eventually led to the discovery of the circulation.

BIBLIOGRAPHY

I. ORIGINAL WORKS. *De re anatomica* (Venice, 1559) was Colombo's only publication. Quite a few reprintings appeared during the later sixteenth and early seventeenth centuries. That of 1593 (Frankfurt) included a commentary by Joannes Posthius. *Anatomia* (Frankfurt, 1609) was a German translation by Johann Schenck.

II. SECONDARY LITERATURE. The most complete summary of what is known of Colombo's life is given in E. D. Coppola, "The Discovery of the Pulmonary Circulation: A New Approach," in *Bulletin of the History of Medicine,* **21** (1957), 44–77, esp. 48–59; his article is a condensed version of his unpublished M.D. thesis, "Realdo Colombo of Cremona (1515?–1559) and the Pulmonary Circulation" (Yale School of Medicine, 1955), which contains additional information and an extensive bibliography. See also G. J. Fisher, "Realdo Colombo," in *Annals of the Anatomical and Surgical Society of Brooklyn,* **2** (1880), 279–284; and H. Tollin, "Matteo Realdo Colombo," in *Pflüger's Archiv für die gesamte Physiologie des Menschen und der Tiere,* **22** (1880), 262–290. For new information about Colombo's studies at Padua, see F. Lucchetta, *Il medico e filosofo Bellunese Andrea Alpago* (Padua, 1964), pp. 60–62. For a general analysis of Colombo's anatomical work, see M. Portal, *Histoire de l'anatomie et de la chirurgie,* II (Paris, 1770), 540–559; on his pathological and anomalous observations, R. J. Moes and C. D. O'Malley, "Realdo Colombo: 'On Those Things Rarely Found in Anatomy,' an Annotated Translation from the *De re anatomica* (1559)," in *Bulletin of the History of Medicine,* **34** (1960), 508–528. For a survey of literature on the priority controversy over the pulmonary circuit, see J. Schacht, "Ibn al-Nafis, Servetus and Colombo," in *Al-Andalus,* **22** (1957), 317–336. For a judicious argument in support of Colombo's independence in this discovery, consult L. G. Wilson, "The Problem of the Discovery of the Pulmonary Circulation," in *Journal of the History of Medicine and Allied Sciences,* **17** (1962), 229–244. Walter Pagel, *William Harvey's Biological Ideas* (Basel–New York, 1964), pp. 154–156, 163–169, 216–218, discusses Colombo's observations on the heartbeat in addition to his work on the pulmonary circulation. The present analysis of Colombo's work is based largely on my unpublished doctoral dissertation, "Cardiovascular Physiology in the Sixteenth and Early Seventeenth Centuries" (Yale Graduate School, 1969), esp. chs. 3, 6, and 7.

JEROME J. BYLEBYL

COLONNA, EGIDIO. See **Giles of Rome.**

COLUMBUS, CHRISTOPHER (*b.* Genoa, Italy, 1451 [?]; *d.* Valladolid, Spain, 20 May 1506), *navigation, exploration.*

For a detailed study of his life and work, see Supplement.

COMAS SOLÁ, JOSÉ (*b.* Barcelona, Spain, 19 December 1868; *d.* Barcelona, 2 December 1937), *astronomy.*

Comas Solá was one of Spain's outstanding scientists. In 1890 he was graduated from the College of Physical and Mathematical Sciences of the University of Barcelona. Shortly after obtaining his degree he began his observations of Mars, which he continued in all its oppositions. In 1894 he made the first Spanish relief map of Mars, in which he incorporated all the latest findings on that planet. Comas Solá was one of the first to theorize that the contours of the so-called Martian canals were more apparent than real. Many of his observations found their way into Flammarion's *La planète Mars.* Comas Solá extended his observations to other planets, notably Jupiter and Saturn. In 1902 he determined the period of rotation of Saturn by using the white tropical spot of Barnard as a point of reference.

In 1915, anticipating the work of foreign scientists better remembered by posterity, Comas Solá published in the *Boletín. Observatorio Fabra* an article entitled "La teoría corpuscular ondulatoria de la radiación," in which he tried to harmonize what had been considered two contradictory theories: the wave and corpuscular theories of radiation.

Comas Solá also did work in seismology, devising a method of ascertaining the depth of seismic epicenters and inventing the stereogoniometer for studying the courses and movements of stars. His constant observations of the heavens resulted in the discovery of two comets, the first to be discovered in Spain for three centuries, and in the discovery of the first eight asteroids in Spain.

He was the founder and director until his death of the Observatorio Fabra, first president of the Sociedad Astronómica de España y América, and editor of *Urania* (Barcelona).

BIBLIOGRAPHY

Comas Solá's books include *Astronomía y ciencia general* (Barcelona, 1906); *El espiritismo ante la ciencia* (Barcelona, 1907); *Astronomía,* 2 vols. (Barcelona, 1910; new ed., 1920, which ran into several editions until 1943); and *El cielo* (Barcelona, 1927). Most of his books reflect Comas Solá's interest in popularizing science. His more important writings are those in the scientific journals. A fairly complete listing of these is in Poggendorff, VI, 1, and VIIb, 2.

There are no works on Comas Solá, and general histories of Spanish science overlook his contributions.

VICENTE R. PILAPIL

COMBES, CHARLES-PIERRE-MATHIEU (*b.* Cahors, France, 26 December 1801; *d.* Paris, France, 11 January 1872), *technology, mechanics.*

Combes was the son of Pierre-Mathieu Combes, a military police chief, and Marie-Salomé-Barbe Beauseigneur. He studied in Paris at the Collège Henri IV, where he won first prize in mathematics and second prize in physics. In 1818 he entered the École Polytechnique, and then spent two years at the École des Mines. He next taught elementary mathematics, design, and draftsmanship at the École des Mineurs de Saint-Étienne. In 1824 he became *ingénieur ordinaire,* then temporarily left the Administration of Mines to manage the silver-bearing lead mines of Sainte-Marie-aux-Mines and Lacroix-aux-Mines. In 1826 he reentered the Administration and became chairman of the departments of mathematics, mechanics, and design in the École des Mines. In addition to these duties, Combes managed the coal fields of Roche-La-Molière and Firminy from 1827 on.

In May 1832 he was named professor of the development of mines at the École des Mines; he taught this subject for sixteen years. He further studied the development of mines in England, traveling there in the summer of 1834.

Combes was especially interested in ventilation. He therefore wanted to measure accurately the flow of air in underground work areas; to this end he constructed an anemometer (modeled on Woltman's winch, used in hydraulics), which he calibrated according to a method that he described in the *Annales des Mines* of 1838. He then obtained a patent for and built a centrifugal ventilator equipped with curved blades inclined away from the direction of air movement, following the rule of hydraulics that maximum output requires that the air be expelled slowly (relative to the forward component of the blade motion). The ventilator was not very efficient, however.

In 1836 Combes became *ingénieur en chef des mines;* until 1842 he directed the regulatory agency for all steam apparatus in the Seine department and also sat on the central committee for steam machines. As secretary of this committee, he drafted the order that required certification for all steam machines. He became inspector general of mines in 1848 and from 1861 was the president of the Central Committee (which, in a decree of 25 January 1865, relaxed Combes's rules somewhat).

From 1839 Combes reported tirelessly to the mechanical arts committee of the Société d'Encouragement pour l'Industrie Nationale. In 1867, he eulogized Théophile Guibal for his invention of the encased-blade ventilator, which was manufactured from 1858 to 1922.

In 1847 Combes was elected to the Académie des Sciences; in 1857 he became director of the École des Mines. From 1869 on he was president of the Conseil Général des Mines.

In 1830 Combes married Louise-Pauline Bousquet, who died in 1841. He had a son who became a judge and two daughters, one of whom married the mineralogist Charles Friedel.

BIBLIOGRAPHY

I. ORIGINAL WORKS. Among Combes's many works are "Note sur les machines à vapeur," in *Annales des mines,* 2nd ser., **9** (1824), 441–462; "Observations sur le travail utile d'une machine d'épuisement . . . et sur les pertes d'eau qui ont lieu dans le jeu des pompes de mines," *ibid.,* 3rd ser., **2** (1832), 73–92; "Épuisement des eaux de quelques mines de houille de Rive-de-Gier," *ibid.,* **3** (1833), 197–208; "Mémoire sur une méthode générale d'évaluer le travail dû au frottement entre les pièces des machines qui se meuvent ensemble en se pressant mutuellement . . .," in Liouville's *Journal de mathématiques pures et appliquées,* **2** (1837), 109–129; "Mémoire sur un nouvel anémomètre . . .," in *Annales des mines,* 3rd ser., **13** (1838), 103–130, extract in *Bulletin de la Société d'encouragement pour l'industrie nationale* (1841), 288–292; "Aérage des mines," in *Annales des mines,* 3rd ser., **15** (1839), 91–308; "Supplément au traité de l'aérage des mines," *ibid.,* **18** (1840), 545–666; *Notice sur les travaux de M. Combes* (Paris, 1843; 2nd ed., 1847); *Traité de l'exploitation des mines,* 3 vols. and atlas (Paris, 1844–1845); *Exposé des principes de la théorie mécanique de la chaleur et de ses applications principales* (Paris, 1863); "Sur l'application de la théorie mécanique de la chaleur aux machines à vapeur . . .," in *Bulletin de la Société d'encouragement pour l'industrie nationale,* 2nd ser., **16** (1869), 13–48; *Études sur la machine à vapeur . . .* (Paris, 1869); and *Deuxième mémoire sur l'application de la théorie mécanique de la chaleur aux machines locomotives dans la marche à contre-vapeur* (Paris, 1869).

Among his collaborative works are "Note sur le fer carbonaté de Lasalle, et sur quelques produits des houillères embrasées des environs d'Aubin (Aveyron)," in *Annales des mines,* 2nd ser., **8** (1823), 431–438, written with T. Lorieux; and "Essais sur l'emploi du coton azotique pour le tirage des rochers," in *Comptes rendus hebdomadaires des séances de l'Académie des sciences,* **23** (1846), 940–944, 1090, extracted in *Bulletin de la Société d'encouragement pour l'industrie nationale* (1846), 621–622, written with Flandin.

II. SECONDARY LITERATURE. See J.-A. Barral, *Éloge biographique de M. Pierre-Charles-Mathieu Combes . . .* (Paris, 1874).

ARTHUR BIREMBAUT

COMBES, RAOUL (*b.* Castelfranc, France, 15 January 1883; *d.* Paris, France, 27 February 1964), *plant physiology, history of science.*

Combes was drawn toward biochemistry and plant physiology very early in life. In 1900 he began his studies in pharmacy under P. Vadam; they were completed in 1910 by a doctorate in science, with the remarkable dissertation "Détermination des intensités lumineuses optima pour les végétaux aux divers stades du développement." In 1905 Combes had published his first work on the quinones found in living cells.

Combes's most important teachers at the Sorbonne were Gaston Bonnier, whose daughter he married in 1910, and Marin Molliard, whom he succeeded at that university.

In 1912 Combes was appointed professor of applied botany at the Institut National Agronomique and at the École Nationale d'Horticulture. He became *chef de travaux* in plant physiology at the Sorbonne in 1921, lecturer in 1931, and full professor of plant physiology in 1937. As president of the Société Botanique de France (1933) he guided the development of the Fontainebleau laboratory, whose director he later became (1937). In 1943 he founded the laboratory of applied biology of the "Station du Froid" at Bellevue and directed the Office de la Recherche Scientifique Coloniale from 1943 to 1956. It was under his direction that the Adiopodoumé and Bondy research institutes were established.

Combes was a member of the Institut de France, the Académie des Sciences d'Outre Mer, and the Académie d'Agriculture. He was the first president of the Société Française de Physiologie Végétale.

At the beginning of his career Combes was interested in the metabolism of anthocyanic tissues; then he began a systematic study of the biochemistry of senescent leaves (transformation of sugars, movement of nitrogen compounds). He generalized his observations to the entire plant and provided the first diagram of the nitrogen metabolism of ligneous plants. Next Combes studied the effects of environment on the form and physiology of plants. The original conclusions of his experiments and those of his students remained classic from then on: in weak lighting or under water, plant cells accumulate nitrates, and at high altitudes carbohydrates are stored.

Combes was also interested in the physiology of flowering and showed for the first time the concentration of carbohydrates and nitrogen in young flowers and the massive movement of nitrogen reserves when hay is cut. His work in histochemistry led him to an analysis of lignin and to the study of the localization of cellular heterosides. Mention should be made of an interesting study, *L'immunité des végétaux vis-à-vis des principes immédiats qu'ils*

élaborent (1919), the conclusions of which were not properly appreciated until much later.

Throughout his career Combes was also interested in problems of a practical nature, including the preservation of fruits in a controlled environment and the biochemistry of forcing. He was an excellent microbiologist; one of his studies in this field was of the dissemination of germs in the atmosphere. From 1917 to 1919 he undertook a series of noteworthy research projects on the typhoid-like afflictions of the horse. Around 1931 Combes became interested in the history and philosophy of science and published *Biologie végétale et vitalisme* (Paris, 1933) and *Histoire de la biologie végétale en France* (Paris, 1933).

BIBLIOGRAPHY

I. ORIGINAL WORKS. Combes's writings include *La vie de la cellule végétale,* 3 vols. (Paris, 1933–1946); *La forme des végétaux et le milieu* (Paris, 1946); and *La physiologie végétale* (Paris, 1948).

II. SECONDARY LITERATURE. On Combes or his work, see *Titres et travaux scientifiques de R. Combes,* 3 vols. (Paris, 1932–1943); and R. Ulrich, "Raoul Combes," in *Compte rendu hebdomadaire des séances de l'Académie d'agriculture de France* (11 Mar. 1964), 430–433.

P. E. PILET

COMENIUS, JOHN AMOS (*b.* Nivnice [near Uherský Brod], Moravia, 28 March 1592; *d.* Amsterdam, Netherlands, 15 November 1670), *theology, pansophy, pedagogy.*

The youngest of five children, Comenius was born into a moderately prosperous family who were devout members of the Bohemian Brethren. His father, Martin, is said to have been a miller. After the death of his parents and two sisters in 1604, presumably from the plague, Comenius lived with relatives and received only a poor education until he entered the Latin school of Přerov, kept by the Brethren, in 1608. Three years later, thanks to the patronage of Count Charles of Žerotín, he matriculated at the Reformed University of Herborn, where he came under the influence of Johann Heinrich Alsted. Significant aspects of Comenius' thought closely resemble Alsted's concerns. Alsted, an anti-Aristotelian and a follower of Ramus, had a profound interest in Ramón Lull and Giordano Bruno, was a chiliast in theology, and worked for the gathering of all knowledge in his famous *Encyclopaedia* (1630). After completing his studies at Heidelberg in 1613 and 1614, Comenius returned to his native land, where he first taught school; but in 1618, two years after his ordination as a priest of the Bohemian Brethren, he became pastor at Fulnek. His first published work, a Latin grammar, dates from these years.

The Thirty Years' War and the battle of the White Mountain in November 1620 had a decisive effect on Comenius' life, since much of his work was directed toward the ultimately unsuccessful effort to have his people's native land and worship restored to them. For the next eight years Comenius led an insecure existence, until the final expulsion of the Brethren from the imperial lands brought him to Leszno, Poland, which he had previously visited to negotiate rights of settlement. During these years his first wife, Magdalena, and their two children died, and he remarried in 1624. He finished *Labyrint Swĕta a Lusthauz Srdce* in 1623, and *Centrum securitatis* in 1625, published in 1631 and 1633, respectively (both in Czech).

From 1628 to 1641 Comenius lived at Leszno as bishop of his flock and rector of the local Gymnasium. He also found time to work on the reformation of knowledge and pedagogy, writing, among other things, his first major work, the *Didactica magna.* Written in Czech, it was not published until 1657, when it appeared in Latin as part of the *Opera didactica omnia,* which contained most of the works he had written since 1627.

In 1633 Comenius suddenly gained European fame with the publication of his *Janua linguarum reserata;* an English version, *The Gate of Tongues Unlocked and Opened,* appeared in the same year. The *Janua* presented a simple introduction to Latin according to a new method based on principles derived from Wolfgang Ratke and from the primers produced by the Spanish Jesuits of Salamanca. The reform of language learning, by making it speedier and easier for all, was characteristic of that general reformation of mankind and the world which all chiliasts sought to bring about in the eleventh hour before the return of Christ to rule on earth.

In England, Comenius gained contact with Samuel Hartlib, to whom he sent the manuscript of his "Christian Pansophy," which Hartlib published at Oxford in 1637, under the title *Conatuum Comenianorum praeludia,* and then again at London in 1639 as *Pansophiae prodromus.* In 1642 Hartlib published an English translation with the title *A Reformation of Schools.* These publications raised such high expectations in certain circles in England that Hartlib found it possible to invite Comenius to London, with the support of Bishop John Williams, John Dury, John Selden, and John Pym. In September 1641 Comenius arrived in London, where he met his supporters as well as such men as John Pell, Theodore Haak, and Sir Cheney Culpeper. He was invited to remain permanently in England, and there were plans for the establishment of a pansophic college. But the Irish Rebellion soon put an end to all these optimistic plans, although Comenius stayed until June 1642. While in London he wrote the *Via lucis,* which circulated in manuscript in England but was not published until 1668 at Amsterdam. In the meantime Comenius had offers from Richelieu to continue his pansophic work in Paris, but he accepted instead an earlier invitation from Louis de Geer to come to Sweden. On his way there he visited Descartes near Leiden, an occasion that can hardly have been a meeting of minds. Descartes thought Comenius confused philosophy with theology, saying: "Beyond the things that appertain to philosophy I go not; mine therefore is that only in part, whereof yours is the whole."

In Sweden, Comenius was to meet difficulty again. The chancellor, Axel Oxenstierna, wanted him to work on useful books for the schools; Comenius, at the urging of his English friends, proposed to work on pansophy. He worked on both, retiring to Elbing, Prussia (then under Swedish rule), between 1642 and 1648. His *Pansophiae diatyposis* was published at Danzig, in 1643, and *Linguarum methodus novissima,* at Leszno, in 1648; in 1651 the *Pansophiae* came out in English as *A Pattern of Universal Knowledge;* and his *Natural Philosophy Reformed by Divine Light: Or a Synopsis of Physics,* a translation of *Physicae ad lumen divinum reformatae synopsis* (Leipzig, 1633), appeared in the same year. In 1648, having returned to Leszno, Comenius became the twentieth—and last—presiding bishop of the Bohemian Brethren (later reconstituted as the Moravian Brethren).

In 1650 Comenius received a call from Prince Sigismund Rákóczy of Transylvania, the younger brother of George II Rákóczy, to come to Sarospatak to give advice on school reform and pansophy. He introduced many reforms into the pansophic school there; but in spite of much hard work, he met with little success, and in 1654 he returned to Leszno. In the meantime Comenius had prepared one of his best-known and most characteristic works, the *Orbis sensualium pictus* (1658), with Latin and German text. Significantly, it opened with an epigraph on Adam's giving of names (Gen. 2:19–20). The first school book consistently to use pictures of things in the learning of languages, it illustrated a principle that was fundamental to Comenius: Words must go with things and cannot properly be learned apart from them. In 1659 Charles Hoole brought out an English version, *Comenius's Visible World, Or a Picture and Nomenclature of All the Chief Things That Are in the World; and of Mens Employments Therein.*

Comenius' lack of success at Sarospatak was probably due in large measure to his acceptance of the fantastic prophecies of the visionary and enthusiast Nicholas Drabik; this was not the first time Comenius had put trust in a latter-day prophet, a weakness he shared with other chiliasts. They were only too willing to listen to optimistic predictions of apocalyptic events and sudden reversals to occur in the near future, such as the fall of the House of Hapsburg or the end of popery and the Roman church. Comenius' publications of these prophecies, with the intent of influencing political events, had a very depressing effect on his reputation.

Soon after Comenius' return to Leszno, war broke out between Poland and Sweden, and in 1656 Leszno was completely destroyed by Polish troops. Comenius lost all his books and manuscripts and was again forced into exile. He was invited to settle at Amsterdam, where he spent the remaining years of his life at the house of Lawrence de Geer, the son of his former patron. During these years he completed the great work that had occupied him for at least twenty years, *De rerum humanarum emendatione consultatio catholica,* a seven-part work summing up his lifelong and all-embracing deliberations on the improvement of human things. The "Pampaedia," directions for universal education, is preceded by the "Pansophia," its foundation, and is followed by the "Panglottia," directions for overcoming the confusion of languages, which alone will make the final reformation possible. Although some parts of the work were published as late as 1702, it was presumed lost until late 1934, when it was found in the Francke Stiftung in Halle. It was first published in its entirety in 1966. Comenius was buried in the Walloon church at Naarden, near Amsterdam. His thought was highly esteemed by German Pietists in the eighteenth century. In his own country Comenius occupies a place of eminence both as a national hero and as a literary artist.

All Comenius' efforts were directed toward the speedy and efficient reformation of all things pertaining to the life of man in the spheres of religion, society, and knowledge. His program was a "way of light" designed to ensure the highest possible enlightenment of man before the imminent return of Christ to reign on earth during the millennium. The universal aims were piety, virtue, and wisdom; to be wise was to excel in all three.

Thus, all Comenius' work found both its beginning and its end in theology. His beliefs and aspirations were shared by many of his contemporaries, but his system was certainly the most comprehensive of the many that were offered in the seventeenth century. It was essentially a prescription for salvation through knowledge raised to the level of universal wisdom, or pansophy, supported by a corresponding program of education. It was in the divine order of things that "modern man" at that time, in what was thought to be the last age of the world, had been made capable of achieving universal reform through the invention of printing and the extension of shipping and commerce, which for the first time in history gave promise of complete universality in the sharing of this new, reforming wisdom.

Since God is hidden behind his work, man must turn to the threefold revelation before him: the visible creation, in which God's power is made manifest; man, created in the image of God and showing forth proof of His divine wisdom; and the Word, with its promise of good will toward man. Thus, all that man needs to know and not know must be learned from three divine books: nature, the mind or spirit of man, and the Scriptures. For the achievement of this education man has been supplied with his senses, his reason, and faith. Since both man and nature are God's creations, they must share the same order, a postulate that guarantees the complete harmony of all things among themselves and with the mind of man.

This familiar macrocosm-microcosm doctrine gives assurance that man is indeed capable of hitherto unrealized wisdom; each individual thus becomes a pansophist, a little god. Heathens, lacking the revealed word, cannot attain this wisdom; even Christians have until recently been lost in a labyrinth of error handed on by tradition and authority in a flood of books that at best contain piecemeal knowledge. Comenius was not a humanist: Man must turn to the divine books alone and begin to learn by direct confrontation with things—by autopsy, as Comenius called it. All learning and knowledge begin with the senses; but according to the correspondence doctrine, it follows that the mind has innate notions, or germs, that make man capable of comprehending the order he confronts. The world and the life of each individual form a school; nature teaches, the teacher is a servant of nature, and naturalists are priests in the temple of nature. Man must know himself (*nosce teipsum*) and nature.

What man needs in order to find his way out of the labyrinth is an Ariadne thread, a method by which he will see the order of things by understanding their causes. This method is to be supplied by the book of pansophy, a book in which the order of nature and the order of the mind will move together stepwise (*per gradatim*) toward wisdom and insight. This book will contain nothing but certain and useful knowledge, thus replacing all other books. A complete record of knowledge arranged according to this method will constitute a true encyclopedia, in much the same sense as Robert Hooke's "repository" of natural curiosities in the Royal Society when arranged according to the categories of John Wilkins' *Essay Towards a Real Character and a Philosophical Language,* a subject on which Comenius also had much to say. By following this natural method, all men will find it easy to gain complete and thorough mastery of all knowledge. From this vast expansion of wisdom true universality will result; and there will again be order, light, and peace. Through this reformation man and the world will return to a state similar to that before the Fall.

This pansophic program led Comenius to take a profound interest in language and in education. From earliest infancy the child must learn to join things and words. His native speech is his first introduction to reality, which must not be clouded by empty words and ill-understood concepts. In school foreign languages—first those of neighboring nations and then Latin—must be learned by reference to the mother tongue, and the school books must follow the pansophic method; the "door to languages" would offer the same material as the "door to things," and both will constitute a little encyclopedia. School books must be graded for each age level, dealing only with things that are already within the child's experience. Latin was best suited for the purpose of wider communication; but Comenius looked forward to the framing of a perfect philosophical language that would reflect the method of pansophy, a language in which nothing false or trivial could be expressed. Language was the mere vehicle of knowledge, but rightly used and taught it was one of the surest means to light and wisdom.

Comenius' concern with didactics was directed not merely toward formal schooling but to all ages of man, on the principle that all of life is a school and a preparation for eternal life. Girls and boys must be educated together; and since all men have an innate desire for knowledge and piety, they must be taught in a spontaneous and playful manner. Corporal punishment must not be applied. Failure in learning was not the learner's fault but evidence of the teacher's inadequacy to perform his role as the servant of nature, or—as Comenius often said—as the obstetrician of knowledge.

Pedagogy is Comenius' most lasting—perhaps his only lasting—contribution to knowledge, and until recently it has generally been considered his primary concern. However, it was only a means toward that universal reformation of mankind of which pansophy was the foundation and theology the single guiding motive. The profusion of scriptural citations in Comenius' works is a constant reminder of this source of inspiration. The books of Daniel and Revelation were the chief texts for the increase of knowledge and the imminence of the millennium. The story of Adam's name-giving in Genesis and The Wisdom of Solomon gave Comenius his conception of man and his conviction of the order that was reflected in pansophy, since God had "ordered all things in measure and number and weight." Comenius relied heavily on elaborate metaphorical and structural uses of the temple of Solomon. To Comenius man was, like Adam, placed in the middle of creation, with the charge of knowing all of nature and thereby controlling and using it. Hence the reformation of man was only a part of the complete reformation of the world, which would restore creation to its initial purity and order and would be the ultimate tribute to its creator.

Comenius made no contribution to natural science, and he was profoundly alienated from the developments in science that occurred during his lifetime. Contrary claims have been made, but only at the cost of ignoring his dependence on a priori postulates and his entire theological orientation. On the other hand, it is also clear that several men who later figured prominently in the Royal Society showed close affinity with much of his thought. The motto of the Royal Society—Nullius in Verba—occupies a significant place in Comenius' *Natural Philosophy Reformed by Divine Light: Or a Synopsis of Physics,* and in both contexts it had the same meaning. It was a reminder that tradition and authority were no longer the arbiters of truth; they had yielded to nature and autopsy as the sole sources of certain knowledge. The much-debated problem of the relationship between Comenius and the early Royal Society is still unsolved, chiefly because discussions of that subject reveal skimpy knowledge of Comenius' writings and nearly total ignorance of his very important correspondence.

The claims that have been made for Comenius' influence on Leibniz would appear to be much exaggerated. Comenius was so typical of certain beliefs, doctrines, and concerns in his age that the same ideas were also expressed by a number of others who occupy more prominent positions in Leibniz' early writings. In addition to the sources already cited, Comenius was influenced by the theology of the Bohemian Brethren (with their strong chiliastic tendencies) as well as by the following figures, to mention only the better known and most important: Johann Valentin Andreae, Jacob Boehme, Nicholas of Cusa, Juan Luis Vives, Bacon, Campanella, Raymond of Sabunde (whose *Theologia naturalis* he published at Amsterdam in 1661 under the title *Oculus fidei*), and Marin Mersenne, whose correspondence gives evidence of a positive attitude toward Comenius and his work.

BIBLIOGRAPHY

I. ORIGINAL WORKS. Comenius' bibliography contains a great many items, published in many places and languages. Two useful guides are Kurt Pilz, *Die Ausgaben des Orbis sensualium pictus* (Nuremberg, 1967); and Emma Urbánková, ed., *Soupis děl J. A. Komenského v Československých Knihovnách, Archivech a Museích* (Prague, 1959). See also Jan Patočka, "L'état présent des études coméniennes," *Historica* (Prague), **1** (1959), 197–240.

Among his major works are *Opera didactica omnia*, repr. of the 1657 ed., 3 vols. (Prague, 1957); and *De rerum humanarum emendatione consultatio catholica*, 2 vols. (Prague, 1966). The great modern ed. is *Veškeré Spisy Jana Amosa Komenského* (Brno, 1914–1938); only 9 of the projected 30 volumes have been published. The correspondence has been published in the following collections: A. Patera, ed., *Jana Amosa Komenského Korrespondence* (Prague, 1892); Jan Kvačala, ed., *Korrespondence Jana Amosa Komenského*, 2 vols. (Prague, 1898–1902); and Johannes Kvačala, ed., *Die pädagogische Reform des Comenius in Deutschland bis zum Ausgange des XVII Jahrhunderts*, 2 vols. (Berlin, 1903–1904), vols. 26 and 32 in the series Monumenta Germaniae Paedagogica. Translations include *The Great Didactic*, M. W. Keatinge, trans. (London, 1896; repr. New York, 1968); and *The Way of Light*, E. T. Campagnac, trans. (Liverpool–London, 1938). Excellent eds., all with good introductions and notes, are *Pampaedia*, Dmitrij Tschiževskij, Heinrich Geissler, and Klaus Schaller, eds. and trans. (Heidelberg, 1960); *Informatorium der Mutterschul*, Joachim Heubach, ed. (Heidelberg, 1962); and *Centrum securitatis*, Klaus Schaller, ed. (Heidelberg, 1964), nos. 5, 16, and 26, respectively, in the series Pädagogische Forschungen, Veröffentlichungen des Comenius-Instituts. *Vorspiele: Prodromus pansophiae, Vorläufer der Pansophie*, Herbert Hornstein, ed. and trans. (Düsseldorf, 1963), with excellent notes and "Nachwort" by Hornstein, forms the best single introduction to Comenius. See also *Janua rerum*, Klaus Schaller, ed. (Munich, 1968), vol. IX of Slavische Propyläen; and *Ausgewählte Werke*, K. Schaller and D. Tschizewskij, eds., 3 vols. (Hildesheim, 1970).

II. SECONDARY LITERATURE. There are two standard biographies: Jan Kvačala, *Johann Amos Comenius, sein Leben und seine Schriften* (Berlin–Leipzig–Vienna, 1892); and Matthew Spinka, *John Amos Comenius* (Chicago, 1943). Anna Heyberger, *Jean Amos Comenius, sa vie et son oeuvre d'éducateur* (Paris, 1928), pp. 243–259, has a convenient list of the works in chronological order. Robert Fitzgibbon Young, *Comenius in England* (London, 1932), contains a useful collection of documents in English translation. See also G. H. Turnbull, *Hartlib, Dury, and Comenius, Gleanings From Hartlib's Papers* (Liverpool, 1947). Three recent German works on Comenius are of unusual distinction: Heinrich Geissler, *Comenius und die Sprache* (Heidelberg, 1959); Klaus Schaller, *Die Pädagogik des Johann Amos Comenius und die Anfänge des pädagogischen Realismus im 17. Jahrhundert*, 2nd ed. (Heidelberg, 1967), with a very full bibliography of primary and secondary literature and relevance far beyond its immediate subject; and Schaller's *Die Pampaedia des Johann Amos Comenius, eine Einführung in sein pädagogisches Hauptwerk*, 3rd ed. (Heidelberg, 1963). These are nos. 10, 21, and 4, respectively, in Pädagogische Forschungen. Also excellent are Herbert Hornstein, *Weisheit und Bildung, Studien zur Bildungslehre des Comenius* (Düsseldorf, 1968); and Charles Webster, ed., *Samuel Hartlib and the Advancement of Learning* (Cambridge, 1970). John Edward Sadler, *J. A. Comenius and the Concept of Universal Education* (London, 1966), may have certain uses but cannot be relied upon for an adequate understanding of the thought of Comenius and his contemporaries. For a characteristic, very unsympathetic early eighteenth-century account of Comenius, see the article on him in *Bayle's Dictionnaire historique et critique*.

HANS AARSLEFF

COMMANDINO, FEDERICO (*b.* Urbino, Italy, 1509; *d.* Urbino, 3 September 1575), *mathematics*.

The little that is known about Commandino's life is derived mainly from a brief biography written by a younger fellow townsman who knew him well for many years toward the end of his career.

Descended from a noble family of Urbino, Commandino studied Latin and Greek for some years with a humanist at Fano. When Rome was sacked on 6 May 1527 by the army of Charles V, the Orsini, a leading noble clan, fled to Urbino. For one of their sons they procured a tutor proficient in mathematics, who also taught Commandino. After this tutor became a bishop on 6 June 1533, he obtained for Commandino an appointment as private secretary to Pope Clement VII. However, the pontiff died on 25 September 1534, and Commandino went to the University of Padua. There he studied philosophy and medicine for ten years, but he took his medical degree from the University of Ferrara.

Returning to his birthplace, Commandino married a local noblewoman, who died after giving birth to two daughters and a son. Commandino resolved not to marry a second time. After his son's death he put the girls in a convent school (and later found husbands for them). Withdrawing from the general practice of medicine, he turned to his true vocation: editing, translating, and commenting on the classics of ancient Greek mathematics. Gaining renown thereby, Commandino was designated the private tutor and medical adviser to the duke of Urbino. The duke, however, was married to the sister of a cardinal; and the latter persuaded Commandino to be his personal physician in his intellectually stimulating household in Rome.

Commandino had been translating into Latin and commenting on Archimedes' *Measurement of the Circle* (with Eutocius' commentary), *Spirals, Quadrature of the Parabola, Conoids and Spheroids,* and *Sand-Reckoner.* Besides the first printed edition of the Greek text of these five works and an earlier Latin translation of them (Basel, 1544), he had access also to a Greek manuscript in Venice, where his patron was residing when Commandino published this Archimedes volume in 1558.

During the previous year Commandino had heard complaints about the difficulty of understanding Ptolemy's *Planisphere,* which showed how circles on the celestial sphere may be stereographically projected onto the plane of the equator. Although the Greek text of the *Planisphere* is lost, it had been translated into Arabic, and from Arabic into Latin. This Latin version, done at Toulouse in 1144, and Jordanus de Nemore's *Planisphere,* both of which had been printed at Basel in 1536, were edited by Commandino and, together with his commentary on Ptolemy's *Planisphere,* were published at Venice in 1558.

Ptolemy's *Analemma* explained how to determine the position of the sun at a given moment in any latitude by an orthogonal projection using three mutually perpendicular planes. Again, as in the case of Ptolemy's *Planisphere,* no Greek text was available to Commandino (a portion was later recovered from a palimpsest); but an Arabic version had been translated into Latin. This was edited from the manuscript by Commandino (Rome, 1562). Besides his customary commentary, he added his own essay *On the Calibration of Sundials* of various types, since he felt that Ptolemy's discussion was theoretical rather than practical.

Commandino's only other original work, dealing with the center of gravity of solid bodies, was published in 1565 at Bologna, of which his patron had become bishop on 17 July of the preceding year. Commandino's interest in this topic was aroused by Archimedes' *Floating Bodies,* of which he had no Greek text, unlike the five other Archimedean works he had previously translated. Since his time a large part of the Greek text of *Floating Bodies* has been recovered, but he had only a printed Latin translation (Venice, 1543, 1565), which he commented on and corrected (Bologna, 1565). In particular the proof of proposition 2 in book II was incomplete, and Commandino filled it out. One step required knowing the location of the center of gravity of any segment of a parabolic conoid. No ancient treatment of such a problem was then known, and Commandino's was the first modern attempt to fill the existing gap.

Archimedes' *Floating Bodies* assumed the truth of some propositions for which Commandino searched in Apollonius' *Conics.* Of the *Conics'* eight books only the first four are extant in Greek, and he had access to them in manuscript. An earlier Latin translation (Venice, 1537) was superseded by his own (Bologna, 1566), to which he added Eutocius' commentary, the relevant discussion in Pappus' *Collection* (book VII), the first complete Latin translation (from a Greek manuscript) of Serenus' *Section of a Cylinder* and *Section of a Cone,* and his own commentary.

Overwork and the death of his patron on 28 October 1565 greatly depressed Commandino; and he returned to Urbino, where he could live quietly, for many months on a salt-free diet. He resumed his former activities, however, after being visited by John Dee, who gave him a manuscript Latin translation of an Arabic work related to Euclid's *On Divisions* (of figures), of which the Greek original is lost. Commandino published this Latin translation and added a short treatise of his own to condense and generalize the discussion in the manuscript (Pesaro, 1570).

At the request of his ruler's son, Commandino translated Euclid's *Elements* into Latin and commented on it extensively (Pesaro, 1572). Also in 1572 he published at Pesaro his Latin translation of and commentary on Aristarchus' *Sizes and Distances of the Sun and Moon,* with Pappus' explanations (*Collection,* book VI, propositions 37–40).

For those of his countrymen who did not know Latin, Commandino supervised a translation of Euclid's *Elements* into Italian by some of his students (Urbino, 1575). His own Latin translation of Hero's *Pneumatics* (Urbino, 1575) was seen through the press by his son-in-law immediately after his death. From a nearly complete manuscript, needing three months' work at most, his faithful pupil Guidobaldo del Monte published Commandino's Latin translation of and commentary on Pappus' *Collection,* books III–VIII (Pesaro, 1588).

In the sixteenth century, Western mathematics emerged swiftly from a millennial decline. This rapid ascent was assisted by Apollonius, Archimedes, Aristarchus, Euclid, Eutocius, Hero, Pappus, Ptolemy, and Serenus—as published by Commandino.

BIBLIOGRAPHY

A list of Commandino's publications is available in Pietro Riccardi, *Biblioteca matematica italiana,* enl. ed., 2 vols. (Milan, 1952), I, cols. 42, 359–365; II, pt. 1, col. 15; II, pt. 2, col. 117; II, pt. 5. cols. 9, 49–50; II, pt. 6, col. 189; II, pt. 7, cols. 25–26. Riccardi omits *Conoids and Spheroids* (I, col. 42); misattributes the Italian translation of Euclid's *Elements* to Commandino himself (I, col. 364); and misdates Pappus' *Collection* as 1558 (correct date, 1588; II, pt. 1, col. 15). To Riccardi's list of writings about Commandino (I, col. 359; II, pt. 1, col. 15) add Edward Rosen, "The Invention of the Reduction Compass," in *Physis*, **10** (1968), 306–308; and "John Dee and Commandino," in *Scripta mathematica,* **28** (1970), 321–326.

Bernardino Baldi's biography of Commandino, completed on 22 November 1587, was first published in *Giornal de' letterati d'Italia,* **19** (1714), 140–185, and reprinted in *Versi e prose scelte di Bernardino Baldi,* F. Ugolino and F.-L. Polidori, eds. (Florence, 1859), pp. 513–537.

EDWARD ROSEN

COMMERSON, PHILIBERT (*b.* Châtillon-les-Dombes, Ain, France, 18 November 1727; *d.* St.-Julien-de-Flacq, Île de France [now Mauritius], 13 March 1773), *natural history*.

Son of Georges-Marie Commerson, notary and adviser to the prince of Dombes, and of Jeanne-Marie Mazuyer, Commerson went to Montpellier in 1747 and received a complete medical education there, obtaining his bachelor's degree on 11 September 1753, his licence on 7 September 1754, and his doctorate on 9 September 1754. His dissertation has been lost. Already deeply interested in botany, he established a notable herbarium. At the request of Linnaeus he drafted a description of the principal species of fish in the Mediterranean. That work, although never published, was later used by Lacepède for his *Histoire naturelle des poissons*.

In 1755 Commerson botanized in Savoy, went to Bern to meet Albrecht von Haller, and visited Voltaire at "Les Délices," near Geneva. In 1756 he returned to Châtillon-les-Dombes and established a botanical garden there in 1758, enriching it through his frequent excursions in the provinces of central France. On 17 October 1760 Commerson married Antoinette-Vivante Beau, daughter of a notary from Genouilly (Saône-et-Loire). She died in 1762, a few days after giving birth to their son, Anne-François-Archambaud. Following the death of his wife Commerson yielded to the urgings of the astronomer Lalande, his friend since childhood, and of Bernard de Jussieu. He left in August 1764 for Paris, where he took lodgings near the Jardin du Roi, which he visited constantly for almost two years.

On the recommendation of Pierre Poissonnier, Commerson was appointed naturalist to Bougainville's expedition. Appointed botanist and naturalist to the king, he boarded the supply ship *Étoile*, accompanied by his maidservant, Jeanne Barré, who was disguised as a man. Setting sail from Rochefort at the beginning of February 1767, they stopped at Rio de Janeiro, where the *Étoile* was joined by the frigate *Boudeuse* (21 June 1767); the Strait of Magellan and Tierra del Fuego (December 1767); Tahiti,

for a brief stay that has become famous due in large part to Diderot's *Supplément au voyage de M. de Bougainville* . . . (5–14 April 1768); the Navigators' Islands (now Samoa) and the New Hebrides (May 1768); the Solomon Islands and New Britain (July 1768); the Moluccas, especially the island of Buru, where they stopped for nearly a week (1–7 September 1768); Java (end of September 1768); and Mauritius (10 November 1768).

Bougainville returned to France, but Pierre Poivre retained Commerson and lodged him in his own quarters. After exploring Mauritius, Commerson went to Madagascar (October 1770–January 1771) to study its flora, fauna, and people. He then went to Île Bourbon (now Réunion; January 1771), where he and Jean-Baptiste Lislet-Geoffroy climbed the Piton de la Fournaise, an active volcano. Returning ill to Port Louis, Mauritius, he put his botanical collections and his notes in order, while Poivre, his patron and friend, went back to France.

Commerson died virtually alone, attended only by his faithful maidservant. Commerson was never a member of the Academy of Sciences. His was the second name proposed on 20 March 1773 for the vacancy created by the promotion of Adanson. But the king chose the candidate named first, Antoine-Laurent de Jussieu. His manuscripts; the beautiful plates executed by his draftsman, Paul Jossigny; his herbarium, containing 3,000 new species and genera; and his collections of every kind, all packed in thirty-two cases, arrived in Paris in 1774 and were turned over to the Jardin du Roi.

BIBLIOGRAPHY

I. ORIGINAL WORKS. Commerson's writings include "Lettre de M. Commerson à M. de Lalande, de l'Île de Bourbon, le 18 avril 1771," in Banks and Solander, *Supplément au voyage de M. de Bougainville ou Journal d'un voyage autour du monde fait par MM. Banks et Solander, Anglois, en 1768, 1769, 1770, 1771,* Anne-François-Joachim de Fréville, trans. (Paris, 1772), pp. 251–286; and *Testament singulier de M. Commerson, docteur en médecine, médecin botaniste et naturaliste du roi, fait le 14 et 15 décembre 1766* (Paris, 1774). Almost all of Commerson's MSS and sketches are preserved at the central library of the Muséum National d'Histoire Naturelle in Paris; a description of them is given by Amédée Boinet in *Catalogue général des manuscrits des bibliothèques publiques de France, Paris,* II, *Muséum d'histoire naturelle* . . . (Paris, 1914), *passim;* and by Yves Laissus in vol. LV of the *Catalogue: Muséum national d'histoire naturelle (supplément)* (Paris, 1965), 32, 45.

II. SECONDARY LITERATURE. On Commerson or his work see the following (listed chronologically): Joseph Jérome de Lalande, "Éloge de M. Commerson," in *Journal de physique,* **5** (1775), 89–119; Paul-Antoine Cap, *Philibert Commerson, naturaliste voyageur. Étude biographique* . . . (Paris, 1861), extract from *Journal de pharmacie et de chimie*

3rd ser., **38** (Dec. 1860), 413–442; F. B. Montessus, *Martyrologe et biographie de Commerson, médecin-botaniste et naturaliste du roi . . .* (Paris, 1889), with portrait; S. Pasfield Oliver, *The Life of Philibert Commerson, D.M., naturaliste du roi. An Oldworld Story of French Travel and Science in the Days of Linnaeus,* G. F. Scott Elliot, ed. (London, 1909); L. Laroche, "Le naturaliste Philibert Commerson," in *Revue périodique de vulgarisation des sciences naturelles et préhistoriques de la Physiophile* (Montceau-les-Mines), 13th year, no. 16 (1 Mar. 1937), 249–253, and no. 17 (1 June 1937), 258–260; Henry Chaumartin, *Philibert Commerson, médecin naturaliste du roy et compagnon de Bougainville* (n.p., 1967), a special publication in the series Petite Histoire de la Médecine; and Étienne Taillemite, "Le séjour de Bougainville à Tahiti, essaie d'étude critique des témoignages," in *Journal de la Société des océanistes,* **24,** no. 24 (Dec. 1968), 3–54. "Collogue Commerson 16–24 octobre 1973," in *Cahiers du Centre universitaire de la Réunion,* spec. no. (St. Denis de la La Réunion, 1974).

YVES LAISSUS

COMMON, ANDREW AINSLIE (*b.* Newcastle-upon-Tyne, England, 7 August 1841; *d.* Ealing, London, England, 2 June 1903), *astronomy.*

Common was noted for his pioneer work in celestial photography and in the design and construction of large telescopes. His professional work as a sanitary engineer prevented his having sufficient time to pursue astronomy effectively until 1875, when he began to experiment with taking photographs through a telescope. The next year he constructed a thirty-six-inch silver-on-glass reflector, with a mirror made by George Calver and a mounting of his own design; the main moving part (the polar axis) floated on mercury to reduce friction. By devising a photographic plate holder that could be moved during exposure to counteract errors in the clock drive of the telescope, Common was the first to make really long exposures successfully. He took the first satisfactory pictures of Jupiter and Saturn and, in 1883, a superb picture of the nebula in Orion that showed the superiority of a photograph over a drawing.

For astronomical colleagues Common constructed many large parabolic mirrors between thirty and sixty inches in diameter and high-quality flat mirrors; in the latter work he was aided by a very sensitive spherometer of his own design. His largest telescope was a sixty-inch reflector, with the polar axis floating in water. Many difficulties attended its construction, and it was not ready for use until 1889. Common had by then become involved in designing gunsights for the Royal Navy—a contemporary claimed these were so successful that they quadrupled fighting efficiency—and this work prevented Common from using his new

instrument before his sudden death. The excellent thirty-six-inch reflector was sold to Edward Crossley of Halifax, England, and presented by him to Lick Observatory, where it was used for photography with great success by James Keeler and others from 1898 to 1900.

Common was awarded the Gold Medal of the Royal Astronomical Society in 1884, was its president from 1885 to 1887, and was elected a fellow of the Royal Society in 1885.

BIBLIOGRAPHY

Significant papers by Common, all in *Monthly Notices of the Royal Astronomical Society,* are "Note on Large Telescopes, With Suggestions for Mounting Reflectors," **39** (1879), 382–386; "Note on Silvering of Large Mirrors," **42** (1882), 77–78; "Note on a Photograph of the Great Nebula in *Orion* and Some New Stars Near Orionis," **43** (1883), 255–257; "Suggestions for Improvements in the Construction of Large Transit Circles," **44** (1884), 288–293; "Note on a Method of Reducing the Friction of the Polar Axis of a Large Telescope," *ibid.,* 366–367; "Note on Stellar Photography," **45** (1885), 25–27; and "Note on an Apparatus for Correcting the Driving of the Motor Clocks of Large Equatorials for Long Photographic Exposures," **49** (1889), 297–300.

A print of a photograph of Jupiter appears in *Observatory,* **3** (1880), pl. III; prints of photographs of Jupiter, Saturn, and the Orion nebula are on the frontispiece and the title page of A. M. Clerke, *A Popular History of Astronomy During the Nineteenth Century* (London, 1893).

Obituaries of Common are in *Monthly Notices of the Royal Astronomical Society,* **64** (1904), 274–278; and *Proceedings of the Royal Society,* **75** (1905), 313–318.

COLIN A. RONAN

COMPTON, ARTHUR HOLLY (*b.* Wooster, Ohio, 10 September 1892; *d.* Berkeley, California, 15 March 1962), *physics.*

Arthur Holly Compton, who shared the Nobel Prize in physics in 1927 for his discovery of the effect that bears his name, was a son of a professor of philosophy and dean of the College of Wooster. His oldest brother, Karl, who became the head of the physics department at Princeton and, later, president of the Massachusetts Institute of Technology, was his close friend and most trusted scientific adviser. His brother Wilson was a distinguished economist and businessman. His sister, Mary, and her husband, C. Herbert Rice, were educators in India. Their mother, Otelia Augspurger, came from a long line of Mennonites and guided the Compton children long after they left Wooster.

The College of Wooster, with its strong missionary tradition, had a decisive influence on Compton throughout his career. His attitudes toward life, science, and the world in general were almost completely determined as he grew to manhood and received his basic education there. The Compton family were interested in many things besides the college: their church, a summer camp in Michigan, Karl's early scientific experiments, Arthur's early interest in paleontology and astronomy and his experiments with gliders, and the boys' athletic activities. All these interests made for an active youth and developed a constructive and buoyant attitude toward life that carried Compton through all the complex and highly demanding responsibilities that his later duties called upon him to assume.

After graduating from the College of Wooster in 1913, Compton entered the graduate school of Princeton, where he earned a master's degree in 1914 and a Ph.D. in 1916. His older brothers had already received the Ph.D. from Princeton. He was an outstanding graduate student, being a "whiz" at problem solving, and conducted research on the quantum nature of specific heats. He also perfected a laboratory method (developed earlier at his home at Wooster) to measure latitude and the earth's rotation (as a vector) independently of astronomical observations. His Ph.D. thesis, begun under O. W. Richardson and completed under H. L. Cooke, began his interest in X-ray diffraction and scattering.

Compton also was active in sports, especially tennis, and in the social life and professional activities of the graduate student community. He became a personal friend and admirer of Henry Norris Russell, whose advice and counsel, along with that of Richardson and William F. Magie, was of great importance in shaping his early scientific career.

Upon completing his graduate studies at Princeton, Compton married his Wooster College classmate Betty Charity McCloskey. She was an enthusiastic and active partner in all of his professional activities, whether these were in the research laboratory or on journeys to the most distant parts of the earth for field measurements. Their children were Arthur Alan, an officer in the U.S. State Department, and John Joseph, professor and head of the philosophy department at Vanderbilt University.

After receiving his Ph.D., Compton taught physics for one year (1916–1917) at the University of Minnesota in the enthusiastic young department that included John T. Tate and Paul D. Foote. He next spent two years as research engineer for the Westinghouse Electric and Manufacturing Company in East Pittsburgh, Pennsylvania. During this period Compton developed aircraft instruments for the Signal Corps, did original work, and was awarded patents on designs for a sodium vapor lamp.

This work led to his later close association, at Nela Park, in Cleveland, Ohio, with the birth of the fluorescent lamp industry in the United States. He worked closely with Zay Jeffries, technical director of the General Electric Company at Nela Park during the years of greatest activity in fluorescent lamp development. In 1934 Compton played a decisive role in providing their engineers with critical information on the work being done at General Electric Ltd., Wembley, England (which had technical exchange relations with the U.S. General Electric Company), which enabled the Nela Park engineers to construct the prototype commercially feasible fluorescent lamp and initiate an extensive research and development program.

During the years (1917–1919) at Westinghouse, Compton's X-ray work continued; two studies are of special interest. In attempting to obtain quantitative agreement between X-ray absorption and scattering data and the classical theory of J. J. Thomson, Compton advanced the hypothesis of an electron of finite size (radius of 1.85×10^{-10} cm.) to account for the observed dependence of intensity on scattering angle. This was the simple beginning that finally led to the concept of a "Compton wavelength" for the electron and other elementary particles. Later his own quantum theory of X-ray scattering and its extension into quantum electrodynamics fully developed the concept.

The second study was initiated with Oswald Rognley in 1917 at the University of Minnesota to determine the effect of magnetization on the intensity of X-ray reflections from magnetic crystals. This work indicated that orbital electron motions were not responsible; and Compton proposed that ferromagnetism was due to an inherent property of the electron itself, which he considered to be an elementary magnet. In 1930, after he had developed the double-crystal X-ray spectrometer, this hypothesis was proved more conclusively by experiments of J. C. Stearns, a graduate student of his at the University of Chicago. Their results on the intensity of X rays diffracted by magnetized and unmagnetized magnetite and silicon steel proved that electronic orbital motions in crystals are not involved in ferromagnetism. This confirmed Compton's prediction, made in 1917, that the electron's magnetic orientation, due to its spin, is the ultimate cause of ferromagnetism.

Following World War I, Compton was awarded one of the first National Research Council fellowships, which enabled him to spend the year 1919–1920 in the Cavendish Laboratory at Cambridge. This was a most hectic and exciting year at that laboratory, which was crowded to capacity with young men from Great

Britain and all parts of the British Empire who had just returned from the war to study with Ernest Rutherford and J. J. Thomson. Compton found this a most inspiring year, not only for his relationship with Rutherford, who provided laboratory space and such support as the crowded conditions at the university afforded, but also for his meetings with Thomson, who formed the highest opinion of Compton's research abilities. The close relationship between Compton and Thomson continued throughout the latter's lifetime.

At Cambridge high-voltage X-ray equipment was not available, so Compton performed a scattering experiment with gamma rays that not only confirmed earlier results of J. A. Gray and others but also started his planning for the more definitive X-ray experiments that would lead him to his greatest discovery.

In attempting to find a classical explanation for these early gamma-ray scattering results, Compton further developed his ideas for an electron of finite size and with various distributions of electricity. However, neither his own efforts nor those of Gray or others could provide a consistent explanation for either the intensity of the scattering or the progressive change in wavelength with scattering angle.

The association with British physicists and their work certainly provided a stimulus for Compton's later great discovery; but since all these early efforts tried to provide a classical explanation for the phenomena, they could hardly have led him to his final correct explanations of the Compton effect as a quantum process, which came later, in his work at St. Louis.

Following the year at Cambridge, Compton returned to the United States in 1920 to begin his tenure as the Wayman Crow professor and head of the physics department at Washington University, in St. Louis, Missouri, where he made his greatest single discovery. This was suggested by his experiments with gamma rays at Cambridge, which provided the first opportunity to test ideas that originated during the Westinghouse period. In St. Louis he used monochromatic X rays, which permitted him, with the help of a Bragg crystal spectrometer, to measure accurately the change in wavelength of the X rays scattered from a target at various angles, the phenomenon since universally known as the Compton effect. Although this discovery can be considered a logical development of his gamma-ray experiments at Cambridge, which showed a systematic reduction in the penetrating power of gamma rays with increasing scattering angle, nevertheless, the Washington University data, determined with great precision, were the principal ones considered in his final development of the Compton quantum theory of scattering.

The definite change in X-ray wavelength on scattering from light elements such as carbon, depending systematically on the scattering angle, could not be reconciled with classical electrodynamics, even when extended by Compton's own suggestion of scattering by electrons of finite size. Neither could his attempts to explain the results as fluorescence be maintained once the changes in wavelength with scattering angle were fully established.

Compton arrived at his revolutionary quantum theory for the scattering process rather suddenly in late 1922, after all his previous attempts at an explanation had failed. He now treated the interaction as a simple collision between a free electron and an X-ray quantum having energy $h\nu$ and momentum $h\nu/c$ and obeying the usual conservation laws. He had previously considered X rays as having linear momentum, but only in the classical sense suggested by the Poynting vector. He was well acquainted with Planck's quantum of energy $h\nu$, principally from his brother Karl's pioneer work on testing Einstein's theory of the photoelectric effect. However, he does not seem to have been aware of Einstein's 1917 paper suggesting that the photon has linear momentum until much later, and then probably only indirectly, through a 1922 paper by Schrödinger on the Doppler effect in scattering in which reference is made to Einstein's paper.

It appears that Compton had been at work on a quantum theory for the scattering process for some time before the Schrödinger paper appeared, but he did not give up the classical concept of the momentum of radiation until he had tried in many ways to bring it into accord with his experimental results.

However, once Compton had the concept of an X-ray photon carrying linear momentum as well as energy, he derived his equations for the Compton effect in the form in which we now know them and found that they agreed perfectly with his data and led to a quantum Compton wavelength h/mc for the electron. However, in 1923 the "old quantum theory" gave no basic significance (other than the numerical agreement) for the Compton wavelength; and it was only after the development of the new quantum mechanics in its relativistic form that the Compton wavelength was seen as having the fundamental significance now ascribed to it.

When Compton reported his discovery at meetings of the American Physical Society, it aroused great interest and strong opposition, especially from William Duane of Harvard, in whose laboratory Compton's results could not be confirmed.

The chief reasons for the reluctance of physicists to accept Compton's experimental results and his theory were that they conflicted with Thomson's theory of X-ray scattering based on classical electrodynamics and that Compton had developed his theory to explain the change in wavelength by using one of the most elegant early applications of the special theory of relativity, which had not received general acceptance at that time.

Compton continued to develop both his theoretical interpretation and the precision of his experiments, however, and reported the results so convincingly in both oral and written presentations that the scientific world at last accepted his data and his theoretical interpretation of them. The last public "debate" on the validity of the Compton effect was held at the Toronto meeting of the British Association for the Advancement of Science in the summer of 1924. Here interest was so great that the president, Sir William H. Bragg, scheduled a special session to consider the Compton effect. Compton's masterful presentation, question answering, and persuasive discussion at this session won over practically everyone to his quantum interpretation of the phenomenon. Duane, in the spirit of a true scientist, however, returned to his laboratory and personally repeated the X-ray experiments that had been in conflict. He discovered some spurious effects in his earlier work and obtained new results which showed clearly that Compton was entirely correct. Duane corrected his stand at the next meeting of the American Physical Society.

The Compton effect, aptly characterized by Karl K. Darrow as one of the most superbly lucid processes in nature, is now part of the fabric of physics; and it is of interest to recall its influence on the development of the quantum theory during the years 1923–1930.

In the first place, it provided conclusive proof that Einstein's concept of a photon as having both energy and directed momentum was essentially correct. Einstein himself brought considerable attention to Compton's discovery by his discussions at the Berlin seminars. Interest was also high at Göttingen, Munich, Zurich, Copenhagen, and other Continental centers where theoretical physics was rapidly developing.

However, the quantitative proof of the photon character of radiation had been established by Compton's use of a Bragg crystal spectrometer, the function of which depended directly on the wave nature of X rays. Thus a more general synthesis was clearly required, in which both the corpuscular photon and the electromagnetic wave would be included and would continue to play the roles demanded by experiment. Later experiments on the Compton effect also proved conclusively that the basic interaction between quantum and electron obeys the conservation laws in individual events and not just statistically, as the theory of Bohr, Kramers, and Slater had proposed in

an effort to avoid the use of photons. However, the Böthe-Geiger and Compton-Simon experiments soon demonstrated the validity of the conservation laws for individual scattering events and showed that the Bohr-Kramers-Slater statistical view of these laws was untenable.

The final great synthesis of quantum mechanics and quantum electrodynamics was forced upon physics by the crucial experiments of the Compton effect, electron diffraction, space quantization, and the existence of sharp spectral lines, which could not be brought into line with classical theory. It required the final relativistic form of quantum mechanics, developed by Paul Dirac, to give a completely quantitative explanation of Compton scattering in regard to both intensity and state of polarization by the formula derived by O. Klein and Nishina from the Dirac relativistic theory of the electron. Compton and Hagenow had shown that the X rays used to discover the Compton effect are completely polarized in the scattering process, thus ruling out fluorescence as an explanation. However, for very high energy the Dirac theory predicted an unpolarized component in the scattered rays; this was confirmed by Eric Rogers in 1936.

One of the most important developments in quantum theory was Heisenberg's uncertainty principle, which is often described as a direct consequence of the recoil of a Compton electron, produced by the high-energy radiation needed to locate its position accurately. This principle of indeterminacy, however, leads to subtler properties for the photon and its interactions than the "old quantum theory" explanations of Einstein and Compton contained. However, the concept of a fundamental length for the electron, the role of the conservation laws in the interaction, and the relationship to experiment in all modern developments are essentially the same as those given by Compton.

The Compton type of interaction is also of basic importance in the inverse Compton effect, in which a high-speed electron imparts great energy to a photon. This process is of central importance both in astrophysics and in high-energy accelerator physics. In the latter one may cite the production of high-energy monochromatic photon beams by the Compton interaction of laser light with the electron beam of the large Stanford linear accelerator.

In 1923 Compton accepted a position as professor of physics at the University of Chicago, a post formerly held by Robert A. Millikan. There he became a colleague of Albert A. Michelson, who was still active in research and whose great work was an inspiration. Compton spent twenty-two years at Chicago, becoming Charles H. Swift distinguished service professor in 1929. His first researches there were to extend his experiments on the Compton effect. These helped clarify and extend the basic discoveries made at Washington University. For his great series of experiments on the Compton effect and their theoretical interpretation he shared the Nobel Prize in physics in 1927. That same year he was elected to the National Academy of Sciences. He was then thirty-five years old.

Other notable work in the physics of X rays that Compton conducted in the early years at Chicago included the extension of his discovery, made at Washington University, of the total reflection of X rays from noncrystalline materials, such as glass and metals, and the first successful application of a ruled diffraction grating to the production of X-ray spectra.

These first X-ray spectra were produced by Richard L. Doan, who carried out a suggestion of Compton's that such spectra might be obtained from a ruled grating by working within the angle of total reflection. Doan designed for this purpose a special grating that was ruled to his specification on Michelson's ruling engine, and with it he photographed the first X-ray grating spectra in 1925.

This X-ray grating work was later developed with great precision by Compton's students J. A. Bearden and N. S. Gingrich. This work alone, extending X rays as a branch of optics, would have assured Compton's reputation as a distinguished physicist.

One of the more important results of this work was that the X-ray grating wavelengths led to values for the Avogadro number and for the electronic charge; the latter was at variance with Millikan's oil-drop determinations, which up to that time had been considered as standard. This work in Compton's laboratory definitively clarified discrepancies that had long existed in the values of fundamental atomic constants.

Other work at the University of Chicago, especially that of E. O. Wollan, used X-ray scattering to determine the density of diffracting matter in crystals and gases. This work required a deep understanding of the X-ray scattering process in order to separate the coherent and incoherent radiations and apply the related theories to the structure determinations. It was reminiscent of Compton's Ph.D. thesis, in which he had first worked out methods to employ X-ray diffraction from crystals to determine the electron distributions in the lattice. This pioneer work, where Fourier analysis of X-ray data was first employed, was developed to a high art by W. L. Bragg and his school at Manchester University.

Compton was an inspiring teacher and research leader. His contagious enthusiasm, friendliness, and great mental powers made his classes and laboratory meetings memorable experiences for all who were privileged to attend them. He always shared most generously all he learned from his many distinguished visitors and from his own travels with his students and younger colleagues.

In the early 1930's Compton changed his main research interest from X rays to cosmic rays. He was led to this because the interaction of high-energy gamma rays and electrons in cosmic rays is an important example of the Compton effect (as today the inverse Compton effect between high-speed electrons and low-energy photons is of central importance for astrophysics). To obtain the essential data needed to determine the nature of the primary cosmic rays, Compton organized and led expeditions to all parts of the world to measure the cosmic ray intensity over a wide range of geomagnetic latitudes and longitudes and at many elevations above sea level. His reports on these measurements often appeared as "Letters to the Editor" of the *Physical Review*.

The first important result of this worldwide study of cosmic rays was to support and greatly extend Jacob Clay's earlier observations showing that the intensity of cosmic rays depends systematically on geomagnetic latitude and on altitude. It proved that, contrary to the generally accepted view held at that time, at least a significant fraction of the primary cosmic rays are charged particles and thus are subject to the influence of the earth's magnetic field.

More refined results that strongly supported this belief were those which demonstrated an east-west asymmetry of the primary cosmic rays, as had been predicted by the theoretical work of Georges Lemaître and M. S. Vallarta and had been discovered experimentally by Compton's student Luis W. Alvarez and, simultaneously, by T. H. Johnson in observations at Mexico City—in the latitude where the theory had predicted the largest effect.

Precise measurements of cosmic rays on the Pacific Ocean, made over long periods of time by Compton and Turner, revealed an anomalous dependence of cosmic ray intensity on atmospheric temperature and barometric pressure. These gave the first indication, as interpreted by P. M. S. Blackett, of the radioactivity of mesons.

Compton's worldwide cosmic ray studies were sponsored in part by the Carnegie Institution of Washington. One of the leading scientists of that organization, Scott E. Forbush, continued the work, making important discoveries, especially the sudden changes in cosmic ray intensity related to conditions on the sun and its magnetic field.

The development of Compton's cosmic ray program was greatly assisted by the collaboration of several of his former students and of refugee physicists from

Europe, for whom he found a haven in Chicago. Their experiments gave additional early evidence on the nature and lifetime of the mu-meson.

The final summary of Compton's cosmic ray work was made at a conference at the University of Chicago in the summer of 1939 which was attended by the leading workers in cosmic rays throughout the world. Compton inspired and led the conference, which was his last peacetime research work.

In spite of his broad sympathies and worldwide interests, Compton had never felt the need to join the pilgrimages to Göttingen, Copenhagen, or Munich, as had so many physicists of his generation. His year (1934–1935) as Eastman professor at Oxford, however, brought him more closely in touch with European scientists and the problems of a troubled world. He became increasingly involved with human problems, which soon led him to his greatest challenge and leadership responsibility.

World War II brought about a complete change in physics research. This was especially true for Compton, who early became involved in the "uranium problem" that led ultimately to the development of nuclear reactors and the atomic bomb. On 6 November 1941, Compton, as chairman of the National Academy of Sciences Committee on Uranium, presented a report on the military potentialities of atomic energy. This report was a masterpiece in setting forth both the scientific and technological possibilities. It had been prepared in close consultation with Ernest O. Lawrence, who had informed Compton of the discovery of plutonium at the Radiation Laboratory at the University of California in Berkeley. This discovery completely changed the long-range prospects for atomic energy, and the initiation of the vast Manhattan Project in the United States was due primarily to the leadership of Compton and Lawrence.

Compton soon gave up all his other activities to organize and direct the work of the Metallurgical Laboratory of the Manhattan District of the Corps of Engineers at the University of Chicago, which was responsible for the production of plutonium. He centralized all the activities in buildings of the University of Chicago and recruited Enrico Fermi, Walter Zinn, Glenn Seaborg, Richard L. Doan, Eugene Wigner, and a large group of young physicists, chemists, and engineers. He was in charge when the first successful nuclear chain reaction with uranium was accomplished by Enrico Fermi and others on 2 December 1942. This first nuclear reactor was designed by Fermi, but its success was possible only through the great support and encouragement given by Compton, who arranged for Crawford H. Greenewalt, an officer and later president of the Du Pont Company, to be present when the reactor first went critical. Greenewalt's witnessing of this historical event was a decisive factor in the Du Pont Company's continued interest in the uranium project. Thereafter, it agreed to build the reactors at Hanford, Washington, without which the whole project could not have succeeded.

Compton had the major role in establishing the Palos Park Laboratory (which became the Argonne National Laboratory) and the Clinton Engineer Works, Oak Ridge, Tennessee, as well as the plutonium production reactor establishment at Hanford, Washington.

As the war reached its end, Compton decided that he should devote himself to university administration and not begin a wholly new physics research program—a very hard decision for him, as he had never lost his keen interests in the laboratory and all new discoveries in physics. Accordingly, he accepted the position of chancellor of Washington University in St. Louis. This challenging assignment to continue the development of a great university was also for Compton a return to the place where he had made his greatest discovery in physics, some twenty-five years before.

When Compton accepted the post of chancellor of Washington University, he had already been offered several university presidencies. Many of his physics colleagues were greatly surprised that he would now make this change, when he could have continued at the University of Chicago and been the leader of the new Institutes for Nuclear Studies, Metals and Microbiology, which he had helped to create. However, his ties with Washington University were very strong, and the tradition of service was equally strong.

Compton's chancellorship at Washington University resulted in many great advances for that distinguished institution. He improved and greatly strengthened the science departments, including the medical school. His own deep interests in the humanities ensured that he would greatly support these activities. He and his wife were able to develop strong support for the university, both in St. Louis and elsewhere in the nation.

In 1954, when Compton had reached the age at which he felt he must retire from the active administrative leadership of the university, he continued to serve with great distinction through his writing, lecturing, teaching, and public service. Perhaps his most notable publication during this period is the book *Atomic Quest,* which gives a complete, clear, and generous account of the activities of all his colleagues in the Manhattan Project during the war. This book and the more technically detailed Smyth Report constitute the basic historical documents on this epoch-making activity.

Although his health was not especially good in the later years, Compton continued to lecture and travel to the very end and accepted the invitation of the University of California to give a series of lectures. He had presented only two when he died.

Arthur Holly Compton will always be remembered as one of the world's great physicists. His discovery of the Compton effect, so vital in the development of quantum physics, has ensured him a secure place among the great scientists.

BIBLIOGRAPHY

I. ORIGINAL WORKS. Compton's books include *X-Rays and Electrons* (Princeton, N.J., 1926); *X-Rays in Theory and Experiment* (Princeton, N.J., 1935), written with S. K. Allison; *Atomic Quest* (New York, 1956); and *The Cosmos of Arthur Holly Compton,* Marjorie Johnston, ed. (New York, 1967), his public papers.

His principal articles in professional journals are "Latitude and Length of the Day," in *Popular Astronomy,* **23** (1915), 199–207; "The Intensity of X-ray Reflection and the Distribution of the Electrons in Atoms," in *Physical Review,* **9** (1917), 29–57, his Ph.D. thesis; "The Softening of Secondary X-Rays," in *Nature,* **108** (1921), 366–367; "Secondary Radiations Produced by X-Rays," in *Bulletin of the National Research Council,* no. 20 (1922), 16–72; "A Quantum Theory of the Scattering of X-Rays by Light Elements," in *Physical Review,* **21** (1923), 483–502; "The Spectrum of Scattered X-Rays," *ibid.,* **22** (1923), 409–413; "The Total Reflection of X-Rays," in *Philosophical Magazine,* **45** (1923), 1121–1131; "Polarization of Secondary X-Rays," in *Journal of the Optical Society of America,* **8** (1924), 487–491, written with C. F. Hagenow; "X-ray Spectra From a Ruled Reflection Grating," in *Proceedings of the National Academy of Sciences of the United States of America,* **11** (1925), 598–601, written with R. L. Doan; "Determination of Electron Distributions From Measurements of Scattered X-Rays," in *Physical Review,* **35** (1930), 925–938; "A Geographic Study of Cosmic Rays," *ibid.,* **43** (1933), 387–403; "A Positively Charged Component of Cosmic Rays," *ibid.,* 835–836, written with L. W. Alvarez; "Cosmic Rays on the Pacific Ocean," *ibid.,* **52** (1937), 799–814, written with R. N. Turner; "Chicago Cosmic Ray Symposium," in *Science Monthly,* **49** (1939), 280–284; and "The Scattering of X-Rays as Particles," in *American Journal of Physics,* **29** (1961), 817–820.

II. SECONDARY LITERATURE. Biographical articles are S. K. Allison, "Arthur Holly Compton, Research Physicist," in *Science,* **138** (1962), 794–799; and "Arthur Holly Compton, 1892–1962," in *Biographical Memoirs. National Academy of Sciences,* **38** (1965), 81–110; and Zay Jeffries, "Arthur Holly Compton," in *Yearbook of the American Philosophical Society* (1962), pp. 122–126.

ROBERT S. SHANKLAND

COMPTON, KARL TAYLOR (*b.* Wooster, Ohio, 14 September 1887; *d.* New York, N.Y., 22 June 1954), *physics.*

Like his brothers, Arthur Holly Compton and Wilson Martindale Compton, Karl Taylor Compton owed much of his later achievements in American science and education to the stimulating intellectual environment of his youth. His father, Elias Compton, was an ordained Presbyterian minister and successively professor of philosophy, dean, and acting president of the College of Wooster. His mother, Otelia Augspurger Compton, was a Mennonite with strong pacifist convictions. Although the American Protestant ethic was a central element in the Compton home, Karl and Arthur were encouraged by their father's broad intellectual outlook to pursue their early interest in science.

Endowed with good health and an excellent physique, Compton excelled in athletics during his high school and college days. Although he had read many of the scientific books in his father's library, his serious interest in the subject was first revealed in 1908, when he was an assistant in the college science laboratory. Attracted by a new X-ray machine there, Compton used it as the basis for a paper that became his master's thesis in 1909 and was the first paper from the College of Wooster to be published in the *Physical Review* (1910).

In further graduate studies at Princeton University, Compton began research on the thermal emission of electrons, under O. W. Richardson, then a leading investigator in that field. His doctoral dissertation in 1912 contributed to Richardson's studies of the emission of photoelectrons from metals. The same year he assisted Richardson in preparing for the *Physical Review* a paper providing experimental evidence for the validity of Einstein's quantum theory of the photoelectric effect. Compton published three more papers on electron physics during his three years as instructor in physics at Reed College in Portland, Oregon (1912–1915).

Compton returned to Princeton as an assistant professor of physics under William F. Magie in 1915. He expanded his earlier research into a general study of electron collisions in ionized gases. When the United States entered World War I in 1917, Augustus Trowbridge recruited Compton to assist in developing sound and flash ranging devices for locating enemy artillery positions, a project organized under the National Research Council. This experience made Compton aware of the importance of scientific research in modern warfare and marked the beginning of an interest that was to dominate his later career.

After the war Compton continued his teaching and

research at Princeton, first as professor and then as chairman of the physics department. From 1918 to 1930 he published almost 100 scientific papers on various aspects of electron physics, including the ionization of gas molecules by electron impacts, the chemical and spectrographic properties of excited atoms, oscillations in low-voltage arcs, and the dissociation of gases by excited atoms. All of these fields were a part of the rapid evolution of atomic physics during the 1920's and 1930's.

Compton's accomplishments in building a graduate department in physics at Princeton led to his appointment as president of the Massachusetts Institute of Technology in 1930. Within a decade, what was essentially an undergraduate engineering college became under his leadership one of the world's leading centers of graduate education in the physical sciences.

With his appointment as chairman of the Science Advisory Board by President Roosevelt in 1933, Compton began to devote an ever-increasing portion of his time to the application of science and technology within the federal government. As a member of the National Defense Research Committee in 1940, he helped to mobilize the nation's scientific resources for World War II. Within NDRC he served as director of Division 14, which developed many of the electronic devices, especially radar, that revolutionized military technology during the war.

In the postwar years Compton continued to have a strong influence on science policy within the federal government. He resigned as president of MIT in 1948 to succeed Vannevar Bush as chairman of the Research and Development Board in the National Military Establishment. From 1948 until his death he served as president of the MIT corporation, as a member of numerous government committees and advisory boards, on the boards of several corporations, and as trustee of educational and research institutions. A member of the National Academy of Sciences, he also served as president of the American Physical Society (1927–1929) and of the American Association for the Advancement of Science (1935–1936).

BIBLIOGRAPHY

I. ORIGINAL WORKS. Compton wrote about 400 articles, of which about half were scientific papers. The others include his writings as president of MIT and as an authority on science and public affairs. For a complete list of his writings see Eleanor R. Bartlett, "The Writings of Karl Taylor Compton," in *Technology Review,* **57** (Dec. 1954), 89–92 ff. For Compton's general interpretation of American scientific research during the 1930's and 1940's, see "The Electron: Its Intellectual and Social Significance," in *Annual Report of the Board of Regents of the Smithsonian Institution—1937* (Washington, D.C., 1938), pp. 205–223; and "The State of Science," in *Annual Report of the Board of Regents of the Smithsonian Institution—1949* (Washington, D.C., 1950), pp. 395–410.

II. SECONDARY LITERATURE. There is as yet no biography of Compton, but the following are useful: James R. Blackwood, *The House on College Avenue: The Comptons at Wooster, 1891–1913* (Cambridge, Mass., 1968); Marjorie Johnston, ed., *The Cosmos of Arthur Holly Compton* (New York, 1967); J. E. Pfieffer, "Top Man in Science," in *New York Times Magazine* (10 Oct. 1948), 10 ff.; and Louis A. Turner, "Karl T. Compton: An Appreciation," in *Bulletin of the Atomic Scientists,* **10** (Sept. 1954), 296.

RICHARD G. HEWLETT

COMRIE, LESLIE JOHN (*b.* Pukekohe, New Zealand, 15 August 1893; *d.* London, England, 11 December 1950), *astronomy, computation.*

A second-generation New Zealander, Comrie had a distinguished academic career in Pukekohe and Auckland, where he specialized in chemistry. After serving with the New Zealand Expeditionary Force in France, where he lost a leg, he revived an early interest in astronomy and entered St. John's College, Cambridge, as a research student. There, and later, he successfully applied new computing techniques to the problems of spherical and positional astronomy. His main contribution was made in the years 1925–1936, when, after teaching astronomy for three years in the United States, at Swarthmore College and Northwestern University, he was deputy and, in 1930, superintendent of the *Nautical Almanac.* He completely revised the *Almanac,* which had been almost unchanged since 1834, and introduced into astronomy the concept of the standard equinox.

After a disagreement with the Admiralty, Comrie left astronomy to found the Scientific Computing Service. There he was free to develop further the techniques of and machines for computation and the calculation and presentation of mathematical tables. He laid a solid foundation for the computational revolution that was to follow the introduction of the electronic computer: he showed how to "program" commercial machines for scientific computation; developed impeccable interpolation techniques; produced mathematical tables of the highest standards of accuracy and presentation; and, in effect, created computational science. For this work he was elected fellow of the Royal Society in 1950, only a few months before his death.

Comrie, with no claim to be a mathematician, had the clarity of mind, tenacity of purpose, scientific courage, and immense energy that enabled him, by using essentially simple and direct methods, to obtain practical solutions to many problems that defied theoretical analysis. But he was inclined to impatience with those who did not share his devotion to perfectionism, and this led to some difficult personal relationships.

BIBLIOGRAPHY

I. Original Works. Among Comrie's writings are "The Use of a Standard Equinox in Astronomy," in *Monthly Notices of the Royal Astronomical Society,* **86** (1926), 618–631; "Explanation," in *The Nautical Almanac and Astronomical Ephemeris* for 1931 (London, 1929); "Interpolation and Allied Tables," in *The Nautical Almanac and Astronomical Ephemeris* for 1937 (London, 1936); *Barlow's Tables of Squares, etc., of Integral Numbers up to 12,500* (London, 1947); and *Chambers's Six-figure Mathematical Tables* (London, 1950).

II. Secondary Literature. Obituaries are W. M. H. Greaves, in *Monthly Notices of the Royal Astronomical Society,* **113** (1953), 294–304; and H. S. W. Massey, in *Obituary Notices of Fellows of the Royal Society,* **8** (1952), 97–107.

D. H. Sadler

COMSTOCK, GEORGE CARY (*b.* Madison, Wisconsin, 12 February 1855; *d.* Madison, 11 May 1934), *astronomy.*

Comstock was the eldest son of Charles Henry and Mercy Bronson Comstock, an itinerant Midwestern family. His education, not atypical for his time, gave him a working knowledge of several fields of endeavor. While a student at the University of Michigan (1873–1877) Comstock was encouraged by James C. Watson to take his course in astronomy. To support his college career Comstock worked each summer, and for a year following his graduation, as recorder and assistant engineer on the U.S. Lake Survey; this practical fieldwork was later put to use in his teaching and in his textbooks. Some time later, to guard against the possibility that astronomy might not provide an adequate living, Comstock studied law, graduating from the University of Wisconsin law school (1883) and being admitted to the bar.

Astronomy, however, did become profitable. In 1879 Comstock moved to Madison, as assistant to Watson, who had recently been appointed the first director of the new Washburn Observatory of the University of Wisconsin; and, except for a brief stint teaching mathematics at Ohio State University (1885–1886) and a summer at the Lick Observatory (1886), Comstock spent the remainder of his scientific career at Wisconsin. He was elected to membership in the National Academy of Sciences (1899), the American Academy of Arts and Sciences, and the Astronomische Gesellschaft; for ten years he was first secretary of the American Astronomical Society and, in 1894, vice-president of the American Association for the Advancement of Science.

By all accounts Comstock was known as an excellent teacher and administrator. His varied teaching duties were directed toward undergraduate and graduate astronomy students, all civil engineering students, and, during World War I, prospective mariners needing knowledge of navigation techniques. In connection with teaching Comstock wrote four textbooks: *An Elementary Treatise Upon the Method of Least Squares* (1890), giving a working rather than an analytical knowledge of the subject; *A Text-book of Astronomy* (1901), for high school students; *A Text-book of Field Astronomy for Engineers* (1902); and *The Sumner Line* (1919). At the University of Wisconsin, Comstock held numerous administrative posts, including chairman, director, and dean of the graduate school.

Comstock's two major scientific researches were primarily visual observational, in the area of precise position measurements; he pursued both problems assiduously for several decades. The first involved a redetermination of the constants of aberration and of atmospheric refraction. Although he did not derive new values for these constants, Comstock was able to verify, or support, Struve's value for aberration and the Pulkovo tables for refraction. He then simplified the formula for atmospheric refraction, producing a formula useful for general work if not for observations of the highest possible precision. Comstock's second investigation, concerning double stars, demonstrated the existence of relatively large proper motions for apparently faint (twelfth magnitude) stars—a question then much in doubt. He finally concluded, as is now well accepted, that there are a large number of intrinsically faint stars in the neighborhood of the sun.

BIBLIOGRAPHY

Comstock's textbooks are *An Elementary Treatise Upon the Method of Least Squares, With Numerical Examples of Its Applications* (Boston, 1890); *A Text-book of Astronomy* (New York, 1901); *A Text-book of Field Astronomy for Engineers* (New York, 1902; 2nd ed., rev. and enl., New York, 1908); and *The Sumner Line; or, Line of Position as an Aid to Navigation* (New York, 1919).

A bibliography of Comstock's writings is in Joel Stebbins, "Biographical Memoir of George Cary Comstock, 1855–1934," in *National Academy of Sciences. Biographical Memoirs,* **20** (1939), 161–182.

Deborah Jean Warner

COMTE, ISIDORE AUGUSTE MARIE FRAN-ÇOIS XAVIER (*b.* Montpellier, France, 19 January 1798; *d.* Paris, France, 5 September 1857), *philosophy, sociology, mathematics.*

The eldest of the three children born to Louis Auguste Xavier Comte and Rosalie Boyer, Comte came from a Catholic and royalist family, his father being a civil servant of reasonable means. An exceptional and rebellious youth, Comte at an early age repudiated the Catholicism of his parents and took up the republican cause in politics. He entered the École Polytechnique in 1814, took part in the disturbances connected with the defense of Paris, and was one of many "subversive" students expelled in the royalist reorganization of the school in 1816. From 1817 to 1823 he was private secretary to Claude Henri de Rouvroy, comte de Saint-Simon, an intellectual association that was profoundly to affect Comte's later development. In 1825 he married a prostitute named Caroline Massin, a most unhappy union that was dissolved in 1842.

Comte's economic position was always an unstable one; he never acquired a university post and survived largely on money earned from the public lectures he gave in Paris, from school examiner's fees, and from the benevolence of admirers (such as Mill and Grote), who were periodically called on to subsidize Comte's researches. In 1830 he founded the Association Polytechnique, a group devoted to education of the working classes. In the early 1840's, he met Mme. Clothilde de Vaux, an intellectual and emotional experience which—even more profoundly than his earlier association with Saint-Simon—was to change his intellectual orientation. In 1848 he founded the Société Positiviste, devoted to the promulgation of the "Cult of Humanity." The last years of his life were spent in developing a godless religion, with all the institutional trappings of the Catholicism that he had repudiated as an adolescent. Abandoned by most of his friends and disciples (usually because of his abuse of them), Comte died in relative poverty and isolation.

Comte's writings exhibit a remarkable scope and breadth, ranging from mathematics to the philosophy of science, from religion and morality to sociology and political economy. What unifies them all is Comte's concern with the problem of knowledge, its nature, its structure, and the method of its acquisition. Positivism, the official name Comte adopted for his philosophy, was primarily a methodological and epistemological doctrine. Traditionally, writers on the theory of knowledge had adopted a psychologistic approach in which the nature and limitations of the human mind and the senses were examined and knowledge treated as a function of certain mental states. Comte's approach to this locus of problems was substantially different. Believing that knowledge could be understood only by examining the growth of knowledge in its historical dimension, he insisted that it is the collective history of thought, rather than the individual psyche, that can illuminate the conditions and limits of human knowledge. It was not knowledge in its static dimension which interested Comte, but the dynamics of man *qua* knower, the progressive development of knowledge. In general outline, this approach was inspired (as Comte acknowledges) by Condorcet and Saint-Simon. What Comte added to this tradition was a firm commitment to studying the history of scientific knowledge, since science was for him (as it was for Whewell) the prototypical instance of knowledge.

The most famous result of this approach is Comte's law of three states. The importance of this law to Comte's theory is crucial, for not only does it provide him with a solution to the problem of the growth of knowledge, but it also serves as an example of the fruitfulness of applying scientific methods to the study of human development. In Comte's eyes the law of three states is as valid—and on the same footing—as the laws of the inorganic world.

Basically, the law (first formulated in 1822) states that human thought, in its historical development, passes successively through three distinct phases: the theological (or fictional) state, the metaphysical (or abstract) state, and the positive (or scientific) state. In the theological state, man explains the world around him in anthropomorphic terms, reducing natural processes to the whims of manlike gods and agencies. Final causes are especially symptomatic of this stage. In the metaphysical state, deities are replaced by powers, potencies, forces, and other imperceptible causal agencies. The positive state repudiates both causal forces and gods and restricts itself to expressing precise, verifiable correlations between observable phenomena. While Comte believes that the theological and metaphysical states are based on a misconception of natural processes, he insists that they were essential preliminaries to the emergence of positive knowledge. Thus, the theological state is a natural one for a civilization which has neither the mathematical nor the experimental techniques for investigating nature, and its importance is that it provides a pattern, however crude, for introducing some element of order into an otherwise capricious world. The metaphysical state, which is purely transitional, contains positive elements which it clothes in the language of powers and forces so as not to offend the sensibilities of theologically inclined minds.

It is not only knowledge in general but every branch of knowledge which evolves through these three states. Different forms of knowledge evolve at different rates, however, and one of Comte's major critical tasks was to assess the degree of progress toward the positive stage in each individual science, a task that occupies most of the six-volume *Cours de philosophie positive* (1830–1842).

This task of assessment led immediately to Comte's hierarchy of sciences. No mere taxonomical exercise, his classification of the sciences is meant to reflect several important characteristics. While most other schemes (for example, those of Aristotle, Bacon, and Ampère) had classified the sciences with respect to their generality or relations of logical inclusion and reduction, Comte arranged the sciences in the hierarchy according to the degree to which they have attained the positive state. On this ranking, the sciences (in order) are (1) mathematics, (2) astronomy, (3) physics, (4) chemistry, (5) biology, and (6) sociology (or "social physics"). Of these, Comte believed that only mathematics and astronomy had reached full positive maturity, while metaphysical and theological modes of thought were still prominent in the others.

Although Comte's classification is based on the "degree of positivity" of the various sciences, it also captures other important characteristics. Neglecting mathematics, the sequence from astronomy to sociology represents an increasing complexity in the phenomena under investigation. Thus, the astronomer is concerned only with motions and positions, the physicist needs forces and charges as well, while the chemist also deals with configurations and structures. The Comtean hierarchy of the sciences reflects moreover important methodological characteristics of each science. Astronomy has only the method of observation, physics can both observe and experiment, and the biologist employs comparison and analogy as well as observation and experiment. It was necessary for Comte to establish the fact that different sciences utilize different methods, since his conception of sociology required a unique method (the "historical method") which none of the other sciences exemplify. Comte was manifestly not a reductionist in the sense of using a classificatory scheme to render one science logically subordinate to another. His scientific beliefs, as well as his insistence on the diversity of methods, made him an outspoken critic of reductionism. The unity of the sciences was not, for Comte, to be found in the identity of concepts, but rather in the positive mentality which he hoped would unite the sciences.

Comte's major impact on his contemporaries was methodological. The *Cours de philosophie positive* is simultaneously a methodological manifesto and an incisive critique of the science of the early nineteenth century. Comte was convinced that a careful analysis of the logic of science would lead him to far-reaching insights into the character of positive (that is, scientific) method. In practice, however, most of his methodological strictures derive from the doctrine of the three states rather than from an objective study of scientific procedures.

Comte claimed that discussion of scientific method had too long been dominated by the naïve division between what he called empiricists and mystics—the former purporting to derive all scientific concepts from experience and the latter from a priori intuition of the mind. He wanted a middle course which recognized the active, acquisitive role of the mind but which at the same time put rigid empirical checks on the conjectures that the mind produces. His approach to this problem was singularly perceptive.

Comte's quarrel with the empiricists had two aspects. By requiring that the scientist must purge his mind of all preconceived ideas and theories in order to study nature objectively, the empiricists demanded the impossible. Every experiment, every observation has as its precondition a hypothesis in the mind of some experimenter. Without theories, Comte insisted, scientific experiments would be impossible. He also urged that the empiricists misunderstood the place of experience in the scientific scheme. The function of experiments is not to generate theories but to test them. It is by subjecting theories to the scrutiny of empirical verification that they are established as scientific.

The stress on verification is a persistent theme in Comte's writings. It is a fundamental principle of the positive philosophy that any idea, concept, or theory that has any meaning must be open to experimental verification. Verification has been the vehicle whereby progress has been made from the metaphysical to the positive state. The forces, powers, and entelechies of the metaphysical epoch were repudiated precisely because they were finally recognized to be unverifiable. Comte's repeated criticisms of many of the physical theories of his own day (for instance, fluid theories of heat and electricity) were grounded largely in his requirement of verifiability. Although Comte was not the earliest writer to stress empirical verification, there is no doubt that it was largely through his influence (especially on such figures as Claude Bernard, J. S. Mill, Pierre Duhem, and C. S. Peirce) that the doctrine of verifiability enjoys the wide currency it has had in recent philosophy and science.

Closely connected with the requirement of verification was another important and influential dogma of Comte's methodology: the unambiguous assertion that the "aim of science is prediction." Since genuinely positive science offers only correlations between phenomena rather than their causes (in the Aristotelian sense of efficient causes), a theory is positively valid only insofar as it permits the scientist to reason from known phenomena to unknown ones. The sole object of the theoretical superstructure of science is to put the scientist in a position to predict what will happen, to substitute ratiocination for direct experimental exploration. The ideal science, for Comte, is one which, given certain empirically determined initial conditions, can deduce all subsequent states of the system. Clearly, it is the science of Laplace and Lagrange rather than that of Buffon or Fourcroy upon which Comte modeled his theory of science.

On other questions of scientific method and the philosophy of science, Comte's views were more traditional. He insisted on the invariability of physical law, argued that scientific knowledge was relative rather than absolute, and believed that scientific laws were approximate rather than precise (a point Duhem was to develop seventy years later). His treatment of the problem of induction, particularly in his *Discours sur l'esprit positif* (1844), is taken largely from Mill's *System of Logic* (1843).

Insofar as Comte identified himself as a natural scientist, it was mathematics which he knew best. Having been a tutor in mathematics for the École Polytechnique in the 1830's, Comte published two straightforward scientific works, the *Traité élémentaire de géométrie* (1843) and the *Traité philosophique d'astronomie populaire* (1844). Both were popular works that grew out of his public lectures in Paris.

Of considerably greater significance was Comte's examination and critique of scientific theories in his *Cours de philosophie positive.* Having laid a solid methodological foundation in the early parts of the *Cours,* Comte devoted the second and third volumes of that work to a scrutiny of the "inorganic" sciences.

Among all the empirical sciences astronomy was closest to the positivist ideal. Concerning themselves exclusively with the position, shape, size, and motion of celestial bodies, astronomers (more from necessity than choice) had restricted themselves to studying the observable properties of the heavens. Astronomy had achieved this positive state because it was not concerned with speculation on the internal constitution of the stars, their elemental composition, or their genesis. (No doubt Comte would have viewed the rise of spectral analysis of stellar objects as a retrograde development, representing the incursion of a metaphysical spirit into an otherwise positive science.) Comte went so far as to assert that astronomy should limit its domain to the solar system—which lends itself to precise mathematical analysis—and should forgo any attempt at the construction of a sidereal astronomy.

If Comte was generally satisfied with the state of contemporary astronomy, his attitude to the physical theories of his day was generally antagonistic. He saw lurking in every elastic fluid and subtle medium a vestige of the metaphysical state. In electricity, heat, light, and magnetism, natural philosophers were attempting to explain observable phenomena by resorting to unobservable, unverifiable, and unintelligible entities. To Comte such entities were not only inconceivable but unnecessary. Since the laws of phenomena are the object of science, only those laws are necessary and causal theories may be dispensed with as Fourier's treatment of heat had demonstrated. Phenomenal laws are valid independently of the theories from which they might be derived. In chemistry, on the other hand, Comte was willing to allow—even to insist on—the use of the atomic hypothesis, which, in many respects, seems as nonphenomenal as an optical ether. The difference between the two is that the properties which the atomic theory attributes to atoms are well-defined and coherent, while the properties attributed to the ether (for example, imponderability) are both inconceivable and unverifiable.

Comte believed chemistry to be in a state of "gross imperfection" with chemists having no clear sense of the aims or the limitations of their science. He believed that organic chemistry was a branch of biology rather than chemistry and that much of what we should now call physical chemistry was in fact physics. The sole function of chemistry, in his eyes, was to study the laws governing the combinations of the various elemental bodies. He likened chemists to the empiricists who were so interested in haphazardly synthesizing new compounds that they completely ignored the rational and theoretical side of the discipline.

Consistent with his classification of the sciences, Comte believed that biological processes (or at least a subset of them known as vital processes) could not be explained by means of physicochemical concepts. Biology, properly conceived, would integrate physiology and anatomy by relating structure to function. Moreover, a legitimate biological theory must study the connection between the organism and its environment, its milieu. He stated the basic problem of biology in the formula, "Given the organ, or the organic modification, find the function or the action, and vice versa."

Methodologically, Comte places biology on a very different footing from the higher sciences. One major difference concerns the amenability of biological phenomena to mathematical treatment. Comte says that life processes are generally too complex to treat quantitatively. This is compensated, however, in that biology has more methods at its disposal than astronomy, physics, or chemistry. Specifically, the biologist can utilize the method of comparison which, in order to understand the life processes in a given organism, successively compares that organism with similar ones of less complexity. If the function of the lungs in man is difficult to ascertain, their function can perhaps be determined by studying the function of lungs or lunglike organs in simpler species. For the method of comparison to have any validity, it is necessary to have a sequence of organisms whose differences from one to the next are very minor. Clearly, an accurate biological taxonomy is crucial to the utilization of this method. Like Cuvier, Comte maintained the fixity of the species, although he confessed that this doctrine was not fully established.

Comte found biology, like physics and chemistry, to be still dominated by the metaphysical mode of thought. Theories of spontaneous generation, mechanistic physiology, materialism, and spiritualism are all manifestations of a prepositive mentality.

The last element in the chain of the sciences is "social physics" or, as he called it after 1840, sociology. Indeed, Comte is often considered the founder of sociology, and his treatment of this topic is his most original and probably his most influential. In his view previous thought about man's social nature had been speculative and a priori, rather than cautious and empirical. Certain moral perspectives and prejudices had stood in the way of an objective philosophy of history, and the subject had been dominated by crude theological and metaphysical perspectives. While social statics had been treated by such writers as Aristotle (especially in the *Politics*) and Montesquieu, social dynamics had been almost completely ignored. The birth, growth, and general life-cycle of a social ensemble were what Comte made the subject matter of sociology. In part, of course, the characteristics of a society (for instance, its family structure, politics, institutions, and so forth) are a function of the biological and physiological characteristics of the men who compose the society. To this extent, sociology is dependent on biology.

But there is another important dimension of sociology that has nothing to do with man's biology—the historical component. What distinguishes the social entity from the physical and biological is that it is uniquely a product of its own past. The structures and institutions of any society—intellectual, political, and economic—are determined by the previous con-

ditions of the society. History thus becomes the heart of sociology, and the aim of the sociologist becomes that of determining the laws of human social progress by an empirical study of the evolution of human institutions. To understand the present and to predict the future, we must know the past. The sociologist seeks to find predictive laws by working in two directions simultaneously—he studies history in order to discover empirical generalizations about social change; while at the same time he attempts to explain these generalizations by deducing them from known laws of human nature, whether biological or psychological. This in essence is the famous "historical method" that Comte advocates in the fourth volume of his *Cours de philosophie positive*.

The basic social unit was, in Comte's view, the family, for the family is the main source of social cohesion. But the basic concept of sociological analysis was that of progress. Comte's theory of progress, while dependent upon the Enlightenment theory of progress, was nonetheless very different from it. He criticized Condorcet for thinking that man was infinitely perfectible and that continuous progress could occur if man would simply decide that he wanted it. Comte, on the contrary, maintained that progress is governed by strict laws, which are inviolate. The churchmen of the middle ages could not have discovered Newtonian astronomy even if they had set their minds to it, for the general social, moral, and intellectual conditions of the Latin West were not capable of embracing such a positive theory. Again, the law of the three states functions as the basic determinant of social change. The primary cause of social change is neither political nor economic, but intellectual. Sophisticated and complex economic and political institutions are possible only when man's intellectual progress has reached a certain level of maturity. To this extent, the sociology of knowledge is the cornerstone of sociological theory.

For most of the last twenty years of Comte's life, he was preoccupied with the problem of formulating the tenets of a "positive religion." Convinced that Christianity was doctrinally bankrupt, he felt nonetheless that formal, organized religion served a vital social and even intellectual function. He believed that egoism must be subordinated to altruism and maintained that this could be achieved only by a "religion of humanity." Such a religion, founded essentially on a utilitarian ethic, dispenses entirely with a deity, substituting mankind in its place. Otherwise, the trappings of traditional religion remain more or less intact. Churches are formed, a priesthood is trained, and sacraments, prayer, and even the saints are preserved, although in a very different guise.

Comte's fanaticism in this matter was a cause of profound dismay to many scientists and men of learning who had been greatly influenced by him in the 1830's and 1840's. It also made it easier for Comte's critics to discredit his earlier ideas by *ad hominem* arguments against the cult of humanity. In spite of his growing estrangement from the intellectual community, however, Comte persisted in his religious speculations, and established more than a hundred positivist congregations in Europe and North America.

The question of Comte's place in history, both as regards the influences on him and his influence on others, is still largely a matter of undocumented conjecture. Certain influences on Comte are virtually undeniable. Comte himself admitted to having learned much from the *philosophes* (especially Condorcet) and from the physiologist Barthez. Of equal authenticity was the role Saint-Simon played in directing Comte's attention to the problem of intellectual progress and its relevance to a philosophy of science. At a less explicit, but probably more pervasive, level of influence was the French tradition of analytic physics. The physics of Laplace and Lagrange, of Fourier and Ampère, was thoroughly positivistic in spirit, with its emphasis on quantitative correlations of phenomena (at the expense of abandoning microreductive theories). Comte himself often suggested that his mission was to extend the methods of mathematical astronomy and physics to the other sciences, especially social physics.

Comte's influence on his contemporaries and successors is a more complicated problem. In his own time, he had numerous disciples, including the literary figures Harriet Martineau, John Stuart Mill, G. H. Lewes, and É. Littré, as well as such scientists as Dumas, Audiffrent, and Claude Bernard, who were sympathetic to Comte's analysis. At the end of the nineteenth century, Comtean positivism became a powerful force in the philosophical critique of the sciences, represented by such figures as J. B. Stallo, Ernst Mach, Pierre Duhem, and A. Cournot. In certain respects, the twentieth-century movement known as logical positivism was a continuation of the philosophical tradition Comte founded.

BIBLIOGRAPHY

1. ORIGINAL WORKS. There is no collected edition of Comte's writings, nor even a full bibliographical list. Among Comte's more important works are *Appel aux conservateurs* (Paris, 1883; English trans. London, 1889); *Calendrier positiviste* (Paris, 1849; English trans. London, 1894); *Catéchisme positiviste* (Paris, 1852; English trans.

London, 1858); *Cours de philosophie positive,* 6 vols. (Paris, 1830–1842; partial English trans. in 2 vols., London, 1853); *Discours sur l'ensemble du positivisme* (Paris, 1848; English trans. London, 1865); *Discours sur l'esprit positif* (Paris, 1844; English trans. London, 1903); *Essais de philosophie mathématique* (Paris, 1878); *Opuscules de philosophie sociale 1819–1828* (Paris, 1883); *Ordre et progrès* (Paris, 1848); *The Philosophy of Mathematics,* W. M. Gillespie, trans. (New York, 1851); *Synthèse subjective* (Paris, 1856; English trans. London, 1891); *Système de politique positive* (Paris, 1824); *Système de politique positive, ou Traité de sociologie,* 4 vols. (Paris, 1851–1854; English trans. London, 1875–1877); *Testament d'Auguste Comte* (Paris, 1884; English trans. Liverpool, 1910); *Traité élémentaire de géométrie analytique* (Paris, 1843); and *Traité philosophique d'astronomie populaire* (Paris, 1844).

Comte was a prolific correspondent and many of his letters have been preserved and published. The most important editions of Comte's correspondence are *Correspondance inédite d'Auguste Comte* (4 vols., Paris, 1903); *Lettres à des positivistes anglais* (London, 1889); *Lettres d'Auguste Comte . . . à Henry Edger et à M. John Metcalf* (Paris, 1889); *Lettres d'Auguste Comte . . . à Richard Congreve* (London, 1889); *Lettres d'Auguste Comte à Henry Dix Hutton* (Dublin, 1890); *Lettres d'Auguste Comte à John Stuart Mill, 1841–1846* (Paris, 1877); *Lettres d'Auguste Comte à M. Valat . . . 1815–1844* (Paris, 1870); *Lettres inédites à C. de Blignières* (Paris, 1932); and *Nouvelles lettres inédites* (Paris, 1939). A chronological list of almost all Comte's correspondence has been compiled by Paul Carneiro and published as a supplement to the *Nouvelles lettres.*

II. SECONDARY LITERATURE. The most important studies of Comte's biography are Henri Gouhier, *La vie d'Auguste Comte* (Paris, 1931) and *La jeunesse d'Auguste Comte et la formation du positivisme,* 3 vols. (Paris, 1933–1941), of which vol. III contains a list of Comte's writings before 1830, 421 ff. Other biographical works include H. Gruber, *Auguste Comte . . . sein Leben und seine Lehre* (Freiburg, 1889); C. Hillemand, *La vie et l'oeuvre d'Auguste Comte* (Paris, 1898); H. Hutton, *Comte's Life and Work* (London, 1892); F. W. Ostwald, *Auguste Comte: Der Mann und sein Werk* (Leipzig, 1914); and B. A. A. L. Seilliera, *Auguste Comte* (Paris, 1924).

The most valuable general works on Comte's philosophy are Jean Delvolvé, *Réflexions sur la pensée comtienne* (Paris, 1932); Pierre Ducassé, *Essai sur les origines intuitives du positivisme* (Paris, 1939); P. Ducassé, *Méthode et intuition chez Auguste Comte* (Paris, 1939); L. Lévy-Bruhl, *The Philosophy of Auguste Comte* (New York, 1903); É. Littré, *Auguste Comte et la philosophie positive* (Paris, 1863); and J. S. Mill, *Auguste Comte and Positivism* (London, 1865).

Specialized studies of various aspects of Comte's works include J. B. G. Audiffrent, *Appel aux médicins* (Paris, 1862); E. Caird, *The Social Philosophy and Religion of Comte* (Glasgow, 1893); G. Dumas, *Psychologie de deux messies positivistes: Saint-Simon et Auguste Comte* (Paris, 1905); F. von Hayek, *The Counter-Revolution of Science* (London, 1964); Jean Lacroix, *La sociologie d'Auguste*

Comte (Paris, 1956); L. Laudan, "Towards a Reassessment of Comte's 'Méthode Positive',," in *Philosophy of Science,* **37** (1970); G. H. Lewes, *Comte's Philosophy of the Sciences* (London, 1853); F. S. Marvin, *Comte: the Founder of Sociology* (New York, 1937); George Sarton, "Auguste Comte, Historian of Science," in *Osiris,* **10** (1952); and Paul Tannery, "Comte et l'Histoire des Sciences," in *Revue générale des sciences,* **16** (1905).

Comte's influence on later science and philosophy has been studied by D. Charlton, *Positivist Thought in France During the Second Empire, 1852–1870* (Oxford, 1959); L. E. Denis, *L'oeuvre d'A. Comte, son influence sur la pensée contemporaine* (Paris, 1901); R. L. Hawkins, *August Comte and the United States, 1816–1853* (Cambridge, Mass., 1936); *Positivism in the United States, 1853–1861* (Cambridge, Mass., 1938); and R. E. Schneider, *Positivism in the United States* (Rosario, Argentina, 1946).

As an intellectual movement, positivism generated numerous journals and periodicals, many of which contain lengthy discussions of various aspects of Comte's work. Chief among these are *Philosophie positive* (Paris, 1867–1883), *El Positivismo* (Buenos Aires, 1876–1877 and 1925–1938), *The Positivist Review* (London, 1893–1923), *Revue occidentale* (Paris, 1878–1914), and *Revue positiviste internationale* (Paris, 1906–1940). Most of Comte's still unpublished manuscripts are kept in the library of the Archives Positivistes in Paris.

LAURENS LAUDAN

CONDILLAC, ÉTIENNE BONNOT, ABBÉ DE (*b.* Grenoble, France, 30 September 1714; *d.* Beaugency, France, 3 August 1780), *philosophy, psychology.*

Condillac's philosophy of science as language occupies a midpoint in the evolution of scientific epistemology between the empiricism of Locke and the positivism of Comte. He was the third son of Gabriel de Bonnot, vicomte de Mably, magistrate and member of the *noblesse de la robe* in Dauphiny. The second son also took orders, styling himself abbé de Mably and achieving as a writer on economic matters the reputation of philosophe somewhat earlier than did Condillac, who took the name by which he is known from another of their father's estates, a village near Romans-sur-Isère.

After the death of his father in 1727, Condillac was brought up by his eldest brother, Jean Bonnot de Mably, in whose household Jean-Jacques Rousseau was for a time a tutor. That acquaintanceship opened literary doors in Paris when in due course Condillac followed his brother the abbé to the seminary at Saint-Sulpice and then to the Sorbonne for his *license.*

Although hampered by bad vision, he was studious and concentrated his mind on philosophy and science. Theology held no interest for him. For many an eighteenth-century abbé, taking holy orders implied nothing special in the way of religious commitment.

Their vocation was for ideas rather than beliefs. They entered the clergy because it was the only profession that accommodated an intellectual career, and they never let their priesthood spoil their interest and pleasure in the world.

Condillac frequented the salons of Mme. du Tencin, Mme. d'Épinay, Mlle. de La Chaux, and Mlle. de Lespinasse. Through Rousseau he met Diderot, and for a time the three dined together weekly. He was soon of the inner circle of the philosophes, and alone among them attained to what in any century would have been regarded as a professional command of the problems of philosophy, whose appellation these men of letters had loosely and somewhat inadvertently taken upon their concerns.

Condillac entered into intellectual life at just the juncture when, for very generalized reasons, French thought was, with some little affectation, disowning the legacy of Descartes and adopting that of Newton and Locke. In consequence of the junction between physics and the empiricist account of the mind, epistemology and psychology became virtually the same subject in the French Enlightenment. Condillac contributed to the synthesis more decisively than did any other writer.

The development of Condillac's thought is recorded in a very coherent way in his main works: *L'essai sur l'origine des connaisances humaines* (1746); *Traité des systèmes* (1749); *Traité des sensations* (1754); *Traité des animaux* (1755); *La logique* (1780); and *La langue des calculs,* which was left incomplete and published posthumously. In the lengthy interval between the last two works and the earlier works, he served the Princess Louise-Elisabeth, daughter of Louis XV and the duchess of Parma, as preceptor to her son, Ferdinand de Bourbon, the heir to the sovereign dukedom. To that end he lived in Parma for more than nine years, and published books on history, education, and political economy drawn up in the course of the young prince's education. These writings are consistent with his philosophy, but they add nothing to it and will not be considered in this article.

On the psychological side, the starting point of Condillac's philosophy was Locke's *Essay Concerning Human Understanding,* which (never having learned English) he knew through the translation by Pierre Coste (1700). "Our first object," he says after Locke, "which we must never lose sight of, is to study the human mind, not to discover its nature, but to know its operations" (*Oeuvres,* I, 4). To that end he insisted on a genetic analysis of knowledge far more single-mindedly than Locke had ever done. "We must," he continued, "retrace our ideas to their origins, exhibit their genesis, follow them to the limits prescribed by nature, thereby establish the extent and boundaries of our knowledge, and thus renew the entire human understanding." It was because philosophers habitually dealt with essences rather than origins that their accounts of knowledge failed—Leibniz with his monads and Descartes with his innate ideas. It ought to be possible, however, to bring to metaphysical and moral problems all the exactness of geometry while employing a method for arriving at the evidence that would be simpler and easier than the synthetic mode of proof.

Instead, Condillac advocated an analytic approach, one that should be as rigorous as classical geometric reasoning and that should also repair a critical defect that in his view Locke the pioneer had left in the empiricist account of the mind. For Locke had admitted a distinction between the process of sensation that imprints direct experiences on consciousness and the process of reflection through which the mind is aware of itself in the act of thinking, doubting, reasoning, or willing. "Internal sense" Locke called this source of ideas in contrast to those deriving from external stimuli, and Condillac would have none of it.

> If, as Locke urges, the soul experiences no perception of which it is not conscious, so that an unconscious perception is a contradiction in terms, then perception and consciousness ought to be taken for a single operation. If the contrary be true, then the two operations are distinct, and it would be in consciousness, and not as I have supposed in perception, that our knowledge originates [Ibid., I, 11].

Locke, in his discursive English way, had failed to notice that this distinction gave all the room required for innate ideas, and Condillac, the heir of Cartesian rationalism in his repudiation of Cartesian metaphysics, set about to eliminate it.

His instrument was a highly original theory of language as the analyst of experience. It is by the mind's capacity to invent and manipulate symbols of uniform and determinate significance that it passes from sensation to reflection and communication and hence to effective knowledge. Approaching all his problems genetically, Condillac adduced the acquisition of language by a baby. In its first years, a baby translates every sensation directly into actions expressing needs or satisfactions. With observation and education, this primitive language of action gives way to French or English. A child who can talk is no longer at the mercy of inexpressible events. The mastery of a symbol or a word for what one needs creates the possibility of securing it. Even if all knowledge originates from outside experience, there-fore, to dispose of a language is to be the master of one's thought. Language, then, the conventionalization of symbols from experience, is the cause of the most complex operations of the mind, those functions usually grouped under the faculties of attention, memory, imagination, intuition, and reflection.

By identifying the operations of language as the cause of intellectual functions, Condillac intended to be making the kind of statement Newton had made when he identified gravity as the cause of the planetary motion—an exact and verifiable generalization of phenomenal effects. In the *Traité des systèmes,* Condillac expounded the Newtonian conception of scientific methodology as the healthy alternative to the metaphysical and necessarily chimerical systems constructed by Descartes, Malebranche, Leibniz, Spinoza, and their followers. Protopositivist strictures about metaphysics were already the fashion, and on the critical side the main distinction of his treatise is that it examined the systems it attacked more carefully and fully than did Voltaire, Diderot, or the more literary writers. On the positive side, his development of inductive empiricism merely systematized the scientific methodology of Newton. The originality of all his work lay in combining that methodology with his theory of language and communication in order to make the science of the human mind a science of experience.

In 1749 Diderot played the part of devil's advocate in several passages of his *Lettre sur les aveugles.* He pointed out that Condillac's restriction of the knowledge that each person has of his own existence to his inner consciousness of sensation could equally well serve the purpose of Bishop Berkeley's idealism, and invited Condillac to distinguish his own account from one that employed strikingly similar arguments to reach entirely opposite conclusions about the reality of the sensible world. Confronted with that challenge, Condillac recognized that his handling of the input from sensible experience was no clearer than Locke's.

He had treated sensations sometimes as images of real objects and sometimes as mere modes of thought, and had never established the objective existence of the world thus represented. Accordingly, in the most comprehensive of his works, the *Traité des sensations,* he set out, faithful to his genetic approach, to trace in principle the process by which a being organized with the capacity of becoming human learns to avail itself of the several senses with which it is provided.

To that end he imagined (not altogether originally) the most famous of eighteenth-century psychological fictions, a statue "organized internally like us, and animated with an intellect devoid of ideas of any sort.

We suppose further that an exterior entirely of marble allows it the use of none of its senses, and we reserve to ourselves the liberty to open them at our will to the different impressions of which they are capable" (*Ibid.*, I, 222).

Condillac chose to begin with the sense of smell since that one appears to contribute the least to human knowledge. To the nose of his statue, he held a rose. What then would its situation be? "Relative to us it will be a statue that is smelling a rose, but relative to itself it will be only the odor of a rose" (*Ibid.*, I, 224). In this state it will be capable of existence and attention, although not yet of desire or choice. Hold a carnation to its nose, however, and the situation changes, for it has comparisons to make, and should the new odor be actually disagreeable, it would come to know the difference between pleasure and pain as a basis for action in its response to external stimuli. Thus did Condillac build up in his statue a schematized sensibility, endowing it with the five senses in succession and with episodes filtered through each one until, having acquired the full syntax of experience, the statue could be imagined receiving the experiences of a varied life.

Not through the experience of smell, taste, hearing, or sight might Berkeley be refuted, however. Only the sense of touch, through which is experienced reaction to the action of putting another body into motion, carries conviction that external objects must exist and be subject to the laws of physics. The point about physical contact with the external world through the sense of touch was critical to Condillac, as indeed it had been in related ways to Voltaire, Descartes, and many another reasoning upon the epistemological significance of a science reducible to mechanics. In the *Traité des animaux* (1755), directed in part against Buffon, Condillac distinguished between the sensitivity of animals and the intellect of men largely on grounds of the superiority of the information conveyed by the human sense of touch. It is not this part of his doctrine that seems the most impressive historically, however. It is rather that his theory of language as the syntax of experience united philosophical empiricism with the account of behavior (later called utilitarian) that explained it by the preference for pleasure over pain.

It was through his last works—*La logique* and, especially, *La langue des calculs*—that Condillac exercised the most decisive influence on the philosophical taste of the generation of scientists immediately following his own. Therein, like his predecessors in the rationalist tradition, he looked to mathematics as the exemplar of knowledge. He parted company with them, however, in developing the preference he had expressed in his early work for the analytic over the synthetic mode of reasoning. Geometry had lent itself to the abuses of the framers of systems, and it was algebra that would exhibit how the operations of any proper science are only those of a "well-made language." Algebraic terms consist of a set of exact symbols. By convention they always mean the same thing. They are combined and manipulated according to rules of a perfectly exact syntax. Algebra, indeed, is at once a language and a method of analysis. By contrast, ordinary language is an inaccurate and clumsy instrument all rusted and corrupted by centuries of sophistry and superstition. To compare it with algebra would reveal the difference between science and the imperfections of life in society.

In that comparison Condillac's philosophy entered into the reforming mission of the Enlightenment, the central imperative of the rationalism then having been to reduce the imperfections of human arrangements by approximating them to the natural and to educate the human understanding in the grammar of nature. In France at least, the congruence of Condillac's philosophy of science with the broader commitments of progressive culture recommended it to scientists themselves as the most authoritative reading of Newtonian methodology. By its canons analysis must first identify the elements of a subject. Once they are made clear, they are to be classified according to the logical (which is the same thing as the natural) relations discerned beneath the complexities of phenomena and the confusion of unanalyzed experience. Analysis dissipates that confusion by finding the science its proper language, a systematic nomenclature chosen to identify the thing by the name, associate the idea with its object, and fasten the memory to nature.

The terminology and symbolism of the modern science of chemistry are examples still alive in science of the practicality of this program. The opening passages of Lavoisier's *Éléments de chimie* (1789) show that the author had taken Condillac's lessons to heart. Other protagonists of the chemical revolution joined him in making the reform of nomenclature one of its central features so that henceforth the identity of a compound should be declared in its name—copper sulfate instead of blue vitriol.

More generally, the taxonomic activity in science at the end of the eighteenth century and its identification with analysis exhibited how widely Condillac's conception of scientific explanation had been adopted—in botany, in zoology, and even in the classification of geometric surfaces according to their mode of generation. The legitimacy with which his philosophy helped invest this program in science proper was certainly a more important legacy than the work of the one school that a generation later identified itself overtly with his influence, that of the *idéologues*—Destutt de Tracy, Cabanis, Volney, Garat, Ginguené, etc. Despite their role in the foundation of the Institut de France, there remains something bookish and disembodied about their psychology. It was not this but rather the historicism that Comte drew from Turgot and Condorcet that developed Condillac's psychological empiricism into the first positivist account of science.

BIBLIOGRAPHY

I. ORIGINAL WORKS. Condillac's works are available in an excellent modern critical edition, edited and with an introduction by Georges Le Roy, *Oeuvres philosophiques de Condillac,* 3 vols. (Paris, 1947–1951). These constitute vols. XXXIII–XXXV of the series "Corpus Général des Philosophes Français," directed by Raymond Bayer. The *Traité des sensations* has been translated into English by Geraldine Carr, *Treatise on Sensations* (Los Angeles, 1930), with a preface by H. Wildon Carr.

II. SECONDARY LITERATURE. The best single book on Condillac is more interested in his psychology than his philosophy of science; it is Georges Le Roy, *La psychologie de Condillac* (Paris, 1937). A recent doctoral dissertation that treats of his place in intellectual history is Isabel F. Knight, *The Geometric Spirit: the Abbé de Condillac and the French Enlightenment* (New Haven–London, 1968).

CHARLES C. GILLISPIE

CONDORCET, MARIE-JEAN-ANTOINE-NICOLAS CARITAT, MARQUIS DE (*b.* Ribemont, France, 17 September 1743; *d.* Bourg-la-Reine, France, 27[?] March 1794), *mathematics, applied mathematics.*

Condorcet's family came originally from the Midi. Although converted at the beginning of the Reformation, the Caritat family renounced Protestantism during the seventeenth century; most of the young men of the family led lives typical of provincial noblemen, becoming either clergymen or soldiers. One of Condorcet's uncles was the bishop of Auxerre and later of Lisieux; his father, Antoine, was a cavalry captain stationed in the tiny Picardy village of Ribemont when he married Mme. de Saint-Félix, a young widow of the local bourgeoisie. It was there that the future marquis was born a few days before his father was killed during the siege of Neuf-Brisach. Raised by an extremely pious mother and tutored in his studies by his uncle the bishop, he was sent to the Jesuits of Rheims and subsequently to the Collège de Navarre in Paris, from which he graduated with a degree in philosophy in 1759, having written a thesis in Latin on mathematics;[1] d'Alembert was a member of the board of examiners.

Condorcet's mathematical ability asserted itself even though his family would have preferred that he pursue a military career. He took up residence in Paris, where he lived on a modest sum provided by his mother. He worked a great deal and became better acquainted with d'Alembert, who introduced him into the salons of Mlle. de Lespinasse and Mme. Helvétius; at the first salon he met Turgot, who subsequently became his close friend. In 1765 Condorcet published a work on integral calculus and followed it with various mathematical *mémoires* that earned him the reputation of a scientist. He was elected to the Académie des Sciences as *adjoint-mécanicien* (adjunct in mechanics), succeeding Bezout (1769), and later as an associate in the same section. He thereafter became closely involved with scientific life. As assistant secretary (1773) and then permanent secretary of the Académie des Sciences (1776), Condorcet published a great many mathematical *mémoires* and, in 1785, the voluminous *Essai sur l'application de l'analyse à la probabilité des décisions rendues à la pluralité des voix.* Simultaneously he wrote his *éloges* of deceased academicians, essays that are often remarkable for his fairness of judgment and broadness of view; he also tried, but in vain, to organize scientific activity in France along rational lines.[2]

In the 1770's, however, another of Condorcet's interests came to the surface, possibly as a result of a meeting with Voltaire at Ferney. The notion of a political, economic, and social reform to be undertaken on scientific bases was its most recurrent subject. From it stemmed, in the theoretical domain, his calculus of probabilities and, in the practical domain, the numerous applied-research projects he had set up or carried out as inspector of the mint (1776) and director of navigation under the ministry of Turgot. To that end he also participated, beginning in 1775, in the movement of political and social dissent from the regime through the publication of various pamphlets. On the eve of the Revolution, Condorcet was inspector of the mint, permanent secretary of the Académie des Sciences, and a member of the Académie Française (1782); in 1786 he married Sophie de Grouchy, whose salon at the Hôtel des Monnaies (the mint) had taken the place of the salons of Mlle. de Lespinasse and Mme. Helvétius. In 1776 he was entrusted with the articles on analysis for Panckoucke's *Supplément* to the *Encyclopédie,* and in 1784 he revised and rewrote, with Lalande and Bossut, the mathematical part of that work, recast as the *Encyclopédie méthodique.* Thus he can be considered one of the most representative personages of the Enlightenment and, so far as France is concerned, one of the most influential.

After 1787 Condorcet's life is scarcely of interest to the historian of science. He then stood forth as an advocate of calling a national assembly that would reform the regime according to the views of the liberal bourgeoisie. After twice failing to be elected to the States-General (from Mantes-Gassicourt, where he had landholdings, and then from Paris), he was elected representative from his quarter of Paris to the Municipal Council, sitting in the Hôtel de Ville, and founded, in association with Emmanuel Sieyès, the Society of 1789. In September 1791 he succeeded in becoming a delegate to the Legislative Assembly and later to the Convention of 1792. In the Assembly he concentrated his efforts mainly on the work of the Commission of Public Instruction, for which he was suited by virtue of the *mémoires* that he had published in the *Bibliothèque de l'homme public* (1791–1792). At the convention he drew up the draft of a constitution (1793), but it was not adopted. A friend of Jacques Brissot and closely linked to the political battle waged by the Girondists, Condorcet came under suspicion following their expulsion from the convention on 2 June 1793; when the draft constitution that had been substituted for his own by the Committee of Public Safety was voted on, he published *Avis aux français*. This pamphlet was the cause of his downfall. An order was issued for his arrest on 8 July, but he managed to escape and found refuge in a house in the present Rue Servandoni. He hid there until 25 March 1794, and there he composed the work that constitutes his philosophical masterpiece, the *Esquisse d'un tableau des progrès de l'esprit humain*. On the latter date, fearful of being discovered, he left on foot for Fontenay-aux-Roses, where he had friends. However, they managed not to be at home when he arrived. On 27 March he was arrested in Clamart under a false name and was taken to the prison of Bourg-la-Reine. The next morning he was found dead in his cell. It has never been possible to verify the rumor, originating as early as 1795, that he committed suicide by poison.

Condorcet seems to have had the character of a systematic and passionate intellectual. Described by d'Alembert as a "volcano covered with snow," he was praised by Mlle. de Lespinasse for his great kindness; and accusations of avarice and social climbing sometimes directed at him have never been corroborated by documentary evidence. He was the typical Encyclopedist and perhaps the last of them. All fields of knowledge fascinated him, as is shown by the equal care and competence that he devoted to his *éloges* of

Euler as well as Buffon, d'Alembert as well as Jean de Witt, Frénicle as well as Pascal. He was thoroughly convinced of the value of science and of the importance of its diffusion as a determining factor in the general progress and well-being of mankind. That is why his interest never flagged in the applications of science,[3] or in the organization of scientific education,[4] or in the establishment of a universal scientific language.[5] His entire concept of scientific knowledge is essentially probabilist: "We give the name of mathematical certitude to probability when it is based on the content of the laws of our understanding. We call physical certitude the probability that further implies the same constancy in an order of phenomena independent of ourselves, and we shall reserve the term probability for judgments that are exposed to other sources of uncertainty beyond that" (*Essai sur l'application de l'analyse*, p. xiv). The concept of a science based on human actions, intrinsically probabilist, therefore seemed to him to be just as natural as a science of nature, and he attempted to establish a portion of such a science by proposing a theory of votes based on the calculus of probability. Thus, despite his reputation as a mathematician, much exaggerated in his lifetime, it is the novelty, the boldness, and the importance of this attempt that today seem to constitute his just claim on the attention of the historian of scientific knowledge.

Let us, nevertheless, briefly examine his work in pure mathematics before summarizing his work on probability and his application of analysis to the theory of votes.

Differential equations. Condorcet's scientific *mémoires* of the type contained in the principal periodicals of the academies have not been collected in book form. To those he did write may be added a few works that have not been republished and several unpublished manuscripts. It must be acknowledged that reading Condorcet's mathematical works is a thankless task and often a disappointing one. The notation is inconsistent, the expression of ideas often imprecise and obscure, and the proofs labored. This is certainly what Lagrange was complaining about when he wrote to d'Alembert on 25 March 1792: "I would like to see our friend Condorcet, who assuredly has great talent and wisdom, express himself in another manner; I have told him this several times, but apparently the nature of his mind compels him to work in this way: we shall have to let him do so. . . ."[6]

Nevertheless, the esteem expressed by good judges, such as Lagrange himself, leaves the contemporary reader perplexed; and one suspects that sometimes friendship and sometimes the respect due to the influential secretary of the Académie des Sciences may have somewhat dulled the critical sense of the mathematicians of his day. Indeed, it now seems that the mathematical part of Condorcet's work contributed nothing original and that it deserves to be considered by the historian merely as evidence of the manner in which a man highly educated in that science could comprehend it and keep abreast of its progress. Wishing to introduce into the theory of differential equations general concepts that would be capable of systematizing it, he prematurely outlined a philosophy of mathematical notions that failed to issue in any coherent or practical organization.

Although he seemed ahead of his time when in an unpublished and incomplete *traité*[7] he defined a function as any relation whatever of corresponding values, he nevertheless thought it possible to propose a systematic and exhaustive classification of all functions. The main idea of the construction of classes of functions was, however, interesting in itself: it was the idea of a procedure of iteration that allows for definitions by recurring algorithms. What he did emphasize was the supposed closed-system character of all analytical entities. He exhibited this intention in an even more dogmatic manner in his *Lettre à d'Alembert* appended to the *Essais d'analyse* of 1768, in which he affirmed that all transcendental functions could be constructed by means of a circle and hyperbola; and he actually expressed himself in many of his *mémoires* as if any nonalgebraic function were of a trigonometric, logarithmic, or exponential nature.

The manner in which Condorcet conceived of the problem of the integration of differential equations or partial differential equations arose from the same tendency to generalization, which appeared a bit hasty even to his contemporaries.[8] Most of his *mémoires* and analytical works deal with this problem: to find conditional equations by relating the coefficients of a differential equation in such a manner as to render it integrable or at least such that its order may be lowered by one degree. Once the existence of solutions was proved, he hoped to be able to determine their form a priori, as well as the nature of the transcendental functions included; a calculation identifying the parameters would thus complete the integration. (See, e.g., "Histoire de l'Académie des sciences de Paris," "Mémoires" [1770], pp. 177 ff.). At least, therefore, he must be credited with having clearly conceived and stated that it is normal for an arbitrary differential equation not to be integrable.

Let us limit ourselves to indicating the approach by which Condorcet proceeded, as set forth in the 1765 text *Du calcul intégral* and resumed in the *traité* of 1786. Given the differential expression of any order $V(x, y, dx, dy, d^2x, d^2y, \cdots, d^nx, d^ny)$, the question is under what conditions it might be the differential of an expression B, such that

$$(1) \qquad V = dB\,(x, y).$$

Let us say $dx = p$, $dy = q$, $d^2y = dq = r$, $d^3y = dr = s$, x being taken as an independent variable, $d^2x = 0 = dp$. The differential of V, dV, has the form $dV = Mdx + Ndy + Pd^2y + Qd^3y + \cdots$, or

$$(2) \qquad dV = Mp + Nq + Pr + Qs + \cdots.$$

The coefficients M, N, P, Q are functions calculable from V.

Condorcet next differentiated the two sides of equation (1):

$$(3) \quad dV = d^2B = d\left(\frac{\partial B}{\partial x}\right)p + d\left(\frac{\partial B}{\partial y}\right)q$$
$$+ \left(\frac{\partial B}{\partial y} + d\left[\frac{\partial B}{\partial q}\right]\right)r + \cdots.$$

The terms of the right-hand side of equations (2) and (3) are then

$$(4) \qquad M = d\left(\frac{\partial B}{\partial x}\right)$$

$$(5) \qquad N = d\left(\frac{\partial B}{\partial y}\right)$$

$$(6) \qquad P = \frac{\partial B}{\partial y} + d\left(\frac{\partial B}{\partial q}\right)$$

$$(7) \qquad Q = \frac{\partial B}{\partial q} + d\left(\frac{\partial B}{\partial r}\right).$$

Obtaining the differential of P, $Q, \ldots$

$$(8) \qquad dP = d\left(\frac{\partial B}{\partial y}\right) + d^2\left(\frac{\partial B}{\partial q}\right)$$

$$(9) \qquad d^2Q = d^2\left(\frac{\partial B}{\partial q}\right) + d^3\left(\frac{\partial B}{\partial r}\right),$$

by successive addition and subtraction of (5), (8), (9), . . . , we obtain

$$(10) \qquad N - dP + d^2Q - \cdots = 0,$$

which is the condition of existence of an integral B.

Condorcet then generalized his example by abandoning the hypothesis $d^2x = 0$, i.e., by supposing that there is no constant difference. He noted that the equations of these conditions are the same ones that determine the extreme values of the integral of V, as established by Euler and Lagrange.

Nowhere in Condorcet's various *mémoires* and other works, whether they deal with differential equations or equations with partial derivatives or equations with finite differences, can any results or methods be found that are truly creative or original relative to the works of Fontaine, d'Alembert, Euler, or Lagrange.

Calculus of probability. In the calculus of probability also, Condorcet did not bring any significant perfection to the resources of mathematics; he did, however, discuss and explain in depth an interpretation of probability that had far-reaching consequences in the applications of the calculus. First of all, he made a very clear distinction between an abstract or "absolute" probability and a subjective probability serving merely as grounds for belief. An example of the former is probability, in throwing an ideal die, that a given side will appear. Condorcet attempted to justify the passage from the first to the second by invoking three axiomatic principles that in effect reduced to the simple proposition "A very great absolute probability gives 'grounds for belief' that are close to certainty." (See "Probabilités," in *Encyclopédie méthodique; Éléments du calcul des probabilités,* art 4.) As for passage from observed frequencies to "grounds for belief," this is effected through an estimated abstract probability; and the instrument for this estimation was Bayes's theorem, which had recently been reformulated by Laplace (*Mémoires par divers savants,* **6**, 1774). Condorcet made a very shrewd study of this. In the fourth "Mémoire sur le calcul des probabilités" (published in 1786), he noted that Bayes assumed the a priori law of probability to be constant, whereas it was actually possible to exhibit experimental variations of that law whether or not it depended on the time factor. Let us examine the latter case by means of an example. In a series of urns everything happens as if the drawings were made each time from an urn selected at random from a group, the numerical order of the drawing having no influence upon the choice. Thus, let $N = m + n + p + q$ decks of mixed cards. The first draw produces m red cards and n black ones from $(m + n)$ decks. We are asked the probability of drawing p red ones and q black ones from the $(p + q)$ remaining decks. Bayes's simple scheme (from a single deck of N cards, $m + n$ drawings have already been made, yielding m red and n black) furnishes the value calculated by Laplace for the probability wanted:

$$\frac{(p + q)! \int_0^1 x^{m+p}(1 - x)^{n+q}\, dx}{p!q! \int_0^1 x^m(1 - x)^n\, dx}.$$

The hypothesis of a variable law led Condorcet to the alternative value

$$\frac{(p + q)! \iiint \cdots \left[\sum \frac{x_i}{N}\right]^{m+p} \left[1 - \sum \frac{x_i}{N}\right]^{n+q} dx_1\, dx_2 \cdots dx_N}{p!q! \iiint \cdots \left[\sum \frac{x_i}{N}\right]^m \left[1 - \sum \frac{x_i}{N}\right]^n dx_1\, dx_2 \cdots dx_N}$$

the multiple integrals being taken, for each variable, between 0 and 1 and x_i being the a priori probability assigned to the first deck of cards.

It is not correct to say, as Todhunter does (p. 404), that such an improvement on Laplace's formula was purely "arbitrary"; it must, nevertheless, be admitted that even though the main idea presupposes a thorough analysis of the requirements of a probabilistic model, it is no less true that the result is a complication offering no real utility.

"Social mathematics." Condorcet's most significant and fruitful endeavor was in a field entirely new at the time. The subject was one that departed from the natural sciences and mathematics but nevertheless showed the way toward a scientific comprehension of human phenomena, taking the empirical approach of natural science as its inspiration and employing mathematics as its tool. Condorcet called this new science "social mathematics." It was apparently intended to comprise, according to the "Tableau général de la science qui a pour objet l'application du calcul aux sciences physiques et morales," *Journal d'instruction sociale* (22 June, 6 July 1795; *Oeuvres,* I, 539–573), a statistical description of society, a theory of political economy inspired by the Physiocrats, and a combinatorial theory of intellectual processes. The great work on the voting process, published in 1785, is related to the latter.

Condorcet there sought to construct a scheme for an electoral body the purpose of which would be to determine the truth about a given subject by the process of voting and in which each elector would have the same chance of voicing the truth. Such a scheme was presented exactly like what is today called a model. Its parameters were the number of voters, the majority required, and the probability that any particular vote voices a correct judgment. Condorcet's entire analysis consisted, then, of calculating different variable functions of these structural parameters. Such, for example, was the probability that a decision reached by majority vote might be correct. An interesting complication of the model is introduced by the assumption that individual votes are not mutually independent. For example, the influence of a leader might intervene; or when several successive polls are taken, the electors' opinions may change during the voting process. On the other hand, the problem of estimating the various parameters on a statistical basis was brought out by Condorcet, whose treatment foreshadowed very closely that employed by modern users of mathematical models in the social sciences. The mathematical apparatus may be reduced to simple theorems of addition and multiplication of probabilities, to binomial distribution, and to the Bayes-Laplace rule.

Here is an example of this analysis. Let v be the individual probability of a correct judgment. The probability that a collective decision having obtained q votes might be correct is

$$\frac{v^q}{v^q + (1 - v)^q}.$$

If one requires a majority of q with n voters, the probability that it will be attained and will furnish a correct decision is given by the sum

$$v^n + \sum_{i=1}^{n-q} C_n^i v^{n-i}(1 - v)^i.$$

In the case of a leader's influence, if a voters out of n shared his opinion in a prior vote, the a posteriori probability of a voter's following the leader is, according to Bayes's rule, $a + 1/n + 2$; if v' represents the probability of any individual's holding a correct opinion and v'' is the same probability for the leader, then the probability of a voter's fortuitously holding the same opinion as the leader is

$$v'v'' + (1 - v')(1 - v'').$$

Condorcet then proposed to measure the magnitude of the leader's influence by the difference between the probabilities:

$$\frac{a + 1}{n + 2} - (v'v'' + [1 - v'][1 - v'']).$$

Along the way he encountered a completely different problem, the decomposition and composition of electoral decisions in the form of elementary propositions on which voters pronounce either "Yes" or "No." He then anticipated, without being aware of it, the logical import of this problem, which was the theory of the sixteen binary sentence connectives, among which he emphasized the conditional. He showed that a complex questionnaire could be reduced to a sequence of dichotomies and that constraints implicitly contained in the complex questionnaire are equivalent to the rejection of certain combinations of "Yes" and "No" in the elementary propositions. This is literally the reduction into normal disjunctive forms as practiced by contemporary logicians (*Essai,* pp. xiv ff.). He therefore brought to light, more completely and more systematically than his predecessor Borda,[9] the possible incoherence of collective judgment in the relative ordering of several candidates.

If there are three candidates to be ranked and sixty voters, the voting may be done on the three elementary propositions p, q, r:

> p: A is preferable to B
> q: A is preferable to C
> r: B is preferable to C.

Since the choice of each voter is assumed to be a coherent one—namely, determining a noncyclical order of the three candidates A, B, and C—an individual choice such as "p and not-q and r" is excluded.

However, let us assume that the results of the vote were as follows:

"p and q and r"	order A-B-C	23	votes
"not-p and not-q and r"	B-C-A	17	votes
"not-p and q and r"	B-A-C	2	votes
"not-p and not-q and not-r"	C-B-A	8	votes
"p and not-q and not-r"	C-A-B	10	votes

If we calculate the votes by opinion, then it is the order A-B-C that wins. But if we calculate the votes according to elementary propositions, the following is obtained: "p," 33 votes; "not-q," 35 votes; "r," 42 votes. These are the majority propositions and define a cyclical order, and therefore an incoherent collective opinion. This is the paradox that Condorcet pointed out, one that poses the problem, taken up again only in modern times, of the conditions of coherence in a collective opinion.

No doubt the results obtained in the *Essai d'application de l'analyse* were modest ones. "In almost all cases," Condorcet said, "the results are in conformity with what simple reason would have dictated; but it is so easy to obscure reason by sophistry and vain subtleties that I should feel rewarded if I had only founded a single useful truth on a mathematical demonstration" (*Essai,* p. ii). One must nevertheless recognize, in this work dated 1785, the first large-scale attempt to apply mathematics to knowledge of human phenomena.

NOTES

1. We believe we can identify a fragment of it with the MS at the Institut de France, fol. 222–223, carton 873.
2. See Baker's excellent "Les débuts de Condorcet . . . ."
3. He became interested in hydraulics together with the Abbé Bossut ("Nouvelles expériences sur la résistance des fluides," in *Mémoires de l'Académie des sciences,* 1778); he became interested in demographic statistics together with Dionis du Séjour and Laplace ("Essai pour connaître la population du royaume," *ibid.,* 1783–1788). The *Mémoires* (1782) contains a curious report on a proposal for a rational distribution of taxes using geographical, economic, and demographic data as a basis; it was formulated in collaboration with two physicists (Bossut and Desmarest), an agronomist (Tillet), and an astronomer (Dionis du Séjour).
4. In the five *mémoires* on public instruction (1791–1792) and in the draft decree of 20 April 1792, he proposed to replace the old literary instruction at the *collèges* with truly modern humanistic subjects including four categories of study: the physical and mathematical sciences, the moral and political sciences, the applications of science, and letters and fine arts.
5. See the fragment of the MSS published in Granger, "Langue universelle . . ."; also *Esquisse,* O. H. Prio, ed. (Paris, 1937), p. 232.
6. *Oeuvres de Lagrange,* XIII (Paris, 1882), 232.
7. Pt. 1, sec. 1. The handwritten MS is at the Institut de France

(cartons 877–879) with the first 152 pages in printed form. Printing was suspended in 1786; the MS was composed in 1778.

8. For example, Lagrange wrote to d'Alembert, on 6 June 1765, that he wanted Condorcet "to explain in more detail the manner in which he arrived at the various integrals to which a single differential equation was susceptible." The reference here is to Condorcet's first work, *Du calcul intégral.* Likewise, on 30 September 1771, Lagrange pointed out to Condorcet that he had tried to apply one of his methods of approximation to an equation already studied, and that the result obtained was inaccurate.

9. Charles de Borda's "Mémoire sur les élections au scrutin" was published in the *Mémoires de l'Académie des sciences* (1781), 657–665, but it had been presented to the Academy in 1770 and Condorcet knew of it.

BIBLIOGRAPHY

The *Oeuvres,* 12 vols., pub. by Mme. Condorcet-O'Connor and François Arago (Paris, 1847–1849), do not contain the scientific writings. A bibliography and a chronology of the latter will be found in Henry, "Sur la vie et les écrits . . ." and in Granger (see below).

Condorcet's MSS are at the library of the Institut de France; the scientific writings are for the most part found in cartons 873–879 and 883–885.

On Condorcet see K. M. Baker, "An Unpublished Essay of Condorcet on Mechanical Methods of Classification," in *Annals of Science,* **18** (1962), 99–123; "Les débuts de Condorcet au Secrétariat de l'Académie royale des sciences," in *Revue d'histoire des sciences,* **20** (1967), 229–280; "Un 'éloge' officieux de Condorcet: Sa notice historique et critique sur Condillac," in *Revue de synthèse,* **88** (1967), 227–251; and "Scientism, Elitism and Liberalism: The Case of Condorcet," in *Studies on Voltaire and the 18th Century* (Geneva, 1967), pp. 129–165; L. Cahen, *Condorcet et la Révolution française* (Paris, 1904), with bibliography; Gilles Granger, "Langue universelle et formalisation des sciences," in *Revue d'histoire des sciences,* **7,** no. 4 (1954), 197–219; and *La mathématique sociale du Marquis de Condorcet* (Paris, 1956), with bibliography; Charles Henry, "Sur la vie et les écrits mathématiques de J. A. N. Caritat Marquis de Condorcet," in *Bollettino di bibliografia e storia delle scienze matematiche,* **16** (1883), 271 ff.; and his ed. of "Des méthodes d'approximation pour les équations différentielles lorsqu'on connaît une première valeur approchée," *ibid.,* 292–324; F. E. Manuel, *The Prophets of Paris* (Cambridge, Mass., 1962); René Taton, "Condorcet et Sylvestre-François Lacroix," in *Revue d'histoire des sciences,* **12** (1959), 127–158, 243–262; and I. Todhunter, *History of the Mathematical Theory of Probability From Pascal to Laplace* (Cambridge, 1865), pp. 351–410.

GILLES GRANGER

CONGREVE, WILLIAM (*b.* Middlesex, England, 20 May 1772; *d.* Toulouse, France, 16 May 1828), *rocketry.*

Congreve was the eldest son of Lieutenant General Sir William Congreve, colonel commandant of the royal artillery and comptroller of the royal arsenal at Woolwich. He was educated at Trinity College, Cambridge (B.A., 1793; M.A., 1795) and became a barrister and, later, the proprietor of the *Royal Standard,* a newspaper. About 1804 he turned his interests to improving and perfecting the rocket as a military weapon. England subsequently adopted "Congreve rockets," which were used with great success against the French at Boulogne, Copenhagen, Leipzig, and elsewhere. They were copied by most European armies by 1830.

Congreve was named an equerry to his friend the prince regent (later George IV) and elected a fellow of the Royal Society in 1811. The same year he was made an honorary lieutenant colonel in the Hanoverian Artillery and subsequently rose to the rank of major general. At the death of his father in 1814, he succeeded as the second baronet (Congreve of Walton) and as comptroller of the royal arsenal.

Congreve married the widow Isabella M'Envoy in 1824 at Wesel, Prussia, and had two sons and a daughter. From 1826 he was compelled to settle permanently on the Continent because of failing health and in order to avoid the scandal of his involvement in a case of fraud. In convalescence, he devised his own wheelchair after losing the use of his legs, and also designed a "wave-wheel"-propelled vessel and a human-muscle-powered aircraft.

Eighteen patents were issued to Congreve in his lifetime. These included new methods of mounting naval ordnance, gunpowder manufacture, printing unforgeable currency, gas lighting, "hydropneumatic" canal locks, several kinds of clocks, a perpetual motion machine, a built-in sprinkler system, and a steam engine.

BIBLIOGRAPHY

I. ORIGINAL WORKS. Congreve wrote over a dozen treatises on his inventions, including *Memoir on the Possibility . . . of the Destruction of the Boulogne Flotilla* (London, 1806); *A Concise Account of the Origin and Progress of the Rocket System* (London, 1814); *A Concise Account of the Origin of the New Class of 24-Pounder Medium Guns* (London, 1814); *Of the Impracticability of Resumption of Cash-Payments* (London, 1819); *Principles on . . . a More Perfect System of Currency* (London, 1819); and *A Treatise on the General Principles . . . of the Congreve Rocket System* (London, 1827).

II. SECONDARY LITERATURE. The best biographical article on Congreve is Col. J. R. J. Jocelyn, "The Connection of the Ordnance Department With National & Royal Fireworks, Including Some Account of . . . Sir William Congreve (2nd Baronet)," in *Journal of the Royal Artillery* (Woolwich), **32** (1905–1906), 481–503. See also Elizabeth M. Harris, *Sir William Congreve and His Compound-Plate*

Printer, Museum of History and Technology Paper no. 71 (Washington, D.C., 1967); Willy Ley, *Rockets, Missiles, and Space Travel* (New York, 1944, and later eds.); Merigon de Montgéry, *Traité des fusées de guerre, nommées . . . fusées à la Congreve* (Paris, 1825); and the unsigned "Obituary, Sir William Congreve, Bart.," in *The Times* (London) (27 Apr. 1828), 2, col. 3.

FRANK H. WINTER

CONKLIN, EDWIN GRANT (*b.* Waldo, Ohio, 24 November, 1863; *d.* Princeton, New Jersey, 21 November, 1952), *biology.*

The son of Dr. Abram V. Conklin and Nancy Maria Hull, Conklin was educated in the public schools of Waldo and received his B.A. from Ohio Wesleyan University (1886) and his Ph.D. from Johns Hopkins (1891). From a prosperous and religious family, he seriously considered entering the ministry before ultimately choosing an academic career. In 1889, while still a graduate student, he married Belle Adkinson, who remained his constant companion until her death in 1940. From 1885 until 1888 Conklin taught at Rust University in Mississippi, from 1891 to 1894 at Ohio Wesleyan, from 1894 to 1896 at Northwestern University, and from 1896 to 1908 at the University of Pennsylvania. In 1908 he accepted a call from Woodrow Wilson, then president of the university, to become professor of biology and chairman of the department at Princeton, a position he held until his retirement in 1933.

Possessed of good health and considerable energy, Conklin was a prodigious investigator and scholar, as well as an indefatigable lecturer, writer, and teacher. Quiet and dignified in manner, he had a warmth and generosity that won the respect not only of his colleagues but also of his students, graduate and undergraduate alike.

Conklin was a member of numerous professional and biological organizations, including the American Philosophical Society, the Academy of Natural Sciences of Philadelphia, the Wistar Institute, the Marine Biological Laboratory and the Woods Hole Oceanographic Institute, the American Academy of Arts and Sciences, and the National Academy of Sciences. He was elected president of the American Society of Zoologists (1899), the American Society of Naturalists (1912), and the American Association for the Advancement of Science (1936). In addition, he was invited to membership in a number of foreign scientific societies, including the Königliche Böhmische Gesellschaft der Wissenschaften, the Royal Society of Edinburgh, the Zoological Society of London, the Académie Royale de Belgique, the Istituto Lombardo (Milan), and the Accademia Nazionale dei Lincei (Rome). He also served on the editorial board of a number of journals, including *Biological Bulletin* (Woods Hole), *Journal of Morphology, Journal of Experimental Zoology, Genetics,* and *Quarterly Review of Biology.*

Outside of his scientific and professional work, Conklin held a long-standing interest in the philosophy of biology and in the relations between science and human values. Many of his lectures, papers, and books dealt with the often hazy but important ground between pure science and the application of science to social problems.

From his rural background Conklin developed an early interest in natural history, which was encouraged by one of his teachers at Ohio Wesleyan, Edwin T. Nelson. After teaching Latin, Greek, English, and most of the sciences at Rust University (a missionary college for black students) from 1885 to 1888, Conklin embarked on graduate study at Johns Hopkins University, taking physiology under H. Newell Martin, geology and paleontology under W. B. Clark, and morphology under W. K. Brooks. Among his fellow students at Hopkins were H. V. Wilson, S. Watasé, T. H. Morgan, and Ross G. Harrison. The intellectual atmosphere of Hopkins was an immense stimulus for scholarly work, as Conklin remarked: "It was as if I had entered a new world with new outlooks on nature, new respect for exact science, new determination to contribute to the best of my ability to 'the increase and diffusion of knowledge among men'" (Harvey, p. 61).

For his dissertation Conklin studied the cell lineage of *Crepidula,* a gasteropod, under the direction of W. K. Brooks. Cell lineage studies were an attempt to follow the fate of daughter cells during embryonic cleavage, from the fertilized egg through later states, such as gastrulation. Conklin felt that homologies, established by comparing early development in various groups of organisms, could provide useful morphological data for understanding evolutionary relationships. This belief was borne out in the summer of 1891 at Woods Hole, when Conklin compared his notes on *Crepidula* with E. B. Wilson's similar work on the annelid *Nereis.* Both Wilson and Conklin were surprised to discover the homologous patterns of cleavage in two groups, which up to that time were thought to be only distantly related.

Working several years later with the naturally pigmented egg of *Cynthia,* a primitive chordate, Conklin followed the distribution of the pigmented particles (originally from one region of the unfertilized egg) in the daughter cells of the developing embryo. The *Cynthia* studies allowed Conklin to test the then-current hypotheses of mosaic versus regulative development. Proponents of the mosaic theory held that specific parts of the egg cell give rise to specific parts of the embryo; proponents of regulative development maintained that specific parts of the egg can give rise to almost any part of the embryo. Conklin showed with *Cynthia* that specific regions of the egg always give rise to specific parts of the later embryo, and thus provided evidence for the mosaic theory. These results brought him into friendly controversy with the outstanding proponent of regulative development, Hans Driesch.

In addition to his cell lineage studies, Conklin also carried out detailed investigations on mitosis (using *Crepidula*) and on the embryology of *Amphioxus.* While most of his investigations were observational, Conklin nonetheless was a strong proponent of experimental biology. Along with T. H. Morgan and Jacques Loeb, Conklin felt that experimental work had been neglected in biology far too long and that, while valuable, observation alone failed to answer many important questions.

Outside of his main area of embryology, Conklin was also interested in evolution and natural selection. He wrote numerous papers on Darwinism, explaining aspects of the theory to lay as well as scientific audiences and defending it against religious and social bigotry. In conjunction with his interest in evolution he also had an interest in the "nature" versus "nurture" controversy in the study of human development. One of his first books, *Heredity and Environment in the Development of Men* (1915), was devoted to an examination of the roles of heredity and environment in shaping human physical, mental, and moral characteristics.

Beyond questions of strictly scientific concern, Conklin was interested in the relationship between science, philosophy, and society. For many years at Princeton he taught a seminar in the philosophy of biology, and he lectured frequently on such topics as "Science and the Future of Man" (1930), "Science and Ethics" (1937), and "The Biological Basis of Democracy" (1938).

Although Conklin was associated with a number of scientific institutions, he was particularly devoted to the affairs of three: Princeton University, the American Philosophical Society, and the Marine Biological Laboratory. He served each of these institutions in various administrative and scholarly capacities. Under his guidance biology at Princeton became a highly popular field of study, and the department a smoothly functioning and stimulating place to work. Conklin twice served as president of the American Philosophical Society (1942–1945, 1948–1952), and from 1897 until 1933 he served the Marine Biological

Laboratory faithfully as a member of its board of trustees. Along with T. H. Morgan and Ross G. Harrison, he was a founder of the *Journal of Experimental Zoology* in 1904 and remained a member of its editorial board until his death.

Conklin's contribution to American biology lay principally in the area of general morphology and embryology. He worked at a time when embryology was emerging from domination by evolutionary questions and was becoming a science with its own set of questions and assumptions. Conklin's cell lineage studies bridged this gap. They were undertaken originally to show evolutionary relationships between groups of invertebrates through homologies in early embryonic development. They came later, however, to shed important light on the embryological question of mosaic versus regulative development.

BIBLIOGRAPHY

I. ORIGINAL WORKS. A complete bibliography of Conklin's published works is given at the end of the biographical sketch by E. Newton Harvey (see below). For an introduction to Conklin's outstanding and detailed embryological work, the following are suggested: "The Embryology of *Crepidula*," in *Journal of Morphology,* **13** (1897), 1–226, his Ph.D. dissertation; "Karyokinesis and Cytokinesis in the Maturation, Fertilization, and Cleavage of *Crepidula* and Other Gasteropoda," in *Journal of the Academy of Natural Sciences of Philadelphia,* **12** (1902), 1–121; "Mosaic Development in Ascidian Eggs," in *Journal of Experimental Zoology,* **11** (1905), 146–223; "The Organization and Cell-Lineage of the Ascidian Egg," in *Journal of the Academy of Natural Sciences of Philadelphia,* **13** (1905), 1–119; and "The Embryology of Amphioxus," in *Journal of Morphology,* **54** (1932), 69–120.

In addition, the following works provide an insight into Conklin's many writings on evolution, ethics, and the human condition: *Heredity and Environment in the Development of Men* (Princeton, 1915), the Norman W. Harris lectures for 1914, delivered at Northwestern University; *The Direction of Human Evolution* (New York, 1921); "Science and the Future of Man," Brown University Papers no. 9 (1930); "Science and Ethics," in *Science,* **86** (1937), 595–603; and "The Biological Basis of Democracy," The Barnwell Address no. 62 (1938).

II. SECONDARY LITERATURE. There is relatively little secondary material on Conklin, or, indeed, on the entire period and subject (embryology) of his greatest scientific achievement. Several sources for more detailed biographical and scientific treatment of Conklin are Garland E. Allen and Dennis M. McCullough, "Notes on Source Materials: The Edwin Grant Conklin Papers at Princeton University," in *Journal of the History of Biology,* **1** (1969), 325–331; F. Bard, "Edwin Grant Conklin," in *Proceedings of the American Philosophical Society,* **208** (1964), 55–56; Elmer G. Butler, "Edwin Grant Conklin (1863–1952)," in

Yearbook. American Philosophical Society (1952), 5–12; and E. N. Harvey, "Edwin Grant Conklin," in *Biographical Memoirs. National Academy of Sciences,* **31** (1958), 54–91.

Particularly interesting is Conklin's brief autobiography, published as a chapter in *Thirteen Americans: Their Spiritual Biographies,* Louis Finkelstein, ed. (New York, 1953).

GARLAND E. ALLEN

CONON OF SAMOS (*b.* Samos; *fl.* Alexandria, 245 B.C.), *mathematics, astronomy, meteorology.*

Conon made astronomical and meteorological observations in Italy and Sicily before settling in Alexandria, where he became court astronomer to Ptolemy III (Ptolemy Euergetes). He became the close friend of Archimedes, who was in the habit of sending him mathematical propositions that he believed to be true but had not yet succeeded in proving. He is famous chiefly for his identification of the constellation *Coma Berenices,* named in honor of Ptolemy's consort. This must have taken place about 245 B.C.; and since he predeceased Archimedes by a considerable number of years, he must have died well before 212. His friend Dositheus took his place as Archimedes' correspondent. To Vergil and Propertius, Conon became a symbolic figure for the astronomer, probably through Callimachus' elegy on the discovery of *Coma Berenices,* which Catullus translated into Latin.

Apollonius relates that Conon, in a work sent to Thrasydaeus, treated the points of intersection of conic sections with each other and with a circle—but inaccurately, so that he was rightly attacked by Nicoteles of Cyrene. Pappus states that Conon enunciated "the theorem on the spiral" proved by Archimedes, but this contradicts what Archimedes himself says. For Archimedes tells Dositheus that he had sent Conon three groups of propositions which were subsequently proved in his treatises *On the Sphere and Cylinder, On Conoids and Spheroids,* and *On Spirals* —as well as two on sections of a sphere that were not correct—and Conon died before he was able to inquire into them sufficiently.

Claudius Ptolemy's *Risings of the Fixed Stars* attributes seventeen "signs of the seasons" to Conon. He clearly played a notable part in the development of the *parapegma,* the Greek astronomical and meteorological calendar. Seneca testifies that Conon diligently studied the records of solar eclipses kept by the Egyptians. The poetic assessment of his work in Catullus (66.1–4)—that he "discerned all the lights of the vast universe, and disclosed the risings and settings of the stars, how the fiery brightness of the sun is darkened, and how the stars retreat at fixed times"—seems, therefore, a fair summary. Probus' commentary on Vergil's *Eclogues* (3.40) ranks Conon

with the great astronomers of antiquity; and the Bern scholiast to this passage calls him *mathematicus, stellarum peritissimus magister* ("a mathematician and most skilled master of the stars").

Conon's chief claim to fame shows, however, his talent as a courtier rather than as an astronomer. Berenice had vowed to dedicate a lock of her hair in the temple of Arsinoë Zephyritis if her newly married husband returned victorious from the Third Syrian War, which had begun in 246 B.C. He quickly returned, and she duly fulfilled her vow. The following day the lock of hair disappeared; and Conon professed to see it in some stars between Virgo, Leo, and Boötes that have been known ever since as Βερενίκης πλόκαμος, *Coma Berenices.*

BIBLIOGRAPHY

According to Probus, Conon left a work in seven books, *De astrologia.* As noted above, Apollonius, *Conics,* IV, pref., records that he wrote a treatise, Πρὸς Θρασυδαῖον, on the points of intersection of conic sections with each other and with a circle. It is a fair deduction from Seneca, *Quaestiones naturales,* VII. iii, 3 that he wrote a book on eclipses of the sun. None of these works has survived.

The fullest modern account is Albert Rehm's article "Konon, 11," in Pauly-Wissowa, XI, 1338–1340. See also Gerald L. Geison, "Did Conon of Samos Transmit Babylonian Observations?," in *Isis,* **58,** pt. 3, no. 193 (1967), 398–401.

IVOR BULMER-THOMAS

CONRAD, TIMOTHY ABBOTT (*b.* near Trenton, New Jersey, 21 June 1803; *d.* Trenton, New Jersey, 8 August 1877), *paleontology, malacology.*

Conrad was one of the first American paleontologists to correlate regional Tertiary strata on the basis of their contained fauna, to compare American fossils with foreign fossils, and to attempt intercontinental correlation. He was the son of Elizabeth Abbott and of Solomon White Conrad, a printer, a minister of the Society of Friends, and professor of botany at the University of Pennsylvania. The family often entertained such eminent naturalists as Say, Nuttall, and Rafinesque. In this environment young Conrad became so interested in science that he was removed from the rolls of the Society of Friends for taking nature walks on the Sabbath. He attended a Quaker school in West Town, Pennsylvania; and although he never went to college, he taught himself Latin, Greek, French, and natural history. He had a talent for drawing and learned lithography in his father's printing shop. These skills served him well in later years, when he prepared many of his own illustrations.

In Conrad's day a naturalist who discovered a new species rarely did more than publish a description of the fossil and place it in his personal cabinet for display. Conrad, however, went on to try to determine its geologic significance. He did this with Tertiary fossils that he collected in the southern and Atlantic states and also with those collected by explorers in the Far West. The resulting papers, together with some on modern marine and freshwater mollusks, the Silurian and Devonian fossils of New York state, the correlation of American and European Cretaceous rocks, and numerous papers on particular genera of mollusks, led to his recognition as an outstanding authority on both paleontology and malacology. He also wrote papers on the general geology of the eastern United States and published the first geologic map of Alabama.

Between 1830 and 1837 Conrad published twenty-two scientific papers, preparing most of the illustrations himself and defraying the cost of publication by subscriptions for collections of specimens and by the sale of his publications. During this time he was often almost destitute. He printed only limited editions of his early papers; and after running off a given number of plates for the first edition, he would grind off the stones used in making its illustrations. If a second edition was needed, the plates were not identical with those in the first edition.

Conrad relied on the hospitality of friends when he was doing fieldwork and financed his expeditions by borrowing from his scientific colleagues, repaying them with collections of specimens. In Alabama he traveled by coach when he had the fare, solicited rides or went afoot when he did not, and made the best of such lodgings as he could find.

In 1837, at the age of thirty-four, Conrad became paleontologist of the New York State Geological Survey and for the first time received a salary. As a result of wise investments in railroad stock he ultimately became financially independent.

Conrad was one of a small group of scientists who in 1840 organized the Association of American Geologists, the predecessor of the American Association for the Advancement of Science. He was an honorary member of many scientific societies at home and abroad; and although he seemed to make light of these honors, he describes himself in one of his papers as "Paleontologist, State of New York; Member Academy of Natural Sciences of Philadelphia; of the Imperial Society of Natural History of Moscow, etc. etc."

His descriptions of fossils are commonly too brief, and some of his illustrations of small shells are unclear because he made it a practice to draw each specimen at its natural size. Conrad wrote letters and labels on odd scraps of paper in an illegible hand and was often careless in giving references and describing localities. He rejected Darwin's theory of evolution, holding instead to the viewpoint of special creation, although he accepted Lyell's uniformitarianism.

Conrad was absent-minded, moody, and often melancholy to the verge of suicide. Although he longed for a home of his own, he never married; and in his later years he was so depressed by failing memory that he sought seclusion. He sometimes wrote poetry, and his letters contained vivid descriptions of the countryside that he so dearly loved. For instance, of South Carolina he wrote:

> The pine forest is here and there varied in the low, moist, richer portions of soil, by a growth of oak, hickory, and a variety of less conspicuous trees, nearly every one of which wears a beard of Spanish moss which a Turk might envy, and on almost every limb, stripped of its panoply of leaves, the mistletoe, as if in pity, hangs its emerald and perennial mantle.

BIBLIOGRAPHY

I. ORIGINAL WORKS. Among Conrad's most important works are *Fossil Shells of the Tertiary Formations of North America* (Philadelphia, 1832–1835), repub. by Harris (1893); *Eocene Fossils of Claiborne, With Observations on This Formation in the United States, and a Geological Map of Alabama* (Philadelphia, 1835), repub. by Harris (1893); and *Fossils of the Medial Tertiary of the United States* (Philadelphia, 1838–1861), repub. by Dall (1893). Numerous papers were published, primarily in *Journal of the Academy of Natural Sciences of Philadelphia,* **6–8**; 2nd ser., **1, 2, 4** (1830–1860); *American Journal of Science,* **23, 39, 41**; 2nd ser., **1, 2, 41** (1833–1866); *Proceedings of the Academy of Natural Sciences of Philadelphia,* **1–3, 6–9, 12, 14, 16, 17, 24, 26–28** (1842–1875); and *American Journal of Conchology,* **1–6** (1865–1871). Bibliographies of Conrad's writings are in Moore and Wheeler (below).

II. SECONDARY LITERATURE. Works that deal with Conrad's career are C. C. Abbott, "Timothy Abbott Conrad," in *Popular Science Monthly,* **47** (1895), 257–263; W. H. Dall, *Republication of Conrad's Fossils of the Medial Tertiary of the United States With an Introduction* (Philadelphia, 1893); G. D. Harris, *Republication of Conrad's Fossil Shells of the Tertiary Formations of North America* (Washington, D.C., 1893), which includes *Eocene Fossils of Claiborne*; J. B. Marcou, 'The Writings of Timothy Conrad,' in "Bibliography of Publications Relating to the Collection of Fossil Invertebrates in the United States National Museum," in *Bulletin of the U.S. National Museum,* no. 30 (1885), 205–222; G. P. Merrill, "Contributions to the History of American Geology," in *Annual Report of the U.S. National*

Museum for 1904 (1906), 189–733, see 306, 320, 354, 355, 357, 368, 396, 397, 693; E. J. Moore, "Conrad's Cenozoic Fossil Marine Mollusk Type Specimens at the Academy of Natural Sciences of Philadelphia," in *Proceedings of the Academy of Natural Sciences of Philadelphia,* **114** (1962), 23–120; and H. E. Wheeler, "Timothy Abbott Conrad (1803–1877), With Particular Reference to His Work in Alabama One Hundred Years Ago," in *Bulletin of American Paleontology,* **23,** no. 77 (1935), i–x, 1–159.

ELLEN J. MOORE

CONSTANTINE THE AFRICAN (*b.* Carthage, North Africa; *d.* Monte Cassino, Italy; *fl.* 1065–1085), *medicine.*

Constantine the African, the first important figure in the transmission of Greco-Arabic science to the West, may have "freed Salerno to speak" (in Karl Sudhoff's phrase), but not to speak about him; and his life and career remain clouded and confused. The most credible account of Constantine's early life is that given by "Matthaeus F." (Ferrarius?), a Salernitan physician of the mid-twelfth century, in the course of a gloss on the *Diete universales* of Isaac Judaeus. According to this, Constantine was a Saracen merchant who, on a visit to the court of the Lombard prince of Salerno in southern Italy, learned from a cleric physician there that Salerno had no Latin medical literature. He immediately returned to North Africa for three years' study and came back to Salerno with a supply of medical texts in Arabic (some of which were damaged in a storm during the crossing), perhaps as early as 1065. Within a few years he had become a Christian and joined the Benedictine community at nearby Monte Cassino. Most of the Latin medical texts bearing his name show signs of having been written at the monastery: two are dedicated to its abbot, Desiderius (later Pope Victor III), and others to Johannes Afflacius, another Muslim turned Christian monk, and Constantine's disciple. It has become traditional to place his death in 1087, but the date seems to rest on no satisfactory evidence.

In a biographical note, Peter Deacon, the untrustworthy historian of the monastery of Monte Cassino, listed some twenty works that the West owed to Constantine. Although this list is clearly incomplete, a precise itemization of the Constantinian writings is not yet possible. He translated a number of books of classical authors from Arabic into Latin (e.g., Hippocrates' *Aphorisms* and *Prognostics* with Galen's commentaries thereupon, and a summary version of Galen's *Megatechne*); but he plainly felt no particular urgency about making such writings available, perhaps because a few Latin versions of classical medicine had remained in use in Europe since late antiq-

uity. The most extensive and important group of texts bearing his name is instead that which for the first time communicated the expanded Arabic medical tradition. We cannot determine the sources of all of these, but it is quite possible that a number of the shorter treatises that now appear to be Constantine's own compositions will prove to be translations. Many of the identifiable translations are of the works of Isaac Judaeus (Isḥāq al-Isrāʿīlī) on urines (perhaps Constantine's first effort), on fevers, and on diets; others are of the works of Ibn al-Jazzār (d. 1009), most notably the text that Constantine entitled *Viaticum.* It is not possible to say much about the order in which the translations were made, but the two Greco-Arabic medical compendia, the *Viaticum* and the *Pantechne* or *Pantegni* (Constantine's version of the *Kitāb al-mālikī* of Haly Abbas [ʿAlī ibn al-ʿAbbās, d. 994]), seem to have been produced relatively late. The *Pantechne,* divided into two ten-chapter sections, one dealing with *theorica* and one with *practica,* was certainly the most ambitious and the most influential of Constantine's productions. As it happens, Constantine was apparently unable to finish the translation of the second half; his copy of the original may have been damaged on his voyage to Salerno. At any rate, internal evidence suggests that he translated only chapters 1–3, part of 9 (the surgery), and perhaps 10, of the *practica,* and that his student Johannes Afflacius completed the translation later.

The *Kitāb al-mālikī* was translated a second time in 1127 by Stephen of Antioch, under the title of *Regalis dispositio;* in passing, Stephen complained that the earlier translation had been incomplete (he had presumably not seen the text in the form completed by Afflacius) and that Constantine had besides suppressed Haly Abbas' name in favor of his own as author. This charge of plagiarism has been leveled repeatedly against Constantine ever since, and the reasons for it lie in his approach to translation. For one thing, he was by no means intent upon exactly reproducing whatever text was in question; rather, as his prefaces reveal, he saw himself as *coadunator,* with the responsibility of summarizing or expanding the substance of the original, perhaps adding material from other sources, in whatever way was best suited to the needs of an essentially ignorant Western audience. It would certainly be wrong to look for any consistent plan, any impulse to systematization or comprehensiveness, in his writings; he was composing primarily to satisfy requests or to fill whatever practical and pedagogical needs arose; and it was this that produced so many explanatory additions. Obviously, therefore, it is nearly impossible to draw a sharp line between a greatly expanded translation, such as the

Antidotarium that seems to derive from the *Pantechne*, and an original collection of material bearing on a particular subject, such as the *Liber de stomacho* that Constantine assembled for his friend Archbishop Alfanus I of Salerno. In a sense, they are equally Constantine's own creations. The fact remains, of course, that Constantine did not always identify his sources. But it should be noted that while he was regularly silent about his indebtedness to Islamic authors (Haly Abbas, Ibn al-Jazzār), he was consistently open about that to Isaac Judaeus—and the suggestion that he was trying not to disturb Christian or Benedictine sensibilities in a land only recently retaken from his former coreligionists is at least a possibility.

Constantine's writings had a very considerable effect upon twelfth-century Salerno. (As the core of the collection entitled *Ars medicine* or *Articella*, which was the foundation of much European medical instruction well into the Renaissance, they exerted a more diffuse influence for centuries.) Johannes Afflacius seems to have fostered their gradual assimilation, continuing the Constantinian program of translation while in association with the medical school at Salerno (there is no evidence that Constantine ever taught there), and by mid-century the Constantinian corpus had become central to Salernitan education. It did not merely enlarge the sphere of practical competence of the Salernitan physicians; it had the added effect of stimulating them to try to organize the new material into a wider, philosophical framework. Constantine had repeatedly insisted that medicine should be treated as a fundamental constituent of natural philosophy, and this attitude was encouraged by the first half of his *Pantechne* (the *theorica*), the first book available to the Salernitans that provided a framework to accommodate all their practical knowledge and to allow them to express and unify it. The achievements of twelfth-century *"Hochsalerno,"* of such writers as Urso of Calabria, mark the eventual triumph of this attitude. In this sense Constantine did indeed "free Salerno to speak."

BIBLIOGRAPHY

I. ORIGINAL WORKS. The study of Constantine's work presents great technical difficulties. His writings were included in two sixteenth-century collections, *Opera omnia Ysaac* (Lyons, 1515) and *Constantini Africani opera* (Basle, 1536), of which the former usually provides the better texts. However, neither is really satisfactory or even adequate; MS versions (the earlier the better) are regularly more coherent. What is really required for each of his works is a study of the complicated MS tradition on which to found a careful edition. A partial list of MSS of Constantine's writings (followed by a short passage in which he summarizes his general aims and expresses his conviction that medicine is the fundamental science) is printed in J. P. Migne, *Patrologia Latina*, CL (Paris, 1880), cols. 1559–1566; a number of such MSS may also be found cited in the articles by Heinrich Schipperges referred to below. Some of Constantine's works have been reprinted relatively recently, unfortunately none critically: his *Chirurgia* (from the *Pantegni*) by J. L. Pagel, "Eine bisher unveröffentlichte Version der Chirurgie der Pantegni nach einer Handschrift der Königliche Bibliothek zu Berlin," in *Archiv für klinische Chirurgie*, **81** (1906), 735–786; the *Microtegni seu de spermate* by V. Tavone Passalacqua (Rome, 1959); and several by Marco T. Malato and Umberto de Martini, *Della melancolia* (Rome, 1959), *Chirurgia* (Rome, 1960), *L'arte universale della medicina* (Rome, 1961), and *Il trattato di fisiologia e igiene sessuale* (Rome, 1962). No one has tried seriously to analyze the authenticity or the sources of the individual works attributed to Constantine since Moritz Steinschneider's "Constantinus Africanus und seine arabischen Quellen," in *Virchows Archiv für pathologische Anatomie*, **37** (1866), 351–410, and a new such study would be valuable.

II. SECONDARY LITERATURE. The account of Constantine's life provided by "Matthaeus F.," summarized in this article, has been printed and analyzed by Rudolf Creutz, "Die Ehrenrettung Konstantins von Afrika," in *Studien und Mitteilungen zur Geschichte des Benediktiner-Ordens*, **49** (1931), 35–44. Two others exist. One, publ. by Charles Singer, "A Legend of Salerno. How Constantine the African Brought the Art of Medicine to the Christians," in *Bulletin of the Johns Hopkins Hospital*, **28** (1917), 64–69, was recognized by Singer as patently false. The other is that given by Peter Deacon, printed in J. P. Migne, *Patrologia Latina*, CLXXIII (Paris, 1894), cols. 766–768, 1034–1035. According to Peter, Constantine left his native Carthage for "Babylon," India, Ethiopia, and Egypt, where in 39 years of study he mastered grammar, dialectic, medicine, and the mathematical sciences of the East. He returned to Africa only to meet with jealousy and hatred, and hurriedly took ship from Carthage to Salerno, where he lived quietly until brought to the attention of Duke Robert (Guiscard) by the brother of the "king of Babylon"; he subsequently became a monk at Monte Cassino and there made his translations and died "full of days." This story, which has become the commonly accepted account of Constantine's life, seems somewhat less realistic than that given by Matthaeus.

The most extensive modern treatments of Constantine's life and work are those of Rudolf Creutz: "Der Arzt Constantinus von Monte Cassino. Sein Leben, sein Werk und seine Bedeutung für die mittelalterliche medizinische Wissenschaft," in *Studien und Mitteilungen zur Geschichte des Benediktiner-Ordens*, **47** (1929), 1–44, and "Additamenta zu Konstantinus Africanus und seinen Schülern Johannes und Atto," *ibid.*, **50** (1932), 420–442, as well as the article previously cited. Lynn Thorndike's discussion in his *History of Magic and Experimental Science*, I (New York, 1923), ch. 32, is also of interest. Constantine's role in the translation movement has been carefully examined by Heinrich Schipperges, "Die übersetzer der arabischer Medizin in chronologischer Sicht," in *Sudhoffs Archiv für Geschichte der Medizin,* **38** (1954), 53–93, and *Die Assimilation der arabischen Medizin durch das lateinische Mittelalter* (*Sudhoffs Archiv für Geschichte der Medizin,* Beiheft 3; Wiesbaden, 1964), pp. 17–54. His role in the development of the Salernitan medical school has been considered by Karl Sudhoff, "Constantin, der erste Vermittler muslimischer Wissenschaft ins Abendland und die beiden Salernitaner Frühscholastiker Maurus und Urso, als Exponenten dieser Vermittlung," in *Archeion,* **14** (1932), 359–369, and more thoroughly by Paul Oskar Kristeller, "The School of Salerno: Its Development and Its Contribution to the History of Learning," in *Bulletin of the History of Medicine,* **17** (1945), 138–194, repr. in *Studies in Renaissance Thought and Letters* (Rome, 1956), pp. 495–551. Few studies of the content of individual works have been made. Rudolf and Walter Creutz, "Die 'Melancholia' des Konstantinus Africanus und seine Quellen," in *Archiv für Psychiatrie,* **97** (1932), 244–269, give a German trans. of the *De melancolia,* discuss its sources, and appraise its psychology.

Michael McVaugh

CONYBEARE, WILLIAM DANIEL (*b.* London, England, June 1787; *d.* Llandaff, Wales, 12 August 1857), *geology.*

Conybeare was the younger son of Rev. William Conybeare, rector of St. Botolph's, Bishopsgate, London. He was educated at Westminster School and Christ Church, Oxford. On marrying in 1814, he took a curacy in Suffolk, became rector of Sully (Glamorganshire) in 1822, took his family living as vicar of Axminster (Devon) in 1836, and became dean of Llandaff in 1845. An early member (1811) of the Geological Society of London, he was elected a fellow of the Royal Society in 1832. In addition to his scientific work he published works on biblical and patristic theology and was Bampton lecturer at Oxford in 1839.

Conybeare was one of the most active early members of the "Oxford school" of geology and was a close associate of William Buckland. He was one of the most able British exponents of the synthesis of progressionism and catastrophism, which dominated geology in the 1820's and 1830's.

His most important single work was his great enlargement and improvement of William Phillips' compilation of English stratigraphy. This created a synopsis of stratigraphical knowledge that was at the time unrivaled in detail and accuracy (1). In the general introduction to the *Outlines,* Conybeare considered the range of "actual causes" but regarded them as inadequate to explain such phenomena as the "diluvium" (glacial drift) and the form of valleys; for these he proposed diluvial explanations, although without stressing any concordance with the scriptural Flood. The *Outlines* described British stratigraphy back to the Carboniferous and was termed "Part I"; Adam Sedgwick was to have assisted Conybeare with a second volume on the earlier strata, but it was never published. Conybeare collaborated with Buckland in a stratigraphical memoir on the coal fields around Bristol (2) that was much admired as a model of clear description and reasoned inference; he also attempted a general correlation with Continental stratigraphy and tectonics (3).

Some fragmentary fossil remains from the Lias of Lyme Regis prompted Conybeare's main work in paleontology. From a detailed comparison of normal reptiles and the highly aberrant *Ichthyosaurus,* he inferred that the new remains were intermediate in anatomy. This reconstruction of the *Plesiosaurus,* which excited great interest, was later confirmed by the discovery of a more complete skeleton (4). His functional anatomy clearly was modeled on Cuvier; but he stressed the interest of intermediate forms as "links in the chain" of organisms, showing, however, by an explicit rejection of Lamarck's transmutation, that the chain was that of an *échelle des êtres,* not an evolutionary series.

Conybeare's exposition of the catastrophist-progressionist synthesis was both more able and more moderate than that of Buckland. He argued in 1829 that the fluvial erosion postulated by Lyell for the valleys of central France was inadequate to explain the form of the valleys of the Thames and other British rivers and suggested that the more powerful agency of a "diluvial" episode was required to account for them (5). He defended the progressionist viewpoint on directional climatic change against the criticisms of John Fleming (6) and later (1830–1831) wrote one of the most important defenses of the whole progressionist synthesis in answer to the more radical attack of Lyell's *Principles of Geology* (7). The moderate and flexible character of his catastrophism is shown, however, by his skepticism about Élie de Beaumont's theory of the parallel and paroxysmal elevation of mountain ranges: here he not only criticized the hasty generalization of the theory and its inapplicability to British geology but also emphasized the slow and gradual nature of many—though not all—tectonic movements (8). His presidency of the Geology Section of the British Association in 1832 gave him an opportunity to review the general progress of the science (9). Here, and later in an important letter to Lyell (10), he expounded his own theoretical viewpoint. This combined an actualistic method in geology with an acceptance of occasional paroxysmal episodes, the overall trend of earth history being one of progressive diminution in the intensity of geological processes coupled with a progressive rise in the complexity of the organic world, culminating in the appearance of man.

BIBLIOGRAPHY

Works are listed in the order cited in the text.

(1) Conybeare and W. Phillips, *Outlines of the Geology of England and Wales, With an Introductory Compendium of the General Principles of That Science, and Comparative Views of the Structure of Foreign Countries. Part I* [all issued] (London, 1822).

(2) W. Buckland and Conybeare, "Observations on the South Western Coal District of England," in *Transactions of the Geological Society of London,* 2nd ser., **1**, pt. 1 (1822), 210–316.

(3) Conybeare, "Memoir Illustrative of a General Geological Map of the Principal Mountain Chains of Europe," in *Annals of Philosophy,* n.s. **5** (1823), 1–16, 135–149, 210–218, 278–289, 356–359; n.s. **6** (1824), 214–219.

(4) Conybeare and H. T. De La Beche, "Notice of a Discovery of a New Fossil Animal, Forming a Link Between the Ichthyosaurus and the Crocodile; Together With General Remarks on the Osteology of the Ichthyosaurus," in *Transactions of the Geological Society of London,* **5** (1821), 558–594; Conybeare, "Additional Notices on the Fossil Genera Ichthyosaurus and Plesiosaurus," *ibid.,* 2nd ser., **1,** pt. 1 (1822), 103–123; and "On the Discovery of an Almost Perfect Skeleton of the Plesiosaurus," *ibid.,* pt. 2 (1824), 381–389.

(5) Conybeare, "On the Hydrographical Basin of the Thames, With a View More Especially to Investigate the Causes Which Have Operated in the Formation of the Valleys of That River, and Its Tributary Streams," in *Proceedings of the Geological Society of London,* **1,** no. 12 (1829), 145–149.

(6) Conybeare, "Answer to Dr Fleming's View of the Evidence From the Animal Kingdom, as to the Former Temperature of the Northern Regions," in *Edinburgh New Philosophical Journal,* **7** (1829), 142–152.

(7) Conybeare, "On Mr Lyell's 'Principles of Geology,'" in *Philosophical Magazine and Annals,* n.s. **8** (1830), 215–219; and "An Examination of Those Phaenomena of Geology, Which Seem to Bear Most Directly on Theoretical Speculations," *ibid.,* 359–362, 401–406; n.s. **9** (1831), 19–23, 111–117, 188–197, 258–270.

(8) Conybeare, "Inquiry How Far the Theory of *M. Élie de Beaumont* Concerning the Parallelism of the Lines of Elevation of the Same Geological Area, Is Agreeable to the Phaenomena as Exhibited in Great Britain," in *Philosophical Magazine and Journal of Science,* **1** (1832), 118–126; **4** (1834), 404–414.

(9) Conybeare, "Report on the Progress, Actual State and Ulterior Prospects of Geological Science," in *Report of the British Association for the Advancement of Science, 1831–2* (1833), pp. 365–414.

(10) M. J. S. Rudwick, "A Critique of Uniformitarian Geology: A Letter From W. D. Conybeare to Charles Lyell, 1841," in *Proceedings of the American Philosophical Society,* **111** (1967), 272–287.

In addition, see F. J. North, "Dean Conybeare, Geologist," in *Transactions of the Cardiff Naturalists' Society,* **66** (1933), 15–68.

M. J. S. RUDWICK

COOK, JAMES (*b.* Marton-in-Cleveland, Yorkshire, England, 27 October 1728; *d.* Kealakekua Bay, Hawaii, 14 February 1779), *maritime discovery.*

The son of James Cook, a farm laborer from Scotland, and his Yorkshire wife, Grace Pace, Cook inherited a vigorous constitution and mind; and although his formal education was only elementary, in 1776 he became a fellow of the Royal Society. He married Elizabeth Batts, of Shadwell, in 1762; only one of their six children survived to maturity.

Cook was apprenticed to John Walker, a Whitby shipowner, at the age of seventeen. His training under the arduous conditions of the North Sea, combined with his natural capacity, made him a first-rate seaman and practical navigator; and in 1755 Walker offered him the command of a ship. Cook, however, preferred to volunteer into the navy as an able seaman. Rapidly promoted master of the sixty-four-gun *Pembroke,* and transferred to the flagship *Northumberland* on the American station in 1759, he was active in surveying the St. Lawrence River before the fall of Quebec and then in further survey work; he learned much from Samuel Holland, the distinguished military surveyor, and assiduously studied the mathematics of navigation. Cook's ability led to his appointment in 1763 to carry out a detailed survey of Newfoundland. His charts were favorably noticed by the Admiralty, and his observation of an eclipse of the sun by the Royal Society; so that when a commander was needed for the expedition to Tahiti to observe the transit of Venus (3 June 1769), Cook seemed an excellent choice.

He sailed in the *Endeavour* in July 1768, entering the Pacific around Cape Horn and arriving in Tahiti in April 1769. Besides observing the transit, he charted the Society and other islands. In August he sailed south to carry out secret instructions: to search for a continent down to latitude 40° south and, if none was found, to go to New Zealand (discovered by Tasman in 1642); he was then to return by whatever route he thought best. The result was a masterly circumnavigation and charting of New Zealand, the discovery and charting of the whole east coast of Australia, and the rediscovery of Torres Strait. After refitting at Batavia, Java, Cook reached England in July 1771. The harvest of this voyage, in geographical, ethnological, and botanical knowledge (Joseph Banks and Daniel Solander were natural history observers), was enormous; its conduct was so brilliant that the Admiralty resolved to send Cook out again on a plan of his own suggestion: to answer finally the question whether or not there was a continent in the Southern Hemisphere.

This Cook did with the ships *Resolution* and *Adventure,* in perhaps the most remarkable voyage ever carried out (July 1772–July 1775). Plunging south from the Cape of Good Hope and sailing east, he circumnavigated the world, utterly destroying the ancient hypothesis of a great southern continent, and reached latitude 71° 10′ south. Using New Zealand and Tahiti as bases for the recruitment of his men, in the warmer latitudes he made new and coordinated old discoveries—among them Easter Island, Tonga, the Marquesas, the New Hebrides, and New Caledonia; finally, he charted part of Tierra del Fuego, South Georgia, and the South Sandwich group. He returned to England without having a single man on the *Resolution* die of the dread scurvy, an astonishing achievement. This voyage was also remarkable for its proof of the chronometer's utility as a navigational instrument.

Cook, promoted post-captain, spent a year writing an account of this voyage and preparing for a third, which he had volunteered to lead. It was to explore the possibility of a northwest passage through America, working from the Pacific coast. He sailed in July 1776 with the *Resolution* and *Discovery,* called at the Cape of Good Hope and New Zealand, and, because of contrary winds, spent some months at Tonga and Tahiti before reaching the North American coast in March 1778—discovering Hawaii on the way. He traced this coast into the Bering Sea and then through Bering Strait, until he was stopped by vast ice fields at 70° 10′N. Before a second attempt he returned to winter in Hawaii. Here he was killed while attempting to recover a stolen ship's boat, an ironic end for a man so humane to the peoples he discovered. Cook was also humane as a commander—his conquest of scurvy at sea would alone have made him famous—a consummate planner of voyages as well as a practical seaman, observer, marine surveyor, and hydrographer. His contribution to knowledge of the Pacific Ocean, in terms of geography, natural history, and ethnology, was correspondingly immense.

BIBLIOGRAPHY

I. ORIGINAL WORKS. The only work that can justly be reckoned as from Cook's own hand is *A Voyage Towards the South Pole, and Round the World. Performed in His Majesty's Ships the Resolution and Adventure, in the Years 1772, 1773, 1774, and 1775,* John Douglas, ed., 2 vols. (London, 1777). The eighteenth-century accounts of Cook's voyages, from his own journals, have been superseded by the Hakluyt Society's ed. of *The Journals of Captain James Cook on His Voyages of Discovery,* J. C. Beaglehole, ed., 3 vols. (Cambridge, 1955–1967).

II. SECONDARY LITERATURE. The standard life is Arthur Kitson, *Captain James Cook* (London, 1907). Smaller good biographies are Hugh Carrington, *The Life of Captain Cook* (London, 1939); and James A. Williamson, *Cook and the Opening of the Pacific* (London, 1946). See also Maurice Holmes, *Captain James Cook, R.N., F.R.S. A Bibliographical Excursion* (London, 1952).

J. C. BEAGLEHOLE

COOKE, JOSIAH PARSONS, JR. (*b.* Boston, Massachusetts, 12 October 1827; *d.* Newport, Rhode Island, 3 September 1894), *chemistry.*

The son of Josiah Parsons Cooke, a prominent lawyer, and Mary Pratt Cooke, Josiah came from a socially prominent and wealthy family. He graduated from Harvard College in 1848 and spent a year in Europe to improve his health. Throughout his life he suffered from poor eyesight and a tremor of his hands. In 1860 he married Mary Hinckley Huntington, who survived him. They had no children.

Cooke was tutor in mathematics at Harvard in 1849 and instructor in chemistry. In 1850 he was appointed Erving professor of chemistry and mineralogy, a position that he held for the rest of his life. Following this appointment he went to Europe to buy apparatus and chemicals, mostly at his own expense, and attended lectures by Regnault and Dumas.

Laboratory instruction in chemistry combined with demonstration experiments during lectures had been developed by Liebig and had been brought to the Lawrence Scientific School of Harvard by E. N. Horsford, a pupil of Liebig's. Cooke became enthusiastic about this method and was assigned a room about twenty by twenty-five feet in the basement of University Hall for a student laboratory. It took seven years of hard fighting, however, before his course was recognized by the college as anything beyond an extra. Finally, in 1871, new accommodations were secured by addition of a story to Boylston Hall.

In Harvard Hall there was a large miscellaneous collection of rocks, minerals, and fossils. Cooke retained Benjamin Silliman, Sr., of Yale to sort this, and the result became the nucleus for the renowned collection of minerals now housed in the University Museum.

Cooke can be considered the founder of Harvard's department of chemistry. All his life he strove for expansion of space, equipment, and personnel, providing much of the equipment from his own funds. His lectures were extremely popular with the students. Cooke was a member of a number of learned societies, notably the American Academy of Arts and Sciences and the Chemical Society of London (foreign honorary member). He received the LL.D. from Cambridge in 1882 and the same degree from Harvard in 1889. Before his death he was so exasperated by the failure of the university to promote Oliver W. Huntington, his wife's nephew and his long-time collaborator, that he canceled a large bequest in its favor.

Cooke's first paper, "The Numerical Relation Between the Atomic Weights and Some Thoughts on the Classification of the Chemical Elements" (1854), attracted wide attention. He maintained that the elements could be arranged in six series, in the manner of organic compounds. In each series atomic weights progressed by integral multiples of an integer peculiar to that series. Physical and chemical properties progressed analogously.

Cooke carried out two important determinations of atomic weights. In his papers on antimony (1877, 1878, 1879, 1882) he identified three crystalline forms of antimony triiodide. He precipitated antimony sulfide, chloride, bromide, and iodide with silver nitrate, avoiding any precipitation of basic salts by addition of tartaric acid. He analyzed the silver salts of the four anions involved. These resulted in four sets of three ratios each: $Ag:S:Sb$, $Ag:Cl:Sb$, $Ag:Br:Sb$, and $Ag:I:Sb$. The results were most satisfactory with the bromide, where $Ag:Br:Sb = 107.66:79.75:119.63$, based on $H = 1.0000$. These are close to modern values.

In 1842 Dumas had determined the very important ratio of oxygen to hydrogen in water by passing an indeterminate amount of hydrogen over heated cupric oxide and weighing the collected water. Cooke and his pupil Theodore William Richards weighed hydrogen in a five-liter bulb and swept it out over heated cupric oxide by a stream of purified air. The hydrogen was prepared by several different reactions. After the first publication of results (1887) John Strutt, the third Baron Rayleigh, pointed out that the volume of hydrogen in the bulb, when evacuated, contracted slightly, so that it displaced less air when weighed. A valid correction was worked out in an "Additional Note" (1888). The final result was an atomic weight of $0 = 15.869 \pm 0.0017$, taking $H = 1.0000$.

Cooke's *Principles of Chemical Philosophy*, which concluded each chapter with many questions and problems, had wide influence. The *London Chemical News* said: "So far as our recollection goes, we do not think that there exists in any language a book on so difficult a subject as this so carefully, clearly and lucidly written." The *American Journal of Science* said: "To Professor Cooke, more than to any American, is due the credit of having made chemistry an exact and disciplinary study in our colleges."

Cooke gave courses of popular lectures in various cities—Lowell and Worcester, Massachusetts, New York, and Washington, D.C. The course at Brooklyn Institute of Arts and Sciences in New York in 1860 was published as *Religion and Chemistry or Proof of God's Plan in the Atmosphere and the Elements* (1864). He maintained that the argument from design in nature is not invalidated by theories of evolution. A similar book, *Credentials of Science and the Warrant of Faith,* appeared in 1888.

BIBLIOGRAPHY

I. ORIGINAL WORKS. A list of Cooke's more important publications is in *Proceedings of the American Academy of Arts and Sciences,* **30** (1895), 544–547. His papers include "The Relation Between the Atomic Weights," in *Memoirs of the American Academy of Arts and Sciences,* n.s. **5** (1854), and in *American Journal of Science,* 2nd ser., **17** (1854), 387; "On the Process of Reverse Filtering," in *Proceedings of the American Academy of Arts and Sciences,* **12** (1877), 124–130; "Revision of the Atomic Weights of Antimony," *ibid.,* **13** (1878), 1–71; "Re-examination of Some of the Haloid Compounds of Antimony," *ibid.,* 72–114; "The Atomic Weight of Antimony," *ibid.,* **15** (1879), 251–255; "Contributions From the Chemical Laboratories of Harvard College," *ibid.,* **17** (1882), 1–22; "The Relative Values of the Atomic Weights of Oxygen and Hydrogen," *ibid.,* **23** (1887), 149–178, written with T. W. Richards; and "Additional Note on the Relative Values of the Atomic Weights of Oxygen and Hydrogen," *ibid.,* **23** (1888), 182, written with T. W. Richards.

Among his books are *Elements of Chemical Physics* (Boston, 1860, 1866, 1877); *Chemical Problems and Reactions to Accompany Stockhard's Elements of Chemistry* (Philadelphia, 1863); *Religion and Chemistry or Proof of God's Plan in the Atmosphere and the Elements* (New York, 1864; new ed., 1880); *Principles of Chemical Philosophy* (Cambridge, Mass., 1868; rev. ed., enl., Boston, 1887); and *The Credentials of Science and the Warrant of Faith* (New York, 1888; new ed., 1893).

II. SECONDARY LITERATURE. There is an unsigned biographical sketch in *Popular Science Monthly* (Feb. 1877). More extensive is *Addresses in Commemoration of Josiah Parsons Cooke. . .* (Cambridge, Mass., 1895), which includes a biographical sketch. The addresses also appeared in

Proceedings of the American Academy of Arts and Sciences, **30** (1895), 513–543.

GEORGE S. FORBES

COOLIDGE, JULIAN LOWELL (*b.* Brookline, Massachusetts, 28 September 1873; *d.* Cambridge, Massachusetts, 5 March 1954), *mathematics.*

Coolidge was the son of John Randolph Coolidge, a lawyer, and his wife, the former Julia Gardner. He received the B.A. at Harvard in 1895 and the B.Sc. at Oxford in 1897. From 1897 to 1899 he taught at the Groton School, then became an instructor at Harvard and in 1902 joined its faculty. From 1902 to 1904 he studied abroad, where his work with Corrado Segre at Turin and Eduard Study at Bonn decisively influenced his scientific career. In 1904 he received the Ph.D. at Bonn with a thesis entitled *Die dual-projektive Geometrie im elliptischen und sphärischen Raume.*

Back at Harvard, Coolidge became assistant professor in 1908 and full professor in 1918. During 1918–1919 he was a liaison officer to the French general staff, and in 1919 he organized courses at the Sorbonne for American servicemen. He returned to the Sorbonne in 1927 as exchange professor. From 1929 until his retirement in 1940 he was master of Lowell House at Harvard. He married Theresa Reynolds; they had two sons and five daughters.

Coolidge's mathematical career can be followed through his books (all, except his thesis, published by the Clarendon Press, Oxford). Four are in the tradition of the Study-Segre school, with many original contributions: *The Elements of Non-Euclidean Geometry* (1909), *A Treatise of the Circle and the Sphere* (1916), *The Geometry of the Complex Domain* (1924), and *A Treatise on Algebraic Plane Curves* (1931). In a class by itself is *Introduction to Mathematical Probability* (1925), one of the first modern English texts on this subject.

The last three books, *A History of Geometrical Methods* (1940), *A History of the Conic Sections and Quadric Surfaces* (1943), and *The Mathematics of Great Amateurs* (1949), reflect the interest that Coolidge, in his later years, showed in the history of mathematics.

Coolidge was an enthusiastic teacher with a flair for witty remarks. He was also a distinguished amateur astronomer.

BIBLIOGRAPHY

I. ORIGINAL WORKS. Coolidge's books are mentioned in the text. Among his papers are "Quadric Surfaces in Hyperbolic Space," in *Transactions of the American Mathe-* matical Society, **4** (1903), 161–170; "A Study of the Circle Cross," *ibid.,* **14** (1913), 149–174; "Congruences and Complexes of Circles," *ibid.,* **15** (1914), 107–134; "Robert Adrain and the Beginnings of American Mathematics," in *American Mathematical Monthly,* **33** (1926), 61–76; "The Heroic Age of Geometry," in *Bulletin of the American Mathematical Society,* **35** (1929), 19–37; and "Analytical Systems of Central Conics in Space," in *Transactions of the American Mathematical Society,* **48** (1940), 354–376.

II. SECONDARY LITERATURE. On Coolidge or his work, see M. Hammond et al., "J. L. Coolidge," in *Harvard University Gazette* (26 February 1955), 136–138; and Dirk J. Struik, "J. L. Coolidge (1873–1954)," in *American Mathematical Monthly,* **62** (1955), 669–682, with bibliography.

DIRK J. STRUIK

COOPER, THOMAS (*b.* London, England, 22 October 1759; *d.* Columbia, South Carolina, 11 May 1839), *chemistry.*

One of the pioneers in bleaching by chlorine in England (*ca.* 1790), Cooper claimed originality in using red lead instead of manganese dioxide (together with common salt and sulfuric acid) for preparing the gas. This, he said, not only gave a purer product but left a residue that was not wasted, since lead could be recovered from it by reduction. He established works at Raikes with Kempe Brydges, C. Teesdale, and Joseph Baker, who had devised the apparatus they used—much simpler and more economical than that in general use. The ingredients, with water, were fed into a large barrel, which was then rotated by hand; after allowing the contents to settle, the liquor was run off and used directly—Cooper says they only used it for the finishing process. The firm proved successful for three years but failed in the depression of 1793. Cooper seems to have lost heavily. He immigrated to America, where he later held professorships in chemistry and mineralogy. He is, however, remembered chiefly for his political agitation and tempestuous personality.

Little is known of Cooper's early life, other than what can be gleaned from his later writings—which are not always perfectly consistent. His parents were apparently wealthy, and Cooper did not lack means. In 1779 he matriculated at Oxford and also married Alice Greenwood. He read law but took no degree; he became a barrister in 1787 but seems to have practiced little. Some years before, he had interested himself in medicine, attending anatomical lectures and veterinary dissections.

It is known that by 1785 Cooper was living near Manchester (he later moved to Bolton); in that year he was elected to membership in the Manchester Literary and Philosophical Society. The papers he read on various subjects displayed erudition and gave expression to radical opinions. He left the society in 1791 in protest at the reticence shown in expressing sympathy for Priestley's losses during the Birmingham riots. He became a member of the Manchester Constitutional Society; and early in 1792, with James Watt, Jr., he visited Paris and read an address pledging the solidarity of the society with the Jacobins. For this Watt and Cooper were bitterly attacked in Parliament by Edmund Burke, who used their action in an attempt to discredit the move for parliamentary reform, against which repressive measures were soon taken.

Cooper sailed for America in August 1793 with two of Priestley's sons and some of his own family (he had five children by his first wife), returning for the remainder the following year. Priestley also emigrated in 1794, and they both settled in Northumberland, Pennsylvania (Cooper lived with Priestley for some time after the latter's wife died in 1796). In 1799 Cooper resumed political activities, embracing the republican cause; and in 1800 he was tried for sedition and libel against the president. He served six months in prison, his wife dying just before his release. He became a close friend of Jefferson after the latter became president, and from 1804 to 1811 he was a member of the state judiciary in Pennsylvania.

In 1802 Cooper became a member of the American Philosophical Society and in 1811 was offered the chair of chemistry at Dickinson College in Carlisle, Pennsylvania. He was professor of applied chemistry and mineralogy at the University of Pennsylvania (1815–1819) and professor of chemistry at South Carolina College (1819–1834); he was elected president of the college in 1821. Much of his time in South Carolina was spent in campaigning vigorously for states' rights and free trade. In retirement he compiled the statute laws of the state.

As a practicing scientist Cooper was not outstanding; his most notable achievement was probably the preparation of potassium in 1810 (almost certainly for the first time in America) by strongly heating potash with iron in a gun barrel, a method originated by Gay-Lussac and Thenard in 1808. Cooper's greatest service to science was undoubtedly the dissemination of information. His biographer, Dumas Malone, wrote (p. 399): "Perhaps no man of Cooper's generation did more than he to advance the cause of science and learning in America."

BIBLIOGRAPHY

I. ORIGINAL WORKS. Cooper brought out American editions of a number of English chemistry textbooks, adding comprehensive notes of his own on recent advances, and wrote *A Practical Treatise on Dyeing and Calicoe Printing* (Philadelphia, 1815). He gave accounts of his bleaching process in *The Emporium of Arts and Sciences* (Philadelphia), n.s. **1** (1813), 158–161 (Cooper edited the first three volumes of the new series, 1813–1815); and in "On Bleaching," in *Transactions of the American Philosophical Society*, n.s. **1** (1818), 317–324. His many works on law, politics, economics, and philosophy are listed by Dumas Malone (see below).

II. SECONDARY LITERATURE. Dumas Malone, *The Public Life of Thomas Cooper, 1783–1839* (New Haven–London, 1926; repr. New York, 1961), gives the fullest available account of his life and a bibliography (not complete). More details on his articles in periodicals and other works are given by M. Kelley, in *Additional Chapters on Thomas Cooper*, University of Maine Studies, 2nd ser., no. 15 (Orono, Me., 1930). Kelley writes of Cooper's science: "For the most part he was rather a theorizing dilettante." A chapter in E. F. Smith, *Chemistry in America* (New York–London, 1914), pp. 128–146, is devoted to Cooper. It includes a long extract from a letter written by Cooper describing his preparation of potassium. A. E. Musson and Eric Robinson, *Science and Technology in the Industrial Revolution* (Manchester, 1969), ch. 3, contains numerous references to Cooper and his firm.

See also E. V. Armstrong, "Thomas Cooper As An Itinerant Chemist," in *Journal of Chemical Education*, **14** (1937), 153–158.

E. L. SCOTT

COPAUX, HIPPOLYTE EUGÈNE (*b.* Paris, France, 9 March 1872; *d.* Etampes, France, 28 August 1934), *chemistry.*

His father died when Copaux was very young, and the boy was brought up by his mother. She sent him to a school run by the Marist Brothers, who encouraged his taste for literature. Nevertheless, Copaux was drawn to chemistry. He graduated first in his class at the École de Physique et de Chimie Industrielles de Paris and later returned to the school, first as head of the department of analytic chemistry (1900), then as professor of general chemistry (1910) and director of studies (1925). In the meantime Copaux was a special assistant to H. Moissan (1895–1897) and received the *licence-ès-sciences physiques* (1896) and the *docteur-ès-sciences physiques* (1905).

Copaux's first studies dealt with the chemistry and crystallography of mineral substances: determination of the physical properties of metallic cobalt (1), which were not accurately known (1902); preparation and study of salts of cobalt sesquioxide: cobaltic selenate, cobaltiacetate of the protoxide of cobalt, cobaltioxalates of alkaline metals (2); and preparation and study of molybdate and tungstate complexes: metallic silicomolybdates (3), including potassium and silver silicomolybdate (4) (triclinic yellow crystals with 30

H$_2$O and triclinic red crystals with 14 H$_2$O), various silicotungstates (5), borotungstates, and metatungstates (6).

During World War I, Copaux was assigned to the Patent Office, as head of the Chemical Department. He developed a process for the rapid preparation of phosphoric acid (7). In 1919 Copaux perfected a method for obtaining beryllium oxide (8) from beryl that, since it greatly facilitates the separation of impurities, is still the basis for the industrial production of beryllium: by treating beryl with sodium fluosilicate at 850°C., one obtains sodium fluoberyllate, which is soluble to the extent of twenty-eight grams per liter in boiling water; sodium fluoaluminate is only slightly hydrolyzed.

From 1925, Copaux continued to direct the laboratory experiments of his young collaborators on active hydrogen (9), beryllium and the heat of formation of beryllium oxide (10), and beryllium chloride, while assuming his new pedagogical and administrative responsibilities.

Several chapters of the *Traité de chimie minérale*, edited by Moissan (1904), were written by Copaux, who also published two works intended especially for students: *Introduction à la chimie générale* (1919) and *Chimie minérale* (1925), the latter with the collaboration of M. H. Perperot.

Copaux belonged to several French chemical and mineralogical societies. He was made a knight of the Legion of Honor in 1923 and an officer in 1933.

BIBLIOGRAPHY

The following abbreviations are used in the listing of Copaux's works cited in the text: *ACP, Annales de chimie et de physique; BSC, Bulletin de la Société chimique de France; BSM, Bulletin de la Société française de minéralogie;* and *CR, Comptes rendus hebdomadaires des séances de l'Académie des sciences.*

1. "Analyse qualitative et quantitative des composés du cobalt," in *BSC,* **29** (1903), 301; "Propriétés physiques comparatives du cobalt et du nickel purs," in *CR,* **140** (1905), 657; and "Recherches expérimentales sur le cobalt et le nickel," in *ACP,* **6** (1905), 508.

2. "Les cobaltioxalates alcalins," in *CR,* **135** (1902), 1214; and "Oxydation des acétates de cobalt et de manganèse par le chlore," *ibid.,* **136** (1903), 373.

3. "Étude chimique et cristallographique des silicomolybdates," in *ACP,* **7** (1906), 118; "De la nature des métatungstates et de l'existence du pouvoir rotatoire dans les cristaux de métatungstate de potassium," in *CR,* **148** (1909), 633; and "Constitution des paramolybdates et des paratungstates," in *BSC,* **13** (1913), 817.

4. "Étude chimique et cristallographique d'un silicomolybdate de potassium et d'argent," in *BSM,* **30** (1907), 292.

5. "Préparation des acides silicotungstiques," in *BSC,* **3** (1908), 101.

6. "Les acides borotungstiques," in *CR,* **147** (1908), 973; "Nouveaux documents sur le dosage du bore," in *BSC,* **5** (1909), 217; and "Recherches sur les tungstates complexes, en particulier sur les borotungstates et les métatungstates," in *ACP,* **17** (1909), 217.

7. "Procédé rapide pour doser l'acide phosphorique," in *CR,* **173** (1921), 656.

8. "Méthode de traitement du béryl pour en extraire l'oxyde du béryllium," in *CR,* **168** (1919), 610.

9. "Quelques expériences sur la production de l'hydrogène actif," in *BSC,* **37** (1925), 141.

10. "Chaleur d'oxydation du béryllium," in *CR,* **171** (1920), 630, and **176** (1923), 579.

ANDRÉ COPAUX

COPE, EDWARD DRINKER (*b.* Philadelphia, Pennsylvania, 28 July 1840; *d.* Philadelphia, 12 April 1897), *paleontology, zoology, natural history.*

For a detailed study of his life and work, see Supplement.

COPERNICUS, NICHOLAS (*b.* Torun, Poland, 19 February 1473; *d.* Frauenburg [Frombork], Poland, 24 May 1543), *astronomy.*

The founder of modern astronomy lost his father in 1483, when he was only a little more than ten years old. Fortunately his maternal uncle stepped into the breach, so that Copernicus was able to enter the University of Cracow in 1491. His own evaluation of his intellectual indebtedness to that institution was publicly reported as follows, at the very time that the end product of his life's work was in the process of being printed:

> The wonderful things he has written in the field of mathematics, as well as the additional things he has undertaken to publish, he first acquired at our university [Cracow] as his source. Not only does he not deny this (in agreement with Pliny's judgment that to name those from whom we have benefited is an act of courtesy and thoroughly honest modesty), but whatever the benefit, he says that he received it all from our university.[1]

Through the influence of his uncle, who had become the bishop of Varmia (Ermland), Copernicus was elected a canon of the cathedral chapter of Frombork (Frauenburg), whose members enjoyed an ample income throughout their lives. In 1496 Copernicus enrolled in the University of Bologna, officially as a student of canon law; but privately he pursued his interest in astronomy, making his earliest recorded observation on 9 March 1497. On 6 November 1500 he observed a lunar eclipse in Rome, where "he lectured on mathematics before a large audience of students and a throng of great men and experts in this branch of knowledge."[2]

On 27 July 1501 he attended a meeting of his chapter, which granted him permission to return to Italy for two more years in order to study medicine: "As a helpful physician he would some day advise our most reverend bishop and also the members of the chapter."[3] For his medical studies Copernicus chose Padua, but he obtained a doctoral degree in canon law from the University of Ferrara on 31 May 1503. Returning soon thereafter to Varmia, he spent the remaining forty years of his life in the service of his chapter.

On 31 March 1513 he bought from the chapter's workshops 800 building stones and a barrel of lime for the purpose of constructing a roofless little tower, in which he deployed three astronomical instruments. He used the parallactic instrument mainly for observing the moon; the quadrant for the sun; and the astrolabe, or armillary sphere, for the stars.

He wrote the first draft of his new astronomical system, *De hypothesibus motuum coelestium a se constitutis commentariolus,* before 1 May 1514 and discreetly circulated a few manuscript copies among trusted friends. The date is that of the catalog of a Cracow professor's books, which included a "manuscript of six leaves expounding the theory of an author who asserts that the earth moves while the sun stands still."[4] This professor was unable to identify the author of this brief geodynamic and heliostatic manuscript because Copernicus, with his customary prudence, had deliberately withheld his name from his *Commentariolus.* But a clue to the process by which his *Commentariolus* found its way into the professor's library is provided by Copernicus' statement that he reduced all his calculations "to the meridian of Cracow, because . . . Frombork . . . where I made most of my observations . . . is on this meridian [actually, Frombork lies about 1/4° west of Cracow], as I infer from lunar and solar eclipses observed at the same time in both places."[5] Furthermore, "as is clear from [lost] letters written with his own hand, Copernicus conferred about eclipses and observations of eclipses with Cracow mathematicians, formerly his fellow students."[6]

In his *Commentariolus,* Copernicus challenged the astronomical system which had dominated Western thought since the days of Aristotle and Ptolemy. Whereas these two revered authorities and their innumerable followers down through the ages insisted on centering the cosmos around the earth, Copernicus proclaimed that "the center of the earth is not the center of the universe,"[7] in which position he stationed the sun. Against the geocentrists' denial of all

motion to the earth, the *Commentariolus* treated "the earth's immobility as due to an appearance."[8] The apparent daily rotation of the heavens results from the real diurnal rotation of the earth. The apparent yearly journey of the sun through the ecliptic is caused by the earth's real annual revolution about the sun. The apparent alternation of retrograde and direct motion in the planets is produced by the earth's orbital travel.

"We revolve about the sun like any other planet."[9] These portentous words in Copernicus' *Commentariolus* assigned to the earth its rightful place in the cosmos. Yet Copernicus laid no claim to priority in this respect (or in any other, since he trod with caution over very dangerous ground). In the compact *Commentariolus* he briefly recalled that in antiquity the Pythagoreans had asserted the motion of the earth. He later identified two of these Pythagoreans when, in June 1542, he wrote that stirring plea for freedom of thought which serves as the dedication of his *De revolutionibus orbium coelestium* (*Revolutions of the Heavenly Spheres*). Therein he named Philolaus as having believed in the earth's revolution (not around the sun, but around an imaginary central fire) and Ecphantus as having attributed to the earth an axial rotation (unaccompanied by orbital revolution).

Copernicus carefully refrained from linking Aristarchus with the earth's motion. He did not hesitate to cite an (unhistorical) determination of the obliquity of the ecliptic by Aristarchus (whom he was misled into confusing with Aristyllus) as 23°51′20″, equal to Ptolemy's.[10] He also reported an equally unhistorical measurement of the length of the tropical year by Aristarchus (again confused with Aristyllus) as exactly 365^d6^h.[11] But the passage in which Copernicus originally associated Aristarchus with Philolaus' advocacy of a moving earth was deleted by Copernicus before he released his *De revolutionibus* for publication.[12] In like manner, Ptolemy's discussion of geodynamism conspicuously omitted the name of Aristarchus, who is nevertheless cited in the *Syntaxis mathematica* in connection with the length of the year.[13] Copernicus had no desire to inform or remind anybody that the fervently religious head of an influential philosophical school had "thought that the Greeks ought to bring charges of impiety against Aristarchus."[14] The latter's superb technical achievements in astronomy were not in question. His geocentric treatise *On the Sizes and Distances of the Sun and Moon* has survived intact; but his account of the heliocentric system has perished, leaving only a trace of the first such statement in the history of mankind.

According to that pioneering declaration, "the sphere of the fixed stars . . . is so great that the circle in which Aristarchus assumes the earth to revolve has the same ratio to the distance of the fixed stars as the center of a sphere has to its surface."[15] Archimedes, who preserved Aristarchus' heliocentric conception by summarizing it in his *Sand-Reckoner,* objected as a mathematician that "since the center of a sphere has no magnitude, neither can it be thought to have any ratio to the surface of the sphere."[16] Accordingly, Archimedes interpreted Aristarchus to mean that the ratio earth : distance earth-sun = distance earth-sun : distance earth-stars. Whatever the defects in Aristarchus' formulation, he unquestionably intended to emphasize the enormous remoteness of the stars.

This fundamental consequence of heliocentrism was expressed in Copernicus' *Commentariolus* by the following inequality: distance earth-sun : distance sun-stars < earth's radius : distance earth-sun. This disproportion is in fact so great that the distance earth-sun is "imperceptible" in comparison with the distance earth-stars or sun-stars.[17] The latter distance measured the size of Copernicus' universe from the sun at its center to the stars at its outermost limit.

Because he abandoned the geocentrism of his predecessors, he likewise had to enlarge the dimensions of their limited cosmos:

Lines drawn from the earth's surface and center [to a point in the firmament] must be distinct. Since, however, their length is immense in relation to the earth, they become like parallel lines. These appear to be a single line by reason of the overwhelming distance of their terminus, the space enclosed by them becoming imperceptible in comparison with their length. . . . This reasoning unquestionably makes it quite clear that, as compared with the earth, the heavens are immense and present the aspect of an infinite magnitude, while on the testimony of the senses the earth is related to the heavens as a point to a body, and a finite to an infinite magnitude.[18]

On the basis of both reason and sense experience, Copernicus' heavens "present the aspect of an infinite magnitude."

But it is not at all certain how far this immensity extends. At the opposite extreme are the smallest, indivisible bodies called "atoms." Being imperceptible, they do not immediately constitute a visible body when they are taken two or a few at a time. But they can be multiplied to such an extent that in the end there are enough of them to combine in a perceptible magnitude. The same may be said also about the position of the earth. Although it is not in the center of the universe, nevertheless its distance therefrom is still insignificant, especially in relation to the sphere of the fixed stars.[19]

When Copernicus' atoms are combined in sufficient quantities, they form a visible object. In like manner, when Copernicus' distance sun-earth is multiplied often enough, the product is Copernicus' distance sun-stars. Whether that distance was finite or infinite, Copernicus declined to say. Regarding the universe's "limit as unknown and unknowable," he preferred to "leave the question whether the universe is finite or infinite to be discussed by the natural philosophers."[20]

Had Copernicus elected to extricate himself from this perennial cosmological dilemma by voting for infinity, he would have had to surrender the sun's centrality, since of course the infinite can have no center. On the other hand, had he retained the limited dimensions of the traditional cosmos, the yearly orbit of his moving earth should have produced an annual parallax of the stars. This perspective displacement is in fact so minute that mankind had to wait nearly three centuries for telescopes sensitive enough to detect it. Copernicus' solution, therefore, was to impale himself on neither horn of the dilemma by declaring the universe to be "similar to the infinite."[21] The qualification "similar" permitted him to regard the universe as capable of possessing a center, while the similarity to the infinite explained the naked eye's inability to perceive annual stellar parallax.

If Copernicus hoped to gain acceptance for his revival of the concept of a moving earth, he had to overcome the ancient objections to such motion. Earth was traditionally regarded as one of the four terrestrial or sublunar elements, the other three being water, air, and fire, whereas the heavenly bodies consisted of a fifth element. Aristotle's theory of the motion of these five elements was summarized by Copernicus as follows:

The motion of a single simple body is simple; of the simple motions, one is straight and the other is circular; of the straight motions, one is upward and the other is downward. Hence every simple motion is either toward the middle, that is, downward; or away from the middle, that is, upward; or around the middle, that is, circular. To be carried downward, that is, to seek the middle, is a property only of earth and water, which are considered heavy; on the other hand, air and fire, which are endowed with lightness, move upward and away from the middle. To these four elements it seems reasonable to assign rectilinear motion, but to the heavenly bodies, circular motion around the middle.[22]

Copernicus had transferred the earth to the category of the heavenly bodies, to which circular motion around the middle could be reasonably assigned. Yet some parts of the earth undeniably "sink of their own weight," while "if any part of the earth is set afire, it is carried from the middle upwards."[23] Such rectilinear motion, however, overtakes things which leave their natural place or are thrust out of it or quit it in any manner whatsoever. . . . Whatever falls moves slowly at first, but increases its speed as it drops. On the other hand, we see this earthly fire . . . after it has been lifted up high, slacken all at once. . . . Circular motion, however, always rolls along uniformly, since it has an unfailing cause. But rectilinear motion has a cause that quickly stops functioning. When rectilinear motion brings bodies to their own place, . . . their motion ends.[24]

Retaining Aristotle's doctrine that every body has its natural place in the universe, Copernicus confined the application of this principle to the displaced parts of the earth, which were subject to the sort of motion classified by Aristotle as violent. Copernicus' planet earth as a whole, on the other hand, possessed perpetual motion, natural to the heavenly bodies. This circular motion was shared by any portion of the earth temporarily detached from it: "The motion of falling and rising bodies in the framework of the universe is twofold, being in every case a compound of straight and circular. . . . Hence, since circular motion belongs to wholes, but parts have rectilinear motion in addition, we can say that circular subsists with rectilinear as animal does with sick."[25] Taken as a whole, earth has only circular motion and no rectilinear motion, just as a healthy animal has no sickness. But a loose portion of the earth has rectilinear motion conjoined with circular motion, just as a diseased beast unites sickness with its animal nature.

The three conventional classes of motion, therefore, do not correspond to entirely separate physical states. "Aristotle's division of simple motion into three types, away from the middle, toward the middle, and around the middle, will be construed as merely an exercise in logic."[26] Similarly, in geometry "we distinguish the point, the line, and the surface, even though one cannot exist without another, and none of them without body."[27]

Besides reinterpreting the traditional theory of motion to fit the requirements of his moving earth, Copernicus endowed the planet earth, as opposed to its disjointed parts, with natural, not violent, motion. Ptolemy had contended that the earth's axial rotation

would have to be exceedingly violent and its speed unsurpassable to carry the entire circumference of the earth around in twenty-four hours. But things which undergo an abrupt rotation seem utterly unsuited to gather bodies to themselves, and seem more likely, if they have been produced by combination, to fly apart unless they are held together by some bond. The earth would long ago . . . have burst asunder . . . and dropped out of the skies.[28]

Ptolemy's anxiety was answered by Copernicus:

What is in accordance with nature produces effects contrary to those resulting from violence. For, things to which force or violence is applied must disintegrate and cannot long endure, whereas that which is brought into existence by nature is well ordered and preserved in its best state. Therefore Ptolemy has no cause to fear that the earth and everything earthly will be disrupted by a rotation created through nature's handiwork, which is quite different from what art or human intelligence can contrive.[29]

Ptolemy was further concerned that "living creatures and any other loose objects would by no means remain unshaken. . . . Moreover, clouds and anything else floating in the air would be seen drifting always westward," since the earth's axial rotation whirls it round swiftly eastward.[30] In reply Copernicus asked:

With regard to the daily rotation, why should we not admit that the appearance is in the heavens and the reality in the earth? . . . Not merely the earth and the watery element joined with it have this motion, but also no small part of the air. . . . [The reason may be] either that the nearby air, mingling with earthy or watery matter, conforms to the same nature as the earth, or that [this] air's motion, acquired from the earth by proximity, shares without resistance in its unceasing rotation.[31]

By contrast with the upper layers of air, the lower layers are firmly attached to the earth and rotate with it. This partnership answers the argument that "objects falling in a straight line would not descend perpendicularly to their appointed place, which would meantime have been withdrawn by so rapid a movement" as the earth's rotation.[32] Pro-Copernicans and anti-Copernicans later conducted experiments to determine whether an object dropped vertically from a height, stationary or moving with respect to the earth's surface, fell precisely at the foot of the height. The divergent results of these numerous trials were variously interpreted; and decisive experimental confirmation of the earth's daily rotation was first provided by Foucault's pendulum in 1851, not long after Bessel, F. G. W. Struve, and T. Henderson published their independent discoveries of annual stellar parallax as direct observational proof of the earth's yearly orbital motion.

In addition to the diurnal rotation and annual revolution, Copernicus felt obliged to ascribe to the earth what he called its "motion in declination."[33] When prolonged, the axis about which our planet rotates daily meets the firmament at the celestial poles. Midway between these poles lies the celestial equator, the intersection of the plane of the earth's equator and the celestial sphere. In performing its annual revolution around the sun, the earth describes what Copernicus termed the "grand circle," the plane of which cuts the celestial sphere in the ecliptic. The poles of the ecliptic are the end points of the axis of the earth's orbital revolution. The plane of that revolution, or ecliptic, is inclined to the celestial equator at an angle known as the obliquity of the ecliptic. As Copernicus said in the *Commentariolus,* "The axis of the daily rotation is not parallel to the axis of the grand circle, but is inclined to it at an angle that intercepts a portion of a circumference, in our time about 23 1/2°."[34]

In Copernicus' time a spherical body revolving in an orbit was considered to be attached inflexibly to the orbit's center, as though from this hub a rigid spoke ran right through the revolving ball. Therefore, if the earth were subject only to the diurnal rotation and annual revolution without the third motion in declination,

no inequality of days and nights would be observed. On the contrary, it would always be either the longest or shortest day or the day of equal daylight and darkness, or summer or winter, or whatever the character of the season, it would remain identical and unchanged. Therefore the third motion in declination is required. . . . [The motion in declination] is also an annual revolution but . . . it occurs in the direction opposite to that of the [orbital] motion of the [earth's] center. Since these two motions are opposite in direction and nearly equal [in period], the result is that the earth's axis and . . . equator face almost the same portion of the heavens, just as if they remained motionless.[35]

The function of Copernicus' third motion in declination was to keep the earth presenting a virtually unchanging aspect to an observer viewing it from a distant star, whereas to a spectator stationed on the sun it would constantly pass through its cyclical seasonal changes. Without the motion in declination Copernicus' earth would always look the same as seen from the sun, while its axis of rotation would describe a huge conical surface in space instead of pointing toward the vicinity of the same star.

The rotational axis, however, is not directed toward precisely the same star because

the annual revolutions of the center and of declination are nearly equal. For if they were exactly equal, the equinoctial and solstitial points as well as the entire obliquity of the ecliptic would have to show no shift at all with reference to the sphere of the fixed stars. But there is a slight variation, which was discovered only as it grew larger with the passage of time.[36]

This slight variation, the precession of the equinoxes, had been explained by Ptolemy as due to a slow eastward rotation of the sphere of the stars. But that sphere had to remain absolutely motionless in the cosmos of Copernicus, who had replaced the apparent daily rotation of the stars by the real axial rotation of the earth.

In like manner, for Ptolemy's motion of the starry sphere in 36,000 years, Copernicus substituted the behavior of the earth:

[Its] two revolutions, I mean, the annual declination and [the orbital motion of] the earth's center, are not exactly equal, the declination being of course completed a little ahead of the period of the center. Hence, as must follow, the equinoxes and solstices seem to move forward. The reason is not that the sphere of the fixed stars moves eastward, but rather that the equator moves westward.[37]

Whereas modern astronomy has adopted Copernicus' account of precession, its rate eluded him. The modern constant value, about 50″ a year, was regarded by him as the mean rate of precession: he was misled by his predecessors' divergent determinations of this minute quantity into believing that it underwent a cyclical variation. He likewise made the same error regarding the obliquity of the ecliptic. The available evidence warranted only the conclusion that the obliquity diminished progressively. Nevertheless, he supposed that after decreasing from a maximum of 23° 52′ before Ptolemy's time to a minimum of 23° 28′ after his own time, it would then reverse itself and increase to its previous maximum, oscillating thereafter in a 24′ cycle of long period.

The sun appears to move with annually recurring variations of speed along its course in the ecliptic, thereby making the four seasons unequal in length. To represent these phenomena, Ptolemy had the sun traverse a circle whose stationary center was separated by some distance from the earth. This eccentric circle's apogee, or point at which the sun attained its greatest distance from the earth, was regarded by Ptolemy as fixed in relation to the stars at 24° 30′ before the summer solstice. Al-Battānī located the apogee only 7° 43′ before the summer solstice.[38] "In the 740 years since Ptolemy it advanced nearly 17°."[39] Al-Zarqālī, however, "put the apogee 12° 10′ before the solstice."[40] Thus,

in 200 years it retrogressed 4° or 5°. Thereafter until our age it moved forward. The entire period [from Ptolemy to Copernicus] has witnessed no other retrogression nor the several stationary points which must intervene at both limits when motions reverse their direction. [The absence of] these features cannot possibly be understood in a regular and cyclical motion. Therefore many astronomers believe that some error occurred in the observations of those men [al-Battānī and al-Zarqālī]. Both were equally skillful and careful astronomers so that it is doubtful which one should be followed. For my part I confess that nowhere is there a greater difficulty than in understanding the solar apogee, where we draw large conclusions from certain minute and barely perceptible quantities. . . . As can be noticed in the general structure of the [apogee's] motion, it is quite probably direct but nonuniform. For after that stationary interval from Hipparchus to Ptolemy the apogee appeared in a continuous, regular, and accelerated progression until the present time. An exception occurred between al-Battānī and al-Zarqālī through a mistake (it is believed), since everything else seems to fit.[41]

Copernicus still believed in the fixity of the earth's aphelion, or—its Ptolemaic counterpart—the solar apogee, when he composed the *Commentariolus* between 15 July 1502 and 1 May 1514. Later, in writing book III of *De revolutionibus,* where he took into account the related work of the Arab astronomers, he made the terrestrial aphelion move. But, the observations of al-Battānī and al-Zarqālī being discordant, he was "doubtful which one should be followed." By the summer of 1539, when his disciple Rheticus drafted the *Narratio prima* (*First Report*) of the Copernican system to be presented in printed form to the reading public, both al-Battānī and al-Zarqālī were suspect in Copernicus' mind. In creating his model for the progressive motion of the earth's aphelion, Copernicus felt justified in lowering al-Battānī's determination by 6° and raising al-Zarqālī's by 4°.

> Now you see [says Rheticus] what great effort my teacher had to put forth to determine the mean motion of the [solar] apogee. For nearly forty years in Italy and here in Frombork, he observed eclipses and the [apparent] motion of the sun. He selected the observation by which he established that in A.D. 1515 the solar apogee was at 6 2/3° of Cancer [= 6 2/3° after the solstice]. Then examining all the eclipses in Ptolemy and comparing them with his own very careful observations, he concluded that the mean annual motion of the apogee with reference to the fixed stars was about 25″. . . .[42]

In his earliest recorded observation, made in Bologna after sunset on 9 March 1497, Copernicus reported an occultation of Aldebaran by the moon. In his *De revolutionibus* he used this observation to support his computation of the lunar parallax.[43] The variation in this quantity and in the length of the moon's apparent diameter was greatly exaggerated in Ptolemy's lunar theory, as Copernicus emphasized in the *Commentariolus*:

> The consequence by mathematical analysis is that when the moon is in quadrature, and at the same time in the lowest part of the epicycle, it should appear nearly four times greater (if the entire disk were luminous) than when new and full, unless its magnitude increases and diminishes in no reasonable way. So too, because the size of the earth is sensible in comparison with its distance from the moon, the lunar parallax should increase very greatly at the quadratures. But if anyone investigates these matters carefully, he will find that in both respects the quadratures differ very little from new and full moon. . . .[44]

Mounting the moon on an epicycle whose deferent was not concentric with the earth, Ptolemy and his followers had the epicycle's center traverse equal arcs in equal times as measured from the earth's center. While Copernicus' predecessors "declare that the motion of the epicycle's center is uniform around the center of the earth, they must also admit that it is nonuniform on its own eccentric, which it describes."[45] Such a model was rejected by the *Commentariolus* as conflicting with "the rule of absolute motion," according to which "everything would move uniformly about its proper center."[46] This principle was violated a second time in the Ptolemaic lunar theory, which had the moon traverse equal arcs on its epicycle, as measured not from the epicycle's center but from a different point known as the equant or the equalizing point.

In order to avoid using an equant, which he regarded as an impermissible device, in his own lunar theory Copernicus obtained an equivalent result by piling on the traditional epicycle a second, smaller epicyclet carrying the moon. This method of adhering to the axiom of uniform motion, at the same time eliminating the equant and the excessive variation in the length of the moon's apparent diameter, had been adopted in the Muslim world by Ibn al-Shāṭir about a century before Copernicus was born. Was Copernicus aware of the work done by his Damascene predecessor? The latter introduced a second epicycle for the sun too, but Copernicus did not follow suit. He used eccentric models, which had been rejected by Ibn al-Shāṭir. His numerical results also differed, being based in part on his own observations. Since he knew no Arabic and Ibn al-Shāṭir's manuscript had not been translated into any language understood by Copernicus, presumably he had no direct acquaintance with the Muslim's thinking. Their conclusions, independently reached, strikingly converged on the same theoretical and practical shortcomings in Ptolemaic astronomy. But there is no inkling of geodynamism in Ibn al-Shāṭir.

The same cannot be said about Ibn al-Shāṭir's contemporary, Nicole Oresme, who around 1377 made the first translation of Aristotle's *De caelo* into a modern language. In his commentary Oresme considered many arguments concerning the diurnal rotation, which should more reasonably, it seemed to him, be assigned to the earth. Yet he admitted that he had discussed this idea "for fun" [47] and, as bishop of Lisieux, he rejected it on the basis of Biblical passages. Oresme's translation-commentary was written in French (which Copernicus did not understand) and was first printed in 1941–1943. Had Copernicus been familiar with it, he would have noticed its complete silence about the earth's orbital revolution. He would surely have been impressed by Oresme's reasoning that the earth benefits from the sun's heat, and in familiar contexts, that what "is roasted at a fire receives the heat of the fire around itself because it is turned and not because the fire is turned around it." [48] That Copernicus had any direct acquaintance with Oresme seems out of the question.

Nevertheless, university teaching may well have been affected by Oresme and even more by his older friend, Jean Buridan. The latter's discussion in Latin of Aristotle's *De caelo* mentioned the idea that "the earth, water, and air in its lower region move jointly with the daily rotation." [49] Buridan also set forth the following argument:

> An arrow shot vertically upward from a bow falls back on the same place on the earth from which it was discharged. This would not happen if the earth moved so fast. In fact, before the arrow fell, the part of the earth from which it was fired would be a mile away. [50]

The absence of the earth's orbital revolution from the thinking of Copernicus' Muslim and Christian predecessors, as well as his use of Arabic observational results, indicate that he did not conceal any intellectual indebtedness to them. On the other hand, with complete openness he expressly acknowledged being inspired by his ancient geodynamic forerunners. Their ideas, however, came down to him as the barest of bones; it was he who first fleshed out the geodynamic astronomy.

Copernicus did away with the stationary earth situated at the center of the Aristotelian-Ptolemaic universe. In his cosmos the earth revolved around the central sun in an annual orbit and at the same time executed its daily rotations. Consequently, the astronomer who inhabits the earth watches the stately celestial ballet from an observatory that is itself both spinning and advancing.

> If any motion is ascribed to the earth, in all things outside it the same motion will appear, but in the opposite direction, as though they were moving past it. This is the nature in particular of the daily rotation, since it seems to involve the entire universe, except the earth and what is around it. However, if you grant that the heavens have no part in this motion but that the earth

rotates from west to east, upon earnest consideration you will find that this is the actual situation, as far as concerns the apparent rising and setting of the sun, moon, stars, and planets. [51]

Three of the planets in Copernicus' cosmos revolve around the sun in orbits larger in size and longer in period than the earth's. Each of these three outer, or superior, planets (Mars, Jupiter, and Saturn in ascending order)

> seems from time to time to retrograde, and often to become stationary. This happens by reason of the motion, not of the planet, but of the earth changing its position in the grand circle. For since the earth moves more rapidly than the planet, the line of sight directed [from the earth] toward [the planet and] the firmament regresses, and the earth more than neutralizes the motion of the planet. This regression is most notable when the earth is nearest to the planet, that is, when it comes between the sun and the planet at the evening rising of the planet. On the other hand, when the planet is setting in the evening or rising in the morning, the earth makes the observed motion greater than the actual. But when the line of sight is moving in the direction opposite to that of the planets and at an equal rate, the planets appear to be stationary, since the opposed motions neutralize each other. [52]

As an outer planet in its normal eastward progression (viewed against the background of the more distant stars) slows down, stops, reverses its direction, stops again, and resumes its direct march, it appears to pass through kinks or loops. These were actual celestial happenings for Ptolemy and his followers. The true nature of these planetary loops was revealed for the first time by Copernicus when he analyzed them in detail as side effects of the observation of the slower planet from the faster earth. The loops are optical illusions, not real itineraries.

> Two entirely different motions in longitude appear in them [the planets]. One is caused by the earth's motion . . . and the other is each [planet's] own proper motion. I have decided without any impropriety to call the first one a parallactic motion, since it is this which makes the stations, direct motions, and retrogressions appear in all of them. These phenomena appear, not because the planet, which always moves forward with its own motion, is erratic in this way, but because a sort of parallax is produced by the earth's motion according as it differs in size from those orbits. [53]

Before Copernicus there was much uncertainty regarding the position of Venus and Mercury in the heavens. But the Copernican system located these two bodies correctly as the inferior, or lower, planets, revolving around the central sun inside the earth's orbit and at a greater speed.

The true places of Saturn, Jupiter, and Mars become visible to us only at their evening rising, which occurs about the middle of their retrogradations. For at that time they coincide with the straight line through the mean place of the sun [and earth], and are unaffected by that parallax. For Venus and Mercury, however, a different relation prevails. For when they are in conjunction with the sun, they are completely blotted out, and are visible only while executing their elongations to either side of the sun, so that they are never found without this parallax.

Consequently each planet has its own individual parallactic revolution, I mean, terrestrial motion in relation to the planet, which these two bodies perform mutually. Combined in this way, the motions of both bodies display themselves interconnected. . . . The motion in parallax, I submit, is nothing but the difference by which the earth's uniform motion exceeds the planets' motion, as in the cases of Saturn, Jupiter, and Mars, or is exceeded by it, as in the cases of Venus and Mercury.[54]

The motion in parallax is smaller, as regards the inner planets, for Venus than for Mercury; and as regards the outer planets, smaller for Mars than for Jupiter and Saturn. Hence,

> the forward and backward arcs appear greater in Jupiter than in Saturn and smaller than in Mars, and on the other hand, greater in Venus than in Mercury. This reversal of direction appears more frequently in Saturn than in Jupiter, and also more rarely in Mars and Venus than in Mercury.[55]

Although the sun was nominally one of the seven Ptolemaic planets, it actually possessed a privileged status in that system. Thus, the center of the epicycle on which Venus was mounted kept exact pace with the sun. This synchronization was accomplished by having the line drawn from the central stationary earth to the annually revolving sun always pass through the center of Venus' epicycle. As a result, Venus' maximum distance to either side of the sun was regulated by the length of the radius of its epicycle. In Ptolemy's words, "the greatest elongations of Venus and Mercury [occur] when the planet reaches the point of contact of the straight line drawn from our eye tangent to the epicycle."[56] This statement applied to Mercury, even though its more irregular motion required a somewhat more complicated arrangement.

In the Ptolemaic theory of the three outer planets the sun again played a special part. As the planet revolved on its epicycle, the radius drawn from the center of the epicycle to the moving planet kept step with the sun revolving around the stationary earth. This coordination was achieved by having the planet's radius vector parallel at all times to the line drawn from the terrestrial observer to the (mean) sun.

Thus, the Ptolemaic theory of each of the three outer and two inner planets introduced the annual revolution. This was imputed by the Ptolemaists to the sun, which they regarded as one of the planets. But they did not explain why the orbital motion of one planet should be so especially privileged as to be an integral part of the theory of five other planets.

In still another respect the sun occupied a privileged position in Ptolemaic astronomy: The sun was placed "between those [planets] which pass through every elongation from it and those which do not so behave, but always move in its vicinity."[57] Copernicus protested that this argument "carried no conviction because its falsity is revealed by the fact that the moon too shows every elongation from the sun."[58] Whatever their other disagreements, Ptolemaists and Copernicans alike separated the moon from the outer planets.

The removal of the sun from the category of the planets was one of Copernicus' most influential contributions to the advancement of astronomy. The limited maximum elongations of Venus and Mercury no longer resulted from the lengths of the radii of their epicycles but were caused by a physical fact: since they were now the inner planets, their orbits lay entirely within the earth's. Therefore, these planets could never be seen from the earth at an angular distance from the sun exceeding 48° for Venus and 28° for Mercury. Hence, these planets could never come to quadrature or opposition, where the difference in geocentric longitude between them and the sun would have to reach 90° or 180°.

In the case of each of the three outer planets, the perpetually parallel orientations of the epicycle's radius directed to the planet and of the line earth-sun were no longer an unexplained coincidence but rather an indication of a physical phenomenon, the earth's orbital revolution around the sun. This "one motion of the earth causes all these phenomena, which the ancient astronomers sought to obtain by means of an epicycle for each" of the three outer planets.[59] By making the earth a planet (or planetizing it, so to say) and deplanetizing the sun, Copernicus took a long step away from previous misconceptions toward the correct understanding of our physical universe:

> Venus seems at times to retrograde, particularly when it is nearest to the earth, like the superior planets, but for the opposite reason. For the regression of the superior planets happens because the motion of the earth is more rapid than theirs, but with Venus, because it is slower; and because the superior planets enclose the grand circle [earth's orbit], whereas Venus is enclosed within it. Hence Venus is never in opposition to the

sun, since the earth cannot come between them, but it moves within fixed distances on either side of the sun. These distances are determined by tangents to the circumference drawn from the center of the earth, and never exceed 48° in our observations.[60]

From the maximum elongation of Venus, Copernicus was able to obtain the first approximately correct planetary distances, which he expressed in terms of the distance earth–sun. This distance, which subsequently became the fundamental astronomical unit, was grossly underestimated by Copernicus, who simply followed the ancient error in this respect. But in computing the distances of the other five planets from the sun as ratios of the distance earth–sun, Copernicus came remarkably close to the values accepted today. For Mars and Venus, he agreed to the second decimal place (1.52, 0.72), and for Jupiter to the first (5.2). For Saturn and Mercury, however, he was less accurate (9.2 as compared with 9.5; 0.376 as compared with 0.387).

In this respect the contrast with the geocentric astronomy is instructive. Ptolemy was familiar with two proposed locations for Venus and Mercury; either below the sun or above it. No transits of the sun by either Venus or Mercury had ever been observed. But the absence of such reports could be explained if the inferior planet's plane did not coincide with the sun's. "Nor can such a determination be reached in any other way, because none of the planets undergoes a perceptible parallax, the only phenomenon from which the [planetary] distances are obtained."[61] Differences in parallax were regarded by Ptolemy as the only method for arranging the planets in the ascending order of distance from the earth. Such parallaxes being unavailable to him, in the *Syntaxis* he virtually renounced the effort to ascertain the distances of the planets. But Copernicus, by using the astronomical unit as his measuring rod, succeeded in establishing the correct order and distance of the known planets with a high degree of accuracy.

Although he did not accept the widespread belief that every planet was moved by a resident angel or spirit, he prudently refrained from explicitly rejecting that popular doctrine. He held instead that, just as physical bodies become spherical when they are unified, so

the motion appropriate to a sphere is rotation in a circle. By this very act the sphere expresses its form as the simplest body, wherein neither beginning nor end can be found, nor can the one be distinguished from the other, while the sphere itself traverses the same points to return upon itself.[62]

Had Copernicus possessed the courage or insight to push this principle to its logical outcome, he would not have left the axial rotation of the sun and planets to be discovered by his followers.

The Copernican celestial spheres, which expressed their form by rotating in a circle, were mainly those which carried either the planets or the planet-carrying spheres. In the former case, the planet was attached to the surface of the sphere at its equator, like a pearl on a ring; however, whereas the pearl was visible, the ring was not. Equally invisible was the rest of the planet-carrying sphere, that is, the sphere of the epicycle. The whole of the deferent sphere, which carried the sphere of the epicycle, was likewise imperceptible. Although Copernicus never explicitly asserted the physical existence of these unseen spheres, he never denied their reality and always implicitly assumed it. Thus, *orbium* in the title of his *De revolutionibus orbium coelestium* referred not to the planetary bodies themselves but to the spheres which carried them or helped to do so. In banishing these spheres from astronomy, Tycho Brahe said,

There really are not any spheres in the heavens. . . . Those which have been devised by the experts to save the appearances exist only in the imagination, for the purpose of enabling the mind to conceive the motion which the heavenly bodies trace in their course and, by the aid of geometry, to determine the motion numerically through the use of arithmetic.[63]

Although Copernicus always proceeded on the assumption that the planetary motions were produced by spheres of one sort or another, he was not unswervingly committed to any particular kind of sphere. Thus, in expounding the motion of the solar apogee, he resorted to an eccentreccentric—that is, an eccentric sphere or circle whose center was carried around by the circumference of a second, smaller eccentric sphere or circle. Then he explained that equivalent results would follow from an epicyclepicyclet—that is, an epicyclet whose center was carried round by an epicycle, whose center in turn revolved on the circumference of a deferent concentric with the sun as the center of the universe. Moreover, mounting an epicycle on an eccentric would serve the purpose as well: "Since so many arrangements lead to the same numerical outcome, I could not readily say which exists, except that on account of the unceasing agreement of the computations and the phenomena I must believe it to be one of them."[64]

In his youthful *Commentariolus* Copernicus located each of the three outer planets on an epicyclet, whose center rode on a larger epicycle carried by a concentric deferent. This device has been called "concentrobi-epicyclic" in contradistinction to the eccentrepicyclic arrangement preferred by Copernicus in his mature *De revolutionibus.* His later shift to the single epicycle mounted on an eccentric deferent was not arbitrary: it was connected with his conclusion that the sun's displacement from the center of the universe was variable and not constant, as he had originally believed on the strength of Ptolemy's statement to that effect.

The center of Copernicus' universe was not the body of the sun, but a nearby unoccupied point. This purely mathematical entity could not fulfill the function served by the center of the pre-Copernican universe. In that cosmos, according to Aristotle, its principal architect, "the earth and the universe happen to have the same center. A heavy body moves also toward the center of the earth, but it does so only incidentally, because the earth has its center at the center of the universe." [65] Having planetized the earth and raised it out of the universe's center to the third circumsolar orbit, Copernicus could not regard his new planet as the collection depot for all the heavy bodies on the move in the universe. On the other hand, he had no reason to deny that heavy terrestrial objects tended toward the earth's center. Hence, he put forward a revised conception of gravity, according to which heavy objects everywhere tended toward their own center—heavy terrestrial objects toward the center of the earth, heavy lunar objects toward the center of the moon, and so on:

> For my part, I think that gravity is nothing but a certain natural striving with which parts have been endowed . . . so that by assembling in the form of a sphere they may join together in their unity and wholeness. This tendency may be believed to be present also in the sun, the moon, and the other bright planets, so that it makes them keep that roundness which they display. [66]

Whereas the pre-Copernican cosmos had known only a single center of gravity or heaviness, the physical universe acquired multiple centers of gravity from Copernicus, who thus opened the road that led to universal gravitation.

This contribution to one of the basic concepts of modern physics and cosmology confirms what we have already witnessed in many other aspects of Copernicus' thought. He was firmly convinced that he was talking about the actual physical world when he transformed the earth from the sluggish dregs of the universe to a satellite spinning about its axis as it whirled around the sun. He would have spurned the doctrine (had he been familiar with it) propounded by Buridan, who said, "For astronomers, it is enough to assume a way of saving the phenomena, whether it is really so or not." [67]

By a quirk of fate, control over the printing of the first edition of Copernicus' *De revolutionibus* passed into the hands of an editor who shared Buridan's fictionalist conception of scientific method in astronomy. Taking advantage of the dying author's remoteness from the printing shop and at the same time concealing his own identity, Andreas Osiander inserted in the most prominent place available, the verso of the title page, an unsigned address "To the Reader, Concerning the Hypotheses of This Work." Therein the reader was not informed that Copernicus used the word "hypothesis" in its strictly etymological sense as equivalent to "fundamental categorical proposition," not in the derivative meaning of "tentative conjecture." Nor was the reader told that in private correspondence with the editor, Copernicus had steadfastly repudiated the principal tenet in the interpolated address: The astronomer's "hypotheses need not be true nor even probable; if they provide a calculus consistent with the observations, that alone is sufficient." [68] Thus it came to pass that Copernicus' *De revolutionibus,* now universally recognized as a classic of science, was first presented to the civilized world in a guise which, however well intentioned, falsified its essential nature and fooled many readers, including J. B. J. Delambre, the renowned nineteenth-century historian of astronomy.

NOTES

1. Albert Caprinus, *Indicium astrologicum* (Cracow, 1542), dedication (cited in Prowe, I, 1, 148).
2. Rosen, p. 111.
3. Prowe, I, 1, 291.
4. Rosen, p. 67.
5. *De revolutionibus,* IV, 7; cf. III, 18, 19.
6. Simon Starowolski's biography of Copernicus (cited in Prowe, I, 1, 149).
7. Rosen, p. 58.
8. *Ibid.,* p. 59.
9. *Ibid.*
10. *De revolutionibus,* III, 2.
11. *Ibid.,* 13.
12. *Gesamtausgabe,* II, 30.
13. III, 1; Heiberg, ed., I, 203:10, 206:5–6, 25.
14. Plutarch, *Face in the Moon,* 923A.
15. Thomas L. Heath, *Aristarchus of Samos* (Oxford, 1959), p. 302 (trans. modified).
16. Heath, *The Works of Archimedes* (Cambridge, 1897; Dover ed., New York, n.d.), p. 222 (trans. modified).
17. Rosen, p. 58.
18. *De revolutionibus,* I, 6.
19. *Ibid.*
20. *Ibid.,* 8.
21. *Gesamtausgabe,* II, 31.
22. *De revolutionibus,* I, 7.
23. *Ibid.,* 8.
24. *Ibid.*
25. *Ibid.*
26. *Ibid.*
27. *Ibid.*
28. *Ibid.,* 7.

29. *Ibid.,* 8.
30. *Ibid.,* 7.
31. *Ibid.,* 8.
32. *Ibid.,* 7.
33. Rosen, p. 63.
34. *Ibid.,* pp. 63–64.
35. *De revolutionibus,* I, 11.
36. *Ibid.*
37. *Ibid.,* III, 1.
38. *Ibid.,* 16.
39. *Ibid.,* 20.
40. *Ibid.,* 16.
41. *Ibid.,* 20.
42. Rosen, p. 125.
43. *De revolutionibus,* IV, 27.
44. Rosen, p. 72.
45. *De revolutionibus,* IV, 2.
46. Rosen, pp. 57–58.
47. Nicole Oresme, "Le livre du ciel et du monde," ed. and with commentary by Albert D. Menut and Alexander J. Denomy, in *Mediaeval Studies,* **4** (1942), 279 (§144c).
48. *Ibid.,* p. 277 (§142b).
49. *Questiones super libris quattuor De caelo et mundo,* E. A. Moody, ed. (Cambridge, Mass., 1942), p. 229.
50. *Ibid.*
51. *De revolutionibus,* I, 5.
52. Rosen, pp. 77–78.
53. *De revolutionibus,* V, 1.
54. *Ibid.*
55. *Ibid.,* I, 10.
56. Ptolemy, *Syntaxis mathematica,* X, 6; Heiberg, ed., II, 317:13–17.
57. *Ibid.,* IX, 1; Heiberg, ed., II, 207:18–20.
58. *De revolutionibus,* I, 10.
59. *Ibid.,* V, 3.
60. Rosen, p. 83.
61. Ptolemy, IX, 1; Heiberg, ed., II, 207:13–16.
62. *De revolutionibus,* I, 4.
63. Tycho Brahe, *Opera omnia,* J. L. E.· Dreyer, ed., 15 vols. (Copenhagen, 1913–1929), IV, 222:24–28.
64. *De revolutionibus,* III, 20.
65. Aristotle, *De caelo,* II, 14; 296b15–18.
66. *De revolutionibus,* I, 9.
67. Buridan, p. 229.
68. Rosen, p. 25.

BIBLIOGRAPHY

A 2nd ed. of Henryk Baranowski's *Bibliografia kopernikowska 1509–1955* (Warsaw, 1958) is being prepared in connection with the celebration in 1973 of the 500th anniversary of the birth of Copernicus. An annotated Copernicus bibliography for 1939–1958 is included in *Three Copernican Treatises,* trans., ed., and with an introduction by Edward Rosen, 2nd ed. (New York, 1959), which also contains an English translation of Copernicus' *Commentariolus* and *Letter Against Werner* and Rheticus' *Narratio prima.* An English translation of Copernicus' *De revolutionibus orbium coelestium* is in Great Books of the Western World, vol. XVI (Chicago, 1952). The Latin text of *De revolutionibus* and a photocopy of Copernicus' autograph manuscript are available in *Nikolaus Kopernikus Gesamtausgabe,* 2 vols. (Berlin–Munich, 1944–1949). The standard biography by Leopold Prowe, *Nicolaus Coppernicus,* 2 vols. (Berlin, 1883–1884; repr., Osnabrück, Germany, 1967), has not yet been superseded, even though it is incomplete (the

planned third volume was never published), nationalistically biased, scientifically inadequate, somewhat inaccurate, and partly obsolete.

EDWARD ROSEN

CORDIER, PIERRE-LOUIS-ANTOINE (*b.* Abbeville, France, 31 March 1777; *d.* Paris, France, 30 March 1861), *geology, mineralogy.*

Cordier was a pioneer in the geological, technical, and economic analysis of French mines, particularly coal mines. He began the use of the polarizing microscope in the study of the constituents of rocks. As a counsellor of state and later a peer during the reign of Louis Philippe, he played an important role in the organization of French railroads, steamboat navigation, and road construction. For three decades he was president of the Conseil des Mines, which afforded him a powerful voice in French mining affairs.

After completing his early education at Abbeville, Cordier went to Paris in 1794 and entered the École des Mines in 1795. He was named engineer in 1797, and in 1798 he was selected by Dolomieu to accompany him to Egypt as a member of the scientific commission of the French expedition. Cordier was an English prisoner for a short time when the venture failed, and returned to France in 1799. He traveled through Belgium, Switzerland, Italy, and Spain until 1803, when he was assigned as engineer in the Department of the Apennines. He became divisional inspector of mines in 1810.

Two of Cordier's early mineral surveys were "Statistique du département du Lot" and "Statistique minéralogique du département des Apennins." These were detailed analyses of the terrain, geology, mineral deposits, and mining and metallurgical works of the two departments. In 1815 Cordier published the memoir "Description technique et économique des mines de houille de Saint-Georges-Chatelaison." This important coal mining property had become the subject of litigation; and Cordier had been appointed as an expert to evaluate the condition of the mines for the court, so that it could judge the rights and interests of the parties involved. The editors of the *Journal des mines* termed Cordier's work one of the most difficult and delicate missions with which an engineer of mines could be charged.

Also in 1815 Cordier published a complete survey of French coal mines and coal production, "Sur les mines de houille de France et l'importation des houilles étrangères." He reported that coal consumption in France had doubled during the period 1789–1812, owing mainly to the substitution of coal for wood, and that French coal production had tripled in the same period because of the cessation of English coal imports. He advocated the continued exploitation of French coal mines to avoid the necessity of imports, to supply domestic needs during wartime, and to aid in gaining a more favorable balance of trade. He recommended a high tariff on imported coal to effect this policy; and, pointing out that transportation costs accounted for 75 percent of the price of coal in France, he urged that the roads, canals, and rivers be improved to lower shipping costs. The publication had important effects, inasmuch as all of Cordier's suggestions became future governmental policy.

In 1816 Cordier published the memoir "Sur les substances minérales, dites en masse, qui servent de base aux roches volcaniques." In it he reported the results of a study of volcanic rocks from active and extinct volcanoes and from volcanic terrains, the origin of which was contested by neptunists and plutonists. He reduced the rocks to powder by compression, separated the particles by flotation, examined the particles microscopically to determine their forms, and employed chemical and magnetic tests to ascertain the pure crystallized minerals of which they were composed. Cordier concluded that the difference between the varieties of modern and ancient lavas were caused only by a very slight modification of their intimate textures and that the lavas of disputed origin were extremely similar in texture and mineralogical composition to modern lavas. His memoir, then, was an important step in the resolution of the basalt controversy.

In 1819 Cordier became professor of geology at the Muséum d'Histoire Naturelle, and in 1822 he was admitted to membership in the Académie des Sciences. In 1827 he published the memoir "Essai sur la température de l'intérieur de la terre." From accumulated temperature observations at various depths in mines, Cordier estimated that the earth's temperature increased one degree centigrade for each thirty to forty meters of depth. He concluded that the earth was fluid beyond a depth of 5,000 meters and that it was a cooled star. It followed that consolidation had taken place from the exterior to the interior and that the layers of primitive terrain closest to the surface were the most ancient, a condition that had not been admitted in geology up to that time but one that geological theory would have to take into account. Further, Cordier attributed the occurrence of volcanoes to the action of the earth's internal high-temperature fluid. This theory remained popular until the discovery of radioactivity.

In 1830 Cordier entered the Conseil d'État of Louis Philippe, and in 1831 he became president of the Conseil des Mines. In this position he presided over all of the permanent and temporary commissions of the Ministry of Public Works having to do with the Corps of Mines, a powerful and time-consuming situation that provoked critics to remark that Cordier could not fulfill all of his responsibilities without adding several days to the week.

By ordinance of 7 November 1839, Cordier was named a peer of France; but he was active only in matters having to do with railways, steamboats, and highways. Throughout the entire early period of railway construction in France, he stoutly maintained the minority opinion that all railroads should be privately owned in perpetuity, that the companies should be freed of all governmental supervision, and that the government should subsidize part of the cost of railway construction. Owing to his heavy administrative duties, Cordier published few scientific memoirs after 1837.

The metamorphic aluminum silicate Cordierite was named for him by Haüy.

BIBLIOGRAPHY

I. ORIGINAL WORKS. Cordier's scientific memoirs appeared principally in the *Journal des mines,* the *Annales des mines,* and the *Journal de physique.* The most important were "Statistique du département du Lot," in *Journal des mines,* **21** (1807), 445–474; **22** (1807), 5–62; "Statistique minéralogique du département des Apennins," *ibid.,* **30** (1811), 81–134; "Description technique et économique des mines de houille de Saint-Georges-Chatelaison," *ibid.,* **37** (1815), 161–214, 257–300; "Sur les mines de houille de France et l'importation des houilles étrangères," in *Journal de physique,* **80** (1815), 272–316; "Sur les substances minérales, dites en masse, qui servent de base aux roches volcaniques," *ibid.,* **84** (1816), 135–161, 285–307, 352–386; and "Essai sur la température de l'intérieur de la terre," in *Mémoires de l'Académie des sciences,* **7** (1827), 473–556.

II. SECONDARY LITERATURE. On Cordier or his work, see H. F. Jaubert, *Notice sur la vie et les travaux de M. Cordier* (Paris, 1862); V. Raulin, *Notice sur les travaux scientifiques de M. Cordier* (Bordeaux, 1862); and C. A. Read, *Notice sur la vie et les travaux de P. L. A. Cordier* (Paris, 1862).

JOHN G. BURKE

CORDUS, EURICIUS (*b.* Simtshausen bei Marburg, Germany, 1486; *d.* Bremen, Germany, 24 December 1535), *medicine, poetry, botany.*

The youngest of thirteen children of a well-to-do farmer, Euricius took as his surname the nickname "Cordus" ("last-born"). He probably attended the schools of Wetter and Frankenberg before beginning his studies in liberal arts at Erfurt (about 1505 to 1507). In 1508 he married Kunigunde Dünnwald, daughter of a Frankenberg pharmacist, who bore him eight children. In 1516 he received a master of arts degree and in 1517 became rector of the Abbey School of St. Mary in Erfurt, where he gave lectures on poetry, rhetoric, and the New Testament.

His income, however, was not nearly adequate; and so in 1519 Cordus decided to study medicine. In 1521 he went to Ferrara, where that autumn Leoniceno conferred a doctor's degree on him. He returned to Erfurt and in 1523 accepted an appointment as municipal physician for Brunswick. Four years later he accepted the offer of the Hessian landgrave Philip the Magnanimous and became professor of medicine at the newly founded (1527) University of Marburg.

In addition to his work at the university and as general practitioner, Cordus devoted himself to botany, laid out a botanical garden, and was the first German university professor to organize excursions for studying plants. Even though Cordus was twice rector of the Philippina and his position seemed assured, there were repeated disputes with colleagues—partially due, no doubt, to his lively temperament. Weary of the continuous hostility, Cordus resigned in 1533 to spend the last years of his life in Bremen as municipal physician and professor at the Gymnasium.

Even from his work it is difficult to be sure whether Cordus should be regarded more as a poet, as a physician interested in the natural sciences, or as a botanist. His work can be divided in two parts: his poetry, particularly the ten eclogues of pastoral poems ("Bucolica") written in the style of Vergil and the ironic-humorous epigrams; and his medical and botanical writings, especially his work on the "English sweat" (sweating sickness), the booklet on the preparation of theriaca, and the *Botanologicon,* as well as the posthumously published *De urinis.* The *Botanologicon* is generally considered to be the first attempt at a scientific systematization of plants.

Cordus may not have been an outstanding physician, but he showed himself to be a highly willful and versatile personality whose progressive ideas characterize the Reformation and the age of humanism.

BIBLIOGRAPHY

I. ORIGINAL WORKS. Cordus' poetry is, for the most part, in *Euricii Cordi Simesusii Germani poetae lepidissimi opera poetica omnia* (n.p., n.d.; Frankfurt, 1550, 1564; Helmstedt, 1614); they include "Bucolicorum eclogae X" and "Epigrammatum libri XIII." The Helmstedt ed. was published by J. Lüder with a biography of Cordus by H. Meibom. The first three books of the epigrams were republished by Carl Krause with a biographical introduction in *Latein-ische Literaturdenkmäler des 15. und 16. Jahrhunderts,* V (Berlin, 1892).

His scientific writings include *Ein Regimennt: Wie man sich vor der newen Plage, der Englische Schweis genant, bewaren . . . sall . . .* (Marburg, 1529), reprinted with commentary by Gunter Mann (Marburg, 1967); *Von der vielfaltigen tugent unnd waren bereitung, dess rechten edlen Theriacs . . .* (Marburg, 1532); *Botanologicon* (Cologne, 1534; Paris, 1551), the Paris ed. including Valerius Cordus' *Annotationes . . . in Dioscoridis de materia medica libros;* and *De urinis* (Magdeburg, 1536; Frankfurt, 1543). The index of the *Botanologicon* may be found in *Pedanii Dioscoridis . . . de medicinali materia libri sex Ioanne Ruellio . . . interprete. . . Per Gualtherum Rivium* (Frankfurt, 1549).

II. SECONDARY LITERATURE. On Cordus or his work see (in chronological order) Wigand Kahler, *Vita D. Euricii Cordi* (Rinteln, 1744), with bibliography; Friedrich Wilhelm Strieder, *Grundlage zu einer hessischen Gelehrten- und Schriftstellergeschichte,* II (Kassel, 1782), 282–294, with bibliography; Carl Krause, "Euricius Cordus. Eine biographische Skizze aus der Reformationszeit," dissertation (Hanau, 1863); F. W. E. Roth, "Euricius Cordus und dessen *Botanologicon* 1534," in *Archiv für die Geschichte der Naturwissenschaften und der Technik,* 1 (1909), 279 ff.; August Schulz, "Euricius Cordus als botanischer Forscher und Lehrer," *Abhandlungen der Naturforschenden Gesellschaft zu Halle,* n.s. 7 (1919); Hans Vogel, "Euricius Cordus in seinen Epigrammen," dissertation (Greifswald, 1932); Helmut Dolezal, "Euricius Cordus," in *Neue deutsche Biographie,* III (1957), 359 f.; and Peter Dilg, "Das *Botanologicon* des Euricius Cordus. Ein Beitrag zur botanischen Literatur des Humanismus," dissertation (Marburg, 1969), with a German translation of the *Botanologicon.*

RUDOLF SCHMITZ

CORDUS, VALERIUS (*b.* Erfurt, Germany, 18 February 1515; *d.* Rome, Italy, 25 September 1544), *botany, pharmacy.*

Valerius Cordus, who appears far more frequently and extensively in literature than his father Euricius, owes his fame chiefly to his authorship of the first official pharmacopoeia in Germany. The importance of his scientific role, however, lies primarily in botany, pharmacognosy, and pharmacy, which he enriched not only by critical plant characterizations but also by new teaching methods based on his own experience and observations. Despite numerous individual publications, so far no comprehensive biography of Cordus exists, perhaps because periods of his short life still show gaps and obscurities. These could be filled in and eliminated only through exhaustive research in the scattered source material.

After spending his childhood and youth at Kassel, then at Erfurt and Brunswick, Cordus went to Marburg, where in 1527 he and his brother Philipp enrolled at the university; he received his bachelor's degree in 1531. During these and the following two years of study, he was under the direct influence of his father, who instructed him in the preparation of medicines as well as in botany. It was with pride that in his *Botanologicon* Euricius Cordus mentioned his son's knowledge, for young Cordus had become very familiar with the science of drugs at an early age. Subsequently Cordus completed this training in the apothecary shop of his uncle Johannes Ralla at Leipzig, where he moved in 1533 and enrolled at the university. He remained there probably until 1539; his enrollment at Wittenberg University can be traced back only to that year.

The only thing known about Cordus' stay at Wittenberg University is that he attended Melanchthon's lectures on the *Alexipharmaka* of Nikander of Colophon and that he himself on three occasions—during the winter semesters of 1539/40 and 1542/43 and the summer semester of 1543—lectured on the *Materia medica* of Dioscorides. This lecturing is important—as evidenced by the reports of his students—because in his research, which was novel at the time, he departed from the purely philological interpretation of and commentary on the text, preferring to rely on his own powers of observation, which he had acquired during walks and longer excursions with students and friends.

During his years at Wittenberg, Cordus also developed close ties to the local apothecary shop of the painter Lucas Cranach. Once again this meant close ties to practical pharmacy. The experience Cordus acquired then found expression in the *Dispensatorium*, which he completed there. During his short visit to Nuremberg in 1542 he submitted that work to the city council, which published it in 1546. The last of the many trips that Cordus took from Wittenberg led him to Italy. Via Venice, Padua, and Bologna he reached Rome, where in 1544—when he was only twenty-nine years old—he died from a severe fever, or possibly an accident. His grave in the Church of Santa Maria dell'Anima, described by his contemporaries, was later lost track of and destroyed during one of several renovations.

A survey of Cordus' work, which appeared in print only after his death and received wide dissemination and recognition particularly through the edition of Conrad Gesner, reveals three principal points of his scientific achievement.

His role in pharmacy is based primarily on the much-praised *Dispensatorium* (1546), which through a limited selection of prescriptions brought order for the first time into the unsystematic corpus of medicaments and soon became the obligatory standard for all of Germany. In addition to describing approxi-

mately 225 medicinal plants and minerals, Cordus also refers, with careful commentary, to the origin and adulteration of drugs. The undated first edition was quickly followed by the second and subsequent editions that made this first official pharmacopoeia known far beyond the borders of Germany. Cordus also is generally called the discoverer of ether, for which—probably based on work by his predecessors—he gave the first method of preparation in *De artificiosis extractionibus liber* (1561).

Cordus' two principal works in botany are *Annotationes in Dioscoridis de materia medica libros* (1549) and *Historiae stirpium libri IV* (1561); the latter was followed by *Stirpium descriptionis liber quintus* (1563). Cordus shows himself to be an observant and critical natural scientist in the *Annotationes,* which served as the basis for his Wittenberg lectures and also ran to several editions, and in the *Historiae,* which contains approximately 500 descriptions of plants, with special emphasis on their smell, taste, and location. In contrast with most of his contemporaries, he attempted to establish distinct differences between species and genus, to make the nomenclature precise, and, above all, to form his own opinion based upon his own observations and to correct by comparison even authors long recognized by tradition.

Cordus' rank as a pharmacognosist is based—apart from the minor publication *De halosantho* (1566)—on his thorough knowledge of the materia medica, as evidenced by his comments in the *Dispensatorium* and the *Annotationes*. In his evaluation of the various remedies he benefited greatly from the experience gained in the apothecary shops in Leipzig and Wittenberg. He is quite justifiably regarded as one of the fathers of pharmacognostics (Tschirch).

Extraordinarily gifted and with an appealing personality—such were descriptions by his friends—he knew how to interest others in science and created new areas in botany with his research. Thus, Valerius Cordus fulfilled in an exemplary way his obligation to extend the legacy of his father. Despite his youth, and even though his works were published only posthumously, he enjoyed during his lifetime the reputation that historians of pharmacy and botany accord him today.

BIBLIOGRAPHY

I. ORIGINAL WORKS. Cordus' works were published after his death, partly from finished MSS and partly from the notes taken by his students. The Swiss naturalist Conrad Gesner deserves special credit for their publication. The *Pharmacorum omnium, quae quidem in usu sunt, conficiendorum ratio. Vulgo vocant Dispensatorium pharmacopolarum*

(Nuremberg, n.d. [1546]) had the title *Pharmacorum conficiendorum ratio. Vulgo vocant Dispensatorium* in its 2nd ed. (Nuremberg, n.d. [1546/1547]) and subsequent ones (Lyons, 1552, 1559; Venice, 1563; Antwerp, 1580; Nuremberg, 1598); there are facsimile eds. of the 1st ed., with introduction by Ludwig Winkler (Mittenwald, 1934), and of the 1598 ed. (Munich, 1969). The *Annotationes . . . in Dioscoridis de materia medica libros* is in *Pedanii Dioscoridis . . . de medicinali materia libri sex Ioanne Ruellio . . . interprete. Per Gualtherum Rivium* (Frankfurt, 1549) and in Euricius Cordus' *Botanologicon* (Paris, 1551).

Gesner published the following, all in one vol.: *Annotationes in Pedanii Dioscoridis Anazarbei de medica materia libros V. longe aliae quam ante hac sunt emulgatae; Historiae stirpium libri IV; Sylva, qua rerum in Germania plurimarum, metallorum, lapidum et stirpium . . . persequitur; Loca medicaminum feracia in Germania; De artificiosis extractionibus liber;* and *Compositiones medicinales* (Strasbourg, 1561). He was also responsible for publication of *Stirpium descriptionis liber quintus . . .* (Strasbourg, 1563), new ed., supplemented and improved, in *Conradi Gesneri opera botanica*, C. C. Schmiedel, ed., I (Nuremberg, 1751); and for *De halosantho, seu spermate ceti . . .* (Zurich, 1566).

II. SECONDARY LITERATURE. On Cordus or his work see the following (listed chronologically): T. Irmisch, "Ueber einige Botaniker des 16. Jahrhunderts," *Gymnasialprogramm* (Sondershausen, 1862), pp. 10–34; and "Einige Mittheilungen ueber Valerius Cordus," in *Botanische Zeitung*, **22** (1864), 315–317; H. Peters, "Die älteste Pharmakopoee in Deutschland," in his *Aus pharmazeutischer Vorzeit* (Berlin, 1886), pp. 129–153; A. Tschirch, "Pharmakohistoria," in *Handbuch der Pharmakognosie*, I (Leipzig, 1908), 775–779, 795–803 and 2nd ed. (Leipzig, 1933), I, sec. 3, 1565–1570; E. L. Greene, "Landmarks of Botanical History," in *Smithsonian Miscellaneous Collections*, **54**, no. 1 (1909), 263–314; A. Schulz, "Valerius Cordus als mitteldeutscher Florist," in *Mitteilungen des Thüringischen botanischen Vereins*, n.s. **33** (1916), 37–66; C. D. Leake, "Valerius Cordus and the Discovery of Ether," in *Isis*, **7** (1925), 14–24; K. Sudhoff, "Valerius Cordus, der Aether und Theophrast von Hohenheim," in *Festschrift für A. Tschirch* (Leipzig, 1926), pp. 203–210; R. Kress, "Valerius Cordus als Botaniker und Pharmakognost," in *Deutsche Apothekerzeitung*, **51** (1936), 1227–1229; T. A. Sprague, "The Herbal of Valerius Cordus," in *Journal of the Linnean Society*, Botany, **52** (1939–1945), 1–113; W. Schneider, "Bemerkungen zum ersten offiziellen deutschen Arzneibuch," in *Süddeutsche Apothekerzeitung*, **89** (1949), 136–137; A. Lutz, "Das Nürnberger *Dispensatorium* des Valerius Cordus von 1546, die erste amtliche Pharmakopoee," in *Festschrift für Ernst Urban* (Stuttgart, 1949), pp. 107–125; K. F. Hoffmann, "Valerius Cordus," in *Münchener medizinische Wochenschrift*, **93** (1951), 181–182; O. Bessler, "Valerius Cordus und der medizinisch-botanische Unterricht," in *450 Jahre Martin-Luther-Universität Halle Wittenberg* (Halle–Wittenberg, 1952), I, 323–333; A. Lutz, "Valerius Cordus und die Pharmakopoeen des 16. Jahrhunderts," in *Schweizer Apothekerzeitung*, **93** (1955), 397 ff.; R. Schmitz, "Zur Bibliographie der Erstausgabe des *Dispensatoriums* Valerii Cordi," in *Sudhoffs Archiv für Geschichte der Medizin und der Naturwissenschaften*, **42** (1958), 260–270; T. Robinson, "On the Nature of Sweet Oil of Vitriol," in *Journal of the History of Medicine*, **14** (1959), 231–233; E. Philipp, *Das Medizinal- und Apothekenrecht in Nürnberg*, no. 3 in the series Quellen und Studien zur Geschichte der Pharmazie (Frankfurt, 1962); R. Schmitz, "Neuere Untersuchungen zur Einführungsgeschichte des *Dispensatoriums* Valerii Cordi," in *Veröffentlichungen der Internationalen Gesellschaft für Geschichte der Pharmazie*, new ed., XXI (Stuttgart, 1963), 85–91; and G. E. Dann, "Cordus-Bildnisse," in *Geschichtsbeilage der Deutschen Apothekerzeitung*, **20**, no. 2 (1968), 9–11; and "Leben und Leistung des Valerius Cordus aus neuer Sicht," in *Pharmazeutische Zeitung*, **113** (1968), 1062–1072.

RUDOLF SCHMITZ

CORI, GERTY THERESA RADNITZ (*b.* Prague, Austria-Hungary [now Czechoslovakia], 15 August 1896; *d.* St. Louis, Missouri, 26 October 1957), *biochemistry.*

The daughter of Otto and Martha Radnitz, Gerty Cori graduated from a school for girls in 1912. Since she wished to study chemistry, she was obliged to prepare for the university entrance examination (*matura*). After passing the examination at the Tetschen Realgymnasium in Prague she entered the medical school of the German University of Prague (Ferdinand University) in 1914. She received the M.D. degree in 1920 and married a fellow student, Carl Ferdinand Cori, in August of the same year. After two years at the Karolinen Children's Hospital in Vienna, where she worked on the problem of temperature regulation in a case of congenital myxedema before and after thyroid therapy, she came to the United States to join her husband at the New York State Institute for the Study of Malignant Diseases in Buffalo, New York. In 1931 the Coris went to the Washington University School of Medicine in St. Louis, Missouri, where Gerty Cori was appointed research associate in the department of pharmacology. In 1946 the Coris moved to the department of biochemistry at the same university, and in 1947 Gerty Cori became professor of biochemistry, the post she occupied at her death. She had one son, C. Thomas Cori, born in 1936.

At Buffalo, in spite of institutional pressure for Gerty Cori to work on selected aspects of cancer, the Coris initiated a close collaboration in research on the metabolism of carbohydrates in animals. Their first joint report on this subject appeared in 1923; and during the succeeding dozen years they described, in a series of important papers, the effects of the hormones epinephrine and insulin on carbohydrate metabolism. During the course of this work the Coris demonstrated that epinephrine increases the rate of conversion of liver glycogen to glucose, an effect counteracted by insulin, and also that epinephrine increases the rate of conversion of muscle glycogen to lactate, with the formation of hexosemonophosphate. A closer study of the hexosemonophosphate led the Coris to discover and to isolate, in 1936, a new phosphorylated intermediate (glucose-1-phosphate) in carbohydrate metabolism. In 1938 they described its enzymatic interconversion with glucose-6-phosphate, already known to be formed by the phosphorylation of glucose in an enzyme-catalyzed reaction involving adenosine triphosphate (ATP). The Coris then demonstrated that the formation of glucose-1-phosphate from glycogen is effected by a new enzyme, phosphorylase, that catalyzes the cleavage and synthesis of polysaccharides. Before these discoveries had been made, it was widely believed that the metabolic breakdown of glycogen involved its hydrolysis to glucose; the Coris showed the existence of an enzymatic mechanism for the phosphorolysis of the glycosidic bonds of a polysaccharide.

The crystallization and characterization of rabbit muscle phosphorylase (fully described in 1943) laid the groundwork for later studies by the Coris and others on the hormonal control of its enzymatic activity. Furthermore, the Coris identified and isolated other enzymes involved in the formation and breakdown of the highly branched glycogen molecule; this knowledge made it possible for them to effect the first synthesis of glycogen in the test tube. For these achievements Carl and Gerty Cori were awarded the 1947 Nobel Prize in physiology or medicine, which they shared with Bernardo A. Houssay of Argentina. Gerty Cori was the third woman to receive a Nobel Prize in science, the other two being Marie Curie and Irène Joliot-Curie.

In subsequent work Gerty Cori used the enzymes involved in the biological cleavage of glycogen as tools for the chemical definition of its molecular structure. This was achieved in 1952, almost exactly 100 years after the discovery of glycogen by Claude Bernard. The insights into the chemistry of glycogen, and of the enzymes concerned with its biological transformations, made it possible for Gerty Cori to illuminate in 1953 the nature of the glycogen storage diseases in children. She recognized two groups of disorders, one involving excessive amounts of normal glycogen and the other characterized by abnormally branched glycogen, and showed them to be a consequence of deficiencies or changes in particular enzymes of the metabolic pathway. Gerty Cori's work thus demonstrated the central importance of the isolation and characterization of individual enzymes,

both for the structural definition of the macromolecules on which they act and for the understanding of dysfunctions of metabolic processes in which these enzymes participate.

BIBLIOGRAPHY

I. ORIGINAL WORKS. Among Gerty Cori's most important papers are "The Formation of Hexosephosphate Esters in Frog Muscle," in *Journal of Biological Chemistry,* **116** (1936), 119–128, written with C. F. Cori; "Crystalline Muscle Phosphorylase. II. Prosthetic Group," *ibid.,* **151** (1943), 31–38, written with A. A. Green; "Crystalline Muscle Phosphorylase. III. Kinetics," *ibid.,* 39–55, written with C. F. Cori and A. A. Green; "The Enzymatic Conversion of Phosphorylase *a* to *b,*" *ibid.,* **158** (1945), 321–332, written with C. F. Cori; "Action of Amylo-1, 6-Glucosidase and Phosphorylase on Glycogen and Amylopectin," *ibid.,* **188** (1951), 17–29, written with J. Larner; "Glucose-6-phosphatase of the Liver in Glycogen Storage Disease," *ibid.,* **199** (1952), 661–667, written with C. F. Cori; and "Glycogen Structure and Enzyme Deficiencies in Glycogen Storage Disease," in *Harvey Lectures,* **48** (1952–1953), 145–171.

II. SECONDARY LITERATURE. On Gerty Theresa Cori or her work, see C. F. Cori, "The Call of Science," in *Annual Review of Biochemistry,* **38** (1969), 1–20; B. A. Houssay, "Carl F. and Gerty T. Cori," in *Biochimica et biophysica acta,* **20** (1956), 11–16; and S. Ochoa and H. M. Kalckar, "Gerty T. Cori, Biochemist," in *Science,* **128** (1958), 16–17.

JOSEPH S. FRUTON

CORIOLIS, GASPARD GUSTAVE DE (*b.* Paris, France, 21 May 1792; *d.* Paris, 17 September 1843), *theoretical and applied mechanics.*

Descended from an old Provençal family of jurists ennobled in the seventeenth century, G. Coriolis (as he signed his name) was born into troubled times. He was the son of a loyalist officer of Louis XVI who had taken refuge in Nancy, where he became an industrialist. Coriolis was naturally drawn to the Napoleonic École Polytechnique, a training ground for civil servants, and was second in the class entering in 1808. He spent several years in the department of Meurthe-et-Moselle and in the Vosges mountains while in active service with the corps of engineers of the Ponts et Chaussées. His already poor health and the need to provide for his family after his father's death led him to accept in 1816 the duties of tutor in analysis at the École Polytechnique on the recommendation of Cauchy, with whom he shared certain political and religious affinities. From then on, his life was dedicated to the teaching of science; it is this teaching that inspired his work.

In 1829 Coriolis assumed the chair of mechanics

at the newly founded École Centrale des Arts et Manufactures; but in 1830, unwilling to assume further duties at the École Polytechnique, he declined the position left vacant by Cauchy's exile. Coriolis had at that time entered into the creative phase of an undertaking that he had developed during the preceding ten years, and he had none too much time to devote to it. However, in 1832 he agreed to assist Navier in applied mechanics at the École des Ponts et Chaussées and succeeded him in 1836. The Academy of Sciences elected him to replace Navier in the mechanics section.

In 1838 Coriolis ended his teaching at the École Polytechnique to become director of studies, a position in which he excelled. His solicitude and attention extended even to working conditions—the water coolers he had installed in the classrooms are still called "Corio's." The unhealthy condition that afflicted him (which also seems to have prevented him from considering marriage) rapidly grew worse during the spring of 1843 and soon overtook him. He was buried in the Montparnasse cemetery on 19 September 1843. Before his death he edited part of the proofs of his last book, which was published the following year.

Coriolis' work is brief and specialized. It belongs to its time, and although it shows no marks of special genius, it was nevertheless innovative. Classical mechanics is indebted to it for fundamental elements necessary to its own complete elaboration.

In 1829 Coriolis published his first book, *Du calcul de l'effet des machines,* begun ten years earlier and inspired by the writings of Lazare Carnot. Coriolis recognized that it was only one item among many others constituting a train of analytical thought addressed to the "economy" of mechanical power, and he modestly declared that his small contribution would be distinctive only in its way of dealing with the subject.

He was right, but his method (formulated while he was teaching at the newly opened École Centrale des Arts et Manufactures) was more important and significant than he was aware. Coriolis was a cultivated man, and for him the word "economy" retained from its Greek etymology a wealth of meaning that was being compromised by the rise of industrialism. While many scientists seemed to favor a radical separation of theory from technology, Coriolis voiced the belief that rational mechanics should be developed as a discipline for the enunciation of general principles applicable to the operation of motors and analysis of the functioning of machinery. The changes in terminology that he proposed, largely as a result of his teaching experience, were in fact conformable to this clearly conceived policy, as they were to the requirements of the theory itself.

The first of these changes consisted in abandoning for the term "force-displacement" the ambiguous designations of mechanical power, quantity of action, and dynamic effect, in all of which was subsumed the consideration that processes occurred in time. The word "work" was in the air following the publication in 1821 of the treatise in which Coulomb had attempted with reference to the limited capacity for activity in men and animals to characterize the notion of the consumption of something in overcoming resistance. The French word—*travail*—conveys the idea particularly well, and it was certainly Coriolis' contribution to assign it a technical meaning and thereby clarify a notion as old as mechanics itself.

Coriolis further proposed the "dynamode" (1,000 kilogram-meters) as a unit of measurement of work (from the Greek *dynamis,* power, and *odos,* path). He based this choice upon a comparison of units related to man, the horse, and the steam engine, and hoped thereby to reach a common denominator that might be applied to all industrial functions.

"Dynamode" did not catch on, but the technical term "work" remained the key to a better formalization of mechanics by eliminating once and for all the ambiguities of the famous principle of *vitesse virtuelle* (virtual velocities). The term itself ultimately disappeared.

The second important innovation made by Coriolis was to apply the term *force vive* (kinetic energy) to one-half the product mV^2. This was a simple matter of coefficient but convenient in the formulation of general equations of dynamics. Coriolis thus expressed the principle of *vis viva* as the "principle of the transmission of work." By development of the applications inherent in this change of viewpoint, Coriolis' "small contribution" marked an important step in the realization of his comprehensive theory.

Coriolis did not delay in producing more. Indeed, he had been led to study the work of internal forces in a material system in order to determine under what conditions this work is nil; he thus discovered the very remarkable characteristic that the value of the work done by a system of forces of which the resultant is equal to zero is independent of the frame of reference in respect to which the changes of position are considered. Wishing to evaluate the work done by fluids in hydraulic machines and steam engines, he found simple expressions that apply to the fixed framework of the machine with respect to which the moving parts are in motion. It was therefore natural that the question of relative motions in machines should occur to Coriolis and that it should entail study of the effects of changes in the system of reference on the fundamental equations of analytical mechanics. But he confined himself at first to the simple problem of comparing two systems of reference in rectilinear translation moving uniformly in respect to each other, for which the work done by inertial forces is identical. On 6 June 1831 Coriolis submitted a memoir to the Academy on the problem of the general case. He envisaged it in a highly characteristic fashion; to the extent that consideration of relative motion in machines was unavoidable either to eliminate or to simplify the work of linking forces, theory has necessarily to deal with the question of inertial forces when the system of reference is changed.

Thus for the first time Coriolis entered into the study of acceleration in composite motions, and the various phases of this study's formalization deserve attention.

In his 1831 paper, Coriolis had limited himself to exhibiting the existence of a term complementary to relative acceleration and to acceleration of the drive. Since his explicit aim was to enrich rational mechanics with a new statement concerning the transmission of work in relative motion, he was satisfied to demonstrate by computation—without interpreting the analytical expressions for complementary acceleration—that the work of connecting inertial forces is nil for real relative displacements; the problem of interpreting this result without calculations disappeared in the result itself.

From the two theorems on the transmission of work—one for absolute motion, the other for relative motion—Coriolis easily deduced the difference in the case of hydraulic wheels in the work absorbed by the frame of the machine. He could feel satisfied to have removed certain of the doubts expressed about the possibility of subjecting these machines to theory.

Poisson's report to the Academy did Coriolis the service of observing that the considerations that had animated him should be studied in greater generality. The memoirs that Coriolis submitted to the Academy after 1833 insured this generality in considering material systems as combinations of molecules with various kinds of connections.

The expression for complementary acceleration, derived from the momentum of relative velocity and the instantaneous rotation of the frame of reference, contained in Coriolis' posthumous work, was the enduring fruit of this effort at generality. It was a major advance.

Poisson recognized Coriolis' great skill in the analytic methods deriving from Lagrange. It is true that he did not succeed in eliminating the formidable difficulties inherent in this legacy. Nor did he succeed

in determining the conditions under which it is legitimate after a change of variables to employ the expression for live force as a function of the new variable in calculating the work done by inertial forces. It is only nowadays that this is an easy operation to perform, and the restriction in no way detracts from the merit of a pioneer whose problems were still those of applied mechanics.

What is noteworthy, given the generality of his approach, is that those memoirs of Coriolis cited above concern phenomena that arise in practice: collision in the presence of friction and perturbations that disturb the conditions of stability.

The statistical treatment that Coriolis proposed for this last phenomenon (brought to his attention by the steam engine) did not survive the crisis of classical mechanics at the end of the nineteenth century, so that history remembers this eminent polytechnician only for the Coriolis force present in a rotating frame of reference. One application of that force is to fluid masses on the earth's surface. Accordingly, in 1963, a French oceanographic research vessel was named for him, thus honoring the scientist—and not the engineer—in a fitting tribute to a career characterized by its union of theory and technical application.

BIBLIOGRAPHY

I. ORIGINAL WORKS. Coriolis' works include *Du calcul de l'effet des machines, ou Considérations sur l'emploi des moteurs et sur leur évaluation pour servir d'introduction à l'étude spéciale des machines* (Paris, 1829); "Sur l'influence du moment d'inertie du balancier d'une machine à vapeur et de sa vitesse moyenne sur la régularité du mouvement de rotation que le mouvement de va et vient du piston communique au volant," in *Journal de l'École Polytechnique,* **13,** pt. 21 (1832), 228–267; "Sur le principe des forces vives dans les mouvements relatifs des machines," a memoir read to the Academy on 6 June 1831 and simultaneously published in *Journal de l'École Polytechnique,* ibid., 265–302, and *Mémoirs des savants étrangers,* **3** (1832), 573–607; "Sur la manière d'étendre les différents principes de mécanique à des systèmes des corps en les considérant comme des assemblages des molécules," in *Comptes rendus de l'Académie des sciences,* **2** (1836), 85, which was repr. in the first part of his *Traité de la mécanique des corps solides et du calcul de l'effet des machines,* 2nd ed. (Paris, 1844). 82–100; and *Théorie mathématique des effets du jeu de billard* (Paris, 1844). In addition, the Bibliothèque Nationale de Paris possesses autograph courses given by Coriolis to the École Centrale, École des Ponts et Chaussées, and École Polytechnique.

II. SECONDARY LITERATURE. For works on Coriolis, see Jacques Binet, *Discours prononcé aux funérailles de Coriolis le 20 septembre 1843* (Paris, 1843); L. S. Freiman, *Gaspard Gustave Coriolis* (Moscow, 1961), which includes an almost complete bibliography; Nicolas Aimé Renard, *Notice historique sur la vie et les travaux de G. Coriolis* (Nancy, 1861); and Henri Aimé Résal, *Traité de mécanique générale*, I (Paris, 1873) pp. 446 ff.

The Secretary's Office at the Archives of the Académie des Sciences possesses a photographic reproduction of Coriolis' portrait.

PIERRE COSTABEL

CORNETS DE GROOT. See **De Groot, Jan Cornets.**

CORNETTE, CLAUDE-MELCHIOR (*b.* Besançon, France, 1 March 1744; *d.* Rome, Italy, 11 May 1794), *chemistry, medicine.*

Cornette, the ninth child of Pierre-Claude Cornette and Claude-Antoine Sauvin, received his early education in the local Jesuit college. In 1760 Cornette began studying pharmacy with a Besançon apothecary named Janson. In 1763 he went to Paris, where he studied chemistry under Macquer and Baumé and pharmacy under Guillaume-François Rouelle until 1768.

Almost nothing is known of Cornette's life between 1768 and 1772. In the latter year he came under the powerful protection of the king's chief physician, Joseph-Marie-François de Lassone. In Lassone's laboratory at Marly-le-Roi he was able to carry on his own research.

He furthered his education by obtaining the title of physician from Montpellier in 1778, after about three years of study. His prominence as a scientist rose when he was named a member of the Academy of Sciences in March of the same year and when, the following year, he joined the Royal Society of Medicine, of which Lassone had been one of the founders. Also in 1779 he became an inspector of the Manufacture Royale des Gobelins. By 1784 he had become physician to the king's aunts, and in 1788, when Lassone died, he replaced him as the king's chief physician. Ironically, his success forced him into exile with the royal family because of the French Revolution, and he died in Rome in 1794.

Cornette's works that deal mainly with chemistry were at first solidly in the Rouellian tradition: he followed Rouelle's theories, as well as several questions that had long interested Rouelle himself. For example, Cornette conceived of chemical union as the *adhérence* (sticking together) of the *latus* (side) of each chemical component, which is what Rouelle also taught. Cornette's several memoirs on salts and their decomposition by mineral acids are strongly reminiscent of Rouelle's important memoirs on this topic. In this study, Cornette was led to suggest corrections to the affinity tables, and his complaint that these were

often too general reflects the staunch empiricism of Rouelle. Finally, his memoirs on the reaction of acids with oils, which were motivated by a prize put up by the Academy of Dijon in 1777, also relied on Rouelle's study on the inflammation of oils.

Later on Cornette turned his attention to chemical drugs (works on soap, mercury, etc.) at the expense of pure chemistry. This shift in emphasis may have been caused by his refusal to espouse the newer theories of Lavoisier, a refusal dictated perhaps more by personal feelings of dislike than by doctrinal convictions.

BIBLIOGRAPHY

Among Cornette's published writings are *Quaestio chemico-medica de diversis saponum generibus, variaque illorum in curandis morbis efficacia . . .* (Montpellier, 1778); and *Mémoire sur la formation du salpêtre et sur les moyens d'augmenter en France la production de ce sel* (Paris, 1779). Many memoirs appeared in the *Mémoires de l'Académie royale des sciences* (1778–1786). Prominent among these are his memoirs on salts, which appeared in the *Mémoires* for 1778, 1779, and 1783; and his memoirs on the reaction between acids and oils, which appeared in the *Mémoires* for 1780 and 1782. Several of his memoirs also appeared in the *Mémoires de la Société royale de médecine,* **3-8** (1782–1788). Many of these were done in collaboration with Lassone. MSS in his hand can be found in the Archives de Seine-et-Oise in Versailles, cataloged as *rubrique* 107, 1 and 2 (E 678–679); *rubrique* 108 (E 689–699); and *rubrique* 109 (E 705).

A more detailed secondary source is André Desormonts, *Claude-Melchior Cornette, apothicaire, chimiste, hygiéniste, médecin, médecin de cour* (Paris, 1933).

JEAN-CLAUDE GUÉDON

CORNU, MARIE ALFRED (*b.* Orléans, France, 6 March 1841; *d.* La Chansonnerie, near Romartin, France, 12 April 1902), *optics.*

Cornu entered the École Polytechnique in 1860, graduated second in his class in 1862, and proceeded to the École des Mines. In 1864 he was appointed *répétiteur* at the École Polytechnique, where he became professor of physics in 1867, the year in which he was awarded his doctorate for a thesis on crystalline reflection. He became a member of the Académie des Sciences in 1878 and was an associate of many foreign scientific bodies, including the Royal Society of London (1884) and the U.S. National Academy of Sciences (1901).

While still a probationer at the École des Mines, Cornu was attracted to experimental optics by an exhaustive study of Félix Billet's celebrated *Traité d'optique,* and he repeated all the experiments in this work in his spare time. Following Jules Jamin, he showed that reflection at metallic surfaces proceeds exactly as in the case of vitreous substances, due allowance being made for ultraviolet frequencies. A series of experiments dating from 1871 led to a re-determination of the velocity of light by Fizeau's method, for which Cornu was awarded the La Caze Prize of the Académie des Sciences and the Rumford Medal of the Royal Society of London in 1878.

Cornu made a number of important contributions to spectrum analysis, including very precise measurements of the wavelengths of certain lines in the hydrogen spectrum. By observing the edges of the solar disk and applying the Doppler-Fizeau principle, he found a means of separating the solar spectra from the terrestrial spectra; in the case of the latter he separated the influence of water vapor from that of air. Following the discovery of the Zeeman effect Cornu showed that the D line of sodium is decomposed, under normal magnetization, into four components, as opposed to three, thus forcing Lorentz to modify his theory of the Zeeman effect.

Cornu's optical researches also included studies of conditions for achromatism in interference phenomena; his work on the measurement of the curvature of lenses; and his explanation of certain observed anomalies in the behavior of diffraction gratings in terms of minute variations in the distance between successive lines. He also engaged in acoustical researches and, with Baille, redetermined the gravitational constant by Cavendish's method.

Apart from his many contributions to experimental physics, Cornu is remembered for the elegant method of the so-called Cornu spiral for the determination of intensities in interference phenomena.

BIBLIOGRAPHY

Cornu's dossier at the Académie des Sciences, Paris, contains a portrait and a number of autograph letters, including six to Fizeau, as well as the orations pronounced at his funeral by Bassot, Mascart, and Poincaré.

Alfred Cornu, 1841–1902 (Rennes, 1904) contains a biographical notice by Henri Poincaré and a complete bibliography of Cornu's published work. Also see *Notice sur les titres scientifiques de M. A. Cornu par H. Fizeau* (Paris, 1873); and the obituary notice in *Proceedings of the Royal Society,* **75** (1905), 184–188.

J. W. HERIVEL

CORONEL, LUIS NUÑEZ (*b.* Segovia, Spain, second half of fifteenth century; *d.* Spain or Canary Islands, 1531), *logic, natural philosophy.*

Coronel received his early education at Salamanca and then went, around 1500, to the Collège de Montaigu at the University of Paris, where he was taught by John Major. He was at the Sorbonne in 1504 as a guest and in 1509 as a fellow. Ordained a priest in 1512, he received the licentiate in theology on 26 January 1514 and the doctorate on 29 May 1514. His brother, Antonio, was John Major's favored disciple and a close friend of Peter Crokaert of Brussels and other Dominicans at St.-Jacques; Antonio is important for his works on logic.

Coronel left Paris around 1517, after Francisco (later Domingo) de Soto's arrival there; in 1520 he was in Flanders as preacher and adviser at the court of Charles V, and in 1521 or 1522 served with the Inquisition at Brussels. Here he came to know Erasmus, who regarded him as an ally and with whom he later corresponded. In 1527 he was secretary to Alfonso Monrique, archbishop of Seville; and in the same year, according to a letter of Juan Luis Vivès to Erasmus, he was made bishop of Las Palmas in the Canary Islands.

At Paris, Coronel published his *Tractatus [de formatione] syllogismorum* in 1507 (or 1508) and his more important *Physicae perscrutationes* in 1511; the latter was based on his lectures at Montaigu and went through three additional known editions (Lyons, 1512, 1530; Alcalá, 1539). Coronel's physical doctrines were influenced by Jean Dullaert of Ghent and Alvaro Thomaz; and he cites approximately the same sources as they among the Oxford "calculators," the Paris "terminists," and various Italian writers. Jean Buridan, Richard Swineshead, Albert of Saxony, and Gregory of Rimini are mentioned by him with greatest frequency.

Coronel's "investigations" (166 folios in the Lyons edition of 1512) are located in the framework of Aristotle's *Physics,* with long digressions on motion (thirty-five folios in book III) and on infinity (thirty-eight folios in book VIII). The tract on motion successively investigates local motion, alteration (including the intensity of forms), augmentation, and the velocities of motions; it contains little original apart from the thesis that impetus is not really distinct from local motion. Coronel treats infinity mainly in relation to God's power, holding that God can produce a syncategorematic, but not a categorematic, infinity; in this he differs from John Major and Juan de Celaya.

BIBLIOGRAPHY

I. ORIGINAL WORKS. Neither of Coronel's works is translated into English; a copy of the *Perscrutationes* is at the University of Wisconsin. Pierre Duhem, *Études sur*

Léonard de Vinci, III (Paris, 1913), gives numerous brief excerpts in French translation. Hubert Élie, ed., *Le traité "De l'infini" de Jean Mair* (Paris, 1938), appendix 5, provides a lengthy French translation from the tract on infinity.

II. Secondary Literature. Coronel's work is discussed in Hubert Élie, "Quelques maîtres de l'université de Paris vers l'an 1500," in *Archives d'histoire doctrinale et littéraire du moyen âge,* **18** (1950–1951), 193–243, esp. 212–213; R. G. Villoslada, *La universidad de Paris durante los estudios de Francisco de Vitoria, O.P. (1507–1522),* vol. XIV in Analecta Gregoriana (Rome, 1938), esp. pp. 386–390; and William A. Wallace, "The Concept of Motion in the Sixteenth Century," in *Proceedings of the American Catholic Philosophical Association,* **41** (1967), 184–195.

WILLIAM A. WALLACE, O. P.

CORONELLI, VINCENZO MARIA (*b.* Venice, Italy, 15 August 1650; *d.* Venice, 9 December 1718), *geography.*

Coronelli was the son of Maffio and Catarina Coronelli. He became a Minorite friar in 1655 and received a doctorate in theology at Rome in 1674. Interested in maps and globes from his early youth, he built his first pair of globes for the duke of Parma about 1678. The workmanship and accuracy of these globes impressed César Cardinal d'Estrées, the French ambassador to the Holy See; and in 1681 Coronelli was invited to Paris, where he spent two years constructing a terrestrial globe and a celestial globe for Louis XIV. These extremely ornate works, 3.90 meters in diameter, remained the largest globes made until the 1920's. They were first displayed at the royal residence of Marly, later the Royal (now National) Library in Paris. Since 1920 these globes have been stored at Versailles.

On his return to Venice in 1684, Coronelli founded the Accademia Cosmografica degli Argonauti, the first geographical society. Named Cosmographer of the Republic of Venice, Coronelli proceeded to design two major atlases, the *Atlante veneto* and the *Isolario;* traveled extensively through Europe; and became known as a civil engineer and geographer. In 1701 he was elected minister-general of the Minorite order, a post he held for three years. In 1705 he returned to Venice and, except for one last journey to Vienna, where he was consulted by the emperor on flood control measures, spent the rest of his life in the Minorite convent there.

Coronelli's work includes, besides the more than 100 large and small globes that have survived, several hundred maps, printed separately and as parts of atlases, and seven volumes of a projected forty-five-volume encyclopedia, the first major encyclopedia to be arranged alphabetically and published in the ver-

nacular. But it is chiefly for his globes that Coronelli is remembered: their accuracy, the wealth and timeliness of the information displayed, and their artistic excellence distinguish their maker as one of the leading geographers and cartographers of the baroque period.

BIBLIOGRAPHY

I. Original Works. Coronelli's writings are *Atlante veneto* (Venice, 1691); *Isolario* (Venice, 1696–1698); and *Biblioteca universale sacro-profana,* 7 vols. (Venice, 1701–1706).

II. Secondary Literature. On Coronelli or his work, see Roberto Almagià, "Vincenzo Coronelli," in *Der Globusfreund,* **1,** no. 1 (1952), 13–27; O.-G. Saarmann Muris, *Der Globus im Wandel der Zeiten* (Berlin-Stuttgart, 1961), pp. 167–173; and Pietro Rigobon, "Biografia e studi del P. Vincenzo Coronelli," in *Archivio veneto,* **3,** no. 1 (1872), 267–271.

GEORGE KISH

CORRENS, CARL FRANZ JOSEPH ERICH (*b.* Munich, Germany, 19 September 1864; *d.* Berlin, Germany, 14 February 1933), *botany, plant genetics.*

Correns was the only child of Erich Correns, a painter and member of the Bavarian Academy of Art. His mother was of Swiss extraction.

Orphaned at seventeen Carl left Munich for St. Gall, Switzerland, where tuberculosis interrupted his schooling and prevented him from obtaining his *Abitur* until 1885. Four years later he graduated from the University of Munich; and after two semesters in Berlin under Schwendener he moved to Tübingen, where he took his *Habilitation* in 1892. In the same year, he married Elizabeth Widmer, a niece of Karl von Naegeli. After ten years as *Privatdozent,* Correns was appointed assistant professor at Pfeffer's institute in Leipzig and full professor at Münster in 1909. When the Kaiser-Wilhelm Institut für Biologie was built in Berlin (Dahlem) in 1913, Correns became its first director. There he remained for the rest of his life, and his unpublished manuscripts were preserved there until their destruction during the bombing of Berlin in 1945.

Correns won fame as a rediscoverer of Mendel's laws. Whereas de Vries and Tschermak, who rediscovered Mendel's work simultaneously, were more concerned with mutation and practical plant breeding, respectively, Correns concentrated on the xenia question: Does foreign pollen have a direct influence on the characteristics of the fruit and seed? This subject was but one of many botanical problems occupying his attention in the 1890's. Correns had studied under

Karl von Naegeli; and his study of cell wall growth, which formed the subject of his *Habilitationsschrift,* was devoted to an attack on E. Strasburger's theory of growth by apposition and a defense of Naegeli's intussusception theory. Other subjects that he studied at this time were the floral morphology of *Dioscorea, Primula vulgaris, Aristolochia, Calceolaria,* and *Salvia;* the physiology of plant sensitivity in *Drosera* and climbing plants; the initiation of leaf primordia; and vegetative reproduction in mosses and liverworts.

The many-sided character of Correns' early research indicates the indecision of one trained in the shadow of giants, Naegeli and Darwin in particular. His work represents a reinvestigation, extension, or criticism of previous work. He studied heterostylism in the primrose because of disbelief in Darwin's and G. G. F. Delpino's views on adaptation to outbreeding. Following his mentor Naegeli, he denied that the large and small pollen grains are an adaptation to long- and short-styled flowers because culture of the grains in a sugar solution failed to yield the required difference in length. In his study of *Drosera rotundifolia* Correns also questioned Darwin's conclusions on the effect of increasing temperature on the sensitivity of leaves.

In 1894 Correns took up the xenia question, studying it by crossing varieties of *Zea, Pisum, Phaseolus, Lilium,* and *Matthiola.* Three years later he concluded that the xenia effect in maize is due either to an enzymatic effect of the embryo on the endosperm tissue or to a genuine hybridization between the secondary embryo sac nucleus and a generative nucleus of the pollen tube. (Mistakenly, he was thinking of a division of the nucleus, which serves as gamete to the egg cell, to form two nuclei, one for the egg cell and one for the endosperm.) While Correns delayed seeking cytological proof, S. G. Nawaschin and L. Guinard showed that in *Lilium* one of the original generative nuclei of the pollen tube fuses with the endosperm nucleus. Correns did not learn of this until late in 1899. Meanwhile, his pea crosses had reached the fourth generation. Here the color change, unlike that in maize, presented no problem to Correns, who, like Mendel, realized that only the cotyledons of the new embryo—not the mother tissue—were involved. His scoring of maize progeny had given him complicated numerical results, but his pea progeny gave him simple ratios; and in October 1899 Correns arrived at the correct explanation. A few weeks later he read Mendel's paper, of which he had learned from a rereading of W. O. Focke's *Die Pflanzenmischlinge.* On 21 April Correns was stimulated to write up his own account by the receipt of de Vries's paper "Sur la loi de disjonction des hybrides." The German Botanical Society received it on 24 April.

Of the three rediscoverers of Mendel's laws, Correns showed the deepest understanding and the most subtle approach. The remainder of his career was devoted to an investigation of the precise extent of the validity of these laws. He was the first to correlate Mendelian segregation with reduction division and to show that Mendelian segregation does not require the dominant-recessive relationship, and that characters, even those due to different physiological agents, can be "coupled" or "conjugated" in heredity and are therefore unable to show independent assortment (Mendel's second law). In 1902 he found in maize a case of coupling between self-sterility and blue coloration of the aleurone layer; here too he found the first evidence of differential fertilization between different gametes. This finding was later supported by his discovery of competition between male and female pollen grains, the former having a selective advantage over the latter. In the same year as Walter Sutton (1902), Correns produced a chromosome theory of heredity that allowed for exchange of genes between homologous chromosomes but not for the block-transfer type later advanced by Morgan. On Correns' scheme, therefore, mapping of the genes from crossover values would be impossible.

In 1903 Correns predicted that sex is inherited in a Mendelian fashion, and in 1907 he proved it by his classic experiments with *Bryonia.* Two years later he obtained the first proof of cytoplasmic inheritance in plants simultaneously and independently of Erwin Baur. From that year until his death Correns studied plant variegation in all its forms, and by crossbreeding experiments he distinguished between nuclear, Mendelian and extranuclear, non-Mendelian inheritance of variegation. Whereas Baur proposed that the plastids themselves are the genetic determinants (Plastom theory) in cases of extranuclear heredity, Correns favored the nonparticulate cytoplasm of the cell considered as a whole (non-Plastom theory). For Baur the change from green to white cells was due to plastid mutation; for Correns in 1909 it was a case of disease, but in 1922 he postulated a labile state of the cytoplasm that could cause the plastids to develop either into normal (green) bodies or into diseased (white) bodies. Although the Plastom theory is now widely accepted both for biparental and for monoparental cytoplasmic inheritance, in Correns' time the failure to detect cells with white and green plastids mixed and to see a clear boundary between mutant and normal tissues seemed to furnish valid objections to Baur's theory.

Correns was unimpressive as a lecturer. He always published in German and rarely traveled. Unlike de Vries, who forced the facts of heredity into an oversimplified Mendelian scheme, Correns stressed the complexities and the exceptions to that scheme. Unfortunately his efforts to introduce more precise terminology were clumsy and did not win acceptance. Yet in 1900 his view of heredity had a sophistication and depth that geneticists in general would not achieve for nearly two decades. He was a hard worker, dedicated to science rather than to success, and willingly gave Mendel full credit, later publishing Mendel's famous letters to Naegeli. He deserves the first place in the history of the rediscovery period of Mendelian genetics.

BIBLIOGRAPHY

I. ORIGINAL WORKS. Nearly all of Correns' 102 papers were reprinted in his *Gesammelte Abhandlungen zur Vererbungswissenschaft aus periodischen Schriften 1899–1924*, F. von Wettstein, ed. (Berlin, 1924). A complete list up to 1923 will be found in Sierp's paper (see below). Although he published no books, Correns did contribute two lengthy monographs to the series Handbuch der Vererbungswissenschaft: "Bestimmung, Vererbung und Verteilung des Geschlechtes bei den höheren Pflanzen," **2c** (1928), 1–138; and "Nicht mendelnde Vererbung," F. von Wettstein, ed., **2h** (1937).

Correns' famous rediscovery paper, "G. Mendel's Regel über das Verhalten der Nachkommenschaft der Rassenbastarde," in *Bericht der Deutschen botanischen Gesellschaft*, **18** (1900), 158–167, is available in translation in C. Stern and E. R. Sherwood, eds., *The Origin of Genetics. A Mendel Sourcebook* (San Francisco–London, 1966) pp. 119–132. Little-known but excellent is Correns' retrospective essay "Die ersten zwanzig Jahre Mendelscher Vererbungslehre," in *Festschrift der Kaiser-Wilhelm Gesellschaft. Förderung der Wissenschaft* (Berlin, 1921), pp. 42–49.

II. SECONDARY LITERATURE. Obituary notices appeared in *Nature*, **131** (1933), 537–538; and *Naturwissenschaften*, **22** (1933), 1–8, with foreword by Max Planck and portrait. Two accounts of Correns' work which complement each other are E. Stein, "Dem Gedächtnis von Carl Erich Correns nach einem halben Jahrhundert der Vererbungswissenschaft," in *Naturwissenschaften*, **37** (1950), 457–463; and H. Sierp, "Die nichtvererbungswissenschaftlichen Arbeiten von Correns," *ibid.*, **12** (1924), 772–780. In the same volume there is a photograph of Correns facing p. 749 and a paper by A. Zimmermann, "Carl Erich Correns," pp. 751–752. For a modern assessment of Correns' work on plastid inheritance see R. Hagemann, *Plasmatische Vererbung* (Jena, 1964); and J. T. O. Kirk and R. A. E. Tilney-Bassett, *The Plastids. Their Chemistry, Structure, Growth and Inheritance* (London–San Francisco, 1967).

ROBERT OLBY

CORTÉS DE ALBACAR, MARTÍN (*b.* Bujaraloz, Spain; *d.* Cádiz, Spain, 1582), *cosmography, navigation.*

The son of Martín Cortés and Martina de Albacar,

Cortés belonged to an ancient and noble Aragon family. It is known that he went to Cádiz before 1530; the only other data on his life are those contained in his *Breve compendio de la esfera*, written in 1545 and reviewed by his contemporaries as possessing such order and clarity as to make it useful to those wishing to learn navigation.

As the fever of discovery continued unabated into the beginning of the sixteenth century, so did the need for works that would expedite the navigation of unexplored oceans. It was truly said that for an enterprise so difficult as guiding a ship by sea and sky alone, those who had faith must turn their eyes to heaven; at the same time, the scanty learning of most sailors did not dispose them to being convinced by scientific arguments. In his letter offering the *Breve compendio* to Emperor Charles V, Cortés included this sentence: ". . . but I would rather call myself the first to have reduced navigation to a brief set of rules."

The *Breve compendio* consists of three parts: the first (twenty chapters) deals with the composition of the world and the universal principles of the art of navigation; the second (twenty chapters) considers the movements of the sun and moon and the effects they produce; and the third (fourteen chapters) deals with the construction and use of instruments and the rules of the art of navigation. In his work, Cortés indicates that the earth has a torrid zone, two temperate zones, and two glacial zones. He describes how there is an atmosphere surrounding the waters of the sea, and in this atmosphere he distinguishes three regions: two hot ones at the extremes and a cold, dark, humid one in the center. He adds that on the earth—which is round, with mountains, valleys, and plains—the sea undergoes oscillations and swellings produced by the moon. The moon, according to him, is the smallest of the planets except for Mercury but appears large because it is near the earth.

He subscribes to the Ptolemaic system of geocentricity and the immobility of the terraqueous globe at the center of the universe; he reasons that the earth is immobile, because a stone thrown into the air comes down on the same spot. He then describes the proper motions of the planets.

Cortés' *Breve compendio* is particularly concerned with practical navigation; a technological section includes rules for the construction and use of cross-staffs, astrolabes, and compasses. He is not satisfied with navigational charts that give only directions and distances (or polar coordinates) and says: "It is also necessary to know the latitudes of the principal headlands and of points and of famous cities." He rejects maps with rectangularly drawn parallels and meridians and states that the farther a point is from the equator, the greater the separation of the parallels and meridians at that point. These ideas were the basis for the separation of the parallels in the cylindrical projection perfected subsequently by the cosmographer Alonso de Santa Cruz.

Cortés accepted Columbus' discovery of the variation of magnetic declination (the declination of the compass to the northeast and northwest) and postulated the existence in the heavens of a magnetic pole. "I conceive of a point beneath the pole of the earth, and this point is outside all the cycles contained under the *primum mobile*. This point or part of the cycle has a power of attraction." Cortés' achievement was to affirm the reality of this point when the existence of magnetic declination was still a matter of doubt; such a point explains precisely and clearly the declination of the compass to the northeast and northwest. Cortés drew upon these theories to account for the variation of magnetic declination at various places on the globe, attributed in other works of that period to the poor quality of magnets and lodestones.

BIBLIOGRAPHY

Cortés' only writing is *Breve compendio de la esfera y de la arte de navegar* (Seville, 1551; repr. Seville, 1556; facs. ed., Zaragoza, 1945). Nine English eds. appeared between 1561 and 1630.

On Cortés or his work, see *Diccionario enciclopédico hispano-americano*, V (Barcelona, 1890), 1172 ff.; M. Fernández de Navarrete, *Disertación sobre la historia de la naútica y de las matemáticas* (Madrid, 1846); and José Gavira Martín, "La ciéncia geográfica española del siglo XVI," in *Boletín. Real. Sociedad geográfica,* **71** (1931), 401–424.

J. M. LÓPEZ DE AZCONA

CORTI, ALFONSO GIACOMO GASPARE (*b.* Gambarana, near Pavia, Italy, 22 June 1822; *d.* Corvino San Quirico, near Casteggio, Italy, 2 October 1876), *microscopic anatomy.*

Corti was one of the biologists who in the middle of the nineteenth century sought to explain the fine structure of organs. Inspired by the color effect of chromic acid solutions, which were used as fixing agents, he introduced carmine staining into microscopic technology. He also described the most important components of the membranous cochlea, and in all his work he looked for the connection between structure and function.

The Cortis, an old noble family of Lombardy, maintained an estate at Gambarana; it was all that remained of their former, more extensive holdings. The location of the estate made them citizens of Sardinia. Alfonso was the oldest son of the Marchese Gaspare Giuseppe Corti di San Stefano Belbo and the Marchesa Beatrice Malaspina di Carbonaro. From youth he was under the influence of his father's scientific interests. Probably he spent his school years in Pavia; from 1841 to 1845 he studied medicine there. Bartolomeo Panizza showed him how to use the microscope, and with Mauro Rusconi he became active in comparative anatomy. Against the opposition of his family he continued his studies, from September 1845, in Vienna. Corti visited clinics there and simultaneously did some anatomical work with Josef Hyrtl. He received his M.D. in Vienna on 6 August 1847; his thesis, *De systemate vasorum Psammosauri grisei,* included his own drawings based on injection specimens that he himself had prepared.

At the end of December 1847, Corti was chosen as Hyrtl's second prosector; he soon had to relinquish this position, however, because war broke out between the Kingdom of Sardinia, as the Piedmontese realm was then styled, and Austria on 23 March 1848. He returned to Turin by way of Zurich. From 2 February to 3 August 1849, Corti was in Bern, where he began his own microscopic studies in collaboration with the physiologist Gustav Gabriel Valentin; Valentin encouraged him to undertake the investigations that were published the following year. Except for a trip to England, where he met some important microscopists (including James Paget, Thomas Wharton Jones, and Richard Owen), Corti spent the rest of 1849 with his relatives in Paris.

In the middle of January 1850, Corti left for Würzburg to work with Albert Kölliker, under whom he mastered normal histology in two months. Some results of his work are in Kölliker's handbooks. In his own investigations on the structure of the retina, Corti succeeded in demonstrating, through isolation, the connection of the nerve cells with the optic nerve fibers; by doing this, he verified A. H. Hassall's disputed, earlier observations. Furthermore, in Würzburg he completed the studies, begun in Bern, on the ciliary epithelium in the digestive organs of the larvae of frogs and toads; he reported on this to the Physikalisch-medizinischen Gesellschaft on 6 July 1850. Corti concluded that the ciliary cells in the digestive organs of these larvae serve, before the formation of muscle layers, to expel the contents of the hollow organs. These observations were unknown to all of Corti's early biographers.

During this period Corti must also have been occupied with his studies on the inner ear, which he had begun entirely on his own; he had not completed these studies when he left Würzburg in the middle of August 1850. He was a member of the wedding of Rudolf Virchow in Berlin, and then he went to the Netherlands. In Utrecht he visited Schroeder van der Kolk and Harting, from whom he learned how to mount moist microscopic preparations in fluid or resin. This technique and the method that he discovered of staining microscopic objects with carmine solutions enabled Corti to distinguish the individual components of the membranous cochlea.

Corti's observations are remarkable since he could use only freshly prepared pieces of the membranous cochlea and not fine sections of it. He described the *ganglion spirale cochleae* with its bipolar cells, the *lamina spiralis membranacea,* the *vas spirale,* the columns, the hair cells, and the *membrana tectoria;* he even saw the vascular epithelium of the *stria vascularis ductus cochlearis* and recognized it as the source of the endolymph. He was unsure only how and where the fibers of the *nervus cochlearis* end. Thus he was unable to clarify completely the function of the individual parts of the membranous cochlea; nevertheless, he expressed the ideas upon which Helmholtz later based his "resonance" theory of hearing.

Corti undertook the completion of his work in Paris, where he had gone toward the end of September 1850. There his efforts in microscopic anatomy were honored: he was elected a corresponding member of the Société de Biologie de Paris (on 20 July 1850); corresponding, and later regular, member of the Verein Deutscher Aerzte und Naturforscher in Paris (17 February and 5 April 1851); and corresponding member of the Paris Medical Society (5 April 1851). Pressing family business forced him to leave at the beginning of 1851 for Turin, where he had to spend several years attending to these matters; the high point of his scientific career had already passed.

The scientific world heard of Corti again with the publication of his letter concerning histological observations on an elephant, sent in April 1853 to Kölliker and published in 1854. The letter testifies to Corti's interest in general science and offers proof that the cells of the various tissues of this large mammal are the same relative size as corresponding cells of other mammals. All later work, which Valentin encouraged him to undertake, remained attempts without conclusive results. A last honor was bestowed upon Corti when he was elected a member of the Imperial Leopoldian-Carolinian Academy (6 January 1854). His letter of appreciation to the president of the academy contains the first reference to his suffering from rheumatoid arthritis. This illness, not dangerous but very protracted, caused Corti—increasingly with the years—to become crippled in his hands and feet.

Earlier, Corti had acquired as part of his inheritance an estate in Mazzolino, which he developed into a model farm for viniculture after his financial situation had improved. This improvement was partially the result of his marriage, on 24 September 1855, to Maria Anna Carlotta Bettinzoli, who died in 1861. The management of his estate and the education of his two children gave substance and meaning to Corti's life until his death at the age of fifty-four. Because of his twenty-two-year silence the scientific world, except for a few friends, had long forgotten him. He is now generally known for the organ in the cochlea, described by him, which at Kölliker's suggestion was designated the organ of Corti—a name familiar to students of medicine and the natural sciences.

BIBLIOGRAPHY

I. ORIGINAL WORKS. Corti's major writings are "Beitrag zur Anatomie der Retina," in *Archiv für Anatomie, Physiologie und wissenschaftliche Medizin* (1850), 273–275, 4 figures in plate 6; "Ueber Flimmerbewegung bei Frosch- und Krötenlarven," in *Verhandlungen der Physikalisch-medizinischen Gesellschaft zu Würzburg,* **1** (1850), 191–192; and "Recherches sur l'organe de l'ouie des mammifères. Première partie: Limaçon," in *Zeitschrift für wissenschaftliche Zoologie,* **3** (1851), 109–169, 12 illustrations in plates 3 and 4.

II. SECONDARY LITERATURE. On Corti or his work, see Alfredo Corti, "Il Marchese Alfonso Corti e le sue ultime ricerche nel Laboratorio di anatomia comparata dell' Università di Torino," in *Rivista di storia delle scienze mediche e naturali,* **47** (1955), 1–28, which includes the Italian translation of the letter from Corti to Kölliker, originally in *Zeitschrift für wissenschaftliche Zoologie,* **5** (1854), 87–93; Erich Hintzsche, *Alfonso Corti (1822–1876),* Berner Beiträge zur Geschichte der Medizin und der Naturwissenschaften, no. 3 (Bern, 1944) with complete bibliography; and "Ein neuer Brief von Alfonso Corti (1822–1876)," in *Gesnerus,* **1** (1944), 137–146; Gennaro Palumbi, "Nel centenario della scoperta dell'organo del Corti," in *Atti della Società italiana di anatomia XIII,* supp. to *Monitore zoologico italiano,* **40** (1952), 22–25; Bruno Pincherle, *La vita e l'opera di Alfonso Corti* (Rome, 1932), which includes a reprint of "Recherches sur l'organe de l'ouie"; and Egon V. Ullman, "Life of Alfonso Corti," in *Archives of Otolaryngology,* **54** (1951), 1–28, more "story" than "history."

ERICH HINTZSCHE

CORTI, BONAVENTURA (*b.* Scandiano, Modena, Italy, 26 February 1729; *d.* Reggio nell' Emilia, Italy, 30 January 1813), *physics, botany.*

A native of the same town as Lazzaro Spallanzani, Corti showed an aptitude for scientific subjects even as a boy. He was sent to Reggio nell' Emilia, where he completed his studies and became a priest.

There, at the age of twenty-five, Corti was appointed professor of metaphysics and geometry at the school he had attended. In 1768 Corti was appointed to the professorship of physics, and the next year he distinguished himself with a valuable publication, *Institutiones physicae.*

At the same time Corti was also rector of the school and of the parish church of SS. Nazario e Celso. In this sensitive office he demonstrated the valuable qualities of poise and shrewdness, thereby winning the esteem of Duchess Maria Teresa Cybo d'Este, who had established her residence there. She chose him as her spiritual adviser and later appointed him her almsgiver and counselor in temporal affairs. She rewarded him for his services with two excellent and extraordinarily valuable microscopes manufactured by Dollond, which enabled Corti to undertake fruitful research.

The Este government had little interest in scientific research or in Corti's scientific achievements. He was appointed director of the Collegio dei Nobili in Modena. Corti headed the *collegio* for about twenty-one years, devoting himself to its reopening and operation. In 1804, at the age of seventy-five, he retired and then was appointed professor of botany and agriculture at the University of Modena.

At eighty Corti relinquished the professorship and retired to Reggio nell' Emilia, where he remained until his death.

Fifteen of Corti's published memoirs and ten unpublished ones, some of which are incomplete, survive. He wrote a two-volume treatise on physics, *Institutiones physicae* (1769), that predates the discoveries of Galvani and Volta. Considerable knowledge is displayed in this work. In 1774 Giuseppe Toaldo wrote of it: "In many courses in physics I have not found subjects that are more appropriate for the schools than in this [treatise], which is a collection made with excellent taste and with a wide selection of doctrines."

From the point of view of originality Corti's *Osservazioni microscopiche sulla Tremella e sulla circolazione del fluido in una pianta acquajuola* (1774), issued in one volume and illustrated with beautiful copperplates, is important. Chapter 21 of part I deserves special mention. On the basis of microscopic findings Corti affirms "that *Tremella* are endowed with movements said to be spontaneous in animals and considered characteristic of animals. And here we have plants that by now are confused with true animals" (p. 66). Further on, he defines the protoplasmic movements: "A certain small dark spot passes from right to left, is lost, and reappears; this is that series of rather dark spots which elsewhere I have said is caused by the elliptical figure of the small rings" (p. 68). This is a clear anticipation of later descriptions of the movements of the protoplasm in the cell.

BIBLIOGRAPHY

I. ORIGINAL WORKS. See *Institutiones physicae,* 2 vols. (Modena, 1769); *Osservazione microscopiche sulla Tremella e sulla circolazione del fluido in una pianta acquajuola* (Lucca, 1774); *Lettera al Signor Conte Agostino Paradisi, che riguarda il movimento della linfa in 38 piante fanerogame* (Modena, 1775); *Storia naturale di quegli insetti che rodono le piante del frumento* (Modena, 1804); "Breve ricerca dei casi, ne' quali il commercio, le ricchezze ed il lusso degli individui, invece di accrescere servono anzi a diminuire le forze e l'autorità di uno stato riguardo alle vicine nazioni," in *Notizie biografiche letterarie degli scrittori dello stato estense* (Reggio, 1834), app. 1, p. 345.

II. SECONDARY LITERATURE. On Corti or his work, see P. Bonizzi, *Intorno alle opere scientifiche di Bonaventura Corti* (Modena, 1883); *Enciclopedia cattolica,* IV (1950), 666; E. Manzini, *Memorie storiche dei Reggiani più illustri* (Reggio nell' Emilia, 1878), pp. 29–35; G. Montalenti, "Storia della biologia e della medicina," in *Storia delle scienze,* III, pt. 1 (Turin, 1962), 368; and G. Venturi, *Storia e teorie dell'ottica,* I (Bologna, 1814), and *Storia di Scandiano* (Modena, 1822).

LORIS PREMUDA

CORVISART, JEAN-NICOLAS (*b.* Dricourt, France, 15 February 1755; *d.* Paris, France, 18 September 1821), *medicine.*

Corvisart was one of the first French physicians to advocate the replacement of the empirical methods of diagnosis (passive observation, cursory examination, brief questioning) with a method of thorough and systematic examination and analytical interpretation of physical symptoms established during the examination. He thus should be considered the true promoter of clinical medicine in France.

Corvisart was born in a modest village in the Ardennes, where his father, an attorney at the *parlement* of Paris, had retired following Louis XV's exile of that unmanageable body. Soon thereafter royal clemency allowed the family to return to the capital, and Jean-Nicolas was enrolled in the Collège Sainte-Barbe. Upon completion of his sometimes tumultuous studies, he received the master of arts degree at the age of eighteen.

His father, wanting his son to follow in his footsteps, urged Corvisart to study law; but the son refused to submit to his father's will and decided to study medicine. Corvisart took courses at the Faculté de Médecine of Paris, where he distinguished himself both by his diligence and by his irreverent nature. On 14 November 1782, after having passed many examinations, he obtained the title of *docteur-régent.*

Wishing to earn a living, Corvisart applied for a position as physician at the Hôpital des Paroisses. Its founder, Mme. Necker, the wife of the minister of finance, rejected his candidacy on the pretext that the applicant refused to wear a wig. He therefore had to resign himself to accepting the position of physician to the poor in Saint-Sulpice, receiving annual fees of 300 francs.

At the same time he continued visiting the Hôpital de la Charité, where he occupied the post of *médecin-expectant,* working with the famous Desbois de Rochefort. In 1788 he became *médecin-en-second.* Even then his talents as a meticulous observer and clever teacher attracted many students, who came from everywhere to accompany him on his rounds. In 1794, when the legislators decided to enact a thorough reform of the teaching of medicine, Corvisart was unanimously appointed to the professorship of internal clinical medicine at La Charité, a position specially created for him.

Corvisart became a professor at the École de Médecine and shortly afterward at the Collège de France. In the morning he taught and during the afternoon he held consultations. He also managed the *Journal de médecine, chirurgie et pharmacie.* He published a translation of Maximilian Stoll's work on fevers and one of Leopold Auenbrugger's on chest percussion, and commentaries on Boerhaave's *Aphorisms.* A long and remarkable study on diseases of the heart constituted his *magnum opus* and can legitimately be considered the first treatise on cardiology.

Of all honors accorded Corvisart, the highest of all was that conferred upon him in 1801, when Napoleon himself, then first consul of the republic, became his patient. Corvisart's common sense, the simplicity of his diagnostics, and the logic of his reasoning triumphed over the skepticism of his new patient, who, after the first interview, declared: "I do not believe in medicine, but I do believe in Corvisart."

Soon Corvisart was appointed surgeon general, and he assumed the duties of first physician of the court when Napoleon became emperor. A few years later he was made a baron and entered the Académie des Sciences. For ten years, until Napoleon's fall, Corvisart remained with the man who had placed such confidence in him. Even though the emperor enjoyed rather good health, he liked to see Corvisart several times a week, to chat or to question him on medical problems. Corvisart enjoyed these conversations but never stooped to obsequiousness. Occasionally, however, he did take advantage of these moments of relaxation to request pay increases, ask for favors, or suggest promotions.

In order to give better service to the court, Corvisart had abandoned all of his hospital and professorial duties. He was, in fact, not only the emperor's physician, but physician to the two empresses and Napoleon's son, the king of Rome. At the time of Napoleon's divorce and remarriage to Archduchess Marie-Louise, he was even summoned to provide advice on the emperor's chances of becoming a father (Napoleon was then forty-one).

After the Bourbons returned to power, Corvisart voluntarily abstained from all political activity. He refused, with dignity, the advances of the new regime, choosing instead to remain loyal to Napoleon. Moreover, his health had become poor. He withdrew from the world, abandoned his profession, and took refuge in silence and solitude. One of Corvisart's last joys was being invited, in 1820, to sit in the Académie Royale de Médecine, the majority of whose members had been his students.

BIBLIOGRAPHY

I. ORIGINAL WORKS. Corvisart's works include *Éloge de Desbois de Rochefort* (Paris, 1789); *Aphorismes sur la connaissance et le caractère des fièvres* (Paris, *an* V [1797]), his trans. of Stoll's work on fevers; *Aphorismi de cognoscendis et curandis morbus chronicis,* his trans. of Boerhaave's *Aphorisms* (Paris, 1802); *Essais sur les maladies organiques du coeur et des gros vaisseaux* (Paris, 1806); and *Nouvelle méthode pour reconnaître les maladies internes de la poitrine par la percussion de cette cavité,* his trans. of Auenbrugger's work, with commentary (Paris, 1808).

II. SECONDARY LITERATURE. On Corvisart or his work, see Isidore Bourdon, *Illustres médecins et naturalistes des temps modernes* (Paris, 1843); Bourguignon, *Corvisart, premier médecin de l'empereur* (Lyons, 1937); Busquet, *Aphorismes de médecine clinique par le baron Corvisart,* collected by P. V. Marat (Paris, 1939); Cabanès, *Au chevet de l'empereur* (Paris, 1924); Thérèse Chemin, *Corvisart et la clinique* (Paris, 1928); Dupuytren, *Discours prononcé à la mémoire de Corvisart à la séance publique de la Faculté de médecine de Paris* (Paris, 1921); Ferrus, *Notice historique sur Corvisart* (Paris, 1921); Paul Ganière, *Corvisart, médecin de Napoléon* (Paris, 1951); Louis Hechemann, *Corvisart et la percussion* (Paris, 1906); Lassus, *Corvisart et la cardiologie* (Paris, 1927); Leroux des Tillets, *Discours prononcé sur le cerceuil de Corvisart* (Paris, 1821); Pariset, *Histoire des membres de l'Académie royale de médecine* (Paris, 1845); and Potain, *Corvisart et son temps* (Paris, 1894).

III. DOCUMENTS. Documentary materials are in Bibliothèque Nationale, manuscrits français; Archives Nationales, files AA I, AA IV, CC 242, F 7, F 15, 02 170, 02 174, 02 186, 02 815, 02 816; Archives de l'Académie des Sciences, Corvisart file; Archives de l'Académie de Médecine, Corvisart file; Bibliothèque Thiers, Fonds Frédéric Masson; Archives Départementales des Ardennes, Corvisart file; and Archives Départementales de la Seine, Corvisart file.

PAUL GANIÈRE

COSSERAT, EUGÈNE MAURICE PIERRE (*b.* Amiens, France, 4 March 1866; *d.* Toulouse, France, 31 May 1931), *geometry, mechanics, astronomy.*

Cosserat first studied at Amiens. When he was seventeen, he was accepted at the École Normale Supérieure. His scientific career was spent in Toulouse: he was assigned to the observatory in 1886, became professor of differential calculus at the Faculty of Sciences in 1896 and of astronomy in 1908, and was director of the observatory from 1908 until his death. A reserved, kindly man and a diligent worker, Cosserat was one of the moving forces of the University of Toulouse for thirty-five years. He was a *membre non-résident* of the Académie des Sciences (1919) and a corresponding member of the Bureau des Longitudes (1923).

For the first ten years Cosserat divided his time between his duties at the observatory, where he made equatorial observations of double stars, planets, and comets, and his research in geometry. His doctoral thesis (1888), which deals with an extension of Plücker's concept of the generation of space by means of straight lines, considers the infinitesimal properties of space created by circles. The congruences and complexes of straight lines are the main subject of his later works, in which he remained a disciple of Darboux.

In studying the deformation of surfaces Cosserat was oriented toward the theory of elasticity and the general problem of continuous mediums. These studies were done between 1885 and 1914 in collaboration with his older brother François, who was chief engineer of the Service des Ponts et Chaussées. François was the main participant in tests on synthesis and philosophical concepts, the mathematical framework of the research being furnished by Eugène.

The most practical results concerning elasticity were the introduction of the systematic use of the movable trihedral and the proposal and resolution, before Fredholm's studies, of the functional equations of the sphere and ellipsoid. Cosserat's theoretical research, designed to include everything in theoretical physics that is directly subject to the laws of mechanics, was founded on the notion of Euclidean action combined with Lagrange's ideas on the principle of extremality and Lie's ideas on invariance in regard to displacement groups. The bearing of this original and coherent conception was diminished in importance because at the time it was proposed, fundamental ideas were already being called into question by both the theory of relativity and progress in physical theory.

The Toulouse observatory participated in the international undertaking of formulating the Carte du Ciel. Having become director, Cosserat organized the important work of meridian observations, photography, and computation of positions in order to make systematic determination of the proper motions of the stars. He personally supervised the details of these operations, including the computations, and was completely occupied with this task for the last fifteen years of his life.

Cosserat was particularly concerned with accuracy: he used his original research, which later appeared as notes in the now-classic works of Darboux, Koenigs, Appell, and Chwolson. Thus, although his name seldom appears in modern works, his influence on them was far-reaching.

BIBLIOGRAPHY

I. ORIGINAL WORKS. In geometry and analysis, Cosserat's main works were published in *Annales de la Faculté des sciences de Toulouse:* "Sur le cercle considéré comme élément générateur de l'espace," **3** (1889), El–E81, his doctoral thesis; "Formes bilinéaires," *ibid.,* M1–M12; "Courbes algébriques dans le voisinage d'un de ses points," **4** (1890), O1–O16; "Congruences de droites et la théorie des surfaces," **7** (1893), N1–N62; "Déformations infinitésimales d'une surface flexible et inextensible," **8** (1894), El–E46; and "Travaux scientifiques de T. J. Stieljes," **9** (1895), [1]–[64]; in *Mémoires de l'Académie des sciences de Toulouse:* "Classe de complexes de droites," **4** (1892), 482–510; and "Théorie des lignes tracées sur une surface," **7** (1895), 366–394; in *Comptes rendus de l'Académie des sciences,* some twenty notes (1888–1908), most notably "Sur les courbes algébriques à torsion constante et sur les surfaces minima . . .," **120** (1895), 1252–1254; and "Sur la théorie des équations aux dérivées partielles du deuxième ordre," in G. Darboux, *Leçons sur la théorie des surfaces,* IV (Paris, 1896), 405–422.

In mechanics he wrote, with his brother François: "Théorie de l'élasticité," in *Annales de la Faculté des sciences de Toulouse,* **10** (1896), Il–Ill6; "Note sur la cinématique d'un milieu continu," in G. Koenigs, *Leçons de cinématique* (Paris, 1897), pp. 391–417; "Note sur la dynamique du point et du corps invariable," in O. Chwolson, *Traité de physique,* I (Paris, 1906), 236–273; "Note sur la théorie de l'action euclidienne," in P. Appell, *Traité de mécanique rationnelle,* III (Paris, 1909), 557–629; and "Note sur la théorie des corps déformables," in O. Chwolson, *Traité de physique,* II (Paris, 1909), 953–1173, also published separately (Paris, 1909).

In astronomy he wrote "Sur quelques étoiles dont le mouvement propre annuel dépasse 0″5," in *Comptes rendus de l'Académie des sciences,* **169** (1919), 414–418; and "Déterminations photographiques de positions d'étoiles," in *Annales de l'Observatoire de Toulouse,* **10** (1933), 1–306.

II. SECONDARY LITERATURE. On Cosserat or his work, see P. Caubet, "E. Cosserat; ses vues générales sur l'astronomie de position," in *Journal des observateurs,* **14** (1931), 139–143; and L. Montangerand, "Éloge de Cosserat," in *Annales de l'Observatoire de Toulouse,* **10** (1933), xx–xxx.

JACQUES R. LÉVY

COSTA, CHRISTOVÃO DA. See **Acosta, Cristóbal.**

COSTA IBN LUCA. See **Qusṭā ibn Lūqā.**

COSTANTIN, JULIEN NOËL (*b.* Paris, France, 16 August 1857; *d.* Paris, 17 November 1936), *botany.*

Although born into a well-to-do commercial family, Costantin prepared for the *baccalauréat* at his own expense and, after studying *mathématiques speciales* at the Lycée Charlemagne for two years, was accepted at the École Polytechnique and at the École Normale Supérieure. He chose to attend the latter, where he eventually became *agrégé-préparateur.* He received his doctorate in 1883 and soon became an *aide-naturaliste* in Van Tieghem's laboratory at the Muséum d'Histoire Naturelle. His next position was as *maître de conférences* at the École Normale. He later returned to the Muséum as a professor, holding the chair of culture (1901) and then the chair of vegetable anatomy and physiology (1919), succeeding Van Tieghem, whose daughter he married. He was a member of the Académie des Sciences (1912) and the Académie d'Agriculture (1923), and he edited the *Annales des sciences naturelles* (botanique).

Costantin was a convinced Lamarckian, and his works were intended first and foremost to establish the influence exerted by environment on the structure of higher plants. For example, he placed the stalk, which is normally aerial, in a subterranean environment and showed, for instance, how a cell that would normally become a lignified fiber could evolve into a parenchymatous cell. Thus function, cellular structure, and the form of the organs are linked. Costantin undertook an analogous study of the action of an aquatic environment. All his life he sought to provide a solid foundation for the theory of inheritance of acquired characters.

For many years mushrooms and their cultivation were his principal interest; then, after 1914, he devoted all his research to the action of symbiosis in the development and maintenance of life. For Costantin and his pupil Joseph Magrou, tuberization of the potato is the result of mycorrhizal symbiosis. Degeneration is caused by an impairment of this symbiosis.

Cultivation and selection in mountainous regions, where the soil is rich in endophytes, appeared to Costantin to be the remedy for this degeneration.

BIBLIOGRAPHY

Costantin's principal works are *Les mucédinées simples* (1888); *Végétaux et milieux cosmiques* (1898); *La nature tropicale* (1899); *Le transformisme appliqué à l'agriculture* (1906); *La vie des orchidées* (1917); *Atlas en couleurs; L'origine de la vie sur le globe* (1923); with Dufour, two *Flores des champignons;* with Faydeau, *Les plantes* (popularization) (1922).

In addition, see *Les orchidées cultivées* (1912); *Atlas des orchidées cultivées* (1912); and *Examen critique du Lamarckisme* (1930). All the above were published in Paris.

L. PLANTEFOL

COSTER, DIRK (*b.* Amsterdam, Netherlands, 5 October 1889; *d.* Groningen, Netherlands, 12 February 1950), *physics.*

Coster is not related to a contemporaneous Dutch writer with the same name. His first post was as teacher in a primary school. Outside support then allowed him to study physics from 1913 to 1916 at Leiden University, where he formed lifelong friendships with his teacher Ehrenfest and his fellow student Kramers. After obtaining his physics degree Coster studied and received a degree in electrical engineering at Delft Technological University because he judged it to have useful experimental applications. His doctoral thesis (Leiden, 1922), on the spectroscopy of X rays, was sponsored by Ehrenfest.

A fellowship enabled Coster to work on precision X-ray spectroscopy from 1920 to 1922 in Manne Siegbahn's laboratory at Lund and from 1922 to 1923 with Niels Bohr in Copenhagen. Here he completely worked out the linking up of X-ray experimental data with Bohr's theory of atomic structure and the periodic table of the elements. His crowning achievement was the discovery in 1923, with G. von Hevesy, of the element hafnium (named after Copenhagen and having the atomic number 72). They worked along lines suggested by Bohr, who expected the element to be a homologue of zirconium (atomic number 40) and rather different in chemical behavior from the rare earth metals (lanthanides, atomic numbers 57–71) that precede it and are very similar to each other because the "inner" shell of fourteen 4f-electrons is gradually filled. Hevesy concentrated the new element from zirconium compounds, in which it occurs naturally, as Bohr had suggested. The enrichment was checked by Coster, using the X-ray spectrum as indicator. This was rendered more difficult by a freak of nature that made the two strongest hafnium lines coincide almost exactly with the second-order reflections of the two zirconium $K\alpha$ lines, thus camouflaging its presence in zirconium compounds.

Following his return to the Netherlands, Coster was assistant to Lorentz at the Teyler Laboratory in Haarlem until 1924, when he was called to the chair of experimental physics at Groningen University; he held the chair until 1949. With a rather limited budget he modernized and extended the physics laboratory, not only introducing X-ray work but also mounting a Rowland grating for studying band spectra, mostly of diatomic molecules. He attracted Dieke and Lochte-Holtgreven as skilled experimenters in this field and R. Kronig as a general modern theorist. Zernike was already doing classical work in interference theory and infrared spectroscopy. The X-ray work was carried on by a staff including Prins, Druyvesteyn, and van der Tuuk, as well as a steadily increasing stream of students (including Wolff, Veldkamp, and Knol) and visitors (including Hanawalt, Nitta, and Smoluchowski). Among the new results were the anomalous dispersion and scattering of X rays in the neighborhood of an absorption edge, leading to differences between the (111) and ($\overline{1}11$) reflections of a zinc sulfide crystal. This effect was later used by Bijvoet to determine the absolute left-hand or right-hand configurations in crystals, which are difficult to obtain from other data.

Another major new point was the elucidation of the fine structure of absorption edges, extending some hundreds of electron volts from the main edge and resulting from the alternation of greater and lesser "ease" with which the crystal lattice accepts the ejected electrons. This explanation arose when the wave theory of the electron was *in statu nascendi* and Kronig (with Sir William Penney) applied it for the first time to the alternation of forbidden and allowed energy ranges in a one-dimensional model of a crystal. The discovery of a special kind of Auger effect, in which a secondary electron is ejected by a primary energy difference between two states having equal principal quantum numbers, also resulted from the collaboration of Coster and Kronig.

In band spectra and ultrasoft X-ray spectra, interesting new results were obtained. Appreciable work was done in neutron and nuclear physics. One gifted pupil, Hugo de Vries, became a specialist in biophysical applications of nuclear and nonnuclear physics, such as carbon-14 dating and the study of the nervous organs of fishes.

Coster was not only an energetic executive but also the moving force of his laboratory, eager to acquire mental pictures of the atomic processes and to explain them in simple terms. He was less interested in mathematical subtleties but never accepted a hazy presentation. The last years of his directorship and teaching were burdened by a progressive spinal disease, resulting ultimately in total paralysis.

BIBLIOGRAPHY

Articles by Coster and his pupils can be found in *Zeitschrift für Physik; Physica* (The Hague); and *Proceedings of the Royal Academy of Amsterdam*.

Coster is discussed in *Jaarboek van de K. Nederlandse Akademie van wetenschappen, gevestigd te Amsterdam* (1951–1952), p. 198; and *Nederlands tijdschrift voor natuurkunde,* **15** (1949), 285.

J. A. Prins

COTES, ROGER (*b.* Burbage, Leicestershire, England, 10 July 1682; *d.* Cambridge, England, 5 June 1716), *mathematics, astronomy.*

Cotes was the second son of the Reverend Robert Cotes, rector of Burbage. His mother was the former Grace Farmer, of Barwell, Leicestershire. He was educated first at Leicester School, where he showed such a flair for mathematics that at the age of twelve his uncle, the Reverend John Smith, took him into his home to supervise his studies personally. Cotes later went to St. Paul's School, London, where he studied mainly classics while keeping up a scientific correspondence with his uncle. He was admitted as a pensioner to Trinity College, Cambridge, in 1699, graduating B.A. in 1702 and M.A. in 1706. He became fellow of his college in 1705 and a fellow of the Royal Society in 1711, and was ordained in 1713. In January 1706, Cotes was named the first Plumian professor of astronomy and natural philosophy at Cambridge on the very strong recommendation of Richard Bentley, master of Trinity. Cotes, who never married, died of a violent fever when only thirty-three. His early death caused Newton to lament: "Had Cotes lived we might have known something."

On his appointment as professor, Cotes opened a subscription list in order to provide an observatory for Trinity. This, with living quarters for the professor, was erected on the leads over King's Gate. Cotes spent the rest of his life here with his cousin Robert Smith, who was his assistant and successor. The observatory was not completed in Cotes's lifetime and was demolished in 1797.

Concerning his astronomical work Cotes supplied, in correspondence with Newton, a description of a heliostat telescope furnished with a mirror revolving by clockwork. He recomputed the solar and planetary tables of Flamsteed and J. D. Cassini and had intended to construct tables of the moon's motion, based on Newtonian principles. According to Halley (1714), he also observed the total solar eclipse of 22 April 1715, noticing the occultation of three spots.

Cotes formed a school of physical sciences at Trinity in collaboration with William Whiston. The two performed a series of experiments beginning in May 1707, the details of which can be found in a posthumous publication, *Hydrostatical and Pneumatical Lectures by Roger Cotes* (1738). These demonstration classes indicate a simple, straightforward style that is both stimulating and thorough. There was no thought of practical work by students at this time.

In 1709 Cotes became heavily involved in the preparation of the second edition of Newton's great work on universal gravitation, the *Philosophiae naturalis principia mathematica*. The first edition of 1687 had few copies printed. In 1694 Newton did further work on his lunar and planetary theories, but illness and a dispute with Flamsteed postponed any further publication. Newton subsequently became master of the mint and had virtually retired from scientific work when Bentley persuaded him to prepare a second edition, suggesting Cotes as supervisor of the work.

Newton at first had a rather casual approach to the revision, but Cotes took the work very seriously. Gradually, Newton was coaxed into a similar enthusiasm; and the two collaborated closely on the revision, which took three and a half years to complete. The edition was limited to only 750 copies, and a pirated version printed in Amsterdam met the total demand. Bentley, who had borne the expense of the printing, took the profits and rewarded Cotes with twelve free copies for his labors. Newton wrote a preface, remarking that in this edition the theory of the moon and the precession of the equinoxes had been more fully deduced from the principles, the theory of comets confirmed by several observations, and the orbits of comets computed more accurately. His debt to Cotes for these improvements cannot be estimated.

Cotes's original contribution to this book was a short preface. He suggested to Newton that he write a description of the scientific methodology used and demonstrate, in particular, the superiority of these principles to the popular idea of vortices presented by Descartes. Cartesian ideas were still vigorous, not only on the Continent but also in England, and continued to be taught at Cambridge until 1730 at least. In particular, Cartesian critics alleged that Newton's idea of action at a distance required the conception of an unexplained, occult force. Newton and Bentley agreed that Cotes should write a preface defending the Newtonian hypothesis against the theory of vortices and the other objections.

Cotes began his preface by considering three possible methods of approaching celestial phenomena. The first, used mainly by the Greeks, was to describe motions without attempting a rational explanation; the second was to make hypotheses and, out of ignorance, to relate them to occult qualities; and the third was to use the method of experiment and observation. He vigorously asserted that Newton's approach belonged only to the third category. Illustrating this by means of the inverse-square law of gravitation, he quoted Newton's discovery that the acceleration of the moon toward the earth confirms this theory and that Kepler's third law of motion, taken in conjunction with Huygens' rule for central forces, implies such a law. He asserted that the paths of comets could be observed as conics with the sun as focus and that in both planetary and cometary motion the theory of vortices conformed neither to reason nor to observation. Cotes concluded that the law of gravitation was confirmed by observation and did not depend on occult qualities.

> But shall gravity be therefore called an occult cause, and thrown out of philosophy, because the cause of gravity is occult and not yet discovered? Those who affirm this, should be careful not to fall into an absurdity that may overturn the foundations of all philosophy. For causes usually proceed in a continued chain from those that are more compounded to those that are more simple; when we are arrived at the most simple cause we can go no farther. . . . These most simple causes will you then call occult and reject them? Then you must reject those that immediately depend on them [*Mathematical Principles,* p. xxvii].

Cotes proceeded positively to imply the principle of action at a distance. "Those who would have the heavens filled with a fluid matter, but suppose it void of any inertia, do indeed in words deny a vacuum, but allow it in fact. For since a fluid matter of that kind can noways be distinguished from empty space, the dispute is now about the names and not the natures of things" (*ibid.,* p. xxxi).

Leibniz later condemned Cotes's preface as "pleine d'aigreur," but it can be seen that Cotes argued powerfully and originally in favor of Newton's hypothesis.

Cotes's major original work was in the field of mathematics, and the decline in British mathematics that followed his untimely death accentuated his being one of the very few British mathematicians capable of following on from Newton's great work.

His only publication during his life was an article entitled "Logometria" (1714). After his death his mathematical papers, then in great confusion, were edited by Robert Smith and published as a book,

Harmonia mensurarum (1722). This work, which includes the "Logometria" as its first part, gives an indication of Cotes's great ability. His style is somewhat obscure, with geometrical arguments preferred to analytical ones, and many results are quoted without explanation. What cannot be obscured is the original, systematic genius of the writer. This is shown most powerfully in his work on integration, in which long sequences of complicated functions are systematically integrated, and the results are applied to the solution of a great variety of problems.

Cotes first demonstrates that the natural base to take for a system of logarithms is the number which he calculates as 2.7182818. He then shows two ingenious methods for computing Briggsian logarithms (with base 10) for any number and interpolating to obtain intermediate values. The rest of part 1 is devoted to the application of integration to the solution of problems involving quadratures, arc lengths, areas of surfaces of revolution, the attraction of bodies, and the density of the atmosphere. His most remarkable discovery in this section (pp. 27–28) occurs when he attempts to evaluate the surface area of an ellipsoid of revolution. He shows that the problem can be solved in two ways, one leading to a result involving logarithms and the other to arc sines, probably an illustration of the harmony of different types of measure. By equating these two results he arrives at the formula $i\phi = \log(\cos\phi + i\sin\phi)$ where $i = \sqrt{-1}$, a discovery preceding similar equations obtained by Moivre (1730) and Euler (1748).

The second and longest part of the *Harmonia mensurarum* is devoted to systematic integration. In a preface to this section Smith explains that shortly before his death Cotes wrote a letter to D. Jones in which he claimed that any fluxion of the form

$$\frac{d\dot{z}z^{\theta n+(\delta/\lambda)n-1}}{e+fz^n}$$

where d, e, f are constants, θ an integer (positive or negative), and δ/λ a fraction, had a fluent that could be expressed in terms of logarithms or trigonometric ratios. He claimed, further, that even fluxions of the forms

$$\frac{d\dot{z}z^{\theta n+(\delta/\lambda)n-1}}{e+fz^n+gz^{2n}}$$

and

$$\frac{d\dot{z}z^{\theta n+(\delta/\lambda)n-1}}{e+fz^n+gz^{2n}+hz^{3n}}$$

had fluents expressible in these terms.

Returning to the text, Cotes then proceeds to evaluate the fluents of no fewer than ninety-four types of such fluxions, working out each individual case as θ takes different values. His calculation was aided by a geometrical result now known as Cotes's theorem, which, expressed in analytical terms, is equivalent to finding all the factors of $x^n - a^n$ where n is a positive integer. The theorem is that if the circumference of a circle is divided into n equal parts OO^1, O^1O^{11}, $\cdots$ and any point P is taken on a radius OC, then $(PC)^n - (OC)^n = PO \times PO^1 \times PO^{11} \times \cdots$ if P is outside the circle and $(OC)^n - (PC)^n = PO \times PO^1 \times PO^{11} \times \cdots$ if P is inside the circle. This result was proved by J. Brinkley (1797).

The third part consists of miscellaneous works, including papers on methods of estimating errors, Newton's differential method, the construction of tables by differences, the descent of heavy bodies, and cycloidal motion. There are two particularly interesting results here. The essay on Newton's differential method describes how, given n points at equidistant abscissae, the area under the curve of nth degree joining these points may be evaluated. Taking A as the sum of the first and last ordinates, B as the sum of the second and last but one, etc., he evaluates the formulas for the areas as

$$\frac{A+4B}{6} \text{ if } n = 3$$

$$\frac{A+3B}{8} \text{ if } n = 4$$

$$\frac{7A+32B+12C}{90} \text{ if } n = 5, \text{ etc.}$$

A modernized form of this result is known as the Newton-Cotes formula.

In describing a method for evaluating the most probable result of a set of observations, Cotes comes very near to the technique known as the method of least squares. He does not state this method as such; but his result, which depends on giving weights to the observations and then calculating their centroid, is equivalent. This anticipates similar discoveries by Gauss (1795) and Legendre (1806).

BIBLIOGRAPHY

I. ORIGINAL WORKS. Cotes's writings are the preface to Isaac Newton, *Philosophiae naturalis principia mathematica*, 2nd ed. (Cambridge, 1713), trans. by Andrew Motte as *Sir Isaac Newton's Mathematical Principles* (London, 1729) and repr. with a historical and explanatory appendix by Florian Cajori (Cambridge, 1934); "Logometria," in *Philosophical Transactions of the Royal Society*, **29** (1714), 5–47; "A Description of the Great Meteor Which Was on the 6th of March 1716 . . .," *ibid.*, **31** (1720), 66; *Epistola ad amicum* *de Cotesii inventis, curvarum ratione, quae cum circulo & hyperbola,* R. Smith, ed. (London, 1722); *Harmonium mensurarum, sive Analysis et synthesis per rationum et angulorum mensuras promotae: Accedunt alia opuscula mathematica per Rogerum Cotesium,* R. Smith, ed. (Cambridge, 1722); and *Hydrostatical and Pneumatical Lectures by Roger Cotes A.M.,* R. Smith, ed. (London, 1738).

II. SECONDARY LITERATURE. On Cotes or his work see the anonymous "An Account of a Book, Intituled, 'Harmonia mensurarum,'" in *Philosophical Transactions of the Royal Society,* **32** (1722), 139–150; D. Brewster, *Memoirs of Sir Isaac Newton,* I (Edinburgh, 1855), 332; J. Brinkley, "A General Demonstration of the Property of the Circle Discovered by Mr. Cotes Deduced From the Circle Only," in *Transactions of the Royal Irish Academy,* **7** (1797), 151–159; A. De Morgan, in *Penny Cyclopaedia,* **8** (1837), 87; and *ibid.,* **13** (1839), 379; J. Edleston, *Correspondence of Sir Isaac Newton and Professor Cotes* (London–Cambridge, 1850); R. T. Gunther, *Early Science in Cambridge* (Oxford, 1937), pp. 78, 161; Edmund Halley, "Observations of the Late Total Eclipse of the Sun on the 22nd of April Last Past . . .," in *Philosophical Transactions of the Royal Society,* **29** (1714), 253–254; A. Kippis, in *Biographia Britannica,* IV (London, 1789), 294–297; A. Koyré, "Attraction, Newton, and Cotes," in *Archives internationales d'histoire des sciences,* **14** (1961), 225–236, repr. in his *Newtonian Studies* (Cambridge, Mass., 1965), pp. 273–282; J. E. Montucla, *Histoire des mathématiques,* III (Paris, 1758), 149, 154; S. P. Rigaud, *Correspondence of Scientific Men of the Seventeenth Century,* I (Oxford, 1841), 257–270; N. Saunderson, *The Method of Fluxions Together With the Demonstration of Mr. Cotes's Forms of Fluents in the Second Part of His Logometria* (London, 1756); C. Walmesley, *Analyse des mesures* (Paris, 1753), a commentary on the *Harmonia mensurarum;* and W. Whiston, *Memoirs of the Life and Writings of Mr. William Whiston* (London, 1749), pp. 133–135.

J. M. DUBBEY

COTTA, CARL BERNHARD VON (*b.* Zillbach, Saxe-Weimar-Eisenach [now German Democratic Republic], 24 October 1808; *d.* Freiberg, Saxony, [now German Democratic Republic], 14 September 1879), *geology.*

Cotta was not only an important geologist in his own time but also, due to his synthesizing abilities, in later eras; he was rightly called the philosopher of geology in a time of increasingly detailed research.

The son of Heinrich Cotta, founder of the Tharandt Forestry Academy near Dresden, and of Christiane Ortmann, Cotta spent his youth (from 1811) in Tharandt, where he attended the local elementary school and received private instruction. From 1822 to 1826 he attended the humanist Gymnasium of the Holy Cross at Dresden. He also received private instruction in mathematics at Dresden from 1826 to 1827 and enrolled at his father's school for the sum-

mer semester of 1827. At the same time his father sought his son's admission to the Freiberg Bergakademie. This, the oldest mining school in Europe, had been made famous by Werner.

Cotta enrolled at the Freiberg Bergakademie in 1827. His teachers included such well-known scholars as the mineralogist F. A. Breithaupt, the physicist Ferdinand Reich, the chemist W. A. Lampadius, and the neptunist geognost K. A. Kühn. After completing his studies in 1831, Cotta spent half a year in the mines and foundries of the Erzgebirge before moving to Heidelberg. There he studied under the volcanist K. C. von Leonhard and graduated at the end of 1832 with the dissertation *Die Dendrolithen in Beziehung auf ihren inneren Bau,* a landmark in the description of the silicified Permian timber of Saxony.

Upon his return to Tharandt, Cotta found employment at the forestry academy. In 1833, however, he agreed to participate in the geological survey of Saxony directed by K. F. Naumann. Cotta mapped a large part of the region, primarily in western Saxony and Lusatia. When this major undertaking was completed in 1845, the critics wrote that this map must be considered "probably the best map of its kind done to date." At the same time the old controversy between the neptunists (followers of Werner) and the volcanists (with L. von Buch as principal representative) over the Lusatia overthrust flared up once again. With a masterful, methodical analysis Cotta clarified the situation and proved that neither theory was correct; rather, the granites and syenites had been thrust above the Cretaceous strata long after their solidification.

In 1839 Cotta received a permanent position at the Tharandt Forestry Academy and became its secretary in the autumn of 1840. When his friend Naumann was called to the University of Leipzig, Cotta succeeded him in 1842 as professor of geognosy and paleontology at the Freiberg Bergakademie. He taught there for thirty-two years, holding his professorship longer than all his predecessors and successors.

Outwardly, Cotta's life after 1842 was without any particular ups and downs. Neither the 1849 revolution in Saxony (he was an active participant), nor the Austro-Prussian War of 1866, nor the Franco-Prussian War altered his position and his duties. Neither the industrialization of Germany nor the conditions at the Bergakademie itself had any major effect on his work.

Cotta's principal efforts were devoted to teaching and research at Freiberg. Aware of the Wernerian heritage, during his many years of teaching he developed a far-reaching geological system. From 1842 he lectured on geognosy, from 1843 on paleontology, and in 1851 he introduced a course in ore deposits. This was probably the first course on this subject anywhere. Among Cotta's students were A. W. Stelzner, C. H. Müller (the "Gang-Müller" of Freiberg), W. Vogelsang, R. Pumpelly, and S. F. Emmons.

Cotta traveled extensively. He was in the Alps and northern Italy in 1843, in Switzerland in 1849, in Bukovina (now divided between Rumania and the U.S.S.R.) in 1854, again in the Alps in 1857 and 1858, in Hungary in 1861, in the Altai in 1868 at the invitation of the Russian czar, and in 1869 in the coal district of southern Russia on the Don. He wrote reports on all trips, some as articles and some as separate works.

Cotta was a founder member of the German Geological Society in December 1848, an honorary member of the Natural Science Society Isis, Dresden, and a member of the Geological Society of London and of the Bergmännische Verein of Freiberg. For the special mapping of Thuringia, done completely on his own during his vacations in 1844–1847, the grand duke of Saxe-Weimar decorated him. He was appointed royal Saxon mining adviser and chief mining adviser in 1862, and from 1869 to 1874 he was a member of the board of trustees, and then of the senate, of the Bergakademie.

In the interval between Werner's formulation of petrography (1787) and Sorby's introduction of petrographical microscopy (1858), Cotta's petrography represented the culmination and end of Werner's macroscopic descriptive petrography. Cotta contributed numerous observations that led to the definition of new rocks, such as porphyrite, felsite, and banatite. He also emphasized the genesis of minerals; and thus his ideas on alkali and lime during metamorphosis, on ultrametamorphism, palingenesis, and hybrid granites (1855) sound quite modern. Cotta attempted to devise a genetic system of petrography, which received its final form in 1866; the principal groupings are still in use. It was broken down into sediments (mechanical, chemical, organogenetic), metamorphites, and magmatites (acid plutonic and volcanic, basic plutonic and volcanic).

Cotta is regarded as an originator of the science of ore deposits. In the series Gangstudien (1850–1862) he and his students dealt with special topics in that science. Cotta summarized the practical experience gained through his numerous travels and his reading in Die Lehre von den Erzlagerstätten (1855). A "Freiberg school" based on Cotta's textbooks became active and later achieved worldwide fame. The degree to which Cotta extended the science of ore deposits from Saxony to the rest of the world was also reflected in the mineral collection at the Freiberg Bergakademie. His science of ore deposits contributed to the fame of the school, which lasted beyond his lifetime.

Cotta was always keenly interested in popularizing scientific findings. At the request of Weigel, a Leipzig publisher, he wrote a commentary (1848–1852) on Kosmos, the brilliant work by Alexander von Humboldt. Between 1852 and 1876 his Geologischen Bilder appeared in six editions.

His deliberate concern with philosophy distinguished Cotta from most of his colleagues. He was committed to the empirical study of nature and regarded the world as knowable, but believed that man's abilities were insufficient for a complete understanding of the world. In his philosophical work Geologie der Gegenwart (1866) he defended certain spontaneous materialistic views. He should not be counted among the materialists, however, for he explained natural law by his concept of God and confined the task of natural science to the exploration of the material world, for which he assumed a comprehensive causal relationship. This outlook places Cotta between Humboldt and Darwin.

Cotta was an ardent defender of the concept of evolution even in the inorganic realm. Thus he defended Lyell's principle of uniformitarianism but took issue with its inherent claim that the history of the earth was merely fluctuation of eternally equal forces, pointing out the earth's historical development. In the twentieth century Hans Stille revived Cotta's concepts and raised them to a new level by connecting Cotta's continuous, but always new and differently determined, variability by means of a tectonic earth history divided into phases. Thus Cotta became the forerunner of Darwin in Germany and, as early as 1848 (eighteen years before Haeckel), expressed the basic biogenetic law.

He retired in 1874 for reasons of health; he suffered a stroke in 1877 and died two years later. Cotta is buried in the cemetery of Donat, as are many famous geologists.

BIBLIOGRAPHY

I. ORIGINAL WORKS. A bibliography of all Cotta's works is in Wagenbreth's "Bernhard von Cotta" (see below), pp. 79–119.

His books are *Die Dendrolithen in Beziehung auf ihren inneren Bau* (Dresden-Leipzig, 1832; 2nd ed., 1850); *Geologisches Glaubensbekenntnis* (Dresden, 1835); *Tharandt und seine Umgebung* (Dresden-Leipzig, 1835); *Geognostische Spezialkarte des Königreiches Sachsen*, 3 vols. (Dresden-Leipzig, 1836–1845), done with K. F. Naumann; *Geognostische Wanderungen* (Dresden-Leipzig, 1836–1838); *Über Thierfährten im bunten Sandstein* (Dresden-Leipzig, 1839); *Anleitung zum Studium der Geognosie* (Dresden-Leipzig, 1842; 2nd ed., 1846; 3rd ed., 1849; 4th ed., 1852); *Geogno-*

stische Karte von Thüringen, 4 vols. (Dresden–Leipzig, 1844–1847); *Winke über Aufsuchen von Braun- und Steinkohlen* (Freiberg, 1846; 2nd ed., 1850); *Briefe über A. von Humboldts Kosmos,* 3 vols. (Leipzig, 1848–1852; 2nd ed., 1849; 3rd ed., 1855), also trans. into Russian (Moscow, 1850–1853); *Die Bergakademie zu Freiberg* (Freiberg, 1849); the series Gangstudien (Freiberg, 1850–1862); *Geognostische Karten unseres Jahrhunderts* (Freiberg, 1850); *Geologische Briefe aus den Alpen* (Leipzig, 1850); *Der innere Bau der Gebirge* (Freiberg, 1851); *Geologische Bilder* (Leipzig, 1852; 2nd ed., 1854; 3rd ed., 1856; 4th ed., 1861; 5th ed., 1871; 6th ed., 1876); *Deutschlands Boden* (Leipzig, 1854; 2nd ed., 1858); *Die Gesteinslehre* (Freiberg, 1855; 2nd ed., 1862; 3rd and 4th eds., entitled *Rocks Classified and Described,* London, 1866 and 1878); *Die Lehre von den Erzlagerstätten* (Freiberg, 1855; 2nd ed., 1859–1861; 3rd ed., entitled *A Treatise on Ore Deposits,* New York, 1870); *Die Lehre von den Flözformationen* (Freiberg, 1856); *Geologische Fragen* (Freiberg, 1858); *Katechismus der Geologie* (Leipzig, 1861; 3rd ed., 1877); *Die Geologie der Gegenwart* (Leipzig, 1866; 2nd ed., 1867; 3rd ed., 1872; 4th ed., 1874; 5th ed., 1878); *Über das Entwicklungsgesetz der Erde* (Leipzig, 1867); *Der Altai* (Leipzig, 1871); and *Beiträge zur Geschichte der Geologie* (Leipzig, 1877).

Cotta also published 101 articles in *Neues Jahrbuch für Mineralogie,* six articles in *Zeitschrift der Deutschen geologischen Gesellschaft,* and 170 articles in the *Berg- und hüttenmännischen Zeitung Freiberg.*

II. SECONDARY LITERATURE. A modern, detailed biography of Cotta has been published only recently. It and other biographies, listed chronologically, are A. W. Stelzner, "Bernhard von Cotta gest.," in *Neues Jahrbuch für Mineralogie* (1879); G. Pethö, "Cotta emlékezete," in *Földtani közlöny* (Budapest) (1880), 90–97; H. C. Sorby, "Obituary," in *Quarterly Journal of the Geological Society of London* (1880); Poggendorff, III (1898), 303; A. von Zittel, "Bernhard von Cotta," in *Allgemeine deutsche Biographie,* XLVII (1903), 538, which is inaccurate; O. Marschall, "Bernhard von Cotta," in *Handwörterbuch Naturwissenschaften,* II (Jena, 1912), 737; K. Lambrecht, W. Quenstedt, and A. Quenstedt, "Cotta," in *Paläontologi catalogus biobibliographicus. Fossilium catalogus* (s'Gravenhage, 1938); E. Krenkel, "Carl Bernhard von Cotta," in *Neue deutsche Biographie,* III (Berlin, 1957), 381; O. Wagenbreth, "Bernhard von Cotta," in *Freiberger Forschungsheft,* **D36** (1965), with complete bibliography; and *Bernhard von Cotta. Sein geologisches und philosophisches Lebenswerk an Hand ausgewählter Zitate,* Berichte Geologische Gesellschaft DDR Sonderheft no. 3 (Berlin, 1965).

HANS PRESCHER

COTTE, LOUIS (*b.* Laon, France, 20 October 1740; *d.* Montmorency, France, 4 October 1815), *meteorology.*

The son of a notary, Cotte was educated in the Oratorian *collèges* of Soissons and Montmorency and entered the Oratorian order upon his graduation in 1758. After teaching for some years in the order's *collèges* in Juilly and Montmorency, Cotte took orders and became vicar (1767), *curé* (1773), and oratory superior (1780) in Montmorency. In 1784 he went to Laon to serve as canon but was left without a post when the Revolution suppressed the canonry and bishopric of Laon. Cotte was then elected *curé* in Montmorency, a title he relinquished upon renouncing the priesthood to marry Antoinette-Marie-Madeleine du Coudray in *an* III (1795). He spent four years (1798–1802) as assistant librarian at the Bibliothèque Sainte-Geneviève in Paris. The Royal Academy of Sciences had elected him correspondent in 1769, and in 1803 he became a member of the Institut de France. He was also a correspondent of the Paris Société d'Agriculture and was associated with over a dozen other societies, both French and foreign.

Cotte's ecclesiastical career afforded him time to cloister himself in his library, where he devoted himself to patient and scholarly collection and assimilation of meteorological data. He traveled little and lived in reclusive style; he did, however, carry on a vast correspondence and routinely recorded meteorological observations several times daily. Cotte made no startling contributions to the science of meteorology but was widely known as a compiler, and he developed a reputation as an advocate of certain periodic meteorological correlations—e.g., among thermometric, barometric, magnetic, and lunar phenomena. Cotte's insistence on the practical applicability of meteorological knowledge, and of scientific information generally, found an outlet in his publications of a popular nature on science and its uses. He believed agricultural utility to be the principal aim of meteorology, with medical application an important secondary purpose.

With a deep Baconian faith in the efficacy of fact-gathering, Cotte was a self-appointed clearinghouse for meteorological information, chronicler of the development of organized study of the weather, and propagandizer of the arrival of meteorology as a distinct science, an arrival that he attributed mainly to the recent development of reliable meteorological instruments. And yet, recognizing that he and his contemporaries had not yet established sound principles that could guide intelligent collection of meteorological data, Cotte contradicted his own wishful declarations by admitting that the efforts of his kind might have to serve the future of meteorology, if not its present. At times appearing almost desperate for universal principles with which to link his assiduously gathered mountains of data and render them meaningful, Cotte nevertheless warned against undue application of the spirit of system, which could produce the illusion of general relationships that do not really exist. He presumed that statistically substantiated relations did not fall into this category.

To Cotte, meteorology encompassed phenomena concerning earthquakes, aurora borealis, terrestrial magnetism, atmospheric electricity, and lunar periodicity, as well as temperature, atmospheric pressure, winds, and precipitation. He assumed regular affiliations within and among all these categories—relationships that merely awaited discovery. He sought, for example, regularities in the variation of barometric pressure with geographical location, as well as with changes in temperature, wind, and weather. Temperature he believed to be subject to correlation with the occurrence of disease. Cotte shared with Giuseppe Toaldo, Lamarck, and others a belief in a periodic lunar influence on the weather. He viewed the alleged nineteen-year cycle of repeating temperature conditions as the clearest case of lunar influence but thought others also were valid. Regarding the essential sources of meteorological change, Cotte opposed the adherents of the "central fire" theory, maintaining instead that the greatest part of terrestrial heat emanates from the sun. When pressed to name the meteorological factor of most significance in alteration of the weather, Cotte upheld the primacy of the winds.

BIBLIOGRAPHY

I. ORIGINAL WORKS. Cotte's principal meteorological works are *Traité de météorologie* (Paris, 1774) and *Mémoires sur la météorologie*, 2 vols. (Paris, 1788). Among over 100 articles on meteorological subjects published in *Journal de physique* between 1774 and 1811, three of the more notable are "Réflexions sur l'application de la période lunaire de dix-neuf ans à la météorologie," **20** (1782), 249–258; "Axiomes météorologiques, ou résultats généraux de mes observations depuis trente ans, & de toutes celles que mes recherches & ma correspondance m'ont fournies," **44** (*an* II [1794]), 231–240; and "Mémoire sur la période lunaire de dix-neuf ans," **61** (*an* XIII [1805]), 129–148. A great number of other articles are in several other journals. Among Cotte's books designed for instruction in natural science and technology are *Leçons élémentaires d'histoire naturelle par demandes et réponses, à l'usage des enfants* (Paris, 1784); *Leçons élémentaires de physique, d'astronomie et de météorologie, par demandes et réponses, à l'usage des enfants* (Paris, 1788); *Leçons élémentaires d'agriculture, par demandes et réponses, à l'usage des enfants* (Paris, 1790); and *Vocabulaire portatif des mécaniques, ou définition, description abrégée et usage des machines, instrumens et outils employés dans les sciences, les arts et les métiers* (Paris, 1801). Books intended for rural use include *Catéchisme à l'usage des habitants de la campagne, sur les dangers auxquels leur santé et leur vie sont exposées et sur les moyens de les prévenir et d'y remédier* (Paris, 1792); and *Leçons élémentaires sur le choix et la conservation des grains, sur les opérations de la meunerie et de la boulangerie et sur la taxe du pain* (Paris, *an* III [1795]).

II. SECONDARY LITERATURE. A contemporary eulogy is Augustin-François Silvestre, "Notices biographiques sur MM. Journu-Auber, Cotte, Allaire, Desmarets et Tenon, membres de la Société Royale et Centrale d'Agriculture de Paris: Lues à la séance publique de la Société, le 28 avril 1816," in *Mémoires d'agriculture, d'économie rurale et domestique, publiés par la Société royale et centrale d'agriculture* (1816), 80–123. More recent biographical sketches are furnished by Parisot and Regnard in Michaud's *Biographie universelle*, LXI (1836), 449–452; by E. Regnard in Hoefer's *Nouvelle biographie générale*, XII (1855), cols. 126–132; and by P. Lemerre in *Dictionnaire de biographie française*, IX (1961), cols. 838–839.

KENNETH L. TAYLOR

COTTRELL, FREDERICK GARDNER (*b*. Oakland, California, 10 January 1877; *d*. Berkeley, California, 16 November 1948), *engineering, chemistry.*

Cottrell was the son of Henry and Cynthia Cottrell. He received a B.S. in chemistry from the University of California at Berkeley in 1896 and a Ph.D. from the University of Leipzig in 1902.

Cottrell was the inventor of electrostatic precipitators for removal of suspended particles from gases. These devices are widely used for abatement of pollution by smoke from power plants and dust from cement kilns and other industrial sources. In 1912 he founded the Research Corporation, a nonprofit foundation for the advancement of science. Patents for the precipitators were assigned to the corporation as an endowment. Cottrell also arranged for the Research Corporation to secure and develop patents for inventors. About 750 patents have been obtained under this arrangement, with assignment of all or a part of the royalties to the corporation.

After receiving a Ph.D. in physical chemistry under Wilhelm Ostwald, Cottrell returned to the University of California as an instructor in chemistry (1902–1906). He continued teaching and development work on the precipitators until 1911, when he resigned from the university to become chief physical chemist of the U.S. Bureau of Mines, where he was successively chief metallurgist (1914), assistant director (1916), and director (1919). He became chairman of the division of chemistry and chemical technology of the National Research Council in 1921. In 1922 he was appointed director of the Fixed Nitrogen Research Laboratory, which was concerned with the development in the United States of the Haber-Bosch process for the catalytic formation of ammonia from atmospheric nitrogen and hydrogen. He resigned in 1930 to return to work associated with the Research Corporation.

Cottrell played a part in development of a process for separation of helium from natural gas. He also had a role in establishing the synthetic ammonia industry in the United States and in a continued attempt to perfect a process for formation of nitric oxide at high temperatures.

BIBLIOGRAPHY

Cottrell's writings are "The Electrical Precipitation of Suspended Particles," in *Journal of Industrial and Engineering Chemistry,* **3** (1911), 542–550; "Recent Progress in Electrical Smoke Precipitation," in *Engineering Mining Journal,* **101** (1916), 385–392; "Oxygen Enrichment of Air in Metallurgy," in *Chemical and Metallurgical Engineering,* **23** (1920), 53–56; "Utilization of Patentable Discoveries of Government Technical Research for the Benefit of Industry," in *Chemical Age,* **28** (1920), 447–450; "The Problem of Nitrogen Fixation I, II, III," *ibid.,* n.s. **11** (1924), 282–284, 310–312, 342–343; "A New Method of Producing and Controlling the Emission of Positive Ions," in *Review of Scientific Instruments,* **1** (1930), 654–661, written with C. H. Kunsman and R. A. Nelson; and "Patent Experience of the Research Corporation," in *Transactions of the American Institute of Chemical Engineers,* **28** (1932), 222–225. His patents include "Recovery of Iodine and Bromine From Brines," U.S. Patent 1,921,563–564 (8 Aug. 1933), and "Electric Filtration of Materials as in Dehydrating Petroleum Emulsions," U.S. Patent 2,116,509 (10 May 1938).

A biography is Frank T. Cameron, *Cottrell—Samaritan of Science* (New York, 1952).

STERLING B. HENDRICKS

COTUGNO, DOMENICO FELICE ANTONIO (*b.* Ruvo di Puglia, Italy, 29 January 1736; *d.* Naples, Italy, 6 October 1822), *anatomy, physiology, medicine.*

Of humble parentage, the son of Michele Cotugno and his second wife, Chiara Assalemme, Cotugno early displayed such intelligence that he was sent to nearby Molfetta for training in Latin; returning to Ruvo for work in logic, metaphysics, mathematics, physics, and the natural sciences, he soon found his natural bent in medicine and continued his studies, often in straitened circumstances, at the University of Naples and the Ospedale degli Incurabili. To these two institutions Cotugno devoted the greater part of his life. In 1765 he made trips to Rome and northern Italy to visit libraries and men of science, including Giovanni Battista Morgagni; and in 1789 he traveled to Austria and Germany as physician to Ferdinand IV, king of Naples. In 1794 Cotugno married Ippolita Ruffo, duchess of Bagnara. In a period of political upheaval in the kingdom of Naples he did not swerve from medicine. An outstanding example of the physician-humanist, he was devoted to books and accumulated a large library; was well versed in art, architecture, numismatics, and antiquities; and had great facility in the Latin language.

Apart from medicine, in which his reputation was such that, the saying went, no one in Naples could die without a passport from him, Cotugno's greatest contributions to science resulted from his fusing of anatomy and physiology to uncover the secrets of the human body. They were made early in his career, when at the Ospedale degli Incurabili he had almost constant opportunities for dissection. In 1761 he published for distribution to friends a plate that traced the course of the nasopalatine nerve, which is responsible for sneezing; Antonio Scarpa acknowledged his priority in knowledge of this nerve. In the same year his anatomical dissertation *De aquaeductibus auris humanae internae,* following the work of Guichard Joseph Duverney and Antonio Maria Valsalva and anticipating that of Hermann Ludwig von Helmholtz, described the vestibule, semicircular canals, and cochlea of the osseous labyrinth of the internal ear, demonstrated the existence of the labyrinthine fluid, and formulated a theory of resonance and hearing. In his commentary *De ischiade nervosa* (1764) Cotugno differentiated between arthritic and nervous sciatica, concluded that the sciatic nerve is responsible for the latter, and in discussing it described extensively for the first time the cerebrospinal fluid. He also described the coagulation of albumin that occurs when the urine of persons afflicted with dropsy is exposed to heat. In addition, Cotugno investigated smallpox, was deeply concerned with controlling pulmonary tuberculosis, and exemplified to many students the true investigative and selfless spirit in anatomy and medicine. Medals were struck in his honor in 1824; in 1931, for the thirty-seventh congress of the Società Italiana di Medicina Interna; and in 1961 (a replica of the 1824 medal), for the tenth International Congress of Rheumatology.

BIBLIOGRAPHY

I. ORIGINAL WORKS. Cotugno's writings are *De aquaeductibus auris humanae internae anatomica dissertatio* (Naples, 1761; repr. Vienna, 1774; Naples–Bologna, 1775), trans. into Italian by Vincenzo Mangano, in the series Collana del "Valsalva" (Rome, 1932) and by L. Ricciardi-Mitolo (Bari, 1951); *De ischiade nervosa commentarius* (Naples, 1764; repr. Carpi, 1768; Vienna, 1770, 1773; Naples–Bologna, 1775, 1789; Naples, 1779; Venice, 1782), trans. into English (London, 1775), into German (Leipzig, 1792), and by Francesco Morlicchio into Italian (Naples, 1860); "Iter italicum anni MDCCLXV," Luigi Belloni, ed., in *Memorie dell'Istituto lombardo di scienze e lettere,* Classe di lettere, scienze morali e storiche, 4th ser., **27** (1960), 3–93, trans. into Italian by Felice Lombardi in the series Scientia Veterum, no. 76 (Naples, 1964); *De sedibus variolarum* σύνταγμα (Naples, 1769; repr. Vienna, 1771; Naples–Bologna, 1775, 1789; Louvain, 1786; "Dello spirito della medicina," in Giovanni Luigi Targioni, *Raccolta di opuscoli medico-pratici,* II (Florence, 1775), also published separately (Naples, 1783; Florence, 1785); "Il viaggio da Napoli a Vienna nel 1790," trans. from his "Iter Neapoli Viennam Austriae anno MDCCXC" by Gennaro de Gemmis in *Fascicoli dell'Archivio provinciale de Gemmis,* **1** (1961), 21–56; *Opuscula medica,* 2 vols. (Naples, 1826–1827); and *Opera posthuma,* Pietro Ruggiero, ed., 4 vols. (Naples, 1830–1833).

Cotugno also published an edition of Pietro Marchetti's *Observationum medico-chirurgicarum rariorum sylloge* under the title *Observationes et tractatus medico-chirurgici* (Naples, 1772).

II. SECONDARY LITERATURE. On Cotugno or his work see the following (listed chronologically): Angelo Scotti, *Elogio storico del cavalier D. Domenico Cotugno* (Naples, 1823); Benedetto Vulpes, *Per la solenne inaugurazione del busto in marmo di Domenico Cotugno nell'Ospedale degl'Incurabili di Napoli avvenuta . . . 10. Maggio dell'anno 1823* (Naples, 1824); Antonio Jatta, *Onoranze cittadine rese a Domenico Cotugno nel giorno 6 ottobre 1891 apponendosi una lapide commemorativa alla casa ove nacque* (Ruvo di Puglia, 1893); Luigi Messedaglia, "L' 'Iter italicum patavinum' di Domenico Cotugno," in *Atti del R. Istituto veneto di scienze, lettere ed arti,* **73,** Pt. 2 (1913–1914), 1691–1803; Guglielmo Bilancioni, "Domenico Cotugno," in Aldo Mieli, *Gli scienziati italiani,* I (Rome, 1923), 164–183 and, with some changes, as the "Proemio" to Vincenzo Mangano's translation of the *De aquaeductibus auris humanae internae anatomica dissertatio* cited above; and "Per la storia dell'anatomia dell'orecchio, lettere inedite di Domenico Cotugno e di Leopoldo Marcantonio Caldani," in Bilancioni's *Sulle rive del Lete* (Rome, 1930), pp. 147–203; Henry R. Viets, "Domenico Cotugno: His Description of the Cerebrospinal Fluid," in *Bulletin of the History of Medicine,* **3** (1935), 701–738; Giuseppe Pezzi, "Ricordo di Cotugno in occasione del rinvenimento del suo sepolcro," in *Riforma medica,* **67** (1953), 1354–1357; Michele Mitolo, "Domenico Cotugno, l'opera anatomo-fisiologica: La sua umanità," in *Puglia chirurgica,* **4** (1961), 289–318; Dorothy M. Schullian, "The Libraries of Rome in the *Iter italicum* (1765) of Domenico Cotugno," in *Journal of the History of Medicine and Allied Sciences,* **17** (1962), 168–181; and "Domenico Cotugno as Humanist," in *Per la storia della neurologia italiana,* no. 6 in the series Studi e Testi issued by the Istituto di Storia della Medicina, Università degli Studi (Milan, 1963), pp. 67–74; and Francesco Aulizio, "Rapporti tra Cotugno e Morgagni . . .," in *Atti del XIX Congresso nazionale di storia della medicina, Aquila, 26–29 settembre 1963* (Rome, 1965), pp. 557–573 and in *Rivista di storia della medicina,* **9** (1965), 34–50.

DOROTHY M. SCHULLIAN

COUES, ELLIOTT (*b.* Portsmouth, New Hampshire, 9 September 1842; *d.* Baltimore, Maryland, 25 December 1899), *ornithology.*

Elliott's father, Samuel Elliott Coues (pronounced Koos), a merchant and admirer of the "mysteries of nature," moved to Washington, D.C., to work for the Patent Office when his son was eleven. There the boy, already interested in natural history—especially birds—became acquainted with Spencer F. Baird and the collections of the Smithsonian Institution. Coues prepared for college at Gonzaga Seminary, then attended Columbian College (now George Washington University), from which he received an A.B. (1861), honorary M.A. (1862), M.D. (1863), and honorary Ph.D. (1869).

During the Civil War and until 1881, Coues served as assistant surgeon in the army. He collected, studied, and published extensively on birds during his peripatetic military assignments, which included Fort Whipple, Arizona; Fort Macon, North Carolina; Fort Randall, Dakota Territory; and, still on army assignment, as naturalist for the Northern Boundary Commission (1873–1876) and for the Hayden survey (1876–1880). From 1877 to 1886 he was professor of anatomy at Columbian College. He devoted seven years to the editorship of natural history subjects for the *Century Dictionary.* Coues married Jane Augusta McKenney in 1867, and after her death he married Mary Emily Bates in 1887. He was elected to the National Academy of Sciences; and, in addition to honorary membership in many scientific societies, he was a founder and very active member of an early conservation group, the American Ornithologists' Union.

Coues's youthful knowledge of eastern ornithology, which first appeared as a significant monograph on sandpipers when he was eighteen, enabled him to make the most of his extensive travels in the unknown western territories, so that as early as 1872 he presented the deservedly popular and immensely useful *Key to North American Birds.* This innovation in identification presented the system of the artificial key, then common only in botany, and also represented a complete taxonomic revision, modified in successive editions to 1903. A correlative of the *Key* was Coues's meticulous *Check-List of North American Birds* (1873, 1882), the first since that of his mentor Baird in 1858. A distinctive feature of the *Check-List* was Coues's corrections in the orthography and pronunciation of original scientific names, a change he was unable to impose on taxonomy. He was a leader in the trend of his era toward reducing the great number of species names to varieties, especially in local forms. In his few but worthy publications on mammals, he did this with less lasting success, in part because considerable anatomical work was yet to be done. A lucid writer with a charming style, second only and successor to Baird in ornithology, Coues presented a great deal of information on the behavior and life histories of birds and even made interesting his detailed collecting manual, *Field Ornithology* (1874), incorporated into later editions of the *Key*. Rather a compulsive compiler, he tried nobly to gather a complete bibliography on ornithology but gave up after publishing the parts of it on North America, Central and South America, and Britain (1878–1880).

In the 1890's Coues checked and annotated manuscripts of several American western explorations, most usefully the journals of Lewis and Clark and of Zebulon Pike, as well as several previously unpublished accounts. In his meticulous fashion, he retraced the explorers' routes and enlarged considerably upon their natural history observations.

BIBLIOGRAPHY

I. ORIGINAL WORKS. Coues's extensive publications—almost 1,000, both technical and popular—are delightfully readable. A bibliography of his principal works, excepting popular ones, is given in Joel A. Allen's "Biographical Memoir" (cited below). The five editions of *Key to North American Birds* stand as Coues's lasting monument. The posthumous 5th ed. (2 vols., Boston, 1903), almost completed at the time of his death, incorporated a history of ornithology, collecting methods, bird anatomy, and nomenclature, as well as identifications and life histories.

Two regional monographs deserve special mention: *Birds of the Northwest: A Hand-book of the Ornithology of the Region Drained by the Mississippi River and Its Tributaries,* U.S. Geological and Geographical Survey of the Territories, Miscellaneous Publication no. 3 (Hayden, Wash., 1874), resulting from the Hayden survey; and *Birds of the Colorado Valley: A Repository of Scientific and Popular Information Concerning North American Ornithology,* U.S. Geological and Geographical Survey of the Territories, Miscellaneous Publication no. 8 (Hayden, Wash., 1878), a remarkably complete treatise on the bird life of the entire Colorado River drainage. The various parts of the intended ornithological bibliography (cited in Allen) were published in 1879 and 1880 in various locations.

Among Coues's publications on mammals, the most worthy, resulting from the Hayden survey, are *Fur-bearing Animals: A Monograph of North American Mustelidae,* U.S. Geological and Geographical Survey of the Territories, Miscellaneous Publication no. 8 (Hayden, Wash., 1877); and *Monographs of North American Rodentia,* U.S. Geological and Geographical Survey of the Territories (Hayden, Wash., 1877), Miscellaneous Publication no. 11.

Many other significant publications are cited in Allen, including the annotated accounts of western explorations.

II. SECONDARY LITERATURE. His colleague Joel A. Allen gives a detailed account of Coues's life and an analysis of his ornithological and other accomplishments in "Biographical Memoir of Elliott Coues," in *National Academy of Sciences. Biographical Memoirs,* **6** (1909), 395–446. Allen also wrote memorials for *Science,* **11** (1900), 161–163; and *Auk* (1900), p. 91. A brief effusive biography of Coues by D. G. Elliot is in the 5th ed. of *Key to North American Birds.* A fine tribute, followed by excerpts from Coues, is in Donald Culross Peattie's *A Gathering of Birds* (New York, 1939), pp. 267–276. Coues's participation in early conservation is referred to in Peter Matthiessen's *Wildlife in America* (New York, 1959), pp. 152–181.

ELIZABETH NOBLE SHOR

COULOMB, CHARLES AUGUSTIN (*b.* Angoulême, France, 14 June 1736; *d.* Paris, France, 23 August 1806), *physics, applied mechanics.*

One of the major figures in the history of physics and engineering, Coulomb's main contributions were in the fields of electricity, magnetism, applied mechanics, friction studies, and torsion. His father, Henry, came from Montpellier, where the family was important in the legal and administrative history of Languedoc. His mother, Catherine Bajet, was related to the wealthy de Sénac family. During Coulomb's youth the family moved from Angoulême to Paris, where he attended lectures at the Collège Mazarin and the Collège de France. An argument with his mother over career plans caused Coulomb to follow his father to Montpellier after the latter became penniless through financial speculations. Coulomb joined the Société des Sciences de Montpellier as an adjoint member in March 1757 and read several papers in astronomy and mathematics there during the next two years.

He went to Paris in the autumn of 1758, seeking the tutoring necessary for him to enter the École du Génie at Mézières. After some months of study he passed the abbé Charles Camus's entrance examination and took up residence at Mézières in February 1760. At about this time he formed lasting friendships with Jean Charles Borda and with the abbé Charles Bossut, his teacher of mathematics at Mézières. Coulomb graduated in November 1761 with the rank of *lieutenant en premier* in the Corps du Génie. His first post was at Brest; but in February 1764 he was ordered suddenly to proceed to Martinique, where he remained until June 1772. Coulomb was put in charge of constructing Fort Bourbon, at a cost of six million livres. He directed several hundred laborers in all phases of the construction, and this experience was important as a foundation for some of his later memoirs in mechanics. Coulomb became seriously ill several times during his stay in Martinique, and these illnesses affected his health to the extent that he was never again a well man.

Following his return to France, Coulomb was posted to Bouchain, where he composed an important memoir in mechanics that earned him the title of Bossut's correspondent to the Paris Academy of Sciences (6 July 1774). Coulomb moved then to duty at Cherbourg, where he began work on a memoir on magnetic compasses that subsequently shared first prize in the Paris Academy's competition for 1777. The importance of this memoir for Coulomb's career is that it contained elements of all of his major physical studies: the quantitative study of magnetism, torsion and the torsion balance, friction and fluid resistance, and the germ of his theories of elasticity and of magnetism.

One other event during his stay at Cherbourg merits attention: his submission in 1776 of a plan for the reorganization of the Corps du Génie. The comte de St.-Germain became minister of war in October 1775, during the administration of Turgot. Coincident with Turgot's reform aims, St.-Germain called for memoirs on the reorganization of the Génie. Coulomb's unpublished "Mémoire sur le service des officiers du Corps du Génie" was organized around two principles, the individual and the state. He sought to define the maximum utility to be obtained for each and to show that the best use of the Génie brought the most to each individual.

Coulomb saw the opportunity for public works in time of peace and favored the establishment of review boards to judge the worth of proposed projects. Most of all, he saw the Corps du Génie and public service as a whole as a "corps à talent," that is, with appointment and advancement based on ability and accomplishment. He stressed not the evils of the state but the potential of the state and individual in balance.

Coulomb was posted to Rochefort in 1779 to aid the notorious marquis de Montalembert in constructing his controversial fort entirely of wood on the nearby Île d'Aix. During this period Coulomb found time to engage in a lengthy series of experiments on friction in the shipyards at Rochefort. The result of these researches won the double first prize at the Academy in Paris in 1781 and gained Coulomb election to the Academy as adjoint *mécanicien*. Membership in the Academy finally assured Coulomb of a Paris residence, after seven different field posts and twenty years' service in the Corps du Génie.

The year 1781 marked a decisive break in Cou-

lomb's life and career. Permanently stationed in Paris, he could find a wife and raise a family. Henceforth, his engineering duties would be only as a consultant, and he was able to devote the major portion of his time to researches in physics. Coulomb the engineer became physicist and public servant. He read twenty-five scientific memoirs at the Academy (and at its successor, the Institut de France) from 1781 to 1806. His most famous memoirs were the series of seven memoirs on electricity and magnetism and the memoirs on torsion and the applications of the torsion balance. In addition to his physics research Coulomb participated in 310 committee reports to the Academy concerning machines, instruments, canals, and engineering and civic projects. In 1787 Coulomb and Jacques René Tenon were sent to England to investigate hospital conditions in London. In 1801 Coulomb was elected to the largely honorary position of president of the Institut de France. Those academicians with whom Coulomb worked most closely were geometers, mechanicians, or astronomers (e.g., Bossut, Leroy, Borda, Prony, Laplace).

Coulomb's most celebrated engineering consulting task was in Brittany in 1783–1784. Here he became involved, against his will, in a commission to recommend canal and harbor improvements. The commission (which included Borda and the abbé Alexis Marie Rochon) submitted a critical report and Coulomb suffered as the scapegoat, being confined to prison for one week in November 1783. Coulomb's excellent reports to the Academy on canals and water supply systems led the comte d'Angiviller to nominate him in July 1784 as intendant of the royal waters and fountains. The task of intendant involved supervising the management of water systems in all royal properties, including a good part of the water supply of Paris. Most biographical sketches of Coulomb mention that he was appointed curator of the large collection of secret military relief maps of French cities and fortresses. Archival records, however, indicate this not to be so.

The Revolution of 1789 caused little outward change in Coulomb's activities. He was in the midst of his great series of memoirs on electricity and magnetism, and his committee reports to the Academy continued as usual. By 1791, however, the National Assembly had overturned or reorganized many of the institutions of the *ancien régime,* and such measures applied to the Corps du Génie led Coulomb to resign from the corps in April 1791. He retired with the rank of lieutenant colonel, was holder of the Croix de St. Louis, and had thirty-one years' service in the corps. He obtained an annual pension of 2,240 livres, which was reduced by two-thirds after the Revolution. Coulomb continued active participation in the Academy until its abolition on 8 August 1793. About the same time he was removed from his position as intendant of waters. He continued work on a committee for standardization of weights and measures until it was "purged" in December 1793. At this time he and Borda retired to La Justinière, some property Coulomb owned near Blois. He returned to his research in Paris in December 1795, upon his election as member for *physique expérimentale* in the new Institut de France. His elder son, Charles Augustin II, was born in Paris on 26 February 1790 and his younger son, Henry Louis, was born there on 30 July 1797. Coulomb legitimized his marriage to Louise Françoise LeProust Desormeaux on 17 brumaire, an XI (1802).

Coulomb's last public service was as inspector general of public instruction from 1802 until his death in 1806, in which office he played a significant role in supervising the establishment of the French system of *lycées*. Coulomb's health, weakened long before during his duty in Martinique, declined precipitously in the early summer of 1806, and he died on the morning of 23 August. Since he had been baptized in the Roman Catholic faith, his final services were held at the Abbaye de St.-Germain-des-Prés. There is little evidence, however, to indicate the extent of his religious convictions. Secondary accounts indicate that the Revolution took most of Coulomb's properties and that he died almost in poverty. Examination of the probate of his estate establishes, however, that Coulomb left over 40,000 francs. (This at a time when a physics professor at a good French university would receive perhaps 6,000 francs per year.) Two decades of field duty in the Corps du Génie must have accustomed Coulomb to a modest style of life. The probate description of his personal belongings accords with this. He was accomplished in history but not a man of letters; his library contained 307 books, 238 of which were volumes issued by the Academy. Coulomb is often referred to as "de Coulomb," implying nobility. He never signed himself as such, and there is no evidence to indicate that any of his family were ennobled.

Applied Mechanics. Generally speaking, Coulomb's studies in mechanics preceded his researches in physics. His mechanics included fundamental memoirs on structural mechanics, rupture of beams and masonry piers, soil mechanics, friction theory, and ergonomics. In these he can be considered one of the great engineers in eighteenth-century Europe. Like Monge, he seemed to apply his talents to whatever was at hand. He took advantage of the peculiarities of each military post and pursued his studies in mechanics accordingly. He was talented but not excep-

tionally gifted in mathematics, although he was one of the first to utilize the variational calculus in practical engineering problems. With the exception of his friction studies, most of Coulomb's mechanics memoirs were little known until utilized by Prony, Thomas Young, and others in the early nineteenth century.

His most important memoir on mechanics was also his first, "Sur une application des règles de *maximis* et *minimis* à quelques problèmes de statique, relatifs à l'architecture" (1773). (The dates given herein for Coulomb's memoirs, unless otherwise indicated, are those dates when he formally presented the memoirs before the Paris Academy of Sciences. The actual dates of publication may be obtained from the bibliography.) The purpose of this memoir, he said, was "to determine, as far as a combination of mathematics and physics will permit, the influence of friction and of cohesion in some problems of statics." Coulomb's statics problems might seem disconnected to the modern reader, but they were at the heart of eighteenth-century engineering mechanics. If one examines the standard early eighteenth-century works (e.g., Bernard Forest de Bélidor's *Science des ingénieurs,* or Amédée François Frézier's *Traité de stéréotomie*), one finds the main engineering topics to be the strength of masonry materials, the design of retaining walls, and the design of arches. These are precisely the problems that Coulomb attacked.

In the beginning of the 1773 essay, Coulomb introduced three propositions relating to equilibrium and resolution of forces. Following this, he considered friction and cohesion, and gave virtually a theory of the flexure of beams and rupture and shear of brittle materials. Coulomb utilized Amonton's law that frictional resistance is proportional to the normal force acting on the surface rather than to the area of the surfaces in contact. He noted, however, that this law is not strictly observed in practice and that the coefficient of friction varies with the materials involved. Following this, he considered cohesion. Friction was seen as resulting from tangential contact between bodies, but cohesion was supposedly due to the effect of close-acting central forces. Cohesion in materials was considered by Coulomb as a mixture of what would today be called shear and tensile strengths. According to him, cohesion is measured by the resistance that solid bodies oppose to direct "disunion" of their parts. In a homogeneous body each part resists rupture with the same degree of resistance. Therefore, total cohesion is proportional to the number of parts to be separated, and thus to the surface area of rupture. Experimenting with stone, mortar, and brick sections, Coulomb found values for ultimate strength under tension and shear. Although his experimentally

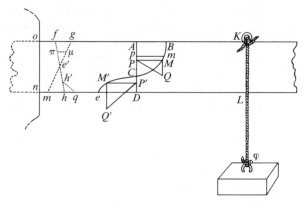

FIGURE 1. Coulomb's analysis of flexure in beams.

determined values varied slightly, he assumed that shear and tensile coefficients were the same.

Having presented these basic propositions and experiments, Coulomb proceeded to a discussion of the flexing of a beam and correctly determined, for the first time, the neutral surface of a beam. Considering a rectangular cantilever beam (Fig. 1) of cross section *AD,* he concluded that the upper portion *AC* will be under tension and the lower portion *CD* under compression. Resolving the forces into horizontal and vertical components, he showed that the sum of horizontal forces along *AD* must equal zero and the sum of vertical forces must equal the load φ. Finally, the moment of the load φ about axis *C* must equal the sums of the internal moments of the beam. Note that although Coulomb took a perfectly elastic beam as an example, he realized the distribution of forces along *BCe* could be *any* sort of curve; and in Figure 1 he drew it as some sort of parabola. In addition, Coulomb recognized that shearing forces could be neglected in long, narrow beams. Following this, he extended his analysis to the rigid, inelastic case.

In this one memoir of 1773 there is almost an embarrassment of riches, for Coulomb proceeded to discuss the theory of compressive rupture of masonry piers, the design of vaulted arches, and the theory of earth pressure. In the latter he developed a generalized sliding wedge theory of soil mechanics that remains in use today in basic engineering practice. A reason, perhaps, for the relative neglect of this portion of Coulomb's work was that he sought to demonstrate the use of variational calculus in formulating methods of approach to fundamental problems in structural mechanics rather than to give numerical solutions to specific problems. It required that group of *Polytechniciens,* teachers and students, in the early nineteenth century to appreciate the importance of this work in the context of the new engineering mechanics. The eighteenth-century engineer preferred to use empirical

design tables, such as those compiled by Jean Rudolph Perronet and Antoine de Chézy.

Coulomb's most celebrated study, one that brought him immediate acclaim, was "Théorie des machines simples," his prize-winning friction study of 1781. He investigated both static and dynamic friction of sliding surfaces and friction in bending of cords and in rolling. From examination of many physical parameters, he developed a series of two-term equations, the first term a constant and the second term varying with time, normal force, velocity, or other parameters. In agreement with Amontons's work of 1699, Coulomb showed that in general there is an approximately linear relationship between friction and normal force; but he extended the investigation considerably to show complex effects due to difference in load, materials, time of repose, lubrication, velocity, and other considerations. Coulomb's work in friction remained a standard of theory and experiment for a century and a half, until the advent of molecular studies of friction in the twentieth century. To quote Kragelsky and Schedrov's recent monograph (p. 52) on the history of friction: "Coulomb's contributions to the science of friction were exceptionally great. Without exaggeration, one can say that he created this science."

Another subject of much interest to Coulomb was the question of efficiency and output in work, and in this field (ergonomics) he made one of the most significant contributions before the studies of F. W. Taylor, a century later. Coulomb began this work in Martinique and read the first of several memoirs on the subject to the Academy in 1778. It was finally published in 1799 as "Résultats de plusieurs expériences destinées à déterminer la quantité d'action que les hommes peuvent fournir par leur travail journalier, suivant les différentes manières dont ils employent leurs forces." Earlier studies tested men or animals only for very brief periods, thus obtaining exaggerated results of productivity. Coulomb investigated various work parameters very realistically and with considerable psychological insight; and he distinguished between useful work and fatigue in work from living "machines," solving to make the ratio of effect to fatigue a maximum. In this he produced the first real study of the practical aspects of labor allocation. Among his findings were that frequent rest periods during certain tasks produce higher overall output, and that maximum daily human work results from seven to eight hours' labor for heavy tasks and ten hours' labor for light tasks. He utilized similar isoperimetric methods to investigate the theory and design of windmills.

Physics. Coulomb's election to the Paris Academy in 1781 and his acquisition of a permanent post in

Paris allowed his research generally to turn from applied mechanics to physics. His physics work, however, is integrally tied to his earlier work in mechanics. His concern with friction and cohesion and his emphasis upon the importance of shear in structural mechanics are continued in his studies of torsion, in his ideas of "coercive force" in electrostatics and magnetism, and in his final studies in magnetism and the properties of matter.

Coulomb's first writings on torsion were presented in his Academy prize-winning memoir of 1777, "Recherches sur la meilleure manière de fabriquer les aiguilles aimantées." He never attacked the general problems of elasticity (these were developed by Navier, Poisson, and Cauchy in the first decades of the nineteenth century), but his simple, elegant solution to the problem of torsion in cylinders and his use of the torsion balance in physical applications were important to numerous physicists in succeeding years. In chapter 3, Coulomb developed the theory of torsion in thin silk and hair threads. Here he was the first to show how the torsion suspension could provide physicists with a method of accurately measuring extremely small forces. He showed that within certain angular limits, torsional oscillation consisted of simple harmonic motion. He examined the parameters relating the angle of twist to the length, diameter, and elastic properties of the torsion thread. In the range of simple harmonic oscillation Coulomb demonstrated that the force of torsion was proportional to the angle of twist. He used this principle in measuring small magnetic forces and also called attention to its use in measuring other forces, notably those of fluids in motion. Eventually he was able to measure forces of less than 9×10^{-4} dynes.

This 1777 memoir contained the design of a torsion suspension declination compass. Adoption of this compass by the Paris Observatory in 1780 and the observatory's subsequent request for a solution to seemingly unresolvable problems in magnetic measurement led Coulomb to undertake a further series of experiments on torsion. His major memoir on torsion, presented 9 September 1784 ("Recherches théoriques et expérimentales sur la force de torsion et sur l'élasticité des fils de métal"), emerged from this latter investigation. This in turn provided him with a means to investigate and determine quantitatively the force relationships in varied physical fields. The torsion balance invented by Coulomb (see note) and the theory of torsion aided him in constructing theories concerning the molecular interaction within fluids and solids and, as is widely known, provided the instrumental foundation for his work in electricity and magnetism.

In his 1784 memoir Coulomb sought (1) to discover the laws of torsion and to determine possible applications of torsion and (2) to investigate the laws of coherence and elasticity of bodies by means of torsion. That is, as he noted, his study was in two regions of the torsion spectrum: the linear and nonlinear regions. Within the first he proposed to determine the linear relationship of force to torsion and to propose practical applications of this phenomenon for use in measuring various small forces. In the nonlinear region he proposed to investigate the mechanism of torsion itself, "in order to determine the laws of coherence and elasticity of metals and of all solid bodies." Coulomb developed both theoretically and experimentally the fundamental equation for torque in thin cylinders to be:

$$M = \frac{\mu B D^4}{L},$$

where M equals the torque, μ equals a constant rigidity coefficient, B is the angle of twist, and D and L are the cylinder diameter and length, respectively. In doing so, he corrected an error of his 1777 memoir (where he supposed the dependence on the diameter to be D^3).

Working with brass and iron wires, Coulomb found that the elasticity limit could be changed by work-hardening. Although the limit of elasticity could be changed, he found that the elasticity itself remained unchanged. This indicated to him the basic dissimilarity of the concepts of cohesion and of elasticity. Above a certain angle of torsion a thin cylinder, for example, either becomes noticeably inelastic or the range of elastic behavior may be shifted (permanent set). Here Coulomb gave the theory that *intra*molecular strains are elastically restored up to a certain limit. Beyond this limit the stresses become great enough to rupture the *inter*molecular bonds, and thus the material fractures or flows along a roughly planar section. After strain beyond the elastic limit but below rupture, the material is rearranged but the *intra*molecular elasticity remains the same. As Coulomb expressed it the *molécules intégrantes* change shape under stress without relative change of place. But when the force of torsion is greater than the force of cohesion, the molecules must separate or slide one over another. Through a certain range the sliding increases the area of mutual contact between the molecules and, therefore, the range of elasticity is increased. Coulomb posited that, within limits, the molecules have a definite shape, and thus there is a maximum possible area of mutual contact between the sides of the molecules. Beyond this point the sliding stops, and outright rupture occurs. His further experiments with regard to magnetism and the various possible mechanical states of iron wires seemed to confirm this idea.

Coulomb's physical theory of torsion here was influenced by his earlier theories concerning soil mechanics and compressive rupture of masonry piers. In harmony with his earlier work he saw permanent set as an *inter*molecular sliding. And, since he believed cohesion to be a positive force acting between bodies, he supposed the molecules would tend to increase their areas of mutual contact. J. T. Desaguliers and others had attributed friction phenomena to a cohesion effect proportional to the surface area of the materials in contact. Similarly, Coulomb attributed the range of permanent set to this sliding or realignment because it gave him definite points for the start of permanent set and of final rupture.

Coulomb's torsion balance was used for many of his quantitative studies in electricity and magnetism. After the French Revolution he continued with his studies of elastic and cohesive properties of matter, particularly with regard to low-velocity fluid motion.

Coulomb's major memoirs in electricity and magnetism are his 1777 memoir on magnetic compasses, the famous series of seven electricity and magnetism memoirs read at the Academy from 1785 to 1791, and several magnetism memoirs prepared after the French Revolution. In his electrical studies Coulomb determined the quantitative force law, gave the notion of electrical mass, and studied charge leakage and the surface distribution of charge on conducting bodies. In magnetism he determined the quantitative force law, created a theory of magnetism based on molecular polarization, and introduced the idea of demagnetization (basically, that combinations of magnetic poles can "cancel" each other).

In the broadest sense Coulomb participated in the articulation and extension of the Newtonian theory of forces to the disciplines of electricity and magnetism. With regard to electricity and magnetism he said, "One must necessarily resort to attractive and repulsive forces of the nature of those which one is obliged to use in order to explain the weight of bodies and celestial physics." And to do this, it would be necessary to obtain exact quantifications of these laws. Particularly in his early writings, Coulomb stressed the importance of destroying the Cartesian notion of vortices, which had again gained favor through an Academy competition of 1746 (in which the winning entries of Leonhard Euler, Daniel and Johann II Bernoulli, and François Dutour had supported the idea of magnetic vortices). It is the attack on these ideas, begun by Franz Aepinus and John Michell and continued by Coulomb, that turned theories of electricity and magnetism toward the idea of action at a distance. Once the boundary conditions could be set for the physical extent of the electric and magnetic "fluids"; once these fluids could be assumed to act as point sources; then regardless of whether one employed the one-fluid or the two-fluid system, the mechanics of the Newtonian system of action at a distance could be applied to electricity and magnetism.

Coulomb worked in both electricity and magnetism throughout the 1780's. Of his seven memoirs in electricity and magnetism the first six are concerned with electricity, and it is to these that we now turn. In the first memoir (1785), Coulomb presented the details of his torsion balance as adapted for electrical studies and demonstrated the inverse-square law of forces for the case of two bodies of opposite electrical charge. This was established statically, using the torsion balance. There were good reasons for Coulomb to limit his early presentation to the case of repulsive forces. The major reason is that the force varies as the inverse square of the distance, while torsion varies as the simple distance. This presents a situation of unstable equilibrium in the use of the torsion balance; and in most instances the charged pith balls under test quickly come together and discharge, nullifying any results.

In the second memoir (1787) Coulomb extended these investigations to the proof of the inverse-square law for electricity and magnetism for the cases of both repulsive and attractive forces. Although he actually succeeded in using the static deflection approach to measure attractive forces, in general Coulomb utilized a dynamic oscillation method to demonstrate the inverse-square law for them. A magnetic needle or charged pith ball was suspended from the torsion balance at a certain distance from another needle or pith ball fixed upon a stand. The method was to deflect the torsion arm and then time the period of the resulting oscillations, repeating this procedure for varying distances between the fixed and the oscillating bodies. This dynamic method requires the assumptions (1) that the electrical or magnetic forces act as if concentrated at a point and (2) that the line of action between the two bodies is along the axis joining their centers, and that the field lines can be considered parallel and equal (that is, the dimensions of the bodies measured must be small compared with the distance between them). If these assumptions hold, the forces responsible for motion will be proportional to the inverse square of the period, and the period will vary directly as the distance between the bodies.

Although Coulomb proved directly by experiment that the electric and magnetic force laws vary inversely as the square of the distance, he never specifically demonstrated that they are also proportional to the product of the respective charges or pole strengths. He simply stated this to be so. That is, Coulomb had demonstrated that $F \propto (1/r^2)$, but only implied that $F \propto q_1q_2$ or $F \propto m_1m_2$. He later introduced the proof plane device. His use of this device and his experiments on magnetizing iron wires show that he indirectly demonstrated the effect of the product of the charges, or pole strengths. Similarly, Coulomb defined "electric mass" and "magnetic density," but only in relative terms. He never defined a unit magnetic pole strength or (unlike Henry Cavendish) a unit electric charge.

In his third to sixth memoirs (1787–1790), Coulomb examined losses due to leakage of electric charge and investigated the distribution of charge on conducting bodies. He determined that charge loss is proportional to the charge, or:

$$\frac{-d\delta}{\delta} = m \; dt,$$

where δ and $-d\delta$ are charge and charge loss, respectively; dt is an element of time; and m is a constant dependent upon humidity and other factors. Coulomb saw charge leakage as taking place by direct contact on a molecular level, through charge-sharing either with adjacent air molecules or across the small *idio-électrique* interval he believed to exist around each molecule in a dielectric. The resistance opposed by each interval recalls his engineering experience with friction and strength of materials, for here is a coercive or passive force that must be overcome.

Experiments with charge leakage and Coulomb's conceptions of material behavior led him to the theory that in electricity there are two classes of substance: perfect conductors and dielectrics. Conduction could then occur in two ways: either through perfect conductors, such as certain metals, gases, and liquids, or through dielectric breakdown. Coulomb believed that in nature there is probably no perfect dielectric; that is, all bodies have a limit above which they cannot resist the passage of electricity. In perfect conductors the electricity can flow freely over the surface of bodies. In dielectrics conduction is resisted by the nature of the dielectric; but if there are "conducting molecules . . . within the imperfect dielectric, or distributed along its surface," then the electricity may flow over the dielectric, provided the electric intensity is sufficient to overcome the coercive force opposed by each *idio-électrique* interval within the dielectric.

Further, Coulomb showed that charge distribution does not depend on chemical affinity or elective attraction, but that it depends "uniquely" on the mutual repulsion of charge of like sign and on the geometry and positioning of the bodies, and that static charge distribution is limited to the surface (and not the interior) of conducting bodies, regardless of the material constituents or geometries of these bodies. He believed that charge could exist within dielectrics as well as on their surfaces, and he proposed to examine this; but the project never materialized. This study of the modes of charge distribution was undertaken partially as a means of preparing for a quantitative study of the effect of body geometry upon the distribution of charge. This became the subject of his fifth (1788) and sixth (1790) memoirs, an experimental investigation of charge distribution between conducting bodies of differing sizes and shapes, both in contact and after separation. These studies made large use of his proof plane to determine the charge density at each point on the charged body.

Following the measurement of charge distributions, Coulomb attempted, with moderate success, to develop analytical support for his results, using various approximative formulations. It was mostly from data presented in these two memoirs that Poisson composed his beautiful theory of electrostatics in 1811.

In the last of his seven memoirs in electricity and magnetism (1791), Coulomb sought to determine the magnetic momenta of magnetic needles and the magnetic intensity at each point as a function of their dimensional parameters. Also in this memoir he presented his fully developed theory of magnetism. In his 1777 memoir on magnetic compasses, Coulomb had leaned to Aepinus' one-fluid theory of magnetism. Although he steadfastly held that the one-fluid and the two-fluid systems were mathematically the same, experimental facts led him to question the basically macroscopic view of a magnet as having an excess of fluid near one pole and a deficiency at the other, or as having positive fluid at one pole and negative fluid at the other. He knew that the magnetic "fluid" could not be physically transferred from one magnet to another. He later discovered that bundles of magnetized wires could produce a more powerful magnet than a single bar of equal weight. The fact that a magnet could be broken into any number of pieces and produce smaller magnets led Coulomb to discard the macroscopic fluid theory and hypothesize that each magnetized particle was in fact a polarized *molécule aimantaire*.

Coulomb's molecular polarization model was amenable to those of both the one-fluid and the two-fluid schools, although he personally preferred the two-fluid model. The molecular model received general approval through the textbooks of René Just Haüy and Jean Baptiste Biot, and it was conceptually important to Biot and Poisson. It was important also to Ampère, although he altered the magnetic polarization idea and suggested that magnetism consisted of molecular electric currents flowing normal to the axis of the molecule. This final memoir in Coulomb's celebrated series of seven was presented just two years before the Academy was dissolved. After the Revolution, Coulomb's studies centered on the magnetic properties of materials as a function of their elastic and thermal history, and on the extent of magnetic properties in all matter.

Coulomb's fundamental researches in electricity and magnetism well represent the extension of Newtonian mechanics to new areas of physics. At the same time they illustrate the emergence of the "empirical" areas of physics from within traditional natural philosophy to positions of sophisticated disciplines in physics.

It may be fitting, finally, to present three statements that define Coulomb's approach to his work. First, both in his view of the Corps du Génie and in public service, he said men should be judged on their abilities and that a public service body was a "corps à talent." Second, in his work in applied mechanics Coulomb called for the use of rational analysis coupled with reality in experiment; for the conduct of research in engineering through use of a "combination of mathematics (*calcul*) and physics." Third, this use of rational analysis and engineering reality, coupled in the pursuit of *physique expérimentale,* led to Coulomb's work in physics and the evaluation of Biot that "It is to Borda and to Coulomb that one owes the renaissance of true physics in France, not a verbose and hypothetical physics, but that ingenious and exact physics which observes and compares all with rigor."

NOTE

Coulomb first mentioned the torsion balance in his 1777 magnetism essay (written 1776, published 1780), both for measuring magnetic declination and in connection with the measurement of fluid resistance. His major memoir on torsion was read at the Academy 9 September 1784 (published 1787). Coulomb claimed to have no knowledge of any predecessors in this work.

Numerous secondary sources, however, cite John Michell as the inventor of the torsion balance. *In no known published* memoir or volume of Michell's is there any mention of a torsion balance. Henry Cavendish ("Experiments to Determine the Density of the Earth," in *Philosophical Transactions,* **88** (1798), 469–70) is usually said to have established Michell's claims, but Cavendish actually says nothing except that Michell privately mentioned his idea of a torsion balance before the publication of Coulomb's experiments. Michell, says Cavendish, did not construct such a balance until a short time before his death (in 1793). Apparently there may have been some fuss about the matter before Coulomb's death in 1806, for at Coulomb's funeral J. J. Lalande, speaking of Coulomb's lack of jealousy, said that "An Englishman seized his idea on the suspension of (magnetic) needles, but Coulomb never bothered to complain."

For a detailed discussion of the invention of the torsion balance, see Gillmor, *Charles Augustin Coulomb.*

BIBLIOGRAPHY

I. ORIGINAL WORKS. Nearly all of Coulomb's published memoirs appear in the publications of the Académie des Sciences (Paris), either in the *Mémoires de l'Académie royale des sciences* (vols. published 1785–1797) and the continuation of these, the *Mémoires de l'Institut national des sciences et arts—Sciences mathématiques et physiques* (vols. published 1799–1806), or in the supplementary series for memoirs of nonmembers, *Mémoires de mathématique et de physique présentés à l'Académie royale des sciences, par divers savans* (vols. published 1776–1785). Extracts of his later memoirs were often published by others in various physical journals; sometimes these bear Coulomb's name, sometimes the name of the reporter who wrote the extract. One paper, "Recherches sur les moyens d'exécuter sous l'eau toutes sortes de travaux hydrauliques sans employer aucun épuisement," was published separately under slightly varying title by C. A. Jombert (Paris, 1779), Du Pont (Paris, 1797), and Bachelier (Paris, 1819, 1846).

There exist collections of Coulomb's major memoirs. The collection of his mechanics memoirs, *Théorie des machines simples* (Paris, 1809, 1821), is rather rare. The collection of his memoirs in electricity and magnetism is more easily located, *Mémoires de Coulomb,* A. Potier, ed., vol. I of *Collection des mémoires relatifs à la physique, publiés par la Société française de Physique,* 5 vols. (Paris, 1884–1891). Potier's edition of the memoirs omits certain important passages and should be checked against the original memoirs as published by the Academy. *Vier Abhandlungen über die Elektricität und den Magnetismus . . . . Uebersetzt und herausgegeben von Walter König* (Leipzig, 1890), no. 13 in Ostwald's *Klassiker der Exakten Wissenschaften,* is a German translation of Coulomb's first four memoirs on electricity and magnetism.

Archival material concerning Coulomb has been located in numerous repositories in France. For a full listing of archival and published sources written by and concerning Coulomb, see Gillmor (below). Most of the extant Coulomb MS material concerns his work in the Corps du Génie and his committee reports to the Academy. Without exception, the location of MS copies of his scientific memoirs is unknown. J. B. J. Delambre and J. B. Biot had a portion of these manuscripts at one time shortly after Coulomb's death. Two MS notebooks compiled by Coulomb shortly before his death are in the Bibliothèque de l'Institut de France, MS1581–82. His unpublished memoir on the reorganization of the Corps du Génie (1776) is in Archives de l'Inspection du Génie, 39, rue de Bellechasse, Paris 7e: art. 3, sect. 10, carton 2, no. 5a.

II. SECONDARY LITERATURE. C. Stewart Gillmor, *Charles Augustin Coulomb: Physics and Engineering in Eighteenth Century France* (in press), presents a full account of Coulomb's life and works; the volume also includes an extensive bibliography. Contemporary short *éloges* of

Coulomb are J. B. J. Delambre, "Éloge historique de M. Coulomb," in *Mémoires de l'Institut national des sciences et arts—Sciences mathématiques et physiques,* **7** (1806), "Histoire," 206–223; and J. B. Biot, "Coulomb," in *Mélanges scientifiques et littéraires de Biot,* 3 vols. (Paris, 1858), III, 99–104. For a discussion of Coulomb's contributions to physics see Edmond Bauer, *L'électromagnétisme—hier et aujourd'hui* (Paris, 1949), pp. 213–235. Coulomb's work in applied mechanics and engineering is discussed in S. B. Hamilton, "Charles Auguste [*sic*] de Coulomb," in *Transactions of the Newcomen Society* (London), **17** (1938), 27–49; and Stephen P. Timoshenko, *History of Strength of Materials* (New York, 1953), pp. 47–54, 61–62, 64–66. For Coulomb's friction studies see I. V. Kragelsky and V. S. Schedrov, *Razvitia nauki o trenii —Sookoi Trenia* ("Development of the Science of Friction—Dry Friction"; Moscow, 1956), pp. 51–69. Hugh Q. Golder, "The History of Earth Pressure Theory," in *Archives internationales d'histoire des sciences,* **32** (1953), 209–219, discusses Coulomb's earth pressure theory.

C. STEWART GILLMOR

COUNCILMAN, WILLIAM THOMAS (*b.* Pikesville, Maryland, 1 January 1854; *d.* York Village, Maine, 26 May 1933), *pathology.*

His father, Dr. John T. Councilman, was a farmer and rural physician, and William Thomas was himself born on a small farm. He attended local schools, St. Johns College in Annapolis, and the school of medicine at the University of Maryland. He graduated with the M.D. degree in 1878 after completing the standard two-year course. Johns Hopkins University had just opened and he made important acquaintances there, in particular the physiologist Henry Newell Martin, who invited him into his department informally for study of problems in elementary experimental physiology.

After short periods of medical service at Baltimore's Marine Hospital and Bay View Asylum, Councilman went abroad for intensive training in pathology in Vienna, Leipzig, Prague, and Strasbourg; he studied with Hans Chiari, Julius Cohnheim, Carl Weigert, and F. D. von Recklinghausen. He returned to Baltimore in 1883 to take a fellowship at Johns Hopkins Medical School. There he advanced rapidly, becoming associate in pathology under William Henry Welch in 1886, and soon attained the rank of associate professor.

His first significant research, after minor publications in other fields, was on malaria. His paper, "Contribution to the Pathology of Malarial Fever," published in 1885 in collaboration with the bacteriologist A. C. Abbott (another Welch associate) enhanced his growing reputation. In 1891—after a period of intensive research, with H. A. Lafleur as coauthor, and under the stimulus of Hopkins' leading light William Osler—he published a monograph on "Amoebic Dysentery," which was immediately recognized as the most authoritative treatise yet printed on the subject. In covering its gross and microscopic anatomical characters meticulously, Councilman and Lafleur established amoebic dysentery as an independent disease entity.

In 1892 Councilman was called to Harvard Medical School as Shattuck professor of pathological anatomy. He soon proved himself a highly competent investigator and an exceptionally able teacher, in both the classroom and the several hospitals with which he became associated. In 1900, together with F. B. Mallory and R. M. Pearce, he brought out a comprehensive monograph on diphtheria. At the same time he developed an intense interest in cerebrospinal meningitis and chronic nephritis and published several important studies on these diseases with Mallory and J. H. Wright.

In 1904, with several associates, Councilman published a monograph on the pathology of smallpox, which was called by his biographer Harvey Cushing the best treatise ever written on the pathology of that disease. Councilman had a unique capacity to work with younger men, many of whom later became leaders in pathology. Much of his best work with them emanated from Boston City, Massachusetts General, and Peter Bent Brigham hospitals.

Councilman was widely honored. He was the principal founder and first president of the American Association of Pathologists and Bacteriologists, and in that capacity greatly stimulated the development of pathology in the United States.

BIBLIOGRAPHY

I. ORIGINAL WORKS. Councilman's major contributions on the pathology of infectious disease include "Amoebic Dysentery," written with H. A. Lafleur, in *Johns Hopkins Hospital Reports,* **2** (1891), 395–548; *Epidemic Cerebrospinal Meningitis and Its Relations With Other Forms of Meningitis,* written with F. B. Mallory and J. H. Wright (Boston, 1898); "A Study of the Bacteriology and Pathology of Two Hundred and Twenty Fatal Cases of Diphtheria," written with F. B. Mallory and R. M. Pearce, in *Journal of the Boston Society of Medical Sciences,* **5** (1900), 139–319; and "Studies on the Pathology and on the Aetiology of Variola and Vaccinia," written with G. B. Magrath et al., in *Journal of Medical Research,* **11** (1904), 1–361.

II. SECONDARY LITERATURE. Biographical sketches of Councilman have been published in several medical journals. See especially Harvey Cushing, "William Thomas Councilman, 1854–1933," in *Science,* **77** (1933), 613–618, repr. in *Biographical Memoirs of the National Academy of Sciences,* **18** (1938), 157–174, with a complete bibliography of Councilman's published papers; and S. B. Wolbach, "William Thomas Councilman, 1854–1933," in *Archives of Pathology,* **16** (1933), 114–119.

ESMOND R. LONG

COUPER, ARCHIBALD SCOTT (*b.* Kirkintilloch, Dumbartonshire, Scotland, 31 March 1831; *d.* Kirkintilloch, 11 March 1892), *chemistry.*

Couper, at the same time as F. A. Kekulé and independently of him, introduced principles that underlie the modern structural theory of organic chemistry. He also introduced into organic chemistry lines to represent valence bonds in chemical formulas. All his papers were published in barely over a year, after which ill health terminated his scientific career. Details of Couper's life were brought to light by Richard Anschütz, Kekulé's biographer and successor at Bonn.

Couper was the only surviving son of Archibald and Helen Dollar Couper. His father was proprietor of a large cotton-weaving business that had been owned by the family for several generations. Couper received most of his early education at home and then went to Glasgow University for study in the humanities and classical languages. The summers of 1851 and 1852 he spent in Germany, and in August 1852 he entered the University of Edinburgh, studying logic and metaphysics under Sir William Hamilton while continuing his language studies. During the year 1854–1855 Couper moved to Berlin, where, sometime before his departure for Paris in 1856, he chose chemistry as his field of major interest. He is known to have attended Sonnenschein's lectures in analytical chemistry at the University of Berlin, and to have worked two months in his laboratory.

After moving to Paris in August 1856, Couper engaged in independent research in the laboratory of Charles Adolphe Wurtz. His first publication—on the bromination of benzene—appeared in August 1857. His other experimental paper, on salicylic acid, was published in 1858. Early in 1858 Couper asked Wurtz to have presented before the French Academy of Sciences a paper entitled "On a New Chemical Theory." Wurtz, not a member of the Academy, procrastinated before finding someone to present it. In the intervening period Kekulé's paper, dated 19 May 1858, appeared in *Liebig's Annalen.*[1] Couper's paper—on essentially the same topic—was presented to the French Academy by J. B. A. Dumas in June and was published soon thereafter. Kekulé attacked the paper almost immediately, claiming priority and greater significance for his own work.[2] He also questioned Couper's researches on salicylic acid and, since Couper never answered, his name was long forgotten.

Couper complained to Wurtz about the delay in presenting his paper and was asked to leave the laboratory. He returned to Scotland in the fall of 1858 and accepted a position as second laboratory assistant to Lyon Playfair, professor of chemistry at Edinburgh University. Soon after beginning his duties he suffered a mental breakdown, underwent treatment, apparently recovered, and went on a fishing expedition, during which, reportedly due to extended exposure to the sun, his illness returned. He never fully recovered and lived in retirement for the remaining thirty-four years of his life.

In Couper's first experimental paper[3] he sought a way to convert benzene, C_6H_6, into the corresponding hydroxy addition compounds, probably C_6H_7OH and $C_6H_6(OH)_2$. Wurtz had just prepared ethylene glycol, $C_2H_4(OH)_2$, from ethylene via the iodide $C_2H_4I_2$ reacted with silver acetate, followed by treatment with potassium hydroxide. Couper reacted benzene with bromine and isolated two new compounds, bromobenzene, C_6H_5Br, and *p*-dibromobenzene, $C_6H_4Br_2$. The first showed only very slight reactivity with silver acetate, while the second led to an explosion.

Couper's other experimental paper dealt with the constitution of the benzene derivative salicylic acid, a subject that a number of chemists, including Kekulé, had been investigating. Treating the acid with phosphorus pentachloride, Couper claimed to have obtained the "trichlorophosphate de salicyle." He proposed detailed structural formulas for the acid and its derivatives. This represented the first time that the relations between the individual carbon atoms of benzene had been depicted in a formula.

Kekulé repeated Couper's experiment more than twenty times without success, but Anschütz later fully confirmed Couper's results. Apparently Kekulé had kept the hot reaction mixture too long before distillation.[4] Couper's work was further confirmed and extended by A. G. Pinkus.[5]

Coming to chemistry from a study of philosophy and classical languages, Couper viewed the complex chemical formulas of his time as if they were words of a foreign language: "To reach the structure of words we must go back, seek out the undecomposable elements, viz. the letters, and study carefully their powers and bearing. Having ascertained these, the composition and structure of every possible word is revealed."[6] By this mode of reasoning he deduced that carbon has a valence, or combining power, of two or four and that it has the power to form valence links with other carbon atoms, thus making possible carbon chains. These two principles are the foundation of the structural theory of organic chemistry. Kekulé had proposed the tetravalence of carbon the previous year.

Couper then devised a pictorial representation of chemical compounds using dotted lines, solid lines, or brackets to show the linkings between atoms in the molecule. These formulas represent the first introduction of the valence line into organic formulas. Since Couper used an atomic weight of eight for oxygen, his formulas always contain a pair of oxygen atoms instead of the single atom used today.

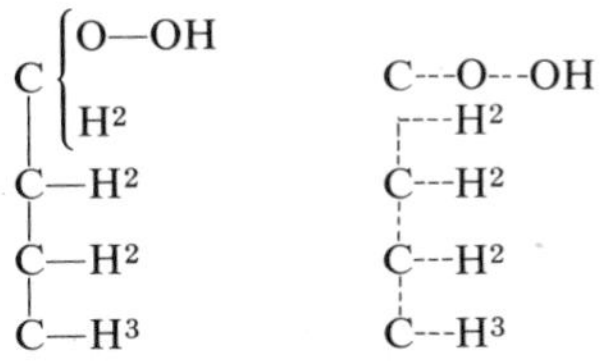

FIGURE 1
Couper's formulas for butyl alcohol, $CH_3CH_2CH_2CH_2OH$

FIGURE 2
Couper's formulas for acetic acid, CH_3COH

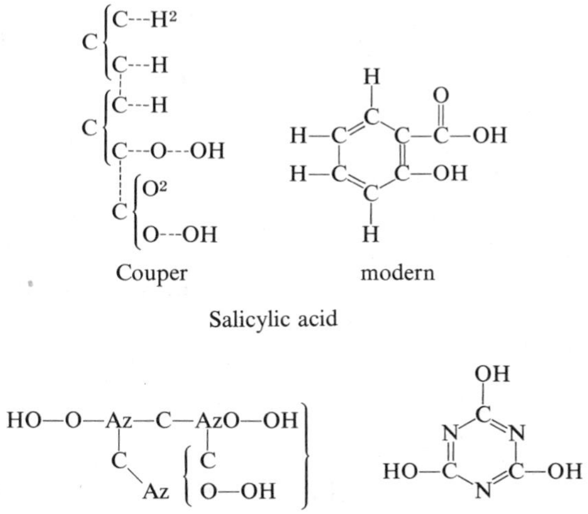

Couper   modern

Salicylic acid

Couper   modern

Cyanuric acid

FIGURE 3

In 1858 Couper introduced the first ring formula into organic chemistry, to represent the structure of cyanuric acid. Kekulé's ring structure for benzene appeared in 1865. Couper's formulas for salicylic and cyanuric acids show some of the carbons as having a valence of two. Couper maintained that several elements, such as carbon, nitrogen, and phosphorus, exhibit multiple valence. Kekulé strongly opposed such a view.

A. Butlerov, who visited Kekulé and Couper during 1858, first strongly criticized Couper's views,[7] but in the well-known paper in which he introduced the term "chemical structure," he credited Couper as the originator of his central ideas.[8]

NOTES

1. "Ueber die Constitution und die Metamorphosen der chemischen Verbindungen und über die chemische Natur des Kohlenstoffs," in *Justus Liebigs Annalen der Chemie,* **106** (1858), 129–159.
2. "Remarques sur l'occasion de 'Couper, sur une nouvelle théorie chimique,'" in *Comptes rendus hebdomadaires des séances de l'Académie des sciences,* **47** (1858), 378–380.
3. Formulas in this paragraph are those now in use.
4. R. Anschütz, "Ueber die Einwirkung von Phosphorpentachlorid auf Salicylsäure," in *Justus Liebigs Annalen der Chemie,* **228** (1885), 308–321.
5. A. G. Pinkus, P. G. Waldrep, and W. J. Collier, "On the Structure of Couper's Compound," in *Journal of Organic Chemistry,* **26** (1961), 682–686.
6. O. T. Benfey, ed., *Classics in the Theory of Chemical Combination,* ·p. 139.
7. "Ueber A. S. Couper's neue chemische Theorie," in *Justus Liebigs Annalen der Chemie,* **110** (1859), 51–66.
8. "Einiges über die chemische Structur der Körper," in *Zeitschrift für Chemie,* **4** (1861), 560.

BIBLIOGRAPHY

I. Original Works. Couper published six papers: "Recherches sur la benzine," in *Comptes rendus hebdomadaires des séances de l'Académie des sciences,* **45** (10 Aug. 1857), 230–232; "Sur une nouvelle théorie chimique," *ibid.,* **46** (14 June 1858), 1157–1160; "On a New Chemical Theory," in *London, Edinburgh, and Dublin Philosophical Magazine and Journal of Science,* 4th ser., **16** (1858), 104–116, a much more extensive treatment than the preceding article; "Sur une nouvelle théorie chimique," in *Annales de chimie et de physique,* 3rd ser., **53** (1858), 469–489, largely a translation of the preceding article but with a different treatment of chemical formulas and with a note on nitrogen compounds, including a ring formula for cyanuric acid; "Recherches sur l'acide salicylique," in *Comptes rendus hebdomadaires des séances de l'Académie des sciences,* **46** (7 June 1858), 1107–1110; and "Researches on Salicylic Acid," in *The Edinburgh New Philosophical Journal,* n.s. **8** (July–Oct. 1858), 213–217, similar to the preceding article but longer and, unlike it, containing structural formulas. R. Anschütz, ed., *"Ueber eine neue chemische Theorie" von Archibald Scott Couper,* no. 183 in Ostwald's Klassiker der Exacten Wissenschaften (Leipzig, 1911), is a German translation of the shorter "Sur une nouvelle théorie chimique" and of the longer article of the same title, with commentary and notes. *On a New Chemical Theory and Researches on Salicylic Acid,* Alembic Club Reprint no. 21 (Edinburgh, 1953), contains English translations of the shorter "Sur une nouvelle théorie chimique" and of

"Recherches sur l'acide salicylique," the two papers published in English, and a short commentary.

II. SECONDARY LITERATURE. On Couper or his work, see R. Anschütz, "Life and Chemical Work of Archibald Scott Couper," in *Proceedings of the Royal Society of Edinburgh,* **29** (Apr. 1909), 193–273, which contains detailed biography and all Couper's papers as originally published—the German version of the biography is R. Anschütz, "Archibald Scott Couper," in *Archiv für die Geschichte der Naturwissenschaften und der Technik,* **1** (1909), 219–261; O. T. Benfey, "Archibald Scott Couper," in Eduard Farber, ed. *Great Chemists* (New York, 1961), pp. 705–715; and O. T. Benfey, ed., *Classics in the Theory of Chemical Combination* (New York, 1963), which includes Couper's two theoretical papers in English as well as Kekulé's 1858 paper; L. Dobbin, "The Couper Quest," in *Journal of Chemical Education,* **11** (1934), 331–338, which details the search for information on Couper and contains further details of his life; and H. S. Mason, "History of the Use of Graphic Formulas in Organic Chemistry," in *Isis,* **34** (1943), 346–354.

OTTO THEODOR BENFEY

COURNOT, ANTOINE-AUGUSTIN (*b.* Gray, France, 28 August 1801; *d.* Paris, France, 31 March 1877), *applied mathematics, philosophy of science.*

Of Franche-Comté peasant stock, Cournot's family had belonged for two generations to the *petite bourgeoisie* of Gray. In his *Souvenirs* he says very little about his parents but a great deal about his paternal uncle, a notary to whom he apparently owed his early education. Cournot was deeply impressed by the conflict that divided the society in which he lived into the adherents of the *ancien régime* and the supporters of new ideas, especially in the realm of religion. One of his uncles was a conformist priest, the other a faithful disciple of the Jesuits, having been educated by them.

Between 1809 and 1816 Cournot received his secondary education at the *collège* of Gray and showed a precocious interest in politics by attending the meetings of a small royalist club. He spent the next four years idling away his time, working "en amateur" in a lawyer's office. Influenced by reading Laplace's *Système du monde* and the Leibniz-Clarke correspondence, he became interested in mathematics and decided to enroll at the École Normale Supérieure in Paris. In preparation, he attended a course in special mathematics at the Collège Royal in Besançon (1820–1821) and was admitted to the École Normale after competitive examinations in August 1821. However, on 6 September 1822 the abbé Frayssinous, newly appointed grand master of the University of France, closed the École Normale. Cournot found himself without a school and with only a modest allowance for twenty months. He remained in Paris, using this free time—which he called the happiest of his life—to prepare at the Sorbonne for the *licence* in mathematics (1822–1823). His teachers at the Sorbonne were Lacroix, a disciple of Condorcet, and Hachette, a former colleague of Monge. A fellow student and friend was Dirichlet.

In October 1823, Cournot was hired by Marshal Gouvion-Saint-Cyr as tutor for his small son. Soon Cournot became his secretary and collaborator in the editing and publishing of his *Mémoires.* Thus, for seven years, until the death of the marshal, Cournot had the opportunity to meet the many important persons around the marshal and to reflect on matters of history and politics. Nevertheless, Cournot was still interested in mathematics. He published eight papers in the baron de Férussac's *Bulletin des sciences,* and in 1829 he defended his thesis for the doctorate in science, "Le mouvement d'un corps rigide soutenu par un plan fixe." The papers attracted the attention of Poisson, who at that time headed the teaching of mathematics in France. When, in the summer of 1833, Cournot left the service of the Gouvion-Saint-Cyr family, Poisson immediately secured him a temporary position with the Academy of Paris. In October 1834 the Faculty of Sciences in Lyons created a chair of analysis, and Poisson saw to it that Cournot was appointed to this post. In between, Cournot translated and adapted John Herschel's *Treatise on Astronomy* and Kater and Lardner's *A Treatise on Mechanics,* both published, with success, in 1834.

From then on, Cournot was a high official of the French university system. He taught in Lyons for a year. In October 1835 he accepted the post of rector at Grenoble, with a professorship in mathematics at the Faculty of Sciences. Subsequently he was appointed acting inspector general of public education. In September 1838, Cournot married and left Grenoble to become inspector general. In 1839 he was appointed chairman of the Jury d'Agrégation in mathematics, an office he held until 1853. He left the post of inspector general to become rector at Dijon in 1854, after the Fortoul reform, and served there until his retirement in 1862.

In the course of his long career as administrator, Cournot, who was extremely scrupulous in fulfilling his duties, was able to exert a strong influence on the teaching of mathematics in the secondary schools and published a work on the institutions of public instruction in France (1864). At the same time he pursued a career as scientist and philosopher. While rector at Grenoble, he published *Recherches sur les principes mathématiques de la théorie des richesses* (1838). Between 1841 and 1875 he published all his mathematical and philosophical works.

Unassuming and shy, Cournot was considered an exemplary civil servant by his contemporaries. His religious opinions seem to have been very conservative. In politics he was an enthusiastic royalist in 1815, only to be disappointed by the restoration of the monarchy. In the presidential elections following the 1848 Revolution, he voted for Louis Eugène Cavaignac, a moderate republican. In 1851, sharply disapproving the organization of public instruction as directed by Louis Napoleon, he decided to become a candidate in the legislative elections in Haute-Saône; this election, however, was prevented by the coup d'état of 2 December.

Cournot's background and his education made him a member of the provincial *petite bourgeoisie* of the *ancien régime*. But as a civil servant of the July monarchy and the Second Empire, he became integrated into the new bourgeoisie of the nineteenth century. Of certainly mediocre talents as far as pure mathematics was concerned, he left behind work on the philosophy of science, remarkably forceful and original for its period, that foreshadowed the application of mathematics to the sciences of mankind. Nobody could express better and more humorously Cournot's importance than he himself when he reported Poisson's appreciation of his first works: "He [Poisson] discovered in them a philosophical depth—and, I must honestly say, he was not altogether wrong. Furthermore, from them he predicted that I would go far in the field of pure mathematical speculation but (as I have always thought and have never hesitated to say) in this he was wrong" (*Souvenirs,* p. 154).

Cournot's mathematical work amounts to very little: some papers on mechanics without much originality, the draft of his course on analysis, and an essay on the relationship between algebra and geometry. Thus, it is mainly the precise idea of a possible application of mathematics to as yet unexplored fields that constitutes his claim to fame. With the publication in 1838 of his *Recherches sur les principes mathématiques de la théorie des richesses* he was a third of a century ahead of Walras and Jevons and must be considered the true founder of mathematical economics. By reducing the problem of price formation in a given market to a question of analysis, he was the first to formulate the data of the diagram of monopolistic competition, thus defining a type of solution that has remained famous as "Cournot's point." Since then, his arguments have of course been criticized and amended within a new perspective. Undoubtedly, he remains the first of the important pioneers in this field.

Cournot's work on the "theory of chance occurrences" contains no mathematical innovation. Nevertheless, it is important in the history of the calculus of probability, since it examines in an original way the interpretation and foundations of this calculus and its applications. According to Cournot, occurrences in our world are always determined by a cause. But in the universe there are independent causal chains. If, at a given point in time and space, two of these chains have a common link, this coincidence constitutes the fortuitous character of the event thus engendered. Consequently, there would be an objective chance occurrence that would nevertheless have a cause. This seeming paradox would be no reflection of our ignorance.

This objective chance occurrence is assigned a certain value in a case where it is possible to enumerate—for a given event—all the possible combinations of circumstances and all those in which the event occurs. This value is to be interpreted as a degree of "physical possibility." However, one must distinguish between a physical possibility that differs from 0 (or 1) only by an infinitely small amount and a strict logical impossibility (or necessity).

On the other hand, Cournot also insisted on the necessary distinction between this physical possibility, or "objective probability," and the "subjective probability" that depends on our ignorance and rests on the consideration of events that are deemed equiprobable[1] since there is not sufficient cause to decide otherwise. Blaise Pascal, Fermat, Huygens, and Leibniz would have seen only this aspect of probability. Jakob I Bernoulli, despite his ambiguous vocabulary, would have been the first to deal with objective probabilities that Cournot was easily able to estimate on the basis of frequencies within a sufficiently large number of series of events.

To these two ideas of probability Cournot added a third that he defined as "philosophical probability." This is the degree of rational, not measurable, belief that we accord a given scientific hypothesis. It "depends mainly upon the idea that we have of the simplicity of the laws of nature, of order, and of the rational succession of phenomena" (*Exposition de la théorie des chances,* p. 440; see also *Essai,* I, 98–99). Of course, Cournot neither solved nor even satisfactorily stated the problem of the logical foundation of the calculus of probability. But he had the distinction of having been the first to dissociate—in a radical way—various ideas that still were obscure, thus opening the way for deeper and more systematic research by more exact mathematicians. He also was able to show clearly the importance of the applications of the calculus of probability to the scientific description and explanation of human acts. He himself—following Condorcet and Poisson—attempted to interpret legal statistics (*Journal de Liouville,* **4** [1838], 257–334; see also *Exposition de la théorie des chances,* chs. 15, 16). But he also warned against "premature and abusive applications" that might discredit this ambitious project.

More than for his mathematical originality, Cournot is known for his views on scientific knowledge. He defined science as logically organized knowledge, comprising both a classification of the objects with which it deals and an ordered concatenation of the propositions it sets forth. It claims neither the eternal nor the absolute: "There can be nothing more inconsistent than the degree of generality of the data with which the sciences deal—data susceptible to the degree of order and the classification that constitute scientific perfection" (*Essai,* II, 189). Therefore, the fundamental characteristic of the scientific object must be defined differently. "What strikes us first of all, what we understand best, is the *form,*" Cournot wrote at the beginning of the *Traité de l'enchaînement des idées,* adding, "Scientifically we shall always know[2] only the form and the order." Thus, it was from this perspective that he interpreted scientific explanation and stressed the privilege of mathematics—the science of form par excellence. Even though establishing himself as forerunner of a completely modern structural concept of the scientific object, Cournot did not go so far as to propose a reduction of the process of knowledge to the application of logical rules. On the contrary, he insisted upon the domination of strictly formal and demonstrative logic by "another logic, much more fruitful, a logic which separates appearance from reality, a logic which connects specific observations and infers general laws from them, a logic which ranks truth and fact" (*Traité,* p. 6).

This discerning and inventive power orients and governs the individual steps of the strictly logical proof; it postulates an order in nature and its realization in the simplest ways.[3] This suggests the opposition Leibniz saw in the laws of logical necessity and the architectonic principles that make their application intelligible (see, e.g., Leibniz' "Specimen dynamicum," in his *Mathematische Schriften,* Gerhardt, ed., VI, 234–246). Cournot also declared himself, on several occasions, a great admirer of Leibniz. But to him the reason that governs the discovery of natural laws was not due to divine wisdom—he was always careful to separate religious beliefs (to which, incidentally, he adhered) from philosophical rationality. Reason, within scientific knowledge, denoted the ineluctable but always hazardous contribution of philosophical speculation. "Everywhere," he assures us, "we must state this twofold fact, that the intervention of the philosophical idea is necessary as a guideline and to give science its dogmatic and regular form; it also must insure that the progress of the positive sciences is not hindered by the indecision of philosophical questions" (*Essai,* II, 252). Thus philosophy, as research on the most "probable" hypotheses regarding the assumption of a maximum of order and a minimum of complexity, becomes an integral part of scientific practice. But if philosophical reason guides the organization of hypotheses, it is the role of logic, obviously, to exhibit consequences and of experience to provide the only evidence that can be decisive in their favor.[4]

From this analysis one must conclude that science cannot be defined as a pure and simple determination of causes. For Cournot the word "cause" meant the generative antecedent of a phenomenon. He wanted science to add to the designation of causes the indication of reasons, i.e., the general traits of the type of order within which the causes act. And since the indication of reasons stems from philosophical speculation, it can only be probable—within a probability that itself is philosophical—that knowledge will advance to the extent that hypotheses are refined and corrected on the basis of experience.

In this sense, Cournot's epistemology is a probabilism. And it is probabilism in another sense, too—since it insists upon the indissoluble connection between the "historic data" and the "theoretical data" in the sciences. Fortuitous facts, in the sense defined above, appear in our experience—by its very nature—and not through our ignorance of causes. These facts appear as knots of contingency within the tissue of theoretical explication and, according to Cournot, cannot be entirely removed from it.

The connection between science and history is defined more precisely by the classification of the sciences proposed in chapter 22 of the *Essai.* According to Cournot, the system of the sciences must show an order that his predecessors had vainly tried to reduce to one dimension. In order to describe this system, we need a double-entry table (Figure 1) that vertically approximates Comte's system of division: mathematical sciences, physical and cosmological sciences, biological and natural sciences, noological and symbolic sciences, political and historical sciences. Horizontally there are three series: theoretical, cosmological, and technical. The technical series gives a special place and autonomous status to certain applied disciplines the importance and development of which "depend upon various peculiarities of the state of civilized nations and are not in proportion to the importance and philosophical standing of the speculative sciences to which they should be linked" (*Essai,* II, 266).

	Theoretical Series	Cosmological Series	Technical Series
Mathematical sciences	Theory of numbers, etc.		Metrology, etc.
Physical and cosmological sciences	Physics Physical chemistry, etc.	Astronomy Geology, etc.	Engineering sciences, etc.
Biological and natural sciences	Anatomy Experimental psychology	Botany Linguistics, etc.	Agronomy Medicine, etc.
Noological and symbolic sciences	Logic Natural theology, etc.	Mythology Ethnography, etc.	Grammar Natural law, etc.
Political and historical sciences	Sociology Economics, etc.	Political geography History, etc.	Judicial sciences, etc.

FIGURE 1

The distinction between the theoretical and the cosmological series corresponds to the separation of a historic and contingent element. This element will always be present in the sciences, even in the theoretical sciences (with the exception, perhaps, of mathematics), and will become more and more dominant as one passes from the physical sciences to the natural sciences (see *Traité*, p. 251). But if the very nature of the process of scientific knowledge demands that the philosophical element cannot be "anatomically" separated, it allows for the establishment of sciences in which the historic element controls the contents and the method of knowing.

Another kind of separation appears in the system of the sciences that Cournot set forth and developed in his works following the *Essai*. This separation is the radical distinction between a realm of physical nature and a realm of life.

For Cournot, the scientific explanation of the phenomena of life requires a specific principle that, in the organism, must control the laws of physics and chemistry. As for man's role among the living beings, it seems that Cournot linked it with the development of community life, for "the superiority of man's instincts and the faculties directly derived from it . . . would not suffice to constitute a distinct realm within Nature, a realm in contrast with the other realms" (*Traité*, p. 365). On the other hand, he adds, "When I see a city of a million inhabitants . . . I understand very well that I am completely separated from the state of Nature . . ." (*ibid.*, p. 366).

This separation from the state of nature is accomplished by man in the course of a development that causes him to cultivate successively the great organizational forms of civilized life: religion, art, history, philosophy, and science. Cournot was careful not to interpret such a development as a straight and continuous march, yet he did not fail to stress that only scientific knowledge could be the sign of great achievement and alone was truly capable of cumulative and indefinitely pursued progress.

NOTES

1. Cournot's definition of an objective probability as the quotient of the number of favorable cases divided by the number of possible cases also entails a hypothesis of equiprobability of these various cases (*Essai*, ch. 2). Cournot does not seem to have noticed this difficulty, which later concerned Keynes and F. P. Ramsey.
2. According to Cournot, order is a fundamental category of scientific thought that can be deduced neither from time nor from space, which it logically precedes. Moreover, it cannot be reduced to the notion of linear succession. Without proceeding to a formal analysis, Cournot very often showed that by "order" he meant any relationship that can be expressed by a multiple-entry table.
3. But Cournot rebelled against the reduction of the principle of order to a maxim postulating the stability of the laws of nature (*Essai*, I, 90).
4. Cournot was always very careful to distinguish between philosophy and science. The following text shows a very rare lucidity, considering when it was written:

> In a century when the sciences have gained so much popularity through their applications, it would be a vain effort to try to pass off philosophy as science or as a science. The public, comparing progress and results, will not be fooled for long. And since philosophy is not—as some would have us believe—a science, one could be led to believe that philosophy is nothing at all, a conclusion fatal to true scientific progress and to the dignity of the human spirit [*Considérations*, II, 222].

BIBLIOGRAPHY

I. ORIGINAL WORKS. Cournot's principal works are *Recherches sur les principes mathématiques de la théorie des*

richesses (Paris, 1838, 1938); *Traité élémentaire de la théorie des fonctions et du calcul infinitésimal,* 2 vols. (Paris, 1841); *Exposition de la théorie des chances et des probabilités* (Paris, 1843); *De l'origine et des limites de la correspondance entre l'algèbre et la géométrie* (Paris, 1847); *Essai sur les fondements de la connaissance et sur les caractères de la critique philosophique,* 2 vols. (Paris, 1861, 1911); *Traité de l'enchaînement des idées fondamentales dans les sciences et dans l'histoire,* 2 vols. (Paris, 1861, 1912); *Considérations sur la marche des idées et des événements dans les temps modernes,* 2 vols. (Paris, 1872, 1934); *Matérialisme, vitalisme, rationalisme* (Paris, 1875); and *Souvenirs,* edited, with intro. and notes, by E. P. Bottinelli (Paris, 1913).

II. SECONDARY LITERATURE. On Cournot or his work, see E. P. Bottinelli, *A. Cournot, métaphysicien de la connaissance* (Paris, 1913), which contains an exhaustive bibliography of Cournot's work; E. Callot, *La philosophie biologique de Cournot* (Paris, 1959); A. Darbon, *Le concept du hasard dans la philosophie de Cournot* (Paris, 1911); F. Mentré, *Cournot et la renaissance du probabilisme au XIXe siècle* (Paris, 1908); and G. Milhaud, *Études sur Cournot* (Paris, 1927). *Revue de métaphysique et de morale* (May 1905) is a special number devoted to Cournot; of special note are Henri Poincaré, "Cournot et les principes du calcul infinitésimal"; G. Milhaud, "Note sur la raison chez Cournot"; and H. L. Moore, "A.-A. Cournot," a biographical study.

G. GRANGER

COURTIVRON, GASPARD LE COMPASSEUR DE CRÉQUY-MONTFORT, MARQUIS DE (*b.* Château de Courtivron, Côte-d'Or, France, 28 February 1715; *d.* Château de Courtivron, 5 October 1785), *mechanics, optics, technology.*

A noble of the sword, Courtivron interrupted his formal education in 1730 by joining the regiment of his maternal uncle. After the armistice of 1735, he began studying science for the purpose of advancing his military career. But he soon became interested in science for its own sake and carefully studied the work of Clairaut. An injury received during the War of the Austrian Succession caused him, in 1743, to give up the military and turn to science as a career.

In mechanics Courtivron hoped to unite statics and dynamics by building on the proposition that a system of rigidly connected bodies acted on by arbitrary forces finds static equilibrium at the position in which it has maximum *vis viva* (mv^2) while moving. By so bringing together Leibniz' concept of "living forces," Johann I Bernoulli's principle of virtual "speeds" (work), and d'Alembert's principle of accelerating forces, he moved toward a formulation of the concept of the conservation of work. Indeed, his principle implied that the change in work equals the change in kinetic energy.

In his treatise on optics Courtivron assailed the Cartesian concept of light and championed the Newtonian. He did not servilely follow Newton, however, for in his view color results from differences in speed rather than from those in weight. Furthermore, in affirming that dense mediums slow down light corpuscles, he made use of Fermat's principle of least time, recently revived by Maupertuis. He burdened the balance of his treatise with translations of Robert Smith's *Compleat System of Opticks* so slavishly that he even contradicted his own theory of colors.

Courtivron contributed to the Academy's *Description des arts et métiers* by directing Étienne Jean Boucher in the writing of a treatise on forges and iron furnaces that incorporated material drawn from the papers and plates of Réaumur and included a translation of Swedenborg's *Minerale de ferro.* Numerous examples of iron smelting practice and an explanation of the properties of iron in terms of Stahl's phlogiston theory offer an intimate view of eighteenth-century scientific practice.

Indeed, Courtivron's career, which covers a broad spectrum of scientific endeavor, reflects the style of Enlightenment science. In 1744 his early research in mechanics earned him a seat in the Academy of Sciences, but in 1765 he became inactive and retired to his country estate.

BIBLIOGRAPHY

I. ORIGINAL WORKS. Courtivron's major scientific works are "Recherches de statique et de dynamique, où l'on donne un nouveau principe général pour la considération des corps animés par des forces variables, suivant une lois quelconque," in *Mémoires de l'Académie des sciences* for 1749 (1753), 15–27; *Traité d'optique, où l'on donne la théorie de la lumière dans le système newtonien, avec de nouvelles solutions des principaux problèmes de dioptrique et de catoptrique* (Paris, 1752); and, with E. J. Boucher, "Art des forges et fourneaux à fer . . .," in Académie des sciences, *Description des arts et métiers,* VII and VIII (Paris, 1761–1762), also sometimes bound with other texts treating the same subject, and also with Boucher, *Observations sur l'art du charbonnier* (Paris, 1767). Minor scientific works are found in the following issues of *Mémoires de l'Académie des sciences* for 1744 (1748), 384–394; *ibid.,* 405–414; for 1745 (1749), 1–8; for 1747 (1752), 287–304; *ibid.,* 449–458; for 1748 (1752), 133–147; *ibid.,* 323–340; for 1755 (1759), 287–292; for 1778 (1781), 10–11. Reports on minor scientific works are given in the following issues of *Histoire de l'Académie des sciences* for 1743 (1746), 120; for 1744 (1748), 13–14; for 1751 (1755), 73–75; for 1757 (1762), 32–33. Courtivron was co-author, with Charles de Beosses de Tournay and Thésut de Verrey, of *Catalogues et armoiries des gentilshommes qui ont assisté à la tenue des États Généraux du Duché de Bourgogne, depuis l'an 1548 jusqu'à l'an 1682 . . .* (Dijon, 1760).

II. Secondary Literature. There is an obituary by Condorcet in *Éloges des académiciens . . .,* **4** (1799), 120–135. E. Jouguet, in *Lectures de mécanique,* I (Paris, 1908), 75, 197–198, discusses Courtivron's mechanics. E. Mach, *The Science of Mechanics* (La Salle, Ill., 1960), p. 84, errs in stating Courtivron's principle. Mention of Courtivron's treatise on optics is found in E. Whittaker, *A History of the Theories of Aether and Electricity,* I (New York, 1960), 99; A. Wolf, *A History of Science, Technology, and Philosophy in the 18th Century,* I (New York, 1961), 162–163; and J. Priestly, *The History and Present State of Discoveries Relating to Vision, Light, and Colours* (London, 1772), pp. 401–403. Courtivron's contributions to iron technology are discussed in T. A. Wertime, *The Coming of the Age of Steel* (Chicago, 1962), *passim;* L. Beck, *Die Geschichte des Eisens in technischer und kulturgeschichtlicher Beziehung,* III (Brunswick, 1897), *passim;* and B. Gille, *Les origines de la grande industrie métallurgique en France* (Paris, 1947), pp. 85–95. J. F. Montucla, *Histoire des mathématiques,* III (Paris, 1802), 61, discusses Courtivron's solution of algebraic equations presented in a paper published in *Mémoires de l'Académie des sciences* for 1744, 405–414.

Robert M. McKeon

COURTOIS, BERNARD (*b.* Dijon, France, 8 February 1777; *d.* Paris, France, 27 September 1838), *chemistry.*

The son of a Dijon saltpeter manufacturer, Courtois studied under Fourcroy at the École Polytechnique and worked as a pharmacist in military hospitals before becoming an assistant to Thénard in 1801. The following year he was assistant to Armand Séguin.

Courtois then returned to Dijon to take over his father's business. The firm used seaweed ash as a source of valuable potassium and sodium salts. By leaching the ashes with water and evaporating the mother liquor, these salts could be precipitated. However, Courtois first had to add strong acid in order to remove the undesirable sulfur-containing compounds. One day late in 1811 he added sulfuric acid in excess and observed clouds of violet vapor evolving from the solution. The vapor, condensing on cold objects, formed dark lustrous crystals. For about six months he investigated the properties of this new substance, preparing its compounds of hydrogen, phosphorus, ammonia, and several metals. Barely able to eke out a living in the saltpeter business, Courtois then decided that he did not have the means to continue work on the substance and abandoned his research.

Sometime in July 1812, Courtois told two Dijon chemists, Charles-Bernard Désormes and Nicolas Clément, of his discovery and urged them to continue the research. With his permission they announced his work to the Institut de France on 29 November 1813.

Before the end of the year both Gay-Lussac and Humphry Davy had examined the substance and independently established it as an element. Gay-Lussac named the new element "iode" because of its violet color.

All of these men acknowledged Courtois as the discoverer of iodine, and in 1831 the Institute awarded him a prize of 6,000 francs for the discovery. By this time Courtois had given up the saltpeter business and, from the 1820's, attempted to make a living by preparing and selling iodine and iodine compounds. This enterprise also failed, and he died in poverty.

BIBLIOGRAPHY

The first publications on iodine are somewhat confusing. Courtois's research is found in a paper attributed to him but actually the work of Clément: "Découverte d'une substance nouvelle dans le Vareck," in *Annales de chimie,* **88** (1813), 304–310. It was followed by an anonymous article, "Sur un nouvel acide formé avec la substance découverte par M. Courtois," *ibid.,* 311–318. Gay-Lussac, who repeated and extended Courtois's work, was responsible for this paper. Courtois himself published nothing. These two articles were immediately followed by short contributions of Gay-Lussac and Humphry Davy on the new element: "Note sur la combinaison de l'iode avec l'oxigène," *ibid.,* 319–321; and "Lettre de M. Humphry Davy sur la nouvelle substance découverte par M. Courtois, dans le sel de Vareck," *ibid.,* 322–329. Gay-Lussac's views (none too flattering to Davy) on the history of the discovery of iodine appeared in his "Mémoire sur l'iode," *ibid.,* **91** (1814), 5–160.

There is a biography of Courtois by L. G. Toraude, *Bernard Courtois, 1777–1838, et la découverte de l'iode, 1811* (Paris, 1921). Among the brief notices on Courtois, the most informative are Paul-Antoine Cap, "Notes historiques sur Bernard Courtois et sur la découverte de l'iode," in *Journal de pharmacie et de chimie,* **20** (1851), 131–138; Paul Richter, "Über die Entdeckung des Jod und ihre Vorgeschichte," in *Archiv für die Geschichte der Naturwissenschaften und der Technik,* **4** (1907), 1–7; and Mary Elvira Weeks, in *Discovery of the Elements,* 7th ed., rev., new material added by Henry M. Leicester (Easton, Pa., 1968), pp. 708–712.

Albert B. Costa

COUTURAT, LOUIS (*b.* Paris, France, 17 January 1868; *d.* between Ris-Orangis and Melun, France, 3 August 1914), *logic, mathematical philosophy, linguistics.*

From early childhood Couturat displayed an exceptional mixture of intellectual and artistic talent, and at the *lycée* his precocity brought him many prizes. He was to become a master of ancient literature, as well as an outstanding critic in the logic of theoretical and applied sciences. Logic was his basic concern, and even his writings on aesthetics show his preoccupation with logical foundations.

When not yet twenty-two Couturat was honored with the *lauréat du concours général* in philosophy and in science. During his fourth year at the École Normale Supérieure he studied mathematics under Jules Tannery and then continued under Picard and Jordan, also taking courses with Poincaré. He received his licentiate in mathematics on 25 July 1892. Thus prepared to handle problems in the philosophy of science, he published a paper on the paradox of Achilles and the tortoise in the *Revue philosophique.*

His Latin thesis for the doctorate was a scientific study of the Platonic myths in the *Dialogues.* For the French thesis he devoted himself to a study of the mathematical infinite. Couturat finished both theses by 12 May 1894, while serving at Toulouse as lecturer on Lucretius and Plato. He defended them at the Sorbonne in June 1896 and again was awarded top honors. In *De l'infini mathématique* he brought to metaphysicians and logicians the theories of the then new mathematics. His treatment of basic concepts served to invalidate the Kantian antinomies, for in the treatment of number and of continuity Couturat adopted a Cantorian stance with respect to an infinite that is defined with logico-mathematical precision. He maintained that a true metaphysics can be founded exclusively on reason. In *De mythis Platonicis,* he showed that the set of mythical passages does not represent the real thought of Plato as represented in the dialectical passages.

A leave of absence enabled Couturat to continue his scientific studies in Paris, where he audited the lectures of Edmond Bouty and Victor Robin. He was called to the University of Caen on 27 October 1897, to lecture on mathematical philosophy.

In October 1899 he returned to Paris on a second leave of absence for research on Leibniz' logic, which, in the various editions, had appeared in fragmentary form only. Couturat believed that Leibniz' metaphysics was a unique product of his logical principles. While in Hannover in 1900–1901 he had access at last to the unpublished works of Leibniz in the Royal Library. His researches resulted in the publication of *La logique de Leibniz* and another volume of more than 200 new Leibnizian fragments, *Opuscules et fragments inédits de Leibniz,* on which he based his theory of Leibniz' logic.

The Leibniz studies brought Couturat into contact with Bertrand Russell and led to his influential edition (1905) of Russell's *Principia mathematica,* with analytical commentary on contemporary works on the subject. Bergson then chose Couturat as his assistant

in the history of logic at the Collège de France (1905–1906).

Influenced by Leibniz' thoughts on the construction of a logical universal language, Couturat became a prime mover in the development of an auxiliary international language. On 1 October 1907 delegates from 310 societies throughout the world met and elected a committee to modify Esperanto. Couturat and Léau were the secretaries. With the collaboration of the Akademie di la Lingue Internaciona Ido, created in 1908, Couturat constructed the complete vocabulary of Ido, a language derived from Esperanto with reforms growing out of scientific linguistic principles. Couturat stood firmly for the application of his own logical principles, despite opposition from many quarters to changes in the already established forms of Esperanto.

Couturat never completed this work. At the age of forty-six and at the height of his intellectual power, he was killed while en route from Ris-Orangis to Melun on the very day Germany declared war on France. A twist of fate brought the speeding automobile carrying the French orders for mobilization into collision with the carriage in which Couturat, a noted pacifist, was riding.

BIBLIOGRAPHY

I. ORIGINAL WORKS. Couturat's books include *De l'infini mathématique* (Paris, 1896); *De mythis Platonicis* (Paris, 1896; Hildesheim, 1961); *La logique de Leibniz* (Paris, 1901); *Opuscules et fragments inédits de Leibniz* (Paris, 1903); *Histoire de la langue universelle* (Paris, 1903), written with Léopold Léau; *L'algèbre de la logique* (Paris, 1905, 1914; Hildesheim, 1965); *Les principes des mathématiques* (Paris, 1905; Hildesheim, 1965), *Étude sur la dérivation en Esperanto* (Coulommiers, 1907); *Les nouvelles langues internationales* (Paris, 1908), sequel to the *Histoire;* and *Étude sur la dérivation dans la langue internationale* (Paris, 1910).

His interest in languages is shown in *Dictionnaire internationale-français* (Paris, 1908), in collaboration with L. de Beaufront; *International-English Dictionary, English-International Dictionary* (London, 1908), in collaboration with L. de Beaufront and P. D. Hugon; *International-deutsches Wörterbuch, Deutsch-internationales Wörterbuch* (Stuttgart, 1908), in collaboration with L. de Beaufront and R. Thomann; *Internaciona matematikal lexiko, en ido, germana, angla, franca ed italiana* (Jena, 1910); and *Dictionnaire français-international* (Paris, 1915), in collaboration with L. de Beaufront.

A complete bibliography of his numerous papers is given in André Lalande, "L'oeuvre de Louis Couturat," in *Revue de métaphysique et de morale,* **22,** supp. (Sept. 1914), 644–688.

II. Secondary Literature. Additional information is in Louis Benaerts, "Louis Couturat," in *Annuaire de l'Association amicale de secours des anciens élèves de l'École normale supérieure* (Paris, 1915); Robert Blanché, "Couturat," in *Encyclopedia of Philosophy* (New York, 1967), II, 248–249; and Ernst Cassirer, "Kant und die moderne Mathematik. Mit Bezug auf Russells und Couturats Werke über die Prinzipien der Mathematik," in *Kantstudien* (1907).

Carolyn Eisele

COWELL, PHILIP HERBERT (*b.* Calcutta, India, 7 August 1870; *d.* Aldeburgh, Suffolk, England, 6 June 1949), *dynamical astronomy*.

The second son of Herbert Cowell, barrister, and of the former Alice Garrett, third daughter of Newson Garrett of Aldeburgh, Cowell had a wide cultural background. After showing unusual mathematical ability at Eton, he went to Trinity College, Cambridge, with an entrance scholarship. He graduated as senior wrangler in 1892 and was awarded the Isaac Newton Studentship in 1894, in which year he was elected fellow of Trinity. In 1896 Cowell was appointed chief assistant at the Royal Observatory at Greenwich, and in 1910 he became superintendent of the *Nautical Almanac*.

Because of the large perturbations by the sun and the extreme complexity of those by the planets, the development of a theory of the motion of the moon is extremely difficult. Cowell, using powerful mathematical and computational techniques, made important contributions to three distinct aspects: the theoretical development of the motion of the node of the orbit; the analysis of observations, from which he deduced the essential correctness, but inadequacy, of the current theories; and the determination of the secular acceleration of the mean longitude from a study of the records of ancient eclipses. For this work he was elected a fellow of the Royal Society in 1906.

Cowell's name is perpetuated in "Cowell's method" of step-by-step numerical integration for the solution of the relatively simple differential equations in rectangular coordinates defining the motions of bodies under their mutual gravitational attraction. He first applied the method to the newly discovered eighth satellite of Jupiter; then, with A. C. D. Crommelin, he used it brilliantly to predict the return of Halley's comet in 1910.

Cowell did not realize his full potential. Neither of his posts provided him with the scope for theoretical research that a Cambridge professorship, which eluded him, would have done; in later years he was frustrated and devoted himself more and more to other pursuits.

BIBLIOGRAPHY

Some fifty of Cowell's research papers are published in *Monthly Notices of the Royal Astronomical Society*, **64–70** (1903–1910). The "Investigation of the Motion of Halley's Comet From 1759 to 1910" is an appendix to *Greenwich Observations* for 1909.

An obituary is Edmund T. Whittaker, in *Obituary Notices of Fellows of the Royal Society of London*, **7** (1949), 375–384.

D. H. Sadler

CRABTREE, WILLIAM (*b.* Broughton, England, June 1610; *d.* Broughton, July 1644 [?]), *astronomy*.

The son of John Crabtree, a Lancashire farmer of comfortable means, and the former Isabel Pendleton, Crabtree established himself as a successful clothier or merchant at Manchester and married Elizabeth Pendleton in 1633. His interest in astronomy led him to undertake a series of precise observations, only a small fraction of which were published some time after his death, and to maintain an active correspondence, most of it now lost, with Samuel Foster, professor of astronomy at Gresham College, London, and the young astronomers Jeremiah Horrocks, William Gascoigne, and Christopher Towneley.

Crabtree's observations convinced him that, despite their errors, Kepler's *Rudolphine Tables* were the best extant; and he became one of the earliest converts to Kepler's new astronomy, accepting the latter's elliptical orbits and concurring in his call for the creation of a celestial physics. By 1637, within a year after they began to correspond, Crabtree convinced Horrocks of the superiority of the Keplerian system over those employing circular components. Both men made many corrections in Kepler's tables, bringing them into better agreement with observation; and Crabtree converted them to decimal form.

As a result of these revisions, Horrocks was able to predict and observe the transit of Venus of 4 December 1639, the first such observation ever made. Having been notified by Horrocks about a month before of the impending transit, Crabtree was the only other astronomer to observe it. Although he failed to record his data as precisely as Horrocks, their general observations were in agreement; and Horrocks obtained improved figures for the diameter of Venus, the elements of its orbit, and the distance of the earth from the sun. Although a capable astronomer in his own right, Crabtree is remembered chiefly for his influence on and relationship to Horrocks.

BIBLIOGRAPHY

I. Original Works. Some of Crabtree's observations were published in Horrocks' *Opera posthuma,* John Wallis, ed. (London, 1672, 1673, 1678), pp. 405–439; and John Flamsteed, *Historia coelestis Britannicae* (London, 1725), I, 4. Extracts from his letters to Gascoigne are printed in William Derham, "Observations Upon the Spots . . . Upon the Sun," in *Philosophical Transactions of the Royal Society,* **27** (1711), 280–290; and "Extracts From Mr. Gascoigne's and Mr. Crabtrie's Letters," *ibid.,* **30** (1717), 603–610.

II. SECONDARY LITERATURE. Horrocks, *Opera posthuma,* pp. 347[247]–338 has extracts of Horrocks' letters to Crabtree. See also Horrocks' *The Transit of Venus Across the Sun,* trans. with "A Memoir of His Life and Labours" by Arundell B. Whatton (London, 1859), *passim;* and John E. Bailey, "Jeremiah Horrox and William Crabtree, Observers of the Transit of Venus, 24 Nov., 1639," in *Palatine Notebook,* **2** (1882), 253–266.

WILBUR APPLEBAUM

CRAIG, JOHN (*b.* Scotland, second half of seventeenth century; *d.* London, England, 1731), *mathematics.*

Little is known of Craig's early life; even the place of his birth is not known with certainty. He was a pupil of David Gregory, who in 1683 had succeeded his uncle, James Gregory, as professor of mathematics at Edinburgh. Most of his life, however, was spent in Cambridge, where he attracted the notice of Newton. He maintained an extensive correspondence with many Scottish mathematicians, including Gregory, the noted mathematician and astronomer Colin Campbell, and later Colin Maclaurin.

Craig lived in an age that was witnessing spectacular advances in the development of mathematics. The Royal Society, of which Craig was elected a fellow in 1711, had already, under the guidance of Newton, established itself as one of the foremost scientific societies in Europe; and its members included many who were to leave their mark upon the progress of mathematics. Living in an age of such intellectual giants, Craig was rarely able to tower above his contemporaries; this is scarcely to be wondered at when it is recalled that they included Leibniz, Johann I and Jakob I Bernoulli, Halley, Moivre, Hooke, and Cotes.

Nevertheless, Craig was unusually gifted, and his writings covered a wide range. He had been received into holy orders, becoming in 1708 prebendary of Salisbury; and he made contributions of value to his adopted profession. It is, however, for his contributions to mathematics that he deserves to be remembered.

Of the vast fields that were thrown open to mathematicians at the close of the seventeenth century, none proved richer than the newly invented calculus; and it was to the extension and application of this that the mathematicians of the period directed their attention. Newton had outlined his discovery in three tracts, the first of which, *De analysi per aequationes numero infinitas,* although it did not appear until 1711, was compiled as early as 1669, and was already known to a number of his contemporaries. Meanwhile, Leibniz had contributed to the *Acta eruditorum* for October 1684 his famous paper "Nova methodus pro maximis et minimis, itemque tangentibus . . . et singulare pro illis calculi genus." For a time the new methods appear to have made surprisingly little impact upon English mathematicians, possibly because when Newton's monumental *Principia* first appeared (1687), there was scarcely any mention of the calculus in its pages; thus, it might well be thought that the calculus was not really necessary. On the Continent, however, Leibniz' great friends, the Bernoullis, lost no opportunity of exploring the new methods. Of the few Englishmen who realized the vast possibilities of the tool that had been placed in their hands, none showed greater zeal than did Craig.

Apart from a number of contributions to the *Philosophical Transactions of the Royal Society,* Craig compiled three major works (the titles are translated):

(1) "Method of Determining the Quadratures of Figures Bounded by Curves and Straight Lines" (1685). In this work Craig paid tribute to the work of Barrow, Newton, and Leibniz. Of great importance is the fact that its pages contain the earliest examples in England of the Leibnizian notation, dy and dx, in place of the "dot" notation of Newton.

(2) "Mathematical Treatise on the Quadratures of Curvilinear Figures" (1693). Here the symbol of integration $\int$ appears.

(3) "On the Calculus of Fluents" (1718), with a supplement, "De optica analytica." Apart from its importance, this work is particularly interesting because on the first page of its preface Craig gives an account of the steps that led to his interest in the fluxional calculus. Translated, it reads:

You have here, kindly reader, my thoughts about the calculus of fluents. About the year 1685, when I was a young man I pondered over the first elements of this. I was then living in Cambridge, and I asked the celebrated Mr. Newton if he would kindly look over them before I committed them to the press. This he willingly did, and to corroborate some objections raised in my pages against D. D. T. [Tschirnhausen] he offered me of his own accord the quadratures of two figures; these were the curves whose equations were $m^2y^2 = x^4 + a^2x^2$, and $my^2 = x^3 + ax^2$. He also informed me that he could exhibit innumerable curves of this kind, which, by breaking off under given conditions, afforded a geometrical squaring of the figures proposed. Later, on returning to my fatherland I became very friendly with Mr. Pitcairne, the celebrated physician, and with Mr. Gregory, to whom I signified that Mr. Newton had a series of such a kind for quadratures, and each of them admitted it to be quite new.

In addition to the above works, Craig contributed a number of papers to the *Philosophical Transactions of the Royal Society*. The titles of the most important of these, translated into English, are (1) "The Quadrature of the Logarithmic Curve" (1698), (2) "On the Curve of Quickest Descent" (1700), (3) "On the Solid of Least Resistance" (1700), (4) "General Method of Determining the Quadrature of Figures" (1703), (5) "Solution of Bernoulli's Problem on Curvature" (1704), (6) "On the Length of Curved Lines" (1708), and (7) "Method of Making Logarithms" (1710).

This is an impressive list and one that bears eloquent testimony to the range and variety of Craig's interests. Nevertheless, he has fared ill at the hands of the historians of mathematics—particularly in his own country—few of whom even mention him and still fewer of whom make any attempt to assess the value of his contributions. French and German historians have treated him more generously.

BIBLIOGRAPHY

An exhaustive account of Craig's contributions to mathematics can be found in Moritz Cantor, *Vorlesungen über Geschichte der Mathematik*, III (Leipzig, 1896), *passim*, esp. pp. 52, 188; and J. P. Montucla, *Histoire des mathématiques*, II (Paris, 1799), 162.

J. F. SCOTT

CRAMER, GABRIEL (*b.* Geneva, Switzerland, 31 July 1704; *d.* Bagnols-sur-Cèze, France, 4 January 1752), *geometry, probability theory.*

Gabriel was one of three sons born to Jean Isaac Cramer, whose family had moved from Holstein to Strasbourg to Geneva in the seventeenth century, and his wife, Anne Mallet. The father and one son, Jean-Antoine, practiced medicine in Geneva. The other two sons, Jean and Gabriel, were professors of law and of mathematics and philosophy, respectively. All three sons were also active in local governmental affairs.

Gabriel Cramer was educated in Geneva and at the age of eighteen defended a thesis dealing with sound. At twenty he competed for the chair of philosophy at the Académie de Calvin in Geneva. The chair was awarded to the oldest of the three contestants, Amédée de la Rive; but the magistrates making the award felt that it was important to attach to their academy two such able young men as Cramer and Giovanni Ludovico Calandrini, the other contestant, who was twenty-one. To do this they split off a chair of mathematics from philosophy and appointed both

young contestants to it. This appointment provided that the men share both the position's duties and its salary. It was also provided that they might take turns traveling for two or three years "to perfect their knowledge," provided the one who remained in Geneva performed all the duties and received all the pay. Calandrini and Cramer, called Castor and Pollux by their friends, secured permission for the innovation of using French rather than Latin, not for courses *ex cathedra* but for recitations, "in order that persons who had a taste for these sciences but no Latin could profit." Calandrini taught algebra and astronomy; Cramer, geometry and mechanics. In 1734 Calandrini was made professor of philosophy and Cramer received the chair of mathematics. In 1750 he was made professor of philosophy when Calandrini entered the government.

Cramer's interests and activities were broad, both academically and in the daily life of his city. He wrote on such topics as the usefulness of philosophy in governing a state and the added reliance that a judge should place on the testimony of two or three witnesses as compared with one. He wrote against the popular idea that wheat sometimes changed to tares and also produced several notes on the history of mathematics. As a citizen Cramer was a member of the Conseil des Deux-Cents (1734) and Conseil des Soixante (1749) and was involved with artillery and fortification. He instructed workers repairing a cathedral and occupied himself with excavations and the search of archives. He was reported to be friendly, good-humored, pleasant in voice and appearance, and possessed of good memory, judgment, and health. He never married.

The encouragement to travel played an important role in Cramer's life. From 1727 to 1729 he traveled, going first to Basel, where he spent five months with Johann I Bernoulli and his students, including Daniel Bernoulli and Leonhard Euler. From Basel he went to England, Leiden, Paris, meeting Nicholas Saunderson, Christopher Middleton, Halley, Sloane, Moivre, James Stirling, s'Gravesande, Fontenelle, Réaumur, Maupertuis, Buffon, Clairaut, and Mairan. In 1747 Cramer visited Paris again with the young prince of Saxe-Gotha, whom he had taught for two years. During the trip he was invited to salons frequented by Réaumur, d'Alembert, and Fontenelle. The friendship with the Bernoullis, formed during the first trip, led to much of Cramer's later editorial work, and the acquaintanceships formed during his travels produced an extended correspondence in which he served as an intermediary for the spread of problems and as a contributor of questions and ideas.

Cramer's major publication, *Introduction à l'analyse*

des lignes courbes algébriques, was published in 1750. During the previous decade he had edited the collected works of Johann I and Jakob I Bernoulli, Christian Wolff's five-volume *Elementa,* and two volumes of correspondence between Johann I Bernoulli and Leibniz. Overwork and a fall from a carriage brought on a decline in health that resulted in his being bedridden for two months. The doctor then prescribed a rest in southern France. Cramer left Geneva on 21 December 1751 and died while traveling.

Cramer received many honors, including membership in the Royal Society of London; the academies of Berlin, Lyons, Montpellier; and the Institute of Bologna. In 1730 he was a contestant for the prize offered by the Paris Academy for a reply to the question "Quelle est la cause de la figure elliptique des planètes et de la mobilité de leur aphélies?" He was the runner-up (*premier accessit*) to Johann I Bernoulli.

This last fact is perhaps typical of Cramer's status in the history of science. He was overshadowed in both mathematics and philosophy by his contemporaries and correspondents. He is best-known for Cramer's rule and Cramer's paradox, which were neither original with him nor completely delineated by him, although he did make contributions to both. His most original contributions are less well-known: the general content and organization of his book on algebraic curves and his concept of mathematical utility.

In the preface to the *Introduction à l'analyse,* Cramer cites Newton's *Enumeration of Curves of the Third Order,* with the commentary by Stirling, as an "excellent model" for the study of curves. He comments particularly on Newton's use of infinite series and of a parallelogram arrangement of the terms of an algebraic equation of degree v in two unknowns. He also refers to a paper by Nicole and one on lines of the fourth order by Christophe de Bragelonne. Cramer gives credit to Abbé Jean Paul de Gua de Malves for making Newton's parallelogram into a triangular arrangement in the book *L'usage de l'analyse de Descartes pour découvrir . . . les propriétés des lignes géométriques de tous les ordres* (1740).

Cramer also says that he would have found Euler's *Introductio in analysin infinitorum* (1748) very useful if he had known of it earlier. That he made little use of Euler's work is supported by the rather surprising fact that throughout his book Cramer makes essentially no use of the infinitesimal calculus in either Leibniz' or Newton's form, although he deals with such topics as tangents, maxima and minima, and curvature, and cites Maclaurin and Taylor in footnotes. One conjectures that he never accepted or mastered the calculus.

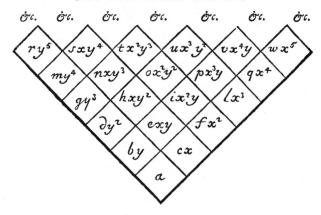

DES DIFFERENS ORDRES

FIGURE 1

The first chapter of the *Introduction* defines regular, irregular, transcendental, mechanical, and irrational curves and discusses some techniques of graphing, including our present convention for the positive directions on coordinate axes. The second chapter deals with transformations of curves, especially those which simplify their equations, and the third chapter develops a classification of algebraic curves by order or degree, abandoning Descartes's classification by genera. Both Cramer's rule and Cramer's paradox develop out of this chapter. The remaining ten chapters include discussions of the graphical solution of equations, diameters, branch points and singular points, tangents, points of inflection, maxima, minima, and curvature. Cramer claims that he gives no example without a reason, and no rule without an example.

The third chapter of Cramer's *Introduction* uses a triangular arrangement of the terms of complete equations of successively higher degree (see Fig. 1) as the basis for deriving the formula $v^2/2 + 3v/2$ for the number of arbitrary constants in the general equation of the vth degree. This is the sum of v terms of the arithmetic progression $2 + 3 + 4 + \cdots$ derived from the rows of the triangle by regarding one coefficient, say a, as reduced to unity by division. From this he concludes that a curve of order v can be made to pass through $v^2/2 + 3v/2$ points, a statement that he says needs only an example for a demonstration. In his example Cramer writes five linear equations in five unknowns by substituting the coordinates of five points into the general second-degree equation. He then states that he has found a general and convenient rule for the solution of a set of v linear equations in v unknowns; but since this is algebra, he has put it into appendix 1. Figure 2 shows the first page of this appendix. The use of raised numerals as indices, not exponents, applied to co-

Soient pluſieurs inconnues z, y, x, v, &c. & autant d'équations

$$A^1 = Z^1z + Y^1y + X^1x + V^1v + \&c.$$
$$A^2 = Z^2z + Y^2y + X^2x + V^2v + \&c.$$
$$A^3 = Z^3z + Y^3y + X^3x + V^3v + \&c.$$
$$A^4 = Z^4z + Y^4y + X^4x + V^4v + \&c.$$
$$\&c.$$

où les lettres A^1, A^2, A^3, A^4, &c. ne marquent pas, comme à l'ordinaire, les puiſſances d'A, mais le prémier membre, ſuppoſé connu, de la prémiére, ſeconde, troiſiéme, quatriéme &c. équation. De même Z^1, Z^2, &c. ſont les coëfficients de z; Y^1, Y^2, &c. ceux de y; X^1, X^2, &c. ceux de x; V^1, V^2, &c. ceux de v; &c. dans la prémiére, ſeconde, &c. équation.

Cette Notation ſuppoſée, s'il n'y a qu'une équation & qu'une inconnue z; on aura $z = \dfrac{A^1}{Z^1}$. S'il y a deux équations & deux inconnues z & y; on trouvera $z = \dfrac{A^1Y^2 - A^2Y^1}{Z^1Y^2 - Z^2Y^1}$, & $y = \dfrac{Z^1A^2 - Z^2A^1}{Z^1Y^2 - Z^2Y^1}$. S'il y a trois équations & trois inconnues z, y, & x; on trouvera

$$z = \frac{A^1Y^2X^3 - A^1Y^3X^2 - A^2Y^1X^3 + A^2Y^3X^1 + A^3Y^1X^2 - A^3Y^2X^1}{Z^1Y^2X^3 - Z^1Y^3X^2 - Z^2Y^1X^3 + Z^2Y^3X^1 + Z^3Y^1X^2 - Z^3Y^2X^1}$$

$$y = \frac{Z^1A^2X^3 - Z^1A^3X^2 - Z^2A^1X^3 + Z^2A^3X^1 + Z^3A^1X^2 - Z^3A^2X^1}{Z^1Y^2X^3 - Z^1Y^3X^2 - Z^2Y^1X^3 + Z^2Y^3X^1 + Z^3Y^1X^2 - Z^3Y^2X^1}$$

$$x = \frac{Z^1Y^2A^3 - Z^1Y^3A^2 - Z^2Y^1A^3 + Z^2Y^3A^1 + Z^3Y^1A^2 - Z^3Y^2A^1}{Z^1Y^2X^3 - Z^1Y^3X^2 - Z^2Y^1X^3 + Z^2Y^3X^1 + Z^3Y^1X^2 - Z^3Y^2X^1}$$

FIGURE 2

efficients represented by capital letters enabled Cramer to state his rule in general terms and to define the signs of the products in terms of the number of inversions of these indices when the factors are arranged in alphabetical order.

Although Leibniz had suggested a method for solving systems of linear equations in a letter to L'Hospital in 1693, and centuries earlier the Chinese had used similar patterns in solving them, Cramer has been given priority in the publication of this rule. However, Boyer has shown recently that an equivalent rule was published in Maclaurin's *Treatise of Algebra* in 1748. He thinks that Cramer's superior notation explains why Maclaurin's statement of this rule was ignored even though his book was popular. Another reason may be that Euler's popular algebra text gave Cramer credit for this "très belle règle."

Cramer's paradox was the outgrowth of combining the formula $v^2/2 + 3v/2$ with the theorem, which Cramer attributes to Maclaurin, that mth- and nth-order curves intersect in mn points. The formula says, for example, that a cubic curve is *uniquely* determined by nine points; the theorem says that two *different* cubic curves would intersect in nine points. Cramer's explanation of the paradox was inadequate. Scott has shown that Maclaurin and Euler anticipated Cramer in formulating the paradox and has outlined later

explanations and extensions by Euler, Plücker, Clebsch, and others. Plücker's explanation, using his abridged notation, appeared in Gergonne's *Annales des mathématiques pures et appliquées* in 1828.

Cramer's work as an editor was significant in the preservation and dissemination of knowledge and reflects both the esteem in which he was held and the results of his early travel. Cantor says he was the first scholar worthy of the name to undertake the thankless task of editing the work of others. Johann I Bernoulli authorized Cramer to collect and publish his works, specifying that there should be no other edition. At the request of Johann, Cramer also produced a posthumous edition of the work of Jakob I Bernoulli, including some unpublished manuscripts and additional material needed to understand them. He also edited a work by Christian Wolff and the correspondence between Johann I Bernoulli and Leibniz.

Throughout his life Cramer carried on an extensive correspondence on mathematical and philosophical topics. His correspondents included Johann I Bernoulli, Charles Bonnet, Georges L. Leclerc, Buffon, Clairaut, Condillac, Moivre, Maclaurin, Maupertuis, and Réaumur. This list shows his range of interests and the acquaintanceships formed in his travels and is further evidence of his function as a stimulator and intermediary in the spread of ideas. For example, a letter from Cramer to Nikolaus I Bernoulli is cited by Savage as evidence of Cramer's priority in defining the concept of utility and proposing that it has upper and lower bounds. This concept is related to mathematical expectation and is a link between mathematical economics and probability theory. Cramer's interest in probability is further revealed in his correspondence with Moivre, in which he at times served as an intermediary between Moivre and various Bernoullis.

Cramer was a proposer of stimulating problems. The Castillon problem is sometimes called the Castillon-Cramer problem. Cramer proposed to Castillon (also called Castiglione, after his birthplace) that the problem of Pappus, to inscribe in a circle a triangle such that the sides pass through three given collinear points, be freed of the collinearity restriction. Castillon's solution was published in 1776. Since then analytic solutions have been presented and the problem has been generalized from a triangle to polygons and from a circle to a conic section.

Gabriel Cramer deserves to have his name preserved in the history of mathematics even though he was outshone by more able and single-minded contemporary mathematicians. He himself would probably have accepted the rule, if not the paradox, as meriting his name but would regret that his major work is less well-known than it merits.

BIBLIOGRAPHY

I. ORIGINAL WORKS. Cramer's chief and only published book is *Introduction à l'analyse des lignes courbes algébriques* (Geneva, 1750). Speziali (see below), pp. 9, 10, refers to two unpublished manuscripts: "Éléments d'arithmétique," written for the young prince of Saxe-Gotha, and "Cours de logique," which was sold in London in 1945 under the mistaken idea that it was by Rousseau.

Two source books have excerpts from Cramer's *Introduction* that present Cramer's rule and Cramer's paradox, respectively: Henrietta O. Midonick, *The Treasury of Mathematics* (New York, 1965), pp. 269–279; and D. J. Struik, *A Source Book in Mathematics* (Cambridge, Mass., 1969), pp. 180–183.

Cramer's minor articles, published chiefly in the *Mémoires* of the academies of Paris (1732) and Berlin (1748, 1750, 1752), include two on geometric problems, four on the history of mathematics, and others on such scattered topics as the aurora borealis (in the *Philosophical Transactions of the Royal Society*), law and philosophy, and the date of Easter. These are listed by Isely, Le Roy, and Speziali.

As editor, Cramer was responsible for the following: Johann I Bernoulli, *Opera omnia,* 4 vols. (Lausanne-Geneva, 1742); Jakob I Bernoulli, *Opera,* 2 vols. (Geneva, 1744), which omits *Ars conjectandi; Virorum celeberr Got. Gul. Leibnitii et Johan. Bernoullii commercium philosophicum et mathematicum* (Lausanne-Geneva, 1745), edited with Castillon; and Christian Wolff, *Elementa matheseos universae,* new ed., 5 vols. (1743–1752). Cantor and Le Roy also give Cramer credit for the 1732–1741 edition.

There is a portrait of Cramer by Gardelle at the Bibliothèque de Genève as well as a collection of 146 letters, many unpublished, according to Speziali. Many of Cramer's letters may be found published in the works of his correspondents. His correspondence with Moivre is listed in Ino Schneider, "Der Mathematiker Abraham de Moivre (1667–1754)," in *Archive for History of Exact Sciences,* **5** (1968/1969), 177–317.

II. SECONDARY LITERATURE. The best and most recent account of Cramer's life and works is M. Pierre Speziali, *Gabriel Cramer (1704–1752) et ses correspondants,* Conférences du Palais de la Découverte, ser. D, no. 59 (Paris, 1959). A short account of his life is to be found in Georges Le Roy, *Condillac, lettres inédites à Gabriel Cramer* (Paris, 1953). L. Isely, *Histoire des sciences mathématiques dans la Suisse française* (Neuchâtel, 1901), gives an extended account of Cramer, beginning on p. 126.

Probably the best account of Cramer's *Introduction* is that in Carl B. Boyer, *History of Analytic Geometry* (New York, 1956), pp. 194–197. Extended discussions are also to be found in Moritz Cantor, *Vorlesungen über Geschichte der Mathematik,* 2nd ed. (Leipzig, 1901), pp. 605–609, 819–842; and Gino Loria, *Storia delle matematiche* (Milan, 1950), pp. 739–741.

See also Carl B. Boyer, "Colin Maclaurin and Cramer's Rule," in *Scripta mathematica,* **27** (Jan. 1966), 377–379; Leonard J. Savage, *The Foundations of Statistics* (New York, 1954), pp. 81, 92–95; and Charlotte Angas Scott, "On the Intersections of Plane Curves," in *Bulletin of the American Mathematical Society,* **4** (1897–1898), 260–273.

PHILLIP S. JONES

CREDNER, (KARL) HERMANN GEORG (*b.* Gotha, Germany, 1 October 1841; *d.* Leipzig, Germany, 21 July 1913), *geology.*

Credner was the son of K. F. H. Credner, a geologist and privy councillor. He studied mining engineering in Clausthal but soon changed his field to geology. He earned the Ph.D. at Göttingen in 1864 with a dissertation on a paleontological problem.

He worked in North America—especially as an expert on gold mining—for about four years and published papers on the geology of the New York and New Brunswick areas. His shorter papers, mostly geographical, dealt with the area to the west of Lake Superior and with the copper deposits of the Keweenaw peninsula. By foot and on horseback he traversed great parts of Missouri, Illinois, Connecticut, and Pennsylvania, as well as New Brunswick and Nova Scotia. He was encouraged in his work by J. D. Dana.

Soon after his return to Germany he succeeded C. F. Naumann in the chair of geology and paleontology at the University of Leipzig. Ferdinand Zirkel was appointed to the chair of mineralogy at the same time. Credner's Leipzig lectures were the basis of his book *Elemente der Geologie,* which was the leading geology text in Germany for some thirty-five years. He was one of the first scientists to recognize the consequences of the European ice ages, while at the same time he approved F. P. W. von Richthofen's interpretation of loess. He was director of the Geologische Landesanstalt; under his guidance a map of the Saxonian granulite mountains, on a scale of 1:100,000, was published, as were many other geological maps of Saxony. At the end of his career he was able to present a complete cumulative survey map of Saxony on the scale of 1:250,000.

Credner did intensive work on Saxon Stegocephalia and was able to classify a series of them as being totally amphibian.

In addition to his academic service, Credner took part in most of the international geological congresses. He was thus personally known to geologists all over the world, who often solicited his advice.

BIBLIOGRAPHY

In addition to the *Elemente der Geologie* (Leipzig, 1871), Credner published treatises, maps, and articles—see especially those in Leipzig journals, listed by Poggendorff.

P. RAMDOHR

CREIGHTON, CHARLES (*b.* Peterhead, Aberdeenshire, Scotland, 22 November 1847; *d.* Upper Boddington, Northamptonshire, England, 18 July 1927), *medical history.*

Creighton, one of Britain's most learned medical historians, incurred the disdain of contemporary physicians by denouncing Jennerian vaccination and by disputing the germ theory of infectious diseases.

Born in a seaport town, Charles was the eldest son and second child of Alexander Creighton, timber merchant, and his wife Agnes Brand, who had five sons and three daughters. He attended the local school and went on to grammar school in Old Aberdeen. In 1864, he won a bursary to King's College, Aberdeen, where he obtained his M.A. in 1867. Enrolling as a medical student at the affiliated Marischal College, he took clinical courses at Edinburgh, and in 1871 passed his M.B. and M.S. examinations at Aberdeen. The university awarded him the M.D. degree in 1878.

After graduation, Creighton studied for a year in Vienna under Karl von Rokitansky and in Berlin under Rudolf Virchow. On his return in 1872, he obtained successive annual appointments as surgical registrar at St. Thomas' Hospital and medical registrar at Charing Cross Hospital, London. He also did part-time research on cancer for the Local Government Board, for whom he worked full-time in 1875 on the physiology and pathology of the breast. These studies, conducted at the Brown Institution, were the subject of his first publications.

Appointed demonstrator of anatomy in the University of Cambridge in 1876, within five years he published the book *Bovine Tuberculosis in Man* and eleven articles on normal and pathological anatomy in the *Journal of Anatomy and Physiology,* of which he became a coeditor in 1879. In 1881 he unaccountably severed these promising academic associations and went to London. After applying unsuccessfully in 1882 for the chair of pathology at Aberdeen, Creighton assumed the mantle of a dedicated, erudite scholar. He lived and worked alone for the next thirty-seven years.

Apart from a three-month visit to India in 1904 to investigate the plague (financed by the Leigh Browne Trust, founded in 1884 "for the promotion of original research in the biological sciences without any recourse to experiments upon living animals of a nature to cause pain"), he resided within walking distance of the British Museum, whose resources were indispensable to him. In 1918 he bought a small house in a Northamptonshire village and lived peacefully there until his death from a cerebral hemorrhage nine years later.

Tall and handsome in his prime, Creighton was abstemious, devout, and fond of music. A kind, gentlemanly, self-contained individualist, he upheld his beliefs inflexibly, regardless of consequences. His dogmatic opinions lost him friends and he became a frugal recluse. His financial anxieties were relieved in later life by a small civil pension granted by Prime Minister Asquith.

Creighton's three-volume translation of August Hirsch's *Handbuch der historisch-geographischen Pathologie,* an outstanding accomplishment, appeared between 1883 and 1886. The next several years were devoted to the great task of compiling *A History of Epidemics in Britain,* whose two volumes, published in 1891 and 1894, earned him lasting distinction. During this period, Creighton also wrote numerous articles for the *Dictionary of National Biography,* besides making notable contributions on medicine and public health to H. D. Traill's *Social England.*

His industry and judgment were not always soundly exercised. A comprehensive article on pathology (1885), commissioned for the ninth edition of the *Encyclopedia Britannica,* cast doubt upon the existence of pathogenic bacteria; another on vaccination (1888) was so reactionary and misleading that it provoked vigorous protests from leading medical journals. Especially condemned was the apparent claim, implicit also in his book *The Natural History of Cowpox and Vaccinal Syphilis* (1887), that vaccination and infantile syphilis were related. Creighton denied this allegation but further blemished his reputation by publishing another polemical volume, *Jenner and Vaccination* (1889).

He maintained an active interest in pathology. For about a decade beginning around 1898, he visited the London Hospital, where his closest friend and fellow-Aberdonian, the bacteriologist William Bulloch, provided access to pathological specimens and records. Creighton prepared and examined the specimens microscopically at home. His subsequent publications on tuberculosis and cancer were unorthodox and had no impact. Likewise his literary studies on Shakespeare gained no following.

Creighton's obdurate rejection of bacteria as causal agents of such diseases as tuberculosis and plague was probably instigated by Rokitansky's humoral theory of pathology and Virchow's early skepticism about the bacteriologists' budding claims. Yet he clung perversely to the least defensible features of these doctrines long after their proponents had modified them. A deep-seated faith in miasmata, soil poisons, and seismic disturbances as instigators of epidemics, which doubtless arose from intimate contact with Hirsch's *Handbuch,* permeates the *History.* His accuracy in citation of rare chronicles is undisputed, although R.

S. Roberts (1968) claims he sometimes selected historical data that fitted favorite theories. The final keys to Creighton's controversial career and anachronistic beliefs are concealed within his enigmatic personality.

Despite many peculiarities, Creighton commands respect for his self-sacrificing industry, rare scholarly insights, linguistic talents, splendid literary style, and especially for his chief work, which will long remain a unique source book on the interrelations of epidemic diseases and social history.

BIBLIOGRAPHY

I. ORIGINAL WORKS. Incomplete bibliographies, listing only Creighton's chief works, have been provided by W. Bulloch and by E. A. Underwood (see below). Among his more noteworthy publications are the following: "Anatomical Research Towards the Aetiology of Cancer," in *Reports of the Medical Officer of the Privy Council and Local Government Board,* no. 4 of n.s. 3 (1874), 95–112; "On the Development of the Mamma and of the Mammary Function," in *Journal of Anatomy and Physiology,* **11** (1877), 1–32; *Contributions to the Physiology and Pathology of the Breast and its Lymphatic Glands* (London, 1878); *Bovine Tuberculosis in Man, an Account of the Pathology of Suspected Cases* (London, 1881); "On the Autonomous Life of the Specific Infections," in *British Medical Journal,* 2 (1883), 218–224; *Dr. Koch's Method of Cultivating the Micro-organism in Tubercle* (London, 1884); *Illustrations of Unconscious Memory in Disease Including a Theory of Alteratives* (London, 1886); *The Natural History of Cow-pox and Vaccinal Syphilis* (London, 1887); *Jenner and Vaccination; a Strange Chapter of Medical History* (London, 1889); *Handbook of Geographical and Historical Pathology,* trans. from 2nd German ed. of Dr. August Hirsch's *Handbuch der historisch-geographischen Pathologie,* 3 vols. (London, 1883–1886); *A History of Epidemics in Britain from A.D. 664 to the Extinction of Plague* (Cambridge, 1891; repr. London, 1965); *A History of Epidemics in Britain, from the Extinction of the Plague to the Present Time* (Cambridge, 1894; repr. London, 1965); *Microscopic Researches on the Formative Property of Glycogen,* 2 vols. (London, 1896–1899); "Plague in India," in *Journal of the Society of Arts,* **53** (1905), 810–827; *Contributions to the Physiological Theory of Tuberculosis* (London, 1908); *Some Conclusions on Cancer* (London, 1920).

Creighton contributed articles on malaria, medicine (synoptical view), Morgagni, pathology, pellagra, and vaccination for the 9th ed. of the *Encyclopedia Britannica* (1885–1888). He wrote forty-seven articles for the *Dictionary of National Biography* (before 1893) and several articles on medical subjects in *Janus.* He also contributed about 30,000 words to H. D. Traill's *Social England,* vols. I–IV (1893–1895). His first and chief Shakespearean work was *Shakespeare's Story of His Life* (London, 1904).

II. SECONDARY LITERATURE. Obituaries include W. Bulloch, "The Late Dr. Charles Creighton," in *Lancet* (1927),

2, 250–251; and "Charles Creighton, M.A., M.D.," in *Aberdeen University Review,* **15** (1928), 112–118; M. Greenwood, "Charles Creighton, M.D.," in *British Medical Journal* (1927), **2,** 240–241.

Other tributes are F. H. Garrison, "A Neglected Medical Scholar," editorial in *Bulletin of the New York Academy of Medicine,* **4** (1928), 469–476; E. A. Underwood, "Charles Creighton, M.A., M.D. (1847–1927): Scholar, Historian and Epidemiologist," in *Proceedings of the Royal Society of Medicine,* **41** (1947), 869–876; and "Charles Creighton, the Man and His Work," in *A History of Epidemics in Britain,* vol. I, repr. ed. (1965), 43–135; D. E. C. Eversley, "Epidemiology as Social History," *ibid.,* 3–39.

Various contemporary editorial reviews of Creighton's publications are cited by E. A. Underwood in a prolix account of his life and work prefacing the repr. ed. of *A History of Epidemics in Britain.* Specific reference is made here to only one editorial, headed "Vaccination Reviewed," *Lancet* (1888), **2,** 1027–1028. Creighton responded in an equivocal letter to the editors, headed "Infantile Syphilis and Vaccinations," *ibid.,* 1096–1097. A later appraisal of Creighton's views on vaccination is M. Greenwood, *Epidemics and Crowd Diseases* (London, 1935), pp. 245–273. A critical review of Creighton's epidemiological theories is R. S. Roberts, "Epidemics and Social History," in *Medical History,* **12** (1968), 305–316.

Useful references to the pathological doctrines of Rokitansky and Virchow are E. R. Ackerknecht, *Rudolf Virchow; Doctor, Statesman, Anthropologist* (Madison, Wis., 1953); L. J. Rather, "Virchow's Review of Rokitansky's *Handbuch* in the *Preussische Medizinal-Zeitung,* Dec. 1846," in *Clio Medica,* **4** (1969), 127–140.

CLAUDE E. DOLMAN

CRELL, LORENZ FLORENZ FRIEDRICH VON (*b.* Helmstedt, Germany, 21 January 1745; *d.* Göttingen, Germany, 7 June 1816), *chemistry.*

His father, Johann Friedrich Crell, professor of medicine at the duchy of Brunswick's university in Helmstedt, died in 1747. Consequently, Lorenz' early education was supervised by his maternal grandfather, Lorenz Heister, also a professor of medicine at Helmstedt and one of Germany's leading surgeons. Crell entered the local university in 1759 and took his M.D. there in 1768. He then spent two and a half years on a study tour to Strasbourg, Paris, Edinburgh, and London. Of the men that he encountered on this trip, William Cullen and Joseph Black, both at Edinburgh, influenced him most. Cullen's ideas about the causes and treatment of fevers fed Crell's desire for medical knowledge, while Black's instruction contributed to a growing interest in chemistry, an interest that his favorite professor at Helmstedt, the alchemist G. C. Beireis, had kindled.

In early 1771, soon after his return to Germany, Crell took a new chair of chemistry at the Collegium

Carolinum, a school for prospective officials in the town of Brunswick. Although medicine apparently remained his first interest during his three years there, he acquired a good knowledge of chemistry's nonmedical uses by teaching at the Collegium. While there he also became a Freemason, making numerous influential friends who assisted him throughout his career.

Crell returned to his native Helmstedt as professor of medical theory and materia medica in 1774. For a while he devoted his spare time to natural theology, which became a lifelong interest, and to short articles on materia medica, which he published in E. G. Baldinger's new medical periodical, *Magazin vor Aerzte*. In 1777, influenced by Baldinger's example, he decided to found and edit a journal for chemistry; it was to become the first successful discipline-oriented journal in science. Besides an almost unquenchable thirst for renown, Crell was motivated by a nationalistic desire to meet the foreign challenge to traditional German leadership in chemistry, the wish to serve the commonweal by spreading information about this useful science, and the hope of promoting intellectual progress by fostering the development of this fundamental discipline. The first volume of his *Chemisches Journal für die Freunde der Naturlehre, Arzneygelahrtheit, Haushaltungskunst und Manufacturen* appeared in 1778. It contained a long foreword in which Crell outlined his goals and invited contributions, a section of original treatises, written mostly by Crell, and a section of articles translated from the memoirs of scientific societies. The reviewers, impressed by Crell's research and aims, gave the journal a favorable reception. More important was the positive response by several young chemists to Crell's call for contributions.

Soon the journal was flourishing. By 1781 it was doing so well that Crell began quarterly publication under a new title that reflected his enthusiasm for the rapid pace of innovation in chemistry: *Die neuesten Entdeckungen in der Chemie*. In 1784 he switched to monthly publication, adopting a title similar to his original one: *Chemische Annalen für die Freunde der Naturlehre, Arzneygelahrtheit, Haushaltungskunst und Manufacturen*. In its first year Crell's monthly had fifty contributors and over 400 subscribers (mostly pharmacists, physicians, professors, and mining officials). Thus, within a few years, Crell had succeeded in creating a forum where German chemists exchanged their findings and views. By putting German chemists in close touch with one another, his journal united them into a German chemical community.

Crell's position in this community enabled him to exert a strong influence on the German reception of the various theoretical systems being developed to accommodate the new phenomena in gas chemistry. By 1784 he had sided with his correspondent R. Kirwan, who built on the notion that inflammable air (hydrogen) was phlogiston. Thanks largely to Crell's backing, Kirwan's views commanded a large following among German chemists in the late 1780's. Lavoisier's ideas received comparatively little attention. Crell did not yet recognize them as a threat to the phlogiston concept, of whose German origin he was very proud. In 1789, however, following the appearance of Lavoisier's *Traité* and the first German converts to the antiphlogistic doctrine, he began to encourage opposition to the French theory.

All his efforts were to no avail. By 1796 the antiphlogistonists had achieved a predominant position within the German chemical community. Crell, who continued his opposition to Lavoisier's theory, only discredited himself. In 1798 A. N. Scherer, a young antiphlogistonist, took advantage of the situation by founding the *Allgemeines Journal der Chemie*. Five years later, the prestigious editorial board consisting of S. F. Hermbstädt, M. H. Klaproth, J. B. Richter, J. B. Trommsdorff, and A. F. Gehlen (the editor in chief) assumed control of Scherer's journal, titling it the *Neues allgemeines Journal der Chemie*. Bested by this competition, Crell stopped publishing the *Chemische Annalen* in 1804 and became an inactive member of his rival's editorial board.

Crell served science well during much of his career as editor. Realizing this, his contemporaries granted him the recognition he so desired: by 1804 he held the title of mining councillor, belonged to the imperial nobility, and was a member of thirty-nine learned societies. His journal provided scientists in other disciplines and nations (for example, the editors of the *Annales de Chimie*) with an excellent model of how the trend to specialization could be served. More important, it enabled German chemists to develop a tradition of interacting as members of a discipline-oriented community, a tradition that Liebig subsequently put to very good use.

When the Napoleonic regime closed the university in Helmstedt in 1810, Crell was transferred to Göttingen. He died there in 1816.

BIBLIOGRAPHY

I. ORIGINAL WORKS. Besides his journal, the successive titles of which are given in the text, Crell published two auxiliary journals, many insignificant pieces on medical and chemical topics, books on natural theology and Freemasonry, and numerous translations. A fairly complete list of his publications is in Johann Stephen Pütter, *Versuch einer academischen Gelehrten-Geschichte von der Georg-*

August-Universität zu Göttingen, III, by J. C. F. Saalfeld (Hannover, 1820), 80–85, which includes a biography of Crell.

II. SECONDARY LITERATURE. In addition to the biography in Pütter, see Johann Friedrich Blumenbach, *Memoria Laurentii de Crell . . .* (Göttingen, 1822); and Karl Hufbauer, "The Formation of the German Chemical Community, 1700–1795," dissertation (University of California, Berkeley, 1970). Pütter, Blumenbach, and others incorrectly give Crell's year of birth as 1744.

KARL HUFBAUER

CRELLE, AUGUST LEOPOLD (*b.* Eichwerder, near Wriezen, Germany, 11 March 1780; *d.* Berlin, Germany, 6 October 1855), *mathematics, civil engineering.*

A son of Christian Gottfried Crelle, an impoverished dike reeve and master builder, Crelle was trained as a civil engineer and became a civil servant with the Prussian building administration. He finally obtained the rank of *Geheimer Oberbaurat* and was made a member of the Oberbaudirektion, under the Prussian Ministry of the Interior. During the years 1816–1826 Crelle was engaged in the planning and construction of many new roads throughout the country. He also worked on the railway line from Berlin to Potsdam, the first to be opened in Prussia, which was built in 1838.

Crelle, who always had been interested in mathematics but lacked the funds to enroll at a university, acquired appreciable knowledge in this field by independent study. At the age of thirty-six he obtained the doctorate from the University of Heidelberg, having submitted a thesis entitled "De calculi variabilium in geometria et arte mechanica usu."

In 1828 Crelle transferred from the Ministry of the Interior to the Ministry of Education, where he was employed as advisor on mathematics, particularly on the teaching of mathematics in high schools, technical high schools, and teachers colleges. During the summer of 1830, on an official tour to France, he studied the French methods of teaching mathematics. In his report to the ministry Crelle praised the organization of mathematical education in France but criticized the heavy emphasis on applied mathematics. In line with the neo-humanistic ideals then current in Germany, he maintained that the true purpose of mathematical teaching lies in the enlightenment of the human mind and the development of rational thinking.

Nevertheless, to the journal for which he is best remembered Crelle gave the name *Journal für die reine und angewandte Mathematik.* He founded it in 1826 and edited fifty-two volumes. From the very beginning it was one of the leading mathematical journals and even today is universally known as *Crelle's Journal.*

Although not a great mathematician himself, Crelle had a unique sensitivity to mathematical genius. He immediately recognized the abilities of such men as Abel, Jacobi, Steiner, Dirichlet, Plücker, Moebius, Eisenstein, Kummer, and Weierstrass and offered to publish their papers in his journal. He also used his influence as ministerial advisor and his acquaintance with Alexander von Humboldt and other important persons to further their careers. It is for this lifelong, unselfish intercession that Crelle deserves a place in the history of science.

Crelle wrote many mathematical and technical papers, textbooks, and mathematical tables and translated works by Lagrange and Legendre. Except for his *Rechentafeln,* these are now mostly forgotten. Also, for many years he published *Journal für die Baukunst.*

Although beginning in the 1830's his health declined until he was hardly able to walk, Crelle continued to further the course of mathematics, even at great personal sacrifice. He was survived by his wife, the former Philippine Dressel.

Crelle was elected full, corresponding, or foreign member of the Prussian Academy of Sciences in Berlin; of the academies of sciences in St. Petersburg, Naples, Brussels, and Stockholm; of the American Philosophical Society; and of the Mathematical Society of Hamburg.

BIBLIOGRAPHY

I. ORIGINAL WORKS. Crelle's works include *Theorie des Windstosses in Anwendung auf Windflügel* (Berlin, 1802); *Rechentafeln, welche alles Multipliciren und Dividiren unter Tausend ganz ersparen,* 2 vols. (Berlin, 1820; latest ed., Berlin, 1954); *Sammlung mathematischer Aufsätze,* 2 vols. (Berlin, 1821–1822); *Lehrbuch der Elemente der Geometrie,* 2 vols. (Berlin, 1825–1827); *Handbuch des Feldmessens und Nivellirens* (Berlin, 1826); and *Encyclopädische Darstellung der Theorie der Zahlen* (Berlin, 1845).

His German translations are A. M. Legendre, *Die Elemente der Geometrie* (Berlin, 1822; 5th ed., 1858); and J. L. Lagrange, *Mathematische Werke,* 3 vols. (Berlin 1823–1824).

He edited *Journal für die reine und angewandte Mathematik,* **1–52** (1826–1856), still being published; and *Journal für die Baukunst,* **1–30** (1829–1851).

His numerous papers on mathematics and engineering were published in these two journals and in the *Abhandlungen* and *Monatsbericht der Berliner Akademie der Wissenschaften* (1828–1853). A selection is in Poggendorff, I, 496–497, and VI, pt. 1, 491.

II. SECONDARY LITERATURE. On Crelle or his work see the following (listed chronologically): Moritz Cantor, in *Allgemeine Deutsche Biographie,* IV (Leipzig, 1876), 589–590; Wilhelm Lorey, "August Leopold Crelle zum

Gedächtnis," in *Journal für die reine und angewandte Mathematik,* **157** (1927), 3–11; Otto Emersleben, "August Leopold Crelle (1780–1855) zum 100. Todestag," in *Wissenschaftliche Annalen . . .,* **4** (1955), 651–656; and Kurt-R. Biermann, "A. L. Crelles Verhältnis zu G. Eisenstein," in *Monatsbericht der Deutschen Akademie der Wissenschaften zu Berlin,* **1** (1959), 67–92; and "Urteile A. L. Crelles über seine Autoren," in *Journal für die reine und angewandte Mathematik,* **203** (1960), 216–220, with a previously unknown portrait of Crelle.

CHRISTOPH J. SCRIBA

CREMONA, ANTONIO LUIGI GAUDENZIO GIUSEPPE (*b.* Pavia, Italy, 7 December 1830; *d.* Rome, Italy, 10 June 1903), *mathematics.*

Cremona was the eldest child of Gaudenzio Cremona and his second wife, Teresa Andereoli. One of his brothers, Tranquillo, attained some fame as an artist.

Luigi was educated at the *ginnasio* in Pavia. When he was eleven, the death of his father threatened to interrupt his schooling; his stepbrothers came to his support, however, and he was enabled to continue his education. He was graduated first in his class with special recognition for his work in Latin and Greek, and entered the University of Pavia.

In 1848 Cremona joined the "Free Italy" battalion in the revolt against Austrian rule and attained the rank of sergeant. He took part in the unsuccessful defense of Venice, which capitulated on 24 August 1849. Because of the gallantry of the defenders, they were permitted to leave the city as a unit, with honors.

When Cremona returned to Pavia, he discovered that his mother had died. With the help of the family he reentered the university. On 27 November 1849 he was granted permission to study civil engineering. Here he came under the influence of A. Bordoni, A. Gabba, and especially Francesco Brioschi. Cremona was always grateful to Brioschi, later writing: "The years that I passed with Brioschi as pupil and later as colleague are a grand part of my life; in the first portion of those years I learned to love science and in the other how to transfer it to a large circle of auditors" (Gino Loria, "Luigi Cremona et son oeuvre mathématique," p. 129).

On 9 May 1853 Cremona received the doctorate in civil engineering and architecture, and on 4 August of the following year he was married. His record of military service against Austrian rule prevented him from obtaining an official teaching post in the educational system, so his first employment was as a private tutor to several families in Pavia.

On 22 November 1855 Cremona was granted permission to teach on a provisional basis at the *ginnasio* of Pavia, with special emphasis on physics. He was already engaged in the mathematical research for which he is so well known. His first paper, "Sulle tangenti sfero-conjugate," had appeared in March 1855.

On 17 December 1856 Cremona was appointed associate teacher at the *ginnasio* in recognition of his good work and his mathematical activity: his second paper, "Intorno al un teorema del Abel" had appeared in May 1856. On 17 January 1857 he was appointed full teacher at the *ginnasio* in Cremona.

He remained in Cremona for nearly three years, during which time he wrote a number of articles. Some were merely answers to problems proposed in the *Nouvelles annales de mathématique,* but at least four contained original results, including his method of examining curves by projective methods: "Sulle linee del terz' ordine a doppia curvatura—nota" (1858); "Sulle linee del terz' ordine a doppia curvatura—teoremi" (1858); "Intorno alle superficie della seconda classe inscritte in una stessa superficie sviluppabile della quarta classe—nota" (1858); and "Intorno alle coniche inscritte in una stessa superficie del quart' ordine e terza classe—nota" (1859).

On 28 November 1859 the Italian government of newly liberated Lombardy appointed Cremona a teacher at the Lycée St. Alexandre in Milan, and on 10 June 1860 he received his first college appointment. A royal decree appointed him ordinary professor at the University of Bologna, where he remained until October 1867.

Cremona's most important original research into transformations in the plane and in space was published while he was at Bologna. His first paper, "Introduzione ad un teoria geometrica della curve piane," appeared in December 1861. The first statement of his general theory for transformations involving plane curves was "Sulle trasformazione geometriche della figure piane" (1863). In March 1866 he published his second paper on transformations, "Mémoire de géométrie pure sur les surfaces du troisième ordre." This earned Cremona half of the Steiner Prize for 1866, the other half going to Richard Sturm. Both papers on transformations were translated into German by Curtze and published as *Grundzüge der allgemeinen Theorie der Oberflächen in synthetischer Behandlung* (1870). It was during this period at Bologna that Cremona developed the theory of birational transformations (Cremona transformations). Besides being a creative mathematician, Cremona was an excellent lecturer: calm, rigorous, yet interesting and even exciting.

In October 1867 Cremona was transferred by royal decree, and on the recommendation of Brioschi, to the Technical Institute of Milan, to be in charge of the courses in higher geometry. He received the title of ordinary professor in 1872. During this period Cremona continued to produce mathematical articles that appeared in many Italian and French journals. His paper "Sulle trasformazione razionale . . . nello spazio . . ." (1871), which extended his transformation theory to space curves, supplemented and completed the main outlines of the theory of birational transformations.

The period at Milan, where he remained until 1873, was the time of Cremona's greatest creativity. He wrote articles on such diverse topics as twisted cubics, developable surfaces, the theory of conics, the theory of plane curves, third- and fourth-degree surfaces, statics, and projective geometry. He also turned out a number of excellent texts, including *Le figure reciproche nella statica grafica* (1872), *Elementi di geometria projettiva* (1873), and *Elementi de calcolo grafico* (1874).

In 1873 Cremona was offered a political post as secretary-general of the new Italian Republic by the minister of agriculture, but he refused. On 9 October of that year Cremona was appointed, by royal decree, director of the newly established Polytechnic School of Engineering in Rome. He was also to be professor of graphic statics. Administrative and supervisory work took up so much of Cremona's time during this period that it effectively ended his creative work in mathematics.

In November 1877 Cremona was appointed to the chair of higher mathematics at the University of Rome, and on 16 March 1879 he was appointed a senator. The duties entailed by this position put a complete stop to his research activities. On 10 June 1903, after leaving a sickbed to act on some legislation, Cremona succumbed to a heart attack.

Cremona's main contributions to mathematics lie in the areas of birational transformations, graphic statics, and projective geometry.

The earliest modern use of one-to-one transformations appears to have been that of Poncelet in 1822. Bobillier used them in 1827–1828, and in 1828 Dandelin used double stereographic projections. The algebraic approach seems to have been used first by Plücker in 1830, and in 1832 Magnus used noninvolutory transformations. Cremona combined all these developments and added much of his own material to create a unified theory. The clarity and polish of his presentation did much to publicize and popularize birational transformations.

The theory of birational transformations is basically as follows: Given a plane curve, $f(x,y) = 0$, which is irreducible and of degree m in x and n in y. Suppose

also that $x' = \varphi_1(x,y)$, $y' = \varphi_2(x,y)$ are rational functions in x and y. The eliminant of the three equations yields a new equation, $F(x',y') = 0$, which may be easier to examine and more revealing than the original, $f(x,y) = 0$.

If we solve the transformation equations $x' = \varphi_1(x,y)$ and $y' = \varphi_2(x,y)$ for x and y thus

$$x = \theta_1(x',y'), \quad y = \theta_2(x',y'),$$

and if these are rational functions in x', y', then we say that the transformation $x' = \varphi_1(x,y)$, $y' = \varphi_2(x,y)$ is birational.

Transformations of this nature are called Cremona transformations when there is a one-to-one reciprocal relation between the sets (x,y) and (x',y'). Geometrically, suppose the curve $f(x,y) = 0$ to be on the plane P and the curve $F(x',y') = 0$ to be on the plane P'. Then we seek a one-to-one correspondence between the sets of points. A very simple example would be that of a curve $f(x,y) = 0$ on P and its projection by a central perspectivity onto the plane P'.

One of the simplest examples of a Cremona transformation is the homographic transformation

$$x' = \frac{ax + by + c}{px + qy + r}, y' = \frac{a'x + b'y + c'}{px + qy + r}.$$

If homogeneous coordinates are used, we may write

$$x':y':z' = (ax+by+c):(a'x+b'y+c'):(px+qy+r).$$

Another example of a birational transformation is the Bertini transformation: $x':y':z' = xy:xz:yz$.

Cremona transformations have been used for studying rational surfaces, for the resolution of singularities of plane and space curves, and for the study of elliptic integrals and Riemann surfaces. They are effective in the reduction of singularities of curves to double points with distinct tangents.

Cremona's main contribution to graphic statics seems to have been the skillful use of the funicular diagram, or the reciprocal figure. This is defined as follows: Let P be a planar polygon with vertices A, B, C, $\cdots$, K, and let V be a point in the plane not on any side. Let VA, VB, VC, $\cdots$, VK be drawn. Construct a polygon whose sides are parallel to VA, VB, VC, $\cdots$, VK. This polygon, P', is called the polygon reciprocal to P, or the funicular diagram.

By a theorem of Maxwell, "If forces represented in magnitude by the lines of a figure be made to act between the corresponding lines of the reciprocal figure, then the points of the reciprocal figure will all be in equilibrium under the action of these forces."

Geometers will probably recognize Maxwell's theorem more readily in the following simplified form.

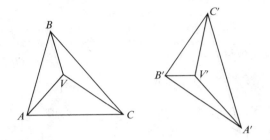

Given △ABC, and V any point in the plane of the triangle; let lines VA, VB, BC be drawn. In △A'B'C', B'C' is parallel to VA, C'A' is parallel to VB, and A'B' is parallel to VC. Now draw a line through A' parallel to BC, through B' parallel to AC, and through C' parallel to AB. These lines will be concurrent.

Note that each diagram is the reciprocal of the other, and that forces applied at any node of one are parallel to and proportional to the sides of the other. Note also that three forces in equilibrium in one figure, and therefore represented by a triangle, have as their images in the reciprocal figure three concurrent lines. It is this property that makes reciprocal figures so useful.

Again, the clarity and elegance of Cremona's presentation helped to disseminate and popularize the theorem and its consequences. Moreover, he collated results obtained by others and made them more readily available. Thus, his *Graphical Statics* contains not only signed lines and signed angles, which are fairly well known, but also the concept of signed and weighted areas. The development of the concepts of centroids of figures is elegant and clear.

The third area to which Cremona contributed was that of projective geometry; in fact, this discipline pervades all his work. Birational transformations arose from the concept of a curve and its projection onto another plane, and graphic statics makes extensive use of projective techniques. It is true that Cremona made no startling discoveries in this area; but he did derive many properties of projectively related figures, and he did present the subject to his classes in a manner calculated to clarify and bring out relationships most simply.

Cremona made use of Euclidean geometry when he thought it most effective, and it may be said that this introduced extraneous factors into projective geometry. It must be remembered, however, that his training and temperament favored the use of intuitive rather than strictly postulational methods.

As an organizer and popularizer of those areas of mathematics in which he did his work, Cremona has

had few peers. His works may still be read with profit and enjoyment.

BIBLIOGRAPHY

I. ORIGINAL WORKS. Cremona's works include "Sulle tangenti sfero-conjugate," in *Annali scienti di matematica,* **6** (Mar. 1855), 382–392; "Intorno al un teorema del Abel," *ibid.,* **7** (May 1856), 97–105; "Sulle linee del terz' ordine a doppia curvatura—nota," in *Annali di matematica pura ed applicata,* **1** (Apr. 1858), 164–174; "Sulle linee del terz' ordine a doppia curvatura—teoremi," *ibid.* (Oct. 1858), 278–295; "Intorno alle superficie della seconda classe inscritte in una stessa superficie sviluppabile della quarta classe—nota," *ibid.,* **2** (Dec. 1858), 65–81; "Intorno alle coniche inscritte in una stessa superficie del quart' ordine e terza classe—nota," *ibid.* (Feb. 1859), 201–207; "Introduzione ad una teoria geometrica della curve piane," in *Memorie della R. Accademia delle scienze dell'Istituto di Bologna,* **12** (Dec. 1861), 305–436; "Sulle trasformazione geometriche della figure piane" (1863); "Mémoire de géométrie pure sur les surfaces du troisième ordre," in *Journal für die reine und angewandte Mathematik,* **68** (Mar. 1866), 1–133; Curtze's translation of the two preceding papers, *Grundzüge der allgemeinen Theorie der Oberflächen in synthetischer Behandlung* (Berlin, 1870); "Sulle trasformazione razionale . . . nello spazio . . .," in *Memorie della R. Accademia delle scienze dell'Istituto di Bologna,* 3rd ser., **1** (1871), 365–386; *Le figure reciproche nella statica grafica* (Milan, 1872); *Elementi di geometria projettiva* (Turin, 1873); and *Elementi de calcolo grafico* (Turin, 1874). English translations of his works include *Graphical Statics,* trans. by Thomas H. Beare (Oxford, 1890); and *Elements of Projective Geometry,* trans. by Charles Leudesdorf (Oxford, 1893). An edition of his works is *Opera matematiche di Luigi Cremona,* Luigi Bertini, ed., 3 vols. (Milan. 1905).

II. SECONDARY LITERATURE. Cremona's work is discussed in P. Appell and E. Goursat, *Théorie des fonctions algébriques,* I (Paris, 1929), 266–292; Hilda P. Hudson, *Cremona Transformations in Plane and Space* (Cambridge, 1927); Gino Loria, "Luigi Cremona et son oeuvre mathématique," in *Bibliotheca mathematica* (1904), 125–195; and Ganesh Prasad, *Some Great Mathematicians of the Nineteenth Century,* II (Benares, 1934), 116–143.

SAMUEL GREITZER

CRESCAS, HASDAI (*b.* Barcelona[?], Spain, *ca.* 1340; *d.* Zaragoza, Spain, 1412), *philosophy, theology.*

Crescas lived in Barcelona and Zaragoza, was one of the most prominent leaders of the Spanish Jewish community, and performed a number of commissions for the monarchs of Aragon. Several scholars name him as a teacher, and there are indications of a circle of students around him. Crescas wrote a Hebrew philosophical-theological work entitled *Or Adonai* ("Light of the Lord"), which offers an unusual constellation of positions, some of them characteristic of the liberals among the medieval philosophers and others characteristic of the conservatives. Thus he is willing, on the one hand, to concede the eternity of the world and in effect sacrifices human free will to considerations of natural causality. But, on the other hand, he rejects the negative theology in vogue among the Neoplatonizing Aristotelians; denies that man's true perfection and, ultimately, human immortality are to be gained by intellectual development; and makes goodness rather than thought the central attribute of the deity.

Of particular interest for the history of natural science is a section in which Crescas examines twenty-five propositions formulated by Maimonides (d. 1204) as a summary of medieval Aristotelian physics and metaphysics. Crescas subjects these propositions to a most exacting analysis and rejects several of the most fundamental. Since at least the physical principles listed by Maimonides were genuinely Aristotelian and since it is especially against them that Crescas directs his criticism, he is in fact attacking much of the foundation of Aristotle's physics. He answers Aristotle's proofs of the impossibility of an infinite magnitude, an infinite place, or a vacuum. He rejects Aristotle's definition of time; his definition of place; the theory that two elements, fire and air, are endowed with an absolute lightness that causes them to rise; and the theory that all four physical elements have their proper, natural place that is the cause of their natural motion. The drift of Crescas' critique is toward a conception of infinite space, with the possibility of infinite worlds, and the uniformity of nature.

The argumentation used by Crescas, as well as the physical theories he advocates in opposition to Aristotle, could have been drawn from his reading; theories and arguments that previously had been raised, only to be answered, are now accepted by Crescas and combined into an overall attack on Aristotelian physics. Crescas' sources were the writings of the Jewish philosophers and the Hebrew translations of Averroës' (Ibn Rushd) commentaries on Aristotle. He knew Aristotle only through Averroës. Recently it has been suggested that Crescas may have been influenced by fourteenth-century Scholastic writers of the Paris school, although no literary connection has been established. Crescas was used and quoted by Pico della Mirandola and also by Spinoza. His position is similar to that of Giordano Bruno.

BIBLIOGRAPHY

An edition and translation of the section of the *Or Adonai* concerned with natural science is H. A. Wolfson, *Crescas' Critique of Aristotle* (Cambridge, Mass., 1929). Wolfson's introduction summarizes Crescas' position, and his masterful notes reveal the sources and history of the problems that Crescas deals with.

A suggestive recent study is Shlomo Pines, *Post Thomistic Scholasticism and the Theories of Hasdai Crescas* (Jerusalem, 1966), in Hebrew.

HERBERT A. DAVIDSON

CROLL, JAMES (*b.* Cargill, Perthshire, Scotland, 2 January 1821; *d.* Perth, Scotland, 15 December 1890), *geology.*

Croll's father, David Croil, was a poor stonemason and crofter who at the age of thirty-seven married Janet Ellis. James, their second son, was too poor to afford the university education he craved; therefore he became first a journeyman millwright, then a house joiner, until an injury forced him to give up manual work. A devout Congregationalist and teetotaler, he gave up smoking following his marriage in 1848 to Isabella Macdonald of Forres. Throughout a varied but universally unsuccessful business career, Croll studied theology and metaphysics, publishing in 1857 *The Philosophy of Theism.* Giving up business in 1858, he wrote for a Glasgow temperance weekly before he found a post as caretaker of Anderson's College and Museum in 1859. Leaving the work to his crippled younger brother David, whom he supported, Croll began to publish papers in chemistry and physics.

A paper in *Philosophical Magazine* (1864), in which he suggested that a change in the eccentricity of the earth's orbit had been responsible for the drastic changes in climate associated with glaciation, brought Croll to the notice of the leaders of British science, with some of whom (like Tyndall and Carey Foster) he began a vigorous correspondence. This led to the post of resident geologist in the newly opened Edinburgh office of the Geological Survey, where Croll remained from 1867 until his retirement in 1880. Here he continued his speculations about the causes of climatic change, publishing papers later brought together in *Climate and Time* (1875). Two more books on related topics, *Discussions on Climate and Cosmology* (1885) and *Stellar Evolution and Its Relations to Geological Time* (1889), appeared after he retired. The latter represents a shift of interest from the physical causes of climatic change to evolution. His final work, *The Philosophical Basis of Evolution* (1890), appeared just before he died.

Croll claimed to have suffered all his life from ill health due to heart disease and strokes; but only the mild stroke in 1880 that led to his retirement and the heart disease that began in 1886 and led to his death seem to have been organic. He was a fellow of the

Royal Society (1876), an LL.D. of the University of St. Andrews (1876), and the recipient of three awards from the Geological Society of London.

Although Croll made some modest contributions to the study of glaciation in Scotland, his chief importance was as a controversialist. In his three scientific books and many papers he never let his lack of training in mathematics and physics keep him from propounding the broadest climatological and cosmological theories and defending them against the mathematically adept, such as George Darwin. A clear and logical thinker whose prodigious reading kept him abreast of the latest observations, Croll had a purely verbal grasp of major scientific issues that drew the admiration of Lord Kelvin, among others. The most important of the many scientific controversies in which he engaged was that on the cause of ocean currents, in which he backed wind stress against the thermohaline, or density, theory of W. B. Carpenter. This argument generated more heat than light, since neither protagonist had the methods or the data to resolve it. Croll's contributions to science are confined to his own time: he stimulated others to develop both better evidence and more quantitative theories, and as a convinced Christian he helped make acceptable the geological ideas that followed from the theory of organic evolution.

BIBLIOGRAPHY

Croll's books are *The Philosophy of Theism* (London, 1857); *Climate and Time* (London, 1875; repr. New York, 1875; Edinburgh, 1885); *Discussions on Climate and Cosmology* (Edinburgh, 1885; repr. New York, 1886); *Stellar Evolution and Its Relations to Geological Time* (London, 1889); and *The Philosophical Basis of Evolution* (London, 1890). His papers include "On the Physical Cause of the Change of Climate During Geological Epochs," in *Philosophical Magazine* (Aug. 1864). William James's annotated copy of *The Philosophical Basis of Evolution* is in the Harvard College library. A list of Croll's published scientific papers is an appendix to the biography cited below.

James Campbell Irons, *Autobiographical Sketch of James Croll With Memoir of His Life and Work* (London, 1896), contains a brief autobiography (1887) followed by a long memoir. The MS materials used by Campbell Irons in preparing the memoir are in the British Museum, B.M. Addl. MSS 41077 (item 189 in Sotheby's sale of 10 April 1924).

HAROLD L. BURSTYN

CROLLIUS, OSWALD (*b.* Wetter, near Marburg, Germany, *ca.* 1560; *d.* Prague, Czechoslovakia, 1609), *medicine, iatrochemistry.*

Crollius was the third son of Johann Crollius, the mayor of Wetter. He received an excellent education at the abbey school in Wetter and at the University of Marburg, which he entered in 1576. Later he studied at Heidelberg, Strasbourg, and Geneva, receiving a doctorate in medicine about 1582. Crollius' first position was with the d'Esnes family in France, as an instructor. During frequent travels he became fluent in Italian and French.

In 1590 Crollius entered the service of Count Maximilian von Pappenheim as a tutor. His close contact with the emperor and high nobility led to Crollius' being entitled to use a coat of arms in 1591. After 1593 Crollius traveled about eastern Europe as a physician. Prague was his permanent residence from 1602. At that time he healed Prince Christian I of Anhalt-Bernburg, who appointed him *Archiat.* Even Emperor Rudolf II often consulted him and said of their relationship "cuius ipsi opera usi fuimus" ("whose work we ourselves did use").

Prince Christian soon formed a close relationship with Crollius, using him as a sort of emissary in Prague and discussing with him questions of politics as well as matters relating to alchemy and iatrochemistry. During these discussions Crollius also received material support for the practical chemical experiments that, in addition to his medical and diplomatic obligations, he conducted to determine the properties of the chemical remedies he had acquired. They became the basis of iatrochemistry.

Crollius recorded his knowledge, his experiments, and his theoretical views in a book entitled *Basilica chymica,* which was printed with the "Tractatus de signaturis." Whether Crollius lived to see the first edition (1609) is uncertain. The *Basilica chymica* became the standard scientific work of iatrochemistry. Johannes Hartmann, the first professor of iatrochemistry, used it as a practical textbook.

The *Basilica* is marked by a peculiar dualism that is typical of Crollius' medical theory. A convinced partisan of the subjective Platonic theory of knowledge, he considered the harmony of microcosm and macrocosm to be the foundation of medicine. In his views on pathology Crollius was entangled in the controversial ideas of Paracelsus. The dualism is also shown in his practical therapy. Healing by purely spiritual means was his ultimate goal, but in daily practice he insisted on the very real and often drastic remedies of iatrochemistry. Crollius also believed in one universal medicine, yet he developed dozens of useful chemical preparations for apothecaries.

In sharp contrast with Paracelsus' vagueness, Crollius describes in detail the individual preparations, their composition, and their application. This explains why the book became such a great success, running to many new editions and occasioning many commentaries.

At Crollius' death iatrochemistry lost one of its great exponents, for he is credited with gaining academic recognition of the medicinal value of many chemicals previously rejected as remedies.

BIBLIOGRAPHY

Crollius' only book is *Basilica chymica continens philosophicam . . . descriptionem, et usum remediorum chymicorum . . .* (Frankfurt, 1609, 1611, 1623; Leipzig, 1634; Geneva, 1642, 1658), published with "Tractatus novus de signaturis internorum rerum."

There are articles on Crollius in *Allgemeine deutsche Biographie,* IV; and *Neue deutsche Biographie,* C. See also J. A. Döderlein, *Historische Nachrichten* (Schwabach, 1739); H. Fränkl, *Zur Geschichte der Medizin in den Anhalt'schen Fürstentümern* (Leipzig, 1858); G. Schröder, "Oswald Crollius," in *Pharmaceutical Industry,* **21** (1959), 405–408; and "Studien zur Geschichte der Chemiatrie," in *Pharmazeutische Zeitung,* **111,** no. 35 (1966), 1246 ff.; and H. Witte, *Diarium biographicum* (Gedern, 1688).

GERALD SCHRÖDER

CROMMELIN, ANDREW CLAUDE DE LA CHEROIS (*b.* Cushendun, Northern Ireland, 6 February 1865; *d.* London, England, 20 September 1939), *astronomy.*

A member of a prominent Huguenot family, Crommelin was educated at Marlborough College and Trinity College, Cambridge, graduating in 1886. From 1891 to 1927 he was additional assistant at the Royal Observatory, Greenwich. He was president of the Royal Astronomical Society during 1929–1931 and of the British Astronomical Association during 1904–1906; he received the latter's Goodacre Medal in 1937.

At Greenwich, Crommelin worked as both observer and computer. He made an accurate determination of the lunar parallax (1911) and prepared the physical ephemerides of the moon and outer planets (1897–1906). His photographs of the Brazilian solar eclipse (1919) helped to establish the relativistic deflection of light.

Crommelin's principal contribution to science, however, involved comets and minor planets. In 1907–1908 he and Cowell traced back the motion of Halley's comet to 240 B.C., identifying observations at almost every perihelion passage. They developed a direct method for studying the very complicated orbit of Jupiter's eighth satellite, the equations being integrated by mechanical quadrature. They also applied the procedure to Halley's comet, due to return in 1910, and for their exact result were awarded the Lindemann Prize of the Astronomische Gesellschaft and honorary D.Sc. degrees from Oxford University. Their predicted perihelion time was three days too early, and their suggestion that the error was due mainly to nongravitational forces has now been confirmed.

From 1891 to 1937 Crommelin wrote the annual reports on minor planets in the *Monthly Notices of the Royal Astronomical Society* and from 1916 the reports on comets as well. He calculated preliminary orbits for many comets, and it is directly due to his indefatigable efforts in providing predictions for the periodic comets that many of these objects have not become lost. From 1935 until his death he was president of the International Astronomical Union's subcommission on periodic comets. He also served as director of the Comet Section of the British Astronomical Association (1897–1901, 1907–1939) and in this capacity produced sequels to Galle's catalogs of cometary orbits. After his retirement from Greenwich he edited the *Circulars* of the association; and the association's *Journal* and *Handbook,* as well as *The Observatory, Nature, Knowledge,* and the *Circulars of the International Astronomical Union,* contain innumerable notes by Crommelin concerning comets, minor planets, and other topics.

In 1929 Crommelin demonstrated that comet Forbes 1928 III was identical with comet Coggia-Winnecke 1873 VII and comet Pons 1818 I, the revolution period being twenty-eight years. He later showed that a comet seen in 1457 was probably (and one in 1625 possibly) the same object. In 1948 the International Astronomical Union changed the name of the comet from Pons-Coggia-Winnecke-Forbes to Crommelin (only the fourth occasion on which a comet has been named after the computer of its orbit, rather than its discoverer), and in 1956 the comet returned to perihelion just four days later than Crommelin had predicted.

BIBLIOGRAPHY

Some of Crommelin's writings are "Essay on the Return of Halley's Comet," *Publikation der Astronomischen Gesellschaft,* no. 23 (1910), written with P. H. Cowell; "Comet Catalogue," *Memoirs of the British Astronomical Association,* **26,** pt. 2 (1925), continued *ibid.,* **30,** pt. 1 (1932); "Tables for Facilitating the Computation of the Perturbations of Periodic Comets by the Planets," in *Memoirs of the Royal Astronomical Society,* **64** (1929), 149–207; and *Comets* (London, 1937), written with Mary Proctor.

An obituary notice is C. R. Davidson, *Monthly Notices of the Royal Astronomical Society,* **100** (1940), 234–236.

BRIAN G. MARSDEN

CRONSTEDT, AXEL FREDRIK (*b.* Turinge, Sweden, 23 December 1722; *d.* Säter, Sweden, 19 August 1765), *chemistry, mineralogy.*

Following in his father's footsteps, Cronstedt studied mathematics at Uppsala and prepared to become a fortifications officer. He changed his mind, however, when, through Johan Wallerius and Sven Rinman, he became interested in mineralogy. He attended the Bergs-Kollegium in 1742 as an observer; there he attracted the attention of Daniel Tilas, a leading figure in mining and metallurgy, who was to have a decisive influence on his further education. An important part of his education included field study in various aspects of mining.

Between 1741 and 1743, when Sweden was at war with Russia, Cronstedt had to serve in the army; part of his service was as secretary to his father, who was a general and head of the Corps of Engineers. It was only toward the end of 1743 that he was able to make the first of many trips to the mines and smelting works, to observe the methods in use. These trips, together with his thorough inquiry into, among other things, the metallurgy of silver, lead, and copper, made Cronstedt a first-rate mining expert. He recorded his experiences and observations in reports and papers that he presented to the Bergs-Kollegium and to the Swedish Academy of Science. To supplement this emphasis on the practical side of his education, in 1746 he took a course in "the art of experiment and chemistry"; he continued the course for several years, with occasional interruptions. His teacher was Georg Brandt, the discoverer of cobalt.

In January 1748, Cronstedt was made a director of the East and West Bergslagen, a large area in the richest ore-bearing region of central Sweden. Among his duties was the supervision of the silver works at the Skiss foundries in southern Dalarna (Kopparbergs län), which had an excellent laboratory for testing metals and minerals. Here Cronstedt found a quiet place where, undisturbed, he could devote himself to his chemical and mineralogical research and experiments. Especially fruitful was his work on "copper-nickel" (niccolite) obtained from the cobalt mines at Los, in Hälsingland. Both Urban Hiärne and Wallerius had declared "copper-nickel" to be a mixture of cobalt, arsenic, copper, and iron. That the ore contained copper seemed evident, since its solution in nitric acid was green, turning to blue upon treatment with ammonia. But when Cronstedt tested copper precipitates with iron or zinc, this did not occur. In another ex-

periment the mineral withstood a humid atmosphere and, after calcination or reduction with the black flux of the lime that was formed, yielded a regulus, yellowish on the outside, with a silver-white metallic broken surface and slight magnetic properties. Since none of the then known metals, alone or in any of its compounds, possessed these properties, Cronstedt felt entitled to claim that he had found a new metal. His results were confirmed by the distinguished director of the Ädelfors placer mines, Henrik Theophilus Scheffer, an assayer and chemist renowned for his knowledge and skill.

Cronstedt further verified his discovery by showing that he could obtain his new metal not only from Swedish ore but also from Freiberg "copper-nickel." His first report on his discovery appeared in the *Kungliga Vetenskapsakademiens Handlingar* for 1751, but not until a paper in the *Handlingar* for 1754 did he give his new metal a name, calling it "nickel," after the material from which it was obtained, called "copparnickel" in Swedish. Most contemporary chemists acknowledged Cronstedt's claim that he had discovered a new metal, but some, such as Balthazar Sage, Antoine Monnet, J. H. von Justi, and G. A. Scopoli, held to the old view represented by E. H. von Henckel and J. A. Cramer: that "copper-nickel" consisted of cobalt, iron, arsenic, and copper and that its regulus did not, in consequence, represent any new metal but was only a compound of previously known ones. It was not until Torbern Bergman's dissertation *De niccolo,* defended in 1775 by Johan Afzelius, that Cronstedt's claim was conclusively accepted.

In a number of other papers appearing in the *Handlingar,* Cronstedt published, among other things, his experiments on new types of iron ore (1751). Here he mentioned a mineral possessing an unusually high specific gravity, which he named "Bastnäs' tungsten"; this led to other important discoveries. In 1803 Berzelius and Hisinger, at the same time as Klaproth, discovered that this substance was a water-containing silicate of a new metal that they named "cerium." The ore, which they called "cerite," was the substance in which Mosander later discovered didymium and lanthanum. Cronstedt reported on other series of experiments: with gypsum (1753), with the mineral that he roasted over a blowpipe flame and named "zeolite" (1756), and with the platina of Pinto (1764).

His discovery of nickel made Cronstedt known in the learned world, and his membership in the Royal Academy of Science and his further work assured his reputation as a scientist. His experiments, oriented toward the chemistry of metals, enabled him to gain an unprecedented insight into the inner structure of minerals. When this insight was combined with a clear understanding of his findings as a basis for a rational mineral system, it revealed the errors in earlier methods of classification. This prompted Cronstedt to publish anonymously his *Försök til mineralogie, eller mineralrikets upställning* (1758). In this essay he sought to lay the foundations for a new mineralogical system. Here, for the first time, he established the correct distinctions between simple minerals and rock minerals consisting of a mixture of several minerals. Since he thereby excluded the rock minerals from mineralogy as such, he also excluded fossil material, concretions, and the like. In an appendix to *Försök* he published a classification of rocks. The work attracted considerable attention; and after it was translated into German in 1760, it quickly became known outside Sweden. Abraham Gottlob Werner, the world-famous mineralogist in Freiberg and the reformer of mineralogy, paid him due homage; and Berzelius declared him to be the founder of chemical mineralogy. In this connection it must be pointed out that even if other investigators had previously used the blowpipe in their tests, nobody before Cronstedt had so methodically applied this tool to the examination of minerals. Thus, he is entitled to be considered the actual founder of systematic blowpipe analysis.

BIBLIOGRAPHY

I. ORIGINAL WORKS. The library of the Academy of Science in Stockholm has a collection of Cronstedt's writings, including 15 vols. of MSS, letters, transcripts, and other material.

His published works include "Rön och försök gjorde med trenne järnmalms arter," in *Kungliga Svenska Vetenskapsakademiens Handlingar,* **12** (1751), 226–232; "Rön och försök gjorde med en malm-art, från Los kobolt-grufvor i Färila socken och Hälsingeland," *ibid.,* 287–292; "Rön och anmärkningar om gips," *ibid.,* **14** (1753), 44–47; "Fortsättning af rön och försök, gjorde med en malm-art från Los kobolt-grufvor," *ibid.,* **15** (1754), 38–45; *Inträdes-tal om medel til mineralogiens vidare förkofran* (Stockholm, 1754); "Rön och beskrifning om en obekant bärg art, som kallas zeolites," in *Kungliga Vetenskapsakademiens Handlingar,* **17** (1756), 120–123; *Försök til mineralogie, eller mineralrikets upställning* (Stockholm, 1758), trans. into German as *Versuch einer neuen Mineralogie,* by G. Wiedemann (Copenhagen, 1760); trans. into English, with notes, by Gustav von Engeström, as *An Essay Towards a System of Mineralogy* (London, 1770), with an appendix by M. T. Brunnich (London, 1772); 2nd ed., rev. and enl. by John Hyacinth de Magellan, 2 vols. (London, 1788); *Cronstedts Versuch einer Mineralogie, vermehret durch Brünnich* (Copenhagen-Leipzig, 1770); *Essai d'une nouvelle minéralogie,* trans. by M. Dreus fils (Paris, 1771); *Saggio per formare un sistema de mineralogia,* trans. by Angelo Talier (Venice, 1775; new ed., 1779); trans. into Russian by Matheus Kardiman (St.

Petersburg, 1776); *Axel von Kronstedts Versuch einer Mineralogie. Aufs neue aus dem Schwedischen übersetzt und vermehret von Abr. Gottlob Werner,* I, pt. 1 (Leipzig, 1780); "Rön och anmärkningar vid Jämtlands mineralhistoria," *ibid.,* **24** (1763), 268–289; "Några rön och anmärkningar vid platina di Pinto," *ibid.,* **25** (1764), 221–228; *Axel Friedrichs von Cronstedt Mineralgeschichte über das westmanländische und dalekarlische Erzgebirge, auf Beobachtungen und Untersuchungen gegründet,* trans. by J. G. Georgi, notes by D. J. C. D. Schreber (Nuremberg, 1781); his autobiography, "Bergmästaren A. F. Cronstedts egenhändiga lefnadsbeskrifning," in *Blad för bergshanteringens vänner inom Örebro län,* **5** (1886–1888), 99–116.

II. SECONDARY LITERATURE. On Cronstedt or his work, see J. Landauer, "Die ersten Anfänger der Löthrohranalyse," in *Berichte der Deutschen Chemischen Gesellschaft,* **26** (1893), 898–908; S. Rinman, *Åminnelsetal öfver Axel Fredrik Cronstedt, hållet den 6 mars 1766* (Stockholm, 1766); and Nils Zenzén, in *Svenskt biografiskt lexikon,* IX (Stockholm, 1929), 279–295.

UNO BOKLUND

CROOKES, WILLIAM (*b.* London, England, 17 June 1832; *d.* London, 4 April 1919), *chemistry, physics.*

Crookes was the eldest son of the sixteen children of Joseph Crookes, a prosperous tailor, by his second wife, Mary Scott. In 1848, after irregular schooling, he received his scholarly introduction to science when he became a student at A. W. Hofmann's Royal College of Chemistry in London. After gaining the Ashburton Scholarship, he served as Hofmann's personal assistant from 1850 to 1854 and came to the attention of Faraday at the Royal Institution. Faraday introduced him to Charles Wheatstone and George Stokes, and together the three men were largely responsible for turning Crookes away from traditional chemical problems and toward chemical physics, exemplified then by the optical problems of photography and later by spectroscopy. There are many indications that Crookes consciously modeled himself on Faraday, with whom he shared a brilliant experimental and lecturing ability, a scrupulous orderliness, and an ignorance of mathematics. This ignorance, however, was often masked in later years through his friendship with Stokes, who privately solved many mathematical problems in physics for him. Nevertheless, the rigorous training in analytical techniques that he received under Hofmann remained the foundation for all of Crookes's subsequent researches and his commercial activities. In 1854, through Wheatstone's influence, he became superintendent of the meteorological department of the Radcliffe (Astronomical) Observatory at Oxford; and in 1855 he taught chemistry at the College of Science at Chester. Finally, in 1856 he settled in London, where, apart from extensive traveling on business, he attempted to bring his name before the scientific community both as a freelance chemical consultant (using a home laboratory) and as an editor of several photographic and scientific journals. He was nominal editor, and proprietor, of the most successful and important of these journals, *Chemical News,* from its founding in December 1859 until his death. He was knighted in 1897 and received the Order of Merit in 1910.

Crookes married Ellen Humphry of Darlington in 1856; and the necessity of supporting ten children helps to explain the amazing diversity and catholicity of his scientific interests, many of which were frankly motivated by the belief that all pure scientific researches would lead to financial rewards. Unlike Faraday, he was intensely ambitious, both in science and in business. In the latter world he drove many hard bargains; but although he was able to make a comfortable living from such commercial ventures as the sodium amalgamation process for gold extraction, the utilization of sewage and animal refuse, and electric lighting, none of them was financially very successful.

Crookes was an experimentalist of consummate skill, and the brilliance of his experimentation was the dominant feature in his scientific career; his success in producing a vacuum of the order of one millionth of an atmosphere made possible the discovery of X rays and the electron. With fame, honors, and responsibilities to various scientific societies, he was fortunate to be able to leave the experimental side of his researches to equally adept assistants, notably Charles H. Gimingham from 1870 to 1882 and James H. Gardiner from 1882 to 1919. To the latter must be attributed the impressive bulk of Crookes's researches in his later years, including work on scandium and on "Crookes lenses," published when he was over eighty.[1] From the 1870's Crookes's papers ceased to be entirely experimental; and with great effect he used various scientific platforms to offer speculations and to make theoretical pronouncements that, although frequently wide of the mark, were nonetheless plausible and stimulating. He was a great syncretist of other scientists' hints and suggestions; and by weaving these together imaginatively with the aid of his literary adviser, Alice Bird, Crookes acquired a well-deserved reputation as a Victorian sage.

Sages often have to face censorious criticism during their lifetimes. The most controversial aspect of Crookes's career, even today, is his investigation "into the Phenomena called Spiritual" during the 1870's. Following Darwinism it fell to Crookes to provoke the last major eruption in the battle between science and religion. Some critics, notably Crookes's archenemy, W. B. Carpenter, argued that there were two Crookeses, one a rational scientist, the other a credulous Spiritualist.[2] Today, with our better understanding of the Victorian mind and of the way in which religion and metaphysics have played a creative role in the scientific lives of men like Newton, it causes no surprise that occult phenomena should have involved the attentions of a major nineteenth-century scientist.

By 1860 Spiritualism was a well-established religion spread, by the grapevine of domestic servants, through all levels of Victorian society. In 1867, following the death of a brother to whom he was devoted, Crookes (who had been brought up with the Christian belief in an afterlife) was persuaded by the electrician and Spiritualist C. F. Varley to attend séances. He became deeply interested in the kinetic, audible, and luminous phenomena that could be witnessed at the fashionable séances of the 1870's and 1880's; and although, like Tyndall, Huxley, and Carpenter, he remained skeptical and was intolerant of the frequent fraudulent practices that he detected, he was persuaded that in the case of a few mediums, notably the Scottish-American D. D. Home, the astounding phenomena were genuine. In 1870 he subjected Home, in the presence of William Huggins, an astronomer, and Edward William ("Sergeant") Cox, a lawyer, to a number of tests; and he became personally convinced that Home possessed a "psychic force" that could modify gravity, produce musical effects, and perform other feats unknown to conjurers or scientists. Crookes published accounts of his investigations in his own *Quarterly Journal of Science,* but the Royal Society rejected his paper on the grounds that the experimental conditions were insufficiently exacting; subsequently, in 1887, when less hysteria surrounded the issue, the Society allowed Crookes to publish some negative observations on the occult force of M. J. Thore.[3]

Even more sensational was Crookes's support in 1874 for the young medium Florence Cook, who materialized a phantom called "Katie King." Was Crookes so infatuated that he lied? Was he the subject of an appalling confidence trick that, if he ever understood, he was too embarrassed to reveal? Did he suffer delusions? Did he in fact observe paranormal phenomena? Although all these questions have been answered affirmatively by various people, there is no doubt of Crookes's sincerity, for he staked his scientific reputation on the validity of the phenomena he described.

Crookes made a prodigious number of investigations of mediums between 1870 and 1900 and was undoubtedly the most experienced, if not always the most critical, of all nineteenth-century psychic investigators. He even built a special room in his house so that investigations could be conducted under rigorous conditions. Not surprisingly for one who claimed to be widely read in "Spiritualism, Demonology, Witchcraft, Animal Magnetism, Spiritual Theology, Magic and Medical Psychology in English, French and Latin," he joined the Society for Psychical Research on its founding in 1882 and was its president in 1897. He also joined Mme. Blavatsky's Theosophy movement in 1883. Until his wife's death in 1916 Crookes never believed that spiritualistic séances presented unequivocal evidence that "spirits" were those of human dead.[4] He is best described, therefore, as an occultist, a man for whom traditional science left huge areas of creation unexplored. These serious investigations into the occult were never entirely divorced from his public science; and for the historian of science they illuminate his purely scientific writings, particularly those on the radiometer and on inorganic evolution, and clarify some of the more purple patches in his deliberately exhibitionistic speeches to the many societies that chose him as their president.

Like Tyndall and Kekulé, Crookes frequently stressed the role played by imagination in discovery, or by an "inward prescience" in shaping the development of his work. The most logical aspect of this was the critical appraisal of "finger posts" that were erected by "anomalous residual phenomena."[5] In this manner Crookes saw a logical sequence running through his work from selenium through thallium, the radiometer, cathode rays, and the rare earth elements to radioactivity.

In 1850, under Hofmann's direction, Crookes made a routine study of the selenocyanides, using selenium ores from Germany. During the next decade, inspired by Wheatstone, he worked enthusiastically on the new subject of photography, investigating the spectral sensitivity of the wet collodion process; attempting to apply photography to the scientific recording of polarization, astronomical objects, and spectra; and, in 1854, devising the first dry collodion process with John Spiller.

His analysis of the optical sensitivity of photographic processes caused Crookes to speculate privately on the origin of spectra; when the solution was published by Bunsen and Kirchhoff in 1860, and they announced the discovery of two new elements, cesium and rubidium, which had been detected by spectrum analysis, Crookes immersed himself in spectroscopy. Everything in his home laboratory was examined through the spectroscope, including the selenium wastes from 1850, which he supposed might show the spectrum of tellurium as an impurity. However, on 5 March 1861 he found an anomalous bright green line in the wastes; and by 30 March he was confident enough to announce the existence of a new element, thallium. He exhibited minute samples of thallium, in the form of a black powder, and of its salts at the International Exhibition in London in 1862, only to find that the element had been isolated simultaneously on a larger, more obviously metallic scale by the French chemist C. A. Lamy. An acrimonious controversy developed, not least because, through an administrative blunder, at first only Lamy received the exhibition's medal. *Chemical News* gave Crookes prestige, but thallium (which he subsequently showed to be a widely distributed element easily extracted from the flue dust of various industrial processes) brought him fame and led to his election to the Royal Society in 1863.

From 1861 to 1871 Crookes patiently developed techniques to determine the atomic weight of thallium, deliberately modeling his standards on the hitherto exacting determinations of atomic weights that had been made by J. S. Stas. Once he had decided to make the determination by the formation of thallium nitrate from the heavy metal, and to check by the precipitation of barium sulfate from thallium sulfate, he took extraordinary pains to purify his reagents, to calibrate his platinum balance weights relative to one another, and to use an extremely sensitive Oertling balance mounted in an iron case that could be exhausted of air. Weighings could then be made at reduced pressures and reduced to a vacuum standard, thus eliminating a major source of previous analytical error, the neglect of barometric variation in gravimetry.[6] The result was an atomic weight for thallium of 203.642 (that of oxygen is 15.960; the difference from the modern value, 204.39, is not a reflection on Crookes's accuracy, but of his use of Stas's inaccurate value for the ratio of nitrogen to oxygen).[7] This seemed to indicate to Crookes that Stas was right to be skeptical of Prout's hypothesis that atomic weights were integral multiples of the atomic weight of hydrogen. Nevertheless, as a spectroscopist faced with the bewildering complexity of spectra, Crookes could not avoid becoming a convinced believer in the complexity of the elements; and he frequently gave space in *Chemical News* to speculations concerning the unity of matter or the relationships between elements (e.g., those of B. C. Brodie, Jr. and J. A. Newlands).

While working with his balance in a vacuum, Crookes noticed another "anomaly": the equilibrium of the balance was disturbed by slight differences in temperature of his samples. In particular he noticed that warmer bodies appeared to be lighter than colder ones; and since this occurred in a strong vacuum, it could not be attributed to either the condensation of vapor on the cooler body or to air currents surrounding the hotter body. At first Crookes believed this was a signpost pointing to a link between heat and gravitation; and because of its bearing on the "psychic force" he was then also investigating, he subjected the anomaly to an intensive examination with the aid of Gimingham, whose skill at glassblowing and vacuum pump technology enabled them to develop a number of delicate and beautiful pieces of apparatus.

They found that if a large mass was brought close to a lighter one suspended in an evacuated space, it was either attracted or repelled. By mounting two pith balls on a pivoted horizontal rod in a tube attached to a mercury pump, they were able to investigate the effect more closely. Since the attraction or repulsion was heightened by a decrease in pressure, Crookes was led to suppose in 1873 that "the movement is due to a repulsive action of radiation." Repulsion was produced not only by heat radiation but also by light, and Crookes concluded—erroneously, as it turned out—that he had found a genuine case of "the pressure of light" postulated by the unfashionable corpuscular theory of light and by Maxwell's as yet unaccepted electromagnetic theory. This belief led him in 1875 to devise the "light-mill," or radiometer, which consisted of "four arms, suspended on a steel point resting in a cup, so that it is capable of revolving horizontally. To the extremity of each arm is fastened a thin disc of pith, lamp-blacked on one side, the black surfaces facing the same way. The whole is enclosed in a glass tube, which is then exhausted to the highest attainable point and hermetically sealed."[8]

The fascinating free rotation of the mill in strong light proved a financial boon to instrument makers, but the explanation of its action proved very difficult and controversial. In the furore, which was exacerbated by his psychic activities, Crookes tried, with difficulty, to maintain an empirical position in the battles between Osborne Reynolds, P. G. Tait and James Dewar, Arthur Schuster, and Johnstone Stoney, and in his personal vendetta with Carpenter. Eventually, in 1876, he accepted Stoney's explanation in terms of the kinetic theory of gases: that the radiometer's motion was due to the internal movements of the molecules in the residual gas. If the mean free path was small compared with the dimensions of the radiometer vessel, molecules that struck the warmer black vane and rebounded with increased velocity would hold back any slower-moving particles advancing toward the surface. Hence a relatively larger number of molecules would hit the cooler surface and prevent any rotation of the vanes. When the pressure was lowered, and the mean free path was large, there was no compensation of the recoil effect, and rotation took place.[9] This theory gave a good qualitative ex-

planation of all Crookes's subsequent radiometer work, although quantitative agreement was the achievement of more mathematically gifted physicists.[10] Even so, in a tedious and exacting series of measurements made between 1877 and 1881, Crookes showed how the radiometer confirmed Maxwell's prediction that the viscosity of a gas was independent of its pressure except at the highest exhaustions.[11]

In 1878 Crookes began a new series of research papers with the conviction "that this dark space coating [the cathode in low-pressure electrical discharges] was in some way related to the layer of molecular pressure causing movement in the radiometer."[12] By attempting to determine the actual paths of "lines of molecular pressure" on the analogy of Faraday's lines of magnetic force, Crookes came to work on the cathode rays, which until then had been the exclusive province of German experimentalists. An electric radiometer whose vanes acted as a cathode showed that the dark space separating the cathode from the cathode glow extended farther from the blackened side of the vane, and that only when the pressure was reduced to a point at which the dark space touched the sides of the radiometer tube did rotation occur. This suggested that the electric discharge in an evacuated tube (a "Crookes tube") was an actual illumination of the lines of molecular pressure. Since the thickness of the dark space (soon called the "Crookes dark space") increased with decrease in pressure, the kinetic theory explanation of the radiometer's action suggested that the dark space was a measure of the length of the mean free path.

In November 1876, Crookes introduced into his work on the molecular physics of high vacua Faraday's beautiful concept of a fourth state of matter, "radiant matter."[13] Lines of molecular pressure were now interpreted as streams of radiant matter in a highly tenuous condition; collisions between molecules were so rare that the familiar physical properties of the gaseous state of matter were modified.

With his thorough grounding in the experimentally difficult art of vacuum physics, Crookes laid the foundation for the fuller investigation by J. J. Thomson of the behavior of radiant matter in the discharge tube, showing, for example, that it induced phosphorescence in minerals like the diamond; that it caused the glass of the discharge tube to phosphoresce; that its stream could be deflected by a magnet; and, most important of all, that since it cast a shadow of an opaque object (for example, a Maltese cross), it traveled in straight lines and was corpuscular in nature.[14] In 1879 he declared:

> In studying this Fourth state of Matter we seem at length to have within our grasp and obedient to our control the little indivisible particles which with good warrant

are supposed to constitute the physical basis of the Universe. . . . We have actually touched the border land where Matter and Force seem to merge into one another, the shadowy realm between Known and Unknown which for me has always had peculiar temptations. I venture to think that the greatest scientific problems of the future will find their solution in this Border Land, and even beyond; here . . . lie Ultimate Realities, subtle, far-reaching, wonderful.[15]

This may seem prescient of the electron, as radiant matter turned out to be; but for Crookes cathode rays were negatively charged molecules, and the above passage is more a hint of the occult. Until Thomson's demonstration of the electron in 1897, Crookes's theory of a fourth state of matter was severely criticized by the German school of physicists, who looked for an etherial wave explanation of cathode rays.

During the 1880's Crookes followed two major research programs: the first, the technology of electric lighting with evacuated tubes, was motivated by the prospect of financial rewards; the second, the examination of the phosphorescent spectra produced by bombarding minerals with radiant matter, led him to speculate about the origin of the elements. Most substances gave continuous phosphorescent spectra, but the beautiful discontinuous spectra produced by rare earth minerals led him to suspect in 1881 that these ores contained many unknown elements. He based this belief principally on the fact that the patterns of discontinuity in the spectra could be enhanced and carried over from one manipulation to another during chemical fractionation. For example, he found that yttria could be separated by fractionation into five or more parts that were differentiated only by their slightly different solubilities in ammonia and by the different prominent lines in their phosphorescent spectra. It turned out later that Crookes had been totally misled by impurities (phosphorogens) in his samples, for, as Lecoq de Boisbaudran, Urbain, and others showed, phosphorescent spectra could be altered simply by adding traces of other earths; and ultrapurified earths were not sensibly phosphorescent. The method, therefore, turned out to be useless for characterizing elements, either known or unknown.

Nevertheless, with the aid of Gardiner, Crookes brilliantly pursued a laborious course of chemical fractionations "repeated day by day, month after month, on long rows of Winchester bottles." Between 1886 and 1890 he concluded erroneously that yttrium, gadolinium, and samarium were complex mixtures of unknown elements; that didymium was more complex than Welsbach's (correct) separation into neodymium and praseodymium in 1885 implied; and that erbia

probably contained four elements. In 1898 he announced that an element, which he called monium or Victorium (in fact a mixture of gadolinium and terbium), could be separated from yttria. Chemists experienced great difficulty in sorting out the relationships between the rare earth elements; Crookes's elegant researches confused matters for a number of years, for he persisted in these increasingly isolated views as late as 1906. Ironically, had he pursued the fractionation of calcium salts, which he began, he undoubtedly would have separated calcium isotopes.[16]

It has happened repeatedly in the history of chemistry that when the number of elements has been experimentally increased, there has been a simultaneous move to decrease their number by the introduction of a simplifying theory of matter. For Crookes, as he explained in superlative addresses to the British Association for the Advancement of Science in 1886 and to the Chemical Society in 1888, the similarity between the rare earth elements was certain evidence that they originated from a common matter.[17] The idea of the chemical element, Crookes felt, had to be expanded to take into account the acceptance of the periodic law of the elements and the speculations of B. C. Brodie, T. S. Hunt, J. N. Lockyer, and others. Crookes rekindled Prout's hypothesis by suggesting that elements were complex bodies that had developed by an inorganic process of Darwinian evolution. Before evolution had commenced, there existed a primary matter that, in an incorrect transcription from the Greek he called "protyle." Prout had originally suggested that hydrogen might be the protyle from which other elements had formed, but sixty years of atomic weight determinations had ruled out this possibility. It was still possible to propose something with less weight than hydrogen; perhaps it was helium, the element then known to exist only in the sun.

With remarkable vision Crookes showed how the elements, with their periodic properties, could be conceived of as arising under the action of an imaginary cosmic pendulum swung by the efforts of two forces, electricity and heat (loss of temperature with time), and dampened according to the laws $x = a \sin (mt)$ (an oscillation: a and m are constants, x is the electrical force, and t is the time) and $y = bt$ (a simple uniform motion at right angles to the oscillation: y is temperature and b is a constant). This effectively produced the curve of the periodic system that had been developed for teaching purposes by J. Emerson Reynolds.[18] But genesis occurred in space. If a third oscillatory motion was conceived, $z = c$

sin $(2mt)$ (an oscillation twice as rapid as the first and at right angles to the two former motions: c is a constant and z is an unspecified force that is probably of electrical origin), then an actual three-dimensional figure-eight model of the genesis of the elements could be constructed such that similar elements were located in a vertical plane on identical parts of the helix.

Crookes held no brief for Kelvin's pessimistic "heat-death" future of the universe; on the contrary, borrowing from Stoney, he believed that the universe was in continual creation. The radiant heat from ponderable elementary matters in the center of the universe flowed toward the periphery, where it was transformed into protyle and genesis recommenced. Definite quantities of electricity were given to each element at its genesis; this electricity determined the element's valence and, hence, its chemical properties. Atomic weight, on the other hand, was only a measure of the cooling conditions that had prevailed at the moment of the element's birth and not, as Mendeleev had implied, a measure of its properties. If the cooling conditions had sometimes been irregular:

. . . elements would originate even more closely related than are nickel and cobalt, and thus we should have formed the nearly allied elements of the cerium, yttrium, and similar groups; in fact the minerals of the class of samarskite and gadolinite may be regarded as the cosmical lumber-room where the elements in a state of arrested development—the unconnected missing-links of inorganic Darwinism—are finally aggregated.[19]

In this way it was conceivable that the atomic weights determined by chemists really represented an average weight of several slightly different atoms.

When we say the atomic weight of, for instance, calcium is 40, we really express the fact that, while the majority of calcium atoms have an actual weight of 40, there are not a few which are represented by 39 or 41, a less number by 38 or 42, and so on.[20]

This explained not only why atomic weights were so persistently and tantalizingly close to integral or half-integral values but also why there was a "close mutual similarity, verging almost into identity" in the rare earth elements. Crookes called these closely related atoms "meta-elements" or "elementoids," and in 1915 he claimed that the concept of isotopes was a vindication of his speculation. However, Crookes's sensational concept was too closely bound up with his imaginative notion of the genesis of the elements and his misinterpreted experiments with the rare earths to be anything more than an inkling of Soddy's isotopes, which were a solution to a different problem upon which it had no influence: that of radioactive

series. In 1902 he identified protyle with the electron, and in 1908 he developed a scheme whereby many elements of high atomic weight were degraded into lower elements.

The theory of genesis was the high-water mark of Crookes's speculative genius, and it would be difficult to find its equal in scientific literature. Even the idea of the noble gases was implicit in the double helix model he built; inevitably, in 1898, he took pains to show how well Ramsay's and Rayleigh's new gases fitted into his system.[21] In 1895 Ramsay had asked Crookes to examine the spectrum of a strange gas extracted from cleveite (uraninite), and in a famous telegram Crookes had announced it was helium.

As a product of the Royal College of Chemistry, which originally had been endowed by agricultural interests, Crookes always maintained a lively interest in agricultural matters. He translated several Continental manuals on fertilizers, and he was a powerful advocate of the use of disinfectants such as phenol in the cattle plague that swept Great Britain from 1865 to 1867. As president of the British Association in 1898 he used his platform not only to summarize his interests in science and the occult but also to warn the world that it would be faced by starvation, because of the failure of wheat supplies to keep abreast of population, unless science found new sources of fertilizers. Many took this as sensationalism, but Crookes had intended to be optimistic that chemists could and would solve the problem of tapping the huge supplies of atmospheric nitrogen for fertilizers. His sensible remarks on the wheat problem caused a furore among economists and politicians and were a factor in the motives that drove O. C. Birkeland and Fritz Haber to develop industrial processes for nitrogen fixation.

In 1900 Crookes was sixty-eight; yet far from retiring from research, he threw himself wholeheartedly into solving the new conundrum of radioactivity, driven in part by the knowledge that a sojourn in South Africa in 1895 had prevented him from beating Roentgen to the discovery of X rays. Despite his speculative powers, Crookes at first took a conservative view of this new science, for he could not believe that radioactive elements decayed spontaneously, since this seemed to imply a violation of the conservation of energy. It was his view, expressed between 1898 and 1900, that the source of activity was external to the radioactive element. He imagined that radium, say, had the ability to act as a Maxwellian demon and select from the atmosphere those air particles which were moving more swiftly than the average, absorb some of their energy, and eject them at a lower speed.

This theory, which never received full publication,

contravened the second law of thermodynamics; and although Crookes thought that he might have experimental support for it, his evidence did not measure up to the critical scrutiny of Stokes.[22] In December 1900 he confided to Stokes that he believed radioactivity was due to "bodies smaller than atoms," and by 1906 he had fully accepted Rutherford's subatomic transmutation theory of radioactivity. Crookes had supplied a fundamental datum for this theory in 1900, when, using photographic plates as indicators of activity, he had shown that if uranium was purified, it could be separated chemically into a nonactive portion and a radioactive portion that he called uranium X. In 1901 Becquerel showed that within a few months the uranium X had lost much of its activity, whereas the previously inactive portion of the separated uranium had regained its radioactivity and would produce further quantities of uranium X. Soddy subsequently showed that Crookes's use of photographic plates masked the fact that uranium emitted α rays, whereas uranium X emitted penetrating β rays. As if in answer to this, in 1903 Crookes developed the spinthariscope, a device whereby the α particles emitted from a sample of a radioactive material were indirectly rendered visible through a simple microscope by their bombardment of a phosphorescent zinc sulfide screen.[23] Even after the invention of more sophisticated electric counting devices, Rutherford and his co-workers used this scintillation-counting method for the estimation of α activity.

Although Crookes produced a dozen more experimental papers between 1910 and 1919, his ability to keep abreast, and ahead, of his contemporaries inevitably declined. His presidency of the Royal Society from 1913 to 1915 was marred not only by the outbreak of war but also by a degree of ill feeling from the younger generation of fellows that he had sowed the wild oats of genius past his allotted time.

NOTES

1. "The Preparation of Eye-preserving Glass for Spectacles," in *Philosophical Transactions of the Royal Society,* **214A** (1914), 1–25.
2. W. B. Carpenter, "The Radiometer and Its Lessons," in *Nineteenth Century,* **1** (1877), 242–256.
3. "On the Supposed 'New Force' of M. J. Thore," in *Philosophical Transactions of the Royal Society,* **178A** (1888), 451–469.
4. E. E. F. d'Albe, *Life of Crookes,* pp. 404–406.
5. Speech at banquet for past presidents of the Chemical Society, in *Chemical News,* **102** (1910), 252.
6. "Researches on the Atomic Weight of Thallium," in *Philosophical Transactions of the Royal Society,* **163** (1875), 277–330. In modern gravimetry reduction of weighings to a vacuum standard is made from calibration tables, and it is not necessary to use a vacuum balance except in the most rigorous investigations.

7. All Stas's weighings were based on a silver standard of 107.93 (modern, 107.880). This gave an atomic weight of nitrogen of 14.04 (modern, 14.008) and helped to lower Crookes's value for thallium, estimated from thallium nitrate, by 0.1 percent. The atomic weight of silver was reappraised by Alexander Scott and Theodore Richards between 1900 and 1910.
8. "On the Attraction and Repulsion Resulting From Radiation," in *Proceedings of the Royal Society,* **23** (1874–1875), 373–378.
9. "On Repulsion Resulting From Radiation, Parts III and IV," in *Philosophical Transactions of the Royal Society,* **166** (1877), 325–376; see 375–376, added 17 January 1877.
10. In fact the simple kinetic theory explanation was later rejected by Maxwell and Reynolds. Unequal heating of the radiometer vanes led to a temperature gradient that caused "thermal creep" or vortex motions of the gas particles. A rise in pressure occurred on the hotter side of the vanes, which consequently moved away from the incident radiation. See J. Clerk Maxwell, "On Stresses in Rarified Gases Arising From Inequalities of Temperature," *ibid.,* **170** (1879), 231–256; and O. Reynolds, "On Certain Dimensional Properties of Gases," *ibid.,* 727–845.
11. "On the Viscosity of Gases at High Temperatures," *ibid.,* **172** (1881), 387–434.
12. "On the Illumination of Lines of Molecular Pressure, and the Trajectory of Molecules," *ibid.,* **170** (1879), 135–164; quotation, 135.
13. "Experimental Contributions to the Theory of the Radiometer," in *Proceedings of the Royal Society,* **25** (1876–1877), 304–314; see 308.
14. See also C. F. Varley, "Some Experiments on the Discharge of Electricity Through Rarified Media and the Atmosphere," *ibid.,* **19** (1870–1871), 236–240.
15. "Radiant Matter," in *Chemical News,* **40** (1879), 91–93, 104–107, 127–131; see 130–131.
16. Crookes, "On the Fractionation of Yttria," *ibid.,* **54** (1886), 158.
17. *Reports of the British Association for the Advancement of Science,* **55** (1886), 558–576; *Journal of the Chemical Society,* **53** (1888), 487–504.
18. J. E. Reynolds, "Note on a Method of Illustrating the Periodic Law," in *Chemical News,* **54** (1886), 1–4.
19. *Reports of the British Association for the Advancement of Science,* **55** (1886), 558–576; see 569.
20. *Ibid.*
21. "On the Position of Helium, Argon, and Krypton in the Scheme of Elements," in *Proceedings of the Royal Society,* **63** (1898), 408–411.
22. "Sur la source de l'energie dans les corps radio-actifs," in *Compte rendu hebdomadaire de l'Académie des sciences,* **128** (1899), 176–178; Larmor, *Stokes Correspondence,* I, 478–481; see also J. Elster and H. Geitel, "Versuche an Becquerelstrahlen," in *Annalen der Physik,* **66** (1898), 735–740.
23. "Certain Properties of the Emanations of Radium," in *Chemical News,* **87** (1903), 241.

BIBLIOGRAPHY

I. Original Works. Crookes published between 250 and 300 papers, many of which are scattered through obscure Victorian newspapers and periodicals. A fairly complete list of his papers may be composed from *Royal Society Catalogue of Scientific Papers (1800–1900); Index to the Proceedings of the Royal Society of London (Old Series) vols. 1–75, 1800–1905* (London, 1913); and *Index to the Proceedings of the Royal Society of London (1905–1930) and to the Philosophical Transactions of the Royal Society of London (1901–1930)* (London, 1932). Articles in *Nature* and *Popular Science Review* may be traced by consulting the not entirely reliable indexes of these journals. A large number of uncataloged papers will be found in the defunct journals that Crookes edited during the years noted: *Liverpool Photographic Journal* (1857–1858), *Photographic News* (1858–1859), *Chemical News* (1859–1919; from 1906 until its failure in October 1932, this was edited by Crookes's assistant, J. H. Gardiner), *Quarterly Journal of Science* (1864–1878), *Journal of Science* (1879–1885), *Electrical News and Telegraphic Reporter* (1875). In addition the following are important articles: T. Belt (Crookes's patent agent), "Improvements in Gold and Silver Amalgamation," in *Mining Journal*, **35** (1865), 407; "Another Lesson From the Radiometer," in *Nineteenth Century*, **1** (1877), 879–887; and "Some Possibilities of Electricity," in *Fortnightly Review*, **51** (1892), 173–181.

Crookes's writings on the paranormal that were not republished in his book of 1874 include "On the Scientific Investigation of Psychic Phenomena," in *The Spiritualist* (15 June 1871), 161; "A Scientific Examination of Mrs. Fay's Mediumship," *ibid.* (12 Mar. 1875), 126–128; his dispute with J. Spiller over Home's mediumship, in *The Echo* (Nov. 1871), *passim*, and in *English Mechanic and World of Science*, **14** (1871), 200, 253, 327; his dispute with W. B. Carpenter over Mrs. Fay, in *Nature*, **17** (1877), 7–8, 43–44, 200; "Notes on Séances With D. D. Home," in *Proceedings of the Society for Psychical Research*, **6** (1889–1890), 98–127; "Differences in 'Psychical Phenomena' With Eusapia Paladino and D. D. Home," in *Journal of the Society for Psychical Research*, **6** (1893–1894), 341–345; "Experiments With D. D. Home and Others," *ibid.*, **9** (1899–1900), 147–148, 324; "Presidential Address," in *Presidential Addresses to the Society for Psychical Research, 1882–1911* (London, 1912), pp. 86–103. These psychic writings, as well as other unpublished letters, are to be published by R. G. Medhurst and Mrs. K. M. Goldney.

Crookes published the following books: *A Handbook to the Waxed Paper Process in Photography* (London, 1857); *On the Application of Disinfectants in Arresting the Spread of the Cattle Plague* (London, 1866), offprinted from *Third Report of the Commissioners Appointed to Inquire Into the Origin of the Cattle Plague; On the Manufacture of Beet-Root Sugar in England and Ireland* (London, 1870); *Select Methods in Chemical Analysis (Chiefly Inorganic)* (London, 1871; 2nd ed., 1886; 3rd ed., 1894; 4th ed., 1905); *Researches in the Phenomena of Spiritualism* (London, 1874; reissued London, 1903, 1953; Manchester, 1926), also in several translations; *A Practical Handbook of Dyeing and Calico Printing* (London, 1874; 2nd ed., 1883); *Dyeing and Tissue Printing* (London, 1882); *London Water Supply* (London, 1882), a report presented to the Local Government Board by Crookes, W. Odling, and C. Meymott Tidy; *The Wheat Problem* (London, 1899; 2nd ed., 1905; 3rd ed., 1917); and *Diamonds* (London, 1909).

Crookes also translated and edited B. Kerl, *A Practical Treatise on Metallurgy* (London, 1868), adapted with the help of E. Röhrig; M. Reimann, *On Aniline and Its Derivatives* (London, 1868); R. Wagner, *A Handbook of Chemical Technology* (London, 1872) and *Manual of Chemical Technology* (London, 1892); G. Auerbach, *Anthracen* (London, 1877); G. Ville, *On Artificial Manures* (London, 1879; 2nd

ed., 1882; 3rd ed., 1909) and *The Perplexed Farmer* (London, 1891); and K. B. Lehmann, *Methods of Practical Hygiene*, 2 vols. (London, 1893).

Crookes edited M. Faraday, *A Course of Six Lectures on the Various Forces of Matter and Their Relations to Each Other* (London, 1860; many reprints) and *Chemical History of A Candle* (London, 1861; many reprints); W. Odling, *A Course of Six Lectures on the Chemical Changes of Carbon* (London, 1869); and J. Mitchell, *A Manual of Practical Assaying* (3rd ed., London, 1868; 4th ed., 1873; 5th ed., 1881; 6th ed., 1888). Patent literature should be consulted for Crookes's several specifications.

Crookes's unpublished material was extensive. E. E. F. d'Albe (below) had some 40,000 documents available to him in 1923. Most of these have disappeared; but some have been dispersed to Science Museum Library, London: laboratory weighing book, 1860–1862; laboratory notebooks, vols. 1–5, 1876–1881, mainly in the hand of Gimingham; and 51 letters from Crookes to Gimingham (1871–1877); Science Museum, London: notebook containing newspaper clippings on thallium, 1861–1864; Royal Institution, London: laboratory notebooks, vols. 6–21, 1881–1919, mainly in the hand of Gardiner; Royal Society, London; correspondence with J. F. W. Herschel, Arthur Schuster, and others; Imperial College archives, London: correspondence with Henry E. Armstrong and S. P. Thompson; University College, London: Sir William Ramsay archive; Wellcome Historical Medical Library, London: correspondence with John Spiller; Psychical Research Society, London: miscellaneous correspondence and Sir Oliver Lodge papers; Cambridge University library: papers of Sir George Stokes; and Britten Memorial Library, The Spiritualists' National Union, Ltd., Tib Lane, Manchester: uncataloged miscellaneous correspondence.

Examples of Crookes's radiometers, discharge tubes, and other apparatus may be seen at the Royal Institution and the Science Museum, London, where a mercury pump by Gimingham should also be studied. Crookes's personal effects may be appraised from John Butler and Co., Ltd., *Auction Sale of Sir William Crookes* (London, n.d.), the sole copy of which is in the Oxford Museum of the History of Science, Gabb papers, item 7.

II. SECONDARY LITERATURE. In view of the wholesale destruction of manuscripts after 1923 the fundamental source is E. E. Fournier d'Albe, *The Life of Sir William Crookes O.M., F.R.S.* (London, 1923); see informative reviews by Sir Oliver Lodge, in *Proceedings of the Society for Psychical Research*, **34** (1924), 310–323, with a "correction" in the *Journal of the Society for Psychical Research*, **22** (1925–1926), 80; and by A. Smithells, in *Nature*, **113** (1924), 227–228. For a complete list of the honors awarded to Crookes, see *Who Was Who*, II (4th ed., London, 1967), 247–248. Obituaries and articles that add significantly to d'Albe are (listed chronologically) P. Zeeman, "Scientific Worthies: Sir William Crookes, F.R.S.," in *Nature*, **77** (1907–1908), 1–3; A. Tilden, obituary, in *Proceedings of the Royal Society*, **96A** (1920), i–ix, reprinted with additions in his *Famous Chemists* (London, 1921), pp. 259–272; Sir W. Barrett, obituary, in *Proceedings of the Society for Psychical*

Research, **31** (1921), 12–29; K. Przibram, "Crookes," in G. Bugge, ed., *Das Buch der grossen Chemiker* (Leipzig, 1930), II, 288–297; Lord Rayleigh, "Some Reminiscences of Scientific Workers of the Past Generation and Their Surroundings," in *Proceedings of the Physical Society,* **48** (1936), 236–242, with photographs of Crookes's laboratories; O. T. Rotini, "Una lettera inedita di J. S. Stas [to Crookes] sopra l'ipotesi di Prout," in *Actes du VIIIe Congrès international d'histoire des sciences (Florence–Milan, 1956)* (Paris, 1958), II, 486–496; F. Greenaway, "A Victorian Scientist: The Experimental Researches of Sir William Crookes (1832–1919)," in *Proceedings of the Royal Institution of Great Britain,* **39** (1962), 172–198; A. E. Woodruff, "William Crookes and the Radiometer," in *Isis,* **57** (1966), 188–198; and J. C. Chaston, "Sir William Crookes. Investigations on Indium Crucibles and the Volatility of the Platinum Metals," in *Platinum Metals Review,* **13** (1969), 68–72.

For Crookes's work with disinfectants, see S. A. Hall, "The Cattle Plague of 1865," in *Medical History,* **6** (1962), 45–58. A useful survey of the context of Crookes's work on rare earths is J. W. Mellor, *A Comprehensive Treatise on Inorganic and Theoretical Chemistry,* 16 vols. (London, 1922–1937), V, 497–503. His work on helium is discussed in M. W. Travers, *A Life of Sir William Ramsay, K.C.B., F.R.S.* (London, 1956), *passim* (index unreliable). On the cooperation with Stokes, consult J. Larmor, ed., *Memoir and Scientific Correspondence of the Late Sir George Gabriel Stokes, Bart.,* 2 vols. (Cambridge, 1907), esp. II. 362–494.

The violent controversies over Crookes's interest in the paranormal are best explored through a critical appraisal of T. H. Hall, *The Spiritualists, the Story of Florence Cook and William Crookes* (London, 1962); R. G. Medhurst and K. M. Goldney, "William Crookes and the Physical Phenomena of Mediumship," in *Proceedings of the Society for Psychical Research,* **54** (1963–1966), 25–157; E. J. Dingwall, *The Critics' Dilemma* (Crowhurst, Sussex, 1966); and A. Gauld, *Founders of Psychical Research* (London, 1968).

W. H. BROCK

CROONE, WILLIAM (*b.* London, England, 15 September 1633; *d.* London, 12 October 1684), *physiology.*

Croone was the fourth child of a London merchant and perhaps the cousin of Robert Boyle. He was educated at Merchant Taylors' School and Emmanuel College, Cambridge, where, after taking his first degree in arts in 1650, he was elected to a fellowship. Twelve years later he was created M.D. by royal mandate, and in 1674 he was made a fellow of the Royal College of Physicians. In 1659 he had been appointed professor of rhetoric at Gresham College, London, where his colleagues were Lawrence Rooke, Christopher Wren, William Petty, and Jonathan Goddard. He was thus drawn into the circle of men out of whose meetings grew the Royal Society. An active interest in experimentation led to his being named "register" and involved him in planning the early program of the new society. When it received its charter in 1662, Croone became one of its first fellows, often served on its council, and remained one of the most active of its early fellows until his death.

Croone saw nature as a whole and often tried to apply his physical experience to the better understanding of living processes. As an experimental physicist his most significant discovery was that water has its maximum density above its freezing point; he was closely associated with Boyle in the latter's classic studies of aerial physics.

Croone's reports to the Royal Society dealt with a variety of subjects, but he had a particular and continuing interest in muscular action and embryology. "On 4 June 1662 Mr. Croune [*sic*] presented two embryos of puppy dogs, which he had kept eight days and were put in spirit of wine in a glass-vial sealed hermetically"[1]—an experiment that is regarded as one of the earliest, if not the first, use of alcohol as a preservative of animal tissues. Croone may have got the idea from Boyle, who was experimenting with alcohol before 1662. There are later reports (1662–1664) recording observations on developing embryos. About that time many authors were at work on the developing chick egg. On 22 February 1672 Malpighi sent to the society a discourse on the chick embryo before and after incubation. A week later Croone stated that he had found the "rudiments of the chick in the egg before incubation" and read the first part of his paper "De formatione pulli in ovo"[2] on 14 March; Cole described it as the "first reasoned attempt, based on observation and illustration, to establish the corporeal existence of preformed foetus in the unincubated egg."[3]

Croone had been the earliest among his contemporaries to concern himself with muscle motion. He published his essay *De ratione motus musculorum* in 1664, and much of the subsequent discussion on this subject was either a commentary on Croone's views or a derivation from them. In the seventeenth century, before Croone, the ideas of muscle motion were derived from Erasistratus and Galen and represented little advance on them. Erasistratus considered that when a muscle is filled with *pneuma,* its breadth increases while its length diminishes; and for this reason it is contracted. Galen did not offer any mechanism to account for muscle contraction but viewed muscle as simply moved by the motor faculty that comes from the brain. To him muscle was made up of fibers and flesh, the fibers being continuous with those of the tendons at either end. Croone, while accepting Galen's concept of the structure of the muscle as tendinous fibers and flesh, describes experiments which show that something necessary to the contraction of muscle does pass along the nerve from the brain. Although he uses the word "spirit" in considering the nature of this force, he makes it a definite material substance, "a rectified and enriched juice." He developed the idea that the nerves or nerve membranes are in a state of tension and that the nerve impulse consists of vibrations transmitted along the nerve in the same way that vibrations are transmitted along the tightened string of a musical instrument. Croone emphasized the role that blood might play in the contraction of the muscle. Not only did he consider that the muscle contraction was brought about by the interaction of the nervous "juice" and the blood, but he also thought that the circulation of the blood within the muscle would be stimulated by its contraction. Since the muscle swelled in contraction, its artery would also necessarily be enlarged and would, therefore, offer less resistance to the inflowing blood.

When the Royal Society suspended its meetings owing to plague in 1665, Croone took the opportunity to visit Montpellier, where he met the young Danish anatomist Steno. Steno had been working at Leiden and Paris with Jan Swammerdam, who had demonstrated experimentally that muscles do not swell during contraction.[4] Steno was thus skeptical of Croone's theory, especially, he argued, since not enough was known about fluids contained within muscles.[5]

On his marriage in 1670 Croone had had to surrender his professorship at Gresham College and succeeded Sir Charles Scarburgh as lecturer in anatomy at Surgeons' Hall. During the years that followed, he evidently continued to ponder the problem of muscle motion. Since his assumption that muscle swelled in contraction had been proved false experimentally, he sought a mechanism that would produce contraction without swelling. He therefore suggested that instead of the whole muscle's forming one large bladder, each individual muscle fiber formed a series of very minute bladders. This modified theory he presented in his lectures in 1674 and 1675,[6] having estimated by a mathematical calculation that the smaller and more numerous the bladders were in the fiber, the smaller would be the outward swelling of the muscle in contraction. This hypothesis was similar to one advanced by Giovanni Borelli in his *De motu animalium* (1680–1681).

Croone made two important contributions to muscle physiology. He altered the concept of "spirits" from that of a vague and ethereal "wind" to that of a definite physical "juice" and thereby made them susceptible to observation and reason, and he introduced the idea that some sort of chemical reaction may be involved in muscle contraction. Furthermore,

he suggested that nerve impulse might be a disturbance or vibration moving along the nerves rather than a flow of substance.

NOTES

1. T. Birch, I, 84.
2. *Ibid.,* III, 30–41.
3. F. J. Cole, *Early Theories of Sexual Generation* (Oxford, 1944), p. 47.
4. Jan Swammerdam, *The Book of Nature* (London, 1758), p. 127.
5. Niels Steno, *Elementorum myologiae specimen* (Florence, 1667), pp. 63–64.
6. *Philosophical Collections,* no. 2, 22–25.

BIBLIOGRAPHY

Among Croone's writings are *De ratione motus musculorum* (London, 1664; 2nd ed., Amsterdam, 1667); "An Hypothesis of the Structure of a Muscle, and the Reason of Its Contraction," in *Philosophical Collections,* Robert Hooke, ed. (1681), no. 2, 22–25; and "De formatione pulli in ovo," in Birch, cited below, appears in trans. in F. J. Cole, "Dr. William Croone on Generation," in *Studies and Essays in the History of Science and Learning Offered in Homage to George Sarton* (New York, 1947), M. F. A. Montagu, ed.

For discussions of Croone's life and theories, see L. M. Payne *et al.,* "William Croone," in *Notes and Records of the Royal Society of London,* 15 (1960), 211–219; and L. G. Wilson, "William Croone's Theory of Muscle Contraction," *ibid.,* 16 (1961), 158–178. See also T. Birch, *History of the Royal Society of London,* 4 vols. (London, 1756–1757), *passim.*

LEONARD M. PAYNE

CROSS, CHARLES WHITMAN (*b.* Amherst, Massachusetts, 1 September 1854; *d.* Chevy Chase, Maryland, 20 April 1949), *petrology.*

Cross was the son of Rev. Moses Kimball Cross and Maria Mason Cross. In 1875 Amherst College awarded him the B.S. degree; he then attended Göttingen and received the Ph.D. at Leipzig in 1880 under Ferdinand Zirkel. On his return to America, Cross joined the U. S. Geological Survey, which he served for forty-five years. He received an honorary D.Sc. degree from Amherst College in 1925. He married Virginia Stevens and had one son, Richard Stevens Cross.

Cross's accurate field investigations, which were carried out primarily in Colorado, constitute his greatest contribution to geology. He and his assistants mapped the rock formations of seven quadrangles in the rugged terrain of the San Juan Mountains, covering over 1,800 square miles on foot and horseback. The region includes the mining camps of Ouray,

Telluride, Silverton, Rico, Needle Mountains, Lake City, and La Plata. About 1,000 square miles were mapped geologically in the Pikes Peak area, which includes the Cripple Creek mining camp, and another 1,000 square miles in the Apishapa quadrangle. In addition to the broad aspects of the rocks as geologic units, Cross tended to the collection, description, and classification of the rocks from these areas and was concerned with the processes that had produced them. Other notable petrologic studies include those of the potash-rich rocks of the Leucite Hills, Wyoming, the first leucite-bearing rocks discovered on the American continent; and the lavas of the Hawaiian Islands, which were found to have a considerable range of composition. Cross discovered an occurrence of trachyte on the island of Hawaii, previously considered generally basaltic in composition.

In addition to his vast fieldwork, Cross collaborated with Joseph P. Iddings, Louis V. Pirsson, and Henry S. Washington in devising a quantitative chemico-mineralogical classification and nomenclature of igneous rocks, commonly referred to as the C.I.P.W. system. From the chemical analysis of a rock, they calculated a "norm" made up of simple theoretical minerals that may be taken to represent, for the most part, the end members of the minerals actually occurring in the igneous rocks. This ingenious scheme, still widely used today, has not only led to an appreciation of the chemical composition of natural igneous rocks but has also served as the basis of much of the experimental investigation of those rocks.

Cross was among the group of geologists who helped persuade the Carnegie Institution of Washington to establish a laboratory (the Geophysical Laboratory since 1906) for the experimental investigation of rocks at high pressures and temperatures. He was a charter member of the Petrologists' Club (organized 1910) along with Bancroft, Bastin, Johannsen, Knopf, Larsen, Lindgren, Paige, and Schaller, among others. The organizational meeting and numerous successive meetings, at which many major petrological concepts were first discussed, took place in his home. As a member of the National Academy of Sciences (elected 1908), Cross served in other ways to advance the progress of geology—for example, by taking an active part in the organization of the National Research Council. He was a member of the Geological Society of America and served as its president in 1918.

Retirement in 1925 did not mark the end of Cross's scientific career; his systematic methods were transferred with equal vigor to the cultivation of roses. He developed new varieties, some of which became available commercially. The outstanding varieties include "Chevy Chase," "Mrs. Whitman Cross," and "Hon. Lady Lindsay." His successes also extended to investment and finance; however, there is no record of his achievements in golf, his chief recreation.

BIBLIOGRAPHY

U.S. Geological Survey folios of Colorado for which Cross was responsible are no. 7, Pikes Peak Sheet (1894); no. 57, Telluride Quadrangle (1899), with C. W. Purington; no. 60, La Plata Quadrangle (1899), with A. C. Spencer and C. W. Purington; no. 120, Silverton Quadrangle (1905), with E. Howe and F. L. Ransome; no. 130, Rico Quadrangle (1905), with F. L. Ransome; no. 131, Needles Mountain Quadrangle (1905), with E. Howe, J. D. Irving, and W. H. Emmonds; no. 153, Ouray Quadrangle (1907), with E. Howe and J. D. Irving; no. 171, Engineer Mountain Quadrangle (1910), with A. D. Hole; and no. 186, Apishapa Quadrangle (1912), with G. W. Stose.

His publications on regional petrology include "Igneous Rocks of the Leucite Hills and Pilot Butte, Wyoming," in *American Journal of Science,* 4th ser., **4** (1897), 115–141; *Lavas of Hawaii and Their Relations,* U.S. Geological Survey Professional Paper no. 88 (Washington, D.C., 1915); and *Geology and Petrology of the San Juan Region of Southwestern Colorado,* U.S. Geological Survey Professional Paper no. 258 (Washington, D.C.), written with E. S. Larsen, Jr.

On the C.I.P.W. system, see "A Quantitative Chemicomineralogical Classification and Nomenclature of Igneous Rocks," in *Journal of Geology,* **10** (1902), 555–690, with J. P. Iddings, L. V. Pirsson, and H. S. Washington; *Quantitative Classification of Igneous Rocks, Based on Chemical and Mineral Characters, With Systematic Nomenclature* (Chicago, 1903), written with Iddings, Pirsson, and Washington; and "The Natural Classification of Igneous Rocks," in *Quarterly Journal of the Geological Society of London,* **66** (1910), 470–506.

H. S. YODER, JR.

CROUSAZ, JEAN-PIERRE DE (*b.* Lausanne, Switzerland, 13 April 1663; *d.* Lausanne, 22 February 1750), *theology, philosophy, mathematics.*

The second son of Abraham de Crousaz and Elisabeth Françoise Mayor, Crousaz belonged to one of the oldest and most noble families of Lausanne, which was then ruled by Bern. Destined at first for a military career by his father, who was a colonel, he eventually obtained permission to study philosophy and Reformed theology at the Academy of Lausanne and then at the Academy of Geneva. At the age of nineteen he went to Leiden and to Rotterdam, where he met the philosopher Pierre Bayle, and from there to Paris. Returning home in 1684, he commenced a theological career, married Louise de Loys, by whom he had three sons and four daughters, and gave private lessons in ancient languages and in the sciences. In 1700 Crousaz became a professor of philosophy and mathematics at the Academy of Lausanne; in his classes he abandoned the traditional use of Latin in favor of French. He was elected rector in 1706 and held this position for three more terms.

Crousaz's existence was arduous: he was a minister; besides his large family, he had in his home boarders, sons of foreign and local noblemen, for whose education and instruction he was responsible; and he wrote books, such as the treatise *Logique* (which first appeared in two volumes, then underwent numerous alterations and had many editions—the fourth, in 1741, reached six volumes) and the *Traité du beau* (first in one volume, then in two). Receptive to all the intellectual currents of the age, a great correspondent, and a virulent polemicist, Crousaz poured into his writings, which he continued to produce throughout his life, all of his immense knowledge and enriched them with reflections and comments that are still of interest.

Mysterious intrigues, tenacious jealousies, and a growing hostility annoyed Crousaz and wounded his pride, although it must be admitted that humility and resignation were certainly not the dominant traits in his character. Having decided to leave Lausanne, he accepted a chair of philosophy and mathematics at the University of Groningen, where he stayed from September 1724 until March 1726 with his wife, two of his daughters, and a large retinue. He then accepted a call from the house of Hesse-Kassel to direct the education of the young heir, Prince Frederick. Crousaz had acquired a solid pedagogical reputation through his *Nouvelles maximes sur l'education des enfants,* published at Amsterdam in 1718, and especially through the *Traité de l'education des enfants,* published in two volumes at The Hague in 1722. (Rousseau had read these two works before writing *Émile.*) His self-esteem flattered by the titles "counselor of the court" and "governor of the prince," Crousaz began a new life that, despite many obligations, left him leisure for his own work—especially since his family had returned to Lausanne. During this period he wrote *Examen du Pyrrhonisme,* a large work in which he refutes the philosophy of Bayle. The stay at Kassel lasted until 1733, when Crousaz was seventy years old; the landgrave of Hesse then granted him a life pension.

He returned to the Academy of Lausanne, where he once more held the chair of philosophy, a post that, thanks to his exceptional health, he did not relinquish until 1749, a year before his death. In this last prolific period he wrote *De l'esprit humain,* a harsh critique

of Leibniz's preestablished harmony, which presupposed a determinism denying free will. (Mme. du Châtelet wrote him a long letter, dated 9 August 1741, defending Leibniz.) Before the end of his long life Crousaz had the satisfaction of following the brilliant scientific career of his grandson Philippe de Loys de Chéseaux (1718–1751). Previously, upon returning from Kassel, he had the pleasure of finding his eldest son, Abraham, as rector of the Academy of Lausanne.

During the first half of his life Crousaz was a follower of Cartesianism, but he later became a defender of the ideas of Newton. It appears, however, that in each case his choice was dictated solely by philosophical and moral arguments, since he placed philosophy, theology, and pedagogy above mathematics and physics. His attacks on Leibniz, moreover, were directed only against the *Théodicée,* for the simple reason that he saw in this system a danger to religion. He had furnished proof of this courageous attitude in his youth while at the Academy of Lausanne, during the period when the teaching of Cartesianism was suspected of undermining orthodoxy.

As for his scientific work, Crousaz was certainly prolific, even prolix, and his treatises on elementary mathematics must have been of some interest and utility in his time. They offer, however, little that is really original; style is often neglected, with a great many repetitions; and the calculations contain errors. In analysis the work that made him famous is the *Commentaire sur l'analyse des infiniment petits,* a large quarto volume of 320 pages, with four plates, which appeared in 1721—three years after the work of l'Hospital, the contents of which Crousaz examines step by step. Long, superfluous passages and excessively elementary detail result in a certain lack of balance in the exposition, in which, nevertheless, one can find new examples. Crousaz had fully assimilated infinitesimal calculus, and he no doubt taught it correctly to his students. Despite its errors and insufficiencies, his book proves this, as does the welcome accorded it by contemporary scientists, Johann I Bernoulli excepted. Indeed, on 15 July 1722 the latter wrote Crousaz a long, very harsh letter filled with criticisms and more or less veiled reproaches. Bernoulli states, however (which is at least curious), that he has read only the preface and the preliminary discourses; he thus allows one to infer that he did not find his name in these pages. The reproach is justified, but if Bernoulli had continued reading, he would have found, on page 303, his name and some kind remarks about himself.

In any case, it is certain that this book contributed in great part to the nomination of Crousaz, in 1725, as associate foreign member of the Académie des

Sciences of Paris. He had already received from this academy, in 1720, first prize in the annual competition for a memoir on the theory of movement. Regarding this work, Johann I Bernoulli criticized him (in the letter mentioned above) for being too Cartesian and for having, like Descartes, made "the essence of bodies consist in extension alone and that of movement in the successive application of their surfaces to the surfaces of contiguous bodies." Crousaz was also three times (1721, 1729, 1735) a laureate of the Académie des Sciences of Bordeaux for his memoirs on the causes of elasticity, on the nature and propagation of fire, and on the different states of matter. In December 1735 he was elected a member of that academy.

Crousaz was also a great letter writer. Some 1,800 letters sent or received by him are preserved at Lausanne, and others are at Geneva, Basel, Paris, London, Leiden, Kassel, and elsewhere; they are almost all unpublished. Among his correspondents were Réaumur, Abbé Nollet, Mairan, Maupertuis, Jacques Cassini, Fontenelle, Bernoulli, and Maclaurin. With Réaumur he obviously discussed natural history, particularly certain shellfish found near Neuchâtel; with Nollet, it was electricity; Mairan was his closest confidant; to Maclaurin he addressed, in 1721, a memoir dealing with optics that, revised, was presented to the Academy of Bordeaux in 1736 under the title "Propagation de la lumière." The secretary of this academy, Sarrau, announced his agreement with the Cartesian conclusions proposed by Crousaz.

Long called "the celebrated professor," Crousaz appears to have had an encyclopedic mind—a mind representative of his age, of his country, and of Europe in the eighteenth century. His merit, if one accepts the *éloge* delivered by Grandjean de Fouchy, "seems to have been less excellence and superiority in a certain genre than universality of knowledge and of literary talents."

BIBLIOGRAPHY

I. ORIGINAL WORKS. Crousaz's scientific writings are *Réflexions sur l'utilité des mathématiques et sur la manière de les étudier, avec un nouvel essai d'arithmétique démontrée* (Amsterdam, 1715); *La géométrie des lignes et des surfaces rectilignes et circulaires*, 2 vols. (Amsterdam, 1718; Lausanne, 1733); *Commentaire sur l'analyse des infiniment petits* (Paris, 1721); *Discours sur le principe, la nature et la communication du mouvement* (Paris, 1721); *De physicae origine, progressibus ejusque tractandae methodo...* (Groningen, 1724); *De physicae utilitate dissertatio philosophica* (Groningen, 1725); *Essai sur le mouvement, où l'on traitte de sa nature, de son origine...* (Groningen, 1726); *Traité de*

l'algèbre (Paris, 1726); *Dissertation sur la nature, l'action et la propagation du feu* (Bordeaux, 1729); and *Dissertation sur la nature et les causes de la liquidité et de la solidité* (Bordeaux, 1735).

II. SECONDARY LITERATURE. The following, listed chronologically, deal with Crousaz or his work: "Éloge de M. de Crousaz," read 14 Nov. 1750, in Grandjean de Fouchy, *Éloges des Académiciens*, I (Paris, 1761), 100–122; Rudolf Wolf, *Biographien zur Kulturgeschichte der Schweiz*, II (Zurich, 1859), 57–70; Eugène Secrétan, *Galerie suisse, Biographies nationales...*, I (Lausanne, 1879), 591–599; Henri Perrochon, "Jean-Pierre de Crousaz," in *Revue historique vaudoise* (1939), 281–298; Suzanne Delorme, *À propos du bicentenaire de la mort de Jean-Pierre de Crousaz: Ses relations avec l'Académie royale des sciences* (Paris, 1954); Jacqueline E. de La Harpe, *Jean-Pierre de Crousaz et le conflit des idées au siècle des lumières* (Geneva, 1955), which is indispensable for the life and literary and philosophical works of Crousaz and contains a complete list of his works; and Marianne Perrenoud, *Inventaire des archives Jean-Pierre de Crousaz* (Lausanne, 1969). A thorough study of his scientific work has yet to be published, and the same is true of his scientific correspondence.

PIERRE SPEZIALI

CRUIKSHANK, WILLIAM CUMBERLAND (*b.* Edinburgh, Scotland, 1745; *d.* London, England, 27 June 1800), *anatomy.*

Cruikshank was the second son of George Cruikshank, supervisor of excise in Glasgow. At the age of fourteen he was sent to the University of Glasgow to study divinity; he matriculated on 14 November 1764 and gave his probationary sermon. Cruikshank put his mastery of Greek, Latin, and French to practical use by giving lessons to other students. On 30 September 1766 he was granted a bursary for two years for the study of theology; it was renewed in 1768 and again in 1770. After receiving the M.A. in 1767, he was appointed tutor in several leading Scottish households, including that of the earl of Dundonald. He became an acquaintance of John Moore, the father of James Carrick Moore, a surgeon; of Sir John Moore, who distinguished himself in the Peninsular War; and of Admiral Sir Graham Moore, the jolly friend of Creevey and the prince regent. Moore encouraged Cruikshank's interest in medicine, particularly in anatomy; and when, in April 1771, William Hunter and William Hewson dissolved their partnership in the Windmill Street School of Anatomy, Cruikshank, on Moore's recommendation, was invited to undertake the care of Hunter's library and museum. His anatomical skill could not fail to recommend him to his new master. To extend the scope of his studies he entered on 31 May 1773 as a surgical pupil at St. George's Hospital, possibly under John Hunter; and before long he took Hewson's place as partner in the anatomy school.

Cruikshank had an unusually good memory and delighted his students not only by his manner of teaching but also by the elegant classical quotations with which he embellished his lectures. In addition he enhanced his reputation by his researches, most of which were suggested and supervised by William Hunter. Reports of his success in injecting the lymphatic vessels in the lungs (and occasionally the thoracic duct) with quicksilver were made in the *Medical and Philosophical Commentaries* in 1774; and two years later, in the same journal, he described "a disease which is not uncommon here"—the presence of loose cartilages in joints—and the method of cure proposed by John Hunter. His "Experiments on Nerves, Particularly on Their Reproduction; and on the Spinal Marrow of Living Animals" was communicated by John Hunter to the Royal Society on 13 June 1776 but was not published until 1795.

In 1778 he carried out a series of experiments to confirm Regnier de Graaf's theories on generation in the rabbit; his observations were reported to the Royal Society by Everard Home on 23 March 1797. In his historical introduction to this paper Cruikshank mentions that in 1672 De Graaf had seen the ovum of a rabbit but "had the fate of Cassandra, to be disbelieved even when he spoke the truth!" Discussing this subject with William Hunter in 1778, Cruikshank remarked that he would like to repeat De Graaf's experiments and was immediately encouraged to pursue the project. Twenty-nine experiments were performed; and from them he was able to state that the ovum is formed in, and comes out of, the ovarium after conception. It then passes down the fallopian tube, in which it is sometimes retained and prevented from entering the uterus. A successful passage to the uterus takes four days. De Graaf did not see the fetus until the tenth day, whereas Cruikshank saw it on the eighth; and De Graaf observed only one ovum, while Cruikshank observed twenty-eight. By means of these experiments Cruikshank was able to throw some light on human reproduction, particularly on the causes of extrauterine pregnancies.

Cruikshank published accounts of three more investigations in 1779: a new method of introducing mercury into the circulation, the absorption of calomel from the internal surface of the mouth, and the insensible perspiration of the human body.

Cruikshank's best-known piece of research was *The Anatomy of the Absorbing Vessels of the Human Body* (1786). Like the Hunters, he maintained that the red veins do not possess any absorbent power and that "the lacteals and lymphatics of the human body are not a trifling appendage of the red veins, but form of themselves a grand system for absorption and, so far as we have yet discovered them, are not only equal in number to the arteries and veins, but actually surpass them" (1st ed., p. 187; 2nd ed., p. 207). William Hunter offered to make this a joint work but died before it was completed. Since he left no provision for the cost of the numerous proposed illustrations, there were only three in the first edition and five in the second. The drawings were the work of Frederick Birnie, Hunter's amanuensis and artist, and the engravings were by Thornthwaite.

After William Hunter's death in March 1783, the anatomy school was left jointly to his nephew, Matthew Baillie, and to Cruikshank; they continued with unabated success the courses of lectures for which it had become famous.

Cruikshank had a large, although not very lucrative, practice. At that time public dispensaries were few, and the sick poor therefore suffered great hardships. This situation led Cruikshank to give treatment free of charge at his own house; but the numbers who took advantage of his charity soon became embarrassing, and after a while he had to curtail this service.

Throughout his life Cruikshank suffered from a nervous affliction that impaired his manual dexterity and prevented him from reaching a high rank in surgery. Nevertheless, recognition of his valuable contributions to medical knowledge came from many sources. In 1783 he received the M.D. at the University of Glasgow; and he was invited to become an honorary member of the Lyceum Medicum Londinense (founded by George Fordyce and John Hunter in 1785), of the Royal Medical Society of Edinburgh, and of the Imperial Academy of Vienna. He was a candidate for the professorship of anatomy at the Royal Academy, to succeed William Hunter. Samuel Johnson wrote to Sir Joshua Reynolds on his behalf, but the choice fell upon John Sheldon.

Cruikshank was one of those who attended Johnson during his last illness without fee; he inherited a book of his own choice—Johnson's copy of Samuel Clark's edition of Homer. On 1 June 1797 he was elected a fellow of the Royal Society; that September he was called upon to attend—at a fee of a guinea a day— probably his most distinguished patient, Admiral Horatio Nelson, whose arm had been amputated two months previously.

After his marriage in 1773 Cruikshank settled in Leicester Square, next door to Sir Joshua Reynolds, whom he attended professionally (his treatment was much criticized). The oldest of his four daughters married Honoratus Leigh Thomas, president of the Royal College of Surgeons in 1829 and 1838.

In anticipation of the time when William Hunter's museum would be transferred to the University of Glasgow, Cruikshank prepared a supplementary collection to illustrate his lectures, sparing neither time nor expense. This had to a large extent been accomplished at the time of his sudden death, the immediate cause of which was apoplexy. His financial affairs were found to be far from flourishing; and his family accepted an offer from the czar to purchase his museum, which was placed in the Medico-Chirurgical Academy in St. Petersburg in 1805.

BIBLIOGRAPHY

I. ORIGINAL WORKS. Cruikshank's writings are a report on his success in injecting quicksilver, in *Medical and Philosophical Commentaries by a Society of Physicians in Edinburgh,* **1** (1774), sec. 3, 430; a report on the presence of loose cartilages in joints, *ibid.,* **4** (1776), sec. 3, 342–347; "Remarks on the Absorption of Calomel From the Internal Surface of the Mouth, With a Preliminary Sketch of the History and Principal Doctrines of Absorption in Human Bodies," in Peter Clare, *Essay on the Cure of Abscesses by Caustic,* 1st ed. (London, 1779), also published separately; *Remarks on the Insensible Perspiration of the Human Body Showing Its Affinity to Respiration* (London, 1779; 2nd ed., 1795), trans. into German by C. F. Michaelis (1798); "Remarks on the New Method of Introducing Mercury Into the Circulation," in Peter Clare, *Essay on the Cure of Abscesses by Caustic,* 1st ed. (London, 1779), pp. 89–154 (introductory remarks by William Hunter); *The Anatomy of the Absorbing Vessels of the Human Body* (London, 1786; 2nd ed., 1790), trans. into French by Petite-Radel as *Anatomie des vaisseaux absorbans du corps humain* (Paris, 1787) and into German by C. F. Ludwig as part of *Geschichte und Beschreibung der einsaügenden Gefässe oder Saugadern des menschlichen Körpers. Mit einigen Anmerkungen und Kupfertafeln vermehrt . . .,* III (Leipzig, 1794); "Experiments on Nerves, Particularly on Their Reproduction; and on the Spinal Marrow of Living Animals," in *Philosophical Transactions of the Royal Society,* **85** (1795), 177–189; and "Experiments in Which, on the Third Day After Impregnation the Ova of Rabbits Was Found in the Fallopian Tubes and on the Fourth Day After Impregnation in the Uterus Itself; With the First Appearance of the Foetus," *ibid.,* **87** (1797), 197–214.

Proof sheets of *Remarks on the Insensible Perspiration of the Human Body . . .,* with MS notes by Cruikshank, are in the library of the Royal College of Surgeons, as are MS notes of Cruikshank's courses in anatomy, physiology, and surgery, taken by various students; the latter provide a valuable source of information on the style and content of his lectures.

II. SECONDARY LITERATURE. On Cruikshank or his work, see James Boswell, *Life of Samuel Johnson, L.L.D.,* IV (London, 1824), 356, 394; Thomas J. Pettigrew, *Medical Portrait Gallery: Biographical Memoirs of the Most Celebrated Physicians, Surgeons, etc. . . .* (London, 1838–1840), II, no. 6; and Honoratus Leigh Thomas, Hunterian Oration (1827), pp. 11–28. Also of value are *Annals of Medicine,* **5** (1800), 497–503; *Annual Register,* XLII (1800), chronicle 62; *Dictionary of National Biography,* XIII (1888), 260–261; and *European Magazine,* **12** (1800), 171.

JESSIE DOBSON

CRUM, WALTER (*b.* Glasgow, Scotland, 1796; *d.* Rouken, Scotland, 5 May 1867), *chemistry.*

For a period of forty years Walter Crum devoted himself to the improvement of calico printing by the application of chemical knowledge. He attended the University of Glasgow and was a student of Thomas Thomson's from 1818 to 1819. Crum utilized his knowledge of chemical analysis in his father's calico-printing business. His analytical interests extended into pure research, and his fourteen published papers, although on subjects connected with calico printing and dyeing, have a value beyond their applicability to this chemical art. Crum became a fellow of the Royal Society in 1844 and was one of the original members of the Chemical Society of London. He joined the Royal Philosophical Society of Glasgow in 1834 and was its president from 1852. Crum also played a leading role in the public life of Glasgow and was president of Anderson's College, founded in 1796 for the instruction of artisans and others unable to attend the University of Glasgow.

Crum's scientific contributions reveal considerable talent and originality. His very first paper (1823) established his European reputation. Crum undertook an analysis of indigo, preparing pure indigo by sublimation and determining its composition in a remarkably accurate piece of organic research even though organic analysis was still imperfect and difficult. Chemists accepted his formula for indigo. He also investigated the sulfonic acids of indigo, obtaining two new derivatives by the action of sulfuric acid.

Crum made several valuable contributions to analytical chemistry. He discovered the lead dioxide test for manganese. In 1847 he developed an ingenious method for the analysis of nitrates, converting them into nitric oxide by the action of sulfuric acid and mercury. He then applied this method to the determination of the amount of nitrogen in guncotton. Among his other contributions was the first preparation of copper peroxide and of colloidal alumina.

In 1844 Crum proposed a theory of dyeing that involved both a mechanical and a chemical aspect. The dyer applied to cotton the mineral basis for the color in a solution of a volatile acid. On drying, the acid volatilized, leaving the metal oxide to adhere to the fiber. This adherence, Crum asserted, was a mechanical action only and was confined to the interior of the tubular structure of the fiber. On dyeing the cotton there was now a purely chemical attraction between the metal oxide inside the cloth and the organic dye, producing a fixed colored compound in the cloth. Crum claimed that his theory resembled the older mechanical theory of dyeing held by Macquer and others more than it did the newer, purely chemical theory held by Berthollet and most contemporary chemists.

In an earlier work (1830) Crum wrote on colors and presented a color theory resembling Goethe's in that the prismatic colors were produced, contrary to Newton, not from white light but from blackness. Crum had made prismatic experiments since 1822 and concluded that white light did not contain any colors and that black was the source of the three primary colors—red, yellow, and blue—as proved, said Crum, by his experiments in which the recomposition of light from these colors produced black, not white.

BIBLIOGRAPHY

I. ORIGINAL WORKS. Crum's analysis of indigo is in "Experiments and Observations on Indigo and on Certain Substances Which Are Produced by Means of Sulphuric Acid," in *Annals of Philosophy,* **5** (1823), 81–100. His analysis of guncotton is in "On a Method for the Analysis of Bodies Containing Nitric Acid, and Its Application to Guncotton," in *Philosophical Magazine,* **30** (1847), 426–431. The theory of dyeing is in "On the Manner in Which Cotton Unites with Colouring Matter," *ibid.,* **24** (1844), 241–246. Crum's color theory is found in *An Experimental Enquiry Into the Number and Properties of the Primary Colours and the Source of Colour in the Prism* (Glasgow, 1830); there is a French translation by Achille Penot in *Bulletin de la Société industrielle de Mulhouse,* **4** (1831), 544–593. All of his papers are listed in the Royal Society's *Catalogue of Scientific Papers,* II, 101.

II. SECONDARY LITERATURE. There are brief obituary notices on Crum in *Proceedings of the Royal Society,* **16** (1867–1868), viii–x and *Journal of the Chemical Society,* **21** (1868), xvii–xviii.

ALBERT B. COSTA

CRUM BROWN, ALEXANDER. See **Brown, Alexander Crum.**

CRUVEILHIER, JEAN (*b.* Limoges, France, 9 February 1791; *d.* Sussac, Haute-Vienne, France, 10 March 1874), *anatomy, pathology.*

Agrégé in surgery at the Faculty of Medicine of Montpellier (1823), Cruveilhier became professor of anatomy at Paris in 1825 and *médecin des hôpitaux*

in 1826. In the same year he reorganized the Société Anatomique and was its founding president until 1866. Elected to the Académie de Médecine in 1836, he was its president in 1839. Cruveilhier devoted himself to his enormous practice, following the rules of a very strict ethic that he condensed in his *Des devoirs et de la moralité du médecin* (1837). A modest and honest physician, he was not gifted with eloquence. He was neither a great clinician nor a great teacher, neither a credulous disciple nor an innovator. Fond of saying that "systems pass and only the facts remain," he was essentially a researcher who owed his reputation more to his books than to his teaching.

Cruveilhier's liking for observation revealed to him, during his service at La Maternité, the importance of the concepts of contagion and isolation. As early as 1821, anticipating Stéphane Tarnier, he called for "the elimination of large maternity hospitals and their replacement by home care, to which might be added a certain number of small hospitals situated outside of Paris, capable of accommodating twelve to twenty women in labor, in which each woman would have a private room."

Cruveilhier was at the same time an experimenter, an anatomist, and a pathologist. In 1836 he became the first holder of the chair of pathological anatomy founded by a bequest of his teacher, Guillaume Dupuytren. At that time he relinquished his chair of normal anatomy to Gilbert Breschet.

His experiments on the formation of callus after bone fractures in pigeons showed Cruveilhier the importance of extraosseous tissues (periosteum, muscles) in the reconstitution of bones of leverage. His injections of mercury into the blood vessels and the bronchial system bore out the theory of phlebitis, which, he said, "dominates the whole of pathology." It made possible the conceptions of embolism and infarction, which were developed by Virchow beginning in 1846. But while Virchow considered the vascular thrombosis to be the primary lesion and the lesion in the venous wall to be secondary, Cruveilhier thought that alteration of the venous wall generated the thrombosis. Later investigations have confirmed his thesis.

Cruveilhier described a valvula at the distal extremity of the lacrimonasal canal that was also known to G. B. Bianchi and Joseph Hasner. He also gave his name to the vertebral nerve in the posterior cervical plexus that issues from the first three cervical pairs. He showed that the styloglossal and palatoglossal nerves can originate in the lingual branch of the facial nerve (Hirschfeld's nerve) which anastamoses with the glossopharyngeal (Haller's ansa).

Cruveilhier's *Cours d'études anatomiques* (1830),

which was expanded into *Anatomie descriptive* (1834–1836), played a major role in the progress of anatomical studies at the École de Médecine at Paris. E. P. Chassaignac was one of the collaborators on this work. Cruveilhier's son, Edouard, brought out a new edition (1862–1867) with the assistance of Marc Sée.

The six-volume *Anatomie pathologique du corps humain* (1828–1842) and the *Traité d'anatomie pathologique générale* (1849–1864) are Cruveilhier's true claims to fame for their illustrations, remarkable even today, and for their conception. In them he describes cysts; gastric *ulcus rodens,* which he isolated from chronic gastritis and from ulcer; colic diverticulosis; progressive muscular paralysis (at the same time as E. Aran and Guillaume Duchenne); "the fibrous bodies of the breast" that, with Velpeau and Sir Astley Cooper, he differentiated from breast cancer; diffuse cerebral sclerosis; the "gelatinous disease" of the peritoneum; and the dilation of the veins of the abdominal wall giving the appearance of a Medusa's head, later described by Clemens von Baumgarten. In 1829 he diagnosed a cerebral tumor with localization of the lesion, which had invaded the acoustic nerve. The observation was so precise that nearly a century later (1927) Harvey Cushing found nothing to add to it. In the diagnosis of tumors Cruveilhier described "cancerous juice" as the criterion of malignancy; it was later replaced by the histological criterion.

Cruveilhier knew little of histology, to which one finds only a few allusions in the fifth volume of the *Traité,* which was edited by his students. Nevertheless, his work has become less dated than some more recent ones that make the most use of the microscope. That is why Virchow called himself Cruveilhier's disciple and why many of his findings remain valid.

BIBLIOGRAPHY

I. ORIGINAL WORKS. Cruveilhier's writings are *Essai sur l'anatomie pathologique en général* (Paris, 1816), his doctoral thesis; *Essai sur l'anatomie pathologique en général et sur les transformations et productions organiques en particulier,* 2 vols. (Paris, 1816); *Médecine pratique éclairée par l'anatomie et la physiologie pathologiques* (Paris, 1821); *An omnis pulmonum exulceratio vel etiam excavatio insanabilis?* (Montpellier, 1824), his *agrégation* thesis; *Discours sur l'histoire de l'anatomie* (Paris, 1825), the opening lecture of his anatomy course; *Anatomie pathologique du corps humain; ou description, avec figures lithographiées et coloriées, des diverses alterations morbides dont le corps humain est susceptible,* 2 vols. (Paris, 1828–1842), illustrated with more than two hundred colored plates; *Cours d'études anatomiques* (Paris, 1830); *Anatomie descriptive,* 4 vols. (Paris, 1834–1836); *Des devoirs et de la moralité du*

médecin, *Discours prononcé dans la séance publique de la Faculté de Médecine, le 3 novembre 1836* (Paris, 1837); *Vie de Dupuytren* (Paris, 1840); "Histoire de l'anatomie pathologique," in *Annales de l'anatomie et de la physiologie pathologiques* (1846), 9–18, 37–46, 75–88; *Traité d'anatomie pathologique générale,* 5 vols. (Paris, 1849–1864); in the *Dictionnaire de médecine et de chirurgie pratique,* 15 vols. (Paris, 1829), the following articles: "Abdomen," "Acéphalocystes," "Adhérences," "Adhésions," "Anatomie chirurgicale médicale," "Anatomie pathologique," "Artères (maladies des)," "Articulations (maladies)," "Entozoaires," "Estomac," "Muscles," and "Phlébites"; *Académie royale de médecine—Rapport fait à cette académie dans la séance du 22 octobre 1839 sur les pièces pathologiques modelées en relief et publiées par le docteur Félix Thibert . . .* (Paris, 1839); "Sur la paralysie progressive atrophique," in *Bulletin de l'Académie de médecine,* **18** (8 Mar. 1853), 490–502 (29 Mar. 1853), 546–584; and *Trois rapports sur un mémoire de M. Jules Guérin relatif aux déviations simulées de la colonne vertébrale, faits à l'Académie royale de médecine* (Paris, 1856).

II. SECONDARY LITERATURE. On Cruveilhier or his work, see P. Astruc, "Les belles pages médicales—Jean Cruveilhier (1791–1874)," in *Informateur médical* (31 Jan. 1932), 2–4; and "Le centenaire de la médecine d'observation," in *Progrès médical,* ill. supp. nos. 10 and 11 (1932), p. 81; de Barante, *Souvenirs,* 8 vols. (Paris, 1901), vol. 7, p. 311; Bardinet, "Discours prononcé aux funérailles de Cruveilhier," in *L'union médicale,* **17** (1874), 436–438; J. Béclard, "M. Cruveilhier—Notice lue dans la séance publique annuelle de l'Académie de médecine—4 mai 1875," in *Mémoires de l'Académie de médecine,* **31** (1875), 21–24 and in *Notices et portraits* (Paris, 1878), pp. 259–289; A. Chéreau, "Cruveilhier," in *Dictionnaire encyclopédique des sciences médicales,* 1st ser., **24,** 10–13; A. Corlieu, in *Centenaire de la Faculté de médecine de Paris (1794–1894)* (Paris, 1896), pp. 249–325; P. Delaunay, "Les médecins, la Restauration et la Révolution de 1830," in *Médecine interne illustrée* (1931–1932), p. 54; L. Delhoume and P. Huard, "Le livre de comptes de Cruveilhier pour les années 1840–43," in *Histoire de la médecine* (1962), 47–52; L. Delhoume, "Une observation de Jean Cruveilhier," in *Concours médical* (1960), 2061–2066; P. Grousset, "Le Docteur Cruveilhier père," in *Le Figaro,* repr. in Labarthe, *Nos médecins contemporains* (Paris, 1868), pp. 131–138; Kelly, Emerson, and Crosly, *Encyclopedia of Medical Sources* (Baltimore, 1948); G. Lacour-Gayet, *Talleyrand, 1754–1838,* III, 366, 374, 384, 386, 389, 392, 397, 408; C. Lasègue, "Cruveilhier—Sa vie scientifique, Ses oeuvres," in *Archives générales de médecine,* **1** (1874), 594–599; Jacques Ménétrier, "Cruveilhier (1791–1874)," in *Progrès médical* (5 Mar. 1927), 357–364; in *Gazette médicale limousine* (July 1927); and in *Aesculape* (July 1927), 182–187 (Aug. 1927), 212–216; Morton T. Leslie, in Garrison and Mottoris, *Medical Bibliography* (London, 1954); J. Rochard, *Histoire de la chirurgie française au XIX° siècle* (Paris, 1874); C. Sachaile, *Les médecins de Paris jugés par leurs oeuvres* (Paris, 1845), pp. 214–217; "La plaque commémorative de Cruveilhier à Limoges," in *Gazette médicale li-*

mousine (July, 1927); "Un appel d'Alfred de Vigny à Cruveilhier," in *Aesculape* (Aug. 1927), 198–199; "Une lettre inédite de Cruveilhier," in *Chronique médicale* (1904), 184; and Monpart, "Cruveilhier," in *Journal de la Santé,* **19** (1902), 2321.

A portrait of Cruveilhier by Apnée is property of Dr. Louis Cruveilhier; one by A. Delbecke is in the Musée Dupuytren, Paris. An unsigned marble bust, at the Faculté de Médecine, Paris, was reproduced as a marble medallion by the sculptor Marcellin Bourgeois for the Faculté de Médecine, Bordeaux, in 1887. Other likenesses are a lithograph by Travies in *Charivari* (19 Oct. 1838); and a lithograph by Maurin (1828), a drawing by Lasnier (1865), and a photograph by Trinquart in *Biographies médicales* (1934), pp. 296, 297, 305. A portrait appears in A. Corlieu, *Centenaire de la Faculté de Médecine (1794–1894)* (Paris, 1896), p. 324.

PIERRE HUARD

CTESIBIUS (KTESIBIOS) (*fl.* Alexandria, 270 B.C.), *invention.*

Ctesibius lived in Alexandria. The date 270 B.C. is fixed by an epigram by Hedylos, quoted by Athenaeus,[1] concerning a singing cornucopia he made for the statue of Arsinoë, the sister and wife of Ptolemy II Philadelphus. Another date, "under Ptolemaeus VII Physkon (145–116 B.C.)," given by Athenaeus[2] from Aristokles, has led Susemihl[3] and others to assume a second Ctesibius at this date; it seems, however, that the manuscripts are at fault and that Ptolemy I Soter is meant.[4]

Ctesibius wrote a book about his inventions,[5] and Vitruvius,[6] who possessed it, tells us that he was the son of a barber. In his father's shop he hung an adjustable mirror with a counterpoise consisting of a ball of lead descending inside a tube; the ball compressed the air, which escaped with a loud noise.

This showed Ctesibius that the air is a body and led to the invention of the cylinder and the plunger.[7] He developed the science of pneumatics, now called hydraulics, probably in collaboration with Strato of Lampsacus,[8] who lived in Alexandria until about 288 B.C. Vitruvius praises Ctesibius' theoretical introduction to the subject.[9]

Ctesibius invented an air pump with valves and connected it to a keyboard and rows of pipes;[10] this organ is known as the water organ because the air vessel was actuated by water. He also invented a force pump for water.[11] Many of the toys described by Philo of Byzantium and Hero of Alexandria in their *Pneumatics* were taken from Ctesibius' book; how many we cannot tell, since the book has been lost.

Another invention of Ctesibius' was the water clock.[12] It depends on a clepsydra with constant flow, i.e., a vessel with a hole in the bottom and an overflow,

which gives it a constant level and a constant flow through the hole. Ctesibius drilled the hole in gold to avoid rust or verdigris, or in a precious stone to guard against wear; the water flowed into a cylindrical container and lifted a float, which carried a pointer to mark the hours. He equipped the float with a rack turning a toothed wheel and made the clock work a number of *parerga:* whistling birds, moving puppets, ringing bells, and the like. An attempt to regulate the flow to suit local hours failed, so he constructed the parastatic clock, in which the pointer, moving at a constant rate, marks hours of different length on a network of lines traced on a vertical cylinder, which was turned a little every day.

Philo of Byzantium records two catapults invented by Ctesibius, one actuated by compressed air[13] and the other by bronze springs;[14] neither seems to have survived him.

Athenaeus the Mechanic[15] attributes to Ctesibius a scaling ladder enclosed in a tube, "a marvellous invention, but of no great use."

Ctesibius was an inventor of the first order; we owe to him the force pumps for air and water and the hydraulic organ with its keyboard and rows of pipes; his water clock has been superseded by the pendulum clock, but his *parerga* still survive in the cuckoo clock.

NOTES

1. Athenaeus, *Deipnosophistae,* bk. 11, p. 497, d–e.
2. *Ibid.,* bk. 4, p. 174, b–e.
3. Franz Susemihl, *Geschichte der griechischen Litteratur in der Alexandrinerzeit* (Leipzig, 1891), 1, 734–736, 775.
4. A. G. Drachmann, "On the Alleged Second Ktesibios," in *Centaurus,* **2** (1951), 1–10.
5. Vitruvius, *De architectura,* bk. 10, ch. 7, v. 5.
6. *Ibid.,* bk. 9, ch. 8, v. 2–4.
7. "Philons *Belopoiika,* Griechisch und Deutsch von H. Diels und E. Schramm," in *Abhandlungen der Preussischen Akademie der Wissenschaften,* Jahrgang 1918, Phil.-hist. Klasse, no. 16(1919), ch. 61.
8. H. Diels, "Ueber das physikalische System des Straton," in *Sitzungsberichte der K. preussischen Akademie der Wissenschaften zu Berlin,* **9** (1893), 106–110.
9. Vitruvius, *De architectura,* bk. 1, ch. 1, v. 7.
10. *Ibid.,* bk. 10, ch. 8.
11. *Ibid.,* ch. 7.
12. *Ibid.,* bk. 9, ch. 8, v. 4–7.
13. "Philons *Belopoiika, . . .,"* chs. 60–62.
14. *Ibid.,* chs. 14, 39–47.
15. Athenaeus Mechanicus, in C. Wescher, *Poliorcétique des Grecs* (Paris, 1867), pp. 29–31.

BIBLIOGRAPHY

The original work of Ctesibius is lost; excerpts are found in Vitruvius, *De architectura,* bk. 9, ch. 8; bk. 10, chs. 7–8.
Secondary works not cited in the notes are A. G. Drachmann, *Ktesibios, Philon and Heron,* no. 4 of the series Acta

Historica Scientiarum Naturalium et Medicinalium (Copenhagen, 1948); Orinsky, "Ktesibios," in Pauly-Wissowa, XI, pt. 2 (1922), col. 2074; and Tittel, "Hydraulis," *ibid.,* IX, pt. 1 (1914), col. 60.

A. G. DRACHMANN

CUDWORTH, RALPH (*b.* Aller, England, 1617; *d.* Cambridge, England, 26 June 1688), *philosophy.*

Cudworth was the most systematic of the Cambridge Platonists, and his attempt to combine Neoplatonism and the mechanical philosophy is important in understanding the background to Newton and the early Newtonians.

At Cambridge he became successively fellow and tutor of Emmanuel College (1639), master of Clare Hall (1644), Regius professor of Hebrew (1645), and master of Christ's College (1654), retaining the mastership and the professorship following the Restoration.

His most important work, *The True Intellectual System of the Universe,* is a fragment of an even larger work he had planned to refute the materialism of Epicurus and of Thomas Hobbes. Cudworth believed that a "rightly understood" mechanical and corpuscular philosophy did not destroy traditional religion but instead offered it new support. If matter was inert and utterly passive, then a spiritual principle was necessary to endow the universe with life and activity. But the principle involved in the ordinary course of nature was not to be equated with God, for then He would be responsible for the "errors and bungles" in nature. Such tasks were performed by a subordinate and unconscious "plastic nature." Cudworth rejected Cartesian dualism and asserted cosmic continuity. His mode of exposition was historical, drawing on classical, patristic, rabbinic, and Renaissance Neoplatonic authorities; and he reaffirmed Henry More's idea of a pristine theology which had been fragmented as it was transmitted from the Hebrews to the Greeks, the scientific part passing to Leucippus and Democritus, who "atheized" it, the theological part alone being taken over by Plato and his successors. Besides the Hobbesian doctrine, Cudworth carefully distinguished other varieties of materialism, including the hylozoism of Spinoza, of whose manuscripts he may have learned from his Dutch correspondents.

BIBLIOGRAPHY

The True Intellectual System of the Universe (London, 1678) was published in a critical Latin ed. by J. Mosheim (Jena, 1733). J. Harrison's useful 3-vol. ed. (London, 1845) incorporates Mosheim's notes.

On Cudworth or his work see the following, listed chronologically: Paul Janet, *Essai sur le méditateur plastique de Cudworth* (Paris, 1860); J. Tulloch, *Rational Theology and Christian Philosophy in England in the Seventeenth Century,* II (London–Edinburgh, 1872), 193–302; E. Cassirer, *Die Platonische Renaissance in England und die Schule von Cambridge* (Leipzig-Berlin, 1932; Eng. tr., London, 1953), ch. 5; J. A. Passmore, *Ralph Cudworth: An Interpretation* (Cambridge, 1951); Rosalie L. Colie, *Light and Enlightenment* (Cambridge, 1957), pp. 117–144; and J. E. McGuire and P. M. Rattansi, "Newton and the 'Pipes of Pan,'" in *Notes and Records. Royal Society of London,* **21** (1966), 108–143.

P. M. RATTANSI

CUÉNOT, LUCIEN (*b.* Paris, France, 21 October 1866; *d.* Nancy, France, 7 January 1951), *zoology, general biology.*

The son of a postal employee, Cuénot was a brilliant student at the Collège Chaptal. After receiving the *baccalauréat ès sciences,* he enrolled in 1883 at the Faculty of Sciences of the Sorbonne, where the courses given by Lacaze-Duthiers led him toward zoology. He placed first on the *licence* examination in 1885 and became *docteur ès sciences naturelles* at twenty-one with a thesis on Asteriidae (1887). Cuénot was appointed *préparateur* in comparative anatomy and physiology at the Faculty of Sciences in Paris in 1888 and began his medical studies. He quickly abandoned them, however, when he was offered a post as lecturer in Lyons. Through a change in plans, however, he was appointed to a course in zoology at the Faculty of Sciences of Nancy in 1890. He spent the rest of his career there, becoming assistant professor in 1896, deputy professor in 1897, and full professor of zoology in 1898; he held that professorship until his retirement in 1937. He consistently refused all appointment offers from Paris.

Knowledgeable in the interpretation of organic structures and able to discern the affinities between living forms and to increase understanding of present species by comparing them with fossil species, Cuénot studied completely different groups, from Protista and Echinodermata to Siphunculata and Tardigrada. He made detailed observations of the fauna of Lorraine and the Arcachon Basin, describing many new species and detailing their ecology and ethology.

Cuénot defined the phases of the sexual cycle of the Gregarina; and using ingenious and sophisticated methods, he analyzed the absorption and excretion processes of Invertebrata. He also studied reflex bleeding, regeneration of antennae, caudal autotomy in rodents, and the homochrome phenomenon. Cuénot became interested in sex determination at a time

(1893) when the chromosome was still unknown; he proved experimentally that sex was not influenced by exterior conditions and acknowledged that determination took place as early as the egg stage. He believed that armadillos from the same litter are true twins born of polyembryony; this was later confirmed.

Cuénot next studied heredity. Through crossbreeding mice he discovered that Mendel's laws were applicable to animals; at almost the same time Bateson conducted similar research on the hen and guinea pig. By extending Mendelism to the animal kingdom, Cuénot became a pioneer in genetics. In analyzing the heredity of pigmentation he discovered that "albinism is a mask" capable of hiding profound genetic differences in coloring; thus he offered an explanation of the phenomenon of atavism, the reappearance of an ancestral characteristic.

Crossbreeding of yellow mice provided Cuénot with the first example of lethality and the discovery of lethal genes (1905); he attributed the impossibility of obtaining pure yellow mice to selective fertilization; this hypothesis was incorrect, but the phenomenon was nevertheless real. Later Kirkham discovered that the pure yellow embryos died in the mother's uterus. Cuénot's genetic studies are still of interest in the heredity of variegation, experimental production of atavism, the intransmissibility of cataract caused by naphthalene, and the heredity of predisposition to cancer. He was one of the first to suggest a comparison between the activity of a gene and that of an enzyme.

After World War I, Cuénot became interested in evolution, especially in adaptation and speciation. He attributed an important role to mutation, but disputed the all-powerful role of natural selection. In an attempt to explain the harmony between form and environment he proposed the theory of preadaptation: the "empty space" or "ecological niche" is populated with mutants already showing characteristics adapted to the conditions of the "empty space."

Cuénot disliked getting involved in the question of the transmission of acquired characteristics. A "dissatisfied mutationist," he would not admit that the randomness of mutations by natural selection could have led to the development of complex organs. He was especially interested in the small but numerous function-specific assemblages of organs in living creatures that, by their structure and function, suggest tools made by man. This principle led Cuénot to a purified neo-finalism, removed from any sort of absolute providentialism; at the end of his life he favored a broad type of pantheistic monism. He was skeptical and independent, refusing to accept illusory solutions.

BIBLIOGRAPHY

I. ORIGINAL WORKS. Cuénot's papers include "Contribution à l'étude anatomique des astérides," *Archives de zoologie expérimentale*, 2nd ser., supp. **5** (1887); "Études physiologiques sur les gastéropodes pulmonés, sur les orthoptères, sur les oligochètes," in *Archives de biologie*, **12** (1892), 683–740; **14** (1895), 293–341; **15** (1897), 79–124; "L'excrétion chez les mollusques," *ibid.*, **16** (1899), 49–96; "Sur la détermination du sexe chez les animaux," in *Bulletin scientifique de la France et de la Belgique*, **32** (1899), 462–535; "Recherches sur l'évolution et la conjugaison des grégarines," in *Archives de biologie*, **17** (1901), 581–652; "La loi de Mendel et l'hérédité de la pigmentation chez les souris," in *Compte rendu hebdomadaire des séances de l'Académie des sciences*, **134** (1902), 779–781, and *Compte rendu des séances de la Société de biologie*, **54** (1902), 395–397; "Sur quelques applications de la loi de Mendel," *ibid.*, 397–398; "Hérédité de la pigmentation chez les souris noires," *ibid.*, **55** (1903), 298–299; "Transmission héréditaire de pigmentation par les souris albinos," *ibid.*, 299–301; "Sur quelques anomalies apparentes des proportions mendéliennes (6ème note)," in *Archives de zoologie expérimentale*, 4th ser., **9** (1908), vii–xvi; "Théorie de la préadaptation," in *Scientia*, **16** (1914), 60–73; "Sipunculiens, échiuriens, priapuliens," in *Faune de France*, **4** (1922); "Tardigrades," *ibid.*, **24** (1932); "Contribution à la faune du Bassin d'Arcachon, IX," in *Bulletin de la station biologique d'Arcachon*, **24** (1927), 229–308; "L'inquiétude métaphysique," in *Mémoires de l'Académie de Stanislas*, 6th ser., **25** (1928); "Génétique des souris," in *Bibliographia genetica*, IV (1928), 179–242; *Echinodermes*, Traité de zoologie no. 11 (Paris, 1948); *Onycophores, tardigrades, pentastomides*, Traité de zoologie no. 6 (Paris, 1949); *Phylogenèse du règne animal*, Traité de zoologie no. 1 (Paris, 1952); *Echinodermes fossiles (hétérostélés et cystidés)*, Traité de paléontologie no. 3 (Paris, 1952); and *Phylogenèse du règne animal*, Traité de paléontologie no. 1 (Paris, 1952), written with A. Tétry.

His books are *Les moyens de défense dans la série animale* (Paris, 1892); *L'influence du milieu sur les animaux* (Paris, 1894); *La genèse des espèces animales* (Paris, 1911, 1932); *L'adaptation* (Paris, 1925); *L'espèce* (Paris, 1936); *Introduction à la génétique* (Paris, 1936), written with J. Rostand; *Invention et finalité en biologie* (Paris, 1941, 1946); and *L'évolution biologique* (Paris, 1951), written with A. Tétry.

II. SECONDARY LITERATURE. An article on Cuénot's life and works was published by R. Courrier in *Académie des sciences* (Paris, 1952); it consists of a biography, complete bibliography, and references to obituaries. A memoriam on the centennial of his birth was published in *Académie des sciences* (Paris, 1967); it consists of texts of speeches by A. Tétry, W. Delafosse, P. Marot, F. Moreau, J. Rostand, P.-P. Grassé, and P. Brien. The following were written by A. Tétry: "Le sort de la notion de préadaptation," in *Proceedings of the XIV International Congress of Zoology, Copenhagen 1953* (Copenhagen, 1956), pp. 159–160; "La place du laboratoire d'Arcachon dans l'oeuvre de L. Cuénot. Colloque international sur l'histoire de la biologie marine," in *Vie et milieu,* supp. **19** (1965), 265–273; and "Les 'incertitudes' de l'évolution biologique," in *Année biologique,* **6** (1967), 573–578.

ANDRÉE TÉTRY

CULLEN, WILLIAM (*b*. Hamilton, Scotland, 15 April 1710; *d*. Kirknewton, near Edinburgh, Scotland, 5 February 1790), *medicine, chemistry.*

Cullen's father was factor to the duke of Hamilton; his mother was the daughter of a younger son of Roberton of Ernock. From Hamilton Grammar School he proceeded to the University of Glasgow, where he could have learned little pertaining to medicine but where, in 1727, he was a member of the mathematics class of Robert Simson, the famous philosophical interpreter of Euclid and Apollonius. After being apprenticed to John Paisley, in whose library he laid the foundation of his wide knowledge, Cullen moved to London in 1729; the following year he was appointed surgeon to a vessel proceeding to the West Indies.

After acting as private physician in Hamilton, Cullen spent the academic years 1734/1735 and 1735/1736 attending classes in the recently founded medical school at the University of Edinburgh, where in all probability the lectures of Andrew Plummer turned his interest to chemistry. Plummer had studied under Boerhaave; but despite statements to the contrary, Cullen himself never studied at Leiden. In 1736 Cullen returned to Hamilton, where for eight years he remained in private practice; during the earlier part of this period William Hunter resided with him as a pupil. In 1740 he graduated M.D. at Glasgow; and in 1741 he married Anna Johnstone, a lady whose charm and intelligence were as much appreciated as were her husband's. To them were born four daughters and seven sons, one of whom, Robert, rose to be a judge of the Court of Session.

In 1744 the removal of the Cullen family to Glasgow marked the beginning of an academic career that remained unbroken until December 1789. Lecturing at first extramurally, Cullen soon reached an agreement with Dr. Johnstone, the professor of medicine, to deliver a course of lectures on that subject—which Johnstone himself had never seen any necessity to do. The year 1747 marked the foundation of the first independent lectureship in chemistry in the British Isles: the university provided a modest sum for the setting up and maintenance of a laboratory. The incumbents were at first William Cullen and John Carrick, assistant to Robert Hamilton, professor of anatomy. The early death of Carrick left Cullen as the first undisputed university teacher of chemistry— not chemistry and medicine, as the earlier chair at

Edinburgh was entitled from its inception until it became the present chair of biochemistry. Cullen was, however, permitted to continue his lectures on the theory and practice of physic, botany, and materia medica.

Cullen's published contribution to the advancement of science (as distinguished from medicine) consists of one paper. His reputation as a teacher is based not merely on skill in imparting information but on a number of innovations, in no one of which he may have been an innovator, but which had never previously been brought together by one man. From the first he lectured in English and freely developed an orderly series of his own notes rather than reading from a stereotyped text; many of his points were illustrated by simple demonstrations, and students were encouraged to take part in laboratory operations at stated times. Above all, Cullen carried the science of chemistry beyond medicine by emphasizing its basic importance both to natural philosophy and to many "arts," such as agriculture, mining, brewing, vinegar manufacture, bleaching, and the manufacture of alkalies. It was his boast, supported by the evidence of records of his lectures extending over many years, that only "when the languor and debility of age shall restrain me . . . shall I cease to make some corrections of my plan or some addition to my course."

On 2 January 1751, after some delay, Cullen was inducted into the chair of medicine at Glasgow, by then vacated by Johnstone, but continued his lectures in chemistry until the summer of 1755, when Plummer was struck down by an illness that rendered him incapable of continuing his chemical lectures at Edinburgh. The intrigues pursuant to the support of rival candidates for succession form a *locus classicus* for the study of the division of university governance between academic *Senatus* and administrative town council: without the strong hand taken by the latter, it is unlikely that Cullen would ever have entered on his illustrious career in Edinburgh. A somewhat similar and even more squalid display occurred when the succession to the chair of practical medicine was raised; on 1 November 1766 Cullen succeeded Robert Whytt in the chair of the institutes (theory) of medicine, later sharing with John Gregory in alternate years the task of lecturing in the practice of physic, in the chair of which he succeeded Gregory in 1773. In these controversies Cullen's greatest pupil, Joseph Black, had been a rival; but no shadow of personal acrimony had ever marred their friendship.

Cullen's sole paper, although appearing in the same volume (II) of *Essays and Observations Physical and Literary Read Before a Society in Edinburgh and Published by Them* (1756), was evidently based on experiments conducted before he left Glasgow. As Cullen himself observed, Richmann had already (1747) given a "very exact account of the *phenomena*" of the cold produced by evaporating fluids: Cullen's achievement was to repeat the experiments on a variety of fluids under a receiver exhausted by the air pump; in the case of "nitrous aether," the surrounding water, contained in a slightly larger vessel, was frozen.

Cullen made no claim to novelty in his exposition of chemistry, although in an earlier (unpublished) paper on "salts" (which then included acids and bases) he was moving toward a classification in terms of the compounds between the "four" acids (to which he later and guardedly seems to add phosphoric, arsenic, and boric) and the "three" alkalies. Strongly influenced by Boyle (rather than by the narrow Newtonianism then fashionable), he gave but nodding respect to atoms, considering them hardly relevant to the understanding of chemical operations; but he was probably the first to give symbolic precision to the "affinity" tables then much in vogue, using reversed arrows to represent what came to be called "double decompositions." He was an uneasy phlogistonist and at one stage suspected that the increase in weight following "calcination" was due to an "acid" in the air; nothing came of it, however.

No adequate appraisal of Cullen's contribution to science can omit reference to his work on nosology or to his deep understanding of the contemporary empirical philosophy, explicit in one of his lectures on the institutes of medicine and everywhere implicit in the orderly deployment of his system of chemistry. He also played a prominent part in the founding of both the Royal Society of Edinburgh and the Royal Medical Society (Edinburgh). He was elected a fellow of the Royal Society of London in 1777 but never signed the roll attesting formal admission.

BIBLIOGRAPHY

Primary printed sources for Cullen's contribution to science (the purely medical are here omitted) comprise only "Of the Cold Produced by Evaporating Fluids and of Some Other Means of Producing Cold," in *Essays and Observations Physical and Literary Read Before a Society in Edinburgh and Published by Them,* II (Edinburgh, 1756), 145 f.; and *Substance of Nine Lectures on Vegetation and Agriculture Delivered to a Private Audience in the Year 1768* (London, 1796). The *Synopsis nosologiae methodicae,* related to the Linnaean system of general systematics, is marginal.

Primary MS sources consist of notes for lectures; a tract on "salts," edited by L. Dobbin, "A Cullen Chemical Manuscript of 1753," in *Annals of Science,* 1 (1936), 138–156; and a short introductory course on the history of chemistry. All of these are in autograph or corrected in Cullen's hand.

Despite its being stigmatized (in the *D.N.B.*) as "diffuse and ponderous," John Thomson's *Account of the Life, Lectures and Writings of William Cullen M.D.* (2 vols., Edinburgh, 1832) is still indispensable. The narrative is soundly based on correspondence and other documents, many of which are cited at length; it is, however, marred by long digressions of doubtful relevance.

Numerous transcripts of Cullen's chemistry courses (probably restricted to those given in Edinburgh) are to be seen at the universities of Glasgow (where most of his autograph miscellanea are), Manchester, and Aberdeen; Clifton College (Bristol); London Medical and Chirurgical Society; Royal College of Physicians of Edinburgh; and Paisley Public Library. The most recent to come to light is the property of Dr. W. A. Smeaton, University College, London. Most of the above are described in articles by W. P. D. Wightman, in *Annals of Science,* 11 (1955), 154–165, and 12 (1957), 192–205, in which a reconstruction of Cullen's first complete course at Glasgow is attempted, together with an appreciation of his views on chemistry.

Other important secondary sources are Andrew Kent, ed., *An Eighteenth Century Lectureship on Chemistry* (Glasgow, 1950), *passim;* and M. P. Crosland, "The Use of Diagrams as Chemical 'Equations' in the Lecture Notes of William Cullen and Joseph Black," in *Annals of Science,* 15 (1959), 75–90; and *Historical Studies in the Language of Chemistry* (London, 1962), pp. 90–91, 99, 119–120.

WILLIAM P. D. WIGHTMAN

CULMANN, KARL (*b.* Bergzabern, Lower Palatinate, Germany, 10 July 1821; *d.* Zurich, Switzerland, 9 December 1881), *graphic statics.*

After preliminary schooling by his father, an evangelical clergyman, Culmann enrolled at the engineering and artillery school in Metz (an uncle was professor there) to prepare for entrance to the École Polytechnique in Paris. He learned of the graphical methods introduced by J. V. Poncelet and studied in Metz before a case of typhoid fever caused him to return home. Following a long convalescence, he attended the Polytechnikum at nearby Karlsruhe. On graduation in 1841 he joined the Bavarian civil service as cadet bridge engineer and was assigned to the Hof railway construction division; he continued to study under the guidance of Schnürlein, a student of Gauss's. To prepare for a study trip abroad, he arranged in 1847 for transfer to the railway bureau in Munich, in order to extend his mathematical training and perfect his English.

Taking leave in 1849, Culmann visited England, Ireland, and the United States, returning in 1851 with detailed observations for two reports. The first dealt with the wooden bridges of the United States, the second with iron bridges in England and America.

These reports present what is probably the most complete story of American bridges of the first half of the nineteenth century: American achievement lay in the largely empirical design of wooden bridges. To compare American and European designs Culmann developed new methods for the analysis of truss systems and approximate procedures for indeterminate structures, such as the Town lattice-truss and Burr arch-truss bridges.

Culmann was called to Zurich in 1855 as professor of engineering sciences at the newly founded polytechnic institute (the present ETH), a post he held until his death twenty-six years later, declining an offer to move to the Munich Polytechnikum in 1868. From 1872 to 1875 he was also director of the institute.

As a teacher Culmann drew high praise for his rich experience, excellent theoretical knowledge, and sympathetic understanding; but it is on his principal work, *Die graphische Statik* (1866), that his fame now rests. His youthful exposure to the developments of Poncelet and other French geometers is reflected in his reports on American bridge practice, in which his independent extensions of graphical methods are already evident. It was the custom to analyze a particular design with equations; but Culmann chose another route to the solution, geometric constructions of a fundamental and widely applicable nature.

Culmann presented the graphical calculus as a symmetrical whole, a systematic introduction of graphical methods into the analysis of all kinds of structures—beams, bridges, roof trusses, arches, and retaining walls. Among other things, he introduced the general use of force and funicular polygons, the method of sections, and the diagram of internal forces based on the equilibrium conditions of successive joints.

The methods, first used by his students, were quickly assimilated by bridge and structural designers, who appreciated the time saving of the graphical methods over the current procedures involving simultaneous equations of analytical mechanics. The contents of the first edition of *Die graphische Statik* was thus known when it was published, and the book was eagerly accepted by a wide circle. For example, graphical analysis was applied to the Eiffel Tower by its structural designer, Maurice Koechlin, a student of Culmann's. Further advances appeared in the first part of the second edition (1875), with more planned for the second volume, which was never published because of Culmann's death. His work was carried forward by a Zurich colleague, W. Ritter (who is not to be confused with his contemporary A. Ritter, a prominent structural engineer who was also professor of mechanics at the Polytechnikum in Aachen). Further developments came from E. Winkler, O. Mohr, L. Cremona, and H. Müller-Breslau, among others.

In his work on beams Culmann showed how the stresses at any point can be analyzed graphically and developed a stress circle for the uniaxial shear state, a particular case of the more general, and later, Mohr circle (1882). With this he was able to draw the stress trajectories of flexural members.

On seeing sections of human bones prepared by Hermann von Meyer, professor of anatomy at the University of Zurich, Culmann noted that in some the trabeculae forming the cancellous tissue (spongiosa) followed the principal stresses on assuming beam loading of the bone, as in the head of the femur. Just how this arrangement is related to the mechanical conditions of loading is not yet clear, even though Meyer's trajectorial theory of bone formation has been the subject of discussion since 1867.

Although modern structural analysis is no longer governed by graphical methods, Culmann's work was fundamental to the present analytical procedures that represent complements and extensions of the older, physical approach.

BIBLIOGRAPHY

I. ORIGINAL WORKS. Culmann's writings are "Der Bau der hölzernen Brücken in den Vereinigten Staaten von Nordamerika," in *Allgemeine Bauzeitung* (Vienna), **16** (1851), 69–129; "Der Bau der eisernen Brücken in England und Amerika," *ibid.*, **17** (1852), 163–222; "Bericht über die Untersuchung der schweizerischen Wildbäche in den Jahren 1858–1863," in *Vierteljahrsschrift der Naturforschenden Gesellschaft in Zürich* (1864); *Die graphische Statik* (Zurich, 1866; 2nd ed., pt. 1, 1875); and "Ueber das Parallelogram und über die Zusammensetzung der Kräfte," in *Vierteljahrsschrift der Naturforschenden Gesellschaft in Zürich* (1870).

II. SECONDARY LITERATURE. On Culmann's work see A. J. du Bois, *Graphical Statics* (New York, 1875); F. G. Evans, *Stress and Strain in Bones* (Springfield, Ill., 1957); H. von Meyer, "Die Architektur der Spongiosa," in *Archiv für Anatomie, Physiologie und wissenschaftliche Medizin,* **47** (1867); W. Ritter, *Anwendungen der graphischen Statik* (Zurich, 1888); and D'Arcy W. Thompson, *On Growth and Form*, repr. of 2nd ed. (Cambridge, 1952), pp. 975–979, 997.

Obituaries are in *Zeitschrift für Bau- und Verkehrswesen* (17 Dec. 1881); and *Vierteljahrsschrift der Naturforschenden Gesellschaft in Zürich,* **27** (1882).

Biographical sketches are C. Matschoss, *Männer der Technik* (Berlin, 1925); W. Ritter, in *Allgemeine deutsche Biographie*, XLVII (1903); M. Rühlmann, *Vorträge über Geschichte der technischen Mechanik* (Berlin, 1885); Fritz Stüssi, in *Neue deutsche Biographie*, III (1957); S. Timoshenko, *History of Strength of Materials* (New

York, 1953); and *Vierteljahrsschrift der Naturforschenden Gesellschaft in Zürich,* **41** (1896), Jubelband 1.

R. S. HARTENBERG

CUMMING, JAMES (*b.* London, England, 24 October 1777; *d.* North Runcton, Norfolk, 10 November 1861), *physics.*

Cumming entered Trinity College, Cambridge, in 1797, graduated tenth wrangler in 1801, became a fellow in 1803, and was elected professor of chemistry in 1815, a position he held until the year before his death. He was apparently excellent as a teacher. Cumming's scientific accomplishments, however, were limited—reportedly by ill health and lack of ambition—to a ten-year period early in his career: his published works span the years 1822–1833.

He became fascinated by the developments in electromagnetism following Oersted's experiments and independently invented the galvanometer, describing it in greater detail than either Schweigger or Poggendorff. P. G. Tait gave Cumming credit for the independent discovery of thermoelectricity, but there seems to be no good reason to accept this. A brief announcement of Thomas Johann Seebeck's discovery appeared in the *Annals of Philosophy* in 1822; Cumming's first description was in the same journal in the following year. On the other hand, Seebeck's detailed paper was not available until 1825. By then Cumming had compared the thermoelectric order of metals with the voltaic and conductivity series, noting the lack of any apparent connection. He also recorded the inversion effect which occurs with certain couples at high temperatures (Seebeck had remarked on this, as had A. C. Becquerel); this anomalous behavior was finally explained by William Thomson on the basis of thermodynamic arguments and the assumption of the "Thomson effect." (Thomson had his attention drawn to this phenomenon by Becquerel's paper but later recognized Cumming's priority; he apparently never realized that it had been described in Seebeck's original publication.)

Cumming was active in the founding of the Cambridge Philosophical Society and from 1824 to 1826 acted as its fourth president.

BIBLIOGRAPHY

I. ORIGINAL WORKS. The galvanometer is described in "On the Connexion of Galvanism and Magnetism," in *Cambridge Philosophical Society, Transactions,* **1** (1821–1822), 268–279, and in "On the Application of Magnetism as a Measure of Electricity," *ibid.,* 281–286. The more important thermoelectric articles are "On the Development of Electromagnetism by Heat," in *Annals of Philosophy,* **21** (1823), 427–429, and "A List of Substances Arranged According to Their Thermoelectric Relations, with a Description of Instruments for Exhibiting Rotation by Thermoelectricity," *ibid.,* **22** (1823), 177–180. The inversion effect was reported in "On Some Anomalous Appearances Occurring in the Thermoelectric Series," *ibid.,* **22** (1823), 321–323.

II. SECONDARY LITERATURE. A biographical sketch appears in the *Dictionary of National Biography.* Reference is made there to some collected papers, but this author has been unable to discover any knowledge of their existence. The galvanometer is discussed in Robert A. Chipman, "The Earliest Electromagnetic Instruments," in *United States National Museum Bulletin,* **240** (1964), 121–136.

BERNARD S. FINN

CURIE, MARIE (MARIA SKLODOWSKA) (*b.* Warsaw, Poland, 7 November 1867; *d.* Sancellemoz, France, 4 July 1934), *physics.*

Maria's parents, descendants of Catholic landowners, were intellectuals held in poor esteem by the Russian authorities. Her father, Wladyslaw, a former student at the University of Saint Petersburg (now Leningrad), taught mathematics and physics in a government secondary school in Warsaw. Her mother, the former Bronislawa Boguska, managed a private boarding school for girls on Freta Street[1] when Maria, her fifth child, was born. The mother subsequently contracted tuberculosis and gave up all professional activity. Misfortune struck the family again in 1876, when Sophia, the eldest child, died of typhus; in 1878 the mother died.

Denied lucrative teaching posts for political reasons, Professor Sklodowski decided, after moving several times, to take in boarders at his home on Leschno Street. Maria, known familiarly as Manya, gave up her room and slept in the living room; she worked there late at night and put everything in order before the boarders had their breakfast. A gold medal for excellence crowned her brilliant high school studies—in Russian—but her health was weakened (1883). A year in the country with her uncle Sklodowski, a notary in Skalbmierz, near the Galician border, restored her. During this period she formed her profound attachment to nature and to country people.

Upon her return Maria gave lessons to earn money. She was a passionate adherent of clandestine movements supporting Polish political positivism and participated in the activities of an underground university—progressive and anticlerical—whose journal, *Pravda,* preached the cult of science. Maria read everything in the original: Dostoevsky and Karl Marx, the French, German, and Polish poets; sometimes she even tried her hand at poetry.

In order that her sister Bronia might study in Paris—for France was the land of liberty of which they both dreamed—Maria became a governess (1 January 1886) in the home of M. Zorawski, administrator of the rich estate of the princes Czartoryski (in Szezuki, near Pzasnysz, Plock district, about sixty miles north of Warsaw).

Moved by the poverty and ignorance of the peasant children, Maria gave them lessons after the seven hours devoted to the education of two of her employer's daughters; she also read many of the books in his scientific library. During the summer Casimir, the Zorawskis' eldest son, a mathematics student at the University of Warsaw, fell in love with her. His family firmly opposed the marriage, however, because Maria was a governess. Although disillusioned, Maria remained with the Zorawskis until the end of her contract (Easter 1889), nearly three years more. Back in Warsaw she again became a governess, dividing her leisure time between her family, the "flying" university, and chemistry. Her cousin Boguski, a former assistant to Mendeleev and director of the modest laboratory of the Museum of Industry and Commerce, entrusted her to Napoleon Milcer, who had studied under Bunsen.

Meanwhile, Bronia, now a medical doctor, had married Casimir Dluski, also a doctor. They insisted that Maria come to Paris and stay with them. She hesitated, then left with her meager savings (1891). She crossed Germany by train, traveling fourth class, seated on a camp stool. Although her relatives (who lived on the Right Bank) welcomed her warmly, Maria, in order to work as she pleased, preferred to live alone in a modest room[2] and content herself with scanty meals. She received an Alexandrovitch Scholarship, which she fully repaid. She passed the *licence* in physics—on 28 July 1893, ranking first, with high honors—and the *licence* in mathematics—on 28 July 1894, with honors, ranking second. Her professors, Paul Appell and Edmond Bouty, took notice of her gifts and her enthusiasm; and Gabriel Lippmann opened his laboratory to her.

Maria met Pierre Curie in April 1894 at the home of a Polish physicist named Kowalski. A lively sympathy brought them together and then a deep affection developed. Pierre proposed to her, but before committing herself, she went to Poland to spend the summer near her friends and family. Their correspondence during the summer was conclusive; Maria returned in October having decided to marry Pierre.

Bronia gave her a room at 39 rue de Châteaudun. There she completed the memoir on her first experimental work, "Sur les propriétés magnétiques des aciers trempés," which Le Chatelier had asked her to write for the Société pour l'Encouragement de l'Industrie Nationale; Pierre advised her on it. She attended Pierre's thesis presentation in March 1895, and on 26 July they were married. On 12 September 1897 their daughter Irène was born in their modest apartment, 24 rue de la Glacière.[3] A little later they moved to 108 boulevard Kellermann, a building since destroyed.[4]

Meanwhile, Marie had placed first on the women's *agrégation* in physics (15 August 1896). She was looking for a thesis topic while visiting Pierre's laboratory at the École de Physique et Chimie and occasionally working with him; the director, Paul Schützenberger, welcomed her warmly.

She already shared in the intense excitement of the scientific world: Roentgen had just discovered invisible rays capable of traversing opaque bodies of varying thicknesses, of exposing photographic plates, and of making the air more conductive. Were they really rays? A respected scientist, Henri Poincaré, had advanced in January 1896 the hypothesis of an emission, called "hyperfluorescence," from the glass wall of a Crookes tube struck by cathode rays. Meanwhile Henri Becquerel, at the Muséum d'Histoire Naturelle, discovered that uranium salts shielded from light for several months spontaneously emit rays related in their effects to Roentgen rays (X rays).[5] Mme. Curie became enthusiastic about this subject filled with the unknown and, as she later acknowledged, involving no bibliographic research.

The first step in the research was to determine whether there existed other elements capable, like uranium, of emitting radiation. Abandoning the idea of hyperfluorescence, couldn't one calculate by electrical measurement the effects on the conductivity of air that were revealed by the gold-leaf electroscope? Pierre Curie and his brother Jacques had constructed an extremely sensitive apparatus to measure weak currents; Mme. Curie employed it in testing both pure substances and various ores. In her first "Note" in the *Comptes rendus . . . de l'Académie des sciences* (12 April 1898) she described the method that she followed throughout her life, the method that enabled her to make comparisons through time and cross-checks with other techniques:

> I employed . . . a plate condenser, one of the plates being covered with a uniform layer of uranium or of another finely pulverized substance [(diameter of the plates, eight centimeters; distance between them, three centimeters). A potential difference of 100 volts was established between the plates.]. The current that traversed the condenser was measured in absolute value by means of an electrometer and a piezoelectric quartz.

In general she preferred the zero method, in which the operator compensates for the current created by the active material by manipulating the quartz. All of her students followed this procedure.

The first results came in 1898: the measurements varied between 83×10^{-12} amperes for pitchblende to less than 0.3×10^{-12} for almost inactive salts, passing through 53×10^{-12} for thorium oxide and for chalcolite (double phosphate of uranium and copper). Thorium would thus be "radioactive" (the term is Mme. Curie's); its radioactive properties were discovered at the same time, independently, by Schmidt in Germany. The same "Note" contained a fundamental observation: "Two uranium ores . . . are much more active than uranium itself. This fact . . . leads one to believe that these ores may contain an element much more active than uranium."

The second stage of the research was to prove that an imponderable mass of an unknown element, too minute to yield an optical spectrum, could be the source of measurable and characteristic effects, whatever the composition of the compound of which it was a part. Mme. Curie showed the strength of her character: She foresaw the immense labor necessary in attempting to concentrate the active substance and the small means at her disposal; and yet she plunged into the adventure. Pierre shared her faith and abandoned—temporarily, he thought—his own research. He participated in the laborious chemical treatments as well as in the physical measurements of the products (of various concentrations), which were then compared with a sample of uranium.

It was already known that natural pitchblende is three or four times more active than uranium: after suitable chemical treatment the product obtained is 400 times more active and undoubtedly contains "a metal not yet determined, similar to bismuth. . . . We propose to call it *polonium,* from the name of the homeland of one of us" ("Note" by M. and P. Curie, *Comptes rendus . . . de l'Académie des sciences* [18 July 1898]). However, the eminent spectroscopist Eugène Demarçay discerned no new lines; and it was necessary to procure more of the ore. Eduard Suess of the University of Vienna, a correspondent of the Institut de France, interceded with the Austrian government; and 100 kilograms were offered to the Curies. A third "Note" followed, signed also by Bémont, Pierre's assistant at the École de Physique et Chimie: "We have found a second radioactive substance, entirely different from the first in its chemical properties, . . . [which are] similar to those of barium" (*Comptes rendus . . . de l'Académie des sciences* [November 1898]). The substance was radium. This time Demarçay observed a new line in the spectrum; confirmation of both the technique of measurement and of the discoveries was in sight.

To make the confirmation irrefutable, still more primary material and treatments were necessary; and André Debierne assisted them. Mme. Curie wrote simply: "I submitted to a fractionated crystallization two kilograms of purified radium-bearing barium chloride that had been extracted from half a [metric] ton of residues of uranium oxide ore." The currents reached 10^{-7} amperes, and the substances obtained were 7,500 times more active than uranium (1899). A few months later they were 100,000 times more active.

The research not only accelerated but also became diversified. Pierre Curie studied the radiations; Marie tried to isolate the polonium, without success, and to determine the atomic weight of radium; Debierne discovered actinium. In Mme. Curie's words:

> None of the new radioactive substances has yet been isolated. To believe in the possibility of their isolation amounts to admitting that they are new elements. It is this opinion that has guided our work from the beginning: it was based on the evident atomic character of the radioactivity of the materials that were the object of our study. . . . This tenacious property, which could not at all be destroyed by the great number of chemical reactions we carried out, which, in comparable reactions, always followed the same path, and manifested itself with an intensity clearly related to the quantity of inactive material retrieved, . . . must be an absolutely essential character of the material itself (1900).

They were reasoning as chemists: the physicist's atom was still in limbo, although the connection between electricity and matter was being revealed, beginning with the electrons, which might be subatomic particles. J. J. Thomson located electrons in a solid sphere, while Jean Perrin imagined that their paths form a sort of miniature solar system (1901).

What slowed the interpretation of the phenomenon of radioactivity, as even Mme. Curie herself acknowledged, was the experimental datum that the radiant activity of uranium, thorium, radium, and probably also of actinium was constant. It is true that the activity of polonium was found to diminish, but Mme. Curie viewed this as an exception (1902). Although since January 1899 the Curies had considered, among other hypotheses, the instability of radioactive substances, their complete faith in experiment prevented them from following Rutherford along the same lines in 1903. If Rutherford and Soddy were right, every radioactive substance would destroy itself according to an exponential law but with a different period of decay. Thus the gaseous emanation of radium would result from the destruction of radium and would destroy itself, producing helium and other substances of a radioactive character. To account for the apparent stability of radium, Rutherford suggested a very long period of decay. This Mme. Curie was not ready to accept until experimental proof had been obtained that radium did not act "on its surroundings (nearby material atoms or the ether in a vacuum) in such a way that it produces atomic transformations. . . . Radium itself would then no longer be an element in the process of being destroyed." However, in 1902 Mme. Curie had isolated a decigram of pure radium and, after great difficulties, determined its atomic weight for the first time, 225 (instead of the presently recognized value, 226). This success brought her the Berthelot Medal of the Académie des Sciences and, for the third time, the Gegner Prize (1902).

But Pierre Curie's meager salary could not support the household and the research; Marie, even before defending her thesis (1903), took a position as lecturer in physics at the École Normale Supérieure in Sèvres (October 1900). There girls who had passed a competitive examination prepared for the *agrégation*. No woman had taught there before, and her lecture experiments assured her success.

Life was hard; then came international recognition of their work. Marie attended Pierre's lecture in London at the Royal Institution in May 1903 and, on 5 November, shared with him the Humphry Davy Medal awarded by the Royal Society. The Nobel Prize for physics was awarded jointly to the Curies and to Henri Becquerel for the discovery of radioactivity (12 December 1903), although, because of weakened health, they did not go to Stockholm to receive it until 6 June 1905. This was followed by many other honors, including the Elliott Cresson Medal in 1909.

Rather than making them happy, this recognition overwhelmed the Curies with solicitations and correspondence that took too much of their time and drained their strength. The Curies denounced "the burden of fame"; their bicycle rides became less frequent and their vacations shorter. On 1 November 1904, a month before the birth of their daughter Eve (6 December), Marie was finally named Pierre's assistant at the Faculté des Sciences, where she had long been working without pay. In 1906 Pierre, at last a member of the Académie des Sciences, presented a "Note" on the period of decay of polonium (140 days), to which Marie applied, for the first time, the Rutherford-Soddy exponential law. She proved, as had these two scientists, the release of helium. Nevertheless, many difficulties remained in interpreting the experimental results: the emanation (radon), induced radioactivity, and radioactive deposits that were more or less short-lived. The situation was further confused, theoretically, by the Curies' observation that "Every atom of a radioactive body functions as a constant source of energy, . . . which implies a revision of the principles of conservation." The scientists pondered;

the press exclaimed, "With radium the Curies have discovered perpetual motion!"

Mme. Curie then stated the policies of her research, to which she always remained faithful; push to the extreme the precision and rigor of measurement; obtain samples that are pure or of maximum concentration, even at the cost of handling enormous quantities of raw materials; and put forth general laws only in the complete absence of exceptions.

A French industrialist, Armet de L'Isle, confident that there was a future for medical and industrial applications of radium, constructed a factory in Nogent-sur-Marne, on the outskirts of Paris, for the extraction of radium from pitchblende residues. In 1904 Debierne installed a section there that prepared the materials needed in the laboratory. The Curies claimed no royalties and refused to take out any patents; they deliberately renounced a fortune, as they had declined a very favorable offer from the University of Geneva in 1900. They remained in France and gave free advice to anyone who asked for it.

Following Pierre's death in April 1906, Marie became a different woman, even with her closest friends; she lost her gaiety and warmth and became distant. Her one thought was to continue: to raise her daughters, even to giving them their daily bath, a task she never entrusted to anyone else; and to go on in the laboratory as if Pierre were still there. Had he not once said to her, "Whatever happens, even if one were to be like a body without a soul, one must work just the same"?

The Ministry of Public Education thought to do her a kindness by offering her a pension, as they had done for Pasteur's widow. She refused; since she was still able to work, why deprive her of that? Surprised, the Faculty Council decided, unanimously, to maintain the chair of physics created for Pierre in 1904 and bestowed it on Marie (1 May 1906). She was confirmed in 1908. For the first time a woman taught at the Sorbonne.

Mme. Curie compelled recognition from her first lecture (5 November 1906), despite her timidity, the emotion she was concealing, her weak voice, and her monotone delivery. She made no introductory remarks, took no notice of the sightseers mingling with the students, and began her lecture with the last sentence that Pierre had spoken in that very place. In every demonstration experiment she watched the result with as much interest as if she did not already know it.

The vocabulary had grown considerably since her thesis; the dissymmetry between the negative charges (electrons), independent of any atomic material, and the positive charges, linked with the material atoms, was obvious. She introduced the terms "disintegration" and "transmutation" and described the advantages of the theory of radioactive transformations. Nevertheless, she included a reservation: "It seems useful to me not to lose sight of the other explanations of radioactivity that may be proposed."

The smallness of the laboratory on the rue Cuvier permitted Marie to have only five or six researchers. Among them were Duane and Stark (two of the first Carnegie Scholars) in 1907 and Ellen Gleditsch in 1908. They fixed standards of measurement, verified that external agents have no effect on radioactivity, and studied the effect of emanation in the formation or condensation of clouds in closed chambers.

Mme. Curie's treatise on radioactivity (1910) admitted, without reservations, the theory of transformations; and in a series of general articles written between 1911 and 1914 she described its consequences. By its nature the phenomenon of radioactivity substantiated the connection between matter and electricity; the causes of the explosion occurring each time an atom emits radiation and is transformed into another product remained to be discovered.[6]

As had Pierre in 1903, Marie Curie declined the Legion of Honor (November 1910), asking only the means to work. Again like Pierre, she yielded to the pleas of their friends and presented herself to the Académie des Sciences. The leading candidate, on the basis of Lippmann's recommendation, she was passed over (23 November 1911) for Branly after a slanderous newspaper campaign. She did not seem at all affected because she had been rejected for extrascientific reasons.

While the creation of a radium institute was in view, through an agreement between the Faculté des Sciences (basic research) and the Institut Pasteur (medical applications), Mme. Curie raised the question of official standards for radium at the Radiology Congress in Brussels in September 1910.[7] A standard was as necessary in research as it was in therapy. She was charged with preparing an ampule containing about twenty milligrams of radium metal, to be deposited with the International Bureau of Weights and Measures in Paris.[8] The same Congress defined the curie (named in honor of Pierre), a new unit corresponding to the quantity of emanation (radon) from or in radioactive equilibrium with one gram of radium. (This definition was redefined in 1953 as the quantity of any radioactive nuclide in which the number of disintegrations per second is 3.700×10^{10}.)

In 1911 the Nobel Prize for chemistry was awarded to Mme. Curie "for her services to the advancement of chemistry by the discovery of the elements radium and polonium," the first time that a scientist had received such an award twice. The major portion of the prize money went directly to research and to friends. At this time her health was impaired; she left the house in Sceaux, 6 rue du Chemin-de-fer, where she had lived since 1906, and moved closer to her laboratory, to 36 quai de Béthune.

In 1913 Mme. Curie had the pleasure of inaugurating the radioactivity pavilion in Warsaw. In January 1914 the Conseil de l'Institut du Radium was formed; she was a member, along with Appell, dean of the Faculté des Sciences; Lippmann, assistant to Appell; and the representatives of the Institut Pasteur, its director, Émile Roux and Prof. Claude Regaud. The building was almost finished by July. Pretending not to know of Mme. Curie's lack of concern with financial matters, the administration constantly made difficulties for her, whether it was a question of dealing with industry or the Service de Mesure, taking inventory of supplies, or importing ores that were subject to taxation.

When World War I broke out, Mme. Curie, after having put her precious gram of radium in safekeeping in Bordeaux (3–4 September), aided the army; the radiological apparatus already used by civilian surgeons was being ignored by military doctors. With the aid of private gifts, she equipped the ambulances that she accompanied to the front lines with portable X-ray apparatus and, on 28 July 1916, she obtained a driver's license in order not to be dependent on a chauffeur. The Red Cross (Union des Femmes Françaises) officially made her the head of its Radiological Service, and the Patronage National des Blessés allocated her funds to increase the number of radiological installations to 140. With her daughter Irène, who became her first experimental assistant, and Marthe Klein (later Mme. Pierre Weiss), she created accelerated courses in radiology for medical orderlies and taught doctors the new methods of locating foreign objects in the human body. Later (1920) she wrote La radiologie et la guerre from her wartime notes.

The Radium Institute began to function. In 1918 Mme. Curie reported to the Committee on Radioactive Substances of the Ministry of Munitions on the radioelements, their role, and their applications. With the return of peace, she finally installed herself at the Institute. Irène, officially named as her préparateur, assisted her particularly in the special courses designed for members of the American Expeditionary Force.

American women were so moved by Mme. Curie's talent and generosity that they opened a national subscription to offer her a gram of radium on the appeal of a journalist, Mrs. Meloney. In May 1921

Mme. Curie made a triumphal visit to the United States with her daughters; she received from President Harding the gold key to the case holding the precious substance. Despite her fatigue and her distaste for display, Mme. Curie was deeply moved by this gesture, and perhaps even more so when it was repeated in 1929 for the benefit of the radium therapy services of Warsaw. The inauguration, on 29 May 1932, of the Maria Sklodowska-Curie Radium Institute (which provided facilities for the treatment of patients) was the occasion for her last trip outside France.

The creation of the Curie Foundation (1920), which was empowered to receive private gifts, and her election to the Académie de Médecine in 1922 assured Mme. Curie contact with the medical profession for two goals that Pierre had cherished: the development of what had come to be called "curietherapy" and the establishment of safety standards for workers. (See her book *Pierre Curie.*)

She was invited by Sir Eric Drummond, secretary general of the League of Nations, to sit on the International Commission for Intellectual Cooperation (17 May 1922) and became its vice-president. She concerned herself with increasing the number of available postgraduate scholarships and with requiring from authors a résumé of their scientific memoirs in order to speed the publication of abstracts. However, the length of discussions on minor subjects, such as standardizing the format of periodicals, sometimes exasperated her.

Her daughter Eve (later Mrs. Henry Labouisse), who devoted herself to literature and music, sometimes took her mother to the theater; Mme. Curie interested herself in all creative endeavors. She followed the work of Sacha Pitoëff (in the 1920's and 1930's) and fostered that of the choreographer Loïe Fuller (in the early 1900's). During vacations she swam and took long nature walks. Mme. Curie closely supervised the work of her collaborators at the Radium Institute. They were of many nationalities, by 1933 numbering seventeen, and their work led from success to success. Irène Curie completed a thesis on the α rays of polonium in 1925. Then Fernand Holweck, using a pump of his own design, studied X rays in the region of maximal absorption and established the relationship between those rays and light. Although it had long been appreciated that X rays and light are of the same nature, no one had succeeded in obtaining a sufficiently complete vacuum to detect X rays of long wavelength. Fernand Holweck's pump, which depended on a technique for soldering glass to metal, made this possible. Salomon Rosenblum, a Curie Scholar, employed Bellevue's powerful electromagnet and discovered the fine structure of the spectrum of the α rays. The discovery of artificial radioactivity by Irène Curie and her husband, Frédéric Joliot, followed in 1933. It was like an echo of Mme. Curie's first ideas on the influence of the radioelements on their environment. At the same time that she was composing a second treatise on radioactivity, Mme. Curie, with the help of Mme. Cotelle and Mlle. Chamié, prepared derivatives of actinium that had not yet been isolated, sometimes staying in the laboratory all night.

Her health declined, but she never spoke of it and grumbled about the interruptions in her work imposed by her doctors. The twenty-fifth anniversary of the discovery of radium was celebrated at the Sorbonne between two of her cataract operations: she had four between 1923 and 1930. She also suffered from lesions on her fingers in 1932, the result of handling radium. She was obliged to enter a nursing home in Paris on 6 June 1934; and then, weakened, she was taken to a sanatorium in the French Alps on 29 June. She did not return.

Of all the honors Mme. Curie received, of all the tributes that were paid her, the one that best fits her personality is a book dedicated to her memory by the French Society of Physics on the centenary of her birth: *Colloquium on Medium and Heavy Nuclei.* The following generations have taken up the torch that she alone had carried for twenty-eight years.

NOTES

1. The house in which Marie was born, in which she lived for only several months, became a laboratory bearing her name.
2. First on rue Flatters, then boulevard de Port-Royal, and finally 11 rue des Feuillantines, in the Latin Quarter.
3. A plaque is affixed there honoring the Curies.
4. A plaque recalls their stay there.
5. "Note," in *Comptes rendus. . . de l'Académie des sciences* (23 Nov. 1896).
6. Communication of Mme. Curie to the Conseil de Physique, Solvay (Brussels, 1911).
7. She had discovered a method of determining the quantity of radium from the radiation emitted (*Le radium,* **7,** 65).
8. Thanks to generous donations, she was compensated for the material she herself had given for this purpose.

BIBLIOGRAPHY

I. Original Works. All of Marie Curie's original memoirs and a few of her general articles have been collected in a single volume by Irène Joliot-Curie: *Oeuvres de Marie Sklodowska-Curie* (Warsaw, 1954). The following works in particular should be mentioned: *Recherches sur les substances radioactives,* 2nd ed. (Paris, 1904), her thesis; *Les théories modernes relatives à l'électricité et à la matière,* the opening lecture of her physics course, 5 Nov. 1906; "Les mesures en radioactivité et l'étalon du radium," in *Journal de physique*, **2** (1912), 715; *Sur les rayonnements des corps radioactifs* (Paris, 1913); and "Les radio-éléments et leur classification," in *Revue du mois* (1914).

See also the first *Traité de radioactivité*, 2 vols. (Paris, 1910); *L'isotope et les éléments isotopes* (Paris, 1922–1923); *Pierre Curie* (Paris, 1924), the American ed. of which (New York, 1923) includes a biography of Marie Curie; and *Radioactivité*, posthumous ed. prepared by Irène and Frédéric Joliot, 2 vols. (Paris, 1935).

II. SECONDARY LITERATURE. Among the innumerable articles and books on Marie Curie are J. Christie, "The Discovery of Radium," in *Journal of the Franklin Institute,* **167,** no. 5 (May 1909); B. Szilard, "Frau Curie und ihr Werk," in *Chemikerzeitung* (1911); B. Harrow, *Eminent Chemists of Our Time* (New York, 1920); Hamilton Foley, "Madame Curie, the Nation's Guest" and "The Sources of Radium," in *Bulletin of the Pan-American Union* (July 1921); Debierne, "Le 25ᵉ anniversaire de la découverte du radium," in *Chimie et industrie* (Mar. 1924); *Le radium, célébration du 25ᵉ anniversaire de sa découverte* (Paris, 1924); C. Regaud, *Marie Sklodowska Curie* (Paris, 1934); Lord Rutherford, "Marie Curie," in *Nature* (21 July 1934); L. Wertenstein, in *Nature,* **141** (18 June 1938), 1079–1081; Eve Curie, *Madame Curie* (Paris, 1939), English trans. by Vincent Sheean (New York–London, 1939); Irène Joliot-Curie, "Marie Curie, ma mère," in *Europe,* **108** (Dec. 1954); and Eugénie Cotton, *Les Curie et la radioactivité* (Paris, 1963).

The ceremonies commemorating the centenary of Mme. Curie's birth have given rise to numerous publications. The catalog of *L'exposition Pierre et Marie Curie* (Paris, 1967) contains an exhaustive chronology of the events at the exposition. In addition, the following may be consulted: the special number of the *Annales de l'Université* (Paris, 1968); *Centenary Lectures, A.I.E.A., Warsaw, 17–20 October 1967* (Vienna, 1968); and the articles of Marcel Guillot in *Nuclear Physics,* **A103** (1967), 1–8; and J. Hurwic, in *Colloque CI,* **29,** supp. to *Journal de physique* (Jan. 1968), followed by the reprinting of Mme. Curie's first "Note."

ADRIENNE R. WEILL

CURIE, PIERRE (*b.* Paris, France, 15 May 1859; *d.* Paris, 19 April 1906), *physics.*

Curie was the son and grandson of physicians. His grandfather came from a Protestant family of Alsatian origin and practiced medicine in Mulhouse. His father, Eugène Curie, had married Sophie-Claire Depoully in 1854; they had two sons, Jacques-Paul, born in 1855, and Pierre. The boys were brought up in a permissive atmosphere. Their father, an enthusiast of science, an idealist, and an ardent republican, opposed any sort of physical or moral servitude. Curie attended neither *école* nor *lycée;* his pensive spirit was unsuited to academic discipline. He worked at home, first with his mother, then with his father and brother; at fourteen he studied with a mathematics professor.

While a boy Curie observed experiments performed by his father and acquired a taste for experimental research. He received his bachelor of science degree on 9 November 1875, and while still a probationary student in pharmacy he enrolled at the Faculty of Sciences in Paris, receiving the *licence* in physical sciences in November 1877. His brother Jacques was a laboratory assistant to Charles Friedel at the Sorbonne mineralogy laboratory. For the two brothers this marked the beginning of a fruitful collaboration in the physics of crystals. When he was eighteen, Pierre and Jacques discovered the phenomenon of piezoelectricity, which proved to be extremely important in its manifold applications. In 1878 Pierre was appointed laboratory assistant to Paul Desains at the Sorbonne physics laboratory; he remained there until 1882, when he became director of laboratory work at the École Municipale de Physique et Chimie, which had just been founded. In 1883 Jacques Curie was appointed professor of mineralogy at Montpellier; this separation ended the scientific collaboration of the two brothers.

Pierre Curie spent twenty-two years at the École de Physique et Chimie, in more isolated work dedicated to theoretical and experimental research. He developed new equipment that all physics laboratories were to use; his studies on magnetism culminated in the publication (1895) of his famous dissertation "Propriétés magnétiques des corps à diverses températures," which was the subject of his doctoral thesis.

In the spring of 1894 Curie met Maria Sklodowska, who was studying the magnetic properties of certain steels at the École de Physique et Chimie. They were married in 1895 and had two daughters: Irène, born in 1897, who had a brilliant scientific career and married Frédéric Joliot, and Eve, born in 1904. It was Marie Curie who, impressed by Roentgen's and Becquerel's discoveries, considered investigating other substances exhibiting the same properties as uranium. For Pierre Curie this was a new period in his scientific career: in close collaboration with his wife, he was to study radioactivity and discover polonium and radium, discoveries that made both of them famous. Until 1905 their work was done in a wretched laboratory at the École de Physique et Chimie. Pierre Curie was appointed professor of physics at the Sorbonne in October 1904 but was unable to realize his dream of working in the new laboratory that he had equipped. He died in his forty-seventh year, after being struck by a truck while crossing the rue Dauphine in Paris.

Serious, reserved, and involved in his own thoughts, Curie was unconcerned with life's material comforts and improvident in his own personal needs; a man of great gentleness and kindly disposition, he had the insatiable need to understand the phenomena of nature and sacrificed all diversions and outside involvements to that end. Aside from his work he liked only outings in the country; he was a superb naturalist and knew where and when various plants, flowers, and insects could be found.

Curie's concept of scientific research was pure and lofty. He never indulged in the sort of hasty publication that is likely to become dated, and his consistently lucid and cogent papers accurately reflected his precise mind. He showed complete unselfishness, was unconcerned with his career, and steadfastly rejected all compromise and complicated diplomacy; in 1903 he firmly refused to be decorated. Lord Kelvin was perhaps one of the first to recognize Curie's great ability. Through him Curie's achievements were acknowledged in British scientific circles more quickly than in France. He maintained a continuing correspondence with Curie from the time Curie discovered piezoelectricity and visited him in his laboratory in October 1893, while Curie was working on magnetism. It was through Kelvin's intervention that Curie was appointed professor at the École de Physique et Chimie in 1894.

Curie's work in radioactivity assured his fame. The Académie des Sciences awarded him the La Caze Prize in physics in 1901 and elected him a member on 3 July 1905; and he was elected a corresponding member of the Academy of Sciences and Letters of Cracow on 15 February 1903. The Royal Society of London awarded the Curies the Davy Medal on 5 November 1903, and on 12 December the Nobel Prize in physics was divided among them and Henri Becquerel. Their honors also included honorary membership in the Royal Institution of Great Britain and in the department of physical sciences of the Academy of Mexico, both in May 1904; the Matteucci Gold Medal of the Italian Society of Sciences in August 1904; and corresponding membership in the Batavian Society of Experimental Philosophy of Rotterdam.

Curie's scientific work and his private life were closely linked. His entire life, basically, was spent in his laboratory; it comprised three periods. The first, begun with his brother Jacques, concerned the physics of crystals, particularly piezoelectricity; then came the years of more isolated but nonetheless fruitful work that mark the complete physicist: he built new apparatus and worked on experimental and theoretical problems of crystallography and magnetism; the third period largely involved the study of radioactivity with his wife.

Piezoelectricity. Jacques Curie, under the direction of Charles Friedel, undertook research on pyroelec-

tricity, a phenomenon that had been known for quite some time and consisted of the appearance of electrical charges in certain crystals when they were heated. This research, conducted in various laboratories, led to experimental observations and interpretations that were often contradictory. Revealing a rare sense of geometry, Pierre and Jacques Curie, by means of simple considerations of symmetry in crystals, discovered the novel phenomenon of piezoelectricity, a property of nonconducting crystals that have no center of symmetry.

These crystals, including zinc sulfide, sodium chlorate, boracite, tourmaline, quartz, calamine, topaz, sugar, and Rochelle salt, were cited in their first publication (1880). These so-called hemihedral crystals may possess axes of symmetry which are polar; in quartz, which they studied extensively, the polar axes are the three binary axes perpendicular to the ternary axis; and in tourmaline it is the ternary axis. By compressing a thin plate cut perpendicular to a binary axis in quartz (still called the electric axis) or perpendicular to the ternary axis in tourmaline, the two faces on which two tin sheets are fastened become charged with equal amounts of electricity of opposite signs, these amounts being proportional to the pressure exerted. For a decrease in pressure of the same value the two faces are charged with the same amounts of electricity but with opposite signs. The amounts of electricity are proportional to the surface of the plates. These measurements were accurate because of the use of Kelvin's electrometer.

As soon as this research was published, Lippmann observed that thermodynamics demanded the existence of the inverse phenomenon, i.e., the strain of the piezoelectric crystal under the action of an electric field.

In 1881 the two brothers proved, with quartz and tourmaline, that the piezoelectric plates of these two substances underwent either contraction or expansion, depending on the direction of the electrical field applied. They showed this extremely slight strain, indirectly at first, by using it to compress another quartz, which exhibited the direct piezoelectric effect, and then directly, with a microscope, amplifying the strain by using a lever. Having established the experimental laws of piezoelectricity, the Curie brothers built a remarkable piece of equipment, the piezoelectric quartz balance, which supplied amounts of electricity proportional to the weights suspended from it. This device, which was immediately used by laboratories engaged in electrical research, was later used by Pierre and Marie Curie in their work on radioactivity.

At first the discovery of piezoelectricity was of only speculative interest. It permitted removal of the contradictions found in pyroelectric observations; thus, in 1882 Friedel and Jacques Curie showed that the electric charges observed in a heated quartz plate originated in internal tensions due to heterogeneous heating. If the temperature remained homogeneous, no electric charge developed; therefore quartz was piezoelectric and not pyroelectric. The industrial uses of piezoelectricity were to come much later. During World War I, Constantin Chilovsky and Paul Langevin, a student of Pierre Curie's, had the idea of placing piezoelectric quartz in an alternating electric field; under the effect of inverse piezoelectricity, predicted by Lippmann and verified by the Curies, the crystal expands and contracts, vibrating with the frequency of the field. This vibration is especially intense when the frequency of the field is the same as that of one of the natural vibration modes of the quartz, i.e., when there is resonance. This is a convenient method of producing high-frequency sound waves, first used to locate enemy submarines and later for underwater soundings. The applications of piezoelectric crystals are innumerable; one of the most important is their use in frequency stabilization of oscillating electromagnetic circuits for radio broadcasting stations. They are used in most piezometers for measuring with great precision either very strong pressure variations, such as those of a cannon at the moment of firing, or very weak ones, such as artery pulsations. These applications have led to the creation of a new industry, the manufacture of large "monocrystals," such as quartz obtained hydrothermally around 500°C. under high water pressures, or crystals such as Rochelle salt, obtained from aqueous solutions. These two substances were mentioned in the Curie brothers' report announcing the discovery of piezoelectricity.

Isolated Work. The study of piezoelectric and pyroelectric phenomena led Pierre Curie to research on crystal symmetry after his brother Jacques left for Montpellier in 1883. As early as 1884 he published several purely geometric reports on symmetry; and in 1885 he published a brief note on the morphology of crystals, in which he used the capillary constants of the different faces of a developing crystal to derive the ideal shape. This theory of Curie's on the growth of crystals is often cited. His ideas on symmetry, which have been fundamental to crystallographers, always gave direction to his experimental research. In 1894 he published an important work on symmetry in physical phenomena, in which he set forth what are today called Curie's laws of symmetry. After discovering radium he returned to this area of study, the results of which he set forth in his course at the Sorbonne.

Curie's laws of symmetry express, in a new and fruitful manner, the principle of causality: When certain causes produce certain effects, the symmetry of the causes reappears, in its entirety, in the effects; if an effect includes an asymmetry, this asymmetry appears, of necessity, in the effective cause.

The important factor in the production of a phenomenon or the realization of a physical state is not the presence, but rather the absence, of certain elements of symmetry. As Curie said, "It is asymmetry that creates the phenomenon." A given phenomenon possesses characteristic symmetry that is the maximum symmetry compatible with the existence of that phenomenon; and this phenomenon cannot appear in an environment that contains an element of symmetry (center, axis, or plane) not included in its characteristic symmetry. The affirmation of this impossibility made possible a considerable economy of research. For example, it establishes that the characteristic symmetry of an electrical field is that of the frustum of a cone of revolution that has neither a plane nor an axis of symmetry perpendicular to the axis of revolution. Which crystals could be pyroelectric and the direction of the electrical field could therefore be predicted. Thus tourmaline is pyroelectric, since its symmetry group (generated by a ternary axis and three planes of reflection passing through this axis) forms a subgroup of the symmetry group of an electrical field; on the other hand, quartz, which has a ternary axis and three perpendicular binary axes, is not. If a crystal is compressed, this compression introduces an asymmetry that homogeneous heating does not introduce; quartz, like all crystals not having a center of symmetry, is piezoelectric. It follows that all pyroelectric crystals are also piezoelectric.

For the magnetic field, the characteristic symmetry is that of a cylinder of revolution turning on its axis, which explains the phenomenon of rotatory magnetic polarization.

Between 1890 and 1895 Curie devoted a great deal of effort to studying the magnetic properties of substances at various temperatures, at that time one of the most obscure areas of physics. The results, presented in a doctoral thesis that he defended on 6 March 1895, form the basis of all modern theories of magnetism. If the work of Curie's first period, relating to crystallography and the principle of symmetry, is essentially of theoretical importance, his studies on magnetism exhibit his experimental ability. Skill and patience were required to measure the force to which a sample was subjected in a nonuniform magnetic field of which the variation in relation to space must be known in its absolute value. The force was measured with a torsion balance and was proportional to the mass of the substance under study, to its magnetic susceptibility, and to the derivative of the square of the field in the direction of the displacement. Heating presented serious difficulties and the influence of convection currents had to be eliminated, lest they interfere with the measurement of extremely small forces; Curie used an electric furnace that allowed him to operate up to 1370°C.

In terms of their magnetic properties the substances investigated by Curie may be divided into three distinct groups: (1) ferromagnetic substances, such as iron, that always magnetize to a very high degree; (2) low magnetic (paramagnetic) substances, such as oxygen, palladium, platinum, manganese, and manganese, iron, nickel, and cobalt salts, which magnetize in the same direction as iron but much more weakly; and (3) diamagnetic substances, which include the largest number of elements and compounds, whose very low magnetization is in the inverse direction of that of iron in the same magnetic field. He had to take great precautions to purify his diamagnetic substances because the very weak forces could be either greatly changed or completely hidden by the presence of traces of a ferromagnetic substance such as iron.

At the beginning of his research Curie stated the problem thus:

At first glance, these three groups are completely separate, but will this separation bear a closer examination? Do transitions between these groups exist? Is this a question of entirely different phenomena, or are we dealing only with a single phenomenon modified in various degrees? These questions were of great concern to Faraday, who often referred to them in his memoirs. He performed one important experiment in this field: it was believed for a long time that iron loses its magnetic properties when it becomes red-hot; Faraday showed that iron remains weakly magnetic when subjected to high temperatures ["Propriétés magnétiques des corps à diverses températures," in *Annales de chimie et de physique,* 7th ser., **5** (1895), 289].

In order to resolve these questions Curie studied, at various temperatures, the diamagnetic substances water, rock salt, potassium chloride, potassium sulfate, potassium nitrate, quartz, sulfur, selenium, tellurium, iodine, phosphorus, antimony, and bismuth; the paramagnetic substances oxygen, palladium, and iron sulfate; and the ferromagnetic substances iron, nickel, magnetite, and cast iron. The large number of measurements taken allowed him to confirm that no parallel can be drawn between the properties of diamagnetic substances and those of paramagnetic substances. The negative susceptibility of diamagnetic substances remains invariable when the temperature varies within wide ranges. This property does not depend on the physical state of the material, since neither fusion (in the case of potassium nitrate) nor allotropic modification (in the case of sulfur) affects the diamagnetic properties of the respective substances. Diamagnetism must therefore be a specific property of atoms. It must result from the action of the magnetic field on the movement of the particles inside the atom, which explains the extreme weakness of the phenomenon and its independence of thermal disturbances or changes of phase, such as fusion and polymorphic transformations. Diamagnetism is thus a property of all matter; it exists also in ferromagnetic or paramagnetic substances but is little apparent there because of its weakness.

Ferromagnetism and paramagnetism, on the other hand, are properties of aggregates of atoms and are closely related. The ferromagnetism of a given substance decreases when the temperature rises and gives way to a weak paramagnetism at a temperature characteristic of the substance and known as its "Curie point." Paramagnetism is inversely proportional to the absolute temperature. This is Curie's law. A little later Paul Langevin, who had been Curie's student at the École de Physique et Chimie, proposed a theory that satisfied these facts by postulating a thermal excitation of the atoms in the phenomena of magnetization. Curie's experimental laws and a quantum mechanical version of Langevin's theory still constitute the basis of modern theories of magnetism.

Curie devoted much of his energy in the laboratory to inventing and perfecting measuring devices, in the design of which his abilities both as a theorist and as a skilled experimentalist are apparent. Among the most notable was the employment of piezoelectric quartz in an instrument for making absolute measurements of very small quantities of electricity. He built a quadrant electrometer that improved upon the one devised by Kelvin by adding an ingenious magnetic damper. Also worthy of note was a standard condenser built of two perfectly parallel glass planes coated with silver and ingeniously mounted so as to eliminate any insulation difficulty. These devices brought Curie, young though he was, to the attention of Lord Kelvin and became the subject of correspondence between the two scientists. In addition he constructed the very sensitive and precise aperiodic balance, with direct reading, which was the first modern balance. All of these devices played an important role in the work on radioactivity that followed.

Radioactivity. It was the research in radioactivity that excited the greatest interest and immortalized the names of Pierre and Marie Curie. The success of their collaboration was assured by the complementary nature of their talents. Pierre Curie appeared to be the complete physicist—theoretician, philosopher, experimenter, and builder of equipment—who was interested in the entire world of nature and had countless ideas in his head; Marie Curie, on the other hand, was trained mainly as a chemist and had a persevering firmness that would not allow for a single moment of discouragement once she attacked a problem. It was Marie who, after having completed a study of the magnetic properties of tempered steel (1898), selected the radiation of uranium for her doctoral thesis.

In 1896 Henri Becquerel had discovered that uranium compounds constantly emitted radiation that was capable of exposing a photographic plate and making air conduct electricity; this radiation had not yet attracted the attention of the scientific world when Marie Curie began her research. On 12 April 1898 she presented, on her own, to the Académie des Sciences a preliminary note that marked the first step in the discovery of radium and opened to mankind the immense world of nuclear physics. Impressed with the importance of this subject, Pierre dropped experiments he had in hand on the growth of crystals (temporarily, he thought) to join in his wife's research. On 18 July 1898 they published their first joint report, "Sur une substance nouvelle radioactive contenue dans la pechblende," in which they announced the discovery of polonium.

The reason for their success may be found in the new method of chemical analysis based on the precise measurement of radiation emitted, a method still in use. It shows the trademark of Pierre Curie. Each product was placed on one of the plates of a condenser, and the conductibility acquired by the air was measured with the aid of the electrometer and piezoelectric quartz he had built. This value is proportional to the quantity of active substance, such as uranium or thorium, present in the product. They examined a great number of compounds of almost all known elements and found that only uranium and thorium were definitely radioactive. Nevertheless, they decided to measure the radiation emitted by the ores from which uranium and thorium were extracted, such as pitchblende, chalcolite (torbernite), and uraninite.

In her first paper Marie Curie had pointed out that emission of radiation seemed to be an atomic property persisting in all physical and chemical states of matter and that in pitchblende it could be even greater than that of known chemical elements, including uranium and thorium. She suggested that a small amount of a very active substance might be found in the ores examined. Pierre and Marie Curie thought that by using the classical methods of analysis and chemical separations, they might be able to isolate new elements that were more active than uranium and thorium.

In that wretched shed at the École Municipale de Physique et Chimie Industrielles they set to work processing huge quantities of pitchblende ore, manhandling the vats and carboys of reagents, carrying out hundreds of fractionations by recrystallization. After each operation they followed up the trace of active fractions remaining in the pitchblende by accurate and rapid measurements of the radioactivity. The importance of this "new method of chemical research," as Pierre Curie called it, is well-known today, with the appearance of radiation detectors such as the Geiger-Müller and scintillation counters.

The pitchblende used recorded, on their plate apparatus, an activity around two and a half times greater than that of uranium. After treating the residue in classical fashion with acid, and the resulting solution with hydrogen sulfide, they discovered that an active substance was present with bismuth in all reactions carried out in solution. Still, they did achieve partial separation by observing that bismuth sulfide was less volatile than the sulfide of the new element, which they called polonium, in honor of Marie Curie's native land.

On 26 December 1898 the Curies and G. Bémont, head of studies at the École Municipale de Physique et Chimie Industrielles, announced in a report to the Académie des Sciences the discovery of a new element, which they called radium, which is present with barium in all chemical separations carried out in solution.

Since radium chloride is less soluble than barium chloride, the Curies were able, by means of fractional crystallizations, to isolate the new element in small quantities; in order to obtain just a few centigrams of pure radium chloride, they had to process two tons of the residue of uranium ore. Almost immediately (October 1899) André Debierne, a student of Pierre Curie's, announced the discovery of a third new radioactive element, actinium. Polonium, and even more so radium, showed enormous radioactivity compared with that of uranium and thorium; for example, the Curies proved that the same effect was produced on a photographic plate by a thirty-second exposure to radium and polonium as could be obtained only after several hours of exposure to uranium and thorium. Further discoveries followed like so many coups de theatre.

First came the announcement in 1899 by Marie Curie of induced radioactivity, brought about by the action of polonium or radium on inactive substances. The induced radioactivity persisted over a considerable period of time, a phenomenon of great concern to Pierre Curie. He took up the question with Debierne, with whom he published two papers in 1901; their experiments could be explained by Rutherford's theory of emanation (radon), a radioactive gas emitted by radium. With J. Danne, Curie measured the diffusion coefficient of radium emanation in the air and proved, as Rutherford had done, that it liquefies at − 150° C. In order to clarify the nature of the emanation he studied the law of diminution of the activity of a solid after having removed it from a chamber in which a radium salt was present. In two notes presented to the Academy on 17 November 1902 and 26 January 1903, Curie showed that this activity diminishes according to an exponential law characterized by a time constant that, for the emanation, is equal to 5,752 days, regardless of the conditions of the experiment. The importance of this discovery, which marks the point of departure for all modern measurements of archaeological and geological dating, did not escape him, for at a meeting of the Société Française de Physique in 1902 he defined a standard for the absolute measurement of time on the basis of radioactivity. Almost immediately Rutherford and Soddy showed that the exponential diminution was caused by the transmutation of radioactive elements.

Curie published two papers on the physiological action of radium rays. In the first, written with Henri Becquerel, the two scientists described burns they had sustained from radioactive barium chloride; in the second, written with the physiologists Charles Bouchard and Victor Balthazard, he participated in experiments on the toxic effect of radium emanation on mice and guinea pigs. On 16 March 1903, in a short paper to the Academy, Curie and his student A. Laborde announced a basic discovery on the heat released spontaneously by radium salts. He showed that a gram atom of radium releases 22,500 calories per hour, a quantity of heat comparable with that of the combustion of a gram atom of hydrogen in oxygen. He concluded from this that the energy involved in the radioactive transformation of atoms is extraordinarily large. This was the first appearance, in human affairs, of atomic energy in the familiar form of heat.

Curie did not fail to reflect upon the misuse that mankind might make of his discoveries. In Stockholm, when he received the Nobel Prize, he said:

> We might still consider that in criminal hands radium might become very dangerous; and here we must ask ourselves if mankind can benefit by knowing the secrets of nature, if man is mature enough to take advantage of them, or if this knowledge will not be harmful to the world.

Nevertheless, he concluded in optimistic fashion: "I am among those who believe that humanity will derive more good than evil from new discoveries."

BIBLIOGRAPHY

I. ORIGINAL WORKS. Many of Curie's writings were brought together in *Oeuvres de Pierre Curie* (Paris, 1908). His papers include "Développement, par pression, de l'électricité polaire dans les cristaux hémièdres à faces inclinées," in *Comptes rendus hebdomadaires des séances de l'Académie des sciences,* **91** (1880), 294, written with Jacques Curie; "Sur l'électricité polaire dans les cristaux hémièdres à faces inclinées," *ibid.,* 383, written with Jacques Curie; "Lois du dégagement de l'électricité par pression dans la tourmaline," *ibid.,* **92** (1881), 186, written with Jacques Curie; "Sur les phénomènes électriques de la tourmaline et des cristaux hémièdres à faces inclinées," *ibid.,* 350, written with Jacques Curie; "Les cristaux hémièdres à faces inclinées, comme sources constantes d'électricité," *ibid.,* **93** (1881), 204, written with Jacques Curie; "Contractions et dilatations produites par des tensions électriques dans les cristaux hémièdres à faces inclinées," *ibid.,* 1137, written with Jacques Curie; "Déformations électriques du quartz," *ibid.,* **95** (1882), 914, written with Jacques Curie; "Symétrie dans les phénomènes physiques, symétrie d'un champ électrique et d'un champ magnétique," in *Journal de physique,* 3 (1894); "Propriétés magnétiques des corps à diverses températures," *ibid.,* 4 (1895); "Nouvelle substance fortement radioactive, contenu dans la pechblende," in *Comptes rendus hebdomadaires des séances de l'Académie des sciences,* **127** (1898), written with Mme. Curie and G. Bémont; "Sur une substance nouvelle radioactive dans la pechblende," *ibid.,* written with Mme. Curie; "Action physiologique des rayons du radium," *ibid.,* **132** (1901), 1289, written with Henri Becquerel; "Conductibilité des diélectriques liquides sous l'influence des rayons du radium et des rayons de Röntgen," *ibid.,* **134** (1902); "Chaleur dégagée spontanément par les sels de radium," *ibid.,* **136** (1903), written with A. Laborde; and "Action physiologique de l'émanation du radium," *ibid.,* **138** (1904), 1384, written with Charles Bouchard and Victor Balthazard.

II. SECONDARY LITERATURE. On Curie or his work see P. Langevin, "Notice sur les travaux de Monsieur P. Curie," in *Revue du mois* (10 July 1906); and *Pierre et Marie Curie* (Paris, 1967), a booklet for visitors to an exhibit held at the Bibliothèque Nationale.

JEAN WYART

CURTIS, HEBER DOUST (*b.* Muskegon, Michigan, 27 June 1872; *d.* Ann Arbor, Michigan, 9 January 1942), *astronomy.*

Curtis was the son of cultivated but somewhat austere parents. While at Detroit High School he showed an aptitude for languages, and at the University of Michigan he devoted himself to Latin, Greek, and other ancient tongues. He studied some mathematics but little science and, apparently, no astronomy whatever.

Curtis obtained his A.B. in 1892 and his A.M. in 1893, then taught Latin at Detroit High School for six months before becoming professor of Latin and Greek at Napa College in California. At Napa his interest was aroused by a small Clark refracting telescope, and when the college merged at San José with the University (now College) of the Pacific, he had the use of a small observatory. In 1896 he transferred to the professorship of mathematics and astronomy.

In 1898 Curtis spent the summer vacation at Lick Observatory, where he was later to carry out his creative work. In 1900 he accepted a Vanderbilt fellowship at the University of Virginia, and he received his Ph.D. there in 1902 with the dissertation "The Definitive Orbit of Comet 1898 I." Meanwhile, he went as a volunteer assistant on the Lick Observatory eclipse expedition to Thomaston, Georgia, in 1900 and on the U.S. Naval Observatory expedition to Solok, Sumatra, in 1901. He so impressed W. W. Campbell that he was invited to join the staff of Lick Observatory as soon as he completed his degree.

Curtis' early years on the Lick staff were devoted mainly to continuing Campbell's studies of the radial velocities of the brighter stars, first at Lick itself and, from 1906, at the station at Santiago, Chile. In 1910 he was recalled by Campbell to work on the program of nebular photography that J. E. Keeler had briefly pioneered at Lick from 1898 to his death in 1900. The outstanding problem of nebular research concerned the nature of the "white" nebulae whose spectrum of light was continuous in a manner consistent with their being large star systems, possible "island universes" comparable with our own star system, or galaxy. Two observational facts were of great weight against this view. First, the vast numbers of white nebulae were not distributed over the whole of the visible sky, as one would expect according to the "island universe" theory, but were concentrated toward the poles of the galaxy and avoided its plane; this "zone of avoidance" suggested that they were subsidiary to the galaxy and not systems of coordinate rank. Second, a bright new star, or nova, had been observed in the great nebula in Andromeda (M31) in 1885; and it had been estimated that, according to the "island universe" theory, this single nova would have to be equivalent in brightness to some 50,000,000 suns.

In the course of his decade of work on nebulae, Curtis steadily became more convinced of the truth of the "island universe" theory. His photographs of white nebulae seen edge-on often showed a dark band of obscuring matter in the plane of the nebula, and he reasoned that a similar band of obscuring matter in the plane of the galaxy would blot out the light of distant nebulae in the direction of the plane and so cause the optical effect of the zone of avoidance. In 1917 Curtis discovered on photographs of nebulae

new stars much fainter than the nova of 1885, and the publication of similar discoveries by G. W. Ritchey and others convinced him that the nova of 1885 was exceptional in brightness and that the inferences drawn from it could not be relied upon.

Other evidence against the "island universe" theory was, however, arising from the comparisons by Adriaan van Maanen of photographs of nebulae taken years apart. The changes revealed in 1916 by his meticulous measurements indicated that the spiral nebula M101 was rotating far too rapidly to be of a size comparable with our galaxy. Curtis himself was skeptical of van Maanen's results, and his skepticism was to prove well-founded. Van Maanen's colleague at Mount Wilson, the brilliant young Harlow Shapley, believed in the alleged rotations; and since Shapley had himself used daring new distance-measuring techniques to argue that the galaxy is far larger than had been thought, he became a leading opponent of the "island universe" theory. On 26 April 1920, Shapley and Curtis presented their contrasting positions before the National Academy of Sciences. The encounter aroused intense professional interest at the time, and Curtis' effectiveness as a public speaker was held to have made his case the more persuasive.

In 1920 Curtis accepted an invitation to become director of Allegheny Observatory of the University of Pittsburgh, and the change marked the end of his major creative contributions to astronomy: his energies were taken up by teaching, administration, and the construction of new instruments. He was content that the main observing program should be a continuation of the stellar parallax investigations initiated by his predecessor. In 1930 Curtis became director of the observatory of the University of Michigan, where he had the promise of money to construct a great telescope. Unfortunately, this gift became a casualty of the depression; and although Curtis persevered in the raising of funds, the telescope was still unconstructed at his death.

BIBLIOGRAPHY

A complete professional bibliography is included in the memoir by Robert G. Aitken cited below. Curtis' most important publications are "Descriptions of 762 Nebulae and Clusters Photographed With the Crossley Reflector," in *Publications of the Lick Observatory,* **31** (1918), 1–42; and "The Nebulae," in *Handbuch der Astrophysik,* V, pt. 2 (Berlin, 1933), 774–936.

On Curtis, see Robert G. Aitken, "Heber Doust Curtis 1872–1942," in *Biographical Memoirs. National Academy of Sciences,* **22** (1943), 275–294.

MICHAEL A. HOSKIN

CURTISS, RALPH HAMILTON (*b.* Derby, Connecticut, 8 February 1880; *d.* Ann Arbor, Michigan, 25 December 1929), *astronomy.*

Curtiss was a skillful and productive spectroscopist who developed experimental techniques and used them to investigate the properties of variable stars.

The youngest of three sons born to Hamilton Burton Curtiss and the former Emily Wheeler, Curtiss grew up in an atmosphere of strict Puritan morality. The family moved in 1892 to California, where he attended high school and the University of California. In the spring of his senior year Curtiss went to Padang, Sumatra, as member of an expedition from Lick Observatory, to observe the solar eclipse of 17–18 May 1901. Returning to Berkeley, he received his B.S. degree in 1901 and his Ph.D. in 1905, while continuing to work at Lick.

In his dissertation ("I. A Method of Measurement and Reduction of Spectrograms for the Determination of Radial Velocities. II. Application to a Study of the Variable Star *W Sagittarii*") Curtiss described a way to get more precise line-of-sight velocities from the Doppler shifts observed in stellar spectra and made the first of his many contributions to the subject of Cepheid variable stars.

During two years at the Allegheny Observatory of the University of Pittsburgh (then called the Western University of Pennsylvania), Curtiss designed his first stellar spectrograph and discovered with it a third component of the star Algol, thereby confirming a suggestion made by Seth Carlo Chandler in 1888. He went in 1907 to the University of Michigan as an assistant professor. Here he initiated an extensive program of observing close binary stars, Cepheid variables, and stars of high surface temperature with bright line spectra ("Class B emission" stars). He became professor in 1918 and married Mary Louise Walton in 1920. Following the death of William Joseph Hussey, he was named director of the Detroit observatory of the University of Michigan in 1927; Heber Doust Curtis succeeded him in that post when he died two years later.

Curtiss was elected fellow of the Royal Astronomical Society of London in 1927. He also served as councilor of the American Astronomical Society and as a member of Commission 29 of the International Astronomical Union.

BIBLIOGRAPHY

I. ORIGINAL WORKS. Curtiss' dissertation, "I. A Method of Measurement and Reduction of Spectrograms for the Determination of Radial Velocities. II. Application to a Study of the Variable Star *W Sagittarii,*" was published in *Astrophysical Journal,* **20** (1904), 149–187; and in *Bulletin of the Lick Observatory,* **3** (1904), 19–40. His work on Algol was "On the Orbital Elements of *Algol,*" in *Astrophysical Journal,* **28** (1908), 150–161; and "The Orbit of Algol from Observations Made in 1906 and 1907" (dated 24 March 1908), in *Publications of the Allegheny Observatory of the University of Pittsburgh,* **1** (1910), 25–32, written with Frank Schlesinger.

What probably constitutes Curtiss' major contribution to astronomy appeared in six memoirs on the Class B emission stars: "The Photographic Spectrum of β Lyrae," in *Publications of the Allegheny Observatory of the University of Pittsburgh,* **2** (1912), 73–120; "The Spectrum of γ Cassiopeiae," in *Publications of the Astronomical Observatory of the University of Michigan,* **2** (1916), 1–35; "Changes in the Spectrum of f¹ Cygni," *ibid.,* 36–38; "Changes in the Spectrum of H.R. 985," *ibid.,* 39–44; "Widths of Hydrogen and Metallic Emission Lines in Class B Stellar Spectra," *ibid.,* **3** (1923), 1–15; and "A Pictorial Study of the Spectrum of Nova Geminorum II," *ibid.,* 253–255, with pls. E–K.

The list of Curtiss' publications in Poggendorff, V (1926), 255, and VI, pt. 1 (1936), 502–503, appears to be complete through 1931—although four articles in vol. 3 of *Publications of the Observatory of the University of Michigan* (also known at that time as Detroit observatory) appear twice (once under each heading). Missing are four posthumous works: Curtiss' excellent historical summary, "Classification and Description of Stellar Spectra," which appeared (in English) as ch. 1 of *Handbuch der Astrophysik,* V, pt. 1 (Berlin, 1932); and three articles prepared for publication by Dean Benjamin McLaughlin: "The Light Curve of R Scuti, 1911–1931," in *Publications of the Observatory of the University of Michigan,* **4** (1932), 129–133; "Variations of Emission Lines in Three B Spectra," *ibid.,* **4** (1932), 163–174; and "Visual Light Curves of Beta Lyrae," *ibid.,* **5** (1934), 177–184. The material on novae mentioned in Rufus' obituary of Curtiss (see below) does not seem to have been published.

II. SECONDARY LITERATURE. The facts of Curtiss' life can be found in several brief obituary notices: two by Edward Arthur Milne, in *Observatory,* **53** (1930), 54–56, and *Monthly Notices of the Royal Astronomical Society,* **90** (1930), 362–363; and two by Joseph Haines Moore, in *Science,* n.s. **72** (1930), 58–60, and *Publications of the Astronomical Society of the Pacific,* **42** (1930), 37–40. The latter includes a portrait, as do two obituaries written by Curtiss' students: Dean Benjamin McLaughlin, in *Journal of the Royal Astronomical Society of Canada,* **24** (1930), 153–158; and Will Carl Rufus, in *Popular Astronomy,* **38** (1930), 190–199.

SALLY H. DIEKE

CURTIUS, THEODOR (*b.* Duisburg, Germany, 27 May 1857; *d.* Heidelberg, Germany, 8 February 1928), *chemistry.*

Most of Curtius' research was on nitrogen-containing organic substances, and two reactions he discovered were named for him: the conversion of an acid into an amine or an aldehyde and the conversion of an azide into an isocyanate.

Curtius was born into a scholarly family originally from Bremen. His grandfather, Friedrich Wilhelm Curtius, was engaged in the manufacture of inorganic chemicals. After graduating from the Gymnasium at Duisburg in 1876, Curtius studied music and science in Leipzig and served in the army from 1877 to 1878, becoming a first lieutenant. He studied in Heidelberg under Bunsen and under Kolbe's direction at Leipzig wrote his doctoral dissertation, "Uber einige neue der Hippursäure analog konstituierte, synthetisch dargestellte Amidosäuren," receiving the degree on 27 July 1882. Curtius worked in Baeyer's laboratories in Munich from 1882 to 1886, when he went to the University of Erlangen as assistant and director of the analytical section of the chemical laboratories. In 1889 he declined an American offer (Worcester Polytechnic) to go to Kiel as professor of chemistry and director of the chemical institute. He succeeded Kekulé at Bonn in 1897 and Victor Meyer (who had succeeded Bunsen in 1889) at Heidelberg in 1898. There he was director of the chemical institute for twenty-eight years and was editor of the *Journal für praktische Chemie.* Curtius entertained himself and his friends by playing the piano, singing, and composing; and he shared his brother's enthusiasm for mountain climbing.

Acting on Kolbe's advice, Curtius investigated the synthesis of hippuric acid by the method of Dessaignes, who had prepared the acid from the zinc salt of glycine and benzoyl chloride in 1853. Curtius improved the method, reporting the results in his doctoral dissertation. Baeyer suggested that he use the ethyl ester of glycine; and when Curtius treated it with sodium nitrite and hydrochloric acid, he discovered the first known aliphatic diazo compound, diazoacetic ester. In 1887 he treated the ethyl diazoacetate with concentrated sodium hydroxide, added hot dilute acid to the product, and discovered hydrazine. The commercial process developed by Friedrich Raschig in 1908 (the oxidation of aqueous ammonia by sodium hypochlorite in the presence of gelatin) was used in Germany during World War II for the production of hydrazine to use in rocket fuel. In 1890 Curtius discovered hydrazoic acid, HN_3, which he called "Stickstoffwasserstoffsäure" (azoimide), by treating benzoyl and hippuryl derivatives of hydrazine with nitrous acid. He reported in 1893 that hydrazoic acid is produced by the action of red fuming nitric acid on a dilute, ice-cold solution of hydrazine. Curtius believed that the structure of hydrazoic acid consisted of a hydrogen atom bonded to one nitrogen atom of a three-nitrogen-atom ring, in line with Emil Fischer's proposed azide structure of 1878. Thiele pointed out in 1911 that a linear structure gives a better explanation of certain reactions, and a linear structure is accepted today.

In 1882 Curtius began a long series of researches into those compounds that Fischer named polypeptides in 1906. Using diazo compounds as intermediates, Curtius synthesized benzoyl derivatives of amino acids; and in 1904 he succeeded in synthesizing a hexapeptide, benzoyl pentaglycine-aminoacetic acid, by fusing ethyl hippurate with glycine. He investigated the rearrangement of aliphatic ketazines and aldazines into pyrazoline derivatives, prepared azides and hydrazides of acids, and obtained benzyl hydrazine from the reduction of benzaldazine. In 1894 Curtius discovered the first of two reactions that were later named for him, the conversion of an acid into the corresponding amine by way of the azide and urethane. By varying the technique, he was able to prepare the corresponding aldehyde in place of the amine in 1906. The second reaction bearing his name is the conversion of an azide into isocyanate, discovered in 1913.

Collaborating with Hartwig Franzen, Curtius began investigating plant chemistry in 1912. They found hexenal but no formaldehyde in green plants. The results of this work caused the abandonment of Baeyer's hypothesis that hexoses are built up from the condensation of hydrated formaldehyde within the plant tissues.

BIBLIOGRAPHY

I. ORIGINAL WORKS. Curtius' doctoral dissertation, *Über einige neue der Hippusäure analog konstituierte, synthetisch dargestellte Amidosäuren,* was published at Leipzig in 1882. *Diazoverbindungen der Fettreihe, ein neue Klasse von organischen Körpern* (Munich, 1886) summarizes his work on diazo compounds. A tribute to his teacher and predecessor at Heidelberg is *Robert Bunsen als Lehrer in Heidelberg* (Heidelberg, 1906). "Über das Diamid (Hydrazin)," in *Berichte der Deutschen chemischen Gesellschaft,* **20** (1887), 1632–1634, describes the discovery of hydrazine; and the discovery of hydrazoic acid is reported in "Über Stickstoffwasserstoffsäure (Azoimid) N_3H," *ibid.,* **23** (1890), 3023–3033, and in "Azoimid aus Hydrazinhydrat und salpetriger Säure," *ibid.,* **26** (1893), 1263. Curtius' early work on polypeptides is described in "Über einige neue der Hippursäure analog constituierte, synthetisch dargestellte Amidosäuren," in *Journal für praktische Chemie,* **26** (1882), 145–208. On the reactions named for Curtius, see "Umlagerung von Säureaziden, $RCON_3$, in Derivate alkylirter Amine (Ersatz von Carboxyl durch Amid)," in *Berichte der Deutschen chemischen Gesellschaft,* **27** (1894), 778–781; "Ersatz von Carboxyl durch Amid in mehrbasischen Säuren," *ibid.,* **29** (1896), 1166–1167; "Umwandlung von Cholalsäure,

$C_{23}H_{39}O_3COOH$, in Cholamin, $C_{23}H_{39}O_3NH_2$," *ibid.,* **39** (1906), 1389–1391; and on conversion of an azide into an isocyanate, "Hippenyl isocyanat, $C_6H_5CONHCH_2CNO$," in *Journal für praktische Chemie,* **87** (1913), 513–541. Two papers by Curtius and Franzen dealing with plant chemistry are "Über die chemischen Bestandteile grüner Pflanzen," in *Justus Liebigs Annalen der Chemie,* **390** (1912), 89–121; and "Das Vorkommen von Formaldehyd in den Pflanzen," in *Berichte der Deutschen chemischen Gesellschaft,* **45** (1912), 1715–1718.

II. SECONDARY LITERATURE. For biographical details and evaluations of Curtius' work see the following (listed chronologically): August Darapsky, "Theodor Curtius zum 70. Geburtstag," in *Zeitschrift für angewandte Chemie und Zentralblatt für technische Chemie,* **40** (1927), 581–583; Heinrich Wieland, "Theodor Curtius," *ibid.,* **41** (1928), 193–194; C. Duisberg, "Theodor Curtius," *ibid.,* **43** (1930), 723–725; and August Darapsky, "Zum Andenken an Theodor Curtius," in *Journal für praktische Chemie,* **125** (1930), 1–22.

A. ALBERT BAKER, JR.

CURTZE, E. L. W. MAXIMILIAN (*b.* Ballenstedt, Harz, Germany, 4 August 1837; *d.* Thorn, Germany [now Torun, Poland], 3 January 1903), *mathematics.*

Curtze was the son of Eduard Curtze, a physician, and of Johanna Nicolai-Curtze. He attended the Gymnasium in Bernburg and from 1857 to 1860 studied in Greifswald, primarily under Johann August Grunert. After he passed the teaching examination in 1861, he taught at the Gymnasium in Thorn, where one of his colleagues was the Copernicus scholar Leopold Prove.

Curtze had an excellent knowledge of the current mathematical literature and an unusual talent for languages; he translated many valuable mathematical works from Italian into German, an outstanding example being Schiaparelli's *Precursori di Copernico nell'antichità* (1876). He did not publish a comprehensive work on his main field—the editing of medieval manuscripts, especially those in the rich collection of the library of Thorn—but he did publish valuable reports on the treasures of this library (1871; 1873–1878). He was also responsible for editions of Oresme's *Algorismus proportionum* (1868) and of his mathematical writings (1870), the *Liber trium fratrum de geometria* (1885), Peter of Dacia's commentary on Sacrobosco's *Algorisms* (1897), and Anaritius' commentaries on Euclid's *Elements* (1899). Other publications are the collection *Urkunden zur Geschichte der Mathematik im Mittelalter und der Renaissance* (1902) and a carefully researched biography of Copernicus (1899).

Curtze contributed many essays to mathematical journals and journals of the history of science. His work was greatly aided by his skill in deciphering hard-to-read handwriting and by visits to libraries in Uppsala and Stockholm (1873) and central Germany (1896). He began corresponding with Moritz Cantor in 1865 but did not meet him until 1896.

In his time Curtze was the outstanding expert on medieval mathematical texts. Through his careful editions he pointed out new paths in a field that was then little investigated.

BIBLIOGRAPHY

I. ORIGINAL WORKS. Curtze's writings include *Die Gymnasialbibliothek zu Thorn* (Thorn, 1871); *Die Handschriften und seltenen Drucke der Gymnasialbibliothek zu Thorn,* 2 vols. (I, Thorn, 1873; II, Leipzig, 1878); a biography of Copernicus (Leipzig, 1899); and *Urkunden zur Geschichte der Mathematik im Mittelalter und der Renaissance* (Leipzig, 1902). Among his translations is that of G. V. Schiaparelli's *Precursori di Copernico nell'antichità* (Leipzig, 1876). Curtze was responsible for editions of Nicole Oresme's *Algorismus proportionum* (Berlin, 1868) and of his mathematical writings (Berlin, 1870); of the *Liber trium fratrum de geometria* (Halle, 1885); of Peter of Dacia's *In algorismum Johannis de Sacro Bosco commentarius* (Copenhagen, 1897); and of Anaritius' *In decem libros elementorum Euclidis commentarii* (Leipzig, 1899).

II. SECONDARY LITERATURE. Obituary notices are Moritz Cantor, "Maximilian Curtze," in *Jahresbericht der Deutschen Mathematiker-Vereinigung,* **12** (1903), 357–368, with portrait, and *Biographisches Jahrbuch und deutscher Nekrolog für das Jahr 1903,* (1904), 90–94; and S. Günther, "Maximilian Curtze," in *Bibliotheca mathematica,* 3rd ser., **4** (1903), 65–81, with portrait and bibliography.

J. E. HOFMANN

CUSA, NICHOLAS, also known as **Nikolaus von Cusa, Nicolaus Cusanus** (*b.* Kues, Moselle, Germany, *ca.* 1401; *d.* Todi, Umbria, Italy, 11 August 1464), *philosophy, mathematics.*

Nicholas was the son of a well-to-do Moselle fisherman, Johann Cryffts (or Krypffs, or Krebs), and his wife Katharina Römer. He may have received his early education at Deventer, in the Netherlands, in the school kept by the Brothers of the Common Life; he would thus have been early influenced by the *devotio moderna* movement. (This reform movement, which strove for a practical, Christocentric religious practice, was spread throughout the low countries and the Rhineland at that period.) Nicholas then entered the University of Heidelberg (he went as companion to Count Ulrich von Manderscheid) and presumably took the introductory course in philosophy there.

In 1417 Nicholas went on to the University of Padua, where he studied canon law with Prosdocimo de Comitibus and others; in 1423 he earned the degree *doctor decretalium*. At Padua, Nicholas met the physician Toscanelli—with whom he attended the lectures on astrology of Prosdocimo Beldomandi—and the humanist educators Guarino da Verona and Vittorino da Feltre. He also attended the penitential sermons of the Franciscan Bernardino of Siena.

In 1425 Nicholas entered the service of Elector Otto von Ziegenhain, archbishop of Trier. In 1426, he was in Cologne, where he had presumably come as a teacher of canon law (although he may have studied theology there in 1425). In Cologne, Nicholas began his independent researches into original source materials, probing deep into the annals of Germanic law; he was thus able to prove that the Donation of Constantine was in fact an eighth-century forgery. He also found copies of Latin writings long believed to be lost; the most important of these were the *Natural History* of Pliny the Elder and twelve comedies of Plautus. His fame spread; in 1428 and again in 1435 he was offered a post as teacher of canon law at the newly established University of Louvain, but he turned it down in each instance.

It was at this period that Nicholas became associated with Heimeric von Campen, who led him into closer acquaintance with the Scholastics as well as introducing him to Proclus' commentary on Plato's *Parmenides* and to the writings of Ramón Lull (whose work Nicholas annotated and, around 1428, copied out in part with his own hand).

Nicholas was ordained priest *ca.* 1430. In this year, too, the archbishopric of Trier was claimed by Ulrich von Manderscheid. In the contest that followed, Nicholas acted as Ulrich's representative in pleading his case before the Council of Basel, convened in 1431. (It was later reconvened in Ferrara and then in Florence.) He did not succeed in getting the council to grant Ulrich's request, but he was recognized as a participant and became one of the leading spokesmen for the conciliar faction. Nicholas dedicated his book *De concordantia catholica* (1433) to his fellow councillors; in it he based his arguments on the thesis that the authority of an ecumenical council was superior to that of a pope and elaborated a comprehensive system of government and society. He stressed the church as the supreme earthly society, and found a divine pattern for priestly concord. (That he could be more realistic as a politician is shown by his 1433 proposal for negotiations with the Hussites.) While in Basel, Nicholas met the humanist Piccolomini, who in 1440 became secretary to Pope Felix V and in 1458 himself became Pope Pius II.

In 1436 the council received an appeal from Christian Byzantium, then sorely pressed by the Turks. The prospect of union of the eastern and western churches triggered stormy arguments in Basel, in which Nicholas sided with the pope's party. In 1437 he took part in an embassy sent to Constantinople to bring the princes of the Byzantine church and the emperor John VIII Palaeologus to the west. Nicholas knew some Greek; he also brought from Constantinople manuscripts containing reports on the councils of Constantinople and Nicaea which shed new light on the doctrinal dispute that centered on the text of the Creed. (Although formal unification of the two churches was achieved in 1437 it was not recognized by Constantinople.)

Through his new alliance with the papal minority Nicholas was given other diplomatic missions. During 1438 and 1439 he was engaged in constant negotiations with the German princes, most of whom either supported the majority conciliar position or wished to remain neutral. It was in part through Nicholas' efforts that these princes agreed to recognize the authority of Pope Eugene IV and his successor Nicholas V. Coincident with his diplomatic work, Nicholas had begun his major philosophical treatise, *De docta ignorantia*, finished in 1440 (by his own account he had struck upon its key notion, the *coincidentia oppositorum*, on his way back from Constantinople).

Nicholas' work puts mathematics and experimental science at the service of philosophy in his attempt to describe the limits of human knowledge. In the *De docta ignorantia* he made new interpretations of those philosophically oriented introductory books of mathematics with which he was acquainted, most notably Boethius' *Institutio geometrica* and Thomas Bradwardine's *Geometria speculativa*.

Trained in the methods of Aristotelian logic, Nicholas found them inadequate to his purpose, since he considered them applicable only to finite phenomena. The Divine, being infinite, is inaccessible to the mind of man, but may be approached through a method of symbolic visualization which resolves apparent antitheses; necessarily, however, at the end of the process man must acknowledge his ignorance. Thus, if the truth is probably inaccessible to the mind of man, man can intellectually get closer and closer to it through the sum of his private knowledge—without ever quite reaching it, since the truth represents an absolute, unchanging maximum limit beyond the scope of man's understanding.

Nicholas made extensive use of geometric figures in his *visio intellectualis;* he chose them because the rational language of demonstration was not suited to explain *intellectus*—the power of knowing that is superior to human reason. Thus one could, for example, increase the number of vertices of a regular polygon until, in infinity, it was transformed into a circle; and while triangles, squares, and circles differed from each other on a finite level, they were resolved beyond it on the infinite scale—although man, of course, could never know such ultimate resolution, being bound by finity to approximations only. Likewise, the contradiction of opposites of a straight line and a circle may be resolved in infinity, since a circle of infinitely long radius has a straight line for its circumference.

As understood by Nicholas, the infinite could take two forms, the infinitely large and the infinitely small. No contradiction was here implied, however, since the infinitely large and the infinitely small could both be contained in the concept of maximum; the largest possible thing was of maximum largeness, while the smallest possible thing was of maximum smallness. The concept of maximum admitted one absolute maximum (God) which could also be seen as unifying absolute maximum and absolute minimum, being infinite and therefore without degree.

Such geometrical examples are the essence of the *coincidentia oppositorum,* by means of which Nicholas hoped to resolve all problems formerly considered insoluble. Since apparent contradictions are united in infinity, the largest possible number must coincide with the smallest possible—one—and since numbers are discrete entities, all are contained in the ultimate unity and can be produced from that unity, which is also the measure of all intermediary quantities. Pursuant to this reasoning Nicholas referred to Anaxagoras and assumed that each entity is present in every other entity. In geometry, continuous forms correspond to numbers; the point stands alone as the smallest initial unit and generates lines, surfaces, and solid forms. The most perfect geometric forms—the infinitely large circle and the infinitely large sphere—are at the same time coincident with their generating point.

In cosmology, the application of the *coincidentia oppositorum* led Nicholas to determine that there could be no cosmic mechanism or center point for the motions of the heavens, since such a point of necessity included the whole universe. The universe mirrors God and is a relative maximum, since it contains all things except God, in Whom all is contained; therefore, the universe has no fixed center and no circumference, being relatively infinite. Therefore, the earth is not the center of the universe, nor is it stationary; it moves, as do all other bodies in space, with a motion that is not absolute but relative to the beholder. Nicholas further suggests by analogy that the earth may not be the only body that supports life. Moreover, the earth is not completely round, there is no

maximum movement of the other heavenly bodies as opposed to the fixity of the earth, and no possible movement of bodies in diametrically opposite directions (such as up and down). Nicholas' cosmological reasoning, although garbed in theological language, here anticipates scientific discovery; his later treatise on the subject, *De figura mundi* (1462) is unfortunately lost.

Also lost is an earlier work, *De conjecturis,* of which only a later, much revised version of some time prior to 1444 exists. In the *De docta ignorantia,* Nicholas makes several specific corrections to the earlier version, from which some indication of its contents may be gained; he draws upon earlier notions of what is knowable, which he restates. It is also clear that the earlier book made the same extensive use of symbolic reasoning as the later, although its purport was more clearly metaphysical. Nicholas' earlier cosmology drew upon the Neoplatonic notions of the hierarchy of God–angel–soul–body and united it with the four elements to produce a metaphysics of unity, to which he assigned analogous mathematical values—i.e., he assigned the numbers 1, 10, 100, and 1,000 to the four elements and maintained the Pythagorean relation $1 + 2 + 3 + 4 = 10$ as a symbol of the *arithmetica universalis.* He paid special attention to the relationship of oneness and otherness, which he represented as two opposed quadratic pyramids, conjoined so that each had its vertex at the center point of the other's base plane. One pyramid stood for light, the other for darkness, one for the male principle, the other for the female, and so on. In the *De docta ignorantia* Nicholas was able to use his newly formulated doctrine of the *coincidentia oppositorum* to resolve these contradictions and to develop his theory of a unity in which each form partakes of and mirrors every other form.

Although the *De docta ignorantia* was both respected and influential, such speculations left Nicholas open to attack from his political enemies for pantheistic teachings and other damnable heresies. Johannes Wenck acted as spokesman for Nicholas' detractors in publishing *De ignota litteratura,* against which Nicholas defended himself with his *Apologia doctae ignorantiae* (1449), quoting at length from Dionysius and Areopagite and the Church Fathers.

Following the publication of the *Apologia doctae ignorantiae* Nicholas undertook further diplomatic missions for the Curia and did not expound or develop his system in more extensive writings. He did, however, dictate a number of short treatises which were copied down and circulated among his friends. In addition, some 300 of his sermons, dating from 1431 on, were recorded in the form of brief notes.

(These sermons, given in both Latin and vernacular German, were at first devotional exercises preached on holidays, but later Nicholas began to introduce his own philosophical tenets into them, and by 1444 the mystical influence of Meister Eckhart is apparent in them.) The sermons provided the bases for Nicholas' later tracts, including *De quaerendo Deum* (1445), an orthodox devotional guide; and *De filiatione Dei* (1445), *De ultimis diebus* (1446), *De genesi* (1447), and *De Deo abscondito* (1450[?]), which derive in part from ideas set forth in *De conjecturis* and *De docta ignorantia*. His *De dato patris tris luminum* (1445 or 1446) and *De visione Dei* (1458) both argue against his alleged pantheism.

Nicholas' services to the Curia were rewarded in 1446 when Pope Eugene IV appointed him a cardinal *in petto;* in 1448 Pope Nicholas V made him a full cardinal, with the titular see of St. Peter in Vincoli. Nicholas received the red hat in 1450 and was named bishop of Brixen, where the cathedral chapter recognized him only reluctantly. At the end of 1450 he was appointed legate for Germany; as such he undertook a journey through Germany, Belgium, and the Netherlands where he preached reform and worked for compromise and conciliation in secular and ecclesiastical disputes. He soon became embroiled in violent quarrels with the nobility; Duke Sigismund of Tirol intervened and threatened to use force.

In 1458 Nicholas went to Rome, where his friend Piccolomini had been elected Pope Pius II. On his return to Germany in 1460, he was immediately locked up in Bruneck castle by Duke Sigismund's mercenaries; eventually Nicholas capitulated to the secular forces and returned to Rome. He was there appointed a papal representative.

During this period, in addition to his sermons, Nicholas undertook brief works on mathematics, to which he sought to apply some of the new insights that he had reached philosophically. His principal aim was to transform a circle into a straight line and a square; he confined himself to approximations. In *Transmutationes geometricae* (1445) he displays familiarity with simpler Euclidean theorems and appears to rely heavily on Bradwardine's *Geometria speculativa,* which he never mentions, however. He further refers to a fragment of the writings of Pappus, and shows considerable knowledge of the practical geometries available to him, drawing upon them for his many examples of applied geometry.

Nicholas' *Complementa arithmetica* may also date from 1445, although the exact year of its first publication is not known and it now exists only in a corrupted form. In it, Nicholas first expressed the idea that the difference between the radii of a series of circles inscribed in a series of regular polygons is proportional to the difference in area between the inscribed circle and its corresponding regular polygon.

In his *De circuli quadratura* (1450) Nicholas used a computational approach similar to that employed in his *Transmutationes geometricae;* it is uncertain what earlier authors may have influenced this work. The *De circuli* is remarkable for its discussion of the disputed intermediate value theorem, derived from Aristotle, and of the angle-of-contingence problem and the problem of exactly squaring a circle. Two of Nicholas' *Idiota* dialogues, *De sapientia* and *De mente* (both 1450) are the philosophical synthesis of his mathematical work at this time: in them, a layman who recognizes God's work in nature is given an opportunity, in the form of Platonic dialogues, to explain Cusan philosophy—especially its mathematicizing tendencies—to an Aristotelian scholar. (A third such dialogue, *Idiota de staticis experimentis,* of the same year, has a more practical bias, and contains numerous methods for determining physical parameters through the use of such apparatus as scales and a water clock—for example, the work tells in detail how to determine the humidity of air by measuring the weight of wool.)

In 1452 Nicholas read Jacob of Cremona's translation of the works of Archimedes, and was much impressed with the latter's indirect method of deduction, which he erroneously associated with his own *coincidentia oppositorum.* He had by now thought through his earlier ideas, and set them down systematically in a series of works, beginning with the *Complementa mathematica* (1453; expanded in 1454), which contained many more approximations than the earlier books (although these dealt largely with special cases, and provided no new insights), and the *Complementum theologicum,* a continuation that pertained primarily to symbolic interpretations. Nicholas presented an important variation of the approximation proportion in the *Perfectio mathematica* (1458)— which, more importantly, also contained the notion that the method of *visio intellectualis* could yield an infinitely small arc of a circle and its corresponding chord (although inadequately expressed by Nicholas, this anticipated an infinitesimal concept of great significance). The *Aurea propositio in mathematicis* (1459) contains the final refinement of these ideas.

During the 1450's, Nicholas continued to apply his doctrine of *coincidentia oppositorum* to religious, as well as mathematical, problems. The shattering event of the year 1453—the fall of the Byzantine empire—led him to write *De pace fidei,* published in that year. In it, he presents a dialogue among seventeen articulate representatives of different nations and faiths and calls for mutual tolerance. Although he maintains the superiority of Christianity, he proposes that the differences among faiths are largely those of ritual and stresses the unifying factor of monotheism. He further undertook a thorough and critical investigation of Islam; his *Cribatio Alkoran* (published in 1461) is, however, based upon a poor translation of the Koran.

Nicholas' later works also include a number of relatively short treatises in which simple physical or mathematical examples are expanded into philosophical symbols. In *De beryllo* (1458) he compares the *visio intellectualis* to the effect of a magnifying glass; and in *De possest* (1460) he uses the example of a circular disc rotating at infinite speed within a stationary ring to show that all points of the circumference of the disc are at all points of the interior of the ring simultaneously—hence, motion and rest are identical, and thus time unfolds from the present and thus the single instant and eternity are the same in infinity. In *De ludo globi* (1463) he moves on to discuss the spiral motion of a partially concave sphere, and in *De non aliud* (1462) he reverts to dialogue form for a critical conversation with Aristotle, Plato, and Proclus. *De venatione sapientiae* (1463) stands in relation to his philosophy as the *Aurea proposito in mathematicis* does to his mathematics; it is a retrospective summation of his earlier attempts to illuminate for others his private intellectual world.

As a philosopher, Nicholas was chiefly concerned with knowing the ways of God; his mystical and symbolic approach was calculated to encourage man to seek the unity of all things and the end of antitheses in light of his insight that the Divine could not be known directly. He rejected the rationalism of the Schoolmen, and revitalized Neoplatonism in his time. That he failed always to be understood and appreciated by his contemporaries is due in part to the peculiarity of his language, neither medieval nor humanistic and flawed by inadequately defined words and concepts.

That he failed also to reach the ideal of conciliation that he preached may be attributed, too, to his character, since he was frequently hot-headed, temperamental, and inclined to arbitrary decisions. Despite his demonstrated skill as an imaginative diplomat, Nicholas spent the last years of his ecclesiastical career in Rome in a series of squabbles with the Italian clique and in ineffectual attempts to reform the clergy, the orders, and the Curia.

Only in the nineteenth century did the importance of Cusan thought begin to become clear, and only in the twentieth century was any thorough study of it begun. Precise study has been made possible through the happy circumstance that the home for the aged in Kues, generously endowed by Nicholas, has survived the ravages of time and war; its library, a chief source for Cusan scholars, has, except for minor losses, remained intact.

BIBLIOGRAPHY

I. ORIGINAL WORKS. Nicholas' *Opera* were widely reprinted, including the editions of Martin Flach (Strasbourg, 1488; repr. in 2 vols. as *Nikolaus von Kues, Werke,* Berlin, 1967); Benedetto Dolcibelli (Milan, 1502); Jodocus Badius (Paris, 1514; three-volume facsimile repr. Frankfurt am Main, 1962); and Henricus Petri (Basel, 1565).

The modern edition prepared by the Heidelberger Akademie der Wissenschaften includes E. Hoffmann and R. Klibansky, eds., *De docta ignorantia* (Leipzig, 1932); R. Klibansky, ed., *Apologia doctae ignorantiae* (Leipzig, 1932); L. Bauer, ed., *Idiota de sapientia, de mente, de staticis experimentis* (Leipzig, 1937); B. Decker and C. Bormann, eds., *Compendium* (Hamburg, 1954); P. Wilpert, ed., *Opuscula I* (Hamburg, 1959); G. Kallen and A. Berger, eds., *De concordantia catholica,* 4 vols. (Hamburg, 1959–1968); and J. Koch, ed., *De conjecturis* (in preparation).

German translations of many individual works are available in the series *Schriften des Nikolaus von Cues* (Leipzig or Hamburg, various dates).

II. SECONDARY LITERATURE. Works on Nicholas include M. de Gandillac, *La philosophie de Nicolas de Cues* (Paris, 1942), which was translated into German by K. Fleischmann (Düsseldorf, 1953); K. Jaspers, *Nikolaus Cusanus* (Munich, 1964); A. Lübke, *Nikolaus von Kues* (Munich, 1968); J. Marx, *Verzeichnis der Handschriftensammlung des Hospitals zu Cues* (Trier, 1907); E. Meuthen, *Nikolaus von Kues 1401–1464. Skizze einer Biographie* (Münster, 1964); P. Rotta, *Il cardinale Niccolò de Cusa* (Milan, 1928); E. Vansteenberghe, *Le cardinal Nicolas de Cues (l'action-la pensée)* (Paris, 1920).

Mitteilungen und Forschungsbeiträge der Cusanus Gesellschaft, Mainz, published since 1961, contains many individual articles on Nicholas and his works as well as a running bibliography of Cusan studies from 1920 on.

J. E. HOFMANN

CUSHING, HARVEY WILLIAMS (*b.* Cleveland, Ohio, 8 April 1869; *d.* New Haven, Connecticut, 7 October 1939), *neurosurgery, neurophysiology.*

Cushing was the sixth son and the tenth child (seven lived to maturity) of Henry Kirke and Betsey Maria Williams Cushing. Henry Kirke Cushing, a third-generation physician, combined a large practice with the professorship of midwifery, diseases of women, and medical jurisprudence at Cleveland Medical College and was also for many years a trustee of Western Reserve University. Reserved with his children, he left much responsibility for their upbringing to his wife, a gracious, highly intelligent woman quite capable of the task. He imposed strict discipline in his household and provided comfortably for physical needs—generously for education. One son entered the law, another geology; two became physicians. They attended the Presbyterian church, public schools in Cleveland, and eastern universities for their college and post-graduate training.

Handsome and of wiry grace, Harvey Cushing maintained his slight figure through life by moderate participation in sports—varsity baseball in college, tennis in later years. His health was good except for an undiagnosed illness during World War I that affected his legs and later caused pain and increasing disability. Four years at Yale nurtured an abiding loyalty to his alma mater, largely through the close friendships formed there and maintained and treasured all his life. Enrolling at Harvard Medical School in 1891, he became the fifth Cushing to enter medicine. He received his M.D. cum laude in 1895 and became a surgical house officer at the Massachusetts General Hospital, Boston, then assistant resident at the newly founded Johns Hopkins Hospital (1889) on the service of William S. Halsted, then pre-eminent among American surgeons.

Stimulated by William H. Welch, William Osler, Howard Kelly, and Halsted, Cushing's restless and inquiring mind and enormous capacity for work found full expression. Their interest in medical history spurred him in his collection of medical books; he was also encouraged by his father, who passed along volumes from his own library that also often carried the signatures of his grandfather and great-grandfather. From 1896 to 1911 Cushing progressed to associate professor of surgery. After a year abroad (1900–1901), he began to move toward neurological surgery. He became the first American to devote full time to its development.

In 1902 Cushing married Katharine Stone Crowell, a Cleveland childhood friend; they had five children. The situation in his father's household was repeated in his: he spent long hours at the hospital, then devoted evenings to writing. Yet many house officers and students remember the warm hospitality of a friendly family and Dr. and Mrs. Cushing as gracious hosts. Their elder son, William, a Yale student, was killed in an automobile accident in 1926, and Cushing's sorrow was deepened because he had only begun to know him.

Several universities, including Yale, offered him professorships; he chose Harvard. Moving to Boston in 1912, he served as Moseley Professor of Surgery and chairman of the department at Harvard Medical School, and as surgeon-in-chief of the Peter Bent Brigham Hospital from 1913—when it opened—until his retirement in 1932. During World War I he was briefly in France in 1915 and again from 1917 to 1919, as chief of Base Hospital No. 5. As he gradually solved the problems of brain surgery, patients and young physicians came to his clinic from all over the world. The results of his labors are recorded in 330 books and papers, some devoted to medical history. He was a talented writer; his biography of William Osler was awarded a Pulitzer Prize in 1926. Cushing returned to Yale as Sterling Professor of Neurology (1933–1937) and there published selections from his war diaries, completed an extensive monograph on the meningiomas, and made plans for leaving his library of some 8,000 items, many of great rarity, to Yale University. He persuaded two friends with fine collections, Dr. Arnold C. Klebs of Switzerland and Dr. John F. Fulton, a Yale physiologist, to join him, together with other friends with smaller collections.

Cushing was awarded honorary degrees from nine American and thirteen European universities; several decorations: Distinguished Service Medal, Companion of the Bath, Officier de la Légion d'Honneur, and Order of El Sol del Perú; and many prizes and awards. He was a member of the American Philosophical Society, the National Academy of Sciences, and the American Academy of Arts and Sciences; a foreign member of the Royal Society; and a member (often honorary or foreign) of more than seventy medical, surgical, and scientific societies in the United States, South America, Europe, and India. Some thirty-five of his young associates formed the Harvey Cushing Society in 1932; it is now called the American Association of Neurological Surgeons and has more than a thousand members.

Macewen of Glasgow accurately diagnosed and located a brain tumor in 1876, and after 1900 Horsley of London performed some successful operations. In America, Cushing's contemporaries, Charles H. Frazier of Philadelphia and Charles A. Elsberg of New York, were starting to study the nervous system, Frazier particularly concerned with relief of pain, Elsberg with surgery of the spinal cord; both worked with neurologists. Cushing's greatest contributions were in the broad field of intracranial tumors; he took full responsibility for diagnosis, localization, treatment, verification of pathology, and follow-up of his patients.

His extraordinary achievement (based on a series of more than 2,000 verified cases of tumor) of reducing mortality from almost 100 percent to less than 10 percent would have been impossible without early and continuing recourse to the experimental laboratory. He was responsible for establishing the Hunterian Laboratory at Johns Hopkins in 1905 and also the Laboratory of Surgical Research at Harvard. Not only did they afford a place for his own investigations, but his course in operative surgery for students, begun in 1902, was basic to another of his important contributions—the training of a generation of surgeons who have extended the boundaries of neurosurgery. From these laboratories came more than 325 papers by his pupils.

Early in his career Cushing employed his considerable talent for drawing—as illustrated in diaries of trips, class notes, and sketches of clinic patients at Harvard—in the recording of important aspects of operations. These were sketched on the patient's record immediately after surgery. During his apprenticeship in general surgery, he studied physiological, bacteriological, and biochemical problems as well. Disturbed by deaths caused by inadequate knowledge of volatile anesthetics, he devised an ether chart on which temperature and respiration were recorded during operations. Later, he experimented with local anesthetics, using them successfully in amputations and hernia operations. He became adept at the surgical handling of perforation of the gut in typhoid cases among soldiers in the Spanish-American War. He carried out experiments on bacterial flora in the abdomen and demonstrated that in the fasting dog they were reduced practically to zero except for a pocket in the cecum harboring colon bacillus. Finding that pure sodium chloride solutions were injurious to nerve-muscle preparations, he worked out a physiologically balanced substitute, which, as Cushing's solution, is still used at the Peter Bent Brigham Hospital. Attempting experimental surgery of the thorax and heart, then clinically unheard-of, he studied chronic valvular lesions in the dog and commented "on the possibilities of future surgical measures in man directed toward the alleviation particularly of the lesion characterizing mitral stenosis." In opening the thorax he discovered the need to maintain lung inflation and used "direct inflation of the lungs by opening the trachea, as commonly used in a physiological laboratory . . .." He thus anticipated the now routine use of positive-pressure endotracheal anesthesia.

Certain technical problems had to be solved before brain surgery could be successful. In Switzerland in 1900–1901 Cushing studied the blood pressure-spinal pressure problem and demonstrated in a classic experiment that as the spinal fluid pressure of a dog is increased, there is initially a vagal effect with bradycardia followed by a high rise in arterial blood pressure. This finding started physiologists such as Walter Cannon on years of further study; for Cushing it made possible safer craniotomies. He took back to America a model of the first blood pressure apparatus and was instrumental in its adoption. The management of hemorrhage was most important, and for this Cushing devised silver clips still used to control bleeding. In 1925 he introduced electrocautery in brain surgery and was able to call back many patients whose tumors he had not dared earlier to attack.

One of his earliest interests, tumors of the hypophysis (pituitary gland), resulted in his most enduring work and offered endocrinologists many provocative ideas for exploration. A contemporary surgeon describes Cushing's work in physiology and endocrinology as that of the natural or intuitive biologist, one who acquires enough general understanding of the natural arrangement of things to sense how things unknown are ordered. The sudden insights of his endocrinological research had this quality of brilliance and enabled him to progress in less than thirty years from the consideration of the peripheral ductless glands to the surgically exposed, diagnosed, examined, and removed pituitary. Cushing early observed that the "hypophysis and the ductless glands in general so influence the function of every organic process that they overlap into every individual specialty." His monograph on the pituitary (1912), based on fifty cases, elucidated this influence, especially the effects of undersecretion (hypopituitarism), oversecretion (hyperpituitarism), and recognition of clinical signs of such malfunction. In 1930 he made the brilliant connection between a reported case of basophile adenoma and his own patients who suffered from painful adiposity of the face and body. In solving this baffling question he recognized a new disease entity that was called pituitary basophilism or, commonly, Cushing's disease.

As his experience in the diagnosis and treatment of intracranial tumors increased, he was able, together with his two associates Percival Bailey and Louise Eisenhardt, to describe the gliomas, blood-vessel tumors, and the meningiomas. He had already completed before World War I an authoritative monograph on tumors of the nervus acusticus. His military contributions included a 50 percent reduction in mortality from compound head wounds and elucidation of surgical principles hardly altered to this day.

Viewed in the perspective of the thirty years since his death, Harvey Cushing remains the dominant figure in neurosurgery. He set an example in the application of the scientific method to clinical problems in this developing field, especially to tumors of the brain; and the influence on succeeding generations of this broad approach will probably be his greatest contribution to medicine.

BIBLIOGRAPHY

I. ORIGINAL WORKS. Cushing's papers, manuscript diaries, and memorabilia are in the Historical Collections of the Yale Medical Library. His vita, honors and awards, bibliography, and the papers from his laboratories by pupils are included in *A Bibliography of the Writings of Harvey Cushing Prepared on the Occasion of his Seventieth Birthday April 8, 1939 by the Harvey Cushing Society* (Springfield, Ill., 1939). His books include *The Pituitary Body and its Disorders* (Philadelphia–London, 1912); *Tumors of the Nervus Acusticus and the Syndrome of the Cerebellopontile Angle* (Philadelphia–London, 1917); *The Life of Sir William Osler* (Oxford, 1925); *A Classification of the Tumors of the Glioma Group on a Histogenetic Basis with a Correlated Study of Prognosis,* written with P. Bailey (Philadelphia, 1926); *Studies in Intracranial Physiology and Surgery* (London, 1926); *Tumors Arising from the Blood-Vessels of the Brain* (Springfield, Ill., 1928); *Consecratio Medici and Other Papers* (Boston, 1928); *Intracranial Tumours. Notes upon a Series of Two Thousand Verified Cases with Surgical-Mortality Percentages Pertaining Thereto* (Springfield, Ill., 1932); *Papers Relating to the Pituitary Body, Hypothalamus and Parasympathetic Nervous System* (Springfield, Ill., 1932); *From a Surgeon's Journal, 1915–1918* (Boston, 1936); and *Meningiomas. Their Classification, Regional Behaviour, Life History, and Surgical End Results,* written with L. Eisenhardt (Springfield, Ill., 1938).

A selection from his papers illustrating the range of his early interests includes "Laparotomy for Intestinal Perforation in Typhoid Fever," in *Johns Hopkins Hospital Bulletin,* **9** (1898), 257–269; "Experimental and Surgical Notes Upon the Bacteriology of the Upper Portion of the Alimentary Canal, with Observations on the Establishment There of an Amicrobic State as a Preliminary to Operative Procedures on the Stomach and Small Intestine," written with L. E. Livingood, in *Johns Hopkins Hospital Reports,* **9** (1900), 543–591; "Concerning the Poisonous Effect of Pure Sodium Chloride Solutions upon the Nerve-Muscle Preparation," in *American Journal of Physiology,* **6** (1901), 77–90; "Some Experimental and Clinical Observations Concerning States of Increased Intracranial Tension," in *American Journal of the Medical Sciences,* **124** (1902), 375–400; "The Blood-pressure Reaction of Acute Cerebral Compression, illustrated by Cases of Intracranial Hemorrhage," in *American Journal of the Medical Sciences,* n.s. **125** (1903), 1017–1044; "Instruction in Operative Medicine. With the Description of a Course Given in the Hunterian Laboratory of Experimental Medicine," in *Johns Hopkins Hospital Bulletin,* **17** (1906), 123–134; "Experimental and Clinical Notes on Chronic Valvular Lesions in the Dog and Their Possible Relation to a Future Surgery of the Cardiac Valves," written with J. R. B. Branch, in *Journal of Medical Research,* n.s. **12** (1908), 471–486; and "Surgery of the Head," in W. W. Keen, ed., *Surgery, Its Principles and Practices,* III (Philadelphia, 1908), 17–276.

II. SECONDARY LITERATURE. The definitive biography is John F. Fulton, *Harvey Cushing: A Biography* (Springfield, Ill., 1946). A shorter biography for the layman is Elizabeth H. Thomson, *Harvey Cushing. Surgeon, Author, Artist* (New York, 1950). *Harvey Cushing. Selected Papers on Neurosurgery,* ed. by Donald D. Matson, William J. German, and a committee of The American Association of Neurological Surgeons (New Haven–London, 1969), is the most recent reprinting of Cushing's papers. The foreword and W. J. German's introductory notes to sections offer useful appraisals, as does Francis D. Moore, "Harvey Cushing: General Surgeon, Biologist, Professor," in *Journal of Neurosurgery,* **31** (1969), 262–270.

E. H. THOMSON

CUSHMAN, JOSEPH AUGUSTINE (*b.* Bridgewater, Massachusetts, 31 January 1881; *d.* Sharon, Massachusetts, 16 April 1949), *micropaleontology.*

Cushman pioneered the use of Foraminifera (an order of microscopic shelled Protozoa) in the search for petroleum. His work stimulated other investigators and initiated a period of unprecedented study of the order.

The son of Darius and Jane Frances Fuller Pratt Cushman, he grew up in the small college town of Bridgewater, where his father sold and repaired shoes. Cushman's early schooling was in Bridgewater, then at Harvard College, where cryptogamic botany was his initial interest. Because of the influence of Robert Tracy Jackson, however, he changed to paleontology. After graduation in 1903 he became a curator at the Museum of the Boston Society of Natural History and spent some summers on botanical collecting trips.

Cushman spent the summers of 1904 and 1905 at the U.S. Fish Commission station at Woods Hole, where Mary J. Rathbun urged him to study rich collections of Foraminifera recently obtained by the commission's steamer *Albatross.* Thus began his systematic identification and description of Holocene Foraminifera. When he started his studies, Foraminifera were regarded as primitive, highly variable organisms having geologic ranges from Cambrian to Holocene and having no value other than as curiosities.

Cushman began his work within the framework of Brady's classification of ten families, which was based on morphology of adult forms. Within a short time he began to formulate a radically different classification that resulted from application of Haeckel's law of recapitulation, in which development of the individual repeats phylogenetic development of the species. Cushman's classification, consisting of forty-five families, was first published in outline form in 1927 and was republished the next year in a textbook, *Foraminifera, Their Classification and Economic Use,* which went through four editions.

Cushman was employed in 1912 by the U.S. Geological Survey as a specialist on Foraminifera and in 1914 made his first diffident attempt at age determination of well samples by Foraminifera (Stephenson, pp. 79–81). By 1918 he was confident enough about Foraminifera to use species as indicators of formations and to include interpretations of their paleoecology. In 1921 he stated that Foraminifera "faunas of the various members of these formations [in the American Gulf Coast plain] are easily recognizable." By that time some economic paleontologists had independently realized the truth of this statement and were using Foraminifera as stratigraphic tools.

In January 1923, Cushman was employed in Mexico by the Marland Oil Company to apply his knowledge of Foraminifera in correlating surface rock outcrops and drill holes. He returned in March to Sharon, Massachusetts, where he built a private laboratory for consulting work. He continued consulting for another year, then retired from commercial work to devote all his time to research. In 1925 Cushman started his privately financed journal, *Contributions From the Cushman Laboratory for Foraminiferal Research,* to contain the numerous papers that came from his prolific pen. He offered his services as teacher without stipend for students from Harvard and Radcliffe and also accepted private students. In 1927 and 1932 he visited Europe to examine type specimens and to collect from type localities.

In the 1920's Cushman's interest expanded from formal description of Foraminifera in the *Albatross* collections to include systematic descriptions of faunas from geologic formations and, in the 1930's, monographic studies of families and genera. In the 1940's Cushman began to realize the potential of planktonic species as a means of stratigraphic zonation, and worldwide zonation by planktonic Foraminifera found wide acceptance shortly after Cushman's death. As his work on Holocene faunas increasingly provided a basis for comparison, paleoecologic interpretation became part of Cushman's work on fossil faunas. During nearly fifty years he produced more than 550 papers. His extensive collections were bequeathed to the U. S. National Museum, where they constitute the world's largest collection of Foraminifera.

During Cushman's lifetime the study of Foraminifera progressed from being an obscure hobby of some few dozen investigators, mostly European, to become a major factor in the oil industry, involving thousands of investigators and large sums of money. Cushman evaluated his own success as being "due to the very thing for which I have been criticized . . . the tendency to split species" (Schuchert, p. 549). His classification, now largely superseded by more detailed ones, nevertheless constituted the basic structure for these later ones. Although additional microfossils and other methods are now used in oil exploration, Foraminifera are still an invaluable tool in biostratigraphy and paleoecology.

BIBLIOGRAPHY

I. Original Works. A complete bibliography of Cushman's works is in Todd *et al.* (see below), pp. 40–68. Cushman's writings include *Some Pliocene and Miocene Foraminifera of the Coastal Plain of the United States,* U.S. Geological Survey Bulletin no. 676 (Washington, D.C., 1918); "Use of Foraminifera in Determining Underground Structure Especially in Petroleum Mining," in *Bulletin of the Geological Society of America,* **33** (1921), 145–146, an abstract; "An Outline of a Re-classification of the Foraminifera," in *Contributions From the Cushman Laboratory for Foraminiferal Research,* **3,** pt. 1 (1927), 1–105, pls. 1–21; and *Foraminifera, Their Classification and Economic Use,* Cushman Laboratory for Foraminiferal Research, special publication no. 1 (Sharon, Mass., 1928).

II. Secondary Literature. Additional information can be found in the following (listed chronologically): L. W. Stephenson, *A Deep Well at Charleston, South Carolina,* U. S. Geological Survey, professional paper 90-H (Washington, D.C., 1914); T. W. Vaughan, "On the Relative Value of Species of Smaller Foraminifera for the Recognition of Stratigraphic Zones," in *Bulletin of the American Association of Petroleum Geologists,* **7,** no. 5 (1923), 517–531; Charles Schuchert, "The Value of Microfossils in Petroleum Exploration," *ibid.,* **8** no. 5 (1924), 539–553; J. A. Waters, "Joseph Augustine Cushman, 1881–1949," *ibid.,* **33,** no. 8 (1949), 1457–1468; Ruth Todd *et al., Memorial Volume* (Sharon, Mass., 1950), final publication from the Cushman Laboratory for Foraminiferal Research; and L. G. Henbest, "Joseph Augustine Cushman and the Contemporary Epoch in Micropaleontology," in *Geological Society of America Proceedings, Annual Report for 1951* (July 1952), 95–101.

Ruth Todd

CUSHNY, ARTHUR ROBERTSON (*b.* Speymouth, Moray, Scotland, 6 March 1866; *d.* Edinburgh, Scotland, 25 February 1926), *pharmacology.*

For a detailed study of his life and work, see Supplement.

CUVIER, FRÉDÉRIC (*b.* Montbéliard, Württemberg, 28 June 1773; *d.* Strasbourg, France, 17 July 1838), *zoology.*

Frédéric Cuvier, four years younger than his famous brother Georges, proved to be a mediocre student and was apprenticed to a watchmaker in Montbéliard. His brother summoned him to Paris in 1797, to help him arrange the comparative anatomy gallery at the Muséum d'Histoire Naturelle. For several years Frédéric completed his education by studying applied science and, in 1801, working on the galvanic battery with Biot. Then his brother entrusted him with preparing the catalog of the comparative anatomy collections at the museum. He did a study of the classification of mammals based on their dentition, *Des dents de mammifères . . .* (1825).

In 1804, at thirty-one, Cuvier's brother appointed him head keeper of the museum menagerie, where he remained until his death. From 1818 to 1837 he published seventy installments of his *Histoire des mammifères,* in which approximately 500 species were described (about 100 were known only slightly); he named important new species and genera, such as *Phacochoerus* and *Spermophilus;* and he published *Histoire naturelle des cétacés* (1838), which has an interesting preface. He was put in charge of the *Dictionnaire des sciences naturelles* edited by Levrault (1816–1830) and wrote many articles for it.

Cuvier's most original work was his scientific study of the behavior of mammals in captivity. He studied the instinctive activities of the young from birth, and the sexual instinct of adults on the basis of their menses and the seasons. He analyzed the remarkable intelligence of a young female orangutan in partial freedom and observed that intelligence apparently diminishes as animals (particularly apes) become adult. With the seal he showed, contrary to the opinion held at that time, that the intelligence of an animal is not a function of the keenness or completeness of its sensory organs.

Cuvier was ahead of his time in the study of mammalian social life. In addition to solitary species, in which individuals do not associate except during the breeding season, he distinguished semipermanent couples, permanent couples, and animals living in groups. Further, he showed that animal groups were hierarchical, with a leader; the young often took their place in the hierarchy only according to the choice made by the females. The individual, once separated from the group, might die or become attached either to an animal of another species or even to man.

Cuvier established an important distinction between the domesticated animal, such as the cat, which accepts our affection but remains solitary by nature, and the truly domestic animal, such as the dog, which, social by instinct, participates in the social life of humans. He felt that certain social mammals, such as the daw, the chigetai, the tapir, and the vicuña, could be domesticated. He claimed that when social animals were hunted, the instinct of self-preservation dominated the social instinct and the group dispersed; but as soon as the species found a peaceful environment again, the social instinct reappeared. Cuvier was able to show this experimentally by causing the gregarious activities and building techniques of beavers to reappear in captivity or semicaptivity. He hoped that his study of mammalian societies would make possible the better understanding of human societies.

Often audacious in his ideas but very modest and gentle, Cuvier had loyal friends. His brother Georges had him appointed inspector of the Academy for Paris (1810), then inspector general of studies (1831); these time-consuming duties were perhaps the cause of his inability to complete his work on the origin of animal behavior. He was elected to the Académie des Sciences in 1826 and to the Royal Society of London; the Muséum d'Histoire Naturelle wanted him to have the chair of animal psychology; when this proved too controversial, it created for him, in December 1837, the chair of comparative physiology. Cuvier died six months later, at almost the same age as his brother and from an illness with the same symptoms. The physiologist Pierre Flourens succeeded him at the museum.

BIBLIOGRAPHY

I. ORIGINAL WORKS. The core of Cuvier's work, in terms of periodical articles (50 titles), is listed in the Royal Society's *Catalogue of Scientific Papers,* II (1868), 112–114; a list of his other works (14 titles) is in the *Catalogue général des livres imprimés de la Bibliothèque Nationale—Auteurs,* XXXIV (1908), cols. 981–983. Of special note is the article "Instinct," in Levrault, ed., *Dictionnaire des sciences naturelles,* XXIII (1822), 528–544. His books include *Des dents de mammifères considérées comme caractères zoologiques* (Paris, 1825); and, with E. Geoffroy Saint-Hilaire, *Histoire naturelle des mammifères . . . figures originales enluminées* (Paris, 1818–1837, 70 installments in folio; supps., 1842).

II. SECONDARY LITERATURE. Flourens's *éloge* of Cuvier, read to the Institute, is in *Mémoires de l'Académie des sciences,* **18** (1842), 1–28; see also *Journal des savants* (1839), 321–333, 461–479, 513–527. He also wrote *De l'instinct et de l'intelligence des animaux. Résumé des observations de Frédéric Cuvier* (3rd ed., Paris, 1858). This book, somewhat superficial despite its scope, also deals with Cuvier's predecessors in animal psychology.

FRANCK BOURDIER

CUVIER, GEORGES (*b.* Montbéliard, Württemberg, 23 August 1769; *d.* Paris, France, 13 May 1832), *zoology, paleontology, history of science.*

Cuvier was born into a poor—but still bourgeois—family; his father, a soldier who had become an officer in the service of France, was married late in life to a woman twenty years his junior and had already retired when the future naturalist was born. Very weak at birth, Cuvier remained in delicate health for a long time. During his childhood he enjoyed drawing and gave evidence of a very precocious intellectual and emotional development. Gifted with an astonishing memory, he mastered the entire works of Buffon. At the age of twelve he began his natural history collections and in the manner of an adolescent prodigy founded a scientific society with some friends.

Montbéliard, geographically French, had been detached from Burgundy in 1397 and rendered subject to the duke of Württemberg; during the sixteenth century its inhabitants adopted Luther's doctrines, while keeping the French language. Cuvier's parents intended for him to become a Lutheran minister like his uncle; but his teachers preferred not to grant him a scholarship to the school of theology at Tübingen. Fortunately for his career, the wife of the governor of Montbéliard recommended him to her brother-in-law, the reigning duke, who was seeking out bright young people to attend the Caroline University (Hohen Karlsschule), which he had founded near Stuttgart.

Cuvier entered that institution in 1784, at the age of fifteen. After two years of general studies, during which he learned German, he decided to specialize in administrative, juridical, and economic sciences, which included a significant portion of natural history. As early as his second year at the university, Cuvier had discovered near Stuttgart some plants that were new to the region. At that time twenty-year-old Karl Friedrich Kielmeyer was the lecturer in zoology. Exceptionally gifted, he became one of the founders of the German school of *Naturphilosophie*. It was Kielmeyer who taught Cuvier the art of dissection and probably comparative anatomy as well. This science was then taught in Tübingen by J. F. Blumenbach, whom Kielmeyer joined in 1786, after having pledged Cuvier eternal friendship in the emotional style of late eighteenth-century Germany.

In 1787 Cuvier received the golden cross of the chevaliers, which allowed him to live with children of noble birth and sometimes with the duke himself. Thus this young man with bright blue eyes, thick red hair, heavy features, and disheveled clothing began his education as a select member of the court. With a few friends he founded a natural history society that awarded decorations to its most active members. Cuvier completed his studies in 1788. There being no vacant positions in the ducal government for this penniless young commoner, he was forced to accept a position as a private tutor in Normandy, with a noble and affluent Protestant family named d'Héricy.

Cuvier traveled through France by stagecoach. The luxury of Paris dazzled him. Revolutionary unrest was beginning, but during the six years he spent in Normandy, Cuvier led a life somewhat outside these dramatic events. His duties as tutor were not very engrossing. During the fall and winter he lived in Caen, where he had rich libraries and a botanical garden at his disposal. In the spring and summer he accompanied the d'Héricy family to the north of Normandy, to the château de Fiquainville, near the sea and the fishing port of Fécamp. This gave him the opportunity to dissect numerous marine organisms and shorebirds. When he was in Stuttgart, Cuvier had begun making notes almost every day and sketching in large notebooks which he called, in the style of Linnaeus, his *Diarium zoologicum* and *Diarium botanicum;* in Normandy he added to them beautiful drawings of fish and of anatomy, accompanied by detailed descriptions.

However, Cuvier missed the Caroline University, where he had left his closest disciple, Christian Heinrich Pfaff. They maintained a correspondence, which kept Cuvier in touch with his university and with the ducal administration (which had knowledge of his letters and the political intelligence they contained). Since he ran the risk that letters would be opened by the French police, Cuvier was forced to feign sympathy to revolutionary ideas. After the Revolution, however, he often expressed his disapproval of this regime—in which, he said, "the populace made the law." He dreaded the "populace" throughout his life.

For the historian of science the Cuvier-Pfaff letters are of double importance. They show that between the ages of nineteen and twenty-three, Cuvier acquired the basic ideas that he developed between 1804 and his death in 1832. These letters also allow one to envisage an influence that Cuvier may have exerted on Lamarck in favor of the theory of the "chain of being." At first Cuvier was hostile toward theories, whether scientific, philosophical, or social. He wrote to Pfaff in 1788: "I wish everything that experience shows us to be carefully disassociated from hypotheses . . . science should be based upon facts, despite systems." In 1791 he explained to his friend that the structure of an animal is, of necessity, in harmony with its mode of life.

Cuvier believed in divine providence and considered himself to be close in spirit to Bernardin de Saint-Pierre. Around 1791 Kielmeyer returned to the Caroline University, and Pfaff sent Kielmeyer's unpublished courses to Cuvier. Pfaff then recalled, in one of his letters, Bonnet's famous theory to the effect that all existing things from the crystal to the man form gradually more complex systems linked through imperceptible transitions, and thus form a continuous chain. Cuvier objected that as many chains must be supposed as there were systems of organs, because in the groups of living beings it was not the same systems of organs that exhibit increasing complexity. Pfaff seems to have replied that this chain might lead in different directions and might be branched like a family tree, for Cuvier replied in 1792: "I believe, I see that aquatic animals were created for the water and the others for the air. But whether it is a question of branches or roots, or even different parts of a single trunk, I say again this is what I am unable to comprehend."

Cuvier seems to have reached soon after a sort of intellectual turning point: in the same year he published in the *Journal d'histoire naturelle* his first work devoted to wood lice. There he suddenly appears to be a proponent of the chain of being: "Here, as elsewhere, nature makes no jumps . . . therefore, the descent is by degrees from crayfish to Squilla, from Squilla to Asellidae, then to wood lice, to Armadilladiidae and to galley worms. All of these genera must be related to a single natural class."

Cuvier, who had become a French citizen upon the annexation of the territory of Montbéliard in 1793, sought recognition in the scientific world of Paris. He wrote to Lacépède and Haüy; and at the suggestion of the agronomist H. A. Tessier he sent a selection of his unpublished works to Étienne Geoffroy Saint-Hilaire, who had been appointed professor at the Muséum d'Histoire Naturelle when only twenty-one. Geoffroy, full of youthful enthusiasm, encouraged Cuvier to come to Paris; he did so at the beginning of 1795. Shortly after his arrival, he took advantage of the numerous dissections that he had performed in Normandy and presented a paper that marked a new stage in the study of invertebrates. "Before me," he wrote in 1829, "modern naturalists divided all nonvertebrate animals into two classes, insects and worms. I was the first . . . to offer another division . . . in which I pointed out the characteristics and limits of mollusks, crustaceans, insects, worms, echinoderms, and zoophytes."

Lamarck, in the introduction to his course of 1796, acknowledged that he was going "to follow to a very great extent [the classification] devised by the learned naturalist Cuvier." Geoffroy had invited Cuvier to work with him, and their collaboration lasted a year. Geoffroy, like his patron Daubenton, was hostile to the idea of the chain of being but changed his mind, probably under the influence of his new friend. In their paper on tarsiers Cuvier and Geoffroy felt that "this genus could be considered as the link uniting quadrumana to Chiroptera or bats." In a paper on orangutans they audaciously proposed the idea of the origin of species from a single type. Lamarck claimed that he owed his theory of the transformation of species to J. J. Barthélemy, who in 1788 had revived the ideas of the Greeks on this subject. Lamarck was very close to Geoffroy, however; and he was certainly influenced in 1795 by Geoffroy's conversion to the theory of the chain of being, a theory that Cuvier may well have borrowed from Kielmeyer through Pfaff.

The rapidity and brilliance of Cuvier's career was the consequence both of the importance of his scientific work and of his ability as a teacher; after only a few minutes of preparation he was able to deliver a logically constructed lecture in a confident manner; without stopping his lecture he illustrated his ideas by means of quick blackboard drawings that were as clear as they were accurate. In Paris, where there was a shortage of zoologists, it was only natural that shortly after his arrival he should be appointed professor of zoology at the *Écoles centrales* (which replaced the former universities for a few years) and assistant professor of animal anatomy at the Muséum d'Histoire Naturelle. Because of his position at the museum, Cuvier was given quarters in the Jardin des Plantes, near the menagerie. He lived there until his death.

In April 1796, Cuvier became a member of the Class of Physical Sciences at the Institut de France. He was only twenty-six at the time. He succeeded Daubenton as a professor at the Collège de France in 1800 and was given the responsibility for organizing the *lycées* in Bordeaux, Nice, and Marseilles. In 1803 he assumed the remunerative duties of permanent secretary of physical sciences at the Institute. The following year the Empire replaced the Consulate. In 1808 Napoleon appointed Cuvier university counselor. He contributed enormously to the organization of the new Sorbonne in Paris. Next he was sent on a mission to Italy, the Netherlands, and southern Germany to reorganize higher education there (1809-1813). As payment for his services he received an endowment and the title of *chevalier* in 1811.

The restored monarchy succeeded the Empire in 1814 and Cuvier—whose political ideal, it was said, had been enlightened despotism—became the devoted servant of the kings. Stendhal wrote: "What servility and baseness has not been shown toward those in power by M. Cuvier!" And indeed, in order to placate those in power, he did not hesitate to contradict his past by associating himself with the adversaries of that liberalism which was so dear to Protestants. On the other hand, he did support the exercise and even the development of Protestantism at a time when the ultraroyalists were hostile to it. He was the director of Protestant universities and, for a while, was also director of non-Catholic religions. E. Trouessart, professor of zoology at the Muséum, believed that the support given by Cuvier to his own coreligionists was as important, in its way, as the work of the zoologist.

Cuvier became councillor of state in 1814, and from 1819 until his death actually presided over the Interior Department of the Council of State. Every day, starting at eleven o'clock, he attended to business of the council of state or the council of public instruction; Monday afternoon was set aside for the Institute. He became a member of the Académie Française in 1818 and was made a baron in 1819, a *grand officier* of the Legion of Honor in 1824, and was nominated peer of France in 1831. Cuvier took advantage of his great influence in higher education by trying to develop the teaching of religion, modern languages, and natural sciences. He obtained positions for his friends and relatives. Numerous applicants wrote to him or visited him, knowing full well that he would accept, with pleasure, the grossest flattery. Cuvier was always in a hurry, easily irritated, and very authoritarian. His loyal associate, C.-L. Laurillard, accused him, above all, of never having told his associates his scientific ideas or goals. He was a very secretive man, and it is not certain whether his writings reflect his true opinions. Nevertheless, he was kind to aspiring young persons, assisting and advising them.

Cuvier was short and during the Revolution he was very thin; he became stouter during the Empire; and he grew enormously fat after the Restoration. He had to walk slowly and did not dare to bend over for fear of apoplexy. His health and his appetite remained excellent, however. Nicknamed "Mammoth" and posing as a kind of bishop of science, he did have a certain majesty when he donned his long, decoration-studded university gown of purple velvet with ermine borders—which, it is said, he himself had designed.

In February 1804, Cuvier married Mme. Davaucelle, a widow and a very devout Protestant, who was kind, outspoken, and energetic. Already the mother of four children, she bore him four more. She saw to everything, and the naturalist's favorite Montbéliard chitterling sausages were never missing from the table. Cuvier had three or four large sources of income, any one of which would have enabled him to enjoy a comfortable life. He had a carriage and servants, visited the Paris salons, and himself received at home on Saturday evenings in the great hall of his library, which contained busts of famous men. His colleagues did not come often; instead, most of his visitors were naturalists from the provinces, foreigners, and even the two great writers, Stendhal and Prosper Mérimée, drawn there by the charms of his daughter and step-daughter. Although he was laden with honors and

money, Cuvier's happiness was clouded by the death of his four children; he was overcome in 1827 by the death of his twenty-two-year-old daughter Clémentine, and he appears to have tried to forget this sorrow by working unceasingly.

One evening in May 1832, Cuvier experienced a slight paralysis and contraction of the esophagus. He grew weaker over the next few days and died on 13 May. Following this rather mysterious illness (it was said to be acute myelitis), the physicians performed a dissection. His brain was found to be exceptional in the bulging configuration of the lobes and to be unusually heavy (1,860 grams). If one believes his widow, he left a fortune insignificant in relation to his enormous income. It was claimed that this reputedly selfish man made large charitable contributions. Was it through the intermediary of his daughter Clémentine, who was known for her generosity? Or did he perhaps have a secret life?

Cuvier's scientific and administrative work was immense, and as he grew older it became greater. This increasing activity may be explained by the power of his memory. His library, arranged according to subject matter, was open to all. At the end of his life it contained 19,000 volumes and thousands of pamphlets. Cuvier had committed almost all of their contents to memory, from which, within a few seconds, he could retrieve the information he needed—whether it was in history, law, natural science, or heraldry. This enormous amount of information seems to have been an obstacle to his ability to synthesize. An opponent of theories, he said, "We know how to limit ourselves to describing." Cuvier sought to do the most possible within the shortest possible period of time, and therefore he rarely sought perfection in form or thought. In order to gain time he surrounded himself with collaborators who lacked his intellectual vigor and would not dare criticize him; thus, the numerous works published under his name contain some weak portions.

In zoology Cuvier's work was, in great part, a result of his dominant position at the Muséum d'Histoire Naturelle, which at the time was the world's largest establishment dedicated to scientific research. The government organized expeditions to distant lands, and all of the collections brought back enriched the museum with prodigious rapidity. Upon his arrival at the museum, Cuvier rearranged the comparative anatomy collections that had been made at the end of the seventeenth century by Claude Perrault and Georges Duverney and at the middle of the eighteenth century by Daubenton; the entire collection consisted of a few hundred skeletons and a few dozen anatomical preparations. In 1804 these collections comprised 3,000 items; and the number had risen to 13,000 by 1832. Cuvier classified the birds and fish in the galleries according to his own system. The huge museum menagerie, founded and directed by Geoffroy Saint-Hilaire, furnished Cuvier with invaluable specimens for dissection and anatomical preparations of mammals and birds. As for fossils and fish, amateurs throughout the world sent him material. Since all the materials for his work were so easily available, Cuvier traveled little for scientific purposes. He took advantage of an administrative mission to Marseilles in 1803 to study marine fauna of the Mediterranean, and he made a few geological excursions in the Paris region with Alexandre Brongniart, starting in 1804. In 1817 he went to England to study fossil remains, but his duties at the Institute and at the Council of State usually prevented him from leaving Paris, even in the summer.

Cuvier published three works of general zoology: *Tableau élémentaire de l'histoire naturelle des animaux* (1797); *Leçons d'anatomie comparée,* in collaboration with C. Duméril (1800) and G. Duverney (1805); and *Le règne animal,* arranged according to his system of classification, in collaboration with Pierre Latreille on insects (1817). Although often a bit hasty and conservative, his classifications, except for fish, did not have the solidity of Lamarck's, Latreille's, or Geoffroy Saint-Hilaire's. His *Historie naturelle des poissons,* begun in Normandy and gradually improved, was published in collaboration with Achille Valenciennes. The first volume appeared in 1828, and the ninth had been edited at the time of his death. Valenciennes stopped with the twenty-second volume in 1849, just short of completing this great work, which constitutes the basis of modern ichthyology; most of the fish families created by Cuvier were so soundly based that they have become orders or suborders in present classification. Before Cuvier, collectors freely described shells of mollusks but often ignored the creatures within; Cuvier dealt with their anatomy, and twenty-three of his papers on this subject were collected in one volume in 1817. Seeking to produce one great *Anatomie comparée,* he spent his life gathering some 13,000 items for the museum's public gallery and collecting drawings and documents; 336 plates made according to his drawings and those of Laurillard appeared between 1849 and 1856, with the title *Anatomie comparée, Recueil de planches de myologie.*

In the history of science Cuvier's work is vast. As permanent secretary of the Academy he had to deliver periodic reports on the progress of research. These reports were bound in four volumes in 1828 and in five volumes in 1833. He also had the responsibility of composing *éloges,* essays on the careers of deceased members of the Academy; these *éloges,* often well documented, make good reading. For Michaud's *Biographie universelle* he wrote articles on Aristotle, Buffon, Daubenton, Linnaeus, Pliny, and Vicq d'Azyr, among others. The first volume of his *Histoire des poissons* traces the development of ichthyology; it contains a great many facts, but they are presented somewhat dryly. This aridity is found again in his last courses at the Collège de France, published as *Histoire des sciences naturelles depuis leur origine;* based on notes taken at his lectures, this was the first great work on this vast subject.

Cuvier was considered by the public to be a bit of a wizard, a man who had brought to life animals that had long since become extinct and of which Buffon, well before him, had understood the scientific importance. But Cuvier knew how to make great strides in studying these creatures and could endow this study with new accuracy. His famous paleontological reconstructions had the living being as their point of departure, a sort of eddy that organized constantly renewed matter (in the manner of Buffon's "organic mold"). This living being "overcomes" physical and chemical laws, Cuvier said, and constitutes "a unique and closed system, all parts of which mutually correspond and concur in the same definitive action through a kind of reciprocal reaction. None of these parts can change without the others changing as well; and consequently each of them taken separately indicates and shows the nature of all the others." Thus a carnivore should have intestines organized for digesting meat, powerful jaws, sharp teeth, and claws. In order to seize prey, these claws should be at the ends of easily moved toes; with musculature appropriate to the osseous structure. Consequently, Cuvier said, every time we have a well-preserved piece of bone, we can determine the class, order, genus, and even the species from which it came as precisely as if we had the entire animal. This method, he states, verified on real skeletons, had been proved infallible. Actually, Cuvier made a great many errors, but he also had spectacular successes. Before witnesses he removed from a stone block the marsupial bones of a small opossum fossil, bones whose existence he had surmised on the basis of the conformation of the visible part of the skeleton. As early as 1804 Cuvier had the idea of reconstructing the musculature of extinct animals from imprints left by the muscles on the bones; then he merely had to imagine the skin over the muscles and the animal was practically brought back to life.

Cuvier conceived the notion of the balance of nature, which was not developed until long after his time. He conceived of living nature as an "immense network" in which the species depended on each other. At first he believed that this network had remained fixed since the six days of Creation, just as the species themselves had remained fixed. But his own paleontological discoveries forced him, as early as 1812, to admit that creation had taken place in several stages. Reptiles, he said, were found on land long before mammals; the species that had become extinct were the first to have appeared; and it is only on the most recent portions of earth that fauna was almost identical to that found there today. Primates, the last beings created, would never have existed in the fossil state, he asserts.

In 1812 Cuvier had brought together his first memoirs on paleontology in his *Recherches sur les ossemens fossiles des quadrupèdes.* The *Discours préliminaire* to the work, printed separately under the title *Discours sur les révolutions du globe,* was in the style of Buffon's *Époques de la nature,* but written less in a philosophical spirit than as a defense of biblical chronology. This essay, often reprinted and translated into many languages, drew its inspiration from the geological concepts of Alexandre Brongniart.

In 1804, wanting to place the Montmartre bed of fossil formations in time, Cuvier, with Brongniart, began research that led to *Géographie minéralogique des environs de Paris* (1808, 1811), which was rewritten and greatly expanded as *Description géologique des environs de Paris* (1822, 1835). In this work, a landmark in the history of geology, Cuvier played the lesser role. Brongniart did the necessary field work, drawing his inspiration from the works of Buffon, Soulavie, Ramond, Palassou, and especially Lamarck; the latter had described the fossil invertebrates of the Paris region and, in his *Hydrogéologie* (1802), had set forth the bases of the theory of "current causes," later developed by Constant Prévost. Cuvier, respecting the short chronology of the Bible, was forced to assume, in addition to "current causes," which act very slowly, rapid catastrophes and global upheavals which had no basis in fact.

Since he considered the theory of the variability of species to be contrary to moral law, to the Bible, and to the progress of natural science itself, Cuvier undertook a battle, which is still famous, against the ideas of his former friends Lamarck and Geoffroy Saint-Hilaire—a battle that was often fought in secret and in which he tested his own political power. Geoffroy Saint-Hilaire, and probably Lamarck, became the object of investigation at a time when religious beliefs were obligatory for all civil servants. In 1792 Cuvier, as we have seen, first disputed and then accepted the theory of the chain of being. Starting in 1802–1804, it appears, he once again rejected this theory for scientific, political, and religious reasons.

Geoffroy Saint-Hilaire had brought back from the Egyptian expedition some 3,000-year-old mummified animals that, when examined in 1802, proved to be similar to present species. For Lamarck this fact merely proved that the transformation of species is so slow as to be imperceptible over a 3,000-year period. To Cuvier such reasoning was absurd; if, over a period of 3,000 years, he said, there is zero modification, one may multiply 3,000 by zero as much as desired, and although that would increase the age of the earth, the modification of species would always remain zero. Political events reinforced Cuvier in his position; in December 1804 Napoleon had himself crowned emperor and restored the official recognition of religion to his own advantage. Thereupon Cuvier in his courses, and much to the astonishment of his audience, attacked Lamarck's materialistic ideas and passed himself off as a defender of the Bible. This return to strict religious orthodoxy was perhaps also connected with his marriage in February 1804, which seems to have reestablished his bonds with Protestant circles in Paris.

On the other hand, Geoffroy Saint-Hilaire remained faithful to the ideas of his aged friend Lamarck and demonstrated that all vertebrates had the same type of body structure, which similarity constituted an argument in favor of their common origin. In 1820 he even claimed to have discovered this unity of structure in invertebrates; Cuvier criticized him with good reason; and Geoffroy, very displeased, sought revenge. In 1824 Cuvier, who worked too quickly, had classified in the crocodile group a reptile of the Jurassic period that was very far removed; Geoffroy was quick to announce the error and claimed that the reptile in question, which he called the Teleosaurus because of its anatomical peculiarities, was a predecessor of mammals of the Tertiary. He thus showed that paleontology, Cuvier's main field, could bring arguments to bear in favor of the chain of being. Geoffroy then developed that part of paleontology whose purpose was to discover the "missing links."

Two of Geoffroy's disciples, Laurencet and Meyranx, through an audacious hypothesis that is still worthy of attention, attempted in 1829 to establish a structural analogy between fish and cephalopods, which made it possible to conceive of a transition between invertebrates and vertebrates. Cuvier sought to prevent an examination of this work by the Academy, and Geoffroy reproached him publicly; Cuvier replied angrily. From 15 February to 5 April 1830, the controversy grew progressively sharper. Cuvier accused Geoffroy and his disciples of being pantheists, a very serious accusation under the reign of Charles X. The press gave this affair extensive coverage, and presented it in different ways, each paper according to its political views. In the July Revolution of 1830 Charles X fled; after a few apprehensive days Cuvier won the favor of the new government. He was then able to resume his attacks against Lamarck and Geoffroy at the Collège de France. In his last lecture, six days before his death, after having pronounced an anathema upon useless scientific theories and upon the pantheism of Kielmeyer, Lamarck, and Geoffroy, he rendered solemn homage to Divine Intelligence, the Creator of all things, before an audience overcome by emotion.

Cuvier's life was one of compensation. Born into a poor family and not a member of the nobility, he became rich and acquired the title of baron. Not particularly handsome, he found consolation in being admired for his intelligence and gave himself a commanding appearance by wearing elaborate attire. His vanity was boundless, as was his hunger for honors and praise, characteristics that dominated his entire career. He had a somewhat Germanic mentality and envisaged society as a sort of organism in which subordination was the rule. Although he was very pliant toward his "superiors," he was authoritarian toward those he deemed his "inferiors"; and he left only second-rate disciples.

In Cuvier's work one must consider separately his still little-known role of support for the Protestant community, a role that would perhaps justify some of his servility toward those in power. His activities on behalf of reorganization of education, often successful, were inspired by some novel ideas that had been popular at the Caroline University when he attended it. His great erudition could have made him the first great historian of natural science if he had had a more precise concept of the role played by theories in scientific research; throughout his life he took great pains to discredit theoretical ideas in favor of what he called positive facts. This proved to be his great error and for a long time was an obstacle to the development of natural science, particularly in France, where the line of "those limited to description" persists to this day.

It is likely that fear of a new revolution played a significant part in the religious revival that Cuvier manifested beginning in 1804; he must not be called to task for this, for thousands of Frenchmen at the time underwent a similar political-religious evolution. Cuvier undoubtedly rediscovered his faith sincerely each time a new period of mourning cast gloom on his existence. But his respect for biblical chronology prevented him from participating in a new form of thought, one that viewed phenomena in four dimensions, the fourth being that of time: time of short duration for physical phenomena, time that was measured in millions or billions of years for the creation of the universe or for that of living species. In this area Cuvier's thought was backward in relation to that of his first teacher, Buffon. His interpretation of the balance of parts of living creatures and the balance of creatures in nature was much too static—even Laplace reproached him for this—but then, by an irony of fate, paleontology, which he advanced in great measure, contributed decisive arguments in favor of the variability of species. When Darwin made the idea of evolution triumphant in 1859, and Christians no longer sought to harmonize the Bible with geology, Cuvier's glory diminished. Nevertheless, for the historian of science as well as for the psychologist who studies the conditions of scientific thought, Cuvier's role, his extraordinary memory, and even his failings remain rich sources of information.

BIBLIOGRAPHY

I. Original Works. There is no complete bibliography of Cuvier's works; one that included all the reprints and translations would probably run to 300 titles. The *Catalogue of Scientific Papers of the Royal Society,* II, VII, IX, XIV, lists works published after 1800 in scientific journals (243 titles); for papers published before 1800 see H. Daudin, *Cuvier et Lamarck* (Paris, 1926), II, 285–286. Basic individual works published in French can be found in the catalog of the Bibliothèque Nationale; translations are listed in catalogs of the larger foreign libraries.

Most of Cuvier's papers are at the library of the Institut de France (very fine analytical catalogs have been published by H. Dehérain) and at the library of the Muséum National d'Histoire Naturelle (mediocre catalog). The archives of the Council of State have disappeared, but those of the ministries probably contain many documents (not cataloged) concerning Cuvier's political and administrative activities. Cuvier's correspondence has not been collected systematically. His library, purchased by the state, was divided between the Muséum d'Histoire Naturelle (works on natural sciences) and the École Normale Supérieure (other works).

II. Secondary Literature. Three of Cuvier's closest associates left behind biographies of him: G. L. Duverney, *Notice historique sur les ouvrages et la vie de M. le Baron Cuvier* (Paris, 1833), chronological bibliography of 213 titles and interesting information on Cuvier's childhood; C. L. Laurillard, *Éloge de M. le Baron Cuvier* (Paris, 1833), reprinted in Cuvier's *Recherches sur les ossemens fossiles des quadrupèdes,* I (1834), 3–78, and in Michaud's *Biographie universelle,* 2nd ed.; and Blainville, *Cuvier et Geoffroy Saint-Hilaire* (Paris, 1890). Cuvier's correspondence with Pfaff is presented in *Georges Cuvier's Briefe an C. H. Pfaff aus den Jahren 1788 bis 1792* (Kiel, 1845), translated by Louis Marchant as *Lettres de Georges Cuvier à C. H. Pfaff* ... (Paris, 1858). There are no adequate works on Cuvier's thought. However, the work by Daudin (*Cuvier et Lamarck*) is of some value, as is W. Coleman, *Georges Cuvier Zoologist* (Cambridge, Mass., 1964). See also G. Petit and J. Théodoridès, in *La biologie médicale,* spec. no. (Mar. 1961). On Cuvier's family see John Viénot, *Georges Cuvier* (Paris, 1932). An iconographic collection by L. Bultingaire is in *Archives du Muséum d'histoire naturelle,* 6th ser., **9** (1932). The biography by Mrs. Lee is an exaggerated and uncritical eulogy. Pierre Flourens's publications on Cuvier's works often substitute the author's personal ideas for those of Cuvier without warning. Other recent publications include *Bicentenaire de la naissance de Georges Cuvier* (Montbéliard, 1969); and Dujarric de La Rivière, *Cuvier, sa vie, son oeuvre. Pages choisies* (Paris, 1969). See also the series in *Réalités* (1969), nos. 280–285, which presents several iconographic documents but otherwise sheds little new light. More interesting is the exposition catalogue *Cuvier und Württemberg* (Stuttgart, 1969), which includes accounts of the documents and contains introductions by Robert Uhland and R. D. Adam.

FRANCK BOURDIER

CYRIAQUE DE MANGIN, CLÉMENT. See Henrion, Denis.

CYSAT, JOHANN BAPTIST (*b.* Lucerne, Switzerland, *ca.* 1586; *d.* Lucerne, 3 March 1657), *astronomy.*

Cysat's father, Renward Cysat, was a man of letters and civic leader in Lucerne, and a vigorous sponsor of the Jesuit college there. Johann Baptist entered the Jesuit order in 1604; and in 1611 he was at the Jesuit college in Ingolstadt as a pupil of Christoph Scheiner, whom he assisted in the observation of sunspots. In 1618 Cysat became professor of mathematics at Ingolstadt, where he made the observations on the comet of 1618–1619 for which he is mainly known. He reports and analyzes these observations in an eighty-page booklet, *Mathemata astronomica . . .* (1619).

The observations cover the interval from 1 December 1618 to 22 January 1619, and form the most nearly continuous series of observations on this much argued-about comet. For each determination of position, Cysat measured the distance of the comet from two fixed stars, using a wooden sextant of six-foot radius. Calculating from Tycho's star catalog, he found that the comet had moved from 9° 24′♍, 11° 37′ N. lat. on 1 December, to 21° 20′♎, 56° 22′ N. lat. on 22 January, at a gradually decreasing rate. He also measured the dimensions of the head and tail and studied the appearance of the comet telescopically.

A major aim of Cysat's analysis was to show that the comet was supralunary. He attempted to determine the true motion from observations on consecutive days when the comet had the same zenith dis-

tance; by comparison of the observed apparent motion in the course of a few hours with the calculated true motion he arrived at a figure for the difference in parallax. His procedure lacked mathematical rigor and was later sharply criticized in Riccioli's *Almagestum novum* (I, 102–109). From the horizontal parallaxes Cysat derived earth-comet distances and then (assuming Tycho's erroneous solar parallax) sun-comet distances. On the basis of these results he proposed two different theories for the comet: a Tychonic-style circular orbit about the sun, located between Venus and Mars, and a straight-line trajectory that fits the observations of 1, 20, and 29 December. The latter theory differs from the rectilinear trajectories espoused by Kepler because Cysat assumed that the earth was immobile.

His telescopic observations led Cysat to picture the nucleus of the comet as breaking into discrete *stellulae* from 8 December on—a phenomenon doubted by Robert Hooke and more recent astronomers. Cysat compared this appearance with that of nebulae in Cancer, Sagittarius, and the sword of Orion; he was once credited with the first reference to the Orion nebula, but this had already been observed in 1610 by Peiresc.

Cysat's astronomical observations continued in later years but apparently did not appear in another published work. He described a lunar eclipse of 1620 in a letter to Kepler; Remus Quietanus and Kepler in 1628–1629 discussed his determinations of apparent planetary diameters; he observed the transit of Mercury of 1631; Riccioli referred critically to his excessive figures for apparent solar diameters, pointing out that they failed to take account of the diffraction of light discovered in the 1650's by Grimaldi (*Astronomia reformata,* pp. 39–41). Meanwhile, Cysat served in many administrative capacities; he was rector of the Lucerne Jesuit college from 1623 to 1627; architect of the Jesuit college church built in Innsbruck in the 1630's; rector of the Innsbruck Jesuit college from 1637 to 1641; and rector of the Eichstadt Jesuit college from 1646 to 1650.

BIBLIOGRAPHY

I. ORIGINAL WORKS. Cysat's cometary work is *Mathemata astronomica de loco, motu, magnitudine, et causis cometae qui sub finem anni 1618 et initium anni 1619 in coelo fulsit . . .* (Ingolstadt, 1619). A *Tabula cosmographica versatilis* by Cysat is listed in Philippe Alegambe, *Bibliotheca scriptorum societatis Iesu* (Antwerp, 1643), p. 223. A "Clavis mathematica," approved by the censors in 1634, apparently was never printed; and a work intended to show the providence of God in the arrangement of the world, mentioned by

Cysat as being in preparation in 1636, also failed to appear in print. A letter of Cysat to Kepler and several letters referring to Cysat's observations are published in Max Caspar, ed., *Johannes Keplers gesammelte Werke*, XVII (Munich, 1955), nos. 838, 840; and XVIII (1959), nos. 910, 1095, and 1103. A collection of Cysat manuscripts at the Universitätsbibliothek in Munich is mentioned by Braunmuhl in *Jahrbuch für Münchener Geschichte* (Munich, 1894), pp. 53 ff.

II. SECONDARY LITERATURE. For details of Cysat's life and work see Bernard Duhr, S.J., *Geschichte der Jesuiten in den Ländern deutsche Zunge* (Freiburg im Breisgau, 1907–1913), *passim;* Rudolf Wolf, *Biographien zur kulturgeschichte der Schweiz*, I (Zurich, 1858), 105–118; and *Geschichte der Astronomie* (Munich, 1877), pp. 319–320, 409, 419; J. H. von Mädler, *Geschichte der Himmelskunde,* I (Brunswick, 1873), 295, 312; and M. W. Burke-Gaffney, S.J., *Kepler and the Jesuits* (Milwaukee, Wis., 1944), pp. 113–119. On the arguments over the 1618 comet, see Stillman Drake and C. D. O'Malley, *The Controversy on the Comets of 1618* (Philadelphia, 1960). On the problems of cometary astronomy before Newton, see James Alan Ruffner, *The Background and Early Development of Newton's Theory of Comets,* Ph.D. dissertation (Indiana University, 1966).

CURTIS WILSON

CZEKANOWSKI, ALEKSANDER PIOTR (*b.* Krzemieniec, Poland [now Kremenets, Ukrainian S.S.R.], 12 February 1833; *d.* St. Petersburg, Russia, 18 October 1876), *geology, exploration.*

Czekanowski's father, Wawrzyniec (Laurenty), worked in the Krzemieniec Academy, a Polish secondary school, and was a keen entomologist and collector of insects. His mother, Joanna Gastell, died during Aleksander's early childhood, and he was brought up by the family of a distinguished naturalist, Willibald Besser, lecturer at Krzemieniec Academy and later professor of botany at the University of Kiev.

Czekanowski finished college in 1850 and entered the Faculty of Medicine of Kiev University. At the end of 1855 he moved to Dorpat University, where he studied in the department of mineralogy up to 1857. Owing to financial difficulties, he returned to Kiev without completing his studies. During the next five years he worked for an electrical engineering firm, Siemens and Halski, which was building a telegraph line to India.

During his studies in Kiev and Dorpat and his later period of employment, Czekanowski carried out studies in the natural sciences, collecting material and making observations. In Podole he gathered together rich collections of Silurian fossils; the armored fish from these collections were later added to by the paleontologist Friedrich Schmidt. Studying in Dorpat

under the guidance of the geologist K. Grewingk, Czekanowski collected rich paleontological material from the Baltic Paleozoic and organized the mineralogical collections of the university. At this time, Czekanowski entered into a friendship and scientific cooperation with Schmidt, as well as with other Polish students at the university: the zoologist B. Dybowski; the paleontologist J. Nieszkowski, and the geologist G. Rupniewski. In Kiev, Czekanowski maintained contact with numerous workers at the university, such as the zoologist K. Jelski, the anthropologist I. Kopernicki, and others. They formed a Polish freedom organization and, after the outbreak of the January insurrection of 1863 in Warsaw, the majority of them were arrested and deported to Siberia.

In 1864 Czekanowski was sentenced to six years in a labor camp. He marched from Kiev with a group of prisoners, including an assistant lecturer in chemistry at Kiev University, Mikolaj Hartung. In spite of the extreme hardship of this trek, Czekanowski did not discontinue his scientific studies. He devised instruments, which he himself constructed, such as a magnifying glass made from a polished glass stopper.

On reaching Tomsk, Czekanowski fell ill with typhus, and only in the spring of 1865 was he able to continue to the place of deportation, in the region of Irkutsk. In April 1866 his sentence was reduced from hard labor to exile in Irkutsk province.

Czekanowski remained in Siberia until 1876. In spite of recurrent illness resulting from improperly treated typhus, he continued to conduct intensive studies in the natural sciences. During the early years of camp labor and later, when he worked for Siberian peasants, he was assisted in his studies only by other Polish deportees, B. Dybowski and W. Godlewski.

In 1868 Czekanowski entered into cooperation with the Russian Geographical Society, through the intervention of Schmidt. He made preliminary observations for the society of the geology of the Lake Baykal basin and afterward worked for three years on the geology of Irkutsk province. The results of these studies were published as a monograph, together with a geological map of the province, in 1874. On completing this work Czekanowski undertood the immense task of investigating the geological structure of the Eastern Siberian Upland, which had not been studied at all until then. In the years 1873–1875 he organized three scientific expeditions along the great Siberian rivers Yenisey, Nizhnyaya Tunguska, Olenek, Lena, and Yana. The routes of these expeditions altogether covered more than 25,000 kilometers. Czekanowski traveled partly by boat and partly on horseback; where conditions were extraordinarily severe he used reindeer teams. He made surveys and maps of the areas investigated, and the observations and geological and paleontological material collected during these expeditions provided a basis for knowledge of the vast region between Mongolia and the Arctic Ocean. Large areas of surface lavas and of Paleozoic and Mesozoic rocks were discovered and investigated. Certain stratigraphic problems were solved, including the assignment of the Siberian coal measures to the Jurassic and not, as had formerly been believed, to the Carboniferous. The rich paleontological, zoological, and botanical collections assembled were found to contain many forms new to science. The material from these collections was later elaborated upon by the following scholars: Jurassic flora by O. Heer and J. Schmalhausen, armored fish by Schmidt, Jurassic fish by J. Rohon and H. Becker, and Triassic Ceratitids by E. Moysinovich.

In 1875, after twelve years of exile, Czekanowski was permitted to leave Siberia. In March 1876 he arrived in St. Petersburg, where he obtained the post of custodian in the Mineralogical Museum of the Academy of Sciences. In connection with work on paleontological material collected in Siberia, he traveled during the summer of that year to Stockholm to acquaint himself with fossils from Spitzbergen. He also received permission to make a brief visit to his home town of Krzemieniec. After returning to St. Petersburg on 30 September 1876, he committed suicide while in a state of depression.

Czekanowski's most extensive work, the diaries of his 1873–1875 expeditions, was not published until 1896. It was prepared for publication by his pupil and successor in Siberian studies, J. Czerski, and his friend of many years, Schmidt.

Czekanowski's work brought him a series of distinctions and awards, such as gold medals from the Russian Geographical Society (1870) for geological studies in Irkutsk province, and from the International Geographical Congress in Paris (1875) for the maps of Eastern Siberia. Several genera and numerous species of plant and animal fossils, as well as four present-day plants, are named after him. A mountain range, about 320 kilometers in length, near the Lower Glensk, as well as one of the peaks of the Chamar-Daban range, Lake Baykal, also bear his name.

BIBLIOGRAPHY

I. ORIGINAL WORKS. A complete list of the publications of A. Czekanowski—manuscripts left by him, as well as biographical works about him—is given in *A. P. Czekanowski, A Collection of the Unpublished Work of A. P. Czekanowski and Dissertations on His Scientific Studies*, published in Russian by the Siberian branch of the U.S.S.R. Academy of Science (Irkutsk, 1962).

II. SECONDARY LITERATURE. Reports from individual studies and of the work of the expeditions organized by Czekanowski were given up to 1876 in periodicals of the Russian Geographical Society. Memoirs and biographical works, all in Polish by B. Dybowski, are "A Czekanowski, an Obituary," in *Reports of the Siberian Branch of the Russian Geographical Society*, 1877; *Siberia and Kamchatka* (Lvov, 1899); *Recollections* (Lvov, 1913); and *Memoirs 1862–1878* (Lvov, 1930). Others are H. Popławska, "The Participation of Poles in Work in the Lake Baykal Region," in *Geographical Review* (in Polish 1922); and R. D. Samoylevitsch, *L'activité scientifique des revolutionaires polonais en Siberie* (Lvov, 1934). Biographical works (in Polish and Russian) are by T. Turkowski, in the *Polish Biographical Dictionary*, vol. IV (Krakow, 1938), and in *Reports of the Earth Museum* (Vilna, 1938); and L. Kleopov and H. B. Lackiy, *A. P. Czekanowski* (Irkutsk, 1962).

STANISŁAW CZARNIECKI

CZERMAK (ČERMAK), JOHANN NEPOMUK (*b.* Prague, Czechoslovakia, 17 June 1828; *d.* Leipzig, Germany, 16 September 1873), *physiology, histology, phonetics, laryngology.*

Czermak is remembered mainly for his contribution to the scientific development of laryngoscopy and its reception as a method of clinical examination. His name is also associated with the slowing of the heart and subsequent loss of consciousness caused by pressure exerted on the neck. (Czermak interpreted this as an effect of the mechanical stimulation of the vagus nerve, but it recently has been shown to originate from the carotid baroreceptors.) Two microscopic features of the teeth are named for him: the Czermak spaces, which are irregular gaps in the dentin that appear in rows, and the Czermak lines, formed by the Czermak spaces, which are arranged in rows and follow the contour of the dentin.

Czermak's father and grandfather were physicians in Prague, and his uncle was professor of advanced anatomy and physiology in Vienna. He studied in Prague, Vienna (1845), Breslau (1847), and Würzburg (1849–1850). Very early he gained the advantage of the advice and sponsorship of Purkinje (then professor of physiology in Breslau), who deeply influenced his scientific interests throughout his life. Indeed, many of Czermak's research subjects were further developments of topics studied by Purkinje (the structure of the teeth, subjective visual phenomena, touch, vertigo, phonetics).

It was his interest in the movements of muscles in speech and the conditions for producing certain unusual sounds (e.g., the Arabic gutturals) that led Czermak to use the laryngeal mirror in his research.

Various instruments had been used previously by Philipp Bozzini (1807), Cagniard de la Tour (1825), Balington (1829), Gerdy (1830), Beaumes (1838), Robert Liston (1840), and Warden (1844) without any wider recognition of their utility. In 1855 Manuel García, a singing teacher, published observations of his own larynx and vocal cords made with a small dental mirror introduced into the throat and using sunlight reflected by another mirror. García was interested in movements connected with the production of the singing voice and did not anticipate the importance of laryngoscopy for medicine. Attempted again two years later by a Vienna neurologist, Ludwig Türck, laryngoscopy did not seem either practical or promising, but Czermak, interested in physiological phonetics, greatly improved the technique. In his lectures in several European countries he brought home to physicians its usefulness and importance, thus opening a new and important field of practical medicine. He was also the first (not Friedrich Voltolini, as is sometimes stated) to use the same means for dorsal rhinoscopy. In phonetics he showed that the voice generated in the larynx does not participate in the production of vowels, but that both the voice and the acoustic conditions of the "joined pipe" (i.e., throat, mouth, and nasal cavities) play an important role in consonants.

In the physiology of sensations Czermak made the first systematic investigation of the spatial localization of skin sensibility, that is, of the two-point threshold in relation to the size of the E. H. Weber sensory circles (1855), postulated a general sense for duration of different specific sensations (*Zeitsinn*), and laid down a program for the investigation of the time sense (1857). He also contributed greatly to physiological experimental techniques, and some of his devices were widely used. He propagated the teaching of physiology by demonstration and designed for the purpose a model institute called a "spectatorium," which was built and opened in Leipzig nine months before his death.

Czermak worked in Prague, Graz, Krakow, Vienna, Budapest, Jena, and Leipzig. Unable to comply with the requirements arising from the natural aspirations of Czechs, Poles, and Hungarians to develop teaching in their own languages at their universities, Czermak had to move several times and worked for some time as an independent scientist. He was afflicted with diabetes in his last years and was apprehensive of an early death—his father, uncle, and other male relatives had died in their forties. Czermak died at the age of forty-five, before he could finish his major theoretical work, "Die Principien der mechanischen Naturauffassung."

BIBLIOGRAPHY

I. ORIGINAL WORKS. Czermak's writings were collected as *Gesammelte Schriften,* 2 pts. in 3 vols. (Leipzig, 1879). Among his books is *Der Kehlkopfspiegel und seine Verwertung in Physiologie und Medizin* (Leipzig, 1860), trans. into French (Paris, 1860) and into English as *On the Laryngoscope and Its Employment in Physiology and Medicine* (London, 1861).

II. SECONDARY LITERATURE. A biographical sketch by Czermak's friend and teacher, the art critic A. Springer, is in *Gesammelte Schriften,* II. His introduction and propagation of laryngoscopy and dorsal rhinoscopy are discussed in all histories of laryngology—see, e.g., F. S. Brodnitz, "One Hundred Years of Laryngoscopy: To the Memory of Garcia, Tuerck and Czermak," in *Transactions of the American Academy of Ophthalmology and Otolaryngology,* **58** (1954), 663–669—but there is no comprehensive discussion of his scientific work. Short biographies are by Durig, in *Wiener medizinische Wochenschrift* (1928), 791–792; and *Münchener medizinische Wochenschrift* (1928), 1509–1510; and Schrutz, in *Praktický lékař* (1928), 527–531. His accomplishments in rhinoscopy are discussed by L. Englert in *Zprávy lékařské,* **4** (1934), 108–114; and by P. Heymann and E. Kronenberg in *Handbuch der Laryngologie und Rhinologie,* I (Vienna, 1898), 1–2, 24–35.

VLADISLAV KRUTA

CZERSKI, JAN (*b.* Svolna, Vitebsk, Russia [now Byelorussia, U.S.S.R.] 15 May 1845; *d.* confluence of Prorva and Kolyma Rivers, Siberia, Russia, 25 June 1892), *geology, zoology, Siberian exploration.*

After the death of his father, Dominik Czerski, a rich landowner, Czerski was brought up by his mother. He attended the Gymnasium and the Institute for Gentry in Vilna. After the outbreak of the Polish insurrection in 1863, Czerski ran away from school to join a rebel regiment, but shortly afterward became ill and was captured by the Russians. For taking part in the insurrection he was sentenced to exile in a penal army regiment stationed in Siberia. After six years of service in a battalion at Omsk he was released from the army in 1869; his health was completely ruined and he was severely neurasthenic.

Czerski stayed in Omsk for the next two years, supporting himself by giving private lessons. During his period of forced army service he had devoted all his free time to the study of science, especially the natural sciences. An engineer named Marczewski and the owner of an extensive library, W. Kwiatkowski—both of whom were Poles living in Omsk—helped him in his self-education, as did the Siberian explorer G. N. Potanin.

In 1871 Czerski obtained permission to settle in Irkutsk, where he was helped by two Polish deportees, the geologist A. Czekanowski and the zoologist

B. Dybowski. With their support he gained the position of custodian of the natural science collections in the Siberian branch of the Russian Geographical Society—the only scientific institution in Siberia. Not long afterward he began organizing several expeditions. Another exile of 1863, the chemist M. Hartung, also took part in them. The purposes of these expeditions were to carry out geological studies, to make use of fossils and archeological materials, and to collect zoological specimens, ethnological observations, and ethnographical materials. From 1872 on Czerski published the results of his studies in the journals of the Russian Geographical Society; later he published in those of the Russian Academy of Sciences as well. Within twenty years he presented some eighty articles, bulletins, and reports drawn from studies that he had carried out, as well as several monographs. In 1879 he published a monograph concerning the remains of Quaternary mammals that he had found in the cave Nizhnyaya Udinska; and in 1891 a comprehensive paper on the remains of Quaternary mammals found during the Novosibirsk expedition of 1885–1886.

Czerski's geological studies of 1877–1880 dealt with Lake Baykal; he published the results of these in several papers. A monograph published in 1886 (with a geological map) synthesizes his studies on Baykal and includes an attempt to explain the origin of that enigmatic lake.

In spite of the excellent results of his scientific work, for which he three times received the Gold Medal of the Russian Geological Society, Czerski was forced to resign from his post in the Siberian branch. In 1885, through the financial aid of J. Zawisza, a Warsaw archaeologist, Czerski was able to move to St. Petersburg. In the course of his journey there he made observations from Baykal to the Urals; these were published in 1888.

In St. Petersburg, Czerski worked in the geological museum of the Russian Academy of Sciences. There he received extensive scientific observations from Siberia—for the most part collected by political deportees—which served him well in writing his addenda to K. Ritter's *The Geography of Asia,* which appeared in two volumes (1893, 1895) after Czerski's death. He was also engaged in preparing for publication the diaries from the expeditions of Czekanowski, as well as in putting in order the geological collections made during these expeditions. This work awakened in him a deep interest in the great Siberian rivers and a desire to continue the studies of Czekanowski. In the summer of 1891 Czerski therefore began his last expedition to the north. With his wife, Marfa, and his twelve-year-old son, Aleksandr, he traveled on horseback from Yakutsk to Vierkhniokolymska.

The severe subarctic winter brought the party to a halt and only at the end of May 1892 was it possible to travel by boat down the Kolyma River. During this winter, however, the state of Czerski's health deteriorated to such an extent that it became clear that he would be unable to lead the expedition to its conclusion. He worked on detailed instructions for his wife on the continuation of his studies and, despite weakness and exhaustion, continued his observations. On 25 June he had a hemorrhage and died; he was buried where the Omolon River flows into the Kolyma, and his wife led the expedition to its end.

Czerski was responsible for the elucidation of the geology of Baykal; for the discovery and elaboration of a rich fauna of Quaternary mammals from Siberia and of Paleolithic occurrences of man in this region; for the assemblage of valuable zoological, geological, and ethnographical collections; and for the first synthetic geological cross section of Siberia from Baykal to the Urals. A mountain chain in the Zabaykalsk region and a range of hills in northern Yakutia, on the upper course of the Kolyma, bear his name; Czerski Peak rises from the northwest shore of Lake Baykal, and the valley of the Kandat River is also named after him.

BIBLIOGRAPHY

I. ORIGINAL WORKS. A list of Czerski's publications is in R. Fleszarowa, *Retrospektywna bibliografia geologiczna polski,* pt. 2 (Warsaw, 1966). A complete bibliography and biographical notice are included in the collective work (in Russian) of the Siberian branch of the Geological Society of Russia, *I. D. Czerski, Unpublished Articles, Letters, and Diaries. Articles on I. D. Czerski and A. I. Czerski* (Irkutsk, 1956).

II. SECONDARY LITERATURE. In addition to the above works, see also B. Dybowski's discussion of Czerski's work in his paper on Siberia and Kamchatka (Lvov, 1899) and in his diaries (Lvov, 1930). In 1892, on the occasion of Czerski's death, a series of memoirs—by I. Kuznietsov, V. Obrushev, S. Nikitin, F. Chernyshev, and others—appeared in a variety of Russian scientific journals. Biographies of Czerski include T. Turkowski, in *Polski słownik biograficzny,* vol. IV (Krakow, 1938) and *Wiadomości Muzeum Ziemi,* vol. I (Warsaw-Vilna, 1938).

STANISŁAW CZARNIECKI

D', DA, DE, DEL', DELLA. Many names with these prefixes are listed under the next word of the name.

DAINELLI, GIOTTO (*b.* Florence, Italy, 19 May 1878; *d.* Florence, 16 December 1968), *geology, geography.*

Dainelli acquired his passion for geology from the

geologist Carlo De Stefani and his love of geography from the geographer Olinto Marinelli. He taught geography at the University of Pisa and geology at the universities of Naples and Florence. His personality as a scientist was characterized by the way in which the two fields blended in his mind. He wrote: "I believe that just as my training as a geologist was the solid foundation of my education as a geographer, so my background in geography has perfected my vision of what the ultimate goal of geology should be: the reconstruction of conditions on earth as they were in former ages." Thus, Dainelli's interest in paleogeography was not accidental. Although much of his work was done in Italy, he was early attracted to foreign exploration, for which the robustness of his body and mind amply fitted him. A great part of his scientific contributions concerned central Asia and eastern Africa.

Of his works on Italy—apart from the geographic studies, among which *Atlante fisico-economico d'Italia* is outstanding—one must mention his paleontologic and tectonic studies of the Friulian Prealps (north of Venezia Giulia). Noteworthy, too, was his paleogeographic study of Italy during the Pliocene. Unfortunately that work, which he began and completed for Tuscany in collaboration with P. Videsott and A. Sestini, was later extended to cover only Campania and Apulia.

In 1913–1914 Dainelli took part in the De Filippi expedition to the western Himalayas and to the Karakoram Range. In 1930 he himself organized an expedition to the eastern Karakoram. Besides his considerable anthropogeographic research, his contributions to geology were of major importance, encompassing terrain series and formations, problems of tectonics, and vast paleogeographic reconstructions. He also made a thorough study of the Ice Age in the same regions. The growth of the ancient ice sheets, which he saw as having occurred in four principal stages, impressed him as a marvel.

Dainelli's work on eastern Africa began in 1905–1906, when, accompanied by Marinelli, he thoroughly explored the Ethiopian plateau and the northern Danakil. He established the fundamental features of the country's geology, studied the volcanoes of the coastal region, and plumbed the geologic history of the Danakil depression. Returning to this field some thirty years later, he organized and led an expedition to Lake Tana (1937). On the basis of his own observations and his critical analysis of the findings of others, he constructed a vast synthesis of the geology of eastern Africa, ranging from stratigraphic and paleontologic observations to paleogeographic reconstructions.

Dainelli's many books, especially the very popular scientific works, show him to have been not only a tireless worker but also a powerful and elegant writer.

BIBLIOGRAPHY

I. ORIGINAL WORKS. Among Dainelli's books are *Risultati scientifici di un viaggio nella Colonia Eritrea* (Florence, 1912), written with O. Marinelli; *L'Eocene friulano* (Florence, 1915); *La struttura delle Prealpi Friulane* (Florence, 1921); *Il mare pliocenico nella Toscana settentrionale* (Florence, 1930), written with P. Videsott; *Atlante fisico-economico d'Italia* (Milan, 1940); and *Geologia dell' Africa orientale* (Rome, 1943). The following volumes, from the series Risultati Geologici e Geografici della Spedizione De Filippi nell'Himalàja e nel Caracorùm, were published in Bologna: I, *Le esplorazioni* (1934); II, *La serie dei terreni* (1934); III, *Studi sul Glaciale* (1922); IV, *Le condizioni fisiche attuali,* written with O. Marinelli (1928); VIII, *Le condizioni delle genti* (1924); IX, *Le condizioni culturali* (1925).

II. SECONDARY LITERATURE. "Giotto Dainelli e la sua opera scientifica," in *Bollettino della Società geografica italiana,* **7** (1954), includes an article by E. Feruglio on Dainelli's geological work and one by R. Riccardi on his geographical work, and a speech by Dainelli with autobiographical notes and a systematic bibliography of his writings up to April 1954, completed in **10** (1969). See also A. Sestini, "L'opera geografica di Giotto Dainelli," in *Rivista geografica italiana,* **76** (1969).

FRANCESCO RODOLICO

DALE, HENRY HALLETT (*b.* London, England, 9 June 1875; *d.* Cambridge, England, 24 July 1968), *medicine, physiology.*

For a detailed study of his life and work, see Supplement.

DALÉCHAMPS, JACQUES (or **Jacobus Dalechampius**) (*b.* Caen, France, 1513; *d.* Lyons, France, 1 March 1588), *botany, medicine.*

Daléchamps, probably best described as a "medical humanist," entered the University of Montpellier in 1545. He received his first degree in medicine the next year, and a doctorate in the same subject in 1547, under Guillaume Rondelet. After a few years in Grenoble and Valence, he moved in 1552 to Lyons, where he spent the remainder of his life. Little is known of his personal life, except that his friends and correspondents included Rondelet, Conrad Gesner, Joseph Justus Scaliger, Robert Constantin, and Jean Fernel.

Daléchamps' most important scientific work is the *Historia generalis plantarum* (1586–1587), the most complete botanical compilation of its time and the first to describe much of the flora peculiar to the region around Lyons. It formed the basis for a later work by Jacobus Antonius Clavenna but was severely criticized in separate writings by Jacobus Pons and Gaspard Bauhin. Daléchamps' other more or less original work is the *Chirurgie françoise* (1570), based largely on Book VI of Paul of Aegina's *De re medica* but also incorporating material from other sources, including Ambroise Paré and Jacques Roy.

Much of Daléchamps' effort, however, was directed toward editing and translating earlier scientific and medical writings. He contributed, in one way or another, to editions of works of Pliny the Elder, the two Senecas, Dioscorides, Paul of Aegina, and Raymond Chalmel de Viviers. In addition, he translated works of Athenaeus into Latin and of Galen into French. Among Daléchamps' unpublished works are a collection of his letters; an "Ornithologie," which consists only of illustrations without descriptions; and a Latin translation of all of the then-known writings of Theophrastus and of some minor writings attributed to Aristotle. He was probably the first to make a complete Latin translation of all of Theophrastus' known writings.

BIBLIOGRAPHY

I. ORIGINAL WORKS. Two of Daléchamps' works are *Chirurgie françoise* (Lyons, 1570; Paris, 1610); and *Historia generalis plantarum,* 2 vols. (Lyons, 1586–1587; French trans., Lyons, 1615, 1653). There are also unpublished works: the Latin translation of Theophrastus (Paris, BN lat. 11,857); the "Ornithologie" (Paris, BN lat. 11,858–11,859); and a collection of his letters (Paris, BN lat. 13,063).

Listings of Daléchamps' works and their various eds. are in Joly and Schmitt (see below).

II. SECONDARY LITERATURE. More on Daléchamps and his work can be found in Georges Grente, ed., *Dictionnaire des lettres françaises: Le seizième siècle* (Paris, 1951), p. 211; E. Gurlt, *Geschichte der Chirurgie* (Berlin, 1898), II, 786–790; F. Hoefer, *Nouvelle biographie générale,* XII, 804–806; Philippe-Louis Joly, *Éloges de quelques auteurs françois* (Dijon, 1742), pp. 350–368; Antoine Magnin, *Prodrome d'une histoire des botanistes lyonnais* (Lyons, 1906), pp. 14–15, also in *Mémoires de la Société Botanique de Lyon,* **31** (1906), 14–15; Ernst H. F. Meyer, *Geschichte der Botanik* (Königsberg, 1857), IV, 395–399; M. Michaud, *Biographie universelle,* new ed., X, 40–41; Jules Roger, *Les médecins normands du XIIe au XIXe siècle* (Paris, 1890–1895), II, 41–42; J.-B. Saint-Lager, *Histoire des herbiers* (Paris, 1885), pp. 47–49; George Sarton, *Appreciation of Ancient and Medieval Science During the Renaissance (1450–1600)* (Philadelphia, 1955), pp. 85–86; Charles B. Schmitt, "Some Notes on Jacobus Dalechampius and His Translation of Theophrastus (Manuscript: BN., Lat. 11,857)," in *Gesnerus,* **26** (1969), 36–53; and Kurt Sprengel,

Geschichte der Botanik, rev. ed., I (Altenburg–Leipzig, 1817), 332–334.

See also H. Christ, "Jacques Dalechamp, un pionnier de la flore des Alpes occidentales au XVI siècle," in *Bulletin de la Société botanique de Genève,* 2nd ser., **9** (1917), 137–164.

CHARLES B. SCHMITT

DALENCÉ, JOACHIM (*b.* Paris, France, *ca.* 1640; *d.* Lille, France, 17 February 1707 [?]), *astronomy, physics.*

Few details are known of Dalencé's life; most information is drawn from archival material and correspondence.

His father, Martin Dalencé, surgeon to the king and a Jansenist, purchased for him the office of royal secretary and counsellor on 15 September 1663.

In 1668, during a trip to England, where he bought a telescope, Joachim formed a friendship with Henry Oldenburg, and in 1675 he is known to have served as an intermediary between Oldenburg and Huygens. He was also in communication with Leibniz and served as liaison between the French Academy and Huygens.

Beginning in 1679 he published anonymously the first six collections of the *Connaisance des temps,* the first French ephemerides of a purely scientific nature.

He gave up this project in 1684 and in 1685 moved to the Low Countries, where for three years he purchased books and art objects for the royal collections.

During this time he published the *Traité de l'aiman* (1687), a well-written discussion of magnets, and the *Traittez des barometres, thermometres et notiometres ou hygrometres* (1688).

His detailed description of the principal meteorological instruments of the period is enriched with several new ideas, such as the calibration of the thermometric scale on the basis of two points of change of state: the point at which water freezes and—a much more contestable point—that at which butter melts.

Dalencé married Geneviève Troisdames, by whom he had at least one son, Denis.

BIBLIOGRAPHY

I. ORIGINAL WORKS. Dalencé's writings include *La connaissance des temps, ou calendrier et ephemerides du lever et du coucher de la lune et des autres planètes, avec les eclipses pour l'année* (Paris, 1679–1684); *Traité de l'aiman,* 2 pts. (Amsterdam, 1687); and *Traittez des barometres, thermometres et notiometres ou hygrometres* (Amsterdam, 1688, 1708, 1724; Liège, 1691; The Hague, 1738).

II. SECONDARY LITERATURE. For Dalencé's life and work see A. Birembaut, "La contribution de Réaumur à la thermométrie," in *La vie et l'oeuvre de Réaumur (1683–1757)* (Paris, 1962), p. 48; M. Daumas, *Les instruments scientifiques au XVIIe et XVIIIe siècles* (Paris, 1953), pp. 77–79; J. Guiffrey, *Comptes des bâtiments du roi sous le règne de Louis XIV,* Vols. II and III (Paris, 1887, 1891), *passim;* A. R. Hall and M. B. Hall, eds., *The Correspondence of Henry Oldenburg,* Vol. V (Madison, Wis., 1968), *passim;* C. Huygens, *Oeuvres complètes, Correspondance,* Vols. VII, IX (The Hague, 1897, 1901); E. Labrousse, *Inventaire critique de la correspondance de Pierre Bayle* (Paris, 1960), *passim;* M. Marie, *Histoire des sciences mathématiques et physiques,* V (Paris, 1884), 130; W. E. R. Middleton, *A History of the Thermometer* (Baltimore, 1966); A. Rivaud, *Catalogue critique des manuscrits de Leibniz,* 2nd *cahier* (Poitiers, 1914–1924), nos. 329, 1105; and the article on Dalancé in *Nouvelle biographie générale,* I (Paris, 1852), col. 786.

RENÉ TATON

DALIBARD, THOMAS FRANÇOIS (*b.* Crannes, France, 1703; *d.* Paris, France, 1779), *natural history.*

Dalibard was the first naturalist in France to adopt Linnaeus' system. In 1749 he published a work entitled *Florae Parisiensis prodomus, ou catalogue des plantes qui naissent dans les environs de Paris,* in which the plants were classified according to Linnaeus' principles. To show his appreciation, Linnaeus named a Canadian bramble for Dalibard.

At the request of Buffon (in whose shadow he developed), Dalibard translated Franklin's *Experiments and Observations on Electricity.* He published this translation in 1752, with a foreword and short history of studies in electricity before Franklin. To a second edition of Franklin's work (1756) Dalibard added a lengthy supplement describing the results he achieved as he reenacted Franklin's experiments (results which he had previously revealed to the Académie des Sciences).

Two of these experiments are especially interesting. The first is the famous experiment performed on 10 May 1752 at Marly-la-Ville, in which Dalibard proved the accuracy of Franklin's theory that the materials of thunder and lightning were similar.

The second experiment concerned the early history of electromagnetism. In a letter to Peter Collinson, Franklin described his experiments in magnetizing needles and reversing their polarities with electricity. Franklin did not assume the identity of magnetic and electrical materials, however. As he repeated the experiment, Dalibard erroneously concluded that he had discovered the electromagnetic relationships necessary to prove this identity. (Franklin himself wrote, erroneously, in 1773 that as to the magnetism apparently produced by electricity, his present opinion was that electricity and magnetic force have no connection with each other and that the production of magnetism is purely accidental.)

BIBLIOGRAPHY

I. ORIGINAL WORKS. Dalibard's works include his trans. of Garcilasso de la Vaga's *Histoire des Incas, rois du Pérou* (Paris, 1774); *Florae Parisiensis prodomus, ou catalogue des plantes qui naissent dans les environs de Paris* (Paris, 1749); and his trans. of Franklin's *Experiments, Expériences et observations sur l'électricité faites à Philadelphie en Amérique par M. Benjamin Franklin; et communiquées dans plusieurs lettres à M. P. Collinson . . .* (Paris, 1752; 2nd ed., Paris, 1756).

II. SECONDARY LITERATURE. On Dalibard, see also articles by Michaud, in *Biographie universelle,* X (Paris, 1855), and Hoeffer, in *Nouvelle biographie générale,* XII (Paris, 1866); and see I. Bernard Cohen, *Franklin and Newton* (Philadelphia, 1956).

PIERRE G. HAMAMDJIAN

DALL, WILLIAM HEALEY (*b.* Boston, Massachusetts, 21 August 1845; *d.* Washington, D.C., 27 March 1927), *malacology, paleontology.*

Dall, a man to whom natural history became a way of life, was the son of Harvard-educated Charles Henry Appleton Dall, a missionary minister, and Caroline Wells Healey, an active writer and lecturer for women's rights whom all, including her son, held in awe. The senior Dall served a variety of missions before becoming the first Unitarian missionary in India; from 1855 he saw his family only at long intervals. In 1862 he took his son, then a student at the English High School in Boston, to meet various Harvard professors, including Louis Agassiz. The young Dall had long been collecting natural history specimens and had begun collecting shells avidly when, at the age of twelve, he found a copy of Augustus A. Gould's beautifully illustrated *Report on the Invertebrata of Massachusetts.* Agassiz directed Dall in studying mollusks and other subjects; Gould helped him in identifications and entered him as a student member of the Boston Society of Natural History.

Although his father urged him to enter the tea business in India, Dall, through the influence of his maternal grandfather, became a clerk for the Illinois Central Railroad in Chicago. Naturally attracted to the museum of the Chicago Academy of Sciences, he was soon acquainted with its director, Robert Kennicott. In 1865, after the first Atlantic cable had broken, he readily accepted Kennicott's invitation to join, as naturalist, the Western Union International Telegraph Expedition to Alaska to find an overland telegraph route to Europe.

Thus was Dall launched on his lifework—the exploration primarily of Alaska but also of the Aleutian chain and the Pacific coast. Dall became the scientific director of the expedition in 1866 upon the sudden death of Kennicott. When the installation of the second Atlantic cable terminated the expedition, Dall continued his explorations at his own expense, traveling up the Yukon River to Fort Yukon. From 1868 to 1870 he studied his collections at the Smithsonian Institution and published *Alaska and Its Resources,* an enlightening book on the new territory. An appointment, obtained through Spencer F. Baird's assistance, as acting assistant with the Coast Survey in 1871 enabled Dall to continue his Alaska studies; he commanded four cruises along the Aleutian Islands and almost to Point Barrow from 1871 to 1880. Just prior to the fourth trip he married Annette Whitney. From 1881 to 1884 he was assistant with the Coast and Geodetic Survey and then became paleontologist for the U.S. Geological Survey, for which he concentrated on Cenozoic mollusks. For the Survey he made six trips to the northwest coast from 1890 to 1910 and in 1899 was a guest on the Harriman Alaska Expedition. Throughout his career he used the facilities of the Smithsonian Institution and from 1881 carried the title of honorary curator there.

Dall more than compensated for his lack of a college education. He was awarded an honorary M.A. by Wesleyan University (1888), D.Sc. by the University of Pennsylvania (1904), and LL.D. by George Washington University (1915). He was an honorary professor of invertebrate paleontology at Wagner Free Institute of Science and received its gold medal. Cataloging shells at the Bishop Museum in Honolulu led to an honorary curatorship there. A founder of the Philosophical Society of Washington and a charter member of the Biological Society of Washington, Dall was also an active member of the American Association for the Advancement of Science. In addition to many other honors, he was elected to the National Academy of Sciences, Phi Beta Kappa, the American Philosophical Society, the American Academy of Arts and Sciences, and the California Academy of Sciences.

Dall was an extraordinarily well-organized scientist. An indefatigable student of natural history, he devoted countless hours to the arranging of his own immense collections and the other conchological material of the U.S. National Museum. This included the magnificent collection of H. Gwyn Jeffreys, the purchase of which was arranged by Dall. He named 5,427 genera, subgenera, and species of mollusks and brachiopods, both recent and fossil. In his extensive revision of the classification of the pelecypods, presented largely in "Contributions to the Tertiary Fauna of Florida"

(1890–1903), he depended primarily on shell features, particularly the hinge. His classification is still a standard reference, although he erred in assigning to the Oligocene some of the Miocene strata of the Caribbean region.

Dall's work in Alaska—including the collection of vast numbers of specimens ranging from coastal to dredged deep-water forms, journal entries and letters, and the compilation of a large library, all at the U.S. National Museum—constitutes a primary source for the area. In addition to *Alaska and Its Resources,* his publications include studies of the birds, land and marine mammals, fishes, climatology, anthropology, currents, and geology of Alaska and the Aleutians. West coast malacologists are indebted to his *Summary of the Marine Shellbearing Mollusks of the Northwest Coast of America from San Diego, California, to the Polar Sea* (1921).

BIBLIOGRAPHY

I. Original Works. Dall's bibliography, compiled by Paul Bartsch, Harald Alfred Rehder, and Beulah E. Shields—"A Bibliography and Short Biographical Sketch of William Healey Dall," in *Smithsonian Miscellaneous Collections,* **104,** no. 15 (1946), 1–96—includes 1,607 entries, many of them reviews and other short items. A remarkably large number, however, are significant scientific publications, with a range of subjects from mollusks and brachiopods to all aspects of natural history. Already referred to is *Alaska and Its Resources* (Boston, 1870). Another valuable Alaska publication was *Pacific Coast Pilot, Coasts and Islands of Alaska* (Washington, D.C., 1883), resulting from his Coast Survey work. Dall's major malacological work was "Contributions to the Tertiary Fauna of Florida," *Transactions of the Wagner Free Institute of Science of Philadelphia,* **3,** pts. 1–6 (1890–1903). In addition, he contributed the sections on Brachiopoda and Pelecypoda and on Gastropoda and Scaphopoda to the reports of the Coast Survey steamer *Blake* in the Gulf of Mexico and the Caribbean Sea—*Bulletin of the Museum of Comparative Zoology at Harvard College,* **12** (1886) and **18** (1889)—and the section on Mollusca and Brachiopoda to the reports of the *Albatross* in Central America and the Gulf of California—*ibid.,* **43,** no. 6 (1908). His summary of northwest coast mollusks was *Bulletin. United States National Museum,* **112** (1921). Descriptions of mollusks from many parts of the world appeared in a variety of publications; land snails were summarized in "Insular Landshell Faunas, Especially as Illustrated by the Data Obtained by Dr. G. Baur in the Galapagos Islands," in *Proceedings of the Academy of Natural Sciences of Philadelphia* (1896), pp. 395–460. A taxonomic catalog of the animals introduced by Dall is Kenneth J. Boss, Joseph Rosewater, and Florence A. Ruhoff, *The Zoological Taxa of William Healey Dall,* U.S.

National Museum Bulletin 287 (Washington, D. C., 1968).

Dall also wrote a number of short biographical memorials and a full-length appreciative one: *Spencer Fullerton Baird: A Biography* (Philadelphia, 1915).

II. SECONDARY LITERATURE. The wealth of memorabilia of Dall at the U.S. National Museum includes personal journals from 1865 to 1927, bound volumes of letters, logbooks of his collections, and autobiographical sketches. W. P. Woodring drew on this material for his memoir on Dall in *Biographical Memoirs. National Academy of Sciences*, **31** (1958), 92–113, which contains a selected bibliography. The bibliography by Bartsch *et al.* (cited above) includes a brief biography. An appreciation by C. H. Merriam appeared in *Science*, **65**, no. 1684 (1927), 345–347.

ELIZABETH NOBLE SHOR

DALTON, JOHN (*b.* Eaglesfield, Cumberland, England, 6 September 1766; *d.* Manchester, England, 27 July 1844), *physics, chemistry, meteorology.*

If the provincial Dissenter of dubiously middle-class background, obscure education, and self-made opportunity is the characteristic figure of late eighteenth-century English natural philosophy, then John Dalton is the classic example of the species. John was the second son of a modest Quaker weaver, Joseph Dalton, and Mary Greenup. The Daltons can be traced back in west Cumberland at least to the late sixteenth century. From that time the family seems to have owned and farmed a small amount of land. Joseph, himself a younger son, had no holding until his elder brother died without issue in 1786. The property Joseph then inherited passed at his death the following year to Jonathan, his elder son. Only when Jonathan, a bachelor, died in 1834 did the then considerably augmented acreage finally pass to John Dalton, who by that time had independently accumulated wealth sufficient to his own frugal and celibate ways.

In the eighteenth century, west Cumberland enjoyed considerable prosperity as a mining and trading area, with an important series of coastal ports engaged in local and overseas commerce. George Fox had earlier seen his first major evangelistic successes in this region, whole villages and families (including the Daltons) undergoing conversion to his doctrines. The area was thus peculiarly important within the developing international life of the Society of Friends. Strong links were forged between these Northern Friends, Quaker manufacturers in the Midlands, London Quaker merchants, and Philadelphia residents. This network of connections, coupled with the sect's strong emphasis on education and the interest in natural philosophy displayed by so many of its members, is the key to understanding the peculiarly favorable context in which Dalton grew and matured as a scientific thinker.

Although his father appears to have been somewhat feckless, his mother came from a more prosperous local family, and John was strongly influenced by her determination and tenacity. He made rapid progress in the village Quaker school, which he himself unsuccessfully took over at the age of twelve. He also quickly attracted the attention of Elihu Robinson, the most prominent of the local Friends and a naturalist of no mean stature. Robinson's encouragement is reflected in the story of how John at the age of thirteen copied out verbatim an issue of the *Ladies' Diary*, a popular but by no means trivial annual devoted to mathematics and philosophy.

At this time Dalton's future seemed uncertain, and he was of necessity put to work as a laborer on the local small-holdings. In 1781 he was rescued by an invitation to replace his elder brother as assistant in a Kendal boarding school, forty miles away.

The school to which Dalton moved was newly built and equipped by the Quakers. The list of benefactors was headed by John Fothergill, the London physician and a personal friend of Robinson, and included such wealthy Midland entrepreneurs as Abraham Darby and Richard Reynolds. More immediately important than the web of contacts the benefactors' list displays was the use that the school's first principal made of the £150 available for the library. George Bewley, himself a distant cousin of Dalton, was quick to purchase not only Newton's *Principia*, but also the supporting texts of 'sGravesande, Pemberton, and Thomas Rutherforth. Later purchases included Musschenbroek's *Natural Philosophy*, the six-volume *Works of the Honourable Robert Boyle*, and Buffon's *Natural History*, among others. The collection was rounded out with various items of apparatus, including a two-foot reflecting telescope, a double microscope, and (for £21) a double-barreled air pump with its subsidiary equipment.

Dalton did not feel such valuable resources as these worth even a mention in the accounts of his early life that he was later to authorize for publication. Nor did he refer to the stimulus available to such a talented and enterprising youth from the continued flow of Quaker visitors. He also forgot to include the public lectures given by itinerant natural philosophers (Kendal being, among other things, an important staging post on the coach route from London to Scotland). Typical of the courses available was that of John Banks in 1782. In a seven-week stay in Kendal, Banks offered "twelve lectures, which include the most useful, interesting and popular parts of philosophy," the lectures being illustrated by extensive apparatus.

Despite his failure to acknowledge their influence, Dalton obviously modeled his own subsequent public courses on lectures such as these.

What Dalton did acknowledge was the presumably still greater stimulus he found in the library, learning, and enthusiasm of another Kendal Quaker, John Gough, the blind natural philosopher of Wordsworth's *Excursion.* As Dalton explained in a 1783 letter to Peter Crosthwaite, a Keswick Friend and fellow naturalist:

> John Gough is . . . a perfect master of the Latin, Greek, and French tongues. . . . Under his tuition, I have since acquired a good knowledge of them. He understands well all the different branches of mathematics. . . . He knows by the touch, taste, and smell, almost every plant within twenty miles of this place. . . . He is a good proficient in astronomy, chemistry, medicine, etc. . . . He has the advantage of all the books he has a mind for. . . . He and I have been for a long time very intimate; as our pursuits are common—viz. mathematical and philosophical. . . .

Under Gough's tuition Dalton made rapid progress in mathematics, meteorology, and botany. Emulating his master, he began to keep a daily meteorological record in 1787, a task he continued steadfastly until the day he died. He also carefully compiled a still-extant eleven-volume *Hortus siccus.* And from 1787 on, this "teacher of the mathematics in Kendal" enjoyed an increasing reputation for his successes in the yearly puzzles and prize competitions of the *Ladies' Diary* and *Gentleman's Diary.*

In 1785 George Bewley withdrew from the Kendal school. Jonathan and John Dalton thereupon took over as joint principals, and their sister Mary moved from Eaglesfield to become housekeeper. Despite his increased responsibilities at the school, John was soon offering his own first series of public lectures in Kendal. The lectures treated mechanics, optics, pneumatics, astronomy, and the use of the globes, with the aid of the school's apparatus. Yet even with such new outlets for his energy and curiosity, John was obviously becoming restless within the confines of a local scientific community whose lessons he had mastered and whose possibilities he had so fully explored. In 1790 he wrote to Bewley, Robinson, and his uncle Thomas Greenup, a London barrister, to seek advice on his prospects in medicine and law (the Society of Friends having no clergy).

Dalton argued that "very few people of middling genius, or capacity for other business" become teachers. His own desire for a profession with "expectation of greater emolument" led to his queries, especially about the feasibility of studying medicine at Edinburgh. The replies were not enthusiastic. Greenup in particular chose to say that medicine and law were "totally out of the reach of a person in thy circumstances" and that Dalton should rather aim at moving in the humbler sphere of apothecary or attorney, where with a little capital and great industry he might perhaps be able to establish himself. Despite this discouragement, such an ambitious and talented young man was not to be confined to a Kendal school.

In 1791 Dalton again offered a public lecture course. In 1792 he paid his first visit to London, ostensibly for the annual meeting of the Society of Friends. Shortly afterwards he was appointed professor of mathematics and natural philosophy in the "New College," which Socinian- and Unitarian-oriented Dissenters had recently established in the dramatically expanding town of Manchester, following the demise of the nearby Warrington Academy, at which Joseph Priestley had once taught.

Initially, Dalton seems to have been well pleased with the Manchester appointment. Reporting on his new situation to Robinson, his early patron, he explained:

> There is in this town a large library [Chetham's], furnished with the best books in every art, science and language, which is open to all gratis; when thou are apprised of this and such like circumstances, thou considerest me in my private apartments, undisturbed, having a good fire, and a philosophical apparatus around me, thou wilt be able to form an opinion whether I spend my time in slothful inactivity of body and mind.

Despite the availability of library and apparatus, teaching duties seem to have absorbed Dalton's energies in his early years in Manchester. Called upon to offer college-level mathematics and natural philosophy for the first time, he soon found himself expected to cover chemistry as well. As he ruefully noted, it was "often expedient to prepare my lectures previously." In addition there was the necessary "attendance upon students 21 hours in the week." Walking tours in the summer vacation, regular local and occasional regional Quaker meetings, return visits to his beloved Lake District, and, in 1796, a further set of Kendal lectures served to more than fill out the remaining time.

On 26 March 1800 Dalton announced his intention to resign his teaching position at the close of the session, for reasons that remain obscure. Perhaps he was dissatisfied with the college's radical posture, perhaps unsettled by its faculty changes and uncertain future, perhaps unhappy that he remained the lowest paid of the three professors (receiving £52.10.0 annual salary, plus approximately £50 in fees), perhaps quite simply confident in his own popularity and teaching abilities. The following September the *Manchester Mercury* advertised the opening of his private "Mathematical Academy," offering tuition in mathematics, experimental philosophy, and chemistry. Success came quickly to the academy. Within two years Dalton could drily observe, "My Academy has done very well for me hitherto. I have about eight or nine day pupils at a medium, at ten guineas per annum, and am now giving upwards of twenty lessons per week, privately, at two shillings each besides. I am not yet rich enough to retire, notwithstanding." Private teaching of this fashion more than adequately supported him for the rest of his days. Far from being regarded as a degrading chore, his activity in this respect was typical of that of a host of lower- to middle-class Dissenters whose academies and popular lectures formed one of the major strengths of English science throughout this period of embryonic professionalization. Self-help, private initiative, technological curiosity, and utilitarian attitudes were characteristics of that Industrial Revolution science which is both exemplified in Dalton's work and encapsulated by his lifetime.

Within five years of leaving the New College Dalton had completed in essential outline the work on which his major and enduring scientific reputation was to rest: the law of gaseous expansion at constant pressure (also called Charles's law after its independent—and earlier—French discoverer); the law of partial pressures in gaseous systems; and the chemical atomic theory (which for the first time gave significance to and provided a technique for calculating the relative weights of the ultimate particles of all known chemicals, whether elements or compounds). Despite this brilliant efflorescence of creative thought, Dalton's achievements can be properly appreciated only when seen against the background of his earlier research and writing.

When he moved from Kendal to Manchester, Dalton also entered a far wider and more demanding scientific world. Indicative of new horizons and new opportunities was his election on 17 October 1794 to membership in the Manchester Literary and Philosophical Society, then in its first great productive epoch, to which Dalton was to contribute so substantially. His sponsors were Thomas Henry, translator of Lavoisier's *Opuscules;* Thomas Percival, pioneer sanitary reformer and medical statistician; and Robert Owen, entrepreneur and visionary socialist. And within a month of his election, Dalton was reading the society his first major paper, on some "Extraordinary Facts Relating to the Vision of Colours, with Observations."

The paper is an excellent example of the careful observation, bold theorizing, and dogmatic belief which together characterize Dalton's work. The paper provided the first systematic notice and attempted explanation of the existence of color blindness, a defect which John shared with his brother Jonathan. Collecting information from other people similarly afflicted, Dalton was able to give a careful account of the actual phenomenon. His explanation of his own failure to see red was in terms of the supposed blue (that is, red-ray absorbing) nature of the aqueous medium of his eye. Characteristically, Dalton refused to entertain Thomas Young's later alternative explanation. He even went so far as to instruct that his own eye be dissected after death to confirm his hypothesis—a dissection duly undertaken, with the opposite result. If the theory now seems inadequate, the meticulous detail and bold speculations on an important and neglected phenomenon were enough to establish the newly elected member's place in the front ranks of Manchester's burgeoning group of natural philosophers.

In contrast to this brilliant early investigation, Dalton's other major scientific achievements present many puzzling problems of chronology and interpretation. His background and scientific formation is ill understood. It is clear that the standard accounts, with their straight line of development from Isaac Newton's speculations on the fundamental particles of matter in the thirty-first query of the *Opticks* to John Dalton's chemical atomic theory a century later, are woefully inadequate. Their replacement by a more careful and convincing narrative is not yet possible, but some tentative outlines may be indicated.

Although not a Newtonian in any simple sense, Dalton was deeply indebted to the British tradition of textbook and popular Newtonianism, pervasive throughout the later eighteenth century. This tradition, at once empirical and speculative, placed great stress on the uniformity (i.e., inertial homogeneity) and "internal structure" of matter and the role in nature of those short-range attractive and repulsive forces everywhere associated with, if not necessarily inherent properties of, that matter. In the hands of more sophisticated thinkers, the path from homogeneity, internal structure, and short-range forces gradually led through the "nutshell" view of matter elaborated by Newton's immediate disciples, to the subtle curves of the Abbé Bošković and the "materialistic" immaterialism of Joseph Priestley. Scottish common-sense philosophy provided one possible answer to the doubts and paradoxes thus arrived at, while some more conservative and evangelical thinkers turned instead to the heterogeneous matter, indivisible atoms, and ethereal fluids of the consciously "revisionist" disciples of John Hutchinson. The shifting political and theological currents of the 1780's, and more especially the 1790's; the association of Priestley's ideas with materialism; and the pressures on chemical theory inherent in the dramatic technological advances of the period have not yet received any sustained investigation. Thus, for instance, the resonances between Dalton's philosophical position on the nature and properties of matter and the teachings of the Hutchinsonians may more easily be noted than explained. It is one of the curiosities of historical exegesis that the intellectual and philosophical context and consequences of what was for a century the dominant scientific theory of matter—the chemical atomic theory—has been so little studied. The situation is better when we turn to the more limited questions of the chronology and logic of the directly scientific and experimental work which fed and helped to fashion Dalton's evolving theoretical concerns.

Besides his mathematical interests, Dalton was early involved in natural history, the compiling of meteorological records, and the construction of barometers, thermometers, rain gauges, and hygrometers. His daily weather records over a five-year period and those of his friends John Gough (also in Kendal) and Peter Crosthwaite (in Keswick) were to form the basis of Dalton's first book, the *Meteorological Observations and Essays* (1793), which well displays the interests, ambitions and energy of the young provincial natural philosopher. The work, which was already with the printer before he left Kendal, provides tables of barometric pressure, temperature, wind, humidity, and rainfall, besides detailing the occurrence of snow, thunder, and the aurora borealis. All these constitute the *Observations.* As such, they testify to Dalton's patience and diligence. Far more interesting are the *Essays,* in which the empirical is made the servant of the speculative.

The essays include a theory of trade winds, anticipated by George Hadley, as Dalton discovered on his move to Manchester with its more adequate libraries; a theory of the aurora borealis, similarly anticipated by Anders Celsius and by Edmund Halley; speculations about variations in barometric pressure, anticipated by Jean Deluc; and ideas on evaporation which include the germs of his own later chemical atomic theory.

Dalton's earliest meteorological researches had not unnaturally awakened a deep and abiding interest in the theory of rain and in the state of water vapor in the atmosphere. The *Meteorological Observations* even went so far as to advance "a theory of the state of vapour in the atmosphere, which, as far as I can discover, is entirely new, and will be found, I believe,

to solve all the phenomena of vapour we are acquainted with." The theory was that "evaporation and the condensation of vapour are not the effects of chymical affinities, but that aqueous vapour always exists as a fluid sui generis, diffused among the rest of the aerial fluids . . . there is no need to suppose a chymical attraction in the case."

In denying the chemical attraction of water for air in which it was "dissolved," Dalton was of course flouting the orthodox and Newtonian view that short-range attractive and repulsive forces were the appropriate media for explaining the process. In support of such a view, chemists could quote no less an authority than Lavoisier. Dalton, with his habit of looking upon all empirical phenomena from a mathematical point of view, was not the one to worry about this. His experiments seemed to show that the absorption of water vapor by air was not pressure dependent, i.e., "that a cubic foot of dry air, whatever its density be, will imbibe the same weight of vapour if the temperature be the same." Such a conclusion (in modern terminology, the vapor pressure of water is constant at constant temperature) could not easily be reconciled with belief in evaporation as a chemical process. Hence Dalton, the mathematically inclined meteorologist, simply abandoned the chemistry.

In the appendix to the work, he went even further, saying that "the vapour of water (and probably of most other liquids) exists at all times in the atmosphere in an independent state." As the quotation shows, Dalton was not afraid to generalize his ideas. The visual nature of his thinking and the essential continuity in his own ideas from before 1793 right down to 1808 is apparent from his supporting statements. Dalton argued that it was an error to assume chemical combination was necessary if water vapor was to exist in the open atmosphere below 212°F. The error arose from assuming that "air pressing upon vapour condenses the vapour equally with vapour pressing upon vapour, a supposition we have no right to assume, and which I apprehend will plainly appear to be contradictory to reason, and unwarranted by facts; for, *when a particle of vapour exists between two particles of air let their equal and opposite pressures upon it be what they may, they cannot bring it nearer to another particle of vapour.*"

The ideas that in a mixture of gases every gas acts as an independent entity (Dalton's law of partial pressures) and that the air is not a vast chemical solvent were thus first stated in the *Meteorological Observations.* The statements brought no immediate reaction. This was only to be expected, Dalton's argument being so tentative and undeveloped, the ideas themselves so curious in a world of all-pervasive chemical forces, and the author and vehicle of publication so comparatively obscure.

Three papers that Dalton read to the Manchester Literary and Philosophical Society in 1799 and 1800 (in which year he became the society's secretary) show how much the question of water vapor continued to exercise him. In the first paper he discussed the balance in nature between rain, dew, river-water runoff, and evaporation. In the course of this discussion, he provided the earliest definition of the dew point. Then followed two competent, but more pedestrian, papers on heat, in which his firm belief in a fluid of heat is well-displayed and his complete acceptance of the particular caloric theory of William Irvine and Adair Crawford is apparent. The really dramatic development came in the summer of 1801. By 14 September, Dalton was confident enough in his ideas to write to William Nicholson's recently established monthly *Journal of Natural Philosophy, Chemistry and the Arts.* It showed no hesitation in publishing his "New Theory of the Constitution of Mixed Aeriform Fluids, and Particularly of the Atmosphere."

That Dalton was convinced of the value of his ideas is apparent. The rough sketch of his theory of mixed gases printed in Nicholson's *Journal* was quickly supplemented in three papers to the Manchester society. These included the first clear statement that "When two elastic fluids, denoted by *A* and *B,* are mixed together, there is no mutual repulsion amongst their particles; that is, the particles of *A* do not repel those of *B,* as they do one another. Consequently, the pressure or whole weight upon any one particle arises solely from those of its own kind." The debt of this generalized "new theory" to his 1793 picture of water vapor in air is obvious. So too is the debt of Dalton's thinking, with its static, particulate gas, to the passage in Newton's *Principia* (bk. II, prop. 23) which discusses the properties that such a gas would have.

Besides this first formal enunciation of the law of gaseous partial pressures, the papers also contained important information on evaporation and on steam pressure, as well as Dalton's independent statement of Charles's law that "all elastic fluids expand the same quantity by heat."

While Dalton's earlier statements had passed unnoticed, the reaction to his 1801 pronouncements was rapid and widespread. The three papers in the *Manchester Memoirs* were abstracted and reprinted on the Continent. Discussion was immediate and lively. C. L. Berthollet, then in the midst of his Newtonian affinity investigations, scornfully dismissed Dalton's diagrammatic representation of the new theory of mixed gases as "un tableau d'imagination," while Humphry Davy quickly sought the judgment of a friend on these "new and very singular" ideas. Even the Literary and Philosophical Society was uncertain what to make of its secretary's dismissal of chemical affinity as a force acting in the atmosphere. More damagingly, the first edition of Thomas Thomson's highly successful *System of Chemistry* (1802) was highly critical. Dalton quickly wrote to the two major monthly scientific journals of the day, rebutting Thomson's criticism; but clearly it was not argument that was needed so much as convincing experimental proof of his beliefs. To provide such proof became Dalton's major aim, and hence the efficient cause of the chemical atomic theory. What began as a particular interest in meteorology thus ended up as a powerful and wide-ranging new approach to the whole of chemistry, although the transition was by no means sudden.

One thing Dalton did in order to provide support for his heavily attacked theory of mixed gases was to begin an experimental inquiry into the proportions of the various gases in the atmosphere. This inquiry accidentally raised the whole question of the solubility of gases in water. By 12 November 1802 he had discovered enough to read to the Manchester Society his paper "On the Proportion of the Several Gases or Elastic Fluids, Constituting the Atmosphere; With an Enquiry into the Circumstances Which Distinguish the *Chymical* and *Mechanical* Absorption of Gases by Liquids." When read, although not when published, it contained the statement that carbon dioxide "is held in water, not by chemical affinity, but merely by the pressure of the gas . . . on the surface, forcing it into the pores of the water." The researches on solubility thus led to an extension of his mechanical ideas.

It seems that it was this extension of Dalton's ideas that provoked his close friend, the Edinburgh-trained chemist William Henry, to begin his own rival and chemically orthodox series of experiments to ascertain the order of affinities of gases for water. Measured with reference to this objective, the experiments were not a success. However, within a month Henry found what Dalton had failed to see—namely, that at a given temperature the mass of gas absorbed by a given volume of water is directly proportional to the pressure of the gas (Henry's law). Aware of this work and quick to see its relevance to his own ideas, Dalton was able to point out faults in Henry's procedure. One consequence was the latter's public admission that "the theory which Mr. Dalton has suggested to me on this subject, and which appears to be confirmed by my experiments, is that the absorption of gases by water is purely a mechanical effect."

In the light of such exciting developments, we can appreciate why Dalton continued to grapple with

"The Absorption of Gases by Water and Other Liquids." A paper with that title, presented to the Manchester Society in October 1803, made it clear that although his theory of mixed gases was much strengthened by the new evidence from solubility studies, "The greatest difficulty attending the mechanical hypothesis arises from different gases observing different laws." Or, to put the problem in its crudest form, Why does water not admit its bulk of every kind of gas alike? To answer this question Dalton proposed that

> ... the circumstance depends upon the weight and number of the ultimate particles of the several gases; Those whose particles are lightest and single being least absorbable and the others more according as they increase in weight and complexity. .... An enquiry into the relative weights of the ultimate particles of bodies is a subject, as far as I know, entirely new; I have lately been prosecuting this enquiry with remarkable success. The principle cannot be entered upon in this paper; but I shall just subjoin the results, as far as they appear to be ascertained by my experiments.

And thus it was that this paper closed with the very first list of what we would now call atomic weights.

Dalton's method of calculating the relative weights of ultimate particles was simplicity itself. Despite their appeal to more orthodoxly Newtonian chemists, the measurement or calculation of interparticle affinity forces held no interest for him. Instead, his mechanistic, visual, and realist view of atoms was joined with prevailing vogue for numerical calculation and with the common assumption of one-to-one combination, in such a way as to yield wholly new insights.

In accord with the postulates of his theory of mixed gases, Dalton assumed that when two elements A and B come together in reaction, it is the mutual repulsion of the atoms of B that is the critical factor in controlling what happens, rather than any attraction between A and B. Thus, assuming spherical atoms of equal size, twelve atoms of B can theoretically come into contact (react) with one atom of A. In practice the most likely outcome is a one-to-one combination of A and B. Two atoms of B combining with one of A is also possible but less likely, since the atoms of B have a mutual repulsion to overcome, even though they will automatically take up positions on opposite sides of A. Three atoms of B to one of A involve greater repulsive forces, a corresponding triangular disposition, and so on. Thus if only one chemical compound of elements A and B is known, it is natural to assume it has the composition AB. If there are two compounds, they are most likely to be AB and AB_2 and so on. In this way Dalton's theoretical views

provided a rationale for deciding on both the formulas of compounds and on their three-dimensional molecular structures. Armed with such a mechanical view of combining ratios, it was a simple matter for him to argue from the knowledge that eight ounces of oxygen combined with one of hydrogen, to the statement that the relative weights of their ultimate particles were as eight to one.

Just how little Dalton or anyone else realized the implications of his work is seen from public reaction to his tables. Although published in the *Manchester Memoirs* and reprinted in the monthly scientific journals, his table of weights—unlike his theory of mixed gases—aroused no reaction at all. When printed by themselves, tables of weight numbers appeared to be just further obscure and unexplained variations on the widely known tables of affinity numbers current at that time. Even when accompanied by an explanation of their significance, a favorable reception was by no means certain. In December 1803, thanks to a slip in his arithmetic, Dalton, in London to lecture at the Royal Institution, was able to show Humphry Davy how the various oxides of nitrogen might be given formulas and particle weights that were in harmony with the latter's own experimental results. Yet Davy, true to his deeper Newtonian vision, simply dismissed these ideas as speculations "rather more ingenious than important."

This lack of enthusiasm for the chemical possibilities of his work must have been a blow to Dalton, for Davy was after all a highly capable and serious chemist. It is not clear, however, that Dalton himself had yet fully grasped the wider implications of his work. In 1804 he did succeed in arriving at formulas for various hydrocarbons which were agreeable both to his calculating system and to his own now rapidly increasing chemical experiments. But 1804 was notable chiefly for controversy over the mixed gases theory and particularly over its denial of weak chemical affinity forces. Continuing criticism of the theory—and the failure of particle weight studies to provide the hoped-for clinching evidence—caused Dalton to revise his ideas on mixed gases during the course of 1805. This revision seems to have strengthened the slowly deepening conviction that his work on particle weights, although not a success in its original purpose, was of fundamental importance as the basis for his *New System of Chemical Philosophy*.

In the syllabuses of the public lectures he gave in London late in 1803, in Manchester early in 1805, and in Edinburgh in April 1807, we can trace the slow shifting of Dalton's interest away from mechanics, meteorology, and mixed gases, toward chemistry. By March 1807 he was writing to Thomas Thomson to offer an Edinburgh lecture course. This was to be on his recent experimental inquiries, including chemical elements or atoms with their various combinations, and would reveal "my latest results, some of which have not yet been published or disclosed in any way, & which I conceive of considerable importance." The lectures duly took place. Dalton's introduction left no doubt as to his awareness that he spoke to the foremost scientific audience of the day. Equally, there was no doubt as to his own estimate of the importance of his ideas. He quickly informed his hearers that he was about to exhibit "a new view of the first principles or elements of bodies and their combinations" and that this view, if established, "as I doubt not it will in time, will produce the most important changes in the system of chemistry, and reduce the whole to a science of great simplicity, and intelligible to the meanest understanding."

The lectures in Edinburgh were little short of a manifesto for the *New System*. That their reception was favorable we know from the dedication attached to the first part of that work when it finally appeared, just over a year later. With the publication of the second part in 1810, and more especially with Thomas Thomson's and W. H. Wollaston's 1808 papers showing the practical power of his approach, the chemical atomic theory was finally launched. The theory was Dalton's last creative piece of scientific thinking, although he continued active work in several fields for more than a quarter century. The main thrust of much of this work was in providing experimental measurements of atomic weights of known chemical compounds. The enormity of this task and Dalton's reluctance to take other people's results on trust are symbolized by his failure ever to complete those later parts of the *New System* which were to embody his results (although vol. II, pt. 1 did belatedly appear in 1827).

Before leaving Dalton's scientific work, mention must be made of his attitude to chemical atomism. The equation of the concepts "atom" and "chemical element" is usually held to be one of the most important aspects of his achievement. It was certainly one that led to controversy and debate throughout the nineteenth century. Dalton's work not only provided a new, fundamental, and enormously fruitful model of reality for the chemist. It also gave focus and rationale to those weight studies that had become of steadily increasing importance to chemistry through the previous two generations. Even so, the systematic utilization and extension of Dalton's ideas on atomic weights was to be plagued by methodological problems, problems only slowly resolved through the work of Gay-Lussac, Avogadro, and Cannizzaro. Dalton's ideas on the real physical existence and actual nature of chemical atoms were to prove even more troublesome, initiating a continuing nineteenth-century debate that was terminated only by the work of Rutherford and Soddy.

The syllabus for his course of lectures at the Royal Institution in December 1803, soon after the first measurement of relative particle weights, speaks in thoroughly orthodox fashion of "Properties of matter. Extension–impenetrability–divisibility–inertia–various species of attraction and repulsion. Motion–forces–composition of forces–collision. Pendulums." Yet by the spring of 1805, under pressure from the continuing success of his chemical investigations, Dalton's public position was changing. The newly discovered syllabus for his Manchester lectures lists "General properties of matter–extension–divisibility–original ideas on the division of matter into elements and their composition–solidity–mobility–inertia." By this time also, one whole lecture was devoted to the elements of bodies and their composition. Tantalizingly, no manuscript survives to enlighten us about the "original ideas on the division of matter into elements" that the syllabus promised. It seems reasonable to suppose, however, that as Dalton slowly came to appreciate the far wider chemical utility of his researches on the relative weights of ultimate particles so he increasingly felt the need to define the nature of these ultimate particles. Because of his background and context, the move to explicit avowal of chemical atoms and heterogeneous matter was a comparatively simple one to make.

It was in the 1807 Edinburgh syllabus that direct mention was first made of indivisible particles or atoms. Lectures 3, 4, and 5 of Dalton's now deliberately propagandizing course were devoted to the chemical elements. The syllabus spoke of elastic fluids, liquids, and solids as consisting of indivisible particles or atoms of matter, surrounded with atmospheres of heat. Even this statement was not as unambiguous as might be supposed. Of the eighteen elastic fluids known to Dalton, fifteen were, by his own reckoning, compounds. Thus Dalton was in part using the word "atom" in the commonplace and acceptable sense of "smallest particle possessing a given nature." In this sense, "atom" was merely a term for a particle which was divisible only with the loss of its distinguishing chemical characteristics. Yet Dalton's position was not so clear-cut. He was also beginning not only to think but to speak in public of chemical atoms in the more radical sense of solid and indivisible particles.

The following year the first part of the *New System* was to say that chemical analysis and synthesis is only the separating and rejoining of existing particles, the actual creation and destruction of matter being beyond the reach of chemical agency. Such a statement was thoroughly orthodox. What was new was the further insistence that "We might as well attempt to introduce a new planet into the solar system, or to annihilate one already in existence, as to create or destroy a particle of hydrogen." In this way Dalton first made formal claim for the privileged status of his chemical atoms. The particle of hydrogen was not to be seen as the complex result of an ordered and intricate internal structure but as the given solid, the planet.

Just why Dalton should have moved to this position of reserving privileged status for his chemical atoms is not fully obvious. No doubt he felt the need of some philosophical justification for his concentration on particle weights at a time when the list of known elements was under renewed, electrochemical, attack. Lavoisier's reforms had by no means settled the question of which substances should be admitted to the status of chemical element. Between 1800 and 1812 no less than fifteen new chemicals were added to the list of eighteen previously known elements. We can therefore appreciate how widely acceptable was Davy's impatient belief that whereas the power of nature was limited the powers of the chemist's analytical instruments were capable of indefinite increase, so that "there is no reason to suppose that any real indestructible principle has yet been discovered."

Having adopted a position, Dalton was not one to settle for half measures. His second lecture series at the Royal Institution, in 1810, was clearly designed to defend and vindicate his now widely known and controversial ideas against Davy and a host of critics. In these lectures he first publicly abandoned the unity of matter. Dalton was prepared to admit how "it has been imagined by some philosophers that all matter, however unlike, is probably the same thing." However, on the excellent principle that attack is the best form of defense, he calmly asserted that "this does not appear to have been [Newton's] idea. Neither is it mine. I should apprehend there are a considerable number of what may be called elementary principles, which can never be metamorphosed, one into another, by any power we can control."

Still on the offensive, Dalton reiterated the same beliefs in print in Nicholson's *Journal* in 1811. He insisted that "atoms of different bodies may be made of matter of different densities." The example he offered was that "mercury, the atom of which weighs almost 170 times as much as that of hidrogen, I should conjecture was larger, but by no means in proportion of the weights." Once again the opposition was disarmed with the bland assertion that Newton

had a better claim to be heard than either Dalton or his critics. And Newton was quoted as saying in the thirty-first query of the *Opticks* that "God is able to create particles of matter of several sizes and figures, and in several proportions to the space they occupy, and perhaps of different densities and forces . . . at least I see nothing of contradiction in all this."

The interesting thing about this quotation is what Dalton chose to omit from it. Newton did not allow that the matter of our own world was heterogeneous, as the missing words—"and thereby to vary the laws of Nature, and make worlds of several sorts in several parts of the universe"—make plain. But Dalton was obviously concerned to utilize Newton in his own defense, not to quote him accurately. It was polemically useful to cite the *Opticks* in favor of elementary principles which could not be metamorphosed and atoms of different bodies made of matter of different densities; however, the roots of such thinking would seem much more complex than Dalton's public defense suggests. Hence the unwillingness of so many chemists to embrace chemical atoms, the utility of which they appreciated, but the ontological base of which they could not understand. This utility, particularly in its visual aspect, was to prove enormous, especially later in the century, when organic chemistry knew its greatest triumphs. Indeed the eventual successes of the chemical atomic theory were so great as to hide from many subsequent chemists and commentators the ambiguities and uncertainties of Dalton's own writings on the subject. The result is that we still do not possess an adequate understanding of the development of Dalton's own thought, the context in which the first debates on chemical atomic theory took place, or the earlier traditions underlying the continuing unease of the later nineteenth century.

While Dalton's creative science stands at the center of his achievement, other facets of his life are of equal interest. The most obvious are his role in the Manchester Literary and Philosophical Society, his activity in other societies, his civil recognition, his public lectures, and his changing place in the mythology of science. Each reflects a different light on the professionalization of the scientific enterprise.

The Manchester Literary and Philosophical Society, England's oldest continuing scientific society apart from the Royal Society of London, was founded in 1781. The first, it was also the foremost of the rash of such societies founded in the growing manufacturing centers of England as the Industrial Revolution progressed. Boldly provincial, utilitarian, and technological in its orientation, it nourished creative science of the highest caliber, of which John Dalton's work is the best-known but by no means the solitary example. While Dalton was ultimately to bring great prestige to it, the society in its turn played an early and critical role in his intellectual development.

The "Lit and Phil" offered legitimation, audience, encouragement, and reward to the scientific practitioner at a time when science still enjoyed little public recognition as a profession. Not only did the society offer an extensive and up-to-date library, a vehicle of publication (the *Manchester Memoirs,* which were eventually to contain twenty-six of the 117 papers Dalton read before the "Lit and Phil"), and, from 1800, a home for Dalton's apparatus and experimental labors. It also offered critical encouragement and personal reward. This last may be seen objectified and institutionalized in Dalton's rise from member, to secretary, to vice-president (1808), and finally to president (1817), in which capacity he ruled the society firmly but efficiently for the remaining twenty-seven years of his life.

If the Manchester group provided the essential environment for the flowering of Dalton's abilities, other scientific societies were more peripheral to his life. Dalton showed considerable reluctance to be a candidate for election to the Royal Society. In 1810 he rebuffed Davy's approaches, and he was finally elected in 1822 only when some friends proposed him without his knowledge. He submitted but four papers to the *Transactions.* (When, in 1839, the last of these papers was rejected for publication, he had it privately printed with the added lament that "Cavendish, Davy, Wollaston and Gilbert are no more.") Although one of the first two recipients of the Royal Medal in 1826, in recognition of his chemical atomic theory, he appears to have been almost completely indifferent to the Society's affairs. This indifference reflects in part the gulf in social class and professional stance between the provincial teacher committed to his science and the still largely amateur, cosmopolitan, and dilettante orientation of the Royal Society. Dalton's attitude may be seen in his comment to Charles Babbage that if the latter's reformist tract on *The Decline of Science* (1830) "should stimulate the officers and other active members of the Royal Society to the performance of their duties it may be of essential service to the promotion of science." Only in 1834, when he himself was at last enjoying widespread social recognition as the archetype of the dedicated and successful man of science, did Dalton finally make his formal bow at the society.

A sharp contrast appears between John Dalton's attitude to the Royal Society and his response to other groups whose socializing functions were more clearly subordinated to the recognition of professional merit and the promotion and dissemination of science.

Thus, in 1816 he willingly accepted his election as a corresponding member of the French Académie des Sciences. In 1822 he even went so far as to visit Paris, where he "had the happiness to know" such preeminent men of science as Laplace, Berthollet, Gay-Lussac, Thénard, Arago, Cuvier, Brequet, Dulong, and Ampère. During this visit he took his seat at a meeting of the Academy, being introduced by Gay-Lussac, then president. He also dined at Arcueil with the members of Berthollet's informal but influential scientific coterie. In 1830 he enjoyed the further honor of being elected one of the eight foreign associates of the Academy, filling the place made available by the death of Davy.

An even clearer case of Dalton's willing involvement with serious scientific endeavor is seen in his attitude to the British Association for the Advancement of Science. One of the few men of scientific distinction present at the 1831 foundation meeting in York, he played an active role in the Association's affairs. He chaired the chemistry, mineralogy, electricity, and magnetism committee in 1832; was vice-president and chairman of the chemistry section in 1833; and in 1834 was deputy chairman, and in 1835 vice-president, of the chemistry and mineralogy committee. In 1836 he was again vice-president of the chemistry section and became a vice-president-elect of the Association. His activity was abruptly halted by two severe paralytic attacks in April 1837. The attacks left Dalton a semi-invalid for the rest of his days, unable to undertake the strenuous traveling necessary to follow the Association in its journeyings. When the annual meeting came to Manchester in 1842, he was too feeble to take on the role of president. The sentiments expressed at that time, however, and the presidential address that followed his death in 1844 leave no doubt as to Dalton's importance to the early life of the Association. His involvement clearly illustrates one way in which England's emerging group of professional scientists was seeking to create and consolidate the institutional forms their professional life demanded.

Dalton's later life also illustrates the growing recognition that society was beginning to offer the man of science. Impeccable scientific credentials, a blameless personal life, and in old age a calm and equable temperament all combined to make Dalton a peculiarly suitable recipient of civil honor. In 1832, in connection with the British Association meeting in the town, the University of Oxford conferred on him the honorary degree of D.C.L. In 1834 he received an Edinburgh LL.D. under similar circumstances. Thanks to the efforts of Charles Babbage, William Henry, and others, he was awarded a government pension of £150 per annum in 1833; the amount was doubled in 1836. Meanwhile, Manchester was not to be outdone in recognition of its adopted savant: A committee raised £2,000 for a statue, and in May 1834 Dalton duly went to London to sit for the fashionable sculptor Chantrey. The esteem newly granted the successful man of science is seen in the way that "the Quaker Doctor" was even presented at Court in the course of this visit.

If Dalton in old age enjoyed widespread recognition and honor, he also knew a rather different public role throughout his life. His earliest public lectures in Kendal were in the tradition of those broadly popular performances by itinerant lecturers on natural philosophy, of which the importance to the developing structure of British science has yet to be fully appreciated. His lectures at the Royal Institution in 1803 were of similar type, although reflecting more closely Dalton's own special concerns. His later courses were often more directly based on his own immediate research interests, reflecting in part the growing sophistication and expertise of the potential audience. The range and importance of this aspect of Dalton's work may be seen from a (probably incomplete) listing of the courses he gave: in Manchester in 1805 and 1806; Edinburgh and Glasgow in 1807; Manchester in 1808; London in 1810; Manchester in 1811 and 1814; Birmingham in 1817; Manchester in 1820; Leeds in 1823; Manchester in 1825, 1827, 1828, 1829, 1834, and 1835. From 1824 he was also lecturer in pharmaceutical chemistry for Thomas Turner's Manchester School of Medicine and Surgery, an association continued for at least six years. Among other things, these varied lecture courses added substantially to Dalton's income—for instance, the Royal Institution paid him eighty guineas, while his first Manchester lectures showed a profit of £58.2.0.

With such an extensive repertoire in addition to his teaching and research, Dalton over the years built up a substantial collection of apparatus. If his predilection was for bold generalizations and elegantly simple experimental tests, he had a considerable range of equipment available, whether in Kendal, at the Manchester New College, or in his laboratory at the Literary and Philosophical Society. In addition, he knew that success as a public lecturer demanded adequate demonstrations. Thus, on one visit to London alone (in 1805) he spent £200 on lecture equipment. The young Benjamin Silliman, while on his first European tour, expressed a suitable awe at the elaborate experiments accompanying Dalton's subsequent Manchester lectures.

Many writers on Dalton have exaggerated the supposed poverty of his training and his lack of experimental equipment. Similarly, the oft-told story of his contempt for books would seem without foundation. His early acquaintance with John Gough, the facilities of the Kendal school, and his eager appreciation of Manchester's libraries all speak of his thirst for, and appreciation of, a wide range of knowledge and information. The sale catalog of his belongings confirms the picture. Despite the excellent libraries so freely available to him, Dalton's own collection eventually contained no less than 700 volumes, largely but by no means solely restricted to science.

In all these ways—lectures, apparatus, books—John Dalton reveals the incipient professionalism of a new class in English science. Without benefit of Oxford, Cambridge, or medicine, from which natural knowledge had previously drawn its devotees, such men as he could not afford a casual or dilettante attitude toward their work. His livelihood and entrée to more rewarding social circles depended too acutely on that mixture of entrepreneurial talent and professional scientific competence which one may also see displayed in the careers of Humphry Davy and Michael Faraday.

If his public activity knew no tumultuous crises, Dalton's private life was even more unruffled. As an active and ambitious youth, he had little free time. His Kendal thoughts on trying medicine or law were encouraged by the knowledge that the emoluments of a Quaker school-master "are not sufficient to support a small family with the decency and reputation I could wish." By 1794 he was saying instead that "my head is too full of triangles, chymical processes, and electrical experiments, etc., to think much of marriage." Lacking a wife and family of his own, he became deeply attached to several relatives and associates. His brother Jonathan, William Henry, Peter Ewart, and Peter Clare were among his closest friends. In addition, his frequent walking tours, lecture trips, and visits to Quaker meetings made him known to a wide circle, although his quiet and reserved manner was often mistaken for indifference or uncouthness by strangers, especially in his later years.

When Dalton died in July 1844, he was accorded a civic funeral with full honors. His body first lay in state in Manchester Town Hall for four days while more than 40,000 people filed past his coffin. The funeral procession included representatives of the city's major civic, commercial, and scientific bodies, and shops and offices were closed for the day as a mark of respect. This attention was in part a measure of Dalton's intellectual stature and in part a display of civic pride by what in his lifetime had become the preeminent provincial city. It was also a recognition of the new and growing importance of the man of science both to the nation at large and, more especially, to its manufactures and commerce.

This recognition unfortunately did not ensure a competent biography. W. C. Henry, Dalton's literary executor, eventually produced a hurried and careless work that has not yet been adequately replaced. The interest displayed in Dalton by members of Manchester University's flourishing school of chemistry late in the nineteenth century did lead to some new studies; these studies, however, systematically undervalued Dalton's youthful experiences. Instead they concentrated heavily on chemistry and on finding an account of the origins of the chemical atomic theory that would emotionally and heuristically satisfy the desire to see in Dalton the lonely pioneer chemist remote from, but also anticipating, Manchester's later glory. The confusion thus generated is not yet dispersed. Even so, we can now recognize that John Dalton is best viewed not as the uncouth and illeducated amateur they saw but as an early provincial prototype of that fateful invention of the nineteenth century, the professional scientist.

BIBLIOGRAPHY

Dalton's three published books are *Meteorological Observations and Essays* (London, 1793; 2nd ed., 1834); *Elements of English Grammar* (Manchester, 1801; 2nd ed., London, 1803); and the classic *New System of Chemical Philosophy* (London, pt. 1, 1808; pt. 2, 1810; vol. II, pt. 1, 1827; 2nd ed. of pt. 1, 1842). His considerable output of papers, notes, etc. is adequately catalogued in A. L. Smyth, *John Dalton 1766–1844. A Bibliography of Works by and About Him* (Manchester, 1966). Smyth also provides a useful guide to the enormous secondary literature and an adequate, but incomplete, list of surviving Dalton manuscripts. The great bulk of the manuscripts was destroyed in World War II, but important extracts from Dalton's scientific notebooks are available in H. E. Roscoe and A. Harden, *New View of the Origins of Dalton's Atomic Theory . . . Now for the First Time Published from Manuscript* (London, 1896), repr. with an intro. by A. Thackray (New York, 1970). A valuable record of current scholarly opinion on Dalton's achievements is available in the bicentennial volume entitled *John Dalton and the Progress of Science,* D. S. L. Cardwell, ed. (Manchester, 1968). Also relevant is the closing section of A. Thackray, *Atoms and Powers: An Essay on Newtonian Matter-Theory and the Development of Chemistry* (Cambridge, Mass., 1970). Two recent popular biographies of Dalton are F. Greenaway, *John Dalton and the Atom* (London, 1966), and E. Patterson, *John Dalton and the Atomic Theory* (New York, 1970).

ARNOLD THACKRAY

DALY, REGINALD ALDWORTH (*b.* Napanee, Ontario, Canada, 18 March 1871; *d.* Cambridge, Massachusetts, 19 September 1957), *geology, geophysics.*

Daly's father, Edward, was a tea merchant and farmer; his mother was the former Jane Marie Jeffers. As a boy Daly displayed no interest in geology or in his early schooling. He graduated from Napanee High School in 1887 and received an A.B. degree from Victoria College of the University of Toronto in 1891, remaining there the next year as instructor in mathematics. While there, according to his own notes, his interest in the science to which he devoted his life was stimulated by A. P. Coleman, professor of geology at the University of Toronto.

In 1892 Daly entered Harvard University for graduate studies, earning the M.A. in 1893 and the Ph.D. in 1896. After two years of postdoctoral studies at Heidelberg and Paris, he was an instructor in geology at Harvard from 1898 to 1901. He then accepted the post of geologist with the Canadian International Boundary Commission and from 1901 to 1907 was engaged in arduous fieldwork in the rugged mountainous area of western Alberta and southern British Columbia. For the next five years he was professor of physical geology at the Massachusetts Institute of Technology, and in 1912 he became the Sturgis Hooper professor of geology at Harvard, a post that he held until his retirement in 1942. He became a citizen of the United States in 1920.

Daly married Louise Porter Haskell of Columbia, South Carolina, in 1903. Their only child, Reginald, Jr., died in his third year. Mrs. Daly's death in 1947 was a severe blow, and Daly's output of scientific papers dwindled thereafter. The robust health that had earlier sustained his strenuous fieldwork deteriorated, and during the last few years of his life he was virtually confined to his home.

Daly was a member or corresponding member of more than a score of scientific organizations and received numerous honorary degrees, awards, and prizes.

A strong advocate of fieldwork, imaginative thinking, and synthesis, Daly practiced what he preached. His earlier fieldwork was chiefly in New England, although he made a trip to Newfoundland and Labrador in 1900. His six field seasons along the Canadian-American boundary in 1901–1907 were followed by field studies that took him to the Scandinavian countries, Hawaii, Samoa, St. Helena, Ascension Island, and South Africa. Everywhere he found problems that involved far-reaching geological principles and often led to new concepts. Thus his fieldwork at Mt. Ascutney, Vermont, stimulated the development of his theory of magmatic stoping during the emplacement of intrusive igneous rocks, and later field studies induced detailed examination of theories of magmatic differentiation to explain the great variety of such rocks. Similarly, Daly's theory of glacial control in the development of coral atolls followed his field studies of Pacific islands. That theory was also involved in his later suggestion that submarine canyons on continental slopes were eroded by turbidity currents. His long-sustained interest in isostasy stemmed from his fieldwork in Labrador, with its elevated postglacial marine shorelines. This in turn led to his geophysical studies, concerned primarily with the strength of the earth's crust and the structure of the interior of the earth.

His seven books and many of his contributions to technical journals testify to Daly's ability to correlate countless observations into coherent genetic syntheses. Thus his impact upon geological thinking was worldwide and long-lasting.

BIBLIOGRAPHY

I. ORIGINAL WORKS. Daly's writings include "The Geology of the Northeast Coast of Labrador," in *Bulletin of the Museum of Comparative Zoology at Harvard College,* **38** (1902), 205–270; "The Geology of Ascutney Mountain, Vermont," in *Bulletin of the United States Geological Survey* (1903), 209; "The Mechanics of Igneous Intrusion," in *American Journal of Science,* 4th ser., **15** (1903), 260–298, and **16** (1903), 107–126; "The Classification of Igneous Intrusive Bodies," in *Journal of Geology,* **13** (1905), 485–508; "The Differentiation of a Secondary Magma Through Gravitative Adjustment," in *Festschrift für Karl Rosenbusch* (Stuttgart, 1906), pp. 203–233; "The Nature of Volcanic Action," in *Proceedings of the American Academy of Arts and Sciences,* **47** (1911), 47–122; *Geology of the North American Cordillera at the Forty-Ninth Parallel,* Geological Survey of Canada, memoir no. 38 (Ottawa, 1912); *Igneous Rocks and Their Origin* (New York, 1914); *Origin of the Iron Ores at Kiruna,* Geology of the Kiruna District, no. 5 (Stockholm, 1915); "The Glacial-Control Theory of Coral Reefs," in *Proceedings of the American Academy of Arts and Sciences,* **51** (1915), 157–251; "Metamorphism and Its Phases," in *Bulletin of the Geological Society of America,* **28** (1917), 375–418; "Genesis of the Alkaline Rocks," in *Journal of Geology,* **26** (1918), 97–134; "The Earth's Crust and Its Stability," in *American Journal of Science,* 5th ser., **5** (1923), 349–371; "The Geology of Ascension Island," in *Proceedings of the American Academy of Arts and Sciences,* **60** (1925), 1–80; "Carbonate Dikes of the Premier Diamond Mine, Transvaal," in *Journal of Geology,* **33** (1925), 659–684; *Our Mobile Earth* (New York, 1926); "The Geology of Saint Helena," in *Proceedings of the American Academy of Arts and Sciences,* **62** (1927), 31–92; "Bushveld Igneous Complex of the Transvaal." in *Bulletin of the Geological Society of America,* **39** (1928), 703–768; *Igneous Rocks and the Depths of the Earth* (New York, 1933); *The Changing World of the Ice Age* (New Haven, 1934); "Origin of 'Submarine Canyons,'" in *American Journal of Science,* 5th ser., **31** (1936), 401–420; *Architecture of the Earth* (New York, 1938); *Strength and Structure of the Earth* (New York,

1940); *The Floor of the Ocean* (Chapel Hill, N.C., 1942); "Meteorites and an Earth Model," in *Bulletin of the Geological Society of America,* **54** (1943), 401–455; and "Origin of 'Land Hemisphere' and Continents," in *American Journal of Science,* **249** (1951), 903–924.

II. SECONDARY LITERATURE. Biographies of Daly are Francis Birch in *Biographical Memoirs. National Academy of Sciences,* **34** (1956), 31–64; and Marland P. Billings, in *Proceedings. Geological Society of America, 1958* (1959), 115–122, which includes a bibliography of 134 titles.

KIRTLEY F. MATHER

AL-DAMĪRĪ, MUḤAMMAD IBN MŪSĀ (*b.* Cairo, Egypt, 1341; *d.* Cairo, 1405), *natural history.*

Al-Damīrī for some years earned his living as a tailor. His fondness for literature led him to study with the leading teachers of the time, among them Ibn ʿAqīl; soon he had mastered traditional science to such an extent that he became a teacher in the country's most distinguished centers of learning, such as al-Azhar University.

His religious convictions led him to join a Sufi brotherhood and to make the pilgrimage to Mecca six times, and the indulgences of his youth gave way to almost daily fasting.

Al-Damīrī's literary fame rests upon only one of his works, the *Ḥayāt al-ḥayawān* ("Life of the Animals"), of which there are three versions: the large (*al-kubrā*), the medium (*al-wusṭā*), and the small (*al-ṣugrā*). In this work he collected as much information as possible on the animals mentioned in the Koran and in Arabic literature. The articles, arranged in alphabetical order, generally give the following information: (*a*) grammatical and lexicographical peculiarities of the name by which the animal is known according to al-Jāḥiẓ, Ibn Sīda, and others; (*b*) description of the animal according to the leading authorities, particularly Aristotle and al-Jāḥiẓ; (*c*) Muslim traditions in which the animal is mentioned; (*d*) juridico-theological considerations regarding the animal, especially whether one may eat its flesh; (*e*) proverbs about the animal, particularly from the *Madjmaʿ al-amthāl* of al-Maydānī; (*f*) the medicinal properties (*khawāṣṣ*), if any, of the various products derived from the animal—in this he generally follows Aristotle, al-Jāḥiẓ, Ibn Sīnā, al-Qazwīnī, and others; (*g*) rules for the interpretation of dreams in which a particular animal appears.

This outline, however, is followed only for the major animals, and one or another of the categories of information is omitted from most of the entries. Al-Damīrī's sources are exclusively Arabic but of course include translations into that language from other tongues, especially Greek. Consequently, the 807 authors that he quotes include representatives of a great many nations.

The literary value of the work, written for the most part in a sober and lucid style, lies in the many observations collected, in the frequent quotations from folklore, and in a series of digressions, like that on the history of the caliphate, under *iwazz* ("goose"), which takes up a thirteenth of the book.

The scientific value of the work is not as great. Al-Damīrī, who simply followed tradition, contributed no observations of his own and compiled solely what he was able to find in other books. The *Hayāt al-hayawān* contains 1,069 articles describing a lesser number of animals, since some are entered under various synonyms and others are purely imaginary, existing only in Muslim tradition. An example of the latter is *al-burāq,* the famous mount with human face, horse's mane, and camel's feet on which Muhammad ascended to heaven. Sometimes al-Damīrī's descriptions provide sufficient data to identify the animals mentioned in the Koran (see 34, 13/14, *al-ʿarada,* the wood borer) or reflect popular legends that may also be found in the fiction of the period, such as *The Thousand and One Nights.* The articles on the lizard and the lion are cases in point.

The *Hayāt al-hayawān* has been republished several times and has been translated into Persian and Turkish.

BIBLIOGRAPHY

I. ORIGINAL WORKS. A list of MSS can be found in C. Brockelmann, *Geschichte der arabischen Literatur,* II (Weimar, 1902), 138, and supp. II (Leiden, 1938), 170. Another list of editions is that included in L. Kopf, "al-Damīrī," in *Encyclopédie de l'Islam,* 2nd ed., II (Paris-Leiden, 1965), 109–110. There is an incomplete English trans. by Col. A. S. G. Jayakar as far as "Abū Firās," a nickname of the lion (Bombay, 1906–1908).

II. SECONDARY LITERATURE. On al-Damīrī or his work, see G. Sarton, *Introduction to the History of Science* (Baltimore, 1948), pp. 1214, 1326, 1639–1641; and many papers by Joseph Somogyi, including "Medicine in ad-Damīrī's *Hayāt al-hayawān,*" in *Journal of Semitic Studies,* **2,** no. 1 (1957), 62–91; "Ad-Damīrī's *Hayāt al-hayawān,* An Arabic Zoological Lexicon," in *Osiris,* **9** (1950), 33 ff.; "Biblical Figures in ad-Damīrī's *Hayāt al-hayawān,*" in *Dissertationes in honorem E. Mahler* (Budapest, 1937), pp. 263–299; "The Interpretation of Dreams in ad-Damīrī's *Hayāt al-hayawān,*" in *Journal of the Royal Asiatic Society* (1940), 1–20; "Chess and Backgammon in ad-Damīrī's *Hayāt al-hayawān,*" in *Études orientales à la mémoire de Paul Hirschler* (Budapest, 1950), pp. 101–110; "A History of the Caliphate in the *Hayāt al-hayawān* of ad-Damīrī," in *Bulletin of the School of Oriental Studies,* **8** (1935), 143–155; and "Index des sources de la *Hayāt al-hayawān* de ad-Damīrī," in *Journal asiatique* (1928), **2,** 5–128.

JUAN VERNET

DANA, JAMES DWIGHT (*b.* Utica, New York, 12 February 1813; *d.* New Haven, Connecticut, 14 April 1895), *geology.*

The son of James Dana, a saddler and hardware merchant, and Harriet Dwight, Dana grew up in a respectable, churchgoing family of modest means. At Yale College, Benjamin Silliman gave direction to his interest in natural history, and under his tutelage Dana developed a consuming interest in geology, particularly mineralogy. Upon graduation, like other college-trained American students of science, he faced the choice of a livelihood with little prospect that it would lie in science, for aside from the occasional post in college teaching, the opportunities for a scientist without independent means to practice his profession were all but nonexistent. Delaying the decision, Dana chose to spend his *Wanderjahr* (1833–1834) as schoolteacher to midshipmen aboard the U.S.S. *Delaware.* On this service he observed his first volcano and made it the subject of his first scientific paper, "On the Condition of Vesuvius in July, 1834."[1] Two years later he was saved for science by Silliman's offer of an assistantship in the chemical laboratory at Yale, then confirmed in his vocation by the invitation to serve with the Wilkes expedition of 1838–1842, having been recommended by his friend and contemporary, the botanist Asa Gray.

An adventure in cultural patriotism, this elaborate enterprise was to be one of the great events in the history of science in the United States. The expedition circled the globe, charting and conducting natural history surveys in Polynesia and confirming along the way the existence of an antarctic continent. To the scientist in a nation committed to the principles of self-help and minimal government, it offered an unprecedented opportunity to practice his profession under government patronage, and it was the making of Dana's career. Although accounting himself a mineralogist (his first book, *A System of Mineralogy,* was published the year before the expedition sailed), Dana shipped as a geologist, and when the conchologist Joseph P. Couthouy was dismissed from the squadron at Sydney, he became marine zoologist as well. As he "served the cruise," sharing all the perils except the coasting of Antarctica (he remained in Australia and New Zealand with the other "scientifics"), for four years the world was his laboratory. The experience gave Dana a competence in natural history matched in his own day only by Darwin's on the *Beagle.* It inspired his lifelong interest in both volcanic and coral phenomena and perhaps his peculiar approach to geology as well, for he ever afterward sought to treat the earth as a geological unit.

Dana's interpretation of these phenomena was profoundly influenced by his religious enthusiasms, for although a model of piety as a child, he had undergone a particularly rigorous religious experience on the eve of sailing. It made him a prissy shipmate. It also predisposed him to comprehend microcosm and macrocosm in a sweeping scientific conception of natural phenomena.

After weathering an unfortunate controversy with Couthouy, whom he too hastily charged with plagiarism, Dana devoted the fourteen years between the expedition's return and his assumption of the duties of Yale's Silliman professorship of natural history in 1856 to writing the expedition reports entitled *Zoophytes, Crustacea,* and *Geology.* The whole subject of zoophytes, the plant-like animals now called coelenterates, and the corals in particular, was new to science. Traditionally the preserve of poets and clergymen, it came laden with conceits and preachments. Although the coral structures actually afforded the most spectacular example of the role of the organism in modifying the earth's crust, it was beyond the comprehension of many that these "animalcules," some of them microscopic, could build islands and reefs, which were ascribed instead to tropical lightning, to "the electric fluid engendered by sub-marine and other volcanoes" (notions which Dana described tartly as "the first and last appeal of ignorance"), and to fishes that labored diligently with their teeth. Many who were willing to credit the polyps saw them as "busy little builders of the deep," exemplifying selfless and patient industry as well as the benevolence of the Creator in providing stone flowers for man's delight and seawalls for his protection. Occupying the border zone between the animal and vegetable kingdoms, they had been transferred by general consent to the animal kingdom only during the past century. Dana early perceived the task that lay before him: disabuse the poets, enlighten the preachers, and put the subject on a scientific basis.

His observations on coral phenomena appeared in *Zoophytes* and *Geology.* The former was a monumental work that cost Dana a vast amount of labor, both because the subject had been so little known (203 of the expedition's 261 actinoid zoophytes and 229 of the 483 coral zoophytes were unknown to science, and even more were unknown in the living state) and because coral animals and the Actiniae (such as the sea anemone, a group that, although related to the coral zoophyte, makes no coral) are among the most difficult groups of the animal kingdom to classify systematically. The work was also a milestone in science, for Dana not only classified the animalcules but also elucidated their physiology, both individual and communal, and their ecology. As Asa Gray predicted in his review, the report long remained the standard authority on the subject and retains its validity after the passage of more than a century, for the large divisions Dana defined and a majority even of his species are still accepted. With the publication of *Zoophytes,* Dana became the source for all that was known of coral zoophytes and the Actiniae.

The popular appeal of these relatively specialized reports was remarkable, and when Dana's observations on coral phenomena were printed as a separate book, *Corals and Coral Islands,* it passed through three editions in his lifetime. While it might be said that the audience for such a topic was forgathered, the book's popular success owed much to Dana's remarkably fine style of writing and, not least perhaps, to the penchant he shared with the preachers and poets for perceiving a design in nature that was flattering to man. Tracing the history of the coral plantation—the shrubs and trees which "stand and wave unhurt in the agitated waters" and are the secretions of the thousands of polyps that "cover the branches, like so many flowers, spreading their tinted petals in the genial sunshine, and quiet seas, but withdrawing when the clouds betoken a storm"—he found that when the polyps die, minute encrusting corals attach themselves to the surface to protect the structure from erosion by the sea until "Finally, the coral becomes subservient to a still higher purpose than the support of polyps and nullipores," and the debris produced by wave action upon the reef settles into the crevices to produce a solid rocky base that gradually becomes one of " 'the sea-girt isles' . . ., the coral polyps now yielding place to the flowers and groves of the land, which fulfill their end in promoting the comfort and happiness of man."[2]

Dana had seen his first coral island, Clermont Tonnerre, when he was twenty-six; and even before the squadron reached Sydney, he had carefully observed the three great types of coral structures: atolls (which Charles Lyell was viewing as merely the crests of submarine volcanoes encrusted with coral), barrier reefs, and fringing reefs. Consequently he was well equipped to address himself to the problem of why there were different and well-defined formations. It was Darwin who showed him the way. Shortly after reaching Sydney in 1839, Dana chanced upon a newspaper statement of the theory Darwin had formed to explain the origin of atolls and barrier reefs. It "threw a flood of light over the subject."[3] Dana had not seen the Gambier Islands on the southeastern fringe of the Tuamotus, which had given Darwin the inspiration for his theory; but the Fijis, where the Americans remained three months, presented similar phenomena in greater variety and on a grander scale and provided the opportunity to test the theory.

Darwin sought to explain the three types of island formations as so many stages in the evolution of the atoll, with subsidence the responsible agent. In areas subject to subsidence, the fringing reef, which lies close to the once-volcanic island and parallel to it, continues to grow upward and also, because the more favorable conditions for coral growth obtain at the reef's outer edge, outward until, separated from the shore by a lagoon, it becomes a barrier reef. As subsidence proceeds and the island is completely submerged, the barrier reef becomes an atoll. Subsidence likewise explained the great depth of many reefs, hitherto something of an anomaly in view of the discovery that the animal could not subsist below a depth of 180 feet. Subsidence granted, the reef could be of any depth so long as its upper portion remained above that critical depth. The theory was simple, ingenious, grand, and, as Darwin noted, it explained "a phenomenon otherwise inexplicable." Unfortunately, it lacked a firm basis without independent confirmation that subsidence had indeed taken place, and Darwin despaired of detecting "a movement the tendency of which is to conceal the parts affected."[4] Nevertheless, he made the attempt by plotting the geographical distribution of coral reefs and volcanic areas, for since he associated barrier reefs and atolls with subsidence, so he associated volcanoes with rising areas and fringing reefs with areas either rising or stable. The result showed that areas of barrier reefs and atolls were widely separated from those of fringing reefs and volcanoes.

Dana's own observations fully bore out Darwin's theory and, although they passed unnoticed for many years, led him to independent confirmation of the idea that barrier reefs were the result of subsidence. In 1839 he had ascended Mount Aorai, Tahiti's second peak and, looking out across the island, realized that the deep gorges that had made the climb so difficult were the result of stream erosion. What effect, he asked himself, would Darwin's subsidence produce here? Deep bays at first, then fjords. Examination of the contours of other islands convinced him that those within barrier reefs generally were embayed. The irregular shoreline was the result not of the action of waves and currents, as Darwin apparently assumed (for such action would tend to obliterate irregularities by filling bays and inlets), but of subaerial erosion of a once-high island. It provided the best evidence yet in support of Darwin's subsidence theory. On reading

through Dana's section on coral structures, Darwin was *"astonished at my own accuracy!!"* and found Dana's support of his own theory "wonderfully satisfactory."[5]

But on the matter of the chart of areas of subsidence and elevation that Darwin had plotted, Dana took him sharply to task. To be sure, coral islands evolved as Darwin said they did, through subsidence, but that did not mean that the vast areas of coral islands or even individual islands were still subsiding, as he had suggested. On those he himself visited, Dana found evidence of elevation (deposits of coral material above the level where they were presently forming, and on islands not coral the presence of sedimentary deposits interstratified among layers of other rocks constituting the hills) varying from one foot at Tahiti to 600 feet at Guam. As Dana saw it, Darwin had placed too much reliance on changes of level in different regions, concluding that areas of active volcanoes were areas of elevation and ascribing to elevation the absence of reefs. Dana pointed out that these were no proofs against subsidence having taken place, for a variety of causes—the heating of the sea by volcanic activity, for example—might delay coral growth or prevent it entirely, as by too rapid subsidence or the flow of continental waters. And the absence of reefs was no proof of elevation. He cited the island of Maui, which had no reefs on its larger part, the scene of recent volcanic action, but did have reefs on the other end, where the fires were long since extinguished, and suggested that one end had been undergoing elevation and the other subsidence.

Still, Dana's differences with Darwin were not great. His observations served to refine and correct particulars of Darwin's theory rather than to supplant it. Having provided independent confirmation of the English naturalist's subsidence as a condition necessary to the existence of barrier reefs and atolls, he could address himself objectively to the matter of their distribution. Dana observed that *"two hundred islands have subsided in the Pacific, which, had there been no corals, would have disappeared without a record."* The coral areas were confined within rather precise limits—the isothermal lines belting the earth that were determined by ocean currents—for the creatures could not live where the water temperature fell below sixty-six degrees. Since each barrier reef or atoll was a sure sign of subsidence, it was clear that a vast area of the Pacific, extending from the Tuamotus to the western Carolines, had once subsided. Dana concluded that subsidence increased from south to north or northeastward and was greatest between the Navigators and the Hawaiian Islands. Since the length of the area amounted to about a quarter of the circumference of the globe and its width to about that of the North American continent, Dana reasoned that the movement had surely constituted "one of the great secular movements of the earth's crust" and must have affected the entire sphere, "for all parts, whether participating or not, must have in some way been in sympathy with it." Probably the downward movement in the tropical Pacific involved the warmer latitudes of all oceanic areas (witness the progressive easterly diminution in the size of the West Indies until they disappear and the Atlantic becomes a blank) and was answered by the elevation of the northern continental lands in the glacial era. Dana's treatment of the subject was nothing if not thorough. Examining it with the eye of both the biologist and the geologist, he linked the tiniest of the coral polyps to the grand movements of the earth's crust.

Darwin's identification of volcanic regions with areas undergoing elevation and his own independent observation of the limits that volcanic activity placed on coral growth whetted Dana's interest in volcanism. He visited many volcanic regions while with the expedition and at the Hawaiian group took up the subject in earnest, becoming the first of a long line of geologists to make detailed observations of Mauna Loa and Kilauea. Recognizing that a history of their activity was essential to an understanding of the processes involved, Dana gathered all the information available since the first recorded eruption in 1789, continued for the rest of his life to collect data by correspondence, and at the age of seventy-four went out to view them once again, then published his book on the subject.[6] Although not particularly successful in the attempt to determine their periodicity, he was able to establish fairly clear-cut steps in the volcanic process.

Audacious by nature and emboldened by his unparalleled opportunities for observation while with the expedition, on his return Dana promptly attacked the theory of Christian Leopold von Buch that volcanic cones were formed by extrusion from below in the form of a dome pushed upward by hot steam. If the dome cooled and solidified, it remained without a crater but, if the steam burst through, the material of the dome fell inward to form the typical crater. Siding with the English geologists Lyell and George J. P. Scrope, Dana denied that craters were formed in this way, insisting, rather, that the cones were produced by the accumulation of ejecta from successive eruptions. He recognized in the gently sloping Hawaiian volcanoes a kind of eruption different from that of steep Vesuvius and ascribed the shape of the former to a quiet outpouring of liquid lava and that of the latter to eruptions of more viscous material.

Dana's ascent of Mount Aorai served him well in volcanology, as it did in his study of corals, and he was able to show that the present rough surface of Tahiti, once a smooth dome like Mauna Loa, was due to stream erosion. At a time when that process was only partially understood, his report, one of the earliest examinations of the erosion of a volcanic cone, called attention to its importance in shaping any land mass.

Dana's chapters on the Hawaiian Islands elucidated many points of volcanology, but to his own way of thinking he had cleared up only a very few sequences in the great and continuing process of the earth's formation. The subject of geographical distribution seemed naturally to lead to macrocosm, and Dana went on to link the geological implications of his study of corals to his volcanology. The ocean bottom in which he detected great subsidence was, he reasoned, at one time the most intensely heated part of a cooling earth and the last to solidify. Consequently it had undergone the greatest contraction. This vastly deepened the ocean basins, thrust up the continents, and formed the great mountain chains along their borders.

The picture that emerged was that of the earth of the present, and indeed Dana's concept of the permanency of the continents and the ocean basins, first presented in his *Geology* of the expedition, was to influence geological thought profoundly. In the emphasis he placed on this concept and the significance he accorded stream erosion, Dana outdid even Lyell in uniformitarianism, the great principle that was liberating geologists from the bonds of miraculous intervention. He did not hesitate to pull the master up sharply when he lapsed from principle, as when, promptly on the expedition's return, he announced his disagreement with Lyell on the source of the heat that metamorphosed sedimentary deposits on the ocean floor.

Yet, like most other uniformitarians, Dana was a catastrophist in biology, committed to the notion that entire species and genera were periodically destroyed and new ones created—man most emphatically excepted. His work in classifying the expedition specimens raised problems in the delineation of species, and time and again in his 1,600-page report *Crustacea* he remarked on the "difficulties in the way of arriving at natural subdivisions. . . . Nature had made her fields without fences" but only "mountain ranges" and "gentle gradations." He considered the problem of geographical distribution of species and noted that the obvious explanation was either migration or local creation. In doubtful cases he even presented a perceptive argument for migration—then rejected it for local creation: "These characteristics are of no climatal origin," he remarked in one such passage. "They are the impress of the Creator's hand. . . ."[7] Like most of his contemporaries, he was simply unable to rid himself of the conventional view of the purposiveness of nature.

As accurate in classification as Darwin and fully as able a systematist, Dana could not help but discern change and, as he interpreted it, progress; and the persistent problems of classification that marked his path through the expedition's *Crustacea* drove him to devise a theory of his own that would describe the progress observed, if it could not explain it without recourse to supernatural intervention. "Cephalization," as he called it, was the release, from lower species to higher, of the organs of locomotion to the service of the brain, and it provided a measure of the intelligence of the species. Increasingly, mind became dominant over body, and reflective and moral faculties appeared.[8] But the concept was no more than a further elaboration of the design Dana perceived in nature. It was his determined attempt to render purposive the succession of catastrophes and new creations that otherwise would have seemed random and idiotic.

Within limits determined at their utmost by extra-scientific considerations, Dana remained remarkably open-minded. In 1856 he opened a vigorous defense of science against the bibliolaters in the pages of the *Bibliotheca sacra*,[9] and when, that year, Darwin wrote him to say, "I am becoming, indeed I should say have become, sceptical as to the permanent immutability of species," and went on to express the hope that Dana would give him credit "for not having come to so heterodox a conclusion without much deliberation,"[10] Dana appended to a long article of his own on the species problem a word of caution: "We should give a high place in our estimate to all investigation tending to elucidate the variation of permanence of species, their mutability or immutability. . . ."[11]

Dana was unable to give any kind of place to the *Origin of Species* when it appeared, for that year his health broke down completely, and it was some years before he read the book. He kept the pages of the *American Journal of Science*, of which he had become an editor in 1846, open to Asa Gray, whose main article was a reply to Agassiz's scathing critique of the *Origin* in the same journal. Gray was rigorously defending the Darwinian doctrine (though with a critical eye), but Dana would have none of it himself. The first public indication of any change in his views appeared in the 1874 edition of his *Manual of Geology*, where, of the concepts he thought "most likely to be sustained by further research," he accorded first place to the proposition that "The evolution of the system of life went forward through derivation of species from species, according to natural methods not yet clearly understood, and with few occasions for supernatural intervention."[12]

Dana was the principal exponent, if not the originator, of the geosynclinal-contraction hypothesis of mountain building. In the later editions of the *Manual of Geology*, he proposed that the forces producing mountains acted laterally and unequally from two opposite directions, that a relatively thin rigid crust overlies a densely viscid or pasty region of great thickness, and that the globe is undergoing continued refrigeration. The contraction beneath the crust was considered to be much greater than the contraction within the crust. The result would be the formation of gigantic linear depressions (the geosynclinals of James Hall) with continued sedimentation in the geosynclinals.

Gravity, together with lateral pressure or thrust of the crust due to the unequal contraction of the subcrustal material, would produce folded mountains. The Appalachian structures described by W. B. and H. D. Rogers would be a typical example.

Dana relied for much of his physical information on the work of G. H. Darwin. Accompanying the formation of the geosynclinal there would necessarily be a parallel geanticlinal. Refusion and partial mobilization of the bottom of the geosynclinal, as discussed theoretically by Herschel, would lead to igneous activity and the action of vapors and liquids, as well as a wide range of other geological processes. Massive evidences from the extensive, regional geologic surveys in the western United States and the Alps were mustered by Dana into a coherent and cohesive geological system which with minor modification remained the guiding theoretical framework of geodynamics through the first quarter of the twentieth century.

A member of the first generation of American specialists in science, a generation that contributed much toward making a profession of science, Dana also belonged to the first generation to be caught up in the warfare between science and revealed religion. Committed to both, he strove to retain a footing in two worlds inexorably drifting apart. But the rest of his life was a progressive surrender to Darwinism, although he continued to insist on those few occasions for supernatural intervention, particularly in the evolution of man, and curiously—but perhaps predictably, from his youthful response to the scenes of savagery in the Pacific—his acceptance of the social Darwinism that was becoming fashionable in the closing years of his life was a good deal more prompt.

Dana retired from Yale in 1890 at the age of seventy-seven. Thereafter the books and scientific papers came somewhat less frequently. In his lectures generations of students and of Americans who took an interest in the things of nature, great and small, had the opportunity to join him on the great adventure of his youth. "If this work gives pleasure to any," he remarked toward the close of his life in the third edition of his book on corals, "it will but prolong in the world the enjoyments of the 'Exploring Expedition.'"[13]

Dana was slight of figure and, from frequent illness during the last three decades of his life, frail in health. He liked to write hymns and love songs for the guitar, and although an active participant in the affairs of the American Association for the Advancement of Science (and its president in 1854) and for many years the commanding figure among the world's geologists, he preferred the tranquillity of his domestic life to public appearance. Of the honors conferred upon him Dana was perhaps most gratified by the remark of ex-President Thiers of France that he had been much strengthened in his faith by the writings of "Monsieur Dana, a professor at New Haven."[14]

NOTES

1. *American Journal of Science,* **27** (1835), 281–288.
2. Dana, *Zoophytes,* pp. 83, 84.
3. Dana, *Corals and Coral Islands* (1890), p. 7.
4. Charles Darwin, *The Structure and Distribution of Coral Reefs* (London, 1896 [first pub. 1842]), p. 125.
5. Gilman, *Life of James Dwight Dana,* pp. 306–407.
6. Dana, *Characteristics of Volcanoes.*
7. Dana, *Crustacea,* I, 45, 158.
8. Dana, "On Cephalization."
9. Dana, "Science and the Bible."
10. Sanford, "Dana and Darwinism," p. 534.
11. Dana, "Thoughts on Species," p. 871.
12. Dana, *Manual of Geology* (1874), pp. 603–604.
13. Dana, *Corals and Coral Islands* (1890), preface.
14. Gilman, *op. cit.,* pp. 357–358.

BIBLIOGRAPHY

I. Original Works. Dana's writings include *A System of Mineralogy* (New Haven, 1837); *United States Exploring Expedition during the Years 1838, 1839, 1840, 1841, 1842, Under the Command of Charles Wilkes, U.S.N. Zoophytes* (Philadelphia, 1846); . . . *Geology* (Philadelphia, 1849); . . . *Crustacea,* 2 vols. (Philadelphia 1852); "Science and the Bible: A Review of 'The Six Days of Creation' of Prof. Tayler Lewis," in *Bibliotheca sacra,* **13** (1856), 80–130, 731–756; **14** (1857), 388–412, 461–524; "Thoughts on Species," *ibid.,* **14** (1857), 854–874; *Manual of Geology* (Philadelphia–London, 1862); "On Cephalization," in *New Englander,* **22** (1863), 495–506; *Corals and Coral Islands* (New York, 1872); and *Characteristics of Volcanoes* (New York, 1890).

The largest collection of Dana's papers is at Yale University. Some fifty pieces are at Harvard, and other letters from his extensive correspondence are to be found at the American Philosophical Society; the University of Rochester, and the University of Illinois. A representative selection, together with a lengthy if not entirely complete bibliography, appears in Gilman's biography.

II. Secondary Literature. See the following, listed in chronological order: Daniel C. Gilman, *The Life of James Dwight Dana* (New York–London, 1899); J. Edward Hoffmeister, "James Dwight Dana's Studies of Volcanoes and of Coral Islands," in *Proceedings of the American Philosophical Society,* **82** (1940), 721–732; Harley Harris Bartlett, "The Reports of the Wilkes Expedition, and the Work of the Specialists in Science," *ibid.,* 691–705; Ann Mozley, "James Dwight Dana in New South Wales, 1839–1840," in *Journal and Proceedings of the Royal Society of New South Wales,* **97** (1964), 185–191; and William F. Sanford, Jr., "Dana and Darwinism," in *Journal of the History of Ideas,* **26** (1965), 531–546.

Willam Stanton

DANDELIN, GERMINAL PIERRE (*b.* Le Bourget, France, 12 April 1794; *d.* Brussels, Belgium, 15 February 1847), *mathematics, military engineering.*

Dandelin's father was French; his mother, Anne-Françoise Botteman, was from Hainaut (now part of Belgium). The father, after the transfer of Belgium to France, occupied administrative functions in that country. Dandelin studied at Ghent, volunteered in 1813 for the defense of Walcheren against the British, and in the same year entered the École Polytechnique in Paris. He was wounded in action at Vincennes on 30 March 1814 and in 1815, during the Hundred Days, was attached to the Ministry of Interior under Carnot. After Waterloo, Dandelin returned to Belgium, where in 1817 he became a citizen of the Netherlands and *sous-lieutenant* in the corps of military engineers. In 1825 he was elected to the Royal Academy of Sciences in Brussels, and from 1825 to 1830 he was professor of mining engineering in Liège. He then served until 1835 as an officer in the Belgian army, taking part in the Revolution of 1830. He afterward held educational and engineering posts (building fortifications) in Namur, Liège, and (from 1843) Brussels. At the time of his death he was *colonel de génie* in the Belgian army.

Dandelin's early work was in geometry, in which he worked in the same spirit as his Belgian colleague Adolphe Quetelet. The theorem named for him, of great use in descriptive geometry, states that when a cone of revolution is intersected by a plane in a conic, its foci (or focus, in the case of a parabola) are (is) the points (point) where this plane is touched by the spheres that are inscribed in the cone. It is published in "Mémoire sur quelques propriétés remarquables de la focale parabolique" (1822). In "Sur l'hyperboloide de révolution et sur les hexagones de Pascal et de Brianchon" (1826) he proved that the theorem also holds for a hyperboloid of revolution and showed the relationship between the Pascal and Brianchon hexagons and the skew hexagon formed by generators of the hyperboloid.

These investigations were closely related to Dandelin's theory of stereographic projection of a sphere on a plane, presented in "Mémoire sur l'emploi des projections stéréographiques en géométrie" (1827). This led him to inversions, by which points P on a line OP connecting them with a fixed pole O are transformed into points P' on OP' such that the product of OP and OP' is constant. He thus found a rational circular cubic curve as the inverse of a conic with the pole on the conic.

Dandelin also wrote on statics, algebra, astronomy, and probability. In his "Recherches sur la résolution des équations numériques" (1823) he outlined in the second supplement a method (already suggested by Edward Waring in 1762) of approximation of the roots of an algebraic equation with roots α_i by determining the coefficients of equations with roots α_i^2, α_i^4, α_i^8, . . . This method, named for Dandelin and C. H. Gräffe, was also proposed by Lobachevski in 1834.

BIBLIOGRAPHY

I. ORIGINAL WORKS. Two books by Dandelin are *Leçons sur la mécanique et les machines* (Liège, 1827) and *Cours de statique* (Paris, 1830). Among his papers are "Mémoire sur quelques propriétés remarquables de la focale parabolique," in *Nouveaux mémoires de l'Académie royale de Bruxelles,* **2** (1822), 171–200; "Recherches sur la résolution des équations numériques," *ibid.,* **3** (1823), 7–71; "Sur l'hyperboloide de révolution et sur les hexagones de Pascal et de Brianchon," *ibid.* (1826), 1–14; "Mémoire sur l'emploi des projections stéréographiques en géométrie," *ibid.,* **4** (1827), 13–47; "Sur la détermination des orbites cométaires," *ibid.,* **13** (1841), 1–23; and "Sur quelques points de métaphysique géométrique," *ibid.,* **17** (1844), 1–44, which misses the geometries of Bolyai and Lobachevski. There are also papers in Quetelet's *Correspondance mathématique et physique* (1825–1839).

II. SECONDARY LITERATURE. A biography is A. Quetelet, "C. P. Dandelin," in *Biographie nationale,* IV (Brussels, 1873), 663–668. On Dandelin's theorem and its generalizations, see M. Chasles, *Aperçu historique sur l'origine et le développement des méthodes en géométrie* (Paris, 1837; 3rd ed., 1889); and E. Kötter, "Entwicklung der synthetischen Geometrie von Monge bis auf von Staudt," in *Jahresbericht der Deutschen Mathematikervereinigung,* **5,** pt. 2 (1901), 60–64. On the Dandelin-Gräffe method, see F. Cajori, *A*

History of Mathematics (New York, 1938), p. 364; and C. Runge, *Praxis der Gleichungen* (Berlin–Leipzig, 1921), pp. 136–158.

DIRK J. STRUIK

DANFORTH, CHARLES HASKELL (*b.* Oxford, Maine, 30 November 1883; *d.* Palo Alto, California, 10 January 1969), *anatomy, genetics.*

Danforth was the son of James and Mary Haskell Danforth. He attended Tufts College, from which he received a B.A. in 1908, an M.A. in 1910, and an honorary D.Sc. in 1941. He took his Ph.D. at Washington University, St. Louis, in 1912. On 24 June 1914 he married Florence Wenonah Garrison, who bore him three sons: Charles Garrison, Alan Haskell, and Donald Reed. Danforth was an instructor in anatomy at Washington University from 1908 to 1914, an instructor at Tufts College from 1910 to 1911, and a teaching fellow at Harvard Medical School from 1910 to 1911. He was an associate at Washington University from 1914 to 1916 and an associate professor at the same school from 1916 to 1922. In 1922 Danforth moved to Stanford University, where he was associate professor of anatomy from 1922 to 1923 and became full professor in 1923. He served as executive head of the department of anatomy at Stanford from 1938 until his retirement in 1949.

Danforth was the author of some 125 papers between 1907 and 1967. His professional career spanned the years from the rediscovery of genetics to the discovery of DNA, but his own work modestly concerned itself with the verification rather than the origination of paradigms.

Much of Danforth's research dealt with problems of inheritance, including such topics as the mechanism and heredity of human twinning, mutation frequency in man, genetic-endocrine balance in birds and mammals, the genetics of mice, human heredity, morphology, and racial differences in man. Indeed, the wide range of his research interests led him to be at home in several different fields, including, besides anatomy and genetics, endocrinology and physical anthropology.

In this last field one of Danforth's major contributions was his help with the measurement of young male Americans being discharged from the army at the end of World War I. Serving as an anthropologist to the surgeon general's office, he, along with others, made basic measurements on some 104,000 soldiers, the results of which were published in *Army Anthropology.* Greulich says: "The data which Professor Danforth and his associates gathered in that survey provide the only reliable information we have" on the subject and provide "a base-line against which subsequent changes in stature and other physical dimensions of our male population can be gauged and evaluated."

The problems of race and evolution were of particular concern to Danforth throughout his career, and he was especially aware of the social consequences of such considerations. "In this country," he wrote in 1926, ". . . it is especially desirable that law-makers and those publicists who would discuss such matters as immigration and the 'melting pot' should have sound views as to what the native stock really is and what influences are affecting it" (*Journal of Heredity*, **17** [1926], 94).

Associates of Danforth, while expressing appreciation for his research accomplishments, stressed his qualities as a human being and his service as a teacher, both in the department of anatomy and in the Stanford medical school, where he instructed students in the dissecting room. Indeed, Greulich says that he "considered teaching to be his primary responsibility to the students and the University and he never permitted his research work to interfere with it."

BIBLIOGRAPHY

No biographical sketches of Danforth or bibliographies of his work have yet appeared in print. In preparing this sketch I was kindly provided with a bibliography of Danforth's publications by B. H. Willier, who is preparing a memoir for the National Academy of Sciences; a copy of the memorial resolution presented to the Academic Council of Stanford University, by Donald J. Gray; and a MS copy of a memoir prepared by W. W. Greulich, for *Yearbook. American Philosophical Society*, for 1969.

CARROLL PURSELL

DANIELL, JOHN FREDERIC (*b.* London, England, 12 March 1790; *d.* London, 13 March 1845), *meteorology, chemistry, electricity.*

The eldest son of George Daniell, a bencher of the Middle Temple, John received an extensive private education in the classics, after the fashion of the prosperous professional classes of late Georgian London. Even so, his predilection for natural and experimental philosophy was apparent from an early age. He eventually found his métier in a blending of useful observation and elegant codification, and in pioneering improvements in instruments and procedures across a wide spectrum of scientific activity. In this respect his life reflects the rapid advances in technical sophistication of the still imperfectly differentiated physical sciences. His teaching, administrative, and governmental work illuminates other facets of the research enterprise in the early nineteenth century. At that time the polite Anglican professionalism which he exemplified was an important mediating force between the declining dilettante tradition of natural philosophy and the more aggressive self-interest of the often provincial and Dissenting autodidacts and entrepreneurs of newly self-conscious science.

As a young man Daniell joined the sugar-refining business of a relative. Improvements in the manufacturing process were one outcome. Of more importance to Daniell was his acquaintance and resulting lifelong friendship with William T. Brande, professor of chemistry at the Royal Institution. Together they made several scientific excursions both in Britain and on the Continent. And in 1816 they revived the defunct journal of the Royal Institution, jointly conducting twenty issues. These contain many of Daniell's early papers, principally on the phenomena of crystallization and other chemical subjects.

In September 1817 Daniell married Charlotte, youngest daughter of Sir William Rule, surveyor of the navy. She bore him two sons and five daughters. Her death in 1834 was a deep blow to Daniell, a man of reserved manner and warm family attachments. As a consequence of his marriage, Daniell became managing director of the Continental Gas Company, formed to profit from the technological lead enjoyed by the new British lighting industry. He visited various French and German cities to suggest plans for their illumination. Characteristically, he was also soon at work on a new process to generate gas by the destructive distillation of resin dissolved in turpentine. This process saw considerable success in places where coal was expensive, notably in New York and some other American cities. Despite his obvious gifts for such manufacturing enterprises, Daniell found administration, teaching, and research in theoretical aspects of science more congenial to his tastes.

In 1820 he published a description of a new dew-point hygrometer. Its operation depended on the rapid indirect cooling and slow reheating of a glass bulb, the mean between the temperatures at which moisture first appeared and last disappeared from the surface of the bulb being taken as the dew point. This hygrometer enjoyed wide popularity and quickly became a standard instrument.

In 1823 Daniell collected and published his various *Meteorological Essays*. Extending and codifying current knowledge, the work had a considerable success, reaching a third edition in 1845. His wide-ranging interests are also revealed in an important 1824 paper to the Horticultural Society, pointing out the necessity of maintaining an appropriate humidity as well as temperature in any hothouse. He was awarded a silver medal for this contribution. In the *Philosophical Transactions of the Royal Society* for 1830 he published an account of a new pyrometer for ascertaining the heats of furnaces. Together with a second paper showing additional applications of the new pyrometer, this work won him the Rumford Medal of the Royal Society (1832).

Daniell's technical ingenuity and love of new instruments also led to the devising of a plan for an elaborate water barometer with which to study more accurately the fluctuations in atmospheric pressure. The Royal Society commissioned him to build such an instrument, which functioned only erratically. Nothing daunted, Daniell moved to new fields and in 1835 began the work with which his name is most frequently connected.

Research in the still new field of current electricity was at this time hindered by the rapid decline in power of the voltaic cell when in use, because of "reverse polarization." Investigating the standard zinc-copper battery, he found that the decline in power was caused by the liberation of hydrogen gas on the copper plate during electrolysis. In an 1836 *Philosophical Transactions* paper he therefore proposed a new form of apparatus (known familiarly as the Daniell cell) to prevent such liberation and thus maintain a continuous and even current. In this apparatus the zinc and its attendant dilute sulfuric acid are separated from the opposite copper electrode and a solution of copper sulfate by a porous barrier, such as a piece of ox gullet. This arrangement effectively prevents the formation of hydrogen gas at the copper electrode and thus enables the battery to function unimpaired for long periods of time.

The use of such cells, often coupled together in long chains to give a large electromotive force, gave great impetus to research in all branches of current electricity and also led to commercial applications in gilding, electroplating, and glyphography. Daniell himself went on to publish a series of papers on various more theoretical aspects of electrochemistry. He investigated the wide-ranging applicability of Ohm's law. He also showed that the ion of the metal, rather than of its oxide, is the carrier of the electric charge when a metal-salt solution is electrolyzed. These researches served to confirm Daniell's skill as an experimenter, but it was the actual construction of the new battery which won him the Copley Medal, the Royal Society's highest award, in 1837.

While active in scientific research and invention all his life (he published more than forty original papers),

Daniell also worked vigorously as a popularizer and teacher. He was one of the originators of the Society for Promoting Useful Knowledge (1827), writing and editing several of the early works of the Society, including their treatise on chemistry. His early papers on crystallization had already established his reputation as a chemist. In combination with his social standing and religious and political views, this reputation made him an obvious choice as first professor of chemistry at the newly established King's College, London, in 1831. For the rest of his days he devoted great effort to his teaching there, developing careful and elaborate experimental illustrations for his lectures. He also supervised a practical class and gave private instruction in laboratory technique. This teaching work led naturally to his elegant *Introduction to the Study of Chemical Philosophy* (London, 1839), dedicated to his admired friend Michael Faraday. Among its other successes, the book was edited and adapted for the more than 5,000 schools of New York State within a year; a second English edition appeared in 1843.

Daniell was continuously active and influential in the affairs of King's College, being largely responsible for the establishment of its department of applied science. A man of such eminent talent and judicious personal views was also much in demand as the calls and pressures on national scientific activity increased and multiplied in the 1830's. He was early involved with the scientific aspects of what became Commander Ross's 1839–1843 Antarctic expedition. He was largely responsible for the meteorological parts of the associated Royal Society report of 1840 on establishing and equipping magnetic and meteorological observatories in the British dominions. In 1839 he was appointed to an Admiralty commission on defending ships from lightning. He also supplied advice on the continual and troubling problem of the corrosion of the copper sheathing of ships on the African station. His report was published in the *Nautical Almanac* for 1841. In that same year Daniell was actively involved in the formation of the Chemical Society of London, serving as a vice-president.

Civil recognition of his abilities came in the form of an Oxford D.C.L. in 1842. The Royal Society itself also recognized the talent of this loyal, active, and scientifically gifted member, and he served as its foreign secretary from 1839. His sudden death, while at a meeting of the Council of the Society, left a considerable gap in the ranks of scientist-administrators. His importance in the changing and professionalizing science of his day was well caught by an obituarist who observed that the "habits of business he acquired in early life, and his extensive intercourse with men in general, added to his natural perspicacity, gave him a clear insight into character, and conferred on him advantages which men of science in general do not possess."

BIBLIOGRAPHY

Listings of Daniell's many publications may be found in the *Royal Society Catalogue of Scientific Papers* and on pp. xxxiii–xxxv of the third edition of his meteorological essays, published as *Elements of Meteorology,* 2 vols. (London, 1845). This latter also contains an interesting obituary (pp. xiii–xxxii), as does *Proceedings of the Royal Society* (London), **5** (1845), 577–580. His hygrometer and water barometer are discussed respectively in W. E. K. Middleton's *Invention of the Meteorological Instruments* (Baltimore, 1969) and *History of the Barometer* (Baltimore, 1964). The Daniell cell and the electrical researches are extensively treated in vol. IV of J. R. Partington's *History of Chemistry* (London, 1964).

ARNOLD THACKRAY

DANTI, EGNATIO (PELLEGRINO RAINALDI) (*b.* Perugia, Italy, April 1536; *d.* Alatri, Italy, 1586), *cosmography, mathematics.*

Danti's father, Giulio, and his grandfather, Pier Vincenzio, were very well-read in Italian literature and in astronomy. The Danti family was originally named Rainaldi, but contemporaries of the grandfather, admiring his talents, began to call him Dante or Danti. The new name was taken by Giovanni Battista, Pier Vincenzio's brother, who was well-known for his work in mathematics and mechanics and who, according to contemporary reports, in 1503 made what appears to have been a successful flight.

At the age of thirteen Danti, having changed his name from Pellegrino to Egnatio, entered the Dominican order. His reputation as a scholar in science and the arts reached Cosimo I de' Medici, perhaps through his brother Vincenzio, a well-known sculptor at the court of the grand duke toward the end of 1562 or early in 1563. Cosimo I ordered him to prepare maps for his collection and a large terrestrial globe, which is still in existence. Cosimo's admiration for his cosmographer later grew to the point that, in 1571, he wrote to the general of the Dominican order, requesting permission for Danti to reside in the palace. While preparing the maps, Danti also did other work; and Cosimo I commissioned him to study reform of the calendar, which was later carried out by Gregory XIII.

Ever since the seventh and eighth centuries chronologists had noted that the length of the year as

determined by Julius Caesar, which was the basis of the calendar then in use (the Julian calendar), did not correspond to the true course of the sun. In 45 B.C., Caesar, at the suggestion of the Alexandrian astronomer Sosigenes, had determined that the year should be 365 days, five hours, forty-eight minutes, and forty-six seconds. The value was eleven minutes greater than the true one, and by the time of Danti the error, which had been increasing for sixteen centuries, amounted to eleven days. The vernal equinox and the summer solstice should have fallen, respectively, on 21–22 March and 21–22 June. In order to establish the displacement in days, Danti, with the permission of the grand duke, constructed on the façade of the church of Santa Maria Novella an astronomical quadrant and an equinoctial armillary. By means of the latter Danti was able to observe the vernal equinox of 1574 and found that it fell on 11 March. Danti had also wanted to construct a large gnomon in Santa Maria Novella, but the death of Cosimo I prevented him from completing the work.

In 1569 Danti published the *Trattato dell'uso et della fabbrica dell'astrolabio,* which was reprinted in 1578. In 1579 he published the *Sphere* of Sacrobosco, translated by his grandfather, which he enlarged with his own comments. His Italian translation of Proclus' *Sphere* appeared in 1573. These were the earliest astronomical treatises in Italian.

Cosimo I had commissioned Danti to give public lectures in the mathematical sciences, but he was obliged to abandon this assignment in 1574, following the death of Cosimo I, when he lost the favor of the new grand duke, Francesco. Within twenty-four hours of the succession he was transferred to Bologna, where a year later he became professor of mathematics. In Bologna he was able to construct a large gnomon in the church of San Petronio, where he made observations on the exact date of the spring equinox.

In 1577 Danti returned to Perugia, where he was commissioned to draw up the topographical map of Perugia and the surrounding countryside. The pope later appointed him to enlarge the map to include all the papal states: Emilia-Romagna, Umbria, Latium, and Sabina. In 1580, Pope Gregory XIII called Danti to Rome to reform the calendar; and at the Vatican Danti constructed a meridian inside the Torre de' Veneti, in the room that later became known as the Calendar Room. In the same year Pope Gregory ordered him to depict in the Belvedere Gallery the various regions of Italy, in thirty-two large panels. On the completion of this work in November 1583, the pope appointed Danti bishop of Alatri. Even with this new office Danti found time to prepare, with corrections and additions, the second edition of Latino

Orsini's *Trattato del radio latino.* In the same year Pope Sixtus V called Danti to Rome, to assist the architect Domenico Fontana in raising the obelisk in St. Peter's Square. On his return from Rome, Danti, although unwell, left Alatri for the transfer of a monastery. He contracted pneumonia, of which he died at the age of forty-nine.

BIBLIOGRAPHY

I. ORIGINAL WORKS. Among Danti's writings, editions, and translations are *Trattato dell'uso et della fabbrica dell' astrolabio con l'aggiunta del planisfero del Rojas* (Florence, 1569), repr. as *Dell'uso et fabbrica dell'astrolabio et del planisfero. Nuovamente ristampato ed accresciuto in molti luoghi con l'aggiunta dell'uso et fabbrica di nove altri instromenti astronomici* (Florence, 1578); *La sfera di Proclo Liceo tradotta da maestro Egnatio Danti. Con le annotazioni et con l'uso della sfera del medesima* (Florence, 1573); *La prospettiva di Euclide . . . tradotta ([dal greco]) da r. p. m. Egnatio Danti. Con alcune sue annotazioni dei luoghi più importanti* (Florence, 1573); *Usus et tractatio gnomonis magni, quem Bononiae ipse in Divi Petroni templo conferit anno d. 1576* (Bologna, n. d.); *Le scienze matematiche ridotte in tavole* (Bologna, 1577); *Anemographia. In anemoscopium verticale instrumentum ostensorem ventorum. His accessit ipsium instrumenti constructio* (Bologna, 1578); *Le due regole della prospettiva pratica di m. Jacomo Berozzi da Vignola con i commentari del R. P. M. Egnatio Danti* (Rome, 1583); and *Trattato del radio latino, instrumento giustissimo et facile più d'ogni altro per prendere qual si voglia misura, et positione di luogo, tanto in cielo come in terra . . . di Orsini Latino. Con i commentari del R. P. M. Egnatio Danti* (Rome, 1586).

II. SECONDARY LITERATURE. On Danti or his work, see J. del Badia, *Egnazio Danti. Cosmografo e matematico e le sue opere in Firenze* (Florence, 1898); Pietro Ferrate, "Recensione e critica di due lettere del Danti in data 23 novembre 1577 e 15 febbraio 1578," in *Giornale di erudizione artistica,* **2** (1873), 174–175; M. Fiorini, *Sfere terrestri e celesti di autori italiani* (Rome 1899), pp. 72 ff.; V. Palmesi, "Ignazio Danti," in *Bollettino della R. deputazione di storia patria per l'Umbria,* **5** (1899); G. Spini, *Annotazioni intorno al trattato dell'astrolabio e del planisfero universale del R. P. Ignazio Danti* (Florence, 1570); and G. B. Vermiglioli, "Elogio di Ignazio Danti detto in Perugia nel giorno 26 Dicembre 1819," in *Opuscoli letterari di Bologna,* III (Bologna, 1820), 1; "Ignatio Danti," in *Biografie degli scrittori perugini e notizie delle opere loro,* I (Perugia, 1829), 366–370.

MARIA LUISA RIGHINI-BONELLI

DARBOUX, JEAN-GASTON (*b.* Nîmes, France, 14 August 1842; *d.* Paris, France, 23 February 1917), *mathematics.*

After having studied at the lycées at Nîmes and Montpellier, Darboux was admitted in 1861 to both the École Normale Supérieure and the École Polytechnique in Paris; in both cases he placed first on the entrance examinations. This—and the fact that he selected the École Normale—brought him a good deal of publicity. While a student at the École Normale, he published his first paper on orthogonal surfaces, which he studied in more detail in his doctoral thesis, *Sur les surfaces orthogonales* (1866).

From 1867 to 1872 Darboux taught in secondary schools. In the latter year his growing fame brought him a teaching position at the École Normale that he held until 1881. From 1873 to 1878 he held the chair of rational mechanics at the Sorbonne as *suppléant* of Liouville. In 1878 he became *suppléant* of Chasles at the Sorbonne, and two years later succeeded Chasles in the chair of higher geometry, which he held until his death. From 1889 to 1903 Darboux served as dean of the Faculté des Sciences. In 1884 he became a member, and in 1900 the *secrétaire perpétuel,* of the Académie des Sciences. A representative figure, Darboux was a member of many scientific, administrative, and educational committees and held honorary membership in many academies and scientific societies: Lebon (see below) lists more than a hundred.

Darboux was primarily a geometer but had the ability to use both analytic and synthetic methods, notably in the theory of differential equations. Conversely, his geometrical way of thinking enabled him to make discoveries in analysis and rational mechanics. Thus he followed in the spirit of Gaspard Monge, and Darboux's spirit can be detected in the work of Élie Cartan. This characteristic of Darboux's approach to geometry is fully displayed in his four-volume *Leçons sur la théorie générale des surfaces* (1887–1896), based on his lectures at the Sorbonne. This collection of elegant essays on the application of analysis to curves and surfaces is held together by the author's deep understanding of the connections of various branches of mathematics. There are many, sometimes unexpected, applications and excursions into differential equations and dynamics. Among the subjects covered are the applicability and deformation of surfaces; the differential equation of Laplace,

$$f_{uv} = A(u,v)f_u + B(u,v)f_v$$

and its applications; and the study of geodesics (these also in connection with dynamic systems) and of minimal surfaces. Typical is the use of the moving trihedral. Relying on the classical results of Monge, Gauss, and Dupin, Darboux fully used, in his own creative way, the results of his colleagues Bertrand, Bonnet, Ribaucour, and others.

In his *Leçons sur les systèmes orthogonaux* (1898), he returned to his early love, with new results: the cyclids, the application of Abel's theorem on algebraic integrals to orthogonal systems in *n* dimensions, and other novel types of orthogonal systems. Earlier than these two books was *Sur une classe remarquable de courbes algébriques* (1873), in which he made an analytic and geometric investigation of the cyclids, of which an early example had been given by Dupin and which can be obtained by inversion from quadrics. Full use is made of imaginary elements, in the tradition of Poncelet and Chasles.

Darboux also did research in function theory, algebra, kinematics, and dynamics. His appreciation of the history of science is shown in numerous addresses, many given as *éloges* before the Academy. He also edited Joseph Fourier's *Oeuvres* (1888–1890).

BIBLIOGRAPHY

I. ORIGINAL WORKS. Darboux's writings include "Remarques sur la théorie des surfaces orthogonales," in *Comptes rendus de l'Académie des sciences,* **59** (1864), 240–242; *Sur une classe remarquable de courbes algébriques et sur la théorie des imaginaires* (Paris, 1873); *Leçons sur la théorie générale des surfaces et les applications géométriques du calcul infinitésimal,* 4 vols. (Paris, 1887–1896; I, 2nd ed., 1914; II, 2nd ed., 1915); his ed. of Fourier's *Oeuvres,* 2 vols. (Paris, 1888–1890); and *Leçons sur les systèmes orthogonaux et les coordonnées curvilignes* (Paris, 1898, 1910). He also wrote many papers pub. in *Comptes rendus de l'Académie des sciences, Annales de l'École normale,* and other periodicals, including *Bulletin des sciences mathématiques,* which he founded in 1870. His *Éloges académiques et discours* (Paris, 1912) contains essays on Bertrand, Hermite, Meusnier, and others. He provided notes for Bourdon's *Applications de l'algèbre à la géométrie* (Paris, 1880; 9th ed., 1906), pp. 449–648.

II. SECONDARY LITERATURE. E. Lebon, *Gaston Darboux* (Paris, 1910, 1913), is a descriptive bibliography. *Éloges académiques et discours* (Paris, 1912) was collected to honor his scientific jubilee. See also Y. Chatelain, *Dictionnaire de biographie française,* X (1962), 159–160; L. P. Eisenhart, "Darboux's Contribution to Geometry," in *Bulletin of the American Mathematical Society,* **24** (1918), 227–237; D. Hilbert, "Gaston Darboux," in *Acta mathematica,* **42** (1919), 269–273; G. Prasad, *Some Great Mathematicians of the Nineteenth Century,* II (Benares, 1934), 144–182, with an analysis of some of Darboux's works; and J. J. Weiss, in *Journal des débats* (20 Nov. 1861).

DIRK J. STRUIK

D'ARCET, JEAN (*b.* Doazit, near St. Sever, Landes, France, 7 September 1725; *d.* Paris, France, 12 February 1801), *chemistry.*

While still very young, d'Arcet was disinherited and left penniless because he decided to devote his life to science instead of following his father in the legal profession. Fortunately, Augustin Roux introduced him to Montesquieu, who took him to Paris in 1742 to be tutor to his son.

Although he became *docteur-régent* of the Faculty of Medicine in Paris on 18 November 1762, d'Arcet never practiced medicine. Rather, he began to attend G. F. Rouelle's courses in chemistry and was so profoundly influenced that he spent the rest of his life in the study of this science. D'Arcet was made professor at the Collège de France in 1774 and was a member of the Paris Académie des Sciences from 1784 until its suppression in 1793. He became a member of the Institut de France at its formation in 1795.

D'Arcet's first major work was a long series of experiments on the action of heat on minerals. The results, read to the Academy in 1766 and 1768, threw new light on the classification of minerals and developed and extended the work of J. H. Pott. It also laid the foundation for the manufacture of true porcelain (*porcelaine dure*) in France. It was quickly followed by work on the action of heat on the diamond and other precious stones (1771, 1773), in which he demonstrated the complete destructibility of the diamond when it is heated in air and distinguished it from other precious stones, such as rubies and emeralds.

During a series of experiments on fusible alloys, d'Arcet made one of lead, bismuth, and tin that was liquid at the temperature of boiling water and found a use in the production of stereotype plates (1775). Further work on alloys, some done in collaboration with Bertrand Pelletier, enabled him to develop a method of separating the copper from church bells and to show how these bells could be melted down to cast cannon (1791, 1794).

Together with the work already discussed, d'Arcet's publication on the geology of the Pyrenees (1776) and his researches on the action of strong heat on calcareous earth (calcium carbonate), published in 1783, indicate that his interest in minerals and their analysis dominated his work throughout his career. That he did not confine himself entirely to this field, however, is shown by his translation into French of the work on viper venom by Felice Fontana and his work on the extraction of gelatin from bones.

Since much of d'Arcet's work had a direct bearing on industrial techniques, he was appointed to several important posts in this field. As inspector at the Gobelins dye works he was able to improve some of the dyeing processes, and he succeeded P. J. Macquer as director of the porcelain works at Sèvres. He also served as inspector general of the mint.

During the last few years of his life d'Arcet did little

original work, but he served on a number of government commissions and contributed to several reports for the Academy. He gradually adopted the new ideas of pneumatic chemistry.

BIBLIOGRAPHY

I. ORIGINAL WORKS. D'Arcet's writings are *Ergo omnes humores corporis tum excremento tum recremento ex fermentatione producuntur* (Paris, 1762), his doctoral thesis; *Analyses comparées des eaux de l'Yvette, de Seine, d'Arcueil, de Ville-d'Avray, de Sainte-Reine et de Bristol* (Paris, 1767), written with Majault, P. I. Poissonnier, La Rivière, and Roux; *Mémoire sur l'action d'un feu égal, violent et continué pendent plusieurs jours sur un grand nombre de terres, de pierres et de chaux métalliques* (Paris, 1766); *Second mémoire sur l'action d'un feu égal, violent et continué pendant plusieurs jours sur un grand nombre de terres, de pierres et de chaux métalliques* (Paris, 1771); *Mémoire sur le diamant et quelques autres pierres précieuses* (Paris, 1771); *Expériences nouvelles sur la destruction du diamant* (Paris, 1773), written with Rouelle; *Rapport pour l'examen des eaux d'Enghien, au dessus de l'étang de Saint-Gratien* (Paris, 1774), written with Bellot, Élie Bertrand, and Roux; *Expériences sur l'alliage fusible de plomb, de bismuth et d'étain* (Paris, 1775); *Discours en forme de dissertation sur l'état actuel des Montagnes des Pyrénées et sur les causes de leur dégradation* (Paris, 1776); *Mémoire sur la maladie de M. Lhéritier* (Paris, 1778); *Traité sur le venin de la vipère,* 2 vols. (Paris, 1781), trans. from the Italian of Felice Fontana; *Mémoire sur la calcination de la pierre calcaire et sur sa vitrification* (Paris, 1783); *Rapport sur l'électricité dans les maladies nerveuses* (Paris, 1785); *Supplément à l'instruction sur l'art de séparer le cuivre du métal des cloches* (Paris, 1791); "Rapport sur les principes de l'art de la verrerie," in *Annales de chimie,* **9** (1791), 113–137, 235–260, written with Antoine de Fourcroy and Claude Berthollet; *Instruction sur l'art de séparer le cuivre du métal des cloches. Publié par ordre du Comité de Salut Public* (Paris, an II [1794], written with Pelletier, which also contains "Rapport sur les derniers essais faits à Romille pour opérer en grand le raffinage du métal des cloches afin d'en séparer le cuivre"; *Rapport sur les divers moyens d'extraire avec avantage le sel de soude du sel marin* (Paris, an II [1794]), written with Claude Lelièvre, Pelletier, and Giroud; *Rapport sur la fabrication des savons* (Paris, an III [1795]), written with Lelièvre and Pelletier; *Rapport sur les couleurs pour la porcelaine . . .* (Paris, an V [1797]), written with Fourcroy and Guyton de Morveau; *Rapport fait à l'Institut nationale sur les résultats des expériences du citoyen Clouet sur les différens états du fer et pour la conversion du fer en acier fondu* (Paris, an VI [1798]); and "Rapport du mémoire de M. Felix sur la teinture et le commerce du coton filé rouge de la Grèce," in *Annals de chimie,* **31** (an VII [1799]), 214–219.

II. SECONDARY LITERATURE. There is an obituary by Antoine de Fourcroy in *Journal de Paris* (28 pluviôse an IX

[1801]). See also Georges Cuvier, "Notice sur Jean d'Arcet," in *Mémoires de l'Institut,* classe des sciences mathématiques et physiques, **4** (an XI [1801]), 74–88; M. J. J. Dizé, *Précis historique sur la vie et les travaux de Jean d'Arcet* (Paris, an X [1802]); and "Éloge historique de Jean Darcet," in *Recueil des éloges historiques lu dans les séances publiques de l'Institut royal de France,* I (Paris, 1819), 165–185.

E. McDONALD

D'ARCY (or **D'ARCI), PATRICK** (*b.* Galway, Ireland, 27 September 1725; *d.* Paris, France, 18 October 1779), *mathematics, astronomy.*

D'Arcy's father, Jean, and his mother, Jeanne Linch, were of noble birth. In 1739, to escape the English persecution of Catholics, he was sent to Paris, where he was cared for by an uncle. He lived in the same quarter as Jean Baptiste Clairaut, a mathematician who tutored him. D'Arcy became a good friend of Clairaut's son, the far more famous Alexis Claude, the pioneer in France of Newtonian mathematical astronomy. D'Arcy was somewhat of a mathematical prodigy and presented two memoirs on dynamics to the Paris Academy of Sciences at about the age of seventeen.

The elder Clairaut normally tutored a number of young military officers. Influenced by this association, d'Arcy entered the army as a captain and served in Germany and Flanders. He participated in an expedition to the coast of Scotland in 1746, was captured, and was later repatriated because of his reputation as a mathematician. D'Arcy reentered the military in 1752 as a colonel and served for the rest of his life, advancing to the rank of *maréchal-de-camp* in 1770. He was admitted to the Academy of Sciences on 12 February 1749 as *adjoint mécanicien* and advanced through its ranks, becoming *pensionnaire géomètre* on 20 February 1771. Profits from mining interests, his military pensions, and a legacy from his uncle left d'Arcy a rather rich man. He married his niece in 1777, just two years before his death from cholera.

A staunch patron of Irish refugees, d'Arcy enjoyed good relations with the English scientific community in spite of his hatred for the English king. His character has been described by Condorcet as "firm, independent, and quick to anger." This is reflected in d'Arcy's polemics with Maupertuis, d'Alembert, and others. He was tall and well built, and his active military career and social interests furthered his tendencies toward being something of a scientific gadfly.

D'Arcy did work in rational mechanics, military technology, and physics. In the *Mémoires* of the Academy of Sciences for 1747 he presented his principle of conservation of areas, stating the moments of bodies with respect to a given axis thus: "The sum of the products of the mass of each body by the area that its radius vector describes around a fixed center, . . . is always proportional to the time." In opposition to Maupertuis's principle of least action, d'Arcy extended his principle to what he called the principle of conservation of action: The sum of the products of the masses, velocities, and perpendiculars (drawn from the center toward the bodies) is a constant. In the extension of his principle to the problem of the precession of the equinoxes, d'Arcy criticized d'Alembert's work and was in turn criticized by both d'Alembert and Lagrange.

In 1751 d'Arcy presented a memoir on the physics and chemistry of gunpowder mixtures, the dimensions and design of cannon, and the placement of the charge in cannon. This work was continued in *Essai d'une théorie d'artillerie* (1760). He had the chemist Antoine Baumé conduct chemical analyses of gunpowder and showed that the physical mixing procedure, rather than the chemical content, was most important in obtaining a good product. In measuring the recoil and power of cannon he invented a momentum pendulum that was adopted by the Régie des Poudres.

In 1749 d'Arcy and Jean Baptiste Le Roy developed a floating electrometer. In it a float in water supported a metal rod and plate. A second, charged plate was brought near it. Weights needed to restore the floating plate to its original level measured the electrostatic force. This device never proved very successful in practice, however. In 1765 d'Arcy presented an interesting memoir, "Sur la durée de la sensation de la vue." His eyesight had been damaged in an accident; and he was forced to use an observer while conducting the experiment, in which he attempted to measure the optical persistence of visual images. For example, in moving a light in a small circle, he found that above seven revolutions per second, the single light gave the appearance of a continuous circle. This hastily performed experiment bears out a further remark of Condorcet's about d'Arcy's tendency to begin a project and leave it unfinished, for although he raised a number of important questions relating to the physiology of vision, d'Arcy never completed the work.

BIBLIOGRAPHY

I. ORIGINAL WORKS. Most of d'Arcy's memoirs are printed in the *Mémoires de l'Académie royale des sciences* for 1747–1765. A fairly complete list of these is given in Poggendorff, I, 58. His first memoir on the principle of conservation of areas is "Principe général de dynamique, qui donne la relation entre les espaces parcourus et les temps, quel que soit le système de corps que l'on considère, et quelles que soient leurs actions les unes sur les autres," in *Mémoires de l'Académie royale des sciences, année 1747* (1752), 348–356. Several of his Academy memoirs were also printed separately, and a partial list of these is given in the entry for d'Arcy in the general catalog of the Bibliothèque Nationale, Paris. His *Essai d'une théorie d'artillerie* was published in Paris (1760) and Dresden (1766).

II. SECONDARY LITERATURE. The best *éloge* of d'Arcy is by Condorcet, in *Histoire de l'Académie royale des sciences, année 1779* (1782), 54–70. J. B. A. Suard utilizes the Condorcet article but adds comments on d'Arcy's relationship with Condorcet in "Patrice d'Arcy," in *Biographie universelle*, II (Paris, 1811), 389. A discussion of the principle of conservation of areas is given in Ernst Mach, *The Science of Mechanics,* 6th Eng. ed. (La Salle, Ill., 1960), pp. 382–395. An excellent critical discussion of this principle and of the evolution of the general law of moment of momentum is C. Truesdell, "Whence the Law of Moment of Momentum?," in *Mélanges Alexandre Koyré, I. L'aventure de la science,* no. 12 of the series Histoire de la Pensée (Paris, 1964), pp. 588–612. The controversy between Maupertuis and d'Arcy over the principle of least action is discussed in P. Brunet, *Maupertuis—l'oeuvre et sa place dans la pensée scientifique et philosophique du XVIIIe siècle* (Paris, 1929), ch. 5.

C. STEWART GILLMOR

DARLINGTON, WILLIAM (*b.* Dilworthtown, Chester County, Pennsylvania, 28 April 1782; *d.* West Chester, Pennsylvania, 23 April 1863), *botany.*

"Nestor of American botany," Asa Gray's wreath, epitomizes this judicious confidant of Gray and recorder of the history of botany. The son of Edward and Hannah Townsend Darlington, he left the farm to apprentice himself to John Vaughan, M.D., of Wilmington, Delaware, and soon enrolled for Benjamin Smith Barton's lectures at the University of Pennsylvania (M.D., 1804). Self-confident then, as throughout his life, he sent a copy of his thesis to Jefferson. Darlington spent seventeen months as ship's surgeon and in 1808 married Catharine Lacey, daughter of General John Lacey, of New Mills, New Jersey; they had eight children.

Darlington volunteered in the War of 1812. He was thrice elected representative to Congress from West Chester, organized the Bank of Chester County, and served as its president for thirty-three years. At the same time he was successfully practicing medicine—he was an advocate of strong purgatives and copious bleeding—but botany always dominated his life. His *Florula Cestrica* (1826), the enlarged *Flora Cestrica* (1837; 2nd ed., 1853), and two books for the farmer, *Agricultural Botany* (1847) and *American Weeds and Useful Plants* (1859), are all readable, often animated books with a strong historic emphasis. Acknowledged the most valued compendium on early American botany is his *Memorials of John Bartram and Humphry Marshall* (1849). His *Reliquiae Baldwinianae* (1843) contains materials on William Baldwin and his contemporaries based on personal acquaintance and unpublished documents.

Darlington was a long-time confidant of Asa Gray; and his voluminous correspondence documents the growth of American botany, including as it does foreign and domestic botanists, both amateur and professional. His antiquarian sympathies saved Rafinesque specimens that another botanist might have discarded. That his works were valued by his contemporaries is witnessed by two honorary degrees, from Yale University (LL.D., 1848) and Dickinson College (Sc.D., 1856). Darlington was something of a classicist: His bookplate reads "Miseris succurrere Disco," and his tombstone in Oaklands Cemetery, West Chester, bears a Latin inscription he composed.

BIBLIOGRAPHY

I. ORIGINAL WORKS. Darlington's notebooks and MSS were dispersed after his death: the most significant materials are now at State College, West Chester, Pennsylvania; the T. J. Fitzpatrick Collection at the University of Kansas; and the Historical Society of Pennsylvania. His extensive correspondence was divided between West Chester and the New-York Historical Society. Some letters in the Conarroe Papers, VII, 42–91, various dates, are in the Historical Society of Pennsylvania. His books, often annotated, are at West Chester; but presentation copies of his own works, freely given, are encountered in botanical libraries here and abroad. Darlington was one of the founders and served as secretary-treasurer of the West Chester Academy, and

was an organizer of the Chester County Cabinet of Natural Sciences and the Medical Society of West Chester. His writings and biographical sketches are listed in Max Meisel, *Bibliography of American Natural History,* 3 vols. (New York, 1924–1929). His *Memorials* and *Reliquiae Baldwinianae* have been reprinted (New York, 1967, 1969) with extensive commentary by J. Ewan.

II. SECONDARY LITERATURE. The most recent biographical accounts are by Dorothy I. Lansing, *William Darlington, M.D.* (West Chester, Pa., 1965) and by Robert H. Leeper, Dorothy I. Lansing, and William R. Overlease in *Dr. William Darlington Commemorative Program Addresses,* issued by the Chester County Medical Society (West Chester, Pa., 1965). Obituaries by those who knew him are T. P. James, "William Darlington," in *Proceedings of the American Philosophical Society,* 9 (1863–1864), 330–343; and Asa Gray, "The Late William Darlington," in *American Journal of Science,* 2nd ser., 36 (1863), 132–139, where an unpublished autobiography is mentioned.

JOSEPH EWAN

DARWIN, CHARLES GALTON (*b.* Cambridge, England, 19 December 1887, *d.* Cambridge, 31 December 1962), *applied mathematics, theoretical and general physics.*

Darwin was the grandson of Charles Darwin and the son of George Darwin. His contributions to science were in three different, although related, areas of activity: (1) theoretical research in optics (particularly X-ray diffraction), atomic structure, and statistical mechanics (in collaboration with R. H. Fowler); (2) educational and scientific administration; (3) world sociological and technical problems, with special reference to population.

Darwin was educated at Cambridge University, where he held a major scholarship in Trinity College

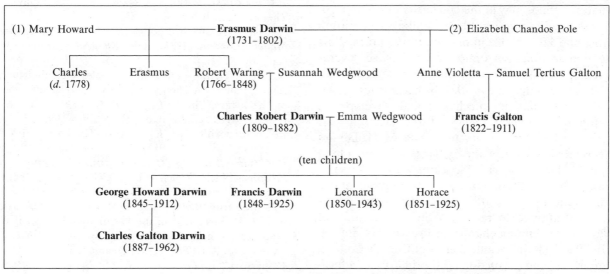

The Darwin family. Names in boldface are discussed in articles.

and took an honors degree in the mathematical tripos in 1910. His training in applied mathematics was particularly strong, although there was little emphasis on contemporary developments in theoretical physics. After leaving Cambridge, Darwin became a postgraduate student with Ernest Rutherford at Manchester. He began research on the absorption of alpha rays and for a time showed interest in the dynamics of Rutherford's nuclear atom model. He soon turned to X-ray diffraction as a subject on which he could exercise his mathematical powers and, after some experimental work with H. G. J. Moseley, produced a series of papers which laid the foundation for all subsequent interpretation of X-ray diffraction by crystals. In these he anticipated by many years the classic work of P. P. Ewald. These researches were probably Darwin's most important contribution to theoretical physics.

After service in World War I, in which he engaged in some early work on acoustic gun ranging, Darwin was appointed fellow and lecturer of Christ's College, Cambridge, a post he held until 1922, when he spent a year in the United States as visiting professor at the California Institute of Technology. The principal fruit of his Cambridge appointment was the collaboration with Fowler on a new method of developing statistical mechanics. Known since this time as the Darwin-Fowler method, it differs from the Maxwell-Boltzmann and Gibbs approaches by calculating directly the averages of physical quantities over assemblies of systems by the method of steepest descents. It served as a particularly effective foundation for the later quantum statistics. For this work and his earlier researches, Darwin was elected a fellow of the Royal Society in 1922.

From 1924 to 1936 Darwin served as Tait professor of natural philosophy at the University of Edinburgh, where his colleagues were E. T. Whittaker in mathematics and J. G. Barkla in physics. Darwin turned his attention to quantum optics and published several papers, particularly in magneto-optics. After a visit to Niels Bohr's institute in 1927, he became interested in the new quantum mechanics and developed a quantum mechanical theory of the electron that proved to be an approximation to P. A. M. Dirac's later relativistic electron theory. This was another high point in Darwin's scientific career. In a sense it marked the termination of his creative investigations in theoretical physics, although during the rest of his life he continued to return to the examination of physical problems that happened to excite his interest.

In 1936 Darwin became master of Christ's College, Cambridge, and devoted himself primarily to educational administration. Presumably he would have been happy to devote the remainder of his professional life to this form of activity. But the approach of war and the resignation of Sir Lawrence Bragg from the directorship of the National Physical Laboratory put pressure on Darwin to take on this important national post. He served in this capacity throughout the war and did not retire until 1949. Darwin's administrative talents were demonstrated by his reorganization of the laboratory both before and after the war. The exigencies of wartime interfered to some extent with his program, since he was engaged in scientific liaison work in the United States during 1941–1942.

The last fifteen years of Darwin's life were devoted largely to the problems of science and society. He paid much attention to genetics and eugenics and to the sociological implications of the population explosion. He became a neo-Malthusian and developed a pessimistic attitude toward man's future on the earth, in spite of obvious technological progress. This view was presented in detail in his well-known book *The Next Million Years*. Here his theory of man as the last "wild" animal is skillfully although rather grimly worked out and has produced much healthy controversy.

Darwin traveled widely, especially in his later years, and showed great interest in international cooperation in science and culture generally. He was a gifted lecturer and knew how to present difficult ideas in simple fashion. Knighted in 1942, he was also honored for his accomplishments by many institutions in Britain and in other countries throughout the world.

BIBLIOGRAPHY

I. ORIGINAL WORKS. A complete bibliography is in Thomson (see below).

Books. Darwin's two books are *The New Conceptions of Matter* (London, 1931); and *The Next Million Years* (London, 1952).

Articles. Of the ninety-three articles the following are representative: "A Theory of the Absorption and Scattering of the α-rays," in *Philosophical Magazine*, 6th ser., **23** (1912), 901; "The Reflexion of the X-Rays," *ibid.,* **26** (1913), 210, written with H. G. J. Moseley; "The Theory of X-Ray Reflexion," *ibid.,* **27** (1914), 315; "The Theory of X-Ray Reflexion," *ibid.,* 675; "The Collisions of α-Particles with Hydrogen Nuclei," *ibid.,* **41** (1921), 486; "On the Reflexion of X-Rays From Imperfect Crystals," *ibid.,* **43** (1922), 800; "On the Partition of Energy," *ibid.,* **44** (1922), 450, written with R. H. Fowler; "On the Partition of Energy, Part II: Statistical Principles and Thermodynamics," *ibid.,* 823, written with R. H. Fowler; "A Quantum Theory of Optical Dispersion," in *Proceedings of the National Academy of Sciences,* 1st ser., **9** (1923), 25–30; "Fluctuations in an

Assembly in Statistical Equilibrium," in *Proceedings of the Cambridge Philosophical Society,* **21** (1923), 4, written with R. H. Fowler; "The Optical Constants of Matter," in *Transactions of the Cambridge Philosophical Society,* **23** (1924), 137–167; "The Intensity of Reflexion of X-Rays by Crystals," in *Philosophical Magazine,* 7th ser., **1** (1926), 897, written with W. L. Bragg and R. W. James; "The Constants of the Magnetic Dispersion of Light," in *Proceedings of the Royal Society,* **114A** (1927), 474, written with W. H. Watson; "The Electron as a Vector Wave," *ibid.,* **116A** (1927), 227; "Free Motion in the Wave Mechanics," *ibid.,* **117A** (1927), 258; "The Wave Equations of the Electron," *ibid.,* **118A** (1928), 654; "The Electromagnetic Equations in the Quantum Theory," in *Nature,* **123** (1929), 203; "Examples of the Uncertainty Principle," in *Proceedings of the Royal Society,* **130A** (1931), 632; "The Diamagnetism of the Free Electron," in *Proceedings of the Cambridge Philosophical Society,* **27** (1931), 1; "Thermodynamics and the Lowest Temperatures," in *Journal of the Institute of Electrical Engineers,* **87** (1940), 528, the thirty-first Kelvin Lecture; "A Discussion on Units and Standards," in *Proceedings of the Royal Society,* **186A** (1946), 149; "Atomic Energy," in *Science Progress,* **135** (1946), 449; "Electron Inertia and Terrestrial Magnetism," in *Proceedings of the Royal Society,* **222A** (1954), 471; "Energy in the Future," in *Eugenics Review,* **46** (1955), 237; "Forecasting the Future," in *New Zealand Science Review,* **14** (1956), 6; "The Value of Unhappiness," in *Eugenics Review,* **49** (1957), 77; "Population Problems," in *Bulletin of the Atomic Scientists,* **14** (1958), 322; "Can Man Control His Numbers?" in *Perspectives in Biology and Medicine* (1960), 252; "The Future Numbers of Mankind," in *Annales Nestle, Humanity and Subsistence Symposium in Vevey* (1960).

II. SECONDARY LITERATURE. A biographical sketch written by Sir George Paget Thomson appears in *Biographical Memoirs of Fellows of the Royal Society,* **9** (1963), 69–85.

See also T. S. Kuhn, J. L. Heilbron, P. Forman, and L. Allen, *Sources for History of Quantum Physics* (Philadelphia, 1960), pp. 30 f. There is an obituary notice by G. B. B. M. Sutherland in *Nature,* **198** (1963), 18.

R. B. LINDSAY

DARWIN, CHARLES ROBERT (*b.* The Mount, Shrewsbury, England, 12 February 1809; *d.* Down House, Downe, Kent, England, 19 April 1882), *natural history, geology, evolution.*

Darwin was the fifth child and second son of Robert Waring Darwin, a successful and respected physician practicing in Shrewsbury, and Susannah Wedgwood, daughter of the potter Josiah Wedgwood I. The Darwin family was characterized by the high intellectual quality of its members, prosperity, industriousness, professional ability, and wide cultural interests.

Darwin's education was begun by his elder sisters after the premature death of his mother. In 1817 he was sent to a day school at Shrewsbury, where he was found to be slow at learning. In 1818 he was entered at Shrewsbury School under Dr. Samuel Butler (grandfather of the author of *Erewhon*). He afterwards complained he was taught nothing but classics, a little ancient history, and geography. "The school as a means of education to me was simply a blank." This statement is perhaps unfair in that his grounding in the classics probably helped him afterwards to think so devastatingly straight. But his headmaster rebuked him publicly for wasting his time on chemical experiments, and his father upbraided him with the remark, "You care for nothing but shooting, dogs, and rat-catching, and you will be a disgrace to yourself and all your family." He was removed from Shrewsbury School in 1825 and sent to Edinburgh University to study medicine.

Darwin stayed at Edinburgh until 1827, attending courses of lectures on materia medica, pharmacy, chemistry, and anatomy, which he found unbearably dull. But worst of all were his experiences attending operations, performed perforce without anesthetics. These repelled him to such an extent that he rushed out and vowed that medicine was no career for him. The chief advantages that he gained from his stay at Edinburgh were his friendship with Robert Grant, zoologist, who accepted Lamarck's teachings on evolution; geological excursions with Robert Jameson; and expeditions onto the Firth of Forth to collect marine animals.

Confronted with the necessity of starting his son on a new line of endeavor, Darwin's father sent him to Cambridge as a preparation for entering the Church of England as a clergyman. At that time, he accepted the Articles of Faith, and perfunctorily attended lectures, but his three years at Cambridge he thought "wasted as far as the academical studies were concerned, as completely as at Edinburgh and at school." But he there made the acquaintance of Adam Sedgwick, who interested him in geology, and, most important of all, John Stevens Henslow, who fired him with a passionate interest in natural history and inspired him with a friendship that gave him confidence in himself, after his discouragement in his own family, at school, and at Edinburgh. Henslow acted as a second father to young Darwin during this critical, formative period in his life, which was climaxed by the voyage of the *Beagle*.

After taking a poor degree at Cambridge in 1831, Darwin was at home when he received an invitation, instigated by Henslow, to join the Admiralty survey ship H.M.S. *Beagle,* under the command of Robert FitzRoy, as unpaid naturalist on a voyage to survey the coasts of Patagonia, Tierra del Fuego, Chile, and Peru; to visit some Pacific islands; and to carry a chain of chronometrical stations around the world.

Darwin wished to accept at once, but his father objected. Darwin's uncle, Josiah Wedgwood II, thought the objection unreasonable and persuaded his brother-in-law to withdraw it. Darwin sailed on the *Beagle* on 27 December 1831.

The five years of the voyage were the most important event in Darwin's intellectual life and in the history of biological science. Darwin sailed with no formal scientific training. He returned a hard-headed man of science, knowing the importance of evidence, almost convinced that species had not always been as they were since the creation but had undergone change. He also developed doubts of the value of the Scriptures as a trustworthy guide to the history of the earth and of man, with the result that he gradually became an agnostic. The experiences of his five years in the *Beagle,* how he dealt with them, and what they led to, built up into a process of epoch-making importance in the history of thought.

On 29 January 1839, Darwin married his first cousin, Emma Wedgwood, daughter of the Josiah Wedgwood who had saved his place in the *Beagle.* At first the young couple lived in London, but ill health began to attack Darwin shortly after his marriage. He found himself increasingly handicapped by great lassitude, nausea, and intestinal discomfort, which made it impossible for him to lead a normal life with social and academic contacts. The Darwins therefore moved to the country, fifteen miles from London, to Down House, in the village of Downe, in Kent. There Darwin had a garden, a small estate about which he could walk, spare rooms for the few friends who were occasionally invited to stay, a study where he could work, and peace and quiet.

His ill health gradually restricted him to a routine of four hours' work a day, walks in the garden, visits to the greenhouses, occasional walks in the neighborhood or rides on a pony, interspersed with periods of rest on a sofa while reading a novel. Dinner was followed by listening to the piano and then early bed.

As his physicians could not discover any organic cause for Darwin's ill health, the idea gained ground that he was a hypochondriac, and in recent years some psychiatrists have attempted to show that he was of a neurotic disposition, but the evidence is not wholly satisfactory. Following the suggestion of Saul Adler, a pathologist expert on South American infections, other commentators note that Darwin was heavily bitten in the Argentine by the Benchuca, the "black bug of the pampas," *Triatoma infestans,* 70 percent of which are now reckoned to be vectors of *Trypanosoma cruzi,* the causative agent of Chagas' disease. As this trypanosome was not discovered until 1909, Darwin's doctors would have been unable to find an organic cause for his trouble; but the clinical picture found in sufferers from Chagas' disease matches Darwin's symptoms in detail. The parasite invades the muscle of the heart causing lassitude; it also invades the nerve cells of Auerbach's plexus in the wall of the intestine, upsetting peristalsis; and it invades the auricular-ventricular bundle of the heart, interference with which may cause heart block. It has unpredictable periods of latency, after which it can be recovered from blood of patients infected many years before. Darwin suffered heart attacks and died from one. Whether in addition Darwin showed neurotic tendencies, accepting the overcare which his wife bestowed on him and taking advantage of his semi-invalidism to protect himself from the distractions of society and amusement, will never be known.

Darwin and his wife had ten children, of whom three died in infancy or childhood. The healths of the survivors (all of whom lived to ripe ages), and their launching in life, gave him anxiety. He once wrote that his chief worries were the discovery of gold in California and in Australia, "beggaring me by making my money on mortgage worthless"; the French coming by the Westerham and Sevenoaks roads, and, therefore, enclosing Down; and, third, "professions for my boys." His fears were groundless. His investments were very profitable, Napoleon III did not invade England, and his sons made careers for themselves of which any father could be proud: Sir George Darwin, mathematician and astronomer; Sir Francis Darwin, botanist; Leonard Darwin, engineer and promoter of eugenics; and Sir Horace Darwin, scientific engineer and founder of the Cambridge Scientific Instrument Co. These men, their children and grandchildren, like their father and ancestors, provide material for Galton's theory of the inheritability of intellectual eminence.

Darwin's contributions to science fall into three main groups: geology, evolution and natural selection, and botany.

Darwin as Geologist. When Darwin sailed in the *Beagle,* the most commonly accepted theory in geology was that of catastrophism, which held that the forces which had produced change during the past history of the earth had been on a far greater scale than any which are active today. When geologists had concluded that stratified rocks had originated as beds of sediment laid down beneath the sea, they had then been obliged to explain how the sedimentary strata had been raised up from the sea bottom to form hills and mountains, and how they had been shifted from the horizontal position, in which they must originally have been deposited, to the inclined or even vertical positions in which they are now so often found. There were two possible explanations: (1) that the sea had formerly stood at a higher level and had in fact extended over the whole earth's surface, or (2) that the strata had been upraised and disturbed. By 1831 most geologists had rejected the idea that the sea had formerly covered the whole earth, but when they tried to imagine forces capable of raising rock strata from the bottom of the sea to form hills or ranges of mountains, it seemed to them that such changes would necessarily involve tremendous convulsions of the earth's surface. Furthermore, they saw a link between periods of convulsion and upheaval in the history of the earth and changes in animal life. They concluded, naturally, that a convulsion of the earth's surface tended to destroy many species of animals and that when the convulsion had subsided new species were created to replace those that had been lost.

In the view of the catastrophists, there had been in the history of the earth a succession of creations of animal and plant life. Each creation had survived during a period of relative tranquillity of the earth's surface, but had later been destroyed in a catastrophic disturbance and upheaval. The history of the earth therefore included periods of tranquillity interrupted by catastrophes, and each catastrophe was followed by a new creation. The catastrophic theory of geological history was reconciled to the account of creation in the Bible by supposing that the most recent catastrophe had been the flood of Noah. Many fossil animals were therefore characterized as antediluvian, or from before the flood. Geologists also supposed that the successive creations were of a progressively higher order which culminated finally in the creation of man.

Since the time of the flood the surface of the earth had been fixed, stable, and unchanging and the whole system of nature tranquil and orderly. The modern order of nature, catastrophists thought, was so very different from that which had existed throughout most of the history of the earth that little could be inferred of past events from present conditions.

The catastrophist viewpoint, firmly held by such English geologists as William Buckland, William Daniel Conybeare, and Adam Sedgwick, was challenged in 1830 by Charles Lyell in his *Principles of Geology.* Lyell had concluded that strata could be elevated from the bottom of the sea by the repeated action of earthquakes continued over long periods of time. In Italy and Sicily he had found recently elevated strata in the neighborhood of active volcanoes and in an area where severe earthquakes had occurred frequently throughout historic time. He showed to his own satisfaction that the strata had been elevated very gradually over an immense period of time, instead of by a single convulsion, and then went on to investigate

how far the phenomena of geology might be explained by the various forces observable on the earth's surface today. With great success he showed that the ordinary action of rain, running water, and the waves of the sea was sufficient to explain the wearing down of land and deposition of sediments, whereas the ordinary, long-continued action of volcanoes and earthquakes was sufficient to elevate continents and mountain ranges.

When Darwin sailed in the *Beagle,* he had with him the first volume of Lyell's *Principles of Geology* (1830), which he had been advised to read but on no account to believe. That is, however, exactly what Darwin did, because he satisfied himself that Lyell's views accorded with the facts: the first step in the making of a man of science possessed of critical judgment.

Santiago in the Cape Verde Islands provided Darwin's first material for geological study. By applying Lyell's principles, he quickly unraveled the history of the island. From the clear exposure of the rocks on the coast he saw that the oldest were crystalline and volcanic. Overlying them was a bed of limestone containing shells of marine organisms of Tertiary date, sixty feet above sea level and twenty feet thick. This bed, Darwin reasoned, had been deposited in a shallow sea and had since been raised to its present height; the island had undergone subsidence and then elevation. Above the limestone was more recent volcanic rock which had altered the uppermost limestone and baked it; evidence of metamorphism *in situ.*

St. Paul's Rocks in mid-Atlantic enabled him to make acquaintance with an island which he correctly recognized as being of neither volcanic nor coralline origin; it is regarded today as an exposed portion of the mantle of the earth. But it was South America that presented itself as a model for his observational and interpretative skill. By comparing lavas of undoubted volcanic origin with igneous rocks of the Andes, he saw that they were closely related. He found that minerals in granites and in lavas were in all respects similar, and he showed a perfect graded series between crystalline granites and glasslike lavas. By observing the direction of strike and the angle of dip of strata, he found that the planes of foliation in schists and gneisses and the planes of cleavage in slates remained constant over areas stretching hundreds of miles, while their planes of inclination could vary greatly. Foliation planes and cleavage planes were parallel to the direction of the great axes along which elevation of the land had taken place. This was directly contrary to the then accepted views put forward by Sedgwick and by Lyell, and Darwin was able to prove that foliation and cleavage are not original phenomena dependent on the deposition of the strata, but had been subsequently superimposed on the beds by pressure, resulting in recrystallization in the case of foliation. He observed clay slate passing over into gneiss as it approached granite, which showed how homogeneous rocks could develop foliation by metamorphosis. Darwin's demonstration of the origin of metamorphic rocks by deformation and of the distinction between cleavage and sedimentary bedding was a major contribution to geology.

The west coast of South America was an equally fruitful geological laboratory for Darwin. He was a witness to the earthquake that devastated Concepción and saw that it had resulted in an elevation of the land by several feet. He was further able to make out a connection between elevation of the land and neighboring volcanic activity because at the time when the land rose near Concepción a number of volcanoes in the Andes renewed their activity, and a new volcano beneath the sea erupted near Juan Fernandez. All these facts were shortly to be woven into an unexpected pattern.

By observing that the shells in beds raised to heights of as much as 1,300 feet belonged to species still living, and in proportions of species similar to those in the ocean, Darwin was able to prove, as Lyell had in Sicily, that such elevation was recent. In the Andes, at a height of 7,000 feet, he found a fossil forest of trees *in situ* (i.e., not drifted from elsewhere) because their trunks projected several feet perpendicularly from the stratum in which they were embedded, which was itself overlaid by thousands of feet of sedimentary deposits alternating with volcanic lavas. This showed that the trees had been buried thousands of feet beneath sea level as the deposits accumulated, and had since been raised to their present height. Here was proof of extensive subsidence as well as elevation.

Beds containing shells that had been raised up were far fewer and much less extensive than low-lying beds, and Darwin found that this was due to the erosion that constantly attacks land surfaces and destroys deposits. This fact had a bearing on estimations of the age of the earth, because while the time taken for a deposit to be formed can be roughly gauged, it is impossible to tell how long an interval of time has gone by while the beds were eroded. This means that the nonconformities between geological formations may represent much longer periods of time than the formation of the beds themselves, and that the age of the earth must be vastly greater than was then imagined.

From these findings, Darwin was led to consider what are the most favorable conditions for the accumulation of beds many thousands of feet thick containing fossils, such as are found elsewhere. He concluded that it could only have been when a slow subsidence continually lowered the bottom of a shallow sea, so that the depth of that sea below sea level remained fairly constant. These conditions further imply that such a sea was near to continental land, from which erosion brought down the material spread out in the deposits. Proof of this was obtained in observation of beds south of Valparaiso, 800 feet thick, containing shells of animals which when alive lived in shallow water, not more than 100 feet deep. The lowest beds of the deposit must have lain 700 feet higher than at present when the first shells became embedded, and as subsidence started the deposits continued and kept pace with the drop of 700 feet. Another bed showed evidence of the same process involving a subsidence and deposits of 6,000 feet. The various conditions required for such accumulations of fossils can only have been met occasionally and at particular localities in the history of the earth, from which it follows that the fossil record must be expected to be erratic and incomplete.

Another line of study to which Darwin was led by his expeditions in the Andes was the change in climate which regions must have undergone, both in geological time and during the life of man. This followed from observation of deserted houses, abandoned by Indians, up to the snowline in places now quite uninhabitable because of lack of water and fertility. This connection between geological phenomena and life was further developed in his studies of coral reefs.

While in South American waters, Darwin had never seen a coral reef, but he knew that atolls in their myriads all over the oceans are about at sea level, and that the coral polyps which build them can live only in waters less than 120 feet deep and at temperatures above 20°C. The theory of the origin of atolls, or lagoon islands, developed by Lyell, was that they were formed on the rims of submarine volcanic craters. Darwin quickly saw that such a hypothesis was untenable, and he put forward his own deductively. There must have been a foundation for the corals at a depth not exceeding 120 feet. It cannot be imagined that there were innumerable submarine mountain ranges with their summits all exactly at this height; therefore, they must have been brought to the required level by subsidence, and he found evidence for it in drowned villages in the Caroline Islands. This has since been confirmed by deep borings at Funafuti, at Bikini, and at Eniwetok, where evidence has been found of over 4,600 feet of coral rock produced by shallow-water organisms during prolonged subsidence.

In addition to atolls or lagoon-islands, there are also barrier reefs, like the Great Barrier Reef of Australia, which is known to have been erected by coral polyps on the down-faulted coastal edge of northeast Australia. Darwin observed a third type of coral reef, the fringing reefs or shore-reefs that surround an island which projects high above the sea, and these reefs are often found lifted above sea level. Fringing reefs are therefore the result of coral action on platforms that are either stationary or have undergone elevation. With all this evidence, Darwin drew a map. He found that large zones of the Pacific Ocean have atolls and barrier reefs and have undergone subsidence, while other zones, parallel to these, have fringing reefs and have undergone elevation. Now the most extraordinary correlation with these facts, bearing in mind Darwin's observations of the connection between land elevation and volcanic activity, was the distribution of active volcanoes in the Pacific: all of them are situated in the zone of fringing reefs, not one of them in the zone of atolls or barrier reefs. Darwin's book on *Coral Reefs,* published in 1842, still remains the accepted explanation, except for a slight addition in the form of R. A. Daly's glacial control theory. This is based on the fact that during the recent Ice Ages, the amount of water immobilized in icecaps reduced the level of the oceans by 150 feet, which would have exposed the atolls to the air and cut down their height by wave-action. But when the Ice Age came to an end and the sea level rose again, the platforms were there for coral polyps to make new atolls.

The evidence which Darwin obtained in South America and in atolls for changes of sea level was later responsible for leading him into error. In 1838 in attempting to explain the Parallel Roads of Glen Roy in Scotland, Darwin imagined that the land had sunk by over 1,000 feet and that the so-called roads were marine beaches. In 1862 it was shown by T. F. Jamieson that the glaciers of Ben Nevis had dammed back the waters of a lake, the level of which dropped twice as outlets were formed, and that the roads were lake beaches. Darwin admitted defeat in words which have an important message for all scientific research workers: "How rash it is in science to argue because any case is not one thing, it must be some second thing, which happens to be known to the writer. . . . My error has been a good lesson to me never to trust in science to the principle of exclusion."

Darwin's trilogy of geological books was completed by the publication in 1844 of *Volcanic Islands,* and at the end of 1846 by *Geological Observations on South America.* The supremacy of the position which geology occupied in Darwin's mind shortly after his return from the voyage of the *Beagle* is shown by the order of subjects in the title of the first edition, published in August 1839, of his *Journal of Researches*

into the Geology and Natural History of the Various Countries Visited by H.M.S. Beagle. In the second edition, published in August 1845, natural history and geology have changed places. The reason for this is to be found in the progressive orientation of Darwin's mind toward the problem of evolution.

Evolution and Natural Selection. When Darwin sailed in the *Beagle,* he had no reason to call in question the accepted view that the species of plants and animals alive on earth were as they had always been since the creation. It was only gradually that doubts began. They arose from four kinds of evidence. The first was that in some areas species had become extinct. Darwin himself had found gigantic fossil armadillos (and other forms) in South America; but armadillos of similar but not identical form also live in South America. This meant that "existing animals have a close relation in form with extinct species." Why was this?

Second, in adjacent areas of South America, Darwin found one species replaced by different, although very similar, species. On the pampas he observed and collected specimens of the South American ostrich (rhea), but when he went farther south into Patagonia, he found a very similar but smaller species (*Rhea darwinii*). But why were the two species of rheas built on the same plan, different from that of the African ostriches? Why were the agoutis and vizcachas of South America built on the same plan as other South American rodents and not on that of North American or European hares and rabbits?

The third doubt was evoked by the fact that inhabitants of oceanic islands tended to resemble species found in the neighboring continents; African-like species in the Cape Verde Islands, South American-like species in the Galápagos archipelago. At the same time, although the geological and physical features of the Cape Verde and Galápagos Islands are very similar, their faunas are quite different. Why was this?

Finally, the different Galápagos islands, identical in climate and physical features, and very close to each other, might be expected to have identical species, but they had not. On different islands he observed that the finches characteristic of the Galápagos group differed in structure and food habits. Some ate insects, others ate seeds, and their sizes and their bills differed in relation to their diets and mode of life. Darwin suspected that they were only varieties of a single species. The local inhabitants could also tell at sight from which island any of the giant tortoises had come. Was all this arbitrary and meaningless or could a pattern of meaning be discerned?

The answer gradually came to Darwin: All these

questions, and many more, could be rationally answered if species did not remain immutable but changed into other species, and diverged, so that one species could have given rise to two or more. It is because they had a common ancestor in South America that fossil armadillos resemble living armadillos, that agoutis resemble capybaras, and that the Galápagos birds resemble South American birds, while the Cape Verde birds resemble African birds with which in each case they had a common ancestor. But why did the finches and other birds of each Galápagos island differ from one another? "One might really fancy that, from an original paucity of birds in this archipelago, one species had been taken and modified for different ends," Darwin wrote. These "ends" are very important; they fit the animals for their mode of life, they are adaptations; and the isolation of each species on each island plays an important part.

Gradually, only a few months after his return to England, these ideas began to crystallize in Darwin's mind: "Animals, our fellow brethren in pain, disease, death, suffering and famine—our slaves in the most laborious works, our companions in our amusements—they may partake of our origin in one common ancestor—we may be all netted together."

In July 1837 Darwin started writing down his ideas at random in his *Notebook on Transmutation of Species*. He soon found that *if* change of species had occurred, there was a ready explanation for a number of otherwise arbitrary and inexplicable facts. Why are the bones of the arm of a man, the foreleg of a dog and a horse, the wing of a bat, and the flipper of a seal built on the same general plan, with the different bones corresponding, each to each? Why are young embryos of lizards, chicks, and rabbits so similar to each other, while their adults are so distinct? Why do some animals have useless rudiments of organs? Why do particular regions of the earth have their characteristic plants and animals? Why do organisms fall into groups, namely, species, that may be arranged in larger groups, namely, genera, that in turn may be arranged in still larger groups, namely, families, and so on? Why is there this apparent relationship of species instead of a random distribution of forms across the whole field of possibilities? Why do more or less similar organisms behave in similar ways? For instance, why do both horses and men yawn; why do both orangutans and men show emotional distress by weeping? Why do fossils in early geological formations differ greatly from those living today, while those from more recent formations differ only slightly?

All these questions could be meaningfully answered if species were related by descent from common ancesters, but not otherwise. It is important to note Darwin's form of argument. He never claimed to demonstrate the change of one species into another, and in his letters constantly repeated that he was tired of impressing this fact on his readers and critics; all that he claimed was that if evolution has occurred, it explains a host of otherwise inexplicable facts. But after he had satisfied himself that evolution had occurred, he kept his views to himself because he saw the great obstacles that lay ahead. Evolution he already saw as the change that species undergo in relation to their adaptation to their environment. A woodpecker differs from other birds in that it has two claws of its feet directed backward, with which it secures a firm grasp on the trunk of a tree, stiff tail feathers with which it supports itself against the tree, a stout bill with which it chisels a hole in the bark, and a very long tongue which it projects through the chiseled hole to extract the grubs beneath the bark. The woodpecker must have evolved these adaptations. How?

In considering this question of how evolution might have occurred, Darwin noticed that the cultivation of plants and the domestication and breeding of animals, both of which man has practiced since the Neolithic period, were the result of selection. Man deliberately chose the parents of the next generation of his plants and animals, to perpetuate and improve the qualities in them which he required. But how did, or could, selection operate in nature, where there had been no man to direct it, since the beginning of life on earth? Darwin had already observed that some species are better adapted than others to life in particular environments, and thus are likely to leave more descendants, while the less well adapted may diminish and become extinct (Notebook 1, MS. p. 37, notebook finished in February 1838). In other words, he had already grasped the principle of natural selection before he saw how it was enforced in nature.

The solution to this problem came to him on 28 September 1838 when he read Thomas Robert Malthus' *Essay on the Principle of Population*. Malthus, whose whole line of thought was tinged with opposition to the principles of egalitarianism displayed by the French Revolution, argued that since the potential rate of increase in man was geometrical, and that his population could double in twenty-five years, while the increase of available foodstuffs could not increase so fast, there was bound to be misery, poverty, starvation, and death among the poor unless the rate of population increase in man was checked by war, famine, disease, or voluntary restraint (which would diminish the incentive to work among the poor).

Malthus, therefore, demonstrated the relentless

pressure which human populations, by their boundless tendency to grow, maintain on their requirements for life whatever the level of their resources may be. If the food supply and other requirements for life can be increased, a population will tend to grow until its growth presses against the limits of the enlarged supply of food. Malthus also observed that in many countries the population did not grow and much of his book is taken up with a study of the means by which the numbers of people were limited in different countries. These ranged from exposure of infants, infanticide, and the intermittent reduction of the population by famine in China, to delayed marriage and strict sexual morality in Norway. But Malthus was able to show that in every country the tendency of a human population to grow was under some form of severe and continued restraint.

Darwin pounced on the argument and applied it, not to man, but to plants and animals in nature, and saw that they are in no position to increase their food supplies, but must die if they outstrip them by their own numbers. Here were the sanctions that made natural selection work.

Like most flashes of genius, it was very simple, and T. H. Huxley said later, "How extremely stupid not to have thought of that." It marked a turning point, not only in the history of science, but in the history of ideas in general, for there is no field of human intellectual endeavor that has not been influenced by the thought and fact of evolution. By good fortune, the note that Darwin scribbled down immediately when the flash struck him has been preserved. It is of such importance that it deserves quotation:

> 28th [September 1838] We ought to be far from wondering of changes in numbers of species, from small changes in nature of locality. Even the energetic language of De Candolle does not convey the warring of species as inference from Malthus. Increase of brutes must be prevented solely by positive checks, excepting that famine may stop desire. In Nature production does not increase, whilst no check prevail, but the positive check of famine and consequently death. I do not doubt every one till he thinks deeply has assumed that increase of animals exactly proportionate to the number that can live.
>
> Population is increase at geometrical ratio in FAR SHORTER time than 25 years, yet until the one sentence of Malthus no one clearly perceived the great check amongst men. There is [a] spring, like food used for other purposes as wheat for making brandy. Even a *few* years plenty makes population in man increase and an *ordinary* crop causes a dearth. Take Europe, on an average every species must have same number killed year with year by hawks, by cold etc. even one species of hawk decreasing in number must affect instantaneously all the

rest. The final cause of all this wedging, must be to sort out proper structure, and adapt it to changes, to do that for form which Malthus shows is the final effect (by means however of volition) of this populousness on the energy of man. One may say there is a force like a hundred thousand wedges trying to force every kind of adapted structure into the gaps in the oeconomy of nature, or rather forming gaps by thrusting out weaker ones.

This passage, in which Darwin's thought penetrates through a jungle of ideas, following on his reading of Candolle's description of the war of nature and Malthus' book, is remarkable for a number of reasons. In the first place it pinpoints the notion that evolution does not take place in a vacuum, but each individual, if it is lucky, lodges in its fortified position in the economy of nature, in what is now called its ecological niche, in dynamic equilibrium not only with the physical factors of the environment, but with the other living organisms of the habitat. It is because Darwin recognized this fact so clearly that he can be reckoned as the founder of the science of ecology.

Second, this passage shows that evolution is not a process that goes on in single individuals, or even in pairs, but which proceeds in populations and consists of some individuals becoming better adapted, under the pressure of natural selection, which is the third lesson of this passage. Ernst Mayr has pointed out that one of the most important recent advances in biology has been the realization that the real units are populations, which have objective existence, and not types (which are only imaginary abstractions), and that Darwin was more responsible than anyone else for the substitution of "population-thinking" for "typological thinking." The change to a different species is nothing but a by-product of the process of a species becoming better adapted, after a certain point of divergence has been reached. The explanation of the fact of divergence itself is foreshadowed in the passage. As Darwin later wrote, "The more diversified the descendants of any one species become in structure, constitution and habits, by so much will they be better enabled to seize on many and widely diversified places in the polity of nature." Again, it is the ecological niches which supply the key to the problem.

Darwin's argument can be formulated as follows: (1) The numbers of individuals in species in nature remain more or less constant. (2) There is an enormous overproduction of pollen, seeds, eggs, larvae. (3) Therefore, there must be high mortality. (4) All individuals in a species are not identical, but show variation and differ from one another in innumerable anatomical, physiological, and behavioristic respects.

(5) Therefore, some will be better adapted than others to their conditions of life and to the ecological niches which they could occupy, will survive more frequently in the competition for existence, will leave more offspring, and will contribute most of the parents that will produce the next generation. (6) Hereditary resemblance between parents and offspring is a fact. (7) Therefore, successive generations will not only maintain but improve their degree of adaptation to their modes of life, i.e., to the conditions of their environments, and as these conditions vary in different places, successive generations will not only come to differ from their parents, but also from each other and give rise to divergent stocks issuing from common ancestors.

This is the formal theory of evolution by natural selection, which recent observation and controlled experiment have proved to be correct in all cases. What modern biology has done is to provide answers to the two questions which, in the state of knowledge of his time, Darwin could not answer: what is the nature of hereditary transmission between parents (and grandparents, and so forth) and offspring? And what is the nature of the origin of heritable variation? Without heritable variation, natural selection could achieve nothing. In Darwin's day, no clear distinction was made between what was heritable and what was not, and few doubted the inheritance of characters acquired by a parent during its own life.

The problem was completely solved (unknown to Darwin, or to the world before 1900) by Gregor Mendel, who proved that the basis of inheritance takes the form of particulate characters, distributed between parents and offspring in accordance with a simple law (since proved by cytology and the behavior of chromosomes) known as Mendel's law of segregation. The inception of a heritable novelty was a sport, now known as a mutation, a random, nonadaptive change in the chemical structure of a particular character, now known as a gene. The application of Mendelian genetics to Darwinian selection was effected by the work of J. B. S. Haldane and Sir Ronald Fisher by 1930. Since then the synthetic theory of evolution by natural selection has been generally accepted by biologists. It is remarkable that even without the knowledge that these advances represent, to say nothing of those achieved in comparative biochemistry, serology, parasitology, and cytogenetics, Darwin was nevertheless able to construct a coherent theory which made general acceptance of evolution possible—and a mechanism to account for it, namely, natural selection. Although long considered inadequate, natural selection is now proved

to be the basis for biological change and for the production of all the diverse structures and functions of living organisms.

Shortly before Darwin died, his son Leonard asked him how long he thought that it would be before positive evidence of natural selection would be forthcoming. Darwin replied, about fifty years. It was a remarkably accurate forecast, for it was in the 1930's that Sir Ronald Fisher showed how expression of a gene, in the form of the bodily character that it controls, is itself the effect of selection. Shortly afterward, E. B. Ford demonstrated that mimetic resemblance in butterflies is adaptive, confers survival value, and originates from heritable variation. The researches of H. B. D. Kettlewell on industrial melanism in moths, P. M. Sheppard and A. J. Cain in snails, T. Dobzhansky on geographical races of fruit flies, and A. C. Allison on sickle cell in man have provided experimental evidence not only of natural selection but of the pressure at which it works, and have permitted actuarial estimates of the longevity of different genetic types. This follows the mathematical studies of Sir Ronald Fisher, Sewall Wright, and J. B. S. Haldane on the effects of selection at different intensities on the composition of populations.

Studies on what B. Rensch has called "ring-races" has shown species in the act of splitting into new species, in gulls, tits, and salamanders, when geographical isolation of portions of populations enables them to evolve independently of each other, so that they become adapted in different directions and eventually no longer interbreed. This speciation is what T. H. Huxley regarded as necessary for the final confirmation of Darwin's views.

Paleontology has been a trap for biologists and others who have neglected David Hume's warning that the existence of a state of affairs does not in itself justify conclusions on how it came about.

Darwin claimed that the fossil record, as known to him, was compatible with evolution; but that if fossils ever provided any evidence contrary to his theory, such, for instance, as to prove that the Cambrian fossils were the first organisms that ever lived, or that a mammal was "created" later than those earliest known mammals of the Stonesfield Slate of the Jurassic period, his theory of evolution must be abandoned. However, he never adduced any evidence from fossils to bear on the problem of the principle of natural selection.

Although the recognition of fossils as former living beings goes back four centuries, it is only recently that paleontology has come into its own as a science, largely as a result of the researches of George G.

Simpson. Fully aware that the mechanism of evolution can be interpreted only in terms of genetics, ecology, selection, and population studies, Simpson has shown that evolutionary sequences previously claimed as "straight" are not straight at all. For example, the evolutionary history of the horse has followed a zigzag course, first in the direction of many-toed browsers on leaves, then in that of many-toed grazers on grass, and finally in that of one-toed grazers on grass. In general the horse increased in size, but some species became smaller. Each lap of the course can be correlated with the environmental conditions of the period: soft or hard ground, leafy or grassy (siliceous) vegetation. At all times the trend has been toward favorable adaptations, to ecological niches which paleontologists are now in a position to describe; in other words, it has been compatible with the requirements of natural selection and has been opportunistic and devoid of any predetermined program. Furthermore, the rate of evolution, now measurable by radioactive dating, is not correlated with variability, nor with fertility (years per generation). In this manner, Simpson has been able to show that natural selection under changing environmental conditions accounts for evolution, and explains why in terrestrial forms evolution has generally been rapid, while in marine forms it has been slow or stationary.

It is regrettable that Darwin in later years allowed himself to be persuaded to accept Herbert Spencer's inappropriate expression "survival of the fittest." This stresses survival when the important thing is the greater fecundity of the better adapted; it emphasizes a superlative "fittest" when it is the slightest comparative superiority of adaptation that confers advantage. It lays the subject open to the taunt of tautology: Who survive? The fittest. Which are the fittest? Those who survive. This is a form of argument which neglects entirely the fundamental fact that forms better adapted to their environment leave more offspring. Finally, the phrase conveys no inkling that the automatic choice exerted by nature in ramming the better adapted variants into their ecological niches is the efficient cause of adaptation. It was not the first (or last) time that so-called philosophers of science have encumbered scientists with their help.

After Darwin had developed his new theory of the origin of species by natural selection in 1838, he did not publish it, or even discuss it with his friends. In 1842 he drew up a rough sketch of his argument, which in 1844 he expanded into an essay but never published. In 1845 when he had finished preparing for publication the results of his geological work on the *Beagle* voyage, he put the species question aside

and started on eight tedious years' study of the structure and classification of living and fossil barnacles. In the course of this work he acquired firsthand knowledge of the amount of variation that is found in nature, and he also made a striking discovery, that of complemental males, small parasitic males found under the mantle of larger hermaphrodite or female individuals.

After his return to England in 1836 Darwin became close friends with Charles Lyell and Joseph Dalton Hooker. They did not accept evolution, which was known to them only in the form of Lamarck's exposition. Lamarck's views were in many ways similar to those of Erasmus Darwin, who posited that there was a "natural tendency" to perfection, and the "inner feelings" of an animal caused it to provide the organs required to meet its needs. In 1844 Robert Chambers published *Vestiges of the Natural History of Creation,* which only brought the subject into disrepute by its amateurish ignorance.

In April 1856 Darwin described his theory of natural selection to Lyell, who urged him to write a book describing his views on species. He began the work in the summer of 1856. On 18 June 1858 Darwin received a letter from Alfred Russel Wallace containing a perfect summary of the views which he had worked out during the preceding twenty years. Thanks to Lyell and to Hooker, Darwin's and Wallace's papers were read together before the Linnean Society of London on 1 July 1858 and published on 20 August of that year.

Darwin then set about writing an "abstract" of his larger work, and *On the Origin of Species* was published on 24 November 1859. The fat was then in the fire. Old-fashioned biologists protested that Darwin indulged in hypotheses that he could not prove (and he did not pretend to prove them). Theologians were aghast at two consequences of Darwin's work; the first was that man and the apes must have had a common ancestor, which dislodged man from his privileged position as created by God in his own image. The second arose from the consideration that if Darwin's views on the origins of plants and animals, including man, by natural selection were true, then much of the argument for the existence of God based on the presence of design in nature was destroyed. The theologian William Paley had argued that highly specialized and coordinated adaptations, such as the tongue, beak, claws, and tail of the woodpecker, or the human hand, manifested a complex design which required one to postulate the presence of a designer in nature. If, however, these same adaptations could be accounted for by natural selection acting on random variations, a designer was no longer needed. Since natural theology, as represented by Paley, had enjoyed enormous influence in England, the destruction of its very foundations was greeted with dismay and outrage.

Matters came to a head at the celebrated Oxford meeting of the British Association for the Advancement of Science on 30 June 1860, when the Reverend Samuel Wilberforce, bishop of Oxford (who knew little of natural history but was coached by the anatomist Richard Owen, who was jealous at what he felt already was Darwin's ascendancy over himself), twitted Darwin's friend Thomas Henry Huxley with the question whether it was through his father or his mother that he claimed descent from an ape. Huxley replied that he was not ashamed to be descended from an ape but he would be ashamed of an ancestor who used great gifts and eloquence in the service of falsehood. Wilberforce was annihilated by Huxley and Hooker, and Darwin's views on evolution started their conquest of the world. In the United States, Darwin's friend, the botanist Asa Gray, had already won a victory over the anatomist and paleontologist Louis Agassiz, who was never able to accept Darwin's theory.

After the *Origin of Species,* Darwin wrote three more books expanding different aspects of the work. *The Variation of Animals and Plants Under Domestication* (1868) took up in detail that subject which had been confined to one chapter of the *Origin.* It contained his hypothesis of pangenesis, by means of which Darwin tried to frame an explanation of hereditary resemblance, inheritance of acquired characters, atavism, and regeneration. It was a brave attempt to account for a number of phenomena which were beyond the bounds of scientific knowledge in his day, such as fertilization by the union of sperm with egg, the mechanism of chromosomal inheritance, and the development of the embryo by successive cell division. His hypothesis of pangenesis could not therefore give a permanently acceptable account of the multitude of phenomena it was designed to explain. It was, however, a point of departure for particulate theories of inheritance in the later nineteenth century.

Darwin's next book, *The Descent of Man* (1871), filled in what was only adumbrated in the *Origin.* T. H. Huxley and W. H. Flower had shown already that the body of man differs less from that of an ape than the latter does from that of a monkey. It is important to remember that Darwin never claimed that man was descended from apes, but that man's ancestor, if alive today, would be classified among the Primates, and would be even lower in the scale than the apes. Man and apes are subject to similar psychological and physiological processes in courtship, reproduction, menstruation, gestation, birth, lactation, and childhood. At early stages of embryonic development the human fetus has a tail, inherited from ancestors who not only had a tail as embryos but also as adults and were quadrupedal. Darwin wrote, "The time will before long come when it will be thought wonderful, that naturalists, who were well acquainted with the comparative structure and development of man and other animals, should have believed that each was the work of a separate act of creation."

Having proved his point with man's body, Darwin then turned to the most difficult aspect, man's mind. Here he showed that the gap, however enormous between man and the highest animal, is not unbridgeable, by the principle of gradation. Man shares with animals the urge to self-preservation, sexual love, maternal affection, paternal protection, and the senses of pleasure, pain, courage, pride, shame, excitement, boredom, wonder, curiosity, imitation, attention, and memory. As for the moral sense, Darwin concluded that it, also, arose gradually, through evolution, and that any animal endowed with social instincts, which man's ancestors undoubtedly possessed, would acquire a moral sense as soon as its intellectual powers had developed (by natural selection) to an extent comparable with those of man and was able to appreciate the survival value of collaboration and mutual affection. Borrowing from Marcus Aurelius the view that social instincts are the prime principles of man's moral constitution, Darwin concluded that these, with the aid of intellectual powers and the effects of new tradition and habits, would lead naturally to the Golden Rule: "As ye would that men should do to you, do ye to them likewise."

In this book Darwin added an essay on a related subject: sexual selection, the preferential chances of mating that some individuals of one sex (usually the male) have over their rivals because of special structures, colors, and types of behavior used in courtship displays, leading them to leave more offspring and accentuate their characters: for example, antlers in deer, trains in peacocks, feathers in birds of paradise, color in sticklebacks. There is no need to assume a power of aesthetic choice in either sex; the characters act as sexual stimulants, real aphrodisiacs, so that mating follows more quickly, occurs more often, and there are more progeny. In man, there can be no doubt that the differences between the sexes in respect of distribution of hair on face and body and of fat in local accumulations under the skin are the result of sexual selection.

Darwin did not explain how the preference of females for particular structures, colors, and courtship displays was related to natural selection, although female preferences themselves must be subject to natural selection. Perhaps Darwin's greatest contribution in this area was to show that secondary sexual characteristics had evolved in relation to a complex pattern of reproductive behavior, which must itself be the product of natural selection.

The last of Darwin's books in this series, *The Expression of the Emotions in Man and Animals* (1872), contains studies of facial muscles and means of expression in man and mammals, emission of sounds, erection of hair, and so forth, and their correlation with suffering, sobbing, anxiety, grief, despair, joy, love, devotion, reflection, meditation, sulking, hatred, anger, pride, disdain, shame, surprise, fear, horror, acceptance (affirmation), and rejection (negation). With this book Darwin founded the study of ethology (animal behavior) and conveyance of information (communication theory) and made a major contribution to psychology.

Botanical Works. From the first Darwin interested himself in adaptations of plants to cross-pollination. He found that trees (with countless flowers) tend to have flowers of one sex, while small plants have hermaphrodite flowers, and he showed that the unisexuality of trees would tend to prevent self-pollination. He concluded that "flowers are adapted to be crossed, at least occasionally, by pollen from a different plant." Here was a general principle which must have a wide meaning, and he set out to study it. His attention was first directed to orchids, which have elaborate adaptations for cross-pollination, making the bees that visit their flowers carry away the pollen sacs with them and pollinate the next flowers that they visit. His book on the *Fertilization of Orchids* (1862) showed that plants are in no way behind animals in the marvels of the adaptations that they show. Darwin further observed that flowers that are pollinated by wind have no colors; it is only those that are pollinated by insects that have bright colors in their petals and sweet-smelling nectaries.

Presently Darwin noticed that in some species flowers differ by the lengths of their anthers and styles, like the primroses, which show two conditions, or loosestrife, which shows three. This is also an adaptation for cross-pollination, and these observations formed the basis of *Different Forms of Flowers on Plants of the Same Species* (1877). The problem continued to fascinate him, and he raised two large beds of seedlings of *Linaria vulgaris,* the one cross-pollinated and the other self-pollinated, all from the same plant. "To my surprise, the crossed plants when fully grown were plainly taller and more vigorous than the self-fertilized ones." Darwin had experi-

mentally discovered and demonstrated the fact of hybrid vigor, or heterosis, which is completely explained by Mendelian genetics. These experiments, conducted over twelve years on fifty-seven species, led to the publication of *Effects of Cross and Self Fertilization* (1876). The demonstration of the advantage that accrues from cross-fertilization explains not only why sexual reproduction (as distinct from asexual budding) increases heritable variation (through recombination of genes), but also reveals the basis for the survival value conferred by the existence of different sexes in a species. This adaptation (for that is what it is) is a very old one, for it was inherited by plants and animals before they diverged from one another.

Darwin's interest in climbing plants was not that of a systematic botanist but an attempt to discover what adaptive value the habit of climbing has. He found that "climbing" is a result of the process of nutation; the apex of the plant's stem bends to one side while it grows and the plane of the bend itself revolves, clockwise or counterclockwise, so that the apex describes circular sweeping movements. In the hop plant—in hot weather, during daylight hours—it takes a little over two hours for each revolution. If the growing stem hits nothing, it continues to circle; if it hits an object it wraps itself around it by twining. Twining thus enables a young and feeble plant, in one season, to raise its growing point and leaves much higher from the ground, with more exposure to sunlight and air, without expending time and energy in the synthesis of woody supporting tissues. There is a further delicate adaptation here; a twining plant will not twine around an object larger than approximately six inches in diameter. This adaptation prevents it from climbing up a large tree, where it would be deprived of air and sun by the tree's own leaves.

Some plants climb by a modification of this method and have stalks and leaves that clasp other objects, and intermediate forms show that leaf-climbers evolved from twiners. In others, such as the Virginia creeper, the leaf tendrils end in little discs full of resinous fluid which anchor the plant to its support. These researches were described in *Climbing Plants* (1875).

The behavior of climbing plants and the bending of their shoots led Darwin to investigate the mechanical cause of such bending. The fact that a stem bends toward the light because it grows faster on the unilluminated than on the illuminated side of the stem had been known for some time. By means of simple but ingenious experiments, Darwin showed that the tip of the shoot was sensitive to light and that the bending was caused by growth of the stem on the side away from the light but some way down from the apex. Furthermore, this growth was due to a substance that comes down from the apex, "some matter in the upper part which is acted upon by light, and which transmits its effects to the lower part." From these researches and experiments (reported in *Power of Movement in Plants,* 1880) has sprung the whole science of growth hormones in plants.

Darwin's chance observation of the number of flies caught on the leaf of the common sundew (*Drosera rotundifolia*) was the starting point for a series of observations and experiments which showed not only how insects are caught, but how their bodies are digested and ingested, and what the significance of this carnivorous habit is for the life of the plant. By feeding one bed of sundew plants with meat and depriving another bed, he found that the fed plants had larger leaves, taller flowerstalks, and more numerous seed capsules. This remarkable adaptation, which insures survival of the plants, explains why sundew plants, which have very few roots, through which only small supplies of nitrogen can be obtained, can live on extremely poor soil. Darwin was particularly impressed by the fact that the living cells of plants possess a capacity for irritability and response similar to that of the nerve and muscle cells of animals. His work on *Insectivorous Plants* was published in 1875.

Darwin's last book connected with plants was *The Formation of Vegetable Mould Through the Action of Worms* (1881). This was published only six months before his death but covers a subject that he had studied for more than fifty years. He showed the services performed by earthworms in eating leaves and grinding earth in their gizzards and turning it into fertile soil, which they constantly sift and turn over down to a depth of twenty inches from the surface, thereby aerating it. He calculated from the weight of worm-castings that on one acre in one year's time eighteen tons of soil are brought up to the surface by worms. This was a pioneer study in quantitative ecology.

Darwin was elected to fellowship of the Royal Society in 1839 at the age of twenty-nine; three foreign universities awarded him honorary doctorates; and fifty-seven leading foreign learned societies elected him to honorary or corresponding membership—the French Academy of Sciences only in 1878, for his views have never been really appreciated in France. The Prussian government awarded him the highly coveted *Ordre pour le mérite,* but from the British sovereign and the British government he never received any recognition. His demonstration of the fact of evolution, and of the automatic mechanism of natural selection which causes it, was unpalatable to the orthodox views of the Church of England. When he died, however, twenty members of Parliament asked the dean of Westminster to allow his burial in Westminster Abbey, an honor which was readily granted. Darwin himself would have been highly amused, for, with his genial sense of humor he once said, "Considering how fiercely I have been attacked by the orthodox, it seems ludicrous that I once intended to be a clergyman." At his funeral there were present not only his friends Hooker, Huxley, and Wallace, but also James Russell Lowell, the American ambassador, as well as diplomats representing France, Germany, Italy, Spain, and Russia.

BIBLIOGRAPHY

I. Books. *Journal of Researches into the Geology and Natural History of the Various Countries Visited by H. M. S. Beagle* (London, 1839); facs. repr. (New York, 1952); 2nd ed., *Journal of Researches into the Natural History and Geology of the Countries Visited During the Voyage of H.M.S. Beagle* (London, 1845); repr., intro. by Sir Gavin de Beer (New York, 1956); *The Structure and Distribution of Coral Reefs* (London, 1842; 2nd ed., 1874); *Geological Observations on the Volcanic Islands Visited During the Voyage of H. M. S. Beagle* (London, 1844); *Geological Observations on South America* (London, 1846); *A Monograph of the Subclass Cirripedia,* 2 vols. (London, 1851, 1854); *A Monograph of the Fossil Lepadidae, or Pedunculated Cirripedes of Great Britain* (London, 1851); *A Monograph of the Fossil Balanidae and Verrucidae* (London, 1854); *On the Origin of Species by Means of Natural Selection, or the Preservation of Favoured Races in the Struggle for Life* (London, 1859; 2nd ed., 1860; 3rd ed., 1861; 4th ed., 1866; 5th ed., 1869; 6th ed., 1872); repr., intro. by Sir Gavin de Beer (London, 1963); 1st American ed. (New York, 1860); variorum text, Morse Peckham, ed. (Philadelphia, 1959); facs. repr. of 1st ed. (Cambridge, Mass., 1964); *On the Various Contrivances by Which British and Foreign Orchids are Fertilized by Insects, and on the Good Effects of Intercrossing* (London, 1862, 2nd ed., 1877); *The Variation of Animals and Plants Under Domestication* (London, 1868; 2nd ed., 1875); *The Descent of Man, and Selection in Relation to Sex* (London, 1871; 2nd ed., 1874); *The Expression of the Emotions in Man and Animals* (London, 1872); *Insectivorous Plants* (London, 1875); *Climbing Plants* (London, 1875); *The Effects of Cross and Self Fertilization in the Vegetable Kingdom* (London, 1876); *The Different Forms of Flowers on Plants of the Same Species* (London, 1877); *The Power of Movement in Plants,* assisted by Francis Darwin (London, 1880); *The Formation of Vegetable Mould, Through the Action of Worms, With Observations on Their Habits* (London, 1881); repr., intro. by Sir Albert Howard (London, 1961); and a preliminary notice in Ernst Krause, *Erasmus Darwin,* W. S. Dallas, trans. (London, 1879).

Posthumously published writings include *Life and Letters of Charles Darwin,* Francis Darwin, ed. (London, 1887); *Autobiography* (London, 1887), unexpurgated ed., Nora Barlow, ed. (London, 1958); *More Letters of Charles Darwin,* Francis Darwin and A. C. Seward, eds. (London, 1903); "Sketch" of 1842 and "Essay" of 1844, in *Evolution by Natural Selection,* foreword by Sir Gavin de Beer (Cambridge, 1958); and the notebooks on transmutation of species, Sir Gavin de Beer, ed., in *Bulletin of the British Museum (Natural History),* Historical Ser., **2** (1960), 23–200, and **3** (1967), 129–176.

II. BIBLIOGRAPHIES. *Handlist of the Darwin Papers at the University Library Cambridge* (Cambridge, 1960); R. B. Freeman, *The Works of Charles Darwin. An Annotated Bibliographical Handlist* (London, 1965).

III. SECONDARY LITERATURE. See Sir Gavin de Beer, "Darwin's Views on the Relations Between Embryology and Evolution," in *Journal of the Linnean Society of London,* **66** (1958), 15–23; "The Origins of Darwin's Ideas on Evolution and Natural Selection," in *Proceedings of the Royal Society,* **155B**, 321–338; "Mendel, Darwin, and Fisher," in *Notes and Records of the Royal Society,* **19** (1964), 192–226; **21** (1966), 64–71; *Atlas of Evolution* (London, 1964); and *Charles Darwin. A Scientific Biography* (New York, 1965).

Also of interest are W. E. Le Gros Clark, *Man-Apes or Ape-Men?* (New York, 1967); A. Cronquist, *The Evolution and Classification of Flowering Plants* (London, 1967); T. Dobzhansky, *Genetics and the Origin of Species* (New York, 1937; 4th ed., 1959); *Mankind Evolving* (New Haven, 1962); R. A. Fisher, *The Genetical Theory of Natural Selection* (Oxford, 1930); E. B. Ford, *Ecological Genetics* (New York, 1964); Bentley Glass, Owsei Temkin, and William R. Strauss, eds., *Forerunners of Darwin* (Baltimore, 1959); Gerald L. Geison, "Darwin and Heredity: the Evolution of His Hypothesis of Pangenesis," in *Journal of the History of Medicine,* **24** (1969), 375–411; Michael T. Ghiselin, *The Triumph of the Darwinian Method* (Berkeley, 1969); John C. Greene, *The Death of Adam. Evolution and Its Impact on Western Thought* (Ames, Iowa, 1959); J. B. S. Haldane, *The Causes of Evolution* (London, 1932); Garett Hardin, *Nature and Man's Fate* (London, 1960); Julian Huxley, A. C. Hardy, and E. B. Ford, eds., *Evolution as a Process* (London, 1954); Julian Huxley, *Evolution. The Modern Synthesis* (London, 1942; repr., 1963); Ernst Mayr, *Animal Species and Evolution* (Cambridge, Mass., 1963); Milton Millhauser, *Just Before Darwin. Robert Chambers and Vestiges* (Middletown, Conn., 1959); Bernard Rensch, *Evolution above the Species Level* (London, 1959); and Anne Roe and George Gaylord Simpson, eds., *Behavior and Evolution* (New Haven, 1958).

See also George Gaylord Simpson, *The Meaning of Evolution. A Study of the History of Life and of its Significance for Man* (London, 1950); *Horses. The Story of the Horse Family in the Modern World and Through Sixty Million Years of History* (New York, 1951); *The Major Features of Evolution* (New York, 1953); *This View of Life. The World of an Evolutionist* (New York, 1964); *The Geography of Evolution* (Philadelphia–New York, 1965); G. Ledyard

Stebbins, *Variation and Evolution in Plants* (New York, 1957); and Sol Tax, ed., *Evolution After Darwin,* 3 vols. (Chicago, 1960).

On Darwin's health see Saul Adler, "Darwin's Illness," in *Nature,* **184** (1959), 1102–1103.

GAVIN DE BEER

DARWIN, ERASMUS (*b.* Elston Hall, near Nottingham, England, 12 December 1731; *d.* Breadsall Priory, near Derby, England, 18 April 1802), *medicine, scientific poetry, botany, technology.*

Erasmus Darwin was the seventh child and fourth and youngest son of Robert Darwin, a retired barrister of independent means. He was educated at Chesterfield School from the age of nine, and from there, at the age of eighteen, he was awarded a Lord Exeter scholarship to St. John's College, Cambridge. At Cambridge Darwin studied classics, mathematics, and medicine, taking the M.B. in 1755. (Although his grandson, Charles Darwin, claimed that Erasmus Darwin took the B.A. at Cambridge in 1754, there is no record of this; nor is there record of his having taken the M.D., although it appears on the title pages of his books, his numerous Royal Society papers, and in many notices, but not on the monument erected by his widow.) He did, however, attend William Hunter's anatomy lectures in London while he was still at Cambridge, and then spent two years, 1754 and 1755, at the Edinburgh Medical School, returning to Cambridge to take the M.B.; later he spent several months in further study at Edinburgh.

In 1756 Darwin started medical practice at Nottingham, but attracted few patients and moved after two months to Lichfield. Shortly after his arrival there, Darwin was called to see a Mr. Inge, who had been pronounced incurable by another physician of the town. Under his care, the patient recovered, and from then on Darwin's fame spread.

As a physician, Darwin dealt with each case upon its merits; he reflected on the symptoms and then individualized treatment. He argued, like Hippocrates, by analogy, which led him to the notion of inoculation for measles, an experiment he tried upon two of his children with singular lack of success. He devised new treatments for illness and disability; recognized the importance of heredity in disease; held enlightened views on the treatment of the mentally afflicted; and, like many of his contemporaries, was interested in public health and urged better ventilation, better sewage disposal, and advised that graveyards be built outside the limits of the town. He was elected fellow of the Royal Society in 1761.

In 1757 Darwin married Mary Howard, with whom he lived happily until her death in 1770. There were three sons of this marriage—Charles, a brilliant medical student at Edinburgh, who died in 1778 at the age of twenty; Erasmus, a lawyer who committed suicide by drowning at the age of forty; and Robert Waring, who practiced medicine at Shrewsbury and was the father of Charles Robert Darwin. Following his wife's death, Darwin fathered two illegitimate daughters, known as the Misses Parker, who remained in his household. (In the 1790's Darwin set up a school for these daughters in Ashbourne, Derbyshire, and in 1797 he wrote for them *A Plan for the Conduct of Female Education in Boarding Schools,* which deals with many topics, not all of them strictly relevant to female education.)

In 1766 Darwin, with Matthew Boulton and Dr. William Small, founded the Lunar Society, which met in Birmingham for scientific discussion, and in the early 1770's founded the Lichfield Botanical Society. The work of the latter led to a bitter quarrel between Darwin and William Withering, a fellow physician and member of the Lunar Society whom Darwin had persuaded to settle in Birmingham. In 1787 the Lichfield Botanical Society's translation, in two volumes, of Linnaeus' *Genera* and *Mantissae Plantarum* appeared. Withering had published his *A Botanical Arrangement of all the Vegetables naturally growing in Great Britain* in 1775, and the Lichfield Linnaeus is filled with sly allusions to Withering (although never naming him) and his work, especially its English nomenclature, such as, "All the uncouth examples of names and terms above mentioned, amongst a multitude of others, equally objectionable . . . are taken from that compilation." It must be recorded that Withering's work has become more important with the passage of time, while the Lichfield text is no longer of any value.

It may be further noted that this was not Darwin's first quarrel with Withering; in 1780 Darwin had translated and published the dissertation and Gold Medal essay of the Aesculapian Society of his dead son Charles, appending a series of case notes on treatment of dropsy by digitalis and thereby apparently claiming priority for this treatment for himself and his son. Withering's classic work *An Account of the Foxglove* was published in 1785, and the preface is dated July of that year. Darwin had already read a paper on the subject to the Royal Society; Withering does not refer to Darwin's work in his book. Nor does Darwin refer to Withering in his appendix of 1780, although the first case reported was seen in consultation with him. Even if priority of publication is granted Darwin, there seems to be little doubt that it was Withering who demonstrated the value of digitalis and inaugurated systematic use

of the drug; in many respects, Darwin seems to have behaved reprehensibly in the course of the wrangle.

In 1777, Darwin bought eight acres of swamp one mile from Lichfield and with great mechanical and botanical skill turned it into a beautiful pleasure garden, comprising lakes and the herboretum that was to serve as his laboratory for the study of plant life. In 1779 he began the first of his major works, *The Botanic Garden,* a science poem with extensive notes.

In 1778, the year in which his son Charles died, Darwin first met Mrs. Elizabeth Chandos Pole, the wife of Colonel Chandos Pole of Radburn Hall, Derbyshire, and the illegitimate daughter of the second Earl of Portmore. Mrs. Pole brought her two sick children to Darwin's house, and remained with them there for a few weeks; Darwin cured them, became enamored of her, and sent her verses and gifts. Shortly thereafter, Mrs. Pole herself became ill; Darwin attended her, and upon her recovery addressed an "Ode to the River Derwent" to her. In 1780 Colonel Pole, many years his wife's senior, died; in 1781 Darwin and Mrs. Pole were married. They had seven children.

Elizabeth Darwin disliked Lichfield, so the Darwins moved to her earlier home, Radburn Hall, and Erasmus Darwin practiced medicine from Derby, four miles away. In 1783 they moved into Derby itself.

Since Derby was thirty-five miles from Birmingham, Darwin rarely attended meetings of the Lunar Society. In 1783, however, he founded the Derby Philosophical Society, a popular center for the discussion of science and applied technology. The Philosophical Society amassed a library for the use of its members, of which part is preserved in the Derby Museum and Library. Darwin further sought to establish a dispensary, and the manuscript relating to this project is preserved in his Commonplace Book, which also contains discussions of a variety of topics of current interest, some of his case notes—including a description of his own gout—and his ideas for and the sketches of his inventions. (Many of these inventions and plans are noteworthy in that they illustrate the protean character of Darwin's imagination; among them are plans for canal locks, a horizontal windmill which was used by Wedgwood to grind colors and flints until the advent of Watt's steam engine, a speaking machine, a telescopic candlestick, a stocking frame, a plan for a water closet of a modern type, a rocket motor that used hydrogen and oxygen as propellants, a mechanical bird that flapped its wings, copying machines, and many other ingenious devices.)

The first of Darwin's four major works, *The Botanic Garden,* an annotated scientific poem in Augustan couplets, appeared in two parts, of which the second, *The Loves of the Plants* (1789), was published before the first, *The Economy of Vegetation* (1791). Darwin decided to publish the second part of the work first because it was better suited "to entertain and charm." The first part of the work is more ambitious than the second, covering all natural philosophy, and embodying many of the researches and inventions of Wedgwood, Watt, Boulton, and others. The design of the totality was, Darwin wrote, "To enlist Imagination under the banner of Science . . . to induce the ingenious to cultivate the knowledge of botany . . . and recommending to their attention the immortal works of the celebrated Swedish naturalist—Linnaeus."

Darwin believed that prose was suited to abstract ideas, but chose to write poetry for its ability to conjure up visual images; he drew upon the Rosicrucian doctrine of Gnomes, Sylphs, Nymphs, and Salamanders, presiding over the four elements, in his personifications of all scientific, technological, and natural phenomena. Although William Cowper and other early critics greeted *The Botanic Garden* with praise, the instinct that led Darwin to anonymous publication was justified; in 1798 Canning (then Foreign Under-Secretary in Pitt's government), in collaboration with Hookham Frere and George Ellis, published a devastating parody of Darwin's work, entitled *The Loves of the Triangles,* and Darwin's reputation as a poet was ruined. One may speculate that Canning's attack was aimed not only at Darwin, but at his fellow members of the Lunar Society who were radical in thought, supporters of the French and American revolutions, and enemies of slavery—in fact, reviled as democrats and atheists.

Darwin published *Zoonomia,* in two large volumes in 1794 and 1796. This, his major treatise on medicine and natural science, embodies the work of twenty years. In it, Darwin sets himself a primary task of definition and explanation of the physiological and indeed psychological bases of, for example, the laws of animal causation, and of the four types—irritative, sensitive, voluntary, and associate—of fibrous motions on which he bases his classification of diseases. He follows Plato in the general concept of disease, but substitutes fibrous motions for the four humors, and regards all diseases as originating in "the exuberance, deficiency, or retrograde action of the faculties of the sensorium, as their proximate cause; and consist in the disordered motions of the fibres of the body, as the proximate effect of the exertions of those disordered faculties." Like his predecessors, Darwin's nomenclatures of "diseases" are, by and large, the Latin translations of signs and symptoms. His classification is indebted to Sauvages de la Croix and

William Cullen; he makes no reference to Morgagni.

The *Zoonomia* also contains a defense of the spontaneous generation of life from the scum of decomposing matter, and gives even clearer expression to the theory of biological evolution that Darwin had begun to develop in *The Botanic Garden.* In his recurrent discussions of evolution and natural selection Darwin rejects the theory of special creation as enunciated by Linnaeus and holds that species are variable and constantly changing. He believes the earth to be millions of years old; in *Zoonomia* he describes the changes occurring in warm-blooded animals in various conditions and asks:

> Would it be too bold to imagine, that all warm-blooded animals have arisen from one living filament . . . with the power of acquiring new parts, attended with new propensities, directed by irritations, sensations, volitions, and associations; and thus possessing the faculty of continuing to improve by its own inherent activity, and of delivering down those improvements to its posterity, world without end!

He extends the argument to include cold-blooded animals and vegetables and concludes, "Shall we conjecture that one and the same kind of living filament is and has been the cause of all organic life," and that "the whole is one family and one parent." As we have seen, Darwin further drew upon the notion of the inheritance of acquired characteristics, as did Lamarck. Both may have been inspired by Buffon; both emphasized evolution as an unconscious striving of the organism for survival and adjustment to environmental conditions, resulting in physical change or modification through the use or disuse of organs. Darwin may be credited, however, with the creation of the first consistent, all-embracing hypothesis of evolution. (It is interesting to note that although Charles Darwin wrote that while he had read *Zoonomia* as a young man, it had had no effect on him; the first draft of *On the Origin of Species* was, however, entitled *Zoonomia.*)

Darwin's third major work, *Phytologia* (1800), is primarily an agricultural treatise. Subtitled *The Philosophy of Agriculture and Gardening,* it illustrates the wide variety of his interests. It was published at the bequest of Sir John Sinclair, President of the London Board of Agriculture. Although the book deals with such practical matters as manures, draining and watering of land, plant diseases and how to cure them, and reform of the Linnaean system of classification, Darwin considered it to be essentially a continuation and application to plants of the discussion of biological evolution that he had begun in *Zoonomia.* Darwin here (as in *The Botanic Garden*) treats plants as inferior animals, endowed with lungs (leaves) and lacteals (because the roots absorb liquids which pass up the stems). Sinclair, who inspired the book, called it, "A most valuable performance . . . but on the whole, it is too philosophical a treatise to be calculated for general use." Nevertheless, a good deal of the less philosophical matter of *Phytologia* is of particular interest; in it Darwin not only recommended a lime-sulfur mixture as an insecticide, but also proposed ecological controls, suggesting that insect pests might be limited by increasing the numbers of their enemies—for example, the larvae of the aphidophorus (ichneumon) fly might be kept down by encouraging the breeding of the hedge birds, larks, and rooks which feed on them.

Darwin's last book, *The Temple of Nature* (published posthumously in 1803), is another long nature poem with copious notes. In it, Darwin clearly affirms his belief that the ancient myths—the Egyptian mysteries, the Greek Eleusinian mysteries, and the old pagan stories—embrace basic natural truths and can thus be united with the world of science. Darwin's extravagant theorizing does not mask his views as an enthusiastic apostle of progress and evolution, however. Indeed, *The Temple of Nature* may be taken as evidence of Darwin's wish to write another *Essay on Man.*

Darwin's enthusiasm for pagan myths and mysteries, his views on evolution, his liberal outlook, and his humanistic principles all led to his being labeled an atheist; the accusation lacks support, however. He is shown by *The Temple of Nature* to be a radical deist who believed no creed superior to another but who did not doubt the divine wisdom of creation or the moral wisdom of the Bible. Perhaps Irwin Primer's comment on Darwin's last book may sum it up: "As a scientific world view, his poem abounds prophetically and forebodingly with the difficulty of reconciling traditional faith in a rational cosmos with the empirical evidence of an expanding and evolving organic nature, a nature that aims at plenitude and seems remarkably careless of individuals."

In 1801, Darwin suffered a serious illness, the symptoms of which were consistent with myocardial infarction, after which he never recovered his energy and zest for life. Early in 1802 he moved to Breadsall Priory, a house bought by his son Erasmus shortly before his suicide, and there had a fatal seizure.

BIBLIOGRAPHY

I. ORIGINAL WORKS. Darwin's papers include "Remarks on the Opinion of Henry Eeles, Esq.; Concerning the Ascent of Vapour," in *Philosophical Transactions of the Royal Society,* **50** (1757), 240–254; "An Uncommon Case of Haemopysis," *ibid.,* **64** (1774), 344–349; "A New Case of Squinting," *ibid.,* **68** (1778), 86–96; "An Account of an Artificial Spring of Water," *ibid.,* **75** (1785), 1–7; "An Account of the successful use of Foxglove, in some Dropsies, and in Pulmonary Consumption," in *Medical Transactions of the College of Physicians,* **3** (1785), 255–286; and "Frigorific Experiments on the Mechanical Expansion of Air, explaining the Cause of the great Degree of Cold on the Summits of high Mountains, the sudden Condensation of aerial vapour, and the perpetual Mutability of atmospheric Heat," in *Philosophical Transactions of the Royal Society,* **78** (1788), 43–52.

Darwin's major works are *The Botanic Garden, Part II, containing The Loves of the Plants, a Poem with Philosophical Notes* (London, 1789); *The Botanic Garden, Part I, containing the Economy of Vegetation, a Poem with Philosophical Notes* (London, 1791); *Zoonomia; or the Laws of Organic Life,* 2 vols. (London, 1794–1796); *Phytologia, or the Philosophy of Agriculture and Gardening, with the theory of draining morasses, and with an improved construction of the drill plough* (London, 1800); and *The Temple of Nature; or the Origin of Society; A Poem with Philosophical Notes* (London, 1803); of which all except *Phytologia* appeared in several editions and were translated into several languages.

In addition Darwin edited his son Charles' *Experiments establishing a Criterion between Mucaginous and Purulent Matter. An account of the Retrograde Motions of the Absorbent Vessels of Animal Bodies in some Diseases* (Lichfield, 1780) and the Lichfield Botanical Society's *The System of Vegetables* (trans. from Linnaeus' *Systema Vegetabilium;* London, 1783) and *The Families of Plants* (trans. from Linnaeus' *Genera Plantarum;* London, 1787). His miscellaneous works include "The Natural History of Buxton and Matlock Waters," in James Pilkington, *A View of the Present State of Derbyshire* (Derby, 1789); *A Plan for the Conduct of Female Education of Boarding Schools* (London, 1797); *Private Families and Public Seminaries, to which is added, "Rudiments of Taste"* in a Series of Letters from a Mother to her Daughter (Philadelphia, 1798); and *Poetical Works* (London, 1806).

Darwin's Commonplace Book is kept in the Erasmus Darwin Room in Down House, Kent, which is now a Charles Darwin museum. Several portraits of Darwin are extant, of which the most notable are the earliest known portrait (*ca.* 1770) by Joseph Wright of Derby, now in the National Portrait Gallery, London; a later portrait by Wright in the Public Art Gallery in Derby; and the last portrait (*ca.* 1802) by James Rawlinson, in the Collection of the Derby Corporation but on loan to the Erasmus Darwin County Secondary School at Chaddersden.

II. SECONDARY LITERATURE. Works on Darwin include Lord Cohen of Birkenhead, "Erasmus Darwin," in *University of Birmingham Historical Journal,* **11** (1967), 17–40; John Dowson, *Erasmus Darwin, Philosopher, Poet and Physician* (London, 1861); Desmond King-Hele, *Erasmus Darwin,* which has extensive bibliographies (London, 1963); Ernst Krause, *Erasmus Darwin,* trans. from the German by

W. S. Dallas, and including a *"Life of Dr. Darwin"* by Charles Darwin (London, 1879); Hesketh Pearson, *Doctor Darwin* (London, 1930); and Anna Seward, *Memoirs of the Life of Dr. Darwin, Chiefly During his Residence at Lichfield* (London, 1804).

Cohen of Birkenhead

DARWIN, FRANCIS (*b.* Down, Kent, England, 16 August 1848; *d.* Cambridge, England, 19 September 1925), *botany.*

Francis Darwin was the third son of Charles and Emma Darwin. He was educated at home and then at Clapham Grammar School before going to Trinity College, Cambridge, in 1866. In 1869, he took a degree in mathematics, and the following year first-class honors in the natural science tripos. He entered St. George's Hospital, London, with the intention of becoming a physician, but his interests turned to research through the influence of Dr. Klein at the Brown Institute, where young Darwin went to learn histology. He presented a thesis and obtained the M.B. degree in 1875.

From 1874, except for a brief spell of research under Julius Sachs in Würzburg and then under H. A. de Bary in Strasbourg, he spent the next eight years working as secretary and botanical assistant to his father. He lived at Down with his parents after the death of his first wife, Amy, in 1876.

Apart from two short papers on the effects of humidity and light on the growth of roots, the results of his collaboration with his father were published in 1880 in *The Movement of Plants.*

In France and Germany, investigations into the causes of plant curvature in response to light were mainly directed toward finding a physicochemical explanation in terms of the direct effect of light on the curving tissues. Francis Darwin, however, working with Sachs, showed that roots grew more slowly in light than in dark.

The Darwins were strongly influenced by the experimental method that had come into biology in the second half of the nineteenth century; but, at the same time, their evolutionary approach left them uncommitted to finding direct relationships between cause and effect.

In *The Movement of Plants* they extended the idea of irregular circumnutatory movements to explain phototropic responses. They showed that geotropisms were similar to phototropisms and demonstrated traumotropisms: the curvature away from an apical wound. Further, they showed that the mechanism of curvature in both root and shoot was the result of differential growth rates. But their most important contribution was to show that there is some matter

in the apex of root and shoot which is acted on by light and gravity and which transmits its effects to other parts of the plant. The effect of the stimulus on curvature was indirect.

Francis Darwin later refined some of the experimental techniques. In 1899, for example, he devised a method of putting the tip of a root in a horizontal glass tube and found that the rest of the root grew in irregular twists in response to the continual signal from the apex to get vertical. He rejected the idea of circumnutation and showed, also in 1899, that gravity affects shoots by apical stimulus in the same way as roots. He took up the suggestion of Němec and Gottlieb Haberlandt that gravi-perception depended on the falling of starch grains. In 1903 he showed that the response was heightened if the root was kept vibrating and, in 1904, found that geotropic excitation occurred only if there was enough force to affect the starch grains.

Subsequent work by Fitting and by Boysen-Jensen proved the Darwins' findings, by showing that a chemical substance moved from the tip to other parts of the shoot and root; in 1928, F. W. Went isolated a growth-promoting substance and showed that light caused lateral transport of the substance from the light to the dark side, so opening up the new era of plant hormones.

After the death of his father, in 1882, Francis Darwin went to live in Cambridge, where, in 1883, he married Ellen Crofts, a lecturer in English literature at Newnham College. He became university lecturer in botany and a fellow of Christ's College and, in 1888, reader in botany. During this time his academic interests were the study of water movements in plants, the teaching of botany, and the editing of his father's letters.

Between 1886 and 1889 the stomatal openings of the undersides of leaves had been seen under the microscope. In 1886 Darwin published his first papers on the distribution of stomata and on the transpiration of plants. In 1897 he introduced the horn hygroscope for the study of stomatal function. A later invention for studying transpiration rate was the porometer, in which air was drawn through a leaf and its velocity measured. Both these simple devices remained in use for many years.

Wiesner had suggested that the effect of light on transpiration was direct, due to its absorption by the tissues and transformation into heat. Darwin again showed cause and effect to be indirect. In 1898 he proved that red rays induced the opening of the stomata and, assuming the intensity of transpiration to be proportional to the degree of opening of the stomata, suggested that it was the stomata that controlled the transpiration rate. In fact, the behavior of stomata has proved more complicated than Darwin realized.

His successful lectures to undergraduates were published in two books: *Practical Physiology of Plants,* written with E. H. Acton, was the first book of its kind in the English language; *The Elements of Botany* provided a stimulating new approach to the basic botanical principles.

Darwin collected and edited his father's letters for publication in 1887. The work was considered one of the best collections of its kind, because of its informative arrangement and tactful annotation, and for its chapter of reminiscences of his father's life. With the later two volumes published in 1903, they have provided the basis for all subsequent work on Charles Darwin.

Francis Darwin wrote several other biographical articles; on his brother George Darwin, Francis Galton, Thomas Hearne, and the botanists Joseph Hooker and Stephen Hales.

For his contributions to science, Francis Darwin received many honors. He became a Fellow of the Royal Society in 1882 and served the society on its council and as foreign secretary. Soon after the death of his second wife in 1903, he retired from all official appointments but continued his botanical research. He was made an honorary Sc.D. of Cambridge in 1909, at the time of the Darwin centenary, and, in 1912, received the Darwin medal of the Royal Society. He received honorary degrees from the universities of Dublin, Liverpool, Sheffield, St. Andrews, Uppsala, Prague, and Brussels. He was knighted in 1913.

In 1913 he married Florence Fisher, the widow of Professor F. W. Maitland, and in 1920 he was once again a widower. His later publications consisted mainly of collections of earlier essays and lectures on a variety of subjects, including musical instruments, which had been a lifelong interest.

BIBLIOGRAPHY

I. ORIGINAL WORKS. Francis Darwin's most important works on movement in plants include "On the Hygroscopic Mechanism by which Certain Seeds are Enabled to Bury Themselves in the Ground," in *Transactions of the Linnean Society of London,* **1** (1876), 149–167; "Über das Wachstum negativ heliotropischer Wurzeln im Licht und im Finstern," in *Arbeiten des botanischen Instituts in Würzburg,* **2** (1880), 521–528; *The Power of Movement in Plants* (London, 1880), written with Charles Darwin; "On Geotropism and the Localization of the Sensitive Region," in *Annals of Botany,* **13** (1899), 567–574; "The Statolith-theory of Geotropism," in *Proceedings of the Royal Society,* **71** (1903), 362–373;

"Notes on the Statolith Theory of Geotropism," in *Proceedings of the Royal Society,* **73** (1904), 477–490; *Lectures on the Physiology of Movement in Plants* (London, 1907).

His most important works on plant physiology are "On the Relation between the 'Bloom' on Leaves and the Distribution of Stomata," in *Journal of the Linnean Society of London,* **22** (1886), 99–116; "On the Transpiration Stream in Cut Branches," in *Proceedings of the Cambridge Philosophical Society,* **5** (1886), 330–367, written with R. W. Phillips; *Practical Physiology of Plants* (Cambridge, 1894), written with E. H. Acton; *The Elements of Botany* (Cambridge, 1895); "Observations on Stomata by a New Method," in *Proceedings of the Cambridge Philosophical Society,* **9** (1897), 303–308; "Observations on Stomata," in *Philosophical Transactions of the Royal Society,* **190B** (1898), 531–621; "On a New Method of Estimating the Aperture of Stomata," in *Proceedings of the Royal Society,* **84B** (1911), 136–154, written with D. F. M. Pertz; "On a Method of Studying Transpiration," in *Proceedings of the Royal Society,* **87B** (1914), 281–299; "On the Relation between Transpiration and Stomatal Aperture," in *Philosophical Transactions of the Royal Society,* **207B** (1916), 413–437.

His biographical works include *The Life and Letters of Charles Darwin,* 3 vols. (London, 1887), which he edited; *More Letters of Charles Darwin,* 2 vols. (London, 1903) edited with A. C. Seward; and *The Foundations of the Origin of Species* (Cambridge, 1909).

Among his many essays are: "Stephen Hales 1677–1761," in *The Makers of Modern Botany,* F. W. Oliver, ed. (Cambridge, 1913), pp. 65–83; "Memoir of Sir George Darwin," in *Scientific Papers of George Darwin* (Cambridge, 1916), pp. ix–xxxiii; biographical and other essays in *Rustic Sounds and Other Studies in Literature and Natural History* (London, 1917); *Springtime and Other Essays* (London, 1920).

II. SECONDARY LITERATURE. See A. C. Seward and F. F. Blackman, "Francis Darwin 1848–1925," in *Proceedings of the Royal Society,* **110B** (1932), i–xxi; W. C. D. Dampier, "Sir Francis Darwin (1848–1925)," in *Dictionary of National Biography: Twentieth Century 1922–1930,* J. R. H. Weaver, ed. (Oxford, 1937), 237–238; P. R. Bell, "The Movement of Plants in Response to Light," in *Darwin's Biological Work,* P. R. Bell, ed. (Cambridge, 1959); H. Meidner and T. A. Mansfield, *Physiology of Stomata* (London, 1968).

WILMA GEORGE

DARWIN, GEORGE HOWARD (*b.* Down House, Kent, England, 9 July 1845; *d.* Cambridge, England, 7 December 1912), *mathematics, astronomy.*

Darwin was the fifth child of Charles Robert Darwin and Emma Wedgwood. His greatgrandfather was Erasmus Darwin; his middle name commemorates Erasmus Darwin's first wife, Mary Howard. The Darwin family was comfortably settled in Kent, where George Howard Darwin began his education at the private school of the Reverend Charles

Pritchard (who was afterward Savillian professor of astronomy at Oxford). Darwin went on to attend Trinity College, Cambridge, from which he graduated second wrangler and Smith's prizeman in 1868. Darwin did not immediately embrace a scientific career, but rather studied law for six years and was admitted to the bar in 1874, although he never practiced that profession.

Darwin was elected fellow of Trinity College in October 1868, but did not return there for good until October 1873. There, in 1875, he began the series of mathematical papers that were eventually to form the four large volumes of his *Scientific Papers*. In 1879 he was elected fellow of the Royal Society and in 1883 he was elected Plumian professor of astronomy and experimental philosophy at Cambridge, to succeed James Challis.

Darwin held the Plumian professorship for the rest of his life. The chair bore no necessary connection with the observatory, and practical astronomy formed no part of his duties. His lectures on theoretical astronomy were poorly attended, but among his students were Ernest W. Brown and Sir James Jeans. During this tenure, Darwin received several honors and distinctions, including, in 1905, a knighthood (knight commander of the Bath) through the offices of his college friend Arthur Balfour. In 1912 he served as president of the International Congress of Mathematicians at Cambridge. This was his last public function; he died of cancer shortly thereafter, and was buried in Trumpington Cemetery, near Cambridge. He was survived by his widow, the former Maud Du Puy of Philadelphia, whom he had married in 1884, and four children, of whom the eldest, Charles Galton Darwin, was also a scientist.

Darwin's paper "On the Influence of Geological Changes on the Earth's Axis of Rotation," published in 1876, marked the beginning of his investigations of essentially geophysical problems. This work was directly inspired by Lord Kelvin, whose great interest in the young Darwin may be said to have been the chief influence in his decision to make science his career. Another group of papers, dated from 1879 to 1880, are concerned with the tides in viscous spheroids, and still show the influence of both Kelvin and Laplace, although their scope is more general. In his paper of this series, "On the Precession of a Viscous Spheroid and on the Remote History of the Earth" (1879), Darwin proposed the "resonance theory" of the origin of the moon, according to which the moon might have originated from the fission of a parent earth as the result of an instability produced by resonant solar tides. His monumental paper "On the Secular Changes in the Elements of the Orbit of a

Satellite Revolving About a Tidally Distorted Planet" was published in 1880.

Following his accession to the Plumian chair Darwin delved even more deeply into the problems of the origin and evolution of the solar system, making numerous investigations of the figures of equilibrium of rotating masses of fluid and, later, making extensive studies of periodic orbits in the restricted problem of three bodies, carried out with special reference to cases obtaining for the particular values of the mass ratio of the two finite bodies of 1:10 and 1:1048 (the latter approximating the mass ratio of Jupiter to that of the sun).

Darwin's most significant contribution to the history of science lies in his pioneering work in the application of detailed dynamical analysis to cosmological and geological problems. That many of his conclusions are now out of date should in no way diminish the historical interest of his experiments, nor the important service that he rendered cosmogony by the example he gave of putting various hypotheses to the test of actual calculations. Darwin's method remains a milestone in the development of cosmogony, and subsequent investigators have favored it over the merely qualitative arguments prevalent until that time.

That Darwin's scientific work is homogeneous is apparent from glancing at the titles of the more than eighty papers collected in the four volumes of his *Scientific Works*. After publishing some short notes on a variety of subjects, he devoted himself steadfastly to the problems of mathematical cosmogony, departing from them only to undertake problems of pressing practical concern (as, for example, in his work on oceanic tides). The greatest part of his work is devoted to the explanation of the various aspects of the history of the double stars, the planetary system, and satellite systems. His papers on viscous spheroids (including those on tidal friction), on rotating homogeneous masses of fluids, and even those on periodic orbits are means to this end.

Darwin's work is further marked by the virtually complete absence of investigations undertaken out of sheer mathematical interest, rather than in the elucidation of some specific problem in physics. Indeed, he was an applied mathematician of the school of Kelvin or Stokes and was content to study physical phenomena by the mathematical methods most convenient to the purpose, regardless of their novelty or elegance. Should the problem fail to yield to analysis, Darwin resorted to computation, never hesitating to embark upon onerous and painstaking numerical work (such as marks his investigations of the stability of pear-shaped figures of equilibrium or of periodic orbits in the restricted problem of three bodies). Indeed, it would seem that he actually preferred quantitative rather than qualitative results, although he seldom carried his calculations beyond pragmatic limits. That this approach was sometimes too blunt is illustrated by Darwin's 1902 investigation of the stability of a rotating pear-shaped figure, which he found to be stable; shortly after publication of Darwin's results, Aleksandr Liapunov announced his proof of the instability of the pear-shaped figure, and several years later Darwin's pupil Jeans showed that Liapunov was indeed correct.

In his speech to the Fifth International Congress of Mathematicians at Cambridge in 1912, Darwin summed up his method, speaking of his work on the problem of three bodies and comparing his technique to that of Poincaré, to whom he had often paid tribute:

> My own work . . . cannot be said to involve any such skill at all, unless you describe as skill the procedure of a housebreaker who blows in a safe door with dynamite instead of picking the lock. It is thus by brutal force that this tantalising problem has been compelled to give up a few of its secrets; and, great as has been the labour involved, I think it has been worth while. . . . To put at their lowest the claims of this clumsy method, which may almost excite the derision of the pure mathematician, it has served to throw light on the celebrated generalisations of Hill and Poincaré.
>
> I appeal, then, for mercy to the applied mathematician, and would ask you to consider in a kindly spirit the difficulties under which he labours. If our methods are often wanting in elegance and do but little to satisfy that aesthetic sense of which I spoke before, yet they constitute honest attempts to unravel the secrets of the universe in which we live.

BIBLIOGRAPHY

I. ORIGINAL WORKS. Darwin's books are *The Tides and Kindred Phenomena in the Solar System* (London, 1898), based on a series of popular lectures delivered in Boston in 1897, and *Scientific Papers* (London, 1907–1916). Among his most important articles are "On the Influence of Geological Changes on the Earth's Axis of Rotation," in *Philosophical Transactions of the Royal Society,* **167A** (1876) 271–312: "On the Precession of a Viscous Spheroid and on the Remote History of the Earth," *ibid.,* **170A** (1879), 447–538; "On the Secular Changes in the Elements of the Orbit of a Satellite Revolving About a Tidally Distorted Planet," *ibid.,* **171A** (1880), 713–891; and "On a Pear-Shaped Figure of Equilibrium," *ibid.,* **200A** (1902), 251–314. Darwin's presidential address presenting the medal to Poincaré is in *Monthly Notices of the Royal Astronomical Society,* **60** (1900), 406–415.

II. SECONDARY LITERATURE. A memoir of Darwin's life by his brother, Sir Francis Darwin, is affixed to vol. V of his *Scientific Papers* (London, 1916).

Obituaries include those by F. J. M. Stratton in *Monthly Notices of the Royal Astronomical Society,* **73** (1913), 204–210; and S. S. Hough, in *Proceedings of the Royal Society,* **89A** (1914), i–xiii.

ZDENĚK KOPAL

DAŚABALA, astronomy.

Daśabala, the son of Vairocana and member of a family from Vallabhī in Saurāṣṭra, was a Buddhist with Śaivite leanings. He wrote two works on astronomy, both of which belong to the *Brāhmapakṣa* (see Essay IV), as modified by the *Rājamṛgāṅka* (written 1042) of his contemporary, the Paramāra monarch Bhojarāja (*fl. ca.* 999–1056).

The *Cintāmaṇisāraṇikā* was written in Śaka 977 (A.D. 1055), while Bhoja was still in power. It contains six sections:

1. On *tithis* (sixty-two verses)
2. On *nakṣatras* (nineteen verses)
3. On *yogas* (twenty-one verses)
4. On diverse subjects (thirty-six verses)
5. On *saṅkrāntis* (four verses)
6. On the sixty-year cycle of Jupiter (sixteen verses).

There is a commentary on it written by Mahādeva, the son of Lūṇiga, also a Gujarātī, in Śaka 1180 (A.D. 1258).

Daśabala's second work, the *Karaṇakamalamārtaṇḍa,* was written in Śaka 980 (A.D. 1058). It contains ten chapters with 270 verses:

1. On mean motions
2. On true longitudes
3. On the three questions relating to the diurnal motion
4. On lunar eclipses
5. On solar eclipses
6. On heliacal risings and settings
7. On the lunar crescent
8. On the *mahāpātas*
9. On conjunctions of the planets
10. On intercalary months and the sixty-year cycle of Jupiter.

BIBLIOGRAPHY

The *Cintāmaṇisāraṇikā* has been edited by D. D. Kosambi in *Journal of Oriental Research,* **19** (1952), supp. The *Karaṇakamalamārtaṇḍa* is known only from the description in Ś. B. Dīkṣita, *Bhāratīya Jyotiḥśāstra* (Poona, 1896; repr., 1931), pp. 239–240.

DAVID PINGREE

DASYPODIUS, CUNRADUS (*b.* Frauenfeld, Thurgau, Switzerland, *ca.* 1530; *d.* Strasbourg, France, 26 April 1600), *mathematics, astronomy.*

Cunradus was the son of the Swiss humanist Petrus Dasypodius; his family name was Rauchfuss (roughfoot, hare). He studied in Strasbourg at the famous academy of Johannes Sturm and became professor there in 1558.

The greater part of Dasypodius' work was destined to be schoolbooks for his students: his editions of Euclid, his *Volumen primum* and *Volumen secundum,* and his *Protheoria* with the *Institutionum mathematicarum erotemata* (in the form of questions and answers) demonstrate his pedagogical interests. These books show that in Strasbourg, under the influence of Sturm, mathematics was studied far more extensively than in many of the universities of the time. Worthy of special mention is his *Analyseis geometricae* (1566). This book, written with his teacher Christian Herlinus, contains the proofs of the first six books of Euclid's *Elements* analyzed as their syllogisms; it was intended to facilitate the study of mathematics for students trained in dialectics.

Confident that the mathematics of his time was far below the Greek level, Dasypodius desired, as did many of his contemporaries (e.g., Commandino and Ramus), publication of all Greek mathematical works. Since he himself owned several manuscripts, he was able to make a beginning in that direction. He edited and translated works of Euclid (partly with and partly without proofs), some fragments of Hero, and (in his *Sphaericae doctrinae propositiones*) the propositions of the works of Theodosius of Bythinia, Autolycus of Pitane, and Barlaamo. His textbooks, too, show his knowledge of Greek mathematics.

Dasypodius' fame is based especially on his construction of an ingenious and accurate astronomical clock in the cathedral of Strasbourg, installed between 1571 and 1574. From his description of this clock it is clear that Dasypodius was influenced by Hero in many details.

BIBLIOGRAPHY

I. ORIGINAL WORKS. See J. G. L. Blumhoff, *Vom alten Mathematiker Dasypodius* (Göttingen, 1796). The following is a short-title bibliography, with location of copies and some additions to Blumhoff (L = Leiden, Univ.; B = Basel, Univ.; P = Paris, Bibl. Nat.). Place of publication is Strasbourg, unless stated otherwise. Dasypodius edited, with a Latin translation, the following works of Euclid: *Catoptrica* (1557, B); *Elementorum primum* (1564, B); *Elementorum II* (1564, B); *Propositiones reliquorum librorum* (1564, B); *Elementorum primum* and *Heronis vocabula geometrica* (1570, B; repr. 1571, B); *Propositiones*

Elementorum 15, Opticorum (1570, B); and *Omnium librorum propositiones* (1571, B). He was also responsible for editions of Caspar Peucer, *Hypotyposes orbium coelestium* (1568, L); *Sphaericae doctrinae propositiones* (1572, P); and *Isaaci Monachi scholia in Euclid* (1579, L). With Christian Herlinus he published *Analyseis geometricae sex librorum Euclidis* (1566, L). He also wrote *Volumen primum* (1567, B), partly repr. in an ed. of M. Psellus, *Compendium mathematicum, aliaque tractatus* (Leiden, 1647); *Volumen II. Mathematicum* (1570, L); *Lexicon* (1573, L); *Scholia in libros apotelesmaticos Cl. Ptolemaei* (Basel, 1578, L); *Brevis doctrina de cometis* (1578, P), also a German ed.; *Oratio de disciplinis mathematicis* and *Heron* and *Lexicon* (1579, P); *Heron mechanicus* and *Horologii astronomici descriptio* (1580, L); *Wahrhafftige Auslegung und Beschreybung des astronomischen Uhrwercks zu Straszburg* (1580, B); *Protheoria mathematica* (1593, B); *Institutionum mathematicarum erotemata* (1593, B); and *Erotematum appendix* (1596, B).

II. SECONDARY LITERATURE. On Dasypodius or his work, see A. G. Kästner, *Geschichte der Mathematik,* I (Göttingen, 1796), 332–345; Wilhelm Schmidt, "Heron von Alexandria, Konrad Dasypodius und die Strassburger astronomische Münsteruhr," in *Abhandlungen zur Geschichte der Mathematik,* **8** (1898), 175–194; and E. Zinner, *Entstehung und Ausbreitung der Coppernicanischen Lehre* (Erlangen, 1943), p. 273.

J. J. VERDONK

DAUBENTON, LOUIS JEAN-MARIE (*b.* Montbard, Côte-d'Or, France, 29 May 1716; *d.* Paris, France, 1 January 1800), *medicine, natural history.*

For a detailed study of his life and work, see Supplement.

DAUBENY, CHARLES GILES BRIDLE (*b.* Stratton, Gloucestershire, England, 11 February 1795; *d.* Oxford, England, 13 December 1867), *chemistry, geology.*

The younger son of Anglican clergyman James Daubeny and of Helena Daubeny, Charles followed the usual classical curriculum at Winchester College and Oxford (B.A. 1814, Chancellor's Latin Essay Prize 1815). A desire to practice medicine led him to attend the university chemical lectures and to meet such Oxford men of science as William Buckland and John and William Conybeare. From 1815 to 1818 he studied medicine in Edinburgh and also attended Robert Jameson's geological lectures. A tour through France preceded his return to a Lay Fellowship at Magdalen, where he graduated M.D. in 1821. Daubeny was active in medical practice until 1829, but his true interests found expression in his 1822 election to the Oxford chemistry professorship. In 1834 he added the chair of botany, and in 1840 that of rural economy. Elected a fellow of the Royal Society in 1822, he was also a member of the Royal Irish Academy and a foreign associate of the Munich Academy of Science. He never married.

Daubeny carried out important research in chemistry, geology, and botany. He also worked vigorously in the British Association for the Advancement of Science, and in the 1840's and 1850's he campaigned to end Oxford's neglect of science. His career well illustrates the growing nineteenth-century involvement of Anglican clerics and the ancient universities in science, and the way that the biological and earth sciences were at the focus of attention.

While in Edinburgh, Daubeny was actively engaged in geological debate. His French tour led first to the important letter "On the Volcanoes of the Auvergnes," then to his masterly *Description of Active and Extinct Volcanoes* (1826). In the latter work he developed a chemical theory of volcanic action, which stated that such action results from penetration of water to the free alkali and alkaline earth metals supposed to exist beneath the earth's crust. A similar theory had been entertained by Humphry Davy and Joseph Louis Gay-Lussac, but Daubeny was the first to develop it in detail and support it with massive factual evidence. In its dependence on chemical ideas the *Description* is typical of all of Daubeny's work. It is no accident that the other book he is also remembered for is a warm defense of the chemical atomic theory, nor that his later botanical work (on the effect of soil condition on vegetation, and of light on plant life) was directly inspired by chemistry.

Daubeny was present at the 1831 founding meeting of the British Association and was instrumental in ensuring the success of its Oxford meeting the next year. Actively involved in arranging the 1836 Bristol meeting, he was vice-president at Oxford in 1847 and president at Cheltenham in 1856. His work for the reform of Oxford science teaching included not only many speeches, pamphlets, and articles but also the practical steps of rearranging and extending the university's botanical garden and building a laboratory for Magdalen at his own expense. His will presented his considerable collection of scientific books, instruments, and specimens to the college, with an endowment for their upkeep.

BIBLIOGRAPHY

I. ORIGINAL WORKS. Daubeny's major works are *Description of Active and Extinct Volcanoes* (London, 1826; 2nd ed., 1848); *An Introduction to the Atomic Theory* (London, 1831; supp., 1840; 2nd ed., 1850); and the literary and scientific essays collected as *Miscellanies*, 2 vols. (London, 1867).

II. SECONDARY LITERATURE. A chronology of Daubeny's life, a full bibliography, and much incidental information may be found in R. T. Günther, *History of the Daubeny Laboratory* (London, 1904). Useful obituaries are in *Proceedings of the Royal Society*, **17** (1868–1869), lxxiv–lxxx; and *Report and Transactions of the Devonshire Association for the Advancement of Science,* **2** (1867–1868), 303–308.

ARNOLD THACKRAY

DAUBRÉE, GABRIEL-AUGUSTE (*b.* Metz, France, 25 June 1814; *d.* Paris, France, 29 May 1896), *geology.*

In 1834 Daubrée entered the University of Strasbourg. His first interest was in tin mining; to study this he visited Norway, Sweden, and England. His membership in an official mission to Cornwall in 1837 resulted in his contribution to Dufrenoy and Élie de Beaumont's *Voyage métallurgique en Angleterre* (1839). His interest in mineralization was shown in the thesis he presented to the Faculty of Sciences in Paris for the doctorate in science: *Thèse sur les températures du globe terrestre . . ..*

Daubrée became engineer for the Department of Bas-Rhin in 1840 and spent the next eight years preparing a geological map of the region, publication of which was followed by a long memoir on the department. He was summoned to Strasbourg to take the chair of mineralogy and geology, and there he established an important experimental laboratory for the study of mineralogical and geological processes.

The pattern of Daubrée's lifework was now established, and he is remembered for his contributions to the understanding of geochemical processes, the application of engineering principles to an understanding of geological structures and mineralization patterns, and the economic exploitation of ore bodies. The first of these interests resulted in *Études et expériences synthétiques . . .* (1859) and the third in his reports for the expositions of London (1862) and of Paris (1867).

In 1861, immediately following an important consultancy on the exploitation of mineral resources in Luxembourg, Daubrée was appointed professor-administrator at the Museum of Natural History in Paris and a member of the Academy of Sciences, where he continued to be an influential researcher and teacher. His interest in experimental geology was reflected in *Rapport sur les progrès de la géologie expérimentale* (1867), which was followed by his work of most lasting significance, *Études synthétiques de géologie expérimentale* (1879). Many of the experiments described in the latter are still referred to, but ironically it was his mechanical rather than his chemical experiments that most influenced later work. In particular, the production of joint patterns associated with folding and torsion proved a very valuable stimulus to the experimental studies of such geologists as Bailey Willis and Ernst and Hans Cloos.

From 1872 until his retirement in 1884 Daubrée was director of the School of Mines in Paris and from 1875 a member of the influential commission charged with the production of the national geological map. His last years, which were dogged by ill health, saw the fruition of two important aspects of his lifework: his studies of meteorites and of the chemical action of underground water on limestone. Daubrée had built up a large collection of meteorites, and his *Météorites et la constitution géologique du globe* (1886) contained significant inferences on the constitution of other extraterrestrial bodies. This work was immediately followed by two works on the action of underground water: *Les eaux souterraines à l'époque actuelle* (1887) and *Les eaux souterraines aux époques anciennes* (1887). His last important work, *Les régions invisibles du globe . . .* (1888), united many of his interests.

At the time of his death Daubrée was a grand officer of the Legion of Honor, the holder of three honorary degrees, and a member of eighty-six learned societies.

BIBLIOGRAPHY

Among Daubrée's writings are his contribution to Dufrenoy and Élie de Beaumont's *Voyage métallurgique en Angleterre,* II (Paris, 1839), 203–257; *Thèse sur les températures du globe terrestre, et sur les principaux phénomènes géologiques qui paraissent être en rapport avec la chaleur propre de la terre; Carte géologique du département du Bas-Rhin, 1:80,000* (Paris, 1849); *Description géologique et minéralogique du département du Bas-Rhin* (Strasbourg, 1852); *Études et expériences synthétiques sur le métamorphisme et sur la formation des roches cristallines* (Paris, 1859); *Produits des mines, des carrières et des usines métallurgiques* (Paris, 1862); *Rapport sur les progrès de la géologie expérimentale* (Paris, 1867); *Substances minérales,* 2nd ed. (Paris, 1868); *Études synthétiques de géologie expérimentale* (Paris, 1879); *Les météorites et la constitution géologique du globe* (Paris, 1886); *Les eaux souterraines à l'époque actuelle, leur régime, leur température, leur composition au point de vue du rôle qui leur revient dans l'économie de l'écorce terrestre* (Paris, 1887); *Les eaux souterraines aux époques anciennes, rôle qui leur revient dans l'origine et les modifications de la substance de l'écorce terrestre* (Paris, 1887); and *Les régions invisibles du globe et des espaces célestes, eaux souterraines, tremblements de terre, météorites* (Paris, 1888).

R. J. CHORLEY

DAVAINE, CASIMIR JOSEPH (*b.* St.-Amand-les-Eaux, France, 19 March 1812; *d.* Garches, France, 14 October 1882), *medicine, biology.*

Davaine was the sixth child of Benjamin-Joseph Davaine, a distiller, and of Catherine Vanautrève. He began his studies at the parochial *collège* at St.-Amand-les-Eaux, entered the *collège* of Tournai (now in Belgium) in 1828, and finished his studies at Lille. At the end of 1830 Davaine started his medical courses in Paris, and in 1834 he competed for a hospital externship. On 1 January 1835 he became extern under Pierre Rayer at La Charité. In December 1837 he presented his doctoral thesis, on the hematocele of the tunica vaginalis, to a committee with Alfred Velpeau as its chairman.

From 1838, Davaine practiced medicine in Paris while carrying on important microbiological, parasitological, pathological, and general biological researches under Rayer. On 23 January 1869 he married an Englishwoman, Maria Georgina Forbes, by whom he had one son, Jules.

Davaine's most important contribution to science was in medical microbiology. As early as 1850 he and Rayer observed small rods, which he later called *bactéridies,* in the blood of a sheep suffering from anthrax. He did not immediately understand the significance of this observation; but from 1863 on, under the influence of Pasteur's work on butyric fermentation, he demonstrated in a series of publications remarkable for their logic and method that the *bactéridie* (*Bacillus anthracis*) is the sole cause of anthrax. Among his findings were the following:

(1) Rabbits and guinea pigs inoculated with blood taken from an animal infected with anthrax always show great numbers of bacilli in their blood, which can be used to inoculate, and thus infect, other animals.

(2) Blood containing anthrax bacilli, when putrefied or heated, no longer transmits the disease because the bacilli have been killed; the same blood, when simply dried, remains infectious.

(3) When the dried blood is mixed with water, the bacilli fall to the bottom of the container. A drop taken from the surface of the liquid will not transmit the disease, but one taken from the bottom of the container will infect the experimental subject.

(4) The blood of a fetus in a female guinea pig suffering from anthrax is not infectious, because the placenta acts as a filter.

(5) The infective power of blood containing anthrax bacilli is very great: a millionth of a drop can still kill a guinea pig.

(6) The period of incubation for the disease corresponds to the time necessary for the bacteria to multiply.

(7) Birds are resistant to anthrax.

(8) Certain types of stinging insects (Diptera) contribute to spreading the disease.

(9) The "malignant pustule" that afflicts man is of anthracic origin, for it contains the same bacteria. Davaine was able to reproduce it experimentally in guinea pigs.

(10) Various chemical substances, such as iodine, can cure anthrax by destroying the bacilli.

(11) Crushed leaves of the walnut (*Juglans regia*) have an antibacterial action. (Today it is known that this plant contains a potent antibiotic substance against the anthrax bacillus.)

During his research on anthrax Davaine distinguished another disease, bovine septicemia, but did not isolate the microbe (1865). In 1869 he stated that (1) the microbes of septicemia are motile, while anthrax bacilli are not; (2) putrefied septicemic blood is no longer virulent, while anthracic blood always is; (3) in septicemia there is neither agglutination of the red blood corpuscles nor any splenomegaly, while anthrax always produces such symptoms.

For all his experimental inoculations Davaine used the recently invented Pravaz syringe rather than the lancet, which presented many inconveniences.

Davaine's contributions to medical and veterinary microbiology were fundamental, for he was the first to recognize the pathogenic role of bacteria. He was not able, however, to elucidate the exact mode of transmission of anthrax because he was unaware that the bacillus had a resistant stage, the spore, that enabled it to survive and to recur in a contaminated region. This stage of the bacillus was described in 1876 by Robert Koch; and later (1877–1881) Pasteur, Émile Roux, and Chamberland definitively proved the role of the bacillus in the etiology of anthrax. Davaine, however, was one of the first medical microbiologists to recognize the role of the bacillus and to differentiate it from bovine septicemia.

Davaine's many disputes at the Paris Academy of Sciences and the Academy of Medicine with the enemies of the germ theory of disease—Leplat, Jaillard, Henri Bouley, André Sanson, Louis Béhier, Alfred Vulpian, and particularly Gabriel Colin—foreshadowed Pasteur's conflicts a few years later. As a matter of fact, Pasteur had a high opinion of Davaine's work and wrote in 1879: "I congratulate myself on having often carried on your clever researches."

Davaine's other scientific contributions had to do with internal parasites of man and domestic animals. His important work on this subject, *Traité des entozoaires,* ran to two editions. As early as 1857 Davaine thought of tracking down intestinal worms by seeking their eggs in the stools, a procedure still followed. Experimentally he specified the mode of development of the Ascaridae (*Ascaris lumbricoides*) and of the Trichocephalus (*Trichuris trichiura*).

Among plant parasites Davaine studied, from 1854 to 1856, the cycle of the wheat worm (*Anguina tritici*) and suggested means of combating this nematode. He was also interested in the mold that causes fruit rot (1866). Thus, he should be considered a pioneer in the study of plant pathology.

Davaine also studied the amoebic movements of the leukocytes, which he noticed as early as 1850. In 1869 he demonstrated that these cells can absorb foreign bodies introduced into the blood and thus observed phagocytosis fourteen years before Élie Metchnikoff did (1883). He was the first to recognize (1852) the protandrous hermaphroditism of oysters. Several of his observations of animal teratology were written up in his "Mémoire sur les anomalies de l'oeuf (1860) and in his article "Monstres, Monstruosités" (1875) for the *Dictionnaire Dechambre*. Davaine's interests further extended to anabiosis among such invertebrates as Protozoa, Nematoda, and Tardigrada (1856); the palatine organ of the Cyprinidae (1850); the thyrohyoid bone of the anoura (1849); and the color mechanism of the tree frog (1849).

In medicine, besides his publications on anthrax and septicemia, Davaine made numerous contributions—some in collaboration with Claude Bernard, Pierre Rayer, or A. Laboulbène—on anatomic-pathological lesions observed in various animals. He also published the important "Mémoire sur la paralysie générale ou partielle des deux nerfs de la septième paire" (1852).

All this research was carried on while Davaine practiced medicine, for he never had a laboratory of his own nor held an official university position. Noted for his humility and modesty, he did not seek honors; the only two he received were the cross of a *chevalier* of the Legion of Honor (1858) and membership in the Academy of Medicine (1868). Under the Second Empire he also was made a *Médecin par quartier de l'Empereur,* a purely honorary title.

During the Franco-Prussian War, while serving in the ambulance corps, Davaine wrote a short philosophical book, *Les éléments du bonheur* (1871). This book summed up in a rather simplistic fashion his inner serenity and his faith in man. His last years were spent at Garches, where his property is now the Fondation Davaine.

BIBLIOGRAPHY

I. ORIGINAL WORKS. Davaine's writings include "Recherches sur les globules blancs du sang," in *Comptes rendus des séances de la Société de biologie,* **2** (1850), 103–105; "Mémoire sur la paralysie générale ou partielle des deux nerfs de la septième paire," in *Mémoires de la*

Société de biologie. **4** (1852), 137–191; "Recherches sur la génération des huîtres," *ibid.,* 297–339; "Recherches sur l'anguillule du blé niellé considérée au point de vue de l'histoire naturelle et de l'agriculture," *ibid.,* 2nd ser., **3** (1856), 201–271; "Sur le diagnostic de la présence des vers dans l'intestin par l'inspection microscopique des matières expulsées," in *Comptes rendus des séances de la Société de biologie,* 2nd ser., **4** (1857), 188–189; *Traité des entozoaires et des maladies vermineuses de l'homme et des animaux domestiques* (Paris, 1860; 2nd ed., enl., 1877), part repr., ed. by W. A. Smith as *On Human Entozoa* (London, 1863); "Mémoire sur les anomalies de l'oeuf," in *Mémoires de la Société de biologie,* 3rd ser., **2** (1860), 183–263; "Nouvelles recherches sur le développement et la propagation de l'ascaride lombricoïde et du trichocéphale de l'homme," *ibid.,* **4** (1862), 261–265; "Recherches sur les infusoires du sang dans la maladie connue sous le nom de sang de rate," in *Comptes rendus hebdomadaires des séances de l'Académie des sciences,* **57** (1863), 220–223; "Nouvelles recherches sur les infusoires du sang dans la maladie connue sous le nom de sang de rate," *ibid.,* 351–353, 386–387, and in *Comptes rendus des séances de la Société de biologie,* 3rd ser., **5** (1863), 149–152; "Nouvelles recherches sur la maladie du sang de rate considérée au point de vue de sa nature," in *Mémoires de la Société de biologie,* 3rd ser., **5** (1863), 193–202; "Nouvelles recherches sur la nature de la maladie charbonneuse connue sous le nom de sang de rate," in *Comptes rendus hebdomadaires des séances de l'Académie des sciences,* **59** (1864), 393–396; "Sur la présence de bactéridies dans la pustule maligne chez l'homme," *ibid.,* 429–431, written with Raimbert; "Recherches sur les vibrioniens," *ibid.,* 629–633; "Recherches sur la nature et la constitution anatomique de la pustule maligne," *ibid.,* **60** (1865), 1296–1299; "Sur la présence constante des bactéridies dans les animaux infectés par la maladie charbonneuse," *ibid.,* **61** (1865), 334–335; "Recherches sur une maladie septique de la vache regardée comme de nature charbonneuse," *ibid.,* 368–370; "Recherches physiologiques et pathologiques sur les bactéries," *ibid.,* **66** (1868), 499–503; "Sur la nature des maladies charbonneuses," in *Archives générales de médecine,* **17** (1868), 144–148; "Rapport sur des recherches de M. Raimbert relatives à la constitution et au diagnostic de la pustule maligne," in *Bulletin de l'Académie de médecine,* **33** (1868), 703–709; "Reproduction expérimentale de la pustule maligne chez les animaux," *ibid.,* 721–722; "Expériences relatives à la durée de l'incubation des maladies charbonneuses et à la quantité de virus nécessaire à la transmission de la maladie," *ibid.,* 816–821; "Bactérie, bactéridie," *in Dictionnaire encyclopédique des sciences médicales,* VIII (1868), 13–39; "Études sur la contagion du charbon chez les animaux domestiques," in *Bulletin de l'Académie de médecine,* **35** (1870), 215–235; "Études sur la genèse et la propagation du charbon," *ibid.,* 471–498; *Les éléments du bonheur* (Paris, 1871); "Recherches sur quelques questions relatives à la septicémie," in *Bulletin de l'Académie de médecine,* 2nd ser., **1** (1872), 907–920, 976–996, 1001–1008, 1095–1105, 1234–1237; "Recherches sur la nature de l'empoisonnement par la saumure," *ibid.,* 1051–1058; "Cas de mort d'une vache par septicémie," *ibid.,*

1058–1062; "Suite des recherches sur quelques questions relatives à la septicémie: La septicémie chez l'homme—recherches expérimentales sur la nature de la fièvre typhoïde," *ibid.,* **2** (1873), 124–144; 487–507; "Rapport sur un mémoire de M. Onimus, relatif à l'influence qu'exercent les organismes inférieurs développés pendant la putréfaction sur l'empoisonnement putride des animaux," *ibid.,* 464–477; "Recherches relatives à l'action de la chaleur sur le virus charbonneux," in *Comptes rendus hebdomadaires des séances de l'Académie des sciences,* **77** (1873), 726–729; "Recherches relatives à l'action des substances dites antiseptiques sur le virus charbonneux," *ibid.,* 821–825; "Réponse à M. Colin sur ses communications relatives à la septicémie," in *Bulletin de l'Académie de médecine,* 2nd ser., **2** (1873), 1272–1281; "Recherches relatives à l'action des substances antiseptiques sur le virus de la septicémie," in *Comptes rendus des séances de la Société de biologie,* 6th ser., **1** (1874), 25–27; "Recherches sur quelques conditions qui favorisent ou qui empêchent le développement de la septicémie," in *Bulletin de l'Académie de médecine,* 2nd ser., **8** (1879), 121–138; and "Recherches sur le traitement des maladies charbonneuses chez l'homme," *ibid.,* **9** (1880), 757–781.

Posthumous publications are "Parasites, parasitisme," in *Dictionnaire encyclopédique des sciences medicales,* 2nd ser., XXI (1885), 66–116, completed by Laboulbène; and A. Davaine, ed., *L'oeuvre de C. J. Davaine* (Paris, 1889).

II. SECONDARY LITERATURE. On Davaine or his work, see P. Huard and J. Théodoridès, "Comment vivait Casimir-Joseph Davaine (1812–1882)," in *Clio-Medica,* **2** (1967), 254–258; A. Laboulbène, *Notice sur C. J. Davaine* (Paris, 1884); and J. Théodoridès, "Une amitié de savants, Claude Bernard et Davaine," in *Histoire de la médecine,* **6** (1956), 35–45; "Un centenaire en parasitologie: Davaine et le diagnostic des helminthiases par l'examen microscopique des selles (1857)," in *Presse médicale,* **65** (1957), 2124; "Casimir Davaine et les débuts de la bactériologie médicale," in *Conférences du Palais de la Découverte,* D95 (Paris, 1964); "Les domiciles parisiens du Docteur Davaine," in *Histoire de la médecine,* spec. no. 4 (1964), 136–147; "Casimir Davaine (1812–1882): A Precursor of Pasteur," in *Medical History,* **10** (1966), 155–165; and "Un grand médecin et biologiste, Casimir-Joseph Davaine (1812–1882)," which constitutes the entirety of *Analecta Medico-historica,* **4** (1968).

JEAN THÉODORIDÈS

DAVENPORT, CHARLES BENEDICT (*b.* Stamford, Connecticut, 1 June 1866; *d.* Cold Spring Harbor, New York, 18 February 1944), *zoology, eugenics.*

A major influence on Davenport was the harsh and puritanical tyranny of his father, Amzi Benedict Davenport, a founder of and teacher in a private academy in Brooklyn, New York, who was later successful in real estate and insurance. The father taught his son to the age of thirteen, as well as demanding almost full-time work in his office. Charles yearned to attend school to escape the drudgery. His mother, Jane Joralemon Dimon Davenport, was of more liberal religious views, enthusiastic about nature, and in favor of college. Work each summer on the family farm near Stamford provided Charles with the opportunity to indulge in nature observation. Although mainly a solitary youth, he did enjoy organizing natural history groups among a few close friends.

At thirteen Davenport entered the Polytechnic Institute of Brooklyn and soon was at the head of his class, despite his previous unorthodox schooling. He received a B.S. in civil engineering in 1886. After nine months as rodman for a railroad survey in Michigan, Davenport entered Harvard College, partly supported by his mother's independent income. Under the guidance of E. L. Mark he majored in zoology, receiving the A.B. in 1889 and the Ph.D. in 1892. He promptly became an instructor at Harvard.

In 1898 Davenport became director of the summer school of the Biological Laboratory of the Brooklyn Institute of Arts and Sciences at Cold Spring Harbor, New York, a commitment he kept until 1923. In 1899 he left Harvard to become assistant professor at the University of Chicago, advancing to associate professor in 1901. Although enthusiastic about the plans of the department chairman, C. O. Whitman, to enlarge the Marine Biological Laboratory at Woods Hole, Massachusetts, and turn it toward intensive study of heredity, Davenport, also keenly independent, in 1904 finally persuaded the Carnegie Institution to support the Station for Experimental Evolution at Cold Spring Harbor, for which he left Chicago to become director. Constantly planning new research, he also established and directed the Eugenics Record Office at the same location from 1910; its funds were acquired from Mrs. E. H. Harriman. The Carnegie Institution assumed the office in 1918. Although the three organizations of which Davenport was director were all adjacent, there was little integration—and frequent confusion of direction. In 1934 Davenport retired from all his commitments, but stayed at Cold Spring Harbor to continue research and writing.

Davenport was a member of sixty-four scientific societies and active in many of them. He was elected to the American Philosophical Society and the National Academy of Sciences. In 1923 he received the gold medal of the National Institute of Social Sciences, and in 1932 he was president of the Third International Congress of Eugenics.

Davenport's research began in the 1890's with statistical studies of populations. With students in his Harvard course in experimental morphology, he published several quantitative studies and pioneered in the use of biometric methods in the United States,

especially through his manual, *Statistical Methods With Special Reference to Biological Variation* (1899), which introduced the methods of Karl Pearson to the United States. The rediscovery of Gregor Mendel's results in 1900 turned many toward similar genetic studies of a variety of organisms; and Davenport's laboratory joined the trend, although he himself was not fully convinced of the validity of Mendelism until many completed experiments and the persuasion of its champion, William Bateson, satisfied him. His station conducted breeding experiments on a variety of animals; but techniques were not always satisfactory, and only the studies of canaries and of chickens produced significant results. Davenport himself was careless and hasty in research, and hypersensitive to criticism.

From 1907 Davenport's interests turned to human heredity and eugenics, a shift sparked at least partly by his wife, Gertrude Crotty Davenport, who for several years was senior author on papers with him. Also, since 1897 Davenport had been acquainted with Francis Galton, the founder of eugenics. A major contribution to early eugenics was *Heredity in Relation to Eugenics* (1911), in which Davenport presented data on inheritance of a great variety of traits, from eye color and certain illnesses to criminality and pauperism, not all with convincing data. This staunch advocate of an improved race of man urged great care in the selection of marriage partners, large families for those who had thus selected, a ban on racial mixing, and the exclusion of undesirable immigrants from the United States. In 1918 Davenport analyzed the physical traits of recruits for the U.S. Army, and he continued along these lines with growth measurements of children in orphanages.

Davenport was a poor lecturer yet an infectiously enthusiastic conversationalist. His major contributions were the introduction of statistical methods into evolutionary studies and the initiation of projects in the laboratories he directed.

BIBLIOGRAPHY

I. ORIGINAL WORKS. Davenport was more at ease writing than speaking and often expressed himself—even at home—in letters to his father. His scientific publications, totaling over 400, began in 1890 and continued to posthumous papers. The great majority were without coauthor. Both biographies cited below contain full bibliographies (except for reviews and short notes).

The manual *Statistical Methods With Special Reference to Biological Variation* went through four editions (New York, 1899, 1904, 1914, 1936). Davenport's animal genetic studies were summarized in *Inheritance in Poultry,* Carnegie Institution of Washington, publication no. 52 (Washington, D.C., 1906); and *Inheritance in Canaries,* Carnegie Institution of Washington, publication no. 95 (Washington, D.C., 1908). His eugenics data and creed were first presented in *Heredity in Relation to Eugenics* (New York, 1911) and were followed by many papers on the subject.

II. SECONDARY LITERATURE. A disarming analysis of Davenport's personality was written by his colleague, E. Carleton MacDowell, in *Bios,* **17,** no. 1 (1946), 2–50, in which the subject's childhood is summarized from personal notes and diaries, and the establishment of the Cold Spring Harbor laboratories is detailed. Oscar Riddle's "Biographical Memoir of Charles Benedict Davenport," in *Biographical Memoirs. National Academy of Sciences,* **25** (1948), 75–110, is derived in great part from MacDowell.

ELIZABETH NOBLE SHOR

DAVID, TANNATT WILLIAM EDGEWORTH (*b.* near Cardiff, Wales, 28 January 1858; *d.* Sydney, Australia, 28 August 1934), *polar exploration, geology.*

David was the son of Rev. William David. He was educated at Magdalen College School, Oxford, and Oxford University, graduating B.A. in classics in 1880. He then became interested in geology and, after a period at the Royal School of Mines in London, went to Australia as a field surveyor.

David vigorously set to work on Australian geological problems and became the greatest contributor of his time to the field. He held the chair of geology at the University of Sydney from 1891 to his retirement in 1924. He received numerous honors from Britain and Australia, including a knighthood in 1921.

As the leading geologist in the Shackleton Expedition of 1907–1909, David carried out pioneering fieldwork in Antarctica. In the course of this work he led expeditions to the summit of Mt. Erebus and to the neighborhood of the south magnetic pole. His party covered 1,200 miles on sledges over broken and splintered sea ice and glacier ice without supporting party, dogs, or mechanical aid. The party was dramatically rescued, near exhaustion, in February 1909.

David's scientific contributions include his early delineation of strata in coalfields of New South Wales; his discoveries in the Hunter River area are a noted classic in Australian geology. His investigations of foundational rocks under the coral atoll of Funafuti (1897), which supplied evidence on Darwin's theory of coral reef formation, won him election to the Royal Society of London in 1900. David also contributed important evidence on past glaciation in Australia and India, and on fossil evidences of life in Precambrian rocks near Adelaide.

After retirement he set out to compile the first comprehensive account of the geology of the whole Australian continent. By 1931 the work had reached the point where David was able to publish a comprehensive geological map of Australia, along with an up-to-date summary of Australian geological field data. A greatly extended version of this work, which he had planned in detail and started before his death, was completed by his colleagues and is a monumental contribution to Australian geology.

BIBLIOGRAPHY

David's publications up to 1920 are listed in W. N. Benson, "Eminent Living Geologists: Professor Sir T. W. Edgeworth David," in *Geological Magazine,* **59** (1922), 4–13. His publications after 1920 are listed in *Obituary Notices of Fellows of the Royal Society,* I (1932–1935), 501.

A preliminary version of David's comprehensive account of the geology of Australia was published as *Geological Map of the Commonwealth of Australia (Scale 1:2,999,000). Explanatory Notes to Accompany a New Geological Map of the Commonwealth of Australia* (Sydney, 1932). A greatly extended version of this work was published sixteen years after his death as *Geological Map of the Commonwealth of Australia,* ed. and supp. by W. R. Browne, 3 vols. (1950).

K. E. BULLEN

DAVIDOV, AUGUST YULEVICH (*b.* Libav, Russia, 15 December 1823; *d.* Moscow, Russia, 22 December 1885), *mechanics, mathematics.*

Davidov, the son of a physician, enrolled in 1841 at Moscow University. He graduated in 1845, and was retained at the university in order to prepare for a teaching career. In 1848 he defended his dissertation, *The Theory of Equilibrium of Bodies Immersed in a Liquid,* and received the master of mathematical sciences degree. Davidov's dissertation was devoted to the exceptionally pressing problem of the equilibrium of floating bodies.

Although Euler, Poisson, and Dupin had worked extensively on the problem they had far from exhausted it, and new, major results were obtained by Davidov. He was the first to give a general analytic method for determining the position of equilibrium of a floating body, applied his method to the determination of positions of equilibrium of bodies, explained the analytical theory by geometric constructions, and investigated the stability of equilibrium of floating bodies.

In 1850 Davidov began teaching at Moscow University, and in 1851 he successfully defended his doctoral dissertation, *The Theory of Capillary Phenomena.* Shortly thereafter he was appointed a professor at Moscow University, where he worked until the end of his life. Both of his dissertations were awarded the Demidovskoy prize by the Petersburg Academy of Sciences.

Of the other valuable mathematics studies done by Davidov mention should be made of those on equations with partial derivatives, elliptical functions, and the application of the theory of probability to statistics. He also compiled a number of excellent texts for secondary schools. Of these, the geometry and algebra textbooks enjoyed special success and were republished many times. Through the next half century the geometry text underwent thirty-nine editions and the algebra text twenty-four.

Davidov conducted much scientific organizational work. For twelve years he was head of the physics and mathematics faculty; for thirty-five years he taught various courses in mathematics and mechanics at the university and prepared two generations of scientific workers and teachers. Along with N.D. Brashman, Davidov was a founder of the Moscow Mathematical Society and was its first president (1866–1885).

BIBLIOGRAPHY

Davidov's works, all in Russian, are *The Theory of Equilibrium of Bodies Immersed in a Liquid* (Moscow, 1848); *The Theory of Capillary Phenomena* (Moscow, 1848); *Elementary Geometry* (Moscow, 1864); *Beginning Algebra* (Moscow, 1866); *Geometry for District Schools* (Moscow, 1873).

See also "Reminiscences of A. Y. Davidov," in *News of the Society of Lovers of Natural Science, Anthropology, and Ethnography,* **51** (1887); and *Reminiscences of A. Y. Davidov. A Speech and Account, Read at the Meeting of Moscow University on 12 January 1886* (Moscow, 1886).

A. T. GRIGORIAN

DAVIDSON, WILLIAM. See **Davison, William.**

DA VIGEVANO. See **Guy Da Vigevano.**

DAVIS, WILLIAM MORRIS (*b.* Philadelphia, Pennsylvania, 12 February 1850; *d.* Pasadena, California, 5 February 1934), *geography, geomorphology, geology, meteorology.*

William Morris Davis was the son of Edward M. Davis and Martha Mott Davis, both members of the Society of Friends. His father, a Philadelphia businessman, was expelled from the society for enlisting in the Union army during the War Between the States. His mother was the daughter of Lucretia Mott, an early and strenuous worker for women's rights and a firm antagonist of slavery. She resigned from the

Society of Friends shortly after the expulsion of her husband from the group.

Davis received a bachelor of science degree from Harvard University in 1869 and a master of engineering degree in 1870. He went directly thereafter to the national observatory at Córdoba, Argentina, as a meteorologist. After three years he returned to the United States and served on the Northern Pacific survey as an assistant to Raphael Pumpelly. In 1877 he was appointed assistant to Nathaniel S. Shaler, professor of geology at Harvard College. From 1879 to 1885 he was instructor in geology at Harvard and in 1885 was appointed assistant professor of physical geography. In 1890 he became full professor of physical geography and in 1898 was appointed Sturgis Hooper professor of geology, becoming emeritus in 1912. He was a visiting professor at Berlin University in 1908 and at the University of Paris in 1911.

After retirement from Harvard, Davis devoted his time to field studies and to writing, and served as visiting lecturer at the University of California at Berkeley (1927–1930), the University of Arizona (1927–1931), Stanford University (1927–1932), the University of Oregon (1930), and the California Institute of Technology (1931–1932). He held honorary degrees from the universities of the Cape of Good Hope, Melbourne, Greifswald, and Christiana (Oslo). Davis was a founding member of the Geological Society of America, its acting president in 1906, and its president in 1911. He was instrumental in founding the Association of American Geographers and served as its president in 1904, 1905, and 1909. He was a member of the American Academy of Arts and Sciences, the American Philosophical Society, the National Academy of Sciences, the Imperial Society of Natural History (Moscow), and the New Zealand Institute. He was an honorary or a corresponding member of more than thirty scientific societies throughout the world and received more than a dozen medals and citations for his work. Daly lists 501 titles in Davis' bibliography, the first entry dated 1880 and the last 1938.

In 1879 Davis married Ellen B. Warner of Springfield, Massachusetts. After her death he married Mary M. Wyman of Cambridge, Massachusetts, and after her death he married Lucy L. Tennant of Milton, Massachusetts, who survived him.

Davis' contributions are in the separate but related fields of meteorology, geology, and geography (now better called geomorphology) and in their teaching at the high school and college levels. He was active in all these fields throughout his life, although the emphasis changed. Geological studies dominated his early publications (1880–1883). During the decade beginning in 1884 he produced some forty meteorological studies. Davis' first published concern with geography, which was one of his few collaborative works, was done with Shaler and focused on the role of glaciers in fashioning the landscape. In 1889, with the publication of "The Rivers and Valleys of Pennsylvania," he laid the cornerstone of his greatest contribution to physical geography: the "Davisian system" of landscape analysis. To this subject he devoted much of his scholarly energy for the rest of his life. From the time he joined the Harvard faculty, Davis was concerned with teaching, particularly of meteorology and physical geography. Not only were his formal courses centered on these subjects, but he was active in promoting them as proper pursuits of learning in both secondary schools and colleges. During the years from 1893 to 1903 he actively expressed this concern in seventeen papers on the subject.

When Davis took his first job with the national observatory in Córdoba, Argentina, the systematic collection of weather data by government was just beginning in the United States. In 1870 a federal service was established in the U.S. Army Signal Corps, a service that later became the U.S. Weather Bureau. It was against this background of developing interest in weather that Davis organized his first formal course shortly after joining the Harvard faculty. He continued to teach this course until 1894, and during this time he organized some of his students and interested amateurs into an informal organization of weather observers in the New England area. The data thus collected allowed early description of such phenomena as the sea breeze, thunderstorms, precipitation, and atmospheric convection. Davis' *Elementary Meteorology* (1894) was an admirable synthesis of the state of meteorological knowledge at that time and was used for more than thirty years as a college text.

Davis' early fieldwork was in geology, as his first reports indicate. His most important contribution was his study, supported by the U.S. Geological Survey, of the Triassic basins of New England and New Jersey. The first of fifteen preliminary reports appeared in 1882, but the most inclusive is his final report, "The Triassic Formation of Connecticut" (1898). This set forth for the first time a comprehensive history of the Triassic volcanic sequence, laid out criteria for distinguishing between intrusive and extrusive rocks, and demonstrated techniques by which subsurface geologic structures could be deduced from the surface topography. Of considerable interest also was his demonstration that the Tertiary "lake beds" in the western states were in large part fluviatile in origin.

Davis' early fieldwork in geology, both in the West and in the northeastern states, coupled with his reading of the reports of the western surveys by Powell, Dutton, and Gilbert, served to develop his interest in the origin and description of landscape. Out of this interest grew his concept of the cycle of erosion and the demonstration of its use to describe the earth's surface features and to decipher earth history. By 1883 Davis was discussing the origin of cross valleys, particularly in the folded belt of Pennsylvania and Virginia. He took exception to the work of Ferdinand Löwl (1882), which held that rivers now cutting across regional (or even local) structures could not maintain their courses by the action of antecedent (pre-existing) rivers flowing across a land slowly rising athwart their paths. Powell (1875, p. 163) had already introduced the idea of an antecedent stream and Davis argued, deductively, that such streams could in fact account for the drainage now cutting across the structures of the Appalachian Mountains.

In "The Rivers and Valleys of Pennsylvania" (1889) and its sequel, "The Rivers and Valleys of Northern New Jersey, With Notes on the Classification of Rivers in General" (1890), Davis firmly established his method of landscape analysis. The method involved assumptions from which he deduced a particular idealized landscape, a summation of the field data, a matching of the idealized form with the actual forms, and finally an explanation of the origin of the landscape.

The cycle of erosion was by far the most influential geographical concept introduced by Davis. For him the present-day landscape resulted from a long-continued and orderly development. To understand it, it was necessary to know the geological structures, the processes which operate on the surface, and the duration of their operation. He held that the present landscape could be understood only by understanding its past and, conversely, that an understanding of present-day landscape was a key to an interpretation of at least some earth history. Davis assumed that a river valley has progressed through one or more cycles, a complete cycle being marked by youth, maturity, and old age as the river valley is worn lower and lower into a landmass. In each stage the river and its valley displayed certain distinctive characteristics. Moreover, a cycle might be interrupted by uplift of the land, which would rejuvenate the river and allow it to impose the beginnings of a new cycle on the remains of an older one. The idea of youth, maturity, and old age in a valley, first presented in the "Rivers and Valleys of Pennsylvania," was extended to embrace large landmasses. The cycle of erosion of a region was also seen as marked by youth, maturity, and old age. The end product of a complete cycle was a nearly featureless plain, called by Davis a peneplain.

The regional cycle of erosion was also subject to interruption and rejuvenation; and various cycles, it was held, could be recognized by erosional surfaces of varying vertical and lateral extent.

Davis was among the vast majority of geologists who accepted uniformitarianism. Writing on the antecedent valleys in the Appalachians (1883, p. 357), he said that he "tests the past by the present." He began "The Rivers and Valleys of Pennsylvania" with "No one now regards a river and its valley as ready-made features of the earth's surface. All are convinced that rivers have come to be what they are by slow processes of natural development. . ." (p. 215). Catastrophe as an explanation of a natural feature was avoided by the uniformitarian whenever another explanation could be offered. Thus Davis wrote: "Perhaps one reason why the explanation [for transverse valleys] has become so popular is that it furnishes an escape from the catastrophic idea that fractures control the location of valleys and is at the same time fully accordant with the ideas of the uniformitarian school that have become current in this half of our century" (*ibid.,* p. 270). Long-continued action of observable processes were demanded by the Davisian approach to landscape, and in this his studies followed in the Hutton-Playfair-Lyell tradition.

Darwin had introduced the idea of organic evolution in 1859, and Davis extended the idea of development of one living form from a pre-existing form to the inorganic world of physical geography. Writing in 1883 about the antecedent streams of the Appalachians, he said, "It seems most probable, that the many pre-existent streams in each river-basin concentrated their water in a single channel of overflow, and that this one channel survives—a fine example of natural selection."

Davis apparently was fully persuaded of the gradual and systematic evolution of topographical forms, and introduced the erosion cycle to "trace the development of . . . river systems . . . from their ancient beginning to the present time" (*ibid.,* p. 183). The erosion cycle was developmental, evolutionary, and in step with the most exciting scientific idea of the time. I. C. Russell in 1904 noted that geomorphology was "vivified by evolution."

Beyond these points Davis approached the explanation of landscape with the eyes of a geologist. His earlier training, in both Montana and the East, and his association with Pumpelly and Shaler stood him in good stead. The cycle of erosion covered long periods of time. He agreed with Powell (1876, p. 196) that mountains cannot remain long as mountains; they are ephemeral topographical forms but, assuming the long perspective of the geologist, he knew that

"a complete cycle [of a river] is a long measure of time" (1889, pp. 203–204).

Davis advanced the cycle of erosion "tentatively," not feeling "by any means absolutely persuaded of the results" (*ibid.,* p. 219). Despite this disclaimer the cycle met with almost immediate acceptance. It appealed most strongly to geologists. This seems to have been true for two reasons. First, the cycle of erosion was set in a framework of long periods of time. That earth history involved immense (if unspecified) amounts of time was inbred in every geologist. Second, geological history had been based on tangible objects, rocks. But what would serve as a basis for history if rocks were missing for a particular time unit? Davis provided a partial answer. In his system the form of the land could, if properly interpreted, provide a history even though no rock record was available.

The cycle of erosion influenced the work of many geologists. Almost immediately after the introduction of the concept, reports analyzing landscape and reconstructing earth history by use of the Davisian analysis of landforms began to appear in the geological journals. Such studies continued to appear throughout Davis' lifetime and until the 1940's.

Davis argued that in the cycle of erosion, uplift of the land was rapid and initiated the erosional process. The land then went through a much longer period in which there was little or no movement and during which erosion wore the land down toward base level and a peneplain. A corollary of this sequence was that erosion was rapid and then slowed as the end of the cycle was approached. With this in mind, one wonders whether Chamberlin, an admiring colleague and close reader of Davis, did not draw some inspiration from the cycle of erosion for his very influential paper "Diastrophism as the Ultimate Basis of Correlation" (1909). The changing nature of the landscape must be reflected in the sedimentary record; and it was to the sedimentary record that Chamberlin appealed, arguing that the diastrophism was not only rapid but worldwide and was part of the sedimentary record. Long periods of stillstand of the land and its accompanying erosion were reflected in the slowly changing nature of the resulting rock types in the sequence of sedimentary rocks, from initially coarse clastics to finer and finer particles and finally to chemical deposits. The Chamberlin model focused on the positive record in depositional basins of the cycle of erosion envisaged by Davis for the continental masses.

When Davis came to the task of the description of landscape, physical geography was an unordered agglomeration of facts and figures. There was no adequate way to describe a landscape in brief, concise, memorable terms. Elevations of mountains, lengths of rivers, angles of slope might represent a landscape; but these facts were not easily retained, nor were the forms they represented impressed upon the mind. Davis came to view the cycle of erosion as representing a series of generalized or idealized landform types. These forms, first deductively arrived at, must be checked against nature and modified accordingly. He sought to develop a store of ideal landscape types correlated with "structure, process, and time" that "shall imitate nature's products" (1902, p. 246). He saw his system as a technique of geographical description, feeling that

> . . . comparing the partial view of the landscape, as seen by the outer sight, with the complete view as seen by the inner sight, [the geographer] determines, with great saving of time and effort just where the next observations should be made in order to decide whether the ideal type he has provisionally selected fully agrees with the actual landscape before him. When the proper type is thus selected, the observed landscape is concisely and effectively named in accordance with it, and description is thus greatly abbreviated [*ibid.*, p. 247].

Davis thought of himself primarily as a geographer, although his influence was probably more profound among geologists. His professional work was in physical geography, which he shortened to "physiography." This he viewed as a description of the land. If the element of time was emphasized—and thus the study became geological—Davis would have used the term "geomorphography." Today the discipline is called "geomorphology." The term generally includes that which Davis called "physiography."

The cycle of erosion was Davis' most important contribution. He at first conceived of it as occurring in a temperate climate—a climate he called "normal." In later years, however, he extended it to the arid and glacial climates.

After his retirement from Harvard, Davis turned his attention to the development of coral reefs and coralline islands. His long monograph *The Coral Reef Problem* (1928) was the culmination of long study during the course of which he traveled to the southwestern Pacific and to the Lesser Antilles. The volume stands as an exhaustive summary of the facts then known, the several hypotheses up to then advanced to explain them, and his own views on the role of ocean floor subsidence and the shifting sea levels of the Pleistocene in the origin of the coral reefs.

In 1930 Davis' paper "Origin of Limestone Caverns" developed the still-accepted interpretation that caves as we now see them have passed through two stages in their development.

Davis' courses in meteorology and physiography were reputed to have been carefully and logically presented. Most American physical geographers and geologists adopted the Davisian approach as it developed in his classroom and was presented in his published papers. He had a strong effect as well on many of the French workers. There was indeed a "Davis" school. Despite this, surprisingly few of his own students carried on directly in his footsteps. Douglas Johnson, as a young instructor at the Massachusetts Institute of Technology, spent much time with Davis and became his most distinguished disciple. He extended Davis' analysis of the landscape history of the northern Appalachians and described the cycle of shoreline development. In meteorology, his student Robert DeCourcy Ward succeeded him as a professor of meteorology and climatology at Harvard and became director of the Blue Hill Observatory, a research institution for meteorology at Harvard.

The cycle of erosion was soon introduced into college and university teaching. Albrecht Penck, in his *Morphologie der Erdoberfläche* (1894), introduced the Davis system of stream classification and discussed the peneplain. Albert de Lapparent did not quote Davis in his third edition of *Traité de géologie* (1897), but he did discuss "aplanissement," which corresponded to peneplanation, and described incised meanders as the inheritance of a stream course through uplift. The same book in its fourth edition (1900) leaned heavily on Davis' physiographical writings, and the *Leçons de géographie physique* (1896) was still more indebted to Davis. In the United States, William Berryman Scott's *Introduction to Geology* (1897) firmly established the Davisian system in American textbooks. Thereafter all American textbooks in introductory geology and physical geography relied upon the Davisian description of landforms. This influence on textbook writers persisted to the mid-twentieth century.

BIBLIOGRAPHY

I. ORIGINAL WORKS. Publications by Davis are listed in Reginald A. Daly, "Biographical Memoir of William Morris Davis," in *Biographical Memoirs. National Academy of Sciences,* **23** (1945), 263–303. A bibliography of meteorological papers compiled and annotated by Lylyan H. Block appears in *Bulletin of the American Meteorological Society,* **15** (1934), 57–61. Twenty-six essays by Davis were collected and edited by Douglas W. Johnson under the title *Geographical Essays* (Boston, 1909; repub. New York, 1954).

Original works by Davis cited in the text are "Origin of Cross Valleys," in *Science,* **1** (1883), 325–327, 356–357; "The Rivers and Valleys of Pennsylvania," in *National Geographic Magazine,* **1** (1889), 183–253; "The Rivers of

Northern New Jersey With Notes on the Classification of Rivers in General," *ibid.*, **2** (1890), 81–110; *Elementary Meteorology* (Boston, 1894); "The Triassic Formation of Connecticut," in *U.S. Geological Survey, Annual Report, 1896–1897,* **18,** pt. 2 (1898), 1–192; *Physical Geography* (Boston, 1898), written with W. H. Snyder; "Systematic Geography," in *Proceedings of the American Philosophical Society,* **41** (1902), 235–259; *The Coral Reef Problem,* American Geographical Society Special Publication no. 9 (New York, 1928); and "Origin of Limestone Caverns," in *Bulletin of the Geological Society of America,* **41** (1930), 475–628.

II. Secondary Literature. Works relevant to the work of Davis cited in the text are: T. C. Chamberlin, "Diastrophism as the Ultimate Basis of Correlation," in *Journal of Geology,* **17** (1909), 685–693; Albert de Lapparent, *Traité de géologie* (3rd ed., Paris, 1893; 4th ed., 1900); *Leçons de géographie physique* (Paris, 1896); Ferdinand Löwl, "Die Enstehung der Durchbruchsthäler," in *Petermanns Mitteilungen aus J. Perthes Geographischer Anstalt,* **28** (1882), 408–416; Albrecht Penck, *Morphologie der Erdoberfläche* (Stuttgart, 1894); J. W. Powell, *Exploration of the Colorado River of the West and Its Tributaries* (Washington, D.C., 1875); *Report on the Geology of the Eastern Portion of the Uinta Mountains and a Region of Country Adjacent Thereto* (Washington, D.C., 1876); I. C. Russell, "Physiographic Problems of Today," in *Journal of Geology,* **12** (1904), 524–550; and William Berryman Scott, *An Introduction to Geology* (New York, 1897).

Accounts of Davis' life and work are given in Daly (see above); Kirk Bryan, "William Morris Davis—Leader in Geomorphology and Geography," in *Annals of the Association of American Geographers,* **25** (1935), 23–31; and Sheldon Judson, "William Morris Davis—an Appraisal," in *Zeitschrift für Geomorphologie,* **4,** no. 3/4 (1960), 193–201.

Sheldon Judson

DAVISON, WILLIAM (*b.* Aberdeenshire, Scotland, 1593; *d.* Paris, France, *ca.* 1669), *medicine, chemistry.*

Davison was the youngest of three sons born to Duncan Davison of Ardmakrone, Aberdeenshire, and Janet Forbes, daughter of William Forbes, baron of Pitsligo. Although the Davisons claimed descent from several branches of the Scottish nobility, William's family was in poor circumstances, particularly after the early death of the father. Supported by the generosity of John, earl of Leslie, he entered the Presbyterian Marischal College, Aberdeen, and graduated Master of Arts in 1617. Shortly afterward Davison migrated to France. Here he appears to have graduated Doctor of Medicine, possibly at Montpellier. He also formed a lifelong friendship with Jean Baptiste Morin, the astrologer and mathematician, and studied chemistry for three years (*ca.* 1620) in the household of Morin's patron, Claude Dormy, bishop of Boulogne.

Later Davison moved to Paris, where he practiced medicine, principally among the British émigré community, and gave private instruction in medical chemistry. In 1648 he was appointed *intendant* of the Jardin Royal des Plantes (now the Muséum d'Histoire Naturelle), Paris, where he introduced public lectures in chemistry, thus becoming the first in the long line of noted chemical teachers at this institution. Following a legal wrangle concerning his position as *intendant,* he resigned in 1651 to become physician to Marie Louise de Gonzague-Nevers (wife of King John II Casimir of Poland) and director of the Royal Botanical Garden in Warsaw. On the death of Queen Marie Louise in 1667 he returned to Paris, where he died about 1669. Davison married a Scottish woman, Charlotte de Thynny, in France; they had one son, Charles.

Davison's principal chemical work, the *Philosophia pyrotechnica* (1633–1635), is characterized by long disquisitions on the metaphysical basis of his chemical theory. He elaborates on the Neoplatonic aspects of Paracelsian theory, with emphasis on the macrocosm-microcosm relationship and the search for incorporeal forces operating in the universe under the veil of observed chemical reactions. In a much revised version, published in French as *Les élémens de la philosophie de l'art du feu ou chemie* (1651, 1657), Davison seeks to incorporate his Neoplatonic world view within the Copernican sun-centered universe. Geometrical analogies are prominent in Davison's theories. He compares the three surfaces required to form a solid angle with the three principles of salt, sulfur, and mercury. He places great emphasis on the five Platonic solids, and in one short treatise of the *Élémens* he describes, with illustrations, the microcosmic manifestations of these solids in the crystal structure of minerals and the morphological structure of plants and insects.

Davison's most ambitious work is his commentary (1660) on the *Idea medicinae philosophicae* of the noted sixteenth-century Paracelsian Peter Severinus. This work marks Davison as a devoted Paracelsian theorist, but by the time of its appearance it was somewhat outdated, since iatrochemical theory had come to be dominated by the work of J. B. van Helmont. Davison's highly theoretical chemical texts seem also to have suffered in comparison with the more practical manuals of such near contemporaries as Jean Beguin, Nicolas Lefèvre, and Christoph Glaser. His works are not frequently cited.

Davison was also the author of two minor treatises: one on the Salic law (1641) and another (1668) on a disease of the scalp known as *Plica polonica,* in which he contested the view that the condition was endemic to Poland.

BIBLIOGRAPHY

I. ORIGINAL WORKS. For a detailed bibliography of Davison's works the reader is referred to the articles by E.-T. Hamy and J. Read cited below. His major works are *Philosophia pyrotechnica seu curriculus chymiatricus,* 4 pts. (Paris, 1633–1635; 1640); *Oblatio Salis sive Gallia Lege Salis condita* (Paris, 1641); *Les élémens de la philosophie de l'art du feu ou chemie* (Paris, 1651; 1657); *Commentariorum in . . . Petri Severini Dani Ideam medicinae philosophicae . . . Prodromus* (The Hague, 1660); and *Plicomastix seu Plicae numero morborum* (Danzig, 1668), published under the pseudonym Theophrastus Scotus.

II. SECONDARY LITERATURE. On Davison or his work see the following (listed chronologically): J. Small, "Notice of William Davidson, M.D.," in *Proceedings of the Society of Antiquaries of Scotland,* **10** (1875), 265–280; E.-T. Hamy, "William Davison, intendant du Jardin du Roi et professeur de chimie (1647–1651)," in *Nouvelles archives du Muséum d'histoire naturelle,* 3rd ser., **10** (1898), 1–38; and J. Read, "William Davidson of Aberdeen, the First British Professor of Chemistry," in *Ambix,* **9** (1961), 70–101.

OWEN HANNAWAY

DAVISSON, CLINTON JOSEPH (*b.* Bloomington, Illinois, 22 October 1881; *d.* Charlottesville, Virginia, 1 February 1958), *physics.*

Davisson's father, Joseph, was a contract painter and his mother, Mary Calvert Davisson, a schoolteacher. After graduating from Bloomington High School in 1902, Davisson entered the University of Chicago. He interrupted his second year to teach physics briefly at Purdue University. In 1905 he moved to Princeton University as instructor in physics, working primarily as research assistant to O. W. Richardson. Returning to Chicago for summer sessions, Davisson earned his B.S. from the university in 1908 and was awarded his Ph.D. in physics (with a minor in mathematics) at Princeton in 1911. On 4 August 1911 he married Charlotte Sara Richardson, sister of O. W. Richardson, and that summer was appointed assistant professor of physics at Carnegie Institute of Technology. While there he published on Bohr's atomic theory.

In 1917, on leave from Carnegie Institute, Davisson moved to the engineering department of the Western Electric Company Laboratories (now Bell Telephone Laboratories) to participate in a project on military telecommunications, intending to return to Carnegie Institute. After the war, however, he decided not to go back to his heavy teaching responsibilities but to remain at the laboratories, where he had been guaranteed freedom to do full-time basic research, an uncommon condition in commercial laboratories at that time.

At Western Electric, Davisson was involved mainly in two different areas of research: thermionics and emission of electrons from metals under electron bombardment. In thermionics one of his more important experiments was the measurement of the work functions of metals, the results of which suggested that the conduction electrons in a metal do not always have the normal thermal energy predicted by classic theory. Davisson's interest in secondary electron emissions started when he and C. H. Kunsman accidentally discovered in 1919 that a few secondary electrons from nickel under electron bombardment have the same energy as the primary electrons. They went on to measure the distribution-in-angle of these secondary electrons and found it to have two maximums. In the years that followed, they repeated these experiments with various metals while Davisson made unsuccessful efforts to understand these results theoretically.

Davisson's investigations on the scattering of electrons entered a new phase when, in April 1925, his target was heavily oxidized by an accidental explosion of a liquid-air bottle. He cleaned the target by prolonged heating and then found the distribution-in-angle of the secondary electrons completely changed, now showing a strong dependence on crystal direction. He traced this change to a recrystallization caused by the heating. Prior to the accident the target had consisted of many tiny crystals, but the heating converted it to several large crystals. Davisson and L. H. Germer, who had replaced Kunsman before the accident, at once began bombarding targets of single crystals.

In the summer of 1926 Davisson attended a meeting of the British Association for the Advancement of Science at Oxford. There he discussed his investigations of the scattering of the electrons with Max Born, James Franck, and others. For the first time he heard in detail about Louis de Broglie's hypothesis that an electron possesses a wave nature and has a wavelength h/mv, where h is Planck's constant, m the electron mass, and v the speed of the electron. The Oxford discussions persuaded Davisson that his experimental results were probably due to the effects of de Broglie waves. This interpretation of the earlier Davisson-Kunsman experiments had already been suggested, a year prior to the meeting, by Walter Elsasser, who carried out experiments himself, trying to confirm his point without success. Davisson knew of Elsasser's suggestion but, failing to find any evidence of a wave phenomenon, dismissed it as irrelevant. Elsasser therefore had no influence on the course of Davisson's experiments, and it was not until the Oxford conference that Davisson became fully aware of what his experimental results might reveal.

When Davisson returned from England, he and Germer began a systematic search for some sort of interference phenomenon, and in January 1927 they observed electron beams resulting from diffraction by a single crystal of nickel. The results were in good agreement with de Broglie's prediction. For his confirmation of electron waves Davisson shared the Nobel Prize in physics in 1937 with G. P. Thomson, who had independently confirmed electron waves by a different method.

In the 1930's Davisson continued to show an interest in electron waves, especially in their application to crystal physics and electron microscopy, developing a technique for electron focusing. He retired from the Bell Telephone Laboratories in 1946. From there he moved to the University of Virginia as a visiting professor of physics and retired in 1954.

BIBLIOGRAPHY

I. ORIGINAL WORKS. Davisson's writings include "Dispersion of Hydrogen and Helium on Bohr's Theory," in *Physical Review,* 2nd ser., **8** (1916), 20–27; "Scattering of Electrons by Nickel," in *Science,* n.s. **54** (1921), 522–524, written with C. H. Kunsman; "Thermionic Work Function of Tungsten," in *Physical Review,* 2nd ser., **20** (1922), 300–330, written with L. H. Germer; "Scattering of Electrons by a Positive Nucleus of Limited Field," *ibid.,* **21** (1923), 637–649; "Scattering of Low Speed Electrons by Platinum and Magnesium," *ibid.,* **22** (1923), 242–258, written with C. H. Kunsman; "Diffraction of Electrons," *ibid.,* **30** (1927), 705–740, written with L. H. Germer; "Scattering of Electrons by a Single Crystal of Nickel," in *Nature,* **119** (1927), 558–560, written with L. H. Germer; "Reflection and Refraction of Electrons by a Crystal of Nickel," in *Proceedings of the National Academy of Sciences of the United States of America,* **14** (1928), 619–627, written with L. H. Germer; and "Reflection of Electrons by a Crystal of Nickel," *ibid.,* 317–322, written with L. H. Germer.

Original MSS are listed in T. S. Kuhn *et al., Sources for History of Quantum Physics* (Philadelphia, 1967), p. 31; and *The National Union Catalog of Manuscript Collections 1963–1964* (Washington, D.C., 1965), p. 253.

II. SECONDARY LITERATURE. On Davisson or his work, see K. K. Darrow, "A Perspective of Davisson's Scientific Work," in *Biographical Memoirs. National Academy of Sciences,* **36** (1962), 64–79; W. Elsasser, "Bemerkungen zur Quantenmechanik freier Elektronen," *Die Naturwissenschaften,* **13** (1925), 711; L. H. Germer, "Low-Energy Electron Diffraction," in *Physics Today,* **17**, no. 7 (July 1964), 19–23; M. Jammer, *The Conceptual Development of Quantum Mechanics* (New York, 1966), pp. 189, 250–253; and M. J. Kelly, "Clinton Joseph Davisson," in *Biographical Memoirs. National Academy of Sciences,* **36** (1962), 51–62. For more extensive bibliographical listings, see Poggendorff, VIIb, pt. 2 (1967), 994; and *Biographical Memoirs. National Academy of Sciences,* **36** (1962), 81–84.

KENKICHIRO KOIZUMI

DAVY, HUMPHRY (*b.* Penzance, England, 17 December 1778; *d.* Geneva, Switzerland, 29 May 1829), *chemistry.*

Humphry Davy was the eldest son of Robert and Grace Millett Davy. His father, of yeoman stock, was a woodcarver but earned little by it and lost money through speculations in farming and tin mining. After his death in 1794 Grace Davy managed a milliner's shop until she inherited a small estate in 1799. After haphazard schooling Davy was apprenticed in 1795 to Bingham Borlase, an able apothecary-surgeon who later qualified as a physician. His scientific career began in 1798, when he was released from his indentures and was appointed superintendent of Thomas Beddoes' Pneumatic Institution at Clifton. He retained a firm, and on occasion fervent, belief in a Supreme Being but does not seem to have been a zealous member of any church. In 1812 he married a wealthy bluestocking widow, Jane Apreece, but the marriage was childless and not happy. In the same year he was knighted and in 1818 was made a baronet. His health was good, aside from a serious illness, probably typhus, at the end of 1807; but in 1826 he suffered a stroke from which he never fully recovered. Most of Davy's working life was spent at the Royal Institution in London; he was elected a fellow of the Royal Society in 1803 and was its president from 1820 to 1827. He was a foreign member of many societies overseas and was acquainted with many Continental scientists.

Davy said of his schooling that he was glad he had not been worked too hard and had been allowed time to think for himself. Schools in Cornwall in the late eighteenth century were not very good, but Davy emerged at fifteen with a fair knowledge of the classics. At about the same time that he began his apprenticeship he drew up a formidable program of self-education, which included theology and geography, seven languages, and a number of science subjects. He read a good deal of philosophy: Locke, Berkeley, Hume, Hartley, Thomas Reid, Condorcet, and even, according to his brother, some transcendentalist writings; and he began to write poetry. Toward the end of 1797 he began the study of chemistry with William Nicholson's *Dictionary of Chemistry* and A. L. Lavoisier's *Traité élémentaire de chimie,* which he read in French. He was fortunate that at this time Gregory Watt, the son of James Watt, lodged with the Davy family, having been advised to spend the winter in a mild climate on account of his health. After Gregory Watt's departure, he and Davy maintained a correspondence. Also during this period Davy became acquainted with Davies Giddy, who later changed his surname to Gilbert and succeeded Davy as president of the Royal Society. Giddy allowed Davy to use his books and helped in the negotiations that led to his going to Clifton. Borlase had quite a good library, and Davy also had access to that of the Tonkins, who were friends of the family.

With this basis of wide but rather undirected reading, Davy began his career as a chemist; and within five years of reading his first chemistry book he had become professor of chemistry at the Royal Institution. His first research, culminating in speculative papers on the role of light, was directed to proving wrong Lavoisier's doctrine of caloric. Lavoisier had placed light and heat, or caloric, at the head of his table of simple bodies and had asked rhetorically whether heat was a modification of light, or vice versa. He believed that caloric combined with other substances and, in particular, that oxygen gas was a compound of a substance or "basis" (so far unknown) and heat. Davy, in a tradition stemming from Newton's *Opticks,* believed that heat was motion but that light was matter. He considered oxygen gas a compound not of oxygen and caloric (for there was no such substance), but of oxygen and light; and he suggested that it be called "phosoxygen."

Davy performed various experiments in support of his hypothesis, of which the most famous was the ice-rubbing experiment of melting ice by friction to prove the kinetic theory of heat. He took precautions against heat flowing into the system from outside, but these apparently were inadequate; there is considerable doubt as to whether Davy could have observed the effects he described. This work was independent of the more quantitative contemporary studies of Benjamin Thompson (Count Rumford). Davy sparked gunlocks *in vacuo* to see if light was emitted in the absence of oxygen and satisfied himself that it was not, even though heat was generated. Light could not, therefore, be a modification of heat. Sir Harold Hartley has cleared up the problem of how Davy obtained the apparatus for these experiments, by uncovering evidence of his friendship with Robert Dunkin, an instrument maker.

Davy wrote up his researches as "An Essay on Heat, Light, and the Combinations of Light." Beddoes declared himself a convert to Davy's views and published the essay in 1799, in a collection entitled *Contributions to Physical and Medical Knowledge, Principally From the West of England.* Davy later repudiated

the essay (*Nicholson's Journal of Natural Philosophy, Chemistry and the Arts,* **3** [1799–1800], 517), but the work is valuable for the light it casts on the working of his mind and for the new observations it contains. Joseph Priestley, among others, was impressed. In place of the theory-loaded term "caloric" Davy proposed "repulsive motion," a phrase that William Thomson later praised for its suggestiveness. In solids, attractive motions exceeded repulsive ones; in fluids the two were balanced; and in gases the repulsive motions exceeded the attractive. Light was another state in which the repulsive motion far exceeded the attractive, and the particles were therefore projected at great speed. This passage foreshadows the idea of a fourth state of matter that is found in the writings of Michael Faraday and the work of William Crookes, who believed the cathode rays to be such a state. Davy's light particles were little affected by gravity and did not contribute perceptible mechanical motion, although they did communicate repulsive motion.

Davy believed that the theory of caloric fluid had been proposed to explain the expansion of bodies with heat, "in conformity to the absurd axiom, *bodies cannot act where they are not*"; but this was to solve a small difficulty by creating a great one. Change of state should not be identified with combination with caloric, for nothing corresponding to chemical change occurred; the quantity of repulsive motion increased, but that was all. Bodies in different states were not compounds with more or less caloric—hydrogen and nitrogen were probably gaseous metals—and it was therefore as inappropriate to talk with the French nomenclators of "hydrogen gas" as it would be to speak of "solid gold." Thus "hydrogen" alone would be the proper term; but oxygen gas was different because it was a compound body containing light, and therefore it should be called "phosoxygen."

Davy was impressed by the importance of light and wrote that the planetary motions seemed to have been designed expressly to supply the solar system with the necessary quantity of light, and that light within and without us was the source of perception, thought, and happiness. He believed that the laws of refraction represented the laws of attraction of diaphanous bodies for light and that the different colors corresponded to different repulsive motions in the light particles. Since the red particles vibrated fastest, they were least attracted by the particles of bodies. Dark bodies contained less light than pale ones. In a passage suppressed in the final text—but referred to elsewhere in the paper—he suggested that electricity might be condensed light, given off at the poles as auroras, and that the phenomena of the mutations of matter probably all arose from the interconversions of the gravitative, mechanical, and repulsive motions. The blue color of air showed that the repulsive motion of light was diminished in passing through it. Phosphorescent bodies contained loosely combined light, whereas in phosoxygen the combination was intimate. In recapitulation, Davy declared that repulsive and attractive motions were the causes of effects uniformly and constantly produced, whereas caloric was imaginary.

As Davy moved into the biological realm, his speculations became more extreme; but he demonstrated photosynthesis in marine vegetation, and he showed the presence of carbon dioxide in venous blood. He wrote that nature had catenated all organic beings, making them mutually dependent and all dependent upon light, and that chemistry, through its connections with the laws of life, promised to become the most sublime and important of the sciences, bringing about the destruction of pain and the increase of pleasure.

In this essay Davy makes rhetorical, but suggestive, cosmological remarks: he attacks Lavoisier for introducing imaginary substances and a theory-loaded nomenclature; appeals to a Newtonian theory of heat, light, and matter; and indulges in speculations concerning light not uncharacteristic of those influenced by the Romantic movement. In his later work Davy kept these strands apart, so that the speculative and cosmological material appeared in his lectures and in his last works, *Salmonia* and *Consolations in Travel,* and in his poetry; in his papers in the *Philosophical Transactions* of the Royal Society, he sought to appear as a Newtonian, using the notion of *vera causa* to demolish "the principle of acidity" as he had used it to expel caloric. Coleridge and Southey praised Davy's poetry and his style of lecturing; it is hard today to work up much enthusiasm for his verses, but his prose works are still pleasing.

Davy met Coleridge and Southey at Clifton, for Beddoes had married an Edgeworth and moved in literary society; and he maintained a correspondence with both poets for some years. He also began a lifelong friendship with Thomas Poole, of Nether Stowey. The object of Beddoes' Pneumatic Institution was to investigate the possibility of using "factitious airs" (synthetic gases) in the cure and prevention of disease; in the course of his researches Davy tried breathing nitrous oxide and discovered its anesthetic properties. Samuel Mitchill, an American, had suggested that this gas was the principle of contagion, which must prove instantly fatal to anyone who respired it. Davy was not convinced by Mitchill's arguments and therefore made the experiment. He suggested that the gas might be employed in minor

surgical operations, but nobody took any notice of this recommendation. Instead, breathing nitrous oxide for the delightful feeling of intoxication became the rage; and Davy, in his book on nitrous oxide and its respiration, published in 1800, gave a series of subjective accounts of nitrous oxide anesthesia, which are among the best on record. But experiments at the Pneumatic Institution failed to reveal any great efficacy of nitrous oxide in curing disease. Davy proceeded to try other gases, including nitric oxide and carbon monoxide; he later warned his brother against perilous experiments of this kind.

Davy's book on nitrous oxide included chemical analyses of the oxides of nitrogen and studies of their reactions. In these, as in his later researches, Davy showed his great facility for qualitative experiment; on the quantitative side, his volumetric analyses were quite good, and he tended afterward to prefer volumetric analyses to more accurate gravimetric methods. Indeed, his laboratory work was characterized by bursts of intense labor, in which he attained his results with great rapidity and showed great ability in adapting existing apparatus to new functions. The accurate and painstaking analyses of Berzelius would not have been possible to one with Davy's temperament; in addition, he lacked formal training and, more important, he always wanted to be original and creative. To review, repeat, or confirm the findings of his contemporaries was much less congenial to him than to press on to new discoveries.

The book on nitrous oxide, which made Davy's reputation, is among his best productions, containing systematic and sustained argument supported by quantitative data. While the researches upon which the book is based were still in progress, Volta's invention of the pile was announced; and before Volta's paper appeared in print, Nicholson and A. Carlisle reported that they had used such a pile to decompose water into oxygen and hydrogen. Davy at once rushed into this new field and published a number of papers on the subject in *Nicholson's Journal of Natural Philosophy, Chemistry and the Arts* in the latter part of 1800. Of all those who worked on the pile, Davy consistently showed himself the clearest-headed. He appears to have realized from the first that the "contact theory"—that electricity was generated by the mere contact of dissimilar metals—was inadequate and that the production of electricity depended upon a chemical reaction's taking place. And in electrolytic cells he remained convinced—and ultimately proved— that the current acted to separate compounds into their components, rather than to synthesize new substances; the latter view enjoyed support in both France and Germany.

Davy found that the pile could not operate if the water between the plates contained no oxygen, and he concluded that the oxidation of the zinc was the cause of the generation of electricity. In accordance with this, he found that nitric acid in the pile was extremely—indeed alarmingly—effective. He established that a battery could be made with poles of charcoal and zinc and also with two fluids and one metallic component, provided one of the fluids would attack the metal. Shortly afterward he invented the carbon arc. He also employed cells in which the poles were in separate vessels connected by moist filaments of amianthus, a woven asbestos fiber.

These researches were interrupted in 1801, when Davy was appointed to a lectureship at the Royal Institution; in the following year he was promoted to professor of chemistry. The Royal Institution had been founded by Rumford, Thomas Bernard, and others largely to provide technical training; but Davy's predecessor, Thomas Garnett, had begun to attract large and fashionable audiences to lectures on science. Under Davy this trend grew stronger, until the Royal Institution became a center for advanced research and for polished demonstration lectures. From the start Davy succeeded in holding large audiences, which Garnett and Thomas Young had begun to lose; they increased until about 1810, when approximately 1,000 people flocked to hear him. The lectures kept the Royal Institution on a tolerably firm financial footing, but Davy seems to have regarded them as an interruption of the research program. Certainly his audiences were able to hear accounts of researches actually in progress and, in particular, to see the latest discoveries in electrochemistry demonstrated before their eyes.

In his first years at the Royal Institution, Davy was set to lecture on a number of technical subjects, and he published a long article on tanning; for this and his other papers he was awarded the Copley Medal in 1805. He gave lectures on a range of sciences and claimed to have given the first public course on geology in London. In 1802 he lectured before the Board of Agriculture on agricultural chemistry; this course was repeated each year until 1812 and was published in 1813. This was the first serious attempt to apply chemistry to agriculture, and it remained a standard work until displaced by Justus von Liebig's publications a generation later. The book is of interest because of its pioneering nature, and it went through a number of editions; but its value must be said to lie more in the impulse it gave toward the application of scientific methods in agriculture than in any of the theories advanced. Through these lectures Davy became acquainted with various great landowners and

attended such functions as the sheepshearing at Woburn.

In 1806 Davy was able to return to electrochemistry, and in November of that year it formed the subject of the Bakerian Lecture, which he delivered before the Royal Society. The first part of the lecture was concerned with the decomposition of water on electrolysis. In 1800 Davy had concluded that oxygen and hydrogen, in the theoretical proportions, were the only products of the electrolysis of pure water; and Berzelius and Hisinger had come to the same conclusion. But other workers had remarked on the presence of acid and alkali around the poles and had noticed that the theoretical yields of oxygen and hydrogen were not attained. Davy established that when pure water, redistilled in silver apparatus, was electrolyzed in gold or agate vessels in an atmosphere of hydrogen (so that nitrogen could not combine with the nascent hydrogen or oxygen), it was decomposed into oxygen and hydrogen only.

In the six years since the experiment of Nicholson and Carlisle, nobody had reasoned and experimented with a clarity approaching this. Davy then proceeded to discuss the use of electrolysis as a method of chemical analysis and the transport of substances during electrolysis, and to propose an electrical theory of chemical affinity. He found that if three vessels connected by filaments were used, with the poles in the outer pair, a neutral salt solution in one of the outer cups, and turmeric or litmus in each vessel, then only near the poles were the indicators affected. But with barium chloride in an outer cup and sulfuric acid in the center one, a precipitate was formed in the center cup. Davy's theory was very general, involving some kind of chain mechanism; Faraday was later able to list a dozen incompatible theories of electrolysis, all of which claimed to derive from Davy's views. Davy found that the electrical condition of a substance can modify its chemical properties; negatively electrified zinc is inert, and positively electrified silver is reactive. This, and that an electric current decomposed compounds, led Davy to propose that chemical affinity is electrical. But it was not his nature to raise his discoveries into systems, which he considered prisons of the mind; and thus he left it to Berzelius to develop the system of dualism from this insight. In "dualism," chemical elements were viewed as electrically positive or negative; they combined to form neutral products, which could be polarized and decomposed by an electric current.

In this lecture Davy did not announce any very important analyses, but in the Bakerian Lecture of 1807 he was able to describe the isolation of potassium and sodium. In 1806 he had remarked that the new methods of investigation promised to lead to a more intimate knowledge of the true elements of bodies. Lavoisier, while laying down the principle that the chemist was concerned not with elements but with those bodies which could not be decomposed, had nevertheless (inconsistently) refused to put the alkalies, soda and potash, on his list of undecomposed substances because of their analogies with ammonia. Davy found that if aqueous solutions of the alkalies were electrolyzed, only the water was decomposed. Dry potash did not conduct; but when, in October 1807, he employed slightly damp fused potash, he found that globules of silvery matter collected at the negative pole. Most of these caught fire, but some could be collected. At the positive pole pure oxygen was liberated. Davy danced about the room in ecstatic delight; he likened the potassium to substances imagined by alchemical visionaries. Because his potassium was impure, containing sodium, it was fluid at room temperature; and its extreme lightness also made it perplexing to classify. Davy at first called it "potagen"; but within a short time of the discovery he had decided that "the analogy between the greater number of properties must always be the foundation of arrangement," and he therefore called it "potasium" (*sic*), a name implying metallic status. By the time he delivered the Bakerian Lecture in November, Davy was able to announce many of the properties of sodium and potassium.

Lavoisier's remarks must have guided Davy in choosing the alkalies to investigate; and after his triumph Davy analyzed the alkaline earths, isolating magnesium, calcium, strontium, and barium; he then obtained boron and silicon. But his discovery that the alkalies were oxides was a puzzle. For Lavoisier oxygen was the principle of acidity; Davy had shown that with equal justice it might be called the principle of alkalinity. His work during the next four years or so was to establish that chemical elements do not behave as "principles" of this kind; he concluded that chemical properties were a function not simply of the components of a substance but also of their relative arrangements. Thus he finally put it beyond doubt that carbon and diamond were chemically identical; that neither all acids nor all alkalies contained oxygen; and that oxygen enjoyed no unique status as the supporter of combustion, but rather that heat was a consequence of any violent chemical change.

The Bakerian Lectures of 1808 and 1809 (Davy was called upon to deliver five of these lectures in succession) show him floundering as he tried one hypothesis after another to account for the ultimate constitution of matter and for the nature of acidity. He believed that the simplicity and harmony of nature demanded that there be very few ultimately distinct forms of matter; it is ironical that one who held that the chemical elements were probably all compounds should have been such a frequent discoverer of new elements. Davy was particularly confused by ammonium amalgam, a pasty material produced when ammonium salts are electrolyzed with a mercury cathode. For Davy and his contempories this proved the metallic nature of "ammonium" and made the compound nature of potassium and sodium, or the metallic nature of nitrogen or hydrogen, almost certain. Davy made the "phlogistic" conjecture that all metals might contain hydrogen, although at the same time he established that experiments of Gay-Lussac and Thenard, which purported to show that potassium was a compound of potash and hydrogen, were misleading.

A change came over Davy's work in 1809, when he embraced the doctrine of definite proportions; however, he never accepted John Dalton's atomic theory, believing it to be speculative. The nearest he came to a definite theory of matter was in a dialogue unfinished at his death, in which he adopted a quasi-Boscovichean atomism; but Davy was not the man to adhere to any theoretical system, believing, as he often wrote, that the only value of hypotheses was that they led to new experiments.

Work on tellurium established that hydrogen telluride, containing no oxygen, was an acid; Davy's first reaction was that hydrogen might be an oxide. But in 1810 the road at last opened in front of him again, and he realized that oxygen was not a constituent of all acids. Muriatic acid, our hydrochloric acid, was the critical case. Lavoisier had declared that it must be an oxide of an unknown substance or "basis," and oxymuriatic acid, our chlorine, a higher oxide. All attempts to extract the "basis" had failed; Davy found that white-hot charcoal extracted no oxygen from oxymuriatic acid and that when tin reacted with the gas and ammonia was added, no oxide of tin was formed. Further, when oxymuriatic acid and ammonia reacted, if both were dry, no water was formed; and when the solid compound of this gas and phosphorus was treated with ammonia, no phosphoric acid was obtained.

Gay-Lussac and Thenard had simultaneously been experimenting on this subject and had concluded that it would be possible to argue for the elementary nature of oxymuriatic acid; but they resisted Davy's arguments when he put them forward. He declared that oxygen was never produced in reactions in which oxymuriatic acid was concerned unless water was present, and concluded that no substance had a better claim to be regarded as undecomposed. Applying Lavoisier's criterion more rigidly than Lavoisier, he placed oxymuriatic acid among the elements as an analog of oxygen and gave it the theory-free name "chlorine." With the demolition of the scaffolding of the oxygen theory of acids, the view that all exothermic reactions were oxidations, and the caloric theory, Lavoisier's revolution was complete; and Davy's chemical career, to all intents and purposes, was over.

In 1812 he published part I of *Elements of Chemistry*, containing a very readable account of that part of the field in which he had worked. Reviewers guessed that part II, which would cover the rest of the science, would never appear, and they were right; Davy could never have produced a systematic treatise. In April 1812, he had been knighted by the Prince Regent, and three days later he married Jane Apreece. In 1813, after receiving from Michael Faraday a fair copy of notes of some of his lectures on chlorine, he offered Faraday a post as laboratory assistant. When he set off for France in the autumn of 1813, to claim the medal established by Napoleon and awarded him by the Institut de France for his electrical discoveries, Faraday accompanied him as his assistant and also acted as his valet. England and France were at war, but Davy was well received by his French colleagues and was elected a corresponding member of the Institute. In a race with Gay-Lussac he elucidated the nature of iodine, using a little case of apparatus he had brought with him, and determined its essential properties, anticipating his rival's more detailed account.

On his return to England in 1815, Davy was asked to turn his attention to the explosions in coal mines, which had recently been the cause of a number of disasters. He visited Newcastle-on-Tyne in August 1815 and arranged for samples of the explosive gas firedamp to be sent to him at the Royal Institution. He confirmed that methane was the main constituent and independently confirmed the observations of Tennant and Wollaston that this gas could be ignited only at a high temperature. Because of this it will not communicate flame through a narrow tube of sufficient length; the cooling effect of the tube is too great. Davy constructed lamps in which the air intake and the chimney were composed of narrow tubes, and found that they did not explode firedamp. He then found that wire gauze was equally efficient; and the Davy lamp, in which the flame is surrounded by wire gauze, was born. The whole research occupied less than three months, and is an example of Davy at his best. It led to research on flame, in which he concluded that luminosity is due to the presence of solid particles. It is sad to have to record that explosions in coal mines did not decrease in number; but coal production rose dramatically in the first half of the

nineteenth century as deeper and more dangerous pits could be worked.

Davy went abroad again in 1818 and attempted to unroll papyruses found at Herculaneum, using chlorine to decompose the gummy matter holding them together. In the summer of 1820 Davy heard of the illness of Sir Joseph Banks, who had become president of the Royal Society a few days before Davy was born; and he returned to London at once. Banks died, and it became clear that the main competitors for the chair would be Davy and W. H. Wollaston, who became acting president for the rest of the year. Davy was determined to win, and Wollaston withdrew before the election.

At this time the council of the Royal Society for the first time contained a majority of active scientists, and Davy hoped for greater endeavors from the fellows and for greater support of science from the government. In 1826 the Royal Medals were endowed by the king, but the government interest in science for which Davy had hoped did not materialize. He also hoped to convert the British Museum into a research institute rather like the Museum of Natural History in Paris; but again nothing came of this. The Royal Society was asked to look into the matter of the corrosion of the copper bottoms of ships, and Davy took this investigation upon himself. He found that if pieces of more electropositive metals, which he called "protectors," were fixed to the copper plates, the copper was not attacked by the seawater; unfortunately, in trials it appeared that marine organisms adhered so tenaciously to the protected copper that the performance of the ships was seriously affected, and the discovery was never taken up. The affair provided more ammunition for Davy's enemies, who considered him arrogant and high-handed.

In this period there were rather inconclusive electrical experiments, following Oersted's discovery of electromagnetic induction, and the experiments with Faraday on the liquefaction of gases in sealed tubes. A sad business was Davy's opposition to Faraday's election as a fellow of the Royal Society in 1824, for he had been generous to Faraday and really liked him. The affair reveals Davy's isolation and unhappiness. In 1826 he delivered his last Bakerian Lecture, adding very little of moment to those of 1806 and 1807.

Soon afterward Davy suffered his first stroke, and thereafter his life consisted of lonely journeys around Europe in search of health, fishing, and shooting. During this period he wrote a paper on volcanoes, based on observations of Vesuvius; Davy had always been interested in geology and, unlike Banks, had encouraged the Geological Society, and later other specialized societies, seeing their role as complementary to that of the Royal Society. He had suggested that volcanoes might have a core of alkali metal, acted on by water to cause eruption; but analyses of lava did not confirm this hypothesis, and he dropped it. It was Davy who persuaded R. I. Murchison to take up geology. Also from his last journeys came *Salmonia* and *Consolations in Travel,* dialogues in which Davy sought to communicate his world view; these went through many editions.

Students have disagreed over the extent to which Davy had a consistent philosophical position; he can be viewed simply as an opportunist, striking at weak points in Lavoisier's chemistry. Others see in the early papers, lectures, and *Consolations* evidence of a Romantic outlook reflected in Davy's friendships and perhaps exerting some influence upon his theorizing, particularly in his attitude toward the atomic theory. But the abiding impression gained from Davy's scientific works is of great experimental facility and of a determination to separate hypotheses from facts as far as possible.

BIBLIOGRAPHY

Davy's works were brought together as *The Collected Works of Sir Humphry Davy, Bart.,* John Davy, ed., 9 vols. (London, 1839–1840). A new edition, with an introduction by R. Siegfried, is being prepared. A bibliography has been compiled by J. Z. Fullmer (Cambridge, Mass., 1969). At the Royal Institution there is a large collection of Davy's MSS, including his notebooks. An edition of these is a desideratum.

Biographies of Davy are John Davy, *Memoirs of the Life of Sir Humphrey Davy, Bart.* (London, 1836); and *Fragmentary Remains, Literary and Scientific, of Sir Humphry Davy* (London, 1858); H. Hartley, *Humphry Davy* (London, 1967); and J. A. Paris, *The Life of Sir Humphry Davy* (London, 1831). J. Z. Fullmer is preparing a full-scale biography of Davy and an edition of his letters.

Recent articles on Davy include J. Z. Fullmer, "Humphry Davy's Adversaries," in *Chymia,* **6** (1960), 102–126; D. M. Knight, "The Scientist as Sage," in *Studies in Romanticism,* **6** (1967), 65–88; W. D. Miles, "Sir Humphry Davie, the Prince of Agricultural Chemists," in *Chymia,* **7** (1961), 126–134; C. A. Russell, "The Electrochemical Theory of Sir Humphry Davy," in *Annals of Science,* **15** (1959), 1–25; **19** (1963), 255–271; and R. Siegfried, "Sir Humphry Davy on the Nature of the Diamond," in *Isis,* **57** (1966), 325–335.

DAVID M. KNIGHT

DAVY, JOHN (*b.* Penzance, England, 24 May 1790; *d.* Ambleside, England, 24 January 1868), *physiology, chemistry, natural history.*

John Davy, youngest child and second son of Robert and Grace Millett Davy, undertook with honor and distinction the medical career originally intended for his brilliant older (by twelve years) brother, Humphry. Left fatherless at the age of four, Davy attended the schools of Penzance until 1808, when he went to London to assist his brother in the laboratory of the Royal Institution. He received no formal instruction there but was encouraged and guided so that he afterward wrote: ". . . in the Laboratory I acquired the habit of research with the love of labour to which whatever little success I may afterwards have attained, I mainly attribute."

During his years at the University of Edinburgh, 1810–1814, Davy successfully defended his brother's recently proposed views on the elemental nature of chlorine against the attacks by John Murray of that university. Out of this work he published the analyses and properties of the metallic chlorides, and first prepared, named, and characterized phosgene gas.

Davy received the M.D. in 1814, submitting a dissertation on the blood, and was soon after commissioned in the army as hospital assistant. Shortly before the battle of Waterloo he was assigned to the hospital at Brussels, and his experience there determined him on a career in the army medical service, in which he spent his entire professional life, eventually attaining the rank of inspector general of hospitals. He served many years abroad, in Ceylon, the Mediterranean islands, and the West Indies. Wherever he was stationed, Davy was a constant observer of the life habits and cultural practices of the native population, which he recorded with sympathy and detachment. He also dosed their bodies, took their temperatures, dissected their domestic and wild animals, and analyzed their minerals.

A diligent researcher and indefatigable writer, Davy derived his major scientific works from his medical service. *Researches Physiological and Anatomical* appeared in 1839, *Diseases of the Army* in 1862, and *Physiological Researches* in 1863. A voluminous output of papers on a broad miscellany of small topics reflects the range of opportunities that his travels presented to his ever alert but rather superficial curiosity.

Davy made a few original observations of significance, especially on the structure of the heart and circulatory system of the amphibians. His most sustained interest was in animal heat, and some of his observations on temperatures within the pulmonary circulation were widely cited at the time. He made thousands of temperature readings of men and beasts in all stages of health and sickness throughout the world; glowworms and elephants, clams and leopards, sharks, dogs, and chickens all felt his thermometer, but he seems not to have established any significant pattern or correlation.

Davy's observations, although more opportune than creative, are not without insight; and his experiments show ingenuity. However, he did not often distinguish between the significant and the merely curious. He avoided speculations, preferring the role of discoverer; and even the most obvious generalizations were put forward with brevity and caution.

Davy was a persistent and oversensitive defender of his brother's reputation. He wrote a two-volume biography in 1836 and edited the *Collected Works of Sir Humphry Davy* in 1839–1840 and a collection of letters and fragmentary works in 1858. He created a trivial quarrel with the gentle Faraday in the 1830's, and as late as the 1860's he was still defending Humphry's conduct as president of the Royal Society against renewed attacks by Charles Babbage.

Although Davy's scientific work is not of first rank, the wealth of original documentation provides a splendid opportunity for a significant study of a man whose professional career and attitudes are not unrepresentative of his times.

BIBLIOGRAPHY

I. ORIGINAL WORKS. John Davy's books, all published in London, are readily classed into four categories:

Professional works are *Researches Physiological and Anatomical,* 2 vols. (1839); *Lectures on the Study of Chemistry and Discourses on Agriculture* (1849); *On Some of the More Important Diseases of the Army, With Contributions to Pathology* (1862); and *Physiological Researches* (1863).

Travel books are *An Account of the Interior of Ceylon* (1821); *Notes and Observations on the Ionian Islands and Malta: With Some Remarks on Constantinople, Turkey, and on the System of Quarantine as at Present Conducted,* 2 vols. (1842); and *The West Indies Before and Since Slave Emancipation* (1854).

Biographical works are *Memoirs of the Life of Sir Humphry Davy, Bart.,* 2 vols. (1836); *Collected Works of Sir Humphry Davy, Bart.,* which he edited, 9 vols. (1839–1840); and *Fragmentary Remains, Literary and Scientific, of Sir Humphry Davy* (1858).

His piscatory colloquies are *The Angler and His Friend* (1855); and *The Angler in the Lake District* (1857).

The *Royal Society Catalogue of Scientific Papers* lists 168 papers by Davy. Several dozen personal notebooks kept by John Davy are in the archives of the Royal Institution in London; six more are in the possession of the Royal Geological Society of Cornwall, and three are in the Morrab Gardens Library, both in Penzance. The library of Keele University, Staffordshire, owns the six notebooks used in the preparation of the *Memoirs* of his brother, along with a few miscellaneous papers including an incomplete autobiography.

II. SECONDARY LITERATURE. No significant study of John Davy has been attempted. Brief accounts are found in the *Dictionary of National Biography; Proceedings of the Royal Society,* **16** (1868), lxxix–lxxxi; *Proceedings of the Royal Society of Edinburgh* (1869), 288–291; and *Medical Times and Gazette* (1868), **1,** 160–161, and (1871), **2,** 390–391.

ROBERT SIEGFRIED

DAWES, WILLIAM RUTTER (*b.* London, England, 19 March 1799; *d.* Haddenham, near Thame, Oxfordshire, England, 15 February 1868), *astronomy.*

Dawes's mother died when he was very young, and since his father was often abroad on colonial service, he was brought up by relatives and friends and his schooling was several times interrupted. At first he intended to become a clergyman, but instead he studied medicine at St. Bartholomew's Hospital and afterward worked a country practice in Berkshire. In 1826, following the death of a sister, Dawes moved to Liverpool, and there he came under the influence of a Dissenting minister who persuaded him to take charge of a small congregation in Ormskirk.

Dawes had been interested in astronomy as a boy, and while at Liverpool he often observed the stars through an open window with a small but excellent refracting telescope. This refractor aroused his interest in double stars, and at Ormskirk he constructed an observatory with a five-foot Dollond refractor that had an aperture of 3.8 inches, which he used to make careful micrometrical measurements of double stars. His measures of 121 double stars made in the period 1830–1833 were published in 1835, and those of 100 double stars in the period 1834–1839 were published in 1851. Chronic ill health forced Dawes to give up his pastoral work, and in 1839 he left Ormskirk to take charge of George Bishop's observatory in Regent's Park. There he continued to devote himself to double stars, and his measurements of about 250 such stars were published in 1852 in Bishop's *Astronomical Observations at South Villa.* His results included the detection of orbital motion in ϵ Hydrae and of third components of γ Andromedae and Σ 3022.

In 1844 Dawes left Bishop's observatory and went to live near Cranbrook, Kent. There he equipped himself with a transit circle by William Simms that was two feet in diameter and an equatorial telescope by Georg Merz of $8\frac{1}{2}$-foot focus and six-inch aperture with a delicate clockwork movement. Unfortunately his headaches and asthma continued, and these forced him for a time to live at the seaside resort of Torquay. In 1850 Dawes felt well enough to move to Maidstone, Kent, and there he observed the crape ring of Saturn on 25 and 29 November, losing priority to the Bonds by only ten days. At this period "the eagle-eyed

Dawes" was establishing himself as a leading observer of Saturn through numerous meticulous observations of the planet and especially of the various rings. These observations and his double-star measurements led to the award of the Gold Medal of the Royal Astronomical Society in 1855.

In 1857 Dawes moved to Haddenham, near Thame, Oxfordshire, where he continued his observations despite rapidly deteriorating health. His second wife, whom he had married in 1842, died in 1860. Dawes was elected a fellow of the Royal Society in 1865 and continued to observe at intervals until 1867. He died at Haddenham, 15 February 1868.

BIBLIOGRAPHY

I. ORIGINAL WORKS. Dawes's scientific publications are contained in George Bishop, *Astronomical Observations at South Villa* (London, 1852), and in numerous papers in the *Memoirs* and *Monthly Notices* of the Royal Astronomical Society. The most important are "Micrometrical Measurements of the Positions and Distances of 121 Double Stars, Taken at Ormskirk, During the Years 1830, 1831, 1832, and 1833," in *Memoirs of the Royal Astronomical Society,* **8** (1835), 61–94; "Micrometrical Measurements of Double Stars, Made at Ormskirk Between 1834.0 and 1839.4," *ibid.,* **19** (1851), 191–212; and "Catalogue of Micrometrical Measurements of Double Stars," *ibid.,* **35** (1867), 137–502.

II. SECONDARY LITERATURE. See the following, listed chronologically: obituary notice in "Report of the Council to the Forty-ninth Annual General Meeting," in *Monthly Notices of the Royal Astronomical Society,* **29** (1868–1869), 116–120; Agnes Clerke, "William Rutter Dawes," in *Dictionary of National Biography,* V, 667–669; and A. F. O'D. Alexander, *The Planet Saturn* (London, 1962), chs. 10–12, 14, *passim.*

MICHAEL HOSKIN

DAWSON, CHARLES (*b.* Fulkeith Hall, Lancashire, England, 11 July 1864; *d.* Uckfield, Sussex, England, 10 August 1916), *paleontology.*

The son of Hugh Dawson, a barrister, Charles Dawson followed his father into the legal profession and became a solicitor upon his graduation from the Royal Academy, Gosport, in 1880. He practiced in Hastings from 1880 to 1890, moving thence to Uckfield, where he remained until his death.

Dawson's avocation, pursued diligently from boyhood, was paleontology. By the age of twenty he had assembled a sizable collection of fossils from the Weald formations around Hastings, which he donated to the British Museum. The Dawson Collection, as it is known, contains several valuable specimens, among them one of the finest extant examples of *Lepidotus mantelli,* the Weald ganoid fish. In recognition of his acumen the museum made Dawson an honorary collector.

He discovered a new species of dinosaur, *Iguanodon dawsoni,* and a new Weald mammal, *Plaugiaulax dawsoni.* At his urging Felix Pelletier and Teilhard de Chardin, then students at the Jesuit college in Hastings, explored the Weald bone beds, finding there *Diprioden valensis,* a second new Weald mammal form.

Dawson was interested in geology as well as paleontology. He reported his discovery of natural gas at Heathfield to the Geological Society of London in 1898 and exhibited zinc blende from the Weald to the same body in 1913. He was elected fellow of the Geology Society at the age of twenty-one, a remarkable achievement for an amateur and one so young.

Archaeology was another field in which Dawson became renowned. He was a fellow of the Society of Antiquaries and a member of the Sussex Archaeological Society. His esteem among the members of the latter group suffered considerably when some excavations he did at their request at the Lavant caves were of exceedingly poor quality. In 1909 Dawson published a two-volume history of Hastings Castle that received wide circulation. Of it Weiner notes, however, that it was "less a product of genuine scholarship than of extensive plagiarism" (p. 176).

Dawson's name is associated, moreover, with one of the blacker events in the history of human paleontology: the Piltdown fraud. It was he who discovered the fragments of *Eoanthropus* in what was supposed to be Red Crag gravel at Barkham Manor between 1909 and 1912, and it was most likely he who had planted them there.

Reconstruction of Piltdown man from Dawson's fragments yielded a creature with a more recent human cranium coupled to a simian mandible. This provoked lively controversy between those who accepted a genetic association of skull and jaw and those who did not. The latter group did not in general challenge the authenticity of the bones, merely their occurrence in the same individual. Dawson had a solid reputation, was respected by many, and was championed by such esteemed scientists as Arthur Smith Woodward and Arthur Keith. Moreover, his claims were reinforced by his discovery of a second Piltdown specimen at Sheffield Park in 1915.

At the time only three forms of early man had been brought to light, and they were in incomplete and imperfect condition: Neanderthal man (1856), Java man (1891), and Heidelberg man (1907). The discovery of fossil forms in China, Java, and Africa in the following decades rendered the Piltdown reconstruction utterly incredible, for in these forms development of jaw preceded development of brain. This left the *Eoanthropus* the sole example of a completely divergent evolutionary line, a bizarre freak rather than an ancestor.

An alternative solution to the puzzle was to reject the assumption that the bones were of great antiquity. By the early 1950's techniques for accurately determining a fossil's age had been developed. Using them, Kenneth P. Oakley, J. S. Weiner, and W. E. Le Gros Clark found that Piltdown man was nothing but an elaborate fraud. Cranium and mandible were from different species: the former from an Upper Pleistocene human, the latter from a modern orangutan. The fragments had been stained with iron sulfate so that they would be the same color and the teeth had been artificially abraded to simulate human wear. Even the "credentials" of *Eoanthropus*—the bone and flint tools and the Villefranchian fauna discovered with it—were shown to be deliberate intrusions.

If Dawson did not actually mastermind this hoax, his complicity is indicated. He could scarcely have been an innocent bystander, for his actions always placed him in circumstances where he could not have avoided knowing what the forger was about to do. He died long before the mystery was solved, so the motives for his involvement could not be ascertained. It is unlikely that they will ever be known.

BIBLIOGRAPHY

I. ORIGINAL WORKS. Dawson's papers include "Discovery of a Large Supply of Natural Gas at Waldron, Sussex," in *Nature,* **57** (1897–1898), 150–151; "Ancient and Modern 'Dene Holes' and Their Makers," in *Geological Magazine,* 4th dec., **5** (1898), 293–302; "On the Discovery of Natural Gas in East Sussex," in *Quarterly Journal of the Geological Society of London,* **54** (1898), 564–571; "List of Wealden and Purbeck-Wealden Fossils," in *Brighton Natural History Society Reports* (1898), 31–37; "On the Discovery of a Palaeolithic Skull and Mandible in a Flint-Bearing Gravel at Piltdown, Fletching (Sussex)," in *Quarterly Journal of the Geological Society,* **69** (1913), 117–151, written with A. S. Woodward; "Supplementary Note on the Discovery of a Palaeolithic Skull and Mandible at Piltdown," *ibid.,* **70** (1914), 82–99, written with A. S. Woodward; and "On a Bone Implement From Piltdown (Sussex)," *ibid.,* **71,** no. 1 (1915), 143–149, written with A. S. Woodward. A book by Dawson is *History of Hastings Castle,* 2 vols. (London, 1909).

II. SECONDARY LITERATURE. On Dawson or his work, see Aleš Hrdlička, "Skeletal Remains of Early Man," in *Smithsonian Miscellaneous Collections,* **81** (1930), 65–90; J. S. Weiner, *The Piltdown Forgery* (London, 1955); and A. S. Woodward, "Charles Dawson," in *Geological Magazine,* 6th dec., **3** (1916), 477–479.

MARTHA B. KENDALL

DAWSON, JOHN WILLIAM (*b.* Pictou, Nova Scotia, 13 October 1820; *d.* Montreal, Quebec, 19 November 1899), *geology.*

Dawson's father, James, was prominent in the business community of Pictou. Dawson obtained the master of arts degree at Edinburgh University in 1841 and returned to Pictou determined to follow geology as a profession. He was fortunate to have met both Charles Lyell, the foremost British geologist, and William Logan, first director of the Geological Survey of Canada, for both of whom he maintained a lifelong admiration and friendship. In 1842 the province of Nova Scotia engaged him to make a geological survey of its coalfields.

Dawson revisited Edinburgh University in 1846 for specialized instruction in natural science, returning in 1847 with his Scottish bride, Margaret Mercer. From 1847 to 1850 he carried out various researches in local geology, conducted courses of lectures and laboratory instruction at Dalhousie College (Halifax, Nova Scotia), and published several scientific papers. In 1850 he was appointed superintendent of education for Nova Scotia, in which capacity he visited every corner of the province; and in his spare time he amassed sufficient geological information to be able to publish his first important work, *Acadian Geology* (1855), which influenced geological thought and work in Nova Scotia until well after the end of the century.

So well did Dawson become known in educational circles that in 1855 he was offered the principalship of McGill University in Montreal, which he accepted, stipulating that he should also occupy the chair of natural history. Arriving at McGill, he was dismayed to find the staff meager, the students few, the buildings wretched, and funds inadequate. Yet almost singlehanded, by dint of perseverance, vision, and devotion, and by appeals to the board of governors and the business community of Montreal, within a score of years he had transformed an academic backwater into a progressive institution; and when he retired in 1893, McGill had become an internationally respected university. Although this accomplishment would have occupied all of the time and energy of an ordinary man, Dawson found time to publish an average of ten scientific papers a year (and probably a like number on educational, social, and religious matters); to take an active part in the Montreal Natural History Society, never missing a meeting or a field excursion, and serving for many years as president; to institute the McGill University library and to serve as its first librarian; to give courses in agriculture, botany, chemistry, geology, paleontology, and zoology, gradually relinquishing all but geology and paleontology; and to help found the Montreal Normal School (he had earlier helped to establish a normal school for Nova Scotia) and for thirteen years to serve as its principal. His interest in higher education for women played a major part in the founding of the Montreal High School for Girls in 1874; and he subsequently arranged, over stubborn opposition, for the admission of women to candidacy for the bachelor of arts degree.

Dawson was brought up in an atmosphere of Presbyterian fundamentalism; and his lifelong, unwavering adherence to this outlook directed and characterized all of his activities. A hundred or more religious articles and a dozen books of a popular nature show his zeal for the propagation of his ideas concerning the relationship between science and religion.

Dawson was elected fellow of the Geological Society of London (1854), Royal Society of London (1862), Academy of Arts and Sciences (Boston), American Philosophical Society, and Geological Society of America (1886; president, 1893). He served as president of the British Association for the Advancement of Science in 1886 and of the American Association for the Advancement of Science in 1882. The latter year he was the first president of the Royal Society of Canada. In 1881 he was awarded the Lyell Gold Medal by the Geological Society of London. He was knighted by Queen Victoria in 1884.

Although in his research Dawson neglected no phase of contemporary geology, it was in paleobotany that his reputation was greatest and most enduring. His treatment of fossil plants in *Acadian Geology* shows that by 1855 he was in full command of contemporary paleobotanical techniques. Moreover, he had already demonstrated in his survey of the Nova Scotia coalfields that the coal beds belonged to three separate stratigraphic units, a concept that showed him to be far ahead of his time. In 1859 he published a description of *Psilophyton,* then the earliest land plant known, which he had found in Devonian strata in Gaspé. Scientists were reluctant to accept so strange a plant until in 1917 Kidston and Lang described similar fossils from Scotland, scarcely improving upon Dawson's original descriptions. In 1884 Dawson began a study of the Cretaceous and Tertiary fossil plants from western Canada for the Geological Survey of Canada; a long series of papers, mostly in the *Transactions of the Royal Society of Canada,* recorded his results. Throughout his career hardly a year passed without several papers on paleobotanical subjects being published; and his *Geological History of Plants* (1888) was a textbook practically without a competitor for several decades.

Dawson's work on paleozoology was less uniform. He interpreted eozoon, now generally agreed to be a product of the metamorphism of limestone, as the fossil of a giant foraminifer—a view which, in the face of mounting and massive opposition, he neither modified nor abandoned. His bibliography contains more than a dozen papers and two books either justifying his stand or attempting to destroy opposing views. His discovery and description of Cambrian(?) sponges at Little Métis, Quebec, was so well done that to date no revision has been necessary. He assembled and described much of the postglacial arctic fauna found in sands and clays of glaciated eastern Canada, embodying this material in *The Canadian Ice Age.* His recognition of the earliest land snails was first recorded in 1853 as a result of a second visit to Joggins, Nova Scotia, with Sir Charles Lyell. This and later discoveries were brought together in *Air Breathers of the Coal Period* (1863). In that publication he also summed up his discoveries of amphibia and reptiles extracted from the fillings of tree stumps exposed along the Joggins shore. Some thirty of his publications were concerned with these vertebrate fossils. Almost immediately upon his arrival at McGill, excavations for new buildings on the college campus revealed artifacts and human bones that Dawson collected and preserved, assigning the site to the village of Hochelaga, visited by Jacques Cartier in 1535. From this he quite naturally progressed to a study of the North American Indians and thence to prehistoric human remains in Europe. His theories and conclusions are contained in his book *Fossil Men* (1880).

Throughout his paleontological writings Dawson stoutly inveighed against the rising tide of evolutionary thought. Chance variations, favored by the Darwinians, left no room for a divine guiding hand and so were rejected as factors in the production of past and present biotas.

In the field of geology, apart from paleontology, Dawson contributed much to the knowledge of the geology of Canada. His first important work, *Acadian Geology,* has already been noted. He never accepted the idea of continental glaciation and lost no opportunity in both popular and scientific writings to combat it; nevertheless, his *Canadian Ice Age* (1893) was, for its time, a remarkable treatment of events succeeding the Pleistocene glaciation. In the handbooks of Canadian geology, of invertebrate zoology, and of agriculture, he compressed vast amounts of information into easily digested form for classroom use.

Critical assessments of Dawson's scientific work too often have emphasized his intransigent attitudes concerning eozoon, organic evolution, and continental glaciation. Glaring though these faults may appear to us today, they were amply compensated for by his positive accomplishments in many fields of science.

BIBLIOGRAPHY

I. ORIGINAL WORKS. Among Dawson's scientific writings are "On the Remains of a Reptile (*Dendrerpeton acadianum* Wyman and Owen) and of a Land Shell Discovered in the Interior of an Erect Fossil Tree in the Coal Measures of Nova Scotia," in *Quarterly Journal of the Geological Society of London,* **9** (1853), 58–63; written with Charles Lyell; *Acadian Geology; an Account of the Geological Structure and Mineral Resources of Nova Scotia and Neighboring Provinces of British America* (Edinburgh, 1855; 2nd ed., 1868; 3rd ed., 1878; 4th ed., 1891); *Air Breathers of the Coal Period* (Montreal, 1863), also in *Canadian Naturalist and Geologist,* **8** (1863), 1–12, 81–88, 159–160, 161–175, 268–295; *First Lessons in Scientific Agriculture for Schools and Private Instruction* (Montreal–Toronto, 1870), new ed. (1897) written with S. P. Robins; *Handbook of Zoology, With Examples From Canadian Species, Recent and Fossil. Invertebrata. Part I* (Montreal, 1870; 3rd ed., 1886); *The Fossil Plants of the Devonian and Upper Silurian Formations of Canada,* Publications of the Geological Survey of Canada, pt. 1 (1871) pp. 1–92; pt. 2 (1882), pp. 93–142; *Handbook of Geology for the Use of Canadian Students* (Montreal, 1871; 1880; 1889); *Notes on the Post-Pliocene Geology of Canada* (Montreal, 1872); "Eozoön canadense," in *Nature,* **10** (1874), 1–103; "On the Results of Recent Explorations of Erect Trees Containing Animal Remains in the Coal Formation of Nova Scotia," in *Philosophical Transactions of the Royal Society,* **173** (1882), 621–659; *On Specimens of Eozoön canadense and Their Geological and Other Relations* (Montreal, 1888); *The Geological History of Plants,* International Scientific Series, no. 61 (New York, 1888); "On Burrows and Tracks of Invertebrate Animals in Paleozoic Rocks, and Other Markings," in *Quarterly Journal of the Geological Society of London,* **46** (1890), 595–617; *The Canadian Ice Age* (Montreal, 1893); and "Additional Notes on Fossil Sponges and Other Organic Remains From the Quebec Group at Little Métis on the Lower St. Lawrence; With Notes on Some of the Specimens by Dr. G. J. Hinde," in *Proceedings and Transactions of the Royal Society of Canada,* 2nd ser., **2,** sec. 4 (1896), 91–121.

In the following popular books Dawson sought to harmonize science with religion: *Archaia, or Studies of the Narrative of the Creation in Genesis* (Montreal, 1857); *The Story of Earth and Man* (London, 1872; Montreal–Toronto, 1873; 9th ed., London, 1887); *The Dawn of Life* (Montreal, 1875); also published as *Life's Dawn on Earth; Being the History of the Oldest Known Fossil Remains* (London, 1875); *The Origin of the World, According to Revelation and Science* (Montreal, 1877; 6th ed., London, 1893); *Fossil Men and Their Modern Representatives* (Montreal, 1880); *The Chain of Life in Geologic Time* (London, 1880; 3rd ed., 1888); *Facts and Fancies in Modern Science* (Philadelphia, 1882); *Egypt and Syria, Their Physical Features in Relation to Bible History* (Oxford, 1885); *Modern Science in Bible Lands* (Montreal–London, 1888; rev. ed., 1892); *Modern Ideas in Evolution as Related to Revelation and Science* (London, 1890); *Some Salient Points in the Science of the Earth* (London, 1893; New York, 1894); *The Meeting Place of Geology and History* (New York–Toronto–Chicago, 1894); and *Relics of Primeval Life* (Chicago–London, 1897).

Dawson's autobiographical notes were published as *Fifty Years of Work in Canada: Scientific and Educational,* Rankine Dawson, ed. (London–Edinburgh, 1901).

II. SECONDARY LITERATURE. On Dawson or his work, see F. D. Adams, "Memoir of Sir J. William Dawson," in *Bulletin of the Geological Society of America,* **11** (1899), 550–557, with portrait; H. M. Ami, "Sir John William Dawson. A Brief Biographical Sketch," in *American Geologist,* **26,** no. 1 (1900), 1–48, with a bibliography of over 500 titles on 19–48; T. H. Clark, "Sir John William Dawson 1820–1899," in *Pioneers of Canadian Science,* G. F. G. Stanley, ed. (Toronto, 1966), pp. 101–113; E. A. Collard, "Lyell and Dawson: A Centenary," in *Dalhousie Review,* **22** (1942), 133–144.

T. H. CLARK

DAY, DAVID TALBOT (*b.* Rockport [now Lakewood], Ohio, 10 September 1859; *d.* Washington, D.C., 15 April 1925), *chemical geology.*

Day's father, Willard Gibson Day, was a not very prosperous Swedenborgian minister. His mother, Caroline Cathcart Day, was remarkably gentle and understanding; her son acquired these traits. The family moved to Baltimore, Maryland, when Day was quite young; and after completing high school there, he attended Johns Hopkins University, where he received the A.B. in 1881. Much interested in chemistry, he studied under Ira C. Remsen, to whom he was an assistant before and after receiving the Ph.D. in 1884. He supplemented a college fellowship by teaching chemistry at a private school. In 1886 he married Elizabeth Eliot Keeler.

From 1884 to 1886 Day was demonstrator in chemistry at the University of Maryland. Already interested in minerals, he wrote reports on manganese, chromium, and tungsten for the U.S. Geological Survey report on resources; and in 1885 he began his long career with that organization. When Albert Williams, Jr., the organizer of the statistical division of the Geological Survey, resigned in 1886, Day was appointed to be his replacement by the director, John Wesley Powell. For the census of 1890 he presented a detailed account of mineral statistics, the first ever gathered.

As the new field of petroleum geology came into prominence, Day took an active interest, soon turning to the study of oil shales. With E. G. Woodruff he completed the first survey of the Green River oil shale beds of northwestern Colorado and northeastern Utah, and he urged the Geological Survey to make detailed maps of other, similar deposits. For the Navy he compiled figures on reserves of petroleum and oil shale. Although the discovery of vast reserves of petroleum has postponed the development of oil-shale reserves, Day's carefully compiled estimates are the basis for their effective development.

His extensive interest in oil shales led to Day's becoming consulting chemist in that subject for the Geological Survey in 1907; in 1914 he was transferred to the Bureau of Mines in the same capacity. In 1920 he left government service to conduct his own research into methods of distilling oil shale and into perfecting the cracking process for converting oil into gasoline and other petroleum distillates. For these studies he left Washington to establish headquarters nearer the subject, at Santa Maria, California. In 1922 he published his invaluable *Handbook of the Petroleum Industry,* a two-volume reference on production and use of oil.

Day was a member of several geological, chemical, and geographic societies and served as vice-president of the American Institute of Mining and Metallurgical Engineers in 1893 and in 1900.

BIBLIOGRAPHY

I. ORIGINAL WORKS. Day's major written contributions were the exhaustive annual reports on mineral resources of the United States for the U.S. Geological Survey (1885–1912), consisting of thousands of pages of detailed summaries. His other scientific papers on individual minerals and resources were few. His culminating work was the compilation of the *Handbook of the Petroleum Industry,* 2 vols. (New York, 1922).

II. SECONDARY LITERATURE. Only two accounts of Day's life seem to be extant: a memorial by M. R. Campbell in *Transactions of the American Institute of Mining and Metallurgical Engineers* (1926), 1371–1373; and one by N. H. Darton in *Proceedings of the Geological Society of America* for 1933 (1934), 185–191. The latter includes a bibliography.

ELIZABETH NOBLE SHOR

IBN AL-DĀYA. See **Aḥmad ibn Yūsuf.**

DEAN, BASHFORD (*b.* New York, N.Y., 28 October 1867; *d.* Battle Creek, Michigan, December 6, 1928), *ichthyology.*

In early childhood Dean showed keen interest in two fields—natural history and ancient armor—in which he excelled equally. His interests were encouraged by his lawyer father, William Dean; by his mother, Emma Frances Bashford; and by their distinguished friends, who enabled the boy to enter the College of the City of New York at the age of fourteen. Dean taught natural history at the college from his graduation in 1886 to 1891, while working for his Ph.D. (1890) at Columbia University, where he became the protégé of John Strong Newberry and heir to his vast collections of Devonian fishes. In 1889 Dean studied oyster culture for the U.S. Fish Commission. In 1890, as the first director of the summer school of biology at Cold Spring Harbor, New York, he helped establish a major center of biological research. He married Alice Dyckman in 1893.

In 1891 Dean began teaching at Columbia University, where he became professor of zoology in 1904. Through his close association with Henry Fairfield Osborn, in 1903 he became curator of reptiles and fishes (honorary from 1910) at the American Museum of Natural History, where he directed the installation of fine habitat groups of primitive fishes and restorations of fossil fishes.

Through a judicious combination of embryology and paleontology, Dean solved several problems in the evolution of primitive fishes. Working especially with the Devonian *Cladoselache newberryi,* he deduced that the pectoral and pelvic fins were derived from continuous fin folds, as opposed to Karl Gegenbaur's theory of feather-like archipterygia as a source. In addition, he concluded that the curious chimaeroid fishes are highly specialized offshoots from the true sharks; he presented in detail the embryology of the Port Jackson shark and the Japanese frilled shark; and he described the embryology, spawning, and nesting habits of several freshwater ganoid fishes. From studies of *Bdellostoma* (now *Eptatretus*) he described the development of hagfishes and their distinction from lampreys. He concluded that the Devonian arthrodires constituted an independent class of chordates, although Erik A. Stensiö later showed them to be related to ancestral sharks.

Dean's meticulous artistry left a wealth of illustrations of fishes and embryology for later workers. His ambidextrous blackboard drawings delighted his students.

His exhaustive three-volume *Bibliography of Fishes* immortalized Dean's name in ichthyology and promptly won for him, in 1923, the Daniel Giraud Elliot Medal of the National Academy of Sciences. In 1910 he received the Lamarck Medal and was made a *chevalier* of the Legion of Honor.

Dean's childhood interest in medieval armor became a lifelong passion and resulted in many scholarly articles and an outstanding collection for the Metropolitan Museum of Art (New York), where he was honorary curator of arms and armor from 1906 to 1927. During World War I, as a major in ordnance, he designed protective armor. Dean's extensive travels and contagious enthusiasm enabled him to collect armor and fishes simultaneously for both museums.

BIBLIOGRAPHY

I. Original Works. Dean's sense of organization led him to index references on fishes from 1890, and he enlisted the aid of the American Museum of Natural History in completing and publishing the *Bibliography of Fishes,* 3 vols. (New York, 1916–1923), with the aid of C. R. Eastman, E. W. Gudger, A. W. Henn, and others.

Dean's writing was characterized by compactness and unusual clarity. An early publication was the textbook *Fishes, Living and Fossil,* no. 3 in Columbia University Biological Series (New York, 1895), which dealt mainly with the lower and older forms. His most valuable paper on hagfishes was "On the Embryology of *Bdellostoma stouti,*" in *Festschrift für Carl von Kuppfer* (Jena, 1899). His work on fin origin was summarized in "Historical Evidence as to the Origin of the Paired Limbs of Vertebrates," in *American Naturalist,* **36** (1902), 767–776. A significant monograph was *Chimaeroid Fishes and Their Development,* Carnegie Institution of Washington, Publication no. 32 (Washington, D. C., 1906). Dean's studies on the development of fish embryos were summarized in "The Plan of Development in Series of Forms of Known Descent and Its Bearing Upon the Doctrine of Preformation," in *Proceedings of the 7th International Zoological Congress* (Boston, 1909). Many other papers represent significant contributions on primitive and archaic fishes. In addition Dean published many articles on medieval armor, mainly in the *Bulletin of the Metropolitan Museum of Art.*

II. Secondary Literature. *The Bashford Dean Memorial Volume,* E. W. Gudger, ed., 2 vols. (New York, 1930), includes many of Dean's drawings. The opening article, an account of his life and accomplishments by W. K. Gregory, includes a complete bibliography. Other memorial sketches on Dean are listed in vol. I, p. 35.

Elizabeth Noble Shor

DE BARY, (HEINRICH) ANTON (*b.* Frankfurt-am-Main, Germany, 26 January 1831; *d.* Strassburg, Germany [now Strasbourg, France], 19 January 1888), *botany.*

Heinrich Anton de Bary's extensive and careful observations of the life histories of the fungi and his contributions on the algae and higher plants were landmarks in the increase of knowledge. Recognized as the foremost mycologist of his day, he is regarded as the founder of that branch of botany.

De Bary was one of ten children born to August Theodor de Bary, a physician, and Emilie Meyer de Bary. His parents encouraged his early propensities toward the study of natural science, and as a youth in Frankfurt he frequently joined the excursions of the active group of naturalists who collected specimens in the nearby countryside. Georg Fresenius, a physician who taught botany at the Senckenberg Institute and was an expert on the thallophytes, guided de Bary's youthful interest in plants and in the examination of fungi and algae. After graduating in 1848 from the Gymnasium in Frankfurt, de Bary began the study of medicine at Heidelberg, continued at Marburg, and from 1850 pursued his studies at Berlin. There he had the opportunity to study under the botanist Alexander Braun, who as teacher and longtime friend communicated to him his own enthusiasm for botany. De Bary was a pupil also of Christian Gottfried Ehrenberg, known for his work on the Infusoria, and of the physiologist and comparative anatomist Johannes Müller, who exercised great influence on the students in his laboratory; de Bary was, some years later, to found laboratories in botany himself and to guide his students in a broad range of investigations. When de Bary received his degree in medicine at Berlin (1853) his dissertation was on a botanical subject and was entitled, *De plantarum generatione sexuali.* The same year he published a book on the fungi that caused rusts and smuts in plants.

De Bary entered the practice of medicine in Frankfurt but soon found himself drawn back to botany. Giving up his medical practice, he became *Privatdozent* in botany at the University of Tübingen. Here Hugo von Mohl, the noted plant anatomist, taught, and de Bary was for a while his assistant. In 1855 de Bary succeeded the botanist Carl von Naegeli at Freiburg im Breisgau.

De Bary established a botanical laboratory at Freiburg; it was the first laboratory of botany and represented a new development in botanical teaching, although its facilities were simple. Here he received the first of the many students who were to gather about him.

De Bary married Antonie Einert in 1861, and four children were born to them. In 1867 he succeeded D. F. L. von Schlechtendal at the University of Halle, where he established another botanical laboratory. Upon the death of Schlechtendal, who with Mohl had founded the *Botanische Zeitung,* de Bary became coeditor of the publication, and he served at times as sole editor. His first paper had appeared in this journal when he was a medical student and through it—both editorially and as a contributor—

he exercised an important influence upon the field of botany.

After the Franco-Prussian War, de Bary was appointed professor of botany at the rechartered University of Strassburg and was appointed its first rector by unanimous vote. The well-appointed botanical institute that he founded there attracted students from Europe and America, and he encouraged them to pursue a diversity of interests. He participated in the activities of various botanical societies while continuing in his editorial capacities. He remained at Strassburg until his death.

De Bary clarified the understanding of the life histories of various fungi at a time when they were still considered by some to arise through spontaneous generation. In 1853, in *Untersuchungen über die Brandpilze,* writing on the fungi that produced rusts and smuts in cereals and other plants, de Bary had already come to the conclusion that these fungi were not the products of the cell contents of the affected plants, nor did they arise from the secretions of sick cells. In his view, they were like other plants, and their developmental histories would bear examination. The question of the origin of plant diseases was most pressing at that time, since fungous diseases, particularly the potato blight, had recently caused sweeping crop devastation with severe economic consequences. Although it was generally acknowledged that plants had parents like themselves, the fungi nevertheless seemed to some observers to present exceptions, and it was noted that they often occurred in association with putrefaction and decay. Much as M. J. Berkeley had insisted in 1846 that the fungus found in potato blight was the cause of the disease, de Bary declared in this early work that the rust and smut fungi were the causes of the pathological changes in diseased plants, and not their results, and that the Uredinales and Ustilaginales were truly parasites.

Until this time the origins and relationships of the fungal stages that had been distinguished were poorly understood, and appearances had been confusing to taxonomists. The importance of following plant life histories was underscored by Louis and Charles Tulasne's recognition of pleomorphism in the fungi; they showed that certain forms which had been classified as separate species actually represented successive stages of development and might even originate from the same mycelium. De Bary continued with a broad range of contributions to morphology.

From his studies of the developmental history of the Myxomycetes, which included the slime molds, de Bary concluded that they ought correctly to be classified among the lower animals and referred to them as the Mycetozoa. He realized the difficulties in clas-sification that arbitrarily separated the animal and vegetable kingdoms, for the Myxomycetes produced flagellate amoeboid swarm spores without walls. Whatever the implications of de Bary's writings on the Myxomycetes (1858) for classification, he pointed out that in one stage they were seen as little more than formless, motile masses of the substance that Dujardin had called *sarcode.* This was the substance of the plasmodium, and protoplasm was the substance of the amoeboid cells as well. De Bary's work served as evidence to his contemporaries of the protoplasmic substratum of life.

Various of de Bary's investigations demonstrated the sexuality of the fungi. He observed the conjugation of *Spirogyra* in 1858 and in 1861 described sexual reproduction in *Peronospora.* He saw the necessity of observing life cycles in continuity and attempted to follow them in the living plant as far as possible. To this end he introduced methods for sowing spores and watching them as they developed.

De Bary's first writing on the fungus that caused the potato blight appeared in 1861, and over the next fifteen or more years he engaged in research on the *Peronosporeae,* which included the potato fungus, then known as *Peronospora infestans.* His researches on this fungus, on *Cystopus* (*Albugo*) and on the Uredinales (the rusts) led him to significant conclusions, which he related in answer to a prize question that the Académie des Sciences proposed in 1861. They were published in the paper "Recherches sur le développement de quelques champignons parasites," in the *Annales des sciences naturelles* in 1863. His direct observations of these parasites of living plants had involved de Bary in the question of spontaneous generation. He had seen numerous methods by which these organisms penetrated healthy plants and developed within their tissues. They had always originated, however, from germs of the same parasitic species, never from the substance of the host plant and never through any predisposition of the host to disease; in fact, they prospered all the more, he claimed, the healthier the plant they invaded.

De Bary outlined the course of the potato blight. He had successfully sown spores of *Peronospora infestans* on healthy potato leaves, and he described the penetration of the leaf and the subsequent growth of the mycelium that affected the tissues, the ensuing formation of the conidia, and the appearance of the characteristic black spots of the potato blight. He had also sown *Peronospora* on potato stalks and tubers, and even had watered conidia into the soil to infect healthy tubers. He believed that the mycelium could survive the winter in the tubers. Certainly there was no spontaneous generation. Some observers might conclude that *Peronospora* arose from within the potato plant because traces of penetration had disappeared, but de Bary maintained that undoubtedly the contagion was due to the numerous germs of parasites and the exterior conditions that favored their invasion of the host plants.

Some years later the Royal Agricultural Society of England asked de Bary to continue his investigations of the potato blight. In 1876 he reported that although he had not yet been able to find the sexual organs of the fungus, which he now named *Phytophthora infestans,* he could account for the known facts of the potato blight, since he could produce the disease in healthy plants by inoculation and reconstruct the fungus' life cycle by experiment and by analogy with others of the *Peronosporeae.*

To de Bary's knowledge, the potato fungus had but a single host, but in his investigations of *Puccinia graminis,* the rust affecting wheat and other grains, he found the phenomenon of heteroecism, for this parasite required two unrelated hosts in order to complete its development. The destruction caused by this rust had been familiar since ancient times, and it was established, but on empirical grounds only, that the occurrence and spread of the rust in wheat fields was somehow related to the presence nearby of the common barberry, *Berberis vulgaris;* in fact, the eradication of barberry plants had long been practiced, and in some instances even required by law, as a preventive measure.

De Bary's observations and experiments extended the investigations of the Tulasne brothers, who had shown the pleomorphism of the Uredinales. He knew that *Puccinia graminis,* the rust of wheat, rye, and other grasses, developed reddish summer spores, the urediospores, and dark winter spores, which he called "teleutospores." Carefully he distinguished the series of reproductive organs that appeared, and he proceeded to actual inoculation experiments. In 1865 de Bary wrote that he had sown sporidia from the winter spores of the wheat rust on the leaves of the common barberry. The sporidia had germinated, and through successive stages had led to the formation of the yellow aecium spores, a familiar sight on the barberry. The aeciospores, and the rest of the reproductive structures of the rust, could reach further development only on cereal plants. In his experiments de Bary had difficulty in obtaining germination of the aecidiospores directly on cereal plants, but in 1866 he could report that he had succeeded in sowing aecidiospores on moisture-retaining slides, then inoculating the leaves of seedling rye plants. In time the reddish summer spores appeared on the leaves, and although his specimens wilted before the formation of the win-

ter-hardy teleutospores, this succession was well known in fields subject to the ravages of the rust. The sporidia from the winter spores germinated, but only upon the barberry; the cycle was then complete.

De Bary thus demonstrated that *Puccinia graminis* required different hosts during the different stages of its development and termed such species heteroecious, as opposed to autoecious species, which passed through their entire development requiring but one host.

His researches clarified other complex relationships as well. He described the formation of the lichen through the associated vegetative processes of the fungus and alga. There were various modes by which parasites lived and affected their hosts; de Bary traced the stages through which they grew and reproduced and the adaptations that enabled them to survive drought and winter. He was careful with his inoculations and cultures, and as he remarked in a popular lecture, *Ueber Schimmel und Hefe* (Berlin, 1869), the organisms termed molds, yeasts, and fungi, despite their peculiarities, exhibited structures and phenomena that were fundamentally those common to plants. To de Bary they were like other plants, only smaller, and were derived ultimately from forms like themselves.

De Bary coined the word "symbiosis" in 1879 in his monograph *Die Erscheinung der Symbiose* (Strasbourg, 1879), using it to mean "the living together of unlike organisms," as a collective term to describe a broad range of relationships. Some organisms, he wrote, existed in conditions of "mutualism"; there was lichenism; there were parasites that could live apart from their hosts at certain periods; there were other parasites completely dependent upon their hosts; between these associations there were infinite gradations.

In his major works on morphology, published between 1866 and 1884, de Bary included the latest botanical researches and the results of his own investigations. His books contain classic descriptions and illustrations, and they were signal contributions to classification and to the systematization of botanical knowledge, establishing mycology as a science.

De Bary's methods and concepts had important applications in the growing field of bacteriology. His lectures on bacteria, *Vorlesungen über Bacterien,* which he published in 1885, presented what was then known of the bacteria. He wrote more than 100 papers on subjects ranging from apogamy in ferns to insect-killing plants. Meanwhile, as a teacher, he continued to encourage his students in their own researches and to accompany them on botanical excursions in the environs of Strasbourg.

At an early period de Bary's views on the question of spontaneous generation were in agreement with those of Pasteur. He was therefore critical of culture procedures and aware of the ease with which "unbidden guests" might intrude into apparatus. His descriptive and experimental work had broad implications for bacteriology as well as for botany, and he was one of the most influential of the nineteenth-century botanists, both in his own time and in ours.

BIBLIOGRAPHY

I. Original Works. De Bary's extensive writings range from comprehensive reference works to numerous contributions to the botanical periodicals. A more complete listing of his publications can be found in the articles by Jost, Reess, and Solms-Laubach (see below). De Bary's *Morphologie und Physiologie der Pilze, Flechten und Myxomyceten* (Leipzig, 1866) was the second volume of Wilhelm Hofmeister's *Handbuch der physiologischen Botanik*. The second edition, *Vergleichende Morphologie und Biologie der Pilze, Mycetozoen und Bacterien* (Leipzig, 1884), in many respects a new work, appeared in an English translation by Henry E. F. Garnsey and Isaac Bayley Balfour as the *Comparative Morphology and Biology of the Fungi, Mycetozoa and Bacteria* (Oxford, 1887). His *Vergleichende Anatomie der Vegetationsorgane der Phanerogamen und Farne* (Leipzig, 1877) constituted the third volume of Hofmeister's handbook; its English edition, F. O. Bower and D. H. Scott, trans., was the *Comparative Anatomy of the Vegetative Organs of the Phanerogams and Ferns* (Oxford, 1884). The second edition of the *Vorlesungen über Bacterien* (Leipzig, 1885, 1887) was translated and revised by Garnsey and Balfour as *Lectures on Bacteria* (Oxford, 1887). The early monograph on the rusts and smuts was *Untersuchungen über die Brandpilze und die durch sie verursachten Krankheiten der Pflanzen* (Berlin, 1853).

The many subjects of de Bary's studies are only indicated by the papers he wrote, including: "Ueber die Myxomyceten," in *Botanische Zeitung*, 16 (1858), 357–358, 361–364, 365–369; "Ueber die Geschlechtsorgane von *Peronospora*," *ibid.*, 19 (1861), 89–91; "Recherches sur le développement de quelques champignons parasites," in *Annales des sciences naturelles, botanique*, 4th ser., 20 (1863), 5–148; "Neue Untersuchungen über die Uredineen, in besonders die Entwicklung der *Puccinia graminis* und den Zusammenhang derselben mit *Aecidium Berberidis*," in *Monatsbericht der Königlicher Preussischen Akademie der Wissenschaften zu Berlin* (1865), 15–50, and "Neue Untersuchungen über Uredineen," *ibid.* (1866), 205–216; and "Researches Into the Nature of the Potato Fungus—*Phytophthora infestans*," in *Journal of the Royal Agricultural Society of England*, 2nd ser., 12 (1876), 239–269.

II. Secondary Literature. The place of de Bary in the history of mycology and the relation of his studies of the life histories of fungi to the problems and questions of his day are discussed by E. C. Large in *The Advance of the Fungi* (New York, 1940, 1962); and aspects of his work by G. W. Martin in "The Contribution of de Bary to Our Knowledge of the Myxomycetes," in *Proceedings of the Iowa Academy of Science*, 65 (1958), 122–127; and W. B. McDougall in "The Classification of Symbiotic Phenomena," in *Plant World*, 21 (1918), 250–256. A contemporary assessment of de Bary's contributions is that of Ferdinand Cohn, himself a noted botanist whose researches were basic to the science of bacteriology, in his obituary notice, "Anton de Bary," in *Deutsche medicinische Wochenschrift*, 14 (1888), 98–99, 118–119. Further sources of biographical information are Ernst Almquist, *Grosse Biologen* (Munich, 1931), pp. 57–60; Ludwig Jost, "Zum hundertsten Geburtstag Anton de Barys," in *Zeitschrift für Botanik*, 24 (1930), 1–74; M. Reess, "Anton de Bary," in *Berichte der Deutschen botanischer Gesellschaft*, 6 (1888), viii–xxvi; H. Graf zu Solms-Laubach, "Anton de Bary," in *Botanische Zeitung*, 47 (1889), 33–49; Marshall Ward, "Anton de Bary," in *Nature*, 37 (1888), 297–299; K. Wilhelm, "Anton de Bary," in *Botanisches Centralblatt*, 34 (1888), 93–94, 156–159, 191–192, 221–224, 252–256; and C. Wunschmann, "Heinrich Anton de Bary," in *Allgemeine deutsche Biographie*, XLVI (Leipzig, 1902), 225–228.

Gloria Robinson

DEBEAUNE (also known as **Beaune**), **FLORIMOND** (*b.* Blois, France, 7 October 1601; *d.* Blois, 18 August 1652), *mathematics*.

Truly representative of a time of intense communication among intellectuals, Debeaune enjoyed great fame although he himself never published anything.

His renown was due entirely to Descartes. The *Notes brèves* that Debeaune wrote on the *Géométrie* were translated and added during his lifetime to the first Latin edition, published by Schooten in 1649. The second Latin edition (1659–1661) also contained two short papers on algebra, edited by Erasmus Bartholin, that are Debeaune's only posthumous publication. The letters published by Clerselier between 1657 and 1667 revealed to a wider public the esteem in which Descartes held his disciple from Blois.

Undoubtedly this was the reason why, in 1682, a chronicler concerned with celebrities of his province wrote a paper on Debeaune, drawing his information from sources close to the family while this was still possible. At the end of the nineteenth century a scholar from Blois confirmed the information by locating various documents in archives, and the great critical edition of Descartes's *Oeuvres* once again brought attention to Debeaune. Paul Tannery had the good fortune to discover a great many handwritten letters in Vienne, which enabled him to gain a great deal of scientific clarity.

On the basis of his interpretation of the signature of these letters, Tannery committed an error in insisting on the spelling "Debeaune," by which he is most frequently cited. Florimond's father, also named Florimond, was undoubtedly the natural son of Jean II de Beaune, brother of the archbishop Renaud de Beaune; but he was legitimized and his titles of nobility assured to his descendants.

A few accurate dates can be furnished for Debeaune. He was baptized on 7 October 1601. He married his first wife, Philiberte Anne Pelluis, on 21 December 1621. She died in August 1622, and he remarried on 15 December 1623. His second wife was Marguerite du Lot, who bore him three sons and one daughter. His burial certificate, designating him as Seigneur de Goulioust, is dated 18 August 1652.

Like Descartes, Debeaune at first did military service, but following a mysterious accident he had to lead a less strenuous life. Taking advantage of his law studies, he bought the office of counselor to the court of justice in Blois. The many years that he divided between this famous city on the banks of the Loire and his nearby country estate, excelling in both jurisprudence and mathematical research, bring Fermat to mind. However, Fermat does not appear on the list of correspondents that the chronicler of 1682 saw among the family papers, a list of which only a small part has been preserved.

Debeaune left his provincial retreat only for business trips to Paris. However, he had many visitors. The first part of Monconis' diary mentions observatory instruments made by him. An inventory made after Debeaune's death confirmed statements in parts of letters that have been preserved: he had built for his own use a shop for grinding lenses. He also had a magnificent library, worthy of a humanist of the preceding century.

Afflicted with various and painful infirmities, particularly gout, Debeaune resigned as counselor around 1648 and withdrew to a town house, the upper floor of which faced due south. There he had—at least for a time—an observatory at his disposal. However, his failing eyesight deteriorated rapidly, and he died shortly after having a foot amputated.

When he was very ill, Debeaune was visited by Erasmus Bartholin, whom he entrusted with arranging for the publication of several of his manuscripts. Despite the intervention of Schooten and Huygens in 1656, Bartholin fulfilled his obligations only partially. Of the manuscripts with which he was entrusted, only "La doctrine de l'angle solide construit sous trois angles plans" was discovered, in 1963. The "Méchaniques" mentioned by Mersenne, and the "Dioptrique" that Schooten knew in 1646 are still missing and may be lost.

This situation is unfortunate, for it deprives us of elements valuable for judging the origin of purely mathematical problems that Debeaune formulated in 1638 and that Paul Tannery analyzed fully according to the correspondence he discovered. According to Beaugrand, the first of these problems—which in the present state of textual study appears to concern itself only with the determination of the tangent to an analytically defined curve—interested Debeaune "in a design touching on dioptrics." As to the second of these problems, the one that has been particularly identified with Debeaune[1] and that ushered in what was called at the end of the seventeenth century the "inverse of tangents"—i.e., the determination of a curve from a property of its tangent—Debeaune told Mersenne on 5 March 1639 that he sought a solution with only one precise aim: to prove that the isochronism of string vibrations and of pendulum oscillations was independent of the amplitude. This statement, which is not easily justified except in the language of differential and integral calculus, was fifty years ahead of scientific developments and—by itself—reveals Debeaune's singular ability to translate physical questions into the abstract language of mathematical analysis, despite the inadequacies of the operative means of his time.

It is not surprising that Debeaune, aware of these inadequacies, eagerly seized upon anything that could possibly be of help in overcoming them. As he had once adopted and assimilated Herigone's algebra (the *Cursus mathematicus,* 1635–1637), he welcomed Descartes's *Géométrie;* and Descartes was right in believing that none of his contemporaries had better understood it. Debeaune's *Notes brèves* clarify and conveniently illustrate some of the difficult passages of the *Géométrie* and played a role in the belated spread of Cartesian mathematics.

As Paul Tannery has shown, Descartes's method for tangents misled Debeaune, at least initially. This purely algebraic method, which consists of determining the subnormal by writing that the equation obtained as a result of an elimination is to have two equal roots, is not susceptible of supporting a process of inversion. But if Debeaune, victim of a misconception that nevertheless bears the stamp of his mathematical genius, could give *his* problem (the first integration problem of a first-order differential equation) only an incorrect solution, he was nevertheless the only one to comprehend the remarkable solution to which it had led Descartes, a solution that anticipated the use of series. This was a remarkable solution that Leibniz, fifty years later, failed to recognize when he replaced it with the aid of new algorithms and the logarithmic function.

Undoubtedly the nature of the various problems

posed by Debeaune becomes clearer when translated into the language of Leibnizian calculus, for Debeaune's language, based on the form of the triangle constructed on the ordinate and the subnormal, is without immediacy. Nevertheless, we should remember the man who dared to pose the inversion problem of tangents at a time when mathematicians had difficulty understanding the direct problem.

The example of Debeaune reminds us that mathematics is sustained more by the perception of profound logical structures than by the invention and use of languages that find acceptance in the structures only with time. As Debeaune wrote to Mersenne (5 March 1639), "I do not think that one could acquire any solid knowledge of nature in physics without geometry, and the best of geometry consists of analysis, of such kind that without the latter it is quite imperfect."[2]

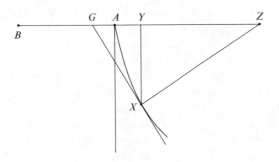

DEBEAUNE'S CURVE

NOTES

1. Find the curve such that ZY is to YX as a given AB is to the difference between YX and AY (letter of 16 October 1638). Descartes presented a solution in letters to Mersenne and Debeaune of 20 February 1639. There is a transformation of the statement by Descartes in a letter to an unknown person dated June 1645. This statement was followed up by Leibniz and Johann I Bernoulli, who formulated it by means of

$$\frac{dy}{dx} = \frac{a}{y - x}.$$

In a letter to Debeaune on 30 April 1639, Descartes congratulates himself for having taken the time to consider the proposed curve lines, stating that he has learned a great deal thereby.

2. "Je ne pense pas qu'on puisse acquérir aulcune cognoissance solide de la nature en physique sans la géométrie, et le plus excellent de la géométrie consiste en l'analyse, en sorte que sans cela elle est fort imparfaicte."

BIBLIOGRAPHY

I. ORIGINAL WORKS. The *Notes brèves* appeared in *Geometria a Renato Descartes anno 1637 edita, nunc autem cum notis Florimundi de Beaune* (Leiden, 1649) and in its 2nd ed., the title of which concludes *Florimundi de Beaune duo tractatus posthumi alter de natura et constitutione, alter de limitibus aequationum, ab Erasmio Bartholino edita* (Amsterdam, 1659–1661). "Notes briefves sur la méthode algébraique de Mr D. C.," the original French text, based on MSS in London and Paris, Charles Adam and G. Milhaud, eds., *Descartes—Correspondance*, III (Paris, 1941), pp. 353–401. "La doctrine de l'angle solide construit sous trois angles plans," handwritten, is at the Secretariat of the Academy of Sciences, Paris. Some of Debeaune's correspondence may be found in *Oeuvres de Descartes*, Charles Adam and Paul Tannery, eds., II and V; and in *Correspondance de Mersenne*, Centre National de la Recherche Scientifique, ed., VIII.

II. SECONDARY LITERATURE. On Debeaune or his work, see Charles Adam and G. Milhaud, eds., *Descartes—Correspondance*, I (Paris, 1931), 436–438; Jean Bernier, *Histoire de Blois* (Paris, 1682), pp. 563–568; Pierre Costabel, "Le traité de l'angle solide de Florimond de Beaune," in *Actes du XIe Congrès international d'histoire des sciences*, III (Warsaw, 1965–1968) pp. 189–194; Paul Tannery, "Pour l'histoire du problème inverse des tangentes," in *Verhandlungen der III Internationalen Mathematiker-Kongresses* (Leipzig, 1904), repub., with additions, in Tannery's *Mémoires scientifiques*, VI (Paris, 1926); and Adrien Thibaut, "Florimond de Beaune," in *Bulletin de la Société des sciences et lettres du Loir et Cher*, **4**, no. 6 (Mar. 1896), 13–29.

PIERRE COSTABEL

DEBENHAM, FRANK (*b.* Bowral, New South Wales, Australia, 26 December 1883; *d.* Cambridge, England, 23 November 1965), *polar exploration, cartography, geography.*

Debenham's early education was at a private school run by his father, Reverend J. W. Debenham, vicar of Bowral. He attended the King's School, Parramatta, New South Wales, and graduated in arts from the University of Sydney in 1904. After being a schoolteacher for three years, he returned to the university and graduated B.Sc. in geology in 1910.

Debenham's research career began soon afterward, when R. F. Scott, while passing through Sydney, selected him as an additional member of the second Scott Antarctic Expedition of 1910–1913. In the Antarctic, Debenham carried out geological surveys and brought to bear the cartographic skills for which he later became famous.

After the return of the expedition Debenham worked in Cambridge for the rest of his life, apart from the interruption caused by World War I, in which he was seriously wounded. In 1919 he was appointed lecturer in surveying and cartography at the University of Cambridge. His early years in this post were devoted mainly to writing up his polar work, including entire responsibility for the account of the survey work of the expedition. At the same time he steadily developed his skills as a cartographer, and his polar writings were followed by several books concerned largely with the surveying and mapping of landforms.

In 1925 Debenham became the first director of the Scott Polar Research Institute at Cambridge and was a central figure in inspiring the surge of British polar exploration that occurred between 1925 and World War II. In 1931 he became the first professor of geography at Cambridge, holding this post until his retirement in 1949.

Debenham was essentially a practical man. Apart from his pioneering work in the Antarctic, he was noted for constantly devising new techniques in cartography that proved to be a great stimulus to geographers. He also made significant contributions concerning the water problems of central Africa. Debenham, an approachable man with a gift for kindly humor, was greatly loved by his colleagues and students, and was an outstanding inspirer and gentle leader of men.

BIBLIOGRAPHY

Debenham's scientific publications include approximately 100 papers, books, and review articles. His most noted works include *The Physiography of the Ross Archipelago. British (Terra Nova) Antarctic Expedition 1910–1913* (London, 1923); *Reports on the Maps and Surveys. British (Terra Nova) Antarctic Expedition 1910–1913* (London, 1923); *Map Making* (London, 1936, 1940, 1945); *The Voyage of Captain Bellingshausen to the Antarctic Seas, 1819–1821,* a translation from the Russian edited by Debenham (London, 1945); *The Use of Geography* (London, 1950); *In the Antarctic: Stories of Scott's Last Expedition* (London, 1952); and *The McGraw-Hill Illustrated World Geography* (New York, 1960), written with W. A. Burns.

K. E. BULLEN

DEBRAY, HENRI JULES (*b.* Amiens, France, 26 July 1827; *d.* Paris, France, 19 July 1888), *chemistry.*

Debray studied in Paris at the École Normale Supérieure. In 1850 he began teaching at the Lycée Charlemagne, and in 1855 he received his doctorate. From 1855 to 1868 he was an assistant at the École Normale, and from 1868 he was lecturer at the École Polytechnique. He held the latter position until his death, serving at the same time as assayer at the mint. In 1877 he was elected to the Institut de France.

Debray was a talented writer of textbooks. His *Cours élémentaire de chimie* appeared in eight editions between 1863 and 1871. He undertook his first scientific work in collaboration with St. Claire Deville, a professor at the École Normale. Their first experi-

ments were in the use of an oxyhydrogen blowpipe to melt platinum. They were commissioned by the Russian government to investigate the applicability of platinum-iridium alloys to coinage and ascertained that these alloys resist corrosion even better than platinum does. At their suggestion the standard meter in Paris was made, under their supervision, from an alloy composed of 90 percent platinum and 10 percent iridium. In further publications Debray reported on investigations of the compounds of tungsten, arsenic, and antimony; on the properties of rhodium; and on the compounds in the platinum metals family and methods of separating them.

Debray also concerned himself with dissociation phenomena. He established that in the dissociation of calcium carbonate the pressure of the carbon dioxide at a given temperature is constant and independent of the degree of dissociation. He made the same finding regarding the water-vapor pressure over hydrated salts. (Wiedemann later raised priority claims in this matter, but Debray obtained his results without knowledge of the results that Wiedemann had published in a little-known journal.) Debray also determined the density of mercurous chloride vapor.

BIBLIOGRAPHY

Debray's books include *Métallurgie du platine et des métaux qui l'accompagnent* (Paris, 1863); and *Cours élémentaire de chimie* (Paris, 1863; 8th ed., 1876). There are about fifty publications in French periodicals.

The only secondary literature is a short notice in Poggendorff, III, 337–338.

F. SZABADVÁRY

DEBYE, PETER JOSEPH WILLIAM (*b.* Maastricht, Netherlands, 24 March 1884; *d.* Ithaca, New York, 2 November 1966), *chemical physics.*

Debye, the son of Wilhelmus and Maria Reumkens Debye, went to school in Maastricht until he left for the nearby Technische Hochschule in Aachen, across the border in Germany. Here he was an assistant from 1904 to 1906, obtaining a degree in electrical engineering in 1905. When Arnold Sommerfeld, the eminent German mathematical physicist, was called from Aachen to the University of Munich in 1906, he took Debye with him. Debye remained in Munich as an assistant for five years, obtaining a Ph.D. in physics in 1908 and serving as *Privatdozent* in his last year. In 1911, at the age of twenty-seven, he succeeded Einstein as professor of theoretical physics at the University of Zurich. After only a year in Switzerland he returned to his native country for a year as professor of theoretical physics at the University of Utrecht, only to leave again for Germany, where he stayed from 1913 to 1920 as professor of theoretical and experimental physics at the University of Göttingen. On 10 April 1913 he married Matilde Alberer. Their son, Peter Paul Ruprecht, who was later to collaborate in some of the light-scattering researches, was born in 1916; their daughter, Mayon M., was born in 1921. The period from 1911 to 1916 was perhaps the most productive for Debye. In spite of holding three professorships in three countries during the first three years of this period, he produced his theory of specific heats, the concept of permanent molecular dipole moments, and the related theory of anomalous dielectric dispersion. With Paul Scherrer he developed the powder method of X-ray analysis.

In 1920 Debye returned to Zurich as professor of experimental physics and director of the Physical Institute at the Eidgenössische Technische Hochschule, where he was surrounded by an able group of students and assistants, one of whom, Erich Hückel, collaborated with him in his next great basic contribution, the Debye-Hückel theory of electrolytes, published in 1923. The work on X-ray scattering and dipole moments continued through the Zurich period along with that on electrolytes. In 1927 Debye moved to the University of Leipzig as professor of experimental physics, a professorship reported to be the most lucrative in Germany. Physical chemists now flocked to Leipzig to Debye's institute as they had to Ostwald's a generation earlier. Although good work continued through this seven-year period at Leipzig, no great basic discoveries were made. In 1934, the second year of the Nazi regime in Germany, Debye moved to the University of Berlin as professor of theoretical physics and supervised the building of the Kaiser Wilhelm Institute of Physics, which he named the Max Planck Institute. Here, as in Leipzig, good work continued; but nothing of outstanding importance emerged. In 1936 Debye received the Nobel Prize in chemistry and, in 1939, had the unusual experience of seeing a bust of himself unveiled in the town hall of his native city.

Debye was not only a brilliant and original scientist but also a wise and shrewd man of the world. In the performance of administrative duties in Berlin, he had to spend a great deal of time dealing with Nazi bureaucrats. He had retained his Dutch citizenship when he came to Berlin, having been told by the minister of education that he would not be required to become a German citizen. However, not long after World War II broke out, he was informed that he could not enter his laboratory if he did not become a German citizen. He refused to do so and soon succeeded in getting

to the United States, where he became a citizen in 1946. He had lectured many times in the United States and had declined offers of professorships at many leading universities; but he now gave the Baker Lectures at Cornell University and was appointed professor of chemistry and head of the chemistry department there, positions which he held from 1940 to 1950, when he became professor emeritus. He continued active in research and consultation until the end of his life. The first ten years at Ithaca produced the work on light scattering, his last great contribution. Debye's many achievements were recognized by his election to membership in some twenty-two academies throughout the world and the award of twelve medals and eighteen honorary degrees.

Debye's physical vigor equaled his vigor of mind and, in middle and old age, his appearance was that of a man at least ten years younger. The extraordinary clarity of his thinking made it possible for him to revise and develop a previously incomplete or inadequate treatment of a phenomenon into an important generalization or new method of investigation. The power to convert this same clarity of thought into words made him a lecturer capable, to a remarkable degree, of making a difficult or obscure subject clear to an audience.

Although much of Debye's work was concerned with the interaction of radiation with matter, it did not include spectroscopy as such. The evolution of his interests and work from mathematical physics to physical chemistry is illustrated by the subjects of his first paper, a theoretical treatment of eddy currents (1), and of one published sixty-one years later, after his death, "Direct Visual Observation of Concentration Fluctuations in a Critical Mixture" (29).

Debye's first major contribution (2, 4, 14) was based on an explanation of the temperature dependence of the dielectric constant. The dielectric constant ϵ of a substance had, for many years, been written in the Clausius-Mosotti equation

$$\frac{\epsilon - 1}{\epsilon + 2} = \frac{4\pi n\alpha}{3}$$

in which n is the number of molecules per cubic centimeter and α is the molecular polarizability, the electric moment induced in a molecule by a unit electric field. Since α was supposed to be a constant characteristic of the substance and n decreases only very slowly with rising temperature as the density decreases, it would appear that ϵ should decrease very slowly with rising temperature. The equation represented exactly the behavior of the dielectric constant of most substances with small dielectric constants; but liquids with large dielectric constants showed a rapid decrease of dielectric constant with temperature, far greater than that predicted by the equation.

The polarization of the substance had been attributed wholly to the induced shift of the electrons within the molecules, giving each molecule a very small electric moment $E\alpha$ in the direction of the electric field E. Debye proposed that the molecules of some substances had permanent electric doublets, or dipoles, in them of moment μ, which would contribute to the total polarization when an external field was applied. The molecule would tend to rotate so as to orient its dipole in the field, but this orientation would be reduced by the thermal motion of the molecules. Using a treatment analogous to that developed by Langevin for magnetic moments, Debye showed that the average moment per molecule in the direction of a unit field would be $\alpha + \mu^2/3kT$. The equation for the dielectric constant was, therefore,

$$\frac{\epsilon - 1}{\epsilon + 2} = \frac{4\pi n}{3} \frac{\alpha + \mu^2}{3kT}$$

in which k is the molecular gas constant and T the absolute temperature. This equation not only represented the behavior of the dielectric constant satisfactorily, but also established the existence of a permanent electric dipole in many molecules and provided a means of determining the moment of the dipole and, from this, the geometry of the molecule. After many years of use in molecular structure investigations, the unit in which the dipole moment was expressed came to be called the "Debye."

In his second outstanding paper (3), following the first dielectric paper by only a few months, Debye treated a solid as a system of vibrating atoms and modified Einstein's theory of specific heats, which had been only partially successful. He showed that the solid could be characterized by a complete spectrum of eigen-frequencies and that the specific heat of a monatomic solid was a universal function of the ratio θ/T, where θ is a temperature characteristic of the particular solid and T is the absolute temperature. Now commonly called the Debye temperature, θ could be calculated from the elastic constants of the solid. The Debye equation, involving the then recently developed quantum theory, gave quantitative agreement with observed specific heat values. Aside from a numerical factor, it differed from the Einstein equation in containing both the compressibility and Poisson's ratio.

Debye (4, 14) showed how the orientation of molecular dipoles in a very high frequency alternating field or in a very viscous medium absorbed energy and gave rise to anomalous dielectric dispersion and dielectric loss. His equations containing the dielectric constant, dielectric loss, frequency, and relaxation time give the classical representation of dielectric behavior, which is often referred to as "Debye behavior." The dependence of the molecular relaxation time upon molecular size and structure, and upon intermolecular forces, makes it of use in the investigation of these properties. The general applicability to liquids of the Debye equations for dielectric constant and loss was improved twenty-five years and more later by the revised treatment of the effect of the internal field in liquids developed by Onsager, Kirkwood, and others.

Within a year of the discovery of X-ray diffraction by crystals by von Laue and the Braggs in 1912, Debye published three papers proving that the thermal movement of the atoms in the crystal affected the X-ray interferences. Here he was examining from a different point of view the atomic lattice treated in his specific heat work. Late in 1913 he sent in for publication a long paper (5) deriving a factor now called the Debye factor, which gave the decrease of intensity of the diffraction spots as a function of wavelength, diffraction angle, and absolute temperature. The best-known paper among Debye's many theoretical investigations of X-ray scattering was that with Scherrer (6) on the X-ray interference patterns of randomly oriented particles, which became the basis for the structure analysis of crystal powders, polycrystalline metals, and colloidal systems by the Debye-Scherrer method, possibly the most powerful tool for the determination of the structures of crystals of high symmetry.

Debye and Scherrer (7) investigated the electron distribution inside the atom by analysis of intensities, introducing the atomic form factor, which was later to prove of importance. Although his long study of X rays had been concerned mainly with the classical wave theory of scattering of radiation by matter, Debye (8) used the quantum theory in 1923 to develop independently a quantitative theory of the Compton effect, which evidenced the dualism of the wave and particle theories of light. When he later extended the work on atomic structure and X-ray scattering (7) to molecules and liquids (15–17), he developed tools for structural investigations that, because of the dualism of the wave and particle theories, provided a foundation for the electron diffraction method, a major method of molecular structure determination.

Physical chemistry could almost be said to date from the quantitative formulation of the theory of ionic conduction by Arrhenius in 1887, but this theory of partial dissociation into ions proved inadequate. A number of investigators proposed a complete dissociation into ions, but it was not until 1923 that Debye and Hückel (9, 10), by mathematical analysis, developed the fundamental thermodynamics of electrolytic solutions and solved the problem of electrolytic conductance. They treated the solution as having a structure somewhat analogous to that of a crystal of sodium chloride, in which each sodium ion is surrounded by six chloride ions and each chloride ion by six sodium ions, as shown ten years earlier by X-ray analysis. However, the extent of this ordering in the solution was determined by the equilibrium between the thermal motion and the interionic attractive forces, which were dependent upon the dielectric constant of the solvent and the concentration of the solute. Each ion, instead of being regarded as in an actual lattice, was treated as surrounded by an ionic cloud whose thickness and relaxation time, reminiscent of that involved in dielectric loss, were important in determining the properties of the solution. The development of the theory of their behavior was a major contribution to our understanding of electrolytes and, in particular, predicted and explained (12, 13) the effect of very high field strengths on conductivity observed by Max Wien. Other publications by Debye and his co-workers during this period extended and applied the basic ideas contained in these papers. Like many of Debye's other pioneering investigations, this work provided the theoretical basis for most of the extensive work subsequently done in the field.

A common thread runs through many of Debye's papers, however diverse the subjects may seem. In an isolated paper (11) on magnetization, the Langevin function, used in the derivation of the equation for the dielectric constant, was shown to be not entirely correct but was employed, nevertheless, in the calculation of the approximate temperature change produced by an adiabatic magnetic process. Debye then raised "the question whether an effort should be made to use such a process in approaching absolute zero" and concluded the paper with a sentence typical of his thinking: "Only experiments can decide, and the above analysis should stimulate the carrying out of these." Such experiments led later to a very close approach to absolute zero.

Having dealt with long electromagnetic waves in his work on dielectric constant and loss, and with short waves in his work on X-ray scattering, Debye developed a suggestion by L. Brillouin and showed both theoretically and experimentally that sound waves in a liquid could form an optical grating, in which the wavelength of the sound waves in the liquid played the same part as the grating constant of a ruled grating (18, 19). Twelve years later Debye was considering light scattering in solution (20), building on the much earlier work of Lord Rayleigh and Einstein on gases and liquids. His previous work on X-ray scattering contributed now to his development of light scattering as a tool for the absolute determination of molecular weights of polymers and the spatial extension of macromolecules in dilute solutions, based essentially on determination of the turbidity of the solutions (20–22). The increase in light-scattering power or turbidity of a solution over that of the pure solvent was found to be proportional both to the number and to the weights of the molecules. The molecular weight was determined by combining measurements of the excess turbidity with the excess of refractive index of the solution over that of the solvent. The averaging process used in calculating the intensity of the scattered light was identical with that used in obtaining the scattering of X rays from a gas molecule (17). In the course of this development of the light-scattering method Debye became so interested in polymers that he worked on their viscosity, diffusion, and sedimentation rate (23); but he also continued the light-scattering attack upon colloidal solutions in the investigation of the size and shape of micelles (24, 25).

In further light-scattering studies (26) Debye investigated visible light scattered by a one-component homogeneous liquid in the vicinity of its critical point and by a homogeneous mixture of two liquids in the vicinity of their critical mixing temperature. He showed that the angular dissymmetry of this scattered light could be used as a measure of the range of molecular forces for ordinary molecules and as a means of measuring the size of coils in polymer molecules too small for measurement by the previously developed light-scattering method. He continued along these lines in his last work (27, 28), exploring the effect of a strong electric field upon critical opalescence by theoretical and experimental methods, the latter involving ingenuity and experimental skill of a high order.

BIBLIOGRAPHY

I. ORIGINAL WORKS. *The Collected Papers of Peter J. W. Debye* (New York, 1954), consisting of fifty-one papers; a brief biography by R. M. Fuoss; and comments by Fuoss, H. Mark, and C. P. Smyth, was published in honor of Debye's seventieth birthday. The papers were selected by Debye himself, translated into English if published originally in another language, and grouped according to subject matter. Many of these papers, together with additional references, are listed below, numbered for reference.

(1) "Wirbelströme in Stäben von rechteckigem Querschnitt," in *Zeitschrift für Mathematik und Physik,* **54** (1907), 418–437.

(2) "Einige Resultate einer kinetischen Theorie der Isolatoren," in *Physikalische Zeitschrift*, **13** (1912), 97–100.

(3) "Zur Theorie der spezifischen Wärmen," in *Annalen der Physik*, **39** (1912), 789–839.

(4) "Zur Theorie der anomalen Dispersion im Gebiete der langwelligen elektrischen Strahlung," in *Berichte der Deutschen physikalischen Gesellschaft*, **15** (1913), 777–793.

(5) "Interferenz von Röntgenstrahlen und Wärmebewegung," in *Annalen der Physik*, **43** (1914), 49–95.

(6) "Interferenzen an regellos orientierten Teilchen im Röntgenlicht. I," in *Physikalische Zeitschrift*, **17** (1916), 277–283, written with P. Scherrer.

(7) "Atombau," *ibid.*, **19** (1918), 474–483, written with P. Scherrer.

(8) "Zerstreuung von Röntgenstrahlen und Quantentheorie," *ibid.*, **24** (1923), 161–166.

(9) "Zur Theorie der Elektrolyte. I. Gefrierpunktserniedrigung und verwandte Erscheinungen," *ibid.*, 185–206, written with E. Hückel.

(10) "Zur Theorie der Elektrolyte. II. Das Grenzgesetz für die elektrische Leitfähigkeit," *ibid.*, 305–325, written with E. Hückel.

(11) "Einige Bemerkungen zur Magnetisierung bei tiefer Temperatur," in *Annalen der Physik*, **81** (1926), 1154–1160.

(12) "Dispersion von Leitfähigkeit und Dielektrizitätskonstante bei starken Elektrolyten," in *Physikalische Zeitschrift*, **29** (1928), 121–132, 401–426, written with H. Falkenhagen.

(13) "Dispersion der Leitfähigkeit starker Elektrolyte," in *Zeitschrift für Elektrochemie*, **34** (1928), 562–565, written with H. Falkenhagen.

(14) *Polar Molecules* (New York, 1929).

(15) "Zerstreuung von Röntgenstrahlen an einzelnen Molekeln," in *Physikalische Zeitschrift*, **30** (1929), 84–87, written with L. Bewilogua and F. Ehrhardt.

(16) "Röntgeninterferenzen und Atomgrösse," *ibid.*, **31** (1930), 419–428.

(17) "Röntgenzerstreuung an Flüssigkeiten und Gasen," *ibid.*, 348–350.

(18) "On the Scattering of Light by Supersonic Waves," in *Proceedings of the National Academy of Sciences*, **18** (1932), 409–414, written with F. W. Sears.

(19) "Zerstreuung von Licht durch Schallwellen," in *Physikalische Zeitschrift*, **33** (1932) 849–856.

(20) "Light Scattering in Solutions," in *Journal of Applied Physics*, **15** (1944), 338–342.

(21) *Angular Dissymmetry of Scattering and Shape of Particles*, Rubber Reserve Company, Technical Report no. 637 (9 Apr. 1945).

(22) "Molecular-Weight Determination by Light Scattering," in *Journal of Physical and Colloid Chemistry*, **51** (1947), 18–32.

(23) "Intrinsic Viscosity, Diffusion, and Sedimentation Rate of Polymers in Solution," in *Journal of Chemical Physics*, **16** (1948), 573–579, written with A. M. Bueche.

(24) "Light Scattering in Soap Solutions," in *Annals of the New York Academy of Sciences*, **51** (1949), 575–592.

(25) "Micelle Shape From Dissymmetry Measurements,"

in *Journal of Physical and Colloid Chemistry*, **55** (1951) 644–655, written with E. W. Anacker.

(26) "Angular Dissymmetry of the Critical Opalescence in Liquid Mixtures," in *Journal of Chemical Physics*, **31** (1959), 680–687.

(27) "Electrical Field Effect on the Critical Opalescence," *ibid.*, **42** (1965), 3155–3162, written with K. Kleboth.

(28) "Electric Field Effect on the Critical Opalescence. II. Relaxation Times of Concentration Fluctuations," *ibid.*, **46** (1967), 2352–2356, written with C. C. Gravat and M. Ieda.

(29) "Direct Visual Observation of Concentration Fluctuations in a Critical Mixture," *ibid.*, **48** (1968), 203–206, written with R. T. Jacobsen.

II. SECONDARY LITERATURE. Articles on Debye are Karl Darrow, in *Annual Year Book of the American Philosophical Society;* and Mansel Davies, in *Journal of Chemical Education*, **45** (1968), 467–473.

CHARLES P. SMYTH

DECHALES, CLAUDE FRANÇOIS MILLIET (*b.* Chambéry, France, 1621; *d.* Turin, Italy, 28 March 1678), *mathematics.*

Not much is known of Dechales's personal life. For some time he was a Jesuit missionary in Turkey. He was well liked in Paris, where for four years he read public mathematics lectures at the Collège de Clermont. After teaching at Lyons and Chambéry, he moved to Marseilles, where he taught the arts of navigation and military engineering and the practical applications of mathematics to science. From Marseilles he went to Turin, where he was appointed professor of mathematics at the university. He died there at the age of fifty-seven.

Although not a first-rate mathematician, Dechales was rather skillful in exposition; Hutton has observed that "his talent rather lay in explaining those sciences [mathematics and mechanics] with ease and accuracy . . . that he made the best use of the production of other men, and that he drew the several parts of the mathematical sciences together with great judgment and perspicuity."

Dechales is best remembered for his *Cursus seu mundus mathematicus,* a complete course of mathematics, including many related subjects that in his day were held to belong to the exact sciences. The first volume opens with a description of mathematical books arranged chronologically that, as De Morgan remarks, is well done and indicates that Dechales had actually read them. This is followed by his edition of Euclid's *Elements* (bks. I–VI, XI, and XII). Arithmetic computation, algebra, spherical trigonometry, and conic sections are of course included. Of the algebraic material, Hutton observes that it is "of a very old-fashioned sort, considering the time when it was written." The algebra of Dechales is imbued with the spirit of Diophantus; as Moritz Cantor points out, Dechales rarely mentions the work of Mydorge, Desargues, Pascal, Fermat, Descartes, or Wallis. Among other subjects included in the *Cursus* are practical geometry, mechanics, statics, geography, magnetism, civil architecture, military architecture, optics, catoptrics, perspective, dioptrics, hydrostatics, hydraulic machinery, navigation, pyrotechnics, gnomonics, astronomy, astrology, meteoritics, the calendar, and music, as well as a section entitled "A Refutation of the Cartesian Hypothesis." Indeed, in his history of mathematics, Cantor gives a detailed description of the *Cursus* both because it was a popular and widely used textbook and because it reflected the totality of mathematical knowledge as possessed by dilettantes or amateur mathematicians of the time who were fairly competent interpreters or expounders of the subject. Thus, while according Dechales due credit for his efforts, Cantor is nevertheless critical of much of the mathematical content of his work, deploring Dechales's failure to make full use of such available contemporary source materials as the first-hand works of mathematicians, their correspondence, and so on.

Dechales's separate edition of Euclid, long a favorite in France and elsewhere on the Continent, never became popular in England.

BIBLIOGRAPHY

I. Original Works. Dechales's works are *Cursus seu mundus mathematicus,* 3 vols. (Lyons, 1674), also ed. by Amati Varcin, 4 vols. (Lyons, 1690); *L'art de fortifier, de défendre et d'attaquer les places, suivant les méthodes françoises, hollandoises, italiennes & espagnoles* (Paris, 1677); *L'art de naviger demontré par principes & confirmé par plusieurs observations tirées de l'expérience* (Paris, 1677); and *Les principes généraux de la géographie* (Paris, 1677). His edition of Euclid, *Les élémens d'Euclide, expliquez d'une manière nouvelle & très facile. Avec l'usage de chaque proposition pour toutes les parties des mathématiques* (Lausanne, 1678, 1683), appeared in revised editions by Ozanam (Paris, 1730) and Audierne (Paris, 1753) and in English translation by Reeve Williams (London, 1685).

II. Secondary Literature. On Dechales or his work, see Moritz Cantor, *Vorlesungen über die Geschichte der Mathematik,* III (Leipzig, 1913), 4–6, 15–19; Augustus De Morgan, *Arithmetical Books From the Invention of Printing to the Present Time* (London, 1847), pp. xv, 53; Charles Hutton, "History of Algebra," in *Tracts on Mathematical Philosophical Subjects,* II (London, 1812), tract no. 33, p. 301; and *Philosophical and Mathematical Dictionary,* I (London, 1815), 395–396; and *The Penny Cyclopaedia of the Society for the Diffusion of Useful Knowledge,* VIII (London, 1837), 343.

William Schaaf

DÉCHELETTE, JOSEPH (*b.* Roanne, Loire, France, 8 January 1862; *d.* Nouron-Vingré, Aisne, France, 4 October 1914), *archaeology.*

The son of rich industrialists, Déchelette studied with the Jesuits. His uncle, Gabriel Bulliot, instilled in him a passionate interest in Gallic archaeology, and one of his cousins, a departmental archivist, introduced him to local history and bibliography. After finishing secondary school, he had to give up entering the École des Chartes in order to sell the fabrics made in his father's factory. He traveled on business in France and abroad, and his visits to museums and his facility in learning foreign and ancient languages stimulated him to undertake archaeological syntheses on a European scale. In 1899, at the age of thirty-seven, Déchelette abandoned industry and devoted himself to archaeology. Next to his magnificent residence he had a large library constructed in which he gathered almost everything that had been published on European archaeology, from prehistory to the Middle Ages.

Methodical organization of his work and the assistance of secretaries and draftsmen enabled Déchelette to publish a synthesis on the Celts in Europe (1901), a work on the Gallic fortified city of Bibracte (1903), and an account of the excavations carried out by himself and his uncle Bulliot on the fortified city of Mont-Beuvray and two volumes on the decorated ceramics of Roman Gaul, with 1,700 drawings to facilitate the dating of the archaeological layers (1904). From 1905 to 1908 he composed the first volume of his famous *Manuel d'archéologie préhistorique, celtique et gallo-romaine,* devoted to prehistory, on which he had the aid of the Abbé Breuil. In 1910 he finished the volume on the Bronze Age. The Iron Age, which was his specialty, required three volumes (1912–1914), and Déchelette was collecting documentation on Gallo-Roman archaeology when World War I began. An ardent patriot, at his own request he took part in an attack. Advancing on horseback at the head of his battalion, he was mortally wounded.

Déchelette's early meticulous researches, within a limited regional framework, gave him a sharp sense of archaeological realities; after having traveled through Europe on business, he continued to travel in order to see for himself the sites, the objects discovered, and the archaeologists. He had the gift of clearly setting forth the essential facts contained in an enormous documentation and stated that "a page of synthesis requires volumes of analyses"; these qualities account for the *Manuel d'archéologie* being still very often consulted, despite a half-century of discoveries and new methods.

BIBLIOGRAPHY

The bibliography of Déchelette's works published by his nephew François Déchelette (see below) contains 183 titles. His principal works are *L'oppidum de Bibracte* (Paris, 1903); *Les fouilles du Mont-Beuvray de 1897–1901* (Paris, 1904); *Les vases céramiques ornés de la Gaule romaine* (Paris, 1904); and *Manuel d'archéologie préhistorique, celtique et gallo-romaine:* I, *Ages de la pierre taillée et polie* (Paris, 1908); II, *Ages du bronze* (Paris, 1910); III, *Le premier âge du fer* (Paris, 1913); IV, *Le second âge du fer et index général* (Paris, 1914). Two appendixes were published as a supp. to vol. II. Two partial reprintings of *Manuel,* including some corrections, were made. A facs. repr. is in preparation. The Gallo-Roman part was later undertaken (but not completed) by Albert Grenier.

Déchelette's widow, carrying out a bequest of her husband, donated to the city of Roanne his residence, made into a museum, and his library, which includes 41 volumes of correspondence received by Déchelette.

François Déchelette, *Livre d'or de Joseph Déchelette* (Roanne, 1962), consists of a rather summary biography, a bibliography, and extracts from a great number of obituary notices.

FRANCK BOURDIER

DECHEN, HEINRICH VON (*b.* Berlin, Germany, 25 March 1800; *d.* Bonn, Germany, 15 February 1889), *mining.*

Dechen was born into a family of Prussian civil servants. At first he was tutored at home with his elder brother; both of them then attended the Gymnasium in Grauen Kloster. Heinrich took his final examination during Easter 1818. He decided to enter the mining profession and began his training at the University of Berlin, then was accepted as a mining *Expektant* by the office of the inspector general of mines and admitted to the Haupt-Bergwerks-Eleven-Institut.

During his student years in Berlin, Dechen served with the military engineers. Following this he began his practical training at Haberbank mine near Sprockhövel in October 1819. Named to the Berg-Eleven on 15 July 1820, he found employment in the mining offices in Bochum and Essen.

At this time Dechen formed a friendship with Karl von Oeynhausen that led first to Dechen's sharing in Oeynhausen's mining studies, and then to a trip together through German, Belgian, and French mining regions. On this expedition, Dechen, through the offices of Leopold von Buch, made the acquaintance of Alexander von Humboldt in Paris.

After the two young friends had prepared a joint report on their journey, Dechen took the examination for *Bergreferendar* in March 1824. After following a tour of the Saxon and Bohemian ore mines, he was named *Oberbergamtsassessor.* He then undertook a second foreign journey with Oeynhausen, this time to England and Scotland. They again reported together in Berlin on their expedition. In 1828 Dechen was transferred to the *Oberbergamt* in Bonn. Here for the first time he came into close contact with Rhineland mining.

After two years Dechen was called back to Berlin, where he became *Oberbergrat* and a councillor in the Ministry of the Interior. He also held the post of assistant professor at the University of Berlin, where he lectured on mining. Dechen became a member of the Prussian Academy of Sciences and presented valuable studies before it. These included papers on material gathered during his journeys, as well as his translation of de la Beche's *Handbook of Geognosy* and his own *Geognostische Übersichtskarte von Deutschland, Frankreich, England und den angrenzenden Ländern* (1838). From 1838 on he also participated in the editing of Karsten's *Archiv für Mineralogie, Geognosie, Bergbau- und Hüttenkunde.*

The philosophical faculty of the University of Bonn honored Dechen's scientific achievements by naming him an honorary doctor. When the position of superintendent of mines in Bonn became open upon the promotion of Count Ernst August von Beust to inspector general of mines, Humboldt and Buch recommended that Dechen be nominated to it. He was named to this post in 1841 and Bonn and the Rhineland became his second home.

In Bonn, Dechen's mining and geological skill fully manifested itself for the first time. He traversed the Rhineland and the neighboring portions of Westphalia, as well as the hard coal areas of the Loire and the Saône regions of France, everywhere occupying himself with the exploration of the earth's interior and the progress of the art of mining. He prepared numerous geological maps, which Humboldt described as models in this field and which for long were standard. The project of a systematic geological survey covering the whole of Prussia—which led in 1873 to the establishment of the Geological Institute in Berlin—was based on Dechen's investigations. Dechen was invited to become a curator of the institute in 1875.

In addition to these major undertakings, Dechen

wrote many individual books and articles for scientific journals and magazines. He also took part in the reform of Prussian mining legislation in Berlin. Indeed, in the winter of 1859–1860 Dechen temporarily assumed the direction of all Prussian mining, metallurgical, and salt-mining operations, but he did not let himself be persuaded to remain permanently in this post in Berlin. He was too much attached to Bonn and the Rhineland, to which he returned. There on 23 May 1860 he obtained the rank of inspector general of mines.

Too much, however, lay on Dechen's shoulders. He wished to devote himself completely to his scientific work and asked to retire from the civil service. His request was granted (after repeated vain appeals to him to change his mind) on 1 January 1864.

Following his retirement from government service Dechen continued to devote himself to research and scientific writing. He further continued to take part in the activities of those service and scientific organizations whose member, honorary member, or chairman he was. Many honors were awarded him, including his appointment to the Prussian Privy Council in 1884. On 10 November 1886 he suffered a stroke from which he never recovered; his condition slowly worsened and he died in 1889.

BIBLIOGRAPHY

I. ORIGINAL WORKS. Dechen wrote about 400 works, most of them published in journals. A complete bibliography is in H. Laspeyres, *Heinrich von Dechen. Ein Lebensbild* (Bonn, 1889).

II. SECONDARY LITERATURE. In addition to Laspeyres's biography, see G. Schmidt, *Die Familie von Dechen* (Merseburg, 1890); Walter Serlo, "Heinrich von Dechen," in *Glückauf!,* **64** (1928), 1517–1519; "Heinrich von Dechen (Lebensbilder zur Geschichte des Bergbaus)," in *Zeitschrift für das Berg-, Hütten- und Salinenwesen,* **82** (1934), 295–297; "Heinrich von Dechen," in Walter Serlo, *Männer des Bergbaus* (Berlin, 1937), pp. 39–40; and von Zittel, "Heinrich von Dechen," in *Allgemeine deutsche Biographie,* XLVII (Leipzig, 1903), 629–631.

M. KOCH

DICTIONARY
OF
SCIENTIFIC BIOGRAPHY

PUBLISHED UNDER THE AUSPICES OF
THE AMERICAN COUNCIL OF LEARNED SOCIETIES

The American Council of Learned Societies, organized in 1919 for the purpose of advancing the study of the humanities and of the humanistic aspects of the social sciences, is a nonprofit federation comprising thirty-four national scholarly groups. The Council represents the humanities in the United States in the International Union of Academies, provides fellowships and grants-in-aid, supports research-and-planning conferences and symposia, and sponsors special projects and scholarly publications.

Member Organizations

AMERICAN PHILOSOPHICAL SOCIETY, 1743

AMERICAN ACADEMY OF ARTS AND SCIENCES, 1780

AMERICAN ANTIQUARIAN SOCIETY, 1812

AMERICAN ORIENTAL SOCIETY, 1842

AMERICAN NUMISMATIC SOCIETY, 1858

AMERICAN PHILOLOGICAL ASSOCIATION, 1869

ARCHAEOLOGICAL INSTITUTE OF AMERICA, 1879

SOCIETY OF BIBLICAL LITERATURE, 1880

MODERN LANGUAGE ASSOCIATION OF AMERICA, 1883

AMERICAN HISTORICAL ASSOCIATION, 1884

AMERICAN ECONOMIC ASSOCIATION, 1885

AMERICAN FOLKLORE SOCIETY, 1888

AMERICAN DIALECT SOCIETY, 1889

ASSOCIATION OF AMERICAN LAW SCHOOLS, 1900

AMERICAN PHILOSOPHICAL ASSOCIATION, 1901

AMERICAN ANTHROPOLOGICAL ASSOCIATION, 1902

AMERICAN POLITICAL SCIENCE ASSOCIATION, 1903

BIBLIOGRAPHICAL SOCIETY OF AMERICA, 1904

ASSOCIATION OF AMERICAN GEOGRAPHERS, 1904

AMERICAN SOCIOLOGICAL ASSOCIATION, 1905

ORGANIZATION OF AMERICAN HISTORIANS, 1907

COLLEGE ART ASSOCIATION OF AMERICA, 1912

HISTORY OF SCIENCE SOCIETY, 1924

LINGUISTIC SOCIETY OF AMERICA, 1924

MEDIAEVAL ACADEMY OF AMERICA, 1925

AMERICAN MUSICOLOGICAL SOCIETY, 1934

SOCIETY OF ARCHITECTURAL HISTORIANS, 1940

ECONOMIC HISTORY ASSOCIATION, 1940

ASSOCIATION FOR ASIAN STUDIES, 1941

AMERICAN SOCIETY FOR AESTHETICS, 1942

METAPHYSICAL SOCIETY OF AMERICA, 1950

AMERICAN STUDIES ASSOCIATION, 1950

RENAISSANCE SOCIETY OF AMERICA, 1954

SOCIETY FOR ETHNOMUSICOLOGY, 1955

DICTIONARY

OF

SCIENTIFIC BIOGRAPHY

CHARLES COULSTON GILLISPIE

Princeton University

EDITOR IN CHIEF

Volume 4

RICHARD DEDEKIND—FIRMICUS MATERNUS

CHARLES SCRIBNER'S SONS · NEW YORK

Editorial Staff

MARSHALL DE BRUHL, *MANAGING EDITOR*

SARAH FERRELL, *Assistant Managing Editor*

LOUISE BILEBOF, *Administrative Editor*

LELAND S. LOWTHER, *Associate Editor*

ROSE MOSELLE, *Editorial Assistant*

JANET JACOBS, *Assistant Editor*

ELIZABETH I. WILSON, *Copy Editor*

DORIS ANNE SULLIVAN, *Proofreader*

JOEL HONIG, *Copy Editor*

Panel of Consultants

Contributors to Volume 4

The following are the contributors to Volume **4.** Each author's name is followed by the institutional affiliation at the time of publication and the names of articles written for this volume. The symbol † indicates that an author is deceased.

A. F. O'D. ALEXANDER †
DREYER

G. C. AMSTUTZ
University of Heidelberg
ESKOLA

R. CHRISTIAN ANDERSON
Brookhaven National Laboratory
DESSAIGNES

COLETTE AVIGNON
University of Orléans-Tours
EVELYN

NANDOR L. BALAZS
State University of New York at Stony Brook
EINSTEIN

GEORGE B. BARBOUR
University of Cincinnati
FENNEMAN

WILLIAM B. BEAN
University of Iowa
FINLAY

ROBERT P. BECKINSALE
University of Oxford
DELUC

SILVIO A. BEDINI
Smithsonian Institution
DIVINI

WHITFIELD BELL, JR.
American Philosophical Society Library
DIXON, J.

LUIGI BELLONI
University of Milan
DUBINI

ALEX BERMAN
University of Cincinnati
DEROSNE; FÉE

PIERRE BERTHON
Archives, Académie des Sciences, Paris
DUNOYER DE SEGONZAC

KURT-R. BIERMANN
German Academy of Sciences
DEDEKIND; EISENSTEIN

ARTHUR BIREMBAUT
ÉLIE DE BEAUMONT

ASIT K. BISWAS
Department of Energy, Mines and Resources, Canada
DUFOUR; FICHOT

A. BLAAUW
European Southern Observatory
EASTON

L. J. BLACHER
Soviet Academy of Sciences
DOGEL; ESCHSCHOLTZ

H. BOERNER
University of Giessen
ENGEL

UNO BOKLUND
Royal Pharmaceutical Institute, Stockholm
EKEBERG

FRANCK BOURDIER
École Pratique des Hautes Études
DEPÉRET

GERT H. BRIEGER
The Johns Hopkins University
EBERTH

W. H. BROCK
University of Leicester
ERDMANN

THEODORE M. BROWN
Princeton University
DESCARTES

STEPHEN G. BRUSH
University of Maryland
ENSKOG

IVOR BULMER-THOMAS
DINOSTRATUS; DIONYSODORUS; DOMNINUS OF LARISSA; EUCLID; EUDEMUS OF RHODES; EUTOCIUS OF ASCALON

JOHN G. BURKE
University of California at Los Angeles
DUFRÉNOY

HAROLD BURSTYN
Carnegie-Mellon University
FERREL

H. L. L. BUSARD
University of Leiden
DEPARCIEUX; DESPAGNET

RONALD S. CALINGER
Rensselaer Polytechnic Institute
DICKSON

ALBERT V. CAROZZI
University of Illinois
DESOR

BERNARDO J. CAYCEDO
ELHUYAR, F; ELHUYAR, J.

JOHN CHALLINOR
FALCONER; FAUJAS DE SAINT-FOND

SEYMOUR L. CHAPIN
Los Angeles State College
DELISLE, J.-N.

ROBERT CHIPMAN
University of Toledo
DEPREZ

EDWIN CLARKE
University College, London
FERRIER

ARCHIBALD CLOW
British Broadcasting Corporation
DUNDONALD

I. BERNARD COHEN
Harvard University
DELAMBRE

ALBERT B. COSTA
University of Notre Dame
DEWAR; DIXON, H. B.; ERLENMEYER

PIERRE COSTABEL
École Pratique des Hautes Études
DU HAMEL, J. B.

CHARLES COURY
University of Paris
DUVAL

C. F. COWAN
EVANS, W. H.

ALISTAIR C. CROMBIE
University of Oxford
DESCARTES

M. P. CROSLAND
University of Leeds
DULONG

L. W. CURRIER †
EMERSON, B. K.

ŽARKO DADIĆ
Yugoslav Academy of Sciences and Arts
DOMINIS

KARL H. DANNENFELDT
Arizona State University
DIOCLES; DIOCLES OF CARYSTUS

MARGARET DEACON
National Institute of Oceanography
DITTMAR

ALLEN G. DEBUS
University of Chicago
DUCHESNE; FLUDD

ALBERT DELAUNAY
Pasteur Institute
DUCLAUX

BERN DIBNER
Burndy Library
DU MONCEL; FERRARIS

D. R. DICKS
University of London
DOSITHEUS; ERATOSTHENES; EUCTEMON

SALLY H. DIEKE
The Johns Hopkins University
DUGAN; DUNCAN; ELKIN

AUBREY DILLER
Indiana University
DICAEARCHUS

CLAUDE E. DOLMAN
University of British Columbia
DOBELL; EHRLICH; ESCHERICH

J. D. H. DONNAY
The Johns Hopkins University
FANKUCHEN

SIGALIA C. DOSTROVSKY
Worcester Polytechnic Institute
DUHAMEL, J.-M.; EUCKEN; EWING,
J. A.; FABRY, C.

A. VIBERT DOUGLAS
Queen's University, Ontario
EDDINGTON

JOHN M. DUBBEY
University College, London
DE MORGAN

K. C. DUNHAM
Institute of Geological Sciences
EVANS, F. J. O.

DAVID R. DYCK
University of Winnipeg
ELLER VON BROCKHAUSEN

JOY P. EASTON
West Virginia University
DEE; DIGGES, L.; DIGGES, T.

SIDNEY M. EDELSTEIN
Dexter Chemical Corporation
DREBBEL

CAROLYN EISELE
*Hunter College, City University of
New York*
ENRIQUES

JON EKLUND
Smithsonian Institution
DUHAMEL DU MONCEAU

VASILY A. ESAKOV
*Institute of the History of Science and
Technology, Moscow*
DOKUCHAEV

JOSEPH EWAN
Tulane University
FERNALD

JOAN M. EYLES
FAREY

V. A. EYLES
DE LA BECHE

EDUARD FARBER †
DÖBEREINER; EDER

W. V. FARRAR
University of Manchester
DONNAN

LUCIENNE FÉLIX
DRACH

E. A. FELLMANN
Institut Platonaeum, Basel
FABRI

EDWIN FELS
DRYGALSKI

BERNARD S. FINN
Smithsonian Institution
EDISON

WALTHER FISCHER
DOELTER

C. S. FISHER
Brandeis University
DEHN; FINE

J. O. FLECKENSTEIN
University of Basel
EMDEN

DONALD FLEMING
Harvard University
DRAPER, J.

MARCEL FLORKIN
University of Liège
DODOENS

PAUL FORMAN
University of Rochester
DUANE

ROBERT FOX
University of Lancaster
DUPRÉ

PIETRO FRANCESCHINI
DELLA TORRE, G.

H. C. FREIESLEBEN
DORNO; ENCKE

DAVID J. FURLEY
Princeton University
EPICURUS

L. K. GABUNA
Soviet Academy of Sciences
DOLLO

GERALD L. GEISON
Princeton University
DUJARDIN

WALTHER GERLACH
University of Munich
ELSTER

CHARLES COULSTON GILLISPIE
Princeton University
DIDEROT

MARIO GLIOZZI
University of Turin
FABBRONI

MARTHA TEACH GNUDI
University of California at Los Angeles
DORN

STANLEY GOLDBERG
Antioch College
DRUDE

J. B. GOUGH
Washington State University
FAHRENHEIT

RAGNAR GRANIT
Nobel Institute for Neurophysiology
FERNEL

NORMAN T. GRIDGEMAN
National Research Council of Canada
DODGSON

M. D. GRMEK
*Archives Internationales d'Histoire des
Sciences*
DODART; ESTIENNE; FERREIN

FRANCISCO GUERRA
DESCOURTILZ; DOMBEY; FEUILLÉE

LAURA GUGGENBUHL
*Hunter College, City University of
New York*
FEUERBACH

HEINRICH GUGGENHEIMER
Polytechnic Institute of Brooklyn
EISENHART

A. RUPERT HALL
*Imperial College of Science and
Technology*
DESAGULIERS

MARIE BOAS HALL
*Imperial College of Science and
Technology*
DIGBY

R. S. HARTENBERG
Northwestern University
EYTELWEIN

JOHN L. HEILBRON
University of California at Berkeley
DUFAY

C. DORIS HELLMAN
*Queens College, City University of
New York*
DÖRFFEL

BROOKE HINDLE
New York University
FARRAR

HEBBEL E. HOFF
Baylor University, College of Medicine
DENIS

J. E. HOFMANN
University of Tübingen
DYCK

S. HOOGERWERF
EINTHOVEN

WŁODZIMIERZ HUBICKI
Marie Curie-Skłodowska University
ERCKER

G. L. HUXLEY
Queen's University, Belfast
EUDOXUS OF CNIDUS

AARON J. IHDE
University of Wisconsin
ELVEHJEM; EULER-CHELPIN

ILSE JAHN
University of Berlin
EHRENBERG

S. A. JAYAWARDENE
Science Museum Library, London
FERRARI

JULIAN JAYNES
Princeton University
FECHNER

SATISH KAPOOR
University of Saskatchewan
DUMAS

GEORGE B. KAUFFMAN
Fresno State College
DELÉPINE

G. B. KERFERD
University of Manchester
DEMOCRITUS

PAUL A. KIRCHVOGEL
Landesmuseum, Kassel
FAULHABER

GEORGE KISH
University of Michigan
DELISLE, G.; DUDLEY

MARTIN J. KLEIN
Yale University
EHRENFEST; EINSTEIN

D. M. KNIGHT
University of Durham
DERHAM

J. KOVALEVSKY
Bureau des Longitudes
DELAUNAY; FAYE

CLAUDIA KREN
University of Missouri
DOMINICUS DE CLAVASIO

VLADISLAV KRUTA
Purkinje University, Brno
DUTROCHET; EDWARDS

LOUIS KUSLAN
Southern Connecticut State College
FAVRE

V. I. KUZNETSOV
Soviet Academy of Sciences
FAVORSKY

RODOLPHINE J. CH. V. TER LAAGE
State University of Utrecht
DONDERS

BENGT-OLOF LANDIN
University of Lund
FABRICIUS

LAURENS LAUDEN
University of Pittsburgh
FERGUSON

P. S. LAURIE
Royal Greenwich Observatory
ELLIS; EVERSHED

GORDON LEFF
University of York
DUNS SCOTUS

HENRY M. LEICESTER
University of the Pacific
DEVILLE

MARTIN LEVEY †
IBN EZRA

JACQUES R. LÉVY
Paris Observatory
ESCLANGON; FABRY, L.

O. LEZHNEVA
Soviet Academy of Sciences
EICHENWALD

G. A. LINDEBOOM
Free University, Amsterdam
EIJKMAN

JAMES LONGRIGG
University of Wisconsin
ERASISTRATUS

SUSAN M. P. McKENNA
University of Michigan
DOWNING

FRANCIS R. MADDISON
Museum of the History of Science, Oxford
DONDI

MUHSIN MAHDI
Harvard University
AL-FĀRĀBĪ

MICHAEL S. MAHONEY
Princeton University
DESCARTES; FERMAT

L. MARTON
National Bureau of Standards
EÖTVÖS

ARNALDO MASOTTI
Politecnico di Milano
FERRO

KIRTLEY F. MATHER
Harvard University
EMMONS, S. F.

ALEXANDER P. D. MAURELATOS
University of Texas
EMPEDOCLES OF ACRAGAS

A. MENIAILOV
Soviet Academy of Sciences
FERSMAN

R. MICHARD
Paris Observatory
DESLANDRES

MIKLÓS MIKOLÁS
Technical University of Budapest
FEJÉR

DONALD G. MILLER
Lawrence Radiation Laboratory
DUHEM

A. M. MONNIER
University of Paris
DEVAUX; ERLANGER; FIESSINGER

JEAN MOTTE
University of Montpellier
DELILE; DRAPARNAUD

JOHN MURDOCH
Harvard University
EUCLID

HENRY NATHAN
OECD, Directorate for Scientific Affairs
FATOU

A. NATUCCI
University of Genoa
FAGNANO DEI TOSCHI, G. C.; FAGNANO DEI TOSCHI, G. F.

AXEL V. NIELSEN †
DUNÉR

LUBOŠ NOVÝ
Czechoslovak Academy of Sciences
DICKSTEIN; DU BOIS-REYMOND, P. D. G.

ROBERT OLBY
University of Leeds
DILLENIUS; FEULGEN

C. D. O'MALLEY †
DUBOIS; EDWARDES; EUSTACHI; FALLOPPIO

JANE OPPENHEIMER
Bryn Mawr College
DRIESCH

OYSTEIN ORE †
DIRICHLET

WALTER PAGEL
Wellcome Institute of the History of Medicine
ERASTUS

A. B. PAPLAUSCAS
Soviet Academy of Sciences
EGOROV

LINUS PAULING
Stanford University
DICKINSON

JACQUES PAYEN
Conservatoire Nationale des Arts et Métiers
DESORMES

KURT MØLLER PEDERSEN
University of Aarhus
DOVE

CONTRIBUTORS TO VOLUME 4

MARGARET R. PEITSCH
Department of Energy, Mines and Resources, Canada
FICHOT

P. E. PILET
University of Lausanne
ERRERA

DAVID PINGREE
University of Chicago
DINAKARA; AL-FAZĀRĪ

A. F. PLAKHOTNIK
Soviet Academy of Sciences
DERYUGIN

ERICH POSNER
Birmingham Regional Hospital Board
DOMAGK

WILLIAM B. PROVINE
Cornell University
EAST

EUGENE RABINOWITCH
State University of New York at Albany
EMERSON, R.

SAMUEL X. RADBILL
College of Physicians of Philadelphia
DUNGLISON

NATHAN REINGOLD
Smithsonian Institution
ESPY

SAMUEL REZNECK
Rensselaer Polytechnic Institute
EATON

JOHN M. RIDDLE
North Carolina State University
DIOSCORIDES

GUGLIELMO RIGHINI
Osservatorio Astrofisico di Arcetri
DEMBOWSKI; EMANUELLI

MARIA LUISA RIGHINI-BONELLI
Istituto e Museo di Storia della Scienza
DONATI

GUENTER B. RISSE
University of Minnesota
DÖLLINGER; EHRET; EICHLER

HANS ROHRBACH
University of Mainz
FEIGL

COLIN A. RONAN
DYSON

PAUL LAWRENCE ROSE
St. John's University, New York
DUDITH

K. E. ROTHSCHUH
University of Münster/Westphalia
DU BOIS-REYMOND, E.; ENGELMANN; FICK

HUNTER ROUSE
University of Iowa
DU BUAT

A. I. SABRA
Warburg Institute
AL-FARGHĀNĪ

MORRIS H. SAFFRON
Rutgers University
DONALDSON; DUGGAR

NORMAN SCHAUMBERGER
Bronx Community College, City University of New York
DOUGLAS, JESSE

EBERHARD SCHMAUDERER
DIELS; EMBDEN

CECIL J. SCHNEER
University of New Hampshire
EMMONS, E.

J. F. SCOTT
St. Mary's College of Education, Middlesex
DELAMAIN

EMILIO SEGRÈ
University of California at Berkeley
FERMI

EDITH SELOW
DINGLER

I. P. SHELDON-WILLIAMS
University College, Dublin
ERIUGENA

ELIZABETH NOBLE SHOR
DEGOLYER; EIGENMANN

DIANA SIMPKINS
Northwestern Polytechnic, London
DIXON, H. H.; ELLIOT SMITH; FARMER

JOSEF SMOLKA
Czechoslovak Academy of Sciences
DIVIŠ

PIERRE SPEZIALI
University of Geneva
DINI

NILS SPJELDNAES
University of Aarhus
EBEL; ESCHOLT

WILLIAM H. STAHL †
FIRMICUS MATERNUS

MAX STECK
University of Munich
DÜRER

WALLACE STEGNER
Stanford University
DUTTON

DIRK J. STRUIK
Massachusetts Institute of Technology
DE GROOT; DUPIN; FANO

CHARLES SÜSSKIND
University of California at Berkeley
DE FOREST; FEDDERSEN; FESSENDEN

JUDITH P. SWAZEY
Harvard University
EGAS MONIZ

FERENC SZABADVÁRY
Technical University of Budapest
ESSON; FARKAS

RENÉ TATON
École Pratique des Hautes Études
DESARGUES; DIONIS DU SÉJOUR

KENNETH L. TAYLOR
University of Oklahoma
DES CLOIZEAUX; DESMAREST; DOLOMIEU; DUPERREY

ANDRÉE TÉTRY
École Pratique des Hautes Études
DELAGE; DUBOSCQ

JEAN THÉODORIDÈS
Centre National de la Recherche Scientifique
FABRE

K. BRYN THOMAS
Reading Pathological Society, Library
DOUGLAS, JAMES

V. V. TIKHOMIROV
Soviet Academy of Sciences
EICHWALD

HEINZ TOBIEN
University of Mainz
DESHAYES; ERMAN; ESCHER VON DER LINTH

VICTOR A. TRIOLO
Temple University
EWING, J.; FINSEN

HENRY S. TROPP
University of Toronto
FIELDS

G. L'E. TURNER
Museum of the History of Science, Oxford
DOLLOND

H. L. VANDERLINDEN
University of Ghent
DELPORTE

J. J. VERDONK
FINK

KURT VOGEL
University of Munich
DIOPHANTUS; FIBONACCI

WILLIAM A. WALLACE, O. P.
Catholic University of America
DIETRICH VON FREIBERG; DULLAERT OF GHENT

CHARLES WEBSTER
University of Oxford
ENT

PIERRE WELANDER
Massachusetts Institute of Technology
EKMAN

GEORGE W. WHITE
University of Illinois
EVANS, L.

CONTRIBUTORS TO VOLUME 4

CHARLES A. WHITNEY
Smithsonian Institution, Astrophysical Observatory
DRAPER, H.

L. PEARCE WILLIAMS
Cornell University
DE LA RUE; FARADAY

WESLEY C. WILLIAMS
Case Western Reserve University
DUVERNEY

CURTIS WILSON
University of California at Berkeley
DOPPELMAYR

J. T. WILSON
University of Toronto
DU TOIT

A. E. WOODRUFF
Yeshiva University
DOPPLER

O. WRIGHT
University of London
AL-FĀRĀBĪ

HATTEN S. YODER, JR.
Geophysical Laboratory, Washington
FENNER

A. P. YOUSCHKEVITCH
Soviet Academy of Sciences
EULER

BRUNO ZANOBIO
University of Pavia
FABRICI

DICTIONARY
OF
SCIENTIFIC BIOGRAPHY

DICTIONARY OF
SCIENTIFIC BIOGRAPHY

DEDEKIND—FIRMICUS MATERNUS

DEDEKIND, (JULIUS WILHELM) RICHARD (*b.* Brunswick, Germany, 6 October 1831; *d.* Brunswick, 12 February 1916), *mathematics.*

Dedekind's ancestors (particularly on his mother's side) had distinguished themselves in services to Hannover and Brunswick. His father, Julius Levin Ulrich Dedekind, the son of a physician and chemist, was a graduate jurist, professor, and corporation lawyer at the Collegium Carolinum in Brunswick. His mother, Caroline Marie Henriette Emperius, was the daughter of a professor at the Carolinum and the granddaughter of an imperial postmaster. Richard Dedekind was the youngest of four children. His only brother, Adolf, became a district court president in Brunswick; one sister, Mathilde, died in 1860, and Dedekind lived with his second sister, Julie, until her death in 1914, neither of them having married. She was a respected writer who received a local literary prize in 1893.

Between the ages of seven and sixteen Dedekind attended the Gymnasium Martino-Catharineum in Brunswick. His interest turned first to chemistry and physics; he considered mathematics only an auxiliary science. He soon occupied himself primarily with it, however, feeling that physics lacked order and a strictly logical structure. In 1848 Dedekind became a student at the Collegium Carolinum, an institute between the academic high school and the university level, which Carl Friedrich Gauss had also attended. There Dedekind mastered the elements of analytic geometry, algebraic analysis, differential and integral calculus, and higher mechanics, and studied the nat-

ural sciences. In 1849–1850, he gave private lessons in mathematics to his later colleague at the Carolinum, Hans Zincke (known as Sommer). Thus, when he matriculated at the University of Göttingen at Easter 1850, Dedekind was far better prepared for his studies than were the majority of graduates from the academic high school. At Göttingen, a seminar in mathematics and physics had just been founded, at the initiative of Moritz Abraham Stern, for the education of instructors for teaching in the academic high school. The direction of the mathematics department was the duty of Stern and Georg Ulrich, while Wilhelm Weber and Johann Benedict Listing directed the physics department. Dedekind was a member of the seminar from its inception and was there first introduced to the elements of the theory of numbers. A year later Bernhard Riemann also began to participate in the seminar, and Dedekind soon developed a close friendship with him. In the first semester, Dedekind attended lectures on differential and integral calculus, which offered him very little new material. He attended Ulrich's seminar on hydraulics but rarely took part in the physics laboratories run by Weber and Listing; Weber's lectures on experimental physics, however, made a very strong impression on him throughout two semesters. Weber had an inspiring effect on Dedekind, who responded with respectful admiration. In the summer semester of 1850, Dedekind attended the course in popular astronomy given by Gauss's observer, Carl Wolfgang Benjamin Goldschmidt; in the winter semester of 1850–1851, he attended Gauss's own lecture on the method of least squares. Although he disliked teaching, Gauss carried out the assignment with his usual conscientiousness; fifty years later Dedekind remembered the lecture as one of the most beautiful he had ever heard, writing that he had followed Gauss with constantly increasing interest and that he could not forget the experience. In the following semester, Dedekind heard Gauss's lecture on advanced geodesy. In the winter semester of 1851–1852, he heard the two lectures given by Quintus Icilius on mathematical geography and on the theory of heat and took part in Icilius' meteorological observations. After only four semesters, he did his doctoral work under Gauss in 1852 with a thesis on the elements of the theory of Eulerian integrals. Gauss certified that he knew a great deal and was independent; in addition, he had prophetically "favorable expectations of his future performance."

Dedekind later determined that this knowledge would have been sufficient for teachers in secondary school service but that it did not satisfy the prerequisite for advanced studies at Göttingen. For instance, Dedekind had not heard lectures on more recent developments in geometry, advanced theory of numbers, division of the circle and advanced algebra, elliptic functions, or mathematical physics, which were then being taught at the University of Berlin by Steiner, Jacobi, and Dirichlet. Therefore, Dedekind spent the two years following his graduation assiduously filling the gaps in his education, attending—among others—Stern's lectures on the solution of numerical equations.

In the summer of 1854, he qualified, a few weeks after Riemann, as a university lecturer; in the winter semester of 1854–1855 he began his teaching activities as *Privatdozent,* with a lecture on the mathematics of probability and one on geometry with parallel treatment of analytic and projective methods.

After Dirichlet succeeded Gauss in Göttingen in 1855, Dedekind attended his lectures on the theory of numbers, potential theory, definite integrals, and partial differential equations. He soon entered into a closer personal relationship with Dirichlet and had many fruitful discussions with him; Dedekind later remembered that Dirichlet had made "a new man" of him and had expanded his scholarly and personal horizons. When the Dirichlets were visited by friends from Berlin (Rebecca Dirichlet was the sister of the composer Felix Mendelssohn-Bartholdy and had a large circle of friends), Dedekind was invited too and enjoyed the pleasant sociability of, for example, the well-known writer and former diplomat, Karl August Varnhagen von Ense, and his niece, the writer Ludmilla Assing.

In the winter semester of 1855–1856 and in the one

following, Dedekind attended Riemann's lectures on Abelian and elliptic functions. Thus, although an instructor, he remained an intensive student as well. His own lectures at that time are noteworthy in that he probably was the first university teacher to lecture on Galois theory, in the course of which the concept of field was introduced. To be sure, few students attended his lectures: only two were present when Dedekind went beyond Galois and replaced the concept of the permutation group by the abstract group concept.

In 1858, Dedekind was called to the Polytechnikum in Zurich (now the Eidgenössische Technische Hochschule) as the successor to Joseph Ludwig Raabe. Thus Dedekind was the first of a long line of German mathematicians for whom Zurich was the first step on the way to a German professorial chair; to mention only a few, there were E. B. Christoffel, H. A. Schwarz, G. Frobenius, A. Hurwitz, F. E. Prym, H. Weber, F. Schottky, and H. Minkowski. The Swiss school counsellor responsible for appointments came to Göttingen at Easter 1858 and decided immediately upon Dedekind—which speaks for his power of judgment. In September 1859, Dedekind traveled to Berlin with Riemann, after Riemann's election as a corresponding member of the academy there. On this occasion, Dedekind met the initiator of that selection, Karl Weierstrass, as well as other leaders of the Berlin school, including Ernst Eduard Kummer, Karl Wilhelm Borchardt, and Leopold Kronecker.

In 1862, he was appointed successor to August Wilhelm Julius Uhde at the Polytechnikum in Brunswick, which had been created from the Collegium Carolinum. He remained in Brunswick until his death, in close association with his brother and sister, ignoring all possibilities of change or attainment of a larger sphere of activity. The small, familiar world in which he lived completely satisfied his demands: in it his relatives completely replaced a wife and children of his own and there he found sufficient leisure and freedom for scientific work in basic mathematical research. He did not feel pressed to have a more marked effect in the outside world; such confirmation of himself was unnecessary.

Although completely averse to administrative responsibility, Dedekind nevertheless considered it his duty to assume the directorship of the Polytechnikum from 1872 to 1875 (to a certain extent he was the successor of his father, who had been a member of the administration of the Collegium Carolinum for many years) and to assume the chairmanship of the school's building commission in the course of the transformation to a technical university. Along with his recreational trips to Austria (the Tyrol), to Switzerland, and through the Black Forest, his visit to the Paris exposition of 1878 should also be mentioned. On 1 April 1894 he was made professor emeritus but continued to give lectures occasionally. Seriously ill in 1872, following the death of his father, he subsequently enjoyed physical and intellectual health until his peaceful death at the age of eighty-four.

A corresponding member of the Göttingen Academy from 1862, Dedekind also became a corresponding member of the Berlin Academy in 1880 upon the initiative of Kronecker. In 1900, he became a correspondent of the Académie des Sciences in Paris and in 1910 was elected as *associé étranger*. He was also a member of the Leopoldino-Carolina Naturae Curiosorum Academia and of the Academy in Rome. He received honorary doctorates in Kristiania (now Oslo), in Zurich, and in Brunswick. In 1902 he received numerous scientific honors on the occasion of the fiftieth anniversary of his doctorate.

Dedekind belonged to those mathematicians with great musical talent. An accomplished pianist and cellist, he composed a chamber opera to his brother's libretto.

In character and principles, in style of living and views, Dedekind had much in common with Gauss, who also came from Brunswick and attended the Gymnasium Martino-Catharineum, the Collegium Carolinum, and the University of Göttingen. Both men had a conservative sense, a rigid will, an unshakable strength of principles, and a refusal to compromise. Each led a strictly regulated, simple life without luxury. Cool and reserved in judgment, both were warm-hearted, helpful people who formed strong bonds of trust with their friends. Both had a distinct sense of humor but also a strictness toward themselves and a conscientious sense of duty. Averse to any excess, neither was quick to express astonishment or admiration. Both were averse to innovations and turned down brilliant offers for other professorial chairs. In their literary tastes, both numbered Walter Scott among their favorite authors. Both impressed by that quality called modest greatness. Thus, it is not astonishing to find their similarity persisting in mathematics in the same preference for the theory of numbers, the same reservations about the algorithm, and the same partiality for "notions" above "notations." Although considerable, significant differences existed between Gauss and Dedekind, what they had in common predominates by far. Their kinship also received a marked visible expression: Dedekind was one of the select few permitted to carry Gauss's casket to the funeral service on the terrace of the Sternwarte.

Aside from Gauss the most enduring influences on Dedekind's scientific work were Dirichlet and Rie-

mann, with both of whom he shared many inclinations and attitudes. Dedekind, Dirichlet, and Riemann were all fully conscious of their worth, but with a modesty bordering on shyness, they never let their associates feel this. Ambition being foreign to them, they were embarrassed when confronted by the brilliance and elegance of their intellect. They loved thinking more than writing and were hardly ever able to satisfy their own demands. Being of absolute integrity, they had in common the same love for plain, certain truth. Dedekind's own statement to Zincke is more revelatory of his character than any description could be: "For what I have accomplished and what I have become, I have to thank my industry much more, my indefatigable working rather than any outstanding talent."

When Dedekind is mentioned today, one of the first associations is the "Dedekind cut," which he introduced in 1872 to use in treating the problem of irrational numbers in a completely new and exact manner.

In 1858, Dedekind had noted the lack of a truly scientific foundation of arithmetic in the course of his Zurich lectures on the elements of differential calculus. (Weierstrass also was stimulated to far-reaching investigations from such observations in the course of preparing lectures.) On 24 October, Dedekind succeeded in producing a purely arithmetic definition of the essence of continuity and, in connection with it, an exact formulation of the concept of the irrational number. Fourteen years later, he published the result of his considerations, *Stetigkeit und irrationale Zahlen* (Brunswick, 1872, and later editions), and explained the real numbers as "cuts" in the realm of rational numbers. He arrived at concepts of outstanding significance for the analysis of number through the theory of order. The property of the real numbers, conceived by him as an ordered continuum, with the conceptual aid of the cut that goes along with this, permitted tracing back the real numbers to the rational numbers: Any rational number a produces a resolution of the system R of all rational numbers into two classes A_1, A_2 in such a way that each number a_1 of the class A_1 is smaller than each number a_2 of the second class A_2. (Today, the term "set" is used instead of "system.") The number a is either the largest number of the class A_1 or the smallest number of the class A_2. A division of the system R into the two classes A_1, A_2, whereby each number a_1 in A_1 is smaller than each number a_2 in A_2 is called a "cut" (A_1, A_2) by Dedekind. In addition, an infinite number of cuts exist that are not produced by rational numbers. The discontinuity or incompleteness of the region R consists in this property. Dedekind wrote,

"Now, in each case when there is a cut (A_1, A_2) which is not produced by any rational number, then we *create* a new, *irrational* number α, which we regard as completely defined by this cut; we will say that this number α corresponds to this cut, or that it produces this cut" (*Stetigkeit,* § 4).

Occasionally Dedekind has been called a "modern Eudoxus" because an impressive similarity has been pointed out between Dedekind's theory of the irrational number and the definition of proportionality in Eudoxus' theory of proportions (Euclid, *Elements,* bk. V, def. 5). Nevertheless, Oskar Becker correctly showed that the Dedekind cut theory and Eudoxus' theory of proportions do not coincide: Dedekind's postulate of existence for all cuts and the real numbers that produce them cannot be found in Eudoxus or in Euclid. With respect to this, Dedekind said that the Euclidean principles alone—without inclusion of the principle of continuity, which they do not contain—are incapable of establishing a complete theory of real numbers as the proportions of the quantities. On the other hand, however, by means of his theory of irrational numbers, the perfect model of a continuous region would be created, which for just that reason would be capable of characterizing any proportion by a certain individual number contained in it (letter to Rudolph Lipschitz, 6 October 1876).

With his publication of 1872, Dedekind had become one of the leading representatives of a new epoch in basic research, along with Weierstrass and Georg Cantor. This was the continuation of work by Cauchy, Gauss, and Bolzano in systematically eliminating the lack of clarity in basic concepts by methods of demonstration on a higher level of rigor. Dedekind's and Weierstrass' definition of the basic arithmetic concepts, as well as Georg Cantor's theory of sets, introduced the modern development, which stands "completely under the sign of number," as David Hilbert expressed it.

Dedekind's book *Was sind und was sollen die Zahlen?* (Brunswick, 1888, and later editions) is along the same lines; in it he presented a logical theory of number and of complete induction, presented his principal conception of the essence of arithmetic, and dealt with the role of the complete system of real numbers in geometry in the problem of the continuity of space. Among other things, he provides a definition independent of the concept of number for the infiniteness or finiteness of a set by using the concept of mapping and treating the recursive definition, which is so important for the theory of ordinal numbers. (Incidentally, Dedekind regarded the ordinal number and not the cardinal number [*Anzahl*] as the original concept of number; in the cardinal number

he saw only an application of the ordinal number [letter to Heinrich Weber, 24 January 1888].) The demonstration of the existence of infinite systems given by Dedekind—similar to a consideration in Bolzano's *Paradoxien des Unendlichen* (Prague, 1851, §13)—is no longer considered valid. Kronecker was critical because Dedekind, agreeing with Gauss, regarded numbers as free creations of the human intellect and defended this viewpoint militantly and stubbornly. Weierstrass complained that his own definition of a complex quantity had not been understood by Dedekind. Hilbert criticized his effort to establish mathematics solely by means of logic. Gottlob Frege and Bertrand Russell criticized Dedekind's opinion that the cuts were not the irrational numbers but that the latter produced the former. However, even his critics and those who preferred Cantor's less abstract procedure for the construction of real numbers agreed that he had exercised a powerful influence on basic research in mathematics.

Just as Kronecker and Weierstrass had edited the mathematical works of those to whom they felt obligated, so Dedekind worked on the erection of literary monuments to those who had stood close to him. Making accessible the posthumous works of Gauss, Dirichlet, and Riemann occupied an important place in his work. In doing this, he gave proof of his congeniality and the rare combination of his productive and receptive intellectual talents.

Publishing the manuscripts of Gauss on the theory of numbers (*Werke,* vol. II [Göttingen, 1863]) gave him the opportunity of not only making available to wider circles the papers of a man he so greatly respected, but also of commenting on them with deep understanding. Dirichlet's *Vorlesungen über Zahlentheorie* (Brunswick, 1863, and later editions) was edited by him. If, as has been said, Dirichlet was the first not only to have completely understood Gauss's *Disquisitiones arithmeticae* but also to have made them accessible to others, then the same is true, to a great extent, of Dedekind's relationship to Dirichlet's lectures on the theory of numbers. Finally, he collaborated in editing the *Werke Bernhard Riemanns* (Leipzig, 1876; 2nd ed., 1892) with his friend Heinrich Weber and with his accustomed modesty placed his name after Weber's.

The editing of Dirichlet's lectures led Dedekind into a profound examination of the theory of generalized complex numbers or of forms that can be resolved into linear factors. In 1871 he provided these lectures with a supplement, in which he established the theory of algebraic number fields, or domains, by giving a general definition of the concept of the ideal—going far beyond Kummer's theory of "ideal numbers"—

that has become fruitful in various arithmetic and algebraic areas. In several papers Dedekind then, independently of Kronecker and with his approval, established the ideal theory that is held to be his masterpiece. Its principal theorem was that each ideal different from the unit ideal R can be represented unambiguously—with the exception of the order of factors—as the product of prime numbers. In his treatises concerning number fields, Dedekind arrived at a determination of the number of ideal classes of a field, penetrated the analysis of the base of a field, provided special studies on the theory of modules, and stimulated further development of ideal theory in which Emmy Noether, Hilbert, and Philipp Furtwängler participated. Paul Bachmann, Adolf Hurwitz, and Heinrich Weber in their publications also disseminated Dedekind's thoughts and expanded them.

That Dedekind did not stand completely apart from the applications of mathematics is shown by a treatise, written with W. Henneberg, which appeared as early as 1851 and concerns the time relationships in the course of plowing fields of various shapes, and also by the completion and publication of a treatise by Dirichlet concerning a hydrodynamic problem (1861).

Finally, we are indebted to Dedekind for such fundamental concepts as *ring* and *unit*.

It was indicative of the great esteem in which Dedekind was held even in foreign countries that, shortly after his death, in the middle of World War I, Camille Jordan, the president of the Académie des Sciences in Paris, warmly praised his theory of algebraic integers as his main work and expressed his sadness concerning the loss.

Although the association of mathematicians' names with concepts and theorems is not always historically justified or generally accepted, the number of such named concepts can provide an indication, albeit a relative one, of a mathematician's lasting accomplishments in extending the science. By this standard, Dedekind belongs among the greatest mathematicians; approximately a dozen designations bear his name.

BIBLIOGRAPHY

I. ORIGINAL WORKS. For a bibliography of Dedekind's writings, see Poggendorff, I (1863), 534, 1555; III (1898), 340; IV (1904), 305; V (1926), 269; and VI (1936), 538. His works include *Gesammelte mathematische Werke,* R. Fricke, E. Noether, and O. Ore, eds., 3 vols. (Brunswick, 1930-1932), with a bibliography in III, 505-507; and *Briefwechsel Cantor-Dedekind,* E. Noether and J. Cavaillès, eds. (Paris, 1937). Extracts from his works appear in Oskar Becker, *Die Grundlagen der Mathematik in Geschichtlicher*

Entwicklung (Freiburg-Munich, 2nd ed., 1964), pp. 224-245, 316.

II. SECONDARY LITERATURE. On Dedekind and his work, see Kurt-R. Biermann, "Richard Dedekind im Urteil der Berliner Akademie," in *Forschungen und Fortschritte,* **40** (1966), 301-302; Edmund Landau, "Richard Dedekind," in *Nachrichten von der Königlichen Gesellschaft der Wissenschaften zu Göttingen, Geschäftliche Mitteilungen* (1917), pp. 50-70, with a bibliography on pp. 66-70; Wilhelm Lorey, *Das Studium der Mathematik an den deutschen Universitäten seit Anfang des 19.Jahrhunderts* (Leipzig-Berlin, 1916), with personal recollections of Dedekind on pp. 81-83; Karl Mollenhauer, "Julie Dedekind," in *Braunschweigisches Magazin,* **21** (1916), 127-130; Richard Müller, "Aus den Ahnentafeln deutscher Mathematiker: Richard Dedekind," in *Familie und Volk,* **4** (1955), 143-145; Nikolaus Stuloff, "Richard Dedekind," in *Neue Deutsche Biographie,* **3** (1957), 552-553; Hans Zincke ("Sommer"), "Erinnerungen an Richard Dedekind," in *Braunschweigisches Magazin,* **22** (1916), 73-81; "Die Akademien zu Paris und Berlin über Richard Dedekind," *ibid.,* 82-84; and "Julius Wilhelm Richard Dedekind," in *Festschrift zur Feier des 25-jährigen Bestehens des Gesellschaft ehemaliger Studierender des Eidgenössischen polytechnischen Schule in Zürich* (1894), pp. 29-30.

KURT-R. BIERMANN

DEE, JOHN (*b.* London, England, 13 July 1527; *d.* Mortlake, Surrey, England, December 1608), *mathematics.*

Dee was the son of Roland Dee, a London mercer, and his wife, Johanna Wilde. He was educated at St. John's College, Cambridge, receiving the B.A. in 1545 and the M.A. in 1548. He was a fellow of St. John's and a foundation fellow of Trinity College (1546). He traveled to Louvain briefly in 1547 and to Louvain and Paris in 1548-1551, studying with Gemma Frisius and Gerhardus Mercator. Throughout his life Dee made extended trips to the Continent and maintained cordial relations with scholars there.

For more than twenty-five years Dee acted as adviser to various English voyages of discovery. His treatises on navigation and navigational instruments were deliberately kept in manuscript; most have not survived, and are known only from his later autobiographical writings. His "fruitfull Praeface" to the Billingsley translation of Euclid (1570), on the relations and applications of mathematics, established his fame among the mathematical practitioners. Although translated by Billingsley, the Euclid is unmistakably edited by Dee, for the body of the work, especially the later books, contains many annotations and additional theorems by him.

Although Dee was a man of undoubted scientific talents, his interests always tended toward the occult.

His favor in court circles was due largely to his practice of judicial astrology. His interest in alchemy and the search for the philosopher's stone led to the gradual abandonment of other work. His last scientific treatise was a reasoned defense of calendar reform (1583). From that time on, he retreated almost wholly into mysticism and psychic research. Dee was certainly duped by his medium, Edward Kelley, but he himself was sincere. He felt that he had been ill rewarded for his many years of serious study and looked for a shortcut to the secrets of the universe through the assistance of angelic spirits.

BIBLIOGRAPHY

I. ORIGINAL WORKS. A more extensive list is in Thompson Cooper's article in *Dictionary of National Biography,* V, 721–729. *Monas hieroglyphica* (Antwerp, 1564) is presented in annotated translation by C. H. Josten in *Ambix,* **12,** nos. 2 and 3 (1964). "A very fruitfull Praeface . . . specifying the chief mathematical sciences," in H. Billingsley, trans., *Euclid* (London, 1570), was reprinted in *Euclid's Elements,* T. Rudd, ed. (London, 1651) and, with additional material by Dee, in *Euclid's Elements,* J. Leeke and G. Serle, eds. (London, 1661). *Parallaticae commentationis praxosque* (London, 1573) contains trigonometric theorems for determining stellar parallax, occasioned by the nova of 1572. "A plain discourse . . . concerning the needful reformation of the Kalendar" (1583) is in the Bodleian Library, Oxford, MS Ashmole, 1789, i. "The Compendious Rehearsal of John Dee" (1592), BM Cotton Vitellius C vii, 1, is available in "Autobiographical Tracts of Dr. John Dee," James Crossley, ed., *Chetham Miscellanies,* I (Manchester, 1851); it is the main source of biographical information but must be read with caution, since it was written as a request for compensation for injury to Dee's library and reputation. *The Private Diary of John Dee, 1577–1601* was edited by J. O. Halliwell as vol. XIX of Camden Society Publications (London, 1842); a corrected version of the Manchester portions, 1595–1601, was edited by John E. Bailey as *Diary, for the Years 1595–1601* (London, 1880).

II. SECONDARY LITERATURE. There is still no adequate biography of Dee. Both Charlotte Fell-Smith, *John Dee* (London, 1909), and Richard Deacon, *John Dee* (London, 1968), stress Dee's nonscientific activities. The latter has revived the theory, originating with Robert Hooke, that Dee's conversations with the angels were intelligence reports in code. Frances A. Yates, *Theatre of the World* (Chicago, 1969), considers Dee as a Renaissance philosopher in the Hermetic tradition. The book contains interesting discussions of Dee's library and of the mathematical preface to Euclid. It is argued that a revival of interest in Vitruvius was spread among the middle classes of London by Dee's Vitruvian references in the preface.

The best assessment of Dee's scientific work may be found in the books of E. G. R. Taylor, especially *Tudor Geography* (London, 1930) and *Mathematical Practitioners of Tudor and Stuart England* (Cambridge, 1954).

Dee had a remarkable library, and many MSS owned by him are extant. M. R. James, "Lists of MSS Formerly Owned by John Dee," a supplement to *Transactions of the Bibliographical Society* (1921), is the basic work, but many others have been located. See, for example, A. G. Watson, *The Library of Sir Simonds D'Ewes* (London, 1966).

JOY B. EASTON

DE FOREST, LEE (*b.* Council Bluffs, Iowa, 26 August 1873; *d.* Hollywood, California, 30 June 1961), *electronics.*

Lee de Forest grew up in Alabama, where his father, Henry Swift De Forest, a Congregational minister of Huguenot stock, was president (1879–1896) of the Negro Talladega College. His mother was Anna Margaret Robbins, the daughter of a Congregational minister.

After preparation at the Mt. Hermon School in Massachusetts, de Forest entered the three-year mechanical engineering course of the Sheffield Scientific School at Yale University. He graduated in 1896 and returned for graduate work under such luminaries as Josiah Willard Gibbs and Henry Bumstead. De Forest enlisted in the army during the Spanish-American War but saw no action, and he received the Ph.D. on schedule in 1899 with a thesis entitled "Reflection of Hertzian Waves From the Ends of Parallel Wires."

From the first de Forest was interested in radiotelegraphy, and his improvements on the early systems enabled him to obtain financing to compete with Marconi and to interest the U.S. Army and Navy in his equipment. Apparatus constructed by de Forest and his associates was used in an attempt to report the America's Cup yacht races of 1903; and in the early part of the Russo-Japanese War of 1904–1905, his equipment was used by European reporters in sending their news dispatches, until the Japanese put an end to the arrangement.

A search for an improved detector (rectifier) of radio signals led de Forest to an invention that was a substantial improvement over the vacuum diode invented in 1904 by J. A. Fleming: the insertion of a third, control electrode between cathode and anode. De Forest applied for a patent on this "device for amplifying feeble electric currents" in 1906, but this first triode initially served only as a superior detector. A careful reading of the contemporary literature on the subject, exegesis of the subsequent claims made by de Forest (partly for purposes of patent litigation), and additional information unearthed by historical research make it clear that he neither fully understood the operation of the triode nor appreciated its possibilities as an amplifier and high-frequency oscillator until years later. In fact, the triode saw very limited use until after 1912, when Fritz Löwenstein invented the negatively "biased grid" circuit for it and H. D. Arnold and Irving Langmuir set about providing the highest attainable vacuum. But that is not to say that this prototype of all electronic amplifiers was not an invention of the greatest significance, matched in importance only by the invention of the transistor some forty years later. The triode made transcontinental telephony possible for the first time and underlies all technological applications in which weak electrical signals are amplified—including the transistor, which is a type of triode.

Following active participation in the development of the infant radio industry in New York, de Forest went to work for the Federal Telegraph Company in Palo Alto, California. He made California his permanent residence and continued to make inventions at an astonishing rate; he was granted more than 300 patents, many quite speculative, in his lifetime. Besides the triode, de Forest's most important contributions were made in the development of sound motion pictures. He was one of the founder members of the Institute of Radio Engineers in 1912, and in 1915 he received its Medal of Honor. He also received the Elliott Cresson Medal of the Franklin Institute and the Cross of the Legion of Honor from the French government.

De Forest was regarded by many as one of the last of the great individualistic inventors; his companies spent fortunes in fighting over patent rights. Several litigations were carried all the way to the U.S. Supreme Court. He had a deep appreciation of the cultural opportunities of broadcasting and deplored its commercialization. "What have you gentlemen done with my child?" he once asked radio executives. "The radio was conceived as a potent instrumentality for culture, fine music, the uplifting of America's mass intelligence. You have debased this child, you have sent him out in the streets in rags of ragtime, tatters of jive and boogie-woogie, to collect money from all and sundry."

Despite his distaste for the uses of radio, de Forest believed strongly in the future of electronics and participated in its development almost to his death; his last work was on the improvement of magnetic tapes and in thermoelectricity. He received his last patent (on an automatic telephone dialing device) at the age of eighty-four.

De Forest was married four times; his last wife was Marie Mosquini, a motion picture actress, whom he married in 1930 and who survived him. When the Foothill Electronics Museum in Los Altos Hills, California, was established in 1969, she donated his papers and many artifacts to it.

BIBLIOGRAPHY

De Forest's autobiography, somewhat immodestly titled *Father of Radio* (Chicago, 1950), contains his 1920 paper on the history of the triode and a list of most of his U.S. patents. Obituaries are in *New York Times* (2 July 1961), pp. 1 ff.; and in *Proceedings of the Institute of Radio Engineers,* **49** (Oct. 1961), 22A. For a well-documented account of the development of the triode, see B. F. Miessner, *The Early History of Radio Guidance* (San Francisco, 1964).

CHARLES SÜSSKIND

DeGOLYER, EVERETTE LEE (*b.* Greensburg, Kansas, 9 October 1886; *d.* Dallas, Texas, 14 December 1956), *geophysics.*

DeGolyer was the eldest child of John William and Narcissa Kagy Huddle DeGolyer, who had homesteaded in Kansas shortly before the boy's birth. Always interested in mining prospects, the senior DeGolyer moved his family to the lead and zinc center of Joplin, Missouri, and then to Oklahoma during the 1901 land opening. The boy finished high school at the University of Oklahoma preparatory school before entering the University of Oklahoma in the mining engineering course, where he was directed by Charles N. Gould and E. G. Woodruff. During the summer he worked for the U.S. Geological Survey, first as a cook but subsequently as a geologist. In the field in 1907 he impressed C. Willard Hayes, chief geologist of the Survey, who hired DeGolyer two years later to head the exploration staff of the Mexican oil company El Águila. DeGolyer's acquaintance with Sir Weetman Pearson (later Lord Cowdray) from his work in Mexico in 1910 later led to financial backing for DeGolyer's early oil companies, Amerada Petroleum Corporation and Rycade Oil Company.

Already holding a high reputation in oil discovery, DeGolyer resumed his interrupted college career at the University of Oklahoma, receiving the B.A. in 1911, and then returned to Mexico. From 1914 he was an independent oil consultant, and he founded and headed an interlocking series of oil exploration and research companies. His ability and unquestioned integrity led to his frequent employment by his own and other governments as adviser on development of oil fields.

DeGolyer received seven honorary doctorates and the Anthony F. Lucas (1941) and John Fritz (1942) medals of the American Institute of Mining and Metallurgical Engineers. He was a member of the

National Academy of Sciences, Sigma Xi, and Phi Beta Kappa and a charter member of the Society of Economic Geologists and of the American Association of Petroleum Geologists, which awarded him the Sidney Powers Memorial Award in 1950.

DeGolyer made his greatest contributions in the application of physical principles to solution of geological problems. He advocated and developed the theory, earlier suggested by European geologists, that salt domes form by the plastic flow of salt upward from deeply buried beds under the pressure of overlying rocks (1918). He then set out to locate salt domes by geophysical methods. Without question, DeGolyer's development and use of these tools gave the search for petroleum its greatest impetus of the twentieth century. He began with the torsion balance, and in 1924 he found the first salt dome by this method. A subsidiary of Amerada, Geophysical Research Corporation, was set up by DeGolyer to perfect and apply the refraction and reflection seismographs to finding oil fields. In 1927–1928 the company found eleven new salt domes by refraction surveys, but it was the discovery of the Edwards oil field in Oklahoma in 1930 by reflection survey that ushered in the modern era of oil exploration.

A knowledgeable collector of books, DeGolyer assembled one of the world's best libraries for the history of science, especially geology. He presented this library to the University of Oklahoma.

DeGolyer took his own life, after having been ill with aplastic anemia for seven years.

BIBLIOGRAPHY

I. ORIGINAL WORKS. Many of DeGolyer's early papers were on the Mexican petroleum industry. As his field enlarged, later papers summarized world petroleum production and methods. His analysis of salt-dome structure was presented in "The Theory of Volcanic Origin of Salt Domes," in *Bulletin of the American Institute of Mining and Metallurgical Engineers*, **137** (1918), 987–1000. The first two sources cited below list DeGolyer's approximately seventy geological writings, and Denison's memorial contains a complete bibliography that indicates DeGolyer's broad range of interests.

II. SECONDARY LITERATURE. Wallace E. Pratt's "Memorial to Everette Lee DeGolyer," in *Proceedings of the Geological Society of America* for 1957 (1958), 95–103, effectively summarizes the life and career of this energetic man. Further comments and indications of DeGolyer's wide interests are in Carl C. Branson's memorial in *Oklahoma Geology Notes*, **17**, no. 2 (1957), 11–21. A. Rodger Denison, in *Biographical Memoirs. National Academy of Sciences*, **33** (1959), 65–86, describes DeGolyer's personality.

ELIZABETH NOBLE SHOR

DE GROOT, JAN CORNETS, also known as **Johan Hugo De Groot** or **Janus Grotius** (*b.* near Delft, Netherlands, 8 March 1554; *d.* Delft, 3 May 1640), *mechanics.*

The son of Hugo Cornelisz and of Elselinge Van Heemskerck, De Groot (or perhaps a namesake) entered the University of Leiden on 5 February 1575, its opening day. He was a master of liberal arts and philosophy at Douai. Belonging to the Delft patriciate, he was a councillor, and from 1591 to 1595 was one of the mayors. From 1594 to 1617 he was a curator of the University of Leiden, which in 1596 awarded him the doctorate of law. After 1617 he served as adviser to the Count of Hohenlohe.

In 1582 De Groot married Alida Borren, from Overschie. One of their five children was the jurist known as Hugo Grotius.

De Groot was a distinguished amateur scientist, acquainted with the best minds in the Netherlands. Stevin, with whom he collaborated in the construction of windmills, praised him as a man well versed in the whole of philosophy, mentioning Euclid, Alhazen (Ibn al-Haytham), Witelo, music, and poetry. De Groot is best known through the experiment he performed with Stevin, in which they proved that lead bodies of different weights falling on a board traverse the same distance in the same time. This anti-Aristotelian experiment, published by Stevin in his *Waterwicht* (1586, p. 66), anticipated Galileo's famous, but apocryphal, experiment at the Leaning Tower of Pisa. De Groot also befriended Ludolph Van Ceulen, on whose behalf he translated Archimedes' *Measurement of the Circle* from the Greek into Dutch and who submitted to him his approximation of π to 20 decimal places (1586). Van Roomen, in his *Ideae mathematicae* (1593), praises De Groot as one of the better mathematicians of his time.

BIBLIOGRAPHY

Only some Latin and Greek poems, correspondence with his son Hugo—*Hugonis epistolae* (Amsterdam, 1607)—and some MS letters in the library of the University of Leiden are extant.

Secondary literature is C. de Waard, "Groot, J. H. de," in *Nieuw Nederlandsch biographisch Woordenboek,* II (1912), cols. 528, 529, with bibliography; and *The Principal Works of Simon Stevin* (Amsterdam, 1955–1966), esp. vols. I, V.

DIRK J. STRUIK

DE HAAS, WANDER J. See **Haas, Wander J. de.**

DEHN, MAX (*b*. Hamburg, Germany, 13 November 1878; *d*. Black Mountain, North Carolina, 27 June 1952), *mathematics.*

Dehn studied at Göttingen under the direction of David Hilbert and received his doctorate in 1900. He then taught at several schools, served in the army, and was a professor of pure and applied mathematics at Frankfurt University from 1921 to 1935, when the Nazi regime forced him to leave. In 1940 he emigrated to the United States, where he taught at the University of Idaho, Illinois Institute of Technology, St. Johns College (Annapolis), and, from 1945 to 1952, at Black Mountain College, North Carolina. He was a member of the Norwegian Academy of Science, the Strassburger Naturforschung Gesellschaft, and the Indian Mathematical Association.

Dehn was an intuitive geometer. Stimulated by Hilbert's work on the axiomatization of geometry, Dehn showed in his dissertation that without assuming the Archimedean postulate, Legendre's theorem that the sum of the angles of a triangle is not greater than two right angles is unprovable, whereas a generalization of Legendre's theorem on the identity of the sums of angles in different triangles is provable. Following this work, Dehn solved the third of the twenty-three unsolved problems that Hilbert had presented in his famous address to the International Congress of Mathematicians in 1900. This problem concerned the congruence of polyhedra, the geometric properties of which Dehn spent much time studying.

In 1907 Dehn and P. Heegaard contributed a report to the *Encyklopädie der mathematischen Wissenschaften* on the topic of analysis situs (now called topology or algebraic topology), which had become prominent as a result of the works of Poincaré. The article was one of the early systematic expositions of this subject. In 1910 Dehn proved an important theorem concerning topological manifolds that became known as Dehn's lemma. In 1928, however, Kneser showed that the proof contained a serious gap; a correct proof was finally given by C. D. Papakyriakopoulos in 1957.

Following Poincaré, Dehn became interested in the groups that are generated in attempts to characterize topological structures. He formulated the central problems of what was to become an important mathematical field: the word, the transformation, and the isomorphism problems. In the case of fundamental groups these have direct topological significance.

Besides his numerous contributions to the field of fundamental groups, Dehn wrote papers on statics, on the algebraic structures derived from differently axiomatized projective planes, and on the history of mathematics. He supervised the work of eight doctoral candidates in Germany and three in the United States.

BIBLIOGRAPHY

A bibliography of Dehn's works may be found in Wilhelm Magnus and Ruth Moufang, "Max Dehn zum Gedächtnis," in *Mathematische Annalen,* **127** (1954), 215–227. See also C. D. Papakyriakopoulos, "Some Problems on 3 Dimensional Manifolds," in *Bulletin of the American Mathematical Society,* **64** (1958), 317–335.

C. S. FISHER

DEINOSTRATUS. See **Dinostratus.**

DE LA BECHE, HENRY THOMAS (*b*. London [?], England, 10 February 1796; *d*. London, 13 April 1855), *geology.*

De la Beche was the son of an army officer, Lt. Col. Thomas Beach, owner of an estate in Jamaica, who had changed his name to de la Beche. He attended school at Ottery St. Mary, Devonshire, for a short time and then moved with his mother to Dorsetshire, residing first at Charmouth and later at Lyme Regis. In these seaside towns Jurassic rocks were well exposed in sea cliffs, and it is most probable that it was while living at Lyme Regis that de la Beche acquired the interest in geology that led him to make it his career. In 1810 he was sent to the Royal Military College at Marlow. The end of the Napoleonic Wars a few years later, however, made the future for an army officer less promising; and he abandoned the idea of a military career.

De la Beche was then a young man of independent means. In 1817 he was elected a member of the Geological Society of London; and his earliest extant diary, for the year 1818, recording geological observations made between Weymouth and Torquay, confirms that he was already keenly interested in field geology. In 1819 he was elected a fellow of the Royal Society. In the same year de la Beche set out on a long tour through France, Switzerland, and Italy, returning through Germany and the Netherlands. During this lengthy residence abroad he made many geological observations that he later published; he also acquired a command of foreign languages that enabled him, after his return, to keep abreast of advances in geology made outside Great Britain. His first scientific paper—on the depth and temperature of Lake Geneva—was, in fact, published in Geneva (1819).

De la Beche spent the year 1824 in Jamaica, visiting the estate he had inherited. During his stay he studied the geology of the island in some detail, and after his return he gave an account of it to the Geological Society (in *Transactions of the Geological Society of London,* 2nd ser., **2** [1827], 143). The first systematic account of the geology of Jamaica published, it included a colored geological map of the eastern half of the island. In 1829 de la Beche made another extensive Continental tour, the results of which are also apparent in later publications. In 1830 he published *Sections and Views Illustrative of Geological Phenomena,* followed in 1831 by his *Geological Manual,* the popularity of which is indicated by its three editions, as well as translations into French and German. In addition to these books he had already contributed a number of papers on British and foreign geology to the *Transactions of the Geological Society of London* and to other periodicals.

De la Beche's pioneer fieldwork was not confined to rocks of any particular kind or age. In conjunction with his extensive reading of the literature, he had acquired a wide general knowledge of geology, and his publications added considerably to the general stock of geological facts. The value of his published work was greatly enhanced by the clear and simple illustrations, based on sketches made in the field—which, incidentally, demonstrate his skill as an artist. In recording facts he was following the original aim of the Geological Society of London, announced on its foundation in 1807: the collection of geological information rather than the promulgation of geological theories. De la Beche was an accurate observer of those details of rock structure that have a bearing on the origin and mode of formation of particular rocks. Although he has not generally been credited with any outstanding contributions to geological theory, it is evident from his writings, particularly *Researches in Theoretical Geology* (1834), that his interests were not confined simply to recording facts. On the other hand, he made it quite clear that theories must always be regarded as tentative until supported by a sufficient body of factual evidence. In this book he applied his knowledge of mineralogy, chemistry, and physics to a discussion of the broader aspects of theoretical geology; but the views advanced are suggestive rather than dogmatic. For example, he suggested that the original solid crust of the earth, formed by cooling, had floated on a fluid interior of molten rock and that the once continuous crust had been broken up by tidal action to form the earliest separate land masses. This was an idea well ahead of its time.

Until about 1832 de la Beche's career had followed a course parallel to that of a number of his contemporaries, such as Charles Lyell and Roderick Murchison, men of independent means who pursued the study of geology for its own sake and were able to travel extensively in furtherance of that study. Changed circumstances, however, resulted in his spending the rest of his life as a professional geologist. This came about partly because the income from his Jamaica estate had diminished, restricting his freedom to travel, and more directly because he was to become closely involved in the formation and development of an official geological survey of Great Britain. He was the first to suggest that such a survey should be undertaken and was one of the prime movers in establishing it on a permanent basis.

Sometime after 1830 de la Beche had conceived the idea of making a geological survey of Devonshire, possibly because new Ordnance Survey topographic maps, on a scale of an inch to the mile, had recently become available for that area (the lack of accurate topographic maps on which to record geological information had previously constituted a serious difficulty for geologists). In 1832 he submitted a memorandum to the master general of the Ordnance, who was then responsible for the primary topographic survey of Great Britain, offering to make an accurate geological survey of the eight sheets covering Devonshire for the sum of £300, mentioning that originally it had been his intention to carry it out at his own expense but that he was no longer able to do so. Senior government officials had already become aware of the potential economic value to the nation of geological information, and his request was granted with little delay. By 1835 he had completed the task.

De la Beche then proposed that a similar survey should be extended to other parts of the country. This request required more serious consideration, and the government sought the advice of an independent committee of geologists. Their report was favorable; and in the same year the official Geological Survey of Great Britain was established, with de la Beche as director at an annual salary of £500. Work commenced in Cornwall, then an important center of metalliferous mining; and when this area was completed, the survey was extended to the coalfields of South Wales. It was later extended to areas other than those in which mining was active. At its commencement the survey was virtually a one-man affair, but during the next fifteen years de la Beche gradually enlarged the field staff and secured the appointment of such specialists as a paleontologist and a chemist. He also assembled large reference collections of fossils and minerals.

The culmination of de la Beche's official career was reached in 1851, when the Museum of Practical Geology, specially built in Jermyn Street, London, with offices for the survey staff, was opened to the public by Prince Albert. Thereafter he had under his direct charge not only the Geological Survey but also the museum, the School of Mines, and the Mining Records Office. Not long after the opening of these establishments the museum was described by Roderick Murchison as "the first Palace ever raised from the ground in Great Britain, which is entirely devoted to the advancement of Science." The expansion of the Geological Survey and the establishment of its important subsidiary activities on a permanent basis were very largely, if not entirely, due to the imaginative drive and administrative ability of de la Beche. What was perhaps of equal importance, in the long run, was that he succeeded in convincing governmental circles of the desirability of considerable state support for scientific research and the teaching of science.

De la Beche's new responsibilities did not prevent him from carrying out geological work himself. In 1839 he published in London his *Report on the Geology of Cornwall, Devon and West Somerset*, a work that remained valuable for many years because of the mass of mining information it contained that otherwise would probably have been lost. It was the first of the long series of official publications issued by the survey, a series that is still continuing. De la Beche contributed to later memoirs, and he continued to publish papers in scientific periodicals. In 1851 he published (also in London) *The Geological Observer*, an expanded version of an earlier work, *How to Observe* (1835).

While de la Beche, over a period of nearly forty years, contributed much to the general stock of geological knowledge through his publications, his wholehearted and determined efforts to advance the then comparatively new science of geology by every means in his power were no less important. Recognition of the value of his contributions to the science came from the government, in the form of a knighthood conferred on him in 1842, and from his fellow geologists, by election to the presidency of the Geological Society of London in 1847 and the award of this society's highest honor, the Wollaston Medal, in 1855. From about 1850 de la Beche suffered from a form of paralysis, but he continued to attend to his duties until two days before his death.

BIBLIOGRAPHY

I. ORIGINAL WORKS. In addition to the works cited above, de la Beche published *A Selection of the Geological Memoirs Contained in the Annales des Mines* (London, 1824; 2nd ed., 1836), an annotated translation, with a correlation of British with French and German strata.

He also contributed a number of papers, mainly geological, to periodicals. The more important were published in either the *Transactions of the Geological Society of London* or their *Proceedings*. They are listed in the Royal Society's *Catalogue of Scientific Papers 1800–1863*, Vol. II (London, 1868).

De la Beche's official publications (in some instances written in collaboration with other authors) include "Report with Reference to the Selection of Stone for Building the New Houses of Parliament," in *Parliamentary Papers for 1839*, vol. XXX (London, 1839); "On the Formation of the Rocks of South Wales and South-western England," in *Memoirs of the Geological Survey of Great Britain*, **1** (1846), 1–296; and "First and Second Reports on the Coals Suited to the Steam Navy," in *Parliamentary Papers* (London, 1848, 1849) and in *Memoirs of the Geological Survey of Great Britain*, **2**, pt. 2 (1948), 539–630. These reports contain a detailed examination of the physical and chemical characteristics of various British coal seams.

De la Beche's "Inaugural Discourse Delivered at the Opening of the School of Mines and of Science Applied to the Arts," in *Records of the School of Mines*, **1**, pt. 1 (1852), 1–22, is of historical interest as an authoritative contemporary account of the formation and objectives of the Geological Survey and its associated institutions.

II. SECONDARY LITERATURE. No biography of de la Beche has been published, although a number of his diaries and letters are now held by the National Museum of Wales, Cardiff. Accounts of his career are included in J. S. Flett, *The First Hundred Years of the Geological Survey of Great Britain* (London, 1937), pp. 23–56; and E. B. Bailey, *Geological Survey of Great Britain* (London, 1952), pp. 21–51. The former deals especially with his official career, and the latter also assesses his capabilities as a geologist. Some additional matter, based partly on unpublished material, is contained in three short articles by F. J. North: "De la Beche and His Activities, as Revealed by His Diaries and Correspondence," in *Abstracts of the Proceedings of the Geological Society of London*, no. 1314 (June 1936), 104–106; "H. T. de la Beche: Geologist and Business Man," in *Nature*, **143** (1939), 254–255; and "The Ordnance Geological Survey: Its First Memoir," *ibid.*, 1052–1053. L. J. Chubb has described in some detail de la Beche's ancestry, his connection with Jamaica, and the geological work he carried out there in the De la Beche Memorial Number of *Geonotes, the Quarterly Newsletter of the Jamaica Group of the Geologists Association* [of London] (Kingston, Jamaica, 1958).

V. A. EYLES

DELAGE, YVES (*b.* Avignon, France, 13 May 1854; *d.* Sceaux, France, 7 October 1920), *zoology, anatomy, physiology, embryogeny, general biology.*

After passing his *baccalauréat ès-sciences* and *baccalauréat ès-lettres*, Delage went to Paris in 1873 to study medicine. Obliged to interrupt his studies, he accepted a post as *répétiteur* at the *lycée* of La Rochelle. In 1875 he was able to resume his medical studies. At the same time he prepared for the *licence* in natural sciences, which he obtained in 1878. In the same year his teacher, Lacaze Duthiers, chose him to direct the zoological station at Roscoff. He defended his doctoral theses in medicine and in science in 1880 and 1881, respectively. He was *chargé des fonctions de maître de conférences* in zoology at the Sorbonne (1880); *chargé de cours* at Caen (1881); titular professor of zoology at Caen and director of the zoological station at Luc (1884); *chargé de cours* at the Sorbonne (1885); titular professor of zoology at the Sorbonne (1886), a post he occupied until his death; and director of the marine laboratory at Roscoff (1902).

Delage's experimental work was drastically curtailed when he suffered a detached retina, and in 1912 he had to give up teaching. Although he became blind, he continued his work.

A disciple of and assistant to Lacaze Duthiers, Delage at first followed the latter's zoological traditions. His works on crustacean circulation display remarkable skill; and the superb accompanying illustrations are considered classics. The two chief qualities of Delage's work were already manifest: experimental ingenuity and technical ability, both guided by a tenacious will. His penetrating critical mind enabled him to establish new truths that were quite different from what was then generally accepted. His discoveries regarding the *Sacculina* and the embryogeny of the sponges are striking evidence of this ability. The inoculation of the kentrogon larva of the *Sacculina* and the inversion of the laminae in the metamorphosis of sponges are major discoveries that have withstood the test of time. Also worthy of mention are his researches on the nervous systems of the *Peltogaster* and of the *Convoluta,* and on the transformation of a leptocephalus into an eel, which provided the key to the metamorphoses of the Anguillidae; an anatomical study of the *Balaenoptera;* and a monograph on the ascidians.

In experimental physiology Delage investigated the function of the otocysts in various animals and of the semicircular canals in man. He established a new conception in the physiology of the inner ear: the semicircular canals stabilize equilibrium.

A crisis of conscience led Delage to abandon pure zoology for general biology and biomechanics. He now investigated the causes of the manifestations of life in the cell, in the individual, and in the species, considering this area more productive of important results. The book *La structure du protoplasma . . .* (1895) marks the turning point of his scientific career.

The qualities evident in Delage's earlier works are also present in those on merogony, fertilization, and artificial parthenogenesis. He would not consider a question without completely settling it. After Jacques Loeb's discovery of artificial fertilization Delage, through theoretical insights, devised a process of chemical fertilization so perfect that for two years he succeeded in raising to the adult stage sea urchins obtained by this process.

Because of the great importance that he accorded to the action of the environment and to the role of acquired characteristics, Delage was a Lamarckian and, therefore, unwilling to accept the ideas of Weismann and Mendel.

Delage founded the *Année biologique* to publish articles on general biology and was its director for fifteen years. He also planned a great zoological treatise based on his belief that each division of the animal kingdom could be reduced to an ideal type that embodied all the fundamental characteristics of that division.

Among his other activities, Delage constructed and experimented with a bathyrheometer designed to measure the speed of ocean currents. After his blindness had condemned him to meditation, Delage concerned himself with the causes and various states of dreaming.

Delage was not merely an incomparable professor who gave illuminating lectures; he was also a philosopher, a novelist, a short-story writer, a poet, and a polemicist. He belonged to many scientific academies and societies and was elected to the Académie des Sciences in 1901. He worked unceasingly at Roscoff to enlarge the biological station, and there is a monument to him there.

BIBLIOGRAPHY

I. Original Works. Among Delage's papers are "Sur l'origine des éléments figurés du sang chez les Vertébrés" (1880), his M.D. thesis; "Contributions à l'étude de l'appareil circulatoire des Crustacés Edriophthalmes marins," in *Archives de zoologie expérimentale et générale,* 9 (1881), 1–176, his thesis for the *doctorat ès-sciences;* "Circulation et respiration chez les Crustacés Schizopodes," *ibid.,* 2nd ser., 1 (1883), 105–131; "Évolution de la Sacculine," *ibid.,* 2 (1884), 417–737; "Sur le système nerveux et sur quelques autres points de l'organisation du Peltogaster," *ibid.,* 4 (1886), 17–37; "Études expérimentales sur les illusions statiques et dynamiques de direction pour servir à déterminer les fonctions des canaux semi-circulaires de l'oreille interne," *ibid.,* 535–695; "Sur une fonction nouvelle des otocystes comme organes d'orientation locomotrice," *ibid.,* 5 (1887), 1–26; "Études sur les Ascidies des côtes de France; les Cyntiadées," in *Mémoires de l'Académie des sciences,* 2nd ser., 45 (1892–1899), 323 pp., written with Lacaze Duthiers; "Embryogénie des Éponges. Développement post-larvaire des Éponges siliceuses et fibreuses," in

Archives de zoologie expérimentale et générale, 2nd ser., **10** (1893), 345–499; "Étude sur la mérogonie," *ibid.,* 3rd ser., **7** (1899), 383–418; and "La parthénogenèse expérimentale," in *Rapport du Congrès zoologique. Graz 1910* (1912), pp. 100–162.

Delage's books include *La structure du protoplasma, les théories sur l'hérédité et les grands problèmes de la biologie générale* (Paris, 1895); *Traité de zoologie concrète,* written with E. Hérouard, composed of the following volumes: *La cellule et les Protozoaires* (Paris, 1896), *Les Vermidiens* (Paris, 1897), *Les Procordés* (Paris, 1898), *Mesozoaires et Spongiaires* (Paris, 1899), *Coelentérés* (Paris, 1901), and *Échinodermes* (Paris, 1903); *Les théories de l'évolution,* in the Bibliothèque de Philosophie Scientifique (Paris, 1909), written with M. Goldsmith; *La parthénogenèse naturelle et expérimentale,* in Bibliothèque de Philosophie Scientifique (Paris, 1913), written with M. Goldsmith; and *Le rêve. Étude psychologique, philosophique et littéraire* (Paris, 1920).

II. SECONDARY LITERATURE. A notice on the life and work of Delage is M. Goldsmith, in *Année biologique,* n.s. **1** (1920–1921), v–xix. L. Joubin's speech at the unveiling of the monument to Delage at Roscoff, a biographical notice, and a bibliography constitute *Académie des sciences* (1924), no. 17. The speeches by both Charles Pérez and L. Joubin at the unveiling of the monument are in *Travaux de la Station biologique de Roscoff,* **5** (1926), 1–30.

ANDRÉE TÉTRY

DELAMAIN, RICHARD (*fl.* London, England, first half of the seventeenth century), *mathematics.*

Delamain was a joiner by trade, and after studying mathematics at Gresham College, London, he supported himself by teaching practical mathematics in London. Later he became mathematical tutor to King Charles I, for whom he fashioned a number of mathematical instruments. He was a pupil of William Oughtred, and in the early days of their association the two men became close friends. Unhappily this did not last, and later they quarreled violently over priority in the invention of the circular slide rule, which Delamain described in his *Grammelogia, or the Mathematicall Ring.*

Delamain's fame rests mainly on this essay, a pamphlet of thirty-two pages. The manuscript was sent to the king in 1629, and the work was published the following year. The king retained Delamain's services as tutor at a salary of £40 per annum. A few years later Delamain petitioned for an engineer's post at a salary of £100 per annum. Following an interview with the king at Greenwich in 1637, he was granted a warrant for making a number of mathematical instruments.

The appearance of the *Grammelogia* was the signal for the beginning of the quarrel. Oughtred had in-

vented the rectilinear slide rule as early as 1622, but his *Circles of Proportion,* which contained a description of the circular slide rule, was not made public until 1632—by which time the *Grammelogia* had been in circulation for two years. William Forster, a friend and pupil of Oughtred, translated from the Latin and published the *Circles of Proportion,* the preface to which contains some ungenerous references to Delamain, who, it states, purloined the design of the circular slide rule from Oughtred. Delamain retaliated vigorously, attacking both Forster and Oughtred; the latter replied with a pamphlet, *The Apologeticall Epistle,* in which he refers to the "slaunderous insimulations of Richard Delamain in a Pamphlet called *Grammelogia, or the Mathematicall Ring*" and maintains that the latter's horizontal quadrant is no other than the horizontal instrument he had invented thirty years earlier.

Delamain perished in the Civil War sometime before 1645. Oughtred lived until 1660, but his last years were embittered by the dispute with his former friend and pupil.

Delamain was a competent mathematician whose genius lay in the practical realm. He excelled in the construction of a number of mathematical instruments. It is thought that the silver sundial which the king always carried with him and, at his execution, entrusted to Mr. Herbert to be given to the young duke of York, was one of Delamain's creations.

> He likewise commanded Mr. *Herbert* to give his son, the Duke of *York,* his large Ring Sundial of silver, a Jewel his Majesty much valued: it was invented and made by *Rich. Delamaine* a very able Mathematician, who projected it, and in a little printed book did shew its excellent use in resolving many questions in Arithmetick and other rare operations to be wrought by it in the Mathematicks [Wood, *Athenae Oxonienses. History of Oxford Writers,* II, 1692, 525].

BIBLIOGRAPHY

I. ORIGINAL WORKS. Delamain's writings are *Grammelogia, or the Mathematicall Ring* (London, 1630); and *The Making, Description and Use of a Small Portable Instrument for the Pocket . . . Called a Horizontal Quadrant* (London, 1632).

II. SECONDARY LITERATURE. On Delamain or his work, see Florian Cajori, *William Oughtred. A Great Seventeenth Century Teacher of Mathematics* (London–Chicago, 1916), *passim; Dictionary of National Biography;* and E. G. R. Taylor, *The Mathematical Practitioners of Tudor and Stuart England* (Cambridge, 1954), p. 201.

J. F. SCOTT

DELAMBRE, JEAN-BAPTISTE JOSEPH (*b.* Amiens, France, 19 September 1749; *d.* Paris, France, 19 August 1822), *astronomy, geodesy, history of astronomy.*

Delambre's early life resembles those novels of the nineteenth century in which industry overcomes hardship and is rewarded with social distinction and financial gain. He began his education in the local schools of Amiens and eventually won a small scholarship that enabled him to move on to Paris and the Collège du Plessis, where he studied literature (chiefly Latin and Greek) and history. He was especially skilled in languages and began to make translations of works in Latin, Greek, Italian, and English. He was apparently so poor that upon graduation he lived for almost a whole year on a diet of bread and water.[1] He then was engaged as a private tutor in Compiègne and undertook private studies in mathematics, presumably so as to be able to teach this subject along with languages, literature, rhetoric, and history—in which he had received schooling. A local doctor seems to have suggested to Delambre that he might eventually learn astronomy. The opportunity to do so did not occur, however, until 1780, when Delambre had been established in Paris for some nine years. He was thus in his early thirties when he first began to study astronomy, the subject in which he was to establish his reputation.

A most fortunate event that helped Delambre in his career occurred in 1771, when he became tutor to the son of Geoffroy d'Assy (*receveur générale des finances*) in Paris. Eventually d'Assy built a small, private, and apparently well-equipped observatory for Delambre's use—acting on the suggestion made by the astronomer Lalande.[2] Delambre had begun to attend Lalande's lectures at the Collège de France in 1780, and he at once attracted the attention of Lalande when the latter made reference to the Greek poet Aratus, of the third century B.C., author of the astronomical poem *Phaenomena.* Delambre, who was endowed with a prodigious memory, thereupon recited the whole passage in question and went on to discuss various explanations and commentaries that had been made by different scholars. Lalande soon learned that Delambre had written a series of annotations and emendations to his own writings, notably his *Astronomie,* the course textbook. After reading Delambre's notes, Lalande made him an assistant, and eventually Delambre became Lalande's scientific collaborator. Lalande fondly referred to Delambre as his "meilleur oeuvre."[3]

The beginning of Delambre's career as an observer is dated (by himself) on a day in 1786 when only he and Messier, of all the astronomers in Paris, had seen the transit of Mercury across the sun. The event occurred three-quarters of an hour later than the time predicted by Lalande; the other observers, too easily discouraged, had given up. Delambre had more faith in Halley's tables, which predicted the occurrence of the transit an hour and a half later, and so he persevered. This particular episode is cited by Jacquinet[4] as an example of the lack of precision in astronomical tables of that time. Delambre's experience with the transit must have provided a strong incentive for making more accurate tables of various major astronomical phenomena.

Before long there was a public challenge to all astronomers to solve the problems of precise planetary motion. The Académie des Sciences announced a general competition for the prize of 1790 on the subject of the motion of the planet Uranus, which had been discovered by William Herschel in 1781. Some idea of the difficulty of the problem may be gained from the simple fact that these eight years of observation of Uranus represent only about one-tenth of its sidereal period. To determine the orbit and motion of Uranus, Delambre had to consider the perturbations produced by Jupiter and by Saturn: in short, he had to combine a skill in computation with a theoretical understanding of applied celestial mechanics. After winning the prize, he went on to establish himself as a foremost expert in positional astronomy. Eventually there followed tables of the sun, of Jupiter, of Saturn, and of the satellites of Jupiter. The high esteem in which his results were held is shown in the statement by Arago: "In perfecting the methods of astronomical calculation he merits, by reason of the variety and elegance of his methods, a distinguished place among the ablest *géomètres* France can boast."

The above-mentioned tables were published by Lalande in a later edition of his *Astronomie* and earned two further honors for Delambre. In 1792 he was again given the annual prize of the Academy, and he was elected *membre associé* in the section of *sciences mathématiques.* This election was a major factor in his being designated to make the fundamental geodetic measurements on which the metric system was to be based.

In 1788 the Academy decided to establish a "uniform system of measures" founded on some "natural and invariable base." The plan for the new system of measures was formally approved by a decree of the Assembly of 8 May 1790, proposed by Talleyrand; it was approved by Louis XVI on the following 22 August. A commission on the metric system, consisting of Borda, Lagrange, Laplace, Monge, and Condorcet, was thereupon appointed by the Academy. In a report submitted on 19 March 1791,[5] the commissioners rejected two proposed bases for the fundamental unit of measure: the length of a seconds pendulum (at 45° latitude), and one-quarter of the terrestrial equator. Instead they chose one-quarter of a terrestrial meridian, the common practical unit to be a ten-millionth part of this quantity. Accordingly, it was proposed to make a careful and accurate measure along an arc of the meridian through Dunkerque (which had in part been measured by the Cassinis in 1718 and in 1740), extending as far south as Barcelona, giving 9.°5 of arc.

Three fundamental tasks were envisaged. First, to determine the exact difference in longitude between Dunkerque and Barcelona (and to make any needed latitude determinations in between); second, to check by new observations and calculations the triangulations used earlier to find the distance between Dunkerque and Perpignan; third, to make new measurements that could serve for successive triangulations. Clearly a major part of this assignment would be to compute carefully the difference in actual lengths (in *toises*) corresponding to the same difference in latitude at various points along the meridian, so as to be able to determine the actual shape of the earth. While these operations were being performed, other scientists would be engaged in establishing a standard of mass. The instruments, chiefly made by Lenoir according to the plans of Borda, were ready by June 1792, and the work was started shortly afterward.

Originally, the geodetic survey was to be entrusted to Méchain, Cassini, and Legendre. The latter two begged off, and Delambre—just made a member of the Academy—was appointed. It was decided that Delambre would be in charge of the survey from Dunkerque to Rodez, leaving the survey from Rodez to Barcelona in the hands of Méchain. An account of the labors and adventures of Méchain and Delambre is available in their joint publication, *Base du système métrique décimal* (3 vols., Paris, 1806, 1807, 1810). This may be supplemented by Delambre's own *Grandeur et figure de la terre* (Paris, 1912), edited and published from Delambre's manuscript about a century later by G. Bigourdan.

Delambre explains the inequality of the assigned distances (Méchain—170,000 *toises* from Rodez to Barcelona; Delambre—380,000 *toises* from Rodez to Dunkerque) as follows: "The reason for this unequal division was that the Spanish part was entirely new, whereas the remainder had already been measured twice; we were agreed that the former would provide many more difficulties." Then he remarks, "We did not know that the greatest difficulties of all would be found at the very gates of Paris." Méchain, the first

to set out, on 25 June 1792, was arrested at his third observational site, at Essonne, by uneasy citizens who were convinced that his activities had some counter-revolutionary aspects. Only by constant explanation and good fortune was Méchain able to continue, and eventually to carry his survey into Spain. Delambre encountered similar difficulties; and, in addition, when he returned to Paris and had to leave again, he had to seek new passports as the government changed. It seems almost incredible that in time of revolution Delambre was able to continue his work as much as he did. In eight months of 1792, however, he had established only four points of triangulation; but in 1793, despite delays in getting his passport, he made better progress. Then, in January 1794, he received an order from the Committee of Public Safety to stop all observations at once. On his return to Paris he learned that as of 23 December 1793 he had been removed from membership in the commission of weights and measures, along with Borda, Lavoisier, Laplace, Coulomb, and Bresson.

Happily, the enterprise was revivified by the law of 18 Germinal *an* III (7 April 1795), and Delambre and Méchain were able to take up their old assignments, now under the title of *astronomes du Dépôt de la Guerre,* serving under the head of that establishment, General Calon, a member of the Convention. Delambre thereupon set out for Orléans on 28 June 1795 and completed his assignment within four years.

Delambre's task was not merely to make a series of correlated astronomical observations and terrestrial measurements; he had also to carry out extremely laborious calculations. The latter were made especially tedious by the need to convert the observations from the new centesimal units of angle-measure (used in Delambre's instruments) to the older units of degrees, on which all tables of logarithms and of trigonometric functions were then based.

The proponents of the metric system succeeded in establishing a decimal-positional system of mass and length (area and volume), but failed in their attempts to introduce similar systems of time or of angle measure. Delambre's instruments were constructed with the new centesimal divisions, in anticipation of their general adoption. On this score he remarks:

> This subdivision is much more convenient for use with the repeating circle, and would be equally convenient for verniers with any sort of instrument. Some people still prefer the old subdivision out of long habit and because they have never tried the new one, but no one who has ever employed them both wants to return to the former system.

Extending the metric system to the subdivision of the circle, however, required the construction of new trigonometric tables. In the year II [1793], M. de Prony was requested to prepare such tables which would leave nothing to be desired in their exactitude and which would constitute the largest and most imposing monument of calculation that had ever been executed or even conceived.[6]

Prony's manuscript, never edited for publication, contained logarithms of sines and tangents to fourteen decimals in tens of centesimal "seconds" and logarithms of numbers from 1 to 100,000 to nineteen decimals. An account was presented to the Institute and was published in 1801 by Lagrange, Laplace, and Delambre, under the title *Notice sur les grandes tables logarithmiques et trigonométriques calculées au bureau du Cadastre sous la direction du citoyen Prony.*

A more usable work was produced by Borda, a set of tables to seven decimals. Completed by Delambre after Borda's death, this work was published in the year IX of the Republic (1801), under the title *Tables trigonométriques décimales, ou tables des logarithmes des sinus, sécantes et tangentes, suivant la division du quart de cercle en 100 degrés, du degré en 100 minutes, et de la minute en 100 secondes; . . . calculées par Ch. Borda, revues, augmentées et publiées par J. B. J. Delambre.*

Méchain died in 1804, and it became Delambre's sole responsibility to complete the computations and to write up the final report. This constituted three volumes containing the history of the enterprise, the observations, and the calculations. The third volume was completed in 1810, some twenty years after the project was begun. When Delambre presented a copy of this work to Napoleon, the emperor responded, "Conquests pass and such works remain."[7]

Delambre's results were put into the hands of a commission of French and foreign scientists, who then determined the unit of length which became the standard meter. Jean Joseph Fourier said that "no other application of science is to be compared with this as regards its character of exactness, utility, and magnitude." The newly constituted Institut de France designated this survey "the most important application of mathematical or physical science which had occurred within ten years" and in 1810 gave Delambre a prize for his share in the great work. The accuracy with which Delambre carried out his task may be seen in a comparison of two base lines: Perpignan and Mélun. Delambre measured both by direct methods. Then, making use of a network of triangulation, he used one to compute the other. According to Fourier, although the distance between the two bases is some 220 leagues, the results of calculation differed from the results of direct measurement by less than three-tenths of a meter, less than one part in 36,000.

By the time of publication of his report on the base of the metric system, Delambre had become a resident member of the newly organized Institut National (*section de mathématiques de la première classe*). He was appointed an inaugural member of the Bureau des Longitudes, founded in 1795. When the Institute was reorganized, he became, on 11 Pluviôse *an* XI (31 January 1803), the first permanent secretary for *les sciences mathématiques.* In 1807, he succeeded Lalande in the chair of astronomy in the Collège de France. In 1813 he published an *Abrégé d'astronomie* and in 1814 a work on *Astronomie théorique et pratique,* based on his lectures. Until 1808, when he moved from the rue de Paradis to the outskirts of the Faubourg Saint-Germain, he continued to make observations from his private observatory, primarily checking stellar positions in the major catalogs from Flamsteed's to Maskelyne's. He also associated himself with Laplace, who was working on the problems of perturbations and other aspects of celestial mechanics, and produced new tables based on these investigations. In the official *éloge,* Fourier wrote:

> Before him astronomical calculations were founded on numerical processes, which were at once indirect and irregular. These he has changed throughout, or ingeniously remodeled. Most of those which astronomers use at the present time belong to him, having been deduced from analytic formulas, which, in their application, have been found alike, sure, uniform, and manageable. The new tables which he has given us of the sun, of Jupiter, of Saturn, of Uranus, and of the satellites of Jupiter, at least some of them, may have been considerably improved by recent labors founded on a greater number of exact observations; yet, in the present state of astronomy, and up to this day, the tables of Delambre just mentioned are those employed in the calculations made for the *Connaissance des temps* and for the nautical and astronomical ephemerides of most nations. In addition, the geodetic operation, for which we are chiefly indebted to him, and of which he bore the greatest share, is the most perfect and extensive which has been executed in any country. It has served as the model of all enterprises of the kind which have been since projected.

As scientist, Delambre is remembered primarily for his improvements in astronomical tables and his contributions to the measurement of the earth (and establishment of the base of the metric system). But he had yet another career, begun in the last decades of his life, as historian. Reference has already been made to Delambre's history of the measurement of the earth (and the historical first volume of the *Base du système métrique décimal*). In 1810 Delambre published a major historical work, *Rapport historique sur les progrès des sciences mathématiques depuis 1789.*

Delambre had at one time intended that his treatise on astronomy would be preceded by a *tableau* of the evolution of this science through the ages. But he found the subject so vast that he decided to devote a separate work to it. He began collecting and organizing his materials in 1812, at the age of sixty-three, and he devoted the remainder of his life to compiling a history of astronomy. By the time he was finished, Delambre had completed six volumes, of which the final one (on the astronomy of the eighteenth century) was published posthumously in 1827 by L. Mathieu. The first two volumes (1817) deal with ancient astronomy, a third (1819) with the astronomy of the Middle Ages, and the fourth and fifth with the astronomy of the Renaissance and the seventeenth century. Delambre also helped the Abbé Halma in his translation of Ptolemy's *Almagest* and he wrote an extensive set of notes of such importance that his name appears along with Halma's on the title page.[8]

Delambre's *Histoire de l'astronomie* is a work without parallel in any of the sciences. It is a technical work, written—as he said—"mainly for astronomers, and mathematicians in general." His aim had been to produce a "tableau complet des différens âges de l'Astronomie," that is, "a repository where could be found all the ideas, all the methods, and all the theorems that have served successively for the calculation of phenomena." There is no synthesis, no generalization, no display of insight into the causes for great progress or decline. Delambre rather presents each major chronological period in a series of discrete analyses of one treatise (or other work) after another. Often, in the case of a long book (as Kepler's *Astronomia nova*), the analysis proceeds chapter by chapter. Thus the reader may readily apprehend what a given astronomical work contains, plus a critical estimate of its worth; and he may follow Delambre as he compares and contrasts several works of a given author. But the method is frustrating in the extreme to anyone who may want to trace a particular topic throughout a whole century or more.

As one would expect, Delambre is especially good on astronomical tables and on methods of observation and calculation. A great virtue is the wealth of information on minor figures, for whom no other account may be available. Above all, Delambre spices his presentation with acerb and delightful comments (including critical remarks about style or errors in Greek and Latin) as his statement that Boulliau's construction is "certainement ingénieuse, mais inutile." Or again, "[We] are writing a history not eulogies." Thus, "The historian owes nothing to the dead save truth. It's not our fault if *in astronomy* Descartes produced nothing but chimeras."[9] Unquestionably the six-volume *Histoire* is the greatest full-scale technical history of any branch of science ever written by a single individual. It sets a standard very few historians of science may ever achieve.

In the course of his geodetic measurements, Delambre gained the service of a young assistant, Leblanc de Pommard. Pommard's mother, then a widow, has been described as "a distinguished Latinist, endowed with solid but not pedantic learning." Delambre married her when he was fifty-five years of age.

Under the Empire, Delambre took on a number of official posts, including *inspecteur général des études* and *trésorier de l'université*. He had the task of establishing a number of *lycées,* including those of Moulins (1802) and Lyons (1803). Despite his "opinions libérales," he became in 1814 a member of the Conseil Royal d'Instruction Publique. He retired from public life in 1815 and was made a *chevalier* of Saint Michel by the royal government. In 1821 he became an *officier* of the Legion of Honor (he had been a *chevalier* since the foundation of the order).

NOTES

1. Mathieu, p. 305*a*.
2. This observatory was situated in the Hotel d'Assy in the rue de Paradis, in the Marais, which is now part of the Archives Nationales. The observatory remained a distinct structure until about 1910 (Jacquinet, p. 195).
3. Lalande listed the date of Delambre's birth in the preface to his *Astronomie,* 3rd ed. (Paris, 1792), p. xxxiii, "parceque, je la regarde comme devant faire époque dans l'histoire de l'astronomie."
4. P. 195.
5. According to the report, "It is apparent here that we are surrendering all claim to the common division of the quarter-meridian into degrees, minutes, and seconds; but this old division could not be kept without harming the unity of the system of measure since decimal division that corresponds to arithmetical gradations is to be preferred for a standard of usage" (Jacquinet, p. 196). See also Delambre's *Grandeur et figure de la terre* (1912).
6. These extracts are taken from Méchain and Delambre, *Base du système métrique.*
7. Delambre wrote this remark in his own copy of this work.
8. In this work, and in his history of ancient astronomy, Delambre adopted a posture which has recently been subject to serious criticism: a "dislike for Ptolemy and the resulting misrepresentation of Hipparchian astronomy as being practically the equivalent of the *Almagest*" (O. Neugebauer, in his preface to the reprint of Delambre's *Histoire de l'astronomie ancienne*).
9. For a critical analysis of Delambre as historian, see I. B. Cohen's introduction to the reprint of Delambre's *Histoire de l'astronomie moderne.*

BIBLIOGRAPHY

I. ORIGINAL WORKS. Delambre published a large number of *mémoires, extraits, notices, éloges,* and other works in the *Connaissance des temps* from 1788 to 1822, and in various publications of scientific societies, including the Académie des Sciences (Paris), and the academies of Berlin and Turin. His major publications, in chronological order, are *Tables de Jupiter et de Saturne* (Paris, 1789); *Tables astronomiques, calculées sur les observations les plus nouvelles, pour servir à la troisième édition de l'Astronomie,* a supp. (with separate pagination) to vol. 1, 3rd ed., of Lalande's *Astronomie* (Paris, 1792)—in the preface Lalande says, "Les tables . . . du Soleil, de Jupiter, de Saturne, et des satellites [de Jupiter], sont de M. de Lambre. . . ."; and *Méthodes analytiques pour la détermination d'un arc de méridien. Précédées d'un mémoire sur le même sujet, par A.M. Legendre* (Paris, an VII [1799]).

The *Tables astronomiques publiées par le Bureau des Longitudes de France,* pt. 1 (Paris, 1806), contains "Tables du Soleil, par M. Delambre"; pt. 2 (Paris, 1808) contains "Nouvelles tables écliptiques des satellites de Jupiter, d'après la théorie de M. Laplace, et la totalité des observations faites depuis 1662, jusqu'à l'an 1802; par M. Delambre"; a separate publication of the latter work, with a slightly different title (*Tables . . . d'après la théorie de M. le Marquis de Laplace . . .*) was issued in Paris, 1817.

Other works are *Base du système métrique décimal, ou Mesure de l'arc du méridien compris entre les parallèles de Dunkerque et Barcelone, exécutée en 1792 et années suivantes, par MM. Méchain et Delambre. Rédigée par M. Delambre. Suite des Mémoires de l'Institut,* 3 vols. (Paris, 1806, 1807, 1810); *Rapport historique sur les progrès des sciences mathématiques depuis 1789, et sur leur état actuel . . .* (Paris, 1810; photo repr., Amsterdam, 1966); *Abrégé d'astronomie, ou leçons élémentaires d'astronomie théorique et pratique* (Paris, 1813); *Astronomie théorique et pratique,* 3 vols. (Paris, 1814); *Histoire de l'astronomie ancienne,* 2 vols. (Paris, 1817); *Histoire de l'astronomie du moyen âge* (Paris, 1819); *Histoire de l'astronomie moderne,* 2 vols. (Paris, 1821); and *Histoire de l'astronomie au dix-huitième siècle.* The six-volume set of histories was repr. in facs., with a preface to vols. I and II by O. Neugebauer, an intro. to vols. IV and V by I. Bernard Cohen, and an intro. to vol. VI by Harry Woolf (New York–London, 1965–1969).

See also *Grandeur et figure de la terre,* with notes and maps (Paris, 1912).

Not included in the above list are certain works edited by Delambre (such as Borda's *Tables trigonométriques décimales . . .*) or reports (such as the one on Prony's tables), even though they are mentioned in the text of the article. But special mention should be made of *Composition mathématique de Claude Ptolémée. Traduite pour la première fois du grec en français, sur les manuscrits originaux de la Bibliothèque Impériale de Paris, par M. [l'Abbé] Halma; et suivie des notes de M. Delambre,* 2 vols. (Paris, 1813–1816; photo repr., Paris 1927).

II. SECONDARY LITERATURE. The major biographical source is Claude Louis Mathieu's article, based on Delambre's manuscript autobiography and other manuscripts, in Michaud's *Biographie universelle* (new ed., Paris, n.d., X, 304–308). Another major source is Pierre Jacquinet, "Commémoration du deux-centième anniversaire de la naissance de J. B. Delambre—son oeuvre astronomique et géodésique," in *Bulletin de la Société Astronomique de France,* 63e annee (1949), 193–207. Jean Joseph Fourier's *Éloge,* prepared for the Académie des Sciences, is available in an English translation by C. A. Alexander in the *Annual report . . . of the Smithsonian Institution . . . for the year 1864* (Washington, 1865), pp. 125–134.

Also available is Joseph Caulle, "Delambre—sa participation à la détermination du mètre," in *Recueil des Publications de la Société Havraise d'Études Diverses,* 103e année (1936), 143–157. Other sources are Charles Dupin, "Notice nécrologique sur M. Delambre," in *Revue encyclopédique,* 16 (1822), 437–460, and the brief account by St. Le Tourneur in the new *Dictionnaire de biographie française,* fasc. 57 (Paris, 1964), p. 675.

As Bigourdan reports, "The Academy of Amiens held a contest for his eulogy; Vulfran Warmé's speech, printed at Amiens in 1824, won for him the *accessit* and a gold medal." A copy of this work, *Éloge historique de M. Delambre, qui a obtenu l'accessit et une medaille d'or au concours de l'Académie d'Amiens* (Amiens, 1824), is in the dossier of Delambre in the library of the Académie des Sciences, Paris, with M. Desboves, *Delambre et Ampère: Discours de réception, suivi de . . . plusieurs lettres inédites de Delambre* (Amiens, 1881). See also David Eugene Smith's *Delambre and Smithson* (New York, 1934).

A summary of the facts of Delambre's life and an evaluation of his work as historian is available in my introduction to the facs. repr. of Delambre's *Histoire de l'astronomie moderne.* In the short biography listed three paragraphs above, Mathieu says of himself: "The author of this article, a student and friend of Delambre and possessor of all his manuscripts, made use of the writings of Delambre, a biographical note written by Delambre himself, and that which he had the opportunity of learning about Delambre during the many years he spent with him. One would have to have heard this modest and sincere man giving an account of his way of life after leaving the Collège du Plessis in order to believe the tiny amount that he spent during one year."

I. BERNARD COHEN

DE LA RUE, WARREN (*b.* Guernsey, 15 January 1815; *d.* London, England, 19 April 1889), *chemistry, invention, astronomy.*

Warren de la Rue was the eldest son of Thomas de la Rue, a printer, and Jane Warren. His early education was at the Collège Sainte-Barbe in Paris. In his teens he entered his father's printing shop and there first came into contact with science and technology. He was one of the first printers to adopt electrotyping and, with a friend, invented the first envelope-making machine. His understanding of machinery and technology was the basis of his contributions to science. De la Rue was not an original thinker but one who perfected instruments and, through these improvements, made accurate observations of theoretical interest.

De la Rue's first scientific contributions were to chemistry, in which he remained interested throughout his life. He made a small improvement of the Daniell constant voltage battery that was announced in his first paper (1836). With August Wilhelm Hofmann, the great German teacher of chemistry in London, he edited an English version of the first two volumes of Liebig and Kopp's *Jahresbericht,* which served to acquaint English chemists with the work of their Continental colleagues. He was an original member of the Chemical Society, serving as its president in the years 1867–1869 and 1879–1880.

De la Rue's major contribution was to astronomy. He was drawn to this science by another inventor-businessman, James Nasmyth, the inventor of the steam pile driver. Nasmyth had been fascinated with the moon for years and had personally drawn some of the best pictures available of our satellite. De la Rue took up astronomy with the purpose of producing more accurate and detailed pictures of the nearby heavenly bodies. He, too, was an excellent draftsman and his drawings of Saturn, the moon, and the sun are superb. His observation of detail was enhanced by improvements he introduced into the polishing and figuring of the thirteen-inch reflecting telescope that he built for his own observatory at Canonbury. His real ability, however, was revealed only when he turned his talents to the application of photography to astronomy. He was able to make stereoscopic plates of the moon's surface, which brought to light details never before noted. He invented a photoheliographic telescope that permitted the sun's surface to be mapped photographically. Applying the stereoscopic methods he had used on the moon, he showed in 1861 that sunspots are depressions in the sun's atmosphere, thus verifying a suggestion made in the eighteenth century by Alexander Wilson of Glasgow.

In later life (1868–1883) De la Rue conducted a series of experiments on electric discharge through gases. They merely multiplied data without leading to any significant theoretical advance.

De la Rue was a fellow of the Royal Society (1850), the Chemical Society, and the Royal Astronomical Society, serving as president of the last-named society from 1864 to 1866. He was also a member and sometime president of the London Institution and member of the Royal Institution and the Royal Microscopical Society.

BIBLIOGRAPHY

De la Rue's published papers are listed in the Royal Society's *Catalogue of Scientific Papers.* The biography that appears in the *Dictionary of National Biography* is generally reliable, although it tends to exaggerate the importance of his contributions to science, particularly chemistry. His correspondence with Michael Faraday, published in *The Selected Correspondence of Michael Faraday,* 2 vols. (London, 1971), reveals his methods and abilities, as well as various aspects of his congenial personality.

L. PEARCE WILLIAMS

DELAUNAY, CHARLES-EUGÈNE (*b.* Lusigny, France, 9 April 1816; *d.* at sea, near Cherbourg, France, 5 August 1872), *celestial mechanics.*

Delaunay's father, Jacques-Hubert, was a surveyor who later bought the office of bailiff. His mother, Catherine, was his confidante all his life, especially after the death of his wife in 1849. A bright pupil in secondary school at Troyes, he showed such a gift for mathematics that he was admitted to the École Polytechnique in 1834 and graduated in 1836, first in his class. He received the newly established Laplace Prize, consisting of the complete works of Laplace, which led to his interest in celestial mechanics.

Although Delaunay was by assignment a mining engineer, he served in various engineering schools and at the University of Paris as professor of mechanics, mathematics, or astronomy. His first researches were on calculus of variations ("De la distinction des maxima et des minima dans les questions qui dépendent de la méthode des variations," 1841, doctor's thesis), on perturbations of Uranus (1842), and on the theory of tides (1844). Delaunay's work in lunar theory started in the 1840's. He published the principle of what is known as the Delaunay method in 1846 and generalized it in 1855.

The Delaunay method, further developed by Anders Lindstedt, Poincaré, and Hugo von Zeipel, was a major contribution to analytical mechanics. It consists of a single procedure permitting elimination from the system of canonical equations, one by one, of all the terms of the disturbing function and hence the building up, term by term, of the solution of the problem. Delaunay applied his method to the moon, computing all the terms up to the seventh order and some additional ones of the eighth and ninth orders. This work was published in 1860 and 1867. It is noteworthy that, in studying the incompatibility between the observed and the computed values of the secular acceleration of the moon, Delaunay suggested that it could be caused by a slowing of the rotation of the earth by tidal friction (1865). This hypothesis is now known to be correct.

There was a long-time rivalry between Delaunay and Le Verrier. Delaunay recognized his colleague's scientific achievements but fought his dictatorial rule over astronomical research. He was appointed director of the Paris observatory in March 1870, after Le Verrier, in a dispute with the staff, was dismissed. But in the two years before his death Delaunay had to devote all his efforts to trying to save the observatory during the Franco-Prussian War and the Commune. He was a member of the Académie des Sciences (1855), the Bureau des Longitudes (1862), and the Royal Society (1869).

BIBLIOGRAPHY

Delaunay's main work is "Théorie du mouvement de la lune," in *Mémoires de l'Académie des sciences,* **28** (1860), entire vol., 883 pages and **29** (1867), entire vol., 931 pages. Most of his findings and scientific discussions were printed in the *Comptes rendus hebdomadaires des séances de l'Académie des sciences* between 1841 and 1872. Among them are "Calcul des inégalités d'Uranus qui sont de l'ordre du carré de la force perturbatrice," *in Comptes rendus hebdomadaires des séances de l'Académie des sciences,* **14** (1842), 371, 406; "Mémoire sur la théorie des marées," *ibid.,* **17** (1843), 344; and "Sur l'existence d'une cause nouvelle ayant une action sensible sur la valeur de l'équation séculaire de la Lune," *ibid.,* **61** (1865), 1023.

The Delaunay method is presented in "Mémoire sur une nouvelle méthode pour la détermination du mouvement de la Lune," *ibid.,* **22** (1846), p. 32; and a modification in "Sur une méthode d'intégration applicable au calcul des perturbations des planètes et de leurs satellites," *ibid.,* **40** (1855), 335.

Further information on Delaunay's life is in Arsène Thévenot, "Biographie de Charles-Eugène Delaunay," in *Mémoires de la Société académique de l'Aube,* **42** (1878), 1–129.

J. KOVALEVSKY

DELÉPINE, STÉPHANE-MARCEL (*b.* St.-Martin-le-Gaillard [Seine Maritime], France, 19 September 1871; *d.* Paris, France, 21 October 1965), *chemistry.*

With his mother's encouragement, Delépine followed in the footsteps of his older brother by studying pharmacy. He received his professional training in Paris, where he studied pharmacy at the École Supérieure de Pharmacie and science at the Sorbonne. From 1892 to 1897 he served as intern in pharmacy and from 1902 to 1927 as pharmacist of the Hospitaux de la Ville in Paris. In 1898 Delépine became *docteur ès sciences physiques.* From 1895 to 1902, he was *preparateur* to the famous chemist and statesman Marcellin Berthelot at the Collège de France, where he worked on thermochemical determinations and remained until Berthelot's death in 1907.

In 1904 Delépine was appointed *agrégé* at the École de Pharmacie, where he was promoted in 1913 to the rank of professor. He remained there until 1930, when he succeeded Charles Moureu as professor at the Collège de France, thus occupying the chair once held by his mentor Berthelot. Although he inspired a number of students, including Raymond Charonnat and Alain Horeau, his successor, Delépine preferred to do most of his own laboratory work. His retirement in 1941 did not decrease his prolific scientific productivity; during his retirement he added some sixty publications to his 200 articles. He continued to work in his laboratory with the aid of an assistant until six weeks before his death at the age of ninety-four. A scientist of international reputation, Delépine was a member of many societies and was the recipient of numerous honors.

Delépine's work encompassed almost all fields of chemistry. Like Alfred Werner, of whose work he was an ardent proponent, he began his long and fruitful career as an organic chemist. The thesis for his degree in pharmacy dealt with the separation of methylamines by formaldehyde, and his doctoral dissertation involved a primarily thermodynamic study of the amines and amides derived from aldehydes. His name is immortalized in the so-called Delépine reaction for the preparation of primary amines. He determined the structure of aldehyde ammonia and demonstrated the reversibility of the formation of acetals. Delépine's voluminous work on organic sulfur compounds included studies of dithiourethanes, the discovery of the monomeric sulfides of ethylene (whose existence had been considered impossible), and the recommendation of dithiocarbamates as analytical reagents. He discovered that certain compounds containing doubly bound sulfur have the property of spontaneous oxidation accompanied by phosphorescence (oxyluminescence). He also made extensive studies of catalytic hydrogenation in the presence of Raney nickel. His organic work also dealt with terpenes, heterocyclic compounds, pyridine compounds, alkaloids, and aminonitriles.

In the field of inorganic chemistry, Delépine immediately adopted Werner's then controversial views, and his numerous studies of coordination compounds, particularly those of the noble metals, confirmed their geometric and optical isomerism and verified Werner's coordination theory. His classical work on iridium, especially the chloro salts, pyridine derivatives, and oxalates, placed the stereochemistry of this element on a firm basis, just as Werner's work had done for the compounds of cobalt. Delépine also perfected a method for preparing pure tungsten for use in electric light filaments. He was also a master of stereo-chemistry and crystallography, and he devised the method of active racemates for resolution of coordination compounds and determination of their configurations. In addition, he published several articles on the history of chemistry.

BIBLIOGRAPHY

I. ORIGINAL WORKS. The majority of Delépine's work was published in the *Bulletin de la Société chimique de France*, the *Annales de chimie*, and *Comptes rendus hebdomadaires des séances de l'Académie des sciences*. His masterly summary of the chemistry of iridium is found in P. Pascal, ed., *Nouveau traité de chimie minérale*, xix (Paris, 1958), 465–575.

II. SECONDARY LITERATURE. R. Oesper, in *Journal of Chemical Education*, **27** (1950), 567–568 gives a very brief description of Delépine's work. A 38-page booklet, *Hommage rendu au Professeur Marcel Delépine par ses amis, ses collègues, ses élèves à l'occasion de sa promotion au grade de Commandeur de la Légion d'Honneur, 23 Novembre 1950* (Paris, 1950), consists of eulogies describing his life and work. Two brief obituaries which discuss his life and work are C. Dufraisse, in *Comptes rendus hebdomadaires des séances de l'Académie des sciences*, **261** (1965), 4931–4935; and A. Horeau, in *Annales de chimie*, **1** (1966), 5–6. A detailed description, evaluation, and bibliography of Delépine's work in the field of inorganic chemistry is found in A. Chrétien, *Revue de chimie minérale*, **3** (1966), 187–200.

GEORGE B. KAUFFMAN

DELILE (or **RAFFENEAU-DELILE**), **ALIRE** (*b.* Versailles, France, 23 January 1778; *d.* Montpellier, France, 5 July 1850), *botany*.

The son of Jean-Baptiste Élie Raffeneau-Delile, *porte-malle ordinaire du Roi*, and Marie Catherine Bar, Delile had barely begun his secondary studies at the Collège de Lisieux in Paris when they were interrupted by the Revolution. He completed them in Versailles after having been forced to return there, then became a nonresident medical student in the charitable institutions of the city. At the same time he was introduced to botany at the Trianon gardens.

On 29 Vendémiaire, *an* IV (21 October 1795) Delile was admitted to the École de Santé in Paris as a result of a competitive examination. At the school he met Desfontaines, who was the deciding factor in his joining the Egyptian expeditionary force organized by Bonaparte. He left France on 19 May 1798 as a botanist attached to the expedition.

Delile was in Egypt until the destruction of the French fleet at Aboukir and then returned to France, with an important herbarium, in November 1801. He sailed for America in April 1803 as an assistant commissioner for commercial relations, representing France in Wilmington, North Carolina. In 1806 Delile left that post to work with Benjamin Smith Barton, a physician at the Pennsylvania Hospital in Philadelphia. He resumed his medical studies there and continued them under David Hosack in New York. On 5 May 1807 he defended a thesis entitled "On the Pulmonary Consumption." Three months later Delile was recalled to France in order to resume the editing of a flora of Egypt.

This work was completed in 1809, and Delile was left without employment. His candidacy to succeed Pierre Broussonnet as professor of botany at the Faculty of Medicine of Montpellier had been unsuccessful—Candolle received the appointment—and Delile had to return to the practice of medicine. On 6 July 1809 he defended his doctoral thesis before the Faculty of Medicine in Paris and treated patients without renouncing his other projects. After the collapse of the Empire, Candolle decided to leave France, and in July 1819 Delile was appointed to take his place. From then on, he remained in Montpellier.

Among Delile's published works those relating to Egypt are the most famous, but they represent no more than eight titles out of the sixty that Joly counted in 1859. At first glance the period spent in America seems to have been less fruitful. Only three papers, including his M.D. thesis, were produced. The period in Montpellier, which was the longest, was the one in which Delile published the most; but the works have no definite orientation. There are essays on culture and acclimatization, descriptions of oriental species, and biological observations; none is without importance, yet none is particularly memorable.

However, Delile's published works represent only one part of his research. At his death he left a great many drawings, observations, and unpublished documents. Joly cites four manuscripts that he himself possessed, but they seem to have been lost since then. However, there is an important piece of evidence of Delile's activities in America, a copy of Michaux's *Flora boreali-americana*, annotated in preparation for a revision. This, together with eight cartons of watercolor drawings representing 332 species of fungi—

done during the Montpellier period—may be found in the archives of the Institut Botanique, Montpellier.

BIBLIOGRAPHY

I. ORIGINAL WORKS. Only Delile's most important works are listed here; a more complete list is provided by Joly: "Description de deux espèces de Séné . . .," in *Mémoires sur l'Égypte,* III (Paris, an X [1801]); "Note critique sur le *Ximenia Aegyptiaca,*" *ibid.;* "Observations sur les lotus d'Égypte," in *Annales du Muséum d'histoire naturelle,* **1** (1802); *An Inaugural Dissertation on the Pulmonary Consumption* (New York, 1807), his M.D. thesis; "Description d'opérations rares et nouvelles d'anévrismes, faites avec succès en Angleterre et en Amérique," in *Journal de médecine* (1809); *Dissertation sur les effets d'un poison de Java appelé Upas Tieuté* (Paris, 1809), his M.D. thesis; "Description et dessin d'une tarière spirale, instrument vulgaire aux États-Unis pour abréger les travaux de charpente," in *Mémoires de la Société d'encouragement de Paris* (1812); "Description du Palmier Doum de la Haute-Égypte," in *Descriptions de l'Égypte,* XIX (Paris, 1824); "Florae Aegyptiacae illustratio," *ibid.;* "Flore d'Égypte, explication des plantes gravées," *ibid.;* "Histoire des plantes cultivées en Égypte," *ibid.;* "Mémoire sur les plantes qui croissent spontanément en Égypte," *ibid.; Centurie de plantes d'Afrique du voyage à Méroé, recueillies par M. Cailliaud et décrites par M. Delile* (Paris, 1826); *Fragments d'une flore de l'Arabie Pétrée* (Paris, 1833); "Nouveaux fragments d'une flore de l'Arabie Pétrée," presented to the Academy of Sciences in April 1834; "Première récolte des fruits du ginkgo du Japon en France," in *Bulletin de la Société d'agriculture de l'Hérault* (1835); "Pomone orientale: Designation d'arbres à fruits à importer de Syrie en France," *ibid.* (1840); and "Correspondance d'Orient: De l'horticulture en Égypte," *ibid.* (1841).

II. SECONDARY LITERATURE. On Delile or his work, see N. Joly, "Éloge historique d'Alyre Raffeneau-Delile," in *Mémoires de l'Académie impériale des sciences, inscriptions et belles-lettres de Toulouse,* 5th ser., **3** (1859); Jean Motte, "Delile l'Égyptien, un botaniste à la suite de Bonaparte," in *Science et nature,* no. 18 (1956); and "Matériaux inédits, préparés par Delile pour une flore de l'Amérique du Nord," in *Les botanistes français en Amérique du Nord avant 1850,* Colloques Internationaux du C.N.R.S. no. 63 (Paris, 1957).

JEAN MOTTE

DELISLE, GUILLAUME (*b.* Paris, France, 28 February 1675; *d.* Paris, 25 January 1726), *geography.*

Delisle was the son of Claude Delisle, historian and geographer, and Nicole-Charlotte Millet de la Croyère. Interested in geography and mapmaking from his early childhood, Delisle was taught both by his father and by Gian Domenico Cassini, the director of the Paris observatory. He published his first important work, a set of maps of the continents, a mappemonde, and a globe, in 1700; these immediately established his fame. He was elected to the French Academy of Sciences in 1702 and ran his own mapmaking establishment in Paris until his death. In 1718 Delisle was given the title *premier géographe du roi,* a distinction no doubt connected with the fact that he tutored the young king, Louis XV, in geography.

Delisle designed and published some ninety maps during his lifetime. These included world maps, maps of continents, and maps of single countries or, in the case of France, provinces. The simple elegance of his work alone would distinguish it from the florid, baroque style affected by his contemporaries and predecessors; but it is the content of his maps that is of importance. Delisle studied under Cassini; and he lived in the era of the great surveys, when a number of places on all continents had their locations accurately determined for the first time, using Cassini's method of observation of the moons of Jupiter. Delisle applied the astronomers' findings to his maps; he omitted guesswork, fantasy, and unnecessary or ornamental detail; he admitted lack of knowledge of unexplored territories; and he insisted on critical use of source materials and dependence on scientifically accurate measurements. He thus acquainted the general public with the results of the work of scientists and became the first modern scientific cartographer.

BIBLIOGRAPHY

Delisle's work can be found in a great many libraries throughout the world, in the form of printed atlases and single maps. Further information can be found in Christian Sandler, *Die Reformation der Kartographie um 1700* (Munich–Berlin, 1905), pp. 14–23.)

GEORGE KISH

DELISLE, JOSEPH-NICOLAS (*b.* Paris, France, 4 April 1688; *d.* Paris, 11 September 1768), *astronomy, geography.*

The ninth child of the historian-geographer Claude Delisle and Nicole-Charlotte Millet de la Croyère, he became known to his contemporaries as Delisle *le cadet* or *le jeune* to distinguish him from his two older brothers. After receiving his early education from his father, he began to develop a taste for mathematics near the end of his formal education in rhetoric at the Collège Mazarin. A solar eclipse in 1706 stimulated this new study and led to instruction in the elements of astronomical calculation under Jacques Lieutaud. He was soon frequenting the Royal Observatory, where he was permitted to copy Jacques Cassini's unfinished lunar and solar tables. When his first attempt to launch his own observational career in the cupola of the Luxembourg Palace was hampered by a lack of instruments, he turned temporarily to the production of various astronomical tables desired by Cassini.

Having equipped his observatory, Delisle began a regular observational program with the lunar eclipse of 23 January 1712. Forced to abandon his observatory in September 1715—when the Duc d'Orleans became regent and installed his eldest daughter, the Duchesse de Berry, in the Luxembourg Palace—he resumed his observations at the end of 1716 in Liouville's former apartment at the Hôtel de Taranne. After almost four years there he moved his instruments to the Royal Observatory and also had some work done in the Luxembourg Palace dome, which he regained in 1722, after the duchess' death in 1719.

This early fulfillment of promise carried Delisle into the Academy of Sciences as a student astronomer attached to Maraldi in 1714, and he quickly began what was to be a long series of contributions—primarily reports of observations of eclipses and occultations—to its *Mémoires*. Since it was also necessary for him to earn a livelihood, he gave mathematics lessons and won a small pension under the regency by drawing up astrological forecasts. An appointment to the chair of mathematics at the Collège Royal in 1718 freed him from such endeavors and also brought him students who aided him in the making and reduction of observations. His best-known students during this period were Godin, Grandjean de Fouchy, and his younger brother, Delisle de la Croyère.

Delisle's growing reputation brought him, in 1721, an offer from Peter the Great to found an observatory and an associated school of astronomy in Russia, an invitation which was transformed into mutually acceptable contractual arrangements in 1725. Planned for four years, Delisle's stay in Russia lasted twenty-two years. There he created an observatory which came to enjoy a good reputation while training many astronomers—with elementary treatises in the preparation of which Delisle participated. Some of these students, as well as his younger brother and the instrument maker who had accompanied him, subsequently engaged in geodetic and cartographic ventures throughout the country, the results of which they communicated to him for a projected, but unrealized, large-scale and accurate map of Russia. To provide corresponding observations for longitude determinations, Delisle published his St. Petersburg observations of eclipses of Jupiter's satellites in each of the first six volumes of the *Commentarii* of Russia's Imperial Academy of Sciences.

Various physical and meteorological data came to

him as well, some of the latter inspired by a "universal thermometer" invented and widely distributed by Delisle. He described this device in a work published in 1738, which also contained his and his brother's numerous observations of aurora borealis in Russia and a record of his own early Paris observations and experiments on light. Furthermore, Delisle returned to an interest in transits of Mercury first manifested in 1723, when he had considered, but failed to demonstrate, that the technique suggested in Halley's famous 1716 paper on the use of transits of Venus to determine the parallax of the sun could be equally utilized in the more frequent transits of Mercury. He now treated this possibility for the 1743 transit of Mercury in a letter to Cassini, which the latter placed in the *Mémoires* of the Paris Academy of Sciences.

That institution took cognizance of the length of Delisle's absence by naming him to veteran status in 1741. This did not change with his return to Paris in 1747, although he resumed attendance at the Academy's meetings. He also returned to his Luxembourg Palace observatory to witness a solar eclipse of July 1748, about which he had prepared an *avertissement* to alert astronomers. In addition he regained his chair at the Collège Royal; most notable among his students in this latter period was Lalande.

One of Delisle's long-standing activities had been the amassing of vast amounts of geographical and astronomical material through an extraordinarily extensive correspondence, through inheritance, and through laborious copying. Because of its great value the French government purchased this collection by giving Delisle the title of *astronome de la marine* and a life annuity of 3,000 livres. Moreover, he obtained a new observatory at the Hôtel de Cluny. It was there, in 1759, that his pupil and assistant, Charles Messier, observed the return of Halley's comet. The place of its reappearance had been the subject of a paper by Delisle in 1757.

In other late works Delisle made some use of his meteorological and cartographic materials from Russia and devoted some attention to longitude determinations. The latter had a significance for transit studies because his perfection of Halley's technique by a simplification of the necessary observations demanded precise longitude information. Having also corrected Halley's planetary tables, Delisle produced an *avertissement* on how to observe the 1753 transit of Mercury and a mappemonde showing the favorable stations. He then determined that the Mercury phenomena were inadequate for parallax determination and concentrated his efforts on the forthcoming transit of Venus in 1761, serving as stimulator and coordinator of its worldwide observation by virtue of another mappemonde and an accompanying memoir distributed through correspondence.

Delisle retired increasingly after this activity. Lalande began to teach in his stead at the Collège Royal in 1761. In 1763 he withdrew to the abbey of Ste.-Geneviève to devote himself to charitable and religious works, although he did publish several maps by his eldest brother, Guillaume. Both the Academy of Sciences and the Collège Royal conferred honors upon him prior to his death from an attack of apoplexy in 1768. His wife, whom he had married before his trip to Russia, died shortly after their return; they had no children.

BIBLIOGRAPHY

I. ORIGINAL WORKS. Delisle's immense collection of observations, maps, and correspondence is scattered among several Paris repositories. He provided a partial description of the collection and some autobiographical materials in a MS preserved in the Bibliothèque Nationale: "Histoire abrégée de ma vie et des mes occupations dans l'astronomie, la géographie et la physique, pour servir d'introduction au catalogue de mes manuscrits et mémoires d'astronomie et de géographie conservés au Dépôt des plans et cartes de la Marine . . .," MS fr. 9678, fols. 24–31. In 1795 the purely astronomical part of the collection was given to the Bureau des Longitudes and became the basis of the archives of the Paris Observatory. That part includes an *abrégé* of his unfinished *Traité complet d'astronomie exposée historiquement et demontrée par les observations*, Archives de l'Observatoire de Paris, A 7 10, and, besides the materials providing the basis for that work, his own early observational journals, which contain a brief autobiographical note on his astronomical beginnings, C 2 14. Only a small part of his correspondence has been printed: E. Doublet, ed., *Correspondance échangée de 1720 à 1739, entre l'astronome J.-N. Delisle et M. de Navarre* (Bordeaux, 1910); and H. Omont, ed., *Lettres de J.-N. Delisle au comte de Maurepas et à l'abbé Bignon sur ses travaux géographiques en Russie (1726–1730)* (Paris, 1919). His contributions to the *Mémoires* of the Paris Academy of Sciences prior to his departure for Russia were as follows: "Sur l'observation des solstices" (1714), pp. 239–246; "Résultat de l'observation de l'éclipse du soleil du 3 mai 1715 au matin, faite au Luxembourg en présence de Madame la Princesse, de M. le Comte de Clermont et de plusieurs autres seigneurs" (1715), pp. 85–86; "Observation de l'éclipse de Vénus par la lune, faite en plein jour au Luxembourg le 28 juin 1715" (1715), pp. 135–137; "Sur l'atmosphère de la lune" (1715), pp. 147–148; "Observation de l'éclipse de Jupiter et de ses satellites par la lune, faite au Luxembourg le 25 juillet 1715 au matin" (1715), pp. 159–160; "Reflexions sur l'expérience que j'ai rapportée à l'Académie d'un anneau lumineux semblable à celui que l'on aperçoit autour de la lune dans les éclipses totales du soleil" (1715), pp. 166–169; "Observation de l'éclipse de lune du 20 septembre 1717 au soir, faite à Montmartre" (1717), pp. 299–301; "Occultation d'Aldebaran par la lune, observée le 9 février 1718 au soir, à l'Hôtel de Taranne" (1718), p. 17; "Sur les projections des éclipses sujettes aux parallaxes; où l'on explique la manière dont les astronomes les considèrent, l'usage qu'ils en font, et où l'on donne l'idée d'une nouvelle projection, qui réduit la détermination géométriques de ces éclipses à une expression plus simple que celle qui se tire des projections ordinaires" (1718), pp. 56–67; "Construction facile et exacte du gnomon pour règler une pendule au soleil par le moyen de son passage au méridien" (1719), pp. 54–58; "Observation de l'éclipse d'Aldebaran par la lune, faite à l'Hôtel de Taranne à Paris le 22 avril 1719, au soir" (1719), p. 318; "Observation de l'éclipse d'Aldebaran par la lune, faite à l'Hôtel de Taranne à Paris, le 30 octobre 1719 au soir" (1719), p. 319; "Détail de l'expérience de la réfraction de l'air dans le vuide" (1719), pp. 330–335; "Sur le dernier passage attendu de Mercure dans le soleil et sur celui de mois de novembre de la présente année 1723" (1723), pp. 105–110; "Observation du passage de Mercure sur le soleil, faite à Paris dans l'Observatoire Royal, le 9 novembre 1723, au soir" (1723), pp. 306–343; "Observations de l'éclipse totale du soleil du 22 mai 1724 au soir, faites à Paris, dans l'Observatoire Royal et au Luxembourg, par MM. Delisle le Cadet et Delisle de la Croyère" (1724), pp. 316–319.

Although some of his observations were reported by himself and others, his only significant contribution to the *Mémoires* during his absence from Paris was the "Extrait d'une lettre de M. Delisle, écrite de Petersbourg le 24 août 1743, et adressée à M. Cassini, servant de supplément au Mémoire de M. Delisle, inséré dans le volume de 1723, p. 105, pour trouver la parallaxe du soleil par le passage de Mercure dans le disque de cet astre" (1743), pp. 419–428.

His contributions to the *Commentarii Academiae imperialis scientiarum petropolitanae* were as follows: "Eclipses satellitum Jovis observatae Petropoli," **1**, 467–474, in collaboration with his brother; "Continuata relatio eclipsium satellitum Jovialium Petropoli," **2**, 491–494; "Observationes altitudinis poli in Observatorio imperiali, quod Petropoli est, habitae," **2**, 495–516; "Tertia series observationum satellitum Jovis in Observatorio imperiali Petropoli factarum," **3**, 425–462; "Continuata relatio eclipsium satellitum Jovis observatarum," **4**, 317–321; "Eclipsium Jovis satellitum in Observatorio petropolitano observatarum continuatio," **5**, 451–457; "Eclipses satellitum Jovis observatae in Imperiali specula astronomica, quae Petropoli est, per integrum annum 1738," **6**, 395–400; "Observationes astronomicae in specula Academiae imperialis scientiarum ab anno 1739–1745, a Josepho Nicolao Delilio cum sociis institutae," **6**, 349–362.

Other publications from this period were a three-part *Abrégé des mathématiques pour l'usage de Sa Majesté impériale de toutes les Russies* (St. Petersburg, 1728), written in collaboration with Jacques Hermann, to which he contributed the second part dealing with astronomy and geography, a *Discours sur cette question: Si l'on peut démontrer, par les seuls faits astronomiques, quel est le vrai système du monde? et si la terre tourne ou non* (St. Petersburg, 1728) which he had read to the Academy, a *Projet de la mesure de la terre en Russie* (St. Petersburg, 1737), and his *Mémoires pour servir à l'histoire et au progrès de l'astronomie, de la géographie et de la physique, recueillis de plusieurs dissertations lues dans les assemblées de l'Académie royale des sciences de Paris et de celle de Saint-Petersbourg, qui n'ont point encore été imprimées, comme aussi de plusieurs pièces nouvelles, observations et réflexions rassemblées pendant plus de 25 années* (St. Petersburg, 1738).

After his return from Russia he placed the following items in the Paris Academy's *Mémoires:* "Observation de l'éclipse du soleil du 25 juillet 1748, faite à Paris au Palais du Luxembourg" (1748), pp. 249–254; "Observations du thermomètre, faites pendant les plus grands froids de la Sibérie" (1749), pp. 1–14; "Observation de l'éclipse de lune du 23 décembre 1749, faite à Paris dans l'Hôtel de Cluny" (1749), pp. 320–321; "Observation de l'éclipse totale de lune du 13 décembre 1750, au matin, faite à Paris dans l'Hôtel de Cluny" (1750), pp. 343–344; "Mémoire sur la longitude de Louisbourg, dans l'Isle Royale" (1751), pp. 36–39; "Observation pour la conjonction de Jupiter avec la lune, du 29 décembre 1751 au soir, faite à Paris dans l'Hôtel de Cluny" (1751), pp. 90–92; "Observation de l'éclipse de lune du 2 décembre 1751 au soir, faite à Paris dans l'Hôtel de Cluny" (1751), pp. 273–274; "Réponse de M. Delisle [à une lettre de M. Bradley]" (1752), pp. 434–439; "Mémoire sur le diamètre apparent de Mercure, et sur le temps qu'il emploie à entrer et à sortir du disque du soleil dans les conjonctions inférieures écliptiques" (1753), pp. 243–249; "Observation de l'occulation de l'étoile ρ du Verseau par la lune, et de la conjonction de l'étoile θ avec la même planète, le 21 novembre 1754 au soir, faites à Paris à l'Hôtel de Cluny" (1754), pp. 382–383; "Détermination de la longitude de l'Isle de Madère, par les éclipses des satellites de Jupiter observées par M. Bory, Lieutenant des Vaisseaux du Roi, comparées avec celles de M. l'abbé de la Caille à l'Isle de France" (1754), pp. 565–571; "Observations des diamètres apparens du soleil, faites à Paris les années 1718 et 1719, avec des lunettes de différentes longueurs; et réflexions sur l'effet de ces lunettes" (1755), pp. 145–171; "Nouvelle théorie des éclipses sujettes aux parallaxes, appliquée à la grande éclipse du soleil qu'on observa le 25 juillet 1748" (1757), pp. 490–515; "Observations du passage de Mercure sur le disque du soleil, le 6 novembre 1756; avec des réflexions qui peuvent servir à perfectionner les calculs de ces passages et les élémens de la théorie de Mercure déduits des observations" (1758), pp. 134–154; "Mémoire sur la comète de 1758" (1759), pp. 154–188; "Mémoire sur la comète de 1759 . . ." (1760), pp. 380–465.

Separately published items of astronomical significance during this later period were his *Avertissement aux astronomes sur l'éclipse de soleil du 25 juillet 1748* (Paris, 1748), the *Lettre de M. Delisle sur les tables astronomiques de M. Halley* (Paris, 1749), the *Avertissement aux astronomes sur le passage de Mercure au devant du soleil, qui doit arriver le 6 mai 1753, avec une mappemonde, où l'on voit les nouvelles découvertes faites au nord de la mer du Sud* (Paris, 1753), and a *Recherche du lieu du ciel où la comète, prédite par M. Halley, doit commencer à paraître* (Paris, 1757).

In the purely geographical realm, he published separately a memoir read to a public assembly of the Academy in 1750 and reported in the *Histoire de l'Académie . . .* of that year (1750), pp. 142–152: *Explication de la carte des nouvelles découvertes au nord de la mer du Sud* (Paris, 1752). His later map publications were noted in *Histoire:* (1763), pp. 112–117; (1764), pp. 158–160; (1766), pp. 114–122.

II. SECONDARY LITERATURE. The laudatory "official" *éloge* for the Academy of Sciences was written by Delisle's student Grandjean de Fouchy and appeared in *Histoire de l'Académie . . .* (1768), 167–183. Equally generous in its praise and somewhat more detailed was Lalande's "Éloge de M. de l'Isle," in *Le nécrologe des hommes célèbres de France, par une société de gens de lettres,* V (Paris, 1770), 1–86. The next treatment of him was quite critical: J. B. J. Delambre, *Histoire de l'astronomie au dix-huitième siècle* (Paris, 1826), pp. 318–327. The rather brief account in J. F. Michaud, ed., *Biographie universelle,* X, 334–335, is inadequate, as is the more recent treatment in Niels Nielsen, *Géomètres français du dix-huitième siècle* (Paris, 1935), pp. 163–166. The first significant attention to his work in Russia was that paid in Petr Pekarski's *Histoire de l'Académie impériale des sciences de Petersbourg* (Paris, 1870), pp. 124–155. The negotiations preceding that trip were treated by J. Marchand in "Le départ en mission de l'astronome J.-N. Delisle pour la Russie (1721–1726)," in *Revue d'histoire diplomatique,* **43** (1929), 1–26; his work there and the maps that he brought back are dealt with in detail in Albert Isnard's "Joseph-Nicolas Delisle, sa biographie et sa collection de cartes géographiques à la Bibliothèque nationale," in *Bulletin de la Section de géographie du Comité des travaux historiques et scientifiques,* **30** (1915), 34–164. On the Delisle materials that went to the Paris observatory, see Guillaume Bigourdan, *Inventaire général et sommaire des manuscrits de la bibliothèque de l'Observatoire de Paris* (Paris, n.d.), taken from *Annales de l'Observatoire de Paris. Mémoires,* **21** (1897), F1–F60. On his Paris observations and observatories, see Bigourdan's *Histoire de l'astronomie d'observation et des observatoires en France,* pt. 2 (Paris, 1930), 20–33, 74–92. On the reputation of the observatory of St. Petersburg under his direction, see Bigourdan's "Lettres de Léonard Euler, en partie inédites," in *Bulletin astronomique,* **34** (1917), 258–319, **35** (1918), 65–96. For brief indications on his teaching at the Collège Royal, see Louis-Amélie Sédillot, *Les professeurs de mathématiques et de physique générale au Collège de France* (Rome, 1869), pp. 128–130. Finally, for an excellent analysis of his contribution to the study of transits and the parallax question, see Harry Woolf, *The Transits of Venus; A Study of Eighteenth-Century Science* (Princeton, 1959), esp. ch. 2.

SEYMOUR L. CHAPIN

DELLA PORTA, GIAMBATTISTA. See **Porta, Giambattista della.**

DELLA TORRE, GIOVANNI MARIA (*b.* Rome, Italy, 12 June 1713; *d.* Naples, Italy, 9 March 1782), *natural sciences.*

Della Torre was born of a noble Genoese family but lived in Naples from his earliest years. He received an ecclesiastical education there and in 1738 was appointed to teach physical sciences in the archiepiscopal lyceum. In 1743 Charles III of Spain, king of the Two Sicilies, named him director of the royal library and royal printing press.

Della Torre was a man of wide culture and of wide scientific curiosity. His studies led him into the history of philosophy, optics, and microscopy (he made several new histological identifications with an excellent compound microscope that he himself had built); in addition, he observed and recorded eruptions of Vesuvius. He was much influenced by the *De rerum natura* of Lucretius.

Della Torre's most important work, however, was as an encyclopedist. His two-volume work, *Scienza della natura* (Naples, 1748–1749; reprinted Venice, 1750—an abridgment, *Institutiones physicae,* was published in Naples in 1753, and a considerably enlarged Neapolitan edition appeared in 1767–1770 as *Elementa physicae*), anticipated the more famous *Encyclopédie,* the publication of which began in 1751.

The *Scienza della natura* is divided into sections, each of which is subdivided into several chapters. The first volume, in five sections, is dedicated to general physics, comprising statics and hydrostatics, dynamics and hydrodynamics (all developed according to the mechanical theories of Galileo and Newton), and includes an entire chapter on thermometry that draws upon the works of Boyle, Perrault, and Fahrenheit. The second volume is also in five sections and deals with the earth (including mineralogy, volcanoes, and earthquakes), the air (including light, sound, and electricity), botany, zoology, and human anatomy. Each volume contains a historical preface and a detailed index. Each is illustrated (there are thirty-one plates in volume I, thirty in volume II); illustrations of particular interest are those of units of measurement, the pendulum, electrostatical machines, the pointing of mortar, the compressed-air gun, the refraction of light rays, and chyliferous vessels in man.

In short, the *Scienza della natura* presents a complete and ordered picture of the state of scientific knowledge in its time. Although Della Torre's work is almost forgotten today, he was strongly influential in establishing the scientific climate of eighteenth-century Italy.

BIBLIOGRAPHY

I. ORIGINAL WORKS. Besides the editions of *Scienza della natura* detailed above, Della Torre published *Storia e fenomeni del Vesuvio col catalogo degli scrittori vesuviani*

(Naples, 1755); *Supplemento alla storia del Vesuvio fino all'anno 1759* (Naples, 1759); and *L'incendio del Vesuvio accaduto il 19 Ottobre 1767* (Naples, 1767).

II. SECONDARY LITERATURE. Further material on Della Torre's life and work may be found in *Biographia degli uomini illustri del Regno di Napoli* (Naples, 1834); *Biographie universelle,* X (Brussels, 1847), 242; G. Bruno, "Giovanni Maria Della Torre istologo napoletano," in *Gazzetta sanitaria,* **20** (1949), 156–159; E. D'Afflitto, *Memorie degli scrittori del Regno di Napoli* (Naples, 1782–1788); S. De Renzi, *Storia della medicina in Italia,* V (Naples, 1848), *passim;* G. De Ruggero, *Sommario di storia della filosofia* (Bari, 1930); and M. Schipa, *Storia del Regno di Napoli al tempo di Carlo di Borbone* (Naples, 1904).

PIETRO FRANCESCHINI

DELLA TORRE, MARCANTONIO. See **Torre, Marcantonio Della.**

DELPORTE, EUGÈNE JOSEPH (*b.* Genappe, Belgium, 10 January 1882; *d.* Uccle, Belgium, 19 October 1955), *astronomy.*

Delporte graduated from Brussels University and obtained the doctorate in physics and mathematics with high honors in 1903, at the age of twenty-one. He immediately joined the Royal Observatory of Belgium in Uccle as a volunteer assistant. In 1904 he became assistant, in 1909 associate astronomer, and in 1923 astronomer. He was appointed to the directorship of the observatory in 1936 and retired from official duty in 1947.

From 1903 to 1919 Delporte was attached to the department of meridian astronomy. He made thousands of transit observations of reference stars, among them 3,533 stars for the zones $+21°$ and $+22°$ of the *Carte du ciel,* and conducted careful investigations of the errors of the divisions of the meridian circle. He determined the difference of longitude between the observatories of Paris and Uccle in 1910 and in 1920, and he supervised the installation of the time service at the observatory.

In 1919 Delporte transferred to the department of equatorials and dedicated himself to systematic observations of comets and asteroids. These observations were first performed visually with the thirty-eight-centimeter Cooke refractor; but about 1925, when first the thirty-centimeter-aperture Zeiss astrograph and then the double Zeiss astrograph with forty-centimeter-aperture objectives were installed, a definite investigation of these bodies was organized. The first discovery was the planet Belgica. It was followed by many more discoveries, including Amor and Adonis, which approach nearest to the earth. Delporte's name is also linked with the independent discovery of the comet Dutoit-Neujmin-Delporte. New techniques for accurate determination of position were investigated; and the precise positions were sent regularly to the Astronomisches Rechen-Institut at Heidelberg, where they were used in the determinations of orbits.

In 1930 Delporte edited two volumes for the International Astronomical Union entitled *Scientific Delimitation of Constellations,* with text, maps, and celestial atlas. These volumes fixed the limits of constellations for the entire sky.

Delporte was actively interested in expanding the work of the observatory. The institution was provided with a reflecting telescope one meter in aperture (this was recently enlarged to 1.20 meters and provided with a Schmidt combination), the double astrograph with a forty-centimeter aperture (already mentioned), and an Askania meridian circle with a nineteen-centimeter aperture. He was also responsible for providing the Cooke visual refractor, which had an aperture of thirty-eight centimeters, with a Zeiss objective forty-five centimeters in diameter.

Delporte was an enthusiastic observer, and he inspired many younger astronomers who are continuing his work. After his official retirement he continued to examine plates of asteroids at the observatory, and it was while pursuing this task that he died suddenly.

Delporte received prizes of the Royal Academy of Sciences of Belgium and was a member of the National Committee on Astronomy from its founding in 1919, its vice-chairman in 1930, and its chairman from 1949 until his death. In addition, he was president of the Commission on the Observation of Planets, Comets and Satellites and Ephemerides of the Internation Astronomical Union, corresponding member of the Bureau des Longitudes in Paris, corresponding member of the Academy of Sciences in Paris, and an associate member of the Royal Astronomical Society. He was secretary-editor of the journal *Ciel et terre* of the Belgian Astronomical Society.

BIBLIOGRAPHY

The following major articles appeared in *Annales de l'observatoire de Belgique:* "Observations méridiennes faites au cercle méridien de Repsold," 2nd ser., **10-12** (1907–1910); "Différence de longitude entre Paris et Uccle," *ibid.,* **14** (1913); "Catalogue de 3533 étoiles de repère de la zone +21°. +22°," 2nd ser., **13** (1914); with H. Philippot, "Étude de la division du cercle méridien de Repsold," *ibid.,* **14** (1918); "Forme des tourillons du cercle méridien de Repsold," 3rd ser., **1** (1921); with Philippot, "Positions moyennes pour 1914 d'étoiles de comparaison de la comète 1913 f (Delavan)," *ibid.,* **1** (1922).
See also "Observations de la lune et de planètes en 1906

et 1907, et comparaison avec la *Connaissance des temps* et le *Nautical Almanach,*" in *Astronomische Nachrichten,* **178** (1908); "Observations du soleil, de la lune et de planètes en 1908, et comparaison avec la *Connaissance des temps* et le *Nautical Almanach,*" ibid., **181** (1909); "Observations photographiques de petites planètes avec l'astrographe Zeiss de l'Observatoire d'Uccle," in *Journal des observateurs,* **7** (1924); "Petites planetes découvertes à l'Observatoire royale de Belgique," in *Bulletin de la classe des sciences de l'Académie royale de Belgique* (1926).

H. L. VANDERLINDEN

DELUC, JEAN ANDRÉ (*b.* Geneva, Switzerland, 8 February 1727; *d.* Windsor, England, 7 November 1817), *geology, meteorology, physics, natural philosophy, theology.*

Deluc was descended from a family who had emigrated from Lucca, Tuscany, and had settled in Geneva, probably in the fifteenth century. His father, François, refuted the satirical ideas of Bernard Mandeville and others in several treatises that were known to Rousseau, who wrote a diverting description of how much they had bored him.

Deluc received an excellent education, particularly in mathematics and natural science. He then took up commerce, which he combined with political activities. In 1768 he went to Paris on a successful embassy to the duke of Choiseul and in 1770 was nominated to the Council of Two Hundred. His travels widened Deluc's knowledge of landscape, but most of his early writings on natural science were based on numerous excursions to the Alps and the Jura. As was then fashionable, he gradually amassed, with the help of his brother, Guillaume Antoine, a collection of minerals and of flora and fauna. Later his nephew, Jean André Deluc, expanded this collection and took on his uncle's role of voluminous discourser on geological topics, trying, for example, to dissuade Buckland and Murchison from accepting any theory regarding glacial action.

Deluc's commercial affairs failed in 1773 and he left Geneva, returning only once, for a few days. However, his decision to migrate to England afforded him greater opportunity for carrying out scientific research and writing, which he did for another forty-four years. In London, soon after his arrival, he was made a fellow of the Royal Society and appointed reader to Queen Charlotte, a post with an income adequate to allow him ample leisure. During this period of his life Deluc undertook several tours on the Continent and lived for six years (1798–1804) in Germany, where he was a nonparticipant honorary professor of philosophy and geology at Göttingen University. He was also a correspondent of the Paris

Academy of Sciences and a member of several other scientific associations.

Deluc's favorite fields were geology, meteorology, and natural philosophy or theology, as one might expect of a Calvinistic Genevan who made many scientific excursions to the Alps. By nature an inveterate discourser, he would write in a moderate tone on anything, including, for example, the history of the solar system before the birth of the sun. His great aim was to reconcile Genesis and geology; and his orthodoxy, versatility, prolixity, productivity, high social standing, and facility in languages earned him an exalted contemporary position. Georges Cuvier ranked him among the first geologists of his age, whereas Zittel (p. 77) affirms that although Deluc was "held in high respect and favour during his lifetime, his papers have no permanent place in [the] literature."

Deluc believed that the six days of the Creation were six epochs that preceded the present state of the globe, which began when cavities in the interior of the earth collapsed and lowered the sea level, thereby exposing the continents. There was thus a distinction between an older creative, or antediluvian, period and a newer, or diluvian, period. Of the former there survived only a few primordial islands, which accounted for the fossils of large animals and the continuity and antiquity of organic life. In the latter period, which started about 2200 B.C., new geological processes were operative but were so ineffectual or incidental that the landscape remained unchanged. To Deluc mountains were the remnants left upstanding when the adjacent areas had collapsed catastrophically, and the large boulders known today as glacial erratics had been blown out when great interior caverns filled with some expansible fluid had collapsed.

In 1790–1791 and later, in many letters, Deluc opposed Hutton's ideas on present erosion, asserting, for example, that soil is not eroded because if it were there would be none left. In his *Elementary Treatise on Geology* (1809) he claimed rather bombastically that he could now demonstrate "the conformity of geological monuments with the sublime account of that series of the operations which took place during the *Six days,* or periods of time, recorded by the inspired penman." This discursive volume contains, *inter alia,* four of his earlier letters refuting the ideas of Hutton and Playfair. In his later geological writings Deluc occasionally proffers an astute minor observation but rarely, if ever, is the originator of a new idea.

Deluc's meteorological researches were of more lasting value but were also hyperbolized by his contemporaries. He is said to have "discovered many facts of considerable importance" relating to atmospheric heat and moisture, but most of his observations had already been developed further by others. For instance, Deluc noticed the disappearance of some heat during the thawing of ice at a time when Joseph Black had already progressed to a hypothesis of latent heat. Deluc, however, probably can claim to be the originator of the theory, later proved more clearly by John Dalton, that the amount of water vapor contained in any space is independent of the density of the air or any other gaseous substance in which it is diffused.

Deluc's early meteorological interest was mainly in measuring heights by barometer, for which he published improved rules (*Philosophical Transactions* [1771], 158) based on many experiments with hygrometers, thermometers, and barometers, and particularly on the fall in the boiling point of water with diminishing atmospheric pressure and increasing altitude. He devised a hygrometer similar to a mercury thermometer but with an ivory bulb that expanded when moistened and thus caused the mercury to descend (*Philosophical Transactions* [1773], 404). Humboldt compared the merits of this with Saussure's hair hygrometer: the latter proved better for measuring altitude on mountains and the former for use at sea level, but Deluc's hygrometer worked so slowly that its readings could seldom be combined with those of other instruments.

Deluc's influence on popular early nineteenth-century British meteorology texts was considerable. J. F. Daniell, in his *Meteorological Essays and Observations,* based his account of atmospheric evaporation and condensation largely on extracts from "the works of Deluc, who was probably one of the most accurate observers of nature that ever existed, and who seldom, indeed, allowed any hypothetical considerations to warp his description of what he observed" (2nd ed. [1827], p. 506). This hyperbole stemmed from Deluc's visual observations on clouds and ground (radiation) fog, which, he stated, can be seen to change shape and evaporate at the same time that they are forming.

The barometric controversy between H. B. de Saussure, professor of philosophy at Geneva, and Deluc is one of lasting scientific interest. In *Essais sur l'hygromètre* (1783, p. 282) Saussure stated that some of Deluc's findings were based on specious reasoning and inadequate experimentation: "Mr. Deluc supposes that pure air is heavier than air mixed with water vapor. . . . This supposition explains well why a lowering of the barometer is a sign of rain. . . ." Saussure, experimenting with closed containers, had found little difference in weight between dry air and humid air, and considered the differences quite inadequate to explain the large variations in barometric pressure that occurred at ground level in Europe.

Modern meteorology has proved that Deluc was right, whereas Saussure was groping toward the influence of air masses and of the passage of cyclonic depressions and anticyclones.

The significance of Deluc's contributions to physics is greatly exaggerated. In 1809 he sent a long article to the Royal Society discussing the mode of action of the galvanic pile and showing that "in Volta's pile, the chemical effects can be separated from the electrical." This, as a biographer in *Philosophical Magazine* (**50**, 393–394) wrote, ". . . led that ingenious philosopher [Deluc] to construct a new meteorological instrument, very desirable for acquiring a knowledge of atmospherical phaenomena, and which he called the Electric column." The ideas expressed differed so much from those prevalent in London that the council of the Royal Society "deemed it inexpedient to admit them into the *Transactions,*" and the article was also published in *Nicholson's Journal* (26 [1810]). This "electric column" (or electroscope) consisted of numerous disks of zinc foil and of paper silvered on one side only, piled horizontally in order of zinc, silver, and paper within a glass tube and firmly screwed together. When the uppermost silver was connected by a wire with the lowest zinc disk, an electric current passed along the wire. Today, however, it is hard to see the importance to meteorology and physics of this electric column, which was later improved by Giuseppe Zamboni. It is claimed as a "very valuable discovery" by Deluc's admirers but its principles, at least, had already been stated clearly by Volta on the Continent and probably also during his visit to England.

Deluc's other ventures into physics and chemistry showed all too clearly his inability to assess truly the quality of progress at home and abroad. He strenuously opposed the new chemical theory associated with Lavoisier and attempted to show in two memoirs on that theory, prefixed to his *Introduction à la physique terrestre par les fluides expansibles,* that meteorological phenomena strongly militate against it and in general that the hypothesis of the composition of water (the fundamental point in the theory) has maintained itself only by numerous other hypotheses which are in contradiction with known facts.

BIBLIOGRAPHY

I. ORIGINAL WORKS. Deluc wrote numerous long articles for periodicals, the chief being *Philosophical Transactions of the Royal Society; The Philosophical Magazine and Journal; British Critic,* especially 1793–1795, for letters addressed to J. F. Blumenbach; *Monthly Magazine; Monthly Review,* especially 1790 and 1791, for letters to Hutton; and *Nicholson's Journal.* Many of these letters or articles were republished later in the following books: *Recherches sur les modifications de l'atmosphère,* 2 vols. (Geneva, 1772), 4 vols. (Paris, 1784); *Lettres physiques et morales sur les montagnes et sur l'histoire de la terre et de l'homme.* (*Adressées à la reine de la Grande Bretagne*), 5 vols. (The Hague, 1779); *Idées sur la météorologie,* 2 vols. (Paris, 1786); *Lettres sur l'histoire physique de la terre* (Paris, 1798), abridged trans. by Henry de la Fite (London, 1831); *Lettres sur l'education religieuse de l'enfance* (Berlin, 1799); *Bacon tel qu'il est* (Berlin, 1800); *Lettres sur le christianisme adressées à M. le pasteur Teller* (Berlin–Hannover, 1801; 1803); *Précis de la philosophie de Bacon,* 2 vols. (Paris, 1802); *Introduction à la physique terrestre par les fluides expansibles,* 2 vols. (Paris, 1803); *Traité élémentaire sur le fluide électricogalvanique,* 2 vols. (Paris, 1804); *Traité élémentaire de géologie* (Paris, 1809), trans. by Henry de la Fite (London, 1809); *Geological Travels in the North of Europe and in England,* 3 vols. (London, 1810–1811); *Geological Travels in Some Parts of France, Switzerland and Germany* (London, 1813).

II. SECONDARY LITERATURE. On Deluc or his work, see R. J. Chorley, A. J. Dunn, and R. P. Beckinsale, *History of the Study of Landforms,* I (London, 1964), *passim; Encyclopaedia Britannica,* 11th ed. (1910–1911); a biography in *Gentleman's Magazine* (1817), pt. 2, 629; Charles C. Gillispie, *Genesis and Geology* (Cambridge–New York, 1951; 1959); *passim;* W. J. Harrison, in *Dictionary of National Biography,* XIV (1888), 328–329; C. Lyell, *Principles of Geology,* 12th ed. (London, 1875), I, 80; II, 506, 507; a biography in *The Philosophical Magazine and Journal,* **50,** no. 1 (Nov. 1817), 393–394; and K. A. von Zittel, *History of Geology and Palaeontology,* M. M. Ogilvie-Gordon, trans. (London, 1901), which mentions Deluc's theoretical articles in *Journal de physique*—this otherwise excellent assessment wrongly suggests that Deluc first proposed the term "geology" in its modern sense.

ROBERT P. BECKINSALE

DEMBOWSKI, ERCOLE (*b.* Milan, Italy, 12 December 1812; *d.* Monte di Albizzate [near Gallarate], Italy, 19 January 1881), *astronomy.*

Dembowski was the son of Jan Dembowski, a general of Napoleon, and Matilde Viscontini, an Italian noblewoman. Until he was thirty-one Dembowski was an officer in the Austrian navy; he made several expeditions to the Orient and participated in some minor battles in which he distinguished himself by gallantry.

Having left the navy Dembowski became interested in astronomy. In 1852 he built his own observatory in the village of San Giorgio a Cremano, near Naples, where he made excellent observations of double stars with only a modest telescope of five-inch aperture.

In 1870 he returned to Lombardy and constructed

at Monte di Albizzate, near Gallarate, a new observatory equipped with a telescope of seven-inch aperture by Merz and a meridian circle by Starke. With these new instruments he continued the revision of Struve's *Dorpat Catalogue* that he had begun in Naples. His energy and perseverance in this work produced an internally consistent series of measurements of the distances and positions of double and multiple stars that extends uninterruptedly over a period of twenty-five years.

Dembowski's first publication (1857) contains measurements of 127 double and triple stars selected from Struve's *Dorpat Catalogue;* each measurement represents the mean of ten observations made on the same night. In 1859 he published a reexamination of all the brightest stars in the *Dorpat Catalogue,* and in 1860 he listed—with great accuracy—the positions of fifty-four double stars. These measurements were used by Argelander in his fundamental work on proper motion of 250 stars.

Dembowski was a very active observer. He regularly published (mainly in *Astronomische Nachrichten*) the results of his observations for the benefit of the other astronomers working in the same field.

Following Dembowski's death Otto Struve credited him with having made about 20,000 observations. At the same time, G. V. Schiaparelli urged the Reale Accademia dei Lincei to undertake the collation and publication of Dembowski's scattered observations. The work, edited by Schiaparelli and Struve, appeared two years later.

Dembowski was elected an associate member of the Royal Astronomical Society (London) on 8 November 1878 and was awarded the Gold Medal of that society.

BIBLIOGRAPHY

I. ORIGINAL WORKS. Dembowski's earliest publication is "Misure micrometriche di 127 stelle doppie e triple di catalogo di Struve," in *Memorie della R. Accademia delle scienze di Napoli,* **2** (1855–1857), which was also published in *Astronomische Nachrichten;* his other observations were published in *Astronomische Nachrichten* volumes for 1859, 1864, 1866, 1869, 1870, 1872, 1873, 1874, 1875, and 1876; his collected observations are G. V. Schiaparelli and Otto Struve, eds., "Misure micrometriche di stelle doppie e multiple fatte negli anni 1852–1878," in *Atti della R. Accademia dei Lincei,* **16** and **17** (1883–1884).

II. SECONDARY LITERATURE. The award to Dembowski of the Gold Medal is noted in *Monthly Notices of the Royal Astronomical Society,* **38** (1878), 249; see also obituaries in *Astronomische Nachrichten* (1881) and in *Monthly Notices of the Royal Astronomical Society,* **42** (1882), 148.

GUGLIELMO RIGHINI

DEMOCRITUS (*b.* Abdera, Thrace, *fl.* late fifth century B.C.), *physics, mathematics.*

There were two main chronologies current in antiquity for Democritus. According to the first, which was followed by Epicurus among others, Democritus was the teacher of the Sophist Protagoras of Abdera and was born soon after 500 B.C. and died about 404 B.C. The other chronology puts his birth about 460 B.C., making him a younger contemporary of Socrates and a generation or more younger than Protagoras; in this case, the tradition that he lived to a great age would bring his death well into the fourth century B.C. According to Democritus' own words, he was a young man when Anaxagoras was old, and he may actually have said that he was younger by forty years. Although there was also more than one ancient chronology for Anaxagoras, this statement probably supports the later dates for Democritus, and these have usually been accepted by modern scholars. The question is an important one for our understanding of the history of thought in the fifth century B.C., and it is unfortunate that the occurrence of the name Democritus, presumably as a magistrate, on a fifth-century tetradrachm of Abdera does not help to settle the question, because we cannot be certain that it is the name of the Democritus here discussed nor can the tetradrachm be dated with certainty earlier than 430 B.C. (this would fit with either chronological scheme).

Most of the stories about Democritus are worthless later inventions, but it is probable that he was well-to-do, and stories of extensive travels may have a foundation in fact. He is reported to have said that he visited Athens, but no one knew him there, and from Cicero and Horace we learn that—at least in their time—he was known as the "laughing philosopher" because of his amusement at the follies of mankind. His only certainly attested teacher was Leucippus. The titles of more than sixty writings are preserved from a catalog that probably represented the holdings of the library at Alexandria. Of these we have only some 300 alleged quotations, many of which may not be genuine. More valuable for the understanding of Democritus' theories are the accounts given by Aristotle, Theophrastus, and the later doxographic tradition. Democritus left pupils who continued the tradition of his teachings and one of them, Nausiphanes, was the teacher of Epicurus. Epicureanism represents a further elaboration of the physical theories of Democritus, and surviving writings of Epicurus and others provide further interpretations and sometimes specific information about earlier atomist doctrines.

According to Posidonius in the first century B.C., the theory of atoms was a very old one and went back to a Phoenician named Mōchus, who lived before the Trojan War, in the second millennium B.C. According to others, Democritus was a pupil of Persian magi and Chaldean astrologers, either as a boy in his native Abdera or later in Egypt. Both stories seem to have originated only in the third century B.C. and to be part of the wholesale attempts to derive Greek thought from Oriental sources that followed the "discovery" of the East resulting from the establishment of Alexander's empire. More intriguing is the fact that certain Indian thinkers arrived at an atomic explanation of the universe, which is expounded in the Vaiśeṣika Sūtra and is interpreted by the aphorisms of Kanada. However, the Vaiśeṣika atoms are not quality-free but correspond to the four elements; nor is soul made from these atoms. Moreover, the date of the first appearance of the doctrine in India is probably subsequent to the founding of the Greek kingdom of Bactria, so that coincidences could be due to Greek influences on Indian thought. There is no early evidence of external sources for Democritus' thought; these are not needed, because the doctrines can be shown to have arisen naturally and almost inevitably as a result of the way in which the problems of explaining the physical universe had been formulated by Democritus' immediate predecessors among the pre-Socratics, who were of course Greeks. Consequently, Aristotle is probably right (*De generatione et corruptione,* 325a23 ff.) in explaining his views as developed in reply to the doctrines of the Eleatics. This need not exclude the possibility that the atomists were also influenced by what is sometimes called Pythagorean number-atomism, although whether this preceded or arose only after the time of Leucippus remains uncertain, and it is clear that Democritus did not invent atomism but received the essentials of the doctrine from Leucippus.

By the middle of the fifth century B.C., it seemed to many thinkers that Parmenides, the founder of the Eleatic school, had proved that nothing can come into being out of that which is not, and that anything which is cannot alter, because that would involve its becoming that which is not. Previous attempts to explain the physical universe as derived from one or more primary substances were thus doomed to failure, as they all involved change in the primary substances and so violated Parmenides' conclusions. Anaxagoras, at least in one view of his doctrine, made a heroic attempt to escape from the difficulty by supposing that all substances were always present in all other substances and that apparent change was simply the emergence of the required substance—which had been present unnoticed all the time. The atomism of Democritus was similar in its approach but went

further in depriving the primary constituents of most, but not all, of the qualities apparent in objects derived from them. Moreover, Leucippus had boldly accepted empty space or void—the existence of which the Eleatics regarded as impossible because it would be that which is not—as necessary to make movement possible.

Atoms and void are the bases of Democritus' system for explaining the universe: solid corporeal atoms, infinite in number and shape, differing in size, but otherwise lacking in sensible qualities, were originally scattered throughout infinite void. In general, the atoms were so small as to be invisible. (They were all invisible for Epicurus, but later sources raise the possibility that for Democritus some exceptional atoms may have been large enough to be seen or even that an individual atom might be as big as the cosmos.) The atoms are physically indivisible—this is the meaning of the name *atomos,* which, while not surviving in the fragments of Democritus, must certainly have been used by him. Whether the atoms were conceptually or mathematically indivisible as well as physically is a matter of dispute. But they were certainly extended and indestructible, so that if he thought about it Democritus ought not to have denied mathematical divisibility, especially as the atoms' variety of shape implied the concept of parts within each physically indivisible atom. They are homogeneous in substance, contain no void and no interstices, and are in perpetual motion in the infinitely extended void, probably moving equally in all directions.

When a group of atoms becomes isolated, a whirl is produced which causes like atoms to tend toward like. Within a kind of membrane or garment, as it were, woven out of hook-shaped atoms, there develops a spherical structure which eventually contains earth, sky, and heavenly bodies—in other words a spherical cosmos. The only detailed description of the process ascribes it to Leucippus (Diogenes Laertius, IX, 30 ff.), but there is no reason to doubt that it was repeated by Democritus. There is no limit to the number of atoms nor to the amount of void, and so not one cosmos but many are formed. Some dissolved again before the formation of our cosmos; others coexist with ours, some larger and some smaller, some without sun or moon, and some without living creatures, plants, or moisture. From time to time a cosmos is destroyed by collision with another.

Our earth and everything in it, like everything elsewhere, is compounded of atoms and void, and there are no other constituents of the universe of any kind. Apart from differences in shape, atoms differ in arrangement and position. As Aristotle says, the letter *A* differs from *N* in shape; *AN* from *NA* in arrangement; and Z from N in position, although both have the same shape. We must add, although Aristotle does not say so here, that the spacings between atoms may vary from the zero space of actual contact through increasing distances apart. Soft and yielding bodies and bodies light in weight contain more void than heavier or harder objects of equal extent. Iron is lighter than lead because it has more void, but it is harder because it is denser than lead at particular points, the void not being distributed evenly throughout, as is the case with lead. It is probable that for Democritus the atoms when entangled do not cease to be in motion (their individual movement is naturally less extensive), but they participate in movements of the object of which they are a part. It appears that atoms were not regarded as possessing weight in their own right; this was Epicurus' innovation. But physical objects possess weight, and according to Aristotle, atoms are heavier in proportion to their excess of bulk. Objects as a whole are heavier the greater the proportion of atoms to void. It may be that weight operates only in a developed world and is the result of a tendency of compound objects to move toward the center of a whirl. For Democritus all movement and all change are due to "necessity," but this is an internal cause and not an agency operating from without: it is the necessary result of the natural movement of the atoms. All events are determined, and if Cicero is right at all in saying Democritus attributed events to chance, this can have meant only that they could not be predicted, not that they were not determined.

The perceived qualitative differences between objects depend upon the nature and arrangement of the relevant atoms and void. The importance and novelty of this doctrine were fully appreciated by Theophrastus, who discussed it at some length in his surviving *De sensibus*. It might have seemed sufficient answer to Parmenides' challenging argument to have said that secondary qualities such as colors and tastes were produced by the appropriate arrangement of atoms in the sense that they were present in any object possessing the appropriate atomic configuration and would be altered or disappear when the configuration changed. But Theophrastus complains that Democritus is inconsistent on this point and that, while explaining sensations causally in terms of configurations, he insists that the perceived qualities depend upon the state of the percipient—for example, his health—to such an extent that the qualities exist not in the object but only in the percipient at the time he is perceiving them. According to Sextus Empiricus, Protagoras, in his "Man is the measure" doctrine, had held that there are present in actual objects multiple qualities which are selectively perceived by different percipients. Democritus is said (fr. 156) to have criticized the doctrine of Protagoras at great length, and it could be that he carried the relativism of Protagoras one step further by supposing that secondary qualities did not exist in the configuration of atoms which constitute a thing but only in the consciousness of the percipient. But not all accept Sextus Empiricus' account of Protagoras on this point.

We lack details of many aspects of Democritus' cosmology. The earth is flat and elongated—twice as long as it is broad. Although earlier it strayed about, it is now stationary at the center of the universe. The angle between zenith and celestial pole is explained by the tilting of the earth because the warmer air to the south—under the earth—offered less support than that in the north. Earthquakes are caused by heavy rain or drought changing the amount of water in the cavities of the earth. While some explanations of meteorological phenomena were offered in terms of the theory of atoms (for example, the attraction of like atoms to like as an explanation of magnetism), in general Democritus seems to have followed traditional explanations drawn from earlier pre-Socratics, above all from Anaximander. Unlike Leucippus, who put the sun's orbit outermost in the heavens, Democritus had the normal order of fixed stars, planets, sun, Venus, moon. The moon, like the earth, contained valleys and glens, and its light was derived from the sun.

Two particularly quick-moving constituents of the universe, fire and soul, were for Democritus composed of spherical atoms. Spherical atoms are not themselves either fire or soul but become such by the suitable aggregation of a number of themselves. Such aggregation cannot be by entanglement, which is not possible with spherical atoms, but only by the principle of the attraction of like to like. Whereas air, water, and perhaps earth, and things containing them, were regarded as conglomerations of atoms of all shapes, only the one shape seems to have occurred in fire and soul. Aristotle more than once speaks as if soul and fire were identical, and he adds that the soul can be fed by breathing in suitable atoms from the air around us. In this way, losses of soul atoms from the body can be replaced. When we can no longer breathe, the pressure from the atmosphere outside continues to squeeze out the soul atoms from the body and death results. A slight excess of loss over replacement produces sleep only and not death. Even when death results, the loss of soul atoms takes time, so that some functions, such as growth of hair and nails, continue for a while in the tomb; a certain degree of sensation may also continue for a time, and in exceptional cases, even resuscitation may be possible. We do not know the contents of the work *On Those in Hades,* attributed to Democritus, except that it included reference to such resuscitations.

Within the living body, soul atoms are distributed throughout the whole in such a way that single atoms of the soul and body alternate, and it has sometimes been said that this involves treating isolated atoms as soul atoms and so reintroducing qualities into individual atoms. But such an alternation could be achieved within a lattice pattern of one kind or another for the soul atoms, so that there is no actual inconsistency. These soul atoms are the immediate source of life, warmth, and motion in a living body. In addition to the soul atoms dispersed throughout the body, there is another part of the soul, the mind, located in one part of the body, namely the head.

Sensation for Democritus was based upon touch and was due to images entering the sense organs from outside and producing alterations in the percipient. Sensation is thus the result of the interaction of image and organ. In the case of flavors, there is always a multitude of configurations of atoms present in what is tasted, but the preponderant configuration exerts the greatest influence and determines the flavor tasted, the result being influenced also by the state of the sense organs. In the case of sight, images continually stream off the objects, which are somehow imprinted —by stamping, as it were—on the intervening air. This imprinted air is then carried to the eyes, where its configuration produces the sensation of color. A similar analysis seems to have been offered for hearing and perhaps for smell. Taste, however, entails direct contact between organ and object: large, rough, polygonal shapes produce astringent flavors, and so on.

Thought, like sensation, is the result of a disturbance of the soul atoms by configurations of atoms from outside; it is what occurs when the soul achieves a fresh balance after the movement which is sensation. The details of the process are obscure, and the text of Theophrastus' description is uncertain. But there is no sure evidence to suggest that Democritus held the later theory of Epicurus that it is possible for certain externally originating images to bypass the senses and secure direct access to the mind in thought. For Democritus, thought follows after sensation, and we may believe that Democritus expressed his real view when he said (fr. 125) that the mind takes its evidence from the senses and then seeks to overthrow them, but that the overthrow is a fall for the mind also. Nonetheless, in an important fragment (fr. 11) Democritus did claim that there were two kinds of knowledge, one genuine or legitimate, and the other bastard. To the bastard belong the senses; genuine

knowledge operates on objects too fine for any sense to grasp. This must surely refer to our knowledge of the atomic theory, including the imperceptible atoms and void of which things are composed, but we do not know what mechanical procedure, if any, Democritus envisaged for the acquisition of such knowledge.

It follows from the above view of the soul and the way it leaves the body at death that there is no survival of the individual soul, although the soul atoms themselves survive because, like all atoms, they are indestructible. It might have been expected that this approach would shed doubt on the existence of gods and spirits, especially since we are told that Democritus attributed early man's fear of the gods to his misunderstanding of natural phenomena such as lightning and eclipses. But he accepted that images of beings both beneficent and maleficent, destructible and yet able to foretell the future while being seen and heard, come to men apparently out of the air itself, without any more ultimate source. We do not know what doctrine lies behind this, but it is likely that there was no external source posited for these images other than the soul atoms at large in the air.

The list of Democritus' writings contains the titles of a number of works on mathematics, and it is clear from the few surviving scattered references that his mathematical interests were not inconsiderable. Protagoras had argued that the tangent touches the circle not at one point but over a distance. Democritus treated the sphere as "all angle," and Simplicius explained this as meaning that what is bent is an angle and the sphere is bent all over. It is inferred that he supposed that the sphere is really a polyhedron with imperceptibly small faces, presumably because a physical sphere involves atoms which cannot be further broken down. In such a case he would be in agreement with Protagoras as to the actual relation between tangent and circle while in disagreement as to the apparent relation. But with atoms in an infinite variety of shapes, there is no reason why Democritus could not have posited a perfect physical sphere made up of atoms of indivisible magnitude but with curved faces. In any case Democritus could probably distinguish a physical from a mathematical sphere well enough.

Of very great interest is Democritus' discussion of the question whether the two contiguous surfaces produced by slicing a cone horizontally are equal or unequal. If equal, it might seem that the cone is a cylinder, while if unequal, the cone becomes steplike and uneven (fr. 155). Chrysippus the Stoic, when discussing Democritus' doctrine, declared that Democritus was unaware of the true answer—namely, so he claimed, that the surfaces are neither equal nor unequal. Unfortunately what Democritus' view was remains in doubt. Some suppose that he argued for a stepped physical cone; others that he regarded the dilemma as genuine; and still others that he considered them equal, at least as far as mathematics was concerned. Archimedes records that Democritus was concerned with the ratios of size between cylinders, pyramids, and prisms of the same base and height. While this is evidence of further interest in problems associated with cones of the kind that were so important for the subsequent history of mathematics, we do not actually know the nature of Democritus' discussions concerning them.

Tantalizing references to individual doctrines and the titles of a number of his writings have suggested to some that Democritus' biological work rivaled Aristotle's in both comprehensiveness and attention to detail. The indications that survive do not for the most part suggest that he made any very particular application of atomic theories to biology, and it is probable that his clearly extensive writings were essentially within the general framework of Ionian speculation. More we cannot say through lack of positive information.

Later writers—as well as some from the fifth and fourth centuries B.C.—preserve details which all seem to come from a single account of the origins and development of human civilization. They have in common not only various particular points but also a basic conception—namely, that civilization developed from lower levels to higher, which contrasted strongly with the dominant view that human history represented a continuous decline from an original golden age. The clearest version of this history of culture survives in the *Bibliotheca historica* of Diodorus (bk. I, ch. 8), written in the age of Cicero. It is clear that Democritus held a similar view, and it is possible, although by no means certain, that he originated the whole tradition. Certain features of it, however, are already in Aeschylus' picture of Prometheus and probably in the writings of Protagoras summarized in Plato's dialogue named after the Sophist. Part of Democritus' treatment of the evolution of culture concerned the origin and development of language, taking the view that names were not natural but conventional.

Special problems affect the reconstruction of Democritus' ethical doctrines, to which a very large part of the surviving fragments relate. Many of these are attributed in the manuscript tradition not to Democritus but to an otherwise unknown Democrates, so that their authority for the reconstruction of the views of Democritus is uncertain. Most of the frag-

ments are extremely commonplace, and hardly any are related to atomic theory. The doxographic tradition does, however, suggest that he had a general theory of *euthymia* ("cheerfulness" or "contentment") as the end of ethics. It was based on a physical state, the actual constitution of the body at any one time, of which the external expression is pleasure or enjoyment when the state itself is satisfactory. Even this much is a matter of conjecture, and we do not know how it was all worked out by Democritus.

Most of the fragments dealing with what we would call political questions are as traditional in content as those dealing with ethics. He seems to have had no doubts about the importance of law, although its function was limited to preventing one man from injuring another. It is inferior to encouragement and persuasion, but "it is right to obey the law, the ruler and the man who is wiser" (fr. 47). Democritus had declared that secondary qualities of perception, such as sweetness, existed only by *nomos,* not in reality, and *nomos,* which means "custom" or "convention," is also the word used for "law." It is perhaps not going too far to say that in ethics and politics, just as in physics, Democritus was searching for a truth and a reality behind or beyond the world of appearances; but at the same time, he wished to reaffirm the importance of changing phenomena as the product of an unchanging reality. It is probable that political obedience to the law was regarded as rooted in the well-being of the soul, just as wrongdoing is not to be justified by the thought that one will escape discovery.

BIBLIOGRAPHY

The fragments and testimonia are collected in H. Diels and W. Kranz, *Die Fragmente der Vorsokratiker,* 6th ed., 3 vols. (Berlin, 1951–1952), vol. II. There is a translation of the fragments by K. Freeman, *Ancilla to the Pre-Socratic Philosophers* (Cambridge, Mass., 1966), and the most important are translated and discussed in G. S. Kirk and J. E. Raven, *The Presocratic Philosophers* (Cambridge, 1957).

For discussions of Democritus, see V. E. Alfieri, *Gli atomisti* (Bari, 1936) and *Atomos idea, l'origine del concetto dell'atomo nel pensiero greco* (Florence, 1953); C. Bailey, *The Greek Atomists and Epicurus* (Oxford, 1928); T. Cole, *Democritus and the Sources of Greek Anthropology,* American Philological Association Monograph (1967); W. K. C. Guthrie, *History of Greek Philosophy,* vol. II (Cambridge, 1965), ch. 8; A. B. Keith, *Indian Logic and Atomism* (Oxford, 1921); H. Langerbeck, *Doxis Epirhysmie, Studien zu Demokrits Ethik und Erkenntnislehre* (Berlin, 1935); S. Luria, *Zur Frage der materialistischen Begründung der Ethik bei Demokrit* (Berlin, 1964); J. Mau, *Zum Problem der Infinitesimalen bei den antiken Atomisten,* 2nd ed.

(Berlin, 1957); P. Natorp, *Die Ethik des Demokritos, Texte und Untersuchungen* (Marburg, 1893); W. Schmid, *Geschichte der griechischen Literatur,* V (Munich, 1948), 236–350; G. Vlastos, "Ethics and Physics in Democritus," in *Philosophical Review,* **54** (1945), 578–592, and **55** (1946), 53–63.

G. B. KERFERD

DE MOIVRE, ABRAHAM. See **Moivre, Abraham de.**

DE MORGAN, AUGUSTUS (*b.* Madura, Madras presidency, India, June 1806; *d.* London, England, 18 March 1871), *mathematics.*

De Morgan's father was a colonel in the Indian Army; and his mother was the daughter of John Dodson, a pupil and friend of Abraham de Moivre, and granddaughter of James Dodson, author of the *Mathematical Canon.* At the age of seven months De Morgan was brought to England, where his family settled first at Worcester and then at Taunton. He attended a succession of private schools at which he acquired a mastery of Latin, Greek, and Hebrew and a strong interest in mathematics before the age of fourteen. He also acquired an intense dislike for cramming, examinations, and orthodox theology.

De Morgan entered Trinity College, Cambridge, in February 1823 and placed first in the first-class division in his second year; he was disappointed, however, to graduate only as fourth wrangler in 1827. After contemplating a career in either medicine or law, De Morgan successfully applied for the chair of mathematics at the newly formed University College, London, in 1828 on the strong recommendation of his former tutors, who included Airy and Peacock. When, in 1831, the college council dismissed the professor of anatomy without giving reasons, he immediately resigned on principle. He resumed in 1836, on the accidental death of his successor, and remained there until a second resignation in 1866.

De Morgan's life was characterized by powerful religious convictions. While admitting a personal faith in Jesus Christ, he abhorred any suspicion of hypocrisy or sectarianism and on these grounds refused an M.A., a fellowship at Cambridge, and ordination. In 1837 he married Sophia Elizabeth Frend, who wrote his biography in 1882. De Morgan was never wealthy; and his researches into all branches of knowledge, together with his prolific output of writing, left little time for social or family life. However, he was well known for his humor, range of knowledge, and sweetness of disposition.

In May 1828 De Morgan became a fellow of the Astronomical Society; he was elected to the council

in 1830, serving as secretary (1831–1838; 1848–1854). He helped to found the London Mathematical Society, becoming its first president and giving the inaugural lecture in 1865. He was also an influential member of the Society for the Diffusion of Useful Knowledge from 1826. De Morgan was a prolific writer, contributing no fewer than 850 articles (one-sixth of the total production) to the *Penny Cyclopaedia* and writing regularly for at least fifteen periodicals.

De Morgan exerted a considerable influence on the development of mathematics in the nineteenth century. As a teacher he sought to demonstrate principles rather than techniques; and his pupils, who included Todhunter, Routh, and Sylvester, acquired from him a great love of the subject. He wrote textbooks on the elements of arithmetic, algebra, trigonometry, calculus, complex numbers, probability, and logic. These books are characterized by meticulous attention to detail, enunciation of fundamental principles, and clear logical presentation.

De Morgan's original contributions to mathematics were mainly in the fields of analysis and logic. In an article written in 1838, he defined and invented the term "mathematical induction" to describe a process that previously had been used —without much clarity—by mathematicians.

In *The Differential and Integral Calculus* (1842) there is a good discussion of fundamental principles with a definition of the limit which is probably the first precise analytical formulation of Cauchy's somewhat intuitive concept. The same work contains a discussion of infinite series with an original rule to determine convergence precisely when simpler tests fail. De Morgan's rule, which is proved rigorously, is that if the series is given by

$$\sum \frac{1}{\phi(n)},$$

then if

$$e = \lim_{n \to \infty} \frac{n\phi'(n)}{\phi(n)},$$

the series converges for $e > 1$ but diverges for $e \leq 1$.

Among his other mathematical work is a system that De Morgan described as "double algebra." This helped to give a complete geometrical interpretation of the properties of complex numbers and, as Sir William Rowan Hamilton acknowledged, suggested the idea of quaternions.

De Morgan's greatest contribution to scientific knowledge undoubtedly lay in his logical researches; and the subsequent development of symbolic logic, with its powerful influences on both philosophy and technology, owes much to his fundamental work. He believed that the traditional method of argument using the Aristotelian syllogism was inadequate in reasoning that involved quantity. As an example De Morgan presented the following argument:

> In a particular company of men,
>   most men have coats
>   most men have waistcoats
> ∴ some men have both coats and waistcoats.

He asserted that it was not possible to demonstrate this true argument by means of any of the normally accepted Aristotelian syllogisms.

The first attempt to extend classical logic by means of quantifying the predicate and reformulating logical statements in mathematical terms was made by George Bentham in 1827. He rephrased the statement "Every X is a Y" into the equation "X in toto = Y ex parte" with the algebraic notation "$tX = pY$." It was more usual at this time, however, for logicians to make more classical attempts to broaden the Aristotelian syllogistic; and De Morgan's work, which commenced in the 1840's, can be seen as the bridge between this older approach and Boole's analytical formulation. Boole acknowledged his debt to De Morgan and Hamilton in the preface to his first logical work, *The Mathematical Analysis of Logic* (1847).

The Scottish philosopher Sir William Hamilton (not to be confused with Sir William Rowan Hamilton) worked out a system for quantifying the predicate a short time before De Morgan did and unjustly accused him of plagiarism. He had no shred of evidence to support his charge, and De Morgan's work was superior to his in both analytical formulation and subsequent development.

De Morgan invented notations, which he sometimes varied, to describe simple propositions. Objects with certain properties were denoted by capital letters $X, Y, Z, \cdots$ and those without this property by the corresponding small letters $x, y, z, \cdots$. One of his notations was

A	Every X is a Y	as	$X)Y$
E	No X is a Y	as	$X.Y$
I	Some X's are Y's	as	XY
O	Some X's are not Y's	as	$X:Y,$

the symbols A, E, I, O, having their usual Aristotelian meaning. He then worked out rules to establish valid syllogistic inferences. Such results were then written in the form

$$X)Y + Y)Z = X)Z$$
$$Y:X + Y)Z = Z:X$$
$$X)Y + Z)Y = xz,$$

and so on. This notation was superseded by Boole's more algebraic one, but it helped De Morgan to establish valid inferences not always obtainable through the traditional rules. Using the notation of Boolean algebra, the two equations $(A \cap B)' = A' \cup B'$ and $(A \cup B)' = A' \cap B'$ are still referred to as the De Morgan formulas.

De Morgan was also the first logician to present a logic of relations. In a paper written in 1860 he used the notation $X..LY$ to represent the statement that X is one of the objects in the relation L to Y, while $X.LY$ meant that X was not any of the L's of Y. He also presented the idea $X..(LM)Y$ as the composition of two relations L,M, and of the inverse relation L^{-1}. This extension of the idea of subject and predicate was not adopted by any of De Morgan's successors, and the idea lapsed until Benjamin Peirce's work of 1883.

De Morgan was steeped in the history of mathematics. He wrote biographies of Newton and Halley and published an index of the correspondence of scientific men of the seventeenth century. He believed that the work of both minor and major mathematicians was essential for an assessment of mathematical development, a principle shown most clearly in his *Arithmetical Books* (1847). This work describes the many arithmetical books in the author's possession, refers to the work of 1,580 arithmeticians, and contains detailed digressions on such subjects as the length of a foot and the authorship of the popular *Cocker's Arithmetick*. De Morgan's book was written at a time when accurate bibliography was in its infancy and was probably the first significant work of scientific bibliography. Despite a lack of means, he collected a library of over 3,000 scientific books, which is now at the London University library.

De Morgan's peripheral mathematical interests included a powerful advocacy of decimal coinage; an almanac giving the dates of the new moon from 2000 B.C. to A.D. 2000; a curious work entitled *Budget of Paradoxes,* which considers, among other things, the work of would-be circle squarers; and a standard work on the theory of probability applied to life contingencies that is highly regarded in insurance literature.

BIBLIOGRAPHY

I. ORIGINAL WORKS. De Morgan's books include *The Elements of Arithmetic* (London, 1830); *Elements of Spherical Trigonometry* (London, 1834); *The Elements of Algebra Preliminary to the Differential Calculus, and Fit for the Higher Classes of Schools etc.* (London, 1835); *The Connexion of Number and Magnitude: An Attempt to Explain the Fifth Book of Euclid* (London, 1836); *Elements of Trig-*

onometry and Trigonometrical Analysis, Preliminary to the Differential Calculus (London, 1837); An Essay on Probabilities, and on Their Application to Life Contingencies and Insurance Offices (London, 1838); First Notions of Logic, Preparatory to the Study of Geometry (London, 1839); Arithmetical Books From the Invention of Printing to the Present Time. Being Brief Notices of a Large Number of Works Drawn up From Actual Inspection (London, 1847), repub. (London, 1967) with a biographical introduction by A. R. Hall; The Differential and Integral Calculus (London, 1842); Formal Logic: or The Calculus of Inference, Necessary and Probable (London, 1847); Trigonometry and Double Algebra (London, 1849); The Book of Almanacs With an Index of Reference, by Which the Almanac May Be Found for Every Year . . . up to A.D. 2000. With Means of Finding the Day of Any New or Full Moon From B.C. 2000 to A.D. 2000 (London, 1851); Syllabus of a Proposed System of Logic (London, 1860); and A Budget of Paradoxes (London, 1872).

Articles by De Morgan can be found in Quarterly Journal of Education (1831–1833); Cambridge Philosophical Transactions (1830–1868); Philosophical Magazine (1835–1852); Cambridge Mathematical Journal (1841–1845); Cambridge and Dublin Mathematical Journal (1846–1853); Quarterly Journal of Mathematics (1857–1858); Central Society of Education (1837–1839); The Mathematician (1850); and British Almanac and Companion (1831–1857). He also contributed to Smith's Classical Dictionary, Dublin Review, Encyclopaedia Metropolitana, and Penny Cyclopaedia.

II. Secondary Literature. On De Morgan or his work, see I. M. Bochenski, Formale Logik (Freiburg–Munich, 1956), pp. 306–307, 345–347, passim; S. De Morgan, Memoir of Augustus De Morgan . . . With Selections From His Letters (London, 1882); G. B. Halsted, "De Morgan as Logician," in Journal of Speculative Philosophy, **18** (1884), 1–9; and an obituary notice in Monthly Notices of the Royal Astronomical Society, **32** (1872), 112–118.

John M. Dubbey

DENIS, JEAN-BAPTISTE (b. Paris, France, 1640 [?]; d. Paris, 3 October 1704), medicine.

Denis was born in Paris, presumably in the 1640's. He was the son of a hydraulic engineer who was Louis XIV's chief engineer in charge of the works distributing the water of the Seine from the pumps at Marly to the fountains at Versailles.

Denis is said to have studied medicine at Montpellier (1), but no records of his inscription as a medical student or of the conferring upon him of a diploma as doctor in medicine can be found in the very complete archives of the Faculty of Medicine. Niceron says that he obtained "un bonnet de Docteur en cette Faculté" and that "il fut aggrégé à la Chambre Royale" (10). On the other hand, Martin de la Martinière, who was a physician in ordinary to the king, in a letter to Denis accuses him of taking the title of "maître" because of a "lettre de Médecine"

that he obtained in Rheims (2). Nothing has yet been found in Rheims indicating that he obtained such a degree. While in Paris he taught philosophy and mathematics, assuming the title of professor, which he placed at the head of most of his works. No evidence for a degree in mathematics or philosophy has yet been found.

Beginning in 1664, Denis gave public lectures in physics, mathematics, and medicine at his home on the quai des Grands-Augustins in Paris, and published these lectures as conference reports (7). He also joined the group surrounding Habert de Montmort, which met to discuss the new philosophy much like the groups in London that preceded the Royal Society. When the Académie des Sciences was established in 1666, the Montmort group did not participate and continued its own meetings independently of that body.

The discovery of the circulation of blood by William Harvey stimulated experiments on the circulation; intravenous injection was begun by Christopher Wren and Clarke in the 1650's. This was followed by the first trial of transfusion of blood in animals. After discussions at the Royal Society as early as its public meeting of 17 May 1665, an account of successful transfusion in dogs was given by Richard Lower in a letter written to Robert Boyle on 6 July 1665 and submitted by Boyle to the Royal Society. This led to another successful transfusion in November 1666 at the Royal Society (9).

When reports of these experiments reached Paris late in 1666 or early in 1667, the Académie des Sciences immediately set about repeating them, appointing a committee including Louis Gayant, an anatomist; Claude Perrault, the physician noted for the east facade (the Colonnade) of the Louvre; and Adrien Auzout, the astronomer. Gayant performed the first transfusion in Paris on 22 January 1667, using dogs. Transfusion also attracted the interest of the Montmort Academy, which apparently appointed Denis and Paul Emmerez, a surgeon from St.-Quentin, to carry out independent studies. On 3 March 1667 Denis performed a transfusion experiment on two dogs (8). On 2 April 1667 various experiments involving transfusion from three calves to three dogs were made. These were published in the Royal Society's Philosophical Transactions (11).

But it was the transfusion of blood in men which was of the greatest interest to Denis, gave him his celebrity, and started the greatest medical controversy of that time. In these experiments he was assisted by Paul Emmerez.

The first transfusion of blood in man was made on 15 June 1667, on a drowsy and feverish young man.

From a lamb he received about twelve ounces of blood, after which he "rapidly recovered from his lethargy, grew fatter and was an object of surprise and astonishment to all who knew him" (4).

The second transfusion was carried out on a forty-five-year-old chair bearer, a robust man who received the blood of a sheep (4). He returned to work the next day as if nothing had happened to him.

The recipient of the third transfusion was Baron Bonde, a young Swedish nobleman who fell ill in Paris while making a grand tour of Europe. He was in such a bad state that he had been abandoned by his physicians; and in despair, having heard of Denis's new cure, his family asked Denis to attempt transfusion of blood as a final recourse. After the first transfusion, which was from a calf, Bonde felt better and began to speak. This improvement lasted only a short time, however, and he died during a second transfusion.

The fourth transfusion patient was a madman, Antoine Mauroy (5), who died during a third transfusion. He may have been poisoned by his wife, who, perhaps to divert suspicion from herself or at the suggestion of the many Paris physicians antagonistic to Denis, accused Denis of having killed her husband. Denis brought the case before the court, and a judgment rendered on 17 April 1668 cleared him of any wrongdoing but forbade the practice of transfusion of blood in man without permission of the Paris Faculty of Medicine. Meanwhile, another transfusion had been made by Denis, on 10 February 1668, on a paralyzed woman. After this, however, the practice of transfusion faded out as suddenly as it had begun.

In 1673 Denis was invited to England by Charles II, who wished to learn about transfusion and other remedies purportedly discovered by Denis. He went to England and successfully treated the French ambassador and several personalities of the court. Despite offers to remain, he became dissatisfied and returned to Paris (10), where he continued his interest in science and mathematics (7) but never practiced medicine or again concerned himself with transfusion. He died suddenly on 3 October 1704.

BIBLIOGRAPHY

1. Jean Astruc, *Mémoires pour servir à l'histoire de la Faculté de Montpellier,* rev. by M. Lorry and P. G. Cavelier, V (Paris, 1767), 378.

2. Martin de la Martinière, *Remonstrances charitables du Sieur de la Martinière à Monsieur Denis* (Paris, 1668).

3. J.-B. Denis, *Lettre à M. L'Abbé Bourdelot . . . pour servir de réponse au Sr. Lamy et confirmer la transfusion du sang par de nouvelles expériences* (Paris, 1667).

4. J.-B. Denis, *Lettre escrite à . . . Montmor . . . touchant une nouvelle manière de guérir plusieurs maladies par la transfusion du sang, confirmée par deux expériences faites sur des hommes* (Paris, 1667).

5. J.-B. Denis, *Lettre escrite à M. . . . touchant une folie invétérée, qui a esté guérie depuis peu par la transfusion du sang* (Paris, 1668).

6. J.-B. Denis, *Lettre écrite à . . . Sorbière . . . touchant l'origine de la transfusion du sang, et la manière de la pratiquer sur les hommes* (Paris, 1668).

7. J.-B. Denis, *Recueil des mémoires et conférences qui ont été présentées à Monseigneur le Dauphin pendant l'année 1672 (1673–1674)* (Paris, 1672–1683).

8. J.-B. Denis, "Extrait d'une lettre de M. Denis, professeur de philosophie et de mathématique, sur la transfusion du sang. De Paris le 9. mars, 1667," in *Journal des sçavans,* **6** (1679).

9. Minutes of the Royal Society (16 Sept. 1663), p. 201.

10. J. P. Niceron, *Mémoires pour servir à l'histoire des hommes illustres de la république des lettres,* XXXVII (Paris, 1727), 77.

11. "An extract of the letter of Mr. Denis . . . touching the transfusion of blood, of April 2. 1667," in *Philosophical Transactions of the Royal Society,* **1,** no. 25 (6 May 1667), 453.

HEBBEL E. HOFF

DEPARCIEUX, ANTOINE (*b.* Clotet-de-Cessous, France, 28 October 1703; *d.* Paris, France, 2 September 1768), *mathematics.*

Deparcieux's father, Jean-Antoine, was a farmer; his mother was Jeanne Donzel. Orphaned in 1715, he was educated by his brother Pierre, who sent him at fifteen to the Jesuit college at Alès. In 1730, after finishing his studies, Deparcieux went to Paris, where he became a maker of sundials. He also investigated problems of hydraulics and conceived a plan for bringing the water of the Yvette River to Paris, which was carried out after his death. In 1746 he was admitted to membership in the Academy of Sciences.

In his *Nouveaux traités de trigonométrie rectiligne et sphérique* (Paris, 1741) Deparcieux gives a table of sines, tangents, and secants calculated to every minute and to seven places, and a table of logarithms of sines and tangents calculated to every ten minutes and to eight places. He also gives the formula for tan $a/2$ in the form of two proportions:

$$\sin s : \sin (s-c) = \sin (s-b) \sin (s-a) : x^2$$
$$\sin (s-a) : r = x : \tan \frac{A}{2},$$

but he did not use the words "cosine" and "cotangent." After long investigations of tontines, individual families, and religious communities, Deparcieux published his results in the famous *Essai sur les probabilités de la durée de la vie humaine* (Paris, 1746; suppl., 1760), one of the first statistical works of its kind. It consists of treatises on annuities, mortality, and life annuities. Deparcieux showed a real progress in his theoretical explanation of the properties of the tables of mortality. However, his tables, which were for a long time the only ones on life expectancies in France, indicated too small a value for the probable life expectancy at every age. He also made further inquiries on the concept of the mean life expectancy.

BIBLIOGRAPHY

On Deparcieux or his work, see J. Bertrand, *L'Académie des sciences et les académiciens de 1666 à 1793* (Paris, 1869), pp. 167, 168, 288, 289; A. von Braunmühl, *Vorlesungen über Geschichte der Trigonometrie*, II (Leipzig, 1903), 90; and G. F. Knapp, *Theorie des Bevölkerungs-Wechsels* (Brunswick, 1874), pp. 68–73.

H. L. L. BUSARD

DEPERET, CHARLES (*b.* Perpignan, France, 25 June 1854; *d.* Lyons, France, 18 May 1929), *paleontology, stratigraphy.*

After submitting a thesis on the Tertiary geology of his native province of Roussillon (1885), Depéret was appointed professor at the Faculté des Sciences of Lyons (1889) and subsequently served as its capable and influential dean, reappointed again and again, for thirty-three years. In 1893 he published, with F. Delafond, a monograph on the Tertiary geology of the Bresse region (between Lyons and Dijon), a work that quickly became a classic. His research on the Tertiary period, especially in the Rhone Valley (where he profited by Fontannes's studies) and in Spain, was accompanied by paleontological studies and often gave rise to detailed geological maps. His *Les transformations du monde animal* (1907), translated into English and German and often reprinted in French, clearly and accurately explains the great problems of paleontology.

Turning to Quaternary geology, until then very obscure, Depéret began in 1906 to present clear and theoretically valid syntheses for the entire world that were based on the theory of eustacy expounded by Eduard Suess and L. de Lamothe. In order to satisfy this theory he conceived of supposed geologic stages called Tyrrhenian and Milazzian and brought about acceptance in France of the notion of alluvial Quaternary terraces of relatively constant altitude (at 20, 30, 60, and 100 meters).

During his lifetime Depéret was considered, both in France and in the Mediterranean countries, to be one of the great masters of science, one whose ideas were adopted without question; extremely powerful

on the administrative level, he had a sense of authority as firm as it was courteous. But after his death it was gradually realized that his paleontological studies were often too hasty; carried away by his theories, which were built upon questionable hypotheses, he had neglected or modified facts inconsistent with those theories. Nevertheless, Depéret did succeed in training loyal disciples and, as a passionate fossil seeker, gathered invaluable paleontological collections for his Lyons laboratory.

BIBLIOGRAPHY

I. ORIGINAL WORKS. Depéret's principal works include *Description géologique du bassin tertiaire du Roussillon* (Paris, 1885); "Recherches sur la succession des faunes de vertébrés miocènes de la vallée du Rhône," in *Archives du Muséum d'histoire naturelle de Lyon,* 4 (1887), 45–313; "Les terrains tertiaires de la Bresse," in *Étude des gîtes minéraux de la France* (Paris, 1893), written with F. Delafond; *Les transformations du monde animal* (Paris, 1907; English trans. New York, 1909; German trans. Stuttgart, 1909); *Notice sur les travaux scientifiques de M. Ch. Depéret* (Lyons, 1913); and "La classification des temps quaternaires et ses rapports avec l'antiquité de l'homme en Europe," in *Revue générale des sciences pures et appliquées* (15 March 1923), 2–8, a résumé of ten notices entitled "Essai de coordination géologique des temps quaternaires" that were published in *Comptes rendus hebdomadaires des séances de l'Académie des sciences* between 1918 and 1922.

II. SECONDARY LITERATURE. On Depéret or his work, see Franck Bourdier, "Origine et succès d'une théorie géologique illusoire: L'eustatisme appliqué aux terrasses alluviales," in *Revue de géomorphologie dynamique,* 10 (1959), 16–29, with 146 references; M. Gignoux, "Charles Depéret," in *Bulletin de la Société géologique de France,* 4th ser., 30 (1930), 1043–1073, with a portrait and a bibliography of 223 titles; and F. Roman, "La vie et l'oeuvre de Charles Depéret," in *Revue de l'Université de Lyon* (July 1929), 304–322.

FRANCK BOURDIER

DEPREZ, MARCEL (*b.* Aillant-sur-Milleron, France, 29 December 1843; *d.* Vincennes, France, 16 October 1918), *engineering.*

Seldom mentioned in English chronologies, Deprez was a major innovator in many fields of technology. After graduating from the National School of Mines, he served as the school's secretary from 1866 to 1872. In this period he invented improved valve and indicator mechanisms for steam engines. During the 1870 siege of Paris he conducted pioneer researches, using instruments of his own creation, on the instantaneous gas pressure, metal strain, projectile velocity, and recoil motion produced in the firing of mortar cannon. For this work and for the invention of a railway dynamometer car he received two prizes from the Académie des Sciences.

Deprez was an early promoter, after 1875, of employing electric power in industry, and he collaborated with d'Arsonval and J. Carpentier in the design and manufacture of a wide variety of direct-current measuring instruments, and adapted small motors to manufacturing and domestic uses. He invented compound winding for voltage and speed stabilization in d.c. machines and showed how the operation of such machines could be fully determined from "open-circuit" and "short-circuit" characteristics.

Convinced of the commercial importance of transmitting power electrically, Deprez presented four dramatic and historic public demonstrations of d.c. electric power transmission, the first at Munich in 1881 and the last in 1886 when he sent seventy-five kilowatts over fifty kilometers of line from Creil to Paris. The 5,800-volt dynamo used was of his own design. He and Carpentier foresaw the advantages of high-voltage a.c. power transmission using transformers and patented the principle in 1881, but they did not develop it commercially.

Deprez was elected to the Académie des Sciences in 1886. In 1890 he was appointed professor of electrotechnology at the Conservatoire des Arts et Métiers. Deprez's work in the areas cited and on other topics, such as the laws of friction, the mechanical equivalent of heat, planimeters, and electric clocks, is described in more than sixty scientific papers, mainly in the *Comptes rendus de l'Académie des sciences.*

BIBLIOGRAPHY

Deprez's works are listed in Poggendorff. *Revue générale de l'électricité* published a special issue (Paris, 1935) commemorating the fiftieth anniversary of Deprez's power-transmission demonstration at Creil. Two obituaries appear in *Comptes rendus de l'Académie des sciences,* **167** (1918), 570–574; and *Electrician* (17 Jan. 1919).

ROBERT A. CHIPMAN

DERHAM, WILLIAM (*b.* Stoughton, Worcestershire, England, 26 November 1657; *d.* Upminster, Essex, England, 5 April 1735), *natural history, natural theology.*

Derham attended Blockley Grammar School and, on 14 May 1675, entered Trinity College, Oxford. He graduated B.A. on 28 January 1679. Ralph Bathurst, the president of the college, recommended him to Bishop Seth Ward, who obtained a chaplaincy for

him. He was ordained a deacon of the Church of England in 1681 and priest in 1682, when he was appointed vicar of Wargrave. In 1689 he became vicar of Upminster, not far from London, where he lived for the rest of his life. In 1702 Derham was elected a fellow of the Royal Society, and in 1711–1712 he delivered the course of Boyle Lectures. On the accession of George I in 1714, he was made chaplain to the Prince of Wales, later George II. In 1716 he became a canon of Windsor, and in 1730 was awarded the degree of doctor of divinity by the University of Oxford. Tall, healthy, and strong, he acted as physician as well as parson in Upminster. His eldest son, William, became president of St. John's College, Oxford.

Derham published a number of papers in the *Philosophical Transactions of the Royal Society* on meteorology, on astronomy, and on natural history—his paper of 1724 on the sexes of wasps was admired. But it is for his editing of works by Robert Hooke and John Ray, and for his books on natural theology, that he is remembered. Ray's *Synopsis methodica avium et piscium* had been sent to a bookseller in 1694; but the latter was in no hurry to publish it, and it remained in manuscript on his shelves until the firm went out of business. On its rediscovery, Derham saw it through the press in 1713. Also in 1713 he supervised a new edition of Ray's *Physico-Theological Discourses* and in 1714, a new edition of his *Wisdom of God.* In 1718 Derham edited Ray's *Philosophical Letters* and wrote a short biography of Ray, which did not appear in print until 1760. After Hooke's death Richard Waller edited some of his papers, publishing them in 1705 as *The Posthumous Works of Robert Hooke.* On Waller's death, Hooke's papers passed to Derham, who in 1726 published them as *Philosophical Experiments . . . of . . . Dr. Robert Hooke.* Also in 1726 he prepared a new edition of *Miscellanea curiosa,* a collection of important scientific papers from various sources.

Of Derham's own works, those of greatest interest are *The Artificial Clockmaker, Physico-Theology* (Boyle Lectures), and *Astro-Theology,* all of which were very successful and went through many editions. *Physico-Theology* was translated into French, Swedish, and German, and *Astro-Theology* into German. None of them shows great originality. *The Artificial Clockmaker* is a useful manual containing some of Hooke's ideas on clockwork, notably on the spiral spring balance, which Hooke claimed to have invented before Christiaan Huygens.

Physico-Theology owes much to Ray's *Wisdom of God,* but it became better known in the eighteenth century than Ray's book and was heavily used by

William Paley. Derham's tone was bland; he sought only to show that this is the best of all possible worlds. Venomous reptiles were difficult to account for, but he reflected that they were mostly to be found in heathen countries. Taken seriously, however, the book abounds in arguments from design to God; and anybody who read it would have acquired a respectable amount of natural history—as one would expect from an author who was himself a naturalist and a friend of John Ray.

Astro-Theology was a similar attempt to argue from astronomy to God, and here again Derham was as well qualified as anybody to do it. He made observations with some of Huygens' telescopes and knew Halley and Newton. The main interest of the book is the distinction between the Copernican system and the new system, in which the universe was infinite and every star a sun, presumably surrounded by populated planets. Although works of natural theology such as these seem tedious in style, they do give a useful glimpse of the background of eighteenth-century science in England.

BIBLIOGRAPHY

I. ORIGINAL WORKS. Derham's writings include *The Artificial Clockmaker* (London, 1696); *Physico-Theology* (London, 1713); and *Astro-Theology* (London, 1714). He edited *Philosophical Experiments and Observations of the Late Eminent Dr. Robert Hooke, and Other Eminent Virtuoso's in His Time* (London, 1726).

II. SECONDARY LITERATURE. On Derham or his work, see A. D. Atkinson, "William Derham, F.R.S. (1657–1735)," in *Annals of Science,* **8** (1952), 368–392; the article in *Biographia britannica,* III; and C. E. Raven, *John Ray, Naturalist* (Cambridge, 1950), pp. xiii–xv.

D. M. KNIGHT

DEROSNE, (LOUIS-)CHARLES (*b.* Paris, France, 23 January 1780; *d.* Paris, 21 September 1846), *chemistry, industrial technology, invention.*

Derosne belonged to a family of pharmacists. His father, François Derosne, was associated with the famous Paris apothecary Louis-Claude Cadet (known as Cadet de Gassicourt). After the death of François Derosne in 1796, the Cadet-Derosne pharmacy on the rue St. Honoré was taken over by Charles's older brother, Jean-François, whose chemical analysis of opium (published in the *Annales de chimie* in 1803) foreshadowed the emergence of alkaloid chemistry as an important field of research. For a time Charles was associated with Jean-François in the practice of pharmacy and in several joint scientific and technological projects. Perhaps the most important result of their collaboration was the investigation in 1807 of the properties of acetone, which they prepared by distilling copper acetate. Both brothers were admitted to the Academy of Medicine in Paris, Jean-François in 1821 and Charles in 1823.

Pharmacy proved too confining, however, for Charles Derosne, who very early in his career demonstrated a remarkable ability for technological innovation. A lifelong interest in improving methods of sugar production led him to introduce new techniques and equipment into sugar technology. In 1808 he refined crude sugar with alcohol, and by 1811 he was able to improve on the methods of beet sugar manufacture described by the contemporary German chemists S. F. Hermbstädt and F. C. Achard. Derosne's innovations and observations were included in his notes to the French translation of Achard's treatise on beet sugar manufacture, published with D. Angar in 1812. Derosne prepared animal charcoal and used it to purify sugar syrup. In 1817 he invented a continuous distillation apparatus and shortly thereafter began to produce other machinery of value in sugar refining at his plant in Chaillot.

Derosne's subsequent partnership with one of his employees, J.-F. Cail, resulted in a rapid expansion and diversification of his business. The Derosne-Cail establishment moved into the manufacture of industrial machinery, locomotives, and railway equipment. By the time of Derosne's death in 1846, an industrial empire had been founded, with factories in Paris, Belgium, Cuba, and Denain, and in 1847 branches were opened in Valenciennes, Douai, and Amsterdam.

BIBLIOGRAPHY

I. ORIGINAL WORKS. Derosne's writings include "Expériences et observations sur la distillation de l'acétate de cuivre et sur ses produits," in *Annales de chimie,* **63** (1807), 267–286, written with J.-F. Derosne; *Traité complet sur le sucre européen de betteraves; culture de cette plante considérée sous le rapport agronomique et manufacturier. Traduction abrégée de M. Achard, par M. D. Angar. Précédé d'une introduction et accompagné de notes . . . par M. Ch. Derosne . . .* (Paris, 1812), the French translation of Achard's work, done with D. Angar; and *De la fabrication du sucre aux colonies et des nouveaux appareils propres à améliorer cette fabrication . . . ,* 2 pts. (Paris, 1843–1844), written with J.-F. Cail.

II. SECONDARY LITERATURE. Brief biographies can be found in *Biographie universelle, ancienne et moderne,* L. G. Michaud and J. F. Michaud, eds., new ed., X, 461–462; *Dictionnaire de biographie française,* X (1965), 1144; *Le grand dictionnaire universel,* Pierre Larousse, ed., VI, 513; *La grande encyclopédie,* XIV, 197; and *Nouvelle biographie générale,* J. C. F. Hoefer, ed., XIII, 718. For an account of the growth of Établissements Derosne et Cail, see Bertrand Gille, *Recherches sur la formation de la grande entreprise capitaliste (1815–1848)* (Paris, 1959), p. 69; and Julien Turgan, *Les grandes usines de France, tableau de l'industrie française au XIXe siècle,* II (Paris, 1860–1868), 1–64. See also "Pharmaciens membres de l'Académie de Médecine," in *Figures pharmaceutiques françaises* (Paris, 1953), p. 263. Additional information about Derosne in connection with his work on phosphorus bottles or tubes ("briquets phosphoriques"), the precursors of phosphorus matches, will be found in Maurice Bouvet, "Les pharmaciens et la découverte des allumettes et briquets," in *Revue d'histoire de la pharmacie,* **11** (Mar. 1954), 230–231.

ALEX BERMAN

DERYUGIN, KONSTANTIN MIKHAILOVICH (*b.* St. Petersburg, Russia, 10 February, 1878; *d.* Moscow, U.S.S.R., 27 December 1938), *earth science, oceanography, zoology.*

From 1896 to 1900 Deryugin was a student in the natural sciences section of St. Petersburg University. In 1899 he took part in a scientific expedition to the White Sea; and from that time on, marine organisms and their habitat became his main scientific interest. After graduation Deryugin remained at the university, in the department of zoology and comparative anatomy, to do research and to teach. He also visited the United States and western Europe, where he became acquainted with the organization of foreign research in oceanography. In 1909 he defended his master's thesis and began to lecture in a course entitled "Life of the Sea."

In 1915 Deryugin defended his dissertation for the doctorate in zoology and comparative anatomy. In 1917 he was made a lecturer at St. Petersburg University and, in 1919, professor. From 1920 he combined teaching with substantial research and administrative responsibilities as deputy director and manager of the oceanic section of the State Hydrological Institute in Leningrad.

Deryugin won fame chiefly as taxonomist of a number of groups of marine organisms (fishes, mollusks, and several others). He also studied the distribution of marine fauna in relation to the environment.

In 1915 Deryugin published his chief work, "Fauna Kolskogo zaliva . . ." ("The Fauna of Kola Bay"), in which he gave the first detailed analysis of the system of zones and biological communities (biocenoses) of the Barents Sea. His full description of the pattern of fauna and flora shows their regular distribution in relation to environmental conditions. This broad approach led Deryugin to consider the history of the fauna of the Barents Sea, and then the history of the sea itself, the structure of the shores and bottom, its geology and petrography, and its hydrology and hydrochemistry.

In 1921 Deryugin reestablished regular hydrological sampling of the Barents Sea (which had been interrupted by the war) along the Kola meridian (33°30′ east longitude) to 75° north latitude and even farther. It soon became clear that the warm currents of the North Cape stream intersected by these samplings frequently change their positions. This discovery had a profound influence on the development of the fishing industry in the Barents Sea, since it showed a connection between the current of warm Atlantic waters and changes of the marine fauna. In particular, as Deryugin demonstrated, these warm currents were responsible for the appearance in the Barents Sea of warm-water forms not previously observed there and for their rapid diffusion to the shores of Novaya Zemlya.

The work of Deryugin and his students, from 1922 on, laid the foundations of present knowledge of the hydrology and biology of the White Sea, a body of water sharply different from other inland seas. Deryugin showed that this comparatively small sea consists of three parts: the basin of the sea itself, the Gorlo Strait, and its funnel. He explained the difference in biological environment between the White Sea and the neighboring Barents Sea by the intensified tidal mingling of waters from the surface to the bottom in the Gorlo Strait, a phenomenon that presented an insurmountable barrier to the dispersion of organisms.

The Pacific Ocean expedition of 1932–1933, which was organized at Deryugin's initiative and worked under his immediate direction, thoroughly investigated the Sea of Okhotsk and the parts of the Sea of Japan, the Bering Sea, and the Chukchi Sea bordering on the Soviet Union. The hydrological samplings were made from six fishing trawlers that collected at depths as great as 3,500 meters. Some work of the expedition was continued in following years. This research, in the course of which the fauna and flora of the Far Eastern seas of the Soviet Union were first seriously studied, brought to light what Deryugin termed a "new world of organisms." In some groups up to 50 percent new forms were found.

Deryugin was responsible for and participated in more than fifty scientific expeditions in twelve bodies of water bordering the Soviet Union. He organized and directed important oceanographic institutions: the Murmansk biological station on the shore of Kola Bay, in Yekaterin Harbor (now Polyarny; 1903–1904) and the Pacific Ocean Scientific Trade Station in Vladivostok (now the Pacific Ocean Scientific Research Institute of Fishing Economy and Oceanography; 1925).

Deryugin gave special attention to the methodological side of oceanographic research. He was responsible for the creation and operation of special methodological stations in the Neva Inlet of the Gulf of Finland (1920), on the White Sea (1931), and on Kamchatka (1932). He organized the design and production of oceanographic instruments in the Soviet Union and spent much of his strength and energy on the planning and building of special ocean research ships.

Deryugin was a member of the Society of Natural Scientists in Leningrad and a life member of the Linnaean Society of Lyons.

BIBLIOGRAPHY

I. ORIGINAL WORKS. The most important of Deryugin's more than 160 published scientific works are "Fauna Kolskogo zaliva i uslovia yeyo sushchestvovania" ("The Fauna of Kola Bay and Its Environment"), in *Zapiski Akademii nauk, fiziko-matematicheskoe otdelenie* ("Notes of the Academy of Sciences, Physics and Mathematics Section"), 8th ser., **34,** no. 1 (1915); "Fauna Belogo morya i uslovia yeyo sushchestvovania" ("The Fauna of the White Sea and Its Environment"), in *Issledovania morey SSSR* ("Investigations of the Seas of the USSR"), no. 7–8 (Leningrad, 1928); "Gidrologia i biologia" ("Hydrology and Biology"), in *Issledovania morey SSSR* ("Investigations of the Seas of the USSR"), no. 11 (Leningrad, 1930), pp. 37–45; "Vlianie prolivov i ikh gidrologicheskogo rezhima na faunu morey i yeyo dalneyshuyu evolyutsiyu" ("The Influence of Straits and Their Hydrological Systems on the Fauna of the Seas and Its Further Evolution"), in *Zapiski Gosudarstvennogo gidrologicheskogo instituta* ("Notes of the State Hydrological Institute"), **10** (1933), 369–374; "Issledovania morey SSSR v biograficheskom otnoshenii" ("Investigations of the Seas of the USSR in Terms of Biogeography"), in *Trudy Pervogo Vsesoyuznogo geograficheskogo sezda (11–18 oktyabrya 1933)* ("Works of the First All-Union Geographical Congress . . ."), pt. 2 (Leningrad, 1934), pp. 36–45; "Uspekhi sovetskoy gidrobiologii v oblasti izuchenia morey" ("The Progress of Soviet Hydrobiology in the Field of Ocean Studies"), in *Uspekhi sovremennoi biologii* ("Progress of Contemporary Biology"), **5,** no. 1 (1936), 9–23; and "Osnovnye cherty sovremennykh faun morey SSSR i veroyatnye puti ikh evolyutsii" ("The Basic Outlines of Contemporary Fauna of the Seas of the USSR and the Probable Course of Their Evolution"), in *Uchenye zapiski Leningradskogo . . . gosudarstvennogo universiteta* ("Scientific Notes of the Leningrad State University"), **3,** no. 17 (1937), 237–248.

II. SECONDARY LITERATURE. Important publications on Deryugin are E. F. Guryanova, "Professor K. M. Deryugin," in *Vestnik Leningradskogo gosudarstvennogo universiteta* ("Leningrad State Herald University"), no. 8 (1949), pp. 81–92; and V. V. Timonov, P. V. Ushakov, and S. Y. Mittelman, "Konstantin Mikhailovich Deryugin kak okeanolog" ("Konstantin Mikhailovich Deryugin as Oceanographer"), in *Trudy Gosudarstvennogo okeanograficheskogo instituta* ("Works of the State Oceanographic Institute"), no. 1, sec. 13 (1947), pp. 9–18.

A. F. PLAKHOTNIK

DESAGULIERS, JOHN THEOPHILUS (*b.* La Rochelle, France, 12 March 1683; *d.* London, England, 10 March 1744), *experimental natural philosophy.*

Desaguliers was taken to Guernsey when he was less than three years old by his Huguenot parents, who in 1694 settled in Islington, where the father taught school and educated his son. After his father's death Desaguliers entered Christ Church, Oxford (28 October 1705), whence he proceeded B.A. in 1709. About this time James Keill abandoned the lectureship in experimental philosophy at Hart Hall that he had held for some ten years; he was succeeded by Desaguliers, who took his M.A. from this college on 3 May 1712. In that year he moved to Channel Row, Westminster, no doubt in the hope of gaining a more remunerative audience. His first book, a translation of *A Treatise on Fortification* from the French of Ozanam, had already appeared (Oxford, 1711).

Continuing in London the style of scientific lecturing he had inherited from Keill, and having taken orders, he was given the living of Whitchurch and Little Stanmore, near Edgeware, to which royal favor later added other benefices. Before long Desaguliers was initiated into No. 4 Lodge of the Freemasons, meeting at the Rummer and Grapes Inn, Channel Row; and by 1719 he had become the third grand master of the recently constituted Grand Lodge of the order. It is said that Desaguliers induced Frederick, prince of Wales, to become a Freemason and also that through him "Freemasonry emerged from its original lowly station and became a fashionable cult" (Stokes). It was at the behest of one such fashionable past grand master, the duke of Wharton (Pope's "scorn and wonder of our days"), that in 1723 Desaguliers (then deputy master) dedicated to the grand master (the duke of Montagu) James Anderson's *Constitutions of the Free-Masons* (in a preface). Others of the Royal Society, to which Desaguliers belonged, joined this distinguished fraternity.

Desaguliers' practical abilities aroused the Royal Society's interest soon after his arrival in London. Late in the winter of 1713/1714, at Newton's suggestion, he was invited to repeat some of Newton's experiments on heat; before long he had become a *de facto* curator of experiments. He was elected a fellow on 29 July 1714, being excused his admission money because of his previous services. Desaguliers continued to furnish the society with experiments until his death. For some time Sir Godfrey Copley's benefaction (1709) of £100 per annum was paid to him; and after the Copley Medal was instituted in 1731, it was awarded to Desaguliers three times as a mark of his experimental ingenuity. Between 1716 and 1742 he contributed no fewer than fifty-two papers to the *Philosophical Transactions,* the earlier ones chiefly on optics and mechanics, the later ones on electricity. In the age of the Bernoullis, Clairaut, Euler, and Maupertuis, Desaguliers' contributions to theoretical mechanics cannot be called outstanding. "Dissertation Concerning the Figure of the Earth" (*Philosophical Transactions,* **33** [1726]) is the most important, yet it follows Keill (correcting his chief error), who followed Huygens (I. Todhunter, *A History of the Mathematical Theories of Attraction . . .* [1873], pp. 103–108). This dissertation was criticized anonymously by Maupertuis in 1741.

In attempting a reconciliation of the measurement of "motion" by the Newtonians (momentum $= mv$) and by the Leibnizians (*vis viva* $= mv^2$), Desaguliers correctly argued that the quantities so expressed were different, and hence the dispute was merely verbal; yet he seems also to have held that the Newtonian concept was better supported by experiment. Desaguliers never employed analytical methods in these papers. In practical mechanics he was highly skilled, being the first English writer to give theoretical analyses of machines on the basis of statics, the ancient treatment of the five simple machines, and elementary dynamics. He was himself a practical improver of various devices, among them Musschenbroek's pyrometer, Stephen Gray's barometric level, Hales's sounding gauge, Joshua Haskins' force pump, and Savery's steam engine. Desaguliers claimed that his improved form of Savery's engine was twice as efficient as the Newcomen pump. He also devised a centrifugal air pump for ventilating rooms, which was employed at the House of Commons. He had the advantage of relying upon Bélidor and his friend Henry Beighton (a Warwickshire engineer) in compiling a very up-to-date account of mechanical practice, including the railroad and steam engine. He clearly understood that a man or a horse could do only a finite amount of work in a given time, no matter what machinery might be used, and understood the fallacy of perpetual motion.

Desaguliers' optical experiments (*Philosophical Transactions,* **29** [1716]) were for the most part repetitions of those described by Newton, made in order to vindicate Newton's accuracy—which had been challenged—and the theoretical conclusions Newton had drawn. Some of them were improved in detail—for example, by the use of a camera obscura. Desaguliers taught that light is a "body" and that reflection, refraction, and diffraction are caused by the varying attractions between light and the media through which it moves. He also hinted at a similarity between the force of electricity and the force of cohesion, which he investigated experimentally. He made little use of the concept of ether (although not wholly avoiding the term), speaking, for instance, of "vacuities" between the particles of matter (the existence of which he thought he could demonstrate experimentally); rather, Desaguliers clung to that part of the Newtonian tradition which emphasized the duality of forces: "There seem to be but two Powers, or general Agents in Nature, which, according to different Circumstances, are concern'd in all the Phaenomena and Changes in Nature; viz. Attraction (meaning the Attraction of Gravity, as well as that of Cohaesion, etc.) and Repulsion" (*Course of Experimental Philosophy,* II, 407). Accordingly, Desaguliers, following Newton's hint, firmly attributed the elasticity of air to a repulsion between its particles. On all such points it might be said that he was "plus Newtonien que Newton," writing of the *Opticks* that it contained a "vast Fund of Philosophy; which (tho' he [Newton] has modestly delivered under the name of *Queries,* as if they were only Conjectures) daily Experiments and Observations confirm," and citing Hales's *Vegetable Staticks* as a book that put several of the "Queries" beyond doubt and showed how well they were founded (*Course of Experimental Philosophy,* preface, pp. vi–vii).

Desaguliers described and demonstrated a great many electrical experiments to the Royal Society (*Philosophical Transactions,* **41, 42;** *Course of Experimental Philosophy,* II, 316–335), although he refrained from so doing until after the death in 1736 of Stephen Gray—who, it is said, lived with Desaguliers and assisted him. This work certainly contributed greatly to the popularization of electrical science. Desaguliers studied charging, conduction, discharge in air, attraction and repulsion, the effects of dryness and moisture, and so forth, using a fragment of thread as detector. He distinguished "electrics per se," which could be charged by friction and so on, from "non-electric bodies," which were incapable of receiving charge directly although they were capable of being electrified indirectly when suitably suspended. Desaguliers did not make a parallel distinction between insulators and conductors, nor did he realize that a "non-electric body" could become an "electric per se" if properly insulated. Nor did he understand the role of leakage to earth in conduction experiments. At a very late stage he commented on the distinction between vitreous and resinous electricity established by Du Fay.

Until the end of his life Desaguliers retained his preeminence as a demonstrative lecturer in the Royal Society, at court, and in his own home (where he took in student boarders). By 1734 he had repeated his course on astronomy, mechanics, hydrostatics, optics, electricity, and machinery more than 120 times. Although he acknowledged that Keill had first "publickly taught Natural Philosophy by Experiments in a mathematical Manner" (the instrument maker Hawksbee had also begun to demonstrate experiments to the public at about the same time), it was Desaguliers who popularized the demonstrative lecture in Britain. (A sample of the scene is provided in two well-known pictures by Joseph Wright of Derby, *ca.* 1760.) "Without Observations and Experiments," he wrote in the preface to the first volume of his *Course of Experimental Philosophy* (1734), "our natural Philosophy would only be a Science of Terms and an unintelligible jargon." By deliberate choice he demonstrated to the eye not only things discovered by experiment but also those "deduc'd by a long Train of mathematical consequences; having contrived Experiments, which Step by Step bring us to the same Conclusions," for he recognized that the Newtonian philosophy was not accessible to all through mathematics. Thus Desaguliers occupies a leading position (along with Keill, Pemberton, and Maclaurin) among those who gave Newtonian science its ascendancy in eighteenth-century England. Not that Desaguliers wholly avoided mathematical reasoning; on the contrary, he employed it continually, but only in simple terms and as an adjunct to empirical evidence. Desaguliers did nothing for serious mathematical physics.

Naturally, Desaguliers was eager to publicize rather than to publish his material. In 1717 he issued as *Physico-Mechanical Lectures* an eighty-page abstract of the twenty-two lectures in the course for the benefit of auditors who did not wish to make their own notes. Two years later one Paul Dawson edited *A System of Experimental Philosophy, Prov'd by Mechanicks . . . As Performed by J. T. Desaguliers.* The lecturer did not produce his own version until 1734 (*A Course of Experimental Philosophy,* Volume I) when he took occasion to denounce this unauthorized version. Meanwhile he was content to print only short syllabi of his lectures: *Mechanical and Experimental Philosophy* (1724) and *Experimental Course of Astronomy* (1725). The former exists in both French and English versions, Desaguliers advertising his willingness to teach in these languages and Latin. At last the long-promised first volume of the *Course* appeared in

1734, containing five long lectures and many additional notes. It is devoted wholly to theoretical and practical mechanics, including both a simple treatment of Newton's system of the world and a description of Mr. Allen's railroad at Bath. Desaguliers attributed the ten-year delay before the appearance of his second tome to his desire to improve the treatment of machines, especially waterwheels; he excused himself for omitting optics altogether, referring the reader to Robert Smith's *Complete System of Opticks.* Continuing with mechanics, in seven lectures he discussed impact and elasticity, *vis viva* and momentum, heat, hydrostatics and hydraulics, pneumatics, meteorology, and more machines. This second volume is even more concerned with applied science and engineering than the first and entitles Desaguliers to be considered a forerunner of the more advanced knowledge of machinery that characterized the Industrial Revolution. Certainly its influence was greater upon practical men and inventors than upon physicists.

Desaguliers was married on 14 October 1712 to Joanna Pudsey, by whom he had several children; the youngest, Thomas, distinguished himself as an artilleryman. About 1739 the construction of the approaches to Westminster Bridge (upon whose design Desaguliers was consulted) necessitated the destruction of Channel Row; he moved his home and classes to the Bedford Coffee House in Covent Garden, where he died. He was buried in the Savoy Chapel, and there is no good reason for supposing him indigent. There are (or were) at least two portraits of Desaguliers: by H. Hysing (1725), engraved by Peter Pelham, and by Thomas Frye (1743), engraved by R. Scaddon.

BIBLIOGRAPHY

I. ORIGINAL WORKS. Desaguliers published translations besides that already mentioned: *Fires Improv'd: Being a New Method of Building Chimneys, so as to Prevent Their Smoking,* from the French by Nicolas Gauger (London, 1715), mostly about an elaborate form of fire grate; *The Motion of Water and Other Fluids,* from the French by Edmé Mariotte (London, 1718); *The Mathematical Elements of Natural Philosophy,* from the Latin by W. J. 'sGravesande (London, 1720); *The Whole Works of Dr. Archibald Pitcairne,* from the Latin (London, 1727), with G. Sewell; and *An Account of the Mechanism of an Automaton,* from the French by J. de Vaucanson (London, 1742). Most were reprinted.

Desaguliers' own writings, in addition to those already discussed, were *The Newtonian System, an Allegorical Poem* (London, 1728), written on the accession of George II; an Appendix on the reflecting telescope, pp. 211–288 in William Brown's translation *Dr. Gregory's Elements of Catoptrics and Dioptrics* (London, 1735), which contains most of the correspondence between Newton and others relating to the development of Newton's form of that instrument in 1668 and subsequently; and *A Dissertation Concerning Electricity* (London, 1742), the French version of which (Bordeaux, 1742) received a prize awarded by the Académie de Bordeaux (*Course of Experimental Philosophy,* II, 335).

There are MSS by Desaguliers in the Sloane and Birch collections of the British Museum and in the archives of the Royal Society.

II. SECONDARY LITERATURE. Modern studies of Desaguliers include Jean Barlais, in *Les archives de Trans en Provence,* **61,** 281–288, for more on his freemasonry; I. Bernard Cohen, *Franklin and Newton* (Philadelphia, 1956), esp. pp. 243–261, 376–384, mainly on electricity; D. C. Lee, *Desaguliers of No. 4 and His Services to Freemasonry* (London, 1932); Paul R. Major, *The Physical Researches of J. T. Desaguliers,* M. Sc. thesis, London University, 1962, with bibliography; and Jean Torlais, *Un Rochelais grand-maître de la Franc-Maçonnerie et physicien au XVIIIe siècle: Le Reverend J.-T. Desaguliers* (La Rochelle, 1937).

A. RUPERT HALL

DESARGUES, GIRARD (*b.* Lyons, France, 21 February 1591; *d.* France, October 1661), *geometry, perspective.*

One of the nine children of Girard Desargues, collector of the tithes on ecclesiastical revenues in the diocese of Lyons, and of Jeanne Croppet, Desargues seems to have studied at Lyons, where the family lived. The first evidence of his scientific activity places him in Paris on 9 September 1626, when, with another Lyonnais, François Villette, he proposed to the municipality that it construct powerful machines to raise the water of the Seine, in order to be able to distribute it in the city. Adrien Baillet, the biographer of Descartes, declares that Desargues participated as an engineer at the siege of La Rochelle in 1628 and that he there made the acquaintance of Descartes, but there is no evidence to confirm this assertion. According to the engraver Abraham Bosse (1602–1676), a fervent disciple of Desargues, the latter obtained a royal license for the publication of several writings in 1630. It was about this time that Desargues, living in Paris, seems to have become friendly with several of the leading mathematicians there: Mersenne, Gassendi, Mydorge, and perhaps Roberval. Although it is not certain that he attended the meetings at Théophraste Renaudot's Bureau d'Adresses (commencing in 1629), Mersenne cites him, in 1635, as one of those who regularly attended the meetings of his Académie Parisienne, in which, besides Mersenne, the following participated more or less regularly: Étienne Pascal, Mydorge, Claude Hardy, Roberval, and soon Carcavi and the young Blaise Pascal.

In 1636 Desargues published two works: "Une méthode aisée pour apprendre et enseigner à lire et escrire la musique," included in Mersenne's *Harmonie universelle* (I, bk. 6), and a twelve-page booklet with one double plate that was devoted to the presentation of his "universal method" of perspective. The latter publication bore a signature that reappeared on several of Desargues's important works: S.G.D.L. (Sieur Girard Desargues Lyonnais).

Moreover, after presenting his rules of practical perspective, Desargues gave some indication of the vast program he had set for himself, a program dominated by two basic themes: on the one hand, the concern to rationalize, to coordinate, and to unify the diverse graphical techniques by his "universal methods" and, on the other, the desire to integrate the projective methods into the body of mathematics by means of a purely geometric study of perspective, several elements of which are presented in an appendix. This publication appears not to have excited a great deal of immediate interest among artists and draftsmen, who were hardly anxious to change their technique; in contrast, Descartes and Fermat, to whom Mersenne had communicated it, were able to discern Desargues's ability.

The publication in 1636 of Jean de Beaugrand's *Geostatice,* then of Descartes's *Discours de la méthode* in May 1637, gave rise to ardent discussions among the principal French thinkers on the various problems mentioned in the two books: the definition of the center of gravity, the theory of optics, the problem of tangents, the principles of analytic geometry, and so on. Desargues participated very actively in these discussions. Although he made Beaugrand his implacable enemy, his sense of moderation, his concern to eliminate all misunderstandings, and his desire to comprehend problems in their most universal aspect won him the esteem and the respect of Descartes and Mersenne, as well as of Fermat, Roberval, and Étienne Pascal. His letter to Mersenne of 4 April 1638, concerning the discussion of the problem of tangents, illustrates the depth of the insights with which he approached such questions and, at the same time, his inclination to synthesis and the universal. Even though Descartes had prepared for him an introduction to his *Géométrie,* designed to "facilitate his understanding" of it, Desargues did not follow Descartes in his parallel attempts to algebraize geometry and to create a new system of explaining all the phenomena of the universe.

Desargues's goal was at once to breathe new life into geometry, to rationalize the various graphical techniques, and, through mechanics, to extend this renewal to several areas of technique. His profound intuition of spatial geometry led him to prefer a thorough renewal of the methods of geometry rather than the Cartesian algebraization; from this preference there resulted a broad extension of the possibilities of geometry. The *Brouillon project* on conics, of which he published fifty copies in 1639, is a daring projective presentation of the theory of conic sections; although considered at first in three-dimensional space, as plane sections of a cone of revolution, these curves are in fact studied as plane perspective figures by means of involution, a transformation that holds a place of distinction in the series of demonstrations. But the use of an original vocabulary and the refusal to resort to Cartesian symbolism make the reading of this essay rather difficult and partially explain its meager success.

Although he praised the unitary conception that inspired Desargues, Descartes doubted that the use of geometry alone could yield results as good as those that a recourse to algebra would provide. As for Fermat, he reserved his judgment, and the only geometer who really comprehended the originality and breadth of Desargues's views was the young Blaise Pascal, who in 1640 published the brief *Essay pour les coniques,* inspired directly by the *Brouillon project.* But since the great *Traité des coniques* that Pascal later wrote has been lost, Desargues's example survived only in certain of the youthful works of Philippe de La Hire and perhaps in a few essays of the young Newton. The rapid success of the Cartesian method of applying algebra to geometry was certainly one of the basic reasons for the poor diffusion of Desargues's ideas. In any case the principles of projective geometry included in the *Brouillon project* were virtually forgotten until the publication in 1820 of the *Traité des propriétés projectives* of J. V. Poncelet—who, moreover, rendered a stirring homage to his precursor, although he knew his work only from a few brief mentions.

In July 1639 Beaugrand criticized Desargues's work, asserting that certain of his demonstrations can be drawn much more directly from Apollonius. Irritated that Desargues, in an appendix to his study of conic sections, had discussed the principles of mechanics and had criticized Beaugrand's conception of geostatics, Beaugrand wrote in July 1640, a few months before his death, another violent pamphlet against the *Brouillon project.*

In August 1640, Desargues published, again under the general title *Brouillon project,* an essay on techniques of stonecutting and on gnomonics. While refining certain points of his method of perspective presented in 1636, he gives an example of a new graphical method whose use he recommends in stonecutting and furnishes several principles that will simplify construction of sundials. He cites the names of a few artists and artisans who have already adopted the graphical methods he advocates: in particular the painter Laurent de La Hire and the engraver Abraham Bosse. In attempting thus to improve the graphical procedures employed by many technicians, Desargues was in fact attacking an area of activity governed by the laws of the trade guilds; he also drew the open hostility of all those who were attached to the old methods and felt they were being injured by his preference for theory rather than practice.

At the end of 1640 Desargues published a brief commentary on the principles of gnomonics presented in his *Brouillon project;* this text is known only through several references, in particular the opinion of Descartes, who found it a "very beautiful invention and so much the more ingenious in that it is so simple." Since 1637 Descartes had conducted an indirect correspondence with Desargues that had been established through Mersenne, and the two men had exchanged ideas on a number of subjects; in this way Desargues took an active part in the discussions that preceded the definite statement and the publication of Descartes's *Méditations.*

At the beginning of 1641 Desargues had Mersenne propose to his mathematical correspondents that they determine circular sections on cones having a conic for a base and any vertex. He himself had a general solution obtained solely by the methods of pure geometry, a solution that is known to us through Mersenne's comments (in *Universae geometriae mixtaeque mathematicae synopsis* [Paris, 1644], the preface to Mydorge's *Coniques,* pp. 330–331). Roberval, Descartes, and Pascal were interested in the problem, which Desargues generalized in his investigation of the plane sections of cones satisfying the above conditions. References in publications of the period seem to indicate that around 1641 Desargues published a second essay on conic sections, cited sometimes under the title of *Leçons de ténèbres.* But since no copy of this work has been found, one may suppose that there may be some confusion here with another work, either the *Brouillon project* of 1639 or with a preliminary edition of certain manuscripts on perspective that were later included in Bosse's *Manière universelle de Mr Desargues pour pratiquer la perspective . . .* (Paris, 1648). Yet a work that appeared later, Grégoire Huret's *Optique de portraiture et de peinture . . .* (Paris, 1670), specifies (pp. 157–158) that the *Leçons de ténèbres* is based on the principle of perspective that inasmuch as the sections of a cone with a circular base and any vertex are, for all cones, circles for two specific orientations of the cutting plane, therefore in general the projective properties of the circle may be extended to various types of conics, considered as perspectives of circles. This systematic recourse to considerations of spatial geometry obviously does not permit the identification of this work with either the *Brouillon project* of 1639 or the geometric texts of 1648 (mentioned below). But, in the absence of the decisive proof that would be provided by the rediscovery of a copy of the *Leçons de ténèbres*, no definite conclusion can be reached.

Desargues strove to spread the use of his graphical methods among practitioners and succeeded in having them experiment with his stonecutting diagrams without encountering very strong resistance. At the beginning of 1642, however, the anonymous publication of the first volume of *La perspective pratique* (written by the Jesuit Jean Dubreuil) gave rise to bitter polemics. Finding that his own method of perspective was both copied and distorted in this book, Desargues had two placards posted in Paris in which he accused the author and the publishers of this treatise of plagiarism and obtuseness. The publishers asserted that they had drawn his so-called "universal" method from a work by Vaulezard (*Abrégé ou raccourcy de la perspective par l'imitation* . . . [Paris, 1631]) and from a manuscript treatise of Jacques Aleaume (1562–1627) that was to be brought out by E. Migon (*La perspective spéculative et pratique . . . de l'invention de feu Jacques Aleaume . . . mise au jour par Estienne Migon* [Paris, 1643]). Desargues having replied with a new attack, Tavernier and l'Anglois, Dubreuil's publishers, brought out in 1642 a collection of anonymous pamphlets against Desargues's various writings on perspective, stonecutting, and gnomonics, to which they added the *Lettre de M. de Beaugrand* . . . of August 1640, which was directed against his projective study of conics.

Desargues, greatly affected by these attacks, which concerned the body of his work and put his competence and his honesty in question, entrusted to his most fervent disciple, the engraver Abraham Bosse, the task of spreading his methods and of defending his work. In 1643 Bosse devoted two treatises to presenting Desargues's methods in stonecutting and in gnomonics: *La pratique du trait á preuves de Mr Desargues, Lyonnois, pour la coupe des pierres en l'architecture* . . . and *La manière universelle de Mr Desargues, Lyonnois, pour poser l'essieu et placer les heures et autres choses aux cadrans au soleil.* Preceded by an "Acknowledgment" in which Desargues states he has given Bosse the responsibility for the spread of his methods, these works are clearly addressed to a less informed audience than the brief essays that Desargues had published on the same subjects. Their

theoretical portion is greatly reduced and more elementary, and numerous examples of applications are handled in a very didactic and often prolix manner. Although only fifty copies of Desargues's essays had been printed, and had been distributed mainly in scientific circles, Bosse's writings were given large printings and were translated into several languages; consequently, they contributed to the diffusion of Desargues's graphical methods among practitioners.

In 1644, however, new attacks were launched against Desargues's work. They originated with a stonecutter, J. Curabelle, who violently criticized his writings on stonecutting, perspective, and gnomonics, as well as the two treatises Bosse published in 1643, claiming to find nothing in them but mediocrity, errors, plagiarism, and information of no practical interest. A very harsh polemic began between the two men, and Desargues published the pamphlet *Récit au vray de ce qui a esté la cause de faire cet escrit,* which contains a number of previously unpublished details on his life and work. He also attempted to sue Curabelle, but the latter seems to have succeeded in evading this action.

Although Desargues apparently gave up publishing, Abraham Bosse wrote an important treatise on his master's method of perspective, commenting in detail on a great many examples of the graphical processes deriving from the "universal method" outlined in 1636. This *Manière universelle de Mr Desargues pour pratiquer la perspective par petit-pied, comme le géométral, ensemble les places et proportions des fortes et foibles touches, teintes ou couleurs* (Paris, 1648) was directly inspired by Desargues and contains, in addition to a reprint of the *Exemple de l'une des manières universelles . . .* of 1636, several elaborations designed "for theoreticians" and others that are purely geometrical. These elaborations, which include the statement and proof of the famous theorem on perspective triangles, should be considered (at least those relating to the theorem should be) as having been written by Desargues. Certain remarks seem to indicate that these theoretical developments may have been the subject of an earlier version published in 1643, under the title of *Livret de perspective,* but no definite proof has yet been established. In 1653 Bosse completed this work with an account of perspective on planes and on irregular surfaces, which included several applications to his favorite technique, copperplate engraving. Desargues's influence is again evident, at least in the first part of this work, but it is less direct than in the *Manière* of 1648.

Meanwhile, relations between the two men had become less close. While continuing his work as an engraver and an artist, Bosse, since 1648, had been teaching perspective according to Desargues's methods at the Académie Royale de Peinture et de Sculpture. He continued to teach there until 1661, when the Academy, following a long and violent polemic in which Desargues intervened personally in 1657, barred him from all his duties, thus implicitly condemning the use and diffusion of the methods of perspective to which it had accorded its patronage for thirteen years. But Bosse continued, through his writings, to conduct a passionate propaganda campaign for his methods.

As for Desargues, after 1644 evidence of his scientific and polemic activity becomes much rarer. Besides the "Acknowledgment" (dated 1 October 1647) and the geometric elaborations inserted in Bosse's 1648 treatise on perspective, Descartes's correspondence (letter to Mersenne of 31 January 1648) alludes to an experiment made by Desargues, toward the end of 1647, in the context of the debates and investigations then being conducted by the Paris physicists on the nature of the barometric space. It seems that while remaining in close contact with the Paris scientists, Desargues had commenced another aspect of his work, that of architect and practitioner. There was no better reply to give to his adversaries, who accused him of wanting to impose arbitrary work rules on disciplines that he understood only superficially and theoretically. Probably, as Baillet states, he had already been technical adviser and engineer in Richelieu's entourage, but he had not yet had any real contact with the graphical techniques he wished to reform. It seems that his new career as an architect, begun in Paris about 1645, was continued in Lyons, to which he returned around 1649–1650, then again in Paris, to which he returned in 1657. He remained there until 1661, the year of his death.

In Paris the authors of the period attribute to Desargues, besides a few houses and mansions, several staircases whose complex structure and spectacular character attest to the exactitude and efficacy of his graphical stonecutting procedures. It also seems that he collaborated, for the realization of certain effects of architectural perspective, with the famous painter Philippe de Champaigne. In the region of Lyons, Desargues's architectural creations were likewise quite numerous; he participated in the planning of several private and public buildings and of rooms whose architecture was particularly delicate. Of Desargues's accomplishments as an engineer, which seem to have been many, only one is well known and is worth mentioning: a system for raising water that he installed near Paris, at the château of Beaulieu. This system, based on the use, until then unknown, of epicycloidal wheels, was described and drawn by

Huygens in 1671 (*Oeuvres de Huygens,* VII, 112), by which time the château had become the property of Charles Perrault. Philippe de La Hire, who had to repair this mechanism, wrote about it (see *Traité de méchanique* [Paris, 1695], pp. 10, 368–374).

To complete this description of Desargues's activity, it is necessary to mention the private instruction he gave at Paris in order to reveal his different graphical procedures. Even before 1640 he had several disciples at Paris, as well as at Lyons, where, Moreri states, he was "of great assistance to the workmen . . . to whom he communicated his diagrams and his knowledge, with no motive other than being useful" (*Le grand dictionnaire historique,* new ed., I [Paris, 1759], 297).

In 1660 Desargues was again active in the intellectual life of Paris, attending meetings at Montmor's Academy, such as one on 9 November 1660, at which Huygens heard him present a report on the problem of the existence of the geometric point and sharply discuss the matter with someone who contradicted him. This is the last trace of his activity; the reading of his will at Lyons on 8 October 1661 revealed only that he had died several days before, without stating the date or place of his death, concerning which no document has yet been found.

A geometer of profoundly original ideas, sustained at the same time by a sense of spatial reality, by a much more precise knowledge of the great classic works than he admitted, and by an exceptional familiarity with the whole range of contemporary techniques, Desargues, in his geometrical work, introduced the principal concepts of projective geometry: the consideration of points and straight lines to infinity, studies of poles and polars, the introduction of projective transformations, the general definition of focuses, the unitary study of conics, and so on. Unfortunately, his work, burdened by a too original vocabulary and the absence of symbolism, and known only in a very limited circle, did not receive the audience it deserved. The disappearance of the essential portion of the work of his chief disciple, Blaise Pascal, and the sudden vogue of analytic geometry and infinitesimal calculus prevented the seventeenth century from witnessing the revival of geometry for which Desargues had laid the foundations. His few known forays into other areas of mathematics and mechanics attest to a perfect mastery of all the problems then under discussion and make us regret the absence of any publication by him. In the field of graphical techniques his contribution is of major importance. Between Dürer and Monge he marks an essential stage in the rationalization of the ensemble of these techniques, as much by the improvements he made in the various procedures then in use as by his concern for unity, for theoretical rigor, and for universality. But in this vast area, too, his innovations were bitterly contested and often rejected with contempt, even though the goal of their author was to reduce the burden of the practitioners through a closer and more trusting collaboration with the theorists.

After the reception, often reserved and sometimes malicious, that it received in his time and the oblivion that it experienced subsequently, Desargues's work was rediscovered and fully appreciated by the geometers of the nineteenth century. Thus, like that of all precursors, his work revealed its fruitfulness much more by its remote extensions than by its immediate repercussions.

BIBLIOGRAPHY

I. ORIGINAL WORKS. Desargues's works, most of them published in editions of a small number of copies, are very rare. N. Poudra republished most of them in *Oeuvres de Desargues réunies et analysées . . .*, 2 vols. (Paris, 1864), but they are imperfect. The purely mathematical texts have been republished in an improved form by René Taton, as *L'oeuvre mathématique de Desargues* (Paris, 1951), pp. 75–212, with a bibliography of Desargues's writings and their editions, pp. 67–73. Aside from the polemic writings, prefaces, acknowledgments, and such, listed in the bibliography mentioned above, Desargues's main works and their most recent editions are the following: "Une méthode aisée pour apprendre et enseigner à lire et escrire la musique," in Mersenne's *Harmonie universelle,* I (Paris, 1638), bk. 6, prop. 1, pp. 332–342, repr. in photocopy (Paris, 1963); *Exemple de l'une des manières universelles du S.G.D.L. touchant la pratique de la perspective . . .* (Paris, 1636), also in Abraham Bosse, *Manière universelle de Mr Desargues pour pratiquer la perspective . . .* (Paris, 1648), pp. 321–334 (incorrect title), and in N. Poudra, *Oeuvres de Desargues,* I, 55–84, which follows the Bosse version; *Brouillon project d'une atteinte aux événemens des rencontres du cone avec un plan,* followed by *Atteinte aux événemens des contrarietez d'entre les actions des puissances ou forces* and an *Avertissement* (errata) (Paris, 1639), also in Poudra, I, 103–230, which includes omissions and errors, and in Taton, pp. 87–184, where the original version is reproduced without the figures; *Brouillon project d'exemple d'une manière universelle du S.G.D.L. touchant la pratique du trait à preuves pour la coupe des pierres en l'architecture . . .* (Paris, 1640), also in Poudra, I, 305–358, with incorrect plates; original plates repub. by W. M. Ivins, Jr., in *Bulletin of the Metropolitan Museum of Art,* new ser., **1** (1942), 33–45; *Leçons de ténèbres* (Paris, 1640 [?]), which has not been found but may exist; *Manière universelle de poser le style aux rayons du soleil . . .* (Paris, 1640), which has not been found but whose existence is definite, partly reconstructed in Poudra, I, 387–392; and *Livret de perspective adressé aux théoriciens* (Paris, 1643 [?]), which has not been found but is perhaps identical with the last part of Abraham Bosse's *Manière universelle de Mr Desargues pour pratiquer la perspective . . .* (Paris, 1648), pp. 313–343.

II. SECONDARY LITERATURE. Desargues directly inspired three works by Abraham Bosse: *La manière universelle de Mr Desargues, Lyonnois, pour poser l'essieu et placer les heures et autres choses aux cadrans au soleil* (Paris, 1643), also trans. into English (London, 1659); *La pratique du trait à preuves de Mr Desargues, Lyonnois, pour la coupe des pierres en l'architecture* (Paris, 1643), also trans. into German (Nuremberg, 1699); and *Manière universelle de Mr Desargues pour pratiquer la perspective . . .*, 2 vols. (Paris, 1648–1653), also trans. into Dutch, 2 vols. (Amsterdam, 1664; 2nd ed., 1686).

A lengthy bibliography of secondary literature is given in Taton's *L'oeuvre mathématique de Desargues.* The most important works cited are the following (listed chronologically): A. Baillet, *La vie de Monsieur des Cartes,* 2 vols. (Paris, 1691), *passim;* R. P. Colonia, *Histoire littéraire de la ville de Lyon . . .,* II (Paris, 1730), 807 f.; L. Moreri, *Le grand dictionnaire historique,* new ed., I (Paris, 1759), 297; J. V. Poncelet, *Traité des propriétés projectives* (Paris, 1822), pp. xxxviii–xxxxiii; M. Chasles, *Aperçu historique sur l'origine et le développement des méthodes en géométrie . . .* (Brussels, 1837), see index; and *Rapport sur les progrès de la géométrie* (Paris, 1870), pp. 303–306; G. Poudra, "Biographie," in *Oeuvres de Desargues . . .,* I, 11–52; and *Histoire de la perspective* (Paris, 1864), pp. 249–270; G. Eneström, "Notice bibliographique sur un traité de perspective publié par Desargues en 1636," in *Bibliotheca mathematica,* **1** (1885), 89–90; "Die 'Leçons de ténèbres' des Desargues," *ibid.,* 3rd ser., **3** (1902), 411; "Über dem französischen Mathematiker Pujos," *ibid.,* **8** (1907–1908), 97; and "Girard Desargues und D.A.L.G.," *ibid.,* **14** (1914), 253–254; S. Chrzaszczewski, "Desargues Verdienste um die Begründung der projectivischen Geometrie," in *Archiv der Mathematik und Physik,* 2nd ser., **16** (1898), 119–149; E. L. G. Charvet, *Lyon artistique. Architectes . . .* (Lyons, 1899), pp. 120–122; F. Amodeo, "Nuovo analisi del trattato delle coniche di Gerard [*sic*] Desargues e cenni da J. B. Chauveau," in *Rendiconti dell'Accademia delle scienze fisiche e matematiche* (Naples), ser. 3a, **12** (1906), 232–262; and *Origine e sviluppo della geometria proiettiva* (Naples, 1939), *passim;* J. L. Coolidge, *The History of Geometrical Methods* (Oxford, 1940), pp. 90, 109; and *The History of Conic Sections and Quadric Surfaces* (Oxford, 1949), pp. 28–33; M. Zacharias, "Desargues Bedeutung für die projektive Geometrie," in *Deutsche Mathematik,* **5** (1941), 446–457; W. M. Ivins, Jr., "Two First Editions of Desargues," in *Bulletin of the Metropolitan Museum of Art,* n.s. **1** (1942), 33–45; "A Note of Girard Desargues," in *Scripta mathematica,* **9** (1943), 33–48; *Art and Geometry. A Study in Space Intuition* (Cambridge, Mass., 1946), pp. 103–112; and "A Note on Desargues's Theorem," in *Scripta mathematica,* **13** (1947), 202–210; and F. Lenger, "La notion d'involution dans l'oeuvre de Desargues," in *IIe Congrès national des sciences, Bruxelles . . .,* I (Liège, 1950), 109–112.

Some more recent studies that should be noted are the following (listed chronologically): P. Moisy, "Textes retrouvés de Desargues," in *XVIIe siècle,* no. 11 (1951), 93–95; R. Taton, "Documents nouveaux concernant Desargues," in *Archives internationales d'histoire des sciences,* **4** (1951), 620–630; and "Sur la naissance de Girard Desargues," in *Revue d'histoire des sciences,* **15** (1962), 165–166; A. Machabey, "Gérard [*sic*] Desargues, géomètre et musicien," in *XVIIe siècle,* no. 21–22 (1954), 346–402; P. Costabel, "Note sur l'annexe du Brouillon-Project de Desargues," in *7° Congrès international d'histoire des sciences. Jérusalem, 1953* (Paris, n.d.), pp. 241–245; A. Birembaut, "Quelques documents nouveaux sur Desargues," in *Revue d'histoire des sciences,* **14** (1961), 193–204; and S. Le Tourneur, in *Dictionnaire de biographie française,* X (1964), 1183–1184.

RENÉ TATON

DESCARTES, RENÉ DU PERRON (*b.* La Haye, Touraine, France, 31 March 1596; *d.* Stockholm, Sweden, 11 February 1650), *natural philosophy, scientific method, mathematics, optics, mechanics, physiology.*

Fontenelle, in the eloquent contrast made in his *Éloge de Newton,* described Descartes as the man who "tried in one bold leap to put himself at the source of everything, to make himself master of the first principles by means of certain clear and fundamental ideas, so that he could then simply descend to the phenomena of nature as to necessary consequences of these principles." This famous characterization of Descartes as the theoretician who "set out from what he knew clearly, in order to find the cause of what he saw," as against Newton the experimenter, who "set out from what he saw, in order to find the cause," has tended to dominate interpretations of both these men who "saw the need to carry geometry into physics."[1]

Descartes was born into the *noblesse de robe,* whose members contributed notably to intellectual life in seventeenth-century France. His father was *conseiller* to the Parlement of Brittany; from his mother he received the name du Perron and financial independence from property in Poitou. From the Jesuits of La Flèche he received a modern education in mathematics and physics—including Galileo's telescopic discoveries—as well as in philosophy and the classics, and there began the twin domination of imagination and geometry over his precocious mind. He described in an early work, the *Olympica,* how he found "in the writings of the poets weightier thoughts than in those of the philosophers. The reason is that the poets wrote through enthusiasm and the power of imagination." The seeds of knowledge in us, "as in a flint," were brought to light by philosophers "through reason; struck out through imagination by poets they shine forth more brightly."[2] Then, after graduating in law from the University of Poitiers, as a gentleman volunteer in the army of Prince Maurice of Nassau in 1618 he met Isaac Beeckman at Breda. Beeckman aroused him to self-discovery as a scientific thinker and mathematician and introduced him to a range of problems, especially in mechanics and acoustics, the subject of his first work, the *Compendium musicae* of 1618; published posthumously in 1650. On 26 March 1619 he reported to Beeckman his first glimpse of "an entirely new science,"[3] which was to become his analytical geometry.

Later in the year, on 10 November, then in the duke of Bavaria's army on the Danube, he had the experience in the famous *poêle* (lit. "stove," "well-heated room"), claimed to have given direction to the rest of his life. He described in the *Discours de la méthode* how, in a day of solitary thought, he reached two radical conclusions: first, that if he were to discover true knowledge he must carry out the whole program himself, just as a perfect work of art or architecture was always the work of one master hand; second, that he must begin by methodically doubting everything taught in current philosophy and look for self-evident, certain principles from which to reconstruct all the sciences. That night, according to his seventeenth-century biographer Adrien Baillet, these resolutions were reinforced by three consecutive dreams. He found himself, first, in a street swept by a fierce wind, unable to stand, as his companions were doing, because of a weakness in his right leg; second, awakened by a clap of thunder in a room full of sparks; and third, with a dictionary, then a book in which he read *Quid vitae sectabor iter?* ("What way of life shall I follow?"), then verses presented by an unknown man beginning *Est et non*; he recognized the Latin as the opening lines of two poems by Ausonius. Before he finally awoke he had interpreted the first dream as a warning against past errors, the second as the descent of the spirit of truth, and the third as the opening to him of the path to true knowledge. However this incident may have been elaborated in the telling, it symbolizes both the strength and the hazards of Descartes's unshakable confidence and resolve to work alone. But he did not make his vision his life's mission for another nine years, during which (either before or after his tour of Italy from 1623 to 1625) he met Mersenne, who was to become his lifelong correspondent, and took part in scientific meetings in Paris. The next decisive incident, according to Baillet, was a public encounter in 1628 in which he demolished the unfortunate Chandoux by using his method to distinguish sharply between true scientific knowledge and mere probability. Among those present was the influential Cardinal de Bérulle, who a few days later charged him to devote his life to working out the application of "his manner of philosophizing . . . to medicine and mechanics. The one would contribute to the restoration and conservation of health, and the other to some diminution and relief in the labours of mankind."[4] To execute this design he withdrew, toward the end of the year, to the solitary life in the Netherlands which he lived until his last journey to Stockholm in 1649, where, as Queen Christina's philosopher, he died in his first winter.

The primarily centrifugal direction of Descartes's thought, moving out into detailed phenomena from a firm central theory (in contrast with the more empirical scientific style of Francis Bacon and Newton), is shown by the sequence of composition of his major writings. He set out his method in the *Rules for the Direction of the Mind,* left unfinished in 1628 and published posthumously, and in the *Discours de la méthode,* written in the Netherlands along with the *Météores, La dioptrique,* and *La géométrie,* which he presented as examples of the method. All were published in one volume in 1637. At the same time his investigation into the true ontology led him to the radical division of created existence into matter as simply extended substance, given motion at the creation, and mind as unextended thinking substance. This conclusion he held to be guaranteed by the perfection of God, who would not deceive true reason. How these two mutually exclusive and collectively exhaustive categories of substance could have any interaction in the embodied soul that was a man was a question discussed between Gassendi, Hobbes, and Descartes in the *Objections and Replies* published with his *Meditations on First Philosophy* in 1641.

It was from these first principles that he had given an account in *Le monde, ou Traité de la lumière* of cosmogony and cosmology as products simply of matter in motion, making the laws of motion the ultimate "laws of nature" and all scientific explanation ultimately mechanistic. This treatise remained unpublished in Descartes's lifetime. So too did the associated treatise *L'homme,* in which he represented animals and the human body as sheer mechanisms, an idea already found in the *Rules.* He withheld these essays, on the brink of publication, at the news of Galileo's condemnation in 1633, and instead published his general system of physics, with its Copernicanism mitigated by the idea that all motion is relative, in the *Principles of Philosophy* in 1644. Finally, he brought physiological psychology within the compass of his system in *Les passions de l'âme* in 1649. This system aimed to be as complete as Aristotle's, which it was designed to replace. It was not by chance that

it dealt in the same order with many of the same phenomena (such as the rainbow), as well as with others more recently investigated (such as magnetism).

A comparison of Descartes's performance with his program of scientific method presents a number of apparent contradictions. He made much of the ideal of a mathematically demonstrated physics, yet his fundamental cosmology was so nearly entirely qualitative that he came to fear that he had produced nothing more than a beautiful "romance of nature."[5] His planetary dynamics was shown by Newton to be quantitatively ridiculous. He wrote in the *Discours,* "I noticed also with respect to experiments [*expériences*] that they become so much the more necessary, the more we advance in knowledge,"[6] yet his fundamental laws of nature, the laws of motion and impact, had to be dismantled by Huygens and Leibniz for their lack of agreement with observation. These apparent contradictions may be resolved in the contrast between Descartes's theoretical ideal of completed scientific knowledge and the actual process and circumstances of acquiring such knowledge. For the modern reader to pay too much attention to his mechanics and to the *Principles,* a premature conception of completed science, can obscure Descartes's firm grasp of the necessity for observation and experiment already expressed in the *Rules* in his criticism "of those philosophers who neglect experiments and expect truth to rise from their own heads like Minerva from Jupiter's."[7]

No other great philosopher, except perhaps Aristotle, can have spent so much time in experimental observation. According to Baillet, over several years he studied anatomy, dissected and vivisected embryos of birds and cattle, and went on to study chemistry. His correspondence from the Netherlands described dissections of dogs, cats, rabbits, cod, and mackerel; eyes, livers, and hearts obtained from an abattoir; experiments on the weight of the air and on vibrating strings; and observations on rainbows, parahelia, and other optical phenomena. Many of his scientific writings reflect these activities and show sound experimental knowledge, although the extreme formalism of his physiological models obscures the question of his actual knowledge of some aspects of anatomy. Attention to the whole range of his scientific thought and practice shows a clear conception not only of completed scientific knowledge but also of the roles of experiment and hypothesis in making discoveries and finding explanations by which the body of scientific knowledge was built up.

Descartes's conception of completed scientific knowledge was essentially that envisaged by Aristotle's true scientific demonstration. It was the geometers' conception of a system deduced from self-evident and certain premises. He wrote,

> In physics I should consider that I knew nothing if I were able to explain only how things might be, without demonstrating that they could not be otherwise. For, having reduced physics to mathematics, this is something possible, and I think that I can do it within the small compass of my knowledge, although I have not done it in my essays.[8]

His optimism about the possibility of achieving such demonstrations seems to have depended on which end of the chain of reasoning he was contemplating. When considering the results of his analysis reducing created existence to extension (with motion) and thought, he seems to have been confident that it would be possible to show that from these "simple natures" the composite observed world must follow. It may be argued that his treatment of motion failed just where his a priori confidence led him to suppose that his analysis (of what was soon seen to be an insufficient range of data) placed his first principles beyond the need for empirical test. But when considering the chain lower down, nearer this complex world, he was more hesitant. He wrote to Mersenne:

> You ask me whether I think what I have written about refraction is a demonstration. I think it is, at least as far as it is possible, without having proved the principles of physics previously by metaphysics, to give any demonstration on this subject . . . as far as any other question of mechanics, optics, or astronomy, or any other question which is not purely geometrical or arithmetical, has ever been demonstrated. But to demand that I should give geometrical demonstrations of matters which depend on physics is to demand that I should do the impossible. If you restrict the use of "demonstration" to geometrical proofs only, you will be obliged to say that Archimedes demonstrated nothing in mechanics, nor Vitellio in optics, nor Ptolemy in astronomy, etc., which is not commonly maintained. For, in such matters, one is satisfied that the writers, having presupposed certain things which are not obviously contradictory to experience, have besides argued consistently and without logical fallacy, even if their assumptions are not exactly true.[9]

The paradox of Descartes as a natural scientist is that his grasp improved the more hopeless he found the immediate possibility of deducing solutions of detailed problems from his general first principles. Standing amidst the broken sections of a chain that he could not cast up to heaven, the experimentalist and constructor of hypothetical models came to life. In Descartes's letter prefaced to the French translation of the *Principles* (1647), he wrote that two, and only two, conditions determined whether the first principles proposed could be accepted as true: "First they must be so clear and evident that the mind of man cannot doubt their truth when it attentively applies itself to consider them"; and secondly, everything else must be deducible from them. But he went on to admit, "It is really only God alone who has perfect wisdom, that is to say, who has a complete knowledge of the truth of all things."[10] To find the truth about complex material phenomena man must experiment, but as the sixth part of the *Discours* shows, the need to experiment was an expression of the failure of the ideal.

As well as being demonstrative, scientific knowledge had to be explanatory; for Descartes the two went together. He wrote, "I have described . . . the whole visible world as if it were only a machine in which there was nothing to consider but the shapes and movements [of its parts]."[11] To such a mechanism it was possible to apply mathematics and calculation, but it was the mechanism that explained. His insistence that even mathematical science without fundamental explanations was insufficient appears in his interestingly similar criticisms of Harvey for starting simply with a beating heart in explaining the circulation of the blood and of Galileo for likewise failing to reduce the mathematical laws of moving bodies to their ultimate mechanisms. He commented on the latter that "without having considered the first causes of nature, he has only looked for the reasons for certain particular effects, and that thus he has built without foundation."[12] By this insistence Descartes here again extracted from the failure of his ideal a fundamental contribution to scientific thinking. He became the first great master to make the hypothetical model, or "conjecture," a systematic tool of research.

Current natural philosophy accepted Aristotle's absolute ontological distinction between naturally generated bodies (inanimate and animate) and artificial things made by man. Hence, in principle no humanly constructed imitation or model could throw real light on the naturally endowed essence and cause of behavior. This distinction had become blurred in the partial mechanization of nature made by some philosophers. Descartes's innovation was to assert the identity of the synthesized artificial construction with the naturally generated product and to make this identification an instrument of scientific research:

> And certainly there are no rules in mechanics that do not hold also in physics, of which mechanics forms a part or species [so that all artificial things are at the same time natural]: for it is not less natural for a clock, made of these or those wheels, to indicate the hours, than for a tree which has sprung from this or that seed to produce a particular fruit. Accordingly, just as those who apply themselves to the consideration of automata, when they

know the use of some machine and see some of its parts, easily infer from these the manner in which others which they have not seen are made, so, from the perceptible effects and parts of natural bodies, I have endeavoured to find out what are their imperceptible causes and parts.[13]

This reduction made the principles of the mechanistic model the only principles operating in nature, thus bringing the objectives of the engineer into the search for the nature of things and throwing the entire world of matter open to the same form of scientific inquiry and explanation. Research, whether into cosmology or physiology, was reduced to the discovery and elucidation of mechanisms. He could construct in distant space the imaginary world of *Le monde* and *L'homme,* and later the world of the *Principles,* as explicitly and unambiguously hypothetical imitations of our actual world, made in accordance with the known laws of mechanics. The heuristic power of the model was that, like any other theory advanced in anticipation of facts, its own properties suggested new questions to put to nature. The main issue in any historical judgment of Descartes here is not whether his own answers were correct but whether his questions were fruitful. In insisting that experiment and observation alone could show whether the model corresponded with actuality, he introduced further precision into his theory of demonstration.

Descartes used the word *demonstrer* to cover both the explanation of the observed facts by the assumed theory and the proof of the truth of the theory. When challenged with the criticism that this might make the argument circular, he replied by contrasting two kinds of hypothesis.[14] In astronomy various geometrical devices, admittedly false in nature, were employed to yield true conclusions only in the sense that they "saved the appearances." But physical theories were proposed as true. He was persuaded of the truth of the assumption that the material world consisted of particles in motion by the number of different effects he could deduce, as diverse as the operation of vision, the properties of salt, the formation of snow, the rainbow, and so on. Thus he made range of application the empirical criterion of truth. He wrote in the *Discours:*

> If some of the matters of which I have spoken in the beginning of the *Dioptrique* and the *Météores* should, at first sight, shock people because I have called them suppositions, and do not seem to bother about their proof, let them have the patience to read them carefully right through, and I hope that they will find themselves satisfied. For it seems to me that the reasonings are so interwoven that as the later ones are demonstrated by the earlier which are their causes, these earlier ones are

reciprocally demonstrated by the later which are their effects. And it must not be thought that in this I commit the fallacy which logicians call arguing in a circle, for, since experience renders the majority of these effects very certain, the causes from which I deduce them do not so much serve to prove them as to explain them; on the other hand, the causes are proved by the effects.[15]

The test implied precisely by the criterion of range of confirmation was the *experimentum crucis.* This is the most obvious feature in common between Descartes's logic of experiment and that of Francis Bacon. Descartes described its function in the *Discours:*

> Reviewing in my mind all the objects that have ever been presented to my senses, I venture truly to say that I have not there observed anything that I could not satisfactorily explain by the principles I had discovered. But I must also confess that the power of nature is so ample and so vast and that these principles are so simple and so general, that I have observed hardly any particular effect that I could not at once recognize to be deducible from them in several different ways, and that my greatest difficulty is usually to discover in which of these ways it depends on them. In such a case, I know no other expedient than to look again for experiments [*expériences*] such that their result is not the same if it has to be explained in one of these ways as it would be if explained in the other.[16]

It was a logician rather than an experimenter who seems to have been uppermost in Descartes's application of this criterion in the same way to very general assumptions, such as the corpuscularian natural philosophy, and to questions as particular as whether the blood left the heart in systole or in diastole. Descartes argued in *La description du corps humain* (1648–1649) that whereas Harvey's theory that the blood was forced out of the heart by a muscular contraction might agree with the facts observed so far, "that does not exclude the possibility that all the same effects might follow from another cause, namely from the dilatation of the blood which I have described. But in order to be able to decide which of these two causes is true, we must consider other observations which cannot agree with both of them."[17] Harvey replied in his *Second Disquisition to Jean Riolan.*

As the great optimist of the scientific movement of the seventeenth century, Descartes habitually wrote as if he had succeeded in discovering the true principles of nature to such an extent that the whole scientific program was within sight of completion. Then, as Seth Ward neatly put it, "when the operations of nature shall be followed up to their staticall (and mechanicall) causes, the use of induction will cease, and syllogisme succeed in place of it." But Descartes would surely have agreed with Ward's qualification that "in the interim we are to desire that men have patience not to lay aside induction before they have reason."[18]

NOTES

1. Fontenelle, *Oeuvres diverses,* new ed., III (The Hague, 1729), 405–406.
2. Part of the *Olympica* incorporated in the *Cogitationes privatae* (1619–1621); see *Oeuvres,* X, 217.
3. *Oeuvres,* X, 156.
4. Baillet, II, 165.
5. *Ibid.,* preface, p. xviii.
6. *Oeuvres,* VI, 63.
7. Rule V; see *Oeuvres,* X, 380.
8. Letter to Mersenne, 11 Mar. 1640; see *Oeuvres,* III, 39. The "Essays" were the volume of 1637.
9. Letter to Mersenne, 27 May 1638; see *Oeuvres,* II, 141–142.
10. *Oeuvres,* IX, pt. 2, 2–3.
11. *Principia philosophiae,* IV, 188; *Oeuvres,* VIII, pt. 1, 315 (Latin); IX, pt. 2, 310 (French, alone with passage in square brackets).
12. Letter to Mersenne, 11 Oct. 1638; see *Oeuvres,* II, 380. For his comments on Harvey, see *Discours V.*
13. *Principia philosophiae,* IV, 203; *Oeuvres,* VIII, pt. 1, 326 (Latin); IX, pt. 2, 321–322 (French, alone with passage in square brackets).
14. Letter to J.-B. Morin, 13 July 1638; see *Oeuvres,* II, 197–202; cf. his letters to Vatier, 22 Feb. 1638, *ibid.,* I, 558–565, and to Mersenne, 1 Mar. 1638, *ibid.,* II, 31–32.
15. *Oeuvres,* VI, 76.
16. *Ibid.,* pp. 64–65; cf. *Principia philosophiae,* III, 46; VIII, pt. 2, 100–101; and IX, pt. 2, 124–125.
17. *Oeuvres,* XI, 241–242; cf. the comments on this controversy by J. B. Duhamel, "Quae sit cordis motus effectrix causa," in *Philosophia vetus et nova.* II, *Physica generalis,* III.ii.2 (Paris, 1684), 628–631.
18. *Vindiciae academiarum* (Oxford, 1654), p. 25.

BIBLIOGRAPHY

Descartes's complete works can be found in *Oeuvres de Descartes,* C. Adam and P. Tannery, eds., 12 vols. (Paris, 1897–1913), together with the revised *Correspondance,* C. Adam and G. Milhaud, eds. (Paris, 1936–   ). Besides these, primary sources for Descartes's life are Adrien Baillet, *La vie de Monsieur Descartes,* 2 vols. (Paris, 1691), which should be read with C. Adam, *Vie et oeuvres de Descartes* (in *Oeuvres,* XII); Isaac Beeckman, *Journal tenu . . . de 1604 à 1634,* C. de Waard, ed., 3 vols. (The Hague, 1939–1953): Marin Mersenne, *Correspondance,* C. de Waard, R. Pintard, B. Rochot, eds. (Paris, 1932–   ).

For Descartes's philosophy and method and their background, see E. Gilson, *Index scolastico-cartésien* (Paris, 1912); *Études sur le rôle de la pensée médiévale dans la formation du système cartésien* (Paris, 1930); *Discours de la méthode: texte et commentaire* (Paris, 1947); Alexandre Koyré, *Entretiens sur Descartes* (Paris–New York, 1944); G. Milhaud, *Descartes savant* (Paris, 1921); L. Roth, *Descartes' Discourse on Method* (Oxford, 1937); H. Scholz, A. Kratzer, and J. E. Hofmann, *Descartes* (Münster, 1951); and Norman Kemp Smith, *New Studies in the Philosophy of Descartes* (London, 1952).

Specific aspects of Descartes's scientific method are discussed in A. Gewirtz, "Experience and the Non-mathematical in the Cartesian Method," in *Journal of the History of Ideas,* **2** (1941), 183–210; and A. C. Crombie, "Some Aspects of Descartes' Attitude to Hypothesis and Experiment," in Académie Internationale d'Histoire des Sciences, *Actes du Symposium International des Sciences Physiques et Mathématiques dans la Première Moitié du XVIIe Siècle: Pise-Vinci, 16–18 Juin 1958* (Paris, 1960), pp. 192–201. An indispensable bibliography is G. Sebba, *Descartes and His Philosophy: A Bibliographical Guide to the Literature, 1800–1958* (Athens, Ga., 1959).

A. C. Crombie

DESCARTES: Mathematics and Physics.

In this section, Descartes's mathematics is discussed separately. The physics is discussed in two subsections: Optics and Mechanics.

Mathematics. The mathematics that served as model and touchstone for Descartes's philosophy was in large part Descartes's own creation and reflected in turn many of his philosophical tenets.[1] Its historical foundations lie in the classical analytical texts of Pappus (*Mathematical Collection*) and Diophantus (*Arithmetica*) and in the cossist algebra exemplified by the works of Peter Rothe and Christoph Clavius. Descartes apparently received the stimulus to study these works from Isaac Beeckman; his earliest recorded thoughts on mathematics are found in the correspondence with Beeckman that followed their meeting in 1618. Descartes's command of cossist algebra (evident throughout his papers of the early 1620's) was perhaps strengthened by his acquaintance during the winter of 1619–1620 with Johann Faulhaber, a leading German cossist in Ulm.[2] Descartes's treatise *De solidorum elementis,* which contains a statement of "Euler's Theorem" for polyhedra ($V + F = E + 2$), was quite likely also a result of their discussions. Whatever the early influences on Descartes's mathematics, it nonetheless followed a relatively independent line of development during the decade preceding the publication of his magnum opus, the *Géométrie* of 1637.[3]

During this decade Descartes sought to realize two programmatic goals. The first stemmed from a belief, first expressed by Petrus Ramus,[4] that cossist algebra represented a "vulgar" form of the analytical method employed by the great Greek mathematicians. As Descartes wrote in his *Rules for the Direction of the Mind* (ca. 1628):

> . . . some traces of this true mathematics [of the ancient Greeks] seem to me to appear still in Pappus and Diophantus. . . . Finally, there have been some most ingenious men who have tried in this century to revive the same [true mathematics]; for it seems to be nothing

other than that art which they call by the barbarous name of "algebra," if only it could be so disentangled from the multiple numbers and inexplicable figures that overwhelm it that it no longer would lack the clarity and simplicity that we suppose should obtain in a true mathematics.[5]

Descartes expressed his second programmatic goal in a letter to Beeckman in 1619; at the time it appeared to him to be unattainable by one man alone. He envisaged "an entirely new science,"

> . . . by which all questions can be resolved that can be proposed for any sort of quantity, either continuous or discrete. Yet each problem will be solved according to its own nature, as, for example, in arithmetic some questions are resolved by rational numbers, others only by irrational numbers, and others finally can be imagined but not solved. So also I hope to show for continuous quantities that some problems can be solved by straight lines and circles alone; others only by other curved lines, which, however, result from a single motion and can therefore be drawn with new forms of compasses, which are no less exact and geometrical, I think, than the common ones used to draw circles; and finally others that can be solved only by curved lines generated by diverse motions not subordinated to one another, which curves are certainly only imaginary (e.g., the rather well-known quadratrix). I cannot imagine anything that could not be solved by such lines at least, though I hope to show which questions can be solved in this or that way and not any other, so that almost nothing will remain to be found in geometry.[6]

Descartes sought, then, from the beginning of his research a symbolic algebra of pure quantity by which problems of any sort could be analyzed and classified in terms of the constructive techniques required for their most efficient solution. He took a large step toward his goal in the *Rules* and achieved it finally in the *Géométrie*.

Descartes began his task of "purifying" algebra by separating its patterns of reasoning from the particular subject matter to which it might be applied. Whereas cossist algebra was basically a technique for solving numerical problems and its symbols therefore denoted numbers, Descartes conceived of his "true mathematics" as the science of magnitude, or quantity, per se. He replaced the old cossist symbols with letters of the alphabet, using at first (in the *Rules*) the capital letters to denote known quantities and the lowercase letters to denote unknowns, and later (in the *Géométrie*) shifting to the *a,b,c; x,y,z* notation still in use today. In a more radical step, he then removed the last vestiges of verbal expression (and the conceptualization that accompanied it) by replacing the words "square," "cube," etc., by numerical

superscripts. These superscripts, he argued (in rule XVI), resolved the serious conceptual difficulty posed by the dimensional connotations of the words they replaced. For the square of a magnitude did not differ from it in kind, as a geometrical square differs from a line; rather, the square, the cube, and all powers differed from the base quantity only in the number of "relations" separating them respectively from a common unit quantity. That is, since

$$1:x = x:x^2 = x^2:x^3 = \cdots$$

(and, by Euclid V, ratios obtain only among homogeneous quantities), x^3 was linked to the unit magnitude by three "relations," while x was linked by only one. The numerical superscript expressed the number of "relations."

While all numbers are homogeneous, the application of algebra to geometry (Descartes's main goal in the *Géométrie*) required the definition of the six basic algebraic operations (addition, subtraction, multiplication, division, raising to a power, and extracting a root) for the realm of geometry in such a way as to preserve the homogeneity of the products. Although the Greek mathematicians had established the correspondence between addition and the geometrical operation of laying line lengths end to end in the same straight line, they had been unable to conceive of multiplication in any way other than that of constructing a rectangle out of multiplier and multiplicand, with the result that the product differed in kind from the elements multiplied. Descartes's concept of "relation" provided his answer to the problem: one chooses a unit length to which all other lengths are referred (if it is not given by the data of the problem, it may be chosen arbitrarily). Then, since $1:a = a:ab$, the product of two lines a and b is con-

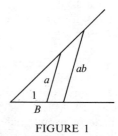

FIGURE 1

structed by drawing a triangle with sides 1 and a; in a similar triangle, of which the side corresponding to 1 is b, the other side will be ab, a line length. Division and the remaining operations are defined analogously. As Descartes emphasized, these operations do not make arithmetic of geometry, but rather make possible an algebra of geometrical line segments.

The above argument opens Descartes's *Géométrie*

and lays the foundation of the new analytic geometry contained therein, to wit, that given a line x and a polynomial $P(x)$ with rational coefficients it is possible to construct another line y such that $y = P(x)$. Algebra thereby becomes for Descartes the symbolic method for realizing the second goal of his "true mathematics," the analysis and classification of problems. The famous "Problem of Pappus," called to Descartes's attention by Jacob Golius in 1631, provides the focus for Descartes's exposition of his new method. The problem states in brief: given n coplanar lines, to find the locus of a point such that, if it is connected to each given line by a line drawn at a fixed angle, the product of $n/2$ of the connecting lines bears a given ratio to the product of the remaining $n/2$ (for even n; for odd n, the product of $(n + 1)/2$ lines bears a given ratio to k times the product of the remaining $(n - 1)/2$, where k is a given line segment). In carrying out the detailed solution for the case $n = 4$, Descartes also achieves the classification of the solutions for other n.

Implicit in Descartes's solution is the analytic geometry that today bears his name. Taking lines AB, AD, EF, GH as the four given lines, Descartes assumes that point C lies on the required locus and draws the connecting lines CB, CD, CF, CH. To apply algebraic analysis, he then takes the length AB, measured from the fixed point A, as his first unknown, x, and length BC as the second unknown, y. He thus imagines the locus to be traced by the endpoint C of a movable ordinate BC maintaining a fixed angle to line AB (the axis) and varying in length as a function[7] of the length AB. Throughout the *Géométrie*, Descartes chooses his axial system to fit the problem; nowhere does the now standard—and misnamed—"Cartesian coordinate system" appear.

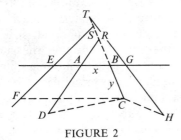

FIGURE 2

The goal of the algebraic derivation that follows this basic construction is to show that every other connecting line may be expressed by a combination of the two basic unknowns in the form $\alpha x + \beta y + \gamma$, where α, β, γ derive from the data. From this last result it follows that for a given number n of fixed lines the power of x in the equation that expresses the ratio of multiplied connecting lines will not exceed

$n/2$ (n even) or $(n - 1)/2$ (n odd); it will often not even be that large. Hence, for successive assumed values of y, the construction of points on the locus requires the solution of a determinate equation in x of degree $n/2$, or $(n - 1)/2$; e.g., for five or fewer lines, one need only be able to solve a quadratic equation, which in turn requires only circle and straightedge for its constructive solution.

Thus Descartes's classification of the various cases of Pappus' problem follows the order of difficulty of solving determinate equations of increasing degree.[8] Solution of such equations carries with it the possibility of constructing any point (and hence all points) of the locus sought. The direct solvability of algebraic equations becomes in book II Descartes's criterion for distinguishing between "geometrical" and "nongeometrical" curves; for the latter (today termed "transcendental curves") by their nature allow the direct construction of only certain of their points. For the construction of the loci that satisfy Pappus' problem for $n \leq 5$, i.e., the conic sections, Descartes relies on the construction theorems of Apollonius' *Conics* and contents himself with showing how the indeterminate equations of the loci contain the necessary parameters.

Descartes goes on to show in book II that the equation of a curve also suffices to determine its geometrical properties, of which the most important is the normal to any point on the curve. His method of normals—from which a method of tangents follows directly—takes as unknown the point of intersection of the desired normal and the axis. Considering a family of circles drawn about that point, Descartes derives an equation $P(x) = 0$, the roots of which are the abscissas of the intersection points of any circle and the curve. The normal is the radius of that circle which has a single intersection point, and Descartes finds that circle on the basis of the theorem that, if $P(x) = 0$ has a repeated root at $x = a$, then $P(x) = (x - a)^2 R(x)$, where $R(a) \neq 0$. Here a is the abscissa of the given point on the curve, and the solution follows from equating the coefficients of like powers of x on either side of the last equation. Descartes's method is formally equivalent to Fermat's method of maxima and minima and, along with the latter, constituted one of the early foundations of the later differential calculus.

The central importance of determinate equations and their solution leads directly to book III of the *Géométrie* with its purely algebraic theory of equations. Entirely novel and original, Descartes's theory begins by writing every equation in the form $P(x) = 0$, where $P(x)$ is an algebraic polynomial with real coefficients.[9] From the assertion, derived inductively, that every such equation may also be expressed in the form

$$(x - a) (x - b) \cdots (x - s) = 0,$$

where $a, b, \cdots, s$ are the roots of the equation, Descartes states and offers an intuitive proof of the fundamental theorem of algebra (first stated by Albert Girard in 1629) that an nth degree equation has exactly n roots. The proof rests simply on the principle that every root must appear in one of the binomial factors of $P(x)$ and that it requires n such factors to achieve x^n as the highest power of x in that polynomial. Descartes is therefore prepared to recognize not only negative roots (he gives as a corollary the law of signs for the number of negative roots) but also "imaginary" solutions to complete the necessary number.[10] In a series of examples, he then shows how to alter the signs of the roots of an equation, to increase them (additively or multiplicatively), or to decrease them. Having derived from the factored form of an equation its elementary symmetric functions,[11] Descartes uses them to eliminate the term containing x^{n-1} in the equation. This step paves the way for the general solution of the cubic and quartic equations (material dating back to Descartes's earliest studies) and leads to a general discussion of the solution of equations, in which the first method outlined is that of testing the various factors of the constant term, and then other means, including approximate solution, are discussed.

The *Géométrie* represented the sum of mathematical knowledge to which Descartes was willing to commit himself in print. The same philosophical concepts that led to the brilliant new method of geometry also prevented him from appreciating the innovative achievements of his contemporaries. His demand for strict a priori deduction caused him to reject Fermat's use of counterfactual assumptions in the latter's method of maxima and minima and rule of tangents.[12] His demand for absolute intuitive clarity in concepts excluded the infinitesimal from his mathematics. His renewed insistence on Aristotle's rigid distinction between "straight" and "curved" led him to reject from the outset any attempt to rectify curved lines.

Despite these hindrances to adventurous speculation, Descartes did discuss in his correspondence some problems that lay outside the realm of his *Géométrie*. In 1638, for example, he discussed with Mersenne, in connection with the law of falling bodies, the curve now expressed by the polar equation $\rho = a\lambda^{\vartheta}$ (logarithmic spiral)[13] and undertook the determination of the normal to, and quadrature of, the cycloid. Also in 1638 he took up a problem posed by Florimond Debeaune: (in modern terms) to construct a curve

satisfying the differential equation $a(dy/dx) = x - y$. Descartes appreciated Debeaune's quadrature of the curve and was himself able to determine the asymptote $y = x - a$ common to the family, but he did not succeed in finding one of the curves itself.[14]

By 1638, however, Descartes had largely completed his career in mathematics. The writing of the *Meditations* (1641), its defense against the critics, and the composition of the magisterial *Principia philosophiae* (1644) left little time to pursue further the mathematical studies begun in 1618.

Optics. In addition to presenting his new method of algebraic geometry, Descartes's *Géométrie* also served in book II to provide rigorous mathematical demonstrations for sections of his *Dioptrique* published at the same time. The mathematical derivations pertain to his theory of lenses and offer, through four "ovals," solutions to a generalized form of the anaclastic problem.[15] The theory of lenses, a topic that had engaged Descartes since reading Kepler's *Dioptrica* in 1619, took its form and direction in turn from Descartes's solution to the more basic problem of a mathematical derivation of the laws of reflection and refraction, with which the *Dioptrique* opens.

Background to these derivations was Descartes's theory of light, an integral part of his overall system of cosmology.[16] For Descartes light was not motion (which takes time) but rather a "tendency to motion," an impulsive force transmitted rectilinearly and instantaneously by the fine particles that fill the interstices between the visible macrobodies of the universe. His model for light itself was the blind man's cane, which instantaneously transmits impulses from the objects it meets and enables the man to "see." To derive the laws of reflection and refraction, however, Descartes required another model more amenable to mathematical description. Arguing that "tendency to motion" could be analyzed in terms of actual motion, he chose the model of a tennis ball striking a flat surface. For the law of reflection the surface was assumed to be perfectly rigid and immobile. He then applied two fundamental principles of his theory of collision: first, that a body in motion will continue to move in the same direction at the same speed unless acted upon by contact with another body; second, that a body can lose some or all of its motion only by transmitting it directly to another. Descartes measured motion by the product of the magnitude of the body and the speed at which it travels. He made a distinction, however, between the speed of a body and its "determination" to move in a certain direction.[17] By this distinction, it might come about that a body impacting with another would lose none of its speed (if the other body remained unmoved) but would

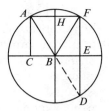

FIGURE 3

receive another determination. Moreover, although Descartes treated speed as a scalar quantity, determination was (operationally at least) always a vector, which could be resolved into components.[18] When one body collided with another, only those components of their determinations that directly opposed one another were subject to alteration.

Imagine, then, says Descartes, a tennis ball leaving the racket at point A and traveling uniformly along line AB to meet the surface CE at B. Resolve its determination into two components, one (AC) perpendicular to the surface and one (AH) parallel to it. Since, when the ball strikes the surface, it imparts none of its motion to the surface (which is immobile), it will continue to move at the same speed and hence after a period of time equal to that required to traverse AB will be somewhere on a circle of radius AB about B. But, since the surface is impenetrable, the ball cannot pass through it (say to D) but must bounce off it, with a resultant change in determination. Only the vertical component of that determination is subject to change, however; the horizontal component remains unaffected. Moreover, since the body has lost none of its motion, the length HF of that component after collision will equal the length AH before. Hence, at the same time the ball reaches the circle it must also be at a distance HF = AH from the normal HB, i.e., somewhere on line FE. Clearly, then, it must be at F, and consideration of similar triangles shows that the angle of incidence ABH is equal to the angle of reflection HBF.

For the law of refraction, Descartes altered the nature of the surface met by the ball; he now imagined it to pass through the surface, but to lose some of its motion (i.e., speed) in doing so. Let the speed before collision be to that after as p : q. Since both

speeds are uniform, the time required for the ball to reach the circle again will be to that required to traverse AB as p : q. To find the precise point at which it meets the circle, Descartes again considered its determination, or rather the horizontal component unaffected by the collision. Since the ball takes longer to reach the circle, the length of that component after collision will be greater than before, to wit, in the ratio of p : q. Hence, if FH : AH = p : q, then the ball must lie on both the circle and line FE. Let I be the common point.

The derivation so far rests on the assumption that the ball's motion is decreased in breaking through the surface. Here again Descartes had to alter his model to fit his theory of light, for that theory implies that light passes more easily through the denser medium. For the model of the tennis ball, this means that, if the medium below the surface is denser than that above, the ball receives added speed at impact, as if it were struck again by the racket. As a result, it will by the same argument as given above be deflected toward the normal as classical experiments with air-water interfaces said it should.

In either case, the ratio p : q of the speeds before and after impact depended, according to Descartes, on the relative density of the media and would therefore be constant for any two given media. Hence, since

$$FH : AH = BE : BC = p : q,$$

it follows that

$$\frac{BE}{BI} : \frac{BC}{AB} = \sin \angle AHB : \sin \angle IBG$$

$$= p : q = n \text{ (constant)},$$

which is the law of refraction.

The vagueness surrounding Descartes's concept of "determination" and its relation to speed makes his derivations difficult to follow. In addition, the assumption in the second that all refraction takes place at the surface lends an ad hoc aura to the proof, which makes it difficult to believe that the derivation represented Descartes's path to the law of refraction (the law of reflection was well known). Shortly after Descartes's death, prominent scientists, including Christian Huygens, accused him of having plagiarized the law itself from Willebrord Snell and then having patched together his proof of it. There is, however, clear evidence that Descartes had the law by 1626, long before Golius uncovered Snell's unpublished memoir.[19] In 1626 Descartes had Claude Mydorge grind a hyperbolic lens that represented an anaclastic derived by Descartes from the sine law of refraction. Where Descartes got the law, or how he got it, remains

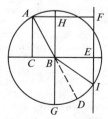

FIGURE 4

a mystery; in the absence of further evidence, one must rest content with the derivations in the *Dioptrique.*

Following those derivations, Descartes devotes the remainder of the *Dioptrique* to an optical analysis of the human eye, moving from the explanation of various distortions of vision to the lenses designed to correct them, or, in the case of the telescope, to increase the power of the normal eye. The laws of reflection and refraction reappear, however, in the third of the *Essais* of 1637, the *Météores.* There Descartes presents a mathematical explanation of both the primary and secondary rainbow in terms of the refraction and internal reflection of the sun's rays in a spherical raindrop.[20] Quantitatively, he succeeded in deriving the angle at which each rainbow is seen with respect to the angle of the sun's elevation. His attempted explanation of the rainbow's colors, however, rested on a general theory of colors that could at the time only be qualitative. Returning to the model of the tennis ball, Descartes explained color in terms of a rotatory motion of the ball, the speed of rotation varying with the color. Upon refraction, as through a prism, those speeds would be altered, leading to a change in colors.

Mechanics. Descartes's contribution to mechanics lay less in solutions to particular problems than in the stimulus that the detailed articulation of his mechanistic cosmology provided for men like Huygens.[21] Concerned with the universe on a grand scale, he had little but criticism for Galileo's efforts at resolving more mundane questions. In particular, Descartes rejected much of Galileo's work, e.g., the laws of free fall and the law of the pendulum, because Galileo considered the phenomena in a vacuum, a vacuum that Descartes's cosmology excluded from the world. For Descartes, the ideal world corresponded to the real one. Mechanical phenomena took place in a plenum and had to be explained in terms of the direct interaction of the bodies that constituted it, whence the central role of his theory of impact.[22]

Two of the basic principles underlying that theory have been mentioned above. The first, the law of inertia, followed from Descartes's concept of motion as a state coequal with rest; change of state required a cause (i.e., the action of another moving body) and in the absence of that cause the state remained constant. That motion continued in a straight line followed from the privileged status of the straight line in Descartes's geometrical universe. The second law, the conservation of the "quantity of motion" in any closed interaction, followed from the immutability of God and his creation. Since bodies acted on each other by transmission of their motion,

the "quantity of motion" (the product of magnitude and speed) served also as Descartes's measure of force or action and led to a third principle that vitiated Descartes's theory of impact. Since as much action was required for motion as for rest, a smaller body moving however fast could never possess sufficient action to move a larger body at rest. As a result of this principle, to which Descartes adhered in the face of both criticism and experience, only the first of the seven laws of impact (of perfectly elastic bodies meeting in the same straight line) is correct. It concerns the impact of two equal bodies approaching each other at equal speeds and is intuitively obvious.

Descartes's concept of force as motive action blocked successful quantitative treatment of the mechanical problems he attacked. His definition of the center of oscillation as the point at which the forces of the particles of the swinging body are balanced out led to quite meager results, and his attempt to explain centrifugal force as the tendency of a body to maintain its determination remained purely qualitative. In all three areas—impact, oscillation, and centrifugal force—it was left to Huygens to push through to a solution, often by discarding Descartes's staunchly defended principles.

Descartes met with more success in the realm of statics. His *Explication des engins par l'aide desquels on peut avec une petite force lever un fardeau fort pesant,* written as a letter to Constantin Huygens in 1637, presents an analysis of the five simple machines on the principle that the force required to lift *a* pounds vertically through *b* feet will also lift *na* pounds *b/n* feet. And a memoir dating from 1618 contains a clear statement of the hydrostatic paradox, later made public by Blaise Pascal.[23]

NOTES

1. Cf. Gaston Milhaud, *Descartes savant* (Paris, 1921), and Jules Vuillemin, *Mathématiques et métaphysique chez Descartes* (Paris, 1960).
2. Cf. Milhaud, pp. 84–87.
3. Defending his originality against critics, Descartes repeatedly denied having read the algebraic works of François Viète or Thomas Harriot prior to the publication of his own *Géométrie.* The pattern of development of his ideas, especially during the late 1620's, lends credence to this denial.
4. In his *Scholarum mathematicarum libri unus et triginta* (Paris, 1569; 3rd. ed., Frankfurt am Main, 1627), bk. I (p. 35 of the 3rd ed.). Descartes quite likely knew of Ramus through Beeckman, who had studied mathematics with Rudolph Snell, a leading Dutch Ramist.
5. *Regulae ad directionem ingenii,* in *Oeuvres de Descartes,* Adam and Tannery, eds., X (Paris, 1908), rule IV, 376–377.
6. Descartes to Beeckman (26 Mar. 1619), *Oeuvres,* X, 156–158. By "imaginary" curve, Descartes seems to mean a curve that can be described verbally but not accurately constructed by geometrical means.

7. Both the term and the concept it denotes are certainly anachronistic. Descartes speaks of the indeterminate equation that links x and y as the "relation [rapport] that all the points of a curve have to all those of a straight line" (Géométrie, p. 341). Strangely, Descartes makes no special mention of one of the most novel aspects of his method, to wit, the establishment of a correspondence between geometrical loci and indeterminate algebraic equations in two unknowns. He does discuss the correspondence further in bk. II, 334–335, but again in a way that belies its novelty. The correspondence between determinate equations and point constructions (i.e., section problems) had been standard for some time.

8. For problems of lower degree, Descartes maintains the classification of Pappus. Plane problems are those that can be constructed with circle and straightedge, and solid problems those that require the aid of the three conic sections. Where, however, Pappus grouped all remaining curves into a class he termed linear, Descartes divides these into distinct classes of order. To do so, he employs in bk. I a construction device that generates the conic sections from a referent triangle and then a new family of higher order from the conic sections, and so on.

9. Two aspects of the symbolism employed here require comment. First, Descartes deals for the most part with specific examples of polynomials, which he always writes in the form $x^n + a_1x^{n-1} + \cdots + a_n = 0$; the symbolism $P(x)$ was unknown to him. Second, instead of the equal sign, $=$, he used the symbol ∞, most probably the inverted ligature of the first two letters of the verb aequare ("to equal").

10. One important by-product of this structural analysis of equations is a new and more refined concept of number. See Jakob Klein, Greek Mathematical Thought and the Origins of Algebra (Cambridge, Mass., 1968).

11. Here again a totally anachronistic term is employed in the interest of brevity.

12. Ironically, Descartes's method of determining the normal to a curve (bk. II, 342 ff.) made implicit use of precisely the same reasoning as Fermat's. This may have become clear to Descartes toward the end of a bitter controversy between the two men over their methods in the spring of 1638.

13. Cf. Vuillemin, pp. 35–55.

14. Ibid., pp. 11–25; Joseph E. Hofmann, Geschichte der Mathematik, II (Berlin, 1957), 13.

15. The anaclastic is a refracting surface that directs parallel rays to a single focus; Descartes had generalized the problem to include surfaces that refract rays emanating from a single point and direct them to another point. Cf. Milhaud, pp. 117–118.

16. The full title of the work Descartes suppressed in 1636 as a result of the condemnation of Galileo was Le monde, ou Traité de la lumière. It contained the basic elements of Descartes's cosmology, later published in the Principia philosophiae (1644). For a detailed analysis of Descartes's work in optics, see A. I. Sabra, Theories of Light From Descartes to Newton (London, 1967), chs. 1–4.

17. "One must note only that the power, whatever it may be, that causes the motion of this ball to continue is different from that which determines it to move more toward one direction than toward another," Dioptrique (Leiden, 1637), p. 94.

18. Cf. Descartes to Mydorge (1 Mar. 1638), "determination cannot be without some speed, although the same speed can have different determinations, and the same determination can be combined with various speeds" (quoted by Sabra, p. 120). A result of this qualification is that Descartes in his proofs treats speed operationally as a vector.

19. See the summary of this issue in Sabra, pp. 100 ff.

20. Cf. Carl B. Boyer, The Rainbow: From Myth to Mathematics (New York, 1959).

21. For a survey of Descartes's work on mechanics, which includes the passages pertinent to the subjects discussed below, see René Dugas, La mécanique au XVIIᵉ siècle (Neuchâtel, 1954), ch. 7.

22. Presented in full in the Principia philosophiae, pt. II, pars. 24–54.

23. Cf. Milhaud, pp. 34–36.

BIBLIOGRAPHY

I. ORIGINAL WORKS. All of Descartes's scientific writings can be found in their original French or Latin in the critical edition of the Oeuvres de Descartes, Charles Adam and Paul Tannery, eds., 12 vols. (Paris, 1897–1913). The Géométrie, originally written in French, was trans. into Latin and published with appendices by Franz van Schooten (Leiden, 1649); this Latin version underwent a total of four eds. The work also exists in an English trans. by Marcia Latham and David Eugene Smith (Chicago, 1925; repr., New York, 1954), and in other languages. For references to eds. of the philosophical treatises containing scientific material, see the bibliography for sec. I.

II. SECONDARY LITERATURE. In addition to the works cited in the notes, see also J. F. Scott, The Scientific Work of René Descartes (London, 1952); Carl B. Boyer, A History of Analytic Geometry (New York, 1956); Alexandre Koyré, Études galiléennes (Paris, 1939); E. J. Dijksterhuis, The Mechanization of the World Picture (Oxford, 1961). See also the various histories of seventeenth-century science or mathematics for additional discussions of Descartes's work.

MICHAEL S. MAHONEY

DESCARTES: Physiology.

Descartes's physiology grew and developed as an integral part of his philosophy. Although grounded at fundamental points in transmitted anatomical knowledge and actually performed dissection procedures, it sprang up largely independently of prior physiological developments and depended instead on the articulation of the Cartesian dualist ontology, was entangled with the vagaries of metaphysical theory, and deliberately put into practice Descartes's precepts on scientific method. Chronologically, too, his physiology grew with his philosophy. Important ideas on animal function occur briefly in the Regulae (1628), form a significant part of the argument in the Discours de le méthode (1637), and lie behind certain parts of the Principia philosophiae (1644) and all of the Passions de l'âme (1649). Throughout his active philosophical life, physiology formed one of Descartes's most central and, sometimes, most plaguing concerns.

Descartes hinted at the most fundamental conceptions of his physiology relatively early in his philosophical development. Already in the twelfth regula, he suggested (without, however, elaborating either more rigorously or more fully) that all animal and subrational human movements are controlled solely by unconscious mechanisms. Just as the quill of a pen moves in a physically necessary pattern determined by the motion of the tip, so too do "all the motions of the animals come about"; thus one can also explain "how ·in ourselves all those operations occur which [we] perform without any aid from the reason." Closely associated in the *Regulae* with this notion of animal automatism was Descartes's belief that human sensation is a two-step process consisting, first, of the mechanical conveyance of physical stimuli from the external organs of sense to a common sensorium located somewhere in the body and, second, of the internal perception of these mechanically conveyed stimuli by a higher "spiritual" principle.

Implicit in these two notions and seeming to tie them together is the assumption, evident in broader compass but in as terse a formulation as elsewhere in the *Regulae,* that all phenomena of the animate and inanimate world, with the sole exception of those directly connected with human will and consciousness, are to be explained in terms of mathematics, matter, configuration, and motion.

The fuller working out of his physiological ideas occupied Descartes in the early 1630's, when he was concerned generally with the development of his ontological and methodological views. In 1632 he several times referred to physiological themes and projects in his correspondence, and in June he informed Mersenne that he had already completed his work on inanimate bodies but still had to finish off "certain things touching on the nature of man." The allusion here was to the *Traité de l'homme,* which with the *Traité de lumière* was meant to form *Le monde.* Along with the *Traité de lumière,* however, the *Traité de l'homme* was suppressed by Descartes after the condemnation of Galileo in 1633, and although it thus had to await posthumous publication in the 1660's, his writing of the *Traité de l'homme* proved extremely important in the further maturation of Descartes's physiological conceptions.

The *Traité de l'homme* begins and ends with a proclamation of literary and philosophical license. In the *Traité,* Descartes writes, we deliberately consider not a real man but a "statue" or *machine de terre* expressly fashioned by God to approximate real men as closely as possible. Like a real man, the *machine de terre* will be imagined to possess an immaterial soul and a physical body, and, also like a real man, its physical body will consist of a heart, brain, stomach, vessels, nerves, *et al.* But since we are considering only an artificial man—a contrivance fashioned more perfectly but on the same principles as a clock or water mill—we will not be tempted to attribute the motions and activities of this man to special sensitive or vegetative souls or principles of life. Nothing more than a contained rational and immaterial soul and "the disposition of organs, no more and no less than

in the movements of a clock or any other automaton" will be needed to comprehend the active functioning of this special contrivance formed by God and operated thereafter by the principles of mechanical action. We are to bear in mind, of course, that the man of the *Traité* is remarkably like men we know, but our literary and philosophical license allows us to hypothesize and analogize freely.

Descartes fully exercises his self-proclaimed license in the rest of the *Traité.* He first surveys various physiological processes, giving for each of them not the traditional or neoclassical account of such recent physiological writers as Fernel or Riolan but mechanistic details by which the particular function is performed automatically in the *homme.* Each of Descartes's explanations borrows something from traditionalist physiological theories, but in each case Descartes wields Ockham's razor to strip away excess souls, faculties, forces, and innate heats from the corpuscular or chemical core of explanation.

Digestion, for example, is for Descartes only a fermentative process in which the particles of food are broken apart and set into agitation by fluids contained in the stomach. Chyle and excremental particles are then separated from one another in a filtration performed merely by a sievelike configuration of the pores and vascular openings in the intestines. Chyle particles go through another filtration and fermentation in the liver, where they thereby—and only thereby—acquire the properties of blood. Blood formed in the liver drips from the vena cava into the right ventricle of the heart, where the purely physical heat implanted there quickly vaporizes the sanguinary mass. The expansion of this sanguinary vapor pushes out the walls of the heart and arteries. Expansion with rarefaction is succeeded by cooling; and, as the vapor condenses, the heart and arteries return to their original size. The heart is fitted with a perfect arrangement of valves, and in addition the *homme* is served by a perpetual circulation of blood. (Descartes had read William Harvey's *De motu cordis,* as is also clear from his prior correspondence, but apparently took seriously only the part on circulatory motion rather than on cardiac action.) Cardiac and arterial pulsation is thus continually and automatically repeated throughout the life of the automaton by mechanical means, not under the control of an active diastolic faculty. And while the sanguinary particles are coursing through the vessels, certain of them separate off into special pores, which accounts for both nutrition and the sievelike production of such secretions as bile and urine.

After this mechanistic survey of general physiology, Descartes moves to the nervous system, which he treats in considerable detail. The nerves are said to be a series of essentially hollow tubes with a filamentous marrow and are similar in operation to the water-filled pipes of those hydraulically controlled puppets and mechanical statues found "in the grottoes and fountains in the gardens of our Kings." Filling the spaces in the nerves is a fine, material substance, the animal spirits. These spirits are actually the most quickly moving particles of the blood that have traveled through the arteries in the shortest, straightest path from the heart to the brain. Once conveyed to the brain according to the laws of mechanics and then mechanically separated from the coarser parts of the blood, these most agile particles become "a wind or very subtle flame."

This spiritual wind can flow into the muscles, which are directly connected with the neural tubes. When a particular muscle is inflated by the influx of animal spirits, its belly distends as if by wind billowing a canvas sail, and the ends or insertions of the muscle are necessarily pulled more closely together. The pulling together of the muscular insertions constitutes muscle contraction. Gross movements in the *homme* (including breathing and swallowing) are produced, therefore, as the necessary mechanical effects of animal spirits discharged to one or another muscle group.

The movement of animal spirits is also controlled, however, by the action of the pineal gland in the brain. The rational soul is itself most closely associated with this gland located centrally in the substance of the cerebral marrow, and by directly causing this gland to move, no matter how slightly, the immaterial, willful soul of man can redirect the animal spirits from one set of nervous channels to another. Redirection of animal spirits results, in turn, in the production of different gross muscle movements. The pineal gland can also be, and often is, directly and unconsciously affected by a whole array of supervening influences, among the most important of which are the sensory. In animals, of course, only the supervening influences operate. Since Descartes has already gotten into a discussion of sensation, he now discusses that subject in great detail. He devotes much attention to the external organs of sense, concentrating to a large degree on the visual apparatus. For his discussion here, Descartes was able to draw upon the prior work of sixteenth-century anatomists and natural philosophers that had culminated in Kepler's fine account of the eye as a camera-like optical system. Yet Descartes inserts his optical account of the eye into his already developed physiological system; for once the image is formed on the retina, Descartes explains, the nerves—and with them Descartes's general physiology—take over. Rays of light focused on particular points of the retina cause specific nerves to jiggle slightly. Since the solid, filamentous part of the nerve is continuous from the sense organ to the brain, the externally caused jiggle is immediately transmitted, like the tug on a bell rope, to the interior of the brain as an internal jiggle. Internal jiggles directly control the streaming of animal spirits in the brain, and the rational soul, operating only at the pineal gland, interprets the patterns of the streams as particular sensations. The soul works "in the dark" this way, too, in all other sensations and in the difficult activity of multilevel perception. The soul "reads" various motions produced in the brain by the nerves and spirits, and interprets particular combinations of motions as taste, odor, color, or even distance.

Descartes moves from this complicated discussion of the five external senses of his *homme* to a consideration of certain internal feelings which it also experiences. The physiology (and psychology) is here, too, based on the manner by which the soul "reads" the messages delivered through the nerves by the spirits, messages which depend, in this instance, upon the internal functioning of the various parts of the body. Thus, when the blood entering the heart is purer and more subtle than usual, it vaporizes very easily in the cardiac chambers, and, as a consequence, it stimulates the nerves placed in the heart in a manner that the soul will associate with joy. All sorts of moods, feelings, and what might be generally labeled chains of somatopsychic effects (Descartes usually calls these "passions") are schematically accounted for in this same manner—by imagining a passive immaterial soul interpreting the varied motions of the spirits as they stream pass the pineal gland.

As a logical extension of his consideration of the "passions," Descartes turns, finally, to a discussion of sleep, dreams, memory, and imagination—other consequences of the interaction of the soul with internal neurophysiology. All these latter psychological phenomena are said to depend on the special motion of animal spirits, sometimes through favored pathways created by habitual or normal daytime activity.

With this discussion (and with his restatement of literary and philosophical license) Descartes terminates the *Traité*. It was obviously written as a full working out of the physiological hints included in the *Regulae* and elaborated in the light of his own philosophical development. A clearly stated dualist ontology runs through the *Traité*, while mechanistic details analogous to those of the *Traité de lumière* are evident at almost every turn.

But the *Traité de l'homme* served not only to clarify and develop Descartes's physiological views; it also quickly became a rich fund of ideas upon which he drew throughout the rest of his intellectual life. In 1637, for example, Descartes published two important works: *Discours de la méthode* and *Dioptrique*. In both he uses physiological ideas from the unpublished *Traité* at important points in his argument. Specifically, in part V of the *Discours,* Descartes employs a summary of his cardiovascular physiology to illustrate how the newly discovered laws by which God orders his universe are sufficient to explicate certain of the most important human functions, and the *Dioptrique* includes a summary of his general theory of sensation as a preliminary to a detailed study of image formation and visual perception. In the 1640's, too, Descartes drew heavily upon the unpublished *Traité*. The complicated arguments of the *Passions de l'âme,* published near the end of that decade, rest firmly on the extensive survey of basic Cartesian physiology incorporated in part I, while Descartes's unruffled assertiveness in his correspondence and later philosophical writings on the "beast-machine" makes full sense only against a background provided by the *Traité*'s automaton.

Descartes, however, had left one major physiological problem untreated in the *Traité:* the reproductive generation of animals and men. He had insisted for reasons of methodological circumspection that the *homme* of the *Traité* was directly contrived by God. The Cartesian program, of course, was to explain all but rational, deliberately willful, or self-conscious behavior in terms of mere mechanism. He had eliminated the souls, principles, faculties, and innate heats of traditional physiology and had systematically replaced them with hypotheses and analogies of purely physical nature. But generation had escaped, and its explanation in mechanistic terms was clearly needed for the logical completion of his system. Recently proposed theories of animal generation had left the subject replete with Galenic faculties and Aristotelian souls, and even William Harvey was soon to show himself content with innate principles and plastic forces as the controlling agents of embryonic development. Descartes, to be consistent, could not accept these explanatory devices and had, therefore, to formulate some alternative.

His correspondence and certain manuscript remains show that Descartes had actually been deeply concerned with the problem of animal generation for a considerable period of time. Earliest references in the former occur in 1629, and snippets of the latter reveal fitful grapplings with the problem, some of them even leading to direct anatomical investigations undertaken, apparently, as a means for providing clues to the processes involved. Descartes's ideas on the subject really seem to have crystallized, however, in the late 1640's, when he triumphantly announced his "solution" to the long-plaguing problem in a series of enthusiastic letters to Princess Elizabeth. The ideas alluded to in these letters appear to be those published as the *De la formation du foetus,* which Descartes completed not long before his death.

First published by Clerselier in 1664, the *Formation* is a curious essay. Unlike the *Traité de l'homme,* which much preceded it in date of composition, the *Formation* consists mainly of bald assertions and only the vaguest mechanisms. Generation commences when the male and female seeds come together and mutually induce a corpuscular fermentation. The motion of certain of the fermenting particles forms the heart, that of others the lungs. Streaming of particles as the process continues furrows out the blood vessels; later, membranes and fibers are formed which ultimately weave together to construct the solid parts. The formation of the bodily parts is described in these vague terms (no mechanism of organ or vascular development is ever made more precise than this), yet Descartes apparently felt satisfied with his results. For by describing generation in chemical and corpuscular, rather than vital or teleological, terms Descartes had, at least in his own mind, completed the mechanistic program for physiology. Everything in the animal's life, from its first formation to its final decay, now had an automatic, mechanical explanation.

The impact of the Cartesian physiological program, once it was publicly known, was enormous. In two ways—philosophically and physiologically—Descartes transformed long-standing beliefs about animals and men. Philosophically, of course, his notions of mind-body dualism and animal automatism had extremely important implications that were not lost on Henry More, Malebranche, Spinoza, and Leibniz, along with many others in the seventeenth century. The "beast–machine" idea also had continuing ramifications in the eighteenth century, leading, at least according to Aram Vartanian, directly to La Mettrie's *L'homme machine*. Also, according to Vartanian, Descartes's posthumously published views on human function and animal generation exerted important philosophical influence, contributing greatly to the eighteenth-century concern with these biological subjects by many of the *philosophes*. But physiologically, too, Descartes's conceptions had an impact that in many ways was even more impressive than the philosophical influence, because it affected the actual course of contemporary science.

Almost as soon as Descartes published his *Discours de la méthode,* a few professors of medicine began to react to specific Cartesian physiological ideas and to the general Cartesian program. Plempius at Louvain and Regius at Utrecht were among the first. Although Plempius proved relatively hostile to Cartesian ideas (his objections were not unlike those William Harvey was to raise a few years later), Regius became so enthusiastic that for a time he entered into something of a student-teacher relationship with Descartes. As their correspondence for 1641 makes clear, Regius would send his students' theses to Descartes for comment and correction. Descartes would then return them with such specific excisions of classical residues as "In the first line of the Thesis I would get rid of these words: vivifying heat."

The intimate contact between Descartes and Regius marked the beginning of the direct influence of Cartesian ideas and modes of thought on seventeenth-century physiology. That influence was deliberately continued later, even more vigorously after Descartes's death, by such influential figures as Thomas Bartholin and Nicholas Steno. These men, especially Steno, tried to wed the Cartesian method of mechanistic explanation to careful anatomical investigation. Steno was particularly highly regarded by contemporaries for his perfection of the mechanical theory of muscular contraction in his *Elementorum myologiae specimen* (1667) and for his defense of the Cartesian physiological methodology in his anatomically sound *Discours sur l'anatomie du cerveau* (1669).

Many other prominent seventeenth-century physiological writers were influenced by the Cartesian program, either directly by reading Descartes's writings or indirectly through such followers as Steno. Among those deeply influenced by Cartesian physiology were Robert Hooke, Thomas Willis, Jan Swammerdam, and Giovanni Alfonso Borelli. These men saw in Cartesian physiology exactly what Descartes had intended it to be: a method of mechanistic formulation by which traditional categories of physiological explanation could be circumvented. Without Descartes, the seventeenth-century mechanization of physiological conceptions would have been inconceivable.

BIBLIOGRAPHY

The main primary sources—letters, MSS, and published works—are all handsomely printed in the Adam-Tannery *Oeuvres de Descartes*; vol. XI contains the largest sample of relevant works.

Useful secondary studies of Descartes's philosophy which seriously consider his physiological writings range from Étienne Gilson's *Études sur le rôle de la pensée médiévale dans la formation du système Cartésien* (Paris, 1930) to Norman Kemp Smith's *New Studies in the Philosophy of Descartes* (New York, 1966). Two older monographic stud-ies are also useful: Bertrand de Saint-Germain's *Descartes considéré comme physiologiste* (Paris, 1869) and Auguste-Georges Berthier's "Le mécanisme Cartésien et la physiologie au XVIIe siècle," in *Isis,* **2** (1914), 37–89, and **3** (1920), 21–58. Some of the background to Descartes's treatment of vision is made clear in A. C. Crombie, "The Mechanistic Hypothesis and the Scientific Study of Vision," in *Proceedings of the Royal Microscopical Society,* **2** (1907), 3–112; fundamental aspects of his mechanistic philosophy are discussed in Georges Canguilhem, *La formation du concept de réflexe aux XVIIe et XVIIIe siècles* (Paris, 1955); while the seventeenth-century impact of Cartesian ideas is studied by Berthier (*op. cit.*) and in two helpful general works, Paul Mouy, *Le développement de la physique Cartésienne* (Paris, 1934) and vol. I of Thomas S. Hall, *Ideas of Life and Matter* (Chicago, 1969).

A sense of the influence of Cartesian ideas and methods can also be gleaned from Michael Foster, *Lectures on the History of Physiology* (Cambridge, 1901); and Gustav Scherz's various studies of Nicholas Steno. Finally, see Aram Vartanian, *Diderot and Descartes* (Princeton, 1953) and *La Mettrie's "L'homme machine"* (Princeton, 1960).

THEODORE M. BROWN

DES CLOIZEAUX, ALFRED-LOUIS-OLIVIER LEGRAND (*b.* Beauvais, France, 17 October 1817; *d.* Paris, France, 6 May 1897), *mineralogy, crystallography.*

Born of a family of the old bourgeoisie with a long tradition in the legal profession, Des Cloizeaux studied at the Lycée Charlemagne in Paris, where he came under the tutelage of Armand Lévy, a teacher of mathematics and mineralogy who instilled in him a fascination with minerals and crystals. Lévy encouraged Des Cloizeaux to attend the courses of Alexandre Brongniart at the Muséum d'Histoire Naturelle and those of Armand Dufrénoy at the École des Mines. Also through Lévy, Des Cloizeaux was introduced into Jean-Baptiste Biot's laboratory at the Collège de France. It was through Biot's influence that Des Cloizeaux was commissioned by the government in 1845 to travel to Iceland to study sources of Iceland spar. This voyage also enabled Des Cloizeaux to visit some British mineralogists (Robert Jameson, Thomas Thomson, and George Bellas Greenough, among others) and to inspect mineralogical collections in Scotland and England. In 1846 he returned to Iceland and joined there with members of a German-Danish expedition (including Bunsen and Wolfgang Sartorius von Waltershausen) in studying geysers and minerals. In succeeding years he traveled widely, especially in the Alpine regions, Scandinavia, and Baltic Russia.

In 1843 Des Cloizeaux had become a tutor at the École Centrale. He defended his thesis for the doctorate in 1857 at the Faculté des Sciences of Paris and was appointed professor in the École Normale Supérieure. In 1876 he was made professor of mineralogy at the Muséum in place of Delafosse, whom he had assisted at the Sorbonne since 1873; he taught at the Muséum until 1892. Des Cloizeaux was elected to the Académie des Sciences in 1869, replacing E. J. A. d'Archiac, having failed of election (to his considerable chagrin) in 1862, when Pasteur was chosen instead. In 1889 he served as president of the Academy.

Des Cloizeaux's main achievements fall into two categories: his studies on the form of crystals, and his investigation of the optical properties of crystalline materials. While questions of morphology occupied him early in his career, after 1855 he devoted himself principally to optical problems in crystallography. In both categories his work was characterized by the broad interest of a naturalist attempting to relate the substance under investigation to its mode of origin.

Des Cloizeaux was able, in part through optical methods, to elucidate the interior structure of minerals that had already been subjected to thorough study (for example, quartz).[1] He set out ambitiously to produce a comprehensive work on the structure (but not in the modern sense) of minerals. The result, the *Manuel de minéralogie,* occupied him for thirty-five years but was never completed beyond two volumes. This project began simply as a plan to translate William Phillips' *An Elementary Introduction to Mineralogy* as extended by Henry James Brooke and William Hallowes Miller, but it grew into a text emphasizing Des Cloizeaux's interest in crystallography and serving to establish the crystallographic notation invented by René-Just Haüy and augmented by Lévy.

The most original aspect of Des Cloizeaux's work lies in the field of optical studies of crystals, which he took up in part through the influence of Henri de Sénarmont. He embarked on the gigantic task of determining the optical characters of all known crystals, and although this proved too large an undertaking for one man he did ascertain the optical properties of nearly 500 substances. He was among the first to perceive the great potential utility of the polarizing microscope for investigating minerals, and with improved polarizing microscopes of his own devising he developed techniques for the determination of significant optical characteristics in crystals (e.g., the angle of the optic axis and the dispersions of the optic axes, indicatrix, and bisectrices). His methods and determinations constituted a part of the foundation of petrology. Among his extensions of the knowledge of polarization in crystals was his demonstration of circular polarization in cinnabar and strychnine sulfate.[2]

Inquiring into the effects of heat on crystalline bodies, he found that prolonged heating beyond a certain temperature permanently alters the positions of the optic axes of certain crystals (notably the orthoclase minerals), thereby providing the geologist with a means of determining whether or not certain rocks have been subjected to high temperatures.[3]

Descloizite, a rare mineral consisting of basic lead and zinc vanadate, was named after Des Cloizeaux by his friend and collaborator Augustin-Alexis Damour.[4]

NOTES

1. *Annales de chimie et de physique,* 3rd ser., **45** (1855), 129–316; *Mémoires présentés par divers savants à l'Académie des sciences. Sciences mathématiques et physiques,* **15** (1858), 404–614.
2. *Comptes rendus hebdomadaires des séances de l'Académie des sciences,* **44** (1857), 876–878, 909–912; *Annales de chimie et de physique,* 3rd ser., **51** (1857), 361–367; *Annalen der Physik und Chemie,* **102** (1857), 471–474.
3. *Comptes rendus,* **53** (1861), 64–68; **55** (1862), 651–654; **62** (1866), 987–990; *Annales de chimie et de physique,* 3rd ser., **68** (1863), 191–203; *Annales des mines,* 6th ser., **2** (1862), 327–328; *Annalen der Physik und Chemie,* **119** (1863), 481–492; **129** (1866), 345–350; *Bulletin de la Société géologique de France,* 2nd ser., **20** (1862–1863), 41–47; *Mémoires présentés par divers savants à l'Académie des sciences. Sciences mathématiques et physiques,* **18** (1868), 511–732.
4. *Annales de chimie et de physique,* 3rd ser., **41** (1854), 72–78.

BIBLIOGRAPHY

I. ORIGINAL WORKS. Des Cloizeaux's major work is *Manuel de minéralogie,* 2 vols. in 3 parts (Paris, 1862–1893). Alfred Lacroix's "Liste bibliographique des travaux de A. Des Cloizeaux," in his *Notice historique sur François-Sulpice Beudant et Alfred-Louis-Olivier Legrand Des Cloizeaux* (Paris, 1930), pp. 91–101, is fairly complete. Des Cloizeaux's own *Notice sur les travaux minéralogiques et géologiques de M. Des Cloizeaux* (Paris, 1869) provides annotations, some of them quite extensive, to a list of many of his publications.

II. SECONDARY LITERATURE. On Des Cloizeaux and his work, see Lacroix's *Notice,* referred to above, also in Lacroix's *Figures de savants,* 4 vols. (Paris, 1932–1938), I, 241–272. Contemporary biographical sources include Charles Barrois's funeral speech in *Bulletin de la Société géologique de France,* **25** (1897), 459–460; the eulogy by Adolphe Chatin, in *Comptes rendus hebdomadaires des séances de l'Académie des sciences,* **124** (1897), 983–984; an obituary by Lazarus Fletcher, in *Proceedings of the Royal Society of London,* **63** (1898), xxv–xxviii; and an anonymous sketch in *L'année scientifique* (1897), p. 409. Conrad Burri provides a recent assessment in "Alfred Des Cloizeaux 1817–1897, Ferdinand Fouqué 1828–1904, Auguste Michel-Lévy 1844–1911," in *Geschichte der Mikroskopie.*

Leben und Werk grosser Forscher, III, Hugo Freund and Alexander Berg, eds. (Frankfurt, 1966), 163–176.

KENNETH L. TAYLOR

DESCOTILS, HIPPOLYTE VICTOR. *See* **Collet-Descotils, Hippolyte Victor.**

DESCOURTILZ, MICHEL ÉTIENNE (*b.* Boiste, near Pithiviers, France, 25 November 1775; *d.* Paris, France, 1836), *medicine, natural history.*

Descourtilz was first trained as a surgeon. Following his marriage to the daughter of Rossignol-Desdunes, who had plantations in Artibonite, he went to Saint-Domingue (Haiti) in 1798, on the way visiting Charleston, South Carolina, and Santiago de Cuba. Descourtilz became involved in the Negro revolution and, in spite of the protection of Toussaint L'Ouverture, was nearly executed by Dessalines. He was forced to join the medical service of the Negro army, but in 1803 he escaped and sailed to Cádiz. After reaching Paris, Descourtilz became a doctor of medicine in 1814; he practiced in Orléans; was for a while physician at the Hôtel-Dieu in Beaumont-en-Gâtinais; and retired to Paris, where he was a member of the Société de Médecine Pratique and became president of the Société Linnéenne. Most of his original drawings and manuscripts, as well as his herbarium, were burned in Haiti; and in writing his books he had to rely on the works of Plumier, Joseph Surian, Alexandre Poiteau, and Turpin. His zoological contributions, particularly those on the caiman, were highly praised.

BIBLIOGRAPHY

In *Voyages d'un naturaliste* (Paris, 1809) Descourtilz narrates his adventures during the Haitian revolution and, at the end, deals with the natural history of the isle. Next to be published was *Code du safranier* (Paris, 1810). Pinel presided over his thesis, *Propositions sur l'anaphrodisie . . .* (Paris, 1814), later expanded into a larger volume, *De l'impuissance et de la stérilité . . .* (Paris, 1831). One of his most popular books was *Guide sanitaire des voyageurs aux colonies* (Paris, 1816). *Manuel indicateur des plantes usuelles aux Antilles* (Paris, 1821) appeared as he began, with one of his eight sons, Jean Théodore, to publish his major work, *Flore pittoresque des Antilles,* 8 vols. (Paris, 1821–1829), arranging the material according to medicinal properties. *Anatomie comparée du grand crocodile . . .* (Paris, 1825) was followed by *Des champignons comestibles* (Paris, 1827) and *Cours d'électricité médicale* (Paris, 1832).

Descourtilz is mentioned in Rulx Léon, *Notes bio-bibliographiques. Médecins et naturalistes de l'ancienne colonie française de Saint-Domingue* (Port-au-Prince, 1933).

FRANCISCO GUERRA

DESHAYES, GERARD PAUL (*b.* Nancy, France, 24 May 1797; *d.* Boran-sur-Oise, France, 9 June 1875), *paleontology, malacology.*

Deshayes was the son of a physics professor at the École Centrale in Nancy. He began studying medicine at the University of Strasbourg but left in 1820 and went to Paris, where in 1821 he became *bachelier-ès-lettres.* He then gave up his medical studies and, as an independent scholar, turned to natural history, particularly geology and malacology. Soon after his arrival in Paris he gave private lectures on geology and led field trips. Those who attended included Élie de Beaumont, d'Archiac, Philippe de Verneuil, Constant Prévost, Desnoyers, and Edmond Hébert. His home soon became a center for exchange of scientific ideas as well as social communications. On behalf of the Paris Academy of Sciences, Deshayes was in Algeria from 1840 to 1842 investigating mollusks, especially those of recent origin. Through his own collecting and through items sent by colleagues in many countries, he amassed a great collection of recent and fossil mollusks that was of exceptional importance because of the many original specimens and types it contained. Circumstances forced Deshayes to sell his collection, together with his comprehensive malacological and paleontological library, to the French government in 1868 for 100,000 francs. Both were turned over to the École des Mines in Paris.

In 1869 Deshayes was appointed professor at the Muséum d'Histoire Naturelle in Paris, occupying the chair of conchology once held by Lamarck, whom he highly regarded. This was Deshayes's first and only public post. Despite his advanced age, poor health, and the Franco-Prussian War, he enlarged the museum's malacological collection with tireless zeal and youthful energy. Deshayes was one of the founders of the Société Géologique de France and was several times its president.

An old and worsening heart ailment weakened Deshayes from 1873. He spent some time in Provence and then returned to Boran-sur-Oise, to the region containing the fossils he had described at the beginning of his career. He was survived by his wife and daughter.

In 1824, with his own modest financial means, Deshayes began the publication of *Description des coquilles fossiles des environs de Paris.* This work, interrupted by descriptions of recent and fossil mollusks of the Peloponnesus and India, was completed in 1837. It is among Deshayes's most important works. Along with the painstaking description and illustration of 1,074 species of mollusks, predominantly from the Eocene of the Paris Basin, including 660 species first described by him, there is a division of the Tertiary into three periods. The basis for this division was the proportion of living species among the total number of Tertiary forms: in the first (oldest) period, 3 percent; in the second (middle), 19 percent; in the third (most recent), 52 percent. Deshayes joined this with the premise that since the beginning of the Tertiary there had been a continuous decrease in temperature. This three-part division of the Tertiary (1831) agreed in methodological principles and main features with the subdivision of the Tertiary into Eocene, Miocene, and Pliocene proposed by Lyell. After Deshayes was brought into contact with Lyell, he provided for the latter's use statistical tables of Tertiary mollusk species. They were published in Lyell's *Principles of Geology* and constitute an essential support for Lyell's very important Tertiary subdivisions.

In the following years Deshayes published several large monographs, collections of articles, and compendia. These included a description of fossil mollusks from the Crimea (1838); a work on mollusks from the Algerian expedition, in twenty-five installments interrupted by the February Revolution of 1848 and the actions of envious colleagues (1844–1848); a new edition, in collaboration with Henri Milne-Edwards, of Lamarck's *Histoire des animaux sans vertèbres* (1833–1858); the continuation of a work begun by J.-B. de Férussac, *Histoire naturelle générale et particulière des mollusques* (1839–1851); "Mollusca," in Cuvier's *Règne animal* (1836–1849); an incomplete *Traité élémentaire de conchyliologie* in three volumes (1839–1857); *Conchyliologie de l'île de la Réunion (Océan Indien)* (1863); and *Description des animaux sans vertèbres découverts dans le bassin de Paris* (1860–1866).

In this last work, Deshayes returned once more to the subject of his first researches. The publications of 1860–1866 and 1824–1837 present his most important contributions.

Deshayes mastered the entire body of the systematic-taxonomic conchology of his time and presented it brilliantly. Even today, despite later publications by other authors, his descriptions of the mollusks of the Tertiary of the Paris Basin form an indispensable basis for further study. His exposition of the stratigraphy of the Paris Basin in the introduction to his *Description des animaux sans vertèbres* (1860) likewise contains valuable data for an understanding of this sedimentary sequence. Also worth reading are his descriptions in this introduction of the delimitation and definition of species, their life spans, and the reasons for their extinction. Finally, his subdivision of the Tertiary, by means of the proportion of living species of mollusks established in Lyell's Eocene, Miocene, and Pliocene stages, belongs to the classic methods of biostratigraphy.

BIBLIOGRAPHY

I. ORIGINAL WORKS. Besides many journal articles, Deshayes wrote the following books: *Description des coquilles fossiles des environs de Paris,* 3 vols. (Paris, 1824–1837); *Traité élémentaire de conchyliologie avec l'application de cette science à la géognosie,* 3 vols. (Paris, 1839–1857); *Exploration scientifique de l'Algérie. Histoire naturelle des mollusques,* 25 pts. (Paris, 1844–1848); *Histoire naturelle générale et particulière des mollusques* (Paris, 1820–1851), with J.-B. de Férussac; and *Description des animaux sans vertèbres découverts dans le bassin de Paris,* 3 vols. (Paris, 1860–1866). Lists of Deshayes's papers may be found in Royal Society of London, *Catalogue of Scientific Papers (1800–1863),* II (London, 1868), 251–254; and *(1864–1873),* VII (London, 1877), 524.

II. SECONDARY LITERATURE. On Deshayes or his work, see H. Crosse and P. Fischer, "Nécrologie. G. P. Deshayes," in *Journal de conchyliologie,* **24** (1876), 123–127; J. Evans, "Obituary Gérard Paul Deshayes," in *Quarterly Journal of the Geological Society of London, Proceedings,* **32** (1876), 80–82; and K. Lambrecht, W. Quenstedt, and A. Quenstedt, "Palaeontologi. Catalogus bio-bibliographicus," in *Fossilium catalogus,* I, *Animalia,* **72** (The Hague, 1938), 112–113.

HEINZ TOBIEN

DESLANDRES, HENRI (*b.* Paris, France, 24 July 1853; *d.* Paris, 15 January 1948), *astronomy, physics.*

Deslandres was born into a family typical of the mid-nineteenth-century French bourgeoisie. He graduated from the École Polytechnique in 1874 and entered the army, in which he served until 1881, when a strong interest in physical sciences led him to resign his military position. He worked first in the physical laboratories of the École Polytechnique and the Sorbonne, devoting himself to ultraviolet spectroscopy under the guidance of M. A. Cornu.

In 1889 Deslandres joined the staff of the Paris observatory, then headed by Admiral Ernest Mouchez, who sought to develop astrophysics in the institution long dedicated to celestial mechanics under Le Verrier. In 1897 he joined Jules Janssen at his astrophysical observatory in Meudon and became its director in 1908, following Janssen's death. In 1926 the Paris and Meudon observatories were united under his management. Deslandres retired in 1929 but pursued an active scientific life almost until his death at the age of ninety-four.

During his long and successful career Deslandres was elected to essentially all the scientific societies of significance, including the Académie des Sciences (1902; president, 1920), the Royal Society, the Royal Astronomical Society, the Accademia dei Lincei, and the National Academy of Sciences of the United States.

In his bearing, his character, and his style of life Deslandres always remained more akin to the soldier (and the officer) than to the scholar. These consequences of his education also appeared in his scientific work: he was more successful in the experimental and technical aspects of physics and astrophysics than in creating new theories (even though he worked at it with tenacity and ultimately with intuition).

In his thesis Deslandres studied the spectra of molecules such as nitrogen, cyanogen, CH, and water and recognized two simple laws in the disconcerting complexity of the numerous bands, each of which is made up of many tens of lines. These laws bear his name.

First, the frequencies (or wave numbers) ν of lines inside a given band can be represented by the parabolic formula

$$\nu = A + 2Bm + Cm^2,$$

where m is an integer ($m \neq 0$). Then, the constants A associated with various bands of a given system can all be fitted to another parabolic formula involving the integers ν' and ν''. These laws were at first useful in the empirical study and classification of molecular spectra; later, with the elaboration of quantum mechanics, it became easy to explain them in terms of the structure of molecules. The integers m, ν', and ν'' were simply the quantum numbers identifying the possible rotational and vibrational energies of the molecule in the initial and final states of a radiative transition; and the various empirical constants in Deslandres's formulas could be related to physical constants that were characteristics of each molecule.

After this splendid success in laboratory spectroscopy, Deslandres turned in 1889 to astrophysics without losing interest in molecules, as is shown by his fine observations of the Zeeman effect on molecular lines. In numerous publications of his late years he sought a unified theoretical interpretation of molecular spectra; he paid little attention to such modernist developments as quantum mechanics.

At the Paris observatory Deslandres attached a spectrograph to the recently built 120-cm. mirror telescope and began observing the spectra of stars and planets, devoting himself to the measurement of line-of-sight velocities through the Doppler-Fizeau effect, the most essential tool of the astronomer in studying the motions and dynamics of celestial bodies. He continued the same type of observations in Meudon with a spectrograph attached to a large re-

fractor with an 83-cm. aperture. Among his most valuable results were the law of rotation of Saturn's ring (1895), which was shown to rotate as a system of independent particles and not as a solid body, and the proof that Uranus (like its known satellites) rotates in the direction opposite to all other planets, a fact of significance to cosmogony.

As an astronomer Deslandres is still better known for his important contributions to the physical study of the sun, particularly for the invention of the spectroheliograph, an instrument that he completed at the Paris observatory in the spring of 1894. It allows one to make photographs of the sun in nearly monochromatic light, the narrow spectral band being selected at will with the help of a dispersing spectrograph. By choosing radiations to which the solar atmosphere is very opaque, one "sees" only its outermost layers, the chromosphere; and a wealth of important phenomena are thus revealed, such as solar plages, prominences, and flares. The spectroheliograph was invented independently by the great American astronomer George Ellery Hale, who actually completed his first instrument more than a year before Deslandres, at Kenwood Observatory in Chicago. However, Deslandres explored with particular tenacity the possibilities of the new method and found useful variants; and for many years he contributed to the description of the structure of the chromosphere and of the complex phenomena of solar activity.

Like Hale, Deslandres believed (and correctly so) that solar activity is dominated by electromagnetic causes; and he attempted imaginative explanations, now of limited interest. However, he showed remarkable intuition in maintaining that the sun produced radio waves, although the crude experiments of Charles Nordmann in 1902 failed to detect them. Solar Hertzian radiation was actually observed only in 1942 by J. S. Hey, and this finding was published shortly after World War II. It is not known whether "the brave soldier" (as Stratton described Deslandres), who then led a rather secluded life, was ever informed that he had been right forty years before.

BIBLIOGRAPHY

I. Original Works. Deslandres's most important publications are "Spectre du pôle négatif de l'azote. Loi générale de répartition des raies dans les spectres de bandes," in *Comptes rendus hebdomadaires des séances de l'Académie des sciences,* **103** (1886), 375–379; "Loi de répartition des raies et des bandes, communes à plusieurs spectres de bandes . . .," *ibid.,* **104** (1887), 972–976; "Spectres de bandes ultra-violets des métalloïdes avec une faible dispersion," his thesis (Faculté des Sciences de Paris, 1888), no.

619; "Étude des gaz et vapeurs du soleil," in *L'astronomie* (Dec. 1894); and "Recherches sur l'atmosphère solaire. Photographie des couches gazeuses supérieures . . .," in *Annales de l'Observatoire d'astronomie physique de Paris,* **4** (1910).

II. SECONDARY LITERATURE. Short biographies of Deslandres are L. d'Azambuja, in *Bulletin de la Société astronomique de France,* **62** (1948), 179–184; and F. J. M. Stratton, in *Monthly Notices of the Royal Astronomical Society,* **109** (1949), 141–144.

R. MICHARD

DESMAREST, NICOLAS (*b.* Soulaines-Dhuys, France, 16 September 1725; *d.* Paris, France, 28 September 1815), *geology, technology.*

Desmarest was the only child of Jean Desmarest, the local schoolteacher, and Marguerite Clement. Practically nothing is known about his youth until 1740, when his father died. In February 1741 Desmarest's mother remarried, and Nicolas, now in the care of a guardian, was placed in the Oratorian *collège* of Troyes. Here he received a sound education, and when he left Troyes for Paris in late 1746 or early 1747, he embarked upon a scientific career that brought him membership in the Academy of Sciences in 1771. His election to that body followed the scientific work with which his name has remained principally associated: his assertion that the basalt columns found in Auvergne and elsewhere are volcanic in origin. The election to the Academy of Sciences crowned with success what had been an often frustrating campaign for membership, begun as early as 1757.

But Desmarest's profession was not exclusively scientific; much of his energy during his best years was consumed by his duties in the government bureaucracy controlling industry, which culminated in his serving as inspector general of manufactures. After the Revolution suppressed this post, Desmarest found himself working for a time in various minor government agencies and was appointed as a teacher in the *écoles centrales.* He married Françoise Tessier, twenty-five years his junior, on an unknown date. They had a son in 1784, and Desmarest's wife died in 1806.

Despite Desmarest's fame for identifying basalt as volcanic, his general geological orientation was by no means solely volcanological. To the extent that he professed any geological doctrine, it reflected the influence of Guillaume-François Rouelle, whose public courses he had attended in Paris. Rouelle expounded a neptunist outlook regarding the origin of terrestrial formations, and in adopting Rouelle's basic tenets Desmarest rejected forever the notion that volcanoes or an internal heat of the globe had been primarily responsible for producing the earth's essential features. Thus, even though Desmarest greatly expanded the extent to which volcanoes were understood to have produced changes in the earth, he cannot be counted as a genuine volcanist; although he was in a sense the founder of the plutonist side of the basalt controversy, he did not represent the volcanists in the general dispute over modes of geological change that emerged from the confrontation of Wernerian and Huttonian theories of the earth. Indeed, Desmarest's geology did not grow out of, and never became altogether adapted to, a "theory of the earth." Instead, his scientific career developed out of the consideration of more narrowly circumscribed problems. His scientific demeanor always remained rather cautious, and in this regard he reflected a typical Enlightenment aversion to the liabilities of *système.*

After his arrival in Paris, Desmarest lived precariously for a time as a private tutor. His climb toward recognition in the learned world began in 1749, when he assisted Pierre-Nicolas Bonamy in editing the *Suite de la clef,* better known as the *Journal de Verdun.* In subsequent years he may have lived in part on fees for editing books; under his guidance there appeared a collection of the thoughts of the Abbé Louis Du Four de Longuerue (1754) and the seventh edition of *Les élémens de géométrie,* by the renowned Oratorian teacher Bernard Lamy (1758). Of far more significance to Desmarest's line of interest, however, was the French edition of Francis Hauksbee's *Physico-Mechanical Experiments* (1754), which Desmarest edited and to which he added voluminous notes and remarks. In 1751 he won the prize competition of the Académie des Sciences, Belles-Lettres et Arts d'Amiens with an essay arguing that England had once been joined to France by an isthmus whose destruction was recent, natural, and noncatastrophic. The published essay (1753) was accompanied by maps and a cross section of the English Channel prepared by the geographer Philippe Buache; the maps, among the first to employ contour lines to show ocean depths, were used by Desmarest to show that the supposed isthmus still existed not far beneath the waves that had reduced it.

After his success in the Amiens competition, Desmarest continued to direct his attention to scientific questions concerning the earth's surface and internal structure. In the 1754 *Expériences physico-méchaniques,* for example, he supported Hauksbee's attack on a peculiar extension of the terraqueous globe doctrine, according to which the earth's primitive formations had been deposited out of the terraqueous fluid in descending order of the specific gravity of the suspended materials. In 1756, just a few months after the Lisbon earthquake, he published an essay on the propagation of earthquake disturbances, presumed to be caused by the combustion of inflammable matter underground and conveyed by means of the worldwide network of interlocking mountain ranges. The seventh volume of the *Encyclopédie* (1757) brought forth two further contributions by Desmarest to his emerging specialty. "Fontaine" deals for the most part with the ancient problem of the origin of springs, or the manner in which rivers are fed and maintained. Desmarest supported the meteoric theory, claiming that rainfall is sufficient to account for the flow of rivers and that the earth can receive and store water in amounts large enough to account for the continued flow of springs. "Géographie physique" is a more general article outlining the aims and methods of investigating the earth's surface features.

By the late 1750's Desmarest had decided to concentrate on physical geography and the study of rocks, and in the course of several years he traveled extensively in France, observing the Paris region, Champagne, Burgundy, Lorraine, Alsace, Franche-Comté, Guienne, and Gascony. His notes from these travels in the late 1750's and early 1760's indicate a commitment to an organizing scheme like that of Rouelle, which called for a division of the earth's surface features into three categories (*ancienne, intermédiaire, nouvelle*) corresponding respectively to a crystalline inner core, and two different sorts of sedimentary layers, all of them deriving from fluid sources.

In 1757 Desmarest began his most significant and extensive professional occupation outside of science. He was appointed by Daniel-Charles Trudaine, the intendant of finances and director of commerce, to make a study of the woolen cloth industry in France. For over three decades thereafter Desmarest was involved in the analysis and control of French industry as an agent of the royal government. He became expert in the technical aspects of numerous industries and developed special proficiency in papermaking technology. His treatises on the manufacture of paper made him a leader in the rationalization of paper technology in the late eighteenth century. As inspector of manufactures in the *généralité* of Limoges under the intendant Anne-Robert-Jacques Turgot, Desmarest devoted himself especially to agricultural matters. He later held the post of inspector of manufactures of the *généralité* of Châlons and ultimately served as inspector general, director of manufactures, for the entire realm (1788–1791).

It was in his capacity as an agent for Trudaine that Desmarest first traveled to Auvergne in 1763. While on a tour to examine industries south of Clermont-Ferrand, Desmarest noticed prismatic basalt columns in association with hardened lavas, such as had been described by Jean-Étienne Guettard after his 1751 visit to the region. Desmarest's excitement at this discovery appears to have been stimulated initially as much by the mere appearance of the columns as by their possibly volcanic origin. At this time there was considerable interest in the finely sculptured basalts, especially in the articulated columns of County Antrim in Northern Ireland, the celebrated Giant's Causeway. Evidently believing that the articulated columns of Auvergne were the first of that type known outside of Ireland, Desmarest momentarily let chauvinistic pride come to the fore but soon set about investigating the geological significance of the columns' presence among lava flows.

Between 1764 and 1766 Desmarest and François Pasumot drew up a map of the main volcanic district of Auvergne, concentrating especially on the southern, Mont-Dore district. Their geological map was published with Desmarest's first long memoir on the Auvergne basalts, which he delivered before the Academy of Sciences in 1771. This map did not fulfill Desmarest's entire cartographic aim, however, and he continued to work intermittently on the Auvergne geological map for the rest of his life. A much larger and more detailed map than the 1771 version was finally published in 1823 by his son, Anselme-Gaëtan Desmarest.

Desmarest's first public statement on the geological significance of the Auvergne basalts came in a paper he delivered to the Academy of Sciences in the summer of 1765, before departing on a year-long journey in Italy in the company of Duke Louis-Alexandre de La Rochefoucauld. The essential contents of this report are included in the sixth volume of plates for the *Encyclopédie* (1768). Here he argued that the presence of prismatic basalts infallibly indicates the former existence of volcanoes in that area. Hypothesizing that "the regular forms of basalt are a result of the uniform contraction undergone by the fused material as it cooled and congealed, shrinking around several centers of activity," he promised a fuller explanation and substantiation of this supposition in a later work. This he never provided, however, and in his own time other voices were heard saying that a rapid, rather than slow, rate of cooling was responsible for the basalt's columnar shape. But Desmarest's attention had not been confined to basalts of prismatic form. Explaining that the basalts of Auvergne also take the form of ellipsoids composed of concentric layers and of large sheets broken into randomly oriented bundles of regular shape, he found himself leading the way to a study of the relation of these materials to their neighboring or enveloping matter.

The notion that prismatic basalts are a volcanic product, rather than rocks of igneous origin, was thus made known to the Academy of Sciences in 1765, and found its way into print in 1768. This opinion became prominent, however, only after 1771, when Desmarest delivered to the Academy of Sciences the lengthy report of his studies on the Auvergne basalts. This was followed by another memoir presented in 1775, and it is in these papers that Desmarest's original contributions to geological science lie. Here he dealt with such problems as the origin of the matter constituting the basalts, the volcanic history of the Auvergne region, and the alterations that the volcanic flows had undergone.

Attempting to reconstruct a mental picture of the former condition of the lava flows whose extremities revealed prismatic basalts, Desmarest was led to a consideration of the destructive effects of flowing water upon the Auvergne terrain. He applied the idea of aqueous degradation, which had become fixed in his mind several years before his first visit in Auvergne, in such a way as virtually to enunciate a principle of uniformity in destructive geological processes. It is noteworthy that he at first took the principle of regularity in cause and rate of degradation as a necessary working hypothesis, but before long began to treat it as a result of his researches. Desmarest utilized his conclusions about Auvergne's physiographical history to hazard a three-stage volcanic history of the region, thus extending historical geology into the field of volcanism.

Desmarest's position on the nature of volcanoes always distinguished between the source of volcanic heat and the molten and solid ejecta. He consistently indicated that volcanoes feed on some combustible or fermenting agent and thus regarded lavas as material heated, as it were, accidentally. Basalt he took to be granite heated moderately—not to an extreme degree. This highly limiting view on the sources of volcanic action, not atypical of his time, naturally placed definite restraints on the possibility of his entertaining truly volcanist ideas, restraints which he always respected. His mature belief, expressed in *Géographie physique,* a large work produced in old age, was that burning beds of underground coal are the most likely cause of volcanic heat, and so he regarded volcanic action as a relatively recent interloper in geological history. Desmarest rejected Hutton's assertion that volcanic heat might be a source of power sufficient to uplift continents. While suggesting that the earth's age must be quite great, he never committed himself to a clear definition of the geological time scale. He retained a faith in the importance of certain catastrophic agencies of terrestrial change and did not regard this as inconsistent with his general adherence to uniformity in degradation.

BIBLIOGRAPHY

I. ORIGINAL WORKS. Desmarest made his scientific debut with *Dissertation sur l'ancienne jonction de l'Angleterre à la France, qui a remporté le prix, au jugement de l'Académie des sciences, belles-lettres & arts d'Amiens, en l'année 1751* (Amiens, 1753), and followed this with *Conjectures physicoméchaniques sur la propagation des secousses dans les tremblemens de terre, et sur la disposition des lieux qui en ont ressenti les effets* (n.p., 1756). Vol. VII of *Encyclopédie, ou dictionnaire raisonné des sciences, des arts et des métiers, par une société de gens de lettres* (Paris, 1757) contains his "Fontaine," pp. 80–101, and "Géographie physique," pp. 613–626. The first published statement of the volcanic origin of prismatic basalt is found in a brief commentary, "Planche VII. Basalte d'Auvergne," in *Recueil de planches, sur les sciences, les arts libéraux, et les arts méchaniques, avec leur explication,* VI (Paris, 1768), pp. 3–4 of section entitled "Histoire naturelle. Règne minéral. Sixième collection. Volcans." Desmarest's detailed studies of the Auvergne basalts were published in three parts, the first two in "Mémoire sur l'origine & la nature du basalte à grandes colonnes polygones, déterminées par l'histoire naturelle de cette pierre, observée en Auvergne," in *Mémoires de l'Académie royale des sciences* for 1771 (1774), 705–775; and the third as "Mémoire sur le basalte. Troisième partie, où l'on traite du basalte des anciens; & où l'on expose l'histoire naturelle des différentes espèces de pierres auxquelles on a donné, en différens temps, le nom de basalte," *ibid.,* for 1773 (1777), 599–670. His general conclusions on the geological history of the Auvergne region were expressed in "Extrait d'un mémoire sur la détermination de quelques époques de la nature par les produits des volcans, & sur l'usage de ces époques dans l'étude des volcans," in *Observations sur la physique, sur l'histoire naturelle et sur les arts,* **13** (1779), 115–126, expanded in "Mémoire sur la détermination de trois époques de la nature par les produits des volcans, et sur l'usage qu'on peut faire de ces époques dans l'étude des volcans," in *Mémoires de l'Institut des sciences, lettres et arts. Sciences mathématiques et physiques,* **6** (1806), 219–289. Desmarest's largest and most comprehensive geological work, which is largely derivative, is a part of the *Encyclopédie méthodique,* entitled *Géographie physique,* 5 vols. (Paris, *an* III [1794–1795]–1828), the last and posthumous volume edited in part by others.

Among the books edited by Desmarest, the most significant is Francis Hauksbee's *Expériences physico-méchaniques sur différens sujets, et principalement sur la lumière et l'électricité, produites par le frottement des corps,* 2 vols. (Paris, 1754).

Foremost among Desmarest's writings in the fields of industry, manufacturing, and agriculture are his essays on papermaking methods: "Premier mémoire sur les principales manipulations qui sont en usage dans les papeteries de Hollande, avec l'explication physique des résultats de ces manipulations," in *Mémoires de l'Académie royale des sciences* for 1771 (1774), 335–364; "Second mémoire sur la papeterie; Dans lequel, en continuant d'exposer la méthode hollandoise, l'on traite de la nature & des qualités des pâtes hollandoises & françoises; De la manière dont elles se comportent dans les procédés de la fabrication & des apprêts: Enfin des différens usages auxquels peuvent être propres les produits de ces pâtes," *ibid.,* for 1774 (1778), 599–687; "Papier. (Art de fabriquer le)," in *Encyclopédie méthodique. Arts et métiers mécaniques,* V (Paris–Liège, 1788), 463–592; and *Art de la papeterie* (Paris, 1789).

The largest and most interesting collections of Desmarest's correspondence are found in the Bibliothèque Nationale, Fonds Français, Nouvelles Acquisitions, MS 803 (letters exchanged by Desmarest and Pierre-Jean Grosley) and MS 10359 (letters from Turgot to Desmarest), and in the Bibliothèque Municipale de Beaune, MS 310 (letters from François Pasumot to Desmarest).

II. SECONDARY LITERATURE. The single most important source of biographical information on Desmarest is a handwritten set of notes by his son, Anselme-Gaëtan Desmarest, "Notes et renseignements sur la vie et les ouvrages de mon père," Bibliothèque de l'Institut de France, Fonds Cuvier, MS 3199. Georges Cuvier depended heavily upon them in preparing his "Éloge historique de Nicolas Desmarets [*sic*], lu le 16 mars 1818," in *Recueil des éloges historiques lus dans les séances publiques de l'Institut royal de France,* II (Strasbourg–Paris, 1819), 339–374. Further references and discussion are found in Sir Archibald Geikie, *The Founders of Geology,* 2nd ed. (London–New York, 1905), pp. 140–175; and Kenneth L. Taylor, "Nicolas Desmarest and Geology in the Eighteenth Century," in Cecil J. Schneer, ed., *Toward a History of Geology* (Cambridge, 1969), pp. 339–356.

KENNETH L. TAYLOR

DESMIER DE SAINT-SIMON, ÉTIENNE J. A. *See* Archiac, Vicomte d'.

DESOR, PIERRE JEAN ÉDOUARD (*b.* Friedrichsdorf, near Frankfurt am Main, Germany, 13 February 1811; *d.* Nice, France, 23 February 1882), *glacial geology, paleontology, stratigraphy.*

Desor was of French origin and studied law at the universities of Giessen and Heidelberg. During a short stay in Paris, he was introduced to geology by Élie de Beaumont. His meeting in 1837 in Switzerland with Louis Agassiz marks the turning point of his career; for almost twenty years he was Agassiz's close friend and chief collaborator. Desor's two volumes on the glacial theory were published for the general public, and his reports of Agassiz's expeditions on the Alpine glaciers led to the formation of numerous Alpine clubs.

Desor followed Agassiz to America in 1846, but—for reasons still not understood—their friendship ended suddenly. In the service of federal and state agencies, Desor studied the fauna of the Atlantic shelf aboard the *Bibb;* took part in the Lake Superior land district survey, directed first by C. T. Jackson and then by J. W. Foster and J. D. Whitney; and undertook a study of the coal basin of Pottsville, Pennsylvania, supervised by H. D. Rogers.

Returning to Neuchâtel in 1852 because of his brother's poor health, Desor was appointed professor of geology at the Academy of Neuchâtel. His brother's death in 1858 left him in a financially advantageous situation and enabled him to resign his professorship and return to the study of fossil echinoderms, the geology of the Jura Mountains, and the archaeology of the Bronze Age lake dwellers. From his home at Combe Varin in the Val des Ponts, well-known as a meeting place for natural philosophers, Desor continued his studies and collaborated for twenty years in the preparation of a geological map of Switzerland.

Desor was naturalized in 1859; his subsequent political career led him to the presidency of the Swiss Federal Assembly in 1874. Forced by gout to spend much of his last years on the French Riviera, he devoted his final works to a study of that region's geology and archaeology.

BIBLIOGRAPHY

I. ORIGINAL WORKS. Desor's writings include *Monographies d'échinodermes vivants et fossiles* (Neuchâtel, 1838–1843), 5 pts. in one vol., written with L. Agassiz and G. Valentin; "Aperçu général de la structure géologique des Alpes," in *Bibliothèque universelle,* XXXVIII (Geneva, 1842), 120–149, written with B. Studer; *Excursions et séjours dans les glaciers et les hautes régions des Alpes, de M. Agassiz et de ses compagnons de voyage* (Neuchâtel, 1844); *Nouvelles excursions et séjours dans les glaciers et les hautes régions des Alpes, de M. Agassiz et de ses compagnons de voyage* (Neuchâtel, 1845); "Catalogue raisonné des familles, des genres, et des espèces de la classe des Échinodermes; précédé d'une introduction sur l'organisation, la classification, et le développement progressif des types dans la série des terrains," in *Annales des sciences naturelles (zoologie),* 3rd ser., **6** (1846), 305–374; **7** (1847), 129–168; **8** (1847), 5–35, 355–381, written with L. Agassiz; "On the Drift of Lake Superior," in *American Journal of Science,* 2nd ser., **13** (1852), 93–109; *Synopsis des échinides fossiles,* 2 vols. and atlas (Paris, 1858); *Études géologiques sur le Jura neuchâtelois* (Neuchâtel, 1859), written with A. Gressly; *Les palafittes ou constructions lacustres du Lac de Neuchâtel* (Paris, 1865); *Description des oursins fossiles de la Suisse. Échinides de la période jurassique,* 16 secs. with atlas (Wiesbaden, 1868–1872), written with P. de Loriol; *Le bel âge du bronze* *lacustre en Suisse* (Paris, 1874), written with L. Favre; and *Le paysage morainique, son origine glaciaire, et ses rapports avec les formations pliocènes d'Italie* (Paris, 1875).

The Royal Society's *Catalogue of Scientific Papers* gives a complete list of Desor's publications in II, 266–269; VII, 525–526; and IX, 688–689.

II. SECONDARY LITERATURE. Notices on Desor are L. Favre, "Notice nécrologique d'Édouard Desor (1811–1882)," in *Bulletin de la Société des sciences naturelles* (Neuchâtel), **12** (1882), 551–576; and J. P. Lesley, "Obituary Notice of Édouard Desor," in *Proceedings of the American Philosophical Society*, **20** (1882), 519–528.

ALBERT V. CAROZZI

DESORMES, CHARLES-BERNARD (*b.* Dijon, Côte-d'Or, France, 3 June 1777; *d.* Verberie, Oise, France, 30 August 1862), *chemistry.*

Desormes entered the École Polytechnique at its founding (1794) and he remained there after his studies were completed, becoming *répétiteur* in chemistry under Guyton de Morveau, a position he held until 1804. From this period dates his relationship with Nicolas Clément, his compatriot who later became his scientific collaborator, industrial associate, son-in-law, and friend. (Detailed information on their important joint scientific works can be found in the article on Clément.) Desormes left Guyton de Morveau only in order to devote himself to the alum factory that he established at Verberie in association with Clément and Joseph Montgolfier.

On 5 July 1819 Desormes was elected a corresponding member of the Académie des Sciences; this honor was refused his self-educated son-in-law. From this time, and especially after 1830, Desormes gradually turned away from science in order to devote his time to politics. He was elected *conseiller général* of the department of Oise in 1830, but was defeated as an opposition candidate for parliament in June 1834. He then founded the *Revue de l'Oise*, which became the *Progrès de l'Oise*. Following two further defeats in November 1837 and in July 1842 (shortly after the death of Clément), Desormes was finally elected to the Constituent Assembly on 23 April 1848. He sat with the republicans and participated in the departmental and communal Committee of Administration.

Besides his works in collaboration with Clément, Desormes's scientific *oeuvre* is slight, consisting of three memoirs dating from 1801 to 1804, the period immediately after the appearance of Volta's pile. Historians of science should certainly pay a bit more attention to the dry piles that Desormes then constructed. They were composed of metallic disks separated by a layer of salt paste. The analogous arrangement of Giuseppe Zamboni dates only from 1812.

BIBLIOGRAPHY

For the works that Desormes published in collaboration with his son-in-law Nicolas Clément, see the article on the latter.

The easiest way to gain an overall idea of Desormes's career is to consult *Procès-verbaux des séances de l'Académie tenues depuis la fondation de l'Institut jusqu'au mois d'août 1835,* 10 vols. (Hendaye, 1910–1924). The most relevant of these are II, 312 (6 ventôse *an* IX [25 Feb. 1801]) and II, 316 (11 ventôse [2 Mar.]), reading of a memoir by Desormes on galvanism; IV, 237 (7 Aug. 1809), Desormes admitted to the session upon having presented at least two memoirs; IV, 315 (22 Jan. 1810), Desormes a candidate following the death of Fourcroy; V, 271 (6 Dec. 1813), Desormes proposed as a correspondent; VI, 113 (25 Nov. 1816), same topic; VI, 118 (2 Dec. 1816), Desormes defeated, having received only two votes; VI, 465 (28 June 1819), Desormes a candidate following the death of Clément; VI, 466 (5 July 1819), Desormes elected corresponding member by forty votes; VI, 483 (13 Sept. 1819), Desormes expresses his appreciation; VIII, 216 (23 May 1825), the role of Desormes in the history of the construction of dry cells is recalled; IX, 284 (3 Aug. 1829), same subject, with remarks by Antoine-César Becquerel.

The announcement of Desormes's death is in *Comptes rendus*, **55** (1862), 418.

I. ORIGINAL WORKS. *Considérations sur les routes en général et sur celles du département de l'Oise* (Senlis, 1834); *Des impôts* (Senlis, 1851); "Expériences et observations sur les phénomènes physiques et chimiques que présente l'appareil électrique de Volta," in *Annales de chimie*, **37** (*an* IX [1801]), 284–321; *Proposition relative à la franchise des lettres, adressée au citoyen président de l'Assemblée nationale, présentée le 12 juillet 1848 par le citoyen Desormes* . . . (Paris, 1849).

Desormes also wrote, in collaboration with Guyton de Morveau, "Essai sur l'analyse et la recomposition de deux alcalis fixes et de quelques-unes des terres reputées simples, lu le six floréal an VIII [26 Apr. 1800]," in *Mémoires de l'Académie des sciences*, 1st. ser., III (*an* XI), 321–336; and, in collaboration with Hachette, "Mémoire pour servir à l'histoire de cette partie de l'électricité qu'on nomme galvanisme," in *Annales de chimie*, **44** (*an* XI), 267–284; and "Du doubleur d'électricité," *ibid.*, **49** (*an* XII), 45–54.

II. SECONDARY LITERATURE. See S.-J. Delmont, *Dictionnaire de biographie française*, X (1965), 1501*a*; and *Grande encyclopédie universelle*, XIV (Paris, n.d. [about 1900]), 263.

JACQUES PAYEN

DESPAGNET, JEAN, *alchemy.*

It is unknown where and when Despagnet was born or where and when he died. Very likely he flourished in the first half of the seventeenth century. We know only that he was president of the Parlement of Bordeaux. His son Étienne, who had the same interests as his father, became a councillor of the same parlement in 1617. Very likely it was the latter whom Christian Huygens mentioned in his letters.

Despagnet acquired a great reputation as a hermetic philosopher and alchemist. Only two of his alchemic works, which are considered classics of their kind, are extant: the *Arcanum Hermeticae philosophiae* and the *Enchiridion physicae restitutae,* both published for the first time in Paris in 1623. The *Arcanum* is also attributed to an unknown author called the Imperial Knight; that attribution was denied in 1664 by Étienne, who, when asked about this by Borrichius, affirmed that his father was the author. In the *Enchiridion,* which is an introduction to the *Arcanum,* nature is regarded as a constant expression of divine will, it being understood that the paradisiacal state is the true nature and its attainment is God's will for humanity. The *Arcanum,* a post-Reformation document, illustrates the deepening sense of the spiritual on the part of physical alchemists.

From Fermat's letter of 22 September 1636 to Roberval, it appears that in 1629 Fermat had visited Despagnet at Bordeaux. Fermat also sent Étienne the original of the second book of his *Loca plana restituta,* which he finished in 1629. From Fermat's letter of February 1638 to Mersenne it appears that he had studied the manuscripts of Viète, which were deposited with Étienne Despagnet and were mentioned in the preface of Van Schooten's edition of Viète's *Opera mathematica* (1646). But, according to Fermat, these manuscripts were so antiquated that it was no longer worthwhile to publish them.

BIBLIOGRAPHY

An English translation of the *Arcanum* can be found in W. W. Westcott, *Collectanea Hermeticae,* I (London, 1893).

The best account of Despagnet's works and their several editions can be found in *Nouvelle biographie universelle,* XV (Paris, 1854), 402–403. Some information can also be found in C. Henry, "Recherches sur les manuscrits de Pierre de Fermat," in *Bullettino di bibliografia e di storia delle scienze matematiche e fisiche,* **12** (1879), 535–537; and A. E. Waite, *The Secret Tradition in Alchemy* (London, 1926), pp. 39, 338, 341.

H. L. L. BUSARD

DESSAIGNES, VICTOR (*b.* Vendôme, France, 30 December 1800; *d.* Paris, France, 5 January 1885), *chemistry.*

Dessaignes, the third son of Jean-Philibert Dessaignes, was born at the Collège of Vendôme, a school his father helped establish. In Paris he studied law, receiving his degree when he was twenty-one. Instead of practicing law, however, he immediately enrolled in medical school, where he was diverted still further by a strong interest in chemistry, undoubtedly stimulated by the then preeminent position in research of French chemists. It was not until he was thirty-five that he received a degree in medicine, defending a thesis on the action of various chemical substances on humans. While still a medical student, he exploited his early investigations into metabolic processes. He was able to pursue his interests by having been appointed tax collector in the city of Vendôme, a sinecure in which his law degree was of service.

At thirty-seven Dessaignes married Mlle. Renou, whose brother became the director of the Observatory at Saint Maur. Mme. Dessaignes died soon after the birth of their only son. Dessaignes then devoted well over ten years of his life—as well as most of his savings and income—to the creation of a private laboratory in which he performed the experiments which brought him to the attention of the scientific world.

Dessaignes was awarded the 1860 Jecker prize, consisting of a citation and a cash award of 2,000 francs, for his elucidation of the structure of a number of important, naturally occurring organic acids, such as hippuric, succinic, butyric, and malic acids, as well as quercitol, a desoxyinositol obtained from acorns. He shared the prize with Berthelot, with whom he evidently had formed a warm relationship. The prize brought him attractive offers to teach, but he elected to continue his research. He was named a *chevalier* of the Legion of Honor in 1863 and, at the request of the Académie des Sciences, became a correspondent for chemistry in 1869. Although he had by then ceased all laboratory work, he remained close to the Paris circle of chemists and was honored at a scientific congress at Blois in September 1884, an occasion marked by a laudatory speech by Friedel on behalf of the Academy. That December he developed bronchitis and, although he promptly recovered, was left in a weakened condition. He died after suffering a respiratory collapse; his son, who had become a professor in the school of medicine of Paris, survived him.

Dessaignes was fortunate to have grown up in a place and at a time when French science—especially French chemistry—had reached a peak. The invigorating intellectual climate of Paris, his father's academic influence, and his freedom from financial worries all combined to give Dessaignes the leisure and opportunity to work in the mainstream of science.

Dessaignes studied, determined the structure, and synthesized several important organic substances, principal among these being hippuric acid. This substance was first isolated by J. Liebig in 1829 from horse urine, thus its name. It occurs in the urine of many herbivorous animals, to a lesser extent in that of humans. It was soon found that by administering benzoic acid by mouth the amount of hippuric acid recoverable from human urine could be increased, a laboratory exercise occasionally still used in undergraduate biochemistry courses. Dessaignes found upon hydrolyzing hippuric acid with either acids or bases that he obtained benzoic acid and glycine. The synthesis followed when Dessaignes reacted benzoylchloride and the zinc salt of glycine and obtained a product identical with naturally occurring hippuric acid. The importance of hippuric acid lay in the realization that the body could eliminate unwanted or dangerous foreign materials by reactions with bodily substances to form compounds which could be excreted readily. Such detoxification mechanisms are observed in almost all vertebrates.

Dessaignes also studied the oxidation and reduction of various compounds, converting malic into succinic acid and tartaric into malic acid, all four carbon acids found widely in plant tissues. A knowledge of the transformations between these acids paved the way for the elucidation of metabolic cycles vital to cellular respiration, which, however, awaited the work of Krebs in the twentieth century.

Dessaignes was the sole author of twenty-five articles and coauthor of three others (two with J. Chautard and one with Schmidt). His papers show marked analytical insight and careful, precise experimentation. His life was that of an enlightened but solitary amateur of science in an age which honored perseverance and modesty.

BIBLIOGRAPHY

I. ORIGINAL WORKS. A complete list of Dessaignes's papers may be found in the *Catalogue of Scientific Papers. Royal Society of London* for 1800–1863. J. C. Poggendorff, III, contains a shorter list. Many of the papers were published in substantially the same form in several journals.

II. SECONDARY LITERATURE. An obituary notice of Dessaignes is in *Comptes rendus hebdomadaires des séances de l'Académie des sciences,* **100** (1885), 18, and was reprinted in *Bulletin. Société chimique de France,* **43** (1885), 145.

R. CHRISTIAN ANDERSON

DEVAUX, HENRI (*b.* Etaules, Charente-Maritime, France, 6 July 1862; *d.* Bordeaux, France, 14 March 1956), *plant physiology, molecular physics.*

Devaux was born into a Protestant family of sailors and farmers. After graduation from the University of Bordeaux he worked for five years under the leading botanists of the University of Paris. His doctoral thesis (1889) concerned the gaseous exchanges in plant tissues. He then returned to the University of Bordeaux, where he held the chair of plant physiology from 1906 until his retirement in 1932. Devaux soon displayed his inclination toward physical chemistry, showing as early as 1896 that aquatic plants accumulate polyvalent metallic ions, such as lead, in their cell membranes, even when the surrounding solution contains only traces of the ion. This accumulation is reversed when a large concentration of a monovalent ion, such as sodium or potassium, is added to the external solution. This was exactly the process that, nearly forty years later, was known as ion exchange, a process with wide scientific and industrial applications.

From 1903 on, Devaux was interested in the physics of surfaces. In 1890 Lord Rayleigh, and shortly afterward Agnes Pockels, had demonstrated that the surface tension of water is reduced when a film of oil, presumably one molecule thick, is spread over the surface. Direct evidence of surface films one molecule thick was presented by Devaux in 1903. He applied this demonstration to a wide range of films, particularly to proteins. The apparatus used by Devaux was of the most elegant simplicity: a photographic tray filled with either water or mercury lightly sprinkled with talcum powder. When a minute amount of film-forming substance is deposited on the liquid surface, the talcum particles are repelled and reassemble in the form of a circle. By a simple calculation involving the diameter of this circle, Devaux obtained the molecular weights of film-forming substances, particularly proteins and heavy organic acids. The results were at first ignored in France, but a decade later they were noticed by Irving Langmuir. The famous American physicist, who was to make such important contributions to the study of surfaces, gave full credit in many of his publications to Devaux for having demonstrated that the behavior of monomolecular films depends essentially on the reactivity of specific, or polar groups of the molecules. Devaux's scientific activity continued until his last years. The "wetting" of solid surfaces, the hydration of molecules in surface films, and the evaporation of odorous substances are among the fields to which he made significant contributions.

BIBLIOGRAPHY

I. ORIGINAL WORKS. Devaux's works appeared mainly in local scientific publications of limited circulation, but some of his most important papers appeared in the following accessible journals: "De l'absorption des poisons métalliques très dilués par les cellules végétales," in *Comptes rendus hebdomadaires des séances de l'Académie des sciences,* **132** (1901), 717–719; "Oil Films on Water and on Mercury," in *Annual Report of the Board of Regents of The Smithsonian Institution* (Washington, D.C., 1913), pp. 261–273; "Action rapide des solutions salines sur les plantes vivantes: déplacement réversible d'une partie des substances basiques contenues dans la plante," in *Comptes rendus hebdomadaires des séances de l'Académie des sciences,* **162** (1916), 561–564; "Ce qu'il suffit d'une souillure pour altérer la mouillabilité d'une surface. Étude sur le contact d'un liquide avec un solide," in *Journal de physique,* **4** (1923), 293–309; "La mouillabilité des substances insolubles et les remarquables puissances d'attraction existant à l'interface des liquides non miscibles," in *Comptes rendus hebdomadaires des séances de l'Académie des sciences,* **197** (1933), 105–108; "L'adsorption de l'ovalbumine à la surface libre de ses solutions, lorsque la concentration de celles-ci varie de 10^{-2} à 10^{-8}," *ibid.,* **200** (1935), 1560–1565; "Action de l'acide carbonique sur l'extension de l'ovalbumine à la surface de l'eau, et variations de l'épaisseur de ces lames en couches monomoléculaires," *ibid.,* **199** (1934), 1352–1354; "Détermination de l'épaisseur de la membrane d'albumine formée entre l'eau et la benzine et propriétés de cette membrane," *ibid.,* **202** (1936), 1957–1960; "Sur une représentation macroscopique des lames monomoléculaires et leur comportement à divers états de compression," *ibid.,* **206** (1938), 1693–1696, written with L. Pallu; "Étude expérimentale des lames formées de graines sur le mercure. Possibilité de déterminer sur les lames minces les 3 dimensions principales des molécules," in *Journal de physique,* **9** (1938), 441–446, written with L. Pallu; "Un rapport remarquable entre la constitution cellulaire et la mouillabilité du corps des mousses," in *Comptes rendus hebdomadaires des séances de l'Académie des sciences,* **208** (1939), 1260–1263; "La mouillabilité des surfaces solides," *ibid.,* **210** (1940), 27–29; "Les lames minces hydrophiles," *ibid.,* **211** (1940), 91–94; "L'adsorption d'une couronne de molécules d'eau autour de chaque molécule d'un sel étendu en lame mince," *ibid.,* **212** (1941), 588–590; "L'adsorption hygroscopique d'une couronne de molécules d'eau autour de chaque molécule des substances étendues en lame monomoléculaire sur le mercure," in *Mémoires de l'Académie des sciences de l'Institut de France,* **66** (1942), 1–28; and "L'arrangement des particules flottant sur du mercure, sous l'influence d'un champ électrique," in *Journal de physique,* **4** (1943), 185–196.

II. SECONDARY LITERATURE. Irving Langmuir gives an extensive account of Devaux's early work in surface physics in "The Constitution and Fundamental Properties of Liquids," in *Journal of the American Chemical Society,* **39** (1917), 1848–1906. A complete bibliography up to 1941 can be found in *Actualités scientifiques et industrielles,* no. 932 (Paris, 1942), pp. 23–36. A detailed obituary was published by Gordin Kaplan, "Henri Devaux, Plant Physiologist, Pioneer of Surface Physics," in *Science,* **124** (1956), 1017–1018.

A. M. MONNIER

DEVILLE, HENRI ÉTIENNE SAINTE-CLAIRE (*b.* St. Thomas, Virgin Islands, 11 March 1818; *d.* Boulogne-sur-Seine, France, 1 July 1881), *chemistry.*

Deville was one of the most prolific and versatile chemists of the nineteenth century, making major contributions in most areas of his science. He and his brother Charles, later a well-known physicist, were sons of the French consul in the Virgin Islands, who was a prominent shipowner. The brothers were educated in Paris, where Henri received his medical degree in 1843. Even before graduation he had been attracted to the study of chemistry by the lectures of Thenard. He established a private laboratory in his own quarters and in 1839 published his first paper, a study of turpentine. Soon after receiving his doctorate in medicine, he followed it with one in science.

In 1845, through the influence of Thenard, Deville was appointed professor of chemistry and dean of the newly established faculty of science of the University of Besançon. Here he established such a reputation that in January 1851 he was chosen as professor of chemistry at the École Normale Supérieure in Paris. The facilities for research were at first poor and the instruction elementary. Deville, however, was an excellent teacher and also was active in research. While always interested in teaching beginning students, he greatly improved the research laboratory of the institution. Many outstanding younger chemists were trained by him. From 1853 to 1866 he gave lectures in chemistry at the Sorbonne. Deville was always close to his brother, whose death in 1876 was a heavy blow to him. His health gradually failed, and he retired in 1880. He died the following year.

Deville was essentially an experimentalist and had little interest in chemical theory. He began his laboratory studies at a time when organic chemistry was developing most actively, and his early work was in this field: investigations of turpentine, toluene, and acid anhydrides. However, his analytical skill and his important synthesis of nitrogen pentoxide in 1849 turned his attention to inorganic chemistry. He worked out a process for producing pure aluminum by reducing its salts with sodium. Deville's methods made both metals readily available and drastically reduced their cost, but he himself did not take much part in their later industrial development. He used the sodium obtained by his method for the preparation of such elements as silicon, boron, and titanium. His investigations of the metallurgy of platinum led to honors from the Russian government. In many of his studies, such as those on the artificial production of natural minerals, Deville employed very high temperatures and became a recognized authority on the use of this technique. His measurements of the vapor densities of compounds at various temperatures helped to confirm Avogadro's hypothesis. These studies led Deville to his most notable discovery, the dissociation of heated chemical compounds and their recombination at lower temperatures. He heated such substances as water, carbon dioxide, and hydrogen chloride and then cooled them suddenly to recover the decomposition products. This work led to a better understanding of the mechanism of chemical reactions and to significant developments in physical chemistry.

BIBLIOGRAPHY

I. ORIGINAL WORKS. There is a complete bibliography of Deville's many individual papers in J. Gay, *Henri Sainte-Claire Deville—sa vie et ses travaux* (Paris, 1889). His first paper on dissociation is in *Comptes rendus hebdomadaires des séances de l'Académie des sciences,* **45** (1857), 857–861.

II. SECONDARY LITERATURE. In addition to the book by Gay mentioned above, accounts of Deville are Maurice Daumas, "Henri Sainte-Claire Deville et les débuts de l'industrie d'aluminium," in *Revue d'histoire des sciences,* **2** (1949), 352–357; the obituary by Louis Pasteur, in *Comptes rendus hebdomadaires des séances de l'Académie des sciences,* **93** (1881), 6–9; the paper by R. E. Oesper and P. Lemay, in *Chymia,* **3** (1950), 205–221; and Sijbren Rienks van der Ley, *Iets over de dissociatetheorie van Deville* (Groningen, 1870).

HENRY M. LEICESTER

DEWAR, JAMES (*b.* Kincardine-on-Forth, Scotland, 20 September 1842; *d.* London, England, 27 March 1923), *chemistry, physics.*

Son of Thomas Dewar, a vintner and innkeeper, and Ann Eadie Dewar, young Dewar attended local schools until he was crippled by rheumatic fever at the age of ten. During his two-year period of convalescence he learned the art of violin making and later said that this was the foundation for his manipulative skills in the laboratory. He entered Edinburgh University in 1858. James David Forbes, professor of natural philosophy, and Lyon Playfair, professor of chemistry, directed his interest to physical science. He was assistant to Playfair (1867–1868) and to Playfair's successor, Alexander Crum Brown (1868–1873). Dewar became lecturer on chemistry in the Royal Veterinary College of Edinburgh (1869) and assistant chemist to the Highland and Agricultural Society (1873). He was elected Jacksonian professor of natural experimental philosophy in Cambridge (1875) and Fullerian professor of chemistry at the Royal Institution (1877) and held both chairs until his death. The Royal Institution was the chief center of his experimental activity.

In 1871 Dewar married Helen Rose Banks, daughter of an Edinburgh printer; they had no children. He was president of the Society of Chemical Industry (1887), the Chemical Society of London (1897–1899), and the British Association (1902). Dewar was knighted in 1904. He also served as consultant to government and industry. He was a member of the government committee on explosives (1888–1891) and with Sir Frederick Abel invented the smokeless propellant cordite, a gelatinized mixture of nitrocellulose in nitroglycerin (1889).

Dewar's earliest work (1867–1877) encompassed a wide variety of subjects in physics, chemistry, and physiology. In 1867 he invented a mechanical device to represent Crum Brown's new graphic notation for organic compounds. Playfair sent the device to Kekulé, and Kekulé invited Dewar to spend the summer in his Ghent laboratory. Dewar suggested seven different structural formulas for benzene, including the diagonal formula

and the formula attributed to Kekulé.

In 1870 he proposed the pyridine ring formula, substituting a nitrogen atom for a CH residue in the benzene ring:

He also suggested that quinoline's structure consisted of fused benzene and pyridine rings.

Dewar's early studies included the heats at formation of the oxides of chlorine, the temperature of the sun and of the electric spark, the atomic volume of solids, and the production of high vacua. In 1872 he determined the physical constants of Thomas Graham's hydrogenium (Graham supposed hydrogen to be the vapor of a volatile metal, "hydrogenium") and first used a vacuum-jacketed insulating vessel. Interspersed with these physical researches were physiological investigations on the constitution and function of cystine, the physiological action of quinoline and pyridine bases, and the changes in the electrical condition of the eye under the influence of light.

At Cambridge and the Royal Institution, Dewar continued his varied interests. There were studies on the coal-tar bases; atomic and molecular weight determinations; the chemical reactions at the temperature of the electric arc, in which he noted the formation of hydrogen cyanide in the carbon arc burning in air (1879); and the determination of the monatomicity of sodium and potassium vapor from gas density studies (1883).

The first area to be thoroughly explored was spectroscopy. He joined George Downing Liveing, professor of chemistry at Cambridge, in an attempt to correlate line and band spectra with atomic and molecular states. They published seventy-eight papers between 1877 and 1904. Dewar's interest in spectroscopy stemmed from a fascination with Henri Sainte-Claire Deville's work on dissociation and reversible interactions and Norman Lockyer's controversial speculations on the dissociation of the elements at high temperatures. He contrasted Deville's exact experimental methods with Lockyer's conjectures, which he felt were based on insufficient evidence. Dewar and Liveing accurately determined the absorption spectra of many elements (especially metallic vapors) and compounds. They studied the general conditions affecting the excitation of spectra and, in particular, the ultraviolet emission spectra of many metals. They noted the contrast between single spectral lines, multiplets, and bands, and they attempted to identify the emitting agents for single, multiplet, and band spectra. They classified great, intermediate, and weak intensities with the spectroscopic series as principal, diffuse, and sharp, respectively. Their studies included the differences between the arc, spark, and flame spectra of metals; the emission spectra of gaseous explosions and of the rare gases; and the effect of temperature and concentration on the absorption spectra of rare-earth salts in solution.

Dewar's coming to the Royal Institution in 1877 marked the beginning of his work in cryogenics, his major field of study. In that year Louis Cailletet and Raoul Pictet liquefied small amounts of oxygen and nitrogen. This achievement interested Dewar because hitherto almost all the work on liquefaction of gases had been done at the Royal Institution, especially by

Michael Faraday, who by 1845 had liquefied all the known gases except the six permanent ones (oxygen, nitrogen, hydrogen, nitric oxide, carbon monoxide, and methane). During a Royal Institution lecture in 1878 Dewar gave the first demonstration in Great Britain of the liquefaction of oxygen. His principal interest was not the liquefaction of gases but the investigation of the properties of matter in the hitherto uninvestigated vicinity of the absolute zero of temperature.

In 1884 the Polish physicists Florenty von Wroblewski and Karol Olszewski improved the refrigerating apparatus, prepared moderate amounts of liquid air and oxygen, and measured their physical properties and critical constants. Dewar further improved the apparatus and methods of the Polish scientists and in 1885 prepared large quantities of liquid air and oxygen by compressing the gases at the temperature of liquid ethylene. In 1891 he discovered that both liquid oxygen and ozone were magnetic.

Dewar hoped to liquefy hydrogen; after a decade of work he had not succeeded. The critical temperature of hydrogen is −241°C. The lowest temperature attainable with liquid air as a refrigerant is about −200°C. His attempts to reach lower temperatures were unsuccessful until 1895, when he took advantage of the Joule-Thomson effect whereby the temperature of a compressed gas decreases with expansion into a vacuum because of the internal work done to overcome molecular attraction. Hydrogen was an exception; its temperature increased slightly. But Dewar showed that hydrogen had a normal Joule-Thomson effect if it was first cooled to −80°C. He cooled hydrogen by means of liquid air at −200°C. and 200 atmospheres pressure and forced it through a fine nozzle. He obtained a jet of gas mixed with a liquid that he could not collect. The temperature of the hydrogen jet was very low, and by spraying it on liquid air or oxygen he transformed them into solids. Dewar was convinced that he could reach still lower temperatures, and he spent a year making a large liquid-air machine. In 1898 his endeavor ended in success. Cooled, compressed hydrogen liquefied on escaping from a nozzle into a vacuum vessel. With liquid hydrogen he reached the lowest temperature then known. Liquid hydrogen boils at −252.5°C. at ordinary pressure. By reducing the pressure he lowered the temperature to −258°C. and solidified the hydrogen. He cooled the solid to −260°C. With liquid hydrogen, every gas except helium could in turn be both liquefied and solidified.

The lowest temperature attainable with hydrogen was still 13° above absolute zero. Could Dewar close this gap? He turned to the recently discovered helium and predicted that if the critical temperature was not below 8°K., then it should be possible to liquefy helium by methods similar to those for hydrogen. As a source of helium he used the gas bubbling from the springs at Bath, which Lord Rayleigh had found to contain the element. He failed in his liquefaction attempts because the Bath spring gas also contained neon, and in the cooling process the neon solidified, blocking the tubes and valves of the apparatus. In 1908 Heike Kamerlingh Onnes at Leiden, using Dewar's methods, succeeded in liquefying the helium isolated from the mineral monazite. Dewar presented Kamerlingh Onnes's work at a British Association meeting and showed that by boiling helium at reduced pressure he could reach a temperature less than 1° from absolute zero.

Dewar's study of the properties of matter at very low temperatures was made possible by his invention in 1892 of the vacuum-jacketed flask, the most important device for preserving and handling materials at low temperatures. The insulating property of the vacuum was well known, and Dewar had used a vacuum flask in 1872 in making specific heat determinations of Graham's hydrogenium. When he wanted to investigate the properties of liquefied gases, the idea of using a vacuum-jacketed vessel suggested itself to him.

Dewar realized that the insulating capacity of the vacuum flask depended on the state of exhaustion of the space between the inner and outer vessels. In 1905 he discovered that charcoal's adsorptive power for gases was enormously increased at −185°C. By putting a small quantity of charcoal in the evacuated space and filling the flask with liquid air, the cooled charcoal adsorbed the remaining traces of air in the space, producing a vacuum of greater tenuity. Furthermore, the charcoal-containing flasks enabled Dewar to substitute metal vessels for glass ones. Metals gave off small quantities of occluded gas, which would impair the vacuum. Since charcoal would adsorb the gas, metal vacuum vessels became feasible. They could be made both larger and stronger than glass ones. (Such vessels are now called Dewar flasks or vessels.)

Dewar used the different condensability of gases on charcoal to separate or concentrate the constituents of a gas mixture. Charcoal preferentially adsorbed oxygen from air passed over it at −185°C. Collecting the liberated gas in fractions as the temperature rose, he obtained air containing eighty-four percent oxygen. Dewar also separated the rare gases from air by this method. In 1908 he used the carbon-adsorption technique in making the first direct measurement of the rate of production of helium from radium.

Dewar examined a wide range of properties in pioneering explorations on the effect of extreme cold on substances. He determined the properties of all the liquefied gases. He measured the decreased chemical reactivity of substances at low temperatures. He studied the effects of extreme cold on phosphorescence, color, strength of materials, the behavior of metal carbonyl compounds, the emanations of radium (with William Crookes), and the gases occluded by radium (with Pierre Curie).

Dewar established that many vigorous chemical reactions did not take place at all at very low temperatures; oxygen, for example, did not react with sodium or potassium. He wanted to test the effect of cold on fluorine, the most reactive element, and in an 1897 collaboration with Henri Moissan, who had isolated the element in 1886, he liquefied fluorine and examined its properties. After Dewar had liquefied hydrogen, they resumed their investigation and solidified fluorine at −233°C. Even when the temperature was reduced to −252.5°C., solid fluorine and liquid hydrogen violently exploded.

Dewar intended to explore the whole field of cryogenics. Between 1892 and 1895 he joined with John A. Fleming, professor of electrical engineering at University College, London, in an investigation of the electrical and magnetic properties of metals and alloys. Their aim was to determine the electrical resistance from 200°C. to the lowest attainable temperature. They obtained temperature-resistance curves and found that the resistance for all pure metals converged downward in such a manner that electrical resistance would vanish at absolute zero. They gathered accurate information on conduction, thermoelectricity, magnetic permeability, and dielectric constants of metals and alloys from 200°C. to −200°C.

Another area of extensive investigation was low-temperature calorimetry (1904–1913). Dewar devised a calorimeter to measure specific and latent heats at low temperatures. He determined the atomic heats of the elements and the molecular heats of compounds between 80°K. and 20°K. He discovered in 1913 that the atomic heats of the solid elements at a mean temperature of 50°K. were a periodic function of the atomic weights.

World War I prohibited continuation of Dewar's costly cryogenic research. He turned to thin films and bubbles, which had been the subject of the first of his nine courses of Christmas lectures for children at the Royal Institution (1878–1879). He studied both solid films, produced by the evaporation of the solvent from amyl acetate solutions of nitrated cotton, and liquid films from soap. He investigated the conditions for the production of long-lived bubbles and of flat films of great size, the distortions in films produced by sound, and the patterns formed by the impact of an air jet on films.

At the time of his death Dewar was engaged in studies with a delicate charcoal-gas thermoscope that he constructed in order to measure infrared radiation. From a skylight in the Royal Institution he measured the radiation from the sky by day and night and under varying weather conditions. Dewar was a superb experimentalist; he published no theoretical papers.

BIBLIOGRAPHY

I. ORIGINAL WORKS. Dewar's papers were reprinted in *Collected Papers of Sir James Dewar,* 2 vols., Lady Dewar, J. D. Hamilton Dickson, H. Munro Ross, and E. C. Scott Dickson, eds. (Cambridge, 1927); and in George Downing Liveing and James Dewar, *Collected Papers on Spectroscopy* (Cambridge, 1915).

Important papers include "On the Oxidation of Phenyl Alcohol, and a Mechanical Arrangement Adopted to Illustrate Structure in Non-Saturated Hydrocarbons," in *Proceedings of the Royal Society of Edinburgh,* **6** (1866–1869), 82–86; "On the Oxidation Products of Picoline," in *Transactions of the Royal Society of Edinburgh,* **26** (1872), 189–196; "On the Liquefaction of Oxygen and the Critical Volumes of Fluids," in *Philosophical Magazine,* 5th ser., **18** (1884), 210–216; "The Electrical Resistance of Metals and Alloys at Temperatures Approaching the Absolute Zero," *ibid.,* **36** (1893), 271–299; and "Thermoelectric Powers of Metals and Alloys Between the Temperatures of the Boiling-Point of Water and the Boiling-Point of Liquid Air," *ibid.,* **40** (1895), 95–119, written with J. A. Fleming.

See also "The Liquefaction of Air and Research at Low Temperatures," in *Proceedings of the Chemical Society,* **11** (1896), 221–234; "Sur la liquéfaction du fluor," in *Comptes rendus hebdomadaires des séances de l'Académie des sciences,* **124** (1897), 1202–1205; "Nouvelles expériences sur la liquéfaction du fluor," *ibid.,* **125** (1897), 505–511; "Sur la solidification du fluor et sur la combinaison à −252.5° du fluor solide et de l'hydrogène liquide," *ibid.,* **136** (1903), 641–643, written with Henri Moissan; "New Researches on Liquid Air," in *Notices of the Proceedings of the Royal Institution of Great Britain,* **15** (1899), 133–146; "Liquid Hydrogen," *ibid.,* **16** (1902), 1–14, 212–217; "Solid Hydrogen," *ibid.,* **16** (1902), 473–480; "Liquid Hydrogen Calorimetry," *ibid.,* **17** (1904), 581–596; "Studies With the Liquid Hydrogen and Air Calorimeters," in *Proceedings of the Royal Society,* **76** (1905), 325–340; "The Rate of Production of Helium From Radium," *ibid.,* **81** (1908), 280–286; "Atomic Specific Heats Between the Boiling Points of Liquid Nitrogen and Hydrogen," *ibid.,* **89** (1913), 158–169; "Studies on Liquid Films," in *Proceedings of the Royal Institution,* **22** (1918), 359–405; and "Soap Films as Detectors: Stream Lines and Sound," *ibid.,* **24** (1923), 197–259.

II. SECONDARY LITERATURE. A bibliography of Dewar's works was compiled by Henry Young, *A Record of the Scientific Work of Sir James Dewar* (London, 1933). Two detailed studies of his accomplishments are Henry E. Armstrong, *James Dewar* (London, 1924) and Alexander Findlay, in Findlay and William Hobson Mills, eds., *British Chemists* (London, 1947), pp. 30–57. Informative accounts include Henry E. Armstrong, "Sir James Dewar, 1842–1923," in *Journal of the Chemical Society,* **131** (1928), 1066–1706, and *Proceedings of the Royal Society,* **111A** (1926), xiii–xxiii; Ralph Cory, "Fifty Years at the Royal Institution," in *Nature,* **166** (1950), 1049–1053, which has many personal remembrances of Dewar by the librarian of the Royal Institution; Sir James Crichton-Browne, "Sir James Dewar, LL.D., F.R.S.," in *Proceedings of the Royal Society of Edinburgh,* **43** (1922–1923), 255–260; and Hugh Munro Ross, in *Dictionary of National Biography, 1922–1930* (London, 1937), pp. 255–257.

Dewar's cryogenic research was analyzed in "Low-Temperature Research at the Royal Institution" by Agnes M. Clerke, in *Proceedings of the Royal Institution,* **16** (1901), 699–718, and Henry E. Armstrong, *ibid.,* **19** (1909), 354–412, and **21** (1916), 735–785. A more recent study is K. Mendelssohn, "Dewar at the Royal Institution," *ibid.,* **41** (1966), 212–233.

ALBERT B. COSTA

DEZALLIER D'ARGENVILLE, ANTOINE-JOSEPH. *See* **Argenville, Antoine-Joseph d'.**

DICAEARCHUS OF MESSINA (*fl.* 310 B.C.).

Dicaearchus was a distinguished disciple of Aristotle and the author of many books in different fields, none of which is preserved, so that our knowledge of them is fragmentary and often problematical.

On the Soul, six books in dialogue form, espoused the view that the soul is "nothing," immaterial, merely a condition of the body, a "harmony of the four elements"—and consequently perishable and not immortal. The doctrine was not new, but this exposition of it was one of the best. *Descent Into the Cave of Trophonius,* also a dialogue, belittled oracles and recognized only dreams and inspirations as valid sources of prophecy. The same view is attributed to Aristotle. *Tripoliticus,* a dialogue on political theory, advocated a composite constitution, combining the three traditional types of government: monarchy, aristocracy, and democracy. It seems that Dicaearchus saw this exemplified in Sparta, a state to which he was partial. The idea of a mixed constitution was taken up by Polybius. *The Life of Greece,* in three books, dealt with anthropological, moral, and cultural history, including the primitive and the Oriental forerunners of Greek civilization. The work was significant for extending history, on Peripatetic principles, over fields other than the political and military. It was the model for a work by Varro on the Romans.

Dicaearchus' only work on natural science was a

geography, *Tour of the Earth,* following the work of Eudoxus of Cnidus. These two were the first geographers to combine the actual knowledge of lands and seas with the theory of the earth as a sphere, which by then was generally accepted. Measuring a long arc north from Syene (Aswan) and observing the zenith points at the ends, they calculated the circumference to be 400,000 or 300,000 stades (stades varied from 148 to 198 meters). The area of the world then known proved to be only a small part of the surface of the sphere, perhaps 45,000 stades by 30,000 stades. Dicaearchus defended the theory of the earth as a sphere by "measuring" the highest mountains, which he found to be only ten or fifteen stades high, showing that they were insignificant in relation to the curvature of the sphere. Within the known world he sought to schematize the masses of land and sea and mountains—without the aid, of course, of specific latitude, much less of longitude. His successor in this field was Eratosthenes, whose great improvements were practical rather than theoretical.

Among the disciples of Aristotle, Dicaearchus and Aristoxenus of Tarentum, another Dorian from the west, seem to have been particularly congenial, while there was some antipathy between Dicaearchus and Theophrastus, the head of the Peripatetic school. Dicaearchus saw the purpose of knowledge as practical action; Theophrastus, as theoretical contemplation. Among posterity Cicero and Atticus were admirers of Dicaearchus and used his works extensively.

BIBLIOGRAPHY

Additional information may be found in Edgar Martini, "Dikaiarchos 3. Peripatetiker," in Pauly-Wissowa, V, pt. 1 (1903), 546–563; and Fritz Wehrli, *Die Schule des Aristoteles, Texte und Kommentar,* Heft I, *Dikaiarchos* (Basel, 1944; 2nd ed., rev. and enl., 1967).

AUBREY DILLER

DICKINSON, ROSCOE GILKEY (*b.* Brewer, Maine, 10 June 1894; *d.* Pasadena, California, 13 July 1945), *physical chemistry, X rays, crystal structure.*

Dickinson's father, George E. M. Dickinson, was a violin teacher and director of music for the Hyde Park, Massachusetts, city schools; his mother's maiden name was Georgie Simmons. He attended grammar school and high school in Hyde Park and then studied chemical engineering at the Massachusetts Institute of Technology, where he received the B.S. degree in 1915. After two years of graduate work there he was appointed instructor at the California Institute of Technology (called Throop College of Technology until 1918). In 1920 he became the first recipient of the Ph.D. degree from this institute. He remained there all his life; at the time of his death he was professor of physical chemistry and acting dean of graduate studies.

As a graduate student Dickinson became familiar with the technique of determining the atomic structure of crystals by the X-ray diffraction method through his contact with C. Lalor Burdick and James H. Ellis, who carried out, in Pasadena, the first crystal-structure determination made in the western hemisphere. At that time the lack of quantitative information about the interaction of X rays and crystals made the task of the crystal-structure investigator a difficult one. The field was, however, especially well suited to Dickinson, whose outstanding characteristics were great clarity of thought, a mastery of the processes of logical deduction, and meticulous care in his experimental work and in the analysis of data. He carried out many crystal-structure determinations, all of which have been found to be reliable to within the limits of error that he assigned. He determined the structures of a number of crystals containing inorganic complexes, including the hexachlorostannates, the tetrachloropalladites and tetrachloroplatinites, and the tetracyanide complexes of zinc and mercury. His determination (with one of his students, A. L. Raymond) of the structure of hexamethylenetetramine was the first structure determination ever made of a molecule of an organic compound. In a decade he developed the leading American school of X rays and crystal structure.

During the last twenty years of his life Dickinson and his students carried on many researches in other fields, including photochemistry, chemical kinetics, Raman spectroscopy, the properties of neutrons, and the use of radioactive indicators in studying chemical reactions. Through this, as well as through his work on crystal structure, he contributed significantly to the development of the California Institute of Technology.

BIBLIOGRAPHY

For a partial list of Dickinson's chemical and physical publications, see Poggendorff, vol. VI.

LINUS PAULING

DICKSON, LEONARD EUGENE (*b.* Independence, Iowa, 22 January 1874; *d.* Harlingen, Texas, 17 January 1954), *mathematics.*

The son of Campbell and Lucy Tracy Dickson, Leonard Eugene Dickson had a distinguished academic career. After graduating with a B.S. in 1893 as class valedictorian from the University of Texas, he became a teaching fellow there. He received his M.S. in 1894. With the grant of a fellowship he then proceeded to the newly founded University of Chicago, where he received its first doctorate in mathematics in 1896. He spent the following year in postgraduate studies at Leipzig and Paris.

Upon his return to the United States, Dickson began his career in mathematics. After a one-year stay at the University of California as instructor in mathematics, in 1899 he accepted an associate professorship at the University of Texas. One year later he returned to the University of Chicago, where he spent the rest of his career, except for his leaves as visiting professor at the University of California in 1914, 1918, and 1922. He was assistant professor from 1900 to 1907, associate professor from 1907 to 1910, and professor from 1910 to 1939. He married Susan Davis on 30 December 1902; their children were Campbell and Eleanor. At the university his students and colleagues regarded him highly as a scholar and a teacher. He supervised the dissertations of at least fifty-five doctoral candidates and helped them obtain a start in research after graduation. In 1928 he was appointed to the Eliakim Hastings Moore distinguished professorship.

Dickson was a prolific mathematician. His eighteen books and hundreds of articles covered many areas in his field. In his study of finite linear groups, he generalized the results of Galois, Jordan, and Serret for groups over the field of p elements to groups over an arbitrary finite field. He gave the first extensive exposition of the theory of finite fields, wherein he stated and proved for $m = 2, 3$ his modified version of the Chevalley theorem: For a finite field it seems to be true that every form of degree m in $m + 1$ variables vanishes for values not all zero in the field. In linear algebra he applied arithmetical concepts and proved that a real Cayley division algebra is actually a division algebra. He also expanded upon the Cartan and Wedderburn theories of linear associative algebras. He studied the relationships between the theory of invariants and number theory.

While he believed that mathematics was the queen of the sciences, he held further that number theory was the queen of mathematics, a belief that resulted in his monumental three-volume *History of the Theory of Numbers,* in which he investigated diophantine equations, perfect numbers, abundant numbers, and Fermat's theorem. In a long series of papers after 1927 on additive number theory, he proved the ideal Waring theorem, using the analytic results of Vinogradov.

Dickson received recognition for his work. The American Mathematical Society, for which he was editor of the *Monthly* from 1902 to 1908 and of the *Transactions* from 1911 to 1916, honored him. He was its president from 1916 to 1918 and received its Cole Prize in 1928 for his book *Algebren und ihre Zahlentheorie.* Earlier, in 1924, the American Association for the Advancement of Science awarded him its thousand-dollar prize for his work on the arithmetic of algebras. Harvard in 1936 and Princeton in 1941 awarded him an honorary Sc.D. In addition to his election to the National Academy of Sciences in 1913, he was a member of the American Philosophical Society, the American Academy of Arts and Sciences, and the London Mathematical Society, and he was a correspondent of the Academy of the French Institute.

BIBLIOGRAPHY

Dickson's books are *Linear Groups With an Exposition of the Galois Field Theory* (Leipzig, 1901); *College Algebra* (New York, 1902); *Introduction to the Theory of Algebraic Equations* (New York, 1903); *Elementary Theory of Equations* (New York, 1914); *Algebraic Invariants* (New York, 1914); *Linear Algebras* (Cambridge, Mass., 1914); *Theory and Applications of Finite Groups* (New York, 1916), written with G. A. Miller and H. F. Blichfeldt; *History of the Theory of Numbers* (Washington, 1919–1923), vol. I, *Divisibility and Primality;* vol. II, *Diophantine Analysis;* vol. III, *Quadratic and Higher Forms* (with a ch. on the class number by G. H. Cresse); *A First Course in the Theory of Equations* (New York, 1922); *Plane Trigonometry With Practical Applications* (Chicago, 1922); *Algebras and Their Arithmetics* (Chicago, 1923); *Modern Algebraic Theories* (Chicago, 1926); *Algebren und ihre Zahlentheorie* (Zurich, 1927); *Introduction to the Theory of Numbers* (Chicago, 1929); *Studies in the Theory of Numbers* (Chicago, 1930); *Minimum Decompositions Into Fifth Powers,* vol. III (London, 1933); *New First Course in the Theory of Equations* (New York, 1939); and *Modern Elementary Theory of Numbers* (Chicago, 1939).

Other writings are *On Invariants and the Theory of Numbers,* American Mathematical Society Colloquium Publications, **4** (1914), 1–110; *Researches on Waring's Problem,* Carnegie Institution of Washington, pub. no. 464 (1935).

A. A. Albert, "Leonard Eugene Dickson 1874–1954," in *Bulletin of the American Mathematical Society,* **61** (1955), 331–346, contains a complete bibliography of Dickson's writings.

RONALD S. CALINGER

DICKSTEIN, SAMUEL (*b.* Warsaw, Poland, 12 May 1851; *d.* Warsaw, 29 September 1939), *mathematics, history of mathematics, science education, scientific organization.*

Dickstein devoted his life to building up the organizational structure for Polish science, especially for mathematics. In the eighteenth century Poland's territory had been divided among Prussia, Austria, and Russia; and thus Polish science education and scientific life depended mostly on personal initiative and not on state support. In his youth Dickstein experienced the escalation of national oppression after the unsuccessful uprising of 1863. There was no Polish university in Warsaw at that time, and higher education was provided in part by the Szkola Główna, which was a teachers' college. From 1866 to 1869 Dickstein studied at the Szkola Główna, which was converted into the Russian University in Warsaw in 1869. After 1870 he continued his studies there and in 1876 received a master's degree in pure mathematics.

From 1870 Dickstein taught in Polish secondary schools, concentrating on mathematics; from 1878 to 1888 he directed his own private school in Warsaw. In 1884, with A. Czajewicz, he founded Biblioteka Matematyczno-Fizyczna, which was intended to be a series of scientific textbooks written in Polish. These books greatly influenced the development of Polish scientific literature. In 1888 Dickstein took part in the founding of the first Polish mathematical-physical magazine, *Prace matematyczno-fizyczne.* Later he founded other publications, such as *Wiadomości matematyczne* and the education journal *Ruch pedagogiczny* (1881).

The Poles' efforts after the creation of the Polish university led in 1906 to the founding of Towarzystwo Kursów Naukowych, which organized the university science courses. Dickstein was the first rector of that society. In 1905 he became a founder of the Warsaw Scientific Society, and he was instrumental in the development of the Society of Polish Mathematicians. After the revival of the Polish university in Warsaw he became professor of mathematics there in 1919. His own mathematical work was concerned mainly with algebra. His main sphere of interest besides education was the history of mathematics, and he published a number of articles on Polish mathematicians that contributed to their recognition throughout the world. Of especial note are the monograph *Hoene Wroński, jego życie i prace* (Cracow, 1896) and the edition of the Leibniz-Kochański correspondence, published in *Prace matematyczno-fizyczne,* **7** (1901), and **8** (1902). Appreciation of his historical works was shown in his election as vice-president of the International Academy of Sciences. The list of his scientific works includes more than 200 titles. Dickstein died during the bombardment of Warsaw and his family perished during the German occupation of Poland.

BIBLIOGRAPHY

I. ORIGINAL WORKS. The list of Dickstein's works up to 1917 is contained in a special issue of the magazine *Wiadomości matematyczne;* works from subsequent years are in the memorial volume *III Polski zjazd matematyczny. Jubileusz 65-lecia działalności naukowej, pedagogicznej i społecznej profesora Samuela Dicksteina* (Warsaw, 1937).

II. SECONDARY LITERATURE. Besides the memorial volume, the basic biographical data and an appreciation are contained in A. Mostowski, "La vie et l'oeuvre de Samuel Dickstein," in *Prace matematyczno-fizyczne,* **47** (1949), 5–12.

LUBOŠ NOVÝ

DIDEROT, DENIS (*b.* Langres, France, 5 October 1713; *d.* Paris, France, 31 July 1784), *letters, technology.*

Diderot's importance in the history of science derives from his having edited the *Encyclopédie*—with the partial collaboration of d'Alembert—and from a sensibility that anticipated and epitomized moral, psychological, and social opportunities and stresses attending the assimilation of science into culture.

Early Life and Work. He came from energetic stock. His father, Didier, was a prosperous master-cutler who aspired to higher spheres for his children. His mother, born Angélique Vigneron, was of a family of tanners with a tendency to the priesthood. Diderot was the eldest child of seven. One sister, Angélique, became a nun and died mad. A brother, Didier-Pierre, took orders and became archdeacon of Langres. It may have been fortunate for the Church that nothing beyond the tonsure at the age of thirteen came of the would-be nepotism of a maternal uncle who thought to make a priest of Diderot in order to leave his nephew his own benefice.

Like many of his fellow *philosophes,* Diderot was well-educated by the Jesuits. Having completed their *collège* at Langres, he was sent to Paris in the winter of 1728–1729. There he enrolled in the Collège d'Harcourt and also followed courses in two other famous establishments, the Collège Louis-le-Grand and the Collège de Beauvais. In 1732 he took his degree *maître-ès-lettres* of the University of Paris. Thereafter his father supposed him to be entering upon legal studies. In fact Diderot was enjoying his freedom, intellectual and amorous, in the literary Bohemia of the capital. Abandoning the pretense of law, he drifted into a catchpenny life, ghost-writing sermons for hard-pressed preachers and missionaries, applying himself to mathematics and teaching it a bit, and perfecting his English and undertaking translations—of which the occasional faithlessness expresses wit, not ignorance. For to the ordinary appetites of Grub Street he added that for information, and unlike the hacks around him, he did study. There have been few writers—perhaps only Voltaire—whose lightness of touch has more gracefully dissembled a capacity for work.

The most considerable of these early commissions planted in his mind the idea that burgeoned in the *Encyclopédie*. In 1744 Diderot, together with three other writers, put in hand for the publisher Briasson the translation of *A Medicinal Dictionary: Including Physic, Surgery, Anatomy, Chemistry, and Botany . . .,* a multivolume work that except for Diderot's connection with it would weigh quite forgotten in both languages upon library shelves.[1] Qualifying himself (for he was not the man to remain ignorant of what he was translating), he attended public courses in anatomy and physiology given by one Verdier, and later those of a certain formidable Mlle. Biher. He thus began an adult self-education in science that he long continued and that put him in the way of the medical humanism which forms a still insufficiently appreciated strain in the naturalistic thought of the Enlightenment and which issued in the philosophy of vitalistic materialism. His marriage in 1743 to Anne-Toinette Champion turned out unhappily almost from the start, although their surviving child, later Mme. de Vandeul, was a comfort to him in old age.

In the 1740's Diderot and his fellow writers began to form a recognizable circle of like-minded free spirits on whom he later drew for contributions to the *Encyclopédie*. He met Rousseau in 1742, Condillac in 1744, and came to know d'Alembert, Grimm, Mably, d'Holbach, and others, and to be known to Voltaire, to whose deistic point of view on science he increasingly opposed the naturalistic standpoint he was developing for himself. The publications by which he made himself known for an original writer went far in the direction of overt skepticism. "I write of God," he announced in the opening sentence of *Pensées philosophiques* (1746), embarking upon a celebration of the passions and identifying them with the creative energies of nature, which, it soon appears, is indistinguishable from God. For although the standpoint from which Diderot was attacking absurdities and inequities in the scriptural tradition was ostensibly that of the deism of Shaftesbury, from whom he borrowed many a theological observation, his inspiration in natural philosophy was actually the pantheism of Spinoza, and never Newton.

The *Lettre sur les aveugles* of 1749, his first truly original work, goes further and undermines deism. In it he initiated a device that he employed more regularly in later writings. A prominent contemporary figure is adopted to be spokesman for views that Diderot was just then trying out. Through the person of Nicholas Saunderson, a blind mathematician who was Lucasian professor in Cambridge, Diderot exhibited how unconvincing it is in the eyes of the sightless to base the existence of God upon the evidence for design in nature. The essay combines humanity with skepticism. In handling this favorite psychological puzzle of eighteenth-century sensationalism—how the world appears to a man deprived of one of his senses, or to whom sight or hearing is suddenly restored—Diderot found himself questioning the artificiality with which the associationists, and notably his friend Condillac, abstracted the operation of the five senses one from another in some mechanical and imaginary sensorium.

He ended by disputing as gratuitous the conclusion of the self-styled empiricists that on regaining sight a blind man, once he learned to use his eyes at all, would not recognize the difference between a sphere and a cube without touching them. There was a psychological shrewdness in Diderot that rejected the notion that touch and sight can be independent even for analytic purposes. His sense of what people are really like, related to a highly personal distrust of all abstractions, animates all his writings.

The most widely read of these pre-Encyclopedic writings was almost certainly *Les bijoux indiscrets* (1747). It is a salacious fantasy and, in certain passages in the vein of *Fanny Hill,* a pornographic one, written gaily rather than grossly, and no doubt mainly for gain. Overtones convey the innocence of sensual enjoyment, and the tale is not out of character. It no longer seems so incongruous coming from a champion of humanity as it did prior to the recent recurrence of a cultural symbiosis, at once libertine and libertarian, between open sensuality—aesthetic, gustatory, sexual—disdain for convention, and the belief that freedom is to be asserted against the corruptions and hypocrisies of society and culture and not merely secured within the operations of law and government. The latter would have satisfied Voltaire, but did not interest Diderot. Nor would he with Rousseau reject society and culture. His yearning was for their transfiguration into a congruence with nature.

Such radicalism made itself felt. Inevitably the authorities thought him dangerous and placed him in detention for more than three months in the summer of 1749 in the confines of the château of Vincennes. There he acknowledged authorship of the three works just mentioned—intemperate thoughts that happened to slip out, he called them—and there he continued preparation of what in an executive sense was the great work of his life, the editing of the *Encyclopédie, ou Dictionnaire raisonné des sciences, des arts, et des métiers,* so fully the signet of the French Enlightenment that the word "Encyclopedists" has become almost a synonym for its exponents. Even his imprisonment was one of those enlightened oppressions that did not prevent a subversive character from reading, writing, or receiving friends.

The *Encyclopédie.* Specifying the importance of the *Encyclopédie* in the history of science does not require following in detail the vexed story of its preparation and publication:[2] its commercial origin at the instance of the publisher Le Breton, who intended a straightforward translation from the English of Chambers' *Cyclopaedia* and John Harris' *Lexicon Technicum;* his enlistment of Diderot and d'Alembert in 1747 after a false start with the Abbé de Gua de Malves as editor, d'Alembert to oversee the mathematical subjects; the beating of bushes for contributors and nagging of contributors for copy; the appearance of the first volume in 1751 to applause from the enlightened and muttering from court and clergy; d'Alembert's desertion in 1758 on the eve of the suspension following volume VII and the subsequent prohibition of the enterprise by the authorities; publication, nevertheless, of the remaining ten volumes with the provisional protection of the chief official charged with censorship, the liberal-minded Malesherbes; the triumphant completion of the text of the work in 1766 consisting finally of seventeen volumes of articles and eleven of splendid plates, largely technical in subject matter. This was not merely the work of reference originally imagined by the publisher, who had had to take three other firms into partnership in order to finance the scale to which Diderot in his energy and enthusiasm had expanded it. It was not merely a place to look things up. A proper dictionary, in Diderot's view, should have "the character of changing the general way of thinking." [3]

The ideological impact of the *Encyclopédie* in its social, economic, political, juridical, and theological aspects is naturally more famous than its technical side. Having largely assimilated the ideology of progress, toleration, and government by consent, the general historical consciousness continues to be titillated by the alarm aroused at the time in traditionalist and privileged quarters over articles like "Certitude," by the Abbé de Prades, an emancipated clergyman who preferred the reasonings of Locke to the obscurities of Revelation; "Fornication," which introduces the word as a term in theology; "Salt," which enlarges on the injustice to the poor of excise taxes on items of subsistence; and "Political Authority," which de-

nied the existence in nature of the right of any man to exercise sovereignty over others.

Diderot's purpose, however, was deeper than unsettling the authorities by purveying tongue-in-cheek reflections on superstition and injustice in the guise of information. The technical contents of the *Encyclopédie* were central to that purpose, which was the dignification of common pursuits over and against the artificiality and pretense of the parasitic encrustations in society and reciprocally the rationalization and perfection of those pursuits in the light of modern knowledge. It would be a mistake to seek the scientific importance of the *Encyclopédie* only or even mainly in articles contributed or commissioned by d'Alembert on topics of mathematics, mechanics, or formal science. True, those articles are often (although not always) valuable summaries of the state and resources of a subject, and were so regarded at the time. But technically the central thrust of the *Encyclopédie* was in its descriptions of the arts and trades, and for that the initiative and responsibility were Diderot's, harking back certainly to his provincial background among the thriving artisans of Langres. The account of the cutlery industry is one of the best and clearest in the work, and in the article "Art" is a passage that may be taken as his credo:

> Let us at last give the artisans their due. The liberal arts have adequately sung their own praises; they must now use their remaining voice to celebrate the mechanical arts. It is for the liberal arts to lift the mechanical arts from the contempt in which prejudice has for so long held them, and it is for the patronage of kings to draw them from the poverty in which they still languish. Artisans have believed themselves contemptible because people have looked down on them; let us teach them to have a better opinion of themselves; that is the only way to obtain more nearly perfect results from them. We need a man to rise up in the academies and go down to the workshops and gather material about the arts to be set out in a book which will persuade artisans to read, philosophers to think on useful lines, and the great to make at least some worthwhile use of their authority and their wealth.

Diderot wrote this himself as he did many of the articles describing particular trades and processes. Not in every case was he able to "go down to the workshops" and base his account on actual observation, and a number are composed from printed sources or other secondary information. It was not on the articles alone, however, but on the illustrative plates to which most of them were keyed that Diderot and his publishers relied to fulfill the promise made to subscribers of a systematic description of eighteenth-century industry in its essential processes and principles. Censorship had interrupted publication of articles in 1759, but there could be no objection to going ahead with technical plates containing no sensitive matter. The series began in 1762 and filled the gap until the remaining ten volumes of text could appear all at once four years later. (The supplementary volume of plates and four of text published by Panckoucke were not edited by Diderot.)

Delicate questions arise about the publication of these plates, not concerning Diderot's treatment of the censorship, but rather his originality and treatment of the rights of others. They were preceded in their appearance by angry charges of having been lifted from engravings prepared for the Academy of Sciences. Since its founding in 1666, during the administration of Colbert, that body had been vested with the responsibility of maintaining a scientific surveillance over French industry. Not a line had appeared of its constitutionally prescribed project for a description of arts and trades, although its most recent director, the naturalist and metallurgist Réaumur, had commissioned a large number of plates before his death in 1757. The Academy rushed one volume of these into print in 1761 in order to forestall Diderot, who himself or through agents must indeed have found, bought, or bribed access to Réaumur's plates during the early stages of preparing the *Encyclopédie*. All that can be said in extenuation is that copyright did not exist in the eighteenth century, that title to artistic and literary property was an amorphous matter, and that whoever engraved the various plates it was through Diderot's deeds and misdeeds that they appeared as a collection, a systematic record of industrial life and methods.

At their best the plates of the *Encyclopédie* are executed with the sweep and style of chefs d'oeuvre of technical illustration, notably in the series devoted to the glass industry, in the coverage afforded to Gobelin tapestries, and in depicting the blast furnace and forge. Typically the reader is given something like an anatomy of machines, a physiology of processes. The technique of illustration might be thought to derive from the anatomy of Vesalius two centuries before. Several of his plates are among those that reappear without acknowledgment as do several of Agricola's depicting sixteenth-century mining. Normally the first plate in each series gives an overall picture of an installation and is followed by sectional views, one lengthwise and one crosswise. Thereafter, cutaway representations penetrate to the intermediate assemblies, sometimes shown in place and sometimes in isolation. Finally, there are drawings of the individual parts, pieces, and tools.

The plates exhibit the state of manufacturing processes just before the industrial revolution, then in its earliest, largely unperceived stages in England. Science is often taken to be the fruitful element in technology, the progenitor of industrialization, and so it appears in the *Encyclopédie,* but only if we limit what we understand by the influence of science to its descriptive role. Basic theory had very little to offer the manufacturer in any industry in the eighteenth century, and it was descriptive science addressed to industry that transformed it by rationalizing procedures and publicizing methods. In effect the *Encyclopédie* turned craftsmanship from lore to science and began replacing the age-old instinct that techniques must be guarded in secret with the concept of uniform industrial method to be adopted by all producers.

Not that the changeover was welcomed by practitioners or easily achieved. Many tradesmen, full of suspicion, resisted inquiries or deliberately misinformed Diderot and his associates after accepting their gratuities. Terminology alone created obstacles. Each trade had its own, often barbarous, jargon. A great many artisans had no desire to understand from a scientific point of view what it was they were doing and preferred working by traditional routine. "It is only an artisan knowing how to reason who can properly expound his work,"[4] exclaimed Diderot in a moment of irritation, a remark that might seem to render somewhat circular his fundamental conception of science and reason as the educators of industry, but that brings out the necessity for its rationalization if the truly popular purpose of the *Encyclopédie* was to be achieved in the deepest sense—that is, in the easing of labor, its liberation from routine, and the summons to pride in its enlightenment.

Moral and Philosophic Position. Particular articles in the *Encyclopédie* exhibit the development of Diderot's scientific sensibility into the psychic materialism of his later years, through his reading of chemistry, natural history, comparative anatomy, and physiology, complemented by the experience and observation of humanity, but it is more satisfactory to follow the writings that he found time and inspiration to leave as his literary legacy. The term legacy is deliberate for it was one of the peculiarities of Diderot's intellectual personality that, prolific writer that he was, it seems to have been more important to him to express his mind than to publish and persuade. Of the major books conveying his philosophy of nature, only *De l'interprétation de la nature* was printed during his lifetime, the first edition in 1753 and a revision in 1754, just when he was winding up his researches on the arts and trades. For the rest (to name the important), *Le rêve de d'Alembert, Entretien* entre Diderot et d'Alembert, and Suite de l'entretien were written in 1769 and published in 1830; the Supplément au voyage de Bougainville was begun in 1772 and published in 1796; and the Éléments de physiologie, begun in 1774 and taken up again in 1778, was published in 1875. (It was the same with his best literary works: the picaresque novel Jacques le fataliste was printed in the completed form in 1796, and the theatrical piece Est-il bon, Est-il méchant? in 1834.

Most extraordinary of all, his masterpiece, Le neveu de Rameau—a complex dialogue that, in shifting the locus of immorality back and forth between the ostensibly degenerate individual and the actually corrupt society, anticipates the diabolism although not the sexual inversion in the writings of Jean Genet—was first printed in a German translation by Goethe from an inexact manuscript and published in French in a largely faithful version only in 1884. Vicissitudes too complicated to follow here led these manuscripts through various minor German courts and the major court of Catherine the Great in St. Petersburg, which Diderot was persuaded to enliven briefly by his presence in later years.

It is possible to identify sources of much of Diderot's scientific inspiration:[5] the chemistry in the lectures of Rouelle at the Jardin du Roi, the natural history in Buffon, the physiology in Haller, the psychology in La Mettrie, and the medical doctrine in his frequent association with Bordeu. But this record of nonpublication, and the consequent implausibility of supposing that his views could have formed a system exerting a coherent influence upon either contemporary or later writers, make it more reasonable to regard his response to the scientific world picture as an anticipation of the program recognizable later, and rather recurrently than consequentially, as that of biological romanticism—an attempt to construct an account of the operations of nature in categories of organism and consciousness rather than impersonal matter in inanimate motion.

Occasionally it has been supposed in Diderot scholarship that he reached this position in a lifelong progression from some solid Newtonian basis in his youth; and it is true that he published in 1748 (the year of Les bijoux indiscrets!) a curious little collection, Mémoires sur différens sujets de mathématiques; and further true that in one memoir, the fifth, he mentions that he had studied Newton formerly, "if not with much success, at least with zeal enough," but that to raise questions about Newton today "is to speak to me of a dream of years gone by."[6] Although very interesting in several respects, the memoirs themselves give no reason beyond the title

for thinking that Diderot had in fact ever been seized of Newtonian mathematical physics.

The first and most considerable is a summary of musical acoustics dressed out in elementary mathematical formalism. Its object is to establish that musical pleasure consists in perception of the relations of sounds as they are propagated in nature and is no mere matter of caprice or culture, although such factors certainly affect the judgment. Ever the good encyclopedist, Diderot reported faithfully the work on vibrating strings and pipes of Taylor, d'Alembert, Mersenne, Sauveur, and Euler. Although in no way original physically, the discussion is, nevertheless, a highly individual approach to the physics of beauty in a manner not to be attempted successfully before the work of Helmholtz. The other essays are much slighter. The second and third are geometrical and concern the design of certain devices that Diderot was proposing to mathematical and musical instrument makers. The fourth is a (virtually computerized) program for enlarging the repertory of barrel organs, and rather wickedly suggests that resistance to the improvement of these popular instruments was a function not of musical taste but of the self-interest and restrictiveness of musicians and music teachers. The last reassures those who have failed to master Newton by taking him to task for an allegedly false assumption about air resistance in the pendulum experiments that he reported in the scholium following the sixth corollary to the laws of motion in the Principia.

Five years later three volumes of the Encyclopédie were in print, and the opening paragraphs of De l'interprétation de la nature predict that mathematics is about to go into a decline, and deservedly so. On all grounds it had exaggerated its claim to be the language of science. Metaphysically it falsifies nature by depriving bodies of the qualities of odor, texture, appearance, or taste through which they appeal to our senses. It impoverishes mechanics by requiring it to operate with the superficial measurements of bodies instead of seeking, as the chemist is said to do, for the activity that animates them. Worst of all, it dries up and blights the sensibilities of those who cultivate it and whom it renders inhuman in their judgments. Such will be the effect of any science that ceases to "instruct and please," for the only thing that will make a science appealing and keep it vital is its capacity to improve the character, understanding, and moral fiber of its possessors. Mathematics leads mainly to arrogance, however, pretending to equip a finite intelligence to plumb the infinite where it has no business. Man being insatiable, we need some criterion not found in mathematics by which to establish bounds between what we need to know and the infinite unknown. So let it be our interests, let it be utility, "which, in a few centuries, will establish boundaries for experimental science, as it is about to do for mathematics."[7]

Apparently a formless rumination, *De l'interprétation de la nature* is actually written in an artful stream of consciousness, a reverie on the Experimental Art, the true road to a science of nature. That road lies through craftsmanship, and here we rejoin the editor of the *Encyclopédie*. For Diderot, it is the common touch that opens up the truth, and genius that is to be distrusted, inclining in its pride to draw a mathematical veil of abstraction and obscurity between nature and the people. It is wrong to say that there are some truths too deep or hard for ordinary understandings. Certainly common men will never attach any value to what cannot be proven useful, and they are right. Only a philosophy derived from actually handling objects is innocent in that it involves no a priori ideas. A kind of intuition in the true craftsman has the quality of inspiration for it derives from genuine participation. Such a man will recognize it in himself, in his solidarity with natural objects. In his hands science and nature are one in the actual operation with materials. Not some mathematical abstraction from nature but manual intimacy with nature, living oneness with nature, is the arm of science.

For nature is the combination of its elements and not just an aggregate. It is continuity that science is to seek in nature, not divisibility. The interesting property of molecules is their transience, not their existence. In genetics the notion of *emboîtement* is unacceptable because of its atomistic implications. For there are no fixed limits in nature. Male and female exist in each other (a fascination with hermaphroditism and the merging of the sexes is another motif in his writings that seems curiously up-to-date in the 1970's). Mineral, vegetable, and animal kingdoms blend, species into species, and only the stream of seminal fluid is permanent, flowing down through time. "Tout change; tout passe; il n'y a que le tout qui reste" ("Everything changes; everything passes; nothing remains but the whole");[8] and in *Le rêve de d'Alembert,* Diderot invokes two models to exhibit this unity. The first is the swarm of bees, for the solidarity of the universe is social. It has the oneness of the social insects, among whom laws of community are laws of nature. In the second the universe is a cosmic polyp, time its life unfolding, space its habitation, gradience its structure, and certainly the two ideas that mattered most to Diderot were those of social naturalism and universal sensibility.

In Diderot's philosophic dialogue with a fictitious d'Alembert, that distinguished mathematical colleague admits that the notion of a sensitive matter containing in itself principles of movement and consciousness is more immediately comprehensible than that of a being which is inextended and yet occupies extension, which differs from matter and yet is united to it, which follows its course and moves it without being moved. "Is it by accident," this straw d'Alembert is made to ask,

> that you would recognize an active sensibility and an inert sensibility, in the way that there is a live force and a dead force?—a live force that manifests itself in motion of translation, a dead force that manifests itself by pressure; an active sensibility that is characterized by certain remarkable actions in animals and perhaps in plants, and an inert sensibility the existence of which is assured by the passage to the state of active sensibility?

"Marvelous," Diderot as interlocutor replies, "you said it."[9]

And in the *rêve* that continues the discussion, d'Alembert, apparently ailing, has been put into a trance—it may be a delirium—in which he speaks truths that would not have come to him from the normal detachment of the mathematical analyst quantifying his inert blocks of matter. Now his interlocutors are his mistress, Mlle. de l'Espinasse, and a doctor, Diderot's mentor in medical vitalism, Théophile de Bordeu, who recognizes these verities for what they are and draws them out in a kind of psychic analysis of the realities of a world alive. He it is who sees nature in the perspective of human nature, and who, therefore, knew the answers all the time. "There is no difference," Diderot makes Mlle. de l'Espinasse observe, "between a doctor keeping watch and a philosopher dreaming."[10]

NOTES

1. Scholarship on Diderot occasionally falls into confusion about the identity of this work, published in 3 vols. (London, 1743–1745), attributing it sometimes to the authorship and sometimes to the firm of Ephraim Chambers, publisher of a famous *Cyclopaedia*. In fact the author was Robert James and the publisher T. Osborne. The French translation was published by Briasson in 6 vols. (1746–1748) and entitled *Dictionnaire universel de médecine, de chirurgie, de chymie, de botanique, d'anatomie, de pharmacie, et d'histoire naturelle*. . . .
2. See Jacques Proust, *Diderot et l'Encyclopédie* (Paris, 1963).
3. Quoted in Arthur Wilson, *Diderot: The Testing Years* (New York, 1957), p. 244.
4. Diderot article, "Encyclopédie," in the *Encyclopédie,* vol. V.
5. Jean Mayer, *Diderot, homme de science* (Rennes, 1959).
6. *Op. cit.,* pp. 202–203.
7. *Op. cit.,* par. 6.
8. "Le rêve de d'Alembert," in Paul Vernière, ed., *Oeuvres philosophiques de Diderot* (Paris, 1956), pp. 299–300.
9. "Entretien entre d'Alembert et Diderot," *ibid.,* p. 260.
10. *Ibid.,* p. 293.

BIBLIOGRAPHY

I. ORIGINAL WORKS. There is no modern edition of the works of Diderot, although one is said to be in preparation. The most recent is *Oeuvres complètes,* J. Assezat and M. Tourneux, eds., 20 vols. (Paris, 1875–1877). A useful selection is *Oeuvres philosophiques de Diderot,* Paul Vernière, ed. (Paris, 1956).

There are a number of selections from the *Encyclopédie* in *L'Encyclopédie* (*Extraits*) (Paris, 1934); *The Encyclopédie of Diderot and d'Alembert, Selected Articles,* John Lough, ed. (Cambridge, 1954); *Textes choisis de l'Encyclopédie,* ed., with commentary, by Albert Soboul (Paris, 1962); and *Encyclopedia Selections,* Nelly S. Hoyt and Thomas Cassirer, eds. (New York, 1965).

The undersigned has edited a selection from the technical plates: *A Diderot Pictorial Encyclopedia of Trades and Industry . . . With Introduction and Notes,* 2 vols. (New York, 1959).

II. SECONDARY LITERATURE. The literature on Diderot is immense. Besides works mentioned in the notes, readers will gain entry into it from Abraham Lerel, *Diderots Naturphilosophie* (Vienna, 1950); Jean-Louis Leutrat, *Diderot* (Paris, 1967); J. Lough, *Essays on the Encyclopédie of Diderot and d'Alembert* (London, 1968); René Pomeau, *Diderot, sa vie, son oeuvre, avec un exposé de sa philosophie* (Paris, 1967); and Franco Venturi, *La jeunesse de Diderot (1713–1753)* (Paris, 1939).

The exposition of *De l'interprétation de la nature* in this article follows closely the discussion of the significance of Diderot's scientific views for the intellectual history of the Enlightenment in the undersigned's *Edge of Objectivity* (Princeton, 1960), ch. 5.

CHARLES COULSTON GILLISPIE

DIELS, OTTO PAUL HERMANN (*b.* Hamburg, Germany, 23 January 1876; *d.* Kiel, Germany, 7 March 1954), *organic chemistry.*

Diels's father, Hermann, was professor of classical philology at the University of Berlin; his brother Paul, professor of Slavic philology at Breslau; and his brother Ludwig, professor of botany at Berlin. His mother, the former Bertha Dübell, was the daughter of a district judge.

As a student, Diels, with his brother Ludwig, eagerly conducted chemical experiments. He studied chemistry in Berlin from 1895 to 1899 and in 1899 obtained the Ph.D., *magna cum laude,* with a dissertation entitled "Zur Kenntnis der Cyanverbindungen." From 1899 he studied under Emil Fischer and served as his assistant until he became a lecturer in 1904. He became department head in 1913 and in 1914 was appointed associate professor at the Chemical Institute of the Royal Friedrich Wilhelm (now Humboldt) University. In 1916 Diels accepted an invitation from Christian Albrecht University, Kiel, where he served as full professor and director of the Chemical Institute until his final retirement in October 1948.

Diels recorded his personal memories in a manuscript entitled "Werden und Wirken eines Chemieprofessors," as well as in an illustrated diary (Diels was a weekend painter) which has been reviewed in detail by Sigurd Olsen. Diels was considered somewhat reserved, yet possessed of a good sense of humor. He was honest, sensitive, an outstanding educator, and a devoted family man. Diels was married in 1909 and had three sons and two daughters. Toward the end of World War II two of his sons were killed at the eastern front, and air raids completely destroyed the Chemical Institute, the library, and his home. Since there was no possibility of carrying on his work and he was suffering from the general privations, Diels filed for retirement in September 1944, to be effective in March 1945. Nevertheless, in 1946 he agreed to resume the directorship of the Chemical Institute, and at the age of seventy he started anew, under the most primitive conditions in makeshift quarters.

Emil Fischer, Diels's teacher, ended a period of chemistry that Willstätter called the age of simple methods and direct observation. During Diels's time the importance of theoretical chemistry, physico-chemical measuring methods, complicated experiments, and teamwork grew to such an extent that in some instances chemical research lost a little of its immediacy.

Diels's work, which was in the field of pure organic chemistry with no significant digressions into biochemistry or into physical chemistry, reveals an outstanding experimenter with original and bold ideas. His *Einführung in die organische Chemie* (1907), which went through nineteen editions by 1962, has a clarity and a precision that have made it one of the most popular textbooks in the field. Diels's lectures, accompanied by experiments, were outstanding and were enthusiastically received by his students.

In 1906 Diels obtained carbon suboxide by dehydrating malonic acid and investigated its properties. In the same year, with E. Abderhalden, he began his research on cholesterol, the structure of which had not yet been determined. He isolated pure cholesterol from gallstones and converted it into "Diels's acid" through cleavage by oxidation. Meanwhile, Windaus had proposed a formula for cholesterol that did not agree with more recent observations. As a result, Diels decided first to establish the aromatic basic structure of cholesterol. Dehydration of cholesterol with sulfur was unsuccessful. Selenium was used next, and in 1927 it yielded a twofold success (Fig. 1): a new, milder, and very effective dehydrating agent had been

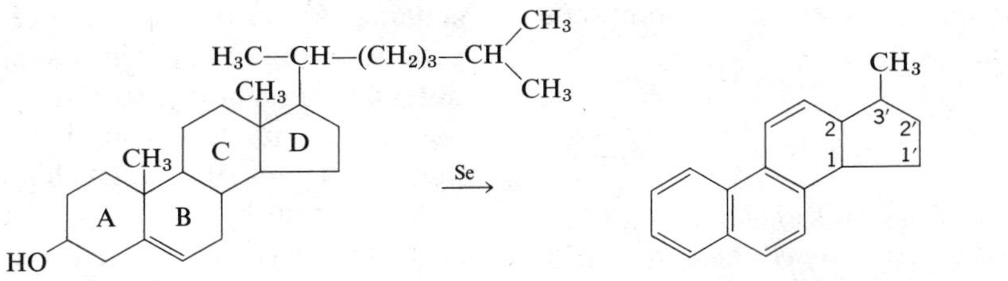

Cholesterol   Diels hydrocarbon
(3'-Methyl-1,2-cyclopentenophenanthrene)

FIGURE 1

discovered; Diels had obtained the aromatic basic structure he had been looking for—3'methyl-1,2 cyclopentenophenanthrene, the structure of which was clarified by Robert Harper, Kon, and Leopold Ruzicka in 1934.

This "Diels hydrocarbon," $C_{18}H_{16}$, the identity of which with the corresponding synthetic product was demonstrated by Diels in 1935, proved to be the basic substance and structure of a number of very important natural products for the chemistry of natural substances. The investigation of these substances corresponded in importance to the discovery of the benzene ring for organic chemistry. The structure and behavior of the sex hormones, the saponins, the cardiac glycosides (digitoxin, strophanthin, etc.), the D vitamins, toad "venom," bile pigments, adrenal-cortex hormones (cortisone), and similar substances could now be clarified.

With Kurt Alder, Diels developed over a period of twenty-two years the diene synthesis, which came to occupy a key position in the theory and practice of organic chemistry. It also yielded new facts concerning the three-dimensional isomerism of the carbon compounds. Starting with Thiele's 1,4-addition theorem and the knowledge of the additive power of azo esters, Diels and Alder attempted in 1928 to combine maleic anhydride with cyclopentadiene (Fig. 2). The dienes (compounds with conjugated carbon double bonds) united with philodienes (compounds with an ethylene radical flanked by carbonyl or carboxyl groups) to form ring-shaped structures. This type of synthesis is not only extraordinarily diverse but also occurs spontaneously, even at room temperature, and in general with good yields, without the use of condensing agents and catalysts. The diene system opens at positions 1 and 4, and the terminal carbons are located at the double bonds of the philodienes.

Diels published thirty-three papers on the practical applications of this new method of synthesis. Windaus, in the field of steroids, used it for the separation of ergosterol and its irradiation products. In a series of important terpenes, such as camphor, dl-santene, butadiene, and α-phellandrene, the structure could be confirmed by synthesis, since these substances are composed of isoprene residues, the building blocks of the diene structure. Also, great progress could be made in the synthesis of heterocyclics.

In 1904 Diels participated in the Louisiana Purchase Exposition in St. Louis and received a gold medal for his exhibit. During the academic year 1925–1926 he served as *rector magnificus* of the University of Kiel. His inaugural address was entitled "Über die Bedeutung von Zufall und Instinkt bei grossen chemischen Entdeckungen." In 1931 the Society of German Chemists awarded Diels the Adolf von Baeyer Memorial Medal, and in 1946 the Medical Faculty of Christian Albrecht University awarded him an honorary doctorate. Diels was a member of the academies of sciences of Göttingen and Halle (Leopoldina) as well as the Bavarian Academy of Sciences. In 1950 he and his pupil Kurt Alder shared the Nobel Prize in chemistry for the development of the diene synthesis. In his Nobel address, "Darstellung und Bedeutung des aromatischen Grundskeletts der

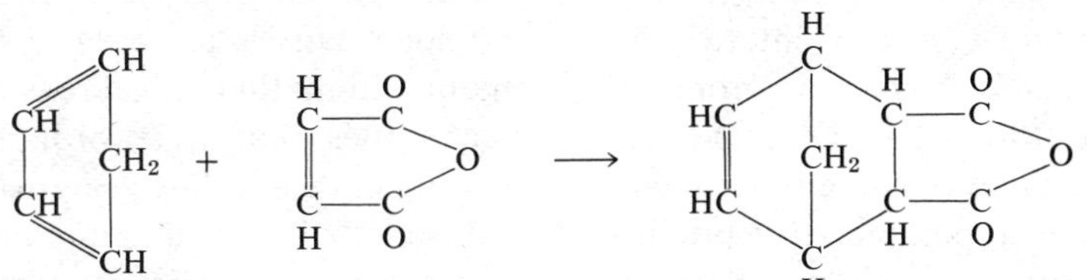

Cyclopentadiene   Maleic anhydride   3,6 Endomethylene-Δ⁴-tetrahydrophthalic
(diene)   (philodiene components)   anhydride (adduced)

FIGURE 2

91

Steroide," Diels compared his research on cholesterol with his work in diene synthesis.

BIBLIOGRAPHY

Diels's writings include "Über das Kohlensuboxyd I," in *Berichte der Deutschen chemischen Gesellschaft,* **39** (1906), 689, written with B. Wolf; *Einführung in die organische Chemie* (Leipzig, 1907; 19th ed., Weinheim, 1962); "Über die Bildung von Chrysen bei der Dehydrierung des Cholesterins," in *Berichte der Deutschen chemischen Gesellschaft,* **60** (1927), 140, written with W. Gädke; "Über Dehydrierungen mit Selen (II. Mitteil)," *ibid.,* 2323, written with A. Karstens; "Synthesen in der hydroaromatischen Reihe, I. Mitteilung, Anlagerungen von 'Di-en'-kohlenwasserstoffen," in *Justus Liebigs Annalen der Chemie,* no. 460 (1928), 98, written with K. Alder; "Die Dien-Synthesen, ein ideales Aufbauprinzip organischer Stoffe. (Vortrag zum 100. Geburtstag von A. Kekulé)," in *Zeitschrift für angewandte Chemie,* **42** (1929), 911; "'Dien-Synthesen' als Aufbauprinzip und Hilfsmittel organisch-chemischer Forschung," in *Jaarboekie van de Natuur-Philosofische Faculteitsvereniging* (Groningen, 1934); "Über organisch-chemischer Entdeckungen und ihre Bedeutung für Gegenwart und Zukunft," in *Chemikerzeitung,* **61** (1937), 7; "Bedeutung der Diensynthese für Bildung, Aufbau und Erforschung von Naturstoffen," in *Fortschritte der Chemie organischer Naturstoffe,* **3** (1939), 1; "Mein Beitrag zur Aufklärung des Sterinproblems," an address summarized in *Angewandte Chemie,* **60** (1948), 78; and "Darstellung und Bedeutung des aromatischen Grundskeletts der Steroide, Nobelvortrag," in *Les Prix Nobel en 1950* (Stockholm, 1950).

A secondary source is Sigurd Olsen, "Otto Diels," in *Chemische Berichte,* **95,** no. 2 (1962), v-xlvi, with bibliography and photographs.

EBERHARD SCHMAUDERER

DIETRICH VON FREIBERG (*b.* Freiberg, Germany, *ca.* 1250; *d. ca.* 1310), *optics, natural philosophy.*

In Latin his name is written Theodoricus Teutonicus de Vriberg (variants: de Vriburgo, de Vribergh, de Vriberch, de Fridiberg, de Frideberch, Vriburgensus). This has been anglicized as Theodoric of Freiberg and rendered into French as Thierry de Fribourg. Which of the many Freibergs or Freiburgs is the place of his birth is not known for certain; Krebs regards Freiberg in Saxony as the most likely. Dietrich entered the Dominican order (province of Teutonia) and probably taught in Germany before studying at the University of Paris, about 1275–1277. He was named provincial of Teutonia in 1293 and was appointed vicar provincial again in 1310. He earned the title of master of theology at St. Jacques in Paris before 1303. In 1304 he was present at the general chapter of his order held in Toulouse, where he was requested by the master

general, Aymeric de Plaisance, to put his investigations on the rainbow into writing. Dietrich is sometimes cited as a disciple of Albertus Magnus; although he is in Albert's tradition, there is no direct evidence that Albert actually taught him.

Apart from his role as a precursor of modern science, Dietrich wrote extensively in philosophy and theology. He is best characterized as an eclectic, although he generally followed the Aristotelian tradition in philosophy and the Augustinian-Neoplatonic tradition in theology. He opposed Thomas Aquinas on key metaphysical theses, including the real distinction between essence and existence. Crombie argues for an influence of Robert Grosseteste on Dietrich from similarities in their optics, but the evidence is meager; Dietrich certainly rejected the "metaphysics of light" taught by Grosseteste and Roger Bacon, and he did not subscribe to the mathematicist view of nature that was common in the Oxford school. Again, Dietrich's interest in Neoplatonism was more theological and mystical than philosophical. He is credited with having influenced the development of speculative mysticism as it was to be taught by Meister Eckhart and Johannes Tauler, both of whom were German Dominicans.

Dietrich's place in the history of science is assured by his *De iride et radialibus impressionibus* ("On the Rainbow and 'Radiant Impressions,'" i.e., phenomena produced in the upper atmosphere by radiation from the sun or other heavenly body), a treatise composed shortly after 1304 and running to over 170 pages in the printed edition (1914). In an age when scientific experimentation was practically unknown, Dietrich investigated thoroughly the paths of light rays through crystalline spheres and flasks filled with water; and he deduced therefrom the main elements of a theory of the rainbow that was to be perfected only centuries later by Descartes and Newton. He also worked out a novel theory of the elements that was related to his optical researches and wrote on the heavenly bodies, although the latter of these contributions is more the work of a philosopher than of a physical scientist in the modern sense.

The anomalous character of Dietrich's contribution poses a problem for the historian of scientific methodology. One is tempted to see in his use of experiment and mathematical reasoning an adumbration of techniques that were brought to perfection in the seventeenth and later centuries. That there is such a foreshadowing is undeniable, and yet the thought context in which Dietrich worked is so different from the Cartesian and empiricist world views that one must be careful not to force too close an identification in method. The mathematical basis for Dietrich's reasoning stems from the *perspectiva,* or geometrical optics, of the Schoolmen and of Arabs such as Ibn al-Haytham (Alhazen); and his measurements are those of medieval astronomy, based on the primitive trigonometry of Ptolemy's *Almagest.* Dietrich does not propose a "theory" in the technical sense, although there is a hypothetical element in his thinking that can be disengaged on careful reading. Rather, he explicitly locates his own method in the framework of Aristotle's *Posterior Analytics,* which puts him on the search for the causes of the rainbow, through discovery of which he hopes to be able to deduce all of the rainbow's properties. This demonstrative ideal of Aristotelian science, it may be noted, did not exclude the use of dialectical (or conjectural) reasoning by its practitioner, although later Scholastics have tended to overlook the latter element. Dietrich's empiricism also derives from the Aristotelian tradition, even though portions of his theory of knowledge are markedly Augustinian. His optics makes implicit use of a method of resolution and composition that was already known to Grosseteste and that was to be refined considerably by the Averroist Aristotelians at Padua, who educated the young Galileo in its use. The general framework of Dietrich's methodology is, thus, far from revolutionary. What characterizes his contribution is his careful application of a method already known in a general way but never hitherto applied with such skill to the detailed explanation of natural phenomena.

Within this setting Dietrich's methodological contribution may be made more precise, as follows. He was not content merely to observe nature but attempted to duplicate nature's operation by isolating the component factors of that operation in a way that permitted their study at close range. Most of his predecessors had regarded the rain cloud as an effective agent in the production of the rainbow; even when they suspected that the individual drop played a significant role, as did Albertus Magnus, they saw no way of isolating it from the collection that produced the bow. When, for example, they compared the colors of the bow with the spectrum resulting from the sun's rays passing through a spherical flask of water, they tended to equate the flask with a cloud or with a collection of drops. It was Dietrich who apparently was the first to see "that a globe of water can be thought of, not as a diminutive spherical cloud, but as a magnified raindrop" (Boyer, p. 112). This insight, coupled with the recognition that the bow is simply the aggregate of effects produced by many individual drops, ultimately led him to the first essentially correct explanation of the primary and secondary bows. Dietrich, of course, used remarkable

experimental acumen in working out all the implications of his discovery. But his genius consisted basically in immobilizing the raindrop, in magnified form, in what approximated a laboratory situation and then studying at leisure and at length the various components involved in the production of the bows.

Dietrich's work represents a great breakthrough in geometrical optics, and yet a simple error in geometry prevented him from giving a correct quantitative theory of the rainbow. In essence, this came about through his using the "meteorological sphere" of Aristotle as his basic frame of reference. Here the observer was regarded as at the center of such a sphere, and the sun and the raindrop (or cloud) were thought to be located on its periphery. Thus, in Figure 1, the observer is at the center, *B,* while the sun is

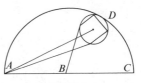

FIGURE 1

behind him at point *A* on the horizon and the much magnified raindrop is elevated at point *D* in front of him. Although this schema permitted Dietrich to use a method of calculation already at hand from medieval astronomy, it automatically committed him to holding that the raindrop and the sun remain always at an equal distance from the observer—an assumption that perforce falsified his calculations.

On the detailed mechanism for the production of the primary, or lower, rainbow (see Figure 2), Dietrich

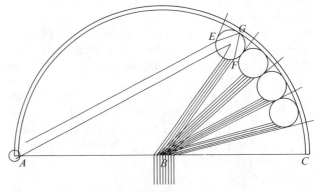

FIGURE 2

was the first to trace correctly the path of the light ray through the raindrop and to see that this involved two refractions at the surface of the drop nearer the observer, i.e., at points *E* and *F,* and one internal reflection at the surface farther from him, i.e., at point *G.* This provided an understanding of why the bow always has a circular form, which was already seen in a rudimentary way by Aristotle; but it also enabled Dietrich to deduce many of the remaining properties of the bow. He was the first to see, for example, that each color in the rainbow is projected to the observer from a different drop or series of drops. He also could deduce, as others before him had merely surmised, that an observer who changes his position sees a different rainbow, in the sense that a completely different series of drops is required for its formation.

This explanation of the primary rainbow alone would have gained for Dietrich a respectable place in the history of optics. He did not stop here, however, but went on to detail the corresponding mechanism for the production of the secondary, or upper, rainbow (see Figure 3). He saw that the light ray, in this case,

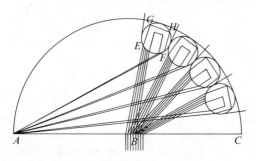

FIGURE 3

follows a path quite different from that in the production of the primary bow, involving as it does two refractions at the surface of the drop nearer the observer, i.e., at points *E* and *F*, and *two* internal reflections at the surface farther from him, i.e., at points *G* and *H*. This insight led immediately to the correct explanation for the inversion of the colors in the secondary bow: the additional internal reflection reverses the ordering of the colors. Thus could Dietrich demolish the competing theories current in his time and go on to deduce other properties of the outer bow: that it is paler in appearance than the inner bow (because of the additional reflection) and that it often fails to appear when the inner bow is clearly seen.

A most interesting part of Dietrich's *De iride*—which led him to compose a companion treatise, *De coloribus* ("On Colors")—is his ingenious but unsuccessful attempt to explain how the colors of the rainbow are generated. It is in these portions of his work, generally passed over rapidly by historians of science, that one can discern in his procedure an interplay between theory and experiment foreshadowing the characteristic methodology of modern science. Dietrich was confident that he had discovered the true "causes" of the bows, and thus he proposed his geometrical explanations of their formation as apo-

dictic demonstrations in the Aristotelian mode. He never was convinced, on the other hand, that he had gotten to the "causes" of radiant color; and thus he had to content himself with the search for the "principles" of such color formation. In this search Dietrich fell back on a Peripatetic argument involving "contraries," the classical paradigm of dialectical reasoning. He used as his analogy the medieval theory of the elements, according to which the four basic contrary qualities of hot-cold and wet-dry, in proper combination, account for the generation of the four elements (fire, air, water, and earth). To employ this, Dietrich had first to establish that there are four colors in the spectrum–and this contrary to Aristotle and almost all of his contemporaries, who held that there are only three. His inductive argument here is superb, and the way in which he employs observation and experiment to overthrow the authority of Aristotle would delight any seventeenth-century thinker. Dietrich was less fortunate in explaining the origin of colors in terms of his two "formal principles" (clear-obscure) and two "material principles" (bounded-unbounded). He did, however, contrive a whole series of experiments, leading to various *ad hoc* assumptions, in his attempt to verify the explanation he proposed. Yet it seems that he was never quite sure of this, and in fact a quite different approach was needed to solve his problem—it was provided by Sir Isaac Newton.

Possibly because of an interest in the elements aroused by his optical studies, Dietrich wrote opuscula entitled *De elementis corporum naturalium* ("On the Elements of Natural Bodies"), *De miscibilibus in mixto* ("On Elements in the Compound"), and *De luce et eius origine* ("On Light and Its Production"). These are neither mathematical nor experimental, but they do shed light on Dietrich's theories of the structure of matter and his analysis of gravitational motion. He also composed treatises relating to astronomy, *De corporibus celestibus quoad naturam eorum corporalem* ("On Heavenly Bodies With Regard to Their Corporeal Nature") and *De intelligenciis et motoribus celorum* ("On Intelligences and the Movers of the Heavens"); the latter has been analyzed by Duhem, who sees it as a retrogression from the astronomical contributions of Albertus Magnus.

For the influence of Dietrich on later optical writers, which was mostly indirect, see the works of Boyer and Crombie cited in the bibliography.

BIBLIOGRAPHY

I. ORIGINAL WORKS. *De iride et radialibus impressionibus*, edited by Joseph Würschmidt in "Dietrich von Freiberg:

Über den Regenbogen und die durch Strahlen erzeugten Eindrücke," in *Beiträge zur Geschichte der Philosophie und Theologie des Mittelalters*, XII, pts. 5–6 (Münster in Westfalen, 1914), contains Latin text, with summaries of chapters in German; an English translation of significant portions of the Latin text with notes, by W. A. Wallace, is to appear in *A Source Book of Medieval Science*, Edward Grant, ed., to be published by the Harvard University Press. The Latin text of *De coloribus, De luce et eius origine, De miscibilibus in mixto,* and portions of *De elementis corporum naturalium* is in W. A. Wallace, *The Scientific Methodology of Theodoric of Freiberg*, Studia Friburgensia, n.s. no. 26 (Fribourg, 1959), pp. 324–376, which contains references to all edited opuscula of Dietrich published before 1959; to these should be added a partial edition of *De visione beatifica* in Richard D. Tétreau, "The Agent Intellect in Meister Dietrich of Freiberg: Study and Text," unpublished Ph.D. thesis for the Pontifical Institute of Mediaeval Studies (Toronto, 1966). A French translation of excerpts from *De intelligenciis et motoribus celorum* is in Pierre Duhem, *Le système du monde*, III (Paris, 1915; repr. 1958), 383–396.

II. Secondary Literature. Biographical material can be found in Engelbert Krebs, "Meister Dietrich, sein Leben, seine Werke, seine Wissenschaft," in *Beiträge zur Geschichte der Philosophie und Theologie des Mittelalters*, V, pts. 5–6 (Münster in Westfalen, 1906). Dietrich's work on the rainbow is well detailed in Carl B. Boyer, *The Rainbow: From Myth to Mathematics* (New York, 1959), pp. 110–124 and *passim;* and in A. C. Crombie, *Robert Grosseteste and the Origins of Experimental Science, 1100–1700* (Oxford, 1953), pp. 233–259. Fuller details on Dietrich's methods are in W. A. Wallace, *Scientific Methodology . . .,* cited above. For an account of Dietrich's other scientific accomplishments, see the following articles by W. A. Wallace: "Gravitational Motion According to Theodoric of Freiberg," in *The Thomist,* **24** (1961), 327–352, repr. in *The Dignity of Science*, J. A. Weisheipl, ed. (Washington, D.C., 1961), pp. 191–216; "Theodoric of Freiberg on the Structure of Matter," in *Proceedings of the Tenth International Congress of History of Science, Ithaca, N.Y., 1962*, I (Paris, 1964), 591–597; and "Elementarity and Reality in Particle Physics," in *Boston Studies in the Philosophy of Science*, III, R. S. Cohen and M. W. Wartofsky, eds. (New York, 1968), 236–271, esp. 243–247.

WILLIAM A. WALLACE, O. P.

DIGBY, KENELM (*b.* Gayhurst, Buckinghamshire, England, 11 July 1603; *d.* London, England, 11 June 1665), *natural philosophy, occult science.*

The son of Sir Everard Digby, executed in 1606 for complicity in the Gunpowder Plot, and of Mary Mulsho of Gayhurst, Digby was brought up a Catholic. In 1617 he accompanied his uncle John Digby (later the first earl of Bristol) on a diplomatic mission to Spain. He was at Oxford, mainly under Thomas Allen, mathematician and astronomer, from 1618 to 1620, after which he set off on a tour of Europe, to France (where he attracted the attention of Marie de' Medici), Italy, and Spain. On his return in 1623 Digby was knighted, presumably for his share, while in Spain, in entertaining Prince Charles and the duke of Buckingham.

In 1625 Digby secretly married the beautiful Venetia Stanley; five children were born, the marriage being made public after the birth of the second in 1627. In this year Digby set off on a privateering mission to the Mediterranean that involved a dramatic but scandalous attack on Venetian shipping off Scanderoon (Alexandretta; now Iskenderun); this won him considerable fame and financial rewards. Probably in the hope of preferment he was converted to Anglicanism in 1630 but returned to Catholicism on the death of his wife in 1633. His wife's death led Digby to give up his gay public life for study. He had already interested himself in literature, alchemy, and religion; and now he turned to writing seriously upon these subjects. He had settled in France, where he met Hobbes and Mersenne, corresponded with Descartes (whom he visited in Holland), and was in close touch with other English Catholics in semiexile. Digby frequently visited England and, aside from a brief imprisonment in 1642, was free to come and go, in spite of his overt royalism. He became chancellor to the widow of Charles I, Queen Henrietta Maria, and undertook a diplomatic mission to the pope on her behalf; he also twice (1648, 1654–1655) tried to negotiate with Cromwell for toleration of Catholics. In the intervals he wrote on natural philosophy (the *Two Treatises*), religion, and literature; collected books and manuscripts; and collected medical, chemical, and household recipes, which he exchanged with others (such as his young relation by marriage, Robert Boyle).

In 1657 his increasingly poor health led Digby to take the waters at Montpellier, where he gave his famous account of the "powder of sympathy," which cured wounds by being rubbed on the weapon that inflicted them. It was a strong solution of vitriol (copper sulfate) in rainwater, which could be improved by drying in the sun and by mixture with gum tragacanth. It worked by a combination of occult and natural powers, that is, by attraction and by the small material particles given off by all objects.

In the same year Digby corresponded with the mathematicians Fermat, Wallis, and Brouncker, serving as intermediary in a dispute concerning Anglo-French priority rather than mathematical fact. After this Digby undertook a long journey through Germany and Scandinavia and thence, at the Restoration, home to England. He was one of those suggested as a member of the new philosophical society that soon became the Royal Society; his *Discourse Concerning the Vegetation of Plants* was read to them on 23 January 1661; he was named to the council in the charters of 1662 and 1663; and his name appears often in the records of their early meetings. In the *Discourse* Digby discusses germination, nutrition, and growth of plants in chemical and mechanical terms; he finds that saltpeter nourishes plants and concludes that, as "the Cosmopolite" (Alexander Seton, a late sixteenth-century alchemist) had said, there is in air a food of life (saltpeter or niter) and an attractive power for this salt in the plant.

Digby was enormously admired in his own day for the fascination of his personality, the flamboyance of his early life, the romance of his love for Venetia, his position in society, and his undoubted intellectual powers. He was at once a lover of the occult and one who appreciated the new trends in natural philosophy. He never completely emancipated himself from traditional Aristotelianism, influenced, perhaps, by his conscious Catholicism and friendship for the English priest and writer Thomas White; yet he read and praised Descartes, Gassendi, and Galileo and could write as scornfully of the "Schoolmen" as they did.

Digby's most important piece of work is the first of the *Two Treatises,* "Of Bodies." Here he displays a clarity and logic of approach that show his appreciation of Descartes. In this work, which deals with both inanimate and animate bodies, he begins with basic definitions. The fundamental properties of bodies are quantity, density, and rarity; and from them motion arises. He discusses motion extensively but qualitatively, although with many admiring references to Galileo's *Two New Sciences* (1638), which not many had read in 1644; he includes Galileo's statement of the law of falling bodies but criticizes Galileo for taking too narrow and strictly functional a view (as Descartes also criticized him). Light is material and in motion; it is in fact fire and can exert pressure, so that when it strikes a body, small particles are carried off with it. Digby's particles, which he sometimes calls atoms, are not fully characterized; they seem neither Epicurean nor Cartesian but certainly are mechanical. The weakness of the work is the lack of precision and definition; this is a general view of natural philosophy, and an interesting one, but Digby had not the ability to explore his subject deeply. Hence, although his book was widely read, it appealed to the virtuoso rather than to the scientist. As a virtuoso himself, Digby may well have intended this, especially in view of the second of *Two Treatises,* "Of Man's Soul."

BIBLIOGRAPHY

I. ORIGINAL WORKS. Digby's writings may be divided into scientific, theological, and personal.

Digby's earliest scientific work is also his most important, *Two Treatises, in One of which, the Nature of Bodies; in the Other, the Nature of Mans Soule, is looked into: in way of discovery, of the Immortality of Reasonable Soules* (Paris, 1644; London, 1645, 1658, 1665, 1669). Best-known is his *Discours fait en une célèbre assemblée, par le Chevalier Digby . . . touchant la guérison des playes par la poudre de sympathie* (Paris, 1658; repr. 1660, 1666, 1669, 1673), English ed., *A late Discourse Made in a Solemne Assembly . . . touching the Cure of Wounds by the Powder of Sympathy* (London, 1658; repr. 1658, 1660, 1664, 1669), and numerous eds. in German, Dutch, and Latin, often appended to other works. His third real scientific work, *A Discourse Concerning the Vegetation of Plants,* read before the nascent Royal Society on 23 January 1661, was first printed at London in 1661 (twice) and, as an appendix to the *Two Treatises,* in 1669; a French trans. was printed at Paris in 1667 and Latin eds. at Amsterdam in 1663, 1669, and 1678. Recipes purporting to come from his MSS were posthumously published in *Choice and Experimented Receipts in Physick and Chirurgery . . . Collected by the Honourable and truly Learned Sir Kenelm Digby* (London, 1668; repr. 1675); these were often selected for inclusion in collections of recipes in other languages. There is also *A Choice Collection of Rare Chymical Secrets and Experiments in Philosophy,* G. Hartman, ed. (London, 1682, 1685). Some of his letters on scientific subjects were printed in John Wallis, *Commercium epistolicum* (Oxford, 1658).

Digby's earliest religious work is *A Conference with a Lady about Choyce of Religion* (Paris, 1638). There followed *Observations upon Religio Medici* (London, 1643; many times repr.); the *Two Treatises* (see above), the theological portion published as *Demonstratio immortalis animae rationalis* (Paris, 1651, 1655; Frankfurt, 1664); *Letters . . . Concerning Religion* (London, 1651); and *A Discourse, Concerning Infallibility in Religion* (Paris, 1652).

The highly miscellaneous personal works include *Articles of Agreement Made Betweene the French King and those of Rochell . . . Also a Relation of a brave and resolute Sea-Fight, made by Sr. Kenelam Digby* (London, 1628); *Sr. Kenelme Digbyes Honour Maintained* (London, 1641); and *Observations on the 22. Stanza in the 9th Canto of the 2d. Book of Spencers Faery Queen* (London, 1643). There are also many posthumously published works, especially *The Closet of the Eminently Learned Sir Kenelme Digbie Kt. Opened* (London, 1669, 1671, 1677, 1910), on food and drink; *Private Memoirs,* Sir N. H. Nicolas, ed. (London, 1827); and *Poems* (London, 1877).

II. SECONDARY LITERATURE. Most important are E. W. Bligh, *Sir Kenelm Digby and His Venetia* (London, 1932), which contains passages not printed in the *Private Memoirs;* and R. T. Petersson, *Sir Kenelm Digby, the Ornament of England 1603–1665* (London, 1956).

MARIE BOAS HALL

DIGGES, LEONARD (*b.* England, *ca.* 1520; *d.* England, 1559 [?]), *mathematics.*

Digges, a member of an ancient family in Kent, was the second son of James Digges of Barham. He was admitted to Lincoln's Inn in 1537 and, if he received the usual education of young gentlemen of the time, may also have attended a university. His works are strongly indebted to contemporary Continental sources, and it is possible that he traveled abroad in 1542.

Digges was interested in elementary practical mathematics, especially surveying, navigation, and gunnery. His almanac and prognostication (1555) contains much material useful to sailors. In 1556 he published an elementary surveying manual, *Tectonicon*. Both of these works went through many editions in the sixteenth century. In 1571 his son Thomas completed and published his more advanced practical geometry *Pantometria,* the first book of which was an up-to-date surveying text. The material in these works is based largely on Peter Apian and Gemma Frisius, but in many cases Digges was the first to describe the instruments and techniques in English.

Digges was a keen experimentalist who gained a reputation, while still quite young, for skill in ballistics. Although his military treatise *Stratioticos* (1579) is largely the work of his son, it is based partly on his notes and the results of his gunnery experiments. The genesis of *Stratioticos* may be found in Digges's association with Sir Thomas Wyatt and others in the preparation of a scheme for an organized militia for Protector Somerset in 1549.

Digges took part in Wyatt's rebellion in 1554. He was attainted and condemned to death but was pardoned for life, probably through the intercession of his kinsman Lord Clinton (later earl of Lincoln), to whom the *Prognostication* was dedicated. He completed payments for the redemption of his property on 7 May 1558 and probably died shortly thereafter.

BIBLIOGRAPHY

I. ORIGINAL WORKS. Digges's writings are *A Prognostication of Right Good Effect* (London, 1555), enl. and retitled *A Prognostication Everlasting* (London, 1556; 11 eds. before 1600); *A Boke Named Tectonicon* (London, 1556; 8 eds. before 1600); *A Geometrical Practise Named Pantometria* (London, 1571, 1591), bk. 1, "Longimetria," repr. by R. T. Gunther as *First Book of Digges Pantometria*, Old Ashmolean Reprints, 4 (Oxford, 1927); and *An Arithmeticall Militare Treatise Named Stratioticos* (London, 1579, 1590).

II. SECONDARY LITERATURE. The *Dictionary of National Biography* article on Leonard Digges is wholly unreliable. Some biographical material can be found in D. M. Loades, *Two Tudor Conspiracies* (London, 1965); and E. G. R. Taylor, *Mathematical Practitioners of Tudor and Stuart England* (Cambridge, 1954). The works on surveying are discussed in E. R. Kiely, *Surveying Instruments* (New York, 1947); and A. W. Richeson, *English Land Measuring to 1800* (Cambridge, Mass., 1967). The *Stratioticos* is discussed in Henry J. Webb, *Elizabethan Military Science* (Madison, Wis., 1965), with reference to Thomas Digges.

JOY B. EASTON

DIGGES, THOMAS (*b.* Kent, England, 1546 [?]; *d.* London, England, August 1595), *mathematics.*

Digges was the son of Leonard Digges of Wotten, Kent, and his wife, Bridget Wilford. He received his mathematical training from his father, who died when Thomas was young, and from John Dee.

Digges was the leader of the English Copernicans. In 1576 he added "A Perfit Description of the Caelestiall Orbes" to his father's *Prognostication.* This contained a translation of parts of book I of Copernicus' *De revolutionibus* and Digges's own addition of a physical, rather than a metaphysical, infinite universe in which the fixed stars were at varying distances in infinite space. He had already published his *Alae seu scalae mathematicae* (1573), containing observations on the new star of 1572 that were second only to those of Tycho Brahe in accuracy. Digges hoped to use these observations to determine whether the Copernican theory was true or needed further modifications, and he called for cooperative observations by astronomers everywhere.

In addition to his astronomical work Digges included a thorough discussion of the Platonic solids and five of the Archimedean solids in his father's *Pantometria* (1571). He also published *Stratioticos* (1579), a treatise on military organization with such arithmetic and algebra as was necessary for a soldier. To this work he appended questions relative to ballistics that were partially answered in the second editions of *Stratioticos* (1590) and *Pantometria* (1591). He was able, on the basis of his own and his father's experiments, to disprove many commonly held erroneous ideas in ballistics but was not able to develop a mathematical theory of his own. These appendixes constitute the first serious ballistic studies in England.

Digges was a member of the parliaments of 1572 (which met off and on for ten years) and 1584 and became increasingly active in public affairs. He was involved with plans for the repair of Dover harbor for several years and served as muster master general of the army in the Low Countries. Apart from his continuing studies in ballistics, his scientific writings cover only a decade; and his promised works on navigation, fortification, and artillery never appeared.

BIBLIOGRAPHY

I. ORIGINAL WORKS. Digges's writings include "A Mathematical Discourse of Geometrical Solids," in Leonard Digges, *A Geometrical Practise Named Pantometria* (London, 1571, 1591), trans. by his grandson Dudley Digges as *Nova corpora regularia* (London, 1634); *Alae seu scalae mathematicae* (London, 1573); "A Perfit Description of the Caelestiall Orbes," in Leonard Digges, *Prognostication Everlastinge* (London, 1576, most later eds.); and *An Arithmeticall Militare Treatise Named Stratioticos* (London, 1579, 1590). For his nonmathematical publications and reports in MS, see the *Dictionary of National Biography,* V, 976–978.

II. SECONDARY LITERATURE. F. R. Johnson, in a letter to the *Times Literary Supplement* (5 Apr. 1934), p. 244, gives information on the dates of Thomas' birth and Leonard's death. The *Dictionary of National Biography* is inaccurate on Thomas' early years, but the account of his later life is useful. For his parliamentary career see J. E. Neale, *Elizabeth I and Her Parliaments* (New York, 1958).

Digges's works are discussed in F. R. Johnson and S. V. Larkey, "Thomas Digges, the Copernican System, and the Idea of the Infinity of the Universe in 1576," in *Huntington Library Bulletin,* no. 5 (Apr. 1934), 69–117; and F. R. Johnson, *Astronomical Thought in Renaissance England* (Baltimore, 1937). For a different interpretation of Digges's infinite universe see A. Koyré, *From the Closed World to the Infinite Universe* (New York, 1957), pp. 34–39. For the ballistics see A. R. Hall, *Ballistics in the Seventeenth Century* (Cambridge, 1952); and for the military treatise H. J. Webb, *Elizabethan Military Science* (Madison, Wis., 1965).

JOY B. EASTON

DILLENIUS, JOHANN JACOB (*b.* Darmstadt, Germany, 1687; *d.* Oxford, England, 2 April 1747), *botany.*

The Dillenius family were civil servants in the state of Hesse who came to Darmstadt at the close of the sixteenth century. Dillenius' grandfather, Justus Dillenius, was a treasury clerk (*Kammerschreiber*), but his father trained as a doctor in the university of Giessen; after several interruptions he completed his studies and was granted a medical licentiate in 1681. Dillenius' mother was the daughter of the clergyman Danile Funk. In 1682 the death of Laurentius Strauss left vacant the chair of medicine in Giessen and Dillenius' father was appointed. In this academic circle the family name, which had already been changed from Dill to Dillen, was altered to Dillenius.

Johann Dillenius followed in his father's footsteps, qualifying in medicine at Giessen in 1713. After a period of practice in Grünberg, Upper Hesse, he was appointed town doctor (*Poliater*) in Giessen. Meanwhile his passion for botany developed and led to his election to the Caesare Leopoldina-Carolina Academia Naturae Curiosorum under the name "Glaucias." About this time he contributed several papers on cryptogams to that academy; these show his concern with the study of cryptogamic sexual organs.

Despite the promise of his work Dillenius was not offered a university post in botany in Germany. It was not until the wealthy English consul at Smyrna, William Sherard, learned of his work that he received an invitation to serve as a full-time botanist, working on Sherard's *Pinax.* Dillenius accepted and by August 1721 he was in England. Apparently it was Sherard's intention to endow the existing chair of botany at Oxford and to see that Dillenius was appointed to it. This could not be realized while Gilbert Trowe occupied the unendowed chair. Dillenius had to wait until Trowe's death in 1734, by which time Sherard had died also. In the thirteen years which remained to him Dillenius completed his magnificent *Historia muscorum* (1741) and continued his study of the fungi with the help of his friend George Deering. Neither the projected book on this subject nor Sherard's ill-fated *Pinax* was completed, however, when Dillenius died after a fit of apoplexy.

Dillenius was elected a fellow of the Royal Society in 1724 and served as its foreign secretary from 1727 to 1747. St. John's College, Oxford, admitted him to the degree of M.D. Oxon. in 1734. His labors in Oxford marked a period of activity in botany there which was not equaled until the appointment of John Sibthorp in 1783.

Botany was passing through an exciting phase in its development in Dillenius' student days. The sexual theory of plant reproduction had been established on the basis of experiments with flowering plants conducted by Rudolph Camerarius, but attempts to determine sexual organs in the flowerless plants had met with virtually no success. With regard to classification the state of affairs was likewise more promising for the students of flowering plants than for those who studied the cryptogams. The best work in the latter field had been that by Samuel Doody, incorporated in the second edition of John Ray's *Synopsis plantarum* (1696). William Sherard was particularly concerned about the inadequate state of such knowledge, and in Dillenius he found an enthusiast for the cryptogams.

Dillenius' failure to make headway in Germany, despite the great interest in the subject there, was undoubtedly due to his unwise criticism of the system of classification of A. Q. Bachmann (Rivinus), which was widely accepted in Germany at the time. He attacked Bachmann's system in his *Catalogus plantarum circa Gissam sponte nascentium* (1719), in which the merits and demerits of the various systems of classification are enumerated with impartiality and justice. Dillenius rightly did not approve of Bachmann's use of the regularity and number of petals as the basis for his classification and preferred the system of Ray to those of both Bachmann and Tournefort. Of course Ray's system was not without its faults—and Dillenius did not escape a harsh reply from Bachmann. Needless to say, Dillenius failed in his role of advocate for Ray's system in Germany.

In England, Dillenius worked on the encyclopedia (or *Pinax*) of all the names that had been given to each plant, on the plan originally conceived by Gaspard Bauhin. Fortunately for science, Dillenius interrupted this work frequently to undertake more fruitful tasks, the first of which was the editing of a third and last edition of Ray's *Synopsis plantarum*. This work brought him into close contact with the small but active circle of British botanists who helped him with it, especially Richard Richardson.

When the *Synopsis* appeared in 1724 the number of flowering plant species in it had been increased to 2,200, and many new species of cryptogams had been added, including 150 moss species. This valuable work served British botanists well until the appearance of Linnaeus' *Species plantarum* in 1761.

The years 1724 to 1732 were largely occupied for Dillenius with illustrating, engraving plates, and describing the plants in William Sherard's brother's garden at Eltham, near London. In this work no love was lost between the proud owner of the garden, James Sherard, and the ardent botanist. James Sherard, who wanted a sumptuous tribute to his glory, was greeted instead with a work of great simplicity, the chief merit of which is its very accurate descriptions and botanical illustrations of exotic plants recently introduced to Europe, especially in the genus *Mesembryanthemum*. Sherard never paid Dillenius for the materials needed to print the book, and Dillenius reckoned that he lost some £200 over the work.

Dillenius began putting together in Oxford the oriental plants collected by Dr. Shaw, the Oxford botanist. In 1736 he was Linnaeus' host in Oxford, and in 1741 he published his most important book, *Historia muscorum,* in which he introduced a new classification of the lower plants (some features of which system are still in use to this day). In his desire to be definitive Dillenius put a prodigious amount of work into this book, which meets the high standards demanded by more modern taxonomy. But it is in the tradition of eighteenth-century British taxonomy and fails to break fresh ground in its approach to the subject or to utilize recent European advances in the knowledge of the sexual organs of cryptogams.

BIBLIOGRAPHY

I. ORIGINAL WORKS. Dillenius' works are *Catalogus plantarum circa Gissam sponte nascentium; cum observationibus botanicis, synonymiis necessariis, tempore & locis, in quibus plantae reperiuntur. Praemittitur praefatio et dissertatio brevis de variis plantarum methodis, ad calcem adjicitur fungorum et muscorum methodica recensio . . .* (Frankfurt am Main, 1718); his ed. of John Ray, *Synopsis methodica stirpium britannicarum . . . Editio tertia multis locis emendata, & quadringentis quinquaginta circiter speciebus noviter detectis aucta* (London, 1724), with illustrations; *Hortus Elthamensis, seu plantarium rariorum, quas in horto suo Elthami in Cantio coluit . . . J. Sherard . . . delineationes et descriptiones* (London, 1732; another ed., Leiden, 1774); and *Historia muscorum inqua circiter sexcentae species veteres et novae ad sua genera relatae describuntur et iconobis genuinis illustrantur: cum appendice et indice synonymorum* (Oxford, 1741), of which another issue of the plates with abbreviated indices is *Historia muscorum: A General History of Land and Water, etc. Mosses and Corals, Containing All the Known Species Exhibited by About 1,000 Figures, on 85 Large Royal Quarto Copper Plates . . . Their Names, Places of Growth, and Seasons, in English and Latin, Referring to Each Figure* (London, 1768).

His correspondence with Linnaeus is included in C. Linnaeus, *Epistolae ineditae C. Linnaei; addita parte commercii litterarii inediti, inprimis circa rem botanicam, J. Burmanni, N. L. Burmanni, Dillenii, . . .* (Groningen, 1830); while Bachmann's reply to Dillenius' criticism of his system of classification of Rivinus is A. Q. Rivinus, *Introductio generalis in rem herbarium. Editio tertia. Accedit . . . Responsio ad J. J. Dillenii objectiones* (Leipzig, 1720).

The Dillenian herbarium is preserved at the Botany School, University of Oxford. Dillenius' MSS are in the Sherard Collection in the Bodleian Library, Oxford. There is a portrait of Dillenius holding a drawing of *Amaryllis formosissima* in the Radcliffe Science Library, Oxford, and a copy of it in the Botany School, Oxford.

II. SECONDARY LITERATURE. For details of Dillenius' early life consult A. J. Schilling, "Johann Jacob Dillenius. 1687–1747. Sein Leben und Wirken," in R. Virchow and F. von Holzendorff, eds., *Sammlungen gemeinverständlicher wissenschaftlicher Vorträge,* 2nd ser., **66** (1889), 1–34.

Dillenius' work in England is well-described in G. C. Druce, *The Flora of Oxford. A Topographical and Historical Account . . . With Biographical Notices of the Botanists Who Have Contributed to Oxfordshire Botany During the Last Four Centuries* (Oxford, 1886), pp. 381–385; and *The Dillenian Herbaria. An Account of the Dillenian Collections in the Herbarium of the University of Oxford, Together With a Biographical Sketch of Dillenius, Selections From His Correspondence, Notes etc.* (Oxford, 1907), ed. and with intro. by S. H. Vines.

For biographical information see R. Pulteney, *Historical and Biographical Sketches of the Progress of Botany in England, From its Origins to the Introduction of the Linnaean System,* 2 vols. (London, 1790); and A. Rees, *The Cyclopaedia; or, Universal Dictionary of Arts, Sciences, and Literature* (London, 1819–1820).

On his botanical work see also J. Reynolds Green, *A History of Botany in the United Kingdom From the Earliest Times to the End of the Nineteenth Century* (London, 1914), pp. 162–173; and M. Moebius, *Geschichte der Botanik von den ersten Anfängen bis zur Gegenwart* (Stuttgart, 1968), where his contributions to cryptogamic botany are discussed critically.

ROBERT OLBY

DINAKARA (*b.* Gujarat, India, *ca.* 1550), *astronomy.*

Dinakara, the son of Rāmeśvara and great-grandson of Dunda, was a resident of Bārejya (probably Bariya [or Devgad Baria] in Rewa Kantha, Gujarat). He belonged to the Moḍha *jñāti* (clan) of the Kauśika *gotra* (lineage). He composed two sets of astronomical tables (see essay in supplement); the epoch of both is Śaka 1500 (A.D. 1578). A third set of tables has as epoch Śaka 1505 (A.D. 1583).

The *Kheṭakasiddhi* contains tables for determining the true longitudes of the five star planets that are based on the *Brahmatulya* of Bhāskara II. The *Candrārkī,* which contains tables of solar and lunar motions and of weekdays, *tithis, nakṣatras,* and *yogas,* was largely influenced by the *Mahādevī* of Mahādeva. Dinakara in turn influenced Haridatta II. There exists an anonymous commentary on the *Candrārkī.* The third set of tables, the *Tithisāraṇī,* is also based on the parameters of the *Brāhmapakṣa;* its purpose is to facilitate the computation of the annual *pañcāṅga* (calendar).

BIBLIOGRAPHY

The *Kheṭakasiddhi* is briefly discussed in Ś. B. Dīkṣita, *Bhāratīya Jyotiḥśāstra* (Poona, 1896; repr. Poona, 1931), p. 277. The *Candrārkī* is described and analyzed in D. Pingree, "Sanskrit Astronomical Tables in the United States," in *Transactions of the American Philosophical Society,* n.s. **58** (1968), 51b–53a. Both the *Kheṭakasiddhi* and the *Tithisāraṇī* are analyzed in the forthcoming *Sanskrit Astronomical Tables in England* by D. Pingree.

DAVID PINGREE

DINGLER, HUGO ALBERT EMIL HERMANN (*b.* Munich, Germany, 7 July 1881; *d.* Munich, 29 June 1954), *philosophy.*

Dingler's mother was Maria Erlenmeyer, daughter of the famous chemist Emil Erlenmeyer; his father, Hermann Dingler, was a professor of botany at the University of Würzburg and a noted scholar. His first wife was Maria Stach von Golzheim, by whom he had one daughter; his second wife was Martha Schmitt.

Dingler passed his *Matura* (school-leaving examination) at the Humanistische Gymnasium in Aschaffenburg and then studied mathematics, physics, and philosophy at Erlangen, Göttingen, and Munich. Among his teachers were David Hilbert, Felix Klein, Edmund Husserl, Hermann Minkowski, Wilhelm Roentgen, and Woldemar Voigt. Dingler received his doctor's degree in mathematics and qualified as lecturer in 1912 at the Technische Hochschule in Munich. In 1920 he became an assistant professor at the University of Munich and remained there until 1932, when he accepted a position at the Technische Hochschule in Darmstadt; two years later he was dismissed from the latter on ideological grounds. Dingler could, however, continue his scientific work and have it published; and in 1935 he participated in a scientific conference at Lund in Sweden, where he gave well-attended lectures and seminars. Difficulties during the Third Reich and privations and adversity after its collapse permanently weakened his health, and in 1954 he succumbed to a heart ailment.

While still a student Dingler, stimulated by John Stuart Mill's *Logic,* had encountered the problem of the validity of axioms, which was to concern him throughout his life. Dingler was an independent, self-willed thinker, one who cannot be classified among those who followed any of the contemporary tendencies, although some influences, especially of Husserl and Henri Poincaré, can be ascertained. He designated himself an antiempiricist and considered himself as holding a position much like Kant's. In more than twenty books and numerous articles written from 1907, he treated the Kantian problem: How is pure science possible? In other words, how is exact research as strict, certain, unambiguous knowledge logically and methodologically possible? Dingler's fundamental investigations were concerned exclusively with the logical and methodological aspect of exact research. He called for a reconstruction of the foundations and the elimination of every presupposition in order to be able really to give an ultimate foundation even to the axioms themselves.

The starting point is the "situation" (*Nullpunkt-Situation*), later also called the "untouched" (*Unberührte*):

That, therefore, which is present in the world at the zero point of all conscious knowledge and tradition, that must be the real world, which enters as a partner into the original relationship between the self and the world. It enters there, so to speak, in an "untouched" condition, that is to say, in a condition untouched by all conscious knowledge and tradition [*Grundriss der methodischen Philosophie*, p. 20].

In the untouched state, that which exists (*das Seiende*) is not yet split into subject and object; there are no concepts, no connection of perceptions. The distinction does not appear until the philosopher gives up his passive attitude and decides to will a first principle (*Dezernismus*): "At the beginning of an ordered structuring of knowledge the philosopher must give up his contemplative attitude and decide on the ultimate principles of a meaningful philosophy" (*Der Zusammenbruch*, 2nd ed., ch. 2, sec. 5). This decision progresses from the will to methodical procedure, to unambiguousness, and to system. The will, which is pure and free of strivings, perceives the way to the goal of knowledge. Dingler's voluntarism is a methodical one, as opposed to Schopenhauer's metaphysical voluntarism or to a completely psychological one.

Starting from the presuppositionless zero-situation, Dingler constructed his "system of pure synthesis." We will the existence of concepts and connections; the concepts must be constant and each new thing that is established must have a sufficient basis. In the construction of the system "pragmatic ordering" (the principle of ordered system-thinking) is determinative, since manual and mental steps cannot occur in just any sequence. The construction takes place according to the principle of simplicity. That is, from among the possible logical forms and steps the simplest are chosen, a principle that also has more or less consciously prevailed in the course of history. Dingler gives a historical survey in order to show that what has come about in consequence of a long period of development may also be assimilated with his "system of pure synthesis."

Dingler followed new paths, building on the ideas of Pierre Duhem, in the concept of the experiment. He wished to refute the belief, which had brought about the dominance of experiment, that one could arrive at general laws of nature through induction. For Dingler an experiment is a willed, intentional action. The geometrical forms, which enter into the measuring apparatus and the measurements required in experiment, are produced according to a priori ideas, their properties being determined from within by the definition of the structure. Dingler speaks of "productive or definitional a priori," which relates to the primary, real world; this differs from Kant's a priori, which refers only to appearances. In experi-

ment the appearances of reality are to be reconstructed by means of suitable, invariable "building stones" (elementary forms and modes of action). At the same time it makes these appearances both dependent upon us and subjected to our will, thus creating mental patterns with which the experimental procedure can be planned and made intellectually manageable.

In numerous publications Dingler presented this foundation for the exact sciences, as he had done for geometry and mechanics in *Das Experiment*. He also derived their axioms and completed them. In the posthumous *Aufbau der exakten Fundamentalwissenschaften* (1964) he brought the foundation of arithmetic and geometry from the preaxiomatic, original basis to the fully established science. In addition, he was convinced that his method was applicable in all other fields, including biology (especially evolution), the philosophy of religion, metaphysics, and ethics.

Dingler's attitude toward non-Euclidean geometry, the theory of relativity, and quantum physics has frequently been misunderstood. In his view only a single, completely determined geometry was demonstrable and demonstrated as a fully defined fundamental science: Euclidean geometry. Nevertheless, non-Euclidean geometries were of great importance in terms of method. In one respect Dingler completely opposed the theory of relativity and quantum physics: the theory of relativity operates in the field of number tables, which are furnished by experiment and within the framework of which any intellectual considerations are permissible. The results of experimental measurements (*Zahlenwolke*) are the domain of theoretical physics, which is obliged further to combine formulas, suggest new experiments, and predict new results. Quantum physics (*Feingebiet*) is therefore open to any theoretical train of thought, but cannot yet be made accessible through measurement and experiment. Thus, physicists should renounce ontological explanations of their mathematical results.

In biology Dingler concerned himself especially with problems of evolution and firmly opposed the vitalistic theses that frequently appeared in philosophical circles. In 1943 he wrote an introduction to a collection edited by Gerhard Heberer, *Die Evolution der Organismen,* which was praised as an original accomplishment by Max Hartmann and also appeared in the second edition of the work (1959). In this work Dingler introduced, completely within his system of pure synthesis, a demonstration of the fact of evolution. What he deduced logically, biological research has confirmed experimentally: i.e., the formation of organic substances that have the property of duplicating themselves, reproduction series representing causal chains of evolutionary theory. Dingler saw in the genes, which he named the "restoration apparatus," the chemical basis for the reproduction of inheritable characteristics. In 1932 he outlined a theory of the factors of evolution that later was supported experimentally.

Relative to the extent of his total work, Dingler paid little attention to logic and rejected the claim that classical logic could be demonstrated by mathematical logic.

The political upheavals in Germany hindered the continuous development of Dingler's work and weakened its influence. Dingler, whose thinking was close to the operationalism of P. W. Bridgman, founded no school but nevertheless had a group of followers scattered far beyond Germany. He did not wish to erect a total system, although he occasionally took a position on ethical and religious problems. His concern, as he states in the foreword to his most famous work, *Der Zusammenbruch,* was to help to achieve the "old, great Greek idea of the unity of the mind."

BIBLIOGRAPHY

I. ORIGINAL WORKS. Among Dingler's writings are *Beiträge zur Kenntnis der infinitesimalen Deformation einer Fläche* (Amorbach, 1907), his dissertation; *Über wohlgeordnete Mengen und zerstreute Mengen im allgemeinen* (Munich, 1912), his *Habilitationsschrift; Die Grundlagen der Naturphilosophie* (Leipzig, 1913); *Das Prinzip der logischen Unabhängigkeit in der Mathematik zugleich als Einführung in die Axiomatik* (Munich, 1915); *Die Grundlagen der Physik. Synthetische Prinzipien der mathematischen Naturphilosophie* (Berlin–Leipzig, 1919; 2nd ed., 1923); *Die Kultur der Juden. Eine Versöhnung zwischen Religion und Wissenschaft* (Leipzig, 1919); *Physik und Hypothese. Versuch einer induktiven Wissenschaftslehre nebst einer kritischen Analyse der Fundamente der Relativitätstheorie* (Berlin–Leipzig, 1921); *Der Zusammenbruch der Wissenschaft und der Primat der Philosophie* (Munich, 1926; 2nd ed., 1931); *Das Experiment. Sein Wesen und seine Geschichte* (Munich, 1928); *Philosophie der Logik und Arithmetik* (Munich, 1931); *Geschichte der Naturphilosophie* (Berlin, 1932); *Die Methode der Physik* (Munich, 1938); "Ist die Entwicklung der Lebewesen eine Idee oder eine Tatsache?," in *Biologe,* **9** (1940), 222–232; *Von der Tierseele zur Menschenseele* (Leipzig, 1941–1943); "Die philosophische Begründung der Deszendenztheorie," in Gerhard Heberer, ed., *Die Evolution der Organismen* (Jena, 1943; Stuttgart, 1959); *Grundriss der methodischen Philosophie* (Füssen, 1949); and *Aufbau der exakten Fundamentalwissenschaften,* P. Lorenzen, ed. (Munich, 1964).

II. SECONDARY LITERATURE. On Dingler or his work, see A. Hubscher, *Denker unserer Zeit* (Munich, 1956), 286–290; W. Krampf, ed., *Hugo Dingler. Gedenkbuch zum 75. Geburtstag* (Munich, 1956), with bibliography; "Über die Philosophie H. Dinglers," in *Zeitschrift für philosophische Forschung,* **10** (1956), 287–299; and *Die Philosophie Hugo Dinglers* (Munich, 1955); and H. C. Sanborn, *Dingler's Methodical Philosophy* (Nashville, Tenn., 1950), also in *Methodos,* **4** (1952), 191–220.

E. SELOW

DINI, ULISSE (*b.* Pisa, Italy, 14 November 1845; *d.* Pisa, 28 October 1918), *mathematics.*

Dini, son of Pietro and Teresa Marchionneschi Dini, came from a very modest background. He studied first in his native city, where, at the age of nineteen, he defended a thesis on applicable surfaces. Having won a competitive examination for study abroad, he left the teachers' college founded by his teacher, Enrico Betti, and went to Paris, where he studied for a year under Joseph Bertrand and Charles Hermite. Seven of his publications on the theory of surfaces date from that brief period.

In 1866 Dini taught higher algebra and theoretical geodesy at the University of Pisa; in 1871 he succeeded Betti (who preferred to direct his efforts to mathematical physics) as professor of analysis and higher geometry and, as early as 1877, also taught infinitesimal analysis. He held these two professorships for the rest of his life. Rector of the university between 1888 and 1890 and director of the teachers' college from 1908 to 1918, Dini was also one of the founders of the School of Applied Engineering in Pisa and was its interim director.

From his youth Dini took an active role in public life; he was a member of the city council of Pisa in 1871 and in various other years until 1895. He was elected to the national parliament in 1880 as a deputy from Pisa and was reelected three times. In 1892 he was appointed a senator of the kingdom.

Dini was an upright, honest, kind man who divided his life between teaching and pure research, on the one hand, and the obligations of a public career completely devoted to the well-being of his native city and his country, on the other.

Two periods of equally intense production may be noted in Dini's scientific activity. The first dealt with infinitesimal geometry and centered on studies of the properties of certain surfaces undertaken by Liouville and Meusnier in France and by Beltrami in Italy. These include surfaces of which the product or the ratio of two principal radii of curvature remains constant (helicoid surfaces to which Dini's name has been given); ruled surfaces for which one of the principal radii of curvature is a function of the other; and the problem suggested by Beltrami, and solved in its entirety by Dini, of representing, point by point, one surface on another in such manner that the geodesic curves of one correspond to the geodesic curves of the other. Dini's complete study of the conformable representation of one surface on another resembles the differential parameters introduced by Beltrami and, generally speaking, equations with partial differential coefficients.

Without losing sight of this geometric research, toward which he guided his best students (such as Luigi Bianchi), Dini preferred to devote himself, after 1871, to analytical studies, in which he was inspired by Weierstrass' and Mittag-Leffler's results on uniform functions and by Dirichlet's on series development of functions of a real variable. He discovered the properties of this development through application of an inversion formula more general than Abel's. Dini of course gave preference to the study of functions of a real variable; but his publication on uniform functions, in which he showed that Weierstrass' and Mittag-Leffler's formulas could be obtained through the method used by Betti in his theory of elliptic functions, proves that he was just as content to develop functions of a complex variable.

Dini devoted a volume to Fourier series and a long chapter of his *Lezioni di analisi infinitesimale* to integral equations, in which many original and fruitful ideas appear. Of his last works in mathematical analysis, the greatest number concern the integration of linear differential equations and equations with partial derivatives of the second order. It must also be mentioned that he discovered a method of solving the linear equation

$$a_0 y^{(n)} + a_1 y^{(n-1)} + \cdots + a_n y = X,$$

in which the *a*'s are given functions of *x*, *X* being a function of *x*. Dini also established a theorem for the upper and lower bounds for the moduli of the roots of an algebraic equation.

BIBLIOGRAPHY

I. ORIGINAL WORKS. Dini's main writings are *Fondamenti per la teoria delle funzioni di variabili reali* (Pisa, 1878), trans. into German by J. Lüroth and A. Schepp as *Grundlagen für eine Theorie der Funktionen einer veränderlichen reellen Grösse* (Leipzig, 1892); *Serie di Fourier e altre rappresentazioni analitiche delle funzioni di una variabile reale* (Pisa, 1880); *Lezioni di analisi infinitesimale*, 2 vols. (Pisa, 1907-1915); and *Lezioni sulla teoria delle funzioni sferiche e delle funzioni di Bessel* (Pisa, 1912). There are articles by Dini in *Annali di matematica pura ed applicata, Atti della Reale Accademia dei Lincei, Comptes rendus hebdomadaires des séances de l'Académie des sciences, Giornale di matematiche,* and *Rendiconti del circolo matematico di Palermo.* The work on uniform functions, "Alcuni teoremi sulle funzioni di una variabile complessa," is in *Collectanea mathematica in memoriam Dominici Chelini* (Milan, 1881), pp. 258-276.

II. SECONDARY LITERATURE. Gino Loria examined the life and works of Dini in "Gli scienziati italiani dall'inizio del medio evo ai nostri giorni," in *Repertorio . . . diretto da Aldo Mieli,* I, pt. 1 (Rome, 1921), pp. 137-150. This work includes a complete bibliography of Dini's works (62 titles), a reproduction of an autograph letter, and several details concerning his political activity. Luigi Bianchi, a student of Dini's, wrote "Commemorazione del socio Ulisse Dini," in *Atti della Reale Accademia dei Lincei,* **28** (1919), 154-163; and the article in the *Enciclopedia Treccani,* XII, 909. See also W. B. Ford, "A Brief Account of the Life and Work of the Late Professor Ulisse Dini," in *Bulletin of the American Mathematical Society,* **26** (1920), 173-177.

PIERRE SPEZIALI

DINOSTRATUS (*fl.* Athens, fourth century B.C.), *mathematics.*

According to Proclus (*Commentary on Euclid, Book I;* Friedlein, ed., 67.8-12), "Amyclas of Heraclea, one of the associates of Plato, and Menaechmus, a pupil of Eudoxus who had also studied with Plato, and his brother Dinostratus made the whole of geometry still more perfect." Dinostratus therefore lived in the middle of the fourth century B.C., and although there is no direct evidence his Platonic associations point to Athens as the scene of his activities. He must have ranged over the whole field of geometry, although only one of his achievements is recorded and the record bristles with difficulties. This is the application of the curve known as the quadratrix to the squaring of the circle.

The evidence rests solely on Pappus (*Collection,* IV. 30; Hultsch ed., 250.33-252.3), whose account is probably derived from Sporus (third century). Pappus says: "For the squaring of the circle there was used by Dinostratus, Nicomedes and certain other later persons a certain curve which took its name from this property; for it is called by them square-forming" (τετραγωνίζουσα sc. γραμμή, quadratrix). The curve was not discovered by Dinostratus but by Hippias, for Proclus, whose account is derived from Eudemus, says: "Nicomedes trisected any rectilineal angle by means of the conchoidal curves, of which he had handed down the origin, order and properties, being himself the discoverer of their special characteristic. Others have done the same thing by means of the quadratrices of Hippias and Nicomedes" (Friedlein, ed., 272.3-10). It has been usual, following Bretschneider, to deduce that Hippias first discovered the curve and that Dinostratus first applied it to finding a square equal in area to a circle, whence it came to be called quadratrix. It is no objection that Proclus writes of the "quadratrix of Hippias," for we regularly speak of Dinostratus' brother Menaechmus as discovering the parabola and hyperbola, although these terms were not employed until Apollonius; nor is there any significance in the plural "quadratrices." It is a more serious objection that Proclus (Friedlein, ed., 356.11) says that different mathematicians have been accustomed to discourse about curves, showing the special property of each kind, as "Hippias with the quadratrices," for this suggests that Hippias may have written a whole treatise on such curves, and he could hardly have failed to omit the circle-squaring aspect; against this may be set the fact that the angle-dividing property of the curve is more fundamental than its circle-squaring property. It is also odd that Proclus does not mention the name of Dinostratus in connection with the quadratrix; nor does Iamblichus, as quoted by Simplicius (*On the Categories of Aristotle*, 7; Kalbfleisch, ed., 192.15–25), who writes of the quadrature of the circle as having been effected by the spiral of Archimedes, the quadratrix of Nicomedes, the "sister of the cochloid" invented by Apollonius, and a curve arising from double motion found by Carpus. Despite all these difficulties, posterity has firmly associated the name of Dinostratus with the quadrature of the circle by means of the quadratrix.

Pappus, IV.30 (Hultsch, ed., 252.5–25), describes how the curve is formed. Let *ABCD* be a square and *BED* a quadrant of a circle with center *A*. If the radius of the circle moves uniformly from *AB* to *AD* and in the same time the line *BC* moves, parallel to its origi-

nal position, from *BC* to *AD*, then at any given time the intersection of the moving radius and the moving straight line will determine a point *F*. The path traced by *F* is the quadratrix. If *G* is the point where it meets *AD*, it can be shown by *reductio per impossibile* (Pappus, IV.31–32; Hultsch, ed., 256.4–258.11) that

$$\text{arc } BED:AB = AB:AG.$$

This gives the circumference of the circle, the area of which may be deduced by using the proposition, later proved by Archimedes, that the area of a circle is equal to a right triangle in which the base is equal to the circumference and the perpendicular to the radius. If Dinostratus rectified the circle in the manner of Pappus' proof, it is one of the earliest examples in Greek mathematics of the indirect proof *per impossibile* so widely employed by Euclid. (Pythagoras before him is said to have used the method to prove the irrationality of $\sqrt{2}$ and Eudoxus must have used it for his proofs by exhaustion.) It is not out of the question that a mathematician of the Platonic school could have proved Archimedes, *Measurement of a Circle*, proposition 1, which is also proved *per impossibile*, but he may only have suspected its truth without a rigorous proof.

According to Pappus, IV.31 (Hultsch, ed., 252.26–256.3), Sporus was displeased with the quadrature because the very thing that the construction was designed to achieve was assumed in the hypothesis. If *G* is known, the circle can indeed be rectified and thence squared, but Sporus asks two questions: How is it possible to make the two points moving from *B* reach their destinations at the same time unless we first know the ratio of the straight line *AB* to the circumference *BED*? Since in the limit the radius and the moving line do not intersect but coincide, how can *G* be found without knowing the ratio of the circumference to the straight line? Pappus endorsed these criticisms. Most modern mathematicians have agreed that the second is valid, for *G* can be found only by closer and closer approximation, but some, such as Hultsch, have thought that modern instrument makers would have no difficulty in making the moving radius and the moving straight line reach *AD* together. It is difficult, however, as Heath argues, to see how this could be done without, at some point, a conversion of circular into rectilinear motion, which assumes a knowledge of the thing sought. Both objections would therefore seem to be valid.

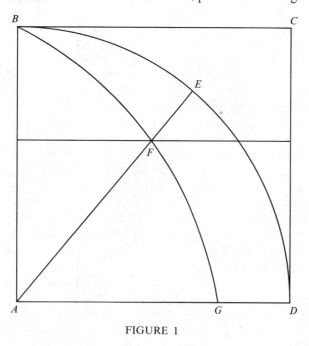

FIGURE 1

BIBLIOGRAPHY

For further reading see the following, listed chronologically: C. A. Bretschneider, *Die Geometrie und die*

Geometer von Euklides (Leipzig, 1870), pp. 95–96, 153–155; Paul Tannery, "Pour l'histoire des lignes et des surfaces courbes dans l'antiquité," in *Bulletin des sciences mathématiques et astronomiques,* 2nd ser., **7**, pt. 1 (1883), 278–284; G. J. Allman, *Greek Geometry From Thales to Euclid* (Dublin, 1889), pp. 180–193; Gino Loria, *Le scienze esatte nell'antica Grecia,* 2nd ed. (Milan, 1914), pp. 160–164; T. L. Heath, *A History of Greek Mathematics,* I (Oxford, 1921), 225–230; Ivor Thomas, *Selections Illustrating the History of Greek Mathematics,* I (London–Cambridge, Mass., 1939), 334–347; B. L. van der Waerden, *Science Awakening* (Groningen, 1954), pp. 191–193; and Robert Böker, in *Der kleine Pauly,* I (Stuttgart, 1964), cols. 1429–1431.

IVOR BULMER-THOMAS

DIOCLES (second century B.C. [?]), *mathematics, physics.*

Nothing is known of the life of this Greek mathematician, but he must have lived after Archimedes (*d.* 212 B.C.) and before Geminus of Rhodes (*fl.* 70 B.C.). Eutocius, a Byzantine mathematician of the fifth and sixth centuries, preserved two fragments of Diocles' work *On Burning Mirrors* in his commentary on Archimedes' *On the Sphere and Cylinder.*

One of these fragments deals with the solution of the problem of the two mean proportionals by means of the cissoid, which Diocles invented. The problem of doubling the cube, the celebrated Delian problem of ancient geometry, had been the subject of mathematical investigation at least as early as the fifth century B.C. Hippocrates of Chios is reported to have discovered that a solution could be found if a way could be devised for finding two mean proportionals in continued proportion between two straight lines, the greater of which line is double the lesser. The question was studied by Plato's Academy and a mechanical solution is even attributed, erroneously, to Plato. Before Diocles, solutions were offered by Archytas, Eudoxus, Menaechmus, Eratosthenes, Nicomedes, Apollonius, Hero, and Philo of Byzantium. All of these, and later solutions, are preserved by Eutocius.

Proposition 4 of Book II of Archimedes' *On the Sphere and Cylinder* presents the problem of how to cut a given sphere by a plane in such a way that the volumes of the segments are in a given ratio to one another. Diocles' solution to the problem, as given in the fragment preserved by Eutocius, was an ingenious geometrical construction that satisfied, by means of the intersection of an ellipse and a hyperbola, the three simultaneous relations which hold in Archimedes' proposition.

Diocles' work *On Burning Mirrors,* judging from the time at which he lived and the work of his predecessors, must have been of considerable scope. It can be assumed that it discussed concave mirrors in the forms of a sphere, a paraboloid, and a surface described by the revolution of an ellipse about its major axis. Apollonius of Perga, a mathematician who was born about 262 B.C., had earlier written a book on burning mirrors, but Arabic tradition associated Diocles with the discovery of the parabolic burning mirror. The Greek *Fragmentum mathematicum Bobiense* contains a fragment of a treatise on the parabolic burning mirror, and some authorities have attributed this work to Diocles. Others consider this attribution very doubtful. William of Moerbeke translated into Latin the fragments of Diocles on mean proportionals and the division of the sphere as a part of his general translation from the Greek of the works of Archimedes and Eutocius' commentaries on them.

BIBLIOGRAPHY

The fragments of Diocles' work can be found in *Archimedis Opera omnia cum commentariis Eutocii iterum,* J. L. Heiberg, ed., III (Leipzig, 1915), 66–70, 160–176.

On Diocles or his work, see Moritz Cantor, *Vorlesungen über Geschichte der Mathematik,* I (Stuttgart, 1907, repr. New York, 1965), 350, 354–355; Thomas Heath, *A History of Greek Mathematics,* 2 vols. (Oxford, 1960), I, 264–266; II, 47–48, 200–203; George Sarton, *Introduction to the History of Science,* I (Baltimore, 1927), 183; Moritz Steinschneider, *Die europäischen Uebersetzungen aus dem arabischen bis Mitte des 17. Jahrhunderts* (Graz, 1965), p. 17; and E. Wiedemann, "Ibn al Haitams Schrift über parabolische Hohlspiegel," in *Bibliotheca mathematica,* 3rd ser., **10** (1909–1910), 202.

KARL H. DANNENFELDT

DIOCLES OF CARYSTUS (*b.* Carystus, Euboea; *fl.* Athens, late fourth century B.C.), *medicine.*

Diocles, the son of Archidamus, also a physician, was still alive shortly after 300 B.C. The Athenians called him a "second Hippocrates" and Pliny the Elder (*Natural History,* XXVI, 10) wrote that Diocles came "next after Hippocrates in time and reputation." Galen and Celsus place him as an equal with Hippocrates, Praxagoras, Herophilus, and Erasistratus. The last three of these physicians were contemporaries of Diocles, and these four raised Greek medicine to a high point in its history. Diocles was a pupil of Aristotle, and he was also a contemporary of such Peripatetics as Theophrastus and Strato. By some, Diocles is considered the leading representative of the dogmatic school, which introduced philosophical speculations into the Hippocratic materials and formalized the medical systems. Diocles saw, however, that philosophical theory could not explain everything, and he is best considered as independent of any school.

Diocles' writings were considerable. The titles of seventeen works are known and more than 190 fragments have been preserved. Unlike the physicians of his time, he wrote in Attic Greek. His writings show a well-polished if simple style, and his language and terminology show the influence of the literary style of Aristotle in scientific writing. The subjects covered in his books range widely.

Diocles' medical writings show the influence of the Aristotelian teleological view of nature. They also indicate that he was the first physician to use a collection of Hippocratic writings, which he may have assembled himself. According to Galen, he was the first to write a book on anatomy and to use that term in the title. While he did not distinguish the nerves from the veins, he did recognize more of the latter than his predecessors. The heart was the source of the blood, which was carried through the aorta and the vena cava. He also described the lungs, ureters, ovaries, fallopian tubes, ileocecal valve, cecum, and the gall bladder with the tube leading to it from the liver. He distinguished between pleurisy and pneumonia and described hepatic and splenic ascites.

In his views on embryology Diocles followed Empedocles. In generation both the man and the woman furnished seed, which contributed to the development of the embryo. The seed, originating in the brain and spinal marrow, was a product of nourishment. Excessive coition was detrimental to the eyes and spinal marrow. In agreement with Empedocles, he felt that the full development of the embryo occurred in forty days, and as the male child grew in the right (i.e., warmer) side of the uterus, it developed quicker than the female. He described human embryos of twenty-seven and forty days. In his studies of sterility, he was especially interested in the mule, and according to Galen, he dissected such animals. Again following Empedocles, he asserted that menstruation occurred during the same period of life for all women, beginning at age fourteen and lasting until sixty. He felt that broad hips, freckles, auburn hair, and manly appearance were certain indicators of fertility. Sterility in the female was attributed to displacement of the uterus.

Diocles' physiology was similar to that of Philistion and was based on the four basic elements of Empedocles—fire, water, air, and earth. The human body also had the four qualities of heat, moisture, cold, and dryness. Health was dependent upon the proper equilibrium of the four elements in the body. Warmth was especially important in the formation of the four humors of blood, phlegm, yellow bile, and black bile. The proper movement of the pneuma, seated in the heart and spreading through the body by means of the veins, had a most important place in health and illness, the latter being independent of outside causes. Fever, disease, or death occurred if the pneuma was hindered by phlegm or bile. Respiration took place through the pores of the skin as well as through the nose and mouth. The Pythagorean number seven was evident in Diocles' view that the seventh, fourteenth, twenty-first, and twenty-eighth days were most critical during illness. Fever was not a disease itself but symptomatic of some morbid condition. He distinguished between continuous and intermittent fevers and also quotidian, tertian, and quartan forms. Like Hippocrates, he stressed practical experience, observation, and the importance of diagnosis and prognosis.

Some indication of Diocles' prominence is seen in the fact that he was known to the rulers of his time. A work on hygiene, written after 300 B.C., was dedicated to the Macedonian prince Pleistarchus, the son of the famous general Antipater. Diocles' letter on hygiene, written between 305 and 301 B.C. and addressed to King Antigone, one of the generals of Alexander the Great, was fortunately preserved by Paul of Aegina, a Greek physician of the late seventh century A.D. Many editions of this work were printed in the sixteenth century in Latin, French, and English.

One of Diocles' works is entitled *Archidamos* in dedication to his dead father. His father had condemned the then current practice of massaging the body with oil, because to do so heated the body too much and made it too dry by rubbing. While refuting his father's arguments, Diocles proposed a compromise: he suggested that in summer a mixture of oil and water be used and in winter pure oil. In the use of oil and water, he is apparently following the idea of a slightly earlier and anonymous work on diet.

Lengthy fragments of Diocles' own work on diet were preserved by Oribasius, physician to Emperor Julian. In this work the Greek physician looked at human life as a whole and by describing the routine of one summer's day prescribed what is suitable and beneficial for men. He made allowances for various ages and changes of seasons. His descriptions are given as ideal standards, dictated by suitable and tasteful behavior—the Aristotelian ethic. He does not describe the various physical exercises, but his whole plan for the day is based on exercise in the morning and the afternoon, revolving around the gymnastics of Greek civilization. His exposition of diet described well the Greek ideals of health, harmony, and balance.

In the history of medical botany or pharmacy, Diocles also deserves recognition. Here, like his colleague Theophrastus, he was probably stimulated to the study of botany by his teacher Aristotle. Diocles was the first scientist to write a herbal on the origin, recognition, nutritional value, and medical use of plants; thus he can be considered the founder of pharmacy. His work was used as a source for all later works until Dioscorides. Two other botanical works, dealing with vegetables and with healing, are practical in nature, but apparently they also advanced the study of plants. Theophrastus, the founder of scientific botany, seems to have made extensive use of the botanical works of Diocles; although he does not name his colleague in his botanical works, in his work *On Stones* he does refer to Diocles as an authority on a certain mineral.

Diocles is credited with two inventions—a bandage for the head and a spoonlike device for the extraction of arrows.

BIBLIOGRAPHY

I. ORIGINAL WORKS. Fragments of Diocles' works are in C. G. Kühn, *Diocles Carystius fragmenta collegit* (Leipzig, 1827); Mauritz Fraenkel, *Dioclis Carystii fragmenta quae supersunt* (Berlin, 1840); Werner Jaeger, "Vergessene Fragmente des Peripatetikers Diokles von Karystos," in *Abhandlungen der Deutschen Akademie der Wissenschaften zu Berlin*, Phil.-hist. Kl., no. 2 (1938); and Max Wellman, "Die Fragmente des sikelischen Aerzte, Akron, Philistion, und des Diokles von Karystos," in *Fragmentsammlung der griechischen Aerzte*, I (Berlin, 1901), 117–207.

II. SECONDARY LITERATURE. On Diocles and his work, see Gustav A. Gerhard, "Ein dogmatischer Arzt des vierten Jahrhunderts vor Christ," in *Sitzungsberichte der Heidelberger Akademie der Wissenschaften*, Phil.-hist. Kl. (1913); W. Haberling, "Die Entdeckung einer kriegschirurgischen Instrumentes des Altertums," in *Deutsche militärärztliche Zeitschrift*, 40 (1912), 658–660; Werner Jaeger, *Paideia, Die Formung des griechischen Menschen*, 3 vols. (Berlin-Leipzig, 1934–1947), trans. into English by G. Highet as *Paideia: The Ideals of Greek Culture*, 3 vols. (New York, 1960), III, 41–44, *passim;* Werner Jaeger, *Diokles von Karystos. Die griechische Medizin und die Schule des Aristoteles* (Berlin, 1938, 1963); George Sarton, *Introduction to the History of Science*, I (Baltimore, 1927), 121; and Max Wellman, *Die pneumatische Schule bis auf Archigenes, in ihrer Entwicklung* (Berlin, 1895); "Das älteste Kräuterbuch der Griechen," in *Festgabe für Franz Susemihl. Zur Geschichte griechischer Wissenschaft und Dichtung* (Leipzig, 1898); and "Diokles von Karystos," in Pauly-Wissowa, *Real-Encyclopädie*.

KARL H. DANNENFELDT

DIONIS DU SÉJOUR, ACHILLE-PIERRE (*b.* Paris, France, 11 January 1734; *d.* Vernou, near Fontainebleau, France, 22 August 1794), *astronomy, mathematics, demography.*

The son of Louis-Achille Dionis du Séjour, counselor at the *cour des aides* in Paris, and of Geneviève-Madeleine Héron, Achille-Pierre studied in Paris at the Collège Louis-le-Grand and then at the Faculté de Droit. A counselor at the Parlement of Paris in 1758, he sat as a member of the Chambre des Enquêtes beginning in 1771 and in 1779 moved to the Grand Chambre, where he was appreciated for his simplicity, his liberalism, and his humanity. He devoted the bulk of his leisure time to mathematical and astronomical research, which brought him election as *associé libre* of the Académie des Sciences on 26 June 1765. (Dionis du Séjour maintained this title at the time of the reorganization of 1785 but resigned it on 14 July 1786 in order to be eligible for election as associate member of the physics section.) His cordiality, his devotion to the cause of scientific research, and his philosophic spirit brought him many friendships; and the quality of his writings earned the respect of Lagrange, Laplace, and Condorcet, among others.

With his friend and future colleague Mathieu-Bernard Goudin, Dionis du Séjour published a treatise on the analytical geometry of plane curves (1756) and a compendium of theoretical astronomy (1761). From 1764 to 1783 he wrote a series of important memoirs on the application of the most recent analytic methods to the study of the principal astronomical phenomena (eclipses, occultations, reductions of observations, determination of planetary orbits, etc.). Revised and coordinated, these memoirs were reprinted in the two-volume *Traité analytique des mouvements apparents des corps célestes* (1786–1789), of which Delambre gives a detailed analysis. The *Traité* is completed by two works, one on comets (1775), in which he demonstrates the near impossibility of a collision of one of these heavenly bodies with the earth, and the other on the varying appearance of the rings of Saturn (1776). All these works are dominated by an obvious concern for rigor and by a great familiarity with analytical methods; if the prolixity of the developments and the complexity of the calculations rendered them of little use at the time, their reexamination in the light of present possibilities of calculation would certainly be fruitful.

In pure mathematics, beyond the study of plane curves, Dionis du Séjour was interested in the theory of the solution of equations, an area where his works have been outclassed by those of his contemporaries Bézout and Lagrange. Finally, in collaboration with Condorcet and Laplace, he undertook a systematic inquiry to determine the population of France. Utilizing the list of communes appearing in the Cassini map of France and the most recent information furnished by the civil registers, this inquiry was based on the empirical hypothesis that the annual number of births in a given population is approximately one twenty-sixth of the total of that population.

The Revolution interrupted Dionis du Séjour's scientific activity. Elected a deputy of the Paris nobility on 10 May 1789, he sat in the National (later Constituent) Assembly until its duties were completed on 30 September 1791. Resigning later from the office of judge of a Paris tribunal, to which post he had been elected on 30 November 1791, he retired to his rich holdings in Argeville, a commune in Vernou, near Fontainebleau, where he died without issue almost a month after 9 Thermidor, having experienced, it seems, a period of difficulties and quite justifiable anxiety.

BIBLIOGRAPHY

I. ORIGINAL WORKS. A list of Dionis du Séjour's papers is in *Table générale des matières contenues dans l'Histoire et dans les Mémoires de l'Académie royale des sciences,* VII-X (Paris, 1768–1809).

His books are *Traité des courbes algébriques* (Paris, 1756), written with Goudin; *Recherches sur la gnomonique, les rétrogradations des planètes et les éclipses du soleil* (Paris, 1761), written with Goudin; *Recueil de problèmes astronomiques résolus analytiquement,* 3 vols. (Paris, 1769–1778), a collection of his papers on astronomy published in the *Histoire de l'Académie royale des sciences; Essai sur les comètes en général; et particulièrement sur celles qui peuvent approcher de l'orbite de la terre* (Paris, 1775); *Essai sur les phénomènes relatifs aux disparitions périodiques de l'anneau de Saturne* (Paris, 1776); *Traité analytique des mouvements apparents des corps célestes,* 2 vols. (Paris, 1786–1789); and *Traité des propriétés communes à toutes les courbes, suivi d'un mémoire sur les éclipses du soleil* (Paris, 1788), written with Goudin.

II. SECONDARY LITERATURE. It should be noted that in any alphabetical listing Dionis du Séjour's name sometimes appears as Dionis, sometimes as Du Séjour, and sometimes as Séjour. On Dionis du Séjour or his work, see the following (listed chronologically): J. S. Bailly, *Histoire de l'astronomie moderne,* III (Paris, 1782), index under Séjour; J. de Lalande, articles in *Magasin encyclopédique ou Journal des sciences, des lettres et des arts,* **1** (1795), 31–34; in *Connaissance des temps... pour l'année sextile VIIe de la République* (May 1797), 312–317; and in *Bibliographie astronomique* (Paris, 1803), pp. 750–752 and index; Nicollet, in Michaud, ed., *Biographie universelle,* XI (1814), 401–403, and in new ed., XI (1855), 90–91; J. B. Delambre, *Histoire de l'astronomie au XVIIIe siècle* (Paris, 1827), pp. xxiii–xxiv, 709–735; R. Grant, *History of Physical Astronomy* (London, 1852), pp. 232, 267; J. Hoefer, in *Nouvelle biographie générale,* XV (Paris, 1858), 295–296; Poggendorff, I, 574–575; A. Maury, *L'ancienne Académie des sciences* (Paris, 1864), see index; J. Bertrand, *L'Académie des sciences et les académiciens de 1666 à 1793* (Paris, 1869), pp. 311–312; J. C. Houzeau and A. Lancaster, *Bibliographie générale de l'astronomie,* 3 vols. (Brussels, 1882–1889; repr. London,

1964) I, pt. 2, 1301, 1313, 1341, II, cols. 385, 483, 1078, 1083, 1150, 1207; J. F. Robinet, A. Robert, and J. le Chapelain, *Dictionnaire historique et biographique de la Révolution et de l'Empire,* I (Paris, 1899), 643–644; F. Matagrin, *Vernou et le château d'Argeville* (Melun, 1905), pp. 128–129; A. Douarche, *Les tribunaux civils de Paris pendant la Révolution,* 2 vols. (Paris, 1905–1907), see index; N. Nielsen, *Géomètres français sous la Révolution* (Copenhagen, 1929), pp. 73–79; and Roman d'Amat, in *Dictionnaire de biographie française,* XI (1967), 390–391.

RENÉ TATON

DIONYSODORUS (*fl.* Caunus [?], Asia Minor, third-second centuries B.C.), *mathematics.*

The Dionysodorus who is the subject of this article is recorded by Eutocius as having solved, by means of the intersection of a parabola and a hyperbola, the cubic equation to which (in effect) Archimedes had reduced the problem of so cutting a sphere by a plane that the volumes of the segments are in a given ratio. Of the many bearers of this name in Greek literature, he has usually been identified with the Dionysodorus who is described by Strabo (XII, 3,16) as a mathematician and is included among the men noteworthy for their learning who were born in the region of Amisene in Pontus, on the shore of the Black Sea. But since Wilhelm Cronert published in 1900 hitherto unknown fragments from the Herculaneum roll no. 1044, and especially since Wilhelm Schmidt commented on them in 1901, it has seemed more probable that he should be identified with Dionysodorus of Caunus, son of a father of the same name, who was probably an Epicurean. One fragment (no. 25) indicates that this Dionysodorus succeeded Eudemus as the teacher of Philonides, and another (no. 7) that Philonides published some lectures by Dionysodorus. Eudemus is obviously the Eudemus of Pergamum to whom Apollonius dedicated the first two books of his *Conics,* and Philonides is the mathematician to whom Apollonius asked Eudemus to show the second book. When we recollect that Caunus in Caria is near Apollonius' birthplace, Perga in Pamphylia, it is clear that this Dionysodorus moved in distinguished mathematical company and would have been capable of the elegant construction that Eutocius has recorded. If this identification is correct, he would have lived in the second half of the third century B.C. If he is to be identified with Dionysodorus of Amisene, all that can be said about his date is that he wrote before Diocles, say before 100 B.C. It is clear that he is not the same person as the geometer Dionysodorus of Melos, who is mentioned by Pliny (*Natural History,* II, 112.248) as having arranged for a message to be put in his tomb saying that he had been to the center of the earth and had found the earth's radius to measure 42,000 stades. Strabo, indeed, specifically distinguishes them.

In the passage quoted by Eutocius, *Commentarii in libros II De sphaera et cylindro* (Archimedes, Heiberg ed., III, 152.28–160.2), Dionysodorus says: Let AB be a diameter of a given sphere which it is required to cut in the given ratio $CD:DE$. Let BA be produced to F so that $AF = AB/2$, let AG be drawn perpendicular to AB so that $FA:AG = CE:ED$, and let H be taken on AG produced so that $AH^2 = FA \cdot AG$. With axis FB let a parabola be drawn having AG as its parameter; it will pass through H. Let it be FHK where BK is perpendicular to AB. Through G let there be drawn a hyperbola having FB and BK as asymptotes. Let it cut the parabola at L—it will, of course, cut at a second point also—and let LM be drawn perpendicular to AB. Then, proves Dionysodorus, a plane drawn through M perpendicular to AB will cut the sphere into segments whose volumes have the ratio $CD:DE$.

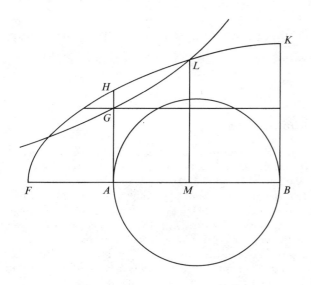

FIGURE 1

It will be more instructive to turn the procedure into modern notation rather than reproduce the prolix geometrical proofs. In his treatise *On the Sphere and Cylinder,* II, 2 and 4, Archimedes proves geometrically that if r be the radius of a sphere and h the height of one of the segments into which it is divided by a plane, the volume of the segment is equal to a cone with the same base as the segment and height

$$h \cdot \frac{3r - h}{2r - h}.$$

If h' is the height of the other segment, and the volumes of the segments stand in the ratio $m:n,$ then

$$nh \cdot \frac{3r - h}{2r - h} = mh' \cdot \frac{3r - h'}{2r - h'}.$$

Eliminating h' by the relationship $h + h' = 2r,$ we obtain the cubic equation in the usual modern form

$$h^3 - 3h^2r + \frac{4m}{m + n}r^3 = 0.$$

If we substitute $x = 2r - h\,(= h')$ we may put the equation in the form solved by Dionysodorus:

$$4r^2:x^2 = (3r - x):\frac{n}{m + n}r.$$

Dionysodorus solves it as the intersection of the parabola

$$y^2 = \frac{n}{m + n}r(3r - x)$$

and the hyperbola

$$xy = \frac{n}{m + n}2r^2.$$

It seems probable (despite Schmidt) that this mathematician is the same Dionysodorus who is mentioned by Hero as the author of the book Περὶ τῆς σπείρας, "On the Tore" (*Heronis opera omnia,* H. Schöne, ed., III, 128.1–130.11), in which he gave a formula for the volume of a torus. If BC is a diameter of the circle $BDCE$ and if BA is perpendicular to the straight line HAG in the same plane, when AB makes a complete revolution around $HAG,$ the circle generates a spire or torus whose volume, says Dionysodorus, bears to the cylinder having HG for its axis and EH for the radius of its base the same ratio as the circle $BDCE$ bears to half the parallelogram $DEHG.$

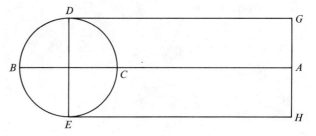

FIGURE 2

That is to say, if r is the radius of the circle and $EH = a,$

$$\frac{\text{Volume of torus}}{\pi a^2 \cdot 2r} = \frac{\pi r^2}{a/2 \cdot 2r},$$

whence

$$\text{Volume of torus} = 2\pi a \cdot \pi r^2.$$

In an example, apparently taken from Dionysodorus, $r = 6$ and $a = 14$, and Hero notes that if the torus be straightened out and treated as a cylinder, it will have 12 as the diameter of its base and 88 as its length, so that its volume is $9956\frac{4}{7}$. This is equivalent to saying that the volume of the torus is equal to the area of the generating circle multiplied by the length of the path traveled by its center of gravity, and it is the earliest example of what we know as Guldin's theorem (although originally enunciated by Pappus).

Among the inventors of different forms of sundials in antiquity Vitruvius (IX, 8; Krohn, ed., 218.8) mentions a Dionysodorus as having left a conical form of sundial—"Dionysodorus conum (reliquit)." It would no doubt, as Frank W. Cousins asserts, stem from the hemispherical sundial of Berossus, and the cup would be a portion of a right cone, with the nodal point of the style on the axis pointing to the celestial pole. Although there can be no certainty, there seems equally no good reason for not attributing this invention to the same Dionysodorus; it would fit in with his known use of conic sections.

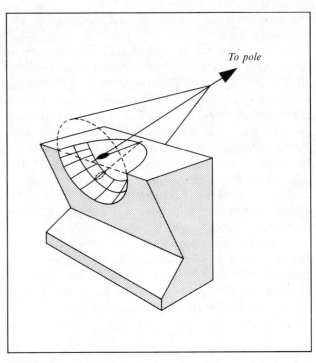

FIGURE 3. Conjectural reconstruction of Dionysodorus' conical sundial.

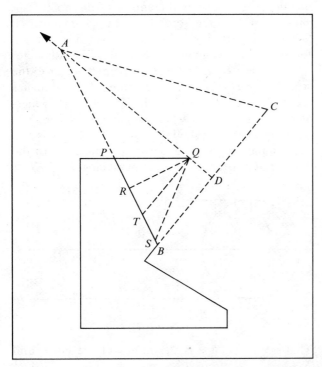

FIGURE 4. Vertical section of Dionysodorus' conical sundial. PQ is the style, its nodal point Q lying on the axis of the cone, which points to the celestial pole. When the sun is overhead at the equator, the nodal point casts a shadow at T, when farthest south at R, and when farthest north at S.

BIBLIOGRAPHY

On Dionysodorus or his work, see Eutocius, *Commentarii in libros II De sphaera et cylindro,* in Archimedes, Heiberg ed., III, 152.27–160.2; Hero of Alexandria, *Metrica,* II, 13—*Heronis opera omnia,* H. Schöne, ed., III, 128.1–130.11; Wilhelm Cronert, "Der Epikur Philonides," in *Sitzungsberichte der K. Preussischen Akademie der Wissenschaften zu Berlin* (1900), 942–959, esp. frag. 7, p. 945, and frag. 25, p. 952. Wilhelm Schmidt, "Über den griechischen Mathematiker Dionysodorus," in *Bibliotheca mathematica,* 3rd ser., **4** (1904), 321–325; Sir Thomas Heath, *A History of Greek Mathematics,* II (Oxford, 1921), 46, 218–219, 334–335; Ivor Thomas, *Selections Illustrating the History of Greek Mathematics,* II (London–Cambridge, Mass., 1941), pp. 135, 163, 364, 481; René R. J. Rohr, *Les cadrans solaires* (Paris, 1965), pp. 31–32, trans. by G. Godin as *Sundials: History, Theory and Practice* (Toronto–Buffalo, 1970), pp. 12, 13; and Frank W. Cousins, *Sundials* (London, 1969), pp. 13, 30 (correcting Cdynus to Caunus).

IVOR BULMER-THOMAS

DIOPHANTUS OF ALEXANDRIA (*fl.* A.D. 250), *mathematics.*

We know virtually nothing about the life of Diophantus. The dating of his activity to the middle of the third century derives exclusively from a letter of

Michael Psellus (eleventh century). The letter reports that Anatolius, the bishop of Laodicea since A.D. 270, had dedicated a treatise on Egyptian computation to his friend Diophantus. The subject was one to which, as Psellus states, Diophantus himself had given close attention.[1] This dating is in accord with the supposition that the Dionysius to whom Diophantus dedicated his masterpiece, *Arithmetica,* is St. Dionysius, who, before he became bishop of Alexandria in A.D. 247, had led the Christian school there since 231.[2] An arithmetical epigram of the *Greek Anthology* provides the only further information (if the data correspond to facts): Diophantus married at the age of thirty-three and had a son who died at forty-two, four years before his father died at the age of eighty-four.[3] That is all we can learn of his life, and relatively few of his writings survive. Of these four are known: *Moriastica, Porismata, Arithmetica,* and *On Polygonal Numbers.*

Moriastica. The *Moriastica,* which must have treated computation with fractions, is mentioned only once, in a scholium to Iamblichus' commentary on Nicomachus' *Arithmetica.*[4] Perhaps the *Moriastica* does not constitute an original treatise but only repeats what Diophantus wrote about the symbols of fractions and how to calculate with them in his *Arithmetica.*

Porismata. In several places in the *Arithmetica* Diophantus refers to propositions which he had proved "in the *Porismata.*" It is not certain whether it was—as seems more probable—an independent work, as Hultsch and Heath assume, or whether such lemmas were contained in the original text of the *Arithmetica* and became lost with the commentators; the latter position is taken by Tannery, to whom we owe the critical edition of Diophantus.

Arithmetica. The *Arithmetica* is not a work of theoretical arithmetic in the sense understood by the Pythagoreans or Nicomachus. It deals, rather, with logistic, the computational arithmetic used in the solution of practical problems. Although Diophantus knew elementary number theory and contributed new theorems to it, his *Arithmetica* is a collection of problems. In the algebraic treatment of the basic equations, Diophantus, by a sagacious choice of suitable auxiliary unknowns and frequently brilliant artifices, succeeded in reducing the degree of the equation (the unknowns reaching as high as the sixth power) and the number of unknowns (as many as ten) and thus in arriving at a solution. The *Arithmetica* is therefore essentially a logistical work, but with the difference that Diophantus' problems are purely numerical with the single exception of problem V, 30.[5] In his solutions Diophantus showed himself a master in the field of indeterminate analysis, and apart from Pappus he was the only great mathematician during the decline of Hellenism.

Extent of the Work. At the close of the introduction, Diophantus speaks of the thirteen books into which he had divided the work; only six, however, survive. The loss of the remaining seven books must have occurred early, since the oldest manuscript (in Madrid), from the thirteenth century, does not contain them. Evidence for this belief may be found in the fact that Hypatia commented on only the first six books (end of the fourth century). A similarity may be found in the *Conics* of Apollonius, of which Eutocius considered only the first four books. But whereas the latter missing material can be supplied in great part from Arabic sources, there are no such sources for the *Arithmetica,* although it is certain that Arabic versions did exist.

Western Europe learned about a Diophantus manuscript for the first time through a letter to Bianchini from Regiomontanus (15 February 1464), who reported that he had found one in Venice; it contained, however, not the announced thirteen books but only six. In his inaugural address at Padua at about the same time, Regiomontanus spoke of the great importance of this find, since it contained the whole "flower of arithmetic, the *ars rei et census,* called algebra by the Arabs."

Reports concerning the supposed existence of the complete *Arithmetica* are untrustworthy.[6] The question, then, is where one should place the gap: after the sixth book or within the existing books? The quadratic equations with one unknown are missing; Diophantus promised in the introduction to treat them, and many examples show that he was familiar with their solution. A section dealing with them seems to be missing between the first and second books. Here and at other places[7] a great deal has fallen into disorder through the commentators or transcription. For example, the first seven problems of the second book fit much better with the problems of the first, as do problems II, 17, and II, 18. As for what else may have been contained in the missing books, there is no precise information, although one notes the absence, for example, of the quadratic equation system (1) $x^2 \pm y^2 = a$; (2) $xy = b$, which had already appeared in Babylonian mathematics. Diophantus could surely solve this as well as the system he treated in problems I, 27, 30: (1) $x \pm y = a$; (2) $xy = b$, a system likewise known by the Babylonians.

Since in one of the manuscripts the six books are apportioned into seven and the writing on polygonal numbers could be counted as the eighth, it has been supposed that the missing portion was not particularly extensive. This is as difficult to determine as how much—considering the above-mentioned problems, which are not always simple—Diophantus could have increased the difficulty of the problems.[8]

Introduction to the Techniques of Algebra. In the introduction, Diophantus first explains for the beginner the structure of the number series and the names of the powers up to n^6. They are as follows:

n^2 is called square number, $\tau\epsilon\tau\rho\acute\alpha\gamma\omega\nu\sigma\varsigma$ ($\dot\alpha\rho\iota\theta\mu\acute\sigma\varsigma$)
n^3 is called cube number, $\kappa\acute\nu\beta\sigma\varsigma$
n^4 is called square-square number, $\delta\nu\nu\alpha\mu\sigma\delta\acute\nu\nu\alpha\mu\iota\varsigma$
n^5 is called square-cube number, $\delta\nu\nu\alpha\mu\acute\sigma\kappa\nu\beta\sigma\varsigma$
n^6 is called cube-cube number, $\kappa\nu\beta\acute\sigma\kappa\nu\beta\sigma\varsigma$

The term n^1, however, is expressed as the side of a square number, $\pi\lambda\epsilon\nu\rho\grave\alpha\ \tauο\hat\nu\ \tau\epsilon\tau\rho\alpha\gamma\acute\omega\nu\sigma\nu$.[9]

Diophantus introduced symbols for these powers; they were also used—with the exception of the second power—for the powers of the unknowns. The symbols are: for x^2, $\Delta^{\mathrm{Y}}(\delta\acute\nu\nu\alpha\mu\iota\varsigma)$; for x^3, K^Y; for x^4, $\Delta^Y\Delta$; for x^5, ΔK^Y; and for x^6, K^YK. The unknown x, "an indeterminate multitude of units," is simply called "number" ($\dot\alpha\rho\iota\theta\mu\acute\sigma\varsigma$); it is reproduced as an s-shaped symbol, similar to the way it appears in the manuscripts.[10] No doubt the symbol originally appeared as a final sigma with a cross line, approximately like this: ς; a similar sign is found (before Diophantus) in a papyrus of the early second century.[11] Numbers which are not coefficients of unknowns are termed "units" ($\mu\sigmaν\alpha\delta\epsilon\varsigma$) and are indicated by $\mathring{\mathrm{M}}$. The symbols for the powers of the unknowns are also employed for the reciprocal values $1/x$, $1/x^2$, etc., in which case an additional index, $^\mathsf{x}$, marks them as fractions. Their names are patterned on those of the ordinals: for example, $1/x$ is the xth ($\dot\alpha\rho\iota\theta\mu\sigma\sigma\tau\acute\sigma\nu$), $1/x^2$ the x^2th ($\delta\nu\nu\alpha\mu\sigma\sigma\tau\acute\sigma\nu$), and so on. All these symbols—among which is one for the "square number," $\square^{os}(\tau\epsilon\tau\rho\acute\alpha\gamma\omega\nu$-$os$)—were read as the full words for which they stand, as is indicated by the added grammatical endings, such as ς^{ol} and $\varsigma\varsigma^{ol}=\dot\alpha\rho\iota\theta\mu\sigmaί$. Diophantus then sets forth in tabular form for the various species ($\epsilon\hatι\delta\sigmaς$) of powers multiplication rules for the operations $x^m \cdot x^n$ and $x^m \cdot x^{1/n}$; thus—as he states—the divisions of the species are also defined. The sign for subtraction, $\wedge$, is also new; it is described in the text as an inverted "psi." The figure is interpreted as the paleographic abbreviation of the verb $\lambda\epsilonί\pi\epsilon\iota\nu$ ("to want").

Since Diophantus did not wish to write a textbook, he gives only general indications for computation: one should become practiced in all operations with the various species and "should know how to add positive ('forthcoming') and negative ('wanting') terms with different coefficients to other terms, themselves either positive or likewise partly positive and partly negative, and how to subtract from a combination of positive and negative terms other terms either positive or likewise partly positive and partly negative."[12] Only two rules are stated explicitly: a "wanting" multiplied by a "wanting" yields a "forthcoming" and a "forthcoming" multiplied by a "wanting" yields a "wanting." Only in the treatment of the linear equations does Diophantus go into more detail: one should "add the negative terms on both sides, until the terms on both sides are positive, and then again . . . subtract like from like until one term only is left on each side."[13] It is at this juncture that he promised that he would later explain the technique to be used if two species remain on one side. There is no doubt that he had in mind here the three forms of the quadratic equation in one unknown.

Diophantus employs the usual Greek system of numerals, which is grouped into myriads; he merely—as the manuscripts show—separates the units place of the myriads from that of the thousands by means of a point. One designation of the fractions, however, is new; it is used if the denominator is a long number or a polynomial. In this case the word $\mu\sigma\rhoί\sigmaν$ (or $\dot\epsilonν\ \mu\sigma\rhoί\omega$), in the sense of "divided by" (literally, "of the part"), is inserted between numerator and denominator. Thus, for example, our expression $(2x^3 + 3x^2 + x)/(x^2 + 2x + 1)$ appears (VI, 19) as

$$K^Y\bar\beta\ \Delta^Y\ \bar\gamma\ \varsigma\ \bar\alpha\ \dot\epsilonν\ \mu\sigma\rhoί\omega\ \Delta^Y\bar\alpha\ \varsigma\ \bar\beta\ \mathring{\mathrm{M}}\ \bar\alpha.$$

One sees that the addends are simply juxtaposed without any plus sign between them. Similarly, since brackets had not yet been invented, the negative members had to be brought together behind the minus symbol: thus, $12 - 1/x - 14x = \mathring{\mathrm{M}}\overline{\iota\beta}\wedge\varsigma^{\mathsf{x}}\ \bar\alpha\varsigma\ \overline{\iota\delta}$ (VI, 22). The symbolism that Diophantus introduced for the first time, and undoubtedly devised himself, provided a short and readily comprehensible means of expressing an equation: for example, $630x^2 + 73x = 6$ appears as $\Delta^Y\overline{\chi\lambda}\ \varsigma\ \overline{\sigma\gamma}\ \ddot\iota\sigma.\ \mathring{\mathrm{M}}\bar\varsigma$ (VI, 8). Since an abbreviation is also employed for the word "equals" ($\ddotι\sigma\sigmaς$),[14] Diophantus took a fundamental step from verbal algebra toward symbolic algebra.

The Problems of the Arithmetica. The six books of the *Arithmetica* present a collection of both determinate and (in particular) indeterminate problems, which are treated by algebraic equations and also by algebraic inequalities. Diophantus generally proceeds from the simple to the more difficult, both in the degree of the equation and in the number of unknowns. However, the books always contain exercises belonging to various groups of problems. Only the sixth book has a unified content. Here all the exercises relate to a right triangle; without regard to dimension, polynomials are formed from the surface, from the sides, and once even from an angle bisector. The first book, with which exercises II, 1–7, ought to be included, contains determinate problems of the first and second degrees. Of the few indeterminate exercises presented there, one (I, 14: $x + y = k \cdot xy$) is transformed into a determinate exercise by choosing numerical values for y and k. The indeterminate exercises I, 22–25, belong to another group; these are the puzzle problems of "giving and taking," such as "one man alone cannot buy"—formulated, to be sure, in numbers without units of measure.[15] The second and all the following books contain only indeterminate problems, beginning with those of the second degree but, from the fourth book on, moving to problems of higher degrees also, which by a clever choice of numerical values can be reduced to a lower degree.[16]

The heterogeneity of the 189 problems treated in the *Arithmetica* makes it impossible to repeat the entire contents here. Many who have worked on it have divided the problems into groups according to the degree of the determinate and indeterminate equations. The compilations of all the problems made by Tannery (II, 287–297), by Loria (pp. 862–874), and especially by Heath (*Diophantus,* pp. 260–266) provide an introductory survey. However, the method of solution that Diophantus adopts often yields new problems that are not immediately evident from the statement of the original problem and that should be placed in a different position by any attempted grouping of the entire contents. Nevertheless, certain groups of exercises clearly stand out, although they do not appear together but are dispersed throughout the work. Among the exercises of indeterminate analysis—Diophantus' own achievements lie in this area—certain groups at least should be cited with individual examples:

I. Polynomials (or other algebraic expressions) to be represented as squares. Among these are:

1. One equation for one unknown:
(II, 23; IV, 31) $ax^2 + bx + c = u^2$.
(VI, 18) $ax^3 + bx^2 + cx + d = u^2$.
(V, 29) $ax^4 + b = u^2$.
(VI, 10) $ax^4 + bx^3 + cx^2 + dx + e$
$= u^2$.
(IV, 18) $x^6 - ax^3 + x + b^2 = u^2$.

One equation for two unknowns:
(V, 7, lemma 1) $xy + x^2 + y^2 = u^2$.

One equation for three unknowns:
(V, 29) $x^4 + y^4 + z^4 = u^2$.

2. Two equations for one unknown ("double equation"):
(II, 11) $a_1x + b_1 = u^2$,
$a_2x + b_2 = v^2$.
(VI, 12) $a_1x^2 + b_1x = u^2$,
$a_2x^2 + b_2x = v^2$.

Two equations for two unknowns:
(II, 24) $(x + y)^2 + x = u^2$,
$(x + y)^2 + y = v^2$.

3. Three equations for three unknowns:
(IV, 19) $xy + 1 = u^2$,
$yz + 1 = v^2$,
$xz + 1 = w^2$.

(II, 34) $x_i^2 + \sum_{k=1}^{3} x_k = u_i^2$ $(i = 1 \cdots 3)$.

(IV, 23) $\prod_{i=1}^{3} x_i - x_k = u_k^2$ $(k = 1 \cdots 3)$.

(V, 21) $\prod_{i=1}^{3} x_i^2 + x_k^2 = u_k^2$ $(k = 1 \cdots 3)$.

4. Four equations for four unknowns:

(III, 19) $\left(\sum_{i=1}^{4} x_i\right)^2 \pm x_k = u_k^2$ $(k = 1 \cdots 4)$.

5. Further variations: In V, 5,[17] to construct six squares for six expressions with three unknowns or six squares for six expressions with four unknowns (IV, 20), etc.

II. Polynomials to be represented as cube numbers.

1. One equation for one unknown:
(VI, 17) $x^2 + 2 = u^3$.
(VI, 1) $x^2 - 4x + 4 = u^3$.

2. Two equations for two unknowns:
(IV, 26) $xy + x = u^3$,
$xy + y = v^3$.

3. Three equations for three unknowns:

(V, 15) $\left(\sum_{i=1}^{3} x_i\right)^3 + x_k = u_k^3$ $(k = 1 \cdots 3)$.

III. To form two polynomials such that one is a square and the other a cube.

1. Two equations for two unknowns:
(IV, 18) $x^3 + y = u^3$, $y^2 + x = v^2$.
(VI, 21) $x^3 + 2x^2 + x = u^3$, $2x^2 + 2x = v^2$.

2. Two equations for three unknowns:
(VI, 17) $xy/2 + z = u^2$, $x + y + z = v^3$.

IV. Given numbers to be decomposed into parts.

1. From the parts to form squares according to certain conditions:

(V, 9) $1 = x + y$; it is required that $x + 6 = u^2$ and $y + 6 = v^2$.

(II, 14) $20 = x + y$; it is required that $x + u^2 = v^2$ and $y + u^2 = w^2$.

(IV, 31) $1 = x + y$; it is required that $(x + 3) \cdot (y + 5) = u^2$.

(V, 11) $1 = x + y + z$; it is required that $x + 3 = u^2$, $y + 3 = v^2$, and $z + 3 = w^2$.

(V, 13) $10 = x + y + z$; it is required that $x + y = u^2$, $y + z = v^2$, and $z + x = w^2$.

(V, 20) $1/4 = \sum_{i=1}^{3} x_i$; it is required that

$$x_i - \left(\sum_{k=1}^{3} x_k \right)^3 = u_i^2 \ (i = 1 \cdots 3).$$

2. From the parts to form cubic numbers:

(IV, 24) $6 = x + y$; it is required that $xy = u^3 - u$.

(IV, 25) $4 = x + y + z$; it is required that $xyz = u^3$, whereby $u = (x - y) + (y - z) + (x - z)$.

V. A number is to be decomposed into squares.

(II, 8) $16 = x^2 + y^2$.

(II, 10) $60 = x^2 - y^2$.

(IV, 29) $12 = x^2 + y^2 + z^2 + u^2 + x + y + z + u$.

(V, 9) $13 = x^2 + y^2$, whereby $x^2 > 6$ and $y^2 > 6$.

In the calculation of the last problem Diophantus arrives at the further exercise of finding two squares that lie in the neighborhood of $(51/20)^2$. He terms such a case an "approximation" ($\pi\alpha\rho\iota\sigma\acute{o}\tau\eta s$) or an "inducement of approximation" ($\grave{\alpha}\gamma\omega\gamma\grave{\eta}$ $\tau\tilde{\eta}s$ $\pi\alpha\rho\iota\sigma\acute{o}$-$\tau\eta\tau os$). Further examples of solution by approximation are:

(V, 10) $9 = x^2 + y^2$, whereby $2 < x^2 < 3$. This is the only instance in which Diophantus represents (as does Euclid) a number by a line segment.

(V, 12) $10 = x + y + z$, where $x > 2$, $y > 3$, and $z > 4$.

(V, 13) $20 = x + y + z$, whereby each part < 10.

(V, 14) $30 = x^2 + y^2 + z^2 + u^2$, whereby each square < 10.

VI. Of the problems formulated in other ways, the following should be mentioned.

(IV, 36) $xy/(x + y) = a$, $yz/(y + z) = b$,   and $xz/(x + z) = c$.

(IV, 38)   The products

$$x_k \cdot \sum_{i=1}^{3} x_i \quad (k = 1 \cdots 3)$$

are to be a triangular number $u(u + 1)/2$, a square v^2, and a cube w^3, in that order.

(IV, 29) $\sum x_i^2 + \sum x_i = 12 \quad (i = 1 \cdots 4)$.

(IV, 30) $\sum x_i^2 - \sum x_i = 4 \quad (i = 1 \cdots 4)$.

(V, 30) This is the only exercise with units of measure attached to the numbers. It concerns a wine mixture composed of x jugs of one type at five drachmas and y jugs of a better type at eight drachmas. The total price should be $5x + 8y = u^2$, given that $(x + y)^2 = u^2 + 60$.

Methods of Problem-solving. In only a few cases can one recognize generally applicable methods of solution in the computations that Diophantus presents, for he considers each case separately, often obtaining an individual solution by means of brilliant stratagems. He is, however, well aware that there are many solutions. When, as in III, 5 and 15, he obtains two solutions by different means, he is satisfied and does not arrange them in a general solution—which, in any case, it was not possible for him to do.[18] Of course a solution could not be negative, since negative numbers did not yet exist for Diophantus. Thus, in V, 2, he says of the equation $4 = 4x + 20$ that it is absurd ($\check{\alpha}\tau o\pi o\nu$). The solution need not be a whole number. Such a solution is therefore not a "Diophantine" solution. The only restriction is that the solution must be rational.[19] In the equation $3x + 18 = 5x^2$ (IV, 31), where such is not the case, Diophantus notes: "The equation is not rational" ($o\grave{v}\kappa$ $\check{\epsilon}\sigma\tau\iota\nu$ $\acute{\eta}$ $\check{\iota}\sigma\omega\sigma\iota s$ $\grave{\rho}\eta\tau\acute{\eta}$); and he ponders how the number 5 could be changed so that the quadratic equation would have a rational solution.

There are two circumstances that from the very beginning hampered or even prevented the achievement of a general solution. First, Diophantus can symbolically represent only one unknown; if the problem contains several, he can carry them through the text as "first, second, etc." or as "large, medium, small," or even express several unknowns by means of one. Mostly, however, definite numbers immediately take the place of the unknowns and particularize the problem. The process of calculation becomes particularly opaque because newly appearing unknowns are again and again designated by the same symbol for x.

Second, Diophantus lacked, above all else, a symbol for the general number n. It is described, for example, as "units, as many as you wish" (V, 7, lemma 1; $\overset{\circ}{\text{M}}$ ὅσων θέλεις). For instance, nx is termed "x, however great" (II, 9; ς ὅσος δήποτε) or "any x" (IV, 39; ἀριθμός τις). Nevertheless, Diophantus did succeed, at least in simple cases, in expressing a general number—in a rather cumbersome way, to be sure. Thus in IV, 39, the equation $3x^2 + 12x + 9 = (3 - nx)^2$ yields $x = (12 + 6n)/(n^2 - 3)$; the description reads "x is a sixfold number increased by twelve, which is divided by the difference by which the square of the number exceeds 3."

Among the paths taken by Diophantus to arrive at his solutions, one can clearly discern several methods:

1. For the determinate linear and quadratic equations there are the usual methods of balancing and completion (see, for example, the introduction and II, 8); in determinate systems, Diophantus solves for one unknown in terms of the other by the first equation and then substitutes this value in the second. For the quadratic equation in two unknowns, he employs the Babylonian normal forms; for the equation in one unknown, the three forms $ax^2 + bx = c$, $ax^2 = bx + c$, and $ax^2 + c = bx$. Moreover, his multiplication of the equation by a can be seen from the criterion for rationality $(b/2)^2 + ac = \square$ or, as the case may be, $(b/2)^2 - ac = \square$.[20]

2. The number of unknowns is reduced. This often happens through the substitution of definite numbers at the beginning, which in linear equations corresponds to the method of "false position." If a sum is to be decomposed into two numbers, for example $x + y = 10$, then Diophantus takes $x = 5 + X$ and $y = 5 - X$. This is also the case with the special cubic equations in IV, 1 and 2.[21]

3. The degree of the equation is reduced. Either a definite number is substituted for one or more unknowns or else a function of the first unknown is substituted.

(V, 7, lemma 1) $xy + x^2 + y^2 = u^2$; y is taken as 1 and u as $x - 2$; this gives $x^2 + x + 1 = (x - 2)^2$; therefore $x = 3/5$ and $y = 1$, or $x = 3$ and $y = 5$.

(V, 29) $x^4 + y^4 + z^4 = u^2$, with $y^2 = 4$ and $z^2 = 9$; therefore $x^4 + 97 = u^2$. With $u = x^2 - 10$ this yields $20x^2 = 3$. Since $20/3$ is not a square, the method of reckoning backward (see below) is employed.

(IV, 37) $60u^3 = v^2$, with $v = 30u$.

(II, 8) $16 - x^2 = (nx - 4)^2$, with $n = 2$. The "cancellation of a species" (see II, 11, solution 2) is possible with $ax^2 + bx + c = u^2$, for example, by substituting $mx + n$ for u and determining the values

of m and n for which like powers of x on either side have the same coefficient. Expressions of higher degree are similarly simplified.

(VI, 18) $x^3 + 2 = u^2$, with $x = (X - 1)$, yields $(X - 1)^3 + 2 = u^2$; if $u = (3X/2) + 1$, then $X^3 - 3X^2 + 3X + 1 = (9X^2/4) + 3X + 1$, and hence a first-degree equation.

(VI, 10) $x^4 + 8x^3 + 18x^2 + 12x + 1 = u^2$, where $u = 6x + 1 - x^2$.

4. The double equation. (II, 11) (1) $x + 3 = u^2$, (2) $x + 2 = v^2$; the difference yields $u^2 - v^2 = 1$. Diophantus now employs the formula for right triangles, $m \cdot n = [(m + n)/2]^2 - [(m - n)/2]^2$ and sets the difference $1 = 4 \cdot 1/4$; thus the following results: $u = 17/8$, $v = 15/8$, and $x = 97/64$. Similarly, in II, 13, the difference 1 is given as $2 \cdot 1/2$; in III, 15, $5x + 5 = 5(x + 1)$; and in III, 13, $16x + 4 = 4(4x + 1)$.

5. Reckoning backward is employed if the computation has resulted in an impasse, as above in V, 29; here Diophantus considers how in $20x^2 = 3$ the numbers 20 and 3 have originated. He sets $20 = 2n$ and $3 = n^2 - (y^4 + z^4)$. With $n = y^2 + 4$ and $z^2 = 4$, $3 = 8y^2$ and $20/3 = (y^2 + 4)/4y^2$. Now only $y^2 + 4$ remains to be evaluated as a square. Similar cases include IV, 31, and IV, 18.

6. Method of approximation to limits (V, 9–14). In V, 9, the problem is $13 = u^2 + v^2$, with $u^2 > 6$ and $v^2 > 6$. First, a square is sought which satisfies these conditions. Diophantus takes $u^2 = 6\frac{1}{2} + (1/x)^2$. The quadruple $26 + 1/y^2$ (with $y = x/2$) should also become a square. Setting $26 + 1/y^2 = (5 + 1/y)^2$ yields $y = 10$, $x^2 = 400$, and $u = 51/20$. Since $13 = 3^2 + 2^2$, Diophantus compares $51/20$ with 3 and 2. Thus, $51/20 = 3 - 9/20$ and $51/20 = 2 + 11/20$. Since the sum of the squares is not 13 (but 13 1/200), Diophantus sets $(3 - 9x)^2 + (2 + 11x)^2 = 13$ and obtains $x = 5/101$. From this the two squares $(257/101)^2$ and $(258/101)^2$ result.

7. Method of limits. An example is V, 30. The conditions are $(x^2 - 60)/8 < x < (x^2 - 60)/5$. From this follow $x^2 < 8x + 60$ or $x^2 = 8x + n\,(n < 60)$, and $x^2 > 5x + 60$ or $x^2 = 5x + n\,(n > 60)$. The values (in part incorrect) assigned according to these limits were no doubt found by trial and error. In IV, 31, the condition is $5/4 < x^2 < 2$. After multiplication by 8^2, the result is $80 < (8x)^2 < 128$; consequently $(8x)^2 = 100$ is immediately apparent as a square; therefore $x^2 = 25/16$. In a similar manner, x^6 is interpolated between 8 and 16 in VI, 21.

8. Other artifices appear in the choice of designated quantities in the exercises. Well-known relations of number theory are employed. For example (in IV, 38), $8 \cdot$ triangular number $+ 1 = \square$, therefore $8[n(n +$

1)/2] + 1 = (2n + 1)². In IV, 29, Diophantus applies the identity $(m + n)^2 = m^2 + 2mn + n^2$ to the problem $x^2 + x + y^2 + y + z^2 + z + u^2 + u = 12$. Since $x^2 + x + 1/4$ is a square, $4 \cdot 1/4$ must be added to 12; whence the problem becomes one of decomposing 13 into four squares. Other identities employed include:

(II, 34) $[(m - n)/2]^2 + m \cdot n = [(m + n)/2]^2$

(VI, 19) $m^2 + [(m^2 - 1)/2]^2 = [(m^2 + 1)/2]^2$

(II, 30) $m^2 + n^2 \pm 2mn = \square$

(III, 19) $(m^2 - n^2)^2 + (2mn)^2 = (m^2 + n^2)^2$

(V, 15)   In this exercise the expressions $(x + y + z)^3 + x$, $(x + y + z)^3 + y$, and $(x + y + z)^3 + z$ are to be transformed into perfect cubes. Hence Diophantus takes $x = 7X^3$ and $(x + y + z) = X$, so the first cube is $(2X)^3$. The other two numbers are $y = 26X^3$ and $z = 63X^3$. From this results $96X^2 = 1$. Here again reckoning backward must be introduced. In I, 22, in the indeterminate equation $2x/3 + z/5 = 3y/4 + x/3 = 4z/5 + y/4$, Diophantus sets $x = 3X$ and $y = 4$. In VI, 16, a rational bisector of an acute angle of a right triangle is to be found; the segments into which the bisector divides one of the sides are set at $3x$ and $3 - 3x$, and the other side is set at $4x$. This gives a hypotenuse of $4 - 4x$, since $3x:4x = (3 - 3x)$: hypotenuse.[22]

In VI, 17, one must find a right triangle for which the area plus the hypotenuse $= u^2$ and the perimeter $= v^3$. Diophantus takes $u = 4$ and the perpendiculars equal to x and 2; therefore, the area is x, the hypotenuse $= 16 - x$, and the perimeter $= 18 = v^3$. By reckoning backward (with $u = m$, rather than $u = 4$) the hypotenuse becomes $m^2 - x$ and the perimeter $m^2 + 2 = v^3$. Diophantus then sets $m = X + 1$ and $v = X - 1$, which yields the cubic equation $X^3 - 3X^2 + 3X - 1 = X^2 + 2X + 3$, the solution of which Diophantus immediately presents (obviously after a factorization): $X = 4$.

It is impossible to give even a partial account of Diophantus' many-sided and often surprising inspirations and artifices. It is impossible, as Hankel has remarked, even after studying the hundredth solution, to predict the form of the hundred-and-first.

On Polygonal Numbers. This work, only fragmentarily preserved and containing little that is original, is immediately differentiated from the *Arithmetica* by its use of geometric proofs. The first section treats several lemmas on polygonal numbers, a subject already long known to the Greeks. The definition of these numbers is new; it is equivalent to that given by Hypsicles, which Diophantus cites. According to this definition, the polygonal number

$$p_a^n = \frac{[(a - 2) \cdot (2n - 1) + 2]^2 - (a - 4)^2}{8 \cdot (a - 2)},$$

where a indicates the number of vertices and n the number of "sides" of the polygon.[23] Diophantus then gives the inverse formula, with which one can calculate n from p and a. The work breaks off during the investigation of how many ways a number p can be a polygonal number.

Porisms and Number-theory Lemmas. Diophantus refers explicitly in the *Arithmetica* to three lemmas in a writing entitled "The Porisms," where they were probably proved. They may be reproduced in the following manner:

1. If $x + a = u^2$, $y + a = v^2$, and $xy + a = w^2$, then $v = u + 1$ (V, 3).[24]

2. If $x = u^2$, $y = (u + 1)^2$, and $z = 2 \cdot (x + y) + 2$, then the six expressions $xy + (x + y)$, $xy + z$, $xz + (x + z)$, $xz + y$, $yz + (y + z)$, and $yz + x$ are perfect squares (V, 5).

3. The differences of two cubes are also the sums of two cubes (V, 16). In this case one cannot say whether the proposition was proved.

In solving his problems Diophantus also employs other, likewise generally applicable propositions, such as the identities cited above (see Methods of Problem-solving, §8). Among these are the proposition (III, 15) $a^2 \cdot (a + 1)^2 + a^2 + (a + 1)^2 = \square$ and the formula (III, 19) $(a^2 + b^2) \cdot (c^2 + d^2) = x^2 + y^2$, where $x = (ac \pm bd)$ and $y = (ad \mp bc)$. The formula is used in order to find four triangles with the same hypotenuse. From the numbers chosen in this instance, $a^2 + b^2 = 5$ and $c^2 + d^2 = 13$, it has been concluded that Diophantus knew that a prime number $4n + 1$ is a hypotenuse.[25] In the examples of the decomposition of numbers into sums of squares, Diophantus demonstrates his knowledge of the following propositions, which were no doubt empirically derived: No number of the form $4n + 3$ is the sum of two square numbers (V, 9), and no number of the form $8n + 7$ is the sum of three square numbers (V, 11). Furthermore, every number is the sum of two (V, 9), three (V, 11), or four (IV, 29, and 30; V, 14) square numbers. Many of these propositions were taken up by mathematicians of the seventeenth century, generalized, and proved, thereby creating modern number theory.

In all his multifarious individual problems, in which the idea of a generalization rarely appears, Diophantus shows himself to be an ingenious and tireless calculator who did not shy away from large numbers and in whose work very few mistakes can be found.[26] One wonders what goals Diophantus had in mind in his *Arithmetica*. There was undoubtedly an irresistible drive to investigate the properties of numbers and to explore the mysteries which had grown up around them. Hence Diophantus appears in the period of decline of Greek mathematics on a lonely height as "a brilliant performer in the art of indeterminate analysis invented by him, but the science has nevertheless been indebted, at least directly, to this brilliant genius for few methods, because he was deficient in the speculative thought which sees in the True more than the Correct."[27]

Diophantus' Sources. Procedures for calculating linear and quadratic problems had been developed long before Diophantus. We find them in Babylonian and Chinese texts, as well as among the Greeks since the Pythagoreans. Diophantus' solution of the quadratic equation in two unknowns corresponds completely to the Babylonian, which reappears in the second book of Euclid's *Elements* in a geometric presentation. The treatment of the second-degree equation in one unknown is also Babylonian, as is the multiplication of the equation by the coefficient of x^2. There are a few Greek algebraic texts that we possess which are more ancient than Diophantus: the older arithmetical epigrams (in which there are indeterminate problems of the first degree), the *Epanthema* of Thymaridas of Paros, and the papyrus (Michigan 620) already mentioned. Moreover, knowledge of number theory was available to Diophantus from the Babylonians and Greeks, concerning, for example, series and polygonal numbers,[28] as well as rules for the formation of Pythagorean number triples. A special case of the decomposition of the product of two sums of squares into other sums of squares (see above, Porisms and Lemmas) had already appeared in a text from Susa.[29] One example of indeterminate analysis in an old Babylonian text corresponds to exercise II, 10, in Diophantus.[30] Diophantus studied special cases of the general Pellian equation with the "side and diagonal numbers" $x^2 - 2y^2 = \pm 1$. The indeterminate Archimedean cattle problem would have required a solution of the form $x^2 - ay^2 = 1$. Consequently, Diophantus certainly was not, as he has often been called, the father of algebra. Nevertheless, his remarkable, if unsystematic, collection of indeterminate problems is a singular achievement that was not fully appreciated and further developed until much later.

Influence. In their endeavor to acquire the knowledge of the Greeks, the Arabs—relatively late, it is true—became acquainted with the *Arithmetica*. Al-Nadīm (987/988) reports in his index of the sciences that Qusṭā ibn Lūqā (*ca.* 900) wrote a *Commentary on Three and One Half Books of Diophantus' Work on Arithmetical Problems* and that Abu'l-Wafā' (940–988) likewise wrote *A Commentary on Diophantus' Algebra,* as well as a *Book on the Proofs of the Propositions Used by Diophantus and of Those That He Himself* [Abu'l-Wafā'] *Has Presented in His Commentary.* These writings, as well as a commentary by

Ibn al-Haytham on the *Arithmetica* (with marginal notations by Ibn Yūnus), have not been preserved. On the other hand, Arab texts do exist that exhibit a concern for indeterminate problems. An anonymous manuscript (written before 972) treats the problem $x^2 + n = u^2, x^2 - n = v^2$; a manuscript of the same period contains a treatise by al-Ḥusain (second half of tenth century) that is concerned with the theory of rational right triangles.[31] But most especially, one recognizes the influence of Diophantus on al-Karajī. In his algebra he took over from Diophantus' treatise a third of the exercises of book I; all those in book II beginning with II, 8; and almost all of book III. What portion of the important knowledge of the Indians in the field of indeterminate analysis is original and what portion they owe to the Greeks is the subject of varying opinions. For example, Hankel's view is that Diophantus was influenced by the Indians, while Cantor and especially Tannery claim just the opposite.

Problems of the type found in the *Arithmetica* first appeared in the West in the *Liber abbaci* of Leonardo of Pisa (1202); he undoubtedly became acquainted with them from Arabic sources during his journeys in the Mediterranean area. A Greek text of Diophantus was available only in Byzantium, where Michael Psellus saw what was perhaps the only copy still in existence.[32] Georgius Pachymeres (1240–1310) wrote a paraphrase with extracts from the first book,[33] and later Maximus Planudes (*ca.* 1255–1310) wrote a commentary to the first two books.[34] Among the manuscripts that Cardinal Bessarion rescued before the fall of Byzantium was that of Diophantus, which Regiomontanus discovered in Venice. His intention to produce a Latin translation was not realized. Then for a century nothing was heard about Diophantus. He was rediscovered by Bombelli, who in his *Algebra* of 1572, which contained 271 problems, took no fewer than 147 from Diophantus, including eighty-one with the same numerical values.[35] Three years later the first Latin translation, by Xylander, appeared in Basel; it was the basis for a free French rendering of the first four books by Simon Stevin (1585). Viète also took thirty-four problems from Diophantus (including thirteen with the same numerical values) for his *Zetetica* (1593); he restricted himself to problems that did not contradict the principle of dimension. Finally, in 1621 the Greek text was prepared for printing by Bachet de Méziriac, who added Xylander's Latin translation, which he was able to improve in many respects. Bachet studied the contents carefully, filled in the lacunae, ascertained and corrected the errors, generalized the solutions, and devised new problems. He, and especially Fermat, who took issue with Bachet's statements,[36] thus became the founders of modern number theory, which then—through Euler, Gauss,[37] and many others—experienced an unexpected development.

NOTES

1. Tannery, *Diophanti opera*, II, 38 f. As an example of "Egyptian analysis" Psellus gives the problem of dividing a number into a determined ratio.
2. Tannery, in his *Mémoires scientifiques,* II, 536 ff., mentions as a possibility that the *Arithmetica* was written as a textbook for the Christian school at the request of Dionysius and that perhaps Diophantus himself was a Christian.
3. Tannery, *Diophanti opera,* II, 60 ff.
4. *Ibid.,* p. 72.
5. V, 30, is exercise 30 of the fifth book, according to Tannery's numbering.
6. Tannery, *Diophanti opera,* II, xxxiv.
7. III, 1–4, belongs to II, 34, 35; and III, 20, 21, is the same as II, 14, 15.
8. Problems such as the "cattle problem" do not appear in Diophantus.
9. Or, as in IV, 1, of a cube number.
10. Tannery, *Diophanti opera,* II, xxxiv.
11. Michigan Papyrus 620, in J. G. Winter, *Papyri in the University of Michigan Collection,* vol. III of *Michigan Papyri* (Ann Arbor, 1936), 26–34.
12. Heath, *Diophantus of Alexandria,* pp. 130–131.
13. *Ibid.*
14. Tannery, *Diophanti opera,* II, xli. There are two parallel strokes joined together.
15. Similar problems exist in Byzantine and in Western arithmetic books since the time of Leonardo of Pisa.
16. Heath (in his *Conspectus*) considers the few determinate problems in bk. II to be spurious. Problems 1, 2, 15, and 33–37 of bk. IV become determinate only through arbitrary assumption of values for one of the unknowns.
17. $x^2y^2 + (x^2 + y^2)$, $y^2z^2 + (y^2 + z^2)$, $z^2x^2 + (z^2 + x^2)$, $x^2y^2 + z^2$, $y^2z^2 + x^2$, $z^2x^2 + y^2$.
18. Sometimes Diophantus mentions infinitely many solutions (VI, 12, lemma 2). In VI, 15, lemma, Diophantus presents, besides a well-known solution of the equation $3x^2 - 11 = y^2$ (namely, $x = 5$ and $y = 8$), a second one: $3 \cdot (5 + z)^2 - 11 = (8 - 2z)^2$.
19. Sometimes, for example in IV, 14, the integer solution is added to the rational solution.
20. For example, in VI, 6; IV, 31; and V, 10.
21. In IV, 1, the system $x^3 + y^3 = 370$, $x + y = 10$, corresponds to the quadratic system $xy = 21$, $x + y = 10$, which was a paradigm in al-Khwārizmī. Tannery (*Mémoires scientifiques,* II, 89) shows how close Diophantus was to a solution to the cubic equation $x^3 = 3px + 2q$.
22. Here one sees the application of algebra to the solution of a geometric problem.
23. The *n*th polygonal number has *n* "sides."
24. This is not a general solution; see Tannery, *Diophanti opera,* I, 317.
25. Heath, p. 107.
26. For example, IV, 25, and V, 30; see Heath, pp. 60, 186.
27. Hankel, p. 165; Heath, p. 55.
28. For example, see Hypsicles' formula used in *On Polygonal Numbers.*
29. See E. M. Bruins and M. Rutten, *Textes mathématiques de Suse,* no. 34 in the series Mémoires de la mission archéologique en Iran (Paris, 1961), p. 117.
30. See S. Gandz, in *Osiris,* **8** (1948), 13 ff.
31. See Dickson, p. 459.
32. Tannery, *Diophanti opera,* II, xviii.
33. *Ibid.,* pp. 78–122. Also in "Quadrivium de Georges Pachymère," in *Studi e testi,* CXIV (Vatican City, 1940), 44–76.
34. *Ibid.,* pp. 125–255.
35. Bombelli and Antonio Maria Pazzi prepared a translation of the first five books, but it was not printed.
36. In his copy of Bachet's edition Fermat wrote numerous critical remarks and filled in missing material. These remarks appeared as a supplement, along with selections from Fermat's letters to Jacques de Billy, in Samuel de Fermat's new edition of Diophantus of 1670.
37. The importance of Diophantus is emphasized by Gauss in the introduction to his *Disquisitiones arithmetice:* "Diophanti opus celebre, quod totum problematis indeterminatis dicatum est, multas quaestiones continet, quae propter difficultatem suam artificiorumque subtilitatem de auctoris ingenio et acumine existimationem haud mediocrem suscitant, praesertim si subsidiorum, quibus illi uti licuit, tenuitatem consideres" ("The famous work of Diophantus, which is totally dedicated to indeterminate problems, contains many questions which arouse a high regard for the genius and penetration of the author, especially when one considers the limited means available to him").

BIBLIOGRAPHY

The first Western presentation of the Diophantine problems was by Raffaele Bombelli, in his *Algebra* (Bologna, 1572; 2nd ed., 1579). A Latin translation was produced by W. Xylander, *Diophanti Alexandrini Rerum arithmeticarum libri sex, quorum duo adjecta habent scholia Maximi Planudis. Item liber de numeris polygonis seu multangulis* (Basel, 1575). The text with a Latin translation was prepared by C.-G. Bachet de Méziriac, *Diophanti Alexandrini Arithmeticorum libri sex, et de numeris multangulis liber unus* (Paris, 1621); a new edition was published by Samuel de Fermat with notes by his father, Pierre de Fermat (Toulouse, 1670). There is also the definitive text with Latin translation by P. Tannery, *Diophanti Alexandrini opera omnia cum Graecis commentariis,* 2 vols. (Leipzig, 1893–1895). An English translation, in a modern rendering, is T. L. Heath, *Diophantus of Alexandria, A Study in the History of Greek Algebra* (Cambridge, 1885); the second edition has a supplement containing an account of Fermat's theorems and problems connected with Diophantine analysis and some solutions of Diophantine problems by Euler (Cambridge, 1910; New York, 1964). German translations are O. Schultz, *Diophantus von Alexandria arithmetische Aufgaben nebst dessen Schrift über die Polygon-Zahlen* (Berlin, 1822); G. Wertheim, *Die Arithmetik und die Schrift über die Polygonalzahlen des Diophantus von Alexandria* (Leipzig, 1890); and A. Czwalina, *Arithmetik des Diophantos von Alexandria* (Göttingen, 1952). A French translation is P. Ver Eecke, *Diophante d'Alexandrie* (Paris, 1959). A Greek text (after Tannery's), with translation into modern Greek, is E. S. Stamatis, Διοφαντου αριθμητικα, η αλγεβρα των αρχαιων ελληνων (Athens, 1963).

On Polygonal Numbers appears in a French trans. by G. Massoutié as *Le traité des nombres polygones* (Mâcon, 1911).

Along with the views of the authors in their text editions and translations, the following general criticisms should be consulted: M. Cantor, *Vorlesungen über Geschichte der Mathematik,* 3rd ed., I (Leipzig, 1907), 463–488; P. Cossali,

Origine, trasporto in Italia, primi progressi in essa dell' *algebra* (Parma, 1797), I, 56–95; T. L. Heath, *History of* *Greek Mathematics* (Oxford, 1921), II, 440–517; B. L. van der Waerden, *Erwachende Wissenschaft,* 2nd ed. (Basel-Stuttgart, 1966), pp. 457–470. See also H. Hankel, *Zur* *Geschichte der Mathematik in Alterthum und Mittelalter* (Leipzig, 1874), 2nd ed. with foreword and index by J. E. Hofmann (Hildesheim, 1965); F. Hultsch, "Diophantus," in Pauly-Wissowa, V, pt. 1, 1051–1073; G. Loria, *Le scienze* *esatte nell'antica Grecia* (Milan, 1914), pp. 845–919; E. Lucas, *Recherches sur l'analyse indéterminée et l'arith-* *métique de Diophante,* with preface by J. Itard (Paris, 1967); G. H. F. Nesselmann, *Die Algebra der Griechen* (Berlin, 1842), pp. 244–476; and G. Sarton, *Introduction* *to the History of Science,* I (Baltimore, 1927), 336 ff.

Special criticism includes P. Tannery, "La perte de sept livres de Diophante," in his *Mémoires scientifiques,* II (Toulouse–Paris, 1912), 73–90; "Étude sur Diophante," *ibid.,* 367–399; and "Sur la religion des derniers mathé-maticiens de l'antiquité," *ibid.,* 527–539.

Historical works include I. G. Bašmakova, "Diofant i Ferma," in *Istoriko-matematicheskie issledovaniya,* **17** (1966), 185–204; L. E. Dickson, *History of the Theory of* *Numbers,* II, *Diophantine Analysis* (Washington, D.C., 1920); T. L. Heath, *Diophantus* (see above), ch. 6 and supplement; K. Reich, "Diophant, Cardano, Bombelli, Viète. Ein Vergleich ihrer Aufgaben," in *Rechenpfen-* *nige, Aufsätze zur Wissenschaftsgeschichte* (Munich, 1968), pp. 131–150; P. Tannery, *Diophanti opera* (see above), II, prolegomena; and P. Ver Eecke, *Diophante* (see above), introduction.

Kurt Vogel

DIOSCORIDES, also known as **Pedanius Dioscorides** **of Anazarbus** (*b.* Anazarbus, near Tarsus in Cilicia; *fl.* A.D. 50–70), *pharmacy, medicine, chemistry, botany.*

A letter attached to Dioscorides' work as a dedica-tory preface reveals almost all that is known of his life. The letter states that Dioscorides lived a soldier's life; this enabled him to learn at first hand the iden-tity, preparation, and uses of medicines. Galen names his birthplace,[1] and some manuscript notations add the name Pedanius. Some authorities believe that Dioscorides studied at Tarsus and Alexandria and later attached himself to the Roman army as a mili-tary physician. These suppositions are based on his statement that he led a "soldier-like life" (οἶσδα γὰρ ἡμῖν στρατιωτικὸν τὸν βίον), his remark that he has "lived" (συνδιάγοντες) with Areius of Tarsus, and the likelihood that his travels would have taken him to Alexandria, where he could have had access to the library. Dioscorides has been dated both by the men-tion of his contemporaries and by Galen's use of Dioscorides' work. Erotian (*fl. ca.* A.D. 60), a com-mentator of the Hippocratic works who lived during the Neronian age, mentions Dioscorides (assuming

that the name is not an interpolation).[2] In his letter to Areius, Dioscorides mentions Laecanius Bassus, presumed to be C. Laecanius Bassus, consul in A.D. 64, who is spoken of by Pliny and Tacitus.[3] Quintus Sextius Niger (*fl.* 25 B.C.) is the latest writer whom Dioscorides cites. Pliny the Elder did not know Dioscorides' works directly, but certain similarities between Pliny's and Dioscorides' texts are explained by their having employed the same written source, Sextius Niger.[4]

Although numerous treatises in Greek and Latin are falsely attributed to Dioscorides—both by virtue of his reputation as a major authority in medicines for around 1,600 years and because of numerous editions of his work—only one treatise, Περὶ ὕλης ἰατρικῆς (*De materia medica*), is now attributed to Dios-corides. The title is taken from book 3 and is the same title as that of Sextius Niger's lost work. Written in five books, the treatise discusses over 600 plants, thirty-five animal products, and ninety minerals in simple, concise Greek; Dioscorides feared that his nonliterary style would hinder recognition of the arduousness of the task he had set himself in collect-ing information. Of the approximately 827 entries, only about 130 substances are included in the Hippo-cratic corpus (since modern subspecies do not always correspond to Dioscorides' varieties [εἴδη], an exact count is difficult). Being the author of by far the largest pharmaceutical guide in antiquity, Dioscorides added considerably to the knowledge of drugs. More important, this procedure for relating information on medicine and his unadorned critical skill determined the general form of later pharmacopoeias, both East-ern and Western. Galen, always a severe critic, ac-knowledged Dioscorides' work to be the best of its kind and showed his respect by numerous citations.[5]

Although an empiricist in method, Dioscorides apparently belonged to no definite philosophical school (his friend Areius was a follower of Asclepi-ades). He cited the need to study each plant in relation to its habitat, to observe rigorously the plants at all seasons, to note all parts from the first shoots to the seeds, to prepare each medicine with precision, and to judge each medicine by its merits.

Dioscorides claimed that his work surpassed that of his predecessors in terms of his industry in collect-ing his information, his unlimited range in finding medicines, and the arrangement of his material. He conceded that the older writers transmitted much accurate information but deplored the fact that recent writers had introduced the element of controversy to medicine by speculating vainly on the causes (αἰτίας) of drugs' powers, while failing to pay proper attention to their experience in the use of drugs. For each item generally he gave a Greek synonym, and the names themselves were often of foreign origin, coming from various languages of Africa, Gaul, Persia, Armenia, Egypt, and the like (for which reason linguists are interested in Dioscorides). There follows a deposition on the substance's origin and physical characteristics. He then gives a discourse on the mode of preparation of the medicine and, finally, a list of its medicinal uses with occasional notations of harmful side effects. Often he relates information about how the simple is compounded in a prescription; further, he gives dietetic hints and even tests for detecting a fraudulent preparation.

Even though he faulted earlier writers for their poor classification, Dioscorides' method is not always clear, although he believed his procedure to be superior. His system is as follows (number of items are approximate): book I (129 items), deals with aromatics, oils, salves, trees, and shrubs (liquids, gums, and fruits); book II (186), with animals, animal parts, animal products, cereals, pot herbs, and sharp herbs; book III (158), with roots, juices, herbs, and seeds; book IV (192), with roots and herbs not previously mentioned; and book V (162), with wines and minerals.

Dioscorides says that whenever possible he saw plants with his own eyes but that he also relied on questioning people in the course of his travels and on consulting previously written works. Dioscorides is credited as being the first to recognize the extensive use of medicines from all three of the natural kingdoms—animal, vegetable, and mineral.[6]

Dioscorides cautioned his readers that knowledge of plants was gained by experience: Differences in climates cause wide variations in living patterns—for example, medicines from plants growing in high-altitude windy areas are stronger than those from plants in marshy, shady locations shielded from the wind. Some medicines, such as the white and black hellebore, retain their power for years; others have a shorter effective lifetime. Herbs that are full of branches, such as abrotonum and absinthium, ought to be gathered at seed time. Seeds are best taken when dry; fruits, when ripe; and flowers, before they fall. When the medicine is from sap or juices, the stem should be cut while at full ripeness, but liquids from roots are to be extracted after the plant has lost its leaves.

He notes that storage of medicines is important: Flowers and aromatics should be placed in drug boxes made of limewood. Some herbs are best kept if wrapped in paper (chartes) or leaves. Moist medicines may be placed in thick vessels made of such things as silver, glass, or horn. Earthenware is satisfactory if it is not too thin. Brass vessels are suitable for eye medicines and for liquids, especially those compounded with vinegar, liquid pitch, or cedar oil; but tin vessels ought to be used for fats and marrows.

In his preface he names as the older authorities upon whom he relied Iolas the Bithynian, Hexaclides the Tarentine, and Andreas the physician. He also cites various more recent writers whom he called Asclepiadeans ('Ασκληπιάδειοι)—namely, Julius Bassus, Niceratur, Petronius, Niger, Diodotus, and, most frequently, Crateuas the rhizotomist—and ten other authorities.

Dioscorides was largely responsible for determining modern plant nomenclature, both popular and scientific, because of the reliance of later authorities on his work. Numerous medicines in Dioscorides' work appear in modern pharmacopoeias, among them almond oil, aloes, ammoniacum, belladonna, calamine, calcium hydrate, cherry syrup, cinnamon, copper oxide, coriander, galbanum, galls, ginger, juniper, lavender, lead acetate, marjoram, mastic, mercury, olive oil, opium, pepper, pine bark, storax, sulfur, terebinth, thyme, and wormwood.

The transmission of Dioscorides' text is as important as what it says. Editors and copyists added or subtracted from Dioscorides' writing as a means of contributing their experiences with various drugs in the context of their needs. The numerous and extensive textual modifications make the problem of arriving at a definite understanding of Dioscorides' own Greek text very difficult. Latin, Arabic, and various European vernacular translations reveal greater variations. The study of its transmission is a veritable introduction to the knowledge of Western pharmacy down to the seventeenth century.

Papyri reveal that as early as the second century A.D. recensions had already appeared. The happy survival of a beautifully illustrated manuscript, written entirely in Greek capitals about A.D. 512, demonstrates that by the sixth century Dioscorides' own order of presentation had been completely redone in favor of an alphabetical order. Produced at Constantinople as a wedding gift for Anicia Juliana, daughter of the emperor Flavius Olybrius, the manuscript was offered for the sale price of one hundred ducats to an ambassador of the emperor Ferdinand I by a Jewish physician of Suleiman. The text includes material from other writers. The lavish botanical illustrations in this manuscript and another of the seventh century are the subject of speculation concerning whether the illustrations are from Dioscorides' original work or from Crateuas, the rhizotomist and physician to Mithridates (120–63 B.C.), who was known to paint herbal illustrations with his own hand.[7] A second-century papyrus fragment of Dioscorides' text has illustrations that are different in at least one instance from those in the Juliana manuscript.[8] There is no direct evidence that Dioscorides himself is responsible for the paintings. Plants are drawn in detailed color showing the entire plant, including the root system, flower, and fruit. Certainly the botanical illustrations became standardized, most copyists being content to draw from precedent rather than from nature. One folio (f. 5v) has a portrait purporting to show Dioscorides writing at a desk on the right, while on the left is a painter thought to be Crateuas, drawing a mandrake plant held by Epindia, or Lady Inventiveness, who is standing at the center of the painting.

In the sixth century, Cassiodorus advised some monks: "If you have not sufficient facility in reading Greek then you can turn to the herbal of Dioscorides, which describes and draws the herbs of the field with wonderful faithfulness."[9] Gargilius Martialis (third century) is the first known Latin author to cite Dioscorides. This reference caused some to believe that Martialis was responsible for the first Latin translation; since, however, the other Latin medical writers living before the sixth century failed to cite Dioscorides, it seems more likely that Dioscorides was translated into Latin later than Martialis. Another consideration is Cassiodorus' use of the word *herbarium* to describe Dioscorides' work—but Dioscorides listed all types of substances, not merely plants. Very popular during the early Middle Ages was a pseudo-Dioscoridean text known as *De herbis femininis,* which described and illustrated some seventy-one plants and herbs. Since Cassiodorus referred only to an illustrated herbal, he might possibly have meant *De herbis femininis,* not the complete Dioscorides, especially since Isidore of Seville is known also to have used *De herbis femininis.* Based on Dioscorides, whom the copyists credited with the authorship, the text is severely edited, with additions from other writers. Twenty-seven known manuscripts of *De herbis femininis,* three of which date from the ninth century, testify to its popularity, especially during the early Middle Ages, when the needs of medical science were simpler than those which the full Dioscorides' work was meant to fill.

The Latin West did not have to await the Arabic-Latin translations of the eleventh to thirteenth centuries in order to possess the complete Dioscorides. A Latin translation made in Italy by the sixth century used Dioscorides' original order rather than that of the Juliana manuscript. Although the Latin spelling and grammar are poor, the translation is fairly accurate, with some omissions. The earliest manuscript is written in the eighth-century Beneventan script;

generally manuscripts of this class possess no illustrations.[10] In addition, there was at least one Old English version of Dioscorides.[11]

Stephanus, son of Basilius (Istifan ibn Basīl), translated Dioscorides into Arabic in the second half of the ninth century, but his translation was corrected by Ḥunayn ibn Isḥāq. Another Arabic translation was by Ibn Juljul in the second half of the tenth century in Cordova.[12] Dioscorides greatly influenced Islamic medical botany and therapy, as witnessed by the reliance on his works by such noted writers as Maimonides, Ibn al-Baytār, Ibn Sīnā (Avicenna), and Yaḥyā ibn Sarāfyūn (Serapion the Elder).

In the late eleventh or early twelfth century, a popular new edition of Dioscorides was produced in Latin that eclipsed the Old Latin translation. The arrangement was alphabetical. The rubric to a Bamberg manuscript (which contains only the wording of Dioscorides' preface in the Old Latin translation) said that Constantine the African (*d. ca.* 1085) was responsible for the alphabetical order.[13] Whoever the editor was, he meant to bring Dioscorides up-to-date by inserting new drugs—for example, ambergris and zeodary. (The inserted items are related to Constantine's translations of other works and thus support his claim to the editorship.) The main body of the text is close to the Old Latin version and is not a translation from the Arabic. In the section on stones, the editor of this version inserted the text of Damigeron (*ca.* second century B.C.), rather than using Dioscorides'; the editor, however, included only those stones that were in Dioscorides' original. Later thirteenth-century writers—Arnold of Saxony and Bartholomew the Englishman, for example—quote extensively a treatise called *Dioscorides on Stones,* which is actually Damigeron's text and not Dioscorides'. But this is not the only treatise falsely ascribed to Dioscorides: two other notable pseudo-Dioscoridean treatises are *De physicis ligaturis*[14] and *Quid pro quo,*[15] the latter being a guide for drug substitutions.

About 1300, Pietro d'Abano lectured and commented on the alphabetical Dioscorides version and said he knew of another, briefer version. When the text of the Latin alphabetical Dioscorides was first published at Colle di Val d'Elsa, Italy, in 1478 and again at Lyons in 1512, the printer included many of d'Abano's comments.

Judged by the number of editions, printings, and translations, Dioscorides was very popular during the Renaissance. The Greek text was in print in 1499. By this time the list of synonyms following each item had grown extensively to include Arabic and European vernacular words. A translation by Jean Ruelle received twenty-five different editions, and by 1544 approximately thirty-five editions of Dioscorides' translations and commentaries had been produced.[16] The most illustrious edition was Mattioli's, first published in Venice in 1554. So many editions and translations were made from Mattioli's critical Dioscorides that it is said that this printing is the basic work for modern botany.

NOTES

1. *On Simple Medicines,* in *Opera omnia,* C. G. Kühn, ed., XI, 794: ὁ δὲ ᾿Αναζαρβεὺς Διοσκουρίδης ἐν πέντε βιβλίοις: cf. XIII, 857: ὡς δὲ Διοσκορίδης ὁ Ταρσεὺς ἔδωκεν ᾿Αρείῳ τῷ ᾿Ασκληπιαδείῳ.
2. *Das Hippokrates-Glossar des Erotianos,* Johannes Ilberg, ed. (Leipzig, 1893), p. 116; also, *Erotiani,* Ernst Nachmanson, ed. (Uppsala, 1918), p. 51.
3. Pliny, *Natural History,* 26. 4. 5; Tacitus, *Annals,* 15.33.
4. Max Wellmann, "Sextius Niger. Eine Quellenuntersuchung zu Dioscorides," in *Hermes,* **23** (1888), 530–569.
5. XI, 795; see also index, vol. XX, for list of citations.
6. Galen, XI, 794.
7. Vienna MS Med. Gr. 1 and Vienna MS Suppl. Gr. 28; see also L. Choulant, "Ueber die HSS. des Dioscorides mit Abbildungen," in *Archiv für die zeichnenden Künste,* **1** (1855), 56–62.
8. J. de M. Johnson, "A Botanical Papyrus With Illustrations," in *Archiv für die Geschichte der Naturwissenschaften und der Technik,* **4** (1913), 403–408.
9. *Institutiones divinarum et humanarum litterarum,* ch. 31.
10. Munich MS lat. 337.
11. London, BM MS Cotton Vitellius C III, eleventh century, pub. by O. Cockayne, *Leechdoms, Wortcunning and Starcraft of Early England,* 3 vols., I (London, 1864–1866; repr., 1961), 251–325; cf. Oxford Hatton MS 76, early eleventh century, ff. 110–124, an Old English version of *De herbis femininis.*
12. See Madrid BN MS Arab. 125 (Gg 147), twelfth century, cited by Hartwig Derenbourg, *Notes critiques sur les manuscrits arabes de la Bibliothèque Nationale de Madrid* (Paris, 1904), pp. 7–8, 19, 30–31.
13. Bamberg, Staatsbibliothek MS med. 6, thirteenth century, f. 28v.
14. London BM MS Sloane 3848, seventeenth century, ff. 36–40; and Cambridge MS Add. 4087, fourteenth century, ff. 244v–254v.
15. Brno MS MK, fifteenth century, 173–174v; Vatican MS lat. 5373, ff. 36–41; and Vienna MS Pal. 5371, fifteenth century, ff. 121–124v.
16. Jerry Stannard, "P. A. Mattioli and Some Renaissance Editions of Dioscorides," in *Books and Libraries at the University of Kansas,* **4,** no. 1 (1966), 1–5.

BIBLIOGRAPHY

(1) Several modern Greek texts and pseudo-Dioscoridean Greek texts exist. There are two critical Greek texts of the Περὶ ὕλης ἰατρικῆς (*De materia medica*), but both contain other treatises formerly attributed to Dioscorides but now regarded as spurious. Definitely outdated is a two-volume ed. by Curtius Sprengel in *Medicorum Graecorum opera quae exstant,* C. G. Kühn, ed. (Leipzig, 1829–1830), vols. XXV–XXVI, which reprints as authentic three pseudo-Dioscoridean texts: Περὶ δηλητηρίων φαρμάκων (*On Poison Drugs*), Περὶ ἰοβόλων (*On Animal and Deadly Poisons*), and

Περὶ ἁπλῶν φαρμάκων (*On Simple Drugs*), sometimes known as Εὐπόριστα. The best critical ed. is Max Wellmann, *Pedanii Dioscuridis Anazarbei De materia medica libri quinque,* 3 vols. (Berlin, 1906–1914; repr., 1958). In "Dioscorides," in Pauly, *Real-Encyclopädie der klassischen Altertumswissenschaft* (Stuttgart, 1903), Wellmann regarded the treatise *On Simple Drugs* as a third- or fourth-century work falsely assigned to Dioscorides. Subsequently he became convinced that it was a legitimate work of Dioscorides and published it in vol. III of his critical ed. of Dioscorides. His attempt to authenticate this work (not convincing to later authorities) was published as *Die Schrift des Dioscorides* Περὶ ἁπλῶν φαρμάκων (Berlin, 1914). A German trans. of *On Simple Drugs* was prepared by Julius Berendes: "Die Hausmittel der Pedanios Dioskurides," in *Janus,* **12** (1907), 10–33. The oldest MS is Florence, Laur. Gr. 74, 10, fourteenth century; no Latin version has been identified. Berendes has also translated the two other pseudo-Dioscoridean Greek texts that Sprengel mistook as authentic: "I. Des Pedanios Dioskurides Schrift über die Gifte und Gegengifte. II. Des Pedanios Dioskurides Schrift über die giftigen Tiere und den tollen Hund," in *Apotheker-Zeitung,* nos. 92–93 (1905), 933–935, 945–954. A pseudo-Dioscoridean lapidary is printed in F. de Mély, *Les lapidaires de l'antiquité et du moyen âge* (Paris, 1902), I, 179–183.

(2) No accurate trans. of Dioscorides has appeared in a modern European language. Berendes prepared a German trans., *Des Pedanios Dioskurides aus Anazarbos Arzneimittellehre in fünf Büchern* (Stuttgart, 1902), but unfortunately it is based on the inadequate Sprengel ed. Woefully inadequate is an English trans. prepared from the Greek by John Goodyer between 1652 and 1655 but not published until much later in Robert T. Gunther, *The Greek Herbal of Dioscorides* (New York, 1934; repr., 1959).

(3) On the basis of a second-century papyrus, Campbell Bonner observed ("A Papyrus of Dioscorides in the University of Michigan Collection," in *Transactions and Proceedings of the American Philological Association,* **53** [1922], 141–168) that Oribasius' citations to Dioscorides in the fourth century resemble the papyrus text more than they do the alphabetical Greek version found in the Juliana MS (Vienna, Gr.1), sixth century. The papyrus text seems close to Escorial MS Gr. III. R.3, eleventh century, which Wellmann, in his critical text, considered more corrupt. A MS containing a section of Dioscorides in red Greek capitals, dating from around A.D. 600, has parts of four chapters in nonalphabetical order from bk. 3 (in order, chs. 82, 83, 78, and 79 of Wellmann ed.). This MS (Naples MS lat. 2; formerly Vienna MS lat. 16) has a Latin treatise written over the Greek Dioscorides text. It is known to have been at the monastery of Bobbio in the eighth century (erroneously reported by Charles Singer, "The Herbal in Antiquity and Its Transmission to Later Ages," in *Journal of Hellenic Studies,* **47** [1927], 34–35; cf. J. V. Eichenfeld, *Wiener Jahrbücher der Literatur,* **25** [1824], 35–37, and Wellmann, "Dioscorides," in Pauly, *Real-Encyclopädie* [1903], col. 1136).

(4) A facsimile repr. of the Vienna MS Med. Gr. 1, with

beautiful color plates, is published as *De materia medica* (Graz, 1966); cf. Otto Waechter, "The 'Vienna Dioskurides' and Its Restoration," in *Libri,* **13** (1963), 107–111; and G. E. Dann, "Ein Faksimile-Druck des Wiener Dioskurides," in *Zur Geschichte der Pharmazie,* **18** (1966), 9–11.

(5) The Latin text is inadequately treated. The old Latin trans. is found in Munich MS 337, eighth century, and Paris BN, lat. 9332, ninth century, pp. 243–321. K. Hoffmann and T. M. Auracher began editing Munich 337 (*Römanische Forschungen,* **1** [1882], 49–105), and the project was continued by H. Stadler (*ibid.,* **10** [1897], 181–247, 369–446; **11** [1899], 1–121; **13** [1902], 161–243; **14** [1903], 601–637). Stadler had the advantage of the discovery of BN 9332, which he used for editing bks. II–V but he did not reedit bk. I. BN 9332 has missing leaves now in Bern MS A 91.7, ff. 1v–2v. Bk. I, using BN 9332, is reedited by H. Mihuescu, *Dioscoride Latino Materia Medica Libro Primo* (Iaşi, Romania, 1938). The alphabetical Latin Dioscorides is perhaps represented best by Cambridge, Jesus Col., MS Q.D. 2 (44), twelfth century, ff. 17–145; Erfurt MS F 41, fourteenth century, ff. 1–62v; and Vatican MS Urb. lat. 1383, twelfth century, ff. 1–116. Pietro d'Abano's commentary, together with the text of Dioscorides, is found in Paris BN lat. 6819, thirteenth century, ff. 1–70v., and BN 6820, fourteenth or fifteenth century, ff. 1–74. It was apparently the MS BN 6820—probably the exact copy—that the printer Johannes de Medemblich mistook part of d'Abano's commentary for Dioscorides' text. The early printings are covered by Jerry Stannard, "P. A. Mattioli. Sixteenth Century Commentator on Dioscorides," in *Bibliographical Contributions, University of Kansas Libraries,* **1** (1969), 59–81.

(6) A text of *De herbis femininis* is published by Heinrich Kästner, "Pseudo-Dioscoridis 'De herbis femininis,'" in *Hermes,* **31** (1896), 578–636, from only three MS sources. This version is faulty; for instance, London BM Sloane MS 1975, twelfth century, ff. 49v–73, adds some herbs, whereas London BM Harley MS 5294, twelfth century, ff. 43v–58, and Oxford Bodl. Ash. MS 1431, late eleventh century, ff. 31v–43, omit many herbs. Kästner's ed. did not use these MSS. *De herbis femininis* is the only version of the Latin MSS always to be illustrated. No study yet has been made of the drawings.

(7) The best overall treatment by modern scholars remains Max Wellmann's preface to his *Pedanti Dioscuridis,* II, v–xxvi; and his articles "Dioscorides" and "Areios," in Pauly, *Real-Encyclopädie.* Charles Singer (*Journal of Hellenic Studies,* **47** [1927], 1–52) studied the transmission problem and worked with the iconographic aspects; see also his "Greek Biology and Its Relation to the Rise of Modern Biology," in *Studies in the History and Method of Science* (Oxford, 1921), II, 1–101. César E. Dubler has written an extensive examination of the various substances mentioned by Dioscorides as they are transmitted through various medical writers, especially writers in Arabic: *La "Materia Médica" de Dioscórides. Transmisión medieval y renacentista,* 5 vols. (Barcelona, 1953–1959). Vol. I has a valuable concordance which traces the translations of the plants and other substances from Greek to Latin, Arabic, and Castilian. With Elias Terés as coauthor, vol. II prints an Arabic

version of Dioscorides, principally from Madrid BN MS 5006 but with other texts consulted. Vol. III reproduces in facsimile the Salamanca 1570 printing of Don Andrés de Laguna's Castilian trans. first printed in Anvers in 1555. The remainder of the volumes are concerned with commentaries, indexes, etc.

(8) Attempts have been made by modern botanists and other scholars to identify the plants in Dioscorides. The best is Berendes' trans. into German with his own commentary (Stuttgart, 1902). Useful are the following: R. Mock, *Pflanzliche Arzneimittel bei Dioskurides die schon in Corpus Hippocratum vorkommen,* diss. (Tübingen, 1919); R. Schmidt, *Die noch gebräulichen Arzneimittel bei Dioskurides,* diss. (Tübingen, 1919); Léon Moulé, "La zoothérapie au temps de Dioscoride et de Pline," in *International Congress for the History of Medicine* (Antwerp, 1920), pp. 451–461; Edmund O. von Lippmann, "Die chemischen Kenntnisse des Dioskorides," in *Abhandlungen und Vorträge zur Geschichte der Naturwissenschaften,* 2 vols. (Leipzig, 1906–1913), I, 47–73; Achille Morricone, "I medicamenti di origine animale recavati dal mare nell'opera di Dioscoride," in *Pagine di storia della medicina,* **7** (1963), 24–28; and Ernst W. Stieb, "Drug Adulteration and Its Detection, in the Writings of Theophrastus, Dioscorides and Pliny," in *Journal mondial de pharmacie,* **2** (1958), 117–134. Attempts to identify some of the plants and animals of the illustrations are made by E. Bonnet, "Étude sur les figures de plantes et d'animaux . . .," in *Janus,* **14** (1909), 294–303, and other of his articles in *Janus,* **8** (1903), 169–177, 225–232, 281–285; and by E. Emmanuel, "Étude comparative sur les plantes dessinées dans le Codex Constantinopolitanus de Dioscoride," in *Schweizerische Apotheker-Zeitung,* **50** (1912), 45–50, 64–72.

JOHN M. RIDDLE

DIRICHLET, GUSTAV PETER LEJEUNE (*b.* Düren, Germany, 13 February 1805; *d.* Göttingen, Germany, 5 May 1859), *mathematics.*

Dirichlet, the son of the town postmaster, first attended public school, then a private school that emphasized Latin. He was precociously interested in mathematics; it is said that before the age of twelve he used his pocket money to buy mathematical books. In 1817 he entered the Gymnasium in Bonn. He is reported to have been an unusually attentive and well-behaved pupil who was particularly interested in modern history as well as in mathematics.

After two years in Bonn, Dirichlet was sent to a Jesuit college in Cologne that his parents preferred. Among his teachers was the physicist Georg Simon Ohm, who gave him a thorough grounding in theoretical physics. Dirichlet completed his *Abitur* examination at the very early age of sixteen. His parents wanted him to study law, but mathematics was already his chosen field. At the time the level of pure mathematics in the German universities was at a low ebb: Except for the formidable Carl Gauss in Göttingen, there were no outstanding mathematicians, while in Paris the firmament was studded by such luminaries as P.-S. Laplace, Adrien Legendre, Joseph Fourier, Siméon Poisson, Sylvestre Lacroix, J.-B. Biot, Jean Hachette, and Francoeur.

Dirichlet arrived in Paris in May 1822. Shortly afterward he suffered an attack of smallpox, but it was not serious enough to interrupt for long his attendance at lectures at the Collège de France and the Faculté des Sciences. In the summer of 1823 he was fortunate in being appointed to a well-paid and pleasant position as tutor to the children of General Maximilien Fay, a national hero of the Napoleonic wars and then the liberal leader of the opposition in the Chamber of Deputies. Dirichlet was treated as a member of the family and met many of the most prominent figures in French intellectual life. Among the mathematicians, he was particularly attracted to Fourier, whose ideas had a strong influence upon his later works on trigonometric series and mathematical physics.

Dirichlet's first interest in mathematics was number theory. This interest had been awakened through an early study of Gauss's famous *Disquisitiones arithmeticae* (1801), until then not completely understood by mathematicians. In June 1825 he presented to the French Academy of Sciences his first mathematical paper, "Mémoire sur l'impossibilité de quelques équations indéterminées du cinquième degré." It dealt with Diophantine equations of the form

$$x^5 + y^5 = A \cdot z^5$$

using algebraic number theory, Dirichlet's favorite field throughout his life. By means of the methods developed in this paper Legendre succeeded, only a few weeks later, in giving a complete proof that Fermat's equation

$$x^n + y^n = z^n$$

has no integral solutions $(x \cdot y \cdot z \neq 0)$ for $n = 5$. Until then only the cases $n = 4$ (Fermat) and $n = 3$ (Euler) had been solved.

General Fay died in November 1825, and the next year Dirichlet decided to return to Germany, a plan strongly supported by Alexander von Humboldt, who worked for the strengthening of the natural sciences in Germany. Dirichlet was permitted to qualify for habilitation as *Privatdozent* at the University of Breslau; since he did not have the required doctorate, this was awarded *honoris causa* by the University of Cologne. His habilitation thesis dealt with polynomials whose prime divisors belong to special arithmetic series. A second paper from this period was inspired

by Gauss's announcements on the biquadratic law of reciprocity.

Dirichlet was appointed extraordinary professor in Breslau, but the conditions for scientific work were not inspiring. In 1828 he moved to Berlin, again with the assistance of Humboldt, to become a teacher of mathematics at the military academy. Shortly afterward, at the age of twenty-three, he was appointed extraordinary (later ordinary) professor at the University of Berlin. In 1831 he became a member of the Berlin Academy of Sciences, and in the same year he married Rebecca Mendelssohn-Bartholdy, granddaughter of the philosopher Moses Mendelssohn.

Dirichlet spent twenty-seven years as a professor in Berlin and exerted a strong influence on the development of German mathematics through his lectures, through his many pupils, and through a series of scientific papers of the highest quality that he published during this period. He was an excellent teacher, always expressing himself with great clarity. His manner was modest; in his later years he was shy and at times reserved. He seldom spoke at meetings and was reluctant to make public appearances. In many ways he was a direct contrast to his lifelong friend, the mathematician Karl Gustav Jacobi.

The two exerted some influence upon each other's work, particularly in number theory. When, in 1843, Jacobi was compelled to seek a milder climate for reasons of health, Dirichlet applied for a leave of absence and moved with his family to Rome. A circle of leading German mathematicians gathered around the two. Dirichlet remained in Italy for a year and a half, visited Sicily, and spent the second winter in Florence.

Dirichlet's first paper dealing with Fermat's equation was inspired by Legendre; he returned only once to this problem, showing the impossibility of the case $n = 14$. The subsequent number theory papers dating from the early years in Berlin were evidently influenced by Gauss and the *Disquisitiones*. Some of them were improvements on Gauss's proofs and presentation, but gradually Dirichlet cut much deeper into the theory. There are papers on quadratic forms, the quadratic and biquadratic laws of reciprocity, and the number theory of fields of quadratic irrationalities, with the extensive discussion of the Gaussian integers $a + ib$, where $i = \sqrt{-1}$ and a and b are integers.

At a meeting of the Academy of Sciences on 27 July 1837, Dirichlet presented his first paper on analytic number theory. In this memoir he gives a proof of the fundamental theorem that bears his name: Any arithmetical series of integers

$$an + b, \; n = 0, \, 1, \, 2, \, \cdots,$$

where a and b are relatively prime, must include an infinite number of primes. This result had long been conjectured and Legendre had expended considerable effort upon finding a proof, but it had been established only for a few special cases, such as

$$\{4n + 1\} = 1, 5, 9, 13, 17, 21, \cdots$$
$$\{4n + 3\} = 3, 7, 11, 15, 19, 23, \cdots.$$

The paper on the primes in arithmetic progressions was followed in 1838 and 1839 by a two-part paper on analytic number theory, "Recherches sur diverses applications de l'analyse infinitésimale à la théorie des nombres." Dirichlet begins with a few general observations on the convergence of the series now called Dirichlet series. The main number theory achievement is the determination of the formula for the class number for quadratic forms with various applications. Also from this period are his studies on Gaussian sums.

These studies on quadratic forms with rational coefficients were continued in 1842 in an analogous paper on forms with coefficients that have Gaussian coefficients. It contains an attempt at a systematic theory of algebraic numbers when the prime factorization is unique, although it is restricted to Gaussian integers. It is of interest to note that here one finds the first application of Dirichlet's *Schubfachprinzip* ("box principle"). This deceptively simple argument, which plays an important role in many arguments in modern number theory, may be stated as follows: If one distributes more than n objects in n boxes, then at least one box must contain more than one object.

It is evident from Dirichlet's papers that he searched very intently for a general algebraic number theory valid for fields of arbitrary degree. He was aware of the fact that in such fields there may not be a unique prime factorization, but he did not succeed in creating a substitute for it: the ideal theory later created by Ernst Kummer and Richard Dedekind or the form theory of Leopold Kronecker.

Dirichlet approached the problem through a generalization of the quadratic forms, using the properties of decomposable forms representable as the product of linear forms, a method closely related to the method later used by Kronecker. One part of algebraic number theory, the theory of units, had its beginning in Dirichlet's work. He had earlier written a number of papers on John Pell's equation

$$x^2 - Dy = N,$$

with particular consideration of the cases in which $N = \pm 1$ corresponds to the units in the quadratic field $K(\sqrt{D})$. But in the paper "Zur Theorie der complexen Einheiten," presented to the Berlin Academy on 30 March 1846, he succeeded in establishing the complete result for the Abelian group of units in an algebraic number field: When the field is defined by an irreducible equation with r real roots and s pairs of complex roots, the number of infinite basis elements is $r + s - 1$; the finite basis element is a root of unity.

After these fundamental papers, the importance of Dirichlet's number theory work declined. He published minor papers on the classes of ternary forms, on the representation of integers as the sum of three squares, and on number theory sums, together with simplifications and new proofs for previous results and theories.

In 1863, Dirichlet's *Vorlesungen über Zahlentheorie* was published by his pupil and friend Richard Dedekind. To the later editions of this work Dedekind most appropriately added several supplements containing his own investigations on algebraic number theory. These addenda are considered one of the most important sources for the creation of the theory of ideals, which has now become the core of algebraic number theory.

Parallel with Dirichlet's investigations on number theory was a series of studies on analysis and applied mathematics. His first papers on these topics appeared during his first years in Berlin and were inspired by the works of the French mathematicians whom he had met during his early years in Paris. His first paper on analysis is rather formal, generalizing certain definite integrals introduced by Laplace and Poisson. This paper was followed in the same year (1829) by a celebrated one published in *Crelle's Journal*, as were most of his mathematical papers: "Sur la convergence des séries trigonométriques qui servent à représenter une fonction arbitraire entre deux limites données." The paper was written under the influence of Fourier's theory of heat conduction as presented in his *Théorie analytique de la chaleur*.

Dirichlet and several other mathematicians had been impressed by the properties of the Fourier series on trigonometric series

$$\tfrac{1}{2}a_0 + (a_1 \cos x + b_1 \sin x)$$
$$+ (a_2 \cos 2x + b_2 \sin 2x) + \cdots,$$

particularly by their ability to represent both continuous and discontinuous functions. Such series, although now commonly named for Fourier, had already been used by Daniel Bernoulli and Leonhard Euler to examine the laws of vibrating strings. The convergence of the series had been investigated shortly before Dirichlet in a paper by Cauchy (1823). In the introduction to his own paper Dirichlet is sharply critical of Cauchy on two accounts: first, he considers Cauchy's reasoning invalid on some points;

second, the results do not cover series for which the convergence had previously been established.

Dirichlet proceeds to express the sum of the first *n* terms in the series corresponding formally to the given function $f(x)$ and examines the case in which the difference between $f(x)$ and the integral tends to zero. In this manner he establishes the convergence to $f(x)$ of the corresponding series, provided $f(x)$ is continuous or has a finite number of discontinuities. Dirichlet's method later became classic; it has served as the basis for many later investigations on the convergence or summation of a trigonometric series to its associated function under much more general conditions.

Dirichlet returned to the same topic a few years later in the article "Über die Darstellung ganz willkürlicher Functionen durch Sinus- und Cosinusreihen," published in the *Repertorium der Physik* (1837), a collection of review articles on mathematical physics on which his friend Jacobi collaborated. An outstanding feature of this article is Dirichlet's abandonment of the until then universally accepted idea of a function as an expression formulated in terms of special mathematical symbols or operations. Instead, he introduces generally the modern concept of a function $y = f(x)$ as a correspondence that associates with each real x in an interval some unique value denoted by $f(x)$. His concept of continuity is, however, still intuitive. For his continuous functions he defines integrals by means of sums of equidistant function values and points out that the ordinary integral properties all remain valid. On this basis the theory of Fourier series is then developed. In a related paper, "Solution d'une question relative à la théorie mathématique de la chaleur" (1830), Dirichlet uses his methods to simplify the treatment of a problem by Fourier: the temperature distribution in a thin bar with given temperatures at the endpoints.

Closely related to these investigations is the paper "Sur les séries dont le terme général dépend de deux angles et qui servent à exprimer des fonctions arbitraires entre des limites données" (1837). The Fourier series can be considered as expansions of functions defined on a circle. In this paper Dirichlet examines analogously the convergence of the expansion in spherical harmonics (*Kugelfunctionen*) of functions defined on a sphere. He later applied these results in several papers on problems in theoretical physics.

Dirichlet's contributions to general mechanics began with three papers published in 1839. All three have nearly the some content; the most elaborate has the title "Über eine neue Methode zur Bestimmung vielfacher Integrale." All deal with methods based upon a so-called discontinuity factor for evaluating multiple integrals, and they are applied particularly to the problem of determining the attraction of an ellipsoid upon an arbitrary mass point outside or inside the ellipsoid.

In the brief article "Über die Stabilität des Gleichgewichts" (1846), Dirichlet considers a general problem inspired by Laplace's analysis of the stability of the solar system. He takes the general point of view that the particles attract or repel each other by forces depending only on the distance and acting along their central line; in addition, the relations connecting the coordinates shall not depend on time. Stability is defined as the property that the deviations of the coordinates and velocities from their initial values remain within fixed, small bounds. Dirichlet criticizes as unsatisfactory the previous analyses of the problem, particularly those by Lagrange and Poisson that depended upon infinite series expansions in which terms above the second order were disregarded without sufficient justification. Dirichlet avoids this pitfall by reasoning directly on the properties of the expression for the energy of the system.

One of Dirichlet's most important papers bears the long title "Über einen neuen Ausdruck zur Bestimmung der Dichtigkeit einer unendlich dünnen Kugelschale, wenn der Werth des Potentials derselben in jedem Punkte ihrer Oberfläche gegeben ist" (1850). Here Dirichlet deals with the boundary value problem, now known as Dirichlet's problem, in which one wishes to determine a potential function $V(x,y,z)$ satisfying Laplace's equation

$$\frac{\delta^2 V}{\delta x^2} + \frac{\delta^2 V}{\delta y^2} + \frac{\delta^2 V}{\delta z^2} = 0$$

and having prescribed values on a given surface, in Dirichlet's case a sphere. This type of problem plays an important role in numerous physical and mathematical theories, such as those of potentials, heat, magnetism, and electricity. Mathematically it can be extended to an arbitrary number of dimensions.

Among the later papers on theoretical mechanics one must mention "Über die Bewegung eines festen Körpers in einem incompressibeln flüssigen Medium" (1852), which deals with the motion of a sphere in an incompressible fluid; it is noteworthy for containing the first exact integration for the hydrodynamic equations. This subject occupied Dirichlet during his last years; in his final paper, "Untersuchungen über ein Problem der Hydrodynamik" (1857), he examines a related topic, but this includes only a minor part of his hydrodynamic theories. After his death, his notes on these subjects were edited and published by Dedekind in an extensive memoir.

In 1855, when Gauss died, the University of Göttingen—which had long enjoyed the reflection of his scientific fame—was anxious to seek a successor of great distinction, and the choice fell upon Dirichlet. His position in Berlin had been relatively modest and onerous, and the teaching schedule at the military academy was very heavy and without scientific appeal. Dirichlet wrote to his pupil Kronecker in 1853 that he had little time for correspondence, for he had thirteen lectures a week and many other duties to attend to. Dirichlet responded to the offer from Göttingen that he would accept unless he was relieved of the military instruction in Berlin. The authorities in Berlin seem not to have taken the threat very seriously, and only after it was too late did the Ministry of Education offer to improve his teaching load and salary.

Dirichlet moved to Göttingen in the fall of 1855, bought a house with a garden, and seemed to enjoy the more quiet life of a prominent university in a small city. He had a number of excellent pupils and relished the increased leisure for research. His work in this period was centered on general problems of mechanics. This new life, however, was not to last long. In the summer of 1858 Dirichlet traveled to a meeting in Montreux, Switzerland, to deliver a memorial speech in honor of Gauss. While there, he suffered a heart attack and was barely able to return to his family in Göttingen. During his illness his wife died of a stroke, and Dirichlet himself died the following spring.

BIBLIOGRAPHY

Many of Dirichlet's works are in L. Kronecker and L. Fuchs, eds., *G. Lejeune Dirichlets Werke, herausgegeben auf Veranlassung der Königlichen Preussischen Akademie der Wissenschaften,* 2 vols. (Berlin, 1889–1897). Included are a portrait; a biography by E. Kummer; correspondence with Gauss, Kronecker, and Alexander von Humboldt; and material from Dirichlet's posthumous papers.

Several of Dirichlet's papers have been reissued in the series Ostwalds Klassiker der exacten Wissenschaften: no. 19, *Über die Anziehung homogener Ellipsoide (Über eine neue Methode zur Bestimmung vielfacher Integrale),* which includes papers by other writers (1890); no. 91, *Untersuchungen über verschiedene Anwendungen der Infinitesimalanalysis auf die Zahlentheorie* (1897); and no. 116, *Die Darstellung ganz willkürlicher Functionen durch Sinus- und Cosinusreihen* (1900).

Dirichlet's lectures have been published in G. Arendt, *Vorlesungen über die Lehre von den einfachen und mehrfachen bestimmten Integrale* (1904); R. Dedekind, *Vorlesungen über Zahlentheorie* (1893); and F. Grube, *Vorle-*

sungen über die im umgekehrten Verhältniss des Quadrats der Entfernung wirkenden Kräfte (1876).

OYSTEIN ORE

DITTMAR, WILLIAM (*b.* Umstadt [near Darmstadt], Germany, 15 April 1833; *d.* Glasgow, Scotland, 9 February 1892), *chemistry.*

Dittmar analyzed samples of seawater collected during the voyage of H.M.S. *Challenger* (1872–1876). He originally qualified in pharmacy, but in 1857 he went to work in Bunsen's laboratory at Heidelberg. There he was befriended by Henry Roscoe, who invited him to England; and when Roscoe became professor of chemistry at Owens College, Manchester, Dittmar accompanied him as his assistant. From 1861 to 1869 he was chief assistant in the chemical laboratory at Edinburgh University.

After spending the next three years in Germany, where he lectured in meteorology at an agricultural college near Bonn, Dittmar returned to Edinburgh in 1872. In 1873 he was appointed assistant lecturer at Owens College but the year after became professor of chemistry at Anderson's College, Glasgow, where he remained until his death.

Dittmar's "Report on Researches Into the Composition of Ocean-Water" (1884) was the result of six years' work on the *Challenger* samples. He improved on existing methods of determining the major constituents of marine salt and confirmed Forchhammer's discovery (1865) that although the salinity of seawater varies from place to place, the ratios of the principal constituents to each other remain almost constant. This showed that calculations of salinity, usually made from measurements of specific gravity, could be done as well, if not better, by determining the weight of chlorine in a sample and multiplying it by a constant factor of 1.8058, the ratio of total salt to chlorine. Dittmar also studied the absorption of gases in seawater and established the proportion of bromine to chlorine.

Dittmar's other work included research on the atomic weight of platinum and on the gravimetric composition of water, for which he was awarded the Graham Medal of the Glasgow Philosophical Society. He was elected a fellow of the Royal Society of Edinburgh in 1863 and of the Royal Society in 1882.

BIBLIOGRAPHY

I. ORIGINAL WORKS. Dittmar's "Report on Researches Into the Composition of Ocean-Water" was published in *Report on the Scientific Results of the Voyage of H.M.S. Challenger: Physics and Chemistry,* I (London, 1884). He also wrote a number of scientific papers and textbooks. See Poggendorff, III, 365–366; IV, 333.

II. SECONDARY LITERATURE. Obituaries of Dittmar are in *Nature*, **45**, no. 1169 (24 Mar. 1892), 493–494; and *Proceedings of the Royal Society of Edinburgh*, **20** (1892–1895), vi–vii. His contribution to oceanography is described in J. P. Riley and G. Skirrow, eds., *Chemical Oceanography*, I (London–New York, 1965), 15, 124–125. His other researches are listed in *Proceedings of the Philosophical Society of Glasgow*, **23** (1891–1892), 310.

MARGARET DEACON

DIVINI, EUSTACHIO (*b*. San Severino delle Marche [near Ancona], Italy, 4 October 1610; *d*. Rome, Italy, 1685), *optical instrumentation.*

Divini was among the first to develop technology for the production of scientifically designed optical instruments. He established himself in Rome about 1646 as a maker of clocks and lenses. In 1648 he constructed an innovative compound microscope with cardboard sliding tubes and convex lenses for the objective and the eyepiece; several years later he developed the doublet lens for microscopes. During the same period he experimented with the construction of telescopes of long focus. It was at this time that Giuseppe Campani of Castel San Felice came to Rome and learned the art of lensmaking. His lenses and instruments were competitive with Divini's, and a bitter rivalry between the two artisans developed into a lasting feud that involved Pope Alexander VII.

Divini constructed long telescopes consisting of wooden tubes with four lenses, of a focal length of seventy-two Roman palms (633 inches). He experimented with the elimination of achromatic aberration in his lenses with some success. He had received some scientific training from Benedetto Castelli, one of Galileo's disciples.

In 1649 Divini published a copper engraving of a map of the moon, based upon his own observations of 1647, 1648, and early 1649, which were made with instruments of his own construction. These instruments incorporated a micrometer of a gridiron design of his own invention.

Divini made a number of astronomical observations, utilizing his instruments. He made observations of the rings of Saturn and the spots and satellites of the planet Jupiter. In his observations he became involved in a controversy with Huygens, in the course of which he published several tracts.

Significant examples of Divini's microscopes and telescopes have survived in such important public collections as those of the Istituto e Museo di Storia della Scienza in Florence, the Osservatorio Astronomico e Museo Copernicano in Rome, the Museo

di Fisica in Padua, and the British Museum in London.

BIBLIOGRAPHY

I. ORIGINAL WORKS. Divini published *Eustachii de Divinis Septempedani brevis adnotatio in systema Saturnium Christiani Hugenii ad Serenissimum Principem Leopoldum Magni Ducis Etruriae fratrem una cum Christiani Hugenii responso ad eumdem Principem* (Rome, 1660); *Eustachius de Divinis Septemedanus pro sua adnotatione in systema Saturnium Christiani Hugenii adversus ejusdem assertionem* (Rome, 1661); *Lettera di Eustachio Divini al conte Carlo Antonio Manzini Si ragguaglia di un nuovo lavoro e compimento di lenti, che servono a occhialoni o simplici o composti* (Rome, 1663); *Lettera di Eustachio Divini con altra del P. Egidio Francesco Gottignes intorno alle macchie nuovamente scoperte nel mese di Luglio 1665, con suoi cannochiali nel pianeta di Giove* (Rome, 1665); and *Lettera sulle ombre delle stelle Medicee nel volto di Giove* (Bologna, 1666).

II. SECONDARY LITERATURE. On Divini and his work see also Silvio A. Bedini, "Seventeenth Century Italian Compound Microscopes," in *Physis*, **5** (1963), 383–397; Giovanni-Carlo Gentili, *Elogio storico di Monsignor Angelo Massarelli di Sanseverino* (Macerata, 1837), pp. 60–86; and Carlo Antonio Manzini, *L'occhiale all'occhio, dioptrica practica* (Bologna, 1660).

SILVIO A. BEDINI

DIVIŠ, PROKOP (also **Procopius Divisch** or **Diwisch**) (*b*. Žamberk, Bohemia, 26 March 1698; *d*. Přímětice, near Znojmo, Moravia, 21 December 1765), *electricity.*

Diviš was one of the most eminent Czech scholars of the mid-eighteenth century. He contributed to the history of science mainly by his studies of atmospheric electricity and by his construction of a lightning conductor. From 1716 he attended secondary school at Znojmo, in southern Moravia. In 1719 he joined the Premonstratensian monastery at Louky, where he completed his study of philosophy and was ordained in 1726. He then taught natural science at the monastery school and introduced practical experiments in the classroom. Diviš also taught theology, becoming doctor of theology at Salzburg in 1733 and doctor of philosophy, probably in 1745 at Olmütz (now Olomouc). In 1736 he was appointed parson in the small village of Přímětice, near Znojmo, where he remained, except for a brief interval, until his death.

Scientific research mattered more to Diviš than his parish duties. Unfortunately, he lived at a great distance from his country's centers of learning (the only universities in Bohemia were those at Prague and Olmütz). He had to keep up with scientific developments through literature, personal contacts, and (chiefly) correspondence. Of the latter, however, only a small portion has been found. Thus circumstance made Diviš more or less self-taught.

During his stay at the monastery of Louky, Diviš concerned himself with practical hydraulics. From his manuscripts it appears that he also worked in chemistry and alchemy. He even constructed a musical instrument called the "denisdor," which resembled a complicated harpsichord and had strings that could be electrified.

Diviš' main interest was electricity. We do not know exactly when he began work in this area, but in 1748 he performed various electrical experiments. He is said to have demonstrated the electric effects of the conductive point in Vienna as early as 1750. By means of unnoticeable pointed wires inserted into his wig, he drew off the electricity from a charged solid body. He had also considered drawing off electricity from clouds, but it is not until early in 1753—only after Dalibard's first experiments—that there is any evidence of Diviš' real interest in and ideas about atmospheric electricity. In that year Diviš also wrote a treatise explaining why the St. Petersburg physicist G. W. Richmann had been killed by a flash of lightning, and sent it to Euler. In a letter dated 24 October 1753, which accompanied the treatise, he hints at his intention of finishing that winter a machine for reducing the severity of thunderstorms and of testing it the following summer. On 15 June 1754, Diviš erected his lightning conductor in the rectory garden of Přímětice.

The basic idea of Diviš' conductor was a consistent application of the point-effect analogy (discovered by Franklin in 1747), that metal points possess the property of allowing electricity to flow away from a charged object. This led Diviš to believe that a thundercloud could be deprived of its charge, thus entirely preventing lightning. Diviš' lightning conductor consisted of a horizontal iron cross with three boxes with twenty-seven points attached to the end of each arm. This complicated *machina meteorologica* was placed on a wooden frame about 108 feet high and connected to the ground by four iron chains. Whether the chains were intended only to increase the stability of so high a structure or whether Diviš was actually aware of the necessity of grounding remains uncertain. At any rate, the grounding made the conductor truly effective, in contrast with the insulated experimental rods of that time. In this respect, Diviš' lightning conductor is considered to be the first to afford actual protection from lightning, although, of course, it did not possess the preventive effects attributed to it by its inventor. The lightning conductor at Přímětice remained standing until 1760,

when the villagers, believing it to be the cause of a great drought, broke the chains and a heavy gale did the rest. Diviš later attached a smaller version to the steeple of his church.

Diviš also was interested in the therapeutic applications of electricity, advocating such treatments and carrying them out. He is said to have cured more than fifty persons. He doubtless was a highly skilled experimenter, but the results he obtained were never beyond the level of the period.

Diviš devoted much attention to the theoretical explanation of electric phenomena. In doing so he referred to Genesis, and in many respects he was influenced by Aristotle and the Scholastics. These influences, together with the general distrust of lightning conductors, seem to have been the main reasons why—despite his continual efforts—Diviš was not more appreciated during the Enlightenment. In 1753 he sought membership in the Berlin Academy of Sciences; in 1755 he took part in a competition at the St. Petersburg Academy; in 1760 he solicited a professorial chair at Vienna University—but all in vain. Likewise, he had great difficulties with his only publication, *Magia naturalis,* which was rejected by the Vienna censors. It was not published until the year of his death, with the help of German Protestants.

BIBLIOGRAPHY

I. ORIGINAL WORKS. Diviš' only published work is *Längst verlangte Theorie von der meteorologischen Electricite, welche er selbst Magiam naturalem benahmet* (Tübingen, 1765; 2nd ed., Frankfurt, 1768). MSS are in National Scientific Library, Olomouc, MSS Dept., sign. III.28. nos. 1–12; National Archives, Brno, Cerroni Collection, no. II.135; and Archives of the Soviet Academy of Sciences, Leningrad: correspondence, fol. 136, op. 2, no. 3; fol. 1, op. 2, no. 39; fol. 1, op. 3, no. 40, 45.

II. SECONDARY LITERATURE. Reports of Diviš' scientific activity are in *Tubingische Berichte von gelehrten Sachen* (1755), pp. 66–73; and *Wochentlicher Intelligenz-Zettel* (Brno, 1758), nos. 3, 4, 7–10.

On Diviš or his work, see E. Albert, "Ein österreichischer Elektrotherapeut aus dem vorigen Jahrhundert," in *Wiener medizinische Presse,* no. 12 (1880), 369–372; I. B. Cohen and R. Schofield, "Did Diviš Erect the First European Protective Lightning Rod, and Was His Invention Independent?," in *Isis,* **43** (1952), 358–364; J. Friess, "Prokop Divisch. Ein Beitrag zur Geschichte der Physik," in *Programm der k. k. Staats-Oberrealschule* (Olomouc, 1884); L. Goovaerts, *Écrivains, artistes et savants de l'Ordre Prémontré,* I (Brussels, 1899), 195–197; K. Hujer, "Father Procopius Diviš—The European Franklin," in *Isis,* **43** (1952), 351–357; F. Pelzel, *Abbildungen böhmischer und mährischer*

Gelehrten und Künstler, III (Prague, 1777), 172–184; and Josef Smolka, "Příspěvky k bádání o Prokopu Divišovi" ("Contribution to the Investigation of Prokop Diviš"), in *Sbornik pro dějiny přirodních věd a techniky,* **3** (1957), 122–152; "Divišova korespondence s L. Eulerem a Petrohradskou akademií ved" ("Diviš' Correspondence With L. Euler and the Petersburg Academy of Sciences"), *ibid.,* **8** (1963), 139–162; "B. Franklin, P. Diviš et la découverte du paratonnerre," in *Actes du dixième congrès international d'histoire des sciences* (Paris, 1964), pp. 763–767; and "Prokop Diviš and His Place in the History of Atmospheric Electricity," in *Acta historiae rerum naturalium necnon technicarum* (Prague), spec. iss. **1** (1965), 149–169.

JOSEF SMOLKA

DIXON, HAROLD BAILY (*b.* London, England, 11 August 1852; *d.* Lytham, England, 18 September 1930), *chemistry.*

Dixon was a leading authority on gaseous explosions and created an international interest in combustion research. The son of William Hepworth Dixon, author of popular historical and travel books and editor of *Athenaeum,* Dixon intended to follow a literary career. He did poorly as a classical student at Oxford (1871–1875), however, and transferred to chemistry. He worked with A. V. Harcourt until 1879 and was a lecturer at Trinity and Balliol colleges. In 1886 he assumed the chair of chemistry at Owens College, Manchester, where he founded the Manchester School of Combustion Research. He was elected to the Royal Society in 1886 and was a president of the Chemical Society of London and the Manchester Literary and Philosophical Society.

The only detailed studies of gaseous explosions had been made by Bunsen, and for twenty years his results were accepted as authoritative. Bunsen claimed that his experiments were inconsistent with Berthollet's law of mass action. In his first researches at Oxford, Dixon found Bunsen's conclusions to be erroneous and that the law of mass action applied to gaseous explosions. In 1880 he discovered the incombustibility of purified and dried gases, proving that prolonged drying of gas mixtures rendered them nonexplosive to electric sparks, whereas wet mixtures exploded readily. His publications began the systematic investigation of the effect of moisture on chemical changes.

Dixon's major contribution to combustion research was his detailed study of the rate and propagation of explosions. On the basis of Bunsen's experiments, scientists believed the rates of gaseous explosions were only a few meters per second; but Dixon proved that very great flame speeds were attained and that Bunsen's speeds applied only to the usually short initial phase of explosions. Dixon determined the velocity

and course of explosions, developing his own photographic methods to detect the moving flame. He named the rapid motion of flame "the detonation wave." Another line of combustion research was his measurement of the ignition temperatures of gases and gas mixtures. He was the first to determine these with accuracy.

BIBLIOGRAPHY

Dixon wrote three long papers that serve as detailed monographs on the subjects of flame and explosions: "Conditions of Chemical Change in Gases: Hydrogen, Carbonic Oxide, and Oxygen," in *Philosophical Transactions of the Royal Society,* **175** (1884), 617–684; "The Rate of Explosions in Gases," *ibid.,* **184A** (1893), 97–188; and "On the Movements of the Flame in the Explosion of Gases," *ibid.,* **200A** (1903), 315–352.

Dixon's life and work are chronicled by his students H. B. Baker and W. A. Bone in "Harold Baily Dixon 1852–1930," in *Journal of the Chemical Society* (1931), 3349–3368. The same essay appears in *Proceedings of the Royal Society,* **134A** (1931–1932), i–xvii; and in Alexander Findlay and William Hobson Mills, eds., *British Chemists* (London, 1947), pp. 126–145.

ALBERT B. COSTA

DIXON, HENRY HORATIO (*b.* Dublin, Ireland, 19 May 1869; *d.* Dublin, 20 December 1953), *botany.*

Dixon was the youngest of seven brothers and two sisters; his father, George Dixon, was owner of a soap works; his mother was the former Rebecca Yeates. In 1887 he entered Trinity College, Dublin, with a classical scholarship and prizes in Italian; but in 1891 he graduated with a first-class degree and gold medal in botany, geology, and zoology. He was stimulated by a very active botany department and also by friendship with the physicist John Joly, with whom he collaborated in fruitful research. He then studied in Bonn with Eduard Strasburger until 1893, when he was appointed assistant to E. Perceval Wright, professor of botany in Dublin, and occupied himself reorganizing the botanical gardens and herbarium. Almost all of Dixon's working life was spent in Dublin: he became professor of botany in 1904 and was active in the Royal Dublin Society, receiving the Boyle Medal in 1916 and serving as president from 1944 to 1947. The Royal Society elected him to fellowship in 1908, and he was the Croonian lecturer of 1937. In 1927 he was visiting professor at the University of California. He resigned his chair in 1949 and was elected an honorary fellow of Trinity College, Dublin. He married Dorothea Mary Franks, a medical student, in 1907; they had three sons.

Dixon's early research was suggestive of his potential rather than significant in botany: work on the cytology of chromosomes and first mitosis of spore mother cells of *Lilium* showed that the appearance of bivalents is due to the approach of chromosomes rather than to splitting of some preexisting structure, thus giving the first indication of reduction division.

He started original work soon after graduating and by 1892 had shown how to grow seedlings in sterile culture, foreshadowing later tissue and root culture. He also suggested mutagenic effects to be expected from cosmic radiation. Work on the resistance of seeds to heat, cold, and poisons developed thermoelectric methods of cryoscopy that he was to use again in work on osmotic pressure.

In 1894 Dixon and Joly read their classic paper "On the Ascent of Sap"; they gave an outline of the theory to the Trinity College Experimental Science Association in March and read the full paper to the Royal Society in October. The tension theory of the ascent of sap in trees arose from the combination of Dixon's knowledge of Strasburger's work on transpiration in high trees and Joly's knowledge of work by François Donny and Berthelot on the tensile strength of columns of sulfuric acid and water. This idea was worked out in experiments on transpiration of various trees; and they showed that since leaves can transpire even against high atmospheric pressures, this tension can be maintained. Transmission of internal stress was found to be due to the internal stability of the liquid when mechanically stretched and to the additional stability of minutely subdivided connective tissue; this is unaffected by dissolved gas, and free gas is restricted by the size of the vessel. The tensile stress in the sap is transmitted to the root, where it establishes in the capillaries of the root surface menisci competent to condense water rapidly from the surrounding soil and thus complete the process of transpiration.

Dixon, working alone, continued to perfect the theory in all details. He published a standard account in 1909 and, in 1914, *Transpiration and the Ascent of Sap in Plants,* a comprehensive monograph bringing together theories and experimental work on the subject in a well-argued account. The transpiration stream is raised either by secretory actions in leaf cells, using energy from respiration in the leaves, or by evaporation from the surfaces of leaves, according to the degree of saturation of surrounding cells. The osmotic pressure of turgid mesophyll cells, calculated from cryoscopic measurements of the freezing point of sap, was shown to be enough to raise sap by tension in even the highest tree, through passive vessels, not living cells.

During World War I, Dixon worked on the micro-scopic identification of different kinds of mahogany. Later he returned to problems of transport of organic substances in the phloem, variation in the permeability of leaf cells, and the mechanisms of transpiration. He wrote a textbook on practical plant biology in 1922 and gave three lectures on the transpiration stream at the University of London in 1924. Some of his manuscript notebooks are held at Trinity College, Dublin.

BIBLIOGRAPHY

I. ORIGINAL WORKS. The classic paper "On the Ascent of Sap," written with J. Joly, was recorded in abstract in *Proceedings of the Royal Society,* **57B** (1894), 3–5, and published in full in *Philosophical Transactions of the Royal Society,* **186B** (1895), 563–576. The standard account of the theory was published as "Transpiration and the Ascent of Sap," in *Progressus rei botanicae,* **3** (1909), 1–66, reprinted in abbreviated form in *Report of the Board of Regents of the Smithsonian Institution, 1910* (1911), 407–425. The work on transpiration was brought together in *Transpiration and the Ascent of Sap in Plants* (London, 1914). The other two volumes published by Dixon are *Practical Plant Biology* (London, 1922; 2nd ed., Dublin, 1943); and *The Transpiration Stream* (London, 1924).

II. SECONDARY LITERATURE. Two assessments of Dixon's scientific importance are by W. R. G. Atkins, who worked with him on the cryoscopy of sap, in *Obituary Notices of Fellows of the Royal Society of London,* **9** (1954), 79–93, with portrait and comprehensive bibliography of Dixon's works; and by T. A. Bennet-Clark, in *Nature,* **173** (1954), 239.

DIANA M. SIMPKINS

DIXON, JEREMIAH (*b.* Bishop Auckland, Durham, England, 27 July 1733; *d.* Cockfield, Durham, England, 22 January 1779), *astronomy.*

The fifth child of George Dixon, a well-to-do Quaker, Dixon was educated at John Kipling's school at Barnard Castle, where he displayed interest in mathematics and astronomy. An acquaintance with John Bird, an instrument maker in London and a native of Bishop Auckland, led to Dixon's appointment as assistant to Charles Mason, whom the Royal Society proposed to send to Sumatra to observe the transit of Venus on 6 June 1761. An encounter with a French man-of-war prevented the party from reaching Bencoolen; they observed the transit instead at the Cape of Good Hope, taking so many other measurements that the astronomer royal declared a few years later: "It is probable that the situation of few places is better known."

In 1763 Mason and Dixon were employed to survey the long-disputed boundary between Pennsylvania and Maryland. They were engaged in the delicate and laborious task for nearly five years, hiring local surveyors to assist in observations and calculations and local laborers to cut "vistoes" and set boundary stones. At Dixon's suggestion, and with the approval of the Royal Society, the surveyors calculated the length of a degree of latitude (363,763 feet, or 470 less than the currently accepted figure). The Mason and Dixon survey put a stop at once to quarrels between the two colonies; in political and social significance the line became and remains the most famous boundary in the United States. Dixon was elected a member of the American Society for Promoting Useful Knowledge (later merged with the American Philosophical Society) in 1768.

He was hardly home when the Royal Society sent him—without Mason—to Hammerfest to observe another transit of Venus on 3 June 1769. While there he prepared "A Chart of the Sea Coast and Islands near the North Cape of Europe." Thereafter Dixon lived comfortably at Cockfield, only occasionally resuming his profession. He never married.

BIBLIOGRAPHY

I. Original Works. Reports from Dixon's three expeditions are in *Philosophical Transactions of the Royal Society,* **52** (1762), 378–394; **58** (1768), 274–323; and **59** (1769), 253–261. The journal of Mason and Dixon in America, 1763–1768, has been edited from the original manuscript in the National Archives, Washington, by A. Hughlett Mason, in *Memoirs of the American Philosophical Society,* **76** (1969).

II. Secondary Literature. See H. W. Robinson, "Jeremiah Dixon (1733–1779)—A Biographical Note," in *Proceedings of the American Philosophical Society,* **94** (1950), 272–274; *Report on the Resurvey of the Maryland-Pennsylvania Boundary Part of the Mason and Dixon Line,* Maryland Geological Survey, vol. VII (1908); and Harry Woolf, *The Transits of Venus* (Princeton, 1959), pp. 84–93.

Whitfield J. Bell, Jr.

DOBELL, CECIL CLIFFORD (*b.* Birkenhead, England, 22 February 1886; *d.* London, England, 23 December 1949), *protozoology.*

Dobell is best-known for his meticulous researches on human intestinal protozoa and for a remarkable monograph on the pioneer microscopist Leeuwenhoek. He was the eldest son and the second of five children of William Blount Dobell, a Birkenhead coal merchant, and his wife, Agnes Thornely. Clifford never used his first Christian name. Owing to his father's unstable business fortunes, his early education was erratic and informal. At thirteen Dobell went to Sandringham School, Southport, as an unhappy

boarder. There he headed his class, developed lasting tastes for classical literature and music, and showed talent for art and biological science. Helped by his headmaster, he entered Trinity College, Cambridge, in 1903. He completed one year of medical studies before his tutor, the zoologist Adam Sedgwick, directed his interests toward protozoology. Obtaining his B.A. degree in the natural sciences tripos with first-class honors in 1906, Dobell took the M.A. in 1910 but disdained the Sc.D. until 1942.

He left Cambridge in 1907 to study at Munich under Richard Hertwig and proceeded thence to the zoological station in Naples, where he began investigating the life history and chromosome cycle of the *Aggregata* of cuttlefish and crabs, on which he published an important monograph in 1925. Returning to Cambridge, he was elected a fellow of Trinity College (1908–1914) and won the Rolleston Prize and the Walsingham Medal for original research in biology. He was also awarded the Balfour Studentship, which enabled him to study parasitic protozoa at Colombo, Ceylon. In 1910 Dobell became lecturer in protistology and cytology at the Imperial College of Science and Technology, London. Moving to the Wellcome Bureau of Scientific Research in 1915, he studied amoebic dysentery for the War Office and directed courses in the diagnosis of intestinal protozoal infections. He was elected a fellow of the Royal Society in 1918. The following year Dobell published his classic monograph, *The Amoebae Living in Man,* and was appointed protistologist to the Medical Research Council. He held this position at the National Institute for Medical Research, Hampstead, until death from cerebral hemorrhage in his sixty-fourth year prevented the retirement he dreaded.

Dobell's fastidious tastes and exacting standards limited the scope but not the depth of his interests and accomplishments. Impatient with inaccuracy and scornful of pretentiousness, he was unpopular for his critical tongue, aloof demeanor, and patrician mien; but he was warmhearted within a restricted circle of friends whom he unstintingly admired, such as Adam Sedgwick, D'Arcy Thompson, and David Bruce. In 1937 Dobell married William Bulloch's stepdaughter, Monica Baker. Their happy relationship relieved the lonely, incessant labors of his last twelve years.

Dobell was a self-demanding perfectionist whose tireless industry, stringent observations, and independent conclusions became legendary. He brooked no technical assistants and abhorred teamwork, made personal pets of his experimental *Macaca* monkeys, and carefully infected himself with all manner of intestinal protozoa. These attributes bore special fruit in the sustained researches, begun in 1928 and nearly completed at his death, which elucidated the *in vitro* life histories and the cross-infectivity of practically all known species of human and simian intestinal amoebae. Many of his publications revealed discriminating scholarship in the history of science. His greatest labor of love was the biographical masterpiece *Antony van Leeuwenhoek and His "Little Animals"* (1932). Dobell had to learn seventeenth-century Dutch to decipher and translate his hero's quaint epistles to the Royal Society. This work will probably outlive recollections of his many contributions to experimental biology and medicine.

BIBLIOGRAPHY

I. ORIGINAL WORKS. The most complete bibliography of Dobell's works (121 items, including four posthumous papers) is provided with the obituary by Doris L. Mackinnon and C. A. Hoare (see below). Among his more outstanding publications, many of them lengthy and illustrated with his own beautifully precise drawings, are "The Structure and Life-History of *Copromonas subtilis,* nov. gen. et nov. spec.: A Contribution to Our Knowledge of the Flagellata," in *Quarterly Journal of Microscopical Science,* **52** (1908), 75–120; "Researches on the Intestinal Protozoa of Frogs and Toads," *ibid.,* **53** (1909), 201–277; "On the So-called 'Sexual' Method of Spore Formation in the Disporic Bacteria," *ibid.,* 579–596; "Contributions to the Cytology of the Bacteria," *ibid.,* **56** (1911), 395–506; "A Commentary on the Genetics of the Ciliate Protozoa," in *Journal of Genetics,* **4** (1914), 131–190; "The Chromosome Cycle in Coccidia and Gregarines," in *Proceedings of the Royal Society,* **89B** (1915), 83–94; "On the Three Common Intestinal *Entamoebae* of Man, and Their Differential Diagnosis," in *British Medical Journal* (1917), **1,** 607–612, written with Margaret W. Jepps; "Experiments on the Therapeutics of Amoebic Dysentery," in *Journal of Pharmacology,* **10** (1917), 399–459, written with H. H. Dale; *A Study of 1,300 Convalescent Cases of Dysentery From Home Hospitals: With Special Reference to the Incidence and Treatment of Amoebic Dysentery Carriers,* Special Report Series, Medical Research Committee, no. 15 (London, 1918), 1–28, written with H. S. Gettings, M. W. Jepps, and J. B. Stephens; "A Revision of the Coccidia Parasitic in Man," in *Parasitology,* **11** (1919), 147–197; *The Amoebae Living in Man. A Zoological Monograph* (London, 1919); *The Intestinal Protozoa of Man* (London, 1921), written with F. W. O'Connor; *A Report on the Occurrence of Intestinal Protozoa in the Inhabitants of Britain, With Special Reference to Entamoeba histolytica,* Special Report Series, Medical Research Council, no. 59 (London, 1921), 1–71, with contributions by A. H. Campbell, T. Goodey, R. C. McLean, Muriel M. Nutt, and A. G. Thacker; "The Life History and Chromosome Cycle of *Aggregata eberthi* (Protozoa: Sporozoa: Coccidia)," in *Parasitology,* **17** (1925), 1–136; "On the Cultivation of *Entamoeba histolytica* and Some Other Entozoic Amoebae," *ibid.,* **18** (1926), 283–318,

written with P. P. Laidlaw; and "Researches on the Intestinal Protozoa of Monkeys and Man. I. General Introduction, and II. Description of the Whole Life-History of *Entamoeba histolytica* in Cultures," *ibid.,* **20** (1928), 357–404. This last item was the first in a series of twelve reports on his researches on the intestinal protozoa of monkeys and man that appeared in *Parasitology.* The last to be completed by Dobell himself was "XI. The Cytology and Life-History of *Endolimax nana,*" *ibid.,* **35** (1943), 134–158. R. A. Neal completed and C. A. Hoare edited a final report from Dobell's notes, and they published it posthumously in his name: "XII. Bacterial Factors Influencing the Life History of *Entamoeba histolytica* in Cultures," *ibid.,* **42** (1952), 16–39.

Dobell's historical writings include tributes to Louis Joblot, C. G. Ehrenberg, T. R. Lewis, and Otto Bütschli. Among several obituaries by him the most notable are "Michał Siedlecki (1873–1940). A Founder of Modern Knowledge of the Sporozoa," in *Parasitology,* **33** (1941), 1–7; and "D'Arcy Wentworth Thompson (1860–1948)," in *Obituary Notices of Fellows of the Royal Society,* **6** (1949), 599–617. His biographical magnum opus was *Antony van Leeuwenhoek and His "Little Animals": Being Some Account of the Father of Protozoology & Bacteriology and His Multifarious Discoveries in These Disciplines* (London, 1932; repr., New York, 1958: paperback ed., New York, 1962).

II. SECONDARY LITERATURE. Obituaries include G. H. Ball, "Clifford Dobell, F.R.S.: 1886–1949," in *Science,* **112** (1950), 294; C. A. Hoare, "Clifford Dobell, M.A., Sc.D., F.R.S.," in *British Medical Journal* (1950), **1,** 129–130; C. A. Hoare and D. L. Mackinnon, "Clifford Dobell (1886–1949)," in *Obituary Notices of Fellows of the Royal Society,* **7** (1950), 35–61; Doris L. Mackinnon and Cecil A. Hoare, "Clifford Dobell (1886–1949), In Memoriam," in *Parasitology,* **42** (1952), 1–15; W. H. van Seters, "In Memoriam: Dr. Clifford Dobell, F.R.S. (1886–1949)," in *Nederlandsch Tijdschrift voor Geneeskunde,* **94** (1950), 1274–1276; and H. E. Shortt, "Dr. Clifford Dobell, F.R.S.," in *Nature,* **165** (1950), 219.

Other references to Dobell's work and character are in C. E. Dolman, "Tidbits of Bacteriological History," in *Canadian Journal of Public Health,* **53** (1962), 269–278; and in Paul de Kruif, *A Sweeping Wind* (New York, 1962), pp. 104–106, 111–113.

CLAUDE E. DOLMAN

DÖBEREINER, JOHANN WOLFGANG (*b.* Hof an der Saale, Germany, 13 December 1780; *d.* Jena, Germany, 24 March 1849), *chemistry.*

Döbereiner was the son of Johann Adam Döbereiner, a farm worker who rose to become an estate manager, and Johanna Susanna Göring. He was self-educated, although his mother supervised his early instruction and in 1784 apprenticed him to an apothecary named Lutz. After three years' service with Lutz, Döbereiner made his journeyman's travels through Germany for five years; then he returned to Hof, married Clara Knab, and started a small business manufacturing white lead, sugar of lead, and other pigments and drugs. Almost simultaneously he began publishing articles on these chemicals in the *Neues Berliner Jahrbüch für die Pharmazie,* of which Adolph Ferdinand Gehlen was editor. Döbereiner's business thrived for a few years, but then declined because of personal intrigue and the Napoleonic Wars.

Döbereiner was almost destitute when he was invited to teach at the technical college of the University of Jena, on the recommendation of Gehlen. He became associate professor of chemistry and pharmacy there in August 1810 and in November of that year received the enabling doctorate. Grand Duke Carl August of Saxony-Weimar, the principal patron of the school, may have expected to turn Döbereiner's work to commercial profit, although Goethe, the chief administrator of the Academy and a close friend of Döbereiner, probably had more purely scientific motives in confirming his appointment.

Some of Döbereiner's work at Jena was indeed practical in design. In 1812 he was engaged in the conversion of starch into sugar by Kirchhoff's process, and at a slightly later date he made experiments with illuminating gas (the grand duke had admired gas lighting during a visit to England in 1814). Döbereiner gave up the latter experiments in 1816, however, following an explosion. He also gave a series of lectures on practical chemistry to a group of technicians, and taught special courses for economists and administrators.

Döbereiner made further experiments with spongy platinum, which he prepared by decomposing platinum salts in solution or by exposing them to direct heat. (The grand duke supported him in his work by generous gifts of the precious metal, obtained from his connections in Russia.) He constructed a pneumatic gas lighter (*Platinfeuerzeug*) which consisted of a hydrogenation device that brought hydrogen to impinge on the finely divided platinum; the ensuing oxidation then brought the metal to white heat. In 1828 Döbereiner wrote that about 20,000 of these lighters were in use in Germany and England but, since he had not taken out a patent, they brought him little profit; he added, "I love science more than money."

In addition to his work with the oxides and complex salts of platinum, Döbereiner investigated that form of the metal that Liebig called "platinum black." He studied the role of this material in the process of oxidation of sulfur dioxide and alcohol and proposed its use to manufacture acetic acid from the latter. He decomposed an alloy of raw platinum black and zinc

with dilute acid and found a black powder that contained platinum, palladium, iridium, ruthenium, and osmium; this black powder was even more intensely reactive in air with acids and alcohols than its parent metal—in a dilute acid it easily oxidized its osmium (which could then be sublimated); and it exploded in a shower of sparks when brought into contact with direct heat in air.

Döbereiner was also involved in stoichiometric studies—for which he suggested the use of simple galvanic cells before Faraday—and wrote a book on the subject in 1816. He studied the action of pyrolusite as a catalyst in the production of oxygen from potassium chlorate and developed a method to separate calcium and magnesium by the use of ammonium oxalate or carbonate in the presence of ammonium chloride.

Döbereiner's chief contribution to chemistry, however, was the result of his examination of the weights of the chemical elements—work which aided in the development of Mendeleev's periodic table of all the known elements. Döbereiner's interest in the relationship of elements to each other began as early as 1817; he based his early studies on analogies within certain groups of elements. He found, for example, that the equivalent weight of strontium is almost exactly equal to the mean weight of calcium and barium, and went on to investigate other such triads in alkalies and halogens. (The first members of a group cannot be fitted into such triads; Döbereiner pointed out in 1829 that fluorine and magnesium stand apart.) He further examined such systems of triads in light of other of their qualities—especially specific gravity and affinity—and found, for instance, that the specific gravity (as well as the atomic weight) of selenium is equal to the mean specific gravity (as well as the mean weight) of sulfur and tellurium. He further found that the intensity of chemical affinity decreases in proportion to increased atomic weight of the salt-forming elements in the triads chlorine-bromine-iodine and sulfur-selenium-tellurium but that it increases with atomic weight of the alkali-forming elements in the triads lithium-sodium-potassium and calcium-strontium-barium.

Döbereiner was less successful in formulating rules for the oxides of what he called "heavy metal alum-forming substances." He hoped to codify these by "a rigorous experimental revision of the specific gravities and atomic weights," but found the principle of grouping into triads doubtful for iron-manganese-cobalt and nickel-copper-zinc (although lead did seem to represent the proper mean for silver and mercury).

Because of his involvement in practical problems (among others, he was concerned with the fermentation of alcohol and developed methods for improving wine, although he did not publish them in any detail) and because of his heavy and diversified teaching schedule, Döbereiner neglected further development of his work on triads. His merit was civilly rewarded, however, when he was made privy councillor and awarded the Cross of the White Falcon.

BIBLIOGRAPHY

I. ORIGINAL WORKS. Döbereiner's writings include *Lehrbuch der allgemeinen Chemie* (Jena, 1811); *Zur Gährungschemie und Anleitung zur Darstellung verschiedener Arten künstlicher Weine, Biere, usw.* (Jena, 1822, 1844); *Chemie für das praktische Leben* (Jena, 1824–1825); "Vermischte chemische Erfahrungen über Platin, Gährungschemie, usw. Ein Schreiben an die Herren Kastner und Schweigger," in *Journal für Chemie und Physik,* **54** (1828), 412–426; "Gruppierung der Elemente," in *Annalen der Physik,* **15** (1829), 301–307; and *Zur Chemie des Platins in wissenschaftlicher und technischer Beziehung. . .* (Stuttgart, 1836).

II. SECONDARY LITERATURE. Wilhelm Prandtl, *Deutsche Chemiker in der ersten Hälfte des neunzehnten Jahrhunderts* (Weinheim, 1956), pp. 37–78. Döbereiner's publication on the triads was reprinted as no. 66 in Ostwald's Klassiker and was translated in H. M. Leicester and H. S. Klickstein, *Source Book in Chemistry 1400–1900* (New York, 1952).

EDUARD FARBER

DODART, DENIS (*b.* Paris, France, 1634; *d.* Paris, 5 November 1707), *botany, physiology.*

Dodart was the son of Jean Dodart, a notary public who loved belles lettres, and of Marie Dubois, a lawyer's daughter. The family was upper middle class, and through his parents' efforts he received a particularly broad and thorough education. After studying medicine he graduated *docteur régent* from the Faculty of Medicine in Paris on 16 October 1660. Even as a student Dodart so distinguished himself by his learning, his eloquence, and his agreeable nature that Gui Patin, the dean of the Faculty of Medicine and a habitually very severe critic, called him, in a letter to a friend, "one of the wisest and most learned men of this century."

In 1666 Dodart became a professor at the School of Pharmacy in Paris. A well-known practitioner, he was first in the service of the duchess of Longueville and then was physician to the house of Conti. A pious man, Dodart spent much of his time helping the poor, both medically and financially. He was sympathetic to Jansenism and became a close friend of Jean Hamon, a monk-doctor, whom he attended at death and whom he succeeded as doctor at Port-Royal.

Dodart had the title, but did not perform the functions, of physician-adviser to the king. Louis XIV disliked Dodart's connections with Port-Royal, but he yielded to Colbert's appeal that he make Dodart a member of the Academy and to that of Mme. de Maintenon that he give Dodart a place at the court (1698) and named him physician at St. Cyr.

Most of Dodart's scientific activity took place within the framework of the Academy of Sciences, of which he became a member (botanist) in 1673. The Perrault brothers regarded Dodart as a man capable of directing the ambitious *Histoire des plantes,* a project which the Academy had proposed as one of the goals of its research since its founding. In 1676 Dodart published *Mémoires pour servir à l'Histoire des plantes,* a preliminary study and an announcement of a large collective work that never appeared. The *Mémoires* contains a methodological introduction and a model showing how to conduct botanical research. Because of Dodart's recommendation of phytochemical analysis, this work marks a new step in botany.

When the Academy of Sciences was reorganized, Dodart was among the first group of *titulaires* named directly by Louis XIV; on 28 January 1699 he was given the title of pensioner-botanist. He justified this nomination by the publication, in 1700, of studies of the influence of gravitation on development of roots and stems and on the fertilization and reproduction of plants. Dodart was an advocate of the theory of encasement or *emboîtement* of seeds, and he strove tirelessly to apply in botany the embryological ideas of N. Andry and other preformationists. He further described several new species of plants; and Tournefort named the genus *Dodartia* for him.

Botany was not the only field in which Dodart excelled. In 1678 he presented to the Academy an important memoir by La Salle "on certain details of the natural history of North America, particularly of the Iroquois territory." He published a good description of ergotism (1676) and several anatomical-pathological and embryological observations; three memoirs on phonation appeared between 1700 and 1707.

Dodart was the first since Aristotle and Galen to present new ideas on the mechanism of phonation. He nursed the idea of writing a history of medicine; but when Daniel Leclerc published one first, he abandoned this project for a history of music. Studies on the human voice and on the nature of tones were to serve as an introduction to this history. Credit must be given to Dodart for pointing out the fundamental role of the vocal cords in phonation. In opposition to the classic theory, which considered the larynx a type of flute, Dodart stated that "the glottis alone makes the voice and all the tones . . .; no wind instrument can explain its functioning . . .; the entire effect of the glottis on the tones depends on the tension of its lips and its various openings."

Finally, Dodart was, with Perrault, one of the few French physicians of the seventeenth century to understand and properly appreciate the experiments and the theories of the Italian iatromechanics. He performed on himself the "static" (i.e., performed with a balance) experiments of S. Santorio, measuring the changes in weight of his body and in particular the quantity of imperceptible perspiration. He demonstrated that perspiration gradually decreases as one grows older and also noted, for example, how much the strictest Lenten fast may decrease weight and how much time is needed to regain this lost weight. As an advocate of mechanical explanations of vital phenomena, Dodart saw in them a proof of the existence of God. His religious convictions found noble expression in his will (an autograph of which is in the archives of the Academy of Sciences in Paris).

Dodart had one son, Claude-Jean-Baptiste (1664–1730), who was chief physician to Louis XV, and one grandson, Denis le Jeune, who was *maître des requêtes* and *intendant*.

BIBLIOGRAPHY

I. ORIGINAL WORKS. Dodart's main work is *Mémoires pour servir à l'Histoire des plantes* (Paris, 1676), which was divided into "Projet de l'Histoire des plantes" and "Description de quelques plantes nouvelles," the latter illustrated with magnificent plates by Abraham Bosse, Nicolas Robert, and L. de Chatillon. A second edition of the "Projet" appeared in 1679.

Dodart published several articles in the *Mémoires de l'Académie royale des sciences:* "Lettre sur le seigle ergoté" (1676), 562; "Sur l'affectation de la perpendiculaire, remarquable dans toutes les tiges" (1700), 47; "Mémoire sur les causes de la voix de l'homme, et de ses différens tons" (1700), 238–287; "Sur la multiplication des corps vivants considérée dans la fécondité des plantes" (1700), 136–160; "Second mémoire sur la fécondité des plantes" (1701), 241–257; and "Supplément au mémoire sur la voix et sur les tons" (1706), 136–148, 388–410; (1707), 66–81.

His physiological experiments on nutrition and on imperceptible transpiration were published posthumously by P. Noguez as *Medicina statica Gallica* (Paris, 1725).

II. SECONDARY LITERATURE. The basic biographical source remains B. de Fontenelle, "Éloge de M. Dodart," in *Histoire de l'Académie royale des sciences pour l'année 1707* (Paris, 1708), pp. 182–192. The text of his will was published in *Chronique médicale,* 5 (1898), 742–750. Remarks on his character are in letters by Gui Patin, *Lettres,* J. H. Reveillé-Parise, ed. (Paris, 1846), III, 231, 277,

293; and in the *éloge* of Dodart's son written by Saint-Simon. Biographical notices include the following (listed chronologically): N. F. J. Eloy, "Dodart," in *Dictionnaire historique de la médecine,* II (Mons, 1778), 64–65; J. A. Hazon, *Notice des hommes les plus célèbres de la Faculté de médecine de l'Université de Paris* (Paris, 1778), pp. 135–138; G. Nomdedeu, *Les médecins de Port-Royal* (Bordeaux, 1931); and Roman d'Amat, "Dodart," in *Dictionnaire de biographie française,* XI (Paris, 1967), cols. 417–418. His work in botany is discussed in Y. Laissus and A. M. Monseigny, "Les plantes du roi," in *Revue d'histoire des sciences,* **22** (1969), 193–236.

M. D. GRMEK

DODGSON, CHARLES LUTWIDGE (*b.* Daresbury, Cheshire, England, 27 January 1832; *d.* Guildford, Surrey, England, 14 January 1898), *mathematics, logic.*

Dodgson was the thirdborn of the eleven offspring of Charles Dodgson, a clergyman, and his wife and cousin, the former Frances Jane Lutwidge. All the children stuttered, and Charles Lutwidge himself is said to have spoken without impediment only to the countless nymphets whom, over decades of adulthood, he befriended, wrote wonderful letters to, entertained, and photographed (often nude) with considerable artistry. The obvious inference from this attraction to young girls seems invalid, for he was strongly undersexed. (Even in the Victorian milieu his puritanism was barely credible: for instance, he nursed a project to bowdlerize Bowdler's Shakespeare, and he demanded assurance from one of his illustrators that none of the work would be done on Sundays.) He was never wholly at ease in the company of grown-ups. Friendship with the three small daughters of Dean Liddell resulted in the celebrated *Alice* books, published under a pseudonym that he had first used in 1856 as a writer of light verse—Lewis Carroll. *Alice* brought him fame, money, and the posthumous honor of becoming the most-quoted litterateur in English discursive scientific writing of the twentieth century.

In our concern here with Dodgson's professional achievements, we must bear in mind that his vocational mathematics and his avocational nonsense commingle in a vein of logic that was his salient characteristic as a thinker. For years it was fashionable to point to the gap between mathematics and ingenious nonsense (and the other nineteenth-century master of nonsense, Edward Lear, was most unmathematical); but today we are aware that, at least in some places, the gap is not that wide. The modern view, that Dodgson was all of a piece, is simpler to sustain. His analytical mind is reflected everywhere in his writings, whose quaintness by no means damages clarity. The pity is that his talents were inhibited by ignorance and introversion, for he made no attempt to keep abreast of contemporary advances in mathematics and logic or to discuss his ideas with other academics.

Dodgson's pedestrian career unfolded without hitch. Graduated from Oxford in 1854, he became master of arts there three years later. Meanwhile, in 1855, he had been appointed lecturer in mathematics at his alma mater, Christ Church College, Oxford. In 1861 he was ordained in the Church of England, although he was never to perform any ecclesiastic duties. As a young man he made a trip to Russia, but later journeyings were restricted to London and quiet seaside vacations. Marriage was unthought of, and Dodgson entered into no close friendships. (Perhaps his acquaintanceship with Ellen Terry, the great actress, most nearly qualified for "close friendship.") He took some part in the administration of his college and was proud of his finicky management of its wine cellars. A part-time inventor of trivia, he devised several aids to writing in the dark—to assuage his chronic insomnia and to help dispel the nameless "unholy thoughts" that occasionally pestered him. But generally speaking his placidity was so well rooted that he was able to make the extraordinary statement, "My life is free from all trial and trouble."

As a lecturer Dodgson was drear; and when he gave up the chore, he noted ruefully that his first lecture had been attended by nine students and his last (twenty-five years later) by two. Away from the classroom he wrote assiduously; and his publications, in book form or pamphlet (a favorite medium), are respectably numerous. His scholarly output falls into four main groups; determinants, geometry, the mathematics of tournaments and elections, and recreational logic. He was modest enough to describe his activities as being "chiefly in the lower branches of Mathematics."

Dodgson's work on determinants opened with a paper in the *Proceedings of the Royal Society* for 1866, and this was expanded into a book that appeared the following year. *An Elementary Treatise on Determinants* is good exposition, but favorable reception was prevented by the author's extensive use of ad hoc terms and symbols.

Dodgson's writings on geometry became well known; it was a subject about which he was almost passionate. His initial contribution was *A Syllabus of Plane Algebraic Geometry* (1860), a textbook whose purpose was to develop analytic geometry along rigorous Euclidean lines. He also published pamphlets on this and related themes, in one of which he introduced an original but not particularly meritorious notation for the trigonometric ratios. His most inter-

esting effort in this genre was a five-act comedy entitled *Euclid and His Modern Rivals,* about a mathematics lecturer, Minos, in whose dreams Euclid debates his original *Elements* with such modernizers as Legendre and J. M. Wilson and, naturally, routs the opposition. The book is an attack on the changing method of teaching classical geometry and not, as is sometimes assumed, on non-Euclidean geometry. Indeed, Dodgson showed himself keenly aware of the infirmity of the fifth postulate (on parallels), and he has his oneiric Euclid admit that "some mysterious flaw lies at the root of the subject." The interesting point here is that Riemann's revolutionary geometry was well established during Dodgson's lifetime, and an English translation of the key paper was available. This is yet another instance of Dodgson's being out of touch with the mathematical research of his day. The Euclid drama (which is most engagingly written) apparently was used as ancillary reading in English schools for a number of years.

Least known but quite praiseworthy is Dodgson's work on tournaments and voting theory. His interest stemmed from two sources: the organization of tennis tournaments and the mechanism of arriving at fair decisions by administrative committees. He decided that both matters needed rethinking. As usual, he did not bother to check the literature and so was unaware that the topic had come in for learned discussion in France before and during the Revolution. However, Dodgson unwittingly improved on existing ideas. His initial publication (a pamphlet, in 1873) reviews different methods of arriving at a fair majority opinion, and he sensibly advocates the use of degrees of preference in voting schedules. His whole approach is fresh and thoughtful, and he was the first to use matrix notation in the handling of multiple decisions.

In contrast with his mathematics, Dodgson's work on logic was written entirely under his pseudonym, which clearly testifies to his view that the subject was essentially recreational. Traditional formal logic had long been a barren and overrated discipline; but during his lifetime a renaissance in technique and significance was taking place, and most of the pioneers were his countrymen. Although he was not ignorant of the new trends, their importance either escaped him or was discountenanced. Dodgson was attracted by the contemporary interest in the diagrammatization of the logic of classes, and he had read and appreciated Venn's seminal contributions. In fact, he modified Venn diagrams by making their boundaries linear and by introducing colored counters that could be moved around to signify class contents—a very simple and effective device. On these foundations Dodgson published a game of logic that featured various forms (some very amusing) of the syllogism. His casual realization of the connections between symbolic logic and mathematics might have become vivid and fruitful had he been properly acquainted with what had already been done in the area. But he did not do the necessary reading—there is, for instance, no indication that he had read Boole's *Laws of Thought,* although he owned a copy! Finally, Dodgson was a prolific composer of innocent-looking problems in logic and paradox, some of which were to engage the attention of professional logicians until well into the twentieth century.

BIBLIOGRAPHY

I. ORIGINAL WORKS. The authoritative conspectus of Dodgson's writings, which included sixteen books (six for children) and hundreds of other items, is S. H. Williams and F. Madan, *A Handbook of the Literature of the Rev. C. L. Dodgson (Lewis Carroll)* (London, 1931; supp., 1935). Two outstanding books are *An Elementary Treatise on Determinants* (London, 1867) and *Euclid and His Modern Rivals* (London, 1879). His initial publication on election theory, *A Discussion of the Various Procedures in Conducting Elections* (Oxford, 1873), is a rare pamphlet, only one copy being known; it is at Princeton. The Morris L. Parrish Collection of Victorian Novelists, in Princeton University library, contains the biggest mass of Dodgsoniana, much of it MS. Warren Weaver, in *Proceedings of the American Philosophical Society,* **98** (1954), 377–381, tells the history of this collection and gives some examples of its mathematical items. Dodgson's two books on recreational mathematics are now available in a 1-vol. paperback: *Pillow Problems and a Tangled Tale* (New York, 1958). Similarly, his two books on logic are bound together in the paperback *Symbolic Logic, and the Game of Logic* (New York, 1958).

II. SECONDARY LITERATURE. Dodgson's nephew, S. D. Collingwood, published the first biography, in the same year as his subject's death: *The Life and Letters of Lewis Carroll* (London, 1898). It remains a primary source book. Among many subsequent biographies and evaluations, Florence Becker Lennon's *Victoria Through the Looking Glass* (New York, 1945), esp. ch. 15, is notable for its perceptive treatment of Dodgson's serious side. R. L. Green, *The Diaries of Lewis Carroll* (London, 1953), is important, although many of the entries on logic and mathematics have been excised or glossed over. Two papers prepared for the centenary of Dodgson's birth are essential reading: R. B. Braithwaite, "Lewis Carroll as Logician," in *Mathematical Gazette,* **16** (1932), 174–178; and D. B. Eperson, "Lewis Carroll—Mathematician," *ibid.,* **17** (1933), 92–100. His work on tournaments and elections is examined in Duncan Black's *The Theory of Committees and Elections* (Cambridge, 1958). Martin Gardner's *New Mathematical Diversions* (New York, 1966), ch. 4, deals with Dodgson's work on games and puzzles. The same author's earlier books, *The Annotated Alice* (New York, 1960) and *The*

Annotated Snark (New York, 1962), provide remarkable insights into the logico-mathematical undercurrents in Dodgson's fantasia.

NORMAN T. GRIDGEMAN

DODOENS (DODONAEUS), REMBERT (*b.* Mechelen, Netherlands [now Malines, Belgium], 29 June 1516; *d.* Leiden, Netherlands, 10 March 1585), *medicine, botany.*

Dodoens' real name was Rembert van Joenckema. He changed it to Dodoens ("son of Dodo"), Dodo being a form of the first name of his father, Denis Van Joenckema, who came from Friesland to Mechelen at the end of the fifteenth century. (The name was latinized into Dodonaeus, from which the French, who were ignorant of its origins, further transformed it into Dodonée.) The year of Dodoens' birth is generally considered to be 1517, although according to Hunger (1923) 1516 is correct. He was married twice (his wives were Catelyne 'sBruijnen and Marie Saerine) and had five children.

Dodoens studied at the municipal college of Mechelen and went from there to the University of Louvain, where he studied medicine under Arnold Noot, Leonard Willemaer, Jean Heems, and Paul Roels. He graduated as licenciate in medicine in 1535. According to the custom of the time, Dodoens then traveled extensively. Between 1535 and 1546 he was in Italy, Germany, and France, where he visited—among others—Gunther of Andernach in Paris. After these *Wanderjahren* he returned to Mechelen.

In 1548 Dodoens published a book on cosmography. In the same year he became one of the three municipal physicians of Mechelen, the other two being Joachim Roelandts (to whom Vesalius wrote his famous letter on the chinaroot) and Jacob De Moor. During this time Dodoens composed a treatise on physiology (published later) and began his botanical works.

In the beginning of the sixteenth century, it was still believed that no plants existed other than those described by Dioscorides in his *Materia medica* of the first century A.D. The great progress of natural sciences in the sixteenth century was helped by the discovery of printing and by the use of wood-block illustrations. In 1530 the *Herbarum* of Otto Brunfels appeared, followed by those of the men Sprengel called (in addition to Brunfels) the "German fathers of botany," Jerome Bock (1539) and Leonhard Fuchs (1542). Dodoens was their follower.

Dodoens' first botanical work—a short treatise on cereals, vegetables, and fodders, *De frugum historia* (1552)—was followed by an extensive herbarium (1554). In 1553 he published a collection of wood-plates with a vocabulary for the use of medical students. Some of the wood-block illustrations were taken from Fuchs's earlier herbarium. In 1554, Dodoens also published, as *Cruydeboek,* a Dutch version of his *De stirpium historia.* This was a national herbarium devoted to species indigenous to the Flemish provinces. The merit of this book was that rather than proceeding by alphabetical order, as Fuchs had done, Dodoens grouped the plants according to their properties and their reciprocal affinities. These earlier works show a tendency toward medical botany; in a later period Dodoens became more inclined toward a more scientific treatment.

In 1557 negotiations were begun in an effort to persuade Dodoens to accept a chair of medicine in Louvain; they were not successful. In 1574 he left Mechelen for Vienna, where he had been appointed physician to the emperor Maximilian II. He remained there in the same capacity in service of Maximilian's successor, Rudolph II. In Vienna, Dodoens met Charles de l'Écluse (Clusius).

In 1580, wishing to return to his native country, Dodoens left Vienna, but the uncertain conditions prevalent in the Low Countries at that time persuaded him to stop for a year in Cologne. There he published in one volume a dissertation on wine and medical observations (1580; these two texts were later reprinted separately) and synoptic tables on physiology (1581). From Cologne he went to Antwerp, where in 1582 he supervised his friend Plantin's printing of his *Stirpium historiae pemptades sex sive libri XXX* (published in full in 1583 and reprinted posthumously in 1616). In this elaborate treatise, Dodoens' most important scientific work, he divided plants into twenty-six groups and introduced many new families, adding a wealth of illustration either original or borrowed from Dioscorides, de l'Écluse, or De Lobel.

In 1582 Dodoens accepted an offer from Leiden University, to which the curators had invited him for the purpose of enhancing the reputation of the Faculty of Medicine by his appointment. He was offered the high salary of 400 florins, with an additional gift of 100 florins and a travel allowance of 50 florins. The conditions of his appointment required his promise not to take part in religious controversies and to limit his activities to scientific questions.

At Leiden, Dodoens was in charge of the lectures on pathology and general therapeutics; there was then no course on botany at Leiden, and the famous botanical garden was not created until 1587. Dodoens remained in Leiden until his death and was buried there in St. Peter's church. Although he was a renowned physician in his time (first as a Galenist and later as a Hippocratian, as shown by his posthumous

Praxis medica), his great fame remains based on his botanical work.

BIBLIOGRAPHY

I. ORIGINAL WORKS. Dodoens' writings are *Cosmographica in astronomiam et geographiam Isagoge* (Antwerp, 1548), of which a second edition is *De sphaera sive de astronomiae et geographiae principiis cosmographica isagoge* (Antwerp–Leiden, 1584); *De frugum historia, liber unus. Ejusdem epistolae duae, una de Fare, Chondro, Trago, Ptisana, Crimno et Alica; altera de Zytho et Cerevisia* (Antwerp, 1552); *Trium priorum de stirpium historia commentariorum imagines ad vivum expressae. Una cum indicibus graeca, latina, officinarum, germanica, brabantica, gallicaque nomina complectentibus* (Antwerp, 1553); *Posteriorum trium . . . de stirpium historia commentariorum imagines ad vivum artificiosissime expressa; una cum marginalibus annotationibus. Item ejusdem annotationes in aliquot prioris tomi imagines, qui trium priorum librorum figuras complectitur* (Antwerp, 1554), which was repr. in two vols. as *Commentariorum de stirpium historia imaginum . . . et stirpium herbarumque complures imagines novae . . .* (Antwerp, 1559); *Cruydeboeck in den welcken de gheheele historie, dat es't gheslacht, 't fatsoen, naem, nature, cracht ende werckinghe, van den cruyden, niet alleen hier te lande wassende, maer oock van den anderen vremden in der medecynen oorboorlyck met grooter neersticheyt begrepen ende verclaert es, med derzelver cruyden natuerlick naer datleren conterfeytsel daer by gestelt* (Antwerp, 1554, 1563), French trans. by Charles de l'Écluse as *Histoire des plantes* (Antwerp, 1557), and English trans. by Henry Lyte (1578, 1586, 1595, 1600, 1619); *Historia frumentorum, leguminum, palustrium et aquatilium herbarum, ac eorum quae pertinent. . . . Additae sunt imagines vivae, exactissimae, jam recens non absque haud vulgari diligentia et fide artificiosissime expressae . . .* (Antwerp, 1565, 1566, 1569); *Florum et coronariorum odoratarumque nonnullarum herbarum historia* (Antwerp, 1568, 1569); *Purgantium aliarumque eo facientium, tam et radicum, convolvulorum ac deletariarum herbarum historiae libri IIII. . . . Accessit appendix variarum et quidem rarissimarum nonnullarum stirpium, ac florum quorumdam peregrinorum elegantissimorumque icones omnino novas nec antea editas, singulorumque breves descriptiones continens . . .* (Antwerp, 1574, 1576); *Historia vitis vinique et stirpium nonnullarum aliarum: item medicinalium observationum exempla rara* (Cologne, 1580, 1583, 1585, 1621); *Physiologiae medicinae partis tabulae expeditae* (Cologne, 1581; Antwerp, 1581, 1585); *Remberti Dodonoei medici caesarei medicinalium observationum exempla rara. Accessere et alia quaedam quorum elenchum pagina post praefationem exhibet: Antonii Beniveni Florentini medici ac philosophi . . . exempla rara ex libris de curandis hominum morbis Valesci Tharantani et Alexandri Benedicti. Historia gestationis foetus mortui in utero, Mathiae Cornacis, Egidii Hertoghii et Achillis Pirminii Gassari* (Cologne, 1581; Harderwijk, 1584; Antwerp–Leiden, 1585; Antwerp, 1586), and rev. and enlarged ed. of text as second part of *Historia vitis vinique et stirpium* . . . (Harderwijk, 1621); *Stirpium historiae pemptades sex sive libri XXX* (Antwerp, 1583, 1616), Dutch trans. as *Herbarius, seu Cruydeboeck van Rembertus Dodonoeus* . . . by Françoys van Ravelingen (Leiden, 1608, 1618; Antwerp, 1644); *Praxis medica* (Amsterdam, 1616), further ed. and annotated by Nicolai Fontani (Amsterdam, 1640), and Dutch trans. with notes by S. Egbertz and Wassenaar (1624); "Remberti Dodonoei consilium medicinale in melancholia per essentiam," in *Opera Laur. Scholzii* (Basel, 1546); and "Remberti Dodonoei ad Balduinum Ronssaeum Epistola de Zytho, Cormi et Cerevisia," in *Balduini Ronssaei medici celeberrimi opuscula medica* (Leiden, 1590).

II. SECONDARY LITERATURE. On Dodoens and his work, see F. W. T. Hunger, "Dodonée comme botaniste," in *Janus,* **22** (1917), 151–162; "Over het geboortejaar van Rembertus Dodonaeus," in *Bijdragen tot de geschiedenis der geneeskunde,* **3** (1923), 116–121; "Een tot dusver onuigegeven brief van Rembertus Dodonaeus," *ibid.,* 306–307; P. C. Molhuysen, *Bronnen tot de geschiedenis der Leidische universiteit, 1e part. 1574–1610* (The Hague, 1913); M. A. Van Andel, "Rembertus Dodoens and his Influence on Flemish and Dutch Folk-Medicine," in *Janus,* **22** (1917), 163; E. C. Van Leersum, "Rembert Dodoens (29 juin 1517–10 mars 1585)," *ibid.,* 141–152; P. J. Van Meerbeeck, *Recherches historiques et critiques sur la vie et les ouvrages de Rembert Dodoens (Dodonaeus)* (Mechelen, 1841); and E. Varenbergh, in *Biographie nationale publiée par l'Académie Royale des Sciences, des Lettres et des Beaux-Arts de Belgique,* IV (Brussels, 1878), cols. 85–112.

MARCEL FLORKIN

DOELTER (CISTERICH Y DE LA TORRE), CORNELIO AUGUST SEVERINUS (*b.* Arroyo, Guayama, Puerto Rico, 16 September 1850; *d.* Kolbnitz, Carinthia, Austria, 8 August 1930), *chemical mineralogy.*

Doelter's father, Carl August (*b.* Emmendingen, Baden, 1818), emigrated to Puerto Rico and there married Francisca de Cisterich y de la Torre, whose plantations he managed before becoming a partner of Aldecoa and Company. With his mother, Doelter moved to Karlsruhe, Baden, in 1855 and entered its lyceum in 1860. In 1865 his father brought him to Paris, where he attended the Lycée St. Louis, transferring in 1866 to the Lycée Bonaparte; in 1869 he received the *bachelier* degree and entered the École Centrale des Arts et Manufactures. In the fall of 1870 Doelter enrolled at the University of Freiburg im Breisgau and transferred in the spring of 1871 to Heidelberg. He studied chemistry, physics, mineralogy, and geology, and his most important teachers were Bunsen, J. F. C. Klein, and E. W. Benecke. Following his receipt of the Ph.D. in 1872 he studied

under Suess, F. von Hochstetter, and Carl Hauer in Vienna. In 1873 Doelter became a laboratory assistant at the Imperial Geological Survey, and in 1875 he qualified as a lecturer at the University of Vienna. On 1 May 1876 he was appointed assistant professor of mineralogy and petrography at the University of Graz; he was named full professor in 1883. In 1907 Doelter succeeded Tschermak at the University of Vienna, and in 1911 he became imperial *Hofrat.* He retired in 1921 and continued to live in Vienna, concerning himself chiefly with the preparation of his *Handbuch der Mineralchemie.*

On 24 August 1876 Doelter married Eleonore Fötterle, who bore him one son and one daughter. The marriage was dissolved in 1915, and on 4 October 1919 he married Maria Theresia Schilgerius. His second wife contributed significantly to the maintenance of his ability to work.

Leitmeier described Doelter thus: "Doelter was a fiery spirit. His ability to make rapid connections was astonishing. Much was not matured, and could scarcely have been so considering the extent of the undertakings. He preferred to direct a staff of assistants, giving only suggestions himself."[1] Grengg elaborated:

> In lectures or in reports, Doelter stood dispassionately above the material, soberly joining one fact to another, illuminating here and there the possibilities of another view, and frequently even leaving to the audience to select for itself what it found most acceptable. Nevertheless, one always had the impression that a man of considerable understanding and wide views was speaking.[2]

In Graz, Doelter was for a long time editor, and in 1892 also president, of the Naturwissenschaftlicher Verein of Styria; in addition, he was a curator of the Landesmuseum Johanneum. The Akademie der Wissenschaften in Vienna named him a corresponding member on 4 May 1902 and a full member on 2 June 1928. His many students dedicated a *Festschrift* to him in 1920.

The question of the origin of dolomite had brought Doelter to Vienna in 1872. He worked in the eruptive regions of Hungary and South Tyrol for the Imperial Geological Survey. He considered microscopic and chemical investigation of igneous rock and the exact observation of geological occurrence to be the foundation of all petrology. Doelter conducted a thorough survey of the Pontine Islands (1874), of the volcanism of Monte Ferru on Sardinia, and of the volcanic areas of the Cape Verde Islands, which he visited in 1880–1881. On the trip to the Cape Verde Islands he also went to Portuguese Guinea, where he convincingly explained laterite formation and also made ethnographic studies. He repeatedly explored the Monzoni region and Predazzo in South Tyrol and worked in the crystalline areas of Styria.

Doelter insisted on the confirmation of all suppositions through experiment. He first devoted himself to synthesis and prepared nepheline and pyroxenes (1884), a sulfide and a sulfosalt (1886), micas (1888), and a zeolite (1890). In this work he improved Charles Friedel's pressure vessels and used liquid and solid carbon dioxide to obtain higher pressure. Experiments on remelting and recrystallization carried out after 1883 had shown that minerals other than the original ones could be separated out from the fused mass. For example, Doelter obtained an augite andesite from a fused mass of eclogite. He likewise investigated the remelting of rocks and the influence of mineralizers on the occurrence of rock-forming minerals. His *Allgemeine chemische Mineralogie* (Leipzig, 1890) presented the knowledge he had gained of these matters.

In 1890 Doelter studied the absorption of water by dehydrated zeolites and the solubility of silicates in water. At the same time he demonstrated that in electrolysis, fused basalt concentrates iron at the cathode and that, therefore, fused silicates behave like electrolytes and not like alloys. Since the application of the physical-chemical laws developed for aqueous solutions to silicate fusions was hampered by the lack of reliable constants, Doelter began in 1899 to determine the melting points of important minerals and, in 1901, those of mixtures. In 1901 he ascertained the volume increase of liquid and solidified rock fusions relative to the solid parent rock and found that minerals which are difficult to fuse are also difficult to dissolve and very hard (1902). In 1902 Doelter presented a viscosity series from liquid basalt to viscous granite, recognizing the influence of mineralizers on viscosity and on the lowering of the melting point, as well as their catalytic and chemical activity—e.g., in the formation of mica and of tourmaline. Moreover, he found that magmatic differentiation "is nothing else than the final result of mineral segregation."[3] For melting-point determinations he constructed the crystallization microscope built by C. Reichert with the heating oven of W. C. Heraeus (described in 1904; improved in 1909). In 1904 he was able to confirm the suspected influence of the inoculation of solution-melts with seed crystals.

With his *Physikalisch-chemische Mineralogie* (Leipzig, 1905), Doelter showed himself to be, along with J. H. L. Vogt, the most important cofounder of this new discipline, even though he held that the direct transfer of the results of physical chemistry was not always possible, because in silicate fusions a restoration of equilibrium is impeded by the subcooling and by reduced diffusion resulting from viscosity. The chief advantage that he saw in physical chemistry was that it had set the direction that experimental work should follow.

In his *Petrogenesis* (Brunswick, 1906), a masterful work for its time, Doelter experimentally verified, on the whole, Rosenbusch's order of crystallization series and related it to solubility, force and velocity of crystal growth, stability, cooling rates, and percentage of mineralizers. Furthermore, he distinguished (1) density differentiation (liquation), which he demonstrated experimentally; (2) crystallization or cooling differentiation by the freezing of basic facies at the boundary of the surfaces of cooling or by the sinking of the earliest segregation products which later form a rock matrix; and (3) isotectic differentiation, in which a definite association of various minerals separates out. In opposition to the views of Vogt, he was able to verify eutectic differentiation only for quartz-orthoclase and sporadically for fayalite fusions.

Doelter next turned to research on estimation of the number of nuclei in fused masses, on the force of crystallization, and on the measurement of viscosity (1905); in 1911, with H. Sirk, he obtained the first measurement of the absolute viscosity of molten diopside at 1300°C. From 1907 to 1910 Doelter studied the dissociation constants, electrical conductivities, and polarization of solid and fused silicates. He discovered that in the transition from the molten to the crystalline state the conductivity changed irregularly, while in the transition to the vitreous state the change was continuous. He held, in opposition to Groth's conception of lattices of atoms in potassium sulfate, that "one could accept it as more probable that such lattices were made of K' and SO_4'' ions."[4] Even solid, nonconducting sodium chloride could, in his opinion, separate completely or largely into ions, which are fixed in the lattice and which, since they are not mobile, cannot exhibit electrical conductivity.

In addition, Doelter concerned himself with gemmology, particularly with the cause of mineral coloration. In 1896 he began to investigate the influence of Roentgen rays on minerals, and later that of radium rays and of ultraviolet light. He also examined the changes in color resulting from heating minerals in oxidizing, reducing, and inert gases. On the basis of his experience with the coloring agents of synthetic corundums and spinels, Doelter warned against considering analytically or spectroscopically demonstrated trace elements as color pigments, without extensive testing of their behavior when heated in gases and under radiation. Unfortunately these warnings were ignored by many later workers in the field, with the result that contradictory assertions are still published. Doelter even drew upon colloid chemistry in order to clarify these questions—without, however, arriving at any definitive conclusions. Nevertheless his extensive collections of material contain very valuable observations.

Doelter's some 200 works and those of his students have provided the essential experimental basis for modern petrology.

NOTES

1. H. Leitmeier, in *Neue deutsche Biographie,* IV (1959), 25.
2. R. Grengg, in *Montanistische Rundschau,* 22, no. 19 (1930), 2.
3. *Physikalisch-chemische Mineralogie,* p. 147.
4. *Sitzungsberichte der Akademie der Wissenschaften in Wien,* sec. 1, 117 (1908), 333.

BIBLIOGRAPHY

I. ORIGINAL WORKS. Doelter's books include *Bestimmung der petrographisch wichtigen Mineralien durch das Mikroskop* (Vienna, 1876); *Die Vulcane der Capverden und ihre Producte* (Graz, 1882); *Über die Capverden nach dem Rio Grande und Futah-Djallon* (Leipzig, 1884); *Edelsteinkunde* (Leipzig, 1893); *Das Radium und die Farben* (Dresden, 1910); *Die Farben der Mineralien, inbesondere der Edelsteine* (Brunswick, 1915); and *Die Mineralschätze der Balkanländer und Kleinasiens* (Stuttgart, 1916). His *Handbuch der Mineralchemie* was edited by C. Doelter and, from III, pt. 2, with H. Leitmeier; it appeared in 4 vols.: II in three pts., III in two pts., and IV in three pts. (Dresden-Leipzig, 1912-1931). About half of the material is by Doelter.

His papers include "Die Vulcangruppe der Pontinischen Inseln," in *Denkschriften der Akademie der Wissenschaften* (Vienna), Math.-nat. Kl., 36, pt. 2 (1876), 141-186; "Der Vulcan Monte Ferru auf Sardinien," *ibid.,* 38, pt. 2 (1878), 113-214; "Die Producte des Vulcans Monte Ferru," *ibid.,* 39, pt. 2 (1879), 41-95; "Chemische Zusammensetzung und Genesis der Monzonigesteine I-III," in *Tschermaks mineralogische und petrographische Mitteilungen,* 21 (1902), 65-76, 97-106, 191-225; "Der Monzoni und seine Gesteine I-II," in *Sitzungsberichte der Akademie der Wissenschaften* (Vienna), Math.-nat. Kl., sec. 1, 111 (1902), 929-986; 112 (1903), 169-236; "Die Silikatschmelzen I-IV," *ibid.,* 113 (1904), 177-249, 495-511; 114 (1905), 529-588; 115 (1906), 723-755; "Über die Dissoziation der Silikatschmelzen I-II," *ibid.,* 116 (1907), 1243-1309; 117 (1908), 299-336; and "Die Elektrizitätsleitung in Krystallen bei höhen Temperaturen," *ibid.,* 119 (1910), 49-111.

An incomplete list of Doelter's works, without his contributions to *Handbuch der Mineralchemie,* is in Poggendorff, III-VI.

II. SECONDARY LITERATURE. On Doelter or his work see E. Dittler, in *Zentralblatt für Mineralogie,* pt. A (1930), 476-477; R. Grengg, in *Montanistische Rundschau;* 22, no. 19 (1930), 1-2; W. Hammer, in *Verhandlungen der Geologischen Bundesanstalt* (Vienna, 1930), 213-214; A. Himmelbauer, in *Almanach der Akademie der Wissenschaften in Wien,* 81 (1931), 314-316; H. Leitmeier, in *Neue deutsche Biographie,* IV (1959), 25-26; and L. J. Spencer, in *Mineralogical Magazine,* 22 (1930), 390-391.

Various biographical information is to be found in the archives at Graz and Vienna.

WALTHER FISCHER

DOGEL, VALENTIN ALEXANDROVICH (*b.* Kazan, Russia, 10 March 1882; *d.* Leningrad, U.S.S.R., 1 June 1955), *zoology.*

His father, A. S. Dogel, was first dissector in the department of histology in Kazan, and then professor in the departments of histology in Tomsk (1888-1894) and in St. Petersburg (1894-1922). V. A. Dogel graduated in 1904 from the natural sciences section of the physics and mathematics faculty of St. Petersburg University and was professor of invertebrate zoology at the university from 1914 to 1955. He occupied the chair of zoology at the (Herzen) Pedagogical Institute for Women from 1908 to 1938, and from 1930 to 1955 was head of the laboratory of fish diseases in the All-Union Institute of Economy for Lake and River Fish and of the laboratory of sea protozoa of the Zoological Institute of the Soviet Academy of Sciences. He was made corresponding member of the Academy in 1939.

Dogel's scientific career was devoted to protozoology and the comparative anatomy of invertebrates. In the area of protozoology he studied the morphology and taxonomy of *Gregarina,* Dinoflagellata, Catenata, Polymastigina, and Hypermastigina (from the intestines of termites) and Infusoria from the stomach of ruminants (Ophryoscolecidae); and established the evolutionary regularity of the development of protozoa—the phenomenon of polymerization. In ecological parasitology, Dogel studied the relation of the parasitofauna of the animal host to the type of diet and to migrations (among fish and migratory birds), and also to hibernation (among bats). In the area of comparative anatomy and evolutionary morphology, in addition to establishing the taxonomic position of pantopods, Dogel formulated the general evolutionary regularities, in particular the regularity of oligomerization of homologous organs and the means by which it comes about (reduction, fusion of organs, change of function). Dogel was the author of several monographs and textbooks and over 250 specialized works. Among his students were

B. E. Bykhovsky, A. P. Markevich, A. V. Ivanov, E. M. Kheysin, Y. I. Polyansky, A. A. Strelkov, and V. L. Vagin.

BIBLIOGRAPHY

I. ORIGINAL WORKS. Dogel's writings include *Catenata. Organizatsia roda Haplozoon i nekotorykh skhodnykh s nim form* ("Catenata. The Organization of the Genus Haplozoon and Certain Forms Similar to Them"), diss. (St. Petersburg, 1910); *Materialy k istorii razvitia Pantopoda* ("Materials for a History of the Development of the Pantopoda"), diss. (1913); "Monographie der Familie Ophryoscolecidae," in *Archiv für Protistenkunde,* **59** (1927), 1–282; *Uchebnik sravnitelnoy anatomii bespozvonochnykh* ("Textbook of Comparative Anatomy of Invertebrates"), 2 pts. (Leningrad, 1938–1940); *Kurs obshchey parazitologii* ("Course in General Parasitology"), 2nd ed. (Moscow, 1947); *Oligomerizatsia gomologichnykh organov* ("Oligomerization of Homologous Organs," Leningrad, 1954); *Allgemeine Parasitologie,* rev. and enl. by Y. I. Polyansky and E. M. Kheysin (Jena, 1963), trans. as *General Parasitology* (Oxford, 1965); and *Obshchaya protozoologia* ("General Protozoology," Moscow–Leningrad, 1962), written with Y. I. Polyansky and E. M. Kheysin.

II. SECONDARY LITERATURE. On Dogel and his work, see (listed chronologically) Y. I. Polyansky, "Professor Valentin Aleksandrovich Dogel," in *Uchenye zapiski Leningradskogo gosudarstvennogo universiteta. Seria biologicheskikh nauk,* no. 2 (Leningrad, 1939); *Materialy k bibliographii uchenykh SSSR. Seria biologicheskikh nauk. Parazitologia,* no. 2 (Leningrad, 1952); and Y. I. Polyansky, *Valentin Aleksandrovich Dogel* (Leningrad, 1969), which contains additions to the list of Dogel's works.

L. J. BLACHER

DOKUCHAEV, VASILY VASILIEVICH (*b.* Milyukovo, Smolensk province, Russia, 1 March 1846; *d.* St. Petersburg, Russia, 8 November 1903), *natural science, soil science, geography.*

Dokuchaev came from the family of a village priest. He received his elementary education at a church school in Vyazma and then studied at the Smolensk seminary. In 1867 he graduated with distinction from the seminary and was accepted at the St. Petersburg Ecclesiastical Academy. In the same year, having decided not to become a priest, he left the academy and entered the physics and mathematics department of St. Petersburg University to study the natural sciences.

In 1871 Dokuchaev graduated from St. Petersburg University with a master's degree. His dissertation was devoted to the study and description of the alluvial deposits of the Kachna River, on the upper reaches of the Volga, near his birthplace. From that

time Dokuchaev's scientific activity was connected with St. Petersburg University, the Society of Natural Scientists, the Free Economic Society, the Mineralogical Society, and the Petersburg Assembly of Agriculturists. With the support of these groups he carried out research on the Russian plains and in the Caucasus. In the fall of 1872 he was made curator of the geological collection of St. Petersburg University, and in 1879 he became *Privatdozent* in geology. Along with his courses in mineralogy and crystallography he began to give a special course, the first anywhere, on Quaternary deposits.

From 1892 to 1895, while remaining a professor at St. Petersburg University, Dokuchaev was occupied with the reorganization and then the direction of the Novo-Aleksandr (now Kharkov) Institute of Agriculture and Forestry (now named for him). There he founded the first department of soil science in Russia and a department of plant physiology that offered courses on microorganisms.

Dokuchaev's first major work, "Sposoby obrazovania rechnykh dolin Evropeyskoy Rossii" ("Methods of Formation of the River Valleys of European Russia"), defended as a doctor's thesis in 1878, was the result of years of profound study of the geological, orographical, and hydrographical peculiarities of the Russian plain, particularly the districts covered by ancient glaciers. He analyzed various earlier hypotheses on the formation of the river plains, particularly those treating valleys as the result of tectonic fissures or receding of the ocean, and he criticized accepted ideas, such as Murchison's "drift theory." Dokuchaev gave a coherent explanation of the genesis of landforms and their relation to specific physical and geographical conditions of the past; on this basis he may be considered one of the founders of geomorphology.

After he began his field studies of Quaternary deposits, Dokuchaev directed his research to the topsoil of European Russia, particularly to chernozem. In 1875 he was invited to write an explanatory note for V. I. Chaslavsky's soil map of European Russia; he then spent several years preparing the map for publication. From 1877 to 1898 he investigated the northern boundary of the chernozem belt: the Ukraine, Moldavia, central Russia, Trans-Volga, the Crimea, and the northern slopes of the Caucasus. His monograph "Russky chernozyom" ("Russian Chernozem," 1883) won numerous honors.

This research attracted the attention of *zemstvos* (village councils) and individual landowners of Saratov and Voronezh provinces; comprehensive research on the natural conditions of these large territories was carried out by a group of young scientists, most of whom were students of Dokuchaev's and worked under his guidance. Their work was important not only for the practice of agriculture but also for the confirmation and development of Dokuchaev's ideas in a new area of natural science—genetic soil science.

Dokuchaev's expeditions to Nizhni Novgorod (1882–1886), Poltava (1888–1896), and other places were conducted according to a special method. In accounts of the expeditions a full description of the natural history of the provinces was given by natural components (geology, soil, water, plant and animal life); and on the basis of an analysis of all the data, an appraisal was made of the agricultural potential. These collections served as the basis for the organization, at Dokuchaev's initiative, of museums of natural history in Nizhni Novgorod and Poltava. Following a plan proposed by Dokuchaev, similar museums were later created in other cities of Russia. Dokuchaev was one of the organizers and leaders of the Eighth Congress of Russian Natural Scientists and Physicians. At his initiative the agronomy and geography sections were separated for the first time. He emphasized the necessity of creating a soil institute and museum and of making a thorough study of the natural history of various areas.

In 1891 there was a severe drought in Russia, and Dokuchaev subordinated his scientific work to the problem of dealing with this disaster. He was commissioned by the Ministry of State Lands to undertake a special expedition that was to devise ways and means of conducting farming, forestry, and water management in the steppe (chernozem) zone. The basis of the work of the expedition was a plan set forth by Dokuchaev in his book *Nashi stepi prezhde i teper* ("Our Steppes Past and Present," 1892), which included preliminary geological, soil, and climatic findings. Three experimental plots in the steppe belt, each about 5,000 hectares, were chosen to survey: Starobelsky, in the watershed between the Don and the Donets; Khrenovsky, between the Volga and the Don; and Veliko-Anadolsky, in the watershed between the Donets and the Dnieper. Of great importance was the network of meteorological stations and rain-gauge points set up on these plots. The careful observations of the climate of the steppe zone made it possible to determine the influence of climatic conditions on agriculture, particularly the role of forests and protective forest belts. Much work was done on artificial forest cultivation, the regulation and use of water resources, and the building of reservoirs.

As head of the Bureau of Soil Science of the Scientific Committee of the Ministry of Agriculture, Dokuchaev also carried out the compilation of a new soil map of European Russia. In 1897, after twenty years of service at St. Petersburg University, he retired for reasons of health. His health having improved slightly, in 1898 Dokuchaev led an expedition to study the soil of Bessarabia; in 1898–1899 and 1900 he studied the Caucasus and the Transcaucasus; and in 1898–1899 he also went to the Trans-Caspian region. Of the results of his research in the Caucasus he wrote that he "not only *predicted* but even *factually proved* the indisputable unusually sharply expressed existence in the whole Caucasus and Trans-Caucasus of *vertical* soil (and, in general, natural history) zones . . . " (*Sobranii sochineny* ["Collected Works"], VIII, 331).

Dokuchaev's last public activity was a series of lectures on soil science and on the results of his three-year study of the soils of the Caucasus, given in Tbilisi in 1900.

Dokuchaev continually sought to popularize the accomplishments of science. He stressed the centralization of all soil work and research and the necessity of creating a soil institute and departments of soil science in universities. He and A. V. Sovetov were responsible for the series *Materialy po izucheniyu russkikh pochv* ("Material for the Study of Russian Soils," from 1885) and *Trudy* ("Works") of the Soil Commission of the Free Economic Society (1889–1899). In 1899, at his initiative, the journal *Pochvovedenie* ("Soil Science") began publication.

A prerequisite for the creation of soil science was Dokuchaev's discovery of soil as a special body that has developed as a result of climate, bedrock, plant and animal life, age of the land, and topography.

Dokuchaev's approach to the evolution of soils allowed him to discover all the complex connections between the soil-forming factors, including the factors of time and human activity. He defined soil as follows:

> It consists essentially of the mineral-organic formations lying on the surface, which are always more or less noticeably colored with humus. These bodies always have their own particular origin; they always and everywhere are the result of the totality of activity of the bedrock, the living and inanimate organisms (plant as well as animal), climate, the age of the country, and the topography of the surroundings. Soils, like every other organism, always have a certain normal structure, normal depth, and normal position and are always related to these in warmth, moisture, and plant growth differently from their bedrock [*ibid.,* II, 260].

Thus, according to Dokuchaev, soils are geobiological formations, the properties of which are closely related to their position on the earth's surface and change regularly as a result of environmental conditions. Starting with the definition of soil and the position that soil depends on soil formers, Dokuchaev created a new classification of soils according to natural history. His methods of classifying soils are now basic for cartographic representation and qualitative appraisal of soils. He distinguished three basic classes of soils: normal, transitional, and abnormal. A more detailed division into sections and types was made after considering the differences in soil formers and their interrelationships. Normal, or zonal, soils are the most typical and most widespread. In the course of time the classification of soils has become more precise and detailed. In Dokuchaev's classifications of 1896 (*ibid.,* VII, 449, inset) and 1898 (*ibid.,* VI, 330, inset) the genetic soil types and soil belts were related to vegetation and climatic belts, which was of great importance "for the accurate and complete *understanding* and *appraisal* of nature and its varied and extremely complex forms" (*ibid.,* VI, 306).

Dokuchaev established the zonality of soil and its coincidence with the zonality of climate, vegetation, and animal life. On this subject he wrote:

> . . . thanks to the known position of our planet relative to the sun, thanks to the rotation of the earth, its spherical shape—climate, vegetation, and animal life are distributed on the earth's surface from north to south, in a strictly determined order . . . which allows the division of the earth's sphere into belts: *polar, temperate, subtropical, equatorial,* and so forth. And since the agents—soil formers, which are subject to known laws in their distribution—are distributed by belts, their results also—the soil—must be distributed on the earth's sphere in the form of definite *zones,* going *more* or *less* (only with certain deviations) parallel to the circles of latitude [*ibid.,* 407].

A view of nature as an entity shaped by a profound mutual interdependence and mutual determination of all its components led Dokuchaev to create the theory of zones of nature. He distinguished five basic geographical zones: boreal, taiga, chernozem (steppe), arid desert, and lateritite (tropical). Tracing the main zones, he noted transitions between them and pointed to the differentiation of natural zones into separate physical-geographical regions.

Dokuchaev stressed that agriculture should be carried out on a zonal basis and defined the main problems of agricultural technology for each zone. The most complete synthesis of his scientific work can be found in his statements on theory of the relationship between "inanimate" and "living" nature. In *Mesto i rol sovremennogo pochvovedenia v nauke i zhizni* ("The Place and Role of Contemporary Soil Science in Science and in Life," 1899) Dokuchaev wrote:

> As is well known, in recent times one of the most interesting *disciplines* in the field of contemporary natural science has developed and become more and more defined, namely, the theory of the multiple and various *relationships* and *interactions,* and equally *laws* which govern their age-old *changes,* which exist between so-called *inanimate* and *living nature,* between (*a*) the surface rocks, (*b*) the plastic layer of the earth, (*c*) soils, (*d*) surface and underground water, (*e*) the climate of the country, (*f*) plant, and (*g*) animal organisms (including, and even chiefly, the lowest) and man, the proud crown of creation [*ibid.,* 416].

Dokuchaev sought to create a unified science that would embrace "the one, whole, and indivisible nature." The core of this new science would be genetic soil science. His death cut off this work; but it has been continued by, among others, Dokuchaev's student Vernadsky, the creator of the theory of the biosphere.

Dokuchaev's work greatly influenced the development of physical geography and geobotany, and made a great contribution to the study of swamps. His "polygenetic" theory of the genesis of loess significantly anticipated contemporary views. His ideas received wide recognition, and his genetic soil classification was applied in making soil maps of England, the United States, Rumania, and other countries.

BIBLIOGRAPHY

Dokuchaev's works have been gathered in *Sobranii sochineny* ("Collected Works"), 9 vols. (Moscow-Leningrad, 1949–1961). Vol. I contains his most important geological and geomorphological works, including *Sposoby obrazovania rechnykh dolin Evropeyskoy Rossii* ("Methods of Formation of the River Valleys of European Russia," St. Petersburg, 1878). Vol. II contains articles and reports on the study of the chernozem (1876–1885) and *Kartografia russkikh pochv* ("Cartography of Russian Soils," St. Petersburg, 1879). Vol. III consists of *Russky chernozyom* ("Russian Chernozem," St. Petersburg, 1883). Vols. IV and V are devoted to *Nizhegorodskie raboty* ("Nizhni Novgorod Works," 1882–1887). Vol. VI contains works on the transformation of the nature of the steppes, on soil research and soil appraisal, and the theory of zonality and classification of soils (1888–1900), including *Nashi stepi prezhde i teper* ("Our Steppes, Past and Present," St. Petersburg, 1892); *K voprosu o pereotsenke zemel Evropeyskoy i Aziatskoy Rossii. S klassifikatsiey pochv* ("Toward the Question of a Reappraisal of the Soils of European and Asiatic Russia. With a Classification of Soils," Moscow, 1898); *K ucheniyu o zonakh prirody* ("Toward a Theory of Zones of Nature," St. Petersburg, 1899); and *Prirodyne pochvennye zony. Selskokhozyaystvennye zony. Pochvy Kavkasa* ("Natural Soil Zones. Agricultural Zones. The Soils of the Caucasus," St. Petersburg, 1900). Vol. VII consists of various articles and reports, material on the organization of soil institutes, and popular lectures (1880–1900). Vol. VIII contains speeches and correspondence. Vol. IX contains S. S.

Sobolev's article on the development of Dokuchaev's basic ideas, a biographical sketch by L. A. Chebotareva, and archival documents for his biography. A substantial part of the volume is devoted to a bibliography of all his works (pp. 165–247) and of literature on him (pp. 248–322), including 49 items in foreign languages (pp. 315–320).

Recent works on Dokuchaev not listed in the bibliography mentioned above are V. A. Esakov and A. I. Soloviev, *Russkie geograficheskie issledovania Evropeyskoy Rossii i Urala v XIX-nachale XX v.* ("Russian Geographical Research on European Russia and the Urals in the Nineteenth and the Beginning of the Twentieth Centuries," Moscow, 1964), pp. 76–90; *Istoria estestvoznania v Rossii* ("History of Natural Sciences in Russia"), III, S. R. Mikulinsky, ed. (Moscow, 1962), 217–238; and G. F. Kiryanov, *Vasily Vasilievich Dokuchaev, 1846–1903* (Moscow, 1966).

VASILY A. ESAKOV

DÖLLINGER, IGNAZ (*b.* Bamberg, Germany, 27 May 1770; *d.* Munich, Germany, 14 January 1841), *physiology, embryology.*

Döllinger should be considered one of the most able and influential German biologists of the early nineteenth century. Although his initial scientific career was associated with the application of Friedrich W. Schelling's philosophical ideas to physiology and embryology, Döllinger simultaneously conducted microscopical observations related to embryonic and vascular structures. His major achievement was to provide enthusiasm, method, and assistance to the new generation of scientists, such as Karl von Baer, Heinrich Christian von Pander, Johann L. Schönlein, Georg Kaltenbrunner, and Eduard d'Alton. Thus Döllinger symbolizes the successful attempt in Germany to proceed beyond the narrow schemes embodied by *Naturphilosophie* and to instruct students in the virtues of observation and experiment.

As the son of the personal physician to Franz Ludwig von Erthal, ruler of the bishopric of Würzburg and Bamberg, Döllinger received an elaborate education. After completing his studies in philosophy and natural sciences at the University of Bamberg, he began his medical career there, transferring later to Würzburg. Under the sponsorship of Erthal, Döllinger next went to Vienna, where he received the clinical training lacking at the time in Germany. In addition he improved his anatomical knowledge and acquired great skill in vascular injection techniques under the direction of Prochaska. The Vienna sojourn was followed by a journey to the University of Pavia, where Döllinger studied under Johann Peter Frank and Antonio Scarpa.

In 1793 Döllinger returned home after the revolutionary wars had forced Pavia to close its doors. In 1794 he received the M.D. from the University of

Bamberg, where he was then named professor of physiology and general pathology; he remained until 1801, when the university was closed as a result of the annexation of the bishopric by Bavaria. In 1803, under the new administration, Döllinger was elected professor of anatomy and physiology at the reorganized University of Würzburg. He remained there for twenty years and was the most important figure at the medical school, attracting a great number of students, especially from Germany and the Russian Empire.

In 1816 Döllinger was accepted as member of the Leopoldinisch-Karolinische Deutsche Akademie der Naturforscher. After being a corresponding member since 1819, he was elected in 1823 to a regular chair at the Bavarian Academy of Sciences. He therefore moved to Munich in order to fill the vacancy left by the anatomist Samuel T. Sömmering. Döllinger's early years in Munich were occupied with the creation and construction of an anatomical amphitheater, which was opened in 1827. Also, after the University of Bavaria had been moved from Landshut to Munich in 1826, he became its professor of human and comparative anatomy. He was elected secretary of the mathematical-physical section of the Bavarian Academy for the years 1827–1830 and 1833–1836, and became a member of the Bavarian Medical Advisory Board in 1836. His health was seriously and permanently affected by the cholera epidemic that ravaged Germany in 1836. Döllinger's last years were spent in virtual seclusion, and he died of a perforated gastric ulcer.

Döllinger's early treatise *Grundriss der Naturlehre des menschlichen Organismus* is an attempt to place the physiological knowledge available at the turn of the nineteenth century within Schelling's philosophical postulates. But in the preface he makes it clear that this formulation will not stifle his spirit of inquiry or hamper further search for knowledge. For him *Naturphilosophie* was pure and absolute philosophical knowledge, which could not determine the finite biological phenomena with which he was dealing.

As a Bavarian academician, Döllinger sought to improve his major research tool, the microscope. With the opticians of the Utzschneider-Fraunhofer Institute at Munich he was instrumental in achieving needed corrections on the aplanatic microscope. He made microscopical observations of the blood circulation, and in his short monograph "Vom Kreislaufe des Blutes" paid special attention to the red and white blood cells and their intravascular and extravascular fates in the different tissues. Döllinger felt that these cells were elementary organic units floating in the serum, and that their chemical transformation was needed for the nutrition and secretion of the different organic structures.

In embryology Döllinger stimulated research on the morphology of the developing chick. He was a believer in the existence of a plastic, organic principle that unfolded its own design within the embryo. He set out to follow this process microscopically and recognized the early stages of embryonic differentiation.

BIBLIOGRAPHY

I. ORIGINAL WORKS. Döllinger's writings include *Grundriss der Naturlehre des menschlichen Organismus, zum Gebrauch bei seinen Vorlesungen* (Bamberg–Würzburg, 1805); *Beyträge zur Entwicklungsgeschichte des menschlichen Gehirns* (Frankfurt, 1814); "Vom Pulse," in *Deutsches Archiv für die Physiologie,* **2** (1816), 356–358; "Versuch einer Geschichte der menschlichen Zeugung," *ibid.,* 388–402; *Was ist Absonderung und wie geschieht sie? Eine akademische Abhandlung* (Würzburg, 1819); "Bemerkungen über die Vertheilung der feinsten Blutgefässe in den beweglichen Teilen des thierischen Körpers," in *Deutsches Archiv für die Physiologie,* **6** (1820), 186–199; and *Vom Kreislaufe des Blutes,* vol. VII in the series Denkschriften der K. Akademie der Wissenschaften (Munich, 1821), pp. 169–228.

II. SECONDARY LITERATURE. On Döllinger or his work, see Burkard Eble, *Curt Sprengel's Versuch einer pragmatischen Geschichte der Arzneikunde* (Vienna, 1837), VI, pt. 1, many refs. between pp. 303 and 618; Albert von Kölliker, *Zur Geschichte der medicinischen Facultät an der Universität Würzburg. Rede zur Feier des Stiftungstages der Julius-Maximilians Universität* (Würzburg, 1871), pp. 31–37; Arthur William Meyer, *Human Generation: Conclusions of Burdach, Doellinger and von Baer* (Stanford, Calif., 1956), pp. 26–32; and Phillip F. von Walther, *Rede zum Andenken an Ignaz Doellinger Dr.* (Munich, 1841).

GUENTER B. RISSE

DOLLO, LOUIS ANTOINE MARIE JOSEPH (*b.* Lille, France, 7 December 1857; *d.* Brussels, Belgium, 19 April 1931), *paleontology.*

Dollo belonged to an old Breton family, several generations of which were sailors and lived in Roscoff. He studied at the University of Lille, where the geologist Jules Gosselet and the zoologist Alfred Giard played major roles in his scientific education. In 1877 he graduated with a degree in civil and mining engineering. In 1879 Dollo moved to Brussels, where he worked first as an engineer in a gas factory. At this time he became acquainted with the works of Kovalevski. Having decided to devote himself to paleontology, in 1882 he became junior naturalist at

the Royal Museum of Natural History in Brussels. From 1891 almost to the end of his life he was curator of the vertebrate section there. He began teaching in 1909 as extraordinary professor at Brussels University, where he delivered lectures on paleontology and animal geography.

Dollo never limited himself to paleontology, and through his enthusiastic lectures in natural science he popularized scientific knowledge. He was seriously interested in linguistics—he attempted to compose a grammar of the Bantu languages of the Congo Basin —and biochemistry, and his love of music was well known to all around him. An active member of the Royal Academy of Belgium and a corresponding member of many foreign academies, Dollo won the Kuhlmann Prize in 1884, the Lyell Medal in 1889, and the Murchison Medal in 1912.

Fossil reptiles occupied the central place in Dollo's scientific work. A long series of brief articles contained careful analyses of the phenomena of adaptation among fossil reptiles, especially iguanodonts of the Lower Cretaceous, to certain conditions of existence. These investigations, which laid the foundations for the ethological study of fossil forms, sought to clarify not only their way of life but also the history of their adaptation. The study of the ethological peculiarities of such dinosaurs as *Triceratops* and *Stegosaurus* enabled Dollo to make a rather unexpected discovery: these four-legged forms must have had functionally biped ancestors. His works on fossil turtles and mosasaurians had great methodological importance. Explaining the ethological type of fossil sea turtles, Dollo showed what a complex and tortuous path their historical development had followed. The evolution of land forms that shifted to sea life did not involve simply the movement from the shores to the open sea; adaptation frequently took place in the opposite direction. Studying the hearing apparatus of the mosasaurian, Dollo found several indications of an ability to dive, which were confirmed by other signs in the structure of its skull. Investigation of the teeth of various representatives of this group led to the identification of several types of adaptation to its food. The precision of observation and the depth of analysis are also striking in the works devoted to rhynchocephalians, crocodiles, and ichthyosaurs.

Among Dollo's most significant works is the monograph on lungfishes, which explains the essential characteristics of the evolution of this very interesting group. The basic conclusions of this and other works on fossil and contemporary fish are still valid.

Dollo successfully applied his ethological method of research not only to fossil vertebrates but also to such invertebrates as the cephalopod mollusks and arthropods. However, probably the greatest achievement of this method was its application to the study of certain animals now living. In particular, the ethological analysis of the properties of the extremities of contemporary marsupials allowed Dollo to establish beyond a doubt their origin from ancient forms.

Like Kovalevski, whom he called his teacher and to whom he dedicated a cycle of remarkable lectures published under the title "La paléontologie éthologique" (1909), Dollo was not satisfied with explaining the functional significance of an organ of an animal being studied. He always tried to explain the historical development of, and the reasons for, its adaptation. Guided in his theoretical research by the ideas of Darwin and Kovalevski, Dollo enriched Darwinism by his famous laws on the irreversibility of evolution.

Limitations can sometimes be noted in Dollo's views in connection with his tendency to delimit mechanically the boundaries between paleontology and geology. Some of his conclusions concerning the mode of life of fossil forms also can be shown—and in fact are being shown—to be imprecise. On the whole, however, his works are of indisputably great methodological and theoretical significance. The most important of them remain unsurpassed models of paleobiological research.

BIBLIOGRAPHY

I. ORIGINAL WORKS. Dollo's writings are "Les lois de l'évolution," in *Bulletin de la Société belge de géologie, de paléontologie et d'hydrologie,* **7** (1893), 164–166; "Sur la phylogénie des dipneustes," in *Bulletin de la Société belge de géologie, de paléontologie et d'hydrologie,* **9** (1895), 79–128; "Les ancêtres des marsupiaux étaient-ils arboricoles?," in *Miscellanées biologiques dédiées au Prof. A. Giard à l'occasion du XXVe anniversaire de la fondation de la station zoologique de Wimereux, 1874–1899* (Paris, 1899), pp. 188–203; "Sur l'évolution des chéloniens marins (considérations bionomiques et phylogéniques)," in *Bulletin de l'Académie royale de Belgique. Classe des sciences,* no. 8 (1903), 801–830; and "La paléontologie éthologique," in *Bulletin de la Société belge de géologie, de paléontologie et d'hydrologie,* **23** (1909), 377–421.

II. SECONDARY LITERATURE. On Dollo or his work, see O. Abel, "Louis Dollo. 7 Dezember 1857–19 April 1931. Ein Rückblick und Abschied," in *Palaeobiologica,* **4** (1931), 321–344; P. Brien, "Notice sur Louis Dollo," in *Annuaire de l'Académie royale de Belgique. Notices biographiques* (1951), pp. 69–138; L. S. Davitashvili, "Lui Dollo," in *Voprosy istorii estestvoznania i tekhniki* ("Problems in the History of Natural Sciences and Technology"), **3** (1937), 108–120; and V. Van Straelen, "Louis Dollo (1857–1931)," in *Bulletin de la Musée d'histoire naturelle de Belgique,* **9,** no. 1 (1933), 1–6.

L. K. GABUNIA

DOLLOND, JOHN (*b.* London, England, 10 June 1706; *d.* London, 30 November 1761), *optics.*

Dollond was born in Spitalfields, London, of French Protestant parents who had originally lived in Normandy. Although he became a silk weaver, his inclination led him to the study of mathematics, astronomy, and the classical languages in whatever spare time he could find. His eldest son, Peter, joined him as a weaver but, stimulated by the knowledge of mathematics and optics learned from his father, subsequently took up the trade of optical instrument maker. This venture was successful, and John Dollond was consequently persuaded to leave the weaving trade. He joined Peter in business in 1752, and the partnership soon became fruitful.

Some ideas for the improvement of the optical arrangement of lenses in the refracting telescope were incorporated in a letter addressed to James Short, a fellow of the Royal Society, who communicated the letter to the society, where it was read on 1 March 1753. Soon after this, on 10 May 1753, another paper was read to the Royal Society, this time on an improved micrometer (heliometer) for the telescope. Dollond had modified the Savery micrometer by using one object glass cut into two equal segments instead of two whole lenses. The micrometer could now be applied to the reflecting telescope, which was immediately done by James Short.

Dollond is popularly known as the inventor of the achromatic telescope; but although he was anticipated in the discovery by about twenty years, it seems that he independently worked out the necessary lens combinations—and he certainly was the first to publish the invention and develop it commercially. Dollond's early successes in optics brought him to the attention of astronomers and mathematicians. He corresponded with many, including Euler, against whom he defended Newton's opinion that no combination of lenses could produce an image free of color, and that in this respect no improvement could be expected in the refracting telescope.

Eventually Dollond conducted (1757–1758) a series of experiments with different kinds of glass to check Newton's findings. The paper incorporating the results, with the conclusion that the objectives of refracting telescopes could be made "without the images formed by them being affected by the different refrangibility of the rays of light," was read to the Royal Society in June 1758. Dollond's composite objective was patented, but the patent was challenged by a group of London optical instrument makers after his death. In 1766 the court upheld Peter Dollond's right to the patent on the grounds that Chester More Hall, the inventor of an achromatic lens combination in the period 1729–1733, did not exploit the invention commercially or publicize his findings.

It seems unlikely that Hall's invention could have been known to anyone capable of realizing its significance, because the Royal Society not only published Dollond's papers but conferred both the Copley Medal (1758) and membership (1761) upon him. The certificate proposing Dollond for membership was signed in February 1761 by ten men, including scientists of the standing of Gowin Knight, John Smeaton, James Short, William Watson, and John Ellicott. The proposal specifically refers to Dollond's invention of "an Object-Glass, consisting of two Spherical Lenses of different densities, so contrived as to correct the Errors arising from the different refrangibility of the Rays of Light."

Early in 1761 Dollond was appointed optician to King George III. Regrettably, he did not enjoy this honor for long; he died of apoplexy later that year. He left three daughters and two sons, Peter and John. The latter joined his elder brother as a partner in the family firm.

BIBLIOGRAPHY

I. ORIGINAL WORKS. The following were first published in *Philosophical Transactions of the Royal Society* and were reprinted in Kelly's *Life* (see below): "A letter from Mr. John Dollond to Mr. James Short, F.R.S. concerning an Improvement of refracting Telescopes," **48**, pt. 1 (1753), 103–107; "A Description of a Contrivance for measuring small Angles," *ibid.,* 178–181; "Letters relating to a Theorem of Mr. Euler, of the Royal Academy of Sciences at Berlin, and F.R.S. for correcting the Aberrations in the Object-Glasses of refracting Telescopes," *ibid.,* 287–296; "An Explanation of an Instrument for measuring Small Angles," **48**, pt. 2 (1754), 551–564; and "An Account of some Experiments concerning the different Refrangibility of Light," **50**, pt. 2 (1758), 733–743.

II. SECONDARY LITERATURE. Reprinted letters and papers by John and Peter Dollond, Short, Euler, and Maskelyne will be found in John Kelly, *The Life of John Dollond, F.R.S. Inventor of the Achromatic Telescope. With a copious Appendix of all the Papers referred to,* 3rd ed. (London, 1808). For an account of the priority of Chester More Hall and the patent litigation, with references, see Thomas H. Court and Moritz von Rohr, "A History of the Development of the Telescope from about 1675 to 1830 based on Documents in the Court Collection," in *Transactions of the Optical Society,* **30** (1928–1929), 207–260 (sec. IV, 228–235).

G. L'E. TURNER

DOLOMIEU, DIEUDONNÉ (called **DÉODAT**) **DE GRATET DE** (*b.* Dolomieu, Dauphiné, France, 23 June 1750; *d.* Châteauneuf, Saône-et-Loire, France, 28 [16? 29?] November 1801), *geology.*

One of eleven children of François de Gratet, marquis de Dolomieu, and of Marie-Françoise de Bérenger, Dolomieu was placed in the Sovereign and Military Order of the Knights of Malta at the age of two. Because of his precocious interest in natural objects, he is supposed to have been sent to Paris for a part of his education before the beginning of his military career. He was a member of the carabiniers in 1764, then served his apprenticeship aboard one of the order's galleys in 1766, achieving the rank of second lieutenant in the same year. He rose to lieutenant in 1774, became a knight in 1778, was promoted to captain in 1779, and commander in 1780.

His career in the Knights of Malta was marked by a long series of difficulties. In 1768, after a duel in which he killed a fellow member of the order, he was sentenced to life imprisonment but was released through the intervention of Pope Clement XIII. In the early 1780's he resigned as *lieutenant du maréchal* of his Langue (Auvergne), following a dispute with the grand master concerning an alleged transgression of the rights of the Langues; he carried on a legal battle for several years thereafter, acquiring some bitter enemies within the order. He finally left Malta in 1791, and his expression of pro-Revolutionary opinion soon began to elicit accusations of his involvement in a plot to destroy the order—which did indeed suffer grave setbacks during the Revolution. Dolomieu eventually played an unwilling role in Napoleon's seizure of Malta in 1798.

Garrisoned in Metz between 1771 and 1774, Dolomieu began to cultivate science under the tutelage of Jean-Baptiste Thyrion, an apothecary who taught chemistry and physics (and who was the mentor of Jean-François Pilâtre de Rozier as well). Dolomieu became the friend of Duke Louis-Alexandre de La Rochefoucauld, who helped secure Dolomieu's election as correspondent of the Royal Academy of Sciences in 1778, and whose mother, the duchess d'Enville, was hostess of a salon where Dolomieu met members of the fashionably learned world. La Rochefoucauld, who had made the grand tour in 1765–1766 in the company of Nicolas Desmarest, helped direct Dolomieu's interests toward mineralogy. In 1775 Dolomieu toured Anjou and Brittany, investigating mines and ironworks and studying the origin of saltpeter. He traveled in the Alps and in Italy in 1776, seeing the region of Vesuvius. His eulogizer Lacépède to the contrary, he probably did not visit Sicily and climb to the summit of Etna.

Dolomieu's determination to pursue geological science became firm in 1778, when he acted as secretary to Prince Camille de Rohan during an embassy of the order to Portugal, where he studied basaltic rocks.

Soon afterward (in 1779 or 1780) he retired from active service with the Knights of Malta and devoted himself to his scientific investigations, although much of his time was taken up in litigation over affairs in the order. He maintained a home in Malta but made frequent journeys for geological and other purposes, traveling in Sicily in 1781, in the Pyrenees (with Philippe-Isidore Picot de Lapeyrouse) in 1782, in Italy and Elba in 1784, Italy and Corsica in 1786, and in the Alps in 1789 and frequently thereafter.

Dolomieu gathered a substantial mineralogical collection, the great bulk of which served him as a pretext for remaining in Malta, despite his ostensible wish to depart because of his political difficulties with the grand master. This collection ultimately became the property of Dolomieu's brother-in-law, Étienne de Drée, part of it later coming to the Muséum d'Histoire Naturelle.

In 1791 Dolomieu made his way to Paris as a strong partisan of the Revolution, to the consternation of his family and many of his noble associates. In that year he wrote to Picot de Lapeyrouse that he refrained from running for public office only on account of the standing it would cost him with relatives. He joined the Club de 1789 and the Club des Feuillants, and styled himself a constitutional monarchist, a liberal position which the march of events soon made conservative. The excesses of the Revolution—most notably the assassination of his friend La Rochefoucauld in August 1792—repelled him, and he turned against these excesses in public rebuke.

During the hard days that followed, from 1792 through part of 1794, Dolomieu lived at La Roche-Guyon, the La Rochefoucauld château. His relatives were incarcerated or executed and his financial resources wiped out. Dolomieu therefore went to work, entering into a contract with the publisher Panckoucke to write the mineralogical portion of the *Encyclopédie méthodique,* a project that Dolomieu never completed. In 1794 he was appointed to teach natural history in the Écoles Centrales of Paris, and he was named *ingénieur* of the Corps des Mines in 1795, which led in 1796 to his assumption of teaching duties in physical geography at the École des Mines. During summer seasons he inspected mines and continued his geological travels in the Alps. When the Academy of Sciences was reconstituted as part of the Institute in 1795, Dolomieu was made a member.

In 1798 Dolomieu joined Napoleon's expedition to Egypt, during which he was maneuvered into taking part in negotiations for the capitulation of Malta. His stay in Egypt was cut short by illness (and perhaps also by his chagrin at having been used by Napoleon); and on the return voyage to France, when a storm forced his ship to put in at Taranto, he and his companions were imprisoned during the Calabrian counterrevolution. Taken to Messina, Dolomieu fell victim to vindictive influences exercised against him at the court of Naples by some of the Knights of Malta. He suffered a trying solitary imprisonment of twenty-one months until his release in March 1801. His return to Paris marked the end of what had become a cause célèbre among French intellectuals, the unconscionable detention of a scientist on the pretext of reasons of war. Having been elected while in prison to Daubenton's former chair, Dolomieu began to teach at the Muséum d'Histoire Naturelle; but his health, seriously affected by his ordeal in prison, failed shortly after his last tour in the Alps. He died a bachelor, in accord with his vow as a Knight of Malta, although he had been known to fancy women.

During his relatively brief career Dolomieu acquired a reputation as one of the most astute geologists. This reputation was not attributable to any remarkable theoretical innovation, although Dolomieu was interested in theory and, indeed, possessed greater theoretical commitments than he readily acknowledged. Instead, he was esteemed as a judicious inquirer within the framework of existing styles of geological research, and it is fitting that he is the eponym of a substance—dolomite—and of the Alpine regions largely composed of it, rather than of a geological principle. He was known primarily for his studies of volcanic substances and regions; among his related interests were earthquakes, the structure of mountain ranges, the classification of rocks, and the fashion in which chemical and mineralogical studies could be applied to historical interpretation of the earth.

Dolomieu ascribed his own interest in volcanoes to the influence of Barthélemy Faujas de Saint-Fond, to whom he wrote from Portugal in 1778 concerning his investigation of the origin of basalt. During the 1770's the new idea that basalt might be of volcanic origin attracted many scientists, especially the French, and Dolomieu rapidly joined the ranks of the adherents of this theory. He accounted for the prismatic form of certain basalts by arguing that they had suffered sudden contraction from the cooling effect of water. His commitment to the proposition that volcanic products are more than casual and accidental creations of the earth, but rather constitute a significant proportion of the earth's features, is reflected in such descriptive writings as *Voyage aux îles de Lipari* (1783) and *Mémoire sur les îles Ponces* (1788), as well as his more analytical papers, most of which were published in *Observations sur la physique.*

Despite his belief in the historical significance of

volcanoes, however, Dolomieu was convinced that aqueous agents were the outstanding causes of geological change. His volcanism was always tempered by this belief. In 1790, at the beginning of the basalt controversy, he declared that "far from extending the empire of subterranean fire, I believe that more than any other [mineralogist] I have circumscribed its true limits and excluded from its domain an infinitude of regions, a multitude of substances that have been attributed to it."[1] He would not join with those who would make volcanoes responsible for the majority of geological events; to his mind, volcanic effects were limited both spatially and temporally. To be sure, volcanic activity was not an ephemeral event and had occurred repeatedly during various stages of the earth's history, but by comparison with aqueous agents it was historically an occasional event of inferior significance. He wrote that "the humid way is the most universal means, the oldest, that which acts quietly in all times and all places, to which almost all of our globe belongs, which everywhere reasserts itself and regains possession of that part of its empire . . . that it yielded momentarily to the dry way." He estimated that no more than one-twentieth of the entire surface of the earth had ever been affected by volcanic action.[2]

In expounding his doctrine of aqueous geological activity, Dolomieu set forth a historical scheme for the earth that contained many of the components of neptunism. He believed that the oldest rocks had been precipitated out of a universal fluid in the earliest epochs from which there remains any evidence. It was perfectly plausible, he acknowledged, that prior to this coagulation the earth may have experienced a history of undeterminable length, but the absence of remnants from this era prevents our knowing anything about it. The history of the earth begins, for our purposes, with a terraqueous globe and the precipitation of matter from the universal fluid.

Dolomieu reasoned that this fluid cannot have been water alone, since so much matter could not have been dissolved without the aid of some other agent, perhaps a principle of fire or light, such as phlogiston (in 1791 Dolomieu had already noted that this principle was being rejected by chemists, but this did not deter him from positing some like essence). Precipitation out of the primordial solvent must have taken place in reverse order of solubility, in a slow and orderly fashion, and upon its completion the solvent material had largely disappeared. The means of its removal posed a problem; perhaps the atmosphere absorbed it.

The next stage in the geological history proposed by Dolomieu was a series of violent upheavals resulting in the rearrangement of the originally crystallized rocks. These catastrophes occurred before the creation of mechanical deposits and determined once and for all the major irregularities of the earth's surface. Whatever consideration Dolomieu gave to geological agents of a regular nature, he did not envision them as capable of accomplishing large alterations in the mountains and basins created by these catastrophes. The source of the violent uplifting force causing this sudden rearrangement was uncertain, but might probably be chosen from among three possibilities—interior force, the loss of interior support (as in the creation of underground caverns), or exterior shock. Dolomieu inclined toward the last possibility. On consulting Laplace, Dolomieu was assured that normal gravitational forces could not account for such upheavals, but that either a passing comet or the "accidental" causes of volcanic eruptions might bring about catastrophic uplift.[3]

Following the process of upheaval, alterations in the rearranged depositions were brought about by "transport," or mechanical deposits deriving from degradation of the mountainous uplands. These beds, however, were also disturbed from time to time by catastrophic currents, probably in the form of immense tides. Dolomieu steadfastly held to a belief in recurrent catastrophic alteration of the fundamentally established order of things, and in the greatly variable intensity of geological forces. His fascination with earthquakes was consistent with this catastrophism. Violent means were the principal causes of change in the earth's surface, he thought, and his geological time scale was accordingly short. He thought 10,000 years to be a generous—even excessive—estimate of the extent of the era following the great catastrophic upheavals. Dolomieu therefore opposed the idea of the action of slow and cumulative forces bringing about geological change over a great period of time. Calling upon force rather than time as the cause of such changes, he wrote that "in fashioning the earth such as we inhabit it, nature has not spent time with as much prodigality as some celebrated writers have supposed."[4] Remarks of this nature and his denial that significant consolidation of rocks occurs in the ocean depths, at least along coasts, suggest that Dolomieu may have intended to make known his opposition to the main tenets of Huttonian theory, although no explicit references to Hutton are known among Dolomieu's writings.

Dolomieu's overall geological scheme, then, was not volcanist, but rather shared much in common with the opponents of doctrinaire volcanism. All the same, he was an authority on volcanoes and volcanic action, and spoke with an influential voice on the subject.

One of his major concerns in studying volcanoes was the nature and source of volcanic ejecta. Dolomieu's view of the mineralogical nature of lava depended on the conception of lavas as being warm and viscous, but never especially hot. An intrinsic source of heat within the lava, he believed, maintains the lava's heat at a relatively even and moderate level. The exact identity of this intrinsic heat source was uncertain; it might be (or contain) sulfur, and it might be the same principle responsible for the binding action within certain rocks (such as granite), which always seem hard when unearthed but often crumble upon exposure to air, presumably because of the release of some substance that had joined the parts.

In any case, Dolomieu rejected any possibility of finding the source of volcanic heat in such fuels as coal or oil. On the other hand, there appears to be little ground for attributing to him a belief in an intense central heat within the globe. His investigations in Auvergne did convince him by 1797 that granites lie above the sources of volcanic eruptions and that the interior of the globe may therefore be fluid, but he did not cast aside the opinion that the modest heat of this viscous core comes from an intrinsic chemical source, such as caloric.[5]

The components of lavas are always traceable to a nonvolcanic origin, according to Dolomieu. In lava these components do not decompose, for the most part, but retain their mineralogical character. Volcanic heat does not destroy the older mineral composition, but merely "dilates" or "disunites" the integrant molecules, allowing them to slip past one another without being disaggregated. Lavas are therefore not vitrifications. The sources (foyers) of volcanoes are quite deep below the surface of the earth and are not limited to one particular type of rock, which explains the variability in composition of lavas. Many of Dolomieu's writings reflect his serious concern to determine which mineral substances have been subject to volcanic action, as well as the effects of volcanic heat on the resulting rock. His deep knowledge of the precise nature of various volcanic ejecta was perhaps the main foundation of the respect accorded him by other scientists.

There is discernible in the development of Dolomieu's thought a drift toward increasing interest in a theory of mineral classification and lithology, accompanied by a growing interest in what German mineralogists were doing. In the early 1780's Dolomieu secured a shipment of Saxon mineral samples through the assistance of Johann Friedrich Charpentier, professor at the Freiberg Bergakademie, and shortly before his death Dolomieu was planning a trip to Freiberg to visit Werner and perhaps take steps toward a reconciliation of German and French mineralogies.

Dolomieu clearly felt a certain affinity with Werner, but his greatest expressions of mineralogical debt and admiration were reserved for Haüy, whom he regarded as the founder of a new and highly fruitful approach to knowledge of mineral substances. Dolomieu came to believe in the importance of the form and constitution of the constituent structural unit, as opposed to the relative quantity of constituent substances, in determining a mineral substance's characteristics. His commitment to the integrant molecule as the basis of a new mineralogy is recorded in *Sur la philosophie minéralogique, et sur l'espèce minéralogique* (1801), which he had begun to compose during his imprisonment and in which he pursued the hope of raising mineralogy to the degree of precision that had recently been achieved, he thought, by chemistry.

Dolomieu was excited by the prospect of reducing the variety of mineral appearances to the chemical and spatial properties of the unique integrant molecule—which by definition established mineral species. Part of his excitement derived from his recognition that by assuming the integrant molecule to exist, one avoided having to deal with mineral species as only a matter of convention.

In light of Dolomieu's avowed empiricism, it is interesting that he saw this promising new development in mineralogy as the outgrowth of conceptual, not observational, investigation; species cannot be definitively distinguished by any mechanical operation, but only in the mind. On the whole, however, Dolomieu regarded himself as a scientist firmly rooted in observational technique. His many trips in the volcanic regions of Italy, in the Alps, and in other areas of geological interest were part of his program for geological investigation, and his mineral collection appears to have been a major focus of his empirical energies, especially early in his career.

Dolomieu was, however, given to making occasional tributes to experience as the fount of scientific knowledge. The scientists he praised most highly were often observationalists like Horace Bénédict de Saussure. He was inclined to see disputes over geological issues as resolvable by recourse to simple observation, if not by reasonable determination to eliminate unnecessary semantic disagreement. In the turmoil of doctrinaire controversy in geology at the end of the eighteenth century, he saw himself as a practical-minded and moderate scientist with excellent credentials as a mediator between rival extremes.

Dolomite was named after Dolomieu by Nicolas Théodore de Saussure, to whom Dolomieu had given samples of the substance after describing it as a calcareous rock from Tirol that was attacked by acid without effervescence.[6]

NOTES

1. "Sur la question de l'origine du basalte," in *Observations sur la physique,* **37** (1790), 193.
2. *Ibid.,* p. 194.
3. Lacroix, *Déodat Dolomieu,* II, 132.
4. "Mémoire sur les pierres composées," in *Observations sur la physique,* **39** (1791), 394.
5. "Rapport," in *Journal de physique,* **3** (1798), 406–411.
6. *Observations sur la physique,* **39** (1791), 3–10; **40** (1792), 161–173.

BIBLIOGRAPHY

I. ORIGINAL WORKS. Dolomieu's books are *Voyage aux îles de Lipari fait en 1781* (Paris, 1783; German trans., Leipzig, 1783); *Mémoire sur les tremblemens de terre de la Calabre pendant l'année 1783* (Rome, 1784; Italian trans., Rome, 1784; German trans., Leipzig, 1789; English trans., as part of John Pinkerton, *General Collection of the Best and Most Interesting Voyages and Travels in All Parts of the World,* V [London, 1809], 273–297); *Mémoire sur les îles Ponces, et catalogue raisonné des produits de l'Etna* (Paris, 1788); and *Sur la philosophie minéralogique, et sur l'espèce minéralogique* (Paris, 1801; German trans., Hamburg–Mainz, 1802).

Among Dolomieu's principal articles are "Mémoire sur les volcans éteints du Val di Noto en Sicile," in *Observations sur la physique, sur l'histoire naturelle et sur les arts,* **25** (1784), 191–205; "Lettre de M. le commandeur Déodat de Dolomieux, à M. le baron de Salis-Masklin, à Coire dans les Grisons: Sur la question de l'origine du basalte," *ibid.,* **37** (1790), 193–202; "Mémoire sur les pierres composées et sur les roches," *ibid.,* **39** (1791), 374–407; **40** (1792), 41–62, 203–218, 372–403; "Mémoire sur la constitution physique de l'Égypte," *ibid.,* **42** (1793), 41–61, 108–126, 194–215; "Distribution méthodique de toutes les matières dont l'accumulation forme les montagnes volcaniques, ou tableau systématique dans lequel peuvent se placer toutes les substances qui ont des relations avec les feux souterrains," in *Journal de physique, de chimie et d'histoire naturelle,* **1** (1794), 102–125; "Mémoire sur les roches en général, & particulièrement sur les pétro-silex, les trapps & les roches de corne, pour servir à la distribution méthodique des produits volcaniques," *ibid.,* **1** (1794), 175–200, 241–263, 406–428; **2** (1794), 81–105; "Discours sur l'étude de la géologie," *ibid.,* **2** (1794), 256–272; "Lettre à M. Pictet, professeur de physique à Genève, et membre de la Société royale de Londres, sur la chaleur des laves, et sur des concrétions quartzeuses," in *Journal des mines,* **4,** no. 22 (1796), 53–72; "Lettre sur la nécessité d'unir les connaissances chimiques à celles du minéralogiste; Avec des observations sur la différente acception que les auteurs allemands et français donnent au mot chrysolithe," *ibid.,* **5,** no. 29 (1797), 365–376; "Rapport fait à l'Institut National, par Dolomieu, sur ses voyages de l'an cinquième & sixième," in *Journal de physique, de chimie, d'histoire naturelle et des arts,* **3** (1798),
401–427, and *Journal des mines,* **7** (1798), 385–402, 405–432.

Dolomieu's "Osservazioni ed annotazioni relative a spiegare ed illustrare la classazione metodica di tutte le produzioni volcaniche" was published with Torbern Bergman's *De' prodotti volcanici considerati chimicamente dissertazione,* Giuseppe Tofani, trans. (Florence, 1789[?]).

Dolomieu's letters to Barthélemy Faujas de Saint-Fond (1778) were published in the latter's *Recherches sur les volcans éteints du Vivarais et du Velay* (Grenoble–Paris, 1778), pp. 440–446. Alfred Lacroix edited several of Dolomieu's previously unpublished papers and notes, including "Une note de Dolomieu sur les basaltes de Lisbonne, adressée en 1779 à l'Académie Royale des Sciences," in *Comptes rendus hebdomadaires des séances de l'Académie des sciences,* **167** (1918), 437–444; "Un voyage géologique en Sicile en 1781," in *Bulletin de la Section de géographie du Comité des travaux historiques et scientifiques* (1918), 29–213; "L'exploration géologique des Pyrénées par Dolomieu en 1782," in *Bulletin de la Société Ramond* (1917–1918), 120–178; and "Vues générales sur le Dauphiné," in *Bulletin de la Société de statistique, des sciences naturelles et des arts industriels du département de l'Isère,* **40** (1919), 237–282. Dolomieu's Egyptian notes, G. Daressy, ed., appeared as "Dolomieu en Égypte (30 juin 1798–10 mars 1799)," in *Mémoires présentés à l'Institut d'Égypte,* **3** (1922).

II. SECONDARY LITERATURE. The fullest biographical treatment is Alfred Lacroix, *Déodat Dolomieu,* 2 vols. (Paris, 1921), consisting largely of selections from Dolomieu's correspondence and other previously unedited material, preceded by a "Notice historique." A contemporary eulogy is Lacépède's "Notice historique sur la vie et les ouvrages de Dolomieu," in *Histoire de la classe des sciences mathématiques et physiques* (Institut National de France), **7** (1806), 117–138, and in *Journal des mines,* **12** (1802), 221–242.

Information on Dolomieu's mineralogical ideas is in Karl Wilhelm Nose, *Beschreibung einer Sammlung von meist vulkanisirten Fossilien die Deodat-Dolomieu im Jahre 1791 von Maltha aus nach Augsburg und Berlin versandte* (Frankfurt-am-Main, 1797). T. C. Bruun-Neergaard's account of Dolomieu's last summer tour is in *Journal du dernier voyage du C^{en}. Dolomieu dans les Alpes* (Paris, 1802).

See also Kenneth L. Taylor, "The Geology of Déodat de Dolomieu," in *Actes,* XIIth International Congress of the History of Science (1968).

KENNETH L. TAYLOR

DOMAGK, GERHARD (*b.* Lagow, Brandenburg, Germany, 30 October 1895; *d.* Burgberg, Germany, 24 April 1964), *medicine, chemistry, pharmacology.*

Domagk, the son of a teacher, decided to study medicine while still at a scientifically oriented grammar school in Liegnitz (now Legnica). During his first term at the University of Kiel, World War I broke out and Domagk volunteered for active service with a German grenadier regiment. After being wounded he transferred to the German army medical corps and received his M.D. at Kiel in 1921. For a short while he worked as an assistant to the chemist Ernest Hoppe-Seyler and in 1924 became reader (*Privatdozent*) in pathology at the University of Greifswald. In 1925 he accepted a similar post at the University of Münster and married Gertrud Strube. They had four children, and his only daughter was one of the first patients to be treated successfully with prontosil rubrum for a severe streptococcal infection.

Domagk became extraordinary professor of general pathology and pathological anatomy at Münster in 1928 and ordinary professor in 1958.

In 1924 Domagk published a paper on the defensive function of the reticuloendothelial system against infections. As a result of that paper and of his well-known interest in chemotherapy the directors of the I. G. Farbenindustrie appointed him—at the age of thirty-two—director of research at their laboratory for experimental pathology and bacteriology at Wuppertal-Elberfeld. It was the turning point of his career.

Since Paul Ehrlich's discovery of arsphenamine in 1909, chemotherapy had advanced in the field of protozoan and tropical diseases, but hardly any progress had been made in regard to bacterial infections of man; and the I. G. Farbenindustrie had decided on further testing of potential antibacterial agents, along the lines laid down by Ehrlich. Domagk's interest centered on the so-called azo dyes, in which one hydrogen atom had been replaced by a sulfonamide group. These dyes, which had been developed as early as 1909 by H. Hörlein and his collaborators, conferred on textiles a high resistance to washing and light, because of their intimate combination with wool proteins.

In 1932 Domagk's colleagues, the chemists Fritz Mietzsch and Josef Klarer, synthesized a new azo dye, hoping that it would prove to be a fast dye for treating leather. It was -4′ sulfonamide-2-4-diaminoazobenzol, which they named prontosil rubrum. Domagk early recognized its protective power against streptococcal infections in mice and its low toxicity, but he withheld publication of his findings until 1935. His paper "Ein Beitrag zur Chemotherapie der bakteriellen Infektionen" has become not only a classic but—measured by strict experimental and statistical yardsticks—a masterpiece of careful and critical evaluation of a new therapeutic agent.

As early as 1933 A. Förster had reported the dramatic recovery of an infant with staphylococcal septicemia after treatment with prontosil rubrum, but Domagk's discovery—after so many years of fruitless searching for specific antibacterial agents—was received with a great deal of skepticism. In 1936

L. Colebrook and M. Kenny of the British Medical Research Council confirmed Domagk's findings and concluded that "the clinical results together with the mouse experiments support the view that . . . there is more hope of controlling these early streptococcal infections by the administration of this or some related chemotherapeutical agent than by any other means at present available." In the first paper on prontosil published in the United States, P. H. Long and E. A. Bliss mentioned Domagk only as one of the investigators of prontosil, but in the same issue of the *Journal of the American Medical Association*, its editor gave Domagk full credit for his significant paper and hoped that "further investigations will disclose a definite group of disorders characterised by high virulence and mortality which can be materially helped by appropriate chemotherapy." Their hopes were indeed fulfilled, but only after workers at the Pasteur Institute in Paris—the Tréfuëls, F. Nitti, D. Bovet, and E. Fourneau—had established that the azo component of prontosil dissociated *in vivo* and that the liberated sulfonamide radical was responsible for the antibacterial effect. This was a very important discovery, because sulfonamide could be produced far more cheaply than prontosil.

Ironically enough, at that very time an agar plate containing an even more powerful antibacterial agent—penicillin—lay forgotten in St. Mary's Hospital in London. Its owner, Alexander Fleming, had become highly interested in prontosil and the sulfonamide derivatives that followed, but in his many papers on antibacterial and antiseptic treatment published between 1938 and 1940 he never mentioned penicillin, the antistaphylococcal action of which he had first observed in 1928.

In 1938 L. E. H. Whitby synthesized sulfapyridine, which soon became the drug of choice in the treatment of the pneumococcal pneumonias. Many other sulfa drugs followed in quick succession. The structural formulas of the most important ones are shown in Figure 1.

Domagk's discovery profoundly changed the prognosis of many dangerous and potentially fatal diseases such as puerperal fever, erysipelas, cerebrospinal meningitis, and pneumonia.

In October 1939, sponsored by American, French, and British scientists, Domagk was awarded the Nobel Prize in physiology or medicine. He duly acknowledged the great honor to the rector of the Caroline Institute; but some weeks later, after having been arrested by the Gestapo, he declined the prize in a letter drafted for him by the German authorities. (After the peace prize had been awarded in 1936 to the German radical and pacifist writer Carl von

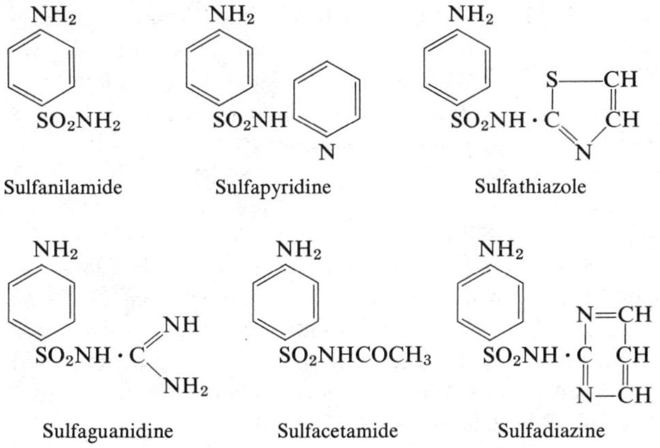

FIGURE 1. Structural formulas of some major sulfonamides.

Ossietzky, who was a prisoner in a concentration camp, Hitler had forbidden any German to accept a Nobel Prize.) Domagk eventually received the Nobel Prize medal (1947), but by then the prize money had been redistributed. He received many other high honors and an honorary professorship of the University of Valencia, and he became doctor *honoris causa* of many European and American universities. He was especially pleased by the award of the Paul Ehrlich Gold Medal and the Cameron Medal of Edinburgh and by his election as a foreign correspondent of the Royal Society of London.

The discovery of the sulfonamides reawakened interest in the sulfones, the derivatives of 4, 4′-diaminodiphenyl-sulfone (DDS) (Fig. 2). Because of their high toxicity the sulfones had never been favored in the treatment of acute bacterial infections, but in 1941, E. V. Cowdry and C. Ruangsiri reported encouraging effects in treating rat leprosy with a DDS derivative, sodium glucosulfone. After G. H. Faget and his co-workers confirmed this in a clinical trial in 1943, the sulfones largely replaced the venerable chaulmoogra oil in the treatment of human leprosy and revolutionized the prognosis of this biblical scourge.

After World War II, with the antibiotics having

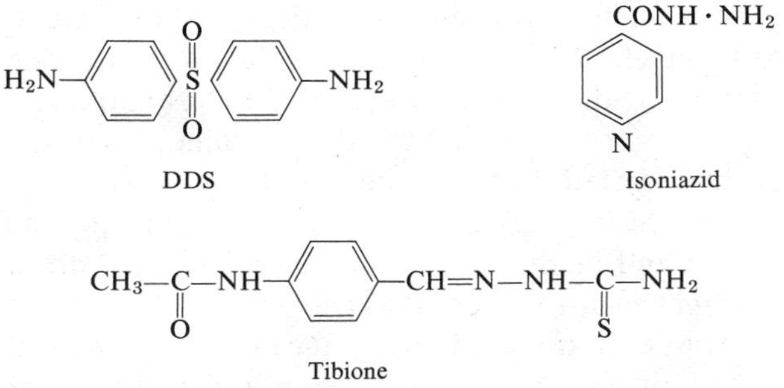

FIGURE 2. Structural formulas of DDS, isoniazid, and tibione.

joined the therapeutic armamentarium against acute bacterial diseases, Domagk's interests shifted to the chemotherapy of tuberculosis. The euphoria induced by the discovery of streptomycin had by that time given way to considerable disillusionment because of the rapidly increasing numbers of streptomycin-resistant strains of the tuberculosis mycobacterium.

Together with R. Behnisch, F. Mietzsch, and H. Schmidt, Domagk reported in 1946 on the tuberculostatic action *in vitro* of the thiosemicarbazones of which the 4′-acetyl-aminobenzaldehyde (Conteben, Tibione) seemed to be the most promising compounds. Because of their many and dangerous toxic side effects the thiosemicarbazones never became popular in clinical medicine, but for many years they were used as "second-line drugs" against mycobacteria that were resistant to one or more of the standard antituberculous drugs.

The work of Domagk and his colleagues with the thiosemicarbazones resulted, however, in a supremely important accidental discovery. In 1945 V. Chorine had reported on the tuberculostatic action of the nicotinamides, and his findings were rediscovered by D. McKenzie, L. Malone, S. Kushner, J. J. Oleson, and Y. Subba Row in 1948. This produced no practical results until it became known that the active agents of the nicotinamides were derivatives of isonicotinic acids. In 1951 H. H. Fox tried to prepare isonicotinaldehyde-thiosemicarbazone. An intermediate product, isonicotinoylhydrazine (isoniazid), was tested in New York's Sea View Hospital in 1952 and has since become one of the most potent and reliable drugs in the treatment of tuberculosis.

Finally Domagk turned to the greatest challenge of all, the chemotherapy of cancer. He experimented mostly with ethylene-iminoquinones, but success evaded him as it did so many other workers in that field. In a letter quoted by Colebrook he wrote: "One should not have too great expectations of the future of cytostatic agents."

It is characteristic of Domagk's intensive scientific curiosity and humane outlook that he wrote at the end of his life: "If I could start again, I would perhaps become a psychiatrist and search for a causal therapy of Mental Disease which is the most terrifying problem of our times."

The dawn of the new chemotherapeutic era in Germany was no accident, if only for the traditional close association between the chemical industry and medical research in that country. Nevertheless, twenty-seven years passed between P. Gelmo's first preparation of a sulfonamide in 1908, during Ehrlich's lifetime, and its recognition by Domagk as an *elixirium magnum sterilisans*. Gelmo's original paper,

"Sulphamides of P-aminobenzene Sulphonic Acid," is in *Journal für praktische Chemie,* **77** (1908), 369–382.

Domagk was fortunate in having adequate chemical help, the lack of which prevented Fleming from advancing with penicillin for eleven years. But here, as in so many similar situations, Pasteur's famous dictum applies: "The lucky chance favors only the prepared mind."

BIBLIOGRAPHY

I. ORIGINAL WORKS. Domagk's writings are "Untersuchungen über die Bedeutung des retikuloendothelialen Systems für die Vernichtung von Infektionserregern und für die Enstehung des Amyloids," in *Virchows Archiv für pathologische Anatomie und Physiologie und für klinische Medizin,* **253** (1924), 594–638; "Ein Beitrag zur Chemotherapie der bakteriellen Infektionen," in *Deutsche medizinische Wochenschrift,* **61** (1935), 250–253; "Über eine neue, gegen Tuberkelbazillen in vitro wirksame Verbindungsklasse," in *Naturwissenschaften,* **33** (1946), 315, written with R. Behnisch, F. Mietzsch, and H. Schmidt; *Pathologische Anatomie und Chemotherapie der Infektions-Krankheiten* (Stuttgart, 1947); "Investigations on the Anti-tuberculous Activity of the Thiosemicarbazones *in vitro* & *in vivo,*" in *American Review of Tuberculosis and Pulmonary Diseases,* **61** (1950), 8–19; "Chemotherapy of Cancer by Ethylenimino-quinones," in *Annals of the New York Academy of Sciences,* **68** (1957–1958), 1197–1204; and "Über 30 Jahre Arzt," in *Therapie der Gegenwart,* **102** (1963), 913–917.

II. SECONDARY LITERATURE. Writings on Domagk or his work are V. Chorine, "Action de l'amide nicotinique sur les bacilles du genre mycobacterium," in *Comptes rendus hebdomadaires des séances de l'Académie des sciences,* **220** (1945), 150–151; L. Colebrook, "Gerhard Domagk," in *Biographical Memoirs of the Royal Society,* **10** (1964), 39–50; L. Colebrook and M. Kenny, "Treatment of Human Puerperal Infections and Experimental Infections in Mice with Prontosil," in *Lancet* (1936), **1**, 1279–1286; E. V. Cowdry and C. Ruangsuri, "Influence of Promin, Starch and Heptaldehyde on Experimental Leprosy in Rats," in *Archives of Pathology,* **32** (1941), 632–640; G. H. Faget *et al.,* "The Promin Treatment of Leprosy, a Progress Report," in *Public Health Reports,* **58** (1943), 1729–1741; E. Fourneau *et al.,* "Chimiothérapie des infections streptococciques par les dérivés du p-aminophénylsulfamide," in *Comptes rendus des séances de la Société de biologie,* **122** (1936), 652–654; H. H. Fox, "The Chemical Attack on Tuberculosis," in *Transactions of the New York Academy of Sciences,* **15** (1952), 234–242; L. S. Goodman and A. Gilman, *The Pharmacological Basis of Therapeutics* (New York, 1960), pp. 1250–1300.

See also H. Hörlein, "Chemotherapy of Infectious Diseases Caused by Protozoa and Bacteria," in *Proceedings of the Royal Society of Medicine,* **29** (1936), 313–324; editorial in *Journal of the American Medical Association,* **108** (1937),

48–49; P. H. Long and E. A. Bliss, "Para-Amino-Benzene-Sulfonamide and Its Derivatives," *ibid.,* pp. 32–37; D. McKenzie *et al.,* "The Effect of Nicotinic Acid-amide on Experimental Tuberculosis of White Mice," in *Journal of Laboratory and Clinical Medicine,* **33** (1948), 1249–1253; I. J. Selikoff *et al.,* "Chemotherapy of Tuberculosis With Hydrazine Derivatives of Isonicotinic Acid," in *Quarterly Bulletin of Sea View Hospital* (New York), **13** (1952), 27; T. L. Sourkes, *Nobel Prize Winners* (New York, 1967), pp. 214–219; O. Warburg, "Gerhard Domagk," in *Deutsche medizinische Wochenschrift,* **90** (1965), 34, 1484–1486; L. E. H. Whitby, "Chemotherapy of Pneumococcal and Other Infections With 2- (p-aminobenzenesulfonamide) pyridine," in *Lancet* (1938), **1**, 1210–1212.

ERICH POSNER

DOMBEY, JOSEPH (*b.* Mâcon, France, 22 February 1742; *d.* Montserrat, West Indies, 18 February 1794), *medicine, botany.*

The tenth of the fourteen children of Jean-Philibert Dombey, a baker, and Marie Carra, Dombey was orphaned at the age of fourteen. After studying with the Jesuits, he studied medicine at Montpellier and graduated in 1767. Antoine Gouan and Pierre Cusson stimulated Dombey to pursue natural history, and he carried out fieldwork in the Pyrenees, the Alps, and the Vosges. He went to Paris in 1772 and took courses under Antoine-Laurent Jussieu and Lemonnier, and in 1776 he was appointed royal botanist for the study of South American plants that could acclimatize to France.

Dombey left Cádiz in 1777 with the Spanish botanists Hipólito Ruiz and José Pavón, reaching Callao, Peru, in 1778. Dombey studied the Peruvian vegetation, particularly the so-called American cinnamon, searched for platinum and saltpeter, analyzed the Chauchín spa, and made archaeological explorations in Chan Chan, Pachacamac, and Tarma; at Huánuco he found cinchona. In 1781 he went to Chile, exploring the mines of Coquimbo and Copiapó, and returned to France in 1785.

Dombey had promised not to publish his observations prior to those of the Spaniards, and his ill health and disappointments caused him to sell his books and to burn all his notes and observations when he retired to Lyons. In 1793, during the Revolution, he became surgeon at the military hospital of Lyons and afterward was commissioned to take the standards of the decimal system to the United States and to arrange the purchase of grain. His ship was captured by the British and he died in prison, but news of his death did not reach France until October 1794.

Dombey introduced the Araucanian pine, named after him, into naval construction and presented to the Jardin des Plantes, Paris, a great number of specimens and his herbarium, containing more than 1,500 new species. He was made a member of the Academia de Medicina, Madrid, and the Académie des Sciences, Paris.

BIBLIOGRAPHY

I. ORIGINAL WORKS. Since he burned his notes and manuscripts, Dombey's only printed works are "Lettre sur la salpêtre du Pérou et la phosphorescence de la mer," published by Lalande in the *Journal de physique,* **15** (1780), 212–214; and certain observations included by Ruiz and Pavón in the *Flora peruviana et chilensis* (4 vols., Madrid, 1798–1802). Dombey's correspondence has been published as E. T. Hamy, *Joseph Dombey, médecin, naturaliste, archéologue, explorateur du Pérou, du Chili et du Brésil (1778–1785), sa vie, son oeuvre, sa correspondance* (Paris, 1905). J. Riquelme Salar, "El Doctor Dombey, médico francés; su labor científica en los reinos del Perú y Chile, año 1777–85," in *Proceedings of the International Congress of the History of Medicine,* **1** (1959), 160–162, also contains the correspondence.

II. SECONDARY LITERATURE. On Dombey or his work, see E. Dubois, *Le naturaliste Joseph Dombey* (Bourg-en-Bresse, 1934); and A. R. Steele, *Flowers for the King. The Expedition of Ruiz and Pavón and the Flora of Perou* (Durham, N.C., 1964).

FRANCISCO GUERRA

DOMINGO DE SOTO. *See* **Soto, Domingo de.**

DOMINIC GUNDISSALINUS. *See* **Gundissalinus, Dominicus.**

DOMINICUS DE CLAVASIO, also known as **Dominicus de Clavagio, Dominicus Parisiensis,** or **Dominic de Chivasso** (*fl.* mid-fourteenth century), *mathematics, medicine, astrology.*

Dominicus de Clavasio's birthdate is unknown, but he was born near Turin and was active in Paris from about the mid-1340's. He taught arts at Paris during 1349–1350, was head of the Collège de Constantinople at Paris in 1349, and was an M.A. by 1350. Dominicus received the M.D. by 1356 and was on the medical faculty at Paris during 1356–1357. He was astrologer at the court of John II and may have died between 1357 and 1362.

Dominicus is the author of a *Practica geometriae* written in 1346; a *questio* on the *Sphere* of Sacrobosco; a *Questiones super perspectivam;* a set of *questiones* on the first two books of the *De caelo* of Aristotle, written before 1357; and possibly a commentary on Aristotle's *Meteorology.* He mentions in the *Practica* his intention to write a *Tractatus de umbris et radiis.* The *questiones* on the *De caelo* have not been edited, although a few that are concerned with physical problems have been examined. They reveal that Dominicus is part of the tradition established at Paris during the fourteenth century by Jean Buridan, Nicole Oresme, and Albert of Saxony. Like these Parisian contemporaries, he adopted the impetus theory as an explanation of projectile motion as well as of acceleration in free fall. Also like his colleagues at Paris, Dominicus considered impetus as a quality. As is true of the *Questiones de caelo* of Albert of Saxony, Dominicus' discussions of impetus reveal the influence of both Oresme and Buridan. If Dominicus were directly familiar with Oresme's conceptions of impetus, he most likely drew them from the latter's early Latin *questiones* on the *De caelo* and obviously not from his much later French *Du ciel.* According to Dominicus, a body in violent motion possessed both impetus and an "actual force" (*virtus actualis*), although the relationship between these factors is unclear. Also, as Nicole Oresme may have done, he may have connected impetus with acceleration rather than velocity.

The *Practica* was a popular work during the Middle Ages and has survived in numerous manuscript versions. It served, for example, as a model for a *Geometria culmensis,* written in both Latin and German near the end of the fourteenth century. The *Practica* is divided into an introduction and three books. The introduction contains arithmetical rules and the description of an instrument, the *quadratum geometricum* of Gerbert. Book I deals with problems of measurement, book II contains geometrical constructions of two-dimensional figures, and book III is concerned with three-dimensional figures. In the course of the *Practica,* Dominicus mentions Ptolemy and the thirteenth-century mathematician and astronomer Campanus of Novara.

The *Questiones super perspectivam* reveal Dominicus' familiarity with the standard authors of the medieval optical tradition, such as Witelo, al-Rāzī (Rhazes), Roger Bacon, Peckham, and Ibn al-Haytham (Alhazen). His work is not based, however, on the influential *Perspectiva communis* of Peckham but is a commentary on the *De aspectibus* of Ibn al-Haytham and the latter's Latin successor, Witelo.

Insofar as his thought has been examined, Dominicus de Clavasio appears not as an innovator but as a fairly conventional continuator of well-established medieval traditions.

BIBLIOGRAPHY

I. ORIGINAL WORKS. H. L. L. Busard, ed., "The *Practica geometriae* of Dominicus de Clavasio," in *Archives for*

History of Exact Sciences, **2** (1962–1966), 520–575, contains the entire text; Graziella Federici Vescovini, "Les questions de 'perspective' de Dominicus de Clivaxo," in *Centaurus,* **10** (1964–1965), 14–28, contains an edition of questions 1 and 6.

II. SECONDARY LITERATURE. On Dominicus or his work, see A. von Braunmühl, *Vorlesungen über Geschichte der Trigonometrie,* I (Leipzig, 1900), 107–110; M. Cantor, *Vorlesungen über Geschichte der Mathematik,* II (Leipzig, 1899), 127, 150–154, 450–452; M. Clagett, *The Science of Mechanics in the Middle Ages* (Madison, Wis., 1959), pp. 635, 636, note; M. Curtze, "Über den Dominicus parisiensis der *Geometria culmensis,*" in *Bibliotheca mathematica,* 2nd ser., **9** (1895), 107–110; and "Über die im Mittelalter zur Feldmessung benutzten Instrumente," *ibid.,* **10** (1896), 69–72; G. Eneström, "Über zwei angebliche mathematische Schulen im christlicher Mittelalter," in *Bibliotheca mathematica,* 3rd ser., **7** (1907), 252–262; Anneliese Maier, *Zwei Grundprobleme der scholastischen Naturphilosophie* (Rome, 1951), pp. 241–243; *An der Grenze von Scholastik und Naturwissenschaft* (Rome, 1952), pp. 121, 209; *Metaphysische Hintergrunde der spätscholastischen Naturphilosophie* (Rome, 1955), p. 365, note; and *Zwischen Philosophie und Mechanik* (Rome, 1958), p. 218; H. Mendthal, ed., *Geometria culmensis. Ein agronomischer Tractat aus der Zeit des Hochmeisters Conrad von Jungingen, 1393–1407* (Leipzig, 1886), which contains the Latin and German texts of the *Geometria culmensis;* K. Michalski, "La physique nouvelle et les différents courants philosophiques au xive siècle," in *Bulletin international de l'Académie polonaise des sciences et des lettres. Classe d'histoire et de philosophie, et de philologie* (Cracow), *année 1927* (1928), 150; P. Tannery, *Mémoires scientifiques,* V, J. L. Heiberg, ed. (Paris, 1922), 329–330, 357–358; Lynn Thorndike, *A History of Magic and Experimental Science,* III (New York, 1934), 587–588; *The Sphere of Sacrobosco and Its Commentators* (Chicago, 1949), p. 37; and E. Wickersheimer, *Dictionnaire biographique des médecins en France au moyen âge* (Paris, 1936), p. 121.

CLAUDIA KREN

DOMINIS, MARKO ANTONIJE (*b.* Rab, Yugoslavia, 1560; *d.* Rome, Italy, 8 September 1626), *physics.*

After finishing his studies in Padua, Dominis lectured on mathematics, logic, and philosophy at Verona, Padua, and Brescia until 1596. Later he was appointed bishop of Senj and then archbishop of Split. The last years of his life were devoted chiefly to theological questions. In his main theological work, *De republica ecclesiastica,* which he wrote during this period and which contributed most to his fame, he urged unity of all Christian churches, their commitment to exclusively spiritual ends, and peace among nations. These beliefs made it necessary for Dominis to flee to England. Soon after his return to Rome he was imprisoned by the Inquisition. He died in a dungeon, and after his death he was found guilty of heresy and his body was burned.

Dominis had written two works on physics by the time he lectured on mathematics in Padua. The first one, *De radiis visus et lucis,* deals with lenses, telescopes, and the rainbow. Dominis knew how light was refracted in its passage from one medium to another, but he was not always consistent in his assertions. He held that it was possible that in some cases light could pass through the border of a medium without being refracted—for instance, into a thin layer of water. In general, his observations on refraction in lenses were correct.

After the invention of the telescope Dominis added its theoretical explanation to his work. His explanation was not entirely satisfactory, however, because his knowledge of the law of refraction was incomplete. He concluded that the image of an object was enlarged by increasing the angle of sight, which he had previously defined correctly. Thus Dominis describes in particular detail the effect on the angle of sight of a lens of greater curvature or of a greater distance between the lens and the object being viewed. With the same thoroughness he examined lens combinations, in particular the combination of a convex object glass and a concave eyepiece. This work led to his discovery of the conditions under which the magnification of an image is possible.

A greater part of *De radiis visus et lucis* is concerned with the rainbow. Dominis held that a rainbow is caused by refraction and reflection in raindrops: upon entering the drop, the light is refracted, then is reflected on the inner side of the drop, and then leaves the drop. This process had already been asserted by Dietrich von Freiberg, but the dependence of Dominis' theory on Dietrich's cannot be proved. Unlike Dietrich, Dominis failed to see that upon leaving the raindrop, the light was refracted again. As for the secondary rainbow, Dominis thought that the light struck on the drops' surfaces more obliquely than in the primary rainbow. His theory does not assert that the secondary rainbow is caused after two reflections on the inner side of the drop. Dominis could not explain the order of the colors in a rainbow because he used the Aristotelian theory of the mixture of darkness and light.

In *Euripus seu de fluxu et refluxu* Dominis is concerned with the tides. The greater part of the book deals with the figure of the earth, advocating sphericity and refuting the opinion of Patricius, who thought that the configuration of the earth was irregular.

Dominis believed the tide was caused by the influence of the moon and the sun on the sea in a manner analogous to the working of a magnet. He adopted Grisogono's theory of a second daily tide caused by the influence of the celestial point opposite the moon or the celestial point opposite the sun. Yielding to objections of his contemporaries, Dominis corrected this theory and introduced the influence of a transpolar circle through the moon and the sun. Dominis summed up the influences of both bodies in any position and thus accounted for all the different elevations of the sea. He was correct in believing that the tide wave was not the result of the lateral transportation of a mass of water, but rather to its rising and falling in depth and place; the wave moves horizontally, but each particle of water vertically. He wrongly ascribed the cause of tide differences in closed seas to characteristics of these seas, such as salinity and warmth.

BIBLIOGRAPHY

I. Original Works. Dominis' scientific works are *De radiis visus et lucis in vitris perspectivis et iride* (Venice, 1611) and *Euripus seu de fluxu et refluxu maris sententia* (Rome, 1624). His main theological work is *De republica ecclesiastica libri X* (London, 1617).

II. Secondary Literature. On Dominis or his work, see R. E. Ockenden, "Marco Antonio de Dominis and His Explanation of the Rainbow," in *Isis,* **26** (1936), 40–49; Carl B. Boyer, *The Rainbow, From Myth to Mathematics* (New York, 1959), pp. 187–192; Josip Torbar, "Ob optici Markantuna de Dominisa" ("On the Optics of Marko Antonije de Dominis"), in *Rad Jugoslavenske akademije znanosti i umjetnosti* (Zagreb), **43** (1878), 196–219; Stanko Hondl, "Marko Antonij de Dominis kao fizičar" ("Marko Antonij de Dominis as Physicist"), in *Vienac,* **36**, no. 2 (1944), 36–48; Žarko Dadić, "Tumačenja pojave plime i oseke mora u djelima autora s područja Hrvatske" ("Explanations of the Tides in the Works of Croatian Authors"), in *Rasprave i gradja za povijest nauka* ("Treatises and Materials for the History of Science"), II (Zagreb, 1965), 87–143. A complete bibliography and discussions of Dominis' life and work (in Croatian) are in *Encyclopaedia moderna,* V–VI (1967), 84–140.

Žarko Dadić

DOMNINUS OF LARISSA (*b.* Larissa, *fl.* fifth century A.D.), *mathematics, philosophy.*

Domninus was a Syrian Jew of Larissa on the Orontes. (In his entry in the *Suda Lexicon* Larissa is regarded as identical with Laodicea, but they appear to have been separate towns.) He became a pupil of Syrianus, head of the Neoplatonic school at Athens, and a fellow student of Proclus. He therefore lived in the fifth century of the Christian era.

Syrianus thought equally highly of Domninus and Proclus, and Marinus relates how he offered to discourse to them on either the Orphic theories or the oracles; but Domninus wanted Orphism, Proclus the oracles, and they had not agreed when Syrianus died. Marinus implies that Domninus succeeded to Syrianus' chair, but if so, he can only have shared it for a short time with Proclus. Their disagreement widened into a controversy over the true interpretation of Platonic doctrine from which Proclus emerged as the victor in the eyes of the Academy. Domninus withdrew to Larissa, and Damascius, the last head of the Neoplatonic school, while admitting Domninus' mathematical competence, accused him of being old-fashioned in philosophical matters. This may have been a partisan judgment. Nor need we pay much attention to an anecdote related by him and intended to show that Domninus lacked the true philosophic mind: When Domninus was troubled by spitting blood and the Aesculapian oracle at Athens prescribed that he should eat swine's flesh, he had no scruples about so doing despite the precept of his Jewish religion. Equally suspect are Damascius' allegations that when advanced in years Domninus loved only the conversation of those who praised his superiority and that he would not admit to his company a young man who argued with him about a point in arithmetic.

Nothing has survived of his philosophical teachings. A reference by Proclus (*In Timaeum,* I.34B; Diehl, ed., I. 109. 30–110.12) indicates that Domninus took some interest in natural science. He held that comets were composed of a dry, vapor-like substance, and he explained the myth of Phaëthon by the assumption that the earth once passed through such a comet, the substance of which was ignited by the sun's rays and which in turn set the earth on fire.

It is on his mathematical work that his claim to remembrance rests. Nothing was known of this work until 1832, when J. F. Boissonade edited from two Paris manuscripts his Ἐγχειρίδιον ἀριθμητικῆς εἰσαγωγῆς ("Manual of Introductory Arithmetic"), and it was not until 1884 that its importance was recognized. Paul Tannery, who in the following year made a critical revision of Boissonade's text, then perceived that this brief work marked a reaction from the arithmetical notions of Nicomachus and a return to the sounder principles of Euclid. It may not be without significance that, according to Marinus, Proclus had become convinced by a dream that he possessed the soul of Nicomachus of Gerasa. Whereas numbers had been represented by Euclid as straight lines, Nicomachus had departed from this convention and, when dealing with unknown quantities, was forced into clumsy circumlocutions; he also introduced highly elaborate classifications of numbers, serving no useful purpose and difficult to justify. Proclus, according to the *Suda Lexicon*, wrote a commentary on Nicomachus' work, but without openly controverting him, Domninus, in the concise and well-ordered text edited by Boissonade, quietly undermines it. As Hultsch recognized, the book is arranged in five parts: an examination of numbers in themselves, an examination of numbers in relation to other numbers, the theory of numbers both in themselves and in relation to others, the theory of means and proportions, and the theory of numbers as figures. In general Domninus follows Euclid in his classification of numbers and departs from Nicomachus. He is content with the arithmetic, geometric, and harmonic means and finds no use for Nicomachus' seven other means. Like Euclid, he admits only plane and solid numbers. In writing this manual Domninus drew not only upon Euclid and Nicomachus but upon Theon of Smyrna and upon a source used by Iamblichus that has since disappeared.

At the end of the manual Domninus avows his intention of setting forth certain subjects more fully in an Ἀριθμητικὴ στοιχείωσις ("Elements of Arithmetic"). Whether he did so is not known, but a tract with the title Πῶς ἔστι λόγον ἐκ λόγου ἀφελεῖν ("How to Take a Ratio out of a Ratio"), edited by C. E. Ruelle in 1883, is almost certainly by Domninus and may have been written as part of the projected *Elements*. "Taking a ratio out of a ratio" does not mean subtraction but manipulation of the ratio so as to get it into other forms.

BIBLIOGRAPHY

Most of what is known about the life of Domninus comes from a long entry in the *Suda Lexicon* (Eva Adler, ed.) that seems to be derived from a lost work by Damascius. There is a short entry in Eudocia, *Violarium* 331, Flach, ed., 239. 7–10. Eudocia's source appears to be the lost *Onomatologos* of Hesychius of Miletus; see Hans Flach, *Untersuchungen ueber Eudoxia und Suidas,* p. 60. A brief notice is given by Marinus, *Vita Procli,* ch. 26, in J. F. Boissonade ed.

Domninus' *Manual of Introductory Arithmetic* may be found in J. F. Boissonade, *Anecdota graeca,* IV (Paris, 1832), 413–429. Paul Tannery commented on it in "Domninos de Larissa," in *Bulletin des sciences mathématiques et astronomiques,* 2nd ser., **8** (1884), 288–298, repr. in his *Mémoires scientifiques,* II (Paris–Toulouse, 1912), 105–117; he revised Boissonade's text at points in "Notes critiques sur Domninos," in *Revue de philologie,* **9** (1885), 129–137, repr. in *Mémoires scientifiques,* II, 211–222; he translated the work into French, with prolegomena, as "Le manuel d'introduction arithmétique du philosophe Domninos de Larissa," in *Revue des études grecques,* **19** (1906), 360–382, repr. in *Mémoires scientifiques,* III (1915), 255–281.

"How to Take a Ratio out of a Ratio" is printed with a French translation, a commentary, and additional notes by O. Riemann, in C. E. Ruelle, "Texte inédit de Domninus de Larissa sur l'arithmétique avec traduction et commentaire," in *Revue de philologie,* **7** (1883), 82–92, with an addendum by J. Dumontier explaining the mathematical import, pp. 92–94.

There is a useful summary by F. Hultsch, "Domninos," in Pauly-Wissowa, *Real-Encyclopädie,* V (1903), cols. 1521–1525.

IVOR BULMER-THOMAS

DONALDSON, HENRY HERBERT (*b.* Yonkers, New York, 12 May 1857; *d.* Philadelphia, Pennsylvania, 23 January 1938), *neurology.*

Donaldson, scion of a banking family, showed an early interest in science, and after studies at Phillips Andover and Yale stayed on an additional year in New Haven to do research in arsenic detection at Sheffield Scientific School (1879–1880) under Russell H. Chittenden. He then received somewhat reluctant parental approval to study medicine at the College of Physicians and Surgeons of New York (1880–1881), but after one year became convinced that his true bent lay in research rather than practice. Donaldson was offered, and accepted, a fellowship at Johns Hopkins, where he spent two years (1881–1883) studying the effects of digitalin on the heart and of cocaine on the nerves controlling temperature. The latter work, done under the supervision of G. Stanley Hall, became the theme of his Ph.D. dissertation (1895). Donaldson next spent almost two years (1886–1887) in Europe at the great neurological centers, studying under such masters as Forel, Gudden, Theodor Meynert, and Golgi. Returning briefly to Johns Hopkins as associate in psychology, he soon followed Hall, who had become president of Clark University, to Worcester, Massachusetts. Here, while assistant professor of neurology (1889–1892), he carried out his classic study on the brain of Laura Bridgman, a blind deaf-mute. This study, characterized as "probably the most thorough study of a single human brain that has ever been carried out," determined the theme that was to dominate all of Donaldson's subsequent research: the growth and development of the human brain from birth to maturity. His early papers were incorporated in a monograph, *The Growth of the Brain: A Study of the Nervous System in Relation to Education* (1895).

In 1892 Donaldson moved to Chicago to join the faculty of the recently opened university. Here he served as professor of neurology and dean of the

Ogden School of Science until 1898, when his teaching career was interrupted by a crippling tubercular infection of the knee which necessitated a prolonged recuperative period in Colorado. From 1891 to 1910 there was a continuous flow of papers concerned with the rate of growth of the brain and spinal cord, and the relationship of their weight and length to that of the entire body.

In 1905, after serious consideration, Donaldson finally decided to accept the distinguished appointment as head of the Wistar Institute of Anatomy and Biology in Philadelphia, a position he was to retain until his death. In 1906 there appeared the first of a long series of papers in which the white or albino rat (rather than the frog) was used as a research tool. Donaldson had already used this animal in his work with Adolf Meyer in Baltimore as early as 1893, but the superiority of the rat over the frog had long been advocated by Shinkishi Hatai, who had joined Donaldson in Philadelphia. He then proceeded to work out the equivalence in age between man and rodent, and as a result of lengthy genetic studies was able to produce the famous Wistar Institute stock of white rats, which have since then figured in innumerable research projects. The fundamental studies on growth, although primarily directed toward the brain and central nervous system, were later expanded to include the muscles, bones, teeth, and viscera.

Among the many distinguished workers associated with Donaldson in Chicago and Philadelphia were Alice Hamilton, John B. Watson, S. W. Ranson, and Frederick S. Hammett. His numerous foreign students included at least twenty Japanese. A member of the American Philosophical Society from 1906, Donaldson served that organization in various important positions until his death. He was also honored with the presidency of the Association of American Anatomists (1916–1918), the American Society of Naturalists (1927), and the American Neurological Society (1937). Both Yale and Clark granted him the honorary D.Sc.

Donaldson was a man of great culture and a true humanist. He loved music, the arts, and literature, but did not neglect to concern himself actively with social problems of the day. From 1888, when he helped to found the Marine Biological Laboratory at Woods Hole, he kept open house each summer for a host of friends and admirers. On his seventy-fifth birthday he was presented with a special volume of the *Journal of Comparative Neurology* (1932), dedicated to him and containing twenty contributions as well as an affectionate eulogy. His first marriage, to Julia Desboro Vaux of New York, produced two sons, one of whom, John C. Donaldson, served as professor of anatomy at the medical school of the University of Pittsburgh.

BIBLIOGRAPHY

See "The Physiology of the Central Nervous System," in *An American Textbook of Physiology* (1898); and *The Rat* (1924). His diaries (1890–1936), in 49 vols. in manuscript, are at the American Philosophical Society Library, Philadelphia.

A biography is Edward G. Conklin, "Henry Herbert Donaldson (1857–1938)," in *Biographical Memoirs. National Academy of Sciences* (1939), 229–243, with a bibliography and portrait.

MORRIS H. SAFFRON

DONATI, GIOVAN BATTISTA (*b.* Pisa, Italy, 16 December 1826; *d.* Florence, Italy, 20 September 1873), *astronomy.*

Donati studied at the University of Pisa, where he was a pupil of Mossotti. In 1852 he went to Florence, where he worked in the observatory, then called "La Specola." It was directed at that time by Amici, whom Donati succeeded in April 1864. In the years 1854–1873 Donati published about 100 works, many of which were devoted to comets (on 2 June 1858 he discovered the comet that is named after him), astrophysics, and atmospheric physics.

His greatest achievement was the development of a new branch of astronomy based on Fraunhofer's famous discovery and leading to the spectroscopic study of the stars. Indeed, after having been in Spain for the eclipse of the sun of June 1860, he devoted himself completely to the application of spectroscopy to the stars and published the results of his studies in *Annali del Regio museo di fisica e storia naturale*. He pointed out the differences between the spectra of fifteen principal stars and that of the sun. He was also the first to obtain the spectrum of a comet and to interpret the observed features. The sun also was an object of his studies, both as an isolated body and in relation to other celestial bodies.

In 1868 he published two papers on the sun, one (in *Nuova antologia,* **8,** 334–353) on determining its distance from the earth and the other (*ibid.,* **9,** 60–93) on its physical structure. In 1869 he noted that certain phenomena heretofore thought to have had an atmospheric source actually originated in higher regions. He thus formulated the basis of a cosmic meteorology.

His experimental work brought forth first of all a spectroscope with five prisms, which was used in Sicily to observe an eclipse in 1870. He also devised a spectroscope of twenty-five prisms; it was exhibited in Vienna and Donati used it to make a series of remarkable observations in 1872. He noted the results that he obtained through the use of these spectroscopes and described them in *Rapporti sulle osservazioni dell'eclisse totale di sole del 22 dicembre 1870* and in *Memorie della Società degli spettroscopisti italiani.*

During the years 1864–1872 Donati was further occupied with the construction of a new observatory at Arcetri, near the house where Galileo died. The observatory was dedicated on 27 October 1872.

Donati's important work "Sul modo con cui si propagarono i fenomeni luminosi della grande aurora boreale osservata nella notte dal 4 al 5 febbraio 1872" was the first number of the intended periodical *Memorie del Regio osservatorio di Firenze ad Arcetri.* (Publication of the *Memorie* ceased with Donati's death; twenty years later A. Abetti started a new series with the title *Osservazioni e memorie dell'osservatorio astrofisico di Arcetri.*)

Donati died of cholera returning from Vienna, where he had taken part in the International Congress of Meteorology.

BIBLIOGRAPHY

Donati's works include "Intorno alle strie degli spettri stellari," in *Annali del Regio museo di fisica e storia naturale di Firenze per il 1865,* n.s. **1** (1866), 1–21; "Intorno alle osservazioni fatte a Torreblanca in Spagna dell'eclisse totale di sole del 18 luglio 1860," *ibid.,* 21–37; "Osservazioni di comete fatte all'Osservatorio del Regio museo di Firenze dall'anno 1854 fino al 1860," *ibid.,* 37–63; "Osservazioni spettroscopiche di macchie solari fatte a Firenze," in *Memorie della Società dei spettroscopisti italiani raccolte e pubblicate per cura del Prof. P. Tacchini,* I (Palermo, 1871), 52–55; *Rapporti sulle osservazioni dell'eclisse totale di sole del 22 Dicembre 1870 eseguite in Sicilia dalla Commissione Italiana* (Palermo, 1872), pp. 31–39; and "Sul modo con cui si propagarono i fenomeni luminosi della grande aurora boreale osservata nella notte dal 4 al 5 febbraio 1872," in *Memorie del Regio osservatorio di Firenze ad Arcetri* (Florence, 1873), pp. 5–31.

MARIA LUISA RIGHINI BONELLI

DONDERS, FRANCISCUS CORNELIS (*b.* Tilburg, Netherlands, 27 May 1818; *d.* Utrecht, Netherlands, 24 March 1889), *physiology, ophthalmology.*

Donders was the youngest of nine children and the only son of Jan Frans Donders and Agnes Elisabeth (Clara?) Hegh. His father (who died when Donders was an infant) was not without means and occupied himself with chemistry, music, and literature, while his wife supervised his business.

Donders therefore passed the earliest years of his life in a family consisting only of women; in 1825 he was sent to a boarding school at Duizel, a village near Eindhoven. He remained there until 1831, working for his tuition as an assistant teacher during the final two years. After a short stay at the French School in Tilburg he attended the Latin School in Boxmeer, from which he graduated *cum laude* on 27 January 1835.

From July 1835 to December 1839 Donders was a student at the military medical school in Utrecht. Since this training conferred the right to treat only military men and their families, he enrolled in the medical faculty of Utrecht University. In February 1840 he was appointed health officer with the garrison in Vlissingen. The same month he passed the doctoral examination in medicine at Leiden and on 13 October 1840 received the M.D. from Leiden University.

Although he was educated as a Roman Catholic, he seems not to have practiced his religion after his time in Utrecht. He pleaded for the separation of religion and science and, according to his statements, was probably a theist or deist.

After he finished his studies Donders remained for some time with the garrison in Vlissingen. In August 1841 he was transferred to The Hague as medical officer. His sojourn there seems to have been important to the development of his career; he read, became familiar with the official and cultural life of that city, and was consulted by the inspector general about the reorganization of the Utrecht military medical school, where he was appointed docent in physiology and anatomy.

In September 1842 he returned to the University of Utrecht, where Gerrit Jan Mulder was intensively occupied with the renovation and expansion of chemistry as a discipline. Mulder requested Donders' cooperation in histological and histochemical research. Important discoveries in embryology and physiology were made in Mulder's laboratory, and Donders was present at the birth of physiological chemistry.

To augment his small income from teaching, Donders translated—among other things—Christian Georg Theodor Ruete's *Lehrbuch der Augenheilkunde.* Indeed, as well as translating the work, he edited it and, where necessary, performed additional experiments. Soon, in addition to his physiological and clinical publications, he began to write articles on ophthalmology. He wrote frequently for the *Holländische Beiträge zu den anatomischen und physiologischen Wissenschaften,* which he published with the physiologists Isaac van Deen and Jacob Moleschott, and for the *Nederlandsch Lancet,* of which he was editor.

His lecture "Blik op de stofwisseling als bron der eigen warmte van planten en dieren" ("Consideration of Metabolism as the Source of Heat in Plants and Animals") was published in 1845. In this work he attributed the regulation of heat mainly to the skin and also mentioned the principle of the law of the conservation of energy.

Donders was appointed extraordinary professor at the University of Utrecht—although there was no vacancy—to retain his services there. He chose to give courses that had not been taught before, including forensic medicine, ophthalmology, and (under the title of general biology) the science of metabolism and histology. He selected the latter term to avoid the word "physiology," in order not to embarrass his beloved teacher, the physiologist J. L. C. Schroeder van der Kolk, with even the appearance of competition.

Because of his courses in the physiology of the eye and its adaptation to pathological problems, Donders was soon consulted as an ophthalmological expert. Although he was urged to establish himself as an ophthalmologist, he hesitated to do so. In 1851 he was invited by Sir James Young Simpson and others to visit the important English eye clinics. This trip was of great significance to him: in England and on his return by way of France he met outstanding English, German, and French physiologists and eye doctors, including Sir William Bowman, Albrecht von Graefe, and Claude Bernard. (In London he also heard about the ophthalmoscope, invented by Helmholtz.) Strengthened by the events of his travels, Donders decided to establish himself as a specialist in diseases of the eye.

In 1852 Donders was appointed ordinary professor at Utrecht and concerned himself especially with ophthalmology. With his own money he opened a polyclinic and managed to obtain the use of the cholera hospital, which, however, soon became too small. A committee of private individuals then raised 40,000 florins to buy a large mansion, which was remodeled as a charity hospital for indigent patients; it opened in 1858. The hospital also functioned as an independent educational institution, primarily at the service of the university, as the university itself provided little opportunity for ophthalmological education. Here Donders established a center for both research and teaching and soon, in addition to university students, many foreign physicians took part in studies on refraction and accommodation anomalies and other ophthalmological problems.

In his autobiography Donders mentions that these years demanded a great deal of his strength. The hospital and the physiological laboratory, his courses in many subjects, research in the laboratory, and work for the press and the university all required his attention. When in 1862 Schroeder van der Kolk died, Donders was offered the professorship in physiology and promised a new laboratory. Donders accepted this offer because, as he said, physiology was his first love. Donders resigned from his ophthalmological practice; he remained, however, as director of the hospital until 1883. The new laboratory, equipped according to his directions, was opened in 1866. Donders' work was not limited to purely scientific studies, and he was often consulted by the university administrators. As dean of the medical faculty he was a capable leader. He was concerned in making science serve the needs of humanity; his publications give an impression of the essential and varied nature of his work.

In 1845 Donders married Ernestine J. A. Zimmerman, the daughter of a Lutheran minister in Utrecht. She died in 1887, after a long illness marked by mental depression. Shortly after he retired in 1888, Donders married the painter Abrahamine Arnolda Louisa Hubrecht, a daughter of his friend, the state councilor P. F. Hubrecht. Although Donders complained about his health in the diaries and correspondence of his last years, he appeared healthy and youthful until the time of his retirement. He died less than a year later, of a progressive brain disease, possibly a tumor.

BIBLIOGRAPHY

I. ORIGINAL WORKS. Donders published more than 340 works, mainly on ophthalmology and physiology, of which about a hundred are of a clinical, pathological, or physiochemical nature. Much of his work was published in French, English, or German. *On the Anomalies of Accommodation and Refraction of the Eye with a Preliminary Essay on Physiological Dioptrics* (London, 1864) has been translated into various languages, but has never appeared in Dutch. Also of importance are his *Handleiding tot de natuurkunde van den gezonden mensch*, written with A. F. Bauduin, 2 vols. (Utrecht–Amsterdam, 1851, 1853); *De voedingsbeginselen. Grondslagen eener algemeene voedingsleer* (Tiel, 1852); and P. B. Bergrath, trans., *Die Nahrungsstoffe, Grundlinien einer allgemeine Nahrungslehre* (Crefeld, 1853), a semipopular work taken from articles in the *Geneeskundige Courant* that was intended to give both the physician and the educated layman an outline of the elements of nutrition as a basis for a rational diet.

II. SECONDARY LITERATURE. On Donders and his work see Sir William Bowman, "In Memoriam F. C. Donders," in *Proceedings of the Royal Society*, **49** (1891), vii–xxiv; F. P. Fischer and G. ten Doesschate, *Franciscus Cornelis Donders* (Assen, 1958), a very extensive monograph in

Dutch, which includes a reprint of the above; M. A. van Herwerden, "Eine Freundschaft von drei Physiologen," in *Janus,* **20** (1915), 174–201, 409–436; "Die Freundschaft zwischen Donders und von Gräfe," *ibid.,* **23** (1918); J. Moleschott, *Franciscus Cornelis Donders, Festgruss zum 27 Mai 1888* (Giessen, 1888); P. J. Nuel, "F. C. Donders et son oeuvre," in *Annales d'oculistique,* 14th ser. (année 52), 5–107, of which pp. 5–45 are biographical and pp. 45–107 contain a systematic bibliography of works by Donders and his students; C. A. Pekelharing *et al.,* Paula Krais, trans., *F. C. Donders Reden gehalten bei der Enthüllung seines Denkmals in Utrecht am 22 Juni 1921* (Leipzig, 1922), which contains a German bibliography of Donders' work based on the one by Nuel; R. A. Pfeiffer, "F. C. Donders," in *Bulletin of the New York Academy of Medicine* (October 1936); H. J. M. Weve and G. ten Doesschate, *Die Briefe Albrecht von Gräfes an F. C. Donders* (Stuttgart, 1936); and *Het Jubileum van Professor F. C. Donders gevierd te Utrecht op 24 en 28 Mei 1888. Gedenkboek uitgegeven door de commissie* (Utrecht, 1889), of which pp. 115–132 contain an autobiographical essay.

RODOLPHINE J. CH. V. TER LAAGE

DONDI, GIOVANNI (*b.* Chioggia, Italy, 1318; *d.* Milan, Italy, 22 June 1389), *horology, astronomy, medicine.*

Dondi was the son of Jacopo (Giacomo) de'Dondi dall'Orologio, municipal physician at Chioggia; his work parallels closely that of his father, with whom he has often been confused. When the family moved to Padua in 1349, Giovanni Dondi became physician to Emperor Charles IV. In 1350 or 1352 he was appointed professor of astronomy at the University of Padua and later was a member of each of the four faculties of medicine, astrology, philosophy, and logic. He lectured on medicine at Florence around 1367–1370. He was ambassador to Venice in 1371; the following year he was a member of a committee of five citizens appointed to establish boundaries between Carrara and the Venetian Republic. Later he was befriended by Gian Galeazzo Visconti of Pavia and from 1379 to 1388 was connected with the University of Pavia; in 1382 he was living in the Visconti palace.

Dondi wrote, probably before 1371, a brief treatise, *De modo vivendi tempore pestilentiali,* mostly concerning diet during times of plague. After 1372 he followed his father's interest in balneology with a treatise describing the hot springs near Padua and methods of salt extraction, *De fontibus calidus agri Patavini consideratio ad magistrum Vicentium.* He also wrote *Quaestiones aliquae in physica et medica.* A friend of Petrarch and Cola di Rienzo, Dondi shared their interest in the ruins and inscriptions of ancient Rome.

Dondi's father, Jacopo, had designed a clock (hence the "dall'Orologio" added to the family name) which was installed in 1344 in the Torre dei Signori of the Palazzo del Capitanio at Padua. Giovanni Dondi's fame rests on an elaborate astronomical clock which he designed and spent sixteen years constructing, completing it in 1364. Petrarch, praising Dondi's astronomical attainments in his will, referred to this "planetarium," and there are early descriptions of it by Dondi's friends Phillippe de Mézières (in his *Songe du vieil pèlerin,* written between 1383 and 1388) and Giovanni Manzini, *podestà* of Pisa (letter to Dondi 11 July 1388). Dondi himself wrote a detailed description of his planetary clock, a treatise known as the *Tractatus astrarii* or *Tractatus planetarii,* copiously illustrated with diagrams of the dials, wheels, and other components. In this treatise, Dondi says, "I derived the first notion of this project and invention from the subtle and ingenious idea propounded by Campanus [of Novara] in his construction of equatoria, which he taught in his *Theorica planetarum*"; the astrarium is one of the earliest geared equatoria, driven by clockwork. A heptagonal frame bears dials for the sun, moon, Mercury, Venus, Mars, Jupiter, and Saturn; below, there are dials or displays showing the twenty-four hours, the times of sunrise and sunset, fixed feasts, movable feasts, and the nodes of the moon's orbit. The astrarium was acquired by Gian Galeazzo Visconti and installed in 1381 in the ducal library in the Castello Visconteo in Pavia. Regiomontanus, who saw the astrarium in 1463 and mentioned it in his introductory lecture on the mathematical sciences at the University of Pavia, said in 1474 that he had such a mechanism under construction in his workshop at Nuremberg. The existence of the astrarium is last recorded when it was offered in 1529 or 1530 to that lover of clocks, Emperor Charles V; it was then so dilapidated that Gianello Torriano (later the emperor's clockmaker) considered it beyond repair and undertook to make a similar device. Two modern replicas of the astrarium have been completed; that now in the Museum of History and Technology, Washington, D.C., is 4'4" in overall height, 1'6" in maximum diameter, and contains 297 parts, of which 107 are wheels and pinions.

Dondi's treatise is of great importance in the history of medieval horology and technology; the only known earlier detailed description of any sort of clock is that by Richard of Wallingford of his astronomical clock installed in St. Albans Abbey (about 1330). Both treatises, written within a century of the invention of the mechanical clock escapement, demonstrate the rapid development of clockwork and the constant concern to mechanize astronomical demonstrational instruments that is so evident in the long tradition of European public astronomical clocks.

BIBLIOGRAPHY

I. ORIGINAL WORKS. Dondi's works include *De modo vivendi tempore pestilentiali*, Karl Sudhoff, ed., in *Archiv für Geschichte der Medizin*, **5** (1911), 351–354; and *De fontibus calidus agri Patavini consideratio ad magistrum Vicentium*, printed in *De Balneis* (Venice, 1553), with Jacopo de'Dondi's brief treatise on the extraction of salt from the hot springs near Padua. Eleven manuscripts have survived of the treatise on the astrarium; these are Padua, Bibl. Capitolare Vescovile, D. 39; Venice, Bibl. Naz. Marciana, 85. Cl. Lat. VIII. 17; Milan, Bibl. Ambrosiana, C. 221 inf. and C. 139 inf.; Padua, Bibl. Civica, CM 631; Oxford, Bodleian, Laud Misc. 620; London, Wellcome Historical Medical Library, 248; Cracow, Bibl. Universytetu Jagiellońskiego, 577 and 589; Eton College, 172. Bi a. l; Salamanca, Universidad, 2621, item 12, 25r–72v (? fragments). A twelfth MS, Turin, Bibl. Naz. XLV, was destroyed by fire in 1904. The MS Padua, Bibl. Capitolare Vescovile, D. 39, has been reproduced in facsimile with transcription and commentary in Dondi's *Tractatus Astrarii*, trans., with introduction and glossary, by Antonio Barzon, Enrico Morpurgo, Armando Petrucci, and Giuseppe Francescato (Vatican City, 1960). Illustrations reproduced from other manuscripts, especially Bodleian, Laud Misc. 620, may be found in works cited below.

II. SECONDARY LITERATURE. On Jacopo and Giovanni de'Dondi and their works, see George Sarton, *Introduction to the History of Science*, III (Baltimore, 1948), 1669–1671, 1672–1677; Lynn Thorndike, *A History of Magic and Experimental Science*, III (New York, 1934), 386–397; and *Giovanni Dondi dall'Orologio, medico, scienziato e letterato* (Padua, 1969). Silvio A. Bedini and Francis Maddison discuss the history of the astrarium and its antecedents, list the MSS, and give a full bibliography in "Mechanical Universe. The Astrarium of Giovanni de'Dondi," in *Transactions of the American Philosophical Society*, n.s. **56**, pt. 5 (1966). Also see G. H. Baillie, "Giovanni de'Dondi and his Planetarium Clock in 1364," in *Horological Journal* (April–May 1934); H. Alan Lloyd, "Giovanni de'Dondi's Horological Masterpiece 1364," in *La Suisse horlogère*, international ed., no. 2 (July 1955), 49–71; and H. A. Lloyd, *Some Outstanding Clocks Over Seven Hundred Years, 1250–1950* (London, 1958), pp. 9–24, which describe the construction and appearance of the astrarium.

FRANCIS MADDISON

DONNAN, FREDERICK GEORGE (*b.* Colombo, Ceylon, 5 September 1870; *d.* Canterbury, England, 16 December 1956), *physical chemistry.*

Donnan, the son of a Belfast merchant, was born while his parents were temporarily abroad. He was educated at Belfast Royal Academy and at Queen's College, Belfast; after graduation he spent three years at Leipzig with Wislicenus and Ostwald, then a year at Berlin under van't Hoff. This unusually long apprenticeship to chemistry was completed with a period spent working with Ramsay at University College, London, to the teaching staff of which he was appointed in 1901. Three years later he became professor of physical chemistry at the University of Liverpool; he was made a fellow of the Royal Society in 1911 and succeeded Ramsay at University College in 1913. Donnan was drawn into industry during World War I, and for several years he worked on problems connected with the manufacture of "synthetic" ammonia and nitric acid, both in London and with the firm of Brunner, Mond in Cheshire. These industrial connections and interests were retained for the rest of his life, and even after his retirement in 1937 he continued to act as a consultant. His London house was destroyed in 1940, and he retired to Kent with his sisters; he was unmarried.

Donnan, having worked with both Ostwald and van't Hoff, was one of the main agents by whom the "new" physical chemistry was introduced into Britain. Van't Hoff had interested him in the problems of colloids, soap solutions, and osmotic pressures, and this interest led to his major paper, "The Theory of Membrane Equilibrium in the Presence of a Nondialyzable Electrolyte" (1911). This examined the effect of confining, by means of a membrane, a mixture of ions, one of which cannot pass through the membrane because of its large size. (In the absence of a membrane, the equilibrium of a protein with a salt solution is a similar case.) The theory of the Donnan membrane equilibrium has important applications in colloid chemistry and in the technologies of leather and gelatin, but above all in the understanding of the living cell, where it can give a quantitative account of ionic equilibria both within the cell and between the cell and its environment.

None of Donnan's subsequent work approaches this in importance. In later years he guided his department in London on a very loose rein, welcoming promising young men and leaving them free to follow their own interests. A wealthy, cultured, and highly articulate man, fond of travel and much given to hospitality, he became interested in the speculative and cosmological aspects of biology, and many of his later publications concern these topics. He wrote no book but was the author of more than 100 papers.

BIBLIOGRAPHY

The most informative obituary notices are those of F. A. Freeth, in *Biographical Memoirs of Fellows of the Royal Society*, **3** (1957), 23–29, which includes a portrait and a complete bibliography; W. E. Garner, in *Proceedings of the Chemical Society* (1957), 362–366, with portrait; and C. F. Goodeve, in *Nature,* **179** (1957), 235–236.

W. V. FARRAR

DOPPELMAYR, JOHANN GABRIEL (*b.* Nuremberg, Germany, 1671 [?]; *d.* Nuremberg, 1 December 1750), *astronomy, mathematics, physics, history of mathematics.*

Doppelmayr's father, Johann Siegmund Doppelmayr, was a merchant who made a hobby of experiments in physics and, according to his son, was the first to introduce into Nuremberg an air pump equipped with a lever and standing upright "like a flower vase."

After graduating from the Aegidien Gymnasium, Doppelmayr entered the University of Altdorf in 1696 with the intention of studying law; but there he heard the lectures on mathematics and physics of Johann Cristoph Sturm, founder of the Collegium Curiosum sive Experimentale and reputedly the most skilled experimenter in Germany. For a brief while in 1700 Doppelmayr attended the University of Halle, but he then decided to give up law for physics and mathematics, and spent two years traveling and studying in Germany, Holland, and England.

After Doppelmayr's return to Nuremberg, he was appointed in 1704 to the professorship of mathematics at the Aegidien Gymnasium, a post he held until his death. His life was devoted to lecturing, writing, astronomical and meteorological observation, and physical experimentation; his reputation was such as to gain him memberships in the Academia Caesarea Leopoldina, the academies of Berlin and St. Petersburg, and the Royal Society of London.

Doppelmayr's writings are not marked by originality; they do, however, provide an index of the scientific interests and information current in Germany, and particularly of the transmission of science from England, Holland, and France into Germany during the first half of the eighteenth century.

Among the astronomical works are *Kurze Erklärung der Copernicanischen Systems* (1707), *Kurze Einleitung zur Astronomie* (1708), and translations of Thomas Streete's astronomy (1705) and of John Wilkins' defense of the Copernican system (1713). His major work, however, is the *Atlas novus coelestis* (1742), a collection of diagrams with explanations intended as an introduction to the fundamentals of astronomy. Besides star charts and a selenographic map, the *Atlas* includes diagrams illustrating the planetary systems of Copernicus, Tycho, and Riccioli; the elliptic theories of Kepler, Boulliau, Seth Ward, and Mercator; the lunar theories of Tycho, Horrocks, and Newton; and Halley's cometary theory.

Doppelmayr's writings on mathematics include *Summa geometricae practicae;* a memoir on spherical trigonometry; an essay on the construction of the sundial; and a translation (with appendices by Doppelmayr) of Nicolas Bion's treatise on mathematical instruments.

Of lasting value for historians is Doppelmayr's *Historische Nachricht* (1730), a 314-page folio volume giving biographical accounts of over 360 mathematicians, artists, and instrument makers of Nuremberg. The biographies are arranged chronologically from the fifteenth to the eighteenth century.

In physics Doppelmayr continued the experimental tradition of Sturm. His *Physica experimentis illustrata* (1731) is a list, in German, of 700 experiments and demonstrations given before the Collegium Curiosum. The procedures are not described in any detail; they are designed to illustrate such topics as the "subtlety" or fineness of subdivision of various materials, electric and magnetic "effluvia," simple machines, the principles of hydrostatics, the optics of the eye, and so on.

More important is the *Neu-entdeckte Phaenomena* (1744), a well-organized and accurate summary of the electrical experiments and theories of Hawksbee, Gray, and Dufay. This work no doubt helped to create and inform the popular interest in electrical phenomena that spread through Germany in the mid-1740's. In the last two chapters Doppelmayr proposes a hypothesis to explain away electrical attraction and repulsion as caused by air movements; Dufay's discovery of the opposite characters of vitreous and resinous electricity is reduced to a difference in electric strength of different materials; and in general Doppelmayr returns to the earlier and less promising theoretical outlook of Hawksbee.

Doppelmayr's electrical investigations continued until his death, which followed a severe shock suffered while experimenting with one of the newly invented condensers.

BIBLIOGRAPHY

I. ORIGINAL WORKS. Doppelmayr's writings include *Eclipsis solis totalis cum mora* (Nuremberg, 1706); *Kurze Erklärung der Copernicanischen Systems* (Nuremberg, 1707); *Kurze Einleitung zur Astronomie* (Nuremberg, 1708); *Neue vermehrte Welperische Gnomonica* (Nuremberg, 1708); *Neu-eroffnete mathematische Werck-Schule* (Nuremberg, 1st ed., 1712, 2nd ed., 1720, 3rd ed., 1741), a trans. of Nicolas Bion's *Traité de la construction et des principaux usages des instruments de mathématique; Johannis Wilkins Vertheidigter Copernicus* (Leipzig, 1713), a trans. of Wilkins' essay of 1640 on the probability of earth's being a planet; *Summa geometricae practicae* (Nuremberg, 1718, 1750); *Grundliche Anweisung zur Verfertigung grossen Sonnenuhren und Beschreibung derselben* (Nuremberg, 1719), also in Latin as *Nova methodus parandi sciaterica solaria;* "Animadversiones nonnullae circa eclipsium observationes," in *Academiae Caesareo-Leopoldinae Carolinae naturae curiosum ephem-*

erides, **2** (1715), Centuriae III et IV, app., 133–136; "Animadversiones circa usum vitrorum planorum in observationibus astronomicis," *ibid.,* **4** (1719), Centuriae VII et VIII, 457–459; "Circa trigonometriam sphaericum," in *Academiae Caesareae Leopoldina-Caroliniae naturae curiosum acta physico-medica . . . exhibentia ephemerides . . .,* **2** (1730), 177–178; *Historische Nachricht von den Nürnbergischen Mathematicis und Künstlern* (Nuremberg, 1730); *Physica experimentis illustrata* (Nuremberg, 1731); *Atlas novus coelestis* (Nuremberg, 1742); and *Neu-entdeckte Phaenomena von bewundernswürdigen Würckungen der Natur, welche bei fast allen Körper zukommenden elektrischen Krafft . . . hervorgebracht werden* (Nuremberg, 1744). J. H. von Mädler mentions as Doppelmayr's first work a Latin trans. (1705) of Streete's *Astronomia Carolina* (1661). George Hadley, in *Philosophical Transactions of the Royal Society,* **42** (1742), 245, refers to published barometrical observations by Doppelmayr for the years 1731–1735.

II. SECONDARY LITERATURE. On Doppelmayr or his work, see *Allgemeine deutsche Biographie,* V, 344–345; J. H. von Mädler, *Geschichte der Himmelskunde,* I (Brunswick, 1873), 129; J. G. Meusel, *Lexikon der vom Jahr 1750 bis 1800 verstorbenen teutschen Schriftsteller* (Leipzig, 1802–1816), II; *Neue deutsche Biographie,* IV, 76; and G. A. Will, *Nürnbergisches Gelehrter-Lexicon,* I (Nuremberg, 1755), 287–290.

CURTIS WILSON

DOPPLER, JOHANN CHRISTIAN (*b.* Salzburg, Austria, 29 November 1803; *d.* Venice, Italy, 17 March 1853), *mathematics, physics, astronomy.*

Christian Doppler was the son of a noted master stonemason. Although he showed talent in this craft, his poor health led his father to plan a career in business for him. Doppler's mathematical abilities were recognized by the astronomer and geodesist Simon Stampfer, at whose advice Doppler attended the Polytechnic Institute in Vienna from 1822 to 1825. Finding the curriculum too one-sided, Doppler returned to Salzburg and pursued his studies privately. He completed the Gymnasium and subsequent philosophical courses in an unusually short time, while tutoring in mathematics and physics. From 1829 to 1833 he was employed as a mathematical assistant in Vienna, and wrote his first papers on mathematics and electricity. In 1835 Doppler was on the point of emigrating to America; he had sold his possessions and had reached Munich when he obtained a position as professor of mathematics and accounting at the State Secondary School in Prague. In 1841 he became professor of elementary mathematics and practical geometry at the State Technical Academy there, during the tenure of which he enunciated his famous principle. He had become an associate member of the Königliche Böhmische Gesellschaft der Wissen-

schaften in Prague in 1840 and was made a full member in 1843. Doppler moved to the Mining Academy at Schemnitz (Banská Štiavnica) in 1847 as *Bergrat* and professor of mathematics, physics, and mechanics. As a result of the turbulence of 1848–1849 he returned to Vienna; there, in 1850, he became director of the new Physical Institute, which was founded for the training of teachers, and full professor of experimental physics at the Royal Imperial University of Vienna, the first such position to exist in Austria. Doppler had suffered from lung disease since his years at Prague. A trip to Venice in 1852 was of no avail, and he died there the following year, survived by his wife and five children.

Doppler's scientific fame rests on his enunciation of the Doppler principle, which relates the observed frequency of a wave to the motion of the source or the observer relative to the medium in which the wave is propagated. This appears in his article "Ueber das farbige Licht der Doppelsterne und einiger anderer Gestirne des Himmels" (read 25 May 1842). The correct elementary formula is derived for motion of source or of observer along the line between them; the extension to the motion of both at the same time appears in an article of 1846. Doppler mentions the application of this result both to acoustics and to optics, particularly to the colored appearance of double stars and to the fluctuations of variable stars and novae. The reasoning in the latter arguments was not always very cogent; for example, he believed that all stars were intrinsically white and emitted only or mainly in the visible spectrum. The colors which he believed to be characteristic of double stars, then, were to have their origin in the Doppler effect. It should be noted that Doppler worked under rather isolated circumstances, being the earliest important physicist in Austria in the nineteenth century. He was unable to justify in his own mind the application of his principle to transverse vibrations of light, an extension performed by B. Bolzano shortly afterwards.

The first experimental verification of the acoustical Doppler effect was performed by Buys Ballot at Utrecht in 1845, using a locomotive drawing an open car with several trumpeters. Buys Ballot also criticized the unsound assumptions upon which Doppler had based his astronomical applications. Doppler replied to these and similar criticisms in a rather stubborn and unconvincing fashion. The acoustical effect was also noted and commented on at the British Association meeting in 1848 by John Scott Russell and by H. Fizeau in the same year, perhaps without knowledge of Doppler's work. Fizeau pointed to the usefulness of observing spectral line shifts in the application to astronomy, a point of such importance that the principle is sometimes called the Doppler-Fizeau principle. Although in 1850 the Italian astronomer Benedict Sestini had published data on star colors apparently supporting Doppler's application of his principle to double stars, its valid astronomical use had to wait until proper spectroscopic instrumentation was available, beginning with the work of the English astronomer William Huggins in 1868. The optical effect was first confirmed terrestrially by Belopolsky in 1901. Modified by relativity theory, the Doppler principle has become a major astronomical tool.

Doppler's principle itself was criticized by the Austrian mathematician Joseph Petzval in 1852, on the basis of an incorrect mathematical argument. Doppler defended himself to good effect in this situation. Doppler also published works of less importance on related optical effects (Bradley's aberration of light; dependence of intensity on the motion of the source; the deviation of waves by a rotating medium, as, for example, an ethereal atmosphere rotating with a star), optical instruments, and topics in mathematics and physics, especially in geometry, optics, and electricity.

BIBLIOGRAPHY

I. Original Works. Doppler's papers on his principle and related topics are found in *Abhandlungen von Christian Doppler,* ed. with notes by H. A. Lorentz (Leipzig, 1907), Ostwald's Klassiker der exakten Wissenschaften, no. 161. A list of most of his publications appears in Poggendorff, I, 594–595. The statement of his principle, "Ueber das farbige Licht der Doppelsterne und einiger anderer Gestirne des Himmels," in *Abhandlungen der Konigl. Böhmischen Gesellschaften der Wissenschaften,* 5th ser., **2** (1842), 465, was also published separately (Prague, 1842). The extension of motion to both source and observer appeared in *Annalen der Physik und Chemie,* **68** (1846), 1–35.

II. Secondary Literature. See the obituary by Anton Schrötter in *Almanach der Kaiserlichen Akademie der Wissenschaften,* **4** (1854), 112–120; further information appears in Julius Scheiner, "Johann Christian Doppler und das nach ihm benannte Prinzip," in *Himmel und Erde,* **8** (1896), 260–271. Some of his ideas and accomplishments are described in B. Bolzano, "Christ. Doppler's neueste Leistungen auf dem Gebiete der physikalischen Apparatenlehre, Akustik, Optik und optischen Astronomie," in *Annalen der Physik und Chemie,* **72** (1847), 530–555.

A. E. Woodruff

DÖRFFEL, GEORG SAMUEL (*b.* Plauen, Vogtland, Germany, 21 October 1643; *d.* Weida, Germany, 16 August 1688), *astronomy, theology.*

Son of Friedrich Dörffel, pastor in Plauen, Georg Dörffel was best known for his representation of the orbit of a comet as a parabola with the sun at the focus. He was married three times and had nine children.

Dörffel studied philosophy, theology, Oriental languages, and mathematics, receiving the degrees of bachelor (Leipzig, 1662), master (Jena, 1663), and bachelor of theology (Leipzig, 1668; dissertation, 1665). He became his father's substitute in Plauen in 1667 and his successor in 1672. In 1684 he became ecclesiastical superintendent in Weida. He had an excellent astronomical library and good eyesight but observed mostly with an old-fashioned astronomical radius. It consisted of a long arm and a short arm at right angles to each other, with a fixed sight at the end of the long arm and two movable sights at the ends of the transversal. His comet observations with it were accurate to one or two minutes. His home was not favorable for observing because of neighboring buildings.

Dörffel found no observable parallax for the comet of 1672 and measured its angular distance from fixed stars. From measurements made with a quadrant on 27 March, he calculated its latitude and longitude and length and breadth. He depicted its apparent path as circular and noted that it moved in the same direction as the planets. Dörffel published his report while the comet was still visible but growing fainter. He realized that it was soon to be lost in the light of the moon. He considered God responsible for comets. In the last half of the seventeenth century, skepticism existed about astrological predictions. Dörffel, although he accepted the ancient belief that comets were evil omens, said of the comet of 1672 that it signified something new that was not good, but that he would not predict specific events which might follow. His next recorded astronomical activity concerns the comet of 1677.

In 1680 Dörffel shifted his attention from the apparent path to the actual path of comets. He noted Apian's statement that comets' tails are directed away from the sun and appreciated Tycho Brahe's observations of the nova of 1572 and the comet of 1577 as forcing abandonment of the concept of the incorruptibility of the heavens. Although not the first to suggest a parabolic path for a comet, Dörffel was the first to describe the path of the comet of 1680, and possibly other comets, as a parabola with the sun at the focus (*Astronomische Betrachtung*). Although he had the comet move around the sun while that body moved around the stationary earth, he was not a confirmed anti-Copernican. Using the radius of the earth's circular orbit around the sun and his observations of the comet's elongation from the sun, he projected the comet's path on the ecliptic. Probably the first (25

August) observer of Halley's comet in 1682, he described it briefly in print and at greater length in correspondence with Gottfried Kirch.

Dörffel's interests extended to the computation of lunar eclipses, to occultations, to meteors, including computation of their paths, and to mock moons. In 1685 he published his discovery, mentioned in 1682, of a new method to determine the distance of a body from the earth, with observations from only one site, utilizing the earth's diurnal rotation and expressing the distance in semidiameters of the earth.

Dörffel's writings, which were in German and anonymous or signed only with initials, were soon superseded by Newton's *Principia*. They received little attention until the mid-eighteenth century, when they began to be of historical interest and appealed to German national pride. In 1791 J. H. Schröter named a lunar mountain range after him.

BIBLIOGRAPHY

I. ORIGINAL WORKS. *De definitione et natura demonstrationis (Dörffel, resp.; Bernhard v. Sanden, praes.)* (Leipzig, June 1662); *Exercitatio philosophica de quantitate motus gravium (Erhard Weidel, praes.)* (Jena, 14 February 1663); *. . . Disputatio philologo-theologica de gloria templi ultimi, ex Hagg. cap. II vers. 6–9, adversus Judaeos asserta . . . sub praesidio Johannis Adami Scherzeri, . . . 1666 . . . eruditorum examini publice subjecit . . .* (Leipzig, n.d.), 2nd ed. in appendix to J. A. Scherzer's *Collegii antisociniani . . . .* 2nd ed. (Leipzig, 1684); *Tirocinium accentuationis ad lectionem Biblicam practice accommodatum* (Plauen, 1670); *Warhafftiger Bericht Von dem Cometen/ Welcher im Mertzen dieses 1672. Jahrs erschienen: Dessen Lauff/ Art und Beschaffenheit/ sampt der Bedeutung/ hiermit fürgestellet wird/ und in Plauen observire worden. Von M.G.S.D.* (1672); *Bericht Von dem neulichsten im Mertzen dieses 1672 Jahres erschienenem Cometen/ Auss einem zu Plauen gedrucktem Bedencken wiederholet und vermehret* (1672); *Extract eines Schreibens aus Plauen im Voigtland an einen guten Freund, von dem neuen Cometen. Welcher im April dieses 1677. Jahrs am Himmel erschienen* [Plauen, 1677]; *Neuer Comet-Stern welcher im November des 1680sten Jahres erschienen, und zu Plauen im Voigtlande dergestalt observiret worden, sampt dessen kurtzer Beschreibung, und darüber habenden Gedanken* [Plauen, 1680?]; *Astronomische Betrachtung des Grossen Cometen/ Welcher im ausgehenden 1680. und angehenden 1681. Jahre höchstverwunderlich und entsetzlich erschienen: Dessen zu Plauen im Voigtlande angestellte tägliche Observationes, Nebenst etlichen sonderbahren Fragen und neuen Denckwürdigkeiten/ sonderlich von Verbesserung der Hevelischen Theoriae Cometarum, . . .* (Plauen, 1681); *De incertitudine salutis aeternae contra Aloysium Richardum,* German title, *Der ärgste Seelen-Gifft Des Trostlosen Pabstthums* (Jena [?], 1682); *Eilfertige Nachricht, von dem itzund am Himmel stehenden neuen Cometen, welcher am 15. Augusti dieses 1682sten Jahres zum erstenmahl, zu Plauen im Voigtlande ist gesehen worden* (Plauen, 1682); "Observatio eclipseos lunae totalis, d. XI. Febr. A. 1682 . . . Symmista Plav. in Variscia, instituta, et cum novissimis tabb. Horroccio-Flamstedianis, vix sensibiliter discrepantibus, collata," in Gottfried Kirch, *Annus III. Ephemeridum motuum coelestium ad . . . M. D. C. LXXXIII* (Leipzig, 1683), recto G_1; *Neues Mond-Wunder/ Wie solches den 24. Jenner dieses angehenden 1684. Jahres/ zu Plauen im Voigtlande/ gesehen worden: Neben einem kurtzen Bedencken/ was hiervon zu halten sey* (Plauen, n.d.); "Calculus Eclipseos Lunaris penumbratilis Anno 1684 d. 17/27 Jun. ex Tabbulis Flamstedianis" in Gottfried Kirch, *Ephemeridum motuum coelestium Annus IV* (Leipzig, 1684), recto D_4; "Methodus nova, phaenomenorum coelestium intervalla a terris facillime determinandi, non variata statione seu loco observationis, neque captis eorundem altitudine vel azimutho," in *Acta eruditorum* (1685), pp. 571–580; and *Gedächtnisspredigt Herrn Hanns George v. Carlowitz gehalten den 19. des Christmonds 1686.*

MS letters from Dörffel to Gottfried Kirch are preserved in the University Library, Leipzig.

II. SECONDARY LITERATURE. See Angus Armitage, "Master Georg Dörffel and the Rise of Cometary Astronomy," in *Annals of Science,* 7 (1951), 303–315; Rudolf Gerlach, in *Neue deutsche Biographie,* IV (1959), 30–31; Norbert Herz, *Geschichte der Bahnbestimmung von Planeten und Kometen,* II (Leipzig, 1894), pp. 252–260; Abraham Gotthelf Kästner, "Nachrichten von Georg Samuel Dörfeln . . .," in Leipzig Gesellschaft der freyen Künste, *Sammlung einiger ausgesuchten Stücke der Gesellschaft,* III (Leipzig, 1756), 252–263; Curt Reinhardt, *Mag. Georg Samuel Dörffel. Ein Beitrag zur Geschichte der Astronomie im 17. Jahrhundert* (Plauen, 1882), a doctoral dissertation presented at Leipzig; and Baron de Zach, *Correspondance astronomique,* VII (1822), 136–138; VIII (1823), 397–399.

C. DORIS HELLMAN

DORN, GERARD (*b.* Belgium [?]; *fl.* Basel, Switzerland, and Frankfurt, Germany, 1566–1584), *medicine, alchemy, chemistry.*

As an early follower of Paracelsus, Dorn contributed significantly, through his translations and commentaries and through his own writings, to the rapid dissemination of Paracelsian doctrines in the late sixteenth century, and his influence has been revived in the twentieth century through its importance to C. G. Jung's studies of alchemy. Yet little appears to be known of his life. His contemporary Michael Toxites referred to him as Belgian (*Belga*).[1] He was a student of the Paracelsist Adam of Bodenstein, to whom he dedicated his first book,[2] but where he received his doctorate remains unclear. For several years he appears to have worked, perhaps on commission, as a translator for the Basel publisher Peter Perna. The dedications of his books indicate that he was still living in Basel in 1577–1578, but by 1581–1584 he apparently had taken up residence in Frankfurt.

Dorn's translations of Paracelsus include works on chemistry, astronomy, astrology, and surgery as well as on therapeutics and pharmacology. His dictionary of Paracelsian terms was influential not only in its Latin editions but also in later Dutch (1614), German (1618), and English translations (1650, 1674). Like Paracelsus he directed attention to the utility of chemistry in medicine, and he defended Paracelsus' new medicine against the attacks of the traditionalist Thomas Erastus. But it was Paracelsus' attempt to build a cosmic philosophy from an amalgamation of hermetic Neoplatonism and chemistry that attracted Dorn most strongly. His objective was not to transmute baser metals into gold, but to change manifest into occult forms, the impure into the pure. Believing that the education of his time was too pagan and scholastic, he presented the operations of alchemy not only as a material procedure but also as a spiritual process of striving toward "sublimity of mind."[3] "Transform yourselves into living philosophical stones," he exhorted.[4]

Dorn was also indebted to John of Rupescissa on the quintessence and to Nicholas of Cusa in his application of numerical symbols[5] to Paracelsian principles. He felt an affinity for the English mathematician and astrologer John Dee, whose symbol of unity (*monas*) he incorporated into the title-page vignette of his *Chymisticum artificium*.[6] His elaborations of Paracelsian concepts and his own meditations, as reprinted in the *Theatrum chemicum*,[7] came to have a special meaning for C. G. Jung, whose later works are based on profound and lengthy researches into philosophical alchemy. Jung's summation of the significance of alchemical symbolism for psychology and religion was notably influenced by Dorn's concept of a unitary world (*unus mundus*) and, like Dorn, he drew a parallel between the alchemical act and the moral-intellectual transformation of man. With this interpretation he placed alchemy in a new perspective in the history of science, medicine, and theology.

NOTES

1. Paracelsus, *Onomastica II,* Michael Toxites, ed. (Strasbourg, 1574), p. 430.
2. *Clavis totius philosophiae chymisticae* (Lyons, 1567), dedication dated Basel, 1566.
3. *Ibid.,* p. 126.
4. Cited by C. G. Jung, in *Psychology and Alchemy,* 2nd ed. (Princeton, 1968), p. 148.
5. *Clavis,* chs. 7–9; *Monarchia triadis in unitate soli deo sacra,* appended to *Aurora thesaurusque philosophorum Paracelsi* (Basel,

1577), pp. 65–127; and his "De spagirico artificio Io. Trithemi sententia," in *Theatrum chemicum,* I (Strasbourg, 1659), 390–391.
6. *Chymisticum artificium naturae theoricum et practicum* (n.p., 1568).
7. *Theatrum chemicum,* I, 192–591.

BIBLIOGRAPHY

I. ORIGINAL WORKS. Bibliographies (listed chronologically) of Dorn's writings are given by K. Sudhoff, "Ein Beitrag zur Bibliographie der Paracelsisten im 16. Jahrhundert," in *Centralblatt für Bibliothekswesen,* **10** (1893), 385–391; J. Ferguson, *Bibliotheca chemica* (Glasgow, 1906), I, 220–222; D. Duveen, *Bibliotheca alchemica et chemica* (London, 1949; repr. London, 1965), 177–179; and J. R. Partington, *A History of Chemistry* (London, 1961), II, 159–160. The *De summis naturae mysteriis* and the *Aurora thesaurusque philosophorum,* presented by Dorn as translations of Paracelsian writings, are considered by Sudhoff to be Dorn's own works: *Versuch einer Kritik der Echtheit der Paracelsischen Schriften* (Berlin, 1894), nos. 125, 177. A. E. Waite's translation of the *Aurora* in *The Hermetic and Alchemical Writings of Paracelsus* (London, 1894; repr. New Hyde Park, N.Y., 1967) does not include, as might be inferred from Partington, either the tract *Monarchia triadis in unitate soli deo sacra* (alternate title of the *Monarchia physica*), which was issued with the *Aurora* (Basel, 1577), pp. 65–127, or the *Anatomia viva,* also issued with *Aurora,* 129–191, on the chemical examination of urine by distillation, which also was probably Dorn's own work and was ascribed to him by Huser in his German translations of the *Aurora* of 1605 and 1618—see Sudhoff, *op. cit.,* nos. 267, 302; see also no. 469. W. Pagel has suggested that, under the name Dominicus Gnosicus Belga, Dorn was the author of a commentary, first printed in 1566, to the *Seven Hermetic Treatises:* "Paracelsus: Traditionalism and Medieval Sources," in L. G. Stevenson, ed., *Medicine, Science and Culture; Historical Essays in Honor of Owsei Temkin* (Baltimore, 1968), pp. 58–60; and "The Eightness of Adam and Related 'Gnostic' Ideas in the Paracelsan Corpus," in *Ambix,* **16** (1969), 131–132.

II. SECONDARY LITERATURE. The fullest accounts of Dorn are given by J. R. Partington, *A History of Chemistry* (London, 1961), II, 159–160; L. Thorndike, *A History of Magic and Experimental Science,* V (New York, 1941; repr. New York, 1954), 630–635; and R. P. Multhauf, *The Origins of Chemistry* (London, 1966), 241–243, 288–289. Other references are in W. Pagel, *Paracelsus; An Introduction to Philosophical Medicine in the Era of the Renaissance* (Basel, 1958), pp. 189–194; and *Das medizinische Weltbild des Paracelsus* (Wiesbaden, 1962), p. 111. The works of C. G. Jung in which Dorn is most frequently cited are vols. XII, XIII, and XIV of his *Collected Works: Psychology and Alchemy,* 2nd ed. (Princeton, 1968); *Alchemical Studies* (Princeton, 1967); and *Mysterium coniunctionis* (New York, 1963). Further references relevant to Dorn's alchemical philosophy and to his influence on Jung are M.-L. von Franz, "The Idea of the Macro- and Microcosmos in the Light of Jungian Psychology," in *Ambix,* **13** (1965), 22–34;

G. H., review of T. Burckhardt's *Alchimie: Sinn und Weltbilt, ibid.,* **8** (1960), 177–180; A. Jaffé, "The Influence of Alchemy on the Work of C. G. Jung," in *Alchemy and the Occult. A Catalogue of Books and Manuscripts From the Collection of Paul and Mary Mellon Given to Yale University Library,* compiled by Ian MacPhail (New Haven, 1968), I, xxi, xxii, xxix; II, 394; W. Pagel, "Jung's Views on Alchemy," in *Isis,* **39** (1948), 44–48; and his review of I. B. Cohen's *Ethan Allen Hitchcock, Discoverer of the True Subject of the Hermetic Art, ibid.,* **43** (1952), 374–375; H. J. Sheppard, "A Survey of Alchemical and Hermetic Symbolism," in *Ambix,* **8** (1960), 35–41; and R. S. Wilkinson, "The Alchemical Library of John Winthrop, Jr. (1606–1676)," *ibid.,* **9** (1963), 33–51; **10** (1966), 139–186.

Five of Dorn's works were in Winthrop's library; the copy that belonged to John Dee bears his initials and the comment in his hand that Dorn used his *monas* without permission. The existence of one of the world's most substantial alchemical libraries in seventeenth-century New England adds another dimension to the Puritan mind.

MARTHA TEACH GNUDI

DORNO, CARL W. M. (*b.* Königsberg, Prussia [now Kaliningrad, U.S.S.R.], 3 August 1865; *d.* Davos, Switzerland, 22 April 1942), *biometeorology.*

The son of Carl Dorno and the former Emma Lehnhard, Dorno came from an old Königsberg merchant family. In 1891 he took over his father's business; and the following year he married Erna Hundt, from Hamburg, who bore him one daughter. When Dorno was thirty-three years old, he began to study chemistry, physics, political economy, and law at Königsberg University, capping his studies with a doctorate in chemistry in 1904. After his daughter contracted tuberculosis, he moved the family to Davos, an alpine town frequented by consumptives; here he became concerned with determining the factors that made the climate beneficial. In order to conduct research he established a private physical-meteorological observatory, financing it with his personal fortune.

From the beginning Dorno sought the exact measurement and recording of solar and celestial radiation, separately and combined, to determine the total energy and the energies of single spectral regions. For this purpose it was necessary to construct better instruments. Aided by R. Thilenius, Dorno constructed a pyrheliograph, based on older pyrheliometer principles, with a continuous intensity scale that permitted its use in atmospheric investigations, and the Davos frigorimeter. With the information recorded by his new instruments Dorno investigated the annual and diurnal variation of radiation, not only for Davos but also for other places at various latitudes and altitudes.

At high altitudes the intensity of solar radiation is nearly constant; this is the most important factor in the beneficial effects of high-altitude air. On the highest peaks radiation is higher in winter than in summer because of the ellipticity of the earth's orbit. For the most part water vapor diminishes radiation, especially ultraviolet radiation. Dorno's investigations on the ultraviolet were so pioneering that the radiation between 2,900 and 3,200 angstrom units is called Dorno's radiation. He also deduced the concept of biological cooling and was thus the founder of bioclimatology.

Dorno's daughter died in 1912; but rather than return to Germany, he decided to devote his life to the work he had begun. The loss of his fortune after World War I brought him many difficulties, but the Swiss government took over his observatory and its financial maintenance, permitting him to observe and publish freely as before. In 1926 Dorno retired as director but remained very active in meteorological-physiological studies. Since the mid-1930's he had suffered from difficulties with his vision, which gradually worsened, so that at last he could no longer continue his scientific work.

Among Dorno's honors were an honorary M.D. from the University of Basel (1922) and the title of "professor" from the Prussian government.

BIBLIOGRAPHY

Dorno's writings are *Studien über Licht und Luft des Hochgebirges* (Brunswick, 1911); *Dämmerungs- und Ringerschein. 1911–1917* (Brunswick, 1917); *Physik der Sonnen- und Himmelsstrahlung* (Brunswick, 1919); *Himmelshelligkeit, Himmelspolarisation und Sonnenintensität, 1911–1918* (Brunswick, 1919); *Klimatologie im Dienste der Medizin* (Brunswick, 1920); "Fortschritte in Strahlungsmessung (Pyrheliograph)," in *Meteorologische Zeitschrift,* **39** (1922), 303–323; *Meteorologische - physikalische - physiologische Studie Muottas-Muraigl* (Davos, 1924); "Über die Verwendbarkeit von Eder's Graukeilphotometer im meteorologische Dienst," in *Meteorologische Zeitschrift,* **42** (1925), 87–97; "Davoser Frigorimeter," *ibid.,* **45** (1928), 401–421; *Assuan, eine meteorologische-physikalische-physiologische Studie* (Davos, 1932); and *Das Klima von Agra, eine dritte und letzte meteorologische-physiologische Studie* (Davos, 1934).

On Dorno, see R. Süring, "Carl Dorno," in *Meteorologische Zeitschrift,* **59** (1942), 202–205.

H. C. FREIESLEBEN

DOSITHEUS (*fl.* Alexandria, second half of the third century B.C.), *mathematics, astronomy.*

He was a friend or pupil of Conon, and on the latter's death, Archimedes, who had been in the habit of sending his mathematical works from Syracuse to

Conon for discussion in the scientific circles of Alexandria, chose Dositheus as the recipient of several treatises, including *On the Quadrature of the Parabola, On Spirals, On the Sphere and the Cylinder* (two books), and *On Conoids and Spheroids*. At the beginning of the first of these Archimedes says, "Having heard that Conon has died, who was a very dear friend of mine, and that you have been an acquaintance of his and are a student of geometry . . . I determined to write and send you some geometrical theorems, as I have been accustomed to write to Conon" (Heiberg, ed., II, 262). The preambles to the other works make it clear that Dositheus on his side often wrote to Archimedes requesting the proofs of particular theorems. Nothing is known about Dositheus' own mathematical work.

His astronomical work seems to have been concerned mainly with the calendar. He is cited four times in the calendar attached to Geminus' *Isagoge* (Manitius, ed., p. 210 ff.) and some forty times in Ptolemy's *Phaseis* for weather prognostications (ἐπισημασίαι) such as formed part of a *parapegma,* a type of almanac, originally engraved on stone or wood and later transmitted in manuscript form—like the two mentioned above—giving astronomical and meteorological phenomena for the days of each month (cf. A. Rehm, "Parapegmastudien," in *Abhandlungen der Bayerischen Akademie der Wissenschaften,* phil.-hist. Abt., n.s., vol. **19** [1941]). Dositheus may have made observations in the island of Cos (*Phaseis,* p. 67. 4, Heiberg, ed.—but the text here is insecure; cf. *proleg.* cliii note 1) as well as in Alexandria.

According to Censorinus (*De die natali,* 18, 5), Dositheus wrote on the *octaëteris* (an eight-year intercalation cycle) of Eudoxus, and he may be the Dositheus Pelusiotes (Pelusium, at the northeastern extremity of the Egyptian delta) mentioned as providing information about the life of Aratus (*Theonis Alexandrini vita Arati,* §2, Maas, ed.—but spelled here Δωσίθεος).

BIBLIOGRAPHY

In addition to the works cited in the text, see F. Hultsch, "Dositheos 9," in Pauly-Wissowa, *Real-Encyclopädie,* X (1905), cols. 1607–1608.

D. R. DICKS

DOUGLAS, JAMES (*b.* Baads, Scotland, 21 March 1675; *d.* London, England, 2 April 1742), *medicine, natural history, letters.*

Douglas was the second son of William Douglas of Baads, near Edinburgh, and his wife, the former Joan Mason. They were an obscure but industrious family of small landowners, a minor branch of the widespread Douglas clan. Of their twelve children, four became fellows of the Royal Society—Walter, James, John, and George—although only James and John, both physicians, produced work of lasting importance. Nothing is known of the early schooling of James Douglas; but on 23 July 1699 he was granted the degree of *docteur* by the Faculty of Medicine of the University of Rheims, and it is likely that he had obtained the M.A. at Edinburgh in 1694. He was working in London by 1700 and early decided on a career in obstetrics and anatomy. In 1705 he read his first paper to the Royal Society and was granted fellowship in 1706. He became fellow of the Royal College of Physicians in 1721. To his contemporaries he was sufficiently outstanding to receive mention in Pope's *Dunciad,* and he was a friend of Sir Hans Sloane, founder of the British Museum; the physician Richard Mead; and William Cheselden, whose lateral operation for removal of bladder stones he described.

Douglas married Martha Wilkes, aunt of John Wilkes, the political reformer and rake. They had two children, but neither married and both died young. Following her husband's death, Martha Douglas gave lodging to William Hunter and to his brother John during their early days in London.

By 1707, when he published his handbook of comparative myology, Douglas had realized the importance of anatomical teaching to the advancement of medicine; and he was among the first to advertise classes, which were well attended. His major publication in this field was that on the peritoneum (1730), an excellent monograph that drew attention to the duplicature of the peritoneal membrane, at that time a controversial subject. In this book there is a short description of the structure later known as the pouch of Douglas and still recognized by that name. There are also the ligament, the line, and the semilunar fold of Douglas. In the Hunterian Library of the University of Glasgow, donated by will of its founder, William Hunter, are no fewer than sixty-four unpublished manuscripts by Douglas, on many aspects of anatomy, natural history, grammar, and orthoepy, and the Blackburn Collection in that library contains an enormous number of documents, drawings, and notes, nearly all in Douglas' hand.

Douglas was also a collector of editions of the works of Horace; and he published a magnificent catalog of his library of Horatiana, containing 557 volumes.

In the Hunterian Library may also be seen an interesting series of case notes, written at the bedside of his patients, dating from 1704 and illustrating the problems of diagnosis and treatment at that time.

They show, too, that he acted as consultant to the London midwives, being called for medical treatment during pregnancy as well as for the complications of labor.

In obstetrical science Douglas carefully studied the anatomy of the female pelvis and of the fetus; in 1735 he attended Anne, princess of Orange, daughter of George II, in Holland. He was also concerned, in 1726, in the exposure of Mary Toft, the "rabbit woman" of Godalming.

In 1719 James assisted his younger brother John, a brilliant but irascible physician, in the promulgation of John's ideas in introducing suprapubic lithotomy, one of the earliest attempts at routine abdominal surgery in England.

Douglas' publications on natural history include a well-produced monograph on the "Guernsay-lilly" (*Nerine sarniensis*) and a paper to the Royal Society on the flamingo (1714). Both demonstrate the care and method of his presentation.

His greatest contribution to the future, however, lay in his encouragement of William Hunter, who came to him as a resident pupil in 1741. The brilliant young student and his shrewd master established an intimate relationship, and on Douglas' death in 1742 Hunter wrote a touching letter to his mother. (The letter is now in the Royal College of Surgeons of England.) During the single year of their contact, Hunter became interested in the anatomical subjects on which Douglas had worked. Thus the anatomy of aneurysms, of the bones, of the "cellular membrane," and above all of the gravid uterus were topics upon which William Hunter elaborated at various later dates. It is not too much to say that the encouragement and training received by Hunter during this formative period was an important factor in those developments in British medical education for which he and his brother John were so largely responsible.

BIBLIOGRAPHY

I. ORIGINAL WORKS. Douglas' writings are *Myographia comparatae specimen, or a Comparative Description of All the Muscles in a Man and a Quadruped . . .* (London, 1707); *Bibliographiae anatomicae specimen sive catalogus omnium pene auctorum qui ab Hippocrate ad Harveium rem anatomicum . . .* (London, 1715); *Index materiae medicae . . .* (London, 1724); *Lilium Sarniense: or a Description of the Guernsay-Lilly . . .* (London, 1725); *The History of the Lateral Operation* (London, 1726); *An Advertisement Occasion'd by Some Passages in Sir R. Manningham's Diary . . .* (London, 1727); *Arbor Yemensis fructum cofé ferens: or a Description and History of the Coffee Tree* (London, 1727); *A Description of the Peritoneum and of . . . the Membrana Cellularis* (London, 1730); and *Catalogus editorum Quinti*

Horatii Flacci ab an. 1476 ad an. 1739 quae in bibliotheca Jacob. Douglas . . . adservantur (London, 1739).

Eleven articles that appeared in the *Philosophical Transactions of the Royal Society* between 1706 and 1731 are listed in K. Bryn Thomas, *James Douglas of the Pouch . . .* (London, 1964), p. 198. Unpublished MSS in the Blackburn Collection at the Hunterian Library of the University of Glasgow are catalogued and annotated in the book by Thomas (above) pp. 85–193; and in J. Young and P. H. Aitken, *A Catalogue of the Manuscripts in the Library of the Hunterian Museum . . .* (Glasgow, 1908), p. 425 ff.

II. SECONDARY LITERATURE. Besides Thomas' book (see above) one may also consult J. C. Carpue, *A Description of the Muscles . . . With the Synonyma of Douglas* (London, 1801); Börje Holmberg, *James Douglas on Pronunciation, c. 1740* (Copenhagen, 1956); and D. Watson, *The Odes, Epodes, and Carmen seculare of Horace . . . With a Catalogue of the Editions of Horace From 1476 to 1739 in the Library of James Douglas* (London, 1747).

K. BRYN THOMAS

DOUGLAS, JESSE (*b.* New York, N.Y., 3 July 1897; *d.* New York, 7 October 1965), *mathematics.*

Douglas became interested in mathematics while he was still a high school student; in his freshman year at the City College of New York he became the youngest person ever to win the college's Belden Medal for excellence in mathematics. He graduated with honors in 1916 and began graduate studies with Edward Kasner at Columbia University. From 1917 to 1920 (in which year he was awarded the doctorate) he also participated in Kasner's seminar in differential geometry; here he developed his love for geometry and first encountered the problem of Plateau.

From 1920 to 1926 Douglas remained at Columbia College, teaching and doing research, primarily in differential geometry. Between 1926 and 1930 he was a National Research fellow at Princeton, Harvard, Chicago, Paris, and Göttingen; during this period he also devised a complete solution to the problem of Plateau, of which the essential features were published in a series of abstracts in the *Bulletin of the American Mathematical Society,* between 1927 and 1930, while a detailed presentation appeared in the *Transactions of the American Mathematical Society* for January 1931. This solution won Douglas the Fields Medal at the International Congress of Mathematicians in Oslo in 1936.

Douglas was appointed to a position at the Massachusetts Institute of Technology in 1930 and taught there until 1936; he was a research fellow at the Institute for Advanced Study at Princeton in the academic year 1938–1939 and received Guggenheim Foundation fellowships for research in analysis and geometry in 1940 and 1941. From 1942 until 1954 he taught at Brooklyn College and at Columbia University, then in 1955 returned to City College, where he spent the rest of his life.

Douglas' work with the problem of Plateau was again rewarded in 1943 when he received the Bôcher Memorial Prize of the American Mathematical Society for his memoirs "Green's Function and the Problem of Plateau" (in *American Journal of Mathematics*, **61** [1939], 545 ff.), "The Most General Form of the Problem of Plateau" (*ibid.*, **61** [1939], 590 ff.), and "Solution of the Inverse Problem of the Calculus of Variations" (in *Transactions of the American Mathematical Society*, **50** [1941], 71–128). The problem of Plateau was apparently first posed by Lagrange about 1760, and had occupied many mathematicians—most notably Riemann, Weierstrass, and Schwarz—in the period from 1860 to 1870. The problem is concerned with proving the existence of a surface of least area bounded by a given contour. Prior to Douglas' solution, mathematicians had succeeded in solving a number of special cases, as when, in the nineteenth century, a solution was obtained for a contour that is a skew quadrilateral having a plane of symmetry. Douglas' solution of 1931 is highly generalized; indeed, it is valid when the contour is any continuous, closed, nonintersecting curve whatever (Jordan curve)—it may even be in space of any number of dimensions. (R. Garnier in 1927 and T. Radó in 1930 had succeeded in solving the problem with less generality by using alternative methods.)

Having disposed of the most fundamental instance of the problem of Plateau—a single given contour and a simply connected minimal surface—Douglas went on to consider surfaces bounded by any finite number of contours and to consider surfaces of higher topological structure—as, for example, one-sided surfaces or spherical surfaces with any number of attached handles or any number of perforations. Between 1931 and 1939 he gave solutions to such problems as these and formulated and solved other general forms of the problem.

The problem of Plateau did not represent Douglas' sole mathematical interest, however. In 1941 he published a complete solution of the inverse problem of the calculus of variations for three-dimensional space—a problem unsolved until then, although in 1894 Darboux had stated and solved the problem for the two-dimensional case. In addition to publishing some fifty papers on geometry and analysis, Douglas' work in group theory is notable; in 1951, he made significant contributions to the problem of determining all finite groups on two generators, A and B, which have the property that every group element can be expressed in the form $A^r B^s$, where r and s are integers.

BIBLIOGRAPHY

I. ORIGINAL WORKS. *Scripta mathematica*, **4** (1936), 89–90, contains a bibliography of Douglas' publications prior to 1936. A complete bibliography of his works is on file in the mathematics department of the City College of New York.

II. SECONDARY LITERATURE. Some information on Douglas' life prior to 1936 may be found in *Scripta mathematica*, **4** (1936), 89–90. In the near future the National Academy of Science will publish a biography of Douglas, including a complete bibliography of his publications. An obituary of Douglas can be found in the *New York Herald Tribune* (8 Oct. 1965).

NORMAN SCHAUMBERGER

DOVE, HEINRICH WILHELM (*b.* Liegnitz, Prussia [now Legnica, Poland], 6 October 1803; *d.* Berlin, Germany, 4 April 1879), *meteorology, physics.*

The son of Wilhelm Benjamin Dove and the former Maria Susanne Sophie Brückner, Dove belonged to a prosperous family of apothecaries and merchants in Liegnitz. His delicate health led him to choose an academic career instead of following the family profession. Dove was an open-minded, communicative person who was interested not only in science but also in politics, history, and philosophy, all of which he studied at the University of Breslau, which he entered in 1821. His doctoral thesis, on climatology, was presented in 1826 to the University of Königsberg, where he lectured until 1829, when he went to Berlin. He became a professor there in 1844. On 26 October 1830 he married Franziska Adelaide Luise Etzel.

Dove devoted much of his time to lecturing, becoming a great popular success. Mainly an experimental physicist, he improved and devised many scientific instruments and illustrated his lectures with elegant experiments. His great interest in education was shown when, in 1837–1849, he edited the *Repertorium der Physik*, for which he wrote many of the articles himself: on the progress of physics, meteorology, theory of heat, optics, magnetism, and electricity. His contribution to physics centered on observations of the earth's magnetism; polarization phenomena, especially the optical properties of rock crystals; and induced electricity.

Dove's principal interest was meteorology. A great step forward in that field had been the extension, in 1780, of a network of thirty-nine observation stations to cover many European countries. This had been organized by the Societas Meteorologica Palatina in Mannheim, Germany, which had demanded uniformity in the mounting and operation of the standard instruments. At the beginning of the nineteenth century, stations were established in Russia and North

America, and Humboldt used their observations to draw maps indicating the distribution of temperature throughout the areas covered. This new field, climatology, also interested Dove, who used the many observations to draw maps showing the monthly mean distribution of heat at the earth's surface. These maps demonstrated how the winds, land masses, and seas influenced heat distribution.

H. W. Brandes emphasized that not only climate but other variables were important in meteorology. In Breslau, Dove studied meteorology with Brandes, and after thorough study of Brandes' meteorological data he formulated his "Drehungsgesetz" ("law of rotation") (1827), which states that the order of meteorological phenomena at a single place on the earth's surface corresponds to what would happen if there were great whirligigs rotating clockwise in the atmosphere. With a southwest wind the temperature rises, and it rains. The wind then moves to the west, the temperature falls, and the barometer rises. When the wind is northeast, the barometric pressure is at a maximum and the temperature at a minimum. Then the pressure begins to fall until the wind again blows from the southwest. It was Dove's great contribution to meteorology to be the first to find a system in weather changes.

Further progress occurred in 1857, when Buys Ballot, one of Dove's students, found that winds arise because of differences in atmospheric pressure. Throughout his life Dove defended his own law, which, in contrast with Buys Ballot's, was founded on the assumption that wind is the primary factor by which weather phenomena should be explained.

Dove was interested in expanding the international collaboration between meteorological institutions that had started with the establishment of the Societas Meteorologica Palatina. He assembled meteorologists from several European countries in Geneva in 1863, but it was not until 1873, with the first international meteorology meeting in Leipzig—in which Dove took no part—that the collaboration was expanded to any great extent.

Dove was director of the Prussian Institute of Meteorology from its founding in 1849, and he was three times elected vice-chancellor of the University of Berlin. In 1853 he received the Copley Medal of the Royal Society, and he was a member of several scientific societies.

BIBLIOGRAPHY

I. ORIGINAL WORKS. Dove's scientific writings are listed in the Royal Society's *Catalogue of Scientific Papers,* **2,** 329–335; **7,** 553–554; and in Neumann's biography (see below), p. 72 ff. Most of them were published in *Abhandlungen* and *Monatsberichte der Preussischen Akademie der Wissenschaft* and J. C. Poggendorff's *Annalen der Physik und Chemie.* The most important meteorological writings are *Meteorologische Untersuchungen* (Berlin, 1837); "Das Gesetz der Stürme," in *Monatsberichte der Preussischen Akademie der Wissenschaft,* **52** (1840), 232–239; and *Die Verbreitung der Wärme auf der Oberfläche der Erde* (Berlin, 1852).

II. SECONDARY LITERATURE. Further information may be found in "Heinrich Wilhelm Dove," in *Nature,* **19** (10 Apr. 1879), 529–530; Hans Neumann, *Heinrich Wilhelm Dove. Eine Naturforscher-Biographie* (Liegnitz, 1925); and *Neue deutsche Biographie,* IV (1959), 92 f.

KURT MØLLER PEDERSEN

DOWNING, ARTHUR MATTHEW WELD (*b.* Carlow, Ireland, 13 April 1850; *d.* London, England, 8 December 1917), *astronomy.*

Downing's chief contribution was the computation of precise positions and movements of astronomical bodies; he was also one of the founders of the British Astronomical Association. The younger son of Arthur Matthew Downing, Esq., of County Carlow, he received his early education under Philip Jones at the Nutgrove School, near Rathfarnham, County Dublin. He thence proceeded in 1866 to Trinity College, Dublin, where he obtained the scholarship in science and graduated B.A. in 1871, gaining the gold medal of his year in mathematics. He took his M.A. degree in 1881, and in 1893 Dublin University granted him the honorary degree of D.Sc.

In 1872 Downing was appointed assistant at the Royal Observatory, Greenwich, to commence 17 January 1873. There he was placed in charge successively of the library and manuscripts, the time department, and the circle computations. Reduction of the circle, altazimuth, and equatorial observations came to constitute his major responsibility, but he also served as one of four regular observers with the transit circle and altazimuth. Following his election as fellow of the Royal Astronomical Society in 1875, he communicated seventy-five papers to the society, dealing principally with the correction of systematic errors in different star catalogs and with the computation of fundamental motions of the heavenly bodies. Among the papers is a calculation done in collaboration with G. Johnstone Stoney of perturbations suffered by the Leonid meteors, which predicted and explained the relative sparseness of the 1899 shower.

Downing's next appointment, as superintendent of the Nautical Almanac Office, extended from 1 January 1892 to his statutory retirement on 13 April 1910. During this period he brought out the *Nautical Almanac* for the years 1896 to 1912, gradually instituting various improvements therein, including increasing the number of ephemeris stars, introducing Besselian coordinates into the eclipse and occultation lists, and expanding the sections on planetary satellites. He also replaced the solar and planetary tables of Le Verrier with those of Simon Newcomb and George Hill, dropped the obsolete Lunar Distance Tables, and introduced into the *Almanac* the physical ephemerides of the sun, moon, and planets.

As one of the founders in 1890 of the British Astronomical Association and subsequent nurturer of its early development, Downing contributed significantly to amateur astronomy. His consistent advance publication of particulars of astronomical occurrences, such as eclipses and occultations, constituted a valuable service to observers in many countries.

In 1896 Downing was elected fellow of the Royal Society and officiated in Paris at the important International Conference of Directors of Ephemerides, which sought to attain uniformity in the adoption of astronomical constants. In 1899 he revised Ernst Becker's *Tafeln zur Berechnung der Precession,* adapting them to a new value of the precessional constant first derived by Simon Newcomb in response to a formal request presented to him at the Paris meeting. The epoch adopted for the tables was 1910, but they were constructed to be useful for at least ten years before and after that year. In 1901 Downing compiled a revised version of Taylor's *Madras Catalogue* of 11,000 stars, reduced without proper motion to the equinox of 1835. His sudden death in 1917, following several years of illness, was from a recurrent heart complaint.

Downing was secretary (1889–1892) and vice-president (1893–1895) of the Royal Astronomical Society, as well as vice-president (1890–1891) and second president (1892–1894) of the British Astronomical Association.

BIBLIOGRAPHY

I. ORIGINAL WORKS. Seventy-five papers communicated to the Royal Astronomical Society are listed in the *General Index to the Monthly Notices of the Royal Astronomical Society* covering vols. **1–52** (1870) and vols. **53–70** (1911). See especially the Downing-Stoney paper "Ephemerides of Two Situations in the Leonid Stream," in *Monthly Notices of the Royal Astronomical Society,* **59** (1898), 539–541. Other joint communications with Stoney on this subject include "Perturbations of the Leonids," in *Proceedings of the Royal Society,* **64A** (1899), 403–409; and two letters to the editor of *Nature:* "Next Week's Leonid Shower," **61** (1899), 28–29, and "The Leonids—a Forecast," **63** (1900), 6. Another paper, by Downing alone, concerning meteors is "The Perturbations of the Bielid Meteors," in *Proceedings*

of the Royal Society, **76A** (1905), 266–270. A report by Downing to the British Astronomical Association on 28 June 1905, concerning his researches on the Bielids, is described in *Journal of the British Astronomical Association,* **15**, no. 9 (1905), 361–363.

Nineteen contributions to *Observatory* (of which Downing was editor of vols. **8–10** [1885–1887]) are in vols. **2–4, 6, 8, 12, 27, 30, 33, 36** and **39**. A determination of the sun's mean equatorial horizontal parallax is in **3**, no. 31 (1879), 189–190. See also *Astronomische Nachrichten,* **96**, no. 2288 (1880), 119–128.

Twenty-one contributions to the *Journal of the British Astronomical Association* are in vols. **1, 3, 8, 10, 15–19,** and **21–22**. Besides accounts of observational phenomena these include "How to Find Easter," **3**, no. 6 (1893), 264–268; and "When the Day Changes," **10**, no. 4 (1900), 176–178, the latter trans. as "Où le jour change-t-il," in *Ciel et terre,* **21** (1900), 84–86. Downing edited the first of the eclipse volumes of the British Astronomical Association dealing with the eclipse of 1896 (he also organized an Association expedition that failed to observe the eclipse because of cloud). He issued in addition several Association circulars giving advance particulars of other current notable eclipses. He was private adviser over many years to the editor of the *Journal of the British Astronomical Association* (see acknowledgment by A. S. D. Maunder, **28**, no. 2 [1917], 67–69).

Ten letters to the editor of *Nature,* two in conjunction with Stoney, see above, are in vols. **17, 59, 61, 63, 65, 71–72, 76, 78,** and **90**. The receipt of advance particulars of various astronomical phenomena from Downing is frequently noted and quoted in the editorial columns of this journal. While superintendent of the Nautical Almanac Office, Downing compiled for the years 1896–1912 the *Nautical Almanac and Astronomical Ephemeris for the Meridian of the Royal Observatory at Greenwich* (for 1896–1901, published London, 1892–1897; for 1902–1912, published Edinburgh, 1898–1909). See seven-page appendix to the *Nautical Almanac* for 1900, which contains a continuation of tables I and III of Damoiseau's *Tables of Jupiter's Satellites for the Years 1900–1910*.

Other contributions are his revision of Becker's *Tafeln . . . Precession: Precession Tables Adopted to Newcomb's Value of the Precessional Constant and Reduced to the Epoch 1910* (Edinburgh, 1899); and his revision of *Taylor's General Catalogue of Stars for the Equinox 1835.0 From Observations Made at the Madras Observatory During the Years 1831 to 1842* (Edinburgh, 1901).

II. SECONDARY LITERATURE. Obituaries are A. C. D. Crommelin, in *Nature,* **100** (1917), 308–309; and A. S. D. Maunder, in *Journal of the British Astronomical Association,* **28**, no. 2 (1917), 67–69. Unsigned obituaries are in *Observatory,* **41**, no. 522 (1918), 70; and *Monthly Notices of the Royal Astronomical Society,* **78**, no. 4 (1918), 241–244. References to his role in founding the British Astronomical Association are made by Howard L. Kelly in *Memoirs of the British Astronomical Association,* **36**, pt. 2 (1948), a Historical Section memoir. Summaries of some of Downing's original papers in *Journal of the British Astronomical*

Association, Monthly Notices of the Royal Astronomical Society, and *Proceedings of the Royal Society* appear in the editor's column of contemporary issues of *Nature.*

SUSAN M. P. McKENNA

DRACH, JULES JOSEPH (*b.* Sainte-Marie-aux-Mines, near Colmar, France, 13 March 1871; *d.* Cavalaire, Var, France, 8 March 1941), *mathematics.*

Drach was born a few months before the Treaty of Frankfurt, by which Alsace ceased to be French. His father, Joseph Louis Drach, and mother, the former Marie-Josèphe Balthazard, modest farmers from the Vosges, took refuge with their three sons at Saint-Dié. From his youth Drach had to work for an architect in order to help his family. He was, however, able to attend elementary school and, at the urging of his teachers, who obtained a scholarship for him, went on to the *collège* in Saint-Dié and then to the lycée in Nancy. Drach was admitted at the age of eighteen to the École Normale Supérieure. Without attempting to make up his failure of the *agrégation,* and encouraged by Jules Tannery, he devoted himself to research and obtained the *doctorat-ès-sciences* in 1898. He taught at the universities of Clermont-Ferrand, Lille, Poitiers (where he married Mathilde Guitton), Toulouse, and Paris (1913), where his courses in analytical mechanics and higher analysis were well received. Drach was elected to the Académie des Sciences in 1929 and was a member of many scientific commissions. His poor health obliged him to reside in Provence for most of the year. His son Pierre entered the École Normale Supérieure in 1926 and had a brilliant career as a biologist.

After retiring to his estate at Cavalaire, Drach pursued his mathematical researches and indulged his love of reading, being interested in the plastic arts as well as in the sciences. Marked by the ordeals of his youth, he always remained close to the poor and was actively involved in the improvement of land held by the peasants.

Drach's mathematical researches display great unity. Galois's algebraic theory, with its extension to linear differential equations just made by Émile Picard (1887), seemed to him to be a model of perfection. He proposed to elucidate the fundamental reason for such a success in order to be able to extend it to the study of differential equations in the most general cases, asserting that the theory of groups is inseparable from the study of the transcendental quantities of integral calculus.

To what he termed the "geometric" point of view, in which one introduces supposedly given functions whose nature is not specified, Drach opposed the "logical" problem of integration, which consists of classifying the transcendental quantities satisfying the rational system verified by the solutions. For this process he introduced the notion of the "rationality group," whose reducibility and primitiveness he investigated. This bold conception, of an absolute character, foreshadowed the axiomatic constructions that were subsequently developed. In discovering regular methods permitting one to foresee the reductions in the difficulties of integration, Drach gave an account of the results obtained before him by Sophus Lie, Émile Picard, and Ernest Vessiot. He completed these and thus extended the special studies concerning, for example, the ballistic equations and those determining families of curves in the geometry of surfaces, such as the "wave surface."

Besides numerous articles in various journals and notes published in the *Comptes rendus . . . de l'Académie des sciences* over some forty years, Drach, with his colleague and friend Émile Borel, prepared for publication a course given at the Faculté des Sciences in Paris by Henri Poincaré—*Leçons sur la théorie de l'élasticité* (Paris, 1892)—and one given by Jules Tannery at the École Normale Supérieure, *Introduction à l'étude de la théorie des nombres et de l'algèbre supérieure* (Paris, 1895). Moreover, he played a large role in preparing for publication, under the auspices of the Académie des Sciences, the works of Henri Poincaré (11 vols., 1916–1956).

BIBLIOGRAPHY

Drach's papers include "Essai sur une théorie générale de l'intégration et sur la classification des transcendantes," in *Annales de l'École normale supérieure,* 3rd ser., **15** (1898), 243–384; "Sur le problème logique de l'intégration des équations différentielles," in *Annales de la Faculté des sciences de Toulouse,* 2nd ser., **10** (1908), 393–472; "Recherches sur certaines déformations remarquables à réseau conjugué persistant," *ibid.,* pp. 125–164; "Le système complètement orthogonal de l'espace euclidien à *n* dimensions," in *Bulletin de la Société mathématique de France,* **36** (1908), 85–126; "Sur l'intégration logique des équations différentielles ordinaires," in *Cambridge International Congress of Mathematicians. 1912,* I (Cambridge, 1913), 438–497; and "Equations différentielles de la balistique extérieure," in *Annales de l'École normale supérieure,* 3rd ser., **37** (1920), 1–94.

LUCIENNE FÉLIX

DRAPARNAUD, JACQUES-PHILIPPE-RAY-MOND (*b.* Montpellier, France, 3 June 1772; *d.* Montpellier, 2 February 1804), *zoology, botany.*

The son of Jacques Draparnaud, a merchant, and Marie-Hélène Toulouse, Draparnaud received his master of arts at the age of fifteen from the École de Droit; his thesis was entitled *De universa philosophia.* After graduating in 1790, he turned to the study of medicine. The Revolution, in which he took an active part, interrupted his studies, sent him to jail, and ruined him financially. He also narrowly escaped the scaffold. Freed in 1794, he withdrew to the École de Sorèze, where he taught physics and chemistry. In 1796 he was assigned the chair of *grammaire générale* at the Écoles Centrales, but he remained there only a short time. In 1802 Chaptal appointed him curator of collections at the Faculté de Médecine of the University of Montpellier and associate to Antoine Gouan, who was then director of the Jardin des Plantes. However, the intrigues of some of his colleagues led Draparnaud to resign in November 1803. He died shortly thereafter of pulmonary tuberculosis, from which he had suffered for many years.

During a scientific career of only fifteen years and despite the obstacles imposed by the Revolution and his illness, Draparnaud published at least forty-five works on politics, philosophy, grammar, physics, mineralogy, zoology (principally malacology), and botany. Besides his observations, he reached certain conclusions in biology and physiology that were remarkably ahead of his time. He introduced in a few words the idea of vital phenomena common to animal and plant life, and attempted to reduce these phenomena to physical and chemical laws.

At his death Draparnaud left numerous manuscripts in various stages of completion. Only one was published: *Histoire naturelle des mollusques terrestres et fluviatiles de la France* (1805); the others have been lost. Among them was one on the history of confervae, to which he attached great importance and on which he had worked for ten years. This work marks him as one of the first algologists. All that is left of it is the herbal of algae, now kept in the Muséum d'Histoire Naturelle in Paris.

BIBLIOGRAPHY

I. ORIGINAL WORKS. A complete list of Draparnaud's works is in Dulieu (1956) and Motte (1964) (see below). The most important include *Discours sur les avantages de l'histoire naturelle* (Montpellier, an IX); *Tableau des mollusques terrestres et fluviatiles de France* (Montpellier, an IX); *Discours sur les moeurs et la manière de vivre des plantes* (Montpellier, an IX); *Discours sur la vie et les fonctions vitales, ou précis de physiologie comparée* (Montpellier-Paris, an X); *Discours sur la philosophie des sciences* (Montpellier-Paris, an X); *Dissertation sur l'utilité de l'histoire naturelle dans la médecine présentée à l'École de Médecine de Montpellier pour obtenir le titre de médecin* (Montpellier, an XI); and *Histoire naturelle des mollusques terrestres et fluviatiles de la France* (Paris-Montpellier, 1805).

II. SECONDARY LITERATURE. On Draparnaud and his work, see J. B. Baumes, *Éloge de Draparnaud* (Montpellier, an XII); L. Dulieu, "Jacques-Philippe-Raymond Draparnaud," in *Revue d'histoire des sciences et de leurs applications,* **9,** no. 3 (1956), 236–358; G. Laissac, "Notes sur la vie et les écrits de Draparnaud," in *Revue du midi,* **2** (1843), 81–112, 239–256; J. Motte, "Vie de Draparnaud," in *Opuscula botanica necnon alia,* **7** (1964), 1–79; and J. Poitevin, *Notice sur la vie et les ouvrages de M. Draparnaud, lue à l'assemblée publique de la Société libre des sciences et belles-lettres de Montpellier, le 13 floréal an XII* (Montpellier, an XIII).

                                        JEAN MOTTE

DRAPER, HENRY (*b.* Prince Edward County, Virginia, 7 March 1837; *d.* New York, N.Y., 20 November 1882), *astronomy.*

Draper's family moved to New York City when he was two years old, and there he maintained his principal residence for the rest of his life. His father, John William Draper, was a distinguished physician and chemist. His mother, the former Antonia Coetana de Paiva Pereira Gardner, was the daughter of the attending physician of the emperor of Brazil.

Draper was stimulated to precocity by his parents and by the intellectual milieu of New York City. He swiftly rose to prominence as a gifted inventor and a deft technician who made most of his own equipment. He pursued excellence and innovation; and his principal fame rests not on his medical profession but on his astronomical avocation, pursued at his own expense. Draper maintained intellectual freedom without slipping into isolation, and he is one of America's outstanding "amateur" scientists.

Educated at the University of the City of New York, where his father was professor of chemistry, Draper entered the medical school at the age of seventeen, after two years in college. In 1857 he had completed all his medical studies; he had written a medical thesis on the spleen, illustrated with daguerreotype microphotographs; and he had passed all of his examinations. He was only twenty, however, and since he had not yet reached the age required for graduation, he went abroad for a year with his older brother, Daniel, to relax and study.

During a visit to the observatory of William Parsons, the third earl of Rosse, in Parsonstown (now Birr), Ireland, Draper conceived the possibility of combining photography and astronomy. Returning to America in the spring of 1858, he received his medical diploma, joined the staff of Bellevue Hospital, and began preparations for grinding and polishing a speculum mirror. This tedious task occupied his spare time until the following summer, and it later ended in frustration when the mirror was split by freezing moisture. Draper's father revealed this tale to Sir John Herschel in June 1860 and was advised that glass was preferable to speculum in ease of figuring, lightness, and brightness when silvered. By November 1861 Draper had completed the first of about 100 glass mirrors, and he installed it in his new observatory on his father's estate at Hastings-on-Hudson, New York. His astronomical career began with daguerreotypes of the sun and the moon.

Early in 1860 Draper was appointed professor of natural sciences at the University of the City of New York, and for the next four years his teaching duties had to compete with experiments in photography and the polishing of glass mirrors. Many of these experiments were discussed at the monthly meetings of the American Photographic Society, of which he was a founding member. His monograph "On the Construction of a Silvered Glass Telescope 15-1/2 Inches in Aperture, and Its Use in Celestial Photography," published in 1864 at the request of Joseph Henry, provided a detailed description of his techniques and became the standard reference on telescope making.

Draper's work was interrupted during the Civil War by service as surgeon of the Twelfth Regiment of the New York State Militia. Because of poor health he was discharged in October 1862, after serving for nine months. During 1863 Draper made 1,500 negatives of the lunar face, and a small number of them bore enlargement to a fifty-inch diameter. His work was severely limited by atmospheric tremor and smoky haze, and he commented that "if the telescope could be transported to the Peruvian plateaus, 15,000 feet above the sea, or somewhere near the equator on the rainless west coast of South America, it would accomplish more." He never reached either of these sites, but both have subsequently been utilized.

The interests of the period 1865–1867 (when he was aged twenty-eight to thirty) were interwoven through his life until his sudden death at forty-five. In this interval he published an article on spectrum analysis, completed a textbook on chemistry, and was appointed professor of physiology and dean of the faculty in the medical department. The department had just lost its building and collections in a fire; and Draper led, and partly financed, its rehabilitation until 1873, when he resigned.

In 1867 Draper married Anna Mary Palmer, the daughter of Courtlandt Palmer. Wealthy and charming, Mrs. Draper proved to be a talented assistant in her husband's laboratory as well as a renowned hostess. The Draper dining table was frequently surrounded by celebrated scientists, politicians, and soldiers; and after Draper's election in 1877 to the National Academy of Sciences, they regularly entertained the members during their New York meetings. During the summer months the Drapers lived in Dobbs Ferry, New York, two miles from the Hastings observatory; and each autumn they returned to their home on Madison Avenue, where Draper maintained an astronomical laboratory that acquired the reputation of being the best-equipped in the world.

Work on a twenty-eight-inch reflecting telescope continued from 1867 to 1872; it was interrupted by the ruling of gratings and by preliminary studies of the spectra of the elements, especially carbon, nitrogen, and hydrogen. As a reference scale for the determination of wavelength, Draper photographed the solar spectrum; and his results far surpassed the best spectra available from 1873 until 1881, attesting to the combined power of photography and his ingenuity.

In May 1872, as soon as the final touches had been added to the twenty-eight-inch reflector, Draper photographed the spectrum of the star Vega, but the initial attempt failed. The low sensitivity of the wet collodion plates and the difficulty of keeping the stellar spectrum motionless on the plate prevented him from obtaining more than a faint continuum without spectrum lines. By August his technique had improved, and he was rewarded with the first photograph of stellar spectrum lines. Later he devised and named a "spectrograph," similar to Huggins' visual spectroscope, employing an entrance slit to purify the resulting spectrum and to permit the impression of reference spectra for the identification of the celestial elements.

When his father-in-law died in 1873, Draper accepted the responsibility of managing the estate; and when he found that this job required several hours a day, he resigned as dean of the medical faculty and accepted the title of professor of analytical chemistry in the academic department. The following year he completely set aside his own research to act as director of the photographic department of the U.S. commission to observe the 1874 transit of Venus. Following the transit the commission asked Congress to order a special gold medal to be struck for Draper, and the following year he was elected to the Astronomische Gesellschaft. In 1876 he attended the Philadelphia Centennial Exposition as a judge of the photographic section.

Draper committed one significant scientific error when, in 1877, he announced the identification of eighteen emission lines of oxygen in the spectrum of the solar disk. His explanation that the emission was due to the "great thickness of ignited oxygen" revealed a failure to appreciate the significance of Kirchhoff's law of radiation. Chemists were delighted by his announcement, but spectroscopists were at once skeptical. Draper repeated the crucial experiment at higher dispersion and carried the new results directly to the Royal Astronomical Society, but he succeeded only in sharpening the debate. After his death his contention that some apparent absorption lines were merely gaps between bright emission lines was definitely refuted by spectroscopy at still higher dispersion.

At the completion of his summer observing, Draper often hunted on horseback in the Rocky Mountains with generals R. B. Marcy and W. D. Whipple, so he was well aware of the clarity of the mountain air. He organized an expedition to observe the summertime solar eclipse of 1878; the party consisted of himself, Mrs. Draper, Thomas Alva Edison (who was able to detect the thermal radiation from the corona), President Henry Morton of the Stevens Institute of Technology, and Professor George F. Barker of the University of Pennsylvania. Their spectroscopic observations during the eclipse revealed the important fact that the corona shines largely by reflected light from the solar disk.

In the spring of 1879 Draper visited Huggins in England and learned that dry photographic plates had become more sensitive than wet collodion. He then called on Lockyer, whose work on the dissociation of the elements had profoundly impressed him. Thus stimulated, he returned to stellar spectroscopy; and in October he wrote to his friend E. S. Holden, "I have had splendid success in stellar spectrum photography this summer, having been able to use a slit only 1/1000 of an inch in width." Draper confirmed Huggins' discovery of hydrogen in Vega, and he noted that the stars so far observed could be placed in Secchi's first two spectrum classes. In his report to the National Academy of Sciences that year he cautioned:

> It is not easy without prolonged study and the assistance of laboratory experiments to interpret the results, and even then it will be necessary to speak with diffidence. . . . It is to be hoped that before long we may be able to investigate photographically the spectra of the gaseous nebulae, for in them the most elementary condition of matter and the simplest spectra are doubtless found.

The convenience of the new dry plates was wonderful; and within the next three years Draper had obtained more than eighty high-quality spectra of bright stars, the moon, Mars and Jupiter, the comet 1881 III, and the Orion nebula. His spectra of the Orion nebula revealed a faint continuous background that he attributed to the scattering of starlight by meteoritic particles.

Draper also achieved splendid and unique results in the direct photography of the moon and of the Orion nebula. The key to this success was the excellent clockwork he devised after rejecting six earlier attempts, and in October of 1880 he wrote to Holden: "The exposure of the Orion Nebula required was 50 minutes; what do you think of that for a test of my driving clock?" The following March he wrote that he had far surpassed the earlier results with an exposure of 140 minutes, so that "the singular proposition is therefore tenable that we are on the point of photographing stars fainter than we can see with the same telescope." Draper had succeeded in making photography the best means of studying the sky, and in May 1882 he wrote: "I think we are by no means at the end of what can be done. If I can stand 6 hours exposure in midwinter another step forward will result." He died before another winter came.

In 1882 Draper received an honorary LL.D. from his alma mater and from the University of Wisconsin. He was a member of the American Philosophical Society and the American Academy of Arts and Sciences, as well as the National Academy of Sciences and the American Association for the Advancement of Science.

In the fall of 1882 Draper resigned his joint professorship of chemistry and physics, to which he had been elected the previous January upon the death of his father. Having thus ensured himself of the free time to pursue his research, he embarked on a two-month hunting trip in the Rocky Mountains. He was exposed one night to severe cold on a mountain slope without shelter, and he returned to New York with less than his usual vigor. But plans for entertaining the National Academy of Sciences were well advanced; and on 15 November 1882 about forty academicians and a few personal friends dined at the Draper home under the novel light of Edison bulbs. Draper retired early with chills and a fever, and he died five days later of double pleurisy.

Mrs. Draper established the Henry Draper Memorial at the Harvard College Observatory to further its research on the photography of stellar spectra.

BIBLIOGRAPHY

I. ORIGINAL WORKS. Draper's published works are "On the Changes of Blood Cells in the Spleen," in *New York Journal of Medicine,* 3rd ser., **5** (1858), 182–189, his M.D. thesis; "On a New Method of Darkening Collodion Negatives," in *American Journal of Photography and the Allied Arts and Sciences,* 2nd ser., **1** (1859), 374–376; "On a Reflecting Telescope for Celestial Photography," in *Report of the British Association for the Advancement of Science,* **2** (1860), 63–64; "On an Improved Photographic Process," in *American Journal of Photography and the Allied Arts and Sciences,* 2nd ser., **5** (1862), 47; "Photography," in *New American Cyclopaedia* (New York, 1863); "On the Construction of a Silvered Glass Telescope 15-1/2 Inches in Aperture, and Its Use in Celestial Photography," in *Smithsonian Contributions to Knowledge,* **14,** pt. 2 (1864); "On the Photographic Use of a Silvered Glass Reflecting Telescope," in *Philosophical Magazine and Journal of Science,* 4th ser., **28** (1864), 249–255; "On a Silvered Glass Telescope and on Celestial Photography in America," in *Quarterly Journal of Science,* **1** (1864), 381–387; "Petroleum: Its Importance, Its History, Boring, Refining," *ibid.,* **2** (1865), 49–59, and in *Dinglers polytechnisches Journal,* **177** (1865), 107–117; "American Contributions to Spectrum Analysis," in *Quarterly Journal of Science,* **2** (1865), 395–401; *A Text Book on Chemistry* (New York, 1866); "Report on the Chemical and Physical Facts Collected From the Deep Sea Researches Made During the Voyage of the School Ship Mercury," in *Report of the Commission on Public Charities* (New York, 1871); "On Diffraction Spectrum Photography," in *American Journal of Science and Arts,* 3rd ser., **6** (1873), 401–409, also in *Philosophical Magazine,* 4th ser., **46** (1873), 417–425; *Annalen der Physik und Chemie,* **151** (1873), 337–350; and *Nature,* **9** (1873), 224–226; "Sur longueurs d'ondes et les caractères de raies violettes et ultraviolettes du soleil," in *Comptes rendus,* **78** (1874), 682–686.

"Photographs of the Spectra of Venus and α Lyrae," in *American Journal of Science and Arts,* 3rd ser., **13** (1877), 95, also in *Philosophical Magazine,* 5th ser., **3** (1877), 238; and *Nature,* **15** (1877), 218; "Astronomical Observations on the Atmosphere of the Rocky Mountains, Made at Elevations of From 4,500 to 11,000 Feet, in Utah and Wyoming Territories and Colorado," in *American Journal of Science and Arts,* 3rd ser., **13** (1877), 89–94; "Discovery of Oxygen in the Sun by Photography and a New Theory of the Solar Spectrum," *ibid.,* **14** (1877), 89–96, also in *Proceedings of the American Philosophical Society* (July 1877), 74–80; *Memorie della Società degli spettroscopisti italiani,* **6** (1877), 69; and *Nature,* **16** (1877), 364–366; "Observations on the Total Eclipse of the Sun of July 29th, 1878," in *American Journal of Science and Arts,* 3rd ser., **16** (1878), 227–230, also in *Philosophical Magazine,* 5th ser., **6** (1878), 318–320; "Speculum," in *New Universal Cyclopedia* (New York, 1878); "On the Coincidence of the Bright Lines of the Oxygen Spectrum with the Bright Lines of the Solar Spectrum," in *American Journal of Science and Arts,* 3rd ser., **18** (1879), 263–277, also in *Monthly Notices of the Royal Astronomical Society,* **39** (1879), 440–447; "On Photographing the Spectra of the Stars and Planets," in *Nature,* **21** (1879), 83–85, also in *American Journal of Science and Arts,* 3rd ser., **18** (1879), 419–425; and *Memorie della Società degli spettroscopisti italiani,* **8** (1879), 81; "Photographs of the Nebula in Orion," in *American Journal of Science and Arts,* 3rd ser., **20** (1880), 433, also in *Philosophical Magazine,* 5th ser., **10** (1880), 388; and *Comptes rendus,* **91** (1880), 688–690; "On a Photograph of Jupiter's Spectrum Showing Evidence of Intrinsic Light From That Planet," in *American Journal of Science and Arts,* 3rd ser., **20** (1880), 118–121, also in *Monthly Notices of the Royal Astronomical Society,* **40** (1880), 433–435; "On Stellar Photography," in *Comptes rendus,* **92** (1881), 964–965; "On Photographs of the Spectrum of Comet b 1881," in *American Journal of Science and Arts,* 3rd ser., **22** (1881), 134–135, also in *Observatory,* **5** (1882), 252–253; "On Photographs of the Nebula in Orion and of Its Spectrum," in *Comptes rendus,* **92** (1881), 173, also in *Monthly Notices of the Royal Astronomical Society,* **42** (1882), 367–368; "On Photographs of the Spectrum of the Nebula in Orion," in *American Journal of Science and Arts,* 3rd ser., **23** (1882), 339–341, also in *Observatory,* **5** (1882), 165–167. An extensive collection of Henry Draper's professional and personal correspondence is on file in the New York Public Library MS collection.

II. Secondary Literature. Principal sources are George F. Barker, "Biographical Memoir of Henry Draper 1837–1882," read before the National Academy (Apr. 1888); and "Researches Upon the Photography of Planetary and Stellar Spectra by the Late Henry Draper, M.D., LL.D., with an introduction by Professor C. A. Young, a List of the Photographic Plates in Mrs. Draper's Possession, and the Results of the Measurement of these Plates by Professor E. C. Pickering," in *Proceedings of the American Academy of Arts and Sciences,* **19** (1884), 231–261. Other sources are "'Minute' on Henry Draper," in *Proceedings of the American Philosophical Society* (Dec. 1882); T. W. Webb, "Draper's Telescope," in *Smithsonian Annual Report* (1864), pp. 62–66, repr. from *Intellectual Observer* (London); and George F. Barker, "On the Use of Carbon Disulphide in Prisms; Being an Account of Experiments Made by the Late Dr. Henry Draper of New York," in *American Journal of Science,* 3rd ser., **29** (1885), 1–10.

Charles A. Whitney

DRAPER, JOHN WILLIAM (*b.* St. Helens, Lancashire, England, 5 May 1811; *d.* Hastings-on-Hudson, New York, 4 January 1882), *chemistry, history.*

Draper was the son of an itinerant Methodist preacher who possessed a Gregorian telescope and who evidently encouraged the boy's scientific interests. The father had purchased two shares in the new London University intended to accommodate scientists, workingmen, and Dissenters, but he died before his son commenced his premedical studies at University College (as it later became) in 1829. There Draper studied chemistry under Edward Turner, an admirer of Berzelius and the author of one of the earliest English textbooks in organic chemistry. Turner interested Draper in the chemical effects of light and thereby gave his career a decisive turn. At a time when Parliament had not yet broken the monopoly of Oxford and Cambridge for granting degrees, Draper had to be contented with a "certificate of honours" in chemistry.

On the urging of maternal relatives, who had gone to America before the Revolution to found a Wesleyan community, Draper immigrated with his mother, his three sisters, and his new wife to Virginia in 1832. He had already collaborated on three minor scientific publications before leaving England and now published eight additional papers between 1834 and 1836 from what he ambitiously described as his "laboratory" in the family farmhouse. The earnings of his sister Dorothy Catharine as a schoolteacher enabled him to take his medical degree at the University of Pennsylvania in 1836. His thesis, "Glandular Action," reflected the interest of his teacher J. K. Mitchell in the researches of Dutrochet on osmosis. His other principal instructor was Robert Hare.

On his return to Virginia, Draper was engaged as chemist and mineralogist to the newly formed Mineralogical Society of Virginia, which had been inspired by the writings of the celebrated pioneer of scientific agriculture in America, the Virginian Edmund Ruffin. A projected school of mineralogy never materialized, but many of the projectors were trustees of Hampden-Sidney College, where Draper served as professor of chemistry and natural philosophy from 1836 to 1839. From 1839 until his death, he was professor of chemistry in the college of New York University. He was also a founding proprietor, in 1841, of the tenuously connected New York University School of Medicine, of which he served as president from 1850. Under his inspiration, the university proper granted the degree of doctor of philosophy five times between 1867 and 1872, to students who had a bachelor's degree in arts or science or a medical degree and had then completed two further years of study in chemistry. This appears to be one of the two earliest attempts in the United States to establish the Ph.D. as a graduate degree. The enterprise petered out with Draper's own advancing years. His other principal institutional exertions were as first president of the American Union Academy of Literature, Science and Art, founded in 1869 as a riposte to the creation of the National Academy of Sciences in 1863. Draper had unaccountably been omitted from the original incorporators of the latter, and the omission was not repaired until 1877. He was, however, elected first president of the American Chemical Society in 1876.

Draper first achieved wide celebrity for his pioneering work in photography. As early as 1837, while still in Virginia, he had followed the example of Wedgwood and Davy in making temporary copies of objects by the action of light on sensitized surfaces. When the details of Daguerre's process for fixing camera images were published in various New York newspapers on 20 September 1839, Draper was ready for the greatest remaining challenge, to take a photographic portrait. A New York mechanic, Alexander S. Wolcott, apparently won the race by 7 October. But if Draper knew of this, he persisted in his own experiments and succeeded in taking a portrait not later than December 1839. His communication to the *Philosophical Magazine,* dated 31 March 1840, was the first report received in Europe of any photographer's success in portraiture. The superb likeness of his sister Dorothy Catharine, taken not later than July 1840, with an exposure of sixty-five seconds, seems to be the oldest surviving photographic portrait.

In the busy winter of 1839–1840, Draper also took the first photograph of the moon and launched, in a very modest way, the age of astronomical photography. He obtained "distinct" representations of the dark spots or lunar maria. He first announced his success to the New York Lyceum of Natural History on 23 March 1840. Fittingly enough, Draper's second son, Henry, became one of the most distinguished astronomical photographers of the nineteenth century. As early as 1850, when Henry was thirteen, Draper enlisted his aid in photographing slides through a microscope to illustrate a projected textbook, *Human Physiology* (1856). In the book they appear as engravings, but the elder Draper was one of the first, if not the very first, to conceive of and execute microphotographs.

Draper's grasp of the uses to which photography could be put and his ingenuity in accomplishing the requisite feats made him a technical innovator of considerable importance; but this was merely incidental to a much deeper concern with the chemical effects of radiant energy in general. By his researches in this field, Draper became easily one of the dozen most important contributors to basic science in the United States before 1870.

He enunciated in 1841 the principle that only absorbed rays produce chemical change—long known as Draper's law but eventually rechristened the Grotthuss or Grotthuss–Draper law from its formulation by the German C. J. D. Grotthuss in 1817. Grotthuss' statement attracted no attention until the close of the nineteenth century, by which time the principle had become well established under Draper's (entirely independent) auspices. In 1843 Draper constructed a "tithonometer," or device for measuring the intensity of light, based on the discovery by Gay-Lussac and Thénard in 1809 that light causes hydrogen and chlorine to combine progressively. The seminal work in this field remained to be done by Bunsen and Roscoe from the mid-1850's onward. Their "actinometer," built after they had rejected Draper's instrument as inaccurate, made use of the identical phenomenon.

With a grating ruled for him by the mechanician of the United States Mint, Joseph Saxton, Draper took, in 1844, what seems to have been the first photograph of the diffraction spectrum. Apparently he was also the first to take with any precision a photograph of the infrared region, and the first to describe three great Fraunhofer lines there (1843). He also photographed lines in the ultraviolet independently of Edmond Becquerel and at about the same time.

In one of his most important memoirs (1847), he proved that all solid substances become incandescent at the same temperature, that thereafter with rising temperature they emit rays of increasing refrangibility, and (a fundamental proposition of astrophysics, later elaborated upon by Kirchhoff) that incandescent solids produce a continuous spectrum. Draper implied in the mid-1840's and stated clearly in 1857 that the maxima of luminosity and of heat in the spectrum coincide. For the sum of his researches on radiant energy, Draper received the Rumford Medal of the American Academy of Arts and Sciences (1875).

By the mid-1850's Draper had become acquainted, directly or indirectly, with the positivism of Auguste Comte and had embraced Comte's law of the three stages of historical development, from the theological through the metaphysical to the "positive," or scientific. Comte had argued for a parallel development of the individual personality from infancy to maturity. Where Draper sharply diverged from Comte was in postulating that the history of mankind had consisted in a succession of dominant nations or cultures, regarded as biological organisms experiencing decrepitude and death as well as birth and development. With this cyclical theory of history, entirely alien to Comte, Draper combined a passionate espousal of environmentalism.

The principal work in which all these notes sounded together was *A History of the Intellectual Development of Europe* (1863). Draper says, however, that the book was sketched by 1856 and completed by 1858. The suspicion that he had been influenced by either Buckle or Darwin is unfounded. Yet he undoubtedly profited from the vogue of both men and blandly assimilated himself to Darwinism whenever he could. Thus he spruced up his totally unamended views on history for the British Association for the Advancement of Science at Oxford in 1860, under the provocative but fraudulent title "On the Intellectual Development of Europe, Considered With Reference to the Views of Mr. Darwin and Others, That the Progression of Organisms Is Determined by Law." This unprofitable paper was the direct occasion for the famous exchange between Bishop Wilberforce and T. H. Huxley. From this time forward, Draper was regarded as a valiant defender of science against religion. His most popular book, widely read in many translations, was a *History of the Conflict Between Religion and Science* (1874), a vigorous polemic against the persecution of scientists by religionists.

BIBLIOGRAPHY

I. ORIGINAL WORKS. Draper's works include *A Treatise on the Forces Which Produce the Organization of Plants* (New York, 1844); *Human Physiology, Statical and Dynamical; or, The Conditions and Course of the Life of Man* (New York, 1856); *A History of the Intellectual Development of Europe* (New York, 1863); *Thoughts on the Future Civil Policy of America* (New York, 1865); *History of the American Civil War*, 3 vols. (New York, 1867–1870); *History of the Conflict Between Religion and Science* (New York, 1874); and *Scientific Memoirs, Being Experimental Contributions to a Knowledge of Radiant Energy* (New York, 1878).

II. SECONDARY LITERATURE. See Donald Fleming, *John William Draper and the Religion of Science* (Philadelphia, 1950), which has an extensive bibliography.

DONALD FLEMING

DREBBEL, CORNELIUS (*b.* Alkmaar, Netherlands, 1572; *d.* London, England, 1633), *mechanics, optics, technology.*

Drebbel's father, Jacob Jansz, a burgher of Alkmaar, was a landowner or farmer. Nothing is known of his mother. He probably had only an elementary education, learning to read and write Latin in his later years. As a young man Drebbel was apprenticed to the famous engraver Hendrik Goltzius and lived in his home in Haarlem. Drebbel proved to be an apt pupil, as is shown by a number of extant engravings from his hand. In addition, he probably acquired an interest in and some knowledge of alchemy from Goltzius, who was an adept.

After his marriage in 1595 to one of Goltzius' younger sisters, Sophia Jansdocther, he settled at Alkmaar, where he devoted himself to engraving and publishing maps and pictures. He soon turned to mechanical inventions, for in 1598 a patent was granted to him for a pump and a clock with a perpetual motion. He is mentioned as having built a fountain for the city of Middelburg, in the province of Zeeland, in 1601; and in 1602 he was granted a patent for a chimney. About 1605 Drebbel moved to London. Apparently some of his mechanical inventions appealed to James I, and he was soon taken into the special service of Henry, prince of Wales, and was installed in the castle at Elpham.

Drebbel's fame as an inventor became well known on the Continent, and he was visited at Elpham by Emperor Rudolf II and by the duke of Württemberg. He was invited to visit Rudolf, and by October 1610 he was in Prague with his family. Drebbel spent his time at the court of Rudolf showing off his "perpetuum mobile" and probably devoting himself to alchemy. After Matthias, Rudolf's brother, had conquered Prague and deposed Rudolf, Drebbel was imprisoned; through the intervention of Prince Henry, however, he was set free to return to England in 1613.

During the next several years Drebbel lived mostly in London, although there are indications that at various times he was on the Continent and again in Prague. About 1620 he began to devote himself to the manufacture of microscopes and to the construction of a submarine. He became acquainted with Abraham and Jacob Kuffler, who with their two other brothers were to become his disciples. Abraham soon married Drebbel's daughter Anna; and Johannes, another brother, married Katherina Drebbel in 1627. The four Kuffler brothers became agents and promoters for the microscopes and other instruments developed by Drebbel, Johannes being the one who did the most to promote Drebbel's inventions after his death.

For the next several years Drebbel was employed by the British navy and was concerned mainly with the famous expedition to La Rochelle. In spite of the failure of the expedition to raise the siege, Drebbel continued to work for the navy for some time at a fairly high salary. From 1629 until his death Drebbel was extremely poor and earned his living by keeping an alehouse. He was also engaged in various schemes for draining land near London, but apparently none was successful.

According to most of Drebbel's contemporaries, he was a light-haired and handsome man of gentle manners, considered to possess good intelligence, to be sharp-witted, and to have many ideas about various inventions. No absolute information on his religion is available; but his biographers have concluded that he was most likely an Anabaptist, since most of his friends and relatives were.

Drebbel left very few writings of his own, and none of them is concerned with his inventions. His most famous work was *Ein kurzer Tractat von der Natur der Elementum* (Leiden, 1608), an alchemical tract on the transmutation of the elements. Later editions contain a description of his "perpetuum mobile." Another treatise, *De quinta essentia* (Hamburg, 1621), is also alchemical in outlook and was written with the help of a friend. In it Drebbel discusses extracts from metals, minerals, plants, and other materials and their use in medicine.

In the strict sense Drebbel was not a scientist but an inventor or practicing technologist. In certain inventions he made use, however, of well-established scientific principles. Unlike many of his predecessors who had been interested in technological inventions, he actually brought his inventions to the practical state, and his finished models worked.

Among Drebbel's best-known inventions are the following.

(1) "Perpetuum mobile." This elaborate toy operated on the basis of changes in atmospheric temperature and pressure. Many models of it were used, and Drebbel extended the basic idea to the operation of clocks. Probably his initial fame in England and Europe rested on this invention, which delighted the people of the time.

(2) Thermostats. Drebbel apparently learned to apply the principles used in the "perpetuum mobile" to temperature regulators for ovens and furnaces. The principle involved was that as the temperature rose, the air expanded and pushed a column of quicksilver to the point at which it would close a damper. As the temperature fell, the damper would be opened. Drebbel applied the same idea to an incubator for hatching duck and chicken eggs.

(3) Optics. Drebbel was an expert lens grinder; and records indicate that his instruments were bought by several well-known persons, including Constantin Huygens. He made compound microscopes as early as 1619; and some of his biographers insist that he was the actual inventor of the microscope with two sets of convex lenses.

(4) The submarine. While living in London, according to many reliable accounts, Drebbel built a submarine that could carry a number of people. It was based on the principle of a diving bell: the bottom was open, and a rower sitting above the water level controlled the submarine. There was apparently no connection between the submarine and the atmosphere. Such reliable authorities as Robert Boyle have said that Drebbel had some means of purifying the air within the submarine.

(5) Chemical technology. Undoubtedly, Drebbel's most important contribution in this field was his discovery of a tin mordant for dyeing scarlet with cochineal. This process was communicated to his son-in-law, Abraham Kuffler, who had a dyehouse in Bow, London; and for many years the scarlet made with tin was known as "color Kufflerianus." The famous scarlets of the Gobelins made use of Drebbel's invention, and the method soon spread to all parts of the Continent. It is said that the discovery was made by accident when some tin dissolved in aqua regia happened to fall into a solution of cochineal that Drebbel

was planning to use for a thermometer. Although not a dyer, he quickly recognized the importance of his fortunate discovery and his family made good use of it. Among other chemical achievements attributed to Drebbel are the introduction into England of the manufacture of sulfuric acid by burning sulfur with saltpeter and the discovery of mercury and silver fulminates.

BIBLIOGRAPHY

I. Original Works. Drebbel's most important books are cited in the text.

II. Secondary Literature. On Drebbel or his work, see (listed chronologically): F. M. Jaeger, *Cornelis Drebbel en zijne tijdgenooten* (Groningen, 1922); Gerrit Tierie, *Cornelius Drebbel (1572–1633)* (Amsterdam, 1932); L. E. Harris, *The Two Netherlanders* (Cambridge, 1961); and J. R. Partington, *A History of Chemistry,* II (London, 1961), 321–324.

Sidney Edelstein

DREYER, JOHANN LOUIS EMIL (*b.* Copenhagen, Denmark, 13 February 1852; *d.* Oxford, England, 14 September 1926), *astronomy.*

Dreyer was the son of Lt. Gen. J. C. F. Dreyer and Ida Nicoline Margarethe Rangrup. He began studying mathematics and astronomy at Copenhagen University in 1869 and won a gold medal in 1874 for an essay entitled "Personal Errors in Observation." He received the M.A. in 1874 and the Ph.D. in 1882. All of Dreyer's astronomical appointments were in Ireland: from 1874 to 1878, assistant at Lord Rosse's observatory, Parsonstown [now Birr], where he observed nebulae through the seventy-two-inch-aperture telescope; from 1878 to 1882, assistant at Dunsink Observatory, Dublin; from 1882 to 1916, director of Armagh Observatory, where he collected and reduced observations made since 1859. This work led to the *Second Armagh Catalogue of 3,300 Stars* (1886). He settled at Oxford after retirement.

Dreyer was a fellow of the Royal Astronomical Society from 1875, gold medalist in 1916, and president from 1923–1926. He married Katherine Hannah Tuthill of Kilmore, Ireland, in 1875; they had three sons and a daughter. Gentle and amiable in disposition, Dreyer was a good astronomical observer and a most erudite scholar, gifted in languages, accurate, painstaking, and devoted to astronomy and astronomical history.

Dreyer enriched astronomy by three monumental works of research, collection, and editing. The first is "New General Catalogue of Nebulae and Clusters of Stars" (1888). The Herschels had discovered and catalogued over 5,000 nebulae and clusters, but by the 1880's numerous discoveries by other astronomers had made Sir John Herschel's catalog out-of-date. Dreyer had tried to remedy this by publishing a supplement listing hundreds of items (1878). Then, at the suggestion of the Royal Astronomical Society, he used the Herschel catalog as a basis for the compilation of the *New General Catalogue,* renumbering all the 7,840 objects discovered up to 1888 and giving their positions and descriptions. During the following twenty years he also published two index catalogs, enumerating all additional discoveries and raising the total to over 13,000. Many galaxies, nebulae, and star clusters are still known by their NGC and IC numbers.

In 1912 Dreyer edited the scientific papers of Sir William Herschel for the Royal Society of London and the Royal Astronomical Society. The two thick volumes contain seventy-one published and some thirty unpublished papers of the elder Herschel. Dreyer prefaced the work with an excellent detailed biography, based mainly on unpublished material.

Interest in a famous figure in his country's history, Tycho Brahe, the last great astronomer of the pretelescope era, had inspired Dreyer from boyhood to become an astronomer. In 1890 his book *Tycho Brahe* provided a scholarly biography rich in information about sixteenth-century astronomy and astronomers. In his last years, Dreyer completed the formidable task of collecting and editing all the works and correspondence of Tycho, including all his observations and those of his assistants. This huge collection (in Latin) was published under the auspices of the Academy of Sciences of Copenhagen between 1913 and 1929.

Among Dreyer's other works are a *History of Planetary Systems* and papers including "Original Form of the Alfonsine Tables" and "A New Determination of the Constant of Precession."

BIBLIOGRAPHY

I. Original Works. Dreyer's 1874 gold medal essay was expanded and published as "On Personal Errors in Astronomical Transit Measurements," in *Proceedings of the Irish Academy,* 2nd ser., **2** (1876), 484–528. Major works edited by Dreyer are "New General Catalogue of Nebulae and Clusters of Stars," in *Memoirs of the Royal Astronomical Society,* **49** (1888), 1–237; "Index Catalogue of Nebulae Found in the Years 1888–1894 (Nos. 1–1529), With Notes and Corrections to NGC," ibid., **51** (1895), 185–228; and "Second Index Catalogue of Nebulae and Clusters of Stars, Containing Objects Found in the Years 1895–1907 (Nos. 1530–5386), With Notes and Corrections to NGC and to First IC," ibid., **59** (1908), 105–198. These three were republished as *New General Catalogue . . .* in 1 vol. by the

Royal Astronomical Society (1953; repr. 1962). Also of considerable importance are *The Scientific Papers of Sir William Herschel*, 2 vols. (London, 1912); and *Omnia opera Tychonis Brahe Dani*, 15 vols. (Copenhagen, 1913–1929). Vols. I–IV contain the observations made by Brahe and his assistants in the following years: Vol. I, 1563–1585; Vol. II, 1586–1589; Vol. III, 1590–1595; Vol. IV, 1596–1601 and observations of comets. Volumes X–XIII are also entitled *Thesaurus observationum*.

Additional works by Dreyer are "A Supplement to Sir John Herschel's *General Catalogue of Nebulae and Clusters of Stars*," in *Transactions of the Royal Irish Academy,* **26** (1878), 381–426, which was included in the NGC; "A New Determination of the Constant of Precession," in *Proceedings of the Royal Irish Academy,* **3** (1883), 617–623, which includes a historical survey of the subject; *Second Armagh Catalogue of 3,300 Stars for Epoch 1875, Deduced From Observations Made at the Armagh Observatory 1859–1883, and Prepared for Publication by J. L. E. Dreyer* (Dublin, 1886); *Tycho Brahe. A Picture of Scientific Life in the Sixteenth Century* (Edinburgh, 1890); *History of the Planetary Systems From Thales to Kepler* (Cambridge, 1906); and "On the Original Form of the Alfonsine Tables," in *Monthly Notices of the Royal Astronomical Society,* **80,** no. 3 (1920), 243–261, a brilliant research based on medieval manuscripts.

II. Secondary Literature. Two works that draw heavily on Dreyer are A. F. O'D. Alexander, *The Planet Saturn* (London, 1962), which draws on Dreyer's works on Brahe and Herschel, and *The Planet Uranus* (London, 1965), which uses the work on Herschel. Obituaries are in *Observatory,* **49** (1926), 293–294; and *Monthly Notices of the Royal Astronomical Society,* **87** (1927), 251–257. There is a biography in *Dictionary of National Biography* for 1922–1930.

A. F. O'D. Alexander

DRIESCH, HANS ADOLF EDUARD (*b.* Bad Kreuznach, Germany, 28 October 1867; *d.* Leipzig, Germany, 16 April 1941), *biology, philosophy.*

Toward the end of the nineteenth century, when Hans Driesch entered upon his unique career, the German universities seemed to be in their prime and members of their scientific faculties were particularly preeminent in the natural sciences, including biology. Yet when Driesch chose his life style, he became an embryologist as an independent rather than as a professional investigator. He was a man of comfortable means, a cosmopolite, and a world traveler. He had no need, and at first no inclination, to join the academic hierarchy; he was not habilitated until twenty years after he had received his doctorate. While working as an amateur, he made numerous contributions of great originality and of far-reaching importance to the new science of experimental embryology. He was always theoretically inclined, and his experimental results turned him eventually toward

philosophical explanations. He had already written, while still an experimental biologist, a number of metaphysical articles and books; and when he finally was habilitated, it was in natural philosophy. He became a professor of philosophy and, as such, the strongest proponent in our times of vitalism.

Driesch was the only child of Paul Driesch, a well-to-do Hamburg merchant who dealt in gold and silver wares, and the former Josefine Raudenkolb. He grew up in Hamburg, where he was first educated at a famous humanistic Gymnasium, Gelehrtenschule des Johanneums, founded by a friend of Martin Luther's. In 1886 he spent two semesters at the University of Freiburg, studying with August Weismann in preparation for becoming a zoologist. He then became a student at the University of Jena, where he received his doctorate under Ernst Heinrich Haeckel in 1889; during the summer of 1888 he interrupted his zoological training to study physics and chemistry at Munich. For ten years after he received his degree he traveled extensively, usually with Curt Herbst, whom he had met at Jena in 1887; many of his scientific ideas were strongly influenced by Herbst's. Driesch performed his most important experiments during these years; they were carried out principally, although not exclusively, at the internationally supported Zoological Station in Naples. He did his last experiments in 1909, the year in which he was habilitated in natural philosophy at the Faculty of Natural Sciences of Heidelberg, where he had settled in 1900. On 23 May 1899 he married Margarete Reifferscheidt; their children were Kurt (*b.* 1904) and Ingeborg (*b.* 1906), both of whom later became musicians.

Driesch became extraordinary professor at Heidelberg in 1912, accepted the ordinary professorship of systematic philosophy at Cologne in 1919, and became professor of philosophy at Leipzig in 1921. He was a visiting professor in China (Nanking and Peking) in 1922–1923, in the United States (University of Wisconsin) in 1926–1927, and in Buenos Aires in 1928. In 1933 he was prematurely placed in emeritus status by the National Socialists because of his lack of sympathy for their regime, but he continued to work until he died in 1941.

Driesch first became attracted to zoology through his mother's collection of exotic birds and animals, maintained in an aviary and in vivaria in her home. He was inspired to specialize in embryology, as were many of his contemporaries, by Haeckel's popular books. Driesch, however, lost interest in phylogeny by 1890. He himself chose the subject for his doctoral dissertation, which played down phylogenetic speculation in favor of an investigation of the laws governing the growth of hydroid colonies (1890, 1891). In the spring of 1891 he performed the experiment for which he is now best remembered, the separation of the blastomeres of the cleaving sea urchin egg.

Shortly before, in 1888, Wilhelm Roux had published the results of experiments showing that when one blastomere of the two-celled frog egg is killed, the remaining cell forms a half-embryo. Roux interpreted this as signifying that each cell is predestined, at the two-cell stage, to form only what it would have formed in the normally developing embryo. Richard and Oskar Hertwig, and later Theodor Boveri, had shaken sea urchin eggs to separate them into nucleated and nonnucleated fragments; Driesch adopted this method to separate the sea urchin blastomeres at the two-cell stage and found that in this egg, in contrast with that of the frog, each blastomere could form a whole, rather than a half, larva. Driesch interpreted these results as signifying that the fate of the cells is not fixed at the two-cell stage and that a cell can form parts that it does not normally form during development. The fate of a cell is a function of its position in the whole; its prospective potency, as Driesch was to put it a few years later (1894), is greater than its prospective significance. Furthermore, the cell not only forms parts that it would not have formed had the experiment not been performed; it also forms an organized, whole individual. It is, as Driesch was to call it, first in 1899, a harmonious equipotential system.

These interpretations were diametrically opposed to those of Roux, who considered that cells self-differentiate independently, forming a sort of mosaic that constitutes the embryo. Driesch's discovery that the development of a cell is not fixed at the two-cell stage and that it can be altered confirmed experimentally that development is epigenetic and opened new paths for exploration by the developing science of experimental embryology. In particular, his emphasis on organic wholeness and his comparison of the prospective potency of a part with its prospective significance provided the conceptual framework for the organizer concept developed by Hans Spemann during the first third of the twentieth century.

Driesch eventually extended his original experiments on the sea urchin eggs by shaking them in calcium-free seawater, according to a method of Herbst's, and separated the cells at later developmental stages. He also performed a corollary experiment, fusing two sea urchin embryos at the blastula stage to produce a single giant larva. He experimented also on eggs of other echinoderms and of ctenophores and ascidians, and he performed a number of experiments on regeneration in adult hydroids and ascidians. The philosophical implications of these experiments were later to influence Driesch's thought profoundly; but in the meantime, during the latter part of the nineteenth century, he performed a number of other strictly embryological experiments of great import and influence in their time, although they are less frequently recalled than the blastomere separation.

Roux and Weismann had postulated that qualitatively unequal nuclear division and subsequent differential distribution of nuclear material to the cells are the prime factors responsible for the formation of particular embryonic parts by particular cells. If this explanation held true, Driesch reasoned, abnormal distribution of the nuclei in the cytoplasm should result in embryonic malformations. He tested this possibility (1892) by compressing sea urchin eggs between glass cover slips at the four-cell stage in order to alter the cleavage pattern. The third cleavage occurred under pressure. The cytoplasm was deformed, with the nuclei displaced and atypically distributed through it. Yet the larvae that developed from the compressed eggs were normal, and Driesch knew that this meant that all the nuclei were equivalent. As Herbst said, "Expressed in terms of modern genetics, this signifies that all the nuclei contain all the genes" ("Hans Driesch als experimenteller und theoretischer Biologe," p. 115).

Driesch had great interest in the mode of nuclear function. He postulated (1894) that factors external to a cell influence its cytoplasm, which in turn influences the nucleus to produce substances that affect the cytoplasm. Furthermore, he postulated that the influence of the nucleus on the cell body is mediated through enzymes (ferments, in the terminology of his times); these concepts were widely disseminated and extremely influential in their day.

Another experiment performed by Driesch on sea urchin eggs, of equal theoretical importance and influence, was one in which he shook sea urchin larvae at the stage when the primary mesenchyme cells were organizing themselves into two clumps, preparatory to developing the larval skeleton. After their displacement by the shaking, the mesenchyme cells returned to their original positions. Driesch believed that they reaggregated there under the influence of the ectoderm and interpreted this in terms of the ability of the mesenchyme cells to react to tactile stimuli from the ectoderm (1896). This is very close to the concept of induction; in 1894 Driesch had written at great length about the possibilities of embryonic induction, including chemical induction and contact induction, in the widely read monograph *Analytische Theorie der organischen Entwicklung.*

Driesch published his first wholly theoretical pamphlet in 1891; in it he expressed his desire to explain development in terms of mechanics and mathematics. In the *Analytische Theorie der organischen Entwicklung* he was still mechanistic in outlook. As early as 1892, however, he had mentioned, in an experimental paper, the possibility that vitalistic interpretation might be compatible with scientific methodology; by 1895, according to his own testimony, he was a convinced vitalist. He despaired of explaining on a mechanistic basis the ability of half of a two-celled egg to form a whole larva, for he could not envisage a machine that could divide itself into two machines, each able to reconstitute itself into a whole. During the first decade of the new century he found himself obliged to invoke an agent-outside-the-machine (for which he borrowed Aristotle's word "entelechy") as a regulator of organic development. Although he believed the entelechy to be a vital agent indefinable in terms of physics and chemistry, he thought its action was somehow brought about by the formation or activation of enzymes (ferments), and he laid strong emphasis on the importance of enzymes as regulatory agents in development.

Driesch was invited to deliver the Gifford lectures in natural theology at the University of Aberdeen in 1907 and 1908. These lectures, first published in 1908 in Driesch's own excellent English under the title *Science and Philosophy of the Organism,* summarized his experiments and the philosophical conclusions to which they led. Driesch later wrote many philosophical articles and books on organic form and organic wholeness, on the mind-body problem, and so forth. As a systematic philosopher he devoted considerable attention to both logic and metaphysics; as a metaphysician, he was strongly influenced by Kant. He also became interested in parapsychology; when he applied for permission to leave Germany to preside at a meeting of the International Society for Psychic Research in Oslo in 1935, he was deprived of his passport by the Nazis.

Driesch received many honors, including honorary degrees from Aberdeen, Hamburg, and Nanking. In honor of his sixtieth birthday, in 1927, eighteen years after he had performed his last experiment, two volumes of *Wilhelm Roux Archiv für Entwicklungsmechanik der Organismen,* including forty-eight articles, were dedicated to him. In his introduction to the *Festschrift,* Spemann wrote: "If Wilhelm Roux's systematic mind discovered and staked out for us a new field of investigation, Driesch's statements of its problems widened its horizons immeasurably" (p. 2). He placed Driesch's work, to which he owed so much, in proper perspective.

Although Driesch's experiments were ingeniously conceived, he was not particularly deft at carrying them out, a deficiency that he recognized and regretted. The biological interpretations of some of his experimental results, particularly those on later cleavage stages of the sea urchin, were subsequently shown to have been erroneous. Nevertheless, the principal conclusions drawn from his early experiments still hold. They made a positive contribution to the ongoing progress of experimental and analytical embryology that cannot be minimized, even though in his later days Driesch chose what was to him the vitalistic imperative.

BIBLIOGRAPHY

I. ORIGINAL WORKS. The article by Herbst (below) includes a bibliography of 130 items by Driesch, classified into two categories (descriptive and experimental, and biotheoretical). The book edited by Wenzl (below) includes a bibliography of 289 items by Driesch, arranged chronologically. The articles and books listed below deal principally with Driesch's experiments and their interpretations: "Tektonische Studien an Hydroidpolypen," in *Jenaische Zeitschrift für Naturwissenschaft,* **24** (1890), 189–226; "Tektonische Studien an Hydroidpolypen. II. Plumularia und Aglaophenia. Die Tubulariden. Nebst allgemeinen Erörterungen über die Natur tierischer Stöcke," *ibid.,* 657–688; "Entwicklungsmechanische Studien. I. Der Werth der beiden ersten Furchungszellen in der Echinodermenentwicklung. Experimentelle Erzeugung von Theil- und Doppelbildungen. II. Ueber die Beziehungen des Lichtes zur ersten Etappe der thierischen Formbildung," in *Zeitschrift für wissenschaftliche Zoologie,* **53** (1891), 160–184; "Heliotropismus bei Hydroidpolypen," in *Zoologische Jahrbücher,* Abteilung für Systematik . . ., **5** (1891), 147–156; *Die mathematisch-mechanische Betrachtung morphologischer Probleme der Biologie. Eine kritische Studie* (Jena, 1891); "Tektonische Studien an Hydroidpolypen. III (Schluss). Antennularia," in *Jenaische Zeitschrift für Naturwissenschaft,* **25** (1891), 467–479; "Entwicklungsmechanische Studien. III. Die Verminderung des Furchungsmaterials und ihre Folgen (Weiteres über Theilbildungen). IV. Experimentelle Veränderungen des Typus der Furchung und ihre Folgen (Wirkungen von Wärmezufuhr und von Druck). V. Von der Furchung doppeltbefruchteter Eier. VI. Ueber einige allgemeine Fragen der theoretischen Morphologie," in *Zeitschrift für wissenschaftliche Zoologie,* **55** (1892), 1–62; *Die Biologie als selbstständige Wissenschaft* (Leipzig, 1893); "Entwicklungsmechanische Studien. VII. Exogastrula und Anenteria (über die Wirkung von Wärmezufuhr auf die Larvenentwicklung der Echiniden). VIII. Ueber Variation der Mikromerenbildung (Wirkung von Verdünnung des Meereswassers). IX. Ueber die Vertretbarkeit der 'Anlagen' von Ektoderm und Entoderm. X. Ueber einige allgemeine entwicklungsmechanische Ergebnisse," in *Mitteilungen aus der Zoologischen Station zu Neapel,* **11** (1893), 221–254; "Zur Theorie der tierischen

Formbildung," in *Biologisches Zentralblatt,* **13** (1893), 296–312; "Zur Verlagerung der Blastomeren des Echinideies," in *Anatomischer Anzeiger,* **8** (1893), 348–357; *Analytische Theorie der organischen Entwicklung* (Leipzig, 1894); "Von der Entwickelung einzelner Ascidienblastomeren," in *Archiv für Entwicklungsmechanik der Organismen,* **1** (1895), 398–413; "Zur Analyse der ersten Entwickelungsstadien des Ctenophoreies. I. Von der Entwickelung einzelner Ctenophorenblastomeren. II. Von der Entwickelung ungefurchter Eier mit Protoplasmadefekten," *ibid.,* **2** (1896), 204–224, in collaboration with T. H. Morgan; "Die taktische Reizbarkeit von Mesenchymzellen von *Echinus microtuberculatus," ibid.,* **3** (1896), 362–380; "Betrachtungen über die Organisation des Eies und ihre Genese," *ibid.,* **4** (1897), 75–124; "Studien über das Regulationsvermögen der Organismen. 1. Von den regulativen Wachstums- und Differenzirungsfähigkeiten der Tubularia," *ibid.,* **5** (1897), 389–418; "Ueber rein-mütterliche Charaktere an Bastardlarven von Echiniden," *ibid.,* **7** (1898), 65–102; *Die Lokalisation morphogenetischer Vorgänge. Ein Beweis vitalistischen Geschehens* (Leipzig, 1899), also in *Archiv für Entwicklungsmechanik der Organismen,* **8** (1899), 35–111; "Studien über das Regulationsvermögen der Organismen. 2. Quantitative Regulationen bei der Reparation der Tubularia. 3. Notizen über die Auflösung und Wiederbildung des Skelets der Echinidenlarven," *ibid.,* **9** (1899), 103–139; "Studien über das Regulationsvermögen der Organismen. 4. Die Verschmelzung der Individualität bei Echinidenkeimen," *ibid.,* **10** (1900), 411–434; "Studien über das Regulationsvermögen der Organismen. 5. Ergänzende Beobachtungen an Tubularia," *ibid.,* **11** (1901), 185–206; "Studien über das Regulationsvermögen der Organismen. 6. Die Restitutionen der *Clavellina lepadiformis," ibid.,* **14** (1902), 247–287; "Studien über das Regulationsvermögen der Organismen. 7. Zwei neue Regulationen bei Tubularia," *ibid.,* 532–538; "Ueber Aenderungen der Regulationsfähigkeit im Verlauf der Entwicklung bei Ascidien," *ibid.,* **17** (1903), 54–63; "Ueber das Mesenchym von unharmonisch zusammengesetzten Keimen des Echiniden," *ibid.,* **19** (1905), 658–679; *Der Vitalismus als Geschichte und als Lehre* (Leipzig, 1905); "Regenerierende Regenerate," in *Archiv für Entwicklungsmechanik der Organismen,* **21** (1906), 754–755; *The Science and Philosophy of the Organism. The Gifford Lectures Delivered Before the University of Aberdeen in the Year 1907* [*and 1908*], 2 vols. (London, 1908; 2nd ed., 1 vol., 1929); "Zwei Mitteilungen zur Restitution der Tubularia," in *Archiv für Entwicklungsmechanik der Organismen,* **26** (1908), 119–129; "Neue Versuche über die Entwicklung verschmolzener Echinidenkeime," *ibid.,* **30,** pt. 1 (1910), 8–23; and *Lebenserinnerungen. Aufzeichnungen eines Forschers und Denkers in entscheidender Zeit* (Basel, 1951).

Later philosophical and theoretical articles are listed in the bibliographies by Herbst and Wenzl.

II. SECONDARY LITERATURE. Selected writings about Hans Driesch, his work, and his thought are Margarete Driesch, "Das Leben von Hans Driesch," in *Hans Driesch, Persönlichkeit und Bedeutung für Biologie und Philosophie von heute . . .,* A. Wenzl, ed. (Basel, 1951), pp. 1–20; Curt

Herbst, "Hans Driesch als experimenteller und theoretischer Biologe," in *Wilhelm Roux Archiv für Entwicklungsmechanik der Organismen,* **141** (1941–1942), 111–153; and E. Ungerer, "Hans Driesch. Der Naturforscher und Naturphilosoph (1867–1941)," in *Naturwissenschaften,* 29 Jahrgang (1941), 457–462.

In addition, the work edited by Wenzl, cited above, includes a bibliography of other articles on Driesch and his work, p. 206. The *Festschrift* referred to in the text, dedicated to Driesch and containing 48 articles in his honor, is *Wilhelm Roux Archiv für Entwicklungsmechanik der Organismen,* **111–112** (1927).

JANE OPPENHEIMER

DRUDE, PAUL KARL LUDWIG (*b.* Brunswick, Germany, 12 July 1863; *d.* Berlin, Germany, 5 July 1906), *physics.*

Drude was the son of a medical doctor in Brunswick. He attended the local Gymnasium and then went on to study at the University of Göttingen. His original ambition was to become a mathematician, and he studied mathematics, first at Göttingen and then at Freiburg and Berlin. In his sixth semester he returned to Göttingen, where he came under the influence of W. Voigt and as a result began to study theoretical physics. Drude's dissertation, under Voigt's direction, was a purely theoretical treatment of the equations governing the reflection and refraction of light at the boundaries of absorbing crystals.

Drude worked with Voigt at Göttingen until 1894. He then moved to Leipzig where he pursued both theoretical and practical researches on the propagation of electromagnetic waves and wireless telegraphy, as well as continuing his work on physical optics. His interest in the physical determinants of optical constants led him toward an attempt to correlate and account for the optical, electrical, thermal, and chemical properties of substances. Drude's interest in these problems was stimulated by his own growing conviction, based on studies begun in 1888, that Maxwell's electromagnetic theory was superior to the older mechanical view of light. This conviction led him to publish *Physik des Äthers* (1894), one of the first German books to base explanations of electrical and optical effects on Maxwell's theory. By 1898 Drude had begun to consider these matters within the structure of the theory of electrons; indeed, he thereby laid the foundation for understanding such phenomena as conduction in metals, thermal conductivity, and optical properties of metals as interactions of the electrical charges of substances with their environment.

In 1894 Drude married Emilie Regelsberger, the daughter of a Göttingen jurist. With the death of Wiedemann in 1889, he assumed the editorship of *Annalen der Physik,* the most prestigious of physics journals. In 1901, shortly after the publication of his *Lehrbuch der Optik*, Drude moved to Giessen where he became director of the Institute of Physics.

In Giessen, where he remained until 1905, Drude continued his work in optics and the electron theory. Having declined other appointments, it was only with some reluctance that he answered the call to Berlin to take over directorship of the physics institute. Almost immediately after Drude assumed this position, the size of the institute's staff was enlarged by a third in order to meet the demands of the increase of both the theoretical and the practical work that he brought with him. He died suddenly and unexpectedly within a year of moving to Berlin.

Drude's chief contributions fall into two categories: his early work in physical optics, in which he concentrated on the relationship between the physical properties and the optical characteristics of crystals, and his later work, in which he attempted to explain both physical and optical properties in a unified theory. In both phases theory and experiment were carefully interwoven; the transition in Drude's orientation is closely correlated with his change from a mechanical to an electromagnetic view of optical phenomena.

Drude may be considered the intellectual descendant of Franz Neumann—the first of Germany's great theoretical physicists, who developed a mechanical theory of light propagation based on the work of Fresnel and closely related to Fresnel's own theory. It was a mechanical theory which assumed that light oscillations were of a mechanical-elastic nature, transmitted through an ether conceived of as an elastic solid. Neumann's theory had its counterparts in England and France but was distinguished by the power and rigor of the mathematical analysis and by the assumption that the density of the ether is the same in all bodies. This leads to the conclusion that the displacement of ether particles in a plane polarized ray is in the plane of polarization. (The Fresnel theory assumed that the elasticity of the ether was the same in all bodies, leading to the conclusion that the displacement of ether particles is perpendicular to the plane of polarization.) Neumann was not only a theoretical physicist. At Königsberg he worked both in the physics department and the department of mineralogy. It was quite natural then that he should do extensive work on the optical properties of crystals. His laboratory was well equipped to investigate the structure of crystals and their elastic properties. Voigt, whose work continued that of Neumann, was particularly interested in magneto- and electro-optics. Drude's dissertation was a direct offshoot of Voigt's work. Voigt then set him the problem of checking his work in the laboratory, using crystals of

bournonite. The experimental difficulties were great, and Drude almost immediately realized that the optical constants of such crystals were not independent of the state of the crystal's surface. He discovered that the index of refraction and the coefficient of reflection of a crystal changed steadily from the time it was freshly cleaved. With characteristic care and thoroughness Drude then undertook a reexamination of the optical constants of a wide variety of absorbing substances, making measurements as difficult and exacting as those of the original experiments. When he was finished the optical constants of a wide variety of substances were known to an accuracy hitherto unthinkable.

This work occupied Drude from 1887 to 1891. During this period, too, he first became interested in Maxwell's work in electrodynamics, stimulated by Hertz's detection of electromagnetic radiation. Maxwell's treatise of 1873 was translated into German in 1882; but Maxwell's views were not widely accepted. The mechanical view of light propagation still held sway, and it had been under the influence of that theory that Drude had been working.

Drude did not become an immediate convert to the electromagnetic point of view. In 1888 he began an intensive four-year study, first immersing himself in the electromagnetic point of view, then reexamining the mechanical theory of light. He did not feel obliged to reject the mechanical theory which—although it presented some difficulty, especially in regard to the propagation of transverse waves through an elastic medium—had served so well.

Finally Drude took a phenomenological approach, attempting to remove nonessential elements from the mechanical formulation of optics. He argued that the differential equations and the imposed boundary conditions must be retained while assumptions about the mechanical nature of light waves and the elasticity of the ether were extraneous. He published the fruits of his investigation in a paper entitled "In wie weit genügen die bisherigen Lichttheorien den Anforderungen der praktischen Physik?" (1892). In this paper Drude pointed out that if the investigator restricted himself to differential equations and necessary boundary conditions, which he designated as the "explanation system" (*Erklärungssystem*), the mechanical and electromagnetic theories were equivalent. For example, to transform the mechanical view to the electromagnetic view, instead of such terms as "density," "elasticity," and "velocity of the ether," one needed only to substitute "magnetic permeability," "dielectric constant," and "magnetic field strength." Drude's paper was much in the spirit of Hertz's own assertions about Maxwell's theory—that the Maxwell theory should be considered as Maxwell differential equations.

Drude gradually took up the electromagnetic viewpoint. In another paper of 1892, "Ueber magnetiooptische Erscheinungen," he developed a system of equations directly from Maxwell's equations to account for Kerr's discovery that the reflectivity of magnetic substances (iron, cobalt, and nickel) is influenced by the state of magnetization and for Kundt's observation that the plane of polarization of light is rotated in passing through thin plates of these substances. Although the ease with which Maxwell's theory allowed such work to be done was important to Drude, he did not yet advocate one theory to the exclusion of the other. Rather, for another two years he lectured at Göttingen on the Maxwell theory; these lectures led to the publication of his first book, *Die Physik des Äthers* (1894). As a result of the heuristic effect that Maxwell's theory had on his own work between 1894 and 1898, Drude became an advocate of the electromagnetic view.

With his move to Leipzig, Drude's work on physical constants and his work on electromagnetic radiation began to merge into one set of coherent concerns. Drude had already hinted in some of his published work that by using the electromagnetic theory one might be able to explain electrical and optical properties of matter as the interaction of electromagnetic fields with electrical charges contained within the body. The publication of Lorentz' electron theory between 1892 and 1895 undoubtedly spurred him in that direction.

Shortly after arriving at Leipzig in 1894, Drude undertook further investigations on the relationship between optical and electrical constants and the constitution of substances. Using seventy to eighty centimeters radiation, he measured coefficients of absorption in a wide variety of solutions and compared these to coefficients of conductivity for the same solutions. According to Maxwellian theory, a close correlation should have existed between electrical conductivity and absorption of light—the higher the conductivity, the greater should be the absorption. Drude found, however, that this was not always the case. For example, the absorption coefficients of amyl alcohol and copper salt solutions might be the same, whereas the conductivity of the copper salt solution might be thousands of times greater than the conductivity of the alcohol. By careful and controlled experimentation, Drude found that, in fact, a whole class of substances absorbed seventy-five centimeters electromagnetic radiation—quite independent of their coefficient of conductivity when it was measured by direct current methods. The same substances also exhibited a marked deviation from expected values of their dielectric constants and also exhibited anomalous dispersion. Drude was able to demonstrate that selective absorption of seventy-five centimeters radiation was directly related to the chemical structure of substances and that it was the hydroxyl radical (OH) that was responsible. Thus he developed a new practical analytic tool for chemists.

The problem presented some theoretical difficulties. The obvious explanation for selective absorption would have to be based on the hypothesis that it represented a resonance phenomenon with the natural period of the molecular constituents of the substance; this suggested to Drude that the natural period at the atomic level was not independent of the particular molecular arrangement since a much higher resonance frequency would be expected of independent considerations.

The organization of Drude's *Lehrbuch der Optik* (1900) reflects his own approach to problems in optics. The first half of the book is devoted almost exclusively to the phenomena and to their mathematical characterization. Then, after a brief outline of the mechanical and electromagnetic theories, Drude gives what he considers to be the advantages of the electromagnetic theory: first, transverse waves are a direct consequence of Maxwell's conception of electromagnetic interaction; second, special boundary conditions are not required in the electromagnetic theory for radiation in the optical region of the spectrum; and third, the velocity of light can be determined directly from electromagnetic experiments. "In fact," Drude wrote, "it is an epoch-making advance in natural science when in this way two originally distinct fields of investigation, like optics and electricity, are brought into relations which can be made the subject of quantitative measurements" (*The Theory of Optics,* C. R. Mann and R. A. Millikan, trans., 2nd ed. [New York, 1959], p. 261).

Drude's move to Giessen thus occurred at a time when he was intent on understanding the optical, thermal, and electrical properties of metals by application of the electron theory. Drude was not the only person interested in such a practical application of the electron theory. Both J. J. Thomson and E. Riecke made substantial though different contributions. In the theory developed by Drude every metal contains a large number of free electrons, which he treated as a gas, the electrons having an average kinetic energy equal to the average kinetic energy of the atoms and molecules of the substance. The essential difference between conductors and nonconductors was that nonconductors contained relatively few free electrons. In early versions of his theory, Drude assumed that both positive and negative electrons were part of the "gas" but in a later simplification assumed that only negative electrons were mobile. Using the temperature of the substance as an index of the average kinetic energy of the particles in the electron gas, the velocity of the electrons should be enormous if it were not for the very small mean free path—due, mainly, to collisions with atomic centers.

Consider a neutral conductor, not under the influence of an electric force. Since the motion of the electron gas is perfectly random, there should be no net charge created at any point and no net transfer of electricity from one point to another. Under the influence of an electric field, however, there should be an increase in the average velocity of electrons in one direction and a decrease in their average velocity in the opposite direction. Such a situation would constitute an electric current whose intensity should be theoretically calculable. Drude arrived at the result that the electrical conductivity would be given by

$$\sigma = \frac{e^2 N l}{2mu},$$

where e is the charge on a single electron, N is the number of electrons per unit volume, l is the mean free path, m is the mass of the electron, and u is the average velocity of the electrons. Since the electrons are treated as a gas, the average kinetic energy of the gas should be proportional to the absolute temperature. The coefficient of conductivity may thus be expressed as absolute temperature:

$$\sigma = \frac{e^2 N l u}{4\alpha T},$$

where T is the temperature and α is a universal constant.

The power of Drude's analysis lies in that when one turns to different phenomena, such as thermal conductivity, the analysis is similar. Suppose the ends of a metal bar be maintained at different temperatures. The conduction of heat in the metal is due to collisions between the free electrons. The mean free path, however, is determined as before by collisions with essentially stationary metal atoms. Based on these assumptions, the coefficient of thermal conductivity for a substance is given by

$$k = \frac{\alpha N l u}{3}.$$

Drude used a scheme of this type to account for such things as thermoelectric and magnetoelectric effects. While agreement with experimental results was never perfect, it was usually within the right order of magnitude. For example, the ratio of Drude's

values for the thermal and electrical conductivity of a substance is proportional to the absolute temperature:

$$k/\sigma = \frac{4}{3}(\alpha/e)^2 T.$$

Since α and e are both universal constants, Drude would have predicted that the ratio of thermal to electric conductivity at a given temperature was the same for all metals. Although this is not precisely true it is a good approximation to what was known to be the case experimentally.

Drude did not make these researches serially; typically, he had several different research projects in progress at the same time—in addition to lecturing, directing doctoral students, heading the various physical institutes, and editing the *Annalen der Physik*. Drude carried this diverse and taxing load with grace and performed his duties with characteristic thoroughness.

Drude died a week after he had written the foreword to the second edition of his *Lehrbuch der Optik* and six days after he had given his inaugural speech at the Berlin Academy—a speech in which he sketched plans for future research.

BIBLIOGRAPHY

I. ORIGINAL WORKS. Drude's writings include "Ueber die Gesetze der Reflexion und Brechung des Lichtes an der Grenze absorbierende Kristalle," in *Annalen der Physik,* **32** (1887), 584–625; "Beobachtungen über die Reflexion des Lichtes am Antimonglanz," *ibid.,* **34** (1888), 489–531; "Ueber Oberflächenschichten," in *Göttingen Nachrichten* (1888), pp. 275–299; and *Annalen der Physik,* **36** (1889), 532–560, 865–897; "Bestimmung der optischen Konstanten der Metalle," *ibid.,* **39** (1890), 481–554; "Ueber die Reflexion und Brechung ebener Lichtwellen beim Durchgang durch eine mit Oberflächenschichten behaftete planparallele Platte," *ibid.,* **43** (1891), 126–157; "In wie weit genügen die bisherigen Lichttheorien den Anforderungen der praktischen Physik?," in *Göttingen Nachrichten* (1892), pp. 366–412.

"Ueber magnetiooptische Erscheinungen," in *Annalen der Physik,* **46** (1892), 353–422; "Ueber die Phasenänderung des Lichtes bei der Reflexion an Metallen," *ibid.,* **50** (1893), 595–624, and **51** (1894), 77–104; "Zum Studium des elektrischen Resonators," in *Göttingen Nachrichten* (1894), pp. 189–223, and *Annalen der Physik,* **53** (1894), 721–768; *Physik des Äthers auf elektro-magnetischer Grundlage* (Stuttgart, 1894); "Die Natur des Lichtes," "Theorie des Lichtes für durchsichtige Medien," "Theorie der anomalen Dispersion," "Doppelbrechung," "Uebergang des Lichtes über die Grenze zweier Medien," "Rotationspolarisation," "Gesetze der Lichtbewegung für absorbierende Medien," and "Polarisation des gebeugten Lichtes," in *Winkelmanns*

Handbuch der Physik, II, pt. 1, "Optik," (Breslau, 1894), 623–840; "Untersuchungen über die elektrische Dispersion," in *Annalen der Physik,* **54** (1894), 352–370.

"Der elektrische Berchungsexponent von Wasser und wässerigen Lösungen," in *Berichte. Sächsische Akademie der Wissenschaften.* Math.-phys. Kl., **48** (1896), 315–360, and *Annalen der Physik,* **59** (1896), 17–63; "Elektrische Anomalie und chemische Konstitution," in *Berichte. Sächsische Akademie der Wissenschaften.* Math.-phys. Klasse, **48** (1896), 431–435, and *Annalen der Physik,* **60** (1897), 500–509; "Neuer physikalischer Beitrag zur Konstitutionsbestimmung," in *Chemische Berichte,* **30** (1897), 930–965; "Ueber Messung der Dielektrizitätskonstanten kleiner Substanzmengen vermittelst elektrischer Drahtwellen," in *Berichte. Sächsische Akademie der Wissenschaften.* Math.-phys. Klasse, **48** (1896), 583–612; "Theorie der anomalen elektrischen Dispersion," in *Annalen der Physik,* **64** (1898), 131–158; "Zur Elektronentheorie. I," *ibid.,* **1** (1900), 566–613; "Zur Elektronentheorie. II," *ibid.,* **3** (1900), 369–402; *Lehrbuch der Optik* (Leipzig, 1900, 1906), C. R. Mann and R. A. Millikan, trans., as *The Theory of Optics* (Chicago, 1902; repr. New York, 1959).

"Zur Elektronentheorie. III," in *Annalen der Physik,* **7** (1902), 687–692; "Zur Konstruktion von Teslatransformatoren," *ibid.,* **9** (1902), 293–610; "Elektrische Eigenschaften und Eigenschwingungen von Drahtspulen mit angehängten geraden Drähten oder Metallplatten," *ibid.,* **11** (1903), 957–995; "Ueber induktive Erregung zweier elektrischer Schwingungskreise mit Anwendung auf Perioden- und Dämpfungsmessung, Teslatransformatoren und drahtlos Telegraphie," *ibid.,* **13** (1904), 512–561; "Optische Eigenschaften und Elektronen-Theorie. I," *ibid.,* **14** (1904), 677–726; "Optische Eigenschaften und Elektronen-Theorie. II," *ibid.,* 936–961; and "Die Natur des Lichtes," "Theorie des Lichtes für durchsichtige, ruhende Medien," "Doppelbrechung," "Uebergang des Lichtes über die Grenze zweier Medien," "Die Gesetze der Lichtbewegung für absorbierende Medien," "Theorie der Dispersion," "Rotationspolarisation," and "Theorie des Lichtes für bewegte Körper," all in *Winkelmanns Handbuch der Physik,* vol. VI (2nd ed., Breslau, 1906).

II. Secondary Literature. On Drude and his work see also M. Plank, "Paul Drude," in *Verhandlungen der Deutschen physikalischen Gesellschaft,* **8** (1906), 599–630; F. Richarz and W. König, *Zur Erinnerung an Paul Drude* (Giessen, 1906); and W. Voigt, "Paul Drude," in *Physikalische Zeitschrift,* **7** (1906), 481–482.

Stanley Goldberg

DRYGALSKI, ERICH VON (*b.* Königsberg, Eastern Prussia [now Kaliningrad, U.S.S.R.], 9 February 1865; *d.* Munich, Germany, 10 January 1949), *geography.*

At the age of seventeen Drygalski began to study mathematics and physics in Königsberg but soon went to Bonn in order to attend the lectures of Ferdinand von Richthofen, whom he followed in 1883 to Leipzig and in 1886 to Berlin. In 1887, while an assistant at the Geodetic Institute in Potsdam, he received his doctorate after completion of a dissertation in geophysics, but Richthofen's strong scientific and personal influence led him to become a geographer. Drygalski qualified as lecturer in geography and geophysics at the University of Berlin in 1898 and became a professor in 1899. In 1906 he accepted a call to the newly established chair of geography at the University of Munich, which he made highly regarded and held until his retirement in 1935.

Ice and oceans figured prominently in Drygalski's lifework. In the summer of 1891 and in 1892–1893 he led the preliminary and main expeditions of the Berlin Geographical Society to western Greenland. This expedition established Drygalski's international reputation. The following years were dedicated to the painstaking preparation of the first German expedition to the South Pole, which was tirelessly advocated and supported by Georg von Neumayer. It was carried out under Drygalski's direction in 1901–1903 on the polar ship *Gauss.* It had little outward publicity in comparison with other South Pole expeditions because of the considerable difficulties of the area allotted to it by international agreement. Of great value, however, were the scientific data, a wealth of scrupulously presented observations of the most varied scientific matters that brought the name "Antarctic University" to the *Gauss* expedition and gave it the highest rank among the South Pole explorations of the "classical" period. Drygalski henceforth was classed among the leading authorities in the fields of polar and oceanic exploration.

Although the expedition's report appeared soon after its return (1904), the scientific conclusions were fully developed only after almost thirty years of indefatigable labor by Drygalski and his co-workers.

Drygalski was a member of many academies, honorary member of numerous geographical societies, and recipient of their medals. In 1944 the Munich Geographical Society, which he had headed for twenty-nine years, established the Erich von Drygalski Medal in his honor.

Drygalski was also an excellent teacher. Thousands of students came to Munich to attend his stimulating classes, which were never confined to his special fields but dealt with many areas of geography, even those in which he had little interest. He emphasized regional geography, especially that of Asia, North America, and Germany. Eighty-four dissertations were written under his guidance; it is characteristic that he did not impose a single one of the subjects, and that none of them was designed to confirm or develop his own views. This absolute scientific freedom was highly appreciated by Drygalski's students and was the reason that the *Festschrift* for his sixtieth birthday was entitled *Freie Wege vergleichender Erdkunde* (1925).

BIBLIOGRAPHY

I. ORIGINAL WORKS. A complete list of Drygalski's writings may be compiled from the following sources: for 1885–1924, his *Festschrift, Freie Wege vergleichender Erdkunde,* pp. 374–386; for 1925–1934, *Zeitschrift für Geopolitik,* **12** (1935), 127–132; and for 1935–1949, *Die Erde. Zeitschrift der Gesellschaft für Erdkunde zu Berlin,* **1** (1949/1950), 69–72.

Among his works are *Grönland-Expedition der Gesellschaft für Erdkunde zu Berlin 1891–1893,* 2 vols. (Berlin, 1897); *Zum Kontinent des eisigen Südens* (Berlin, 1904); *Deutsche Südpolar-Expedition 1901–1903,* 20 vols. text and 2 vols. maps (Berlin–Leipzig, 1905–1931); and "Gletscherkunde," in *Enzyklopädie der Erdkunde,* VIII (Vienna, 1942), written with Fritz Machatschek.

II. SECONDARY LITERATURE. The *Festschrift* issued for Drygalski's sixtieth birthday was L. Distel and E. Fels, eds., *Freie Wege vergleichender Erdkunde* (Munich, 1925). Obituaries are N. Greutzburg, in *Erdkunde,* **3** (1949), 65–68; H. Fehn, in *Berichte zur deutschen Landeskunde,* **8** (1950), 46–48; E. Fels, in *Forschungen und Fortschritte,* **25** (1949), 190–191, and *Die Erde. Zeitschrift der Gesellschaft für Erdkunde zu Berlin,* **1** (1949/1950), 66–72; O. Jessen, in *Geographische Rundschau,* **1** (1949), 116–117, and *Jahrbuch der Bayerischen Akademie der Wissenschaften* (1949/1950), pp. 133–136; W. L. G. Joerg, in *Geographical Review,* **40** (1950), 489–491; W. Meinardus, in *Petermanns geographische Mitteilungen,* **93** (1949), 177–180; and S. Passarge, in *Mitteilungen der Geographischen Gesellschaft in München,* **35** (1949/1950), 105–107.

EDWIN FELS

DUANE, WILLIAM (*b.* Philadelphia, Pennsylvania, 17 February 1872; *d.* Devon, Pennsylvania, 7 March 1935), *physics, radiology.*

William Duane was the younger son, by his second wife, of Charles William Duane, an Episcopalian minister. On his father's side he was a direct descendant of Benjamin Franklin and of several Duanes who had played prominent political roles in the early republic; through his mother he held a good Bostonian pedigree and was distantly related to Charles W. Eliot, president of Harvard. From 1882 to 1890 his father was rector of St. Andrew's Church in West Philadelphia; young William attended private schools in Philadelphia and then the University of Pennsylvania, where he studied mathematics (including quaternions), wrote papers on the Sophists and on the "silver question," and was graduated A.B. in 1892 as valedictorian of his class. Moving on to Harvard—his father was rector of Christ Church (Old North

Church), Boston, from 1893 to 1910—he received an A.B. in 1893; was then two years assistant in physics, aiding John Trowbridge in experiments on the velocity of Hertzian waves; and received an A.M. in 1895.

With a Tyndall fellowship in physics from Harvard, Duane spent the next two or three years in Germany. At the University of Berlin he continued experimental work on electromagnetism under Emil Warburg (discovering an unforeseen effect and then showing it to be a "dirt effect"); heard the philosopher Wilhelm Dilthey; and studied physical chemistry with Landolt, mineralogy with Klein, experimental physics with Neesen, and especially theoretical physics with Planck, who testified in June 1898 that Duane had attended his lectures and exercise sessions for several semesters, displaying genuine aptitude and conscientious application. At Göttingen he heard the organic chemist Wallach but worked especially with Nernst. Duane's experimental and theoretical investigation, *Über elektrolytische Thermoketten,* suggested and guided by Nernst, was accepted by Planck in December 1897 as a University of Berlin doctoral dissertation.

In 1898 Duane was appointed professor of physics at the University of Colorado. He married in 1899 (and eventually had four children) and began to accumulate apparatus for experimental work in physical chemistry (1900–1901). By 1902, however, Duane had lost interest in physical chemistry and turned back to electromagnetism—perhaps because of his teaching in this field. His attention was increasingly to applications, with much effort devoted to a multiplex telegraph based upon synchronous motors at the two stations. It was the sabbatical year 1904–1905 which brought Duane back to fundamental problems; the winter was spent in the Curies' laboratory, where he learned techniques of research in radioactivity, and the spring with J. J. Thomson. In Paris he had determined the total ionization produced by a radioactive source of given intensity; back at the University of Colorado in the winter of 1905–1906 Duane was determining the total charge carried by the α and β rays rather than the ionization they produced.

Duane liked Paris, and the Curies had been impressed by him. Late in 1905 Pierre Curie asked Andrew Carnegie for a fellowship for Duane to continue work in the Laboratoire Curie. At the end of 1906, after Pierre Curie's death, Carnegie provided Mme. Curie with 12,500 francs per year for two or three fellowships. Duane was granted 7,500 francs from that sum for each of three years; in fact, he stayed six, and remained thereafter a member of the Société Française de Physique. His work in this period, 1907–1913, was very solid but not truly outstanding, either quantitatively or qualitatively. Perhaps the most difficult and most important of these researches was the measurement of the rate of evolution of heat from minute samples of radioactive substances.

In 1913 the physics department at Harvard University had a vacancy due to the retirement of John Trowbridge; and the newly founded Harvard Cancer Commission, at the newly constructed Huntington Hospital, wished the services of a physicist experienced in handling radioactive substances to initiate there the treatment of cancer by implantation of sources of intense radiation. Duane was appointed assistant professor at the Jefferson Physical Laboratory in Cambridge and research fellow in physics at Huntington Hospital in Boston; his time and his salary were thenceforth divided between these two institutions. The techniques of collecting the radium emanation (radon) continually evolving from a dissolved radium salt, purifying it, compressing it, and sealing it into a tube whose volume was a fraction of a cubic millimeter were all familiar to Duane from his Paris period; he now designed a far more efficient apparatus for manufacturing such "radioactive lamps" and himself applied the "lamps" to the patients. In 1917 Duane was promoted to professor of biophysics, a title created for him; he proudly asserted it to be the first such in America.

Besides radioactivity there was a second area of physical research with direct applications to cancer therapy: X rays. Duane had had no experience in this field, but his new position and responsibilities demanded that he make himself thoroughly familiar with it; and so arose his truly important contributions to physics. The time and place were opportune: Laue and the Braggs had just opened the field of X-ray spectroscopy; W. D. Coolidge's high-vacuum, high-voltage, heated cathode X-ray tubes were just becoming available; and Duane had inherited Trowbridge's unique 45,000-volt storage battery—the ideal power supply to exploit the Coolidge tube. Duane's initial concern was with the therapeutically important "Relation Between the Wave-Length and Absorption of X-Rays," the title of a paper read to the American Physical Society at the end of October 1914 but not printed.

Duane's attention soon turned, however, to the relation between the energy of the cathode rays and the frequency of the X rays produced by them, and at the end of December 1914 he described to the American Physical Society experiments showing that the ratio of these two quantities was equal, within a factor of two, to Planck's constant, h. The real advance, reported at the annual meeting of the society in April 1915, was made in collaboration with Franklin L. Hunt,[1] the first of the many graduate students and postdoctoral assistants with whose aid almost all of Duane's subsequent physical researches were carried out. Now for the first time Duane drove his Coolidge tube with the high-tension battery—i.e., with a constant voltage—and was impressed by the fact that "a constant difference of potential . . . does not produce homogeneous X-rays," i.e., X rays whose fractional absorption per unit of path length is independent of the thickness of the (homogeneous) absorber. From this naïve discovery Duane and Hunt jumped to a sophisticated question: "We therefore set ourselves the problem of determining the minimum wave length that can be produced by a given difference of potential."

From David L. Webster, then a young instructor at Harvard, they borrowed his newly constructed X-ray spectrometer. Fixing it at a given angle (i.e., wavelength), they observed the intensity of the X rays as a function of the voltage applied to the tube—possible because, and only because, of the Coolidge-Trowbridge apparatus. With decreasing voltage the intensity plunged to zero, thus showing dramatically that there was indeed a maximum frequency in the radiation produced by electrons of a given energy, and for this frequency the equation $E = h\nu$ held to within a few tenths of a percent. In the next two years Duane developed the "Duane-Hunt law" into a precision method for determining h, and it soon came to be regarded as the most accurate method available.

The war brought only a slight dip in Duane's productivity, and with the aid of Chinese and Japanese students he turned to accurate measurements, on a variety of elements, of the critical potentials and wavelengths for excitation of their characteristic X rays and, again, accurate measurements of the wavelengths of these X-ray spectra. Theoretical atomic physicists, notably Sommerfeld and Bohr, craved data of this sort to fix the number of electron shells and subshells in the various atoms, their energies, their quantum numbers, and the number of electrons in each shell. In the period 1918–1921 Manne Siegbahn's laboratory in Lund—using photographic recording in a closed, evacuated spectrometer—and Duane's laboratory—where the measurement of the intensity of the diffracted X rays at each angle by means of an ionization chamber achieved its highest development—were the two principal reliable sources of this vital data. Moreover, during these and especially the following years Duane gave much effort to the introduction and promotion of the treatment of cancer with high-voltage X rays—designing apparatus, supervising its installation at Huntington Hospital[2] and then at other institutions, developing the technique of measuring X-ray dosage in terms of the ionization of air, and securing the official adoption of this standard in the United States and then internationally in 1928.

In the year 1922–1923 Duane's career reached its zenith—and then fell precipitously. In April 1920 he had been elected to the National Academy of Sciences; in the fall of 1922 he was selected to receive the 1923 Comstock Prize—an award made by the academy at five-year intervals—for having established through his X-ray researches "relations which are of fundamental significance, particularly in their bearings upon modern theories of the structure of matter and of the mechanism of radiation."[3] In 1922–1923 Duane was chairman of the Division of Physical Sciences of the National Research Council and in 1923 president of the Society for Cancer Research. All must have supposed that many more laurels would in the following years be laid upon the brow of this late-blooming, quiet, outwardly modest, unexcitable Episcopalian; this competent pianist and organist; this Beacon Street Bostonian who sought recreation in bridge whist with his friends at the Somerset Club.

At the end of 1921 George L. Clark came into Duane's Cambridge laboratory as a postdoctoral National Research fellow. Three and a half years later it became clear that the several wholly new discoveries to which their fourteen collaborative papers had been devoted were so many pseudo phenomena. Clark was a chemist, then interested in crystal structure. The original goal of their research was a new method of crystal analysis using the Duane-Hunt limit to determine directly the wavelength of the X ray reflected at a given angle, and thus the interatomic distances in the reflecting crystal. With potassium iodide they found, along with other anomalous phenomena, intense reflected X rays not merely with the wavelengths of the characteristic X rays of the anticathode of the X-ray tube but also with the wavelengths of the K series of iodine. They had thus "discovered" the diffraction of the characteristic X radiation emitted by the atoms of the crystal—the very phenomenon which in 1912 Laue, Walter Friedrich, and Paul Knipping had unwarrantably expected to find when they in fact discovered diffraction of the incident X rays.[4] Bohr, among many others, was especially interested in this "discovery," for he felt it had to be explicable from the general viewpoint on the interaction of radiation and matter for which he had been groping and which emerged at the end of 1923 as the Bohr-Kramers-Slater theory.

This "discovery" also led Duane himself to some theoretical considerations of astonishing simplic-

ity—and novelty. Reasonably well grounded in classical mathematical physics (as most of the best American experimentalists were) but for want of personal contact with the contemporary European theoretical physicists typically unaware of the extent of his naïveté, Duane had published a number of theoretical papers in the preceding years: on magnetism as the nuclear force (1915); on a new derivation of Planck's law (1916); and on modifications of the electron ring positions, quantum numbers, and populations in the Bohr model (1921). These, quite properly, had been ignored by the theorists.

Now, however, stimulated by the fact that the selective reflection of the characteristic X radiations of the atoms of the diffracting crystal "does not appear to be explicable in a simple manner by the theory of interference of waves," Duane suggested in February 1923 an interpretation of diffraction by a grating or crystal "based on quantum ideas without reference to interference laws."[5] In much the same way that A. H. Compton was simultaneously explaining the increased wavelength of scattered X rays, Duane pictured diffraction as a collision of a light quantum with a grating. He pointed out that if one applies the familiar quantum conditions to the grating, the periodicity of its structure restricts the momentum it can take up from, or impart to, the light quantum, with the result that all the equations expressing the conditions for constructive interference of waves (e.g., the Bragg law, $n\lambda = 2\,d \sin \theta$) are retrieved if the energy transferred to the grating can be neglected.

This reinterpretation came as a revelation to many theorists puzzling over the wave versus quantum theory of light (Gregory Breit, Paul Ehrenfest, Paul Epstein, Adolf Smekal, Gregor Wentzel—but not Bohr), although they of course disregarded the *ad hoc* extensions of this mechanism by means of which Duane claimed to have explained the selective reflection. A curious consequence of the theorists' efforts to achieve a more general statement of Duane's reinterpretation of diffraction was to bring forward once again the representation of a grating (or crystal) by a Fourier series or integral. Duane himself then pointed out that since the distribution of intensity in the spectrum is essentially the Fourier transform of the grating producing it, in principle it should be possible to invert the transform and from intensity measurements determine the distribution of electrons in the diffracting crystal.[6] Duane put a National Research fellow to work on this, but the idea, which is the basis of all the subsequent analyses of the structure of biologically important molecules, caught on only through its adoption and advocacy by W. L. Bragg in the following years.

Perhaps even more damaging to Duane's reputation than the "discovery" of selective reflection was his opposition to the Compton effect. The most eminent American X-ray spectroscopist was the last to reproduce Compton's observation and the most vociferous in denying and explaining away his younger compatriot's discovery—even to the point of claiming as his own discovery the tertiary radiation that he had originally advanced as the probable source of Compton's shifted wavelengths.[7] It was not, of course, that Duane was opposed to the notion of light quanta; rather, there was an element of competition between his own and Compton's light quantum theories, which probably disposed Duane, who had in those years so little time for laboratory work, to accept uncritically the various new effects and negative observations with which Clark was plying him.

These disastrous episodes were scarcely behind Duane when, late in 1925, his capacity for work was severely reduced by the onset of acute diabetes, and in 1926 he was obliged to take a year's leave. By 1927 his sight had so deteriorated that he was compelled to do much of his reading and writing through his secretary. In 1931 Duane suffered a paralytic stroke, recovering only slowly and incompletely; in the fall of 1933 he took a leave of absence and retired in the fall of 1934. Six months later he died of a second stroke.

NOTES

1. Duane and Hunt, "On X-Ray Wave-Lengths," in *Physical Review,* **6** (1915), 166–171.
2. "Improved X-Rays for Cancer Work," in *New York Times* (14 Feb. 1921), 7, col. 2.
3. *Report of the National Academy of Sciences* for 1923 (Washington, D.C., 1924), pp. 5, 20.
4. P. Forman, "The Discovery of the Diffraction of X-Rays by Crystals: A Critique of the Myths," in *Archive for History of Exact Sciences,* **6** (1969), 38–71.
5. Duane, "The Transfer in Quanta of Radiation Momentum to Matter," in *Proceedings of the National Academy of Sciences,* **9** (1923), 158–164.
6. Duane, "An Application of Certain Quantum Laws to the Analysis of Crystals," in *Physical Review,* **25** (1925), 881; R. J. Havighurst, "The Application of Fourier's Series to Crystal Analysis," *ibid.;* Duane, "The Calculation of the X-Ray Diffracting Power at Points in a Crystal," in *Proceedings of the National Academy of Sciences,* **11** (1925), 489–493; R. J. Havighurst, "The Distribution of Diffracting Power in Sodium Chloride," *ibid.,* 502–507.
7. Clark and Duane, "On the Theory of the Tertiary Radiation Produced by Impacts of Photoelectrons," in *Proceedings of the National Academy of Sciences,* **10** (1924), 191–196.

BIBLIOGRAPHY

I. ORIGINAL WORKS. The best bibliography of Duane's publications, although still woefully incomplete in both form and content, is in Bridgman (below). A few letters from Duane to Bohr and Sommerfeld in 1924 are in the Archive for History of Quantum Physics, for which see T. S. Kuhn, *et al., Sources for History of Quantum Physics* (Philadelphia, 1967). The Niels Bohr Library, American Institute of Physics, New York, holds some twelve letters to Duane, notably a testimonial by Planck (1898), two letters from J. J. Thomson (1905), and four letters from Marie Curie (1905–1907).

II. SECONDARY LITERATURE. Biographical articles are P. W. Bridgman, "Biographical Memoir of William Duane, 1872–1935," in *Biographical Memoirs. National Academy of Sciences,* **18** (1937), 23–41; and G. W. Pierce, P. W. Bridgman, and F. H. Crawford, "Minute on the Life and Services of William Duane, Professor of Bio-Physics, *Emeritus,*" in *Harvard University Gazette* (11 May 1935). There is also "Charles William Duane," in *National Cyclopaedia of American Biography,* XVIII (New York, 1922), 403–404. Useful information is contained in the *vita* of Duane's dissertation, *Über elektrolytische Thermoketten* (Berlin, 1897). Recollections of the controversy with Compton by one of Duane's postdoctoral students—the one who finally found the effect—may be found in Samuel K. Allison, "Arthur Holly Compton, Research Physicist," in *Science,* **138** (1962), 794–797.

PAUL FORMAN

DUBINI, ANGELO (*b.* Milan, Italy, 8 December 1813; *d.* Milan, 28 March 1902), *medicine, helminthology.*

Dubini took the M.D. at the University of Pavia in 1837. He began his practice at the Ospedale Maggiore in Milan, then returned to Pavia for the academic biennium 1839–1841 as an assistant at the Clinica Medica, where he gave a free course in auscultation. In November 1841 he began a postgraduate trip to France, England, and Germany; in Paris he attended the courses of Gabriel Andral. At the end of 1842 Dubini returned to Milan and resumed his work as a medical assistant at the Ospedale Maggiore. In 1865 he was nominated as both head physician and director of the newly established dermatology department.

At the Ospedale Maggiore Dubini was the most noteworthy exponent of the *médecine d'observation,* which, proceeding along the lines indicated by G. B. Morgagni, had been developed in Parisian hospitals by J. N. Corvisart and R. T. H. Laënnec, among others. This new clinical medicine attempted to formulate in the living being a diagnosis that is substantially anatomical, by means of a continuous dialogue—possible only in the hospital—between clinical medicine and anatomical pathology, with a common purpose and subject.

Indeed, Dubini's most important discovery was precisely the result of his anatomical-pathological work within the hospital and, in particular, of the diligence with which he systematically opened the intestine in accordance with recent studies made by French physicians on typhic and tubercular ulcers. In May 1858 Dubini recorded a "new human intestinal worm" following the dissection of the corpse of a peasant woman who had "died of croupous pneumonia." He confirmed this discovery in November 1842 and published his description of it in April 1843, describing the new worm as *Anchylostoma duodenale,* derived from the hooked mouth of the organism and from its habitat in the human host. Dubini's helminthological description is highly accurate and was further developed in his *Entozoografia* (1850).

As early as his work of 1843 Dubini had stressed the high frequency of occurrence of the worm, "which, although it had not yet been seen by others, nor described, is nevertheless found in *twenty out of one hundred corpses* that are dissected with the aim of finding it." This affirmation—as well as testifying to the high incidence of ancylostomiasis in the countryside around Milan—demonstrates that Dubini (who had also noted that the worm seemed to be hematotrophic) was unwilling to attribute any particular pathogenicity to the duodenal Ancylostoma. (The pathogenicity of Ancylostoma was eventually confirmed in the course of studies on Egyptian chlorosis made by F. Pruner, W. Griesinger, and T. Bilharz and in D. Wucherer's work on tropical chlorosis; it was proven beyond doubt in 1882 in the research of B. Grassi, C. and E. Parona, E. Perroncito, C. Bozzolo, and L. Pagliani on the serious epidemic of miner's cachexia that spread among the miners of the St. Gotthard tunnel.)

Dubini's name is given to the electric chorea that he diagnosed and described, thereby ensuring himself a place in the history of lethargic encephalitis, of which such chorea is a mark.

BIBLIOGRAPHY

I. ORIGINAL WORKS. Dubini's major work is *Entozoografia umana per servire di complemento agli studi d'anatomia patologica* (Milan, 1850). His earlier works are "Nuovo verme intestinale umano (Anchylostoma duodenale), costituente un sesto genere di nematoidei proprii dell'uomo," in *Annali universali di medicina,* **106** (1843), 4–13; and "Primi cenni sulla corea elettrica," *ibid.,* **117** (1846), 5–50.

II. SECONDARY LITERATURE. Works on Dubini are Luigi Belloni, "Per la storia del cuore tigrato," in *L'Ospedale Maggiore,* **44** (1956), 252–258; "La medicina a Milano dal settecento al 1915," in *Storia di Milano,* XVI (Milan, 1962), 991–997; "La scoperta dell'Ankylostoma duodenale," in *Gesnerus,* **19** (1962), 101–118; "Dalla scoperta dell'Anky-

lostoma duodenale alla vittoria sull'anemia dei minatori," in *Folia medica,* **48** (1965), 836–855, and in *Minerva medica,* **57** (1966), 3215–3233; and Ambrogio Bertarelli, "Angelo Dubini (8 dicembre 1813–28 marzo 1902)," in *Bollettino della Associazione Sanitaria Milanese,* **4** (1902), 115–119.

LUIGI BELLONI

DUBOIS, FRANÇOIS. *See* **Sylvius, Franciscus.**

DUBOIS, JACQUES (Latin, **JACOBUS SYLVIUS**) (*b.* Amiens, France, 1478; *d.* Paris, France, 13 January 1555), *medicine.*

Jacques Dubois, hereinafter referred to as Sylvius, came to Paris at the invitation of his brother François, professor at and principal of the Collège de Tournai. Sylvius acquired a good command of Greek and Latin and was particularly attracted to the medical writings of Hippocrates and Galen. He studied medicine informally with members of the Paris Faculty of Medicine and particularly anatomy with Jean Tagault, whom he later described as "mihi in re medica praeceptor." Prevented from having any sort of medical career by lack of a degree, Sylvius went to Montpellier, where he was graduated M.B. in 1529 and M.D. in 1530. Upon returning to Paris, he was incorporated M.B. in 1531, permitted to take the examinations for the degree of licenciate, and thus allowed to teach at the Collège de Tréguier. In 1536 the Faculty of Medicine gave recognition to his course by permitting him to lecture in the Faculty and to receive students' fees.

Sylvius was a very popular teacher of anatomy who, unlike many of his contemporaries, was not unwilling to perform his own dissections. His most distinguished student was Andreas Vesalius; but since Sylvius was the arch-Galenist of Paris, wholly confident of Galen's medical omniscience and determined at all costs to defend him against open, critical attack, he became intensely hostile to his former student upon publication of Vesalius' *Fabrica* (1543). Sylvius' most bitter attack, which appeared under the title of *Vaesani cuiusdam calumniarum in Hippocratis Galenique rem anatomicam depulsio* (1551), was so unrestrainedly abusive that Renatus Henerus, in his later defense of Vesalius, *Adversus Jacobi Sylvii depulsionum anatomicarum calumnias pro Andrea Vesalio apologia* (1555), declared that Sylvius' invective "wearied our ears and aroused the indignation of many of us." Despite such irascibility, Sylvius was genuinely concerned over the welfare of his more orthodox students, for whom he wrote *Victus ratio scholasticis pauperibus partu facilis & salubris* (1540) and *Conseil tresutile contre la famine: & remedes d'icelle* (1546).

Sylvius was a prolific writer of commentaries, of

which the following were the most frequently reprinted and the most influential: *Methodus sex librorum Galeni in differentiis et causis morborum et symptomatum* (1539), *Methodus medicamenta componendi* (1541), *Morborum internorum prope omnium curatio ex Galeno et Marco Gattinaria* (1548), and *De febribus commentarius ex Hippocrate et Galeno* (1555). His major contribution to anatomy is represented by the posthumous *In Hippocratis et Galeni physiologiae partem anatomicam isagoge* (1555). It is a systematic account of anatomy, written at some time after 1536 (possibly in 1542) and based on the writings of Galen, on a certain amount of human anatomical dissection, and, as Sylvius admitted, on the *Anatomiae liber introductorius* (1536) of Niccolo Massa, a Venetian physician and anatomist.

As a self-appointed defender of Galenic anatomy Sylvius could not, like Vesalius, call attention openly to Galen's errors in the course of presenting more nearly correct anatomical descriptions in his *Isagoge*. His procedure was therefore (1) to acknowledge the best of Galenic anatomy; (2) to describe without critical comment such anatomical structures as Galen had overlooked or, where Galen had permitted an alternative, to make a better choice; and (3) if necessary, to criticize not Galen but the human structure, which Sylvius declared to have degenerated and thus to have betrayed Galen's earlier, correct descriptions. In general, Sylvius' systematic presentation is worthy of commendation, as is his relatively modern method of numbering branches of vessels, structures, and relationships. Notably, he provided a clear scheme for the identification of muscles, based, like that of Galen, on their attachments. It has been called the foundation of modern muscle nomenclature. Relative to this contribution, Sylvius introduced and popularized a number of other anatomical terms that have persisted, such as crural, cystic, gastric, popliteal, iliac, and mesentery.

Further examples of his method and contributions are to be found in his description of the heart, where, perhaps influenced by Massa as well as by his own dissections, Sylvius describes the passage of blood by the pulmonary artery from the right ventricle to the lungs and thence to the left ventricle (ed. Venice, 1556, fol. 89v). It is true that Galen had described this route, although he considered it of lesser importance than the one that he proposed through "pores" in the cardiac septum. Sylvius, however, does not refer to the latter route or to the implications of his silence—which perhaps he did not realize, for in effect they denied Galenic cardiovascular physiology. Furthermore, he did not accept the standard existence of the *rete mirabile* in the human brain: "This plexus seen by Galen under the gland still appears today in brutes" (fol. 57r). Thus he suggests that through degeneration the *rete mirabile* had disappeared from the human structure. This attitude is clearly expressed in a further statement relative to thoracic structures: "The azygos vein [was] always observed under the heart by Galen in those in whom the sternum formed of seven bones made a longer thorax, but in our bodies, because of the shortness of the sternum and thorax, it arises more or less above the heart and pericardium" (fols. 46v–47r). In summation, the *Isagoge* may be described as an introduction to human anatomy based on an attempt to reconcile the best of classical teachings with the results of observation, direct or at second hand, of human dissection. Despite such contributions as were mentioned above, and others, the work retains the defects of compromise.

Sylvius died in Paris and was interred in the Cemetery of the Poor Scholars.

BIBLIOGRAPHY

I. Original Works. Of Sylvius' many publications the following list represents a selection of the most important and representative: *Methodus sex librorum Galeni in differentiis et causis morborum et symptomatum* (Paris, 1539); *Ordo et ordinis ratio in legendis Hippocratis et Galeni libris* (Paris, 1539); *Methodus medicamenta componendi ex simplicibus judicio summo delectis, et arte certa paratis* (Paris, 1541); *Victus ratio scholasticis pauperibus paratu facilis & salubris* (Paris, 1542); *Morborum internorum prope omnium curatio brevi methodo comprehensa ex Galeno praecipue & Marco Gattinario* (Paris, 1545); *Vaesani cuiusdam calumniarum in Hippocratis Galenique rem anatomicam depulsio* (Paris, 1551); *De febribus commentarius ex libris aliquot Hippocratis & Galeni* (Paris, 1554); *Commentarius in Claudii Galeni duos libros de differentiis febrium* (Paris, 1555); *In Hippocratis et Galeni physiologiae partem anatomicam isagoge* (Paris, 1555); *Commentarius in Claudii Galeni de ossibus ad tyrones Libellum* (Paris, 1556); and *Iacobi Sylvii Opera medica*, René Moreau, ed. (Geneva, 1634).

II. Secondary Literature. The fullest biography of Sylvius is the "Vita" prefixed to René Moreau's edition of *Iacobi Sylvii Opera medica*, cited above; corrections will be found in Louis Thuasne, "Rabelaesian: Le Sylvius Ocreatus," in *Revue des bibliothèques*, **15** (1905), 268–283. More specialized topics are dealt with in Curt Elze, "Jacobus Sylvius, der Lehrer Vesals, als Begründer der anatomischen Nomenklatur," in *Zeitschrift für Anatomie und Entwicklungsgeschichte*, **114** (1949), 242–250; C. E. Kellett, "Sylvius and the Reform of Anatomy," in *Medical History*, **5** (1961), 101–116; and C. D. O'Malley, "Jacobus Sylvius' Advice for Poor Medical Students," in *Journal of the History of Medicine*, **17** (1962), 141–151.

C. D. O'Malley

DU BOIS-REYMOND, EMIL HEINRICH (*b.* Berlin, Germany, 7 November 1818; *d.* Berlin, 26 December 1896), *electrophysiology.*

Emil's father, Felix Henri du Bois-Reymond, moved from Neuchâtel, Switzerland (then part of Prussia), to Berlin in 1804 and became a teacher at the Kadettenhaus. Later he was the representative from Neuchâtel to the Prussian government, and in 1832 he published a fundamental work on linguistics. His orthodox Pietism and authoritarian manner soon aroused his son's spirit of resistance. Emil's mother, the former Minette Henry, was the daughter of the minister who served the French colony in Berlin and the granddaughter of Daniel Chodowiecki, a well-known artist. Emil had two sisters, Julie and Felicie, and two brothers; his brother Paul became a distinguished mathematician. The family's background made them feel that they belonged to the French colony in Berlin. They usually spoke French at home; and Emil attended the French academic high school in Berlin, except for a year in Neuchâtel.

In 1837 du Bois-Reymond began his studies at the University of Berlin, where, at first, somewhat undecided about his future, he attended lectures in theology, philosophy, and psychology. During a short period at the University of Bonn (1838–1839), he studied logic, metaphysics, and anthropology, in addition to botany, geology, geography, and meteorology. In the winter semester of 1839, having returned to Berlin, he was inspired by Eduard Hallmann, assistant to the anatomist Johannes Müller, to study medicine. It was also Hallmann who taught him the basic principles of osteology and botany and worked out a schedule of the lectures he should attend. His letters to Hallmann are splendid proof of du Bois-Reymond's intellectual liveliness, but they also show his initial uncertainty about his course of study and his own talents. He was easily influenced and was able only slowly to eradicate the prejudice for Müller that he had acquired from Hallmann.

Du Bois-Reymond was soon acquainted with such teachers and researchers as Heinrich Dove, Theodor Schwann, and Matthias Schleiden, and became a close friend of Ernst Brücke, Hermann Helmholtz, Carl Reichert, and Carl Ludwig. In 1840 he worked more closely with Müller, concerning himself with anatomical preparations, comparative anatomy, physiology, and microscopy. He was still very interested in morphology. During this period he also studied the philosophical writings of Hegel and Schelling with great interest and attended the clinical lectures of Dieffenbach and Johann Schönlein, among others.

On 10 February 1843 du Bois-Reymond received his degree with a historical-literary paper on electric fishes. This was a subsidiary result of his interest in the history of animal electricity and also a preliminary study for the experimental verification, recommended to him by Müller, of the new papers of Carlo Matteucci, who in 1840 had published his *Essai sur les phénomènes électriques des animaux.* This marked the start of his lifelong, almost monomaniacal experimental analysis of animal electricity. It occupied him constantly from 1840 to 1850; and in the course of his work he developed a strong preference for experimental physics, particularly the application of physical principles and methods of measurement to the problems of physiology. He was encouraged greatly in this by Brücke and later by Helmholtz. On 14 January 1845 he founded the Physikalische Gesellschaft with Brücke, Dove, and others in Berlin. In December 1845 he met Helmholtz and was deeply impressed by him.

Du Bois-Reymond's first experimental and theoretical investigations of animal electricity produced definite conclusions in November 1842. Upon Müller's advice the results were hurriedly submitted for publication in Poggendorff's *Annalen der Physik und Chemie,* appearing as "Abriss einer Untersuchung über den sogenannten Froschstrom und über die electrischen Fische" in January 1843. Through Humboldt he also sent an extract to the Académie des Sciences in Paris.

On 6 July 1846 du Bois-Reymond qualified as a university lecturer with the paper "Über saure Reaktion des Muskels nach dem Tode." He had already completed a great deal of the manuscript of *Untersuchungen über thierische Elektrizität,* mainly the preface of volume I (which became famous), the historical introduction, and the techniques of electrophysiology. From 1848 to 1853 du Bois-Reymond was instructor in anatomy at the Berlin Academy of Art. He did not lecture at the university until 1854, when Müller asked him to lecture on physiology with him. Du Bois-Reymond was now concerned exclusively with physiology, and with his friends Brücke, Ludwig, and Helmholtz he became a pioneer in the new physical orientation of the field, which sought to explain all processes in an organism by means of physical, molecular, and atomic mechanisms, without drawing upon hypothetical vital forces.

Thanks to the great interest which Humboldt had had in galvanism since his youth, du Bois-Reymond was elected to membership in the Prussian Academy of Sciences in 1851, at the age of thirty-three. From 1876 he was one of the permanent secretaries of the academy, and a great part of his work was dedicated to preparing the meetings and the official speeches for the annual celebrations in memory of Leibniz, its founder, and of its great patron, Frederick II.

In 1853 du Bois-Reymond married Jeanette Claude. They had four sons and five daughters. Of the four sons, René became a physician and a physiologist, Claude an ophthalmologist, and Allard and Felix mathematicians and engineers. Of the daughters, Estelle gained fame by editing her father's posthumously published works.

In 1855 du Bois-Reymond was named associate professor. When Müller died suddenly in 1858, the chair was divided. Reichert received the professorship of anatomy and du Bois-Reymond that of physiology. Now he had to carry the entire burden of lecturing as well as a heavy schedule of academy duties. At this time the physiology department was located in the west wing of the university building on Unter den Linden; along with the anatomy department and the museum, there were a few shabby rooms, inadequate for the needs of the physiology department. The conditions for experimentation were so unsuitable that du Bois-Reymond had to conduct the greater part of his experiments in his own apartment. Only after long efforts was a new institute for physiology, located on the Dorotheenstrasse, completed in 1877; after Carl Ludwig's institute in Leipzig it was the largest and most modern in Germany. There were four departments: physiological chemistry (Eugen Baumann, Kossel), physiological histology (Gustav Fritsch), physiological physics (Arthur Christiani), and a special department for experimentation with animals (Karl Hugo Kronecker, J. Gad). Also at this time the *Archiv für Anatomie, Physiologie und wissenschaftliche Medizin,* which had been taken over from Müller and since then edited by Reichert and du Bois-Reymond, was divided into an anatomical section and a physiological section, the latter being edited by du Bois-Reymond. Most of the papers by his colleagues and pupils appeared in this journal, as did most of the publications from Ludwig's institute in Leipzig. Among du Bois-Reymond's pupils were Eduard Pflüger, Ludimar Hermann, Isidor Rosenthal, Hermann Munk, F. Boll, Carl Sachs, and Gad, all of whom became prominent physiologists. A great many Russians worked in the institute also. A list of his colleagues and pupils, along with their publications, is in the dissertation by J. Marseille (1967). Du Bois-Reymond gave his colleagues and pupils great personal freedom and latitude to develop on their own, occupying himself almost exclusively with the problems of electrophysiology. After 1877 his publications are dominated by public speeches at the academy and his investigations of electric fishes.

By heritage du Bois-Reymond was particularly open to things French; he sought contact with French colleagues and visited Paris as early as 1850 in order to present the results of his experiments. He also met Claude Bernard there. However, he obviously felt that he did not receive the recognition he had expected. He reproached French researchers for reading only French publications and later leveled harsh criticism against his French neighbors, particularly during the Franco-Prussian War. He got along better with his English colleagues, traveling to England in 1852, 1855, and 1866 to visit or to attend congresses. H. Bence-Jones, who became a good friend and colleague, published a short version of du Bois-Reymond's papers in 1852.

Du Bois-Reymond was twice rector of the University of Berlin, in 1869–1870 and in 1882–1883. He was of course a member of almost all noteworthy scientific academies. On 11 February 1893 he celebrated the fiftieth anniversary of obtaining his doctorate. The formal address was delivered by Virchow. On 26 December 1896 du Bois-Reymond died of senile heart disease. The eulogy given by his pupil Rosenthal was used as a preface to the second edition of du Bois-Reymond's speeches (1912). The best biography is by his pupil E. Boruttau but it lacks a bibliography. With du Bois-Reymond, the last of that group died which had led German physiology to its position of uncontested leadership at the end of the nineteenth century.

From the beginning du Bois-Reymond directed his research to electrical phenomena that had been thought to be involved in various life processes since the time of Galvani but were long known to exist with certainty only in the discharges of electric fishes. Along with Galvani and Volta, Humboldt (1797) had concerned himself with these phenomena; this was the reason for his great interest in du Bois-Reymond's research. Interest in the phenomenon of animal electricity had generally receded since about 1820; but Aldini, Nobili, and particularly Matteucci continued to concern themselves with its explanation and measurement. In 1828 Nobili, using an improved Schweigger multiplier, was able to demonstrate the presence of an electric current (*courant propre*) on an intact but skinned frog trunk. From 1836 Matteucci concerned himself with the shock of electric fishes and confirmed the existence of the *courant propre*. In 1842 he demonstrated the existence of a *courant musculaire,* the demarcation, or injury, current between the uninjured surface and the cross section of a muscle. Matteucci also observed the "induced contraction" that a nerve-muscle preparation shows when its nerve is laid over the thigh muscle of a second, contracting preparation. In addition, he was the first to observe deflections in the galvanometer when a muscle contracted in strychnine tetanus. His later investigations were published in *Untersuchungen über die thierische Elektrizität* (1848) and in *Gesam-*

melte Abhandlungen zur allgemeinen Muskel- und Nervenphysik (1875–1877).

Du Bois-Reymond's most significant achievement was introducing clear physical methods and concepts into electrophysiology. In 1842–1843 he described (incorrectly at first) an autogenous current from the intact surface of the muscle to the tendon and (correctly) the injury current between the surface and the cross section; he used a multiplier that he had coiled and improved himself. He found this current even in the smallest pieces of muscle and traced it correctly to the individual muscle fibers, the interior of which is negative with respect to the surface of the fibers. The contracting muscle thus reveals a change, the so-called "negative fluctuation" of the injury current (1849). It occurs during every muscle contraction; but during tetanus, which arises from summation of many individual contractions, it becomes much clearer. Du Bois-Reymond confirmed induced contraction and identified it correctly as "secondary contraction" caused by the stimulus that the electric current of a contracting muscle in one nerve-muscle preparation gives to the nerve of a second preparation.

Du Bois-Reymond's interpretation of the basic molecular processes was analogous to Ampère's interpretation of the magnet. He believed that the muscle fiber was made up of numerous peripolar electromotive molecules, each thought to consist of a positive equatorial zone and two negative polar zones. For the tendons he assumed electrically neutral parelectronomic molecules. When Hermann was able to demonstrate the lack of current in intact muscle fibers in 1867–1868, du Bois-Reymond tried in vain to save his theory of preexistence by means of additional assumptions. In any case, he interpreted the currents from intact, injured, and contracting muscles as having a single cause.

Because of the extremely high sensibility of his multiplier, he was able in 1849 to show the injury current also in the nerve. He succeeded further in demonstrating the "negative fluctuation" in tetanized nerves and thus proved the electric nature of the *Nervenprinzip*.

Du Bois-Reymond also discovered that polarization occurs at the points of entry and exit during the flow of direct current through a nerve. It is shown in a change of charge at and near the positive and negative poles of the section through which the current flows. This "electrotonus" was the subject of his work for many years, and his pupil Pflüger continued to investigate the subject intensively. In connection with this du Bois-Reymond proposed the thesis that the effect of the electrical stimulation, apart from the polarization, depends upon the slope of the change

in intensity of the current at the point of stimulation, and not upon the duration or the absolute intensity of the current. After 1869, when Wilhelm Krause developed the theory that the transmission of stimulation from nerves to the muscle fiber is the result of an electrical discharge of the end plate, du Bois-Reymond also pursued these questions. He considered it possible for a chemical mechanism for transmitting stimulation to exist along with the electrical mechanism.

Quite a large part of du Bois-Reymond's research concerned the explanation of the nature and origin of the shock given by electric fishes. Many papers, particularly after 1877, written with Sachs and Fritsch are concerned with the anatomy and the production of electricity in these creatures. From 1857 he studied living examples of the *Malapterurus electricus* (electric catfish), the torpedo, and the *Gymnotus electricus.*

Du Bois-Reymond owed his great scientific success to the development of new electrophysiological methods of deriving and measuring current. He was the first to avoid the many difficulties and sources of error that make it very hard to obtain clear results in electrophysiology, and he created much of the apparatus of electrophysiology. His first multiplier (1842) had almost twice as many wire coils as previous ones, which made it unusually sensitive. In 1849 he once more increased them several times for the measurement of the nerve current. He was also the first to develop a deriving electrode that could not be polarized, by using glass tubes closed with a clay stopper and including a combination of zinc slabs in a solution of zinc sulfate (1859); it remained in use until about 1940. Du Bois-Reymond also was the first to develop the procedure of measuring weak bioelectric currents without loss by means of compensation with a rheochord bus-bar or a handy round compensator (1861). For decades physiologists used the du Bois-Reymond sliding-carriage induction coil for many purposes. The principle of induction had been known since the time of Oerstedt and Faraday. Du Bois-Reymond built an apparatus with a secondary coil that could be moved on a sliding carriage (1849). This made it possible to graduate and calibrate the intensity of the first and last induction shocks. This instrument was the starting point for many devices used in the medical applications of Faradic stimulations in electrotherapy.

Many very useful aids in electrophysiology were developed or adapted to the needs of animal experimentation by du Bois-Reymond. He constructed "simple" switches that guaranteed a definite electrical contact, as well as the mercury switch, the rocker, the electrode holder, stands, and clamps. He was the first to succeed in clarifying, eliminating, or avoiding the many sources of error in electrophysiological procedure, such as losses due to leakage, the polarization phenomena on metal electrodes, the deflection caused by nonparallel muscle fibers, the temporal change in the injury current following death, and the influence of extension on the voltage derived. Du Bois-Reymond also investigated the currents which occur in secretory glands (1851) and believed that he could derive electric currents from the human arm during voluntary contraction.

Most of du Bois-Reymond's experimental findings and technical procedures have remained valid. Some of his theories and several of the conceptions derived from his incorrect molecular hypothesis (such as denying the absence of current in intact muscles) did not last. Students, co-workers, and visitors from all over the world took up electrophysiology; improved the measuring devices; and demonstrated electrical phenomena in glands, the eye, the heart, and the brain. In this way, new areas of physiology originated from his preliminary work.

Du Bois-Reymond's interest in molecular physics led him to lecture during almost every summer semester from 1856, on the "physics of organic metabolism." This involved such processes as diffusion of gases and liquids, diffusion through pores, adsorption, the theory of solutions, capillarity, surface tension, swelling, osmosis, and secretion. The lectures, which were edited by his son René (1900), show du Bois-Reymond's interest in the subject and make one aware that his intellectual effort belonged completely to the period up to the end of the 1860's. His last decades were, for the most part, filled with other work and problems that were much less those of physiology than general problems of scientific knowledge, problems of methods and limitations, and historical questions. After about 1870 du Bois-Reymond became increasingly active in the public discussion of the relation between the natural sciences and the humanities, particularly philosophy, theology, and history. He had uncommonly wide knowledge and judgment in both science and various branches of the humanities. This is shown especially by the two-volume *Reden*.

In France and Germany in the second half of the eighteenth century, the idea had prevailed that the processes of formation, conservation, irritability, sensitivity, and such could not be explained by the laws of inorganic nature. Something like a vital force, a conservative force, or an educative force—analogous to the force of gravity—was supposed to direct the vital processes. Even Müller, du Bois-Reymond's teacher, subscribed to this belief; thus it was significant when Müller's young assistant eloquently demonstrated the inconsistencies in this theory: "Matter is not a wagon to which forces can be arbitrarily hitched or unhitched like horses." Forces do not exist independently of matter; and where they are expressed, they are the same in the living and the dead. At that time Schwann, Hermann Lotze, Brücke, Ludwig, and Helmholtz thought similarly.

Du Bois-Reymond's detailed, well-documented memorial speech for Müller (1858) is of the greatest historical interest. The same is true of his speech for Helmholtz (1895). His speech at the opening of the new institute for physiology (1877) is also an important document. His speeches on Voltaire (1868), La Mettrie (1875), and Maupertuis (1892) are prime examples of analysis of intellectual history. To him Voltaire was a fighter for intellectual freedom, human dignity, and justice, who had disseminated the significance of Newtonian thought. Du Bois-Reymond also portrayed the astonishing gifts of Diderot (1884), who was equally productive in treatises and novels, in art and science, ethics, metaphysics, philology, and philosophy. He demonstrated in several speeches his excellent knowledge of Leibniz, his philosophy, and his scientific significance. His lively feeling for the history of science was expressed in his address of 1872. He praised the charisma of the master scientists of the past and considered the history of science the most important, but most neglected, part of cultural history.

Du Bois-Reymond's lecture "Kulturgeschichte und Naturwissenschaft" (1876) contains a complete analysis of Western cultural history in relation to the inductive sciences. He saw the absolute organ of culture in the natural sciences and the true history of mankind in the history of the natural sciences. Man had become a "rational animal who travels with steam, writes with lightning, and paints with sunbeams." He portrayed the weaknesses of the contemporary schools, which provide classical languages but are deficient in mathematics and the theory of conic sections. His political speeches "Der deutsche Krieg" (1870), "Das Kaiserreich und der Friede" (1871), and "Über das Nationalgefühl" (1878) are not without exaggerated complaints about the self-praise and chauvinistic feelings of superiority of the French.

In any case, du Bois-Reymond was not afraid to express unpopular thoughts. His inaugural speech as rector, "Goethe und sein Ende" (1882), annoyed a great many intellectuals. He mercilessly portrayed the weaknesses of Goethe's concept of nature, his inclination to deduction, the deficiencies of his theory of color, and even the curious contradictions in his *Faust*. Du Bois-Reymond maintained that natural science had come as far as it had without Goethe's scientific writings: one should leave Goethe alone as a scientist.

Other speeches show how well read du Bois-

Reymond was and his ability to judge questions of art. The greatest excitement and the most bitter opposition were caused by the two speeches "Über die Grenzen des Naturerkennens" (1872) and "Die sieben Welträtsel" (1880). For him there were two insoluble questions for natural science, that of the essence of matter and force and that of the occurrence of consciousness in connection with molecular processes in the brain. Even an intellect like Laplace would not be able to know all of the factors involved in these questions. The seven riddles of the world are, according to du Bois-Reymond, those questions which science can answer only with the words *ignoramus* or *ignorabimus:* (1) the essence of force and matter, (2) the origin of movement, (3) the origin of life, (4) the teleology of nature, (5) the origin of sense perception, (6) the origin of thought, and (7) free will. He considered these questions transcendental.

Du Bois-Reymond's views annoyed both extreme natural scientists, like Ernst Haeckel, and theologians. The controversy he stimulated filled the daily press as well as the scientific literature. Accordingly, du Bois-Reymond significantly affected his own time and posterity on two levels. In electrophysiology he laid the foundations of the methods that were used for a century. In his treatment of problems of scientific boundaries and principles, he developed such brilliant formulations that his arguments still arouse great interest.

Du Bois-Reymond had an unusual gift for language and a finely developed sense of beauty, and he chose his words carefully. He loved figurative comparisons; those he used were sometimes audacious but never dull. His language was clear, his thought structure logical. He loved to introduce quotations from both classical and contemporary poets. French heritage blended with German thoroughness, eloquence with awareness of problems. From his youth du Bois-Reymond was receptive to philosophy and religion; but in protest against his Kantian and pious father, he tended very early toward cognitive-theoretical empiricism and free-thinking atheism. His study of La Mettrie had played not a slight role. He had only little love for nineteenth-century Christianity. Metaphysics, he thought, should not be mixed with natural science: the idea of the vital force was a mistake of this kind because the law of the conservation of energy, the framework for all transformations of energy, forbids such a hypothesis.

The neovitalism of Hans Driesch and Gustav von Bunge drew du Bois-Reymond's sharp condemnation. Thus, as he wrote in 1875, he found in himself a union of "intellectual inclinations which drive me with almost equal intensity in very different directions of

perceiving nature." Du Bois-Reymond, for all his modesty, was self-confident and certain. His intellectual vitality and his many talents allowed him to make friends rapidly. He was uncommonly devoted to his friends, such as Hallmann and Ludwig, and did not allow his co-workers, such as Sachs and Fritsch, to go unrecognized. Understandably, he found it difficult in his old age to encounter much hostility and many refutations of his molecular theory of animal electricity. The opposition of his talented student Hermann was a bitter blow, but the scientific world of the nineteenth century never lacked polemics.

BIBLIOGRAPHY

I. ORIGINAL WORKS. Du Bois-Reymond's writings include "Vorläufiger Abriss einer Untersuchung über den sogenannten Froschstrom und über die elektrischen Fische," in Poggendorff's *Annalen der Physik und Chemie,* **58** (1843), 1–30; *Untersuchungen über thierische Elektrizität,* 2 vols. (Berlin, 1849–1884), vol. II in 2 secs., sec. 2 in 2 pts.; and *Abhandlungen zur allgemeinen Muskel- und Nervenphysik,* 2 vols. (Leipzig, 1875–1877). The books contain many papers that had already appeared in the *Monatsberichten . . ., Sitzungsberichten . . .,* and *Abhandlungen der Preussischen Akademie der Wissenschaften* and also those that had been published in Poggendorff's *Annalen* and in *Archiv für Anatomie, Physiologie und wissenschaftliche Medizin.* A complete bibliography of du Bois-Reymond's publications is in the dissertation by J. Marseille (below). The list of academic speeches (incomplete) is in A. von Harnack, *Geschichte der Königlichen Preussischen Akademie der Wissenschaften zu Berlin,* III; and in Otto Köhncke, *Gesamtregister über die in den Schriften der Akademie von 1700–1890 erschienen wiss. Abhandlungen und Festreden* (Berlin, 1900). The 2-vol. 2nd ed. of his *Reden,* Estelle du Bois-Reymond, ed. (Leipzig, 1912), contains almost all of his public speeches. Further sources are *Emil du Bois-Reymond. Jugendbriefe an Eduard Hallmann,* Estelle du Bois-Reymond, ed. (Berlin, 1918); and *Zwei grosse Naturforscher des 19. Jahrhunderts. Ein Briefwechsel zwischen Emil du Bois-Reymond und Karl Ludwig,* Estelle du Bois-Reymond and Paul Diepgen, eds. (Leipzig, 1927).

II. SECONDARY LITERATURE. Accounts based on original sources are H. Bence-Jones, *On Animal Electricity, Being an Abstract of the Discoveries of Emil du Bois-Reymond* (London, 1852); and Ilse Jahn, "Die Anfänge der instrumentellen Elektrobiologie in den Briefen Humboldts an Emil du Bois-Reymond," in *Medizin historisches Journal,* **2** (1967), 135–156. On the history of electrophysiology, see the following by K. E. Rothschuh: "Die neurophysiologischen Beiträge von Galvani und Volta," in L. Belloni, ed., *Per la storia della neurologia italiana* (Milan, 1963), pp. 117–130. In addition see "Alexander von Humboldt und die Physiologie seiner Zeit," in *Archiv für Geschichte der*

Medizin, **43** (1959), 97–113; "Von der Idee bis zum Nachweis der tierischen Elektrizität," *ibid.,* **44** (1960), 25–44; and "Emil du Bois-Reymond und die Elektrophysiologie der Nerven," in K. E. Rothschuh, ed., *Von Boerhaave bis Berger* (Stuttgart, 1964), pp. 85–105. Also of value are Giuseppe Moruzzi, "L'opera elettrofisiologica di Carlo Matteucci," in *Physis,* **4** (1964), 101–140, with a bibliography of works by Matteucci; and J. Marseille, "Das physiologische Lebenswerk von E. du Bois-Reymond mit besonderer Berücksichtigung seiner Schüler," an M.D. dissertation (Münster, 1967), with bibliography.

Biographies, memorials, and obituaries are Heinrich Boruttau, *Emil du Bois-Reymond,* vol. III in the series Meister der Heilkunde (Vienna-Leipzig-Munich, 1922); and I. Munk, "Zur Erinnerung an Emil du Bois-Reymond," in *Deutsche medizinische Wochenschrift,* **23** (1897), 17–19, with portrait. Further references to obituaries are in *Index medicus,* 2nd ser., **2** (1897), 521–522; 3rd ser., **3** (1922), 178.

Further secondary literature and historical evaluations are Erich Metze, *Emil du Bois-Reymond, sein Wirken und seine Weltanschauung,* 3rd ed. (Bielefeld, 1918); Paul Grützner, in *Allgemeine deutsche Biographie,* XLVIII (1903), 118–126; Friedrich Harnack, *Emil du Bois-Reymond und die Grenzen der mechanistischen Naturauffassung* (Festschrift zur 150-Jahr-Feier der Humboldt-Universität Berlin), I (Berlin, 1960), 229–251; F. Dannemann, "Aus Emil du Bois-Reymond's Briefwechsel über die Geschichte der Naturwissenschaften," in *Mitteilungen zur Geschichte der Medizin und der Naturwissenschaften und der Technik,* **18** (1919), 274 ff.; K. E. Rothschuh, *Geschichte der Physiologie* (Berlin-Göttingen-Heidelberg, 1953), pp. 130–139, with portrait; Wolfgang Kloppe, "Du Bois-Reymond's Rhetorik im Urteil einiger seiner Zeitgenossen," in *Deutsches Medizin historisches Journal,* **9** (1958), 80–82; and Rudolf Virchow, "Ansprache zum 50-jährigen Dr.-Jubiläum von Emil du Bois-Reymond," in *Berliner klinische Wochenschrift,* **30** (1893), 198–199.

K. E. ROTHSCHUH

DU BOIS-REYMOND, PAUL DAVID GUSTAV (*b.* Berlin, Germany, 2 December 1831; *d.* Freiburg, Germany, 7 April 1889), *mathematics.*

Paul du Bois-Reymond was the younger brother of the famous physiologist Emil du Bois-Reymond. He studied first at the French Gymnasium in Berlin, then at the *collège* in Neuchâtel and the Gymnasium in Naumburg. Following the example of his brother, he began to study medicine at the University of Zurich in 1853 and by the next year had published four articles that dealt basically with physiological problems. But soon du Bois-Reymond began to apply his talents to the mathematical and physical sciences. He continued his studies at the University of Königsberg, where, mainly through the influence of Franz Neumann, he turned to the study of mathematical physics, joining his talent for observation with that for making theoretical analyses. He specialized in the study of liquids, especially the areas of liquidity and capillarity. In 1859 he received his doctorate at the University of Berlin on the basis of his dissertation, "De aequilibrio fluidorum." Du Bois-Reymond then became a professor of mathematics and physics at a secondary school in Berlin and continued to devote himself systematically to mathematics until his appointment at the University of Heidelberg in 1865. In 1870 he went from Heidelberg to Freiburg as a professor, and thence to the University of Tübingen, in 1874, as the successor to H. Hankel. From 1884 until the end of his scientific career he occupied a chair of mathematics at a technical college in Berlin.

Du Bois-Reymond worked almost exclusively in the field of infinitesimal calculus, concentrating on two aspects: the theory of differential equations and the theory of the functions of real variables.

Studying the problems of mathematical physics led du Bois-Reymond to the theory of differential equations. He was concerned with these problems at the start of his scientific career and returned to them in the last years of his life. His basic study, *Beiträge zur Interpretation der partiellen Differentialgleichungen mit drei Variablen* (part 1, *Die Theorie der Charakteristiken* [Leipzig 1864]), was one of the first to follow up Monge's idea of the "characteristic" of a partial differential equation. This idea, expressed by Monge for equations of the second order as early as 1784, depended on the geometric expression of the integral of a differential equation as the surface defined by a system of curves. Du Bois-Reymond generalized this for partial differential equations of the nth order. These ideas, such as the simple case of a study of contact transformations, led in generalized form to the studies by Lie and Scheffers.

The chief means of solving partial differential equations at that time was by Fourier series. As early as the 1820's Cauchy, Abel, and Dirichlet had pointed out some of the difficulties of the expansion of "arbitrary" function in a Fourier series and of the convergence of this series. These problems contributed substantially to the rebuilding of the foundations of mathematical analysis. One of the first to deal with them systematically was du Bois-Reymond. He published his main results toward the end of the 1860's and in the 1870's—at first under the influence of Riemann's ideas and at a time when the results of Weierstrass' work had not been published and were little known.

He achieved a number of outstanding results. As early as 1868, when he was studying some properties of integrals, du Bois-Reymond expressed both precisely and generally and demonstrated the mean-value theorem for definite integrals, which was then an important aid in the study of Fourier series. This theorem was later expressed independently by Dini (1878), who ascribed it to Weierstrass. The latter, however, developed a similar but more specialized proposition and made no claim to du Bois-Reymond's discovery.

Like other mathematicians who relied on Dirichlet, du Bois-Reymond also originally tried to show that each continuous function in a given interval is necessarily representable by its Fourier series (or another series analogous to a trigonometric one). A decisive turn came in 1873, when du Bois-Reymond published "Über die Fourier'schen Reihen." This article contains an exposition of the chief idea of a construction of a continuous function with a divergent Fourier series at any point. Later he also attempted to show the properties of this continuous function (which has a very complicated construction) using considerations, difficult to comprehend, that concern the infinitely small and the infinitely large (the so-called *Infinitär-calcul*).

Two of the other results of du Bois-Reymond's work should be mentioned. First and foremost is the solution of the problem of the integrability of Fourier series, which he proposed (but did not publish until 1883) and for which he demonstrated certain conditions that made it possible to distinguish Fourier's from other trigonometric series. The second is the solution to a question that concerned mathematicians of that time: the publication (1873) of an example and the precise demonstration of the properties of the function that is continuous in a given interval but without derivatives. This achievement was inspired by Weierstrass.

Du Bois-Reymond was then led to attempt a general exposition of the fundamental concepts of the theory of functions in his book *Die allgemeine Functionentheorie,* the first part of which was published in 1882. Among other things, this work shows that its author was aware that a precise theory of real numbers was needed for the further progress of the theory of functions, but he did not make any real contributions to that progress. Instead, he wrote of the problems of the philosophy of mathematics, recognizing the advantages of different approaches and expressing grave doubts about the usefulness of formalism.

Du Bois-Reymond's work was directed at the basic questions of the mathematical analysis of the time and is marked by both the personality of the author and the state of the mathematics of the period. It appeared before completion of the revision of the foundations of mathematical analysis for which he was striving. Led by sheer mathematical intuition, he did not hesitate to publish even vague considerations and assertions that were later shown to be false. Further developments, some while du Bois-Reymond was still alive, disclosed these weaknesses (e.g., Pringsheim's criticism) and also rapidly outdated his results, even on the main questions. Among them is his attempt to give a general theory of convergence tests. This meant that his work, which had been greatly appreciated by his contemporaries, soon sank into oblivion, although it had included very important questions and notions that were later reflected in the work of such mathematicians as W. H. Young, A. Denjoy, and H. Lebesgue.

BIBLIOGRAPHY

Du Bois-Reymond's "Über die Fourier'schen Reihen" appeared in *Nachrichten von der Gesellschaft der Wissenschaften zu Göttingen* (1873), 571–584. *Die allgemeine Functionentheorie* was translated by G. Milhaud and A. Girot as *Théorie générale des fonctions* (Paris, 1887). On du Bois-Reymond and his work, see (in chronological order), L. Kronecker, "Paul du Bois-Reymond," in *Journal für die reine und angewandte Mathematik,* **104** (1889), 352–354; "P. du Bois-Reymond's literarische Publicationen," in *Mathematische Annalen,* **35** (1890), 463–469; H. Weber, "Paul du Bois-Reymond," *ibid.,* 457–462; *Paul du Bois-Reymond, Zwei Abhandlungen über unendliche* (*1871*) *und trigonometrische Reihen* (*1874*), Ostwalds Klassiker, no. 185 (Leipzig, 1912); *Paul du Bois-Reymond, Abhandlungen über die Darstellung der Funktionen durch trigonometrische Reihen* (*1876*), Ostwalds Klassiker, no. 186 (Leipzig, 1913); and A. B. Paplauskas, *Trigonometricheskie ryady ot Eylera do Lebega* ("Trigonometrical Series from Euler to Lesbesgue," Moscow, 1966).

LUBOŠ NOVÝ

DUBOSCQ, OCTAVE (*b.* Rouen, France, 1 October 1868; *d.* Nice, France, 18 February 1943), *protistology, cytology.*

The son of a minor railroad employee who was crushed between two cars, Duboscq was orphaned at a very young age; his mother died of grief only a few months later. His aunt took him in and cared for him. He was a brilliant student at the lycée in Coutances. In 1886 he began to study medicine at Caen while preparing for his *licence ès sciences naturelles,* which he obtained in 1889. He went to Paris to complete his medical studies and in 1894 defended his doctoral thesis in medicine.

However, scientific research attracted Duboscq, and he abandoned his medical career. He was named *préparateur* at Caen, then *chef de travaux* at Grenoble. He defended his doctoral thesis in science in 1899 and became lecturer in zoology at Caen (1900), professor of zoology at Montpellier (1904) and then of marine biology at Paris, and director of the Arago Laboratory at Banyuls-sur-Mer (1923), where he remained until his retirement (1 October 1937). He then took up residence in Paris and he spent three months each winter in Nice with his brother, who died of a heart attack in 1942. Grief over this death and malnutrition as a result of the war led to his own death.

Louis Joyeux-Laffuie, professor of zoology at Caen, Georges Pruvot of Grenoble, Louis Léger of Grenoble, and Yves Delage of Paris profoundly influenced Duboscq's work as well as his teaching. His collaboration with Léger lasted nearly a quarter of a century. A brilliant teacher, around 1912 Duboscq created a certificate in cytology and protistology and organized a teaching method that accorded a large place to practical topics.

Duboscq's work, at once original and substantial, began with researches on the microanatomy of the arthropods (venom glands, nervous system of the chilopods, the digestive duct of the insects and crustaceans, spermatogenesis of the *Sacculina*). His essential work, however, represents an effort very important to the knowledge of several classes of the Protista. His investigations concerned the structure and cycle of the schizophytes and the intestinal and sanguicolous spirochetes of saltwater fishes. With Léger he studied the eccrinids, which are filamentous parasitic protophytes of the arthropods, and determined their development cycles; they also studied the sporozoans, establishing their general cycle and that of the gregarines in particular. They also discovered the cycle of the *Porospora*. They studied the coccidiomorphs and made a valuable contribution to the knowledge of the *Pseudoklossia*, of the *Selenococcidium*, and of the *Aggregata*. Moreover, they offered original ideas on the phylogeny and classification of the sporozoans.

Duboscq was also interested in the flagellates, especially in the sexual reproduction of the *Peridinia* and in the sexuality of the flagellate organisms that live in termites. The sporozoans represent the major work of his maturity. In his old age he devoted himself to the sponges. James Brontë Gatenby discovered the indirect fertilization of the sponges, and Duboscq and his student O. Tuzet specified the details and variations of the process.

BIBLIOGRAPHY

Among Duboscq's numerous writings are "Recherches sur les chilopodes," his thesis for the doctorate in science, in *Archives de zoologie expérimentale*, 3rd ser., **7** (1899);

"Les éléments sexuels et la fécondation chez *Pterocephalus*," in *Comptes rendus hebdomadaires des séances de l'Académie des sciences*, **134** (1902), 1148–1149; "Aggregata vagans, n. sp. grégarine gymnosporée parasite des pagures," in *Archives de zoologie expérimentale*, n. et r., 4th ser., **1** (1903), 147–151; "Notes sur les infusoires endoparasites. II. *Anoplophrya brasili*. III. *Opalina saturnalis*," *ibid.*, **2** (1904), 337–356; "Selenococcidium intermedium et la systématique des sporozoaires," *ibid.*, 5th ser., **5** (1910), 187–238; "Deux nouvelles espèces de grégarines appartenant au genre *Porospora*," in *Annales de l'Université de Grenoble*, **23** (1911), 399–404; "Selysina perforans, description des stades connus de sporozoaire de *Stolonica* avec quelques remarques sur le pseudovitellus des statoblastes et sur les cellules géantes," in *Archives de zoologie expérimentale*, **58** (1918), 1–53; "L'appareil parabasal des flagellés et sa signification," in *Comptes rendus hebdomadaires des séances de l'Académie des sciences*, **180** (1925), 477–480; "Les porosporides et leur évolution," in *Travaux de la Station zoologique de Wimereux*, **9** (1925), 126–139; "L'évolution des *Paramoebidium*, nouveau genre d'eccrinides, parasites des larves aquatiques d'insectes," in *Comptes rendus hebdomadaires des séances de l'Académie des sciences*, **189** (1929), 75–77; "L'appareil parabasal et les constituants cytoplasmiques des zooflagellés," *ibid.*, **193** (1931), 604–605; "L'appareil parabasal des flagellés avec des remarques sur le trochosponge, l'appareil de Golgi, les mitochondries et le vacuome," in *Archives de zoologie expérimentale*, **73** (1933), 381–621; "L'ovogenèse, la fécondation et les premiers stades du développement des éponges calcaires," *ibid.*, **79** (1937), 157–316; and "Recherches complémentaires sur l'ovogenèse, la fécondation et les premiers stades du développement des éponges calcaires," *ibid.*, **81** (1942), 395–466.

A biographical notice with a chronological list of Duboscq's scientific works is P. P. Grassé, in *Archives de zoologie expérimentale*, **84** (1944), 1–46.

Andrée Tétry

DU BUAT, PIERRE-LOUIS-GEORGES (*b.* Tortizambert, Normandy, France, 23 April 1734; *d.* Vieux-Condé, Flanders [now part of Nord, France], 17 October 1809), *hydraulics*.

Born in the manor of Buttenval, the second son of a minor nobleman, Du Buat was in all probability educated at Paris, where he became a military engineer at the age of seventeen. By 1787 he had risen to the rank of colonel, which he then resigned to accept appointment as *lieutenant du roi*. His earliest assignments were canal, coastal, harbor, and fortification works in the north of France.

In 1758 Du Buat married a native of Condé (near Valenciennes, on the Belgian border), by whom he eventually had eleven children. On the death of his older brother in 1787, he inherited their late father's title of count; but with the advent of the Revolution he lost titles and properties and was forced in 1793 to flee with his family to Belgium, then Holland, and finally Germany. In 1802 he returned to Vieux-Condé, and a portion of his estate was restored to him.

Du Buat began his hydraulic studies in 1776, and by 1779 he had published the first edition of his major work, *Principes d'hydraulique,* copies of which are now quite rare. This was enlarged in 1786 to two volumes, the first of which was analytical and the second experimental; it is supposed to have been translated into English (no copy can be found) as well as German. A posthumous edition of three volumes appeared in 1816, the third volume (*Pyrodynamique*) having been written during his exile. All three editions carried essentially the same prefatory remarks reviewing the state of the art—in particular those many important topics about which little or nothing was known; portions of the discourse are often quoted in subsequent works because of their continued relevance.

The analytical part of Du Buat's writings was perhaps more effective than that of such contemporaries as Jean Charles Borda and Charles Bossut, for he dealt extensively with matters of boundary resistance, velocity distribution, underflow, overflow, and backwater. It was in experimental work that he excelled. The results of his 200 separate tests on flow in pipes, artificial channels, and natural streams were to be used by engineers for generations. Even more original was his treatment of immersed bodies. He showed that tests in air and in water could be correlated in terms of the relative density of the resisting medium, and calculated, for example, the size of a parachute required to break the fall of a man of a given weight. He was also the first to demonstrate that the shape of the rear of a body is as important in controlling its resistance as is that of the front.

Du Buat also made 100 measurements on the distribution of pressure around bodies and sought through his findings to develop a new form of Henri Pitot's "machine" for the measurement of velocity. The basis for his conclusion that the force exerted upon a stationary body by running water is greater than that required to move the same body at the same relative speed through still water (Du Buat's paradox) is not clear from his writings—i.e., whether the cause is the uneven velocity distribution or the turbulence of the flow. Du Buat is called by some the father of French hydraulics, although at least partial credit should go to several of his contemporaries—not to mention Edmé Mariotte, more than a century his senior.

BIBLIOGRAPHY

Du Buat's major work was *Principes d'hydraulique, ouvrage* . . . (Paris, 1779); 2nd ed. entitled *Principes d'hydraulique vérifiés* . . ., 2 vols. (Paris, 1786); 3rd ed. entitled *Principes d'hydraulique et de pyrodynamique,* 3 vols. (Paris, 1816).

On Du Buat and his work, see H. Rouse and S. Ince, *History of Hydraulics* (New York, 1963), pp. 129–134; and B. de Saint-Venant, "Notice sur la vie et les ouvrages de Pierre-Louis-Georges, comte du Buat," in *Mémoires de la Société impériale des sciences de Lille,* 3rd ser., **2** (1865), 609–692.

HUNTER ROUSE

DUCHESNE, JOSEPH, also known as **Josephus Quercetanus** (*b.* L'Esture, Armagnac, Gascony, France, *ca.* 1544; *d.* Paris, France, 1609), *chemistry, medicine.*

The son of a physician, Duchesne (occasionally referred to as Sieur de la Violette) studied first at Montpellier. He married a granddaughter of the humanist Guillaume Budé and, because of persecution of the French Protestants, spent many years away from his homeland. Duchesne received his medical degree at Basel in 1573 and for some time was settled at Kassel, the capital of the grand duchy of Hesse. At this time and later the grand dukes were noted for their patronage of the new Paracelsian-Hermetic medicine. Later Duchesne moved to Geneva where he was received as a citizen in 1584. After election to the Council of Two Hundred (1587), he was sent on several diplomatic missions. In 1592 he helped determine the peace terms which the Republic of Geneva made with its neighbors. The following year Duchesne returned to Paris, where he was appointed physician in ordinary to King Henry IV.

Duchesne is a figure of some importance in French literature as well as science and medicine. His *La morocosmie* (1583, 1601) and *Poesies chrestiennes* (1594) have been commented on favorably by literary historians while his other poetical work, *Le grand miroir du monde* (1584, 1595), is important for Duchesne's concept of the elements. In addition, he ventured into tragicomedy with *L'ombre de Garnier Stauffacher* (1583), a work which took as its theme the alliance between Zurich, Berne, and Geneva.

Duchesne's medicoscientific works are best seen as part of the late sixteenth- and early seventeenth-century debate on the place of chemistry in medicine and natural philosophy. The flood of Paracelsian texts published in the third quarter of the sixteenth century had gained many adherents to the new medicine, but at the same time it had brought forth strong opposition from the medical establishment. Peter Severinus had attempted to systematize the works of Paracelsus in 1571, and Guinther von Andernach had written in defense of the new chemically prepared medicines in the same year, but Thomas Erastus at Basel had prepared a lengthy and detailed attack on Paracelsus and his views (1572–1573). Alarmed by the increasing internal use of minerals and metals, the Faculty of Medicine at Paris forbade the further prescription of antimony in this fashion (*ca.* 1575).

The strong critique of the views of Paracelsus on chemical medicines and the origin of metals written by Jacques Aubert in 1575 was the occasion for Duchesne's first publication. His *Responsio* to Aubert (1575) was a strong defense of the iatrochemical position, and although it was a short work, it was reprinted often and attracted considerable attention.

In the *Responsio* and many other works Duchesne offered a large number of pharmaceutical preparations. His *Sclopetarius* (1576), which dealt with the cure of gunshot wounds, and his *Pharmacopoea dogmaticorum* (1607) are only two of many works by him that were translated into several languages and went through numerous editions. These works offer a large number of remedies prepared from substances of mineral, vegetable, and animal origin. In all of his practical texts Duchesne placed strong emphasis on chemical procedures and his works contain the first printed directions for the preparation of turpeth mineral (basic mercuric sulfate), antimony sulfide, urea, and—possibly—calomel as medicines. Devaux has pointed to Duchesne's use of sulfur for respiratory problems and iodated substances (calcinated sea sponges) for the goiter.

A series of polemical works debating the value of the new medicine and the extent to which chemistry might be employed by physicians were printed in the last quarter of the century. In France the matter reached a climax when Duchesne published his *De priscorum philosophorum verae medicinae materia . . .* (1603). This work was immediately answered by the elder Jean Riolan who accused him of wishing to sweep away the venerable medicine of the ancients in his *Apologia pro Hippocratis Galeni medicina* (1603). In his reply to Riolan, published the following year, Duchesne denied this charge and answered that he wished only to use the best of the old medicine along with the new chemistry. These works were followed by a series of other works in which both Riolans, Israel Harvet, Theodore Turquet de Mayerne, Andreas Libavius, and many other authors participated.

In the course of this debate it became clear that there was more at stake than the simple acceptance or rejection of chemical remedies. For Duchesne—as for other iatrochemists—chemistry was to serve as a key to all nature. His cosmology was based on the biblical story of the Creation, and in his discussion he pictured the Creator as an alchemist separating the elements from the unformed chaos. In the fifteenth chapter of the *Ad veritatem hermeticae medicinae ex*

Hippocratis veterumque decretis ac therapeusi . . . (1604), Genesis is clearly interpreted in terms of the three Paracelsian principles of salt, sulfur, and mercury. In the earlier *Le grand miroir du monde* (1584) Duchesne had also accepted the Aristotelian water and earth as elementary substances. This five-element principle system has much in common with the five-element descriptions so common in the works of later seventeenth-century chemists.

Duchesne rejected the four humors of the ancients and when discussing the vascular system specifically spoke of the "circulation" of the blood. By this, however, he meant a series of local circulations in different organs, analogous to the heating of liquids in distillation flasks. His was a world view based on a close relation of the macrocosmic and microcosmic worlds. An integral part of this was his sincere belief in the doctrine of signatures, which for him were an important guide to divine gifts existing here on earth. Duchesne wrote of the need of "experientia" and new observations for a proper understanding of nature, and although he objected to being called a Paracelsian, his views were similar to those of Paracelsus in many respects.

Much of Duchesne's influence derives from the debate his work had initiated at Paris. His publications of 1603 and 1604 went through numerous editions in several languages and did much to publicize his version of the chemical philosophy. In these works Duchesne had discussed at length the Creation and the three principles. More than half a century later Robert Boyle still found it necessary to comment on these views in *The Sceptical Chymist* (1661).

In addition, the Parisian debate was influential in bringing about a more general acceptance of chemically prepared medicines. The chemists had been generally agreed that their aim was not to destroy all of the old medicine, but rather to apply what they found valuable in the works of the ancients along with the best of the new chemical medicine. This surely had been the view taken by Duchesne and it was also that of Mayerne, who had been the first of Duchesne's colleagues to support him in 1603. Mayerne was later to become chief physician to King James I of England, and he advocated this compromise position in the publication of the important *Pharmacopoeia* (1618) of the Royal College of Physicians. This work is notable both for its prominent inclusion of chemicals alongside the traditional Galenicals and also for its prefatory defense of the new methods of cure.

BIBLIOGRAPHY

I. ORIGINAL WORKS. There is no complete list of Duchesne's iatrochemical books. Used together, J. Ferguson's

Bibliotheca chemica, 2 vols. (Glasgow, 1906), and J. R. Partington's *A History of Chemistry,* II (London, 1961), will furnish most titles if not all of the editions.

Among his many works, Duchesne's first publication, the *Ad Iacobo Auberti vindonis de ortu et causis metallorum contra chymicos explicationem Iosephi Armeniaci, D. Medici breuis responsio* (Lyons, 1575), was considered a major work in support of the iatrochemical position. It was reprinted often in Latin, French, and German, and an English translation by John Hester was printed in London in 1591. The next year Duchesne published his *Sclopetarius, sive de curandis vulneribus quae sclopetarum ictibus acciderunt* (Lyons, 1576), a work that appeared in French translation in the same year. John Hester made this text available in English in 1590. The *De priscorum philosophorum verae medicinae materia . . .* (St. Gervais, 1603) initiated the debate over the chemical medicine at Paris. Duchesne's reply to Jean Riolan, the *Ad veritatem hermeticae medicinae ex Hippocratis veterumque decretis ac therapeusi . . .* (Paris, 1604), is also a major theoretical statement. The first of these appeared in French translation (Paris, 1626) and selections from both were translated into English by Thomas Timme as *The Practise of Chymicall and Hermeticall Physicke for the Preseruation of Health* (London, 1605). A final—and much less well known—*Ad brevem Riolani excursum brevis incursio* (Marburg, 1605) concluded Duchesne's contributions to this debate.

There is little question that Duchesne's most popular work was the *Pharmacopoea dogmaticorum restituta pretiosis selectisque hermeticorum floribus abunde illustrata* (Paris, 1607). There are twenty-five known editions of this work from the first half of the seventeenth century. The *Opera medica* includes the *Responsio* to Aubert, the *Sclopetarius,* and the *De exquisita mineralium animalium, et vegetabilium medicamentorum spagyrica preparatione et vsu, perspicua tractatio.* This went through at least two Latin editions (Frankfurt am Main, 1602; Leipzig, 1614) and one German edition (Strasbourg, 1631). The most extensive collection was the three-volume *Quercetanus redivivus* prepared by Johann Schröder, a massive text that went through three editions (Frankfurt am Main, 1648, 1667, 1679).

II. SECONDARY SOURCES. Pierre Lordez's *Joseph du Chesne, sieur de La Violette, médecin du roi Henri IV, chimiste, diplomate et poète* (Paris, 1944) contains useful information.

Guy Devaux's "Quelques aspects de la médecine et de la pharmacie au XVIe siècle à travers la 'Pharmacopée des dogmatiques' de Joseph Du Chesne, Sieur de la Violette, conseiller et médecin du roy," in *Revue d'histoire de la pharmacie,* **19** (1969), 271–284, is a study of the pharmaceutical preparations in Duchesne's most popular work, while the discussion of Duchesne in Partington's *History of Chemistry,* II (London, 1961), 167–170, centers on the chemical preparations known to Partington.

W. P. D. Wightman, *Science and the Renaissance,* I (Edinburgh–London–New York, 1962), 256–263, offers a helpful discussion of the complex debate at Paris in the early years of the seventeenth century.

For element theory in Duchesne, see R. Hooykaas, "Die Elementenlehre der Iatrochemiker," in *Janus,* **41** (1937), 1–18; and Allen G. Debus, *The English Paracelsians* (London, 1965), pp. 87–101. Duchesne's views on the circulation of the blood are discussed in Allen G. Debus, "Robert Fludd and the Circulation of the Blood," in *Journal of the History of Medicine and Allied Sciences,* **16** (1961), 374–393.

Finally, Pagel has pointed to a Paracelsian tract (1635) by Fabius Violet (possibly Duchesne) in which the digestive factor in the stomach is identified with the "hungry acid" of Paracelsus. This is a statement that comes close to van Helmont's position first printed in 1648. The influence of Violet on van Helmont is surely possible although the former did not go on to identify this acid with hydrochloric acid as did the latter. On this see Walter Pagel, *Paracelsus. An Introduction to Philosophical Medicine in the Renaissance* (Basel, 1958), 161–164.

ALLEN G. DEBUS

DUCLAUX, ÉMILE (*b.* Aurillac, Cantal, France, 24 June 1840; *d.* Paris, France, 2 May 1904), *biochemistry.*

Duclaux belonged to that group of physicists and chemists, still limited in the second half of the nineteenth century, who through their work increased our knowledge of living matter. It appears, however, that his lasting fame derives less from his discoveries than from the close ties that bound him to Pasteur and his followers throughout his adult life. He was, in fact, one of the first to believe in microbes, and the books he devoted to them have long remained the "gospel" of the new doctrine.

His father was bailiff of the court at Aurillac, where his mother ran a small grocery. As a child, his long walks through the beautiful Auvergne countryside gave Duclaux a taste for nature and poetry; and his parents' example revealed to him the value of sincerity, simplicity, and perseverance.

In 1857, upon completing his classical education at the local *collège,* Duclaux left Aurillac and went to Paris to attend the special mathematics course at the Lycée St. Louis. Two years later he was accepted at both the École Polytechnique and the École Normale Supérieure; he chose the latter. In 1862 he became *agrégé* in the physical sciences and was then retained by Pasteur as his laboratory assistant (*agrégé-préparateur*) at the school. It was during this period that the discussions of the possibility or impossibility of spontaneous generation were at their liveliest. Pasteur maintained that the microscopic creatures responsible for fermentation came from parents similar to themselves. Nicolas Joly, Pouchet, and Musset asserted that, on the contrary, these creatures were born spontaneously in organic fluids. From time to time Dumas and Balard, members of the commission named by the Académie des Sciences to settle the question, came to the École Normale. Duclaux, who had already participated in the experiments of his mentor, now attended the debates. The impression they made on him showed him his true course in life.

Dissociated from Pasteur's laboratory, an *agrégé-préparateur* faced an uncertain future. After defending his doctoral thesis in physical sciences in 1865, Duclaux decided to leave Paris. He became a teacher first at the *lycée* in Tours, then at the Faculty of Sciences at Clermont-Ferrand, in which city his mother, a widow since 1860, joined him. He was able to renew his collaboration with Pasteur, first at Pont-Gisquet, Gard, where the master was pursuing his work on silkworm diseases, and a little later at Clermont-Ferrand. The experiments—on fermentation—began in a makeshift laboratory set up by Duclaux and were repeated on a much greater scale at the Kuhn brewery in Chamalières, which is between Clermont-Ferrand and Royat. It is well known that these experiments were requested in order to revive the brewing industry.

New professional duties brought Duclaux to Lyons in 1873 and finally to Paris in 1878. In Paris he won a competition for the professorship of meteorology at the Institut Agronomique, and he was also given a lectureship in biological chemistry at the Sorbonne. He immediately used this opportunity to give a course in microbiology, the first of its kind anywhere.

His young wife, the former Mathilde Briot, succumbed suddenly to puerperal fever following the birth of their second son. To forget his grief, Duclaux threw himself into his work with even greater energy. He taught, experimented, and wrote; and he followed, day after day, Pasteur's extraordinary series of accomplishments. These included the development of vaccines against fowl cholera, anthrax, swine fever, and, in 1885, against rabies. In 1888 the Institut Pasteur was founded in Paris on rue Dutot. Duclaux, who meanwhile had become titular professor at the Sorbonne, transferred his teaching activities to the Institut Pasteur. A little earlier, through his efforts a new monthly journal, the *Annales de l'Institut Pasteur,* was created to publish research in microbiology.

Beginning with this period, one may say that Duclaux's life was almost inseparable from that of the Institut Pasteur. At the death of its brilliant founder in 1895, he took over its direction and in a few years made it into a sort of "scientific cooperative," in which each scientist, while preserving the independence of his own ideas, worked toward a common goal. To the original buildings were added, at the beginning of the century, the Institut de Chimie Biologique and a hospital.

Duclaux became a member of the Académie des Sciences in 1888, of the Société Nationale d'Agriculture in 1890, and of the Académie de Médecine in 1894. In 1901 he married Mme. James Darmesteter (the former Mary Robinson), a woman remarkable for both her intelligence and her warmth. He had finally found familial happiness again, but this happiness did not last. In January 1902 he suffered his first stroke. Scarcely recovered, he began to write again for the *Annales* and in the spring of 1903 recommenced his lectures. But this was too much to demand of an overtaxed body. On the evening of 2 May 1904 Duclaux suddenly lost consciousness and died in the night. His place as director of the Institut Pasteur was assumed by one of his pupils from Clermont-Ferrand, Émile Roux. The latter had become well known for his research with Pasteur, his discovery of the diphtheria bacillus, and his development of a specific diphtheria antitoxin.

Duclaux's scientific work is at once that of a physicist and that of a chemist. As a physicist he studied the phenomena of osmosis, of molecular adhesion, and of surface tension. As a chemist he concentrated especially on fermentation processes. In this area he was to some extent following up the work of Pasteur. As the years passed he was led to accord to enzymes (then called *diastases*) an increasingly important role in the phenomena of life. He devoted a long series of investigations to the respective roles played in the intestinal tract of men and animals by enzymes issuing from glands and by those liberated by microbes. He recognized that the microbes had no role in gastric and pancreatic digestion, which involve only juices released from the tissues. Microbial digestion does not begin until the intestine, but then rapidly becomes important. In a related area, Duclaux realized that microbes are indispensable in the formation in the soil of plant nutrients. Without microbes the earth is infertile, because the enzymes in the plant cells cannot leave the cells and thus cannot act outside the plant.

Milk provided Duclaux with a material ideally suited to the study of enzymes. In the first stage, through a great number of analyses, he was able to develop methods permitting the determination of the proportions of its constituents. In the second stage he studied the enzymes capable of modifying the constituents. The great importance of enzymes was shown in the transformation of milk into cheese. In this case, however, the active agents are of external origin. A cheese is in fact the result of microbial cooperation: "Each of the microscopic workers must act in its turn and stop at the right moment. Such a workshop is difficult to direct, and one may say that it has required the experience of centuries to obtain products whose taste and appearance are always the same" (Émile Roux, in *Annales de l'Institut Pasteur,* **18** [1904], 337). Duclaux studied several types of cheese, but undoubtedly with particular relish the cheese from Cantal, one of the riches of his native area.

Duclaux the teacher was no less remarkable than Duclaux the researcher. His pupil Roux, recalling his days as a medical student at Clermont-Ferrand, wrote: "Duclaux presented a subject so clearly that everyone understood. His words were those of a scientist burning with the 'sacred fire.' He set thinking, to the point that when one had finished his course, he seemed to be there still" (*ibid.*).

In addition to his research papers Duclaux wrote a great many didactic works; and the critical reviews he published in the *Annales* remain models. It has been said of them that they display "the logic of the scientist and the style of the poet. . . . He could extract from a memoir . . . possible consequences that the author himself had not always suspected. How many ideas he explored; what new insights he lavishly bestowed. Duclaux sowed the high wind . . ." (Émile Roux, in *Bulletin de l'Institut Pasteur,* **2** [1904], 369).

Duclaux was captivating, full of wit and verve. He was also a just man with a passionate soul. He dreamed of a universal brotherhood under the banner of science—"the common fatherland," as he used to say, "where one could have passions without having hatreds." But he was not oblivious to what was happening outside his laboratory. On several occasions his devotion to the truth led him to enter into political conflicts. In particular he took a very active part in the campaign that finally forced the reinstatement of Captain Dreyfus.

BIBLIOGRAPHY

I. ORIGINAL WORKS. A complete list of Duclaux's scientific publications appears in *Annales de l'Institut Pasteur,* **18** (1904), 354–362.

Among his most important books are *Ferments et maladies* (Paris, 1882); *Le microbe et la maladie* (Paris, 1886); *Cours de physique et de météorologie* (Paris, 1891), a published version of his course at the Institut Agronomique; *Traité de microbiologie,* 4 vols. (Paris, 1891–1901); *Principes de laiterie* (Paris, 1892); *Le lait, études chimiques et microbiologiques* (Paris, 1894); *Pasteur, histoire d'un esprit* (Paris, 1896), which contains the frequently cited phrase, "Chemistry has taken possession of medicine, and will not let go"; and *L'hygiène sociale* (Paris, 1902).

II. SECONDARY LITERATURE. On the life and works of Duclaux, two articles by Émile Roux are classics: "Notice sur la vie et les travaux d'Émile Duclaux," in *Annales de l'Institut Pasteur,* **18** (1904), 337–362; and "Émile Duclaux," in *Bulletin de l'Institut Pasteur,* **2** (1904), 369–370. A biography, *La vie d'Émile Duclaux,* was written by his second wife, Mary Darmesteter Duclaux (Paris, 1906).

ALBERT DELAUNAY

DUCROTAY DE BLAINVILLE, HENRI MARIE. *See* **Blainville, Henri Marie Ducrotay de.**

DUDITH (DUDITIUS), ANDREAS (*b.* Buda [now Budapest], Hungary, 16 February 1533; *d.* Breslau, Germany [now Wrocław, Poland], 23 February 1589), *astronomy, astrology, mathematics.*

Andreas Dudith combined political and religious activity with humanist and scientific interests in a manner fairly common in the sixteenth century. Of mixed Hungarian and Italian descent, Dudith was educated in the Hungarian tradition of Erasmian humanism. He traveled widely in Italy, France, and England from 1550 to 1560, serving for a time as secretary to Cardinal Reginald Pole. After attending the Council of Trent in 1562–1563, Dudith received the bishopric of Pécs and performed various diplomatic missions to Poland for the emperors Sigismund II and Maximilian II between 1563 and 1576. Dudith's first marriage in 1567 to Regina Strass, a Polish noblewoman, and his subsequent adherence to Lutheranism brought upon him the condemnation of Rome and weakened his position at the Viennese court. (His second marriage, in 1574, was to Elisabeth Zborowski.) After some political reverses he retired from affairs of state in 1576 and later devoted himself to scientific and theological matters at Breslau, inclining to Calvinist and Socinian doctrines.

Dudith was familiar with the leading intellectual movements of his day; his visits to Italy had acquainted him with humanists and bibliophiles like Paulus Manutius and Giovanni-Vincenzo Pinelli and also with the works of Pietro Pomponazzi and the Paduan Averroists. In the 1570's Dudith took up the study of mathematics and cultivated the friendship of the Englishmen Henry and Thomas Savile and the German Johann Praetorius. Medicine also interested Dudith; he studied Galen and corresponded with many physicians, including the imperial physician Crato. In the breadth of his intellectual interest Dudith was typical of Renaissance humanists, and his library of printed books and manuscripts reflects this encyclopedism. Like many Italians, notably his friend Pinelli, he collected Greek mathematical manuscripts for both their philological and scientific interest. Among his manuscripts were the *Arithmetic* of Diophantus (which he loaned to Xylander to use as the text for the first Latin translation published at Basel in 1575); the *Mathematical Collections* of Pappus; the

Elementa astronomiae of Geminos (used for the *editio princeps* of 1590); and his own transcription of the *Tetrabiblos* of Ptolemy. Several of these manuscripts were lost following the dispersal of his library, but many of his manuscripts and his 5,000 printed books are now in the Vatican, Paris, Leiden, and various Swedish libraries.

Dudith is known mainly for his contribution to the controversy over the comet of 1577. (Hellman lists more than 100 publications on this comet.) He knew several of the personalities involved in the dispute, including Thomas Erastus, Thaddaeus Hagecius (Hayck), and Tycho Brahe; and a collection of tracts on the topic was dedicated to him in 1580. Although originally interested in the astrology of the *Tetrabiblos* of Ptolemy, Dudith became an opponent of the astrologers. His *De cometarum significatione* shows the influence of Erastus in its rejection of astrology as a vain pseudoscience. Both Dudith and Erastus argued that comets could appear without causing or portending natural or political calamities. (In his first letter to Hagecius, Dudith remarked that astrology was condemned by Christian authorities and, despite his own Calvinist leanings, that astrology infringed upon free will.)

Dudith accepted, however, Aristotle's physical explanation of comets as accidental exhalations of hot air from the earth that rise in the sublunar sphere. But an insistence on mathematical astronomy rather than astrology soon led Dudith to the rejection of Aristotelian physical doctrine. In 1581 Dudith learned in a letter from Hagecius, a believer in astrology, of the latter's observation that the parallax of another comet indicated that the comet was beyond the moon. In his letter of 19 January 1581 to Rafanus, Dudith argued that this observation proved the Aristotelian explanation fallacious. If the comets were terrestrial in origin they could not penetrate beyond the sublunar sphere; if, however, they originated in the immanent heavens comets could not be classified as accidental phenomena. Dudith remarked that many recently observed comets seemed to form and dissolve in the region of permanent things. This fact suggested serious flaws in the Aristotelian system. (Tycho arrived at a similar conclusion from his observation that the 1577 comet had no parallax and must therefore be farther from the earth than was the moon. Tycho also attempted to calculate the orbit of that comet.)

Dudith's use of a mathematically precise observation to criticize a general physical theory of Aristotle's betokens the same kind of dissatisfaction with Aristotelian physical doctrines that was most eloquently expounded in the works of Galileo fifty years later.

BIBLIOGRAPHY

I. ORIGINAL WORKS. Dudith's main scientific work is *De cometarum significatione commentariolus* . . . (Basel, 1579), repr. in the 2nd pt. of *De cometis dissertationes novae clariss. Virorum Th. Erasti, Andr. Duditii* . . . (Basel, 1580). Dedicated to Dudith, it includes Dudith's letter to Erastus of 1 Feb. 1579. Subsequent eds. are: Breslau, 1619; Jena, 1665; Utrecht, 1665. The first letter to Hagecius (26 Sept. 1580) is in J. E. Scheibel, *Astronomische Bibliographie,* II (Breslau, 1786), 160–182, with other materials on the comet of 1577. The second letter (1 Feb. 1581), congratulating Hagecius on his observation of a comet's parallax, appears at the beginning of Thaddaeus Hagecius, *Apodixis physica et mathematica de cometis* (Görlitz, 1581). The letter to Rafanus is in Lorenz Scholtz, *Epistolarum philosopharum medicinalium, ac chymicarum volumen* (Frankfurt, 1598), letter 28. Details of Dudith's voluminous correspondence are given in the Costil biography cited below.

For Dudith's writings on the marriage of priests see Q. Reuter, *Andreae Dudithii orationes in concil. Trident. Habitae*. . . (Offenbach am Main, 1610). See also the references to Dudith in J. L. E. Dreyer, ed., *Tychonis Brahe opera omnia,* 15 vols. (Copenhagen, 1913-1929), IV, 453, 455; VI, 327-328; VII, 63, 123, 182, 214; VIII, 455.

II. SECONDARY LITERATURE. An excellent biography and bibliography is Pierre Costil, *André Dudith: humaniste hongrois 1533–1589, sa vie, son oeuvre et ses manuscrits grecs* (Paris, 1935). For the controversy on the comet of 1577, see Dreyer, *op. cit.,* IV, 509; C. Doris Hellman, *The Comet of 1577: Its Place in the History of Astronomy* (New York, 1944); and Lynn Thorndike, *A History of Magic and Experimental Science,* 8 vols. (New York, 1923-1958), VI, 67-98, 183-186.

PAUL LAWRENCE ROSE

DUDLEY, ROBERT (*b.* Sheen House, Surrey, England, 7 August 1573; *d.* Villa di Castello, Florence, Italy, 6 September 1649), *navigation.*

Dudley was the son of Robert Dudley, earl of Leicester, and Lady Douglas Sheffield. The legitimacy of Dudley's birth was questioned in his lifetime, yet he was given every advantage commensurate with his father's position in Elizabethan England. He was a student at Christ Church, Oxford, and at the age of twenty-one sailed in command of two ships to the West Indies. In 1596 he was in the battle of Cádiz with the earl of Essex and was knighted for his bravery. In 1605 Dudley left his wife and children in England and traveled to Italy, accompanied by one of the beauties of the day, Elizabeth Southwell. He established himself in Florence, became a Catholic, married Miss Southwell, and entered the service of the grand duke of Tuscany. He was put in charge of several major engineering projects, including the building of the port of Leghorn, and the beginnings of land reclamation near Pisa. He never returned to

England; his assumed titles, earl of Warwick and duke of Northumberland, invalid in England, were confirmed by the Holy Roman Emperor Ferdinand II in recognition of his services.

Dudley's first work, an account of his voyage to the West Indies, was printed by Hakluyt in the second edition of his *Voyages* under the title "A Voyage . . . to the Isle of Trinidad and the Coast of Paria." He had become interested in navigation while at Oxford, and the interest had been further stimulated by his close association with the great sea captain Thomas Cavendish, brother of his first wife. He continued to work on the pressing problems of navigation, including the determination of longitude; made a collection of the best and most advanced navigational instruments, now in the Florence Museum of Science; and at the age of seventy-three published his great work, *Dell'arcano del mare* (three volumes, Florence, 1646–1647). It is one of the great sea atlases of all time, magnificently engraved, and may justly be regarded as an encyclopedia of knowledge regarding the sea. It contains a treatise on naval strategy; a manual of shipbuilding; directions on building coastal fortifications; instructions to navigators, including the essential elements of nautical almanacs; and a set of maps of the entire world. It is these maps that give Dudley's work special significance; *Dell'arcano del mare* is the first sea atlas with all maps drawn on Mercator's projections, as modified by Edward Wright. The maps, virtually without ornamentation and restricted to the information essential to the seaman, are, in spite of errors and imperfections, among the milestones of naval cartography.

BIBLIOGRAPHY

See G. F. Warner's biographical sketch and preface to the Hakluyt Society's edition of Dudley's *Voyage* (London, 1899); and Vaughan Thomas, *The Italian Biography of Sir Robert Dudley* (Oxford, 1861?).

GEORGE KISH

DUFAY (DU FAY), CHARLES-FRANÇOIS DE CISTERNAI (*b.* Paris, France, 14 September 1698; *d.* Paris, 16 July 1739), *physics.*

Dufay came from a family that had followed military careers for over a century. He himself joined the Régiment de Picardie as a lieutenant in 1712, at the warlike age of fourteen; apparently he missed the closing battles of the War of the Spanish Succession, but he participated in the successful siege of Fuenterrabia (1718/1719), which helped force Philip V to abandon his adventures in Italy. Shortly after the campaign, Dufay accompanied his father and Cardinal de Rohan, the leading churchman in France, on an extended visit to Rome (1721). This marked the end of his military service. On his return to France in 1722 he became a candidate for the position of "adjunct chemist" in the Académie des Sciences, Paris.

This step did less violence to family tradition than might appear. Dufay's grandfather, an amateur alchemist who appreciated the value of education, had sent his son, Dufay's father, to the Jesuits at the Collège de Clermont. There he met the future cardinal and contracted a bibliomania that dominated his life after the loss of a leg ended his soldiering in 1695. Dufay grew up among his father's books and erudite friends, "raised, like an ancient Roman, equally for arms and for letters" (Fontenelle). It was very likely Cardinal de Rohan who directed the attention of the scientific establishment toward the unknown young officer. The Academy's leading scientist, Réaumur, and its titular head, the Abbé Bignon, managed Dufay's candidacy, which terminated successfully in May 1723. He became associate chemist in 1724, pensionary in 1731, and director in 1733 and 1738.

Dufay very quickly justified the influence exercised in his favor. His first academic paper (1723), on the mercurial phosphorus, already displayed the characteristics which distinguished his later work: full command of earlier writings, clear prescriptions for producing the phenomena under study, general rules or regularities of their action, thorough study of possible complications or exceptions, and cautious mechanical explanations of a Cartesian flavor. This "phosphor"—the light sometimes visible in the Torricelli space when a barometer is jostled—much perplexed the physicists of the era, primarily because it did not always occur under apparently identical conditions. Dufay found that traces of air or water vapor occasioned the failures, which could be entirely eliminated with a technique of purification taught him by a German glassmaker. He explained the light in terms of Cartesian subtle matter squeezed from the agitated mercury; although he knew the work of Francis Hauksbee (the elder), he suggested no connection with electricity.

This maiden effort, however useful for the development of technique, did not provide a continuing line of research. For several years Dufay flitted from one subject to another: he studied the heat of slaked lime (1724), invented a fire pump (1725), touched on optics (1726), plane geometry (1727), the solubility of glass (1727), and the coloring of artificial gems (1728). He eventually published at least one paper in each of the branches of science recognized by the Academy, the only man, perhaps, who has ever done so. In 1728 he took up magnetism, the first subject to enlist his interest for an extended period. In the first of three memoirs he attacked the vexed question of natural magnetism: Under what conditions, and in what positions, do iron tools acquire a magnetic virtue? The apparent answer—oriented vertically—suggested an easy Cartesian model, for the "hairs" which determine the direction in which the magnetic effluvia pass through the pores of iron might be expected to line up under their gravity when the bodies containing them stand upright. The last two memoirs (1730, 1731), which attempt to measure the force of magnetic poles, are most instructive. Although Dufay took the greatest pains over the experiments, varying sizes, shapes, and measuring devices, he failed to find any simple relation between force and distance; the apparently straightforward procedures of Coulomb in fact are far from obvious.

In 1730 Dufay returned to his original subject, phosphorescence, with a memoir of great importance in the development of his method. Chemists had long been acquainted with a few minerals which, like the Bologna stone (BaS) and Balduin's hermetic phosphor (CaS), glowed after exposure to light. Great mystery surrounded these expensive and supposedly rare substances. Dufay detested mysteries and held as a guiding principle that a given physical property, however bizarre, must be assumed characteristic of a large class of bodies, not of isolated species. He set about calcining precious stones, egg and oyster shells, animal bones, etc., most of which became phosphorescent; indeed, he found that almost everything except metals and very hard gems could be made to shine like Bologna stones. He gave clear recipes for producing the phosphors and patiently examined the endless variations in their colors and intensities: "How differently bodies behave which seemed so similar, and how many varieties there are in effects which seemed identical!" This line of work ended in 1735, with a study of the luminescence of gems. Dufay distinguished excitation by friction, by heat, and by light, and tried to find some general rules of their operation; but the phenomena proved altogether too complex, and he established little more than that diamonds usually can be excited in more ways than lesser stones.

In 1732 Dufay at last found a subject ripe for his practiced talents. A year earlier Stephen Gray had published an account of his discovery that "electricity"—the attractive and repulsive "virtue" of rubbed glass, resins, precious stones, etc.—could be communicated to bodies, like metals or human flesh, which could not be electrified by friction. Gray had also succeeded in transmitting the virtue of a glass tube through lengths of stout cord suspended by silk threads. It appeared to Dufay that electricity, far from

being the parochial, effete effect discussed by earlier writers, was one of nature's favorite phenomena. He proceeded as with the phosphors: first a survey of the existing literature, which became his initial memoir on electricity; next, an attempt to electrify every natural object accessible to experiment. As he expected, all substances properly treated—save metals, animals, and liquids—could be electrified by friction; while all bodies whatsoever could be made so by communication. In the process he distinguished insulators from conductors more sharply than Gray had done and ended the desultory search for new electrics which had characterized the study of electricity since the time of Gilbert.

Dufay's most notable discoveries (1733) resulted from an attempt to clarify the connection between electrostatic attraction and repulsion. Ever since Hauksbee had found that light objects drawn to a glass tube are sometimes forcibly driven from it, physicists had tried to understand the relation between motions toward and away from an excited electric. Hauksbee had given incompatible theories; others, like the Dutch Newtonian W. J. 'sGravesande, taught that the tube possessed an electrical "atmosphere" whose pulsations caused alternate "attractions" and "repulsions," an elegant theory which, however, misrepresents the facts; and still others suspected, as Dufay did initially, that repulsion did not exist at all, an object apparently repelled by an excited electric in fact being drawn away by neighboring bodies electrified by communication. Further experiment suggested another possibility to him: Since substances the least excitable by friction, like the metals, respond most vigorously to the pull of the tube, might not "an electric body attract all those that are not so, and repel all those that become electric by its approach, and by the communication of its virtue?" The apparent confirmation of this capital insight—bits of metal electrified by the tube were found to repel one another—may be regarded as the decisive step in the recognition of electrostatic repulsion, the uncovering of the phenomenological connection between motions toward and away from the tube. It also prepared the way for a detection still more surprising.

Experience had taught Dufay not to draw general conclusions without examining a wide range of substances. Accordingly he tried electrifying one of the two metal bits by a rod of gum copal; the resultant attraction, which flabbergasted him, soon forced him to recognize the existence of two distinct "electricities," and to determine their basic rule of operation. This "bizarrie" (as Dufay called the double electricities) proved a great difficulty for the usual theories of electricity, which relied solely on matter in motion. Dufay expected that a representation in terms of vortices might someday be found, but he did not insist; he was concerned first to establish the regularities and only later to add the mechanical pictures. In this point of method Benjamin Franklin—and not Dufay's protégé the Abbé Nollet—was his lineal descendant.

From his classic researches on the two electricities Dufay turned to Gray's quixotic experiment of the charity boy, the electrification of a small insulated orphan. Playing the leading part himself, Dufay received a sharp shock when an assistant tried to touch him, the stroke penetrating even his waistcoat and shirt; and both he and Nollet noticed that a spark passed just before contact when the experiment was repeated at night. These phenomena utterly astounded him, inured though he was to "meeting the marvelous at every turn." He devoted great effort to studying the electric light, to which his earlier research on phosphors naturally inclined him. Here again he was stopped by the vast complexity of the phenomena, which gave no intelligible clue to the advancement of electrical theory.

Dufay's substantial electrical discoveries—the relation between attraction and repulsion, the two electricities, shocks, and sparking—are but one aspect, and perhaps not the most significant, of his achievement. His insistence on the importance of the subject, on the universal character of electricity, on the necessity of organizing, digesting, and regularizing the known facts before grasping for more, all this helped to introduce order and professional standards into the study of electricity at precisely the moment when the accumulation of data began to require them. He found the subject a hodgepodge of often capricious, disconnected phenomena; he reduced the apparent caprice to rule; and he left electricity in a state where, for the first time, it invited prolonged scrutiny from serious physicists.

Electricity by no means exhausted Dufay's energy or talent; between his sixth and seventh memoirs on the subject, he published papers on parhelia (1735), on fluid mechanics (1736), on dew (1736), on sensitive plants (1736), and on dyestuffs (1737). The last two studies were by-products of still another side of Dufay's ceaseless activity. That on dyestuffs related to an onerous charge he had received from the government, namely the revision of standards for the closely regulated dye industry. The botanical paper grew out of an even larger job, the administration of the Jardin Royal des Plantes.

The Jardin, founded in 1635 as a medical garden and a school of pharmacy and medicine, had extended its functions under the inspired direction of the royal physician, Guy Crescent Fagon, who encouraged the study of chemistry and the expansion of the botanical collections. Regrettably Fagon's successor, Pierre Chirac, a much more limited doctor who cared only for his own profession, neglected the garden and alienated the professors. A nonmedical man was needed to repair the damage. Dufay's industry, wide interests, and practical good sense, not to mention his ministerial connections, made him an ideal administrator. With the advice of his friends the brothers Jussieu, who held chairs at the Jardin, Dufay replanted, built new greenhouses for foreign flora, and established close relations, including exchanges of specimens, with the directors of similar institutions elsewhere in Europe. His official visits to Holland and England (1733/1734), accompanied by Bernard de Jussieu and Nollet, advanced not only French botany, but—through connections formed by Nollet—French experimental physics as well. In the seven years of his intendancy Dufay transformed Chirac's collection of weeds into "the most beautiful garden in Europe" (Fontenelle), providing the basis for the great expansion effected by his successor, the comte de Buffon.

Dufay's diverse activities made him a careful economist of his time. Although his position, acquaintance, and good humor opened endless opportunities for social engagements, he preferred to live quietly, finding relaxation in the satires of Swift, in the small circle of his mother's friends, or at the home of a kindred soul like the marquise du Châtelet. He never married. It was a great blow to French science (and a measure of its incompetence) when, at the age of forty, Dufay succumbed to the smallpox.

BIBLIOGRAPHY

Dufay's chief papers were published in the *Mémoires de l'Académie des sciences* (Paris). Among the most important are "Mémoire sur les baromètres lumineux" (1723), 295–306; "Observations sur quelques expériences de l'aimant" (1728), 355–369; "Suite des observations sur l'aimant" (1730), 142–157; "Mémoire sur un grand nombre de phosphores nouveaux" (1730), 524–535; "Troisième mémoire sur l'aimant" (1731), 417–432; "Mémoires sur l'électricité" (1733), 23–35, 73–84, 233–254, 457–476; (1734), 341–361, 503–526; (1737), 86–100, 307–325; and "Recherches sur la lumière des diamants et de plusieurs autres matières" (1735), 347–372. Dufay summarized his first electrical memoirs in "A Letter . . . Concerning Electricity," in *Philosophical Transactions of the Royal Society,* **38** (1733/1734), 258–266.

A full bibliography of the French papers is given in *Nouvelle table des articles contenus dans les volumes de l'Académie royale des sciences de Paris depuis 1666 jusqu'en 1770* (Paris, 1775/1776), and in P. Brunet, "L'oeuvre scientifique de Charles-François Du Fay (1698–1739)," in *Petrus nonius,* **3,** no. 2 (1940), 1–19. Poggendorff omits several items published after 1735. I. B. Cohen, *Franklin and Newton* (Philadelphia, 1956), p. 616, gives complete titles of the memoirs on electricity. Dufay autographs are quite rare. There are a few unimportant letters among the Sloane Manuscripts at the British Museum; a dossier including notes of Hauksbee's work and correspondence with Gray's collaborator, Granville Wheler, at the Institut de France; several letters to Réaumur published in *La correspondance historique et archéologique,* **5** (1898), 306–309; and a few administrative documents noticed in A.-M. Bidal, "Inventaire des archives du Muséum national d'histoire naturelle," in *Archives du Muséum,* **11** (1934), 175–230.

The biographical sources are surprisingly meager. The most important is Fontenelle's "Éloge de M. Du Fay," in *Histoire de l'Académie des sciences* (1739), 73–83; scattered data appear in G. Martin, *Bibliotheca fayana* (Paris, 1725); *Les lettres da la marquise du Châtelet,* T. Besterman, ed. (Geneva, 1958); *Correspondence of Voltaire,* T. Besterman, ed. (Geneva, 1953–1965); and J. Torlais, *Un esprit encyclopédique en dehors de l'Encyclopédie. Réaumur,* 2nd ed. (Paris, 1961).

For assessments of Dufay's scientific work see the publications of Brunet and Cohen cited above; H. Becquerel, "Notice sur Charles François de Cisternai du Fay . . . ," in *Centenaire de la fondation du Muséum d'histoire naturelle* (Paris, 1893), pp. 163–185; J. Daujat, *Origines et formation de la théorie des phénomènes électriques et magnétiques* (Paris, 1945); and E. N. Harvey, *A History of Luminescence* (Philadelphia, 1957). For Dufay's administrative accomplishments see A.-L. de Jussieu, "Quatrième notice historique sur le Muséum d'histoire naturelle," in *Annales du Muséum,* **4** (1804), 1–19; and the bibliography in Y. Laissus, "Le Jardin du Roi," in R. Taton, ed., *Enseignement et diffusion des sciences en France au XVIIIe siècle* (Paris, 1964), pp. 287–341.

JOHN L. HEILBRON

DUFOUR, GUILLAUME-HENRI (*b.* Constance, Switzerland, 15 September 1787; *d.* Les Contamines [near Geneva], Switzerland, 14 July 1875), *technology.*

Dufour's family came from Geneva, where he studied military engineering. He also studied at the École Polytechnique in Paris and the École de Génie at Metz. When Geneva was incorporated into the French territory, he served as a sublieutenant in the army of his new country. In 1813, he was in Napoleon's army defending Corfu and had become a captain by the time of the fall of the Empire. In 1817 he returned to Switzerland and was appointed *ingénieur cantonal.* His work on fortification at Grenoble and his construction works in Geneva—which greatly improved the city—made him well-known. In 1818, he became the chief instructor of the military school that he had helped to establish at Thun.

When Geneva was reintegrated into Switzerland, Dufour joined the Swiss army. He was made a colonel in 1827. In 1831 he became chief of staff, and in 1833 he commanded a division that restored order in Basel. He began his pioneering work in triangulation the same year in order to prepare topographical maps of the Confederation. The maps were later published (1842–1864). He was elected a general of the Swiss army in 1847, during the Sonderbund War, again in 1849 to preserve Swiss neutrality, in 1856 during the conflict with Prussia over Neuchâtel, and finally—for the fourth time—in 1859, when the French threatened to annex Savoy. He was a conservative member of the federal assembly, and in 1864 presided over the Geneva international congress which drew up the rules for treatment of the wounded in wartime and resulted in the creation of the Red Cross.

A bronze equestrian statue of Dufour stands in the Place Neuve in Geneva.

BIBLIOGRAPHY

I. ORIGINAL WORKS. Dufour's works include *Cours de tactique* (Paris, 1840); *De la fortification permanente* (Geneva, 1822); *Mémoire sur l'artillerie des anciens et sur celle du moyen âge* (Paris, 1840); and *Mémorial pour les travaux de guerre* (Geneva, 1820).

II. SECONDARY LITERATURE. See also E. Chapuisat, *Le Général Dufour,* 2nd ed. (Lausanne, 1942); and T. Stark, *La famille du général Dufour et les Polonais* (Geneva, 1955).

ASIT K. BISWAS

DUFRÉNOY, OURS-PIERRE-ARMAND (*b.* Sevran, Seine-et-Oise, France, 5 September 1792; *d.* Paris, France, 20 March 1857), *geology, mineralogy.*

With Élie de Beaumont and under the direction of Brochant de Villiers, Dufrénoy prepared the first modern geological map of France. He published over sixty memoirs on geology and crystallographic and chemical mineralogy and an important mineralogical work entitled *Traité de minéralogie.* As the inspector of courses and later director of the École des Mines, he instituted important changes in the curriculum and teaching methods.

Dufrénoy's father had been Voltaire's literary agent, and his mother, Adélaïde, was a scholar of the classics and an accomplished poet. Thus the boy grew up in an intellectual atmosphere despite the family's extreme poverty, occasioned by the French Revolution and the father's loss of eyesight. Dufrénoy first attended the lycée at Rouen and then the Lycée Louis-le-Grand in Paris. He won the first prize in mathematics at the general examinations in 1810 and entered the École Polytechnique in 1811. In 1813 he was admitted to the École des Mines, at that time located at Peisey in Tarentaise (Savoy), and he returned with it when the school was moved to Paris in 1816. In 1818 he was named engineer of mines and head of the collections at the École des Mines. Stimulated by the publication of Greenough's geological map of England, Brochant de Villiers was successful in gaining authorization for the preparation of a similar map of France. Dufrénoy and Élie de Beaumont were selected to carry out the necessary fieldwork and in 1822 were sent to England for two years to learn Greenough's procedures. As a result of this visit, Dufrénoy and Élie de Beaumont coauthored a work entitled *Voyage métallurgique en Angleterre,* which described in detail the mining and metallurgical industries of England.

For the French geological map Dufrénoy was assigned the area south and west of a line running from Honfleur, Alençon, Avallon, Chalons-sur-Saône, and the Rhone. Each summer from 1825 to 1829 Dufrénoy and Élie de Beaumont made field trips, surveying and taking notes on their respective areas; from 1830 to 1834 they traveled together in order to resolve difficulties and coordinate their work. They published their notes between 1830 and 1838 and the explanation of their map in 1841. The map was on a scale of 1:500,000, and the explanation was essentially a physical description of France, together with the history, composition, and disposition of its terrain. They did not perform any detailed stratigraphic work. The regions of the Pyrenees and Britanny had hardly been explored geologically at that time, and Dufrénoy's work in these areas was excellent. He was admitted to membership in the Académie des Sciences in 1840, and in 1843 the Geological Society of London presented the Wollaston Medal to Dufrénoy and Élie de Beaumont jointly for their work.

While preparing the geological map, Dufrénoy was also engaged in other important studies. In 1830 he determined that the large coal deposits in the department of Aveyron could be used directly in various metallurgical processes, as the English had been doing; his study resulted in the establishment of industries at Decazeville. In 1832 he was sent to Scotland to study the use of high-temperature air in iron blast furnaces, and his detailed report caused the French iron industry to adopt this practice immediately. He also studied the thermal springs at Vichy and drew up plans for their exploitation.

Dufrénoy served as assistant professor of mineralogy at the École des Mines from 1827 to 1835 and as professor from 1835 to 1848. In 1836 he became inspector of courses there and director of the school in 1846, so that, in effect, he governed instruction in

the institution for twenty years, until his death in 1857. During his tenure Dufrénoy introduced to courses the use of specimens from the rich mineralogical collection, and he modernized the curriculum by adding courses in railway construction, law, economics, and paleontology. He entered the Conseil des Mines in 1846 and was promoted to the superior grade in 1851. After Brongniart's death in 1847, Dufrénoy was named professor of mineralogy at the Muséum d'Histoire Naturelle.

BIBLIOGRAPHY

I. ORIGINAL WORKS. With Élie de Beaumont, Dufrénoy published the following works: *Voyage métallurgique en Angleterre,* 2 vols. (Paris, 1827); *Mémoires pour servir à une description géologique de la France,* 4 vols. (Paris, 1830–1838); and *Explication de la carte géologique de la France, rédigée sous la direction de M. Brochant de Villiers,* 3 vols. (Paris, 1841). In addition, Dufrénoy published *Mémoire sur la position géologique des principales mines de fer de la partie orientale des Pyrénées* (Paris, 1834); and *Traité de minéralogie,* 4 vols. (Paris, 1844–1847). His sixty-five memoirs were published principally in the *Journal des mines,* the *Annales des mines,* and the *Annales de chimie et de physique.*

II. SECONDARY WORKS. On Dufrénoy or his work, see A. d'Archiac, *Notice sur la vie et les travaux de P. A. Dufrénoy* (Paris, 1860); A. Daubrée, "Dufrénoy," in *École polytechnique: Livre du centenaire 1794–1894,* I (Paris, 1895), 375–381; and A. Lacroix, "Notice historique sur le troisième fauteuil de la section de minéralogie," in *Académie des sciences—séance publique annuelle du lundi 17 décembre 1928* (Paris, 1928), pp. 24–33.

JOHN G. BURKE

DUGAN, RAYMOND SMITH (*b.* Montague, Massachusetts, 30 May 1878; *d.* Philadelphia, Pennsylvania, 31 August 1940), *astronomy.*

Dugan devoted over half his life to the study of eclipsing variables—those pairs of stars which by chance have orbits almost edge on to us and therefore appear to vary in brightness because of repeated mutual eclipses. His prolonged and careful observations resulted in a wealth of information about many fundamental properties of stars. He also collaborated with Henry Norris Russell in writing one of the best elementary astronomy texts of this century.

His mother, Mary Evelyn Smith Dugan, was of Puritan stock, and his father, Jeremiah Welby Dugan, was but one generation removed from Ireland. After receiving a B.A. degree from Amherst College in 1899, Dugan went to Beirut (then Syria) to serve both as instructor in mathematics and astronomy and as acting director of the observatory in the Syrian Protestant College (now American University). Here he remained until 1902, when Amherst granted him an M.A. degree and he transferred to the University of Heidelberg to work and study for three years under Max Wolf. During this period he discovered eighteen asteroids, found two new variable stars, and earned a Ph.D. degree with a dissertation on the star cluster known as the Pleiades.

Having taken part in the Lick Observatory expedition to Alhama de Aragón, Spain, for the solar eclipse of 30 August 1905, Dugan returned to the United States to become instructor in astronomy at Princeton University. In 1909 he married Annette Rumford Odiorne, and in 1920 he was named professor, which rank he held until death terminated his thirty-five-year association with Princeton.

Endowed with what a contemporary referred to as "the world's most accurate photometric eyes," Dugan used Princeton's twenty-three-inch telescope with a polarizing photometer to make long series of observations on a selected few eclipsing variables; for several of them this involved more than 18,000 individual settings. The resulting light curves, when analyzed according to methods developed in large part by his colleague Russell, revealed the relative sizes of the two stars, how far apart they were, their individual surface brightnesses, mutual tidally induced distortions into ellipsoids, and even how their densities varied from center to periphery. Dugan was the first to detect the so-called reflection effect, a brightening of the fainter member on the side facing its more luminous partner. He also found evidence that even widely separated pairs could be tidally distorted; furthermore he was able to explain some of the gradually lengthening periods of revolution that he observed as resulting from tidal evolution.

Dugan was elected a member of the American Philosophical Society in 1931 and served from 1935 until his death as chairman of the Commission on Variable Stars of the International Astronomical Union. He was also secretary of the American Astronomical Society from 1927 to 1936 and vice-president from 1936 to 1938.

BIBLIOGRAPHY

I. ORIGINAL WORKS. The textbook referred to above was by Henry Norris Russell, Raymond Smith Dugan, and John Quincy Stewart, *Astronomy,* I. *The Solar System* (Boston, 1926; rev. ed., 1945), II. *Astrophysics and Stellar Astronomy* (Boston, 1927; repr. with supp. material, 1938).

Dugan's discoveries of asteroids were reported in *Astronomische Nachrichten,* **160** (1902–1903), cols. 183, 216; **161** (1903), col. 143; **162** (1903), cols. 15, 111, 160; **163** (1903), cols. 255, 285, 379; **164** (1903–1904), cols. 15, 191, 224; **165**

(1904), cols. 110, 191; his discovery of two variable stars, *ibid.,* col. 43. A shortened version of his dissertation also appeared in *Astronomische Nachrichten,* **166** (1904), 49–56; the full text, "Helligkeiten und mittlere Örter von 359 Sternen der Plejaden-Gruppe," is in *Publikationen des Astrophysikalischen Observatoriums Königstuhl-Heidelberg,* **2** (1906), 29–55. His work on eclipsing variables, under the general title "Photometric Researches," makes up (if four papers by his students are included) 17 of the 19 *Contributions From the Princeton University Observatory* that appeared between 1911 and 1940: details concerning equipment and observing techniques are in "The Algol System RT Persei," in no. 1 (1911), 1–47, with a shortened version published in *Monthly Notices of the Royal Astronomical Society,* **75** (1915), 692–702; his first detection of the reflection effect is described in "The Algol System Z Draconis," in no. 2 (1912), 1–44, with a shortened version in *Monthly Notices of the Royal Astronomical Society,* **75** (1915), 702–710; confirmation of the reflection effect and observation of ellipsoidal shapes in widely separated pairs appeared in "The Eclipsing Variables RV Ophiuchi and RZ Cassiopeiae," in no. 4 (1916), 1–38, also covered in *Astrophysical Journal,* **43** (1916), 130–144, and **44** (1916), 117–123, and for RZ Cassiopeiae alone in *Monthly Notices of the Royal Astronomical Society,* **76** (1916), 729–739; his first evidence for a lengthening period, indicating tidal evolution, is in "The Eclipsing Variable U Cephei," in no. 5 (1920), 1–34, with a shortened version in *Astrophysical Journal,* **52** (1920), 154–161; the first evidence for radial variation in density appeared in "The Eclipsing Variable Y Cygni," in no. 12 (1931), 1–50; and his critical summary of the first twenty-nine years of his work at Princeton is in the introduction to "A Finding List for Observers of Eclipsing Variables," in no. 15 (1934), 1–33.

For insight into the conditions under which Dugan worked (and a taste of his dry wit), see "The Old Princeton Observatory and the New," in *Popular Astronomy,* **43** (1935), 146–151, repr. from *Princeton Alumni Weekly* (2 Oct. 1934).

The list of Dugan's publications in Poggendorff, VI, 612, and VIIb, 1149–1150, is complete except for some early items (included above).

II. SECONDARY LITERATURE. Obituary notices on Dugan by Henry Norris Russell appeared in *Popular Astronomy,* **48** (1940), 466–469; *Science,* n.s. **92** (1940), 231; and *Yearbook of the American Philosophical Society* (1940), 419–420. Other brief notices are listed in Poggendorff, VIIb (see above). Dugan's entry in *Who Was Who in America,* I (Chicago, 1943) appears on p. 344.

SALLY H. DIEKE

DUGGAR, BENJAMIN MINGE (*b.* Gallion, Alabama, 1 September 1872; *d.* New Haven, Connecticut, 10 September 1956), *plant pathology.*

The fourth of six sons of a country practitioner, Duggar entered the University of Alabama shortly before his fifteenth birthday, but a compelling interest in agricultural science led him to transfer after two years to the Mississippi Agricultural and Mechanical College (now Mississippi State College). Shortly after graduation (1891) Duggar found a sympathetic mentor in George F. Atkinson at Alabama Polytechnic Institute, where he served one year as assistant in mycology and plant physiology and received his M.Sc. for the carefully documented thesis "Germination of Teleutospores of Ravenelia Cassiaecola." He spent an additional year as assistant director of the Agricultural Experiment Station at Uniontown, Alabama, before transferring to Harvard. Here, from 1893 to 1895, he worked under W. G. Farlow and Roland Thaxter, taught botany at Radcliffe, and completed required courses in the humanities that finally brought him a highly cherished Harvard M.A. After another year of fieldwork—this one concerned with the wheat-devastating chinch bug in Illinois—Duggar rejoined Atkinson, at Cornell. Here he concentrated on chemistry and fungus spore germination and completed the work required for the Ph.D.

Traveling in Europe for a year (1899–1900) Duggar studied with such eminent authorities as Wilhelm Pfeffer at Leipzig and Georg Klebs and Julius Kühn, both at Halle. Returning to America, he spent one year as plant physiologist with the Bureau of Plant Industry of the U. S. Department of Agriculture. There he developed a lasting interest in cotton diseases and mushroom culture. While continuing to serve as consultant to the bureau, Duggar accepted his first major educational post, as professor of botany and head of the department at the University of Missouri (1902). For an exhibit of mushrooms and other fungi he was awarded a grand prize at the St. Louis Fair (1904). Another foreign tour (1905–1906) took him first to Munich, where he worked with Karl von Goebel at the Botanical Institute, and later to Bonn, Montpellier, and Algiers. Returning to Cornell as professor of plant physiology, Duggar completed two major works: *Fungus Diseases of Plants* (New York, 1909), the first monograph in any language devoted exclusively to the subject of plant pathology, which remained a standard text for many years, and *Plant Physiology With Special Reference to Plant Production* (New York, 1911).

Duggar next returned to Missouri as research professor of plant physiology at Washington University and director of the Missouri Botanical Garden. Studies of this period include those on red pigment formation in the tomato, enzymes in red algae, nitrogen fixation, and methods for determining hydrogen concentration in biological fluids. During World War I, Duggar contributed valuable data on the salt requirements of higher plants. He later (1920) turned his attention to the serious problem of the tobacco mosaic virus, becoming a leading investigator in this field.

After fifteen years at Washington University, Duggar accepted his last university assignment, the professorship of plant physiology and economic botany at the University of Wisconsin, a post he held from 1927 to 1943, when he retired emeritus. But the most rewarding period of his career was yet to come. During World War II he served as adviser to the National Economic Council, and in 1944 he accepted a position as consultant in mycological research with the Lederle Division of the American Cyanamid Company. After a short period devoted to the investigation of antimalarial drugs, Duggar turned his attention to the quest for new antibiotic-producing organisms. A three-year study of *Streptomyces aureofaciens* led to the introduction of chlortetracycline (Aureomycin) (1948), another milestone in the story of man's attempt to conquer pathogenic bacteria.

Duggar, a man of great enthusiasm and physical vitality, played a dynamic role in organizing plant scientists in America. A founder of the American Society of Agronomy (1907) and the American Phytopathological Society (1908), he was also active in the American Association for the Advancement of Science, the American Society of Naturalists, the American Botanical Society (president, 1923), and the American Society of Plant Physiologists (president, 1947). A voluminous writer, he also was editor of several important publications, including *Proceedings of the International Congress of Plant Sciences* (1926); *Biological Effects of Radiation* (1936), for which he wrote a chapter, "Effects of Radiation on Bacteria"; and *Botanical Abstracts for Physiology* (1917–1926). When the latter publication was absorbed by *Biological Abstracts,* he continued as editor of the plant physiology section (1926–1933).

Among the many honors bestowed on Duggar were membership in the American Philosophical Society (1921) and the National Academy of Sciences (1927), as well as honorary degrees from Missouri (LL.D., 1944), Washington (Sc.D., 1953), and Wisconsin (D.Sc., 1956). A modest individual, although always a perfectionist, Duggar enjoyed many aspects of life to the fullest. His marriage in 1901 to Marie L. Robertson (*d.* 1922) produced two sons and three daughters. A second marriage to Elsie Rist (1927) produced one daughter.

As a scientist Duggar helped to advance research in his chosen field from the era of morphology to the modern period, with its emphasis on physiology. Certainly his pioneer work of 1909 has already won a secure place among the classics of American science, and the discovery of Aureomycin assures Duggar an honored position in the history of medicine.

BIBLIOGRAPHY

A biography is J. C. Walker, "Benjamin Minge Duggar," in *Biographical Memoirs. National Academy of Sciences,* **32** (1958), 113–131, with bibliography and photograph.

See also "Benjamin Minge Duggar," in *Current Biography 1952* (1953), pp. 166–169.

MORRIS H. SAFFRON

DU HAMEL, JEAN-BAPTISTE (*b.* Vire, Normandy, France, 11 June 1623; *d.* Paris, France, 6 August 1706), *institutional history.*

Although usually designated an anatomist, this distinguished priest and humanist had in reality no such specialized scientific interests and indeed owes his fame primarily to the high office that he held from 1666 to 1697 in the first great French institution. His successor Fontenelle, who knew him well, suggested that du Hamel inherited from his lawyer father a talent for conciliation. Certainly he inherited from his family a sensitivity to social relations in the legal milieu in general, and the manner in which he exploited his heritage has earned him a place in the history of science.

In Paris, du Hamel completed the studies in rhetoric and philosophy that he had begun in Caen. He immediately applied his talents to mathematics at the scholarly institution called the Académie Royale, which was being enlivened by the Jesuits. His short treatise *Elementa astronomica* (1643), intended as a primer on astronomy, testifies to his ability. Having already taken minor orders, he was admitted to the Institution de l'Oratoire in Paris on Christmas day of 1643. In September 1644 he was sent to the Collège Université of Angers, where he taught philosophy with great success and where he was ordained a priest in 1649. In October 1652 he was recalled to the house in the rue St. Honoré to instruct the young Oratorians in positive theology. But at the request of his lawyer brother, who sought his aid in ecclesiastical matters, he left the Oratory in 1653. He became a curé in Neuilly-sur-Marne until 1663. During these ten years he was both a zealous pastor and an industrious intellectual. The works that he published in 1660 and 1663 assure his reputation and reflect perfectly his scholarly personality.

Directed to a lay audience, these works outlined the then current state of physics and of philosophical disputes. Their originality lies in the effort to emphasize what is valuable in the ancients for the moderns, in an interesting compilation of knowledge in the era following the death of Descartes.

Appointed royal chaplain in 1656, du Hamel relinquished the vicariate of Neuilly in 1663 to assist

the bishop of Bayeux in Paris as chancellor. In 1666 the founding of the Académie Royale des Sciences brought him another office. As Fontenelle said, he had given, without intending to, evidence of all the qualities necessary in a secretary of the new organization. The choice proved to be judicious.

From 1668 to 1670 he attended the marquis of Croissy at the negotiations of Aix-la-Chapelle, at which an expert Latinist was required, and then accompanied him to England and the Netherlands on diplomatic missions. Du Hamel resumed his position at the academy, enriched by his contact with foreign scholars, notably Boyle and Oldenburg.

His zeal in the service of secular knowledge was soon tempered, however. He dedicated himself increasingly to important publications for the sacred sciences. Without doubting the sincerity of the scruples that he expressed with regard to his sacerdotal state, one may suppose that with time, and for a variety of reasons, his responsibilities at the Academy became more trying. The Academy had not maintained its original distinction and in the last decade of the seventeenth century was afflicted with various ills that threatened to hasten its demise. Surrounded by the debates of an advancing science and suffering from faults in its administrative organization, the institution was in need of reform as well as protection. By securing his position for as long as possible and by passing it on to Fontenelle in 1697, du Hamel certainly assisted in preserving the Academy, at the same time that he published the first printed summary of its history.

A pensionary anatomist in the revived Academy of 1699, du Hamel once again saw how to make way for a qualified member, Littré Alenis. In the larger interest of science he permitted Varignon, in 1701, to assume the chair of Greek and Latin philosophy, which he had held since 1682, at the Collège de France.

A man of the Church who had a deep inner life, du Hamel put the advantages of his position and background in the service of the broadest scientific progress.

BIBLIOGRAPHY

I. ORIGINAL WORKS. The following list includes only those works concerning science and its history: *Elementa astronomica ubi Theodosii Tripolitae sphaericorum libri tres cum universa triangulorum resolutione nova succincta et facillime arte demonstrantur* (Paris, 1643), some eds. with the commentary on Euclid by P. Georges Fournier, S.J., one of which eds. was repr. (London, 1654); *Astronomia physica, seu de luce, natura et motibus corporum caelestium, libri duo . . . Accessere P.Petiti observationes aliquot eclypsium solis et lunae* (Paris, 1660); *De meteoribus et fossilibus libri duo* (Paris, 1660); *De consensu veteris et novae philosophiae ubi Platonis, Aristotelis, Epicuri, Cartesii aliorumque placita de principiis rerum excutiuntur et de principiis chymicis* (Paris, 1663); *De corporum affectionibus tum manifestis tum occultis libri duo* (Paris, 1670); *De mente humana libri quatuor* (Paris, 1672); *De corpore animato libri quatuor* (Paris, 1673); and *Regiae scientiarum academiae historia* (Paris, 1698), 2nd augmented ed. (Paris, 1701).

II. SECONDARY LITERATURE. On du Hamel and his work, see "Mémoire sur la vie et les écrits de J. B. du Hamel, prieur de St. Lambert," in *Journal des sçavans,* supp. (Feb. 1707), pp. 88–94 (author anon.); Fontenelle, "Éloge de Mr. du Hamel," in *Histoire et mémoires de l'Académie royale des sciences,* pt. 1 (Paris, 1707), pp. 142–153; Louis Batterel, *Mémoires domestiques pour servir à l'histoire de l'Oratoire,* Bonnardet-Ingold, ed., III (Paris, 1904), 142–155; and a medallion in *Museum Mazzuchelianum seu numismata virorum doctrina praestantium,* II (Venice, 1761), 89 and pl. 120, no. 4.

PIERRE COSTABEL

DUHAMEL, JEAN-MARIE-CONSTANT (*b.* St.-Malo, France, 5 February 1797; *d.* Paris, France, 29 April 1872), *mathematics, physics.*

Duhamel's scientific contributions were minor but numerous and pertinent; his teaching and involvement in academic administration were probably the primary sources of his influence. Today his name is best remembered for Duhamel's principle in partial differential equations;[1] Duhamel obtained this theorem in the context of his work on the mathematical theory of heat. Using the techniques of mathematical physics he also studied topics of acoustics. In his interests and approaches Duhamel was clearly in the tradition of the French *géomètres.*

Duhamel entered the École Polytechnique in Paris in 1814, after studying at the lycée of Rennes. The political events of 1816, which caused a reorganization of the school, obliged him to return to Rennes, where he studied law. He then taught in Paris at the Institution Massin and the Collège Louis-le-Grand (probably mathematics and physics) and founded a preparatory school, later known as the École Sainte-Barbe. The subject of his first memoir, presented in 1828,[2] indicates that by this time he was quite involved in current problems of mathematical physics.

Except for one year, Duhamel taught continuously at the École Polytechnique from 1830 to 1869. He was first given provisional charge of the analysis course, replacing Coriolis. He was made assistant lecturer in geodesy in 1831, entrance examiner in 1835, professor of analysis and mechanics in 1836, permanent examiner in 1840, and director of studies in 1844. The

commission of 1850 demanded his removal because he resisted a program for change, but he returned as professor of analysis in 1851, replacing Liouville. Duhamel also taught at the École Normale Supérieure and at the Sorbonne. He was known as a good teacher, and students commented especially on his ability to clarify the concept of the infinitesimal, a topic also emphasized in his text.[3] Duhamel's most famous student was his nephew by marriage, J. L. F. Bertrand.

Duhamel's earliest research dealt with the mathematical theory of heat and was based on the work of Fourier and Poisson. It was the subject of the theses, accepted in 1834,[4] that he submitted to the Faculty of Sciences. His first memoir treated heat propagation in solids of nonisotropic conductivity, and the laws that he obtained were later verified experimentally by Henri de Sénarmont.[5] In 1833 Duhamel published a solution for the temperature distribution in a solid with variable boundary temperature. He was considering the situations in which the surface radiates into a medium and in which the temperature of the medium changes according to a known law. His object was to reduce these cases to those of a surface at constant temperature. His method, based on the principle of superposition, generalized a solution by Fourier and substituted, in place of the original temperature function, the sum of a constant temperature term and an integral term (an integral of the rate of change of the temperature function).[6] This method generalizes to Duhamel's principle.

In acoustics Duhamel studied the vibrations of strings, the vibrations of air in cylindrical and conical pipes, and harmonic overtones. For an experimental check on his analysis of a weighted string Duhamel used a novel method whereby a pointer attached to the string leaves a track on a moving plane. His study of the excitation of vibration by the violin bow, based on Poisson's *Mécanique,* used the expression for friction force that had been experimentally determined by Coulomb and Morin. Duhamel was intrigued by harmonic overtones and suggested, independently of Ohm,[7] that one perceives a complex sound as the group of simultaneous sounds into which its vibrations can be decomposed.[8]

NOTES

1. For example, R. Courant and D. Hilbert, *Methods of Mathematical Physics,* II (New York, 1966), 202–204.
2. Published in 1832.
3. Commenting on the disagreement among mathematicians over definitions of the differential, Duhamel pointed out in an introductory note to the second edition of his *Cours d'analyse* (Paris, 1847) that he had changed his own approach. Instead of considering the differential as an infinitely small addition to the

variable, as he had done in the first edition, he was now considering differentials as quantities whose ratios in the limit are the same as the ratios of the variables.

4. *Théorie mathématique de la chaleur* (Paris, 1834) and *De l'influence du double mouvement des planètes sur les températures de leur différents points* (Paris, 1834).

5. "Sur la conductibilité des corps cristallisés pour la chaleur," in *Annales de chimie*, **21** (1847), 457–476; "Second mémoire sur la conductibilité des corps cristallisés pour la chaleur," in *Comptes rendus hebdomadaires des séances de l'Académie des sciences*, **25** (1847), 459–461, 707–710; *Annales de chimie*, **22** (1848), 179–211.

6. For a modern discussion of Duhamel's principle and its use in solving heat problems, see H. S. Carslaw and J. C. Jaeger, *Conduction of Heat in Solids*, 2nd ed. (Oxford, 1959), 30–32.

7. G. S. Ohm, "Ueber die Definition des Tons, nebst daran geknüpfter Theorie der Sirene und ähnlicher tonbildender Vorrichtungen," in *Annalen der Physik*, **59** (1843), 513–566.

8. *Comptes rendus*, **27** (1848), 463.

BIBLIOGRAPHY

I. ORIGINAL WORKS. A partial listing of Duhamel's textbooks includes *Cours d'analyse*, 2 vols. (Paris, 1840–1841; 2nd ed., 1847); *Cours de mécanique*, 2 vols. (Paris, 1845–1846; 2nd ed., 1853–1854); *Élements de calcul infinitésimal*, 2 vols. (Paris, 1856; 2nd ed., 1860–1861); *Des méthodes dans les sciences de raisonnement*, 3 vols. (Paris, 1865–1873).

His articles include "Sur les équations générales de la propagation de la chaleur dans les corps solides dont la conductibilité n'est pas la même dans tous les sens," in *Journal de l'École polytechnique*, **13** (1832), *cahier* 21, 356–399; "Sur la méthode générale relative au mouvement de la chaleur dans les corps solides plongés dans des milieux dont la température varie avec le temps," in *Journal de l'École polytechnique*, **14** (1833), *cahier* 22, 20–77; "De l'action de l'archet sur les cordes," in *Comptes rendus hebdomadaires des séances de l'Académie des sciences*, **9** (1839), 567–569; **10** (1840), 855–861; *Mémoires présentés par divers savants à l'Académie des sciences*, **8** (1843) 131–162; "Mémoire sur les vibrations des gaz dans les tuyaux de diverses formes," in *Comptes rendus hebdomadaires des séances de l'Académie des sciences*, **8** (1839), 542–543; *Journal de mathématiques pures et appliquées*, **14** (1849), 49–110; "Sur les vibrations d'une corde flexible chargée d'un ou de plusieurs curseurs," in *Journal de l'École polytechnique*, **17** (1843), *cahier* 29, 1–36; "Sur la résonnance multiple des corps," in *Comptes rendus hebdomadaires des séances de l'Académie des sciences*, **27** (1848), 457–463; and "Sur la propagation de la chaleur dans les cristaux," in *Journal de l'École polytechnique*, **19** (1848), *cahier* 32, 155–188.

For an extended list of Duhamel's papers, see the *Royal Society Catalogue of Scientific Papers, 1800–1863*, II (1868), 376–377; and *1864–1873*, VII (1877), 569.

II. SECONDARY LITERATURE. Not much material is available on Duhamel. For short accounts, see Louis Figuier, *L'année scientifique et industrielle* (*1872*), XVI (Paris, 1873), 537–540; M. Maximilien Marie, *Histoire des sciences mathématiques et physiques*, XII (Paris, 1888), 220–224; and E. Sarrau, "Duhamel," in *Ecole polytechnique, livre du centenaire 1794–1894*, I (Paris, 1894–1897), 126–130.

SIGALIA DOSTROVSKY

DUHAMEL DU MONCEAU, HENRI-LOUIS (*b.* Paris, France, 1700; *d.* Paris, 23 August 1782), *agronomy, chemistry, botany, naval technology.*

The son of Alexandre Duhamel (the name can be found variously listed as Hamel, du Hamel, or Monceau), *seigneur* of the estate of Denainvilliers in Gâtinais, Duhamel was a wealthy but minor member of the French landed gentry. Although the stories about Duhamel's problems at the Collège d'Harcourt have been rather exaggerated for effect, there is little doubt that he began to apply himself seriously only after hearing science lectures at the Jardin du Roi in the 1720's. During this time he became acquainted with the younger group of French scientists, such as the chemist Charles-François Dufay and the botanist Bernard de Jussieu, as well as with such established members of the scientific community as Louis Lémery and Étienne-François Geoffroy.

Duhamel first achieved scientific recognition with his explanation of the cause of a blight which attacked the saffron plant with particular ferocity in the 1720's. This study, unusual from its inception because it was given by the Academy to a nonmember, showed that the disease was caused by a plant parasite which spread underground from one saffron bulb to another. The work, read to the Academy in April 1728, was well-conceived, thorough, and conclusive, and led to his election as *adjoint chimiste* in the same year.

The choice of position was partly due to a lack of openings in botany, but it was not unreasonable in absolute terms since Duhamel's breadth of interests was somewhat surprising even for an eighteenth-century polymath. Moreover, although his early researches for the Academy were devoted to botanical subjects, he undertook a series of chemical investigations in the early 1730's in collaboration with the chemist Jean Grosse. These included an attempt to make tartar soluble and the well-known study of Frobenius' ether, which, if one can accept Duhamel's testimony, was carried out largely by Grosse. In the mid-1730's Duhamel investigated contrasting claims for the synthesis of sal ammoniac and examined the nature of the purple dye commonly obtained from shellfish. His most important work in chemistry during this period was "Sur la base du sel marin," read in January 1737, in which he argued that there were different fixed alkaline components of salts and that these were essentially soda and potash. Although challenged by the German chemist J. H. Pott, Duhamel made this important idea credible. He occasionally undertook chemical projects in later years, but other interests, primarily botany and agronomy, occupied most of his remaining working life. One of these interests was in the cultivation and use of timber. With his primary interest in plants, Duhamel was made *associé botaniste* of the Academy in 1730; and in 1732 he was appointed *inspecteur général de la marine,* with the understanding that his duties would include supervising the timber to be used in the French navy. His experiences in this position led to several studies published in the 1740's on the structural properties of wood and on management of tree stands, and to his first book (1747), a treatise on the rigging of ships.

But Duhamel's major interest and contribution to technology and society was in agriculture. The first half of the eighteenth century had witnessed the beginning of a technological renaissance in agriculture, chiefly in England, where it was notably celebrated in the writings of Jethro Tull, whose major work, *Horse-Hoeing Husbandry,* was published in 1733. As the work of the progressive English landed gentry began to bear fruit, traveling French savants were quick to publish critical comparisons of French and English practices. In 1750, stimulated by a trip to England and by wide reading in agronomy, Duhamel published the first volume of his *Traité de la culture des terres.* . . . This was an exposition, rather than a translation, of Tull's writings. Moreover, he adapted Tull's system to France based on his own wide reading in French agronomy and on original experiments. Although a supporter and admirer of Tull's system, he was not a slavish disciple: not only was he critical of Tull's experiments and ideas, but he refused to accept one of Tull's central principles, a doctrinaire rejection of the use of manures. Later volumes of the *Traité* were devoted to clarifications of and additions to the original work and, most important, to case histories—most of which were gleaned from Duhamel's extensive correspondence—of successful applications of the *nouveau système* to support its adherents.

Although the ideas of Tull and Duhamel enjoyed substantial popularity among a progressive group of French landowners, the opposition—either in the form of active criticism or in the passive inertia of an almost medieval agrarian society—was too strong for France to enjoy the rapid agricultural changes which occurred in England and, shortly after, in Scotland. However, enough progress was made for Duhamel to receive recognition for his pioneering work during his lifetime.

Duhamel never married and, according to one biographer, never planned or desired to do so. He divided his time between Paris and Denainvilliers, where his brother carried out many of the agricultural projects which Duhamel designed, and managed the family fortune that allowed Duhamel to pursue various experiments of his own. When his brother died,

Duhamel was looked after by a niece and particularly by his nephew and protégé, Fougeroux de Bondaroy. Straightforward in speech and thought, Duhamel consciously wrote for an audience of technicians rather than scientists. Although his later writings reflect the professed distaste for theory that Condorcet attributed to him, his papers as a whole were not as barren of interpretation as one might suppose from his biographers. Indeed, that his works include ideas as well as simple techniques is amply attested by those who disagreed with his papers and treatises, as well as by the honors he received from more than a dozen learned societies.

BIBLIOGRAPHY

Duhamel had the habit of writing supplements to his books, sometimes years after the original publication. As the supplements were book-length themselves, they were frequently issued as "Vol. II" of the original and sometimes there were minor changes in the titles. The *Traité . . . des terres* thus grew from a one-volume work to a six-volume work and generally bears the dates 1753 (the year of publication of the second supplement as part of a set that included the original and first supplement) and 1761 (the year of the final supplement). I have, however, given original publication dates and not the dates of reissues; that is, all dates in the text and below are those years in which Duhamel's works were first made available to the public.

I. ORIGINAL WORKS. Duhamel is credited with over a hundred entries in the *Histoires de l'Académie royale des sciences, avec des mémoires de mathématique et de physique.* Although he was certainly prolific, this figure is somewhat misleading. More than a third of the *mémoires* under his name are his "Observations botanico-météorologiques," an intellectually routine annual series (1740–1780) of recorded daily weather conditions, including temperature and barometric pressure, with additional data on crops, floods, and plant growth in general. But even subtracting *histoires,* reviews of his books, and the "Observations," Duhamel contributed some fifty-five papers to the French Academy. In addition, he wrote a score of articles for the *Descriptions des arts et métiers* (1760–1775). A few contributions to other journals are also known.

His impressive output of separate works includes *Traité de la fabrique des manoeuvres pour les vaisseaux ou l'art de la corderie perfectionné,* 2 vols. (Paris, 1747–1769); *Traité de la culture des terres suivant les principes de M. Tull,* 6 vols. (Paris, 1750–1761); *Eléments de l'architecture navale* (Paris, 1752); *Avis pour le tránsport par mer des arbres, des plants vivaces, des semences, et de diverses autres curiosités d'histoire naturelle* (Paris, 1753); *Traité de la conservation des grains et en particulier du froment* (Paris, 1753); *Traité des arbres et arbustes qui se cultivent en France en pleine terre,* 2 vols. (Paris, 1755); *Mémoires sur la garance et sa culture* (Paris, 1757); *La physique des arbres,* 2 vols. (Paris, 1758); *Moyens de conserver la santé aux équipages des vaisseaux* (Paris, 1759); *Des semis et plantations des arbres et de leur culture* (Paris, 1760); *Éléments d'agriculture*, 2 vols. (Paris, 1762); *Histoire d'un insecte qui dévore les grains dans l'Angoumois* (Paris, 1762); *De l'exploitation des bois*, 2 vols. (Paris, 1764); *Du transport, de la conservation et de la force des bois* (Paris, 1767); *Traité des arbres fruitiers*, 2 vols. (Paris, 1768); *Traité général des peches, et histoire des poissons qu'elles fournissent,* 3 vols. (Paris, 1769–1777). There are various translations in English, German, Spanish, and Italian.

II. Secondary Literature. Most biographies rely heavily on the *éloge* by Condorcet in the *Histoire de l'Académie royale des sciences . . . 1782* (1785), 131–155.

Other material may be found in *Biographie universelle, ancienne et moderne,* XII (Paris, 1814), 185–190; *Dictionnaire de biographie française* (Paris, 1968), p. 22; *Dictionnaire historique de la médecine, ancienne et moderne,* II, pt. 1 (Paris, 1834), 147–149; *Dictionnaire des sciences médicales,* III (Paris, 1821), 538–541; *Nouvelle biographie générale,* XV (Paris, 1868), 106–107.

Jon Eklund

DUHEM, PIERRE-MAURICE-MARIE (*b.* Paris, France, 10 June 1861; *d.* Cabrespine, France, 14 September 1916), *physics, rational mechanics, physical chemistry, history of science, philosophy of science.*

Duhem was that rare, not to say unique, scientist whose contributions to the philosophy of science, the historiography of science, and science itself (in thermodynamics, hydrodynamics, elasticity, and physical chemistry) were of profound importance on a fully professional level in all three disciplines. Much of the purely scientific work was forgotten until recently. His apparent versatility was animated by a single-mindedness about the nature of scientific theories that was compatible with a rigidly ultra-Catholic point of view, an outlook unusual among historians, philosophers, or practitioners of science—Cauchy is the only other example that comes to mind.

Duhem's historical work, the major part of which traces the development of cosmology from antiquity to the Renaissance, was meant partly to redeem the centuries of Scholasticism, the great age for his church, from the reputation of scientific nullity, but mainly to exemplify the central epistemological position of his philosophy. This assigned to scientific theories the role of economizers of experimental laws which approach asymptotically some sort of reality, rather than that of models of reality itself or bearers of truth. Thus would the truth be independent of science and reserved for theology. This position coincided in important, although not all, respects with that of contemporary positivists, who came to it from the other extreme ideologically and without concern for defending theology.

Among the areas of agreement between Duhem,

Ernst Mach, and Wilhelm Ostwald was a common predilection for the energeticist over the mechanistic position in physics itself, involving skepticism about the reality of known physical entities, although he differed from them in allowing the existence of real entities in principle, however unknowable. A similar skeptical view was held by Henri Poincaré.

Duhem's father, Pierre-Joseph, was a commercial traveler from Roubaix in the industrialized north of France. His mother, born Alexandrine Fabre, was of a bourgeois family originally from Cabrespine, a town in Languedoc, near Carcassonne. They settled in Paris and sent Duhem, the eldest of their four children, to the Collège Stanislas from his eleventh year. He was a brilliant student and there acquired the firm grasp of Latin and Greek that he would need in his historical scholarship, while being attracted primarily to scientific studies and especially thermodynamics by a gifted teacher, Jules Moutier. His father hoped that for his higher education he would enter the École Polytechnique, where the training and tradition assured most graduates eminent technical careers in the service of the state. His mother, on the other hand, fearful that science or engineering would diminish his religious faith, urged him to study humanities at the École Normale Supérieure. Having placed first in the entrance examinations, he chose the middle ground of science at the École Normale, indicating his desire for an academic career. He published his first paper, on the application of the thermodynamic potential to electrochemical cells, in 1884, while still a student.

He proceeded with distinction through the *licence* and *agrégation,* after meeting a setback with a thesis for the doctorate that he presented in 1884 (prior to receiving the *licence,* an uncommon event). The subject concerned the concept of thermodynamic potential in chemistry and physics, and the argument included an attack on Marcellin Berthelot's twenty-year-old principle of maximum work, whereby the heat of reaction defines the criterion for the spontaneity of chemical reactions. This principle is false. Duhem, following J. W. Gibbs and Hermann von Helmholtz, properly defined the criterion in terms of free energy. Berthelot was extremely influential, resented the neophyte challenge, and was able to get the thesis refused. At risk to his career, Duhem later published the thesis as a book, *Le potentiel thermodynamique* (1886). Duhem was placed under the necessity of preparing another subject for the doctorate. He received the degree in 1888 for a thesis on the theory of magnetism, this one falling within the area of mathematics.

Unfortunately the enmity between Berthelot and Duhem was not dissipated until after 1900. Moreover, Duhem was of a contentious and acrimonious disposition, with a talent for making personal enemies over scientific matters. He blamed Berthelot, who was minister of education from 1886 to 1887, together with the circle of liberal and free-thinking scientists who advised successive ministers, for preventing him from ever receiving the expected call to a professorship in Paris. Aside from the hearsay evidence of anecdotes from the personalities involved, it must be admitted that there is no other instance in modern French history of a scientist of equivalent productivity, depth, and originality remaining relegated to the provinces throughout his entire postdoctoral career. Duhem taught at Lille (1887–1893), Rennes (1893–1894), and Bordeaux (1894–1916). He spurned an offer of a professorship in the history of science at the Collège de France shortly before his death, on the grounds that he was a physicist and would not enter Paris by the back door of history. In 1900 he was elected to corresponding membership in the Academy of Sciences. In 1913 he was elected one of the first six nonresident members of the Academy, a recognition that, together with various honorary degrees and foreign academic memberships received earlier, mollified his feelings to some degree.

Duhem had few qualified students, but those he did have considered him an extraordinary teacher. His personal friendships were as warm as his professional enmities were bitter. In October 1890 while at Lille he married Adèle Chayet. She died only two years later while giving birth to their second daughter, who also died. Duhem made his home thereafter with the surviving daughter, Hélène. She saw to the publication of the final five volumes (1954–1959) of his historical masterpiece, Le système du monde, left in manuscript after his death. He died at fifty-five of a heart attack brought on by a walking expedition during vacation days at Cabrespine. His health had never been vigorous.

Duhem's interests fell roughly into periods. Thermodynamics and electromagnetism predominated between 1884 and 1900, although he returned to them in 1913–1916. He concentrated on hydrodynamics from 1900 to 1906. His interest in the philosophy of science was mostly in the period 1892–1906, and in the history of science from 1904 to 1916, although his earliest historical papers date from 1895. The extraordinary volume of Duhem's production is impressive—nearly 400 papers and some twenty-two books. Among them, certain wartime writings (La science allemande and La chimie est-elle une science française?) express, as do his philosophical judgments of the style of British science, a certain chauvinism that remains the only unattractive characteristic of his nonscientific writings. It will be best to consider his most important work in the order of philosophy, history, and physics; to do so will reverse its chronology but will respect its intellectual structure.

Philosophy of Science. Duhem published his major philosophical work, La théorie physique, son objet et sa structure, in 1906, after having largely completed his researches in physical science. "A physical theory," he held there, ". . . is a system of mathematical propositions, deduced from a small number of principles, which has the object of representing a set of experimental laws as simply, as completely, and as exactly as possible." In adopting this position, he was explicitly rejecting what he considered to be the two alternatives to which any serious existing or previous account might be reduced.

According to epistemologies of the first sort, proper physical theories have the aim of accounting for observed phenomena by proposing hypotheses about, and preferably by actually revealing, the nature of the ultimate entities underlying the phenomena in question. Duhem rejected this view as illusory because experience showed that acting upon it had had the effect historically of subordinating theoretical physics to metaphysics, thereby encumbering and distracting it with all the difficulties and disputes afflicting that subject. He allowed that physicists may appropriately hope to form theories of which the structure "reflects" reality. It may be thought of such theories that their mode of interrelating empirical laws somehow fits the way in which the real events that give rise to the observations are interrelated. This hope can be based only on faith, however. There is and can be no evidence to support it.

Little in Duhem's philosophical writings clarifies the idea of such a fit, beyond the notion that the evolution of physical theories caused by successive adjustments to conform to experiment should lead asymptotically to a "natural classification" which somehow reflects reality. But his historical writings allude to numerous examples of what he had in mind, and his Notice (1913) indicates that they were in part originally motivated by it. It is no doubt for this reason that, despite his enthusiastic discovery of Scholastic mechanics in the Middle Ages, his favorite philosopher of antiquity was Plato, to whom he attributed the origin of the view (clearly akin to his own) that the healthy role of astronomical or other mathematical theory is to "save the phenomena." At the same time he had great faith in the syllogism as a logical instrument. He believed that mathematical reasoning could in principle be replaced with syllogistic reasoning, and he went so far as to reject Poincaré's argument that mathematical induction involves nonsyllogistic elements.[1]

The second category of philosophies or methodologies of physics that Duhem found unacceptable were those in which theories were expected to provide models in the form of mechanical analogies or constructs that permit visualizing the phenomena and offer handles for thought. He rejected this alternative partly on utilitarian and partly on aesthetic grounds. He felt that physical theories should have practical value, and he preferred the analytic to the geometric mode in mathematical thinking. Theories of the kind he advocated permit deducing many laws from a few principles and thus dispense the physicist from the necessity of trying to remember all the laws. Duhem evidently considered reason a higher faculty than memory. Complex models are distracting to people who can reason but cannot remember a mass of concrete detail. They are not, he believed, likely to lead to discovery of new laws. Merely artificial constructs, they never attain to the status of natural classifications. Duhem was highly critical of British physics for its reliance upon the use of just such mechanistic models. In his view this national habit resulted from a defect of cultural temperament. He described the British mind in science as wide and shallow, the French as narrow and deep. As will appear in the discussion of his electrodynamics, Maxwell was his bête noire in this respect. It must be acknowledged that a certain rigidity in his opinions accorded ill with the subtle nature of his philosophy.

Duhem's philosophy was certainly empiricist but never naïvely so. He showed very beautifully that there can be no such thing as simply observing and reporting an experiment. The phenomenon observed must be construed—must be seen—in the light of some theory and must be described in the terms of that theory. Laws arrived at experimentally must be expressed by means of abstract concepts that allow them to be formulated mathematically and incorporated in a theory. At their best they can merely approximate experimental observations. It is quite impossible to test or verify the fundamental hypotheses of a theory one by one. Thus there cannot be a crucial experiment, and induction from laws can never determine a unique set of hypotheses. Thus data and logic leave much to the discretion of the theorist. He must supplement their resources with good sense and historical perspective on his problems and his science.

It is an aspect of Duhem's recognition of the role of taste in scientific research that he never insisted that his philosophy require the adoption of an energeticist, to the exclusion of a mechanistic, point of view. That was an empirical, not a philosophical, issue. What his philosophy purported to establish was that an energeticist approach was no less legitimate

than a mechanistic one. The discussion explains how theories are to be judged and looked at merely in point of preference or policy; and in the absence of concrete facts, either type of theory would in principle be acceptable, so long as no metaphysical import be loaded into the choice. The issue was one that Duhem discussed in *L'évolution de la mécanique* (1902) and also in the essay "Physique de croyant," included in later editions of *La théorie physique.*

History of Science. Like Ernst Mach, his contemporary in the positivist school, Duhem relied heavily on historical examples in presenting his philosophy of science. *L'évolution de la mécanique* may be compared to Mach's famous *Die Mechanik in ihrer Entwicklung, historisch-kritisch dargestellt* (1883) as a philosophical critique of a science based upon its history, although Duhem was by far the more faithful to the original texts and the intentions of their authors. A history of the concept of chemical combination appeared in 1902 and a two-volume study of early statics in 1905–1906.

The object of historical examples was to attempt to see the trend toward the "natural classification," which requires the examination of preceding theories. Duhem was primarily led into his historical studies by following such theories backwards. Thus he always claimed that his conception of physical theory was justified by the history of physics, not because it corresponded to views shared by all, or most, or even (as Mach had tended to imply of his own position) by the best physicists, but because it did yield an analysis of the nature of the evolution of physics and of the dialectic responsible for that process.

The most impressive monument to the scholarly fertility of that claim remains his massive contribution to the knowledge of medieval science in his three-volume *Études sur Léonard de Vinci* (1906–1913) and the ten-volume *Système du monde* (1913–1959). These works contain a detailed exposition of two theses: (1) a creative and unbroken tradition of physics, cosmology, and natural philosophy was carried on in the Latin West from about 1200 to the Renaissance, and (2) the results of this medieval activity were known to Leonardo da Vinci and Galileo, and played a seminal role in the latter's transformation of physics. Duhem was led to his theses, and to the almost single-handed discovery of this medieval activity, by recognizing in Leonardo's notebooks statements by earlier writers and references to works fortunately available in manuscript in the Bibliothèque Nationale. Pursuing these citations and references still further he found wholly unsuspected "schools of science." He emphasized the significance of Paris: particularly important was a series of Parisian masters who were relatively unknown before Duhem's researches—Jordanus de Nemore, Jean Buridan, Francis of Méyronnes, Albert of Saxony, and Nicole Oresme. Duhem also brought out of obscurity the contributions of Mersenne and Malebranche. Expressed in dramatic form and supported by extensive quotation from the original texts (particularly in *Le système du monde*), Duhem's discoveries revolutionized, if they did not completely create, the study of medieval physics. While it is true that recent studies have seriously modified and qualified some of his conclusions, Duhem's studies remain the indisputable starting point for the study of medieval natural philosophy.[2]

Scientific Thought and Work. It must be recalled that Duhem's scientific formation took place in the period 1880–1890, well before the discovery of radioactivity and the experiments of Jean Perrin and, later, Henry G. F. Moseley. Discontent with the notion of reducing all physical concepts to classical mechanics or to mechanical models was growing. It was fed by the necessity to modify ad hoc the often contradictory properties of supposedly fundamental atomic or molecular particles in order to maintain the applicability of the model to newly determined phenomena, particularly in chemical dynamics and in the physics of heat and gases. Duhem early became convinced that rather than try to reduce all of physics and chemistry to classical mechanics, the wiser policy would be to see classical mechanics itself as a special case of a more general continuum theory. He believed that such underlying descriptive theory for all of physics and chemistry would emerge from a generalized thermodynamics. The central commitment of his scientific life was the building up of such a science, one that would include electricity and magnetism as well as mechanics. His attempts culminated in the *Traité d'énergétique* (1911), in which valuable work there is not a single word about atoms or molecules. Duhem always considered that it was his most important—and would prove to be his most lasting—contribution to science. He had not succeeded, however, in his goal of including electricity and magnetism in its purview.

His conception of the nature of physical theory had in fact influenced both the direction of his work and the form of his writings. His contemporaries (see Secondary Literature) often remarked that many of his papers opened with the barest of assumptions followed by a series of theorems. In his mode of posing "axioms," he gave little motivation, made hardly any appeal to experiment, and of course made no use whatever of atomic or molecular models. In his concern over extracting the logical consequences of a set of axioms for a portion of physics or chemis-

try, Duhem was a pioneer. Today a flourishing school of continuum mechanics follows a similar path, with a strong interest in foundations and in finding general theorems about more general fluids or elastic bodies with nonlinear constitutive equations or with fading memory. They often cite Duhem and his more famous predecessors such as Euler and Cauchy. However, because of the special hypotheses and restricted constitutive equations built into Duhem's thermodynamics from the beginning, modern workers no longer view his generalized thermodynamics as the best way to approach continuum mechanics.[3,4]

Duhem began his scientific work with the generalization and application of thermodynamics. While still at the Collège Stanislas and under Moutier's guidance, he had read G. Lemoine's description of J. W. Gibbs's work[5] and the first part of Hermann von Helmholtz' "Die Thermodynamik chemischer Vorgänge."[6] These papers emphasized the characteristic functions, closely related to those invented by F. J. D. Massieu,[7] now called the Gibbs and Helmholtz free energies—*G* and *A,* respectively. These functions play a role for thermodynamics directly analogous to the one played by the potential of classical mechanics. Duhem was one of the first to see real promise in this, calling Massieu's functions "thermodynamic potentials." Using this idea together with the principle of virtual work, he treated a number of topics in physics and chemistry.

Among the subjects treated systematically were thermoelectricity, pyroelectricity, capillarity and surface tension, mixtures of perfect gases, mixtures of liquids, heats of solution and dilution, saturated vapors, solutions in gravitational and magnetic fields, osmotic pressure, freezing points, dissociation, continuity between liquid and gas states, stability of equilibrium, and the generalization of Le Chatelier's principle. The Duhem-Margules equation was first obtained by Duhem in the course of this work. His success with these problems in the period 1884–1900 rank him with J. H. van't Hoff, Ostwald, Svante Arrhenius, and Henry Le Chatelier as one of the founders of modern physical chemistry.

Duhem's results are of course an extension and elaboration of the pioneer work of Gibbs and Helmholtz. But Duhem's elaboration, explanation, and application of their suggestions in his *Traité de mécanique chimique* (1897–1899) and *Thermodynamique et chimie* (1902) provided a whole generation of French physicists and chemists with their knowledge of chemical thermodynamics.

Duhem made a number of other contributions to thermodynamics. In the first part of his rejected thesis, *Le potentiel thermodynamique* (1886), Duhem presented or rederived by means of the thermodynamic potential a number of known results on vapor pressure of pure liquids and solutions, dissociation of gases and of heterogeneous systems, and the heat effects in voltaic cells. In the second and third parts he obtained new results on solubility and freezing points of complex salt solutions and on electrified systems. There is also the first application of Euler's homogeneous-function theorem to the extensive properties of solutions. This technique, now common, reduces the derivation of relations among the partial molal properties of a solution to the repeated application of this theorem. One of the equations so derived is the Gibbs-Duhem equation. Also included is a discussion of electrified systems which contains an expression equivalent to the electrochemical potential. This book, popular enough to be reprinted in 1896, is historically important for the systematic use of thermodynamic potentials, when others were still using osmotic pressure as a measure of chemical affinity and using artificial cycles to prove theorems.

Duhem was the first (1887) to publish a critical analysis[8] of Gibbs's "Equilibrium of Heterogeneous Substances."[9] In Duhem's paper is the first precise definition of a reversible process; earlier versions by others (unfortunately often preserved in today's textbooks) are too vague. Duhem emphasizes that the reversible process between two thermodynamic states A and B of a system is an unrealizable limiting process. The limit of the set of real processes for getting from A to B is obtained by letting the imbalance of forces between the system and the surroundings at each step tend toward zero. Each member of this set of real processes must pass through nonequilibrium states, or else nothing would happen. However, the limit of this set, where the forces balance at every step, is a set of equilibrium states. Since once the system is in equilibrium nothing can happen, this limit is thus in principle an unrealizable process. This limiting process is now called a "quasi-static" process. If a similar set of realizable processes for getting from B and A has the same (unrealizable) limit, then the common sequence of equilibrium states is defined by Duhem as a reversible process.

Duhem later pointed out in the "Commentaire aux principes de la thermodynamique" (pt. 2, 1893) that there exist situations such as hysteresis where the limiting set of equilibrium states for the direction AB is not the same as that for the direction BA. Therefore, it is possible to go from A to B and back by quasi-static processes, but not reversibly. This distinction was noted fifteen years before the celebrated paper of Carathéodory.[10]

Duhem believed that the "Commentaire" (1892–1894) was one of his more significant contributions.

It contains a very detailed analysis of the steps leading from the statement of the second law of thermodynamics to the definitions of entropy and thermodynamic potential. It also contains an axiomatic treatment of the first law of thermodynamics which is surprisingly good by present-day standards. (A different version is given in the *Traité d'énergétique* [1911].) The concepts of *oeuvre* (total energy including kinetic energy) and *travail* (work) are taken as undefined ideas. Axioms about *oeuvre* include independence of path, additivity along a path, commutativity, associativity, conservation, plus other matters often left implicit. Important to note is that the concept "quantity of heat" was not assumed but was defined in terms of energy and work. Consequently the definition, although more diffusely stated, was equivalent to and preceded that of C. Carathéodory (1909)[10] and Max Born (1921),[11] and should be called Duhem's definition. Duhem's axiomatic outlook which characterized this discussion of the first law was indeed pioneering for physics and to some extent anticipated the major axiomatic research in mathematics. Thus, although the axiomatization of arithmetic began in the first half of the nineteenth century, the research for axiomatic foundations for other branches of mathematics (Euclidean geometry, fields, groups, Boolean algebra) did not begin in earnest until 1897–1900.

In "Sur les déformations permanentes et l'hystérésis" (1896–1902), Duhem considered in some detail the thermodynamics of nonreversible but quasi-static processes and some irreversible processes, including hysteresis and creep. The results were mostly qualitative, not entirely satisfactory, and of little influence. As of this writing there is no really adequate thermodynamic theory of such systems, although interest in this subject has recently been revived.

Duhem provided the first explicit unrestricted proof of the Gibbs phase rule, based on Gibbs's suggestions, in "On the General Problem of Chemical Statics" (1898). At the same time he extended it beyond the consideration of just the intensive variables, giving the conditions necessary to specify the masses of the phases as well. The conditions are different for the pairs of variables pressure–temperature and volume–temperature, and their statement is called Duhem's theorem.[12] In addition the properties of "indifferent" systems, of which azeotropes are a simple special case, were discussed in some detail.

Duhem attached great importance to his thermodynamics of false equilibrium and friction.[13] According to Duhem, false equilibria can be divided into two classes: *apparent,* as for example a supersaturated solution, which, as a result of a small perturbation, returns instantly to thermodynamic equilibrium; and *real,* as for example organic compounds, such as diamond or petroleum constituents. Such compounds are unstable thermodynamically with respect to other substances but have remained unchanged for large perturbations throughout geological periods of time. Yet they will transform into the stable products if the perturbations are large enough (diamond to graphite by heating). A similar view was held by Gibbs (his passive resistances). The false equilibrium viewpoint was very useful to E. Jouguet, a major contributor to explosives theory and one of Duhem's disciples.[14] However, real false equilibria can also be considered as instances of extremely slow reaction rates. A violent polemic over this issue took place between 1896 and 1910. Most, but by no means all, of those interested in such questions today prefer the infinitely slow reaction rate view. Since the results are the same from either view, the choice is a personal one.

A major portion of Duhem's interest was focused on hydrodynamics and elasticity. His second book, *Hydrodynamique, élasticité, acoustique* (1891), had an important influence on mathematicians and physicists because it called attention to Hugoniot's work on waves. Jacques Hadamard, a colleague for one year and lifelong friend, remarked that this book and later conversations with Duhem led him into a major portion of his own work in wave propagation, Huygens' principle, calculus of variations, and hyperbolic differential equations. Duhem was both a pioneer and almost alone for years in trying to prove rigorous general theorems for Navier-Stokes fluids and for finite elasticity in Kelvin-Kirchhoff-Neumann bodies. His results are important and of sufficient interest later that his *Recherches sur l'hydrodynamique* (1903–1904) was reprinted in 1961.

In hydrodynamics Duhem was the first to study wave propagation in viscous, compressible, heat-conducting fluids using stability conditions and the full resources of thermodynamics (*Recherches sur l'hydrodynamique*). He showed the then startling result that no true shock waves (i.e., discontinuities of density and velocity) or higher order discontinuities can be propagated through a viscous fluid. This is contrary to the result for rigorously nonviscous fluids. The only discontinuities that can persist are transversal, which always separate the same particles; these Duhem identified with the "cells," observed by Bénard, formed when a liquid is heated from below. Since real fluids are both viscous and heat conducting, how is it possible to have sound waves propagated, as in air? Duhem's answer was that while true waves are not possible, "quasi waves" are. A quasi wave is a thin layer whose properties, including velocity, change smoothly but rapidly. If we consider a series of similar fluids whose values of the heat conductivity k and viscosity η approach zero, then the thickness of the associated quasi wave also approaches zero and the smooth change of properties approaches a discontinuity. When k and η are small, as in air, such quasi waves behave exactly as a true longitudinal shock wave in a perfect fluid with $k = \eta = 0$, i.e, propagating with the Laplace velocity. Duhem's concept and theory of the quasi wave is more general and more precise than the later ideas of Prandtl (1906) about the "shock layer." Some of Duhem's theorems on shock waves have been improved recently. For perspective, it should also be noted that Duhem considered only the then universally accepted Navier-Stokes fluid. There are more general concepts of a fluid with viscosity which do allow wave propagation.[4]

Duhem generalized and completed earlier results on the stability of floating bodies (including those containing a liquid). He showed that while some earlier methods were incorrect, certain results (in particular the famous rule of metacenters) were still correct. Finally, the article "Potentiel thermodynamique et pression hydrostatique" (1893) contains, but does not develop, the idea of an oriented body that consists not only of points but of directions associated with the points. Such an oriented body can represent liquid crystals or materials whose molecules have internal structure. Eugène and François Cosserat adapted this idea to represent the twisting of rods and shells in one and two dimensions (1907–1909). This concept has also been useful for some recent theories of bodies with "dislocations."

In elasticity Duhem was again interested in rigorous general theorems (*Recherches sur l'élasticité* [1906]). He kept a correct finite elasticity alive and inspired other workers. He was the first to study waves in elastic, heat-conducting, viscous, finitely deformed systems. The results are similar to that for fluids; namely, in any finitely deformed viscous elastic system, whether crystalline or vitreous, no true waves can be propagated and the only possible discontinuities always separate the same particles (as in the Bénard problem). Quasi waves are expected in viscous solids, but Duhem did not carry his analysis that far. Duhem was also the first to study the relationships between waves in isothermal (heat-conducting) and adiabatic (nonconducting) finitely deformed systems without viscosity. Duhem was also interested in the general conditions for solids (vitreous or crystalline) to be stable. He had to choose special conditions of stress or strain, but he was able to prove some general theorems. All this was based on the then universally accepted Kelvin-Kirchhoff-

Neumann elastic body. At the present writing, more general concepts of elastic bodies are being considered.

After Gibbs, Duhem was among the few who were concerned about stability of thermodynamic systems. His techniques were a natural consequence of his interest in thermodynamic potentials. He was the first to consider solutions ("Dissolutions et mélanges" [1893]); and he often returned to stability questions ("Commentaire aux principes de la thermodynamique" [1894]; "On the General Problem of Chemical Statics" [1898]; *Recherches sur l'élasticité* [1906]; *Traité d'énergétique* [1911]). Because he tried to be more explicit and more general than Gibbs and because he often took a global point of view, he had to face more difficult problems than did Gibbs. He succeeded fairly well with sufficient conditions but was less successful with necessary ones. In his *Énergétique* he showed familiarity with Liapounoff's work, but his own previous results were based on more special hypotheses. As a result, there is some confusion in Duhem's results over what are the proper necessary and sufficient conditions for thermodynamic stability. Such questions have only recently been rigorously resolved.

Electricity and magnetism and his attempts to bring them into the framework of his *Énergétique* (which was not the same as the philosophical school of "energetics") were important to Duhem. If a system's currents are zero or constant, then its electrodynamic energy is zero or constant. In this case, the total energy divides neatly into internal and kinetic energies, and energetics can be successfully applied. Thus Duhem was able to treat pyroelectricity and piezoelectricity in a general way without needing the special hypotheses of F. Pockels and W. Voigt. However, if currents are not constant, then matters are much more complex, and the electrodynamic energy must be accounted for using some electromagnetic theory.

Although Duhem recognized J. Clerk Maxwell's ingenuity, he could not appreciate Maxwell's theory at its real value because of its contradictions and unrigorous development, its mistakes in sign, and its lack of experimental foundation. Duhem preferred an electromagnetic theory due to Helmholtz, since it could be logically derived from the classical experiments. This theory, which Duhem helped to elaborate—and improve—is more general than Maxwell's because it contains two additional arbitrary parameters. By an appropriate choice of values for these parameters, it can be shown that Maxwell's equations appear as special cases of Helmholtz' theory. In particular, if the Faraday-Mossotti hypothesis is adopted

(equivalent to one parameter being infinity), then transverse fluxes propagate with the velocity of light. This results in an electromagnetic theory of light and an explanation of Heinrich Hertz's experiments. If the other parameter (Helmholtz') is chosen to be zero, then no longitudinal fluxes can be propagated, which circumstance is in agreement with Maxwell's equations. Duhem, however, believed that there were experiments showing that such longitudinal fluxes exist and are also propagated at the velocity of light. He suggested (1902) that perhaps the recently discovered X rays might be identified with these longitudinal fluxes.

Duhem was a pitiless critic of Maxwell's theory, claiming that it not only lacked rigorous foundation but was not sufficiently general to explain the existence of permanent magnets (*Les théories électriques de J. Clerk Maxwell* [1902]). Similar reservations about lack of rigor were expressed by many Continental physicists (e.g., Poincaré), and Helmholtz worked out his own electromagnetic theory because of his dissatisfaction with Maxwell's approach. Duhem later admitted that not only had his criticisms not been accepted, they had not even been read or discussed; and of course Maxwell's theory has triumphed. However, both L. Roy[15] and A. O'Rahilly[16] have contended that the logical derivation of Maxwell's equations from a continuum viewpoint comes best through the Helmholtz-Duhem theory with the proper choice of constants.

The foregoing discussion covers an extraordinary output of purely scientific work. It is curious that until recently working scientists were almost completely unaware of these contributions, with the exception of the Gibbs-Duhem and Duhem-Margules equations, which have been well known to physical chemists. The reason for the neglect of Duhem's scientific work, the failure to call him to Paris, and the long delay in his election to the Academy—despite the high quality of his work and the foreign honors accorded him—are interesting and are summarized below. They involve aspects of Duhem's personality as well as differences between competing scientific schools of the period. (A more complete account of the antagonisms and suppression, interwoven with a biography, may be found in Miller, *Physics Today,* **19,** no. 12 [1966], 47–53.)

Duhem's contentious characteristics have already been noted. On the one hand, his extremely conservative religious and political views conflicted sharply with those of the freethinkers and liberals who then dominated French science. On the other hand, the polemical nature of his writings on such controversies as energetics vs. atomism, Maxwell's theory vs. Helmholtz', relativity, false equilibrium, and the maximum work principle made personal enemies of many of his scientific contemporaries. Their combined opposition blocked his career and resulted in partial suppression of his work or in its being taken over without citation.

In part, however, the neglect of his work is to be explained by the triumph of views that he bitterly opposed, such as atomic theories and Maxwell's theory. His objection to relativity derived from its "mutilation" of classical mechanics in order to leave unaltered Maxwell's theory and atomic theories of electrons.

With the crystal clarity of a half century of hindsight, it would seem that Duhem should not have opposed corpuscular models so strongly. Since he had based his whole philosophy on the deliberate avoidance of such aids and given the rigid nature of his personality, he could not change his views as the evidence mounted and the use of such models became more plausible. It is essential to recall, however, that Duhem was not alone in his objection to corpuscular models, Maxwell, and relativity. At the time he was in the company of many eminent scientists.

Pierre Duhem is a fascinating example of a brilliant scientist caught up in historical and personal circumstances that blocked his career and partially suppressed his scientific work. Right-wing, royalist, anti-Semitic, anti-Dreyfus, anti-Republican, and a religious extremist, he was exiled to the provinces and his scientific work was almost systematically ignored in France.

Nevertheless, Duhem's scientific ideas and outlook had a major influence on French physical chemistry and particularly on Hadamard, Jouguet, and the Cosserats. He was a pioneer in attempting to prove rigorous general theorems about thermodynamics, physical chemistry, Navier-Stokes fluids, finite elasticity, and wave propagation. His purely scientific investigations and results in these fields are important, useful, and significant today, although the ascendancy of atomic theories has diminished the relative importance of his contributions to science as a whole.

By midcentury Duhem's scientific work had been almost completely forgotten. Since then, his contributions have been rediscovered, and are being increasingly cited and given the recognition they deserve.[3,4,12] There has never been, of course, any question about the importance of his work in the philosophy and history of science. Since his contributions to any one of the fields of pure science, philosophy, or history would have done credit to one person, the ensemble from the pen of a single man marks

Duhem as one of the most powerful intellects of his period.

NOTES

1. "La nature du raisonnement mathématique," in *Revue de philosophie,* **21** (1912), 531–543.
2. M. Clagett, *Science of Mechanics in the Middle Ages* (Madison, Wis., 1959).
3. C. Truesdell and R. Toupin, "The Classical Field Theories," in S. Flügge, ed., *Encyclopedia of Physics,* III, pt. 1 (Berlin, 1960); C. Truesdell and W. Noll, "The Non-Linear Field Theories of Mechanics," *ibid.,* III, pt. 3 (Berlin, 1965).
4. B. D. Coleman, M. E. Gurtin, I. Herrara, and C. Truesdell, *Wave Propagation in Dissipative Materials* (New York, 1965).
5. G. Lemoine, *Études sur les équilibres chimiques* (Paris, 1882).
6. H. von Helmholtz, "Die Thermodynamik chemischer Vorgänge," in *Sitzungsberichte der Deutschen Akademie der Wissenschaften zu Berlin,* **1** (1882), 22–39.
7. F. J. D. Massieu, "Sur les fonctions caractéristiques des divers fluides," in *Comptes rendus hebdomadaires des séances de l'Académie des sciences,* **69** (1869), 858–864, 1057–1061.
8. P. Duhem, "Étude sur les travaux thermodynamiques de J. Willard Gibbs," in *Bulletin des sciences mathématiques,* 2nd ser., **11** (1887), 122–148, 159–176.
9. J. W. Gibbs, "On the Equilibrium of Heterogeneous Substances," in *Transactions of the Connecticut Academy of Arts and Sciences,* **3**, pt. 1 (1876), 108–248; **3**, pt. 2 (1878), 343–520.
10. C. Carathéodory, "Untersuchungen über die Grundlagen der Thermodynamik," in *Mathematische Annalen,* **67** (1909), 355–386.
11. M. Born, "Kritische Betrachtungen zur traditionellen Darstellung der Thermodynamik," in *Physikalische Zeitschrift,* **22** (1921), 218–224, 249–254, 282–286.
12. I. Prigogine and R. Defay, *Chemical Thermodynamics* (New York, 1954), ch. 13.
13. P. Duhem, "Théorie thermodynamique de la viscosité, du frottement, et des faux équilibres chimiques," in *Mémoires de la Société des sciences physiques et naturelles de Bordeaux,* 5th ser., **2** (1896), 1–208; *Thermodynamique et chimie* (Paris, 1902; 2nd ed., 1910).
14. E. Jouguet, *Mécanique des explosifs, étude de dynamique chimique* (Paris, 1917).
15. L. Roy, *L'électrodynamique des milieux isotropes en repos d'après Helmholtz et Duhem* (Paris, 1923).
16. A. O'Rahilly, *Electromagnetics* (London, 1938), ch. 5; repr. as *Electromagnetic Theory,* 2 vols. (New York, 1965).

BIBLIOGRAPHY

I. ORIGINAL WORKS. Duhem published twenty-two books in forty-five volumes, as well as nearly 400 articles and book reviews in scientific and philosophical journals. An extensive bibliography (although lacking some twenty-five articles and more than fifty book reviews) is given by O. Manville, in *Mémoires de la Société des sciences physiques et naturelles de Bordeaux,* 7th ser., **1,** pt. 2 (1927), 437–464.

Duhem's correspondence consists of letters to him from some 500 correspondents and is being copied with the permission of Duhem's daughter, Mlle. Hélène Pierre-Duhem. Copies will ultimately be deposited in the University of California, Berkeley, and University of California, San Diego, libraries. Few letters by Duhem survive. Little of the correspondence seems to have major scientific value, although there are a few interesting historical items.

Duhem's major scientific books are *Le potentiel thermodynamique et ses applications à la mécanique chimique et à la théorie des phénomènes électriques* (Paris, 1886); *Hydrodynamique, élasticité, acoustique,* 2 vols. (Paris, 1891); *Leçons sur l'électricité et le magnétisme,* 3 vols. (Paris, 1891–1892); *Traité élémentaire de la mécanique chimique,* 4 vols. (Paris, 1897–1899); *Les théories électriques de J. Clerk Maxwell: Étude historique et critique* (Paris, 1902); *Thermodynamique et chimie* (Paris, 1902; 2nd ed., 1910), English trans. by G. Burgess (New York, 1903); *Recherches sur l'hydrodynamique,* 2 vols. (Paris, 1903–1904; repr., 1961); *Recherches sur l'élasticité* (Paris, 1906); and *Traité d'énergétique,* 2 vols. (Paris, 1911).

His major historical books are *Le mixte et la combinaison chimique. Essai sur l'évolution d'une idée* (Paris, 1902); *L'évolution de la mécanique* (Paris, 1902); *Les origines de la statique,* 2 vols. (Paris, 1905–1906); *Études sur Léonard de Vinci, ceux qu'il a lus et ceux qui l'ont lu,* 3 vols. (Paris, 1906–1913); and *Le système du monde. Histoire des doctrines cosmologiques de Platon à Copernic,* 10 vols. (Paris, 1913–1959).

His philosophy of science is stated in *La théorie physique, son objet et sa structure* (Paris, 1906; 2nd ed., 1914; 3rd ed., 1933), German trans. by F. Adler, with foreword by E. Mach (Leipzig, 1908); English trans. by Philip P. Wiener as *The Aim and Structure of Physical Theory* (Princeton, 1954; repr. New York, 1963). See also "Notation atomique et hypothèse atomistique," in *Revue des questions scientifiques,* 2nd ser., **31** (1892), 391; *Les théories électriques de J. Clerk Maxwell* (Paris, 1902); "Analyse de l'ouvrage de Ernst Mach," in *Bulletin des sciences mathématiques,* 2nd ser., **27** (1903), 261; and Σωξειν τα φαινομενα (Paris, 1908).

Duhem's most important scientific papers include "Étude sur les travaux thermodynamiques de J. Willard Gibbs," in *Bulletin des sciences mathématiques,* 2nd ser., **11** (1887), 122, 159; "Commentaire aux principes de la thermodynamique," in *Journal de mathématiques pures et appliquées,* **8** (1892), 269; **9** (1893), 293; and **10** (1894), 207; "Le potentiel thermodynamique et la pression hydrostatique," in *Annales scientifiques de l'École normale supérieure,* **10** (1893), 183; "Dissolutions et mélanges," in *Travaux et mémoires des Facultés de Lille,* **3,** no. 11 (1893), no. 12 (1893), and no. 13 (1894); "Sur les déformations permanentes et l'hystérésis," in *Mémoires de l'Académie royale de Belgique. Classe des sciences,* **54,** nos. 4, 5, and 6 (1896); **56,** no. 6 (1898); and **62,** no. 1 (1902); "Théorie thermodynamique de la viscosité, du frottement, et des faux équilibres chimiques," in *Mémoires de la Société des sciences physiques et naturelles de Bordeaux,* 5th ser., **2** (1896), 1; and "On the General Problem of Chemical Statics," in *Journal of Physical Chemistry,* **2** (1898), 91.

Some of Duhem's papers on electrodynamics may be found in *Annales de la Faculté des sciences de l'Université de Toulouse,* **7,** B, G (1893); **10,** *B* (1896); 3rd ser., **6** (1914), 177; *American Journal of Mathematics,* **17** (1895), 117; *L'éclairage électrique,* **4** (1895), 494; and *Archives néer-*

landaises des sciences exactes et naturelles, 2nd ser., **5** (1901), 227. His principal papers on floating bodies are found in *Journal de mathématiques pures et appliquées,* 5th ser., **1** (1895), 91; **2** (1896), 23; **3** (1897), 389; 6th ser., **7** (1911), 1.

His major historical papers are "Les théories de la chaleur," in *Revue des deux-mondes,* **129** (1895), 869; and **130** (1895), 380, 851.

For Duhem's own assessment of his work to 1913, see *Notice sur les titres et travaux scientifiques de Pierre Duhem* (Bordeaux, 1913), prepared by him for his candidacy at the Académie des Sciences. There remains an unpublished work on capillarity written several years before his death.

II. SECONDARY LITERATURE. *Mémoires de la Société des sciences physiques et naturelles de Bordeaux,* 7th ser., **1,** pt. 2 (1927) is a special issue entitled "L'oeuvre scientifique de Pierre Duhem"; in addition to the bibliography by O. Manville, cited above, it contains Manville's detailed discussion of Duhem's physics, pp. 1–435, and shorter discussions of his mathematical work by J. Hadamard, pp. 465–495, and of his historical work by A. Darbon, pp. 497–548.

See also E. le Roy, "Science et philosophie," in *Revue de métaphysique et de morale,* **7** (1899), 503; and "Un positivisme nouveau," *ibid.,* **9** (1901), 143–144; A. Rey, "La philosophie scientifique de M. Duhem," *ibid.,* **12** (1904), 699–744; a short review of Duhem's scientific work by his best-known disciple, E. Jouguet, in *Revue générale des sciences,* **28** (1917), 40; L. Roy, *L'électrodynamique des milieux isotropes en repos d'après Helmholtz et Duhem* (Paris, 1923); and A. Lowinger, *The Methodology of Pierre Duhem* (New York, 1941).

Biographical sources include P. Humbert, *Pierre Duhem* (Paris, 1932); E. Jordan, in *Annuaire de l'Association des anciens élèves de l'École normale supérieure* (1917), pp. 158–173, and *Mémoires de la Société des sciences physiques et naturelles de Bordeaux,* 7th ser., **1,** pt. 1 (1917); D. Miller, in *Physics Today,* **19,** no. 12 (1966), 47–53, based in part on several interviews with Hélène Pierre-Duhem; E. Picard, *La vie et l'oeuvre de Pierre Duhem* (Paris, 1921), which also includes a summary review of all his work; and Hélène Pierre-Duhem, *Un savant français: Pierre Duhem* (Paris, 1936).

DONALD G. MILLER

DUJARDIN, FÉLIX (*b.* Tours, France, 5 April 1801; *d.* Rennes, France, 8 April 1860), *protozoology.*

Both Dujardin's father and grandfather were skilled watchmakers, originally in Lille, and Félix, who for a time trained in the trade, seems to have acquired some of his interests—as well as his remarkable manual dexterity—from them.

With his two brothers, Dujardin attended the classes of the Collège de Tours as a day pupil. He was originally attracted to art, especially drawing and design. His interest in science was apparently first aroused by a surgeon who was a friend of the family and who lent him some books on anatomy and natural history as well as Fourcroy's *Chimie.* Chemistry became for a time Dujardin's chief interest and, using a textbook by Thénard and some basic chemical reagents, he conducted simple experiments at home. Intending to study chemistry in the laboratories of Thénard and Gay-Lussac at Paris, he began to prepare himself for the entrance examination at the École Polytechnique. He persuaded his older brother to join him in these studies—particularly mathematics—and they both presented themselves for the examination in 1818. His brother succeeded, but Dujardin failed.

Discouraged by this failure, Dujardin went to Paris to study painting in the studio of Gérard, although he did not entirely forsake his scientific studies. In order to make a living, however, he soon accepted a position as a hydraulic engineer in the city of Sedan. He was married to Clémentine Grégoire there in 1823. Still restless, he returned to Tours, where he was placed in charge of a library. He began simultaneously to teach, especially mathematics and literature, and soon achieved sufficient success to give up his duties at the library. In his leisure, he pursued scientific studies of various kinds. His earliest publication, on the Tertiary strata and fossils of the Touraine area, were valuable enough to attract the attention of Charles Lyell.[1]

When in 1826 the city of Tours decided to inaugurate courses in applied science, Dujardin was assigned to teach geometry. In 1829 he was asked to teach chemistry as well and was provided with liberal funds for the establishment of a laboratory. This gave Dujardin the opportunity to return to his initial interest in chemical research. He also pursued studies in optics and crystallography and found time for botanical excursions, which led in 1833 to the publication (with two collaborators) of *Flore complète d'Indre-et-Loire*.

About this time, the diversity of his interests began to trouble Dujardin. On the advice of Henri Dutrochet, he decided to specialize in zoology and left Tours for Paris in pursuit of this goal. For the next several years, he apparently supported himself and his family by writing for scientific journals and encyclopedias.

In 1839, on the strength of his work in geology, Dujardin was appointed to the chair in geology and mineralogy at the Faculty of Sciences at Toulouse. In November 1840 he was called to the newly established Faculty of Sciences at Rennes as professor of zoology and botany and dean of the faculty—a position that for several years embroiled him in disputes with his colleagues. The intensity of these disputes diminished somewhat after he gave up the deanship in 1842. Although he was nominated several times for more important positions in Paris, he seemed always to end up second in the voting. Convinced, with some justice, that he was being persecuted from all sides (his colleagues sought to undermine his authority by such tactics as spreading rumors about his sex life), Dujardin became almost a recluse and spent his final years at Rennes in quiet obscurity. Shortly before his death, he was elected corresponding member of the Académie des Sciences, twelve years after his name was first proposed.

From the beginning of his career in zoology, Dujardin seems to have perceived the importance of observing organisms in the living state. Having already traveled widely during his geological and botanical studies, he expanded his excursions in pursuit of living animal specimens. Some of this spirit is reflected in his rare but charming little book *Promenades d'un naturaliste* (Paris, 1838).

In the autumn of 1834 Dujardin went to the Mediterranean coast to study microscopic marine animals. It was this work that led him to suggest the existence of a new family, the Rhizopods (literally, "rootfeet"). This suggestion was based primarily on his careful examination of several living species belonging to a widely distributed group long known as the Foraminifera. The most obvious feature of these tiny organisms (especially in the fossil state) is a delicate multichambered shell, outwardly similar to the shell of such mollusks as the Nautilus, and they had consequently been classified as "microscopic cephalopods" by Alcide d'Orbigny in 1825. Although d'Orbigny's classification was subsequently supported by the authority of Georges Cuvier, Dujardin rejected it because he was unable to see in the Foraminifera any evidence of the internal structure one ought to find in a mollusk. He perceived that the shell was only a secondary, external structure. By carefully crushing or decalcifying these delicate shells, he exposed a semifluid internal substance having no apparent structure.

As Dujardin observed the Foraminifera in their living state, he was struck by the activity of this contractile internal substance, which exuded spontaneously through pores in the calcareous shells to form pseudopodic rootlets. With equal spontaneity, these rootlets might then retract within the shell again. Dujardin became convinced that he was observing a special sort of amoeboid movement, in effect an amoeba within a porous shell. But pseudopodic rootlets could also be seen in microscopic animals having a less distinct casing than that of the Foraminifera, and Dujardin suggested that all such organisms should be joined in a new family to be called the Rhizopoda. According to this view, the Foraminifera, d'Orbigny's so-called "microscopic cephalopods," were in truth merely rhizopods with shells (*Rhizopodes à coquilles*).

This work in systematics led Dujardin to conclusions of far greater significance. In particular he now denied the famous "polygastric hypothesis" of Christian Ehrenberg, the foremost protozoologist of the era. Ehrenberg had recently revived Leeuwenhoek's view that infusoria were "complete organisms"; more specifically, that they possessed organ systems that imitated in miniature the general features of the organ systems of far more complex organisms, including the vertebrates. Like d'Orbigny, Ehrenberg enjoyed the support of Cuvier, and his theory was generally accepted. In his classificatory scheme, Ehrenberg placed several hundred species of infusoria in a new class, the Polygastrica (literally, "many stomachs"), in conformity with his belief that the globules or vacuoles which appear in most infusoria are tiny stomachs (as many as 200) connected together by an intestine. The strongest evidence for this belief came from experiments in which Ehrenberg had fed infusoria with various dyes (indigo and carmine, for example) and had then observed coloration of the "stomachs."

Dujardin reported that this conception had troubled him for some time. Although he could see neither the intestine nor the anal and oral orifices that Ehrenberg had posited, the "stomachs" were clearly visible. "I would," he wrote, "probably have lost courage and abandoned this research . . . if I had not fortunately found the solution to my problem in the discovery of the properties of *sarcode*."

"Sarcode" (from the Greek word for flesh) was the name Dujardin gave to the structureless substance he had found within the Foraminifera and other rhizopods and that he had found to be in every sense comparable to the substance of the amoeba and other Polygastrica. "The strangest property of *sarcode*," wrote Dujardin, "is the spontaneous production, in its mass, of vacuoles or little spherical cavities, filled with the environing fluid." It was these spontaneously produced vacuoles (*vacuoles adventives*) that Ehrenberg had mistaken for stomachs. Far from being complex organs, they were a natural result of the physical properties of sarcode; vacuoles could be formed at any time, by a spontaneous separating out of a part of the water present in living sarcode.

Ehrenberg's feeding experiments did not prove the existence of true stomachs, since the vacuoles did not become distended upon ingestion as might be expected of walled stomachs and only some of the vacuoles took on color, while others remained colorless. If they were stomachs, how could one explain "this choice of different aliments for different stomachs?" Dujardin thus rejected Ehrenberg's theory "with complete conviction," finding no reason to believe that his microscope and his sight were inferior to Ehrenberg's, especially since in several infusoria he had seen essential details which had escaped the German observer.

Dujardin presented all this work in a memoir of 1835. Ehrenberg did not retract, however. When in 1838 he published his monumental work on the infusoria as complete animals, he took every opportunity to ridicule Dujardin. In 1841 Dujardin gathered his work together in a large but less pretentious treatise on the infusoria. In this work, which became the starting point for later attempts to classify the protozoa, Dujardin reasserted his views but treated Ehrenberg rather more fairly than Ehrenberg had treated him. The polemic between Dujardin and Ehrenberg stimulated great interest in the microscopic animals and focused attention on one of the most important and recurrent issues in the history of biology—the relation between structure and function. By 1870 this issue had been resolved at one level by the general acceptance of the protoplasmic theory of life, according to which the basic attributes of life resided in a semifluid, largely homogeneous ground substance (protoplasm) having no apparent structure.

Dujardin's description of sarcode represents an important step toward this view. In his memoir of 1835, he wrote: "I propose to name *sarcode* that which other observers have called living jelly [*gelée vivante*], this diaphanous, glutinous substance, insoluble in water, contracting into globular masses, attaching itself to dissecting-needles and allowing itself to be drawn out like mucus; lastly, occurring in all the lower animals interposed between the other elements of structure." Dujardin went on to describe the behavior of sarcode when subjected to various chemicals. Potash seemed to hasten its decomposition by water, while nitric acid and alcohol caused it to coagulate suddenly, turning it white and opaque. "Its properties," wrote Dujardin, "are thus quite distinct from those of substances with which it might be confused, for its insolubility in water distinguishes it from albumen (which it resembles in its mode of coagulation), while at the same time its insolubility in potash distinguishes it from mucus, gelatin, etc."

Because this is such a remarkably complete and accurate description of what would later be called protoplasm, some of Dujardin's admirers have insisted that the German-directed (most especially by the histologist Max Schultze) substitution of "protoplasm" for "sarcode" represents "a violation of all good rules of nomenclature and justice."[2] If this attitude is meant to suggest that Dujardin was the rightful discoverer of the substance of life, one major objection can be raised; namely, that it ascribes to Dujardin's work a broader interpretation than he himself seems to have given it. He did suggest, even in 1835, that sarcode was present in a number of animals more complicated than the infusoria (worms and insects, for example), and he did soon after recognize that the white blood corpuscles were also composed of sarcode. The identity between plant protoplasm and animal sarcode seems to have escaped him, however, and was emphasized instead by German workers, most notably Ferdinand Cohn and Max Schultze. Until this identity was recognized, the notion of a substance of life had little meaning. Perhaps Dujardin missed the identity because he never integrated his notion of sarcode with the concept of the cell.

Dujardin published memoirs on a variety of animals other than the infusoria, particularly the coelenterates, intestinal worms, and insects. In 1838 he described a rare species of spiculeless sponge, to which his name was later attached. He also considered the then disputed question whether sponges were animals or plants, and concluded that they were animals. In 1844, he published a major treatise on the intestinal worms, which laid the basis for much of the work done since in helminthology and parasitology.

At the time of his death, Dujardin was engaged in a major study of the echinoderms, although he was by then more interested in questions of broader biological significance. He regretted that this work on the echinoderms kept him from a proper investigation of the "division of germs," of the species problem, and particularly from a new study on sarcode. This last point is especially interesting because by 1852 at least, Dujardin clearly recognized that the properties of sarcode led to an idea of great biological significance—the idea of "life as anterior to organization, as independent of the permanence of forms, as capable of making and defying organization itself."[3] It should be emphasized that Dujardin did not really deny all organization whatever to sarcode. Rather, he argued that its organization could not be compared to the definite structures observable in higher organisms. He seems to have had an almost prophetic vision of the importance of organization at the more subtle molecular level, and with the benefit of hindsight, E. Fauré-Fremiet makes a persuasive case for considering Dujardin a pioneer in the colloidal chemistry of protoplasm.[4]

Apart from this prophetic vision, perhaps the most appealing feature of Dujardin's work is his consistent modesty and rigorous attention to methodology. He always recognized that his work might undergo significant modification through the efforts of later workers and rarely made a claim that was not supported by his own direct observations. In placing the bacteria among the animals rather than the plants, in failing to recognize the significance of the nucleus, and in considering spontaneous generation possible, Dujardin was in the company of most of his contemporaries. His close attention to microscopic method is particularly apparent in his *Manuel de l'observateur au microscope* (1843), but it also informs his major treatise on the infusoria, which contains a brief but suggestive sketch of the historical interrelationship between developments in microscopic technique and developments in knowledge about the microscopic animals.

The breadth of Dujardin's early interests was crucial to his later success in protozoology. His artistic talent and training is evident in the many careful and beautiful plates with which his works are illustrated. His knowledge of optics allowed him to develop an improved method of microscopic illumination which bore his name and which can be considered an ancestor of the present condenser. Finally, his knowledge of physics and chemistry was important in enabling him to describe so completely and so accurately the properties of sarcode. It is easy to agree with Dujardin's admirers that his work was improperly appreciated during his lifetime, and easy to understand why protozoologists still cite his work with admiration today.[5]

NOTES

1. Charles Lyell, "On the Occurrence of Two Species of Shells of the Genus *Conus* in the Lias, or Inferior Oolite, near Caen in Normandy," in *Annals of Natural History*, **6** (1840), 293; and *Principles of Geology* (9th ed., London, 1853), p. 236.
2. Yves Delage, *La structure du protoplasma et les théories sur l'hérédité et les problèmes grands de la biologie générale* (Paris, 1895), p. 19. See also L. Joubin, p. 10.
3. E. Fauré-Fremiet, pp. 261–262.
4. *Ibid.,* 266–268.
5. See, e.g., Reginald D. Manwell, *Introduction to Protozoology* (New York, 1968).

BIBLIOGRAPHY

I. ORIGINAL WORKS. Dujardin's major works are "Recherches sur les organismes inférieurs," in *Annales des sciences naturelles* (*zoologie*), 2nd ser., **4** (1835), 343–377; *Histoire naturelle des zoophytes. Infusoires, comprenant la physiologie et la classification de ces animaux et la manière de les étudier à l'aide du microscope* (Paris, 1841); and *Histoire naturelle des Helminthes ou vers intestinaux* (Paris, 1845).

A complete bibliography of Dujardin's ninety-six published works can be found in Joubin (see below), pp. 52–57, while sixty-four of his papers are cited in the *Royal Society Catalogue of Scientific Papers,* II, 378–380.

Dujardin's rich collection of manuscripts, including laboratory notes and more than 500 letters, many of which are from the leading scientists of the day, is preserved at the Faculty of Sciences in Rennes. This probably important collection remains largely untapped, although Joubin and E. Fauré-Fremiet have made some use of it.

II. SECONDARY LITERATURE. The basic source is L. Joubin, "Félix Dujardin," in *Archives de parasitologie,* **4** (1901), 5–57. At the time he wrote this paper, Joubin held the chair at Rennes once occupied by Dujardin, and it was his clear intention to bestow on his predecessor all the honor he had been denied in life. The attempt was marred by Joubin's consistent and uncritical tendency to give Dujardin's work an importance that only hindsight can provide.

Also on Dujardin, see Enrique Beltrán, "Felix Dujardin y su *Histoire naturelle des zoophytes. Infusoires,* 1841," in *Revista de la Sociedad mexicana de historia natural,* **2** (1941), 221–232; "Notas de historia protozoologica. I. El descubrimiento de los sarcodarios y los trabajos de F. Dujardin," *ibid.,* **9** (1948), 341–345; and E. Fauré-Fremiet, "L'oeuvre de Félix Dujardin et la notion du protoplasma," in *Protoplasma,* **23** (1935), 250–269.

More generally, see J. R. Baker, "The Cell Theory: A Restatement, History, and Critique. Part II," in *Quarterly Journal of the Microscopical Sciences,* **90** (1949), 87–107; F. J. Cole, *The History of Protozoology* (London, 1926); G. L. Geison, "The Protoplasmic Theory of Life and the Vitalist-Mechanist Debate," in *Isis,* **60** (1969), 273–292; *Toward a Substance of Life: Concepts of Protoplasm, 1835–1870* (unpublished M.A. thesis, Yale University, 1967); and Arthur Hughes, *A History of Cytology* (London, 1959).

GERALD L. GEISON

DULLAERT OF GHENT, JEAN (*b.* Ghent, Belgium, *ca.* 1470; *d.* Paris, France, 19 September 1513), *logic, natural philosophy.*

Dullaert is sometimes confused with John of Jandun, owing to the Latin form of his name, Joannes de Gandavo. At the age of fourteen Dullaert was sent to Paris to study. He was a pupil of, and later taught with, John Major at the Collège de Montaigu; in 1510 he became a master at the Collège de Beauvais. Among his students at Montaigu were the Spaniards Juan de Celaya and Juan Martínez Silíceo, both important for their contributions to the rise of mathematical physics, and the humanist Juan Luis Vives.

At Paris, Dullaert published his questions on the *Physics* and the *De caelo* of Aristotle (1506), which appeared in at least two subsequent editions (1511; Lyons, 1512); a commentary on the *De interpretatione* of Aristotle (1509; Salamanca, 1517, edited by Silíceo); and an exposition of Aristotle's *Meteorology* (1512). The last title was reissued posthumously, "with Dullaert's questions," in 1514 by Vives, who prefaced

a brief biography wherein he states that Dullaert had left unfinished a commentary on the *Prior Analytics;* this apparently was prepared for publication in 1520/1521 by Jean Drabbe, also of Ghent. At the time of his death Dullaert was also working on a general edition of the works of Albertus Magnus that was based on previously unedited manuscripts which he himself had discovered. Earlier he had edited and revised for publication Jean Buridan's questions on Aristotle's *Physics* (1509) and Paul of Venice's *De compositione mundi* (*ca.* 1512) and *Summa philosophiae naturalis* (1513).

Like Paul of Venice, Dullaert was an Augustinian friar, and he showed a predilection for the realist views of this confrere while being strongly attracted also to the nominalist teaching then current at Paris. Perhaps for this reason his questions on the *Physics* are eclectic and, in many passages, inconclusive. At the same time they summarize in great detail (and usually with hopelessly involved logical argument) the teachings of Oxford "calculatores" such as Thomas Bradwardine, William Heytesbury, and Richard Swineshead; of Paris "terminists" such as Jean Buridan, Albert of Saxony, and Nicole Oresme; and of Italian authors such as James of Forlì, Simon of Lendenaria, and Peter of Mantua—while not neglecting the more realist positions of Walter Burley and Paul of Venice. The logical subtlety of Dullaert's endless dialectics provoked considerable adverse criticism from Vives and other humanists, but otherwise his teachings were appreciated and frequently cited during the early sixteenth century.

The structure of Dullaert's treatment of motion, which covers sixty-nine of the 175 folios constituting the questions on the *Physics* and *De caelo,* shows the strong influence of Heytesbury's *Tractatus de tribus praedicamentis,* with some accommodation along lines suggested by Albert of Saxony's *Tractatus proportionum.* Dullaert treats successively the entitative status of local motion, the velocity of local motion (both rectilinear and curvilinear), the velocity of augmentation, and the velocity of alteration, digressing in the latter tract to take up the intension and remission of forms. He is ambiguous in discussing the reality of local motion but holds for the impetus theory of Jean Buridan, regarding impetus as a kind of accidental gravity in the projectile that is corrupted by the projectile's own natural tendencies. He raises the question whether the impetus acquired by a falling body is proportional to the weight of the body but declines an answer. Following John Major, he teaches that God has the power to produce an actual (as opposed to a potential) infinity, and he sees no difficulty in the existence of a void. His views were generally taken over and clarified by Juan de Celaya, Luis Coronel, and others at Paris and in Spanish universities.

BIBLIOGRAPHY

None of Dullaert's works is available in English. A copy of the *Quaestiones super octo libros Aristotelis physicorum necnon super libros de caelo et mundo* (Lyons, 1512) is in Houghton Library of Harvard University. Pierre Duhem, *Études sur Léonard de Vinci*, III (Paris, 1913), gives numerous brief excerpts from this in French translation.

On Dullaert or his work, see Hubert Élie, "Quelques maîtres de l'université de Paris vers l'an 1500," in *Archives d'histoire doctrinale et littéraire du moyen âge*, **18** (1950–1951), 193–243, esp. 222–223; R. G. Villoslada, *La universidad de Paris durante los estudios de Francisco de Vitoria, O.P., (1507–1522)*, Analecta Gregoriana XIV (Rome, 1938); and William A. Wallace, "The Concept of Motion in the Sixteenth Century," in *Proceedings of the American Catholic Philosophical Association,* **41** (1967), 184–195; and "The Enigma of Domingo de Soto: *Uniformiter difformis* and Falling Bodies in Late Medieval Physics," in *Isis,* **59** (1968), 384–401.

WILLIAM A. WALLACE, O.P.

DULONG, PIERRE LOUIS (*b.* Rouen, France, 12/13 February 1785; *d.* Paris, France, 19 July 1838), *chemistry, physics.*

Dulong's father died shortly after his birth; and when he was four and a half, his mother died. An aunt, Mme. Fauraux, assumed responsibility for the young orphan and took him into her home at Auxerre, where he attended the local *collège.* His teachers encouraged his mathematical ability, and he was able to study some science at the *école centrale* founded in Auxerre in 1796. He succeeded in the competition for the École Polytechnique in Paris, which he entered in 1801 at the minimum age of sixteen. Excessive study ruined Dulong's health, and in his second year he withdrew from the school. He then turned to medicine as a career. He could do this because formal qualifications were not required in post-Revolutionary France. Dulong practiced in one of the poorer districts of Paris. Although the number of patients rapidly increased, his small capital ran out, for the tenderhearted doctor not only treated his patients without charge but offered to pay for their prescriptions.

After leaving medicine, Dulong turned first to botany and then to chemistry, hoping to make a name for himself in this newly famous science. In chemistry he found his métier, although his subsequent work in physics and physical chemistry exhibited the value of his mathematical training at the École Polytechnique. First Dulong obtained a position as an assistant in Louis Jacques Thenard's laboratory, and then Berthollet offered him a place in his private laboratory at Arcueil.

Dulong married Émélie Augustine Rivière on 29 October 1803. They had four children, one of whom died in infancy. Dulong's mature years were dominated by a conflict between his desire to do research and the necessity of accepting numerous teaching, examining, and administrative positions in order to buy apparatus and provide for his family. This conflict often developed into a crisis because of his persistent bad health. His first teaching post (1811) was at the École Normale, and in 1813 he was appointed to teach chemistry and physics at the École Vétérinaire d'Alfort. In 1820 he was appointed professor of chemistry at the Faculté des Sciences in Paris. At the École Polytechnique he was first appointed as examiner (1813–1820) and then professor of physics (1820–1830). Of the latter appointment Dulong wrote:

> Through a weakness of character for which I reproach myself incessantly, I have consented to accept the professorship of physics at the École Polytechnique, which the death of my unfortunate friend [Petit] has left vacant. Even with good health the duties of this post would have left me with little free time, so judge how much of this I have had, sick as I have been for the past eighteen months.[1]

In 1831 Dulong's friend Arago succeeded in relieving him of his teaching duties at the École Polytechnique by securing for him the post of director of studies, an office that was purely administrative. Dulong's health subsequently improved, but he bitterly regretted that he no longer had the laboratory facilities of his teaching post.

Dulong was a member of the Société d'Arcueil (1811) and the Société Philomatique (1812). He was elected to the physics section of the Académie des Sciences on 27 January 1823. In 1828 he became president of the Academy for one year, and later ill-advisedly accepted the position of permanent secretary for a short time (1832–1833).

Dulong's first publication, a report on work that he had carried out at Arcueil, reflects very clearly the influence of his patron, Berthollet. This memoir, read at a meeting of the First Class of the Institute in July 1811, provided a striking confirmation of Berthollet's thesis that chemical affinities were not fixed and that chemical reactions could often be reversed. Dulong showed that when barium sulfate, a notoriously "insoluble" salt, was boiled with a solution containing an equivalent quantity of potassium carbonate, it was partly decomposed. In another experiment, in which "insoluble" barium carbonate was added to a boiling solution of potassium sulfate, a partial exchange took place. From the small amount of barium sulfate formed, Dulong considered that a state of equilibrium had been reached.

It was his work on "insoluble" salts that led Dulong to make a special study of the oxalates of barium, strontium, and calcium. He went on to study the action of heat on oxalates and reached the conclusion that oxalic acid consisted of hydrogen and carbon dioxide;[2] he suggested that it might accordingly be renamed "hydrocarbonic acid." To Dulong, a metal oxalate was a simple compound of the metal and carbonic acid, although this was contrary to current chemical theory, according to which metals could combine with acids only in the form of oxides. Dulong's work therefore contributed to the post-Lavoisier conception of an acid as a compound containing hydrogen that is replaceable by a metal. Dulong compared the carbon dioxide in a metal oxalate to cyanogen in a cyanide or chlorine in hydrochloric acid. He thus deserves mention as a precursor of the radical theory in organic chemistry.

Better known, and certainly more spectacular, was Dulong's discovery of the spontaneously explosive oil nitrogen trichloride. Not realizing the danger involved when he first prepared a small quantity of the substance in October 1811, he lost a finger and the sight of one eye. He began his memoir (1813) by pointing out that only three elements—nitrogen, carbon, and boron—were apparently unable to combine with "oxymuriatic acid" (chlorine). He presented his research as an attempt to make nitrogen and chlorine combine. Although the gases did not combine directly, Dulong was more successful when he passed a current of chlorine through a fairly concentrated solution of ammonium chloride. After the reaction had been going on for two hours at a temperature of between 7° and 8° C., a yellow oil began to form. This was what exploded.

By February 1812 Dulong had sufficiently recovered from his injuries to resume his investigation. He now realized that he could avoid an explosion only if he kept the temperature low. Accordingly, he waited until the following October to carry out further research. He succeeded in determining the qualitative composition of the oil by allowing it to come into contact with copper; the only products were copper chloride and nitrogen. After receiving further injuries, Dulong abandoned this research. His memoir, however, contains the following remarks about the explosive properties of the new compound. They are significant in the light of his later thermochemical studies:

It seems to me that this compound contains a certain amount of combined heat which, when its elements separate from each other, raises their temperature and imparts to them a great elastic force.[3]

In 1816 Dulong carried out two pieces of chemical research prompted by the interest of Berthollet's circle at Arcueil in combining proportions. There was very great divergence between the analyses of phosphoric acid published by various chemists, and there was also some difference of opinion between Gay-Lussac and Thenard on the composition of phosphorous acid. It was in an attempt to clarify this situation that Dulong found that there were at least four acids of phosphorus. In addition to the two mentioned above, the composition of which he determined accurately, Dulong was able to confirm the existence of hypophosphoric acid and discovered a fourth acid that was obtained from the solution remaining after the action of water on barium phosphide. Dulong named this syrupy acid "hypophosphorous acid." His memoir (1817) contained a careful analysis of the new acid and a discussion of its salts.

The various oxides of nitrogen had created some confusion in the early nineteenth century. In 1816 Gay-Lussac had succeeded in distinguishing five oxides of nitrogen and had given their correct chemical composition. Dulong repeated Gay-Lussac's preparation of dinitrogen tetroxide. By heating *dry* crystals of lead nitrate, he excluded water from the product, which was collected as a liquid in a tube surrounded by a freezing mixture. He was the first to make a study of the color changes undergone by this interesting compound over a wide range of temperature, from a colorless solid at $-20°$C. to a deep red vapor when heated.

In 1819, when Berzelius was in Paris, Dulong collaborated with him in determining the gravimetric composition of water. This was a fundamental datum of chemistry, and it was important that it should be determined with great precision. They passed pure hydrogen over heated copper oxide and absorbed the water formed with anhydrous calcium chloride. The mean of their results gave the ratio $H:O = 11.1:88.9$, or, as Dumas later represented it, $1:8.008$. Dulong and Berzelius had been able to work only to an accuracy of approximately 1/60, and the remarkable accuracy of their result was due largely to the canceling out of errors. Dumas did not carry out his classic redetermination of this ratio until 1842.

In 1815 Dulong's famous collaboration with the mathematical physicist Alexis Thérèse Petit began; it produced three important memoirs on heat. The best-known part of this work is the statement of the law of constant atomic heats that bears their names, which is discussed further below. They began with the fundamental problem of measuring quantities of heat, which involved a critical analysis of thermometric scales. In 1804–1805 Gay-Lussac had carried out a comparison of mercury and air thermometers between 0°C. and 100°C. Dulong and Petit extended the range of comparison up to 300°C. and found an increasing discrepancy between the two scales at higher temperatures.

Dulong and Petit continued their researches in 1817, stimulated by the subjects of the prize to be awarded by the Académie des Sciences in 1818. The first of the three subjects for this prize was to determine the movement of the mercury thermometer as compared with an air thermometer from −20°C. to +200°C. They approached the subject by determining the *absolute* coefficient of expansion of mercury, and to do this they introduced the now classic method of balancing columns. Two vertical columns of mercury, one hot and the other cold, were connected by a thin horizontal tube. Since the columns balanced, the two pressures were equal, or

$$h:d = h':d',$$

where h is the height of the column and d is the density. Since density is inversely proportional to volume, a simple method was now available for direct measurement of the expansion of the mercury without reference to the material of the vessel. A refined version of this apparatus, introduced later by Regnault, became the standard apparatus for determining a liquid's coefficient of absolute expansion.

The second part of the subject for the Academy's prize was to determine the laws of cooling in a vacuum; Dulong and Petit accordingly undertook a complete re-examination of Newton's law of cooling. The most remarkable feature of their work was the way they broke down a complex phenomenon into its constituent parts and dealt with each factor separately. For example, they distinguished losses due to radiation from those due to contact with particles of a gas. They thus arrived at a series of laws relating to different special cases. It was not until 1879 that Stefan was able to reduce the phenomenon of radiation to a simple law. The memoir in which he did this took the work of Dulong and Petit as its starting point, and Stefan was at pains to show that his law agreed with their experimental results.[4] There is no doubt that the young Frenchmen deserved to win the 3,000-franc prize offered by the Academy.

The third and last joint memoir of Dulong and Petit was also on heat. Among its far-reaching implications was a new approach to Dalton's atomic theory, which had been received in France with deep skepticism. In considering some of the implications of the atomic theory, Dulong and Petit therefore began on a defensive note:

> Convinced . . . that certain properties of matter would present themselves under more simple forms, and could be expressed by more regular and less complicated laws, if we could refer them to the elements upon which they immediately depend, we have endeavoured to introduce the most certain results of the atomic theory into the study of some of the properties which appear most intimately connected with the individual action of the material molecules.[5]

They were concerned with the specific heats of elements; but if these elements really existed as atoms, it seemed possible that there might be a connection between the weight of the atom and the amount of heat required to raise the temperature of a given weight of that element by a certain amount.

Dulong and Petit first had to develop a reliable method of determining specific heats, for the published data were quite unreliable. They adopted the method of cooling that used the finely powdered solid packed round the bulb of a thermometer (since the rates of cooling are directly proportional to the thermal capacities and hence to the specific heats). They recorded the specific heats of a dozen metals and sulfur, and then multiplied each by the element's atomic weight. The following table shows the remarkably constant value obtained.

	Specific heat (water = 1)	Atomic weight (oxygen = 1)	Product of specific heat and atomic weight
Bismuth	0.0288	13.30	0.3830
Lead	0.0293	12.95	0.3794
Gold	0.0298	12.43	0.3704
Platinum	0.0314	11.16	0.3740
Tin	0.0514	7.35	0.3779
Silver	0.0557	6.75	0.3759
Zinc	0.0927	4.03	0.3736
Tellurium	0.0912	4.03	0.3675
Copper	0.0949	3.957	0.3755
Nickel	0.1035	3.69	0.3819
Iron	0.1100	3.392	0.3731
Cobalt	0.1498	2.46	0.3685
Sulfur	0.1880	2.011	0.3781

Dulong and Petit said that inspection of this table showed "the existence of a physical law susceptible of being generalized and extended to all elementary substances." Certainly it showed that the specific heats of the elements tested were inversely proportional to their atomic weights. Their interest in the atomic structure of matter is revealed by their conclusion, "The atoms of all simple bodies have exactly the same capacity for heat," and also by their suggestion that the actual distances between the atoms might be calculated from thermal expansion data. Of more immediate concern to chemists, however, was the use of their law of atomic heats in settling disputed values of atomic weights. On the basis of their law, Dulong and Petit changed some of Berzelius' atomic weights—for example, they halved his values for silver and sulfur. Following his success in relating atomic weights to specific heats, Dulong investigated the possibility of a relation between the atomic (or molecular) weights of gases and their refractive indexes, but with little success.

After Petit's death in 1820, Dulong carried out further research on heat by himself and published a memoir on the specific heats of gases (1829). He determined the relation of the specific heats of various gases at constant pressure and constant volume by measuring the effect of change in temperature on the tone produced when the respective gases were passed through a flute. His method was an extension of one used by Chladni in 1807. He concluded that (1) equal volumes of all gases under the same conditions of temperature and pressure, when suddenly compressed or expanded to the same fraction of their original volume, give off or absorb the same quantity of heat; (2) the resulting temperature changes are inversely proportional to the specific heats of the respective gases at constant volume.

Dulong also worked on animal heat, taking up the subject when it was chosen in 1821 for the prize of the Académie des Sciences. He devised a respiration calorimeter in which the heat was absorbed by water rather than by ice, as in the classic apparatus of Lavoisier and Laplace. Because of unsatisfactory agreement between the theoretical and actual quantities of heat obtained (due largely to incorrect data), Dulong was not satisfied with his work and it was not published until after his death. The situation was similar in the case of his work on thermochemistry, where he measured several heats of combustion. In conversations with Hess in 1837, Dulong made such generalizations as "The quantities of heat evolved are approximately the same for the same substances, combining at different temperatures," and "Equal volumes of all gases give out the same quantity of heat." Dulong's concern with generalizing about heats of reaction may have inspired Hess to formulate his law of constant heat summation (1840).

Dulong collaborated with Thenard on a study of catalytic phenomena. They confirmed Döbereiner's discovery that a jet of hydrogen could be kindled by allowing it to impinge in air on spongy platinum. They found that palladium, rhodium, and iridium are active at room temperature and other metals at higher temperatures. They realized that the metal's activity was dependent on its physical state but offered no explanation of the action of these substances, which Berzelius later called *catalysts*.

Dulong collaborated with Arago on a long and perilous study of the pressure of steam at high temperatures. This work was prompted by the French government's concern about the safety of boilers. They first confirmed the validity of Boyle's law at pressures up to twenty-seven atmospheres. They were then able to measure the steam pressure in boilers by means of a manometer, and afterward they calculated the temperatures corresponding to these pressures.

NOTES

1. Letter to Berzelius, 21 Aug. 1821, in Berzelius, *Bref* (Uppsala, 1912–1925), II, pt. 4, 29.
2. Although this was never published by Dulong, Cuvier gave an account of it in *Mémoires de l'Institut* (1813–1815), Histoire, 198–200. See also Gay-Lussac, in *Annales de chimie et de physique*, **1** (1816), 157; and Ampère, *ibid.*, p. 298.
3. *Mémoires de physique et de chimie de la Société d'Arcueil*, **3** (1817), 62.
4. "Über die Bezeihung zwischen der Wärmestrahlung und der Temperatur. I. Über die Versuche von Dulong und Petit," in *Sitzungsberichte der kaiserlichen Akademie der Wissenschaften.* Mathematisch-naturwissenschaftliche Classe, **59**, Abt. 2 (1879), 391–410.
5. *Annales de chimie et de physique*, **10** (1819), 395.

BIBLIOGRAPHY

I. ORIGINAL WORKS. Dulong wrote both alone and in collaboration. Works of which he was the sole author include "Recherches sur la décomposition mutuelle des sels solubles et insolubles," in *Annales de chimie et de physique,* **82** (1812), 273–308; "Mémoire sur une nouvelle substance détonnante," *ibid.*, **86** (1813), 37–43, and in *Mémoires de physique et de chimie de la Société d'Arcueil,* **3** (1817), 48–63; "Observations sur quelques combinaisons de l'azote avec l'oxigène," in *Annales de chimie et de physique,* **2** (1816), 317–328; "Mémoire sur les combinaisons du phosphore avec l'oxigène," in *Mémoires de physique et de chimie de la Société d'Arcueil,* **3** (1817), 405–452; "Recherches sur les pouvoirs réfringents des fluides élastiques," in *Annales de chimie et de physique,* **31** (1826), 154–181; "Recherches sur la chaleur specifique des fluides élastiques," *ibid.,* **41** (1829), 113–158; "Sur la chaleur dégagée pendant la combustion de diverses substances simples ou composées," in *Comptes rendus de l'Académie des sciences,* **7** (1838), 871–877; and "De la chaleur animale" (1822–1823), in *Annales de chimie et de physique,* 3rd ser., **1** (1841), 440–455.

With Petit, he wrote "Lois de la dilatation des solides, des liquides et des fluides élastiques à de hautes températures," in *Annales de chimie et de physique,* **2** (1816), 240–264; "Recherches sur la mesure des températures, et sur les lois de la communication de la chaleur," *ibid.,* **7** (1817), 113–154, 225–264, 337–367; and "Recherches sur quelques points importants de la théorie de la chaleur," *ibid.,* **10** (1819), 395–413.

Berzelius collaborated with him on "Nouvelles déterminations des proportions de l'eau et de la densité de quelques fluides élastiques," in *Annales de chimie et de physique,* **15** (1820), 386–395.

Dulong and Thenard's researches produced "Note sur la propriété que possèdent quelques métaux de faciliter la combinaison des fluides élastiques," in *Annales de chimie et de physique,* **23** (1823), 440–444; and "Nouvelles observations sur la propriété dont jouissent certains corps de favoriser la combinaison des fluides élastiques," *ibid.,* **24** (1823), 380–387.

Arago and Dulong wrote "Exposé des recherches faites par ordre de l'Académie Royale des Sciences, pour déterminer les forces élastiques de la vapeur d'eau à de hautes températures," in *Annales de chimie et de physique,* **43** (1830), 74–110.

II. SECONDARY LITERATURE. Works on Dulong or his contributions are F. Arago, "Dulong," in *Notices biographiques,* 2nd ed. (Paris, 1865), III, 581–584; M. P. Crosland, *The Society of Arcueil. A View of French Science at the Time of Napoleon I* (Cambridge, Mass., 1967); R. Fox, "The Background to the Discovery of Dulong and Petit's Law," in *British Journal for the History of Science,* **4** (1968–1969), 1–22; J. Girardin and C. Laurens, *Dulong de Rouen. Sa vie et ses ouvrages* (Rouen, 1854); J. Jamin, "Études sur la chaleur statique et la vapeur, travaux de Dulong et Petit," in *Revue des deux mondes,* **11** (1855), 377–397; P. Lemay and R. E. Oesper, "Pierre Louis Dulong, His Life and Work," in *Chymia,* **1** (1948), 171–190; and G. Lemoine, "Dulong," in *Livre centenaire de l'École polytechnique* (Paris, 1895), I, 269–278.

M. P. CROSLAND

DUMAS, JEAN-BAPTISTE-ANDRÉ (*b.* Alès [formerly Alais], Gard, France, 14 July 1800; *d.* Cannes, France, 11 April 1884), *chemistry.*

Dumas, son of the town clerk of Alès, was educated at the classical *collège* in that southern town and then was apprenticed to an apothecary. In 1816 he emigrated to Geneva, where he studied pharmacy and was taught chemistry by Gaspard de La Rive, physics by Marc Pictet, and botany by Augustin de Candolle. He was given permission to conduct experiments in the chemical laboratory of Le Royer, a local pharmaceutical firm.

Dumas's earliest researches were in medicine and physiology. In 1823 he returned to France and was appointed *répétiteur* in chemistry at the École Poly-

technique. Shortly afterward he succeeded to Robiquet's chair of chemistry at the Athenaeum, where evening classes were held for adults.

In 1824, with Adolphe Brongniart and J. V. Audouin, Dumas founded *Annales des sciences naturelles.* Two years later he married Hermine Brongniart, daughter of Alexandre Brongniart, director of the royal porcelain works at Sèvres. In 1828 he published the first volume of his *Traité de chimie appliquée aux arts,* and the following year he was a cofounder of the École Centrale des Arts et Manufactures. Dumas was appointed assistant professor at the Sorbonne and became full professor in 1841, a position he held until his retirement in 1868. Since the contemporary practice was to hold several academic appointments at once, Dumas also occupied a chair at the École Polytechnique (from 1835) and in 1839 became professor of organic chemistry at the École de Médecine. He lectured occasionally at the Collège de France and gave instruction in experimental chemistry at his private laboratory from 1832 to 1848. From 1840 he was an editor of *Annales de chimie et de physique.*

Dumas, a moderate conservative, became actively involved in politics after the February Revolution of 1848 and was elected to the legislative assembly from Valenciennes immediately after the fall of Louis Philippe. He was minister of agriculture from 1850 to 1851, and when Napoleon III became emperor he was made a senator. He was also a member, vice-president (1855), and president (1859) of the Paris Municipal Council. With Haussmann, Dumas undertook the transformation and modernization of the capital, supervising the installation of modern drainage systems, water supply, and electrical systems. He became permanent secretary of the Academy of Sciences in 1868.

Dumas was a brilliant teacher and trained a galaxy of chemists, including Laurent, Stas, Leblanc, and Louis Melsens. The iniquitous system of multiple professorships was responsible for a great deal of bitterness directed against him by some of the younger chemists, a few of them his former pupils. Dumas, however, refrained from indulging in retaliatory measures, even though he was repeatedly subjected to unfounded attacks.

Dumas's work is notable for its wide range rather than for its depth and insight. His most original contributions stemmed from the adaptation of existing ideas and not from the desire to make revolutionary breakthroughs. This was partly the result of his eminently practical personality, always willing to compromise; but it was chiefly the result of his familiarity with the historical tradition of chemistry, which enabled him to situate every problem within a broader perspective. His historically oriented *Leçons sur la philosophie chimique* was very influential upon subsequent studies in the history of chemistry.

Dumas's practical interests resulted in numerous contributions to applied chemistry, including the publication of *Traité de chimie appliquée aux arts*. He investigated problems in metallurgy, such as the preparation of calcium and the treatment of iron ores; he studied the nature and properties of different kinds of commercial glass; and he was interested in questions as diverse as the materials used in thirteenth-century frescoes and the nature of the compounds of phosphorus and of minium. Dumas's researches on dyes were probably his most lasting contributions to industry: he analyzed indigo and established the relationship between the colorless and blue types. He was also the first chemist to show that picric acid, the yellow organic compound commonly used for dyeing during the period, was a derivative of phenol. Dumas made extensive studies in pharmaceutics and established the correct formulas for several alkaloids, chloroform, and other substances. His interest in animal and plant physiology led him to suggest numerous improvements in those fields.

He investigated the mechanism involved in the formation of animal fat and attempted to establish that it was utilized in the maintenance of body heat and combustion while it formed a reserve, stored in the body tissues, which could be released for metabolism whenever required. He also showed that there was a close analogy between vegetable and animal metabolism. Because of the growing exasperation of German scientists with the dominant position of French science, this period saw an increasing number of violent diatribes against any major discoveries made in France. J. Liebig was the undisputed champion of this growing and squalid German nationalism in scientific affairs. Along with the discoveries of most other major French chemists—including Laurent, Gerhardt, and Chevreul—he abusively attacked the physiological discoveries of Dumas in the most violent and unjustified manner.

The most important problem with which Dumas was concerned throughout his career was the classification of chemical substances. He sought to devise comprehensive classificatory schemes for organic compounds and for the elements. Dumas's earliest contribution to organic chemistry was his study of nine alkaloids, published in 1823, jointly with Pierre Pelletier.[1] He analyzed the elemental constituents of these organic "bases" and attempted to prove that their relative proportions of oxygen followed Dalton's law of multiple proportions. He had embraced the ideas of the two reigning theories in contemporary chemistry: dualism, with its division of substances into electronegative (acid) and electropositive (alkaline); and atomism, which Dalton had used to explain his law. Dumas spent the next few years attempting to create an adequate system of classification of organic compounds based upon these two theories.

In 1826 Dumas developed a new method for directly measuring the vapor densities, and indirectly (by calculation) the relative molecular weights of different substances in the gaseous state. His method, which had the merit of being both precise and simple, is still used in chemical analysis. Dumas used the method himself to determine the molecular weights of phosphorus, arsenic, and boron.

Although he explicitly referred to Avogadro and Ampère, Dumas nevertheless failed to make a clear distinction between molecules and atoms. He thought that the atomic weights of gaseous substances could be derived directly by measuring their densities. Dumas circumvented the limitation imposed upon the application of this principle by the small number of elements observed in a gaseous state with the help of Gay-Lussac's law of combining volumes. Since those elements formed gaseous compounds, it was relatively easy to determine the simple volumetric proportions in which they combined. Atomic weights could then be indirectly calculated from the measurement of the density and the application of Boyle's law and Gay-Lussac's law.

However, Dumas's original enthusiasm for this method was soon tempered by his realization of its obvious inadequacies, which could have been removed only by a clearer recognition of Avogadro's distinction between a molecule and an atom. Dumas pointed out several anomalies. For instance, a liter of chlorine and a liter of hydrogen contained the same number of atoms—say 1,000—at a given temperature and pressure. Upon combination, one atom of either element united with an atom of the other to form a single atom of hydrochloric acid gas. If it were true that all gases contained the same number of atoms under the same conditions, hydrogen and chlorine in the example above would have combined to produce one liter of hydrochloric acid gas containing 1,000 atoms. But this was not the case: two liters of the gas resulted. Therefore chlorine and hydrogen atoms could not be indivisible: they must have divided in two before combination in order to produce as many atoms of the compound gas as of the two elemental gases taken together—2,000—assuming that a liter of any gas contained 1,000 atoms. Dumas's initial hypothesis that the vapor density of a gas could give a direct measurement of its relative atomic weight was thus disproved.

Dumas tried to save the situation by postulating a distinction between two types of particles: those corresponding to molecules, which could not be split any further by purely physical means (such as heat), and the true chemical atoms, which were the smallest units entering into any chemical reaction. It was only the former whose relative weights could be determined by comparing vapor densities. In spite of this classification, Dumas's ideas on the subject were not always consistent; he accepted the concept of an atom grudgingly as his career advanced. He cited the particles found in identical numbers in all gases under similar conditions as examples of physical atoms. He found it impossible, however, to ascertain that the smallest particles involved in any chemical reaction were genuine examples of chemical atoms because there was always the chance that reactions were possible only with aggregates of chemical atoms rather than with single atoms.

He was, however, so far from rejecting atomism that from 1840 onward he carried out an important revision of the atomic weights of thirty elements. His most valuable contribution in this field was his very precise determination of the atomic weight of carbon (jointly with his pupil Stas) in 1840.[2] A previously accepted weight, determined by Berzelius as $C = 12.20$ ($O = 16$), was shown to be incorrect. Dumas proved that $C = 12 \pm .002$ ($O = 16$) or $C = 75$ ($O = 100$). The analysis was made by burning diamond and artificial and natural graphite in oxygen; the carbon dioxide formed was weighed in potash solution. The results were in close agreement. The "new" weight of carbon had a great effect on the progress of organic chemistry.

Dumas never doubted that organic compounds were to be classified according to their structures, which depended upon their having an atomic (or particulate) constitution. His first important contribution to classificatory organic chemistry came in 1827, when he and his assistant Polydore Boullay published the first part of a study of ether; the final part appeared the following year. This paper assumed that the composition of organic compounds was dualistic, consisting of two parts corresponding to the acidic and basic constituents of an inorganic salt. First, the composition of alcohol and ether was determined by analysis and vapor density measurements. It was concluded that they were both hydrates of ethylene ("hydrogène bicarbone"); alcohol contained twice as much water combined with the hydrocarbon as ether did. Extending the analysis to "compound ethers" (i.e., esters) of nitrous, benzoic, oxalic, acetic,

and other acids with alcohol, it was shown that "compound ethers" could be divided into two kinds: those that were formed by oxyacids, which contained no water and were the salts of ether and anhydrous acids; and those that were formed by hydracids, which contained water and were salts of ether and hydrated acids.

Dumas affirmed that this dualistic interpretation should be related to the nature of ethylene rather than to that of ether. In all these cases it was the hydrocarbon that played the role of a powerful base, having a saturation capacity equal to that of ammonia. If it did not act upon litmus paper and other indicators, this was due to its insolubility in water. (The suggestion that ethylene was a strong base had originally been made by Chevreul.) This view was extended by Dumas to cover a large number of other cases. For instance, he suggested that cane sugar and grape sugar were salts formed by carbonic acid, ethylene, and water.

The dualistic theory was interpreted by Dumas in electrochemical terms in 1828; and he maintained this view for another ten years, although with diminishing enthusiasm as the discrepancies accumulated over the years. Until 1835 he was convinced that the electrochemical theory had been established with almost complete certainty. "All present-day chemistry is based upon the view that there is opposition between substances, which is admirably borne out by the evidence from electrical phenomena."[3] The constitution of oxamide (analogous to ethers), also investigated by Dumas, was explained by postulating the existence of the amide radical (N_2H_4) which remained after ammonia (N_2H_6) had lost hydrogen at the negative pole. Oxamide was a binary compound formed by the combination of carbon [mon]oxide (C_2O) with the amide radical. Even though the latter had not been isolated, its presence was to be assumed because it helped to explain, predict, and classify a large number of phenomena. For example, urea was best understood as being made up of carbon [mon]oxide (C_2O) combined with two amide radicals: $C_2O + 2N_2H_4$ (1830). Similarly, the strongly electropositive alkaline metals, sodium and potassium, formed amides, with the amide radical functioning electronegatively, like chlorine in metallic chlorides.

In 1835, through his investigations into "spirit of wood" (methyl alcohol), Dumas showed how the presence of a radical gave rise to a whole series of compounds. Something new was added to the earlier dualistic theories in this conception: isomerism. He had found that in various hydrocarbons, such as naphthalene and anthracene, or ethylene and isobutylene, carbon and hydrogen were combined in the same relative proportions but were "more closely packed together in one member than in the other." This discovery led Dumas to postulate the existence of a third hydrocarbon analogous to ethylene and isobutylene, where hydrogen and carbon were combined in a 1:1 ratio, although in different states of condensation. He suggested that the three hydrocarbons would constitute a series such that the condensation for each successive term was twice that of its immediate predecessor. In other words, ethylene was C_2H_2 ($C = 6$) and isobutylene C_4H_4; the new member would be CH, the immediate predecessor of ethylene. By this reasoning, based purely upon analogy, Dumas predicted and succeeded in discovering the whole methyl series. The first member of the series (CH) was called "methyl." In this way Dumas not only established a link between ethyl and methyl alcohols but also discovered the radical of cetyl alcohol, which had been known from Chevreul's earlier investigations.

From the known constitution of ethyl alcohol—interpreted as being composed of a hydrocarbon, ethylene (C_8H_8), and water—Dumas reasoned that there must be similar hydrocarbons to be found in other alcohols if their water could be extracted. Thus he succeeded in discovering, although not necessarily isolating, hydrocarbons combined with water in methyl alcohol, cetyl alcohol, etc.

A second set of analogies, worked out in conjunction with the hydrocarbon contained in methyl alcohol, indicated that this hydrocarbon could be made to combine with a host of substances—nitric acid, ammonia, chlorine, etc.—and give rise to a complete series of compounds in which the hydrocarbon is transferred from one combination to another. At the same time he realized that in certain hydrocarbons, carbon and hydrogen were contained in the same relative proportions, although not in the same relative quantities.

The Theory of Substitutions. The theory of substitutions (or "métalepsie") was stated by Dumas in 1834. Its main assertion was that the hydrogen in any compound could be replaced by an equivalent amount of a halogen, oxygen, or other element. Furthermore, in order to explain the action of chlorine on a substance containing hydrogen linked to oxygen (rather than to carbon), the theory maintained that "if the hydrogenized compound contains water (i.e., hydrogen linked to oxygen), the hydrogen is eliminated without replacement; but if a further quantity of hydrogen in subsequently removed, then it is replaced by an equivalent amount of chlorine, etc."

The importance of this law, which was the first to explain the mechanism of substitution reactions in organic chemistry, cannot be overestimated. Its historical origin has been explained in various fashions. The most likely explanation is the one offered by Dumas himself, which the context renders probable: he was interested in testing the correctness of his dualistic theory of the constitution of alcohols by examining the action of the halogens on these compounds. Both ethyl and methyl alcohol produced chloroform when subjected to the action of chlorine. This led him indirectly to the theory of substitutions.

In January 1834, Dumas had read to the Académie des Sciences the results of significant research[4] in which the correct molecular formulas for chloroform, bromoform, iodoform, and chloral were given for the first time. He had also observed during his investigations of the action of chlorine on alcohol to give chloral that ten volumes of hydrogen ($C = 6$) were removed from alcohol but were replaced by only six volumes of chlorine. This was contrary to the evidence he had obtained earlier when studying the action of chlorine on essence of turpentine, where each atom of hydrogen had been replaced by one of chlorine. However, the reaction of alcohol and chlorine was explicable if it were assumed that an atom of hydrogen directly combined with oxygen behaved differently from hydrogen atoms combined with carbon in an organic compound. In other words, if Dumas's theory of the constitution of alcohol were correct, it followed that the water molecule would react differently from the ethylene molecule. The action of chlorine on alcohol thus appeared to constitute a direct proof of the correctness of Dumas's dualistic theory: the hydrogen atoms lost by ethylene were replaced by chlorine; those eliminated from water were not, as was shown by the following equation:

$$C_8H_8,H_4O_2 + Cl_{16} = C_8H_2Cl_6O_2 + 5H_2Cl_2$$

Alcohol   Chlorine   Chloral   Hydrochloric Acid

Dumas continued his researches on the theory of substitution in order to seek further proof for his view of the constitution of alcohol and the ethers. Paradoxically, the theory was correct only for that part which Dumas had thought was subsidiary to his main proof, a proof based upon erroneous assumptions.

The Theory of Types. From 1837 he became progressively dissatisfied with the electrochemical dualistic theory because of the numerous difficulties that it could not resolve. Encouraged by the example of several young chemists who had developed alternative modes of explanation and classification in organic chemistry while ignoring electrochemical ideas, Dumas was also progressively led to abandon the dualistic theory in favor of a unitary view in which the whole molecule was conceived of as a single structure without polarization into negative and positive parts. Laurent had pointed out (1837) that within a series generated by a hydrocarbon, all the hydrogen molecules could be replaced by their equivalents of the halogens, oxygen, or other substance without the fundamental chemical characteristics of the compound being markedly affected. He therefore assumed that all molecules were unitary structures whose properties were dependent upon the position and arrangement of their component elements and not upon the intrinsic natures of the latter, whether electropositive or electronegative.

In 1839 Dumas discovered that the action of chlorine upon acetic acid formed a new compound (trichloroacetic acid) in which the hydrogen atoms of the acetic acid had been replaced by chlorine. But the new compound had virtually the same physical and chemical characteristics as the acetic acid, even though electronegative chlorine had replaced the strongly electropositive hydrogen. Dumas was converted to the unitary view by this experiment.

The role of his younger contemporaries in his adoption of the unitary theory was admitted by Dumas. He explicitly recognized the contributions of Laurent, Regnault, Faustino Malaguti, Rafaelle Piria, and J. P. Couerbe in his earliest papers, before a rather distasteful set of accusations about priorities were made against him in print by Laurent, Baudrimont, and several others. Dumas's fairness is demonstrated by his reference to Couerbe, who had emphasized the role of arrangement and position of atoms within a molecule in determining its properties: "I attribute the properties of alkaline compounds to the physical form of the molecule, a form produced by the grouping of the elementary atoms of this molecule. This idea, which I have generalized, is the cause, if not primary, at least the secondary cause, of its properties."[5] Dumas affiliated his views on unitary structures with those of his predecessors. In fact, Dumas was so generous that he did not even claim to have discovered the law of substitutions entirely by himself and contented himself with the modest role of having generalized a discovery made by a group of contemporary chemists: "I do not claim to have discovered it (the law of substitutions), for it does no more than reproduce more precisely and in a more generalized form, opinions that could be found in the writings of a large number of chemists. . . ."[6]

It is, however, of interest to see the gradual evolution of Dumas's thought on this subject, even during the period when he was convinced of the quasi certainty of the electrochemical theory. In 1828 he had already declared that the electrochemical theory was

powerless to account for the dual behavior of certain elements that were negative in some combinations and positive in others. This implied the contradictory assumption that some molecules were both negatively and positively charged. For example, the halogens—chlorine, bromine, and iodine—were positive toward oxygen and negative toward hydrogen. Even more difficult to explain was the fact that while chlorine was positive toward oxygen and both chlorine and oxygen were negative in their compounds with calcium, chlorine displaced oxygen from calcium oxide.

In order to avoid a complete impasse, Dumas hinted at another mode of explanation: "It must be admitted that electrical relations are not alone in determining chemical reactions; in certain cases, the *number* of molecules, their *relative positions*"[7] were perhaps equally influential in modifying the outcome.

By 1834 these anomalous cases had vastly increased because of Dumas's interpretation that frequently, in binary organic compounds, carbon acted both electronegatively and electropositively. Often it was electropositive in an organic acid and negative in the corresponding base. For example, oxalic ester was composed of an acid (oxalic acid), a base (ethylene), and water; in it carbon functioned positively in the first constituent and negatively in the second. It is strange that the replacement of hydrogen by the electronegative halogens, in alcohols and other compounds, did not appear anomalous to Dumas when he formulated the law of substitution. In fact, he was still persuaded that hydrogen was the only absolutely electropositive element. This is all the more difficult to reconcile with his later (1838) remark to Berzelius that his theory of substitution was a simple empirical rule that described but did not explain phenomena, especially since almost immediately afterward he abandoned electrochemistry because of the anomalous role of hydrogen in substitution reactions. In fact, it is closer to the truth to say that Laurent and Baudrimont's conclusions about the unitary structure of molecules had been associated with the discovery of substitution reactions by Dumas as early as 1836, when a new note of caution crept into the latter's attitude toward electrochemistry. The dogmatic certainty of this theory had been replaced by Dumas's admission that the electrochemical theory was nothing more than a series of hypotheses for which no final proof was forthcoming.

After 1840 Dumas developed the type theory, in which he classified compounds according to two types: chemical and mechanical. The former were substances like acetic and chloroacetic acids, which have similar chemical properties, while the latter had more obscure analogies, basically of a physical kind.

Dumas's mechanical type, whose origin he attributed to Regnault's work on the ethers, was shown to be untenable by Laurent.

Whereas Laurent had adopted a static model for his fundamental types, based upon an analogy with crystalline structures, Dumas had adopted a dynamic planetary model in which the atoms in a molecule were seen as analogous to the planets in the solar system. Laurent's model was ultimately derived from Haüy, while Dumas was influenced by Berthollet.

Dumas and the Classification of Elements. In 1831, after the discovery of isomerism in compounds, Dumas had been led to speculate upon the possibility of isomerism among the elements: different elements might in fact be nothing but multiple structures in which the same fundamental element was duplicated or "condensed." This was supported by the comparison of atomic weights, since several elements had atomic weights which were whole-number multiples of one another, as was shown by the following table[8] drawn up at the time:

	Zinc	403.22
	Yttrium	401.84
1/2	Antimony	403.22
1/2	Tellurium	403.22
1/2	Sulfur	402.33
	Platinum	1233.26
	Indium	1233.26
	Osmium	1244.21
	Gold	1243.01
	Bismuth	1330.37
2	Palladium	1331.68
	Cobalt	369.99
	Nickel	369.67
1/2	Tin	367.64
	Cerium	574.7
	Tantalum	576.8
	Copper	395.7
1/2	Iodine	394.6
	Molybdenum	598.5
1/2	Tungsten	596.5
	Silicium	277.4
2	Boron	271.9

After his revision of atomic weights in the 1840's, Dumas had wanted to revive the speculation about a *materia prima* in conjunction with Prout's hypothesis that all elements were multiples of the hydrogen atom. In 1851 he read a paper to the British Association in which he attempted to establish how certain regular patterns might be found in arranging elements, such that the heavier atoms were derived from combinations of lighter ones. He also published two papers[9] in which he tried to develop the view that for the classification of the elements it was possible to discover "generating" relations similar to those defining the series of organic compounds. The elements could be divided into "natural families." The atomic weights of all the members of the same family were linked by a simple arithmetic relationship; they increased by multiples of sixteen:

Li	7
Na	$7 + (1 \times 16) = 23$
K	$7 + (2 \times 16) = 39$
O	16
S	$16 + (1 \times 16) = 32$
Şe	$16 + (4 \times 16) = 80$
Te	$16 + (7 \times 16) = 128$
Mg	24
Ca	$24 + (1 \times 16) = 40$
Sr	$24 + (4 \times 16) = 88$
Ba	$24 + (7 \times 16) = 136$

NOTES

1. "Recherches sur la composition eléméntaire et sur quelques propriétés caractéristiques des bases salifiables," in *Annales de chimie et de physique,* **24** (1823), 163–191.
2. "Sur le véritable poids atomique du carbone," *ibid.,* **1** (1841), 5–55, written with J. S. Stas; also in *Comptes rendus hebdomadaires des séances de l'Académie des sciences,* **11** (1840), 991–1008.
3. *Journal de pharmacie,* **20** (1834), 262.
4. "Recherches de chimie organique," in *Annales de chimie et de physique,* **56** (1854), 113–154; repr., with a few adds., as "Recherches de chimie organique, relative à l'action du chlore sur l'alcool," in *Mémoires de l'Académie des sciences,* **15** (1838), 519–556.
5. J. P. Couerbe, "Du cerveau considéré sous le point de vue chimique et physique," in *Annales de chimie et de physique,* **56** (1834), 189 n.
6. "Mémoire sur la loi des substitutions et la théorie des types," in *Comptes rendus hebdomadaires des séances de l'Académie des sciences,* **10** (1840), 178.
7. See the intro. to the *Traité de chimie appliquée aux arts,* I (Paris, 1828), lx; the italics are the author's.
8. "Lettre de M. Dumas à M. Ampère sur l'isomérie," in *Annales de chimie et de physique,* **47** (1831), 335.
9. "Sur les equivalents des corps simples," in *Comptes rendus hebdomadaires des séances de l'Académie des sciences,* **45** (1857), 709–731; **46** (1858), 951–953; and **47** (1858), 1026–1034; also in *Annales de chimie et de physique,* **55** (1859), 129–210.

BIBLIOGRAPHY

I. ORIGINAL WORKS. Dumas published most of his work in the *Annales de chimie et de physique* and in the *Mémoires* and the *Comptes rendus* of the Académie des Sciences. See the indexes for titles.

His books are *Phénomènes qui accompagnent la contraction de la fibre musculaire* (Paris, 1823); *Traité de chimie appliquée aux arts,* 8 vols. (Paris, 1828); *Leçons sur la philosophie chimique* (Paris, 1837); *Thèse sur la question de l'action du calorique sur les corps organiques* (Paris, 1838); *Essai sur la statique chimique des êtres organisés* (Paris, 1841).

II. SECONDARY LITERATURE. On Dumas's life and work see J.-B. Dumas, *La vie de J.-B. Dumas, par le général J.-B. Dumas son fils* (Paris, 1924), 230 mimeographed pp.; S. C. Kapoor, "Dumas and Organic Classification," in *Ambix,* **16** (1969), 1–65; and E. Maindron, *L'oeuvre de J.-B. Dumas* (Paris, 1886).

SATISH C. KAPOOR

DUMBLETON. *See* **John of Dumbleton.**

DU MONCEL, THÉODOSE ACHILLE LOUIS (*b.* Paris, France, 6 March 1821; *d.* Paris, 16 February 1884), *electricity, magnetism.*

Du Moncel studied at the *collège* of Caen and at the age of eighteen published two works on perspective. After graduation he traveled through Turkey and Greece and later published an elaborate account of his travels, for which he drew the lithograph illustrations. His interest in electricity began in 1852, following the publication in Cherbourg the previous year of a work on meteorology. Some sixty-five books and papers on electricity and magnetism followed during thirty years of active writing, his works being translated into English, German, Portuguese, and Italian.

Du Moncel's interest in electricity spanned the most fertile period of its development, from Faraday to Edison, and his publications analyzed each discovery and invention in the framework of the entire science. His early work dealt with the determination of the characteristics of electromagnets and their application to motor design, and the mutual interaction of magnets and energized conductors.

His first popular work, *Exposé des applications de l'électricité,* appeared in Paris in two volumes in 1853–1854 and was expanded to five volumes of 2,870 pages in 1856–1862, making it a valuable reference encyclopedia of electrical development up to that time. In it Du Moncel reviewed Charles Bourseul's proposal for the electric transmission of speech, the earliest approach to practical telephony. Du Moncel wrote of this, "I thought it incredible," yet it was held to have contained the germ of later Bell and Gray inventions. In the contests among telephone inventions, Du Moncel soon differentiated between those devices capable of transmitting only music and those

which could transmit the more complex articulations of the human voice. He gave maximum praise to Bell. Du Moncel also described electromagnetic equipment and its widening use in telegraphy, mechanics, and medicine. His most popular work, *Le téléphone, le microphone et le phonographe,* was first published in Paris in 1878 and was translated into English the following year. He collaborated with Sir William Henry Preece in publishing on electric illumination in 1882 and with Frank Geraldy on electric motors in the following year.

Du Moncel's publications also dealt with the printing telegraph, electromagnetic applications, especially electric motors and railway signals, the Ruhmkorff induction coil, lightning theory and lightning protection, the effect of the sun's passage over telegraph lines, and the forms and operation of electric batteries, clocks, and lamps. He then turned to the Atlantic cable, mathematical analysis of electromagnets, and grounded telegraph circuits. His final work, which concerned the electric motors of P. Elias and the Pacinotti dynamo, was published in 1883.

Although Du Moncel contributed no great discovery or invention of his own, his clear, widely read books and papers spread the advances in electrical science, which was rapidly expanding. His concern was less with electrical theory than with its devices and practical applications. His assiduous experiments and interpretations of the work of his colleagues helped organize the electrical innovations from the 1850's to the 1880's. He accepted tasks on behalf of his nation and profession and was accorded many honors. He became a member of the Technical Committee of the Administration of Telegraphs of France in 1860 and in 1866 was named a *chevalier* of the Legion of Honor. He was made a member of the Institute of France, elected to the Academy of Sciences, and in 1879 became editor of *Lumière électrique.* He was an early member of the Society of Telegraph Engineers and Electricians of London and was awarded the Order of St. Vladimir of Russia.

BIBLIOGRAPHY

Among Du Moncel's many books and monographs on electrical subjects are *Considérations nouvelles sur l'électromagnétisme* (Paris, 1853); *Exposé sommaire des principes et des lois de l'électricité* (Cherbourg, 1853); *Exposé des applications de l'électricité,* 2 vols. (Paris, 1853–1854; 2nd ed., 5 vols., 1856–1862; rev. 3rd. ed., 1872–1878); *Notice sur l'appareil d'inductions électrique de Ruhmkorff* (Paris, 1855; 4th ed., 1859); *Notices historiques et théoriques sur le tonnerre et les éclairs* (Paris, 1857); *Étude du magnétisme et de l'électro-magnétisme* (Paris, 1858); *Revue des applications de l'électricité en 1857 et 1858* (Paris, 1859); *Étude des lois des courants électriques* (Paris, 1860); *Traité théorique et pratique de télégraphie électrique* (Paris, 1864); *Le téléphone, le microphone et le phonographe* (Paris, 1878), Eng. trans. (New York, 1879); *L'éclairage électrique* (Paris, 1880), Eng. trans. (London, 1882), Italian trans. (Turin, 1885–1887); and *L'électricité comme force motrice* (Paris, 1883), Eng. trans. (London, 1883).

BERN DIBNER

DUNCAN, JOHN CHARLES (*b.* Duncan's Mill, near Knightstown, Indiana, 8 February 1882; *d.* Chula Vista, California, 10 September 1967), *astronomy.*

Duncan's chief contribution to astronomy was his photographic demonstration of expansion in the Crab nebula. He is perhaps better known, however, as the author of *Astronomy,* a standard college textbook for over thirty years, which was illustrated with many of his own excellent photographs of nebulae and galaxies.

The son of Daniel Davidson Duncan and his wife, Naomi Jessup, Duncan grew up in Indiana and taught at a country school there from 1901 to 1903 while an undergraduate at Indiana University in Bloomington. Between receiving his B.A. in 1905 and his M.A. in 1906, both from Indiana University, he was a fellow of Lowell Observatory in Flagstaff, Arizona. In 1907, following his marriage to Katharine Armington Bullard the previous year, he enrolled at the University of California, where he received a Ph.D. in 1909; his dissertation, on Cepheid variables, was written under the direction of William Wallace Campbell.

Returning to the East, he served as instructor in astronomy at Harvard University from 1909 to 1916, before becoming professor of astronomy and director of Whitin Observatory at Wellesley College. Upon his retirement from these posts in 1950—at the age of sixty-eight—he spent the next twelve years as visiting professor at the University of Arizona and visiting astronomer at Steward Observatory.

The Crab nebula, located in the constellation of Taurus, is still today a fruitful subject for investigation because of its association with the pulsar NP 0532; it is believed to be the remnant of a supernova observed in Japan and China in A.D. 1054. By comparing a photograph taken with the sixty-inch telescope at Mount Wilson in 1909 by George Willis Ritchey with one he took himself in 1921 with the same instrument, Duncan was able to demonstrate outward motions in the filaments of the Crab. He later confirmed these motions with another photograph taken in 1938, thus showing that it was indeed an expanding envelope such as has been observed around other novae.

Duncan also investigated comets, spectroscopic binary stars, and novae. His long-exposure photographs of nebulae and galaxies were taken during an appointment as astronomer at Mount Wilson in 1920 and during many subsequent summers spent there as a voluntary research assistant.

BIBLIOGRAPHY

I. ORIGINAL WORKS. Duncan's textbook, *Astronomy* (New York, 1926; 5th ed., 1955), also appeared in an abridged version, *Essentials of Astronomy* (New York, 1942).

His dissertation was published as "The Orbits of the Cepheid Variables Y *Sagittarii* and RT *Aurigae;* with a Discussion of the Possible Causes of This Type of Stellar Variation," in *Lick Observatory Bulletin,* **5** (1908–1910), 82–94. His work on the Crab nebula appeared as "Changes Observed in the Crab Nebula in Taurus," in *Proceedings of the National Academy of Sciences,* **7** (1921), 179–180; and as "Second Report on the Expansion of the Crab Nebula," in *Astrophysical Journal,* **89** (1939), 482–485.

Reproductions and descriptions of the best of Duncan's photographs are contained in six papers: "Bright Nebulae and Star Clusters in Sagittarius and Scutum," in *Astrophysical Journal,* **51** (1920), 4–12, with 4 plates; "Bright and Dark Nebulae near ζ Orionis, Photographed with the 100-inch Hooker Telescope," *ibid.,* **53** (1921), 392–396, with 2 plates; "Photographic Studies of Nebulae, Third Paper," *ibid.,* **57** (1923), 137–148, with 11 plates; "Photographic Studies of Nebulae, Fourth Paper," *ibid.,* **63** (1926), 122–126, with 4 plates; "Photographic Studies of Nebulae, Fifth Paper," *ibid.,* **86** (1937), 496–498, with 6 plates; and "Photographic Studies of Nebulae VI. The Great Nebulous Region in Cygnus Photographed in Red Light," *ibid.,* **109** (1949), 479, with 2 plates.

There are 38 articles by Duncan listed in Poggendorff, VI, pt. 1, 615, and VIIb, pt. 2, 1155–1156, which include all those mentioned above except the second.

II. SECONDARY LITERATURE. Joseph Ashbrook's brief, unsigned obituary notice, with photograph, appeared in *Sky and Telescope,* **34** (1967), 283. Other facts about Duncan's life can be found in *Who's Who in America,* XXVII (Chicago, 1952), 690–691, and XXVIII (Chicago, 1954), 746; and in *American Men of Science,* 11th ed., The Physical and Biological Sciences, D–G (New York, 1965), 1313.

SALLY H. DIEKE

DUNDONALD, ARCHIBALD COCHRANE, EARL OF (*b.* Culross Abbey [?], Scotland, 1 January 1749; *d.* Paris, France, 1 July 1831), *chemistry.*

Dundonald was the eldest son of Thomas Cochrane of Culross and Ochiltree, eighth Earl of Dundonald, and his second wife, Jane Stuart. Following family tradition he entered on a military career but subsequently transferred in turn to the navy and back to the army. He inherited his title in 1778 but little else other than saltpans and mineral rights on the Culross

Abbey estate on the north shore of the River Forth. Dundonald spent most of his long life attempting to apply science to the art of manufactures; he achieved considerable technical but little commercial success.

In 1781 he returned to Culross Abbey, where he associated with such Edinburgh intellectuals as Joseph Black, James Hutton, and John Hope. By this time Dundonald had conceived the idea of a substitute for wood tar made from coal, and he built kilns at Culross Abbey. In 1781 he was granted a patent (B.P. 1781 No. 1291) covering not only coal tar but "essential oils, volatile alkali, mineral acids, salts and cinders (coke)." The kilns are described in *The Statistical Account of Scotland,* and there is a drawing, probably by Dundonald, in the Boulton and Watt Collection (Reference Library, Birmingham). In 1782 Dundonald founded the British Tar Company to operate the patent and build kilns associated with various ironworks. He failed, however, to interest the British Admiralty in coal tar. Coal-gas lighting was almost a by-product of the same experiments, but Dundonald missed the possibility.

Failure with coal tar led to interest in other materials: first alum, a mordant used by dyers and by silk and calico printers (B.P. 1794 No. 2015). His chief contribution to late eighteenth-century industrial chemistry was the production of soda from common salt (B.P. 1795 No. 2043), which solved one of the major technical problems of the late eighteenth century: to find a synthetic substitute for the dwindling supplies of barilla, kelp, wood ash, and weed ash that were essential to the soap, glass, and textile industries. In 1790 Dundonald had gone to Newcastle-upon-Tyne, where William Losh and Thomas Doubleday were trying to make alkali from the ash of marine plants by a LeBlanc-like process. Losh was sent to Paris and in 1796, following his return, The Walker Chemical Company, at Walker-on-Tyne, County Durham, was established to operate Dundonald's patent. Similar works were subsequently established near Newcastle and Glasgow. His other patents cover the manufacture of white lead (B.P. 1779 No. 2189); a variety of heavy chemicals (B.P. 1798 No. 2211) including soda, saltpeter, sal ammoniac, alum, Epsom salts, potassium chloride and sulfate, and sodium phosphate, and the production of alkali from vegetable sources (B.P. 1812 No. 3547).

Dundonald's other interests included making bread from potatoes; the substitution of potatoes for grain in alcohol production; finding a substitute for gum senegal; paint, pottery, and textile production; and iron and coal mining. His treatise on the connection between agriculture and chemistry foreshadowed much of Humphry Davy's *Elements of Agricultural Chemistry,* including the recognition of phosphorus as an essential plant nutrient.

So speculative and widespread were his enterprises that he was known in Scotland as "Daft Dundonald." Unhappily none of them helped his family fortunes; he died in poverty in Paris in 1831.

BIBLIOGRAPHY

I. ORIGINAL WORKS. Dundonald's writings include *Account of the Quality and Uses of Coal Tar and Coal Varnish* (London, 1785); *The Present State of Manufacture of Salt Explained* (London, 1785); *Letters of the Earl of Dundonald on Making Bread from Potatoes* (Edinburgh, 1791); *A Treatise Showing the Intimate Connection That Subsists Between Agriculture and Chemistry* (London, 1795); and *Directions by Lord Dundonald for Extracting Gum From Lichen and Tree Moss* (Glasgow, 1801).

See also "Dundonald Papers Concerning The British Tar Company," National Library of Scotland (Edinburgh); "Boulton and Watt Papers," Assay Office Library (Birmingham, England); "Session Papers 241/25," Library of the Writers to H. M. Signet (Edinburgh); *Abridgments of Specifications Relating to Acids, Alkalis . . .,* Patents Office Library (London); and "Newcastle: Chemical Manufacturers in the District," British Association Report (1863), p. 701.

II. SECONDARY LITERATURE. On Dundonald and his work, see W. G. Armstrong, ed., *The Industrial Relations of the Three Northern Rivers, Tyne, Wear, and Tees* (London, 1864); Archibald and Nan L. Clow, "Lord Dundonald," in *Economic History Review,* **12** (1942), 47; "Archibald Cochrane, 9th Earl of Dundonald," in *Chemistry and Industry,* **24** (1944), 217; *The Chemical Revolution* (London, 1952); Thomas Cochrane, 10th Earl of Dundonald, *Autobiography of a Seaman* (London, 1860); and John Sinclair, ed., *The Statistical Account of Scotland,* X (Edinburgh, 1791–1799), 412.

ARCHIBALD CLOW

DUNÉR, NILS CHRISTOFER (*b.* Billeberga, Sweden, 21 May 1839; *d.* Stockholm, Sweden, 10 November 1914), *astronomy.*

Dunér studied astronomy at the University of Lund and obtained his doctor's degree in 1862. From 1864 to 1888 he was senior astronomer at the Lund Observatory. In 1888 he was appointed professor of astronomy at Uppsala University and director of the observatory. He retired in 1909.

Dunér's dissertation of 1862 deals with the determination of the orbit of the planetoid Panopea, which had been discovered the previous year. Swedish astronomy at that time was of necessity strictly nonobservational, as the observatories were obsolete. During the next few years a new observatory was erected at Lund, and Dunér became an observing astronomer, "the strict empiricist," who introduced the "new astronomy" to Sweden. His work covered measurement of visual double stars and discussion of their relative movements; description and measurement of the spectra of red stars (the Vogel spectral type III), of which he discovered more than 100; spectroscopic determination of the rotation of the sun (a spectroscope with the largest grating of the time had been constructed); and observation and reduction of the star positions of the Lund Zone (declination $+35°$ to $+40°$) of the meridian circle survey until the declination $-23°$ of the *Astronomische Gesellschaft*.

Having transferred to Uppsala Observatory, Dunér again had to wait for improved equipment, but he succeeded in developing an efficient observatory. He revived his measurements of solar rotation and obtained further evidence for the decrease in the velocity of rotation from the equator to the latitudes $±75°$ (the results of his Lund measurements had also contributed to the waning discussion of the reliability of the Doppler principle). He continued his observations of the red stars; about twenty years later Hale and Ellerman pointed out how Dunér's results, obtained visually and with small and primitive instruments, were confirmed photographically. Also at Uppsala he found the solution to the special problems of the eclipsing binary Y Cygni, pointing out that this system consists of two similar suns moving around their common center of gravity in elliptic orbits, the common line of apsides of which simultaneously rotates in the plane of the orbits. After summing up these results in a few lines Dunér added modestly: "These investigations may well claim some interest."

In 1887 Dunér went to Paris as a Swedish delegate to the meeting concerning the gigantic *Carte photographique du ciel* project, and he was a member of the commission appointed to plan and supervise its effectuation. His foresight is apparent in his suggestion to postpone the project for a quarter of a century, in view of the rapid development of instrumental and photographic facilities that was expected. But his observatory did participate in the photographic campaign of the years 1900–1901 to determine the solar parallax by means of observations of the planetoid Eros.

In 1861 and 1864 Dunér was a member of expeditions to the Spitsbergen Islands as geographer and physicist, and his experiences were later taken into account by a joint Swedish-Russian geodetic survey of these northern islands about 1900.

In characterizing Dunér's qualities as a scientist, Ångström, in his obituary, praised his clear mind for inductive reasoning and his great experimental genius.

BIBLIOGRAPHY

I. ORIGINAL WORKS. A bibliography of Dunér's published papers is in von Zeipel's obituary. Dunér's more important works are "Mesures micrométriques d'étoiles doubles, faites à l'Observatoire de Lund, suivies de notes sur leurs mouvements relatifs," in *Acta Universitatis lundensis*, **12**, no. 2, pt. 1 (1876), 1–266; "Sur les étoiles à spectres de la troisième classe," in *Kungliga Svenska vetenskapsakademiens handlingar*, **21**, pt. 2 (1884), 1–137; "Recherches sur la rotation du soleil," in *Nova acta Regiae Societatis scientiarum upsaliensis*, 3rd ser., **14**, pt. 13 (1891), 1–78; "On the Spectra of Stars of Class III *b*," in *Astrophysical Journal*, **9** (1899), 119–132; "Calculation of Elliptic Elements of the System of *Y* Cygni," ibid., **11** (1900), 175–191; "Om den på fotografisk väg framställda stjernkatalogen," in *Öfversigt af Kungliga Vetenskapsakademiens förhandlingar* (1900), 399–407; and "Über die Rotation der Sonne," in *Nova acta Regiae Societatis scientiarum upsaliensis*, 4th ser., **1**, pt. 6 (1907), 1–64.

II. SECONDARY LITERATURE. An article on Dunér by Ö. Bergstrand is in *Svenskt biografiskt lexicon*, XI (1945), 528–535. Obituaries are A. Ångström in *Astrophysical Journal*, **41** (1915), 81–85; Ö. Bergstrand in *Astronomische Nachrichten*, **199** (1914), 391–392; A. Fowler in *Monthly Notices of the Royal Astronomical Society*, **75** (1915), 256–258; B. Hasselberg in *Vierteljahrsschrift der Astronomischen Gesellschaft*, **52** (1917), 2–31; and H. von Zeipel in *Kungliga Svenska vetenskapsakademiens Årsbok för År 1916* (Uppsala), 291–312.

AXEL V. NIELSEN

DUNGLISON, ROBLEY (*b.* Keswick, England, 4 January 1788; *d.* Philadelphia, Pennsylvania, 1 April 1869), *medical education, lexicography, physiology.*

His father, William, and his maternal grandfathers were wool manufacturers; his mother was Elizabeth Jackson, and his maternal grandmother was a Robley, hence his first name. Orphaned as a child, Dunglison received a classical education at Green Row Academy, Abbey Holme, through a legacy from a rich uncle. There he obtained an excellent knowledge of Greek and Latin as well as a fluent pen in English; later he was also to become well-versed in French and German. Having decided upon a medical career, he took a preceptorship with a surgeon at Keswick and went to Edinburgh, Paris, and London for his formal medical education; he obtained his degree by examination from Erlangen. In London he assisted the ailing Dr. Charles Thomas Haden, a prominent practitioner, who greatly influenced the development of Dunglison's professional and social character. Dunglison passed the examinations of the Royal College of Surgeons and of the Society of Apothecaries and commenced practice in 1819 at London, where he was appointed physician-accoucheur to the Eastern

Dispensary. His pen, however, was busier than his lancet, and by 1824 he had published articles on the English Lake Region, belladonna, malaria, and meningitis; a book on the bowel complaints of children; numerous book reviews; translations of Félix-Hippolyte Larrey's *Moxa* and of François Magendie's *Formulary;* and an edition of Robert Hooper's *Vade-Mecum;* and had served on the editorial boards of two medical journals. In 1824 he married Harriet Leadam, daughter of a London apothecary; they had seven children. Shortly after their marriage they went to the University of Virginia where, at the behest of Thomas Jefferson, Dunglison was appointed to the chair of medicine. Responsible only for teaching, Dunglison was able to prepare textbooks on those subjects he taught. (Medical instructors, who had, until then, been actively engaged in practice, relied chiefly upon the European literature for information on current advances.) Dunglison thus became the first full-time professor of medicine in the United States and the first American author of a book on physiology, a medical dictionary, and a history of medicine, as well as a pioneer in the publication of works on public health (or hygiene, as he called it), materia medica and therapeutics, medical jurisprudence and toxicology, medical education, and internal medicine. (He abhorred the knife and completely avoided surgery.) Dunglison also made important contributions to William Beaumont's classic work on the physiology of digestion.

After eight years at the University of Virginia, Dunglison moved to the University of Maryland and then, after three years, to the Jefferson Medical College at Philadelphia, where he taught for the next thirty-two years. When he arrived, faculty dissension and rivalry with another Philadelphia medical school were threatening to destroy the college, but Dunglison's skillful reorganization of the faculty welded it into a coherent, cooperative teaching group, and he succeeded in establishing the school as one of the country's best medical centers. Fluent, lucid, elegant, entertaining, instructive, and stimulating as a lecturer, he attracted many students: more than 5,000 nineteenth-century physicians proudly displayed his signature on their diplomas. Elected to many organizations, he was especially active in the American Philosophical Society and in the Musical Fund Society of Philadelphia. As a member of the Pennsylvania Institution for the Blind he was an early advocate of raised type for the blind. He was also interested in the Elwyn School for the mentally retarded and worked for improved care of the insane poor, preparing several reports that led to reforms in asylums. An Episcopalian, he was a member of the vestry of St. Stephen's Church, Philadelphia. Two of his sons, Richard James and Thomas Randolph, were physicians. Although medical practice was not to his taste, he attended Thomas Jefferson in his last illness and was consulted by presidents Monroe, Madison, and Jackson and by families connected with the University of Virginia. His importance to American medical history rests in his extraordinary success in sifting from the world literature information of importance to medical students and physicians, and in his ability to present this information effectively. In addition we owe to him the firm establishment of two great medical institutions, the University of Virginia School of Medicine and the Jefferson Medical College.

BIBLIOGRAPHY

I. ORIGINAL WORKS. A fairly complete list of Dunglison's writings appears in "The Autobiographical Ana of Robley Dunglison," ed. with notes and an intro. by Samuel X. Radbill, in *Transactions of the American Philosophical Society*, n.s., **53** (1963), 196–199. His most significant medical publications are *Commentaries on the Diseases of the Stomach and Bowels of Children* (London, 1824); *Syllabus of Lectures on Medical Jurisprudence, and on the Treatment of Poisoning and Suspended Animation* (Charlottesville, Va., 1827); and *Human Physiology*, 2 vols. (Philadelphia, 1832; 8th ed., 1856). *A New Dictionary of Medical Science and Literature*, 2 vols. (Boston, 1833) appeared in 1 vol. in its 2nd and subsequent eds.; the 19th ed. (Philadelphia, 1868) was the last published in Robley Dunglison's lifetime; his son Richard edited several subsequent eds.; in 1911 Thomas Lathrop Stedman continued it as *Stedman's Practical Medical Dictionary*, and it is still appearing, a century after Dunglison's death. Other works are *Elements of Hygiene* (Philadelphia, 1835); *General Therapeutics, or Principles of Medical Practice* (Philadelphia, 1836); *The Medical Student* (Philadelphia, 1837); *New Remedies* (Philadelphia, 1839); and *The Practice of Medicine*, 2 vols. (Philadelphia, 1842). The *Dictionary for the Blind in Tangible Type*, 3 vols. (Philadelphia, 1860) was prepared by W. Chapin under the supervision of Dunglison. Dunglison also edited the following journals: *London Medical Repository* (1823–1824), *Medical Intelligencer* (1823), *Virginia Literary Museum and Journal of Belles Lettres, Arts, Sciences, etc.* (1830), and *American Medical Library and Intelligencer* (1837–1842). Dunglison's interest in the mentally retarded is reflected in *Appeal to the People of Pennsylvania on the Subject of an Asylum for the Poor of the Commonwealth* (Philadelphia, 1838); a *Second Appeal* was pub. in 1840. A posthumous work, *History of Medicine*, was arranged and ed. by his son Richard J. Dunglison, M.D. (Philadelphia, 1872).

II. SECONDARY LITERATURE. On Dunglison and his work, see William B. Bean, "Mr. Jefferson's Influence on American Medical Education," in *Virginia Medical Monthly*, **87** (1960), 669–680; John M. Dorsey, *Jefferson-Dunglison Let-*

ters (Charlottesville, Va., 1960); Chalmers L. Gemmill, "Educational Work of Robley Dunglison, M.D. at the University of Virginia," in *Virginia Medical Monthly,* **87** (1960), 307–309; Chalmers L. Gemmill and Mary Jeanne Jones, *Pharmacology at the University of Virginia School of Medicine* (Charlottesville, Va., 1966), pp. 9–23; Samuel D. Gross, "Memoir of Robley Dunglison," in *Transactions of the College of Physicians of Philadelphia,* n.s., **4** (1874), 294–313, and *Autobiography,* II (Philadelphia, 1887), 334; Mary Jeanne Jones and Chalmers L. Gemmill, "The Notebook of Robley Dunglison, Student of Clinical Medicine in Edinburgh, 1816–1818," in *Journal of the History of Medicine and Allied Sciences,* **22** (1967), 261–273; Henry Lonsdale, *Worthies of Cumberland,* VI (London, 1875), 262–279; Samuel X. Radbill, "Robley Dunglison, M.D., 1788–1869: American Medical Educator," in *Journal of Medical Education,* **34** (1959), 84–94; and "Dr. Robley Dunglison and Jefferson," in *Transactions and Studies of the College of Physicians of Philadelphia,* 4th ser., **27** (1959), 40–44.

SAMUEL X. RADBILL

DUNOYER DE SEGONZAC, LOUIS DOMINIQUE JOSEPH ARMAND (*b.* Versailles, France, 14 November 1880; *d.* Versailles, 27 August 1963), *physics.*

Dunoyer was the son of Anatole Dunoyer, a founder of the École des Sciences Politiques in Paris, and Jeanine Roquet. He married Jeanne Picard, daughter of Émile Picard, on 4 June 1907. They had two sons, whose studies Dunoyer supervised himself.

As a youth, Dunoyer placed first in the general physics competition and was second on the admissions list of the École Polytechnique and first on that of the École Normale Supérieure. He chose to attend the latter (1902–1905), which oriented him toward teaching and research. He placed first in the physics *agrégation* in 1905 and in that year was an assistant to P. Langevin at the Collège de France. His first research concerned the difficulties of compensating compasses in iron and iron-clad ships. This work furnished the subject for his doctoral thesis[1] and took concrete form in the dygograph and the type of electromagnetic compass[2] that was mounted in Lindbergh's *Spirit of St. Louis.* In 1908 he won the Prix Extraordinaire de la Marine[3] for his research in magnetism.

A Carnegie scholar in the laboratory of Marie Curie in 1909, Dunoyer conducted the fundamental experiment on molecular beams in 1912.[4] Originally designed to verify the kinetic theory, the experiment also resulted in the preparation of thin films of alkali metals. "He showed that in a good vacuum one could obtain a linear beam of molecules, but that if the vacuum degenerated, the impacts of the molecules against each other produced a broadening and a

disintegration of the beam." [5] (This work was followed by the studies of the properties of molecules, without perturbation, by Otto Stern and others.) This experimental demonstration of the kinetic theory of gases was the origin of the preparation of thin films by thermal vaporization and of the studies of the properties of atoms and molecules by the so-called molecular ray method.[6] In 1912 Dunoyer was awarded the Subvention Bonaparte[7] for his work on the fluorescence of pure sodium vapor and for the complete investigation of the fluorescence and absorption spectra of the alkaline metals. In 1913 he won the Prix Becquerel for his research on the electrical and optical properties of metallic vapors, notably of sodium vapor.[8] In the same year he was appointed deputy professor at the Conservatoire National des Arts et Métiers. He studied the surface resonance of sodium vapor with R. W. Wood in 1914.

An aviation officer and inspector, Dunoyer was wounded and became *chevalier* of the Legion of Honor, receiving the Croix de Guerre in 1915 as well. He became interested in meteorology and aerial navigation and invented a bombsight.[9] In 1918 he was awarded the Prix Danton for his work on radiant phenomena.[10]

Dunoyer became a lecturer at the Institut d'Optique in 1919. He was physicist at the observatory of Meudon from 1927 to 1929 and professor at the Institut d'Optique from 1921 to 1941. He participated in the founding of the Société de Recherches et de Perfectionnements Industriels, and while secretary-general of the Société Française de Physique he devised a special lens for the illumination of atomic beams. A glassblower and remarkable technician, he improved the procedures of Wolfgang Gaede (1913) and of Langmuir (1916) and developed various diffusion pumps and devices for measuring very low temperatures. These accomplishments brought him the Subvention Loutreuil in 1925.[11]

Simultaneously Dunoyer pursued his research on photoelectricity and the construction of photoelectric cells; the first application (1925) was to talking movies, where a potassium cell was employed. He won the Prix Valz in 1929 for his research on the spirit level and on photoelectric cells.[12] In 1930 he was awarded the Subvention Loutreuil for the continuation of his research on photoelectric cells.[13]

In 1935 Dunoyer's studies on thermal vaporization in a vacuum enabled him to construct the first aluminized mirrors.[14] On 10 February 1937 he was elected artist member of the Bureau des Longitudes, replacing Louis Jolly. From 1941 to 1945 he was titular professor at the Sorbonne and director of the Institut de Chimie Physique, where he taught a remarkable course on the kinetic theory of gases. Dunoyer was named honorary president of the Société des Ingénieurs du Vide, in whose journal, *Le vide,* he published seven articles between 1949 and 1956. The society dubbed him "Grandfather of the Vacuum."

NOTES

1. The thesis was under the direction of E. E. N. Mascart and P. Langevin and was entitled *Étude sur les compas de marine et leurs méthodes de compensation. Un nouveau compas électromagnétique* (Paris, 1909). It is *thèse* Fac. Sciences Paris, no. 1336.
2. See *Comptes rendus hebdomadaires des séances de l'Académie des sciences,* **145** (1907), 1142–1147, 1323–1325; **147** (1908), 834–837, 1275–1277; *Bulletin de la Société française de physique,* **295** (1909–1910); *Revue maritime,* **315** (1910). The *Comptes rendus* are hereafter cited as *CR.*
3. See *CR,* **147** (1908), 1111, 1113–1117, for his electromagnetic compass tested on the battleship *Patrie* and the dygograph placed on the battleship *Danton.* It replaced the ordinary compass, which was rendered useless by rarefaction of the magnetism, resulting from the ship's armor.
4. See *CR,* **152** (1911), 592–595; **153** (1911), 333–336; **154** (1912), 815–818, 1344–1346; **155** (1912), 144–147, 270–273; **157** (1913), 1068–1070; **158** (1914), 1068–1071, 1265–1267, written with R. W. Wood; *Bulletin de la Société française de physique,* four memoirs between 1912 and 1914; *Journal de physique et radium,* **185** (1913); *Collection de mémoirs relatifs à la physique* (1912); *Radium,* seven memoirs between 1910 and 1914.
5. Robert Champeix, *Le vide* (Paris, 1965).
6. *Le vide,* **106** (July-Aug. 1963).
7. See *CR,* **155** (1912), 93, 1407.
8. See *CR,* **157** (1913), 1287.
9. See *CR,* **165** (1917), 1068–1071, written with G. Reboul; **166** (1918), 293–295; **168** (1919), 47, 138, 457–459 (with Reboul), 785–787 (with Reboul), 726–729, 1102–1105; **169** (1919), 762, 78–79, 191–193 (the last two written with Reboul); **170** (1920), 744–747 (with Reboul); **173** (1921), 1101–1104; *Bulletin de la Société française de physique* (1920); *Technique aéronautique* (1921).
10. See *CR,* **167** (1918), 829.
11. See *CR,* **181** (1925), 1012, 1016. At the same time he was given money from the research fund of the secretary-general of the Société Française de Physique in order to pursue his investigations of certain problems concerning modern methods of measuring high vacuums.
12. See *CR,* **189** (1929), 1123.
13. See *CR,* **191** (1930), 1245. See also *CR,* **174** (1922), 1615–1617 (written with P. Toulon); **176** (1923), 953–955, 1213; **179** (1924), 148–151, 461–464, 522–575 (all written with P. Toulon); **182** (1926), 686–688; **185** (1927), 271–273; **196** (1933), 684–686 (written with Paounoff); **198** (1934), 909–911; **200** (1935), 1835–1838; *Bulletin de la Société française de physique,* 13 memoirs between 1922 and 1930; *Revue d'optique théorique et instrumentale,* 11 memoirs between 1922 and 1948; *Journal de physique et radium,* 2 memoirs.
14. See *CR,* **202** (1936), 474–476; **220** (1945), 520–522, 686–688, 816–817, 907–909; **221** (1945), 97–99; **230** (1950), 57–58; **232** (1951), 1080–1082; **233** (1951), 125, 919–921.

BIBLIOGRAPHY

In addition to the articles cited in the notes, see *La technique du vide* (Paris, 1924); *Les émissions électroniques des couches minces* (Paris, 1932); *Les radiations monochromatiques* (Paris, 1935); "Les cellules photoélectriques," in

Comptes rendus. 2° Congrès international d'électricité (Paris, 1936); *Allocution pour le vingtième anniversaire de la mécanique ondulatoire* (Paris, 1944); *Le vide et ses applications,* in the series Que Sais-je?, no. 430 (Paris, 1950).

His articles published in *Le vide* are "Étude d'un micromanomètre thermique, précédée de quelques remarques générales sur ce type d'instrument," **4** (1949), 571–581, 603–618, 643–660; "Sur certaines phénomènes de dégagement gazeux observés pendant le pompage de lampes à incandescence à basse tension," **5** (1950), 793–806; "Quelques remarques sur les formules de l'écoulement des gaz raréfiés dans les canalisations," *ibid.,* 881–886; "Bases théoriques de la dessication dans le vide," **6** (1951), 1025–1040, 2077–2090; "Expériences sur l'évaporation de l'eau dans le vide et comparaison avec la théorie," **8** (1953), 1280–1294; "Fonctionnement des condensateurs de vapeur d'eau dans les appareils de dessication dans le vide," **10** (1955), 165–184; and "Quelques appareils pour la production de faisceaux moléculaires et quelques expériences sur ceux de sodium en resonance optique," **11** (1956), 172–189.

P. BERTHON

DUNS SCOTUS, JOHN (*b.* Roxburghshire, Scotland, *ca.* 1266; *d.* Cologne, Germany, November 1308), *philosophy.*

Little is known of the life of John Duns Scotus, who was among the outstanding thinkers of the later Middle Ages. He entered the Franciscan order probably in 1279 or 1280 and was ordained in 1291. He studied first at Oxford University and then at Paris University, returning to Oxford in 1300 to complete the requirement for his doctorate. Before he could take his degree, however, he was once again sent by his superiors to Paris, where he finally became a doctor of theology in 1305, having been temporarily banished from France in 1303, together with about seventy other friars, for supporting Pope Boniface VIII in his quarrel with the French king, Philip the Fair. We last hear of him at Cologne in 1308, teaching in the Franciscan house there.

Duns Scotus' premature death together with the vicissitudes of his career have combined to make his writings more than usually problematical. Only gradually is the correct relation between his lectures at Paris and those at Oxford being established, while the authority of other works ascribed to him has still to be definitively established. His major writings are his two commentaries on the *Sentences* of Peter Lombard, a compendium of theology, which constituted one of the main exercises for a degree in that subject. Because of his studies in the theological faculties of both Oxford and Paris, Duns Scotus wrote two such commentaries, *Opus Oxoniense* and *Reportata Parisiensis.* Each was left in an unfinished state, as were all his other main works; the unraveling of the correct relation of the two commentaries to each other has been one of the preoccupations of the Scotist editorial commission over the past thirty years and is still not complete. Even when it is, Duns Scotus' thought will always be incompletely understood. Within a few years of his death his teaching had been developed by his followers into a definite set of tenets, from which it is sometimes difficult to disentangle his own positions. Scotism became one of the dominant schools of later medieval thought, and much in Duns Scotus' teaching formed the point of departure for William of Ockham's own, more far-reaching radicalism.

Like the majority of medieval thinkers, Duns Scotus was primarily a theologian. He sought to provide a new, metaphysical basis for a natural theology, which would thereby free such discourse from dependence upon natural phenomena. Duns Scotus was writing in the aftermath of the great 1277 condemnations at Paris and at Oxford of over 200 theses that had applied criteria drawn from the sensory world to the articles of Christian faith. The condemnations had crystallized the danger inherent in employing the categories of nature in seeking knowledge of the divine. As a consequence, many theologians in the years immediately before Duns Scotus had sought a return to the older, traditional stress upon inner, nonsensory awareness as the source of higher knowledge. Duns Scotus, however, denied the human mind any but a sensory source for its knowledge. Accordingly, the problem was how to arrive at concepts that could be held independently of sensory experience. Scotus found the answer in metaphysics—the study of being in itself—and more specifically in the notion of being. As a concept, being was the most universal of all categories, under which every other concept fell. In this most generalized form, being was univocal: it applied indifferently to all that is, regardless of different kinds of being. It therefore transcended the physical properties of specific beings known through the senses; thus, if it could be applied to God, it would free any discussion of him from reliance upon physical categories. In that way, God could be the object of metaphysical, as opposed to physical, discourse. Duns Scotus held that the way to this lay in considering being in its two main modes, infinite and finite. Infinite being was by definition necessary and uncaused, while finite being was dependent upon another for its existence and, so, contingent. Accordingly, metaphysics could adduce God's existence as necessary being and that of his creatures as finite. But that was as far as it could go. Beyond saying God was first being, one could know his nature only when one turned from metaphysics to theology; in like manner, what he had

ordained for creation belonged to the articles of faith, not to natural reason.

The effects of Duns Scotus' reorientation of metaphysics were to put a new stress upon infinity and contingency. On the one hand, only God was infinite and, so, beyond the compass of human discourse; once having established God as the first infinite being, metaphysics could offer no analogies between the divine and the created. There was no place for Aquinas' five proofs of God's existence drawn from knowledge of this world, just as Duns Scotus allowed none to the older Augustinian doctrine of divine illumination of the soul, by which the soul was enabled to know eternal truths. On the other hand, creatures, since they were merely contingent, had no other *raison d'être* than God's having willed them. God's will was the only reason for the existence of that which was finite and need not have been. Moreover, God was absolutely free to do anything save contradict himself, which would limit him. Duns Scotus gave a renewed emphasis to God's omnipotence by reviving the distinction between God's ordained power as applied to this world and his absolute power by which he could do anything. Whereas by his ordained power he had decreed the unchanging laws that govern creation, by his absolute power God could supersede those laws and thus, for example, reward a man without first having infused him with grace. Duns Scotus does not appear to have pressed very far the contrast between these two aspects of God's power, but in this, as in stressing God's infinity, he opened the way to a much more radical application by William of Ockham and his followers.

The significance of Duns Scotus in the history of thought is that he broke away from the previous ways of establishing a natural theology. In doing so, he limited the area of meaningful natural discourse about the divine and gave new force to the contingent nature of creation. He thereby took an important step in separating natural experience and reason from revealed theological truth and from the preordained determinism against which the condemnations of 1277 had been especially directed. Those of the next generation, above all William of Ockham, were to make unbridgeable the gulf thus opened between knowledge and faith and to arrive at new and fruitful ways of interpreting natural phenomena.

BIBLIOGRAPHY

Duns Scotus' works were collected as *Opera omnia*, 12 vols. (Lyons, 1639; repr., Paris, 1891–1895). A new critical edition by the Scotist Commission, under C. Balíc, at Rome is in progress.

Modern editions of individual works include *Tractatus de primo principio,* ed. and with English trans. by E. Roche (New York, 1949). Selections from Duns Scotus in English translation are contained in *John Duns Scotus: Philosophical Writings,* A. Wolter, ed. and trans., which also provides a selected bibliography. A fuller bibliography is to be found in E. Gilson, *History of Christian Philosophy* (London, 1955), pp. 763–764.

GORDON LEFF

DUPERREY, LOUIS-ISIDORE (*b.* Paris, France, 21 [22?] October 1786; *d.* Paris, 25 August [10 September?] 1865), *navigation, hydrography, terrestrial magnetism.*

A sailor in the French navy from 1803 onward, Duperrey rose rapidly through the ranks, was assigned to carry out a hydrographic mission off the coast of Tuscany in 1809, and received his first command in 1814. He was in charge of hydrographic activities on Louis de Freycinet's expedition around the world in 1817–1820 and produced valuable observations on the earth's shape and on terrestrial magnetism, together with numerous charts. About one year following his return Duperrey presented to the naval minister a plan for another circumnavigating expedition, and in 1822 he embarked in command of the *Coquille.* Second in command was Jules-Sébastien-César Dumont d'Urville, and the expedition was joined by two naturalists, René-Primevère Lesson and Prosper Garnot.

Before his return to Marseilles in 1825 (accomplished without the loss of a single man), Duperrey and his company discovered a number of unknown islands, prepared charts of previously little-known areas of the South Pacific (especially in the Caroline Archipelago), studied ocean currents, gathered new information on geomagnetic and meteorological phenomena, and collected an impressive array of geological, botanical, and zoological specimens for the Muséum d'Histoire Naturelle. After the return of the expedition Duperrey and his collaborators worked assiduously to prepare the results of the journey for publication. The expedition particularly distinguished itself in producing new knowledge of the behavior of ocean currents in the Atlantic and Pacific oceans (there was a certain fascination with asymmetry in the weather of the Northern and Southern hemispheres) and in its contributions to knowledge of variations in intensity and direction of terrestrial magnetism. Duperrey himself was especially concerned with the determination of the earth's magnetic equator.

Duperrey was elected to membership in the Academy of Sciences (section for geography and navigation) in 1842, became vice-president in 1849, and served as president in 1850.

BIBLIOGRAPHY

I. ORIGINAL WORKS. The results of the voyage of the *Coquille* were published as *Voyage autour du monde, exécuté par ordre du Roi, sur la corvette de Sa Majesté, la Coquille, pendant les années 1822, 1823, 1824 et 1825. . .* (7 vols. plus 4 vols. of plates and maps, Paris, 1825–1830). Duperrey himself prepared the *Histoire du voyage* (1825), *Hydrographie et physique* (1829), *Hydrographie* (1829), and *Physique* (1830); *Botanique* (2 vols., 1828) was prepared by Dumont d'Urville, J. B. Bory de Saint-Vincent, and Adolphe Brongniart; while Lesson, Garnot, and Félix-Édouard Guérin-Méneville collaborated on *Zoologie* (2 vols., 1826–1830). Duperrey also published separately *Mémoire sur les opérations géographiques faites dans la campagne de la corvette de S. M. la Coquille, pendant les années 1822, 1823, 1824 et 1825* (Paris, n.d. [1827]).

Duperrey's publications of the scientific results of his expeditions also appeared in the form of articles, such as "Résumé des observations de l'inclinaison et de la déclinaison de l'aiguille aimantée faites dans la campagne de la corvette de S. M. la *Coquille,* pendant les années 1822, 1823, 1824, et 1825," in *Annales de chimie et de physique,* **34** (1827), 298–320; "Notice sur la configuration de l'équateur magnétique, conclue des observations faites dans la campagne de la corvette la *Coquille,*" ibid., **45** (1830), 371–386; "Notice sur la position des pôles magnétiques de la terre," in *Comptes rendus hebdomadaires des séances de l'Académie des Sciences,* **13** (1841), 1104–1111; and "Réduction des observations de l'intensité du magnétisme terrestre faites par M. de Freycinet et ses collaborateurs durant le cours du voyage de la corvette l'*Uranie,*" ibid., **19** (1844), 445–455. Other articles were published in *Additions à la connaissance des temps* and *Annales maritimes et coloniales.* Information gathered by Duperrey on terrestrial magnetism is set forth extensively in Antoine César Becquerel, *Traité expérimental de l'électricité et du magnétisme, et de leurs rapports avec les phénomènes naturels,* vol. VII (Paris, 1840).

II. SECONDARY LITERATURE. On Duperrey and his work see Dominique F. J. Arago and others, "Rapport fait à l'Académie des Sciences, le lundi 22 août 1825, sur le voyage de découvertes, exécuté dans les années 1822, 1823, 1824 et 1825, sous le commandement de M. Duperrey, lieutenant de vaisseau," in *Additions à la connaissance des temps,* année 1828, pp. 240–272; this appears to be the principal source for a long biographical article by P. Levot in Hoefer's *Nouvelle biographie générale,* vol. XV (Paris, 1856), cols. 278–286. Duperrey published a *Notice sur les travaux de M. L.-I. Duperrey, ancien officer supérieur de la marine* (Paris, 1842). Contemporary biographical notices appear in Figuier's *L'année scientifique et industrielle* (1866), pp. 477–478; and in Edouard Goepp and Henri de Mannoury d'Ectot, *La France biographique illustrée: Les marins,* 2 vols. (Paris, 1877), II, 227–228. A recent sketch by É. Franceschini is in *Dictionnaire de biographie française,* fasc. LXVIII (1968), cols. 338–339. Information on Duperrey's circumnavigating expedition is found in Paul Chack, *Croisières merveilleuses* (Paris, 1937), pp. 131–147;

and in Robert J. Garry, "Geographical Exploration by the French," in *The Pacific Basin: A History of Its Geographical Exploration,* Herman R. Friis, ed. (New York, 1967), pp. 201–220.

KENNETH L. TAYLOR

DUPIN, PIERRE-CHARLES-FRANÇOIS (*b.* Varzy, France, 6 October 1784; *d.* Paris, France, 18 January 1873), *mathematics, economics, education.*

Dupin grew up in his native Nivernais, where his father, Charles-André Dupin, was a lawyer and legislator. His mother was Cathérine Agnès Dupin (her maiden name was also Dupin). The second of three sons, Dupin graduated in 1803 from the École Polytechnique in Paris as a naval engineer. In 1801, under the guidance of his teacher Gaspard Monge, he had made his first discovery, the cyclid (of Dupin). After assignments in Antwerp, Genoa, and Toulon, he was placed in charge of the damaged naval arsenal on Corfu in 1807. He restored the port, did fundamental research on the resistance of materials and the differential geometry of surfaces, and became secretary of the newly founded Ionian Academy. In 1810, on his way back to France, he was detained by illness at Pisa; and during his convalescence he edited a posthumous book by his friend Leopold Vacca Berlinghierri, *Examen des travaux de César au siège d'Alexia* (Paris, 1812). At the Toulon shipyard in 1813, Dupin founded a maritime museum that became a model for others, such as that at the Louvre. That year he published his *Développements de géométrie.*

In 1816, after some difficulty, Dupin was allowed to visit Great Britain to study its arsenals and other technical installations. The results were published in *Voyages dans la Grande Bretagne entrepris relativement aux services publics de la guerre, de la marine . . . depuis 1816* (1820–1824).

Settling down to a life of teaching and public service, Dupin accepted the position of professor of mechanics at the Paris Conservatoire des Arts et Métiers, a position he held until 1854. His free public lectures, dealing with mathematics and mechanics and their industrial applications, became very popular. His *Applications de géométrie et de mécanique* (1822) was a continuation of the *Développements* but placed greater stress on applications. Many of Dupin's lectures on industry and the arts were published in *Géométrie et mécanique des arts et métiers et des beaux arts* (1825); his *Sur les forces productives et commerciales de France* appeared two years later. In 1824 the king made him a baron.

The *Développements* contains many contributions to differential geometry, notably the introduction of conjugate and asymptotic lines on a surface, the so-

called indicatrix of Dupin, and "Dupin's theorem," that three families of orthogonal surfaces intersect in the lines of curvature. A particular case Dupin investigated consisted of confocal quadrics. In the *Applications* we find an elaboration of Monge's theory of *déblais et remblais*—and, hence, of congruences of straight lines, with applications to geometrical optics. Here Dupin, improving on a theorem of Malus's (1807), stated that a normal congruence remains normal after reflection and refraction. He also gave a more complete theory of the cyclids as the envelopes of the spheres tangent to three given spheres and discussed floating bodies. In 1840 he introduced what is now called the affine normal of a surface at a point.

In 1828 Dupin was elected deputy for Tarn, and he continued in politics until 1870. In 1834 he was minister of marine affairs, in 1838 he became a peer, and in 1852 he was appointed to the senate. He tirelessly encouraged the establishment of schools and libraries, the founding of savings banks, the construction of roads and canals, and the use of steam power. In 1855 he reported on the progress of the arts and sciences, as represented at the Paris World Exhibition; the part of the report dealing with Massachusetts was published in English (1865).

Dupin married Rosalie Anne Joubert in 1830. He was a correspondent of the Institut de France (1813) and a member of both the Académie des Sciences (1818) and the Académie des Sciences Morales et Politiques (1832). His older brother, André, known as Dupin *aîné,* was a prominent lawyer and politician.

BIBLIOGRAPHY

I. ORIGINAL WORKS. Among Dupin's writings are his ed. of Berlinghierri's *Examen des travaux de César* (Paris, 1812); *Développements de géométrie* (Paris, 1813); *Voyages dans la Grande Bretagne,* 3 vols. (Paris, 1820–1824); *Applications de géométrie et de mécanique des arts et métiers,* 3 vols. (Brussels, 1825); *Sur les forces productives,* 2 vols. (Paris, 1827); "Mémoire sur les éléments du troisième ordre de la courbure des lignes," in *Comptes rendus de l'Académie des sciences,* **26** (1848), 321–325, 393–398; and *Forces productives des nations de 1800 à 1851* (Paris, 1851). For his many economic and technical writings, see A. Legoyt, in *Nouvelle biographie générale,* XIV (1868), 315–326.

II. SECONDARY LITERATURE. On Dupin's life, see J. Bertrand, *Éloges académiques* (Paris, 1890), pp. 221–246; and A. Morin, *Discours funéraires de l'Institut de France* (Paris, 1873). Dupin's mathematical work is discussed in J. G. Darboux, *Leçons sur la théorie générale des surfaces* (Paris, 1887–1896, see index), and *Leçons sur les systèmes orthogonaux et les coordonnées curvilignes* (Paris, 1898; 2nd ed., 1910), ch. 1.

For information on the Dupin family, I am indebted to M. Baron Romain, Corvol d'Embernard (Nièvre).

DIRK J. STRUIK

DUPRÉ, ATHANASE LOUIS VICTOIRE (*b.* Cerisiers, France, 28 December 1808; *d.* Rennes, France, 10 August 1869), *physics, mathematics.*

After early education at the Collège of Auxerre, Dupré entered the École Normale Supérieure in Paris in 1826, gained first place in science in the *agrégation* of 1829, and immediately took a post at the Collège Royal in Rennes. There he taught mathematics and physical science until, in 1847, he was appointed to the chair of mathematics in the Faculty of Science in Rennes. His last post, from 1866, was as dean of the faculty there. He received the Legion of Honor in 1863 but won no other major honor and was never a member of the Académie des Sciences. He was an ardent Catholic throughout his life.

Dupré's scientific career fell into two parts. In the first, which lasted from his years at the École Normale until about 1859, he contributed to several branches of mathematics and physics. The most important of his papers from this period was his entry of 1858 for the competition in mathematics set by the Académie des Sciences. The paper, a study of an outstanding problem in Legendre's theory of numbers, earned Dupré an honorable mention but only half the prize of 3,000 francs.

During the second period, which covered the remaining ten years of his life, Dupré concerned himself exclusively with the mechanical theory of heat, his main interest being the implications of the theory for matter on the molecular scale. He made an important contribution to the dissemination in France of the newly discovered principles of thermodynamics in nearly forty communications to the Academy, in an entry for the Academy's Prix Bordin in 1866 (which again won him only an honorable mention and half the prize money), and in a successful advanced textbook, *Théorie mécanique de la chaleur* (1869).

BIBLIOGRAPHY

I. ORIGINAL WORKS. The most important of Dupré's papers were published in the *Annales de chimie et de physique;* those on the mechanical theory of heat were summarized in the *Théorie mécanique de la chaleur* (Paris, 1869). His entry for the 1858 competition was published as *Examen d'une proposition de Legendre relative à la théorie des nombres* (Paris, 1859).

II. SECONDARY LITERATURE. The only full biographical sketch, by Simon Sirodot, Dupré's successor as dean of the Faculty of Sciences in Rennes, appeared in the annual publication *Université de France. Académie de Rennes. Rentrée solennelle des facultés des écoles préparatoires de médecine et de pharmacie et des écoles préparatoires à l'enseignement supérieur des sciences et des lettres de l'Académie de Rennes* (Rennes, 1869), pp. 52–57.

ROBERT FOX

DÜRER, ALBRECHT (*b.* Nuremberg, Germany, 21 May 1471; *d.* Nuremberg, 6 April 1528), *mathematics, painting, theory of art.*

Dürer was the son of Albrecht Dürer (or Türer, as he called and signed himself) the Elder. The elder Dürer was the son of a Hungarian goldsmith and practiced that craft himself. He left Hungary, traveled through the Netherlands, and finally settled in Nuremberg, where he perfected his craft with Hieronymus Holper. He married Holper's daughter Barbara. The printer and publisher Anton Koberger stood godfather to the younger Dürer.

Dürer attended the *Lateinschule* in St. Lorenz and learned goldsmithing from his father. From 1486 to 1489 he studied painting with Michael Wolgemut (then the leading church painter of Nuremberg); in Wolgemut's workshop he was able to learn not only all the standard painting techniques but also wood- and copper-engraving. In 1490, in accordance with the custom of the painter's guild, Dürer went on his *Wanderjahre.* Until 1494 he traveled through the Upper Rhine and to Colmar, Basel, and Strasbourg, presumably making his living as a draftsman.

Dürer returned to Nuremberg and on 7 July 1494 married Agnes Frey, the daughter of Hans Frey. Frey, who had been a coppersmith, had become prosperous as a mechanician and instrument maker. He belonged to that school of craftsmen in metals for which Nuremberg was famous. The marriage brought Dürer's family increased social standing and brought Dürer a generous dowry.

As early as his *Wanderjahre* Dürer had come to appreciate the works of Mantegna and other Italian artists. He wished to learn more of the artistic and philosophical rediscoveries of the Italian Renaissance (he knew from books about the Academy of Florence, modeled on Plato's Academy). Moreover, he had become convinced that the new art must be based upon science—in particular, upon mathematics, as the most exact, logical, and graphically constructive of the sciences. It was this realization that led him to the scientific work for which he was, in his lifetime, as celebrated as for his art. He decided to travel to Italy and in 1494 left his wife in Nuremberg and set off on foot to visit Venice.

On his return to Nuremberg in 1495, Dürer began serious study of mathematics and of the theory of art as derived from works handed down from antiquity, especially Euclid's *Elements* and the *De architectura* of Vitruvius. These years were highly productive for Dürer; in 1497 he adopted his famous monogram 𝔸 to protect his work against being counterfeited. At about the same time he formed an important and lasting friendship with Willibald Pirckheimer, subject of one of his most famous portraits. (Dürer was fortunate in his patrons and friends; besides Pirckheimer these included such humanists as Johannes Werner, the mathematician; Johann Tscherte, the imperial architect; and Nicholas Kratzer, court astronomer to the English King Henry VIII.)

Most important, however, this period marked the beginning of Dürer's experiments with scientific perspective and mathematical proportion. The mathematical formulations of Dürer's anatomical proportion are derived both from antiquity and from the Italian rediscoveries; he drew upon both Polyclitus the Elder and Alberti, and to these he added the notion of plastic harmony after the mode of musical harmony taken from Boethius and Augustine. The earliest of Dürer's documented figure studies to be constructed in accordance with one or several strictly codified canons of proportion date from 1500 and include the study of a female nude (now in London). In addition, critics have pointed out that the head of the famous Munich self-portrait may be shown to have been constructed proportionally.

Throughout the years 1501–1504 Dürer continued to work with the problem of proportion, making numerous studies of men and horses. His copper engraving *Adam and Eve* (1504) marks the high point of his theoretical mastery—the figures were methodically constructed, he wrote, with a compass and a ruler. The preliminary studies for the *Adam and Eve* (now in Vienna) reveal Dürer's method. During this time he also mastered the techniques of linear perspective, as may be seen in his series of woodcuts, *The Life of the Virgin*.

In 1505–1507 Dürer returned to Venice. He extended his Italian travels to Bologna on this occasion, "on account," he wrote, "of secret [knowledge of] perspective." He most probably made the journey to meet with Luca Pacioli, a mathematician and theorist of art. Pacioli's book, *Divina proportione* (in which Leonardo da Vinci collaborated), propounded the notion that the *sectio aurea* (the famous "golden mean" of classical sculpture and architecture), being mathematical in nature, related art to that science exclusively. In Venice, at the close of his second Italian trip, Dürer bought Tacinus' 1505 edition of Euclid, which was henceforth to be his model for mathematical formulation. This period of Dürer's life marks

the full maturity of his mathematical, philosophical, and aesthetic theory; in his painting he had begun to realize the full synthesis of late German Gothic and Italian Renaissance painting.

Between 1506 and 1512 Dürer devoted himself to the rigorous study of the problem of form, which presented itself to him in three aspects: true, mathematical form; beautiful, proportional form; and compositional form, used in an actual work of art, ideally the fusion of the preceding. In solving these problems Dürer drew upon the resources of arithmetic and geometry; it was in his achievement as a painter that his formal solutions were meaningful and expressive.

From about 1508 Dürer sketched and wrote down the substance of his theoretical studies (fragments of these notes and drawings are preserved in the notebooks in London, Nuremberg, and Dresden). Some of these fragments may have been intended for inclusion in the encyclopedic *Speis' der Malerknaben* that Dürer had planned to publish; this *Malerbuch* was to have presented his mathematical solutions to all formal problems in the plastic arts. Although the *Malerbuch* was never completed, Dürer extracted a part of it for his major "Treatise on Proportion" (*Proportionslehre*).

In 1520–1521 Dürer traveled to the Netherlands, particularly Bruges and Ghent, where he saw the works of the early Flemish masters. He returned to Nuremberg ill with malaria; henceforth he devoted himself primarily to the composition and printing of his three major theoretical books. (He continued to paint, however; his pictures from this period include several notable portraits of his friends as well as the important diptych of the *Four Apostles,* given to the city council of Nuremberg by Dürer in 1526 and now in Munich.)

Dürer had completed the manuscript of the "Treatise on Proportion" by 1523, but he realized that a more basic mathematical text was necessary to its full comprehension. For that reason, in 1524–1525, he wrote such a text, the *Underweysung der Messung mit Zirckel und Richtscheyt in Linien, Ebnen und gantzen Corporen* ("Treatise on Mensuration With the Compass and Ruler in Lines, Planes, and Whole Bodies"), which was published by his own firm in Nuremberg in 1525.

The *Underweysung der Messung* is in four books. In the first, Dürer treats of the construction of plane curves (including the spiral of Archimedes, the logarithmic spiral, tangential spirals, conchoids, and so forth) and of helices according to the methods of descriptive geometry. In addition he includes a method for the construction of "Dürer's leaf" (the *folium Dureri*), presents the notion of affinity by the example of the ellipse as a related representation of the circle, and, most important, describes the conic sections in top and front views as well as demonstrating their construction.

In book II Dürer develops a morphological theory of regular polygons and their exact or approximate constructions. He shows how to make use of such constructions as architectural ornaments, in parquet floors, tesellated pavements, and even bull's-eye window panes. The book concludes with theoretical investigations (culminating in the Vitruvian approximation for squaring the circle, a process which had already been noted by Dürer in a proportional study made in Nuremberg in 1504 or 1505) and with the computation of π (as 3.141).

The first part of the third book includes bird's-eye and profile elevations of pyramids, cylinders, and columns of various sorts (in 1510, in Nuremberg, Dürer had already sketched a spiral column with spherical processes). The second part of the book deals with sundials and astronomical instruments; Dürer had a small observatory at his disposal in the house that he had acquired from Bernhard Walther, a student of Regiomontanus, and could also make use of Walther's scientific library, part of which he bought. In the third part of the third book Dürer is concerned with the design of letters and illustrates the construction in a printer's quad of capitals of the Roman typeface named after him as well as an upper- and lowercase *fraktur* alphabet.

In book IV Dürer presents the development of the five Platonic solids (polyhedra) and of several semiregular (Archimedean) solids. He additionally shows how to construct the surfaces of several mixed bodies and, of particular importance, presents an approximate development of the sphere (he had begun work on the last for the construction of the first globe in Nuremberg in 1490–1492; his work on other globes, celestial charts, and armillary spheres is well known). He also shows how to duplicate the cube (the Delian problem) and related bodies, demonstrates the construction of the shadows of illuminated bodies, and finally summarizes the theory of perspective.

Except for the *Geometria Deutsch* (*ca.* 1486–1487), a book of arithmetical rules for builders which Dürer knew and used, the *Underweysung der Messung* is the first mathematics book in German. With its publication Dürer could claim a place in the front ranks of Renaissance mathematicians.

Dürer's next technical publication, the *Befestigungslehre* ("Treatise on Fortifications"), was a practical work dictated by the fear of invasion by the Turks, which gripped all of central Europe. This book was published in Nuremberg in 1527; as well as summarizing the science of fortification it contains some of Dürer's chief architectural work (various other architectural drawings and models are extant). Many of his ideas were put to use; the city of Nuremberg was strengthened according to his plan (in particular the watchtowers were fortified), similar work was undertaken at Strasbourg, and the Swiss town of Schaffhausen built what might be considered a model of Dürer's design with small vaults above and below ground, casemates, and ramparts that still survive intact.

Dürer's third book, his "Treatise on Proportion," *Vier Bücher von menschlicher Proportion,* was published posthumously in 1528; Dürer himself saw the first proof sheets (there are no other details of his last illness and death) and his friends saw to the final stages of publication. This book is the synthesis of Dürer's solutions to his self-imposed formal problems; in it, he sets forth his formal aesthetic. In its simplest terms, true form is the primary mathematical figure (the straight line, the circle, conic sections, curves, surfaces, solids, and so forth), constructed geometrically or arithmetically, and made beautiful by the application of some canon of proportion. The resulting beautiful form may be varied within limits of similarity. (In the instance of human form, there should be sufficient variation to differentiate one figure from another, but never so much as that the figure becomes deformed or nonhuman.) Dürer's Platonic idea of form figures in his larger aesthetic; for him beauty was the aggregate of symmetrical, proportionate, and harmonious forms in a more highly symmetrical, proportionate, and harmonious work of art.

Dürer's aesthetic rules are firmly based in the laws of optics—indeed, he even designed special mechanical instruments to aid in the attainment of beautiful form. He used the height of the human body as the basic unit of measurement and subdivided it linearly to reach a common denominator for construction of a unified artistic plan. This canon was not inviolable; Dürer himself modified it continually in an attempt to approximate more closely the canon of Vitruvius (which was also the canon most favored by Leonardo). Thus the artist retains freedom in the act of selecting his canon. In books I and II of the *Vier Bücher* Dürer deals, once again, with the arithmetic and geometrical construction of forms; in books III and IV he considers the problems of variation and movement.

The last of the *Vier Bücher* is perhaps of greatest mathematical interest since in treating of the movement of bodies in space Dürer was forced to present new, difficult, and intricate considerations of descriptive spatial geometry; indeed, he may be considered the first to have done so. At the end of this book he summarizes and illustrates his theories in the construction of his famous "cube man."

Dürer's chief accomplishment as outlined in the *Vier Bücher* is that in rendering figures (and by extension, in the composition of the total work of art) he first solved the problem of establishing a canon, then considered the transformations of forms within that canon, altering them in accordance with a consistent idea of proportion. In so doing he considered the spatial relations of form and the motions of form within space. His triumph as a painter lay in his disposition of carefully proportioned figures in surrounding space; he thereby elevated what had been hit-or-miss solutions of an essential problem of plastic composition to a carefully worked out mathematical theory. No earlier method had been so successful, and Dürer's theoretical work was widely influential in following centuries.

BIBLIOGRAPHY

I. ORIGINAL WORKS. Editions of Dürer's works include *Underweysung der Messung mit Zirckel und Richtscheyt in Linien, Ebnen, und gantzen Corporen* (Nuremberg, 1525; 2nd ed., Nuremberg, 1538; facsimile ed. by Alvin Jaeggli and Christine Papesch, Zurich, 1966); *Etliche Underricht zu Befestigung der Stett, Schloss und Flecken* (Nuremberg, 1927), repr. as W. Waetzoldt, ed., *Dürer's Befestigungslehre* (Berlin, 1917); and *Vier Bücher von menschlicher Proportion* (Nuremberg, 1528), of which there is a facsimile ed. in 2 vols. with text and commentary, Max Steck, ed. (Zurich, 1969).

II. SECONDARY LITERATURE. Max Steck, *Dürer's Gestaltlehre der Mathematik und der bildenden Künste* (Halle–Tübingen, 1948), contains an extensive bibliography of works by and about Dürer as well as a scientific analysis of the sources of the *Underweysung der Messung;* see also Hans Rupprich, *Dürer-Schriftlicher Nachlass*, vols. I and II (Berlin, 1956–1966), vol. III (in preparation); and Max Steck, "Albrecht Dürer als Mathematiker und Kunsttheoretiker," in *Nova Acta Leopoldina*, **16** (1954), 425–434; "Albrecht Dürer als Schrifsteller," in *Forschungen und Fortschritte*, **30** (1956), 344–347; *Dürer: Eine Bildbiographie* (Munich, 1957; 2nd ed. Munich, 1958; other German eds.), trans. into English as *Dürer and His World* (London–New York, 1964); "Ein neuer Fund zum literarischen Bild Albrecht Dürer's im Schrifttum des 16. Jahrhunderts," in *Forschungen und Fortschritte*, **31** (1957), 253–255; "Drei neue Dürer-Urkunden," *ibid.*, **32** (1958), 56–58; "Grundlagen der Kunst Albrechts Dürers," in *Universitas* (1958), pp. 41–48, also trans. for English and Spanish eds.; "Theoretische Beiträge zu Dürers Kupferstich 'Melancolia I' von 1514," in *Forschungen und Fortschritte,* **32** (1958), 246–251; *Albrecht Dürer, Schriften—Tagebücher—Briefe* (Stuttgart, 1961); "Albrecht Dürer as a Mathematician," in *Proceedings of the Tenth International Congress of the History of Science*, II (Paris, 1964), 655–658; "Albrecht Dürer als Mathematiker und Kunsttheoretiker," in *Der Architekt und der Bauingenieur* (Munich, 1965), pp. 1–6; and "Albrecht Dürer: Die exaktwissenschaftlichen Grundlagen seiner Kunst," in *Scientia, Milano,* **1C** (1966), 15–20, with French trans. as "Albert Dürer: Les sciences exactes sont les bases de son art," *ibid.*, pp. 13–17.

MAX STECK

DU TOIT, ALEXANDER LOGIE (*b.* Rondebosch, near Cape Town, South Africa, 14 March 1878; *d.* Cape Town, 25 February 1948), *geology.*

Du Toit was not only the most honored of South African geologists but also, in the words of R. A. Daly, the "world's greatest field geologist." To a remarkable degree he combined two traits not often found together: an extremely careful observer, he noted and drew deductions from details that escaped others, and at the same time he was able to synthesize information in broad fashion. Toward the end of his life he supported the hypothesis of continental drift with arguments drawn from all parts of the world and considerations about the underlying mantle.

Du Toit's versatility of mind is demonstrated by the important contributions he made to such varied subjects as the stratigraphy of both Precambrian and Karroo beds, paleobotany, petrology, hydrogeology, geomorphology, and the economic geology of base metal, nonmetallic, and diamond deposits. He was very active and mapped the geology of more than 100,000 square miles, much of it in detail, using a plane table with a bicycle for transport and a donkey cart as a base.

Those who remember du Toit have a deep respect for his intellect, knowledge, and activity but an even greater regard for his modesty, frugality, and kindness to all. These qualities, coupled with a strong character, made him an outstanding leader of men and a dominant figure in any company.

Du Toit was born on his family's estate near Cape Town. His father's family, one of the largest and most distinguished in South Africa, was of Huguenot descent and had been in the Cape since 1687. His mother was Anna Logie, daughter of a Scottish immigrant. He went to school at the local diocesan college and graduated from South Africa College (now the University of Cape Town) before spending two years qualifying in mining engineering at Royal Technical College, Glasgow, and studying geology at the Royal College of Science, London.

While in Glasgow, du Toit married Adelaide Walker. They had one child, Alexander Robert. Du Toit's wife died in 1923, and two years later he married Evelyn Harvey. At this period of his life he became a proficient musician, his favorite instrument being the oboe, and did some motorcycle racing.

In 1901 du Toit became lecturer at both the Royal Technical College, Glasgow, and the University of Glasgow. In 1903 he returned to South Africa to join the Geological Commission of the Cape of Good Hope. He spent the next seventeen years almost continuously in the field, mapping, at times accompanied by his wife and child. This period laid the foundations for his broad understanding and unrivaled knowledge of the details of South African geology.

His first season, spent with A. W. Rogers in the western Karroo, determined many of what were to become du Toit's abiding interests. Together they established the stratigraphy of the Lower and Middle Karroo system noting the glacial origin of the Dwyka tillite. They recorded systematic phase changes from place to place in the Karroo and Cape systems. They mapped numerous dolerite intrusives, their acid phases, and their metamorphic aureoles.

From 1903 to 1905 du Toit was in the rugged Stormberg area, mapping so well that his accounts of the paleobotany of the coal-bearing Molteno beds and of the volcanicity have remained classics. From 1905 to 1910 he worked in the northern part of the old Cape Colony, mapping nonfossiliferous rocks of early Precambrian to Permian age. He became interested in geomorphology and hydrogeology and collaborated with Rogers in a new edition of the book *Introduction to the Geology of Cape Colony*.

Between 1910 and 1913 du Toit was near the Indian Ocean, mapping Karroo coal deposits, the flexure of the Lebombo Range along the coast, and an immense number of basic intrusions and charnockite rocks that he discovered there. In 1910 he received the D.Sc. degree from the University of Glasgow for his report on the copper-nickel deposits of the Insiza Range. His "Underground Water in South-East Bechuanaland" (1906) and "The Geology of Underground Water Supply" (1913) were important monographs which served to establish him as the leading authority on groundwater in South Africa. In 1914 du Toit visited Australia to study the rocks equivalent to the Karroo System and the groundwater geology of the Great Artesian Basin. From the outbreak of World War I until the campaign in South West Africa was over in 1915, he was hydrogeologist to the South African forces, holding the rank of captain.

Returning to Natal, du Toit became increasingly involved in work for the irrigation department and transferred to it in 1920. The relief from continuous fieldwork that this provided enabled him to produce a series of important papers and books on the Karroo System (1918), Karroo dolerites (1920), Carboniferous glaciation (1921), past land connections with other continents (1921), the South African coastline (1922), and the geology of South Africa (1926).

In 1923 a grant-in-aid from the Carnegie Institution of Washington enabled du Toit to make a trip to South America for the purpose of comparing the geology of that continent with that of Africa. He left Cape Town on 12 June and spent five months in Brazil, Paraguay, and Argentina. He described this visit in *A Geological Comparison of South America with South Africa* (1927), in which he also outlined points of similarity between the two continents.

He found the two continents to be alike in (1) Precambrian crystalline basement with infolded pre-Devonian sediments; (2) in the far north, a gentle syncline of marine Silurian and Devonian strata; (3) farther south, gently dipping Proterozoic and Lower Paleozoic strata cut by granites; (4) an area of flat-lying Devonian strata; (5) in the extreme south, conformable Devonian to Permian strata including Carboniferous tillites crumpled by later mountain building; (6) tillites extending northward transgressing across the Devonian on to the Precambrian basement before dying out; (7) glacial deposits overlain by continental Permian and Triassic strata with *Glossopteris* flora, followed by extensive basalt flows and dolerite intrusives; (8) Gondwana beds extending northward continuously from the southern Karroo to the Kaokoveld in Africa and from Uruguay to Minas Geraes in South America, with further great detached areas in the north, in each instance some distance inland, in the Angola-Congo and Piauhy-Maranhão regions; (9) an intraformational break occurring commonly below the late Triassic; (10) tilted Cretaceous beds occurring only along the coast; (11) widespread horizontal Cretaceous and Tertiary strata; (12) a succession in the Falkland Islands closely resembling that of the Cape, but distinct from that of Patagonia; (13) seven corresponding faunal assemblages in the similar strata; and (14) the geographical outline of the continent.

From 1927 to 1941 du Toit was consulting geologist to De Beers Consolidated Mines but continued to write on many topics; in 1937 he published his well-known book *Our Wandering Continents*. In 1932 he visited North America, in 1937 the Soviet Union, and in 1938 India. From his retirement until his death he lived at Cape Town and maintained his varied interests, extending them to include archaeology and vertebrate zoology.

Du Toit received many honors and awards, including five honorary degrees. He was twice president of the Geological Society of South Africa, a corresponding member of the Geological Society of

America, and a fellow of the Royal Society of London.

More than most scientists, du Toit's reputation, already high, has continued to grow, because in many of his deductions he was ahead of his time. His forte was meticulous and extensive fieldwork which enabled him to grasp virtually all aspects of the geology of South Africa. Many of the ideas that he espoused concerning groundwater, economic deposits of copper and nickel, and geomorphology and stratigraphy of rocks were original but not remarkably different from those held by others. The most significant factor of du Toit's work was his early espousal of the theory of continental drift; he was the first to realize that the southern continents had at one time formed the supercontinent of Gondwanaland, which was distinct from the northern supercontinent of Laurasia. His championship of continental drift, unpopular at the time, is now widely hailed as having been correct.

BIBLIOGRAPHY

I. ORIGINAL WORKS. Du Toit's works include "The Stormberg Formation in Cape Colony," in *Report and Papers, South African Association for the Advancement of Science,* **2** (1905), 47; "Underground Water in South-East Bechuanaland," in *Transactions of the South African Philosophical Society,* **16** (1906), 251–262; "Report on the Copper-Nickel Deposits of the Insizwa, Mount Ayliff, East Griqualand," in *Annual Report of the Geological Commission for the Cape of Good Hope for 1910* (1911), pp. 69–110; "The Geology of Underground Water Supply," in *Mining Proceedings of the South African Society of Civil Engineers for 1913* (1913), pp. 7–31; "The Problem of the Great Australian Artesian Basin," in *Journal and Proceedings of the Royal Society of New South Wales,* **51** (1917), 135–208; "The Zones of the Karroo System and Their Distribution," in *Proceedings of the Geological Society of South Africa,* **21** (1918), 17–36; "The Karroo Dolerites of South Africa: A Study in Hypabyssal Injection," in *Transactions of the Geological Society of South Africa,* **23** (1920), 1–42; "The Carboniferous Glaciation of South Africa," *ibid.,* **24** (1921), 188–227; "Land Connections Between the Other Continents and South Africa in the Past," in *South African Journal of Science,* **18** (1921), 120–140; "The Evolution of the South African Coastline," in *South African Geographical Journal,* **5** (Dec. 1922), 5–12; *The Geology of South Africa* (Edinburgh, 1926); *A Geological Comparison of South America with South Africa,* Carnegie Institution Publication no. 381 (Washington, D. C., 1927); *Our Wandering Continents: An Hypothesis of Continental Drifting* (Edinburgh, 1937); "Tertiary Mammals and Continental Drift," in *American Journal of Science,* **242** (1944), 145–163; and "Palaeolithic Environments in Kenya and the Union—A Contrast," in *South African Archeological Bulletin,* **2** (1947), 28–40.

II. SECONDARY LITERATURE. On du Toit and his work see T. W. Gevers, "The Life and Work of Dr. Alexander L. du Toit," Alexander L. du Toit Memorial Lecture no. 1, in *Proceedings of the Geological Society of South Africa,* **52** (1949), annexure 1–109; S. H. Haughton, "Alexander Logie du Toit, 1878–1948," in *Biographical Memoirs of Fellows of the Royal Society,* **6** (1948), 385–395; and "Memorial to Alexander Logie du Toit," in *Proceedings. Geological Society of America,* Annual Report for 1949 (1950), pp. 141–149.

J. T. WILSON

DUTROCHET, RENÉ-JOACHIM-HENRI (*b.* Néon, France, 14 November 1776; *d.* Paris, France, 4 February 1847), *animal and plant physiology, embryology, physics, phonetics.*

Dutrochet was born at a seignorial mansion near Poitiers; he was the eldest son of wealthy and noble parents. His early childhood was spoiled by clubfoot, which—after unsuccessful medical consultations—was completely healed by a renowned healer, who was also a hangman.

The Revolution brought expropriation of family property after his father emigrated; Dutrochet was therefore forced to rely on his own resources. He volunteered for the navy in 1799, then deserted and joined his brothers in the last Royalist units. Following 18 Brumaire, which with the accession of Bonaparte to the Consulate brought an end to the Royalist resistance and a general amnesty, Dutrochet spent two peaceful years with his family. He was, however, dissatisfied with such an empty and useless life, so in 1802, at the age of twenty-six, he went to Paris to study medicine. Graduating in 1806, he qualified as a military medical officer in 1808 and was sent to Spain. With great devotion and sacrifice and lacking adequate material means, he dealt with a severe outbreak of typhoid in Burgos. He believed that he had found his true vocation until, having contracted the disease himself, he was so weakened that he had to return to France for a long convalescence. He joined his mother in a country house near Château-Renault in Touraine, where he lived in seclusion from society and decided, at the age of thirty-four, to abandon medical practice and devote all his efforts to natural science. In 1819 he was elected corresponding member of the Académie des Sciences; in 1831 he was elected to full membership. After his mother's death he lived in Paris during the winter months and returned in summer to his country residence. In 1845 he suffered a blow on the head which, after a long illness marked by severe headaches, led to his death.

Dutrochet's first scientific interest, which he pursued even before he finished his medical studies, was phonetics. In 1806 he repeated Ferrein's experiments on the larynx and tried to establish the relationship between pitch and tension of the vocal cords under different loads—these experiments, however, have been overshadowed by his subsequent studies and are almost forgotten.

Dutrochet's investigations into the development of birds, reptiles, batrachians, and mammals, published in 1814, are more important. In them he paid special attention to the hitherto neglected early stages of development of the egg within the ovary, to its detachment, and to the fetal membranes. (One of them, the external yolk membrane of the bird's egg, whose fibers are continuous with the chalazae, is called "Dutrochet's membrane.") He also made several original observations on fetal development, but it was his demonstration of the analogy of fetal envelopes in ovipara and vivipara that suggested a unity of the main features in the development of animals and proved extremely valuable for further studies.

The principal field of Dutrochet's studies was plant physiology, although he also studied that of animals. He further explored the areas that were common to both, especially the exchange of gases between the atmosphere and plant or animal tissues—the key to the life processes. He asserted that respiration is of the same nature in both plants and animals. Active breathing in animals had been evident to observers since very early times; in plants, however, the existence of respiration was brought to light much later. In 1832 Dutrochet showed that the minute openings on the surface of leaves (the stomata) communicate with lacunae in deeper tissue. He further demonstrated that only the green parts of the plant can absorb carbon dioxide and thus transform light energy into chemical energy that can then serve to accomplish all kinds of syntheses.

In his studies of excitability and motility Dutrochet tried to demonstrate that these widespread phenomena are essentially the same in both plants and animals, since they utilize the same organs and mechanisms. In contrast to the then current explanation of these phenomena—which was based on *Naturphilosophie* and depended on intervention of the "vital force"—he stressed anatomical and mechanical arguments. For example, he emphasized the importance to plant motility of the turgor of the hinge cells, the passage of water out of the cells on one side into the intercellular spaces, and so forth.

Dutrochet's research on the phenomena of osmosis and diffusion (or endosmosis and exosmosis, as he not very aptly called them) and their applications to the study of previously unexplained vital phenomena attracted general attention. His chief observation was that certain organic membranes allow the passage of water but stop the molecules dissolved in it, so that between two solutions of different concentration, separated by such a membrane, water passes from the less concentrated to the more concentrated, even against gravity. Although the conditions of Dutrochet's experiments were rather simple and did not allow of great accuracy, he made the first important steps toward the study of osmosis and diffusion. He constructed an osmometer for measurements of osmotic pressure and pointed to such pressure as the possible cause of circulation and rise of sap in plants, absorption of nutrients in plants and animals. His experiments were developed by many of his younger colleagues, and his ideas played an important role in their thinking, for example, in Carl Ludwig's hypothesis of the formation of urine in the kidneys (1842).

Among Dutrochet's other discoveries, it may be noted that in 1831 he demonstrated that mushrooms are in fact the fruiting bodies of the mycelium; they had been previously considered to be a particular genus (called *byssus*). He was also the first (1840) to detect, by a thermoelectric technique, the production of heat in an individual plant and in an insect muscle during activity.

Dutrochet is also considered to be the founder of the cell theory; but his ideas are actually more in the nature of shrewd, intuitive anticipations rather than conclusions based on his own microscopic observations. His illustrations are not convincing and it seems that, at least in some cases, Dutrochet's "globules" (cells) were optical artifacts produced by poor lenses and bad illumination. Although he expressed his ideas in language similar to that of the cell theory, these ideas and the observations upon which they were based were not equivalent to it. Dutrochet himself made no claim to priority when Schwann's book was published in 1839.

Dutrochet's observations and experiments were often unsatisfactory; his means were largely inadequate. He was often mistaken in his conclusions or made false parallels between plants and animals. His importance lies more in the systematic endeavor to demonstrate that vital phenomena can be explained on the basis of physics and chemistry, that living organisms use physical and chemical forces, and that there is no reason to suppose the existence of some intervention of a "vital force." He strove to generalize and to show the unity of basic processes in all living things, both plant and animal. His experimental studies and observations led him to the conclusion that there is only one physiology, only one general science of the function of living bodies. His attempt to apply physicochemical forces and phenomena in explanation of physiological processes overcame that mysticism which had been introduced into physiology by teleologically minded physiologists.

A convinced antivitalist, Dutrochet developed a unitary conception of a nature—animate and inanimate, organic and inorganic, all subject to the laws of physics and chemistry. For this reason he had a great influence on his younger colleagues; for example, in 1841 du Bois-Reymond wrote, "I am gradually returning to Dutrochet's view. The more one advances in the knowledge of physiology, the more one will have reason for ceasing to believe that the phenomena of life are essentially different from physical phenomena." Dutrochet's work was also greatly appreciated by the distinguished plant physiologist Julius Sachs.

BIBLIOGRAPHY

I. ORIGINAL WORKS. Dutrochet's writings include *Essai sur une nouvelle théorie de la voix, avec l'exposé des divers systèmes qui ont paru jusqu'à ce jour sur cet objet* (Paris, 1806); *Mémoire sur une nouvelle théorie de la voix* (Paris, 1809); *Mémoire sur une nouvelle théorie de l'harmonie dans lequel on démontre l'existence de trois modes nouveaux, qui faisaient partie du système musical des Grecs* (Paris, 1810); *Recherches anatomiques et physiologiques sur la structure intime des animaux et végétaux* (Paris, 1824); *L'agent immédiat du mouvement vital dévoilé dans sa nature et dans son mode d'action chez les végétaux et animaux* (London–Paris, 1826); and *Nouvelles recherches sur l'endosmose et l'exosmose, suivies de l'application expérimentale de ces actions physiques à la solution du problème de l'irritabilité végétale, et à la détermination de la cause de l'ascension des tiges et de la descente des racines* (Paris, 1828). Further observations on the osmotic phenomena were published as "Nouvelles observations sur l'endosmose et l'exosmose, et sur la cause de ce double phénomène," in *Annales de chimie et de physique,* **35** (1827), 393–400; "Nouvelles recherches sur l'endosmose et l'exosmose," *ibid.,* **37** (1828), 191–201; "Recherches sur l'endosmose et sur la cause physique de ce phénomène," *ibid.,* **49** (1832), 411–437; "Du pouvoir d'endosmose considéré comparativement dans quelques liquides organiques," *ibid.,* **51** (1832), 159–166; and "De l'endosmose des acides," *ibid.,* **60** (1835), 337–368.

II. SECONDARY LITERATURE. On Dutrochet and his work, see *Notice analytique sur les travaux de M. Henri Dutrochet* (Paris, 1832); A. Brongniart, *Notice sur Henri Dutrochet* (Paris, 1852); J. J. Coste, *Éloge historique de Henri Dutrochet* (Paris, 1866); *Gazette médicale de Paris,* no. 11 (1866); and I. Geoffroy Saint-Hilaire, in *Biographie universelle,* XII (Paris, 1855). For Dutrochet's place in the history of the cell theory see A. R. Rich, "The Place of R. J. H. Dutrochet in the Development of the Cell Theory," in *Bulletin of the Johns Hopkins Hospital,* **39** (1926), 330–365; F. K. Studnička, "Aus der Vorgeschichte der Zellentheorie. H. Milne Edwards, H. Dutrochet, F. Raspail, J. E. Pur-

kinje," in *Anatomischer Anzeiger,* **73** (1931), 390–416; and J. W. Wilson, "Dutrochet and the Cell Theory," in *Isis,* **37** (1947), 14–21. His work in plant physiology is discussed by J. Sachs in *Geschichte der Botanik* (Munich, 1875), English trans. by H. E. Garney (Oxford, 1890).

VLADISLAV KRUTA

DUTTON, CLARENCE EDWARD (*b.* Wallingford, Connecticut, 15 May 1841; *d.* Englewood, New Jersey, 4 January 1912), *geology.*

Prepared for college early (he was ready to matriculate at thirteen), Dutton was held back by his parents, Samuel and Emily Curtis Dutton, as too young. Entering Yale at fifteen, he at first showed more literary than scientific aptitude. After graduation in the class of 1860, he entered Yale Theological Seminary. His studies were interrupted by the Civil War; he entered the army, where he quickly discovered a liking for mathematics that led him to make the army his career. Emerging from the war a captain of ordnance, he was stationed first at Watervliet Arsenal, near the Bessemer steelworks in West Troy, New York. His first scientific paper was on the chemistry of the Bessemer process.

Later Dutton was transferred to the Washington Arsenal; here he was led toward geology by the brilliant group of men who were then creating a structure of government-supported science. At Washington Philosophical Society meetings he met S. F. Baird and Joseph Henry of the Smithsonian Institution; E. W. Hilgard of the Coast and Geodetic Survey; Simon Newcomb, Asaph Hall, and William Harkness of the Naval Observatory; and F. V. Hayden and J. W. Powell of the western surveys. By 1875 Powell had enough confidence in Dutton as a geologist to ask that he be assigned to special duty with his geographical and geological survey of the Rocky Mountain region. When the western surveys of Powell, Clarence King, Hayden, and G. M. Wheeler were consolidated in 1879, Dutton continued with the United States Geological Survey until 1890, when he returned to regular army duty.

From the beginning of his geological studies, Dutton was interested in orogenic problems, and during his years of work in the plateau region of Utah, Arizona, and New Mexico he had the opportunity to study not only the faults and monoclines along which uplift and subsidence had taken place but also the extensive volcanism that had accompanied these earth movements. In a number of reports and papers and in two major monographs (*Report on the Geology of the High Plateaus of Utah,* 1880, and *The Tertiary History of the Grand Canyon District,* 1882) he established himself as a brilliant interpreter of physical features. With Powell and Grove Karl Gilbert, his close collaborators, he established some of the basic principles of structural geology—in particular the theory of isostasy, which he developed to explain crustal movements. His geological writings, moreover, are marked by great charm of style; and he virtually formulated a new aesthetic for the startling scenery of the Grand Canyon country.

In 1882 Dutton studied live Hawaiian volcanoes, intending thereafter to examine the extinct volcanoes and lava beds of Oregon. He was diverted by the Charleston earthquake of 1886, on which he prepared a monograph; and from 1888 to 1890 he directed the hydrographic work of the irrigation surveys under Powell. After returning to army duty in 1890, he wrote further papers on volcanism and earthquakes, including several on the possibilities of earthquakes along the route of the proposed Nicaragua Canal. His last major contribution was *Earthquakes in the Light of the New Seismology,* which linked volcanism to radioactivity.

BIBLIOGRAPHY

I. ORIGINAL WORKS. A full listing of Dutton's papers is appended to Wallace Stegner, *Clarence Edward Dutton: An Appraisal* (Salt Lake City, n.d. [1936]). The most important are *Report on the Geology of the High Plateaus of Utah,* vol. XXXII, U.S. Geographical and Geological Survey of the Rocky Mountain Region (Washington, D.C., 1880); *The Tertiary History of the Grand Canyon District,* U.S. Geological Survey Monograph no. 2 (Washington, D.C., 1882); "Hawaiian Volcanoes," in *Report of the United States Geological Survey,* **4** (1884), 75–219; "Mount Taylor and the Zuñi Plateau," *ibid.,* **6** (1885), 105–198; "The Charleston Earthquake of August 31, 1886," *ibid.,* **9** (1889), 203–528; "General Description of the Volcanic Phenomena Found in That Portion of Central America Traversed by the Nicaragua Canal," United States Senate Document no. 357, 57th Congress, 1st Session (1901–1902), XXVI, 55–62; and *Earthquakes in the Light of the New Seismology* (New York, 1904).

II. SECONDARY LITERATURE. In addition to Stegner, above, brief discussions of Dutton's life and work may be found in *Biographical Record, Class of Sixty* (Boston, 1906); and in G. P. Merrill, *The First Hundred Years of American Geology* (New Haven, 1904); as well as in the administrative reports of the Powell Survey and the United States Geological Survey.

WALLACE STEGNER

DUVAL, MATHIAS MARIE (*b.* Grasse, France, 7 February 1844; *d.* Paris, France, 28 February 1907), *histology, physiology, comparative anatomy, embryology.*

The son of Joseph Duval, the botanist and naturalist, and Marie Jouve, he spent his childhood in Grasse, Algiers, and then Strasbourg, where his father had been appointed school inspector. Duval studied medicine there from 1863 to 1869. His doctoral thesis was entitled "Étude sur la valeur relative de la section du maxillaire supérieur." He was a student of Joseph Alexis Stolz, Charles Basile Morel, and Émile Küss; he became an anatomy assistant in 1866 and a prosector in 1869. During the Franco-Prussian War he served under General Charles Bourbaki. The loss of Alsace forced him to pursue his career first in Montpellier and finally in Paris. After defending a thesis entitled "La rétine; structure et usages," he became *agrégé* in anatomy and physiology in 1873.

From 1873 until 1899 he taught—among other subjects—anatomy for artists at the École Nationale Supérieure des Beaux-Arts. His course, greatly influenced by Guillaume Duchenne, dealt mainly with the physiology of the face, the muscles of physiognomy, and the expression of strong emotions. In 1880 Duval succeeded Pierre Paul Broca as professor of zoological anthropology and became director of the anthropology laboratory at the École Pratique des Hautes Études. In December 1885 he was appointed professor of histology at the Faculty of Medicine of the University of Paris. His clear, precise, and efficacious teaching method was highly appreciated by students. In 1882 Duval was elected to the Académie Nationale de Médecine. In 1889 he was president of the Société d'Anthropologie de Paris. Blinded by bilateral cataract, he was forced to reduce and finally to discontinue his teaching activity. He died probably of cancer and was buried in Neuville-les-Dames, near Dieppe.

Duval's work was strongly influenced by that of Charles Darwin and the French histologist Charles Robin. In histology his most important original works involve the microscopic structure of the central nervous system and sensory organs, as well as the true origin of the cranial nerves. Duval declared: "Embryological studies can have no guiding hypotheses other than those expressed in transformist doctrine." He did a great deal to disseminate Darwin's theory of evolution. His research in this area reflects his training as a histologist and physiologist: it concerned the formation of the gastrula, development of the blastoderm, the three primitive germ layers and their derivatives in the various species, segmentation of the egg, and embryonic appendages of birds and mammals. Extremely well-versed in anatomy, he defended the concepts of "animal colonies" and of the invertebrate origin of higher forms of animal life; among other homologies he established relationships between the "primitive lineage" of birds and the "Rusconian orifice" of batrachians. He was also interested in teratology based on fertilization anomalies. His written didactic works exerted tremendous influence both in France and abroad.

BIBLIOGRAPHY

I. ORIGINAL WORKS. Between 1868 and 1900 Duval published more than 250 papers, works, and articles in various dictionaries and journals of anatomy, physiology, anthropology, general biology, and even history and ethnography. A complete list may be found in the two *Notices sur les titres et travaux scientifiques de M. Mathias Duval* (Paris, 1885, 1896) and in E. Retterer, "Mathias Duval; sa vie et son oeuvre," in *Journal de l'anatomie et de la physiologie de l'homme et des animaux,* **43,** no. 3 (1907), 241–331.

His main didactic works are *Cours de physiologie,* which appeared in eight eds. from 1872 to 1897 and was translated into English (Boston, 1875), Spanish (Madrid, 1876, 1884), Greek (Athens, 1887), and Russian (St. Petersburg, 1893); *Manuel du microscope dans ses applications au diagnostic et à la clinique* (with L. Lereboullet, 1873, two eds.); *Précis de technique microscopique et histologique* (1878); *Manuel de l'anatomiste* (with C. Morel, 1882); *Leçons sur la physiologie du système nerveux* (1883); *Précis d'anatomie à l'usage des artistes* (1882, three eds.; English trans., London, 1884 and 1905); *Dictionnaire usuel des sciences médicales* (with Dechambre and Lereboullet, 1885, two eds.); *Le Darwinisme* (1886); *Atlas d'embryologie* (1889); *L'anatomie des maîtres* (with A. Bical, 1890); *Anatomie et physiologie animales* (with P. Constantin, 1892, two eds.); *Le placenta des carnassiers* (1895); *Précis d'histologie* (1897; two eds.; Italian translation by Fusari, Turin, 1899); and *Histoire de l'anatomie plastique* (with E. Cuyer, 1899).

II. SECONDARY LITERATURE. On Duval and his work see A. Gautier, "Notice nécrologique sur M. Duval," in *Bulletin de l'Académie nationale de médecine,* **57** (1907), 343–344; G. Hervé, "Mathias Duval," in *Revue de l'École d'anthropologie,* **17** (1907), 69–74; and H. Roger, "Mathias Duval," in *Presse médicale,* **15,** no. 19 (1907), 145–146.

CHARLES COURY

DUVERNEY, JOSEPH-GUICHARD (*b.* Feurs, France, 5 August 1648; *d.* Paris, France, 10 September 1730), *anatomy.*

Duverney, the son of the village doctor, went to Avignon when he was fourteen years old to study medicine, receiving his medical degree there in 1667. Shortly thereafter he went to Paris, where he soon began to attend the weekly scientific meetings at the house of the Abbé Bourdelot. At these meetings Duverney often spoke on anatomical subjects. Here, too, he probably met Claude Perrault, who asked him to assist in dissections.

Perrault was the leader of a group of anatomists, who came to be known as the "Parisians," who collaborated with one another to an uncommon degree, regularly performing dissections as a group and collectively reviewing both the text and plates before publishing their collaborative work. Individually and together they dissected a wide variety of animals, many of which came from the royal menagerie at Versailles. (Duverney performed the dissection of an elephant in the presence of Louis XIV.) These anatomists considered most zoological writings inadequate and wished to assemble a large series of observations to constitute a new *Historia animalium* to replace that of Aristotle—one that would be worthy of their monarch.

The Paris group concentrated on describing unusual species and distinctive anatomical features, barely more than cataloging the commonplace. They used the human body as a standard of reference, not because of its assumed perfection, but for their readers' familiarity with it. When appropriate, domestic animals were used for purposes of comparison, although these were not described in any detail. They published their comparative anatomical studies anonymously at first. Although individual contributions to these earlier papers can only occasionally be determined, there is little question that Duverney contributed to them, probably heavily, presumably beginning with the *Description anatomique . . .* of 1669, which contained descriptions of a chameleon, beaver, dromedary, bear, and gazelle.

Duverney's connection with the Académie des Sciences began in 1674 when he was enlisted to assist in the completion of the two sumptuous elephant-folio volumes of the *Mémoires* that were published anonymously at the king's expense. The work had been begun by Perrault, Louis Gayant, and Jean Pequet but had been interrupted by the deaths of the latter two. The same year, the Academy sent Duverney to Bayonne and lower Brittany to dissect fishes; Phillipe de la Hire accompanied him as his illustrator. Duverney was elected to full membership in 1676.

In 1679 Duverney was appointed to the chair of anatomy at the Jardin du Roi. He was a highly successful lecturer, and his auditors included the curious and the fashionable as well as serious students of anatomy. Jacques-Bénigne Winslow, F. P. du Petit, and J.-B. Senac—who edited two of Duverney's posthumous works—were among his students.

In 1688 Perrault died, and Duverney became responsible for all the comparative-anatomical work sponsored by the Academy. He also inherited Perrault's manuscripts relating to such work, including descriptions of sixteen animals that needed only editing for publication, but these did not appear until after Duverney's death. Duverney had a certain reluctance to publish—for example, he bought the manuscript of Jan Swammerdam's *Biblia natura* with the intention of publishing it, but the book did not appear until Hermann Boerhaave bought it from him. Nor did he ever produce the new edition of the *Mémoires,* despite the urgings of the Academy.

The only major work written by Duverney alone and published during his lifetime was, in fact, his *Traité de l'organe de l'ouie . . .* (1683), the first thorough, scientific treatise on the human ear. In it he describes the structure, functions, and diseases of the ear and includes a further description of the fetal ear, noting its differences from the adult structure. Duverney based his study of the ear on a study of its sensory innervation; to this end he had a new plate engraved to illustrate the base of the brain and the origin of these nerves, since he had found no adequate figure of this region. His interpretation of aural function was mechanical; for example, in the *Traité* he states that sound is transmitted within the ear as vibrations carried by the enclosed air and by the malleus, incus, and stapes. These vibrations reach the end of the nerves and set up a flow of spirits to the brain; the muscles (except for the muscles of the neck that turn the head) are motivated by another flow of spirits from the brain. Duverney believed that there is a direct communication by the nerves from the outer ear to the neck muscles, so that a flow of spirits along this route is responsible for the turning of the head when a noise is heard.

In addition, Duverney read numerous papers to the Academy, of which the most important are a group dealing with the circulatory and respiratory systems in cold-blooded vertebrates. In 1699 he presented a paper on these subjects, especially in the tortoise but also in the carp, frog, and viper. He presented a highly accurate description of the heart of the tortoise, demonstrating the single ventricle and its three cavities, the flow pattern of the blood, and the mixing of the arterial and venous bloods. He noted that the pulmonary artery carries venous blood, and he recognized the respiratory function of the gills. He here displays a knowledge of the piscine circulatory system that surpasses that of any other seventeenth-century work.

In a paper of 1701 Duverney limited himself to fishes with gills, but did not go significantly beyond his earlier work. He did describe the role of the gills in greater detail, however, in particular the diffusion of blood in the gills to provide a greater respiratory surface. He recognized that the change of color in the gills marked the conversion of venous to arterial blood. Knowledge of the cold-blooded vertebrates' circulatory state had been chaotic and disorganized prior to Duverney's work; he systematized it and advanced it considerably.

Three anatomical structures are sometimes given Duverney's name. The first, an incisura in the cartilage of the external auditory meatus, was described in his *Traité;* the second, the *pars lacrimalis musculus orbicularis oculi,* was described in the posthumously published (1749) *L'art de disséquer;* while the third is commonly known as Bartholin's glands, after Caspar Bartholin (1655–1738), who first described them in humans—Duverney had previously observed them in the cow.

BIBLIOGRAPHY

I. ORIGINAL WORKS. The collected works of the Parisians, to which Duverney would have contributed, were the *Mémoires pour servir à l'histoire naturelle des animaux,* 2 vols. (Paris, 1671–1676), trans. into English by Alexander Pitfield (London, 1688; later eds., 1701, 1702). An expanded ed. appeared after Duverney's death but included much of his material, 3 vols. (Paris, 1732–1734). Duverney's principal individual study was the *Traité de l'organe de l'ouie, contenant la structure, les usages & les maladies de toutes les parties de l'oreille* (Paris, 1683). A Latin ed. appeared (Nuremberg, 1684), and there were two English eds., trans. by J. Marshall (London, 1737, 1748).

Individual papers cited in the text are "Observations sur la circulation du sang dans le foetus et description du coeur de la tortue et de quelques autres animaux (du coeur de la grenouille; de la vipère; de la carpe)," in *Mémoires de l'Académie des Sciences de Paris* (1699); and "Mémoire sur la circulation du sang des poissons qui ont des ouyes et sur respiration," *ibid.* (1701).

Duverney's posthumous writings include *L'art de disséquer méthodiquement les muscles du corps humain* (Paris, 1749); *Traité des maladies des os,* J.-B. Senac, ed. (Paris, 1751), English trans., *The Diseases of the Bones,* S. Ingham, trans. (London, 1762); and *Oeuvres anatomiques,* 2 vols., J.-B. Senac, ed. (Paris, 1761), which contains most of his papers published in the *Mémoires de l'Académie des sciences,* including the important ones on the circulatory system of the cold-blooded vertebrates.

II. SECONDARY LITERATURE. Principal biographical sources are the article in *Biographie universelle (Michaud) ancienne et moderne,* vol. XII (Paris, 1855); and Bernard Le Bovyer de Fontenelle, "Eloge," in *Oeuvres,* new ed., vol. VI (Paris, 1742).

WESLEY C. WILLIAMS

DYCK, WALTHER FRANZ ANTON VON (*b.* Munich, Germany, 6 December 1856; *d.* Munich, 5 November 1934), *mathematics.*

Dyck was the son of Hermann Dyck, a painter and the director of the Munich Kunstgewerbeschule, and

Marie Royko. He married Auguste Müller in 1886; they had two daughters.

Dyck studied mathematics in Munich, Berlin, and Leipzig. He qualified as a university lecturer in Leipzig in 1882 and was an assistant of F. Klein. In 1884 he became a professor at the Munich Polytechnikum. He made noteworthy contributions to function theory, group theory, topology, potential theory, and the formative discussion on integral curves of differential equations. He was also one of the founders of the *Encyclopädie der mathematischen Wissenschaften.* Appointed director of the Polytechnikum in 1900, he brought about its rise to university standing as the Technische Hochschule; and as rector (1903–1906, 1919–1925) he carried out a major building expansion. In 1903 he was enlisted, along with Carl von Linde, by Oskar von Miller to aid in the establishment and early development of the Deutsches Museum; he also served as its second chairman from 1906. As a dedicated member of the Bayerische Akademie der Wissenschaften (and a class secretary in 1924), he prepared the plan and organization of the complete edition of the writings and letters of Kepler, including the posthumous works (for the most part in Pulkovo, near Leningrad). Moreover, as a founder (along with F. Schmitt-Ott) of the Notgemeinschaft der Deutschen Wissenschaften, he concerned himself with assuring financial support for the edition.

Linguistically gifted and a warm, kind-hearted man of wide-ranging and liberal interests, including art and music, Dyck was an outstanding scholar and organizer and an enthusiastic and inspiring teacher.

BIBLIOGRAPHY

I. ORIGINAL WORKS. Dyck's writings include *Uber regulär verzweigte Riemannsche Flächen und die durch sie bestimmten Irrationalitäten* (Munich, 1879), his doctoral dissertation; "Gruppentheoretische Studien," in *Mathematische Annalen,* **20** (1882), 1–44; **22** (1882), 70–108; "Beiträge zur Analysis situs," in *Sitzungsberichte der Sächsischen Akademie der Wissenschaften zu Leipzig* (1885), 314–325; (1886), 53–69; (1888), 40–52; and in *Mathematische Annalen,* **32** (1888), 457–512; **37** (1890), 273–316; *Katalog math.-physik. Modelle . . .* (Munich, 1892; supp., 1893); "Beiträge zur Potentialtheorie," in *Sitzungsberichte der Bayerischen Akademie der Wissenschaften zu München* (1895), 261–277, 447–500; (1898), 203–224; *Spezialkatalog der mathematischen Austellung, Deutsche Unterrichtungsabteilung in Chicago* (Munich, 1897); L. O. Hesse, *Gesammelte Werke,* ed. with S. Gundelfinger, J. Lüroth, and M. Noether (Munich, 1897); "Nova Kepleriana," in *Abhandlungen der Bayerischen Akademie der Wissenschaften,* **25** (1910), 1–61; **26** (1912), 1–45; **28** (1915), 1–17; **31** (1927), 1–114, written with M. Caspar; n.s. **17** (1933), 1–58, written with M. Caspar; **18** (1933), 1–58; **23** (1934), 1–88; *G. von Reichenbach* (Munich, 1912), a biography; and *J. Kepler in seinen Briefen,* 2 vols. (Munich, 1930), written with M. Caspar.

II. SECONDARY LITERATURE. Obituary notices by G. Faber are in *Forschungen und Fortschritte,* **34** (1934), 423–424, *Jahresbericht der Deutschen Mathematikervereinigung,* **45** (1935), 89–98, with portrait and bibliography, and *Jahrbuch der Bayerischen Akademie der Wissenschaften* (1934); an anonymous obituary is in *Almanach. Österreichische Akademie der Wissenschaften,* **85** (1935), 269–272; see also J. E. Hofmann, in *Natur und Kultur,* **32** (1935), 61–63, with portrait; J. Zenneck, in *Mitteilungen der Gesellschaft deutscher Naturforscher und Arzte,* **11** (1935), 2–3. On his seventieth birthday see H. Schmidt, in *Denkschriften der Technische Hochschule München,* **1** (1926), 3–4, with portrait as a youth. On the centenary of his birth see R. Sauer, in *Wissenschaftliche Vorträge, gehalten bei der akademischen Jahresfeier der Technischen Hochschule München* (1957), 10–11. A short biography by G. Faber is in *Neue Deutsche Biographie,* IV (Berlin, 1959), 210. A bronze bust by Hermann Hahn, at the Technische Hochschule in Munich, was unveiled in 1926.

J. E. HOFMANN

DYSON, FRANK WATSON (*b.* Measham, near Ashby-de-la-Zouch, Leicestershire, England, 8 January 1868; *d.* on board ship near Cape Town, South Africa, 25 May 1939), *astronomy.*

On graduating in the mathematical tripos at Cambridge, England, as second wrangler in 1889, Dyson began research on gravitational problems. He was appointed chief assistant at the Royal Observatory at Greenwich in 1894; astronomer royal for Scotland in 1905; and in 1910 he returned to Greenwich to become the eleventh astronomer royal. It was by cooperation in, and direction of, the preparation of fundamental astronomical measurements that he contributed significantly to the progress of astronomy.

From 1894 Dyson improved the methods used at Greenwich for reduction of the measurement of star positions from photographs, and during the opposition of Eros in 1900–1901 he organized the observations and reduction of the data compiled from them to provide new standards of accuracy. With W. G. Thackeray he reobserved the 4,239 stars that had been cataloged by Stephen Groombridge from 1806 to 1816 and compared positions so that proper motions of the stars could be determined over an eighty-year interval. Following this, Dyson extended J. C. Kapteyn's hypothesis of two star streams to fainter stars and rephotographed the stars in the internationally delimited Greenwich astrographic zone to allow more proper motions to be determined.

In observing the total solar eclipses of 1900, 1901, and 1905, Dyson measured the wavelengths of 1,200 lines in the spectrum of the chromosphere and compared the strengths with those for which laboratory evidence was available. His intensity measurements confirmed the work of J. N. Lockyer and A. Fowler, which was then subject to much criticism. It was due to Dyson that two Greenwich expeditions, one to Principe Island off Spanish Guinea and one to Sobral in Brazil, were sent to observe the 1919 solar eclipse. They verified the deflection of starlight by the sun's gravitational field to the degree predicted by relativity theory.

Dyson developed geophysical work at Greenwich by bringing up to date the observatory's equipment and techniques for measurements of terrestrial magnetism and by moving the department to the country when local railroad electrification made this necessary. He also reorganized accurate latitude measurement, developing the methods for using the floating zenith telescope with great success. Dyson took much interest in time determination: he installed the Shortt synchronome free-pendulum clocks in 1924 and arranged for public radio time signals that were soon extended to give worldwide coverage.

Dyson had a genius for collaboration, and the majority of his published work was as joint author. A strong supporter of the International Astronomical Union, he attended every meeting from 1922 to 1935; his charming hospitality at Greenwich was a byword among astronomers from every country. In 1931 W. J. Yapp, a wealthy manufacturer, donated a thirty-six-inch reflector to commemorate Dyson's tenure as astronomer royal; it was first used in April 1934. Dyson retired on 28 February 1933 but continued to offer advice, particularly on the removal of the Radcliffe Observatory from Oxford to Pretoria, South Africa.

Dyson received many honors, including four gold medals from learned societies. He was elected a fellow of the Royal Society in 1901, was president of the Royal Astronomical Society from 1911 to 1913, was created a knight bachelor in 1915, and received the K.C.B. in 1926.

BIBLIOGRAPHY

Dyson's important writings (alone or in collaboration) include "On the Determination of Positions of Stars for the Astrographic Catalogue at the Royal Observatory," *Monthly Notices of the Royal Astronomical Society,* **51** (1896), 114–134; "Determination of the Constant of Precession and the Direction of the Solar Motion From a Comparison of Groombridge's Catalogue (1810) With Modern Greenwich Observations," *ibid.,* **65** (1905), 428–457; "Determinations of Wavelengths From Spectra Obtained at the Total Solar Eclipses of 1900, 1901 and 1905," in *Philosophical Transactions of the Royal Society,* **206A** (1906), 403–452; "A Statistical Discussion of the Proper Motions of the Stars in the Greenwich Catalogue for 1910," in *Monthly Notices of the Royal Astronomical Society,* **77** (1917), 212–219; "A Determination of the Deflection of Light by the Sun's Gravitational Field, From Observations Made at the Total Eclipse of 1919 May 29," in *Philosophical Transactions of the Royal Society,* **220A** (1920), 291–333; "Variability of the Earth's Rotations," in *Monthly Notices of the Royal Astronomical Society,* **89** (1929), 549–557; and *Eclipses of the Sun and Moon* (Oxford, 1937). Manuscript material is in the archives of the Royal Greenwich Observatory, Herstmonceux, Sussex, England.

Obituaries of Dyson are A. S. Eddington, in *Obituary Notices of Fellows of the Royal Society of London,* **3** (1940), 159–172; and J. Jackson, in *Monthly Notices of the Royal Astronomical Society,* **100** (1940), 236–246. A biography is M. Wilson, *Ninth Astronomer Royal* (Cambridge, 1951).

COLIN A. RONAN

EAST, EDWARD MURRAY (*b*. Du Quoin, Illinois, 4 October 1879; *d*. Boston, Massachusetts, 9 November 1938), *genetics.*

East was the only son of William Harvey East, a mechanical engineer, and Sarah Granger Woodruff. The family on both sides had a long history of scholarly pursuits. After graduating from high school at fifteen, he worked for two years in a machine shop before entering the Case School of Applied Science (now Case Western Reserve University) in Cleveland in 1897. He found Case intellectually too narrow and left the next year for the University of Illinois, where he took his bachelor's (1900), master's (1904), and doctorate (1907) degrees. In 1903 he married Mary Lawrence Boggs; they had two daughters. East was ill for much of his life, and this may account for his irascibility with some of his students and colleagues.

East was trained as a chemist at Illinois. While a student there he assisted C. G. Hopkins of the Illinois Agricultural Experiment Station on selection experiments to alter the protein and fat content of corn. His job was to conduct the chemical analysis of the kernels. He soon became dissatisfied with this perfunctory job and wanted to elucidate the genetic mechanisms involved. East was especially intrigued by the decrease in yield which accompanied the success of the selection experiments. He wondered if increased inbreeding in the selected stock caused the decrease in yield, and if so, why. He began experiments on inbreeding in corn before he left the University of Illinois in October 1905 for a position at the Connecticut Agricultural Experiment Station. He remained there four years and conducted numerous experiments on inbreeding and outbreeding in tobacco, potatoes, and corn. These four years of intense experimentation were crucial for East's development as a scientist and determined to a great extent his later scientific interests. In 1909 he received an offer from the Bussey Institution of Harvard University. He accepted, became a full professor in 1914, and remained there until his death.

While conducting experiments on inbreeding and outbreeding between 1905 and 1908, East was struck by the problem of accounting for the inheritance of continuously varying characters. It was widely believed at this time that such blending inheritance could not be accounted for by Mendelian inheritance. East, along with H. Nilsson-Ehle of the Agricultural Research Station at Svalöf, Sweden, showed experimentally that some cases of blending inheritance could indeed be interpreted in terms of Mendelian inheritance. This experimental result was extremely important because Mendelian inheritance could then be seen to cover the entire spectrum of inherited characters, whereas before it was generally known to apply only to phenotypic characters inherited as a unit. East's 1910 paper on this topic, "A Mendelian Interpretation of Variation That Is Apparently Continuous," was particularly influential in America because Nilsson-Ehle's papers were written in German and published in a Swedish journal generally unavailable in the United States. After 1910 East published other papers on the inheritance of quantitatively varying characters. Perhaps the most important of these was his 1916 paper on the inheritance of size in *Nicotiana*, published in the first volume of the new journal *Genetics*. East's work on the problem of quantitative inheritance was widely known and hailed by other geneticists.

Another result of East's experiments in inbreeding and outbreeding was his interpretation of the role of sexual reproduction in the production of heritable variation. An understanding of heritable variation was of course crucial for an understanding of evolution. East popularized the now well-known view that sexual reproduction leads to recombination in the germ plasm and thus to vastly increased numbers of heritable variations. His 1918 article, "The Role of Reproduction in Evolution," came at a time when geneticists were just beginning to understand the importance of sexual reproduction as an immense source of heritable variation upon which selection could act. Thus he contributed significantly to a major tenet in all modern genetic theories of evolution.

By 1912 East had begun planning a book proposing a general theory of the effects of inbreeding and outbreeding. This book, *Inbreeding and Outbreeding: Their Genetic and Sociological Significance,* was finally published in 1919 with the collaboration of Donald F. Jones. The basic theory proposed by East was that inbreeding in a genetically diverse stock caused increased homozygosity. Believing with the *Drosophila* workers (Thomas Hunt Morgan *et al.*) that most mutations were recessive and deleterious, he concluded that the increased homozygosity caused by inbreeding should generally be accompanied by detrimental effects. The theory also explained why inbreeding was not necessarily deleterious in all cases. Unless deleterious recessives were present, inbreeding caused no ill effects. Outbreeding of course had the opposite effect of increasing heterozygosity and was often accompanied by heterosis, or hybrid vigor. In later years East continued to work on the physiological interpretation of heterosis and published a long paper on the subject only two years before his death. The theory of inbreeding and outbreeding was not conceived by East alone, but the 1919 book was well conceived and the theory presented with a wealth of evidence. It was widely read and cited by geneticists.

From the beginning of his work at the Connecticut Agricultural Experiment Station, East was concerned with the problems of the commercial production of agricultural products. With G. H. Shull he pioneered in developing a new method of corn breeding which revolutionized the production of corn in America and elsewhere. He also published papers on the improvement of potato and tobacco yields.

During World War I, East served as a chairman of the Botanical Raw Products Committee of the National Research Council and as acting chief of the Statistical Division of the U.S. Food Administration. While serving in these capacities he became immersed in the problems of world food production, overpopulation, and eugenic improvement of mankind. The gravity of these problems was brought home to him by the effects of World War I and by reading Malthus' *Essay on Population.* He became convinced that biologists had an obligation to speak out on the social implications of their science, and after 1920 his major efforts went in this direction. East believed that there was a genetic aspect to nearly all the problems of society; thus geneticists were qualified to speak on social problems. He himself was outspoken on the major social issues of the time. In 1923 he published *Mankind at the Crossroads,* in 1927 *Heredity and Human Affairs,* and in 1931 *Biology in Human Affairs,* a book which he edited and to which he contributed two chapters.

East was very concerned with the problems of overpopulation because he believed that mankind was reproducing faster than the food supply was increasing. He predicted that the world would soon be faced with mass starvation. Believing in the overwhelming importance of heredity as compared with environment, he proposed a eugenic plan of birth control in which the less desirable elements of society would be prevented from having children. This would solve the overpopulation problem and improve mankind at the same time.

Along with many other geneticists East believed that human races differed in their inherent capacities, both physical and mental, and that crosses between divergent races were biologically detrimental. He was firmly convinced that the racial crossing of whites and blacks in the United States should be prevented at all costs because the Negro was an inferior race: "In reality the negro is inferior to the white. This is not hypothesis or supposition; it is a crude statement of actual fact. The negro has given the world no original contribution of high merit" (*Inbreeding and Outbreeding*, p. 253). East justified his belief in the genetic inferiority of the Negro by pointing to the results of psychological testing of U.S. recruits during World War I. Negroes had scored consistently lower than whites on I.Q. tests. He was unafraid of the prospect that the United States or the world would be inundated by blacks, because he claimed their natural rate of increase was low in comparison with that of whites. He believed the blacks' only chance for extended survival in the United States was amalgamation with the whites, a possibility he clearly wanted to prevent.

One of East's major contributions to biology was his influence as a teacher of geneticists. He and William Castle worked together at the Bussey Institution to produce many of the best-known geneticists in the world, including D. F. Jones, Karl Sax, L. J. Stadler, R. A. Brink, L. C. Dunn, and Sewall Wright. East was known to be harsh and unduly critical of his students at times, but clearly his students were successful.

Another way he helped shape genetic research was by his active participation in professional organizations. He was a member of nearly every group of biologists concerned with genetics and its social import and served as an officer on many occasions. East helped found the journal *Genetics* in 1916 and was on its editorial board for many years. He not only helped direct the progress of genetic research through participation in these activities but also helped direct the interests of other geneticists toward the social implications of their scientific work. Some geneticists were stirred to write about the social im-

plications of genetics because of their opposition to East's ideas.

East contributed significantly to genetic research. His attempts to portray its social implications were less successful, and on some issues, particularly race, he now seems totally misguided. He appears to have examined his evidence much less carefully when analyzing the social implications of genetics than when analyzing a problem within genetics.

BIBLIOGRAPHY

A complete bibliography of East's work may be found in Donald F. Jones, "Edward Murray East," in *Biographical Memoirs. National Academy of Sciences*, **22** (1944), 217–242. This memoir is the only substantial secondary source dealing with the life and work of East.

WILLIAM B. PROVINE

EASTON, CORNELIS (*b.* Dordrecht, Netherlands, 10 September 1864; *d.* The Hague, Netherlands, 3 June 1929), *astronomy, climatology.*

The son of J. J. Easton, a sailor, and M. W. Ridderhof, Easton was principally a journalist. After graduating from high school in 1881, he first attended courses for those wishing to become government employees in Indonesia (then under Dutch rule) and subsequently studied French at the Sorbonne until 1886. After a short period of teaching, he became associated with various newspapers, notably the *Nieuwe Rotterdamsche Courant* (1895–1906), the *Nieuws van den Dag* of Amsterdam (1906–1923), and the *Haagsche Post* (from 1923).

Easton's contributions to astronomy deal mostly with the description and interpretation of the Milky Way. At the age of seventeen, as an amateur astronomer, he made his first drawings of the distribution of its brightness. The subsequent perfection and Easton's interpretation of these drawings gained him international fame and, in 1903, an honorary doctorate in physical sciences from the University of Groningen, at the proposal of the famous astronomer J. C. Kapteyn. The drawings aimed, first of all, at the representation of the northern Milky Way as a whole; detailed descriptions of certain regions of the sky published by such authors as Heis and Otto Boeddicker did not allow the construction of a homogeneous overall picture. The drawings were first published at Paris in 1893, under the title *La Voie Lactée, dans l'hémisphère boréal.*

Subsequent work deals with the comparison of those drawings with the distribution of the stars, and with the problem of the structure of the Milky Way stellar system. Counts of the faint stars in the *Bonner Durchmusterung* (around ninth-magnitude) revealed close correlation between their distribution in the sky and the drawings. In his attempts to interpret these findings, Easton adopted the hypothesis that the Milky Way system resembles other celestial objects showing spiral structure, and he proposed various solutions putting the center of the galaxy in the direction of the constellation Cygnus. The work is synthesized in "A Photographic Chart of the Milky Way and the Spiral Theory of the Galactic System," in *The Astrophysical Journal* (**37** [March 1913]). This concept of the galactic spiral structure has not survived subsequent research, which has led to the establishment of the galactic center in the direction of Sagittarius. But Easton's work inspired, and was highly esteemed by, such contemporary professional astronomers as Kapteyn, Pannekoek, and Seeliger.

Easton was active in many other fields of science besides astronomy. Particular mention should be made of his efforts in climatology. His monumental work *Les hivers dans l'Europe occidentale* (Leiden, 1928) contains a statistical-historical study of the climatological conditions in western Europe that attempts to connect data as far back as the thirteenth century with modern ones and critically studies suggested periodic variations. His accomplishments in this field led to Easton's appointment to the Board of Curators of the Netherlands Meteorological Institute in 1923.

BIBLIOGRAPHY

Easton's principal communications or astronomical studies, apart from those cited in the text, are in the publications of the Royal Netherlands Academy of Sciences (Amsterdam), *The Astrophysical Journal, Astronomische Nachrichten,* and *Monthly Notices of the Royal Astronomical Society.*

For descriptions of Easton's life and works see an article by J. J. Beyermann, in *Nieuwe Rotterdamsche Courant* (10 Sept. 1964); and J. Stein, "C. Easton in Memoriam," in *Hemel en Dampkring* (July, Aug.–Sept. 1929).

A. BLAAUW

EATON, AMOS (*b.* Chatham, New York, 17 May 1776; *d.* Troy, New York, 10 May 1842), *geology, botany, scientific and applied education.*

The son of a farmer, Abel Eaton, of old New England stock, Amos was born in eastern New York, just over the border from Massachusetts. He had a conventional education that culminated in graduation from Williams College in 1799. He taught briefly in a country school, but primarily he read law in New York City and was admitted to the state bar. Thereafter he established himself in Catskill, New York, as a lawyer and land agent. To these roles he added, as was then usual, practice in surveying. Even in those early years, however, Eaton gave evidence of an interest in popular science and education. In 1802 he published a pamphlet on surveying, *Art Without Science.* In Catskill he offered popular lectures in botany and wrote a manual on the subject that won the approval of David Hosack, an eminent authority in the field.

A tragic turn of events abruptly terminated Eaton's legal career in 1810. He was convicted of an alleged forgery during the Hudson Valley land disputes. Eaton and many others always maintained his innocence, but he spent five years in the Greenwich jail in New York City. There he turned to scientific studies, aided by John Torrey, son of the warden and subsequently a distinguished botanist. On release from jail, almost forty years of age, already married twice and a father, Eaton spent a year at Yale College, studying science under Benjamin Silliman and Eli Ives. Then he returned to Williams College, where he introduced a course of very popular and successful scientific lectures in 1817. In this year too appeared his first ventures in scientific publication, *A Botanical Dictionary* and the first edition of *Manual of Botany for the Northern States.*

Eaton was deeply grateful to the academic communities of Yale and Williams for admitting him to their company after his earlier humiliation. An important effect of his imprisonment was undoubtedly a humble and self-deprecating manner, which expressed itself in an effort at once to suppress and to surmount the resulting handicap. Eaton was gifted with an articulate and even voluble style, which fitted him well for his new career as a popular lecturer. It was an age and an environment lacking a truly professional tradition, and he was able to range widely as an amateur over the whole of science, from botany to chemistry, zoology, and geology. In 1818 he moved westward into the Troy–Albany area, in which he had been born and raised. This was then an active center of growth and internal improvement, as manifested particularly by the Erie Canal. Here he became associated with Governor De Witt Clinton and Stephen Van Rensselaer, both patrons and promoters of science and public improvement, who were convinced that the geological study of western New York could not fail to uncover coal and other mineral resources.

For the next half-dozen years Eaton was busy in several capacities. He was an itinerant lecturer in village and school, from West Point in the lower Hudson Valley to the Castleton Medical Academy in

Vermont. He sponsored the formation of the Troy Lyceum of Natural History, and he compiled textbooks in chemistry, zoology, and geology. Most important, under Stephen Van Rensselaer's patronage he executed geological and agricultural surveys of the local counties and across New York State along the Erie Canal route. He was thus drawn into a kind of specialization in geology, and he described himself as "the only person in North America capable of judging strata." His published reports of these surveys, bridging the earlier surveys of Maclure and the classical stratigraphy of the New York State Survey, earned him recognition in American geology, and the decade of the 1820's has been designated as the "Eatonian era."

His persistent efforts to devise and develop an American nomenclature for New York stratigraphy often led Eaton into opinionated and extravagant theorizing. Basically he was a Wernerian, following Abraham Werner's fivefold classification of Primitive, Transition, Secondary (or *Floetz*), Volcanic, and Alluvial. He thought of his task as primarily one of correlation of American, and particularly New York, strata with their English and Continental equivalents, recognizing at the same time the differences in the lithologic sequence. Although Eaton described fossils, his stratigraphic distinctions were drawn, as were those of Maclure before him, on the basis of lithology and the structural attitude of beds. This led him to repeat Maclure's error in correlating the New York plateau sediments with the English Secondary, although he correctly associated the Catskill brownstones with the Old Red Sandstone.

A characteristic product of the American frontier, Eaton was a kind of "jack-of-all-sciences," opening new vistas and stressing simplicity and practicality. In botany, too, he was very prolific, issuing eight editions of *Manual of Botany for the Northern States*. Asa Gray, like Eaton a product of northern New York, who became America's greatest botanist in the nineteenth century, began his studies with Eaton's *Manual*. In later years, however, he severely criticized Eaton's deficiencies as a botanist.

Eaton's final and most noteworthy contribution was to scientific education. He evolved a pedagogical theory emphasizing "the application of science to the common purposes of life"; students were to learn by doing, in sharp contrast with the conventional method of learning by rote. They were to perform experiments in the laboratory, collect specimens in the field, and even prepare their own lectures, leaving to the instructor and fellow students the role of critic. For the implementation of this then novel theory, Eaton persuaded Stephen Van Rensselaer to establish the Rensselaer School in Troy, New York, in 1824. Here, for the rest of his life, Eaton served as senior professor, struggling to realize his concept of an all-scientific and practical course of education. It was virtually a one-man institution, and Eaton's zeal and dedication were unflagging. In this school he trained a small but significant band of scientists who carried on and diffused his influence widely. Chief among his disciples were James Hall, J. C. Booth, Asa Fitch, Ebenezer Emmons, G. H. Cook, Abram Sager, E. S. Carr, Douglass Houghton, and E. N. Horsford. Although most were concerned primarily with geology, some also acquired an interest in chemistry and botany, as well as mineralogy and zoology.

In 1835 Eaton expanded his program and gave it greater practicality. Renamed the Rensselaer Institute, it was divided into two departments, one for science and one for engineering. Eaton created two degrees new to American education: bachelor of natural science, and civil engineer. In the pragmatic environment of nineteenth-century America, engineering gained headway, and the role of science was eventually subordinated in what became known after the middle of the century as the Rensselaer Polytechnic Institute.

Amos Eaton transmitted his zeal for science to his children. A daughter taught science in a girls' academy. Two of his sons were educated and taught at the Rensselaer School, and a third became a professor of natural science at Transylvania University in Kentucky. All three died young. A grandson, Daniel Cady Eaton, was for many years professor of paleobotany at Yale University. Thus, largely self-taught, Amos Eaton was a zealous and pioneering explorer and teacher of natural science in early America. Above all, he laid the foundations of a novel school and course of scientific and technological education. Perhaps he was also responsible for introducing a basic dichotomy between the traditional and the new technical types of education, a dichotomy that has not been easy to resolve.

BIBLIOGRAPHY

I. ORIGINAL WORKS. For more than a quarter of a century Amos Eaton was a prolific writer of texts, manuals, reports, and articles on many subjects. A full bibliography is in E. McAllister's *Amos Eaton* (see below). Only a few titles need be listed here, chiefly to illustrate the broad scope of his scientific interests: *Art Without Science* (Hudson, 1802; Albany, 1830); *The Young Botanists' Tablet of Memory* (Catskill, 1810); *A Botanical Dictionary* (New Haven, 1817); *Manual of Botany for the Northern States* (Albany, 1817; 8th ed., 1840); *An Index to the Geology of the Northern States* (Albany, 1818); *Chemical Instructor* (Albany, 1822; several eds. to 1836); *A Geological and Agricultural Survey of the District Adjoining the Erie Canal* (Albany, 1824); *Zoological Text-Book* (Albany, 1826); *Geological Nomenclature for North America* (Albany, 1828); *Prodromus of a Practical Treatise on the Mathematical Arts* (Troy, 1838).

II. SECONDARY LITERATURE. Aside from numerous biographical sketches, in the *Dictionary of American Biography* and elsewhere, the sole full-length life of Eaton is Ethel M. McAllister, *Amos Eaton, Scientist and Educator* (Philadelphia, 1941). Other references deal with the various aspects of Eaton's scientific and educational career. To be mentioned are P. C. Ricketts, *History of the Rensselaer Polytechnic Institute* (New York, 1895, 1914, 1934); and Samuel Rezneck, *Education for a Technological Society: A Sesquicentennial History of Rensselaer Polytechnic Institute* (Troy, N.Y., 1968). Other special topics include G. P. Merrill, *The First One Hundred Years of American Geology* (New Haven, 1924); Samuel Rezneck, "Amos Eaton: A Pioneer Teacher of Science in Early America," in *Journal of Geological Education,* **13** (Dec. 1965), 131 ff.; and "Amos Eaton the Old Schoolmaster," in *New York History,* **39** (Apr. 1958), 165 ff.; W. M. Smallwood, "Amos Eaton, Naturalist," *ibid.,* **18** (Apr. 1937), 167 ff.; H. S. Van Klooster, *Amos Eaton as a Chemist,* Rensselaer Science and Engineering Series, no. 56 (Troy, N.Y., 1938); and John W. Wells, *Early Investigations of the Devonian System in New York,* Special Papers, Geological Society of America, no. 74 (New York, 1963), esp. pp. 25–64, "The Eatonian Era." For Eaton's geological bibliography, see J. M. Nickles, *Geological Literature on North America, 1785–1918,* U.S. Geological Survey Bulletin 746 (Washington, D.C., 1923).

SAMUEL REZNECK

EBEL, JOHANN GOTTFRIED (*b.* Züllichau, Germany, 6 October 1764; *d.* Zurich, Switzerland, 8 October 1830), *medicine, geography.*

Ebel's father was a wealthy merchant, and his mother died when he was still young. He went to high school in Neuruppin and in 1780 entered the University of Frankfurt-an-der-Oder to study medicine. After studying there and in Vienna, he received his M.D. degree in 1788 with a thesis on the comparative anatomy of the brain. He continued his scientific work but also entered politics. Ebel lived for some years in Paris, where he came to know several of the leading figures of the French Revolution. During his extensive travels he became fascinated by Switzerland and wrote his famous *Anleitung* (1793), which is essentially a geological and historical guide to the country. During the wars of the French Revolution and the Napoleonic era, Ebel used his influence with the French leaders to improve conditions for the Swiss population and was rewarded with Swiss citizenship (1801). He settled there in 1810.

Like many intellectuals of his time, Ebel worked in many fields. He translated the political works of Sieyès into German, and he worked extensively in ethnology, statistics, and comparative anatomy. His reputation was such that Goethe suggested him for the chair of surgery and anatomy at the University of Jena (1803). Ebel loved his adopted country and wrote a number of enthusiastic books about it. They contributed to the image of Switzerland, and because of their popular style, especially that of the *Anleitung,* they attracted many visitors to the country. He also participated in establishing new hotels and set up the first lookout point in Switzerland, at Rigi. The purpose of his books was to spread knowledge of Switzerland, and he can thus be considered one of the pioneers of the Swiss tourist industry.

BIBLIOGRAPHY

Ebel's works in medicine are almost forgotten, and his fame rests on his books about Switzerland, the most important of which is *Anleitung auf die nützlichste und genüssvollste Art die Schweiz zu bereisen,* 2 vols. (Zurich, 1793; 2nd ed., 4 vols., 1804–1805; 3rd ed., 4 vols., 1809–1810). The last five eds. (8th ed., 1843) were in 1 vol. This popular work was translated into several languages and also appeared in pirated editions. Ebel's other works include *Schilderungen der Gebirgsvölker der Schwiz,* 2 vols. (Tübingen, 1798–1802; 2nd ed., Leipzig, 1802–1803), an ethnological description of the population in the cantons of Glarus and Appenzell. His most important geological work is *Über den Bau der Erde in Alpengebirge,* 2 vols. (Zurich, 1808). He also wrote a number of popular, descriptive books about Switzerland, such as *Malerische Reise durch Graubünden* (Zurich, 1825).

As a result of his fame and popularity, many (mostly panegyric) biographies of Ebel were written just after his death. Among the later ones is H. Escher, "J. G. Ebel," in *80. Neujahrblatt zum Besten des Waisenhaus in Zürich* (1917).

NILS SPJELDNAES

EBERTH, CARL JOSEPH (*b.* Würzburg, Germany, 21 September 1835; *d.* Berlin, Germany, 2 December 1926), *comparative anatomy, pathology, bacteriology.*

Eberth was the son of an artist, who died when Carl Joseph was still young. The boy helped his mother support the family by cutting out silhouette pictures. Nevertheless, he was able to attend the University of Würzburg, where he was drawn to biology and medicine by some of Germany's foremost teachers: Kölliker, Heinrich Müller, Leydig, and Virchow.

From 1856 to 1859 Eberth worked as an assistant in the Pathological Institute in Würzburg. In the latter year he completed a dissertation on the biology and parasitic characteristics of whipworms and was

275

granted the M.D. degree. He then became a prosector under Heinrich Müller at the Institute of Comparative Anatomy. Here he concentrated on histology, both normal and abnormal. He passed the *Habilitation* in 1863 and two years later moved to Zurich as *extraordinarius* in pathology, becoming *ordinarius* in 1869. From 1874, until his call to Halle in 1881, Eberth also taught histology and embryology in the school of veterinary medicine at the University of Zurich.

In Halle, Eberth was professor of comparative anatomy and histology until 1895, when he assumed directorship of the Pathological Institute. He held the latter position until his retirement at age seventy-five in 1911. As a teacher Eberth was patient and much admired. As a scientist he was thorough, meticulous, and humble, despite wide acclaim for his work.

Eberth married Elisabeth Hohensteiner, a minister's daughter, in Zurich in 1870. They had three daughters. Eberth was an avid naturalist and mountain climber, activities he continued into his seventies. After becoming emeritus, Eberth lived near Berlin with a daughter. He continued to be in excellent health until shortly before his death.

Eberth is best known for his discovery of the typhoid bacillus (*Salmonella typhosa,* earlier known as *Eberthella typhosa*), but this was only one of many important contributions he made in a fifty-year-long career in biological and medical science.

In his earlier scientific papers, many of which were published in *Virchows Archiv für pathologische Anatomie,* Eberth dealt with the histological structure of various parts of human and animal bodies. Particularly noteworthy were his descriptions of the ciliated epithelium and its function and several papers describing the normal and abnormal microscopic anatomy of the liver. As was true of many comparative anatomists and pathologists about 1870, Eberth became interested in the process of inflammation. He clearly differentiated between epithelial degeneration and regeneration in the cornea, and he was drawn to the study of inflammations caused by microorganisms. Along with Edwin Klebs and very few others, Eberth was instrumental in bringing the studies of bacteria and their actions, in which Davaine and Pasteur in France had pioneered, to the attention of German scientists.

In a remarkable small monograph, *Zur Kentniss der bacteritischen Mycose* (1872), Eberth set forth the results of his thorough observational and experimental techniques. Especially noteworthy is that this work was carried out four years before Koch dramatically demonstrated the isolation and cultivation of anthrax bacilli. The first part of Eberth's monograph described his studies of tissues from patients who had died of diphtheria, then a prevalent disease. He saw organisms (not clearly identified as diphtheria bacilli until 1884 by Klebs and Loeffler) that were most plentiful in the exudate covering the tonsils and the necrotic membrane in the pharynx. As a result of his investigations Eberth concluded that the organisms associated with diphtheria appeared first on the mucous membrane or on the edges of wounds. Further growth of the bacteria led to marked tissue destruction. All these conclusions are now known to be essentially correct. He went even further, saying that without these organisms there is no diphtheria ("Ohne diese Pilze keine Diphtherie . . .").

In the case of a newborn baby dying of respiratory failure, Eberth described a gelatinous exudate, rich in bacteria, filling the alveoli of the lungs. He did not clearly identify the organisms, but he stained them with iodine and hematoxylin and showed their existence in the heart and spleen as well as in the lungs. In the final section of the monograph Eberth confirmed experimentally what Davaine in France had shown before: that rod-shaped bodies in the blood of animals sick with anthrax were the cause of the disease. He mixed anthrax-infected blood with large volumes of water and allowed the mixture to settle. When he inoculated experimental animals with the supernatant fluid, no infection resulted. The sediment, however, was capable of producing anthrax. These techniques were to become commonplace in the laboratories of Europe during the next decade, but Eberth's work and his observations made him one of the earliest laboratory bacteriologists. He was thus one of the first of many pathologists seriously to take up bacteriological investigations.

In 1879 Eberth studied twenty-three cases of typhoid fever and reasoned that the characteristic changes found in the spleen and lymph nodes of the abdomen occurred because bacterial activity was most intense in these areas. He found rod-shaped organisms in twelve of his cases and published his results in *Virchows Archiv* in 1880. While he is, therefore, given credit for discovering the typhoid bacillus, he did so by histopathological techniques. The bacillus was not actually isolated and cultivated until 1884, when Gaffky, a student of Koch's, was able to grow it. Eberth, along with Koch and others, demonstrated the pneumonia diplococcus microscopically, but he did not cultivate that organism either.

Eberth contributed many papers describing important techniques and discoveries. He described the process of amyloid deposition in tissues and clearly showed that this substance came from outside the cells and was not a product of the cells in the affected areas. Thus, it was not necessarily a degenerative process of the cells that caused the amyloid to appear; rather, the cells were damaged by the amyloid deposited in the spaces between them.

Perhaps Eberth's major work in pathology was his contribution to the understanding of thrombosis, one of the most common pathological findings. Thrombosis is the process through which clots form in blood vessels during life. Because of its frequency and importance it had received much attention since the earlier part of the century. Virchow, in the 1840's, and others studied the problem, and most thought it was merely a blood coagulum. In the 1870's Zahn carried out systematic studies of thrombosis in the frog's mesentery. By direct observation he noticed that blood cells were deposited on the inner wall of the blood vessels and continued to accumulate in layers until the lumen became completely occluded. Zahn thought the cells were mainly white blood cells. Georges Hayem and Bizzozero in the early 1880's implicated blood platelets. In the mid-1880's Eberth, with his pupil and assistant Curt Schimmelbusch, who later became instrumental in perfecting the aseptic technique for surgery, carried out a thorough study of the role of the platelets.

Eberth and Schimmelbusch, by means of meticulous microscopic studies of experimentally induced thrombi, concluded that slowing of the flow of blood or injury to the inner lining of the vessel caused platelets to adhere to the wall, forming the beginning of a plug. By a process of viscous metamorphosis, now better understood, they believed the platelets adhered to one another and attracted red and white cells as formation of the thrombus continued. Eberth and Schimmelbusch called this process conglutination and were careful to distinguish it from coagulation, which they regarded as a later event in the development of the thrombus.

While some of the details of their explanation were disputed, Eberth and Schimmelbusch's major conclusions—that it was the platelets that were first involved, and that a combination of injury to the vessel and slowing of the blood flow were necessary for thrombosis to occur—have essentially stood the test of time. Their papers and subsequent monograph of 1888 do not give them priority of discovery, yet they deserve major credit for summarizing and elucidating the process in modern terms.

In Zurich and Halle, Eberth had many students. As an aid to them and students everywhere he undertook in 1889 to bring out a new edition of a widely used manual of techniques for pathological studies written by Carl Friedländer. Eberth contributed substantially toward making this popular book even more useful in the fourth and fifth editions. He nearly doubled the text, added many illustrations, and provided an index. Thus, Eberth was able to communicate to others the methods of microscopic investigation of tissues and cells that he had so successfully used himself.

BIBLIOGRAPHY

I. ORIGINAL WORKS. Much of Eberth's work was reported in the major German and Swiss medical journals. He was a frequent contributor to *Virchows Archiv für pathologische Anatomie,* where some of his major discoveries appeared, including "Untersuchungen über die normale und pathologische Leber," **39** (1867), 70–89; **40** (1867), 305–325; "Die amyloide Entartung," **80** (1880), 138–172; "Die Organismen in den Organen bei Typhus abdominalis," **81** (1880), 58–74; and "Neue Untersuchungen über den Bacillus des Abdominaltyphus," **83** (1881), 486–501.

The major monographs were *Zur Kentniss der bacteritischen Mycose* (Leipzig, 1872); *Die Thrombose nach Versuchen und Leichenbefunden* (Stuttgart, 1888), written with C. Schimmelbusch; and new eds. of Carl Friedländer, *Microscopische Technik zum Gebrauch bei medicinischen und pathologisch-anatomischen Untersuchungen* (4th ed., Berlin, 1889; 5th ed., 1894).

II. SECONDARY LITERATURE. See the following, listed chronologically: H. Ribbert, "Karl Joseph Eberth zum 70. Geburtstag," in *Deutsche medizinische Wochenschrift,* **31** (1905), 1511–1512; R. Beneke, "Zu Carl Josef Eberth's 80. Geburtstag," in *Berliner klinische Wochenschrift,* **52** (1915), 1010–1013; and "Carl Josef Eberth," in *Zentralblatt für allgemeine Pathologie und pathologische Anatomie,* **39** (1927), 226–228; W. Wachter, "Carl Joseph Eberth," in *Apothekerzeitung,* **42** (1927), 310–313; R. Beneke, "Zur Erinnerung an Karl Joseph Eberth," in *Münchener medizinische Wochenschrift,* **82** (1935), 1536–1537; Ernst Galgiardi, Hans Nabholz, and Jean Strohl, *Die Universität Zurich 1833–1933 und ihre Vorläufer* (Zurich, 1938), pp. 564–565; Heinrich Buess, "Carl Joseph Eberth," in *Les médecins célèbres,* R. Dumesnil and F. Bonnet-Roy, eds. (Geneva, 1947), pp. 196–197, trans. into German in *Die berühmten Ärzte,* R. Dumesnil and H. Schadewaldt, eds. (Cologne, 1966), pp. 235–236; and H. von Meyenburg, "Geschichte des pathologischen Instituts," in *Zürcher Spitalgeschichte,* 2 vols. (Zurich, 1951), II, 559–580, esp. 565–566.

GERT H. BRIEGER

EDDINGTON, ARTHUR STANLEY (*b.* Kendal, England, 28 December 1882; *d.* Cambridge, England, 22 November 1944), *astronomy, relativity.*

Eddington was the son of a Somerset Quaker, Arthur Henry Eddington, headmaster of Stramongate School in Kendal from 1878 until his death in 1884, and of Sarah Ann Shout, whose forebears for seven generations had been north-country Quakers. Following the death of her husband, Mrs. Eddington took Arthur Stanley, not yet two, and her daughter Winifred, age six, back to Somerset, where they made their home at Weston-super-Mare. In the atmosphere of this quiet Quaker home, the boy grew up. He remained a Quaker throughout his life.

Eddington's schooling was fortunate. Brynmelyn School, to which he went as a day boy, had three exceptionally gifted teachers who imparted to him a keen interest in natural history, a love of good literature, and a splendid foundation in mathematics. Reserved and studious by nature, Eddington was also physically active, playing on the first eleven at both cricket and football and enjoying long bicycle rides through the Mendip Hills. Before he was sixteen, he won an entrance scholarship to Owen's College (now the University of Manchester), where again he was fortunate in his teachers—Arthur Schuster in physics and Horace Lamb in mathematics. In the autumn of 1902, with an entrance scholarship, Eddington went into residence at Trinity College, Cambridge.

After two years of intensive concentration on mathematics under the guidance of the distinguished coach R. A. Herman, who stressed both the logic and the elegance of mathematical reasoning, Eddington sat the fourteen papers of the tripos examinations in 1904. He won the coveted position of first wrangler, the first time that a second-year man had attained this distinction. In 1905 he gained his degree and proceeded to coach pupils in applied mathematics and to lecture in trigonometry during the following term.

In February 1906 Eddington took an appointment as chief assistant at the Royal Observatory, Greenwich, where he remained until 1913. Here he obtained thorough training in practical astronomy and began the pioneer theoretical investigations that placed him in the forefront of astronomical research in a very few years. Besides his participation in the regular observing programs, Eddington had two special assignments: he went to Malta in 1909 to determine the longitude of the geodetic station there, and to Brazil in 1912 as leader of an eclipse expedition. Two further tests of his ability as a practical astronomer came after his return to Cambridge as Plumian professor of astronomy and director of the observatory. During the war years Eddington completed single-handed the transit observations for the zodiacal catalog. In 1919 he organized the two eclipse expeditions that provided the first confirmation of the Einstein relativity formula for the deflection of light in a gravitational field.

During these years Eddington was elected to fellowships in the Royal Astronomical Society (1906) and the Royal Society (1914). He was knighted in 1930, and his greatest honor, the Order of Merit, was conferred on him eight years later.

Eddington was president of the Royal Astronomical Society from 1921 to 1923 and of the Physical Society and the Mathematical Association from 1930 to 1932. In 1938 he became president of the International Astronomical Union. After his death an annual Eddington Memorial Lectureship was established and the Eddington Medal was struck for annual award, the first recipient being a former pupil of Eddington's, Canon Georges Lemaître of Louvain.

Eddington never married. After his appointment in 1913 to the Plumian professorship in Cambridge, he moved into Observatory House as director of the observatory and brought his mother and sister to live with him. Here he remained until the autumn of 1944, when he underwent a major operation from which he did not recover.

Of Eddington's scientific work, particularly in the field of stellar structure, E. A. Milne wrote in 1945 that he "brought it all to life, infusing it with his sense of real physics and endowing it with aspects of splendid beauty. . . . Eddington will always be our incomparable pioneer." His intuitive insight into the profound problems of nature, coupled with his mastery of the mathematical tools, led him to illuminating results in a wide range of problems: the motions and distribution of the stars, the internal constitution of the stars, the role of radiation pressure, the nature of white dwarfs, the dynamics of pulsating stars and of globular clusters, the sources of stellar energy, and the physical state of interstellar matter. In addition, he was the first interpreter of Einstein's relativity theory in English, and made his own contributions to its development; and he formulated relationships between all the principal constants of nature, attempting a vast synthesis in his provocative but uncompleted *Fundamental Theory.*

It is important to remember how rudimentary was much of our knowledge of astrophysics and of stellar movements at the beginning of this century. Proper motion or transverse motion had been known since the time of Halley and radial velocity since Doppler, but the assumption of William Herschel of random motion of the stars relative to the sun had been abandoned of necessity by Kapteyn in 1904. Schwarzschild attempted to show that the radial velocity vectors could be represented as forming an ellipsoid. This problem of the systematic motions of the stars was the subject of Eddington's first theoretical investigations. He chose to work with proper motions and isolated two star streams or drifts. In 1917 he compared the two theories thus:

The apparent antagonism between the two-drift and the ellipsoidal hypotheses disappears if we remember that the purpose of both is descriptive. Whilst the two-drift theory has often been preferred in the ordinary proper motion investigations on account of an additional constant in the formulae which gives it a somewhat greater flexibility, the ellipsoidal theory has been found more suitable for discussions of radial velocities and the dynamical theory of the stellar system [*Monthly Notices of the Royal Astronomical Society,* **77**, 314].

Eddington's remarkable statistical analyses of proper-motion data fully confirmed the existence of the two star streams, and he was able to determine their directions and relative numbers. He went on to other problems, such as the distribution of stars of different spectral classes, planetary nebulae, open clusters, gaseous nebulae, and the dynamics of globular clusters. In his first book, *Stellar Movements and the Structure of the Universe* (1914), Eddington brought together all the material of some fifteen papers, most of which had been published in the *Monthly Notices of the Royal Astronomical Society* between 1906 and 1914. The cosmological knowledge of the period was summarized and the most challenging problems were delineated, and he clearly declared his preference for the speculation that the spiral nebulae were other galaxies beyond our Milky Way, which was itself a spiral galaxy.

Eddington's great pioneer work in astrophysics began in 1916. His first problem was radiation pressure, the importance of which had been pointed out a decade earlier by R. A. Sampson. A theory of the radiative equilibrium of the outer atmosphere of a star was subsequently developed by Schwarzschild in Germany. Eddington delved deeper, in fact to the very center of a star, showing that the equation of equilibrium must take account of three forces—gravitation, gas pressure, and radiation pressure. Replacing the assumption of convective equilibrium of Lane, Ritter, and Emden with radiative equilibrium, he developed the equation that is still in general use. At that time he felt justified in assuming that perfect gas conditions existed in a giant star, and he adopted Emden's equation for a polytropic sphere with index $n = 3$. This is still referred to as Eddington's model of a star. Not until 1924 did he realize that this assumption and, therefore, this model were also applicable to dwarf stars.

That matter under stellar conditions would be highly ionized had been recognized by several astronomers, but it was Eddington who first incorporated this into the theory of stellar equilibrium by showing that high ionization of a gas reduced the average molecular weight almost to 2 for all elements except hydrogen.

Finding that the force of radiation pressure rose with the mass of the star, and with startling rapidity as the mass exceeded that of our sun, Eddington concluded that relatively few stars would exceed ten times the sun's mass and that a star of fifty times the solar mass would be exceedingly rare. To obtain a theoretical relation between mass and luminosity of a star, some assumption was necessary about internal opacity. At first he regarded opacity as mainly a photoelectric phenomenon, a view that drew strong criticism; but when Kramers' theory of the absorption coefficient became available, Eddington adapted it to the stellar problem, introducing his "guillotine" factor, and obtained his important mass-luminosity relation, announced in March 1924. Since the observational data for dwarf stars, as well as for giant stars, closely fitted the theoretical curve, he announced that dwarfs also must be regarded as gaseous throughout, in spite of their densities exceeding unity. He realized that the effective volume of a highly or fully ionized atom is very small, and hence deviations from perfect gas behavior will occur only in stars of relatively high densities. The mass-luminosity relation has been widely used and is still of immense value, although its applicability has been somewhat limited in recent years by the more detailed classification of both giants and dwarfs and by the recognition of the distinctive characteristics, for example, of subdwarfs, which do not conform to the mass-luminosity relation.

Eddington had calculated the diameters of several giant red stars as early as the summer of 1920. In December, G. E. Hale wrote him of the Pease and Anderson interferometer measurement of α Orionis on 13 December "in close agreement with your theoretical value and probably correct within about 10 per cent." Later Eddington applied his calculations to the dwarf companion of Sirius, obtaining a diameter so small that the star's density came out to 50,000 gm./cc., a deduction to which he said most people had mentally added "which is absurd!" However, in the light of his 1924 realization of the effects of high ionization, he claimed these great densities to be possible and probably actually to exist in the white dwarf stars. He therefore wrote W. S. Adams, asking him to measure the red shift in the Mount Wilson spectra of Sirius B, since, if a density of 50,000 or mo did exist, then a measurable Einstein relativity to the red would result. Adams hastened to c and wrote Eddington that the measured shi confirmed the calculated shift and, henc both the third test of relativity theor mense densities that Eddington had exchange of historic letters in 1 corded in *Arthur Stanley Eddi*

A direct consequence of this work was the challenge it presented to physicists, a challenge taken up in 1926 by R. H. Fowler, who achieved a brilliant investigation of the physics of super-dense gas, afterward called "degenerate" gas, by employing the newly developed wave mechanics of Schrödinger.

A consequence of Eddington's mass-luminosity relation was his realization that a time scale of several trillion (i.e., 10^{12}) years was essential for the age of stars if the then current Russell-Hertzsprung sequence of stellar evolution was to be retained. Except in the rare case of a nova or supernova that hurls out much of its matter, the loss of mass by a star is due to radiation. For a massive O or B class star to radiate itself down to a white dwarf, at least a trillion years would be required. This brought into the limelight the theory of conversion of matter into radiation by annihilation of electrons and protons, a hypothesis that appears to have been first suggested by Eddington in 1917. For seven years, in spite of severe criticism in Great Britain, he defended the general idea that the chief source of stellar energy must be subatomic. After 1924 many astronomers and physicists turned their attention to this. In 1934, after the discovery of the positron, Eddington urged abandonment of the electron-proton annihilation theory, on the ground that electron-positron annihilation was not only a more logical supposition but also an observed fact. In 1938 came the famous carbon-nitrogen-oxygen-carbon cycle of Hans Bethe, elegantly solving some of the problems of stellar energy and invoking the electron-positron annihilation hypothesis.

In 1926 Eddington published his great compendium, *The Internal Constitution of the Stars* (reprinted in 1930). In this book he drew attention to the unsolved problems partially treated in his investigations, among them the problem of opacity and the source or sources of stellar energy, which he called "two clouds obscuring the theory." Another obstinate problem was the phase relation of the light curve and the velocity curve of a Cepheid variable. In 1918 and 1919 he had published papers on the mathematical theory of pulsating stars, explaining many observed features of Cepheid variables but not the phase relation. He returned to this problem in 1941, when more was known about the convective layer and he could apply the physics of ionization equilibrium within this layer with encouraging results.

Other problems dealt with in these years were the central temperatures and densities of stars and the great cosmic abundance of hydrogen (recognized independently by Strömgren). Eddington developed a theory of the absorption lines in stellar atmospheres, extending earlier work of Schuster and Schwarzschild.

This made possible the interpretation of many observed line intensities. When the "nebulium" lines were identified by Bowen in 1927 as the result of so-called forbidden transitions in ionized nitrogen and oxygen atoms, Eddington explained how and why these emission lines can be produced within the highly rarefied gases that constitute a nebula. Another line of adventurous thinking concerned the existence, composition, and absorptive and radiative properties of interstellar matter. He calculated the density and temperature and showed that calcium would be doubly ionized, with only about one atom in 800 being singly ionized. He discussed the rough measurement of the distance of a star by the intensity of its interstellar absorption lines, a relation soon confirmed by O. Struve and by J. S. Plaskett.

In the field of astrophysics Eddington undoubtedly made his greatest—but by no means his only—contributions to knowledge. Here he fashioned powerful mathematical tools and applied them with imagination and consummate skill. But during these same years his mind was active along other lines; thus we have his profound studies on relativity and cosmology, his herculean but unsuccessful efforts to formulate his *Fundamental Theory,* and his brilliant, provocative attempts to portray the meaning and significance of the latest physical and metaphysical thinking in science.

Einstein's famous 1915 paper on the general theory of relativity came to England in 1916, when deSitter, in Holland, sent a copy to Eddington, who was secretary of the Royal Astronomical Society. Immediately recognizing its importance and the revolutionary character of its implications, Eddington threw himself into a study of the new mathematics involved, the absolute differential calculus of Ricci and Levi-Civita. He was soon a master of the use of tensors and began developing his own contributions to relativity theory. At the request of the Physical Society of London, he prepared his *Report on the Relativity Theory of Gravitation* (1918), the first complete account of general relativity in English. He called it a revolution of thought, profoundly affecting astronomy, physics, and philosophy, setting them on a new path from which there could be no turning back. A second edition (1920) contained the results of the eclipse expeditions of 1919, which had appeared to confirm the bending of light in a gravitational field, as predicted by Einstein's theory; it also contained a warning that the theory must meet the test of the reddening of light emitted from a star of sufficient density. This test was met when the measurements on Sirius B made by W. S. Adams at Eddington's request were announced in 1924.

Eddington published a less technical account of relativity theory, *Space, Time and Gravitation,* in 1920. This book brought to many readers at least some idea of what relativity theory was and where it was leading in cosmological speculation. It showed, too, how Eddington's mind had already entered philosophical grooves in which it continued to run—his selective subjectivism, almost universally repudiated, and his logical theory of structure, "a guiding illumination," in the words of Martin Johnson, who added, "As elucidator of the logical status of physics, Eddington led well his generation."

In 1923 came Eddington's great book, *Mathematical Theory of Relativity.* Einstein said in 1954 that he considered this book the finest presentation of the subject in any language, and of its author he said, "He was one of the first to recognize that the displacement field was the most important concept of general relativity theory, for this concept allowed us to do without the inertial system."

In this book Eddington gave the substance of the original papers of Einstein, deSitter, and Weyl but departed from their presentations to give a "continuous chain of deduction," including many contributions of his own, both in interpretation and in derivation of equations. With intuitive brilliance he modified Weyl's affine geometry of world structure by means of a new mathematical procedure, parallel displacement, which in itself was a not unimportant contribution to geometry. This led to his explanation of the law of gravitation ($G_{\mu\nu} = \lambda g_{\mu\nu}$) as implying that our practical unit of length at any point and in any direction is a definite fraction of the radius of curvature for that point and direction, so that the law of gravitation is simply the statement of the fact that the world radius of curvature everywhere supplies the standard with which our measure lengths are compared. This led subsequently to his theoretical determination of the cosmic constant λ. Assuming the principle that the wave equation determining the linear dimensions of an atomic system must give these dimensions in terms of the standard world radius, he obtained a value for λ in terms of the atomic constants that appear in the ordinary form of the wave equation.

This fascination with the fundamental constants of nature—the gravitation constant, the velocity of light, the Planck and Rydberg constants, the mass and charge of the electron, for example—and the basic problem of atomicity had driven Eddington to seek this bridge between quantum theory and relativity. Having found it, he eventually established relationships between all these and many more constants, showing their values to be logically inevitable. From seven basic constants Eddington derived four pure numbers, including the famous 137 forever associated with his name. This is the fine structure constant. He evolved the equation $10m^2 - 136m + 1 = 0$, the coefficients of which are in accordance with the theory of the degrees of freedom associated with the displacement relation between two charges and the roots of which give the ratio of masses of proton and electron as 1847.60. He showed that the packing factor for helium should be 136/137. Later Eddington identified the total number of protons and electrons in the universe with the number of independent quadruple wave functions at a point; he evaluated this constant as $3/2 \times 136 \times 2^{256}$, which is a number of the order of 10^{79}. In all, he evaluated some twenty-seven physical constants.

As all this work proceeded, Eddington published a succession of books, both technical and nontechnical, dealing with the above problems and also with the new problems that were arising in cosmological theories. The spherical Einstein universe was found to be unstable, and in 1927 Georges Lemaître published in an obscure journal his cosmology of an expanding universe, the result of the catastrophic explosion of a primeval atom containing all the matter of the universe. He sent a reprint to Eddington only in 1930. Immediately his own modification of this became the basis of all of Eddington's further work in this field. In 1928 Dirac published his new interpretation of the Heisenberg symbols q and p, an approach to a recondite subject that sent Eddington's mind racing off in a new direction. He developed a theory of matrices providing "a simple derivation of the first order wave equation, equivalent to Dirac's but expressed in symmetrical form" and also "a wave equation which we can identify as relating to a system containing electrons with opposite spin." He then developed his *E*-number theory, which proved to be a powerful tool in much subsequent work.

The Nature of the Physical World (1928) and *The Expanding Universe* (1933) deal with the above ideas and his epistemological interpretation. *New Pathways in Science* (1935) and *The Philosophy of Physical Science* (1939) carried his ideas further. All these books are rich in literary excellence and in the sparkle of his imagination and humor, as well as being gateways to new ideas and adventures in thinking.

His technical book *The Relativity Theory of Protons and Electrons* (1936), based almost wholly on the spin extension of relativity, spurred Eddington to evolve a statistical extension. Thus, during his last years he worked indomitably toward his dream—"Bottom's dream," he called it—his vision of a harmonization of quantum physics and relativity. The difficulties were immense and, as we now know, the greatest complexities of nuclear physics and subatomic particles were not yet discovered. But he took hurdle after hurdle as he saw them, with daring leaps, always landing, as he believed, surefootedly on the far side, even though he could not demonstrate his trajectories with mathematical rigor.

The obscurities and gaps in logical deduction in *Fundamental Theory* have discouraged most scientists from taking it seriously, but a few able men—Whittaker, Lemaître, Bastin, Kilmister, Slater—have seen Eddington's vision and have felt it worthwhile to explore further. Slater isolated an erroneous numerical factor in Eddington's work, a factor of 9/4 which modified the calculated recessional constant that had agreed reasonably well with the Mount Wilson observed value. Thus, in 1944, although he did not realize it himself, Eddington's theory had really demanded the change in the distance scale of the universe that Baade announced in 1952 from observational studies of the Cepheid variables in the Andromeda galaxy.

Eddington's biographer has referred to *Fundamental Theory* as an "unfinished symphony" standing as a challenge to "the musicians among natural philosophers of the future." His mystical approach to all experience necessarily embraced the sensual, the mental, and the spiritual. He believed that truth in the spiritual realm must be directly apprehended, not deduced from scientific theories. His Swarthmore Lecture to the Society of Friends, published as *Science and the Unseen World* (1929), and his chapter entitled "The Domain of Physical Science" in *Science, Religion and Reality* (1925), as well as passages throughout his books, reveal a deeply sincere, mystical, yet essentially simple, approach to consideration of the things of the spirit. In the search for truth, whether it be measurable or immeasurable, "It is the search that matters," he wrote. "You will understand the true spirit neither of science nor of religion unless seeking is placed in the forefront."

BIBLIOGRAPHY

I. ORIGINAL WORKS. Eddington's books are *Stellar Movements and the Structure of the Universe* (London, 1914); *Report on the Relativity Theory of Gravitation* (London, 1918); *Space, Time, and Gravitation* (Cambridge, 1920); *Mathematical Theory of Relativity* (Cambridge, 1923); *The Internal Constitution of the Stars* (Cambridge, 1926); *Stars and Atoms* (Oxford, 1927), Eddington's only popular account of astrophysical researches; *The Nature of the Physical World* (Cambridge, 1928); *Science and the Unseen World* (London, 1929); *The Expanding Universe* (Cambridge, 1933, 1940); *New Pathways in Science* (Cambridge, 1935); *Relativity Theory of Protons and Electrons* (Cambridge, 1936); *The Philosophy of Physical Science* (Cambridge, 1939); and *Fundamental Theory*, Edmund T. Whittaker, ed. (Cambridge, 1946), published posthumously.

II. SECONDARY LITERATURE. Writings on Eddington or his work include Herbert Dingle, *The Sources of Eddington's Philosophy* (Cambridge, 1954), an Eddington Memorial Lecture; A. Vibert Douglas, *Arthur Stanley Eddington* (Edinburgh and New York, 1956), a biography that includes a comprehensive list of Eddington's books and more than 150 scientific papers on pp. 193–198 and a genealogical table, pp. 200–201; Martin Johnson, *Time and Universe for the Scientific Conscience* (Cambridge, 1952), an Eddington Memorial Lecture; C. W. Kilmister, *Sir Arthur Eddington*, in Selected Readings in Physics series (London, 1966); C. W. Kilmister and B. O. J. Topper, *Eddington's Statistical Theory* (Oxford, 1962); S. R. Milner, *Generalized Electrodynamics and the Structure of Matter* (Sheffield, 1963); J. R. Newman, *Science and Sensibility*, I (London, 1961); A. D. Ritchie, *Reflections on the Philosophy of Sir Arthur Eddington* (Cambridge, 1947), an Eddington Memorial Lecture; Noel B. Slater, *Eddington's Fundamental Theory* (Cambridge, 1957); Edmund Whittaker, *From Euclid to Eddington* (Cambridge, 1949), and *Eddington's Principle in the Philosophy of Science* (Cambridge, 1951), an Eddington Memorial Lecture; and J. W. Yolton, *The Philosophy of Science of A. S. Eddington* (The Hague, 1960).

A. VIBERT DOUGLAS

EDER, JOSEF MARIA (*b.* Krems, Austria, 16 March 1855; *d.* Kitzbühel, Austria, 18 October 1944), *chemistry, photography.*

Eder went to Vienna to study chemistry and remained there for his entire professional life. His first independent research was carried out in a competition to explain the reaction between chromic acid or chromates and organic substances, especially gelatin, which in chromate becomes insoluble when exposed to light. He found in 1878 that a labile chromium dioxide is formed and easily decomposes into the lower and higher oxides. This study earned Eder first prize, and he decided to devote his life to photochemistry and photography. He became assistant professor at the Technische Hochschule in 1880 and professor of chemistry at the state vocational high school in 1882. There he founded an institute for education and research in the graphic arts, directing it from its beginning in 1889 until he retired in 1923.

In 1881 Eder extended and improved the use of silver chloride in gelatin emulsion by the method of precipitation, washing, and developing by ferrous citrate or hydroquinone. Such emulsions and techniques became important in the production of transparencies and copying papers. Eder became interested in sensitometry in 1884. He measured the effect of adding the dye eosin to silver chloride or bromide in gelatin or collodion emulsion by determining the depth of "blackening" after exposure for various lengths of time in specific regions of the solar spectrum. He found advantages in using the iodine derivative erythrosin with eosin. With monobromofluorescein and the methyl violet that others had recommended, Eder obtained sensitivities of silver bromide-collodion emulsions that made them suitable for the autotype process of printing in three colors. He recommended mercury oxalate for measurements in the ultraviolet. From spectrophotometry he went to the use of a concave Rowland grating with 13,000 lines per inch for the spectra of many elements, including those of the rare earths.

In 1884 Eder published the first volume of an extensive handbook of photography that grew to four volumes and appeared in a second edition in 1892. Eder was responsible for subsequent editions and enlargements. He also started a yearbook of photography and reproduction techniques in 1887; its thirtieth volume appeared in 1928. His history of photography saw several German editions and was translated into English by Edward Epstein in 1945.

BIBLIOGRAPHY

I. ORIGINAL WORKS. Eder's books were published in Halle. In later editions, the four volumes of the *Handbuch* were divided into parts, as follows: I, pt. 1, *Geschichte der Photographie* (4th ed., 1931); pt. 2, *Photochemie* (1906); pt. 3, *Photographie bei künstlichem Licht, Spektrumphotographie, Aktinometrie* (1912); pt. 4, *Photographische Objektive* (1911); II, pt. 1, *Die Grundlagen der Negativ-Verfahren* (1927), written with Lüppo-Cramer; pt. 2, *Photographie mit Kollodium Verfahren* (1927); pt. 3, *Daguerrotypie, Talbotypie, und Niepcotypie* (1927), written with E. Kuchinka; pt. 4, *Autotypie* (1928), written with A. Hay; III, pt. 1, *Fabrikation photographischer Platten, Filme, und Papiere* (1930), written with F. Wentzel; pt. 2, *Verarbeitung der photographischen Platten, Filme, und Papiere* (6th ed., 1930), written with Lüppo-Cramer, M. Andresen, and Tanzen; pt. 3, *Sensibilisierung und Desensibilisierung* (1930), written with Lüppo-Cramer, R. Schuloff, G. Sachs, J. Eggert, W. Ditterle, and M. Biltz; pt. 4, *Sensitometrie und Spektroskopie* (1930); IV, pt. 1, *Die photographischen Kopierverfahren mit Silbersalzen und photographische Rohpapiere* (1928), written with F. Wentzel; pt. 2, *Pigmentverfahren usw.* (4th ed., 1926); pt. 3, *Heliogravüre und Rotationstiefdruck* (1922); pt. 4, *Lichtpausverfahren und Kopierverfahren ohne Silbersalze* (1929), written with A. Trumm. His other works include *Atlas typischer Spektren* (1911); and *Rezepte, Tabellen, und Arbeitsvorschriften für Photographen und Reproduktionstechniker* (13th ed., 1927).

II. SECONDARY LITERATURE. Obituaries and biographies include W. Greenwood, in *Photographic Journal,* **86A** (1946), 266 ff.; O. Kempel, in *Österreichische Naturforscher und Techniker* (Vienna, 1951), pp. 125–127; and Erich Stenger, in *Zeitschrift für wissenschaftliche Photographie,* **8** (1948), 255–256.

EDUARD FARBER

EDISON, THOMAS ALVA (*b.* Milan, Ohio, 11 February 1847; *d.* West Orange, New Jersey, 18 October 1931), *technology.*

Edison's parents emigrated from Canada to Milan, Ohio, after his father joined an unsuccessful insurrection in 1837. The elder Edison prospered as a shingle manufacturer until the railroad bypassed the town, and in 1854 the family moved to Port Huron, Michigan, where the father conducted a less profitable grain and lumber business.

Edison's formal schooling was limited to about three months, followed by four years of instruction by his mother. He was an entrepreneur at age twelve, riding the trains to sell newspapers and food and to pick up odd jobs. He had an early and avid interest in chemistry and electricity, performing experiments at home and on the train. He acquired the habit of going for long periods with little sleep—an idiosyncrasy he kept throughout his life. At about age twelve Edison began to grow deaf, to the point where he could hear nothing below a shout. One result of this was to shut him further into himself and to encourage him in a vast self-directed program of reading. A bout with Newton's *Principia* at age fifteen "gave me a distaste for mathematics from which I have never recovered." He was, however, fascinated by various more elementary practical treatises.

In 1863 Edison became a telegraph operator, and this was his main source of income as he moved from city to city, ending up in Boston in 1868. His resolve to become an inventor became dominant, even though some initial attempts proved financially disastrous. He went to New York in 1869 to seek better fortune. In 1870, at age twenty-three, Edison received $40,000 for improving the stock-ticker system and used the money to set up a private fifty-man laboratory. In 1876 this laboratory was moved to Menlo Park, New Jersey, where his most concentrated and productive work was done. Eleven years later he moved to enlarged facilities at West Orange, New Jersey.

Edison was the epitome of the technologist-inventor. He was not unlearned in science—his prodigious reading had carried him through countless semipopular works, and during the year in Boston he obtained the first two volumes of Faraday's *Experimental Researches,* which he later claimed was a source of considerable inspiration to him; certainly the ability of Faraday to get along without mathematics must have been appealing. But his purposes were practical; he invented by design. He would see a gap in the economy, then invent to fill it; and at this he was very good. Examples include his work on stock tickers, multiplex telegraphy, incandescent lighting, magnetic iron-ore separation, and the storage battery. Some items were developed on very short notice to protect a patent position. Edison's chalk-drum telegraph relay and loudspeaking telephone receiver are especially good examples of this. His method in virtually all cases was to try the hundreds or thousands of possibilities that seemed plausible. This was not done in completely haphazard fashion, since he often obtained detailed knowledge of materials before testing them; but his procedure is rightly considered close to the ultimate in "cut-and-try."

The "Edison effect" (emission of electrons from a hot cathode) is often cited as his sole scientific discovery. In 1883 Edison performed a series of experiments to investigate the dark shadow that formed on the inside of a light bulb. He placed a second electrode inside the bulb and found that negatively charged carbon particles were emitted from the filament. He patented the device as a possible meter and then abandoned it. John A. Fleming, a British consultant to Edison, performed some further experiments, and the matter was still in his mind twenty years later when, as a consultant to Marconi, he saw the possibility of using the rectifying properties of a two-element bulb as a radiowave detector.

One product of his practical motivation was that Edison approached certain problems with a point of view different from that of a scientist. Thus some of the latter, contemplating the possibilities of incandescent lighting in the late 1870's, used available information—including indications that the successful lamp (as yet not invented) would have a low resistance—to prove that a system of independently controlled lights was infeasible. Edison changed the parameters by developing a high-resistance lamp and constructed a system that worked. Similarly the experts extolled the value of generators in which the internal and external resistances were equal, hence producing an efficiency of 50 percent. This was the condition for maximum energy transfer. Edison recognized that he did not need maximum energy, and that therefore he could use machines of low internal resistance to obtain much higher efficiencies. Edison may not have been unique in either of these realizations, but he was certainly the first and most successful in putting them together into a practical lighting system.

Edison's laboratories are considered prototypes of the modern industrial research laboratories in terms of the support they gave to manufacturing operations and the training they gave to staff members. The centralization of effort around the ideas of one man, however, was much greater than in later organizations.

In 1915 a consulting board, with Edison as its president, was established to advise the U.S. Navy on the possibilities of using new scientific and technical devices in war. Tangible results were limited, but one of Edison's early suggestions—a permanent scientific laboratory within the Navy—eventually found fruit in the establishment of the Naval Research Laboratory.

Edison was elected to membership in the National Academy of Sciences in 1927.

BIBLIOGRAPHY

I. Original Works. A large body of notebooks, photographs, and other MS materials is preserved at the Edison Laboratory National Monument at West Orange, New Jersey. Other miscellaneous sources can be identified in the Josephson work cited below. Some original apparatus has been saved: in the Menlo Park laboratory building, which has been restored and moved to Greenfield Village in Dearborn, Michigan; in the West Orange laboratory, which has been preserved at its original site; and at the Smithsonian Institution in Washington.

II. Secondary Literature. The best of the Edison biographies is M. Josephson, *Edison* (New York, 1959), although technical details are generally lacking. The Menlo Park years are treated in some depth in F. Jehl, *Menlo Park Reminiscences* (Dearborn, Mich., 1938).

See also H. C. Passer, "Electrical Science and the Early Development of the Electrical Manufacturing Industry in the United States," in *Annals of Science,* 7 (1951), 382–392.

Bernard S. Finn

EDWARDES (or **EDGUARDUS**), **DAVID** (*b.* Northamptonshire, England, 1502; *d.* Cambridge [?], England, *ca.* 1542), *medicine.*

Edwardes was admitted on 9 August 1517 as a scholar to Corpus Christi College, Oxford, where at different times he seems temporarily to have held the readership in Greek, and became B.A. in 1521 and M.A. in 1525. Additionally it appears that he had "seven years study of medicine" at Oxford, and at some undetermined time practiced that profession at Bristol. In 1528–1529 he continued his medical studies at Cambridge, where he became a member of the medical faculty, a position that he retained until his death.

Edwardes produced a small book of two treatises (London, 1532), the first entitled *De indiciis et praecognitionibus,* dealing with uroscopy and medical prognostication; the second, *In anatomicen introductio luculenta et brevis,* dedicated to the earl of Surrey, was the first work published in England to be devoted solely to anatomy. The latter treatise contains reference to Edwardes' dissection of a human body in 1531 at or near Cambridge, the first recorded, although legally unsanctioned, human dissection in England. The work is brief, occupying only fifteen printed pages, and follows the pattern of exposition popularized by Mondino, that is, progressing from the most to the least corruptible parts. It displays a scorn of medieval anatomy and, in contrast, reflects the new, humanistic Greek anatomical nomenclature. In fact, Edwardes' occasional use of words and phrases in Greek letters is one of the first such instances in England. The anatomical content of the treatise is mostly Galenic in character, although the author was sufficiently under Aristotelian and Avicennan influence to describe a three-chambered heart. Astonishingly for the time, he referred to the left kidney as being higher than the right, a statement that ran counter to the accepted Galenic doctrine and clearly demonstrated the use of independent observation and judgment.

Edwardes' treatise on anatomy antedated the development of anatomical studies in England by almost a generation and, perhaps for this reason, appears to have been utterly without influence.

BIBLIOGRAPHY

The combination of Edwardes' two treatises is known today only in the copy in the British Museum Library. There is a second copy of *De indiciis et praecognitionibus* in the library of the Royal Society of Medicine, London. A facsimile of *In anatomicen introductio luculenta et brevis,* with an English translation and such slight biographical information about Edwardes as exists, is to be found in *David Edwardes. Introduction to Anatomy 1532,* C. D. O'Malley and K. F. Russell, eds. (London, 1961).

C. D. O'MALLEY

EDWARDS, WILLIAM FRÉDÉRIC (*b.* Jamaica, West Indies, 6 April 1776; *d.* Versailles, France, 23 July 1842), *physiology, ecology, anthropology, ethnology, linguistics.*

Son of a wealthy planter of English origin, Edwards, like his brother, Henri Milne-Edwards, grew up and was educated in Bruges, where his family had moved. He became keeper of the Bruges Public Library and, interested in natural sciences, began to study medicine. Since Flanders was then a part of France, the family acquired French citizenship. In 1808 Edwards went to Paris to complete his medical studies. He did not graduate until 1814, at the age of thirty-eight, with a dissertation on the inflammation of the iris and black cataract. He worked at that time with the physiologist Magendie, who in his two-volume *Précis élémentaire de physiologie* (1816–1817) acknowledged Edwards' constant assistance in his experiments and in the preparation of the book.

After a short excursion into mineralogy, Edwards devoted much time to the study of the influence of environmental factors on the "animal economy." His early results were honored by the Prix Montyon of the Académie des Sciences (1820), and in 1824 he published his findings in a book. His main idea was that vital processes depend on external physical and chemical forces but are not entirely controlled by them. Life is different from heat, light, or electricity, forces which, however, contribute to the production of vital phenomena. Edwards systematically examined all principal functions, mostly of vertebrate species; and by varying the external conditions, he determined the nature and degree of their modification. Among the phenomena studied were the minimum and maximum temperatures compatible with life; heat production in young and adult animals; resistance of young animals to cold and to lack of oxygen; the importance of humidity, pressure, and movement of air in the loss of heat by transpiration; the role of light in the development of batrachians; and expiration of carbon dioxide by animals deprived of oxygen. Important was his finding that some warm-blooded animals (carnivores, rodents, some birds) are born less developed and have a much smaller capacity for heat production than those not born helpless. The former need external heat and cannot live without it. Body temperature of newborn carnivores and rodents drops by 10–12°C. as soon as they move away from their mother, but in contact with her it differs by only 1–2°C. Similarly, eight-day-old birds (starlings) in the nest maintain a body temperature of 35–37°C., while outside the nest (at 17°) body temperature drops within one hour to 19°C. These findings proved important in the prevention of infant mortality. Adult animals are, according to the conditions of life of their species, adapted to certain external temperature and thus to certain geographical distribution. Edwards' book is a classic pioneer work on animal ecology.

Soon afterward (*ca.* 1826) Edwards turned to some linguistic problems (etymology in Indo-Germanic languages, Celtic idioms) and was impressed during a journey to southern France and northern Italy by the problem of human types (races). In his opinion human races—in spite of their mixing—have fixed features and persist in their original type for centuries, so that descendants of all known great nations of antiquity could still be found among contemporary peoples (he gave several examples). His view, backed by J.-A. Colladon's early mice hybridization experiments, agrees with the more recent views of geneticists.

In 1832 Edwards was elected to the Académie des Sciences. His last studies led him to found the Société Ethnologique de Paris in 1839 (followed soon in England and the United States). Publication of its *Mémoires* drew attention to a field hitherto rather neglected. The word "ethnologie," introduced by Edwards, designates matters later included in the scope of anthropology.

BIBLIOGRAPHY

I. ORIGINAL WORKS. Edwards' writings include *Dissertation sur l'inflammation de l'iris et de la cataracte noire* (Paris, 1814); *De l'influence des agents physiques sur la vie* (Paris, 1824), trans. by Hodgkin and Fisher as *On the Influence of Physical Agents on Life* (London, 1832; Philadelphia, 1838), with observations on electricity and notes to the work of Edwards by Hodgkin; *Des caractères physiologiques des races humaines, considérés dans leurs rapports avec l'histoire* (Paris, 1829); "Animal Heat," in R. B. Todd, ed., *Cyclopaedia of Anatomy and Physiology,* II (London, 1836–1839), 648–684; *Mémoires de la Société ethnologique de Paris,* **1, 2** (1841–1842); and *Recherches sur les langues celtiques,* H. Milne-Edwards, ed. (Paris, 1844).

II. SECONDARY LITERATURE. See *Analyse succincte des principaux travaux de William Edwards, docteur en médecine* (Paris, n.d.); *Funérailles de William Edwards. Discours de Beriat-Saint-Prix* (Paris, 1842); and A. Quatrefages, in Michaud, *Biographie universelle,* XII (Paris, 1855), 280–282.

VLADISLAV KRUTA

EGAS MONIZ, ANTONIO CAETANO DE ABREU FREIRE (*b.* Avança, Portugal, 29 November 1874; *d.* Lisbon, Portugal, 13 December 1955), *neurology.*

Egas Moniz, the son of Fernando de Pina Rezende Abreu and Maria do Rosario de Almeida e Sousa, was educated by his uncle, an *abbé,* before entering the University of Coimbra in 1891. He studied mathematics and considered a career in engineering before deciding to enter medicine; he received his M.D. degree in 1899. Selecting neurology as his specialty, Egas Moniz went to Paris and Bordeaux to study with the leading figures in neurology and psychiatry, such as J. F. F. Babinski, J. J. Dejerine, Pierre Marie, and J. A. Sicard. In 1902 he became professor at

Coimbra and married Elvira de Macedo Dias. Egas Moniz was appointed to the chair of neurology at the new University of Lisbon in 1911, a position he held until his retirement in 1945. His honors included the Nobel Prize in physiology or medicine in 1949, honorary degrees from the universities of Bordeaux and Lyons, and awards from the Portuguese, Spanish, Italian, and French governments. He served as president of the Lisbon Academy of Sciences and was a member or honorary member of many other scientific societies, including the Royal Society of Medicine, the Academy of Medicine in Paris, and the American Neurological Association.

In addition to his scientific achievements, Egas Moniz was an accomplished historian, literary critic, and composer and had a distinguished career in politics. He was a deputy in the Portuguese Parliament from 1900 until his appointment as ambassador to Spain in 1917; he became foreign minister in 1918 and led Portugal's delegation to the 1919 Paris Peace Conference. Egas Moniz retired from politics in 1919 after a political quarrel involved him in a duel.

Egas Moniz' two most outstanding contributions to medicine were the diagnostic technique of cerebral angiography to locate brain tumors, and the first clinical use of psychosurgery. When he entered neurology, the method by which physicians attempted to use the still new technique of X-raying to locate intracranial tumors was the one developed by W. E. Dandy, involving the injection of air into the brain cavities. Seeking a more exact as well as a less hazardous technique, Egas Moniz began a series of cadaver experiments in which he injected various radiopaque solutions into the brain's arteries. After mapping the normal distribution of the intracranial blood vessels, he introduced his method clinically in 1927, outlining with X rays the location and size of a patient's brain tumor by the tumor's displacement of injected arteries. Egas Moniz and his colleagues published over 200 papers and monographs on normal and abnormal cerebral angiography, and the technique has been refined and elaborated for the localization of tumors and vascular disorders throughout the body.

The Nobel Prize went to Egas Moniz "for his discovery of the therapeutic value of leucotomy in certain psychoses." Early in his career he had worked with F. Regis on the problem of toxic psychoses and had become convinced that "only by an organic orientation can psychiatry make real progress." In 1935, at the Second International Neurological Congress in London, he heard J. F. Fulton and G. F. Jacobsen discuss the effects of frontal leucotomy (surgical division of the nerves connecting the frontal lobes to the rest of the brain) on the behavior of two chimpanzees: the animals remained friendly, alert, and intelligent but were no longer subject to temper tantrums or other symptoms of the experimental neuroses that had been successfully induced prior to surgery.

On the basis of this work Egas Moniz and his young surgical colleague, Almeida Lima, worked out a frontal leucotomy technique that they felt might alleviate certain psychiatric conditions, particularly those dominated by emotional tensions. The report of their first clinical trials on mental hospital patients—no operative deaths and fourteen out of twenty patients "cured" or "improved"—created worldwide interest and debate over the possibility that mental illness could be corrected by operating on brains that are not organically diseased. Modifications of their psychosurgical procedure were employed widely for two decades, then declined in use with the advent of psychopharmacology. At the Nobel presentations in 1949, Herbert Olivecrona captured the significance of Egas Moniz' work when he said:

> Frontal leucotomy, despite certain limitations of the operative method, must be considered one of the most important discoveries ever made in psychiatric therapy, because through its use a great number of suffering people and total invalids have recovered and have been socially rehabilitated.

NOTE

The presentation speech by Olivecrona at the Nobel awards, 1949, is in *Nobel Lectures. Physiology or Medicine, 1942–1962* (Amsterdam–New York, 1964), p. 246.

The techniques of frontal leucotomy or lobotomy pioneered by Egas Moniz in treating mental conditions were applied to the alleviation of intractable pain by W. H. Freeman and his colleagues in the mid-1940's. Variations of the surgical procedure have been devised in an effort to minimize some of the undesirable personality changes that were found to follow leucotomy. Use of the procedure to treat conditions such as schizophrenia has declined greatly since the advent of tranquilizers and other psychopharmacological agents.

BIBLIOGRAPHY

I. ORIGINAL WORKS. Egas Moniz' writings include *A vida sexual (fisiologia e patologia)* (Coimbra, 1901); *A neurologia na guerra* (Lisbon, 1917); *Um ano de politico* (Lisbon, 1920); "L'encéphalographie artérielle: Son importance dans la localisation des tumeurs cérébrales," in *Revue neurologique*, **2** (1927), 72–90; *Diagnostic des tumeurs cérébrales et épreuve de l'encéphalographie artérielle* (Paris, 1931); *L'angiographie cérébrale, ses applications et résultats en anatomie, physiologie et clinique* (Paris, 1934); "Essai d'un traitement chirurgical de certaines psychoses," in *Bulletin de l'Académie de médecine*, **115** (1936), 385–392; *Tentatives opératoires dans le traitement de certaines psychoses* (Paris, 1936); *Ao lado da medicina* (Lisbon, 1940); *Como cheguei a realizar a leucotomia pré-frontal* (Lisbon, 1948).

II. SECONDARY LITERATURE. See "Obituary. Antonio Egas Moniz, M.D.," in *Lancet* (1955), **2**, 1345; and F. R. Perino, "Egas Moniz, 1874–1955," in *Journal of the International College of Surgeons,* **36** (1961), 261–271.

JUDITH P. SWAZEY

EGOROV, DIMITRY FEDOROVICH (*b.* Moscow, Russia, 22 December 1869; *d.* Kazan, U.S.S.R., 10 September 1931), *mathematics.*

After graduating from the Gymnasium in Moscow, Egorov entered the division of physics and mathematics of Moscow University, from which he received a diploma in 1891. After obtaining his master's degree he remained at the university to prepare for a professorship, and in 1894 he became *Privatdozent* there. In 1901 he received his doctorate with a dissertation entitled "Ob odnom klasse ortogonalnykh sistem," and two years later he was appointed extraordinary professor in the division of physics and mathematics at the University of Moscow. In 1909 Egorov was made ordinary professor and was appointed director of the Mathematical Scientific Research Institute. He was elected corresponding member of the USSR Academy titled "Ob odnom klasse ortogonalnykh sistem," and member. Egorov was a member of the Moscow Mathematical Society; in 1902 he was elected to the French Mathematical Society and the Mathematical Society of Berlin University. From 1922 almost until his death, he was president of the Moscow Mathematical Society, and from 1922 he was editor-in-chief of *Matematicheskii sbornik.*

Egorov's investigations on triply orthogonal systems and potential surfaces, i.e., surfaces *E,* contributed greatly to differential geometry. The results of these investigations were presented by Darboux in his monograph *Leçons sur les systèmes orthogonaux et les coordonnées curvilignes* (2nd ed., Paris, 1910).

Egorov considerably advanced the solution of Peterson's problem on the bending on the principal basis. In the theory of functions of a real variable, wide use is made of Egorov's theorem: Any almost-everywhere converging sequence of measurable functions converges uniformly on a closed set, the complement of which has an infinitely small measure. This theorem, as well as Egorov's scholarship in new trends, led to the creation of the Moscow school dealing with the theory of functions of a real variable. Among the mathematicians belonging to Egorov's school are the well-known Soviet mathematicians N. N. Lusin, V. V. Golubev, and V. V. Stepanov.

Egorov also worked in other areas; for instance,

he initiated an investigation into the theory of integral equations. A brilliant lecturer and scholar, Egorov wrote some college textbooks on the theory of numbers, on the calculus of variations, and on differential geometry.

BIBLIOGRAPHY

Egorov's writings include "Uravnenia s chastnymi proisvodnymi vtorogo poriadka po dvum nezavisimym peremenym," in *Uchenye zapiski Moskovskogo universiteta,* **15** (1899), i–xix, 1–392; "Ob odnom klasse ortogonalnykh sistem," *ibid.,* **18** (1901), i–vi, 1–239, his doctoral diss.; "Ob izgibanii na glavnom osnovanii pri odnom semeystve ploskikh ili konicheskikh linii," in *Matematicheskii sbornik,* **28** (1911), 167–187; "Sur les suites de fonctions mesurables," in *Comptes rendus hebdomadaires des séances de l'Académie des sciences,* **152** (1911), 244–246; and "Sur l'intégration des fonctions mesurables," *ibid.,* **155** (1912), 1474–1475; *Elementy teorii chisel* (Moscow, 1923); *Differentsialnaya geometria* (Moscow–Petrograd, 1924); "Sur les surfaces, engendrées par la distribution des lignes d'une famille donnée," in *Matematicheskii sbornik,* **31** (1924), 153–184; "Sur la théorie des équations intégrales au noyau symétrique," *ibid.,* **35** (1928), 293–310; "Sur quelques points de la théorie des équations intégrales à limites fixes," in *Comptes rendus hebdomadaires des séances de l'Académie des sciences,* **186** (1928), 1703–1705.

A secondary source is V. Steklov, P. Lazarev, and A. Belopolsky, "Zapiska ob uchenykh trudakh D. F. Egorova," in *Izvestiya Rossiiskoi akademii nauk,* **18,** no. 12–18 (1924), 445–446.

A. B. PAPLAUSKAS

EHRENBERG, CHRISTIAN GOTTFRIED (*b.* Delitzsch, near Leipzig, Germany, 19 April 1795; *d.* Berlin, Germany, 27 June 1876), *biology, micropaleontology.*

Ehrenberg's father, Johann Gottfried Ehrenberg, was a municipal magistrate in the small city of Delitzsch; his mother, Christiane Dorothea Becker, was the daughter of an innkeeper. She died when Ehrenberg was thirteen years old. At the age of fourteen he entered school in Schulpforta, near Naumburg, a Protestant boarding school with a high level of instruction and a classical-philological orientation. In 1815, after passing his final examination, Ehrenberg, at his father's request, began to study theology in Leipzig. He then changed to medicine, however, and during his five semesters in Leipzig also attended J. C. Rosenmüller's lectures on anatomy and Schwägrichen's lectures on botany and zoology.

In 1817 Ehrenberg continued his medical studies at the University of Berlin, where in 1818 he passed the state medical examination. He studied under Christoph W. Hufeland, K. A. W. Berends, and K. F. von Graefe and was especially influenced by Karl Rudolphi and Heinrich Link, who encouraged his botanical and zoological studies and stimulated his interest in microscopical technique. With the presentation of a botanical dissertation, "Sylvae mycologicae Berolinensis," he was made a doctor of medicine in November 1818. This work not only depicts some 250 species of fungi found in the vicinity of Berlin (including sixty-two described for the first time) but it also demonstrates the constancy of the fungi species and their origin from seeds, which was still disputed. In a specialized work on the *Syzygites megalocarpus* (1819) he described for the first time the copulatory process among the molds and provided still further proofs of the sexual generation of the mushrooms in the essay *De mycetogenesis epistola* (1821).

These fundamental microscopical investigations led Ehrenberg to reject the then dominant view that spontaneous generation (*generatio aequivoca*) was possible in principle and that "lower" organisms, among which algae and fungi were grouped, could originate directly out of a basic inorganic substance, that is, water and slime. A major portion of Ehrenberg's lifework, above all the later microscopical research, was directed toward the clarification of this question. In the context of this problem, it becomes understandable why throughout his life Ehrenberg placed so much weight on demonstrating the constancy of species even among the lower organisms. Through his research on fungi, Ehrenberg became acquainted with Nees von Esenbeck, the president of the Deutsche Akademie der Naturforscher Leopoldina, of which Ehrenberg had become a member in 1818. His acquaintance with Adelbert von Chamisso was also a result of his botanical studies. While preparing the material from his expeditions, Ehrenberg was able to observe the generation of the lichen.

Ehrenberg's scientific career took a decisive turn in 1820 when he was presented with the opportunity to participate in the archaeological expedition of Count Heinrich von Menu von Minutoli, who planned to travel in Egypt. With a friend from his student days, Wilhelm Hemprich, Ehrenberg joined this expedition, which was advocated by Alexander von Humboldt (then residing in Paris) and by Heinrich Lichtenstein, the director of the university's zoological museum. The expenses of Ehrenberg and Hemprich were financed primarily by the Prussian Academy of Sciences, which also gave the instructions for the expedition. The two men were first to study thoroughly the natural history collections in Vienna and to consult scientists residing in that city. Therefore, in June 1820, Ehrenberg traveled from Berlin to Vienna and then to Trieste, where the party boarded ship for Alexandria on 5 August. The journey led through the Libyan desert to Cyrenaica, to Fayum in 1821, toward the Nile to Dongola, and to the shores of the Red Sea (in 1823) and yielded an unexpectedly large body of scientific results. Of the animal species alone, 3,987 (34,000 individual zoological objects) were sent to the collections of the Berlin Zoological Museum, which in that period were under the supervision of Heinrich Lichtenstein. As for botany, 46,000 plant specimens representing approximately 3,000 species were collected. The poor organization of the journey and its insufficient financing resulted in a considerable loss of time and many privations. Many members of the expedition became seriously ill. Ehrenberg's companion Hemprich fell victim to a fever in Massawa shortly before the end of the trip in 1825. Ehrenberg was forced to halt the journey ahead of schedule, without being able to explore the interior of Abyssinia; however, he had already made the littoral observations that were to make him famous as a zoologist. In his work on the coral polyps of the Red Sea (1834), he presented the first exact investigations on the anatomy, nourishment, and growth of the corals, and in individual works (published mostly in Poggendorf's *Annalen der Physik*) he explained the causes of the coloring of the Red Sea and the composition of the dust of trade winds. He also made drawings, based on microscopical studies, of numerous marine animals (mollusks, echinoderms, medusae, and electric rays) *in situ;* examined vertebrates for endoparasites; and made anatomical and embryological observations on insects, crabs, and spiders. In accordance with the scientific instructions given by the Berlin Academy, the collecting was not confined to plants and animals but also included rocks, fossils, and geographic measurements, as well as historical and ethnographic data and materials, among which were six manuscripts of ancient Arab physicians.

At the beginning of December 1825, Ehrenberg landed again in Trieste, the only survivor of an expedition that had lost nine members, including Hemprich. After his return Ehrenberg, originally humorous and joyful, experienced further disappointments that spoiled his enjoyment of the scientific utilization of this unusual and very productive journey. Even before his return his collections were decimated through the sale of the duplicates, and some of the boxes were damaged during quarantine, resulting in the destruction of their contents; consequently, of the originally immense series of forms of each species, only a fraction was still usable. Moreover, labels and sketches were missing, and Ehrenberg could hardly put the collections in order by himself.

Although in 1826 he published the plan for a comprehensive work (in the journal *Hertha,* pp. 92–94), the first section of which was to contain an account of the expedition, and the second, illustrated descriptions of the individual plants and animals, neither portion was completed, even though draftsmen paid by the state aided in preparing the material. The description of the journey and the first parts of the *Symbolae physicae,* dealing with animal and bird descriptions, appeared in 1828. Moreover, individual results obtained on the expedition had been reported since 1826 in the sessions of the Berlin Academy of Sciences by its members Alexander von Humboldt, Heinrich Lichtenstein, Heinrich Link, Karl Rudolphi, and Christian Weiss.

In July 1827, Ehrenberg was elected a member of the Berlin Academy of Sciences, having already become, on 24 March of that year, an assistant professor at the University of Berlin. In these years Ehrenberg obtained financial aid and scientific support through the influential Alexander von Humboldt, who in May 1827 left Paris to settle permanently in Berlin. He persuaded Ehrenberg to participate in a scientific expedition to Siberia financed by Czar Nicolas I. Ehrenberg used this opportunity not only to collect botanical and zoological specimens but also to undertake geological and paleontological studies, and to make microscopical observations on the Infusoria (or, as they were then called, animalcules). Humboldt's eight-month journey, in which, besides Ehrenberg, the Berlin mineralogist Gustav Rose took part, went from St. Petersburg and Moscow to Nizhni Novgorod (now Gorki), up the Volga to Kazan and to Ekaterinburg (now Sverdlovsk), and from there into the northern Urals. On the return journey, based in Astrakhan, they navigated the Volga and the Caspian Sea, where Ehrenberg, in addition to gathering fish for the St. Petersburg, Paris, and Berlin museums, made observations on living plankton.

Following his return home in 1831, Ehrenberg married Julie Rose, the niece of Gustav Rose and the daughter of Johannes Rose, a businessman and the consul in Wismar. After her early death (1848) he was married again, in 1852, this time to Karoline Friederike Friccius, sister-in-law of the chemist Eilhard Mitscherlich. From his first marriage he had one son, Hermann Alexander, and four daughters. The eldest daughter, Helene, married the botanist Johannes Hanstein, and another married the chemist Karl Friedrich Rammelsberg; the youngest daughter, Clara, aided her father in his scientific research.

When in 1833 Johannes Müller was appointed

successor to the late Karl Rudolphi, Ehrenberg was disappointed in his hope of being named to the chair of comparative anatomy, which would have allowed him to use the zootomical collections (including the material he himself had contributed) unhindered by questions of competence. As a result of Alexander von Humboldt's vigorous intercession, a strengthening of Ehrenberg's academic standing was sought, and in 1839 he was given a full professorship in "Methodologie, Enzyklopädie und Geschichte der Medizin." Ehrenberg did not, however, view this teaching post in terms of an active role. Generally, he hardly functioned as a university teacher, although as a member of the Berlin Academy of Sciences he had the right to lecture at the university even without a teaching appointment.

Ehrenberg's lifework consisted mainly of specific research, which he carried out at the Berlin Academy. It was recognized through his election to the Académie des Sciences of Paris (1841), his appointment as secretary of the Mathematics-Physics Section of the Berlin Academy of Sciences (1842), and by the award of the Order of Merit (1842), in which Alexander von Humboldt's opinion was important.

The scientific work that Ehrenberg undertook in Berlin following his return from the Russo-Siberian expedition was at first still related to the observations he had made on the Middle Eastern journey, especially those from the Red Sea. Included in this category are the publications on Hydrozoa and mollusks, especially the works on the coral polyps of the Red Sea (1831–1834); contributions to the knowledge of the physiology of the coral polyps in general and those of the Red Sea in particular, including an essay on the physiological systematics of these animals (1832); the medusae of the Red Sea (1834, 1835); marine phosphorescence (1835); the development and structure of the gastrotricha and rotatoria (1832); and the first reports on the so-called animalcules (Infusoria), the group that later absorbed Ehrenberg's interest. Already in 1828 and in 1830 he had given lectures at the Berlin Academy on the organization, systematics, and geographical relations of the Infusoria, which were soon followed by a series of papers entitled "Zur Erkenntnis der Organisation in der Richtung des kleinsten Raumes" (from 1832). Ehrenberg first treated this theme in monograph form in 1838, in *Die Infusionsthierchen als vollkommene Organismen,* in which he also presented a detailed historical sketch on the investigation and significance of this heterogeneous class of animals and elucidated the method of study underlying his microscopical researches. Although in all these works Ehrenberg utilized primarily observations and specimens from his Middle East expedition, he nevertheless almost always completed them with comparative material from the Baltic Sea and North Sea or from the Russo-Siberian expedition.

All Ehrenberg's individual observations were viewed in the light of a fundamental conception held consistently since his student years: this consisted of examining the theory, revived by Leibniz in the eighteenth century, of the existence of a "chain of being" (*scala naturae*) in nature. One of the bases for this theory was the various levels of organization among organisms, combined with the ideas that there exist gradual transitions in structure and performance from the mineral kingdom to man, that transformations from "lower" into more highly organized creatures still take place everywhere in nature, and that the lowest organisms can emerge spontaneously out of inorganic matter. These theories, which were earnestly discussed until the middle of the nineteenth century, were based mainly on that little-investigated group of organisms that Cuvier had united in the fourth class of his "radial animals." The five classes of the "radial animals" were, according to Cuvier, echinoderms, entozoa, medusae, polyps, and Infusoria; and they were considered to be simple organized animals, as opposed to the divisions within the vertebrates, mollusks, and arthropods. From the beginning of his research Ehrenberg strove to investigate the inner organization of the animals of these five classes and particularly to provide a new, systematic grouping for the Infusoria, a task to which he was led in large part by the use of the microscope.

In his opinion, all animals possess with an equal degree of completeness the important organs of life, e.g., nervous and vascular systems, muscles, and digestive and sexual organs. Through comparative anatomical investigations Ehrenberg examined Cuvier's five classes for the presence of these organs. In order to ascertain the nervous capacity of the echinoderms and medusae, he employed, on the advice of Alexander von Humboldt, galvanic currents as stimuli. In exploring the structure of the digestive organs he utilized, beginning in 1833, food colored with indigo or carmine. He carried out an extensive series of studies of this type with the Infusoria in particular. At the time this group still included such heterogeneous organisms as bacteria, all single-celled animals, the many-celled rotatoria, and several worms. Ehrenberg did not yet separate the many-celled animals from the single-celled ones, a concept that became current in systematic zoology only after 1850; rather, he believed that he could demonstrate the presence of complete organ systems in single-celled animals. This was for him an important argument against spontaneous generation and the "chain of being."

In spite of a critical, inductive research method, Ehrenberg succumbed to an optical error, especially when he consciously renounced microscopical magnifications of greater than 300. His error was similar to that of the pioneers of the microscope in the seventeenth century, like Leeuwenhoek and Swammerdam, who considered the indistinct structures in the egg and the sperm to be complete organisms and on these grounds derived the preformation theory. The basis for the correction of Ehrenberg's errors was set out only in 1863, with the union of protozoology and cytology. In his later years Ehrenberg could no longer accept the more correct perceptions of Felix Dujardin (his most vehement opponent), Theodor von Siebold, Max Schultze, and others, since they did not concern an individual error but called into question the entire conception of his system of the animal kingdom.

As a systematist Ehrenberg proceeded from Cuvier, but he had rejected both the latter's graduated hierarchy of the more highly and thoroughly organized animals and his classification of man among the mammals. His own system, which he proposed in 1836 in the sketch of the animal kingdom according to the principle of a single type reaching down to the monad, set man (*Kreis der Völker*) as a systematic category in contrast with the animals (*Kreis der Tiere*). He based this procedure on the "capacity for mental development" (*geistigen Entwicklungsfähigkeit*) of the human race, and even for the classification of the animals he employed social behavior as the most important taxonomic characteristic. This conception remained limited to its time and later hindered Ehrenberg's acceptance of the Darwinian theory of evolution.

So much the greater, then, is the importance of the continuation of the studies on single-celled marine and fossil animals, with which Ehrenberg completed his pioneering achievements. With the microscope he discovered single-celled fossils that built up geological strata; he gave exact descriptions of and discriminated among the shells and skeletons of freshwater and marine animals, thereby becoming the founder of microgeology and micropaleontology in Germany. His collection of samples, containing many types, along with his manuscripts and correspondence, are still available for study in the Museum für Naturkunde in Berlin.

Through his worldwide marine investigations Ehrenberg was invited to participate in oceanographic research projects of international importance. Thus he influenced the instructions, drawn up by Humboldt, for James Ross's Antarctic expedition (1839–

1843) and for the Novara expedition (1857–1859) and worked on the deep-bottom samples of the American researchers Silliman, L. W. Bailey, and M. F. Maury; the latter, beginning in 1853, provided him with material taken from depths of 10,000–12,000 feet. These results made possible L. Brooke's invention of the deep-sea lead, for the employment of which on German ships such as the *Arkona* and the *Thetis* (1860–1862) and the *Nymphe* (1865–1868) Ehrenberg tirelessly campaigned. Finally, the *Gazelle* expedition (1874–1876) was equipped with this device, but Ehrenberg did not live to see the results. Even the investigation of the sea bottom served Ehrenberg in his refutation of the theory of spontaneous generation, which had again come under discussion from the standpoint of the theory of evolution, through the hypothetical prehistoric organism *Bathybius haeckeli*.

In his lifetime and until the present, Ehrenberg has been reproached for not completing the utilization of the collections assembled on his Middle East expedition, for not accepting the findings of cytology and of the theory of evolution, and for not correcting his errors. The first stemmed both from technical problems, extending to a lack of scientific organizational ability, and from his very exact and laborious method of working, which aimed at comprehensive analysis. Moreover, the results were embedded in a philosophical system that presupposed the spiritual origin and constancy of the world order and therefore resisted materialistic and evolutionary interpretations. His method was based on the comparative anatomy and morphology of the first half of the nineteenth century; in these areas, as a pioneer of microscopy, he employed polarized light and pursued a comparative microscopical anatomy—with outmoded optical means. He expressed the program of his life and his research in a youthful letter (1821) to Nees von Esenbeck: "Until now my favorite pursuit has been neither naked systematizing nor unsystematic observation, and whenever time and circumstances, together with my ability, allow it, I prefer getting down to the grass-roots level" (Stresemann, "Hemprich und Ehrenberg. Reisen zweier naturforschender Freunde . . .," p. 42). In later years he considered only purely empirical knowledge to be valid; the lasting merit of his description and classification of the fossil protozoans stems from this position.

BIBLIOGRAPHY

I. ORIGINAL WORKS. Ehrenberg's writings include *Reisen in Aegypten, Libyen, Nubien und Dongola*, I, pt. 1 (Berlin, 1828), not completed; *Symbolae physicae* (Berlin,

1828–1845): *Aves,* pts. 1–2 (1828–1829), *Mammalia,* pts. 1–2 (1828–1832), *Evertebrata excl. Insecta,* pts. 1–2 (1829–1831), *Insecta,* pts. 1–5 (1829–1845), none of the divisions completed; "Ueber die Natur und Bildung der Corallenbänke des rothen Meeres und über einen neuen Fortschritt in der Kenntnis der Organisation im kleinsten Raume durch Verbesserung des Mikroskops von Pistor und Schiek," in *Abhandlungen der Preussischen Akademie der Wissenschaften zu Berlin* (1832), pp. 381–438; "Ueber den Mangel des Nervenmarks im Gehirn der Menschen und Thiere, den gegliederten röhrigen Bau des Gehirns und über normale Krystallbildung im lebenden Tierkörper," in *Annalen der Physik,* **28,** no. 3 (1833); *Ueber die Natur und Bildung der Coralleninseln und Corallenbänke im rothen Meere* (Berlin, 1834); *Das Leuchten des Meeres. Neue Beobachtungen nebst Übersicht der Hauptmomente der geschichtlichen Entwicklung dieses merkwürdigen Phänomens* (Berlin, 1835); *Die Akalephen des rothe Meeres und den Organismus der Medusen der Ostsee erläutert und auf Systematik angewendet* (Berlin, 1836), which contains the first draft and a review of his system; *Beobachtung einer auffallenden bisher unerkannten Structur des Seelenorgans bei Menschen und Tieren* (Berlin, 1836); "Ueber das Massen verhältnis der jetzt lebenden Kiesel-Infusorien...," in *Abhandlungen der Preussischen Akademie der Wissenschaften zu Berlin* (1836), pp. 109–136; *Die Infusionsthierchen als vollkommene Organismen. Ein Blick in das tiefere organische Leben der Natur* (Leipzig, 1838), with an atlas and 64 colored copperplates; "Über noch zahlreich jetzt lebende Tierarten der Kreidebildung," in *Abhandlungen der Preussischen Akademie der Wissenschaften zu Berlin* (1839), pp. 81–174; "Kieselschaligen Süsswasserformen am Wasserfall-Fluss im Oregon" and "Mikroskopisches Leben in Texas," in *Monatsberichte der Akademie der Wissenschaften zu Berlin* (1848–1849), pp. 76–98; "Über das mikroskopische Leben der Galapagos-Inseln," *ibid.* (1853), pp. 178–194; "Über die erfreuliche im Grossen fördernde Teilnahme an mikroskopischen Forschungen in Nord-Amerika," *ibid.,* pp. 203–220; "Über die seit 27 Jahren noch wohl erhaltenen Organisations-Praparate des mikroskopischen Lebens," in *Abhandlungen der Preussischen Akademie der Wissenschaften zu Berlin* (1862), pp. 39–74, with three color plates; "Über die wachsende Kenntnis des unsichtbaren Lebens als felsbildende Bacillarien in Californien," *ibid.* (1870), pp. 1–74, with three plates; "Uebersicht... über das von der Atmosphäre getragene organische Leben," *ibid.* (1871), pp. 1–150; "Nachtrag zur Übersicht der organischen Atmosphärilien," *ibid.* (1871), pp. 233–275, with three plates; "Mikrogeologische Studien über das kleinste Leben der Meeres-Tiefgründe aller Zonen und dessen geologischen Einfluss," *ibid.* (1873), pp. 131–398, with twelve plates; and "Fortsetzung der mikrogeologischen Studien als Gesammt-Übersicht der mikroskopischen Paläontologie gleichartig analysierter Gebirgsarten der Erde, mit specieller Rücksicht auf den Polycistinen-Mergel von Barbados," *ibid.* (1875), pp. 1–225.

Unpublished travel diaries, MSS, and letters on his journeys to the Middle East and Siberia are in the archives of the Berlin Academy of Sciences, and original drawings are in the library of the zoological museum (Museum für Naturkunde) of Humboldt University, Berlin.

II. Secondary Literature. Biographies of Ehrenberg are Clara Ehrenberg, *Unser Elternhaus. Ein Familienbuch* (Berlin, 1905); Johannes von Hanstein, *Christian Gottfried Ehrenberg* (Bonn, 1877); and Max von Laue, *Christian Gottfried Ehrenberg. Ein Vertreter deutscher Naturforschung im neunzehnten Jahrhundert* (Berlin, 1895), with portrait and bibliography.

On specific aspects of Ehrenberg's work, see H. Engel, "Het Levenswerk van Christian Gottfried Ehrenberg," in *Microwereld,* **15** (1960), 19–32; Gerhard Engelmann, "Christian Gottfried Ehrenberg, ein Wegbereiter der deutschen Tiefseeforschung," in *Deutsche hydrographische Zeitschrift,* **22** (1969), 145–157; Siegmund Günther, "Chr. G. Ehrenberg und die wissenschaftliche Erdkunde," in *Deutsche Rundschau für geographische Statistik,* **17** (1895), 529–538; A. von Humboldt, "Bericht über die naturhistorischen Reisen der Herren Ehrenberg und Hemprich," in *Hertha* (1827), 73–92; Otto Koehler, "Christian Gottfried Ehrenberg," in H. Gehrig, ed., *Schulpforta und das deutsche Geistesleben* (Darmstadt, 1943), pp. 58–68; Erwin Stresemann, "Hemprich und Ehrenberg. Reisen zweier naturforschender Freunde im Orient, geschildert in ihren Briefen aus den Jahren 1819–1826," which constitutes *Abhandlungen der Deutschen Akademie der Wissenschaften zu Berlin, Klasse für Math. und allg. Naturwiss.* (1954), no. 1, with a portrait; "Hemprich und Ehrenberg zum Gedenken. Ihre Reise zum Libanon im Sommer 1824 und deren ornithologische Ergebnisse," in *Journal für Ornithologie,* **103** (1962), 380–388; and Sigurd Locker, "Mikrofossilien aus der Sammlung Christian Gottfried Ehrenberg," in *Wissenschaftliche Zeitschrift der Humboldt-Universität zu Berlin,* Math.-Naturwiss. Reihe., **19,** no. 2–3 (1970), 186–189.

There is no comprehensive assessment of Ehrenberg's importance in micropaleontology and geology. Unpublished letters and MSS in these areas are included in the micropaleontological collection of the Museum für Naturkunde, Berlin.

Ilse Jahn

EHRENFEST, PAUL (*b.* Vienna, Austria, 18 January 1880; *d.* Amsterdam, Netherlands, 25 September 1933), *theoretical physics.*

Paul Ehrenfest was the youngest of the five sons of Sigmund and Johanna Jellinek Ehrenfest. His childhood was spent in a working-class district of Vienna, where his father ran a successful grocery business. He grew up surrounded by the crowded and varied life of the many nationality groups in the Austro-Hungarian capital, constantly reminded by the ugly, widespread anti-Semitism that he was a Jew. Ehrenfest's early interest in mathematics and science was stimulated by his oldest brother, Arthur, and this fascination with science helped him through a difficult adolescence. He studied theoretical physics at Vienna, where he received his doctorate in 1904 for a dissertation on the extension of Hertz's mechanics to problems in hydrodynamics. The dissertation was supervised by Ludwig Boltzmann, whose work and style greatly influenced Ehrenfest.

On 21 December 1904 Ehrenfest married Tatyana Alexeyevna Afanassjewa, a Russian student of mathematics whom he had met at Göttingen in 1902, during a year of study there. According to Austro-Hungarian law, the marriage of a Christian to a Jew could occur only if both partners officially renounced their religions, which Ehrenfest and his Russian Orthodox bride therefore did. During the early years of their marriage the Ehrenfests collaborated on several papers that clarified some of the obscurities in the statistical mechanics of Boltzmann and Josiah Willard Gibbs. As a result they were invited by Felix Klein to prepare a monograph on the foundations of statistical mechanics for the *Encyklopädie der mathematischen Wissenschaften*. After their marriage the Ehrenfests lived first in Vienna and Göttingen and in 1907 moved to St. Petersburg, hoping to settle in Russia. Ehrenfest had no regular employment in any of these cities, but they were able to manage on their small inherited incomes. In 1911 he began a difficult and depressing search for an academic position, a search complicated by his anomalous religious status. This search came to an unexpectedly successful conclusion in 1912, when Ehrenfest was appointed to the chair of theoretical physics at Leiden as the successor to H. A. Lorentz, to whom he became deeply attached.

Ehrenfest moved to the Netherlands in October 1912 and immediately brought new vitality to the scientific life of Leiden. He started a weekly colloquium, established a reading room for physics students, revived a student science club, and generally devoted his efforts to maintaining real intellectual and human contact among all members of Leiden's scientific community. As a teacher Ehrenfest was unique. Albert Einstein described him as "peerless" and "the best teacher in our profession whom I have ever known." His lectures always brought out the basic concepts of a physical theory, carefully extracting them from the accompanying mathematical formalism. He worked closely with his students, doing everything in his power to help them develop their own talents. His nickname among the students, "Uncle Socrates," captures his probing questioning, the force of a personality that could sometimes be overwhelming, and the infectious warmth of his humor.

Ehrenfest's special gift as a theoretical physicist was his critical ability, rather than his creative power or his calculational skill. This ability is particularly evident in his writings on statistical mechanics. Boltzmann had developed this subject over a thirty-year period, and his ideas had changed a good deal during this time as he responded to a variety of difficulties pointed out to him by others. As a consequence, there was a certain amount of confusion about what his theory asserted, which assertions had been proved, and how much of the theory had survived the various attacks on it. In 1907 Ehrenfest proposed a simple theoretical model (the Ehrenfest urn model) that showed how the laws of probability could produce an average trend toward equilibrium, even though the behavior of the model was reversible in time and every one of its states would eventually recur. This meant that Boltzmann's *H*-theorem (showing that molecular collisions will produce an approach to equilibrium with the entropy increasing monotonically in time), if interpreted in a suitable statistical way, did not necessarily contradict the reversible laws of mechanics, as Loschmidt had argued, or Poincaré's recurrence theorem, as Zermelo had argued.

In their *Encyklopädie* article, which appeared in 1911, the Ehrenfests brought out both the logical structure and the remaining difficulties of this theory. They made a clear distinction between the older approach (before 1877), which treated the molecules statistically but tried to make universally valid statements about the gas as a whole, and the later work, in which the gas itself was treated by statistical methods. The role of the ergodic hypothesis in relating time averages to averages over an ensemble was brought out clearly; so clearly, in fact, that the attention of mathematicians was drawn to the ergodic problem. The Ehrenfests formulated the sequence of theorems that still needed proving before the statistical foundation of the second law of thermodynamics could be said to be firmly established. Their analysis of Gibbs's approach to the subject was less sympathetic. They found fault with his treatment of irreversibility and underestimated the importance of Gibbs's powerful ensemble methods in dealing with complex systems.

The critical approach also led Ehrenfest to his greatest positive contribution to physics: the adiabatic principle. Ehrenfest was one of the first to try to understand the significance of the strange new concept of energy quanta that Max Planck had introduced into physics in 1900 in his theory of blackbody radiation. In a series of papers culminating in his major study of 1911, "Which Features of the Quantum Hypothesis Play an Essential Role in the Theory of Heat Radiation?," Ehrenfest picked out the essentials of the early quantum theory and showed how they fit together. He proved rigorously that the energy of electromagnetic vibrations cannot take on all values—cannot vary continuously—if the total energy of the blackbody radiation in an enclosure is to be finite: Planck's assumption that energy is a discrete variable was, therefore, logically necessary and not just sufficient. Ehrenfest also showed, by an analysis of Wien's displacement law, that the ratio of energy to frequency was the only variable that could be quantized for a harmonic oscillator, if one wanted to maintain the statistical interpretation of entropy. Planck's quantum condition for the energy E of an oscillator of frequency ν,

$$E/\nu = nh,$$

where n is a nonnegative integer and h is Planck's quantum of action, no longer seemed arbitrary. The quantity E/ν was the only one that kept a constant value if one varied the frequency-determining parameters sufficiently slowly, that is, adiabatically.

In a series of papers that appeared from 1913 to 1916, Ehrenfest studied the possibility of generalizing the notion of quantization, previously applied only to oscillators. He showed that every periodic system possesses a property invariant under slow (adiabatic) changes in its parameters: the ratio of the kinetic energy, averaged over one period, to the frequency. Ehrenfest proposed that only such adiabatic invariants could properly be quantized and also that when the parameters of a quantized system are changed adiabatically, the allowed quantum states of the original system continue to be the allowed quantum states. Ehrenfest and his student J. M. Burgers showed that this adiabatic principle encompassed the various quantization methods introduced independently by a variety of physicists. The adiabatic principle was widely used and highly prized as one of the few reliable guides to progress during the difficult years of the "old quantum theory," when even the laws of conservation of energy and momentum were suspect.

During the late 1920's and early 1930's Ehrenfest did his utmost to try to insure the intelligibility of the new quantum mechanics and to stress its relationships with classical physics. He proved the result, still known as Ehrenfest's theorem, that quantum mechanical expectation values of coordinates and momentum obey the classical equations of motion. One of his last papers consisted entirely of a series of fundamental questions on the physical and mathematical aspects of quantum mechanics. These questions were probably troubling many physicists, but only Ehrenfest was willing to risk the odium of asking questions that might be put aside as "meaningless." They were far from that, as Wolfgang Pauli soon demonstrated by writing a paper answering some of Ehrenfest's questions.

Ehrenfest affected the development of physics even more by his personal influence on other physicists than by his writings, particularly in the last decade of his life. He traveled widely, lecturing or attending conferences, and was a welcome visitor at universities from Moscow to Pasadena. When he was at home, there were always visiting physicists—older colleagues like Planck, contemporaries like Abram Fedorovitch Joffe, or young men like Enrico Fermi and Robert Oppenheimer. Ehrenfest's way of living his physics had its effect on all who knew him, and on others through them. Both Albert Einstein and Niels Bohr were his close friends, and Ehrenfest arranged a number of the historic conversations between them on the fundamental ideas of quantum physics.

All his life Paul Ehrenfest suffered from feelings of inadequacy and inferiority. They persisted despite his extraordinary success as physicist and teacher, despite his close and warm ties to many people of many kinds. They were accentuated by the growing difficulty Ehrenfest felt in keeping up with the latest developments in his science. Finally, in September 1933, depressed by the plight of his Jewish colleagues in Nazi Germany (on whose behalf he had been exerting himself to the limit of his powers) and faced with a multitude of personal problems that seemed insuperable, Ehrenfest took his own life.

BIBLIOGRAPHY

I. ORIGINAL WORKS. Ehrenfest's scientific papers, including his unpublished dissertation, the *Encyklopädie* article, and several lectures, are reprinted in his *Collected Scientific Papers* (Amsterdam, 1959). An English translation by M. J. Moravcsik of the *Encyklopädie* article appeared under the title *The Conceptual Foundations of the Statistical Approach in Mechanics* (Ithaca, N.Y., 1959).

Ehrenfest's MSS, notebooks, and scientific correspondence are in the National Museum for the History of Science at Leiden. For further information, see T. S. Kuhn, J. L. Heilbron, P. Forman, and L. Allen, *Sources for History of Quantum Physics* (Philadelphia, 1967), pp. 33–35.

II. SECONDARY LITERATURE. For a full discussion of Ehrenfest's life and work through the period of World War I, see Martin J. Klein, *Paul Ehrenfest,* I, *The Making of a Theoretical Physicist* (Amsterdam, 1970). The second volume of this biography is in preparation. A biography in Russian by Viktor J. Frenkel has been announced but has not yet appeared.

Valuable information can be found in Albert Einstein's essay, "Paul Ehrenfest in Memoriam," repr. in his *Out of My Later Years* (New York, 1950), pp. 214–217; H. A. Kramers, "Physiker als Stilisten," in *Naturwissenschaften,* **23** (1935), 297–301; Wolfgang Pauli, "Paul Ehrenfest†," *ibid.,* **21** (1933), 841–843; and George E. Uhlenbeck, "Reminiscences of Professor Paul Ehrenfest," in *American Journal of Physics,* **24** (1956), 431–433.

MARTIN J. KLEIN

EHRET, GEORG DIONYSIUS (*b.* Heidelberg, Germany, 30 January 1708; *d.* London, England, 9 September 1770), *botany.*

Ehret was a gifted artist and teacher whose skillfully executed botanical drawings significantly advanced the knowledge of many new and exotic plants. His work and teaching contributed to the successful introduction of the Linnaean system into England.

Ehret was the son of a poor gardener to the margrave of Baden-Durlach. His father's early death forced him to leave school and begin an apprenticeship as a gardener with his uncle. During the ensuing years he worked as a journeyman gardener in several cities of Germany, executing at the same time a large number of drawings and paintings of the plants under his care. The turning point of his career came about 1732, when Ehret made the acquaintance of the German botanist and physician Christoph Jacob Trew. The financial good fortune resulting from Trew's patronage enabled Ehret to travel across Europe. He visited the most celebrated gardens in France and Holland, collecting and drawing many rare plants. In 1734 he met the French botanist Bernard de Jussieu in Paris, and shortly thereafter the English physician and botanist Hans Sloane in London. His travels culminated with a visit to the great Swedish botanist Linnaeus in 1737, at Haarlem. There he completed all the illustrations for Linnaeus' book *Hortus Cliffortianus,* which was published in the same year. Ehret returned to England in 1740 and began his successful career as an artist and teacher. Among his patrons were the English naturalists Sir Joseph Banks and Griffith Hughes, the physicians Richard Mead and John Fothergill, and the duchess of Portland. He was also temporarily employed at the botanical garden of the University of Oxford in 1750–1751. Ehret was elected fellow of the Royal Society on 19 May 1757 and read several botanical papers before the group. His name was immortalized in the genus *Ehretia,* an honor proposed by his lifelong friend and mentor Trew and confirmed by Linnaeus.

BIBLIOGRAPHY

I. Original Works. "A Memoir of Georg Dionysius Ehret," translated into English, with notes, by E. S. Barton, is in *Proceedings of the Linnean Society of London* (1894–1895), pp. 41–58; Ehret's original German MS is preserved in the botanical department of the British Museum. With Christoph J. Trew he published *Plantae selectae quarum imagines ad exemplaria naturalia Londini in hortis curiosorum nutrita* (Nuremberg, 1750–1773). Some of his paintings were collected in *Twelve Coloured Reproductions From the Original Paintings on Vellum,* with an introduction and descriptive text by Wilfrid Blunt (Guildford, England, 1953). Among Ehret's articles the following should be mentioned: "An Account of a Species of Ophris," in *Philosophical Transactions of the Royal Society,* **53** (1763), 81–83; "An Account of a New Peruvian Plant Lately Introduced Into the English Gardens," *ibid.,* 130–132; and "A Description of the Andrachne With Its Botanical Characters," *ibid.,* **57** (1767), 114–117.

II. Secondary Literature. For a short and almost contemporary account of Ehret's life see Richard Pulteney, *Historical and Biographical Sketches of the Progress of Botany in England* (London, 1790), II, 284–293. Other biographies appeared in *Dictionary of National Biography,* VI, 585; *Proceedings of the Linnean Society of London* (1883–1884), 42–43; and in *Journal of Botany, British and Foreign,* **34** (1896), 316–317. Seven letters written to Ehret by Linnaeus between 1736 and 1769 were published in *Proceedings of the Linnean Society of London* (1883–1884), 44–51.

Guenter B. Risse

EHRLICH, PAUL (*b.* Strehlen, Germany [now Strzelin, Poland], 14 March 1854; *d.* Bad Homburg, Germany, 20 August 1915), *hematology, immunology, chemotherapy.*

Among medical scientists of his generation Ehrlich was probably the most original, stimulating, and successful. The fruitfulness of his concepts initiated advances in all fields of biomedical research to which they were applied. Hematology became a recognized discipline through his pioneering studies of dye reactions on red and white blood cells. In exhaustive experiments on the production of high-potency diphtheria antitoxin and on methods of assaying and standardizing such products, he developed techniques and established fundamental principles of immunity. His crowning achievement was the synthesis of Salvarsan and the demonstration of its therapeutic efficacy in syphilis and allied diseases.

Paul Ehrlich was born into a comfortable, lively household in a country town in Prussian Silesia, about twenty miles south of Breslau (now Wrocław, Poland). He was the only son and fourth child of Ismar Ehrlich, a respected but somewhat eccentric Jewish distiller, innkeeper, and lottery collector, and his wife Rosa Weigert, an industrious woman of notable intelligence, charm, and organizational talent. Her cousin Carl Weigert, the distinguished pathologist, was only nine years older than Paul, and the two became close friends. Besides many of his mother's characteristics the boy had his father's excitability and interjection-ridden manner of speech, and perhaps inherited certain aptitudes from his paternal grandfather, Heimann Ehrlich, a prosperous liqueur merchant, who collected an extensive private library and late in life gave lectures on science to fellow citizens of Strehlen.

In 1860, when he was six years old, Ehrlich entered the local primary school. At age ten he went to the St. Maria Magdalena Humanistic Gymnasium in Breslau and boarded with a professor's family. He accepted Spartan living and classroom conditions; was unobtrusive and conscientious; and though not outstanding, was often near the top of his class. He disliked all examinations, however. His favorite subjects were mathematics and Latin; his weakest was German composition.

After matriculating in 1872, Ehrlich took a disappointing introductory course in natural sciences at Breslau University and then spent three semesters at Strasbourg, which largely determined his life's course. He was impressed by the anatomist Wilhelm von Waldeyer's broad comprehension of medicine, and the professor in turn noted the many extra hours this unusual student devoted to making excellent histological preparations with his own modifications of new aniline dyes. Ehrlich visited the Waldeyer household, and a lasting friendship was established.

Although lacking formal courses in chemistry, Ehrlich became fascinated with the subject while studying for his *Physikum* at Strasbourg. Having passed this examination, he returned in 1874 to Breslau, where he completed studies for his medical degree, except for one semester in 1876 at the Physiology Institute of Freiburg im Breisgau and a final term at Leipzig in 1878. In Breslau he was influenced by the pathologists Julius Cohnheim and Carl Weigert, the physiologist Rudolf Heidenhain, and the botanist Ferdinand Cohn, sponsor of Robert Koch's researches on anthrax bacilli. At the Pathology Institute, Ehrlich became friendly with such outstanding visitors as W. H. Welch, the American pathologist, and C. J. Salomonsen, the Danish bacteriologist. Weigert had introduced aniline dyes into microscopic technique, and in his cousin's laboratory Ehrlich studied their selective action on cells and tissues. His first paper on the properties of these dyes appeared in 1877, in which year he passed the state medical examination. His doctoral dissertation, "Beiträge zur Theorie und Praxis der histologischen Färbung," was approved at Leipzig University in 1878. These two works included descriptions of large, distinctively stained cells containing basophilic granules, for which Ehrlich coined the term "mast cells," differentiating them from the rounded "plasma cells" observed in connective tissue by Waldeyer. In 1879 he defined and named the eosinophil cells of the blood.

Upon graduation Ehrlich was appointed head physician (*Oberarzt*) in Friedrich von Frerichs' renowned medical clinic at the Charité Hospital in Berlin. Frerichs, an imaginative clinician with deep interests in experimental pathology, encouraged Ehrlich's histological and biochemical researches, and the latter thereby gained lasting insights into diagnostic and therapeutic problems. His reports on the morphology, physiology, and pathology of the blood cells advanced hematology into a new era by establishing methods of detecting and differentiating the leukemias and anemias. Further, the observations that basic, acidic, and neutral dyes reacted specifically with such cellular components as leukocyte granules and nuclei implanted in Ehrlich's mind the fundamental concept underlying his future work: that chemical affinities govern all biological processes. He extended comparable staining methods to bacteria and protozoa and rendered Koch's discovery of the tubercle bacillus immediately more important by showing that its failure to stain in aqueous dye solutions could be circumvented by use of basic dyes in an aqueous-aniline oil solution, which penetrated the bacillary coating and then remained acid-fast.

Ehrlich was determined to explore the avidity of living tissues for certain dyes. In 1885 a remarkable monograph, *Das Sauerstoffbedürfnis des Organismus,* reporting his investigations into the distribution of oxygen in animal tissues and organs, gained widespread attention from medical scientists. Using two vital-staining dyes, alizarin blue (reducible to a leuko form with difficulty) and indophenol blue (readily reducible), he demonstrated that while living protoplasm in general has potent reducing properties, bodily organs are classifiable into three categories according to their oxygen avidity. Challenging Pflüger's assertion that tissue oxidation and reduction entail direct entry and exit of oxygen, he contended that these processes involve withdrawal and insertion of hydrogen atoms. Two years later the monograph won the Tiedemann Prize and served as Ehrlich's *Habilitation* thesis before he became *Privatdozent* in internal medicine at Berlin University. In 1886 he described methylene blue as a selective vital stain for ganglionic cells, axis cylinders, and nerve endings. Later, with A. Leppmann, he used this dye therapeutically to kill pain in neuralgias; and in 1891, with P. Guttmann, he pursued to its logical conclusion the finding that malaria parasites stain well with methylene blue, administering the dye to two malarial patients with apparent success.

Further by-products of Ehrlich's ingenuity with dyestuffs were the use of fluorescein to observe the streaming of the optic humors (1882) and his diazo reaction, a color test for the presence of bilirubin in the urine, regarded long afterward as a useful prognostic test in severe acute infections, such as typhoid fever (1883). His other Charité investigations that strengthened the developing conviction that chemical composition, distribution within the body, and pharmacological effect of biologically active substances were interrelated included the treatment of iodine poisoning by detoxification with sulfanilic acid (1885); the lipotropism of thalline and its homologues, and the dependence of thalline's antipyretic action on the ortho-position of the methoxyl group in the molecule (1886); the correlation between lipotropism and neurotropism, as displayed in rabbits inoculated with certain dyes of the basic and the nitrated, but not the sulfonic acid, groups (1887); and the demonstration that liver degeneration in cocaine-poisoned mice was not caused by the benzoyl radical responsible for the drug's anesthetizing properties (1890).

In 1883 Ehrlich had married Hedwig Pinkus, daughter of a prosperous textile industrialist of Neustadt, Upper Silesia, whom he had met during a visit to Strehlen. Ten years his junior, she proved an understanding, faithful companion, and their marriage was happy. They had two daughters—Stephanie, born in 1884, and Marianne, born in 1886—to whom he was greatly attached. One year after his marriage Ehrlich was made a titular professor at Berlin, on Frerichs' recommendation. When Frerichs died suddenly in 1885 and the more conservative Karl Gerhardt succeeded him, Ehrlich found his researches disturbingly impeded. In 1888, discovering tubercle bacilli (presumably of laboratory origin) in his sputum, he ended a decade of fruitful association with the clinic and journeyed with his young wife to Egypt, where he stayed over a year. In 1889 he returned to Berlin apparently cured of pulmonary tuberculosis, received Koch's newly discovered tuberculin treatment, and never had a recurrence.

Now without appointment, Ehrlich set up a small private laboratory in a rented flat and launched a series of fundamental studies in immunity that captured attention for many years. Using as antigens the toxic plant proteins ricin and abrin, he demonstrated that young mice could be protected against these agents if fed or injected with them in initially minute but increasing dosages. Such "actively" immunized mice developed high levels of specific antibodies in their blood. After describing these observations in two papers entitled "Experimentelle Untersuchungen über Immunität" (1891), Ehrlich showed that the progeny of a ricin- or abrin-immunized mother inherited a specific transient immunity, sustainable at higher levels by sucklings through absorption of antitoxin in the maternal milk. A similar state of "pas-

sive" immunity was induced in the progeny of a nonimmune mother that were suckled by an actively immunized mouse. Further, a normal lactating mouse injected with antiserum from an animal highly immunized against abrin, ricin, or tetanus toxin conferred specific passive immunity upon her offspring. These "wet nurse" and related experiments were reported in 1892.

Some of this work was carried out during Ehrlich's brief appointment (arranged by Koch in 1890) as clinical supervisor at the Moabit Municipal Hospital in Berlin. There he and P. Guttmann found that small doses of tuberculin were valuable in pulmonary and laryngeal tuberculosis. Ehrlich reported this finding at the Seventh International Congress for Hygiene and Demography at London in 1891. Thereafter he performed his immunological studies in a small laboratory at the newly founded Institute for Infectious Diseases in Berlin, of which Koch had become director. Ehrlich worked here for more than three years without salary, despite his appointment as extraordinary professor at Berlin University in 1891.

The institute's dedication to problems of infection, his own experiences with tuberculin, and Emil von Behring's discoveries of diphtheria and tetanus antitoxins led Ehrlich to investigate bacterial toxins and antitoxins by methods comparable with those employed in his plant protein studies. With L. Brieger he produced potent antitoxic serums in actively immunized large animals and demonstrated that these substances could be concentrated and partially purified. In 1894 he reported, with H. Kossel and A. von Wassermann, on 220 unselected diphtheritic children treated with antitoxin, stressing the importance of early, liberal dosages. Meanwhile, Behring had overcome serious difficulties in diphtheria antitoxin production by exploiting Ehrlich's assistance, procuring for himself a remunerative contract for supervising commercial manufacture of antitoxin.

Early in 1895, on the initiative of the director of the Prussian Ministry of Educational and Medical Affairs, Friedrich Althoff, an enlightened public servant who admired Ehrlich's ability, an antitoxin control station was established at Koch's institute under the supervision of Ehrlich, assisted by Kossel and Wassermann. This function was transferred in 1896 to a center for serum research and testing at Steglitz, a Berlin suburb. Ehrlich was appointed director, with Wilhelm Dönitz, and later Julius Morgenroth and Max Neisser, as his associates. The Institut für Serumforschung und Serumprüfung consisted of a one-story ramshackle building, variously described as a former almshouse or disused bakery, with an adjacent stable for laboratory animals. Nevertheless,

Ehrlich took pride in his unpretentious establishment, and excellent work was done in it.

After months of arduous work involving "hecatombs" of guinea pigs, he concluded that serum samples should be assayed in terms of a relatively stable international unit of antitoxin, distributable in dried form in vacuum tubes. Moreover, in titration the "test dose" of toxin should be the minimum amount that, added to one standard unit of antitoxin, kills within four days a 250-gram guinea pig injected therewith. These recommendations were widely adopted, and Ehrlich's L†, or *Limes-Tod,* designation for the test dose survives among his striking legacy of biomedical terms. Besides such practical accomplishments he sought theoretical explanations for the instability of diphtheria toxins that involved their lethality for guinea pigs and their ratio of lethality to antitoxin-binding power. He considered the interaction between diphtheria toxin and antitoxin a chemical process in which the reagents combine in constant proportion, as did abrin and ricin with their respective antiserums.

Ehrlich also surmised arbitrarily that one standard unit of antitoxin should fully neutralize exactly 200 minimal doses of pure toxin. When unpredictable rates of toxin degradation and varying avidities among antitoxin samples challenged this oversimplified view, he postulated the formation of toxoids (with combining power intact but toxicity absent) and of epitoxoids or toxones (with lessened combining power and altered toxicity). According to Ehrlich, each preparation of crude toxin had its own "spectrum" (*Giftspektrum*), divided into 200 segments, in which toxin, toxoid, and other designated components showed simple quantitative interrelationships.

Although certain of these proposals, set forth in the papers "Die Wertbemessung des Diphtherieheilserums und deren theoretische Grundlagen" (1897), and "Ueber die Constitution des Diphtheriegiftes" (1898), mystified some readers and aroused opposition from others, in the main they won acceptance and brought their author international recognition. He was appointed *Geheimer Medizinalrat* in 1897. Althoff realized that Ehrlich's genius deserved better facilities, and with the lord mayor of Frankfurt am Main, Franz Adickes, arranged for construction of a suitable building near the city hospital. Opened in 1899, the Royal Prussian Institute for Experimental Therapy was directed by Ehrlich until his death sixteen years later.

The new "Serum Institute" was not only responsible for routine state control of immunotherapeutic agents, such as tuberculin and diphtheria antitoxin, but also for research and training in experimental therapy. To this latter function Ehrlich devoted himself and his disciples, including Dönitz, Neisser, and Morgenroth, who followed him from Steglitz, and such subsequent staff members as Hans Sachs, E. von Dungen, E. Marx, Hugo Apolant, and Alfred Bertheim. In 1906 the adjacent Research Institute for Chemotherapy (designated the Georg-Speyer-Haus) was erected and endowed by Franziska Speyer in memory of her late husband. She did so on the advice of her brother, L. Darmstädter, to whom the promising possibilities of the specific chemotherapy of infectious diseases had been expounded by Ehrlich early in 1905. His spreading fame brought numerous visitors from abroad to work in the combined institutes, including Reid Hunt, Christian Herter, and Preston Keyes from the United States, Carl Browning and H. H. (later Sir Henry) Dale from Britain, and Kiyoshi Shiga and Sahachiro Hata from Japan.

Ehrlich's activities in Frankfurt fall into three periods. The first, 1899–1906, was marked by the emergence and elaboration of his side-chain theory, the conclusion of his work on diphtheria, extensive researches into the mechanisms of hemolytic reactions (with Morgenroth), and his cancer investigations (with Apolant). The second period dates from an address at the ceremonial opening of the Georg-Speyer-Haus in September 1906, in which Ehrlich prophesied the creation of substances "in the chemist's retort" that would "be able to exert their full action exclusively on the parasite harbored within the organism and would represent, so to speak, magic bullets which seek their target of their own accord." It culminated in his announcement before the Congress for Internal Medicine at Wiesbaden, in April 1910, that a synthetic arsenical compound, which he called dioxy-diamidoarsenobenzol (Salvarsan), had shown curative properties in rabbit syphilis and fowl spirillosis, and also in clinical trials on syphilitic patients. The third period, 1910–1915, covered Ehrlich's gallant struggle to handle the multiplex problems that followed the discovery of Salvarsan. The highlights of these periods will be reviewed consecutively.

In the final publications begun at Steglitz, Ehrlich summarized his doctrine of the interrelationship of "composition, distribution, and effect" and outlined his side-chain theory. This theory, presaged in his *Sauerstoffbedürfnis* (1885), was brought into focus mainly to account for diphtheria toxin's two distinct attributes, toxicity and antitoxin-binding power. It postulated two different chemical groups in the toxin molecule, one designated haptophore and the other toxophore. The former "anchors" the toxin molecule to the side chains (later termed "receptors") of a cell for which it has chemical affinity, by a process akin to the "lock and key" simile of the organic chemist

Emil Fischer, thus exposing the cell to damage or destruction by the toxophore group. If the cell survives the attack, the receptors rendered inert by combination with the haptophore group are replicated to excess, following Weigert's theory that tissue injury incites proliferative regeneration. Some of these surplus receptors, adapted to absorbing and neutralizing the toxin molecules, are shed and appear as circulating antitoxin—in Ehrlich's words, "handed over as superfluous ballast to the blood."

The theory was expounded by Ehrlich in his Croonian lecture, "On Immunity With Special Reference to Cell Life," delivered before the Royal Society in 1900. This fertile, heuristic hypothesis was a bold attempt to integrate the newer knowledge of nutrition, immunology, and pharmacology, but the ingenious arguments advanced by Ehrlich to bring fresh data within its purview were sometimes farfetched or obscure. He investigated the hemolytic reactions of animal serums reported by Jules Bordet in 1898 because they showed analogies to bacteriolytic phenomena and could be studied precisely *in vitro*. Bordet's observation—that the heterolysin produced by injecting an animal with red blood cells from an alien species became manifest only in the presence of a heat-labile factor (designated "alexine" by Bordet and "complement" by Ehrlich), found in most fresh normal serums—was confirmed. Whereas Bordet contended that alexine destroyed the red cells after their sensitization by a single immune body (*substance sensibilatrice*), Ehrlich visualized a far more complex situation. In several papers written with Morgenroth (1899–1901) he postulated two haptophore components in the immune body of an active hemolysin, one having strong affinity for the corresponding red blood cell receptor, the other combining with complement. Later he compared the immune body (amboceptor) and complement to the haptophore and toxophore groups of a toxin and presupposed an "extraordinary multiplicity" of hemolysins and a plurality of complements.

From Steglitz, in the midst of illuminating and practically unchallenged toxin-antitoxin titrations, Ehrlich had confided his perplexity and disenchantment to Carl Weigert. The situation now was different. In 1901 Max von Gruber launched a two-year polemic, which became inexcusably insulting, against the side-chain theory. Moreover, Svante Arrhenius and Thorvald Madsen, and Bordet as well, constructively criticized Ehrlich's views on the strictly chemical nature of the union between diphtheria toxin and antitoxin. At Frankfurt pertinacious efforts to clarify the mechanisms of hemolytic and toxin-antitoxin reactions continued. When J. Bang and

J. Forssman criticized the side-chain theory anew in 1909, Ehrlich and Sachs defended it in two final papers. To confound contemporaries who proclaimed the theory without practical value and its creator a "theoretician," Wassermann testified that the complement-fixation test for syphilis could not have been developed without Ehrlich's teaching.

In 1901 Adickes and Althoff persuaded the Theodor Stern Foundation to finance a cancer research station at the Serum Institute. After two rather unproductive years, C. J. Jensen's discovery that mouse mammary tumors are malignant and transplantable incited Ehrlich and Apolant to perform thousands of tumor-grafting experiments. Applying familiar techniques to this new field, they increased the tumor virulence for mice tenfold, until 100 percent of grafts took; and with single injections of slightly virulent cell suspensions they induced high degrees of immunity against virulent transplanted tumors. While closely following over many generations the structural changes that accompanied increased virulence, they observed a strain of mouse carcinoma apparently transforming into sarcoma. To explain the failure of a second graft to grow in an animal already carrying a tumor, whereas after resection of the first tumor a subsequent graft would take, Ehrlich coined the term "athreptic immunity." "Athrepsia," derived from the Greek τρεφω, "to nourish," signified exhaustion of the host's supply of nutrients essential for tumor growth. In his second Harben lecture (1907), Ehrlich suggested broader applications of the term—which, however, found little acceptance. This cancer work represented an unsought digression from his main course, and by 1909 chemotherapeutic researches had entirely superseded it.

In his long-standing aim to discover synthetic chemicals that act specifically upon pathogenic microorganisms, Ehrlich was aided by Arthur Weinberg and Ludwig Benda, director and chemist, respectively, of the Farbwerke Cassella & Co. near Frankfurt, who made compounds to his specifications even before the Georg-Speyer-Haus was established. In 1904 he reported with Shiga that one such substance, trypan red, cured mice experimentally infected with *Trypanosoma equinum*, causal parasite of *mal de caderas*. When he and Bechhold investigated the relationship between molecular constitution and disinfectant action of phenolic compounds upon bacterial suspensions, they found these effects inhibited by serum; moreover, the agents proved toxic and failed to produce "internal antisepsis" when injected into artificially infected animals. Hence Ehrlich pursued his earlier chemotherapeutic studies on trypanosome-infected mice and rats.

Recurrent infections in treated animals were ascribed to specific resistance to trypan red and related dyes acquired by the surviving parasites. However, such resistant strains were susceptible to atoxyl, an arsenical compound reported by H. W. Thomas and A. Breinl in 1905 to cure trypanosome-infected rodents. Ehrlich therefore postulated sessile "chemoceptors" (including an "arsenoceptor") in the parasite's protoplasm that were not released into the blood like antitoxin but had anchoring facilities for certain specific radicals. In 1907, having discovered that atoxyl was the sodium salt of *p*-aminophenylarsonic acid, or arsanilic acid, he and Bertheim synthesized and tested several hundred derivative compounds. By tailoring molecular appendages to fit the receptors of broadly resistant trypanosomal strains, they hoped to create drugs of maximum "parasitotropism" and minimum "organotropism."

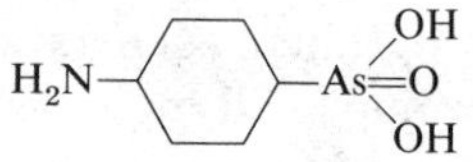

FIGURE 1. *p*-Aminophenylarsonic Acid
(Arsanilic Acid)

Meanwhile, Paul Uhlenhuth and others, stimulated by E. Roux and Elie Metchnikoff's successful transfer of syphilis to apes (1903), Fritz Schaudinn's discovery of the spirochete of syphilis (1905), and certain parallels between spirochetal and trypanosomal infections in animals and man, reported beneficial effects from atoxyl treatment of dourine, fowl spirillosis, and syphilis in rabbits, apes, and man. Since blindness sometimes followed treatment of human sleeping sickness with this agent, Ehrlich sought safer and more effective remedies. For example, arsacetin, prepared by introducing the acetyl radical into the amino group of atoxyl, was less poisonous and cured mice a few hours away from death, but it was still too toxic for clinical use.

Late in 1908, lecturing before the German Chemical Society, Ehrlich described a trivalent arsenobenzene compound of low toxicity for mice that was derived from atoxyl by two-stage reduction. This was arsenophenylglycine, number 418 in the series under test. Its high trypanocidal effectiveness inspired Ehrlich to introduce one of his favorite and best-known Latin tags, "therapia sterilisans magna," denoting "complete sterilization of a highly infected host at one blow." Six weeks later, in his Nobel lecture, "Ueber Partialfunctionen der Zelle," he asserted that through this substance "one can actually, with all kinds of animals and with every kind of trypanosome infection, achieve a complete cure by a single injection." In trials elsewhere, particularly by his friend

from earliest school days, the Breslau dermatologist Albert Neisser, arsenophenylglycine gave excellent results in the treatment of dourine and other treponemal diseases of animals but was less satisfactory in fowl spirillosis and in simian and human syphilis. Moreover, it was unstable, forming toxic oxidation products.

The search for an agent whose therapeutic index (ratio of curative to tolerated dose) was very small halted in 1909. Hata arrived that spring from Tokyo to work with Ehrlich. He was familiar with rabbit syphilis, and the emphasis switched to this and fowl spirillosis for appraisal of the many new compounds now on hand. Hata found number 606, dihydroxy-diamino-arsenobenzene-dihydrochloride (distantly related to arsanilic acid through a three-stage reduction process), had a "dosis curativa" to "dosis tolerata" ratio for fowl spirillosis of only 1:58. Intensive trials on rabbit syphilis confirmed the outstanding spirocheticidal properties of this compound. Ehrlich released limited supplies to selected specialists for clinical trials. Paralytic syphilis cases showed little improvement, but in relapsing fever and early syphilis the results were excellent. After additional favorable trials, Ehrlich, Hata, and several clinicians announced their findings in April 1910, before the Congress for Internal Medicine at Wiesbaden.

The rush for the new remedy was uncontrollable. Ehrlich tried to restrict its distribution to qualified acquaintances in various countries but was importuned by mail and by physicians who flocked to Frankfurt. Five months later, at another congress in Königsberg, he announced that "606" would not be generally available until 10,000–20,000 cases had received treatment, but further enthusiastic reports increased the demand. By the year's end, when the full resources of the Georg-Speyer-Haus had provided about 65,000 doses gratis, large-scale facilities at the nearby Höchst Chemical Works were enlisted and the product patented under the name Salvarsan. In the United States it later became known as arsphenamine.

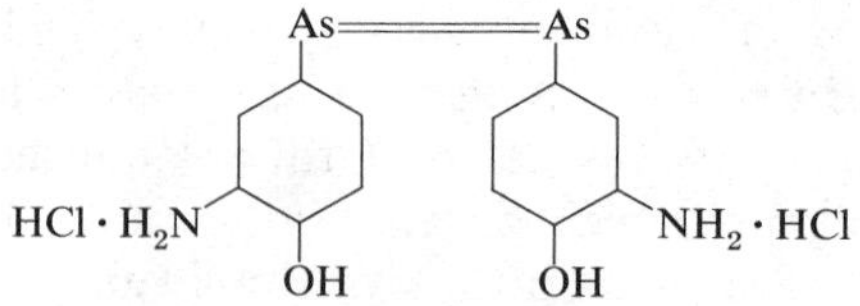

FIGURE 2. Dihydroxydiamino-arsenobenzene-dihydrochloride
(Salvarsan, Arsphenamine)

The invention of Salvarsan brought Ehrlich four years of both tragedy and triumph. He battled problems that stemmed from the drug's imperfections, from the complex pathology of syphilis, and from human carelessness, cupidity, and malice. The tricky manufacturing process and rigid biological tests on every batch came under his scrutiny. The best method of administration for counteracting the product's oxidizability and acidity and for reducing reactions remained uncertain; and although Ehrlich emphasized the therapeutic principle "frapper fort et frapper vite," routes of injection and dosages were still largely empirical. His ideal, "sterilisatio magna," was apparently feasible for relapsing fever, yaws, and certain animal diseases, but it seemed elusive or unattainable in syphilis. Neurological recurrences in undertreated cases and Jarisch-Herxheimer reactions (from hypersensitivity to massively destroyed spirochetes) were alarming, despite Ehrlich's explanations. Again, on every possible patient serological reports on Wassermann's complement-fixation test were correlated with clinical progress. As each complaint or complication was pursued, Ehrlich's correspondence reached staggering proportions and the institute overflowed with visiting physicians and would-be patients. Meanwhile, he published several reviews and edited collections of reports on Salvarsan and chemotherapy. Despite all the turmoil, he devised an arsenical derivative, number 914, which went into neutral solution without loss of effectiveness. It was introduced for clinical use in 1912 as Neosalvarsan. With Paul Karrer, his last collaborator, Ehrlich attempted further improvements by combining Salvarsan with such metals as copper, silver, bismuth, and mercury.

Such burdens would have daunted and overtaxed any man. Ehrlich's frail health began to crumble, and his peace of mind was disturbed by calumnies. Fanatic sensationalists accused him of charlatanism, profiteering, and ruthless experimentalism. The slander continued, led by the Berlin police doctor, until in March 1914 the Reichstag, forced to debate the merits of Salvarsan, endorsed it as "a very valuable enrichment of the remedies against syphilis." Three months later Ehrlich was defense witness for the Frankfurt Hospital when a local newspaper brought suit alleging that prostitutes were being forcibly subjected to dangerous treatment with Salvarsan. The complainant was sentenced to one year in jail. The outbreak of World War I drew public attention elsewhere, and Ehrlich suffered no further indignities.

Ardently although quietly patriotic and on friendly terms at court, Ehrlich was grievously distressed by the war; he brooded over his isolation from scientific friends abroad and was disconcerted by the enforced diversion of the institute's activities. In December 1914 he suffered a slight stroke. The arteriosclerotic and diabetic manifestations were treated by banning the strong cigars that he habitually smoked to excess

and by regimenting his diet, but he regained neither health nor sanguine temperament. Persuaded early in August 1915 to enter a sanatorium for treatment and rest, he shortly had a second, peacefully terminal stroke. He was buried in the Frankfurt Jewish Cemetery.

Many honors came his way. After sharing the 1908 Nobel Prize with Metchnikoff, awarded in recognition of their work on immunity, Ehrlich was renominated in 1912 and 1913 for his contributions to chemotherapy. The value of Salvarsan was considered still too disputed; and before the question was settled, Ehrlich had died. He received the Prussian Great Gold Medal for Science (1903), the Liebig Medal (1911), and the Cameron Prize (1914). Twelve orders (ten from foreign governments) and five honorary doctorates were conferred on him. He was granted the title of *Geheimer Obermedizinalrat* in 1907 and of *Wirklicher Geheimer Rat,* with the predicate *Excellenz,* in 1911. From 1904 he was honorary ordinary professor at Göttingen, and in 1914 he became ordinary professor at the new Frankfurt University. He held honorary or foreign memberships in about eighty scientific and medical societies. In 1912 he received the freedom of the city of Frankfurt, and the street containing his institutes was renamed Paul-Ehrlich-Strasse. His friends and disciples celebrated his sixtieth birthday in 1914 by preparing a remarkable *Festschrift,* each of whose thirty-seven chapters commemorated one aspect of his manifold accomplishments. The Paul Ehrlich Prize for outstanding achievement in one of his fields of research is given biennially by the Paul Ehrlich Institut as a living memorial to him.

Despite the varied nature of his investigations, a unifying principle is discernible throughout. As a student Ehrlich was fascinated by E. H. Heubel's observation (1871) that in chronic lead poisoning the organs showed wide differences in content of the toxic element, differences that were paralleled in organs from normal animals immersed in dilute lead solutions. Thus the fruitful doctrine was initiated that biological activities are determined by specific chemical affinities and are quantitatively measurable. Ehrlich's early work on dyes, on the oxygen need of the tissues, and on methylene blue treatment of malaria strengthened this belief, which also animated his chemotherapeutic strivings. The adapted aphorism "Corpora non agunt nisi fixata," introduced in his address on chemotherapy before the 1913 International Congress of Medicine in London, epitomized a concept that is still valid and fruitful, particularly in cytochemistry.

To the creative momentum of a sound original

principle of broad applicability, Ehrlich harnessed brilliant talents: a darting intelligence linked to untrammeled imagination; compulsive industriousness; the faculty of stereognostically visualizing benzene rings and structural chemical formulas; technical ingenuity and punctiliousness, and unique virtuosity in "test-tube" chemistry; the capacity to direct several lines of research simultaneously, through a system of daily "blocks" carrying written instructions to every co-worker; and the foresight to abandon paths that were unpromising. An autodidact, he was nobody's disciple. His gift for coining words, phrases, and metaphors enriched the common vocabulary of science. Ehrlich conversed and lectured in German only, but he could read English and French and perused relevant scientific publications avidly and rapidly (reading "diagonally"). His tastes in general literature aspired no higher than Conan Doyle and he lacked feeling for art, but he was refreshed by simple music.

By nature Ehrlich was enthusiastic, good-humored, at times even bantering; but meanness, unfair criticism, or false claims to priority aroused fierce indignation. Although genuinely modest, he knew the importance of his work. He never lobbied for his own ends, was devoid of mercenary instinct, and was completely honorable in all his dealings. Lovably loyal to his family and countless friends, he was the very embodiment of minor eccentricity and true genius. As Sir Robert Muir wrote, "Ehrlich must be with the greatest, however small that company may be."

BIBLIOGRAPHY

I. ORIGINAL WORKS. The most complete edition is *The Collected Papers of Paul Ehrlich,* compiled and edited by F. Himmelweit, assisted by Martha Marquardt, under the general direction of Sir Henry Dale (London–New York, 1956–1960). The first 3 vols. contain all his important papers, including a few hitherto unpublished. These are grouped according to topics: vol. I, *Histology, Biochemistry, and Pathology;* vol. II, *Immunology and Cancer Research;* vol. III, *Chemotherapy.* Vol. IV (not yet published) will include his letters and a complete bibliography. Of 158 publications reproduced in this edition, all are in German except 11 articles which originally appeared in English or French. Fresh English translations are appended to 15 of the remaining 147 items.

Ehrlich's pupil and co-worker Hans Sachs compiled a bibliography in 1914 as an appendix to Ehrlich's sixtieth birthday *Festschrift* (see below). This listed 212 separate publications in well-known journals, as well as several monographs, over the period 1877–1913. Ehrlich was sole author of roughly three-quarters of these items, many of which underwent multiple publication. Sachs also ap-pended a bibliography of 400 reports by Ehrlich's disciples.

Apart from the special translations in the *Collected Papers,* Ehrlich's Croonian (1900), Herter (1904), and Harben (1907) lectures were first published in English-language journals (see below). The Harben lectures were republished as *Experimental Researches on Specific Therapy* (London, 1908). *Collected Studies in Immunity* (New York, 1906), C. Bolduan, trans., includes 41 reports by Ehrlich and his co-workers between 1899 and early 1906, of which 38 had appeared previously in *Gesammelte Arbeiten zur Immunitätsforschung* (Berlin, 1904). A later edition of *Collected Studies* (1910) contains Bolduan's translations of Ehrlich's Nobel Prize address and of seven additional papers by his pupils.

Other collections of Ehrlich's papers in book form are *Farbenanalytische Untersuchungen zur Histologie und Klinik des Blutes* (Berlin, 1891); *Constitution, Vertheilung und Wirkung chemischer Körper; ältere und neuere Arbeiten* (Leipzig, 1893); and *Beiträge zur experimentellen Pathologie und Chemotherapie* (Leipzig, 1909). He also published several monographs, including his *Habilitation* thesis, *Das Sauerstoffbedürfnis des Organismus. Eine farbenanalytische Studie* (Berlin, 1885); *Die Anaemie* (Vienna, 1898–1900), written with A. Lazarus; *Leukaemie. Pseudoleukaemie. Haemoglobinaemie* (Vienna, 1901), written with A. Lazarus and F. Pinkus; *Die experimentelle Chemotherapie der Spirillosen* (Berlin, 1910), written with S. Hata; *Aus Theorie und Praxis der Chemotherapie* (Leipzig, 1911); and *Grundlagen und Erfolge der Chemotherapie* (Stuttgart, 1911). Ehrlich was coeditor, with R. Krause, M. Mosse, H. Rosin, and C. Weigert, of *Enzyklopädie der mikroskopischen Technik,* 3 vols. (Vienna, 1902–1903; 2nd ed., 1910; 3rd ed., 1926–1927); he was sole editor of *Abhandlungen über Salvarsan,* 4 vols. (Munich, 1911–1914); contributed chapters and forewords to several monographs; and wrote obituaries of or tributes to E. Albrecht, H. Apolant, A. Bertheim, R. Koch, and C. Weigert.

Among his more important and characteristic reports in scientific and medical journals are "Beiträge zur Kenntnis der Anilinfärbungen und ihrer Verwendung in der mikroskopischen Technik," in *Archiv für mikroskopische Anatomie,* **13** (1877), 263–277, published while he was still a medical student; "Beiträge zur Theorie und Praxis der histologischen Färbung," his inaugural dissertation at Leipzig University (1878); "Beiträge zur Kenntnis der granulirten Bindegewebszellen und der eosinophilen Leukocythen," in *Archiv für Anatomie und Physiologie,* Physiologische Abteilung (1879), 166–169; "Ueber die spezifischen Granulationen des Blutes," *ibid.,* 571–579; "Methodologische Beiträge zur Physiologie und Pathologie der verschiedenen Formen der Leukocyten," in *Zeitschrift für klinische Medizin,* **1** (1880), 553–560; "Ueber paroxysmale Hämoglobinurie," in *Deutsche medizinische Wochenschrift,* **7** (1881), 224–225; "Ueber provocirte Fluorescenzerscheinungen am Auge," *ibid.,* **8** (1882), 21–22, 35–37, 54–55; "Ueber eine neue Methode zur Färbung von Tuberkelbacillen," in *Berliner klinische Wochenschrift,* **20** (1883), 13; "Ueber eine neue Harnprobe," in *Charité-Annalen,* **8** (1883), 140–166; and "Zur biologischen Verwertung des Methylenblaus," in *Zentralblatt für die medizinische Wissenschaft*, **23** (1885), 113–117.

Other papers include "Ueber Wesen und Behandlung des Jodismus," in *Charité-Annalen*, **10** (1885), 129–135; "Zur Physiologie und Pathologie der Blutscheiben," *ibid.*, 136–146; "Ueber die Methylenblaureaction der lebenden Nervensubstanz," in *Deutsche medizinische Wochenschrift*, **12** (1886), 49–52; "Beiträge zur Theorie der Bacillen-färbung," in *Charité-Annalen,* **11** (1886), 123–138; "Experimentelles und Klinisches über Thallin," in *Deutsche medizinische Wochenschrift*, **12** (1886), 849–851, 889–891; "Zur therapeutischen Bedeutung der substitutierenden Schwefelsäuregruppe," in *Therapeutische Monatsheft*, **1** (1887), 88–90; "Ueber die Bedeutung der neutrophilen Körnung," in *Charité-Annalen*, **12** (1887), 288–295; "Ueber schmerzstillende Wirkung des Methylenblau," in *Deutsche medizinische Wochenschrift*, **16** (1890), 493–494, written with A. Leppmann; "Studien in der Cocainreihe," *ibid.*, 717–719; "Die Wirksamkeit kleiner Tuberkulindosen gegen Lungenschwindsucht," *ibid.*, 793–795, written with P. Guttmann; "Recent Experiences in the Treatment of Tuberculosis (With Special Reference to Pulmonary Consumption) by Koch's Method," in *Lancet* (1891), **2**, 917–920, trans. by T. W. Hime from a paper presented by Ehrlich at the Seventh International Congress of Hygiene and Demography, London; "Ueber die Wirkung des Methylenblau bei Malaria," in *Berliner klinische Wochenschrift*, **28** (1891), 953–956, written with P. Guttmann; "Experimentelle Untersuchungen über Immunität. I. Ueber Ricin. II. Ueber Abrin," in *Deutsche medizinische Wochenschrift*, **17** (1891), 976–979, 1218–1219; and "Ueber Immunität durch Vererbung und Säugung," in *Zeitschrift für Hygiene*, **12** (1892), 183–203.

See also "Ueber die Uebertragung von Immunität durch Milch," in *Deutsche medizinische Wochenschrift*, **18** (1892), 393–394, written with L. Brieger; "Ueber Gewinnung und Verwendung des Diphtherieheilserums," *ibid.*, **20** (1894), 353–355, written with H. Kossel and A. Wassermann; "Die staatliche Kontrolle des Diphtherieserums," in *Berliner klinische Wochenschrift*, **33** (1896), 441–443; "Die Wertbemessung des Diphtherieheilserums und deren theoretische Grundlagen," in *Klinische Jahrbuch*, **6** (1897), 299–326; "Zur Kenntnis der Antitoxinwirkung," in *Fortschritte der Medizin*, **15** (1897), 41–43; "Ueber die Constitution des Diphtheriegiftes," in *Deutsche medizinische Wochenschrift*, **24** (1898), 597–600; "Observations Upon the Constitution of the Diphtheria Toxin," in *Transactions of the Jenner Institute of Preventive Medicine*, **2** (1899), 1–16; "Zur Theorie der Lysinwirkung," in *Berliner klinische Wochenschrift*, **36** (1899), 6–9, written with J. Morgenroth, as was "Ueber Haemolysine," *ibid.*, **36** (1899), 481–486; **37** (1900), 453–458; **38** (1901), 251–257, 569–574, 598–604.

Other papers published after 1900 include "On Immunity, With Special Reference to Cell Life," in *Proceedings of the Royal Society*, **66** (1900), 424–448, the Croonian lecture; "Die Schutzstoffe des Blutes," in *Deutsche medizinische Wochenschrift*, **27** (1901), 865–867, 888–891, 913–916; "Ueber die Beziehungen von chemischer Constitution, Vertheilung, und pharmakologischen Wirkung,"

in *Ernst von Leyden-Festschrift,* I (Berlin, 1902), 645–679, address delivered to Verein für innere Medicin, Berlin, 12 December 1898, trans. by C. Bolduan as "The Relations Existing Between Chemical Constitution, Distribution, and Pharmacological Action," in *Studies on Immunity* (1906), pp. 404–442; "Ueber die Vielheit der Complemente des Serums," in *Berliner klinische Wochenschrift*, **39** (1902), 297–299, 335–338, written with H. Sachs, as was "Ueber den Mechanismus der Amboceptorenwirkung," *ibid.*, 492–496; "Ueber die complementophilen Gruppen der Amboceptoren," *ibid.*, 585–587, written with H. T. Marshall; "Ueber die Giftcomponenten des Diphtherie-Toxins," *ibid.*, **40** (1903), 793–797, 825–829, 848–851; "Toxin und Antitoxin. Entgegnung auf den neuesten Angriff Grubers," in *Münchener medizinische Wochenschrift*, **50** (1903), 1428–1432, 1465–1469; "Toxin und Antitoxin. Entgegnung auf Grubers Replik," *ibid.*, 2295–2297; "Vorläufige Bemerkungen zur Mittheilungen von Arrhenius zur Theorie der Absättigung von Toxin und Antitoxin," in *Berliner klinische Wochenschrift*, **41** (1904), 221–223; "The Mutual Relations Between Toxin and Antitoxin," in *Boston Medical and Surgical Journal*, **150** (1904), 443–445; "Physical Chemistry v. Biology in the Doctrine of Immunity," *ibid.*, 445–448; "Cytotoxins and Cytotoxic Immunity," *ibid.*, 448–450, the Herter lectures; "Farbentherapeutische Versuche bei Trypanosomenerkrankungen," in *Berliner klinische Wochenschrift*, **41** (1904), 329–332, 362–365, written with K. Shiga; and "Ueber den Mechanismus der Antiamboceptorenwirkung," *ibid.*, **42** (1905), 557–558, 609–612, written with H. Sachs.

Also of interest are "Beobachtungen über maligne Mäusetumoren," *ibid.*, 871–874, written with H. Apolant; "Beziehungen zwischen chemischer Konstitution und Desinfektionswirkung. Ein Beitrag zum Studium der 'innern Antisepsis,'" in *Hoppe-Seyler's Zeitschrift für physiologische Chemie*, **47** (1906), 173–199, written with H. Bechhold; "On Immunity With Special Reference to the Relationship Between Distribution and Action of Antigens," in *Journal of the Royal Institute of Public Health*, **15** (1907), 321–340; "On Athreptic Functions," *ibid.*, 385–403; "Chemotherapeutic Studies on Trypanosomes," *ibid.*, 449–456, the Harben lectures; "Ueber *p*-Aminophenylarsinsäure. Erste Mitteilung," in *Berichte der Deutschen chemischen Gesellschaft*, **40** (1907), 3292–3297, written with A. Bertheim; "Ueber spontane Mischtumoren der Maus," in *Berliner klinische Wochenschrift*, **44** (1907), 1399–1401, written with H. Apolant; "Ueber den jetzigen Stand der Chemotherapie," in *Berichte der Deutschen chemischen Gesellschaft*, **42** (1909), 17–47; "Ueber Partialfunktionen der Zelle," in *Münchener medizinische Wochenschrift*, **56** (1909), 217–222, Nobel Prize address, 11 December 1908; "Ueber serumfeste Trypanosomenstämme," in *Zeitschrift für Immunitätsforschung*, **3** (1909), 296–299, written with W. Roehl and R. Gulbransen; "Kritiker der Seitenkettentheorie im Lichte ihrer experimentellen und literarischen Forschung. Ein Kommentar zu den Arbeiten von Bang und Forssman," in *Münchener medizinische Wochenschrift*, **56** (1909), 2529–2532, written with H. Sachs, as was "Ist die Ehrlichsche Seitenkettentheorie mit den tatsächlichen

Verhältnissen vereinbar? Bemerkungen zu der II Mittheilung von Bang und Forssman," *ibid.,* **57** (1910), 1287–1289; "Reduktionsprodukte der Arsanilsäure und ihre Derivate. Zweite Mittheilung: Ueber *p, p'*-Diamino-arsenobenzol," in *Berichte der Deutschen chemischen Gesellschaft,* **43** (1910), 917–927, written with A. Bertheim; "Allgemeines über Chemotherapie," in *Verhandlungen des 27. Kongresses für innere Medizin, Wiesbaden,* **27** (1910), 226–234; "Die Behandlung der Syphilis mit dem Ehrlichschen Präparat 606," in *Deutsche medizinische Wochenschrift,* **36** (1910), 1893–1896; "Die Salvarsantherapie. Rückblicke und Ausblicke," in *Münchener medizinische Wochenschrift,* **58** (1911), 1–10; "Ueber Salvarsan," *ibid.,* 2481–2486; "Chemotherapeutics: Scientific Principles, Methods, and Results," in *Lancet* (1913), **2,** 445–451, address in pathology to 17th International Medical Congress, London, 1913; and "Deaths After Salvarsan," in *British Medical Journal* (1914), **1,** 1044–1045.

Ehrlich's former institute at Frankfurt contains a small collection of memorabilia (group photographs and laboratory notebooks). Other relics were donated by his family to the New York Academy of Medicine. His surviving papers, retrieved from the bomb-damaged institute after World War II and from a village hiding place in the Taunus Mountains, were placed by Ehrlich's executor, his grandson, on indefinite loan in the custody of the Wellcome Institute of the History of Medicine, London. These include most of his copybooks of handwritten letters, from late 1898 to early 1903, and of typed ones thereafter to 1915, as well as copies of his "blocks" to collaborators, 1906–1915.

II. SECONDARY LITERATURE. Obituaries in German include G. Joannovics, "Paul Ehrlich 1854–1915," in *Wiener klinische Wochenschrift,* **28** (1915), 937–942; M. Kirchner, "Paul Ehrlich†," in *Zeitschrift für ärtzliche Fortbildung,* **12** (1915), 513–515; A. Neisser, "Paul Ehrlich, Gestorben den 20 August 1915," in *Archiv für Dermatologie und Syphilis,* **121** (1919), 557–578; F. Pinkus, "Paul Ehrlich geboren 14 März 1854, gestorben 20 August 1915," in *Medizinische Klinik,* **11** (1915), 985, 1116–1117, 1143–1145; H. Sachs, "Paul Ehrlich†," in *Münchener medizinische Wochenschrift,* **62** (1915), 1357–1361; and "Paul Ehrlich, geb. 14.III.1854, †20.VIII.1915," in *Natur und Volk,* **46** (1916), 139–152; A. von Wassermann, "Paul Ehrlich†," in *Deutsche medizinische Wochenschrift,* **41** (1915), 1103–1106, 1135–1136; and A. von Weinberg, "Paul Ehrlich," in *Berichte der deutschen chemischen Gesellschaft,* **49** (1916), 1223–1248.

Obituaries in English include C. H. Browning, "Professor Paul Ehrlich," in *British Medical Journal* (1915), **2,** 349–350; R. Muir, "Paul Ehrlich. 1854–1915," in *Journal of Pathology and Bacteriology,* **20** (1915), 349–360; and the following unsigned tributes: "Wirkl.-Geheimrat Paul Ehrlich," in *Lancet* (1915), **2,** 525–526; "Professor Paul Ehrlich," in *British Medical Journal* (1915), **2,** 349; "Professor Paul Ehrlich," in *Boston Medical and Surgical Journal,* **173** (1915), 637–640; and "Death of Paul Ehrlich," in *Journal of the American Medical Association,* **65** (1915), 814, 1123.

Other references in German to Ehrlich's life and work are H. Apolant, L. Benda, A. Bertheim, et al., *Paul Ehrlich. Eine Darstellung seines wissenschaftlichen Wirkens* (Jena, 1914), the *Festschrift* for his sixtieth birthday; S. Arrhenius and T. Madsen, "Anwendung der physikalischen Chemie auf des Studium der Toxine und Antitoxine," in *Zeitschrift für physikalische Chemie,* **44** (1903), 7–62; L. Aschoff, *Ehrlich's Seitenkettentheorie und ihre Anwendung auf die künstlichen Immunisierungsprozesse* (Jena, 1902); A. Beyer, *Paul Ehrlich und Emil v. Behring* (Berlin, 1954); W. Dönitz, "Bericht über die Thätigkeit des königlichen Instituts für Serumforschung und Serumprüfung zu Steglitz. Juni 1896–September 1899," in *Klinische Jahrbuch,* **7** (1899), 1–26; A. von Engelhardt, *Hundertjahrfeier der Geburtstage von Paul Ehrlich und Emil von Behring in Frankfurt-Main, Marburg-Lahn und Hoechst,"* (Marburg, 1954); I. Fischer, "Paul Ehrlich," in *Biographisches Lexikon der hervorragender Aerzte,* I (1932), 352–353.

See also W. Greuling, *Paul Ehrlich. Leben und Werk* (Düsseldorf, 1954); M. Gruber, "Zur Theorie der Antikörper. I. Ueber die Antitoxin-Immunität," in *Münchener medizinische Wochenschrift,* **48** (1901), 1827–1830, 1880–1884; "II. Ueber Bakteriolyse und Haemolyse," *ibid.,* 1924–1927, 1965–1968; "Neue Früchte der Ehrlichschen Toxinlehre," in *Wiener klinische Wochenschrift,* **16** (1903), 791–793; "Toxin und Antitoxin. Eine Replik auf Herrn Ehrlichs Entgegnung," in *Münchener medizinische Wochenschrift,* **50** (1903), 1825–1828; and "Bemerkungen zu Ehrlich's 'Entgegnung auf Grubers Replik,'" *ibid.,* 2297; M. Gruber and C. von Pirquet, "Toxin und Antitoxin," *ibid.,* 1193–1196, 1259–1263; B. Heymann, "Zur Geschichte der Seitenkettentheorie Paul Ehrlichs," in *Klinische Wochenschrift,* **7** (1928), 1257–1260, 1305–1309; Janina Hurwitz, "Paul Ehrlich als Krebsforscher," in *Zürcher medizingeschichtliche Abhandlungen,* n.s. no. 7 (1962); G. Joannovics, "Referat: Paul Ehrlich. Eine Darstellung seines wissenschaftlichen Wirkens. Festschrift zum 60. Geburtstage des Forschers," in *Wiener klinische Wochenschrift,* **28** (1915), 93–95; A. Lazarus, "Paul Ehrlich," in M. Neuberger, ed., *Meister der Heilkunde,* I (Vienna, 1921), 9–88; W. Leibbrand, "Paul Ehrlich," in *Neue deutsche Biographie,* IV (1957), 364–365; and "Paul Ehrlich und Emil von Behring zum hundertsten Geburtstag am 14. und 15. März," in *Münchener medizinische Wochenschrift,* **96** (1954), 298–299; H. Loewe, "Paul Ehrlich. Schöpfer der Chemotherapie," in *Grosse Naturforscher,* VIII (Stuttgart, 1950); Leonor Michaelis, "Zur Erinnerung am Paul Ehrlich: Seine wiedergefundene Doktor-Dissertation," in *Die Naturwissenschaften,* **7** (1919), 165–168; A. Neisser, "Ueber das Arsenophenylglyzin und seine Verwendung bei der Syphilisbehandlung," in *Archiv für Dermatologie und Syphilis,* **121** (1916), 579–612; A. Neisser *et al.,* "Die Behandlung der Syphilis mit den Ehrlichschen Präparat 606," in *Deutsche medizinische Wochenschrift,* **36** (1910), 1889–1924; and *Beitrage zur Pathologie und Therapie der Syphilis* (Berlin, 1911); H. Satter, *Paul Ehrlich, Begründer der Chemotherapie* (Munich, 1962); P. Uhlenhuth, *Experimentelle Grundlagen der Chemotherapie der Spirochaetenkrankheiten mit besondere Berücksichtigung der Syphilis* (Berlin-Vienna, 1911); O. H. Warburg, "Paul Ehrlich, 1854–1915," in *Die*

grossen Deutschen, IV (1957), 186–192; and A. Wassermann, "Paul Ehrlich," in *Münchener medizinische Wochenschrift*, **56** (1909), 245–247.

English and French references to Ehrlich's life and work include J. Almkvist, "Reminiscences of Paul Ehrlich," in *Urologic and Cutaneous Review*, **42** (1938), 214–220; H. Bauer, "Paul Ehrlich's Influence on Chemistry and Biochemistry," in *Annals of the New York Academy of Sciences*, **59** (1954), 150–167; J. Bordet, "Sur le mode d'action des antitoxines sur les toxines," in *Annales de l'Institut Pasteur*, **17** (1903), 161–186; C. H. Browning, "Paul Ehrlich—Memories of 1905–1907," in *British Medical Journal* (1954), **1**, 664–665; and "Emil Behring and Paul Ehrlich. Their Contributions to Science," in *Nature*, **175** (1955), 570–575, 616–619; Henry Dale, "Paul Ehrlich, Born March 14, 1854," in *British Medical Journal* (1954), **1**, 659–663; C. E. Dolman, "A Fiftieth Anniversary Commemorative Tribute to Paul Ehrlich, With Two Letters to American Friends," in *Clio Medica*, **1** (1966), 223–234; and "Paul Ehrlich and William Bulloch: A Correspondence and Friendship (1896–1914)," *ibid.*, **3** (1968), 65–84; H. Goodman, "Paul Ehrlich. 'A Man of Genius and an Inspiration to Humanitarians,'" in *American Medicine*, **42** (1936), 73–78; L. W. Harrison, "Paul Ehrlich and the Development of 606," in *St. Thomas's Hospital Gazette*, **52** (1954), 37–42; E. Jokl, "Paul Ehrlich—Man and Scientist," in *Bulletin of the New York Academy of Medicine*, **30** (1954), 968–975; Martha Marquardt, *Paul Ehrlich* (London, 1949); and "Paul Ehrlich, Some Reminiscences," in *British Medical Journal* (1954), **1**, 665–667; G. H. F. Nuttall, "Biographical Notes Bearing on Koch, Ehrlich, Behring and Loeffler, With Their Portraits and Letters From Three of Them," in *Parasitology*, **16** (1924), 214–238 ("Paul Ehrlich, 1854–1915", pp. 224–229); H. G. Plimmer, "A Critical Summary of Ehrlich's Recent Work on Toxins and Antitoxins," in *Journal of Pathology and Bacteriology*, **5** (1897), 489–498; C. P. Rhoads, "Paul Ehrlich and the Cancer Problem," in *Annals of the New York Academy of Sciences*, **59** (1954), 190–197; and "Paul Ehrlich in Contemporary Science," in *Bulletin of the New York Academy of Medicine*, **30** (1954), 976–987; L. Vogel, "Paul Ehrlich (1854–1915)," in *Revue d'histoire de la médecine hebraïque*, **22** (1969), 75–85, 107–117; W. Wechselmann, "Beobachtungen an 503 mit Dioxydiamido-Arsenobenzol behandelten Krankheitsfällen," in *Deutsche medizinische Wochenschrift*, **36** (1910), 1478–1481; and *The Treatment of Syphilis With Salvarsan* (New York-London, 1911), A. L. Wolbarst trans.; and E. Witebsky, "Ehrlich's Side-Chain Theory in the Light of Present Immunology," in *Annals of the New York Academy of Sciences*, **59** (1954), 168–181.

CLAUDE E. DOLMAN

EICHENWALD, ALEKSANDR ALEKSANDRO-VICH (*b.* St. Petersburg, Russia, 4 January 1864; *d.* Milan[?], Italy, 1944), *physics, engineering.*

Eichenwald's father was a photographer and artist; his mother, a professor of harp at the St. Petersburg Conservatory and, later, a soloist of the Bolshoi Theater orchestra in Moscow. His sisters and brother were also professional musicians, and he himself was a pianist and a connoisseur of music, which stimulated his interest in acoustics.

While in high school, from which he graduated in 1883, he formed his friendship with the future physicist P. N. Lebedev. After completing two years of study at the Faculty of Physics and Mathematics of Moscow University, Eichenwald entered the St. Petersburg Railway Institute, from which he graduated in 1888. After working for seven years as an engineer, he went to Strasbourg to continue his education and devoted himself to physics. K. F. Braun was his instructor in experimental physics, and Emil Cohn in theoretical physics. His Ph.D. dissertation was entitled "Absorption elektrischer Wellen bei Elektrolyten" (1897).

From 1897 to 1921 Eichenwald worked at the Moscow Engineering College (now the Moscow Institute of Railway Engineers). In the excellent scientific laboratory that he organized there, he carried out the fundamental experiments described in his dissertation for a Russian doctorate, *O magnitnom deystvii tel, dvizhushchikhsya v elektrostaticheskom pole* ("On the Magnetic Action of Bodies Moving in an Electrostatic Field," 1904), and undertook investigations of the propagation of electromagnetic and sound waves. In 1905–1908 Eichenwald was director of the Institute of Railway Engineers and from 1901 was also an instructor at the Higher Women's Courses and, in 1906–1911, at Moscow University. After Lebedev's death he headed the Moscow Physics Society, which Lebedev had founded. In 1917–1920 he participated in the reorganization of higher education. After two operations in Moscow and Berlin for cancer, Eichenwald moved to Milan, where he wrote textbooks that were published in the Soviet Union. His textbook on electricity saw eight editions from 1911 to 1933. In 1926–1932 the first three and the sixth volumes of *Theoretical Physics* were published.

Eichenwald was simultaneously a keen experimenter, a serious theoretician, a brilliant lecturer and methodologist, and an inventor of demonstration apparatus. He won world fame by his unquestionable proof that the motion of an electrically charged body produces an electric field, by his exact proof of the equivalence of convection and conduction currents, and by the first proof, based on direct measurements, of the existence of a magnetic field when the polarization of a dielectric changes, i.e., a magnetic field of a displacement current (1901–1904).

By his direct and accurate experiments on the detection and measurement of the magnetic field of convection currents, Eichenwald completed the final step in a series of experiments with contradictory results that had been started by H. A. Rowland (1876) and continued by V. Grémieu, Ernst Lecher, and Harold Pender (1902).

Besides the magnetic field created by the motion of charged conductors, Eichenwald measured the currents produced by the motion of a dielectric in a nonuniform electric field (they had been discovered by Roentgen in 1888) and organized a new type of experiment (the Eichenwald experiment), by means of which the existence of the magnetic field of the displacement current in dielectrics was established and its magnitude was measured for the first time.

Half of a disk made from a dielectric and rotating about its axis passed constantly between the plates of one capacitor, and the other half between those of another capacitor. The electric fields in the capacitors were oppositely directed. When each element of the dielectric passed from the zone of one capacitor to that of the other, the polarization of this element became the opposite of what it had been. Eichenwald discovered the magnetic field of the displacement current appearing in the dielectric by observing the change in the oscillations of a small magnetic needle when the disk was stationary and in motion. The needle was arranged so as not to react to convection currents. The experiment was organized in connection with the question of the conduction of ether by moving bodies. The result conformed with the theories of H. A. Lorentz and E. Cohn, in which motionless ether was assumed, but after the appearance of the theory of relativity Eichenwald proved that his experiment could be interpreted in accordance with the new concept.

In "O dvizhenii energii pri polnom vnutrennem otrazhenii sveta" ("On the Motion of Energy With Complete Internal Reflection of Light," 1908), Eichenwald completely explained this phenomenon from the standpoint of J. C. Maxwell's electromagnetic theory of light, indicating the reason for Drude's error. (According to Drude's theory of the motion of light energy along a reflecting surface, the direction of the vector of the electric field coincides with the direction of this motion, instead of being perpendicular to it.)

The equations deduced in this work, reflecting the curvilinear nature of the propagation of light in a reflecting medium, were also applicable in other cases and were published in a generalized form in "Das Feld der Lichtwellen bei Reflexion und Brechung" (1912).

In the investigation "Akusticheskie volny bolshoy amplitudy" ("Sound Waves of Large Amplitude"), a different, simpler, and physically more illustrative method of calculation than that of Riemann was proposed for strong sounds, when the approximate equations of wave propagation cannot be used.

BIBLIOGRAPHY

I. ORIGINAL WORKS. Some of Eichenwald's works were collected as *Izbrannye trudi* ("Selected Works"), A. B. Mlodzeevsky, ed. (Moscow, 1956), with remarks and a biographical essay. Among his writings are "Absorption elektrischer Wellen bei Elektrolyten," in *Annalen der Physik und Chemie,* **62** (1897), 571–587; "Über die magnetischen Wirkungen elektrischer Konvektion," in *Jahrbuch der Radioaktivität und Elektronik,* **5,** no. 1 (1908), 82–98—see also L. Graetz, *Handbuch der Elektrizität und des Magnetismus,* II (Leipzig, 1914), 337–365; "O dvizhenii energii pri polnom vnutrennem otrazhenii sveta" ("On the Motion of Energy With Complete Internal Reflection of Light"), in *Izvestiya Moskovskago inzhenernogo uchilishcha* (Apr. 1908), 15–41—see also *Annalen der Physik,* **35** (1911), 1037–1040; "Das Feld der Lichtwellen bei Reflexion und Brechung," in *Festschrift Heinrich Weber* (Leipzig, 1912), 37–56; and "Akusticheskie volny bolshoy amplitudy" ("Sound Waves of Large Amplitude"), in *Uspekhi fizicheskikh nauk,* **14,** no. 5 (1934), 552–585—see also *Rendiconti del Seminario matematico e fisico di Milano,* **6,** no. 10 (1932), 1–28.

II. SECONDARY LITERATURE. On Eichenwald and his work, see N. A. Kaptsov, "Aleksandr Aleksandrovich Eichenwald," in *Uchenye zapiski Moskovskogo gosudarstvennogo universiteta,* Jubilee ser., no. 52 (1940), 166–171; G. Mie, *Elektrodynamik* (Leipzig, 1952), pp. 51–54, 60–62; and A. B. Mlodzeevsky, "A. A. Eichenwald," in *Ocherki po istorii fiziki v Rossii* ("Essays on the History of Physics in Russia," Moscow, 1949).

O. LEZHNEVA

EICHLER, AUGUST WILHELM (*b.* Neukirchen, Germany, 22 April 1839; *d.* Berlin, Germany, 2 March 1887), *botany.*

Eichler has been considered one of the most prominent systematic and morphological botanists of his time. He was an enemy of dogmas and philosophical speculations; his main contributions concern the symmetry of flowers and the taxonomy of higher plants. These accomplishments followed in the tradition of another great German botanist, Alexander Braun, and culminated in the introduction of a widely adopted system of plant classification.

The eldest son of a cantor who also taught natural sciences, Eichler demonstrated an early interest in nature, collecting minerals and flowers and becoming skilled in mountain climbing. After his early education in Eschwege and Hersfeld, he studied mathematics and natural science at the University of

Marburg from 1857 to 1861. There Eichler developed a close personal relationship with one of his teachers, the botanist Albert Wigand, an association that proved decisive for his future career. His main interest turned to the study of flowers and their basic structure and, in opposition to Wigand's ideas, he became a vigorous defender of Darwin. Warmly recommended by his teachers, Eichler went to Munich after his graduation, as private assistant to the naturalist Karl Friedrich von Martius. There he assisted Martius in editing the monumental *Flora Brasiliensis.* After becoming a lecturer at the University of Munich, Eichler assumed the sole editorship of this ambitious project upon the death of Martius in 1868. Three years later, in 1871, he accepted an offer to become professor of botany at the Technische Hochschule of Graz and to supervise the local botanical gardens. Eichler occupied this post for only one year, until he heeded a call from the Prussian government to assume the chair of botany at Kiel. There he completed the first part of his most famous work on the comparative structure of flowers, *Blüthendiagramme.* This publication was based on meticulous and repeated observations, and Eichler himself executed some of the woodcuts that illustrated the book.

After the death of Alexander Braun in 1877, Eichler was appointed professor of systematic and morphologic botany at the University of Berlin. In addition he assumed the direction of the university's herbarium and the Royal Botanical Gardens at Schoeneberg. Eichler's contemporaries were impressed with his ability as a teacher as well as with the combination of scholarship, strict objectivity, and personal modesty that he constantly exhibited. What had begun as an eye complaint during the years at Kiel gradually became a disabling systemic disease, diagnosed before his death as leukemia. It greatly hampered his research and teaching during the years in Berlin.

In 1867 Eichler was elected secretary of the International Botanical Congress held in Paris on 16–23 August of that year. After his call to the German capital, he was elected a member of the Berlin Academy of Sciences in 1880. He also became a corresponding or honorary member of several other societies, such as the French Academy of Sciences, the Royal Academy of Belgium, the Academy of Sciences of Munich, and, in 1881, the Linnean Society of London.

BIBLIOGRAPHY

I. ORIGINAL WORKS. A complete list of Eichler's publications can be found in *Annals of Botany,* **1** (1887–1888), 400–403; and in I. Urban, "A. W. Eichler's botanische

Arbeiten," in *Botanisches Centralblatt,* **32** (1887), 123–127.

Among his best-known books are *Zur Entwicklungsgeschichte des Blattes mit besonderer Beruecksichtigung der Nebenblattbildungen* (Marburg, 1861); *Bewegung im Pflanzenreiche* (Munich, 1864); *Blüthendiagramme,* 2 vols. (Leipzig, 1875–1878); *Beitraege zur Morphologie und Systematik der Marantaceen* (Berlin, 1883); and *Syllabus der Vorlesungen ueber specielle und medicinisch-pharmaceutische Botanik,* 4th ed., rev. (Berlin, 1886).

Eichler was also one of the editors of *Flora Brasiliensis,* 15 vols. (Munich–Leipzig, 1840–1906).

Some of his best articles are "On the Formation of the Flower in Gymnosperms." T. Thomson, trans., in *Natural History Review,* **4** (1864), 270–290; "Bemerkungen ueber die Structur des Holzes von Drimys und Trochodendron, sowie ueber die systematische Stellung der letzteren Gattung," in *Flora, oder allgemeine botanische Zeitung,* **47** (1864), 449–458 and **48** (1865), 12–15, partially trans. into English in *Journal of Botany, British and Foreign,* **3** (1865), 150–154; and "Einige Bemerkungen ueber den Bau der Cruciferenbluethe und das Dédoublement," in *Flora, oder allgemeine botanische Zeitung,* **52** (1869), 97–109.

II. SECONDARY LITERATURE. An extensive biography of Eichler is Carl Mueller, "August Wilhelm Eichler, ein Nachruf," in *Botanisches Centralblatt,* **31** (1887), 61–63, 120–128, 155–160, 188–191, 229–232, 261–263, 294–296, 325–327, 357–360; **32** (1887), 27–32, 61–63, 121–123. This biography was also published as a book, together with I. Urban's bibliography, under the same title (Kassel, 1887). Other biographical sketches appeared in *Proceedings of the Linnean Society of London* (1886–1887), 38–39; *Berichte der Deutschen botanischen Gesellschaft,* **5** (1887), xxxiii–xxxvii; and *Flora, oder allgemeine botanische Zeitung,* **70** (1887), 243–249.

GUENTER B. RISSE

EICHWALD, KARL EDUARD IVANOVICH (*b.* Mitau, Latvia [now Jelgava, Latvian S.S.R.], 4 July 1795; *d.* St. Petersburg, Russia, 16 November 1876), *geology, paleontology.*

Eichwald's father, Johann Christian Eichwald, was a private tutor to the family of a baron of Courland and later a lecturer in modern languages and natural history. His mother, Charlotte Elizabeth Louis, was the daughter of the court hairdresser. Eichwald was tutored at home before attending the Gymnasium. He began his studies at Dorpat University in 1814 but soon transferred to Berlin University, where he studied medicine and natural sciences. To expand the range of his knowledge he traveled to Germany, Switzerland, France, and England in 1817, taking specialized courses at the universities of Vienna and Paris. In 1819, upon his return to Russia, Eichwald defended his dissertation at Vilna University and was awarded the M.D. degree; he then worked as a physician for two years.

In 1821 Eichwald became assistant professor at

Dorpat University and lectured on geology and paleontology. Two years later he became ordinary professor of obstetrics and zoology at Kazan University, where he also lectured on botany and mineralogy, directed the botanical garden, founded a laboratory of comparative anatomy, and furthered the development of the department of natural history. While in Kazan he married Sofia Ivanovna Finke, the daughter of a professor at the university.

In 1827 Eichwald moved to Vilna and was given the chair of zoology, comparative anatomy, and obstetrics at the university; from 1831 to 1837 he held the chair of zoology, mineralogy, and anatomy at the Medical-Surgical Academy in Vilna. In 1838 Eichwald moved to St. Petersburg, where he received the same chair in a similar type of academy. Until his retirement in 1855 he simultaneously lectured on paleontology at the Mining Institute and on mineralogy and geognosy at the Engineering Academy. In 1826 the Academy of Sciences in St. Petersburg elected him a corresponding member. In 1846 the Medical-Surgical Academy conferred the doctorate of surgery on him, and the University of Breslau awarded him the Ph.D.

Eichwald was a naturalist of wide interests. At different periods of his life he successfully devoted himself to medicine and zoology, as well as to botany, geology, paleontology, anthropology, ethnography, and archaeology. His major fields of concentration were geology and especially paleontology.

During the first years of his scientific activity Eichwald traveled a great deal, which opportunity enabled him to collect abundant and varied material in natural history. In addition to visiting western Europe, he traveled extensively throughout European Russia, concentrating on the Baltic provinces. He also visited Scandinavia and in 1846 made excursions to Italy, Sicily, and Algeria. The results of his trip in 1825–1826, across the Caspian Sea, the Caucasus, Persia, and western Turkmenistan, were extremely fruitful: he collected extensive material on the existing flora and fauna and on the geology, paleontology, and geography of a region that was virtually unstudied at the time.

In 1829–1831 Eichwald published a three-volume monograph on zoology, in which he gave a classification of animals and supplied comparative anatomical, physiological, and paleontological data. By a natural system of classification Eichwald understood not merely a simple and definite grouping of animals by criteria of similarity in organization, but also an expression of the genetic relations actually existing between them. In his researches on present-day fauna Eichwald described many previously unknown forms of mollusks and fishes of the Caspian, brackish-water animals of the Black Sea area, and reptilians of certain regions of the Caucasus.

In the 1830's Eichwald became increasingly interested in the study of fossil organisms. He soon received general recognition as the leading paleontologist of Russia and retained this reputation for more than thirty years. He demonstrated exceptional erudition in a wide range of areas concerning fossil organisms and studied both the flora and fauna of all orders and classes throughout the entire geologic sequence from Cambrian to Recent deposits.

His researches were not limited to the fossils that he collected but also included numerous collections regularly sent to him by geologists from the most diverse regions of Russia. Amid this vast quantity of material Eichwald discovered and described for the first time a great number of previously unknown forms. Endeavoring to summarize and systematize the accumulated data, he began the compilation of an extensive summary of the paleontology of Russia. As a result, during the period from 1853 to 1868 he published (simultaneously in Russian and French) three volumes, in separate sections, of his fundamental monograph, *Lethaea Rossica*. The total work consists of about 3,500 pages of text and three atlases that contain 133 plates representing more than 2,000 different fossil organisms.

While preparing this exceptionally voluminous work, Eichwald worked alone and tried to do everything without assistants. Having undertaken a task too great for one man, he inevitably produced a number of inaccurate descriptions and made certain errors in determining the systematic position and geologic age of the fossils. Nevertheless, his paleontological summary was a valuable scientific contribution; it was very widely used, and many of the new species that he described have retained his nomenclature.

In paleontology Eichwald sought to study a fossil without separating it from geology, a practice that enabled him to draw conclusions on the age of the strata and on the physicogeographical environment that the extinct organisms had inhabited. His stratigraphic deductions formed the basis for most of the geologic research carried out in Russia during that period. In the early 1830's Eichwald was the first to divide the entire geologic column over the distance from Lithuania to the Black Sea, distinguishing detailed units from Transition beds, i.e., Lower Paleozoic, up to Quaternary deposits, inclusive. He also confirmed the wide development of Silurian deposits in the Baltic provinces and supplied the earliest information on the fauna of Pliocene and Quaternary terraces on the coast of the Caspian Sea. In addition, he correctly resolved the question of the geologic age of deposits found throughout the wide expanses of central Russia.

In studying the lithologic features of the rocks in which the fossils were found, Eichwald, outstripping his mid-nineteenth-century contemporaries, began to draw conclusions on the paleogeographic conditions that had existed in the distant geologic past and on the ecological environments that the organisms had inhabited. On a number of geologic questions he adhered basically to opinions that were progressive for his time. For instance, contrary to the majority of his contemporaries, who in conformity with the views of the catastrophist school considered orogeny to be a rapid process, Eichwald wrote in the 1830's that mountains originate from repeated slow uplifts that do not occur simultaneously in different parts of the mountain range.

Regarding the nature of island arcs of the Aleutian type, Eichwald expressed an opinion congruent with the present concepts of the formation of such arcs in weakened zones of the earth's crust. He also stated that prominent faults exist on the boundaries between the continents and the oceans and that they are marked by volcanoes.

In 1827—before the contraction hypothesis was formulated—Eichwald wrote that folding results from the combined effect of gravity and lateral compression, in other words, that it is caused by a combination of vertical and tangential strains.

Eichwald several times changed his ideas on the development of the organic world. Originally, in the 1820's, his views were transformist: he thought that all types and classes of animals originated from a primal protoplasm. Later, under the influence of proponents of the catastrophist theory, Eichwald wrote of the periodic destruction of every living thing and the subsequent appearance of a completely new fauna and flora as the result of an act of divine creation, this new life being more highly organized than the one that had previously existed. Following the publication of Darwin's *Origin of Species,* however, Eichwald renounced his catastrophist and creativist concepts and became an adherent of evolutionary theory.

Eichwald was interested in the problems of the origin of certain rocks and minerals. He believed, among other things, that dolomite was formed as a result of ordinary sedimentation from water, despite the dominance at that time of Buch's hypothesis, which asserted that this rock was formed from limestones under the influence of "magnesian vapors." Eichwald divided hard coals according to their origin into two types that correspond to the present allochthonous and autochthonous coals. He thought that the process of coal formation takes place at a great depth as a result of the pressure exerted by the rock mass above and the action of heat from below.

Eichwald's textbooks on mineralogy (1844) and on geology (1846) were based on the specific features of the geologic structure of Russia. He was one of the first (1821–1823) to lecture systematically on paleontology, thus laying the foundation for the creation of chairs of paleontology in Russian universities and institutes.

A member of many Russian and foreign scientific societies, Eichwald was especially active in the work of the Free Economic Society of Russia.

BIBLIOGRAPHY

I. ORIGINAL WORKS. Eichwald's major works include *Ideen zu einer systematischen Oryktozoologie oder über verändert und unverändert ausgegrabene Thiere* (Mitau [Jelgava], 1821); *Zoologia specialis quam exposites animalibus, tam fossilibus potissimum Rossiae in universum et Poloniae in specie,* 3 vols. (Vilna, 1829–1831); "Fauna Caspio-Caucasia nonnulis observationibus novis illustravit," in *Nouveaux mémoires de la Société des naturalistes de Moscou,* **7** (1842); *Polny kurs geologicheskikh nauk preimuschestvenno po otnosheniyu k Rossii:* I. *Oriktognozia preimushchestvenno po otnosheniyu k Rossii i s prisovokupleniem upotreblenia mineralov* (St. Petersburg, 1844), and II. *Geognozia, preimushchestvenno po otnosheniyu k Rossii* (St. Petersburg, 1846); and *Lethaea Rossica ou Paléontologie de la Russie,* 3 vols. (Stuttgart, 1853–1868).

II. SECONDARY LITERATURE. On Eichwald and his work, see E. Lindemann, "Das fünfzigjährige Doktorjubiläum Eduard von Eichwald's, Dr. der Philosophie, Medizin und Chirurgie," in *Verhandlungen der Russisch-kaiserlichen mineralogischen Gesellschaft zu St. Petersburg,* 2nd ser., **5** (1870), 278–358.

V. V. TIKHOMIROV

EIGENMANN, CARL H. (*b.* Flehingen, Germany, 9 March 1863; *d.* Chula Vista, California, 24 April 1927), *ichthyology.*

The son of Philip and Margaretha Lieb Eigenmann, Carl intended to study law when he entered Indiana University in 1879, two years after his arrival in the United States with an uncle. A course in biology under David Starr Jordan turned him to an extremely productive career in ichthyology. After receiving his bachelor's degree (1886), he studied the South American fish collections at Harvard University for one year and then became curator at the Natural History Society in San Diego, the home of his wife, Rosa (also an ichthyologist). At Jordan's departure in 1891, Eigenmann replaced him as professor of zoology at Indiana University, having received his Ph.D. there in 1889. In 1892 he was made director of the Biological Survey of Indiana; in 1908 he became the first dean of the graduate school at Indiana; and from 1909 to 1918 he was honorary curator of fishes at the Carnegie Museum in Pittsburgh.

Eigenmann was a member of the National Academy of Sciences, of Sigma Xi and Phi Beta Kappa, a fellow of the American Association for the Advancement of Science, and an honorary member of the California Academy of Sciences and of the Sociedad de Ciencias Naturales of Bogotá, Colombia.

Although he conducted other researches, Eigenmann repeatedly turned to painstaking analyses of the classification, distribution, and evolution of the freshwater fishes of South America, based on studies of many museum collections and the results of his own expeditions. From comparisons of the African and South American Cichlidae and Characidae, he concluded that a pre-Tertiary land connection between the two continents must have existed. In a number of monographs he presented the classification of families of South American fishes, climaxed by the exhaustive five-part "American Characidae" (1917–1925). His brief stay on the West Coast resulted in valuable papers on the taxonomy, variation, and habits of the fishes of San Diego (1892), and at Indiana he published considerably on the fish there. Curiosity about blindness of cave animals led Eigenmann to detailed study of specimens from Indiana, Missouri, Texas, Kentucky, and Cuba. He concluded that the degenerative characteristics of subdued coloration and of blindness become inherited when they have adaptive environmental value.

A significant participant in the "golden age of descriptive ichthyology in the United States," Eigenmann left a legacy of meticulous classification and many grateful students who credited him especially with teaching them self-reliance.

BIBLIOGRAPHY

I. ORIGINAL WORKS. Besides many monographs on fish families and single papers, Eigenmann's major publication was "The American Characidae," in vol. **43** of the *Memoirs of the Museum of Comparative Zoology of Harvard College:* pt. 1 (1917), 1–102; pt. 2 (1918), 103–208; pt. 3 (1921), 209–310; pt. 4 (1927), 311–428; and pt. 5 (1929), 429–558, written with G. S. Myers. The geographic studies of South American fishes appeared in "The Fresh-Water Fishes of Patagonia and an Examination of the Archiplata-Archhelenis Theory," in *Reports of the Princeton University Expedition to Patagonia, 1896 to 1899,* Zoology, III (Princeton, 1909), 225–374. The studies of cave fauna are summarized in *Cave Vertebrates of North America: A Study in Degenerative Evolution,* Publications of the Carnegie Institution, no. 104 (Washington, D.C., 1909), pp. 1–341.

II. SECONDARY LITERATURE. Little is known of Eigenmann's early life, but his professional career is well summarized in Leonard Stejneger, "Carl H. Eigenmann," in *Biographical Memoirs. National Academy of Sciences,* **18** (1937), 305–336, where it is commented that the "middle initial did not stand for a name." A full bibliography accompanies the Stejneger memoir. For a lively account of the "golden age of descriptive ichthyology," see Carl L. Hubbs, "History of Ichthyology in the United States After 1850," in *Copeia,* no. 1 (1964), pp. 42–60.

ELIZABETH NOBLE SHOR

EIJKMAN, CHRISTIAAN (*b.* Nijkerk, Netherlands, 11 August 1858; *d.* Utrecht, Netherlands, 5 November 1930), *medicine, physiology, nutrition.*

Eijkman, who in 1929 shared the Nobel Prize in physiology or medicine with F. G. Hopkins, was the seventh child of a boarding-school proprietor in the small Gelderland town of Nijkerk, situated at the northern border of the Veluwe. His parents, Christiaan Eijkman and Johanna Alida Pool, had several gifted sons: one brother became a chemist and a professor at Tokyo and Groningen; another was a linguist; and a third was one of the first roentgenologists in the Netherlands.

When Eijkman was only three years old the family moved to Zaandam, where he received sufficient instruction to pass the examination that enabled him to enter the university (1875). The costs of his study were defrayed by the government because he enrolled for later service as an army physician. His ability soon became apparent; he passed three examinations cum laude or magna cum laude.

As a student, for two years Eijkman was assistant to the professor of physiology, Thomas Place. In 1883 he qualified as physician and took his medical degree after defending a thesis on polarization in the nerves ("Over Polarisatie in de Zenuwen"). He was immediately sent as medical officer to the Dutch East Indies, where he worked for two years on Java and Sumatra. A severe attack of malaria forced him to repatriate on sick leave in November 1885. Two months later his young wife of three years, Aaltje Wigeri van Edema, died. After his recovery Eijkman decided to train himself in bacteriology, then a new and rising science. After studying under Josef Förster at Amsterdam, he went to work with Robert Koch at Berlin. Here he made some acquaintances that were decisive for his future career.

In the Dutch East Indies and other eastern countries a disease called beriberi was spreading, especially in closed communities—the army, the navy, prisons, and so on. In some cases cardiac insufficiency with massive edema of the legs dominated the clinical picture, in others a progressive paralysis of the legs (hydropic and "dry" forms). In view of its apparent epidemic character, a bacteriological origin seemed obvious. The Dutch government appointed a committee to study the disease on the spot. The committee consisted of C. A. Pekelharing and C. Winkler, then a young reader and later a well-known neurologist. Before undertaking their difficult mission, both men went to Koch at Berlin to learn something of bacteriology. The result of their meeting there with Eijkman was that the latter, at his own request, was added to the committee that departed for the East in October 1886.

After some two years the committee had shown that in beriberi a polyneuritis could be proved by clinical and microscopic examination, and it believed it had isolated from the blood of beriberi patients the causative agent: a micrococcus, the toxins of which caused the polyneuritis. Pekelharing and Winkler returned to Europe in 1887, leaving Eijkman as head of a small laboratory built for the purpose of continuing the research. Eijkman was also director of the Javanese Medical School.

At that time Eijkman also considered beriberi to be an infectious disease, but he did not succeed in producing the disease in animals by inoculation with the micrococci. He then had the good fortune to see a similar disease develop spontaneously in fowls. The animals showed paresis or paralysis of the legs with dyspnea and cyanosis; microscopic examination of the nerves confirmed the presence of a polyneuritis. Eijkman considered this *polyneuritis gallinarum* to be the equivalent of the polyneuritis in beriberi.

In order to extend the observations, the fowls were removed to another place; but then, unexpectedly, the disease inexplicably disappeared. Eijkman noticed that at the same time a slight change had occurred in the food of the fowls: the original food was obtained from the leavings of boiled rice from the officers' table in the military hospital; later they received unpolished rice, or "paddy," because a new cook had refused military rice to "civilian" fowls. Eijkman now supposed that the causative factor must be sought in the food, especially the polished and boiled rice.

On the basis of extensive food experiments, Eijkman proved that unpolished rice had both a preventive and curative effect on polyneuritis in fowls; but he did not perceive the correct explanation and continued to believe that some chemical agent was causing the polyneuritis, for example, a toxic substance originating from the action of intestinal microorganisms on boiled rice. He even adhered to such a hypothesis for some years after Grijns, in 1901, had advanced the idea of a nutritional deficiency. Nevertheless, Eijkman's observations were the starting point for a line of scientific research that led to the discovery of thiamine (vitamin B_1), found in the pericarp of the unpolished rice grains, as the substance protecting against beriberi.

In 1896 Eijkman, who had married Berthe Julie Louise van der Kemp in 1888, returned home again on sick leave. He made statistical studies on beriberi, on osmosis in the blood, and on the influence of summer and winter on metabolism. In comparing the metabolisms of Europeans in the tropics and natives, he had found no disparities in respiratory metabolism, perspiration, and temperature metabolism. In 1898 he was appointed professor of public health and forensic medicine at the University of Utrecht. He took office with a formal address on health and diseases in the tropics, *Gezondheid en Ziekte in Heete Gewesten* (Utrecht, 1898). During the thirty years of his professorship Eijkman guided many research projects in his laboratory. In the academic year 1912–1913 he acted as rector magnificus, leaving this office with a rectorial oration entitled "Simplex non veri sigillum." He was also a member of several governmental committees in the field of public health and of many national and foreign societies, including the Royal Academy of Sciences of the Netherlands, to which he was appointed in 1907. He was a recipient of the John Scott Medal and a foreign associate of the National Academy of Sciences, Washington, D.C.

In 1928, seventy years old, Eijkman retired; the following year the state of his health did not allow him to accept the Nobel Prize personally, but the address he had intended to deliver was published in *Les Prix Nobel*. During his long life he had performed research in various fields, but his discovery of the role of polished rice in causing *polyneuritis gallinarum* remains his claim to fame because it was the foundation of the later doctrine of the role of vitamins in human nutrition.

About the man himself little information is obtainable. By his second marriage he had one son, Pieter Hendrik, who became a physician.

BIBLIOGRAPHY

I. ORIGINAL WORKS. Except for two textbooks on physiology and chemistry and his orations, nearly all of Eijkman's publications appeared in annual reports and periodicals, most of them in Dutch. A full list is given by Jansen.

Of special note are *Specifieke Antistoffen* (Haarlem, 1901); *Onzichtbare Smetstoffen* (Haarlem, 1904); *Een en Ander over Voeding* (Haarlem, 1906); and *Hygiënische Strijdvragen* (Rotterdam, 1907). See also *Nobel Lectures. Physiology or Medicine*, II (Amsterdam–New York, 1965), 199–207.

II. SECONDARY LITERATURE. See J. M. Baart de la Faille, "Christiaan Eijkman," in T. P. Sevensma, ed., *Nederlandsche Helden der Wetenschap* (Amsterdam, 1946), pp. 299–333, with portrait; and B. C. P. Jansen, *Het Levenswerk van Christiaan Eijkman 1858–1930* (Haarlem, 1959).

G. A. LINDEBOOM

EINSTEIN, ALBERT (*b.* Ulm, Germany, 14 March 1879; *d.* Princeton, New Jersey, 18 April 1955), *physics.*

Albert Einstein was the only son of Hermann and Pauline (Koch) Einstein. He grew up in Munich, where his father and his uncle ran a small electrochemical plant. Einstein was a slow child and disliked the regimentation of school. His scientific interests were awakened early and at home—by the mysterious compass his father gave him when he was about four; by the algebra he learned from his uncle; and by the books he read, mostly popular scientific works of the day. A geometry text which he devoured at the age of twelve made a particularly strong impression.

When his family moved to Milan after a business failure, leaving the fifteen-year-old boy behind in Munich to continue his studies, Einstein quit the school he disliked and spent most of a year enjoying life in Italy. Persuaded that he would have to acquire a profession to support himself, he finished the Gymnasium in Aarau, Switzerland, and then studied physics and mathematics at the Eidgenössische Technische Hochschule (the Polytechnic) in Zurich, with a view toward teaching.

After graduation Einstein was unable to obtain a regular position for two years and did occasional tutoring and substitute teaching, until he was appointed an examiner in the Swiss Patent Office at Berne. The seven years Einstein spent at this job, with only evenings and Sundays free for his own scientific work, were years in which he laid the foundations of large parts of twentieth-century physics. They were probably also the happiest years of his life. He liked the fact that his job was quite separate from his thoughts about physics, so that he could pursue these freely and independently, and he often recommended such an arrangement to others later on. In 1903 Einstein married Mileva Marić, a Serbian girl who had been a fellow student in Zurich. Their two sons were born in Switzerland.

Einstein received his doctorate in 1905 from the University of Zurich for a dissertation entitled, "Eine neue Bestimmung der Moleküldimensionen" ("A New Determination of Molecular Dimensions"), a work closely related to his studies of Brownian motion, discussed below. It took only a few years until he received academic recognition for his work, and then he had a wide choice of positions. His first appointment, in 1909, was as associate professor (*extraordinarius*) of physics at the University of Zurich. This was followed quickly by professorships at the German University in Prague, in 1911, and at the Polytechnic in Zurich, in 1912. Then, in the spring of 1914, Einstein moved to Berlin as a member of the Prussian Academy of Sciences and director of the Kaiser Wilhelm Institute for Physics, free to lecture at the university or not as he chose. He had mixed feelings about accepting this appointment, partly because he disliked Prussian rigidity and partly because he was unhappy about the implied obligation to produce one successful theory after another. As it turned out he found the scientific atmosphere in Berlin very stimulating, and he greatly enjoyed having colleagues like Max Planck, Walther Nernst, and, later, Erwin Schrödinger and Max von Laue.

During World War I, Einstein's scientific work reached a culmination in the general theory of relativity, but in most other ways his life did not go well. He would not join in the widespread support given to the German cause by German intellectuals and did what he could to preserve a rational, international spirit and to urge the immediate end of the war. His feeling of isolation was deepened by the end of his marriage. Mileva Einstein and their two sons spent the war years in Switzerland and the Einsteins were divorced soon after the end of the war. Einstein then married his cousin Elsa, a widow with two daughters. Einstein's health suffered, too. One of his few consolations was his continued correspondence and occasional visits with his friends in the Netherlands—Paul Ehrenfest and H. A. Lorentz, especially the latter, whom Einstein described as having "meant more to me personally than anybody else I have met in my lifetime"[1] and as "the greatest and noblest man of our times."[2]

Einstein became suddenly famous to the world at large when the deviation of light passing near the sun, as predicted by his general theory of relativity, was observed during the solar eclipse of 1919. His name and the term *relativity* became household words. The publicity, even notoriety, that ensued changed the pattern of Einstein's life. He was now able to put the weight of his name behind causes that he believed in, and he did this, always bravely but taking care not to misuse the influence his scientific fame had given him. The two movements he backed most forcefully in the 1920's were pacifism and Zionism, particularly the creation of the Hebrew University in Jerusalem. He also took an active part for a few years in the work of the Committee on Intellectual Cooperation of the League of Nations.

Soon after the end of the war, Einstein and relativity became targets of the anti-Semitic extreme right wing. He was viciously attacked in speeches and articles, and his life was threatened. Despite this treatment Einstein stayed in Berlin, declining many offers to go elsewhere. He did accept an appointment as special professor at Leiden and went there regularly for periods of a week or two to lecture and to discuss current problems in physics. In 1933 Einstein was considering an arrangement that would have allowed him to divide his year between Berlin and the new Institute for Advanced Study at Princeton. But when Hitler came to power in Germany, he promptly resigned his position at the Prussian Academy and joined the Institute. Princeton became his home for the remaining twenty-two years of his life. He became an American citizen in 1940.

During the 1930's Einstein renounced his former pacifist stand, since he was now convinced that the menace to civilization embodied in Hitler's regime could be put down only by force. In 1939, at the request of Leo Szilard, Edward Teller, and Eugene Wigner, he wrote a letter to President Franklin D. Roosevelt pointing out the dangerous military potentialities offered by nuclear fission and warning him of the possibility that Germany might be developing these potentialities. This letter helped to initiate the American efforts that eventually produced the nuclear reactor and the fission bomb, but Einstein neither participated in nor knew anything about these efforts. After the bomb was used and the war had ended, Einstein devoted his energies to the attempt to achieve a world government and to abolish war once and for all. He also spoke out against repression, urging that intellectuals must be prepared to risk everything to preserve freedom of expression.

Einstein received a variety of honors in his lifetime—from the 1921 Nobel Prize in physics to an offer (which he did not accept) of the presidency of Israel after Chaim Weizmann's death in 1952.

One of Einstein's last acts was his signing of a plea, initiated by Bertrand Russell, for the renunciation of nuclear weapons and the abolition of war. He was drafting a speech on the current tensions between Israel and Egypt when he suffered an attack due to an aortic aneurysm; he died a few days later. But despite his concern with world problems and his willingness to do whatever he could to alleviate them, his ultimate loyalty was to his science. As he said once

with a sigh to an assistant during a discussion of political activities: "Yes, time has to be divided this way between politics and our equations. But our equations are much more important to me, because politics is for the present, but an equation like that is something for eternity."[3]

Early Scientific Interests. Albert Einstein started his scientific work at the beginning of the twentieth century. It was a time of startling experimental discoveries, but the problems that drew his attention and forced him to produce the boldly original ideas of a new physics had developed gradually and involved the very foundations of the subject. The closing decades of the nineteenth century were the period when the long-established goal of physical theory—the explanation of all natural phenomena in terms of mechanics—came under serious scrutiny and was directly challenged. Mechanical explanation had had great successes, particularly in the theory of heat and in various aspects of optics and electromagnetism; but even the successful mechanical theory of heat had its serious failures and unresolved paradoxes, and physicists had not been able to provide a really satisfactory mechanical foundation for electromagnetic theory. Many were questioning the whole program of mechanism, and alternatives ranging from the energetics of Wilhelm Ostwald to the electromagnetic world view of Wilhelm Wien were widely considered and vigorously debated.

To a young man who looked to science for nothing less than an insight into the "great eternal riddle"[4] of the universe, these basic questions were the most challenging and also the most fascinating. Einstein was impressed by both the successes and the failures of mechanical physics and was attracted to what he later called the "revolutionary" ideas of James Clerk Maxwell's field theory of electromagnetism. His study of the writings of the nineteenth-century masters received a new direction when he read Ernst Mach's *Science of Mechanics*. This concern with general principles required something else to make it fruitful, however, and Einstein himself described what it was. He realized that each of the separate fields of physics "could devour a short working life without having satisfied the hunger for deeper knowledge," but he had an unmatched ability "to scent out the paths that led to the depths, and to disregard everything else, all the many things that clutter up the mind and divert it from the essential."[5] This ability to grasp precisely the particular simple physical situation that could throw light on obscure questions of general principle characterized much of Einstein's thinking.

His earliest papers—"my two worthless beginner's works,"[6] as he referred to them a few years later—

were an attempt to learn something from experimental materials about intermolecular forces with a view toward their possible relationship with long-range gravitational force, a problem going back to Newton's time. This work led nowhere, and Einstein's next series of three articles, published during the years 1902 to 1904, dealt with quite another set of ideas and was clearly the work of a mature scientist. In these articles Einstein rederived by his own methods the basic results of statistical mechanics: the canonical distribution of energy for a system in contact with a heat bath, the equipartition theorem, and the physical interpretations of entropy and temperature. He also emphasized that the probabilities that appear in the theory are to be understood as having a very definite physical meaning. The probability of a macroscopically identifiable state of a system is the fraction of any sufficiently long time interval that the system spends in this state. Equilibrium is dynamic, with the system passing through all its possible states in an irregular sequence. Ludwig Boltzmann had introduced this point of view years before, but Einstein made it very much his own.

It was in the last of this early series of papers, however, that Einstein introduced a new theme. There is one fundamental constant in statistical mechanics, the constant now known as Boltzmann's constant, k. It appears in the typical exponential factor of the distribution law, $\exp(-E/kT)$, where E is the energy of the system and T is its absolute temperature. It appears too in the relation between the entropy S and the probability W of a state

$$S = k \ln W. \tag{1}$$

Einstein asked for the physical significance of this constant k. It was already well-known from the theory of the ideal gas that k was simply related to the gas constant R and to Avogadro's number, N_0, the number of molecules in a gram-molecular weight of any substance,

$$k = \frac{R}{N_0}. \tag{2}$$

Einstein showed that k entered into still another basic equation of the statistical theory, the expression for the mean square fluctuation $\langle \Delta^2 \rangle$ of the energy E about its average value $\langle E \rangle$:

$$\langle \Delta^2 \rangle = \langle (E - \langle E \rangle)^2 \rangle = kT^2 \frac{d\langle E \rangle}{dT}. \tag{3}$$

This meant that k defines the scale of fluctuation phenomena or, as Einstein put it, that it determines the thermal stability of a system. This result shows that fluctuations are normally negligibly small so that the average or thermodynamic value of the energy is a very good measure of this quantity, but Einstein was more interested in its other implications. If one could actually measure the energy fluctuations of any system, then k could be determined and with it Avogadro's number and the mass of an individual atom. None of these quantities was known with any precision, and previous determinations involved very indirect theoretical arguments. Einstein could not refer to any measurements of fluctuations, but he did give a very plausible analysis of the energy fluctuations in black-body radiation showing how k was related to the constant in Wien's displacement law.

This 1904 paper made little if any impression on Einstein's contemporaries, but it contained the seeds of much of his later work. No one before Einstein had taken seriously the fluctuation phenomena predicted by statistical mechanics, but he saw that the existence of such fluctuations could be used to demonstrate the correctness of the whole molecular theory of heat. The problem was to find a situation in which fluctuations could be observed, and Einstein found a solution to this problem in 1905, in his paper "Die von der molekularkinetischen Theorie der Wärme geforderte Bewegung von in ruhenden Flüssigkeiten suspendierten Teilchen" ("On the Movement of Small Particles Suspended in a Stationary Liquid Demanded by the Molecular-Kinetic Theory of Heat"). This predicted motion of colloidal particles was already widely known as Brownian motion, but at the time Einstein wrote this paper he knew virtually nothing about what had been observed and hesitated to identify the two motions. He was not trying to explain an old and puzzling phenomenon, but rather to deduce a result that could be used to test the atomic hypothesis and to determine the basic scale of atomic dimensions.

One essential assumption Einstein made was that a colloidal particle will come into thermodynamic equilibrium with the molecules of the fluid in which it is suspended, so that the average kinetic energy of the particle associated with its motion in any one direction is just the equipartition value, $kT/2$. The quantity that Einstein calculated for this random motion of colloidal particles was not the velocity, which is unmeasurable even in principle, but rather the mean square displacement $\langle \delta_x^2 \rangle$ in some particular direction x during the time interval τ. For spherical particles of radius a, satisfying the same law of resistance that a macroscopic sphere would obey in this fluid of viscosity η, he obtained the result

$$\langle \delta_x^2 \rangle = \frac{kT}{3\pi\eta a}\tau. \tag{4}$$

The hope Einstein expressed at the end of his paper, that "some enquirer" undertake an experimental test of his predictions, was fulfilled several years later when Jean Perrin's experiments confirmed the correctness of all features of the Brownian motion equation and provided a new determination of Avogadro's number. These results helped to convince the remaining skeptics, such as Wilhelm Ostwald, that molecules were real and not just a convenient hypothesis. The theory of Brownian motion was developed further by both Einstein and Maryan von Smoluchowski. Several years later both men worked on the theory of another fluctuation phenomenon—the opalescence exhibited by a fluid in the immediate neighborhood of its critical point. Einstein's work, published in 1910, was especially notable for its generalization of fluctuation theory in a form independent of the mechanical foundations of the theory, an old idea of his and one that later proved to be of considerable influence.

All the work discussed thus far, significant as it was, does not represent the predominant concern of Albert Einstein throughout his career—the search for a unified foundation for all of physics. Neither the attempts at a mechanical theory of the electromagnetic field nor the recent efforts to base mechanics on electromagnetism had been successful. The disparity between the discrete particles of matter and the continuously distributed electromagnetic field came out most clearly in Lorentz' electron theory, where matter and field were sharply separated for the first time. This theory strongly influenced Einstein, who often referred to the basic electromagnetic equations as the Maxwell-Lorentz equations. The problems generated by the incompatibility between mechanics and electromagnetic theory at several crucial points claimed Einstein's attention. His struggles with these problems led to his most important early work—the special theory of relativity and the theory of quanta.

For the sake of clarity and convenience, Einstein's development of relativity theory is treated in a separate article following the discussion of his contribution to quantum mechanics that occupies the remainder of the present article. It must be pointed out, however, that separating these two main themes in Einstein's work does an injustice to the unity of his fundamental purpose.

Quantum Theory and Statistical Mechanics. Einstein once described his first paper of 1905, "Über einen die Erzeugung und Verwandlung des Lichtes betreffenden heuristischen Gesichtspunkt" ("On a Heuristic Viewpoint Concerning the Production and Transformation of Light"), as "very revolutionary." He was not exaggerating. The heuristic viewpoint of

the title was nothing less than the suggestion that light be considered a collection of independent particles of energy, which he called light quanta. Einstein had his reasons for advancing such a bold suggestion, one that seemed to dismiss a century of evidence supporting the wave theory of light. First among these reasons was a negative result: The combination of the electromagnetic theory of light with the (statistical) mechanics of particles was incapable of dealing with the problem of black-body radiation. It predicted that radiation in thermodynamic equilibrium within an enclosure would have a frequency distribution corresponding to an infinite amount of energy at the high-frequency end of the spectrum. This was incompatible with the experimental results, but, worse than that, it meant that the theory did not give an acceptable answer to the problem. Einstein was the first to point to this result, known later on as the "ultraviolet catastrophe," as a fundamental failure of the combined classical theories, although Lord Rayleigh had hinted at this in a paper in 1900.

Although he was convinced that a new unified fundamental theory was needed for an adequate treatment of the radiation problem, Einstein had no such theory to offer. What he did instead was to analyze the implications of the observed radiation spectrum, well-described, except at low frequencies, by Wien's distribution law. To carry out his analysis, Einstein used the methods of thermodynamics ("the only physical theory of universal content concerning which I am convinced that, within the framework of the applicability of its basic concepts, it will never be overthrown") and statistical mechanics. What he found was that the entropy of black-body radiation in a given frequency interval depends on the volume of the enclosure in the same way that the entropy of a gas depends on its volume. And because the latter dependence has its origin in the independence of the gas molecules rather than the details of their dynamics, Einstein leaped to the conclusion that the radiation, too, must consist of independent particles of energy. This identification required the energy E of the particles to be proportional to the frequency ν of the radiation,

$$E = h\nu, \qquad (5)$$

where the universal proportionality constant h was the product of k and one of the constants in Wien's distribution law.

Einstein showed that his strange proposal of light quanta could immediately account for several puzzling properties of fluorescence, photoionization, and especially of the photoelectric effect. His quantitative prediction of the relationship between the maximum energy of the photoelectrons and the frequency of the incident light was not verified experimentally for a decade. The light quantum hypothesis itself attracted only one or two adherents; it represented too great a departure from accepted ideas. It went far beyond the work Max Planck had done in 1900, in which the energy of certain material oscillators was treated as a discrete variable, capable only of values that were integral multiples of a natural unit proportional to the frequency. Planck's quantum hypothesis had been introduced as a way of deriving the complete distribution law for black-body radiation, but in 1905 it was only just starting to receive critical study.

During the years between 1905 and 1913 it was Einstein who took the lead in probing the significance of the new ideas on quanta. He soon decided that Planck's work was complementary to his own and not in conflict with it, as he had first thought. Einstein then realized that if Planck had been right in restricting the energies of his oscillators to integral multiples of $h\nu$, in discussing the interaction of molecular oscillators with black-body radiation, then this same restriction should also apply to all oscillations on the molecular scale. The success of Planck's work had to be looked upon as demonstrating the need for a quantum theory of matter, even as his own 1905 paper demonstrated the need for a quantum theory of radiation.

In 1907 Einstein pointed out how one could use the quantized energy of the oscillations of atoms in solids to account for departures from the rule of Dulong and Petit. This empirical rule, that the specific heat is the same for one mole of any element in solid form, was understood as a consequence of the theorem of equipartition of energy. Many light elements, however, had specific heats at room temperature that were much smaller than the Dulong-Petit value. Einstein showed how one could easily calculate the specific heat of a solid all of whose atoms vibrated with the same frequency (an assumption he made only as a convenient simplification) and obtain a universal curve for the variation of specific heat with temperature. The only parameter in the theory was the frequency of the quantized vibrations. This specific heat curve approached the Dulong-Petit value at high temperatures; accounted qualitatively for all the departures from the equipartition result, including the absence of electronic contributions to the specific heat; and predicted a new and general law: The specific heats of all solids should approach zero as the absolute temperature approaches zero. Einstein indicated how the vibration frequencies could be determined from infrared absorption measurements in

many cases; several years later he suggested another way of determining these frequencies using their relationship to the elastic constants of the solid.

As it turned out, Einstein's quantum theory of specific heats appeared at a time when the behavior of specific heats at low temperatures had just become of interest for very different reasons. Walther Nernst was planning a program of such measurements to establish his own new heat theorem, later known as the third law of thermodynamics. Nernst's results matched the predictions of Einstein's theory in all essential respects and convinced him that there was something really significant in this "odd" and "grotesque" theory,[7] as he called it. The success of Einstein's theory of specific heats in explaining old difficulties, predicting new laws, and establishing unexpected connections among thermal, optical, and elastic properties of crystals was the single most important element in awakening the interest of physicists in the quantum theory.

Wave-Particle Duality. For Einstein, however, the central problem continued to be the nature of radiation. In 1909, speaking in Salzburg at his first major scientific meeting, he argued that the future theory of light which would have to be constructed would be "a kind of fusion of the wave and emission theories."[8] Einstein's prediction was based on the results of his continued probings into the implications of Planck's distribution law for black-body radiation. He had calculated the energy fluctuations of the radiation in a small frequency interval with the help of equation (3) and had found that the fluctuations were the sum of two terms, indicating two apparently independent mechanisms for energy fluctuations. One term was readily intelligible as due to interfering waves, the other as due to variations in the number of light quanta present in the subvolume under study. Neither a wave nor a particle theory could account for the presence of both terms. Einstein confirmed this result by a completely independent calculation of the Brownian motion that a mirror would have to undergo if it were suspended in an enclosure containing a gas and black-body radiation in thermodynamic equilibrium. Once again there were wave and particle contributions to the fluctuations in momentum of the suspended mirror.

Einstein saw this wave-particle duality in radiation as concrete evidence for his conviction that physics needed a new, unified foundation. His view of the role of light quanta in this new fundamental theory had evolved since he put forward the heuristic suggestion of a corpuscular approach to radiation in 1905. Einstein now envisaged a field theory, based on appropriate partial differential equations, probably nonlinear, from which quanta would emerge as singular solutions, along the lines of the electric charges in electrostatics. He found some support for this parallel in the fact that Planck's constant, h, characteristic for light quanta, was dimensionally equivalent to e^2/c, where e is the unit electric charge and c is the velocity of light. To Einstein this suggested that the discreteness of energy and the discreteness of charge might be explained together by the new fundamental theory.

There was unfortunately very little to go on in the search for this new theory. It would have to be consistent with the special theory of relativity, but Einstein saw that theory as only a universal formal principle, analogous to the laws of thermodynamics, which gave no clue to the structure of matter or radiation. The fluctuation properties of radiation, which he had established, "presented small foothold for setting up a theory."[9] We know from Einstein's correspondence as well as from the brief remarks in his papers of this period that he devoted much of his effort to this problem in the years 1908 to 1911, using Lorentz' theory of electrons as one of his points of departure. His efforts along this line seem to have been comparable in their intensity, although not in their fruitfulness, to his efforts during the following years to create the new gravitational theory—the general theory of relativity.

When in 1911 Einstein put aside his intense work on the problem of developing a theory from which he could "construct" quanta—"because I now know that my brain is incapable of accomplishing such a thing"[10]—he did not give up his interest in quanta. He continued to reflect on the questions surrounding the quantum theory. In a paper in 1914, for example, he used familiar thermodynamic arguments to give a new derivation of Planck's expression for the average energy of an oscillator. This work led him to suggest the identity of physical and chemical changes at the molecular level: "A quantum type of change in the physical state of a molecule seems to be no different in principle from a chemical change."[11]

Relation to Bohr's Early Work. When Einstein returned to the radiation problem in 1916, the quantum theory had undergone a major change. Niels Bohr's papers had opened a new and fertile domain for the application of quantum concepts—the explanation of atomic structure and atomic spectra. In addition Bohr's work and its generalizations by Arnold Sommerfeld and others constituted a fresh approach to the foundations of the quantum theory of matter. Einstein's new work showed the influence of these ideas. He had found still another derivation of Planck's black-body radiation law, an "astonishingly

simple and general" one which, he thought, might properly be called "*the* derivation"[12] of this important law. It was based on statistical assumptions about the processes of absorption and emission of radiation and on Bohr's basic quantum hypothesis that atomic systems have a discrete set of possible stationary states. The proof turned on the requirement that absorption and emission of radiation, both spontaneous and stimulated, suffice to keep a gas of atoms in thermodynamic equilibrium. (This paper introduced the concept of stimulated emission into the quantum theory and is therefore often described as the basis of laser physics.) Einstein himself considered the most important contribution of this work to be not the new derivation of the distribution law but rather the arguments he presented for the directional character of energy quanta. Each quantum of frequency ν emitted by an atom must carry away momentum $h\nu/c$ in a definite direction; spherical waves would simply not exist.

Although Einstein put particular emphasis on the directionality of light quanta, there was no direct evidence for it until 1923 when Arthur Compton explained his experiments on the increase in X-ray wavelength after scattering from free electrons. Compton simply treated the process as a collision, obeying the conservation laws, between the electron and a quantum of energy $h\nu$ and momentum $h\nu/c$ in the direction of the incident X-ray beam. Even before this, however, Einstein was trying to devise a crucial experiment to settle the question of the nature of radiation. He held fast to his view that light quanta were indispensable since they described the particle properties really manifested by radiation. Light quanta did not have many other supporters until after the Compton effect, and they were particularly unpopular with Bohr and his co-workers. Bohr saw no good way of reconciling them with the correspondence principle and was willing to give up the exact validity of the conservation laws in order to avoid quanta. Experiments to check Bohr's proposals early in 1925 vindicated Einstein's belief in both the conservation laws and the validity of light quanta.

Bose-Einstein Statistics and Wave Mechanics. In 1924 Einstein received a paper from a young Indian physicist, S. N. Bose, setting forth a theory in which radiation was treated as a gas of light quanta. By changing the statistical procedure for counting the states of the gas, Bose had arrived at an equilibrium distribution which was identical with Planck's radiation law. Einstein was much taken with this extension of his old idea. He not only translated Bose's paper into German and saw to its publication, but he also applied Bose's new statistical idea to develop an analogous theory for an ideal gas of material particles. A gas obeying the Bose-Einstein statistics, as the new counting procedure was later called, showed a variety of interesting properties. Even though the particles exerted no forces on each other the gas showed a peculiar "condensation" phenomenon: Below a certain temperature a disproportionately large fraction of the total number of particles are found in the state of lowest energy.

Einstein's interest in the parallel between the gas of particles and the gas of light quanta deepened when he read Louis de Broglie's Paris thesis late in 1924. De Broglie, inspired by Einstein's earlier work on the wave-particle duality, had become convinced that this duality must hold for matter as well as radiation. In his thesis he developed the idea that every material particle has a wave associated with it, the frequency ν and wavelength λ of the wave being related to the energy E and momentum p of the particle by the equations

$$E = h\nu, \quad p = h/\lambda. \tag{6}$$

De Broglie had no experimental evidence to support his idea and deduced no experimentally testable conclusions from it, so it aroused very little interest. Einstein, however, was immediately attracted to the idea of matter waves because he saw its relationship to his new theory of the ideal gas. He found a confirmation of de Broglie's wave-particle duality for matter in the results of his calculation of the density fluctuations of this ideal gas. These fluctuations showed the same structure as had the energy fluctuations of black-body radiation; only now it was the particle term that would have been the only one present in the classical gas theory. Einstein saw the wave term in the fluctuations as a manifestation of the de Broglie waves, and he was sure he was not dealing with a "mere analogy." He proposed several kinds of experiments which might detect the diffraction of de Broglie waves.

Einstein's support for de Broglie's work brought it the attention it deserved, particularly from Erwin Schrödinger. In describing the origins of his wave mechanics a few years later, Schrödinger wrote: "My theory was stimulated by de Broglie's thesis and by short but infinitely far-seeing remarks by Einstein."[13] Those remarks were the ones linking de Broglie's ideas to the properties of the Bose-Einstein gas.

When the new matrix mechanics appeared, in the papers of Werner Heisenberg, Max Born, and Pascual Jordan, Einstein was interested but not convinced. "An inner voice tells me that it is still not the true Jacob,"[14] he wrote to Born in 1926. He looked more favorably on Schrödinger's wave mechanics: "I am convinced that you have made a decisive advance with your formulation of the quantum condition, just as I am equally convinced that the Heisenberg-Born route is off the track."[15]

Discontent With Quantum Mechanics. In 1927 the synthesis that constituted the new quantum mechanics was worked out. One of its key features was Born's statistical interpretation of Schrödinger's wave function. This meant that a full quantum mechanical description of the state of a system would generally specify only probabilities rather than definite values of the dynamical variables of the system. The new theory was intrinsically statistical and renounced as meaningless in principle any attempt to go beyond the probabilities to arrive at a deterministic theory. Bohr expressed what became the generally accepted viewpoint when he described quantum mechanics as a "rational generalization of classical physics," the result of "a singularly fruitful cooperation of a whole generation of physicists."[16]

Einstein dissented from this majority opinion. He never accepted the finality of the quantum mechanical renunciation of causality or its limitation of physical theory to the unambiguous description of the outcome of fully defined experiments. From the Solvay Congress of 1927, when the quantum mechanical synthesis was first discussed, to the end of his life, Einstein never stopped raising questions about the new physics to which he had contributed so much. He tried at first to propose conceptual experiments that would prove the logical inconsistency of quantum mechanics, but these arguments were all successfully refuted by Bohr. In 1935 Einstein began to stress another objection to quantum mechanics, arguing that its description of physical reality was essentially incomplete, that there were elements of physical reality which did not have counterparts in the theory. Bohr answered this argument, saying that Einstein's criterion of physical reality was ambiguous and that from Bohr's own complementarity standpoint the theory satisfied any reasonable standard of completeness.

Einstein never abandoned his opposition to the prevailing mode of thought despite the enormous success of quantum mechanics. He was convinced that a fundamental theory could not be statistical, "that *He* doesn't play dice."[17] Even more serious in Einstein's view was the incompleteness of the theory. He would not give up the idea that there was such a thing as "the real state of a physical system, something that objectively exists independently of observation or measurement, and which can, in principle, be described in physical terms."[18] The search for a theory that could provide such a description of reality was Einstein's program. He never lost his hope that a field theory of the right kind might eventually reach this goal.

That Einstein, without whom twentieth-century physics would be unthinkable, should have chosen to follow a separate path was a source of great regret to his colleagues. In Max Born's words: "Many of us regard this as a tragedy—for him, as he gropes his way in loneliness, and for us who miss our leader and standard-bearer."[19] But to Einstein himself his choice was inevitable; it was the natural outgrowth of all his years of striving to find a unified foundation for physics. This was what he meant when he ended his scientific autobiography by writing that he had tried to show "how the efforts of a lifetime hang together and why they have led to expectations of a definite form."[20]

MARTIN J. KLEIN

EINSTEIN: Theory of Relativity.

Einstein first wrote on radiation in his statistical mechanical discussion of Wien's law in 1904. He had already long thought about the fundamental problems of radiation; at age sixteen he had puzzled deeply over the question of what light would look like to an observer moving with it. As a student, one of the extramechanical applications of mechanics that had most fascinated him was the theory of light as a wave motion in a quasi-rigid elastic ether. In 1901 he was absorbed in unpublished, independent investigations of the critical problem of the motion of matter through the light ether.

At the turn of the century, the focus of discussion of the light ether was the electron theory. The electron theory had not been taught at the Zurich Polytechnic, and Einstein had had to instruct himself in it. Early in 1903 he began an intensive study of the theory, especially H. A. Lorentz' formulation of it. Lorentz' theory was founded on the concept of an absolutely stationary light ether. The ether completely permeated matter, with the consequence that bodies moving through it were not impeded and did not drag the ether with them. The ether was a dynamical substance but clearly not a mechanical one. Its dynamical properties were described precisely by Maxwell's electromagnetic field equations. The sole connection between the ether and matter occurred through the electrons that Lorentz assumed were contained in all ponderable molecules. The two kinds of physical entities in Lorentz' theory were the continuous ether and the discrete electrons; Maxwell's partial differential equations for the continuous field described the

state of ether, and the ordinary differential equations of Newtonian mechanics described the motion of the electrons.

The two entities together with their respective formalisms—continuous field theory and particle mechanics—constituted the characteristic dualism of Lorentz' theory. This dualism, which pervaded late nineteenth-century physics, was most clearly defined and confronted in Lorentz' theory. From 1900 on there was increasing concern to eliminate the dualism by recognizing the mass concept and the laws of Newtonian mechanics as consequences of the more fundamental laws of electron dynamics. This reduction was the program of the electromagnetic view of nature, which was advocated by W. Wien, M. Abraham, and others.

Like the more influential of his contemporaries, Einstein regarded the separateness of the concepts of electromagnetism and particle mechanics as the outstanding fault of physical theory. He did not, however, subscribe to the electromagnetic program but originated new strategies for unifying the parts of physics; his 1905 light-quantum hypothesis and relativity theory were the fruits of such strategies. In his light-quantum study, he attacked the dualism of field and particle concepts by showing reasons to conceive radiation not as a continuous wave phenomenon, but as a finite collection of discrete, independent energy particles, or quanta. Quanta were foreign to the Maxwell-Lorentz theory, and Einstein was convinced that that theory had to be changed—one reason why he could not accept the electromagnetic program, since it posited the existing electromagnetic theory as exact.

Einstein wrote his theory of relativity in full awareness of its relation to his work on light quanta earlier that year; relativity did not depend on the exactness of Maxwell's theory, a fact that was important to Einstein. He recognized that electromagnetism, no less than mechanics, had to be reformed; and he retained certain concepts of both sciences in seeking the synthesis that removed the dualism from physical theory. He introduced the particle concept from mechanics into the theory of light in his light-quantum study, and he introduced the mechanical concept of relativity into field theory in his relativity study. In his light-quantum study he concluded that light is discontinuous; in his relativity study he rejected the concept of the ether outright as being superfluous in a consistent electromagnetic theory. He saw the stationary, continuous ether of the electron theory as the chief impediment to a unified physics, and in 1905 he put forward two distinct arguments against its admissibility.

Special Relativity. The stated purpose of Einstein's first paper on relativity in 1905, "Zur Elektrodynamik bewegter Körper"[21] ("On the Electrodynamics of Moving Bodies"), was to produce a "simple and consistent theory of the electrodynamics of moving bodies based on Maxwell's theory for stationary bodies." Until the publication of his paper, the current theory had been neither simple nor consistent. As an example of what struck Einstein as undesirable complications, there was the sharp theoretical distinction made between the two ways in which the interaction between a magnet and a conductor was supposed to produce a current in the latter. In one case the magnet was assumed to be at rest, with the conductor in motion; in the other case the conductor was assumed to be at rest, with the magnet in motion. Although the resulting current was the same in both cases, the respective explanations differed from one another and invoked different concepts. Given Einstein's strong conviction that the logical simplicity of a scientific theory was an important token of its validity, the foregoing example (a commonplace experiment in elementary physics) suggested the desirability of finding a point of view from which the phenomena could be accounted for more simply. It suggested to him the possibility that, as was already the case in mechanics, a theory of the electrodynamics of moving bodies should specify only relative motions, there being no phenomenological basis for defining absolute motions.

The validity of this point of view was further confirmed for Einstein by the failure of various "ether-drift" experiments designed to detect the "absolute motion" of the earth, through variations in the velocity of light or other optical or electromagnetic effects of such motion. These experiments were undertaken in the expectation that the laws of electrodynamics and optics for a stationary reference system must take different forms in a moving system.

A variety of attempts were made by contemporary physicists to remove the conflicts with accepted theory produced by such experiments. One of the most famous of these expedients was the hypothesis of the Fitzgerald contraction proposed by G. F. Fitzgerald—and, independently, by Lorentz—to account for the failure of the Michelson-Morley experiment. The interferometer employed in this particular attempt to measure the absolute speed, v, of the earth's motion through the ether, was of sufficient sensitivity to detect variations in the speed of light, c, to the second order of the magnitude v/c. The conclusively negative result was explained by Fitzgerald as having been caused by a contraction of the arm of the interferometer in the direction of its motion through the ether, by the factor $(1 - v^2/c^2)^{1/2}$, this contraction being just sufficient to offset the expected variation in the velocity of light.

This and other similar supplementary hypotheses added further complications to a theory of electrodynamics which in Einstein's opinion was already unnecessarily complex. By his own testimony the failure of the ether-drift experiments did not play a determinative role in his thinking but merely provided additional evidence in favor of his belief that inasmuch as the phenomena of electrodynamics were "relativistic," the theory would have to be reconstructed accordingly.

Another critic of "arbitrary hypotheses" such as the Fitzgerald contraction was the notable mathematician Henri Poincaré, who as early as 1895 had perceived the operation of a general law in the repeated failures of experiments designed to detect the absolute motion of the earth. Poincaré complained about there being certain explanations for the absence of first-order effects and other explanations for the absence of second-order effects,[22] and he was prepared to postulate that no physical experiment, regardless of its degree of accuracy, could detect the earth's absolute motion. He called this postulate the principle of relativity,[23] perhaps borrowing from Maxwell, who had referred to the "doctrine of the relativity of all physical phenomena" in his *Matter and Motion*.

Poincaré anticipated Einstein in asserting that all laws of nature, optical and electrodynamical as well as mechanical, should be brought within the scope of the principle of relativity. But their points of view and their programs were not identical. Poincaré remained in many respects committed to the traditional theory of electrodynamics. He adhered to the concept of the ether; and, while he appealed to the principle of relativity in deducing important results in electrodynamics, he appears to have imagined that the principle of relativity itself might be accounted for by an appropriate modification of the ether theory.

Einstein's approach was both more radical and more consistent. In his "Autobiographical Notes," Einstein recalled that he had long attempted to correct the dualistic fault of Lorentz' electron theory by direct, constructive approaches. But by the middle of 1905 he had come to see that to succeed he must proceed indirectly, by means of some universal principle. The model he had before him was thermodynamics, the science that had already guided his thought in statistical mechanics and radiation theory. He characterized thermodynamics as a theory of principle, one based on statements such as that of the impossibility of perpetual motion. He contrasted thermodynamics with the more common constructive theory built up from hypothetical statements, notably the theory of the continuous ether and the kinetic-molecular theory of gases.

In 1905 Einstein refounded the Maxwell-Lorentz theory on a new kinematics based on two universal postulates. The first postulate, or the "principle of relativity," pointed directly to Einstein's goal of unifying mechanics and electromagnetism; the postulate stipulated that the "same laws of electrodynamics and optics will be valid for all frames of reference for which the equations of mechanics hold good." The second postulate stipulated that light always moves with the same velocity in free space, regardless of the motion of the source. The second postulate was later described by W. Pauli as "the true essence of the old ether point of view."[24]

Both of Einstein's postulates, taken separately, had considerable experimental support, but in the ordinary view they appeared to be irreconcilable. For if the second postulate held true in one inertial system K (as it was assumed to hold true in the ether), then by virtue of the first postulate it would have to hold true in all reference systems in uniform translatory motion relative to K. But if the velocity of light were measured as c in terms of the space and time coordinates x, y, z, t of the reference system K, then it could not in current theory take the same value c in terms of the coordinates x', y', z', t' of another reference system K' in uniform motion v relative to K. For it was taken for granted that the coordinates of the two systems K and K' had to be related by the transformation equations:

$$
\begin{aligned}
x' &= x - vt, \\
y' &= y, \\
z' &= z, \\
t' &= t.
\end{aligned}
\tag{1}
$$

The correctness of the transformation (1) had been held virtually above suspicion throughout the history of modern science. It was with respect to this transformation that the laws of classical mechanics remained invariant for all inertial systems. The equations of (1) were also obviously true in the common-sense view of space and time.

It was Einstein's fundamental insight into the problems of electrodynamics to perceive that the transformation (1) could not be assumed to be true a priori. Their form was an assumption made plausible by the invariance properties of Newtonian dynamics, but it was by no means necessary. Indeed, the transformation equations relating the space and time measurements between two coordinate systems are part of a physical theory and have to be consistent with experience. Accordingly Einstein was led to undertake a profound analysis of the appropriate procedures by which space and time coordinates are established within a reference system. His object was to provide a physically meaningful and justifiable basis for the derivation of an alternative set of transformation equations consistent with the joint validity of his two postulates.

One step in that direction had been taken ten years earlier by Lorentz. In a treatise on "electrical and optical phenomena in moving bodies," Lorentz introduced a new concept which he called *Ortszeit* ("local time").[25] He used this primarily as a mathematical shortcut to simplify the form of Maxwell's equations in a system K' assumed to be in uniform motion relative to the unique stationary system K in which these equations held true exactly. Local time involved a departure from the transformation equation $t' = t$. Although Lorentz appears to have viewed local time as a mathematical artifice, it represented in embryo a concept of time that Einstein would later justify adopting for the whole of physics.

The "Kinematical Part" of Einstein's paper "On the Electrodynamics of Moving Bodies" begins with an analysis of the meaning of time in physics. This had been a subject relatively exempt from fundamental scrutiny because of the extraordinary strength of traditional intuitive beliefs. Einstein later acknowledged that his familiarity with the writings of David Hume and Ernst Mach had fostered the kind of critical reasoning underlying this part of his work.

The formidable psychological obstacle to revising the transformation equations (1) had been the concept of absolute simultaneity "rooted in the unconscious."[26] But to be physically meaningful, the synchronization of spatially separated clocks must be defined in terms of an actual physical process. Although, as Einstein later emphasized, one is not in principle restricted to light signals as the standard process for coordinating the clock settings, he chose that particular method on the grounds that the propagation of electromagnetic waves was a process about which most was known.[27] Accordingly, he proposed that within an inertial system the clocks at any two points A and B could be synchronized by stipulating that the time interval in which light travels from A to B is the same as from B to A.

Einstein then proceeded to deduce on the basis of his two postulates the transformation equations relating the four coordinates x', y', z', t' in the system K' to the coordinates x, y, z, t in the inertial system K, with respect to which K' was in uniform translatory velocity v along the x-axis. (For simplicity the two x-axes were assumed to coincide, the other pairs of axes remaining parallel.) Fundamental to the derivation and meaning of this transformation was its dependence upon the synchronizing operation he had already defined for establishing the time coordinates in each system.

The following equations were deduced:

$$x' = \frac{x - vt}{(1 - v^2/c^2)^{1/2}}$$
$$y' = y$$
$$z' = z \qquad (2)$$
$$t' = \frac{t - vx/c^2}{(1 - v^2/c^2)^{1/2}}$$

It should be noted here that, unknown to Einstein, the transformation equations (2) had already appeared in a paper published in 1904 by Lorentz, "Electromagnetic Phenomena in a System Moving With Any Velocity Smaller Than That of Light."[28] Therefore, these equations were called the Lorentz transformation by Poincaré. (Einstein did not use this name, either in 1905 or in his 1907 review paper.)

In Einstein's theory the transformation equations (2) express the kinematical content of his two postulates. Solving all four equations (2) for x, y, z, t in terms of x', y', z', t', their symmetry becomes perspicuous, the only change between the two sets being the sign (i.e., the direction) of v. As Einstein perceived, the new transformation presented a revolutionary theory of space and time. Consider any two inertial systems K and K' moving with relative velocities $\pm v$, respectively. By application of the equations (2), it follows that a rigid body of length l as measured in K, in which it is at rest, measures $l(1 - v^2/c^2)^{1/2}$ in K'; and a rigid body of length l as measured in K', in which it is at rest, measures $l(1 - v^2/c^2)^{1/2}$ in K. A clock at rest in K runs slow by $1 - (1 - v^2/c^2)^{1/2}$ seconds per second when timed by the clocks in K'; and reciprocally an identical retardation occurs for the rate of a clock at rest in K' as measured in K. Thus, lengths and time intervals are magnitudes relative to the inertial systems in which they are measured. The reciprocity of length contraction and time dilation between any two inertial systems renders physically meaningless questions as to whether such effects are "apparent" in one system and "real" in the other, or vice versa. For those contemporaries of Einstein who were committed to the ether theory or the concept of absolute simultaneity this conclusion was difficult to accept.

Another important kinematical theorem of "On the Electrodynamics of Moving Bodies" was Einstein's revised law for the addition of velocities. For the simplest case (in which the velocities v and w are in the direction of the x-axis of the inertial system in which they are composed) this law takes the form:

$$V = \frac{v + w}{1 + vw/c^2}. \qquad (3)$$

From this equation the limiting value of the velocity of light, c, can also be deduced. The composition of no two velocities v and w, each of which is less than c, can equal c; and the composition of any velocity less than c with c equals c.

Because of the complete generality of Einstein's first postulate, it follows that the fundamental principle of the special theory of relativity can be expressed by stating that the laws of physics are invariant with respect to the Lorentz transformation. The imposition of this formal restriction on all possible laws immediately facilitated the development of a greatly simplified theory of electrodynamics for both stationary and moving bodies. That, of course, had been Einstein's original objective.

Applications of Special Relativity. In the concluding or "Electrodynamical Part" of Einstein's original paper on relativity, he presented applications of his theory to various phenomena of electrodynamics. He proved that the Maxwell-Hertz equations for the electromagnetic field, both in empty space and when convection currents are taken into account, were invariant under the Lorentz transformation. He also showed how the force acting upon a point charge in motion in an electromagnetic field could be calculated simply by a transformation of the field to a system of coordinates at rest relative to the charge. In the new theory, electric and magnetic forces did not exist independently of the motion of the system of coordinates. From this point of view, the explanation of the currents produced by the relative motion of a magnet and a conductor was not complicated by theoretical distinctions based on their absolute motions.

Einstein derived several other theorems in the optics and electrodynamics of moving bodies on the basis of the theory for stationary bodies. His method was to choose the appropriate coordinate systems in each case and then apply the transformation equations (2). In this way he deduced relativistic (i.e., Lorentz-invariant) laws for Doppler's principle, for aberration, for the energy of light, and for the pressure of radiation on perfect reflectors. In what was a more difficult problem, he derived the three relativistic laws describing the motion of an electron in an electromagnetic field.

The requirement of Lorentz invariance for the laws of physics led Einstein and fellow physicists, including Planck, to the revision of a number of the laws of classical mechanics. This revision had already been initiated by Einstein in his laws of motion for electrons. In their relativistic formulations, masses and, correspondingly, forces could no longer have absolute magnitudes independent of the coordinate systems in which they were measured. Thus the expressions for momentum and energy also took new relativistic forms. Einstein later claimed that one of the foremost achievements of the special theory was its unification of the conservation laws for momentum and energy.[29]

Another demonstration of the heuristic power of the principle of Lorentz invariance was provided by a second paper published by Einstein in 1905, "Ist die Trägheit eines Körpers von seinem Energiegehalt abhängig?" ("Does the Inertia of a Body Depend Upon Its Energy-Content?").[30] In calculating, by means of the Lorentz transformation, the loss of kinetic energy $(K_0 - K_1)$ for a body emitting radiation energy in the amount L, Einstein was able to deduce the equation

$$K_0 - K_1 = Lv^2/2c^2.$$

This expression revealed (in view of the definition of kinetic energy: $mv^2/2$) that as the result of its radiation the mass of the body had been diminished by L/c^2. Arguing that the particular form in which the body lost some of its energy did not affect the calculated diminishment of its mass, Einstein concluded that the mass of a body is a measurement of its energy content. Differences in its energy content equal differences in its mass in accordance with the equation

$$\Delta E = \Delta mc^2.$$

In accordance with his frequent practice of suggesting appropriate experimental research, Einstein proposed that this law might be tested by experiments with radium salts. He observed that the exchange of radiation between bodies should involve an exchange of mass; light quanta have mass exactly as do ordinary molecules, and thus a bridge was established between the concepts of electromagnetism and mechanics. The mechanical concept of mass lost its isolation, becoming a form of energy, as characteristic of radiation as of ordinary matter. The chief value of the mass-energy law for Einstein lay in its contribution to the problem of the dualism in physical theory. In 1907 he carried this viewpoint one step further, assuming that mass and energy are completely equivalent concepts, the rest mass of a body being a measure of its "latent" energy content in accordance with the famous equation[31]

$$E = mc^2.$$

The electromagnetic mass question and other questions central to the work of Lorentz and his contemporaries remained the focus of German electromagnetic research for several years after Einstein's 1905 relativity paper. The universal significance of relativity was not generally recognized at first; Einstein's theory was regarded as merely another statement of Lorentz' electron theory, not as an important statement in its own right.

Geometric Significance of Relativity. The universal implications of Einstein's theory were first clearly revealed by the Göttingen mathematician Hermann Minkowski in 1907 and 1908. Minkowski argued that relativity implied a complete revision of our conception of space and time, and that this revision applied throughout physics and not just to electrodynamics where it originated. His four-dimensional formulation of the theory, his application of it to mechanics, and his advocacy generally had a decisive historical importance in winning physicists to the new theory and in clarifying its revolutionary significance. By 1910 relativity was fairly well understood in its full generality, and it began to be widely accepted, especially in Germany.

Minkowski recast the special theory of relativity in a form which had a decisive influence in the geometrization of physics. (The memoir was published in 1908.) As David Hilbert expressed it in his memorial lecture, Minkowski, in the formalism which he developed, "was able to reveal the inner simplicity and the true essence of the Laws of Nature." Minkowski considered the world as described by the special theory of relativity to be a four-dimensional flat space-time in which the events are points, the histories of particles represented by curves (world lines), and the inertial frames correspond to Cartesian coordinates spanning this space-time. The history of a particle moving in the absence of an external force is the straightest possible, it is a geodesic. This space-time is flat. Given a geodesic it is always possible to find another which does not intersect the first one. (The Euclidean space of elementary geometry is also flat. There the geodesics are straight lines, and parallel straight lines do not intersect.) Going from one inertial frame to another and using the transcription of data as given by the Lorentz transformations (2) corresponds to relabeling the events by changing the coordinate system in space-time. This very strongly geometrical point of view exhibited the fundamental features of the special theory clearly, and ultimately led to Einstein's belief that all laws of nature should be geometrical propositions concerning space-time.

Gravitational Theory. Einstein understood the universal significance of his relativity theory from the start. In the years immediately following 1905, he continued to work and publish on problems in relativistic electron theory; at the same time he tried to frame a relativistic theory of gravitation, a branch of physics that belonged to mechanics, not electromagnetism. In 1905 he had already freed gravitation from its exclusively mechanical context by his law of mass-energy equivalence; radiation too has mass and should gravitate. He tried to revise the Newtonian gravitational law so that it agreed with the demands of relativistic kinematics.

Any theory of gravitation contains three major parts: (a) the field equations relating the gravitational field to its sources (in Newtonian theory, there is just one equation, Poisson's equation for the gravitational potential using the mass density as the source); (b) the equations of motion of material bodies in this gravitational field (the Newtonian equation of motion in the Newtonian theory); and (c) the equations of motion of the electromagnetic field in the presence of gravitational fields (in the classical theory these are Maxwell's equations *uninfluenced* by the gravitational field). (See Table I for an outline of the development of the theory in these terms.)

Einstein, in his first attempt, retained the scalar potential of classical gravitation theory, generalizing the potential equation by adding a second time-derivative term. The field equation of gravitation then transforms correctly, and gravitation becomes, like electromagnetism, a finitely propagated action. He did not get far with this approach and did not publish it. The difficulty was that, according to the mass-energy law, the inertial mass of a body varies with its internal and kinetic energies, so that the acceleration of free fall might depend on these energies; this would contradict the notion, suggested by experience and adopted by Einstein as a premise, that all bodies have the same gravitational acceleration regardless of their velocities and internal states. This persuaded him that his 1905 principle of relativity was an inadequate basis for a gravitational theory and, hence, for a unified, nondualistic physics.

There was another way to approach the gravitational problem, one based on the recognition that the

Date	Equations of Motion in the Presence of Gravitation	Field Equations for Gravitation
1907	Equivalence principle using uniformly accelerated frames; Maxwell's equations in the presence of gravitation	
1911	Equivalence principle rediscussed	Scalar field theory
1912		Nonlinearity of field equations Gravitational induction (leading to Mach's principle of 1918)
1913	Equivalence of all frames in which conservation laws hold History of particle as a geodesic in space-time	Introduction of metric in space-time as seat of gravitation and tensorial formulation of laws
1914	Maxwell's equations correctly given in presence of gravitation	
1915	Final formulation of general theory (announced 25 March)	
1916	Equations of motions of particles and of the electromagnetic field	Field equations of gravitation expressed through curvature of space-time

TABLE I. Evolution of the General Theory

EINSTEIN EINSTEIN

free fall of a body is independent of its energy if its
gravitational mass varies with energy in the same way
as its inertial mass. Although there was no theoretical
reason why the two kinds of mass should behave in
the same way, Einstein did not doubt that they did
so. He made the strict equivalence of inertial and
gravitational mass the key to a proper understanding
of gravitation, and he developed this understanding
in his first published statement on gravitational theory. In the same survey article on relativity, which
contained the energy-mass equation, he elevated the
equality of inertial and gravitational mass, or, equiv-
alently, the equality of the acceleration of the free
fall of all bodies, to the status of an equivalence
principle.

Einstein explained that the principle of relativity
must be extended to accelerated coordinate systems;
a coordinate system accelerated relative to an inertial
frame is equivalent, in a sufficiently small spatial
region, to a frame which is not accelerated relative
to an inertial one but in which a gravitational field
is present. This comes about in the following way:
In an inertial frame let bodies be at rest, or move
with a uniform motion. From the point of view of
an observer at rest in the accelerated frame, these
bodies appear to have an acceleration; this acceleration is the same for all bodies independent of their
mass (being equal and opposite to the acceleration
of the frame accelerated relative to the inertial frame).
This acceleration naturally can be transformed away
by simply using the original inertial frame as the
frame of reference.

Now Einstein observed that a gravitational field
locally generates a physical situation which is identical to the one described by the accelerated observer.
For all bodies undergo the same acceleration in a
gravitational field independent of their mass, and the
acceleration of a body can be transformed away using
a frame of reference which falls freely with the body
whose acceleration we wish to transform away. If this
be the case, one can immediately discuss, at least in
a heuristic fashion, the influence of gravitational fields
on phenomena, by solving another problem, the description of the same phenomena in the absence of
a gravitational field but viewed from an accelerated
frame. In this way Einstein showed that gravitational
fields influence the motion of clocks. Since the frequency of an emitted spectral line can be used as
a clock, it follows that there is a frequency shift between an emitted and an observed spectral line, if
the gravitational potential at the location of the observer is different from that at the emitter. He also
showed that all electromagnetic phenomena are influenced by a gravitational field; for example, the

light rays are bent if they pass in the vicinity of
gravitating bodies. It had long been suspected that
there might be an interaction between electromagnetic and gravitational fields (see, for example,
Faraday's experiments); however, this was the first
concrete suggestion as to what this interaction should
be, how it arises, and what the order of the magnitude
of these effects are.

Program for General Relativity. The year 1907 was
a turning point in relativity. From then on Einstein
(then twenty-eight years old) was interested less in
the special theory of relativity and more in its possible
generalization. From the point of view of the special
theory, space-time was a given framework, within
which the natural phenomena took place. These were
still described as in classical physics; there were fields
of force and material bodies which acted on each
other. The next generalization consisted in eliminating the gravitational fields of force by allowing the
structure of space-time to change in such a way that
the free motion in this altered space-time should in
some way correspond to the motion under the influence of gravitational fields in the space-time of the
special theory. The final generalization, which was
never successfully attained, would have consisted in
eliminating the electromagnetic fields of force as well
by altering the geometry of space-time in a suitable
way. The first generalization, the geometrization of
gravitation, led eventually to the general theory of
relativity; the additional geometrization of the electromagnetic fields of force led to the invention of the
unified field theories.

In 1907 the details of this program were still obscure. The equivalence principle, relating the gravitational field and accelerated frames of reference, was
already there. This helped Einstein to discuss some
of the effects of a gravitational field on the electromagnetic field. The geometrization of this principle,
the mathematical characterization of the gravitational
field, its sources, and the relation between the field
and its sources, i.e., the gravitational field equations,
were still missing. In the next twelve years, Einstein
was occupied with building a complete theory of
gravitation rising out of his heuristic principle.

Between 1907 and 1911 Einstein published nothing
more on gravitation. He was preoccupied with finding
a reformed electron theory that incorporated both
electric charges and light quanta; his chief heuristic
guides in this search were thermodynamics, his fluctuation method in statistical mechanics, and his special theory of relativity. In the context of this largely
unpublished work, he came to understand that the
solution to the dualism problem was to write physics
in terms of continuous field quantities and nonlinear

partial differential equations that yield singularity-free particle solutions. The field equations were to account for particles and their interaction; they were to contain the laws of motion of particles, eliminating the dualistic need for particle mechanics in addition to field theory. He was soon to find additional support for this understanding in his general relativity theory, which required that field equations have just those mathematical properties he had decided upon on physical grounds.

For Einstein the connection of the particle and the field was the central problem of physics, and he saw the whole of the problem as contained in the connection of the electron and the electromagnetic field. A cardinal point of his unification objective in physics was to deduce the electron and its motion from the field equations. The whole difficulty was that it proved impossible to find the proper modification of Maxwell's equations that would permit such a deduction; any modification seemed arbitrary without a universal principle to determine its selection. In the years prior to 1911, he believed that the special relativity principle, together with statistical mechanical considerations, was an adequate guide for finding the new electromagnetic equations capable of describing particles. After several years of arduous effort, he recognized that he had been mistaken and shifted his expectations to the more powerful universal principle—the postulate of general relativity—as offering the possibility of avoiding arbitrariness in constructing field theories with particle solutions. The new postulate restricted much more severely the mathematical form the field equations could take.

By 1911 Einstein had become deeply pessimistic over the prospect of soon finding a new electron theory that incorporated quanta in a natural way. That year he returned to his 1907 gravitational theory, and for the next several years he looked to gravitation rather than electromagnetism as the starting point for the reform of physical theory. In 1911 the first Solvay Congress met to discuss the crisis in physics signaled by the quantum theory. Two years later, in 1913, Niels Bohr published his quantum theory of atoms and molecules. Just when the physics community began seriously to reorient itself toward quantum problems, Einstein seemed to move away from them. By 1911 he had come to regard the particle or quantum aspect of nature as secondary to the field aspect. He thought it was futile to attempt a fundamental understanding of the microscopic structure of nature until the macroscopic structure of the field was understood; when it was, quantum phenomena would be deduced from it. He never ceased to struggle with quanta, but his concern was less obvious than it had

once been. His direct contribution to the later development of the quantum theory tended to be in the nature of criticism and suggestion; all the while he struggled to vindicate this way of developing physics by seeking a theory of the total field that would finally clarify the quantum problem.

Following a 1911 paper on the influence of gravitation on the propagation of light,[32] Einstein published two remarkable memoirs in 1912[33] which were efforts to construct a complete theory of gravitation incorporating the equivalence principle. In these memoirs Einstein supposed that the gravitational field can be characterized completely by one function, the local speed of light, analogous to the Newtonian description, where only the gravitational potential appears. The equivalence principle gives no clue to how the field equation describing the gravitational field should be constructed. By an extraordinary argument he extended the potential equation of Newton, which determines in that theory the gravitational potential, and came to the conclusion (a) that the equation must be nonlinear and (b) that this nonlinearity can be interpreted to show that the source of the gravitational field is not only the energy associated with the rest mass of bodies but also depends on the energy residing in the gravitational field itself. This was the first appearance of a nonlinear field equation for gravitation. In the first of his 1912 memoirs, Einstein wrote the differential equation of the static gravitational field as

$$\Delta c = kc\rho,$$

where c is the local velocity of light, k is a universal gravitational constant, and ρ is the density of matter. Next he derived the equations of motion of a body, using the equivalence principle: the force on a material point of mass m at rest in a gravitational field is

$$-m \operatorname{grad} c.$$

In his second memoir in 1912, he used the equivalence principle to show the influence of a static gravitational field on electromagnetic and thermal processes. Also in the same year, Einstein pointed out that from the energy expression which followed from his theory as it then was, one could conclude that if a body is enclosed in an envelope of gravitating matter its mass might be expected to increase. This he considered as a suggestion that perhaps the whole mass of a body could be conceived as arising from the gravitational interaction of this body with all the other bodies in the universe. He further pointed out that a similar point of view, without any theory, had already been advocated by Mach. This suggestion, that the mass of a body is the manifestation of the presence of other bodies in the universe, Einstein later (1918) called Mach's principle; the first appearance of an idea that a theory of gravitation should discuss the problem raised by Mach was in 1912. During this same period, others were more reluctant to take these far-reaching steps and there were several attempts to invent a gravitational theory in which there are preferred frames of reference such that the velocity of light is constant and has the same value as in empty space. Such a theory is, however, in conflict with the equivalence principle. The most comprehensive attempt was G. Nordström's in 1913.

Einstein himself came to the opposite conclusion. Instead of abandoning the investigation of uniformly accelerated frames of reference, he decided that the approach was too narrow and that in the generalization of the special theory not only these special transformations should be permitted, but more general ones. He was as yet unwilling to introduce general coordinate transformations. In his 1913 review article and in his memoir with Grossman, he insisted that only those frames of reference are admissible in which the conservation laws hold true. In fact these conservation laws were used by him to invent the field equations for gravitation. The use of more general frames of reference brought two important aspects into the theory. One was the use of more general mathematical tools which practically forced Einstein toward the final answer; the other was the observation that if more general transformations are permitted the gravitational field must be characterized not by one function but by ten functions. Moreover these ten functions have a simple geometric significance; they characterize the metric properties of space-time at every point.

This step was immense in its implications: (*a*) it forced the abandonment of the Newtonian notion that the gravitational field could be characterized by one scalar function, the gravitational potential; (*b*) it forced on Einstein the notion that gravitation is explicitly related to the geometrical structure of space-time. In his 1913 paper with Grossman,[34] Einstein adopted as the fundamental invariant of this theory the generalization of the four-dimensional line element *ds* originally introduced by Minkowski for a flat space-time. If space-time is not flat, *ds* must be expressed in terms of a general coordinate frame. Then two events, labeled

$$x^1, x^2, x^3, x^4$$

and

$$x^1 + dx^1, \quad x^2 + dx^2, \quad x^3 + dx^3, \quad x^4 + dx^4,$$

have a separation *ds* whose square is given by

$$ds^2 = \sum_{\mu=1}^{4} \sum_{\nu=1}^{4} g_{\mu\nu} \, dx^\mu \, dx^\nu.$$

The sixteen functions $g_{\mu\nu}$ form a symmetric tensor field (thus ten $g_{\mu\nu}$ are independent). Einstein considered these functions as the basic objects of his theory describing the manifestations of gravitation. In particular he assumed that as in the special theory the history of a body will be a geodesic; thus the history will be that curve in space-time for which $\int ds$ is a minimum, $\delta \int ds = 0$. Since the $g_{\mu\nu}$ appear in *ds* this principle will now determine the motion of a body influenced by gravity. Einstein regarded the $g_{\mu\nu}$ as the gravitational potentials, which replaced the single scalar *c* of his 1912 theory. The principal problem of his gravitational theory was, then, to determine the $g_{\mu\nu}$. For this purpose, Einstein sought an extension of Poisson's equation for the gravitational potential ϕ,

$$\Delta\phi = 4\pi\rho,$$

which he wrote

$$\Gamma_{\mu\nu} = X\Theta_{\mu\nu}.$$

$\Gamma_{\mu\nu}$ is constructed from derivatives of the $g_{\mu\nu}$ and is the analogue of $\Delta\phi$; $\Theta_{\mu\nu}$ contains the material sources of the field and is the analogue of ρ; X is a gravitational constant. In any given physical situation, the $\Theta_{\mu\nu}$ may be assumed known. The problem was to determine the $\Gamma_{\mu\nu}$; Einstein believed that the principle of the conservation of energy and momentum is sufficient for this purpose.

Enunciation of General Relativity. Thus in 1913 the situation was as follows: The equation of motion of a particle in a gravitational field had been given; the equations of motion of the electromagnetic field in the presence of a gravitational field were incomplete; the field equations of gravity were incomplete; Newtonian concepts had been retained in order to save the conservation laws together with preferred frames of reference, to wit, precisely those where the conservation laws still held true as laws. This last stage was soon passed. In 1914 the field equations for the electromagnetic field (Maxwell's equations) were given correctly in the presence of a gravitational field. On 25 March 1915 Einstein announced in the Prussian Academy of Sciences that he also had the field equations of gravitation in hand. The results were published in several articles later in the same year,[35] and on 20 November, David Hilbert, in Göttingen, independently found the same field equations. A greatly expanded and detailed memoir appeared in 1916 which, so to speak, became the Authorized Version.[36] Einstein's new understanding in 1915 and 1916 was that the gravitational field can be characterized by Riemann's curvature tensor $G_{\mu\nu}$, a tensor obtained from the $g_{\mu\nu}$ by differentiation. He wrote the gravitational field equations as

$$G_{\mu\nu} = -K(T_{\mu\nu} - \tfrac{1}{2} g_{\mu\nu} \, T),$$

where T is the scalar of the material energy tensor $T_{\mu\nu}$ and K is a gravitational constant. This was his new analogue of Poisson's equation. The $g_{\mu\nu}$ that are determined by the field equations determine, through the separate equations of motion, the history of a body in the gravitational field.

The final theory now presented was of immense sweep and great conceptual simplicity. All frames of reference are equally good; the classical conservation laws fade away—they are no longer laws but mere identities and lose the significance they had before. There are no gravitational forces present in the sense that the theory contains electromagnetic ones, or elastic ones. Gravitation appears in a different way. The fixed, given space-time of the special theory has gone; what before had erroneously been labeled as the influence of a body by gravitation on the motion of another is now given as the influence of one body on the geometry of space-time in which the free motion of the other body occurs. This free motion in the altered space-time is what was mistaken as the forced motion (forced by a gravitational field) in an unaltered space-time. The laws of nature are now geometrical propositions concerning space-time. Space-time is a metric space, which means that we have a rule how to compute the separation of any two points in the space. This can be done if we know ten numbers at any point and thus have ten functions in space-time. Once these ten functions are given, everything that can be known can be computed. As stated before, three questions should be answered by the theory:

(1) *What corresponds to the field equations of gravitation?* The presence of matter alters the metric properties of space-time; in particular, the curvature of space-time at a point is determined by the amount of matter and electromagnetic field and their motion at that point. This alteration of the metric properties has an effect on the history of the motion of a body and on the history of the development of the electromagnetic field. These are embodied in the equations of motion.

(2) *What are the equations of motion of a body?* The equation of motion of a body is given by the statement that the history of a body always be a geodesic. But, of course, a geodesic in a curved space is quite different from a flat one; altering the curvature alters the motion.

(3) *What are the equations of motion of the electromagnetic field?* The equations of motion of the elec-

tromagnetic field are not geometrical propositions. Here we take over Maxwell's equations from the special theory where it was specified for a flat space-time, and simply transcribe them for a curved one. (The fact that these equations do not refer to anything geometrical sent Einstein in search of a more general theory, a unified field theory, where all the laws would have a geometrical significance.)

The equivalence principle is contained in the theory in the following fashion. According to this principle, one can make the effects of a gravitational field, on the motion of a particle or on an electromagnetic field, disappear locally by a transformation to a frame of reference which has the same acceleration as the gravitational field at that point. In this frame one can describe locally all events as in the special theory. Since the effects of the gravitational field are represented in the general theory by the fact that space-time is curved, this must mean that the equations of motion, on the one hand, and space-time, on the other, must be of such nature that the effects of this curvature can be transformed away locally. This arises because (*a*) the equations of motion of a particle, and of the electromagnetic field, are such that the curvature does not appear in them explicitly, and (*b*) space-time can be approximated locally by its flat tangential space, in the same way that a curved surface can be approximated by a flat tangential plane in the immediate vicinity of the point of contact.

Experimental Predictions of General Relativity. What were the experimental predictions? The equivalence principle had already predicted that the gravitational field must influence the electromagnetic field, and this result was also obtained in the developed theory. In addition Einstein in 1915 had already pointed out that according to Newtonian theory a solitary planet moving around the sun describes an elliptic orbit. Because of the presence of the other planets, this motion is perturbed and the axis of the ellipse slowly rotates relative to the fixed stars; i.e., it precesses. If these effects are computed on a Newtonian basis for the planet Mercury, it will be found that the experimentally observed precession is larger than the computed one. Einstein showed that if the motion of the single planet around the sun is calculated from the general theory, there is already a small precession. If this value is added to the Newtonian value of the precession caused by the other planets, the resulting total precisely fits the experimental results. This was the first unexpected success of the theory. The two other predictions concerned the effect of the gravity on light. In testing the first, the sun is treated as the body causing the bending. In order that the effect on the light coming from other stars might be observed, an eclipse was requisite. Although Einstein suggested such an experiment before the 1914 eclipse, the first observations were made in 1919. The values observed scatter around the theoretically predicted value with probable errors large enough to include it. The second prediction concerned the shift of spectral lines in the presence of a gravitating body and, for some time, the astrophysical observations were inaccurate. Not until 1960 was the prediction accurately verified in the laboratory.

Subsequent Investigations. For the rest of his life Einstein's investigations on relativity centered on the following points: (1) mathematical investigations into the structure of the theory; (2) approximate solutions of the general theory and their physical implications; (3) the application of the theory to the universe as a whole, to cosmology, and to Mach's principle; (4) general discussions of relativity and popular expositions; and (5) efforts to incorporate the electromagnetic field into the geometry of space-time. We shall take up the discussion of each point in chronological order.

Mathematical Investigations. As far as the mathematical structure of general relativity was concerned there were three classes of problems which concerned Einstein particularly: the role and meaning of conservation laws; the relation of the equations of motion of bodies to the field equations of gravitation; and the role and nature of singularities in the theory.

During the development of the general theory, Einstein had intended to hold fast to the conservation of energy and momentum in the usual (special relativistic) sense as far as possible. At the same time he was driven by other considerations toward the idea that the laws should be generally covariant, i.e., that the laws should have the same form for all observers in space-time, irrespective of their states of motion. These two desires, the maintenance of these conservation laws and of general covariance, proved mutually incompatible. The final theory is generally covariant; it has conservation identities and not conservation laws in the usual sense, although certain covariant laws do exist for special cases in general relativity.

The problem of the equation of motion of bodies is the following. The 1916 theory had a classical structure in the sense that there were both field equations (the curvature of space-time is determined by the mass and motion of bodies in space-time) and equations of motion of bodies (the world line of small mass is a geodesic). Are these two statements really separate? If the field equations were linear, they indeed would be. They are not linear, however, and Einstein showed that if matter is represented by a

point singularity of the metric field, these singularities are located on world lines that are geodesics of space-time, provided its metric satisfies the equation of general relativity.[37]

The role and nature of singularities in the solutions greatly troubled Einstein both for mathematical and for physical reasons, and the question influenced his thinking on Mach's principle and on the necessity of unified field theories (the latter are discussed below).

Approximate Solutions. Experimentally, we know that there are observers such that the effects of gravitation are quite small in extended portions of space-time. This enables one to solve the equations of the general theory approximately. These approximate solutions show that, contrary to the Newtonian theory, there are gravitational disturbances that travel as waves. Recent experimental observations by Joseph Weber suggest that these waves may occur in nature. If these waves are compared with electromagnetic waves in empty space, significant differences are noted. Their polarization properties are different. They are associated with more complicated modes of motion of the source. They are thus less efficient in carrying away the kinetic energy of the motion of the source. Both types of waves propagate with the same velocity, however—that of light.

Application of the Theory. If we consider the universe as a whole, we cannot consider gravitation as a small effect, a small deformation of an otherwise flat space-time. In particular the following predicament arises. If inertia and gravitation are inseparable, how is it possible to have a situation in which the effects of gravitation are small? After all, that would mean that in the absence of any such small effect, we still would have a flat space-time with an inertial motion possible in it, while gravitation would be entirely absent. Previously (1912),[38] Einstein had considered the possibility that there should be a gravitational induction effect, according to which the presence of other masses alters the value of the mass of a given body. From this consideration he erected the hypothesis that, conceivably, the whole mass of a body is generated by the presence of other masses.[39] If this be true within the general theory, two things should follow:

(*a*) There should be no solution of the field equations applicable to the whole universe which can describe an empty space-time (since then in this empty space-time geodesics would exist, which could be taken as giving the history of the inertial motion of a particle).

(*b*) The value of the mass of a body should be determined by the presence and amount of other masses in the universe. (It is now also believed that

Mach's principle should contain an explanation of why the gravitational interaction is always attractive.)

It was the notion that the existence of inertia and gravitation must be explained along these lines that Einstein called Mach's principle (1918). In 1917 he grappled with the first half of this problem.[40] If matter in the universe is generally distributed uniformly, is it possible to find a time-independent solution of the field equations that describe a spatially finite (closed) space-time, and will this space-time vanish if the total mass of the universe is spatially infinite? He found that no such solution existed unless he modified the field equations, adding to them the so-called cosmological term. This term has no observable effect on any of the local solutions used in the experimental tests but alters the solution as a whole. With this modification, however, a solution does exist. Thereby, the first aspect of Mach's principle would be satisfied. But Einstein was dissatisfied with this answer because an arbitrary modification had had to be introduced into the field equations.

In 1922 A. Friedmann found that even without the cosmological term there are still solutions of the field equations where matter has a finite density everywhere in space, provided this density is not time-independent. In 1929 Hubble announced his discovery that the red shift of spectral lines coming from distant sources increases uniformly with the distance. This phenomenon can be interpreted as evidence for a uniform expansion of the universe as can be described by the Friedmann solutions.

General Discussions and Expositions. During these years Einstein was also concerned to clarify misconceptions about the theory of relativity and to present his views on natural sciences on a less abstract level. Among his efforts in this direction, one particularly beautiful lecture must be mentioned. In 1921, at the Prussian Academy's commemorative session honoring Frederick the Great, Einstein delivered a lecture on geometry and experience in which he summed up his views on the geometrization of physics and relativity and the relation of mathematics to the external world.[41] Here he gave his famous answer to the puzzling question of why mathematics should be so well adapted to describing the external world: "Insofar as the Laws of Mathematics refer to the external world, they are not certain; and insofar as they are certain, they do not refer to reality."

The Electromagnetic Field and the Geometry of Space-Time. There were two main reasons for Einstein's dissatisfaction with the general theory. One was the seemingly still-inadequate geometrization of physics. He felt that not only gravitational but also electromagnetic effects should be manifestations of the geometry of space-time. The other interactions, such as nuclear forces and the forces responsible for beta decay, were not yet known. Einstein never considered the geometrization of the other interactions, although in the 1940's E. Schrödinger made an attempt to invent a unified field theory incorporating gravitational, electromagnetic, and nuclear interactions.

The other problem was the relation of matter to the singularities of the gravitational field. Einstein felt that a complete and correct field theory should be without singularities while in the general theory the field equations are in general singular.[42] This, he believed, is due to the inadequate description of matter as handled in the general theory. Thus the stage was set for a search for a more extended theory. This new theory should have two basic features. It should enlarge the geometry of space-time in such a manner that new geometrical objects could be introduced which can be associated with the electromagnetic field; the physically relevant solutions (whatever that may mean) should be nonsingular. Although the initial steps in this direction were not taken by Einstein himself, he did become more and more preoccupied with this problem and in his later years the construction of such a theory was his main concern. This was Einstein's ultimate response to the mechanical-electromagnetic crisis in physical theory he had first talked about in the opening of his 1905 light-quantum paper. (In 1953 Einstein said to the author that although it is doubtful that a unified field theory of the type he was seeking could exist, even its nonexistence would be of sufficient interest to be worth establishing it. If he did not do it, Einstein said, perhaps nobody ever would.)

How might the geometry of space-time be enriched with new geometrical objects which then could be considered as candidates for the description of electromagnetic phenomena? Since no clear guiding principle stemming from physics existed (or exists even today), we must rely on geometrical intuition, which necessity would also serve as a motivation.

Practically all the work in this direction fell into one of two categories. Either the dimensionality of space-time would be preserved and the geometry altered in a formal fashion, or the dimensionality of space-time would be enlarged in a formal fashion and the metric geometry preserved. Hermann Weyl initiated the first line of thought in 1918; Kaluza the second in 1921.

Weyl's unified field theory considered space-time to be endowed with a more general geometry. This approach enabled him to introduce four extra functions in space-time, in terms of which the electromagnetic field can be expressed. Einstein immediately noticed (1918), however, that if the same physical interpretation for the geometry be maintained as in the general theory, Weyl's theory leads in its original form to results that contradict experience. In 1921 A. S. Eddington observed that Weyl's geometry of space-time is a special case of a much more general class of geometries, usually called affine geometries, which depend on a profound generalization of the notion of parallelism. Einstein's first investigation of these ideas (1923) introduced the notion of distant parallelism. In 1930, however, he found that the new theory admitted solutions that describe gravitating masses represented by singularities at rest relative to each other under the sole influence of their gravitational interaction. Experience clearly contradicts this consequence. Einstein originally rejected these solutions on the grounds that they contain singularities, and later he rejected the theory itself.

In 1931 Einstein and Walter Mayer reformulated Kaluza's five-dimensional theory retaining a four-dimensional space-time. In 1938 and 1941 Einstein again discussed theories of this type, before returning to the notion that space-time may be endowed with an affine geometry. Several different geometries were envisaged in papers written in collaboration with V. Bargmann (1944), E. G. Strauss (1945–1946), and Bruria Kaufmann (1955). The last was his final published memoir.

Summary. If we turn to summarize Einstein's achievements in relativity what we see is a new point of view and innumerable consequences. The basic new point of view was the explicit recognition that the invariance properties of the laws of physics are of fundamental importance and that these invariance properties stem from immediate physical facts and are required by them. The physical notion of invariance arises from the experimental fact that the descriptions change in a specific way if the arrangement of measuring devices is altered in a specific manner. This alteration may be a simple spatial rotation, or a simple transfer of the origin of the coordinate system, or something more complicated that endows the whole laboratory with a uniform motion, or with a motion with a uniform acceleration. That the results so obtained can be linked together implies that these observations, and hence the physical quantities they describe, transform in a given way as the measuring devices are altered.

The special theory concerned itself with that transformation of labeling of space-time points which corresponds physically to uniform translation of the whole laboratory. Einstein's great achievement was the explicit realization that this transformation of

labels cannot be specified without a specific assumption about the operational meaning of simultaneity with respect to events that are spatially separate. From this realization Einstein was led to the only consistent definition of simultaneity, and thus to the correct understanding of the Lorentz transformations and to an appreciation of their great generality.

If invariance properties are of such importance then a strong geometrical interpretation of the laws of nature becomes highly desirable, since a thing that has no definite invariance properties cannot be even thought of geometrically. This consideration led Einstein to accept Minkowski's point of view that space and time should be considered as forming one geometrical object, space-time, a four-dimensional flat space.

The next step was to analyze the relations between the descriptions of phenomena in frames accelerated relative to each other. From these relations Einstein drew the conclusion that this four-dimensional space-time cannot be flat, and that gravitation is the name given to those phenomena that appear because space-time is not flat. The curvature of space-time is due to its energy and mass content. The remarkable success of this theory derived from its automatic explanation of two features of gravitation: Why is the inertial mass equal to the gravitational mass? And why is gravitation a universal property acting on everything in the universe? The answer is that the two masses are equal because they are one and the same, since they appear in the theory uniquely as the cause of the curvature of space-time. Gravitation is a universal manifestation because it is the property of space-time, and hence everything that is in space-time (which is, literally, every thing) must experience it.

The last efforts of Einstein on unified field theories were a logical continuation of his previous efforts. The chain of argument may be said to run as follows. If invariance properties are of utmost importance, physics should be thought of as a geometry, because thought is then occupied only with objects that have invariant properties. This led to a theory according to which the structure of space-time is the seat of gravitation interaction. Is this structure perhaps so rich that not only gravitational interaction but other interactions are also determined by it? That investigation proved to be unsatisfactory, perhaps because only gravitation is a universal interaction. Nevertheless even this effort turned out to be prophetic; in modern physics it is more and more the practice to proceed with a formal guessing at the laws of nature, the guess being based on formal simplicity and on invariance. The interpretation of the theory often emerges only after the structure of the equations guessed at are better understood. In this, Einstein was a forerunner.

When Einstein's total work in physics is considered, it can be said that his achievements are rivaled only by those of Isaac Newton. Both scientists were guided in their work by unique insights into the nature of physical reality and both represent the utmost fulfillment of the creative imagination in science.

NANDOR L. BALAZS

NOTES

1. A. Einstein, "H. A. Lorentz, His Creative Genius and His Personality," in G. L. de Haas-Lorentz, ed., *H. A. Lorentz. Impressions of His Life and Work* (Amsterdam, 1957), p. 5.
2. A. Einstein, *The World As I See It* (New York, 1934), p. 250.
3. Ernst Straus, "Assistant bei Albert Einstein," in C. Seelig, ed., *Helle Zeit–Dunkle Zeit* (Zurich, 1956), p. 71.
4. A. Einstein, "Autobiographical Notes," in P. A. Schilpp, ed., *Albert Einstein: Philosopher-Scientist* (Evanston, Ill., 1949, 1951), p. 5.
5. *Ibid.,* p. 16.
6. A. Einstein to Johannes Stark (7 Dec. 1907); quoted in A. Hermann, "Albert Einstein und Johannes Stark," in *Sudhoffs Archiv,* **50** (1966), 272.
   Einstein's papers can be readily located by consulting the years of publication in the bibliographies listed below.
7. Walther Nernst, "Über neuere Probleme der Wärmetheorie," in *Sitzungsberichte der Preussischen Akademie der Wissenschaften zu Berlin* (1911), p. 86.
8. A. Einstein, "Über die Entwicklung unserer Anschauungen über das Wesen und die Konstitution der Strahlung," in *Physikalische Zeitschrift,* **10** (1909), 817.
9. *Ibid.,* p. 824.
10. A. Einstein to Michele Besso (13 May 1911).
11. A. Einstein, "Beiträge zur Quantentheorie," in *Verhandlungen der Deutschen physikalischen Gesellschaft,* **16** (1914), 823.
12. A. Einstein to M. Besso (11 Aug. 1916).
13. Erwin Schrödinger, "Über das Verhältnis der Heisenberg-Born-Jordanschen Quantenmechanik zu der meinen," in *Annalen der Physik,* **79** (1926), 735.
14. A. Einstein to Max Born (4 Dec. 1926).
15. A. Einstein to E. Schrödinger (26 Apr. 1926).
16. Niels Bohr, *Atomic Physics and Human Knowledge* (New York, 1958), pp. 66, 71.
17. A. Einstein to M. Born (4 Dec. 1926).
18. A. Einstein, "Einleitende Bemerkungen über Grundbegriffe," in *Louis de Broglie, Physicien et penseur* (Paris, 1953), p. 6.
19. M. Born, "Einstein's Statistical Theories," in Schilpp, p. 163.
20. A. Einstein, "Autobiographical Notes," in Schilpp, p. 94.
21. A. Einstein, *Annalen der Physik,* **17** (1905), 891–921.
22. H. Poincaré, *Rapports présentés au Congrès International de Physique de 1900* (Paris, 1900), I, 22–23.
23. H. Poincaré, "L'état actuel et l'avenir de la physique mathématique," in *Bulletin des sciences mathématiques,* **28** (1904), 306; repr. in English as "The Principles of Mathematical Physics," in *Monist,* **15** (1905), 5.
24. W. Pauli, *Theory of Relativity,* G. Field, trans. (New York, 1958), p. 5.
25. H. A. Lorentz, *Versuch einer Theorie der electrischen und optischen Erscheinungen in bewegten Körpen* (Leiden, 1895), pp. 49–50.
26. A. Einstein, "Autobiographical Notes," in Schilpp, p. 52.
27. A. Einstein, *The Meaning of Relativity* (Princeton, N. J., 1922), p. 28.
28. H. A. Lorentz, in *Proceedings of the Royal Academy of Sciences of Amsterdam,* **6** (1904), 809.
29. A. Einstein, "Autobiographical Notes," in Schilpp, p. 60.
30. A. Einstein, in *Annalen der Physik,* **18** (1905), 639–641.
31. A. Einstein, "Über das Relativitätsprinzip und die aus demselben gezogenen Folgerungen," in *Jahrbuch der Radioaktivität,* **4** (1907), 442.
32. A. Einstein, "Einfluss der Schwerkraft auf die Ausbreitung des Lichtes," in *Annalen der Physik,* 4th ser., **35** (1911), 898.
33. A. Einstein, "Lichtgeschwindigkeit und Statik des Gravitationsfeldes," *ibid.,* **38** (1912), 355; "Theorie des statischen Gravitationsfeldes," *ibid.,* p. 443.
34. A. Einstein, "Entwurf einer verallgemeinerten Relativitätstheorie und eine Theorie der Gravitation," in *Zeitschrift für Mathematik und Physik,* **62** (1913), 225, pt. 2 written with M. Grossman.
35. A. Einstein, "Zur allgemeinen Relativitätstheorie," in *Sitzungsberichte der Preussischen Akademie der Wissenschaften zu Berlin,* pt. 2 (1915), pp. 778, 799; "Erklärung der Perihelbewegung des Merkur aus der allgemeinen Relativitätstheorie," *ibid.,* p. 831; "Feldgleichungen der Gravitation," *ibid.,* p. 844.
36. A. Einstein, "Grundlagen der allgemeinen Relativitätstheorie," in *Annalen der Physik,* 4th ser., **49** (1916), 769.
37. A. Einstein, "Allgemeine Relativitätstheorie und Bewegungsgesetz," in *Sitzungsberichte der Preussischen Akademie der Wissenschaften zu Berlin,* Phys.-math. Kl. (1927), pp. 2, 235, pt. 1 written with J. Grommer; "Gravitational Equations and the Problems of Motion," in *Annals of Mathematics,* 2nd ser., **39** (1938), 65, written with L. Infeld and B. Hoffmann; "Gravitational Equations and the Problems of Motion. II," *ibid.,* **41** (1940), 455, written with L. Infeld.
38. A. Einstein, "Gibt es eine Gravitationswirkung die der elektrodynamischen Induktionswirkung analog ist?," in *Vierteljahrschrift für gerichtliche Medizin und öffentliches Sanitätswesen,* 3rd ser., **44** (1912), 37.
39. A. Einstein, "Prinzipielles zur allgemeinen Relativitätstheorie," in *Annalen der Physik,* 4th ser., **55** (1918), 241.
40. A. Einstein, "Kosmologische Betrachtungen zur allgemeinen Relativitätstheorie," in *Sitzungsberichte der Preussischen Akademie der Wissenschaften zu Berlin,* pt. 1 (1917), p. 142.
41. A. Einstein, *Geometrie und Erfahrung* (Berlin, 1921).
42. A. Einstein, "Demonstration of the Non-existence of Gravitational Fields With a Non-vanishing Total Mass Free of Singularities," in *Revista. Instituto de física, Universidad nacional de Tucumán,* **2A** (1941), 11.

BIBLIOGRAPHY

I. There are three principal bibliographies of Einstein's writings. The first, compiled by Margaret C. Shields, covers his writings to May 1951 and includes general works as well as scientific articles and books. It is to be found in P. A. Schilpp, ed., *Albert Einstein: Philosopher-Scientist* (Evanston, Illinois, 1949, 1951), pp. 689–760. This important book also contains Einstein's "Autobiographical Notes," a series of essays on his work by physicists, mathematicians, and philosophers, and Einstein's "Remarks" concerning these essays. The second bibliography, containing only the scientific writings, is E. Weil, *Albert Einstein. A Bibliography of His Scientific Papers 1901–1954* (London, 1960). The third is Nell Boni, Monique Ross, and Dan H. Laurence, *A Bibliographical Checklist and Index to the Published Writings of Albert Einstein* (New York, 1960).

No regular edition of Einstein's scientific papers has

appeared yet. There is a microfilm edition (Readex Micro-print, New York).

II. Einstein's books include *Relativity, the Special and the General Theory: A Popular Exposition* (London, 1920); *The Meaning of Relativity* (Princeton, N. J., 1921); *Investigations on the Theory of the Brownian Movement*, R. Fürth, ed. (London, 1926); *The World As I See It* (New York, 1934); *The Evolution of Physics* (New York, 1938), written with Leopold Infeld; *Out of My Later Years* (New York, 1950); and *Ideas and Opinions* (New York, 1954).

III. Portions of Einstein's correspondence have been published: *Albert Einstein, Hedwig und Max Born. Briefwechsel 1916–1955*, M. Born, ed. (Munich, 1969); *Albert Einstein/Arnold Sommerfeld Briefwechsel*, A. Hermann, ed. (Basel, 1968); Albert Einstein, Erwin Schrödinger, Max Planck, H. A. Lorentz, *Letters on Wave Mechanics*, K. Przibram, ed., M. J. Klein, trans. (New York, 1967); Albert Einstein, *Lettres à Maurice Solovine* (Paris, 1956).

Also see O. Nathan and H. Norden, eds., *Einstein on Peace* (New York, 1960), and M. J. Klein, *Paul Ehrenfest. Volume 1. The Making of a Theoretical Physicist* (Amsterdam, 1970). Both books quote extensively from Einstein's correspondence.

IV. Many biographies of Einstein have appeared, but nothing like a definitive study of either the man or his work yet exists. Philipp Frank, *Einstein. His Life and Times*, G. Rosen, trans. (New York, 1947), written by a physicist and philosopher of science who knew Einstein for over forty years, is the most thorough work. It does not, however, discuss Einstein's work in any detail. It suffers from having been written during Einstein's lifetime and without the use of manuscript sources. Carl Seelig, *Albert Einstein. A Documentary Biography*, M. Savill, trans. (London, 1956), quotes extensively from Einstein's correspondence and is particularly good on the earlier part of his life.

Another biography of particular interest is that by Rudolf Kayser, Einstein's son-in-law, *Albert Einstein. A Biographical Portrait* (New York, 1930); this was actually written under the pseudonym Anton Reiser.

A recent biography that presents interesting ideas on Einstein's thought is Boris Kuznetsov, *Einstein*, V. Talmy, trans. (Moscow, 1965).

V. Some articles of particular interest are Robert S. Shankland, "Conversations With Albert Einstein," in *American Journal of Physics*, **31** (1963), 37–47; Gerald Holton, "On the Origins of the Special Theory of Relativity," *ibid.*, **28** (1960), 627–636; "Influences on Einstein's Early Work in Relativity Theory," in *American Scholar*, **37** (winter 1968), 59–79; "Mach, Einstein, and the Search for Reality," in *Daedalus*, **97** (1968), 636–673; and "Einstein, Michelson, and the 'Crucial' Experiment," in *Isis*, **60** (1969), 133–197; Tetu Hirosige, "Theory of Relativity and the Ether," in *Japanese Studies in the History of Science* (1968), pp. 37–53; Martin J. Klein, "Einstein's First Paper on Quanta," in *The Natural Philosopher*, **2** (1963), 57–86; "Einstein and the Wave-Particle Duality," *ibid.*, **3** (1964), 1–49; "Einstein, Specific Heats, and the Early Quantum Theory," in *Science*, **148** (1965), 173–180; "Thermodynamics in Einstein's Thought," *ibid.*, **157** (1967), 509–516; and "The First Phase

of the Bohr-Einstein Dialogue," in *Historical Studies in the Physical Sciences*, **2** (1970), 1–39; R. McCormmach, "Einstein, Lorentz, and the Electron Theory," *ibid.*, 41–87.

VI. Einstein's manuscripts, notes, and correspondence have been collected by the estate of Albert Einstein and are kept at present at the Institute for Advanced Study, Princeton, N. J. Information on certain other Einstein manuscripts may be found in T. S. Kuhn, J. L. Heilbron, P. L. Forman, and L. Allen, *Sources for History of Quantum Physics* (Philadelphia, 1967).

EINTHOVEN, WILLEM (*b.* Semarang, Java, 21 May 1860; *d.* Leiden, Netherlands, 28 September 1927), *physiology.*

Einthoven's father was municipal physician of Semarang; he married Louise M. M. C. de Vogel. He died in 1866, and four years later his widow settled in Utrecht with their six children. There Willem Einthoven graduated from high school and registered as a medical student in 1879. In 1886 he married his cousin Frédérique Jeanne Louise de Vogel; they had three daughters and a son.

While a student, Einthoven was active in sports; when he broke his wrist in a fall, he made it the occasion to publish a study on the pronation and supination of the forearm (1882). On 4 July 1885 he received the Ph.D in medicine *cum laude* with a thesis on stereoscopy through color differentiation. The following December he was appointed professor of physiology at Leiden.

In 1895, after the London physiologist A. D. Waller had published the curve for the action current of the heart as deduced from the body surface and had announced that he was unable to calculate its true shape (as recorded with Lippmann's capillary electrometer), Einthoven repeated this experiment. He defined the physical constants of the capillary electrometer and calculated the true curve, which he called the electrocardiogram. Einthoven considered direct registration of the curve's true shape a necessity. Starting from the mirror galvanometer of Deprez-d'Arsonval, he arrived at his brilliant conception of the string galvanometer. In 1896, while working on the construction of this instrument and developing the necessary photographic equipment, he registered electrocardiograms with the capillary electrometer as well as heart sounds of humans and animals.

For making electrocardiograms Einthoven chose the ordinate and abscissa in such a way that all details of the electrocardiogram would appear as clearly as possible. In 1903 he defined the standard measures for general use—one centimeter movement of the ordinate for one millivolt tension difference and a shutter speed of twenty-five millimeters per second, so that one centimeter of the abscissa represented 0.4 second. He indicated the various extremes by the random letters *P, Q, R, S,* and *T* and chose both hands and the left foot as contact points. This gave three possible combinations for contact which he labeled I (both hands); II (right hand–left foot); and III (left hand–left foot).

In 1912 Einthoven's research on the explanation of the respiratory changes in the electrocardiogram led him to the scheme of the equilateral triangle, considering the extremities as elongations of the electrodes. The information received from the contacts thus represents the projection of what takes place in the heart. With simultaneous registration of the three contacts, the size and direction of the resultant of all potential differences in the heart could be calculated minute by minute. Einthoven referred to this as the manifest size and direction of the electrical axis. He indicated the direction by the angle α of this axis with the horizontal and called it positive when it turned clockwise, negative when counterclockwise. Clinical electrocardiograms were studied by connecting patients with heart disease in the academic hospital to the instrument in Einthoven's laboratory by means of a cable 1.5 kilometers long (1906).

These "telecardiograms" acquainted Einthoven with many forms of heart disease. In addition he deepened his insight by registering heart sounds and murmurs simultaneous to the electrocardiogram by means of a second string galvanometer. The construction of a string recorder and a string myograph, both based on the torsion principle, enabled him to prove that the electrocardiogram and muscle contraction are inseparably connected.

While visiting America to give the Dungham lectures (1924) Einthoven was awarded the Nobel Prize for physiology or medicine. Upon his return to Leiden he found two foreign requests to register the action currents of the cervical sympathetic nerve. With the newly constructed vacuum string galvanometer he succeeded, on 28 April 1926, in registering the tonus action current and, after irritation of the organ, the thereupon induced action current of the cervical sympathetic nerve. His last major physical experiment, which he carried out in company with his son, was concerned with the reception of radiotelegrams broadcast by the machine transmitter "Malabar" in Java. In this case the string of 0.1 micron diameter and six millimeters length had to be synchronized with the 40,000 vibrations of the transmitting wave. Einthoven and his son found the resonance point after they achieved a variation in tension of one micromicron, after which telegrams from the machine

transmitter, working at top speed, were perfectly photographed on paper one centimeter wide.

Einthoven's last work was his treatise on the action current of the heart, which appeared posthumously in *Bethe's Handbuch der normalen und pathologischen Physiologie.*

BIBLIOGRAPHY

I. ORIGINAL WORKS. Einthoven's works include "Quelques remarques sur le mécanisme de l'articulation du coude," in *Archives néerlandaises des sciences exactes et naturelles,* **17** (1882), 289–298; "Stéréoscopie dépendant d'une différence de couleur," *ibid.,* **20** (1886), 361–387; "Lippmann's Capillarelektrometer zur Messung schnellwechseln der Potentialunterschiede," in *Pflügers Archiv für die gesamte Physiologie des Menschen und der Tiere,* **56** (1894), 528–540; "Die Registrierung der Herztöne," *ibid.,* **57** (1894), 617–639, written with M. A. J. Geluk; "Über den Einflusz des Leitungswiderstandes auf die Geschwindigkeit der Quecksilberbewegungen in Lippmann's Capillarelektrometer," *ibid.,* **60** (1895), 91–100; "Über die Form des menschlichen Elektrocardiogramms," *ibid.,* 101–123; "Beitrag zur Theorie des Capillarelektrometers," *ibid.,* **79** (1900), 1–25; "Eine Vorrichtung zum Registrieren der Ausschläge des Lippmann'schen Capillarelektrometers," *ibid.,* 25–38; "Über das normale menschliche Elektrokardiogramm und die capillarelektrometrische Untersuchung einiger Herzkranken," *ibid.,* **80** (1900), 139–160, written with K. de Lint; "Un nouveau galvanomètre," in *Archives néerlandaises des sciences exactes et naturelles,* **6** (1901), 625–633; "Die galvanometrische Registrierung des menschlichen Elektrokardiogramms, zugleich eine Beurteilung der Anwendung des Capillarelektrometers in der Physiologie," in *Pflügers Archiv für die gesamte Physiologie des Menschen und der Tiere,* **99** (1903), 472–480.

See also "Über einige Anwendungen des Saitengalvanometers," in *Annalen der Physik,* **14** (1904), 182–191; "Über eine neue Methode zur Dämpfung oszillierender Galvanometerausschläge," *ibid.,* **16** (1904), 20–32; "Weitere Mitteilungen über das Saitengalvanometer. Analyse der saitengalvanometrischen Kurven. Masse und Spannung des Quarzfadens und Widerstand gegen die Fadenbewegung," *ibid.,* **21** (1906), 483–514, 665–701; "Le télécardiogramme," in *Archives internationales de physiologie,* **4** (1906), 132–165; "Die Registrierung der menschlichen Herztöne mittels des Saitengalvanometers," in *Pflügers Archiv für die gesamte Physiologie des Menschen und der Tiere,* **117** (1907), 461–472, written with A. Flohil and P. J. J. A. Battaerd; "Ein dritter Herzton," *ibid.,* **120** (1907), 31–43, written with J. H. Wieringa and E. P. Snijders; "Weiteres über das Elektrokardiogramm," *ibid.,* **122** (1908), 517–585, written with B. Vaandrager; "Die Konstruktion des Saitengalvanometers, *ibid.,* **130** (1909), 287–321; "Über die Deutung des Elektrokardiogramms," *ibid.,* **149** (1913),

65–86; "Eine Vorrichtung zur photographischen Registrierung der Zeit," in *Zeitschrift für biologische Technik und Methodik,* **3** (1912), 1–8; and "Über die Richtung und die manifeste Grösse der Potentialschwankungen im menschlichen Herzen und über den Einfluss der Herzlage auf die Form des Elektrokardiogramms," in *Pflügers Archiv für die gesamte Physiologie des Menschen und der Tiere,* **150** (1913), 275–315, written with G. Fahr and A. de Waart.

Subsequent works are "On the Variability of the Size of the Pulse in Cases of Auricular Fibrillation," in *Heart,* **6** (1915), 107–121, written with A. J. Korteweg; "Die gleichzeitige Registrierung elektrischer Erscheinungen mittels zwei oder mehr Galvanometer und ihre Anwendung auf die Elektrokardiographie," in *Pflügers Archiv für die gesamte Physiologie des Menschen und der Tiere,* **164** (1916), 167–198, written with L. Bergansius and J. Bijtel; and "Über den Zusammenhang zwischen Elektro- und Mechanokardiogramm," in *Berichte über die gesamte Physiologie und experimentelle Pharmakologie,* **2** (1920), 178.

His last works include "L'électrocardiogramme tracé dans le cas où il n'y a pas de contraction visible du coeur," in *Archives néerlandaises de physiologie de l'homme et des animaux,* **5** (1921), 174–183, written with F. W. N. Hugenholtz; "Über die Beobachtung und Abbildung dünner Fäden," in *Pflügers Archiv für die gesamte Physiologie des Menschen und der Tiere,* **191** (1921), 60–98; "Über Stromleitung durch den menschlichen Körper," *ibid.,* **198** (1923), 439–483, written with J. Bijtel; "Functions of the Cervical Sympathetic Manifested by Its Action Currents," in *American Journal of Physiology,* **65** (1923), 350–362, written with Joseph Byrne; "The Relation of Mechanical and Electrical Phenomena of Muscular Contraction, With Special Reference to the Cardiac Muscle," in *The Harvey Society Lectures* (Philadelphia–London, 1924–1925), pp. 111–131; "Das Saitengalvanometer und die Messung der Aktionsströme des Herzens," in *Les Prix Nobel 1924–1925* (Stockholm, 1926), p. 18, his Nobel Prize acceptance speech; "Gehirn und Sympathicus, die Aktionsströme des Hallssympathicus," in *Pflügers Archiv für die gesamte Physiologie des Menschen und der Tiere,* **215** (1927), 443–453, written with S. Hoogerwerf, J. P. Karplus, and A. Kreidl; and "Die Aktionsströme des Herzens," in *Bethe's Handbuch der normalen und pathologischen Physiologie,* **8** (1928), 785–862.

II. SECONDARY LITERATURE. On Einthoven and his work see S. L. Barron, *Willem Einthoven, Biographical Notes,* Cambridge Monograph no. 5 (London, 1952), pp. 1–26; F. L. Bergansius, "Willem Einthoven," in *Wetenschappelijke bladen,* **1** (1925), 257; A. V. Hill, "Obituary. Prof. W. Einthoven," in *Nature,* **120** (1927), 591–592; Leonard Hill, "Willem Einthoven," in *British Medical Journal* (1927), **2**, 665; S. Hoogerwerf, *Leven en Werken van Willem Einthoven* (Hoorn, 1925), 9–93; and "Willem Einthoven," in T. P. Sevensma, ed., *Nederlandsche Helden der Wetenschap* (Amsterdam, 1946), 239–297; J. E. Johansson, "W. Einthoven (1924–1925)," in *Les Prix Nobel 1924–1925* (Stockholm, 1926); C. L. de Jongh, "Het levenswerk van

Einthoven," in *Nederlandsch tijdschrift voor geneeskunde,* **98** (1954), 270–273; T. Lewis, "Willem Einthoven," in *British Medical Journal* (1927), **2,** 664–665; G. van Rijnberk, "Willem Einthoven," in *Nederlandsch tijdschrift voor geneeskunde,* **68** (1924), 2424–2430; "In Memoriam," *ibid.,* **71** (1927), 1502–1503; E. Schott, "Willem Einthoven und die Fortschritte, welche wir der Erfindung des Saitengalvanometers verdanken," in *Münchener medizinische Wochenschrift,* **72** (1925), 391–392; A. Sikkel, "In Memoriam W. Einthoven," in *Geneeskundige gids,* **5** (1927), 925; "Necrologie Einthoven," in University of Leiden, *Jaarboek, 1928;* A. de Waart, *Einthoven* (Haarlem, 1957), with a complete list of his works; K. F. Wenckebach, "W. Einthoven," in *Deutsche medizinische Wochenschrift,* **51** (1927), 2176; F. A. F. C. Went, "Herdenkingsrede," in *Verslagen van de gewone vergadering van de Koninklijke Nederlandsche Academie van Wetenschappen. Afdeling Natuurkunde,* **8** (1927), 936–938; and H. Winterberg, "W. Einthoven," in *Wiener klinische Wochenschrift,* **40** (1927), 1460–1461.

S. HOOGERWERF

EISENHART, LUTHER PFAHLER (*b.* York, Pennsylvania, 13 January 1876; *d.* Princeton, New Jersey, 28 October 1965), *mathematics.*

Eisenhart was the second son of Charles Augustus Eisenhart and the former Emma Pfahler. His father was a dentist, a founder of the Edison Electric Light and York Telephone companies, and secretary of the Sunday school of St. Paul's Lutheran Church. Eisenhart was taught by his mother before he entered school and completed grade school in three years. He then attended York High School until, in his junior year, he was encouraged by the principal to withdraw and devote his time to the independent study of Latin and Greek for early admission to Gettysburg College, which he attended from 1892 to 1896. Being the only upper-division mathematics student, during the last two years of college Eisenhart studied mathematics through independent guided reading.

After teaching for a year at the preparatory school of the college, he began graduate study at Johns Hopkins University in 1897 and obtained the Ph.D. in 1900 with a thesis whose topic, "Infinitesimal Deformations of Surfaces," he had chosen himself. He was introduced to differential geometry through a lecture by Thomas Craig and studied the subject through the treatises of Gaston Darboux. According to his own testimony, the experience of independent study led Eisenhart to propose the four-course plan of study adopted at Princeton in 1923, which provides for independent study and the preparation of a thesis. Eisenhart's scientific career was spent at Princeton; he retired in 1945.

In 1908 Eisenhart married Anna Maria Dandridge Mitchell of Charles Town, West Virginia; she died in 1913. In 1918 he married Katharine Riely Schmidt of York, Pennsylvania. He had one son, Churchill, by his first marriage and two daughters, Anna and Katharine, by his second.

Eisenhart's work in differential geometry covers two distinct periods and fields. The first period, to about 1920, was devoted mainly to the theory of deformations of surfaces and systems of surfaces.

Modern differential geometry was founded by Gaston Darboux as a field of applications of partial differential equations. His methods were taken up by Luigi Bianchi, who created an extensive theory of the deformations of surfaces of constant negative curvature. In another direction, Claude Guichard showed between 1897 and 1899 how the partial differential equations of the deformations of triply orthogonal systems of surfaces can be interpreted in terms of the systems of lines connecting a point and its image point. These discoveries made the theory of deformations of surfaces one of the focal points of geometric research in Europe at the turn of the century. Although there were quite a number of able mathematicians working in America in the field of geometry at that time, Eisenhart was the only one to turn to the topic of deformations. His main contribution to the theory was a unifying principle: The deformation of a surface defines the congruence (two-parameter family) of lines connecting a point and its image (following Guichard). In general, a congruence contains two families of developable surfaces (a developable surface is formed by the tangent to a space curve). Eisenhart recognized that in all known cases, the intersections of these surfaces with the given surface and its image form a net of curves with special properties. This allows not only a unified treatment of many different subjects and a replacement of tricks by methods, but also leads to many new results that round off the theory. Eisenhart gave a coherent account of the theory in *Transformations of Surfaces* (1923). The book also contains most of Eisenhart's previous results either in the text or in the exercises, with references. Some aspects of the theory were taken up later in the projective setting by Eduard Čech and his students. All these investigations deal with small neighborhoods for which existence theorems for solutions of differential equations are available.

Of the few papers not dealing with deformations dating from this period, a noteworthy one is "Surfaces Whose First and Second Forms Are Respectively the Second and First Forms of Another Surface" (1901), one of the first differential geometric characterizations of the sphere, a topic started by Heinrich Liebmann in 1899. Eisenhart proved that the unit sphere is the only surface whose first and second fundamental forms are, respectively, the second and first fundamental forms of another surface.

Einstein's general theory of relativity (1916) made Riemannian geometry the center of geometric research. The analytic tools that turned Riemannian geometry from an idea into an effective instrument were Ricci's covariant differential calculus and the related notion of Levi-Civita's parallelism. These tools had been thoroughly explored in Luigi Bianchi's *Lezioni di geometria differenziale*. As a consequence, the attention of geometers immediately turned to the generalization of Riemannian geometry. Most of Eisenhart's work after 1921 was in this direction. The colloquium lectures *Non-Riemannian Geometry* (1927) contain his account of the main results obtained by him and his students and collaborators. An almost complete coverage of Eisenhart's results, with very good references, is given in Schouten's *Ricci Calculus*. Three directions of generalization of Riemannian geometry were developed in the years after 1920. They are connected with the names of Élie Cartan, Hermann Weyl, and Eisenhart. Cartan considered geometries that induce a geometry of a transitive transformation group in any tangent space. Weyl gave an axiomatic approach to the maps of tangent spaces by parallelism along any smooth curve. Eisenhart's approach, inspired by Oswald Veblen's work on the foundations of projective geometry and started in cooperation with Veblen, is the only one to deal directly with the given space. In Riemannian geometry, the measure of length is prescribed and the geodesic lines are determined as the shortest connections between nearby points. In Eisenhart's approach, the geodesics are given as the solution of a prescribed system of second-order differential equations and the non-Riemannian geometries are obtained by asking that there should exist a Levi-Civita parallelism for which the tangents are covariant constant.

While Cartan's and Weyl's generalizations have become the foundations of the fiber space theory of differentiable manifolds, Eisenhart's theory does not fit the framework of these topological theories. The reason is that the geometric objects intrinsically derived from the "paths" of the geometry, the projective parameters of Tracy Y. Thomas, have a more complicated transformation law than the generalized Christoffel symbols of Cartan and Weyl. However, there are a number of modern developments, such as the theory of Finsler spaces and the general theory of the geometric object, that fit Eisenhart's framework but not that of the algebraic-topological approach.

As far as metric geometry is concerned, the most fruitful approach seems to be to give the geodesics directly as point sets and to throw out all differential equations and analytical apparatus. On the other hand, for nonmetric geometries Eisenhart proved (in "Spaces With Corresponding Paths" [1922]) that for every one of his geometries there exists a unique geometry with the same paths and for which the mapping of tangent spaces induced by the flow of tangent vectors with unit speed along the paths is volume-preserving. For the latter geometry, which would appear to give a natural setting for topological dynamics, the Cartan, Weyl, and Eisenhart approaches are equivalent.

A number of interesting avenues of development of Riemannian geometry were opened by Eisenhart. The papers "Fields of Parallel Vectors in the Geometry of Paths" (1922) and "Fields of Parallel Vectors in a Riemannian Geometry" (1925) started the topic of recurrent fields and harmonic spaces (for a report with later references, see T. J. Willmore, *An Introduction to Differential Geometry,* ch. 7, sec. 13). The so-called Eisenhart's theorem appears in "Symmetric Tensors of the Second Order Whose First Covariant Derivatives Are Zero" (1923): If a Riemannian geometry admits a second-order, symmetric, covariant constant tensor other than the metric, the space behaves locally like the product of two lower-dimensional spaces. Together with a theorem of Georges de Rham to the effect that a simply connected, locally product Riemannian space is in fact a Cartesian product of two spaces, the theorem is an important tool in global differential geometry. An extension of the theorem is given in "Parallel Vectors in Riemannian Space" (1938).

The basic equations for the vectors of a group of motions in a Riemannian space had been given by Killing in 1892. Eisenhart developed a very powerful analytical apparatus for these questions; the results are summarized in *Riemannian Geometry* (1926; ch. 6) and *Continuous Groups of Transformations* (1933). The later developments are summarized in Kentaro Yano's *Groups of Transformations in Generalized Spaces* (1949).

Eisenhart's interest in mathematical instruction found its expression in a number of influential textbooks—such as *Differential Geometry of Curves and Surfaces* (1909), *Riemannian Geometry* (1926), *Continuous Groups of Transformations* (1933), *Coordinate Geometry* (1939), *An Introduction to Differential Geometry With Use of the Tensor Calculus* (1940)—some in fields that until then had been dependent upon European monographs devoid of exercises and other student aids. His interest in history resulted in several papers: "Lives of Princeton Mathematicians" (1931), "Plan for a University of Discoverers" (1947), "Walter Minto and the Earl of Buchan" (1950), and the preface to "Historic Philadelphia" (1953).

BIBLIOGRAPHY

I. ORIGINAL WORKS. Eisenhart's works published between 1901 and 1909 are "A Demonstration of the Impossibility of a Triply Asymptotic System of Surfaces," in *Bulletin of the American Mathematical Society,* **7** (1901), 184–186; "Possible Triply Asymptotic Systems of Surfaces," *ibid.,* 303–305; "Surfaces Whose First and Second Forms Are Respectively the Second and First Forms of Another Surface," *ibid.,* 417–423; "Lines of Length Zero on Surfaces," *ibid.,* **9** (1902), 241–243; "Note on Isotropic Congruences," *ibid.,* 301–303; "Infinitesimal Deformation of Surfaces," in *American Journal of Mathematics,* **24** (1902), 173–204; "Conjugate Rectilinear Congruences," in *Transactions of the American Mathematical Society,* **3** (1902), 354–371; "Infinitesimal Deformation of the Skew Helicoid," in *Bulletin of the American Mathematical Society,* **9** (1903), 148–152; "Surfaces Referred to Their Lines of Length Zero," *ibid.,* 242–245; "Isothermal-Conjugate Systems of Lines on Surfaces," in *American Journal of Mathematics,* **25** (1903), 213–248; "Surfaces Whose Lines of Curvature in One System Are Represented on the Sphere by Great Circles," *ibid.,* 349–364; "Surfaces of Constant Mean Curvature," *ibid.,* 383–396; "Congruences of Curves," in *Transactions of the American Mathematical Society,* **4** (1903), 470–488; "Congruences of Tangents to a Surface and Derived Congruences," in *American Journal of Mathematics,* **26** (1904), 180–208; "Three Particular Systems of Lines on a Surface," in *Transactions of the American Mathematical Society,* **5** (1904), 421–437; "Surfaces With the Same Spherical Representation of Their Lines of Curvature as Pseudospherical Surfaces," in *American Journal of Mathematics,* **27** (1905), 113–172; "On the Deformation of Surfaces of Translation," in *Bulletin of the American Mathematical Society,* **11** (1905), 486–494; "Surfaces of Constant Curvature and Their Transformations," in *Transactions of the American Mathematical Society,* **6** (1905), 473–485; "Surfaces Analogous to the Surfaces of Bianchi," in *Annali di matematica pura ed applicata,* 3rd ser., **12** (1905), 113–143; "Certain Surfaces With Plane or Spherical Lines of Curvature," in *American Journal of Mathematics,* **28** (1906), 47–70; "Associate Surfaces," in *Mathematische Annalen,* **62** (1906), 504–538; "Transformations of Minimal Surfaces," in *Annali di matematica pura ed applicata,* 3rd ser., **13** (1907), 249–262; "Applicable Surfaces With Asymptotic Lines of One Surface Corresponding to a Conjugate System of Another," in *Transactions of the American Mathematical Society,* **8** (1907), 113–134; "Certain Triply Orthogonal Systems of Surfaces," in *American Journal of Mathematics,* **29** (1907), 168–212; "Surfaces With Isothermal Representation of Their Lines of Curvature and Their Transformations (I)," in *Transactions of the American Mathematical Society,* **9** (1908), 149–

177; "Surfaces With the Same Spherical Representation of Their Lines of Curvature as Spherical Surfaces," in *American Journal of Mathematics,* **30** (1908), 19–42; and *A Treatise on the Differential Geometry of Curves and Surfaces* (Boston, 1909; repub. New York, 1960).

Between 1910 and 1919 he published "The Twelve Surfaces of Darboux and the Transformation of Moutard," in *American Journal of Mathematics,* **32** (1910), 17–36; "Congruences of the Elliptic Type," in *Transactions of the American Mathematical Society,* **11** (1910), 351–372; "Surfaces With Isothermal Representation of Their Lines of Curvature and Their Transformations (II)," *ibid.,* 475–486; "A Fundamental Parametric Representation of Space Curves," in *Annals of Mathematics,* 2nd ser., **13** (1911), 17–35; "Sopra le deformazioni continue delle superficie reali applicabili sul paraboloide a parametro puramente immaginario," in *Atti dell'Accademia nazionale dei Lincei. Rendiconti,* Classe di scienze fisiche, matematiche e naturali, 5th ser., **21¹** (1912), 458–462; "Ruled Surfaces With Isotropic Generators," in *Rendiconti del Circolo matematico di Palermo,* **34** (1912), 29–40; "Minimal Surfaces in Euclidean Four-Space," in *American Journal of Mathematics,* **34** (1912), 215–236; "Certain Continuous Deformations of Surfaces Applicable to the Quadrics," in *Transactions of the American Mathematical Society,* **14** (1913), 365–402; "Transformations of Surfaces of Guichard and Surfaces Applicable to Quadrics," in *Annali di matematica pura ed applicata,* 3rd ser., **22** (1914), 191–248; "Transformations of Surfaces of Voss," in *Transactions of the American Mathematical Society,* **15** (1914), 245–265; "Transformations of Conjugate Systems With Equal Point Invariants," *ibid.,* 397–430; "Conjugate Systems With Equal Tangential Invariants and the Transformation of Moutard," in *Rendiconti del Circolo matematico di Palermo,* **39** (1915), 153–176; "Transformations of Surfaces Ω," in *Proceedings of the National Academy of Sciences,* **1** (1915), 62–65; "One-Parameter Families of Curves," in *American Journal of Mathematics,* **37** (1915), 179–191; "Transformations of Conjugate Systems With Equal Invariants," in *Proceedings of the National Academy of Sciences,* **1** (1915), 290–295; "Surfaces Ω and Their Transformations," in *Transactions of the American Mathematical Society,* **16** (1915), 275–310; "Sulle superficie di rotolamento e le trasformazioni di Ribaucour," in *Atti dell'Accademia nazionale dei Lincei. Rendiconti,* Classe di scienze fisiche, matematiche e naturali, 5th ser., **24²** (1915), 349–352; "Surfaces With Isothermal Representation of Their Lines of Curvature as Envelopes of Rolling," in *Annals of Mathematics,* 2nd ser., **17** (1915), 63–71; "Transformations of Surfaces Ω," in *Transactions of the American Mathematical Society,* **17** (1916), 53–99; "Deformations of Transformations of Ribaucour," in *Proceedings of the National Academy of Sciences,* **2** (1916), 173–177; "Conjugate Systems With Equal Point Invariants," in *Annals of Mathematics,* 2nd ser., **18** (1916), 7–17; "Surfaces Generated by the Motion of an Invariable Curve Whose Points Describe Straight Lines," in *Rendiconti del Circolo matematico di Palermo,* **41** (1916), 94–102; "Deformable Transformations of Ribaucour," in *Transactions of the American Mathematical Society,* **17** (1916), 437–458;

"Certain Surfaces of Voss and Surfaces Associated With Them," in *Rendiconti del Circolo matematico de Palermo,* **42** (1917), 145–166; "Transformations *T* of Conjugate Systems of Curves on a Surface," in *Transactions of the American Mathematical Society,* **18** (1917), 97–124; "Triads of Transformations of Conjugate Systems of Curves," in *Proceedings of the National Academy of Sciences,* **3** (1917), 453–457; "Conjugate Planar Nets With Equal Invariants," in *Annals of Mathematics,* 2nd ser., **18** (1917), 221–225; "Transformations of Applicable Conjugate Nets of Curves on Surfaces," in *Proceedings of the National Academy of Sciences,* **3** (1917), 637–640; "Darboux's Contribution to Geometry," in *Bulletin of the American Mathematical Society,* **24** (1918), 227–237; "Surfaces Which Can Be Generated in More Than One Way by the Motion of an Invariable Curve," in *Annals of Mathematics,* 2nd ser., **19** (1918), 217–230; "Transformations of Planar Nets," in *American Journal of Mathematics,* **40** (1918), 127–144; "Transformations of Applicable Conjugate Nets of Curves on Surfaces," in *Transactions of the American Mathematical Society,* **19** (1918), 167–185; "Triply Conjugate Systems With Equal Point Invariants," in *Annals of Mathematics,* 2nd ser., **20** (1919), 262–273; "Transformations of Surfaces Applicable to a Quadric," in *Transactions of the American Mathematical Society,* **20** (1919), 323–338; and "Transformations of Cyclic Systems of Circles," in *Proceedings of the National Academy of Sciences,* **5** (1919), 555–557.

Eisenhart's works published between 1920 and 1929 are "The Permanent Gravitational Field in the Einstein Theory," in *Annals of Mathematics,* 2nd ser., **22** (1920), 86–94; "The Permanent Gravitational Field in the Einstein Theory," in *Proceedings of the National Academy of Sciences,* **6** (1920), 678–682; "Sulle congruenze di sfere di Ribaucour che ammettono una deformazione finita," in *Atti dell'Accademia nazionale dei Lincei. Rendiconti,* Classe di scienze fisiche, matematiche e naturali, 5th ser., **29²** (1920), 31–33; "Conjugate Systems of Curves *R* and Their Transformations," in *Comptes rendus du sixième Congrès international des mathématiciens* (Strasbourg, 1920), pp. 407–409; "Darboux's Anteil an der Geometrie," in *Acta mathematica,* **42** (1920), 275–284; "Transformations of Surfaces Applicable to a Quadric," in *Journal de mathématiques pures et appliquées,* 8th ser., **4** (1921), 37–66; "Conjugate Nets *R* and Their Transformations," in *Annals of Mathematics,* 2nd ser., **22** (1921), 161–181; "A Geometric Characterization of the Paths of Particles in the Gravitational Field of a Mass at Rest" (abstract), in *Bulletin of the American Mathematical Society,* **27** (1921), 350; "The Einstein Solar Field," *ibid.,* 432–434; "Sulle trasformazioni *T* dei sistemi tripli coniugati di superficie," in *Atti dell'Accademia nazionale dei Lincei. Rendiconti,* Classe di scienze fisiche, matematiche e naturali, **30²** (1921), 399–401; "Einstein Static Fields Admitting a Group G_2 of Continuous Transformations Into Themselves," in *Proceedings of the National Academy of Sciences,* **7** (1921), 328–334, abstract in *Bulletin of the American Mathematical Society,* **28** (1922), 34; "The Riemann Geometry and Its Generalization," in *Proceedings of the National Academy of Sciences,* **8** (1922), 19–23, abstract in *Bulletin of the American Mathematical*

Society, **28** (1922), 154, written with Oswald Veblen; "Ricci's Principal Directions for a Riemann Space and the Einstein Theory," in *Proceedings of the National Academy of Sciences,* **8** (1922), 24–26, abstract in *Bulletin of the American Mathematical Society,* **28** (1922), 238; "The Einstein Equations for the Solar Field From the Newtonian Point of View," in *Science,* n.s. **55** (1922), 570–572; "Fields of Parallel Vectors in the Geometry of Paths," in *Proceedings of the National Academy of Sciences,* **8** (1922), 207–212; "Spaces With Corresponding Paths," *ibid.,* 233–238; "Condition That a Tensor Be the Curl of a Vector," in *Bulletin of the American Mathematical Society,* **28** (1922), 425–427; "Affine Geometries of Paths Possessing an Invariant Integral," in *Proceedings of the National Academy of Sciences,* **9** (1923), 4–7; "Another Interpretation of the Fundamental Gauge-Vectors of Weyl's Theory of Relativity," *ibid.,* 175–178; "Orthogonal Systems of Hypersurfaces in a General Riemann Space," in *Transactions of the American Mathematical Society,* **25** (1923), 259–280, abstract in *Bulletin of the American Mathematical Society,* **29** (1923), 212; "Symmetric Tensors of the Second Order Whose First Covariant Derivatives Are Zero," in *Transactions of the American Mathematical Society,* **25** (1923), 297–306, abstract in *Bulletin of the American Mathematical Society,* **29** (1923), 213; "Einstein and Soldner," in *Science,* n.s. **58** (1923), 516–517; "The Geometry of Paths and General Relativity," in *Annals of Mathematics,* 2nd ser., **24** (1923), 367–393; *Transformations of Surfaces* (Princeton, 1923; corr. reiss. New York, 1962); "Space-Time Continua of Perfect Fluids in General Relativity," in *Transactions of the American Mathematical Society,* **26** (1924), 205–220; "Spaces of Continuous Matter in General Relativity," abstract in *Bulletin of the American Mathematical Society,* **30** (1924), 7; "Geometries of Paths for Which the Equations of the Paths Admit a Quadratic First Integral," in *Transactions of the American Mathematical Society,* **26** (1924), 378–384, abstract in *Bulletin of the American Mathematical Society,* **30** (1924), 297; "Linear Connections of a Space Which Are Determined by Simply Transitive Continuous Groups," in *Proceedings of the National Academy of Sciences,* **11** (1925), 243–250; "Fields of Parallel Vectors in a Riemannian Geometry," in *Transactions of the American Mathematical Society,* **27** (1925), 563–573, abstract in *Bulletin of the American Mathematical Society,* **31** (1925), 292; "Einstein's Recent Theory of Gravitation and Electricity," in *Proceedings of the National Academy of Sciences,* **12** (1926), 125–129; *Riemannian Geometry* (Princeton, 1926); "Geometries of Paths for Which the Equations of the Path Admit $n(n + 1)/2$ Independent Linear First Integrals," in *Transactions of the American Mathematical Society,* **28** (1926), 330–338, abstract in *Bulletin of the American Mathematical Society,* **32** (1926), 197; "Congruences of Parallelism of a Field of Vectors," in *Proceedings of the National Academy of Sciences,* **12** (1926), 757–760; "Displacements in a Geometry of Paths Which Carry Paths Into Paths," in *Proceedings of the National Academy of Sciences,* **13** (1927), 38–42, written with M. S. Knebelman; *Non-Riemannian Geometry* (New York, 1927; 6th pr., 1968); "Affine Geometry," in *Encyclopaedia Britannica,* 14th ed. (1929), I, 279–

280; "Differential Geometry," *ibid.,* VII, 366–367; "Contact Transformations," in *Annals of Mathematics,* 2nd ser., **30** (1929), 211–249; and "Dynamical Trajectories and Geodesics," *ibid.,* 591–606.

Between 1930 and 1939 Eisenhart published "Projective Normal Coordinates," in *Proceedings of the National Academy of Sciences,* **16** (1930), 731–740; "Lives of Princeton Mathematicians," in *Scientific Monthly,* **33** (1931), 565–568; "Intransitive Groups of Motions," in *Proceedings of the National Academy of Sciences,* **18** (1932), 195–202; "Equivalent Continuous Groups," in *Annals of Mathematics,* 2nd ser., **33** (1932), 665–676; "Spaces Admitting Complete Absolute Parallelism," in *Bulletin of the American Mathematical Society,* **39** (1933), 217–226; *Continuous Groups of Transformations* (Princeton, 1933; repr. New York, 1961); "Separable Systems in Euclidean 3-Space," in *Physical Review,* 2nd ser., **45** (1934), 427–428; "Separable Systems of Stäckel," in *Annals of Mathematics,* 2nd ser., **35** (1934), 284–305; "Stäckel Systems in Conformal Euclidean Space," *ibid.,* **36** (1935), 57–70; "Groups of Motions and Ricci Directions," *ibid.,* 823–832; "Simply Transitive Groups of Motions," in *Monatshefte für Mathematik und Physik,* **43** (1936), 448–452; "Invariant Theory of Homogeneous Contact Transformations," in *Annals of Mathematics,* 2nd ser., **37** (1936), 747–765, written with M. S. Knebelman; "Graduate Study and Research," in *Science,* **83** (1936), 147–150; "Riemannian Spaces of Class Greater Than Unity," in *Annals of Mathematics,* 2nd ser., **38** (1937), 794–808; "Parallel Vectors in Riemannian Space," in *Annals of Mathematics,* 2nd ser., **39** (1938), 316–321; and *Coordinate Geometry* (Boston, 1939; repr. New York, 1960).

In the 1940's Eisenhart published *An Introduction to Differential Geometry With Use of the Tensor Calculus* (Princeton, 1940); *The Educational Process* (Princeton, 1945); "The Far-Seeing Wilson," in William Starr Myers, ed., *Woodrow Wilson, Some Princeton Memories* (Princeton, 1946), pp. 62–68; "Plan for a University of Discoverers," in *The Princeton University Library Chronicle,* **8** (1947), 123–139; "Enumeration of Potentials for Which One-Particle Schrödinger Equations Are Separable," in *Physical Review,* 2nd ser., **74** (1948), 87–89; "Finsler Spaces Derived From Riemann Spaces by Contact Transformations," in *Annals of Mathematics,* 2nd ser., **49** (1948), 227–254; "Separation of the Variables in the One-Particle Schrödinger Equation in 3-Space," in *Proceedings of the National Academy of Sciences,* **35** (1949), 412–418; and "Separation of the Variables of the Two-Particle Wave Equation," *ibid.,* 490–494.

Eisenhart's publications of the 1950's are "Homogeneous Contact Transformations," in *Proceedings of the National Academy of Sciences,* **36** (1950), 25–30; "Walter Minto and the Earl of Buchan," in *Proceedings of the American Philosophical Society,* **94,** no. 3 (1950), 282–294; "Generalized Riemann Spaces," in *Proceedings of the National Academy of Sciences,* **37** (1951), 311–315; *Uvod u diferentsijalnu geometriiu* (Belgrade, 1951), translation of *Introduction to Differential Geometry* . . .; "Generalized Riemann Spaces, II," in *Proceedings of the National Academy of Sciences,* **38** (1952), 506–508; "Generalized Riemann Spaces and

General Relativity," *ibid.,* **39** (1953), 546–550; Preface to "Historic Philadelphia," in *Transactions of the American Philosophical Society,* **43,** no. 1 (1953), 3; "Generalized Riemann Spaces and General Relativity, II," in *Proceedings of the National Academy of Sciences,* **40** (1954), 463–466; "A Unified Theory of General Relativity of Gravitation and Electromagnetism. I," *ibid.,* **42** (1956), 249–251; II, *ibid.,* 646–650; III, *ibid.,* 878–881; IV, *ibid.,* **43** (1957), 333–336; "Spaces for Which the Ricci Scalar *R* Is Equal to Zero," *ibid.,* **44** (1958), 695–698; "Spaces for Which the Ricci Scalar *R* Is Equal to Zero," *ibid.,* **45** (1959), 226–229; and "Generalized Spaces of General Relativity," *ibid.,* 1759–1762.

The early 1960's saw publication of the following: "The Cosmology Problem in General Relativity," in *Annals of Mathematics,* 2nd ser., **71** (1960), 384–391; "The Paths of Rays of Light in General Relativity," in *Proceedings of the National Academy of Sciences,* **46** (1960), 1093–1097; "Fields of Unit Vectors in the Four-Space of General Relativity," *ibid.,* 1589–1601; "Generalized Spaces of General Relativity II," *ibid.,* 1602–1604; "Spaces Which Admit Fields of Normal Null Vectors," *ibid.,* 1605–1608; "The Paths of Rays of Light in General Relativity of the Nonsymmetric Field V_4," *ibid.,* **47** (1961), 1822–1823; "Spaces With Minimal Geodesics," in *Calcutta Mathematical Society Golden Jubilee Commemorative Volume* (Calcutta, 1961), pp. 249–254; "Spaces in Which the Geodesics Are Minimal Curves," in *Proceedings of the National Academy of Sciences,* **48** (1962), 22; "The Paths of Rays of Light in Generalized General Relativity of the Nonsymmetric Field V_4," *ibid.,* 773–775; "Generalized Riemannian Geometry II," *ibid.,* **49** (1963), 18–19; and "The Einstein Generalized Riemannian Geometry," **50** (1963), 190–193.

II. SECONDARY LITERATURE. Biographical memoirs are Gilbert Chinard, Harry Levy, and George W. Corner, "Luther Pfahler Eisenhart (1876–1965)," in *Year Book of the American Philosophical Society* for 1966 (Philadelphia, 1967), pp. 127–134; and Solomon Lefschetz, "Luther Pfahler Eisenhart," in *Biographical Memoirs. National Academy of Sciences,* **40** (1969), 69–90.

Eisenhart's work is discussed in Luigi Bianchi, *Lezioni di geometria differenziale,* Nichola Zanichelli, ed., II, pt. 2 (Bologna, 1930); Herbert Busemann, *The Geometry of Geodesics* (New York, 1955); J. A. Schouten, *Ricci Calculus,* 2nd ed. (Berlin–Göttingen–Heidelberg, 1954); T. Y. Thomas, "On the Projective and Equi-projective Geometries of Paths," in *Proceedings of the National Academy of Sciences,* **11** (1925), 198–203; T. J. Willmore, *An Introduction to Differential Geometry* (London, 1959); and Kentaro Yano, *Groups of Transformations in Generalized Spaces* (Tokyo, 1949).

H. GUGGENHEIMER

EISENSTEIN, FERDINAND GOTTHOLD MAX

(*b.* Berlin, Germany, 16 April 1823; *d.* Berlin, 11 October 1852), *mathematics.*

Eisenstein's father, Johann Konstantin Eisenstein, and his mother, the former Helene Pollack, had con-

verted from Judaism to Protestantism before Gotthold was born. His father, who had served eight years in the Prussian army, tried his hand at various commercial enterprises, including manufacturing, but without financial success. Not until late in life did he begin to make a decent livelihood. Eisenstein's five brothers and sisters, born after him, died in childhood, nearly all of meningitis, which he also contracted. His interest in mathematics, awakened and encouraged by a family acquaintance, began when he was about six. "As a boy of six I could understand the proof of a mathematical theorem more readily than that meat had to be cut with one's knife, not one's fork" ("Curriculum vitae," p. 150). Early, too, Eisenstein showed musical inclinations that continued throughout his life and that found expression in playing the piano and composing.

Even while he was in elementary school, his persistently poor health prompted his parents to send him for a time to board in the country. From about 1833 to 1837 he was a resident student at the Cauer academy in Charlottenburg (near Berlin), where the quasi-military discipline was little to his taste. The effects upon him of its Spartan pedagogical methods were manifested in frequent, often feverish illnesses and depression. From September 1837 to July 1842 he attended the Friedrich Wilhelm Gymnasium and then, as a senior, the Friedrich Werder Gymnasium in Berlin. In addition, he went to hear Dirichlet and others lecture at the university.

> What attracted me so strongly and exclusively to mathematics, apart from its actual content, was especially the specific nature of the mental operation by which mathematical things are dealt with. This way of deducing and discovering new truths from old ones, and the extraordinary clarity and self-evidence of the theorems, the ingeniousness of the ideas . . . had an irresistible fascination for me. . . . Starting from the individual theorems, I soon grew accustomed to pierce more deeply into their relationships and to grasp whole theories as a single entity. That is how I conceived the idea of mathematical beauty. . . . And there is such a thing as a mathematical sense or instinct that enables one to see immediately whether an investigation will bear fruit, and to direct one's thoughts and efforts accordingly ["Curriculum vitae," pp. 156–157].

Eisenstein had the good fortune to find in the meteorologist Heinrich W. Dove and the mathematician Karl Schellbach teachers who understood and encouraged him. What he learned in class and at lectures led him to deeper, independent study of the works of Euler, Lagrange, and Gauss, although it was the last who influenced him most. In the summer of 1842, before completing school, he accompanied his

340

mother to England to join his father, who had gone there two years earlier in search of a better livelihood. In neither England, Wales, nor Ireland could the family gain a firm footing. Eisenstein used the time to steep himself in Gauss's *Disquisitiones arithmeticae* and started on his own to study forms of the third degree and the theory of elliptic functions. In Dublin in early 1843 he made the acquaintance of W. R. Hamilton, who gave him a copy of his work "On the Argument of Abel, Respecting the Impossibility of Expressing a Root of Any General Equation Above the Fourth Degree," to be presented to the Berlin Academy.

By around mid-June 1843 Eisenstein and his mother were back in Berlin. His parents were now living apart, and from then until his death Eisenstein stayed with his mother only briefly from time to time. In August 1843 he applied to the Friedrich Wilhelm Gymnasium in Berlin for permission, as a nonstudent, to take their final examinations (a prerequisite for admission to regular university study). In the brief autobiography appended to his application he mentioned (at age twenty) the "hypochondria that has been plaguing me for two years." On 22 September 1843 Eisenstein passed his final secondary school examination, and Schellbach wrote of him in his report: "His knowledge of mathematics goes far beyond the scope of the secondary-school curriculum. His talent and zeal lead one to expect that some day he will make an important contribution to the development and expansion of science" (a remarkable opinion, compared with the wrong ones put forth by other teachers, Galois for example).

Immediately after passing his examinations, Eisenstein enrolled at the University of Berlin. In January 1844 he delivered to the Berlin Academy the copy of Hamilton's study that he had received in Dublin, using the occasion to submit a treatise of his own on cubic forms with two variables. A. L. Crelle, whom the Academy had commissioned to evaluate Eisenstein's work and make appropriate reply to him on its behalf, accepted the treatise for publication in his *Journal für die reine und angewandte Mathematik,* thus again demonstrating Crelle's keen eye for mathematical genius, which had earlier spotted Abel, Jacobi, Steiner, and, later, Weierstrass. At the same time, Crelle introduced the young author to Alexander von Humboldt, who immediately took an interest in him. Time and again Humboldt requested financial support for Eisenstein from the Prussian ministry of education, the king, and the Berlin Academy, and often helped him out of his own pocket. Eisenstein had no feeling of economic security, since these official grants were awarded only for short pe-

riods and always had to be reapplied for, with the approved extensions often arriving late and the sums involved being quite modest and certainly not owed to the recipient. His constant dependence on gifts and charity weighed heavily on him, yet he had found in Humboldt a tireless mentor and protector, the like of which few young talents are ever blessed with. And Humboldt made it clear that he valued Eisenstein not only as a promising young scholar but also as a human being, and with tact and sensitivity he tried (albeit in vain) to divert and cheer him.

The twenty-seventh and twenty-eighth volumes of Crelle's *Journal,* published in 1844, contained twenty-five contributions by Eisenstein. These testimonials to his almost unbelievable, explosively dynamic productivity rocketed him to fame throughout the mathematical world. They dealt primarily with quadratic and cubic forms, the reciprocity theorem for cubic residues, fundamental theorems for quadratic and biquadratic residues, cyclotomy and forms of the third degree, plus some notes on elliptic and Abelian transcendentals. Gauss, to whom he had sent some of his writings, praised them very highly and looked forward with pleasure to an announced visit. In June 1844, carrying a glowing letter of recommendation from Humboldt, Eisenstein went off to see Gauss. He stayed in Göttingen fourteen days. In the course of the visit he won the high respect of the "prince of mathematicians," whom he had revered all his life. The sojourn in Göttingen was important to Eisenstein for another reason: he became friends with Moritz A. Stern—the only lasting friendship he ever made. While the two were in continual correspondence on scientific matters, even Stern proved unable to dispel the melancholy that increasingly held Eisenstein in its grip. Even the sensational recognition that came to him while he was still only a third-semester student failed to brighten Eisenstein's spirits more than fleetingly. In February 1845, at the instance of Ernst E. Kummer, who was acting on a suggestion from Jacobi (possibly inspired by Humboldt), Eisenstein was awarded an honorary doctorate in philosophy by the School of Philosophy of the University of Breslau.

The year 1846 found Eisenstein suddenly involved in an unpleasant priority dispute with Jacobi, who accused him of plagiarism and of misrepresenting known results. Writing to Stern on 20 April 1846, Eisenstein explained that "the whole trouble is that, when I learned of his work on cyclotomy, I did not immediately and publicly acknowledge him as the originator, while I frequently have done this in the case of Gauss. That I omitted to do so in this instance is merely the fault of my naïve innocence."

Jacobi charged him with scientific frivolity and appropriating as his own the ideas imparted to him by others, and he maintained that Eisenstein had no original achievements to his credit but had merely cleverly proved certain theorems stated by others and carried out ideas conceived by others. This was in curious contrast with Jacobi's attitude in 1845, when he had recommended Eisenstein for the honorary doctorate.

In 1846–1847 Eisenstein published various writings, mainly on the theory of elliptic functions. Humboldt, who had tried in vain in 1846 to draw the attention of Crown Prince Maximilian of Bavaria to Eisenstein, early in 1847 recommended him for a professorship at Heidelberg—even before he had earned his teaching credentials at the University of Berlin—but again without success. During the summer semester of 1847 Riemann was among those who attended Eisenstein's lecture on elliptic functions. In September 1847 a great honor came to Eisenstein: Gauss wrote the preface to a volume of his collected treatises. No longer extant, unfortunately, is the letter from Gauss to Eisenstein in which, the latter reported to Riemann, Gauss set down the essentials of his proof of the biquadratic reciprocity law with the aid of cyclotomy.

Early in 1848 Eisenstein had attended meetings of certain democratically oriented clubs, although he took no active part in the pre-March political ferment. During the street battles on 19 March, however, he was forcibly removed from a house from which shots had been fired and was taken with other prisoners to the Citadel at Spandau, suffering severe mistreatment en route. Although he was released the next day, the experience gravely affected his health. Moreover, when word spread that he was a "republican," financial support for him dwindled, and it took Humboldt's most strenuous efforts to keep it from drying up altogether. Eisenstein's situation visibly worsened. Alienated from his family and without close friends or any real contact with other Berlin mathematicians, he vegetated. Only occasionally did he feel able to deliver his lectures as *Privatdozent,* from his bed, if he managed to lecture at all. Yet all this time he was publishing one treatise after another in Crelle's *Journal,* especially on the quadratic partition of prime numbers, on reciprocity laws, and on the theory of forms. In August 1851, on Gauss's recommendation, both Eisenstein and Kummer were elected corresponding members of the Göttingen Society, and in March 1852 Dirichlet managed his election to membership in the Berlin Academy. In late July of that year Eisenstein suffered a severe hemorrhage. Funds raised by Humboldt so that Eisenstein could spend a year convalescing in Sicily came too late: on 11 October he died of pulmonary tuberculosis. Despite all the public recognition, he ended his days in forlorn solitude. The eighty-three-year-old Humboldt accompanied the coffin to the graveside.

Eisenstein soon became the subject of legend, and the early literature about him is full of errors. Only latter-day research has illumined the tragic course of his life. For instance, no evidence at all has been found of the dissolute existence that he was frequently rumored to have led. His lectures were usually attended by more than half of Berlin's mathematics students, which was the more remarkable since Dirichlet, Jacobi, and Steiner were then teaching at Berlin. Eisenstein was ever at pains, as he himself emphasized, to bring home to his listeners the most recent research results.

His treatises were written at a time when only Gauss, Cauchy, and Dirichlet had any conception of what a completely rigorous mathematical proof was. Even a man like Jacobi often admitted that his own work sometimes lacked the necessary rigor and self-evidence of methods and proofs. Thus it is not surprising that, as Leo Koenigsberger tells us, Eisenstein's "Study of the Infinite Double Products, of Which Elliptic Functions Are Composed as Quotients" should have been criticized by Weierstrass, who, in representing his own functions in terms of infinite products, was not picking up the torch from his forerunner, Eisenstein, but was drawing directly upon Gauss. Weierstrass correctly rated Riemann over Eisenstein, who was unable to grasp Riemann's general ideas about functions of complex variables. While Klein did concede that the simplest elliptic functions are defined by Eisenstein's everywhere absolutely convergent series, he called Eisenstein a "walking formula who starts out with a calculation and then finds in it the roots of all his knowledge." Unjustly Klein attributed to him a persecution complex and megalomania. Eisenstein's oft-quoted statement to the effect that through his contributions to the theory of forms (including his finding the simplest covariant for the binary cubic form) he hoped "to become a second Newton" (letter to Humboldt, July 1847) is nothing more than a bad joke.

The development that led to the reciprocity law of nth-power residues will be permanently associated with Eisenstein's work on cubic and biquadratic reciprocity laws. The Eisenstein series have become an integral part of the theory of modular forms and modular functions. They and the Eisenstein irreducibility law (along with the Eisenstein polynomial and the Eisenstein equation) continue to bear his name and to assure him a position about halfway between that contemptuous assessment by Klein and the verdict of Gauss (expressed, of course, in a letter intended for display), who held Eisenstein's talents to be such as "nature bestows on only a few in each century" (letter to Humboldt, 14 April 1846).

BIBLIOGRAPHY

I. ORIGINAL WORKS. Nearly all of Eisenstein's scientific writings were published in the *Journal für die reine und angewandte Mathematik,* specifically in vols. **27** (1844) to **44** (1852); see the bibliography by Kurt-R. Biermann in *Journal für die reine und angewandte Mathematik,* **214/215** (1964), 29–30. Selected *Mathematische Abhandlungen besonders aus dem Gebiete der höhern Mathematik und der elliptischen Functionen* (Berlin, 1847) were published with a preface by Gauss; repr., with intro. by Kurt-R. Biermann (Hildesheim, 1967). An autobiography, "Curriculum vitae des Gotth. Ferdinand Eisenstein," ed. and with intro. by F. Rudio, was published in *Zeitschrift für Mathematik und Physik,* **40** (1895), supp., 143–168. The letters from Eisenstein to M. A. Stern were published by A. Hurwitz and F. Rudio in *Zeitschrift für Mathematik und Physik,* **40** (1895), supp., 169–203. A report by Eisenstein on his imprisonment is found in Adalbert Roerdansz, *Ein Freiheits-Martyrium. Gefangene Berliner auf dem Transport nach Spandau am Morgen des 19. März 1848* (Berlin, 1848), pp. 130–135. A bibliography of Eisenstein's writings is given by Kurt-R. Biermann in *Istoriko-mathematicheskie issledovaniya,* **12** (1959), 493–502.

Historical records are available primarily at the following institutions: Archiv der Deutschen Akademie der Wissenschaften zu Berlin; Archiv der Humboldt-Universität zu Berlin; Deutsche Staatsbibliothek, Berlin; Niedersächsische Staats- und Universitäts-Bibliothek, Göttingen; Archiv der Akademie der Wissenschaften, Göttingen; and Deutsches Zentralarchiv, Historische Abteilung II, Merseburg. See also the survey by Kurt-R. Biermann in *Journal für die reine und angewandte Mathematik,* **214/215** (1964), 28.

II. SECONDARY LITERATURE. See the bibliography by Kurt-R. Biermann in *Journal für die reine und angewandte Mathematik,* **214/215** (1964), 28–29. Only the literature devoted directly to Eisenstein will be cited here. See Wilhelm Ahrens, "Gotthold Eisenstein," in *Deutsche allgemeine Zeitung,* no. 177 (17 April 1923); Moritz Cantor, "Eisenstein," in *Allgemeine deutsche Biographie,* V (1877), 774, which contains errors; J. Loewenberg, "A. v. Humboldt und G. Eisenstein," in *Allgemeine Zeitung des Judenthums,* **55** (1891), 246–248; and Julius Schuster, "A. v. Humboldt und F. G. Eisenstein," in *Janus,* **26** (1922), 99.

See also the following works by Kurt-R. Biermann: "A. v. Humboldt als Protektor G. Eisensteins und dessen Wahl in die Berliner Akademie," in *Forschungen und Fortschritte,* **32** (1958), 78–81; "Zur Geschichte der Ehrenpromotion G. Eisensteins," *ibid.,* 332–335; "Die Briefe A. v. Humboldts an F. G. M. Eisenstein," in *Alexander von Humboldt, Gedenkschrift* (Berlin, 1959), 117–159;

"A. L. Crelles Verhältnis zu G. Eisenstein," in *Monatsbericht der Deutschen Akademie der Wissenschaften zu Berlin,* **1** (1959), 67–72; "Eisenstein," in *Neue deutsche Biographie,* IV (1959), 420–421; "Einige neue Ergebnisse der Eisenstein-Forschung," in *Zeitschrift für Geschichte der Naturwissenschaften, Technik und Medizin,* **1,** no. 2 (1961), 1–12; and "G. Eisenstein, Die wichtigsten Daten seines Lebens und Wirkens," in *Journal für die reine und angewandte Mathematik,* **214/215** (1964), 19–30.

KURT-R. BIERMANN

EKEBERG, ANDERS GUSTAF (*b.* Stockholm, Sweden, 15 January 1767; *d.* Uppsala, Sweden, 11 February 1813), *chemistry, mineralogy.*

Ekeberg studied in Uppsala, Greifswald, and Berlin from 1784 to 1790. He worked at the Council of Mining in 1794 and in the same year became assistant professor of chemistry at the University of Uppsala. In 1799 he became associate professor and a member of the Royal Swedish Academy of Science.

While in Greifswald, Ekeberg had studied under Christian Ehrenfried Weigel, the follower and German translator of Lavoisier. That the New Chemistry was introduced into Germany at such an early date was undoubtedly due to Weigel; as Weigel's enthusiastic pupil, Ekeberg in his turn helped to spread it northward.

Ekeberg's article "Om Chemiska Vetenskapens närvarande skick" ("On the Present State of the Chemical Science"), published in 1795, was the first attempt to present the antiphlogiston theory in Sweden. Ekeberg consolidated his position with a pamphlet, published the same year, entitled *Försök till Svensk nomenklatur för Chemien* . . . ("An Attempt Toward a Swedish Nomenclature for Chemistry"), in which the terminology introduced was Lavoisier's. Both of these outspokenly antiphlogistic works were published anonymously because Ekeberg was anxious to avoid conflicts with Johan Afzelius, his superior in Uppsala, who distrusted the new theories.

Ekeberg was an extraordinarily capable analytic chemist. Shortly after taking up his duties at Uppsala, probably about 1795–1796, he became interested in a remarkable mineral quarried in Ytterby in Sweden; he made a thorough investigation of it and was thus able in 1797 to confirm Gadolin's earlier discovery of yttria. After further prolonged research he announced in 1802 that he had found yttria in a new mineral from Ytterby which also contained a hitherto unknown heavy metal. Ekeberg was the first to define this heavy metal precisely. On the basis of the inability of its oxide to combine with even the smallest particle of acid—even when it was submerged in it—Ekeberg compared the new metal to Tantalus and called it "tantalum."

In addition to his scientific ability Ekeberg possessed a considerable literary talent which he demonstrated in his younger years. He suffered poor health throughout his life, however, and when this was aggravated by an impairment of vision and hearing, his vitality decreased and his promising scientific career came to a premature end.

BIBLIOGRAPHY

I. Original Works. Ekeberg's publications include "Om Chemiska Vetenskapens närvarande skick," in *Litteratur tidning för år 1795,* **1** (1795), 91–104; *Försök till Svensk Nomenklatur för Chemien, lämpad efter de sednaste uptäckterne* (Uppsala, 1795); "Ytterligare undersökningar af den svarta stenarten från Ytterby och den däri fundne egna jord," in *Kongliga Vetenskaps Academiens nya Handlingar,* **18** (1797), 156–164; "Uplysning om ytterjordens egenskaper, i synnerhet i jämförelse med berylljorden: Om de fossilier hvari förstnämnde jord innehålles, samt om en ny uptäckt kropp af metallisk natur. Tantalum," *ibid.,* **23** (1802), 68–73; "Chemisk undersökning af et hårdt oktaedriskt kristalliseradt fossil ifrån Fahlun," in *Afhandlingar i fysik, kemi och mineralogi,* **1** (1806), 84–90; and "Undersökning af ett natronhaltigt fossil ifrån Hesselkulla," *ibid.* (1807), 144–153.

II. Secondary Literature. On Ekeberg's life and work see "Anders Gustaf Ekebergs biographie," in *Kongliga Vetenskaps Academiens Handlingar,* 3rd ser. (1813), 276–279; and Arne Westgren, "Anders Gustaf Ekebergs föreläsningar 1805–1811," in *The Svedberg 1884 $\frac{30}{8}$ 1944* (Uppsala, 1944).

Uno Boklund

EKMAN, VAGN WALFRID (*b.* Stockholm, Sweden, 3 May 1874; *d.* Gostad, Stockaryd, Sweden, 9 March 1954), *oceanography.*

Ekman belonged to the group of Scandinavian oceanographers who, at the beginning of this century, started a new and very fruitful line in physical oceanography. He worked with Fridtjof Nansen, Vilhelm Bjerknes, and B. Helland-Hansen and, like them, must be ranked as one of the great oceanographers.

Ekman was the youngest son of Fredrik Laurentz Ekman, a professor who also worked in oceanography. He went to school in Stockholm and then studied at the University of Uppsala. His interest in hydrodynamics and oceanography came through his contact with Bjerknes, who introduced him to the theoretical problem of the "wind spiral," which became the subject of his thesis. The original idea of the spiral came from Nansen, who observed a systematic drift of the ice to the right of the wind direc-

tion during his famous *Fram* expedition. Nansen suggested that Ekman investigate the problem mathematically, reasoning that each layer of the sea must be set in motion by the layer immediately above and be successively more deflected to the right by the Coriolis force (on the northern hemisphere). Ekman's thesis was a short paper in Swedish published in 1902, and it did not immediately attract any attention. Through an enlarged paper, "On the Influence of the Earth's Rotation on Ocean Currents," published in 1905 in *Arkiv för matematik, astronomi och fysik,* his theory became known to the international scientific community. Its importance has since been well established. The "Ekman layer" is one of the cornerstones of modern theories of oceanic circulation, and it plays an important role in practically all theoretical and experimental works on rotating fluids.

Another important theoretical contribution was given in a paper published in 1923 in the same journal. Here Ekman developed, for the first time, a complete mathematical theory for the wind-driven circulation in an oceanic basin. It took almost thirty years before a new generation of theoretical oceanographers was able to catch up with these ideas and build further on Ekman's theory. Both the 1905 and the 1923 papers are masterpieces of clarity and elegance and can certainly be enjoyed by all those interested in the subject, whether professionally or not. The first paper has been reprinted in booklet form by the Royal Swedish Academy of Sciences, Stockholm, and there are plans to print an English edition of the second paper.

Ekman was a very good experimentalist. He made determinations of the equation state for seawater and studied several important hydrodynamic phenomena both in the laboratory and in the sea. Probably the most widely known of his studies is that devoted to dead water, the strong resistance experienced by ships in the Norwegian fjords because of a particular stratification of the water.

Ekman also constructed oceanographic instruments. His current meter, which gives speed and direction by use of a purely mechanical system, is still a standard tool in oceanographic studies. With Helland-Hansen, Ekman made a number of cruises in the Norwegian Sea to test his current meter. Later, in 1930, he cruised to the trade-wind belt in the Atlantic to carry out systematic current measurements. Unfortunately some of the material was lost during World War II, and it was not until 1953, when Ekman was seventy-nine, that the results were finally published. As usual, Ekman had perfected his work in every detail.

While carrying out his oceanographic research,

Ekman was assistant at the International Oceanographic Laboratory in Oslo from 1902 to 1908, then lecturer in mechanics and mathematical physics in Lund, where he received a full professorship in 1910. He became a member of several learned societies and received the Agassiz Medal in 1928 and the Vega Medal in 1939. Ekman published a total of more than 100 scientific articles as well as several articles on philosophical and religious subjects. He was an active member of the Lutheran church and published in its newspaper, *Kyrkobröderna*.

Ekman was extraordinary in his requirements for truth and exactness in every detail, both in his scientific work and in his private life. He believed that no human being has the right to cause injustice or harm to anyone else, and he certainly lived up to his ideals. Governed rigorously by his principles, Ekman may have appeared impersonal to some, but he had a warmth and spontaneity that was revealed to his family and close friends.

BIBLIOGRAPHY

A complete bibliography, with biography by B. Kullenberg, is given in a repr. of "On the Influence of the Earth's Rotation on Ocean Currents" (Uppsala, 1963). Among his works are "On Dead Water," in *The Norwegian North Polar Expedition 1893–1896. Scientific Results*, V (1904), 15; "On the Influence of the Earth's Rotation on Ocean Currents," in *Arkiv för matematik, astronomi och fysik*, **2** (1905), 11; *Tables for Sea Water Under Pressure*, Conseil Permanent International pour l'Exploration de la Mer, Publication de Circonstance no. 49 (1910); "Über Horizontalzirkulation bei winderzeugten Meeresströmungen," in *Arkiv för matematik, astronomi och fysik*, **17** (1923), 26; *On a New Repeating Current Meter*, Conseil Permanent International pour l'Exploration de la Mer, Publication de Circonstance, no. 91 (1926); and "Turbulent, Periodic and Mean Motions. Some Measurements in the Atlantic," in *Procès-verbaux. Association d'océanographie physique*, **4** (1949), written with B. Helland-Hansen.

PIERRE WELANDER

ELHUYAR (or **ELHUYART**), **FAUSTO D'** (*b.* Logroño, Spain, 11 October 1755; *d.* Madrid, Spain, 6 January 1833), *chemistry, mineralogy, assaying.*

The younger brother of Juan José D'Elhuyar, Fausto shared his brother's studies, profession, and travels until 8 October 1781, when they separated in Vienna and Fausto returned to Spain. There he taught mineralogy and structural, or geotectonic, geology at the Real Seminario Patriótico in Vergara and, with François Chabaneau, lecturer in physics and chemistry, founded the Real Escuela Metalúrgica. There too D'Elhuyar collaborated in the experiments conducted by his brother Juan José, the discoverer of metallic tungsten, and in 1784 he distributed the monograph that had been published under both their names and made it well known when Juan José left for South America. Thus Fausto is the better known of the two and often is erroneously credited with having made the larger contribution to the discovery of metallic tungsten.

In 1785 D'Elhuyar abandoned the teaching of mineralogy and worked with Chabaneau on separating platinum and rendering it malleable. On this subject he had his own ideas and methods, which were opposed to those of Carl von Sickingen. He was unable to develop them, however, for the Spanish government commissioned him to visit his friend Ignaz von Born in Hungary, to see at firsthand Born's new method of amalgamation for the treatment of gold and silver ores. Fausto sent this information to his brother, as well as drawings and models of apparatus to assist Juan José in Nueva Granada (now Colombia). The method was regarded as a revolutionary one, but Born conceded that its origins were in Alvaro Alonso Barba's *El arte de los metales*. D'Elhuyar improved it, however, by adding salt and lime and showed that, contrary to the prevailing view, gold and silver ores could be roasted. The French chemist Joseph-Louis Proust extolled his work in his "Extracto de los descubrimientos de Don Fausto D'Elhuyar" (published in Spanish in *Anales del real laboratorio quimico de Segovia*, **1** [1791]). D'Elhuyar was an avid researcher. In the first of his *Disertaciones metalúrgicas* (Madrid, 1933) he asserted that there were no essential differences between chemical substances, and he adopted the phlogiston theory only in order to make himself easily understood. From his experiments he concluded that "there is no other difference between metals than that which distinguishes each in kind"—there are only gradations. D'Elhuyar also discovered chloroargentic acid and obtained new results in combining sulfur and metals.

The Spanish government put him in charge of organizing missions of metallurgists to go to Mexico and Peru. D'Elhuyar selected Friedrich Sonneschmidt to head the first and Baron Nordenflicht to head the second. The development and prosperity of Mexican mining was due largely to D'Elhuyar's efforts, for when he arrived it was on the decline. After Alexander von Humboldt visited Mexico, he wrote in his *Political Essay on the Kingdom of New Spain:* "No city on the New Continent, not even in the United States, offers scientific establishments so vast and so solid as does the capital of Mexico. It is enough to cite here the School of Mines, of which the scholar D'Elhuyar is Director."

With the start of the War of Independence in 1810, the work of the College of Mines was interrupted and it went into decline, as did mining throughout Mexico. During the conflict D'Elhuyar remained loyal to Spain. In 1821 he resigned his post and returned to Madrid, where he became a member of the Directorate General of Public Credit and of the Development Board. His *Memoria para la formación de una ley orgánica para el gobierno de la minería en España* was the basis for the mining law enacted in 1825. Although he was appointed Director General of Mines, he was assigned to other duties, and his learning and research skills were not fully utilized.

BIBLIOGRAPHY

On D'Elhuyar or his work see the following: *Biografía del ilustrísimo Señor Don Fausto D'Elhuyar y de Subice* (sic) (Madrid, 1853); Juan Fages y Virgili, *Los químicos de Vergara y sus obras* (Madrid, 1909); A. de Gálvez-Cañero y Alzola, *Apuntes biográficos de Don Fausto D'Elhuyar* (Madrid, 1933); A. Federico Gredilla, *Biografía de José Celestino Mutis* (Madrid, 1911); J. Guzmán, *Las disertaciones metalúrgicas de Fausto D'Elhuyar* (Madrid, 1933); C. López-Sánchez, *Elhuyar, minero-metalúrgico* (Madrid, 1933); E. Moles, "Elhuyar, químico," in *Anales de la Sociedad española de física y química* (Feb. 1933); Nicolás de Soroluce, *Real sociedad bascongada de amigos del país* (San Sebastián, Spain, 1880); Mary Elvira Weeks, *Discovery of the Elements,* 6th ed. (Detroit, 1956); and Arthur P. Whitaker, "The Elhuyar Mining Missions and the Enlightenment," in *Hispanic American Review* (Nov. 1961); and *Latin America and the Enlightenment* (Ithaca, N.Y., 1961).

Manuscript sources include the author's library, which contains some of D'Elhuyar's correspondence; and the parish records of Logroño, Spain, and Bayonne and Saint Jean-de-Luz, France.

BERNARDO J. CAYCEDO

ELHUYAR (or **ELHUYART**), **JUAN JOSÉ D'** (*b.* Logroño, Spain, 15 June 1754; *d.* Bogotá, Nueva Granada [now Colombia], 20 September 1796), *chemistry, mineralogy, metallurgy.*

D'Elhuyar's father, Juan, a well-known surgeon, and his mother, Ursula Lubice, were French Basques who moved to Spain and settled in Logroño. Juan José received his elementary education at Oyón (Navarre) and Logroño. Sent to Paris with his brother Fausto, he studied medicine for five years (1772–1777) and was a pupil of the chemist and mineralogist Hilaire-Marin Rouelle. Upon returning to Spain in 1777 he joined the Sociedad Económica de Amigos del País, founded by the count of Peñaflorida, whose son Antonio had been a schoolmate of D'Elhuyar's. The latter had to give up his medical career, however,

because the minister of the navy, concerned with preparing Spain for war with Great Britain, sent him and his brother Fausto to study geology, mineralogy, and metallurgy at the Mining Academy in Freiberg, Saxony, so that their knowledge could be applied to the treatment of iron and steel in the manufacture of cannon.

In Freiberg, D'Elhuyar attended lectures by Johann W. Charpentier on structural, or geotectonic, geology, by Gellert on metallurgical chemistry, and especially those by Werner, on geology and petrography. Both the lectures at the academy and the fieldwork contributed to his earning in later years the cognomen "sabio" (scholar), which was applied to him by his friend José Celestino Mutis, a physician and botanist. After visiting the mercury deposits in Idrija and the mining districts of Rosenau, Hungary, Bohemia, the duchy of Zweibrücken, the Rhenish Palatinate, and Austria with his brother, D'Elhuyar proceeded by himself in 1781 to Sweden, Norway, and Denmark. At the University of Uppsala he attended lectures by Peter Jacob Hjelm and especially those by Bergman. He also met Scheele. Bergman and Scheele were seeking a new element but got no further than obtaining tungstic acid. D'Elhuyar participated in the experiments, and after his return to Spain, he succeeded in isolating the metallic element tungsten.

In 1783 D'Elhuyar and his brother Fausto published *Análisis químico del wolfram y examen de un nuevo metal que entra en su composición,* which attracted great interest in scientific circles and was translated into several languages. In December 1783 he was appointed director of mines of Nueva Granada (now Colombia); his principal occupation consisted of managing the Mariquita silver mines. In 1796 Viceroy Ezpeleta ordered the exploitation of the silver mines stopped, and D'Elhuyar moved to Bogotá, where he died, leaving a widow (the former Josefa Bastida-Lee) and three children. He wrote many reports to and carried on considerable correspondence with three viceroys and other officials of the Spanish Crown. He perfected Born's method for amalgamating silver and mercury and recommended a new process for the isolation of platinum. The colonial authorities interfered with D'Elhuyar's work and did not utilize his abilities as a scientific researcher.

BIBLIOGRAPHY

I. ORIGINAL WORKS. D'Elhuyar's published work is *Análisis químico del wolfram y examen de un nuevo metal que entra en su composición* (Vitoria, Spain, 1783), written with his brother Fausto. Notes taken by D'Elhuyar while attending Bergman's course in special chemistry at the University of Uppsala in 1782, concerning the new discoveries in chemistry, are in *Lychnos* (1959), pp. 162–207, with an introduction by Stig Rydén and Arne Fredga.

II. SECONDARY LITERATURE. On D'Elhuyar or his work, see *Archivo epistolar del sabio naturalista José Celestino Mutis*, I (Bogotá, 1947), 82–163; Juan Fages y Virgili, *Los químicos de Vergara y sus obras* (Madrid, 1909); Federico Gredilla, *Biografía de Don José Celestine Mutis* (Madrid, 1911), 230–246, 301–316; Vicente Restrepo, "Biografía de Juan José D'Elhuyar," in *Estudio sobre las minas de oro y plata de Colombia* (Bogotá, 1888), pp. 230–246; Stig Rydén, *Don Juan José D'Elhuyar en Suecia y el descubrimiento del tungsteno* (Madrid, 1962); Mary Elvira Weeks, *Discovery of the Elements,* 6th ed. (Detroit, 1956), pp. 113–114, 132–147; Arthur P. Whitaker, "The Elhuyar Mining Missions and the Enlightenment," in *Hispanic American Historical Review,* **31,** no. 4 (Nov. 1951), 558–585; and José Zamora Mendoza, *Don Juan D'Elhuyar, prestigioso cirujano del Hospital de Logroño* (Logroño, 1956), and Marcelino Menéndez Pelayo, *La ciencia española,* vol. II.

Manuscript sources are to be found in the following locations: National Archives, Bogotá, Cataloging Room, no. 169, notarial registries; the author's archives, which contain some of D'Elhuyar's correspondence, as well as many documents relating to both his private and his public life; the parish records of Logroño, Spain; Bayonne, France; and Bogotá, Colombia; and the records of the Real Expedición Botánica del Nuevo Reino de Granada, at the Botanical Garden in Madrid, file 24.

BERNARDO J. CAYCEDO

ÉLIE DE BEAUMONT, JEAN-BAPTISTE-ARMAND-LOUIS-LÉONCE (*b.* Canon, Calvados, France, 25 September 1798; *d.* Canon, 21 September 1874), *geology.*

Élie de Beaumont was the elder son of Armand-Jean-Baptiste-Anne-Robert Élie de Beaumont and Marie-Charlotte-Eléonore Mercier Dupaty; their marriage united the families of two jurists who had achieved fame under the *ancien régime.*

Following a widespread custom of the bourgeoisie in the eighteenth century, Élie de Beaumont's grandfather, Jean-Baptiste, a Norman lawyer, joined to his patronymic the name of an estate, in order to distinguish himself from his brother Jean-Antoine, *docteur régent* of the Faculté de Médecine of Paris.

Under the Empire, Léonce's parents, who were living at Canon, engaged a Benedictine monk, Dom Raphaël de Hérino, to serve as tutor to him and his brother, Charles-Adolphe-Eugène. At the beginning of the Restoration, when the children had nearly completed their secondary studies, the family moved to the rue de la Muette, in Faubourg Saint-Antoine, a quarter of Paris not far from the Collège Royal Charlemagne. At this *collège,* where he was called

Élie-Debeaumont, Léonce obtained fourth honorable mention in elementary mathematics in the general competition of 1816. In 1817, after a year of higher mathematics at the Collège Henri IV, he won first prize in mathematics and physics in the general competition and was second on the admissions list of the École Polytechnique, which methodically collected the country's most gifted students for intensive training in mathematics. Graduated first in his class from the Polytechnique, he chose the Corps des Mines, which had not received any engineering students from the last three graduating classes. He entered the École Royale des Mines on 15 November 1819.

The École des Mines offered four two-year courses. Baillet du Belloy, in his course on the working of mines, also discussed hydraulic engines, steam engines, and subterranean topography; Berthier taught docimasy (analytic mineral chemistry); Hassenfratz described the primary treatment processes for all kinds of ores (that is, ore dressing, or beneficiation of ore); and Brochant de Villiers alternately taught one year of mineralogy and one of geology. Between the first and second years the engineering students worked in the laboratory, practiced drafting, and made plans of the catacombs. After the second year they undertook study trips of several months' duration in a mining or metallurgical region, following an itinerary outlined in detail by the council of the École des Mines. Their studies were judged complete when they had attained a certain level in each subject area. In general, students had to spend a third year at the École des Mines in order to reach that level and then were required to take a second study trip.

This was the case with Élie de Beaumont, who, beginning in November 1820, also attended the Faculté des Sciences in Paris. During the summer of 1821 he devoted his first study trip to visiting the iron mines and forges in eastern France and began making geological observations in the Vosges, where his guide was Philippe Voltz, a mining engineer in Strasbourg who was particularly interested in paleontology. In 1822 a second study trip took Élie de Beaumont to Switzerland, where his guide was Jean de Charpentier, director of the salt mines of Bex and a former student of Werner's at the Freiberg Bergakademie. Charpentier also acquainted Élie de Beaumont with his observations on glaciers: he was the first to propose that they had transported erratic boulders deposited on the Swiss Jura. Élie de Beaumont returned on foot to Paris, passing through Auvergne. The main portions of his journals of the two trips, which were considered useful in teaching others, were published in the *Annales des mines* in 1822 and 1824.

On 1 January 1823 Élie de Beaumont was named an engineering cadet at the same time as Charles Combes, who had been in the class after his. All the mining engineers, including the older ones like Hassenfratz and Héron de Villefosse, were concerned about France's underdeveloped industry, especially vis-à-vis England. They thought that first priority should be given to putting France on an equal footing with its rival, especially in the use of steam engines, the development of collieries and mines, and in metallurgy. At the Ministry of the Interior their point of view was shared by the director general of bridges, highways, and mines, Louis Becquey.

The absence of outcroppings and the lack of knowledge of the substrata made prospecting for sedimentary deposits, particularly the search for coal beds, extremely uncertain. To all those concerned with the development of a mineral industry in France, the importance of preparing a geological map had become evident. The idea of representing the nature of the terrain on a topographical map was not new in France. In 1664 the monograph of the Abbé Louis Coulon, *Les rivières de France, ou Description géographique et historique du cours et débordement des fleuves, rivières . . . de France, avec un dénombrement des villes, ponts, passages, . . .*, had been reprinted, accompanied by a map indicating the boundaries of the granite and the sedimentary formations.

In the middle of the eighteenth century, when Guettard had surveyed northern France, he had perceived that the different formations that constitute its soil form large concentric bands about Paris. In 1746 he published in the *Mémoires de l'Académie royale des sciences* a map in which he proposed "to show that there is a certain regularity in the distribution of the rocks, the metals, and most of the fossil substances." Subsequently he was commissioned by Bertin to explore the whole of France from the point of view of mineralogy and to publish descriptions and maps of the provinces.

Originally it had been planned to cover all of France by means of 214 mineralogical maps, but a lack of funds permitted only forty-five maps to be made. Monnet published them in 1780 in the *Atlas et description minéralogiques de la France, entrepris par ordre du Roi . . . 1re partie, comprenant le Beauvoisis, la Picardie, le Boulonnais, la Flandre Française, le pays Messin, et une partie de la Champagne.*

On 6 July 1794 the Committee of Public Safety issued a decree (the text of which Hassenfratz had composed) that the engineers of the Corps des Mines were to search for mineral substances in their districts and to map their discoveries. The *Journal des mines* published several geological memoirs written by mining engineers, as well as mineralogical reports on several departments by Coquebert de Montbret. In collaboration with the latter, the geologist Omalius d'Halloy prepared, between 1810 and 1813, "Essai d'une carte géologique de la France, des Pays-Bas et de quelques contrées voisines." It was published in the *Annales des mines,* but not until 1822, when the subject had become of current interest. This map, drawn on a scale of about 1:3,600,000, distinguished six great rock systems: primordial, penean, ammonean, cretaceous, mastozooic, and pyroidic.

Brochant de Villiers, commissioned in 1802 to teach geology at the École des Mines, had seen the importance of preparing a more detailed geological map. In 1811 he had presented to the director-general of mines, Count Laumond, a project for the execution of such a map. But Napoleon and his ministers, who did not understand the revolutionary economic role of the steam engine and were thus unable to grasp the importance of developing the coal industry, considered the establishment of a geological map to be one of those academic exercises that can be postponed indefinitely.

The Corps des Mines had to wait for the fall of the Empire to triumph over Napoleonic obtuseness. The royal order of 5 December 1816 instructed the council of the École des Mines, reestablished in Paris at the Hôtel Vendôme, to assemble all the materials necessary to complete the mineralogical description of France, and subsequently commissioned it to amass mineral collections and to publish geological and mining maps.

In 1820 the English geologist George Bellas Greenough published a geological map of England in six sheets, a copy of which he sent to the director-general of bridges, highways, and mines. On 11 June 1822 the council of the École des Mines, judging the occasion favorable, repeated its intention to produce a geological map of France and invited Brochant de Villiers to present a new report on this subject to Becquey. On 15 June, Brochant proposed that a mission composed of himself and two young mining engineers go the following year to England in order to confer with and study the methods of the English geologists. Dufrénoy, promoted to mining engineer on 1 June 1821, was the first collaborator that Brochant thought of, and Élie de Beaumont's good fortune was to complete his studies at the moment when a second assistant was being sought.

On 25 April 1823 Élie de Beaumont was invited by Becquey to leave immediately for Dover, "in order to improve his knowledge of the English language by speaking with Englishmen," and to wait there for Brochant and Dufrénoy. Their mission lasted six months. They inspected, in particular, tin and copper mines in Cornwall, lead mines in Cumberland and Derbyshire, and coal mines and ironworks in Wales. Upon their return Becquey complimented them for their work and for the smallness of their expenses. Dufrénoy and Élie de Beaumont devoted 1824 to composing memoirs on their mission, which were published in the *Annales des mines.*

Élie de Beaumont, appointed as mining engineer in May 1824, was assigned on 1 September 1824 to the Service des Mines at Rouen and placed in charge of the Seine-Inférieure (now Maritime) and Eure departments. On 29 June 1825 Becquey wrote to Brochant:

> I have reread your report of 15 June 1822 and the opinion of the council of the École des Mines. All these documents have strengthened me in the resolution that I announced to you in 1820: having a geological map of France made. This work is important: its results should be eminently useful and its execution can only bring honor to the Royal Corps of Mining Engineers. Consequently I approve the various measures that you proposed to me in concert with the council of the École des Mines. The mineralogical description of France ought therefore to include all the information relative to the nature of the terrain that could be of interest at once to *geology, the art of mining, and all the other arts practiced on mineral substances,* as well as information on the different kinds of formations, their relationships, their boundaries, exploited and unexploited deposits of useful minerals, and finally, the location of mine mills. In order to present without confusion such manifold and very often closely related data, one should draw up: (1) *a general geological map of France* including all information on the nature of the formations; (2) *departmental geological and mineralogical maps.* The work will commence with the general geological map. . . . Your proposed geological division of the work among the engineers seems to me clear and precise. I easily follow the twisting diagonal line by which you divide France, which goes from the northwest to the southeast, running along the boundary of the great chalk massif from Calvados to the vicinity of Saumur, Châtellerault, and Auxerre, then turning back toward Avallon and Chalon and, beyond that, following the course of the Saône and the Rhone. . . . Each year please inform me in advance of the probable expense, since funds are so limited that I must be sure of being able to meet expenses before authorizing them. According to the information that you have presented, the total expense for 1825 will be approximately 3,000 francs, including everything.

Dufrénoy, in charge of the western division, and Élie de Beaumont, in charge of the eastern division, spent the first five years, from 1825 to 1829, exploring their sectors on foot during the summer months,

noting their observations on Cassini de Thury's map, drawn up on 180 sheets on a scale of 1 line for 100 *toises* (1:86,400). From 1826 to 1828 Dufrénoy was assisted by E. de Billy, and Élie de Beaumont by Fénéon. In 1830 they went to the Alps with Brochant to verify the abnormal contact, observed by Élie de Beaumont, of granite and an underlying calcareous layer; this inaugurated their joint expeditions, which were continued until 1836. On 20 December 1841, they presented to the Académie des Sciences the *Carte géologique générale de la France,* drawn on six sheets on a scale of 1:500,000, and the first volume of the *Explication.* The 100-page introduction to the latter is still the best that can be placed at the beginning of modern treatises on physical geography.   .

Dufrénoy and Élie de Beaumont, appointed mining engineers first class in May 1832, henceforth received the same promotions at the same time: chief engineer in May 1833 and inspector general in March 1848.

From 1825 Élie de Beaumont devoted himself almost exclusively to geology, on which he started to lecture in 1827 at the École des Mines and in 1832 at the Collège de France, where in 1848 he had as auditors the students of the first École d'Administration.

On 22 June 1829 Élie de Beaumont presented to the Académie des Sciences his first ideas on tectonics, showing that the various mountain chains are of different ages. In this exposition, to which Arago devoted seventeen enthusiastic pages in the *Annuaire pour 1830* of the Bureau des Longitudes, Élie de Beaumont distinguished six systems of uplift, each characterized by one direction. In 1833, in a note inserted in the translation of Henry De la Beche's *Manual of Geology,* he increased the number of systems to twelve. In 1834, at the Sociéte Géologique de France, of which he had been one of the founders in March 1830, his hypotheses were sharply criticized by Ami Boué.

In 1833, in a memoir entitled *Sur les groupes du Cantal, du Mont-Dore et sur les soulèvements auxquels ces montagnes doivent leur relief actuel,* Dufrénoy and Élie de Beaumont unfortunately borrowed from Leopold von Buch the theory of elevation craters. They thought that great lava flows can spread only over surfaces that are almost horizontal. Occasionally a force acting upward from below, which they supposed was the upheaval of a plug of solid lava, would raise the flows thus formed and build what they called an elevation cone. At the point of application of this force, they stated, a crack with divergent fissures was produced; this divided into triangular sectors the fragments of lava that henceforth constituted the sides of the cone. When flows or thrusts produce a gap at the summit, this constitutes an elevation crater. Since there are spaces between the lava sectors thus formed, Dufrénoy and Élie de Beaumont developed formulas for the calculation of their area as a function of the diameter of the base of the cone and the slope of its sides. After stating these formulas, they believed, contrary to the evidence, that they were verified by observation in central France and then at Vesuvius and Etna, which they inspected the following year with Buch.

In 1846–1847 Élie de Beaumont devoted his course at the Collège de France to volcanic and metalliferous emanations; he continued it on 5 July 1847 at the Société Géologique de France in a statement that was the first complete theory of metalliferous veins. It is his most solid scientific work besides his surveys for the geological map.

On 5 May 1838, at the Société Philomathique, Élie de Beaumont appeared like a mathematician lost among the natural sciences, applying to sedimentary folds considerations borrowed from the theory of ruled surfaces; and these considerations, moreover, were not original, since, as Babinet observed, they had already been stated by Monge in another form.

A growing detachment from observation and a love of calculation led Élie de Beaumont, beginning in 1850, to connect the stylized directions of the mountain chains to a system of terrestrial great circles forming regular pentagons in gnomonic projection. This was the point of departure for his delusory theory of the pentagonal grid, which he represented in 1866 on the *Carte géologique générale de la France* and which was taught for some thirty years at the École des Mines.

On 21 December 1835 Élie de Beaumont was elected to the Académie des Sciences, in the mineralogy section; on 19 December 1853 he was named perpetual secretary for the mathematical sciences, replacing Arago. In December 1859 he married Thérèse-Marie-Augusta de Quélen, the widow of the marquis du Bouchet; she died childless in 1866.

Élie de Beaumont was director of the Service de la Carte Géologique from its organization in 1865 until 1868, when, having reached the age of retirement, he had to cede to Combes the chairmanship of the Conseil Général des Mines, an office which he had held since 1861.

A dogmatic, cold, and distant mathematician, Élie de Beaumont appeared to his contemporaries as a misguided pundit—an impression not lessened by his attachment to the theory of elevation craters and to the pentagonal grid, his noncomprehension of the discoveries of Boucher de Perthes, and his unreserved support of the gullible geometer Michel Chasles during the scandal that developed from the latter's presentation to the Academy of forged manuscripts which he had purchased.

BIBLIOGRAPHY

I. ORIGINAL WORKS. Élie de Beaumont's papers are too numerous to cite individually. A nearly complete list is in the Royal Society of London *Catalogue of Scientific Papers,* I (1868), 476–479; VI (1872), 648; VII (1877), 607; and IX (1891), 787. Poggendorff, I (1863), 657–658, and III (1898), 404, is incomplete.

The major works were written with Dufrénoy: *Voyage métallurgique en Angleterre, ou Recueil de mémoires sur le gisement, l'exploitation et le traitement des minerais d'étain, de cuivre, de plomb, de zinc et de fer dans la Grande-Bretagne,* 2 vols. (Paris, 1827); *Mémoires pour servir à une description géologique de la France,* 4 vols. (Paris, 1830–1838); *Carte géologique de la France (6 feuilles), Tableau d'assemblage des six feuilles de la carte géologique* (Paris, 1841); *Explication de la carte géologique de la France,* 2 vols. (Paris, 1841–1848); *Description du terrain houiller de la France* (Paris, 1842).

II. SECONDARY LITERATURE. Works on Élie de Beaumont are J. Bertrand, *Éloge historique de Élie de Beaumont* (Paris, 1875); P. Fallot, "Élie de Beaumont et l'évolution des sciences géologiques au Collège de France," in *Annales des mines,* 13th ser., *Mémoires,* **15** (1939), 75–107.

Additional materials may be found in the Archives Nationales (Paris), F14.2723[1], and in the library of the École des Mines de Paris (Élie de Beaumont's papers).

ARTHUR BIREMBAUT

ELKIN, WILLIAM LEWIS (*b.* New Orleans, Louisiana, 29 April 1855; *d.* New Haven, Connecticut, 30 May 1933), *positional astronomy, meteoritics.*

At the time of Elkin's birth his father, Lewis, was inspector of public schools for the city of New Orleans, but he later became a manufacturer of carpets and prospered financially. In 1867 he was appointed representative of the state of Louisiana at the forthcoming Paris Exposition but died unexpectedly just before the family's departure for Europe. All arrangements had been made and his widow, Jane Fitch Elkin, was persuaded to go as scheduled, taking with her their twelve-year-old son William, the only survivor of her five children. Mother and son became expatriates, leading an interesting but peripatetic life in various countries, so that Elkin's secondary education was unorthodox. He did, however, attend private schools in Switzerland but at the age of fifteen had a serious illness, which left his health permanently impaired.

Nevertheless, Elkin decided upon a career as an

engineer and enrolled in the Royal Polytechnic School at Stuttgart, from which he graduated in 1876 with a C.E. degree. But by now his interest had changed to astronomy; so he went to Strasbourg to study under Friedrich August Theodor Winnecke, then considered the foremost teacher of astronomy in Europe. He was a third-year graduate student, preparing a dissertation on the star α Centauri, when a chance meeting with David Gill, newly appointed royal astronomer at the Cape of Good Hope, led to an invitation to South Africa for a working (but personally financed) visit. After receiving his Ph.D. in 1880, Elkin joined Gill and his wife at the Cape and lived with them like a son for more than two years. Together Gill and Elkin set up and shared an ambitious observing program with a four-inch heliometer, to determine the parallaxes of nine stars: three observed by each separately and three by both of them. One of the latter was α Centauri, for which they published a value of 0.75″, with a probable error of 0.01″ (compared to today's value of 0.760″ ± 0.005″).

Meanwhile, a six-inch heliometer had been installed in the Yale University Observatory, and Elkin was invited to come there, as "astronomer in charge of the heliometer." He arrived in 1884 and remained in New Haven for the rest of his life, at first actively observing, then serving as director of the Yale observatory from 1896 (the year he married Catharine Adams) until increasing disability forced his retirement in 1910, at the age of fifty-five. The remaining twenty-three years of his life were spent as an invalid, with no further astronomical activities.

During his productive years at Yale, Elkin performed a formidable amount of work with the heliometer. The labor of measuring parallactic angles is great enough with today's photographic techniques but was many times greater with the heliometer, which in addition was an exhausting instrument to use. Elkin, together with his students, used the Yale heliometer to determine the parallaxes of 238 stars; this was, in the words of Frank Schlesinger (Elkin's successor at Yale), "by far the most important single contribution to our knowledge of stellar distances up to that time."

Concurrently Elkin was cooperating with Gill in an attempt to get an improved value for the solar parallax, by making simultaneous observations, from their widely separated locations, of the minor planets Iris, Victoria, and Sappho. Other observatories with heliometers later joined in this program. Their value, published in 1896–1897, was 8.812″ ± 0.009″, close to the 8.790″ obtained in 1930–1931 with photographic techniques; however, both have now been superseded by a value of 8.79415″, derived from 1961 radar observations of the planet Venus.

Elkin also undertook a program for determining the positions of stars near the north celestial pole, at the request of E. C. Pickering (then director of the Harvard College Observatory), who needed good reference points for a projected photographic survey of that part of the sky. Elkin himself never used photography for determining stellar positions, although in papers published in 1889 and 1892 he compared photographic data on stars in the Pleiades cluster with his own heliometer measurements and declared that the new method showed promise. He did, however, apply photographic techniques to the study of meteors: using two "meteorographs," each consisting of a number of cameras mounted on a motor-driven polar axis and placed several miles apart, he was able to determine the heights at which one Leonid meteor appeared and disappeared—many more were observed visually, but apparently his photographic emulsion was not fast enough to record them. Of greater significance was his use of a rotating bicycle wheel with occulting segments placed in front of his cameras to produce the first successful interrupted photographs of meteor trails, thereby providing information on their velocities.

Recognition for his work was international. Elkin was elected a foreign associate of the Royal Astronomical Society in 1892, received an honorary M.A. from Yale in 1893, and became a member of the American Academy of Sciences in 1895. He shared with his student Frederick Lincoln Chase the Prix Lalande of the French Academy of Sciences in 1908 (for work on stellar parallaxes) and was awarded an honorary Ph.D. by the University of Christiania (now Oslo) in 1911.

BIBLIOGRAPHY

I. Original Works. Elkin's dissertation was published as *Parallaxe von α-Centauri* (Karlsruhe, 1880). His work at the Cape observatory appeared as "I. Heliometer-Determinations of Stellar Parallax in the Southern Hemisphere," in *Memoirs of the Royal Astronomical Society*, **48** (1884), 1–194, written with David Gill. His heliometer work at Yale was published under the general title "Researches With the Heliometer," in *Transactions of the Astronomical Observatory of Yale University,* with those papers mentioned in the text being "Determination of the Relative Positions of the Principal Stars in the Group of the Pleiades," **1**, pt. 1 (1887), 1–105; "Triangulation of Stars in the Vicinity of the North Pole," **1**, pt. 3 (1893), 149–182; and "Catalogue of Yale Parallax Results," **2**, pt. 4 (1912), 385–400, written with Frederick L. Chase and Mason F. Smith.

The solar parallax program is described in *Annals of the Royal Observatory, Cape Town,* **6** (1897) and **7** (1896), "planned and discussed by David Gill . . . with the cooperation of Arthur Auwers and W. L. Elkin"; "Discussion of the Observations of Iris," **6,** pt. 4 (1)–(169), is the only part signed by Elkin.

Elkin's critiques of photographic methods were "Comparison of Dr. Gould's Reductions of Mr. Rutherfurd's *Pleiades* Photographs With the Heliometer-Results," in *Astronomical Journal,* **9** (1889), 33–35; and "The Rutherfurd Photographic Measures of the *Pleiades,*" in *Publications of the Astronomical Society of the Pacific,* **4** (1892), 134–138.

Elkin published only four brief papers on meteors (but see paper by Olivier listed below): "Photography of Meteors," in *Astronomical Journal,* **13** (1893), 132; "Photographic Observations of the Leonids at the Yale Observatory," in *Astrophysical Journal,* **9** (1899), 20–22; "Results of the Photographic Observations of the Leonids, November 14–15, 1898," *ibid.,* **10** (1899), 25–28; and "The Velocity of Meteors as Deduced From Photographs at the Yale Observatory," *ibid.,* **12** (1900), 4–7.

A list of thirty-one publications by Elkin appears in Frank Schlesinger's biographical memoir (see below).

II. SECONDARY LITERATURE. Gill's account of his first encounter with Elkin and their subsequent association at the Cape observatory constitutes the preface to their joint article in *Memoirs of the Royal Astronomical Society* (see above). Further facts about Elkin appear in *Monthly Notices of the Royal Astronomical Society,* **94** (1934), 285–289, with photograph (written by H. Spencer Jones, then astronomer royal); and in Frank Schlesinger, *Biographical Memoirs. National Academy of Sciences,* **18** (1938), 175–188, with portrait and list of publications. Elkin is listed in *Who Was Who in America,* I (Chicago, 1943), 365.

An account of the awarding of the Prix Lalande for 1908 occurs in *Comptes rendus hebdomadaires des séances de l'Académie des sciences,* **147** (1908), 1123. Elkin's unpublished work on meteors was presented and analyzed by Charles P. Olivier as "Results of the Yale Photographic Meteor Work, 1893–1909," in *Astronomical Journal,* **46** (1937–1938), 41–57.

SALLY H. DIEKE

ELLER VON BROCKHAUSEN, JOHANN THEODOR (*b.* Plötzkau, Germany, 29 November 1689; *d.* Berlin, Germany, 13 September 1760), *medicine, chemistry.*

Eller's father, Jobst Hermann Eller, was an eminent military man under the prince of Anhalt; his mother belonged to the Behm family, an ancient family in Livonia. He had an excellent education in law at Quedlinburg College and Jena University, but he changed to the study of medicine while at Jena. In 1711 he left Jena to search for better instruction in anatomy, going to Halle, then Leiden, and finally in 1712 to Amsterdam, where he found the most capable

anatomists in Europe, Rau and Ruysch. When Rau moved to Leiden, Eller went with him and performed public dissections for him until 1716. Eller was not prepared at this point to settle down to the practice of medicine, but turned to the study of mineralogy and chemistry, first with Lemery and Homberg in Paris, and then with Hauksbee and Desaguliers in London.

Upon his return to Anhalt-Bernberg in 1721, Eller was made court physician by the prince. He married Catherine Elizabeth Burckhard in October of that year. In 1724 King Frederick William I called Eller to Berlin and made him professor of anatomy and permanent dean of the Medical College, as well as physician to the army. When Frederick the Great became king in 1740, he made Eller his personal physician and appointed him director of the Berlin Academy of Sciences. Eller's wife died in 1751, and in 1753 he married Henrietta Catherine Rosen. In 1755 Frederick made Eller a privy councillor, a position he held until his death.

Eller held the highest medical positions in Prussia during his lifetime. He was a very competent doctor and administrator. In administration, for example, he was responsible, together with Georg Ernst Stahl, for laying the foundation for all subsequent developments in medical services in Prussia. Eller's writings reflect his medical knowledge and consist largely of compilations of case histories, like the one he published in Berlin in 1730, based on his experiences in the Charité Hospital in Berlin. He carried out some studies on human blood, examined human calculi, and warned about the dangers of using copper kitchen utensils, but he is not noted for any significant development in medicine.

Eller's theoretical chemistry is characterized by the central role given to heat (fire). His writings clearly place him in the Continental tradition that considered heat to be material in nature. With Eller, the Stahlian principle of phlogiston became simply another name for the primary element fire in the fixed state. In this state, fire is chemically combined with most substances and generally is released during chemical reactions. In the active state fire is the sole cause of the fluidity of water and fluids in general (air included). This principle is also integrated into his main practical chemical work, which dealt with the solubility of salts in water. Eller argued that since solubility generally increased with temperature, it must be because the fire in the water, whose activity increased with temperature, could more readily break up (dissolve) the salts.

Eller's influence on chemical theory has never been considered of great consequence, yet it is known that

Lavoisier read Eller's papers, especially those dealing with the elements, and his early views have an interesting resemblance to Eller's ideas about heat and fluidity.

BIBLIOGRAPHY

I. ORIGINAL WORKS. Eller's main work in medicine is *Nützliche und auserlesene medicinische und chirurgische Anmerkungen so wohl von innerlichen, als auch äusserlichen Krankheiten, . . . ; nebst einer vorangegebenen kurtzen Beschreibung der Stiftung, Anwachs und jetzigen Beschaffenheit dieses Hauses, usw.* (Berlin, 1730). All of Eller's papers, which were originally published in the Berlin Academy *Memoirs* and *Miscellanea Berolinensia,* are found in German translation in his collected papers, *Physikalisch-chymisch-medicinische Abhandlungen, aus den Gedenkschriften der königlichen Akademie der Wissenschaften,* D. Carl Abraham Gerhard, ed. (Berlin, 1764).

II. SECONDARY LITERATURE. There is no biography of Eller. Some information may be obtained from the *éloge* in the Berlin Academy *Memoirs* of 1761 and the *Allgemeine deutsche Biographie,* VI (Leipzig, 1877), 52–53. Only the larger histories of chemistry say anything about his scientific work.

DAVID R. DYCK

ELLIOT SMITH, GRAFTON (*b.* Grafton, Australia, 15 August 1871; *d.* Broadstairs, England, 1 January 1937), *anatomy.*

Elliot Smith was the son of Stephen Sheldrick Smith, a schoolteacher who had emigrated from England to Australia, and of his wife, Mary Jane Evans. At school he was interested in both physics and medicine, and he dated his interest in the brain to the age of ten, when he dissected a shark. He read medicine at the University of Sydney, graduated M.B., Ch.M. in 1892, and took some clinical posts while beginning research on brains. In 1895 Elliot Smith was awarded an M.D. and gold medal for a thesis on the anatomy and histology of the cerebrum of the nonplacental mammal. In 1896 he went to England, where he continued research for a Ph.D. at Cambridge and in 1899 was elected a fellow of St. John's College. He was also asked to help in preparing a catalog of brains of Reptilia and Mammalia in the Museum of the Royal College of Surgeons; this was published in 1902.

Elliot Smith was invited to be professor of anatomy in the new Government School of Medicine in Cairo and went there in 1900 to create an active department and continue his neurological work. In spite of an early determination to resist the lure of Egyptology, he became interested when he was asked to make anatomical investigations of old skeletons and mummies, particularly when there were remains of soft parts, including brains. These investigations led to the *Catalogue of the Royal Mummies in the Cairo Museum,* published in 1912. In 1907 he was elected to fellowship of the Royal Society and was appointed anatomical adviser to the Archaeological Survey of Nubia, which involved examination and description of thousands of skeletons excavated before the Aswan Dam was raised. The report was published in 1910.

Returning to England in 1909 to occupy the chair of anatomy at Manchester, Elliot Smith continued to work on both neurology and the Nubian remains and developed his theory of the diffusion of culture, which has never been generally accepted by anthropologists. He was one of several experts deceived by the Piltdown skull. During World War I, Elliot Smith worked for short periods in hospitals and did research on shell shock, and in 1919 he transferred to the chair of anatomy at University College, London, where he emphasized the importance of studying human biology with its psychological and cultural aspects, and also the history of medicine. He traveled frequently to the United States, China, and Australia, mainly on anthropological work, and trained several assistants who were later prominent anthropologists. He married Kathleen Macredie in 1900, and they had three sons. He was knighted in 1934 and retired in 1936.

The weight of Elliot Smith's work lies in his anatomical studies. His detailed comparative anatomical descriptions of the brains of reptiles and nonplacental and placental mammals contributed to the study of evolution as well as to neurology, and he related the development of the visual area of the brain to arboreal life in primates. His descriptions of Egyptian mummies were the first to be so comprehensive and so detailed; many of them are not yet superseded.

Some of his manuscripts are at the University of Manchester, and a bronze head, done by A. H. Gerrard in 1937, is in the Medical Sciences Library of University College, London.

BIBLIOGRAPHY

The most comprehensive biography, by Warren R. Dawson, forms the longest section of Warren R. Dawson, ed., *Sir Grafton Elliot Smith: A Biographical Record,* by his colleagues (London, 1938); the volume also contains a 434-item bibliography of Elliot Smith's publications and refers to a collection of his letters held by Dawson.

There is an entry by H. A. Harris in the *Dictionary of National Biography, Supplement 1931–1940* (London, 1949), 816–817, with a list of obituaries, including that by J. T. Wilson in *Obituary Notices of Fellows of the Royal Society of London,* **2** (1938), 323–333. Volume **71** of the *Journal of Anatomy* is a memorial volume to Elliot Smith, who had for a time assisted in editing it; included is a bibliography of his anatomical writings. A more recent assessment is A. A. Abbie, "Sir Grafton Elliot Smith," in *Bulletin of the Post Graduate Committee in Medicine, University of Sydney,* **15** (1959), 101–150. Abbie includes a bibliography strong in biographical material and refers to a valuable collection on Elliot Smith in the Sydney University department of anatomy.

DIANA M. SIMPKINS

ELLIS, WILLIAM (*b.* Greenwich, England, 20 February 1828; *d.* Greenwich, 11 December 1916), *geomagnetism, meteorology, astronomy.*

Ellis' father, Thomas, joined the staff of the Royal Observatory at Greenwich in 1825 and obtained employment there for his son when the boy was thirteen.

Twice married but childless, Ellis occupied himself in his spare time with local church affairs and contributed some 100 articles on a wide range of subjects to scientific journals. He was elected fellow of the Royal Astronomical Society in 1864, honorary member of the British Horological Institution in 1865, member of the Institute of Electrical Engineers in 1873, fellow of the Royal Meteorological Society in 1875 (president 1886–1887), and fellow of the Royal Society in 1893.

Ellis joined the Royal Observatory as a temporary computer on 2 August 1841 and was employed on lunar reductions and the comparison of standards of length during the restoration of the British standards. He left the Royal Observatory in March 1852 to become astronomical observer at Durham University but returned to Greenwich on 13 May 1853 to become a second-class assistant on the permanent staff, which then numbered only nine, including the astronomer royal, George Biddell Airy.

Ellis remained a transit circle observer for more than twenty years, took part in the Harton Colliery geodetic experiments in 1854, and assumed charge of the chronometric galvanic department in 1856. His duties included the care and rating of Royal Navy chronometers and supervision of the hourly galvanic time signals (instituted by Airy in 1853) to a central London telegraph office for distribution throughout the country; he was also in charge of arrangements for the telegraphic determination of longitudes.

Promoted to first-class assistant on 1 February 1871, Ellis was at his own request transferred to the magnetic and meteorological department as superintendent in 1875, a position he held until his retirement on 31 December 1893. It was during this later period that he carried out the work for which he is best known in geomagnetism and meteorology. His paper "On the Relation Between the Diurnal Range of Magnetic Declination and Horizontal Force as Observed at Greenwich During the Years 1841–1877 and the Period of Solar Spot Frequency" (1880) was accepted by most people as proof of the relationship between terrestrial magnetism and sunspots suggested in 1852 by Sabine and others. In his eighty-eighth year Ellis contributed his last paper, "Sunspots and Terrestrial Magnetism," to *Observatory* magazine.

BIBLIOGRAPHY

Ellis' own publications include many regular contributions on meteorology to the *Quarterly Journal of the Royal Meteorological Society* from 1877 on. Among his papers on magnetism are "Account of Some Experiments Showing the Change of Rate in a Clock by a Particular Case of Magnetic Action," in *Philosophical Magazine,* **25** (May 1863), 325–331; "On the Relation Between the Diurnal Range of Magnetic Declination and Horizontal Force as Observed at Greenwich During the Years 1841–1877 and the Period of Solar Spot Frequency," in *Philosophical Transactions of the Royal Society,* **171** (1880); "Earth Currents and the Electric Railway," in *Nature,* **44** (June 1891), 127–128; "On the Simultaneity of Magnetic Variations at Different Places on Occasions of Magnetic Disturbance, and on the Relation Between Magnetic and Earth Current Phenomena," in *Proceedings of the Royal Society,* **52** (1892), 191–212; "On the Relation Between the Diurnal Range of Magnetic Declination and Horizontal Force, and the Period of Solar Spot Frequency," *ibid.,* **63** (1898), 64–78; "Magnetic Results at Greenwich and Kew Discussed and Compared, 1889 to 1896," in *Report of the British Association for the Advancement of Science* (1898), 80–108; and "Sunspots and Terrestrial Magnetism," in *Observatory,* **39** (Jan. 1916), 54–59.

P. S. LAURIE

ELSTER, JOHANN PHILIPP LUDWIG JULIUS (*b.* Bad Blankenburg, Germany, 24 December 1854; *d.* Bad Harzburg, Germany, 6 April 1920), *experimental physics.*

Elster's scientific work can be discussed only with that of Hans Geitel; they jointly carried out and published almost all of their investigations from 1884 to 1920. They were teachers of mathematics and physics at the Herzoglich Gymnasium in Wolfenbüttel, near Brunswick.

Elster and Geitel studied together from 1875 to 1877 in Heidelberg, and until 1878 in Berlin. Elster then returned to Heidelberg, to study with Georg Quincke, under whom he received the doctorate in 1879 for his dissertation "Die in freien Wasserstrahlen auftretenden elektromotorischen Kräfte." After taking the examination to become a teacher he went in 1881 to Wolfenbüttel, where Geitel had been teaching since 1880. In 1884 they began collaborating on scientific works, which eventually totaled almost 150. They were especially concerned with the following problems, which were then new: electrical phenomena in the atmosphere, the photoelectric effect and thermal electron emission, photocells and their use in photometry, various aspects of radioactivity, and the development of apparatus and methods for the measurement of electrical phenomena in gases. The vast scope of Elster and Geitel's pioneering work in all these areas can be determined from the contemporary literature and is emphasized in textbooks of both the nineteenth and the twentieth centuries. Many results of their investigations are now part of the accepted foundations of the areas covered. Today, in the age of Geiger and Müller counters, of cloud and bubble chambers, and of electronics, their methods of measurement are no longer employed; but until 1920 they were crucial to research in the respective areas.

Elster and Geitel's first joint work was concerned with the electrification of flames (1884). This was followed by the first investigations of electrical processes in thunderclouds,[1] the development of electricity in rain, and the "dispersion of electricity" in the atmosphere and its dependence on the electric field of the earth and the measurement of that field. The ionization of gas was not yet known; it arose from J. J. Thomson's discovery of the conductivity of air induced by roentgen rays (1896). With the theory of gas ions, a rational treatment of the phenomena of atmospheric electricity was finally possible; a comprehensive presentation of the results obtained up to 1901 is in a report by Geitel[2] (the investigations continued until 1905). Measurements of atmospheric electricity were made in the Austrian Alps in 1891–1893; on Mt. Brocken, Mt. Säntis, and Mt. Gornergrat in 1900; during the total solar eclipses in Algeria in 1900 and on Majorca in 1905; and on Spitsbergen and Capri in 1902.

A series of twenty investigations on the photoelectric effect began in 1889 with the discovery that negatively charged magnesium filaments, freshly ground with emery, are discharged not only by ultraviolet light but even by "dispersed evening daylight."[3] The investigations of the sensitivity of the photocathodes to visible light led to the actinoelectric series—rubidium, potassium, sodium, magnesium, thallium, zinc—and to the discovery (1910) of the sensitivity of the hydrogenized potassium cathode, which was found to extend into the infrared range.[4] By use of a photocathode of a fluid, absolutely smooth potassium-sodium alloy, the dependence of the photoelectric current on the polarization of the light was discovered,[5] as was the existence of a "normal" and a "selective" photoelectric effect, which later became of decisive importance in the electron theory of metals. The Elster-Geitel photocell was for decades the photometric instrument of physics and astronomy.

In 1887 Elster and Geitel discovered the "electrification of gases by means of incandescent bodies,"[6] a finding that later was very significant in thermionics. Their finding of the emission of negative electricity from incandescent filaments was decisive in the proof that Thomson's "corpuscles" (electrons) are constituents of all matter.

Soon after the discovery (1896) of radioactivity Elster and Geitel began to study Becquerel rays,[7] in order to determine the origin of the energy of these rays. Crookes had proposed the hypothesis that the air molecules with the greatest velocity stimulated the rays; energy was therefore extracted from the surrounding air. Elster and Geitel placed uranium in a glass vessel that was then evacuated: even at the highest vacuum the radiation remained constant. They also placed uranium and a photographic plate in a container: the blackening of the plate was independent of the pressure. Therefore the radiation could not be stimulated by the air.

Mme. Curie suggested another hypothesis: the radioactive emission was a fluorescence of the uranium, which was excited by a very penetrating radiation that fills all of space. She therefore named the new phenomenon la radioactivité, i.e., "activated by radiation." Elster and Geitel showed, however, that the intensity of the uranium radiation above the earth is the same as it is in a mine 852 meters below the surface. They also investigated whether uranium emitted stronger Becquerel radiation when under the influence of cathode rays. For this purpose they developed a new Lenard cathode-ray tube, which let pass into the atmosphere an intense electron beam with a cross section of several square centimeters. (They closed off the discharge tube with a copper net covered with a very thin aluminum foil; the cathode rays escaped through the net's interstices.) The result was negative. They also demonstrated that Becquerel radiation is independent of the temperature of the uranium and of the compound in which it occurs. They concluded from these and other experiments that the radioactive emission is not the consequence of an external influence, but can only be a spontaneous release of energy by the atom. They inferred "that the atom of a radioactive element behaves like an unstable compound that becomes stable upon the release of energy. To be sure, this conception would require the acceptance of a gradual transformation of an active substance into an inactive one and also, logically, of the alteration of its elementary properties."[8] With this statement radioactivity was defined for the first time as a natural, spontaneous transformation of an element attendant upon the release of energy.

A magnetic deflection of the Becquerel rays was sought, but it could not be demonstrated unequivocally; nevertheless, the question raised here for the first time was answered definitively by Giessel (1899) and Mme. Curie (1900). Elster and Geitel did not succeed because in their positive tests beta rays were measured, while in their negative tests gamma rays were measured—this distinction, however, was not yet known. Elster and Geitel had already announced in 1898 that they had obtained new, highly radiant substances from the chemical treatment of Joachimstal pitchblende. Polonium, just discovered by the Curies, was immediately prepared by their methods. A short while later Mme. Curie and G. Bémont made known their discovery of radium. At the same time Elster and Geitel communicated their finding that lead extracted from pitchblende is highly radioactive; the radioactivity of ordinary lead and the amount of radium D, E, and F that it contains were repeatedly investigated in later studies. In 1899 the ionization of air by Becquerel rays was examined for the first time: the radiation produced equal numbers of positive and negative ions in the air mass between the electrodes, not at the electrodes themselves, and thus resembled the effect of roentgen rays, which J. J. Thomson and Ernest Rutherford had demonstrated for the first time in 1896. The influence of the Becquerel rays on spark and brush discharge in the air under various pressures was also described.

In 1899 Elster and Geitel had observed and demonstrated that uranium potassium sulfate glows with constant intensity, completely independently of all external influences. They then investigated the visible fluorescence that Becquerel rays excite in many crystals ("radioactive luminous paint"). After the three types of radioactive Becquerel rays (alpha, beta, gamma) were discovered, such experiments could be carried out separately with each of these types. In this process they discovered[9] (at exactly the same time as Crookes) the scintillation of zinc sulfide by alpha rays: the appearance of a flash of light as each alpha particle enters the crystal. The scintillation method was important in radioactivity research until 1920.

With the experience they had gained in radioactivity investigations, Elster and Geitel set themselves the question[10] of whether the ionization of the atmosphere results from radioactive material within it.

Geitel had shown that the ion content of a quantity of air hermetically sealed off from the outside becomes constant after some time; since both positive and negative ions disappear from the air, for example, through recombination to neutral molecules, an ionizing source must be present. Hence a wire one meter long was suspended in the air at a potential of $-2,000$ volts against earth; after several hours it was radioactive. Under definite, accurately determined experimental and measurement conditions, its activity was found to be proportional to the concentration of the radium emanation (radon) of the free atmosphere. (This is known as the Elster-Geitel activation number.)[11] This simple method provided information on the distribution of the emanation in the atmosphere over land and water, its dependence upon the height, upon meteorological data, and upon the earth's local electric field and its high concentration in narrow valleys and caves. Next came extensive measurements of the radioactivity of rocks, lakes, and spring waters and spring sediments, especially at health spas.[12] In 1913 Ernest Rutherford wrote: "The pioneers in this important field of investigation were Elster and Geitel and no researcher has contributed more to our knowledge of the radioactivity of the earth and the atmosphere than they have."[13]

Elster and Geitel, as inseparable in their life as in their work, were called "the Castor and Pollux of physics." They set up their physics laboratory in their residence, which also contained an astronomical telescope, terraria with tropical animals, and all kinds of natural history collections. They took vacation trips together to investigate the electricity in mountain and sea air and to measure the radioactivity of rocks, springs, and spas. When Geitel received a call to the University of Breslau in 1899, it was obvious that he would go there only with Elster. Elster was also considered for a professorship at Breslau, but they both finally rejected the offers: they feared that in Breslau they would not have the independence and quiet for their research that they had in Wolfenbüttel.

Skilled in designing and making equipment, they constructed all of their apparatus but refused to take out patents on their inventions (for example, the electrometer and the photocell). In 1889 they obtained a great deal of financial support from the Elizabeth Thompson Science Fund, of Boston, Massachusetts.

Both were good teachers, beloved by their students; Geitel especially understood how to train students to think independently. Elster's absent-mindedness was almost legendary: When told that he had placed a stamp of too high a value on a letter, he crossed it out and stuck one of the correct value next to it. And the door of his apartment had a large opening and a small one cut out at the bottom because he had a large dog and a small dog.

The general respect in which Elster and Geitel were held is shown by the large (719 pages) *Festschrift* that was dedicated to them in 1914–1915, a gift from older and younger physicists on their sixtieth birthdays. Among the authors who contributed original works were Max Born, Laue, Lenard, Gustav Mie, Planck, R. W. Pohl, Regener, and Sommerfeld.

See also Geitel article in Vol. 5.

NOTES

1. "Observations on the Electrical Processes in Thunder-Clouds," in *Philosophical Magazine*, **20** (1885).
2. H. Geitel, *Anwendung der Lehre von den Gasionen auf die Erscheinungen der atmosphärischen Elektrizität* (Brunswick, 1901), 27 pp.
3. "Entladung negativ elektrisierter Körper durch Sonnen- und Tageslicht," in *Annalen der Physik*, **38** (1889), 497.
4. "Über gefärbte Hydride der Alkalimetalle und ihre photoelektrische Empfindlichkeit," in *Physikalische Zeitschrift*, **11** (1910), 257; "Über den lichtelektrischen Effekt im Infrarot und einige Anwendungen hochempfindlicher Kaliumzellen," in *Physikalische Zeitschrift*, **12** (1911), 758.
5. "Abhängigkeit der Intensität des photoelektrischen Stromes von der Lage der Polarisationsebene des erregenden Lichtes zu der Oberfläche der Kathode," in *Sitzungsberichte der Berliner Akademie der Wissenschaften* (1894); *Annalen der Physik*, **55** (1895), 684, and **61** (1897), 445; *Physikalische Zeitschrift*, **10** (1909), 457.
6. "Elektrisierung der Gase durch glühende Körper," in *Annalen der Physik*, **31** (1887), 109, and **37** (1889), 315; *Sitzungsberichte der Akademie der Wissenschaften in Wien*, **97** (1888), IIa, 1175.
7. "Versuche an Becquerel-Strahlen," in *Annalen der Physik*, **66** (1898), 735, and **69** (1889), 83.
8. *Jahresberichte des Vereins für Naturwissenschaft zu Braunschweig*, **10/12** (1902), 39; *Annalen der Physik*, **69** (1899), 83.
9. "Über die durch radioaktive Emanation erregte szintillierende Phosphoreszenz der Sidotblende," in *Physikalische Zeitschrift*, **4** (1903), 439.
10. "Analogie im elektrischen Verhalten der natürlichen Luft und der durch Becquerel-Strahlen leitend gemachten," *ibid.*, **2** (1901), 590; "Radioaktivität der im Erdboden enthaltenen Luft," *ibid.*, **3** (1902), 574.
11. "Radioaktive Emanation in der atmosphärischen Luft," in *Sitzungsberichte der Bayerischen Akademie der Wissenschaften*, **33** (1903), 301–323; and in *Physikalische Zeitschrift*, **4** (1903), 522.
12. "Sédiments radioactives des sources thermales," in *Archives des sciences physiques et naturelles*, 4th ser., **19** (1905), 5.
13. *Handbuch der Radiologie*, II (Leipzig, 1913), 563.

BIBLIOGRAPHY

In addition to the works cited in the text and the notes see the following.

I. ORIGINAL WORKS. Among the writings published by Elster and Geitel are *Ziele und Methoden der luftelektrischen Untersuchungen* (Wolfenbüttel, 1891); and *Ergebnisse neuer Arbeiten über atmosphärische Elektrizität* (Wolfenbüttel, 1897).

II. SECONDARY LITERATURE. Elster's work is discussed in *Handbuch der Radiologie*, Vol. I (Leipzig, 1920), Vol. II (1913); and R. Pohl and P. Pringsheim, *Die lichtelektrischen Erscheinungen* (Brunswick, 1914). An obituary is E. Wiechert, in *Nachrichten der Akademie der Wissenschaften zu Göttingen* (1921), 53–60.

WALTHER GERLACH

ELVEHJEM, CONRAD ARNOLD (*b.* McFarland, Wisconsin, 27 May 1901; *d.* Madison, Wisconsin, 27 July 1962), *biochemistry.*

Elvehjem, the son of Ole and Christine Lewis Elvehjem, grew up on the family farm and was educated in local schools before entering the University of Wisconsin. He received his B.S. in agricultural chemistry in 1923, the M.S. in 1924, and the Ph.D. in 1927. Upon graduation he became an instructor in agricultural chemistry, rising to the rank of full professor in 1936. In 1944 he became chairman of the department (renamed biochemistry), a position he held until he became the university's thirteenth president in 1958. From 1946 to 1958 he served as dean of the graduate school.

Elvehjem married Constance Waltz in 1926. They had two children, Peggy Ann and Robert Stuart. He died unexpectedly of a heart attack suffered while at work in the presidential office. His only extended absence from the university was a year (1929–1930) spent at Cambridge University studying catalytic oxidations in the laboratory of F. G. Hopkins as a National Research Council Fellow. Elvehjem was elected to the National Academy of Sciences in 1942 and to the National Academy of Arts and Sciences in 1953. He received numerous other honors—the Willard Gibbs Medal, the Osborne and Mendel Award, the Nicolas Appert Award, and the Lasker Award—and was a member of numerous national committees.

Elvehjem's entire scientific career dealt with animal nutrition, particularly the role of trace elements and vitamins. He ranged widely in his interests, yet there was an overall interrelationship in the more than 800 research papers published during his career. His first work, undertaken with E. B. Hart and H. Steenbock, dealt with the influence of light on metabolism of calcium and phosphorus in lactating animals. During the next several years, in association with these same investigators, he studied milk-induced anemia and showed that traces of copper are essential for satisfactory iron uptake in hemoglobin formation. Mineral metabolism in animals remained a matter of primary concern and ultimately included work on the roles of zinc, cobalt, manganese, molybdenum, boron, potassium, aluminum, fluorine, and arsenic.

Elvehjem contributed in many ways to the growth of understanding of members of the vitamin B complex. In 1937, after Euler-Chelpin and Otto Warburg showed that Harden's coenzyme I and related coenzymes contained nicotinic acid, Elvehjem and his associates showed that nicotinic acid cured blacktongue in dogs. Goldberger had shown in 1926 that blacktongue is the canine equivalent of human pellagra, and medical investigators soon showed that nicotinic acid cured pellagra in human beings. Elvehjem's group showed the vitamin to be present as the amide in active concentrates of the vitamin prepared from liver.

When it later became evident that most cereal grains are like corn in being low in nicotinic acid yet—unlike corn—are protective or curative against pellagra, the reason for the anomaly was sought. The proteins of corn are lower in tryptophan than are those of other cereals, and corn has traditionally been present in human diets everywhere in the world that pellagra is endemic. Elvehjem and his associates showed that tryptophan can serve as a substitute for nicotinic acid in diets low in that vitamin, and other investigators established the metabolic conversion of tryptophan to nicotinic acid.

Elvehjem and his associates carried out studies on biotin, pantothenic acid, para-aminobenzoic acid, folic acid, and inositol as these substances became available in pure form, and helped to clarify their role in the nutrition of many species, particularly rats, chickens, dogs, and monkeys. All of these vitamins were originally recognized in connection with the growth of bacterial species; Elvehjem's laboratory pioneered in testing their role in the nutrition of higher animals. His laboratory also did extensive work in clarifying the role of intestinal bacteria in the synthesis of various trace nutrients. This work revealed that certain species do not require a dietary source of a particular vitamin, because it is synthesized by their normal intestinal flora. When the normal flora is inhibited by such drugs as sulfaguanidine or succinylsulfathiazole, the animal becomes dependent upon dietary supplements.

Elvehjem's quiet but forceful, perceptive, hardworking example made him a natural leader as a research director. Many of the eighty-eight students who took the Ph.D. under him went on to make significant contributions to the field. This same intensity of effort in the laboratory was evident in his administrative work. As president of the university he encouraged the growth of research in the humanities and social studies while maintaining the strength of the sciences.

BIBLIOGRAPHY

I. ORIGINAL WORKS. The archives at the University of Wisconsin have extensive holdings of Elvehjem's papers: the material associated with his presidency in the presidential papers, that with his deanship in the graduate school papers, that with his research and instruction in the papers of the biochemistry department. More than a third of his research papers were published in the *Journal of Biological Chemistry*. Other journals containing numerous papers are *Journal of Nutrition, American Journal of Physiology,* and *Proceedings of the Society for Experimental Biology and Medicine*. There is no published bibliography of his works other than the incomplete one in Poggendorff and the listings in the author indexes of *Chemical Abstracts*. The University of Wisconsin biochemistry department holds a bound set of his collected works.

The principal paper dealing with copper and its role in anemia is "Iron in Nutrition, VII. Copper as a Supplement to Iron for Hemoglobin Building in the Rat," in *Journal of Biological Chemistry,* **77** (1928), 797–812, written with E. B. Hart, H. Steenbock, and J. Waddell; see also "Mineral Metabolism," in *Annual Review of Biochemistry,* **5** (1936), 271–294, written with E. B. Hart; and "The Biological Significance of Copper and Its Relation to Iron Metabolism," in *Physiological Reviews,* **15** (1935), 471–507. The curative effect of nicotinic acid for blacktongue in dogs was announced in "Relation of Nicotinic Acid and Nicotinic Acid Amide to Canine Black Tongue," in *Journal of the American Chemical Society,* **59** (1937), 1767–1768, and in more detail in "The Isolation and Identification of the Anti-Black Tongue Factor," in *Journal of Biological Chemistry,* **123** (1938), 137–147; both articles were written with R. J. Madden, F. M. Strong, and D. W. Woolley. See also "Relation of Nicotinic Acid to Pellagra," in *Physiological Reviews,* **20** (1940), 249–271; and "The Biological Significance of Nicotinic Acid," in *Bulletin of the New York Academy of Medicine,* **16** (1940), 173–189. On the role of tryptophan see two articles written with W. A. Krehl, L. J. Teply, and P. S. Sarma: "Growth-Retarding Effect of Corn in Nicotinic Acid-Low Rations and Its Counteraction by Tryptophane," in *Science,* **101** (1945), 489–490; and "Factors Affecting the Dietary Niacin and Tryptophane Requirement of the Growing Rat," in *Journal of Nutrition,* **31** (1946), 85–106.

The role of Elvehjem and his associates in nutritional research is brought out in several review articles which place the work of his laboratory into perspective with overall activities in the field. See, in addition to those cited above, the following papers: "The Water Soluble Vitamins," in *Journal of the American Medical Association,* **120** (1942), 1388–1397; "Recent Advances in Our Knowledge of the Vitamins," in *Scientific Monthly,* **56** (1943), 99–104; "Present Status of the Vitamin B Complex," in *American Scientist,* **32** (1944), 25–38; and "Recent Progress in Nutrition and Its Relation to Drug Therapy," in *Journal of the American Medical Association,* **136** (1948), 915–918.

Elvehjem's Willard Gibbs Medal address, "Newer

Members of the Vitamin B Complex. Their Nutritional Significance," in *Chemical and Engineering News,* **21** (1943), 853–857, is somewhat autobiographical, as is "Early Experiences With Niacin—a Retrospect," in *Nutrition Reviews,* **11** (1953), 289–292. His education philosophy is reflected in his inaugural address, "Essentials of Progress," delivered 9 October 1958 (Madison, 1958).

II. SECONDARY LITERATURE. There are no biographies of Elvehjem other than short journalistic pieces. There are lengthy obituary notices in *The Capital Times* (Madison, 27 July 1962) and *The Milwaukee Journal* (27 July 1962); *The Wisconsin State Journal* (28 July 1962); and *Wisconsin Alumnus,* **64,** no. 1 (1962), 9–12.

AARON J. IHDE

EMANUELLI, PIO (*b.* Rome, Italy, 3 November 1888; *d.* Rome, 2 July 1946), *astronomy.*

The son of a clerk at the Vatican, Emanuelli became interested in astronomy when only ten years old. His first astronomical work was the observation of the solar eclipse of 20 May 1900, on which he made a report to the French astronomer Camille Flammarion. Emanuelli studied in Rome, where he served as a volunteer at the Collegio Romano Observatory until, at the age of twenty-two, he became astronomer at the Vatican Observatory.

At that time the great international astronomical enterprise was the *Astrographic Catalogue,* to which the Vatican Observatory contributed, and Emanuelli worked actively in that area. Even during World War I, when he was called into the army, he maintained his scientific contacts and offered advice and suggestions for the continuation of this important work.

Emanuelli computed several orbits of small planets, elements of many solar eclipses, and was particularly interested in the problem of the relativistic deflections of the stars in the neighborhood of the eclipsed sun. For some eclipses, he published celestial maps and tables with data concerning the amount of the expected displacements. Well known to all astronomers are Emanuelli's tables for the conversion of the equatorial in galactic coordinates, which were used until the new position of the galactic poles was established.

Emanuelli had, besides his professional work in astronomy, a deep interest in the history of science and its dissemination. In these two fields he published many writings in different journals. At his death he left a large number of unpublished manuscripts, which are now deposited at the Domus Galileiana in Pisa.

BIBLIOGRAPHY

Emanuelli's writings include "Eclisse Solare del 17 aprile 1912," in *Memorie della Società degli spettroscopisti italiani,*

40 (1912), 123; observations and computations of the elements of planetary orbits, often untitled, in *Astronomische Nachrichten,* **178** (1908), 319; **181** (1919), 209; **215** (1922), 211; **216** (1922), 137, 419; **219** (1923), 161, 219; **223** (1925), 119; **234** (1929), 357; and **237** (1930), 237; "Gli eclissi di Sole totali non centrali," in *Atti dell'Accademia nazionale dei Lincei. Rendiconti della classe di scienze fisiche, matematiche e naturali,* 6th ser., **8** (1928), 214; and "Il polo galattico e la regione circumpolare galattica," *ibid.,* **9** (1929), 1096.

See also "Posizione di Venere nel 25 marzo del 1300, determinazione dell'ora in cui sono sorti Venere ed il Sole il 25 marzo del 1300 nella montagna del Purgatorio," in *Memorie della R. Accademia d'Italia,* **14** (1943), 193. A list of Emanuelli's MSS is P. Maffei, "Gli scritti inediti di Pio Emanuelli," in *Memorie della Società astronomica italiana,* **37** (1966), 803.

G. RIGHINI

EMBDEN, GUSTAV (*b.* Hamburg, Germany, 10 November 1874; *d.* Frankfurt, Germany, 25 July 1933), *physiological chemistry.*

Embden, son of a Hamburg lawyer, studied medicine at the universities of Freiburg im Breisgau, Munich, Berlin, and Strasbourg. His teachers, Johannes von Vries in Freiburg and Franz Hofmeister in Strasbourg, directed his interest toward physiology. In 1903 Embden was appointed assistant at the Physiological Institute, having worked in Hofmeister's laboratory since his graduation in 1899. He also worked for short periods with Gaule in Zurich, Paul Ehrlich in Frankfurt, and Ernst Ewald in Strasbourg. These brief stints increased Embden's skill in experimentation. But it was Hofmeister's influence that was decisive in directing him toward chemicophysiological research and in increasing his ability to think in a biologically oriented way.

In 1904 Carl von Noorden made Embden director of the newly organized chemistry laboratory of the medical clinic at the municipal hospital of Frankfurt-Sachsenhausen. Embden helped to create such a fine reputation for this laboratory that in 1907 it was expanded into the Physiological Institute. In 1909 this institute became autonomous, and in 1914, with the founding of the university, it was renamed the University Institute for Vegetative Physiology, with Embden as director and full professor. In 1907 Embden had qualified as lecturer at the University of Bonn on the basis of his work in Frankfurt. Two years later he was appointed professor. He married Hanni Fellner, granddaughter of Frankfurt's former lord mayor. From 1925 to 1926 he served as rector of the university.

In his lectures, as in his research, Embden preferred

a presentation of the deeper relationships of chemico-physiological processes to a collection of individual facts. At the Physiological Institute he created an atmosphere of dedicated teamwork through his sympathy, helpfulness, and ability to foster close personal contacts and inspire his assistants to cooperate in the solution of common problems. Embden was known as a researcher who made absolutely sure of his results, without impeding his bold conclusions and theories.

At the time Embden began his biochemical research, the physiology of metabolism was still dominated by the energy principle, and pertinent research was concentrated mainly on the initial and final stages of the metabolic processes. Embden focused on the biologically specialized position of the chemical processes in the living organism and particularly investigated the several stages of intermediate metabolism. His works form an important part of the transition from calorimetric investigations to the physiology of the metabolism of the living cell. Embden's scientific works are divided into two large, clearly distinct, and logically related groups. Initially he investigated intermediate metabolic processes in the liver in order to consider the physiological-chemical processes involved in muscular exertion.

Embden recognized that experiments on the undissected animal produced unclear results, while dissected tissue was unsatisfactory because of breakdown of the organic structure and cell damage. Therefore he developed a new method by using the livers of warm-blooded animals, kept in good condition by a special perfusion technique. With the help of this method he recognized oxidative deamination as a way to break down amino acids, the synthesis of sugar from lactic acid, and—in connection with the β-oxidation of fatty acids discovered by Franz Knoop—that acetoacetic acid and acetone are the products of pathological sugar metabolism. This last discovery formed a basis for research into sugar metabolism and diabetes. In their entirety, these investigations showed that the liver is the most important metabolic-physiological organ of the body.

Embden's research work on intermediate carbohydrate metabolism turned his attention to the chemical processes involved in muscular activity. At first he selected the fluid pressed from a muscle—analogous to fluid pressed from yeast—as a cell-free research medium. In 1924 he succeeded in isolating a hexose diphosphate as an intermediate product, naming it lactacidogen. It showed him that—analogous to the processes in the yeast cell—glucose must be esterified with phosphoric acid before it can be broken down further. In 1927 he discovered hexose

monophosphate, the so-called "Embden ester," in the muscle cells. During twenty years of tenacious work Embden and his assistants isolated important phosphor-containing intermediate products of carbohydrate metabolism in the muscle. This led to his discovery of adenyl phosphoric acid in the muscle, thus opening a new, large field in biochemistry. Embden was the first to recognize the rapid reversibility of chemical processes in muscle contraction. Finally, in the course of his last experiments, in 1932–1933, he and his assistants succeeded in tracing all stages of the breakdown of glycogen in the muscle to lactic acid. In these biochemical investigations Embden never lost sight of the problem of general cell physiology, always endeavoring to integrate his discoveries of phosphorylation and metabolic processes with the relationships between activity, fatigue, and training on the one hand and the colloidal state of protoplasm on the other.

BIBLIOGRAPHY

I. ORIGINAL WORKS. Embden's inaugural dissertation is *Anatomische Untersuchung eines Falles von Elephantiasis fibromatosa* (Strasbourg, 1899). Posthumously there appeared "Gustav Embdens und seiner Mitarbeiter letzte Arbeiten," in *Hoppe-Seyler's Zeitschrift für physiologische Chemie,* **230** (1934), 1–108, with a biography (not by Embden).

II. SECONDARY LITERATURE. On Embden or his work see H. J. Denticke, "Gustav Embden," in *Ergebnisse der Physiologie (biologischen Chemie und experimentellen Pharmakologie),* **35** (1933), 32–49, with a bibliography; E. Lehnartz, "Gustav Embden," in *Arbeitsphysiologie,* **7,** no. 5 (1934), 475–483; and J. C. Poggendorff, VI, pt. 1, 660; VIIa, pt. 1, 500.

EBERHARD SCHMAUDERER

EMDEN, ROBERT (*b.* St. Gallen, Switzerland, 4 March 1862; *d.* Zurich, Switzerland, 8 October 1940), *astrophysics.*

Emden received his doctorate at Strasbourg in 1887 as a student of the physicist August Kundt. In 1907 he became an assistant professor (*extraordinarius*) of physics and meteorology at the Technische Hochschule in Munich. He held that position until 1928, when he became an assistant professor for astrophysics at the University of Munich. His work *Gaskugeln, Anwendungen der mechanischen Wärmetheorie auf kosmologische und meteorologische Probleme* (Leipzig, 1907) was epoch-making; all subsequent textbooks on astrophysics have been based on it. He was therefore entrusted with writing the article on the thermodynamics of celestial bodies in the *Enzyklopädie der mathematischen Wissenschaften.*

Emden introduced the concept of "polytropic change of state," although he did not need to display the radiation pressure explicit in his calculations. Nevertheless, this pressure appeared mathematically in the exponents of the polytropic lines. Emden was also the first to give a derivation for radiative equilibrium for nondiscernible particles, that is, photon statistics. He thereby became a precursor in the use of the Bose-Einstein statistics. In further works he dealt with astronomical refraction, the thermodynamics of the atmosphere, and propagation of sound in the atmosphere. In addition, he contributed significantly to the development of the theory of balloon flight.

In 1920 he became a member of the Bavarian Academy of Sciences. From 1916 on, formal difficulties kept him from obtaining German citizenship. Consequently, he at least avoided financial losses upon his dismissal when the Nazis came into power in 1933. He died in Zurich, in his native Switzerland, on 8 October 1940.

BIBLIOGRAPHY

Emden's works include, in addition to those cited in the text, "Theoretische Grundlagen der Ballonführung," in *Illustrierte aeronautische Mitteilungen* (1901); *Grundlagen der Ballonführung* (Leipzig, 1910); "Abnorme Hörbarkeit," in *Sitzungsberichte der Bayerischen Akademie der Wissenschaften zu München* (1916), 113–123; "Zur Thermodynamik der Atmosphäre," in *Meteorologische Zeitschrift,* **33** (1916), 351–360; **35** (1918), 13–29, 74–81, 114–123; **40** (1923), 171–177; "Lichtquanten," in *Physikalische Zeitschrift,* **22** (1921), 513–517; "Astronomische Refraktion," in *Astronomische Nachrichten,* **219** (1923), 45–56; "Strahlungsgleichgewicht," in *Zeitschrift für Physik,* **23** (1924), 176–213; and "Freiballon," in *Handbuch der Experimentalphysik,* IV, pt. 3 (1930), 115–131.

J. O. FLECKENSTEIN

EMERSON, BENJAMIN KENDALL (*b.* Nashua, New Hampshire, 20 December 1843; *d.* Amherst, Massachusetts, 7 April 1932), *geology.*

Emerson belonged to a distinguished New England family that was eminent in the educational world. His father, Benjamin F. Emerson, was a lawyer; his mother was Elizabeth Kendall. Emerson received his secondary education at Tilton (Vermont) Academy. Inspired by the work of the famous New England geologist Edward Hitchcock, he went to Amherst College, from which he graduated with distinction. After a period of teaching sciences at the old Groton Academy, he studied geology at Berlin and Göttingen, receiving the Ph.D. at the latter in 1870.

Emerson returned to Amherst as instructor in geology and zoology, became professor of geology in 1872, and of geology and theology in 1881. That he was an inspiring teacher is shown by the number of renowned geologists who had been his students.

Emerson was also distinguished as a field geologist. For thirty years he was on the staff of the U.S. Geological Survey. He was a member of the Harriman expedition to Alaska in 1899 and wrote the geologic section of its report. His outstanding contributions were on the geology of the Connecticut Valley and bordering plateaus of central and southern New England; his *Geology of Old Hampshire County* is a classic and, like his *Geology of Massachusetts and Rhode Island,* still is an important source for field geologists. Many of his interpretations of stratigraphic, petrologic, and metamorphic geology are a marked advance over earlier works and have served as a progressive link between nineteenth-century and early twentieth-century geology. Besides the newer terminology, he brought in modern concepts of petrogenesis and stratigraphy, drastically modifying and adding to the earlier interpretations of Edward and Charles H. Hitchcock and others. He recognized transitional metamorphic facies and the metasomatic effects of granitic solutions, giving some emphasis to their role in promoting regional metamorphism. In *Geology of Old Hampshire County* there is a detailed description of the Bernardston formation (Devonian), basically important in regional correlations; this monograph also provides the first detailed treatment of the igneous and sedimentary rocks of the Triassic basin and of the Pleistocene—chiefly glacial—deposits of the Connecticut Valley. His *Mineralogical Lexicon* . . . provides a remarkable catalog of mineral occurrences in south-central New England.

Emerson was a founder and original fellow of the Geological Society of America and one of its early presidents (1899). He was also a fellow of the American Academy of Arts and Sciences, the Washington Academy of Sciences, the American Philosophical Society, and the American Geographical Society. He was also a member of the Deutsche Geologische Gesellschaft and several other learned societies.

BIBLIOGRAPHY

I. ORIGINAL WORKS. Fifty-one titles are recorded in *Bibliography of Geologic Literature on North America 1785–1918,* U.S. Geological Survey Bulletin no. 746 (Washington, D.C., 1923), pp. 343–344. The following may be considered Emerson's principal contributions: "A Description of the 'Bernardston Series' of Metamorphic Upper Devonian Rocks," in *American Journal of Science,* 3rd ser.,

40 (1890), 263–275, 362–374; *Mineralogical Lexicon of Franklin, Hampshire, and Hampden Counties, Massachusetts,* U.S. Geological Survey Bulletin no. 126 (Washington, D.C., 1895); *Geology of Old Hampshire County, Massachusetts, Comprising Franklin, Hampshire, and Hampden Counties,* U.S. Geological Survey Monograph no. 29 (Washington, D.C., 1898); *Outlines of the Geology of Western Massachusetts: Description of the Holyoke Quadrangle,* U.S. Geological Survey Geologic Atlas, Holyoke Folio, no. 50 (Washington, D.C., 1898); *The Geology of Eastern Berkshire County, Massachusetts,* U.S. Geological Survey Bulletin no. 159 (Washington, D.C., 1899); and *Geology of Massachusetts and Rhode Island,* U.S. Geological Survey Bulletin no. 597 (Washington, D.C., 1917).

Of the remaining forty-five titles in the cited bibliography two major publications were written with J. H. Perry: *The Geology of Worcester, Massachusetts* (Worcester, 1903); and *The Green Schists and Associated Granites and Porphyries of Rhode Island,* U.S. Geological Survey Bulletin no. 311 (Washington, D.C., 1907).

Other titles, short papers, relate mostly to the tetrahedral theory of the earth (discussional); forms, distribution, and mineralogy of Triassic traprocks of Massachusetts; and glacial and postglacial features of the Connecticut Valley. Several short notes on mineralogic and petrologic subjects appeared in various scientific journals.

II. SECONDARY LITERATURE. The principal biographical sketch is by one of Emerson's former students, F. B. Loomis, "Memorial of Benjamin Kendall Emerson," in *Bulletin of the Geological Society of America,* 44, pt. 2 (1933), 317–325. Briefer sketches are Charles R. Keyes, in *Pan-American Geologist,* 58, no. 1 (1932), 1–6; and A. C. Lane, in *Proceedings of the American Academy of Arts and Sciences,* 68, no. 13 (1933), 625–627.

L. W. CURRIER

EMERSON, ROBERT (*b.* New York, N.Y., 4 November 1903; *d.* New York, 4 February 1959), *plant physiology.*

After studying at Harvard, Emerson received his doctorate under Otto Warburg at Berlin in 1927 and joined the California Institute of Technology biology department in 1930. From 1937 to 1940 he worked at the Carnegie Laboratory of Plant Biology at Stanford, California. After returning to Cal Tech, Emerson spent the World War II years working with Japanese deportees from the West Coast, attempting to develop rubber production from guayule, a Mexican desert shrub. In 1947 he became director of the newly founded photosynthesis research laboratory associated with the botany department of the University of Illinois in Urbana, which he built into one of the leading research laboratories in this field. He died, at the height of his research career, in a plane crash in the East River, off New York City. In 1949 Emerson received the Stephen Hales Award of the American Society of Plant Physiologists, and in 1950 he was elected to membership in the National Academy of Sciences.

In appearance and character Emerson was a typical New Englander: tall, lean, long-headed, self-denying, hard-working, expecting (and appreciating) hard work in others. He exerted great influence on his co-workers and students. A pacifist and believer in democratic socialism, he was always defending the underdog—working with deported Japanese, fighting housing discrimination, befriending students from Africa and India. Emerson was a perfectionist in experimental research and skillful in manual work, including cabinetmaking and gardening. He combined pride in the quality of his own work and critical rejection of less careful work with great modesty and deep respect for the achievements of others.

Emerson's lifelong concern was the precise, quantitative study of photosynthesis—the basic process of life on earth by which organic matter is synthesized by plants from water and carbon dioxide with the aid of light absorbed by plant pigments (of which the green pigment chlorophyll is the most important and ubiquitous).

In 1937 Emerson set out to check the conclusions of his teacher Otto Warburg that plants can synthesize sugar (glucose), using only four light quanta (photons) for each molecule of carbon dioxide utilized and of oxygen liberated. This suggested a remarkable efficiency of the process—conversion of up to 30 percent of the absorbed light energy into chemical energy of the products. Steadily improving the measurement techniques and systematically determining the quantum requirement of photosynthesis in green, brown, red, and blue-green algal cells, in monochromatic light of widely different wavelengths, Emerson arrived at a number of important conclusions.

1. The minimum quantum requirement of photosynthesis for all plants is not four but eight. This conclusion led to a drawn-out controversy with Warburg, in which Emerson's conclusions were gradually accepted as correct by most workers in the field, although not by Warburg himself.

2. Quanta absorbed in chlorophyll *a,* chlorophyll *b,* and the red and blue phycobilin pigments of certain algae are about equally effective in producing photosynthesis. Light quanta absorbed by the yellow pigments (the carotenoids) have a much smaller efficiency—with the exception of a special carotenoid, fucoxanthol, present in brown algae and diatoms.

3. At the longwave end of the absorption band of chlorophyll *a* (above 680 nm in green cells and above 650 nm in red algae) the yield of photosynthesis drops sharply ("red drop"); it can be restored to normal

by additional illumination with shortwave light (the "Emerson effect").

This last result has become one of two main foundations of the now widely accepted theory according to which photosynthesis involves two successive photochemical processes brought about by two pigment systems. Light absorption in one system oxidizes water, liberating oxygen and reducing an intermediate product (perhaps a cytochrome); light absorbed in the other system reduces carbon dioxide (sugar is the ultimate product), oxidizing the intermediate product that had been reduced by the first system. Each of the two steps requires four quanta (to move four hydrogen atoms "uphill," that is, with an increase in chemical energy), which explains the total quantum requirement of eight. In the region of the red drop, too many quanta are absorbed in one pigment system and not enough in the other; this can be corrected by supplementary shortwave illumination.

Another important work of Emerson's (together with William Arnold) dealt with photosynthesis in flashing light. Since photosynthesis involves one or two photochemical steps, preceded and followed (and also separated) by nonphotochemical, enzyme-catalyzed "dark" reactions, the study of photosynthesis in flashing light permits the separation of the light stage from the dark stage. By varying the dark interval between flashes, Emerson proved that the dark stage needs about 0.01 second for its completion at room temperature. Another important result was to show that a single, intense flash of light can produce, in normal healthy plants, only one molecule of oxygen (and reduce one molecule of carbon dioxide) per approximately 2,500 chlorophyll molecules present. This finding became the starting point of the theory of the photosynthetic unit, which postulates the association in plant cells of about 300 chlorophyll molecules (2,500 divided by 8) with a single reaction center (an enzyme molecule) to which the light energy absorbed in any one of the 300 associated pigment molecules is conveyed by a special physical mechanism (resonance energy migration). This is one of the basic concepts of the present-day theory of photosynthesis.

Emerson carried out all his experiments himself, alone or with a trusted assistant. Among his students and co-workers were William Arnold, Charleton Lewis, Shimpe Nishimura, Mrs. Marcia Brody, Carl Cederstrand, and Mrs. Ruth Chalmers.

BIBLIOGRAPHY

I. ORIGINAL WORKS. Emerson's principal writings include "A Separation of the Reactions in Photosynthesis by Means of Intermittent Light," in *Journal of General Physiology,* **15,** no. 4 (1932), 391–420, written with W. Arnold; "The Photochemical Reaction in Photosynthesis," *ibid.,* **16,** no. 2 (1932), 191–205, with W. Arnold; "Photosynthesis," in *Annual Review of Biochemistry,* **6** (1937), 535–556; "Carbon Dioxide Exchange and the Measurement of the Quantum Yield of Photosynthesis," in *American Journal of Botany,* **28,** no. 9 (1941), 789–804, with C. M. Lewis; "The Dependence of the Quantum Yield of Chlorella Photosynthesis on Wave Length of Light," *ibid.,* **30,** no. 3 (1943), 165–178, with C. M. Lewis; "Some Factors Influencing the Long-wave Limit of Photosynthesis," in *Proceedings of the National Academy of Sciences of the United States of America,* **43** (1957), 133–143, with R. Chalmers and C. Cederstrand; "The Quantum Yield of Photosynthesis," in *Annual Review of Plant Physiology,* **9** (1958), 1–24; and "Red Drop and Role of Auxiliary Pigments in Photosynthesis," in *Plant Physiology,* **35** (1960), 377–485, with E. Rabinowitch.

II. SECONDARY LITERATURE. On Emerson and his work, see E. Rabinowitch and Govindjee, *Photosynthesis* (New York, 1969).

EUGENE RABINOWITCH

EMMONS, EBENEZER (*b.* Middlefield, Massachusetts, 16 May 1799; *d.* Brunswick County, North Carolina, 1 October 1863), *geology.*

Ebenezer Emmons was the focus of the great Taconic controversy, the gravest dispute ever to divide American geology. A principal figure in the first geological survey of New York between 1837 and 1843, he played a leading role in the establishment of a geological column for America and a stratigraphy independent of the Anglo-Continental model. Before the New York survey, correlation with the succession of English strata was the principal objective of American geologists from Maclure and Eaton through Edward Hitchcock and David Dale Owen. By 1842 the Transition strata of New York between the Carboniferous of Pennsylvania and the Primary had been separated into the Catskill, Erie, Helderberg, Ontario, and Champlain groups of a New-York system and a separate Taconic system. The Potsdam sandstone, the Chazy limestone, the black marble of Isle La Motte, and the Lorraine shales, as well as the major group names and the two system designations (New-York and Taconic) were bestowed by Emmons. The strata were characterized both paleontologically and lithologically, and type sections were described by Emmons and his colleagues. The system of nomenclature by geographic reference, adopted for the first time in North America, was specifically the work of Emmons. He also named the Adirondack Mountains,[1] and it was at a meeting of the New York Board of Geologists at Emmons' home in Albany in 1838 that the Association of American Geologists, which developed into the American Association for the Advancement of Science, was planned.

Emmons studied at Williams College and later the Berkshire Medical School, in the midst of the Taconic country. At Williams he was protégé and assistant of Chester Dewey, with whom he began his geological surveys. He joined Dewey in 1828 as instructor and later succeeded him as professor of natural history. In 1826 he was studying geology and assisting Amos Eaton at the newly organized Rensselaer School (now Rensselaer Polytechnic Institute) as well as lecturing in chemistry at the Albany Medical College. Emmons achieved a certain financial independence through the practice of medicine, chiefly obstetrics, which he continued throughout his life.

James Hall was a student at Rensselaer from 1830, the year in which Emmons became junior professor, to 1832. Emmons chose Hall as his assistant in the first field season of his survey of the second New York district. W. W. Mather, Lardner Vanuxem, and T. A. Conrad were initially responsible for the first, third, and fourth of the four districts into which the state had been divided. Hall later succeeded Conrad. Lewis C. Beck was attached to the survey as mineralogist.

Emmons began by tracing an orderly succession of scarcely disturbed fossiliferous strata, beginning with the Potsdam sandstone dipping gently away from the crystalline rocks of the Adirondack outlier. Across the Hudson and Champlain valleys to the east, he found the abrupt front of the Taconic Range, consisting of broken slates with intercalated carbonates, generally dipping steeply toward the crystalline rocks of the New England mountains. In spite of the reverse dip, Emmons had always considered these rocks to be older than the less disturbed shales and carbonates of the Hudson and Champlain valleys[2] and now proposed that they constituted a vast new sedimentary series between the Potsdam and the primary, the true primordial system and the base of the sedimentary column. Vanuxem and Conrad, and apparently even Hall at first, agreed. Mather disagreed, as did William and Henry Barton Rogers, whose observations of the merging of the flat rocks of the Allegheny Plateau into the folded strata of the valley and ridge province persuaded Mather that the Taconic strata are an overfolded extension of the Champlain group.

As the original survey was completed, Hall and Emmons were in competition for continuing geological positions with the state, Hall maneuvering to obtain the post of state paleontologist. Although aspiring to the title of state geologist, Emmons was diverted into the position of state agriculturalist.

His Taconic system came under direct attack from the Rogers brothers, who maintained that the entire folded Appalachian chain contained neither faults nor unconformities. In 1844 and again in 1846, in *Agriculture of New York,* Emmons published a vigorous exposition of the Taconic system, by then supported by paleontological evidence.

James T. Foster, a New York schoolteacher, relying on Emmons' expertise, prepared a geological map that he succeeded in selling to the New York Legislature in 1850 for use in the schools of the state. Louis Agassiz and Hall (who hurriedly prepared his own rival map) induced the state to cancel the order—Hall going so far, according to his assistant and, later, successor, J. M. Clarke, as to dump Foster's maps into the Hudson River. Thereupon, Foster sued Agassiz and Hall for libel. The case against Agassiz was heard first, with Emmons the sole witness to the scientific reputability of Foster's chart. Joseph Henry, James Dwight Dana (who later wrote fifteen papers disproving the Taconic system), J. D. Whitney, the Rogers brothers, Eben Horsford, Mather, Hitchcock, and even Sir Charles Lyell volunteered their testimony in Agassiz's behalf. The case, which ended in dismissal, amounted to the excommunication of Emmons from the ranks of American science. The following year he accepted the post of state geologist for North Carolina and moved south, there to be caught by the Civil War. He continued to assemble evidence of the wide extension of his pre-Potsdam system, and before his death in 1863 he had the satisfaction of seeing his claim to be discoverer of the "true primordial fauna" and the base of the sedimentary column supported by the contentious Jules Marcou; Joachim Barrande, by then the acknowledged authority on the Lower Paleozoic; the Canadian geologists William Logan and Elkanah Billings; and T. S. Hunt.

A man of stern appearance and strict religious observance, Emmons was nevertheless a popular teacher and greatly in demand as an obstetrician. His controversial views made him an underground favorite among students and younger faculty in the strongholds of his opponents. Successive editions of his textbooks attest to his extended influence. Perhaps more significantly, the *Manual of Geology* by Dana, his most uncompromising foe,[3] closely follows the organization and basic pedagogy of Emmons' *Manual of Geology* more than any of the available English examples.

The Taconic controversy, continuing in some aspects to the present day, came to overshadow Emmons' major contribution to world geology, which was the extension in New York of the method begun by William Smith, Georges Cuvier, and Alexandre Brongniart. It was the New York survey, but especially the New-York system classification and Emmons' nomenclature, that set the model in America, as Sedgwick and Murchison were setting it in England, for the subsequent development of stratigraphy and the geological time scale in the next century.

NOTES

1. Wherever possible Emmons took his designations from the names for the original Indian inhabitants. The precedent may have been Murchison and Sedgwick's Silurian system, first used in 1835, when the New York survey was beginning.
2. He had convinced Hitchcock of this in 1833, but by 1842 Hitchcock had decided that the Taconic rocks were simply metamorphosed Champlain strata (E. Hitchcock, *Geology of Massachusetts* [Amherst, 1833], p. 300; W. W. Mather, "Geology of the First District," in *Geology*, pt. 4 of the *Natural History of New York* [Albany, 1843]).
3. In 1888, when the evidence of extensive Lower Cambrian fossiliferous strata was incontrovertible, the American Committee of the International Geological Congress recommended the denomination "Taconic system" for the first group of strata above the Archean. The aged and mellowed Hall agreed, but Dana and C. D. Walcott led a successful fight to reject the recommendation.

BIBLIOGRAPHY

I. ORIGINAL WORKS. Emmons' most significant publication was the now almost unobtainable final report, "Geology of the Second District," in *Geology*, pt. 4 of the *Natural History of New York* (Albany, 1842). *Agriculture of New York*, pt. 5 of *Natural History of New York* (Albany, 1846), contains his geology of New York, the only summary publication of the results of the New York survey. It is profusely illustrated with exceptional lithographs by his son, Ebenezer Jr., and engraved sections, some hand-colored. The text speaks of a map but, according to Marcou, although the map was printed, Hall succeeded in suppressing it. Emmons' several textbooks made free and presumably profitable use of the fossil illustrations prepared under the direction of the New York Board of Geologists, thereby infuriating Hall.

Emmons' *Manual of Mineralogy and Geology* (Albany, 1826; 2nd ed., 1832) should be compared with his last textbook, *Manual of Geology* (Philadelphia, 1860), to illustrate his lack of concern with theory and a characteristic pragmatic concentration on historical geology. For a complete bibliography, consult John M. Nickles, "Geologic Literature on North America 1785-1918," *U.S. Geological Survey Bulletin* no. 746 (1922), 345-346.

II. SECONDARY LITERATURE. A major source of information about Emmons is J. M. Clarke, *James Hall of Albany* (Albany, 1923), by Hall's assistant and successor as New York State paleontologist; see pp. 30, 40, 42, 53, 57, 99, 206. The Taconic controversy is extensively treated in G. P. Merrill, *The First One Hundred Years of American*

Geology (New Haven, 1924), pp. 594-614; a biographical essay by Jules Marcou, *American Geologist*, 7 (1891), 1-23, is somewhat overblown; the article by G. P. Merrill in the *Dictionary of American Biography*, III, 149, is unduly deflating. Emmons' own records apparently were lost in the disorders attendant on his wartime death. Any official records of the *Foster* v. *Agassiz* case, beyond the bare fact of its dismissal, were lost in a fire. See C. J. Schneer, "Ebenezer Emmons and the Foundations of American Geology," in *Isis, 60*, pt. 4 (1970), 439-450.

CECIL J. SCHNEER

EMMONS, SAMUEL FRANKLIN (*b.* Boston, Massachusetts, 29 March 1841; *d.* Washington, D.C., 28 March 1911), *geology, mining.*

Emmons was a descendant of Thomas Emmons, one of the founders of the Rhode Island Colony, who was "admitted to be an inhabitant of Boston" in 1648. His father was Nathaniel Henry Emmons, a highly respected and affluent Boston merchant engaged in the East India and China trade. His mother, Elizabeth Wales, was a descendant of Nathaniel Wales, who emigrated from Yorkshire to Boston in 1635. He was named for a great-grandfather on his father's side, Samuel Franklin, a cousin of Benjamin Franklin.

During his boyhood Emmons attended private schools, including the Dixwell Latin School, where he had rigorous training in English composition and some instruction in physical geography and map-making, all of which stood him in good stead in later years. He entered Harvard College at the age of seventeen and graduated with an A.B. degree in 1861. The next five years were spent in Europe, climbing in the Alps in the summer of 1861 and then studying with private tutors in Paris to gain admission to the École Impériale des Mines, where he was a student during the academic years 1862-1863 and 1863-1864. He then enrolled in the Bergakademie at Freiberg, Saxony, where he remained until midsummer 1865, after which he visited many of the important European mining centers, finally returning to Boston from Rome in June 1866, at which time Harvard awarded him an A.M. degree.

In late 1866 and early 1867 the geological exploration of the fortieth parallel, under the direction of the chief of engineers, U.S. Army, was authorized by Congress and organized by Clarence King, who was chief geologist. Arnold Hague, who had become acquainted with Emmons when they were students at Freiberg, was appointed one of King's assistant geologists. Through Hague's influence King accepted Emmons as a volunteer assistant for the 1867 field season. The following winter Emmons received an official appointment as assistant geologist, a position he held until the completion of his reports in 1877, when he resigned to engage in cattle ranching in Wyoming. The U.S. Geological Survey was created by act of Congress on 3 March 1879. Clarence King was its first director, and one of his first official acts was to appoint Emmons as geologist in charge of the Rocky Mountain Division. Later he was placed in charge of the Division of Economic Geology, where he remained until his death.

Emmons was married three times: on 5 August 1876 to Waltha Anita Steeves of New York, from whom he was subsequently divorced; on 14 February 1889 to Sophie Dallas Markoe of Washington, who died on 19 June 1896; and on 4 August 1903 to Suzanne Earle Ogden-Jones of Dinard, France, who survived him. He left no children.

Emmons was active in many scientific organizations. He was a member of the National Academy of Sciences and its treasurer from 1902 to 1910. He was one of the founders of the Geological Society of America in 1888 and its president in 1903. Earlier he had helped to establish the Colorado Scientific Society and was its first president in 1882. Emmons was a member or fellow of the American Academy of Arts and Sciences, the American Philosophical Society, the American Association for the Advancement of Science, the Washington Academy of Sciences, the Geological Society of Washington (of which he was president for one term), and the American Institute of Mining Engineers, of which he was vice-president in 1882 and again in 1890 and 1891. When the International Geological Congress met in Washington in 1871, he was its general secretary, and at meetings in other countries in 1897, 1903, and 1910 he was one of its vice-presidents. Emmons became a fellow of the Geological Society of London in 1874 and later an honorary member of the Société Helvétique des Sciences Naturelles. Columbia and Harvard conferred honorary Sc.D. degrees upon him in 1909.

Although mining geology, especially the origin of ore deposits, was Emmons' major interest, his contributions to regional and structural geology were notable. His assignment with the fortieth-parallel exploration involved the survey of an area, about 100 miles in width, extending from the Sierra Nevada to the Great Plains. Much of the region was then so little settled that detachments of the U.S. Army accompanied the field parties to protect them from hostile Indians. The report on this pioneer work (1877) promptly became a model for regional studies and was emulated by many geologists.

The directives for Emmons' early work in the U.S. Geological Survey were conflicting. Most desirable from his point of view were the instructions to prepare a monograph on the region of Leadville, Colorado, at that time one of the most productive mining localities in the Rocky Mountains. But this was in 1879, and the Geological Survey had undertaken the collection of statistics on precious metals for the tenth census (1880). That task was also given to Emmons and G. F. Becker. With characteristic energy Emmons fulfilled both duties simultaneously. Volume XIII of the *Tenth Census Reports,* by Emmons and Becker, with its geological descriptions of mining regions, was published in 1885. A preliminary report on the geology and mining industry of the Leadville region was published in 1882 and the definitive monograph in 1886. This immediately attracted widespread attention and stimulated the investigation of the origin of ore deposits in other mining regions.

Emmons' conclusion that the ores had been derived mainly from the intruded igneous rocks and deposited in adjacent sedimentary rocks by hot aqueous solutions led to the classification of many ore bodies all over the world as results of contact metamorphism. His suggestion that the hot aqueous solutions were of meteoric origin, heated at depth by contact with hot igneous rock, provoked long-continuing discussion, and in his later work he modified his original theory to include the idea that they were partly of magmatic origin. This development of Emmons' concepts of ore genesis, as well as of secondary enrichment, appeared in his later publications and profoundly influenced the work of many geologists, including those who studied mining regions under his supervision when he was in charge of the Division of Economic Geology in the U.S. Geological Survey.

BIBLIOGRAPHY

I. ORIGINAL WORKS. Emmons' writings include "Geology of Toyabe Range," in *Geological Exploration of the 40th Parallel,* III (Washington, D.C., 1870), 320–348; "Descriptive Geology of the 40th Parallel," *ibid.,* II (Washington, D.C., 1877), 1–890, written with Arnold Hague; "Abstract of a Report Upon the Geology and Mining Industry of Leadville, Colorado," in *Report of the United States Geological Survey* (Washington, D.C., 1882), pp. 203–290; "Statistics and Technology of the Precious Metals," in *Tenth Census Reports,* XIII (Washington, D.C., 1885), 1–540, written with G. F. Becker; "Geology and Mining Industry of Leadville, Colorado," in *Monographs of the U.S. Geological Survey,* 12 (1886), 1–770; "Structural Relations of Ore Deposits," in *Transactions of the American Institute of Mining Engineers,* 16 (1888), 804–839; "Orographic Movements in the Rocky Mountains," in *Bulletin of the Geological Society of America,* 1 (1890), 245–286; *Geology and Mineral Resources of the Elk Mountains,*

Colorado, U.S. Geological Survey folio no. 9 (Washington, D.C., 1894); "Geology of the Denver Basin in Colorado," in *Monographs of the U.S. Geological Survey,* **27** (1896), 1–556, written with Whitman Cross and G. E. Eldridge; "The Mines of Custer County, Colorado," in U.S. Geological Survey, *Seventeenth Annual Report* (Washington, D.C., 1896), pt. 2, 411–472; *Geology of the Ten-Mile District, Colorado,* U.S. Geological Survey folio no. 48 (Washington, D.C., 1898); "Secondary Enrichment of Ore Deposits," in *Transactions of the American Institute of Mining Engineers,* **30** (1901), 177–217; "Theories of Ore Deposition, Historically Considered," in *Bulletin of the Geological Society of America,* **15** (1904), 1–28; "Development of Modern Theories of Ore Deposition," in *Mining and Scientific Press,* **99** (1909), 400–403; "The Downtown District of Leadville, Colorado," in *Bulletin of the United States Geological Survey,* no. 320 (1907), 1–75, written with J. D. Irving; and "Cananea Mining District of Sonora, Mexico," in *Economic Geology,* **5** (1910), 312–366.

II. Secondary Literature. Biographies of Emmons are George F. Becker, in *Transactions of the American Institute of Mining Engineers,* **42** (1911), 643–661, with a bibliography of 93 titles; Arnold Hague, in *Bulletin of the Geological Society of America,* **23** (1912), 12–28, with a bibliography of 94 titles; and in *Biographical Memoirs. National Academy of Sciences,* **7** (1913), 307–334, with a bibliography of 98 titles.

Kirtley F. Mather

EMPEDOCLES OF ACRAGAS (*b.* Acragas [now Agrigento, Sicily], *ca.* 492 b.c.; *d. ca.* 432 b.c.), *natural philosophy.*

The originator of the four-element theory of matter, Empedocles was the author of two hexameter poems, a physical-cosmological one traditionally entitled "On Nature" (estimated to have been 2,000–3,000 lines in length) and a religious-mystical one, "Purifications," on themes of personal salvation (including lists of taboos), metempsychosis, and eschatology. A total of 450 lines from the two poems, the largest amount of text available to us from any of the pre-Socratics, have been preserved in the form of quotations by later authors (Simplicius, Aristotle, Plutarch, and others). Also attributed to him in antiquity were a treatise on medicine, tragedies, and other works, but the sources tell us nothing about the contents of any of these, and modern scholars are generally of the opinion that the attributions were spurious or confused. It is noteworthy, nevertheless, that Empedocles is often mentioned as a physician by the medical writers (Galen refers to him as founder of the Sicilian school of medicine) and that Aristotle called him the inventor of rhetoric.

The ancient biographical tradition concerning Empedocles is overlaid with legend. Most of the stories told about him can be seen to be fanciful elabora-

tions of personal remarks made in his poems. He is supposed to have stopped an epidemic by diverting and mixing river streams, to have improved climate by erecting a windbreak across a gorge, to have moderated the etesian winds by drawing them into sacks, and to have revived a woman who had had neither breath nor pulse for thirty days. Of his death it was said that, convinced of his immortality, he jumped into one of the craters of Etna. An alternative version (this one not tinted with sarcasm) has him ascending to the sky. An ancient tradition, enthusiastically revived by the Sicilians in the days of Garibaldi, was that Empedocles, although born an aristocrat, became a champion and hero of democratic politics.

The contents and style of his poetry reveal a man of fervid imagination, versatility, and eloquence, with a touch of theatricality. Perhaps some of the traits of the historical Empedocles have indeed been captured in the colorful portrait of the biographical tradition. The legend-making has continued in modern times: Empedocles has been the hero of Romantic tragedy-poems by Hölderlin and Matthew Arnold, and of other literary works (a French play as recently as 1950).

While the religious poem betrays the influence of Pythagoreanism and kindred strands of what has been called Orphism by some scholars and the Greek "puritan psychology" by others, the cosmological poem is unmistakably a development, with crucial modifications, of Parmenidean metaphysics. Parmenides of Elea had deduced that the real must be (a) unborn and imperishable, (b) one and indivisible, (c) immobile, (d) a complete actuality. Since familiar entities of the world of sense fail to conform to these criteria, these entities are a man-made illusion.

Empedocles moderates the extreme transcendent rationalism of the Parmenidean deduction. He postulates four eternal and unchanging elements, earth, water, air, and fire (he actually calls them "roots of all things" and also refers to them by mythological names) and two forces, Love and Hate. Viewed distributively, all six conform, in some sense that Empedocles considers appropriate, to the Parmenidean criteria. The familiar entities of the manifest world (animals, plants, minerals) result from the mixture, in various degrees of combination and according to various proportions of components, of the four elements ("So much does the mixing alter their [the elements'] look," fragment 21.14). Generation and destruction—change in general—are nothing but aggregation and dispersal of the elements by the two cosmic forces acting externally upon them.

Parmenides' requirement for unity and total integration of the real, expressed by him in a comparison with a "sphere well-rounded from all sides," is also fulfilled in a collective sense in Empedocles' system. The latter postulates a cosmic cycle involving four phases: (1) complete mixture of the four elements in a homogeneous sphere; (2) partial and increasing separation owing to the ascendancy of Hate; (3) an interval of total separation; (4) partial and increasing integration owing to the ascendancy of Love. A cosmos like ours can exist, it would seem, only in phases 2 and 4. (What is given here is the traditional interpretation, most recently defended by O'Brien. Bollack and others have argued that the ancient evidence does not support the ascription of a cycle, in the sense of chronological repetition of these phases.)

To explain the origins of animals in phase 4, Empedocles invokes chance and natural selection. From random combinations of stray limbs and organs, monsters emerge at the early stages. Since these are not adapted for survival and reproduction, they perish. Eventually viable organisms come to be assembled; they succeed in producing offspring, and they proliferate. Darwin mentions this Empedoclean theory (indirectly, by citing a passage from Aristotle in which the views of Empedocles are being discussed) in the first note of the historical preface to *The Origin of Species*. But it should be stressed that in Empedocles the mechanism of selection ceases to function precisely where Darwinian evolution starts: when the mechanism of heredity begins to play a role.

Closely related to the cosmic cycle that culminates in the sphere is the Empedoclean picture of our universe as a spherical (or perhaps egg-shaped) plenum, with an encompassing crystalline firmament. Both fixed stars and planets are islands or pockets of fire, the former rigidly attached to the firmament. Empedocles may well have been the first Greek to articulate this influential picture. (Some scholars credit this to Anaxagoras, on the assumption that his date is earlier; others to Parmenides, or to the Pythagoreans, or to even older thinkers, but on evidence that is less firm than what we possess for Empedocles.) That he regarded the earth as spherical is open to doubt, since he explained the inclination of the celestial axis as the result of "tilting" caused by air pressure. Three more doctrines of his astronomy are worth mention. He adopted Parmenides' account of moonlight as a reflection from the sun and gave the correct explanation of solar eclipses. But he was not content to limit the hypothesis of reflection to the moon; he considered the sun itself to be an image of the whole daytime sky as the latter is reflected from the earth's surface.

The macrocosmic cycle reverberates at all levels of the universe. In fragment 100 Empedocles explains respiration and the movement of blood in terms of ebbing and flowing, and gives as illustration the movement of a liquid in the clepsydra, "the water snatcher"—essentially our pipette but wider and with multiple holes at the lower end, suitable for drawing and serving wine from a deep jar. The illustration has often been extravagantly hailed by modern scholars as an experiment. More significantly, the passage represents the earliest statement of the tidal theory of blood movement that remained standard until the time of William Harvey (1628).

The theory of the four elements was adopted by Plato and Aristotle, although both postulated sub-elemental principles and allowed for transmutation. The Empedoclean theory also inspired or influenced the similar doctrine of four elements and four humors in the Hippocratic school of medicine. Through these three avenues of Platonism, Aristotelianism, and the medical tradition, Empedocles' theory of matter remained the dominant one until the revival of atomism (by Gassendi and Boyle) in the seventeenth century.

While critical of the four-element theory and of Empedocles' conception of the world as a plenum, the ancient atomists nevertheless drew heavily on him. As can be seen very clearly in Lucretius, they adapted Empedoclean ideas not only in the areas of cosmology and zoogony but also in explanations of the phenomena of perception. Empedocles' theory of filmlike "effluences" that are emitted by all things and of corresponding "pores" that serve as selective receptors for these emissions was especially influential.

To trace the various types of Love-Hate metaphysics, speculative physics, and *Naturphilosophie* to Empedocles would be gratuitous, of course, since that particular pair of forces has the universality of a psychological archetype. Freud, who was struck by the resemblance of his later theory of Eros-Destructiveness to the Empedoclean scheme, mused that his own theory might be a case of cryptomnesia of his early readings in pre-Socratic philosophy (*The Standard Edition of the Complete Psychological Works of Sigmund Freud,* James Strachey, trans. [London, 1964], XXIII, 244–246).

Many of Empedocles' ideas that did not have historical influence turned out to have been prophetically right. Most often mentioned in this connection is his evolutionary paleontology—a prime example, according to Hans Reichenbach, of how "a good idea stated within an insufficient theoretical frame loses its explanatory power and is forgotten" (*The Rise of Scientific Philosophy* [Berkeley, 1957], p. 197). The same holds for Empedocles' distinction between matter and mechanical force, his ultimate dualism of attractive and repulsive forces, his postulate of the conservation of energy and matter, his doctrine of constant proportions in chemical reactions, and his assumption that light is corporeal and has finite velocity. If one of the two presently competing theories of cosmology (the "big bang" theory) turns out to be right, even his vision of the universe under the influence of Strife will have found a counterpart in modern physics.

The question of the relation between "On Nature" and "Purifications" remains an enigma. The contrast is not only one of mood; the doctrines of personal salvation and metempsychosis cannot easily be reconciled with the essentially materialist metaphysics of the physical poem. Most modern interpreters have despaired of finding more than a psychological or biographical solution. They see the antinomy implicit in the extant fragments as significantly connected with Empedocles' dual reputation as philosopher-scientist and miracle worker. "The last of the Greek shamans," "a Faust," "a Greek Paracelsus" are some of the more suggestive characterizations that have been proposed.

BIBLIOGRAPHY

For the fragments and ancient reports, see Hermann Diels, *Die Fragmente der Vorsokratiker,* Walther Kranz, ed., 6th ed., 3 vols. (Zurich–Berlin, 1951), ch. 31. Selections with English translation and commentary are in G. S. Kirk and J. E. Raven, *The Presocratic Philosophers* (Cambridge, 1957), pp. 320–361. The most recent lengthy accounts are Jean Bollack, *Empédocle:* I, *Introduction à l'ancienne physique* (Paris, 1965); II and III, *Les origines* (Paris, 1969); W. K. C. Guthrie, *A History of Greek Philosophy,* II (Cambridge, 1965), 122–265; and D. O'Brien, *Empedocles' Cosmic Cycle* (Cambridge, 1969), which includes an annotated bibliography of all publications on Empedocles from 1805 to 1965.

ALEXANDER P. D. MOURELATOS

ENCKE, JOHANN FRANZ (*b.* Hamburg, Germany, 23 September 1791; *d.* Spandau [near Berlin], Germany, 28 August 1865), *astronomy.*

The eighth child of a Lutheran preacher, J. Michael Encke, and his wife, M. Elisabeth Misler, Encke displayed an early interest in mathematics but did not enter the University of Göttingen until the autumn of 1811. During his years of study, which were twice interrupted by his military service in the Wars of Liberation, he was greatly impressed and guided by Gauss, who reciprocated his esteem and in May 1816 procured a post for him at the small Seeberg

observatory, near Gotha. After serving as assistant, Encke qualified as professor at and director of this observatory through his theoretical work, of which the computation of the orbit of a comet discovered by Pons is the most essential. This comet was later called Encke's comet. Encke demonstrated that this comet had a period of scarcely four years and that it had been observed repeatedly. Prior to this only a few comets with elliptical orbits and much longer periods had been known. In 1825 Encke—already famous—was offered a professorship at the Academy of Sciences in Berlin and the directorship of the Berlin observatory.

As member of the Academy Encke directed his attention to creating new star charts. This work, only partly based on new observations, was done by many observatories. Those parts finished by the middle of the 1840's led to the discovery of several minor planets and of Neptune in 1846. In 1859 the task was completed, but the charts were soon excelled by those of Argelander.

Encke continued his theoretical work on comets in Berlin, and his investigations into special perturbations are worth attention even today. As a disciple of Gauss, he had already computed the perturbations of the four oldest planetoids, using Gauss's method, which he subsequently improved considerably. As academician Encke was entitled to lecture without passing the *Habilitation* or receiving a doctorate. Upon request by the ministry he lectured until 1862, but without much pleasure.

Encke's lectures were nevertheless influential, since a whole generation of astronomers, including Galle, F. F. E. Brünnow, B. A. Gould, K. N. A. Kruseger, W. J. Förster, Friedrich Tietjen, and K. C. Bruhns were among his disciples. His lectures covered all areas of astrometry and included practical training in the use of measuring instruments, in determining orbits, and in computing perturbations, as well as in fields now considered part of applied mathematics. Encke also lectured on the history of astronomy. He became ordinary professor at the University of Berlin in 1844.

In 1825 the Berlin observatory was obsolete. With the strong support of Alexander von Humboldt, Encke soon succeeded in obtaining funds for a better and more suitably located structure, which began operation in 1835. Besides a meridian circle and a large Fraunhofer refractor, it was equipped with several special-purpose instruments, such as a heliometer. Particular attention was given to observing the positions of stars, particularly of movable stars. Physical observations of planets were of minor interest to Encke. An eager observer himself, he guided his assistants in observing without interfering too much with their work.

After his appointment at Berlin, Encke undertook the editing of the *Berliner astronomisches Jahrbuch.* With the support of his assistants, especially J. P. Wolfers and Bremiker, he issued the yearbooks for 1830–1866. For coverage of minor planets, which was requiring more and more space in the books, several of Encke's disciples were engaged. Apart from 1844–1851, when the yearbook appeared together with the nautical ephemerides as an official publication, its issuance was a private matter, supported by the state but not without economic risk. The opportunity to publish made possible the appearance of Encke's treatises in the yearbooks, dealing particularly with orbit determinations and perturbation computations.

Encke published several of his papers in *Astronomische Nachrichten;* they referred almost exclusively to bodies of our solar system and only to a small extent to fixed stars. In 1823 he married Amalie Becker, who bore him three sons and two daughters. He died peacefully in 1865, after suffering three strokes, and was buried in Berlin.

BIBLIOGRAPHY

Many of Encke's writings were brought together as *Astronomische Abhandlungen,* 3 vols. (Berlin, 1868). Among his papers are "Polhöhe der neuen Berliner Sternwarte," in *Abhandlungen der Preussischen Akademie der Wissenschaften* (1845); "Entwicklung der allgemeinen Störungen der Flora durch Jupiter und Saturn," in *Berichte der Preussischen Akademie der Wissenschaften* (1853); "Berechnung der Pallas-Störungen," *ibid.* (1855); "Telegraphische Bestimmung der Längenunterschiedes Brüssel-Berlin," in *Abhandlungen der Preussischen Akademie der Wissenschaften* (1858); and "Der Comet von Pons," *ibid.* (1859), composed of eight papers. *Vorlesungen über Geschichte der Astronomie im Altertum* was edited by K. C. Bruhns and published after Encke's death (Altona, 1869). Many notices appeared in *Astronomische Nachrichten.*

A biography is K. C. Bruhns, *Johann Franz Encke* (Leipzig, 1869).

H. C. FREIESLEBEN

ENGEL, FRIEDRICH (*b.* Lugau, near Chemnitz [now Karl-Marx-Stadt], Germany, 26 December 1861; *d.* Giessen, Germany, 29 September 1941), *mathematics.*

The son of a Lutheran pastor, Engel attended the Gymnasium at Greiz from 1872 to 1879, studied mathematics in Leipzig and Berlin from 1879 to 1883, and received his doctorate in Leipzig in 1883 under Adolph Mayer. In 1884 and 1885 he studied with Sophus Lie in Christiania (now Oslo). In 1885 Engel qualified as a lecturer in pure mathematics at Leipzig and became an assistant professor there in 1889 and an associate professor in 1899. In 1904 he succeeded his friend Eduard Study as full professor at Greifswald, and in 1913 he went in the same capacity to Giessen, where, after his retirement in 1931, he continued to work until his death.

Although Engel was himself an important and productive mathematician he has found his place in the history of mathematics mainly because he was the closest student and the indispensable assistant of a greater figure: Sophus Lie, after N. H. Abel the greatest Norwegian mathematician. Lie was not capable of giving to the ideas that flowed inexhaustibly from his geometrical intuition the overall coherence and precise analytical form they needed in order to become accessible to the mathematical world. It was no less a mathematician than Felix Klein who recognized that the twenty-two-year-old Engel was the right man to assist Lie and who sent him to Christiania.

Shortly after Engel's return to Leipzig in 1886, Lie succeeded Klein there, and the fruitful collaboration was continued. The result was the *Theorie der Transformationsgruppen,* which appeared from 1888 to 1893 in three volumes "prepared by S. Lie with the cooperation of F. Engel."

Engel performed two further services for the great man long after the latter's death in 1899. In 1932 there appeared Engel's lectures *Die Liesche Theorie der partiellen Differentialgleichungen: Erster Ordnung,* prepared for publication by Karl Faber. For Lie the transformation groups had only been an important aid in handling differential equations; however, he never succeeded in composing a work on his theory of these differential equations. In Faber the seventy-year-old Engel had found the right person to help him in completing this work of his teacher.

Between 1922 and 1937, Engel published six volumes and prepared the seventh of the seven-volume edition of Lie's collected papers, an exceptional service to mathematics in particular and scholarship in general. Lie's peculiar nature made it necessary for his works to be elucidated by one who knew them intimately, and thus Engel's *Anmerkungen* ("Annotations") competed in scope with the text itself. The seventh volume finally appeared in 1960.

Engel's numerous independent works also are concerned primarily with topics in the fields of continuous groups and of partial differential equations: contact transformations (in his dissertation, before his meeting with Lie), Pfaffian equations, Lie's element sets and higher differential quotients, and many others. Lie's ideas were also applied to the *n*-body problem in mechanics (the ten general integrals).

Engel also edited the collected works of Hermann Grassmann, thus bringing posthumous fame to this great mathematician. In addition, with his friend P. Stäckel, Engel investigated the history of non-Euclidean geometry; along with this study he translated the essential works of N. I. Lobachevsky from Russian into German, their first appearance in a Western language.

Engel was a member of the Saxon, Russian, Norwegian, and Prussian academies. He received the Lobachevsky Gold Medal and the Norwegian Order of St. Olaf and was an honorary doctor of the University of Oslo. In 1899 he married Lina Ibbeken, the daughter of a Lutheran pastor. Their only child died very young.

BIBLIOGRAPHY

On Engel's work, see "Friedrich Engel," in *Deutsche Mathematik,* **3** (1938), 701–719, which includes a detailed bibliography with a short summary of each item by Engel himself; G. Kowaleski, "Friedrich Engel zum 70sten Geburtstag," in *Forschungen und Fortschritte* (1931), p. 466; E. Ullrich, "Ein Nachruf auf Friedrich Engel," in *Mitteilungen des Mathematischen Seminars der Universität Giessen,* no. 34 (1945), which contains a supplement to the bibliography in *Deutsche Mathematik* and, with nos. 35 and 36, containing two previously unpublished works of Engel's, is bound to form *Gedenkband für Friedrich Engel;* and "Friedrich Engel, ein Nachruf," in *Nachrichten der Giessener Hochschulgesellschaft,* **20** (1951), 139–154, and in *Mitteilungen des Mathematischen Seminars der Universität Giessen,* no. 40 (1951), which also contains the bibliographical supplement that appeared in his earlier article. Also see H. Boerner's article in *Neue deutsche Biographie,* IV (1959), 501–502.

H. BOERNER

ENGEL, JOHANN. *See* **Angelus, Johannes.**

ENGELMANN, THEODOR WILHELM (*b.* Leipzig, Germany, 14 November 1843; *d.* Berlin, Germany, 20 May 1909), *physiology.*

Engelmann was the son of the well-known bibliographer and publisher Wilhelm Engelmann and his wife, Christiane Therese Hasse, daughter of the Leipzig historian Friedrich Christian August Hasse. He was very musical and graduated from the Thomas Schule in Leipzig. In the winter semester of 1861–1862 he began his studies in the natural sciences and medicine at Jena, where he was introduced to comparative anatomy by Gegenbaur, to physiology by

his brother-in-law, Adalbert von Bezold, and to botany by Schleiden. He continued his studies at Heidelberg and Göttingen, but it was at Leipzig that he finished his studies and in 1867 took his doctorate under the ophthalmologist Theodor Ruete, with a dissertation on the cornea.

At the beginning of 1867 Engelmann went to Utrecht, at Ruete's recommendation, as assistant to the physiologist and ophthalmologist Franz Cornelis Donders. In 1869 he married Donders' daughter, Marie. Following her early death he married Emma Vick, a well-known pianist whose professional name was Emma Brandes. Engelmann became associate professor of general biology and histology at Utrecht in 1871. In 1888 he was Donders' successor in the chair of physiology, having declined offers of appointment from Freiburg im Breisgau, Zurich, and Jena, primarily because he suffered greatly from migraine headaches. In 1877 he was rector at Utrecht. Oxford conferred an honorary doctorate on him in 1894.

In the winter semester of 1897 Engelmann became professor of physiology at Berlin, succeeding Emil du Bois-Reymond. He not unhesitatingly exchanged his contemplative existence at Utrecht for the activity of the cosmopolitan city of Berlin. Senile diabetes and failing strength forced him to retire on 14 October 1908. He died the following year from progressive arteriosclerosis.

Engelmann became scientifically oriented at an extremely early age. He was a passionate botanist and microscopist while still a schoolboy. Before his dissertation (1867) he had already published eight papers in zoology and biology, especially on Infusoria, the connection between the nerves and the muscle fibers, and the excitability of nerves and muscles under the influence of induction currents. The microscope was also his most important research tool in Utrecht. Engelmann was one of the founders of the cell physiology that was widespread in the second half of the nineteenth century. In Utrecht (1869) he investigated the flagellating movements of protozoa in great detail and described the *Flimmermühle* and the *Flimmeruhr,* physiological devices for measuring oscillatory motion.

At the same time Engelmann began his studies on the transmission of stimuli in the muscles of the ureter and on the physiology of peristalsis. He was an energetic advocate of myogenic formation and conduction of stimuli (1869). He claimed the same thing for the heart and proved his claim with the famous "zig-zag experiment," in which the heart of a frog was dissected spirally. In spite of its nerves being cut, the strip remained capable of forming and conducting stimuli (1875). Engelmann perfected the method of lever suspension with the frog heart and analyzed the laws of extrasystoles, the refractory phase, and the compensatory pause (1892–1895). He was the first to prove the lack of current in the intact and resting heart (1878) in opposition to du Bois-Reymond's theory of the preexistence of the electrical charge in the intact and resting muscle fiber. Earlier he had determined the velocity of the conduction of stimuli in cardiac muscle (1875). He formulated the law of conservation of the physiological stimulus period (1895). Engelmann was the first to distinguish the four types of activity of the heart nerves: inotropic, bathmotropic, chronotropic, and dromotropic (1896). This was the final prerequisite for an improved understanding of cardiac function in general and of the excitatory processes in particular.

Engelmann's second main area of work was the physiology of muscle contraction, in which he made much use of the microscope. He described the diminution of double refraction in the contracted muscle fiber in polarized light (1873) and believed the cause of contraction to be a shifting of fluid from the isotropic to the anisotropic substance, suspecting swelling processes to be the cause. He constructed an artificial model of the muscle fiber (a birefringent violin string) in order to elucidate the contraction process, and believed that he was able to demonstrate that heat was directly transformed into mechanical work in the course of contraction (1893). A lively conflict of scientific opinion arose from this, a battle he finally lost.

A remarkable investigation with Genderen furnished the microscopic proof that the retinal cones of the frog shift in the course of the change from light to darkness (1884) and that such movements are binocular even if only one of the two eyes is illuminated. Finally, Engelmann analyzed the sensitivity of protozoa to light and color and chemotaxis in bacteria. He had a difference of opinion with Ranvier concerning the structure of the axis cylinder of peripheral nerves, since Engelmann (1880) had incorrectly believed that the nodes of Ranvier represented a discontinuity of the axis cylinder.

Engelmann's interests were directed very early to microscopy and cellular physiology. At first, therefore, his subjects were more biological than physiological. Only with cardiac physiology did he enter the central area of experimental animal physiology. Engelmann never expressed himself on questions of theoretical biology or natural philosophy.

He lived a simple, modest, and retiring life. He loved music and musicians. His house in Utrecht and later in Berlin was frequently a meeting place for well-known musicians. He was an avid cellist and a close friend of Johannes Brahms, who dedicated the Quartet in B Major op. 67 to him. His correspondence with Brahms was published in 1918.

BIBLIOGRAPHY

I. ORIGINAL WORKS. His publications usually appeared both in Dutch, in the archives in the Netherlands, and in German, particularly in *Pflügers Archiv für die gesamte Physiologie* and later in *Archiv für Anatomie und Physiologie,* of which he was editor from 1900 to 1909. Of his 245 publications (see the bibliography in Kingreen) only the most important can be listed here.

Engelmann's books and surveys include *Zur Naturgeschichte der Infusionsthiere* (Leipzig, 1862); *Über die Hornhaut des Auges* (Leipzig, 1867), his inaugural dissertation; *Über die Flimmerbewegung* (Leipzig, 1868); "Physiologie des Protoplasma und der Flimmerbewegung," in *Hermanns Handbuch der Physiologie,* I (Leipzig, 1879), 343–408; *Über den Ursprung der Muskelkraft* (Leipzig, 1892); *Tafeln und Tabellen zur Darstellung der Ergebnisse spektroskopischer Beobachtungen* (Leipzig, 1897); and *Das Herz und seine Thätigkeit im Lichte neuerer Forschung* (Leipzig, 1904).

His journal articles include the following, all in *Pflügers Archiv für die gesamte Physiologie:* "Zur Physiologie der Ureter," **2** (1869), 243–293; "Die Hautdrüsen des Frosches," **6** (1872), 97–157; "Mikroskopische Untersuchungen über die quergestreifte Muskelsubstanz," **7** (1872), 33–71, 155–188; "Contractilität und Doppelbrechung," **11** (1875), 432–464; "Über die Leitung der Erregung im Herzmuskel," *ibid.,* 465–480; "Flimmeruhr und Flimmermühle. Zwei Apparate zum Registrieren der Flimmerbewegung," **15** (1877), 493–510; "Über das elektrische Verhalten des thätigen Herzens," **17** (1878), 68–99; "Über die Discontinuität des Axenzylinders und den fibrillären Bau der Nervenfasern," **22** (1880), 1–30; "Zur Anatomie und Physiologie der Flimmerzellen," **23** (1880), 505–535; "Neue Methode zur Untersuchung der Sauerstoffausscheidung pflanzlicher und thierischer Organismen," **25** (1881), 285–292; "Über Licht- und Farbenperception niederster Organismen," **29** (1882), 387–400; "Über Bewegungen der Zapfen und Pigmentzellen der Netzhaut unter dem Einfluss des Lichts und des Nervensystems," **35** (1885), 498–508; "Beobachtungen und Versuche am suspendirten Herzen. I," **52** (1892), 357–393 (Suspensionsmethode); II, **56** (1894), 149–202 (Erregungsleitung); III, **59** (1895), 309–349 (Physiol. Reizperiode); and "Über den Ursprung der Herzbewegung . . .," **65** (1897), 109–214.

Also see "Über die Wirkungen der Nerven auf das Herz," in *Archiv für Anatomie und Physiologie* (1900), pp. 315–361; "Über die bathmotropen Wirkungen der Herznerven," *ibid.* (1902), supp. 1–26; and "Über den causalen Zusammenhang zwischen Kontraktilität und Doppelbrechung (und ein neues Muskelmodell)," in *Sitzungsberichte der Preussischen Akademie der Wissenschaften zu Berlin* (1906).

II. Secondary Literature. Obituaries include *Deutsche medizinische Wochenschrift,* **35** (1909), 1110; R. du Bois-Reymond, in *Berliner klinische Wochenschrift,* **46** (1909), 1097–1099; *Nederlands Tijdschrift voor Geneeskunde,* **22** (1909), 1786–1790; H. Piper, in *Münchener medizinische Wochenschrift,* **56** (1909), 1797–1800; M. Rubner, in *Verhandlungen der Physiologischen Gesellschaft zu Berlin,* **34** (1910), 84–90; and M. Verworn, in *Zeitschrift für allgemeine Physiologie,* **10** (1910), i–vi.

Details of his life and assessments of his work may be found in *Biographisches Lexikon der hervorragendsten Arzte . . .,* I (Vienna-Berlin, 1932), 367; H. Kingreen, "Theodor Wilhelm Engelmann (Biobibliographie)," unpub., inaugural diss. (Münster, 1969); K. E. Rothschuh, *Geschichte der Physiologie* (Berlin-Göttingen-Heidelberg, 1953), pp. 213–214; and M. Stürzbecher, "Beitrag zur Biographie von Th. W. Engelmann," in *Berliner medizinische Wochenschrift,* **27** (1958), 470–474; and the article on Engelmann in *Neue deutsche Biographie,* IV (Berlin, 1959), 517–518.

K. E. ROTHSCHUH

ENGELS, FRIEDRICH (*b.* Barmen, Germany, 28 September 1820; *d.* London, England, 5 August 1895), *philosophy.*

For a detailed study of his life and work, see Supplement.

ENRIQUES, FEDERIGO (*b.* Leghorn, Italy, 5 January 1871; *d.* Rome, Italy, 14 June 1946), *mathematics, philosophy and history of mathematics and science.*

Enriques, the son of S. Giacomo and Matilda Enriques, was educated in Pisa, where the family moved during his childhood. He attended the university and the Scuola Normale with a brilliant record in mathematics and took his degree in 1891. After a year of graduate study in Pisa, a second one in Rome, and some further work in Turin with Corrado Segre, Enriques undertook the teaching of projective and descriptive geometry at the University of Bologna, where in 1896 he was elevated to a professorship in those subjects. He remained there until 1923. He was honored by the University of St. Andrews with an honorary doctorate.

Guido Castelnuovo speaks of the happy years spent at Bologna as being perhaps the most fruitful of Enriques' entire life. His intense interest in all fields of knowledge was nurtured by close contact with professors from all the faculties, and in the period 1907–1913 he served as president of the Italian Philosophical Society. In this capacity he organized the Fourth International Congress of Philosophy, held at Bologna in 1911.

In 1923 Enriques accepted the offer of the chair of higher geometry at the University of Rome. While there he founded the National Institute for the History of Science and a school dedicated to that discipline. Since his way of life and his philosophy made it impossible for him to cooperate with the dictates of a fascist regime, Enriques retired from teaching during the years 1938–1944.

As a young man Enriques studied under Betti, Dini, Bianchi, and Volterra and was influenced in his views on algebraic geometry by Segre. In 1892 he turned to Castelnuovo in Rome for advice on the direction of his work, and their many consultations led to Enriques' specialization in the theory of algebraic surfaces and to their collaboration in the field. The Turin Academy of Sciences published Enriques' first paper on the subject in June 1893.

A short summary of Enriques' contributions to this field—relating them to those of Castelnuovo, Picard, Severi, Humbert, and Baker—may be found in F. Cajori's *A History of Mathematics.*[1] Greater detail is given in each of two other accounts, both by Castelnuovo and Enriques in collaboration. The first, entitled "Sur quelques résultats nouveaux dans la théorie des surfaces algébriques,"[2] summarizes the Italian contribution up to 1906. The second, an earlier paper, carries the title "Sur quelques récents résultats dans la théorie des surfaces algébriques."[3] H. F. Baker's presidential address to the International Congress in Cambridge (12 December 1912), published as "On Some Recent Advances in the Theory of Algebraic Surfaces,"[4] also serves to highlight the contributions in that field and in so doing details Enriques' major contributions.

Enriques also contributed to the differential geometry of hyperspace. In 1907 he and Severi received the Bordin Award of the Paris Academy of Sciences for their work on hyperelliptical surfaces. The French honored him again in 1937 by making him a corresponding member of the Académie des Sciences Morales et Politiques.

As early as 1898 Enriques' interest in foundations of mathematics was reflected in his use of a system of axioms in his textbook writings. Having written, at Felix Klein's request, the article on the foundations of geometry ("Principien der Geometrie") for the *Encyklopädie der mathematischen Wissenschaften* (III, 1–129), he became instrumental in the writing of textbooks for both elementary and high schools that greatly influenced teaching in Italy. He was responsible for the publication, in Italian, of Euclid's *Elements* with historical notes and commentary, and he encouraged the publication of historical and didactic articles in *Periodico di matematiche,* which he headed for twenty years. His interest in teaching and in teachers is well reflected in his service as president of the National Association of University Professors.

By 1895 Enriques had concluded that besides the logical criteria of independence and compatibility, a psychological criterion involving the sensations and experiences that lead to the formulation of the postulates must be considered. In an 1898 paper he set up conditions justifying the introduction of coordinates on surfaces, thus supplementing Riemann's a priori approach in the assumption of such an existence. His interest in physiological psychology led to his writing studies for the *Rivista filosofica* that were later expanded into his *Problemi della scienza* (1906). Castelnuovo describes Enriques' thesis as being that topology and metrical and projective geometry are linked, respectively, to three different orders of sensations: to the general tactile-muscular, to those of the special sense of touch, and to those of vision. In the second part of the *Problemi,* a critical examination is made of the principles of mathematical, physical, and biological sciences. In the treatment of the principles of mechanics Enriques anticipated some of the foundations of Einstein's theory of relativity. His views on structure are given in *Causalité et déterminisme* (1940), and his philosophical thought is found in *Scienza e razionalismo* (1912). A causal explanation involves a "why" as well as a "how" and links effect to cause. Theory should be "plausible in itself" and "satisfy the principle of sufficient reason which is the mental aspect of causality." Determinism thus becomes a premise of scientific research. Enriques' philosophical and historical beliefs pervade *Per la storia della logica* (1922).

In the introductory note to the English translation of *Problemi della scienza* (1914), Josiah Royce writes of the pragmatistic element in Enriques' thought that brings to the thinking process an adjustment to situations; of his stress on the unifying aspect of scientific theory, the association of concepts and of scientific representation. Enriques' philosophical stance differs from that of the Comtean school. He disagrees with Mach and Pearson in their limitation of science to a simple description of physical phenomena, yet writes: "In the formation of concepts, we shall see not only an economy of thought in accordance with the views of Mach, but also a somewhat determinate mental process. . . ." He maintains a positivistic position toward the transcendental and the absolute in his emphasis on the tentative and relative character of scientific theory; yet his theory progresses toward a comprehension of the essential core concealed in every question. Enriques maintained that "It is plainly seen that scientific questions include something essential, apart from the special way in which

they are conceived in a particular epoch by the scholars who study such problems."

NOTES

1. (New York, 1961), p. 316.
2. Émile Picard and George Simart, *Théorie des fonctions algébriques de deux variables indépendentes,* II (Paris, 1906), 485–522.
3. *Mathematische Annalen,* **48** (1897), 241–316.
4. *Proceedings of the London Mathematical Society,* **12** (1913), 1–40.

BIBLIOGRAPHY

I. ORIGINAL WORKS. *Federigo Enriques: Memorie scelte di geometria,* 3 vols. (Bologna, 1956–1966), is a collection of 74 papers written between 1893 and 1940 and contains a bibliography of his works. Among his writings are *Lezioni di geometria descrittiva,* J. Schimaglia, ed. (Bologna, 1893–1894; 2nd ed., 1894–1895); republished in a new ed., U. Concina, ed. (Bologna, 1902; 2nd ed., 1908); *Lezioni di geometria proiettiva,* C. Pedretti, ed. (Bologna, 1893–1894; 2nd ed., G. Serrazanetti, ed., 1894–1895); republished in a new ed. (Bologna, 1898; 4th ed., 1920); *Conferenze di geometria: Fondamenti di una geometria iperspaziale* (Bologna, 1894–1895); *Elementi di geometria ad uso delle scuole normali* (Bologna, 1903), written with U. Amaldi; *Elementi di geometria ad uso delle scuole secondarie superiori* (Bologna, 1903), written with U. Amaldi; *Problemi della scienza* (Bologna, 1906; 2nd ed., 1908; repr. 1926), trans. into English by K. Royce with introductory note by J. Royce (Chicago, 1914); *Elementi di geometria ad uso delle scuole tecniche* (Bologna, 1909), written with U. Amaldi; *Nozioni di geometria ad uso delle scuole complementari* (Bologna, 1910), written with U. Amaldi; *Nozioni di geometria ad uso dei ginnasi inferiori* (Bologna, 1910), written with U. Amaldi; *Scienza e razionalismo* (Bologna, 1912); *Nozioni di matematica ad uso dei licei moderni,* 2 vols. (Bologna, 1914–1915); *Lezioni sulla teoria geometrica delle equazioni e delle funzioni algebriche,* 4 vols. (Bologna, 1915–1934; new ed. of vol. I, 1929), written with O. Chisini; *Conferenze sulla geometria non-euclidea,* O. Fernandez, ed. (Bologna, 1918); *Per la storia della logica. I principii e l'ordine della scienza nel concetto dei pensatori matematici* (Bologna, 1922), trans. into English by J. Rosenthal (New York, 1929); *Algebra elementare,* 2 vols. (Bologna, 1931–1932), written with U. Amaldi; *Nozioni di geometria ad uso delle scuole di avviamento al lavoro* (Bologna, 1931), written with U. Amaldi; *Nozioni intuitive di geometria ad uso degli istituti magistrali inferiori* (Bologna, 1931), written with U. Amaldi; *Lezioni sulla teoria delle superficie algebriche,* pt. 1 (Padua, 1932), written with L. Campedelli; pt. 2 was published in *Rendiconti del seminario matematico della reale università di Roma* (1934) as "Sulla classificazione delle superficie algebriche particolarmente di genere zero"; both parts were reorganized by G. Castelnuovo and published as *Le superficie algebriche* (Bologna, 1949).

See also *Storia del pensiero scientifico*, I, *Il mondo antico* (Bologna, 1932), written with G. de Santillana; *Nozioni di geometria ad uso delle scuole di avviamento professionale* (Bologna, 1934), written with U. Amaldi; *Il significato della storia del pensiero scientifico* (Bologna, 1936); *Compendio di storia del pensiero scientifico dall'antichità fino ai tempi moderni* (Bologna, 1937), written with G. de Santillana; *Le matematiche nella storia e nella cultura*, A. Frajese, ed. (Bologna, 1938); *La théorie de la connaissance scientifique de Kant à nos jours*, Actualités Scientifiques et Industrielles no. 638 (Paris, 1938); *Le superficie razionali* (Bologna, 1939), written with F. Conforto; *Causalité et déterminisme dans la philosophie et l'histoire des sciences*, Actualités Scientifiques et Industrielles no. 899 (Paris, 1940); *Elementi di trigonometria piana ad uso dei licei* (Bologna, 1947), written with U. Amaldi; *Le dottrine di Democrito d'Abdera, testi e commenti* (Bologna, 1948), written with M. Mazziotti; and *Natura, ragione e storia*, L. Lombardo Radice, ed. (Turin, 1958), a collection of his philosophical writings with a bibliography.

See also *Questioni riguardanti la geometria elementare* (Bologna, 1900), which Enriques collected and arranged; *Questioni riguardanti le matematiche elementari*: I, *Critica dei principii* (Bologna, 1912); II, *Problemi classici della geometria. Numeri primi e analisi indeterminata. Massimi e minimi* (Bologna, 1914); pt. 2, *I problemi classici della geometria e le equazioni algebriche*, 3rd ed. (Bologna, 1926); pt. 3, *Numeri primi e analisi indeterminata. Massimi e minimi,* 3rd ed. (Bologna, 1927), all collected and arranged by Enriques; and *Gli Elementi d'Euclide e la critica antica e moderna,* ed. by Enriques and many others: bks. I–IV (Rome-Bologna, 1925); bks. V–IX (Bologna, 1930); bk. X (Bologna, 1932); bks. XI–XIII (Bologna, 1935).

II. SECONDARY LITERATURE. On Enriques or his work, see H. F. Baker, "On Some Recent Advances in the Theory of Algebraic Surfaces," in *Proceedings of the London Mathematical Society,* 2nd ser., **12** (1913), 1–40; F. Baron, "Enriques, Federigo," in *Enciclopedia filosofica* (Venice-Rome, 1957), cols. 1916–1917; Guido Castelnuovo, "Commemorazione di Federigo Enriques," in *Federigo Enriques: Memorie scelte di geometria,* pp. x–xxii; Poggendorff, IV, 388–389; *Proceedings of the Fifth International Congress of Mathematicians* (Cambridge, 1912), I, 40; II, 22; Ferruccio Rossi-Landi, "Enriques, Federigo," in *Encyclopedia of Philosophy* (New York, 1967), III, 525–526; Ferruccio Rossi-Landi and Vittorio Somenzi, "La filosofia della scienza in Italia," in *La filosofia contemporanea in Italia* (Rome, 1958), pp. 407–432; and Antonio Santucci, *Il pragmatismo in Italia* (Bologna, 1963), pp. 302–322.

CAROLYN EISELE

ENSKOG, DAVID (*b.* Västra Ämtervik, Värmland, Sweden, 22 April 1884; *d.* Stockholm, Sweden, 1 June 1947), *physics.*

Enskog's father, Nils Olsson, was a preacher; his mother was Karolina Jonasdotter. He was educated at the Karlstads Läroverk (high school) and at Uppsala University, where he received the Ph.D. in 1917. After teaching in secondary schools and colleges for several years, he was appointed professor in mathematics and mechanics at the Royal Institute of Technology, Stockholm, in 1930. In 1913 he married Anna Aurora Jönsson.

Enskog is best-known for his development of a method for solving the Maxwell-Boltzmann transport equations in the kinetic theory of gases. These equations, describing the effect of molecular collisions and external variables (such as temperature gradients) on the flow of molecules, momentum, and energy in a gas, had originally been formulated by James Clerk Maxwell in 1867. The solution of the equations, however, depends in general on a determination of the velocity distribution function in a nonequilibrium gas. Maxwell was unable to determine this function. However, he did show that for a special molecular model—point centers of force with repulsive forces varying inversely as the fifth power of the distance between the molecules—the transport coefficients such as viscosity and thermal conductivity can be calculated even if the velocity distribution function is unknown. In 1872 Ludwig Boltzmann reformulated Maxwell's equations as a single integrodifferential equation for the velocity distribution function, but he was not able to find an exact solution except for the same special model that Maxwell had introduced.

Little progress was made toward solving the Maxwell-Boltzmann equations until 1911, when Enskog in Sweden and Sydney Chapman in England began their researches. In that year, Enskog obtained his philosophy licentiate (equivalent to the M.A.), partly for experimental work on gas diffusion, and also published two papers on a generalization of the Maxwell-Boltzmann kinetic theory. In the second of these papers he noted briefly the existence of a term proportional to the temperature gradient in the theoretical formula for the rate of diffusion in a mixture of two gases. Although it was not followed up at the time, this was later recognized as the first theoretical prediction of the important phenomenon known as thermal diffusion, established experimentally by Chapman and F. W. Dootson in 1917.

The calculations of Enskog and of Chapman, published in 1911–1912, while representing an advance over earlier work, were not yet satisfactory, since they depended on arbitrary assumptions about the nonequilibrium velocity distribution function. In 1912, David Hilbert published a short paper on the Maxwell-Boltzmann equation, in which he applied methods developed earlier in his general theory of integral equations. Enskog immediately saw the value of Hilbert's approach and used it with some modifications to work out a systematic series expansion of the velocity distribution function. The results were presented in his 1917 dissertation at Uppsala, shortly after Chapman published his own calculations based on an equivalent method. Chapman was one of the first scientists to recognize the value of Enskog's work, and in the monograph on *The Mathematical Theory of Non-Uniform Gases,* now the standard work on the subject, he used Enskog's procedure for solving the Maxwell-Boltzmann equations in preference to his own. It was also partly on Chapman's recommendation that Enskog obtained his chair at Stockholm in 1930, although according to Chapman, "His transfer to a university chair seemed rather to bring him new duties than increased leisure, and this, with renewed ill-health, reduced his productivity in later years."

With their new methods Chapman and Enskog were able to calculate accurately a number of the transport properties of gases, such as the coefficients of viscosity, thermal conduction, and diffusion, without having to rely on the mean-free-path approximation introduced by Clausius in 1858 and used by Maxwell in his early work. (Although Maxwell himself abandoned the mean-free-path method, it continued to be used by other scientists and is still discussed in most modern textbooks.) Whereas earlier kinetic theories had been limited to the use of very special molecular models, such as elastic spheres or the Maxwellian inverse fifth-power repulsive force, the Chapman-Enskog theory now made it possible to do calculations with a much larger class of models, including both attractive and repulsive forces varying with any power of the distance.

The Chapman-Enskog theory would have been quickly taken up and exploited by many other scientists during the 1920's if that had not been the time when quantum theory was being vigorously developed. It did not become clear until the 1930's that the classical kinetic theory of gases is still valid over a large range of temperatures and densities, even though the nature of the intermolecular force law is determined by quantum-mechanical considerations. Thus the main impact of the work of Enskog and Chapman before 1920 was not apparent until after 1945, when it provided the basis for a revival of activity in kinetic theory, including applications to phenomena such as sound propagation, shock waves, aerodynamics, the behavior of electrons in metals, and the diffusion of neutrons in nuclear reactors.

One other contribution by Enskog played an important role in the postwar development of kinetic theory. In 1922, he proposed a generalization of the Maxwell-Boltzmann equations to higher densities, taking account of the effect of finite diameter of the molecules. He assumed that the frequency of collisions would be changed by an amount that could be related to the equation of state (equilibrium pressure-volume-temperature relation), since that also depends on collision rate. In this way he obtained formulas for the transport coefficients in which the variation with density can be determined empirically if the equation of state is known from experimental measurements, or theoretically for simple models if the equation of state itself can be calculated theoretically. Until 1965 the Enskog theory of dense gases remained the only accepted theory that had been sufficiently worked out to permit experimental verification, although many more elaborate theories had been attempted. It is only in the last few years, as a result of the work of J. Weinstock, J. R. Dorfman, E. G. D. Cohen, R. Goldman, E. A. Frieman, and J. V. Sengers, that the kinetic theory of dense gases has definitely progressed beyond the point reached by Enskog from a fundamental basis; and according to the most recent calculations of Sengers, the numerical results for the transport coefficients computed from Enskog's theory are probably accurate to within 5 percent as compared with the exact theory.

BIBLIOGRAPHY

I. Original Works. Enskog's major works on kinetic theory include *Kinetische Theorie der Vorgänge in mässig verdünnten Gasen* (Uppsala, 1917) and "Kinetische Theorie der Wärmeleitung, Reibung und Selbstdiffusion in gewissen verdichteten Gasen und Flüssigkeiten," in *Kungliga Svenska vetenskapsakademiens handlingar,* n.s. **63**, no. 4 (1922). Translations of these works and references to others will be found in the book by S. G. Brush cited below.

II. Secondary Literature. On Enskog's kinetic theory of gases, see Sydney Chapman and T. G. Cowling, *The Mathematical Theory of Non-Uniform Gases* (Cambridge, 1939; 2nd ed., 1952). On the history of kinetic theory, see S. G. Brush, *Kinetic Theory,* 3 vols. (Oxford, 1965–   ); vol. III (in press) includes translations of the two major works of Enskog listed above and a discussion of more recent work.

For further biographical details, see Hilding Faxén, *Svenskt biografiskt lexikon,* XIII (Stockholm, 1950), 765–767.

Stephen G. Brush

ENT, GEORGE (*b.* Sandwich, Kent, England, 6 November 1604; *d.* London, England, 13 October 1689), *medicine.*

Ent was the son of Josias Ent, a merchant from

the Low Countries. His early education was at Rotterdam; he then studied at Sidney Sussex College, Cambridge, from 1624 to 1631, obtaining his M.A. in 1631. Shortly afterward he probably settled in Padua as a student of medicine. He obtained the M.D. there in 1636, the event being celebrated by a volume of poems, *Laureae Apollinari* (Padua, 1636), contributed by his friends—including P. M. Slegel, J. Rhode, and J. Greaves, all of whom became friends and defenders of Harvey.

The rest of Ent's life was spent in London as a successful and moderate medical practitioner. He rapidly reached a position of esteem among his professional and scientific colleagues. Although he probably had royalist sympathies he was not subject to recriminations during the Cromwellian period. During these years he came into prominence at the College of Physicians, in which he was elected to the most important executive positions, serving as president for seven years between 1670 and 1684. He was married to Sarah Meverall, daughter of the treasurer of the college. The Royal Society provided an outlet for his wider scientific interests; he was a founder fellow and member of its council, although playing a relatively small part in its scientific affairs. Ent was one of the last to give the annual anatomy lectures at the College of Physicians. At these lectures in 1665 he was granted a knighthood by Charles II.

Ent owed much of his scientific reputation to his friendship with William Harvey, which dated from their chance meeting in Rome in 1636. In reaction to the mounting published criticism of *De motu cordis*, Ent became one of the first writers to compose a detailed defense of Harvey, *Apologia pro circulatione sanguinis* (1641). This counteracted the criticisms of Emilius Parisianus; Ent quoted primarily from Harvey but also displayed a wide familiarity with ancient and modern authorities. In a series of digressions he showed a distinctive approach, being more receptive to hermetic authors than Harvey. This is particularly obvious in the sections on innate heat and respiration, which point toward the theories of John Mayow and suggest that nitrous particles from the air are absorbed by the lungs or gills, to support the physiological flame burning in the heart—the source of innate heat. Ent further proposed a less fortunate, but popular, theory whereby a highly nutritive fluid is dispensed through the nerves. Accordingly, the nervous role is reduced; Ent emphasized the role of tissue irritability and natural movement. His ideas on irritability are particularly prominent in his unpublished anatomy lecture notes.

Ent's association with Harvey continued. In about 1648 he persuaded the elderly Harvey to release the manuscript of *De generatione*, which Ent edited and published with a commendatory preface in 1651. His transcript of Harvey's correspondence was used in the College of Physicians edition of Harvey's works in 1766. Harvey's gratitude was indicated in the terms of his will, in which Ent was charged with dispersing his library, that is, selling worthless books and buying better ones to be deposited, with the rest of the library, in the College of Physicians. Ent was also given five pounds to purchase a ring in remembrance of Harvey.

Ent's other writings are less important. His proficient studies of the anatomy of *Lophius, Galeus,* and *Rana,* entitled "Mantissa anatomica," were published in one of Charleton's lesser works. This appears to have been the meager outcome of an elaborate mutual comparative anatomy project, conceived in the 1650's. Finally, Ent published a critique of Malachi Thruston's ideas on respiration, which showed little advance on the *Apologia* and lacked the empirical foundation of Mayow's writings on the same theme.

BIBLIOGRAPHY

I. ORIGINAL WORKS. Ent's writings are *Apologia pro circulatione sanguinis: Qua respondetur Aemilio Parisano* (London, 1641; 2nd ed., with some additions, 1685); *ΑΝΤΙΔΙΑΤΡΙΒΗ. Sive animadversiones in Malachiae Thrustoni, M.D. Diatribam de respirationis usu primario* (London, 1679, 1685); and *Opera omnia medico-physica* (Leiden, 1687), which contains essays on tides. See also "Mantissa anatomica," in Walter Charleton, *Onomasticon zoicum,* 2nd ed. (Oxford, 1677). Ent's anatomy lectures are in Bodleian Library, Ashmolean MS 1476, and Royal College of Physicians, London, MS 110.

II. SECONDARY LITERATURE. On Ent's life and work, see Sir George Clark, *History of the Royal College of Physicians,* vol. I (London, 1964); Sir Geoffrey Keynes, *The Life of William Harvey* (Oxford, 1966); William Munk, *Roll of the Royal College of Physicians,* I (London, 1878), 223–227; J. R. Partington, *History of Chemistry,* II (London, 1962), 564, 573–574; J. and J. A. Venn, *Alumni cantabrigenses,* II (Cambridge, 1924), 104; and C. Webster, "The College of Physicians 'Solomon's House' in Commonwealth England," in *Bulletin of the History of Medicine,* **41** (1967), 393–412.

CHARLES WEBSTER

EÖTVÖS, ROLAND, BARON VON (*b.* Budapest, Hungary, 27 July 1848; *d.* Budapest, 8 April 1919), *physics.*

Most of the world literature lists Eötvös (pronounced ut' vûsh) in the fashion given in the heading, which is the German version of the Hungarian name. The reason for this is that he published most of his major papers in both Hungarian and German. Roland is a translation of the Hungarian name Loránd, and his full name in Hungarian is Vásárosnaményi Báro Eötvös Loránd. The correct full name in English translation is Roland, Baron Eötvös of Vásárosnamény.

Eötvös was the scion of an aristocratic and intellectual family. His father, Joseph, Baron Eötvös of Vásárosnamény, at the time of Roland's birth held the portfolio of public instruction and religious affairs in the first, short-lived, responsible Hungarian cabinet; his mother was the former Agnes Rosty. The family had a long background of public service (the barony was conferred on his great-grandfather in the eighteenth century), but its intellectual tendencies came to full bloom only in his father, who became Hungary's foremost writer and political philosopher of the nineteenth century. Young Roland thus grew up in an environment leading more or less toward a study of law and government (his family, by hereditary right, belonged to the upper house of parliament). He entered the University of Budapest in 1865 as a law student but, already interested in the mathematical and physical sciences, took private lessons in mathematics from Otto Petzval.

At his father's request Joseph Krenner, the future professor of mineralogy at the university, introduced Eötvös to the study of physical sciences; at the same time he worked in the chemistry laboratory of Charles Than.

In 1867 Eötvös definitely abandoned the study of law and entered the University of Heidelberg. His studies included mathematics, physics, and chemistry, taught there by such outstanding teachers as Kirchhoff, Helmholtz, and Bunsen. After three semesters he went to the University of Königsberg but found the lectures of the theoretical physicist Franz Neumann and of the mathematician Friedrich Richelot less to his taste. For a while Eötvös toyed with the idea of joining the arctic expedition headed by August Petermann; but he finally decided, on his father's advice, to return to Heidelberg, where he obtained his doctorate *summa cum laude* in the summer of 1870. Apparently the subject of his doctoral thesis was identical with the subjects of three papers published by him in 1871, 1874, and 1875; they dealt with a problem formulated by Fizeau. The question was raised whether the relative motion of a light source, with respect to an immobile ether, can be detected by measuring the light intensities in both the same and the opposite directions of the motion. Eötvös generalized the calculations for both the emitter and the detector being in motion and extended it to astronomical observations. This purely theoretical

work became, decades later, the object of many important papers, leading ultimately to the theory of relativity.

His professors profoundly influenced Eötvös's working habits. Kirchhoff taught him the importance of accuracy in measurements. Helmholtz liked to spend as much time as possible with his students and showed Eötvös the value of individual discussion. His knowledge of theoretical physics, and in particular of potential theory, came from Franz Neumann.

At the end of his studies in 1870, Eötvös returned to Hungary and in 1871 became *Privatdozent* at the University of Budapest. In 1872 he was promoted to full professor at the same university. At first he taught theoretical physics; in 1874 he added experimental physics to his duties; and in 1878, at the retirement of Ányos Jedlik, professor of experimental physics, he took over that chair.

In 1876 Eötvös married Gizella Horváth, the daughter of the minister of justice, Boldizsár Horváth. They had two daughters, Ilona and Rolanda.

A few years earlier, while he was still a student at the University of Königsberg, Eötvös designed a simple optical method for determining the constant of capillarity (surface tension). He presented the subject at a physics colloquium, and Franz Neumann found the idea quite praiseworthy. Capillarity thus became the first research subject he attacked and led him to his first important discovery. He showed that the temperature coefficient of the molecular surface energy of a liquid—expressible as $\frac{d}{dT}[\gamma(Mv)^{2/3}]$, where M is the molecular weight, v is the specific volume of the liquid, and γ is the capillarity constant or surface tension—is independent of the nature of simple unassociated liquids. The integral form of the law of Eötvös is usually written as $\gamma v^{2/3} = k(T_0 - T)$, where k is a constant for all simple liquids ($k \approx 2.12$) and T_0 is (approximately) the critical temperature.

His investigations on capillarity were published in a few papers between 1876 and 1886. After 1886 there were no further communications by Eötvös on this subject, although the law of Eötvös attracted wide attention and during the next few decades a considerable number of papers appeared, examining and extending the concepts introduced by him.

After 1886 practically all of Eötvös's scientific papers concentrated on his lifework: gravitation. He was interested in this subject on and off before then, and there is some evidence of a gradually awakening interest in earlier papers and speeches. The exact year when his interest swung from phenomena involving van der Waals forces to the weakest known forces in the universe cannot be ascertained. A partial moti-

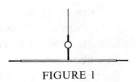

FIGURE 1

vation may have been a request by the Természet-tudományi Társulat (Hungarian Society for Natural Sciences) in 1881 for the determination of the gravitational acceleration in different parts of Hungary.

Eötvös's first short Hungarian-language publication on gravitational phenomena appeared in 1888. In January 1889 he presented a short paper to the Hungarian Academy of Sciences concerning his search for a difference in gravitational attraction exerted by the earth on different substances. This short paper, published in 1890, reported that within the accuracy achieved with his torsion balance, all substances investigated experienced the same force of attraction per unit of mass.

From the beginning of his gravitational researches Eötvös concentrated on the use of the instrument he called in a later paper the Coulomb balance, in recognition of the invention and use of torsion balance by Charles Coulomb. Actually, the torsion balance had been invented earlier (and independently) by Rev. John Michell, who had applied the principle upon which it is based as early as 1768 and, shortly before his death, completed construction of the particular torsion balance that in the hands of Henry Cavendish became an outstanding instrument for the determination of the attraction between two masses. The original Michell-Cavendish instrument places two masses at the ends of a horizontal bar suspended to allow a horizontal displacement of the masses around the torsion axis (see Figure 1). If the gravitational potential U is a function of the Cartesian coordinates x, y, z, it is possible to determine with that instrument the curvature of the gravitational field in a horizontal plane, i.e.,

$$\left(\frac{\partial^2 U}{\partial x \partial y}\right) \text{ and } \left(\frac{\partial^2 U}{\partial y^2} = \frac{\partial^2 U}{\partial x^2}\right).$$

For this reason Eötvös called this instrument the curvature variometer, to distinguish it from his horizontal variometer. The latter (see Figure 2) still supports two masses on a horizontally suspended bar, but the masses are offset both horizontally and vertically. By using the new geometry, $\left(\frac{\partial^2 U}{\partial x \partial z}\right)$ and $\left(\frac{\partial^2 U}{\partial y \partial z}\right)$ can be measured in addition to the two components measured by the earlier instrument. An added refinement was measurement of the oscillation period of the

torsion pendulum, instead of the static deviation of the suspended bar, thus gaining an added sensitivity for the instrument.

The achievements of Eötvös in the use of his instrument are threefold. By developing the complete theory of the Eötvös balance, he was able to push its sensitivity to such a point that it took decades to devise methods for exceeding his precision. It is only proper to mention that the high degree of precision he achieved was not due solely to the design of the instrument but depended also on the unparalleled skill he displayed in using it.

The other two accomplishments encompassed the clear recognition of the very important applications of the balance: geophysical exploration and the equivalence of gravitational and inertial mass. In both cases the recognition was followed by intense work proving his insight.

Prospecting by gravitational methods is the technique of measuring the gravitational field at the earth's surface and predicting, from the data obtained, the structure beneath the surface. In principle this information can be derived from direct measurements of gravity by means of a gravimeter, or from gradient measurements by means of the Eötvös balance. While the gravimeter can give faster results (the time required for a single observation is a small fraction of that required by the torsion balance) the signal-to-noise ratio was originally more favorable for the gradient measurement. As a consequence the Eötvös balance was, until good gravimeters were developed, the leading instrument for geophysical prospecting.

Between 1888 and 1922 Eötvös, together with his collaborators, published a number of papers on his investigations. These included the theory and design

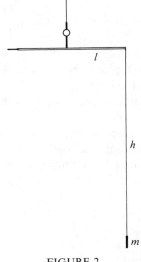

FIGURE 2

of the instrument and the results of its widespread application in Hungary and abroad.

The second extremely important application of the Eötvös balance involved a redetermination of the rate of gravitational acceleration for different bodies. It had been known from earlier work that all bodies fall with the same acceleration (in a vacuum), but the best previous determinations yielded only a limited accuracy. In response to a prize announcement by the University of Göttingen, Eötvös and his collaborators followed up his early measurements on this subject. The new measurements provided not merely a more accurate proof of a principle believed right until then, but much more: his results, proving that gravitational mass and inertial mass are equivalent, the possible deviation being about five parts in 10^9, became one of the building stones of the theory of general relativity. The experiment proves the "weak" form of the principle of equivalence, which states that the trajectory of a test particle, under the influence of gravitational fields only, depends only on its initial position and velocity, not on its mass and nature. Later confirmation of his results (during the last fifty years) reduced the possible deviation from perfect equivalence by a factor of 1,000.

Late in life Eötvös became interested in the variation of the gravitational acceleration caused by the relative motion of a body with respect to the earth. The experimental proof of this effect was the subject of two posthumous papers.

Parallel to the geophysical application of the torsion balance, he pursued an investigation of the magnetic anomalies accompanying the gravitational effects. His interest in magnetism led him to paleomagnetic work on bricks and other ceramic objects that covered a period of about 2,000 years. Another side issue attracting his interest was the shape of the earth.

Eötvös's intensive research efforts did not prevent him from pursuing other interests. Shortly after his appointment as professor of physics, he became aware of the shortcomings of both high school and university instruction in Hungary, and from then on he devoted considerable effort to improving both. In 1881 the minister of instruction requested him to make a trip to Paris for the purpose of studying the French system of higher education. These efforts led to a short appointment (June 1894–January 1895) as minister of public instruction and religious affairs (the cabinet post held twice by his father). One of the highlights of this period was the founding of the Eötvös Collegium, patterned after the French École Normale Supérieure and named for his father. Its stated purpose was to improve the training of high school teachers. The effort was quite successful; the surprising increase in the number of outstanding Hungarian scientists during the twentieth century may be indirect proof of the effectiveness of the new school.

Eötvös served one year as rector of the University of Budapest. As professor of physics he devoted much care and time to the preparation of his lectures; he also invented many good demonstration experiments and insisted that his students understand the basic principles underlying them. He became a corresponding member of the Hungarian Academy of Sciences in 1873 and a full member ten years later; in 1889 he was elected its president. Although the usual term for the presidency was three years, he was reelected until his resignation in 1905. Many honors were bestowed upon him, including election to a number of foreign academies, as well as prizes and decorations.

In 1885 Eötvös and four friends founded the (Hungarian) Society for Mathematics. A little later physicists were attracted to the new society. In 1891 the Mathematical and Physical Society was founded, and Eötvös was elected its first president, a post he held until his death.

His main relaxation was mountain climbing. For quite a long time he was well-known as one of Europe's foremost climbers: a peak in the Dolomites is named for him, and the mountaineering handbooks record a number of "first climbs" he made either alone or with his daughters, who became his steady climbing companions.

BIBLIOGRAPHY

I. ORIGINAL WORKS. Eötvös's most important papers include "Über den Zusammenhang der Oberflächenspannung der Flüssigkeiten mit ihrem Molekularvolumen," in *Annalen der Physik und Chemie,* **27** (1886), 448–459; "Untersuchungen über Gravitation and Erdmagnetismus," *ibid.,* **59** (1896), 354–400; "Étude sur les surfaces de niveau et la variation de la pesanteur et de la force magnétique," in C.-E. Guillaume and L. Poincaré, eds., *Rapports présentés au Congrès international de physique réuni à Paris en 1900,* III (Paris, 1900), 371–393; "Programme des recherches gravimétriques dans les régions vésuviennes," in *Comptes rendus des séances de la première réunion de la commission permanente de l'Association internationale de sismologie réunie à Rome* (1906), pp. 177–179; "Bericht über Arbeiten mit der Drehwage ausgeführt im Auftrage der königlich ungarischen Regierung in den Jahren 1909–1911," in *Verhandlungen der XVII. allgemeinen Konferenz der internat. Erdmessung in Hamburg,* I (1912), 427–438; and "Beiträge zum Gesetze der Proportionalität von Trägheit und Gravität," in *Annalen der*

Physik, **68** (1922), 11–66, written with D. Pekár and J. Fekete.

Most of these, together with a few other papers, were republished by the Hungarian Academy of Sciences under the title *Roland Eötvös Gesammelte Arbeiten,* P. Selényi, ed. (Budapest, 1953). A complete bibliography, except for posthumous papers, is given (in Hungarian) in the special Eötvös issue of *Matematikai és physikai lapok,* **27** (1918), 284–290. A reproduction of the same bibliography, with the posthumous papers included, is to be found in Elek Környei, *Eötvös Loránd, A tudós és müvelödéspolitikus irásaiból* (Budapest, 1964).

II. SECONDARY LITERATURE. The book by Környei listed above and the special issue of *Matematikai és physikai lapok* contain discussions (in Hungarian) of his life and theories. His contributions to physics are the subject of an excellent paper by R. H. Dicke, "The Eötvös Experiment," in *Scientific American,* **205** (Dec. 1961), 84–94, as well as R. H. Dicke, "Some Remarks on Equivalence Principles," in *Annales Universitatis scientiarum budapestinensis de Rolando Eötvös Nominatae,* Geological Section, **7** (1964). See also R. H. Dicke, "Remarks on the Observational Basis of General Relativity," in H. Y. Chiu and W. F. Hoffmann, eds., *Gravitation and Relativity* (New York, 1964), pp. 1–16; and P. G. Roll, R. Krotkov, and R. H. Dicke, "The Equivalence of Inertial and Passive Gravitational Mass," in *Annals of Physics,* **26** (1964), 442–517. The latter paper emphasizes also the importance of the Eötvös experiment for the theory of general relativity. An appreciation of the geophysical aspects is given by A. H. Miller, in "The Theory and Operation of the Eötvös Torsion Balance," in *Journal of the Royal Astronomical Society of Canada,* **28** (1934), 1–31, with the most important relevant papers by Eötvös and his collaborators listed in the bibliography (unfortunately with many misprints). Another useful source of information on the geophysical importance of the torsion balance can be found in Donald C. Barton, "Gravity Measurements With the Eötvös Torsion Balance, Physics of the Earth II., The Figure of the Earth," in *Bulletin of the National Research Council. Washington,* **78** (1931), 167–190.

L. MARTON

EPICURUS (*b.* Samos, 341 B.C.; *d.* Athens, 270 B.C.), *moral and natural philosophy.*

Epicurus' father, Neocles, a schoolmaster, was an Athenian of the deme Gargettus who emigrated to the Athenian colony in Samos. At eighteen Epicurus was required to go to Athens to do his military service, after which he rejoined his family, who had by then moved to the Ionian mainland town of Colophon. When he was thirty-two he moved to Mytilene, on Lesbos, and then to Lampsacus on the Hellespont; in both places he set up a school. He returned to Athens about 307/306 B.C. and bought a house, with a garden that became the eponymous headquarters of his school of philosophy. His extant writings, apart from fragments of lost works, consist of *Letter to Herodotus,* which is a summary of his philosophy of nature; *Letter to Pythocles,* on celestial phenomena (possibly the work of a pupil); *Letter to Menoeceus,* on morality; and two collections of aphorisms, one called *Kyriai doxai* (*Principal Doctrines*), the other now known as *The Vatican Collection.*

Epicurus' main concern was to teach an attitude toward life that would lead to personal happiness. He rejected the philosophical ideals of the good life propounded by the Platonists and Aristotelians and substituted a moderate hedonism. Pleasure is the good. Pain is the obstacle to be removed or avoided. Unsatisfied desires are painful, so the wise man learns to limit his desires to things that can easily be obtained. The good Epicurean seeks a quiet life with a few like-minded friends and avoids becoming deeply involved in the affairs of the world.

The moral message was reinforced by a cosmology, and it was this that gave Epicurus whatever importance he has for the history of science. Peace of mind, he thought, was threatened by ignorance about the natural world, by certain widespread beliefs in the intervention by supernatural powers in man's environment, and by belief in rewards and punishments in a life after death: "If we were not troubled by doubts about the heavens, and about the possible meaning of death, and by failure to understand the limits of pain and desire, then we should have no need of natural philosophy [$\phi v\sigma\iota o\lambda o\gamma i\alpha$]" (*Kyriai doxai,* 11).

Epicurus found a world view that suited his moral purpose in the atomism of Leucippus and Democritus, which he first learned from his teacher Nausiphanes. The historian is in no position to make an accurate assessment of Epicurus' originality, since information about Democritus is scanty and biased. It is certain that the main framework of the atomist system was completed by Democritus. All phenomena were explained on the assumption that the whole natural world consists of imperceptibly small, indestructible, and changeless atoms, made of a single common substance, differing only in shape and size, moving in the infinite void. Democritus explained how perceptible qualities were generated in compounds according to the shapes and sizes of the component atoms and the quantity of void between them. He gave some account of the origin and destruction of worlds in the infinite universe, brought about by random collisions of atoms moving through the void. He wrote about the natural origin of living forms and the natural development of human society and culture.

All of this was taken over by Epicurus. Several

modifications in the system can be observed, however, and no doubt more would be revealed if the evidence were more complete. Some of the modifications can be seen to be attempts to meet criticisms brought against Democritus by Aristotle. For example, Aristotle's criticism of "indivisible magnitudes" (especially in *Physics*, Z) appears to be the reason for Epicurus' contradicting Democritus about the indivisibility of the atoms; the Epicurean atom has "minimal parts" that can be distinguished theoretically but not split off physically (*Letter to Herodotus*, 56–59). Aristotle's analysis of "the voluntary" (*Nicomachean Ethics*, III, 1–5) was one of the factors that led to the notorious "swerve" of atoms in Epicurean theory. Democritus' theory of motion was thought not to allow human beings to initiate motion, since all the motions of the atoms that constitute a mind could be explained by their own previous motions and their interaction with the environment. Epicurus said that atoms deviated unpredictably from time to time, and thus he provided for breaks in the chains of causation. He also modified Democritus' theory of motion in another way: instead of taking basic atomic motion as an unexplained assumption of the theory, he said that all atoms have a natural motion "downwards," because of weight. The swerve was therefore needed for another purpose, since without it the theory could not explain why atoms do not all drop in parallel straight lines through the infinite void, without colliding.

Some of Epicurus' views about the natural world were extremely naïve and reactionary. His avowed purpose was to pursue the inquiry only as far as was necessary to remove anxiety. His "canonic," or rules of procedure, held that any view not in conflict with the evidence of the senses could be regarded as true. Thus the hypothesis that the cosmos was created by an intelligent deity was ruled out as being in conflict with the observed facts of the world's imperfections and with the true conception of what it is to be a god (*Letter to Herodotus*, 76–77; see also Lucretius, *De rerum natura*, V, 55–234). But the sun's motion in the ecliptic may be due to the tilting of the heavens, or to winds, or to some other cause (*Letter to Pythocles*, 93). *De rerum natura*, book VI, and *Letter to Pythocles* contain many cases in which multiple explanations, ranging from the more or less correct to the ridiculous, are offered for natural phenomena.

The main importance of Epicurus for the history of science is that he reasserted the principles of Democritus' atomic theory in opposition to the teleological natural philosophy of Plato and Aristotle. His own major work, *On Nature*, did not survive long enough to be very influential; but the essentials of his theory were preserved in the letters that Diogenes

Laertius included in book X of his *Lives and Opinions of the Philosophers*, in some of the philosophical works of Cicero, and especially in the poem *De rerum natura* of the devoted Roman Epicurean, Lucretius. These were the main sources from which post-Renaissance philosophers drew their knowledge of ancient atomism, when Aristotelianism began at last to lose its dominant position.

BIBLIOGRAPHY

Text with English translation and commentary is in Cyril Bailey, *Epicurus* (Oxford, 1926; repr. New York, 1970). The most recent critical edition is G. Arrighetti, *Epicuro* (Turin, 1960), with Italian trans. and commentary; this also includes the papyrus fragments of *On Nature*. English translation is in Russel M. Geer, *Epicurus: Letters, Principal Doctrines and Vatican Sayings* (New York, 1964). Text with ancient testimonia is in H. Usener, *Epicurea* (Leipzig, 1887; repr. Stuttgart, 1966).

Studies of Epicureanism include Cyril Bailey, *The Greek Atomists and Epicurus* (Oxford, 1928; repr. New York, 1964); Benjamin Farrington, *The Faith of Epicurus* (New York, 1967); David J. Furley, *Two Studies in the Greek Atomists* (Princeton, 1967); Jürgen Mau, *Zum Problem des Infinitesimalen bei den antiken Atomisten* (Berlin, 1954); W. Schmid, *Epikurs Kritik der platonischen Elementenlehre* (Leipzig, 1936); "Epikur," in *Reallexikon für Antike und Christentum* (Stuttgart, 1961); and Gregory Vlastos, "Minimal Parts in Epicurean Atomism," *Isis,* **56** (1965), 121–147.

A conference on Epicureanism is recorded by Association Guillaume Budé, *Actes du VIIIᵉ congrés, Paris, 5–10 avril 1968* (Paris, 1969).

Later history of Epicureanism is discussed in Marie Boas, "The Establishment of the Mechanical Philosophy," in *Osiris,* **10** (1952), 412–541; Robert H. Kargon, *Atomism in England From Hariot to Newton* (Oxford, 1966); and Kurd Lasswitz, *Geschichte der Atomistik vom Mittelalter bis Newton* (Hamburg, 1890; repr. Hildesheim, 1963).

For fuller bibliography, see Bursian's *Jahresbericht über die Fortschritte der klassischen Altertumswissenschaft,* no. 281 (1943), pp. 1–194; and P. DeLacy, "Some Recent Publications on Epicurus and Epicureanism, 1937–1954," in *Classical Weekly,* **48** (1955), 169 ff.

DAVID J. FURLEY

ERASISTRATUS (*b.* Iulis, Ceos, *ca.* 304 B.C.; *d.* Mycale), *anatomy, physiology.*

Erasistratus was born into a medical family. His father, Cleombrotus, was a doctor, and his mother, Cretoxene, was the sister of the doctor Medios. Like his brother Cleophantus, Erasistratus entered the family profession. He studied medicine first in Athens as a pupil of Metrodorus, the third husband of Aristotle's daughter, Pythias; and it was probably, at least in part, through him that he became so strongly influenced by Peripatetic thought.[1] About 280 B.C. he entered the university in Cos, where the medical school of Praxagoras flourished. Cos had strong political and cultural ties with Alexandria, and in that city Erasistratus came under the influence of Chrysippus the Younger (the palace doctor of Ptolemy Philadelphus), especially in the fields of anatomy, physiology, and pathology. In his old age Erasistratus gave up medical practice and entered the Museum at Alexandria, where the unrivaled facilities afforded by the Ptolemies allowed him to devote himself to his researches. According to later tradition, an incurable ulcer on his foot caused him to commit suicide by drinking hemlock.

Erasistratus wrote a large number of works, notably on anatomy, abdominal pathology, hemoptysis, fevers, gout, dropsy, and hygiene. None of these works has survived. He is best-known for his anatomical and physiological researches. In these fields distinguished work had already been done by his immediate predecessors, Diocles and Praxagoras, and by his teachers Chrysippus and Herophilus, and the Museum itself provided highly advantageous conditions for anatomical research. The public dissection of human bodies was an innovation of the Ptolemies and remained almost exclusively the preserve of the Alexandrians. It is possible that familiarity with the Egyptian practice of embalming bodies contributed to the creation of an environment in which the dissection of the human body was not viewed with misgivings. Certain ancient authorities, notably Celsus, followed by Tertullian and St. Augustine, even accused Erasistratus of vivisecting criminals taken out of the royal prisons. Scholarly opinion is divided in its acceptance of this tradition, and Galen's silence in the matter cannot be taken as conclusive evidence against it. It is clear from Galen that Erasistratus vivisected animals, and these dissections enabled him to draw parallels between men and beasts. For example, he made a detailed investigation of the cavities and convolutions in the brains of man, hare, and stag and correctly inferred that the number of convolutions varied with the degree of intellectual development. In addition to his ventures in comparative anatomy, Erasistratus was a pioneer in the field of pathological anatomy, conducting postmortem examinations of the bodies of men who had just died, in order to study structural changes due to morbid conditions.

As the basis for his physiology Erasistratus combined a corpuscular theory with the doctrine of the pneuma. In both these respects it seems likely that he reveals the influence of the Lyceum, where these theories played an important role. A very important factor in persuading Erasistratus to adopt the latter theory must have been the influence of such doctors as Diocles, Praxagoras, and Herophilus, in whose medical theory the pneuma doctrine plays so fundamental a role and who were themselves either directly or at least indirectly influenced by Peripatetic teachings. It is probable, too, that Erasistratus was induced to adopt his corpuscular theory by the direct influence of Strato of Lampsacus, the third head of the Lyceum and the teacher of Ptolemy Philadelphus in Alexandria shortly before Theophrastus' death. It is perhaps worth recording that Strato also seems to have subscribed to the pneuma doctrine.[2] Like Strato, Erasistratus conceived of his particles as very small, imperceptible, corporeal entities surrounded by a vacuum in a finely divided or discontinuous condition.

Upon the basis of these two theories he sought to assign natural causes to all phenomena and rejected the idea that there were hidden forces such as the power of attraction of certain organs, which many medical authors had postulated in order to explain such physiological processes as the assimilation of food and the secretion of humors. For this idea Erasistratus substituted the theory of πρὸς τὸ κενούμενον ἀκολουθία, the *horror vacui*, derived from Strato, whereby those empty spaces which suddenly form in the living body are continually filled.

Erasistratus may also have derived his experimental method from Strato. The *Anonymus Londinensis* (col. xxxiii) preserves some evidence of his methodology. It describes an experiment of his which anticipates another experiment, generally considered to be the beginning of the modern study of metabolism, performed in the seventeenth century by Santorio. In order to prove that animals give off certain emanations, Erasistratus recommended that a bird or similar creature should be kept in a vessel without food for some time and then weighed together with any excrement that had been passed. A great loss of weight would then be discovered.

To repair the bodily wastage which Erasistratus had so strikingly demonstrated took place not only visibly but in some part invisibly, Galen tells us that he held that Nature had provided mechanism in the form of appetites (ὀρέξεις), forces (δυνάμεις), and substances (ὕλαι), i.e., he believed that part of the purposive activity of the pneuma was the absorption of food into the body. He had made the striking discovery that all organic parts of the living creature were a tissue composed of vein, artery, and nerve (the τρυπλοκία τῶν ἀγγείων), bodies so fine that they were knowable only by reason. The vein carried the food, the artery the pneuma, and the nerve the psychic pneuma. There was no need for Erasistratus to postulate a hidden power of attraction. He held that the supply of nourishment to each particular organ took place by a process of absorption (διάδοσις) through the extremely fine pores (κενώματα) in the walls of the veins contained in it. The particles of nourishment contained in the blood were able to pass through the veins, in accordance with the principle of πρὸς τὸ κενούμενον ἀκολουθία, to fill those spaces left empty by the evacuations and emanations. Growth, then, proceeded on the principle of the accretion of like to like. Galen tells us that Erasistratus likened the growth of an animal to that of a sieve, rope, bag, or basket, since each of these grows by the addition to it of materials similar to those out of which it began to be made. The two materials which serve for the preservation of the creature, then, are blood and pneuma, the former providing the nutriment and the latter helping to transmit the natural activity.

Erasistratus rejected the widely held beliefs that digestion was a process analogous either to cooking (as Aristotle had held) or to fermentation (as Diocles had believed). He was well aware that the epiglottis closes the larynx during swallowing, thereby preventing liquids and solids from entering the trachea. He thus rejected the old and much debated belief held by Plato, Philistion, Diocles, and Dexippus that it was possible for drink to enter the lungs. He held that the food, once in the stomach, was torn to pieces by the peristaltic motion of the gastric muscles under the influence of pneuma. He gave an accurate description of the structure and function of these muscles. Unlike Diocles, he was not of the opinion that the pneuma arrived in the stomach with the food but held that it was introduced there by the gastric arteries. In accordance with this theory he sought to explain why digestion is impaired during a fever. This was due, he thought, to the impediment of the free motion of the pneuma by the blood that had penetrated into the arteries.

A portion of the food broken into pulp in the stomach was subsequently conveyed in the blood vessels to the liver, where, he believed, it was transformed into blood. Galen complains that Erasistratus did not reveal how this transformation takes place. During this process the biliary constituents are separated off and pass to the gallbladder while the pure blood from the liver is conveyed via the vena cava (κοίλη φλέψ) to the heart.

Erasistratus' description of the vascular system and his views on the significance and structure of the heart represent a great advance over his predecessors. Some scholars have claimed that Erasistratus came near to anticipating Harvey's discovery of the circulation of the blood, but such a claim is unwarranted. He did,

however, conceive the function of the heart to be that of a pump and compared it to a blacksmith's bellows actively dilating and contracting by its own innate force. The arteries, on the other hand, which he likened to a skin bag, were passively dilated by the pneuma forced into them by the heart's contraction. In this respect Erasistratus rejected Herophilus' claim, subsequently revived by Galen, that the arteries were subject to dilations and contractions synchronous with those of the heart.

Erasistratus rightly believed that both veins and arteries originated from the heart and that both of these vessels extended throughout the body, dividing into extremely fine capillaries. But, following Praxagoras, he held that only the veins contained blood, the arteries being full of pneuma. His dissections of dead animals would have confirmed this belief. He had not, however, failed to observe that the arteries of a wounded animal spurt blood. To account for this phenomenon he maintained that when an artery was severed, the escaping pneuma caused a vacuum and the pull of this vacuum (the πρὸς τὸ κενούμενον ἀκολουθία) drew blood from the veins through certain fine capillaries which were usually closed (the synanastomoses, συναναστομώσεις). The blood then spurted out of the arteries after the escaping pneuma. This ingenious hypothesis was not disproved until four and a half centuries later, when Galen, who was otherwise greatly influenced by Erasistratus' theory, showed by careful experiments in vivisection that the arteries of living beings carry blood continuously.

Erasistratus rejected the view that the pneuma was innate in the body. He believed that it was ἐπικτητόν, i.e., drawn in from outside through the nose and the mouth in the process of inhalation, that it passed via the bronchi to the lungs and thence through the pulmonary vein to the left ventricle of the heart. From the heart the vital pneuma was carried through the aorta ascendens and aorta descendens to the brain and to the whole body. In the brain the vital pneuma was transformed into psychic pneuma and was carried from the brain to various parts of the body by the nervous system, there to cause muscular movements by its effect upon the muscles. In order to perform this function the inspired pneuma must have a certain density. If it were too fine, it would presumably escape through pores.

The respiratory process Erasistratus held to be due to the muscular activity in the thorax. When the chest is expanded, so are the lungs, which, like the arteries, possess no motion of their own. Into the resulting empty space the outer air streams in accordance with the principle of *horror vacui*. The contraction of the thorax also brings about the contraction of the lungs "like a sponge compressed by the hands," and the air is expired. The purpose of respiration, according to Erasistratus, is not to cool the innate heat, as Philistion and Diocles had believed, but to fill the arteries with pneuma. Galen provides the further information that Erasistratus declared that if the activity of the thorax were to stop, the heart would not be able to draw in any air from the lungs and suffocation would ensue.

As has already been seen, the pneuma passes from the lungs through the so-called ἀρτηρία φλεβώδης, "the artery resembling a vein," i.e., the pulmonary vein, and, with the expansion of the heart, into the left ventricle in accordance with the principle πρὸς τὸ κενούμενον ἀκολουθία. When the heart contracts, the pneuma is then driven, via the aorta, through the arteries all over the body. The return of the pneuma from the heart into the lungs is prevented by the bicuspid valve (mitral valve) and from the aorta into the heart by the semilunar (sigmoid) valves.

Since there is no possibility of any return to the lungs, any superfluous pneuma distributed throughout the arteries is, presumably, ultimately given off from the body by that process of skin respiration which Erasistratus had earlier sought to demonstrate by the experiment with the bird.

Just as the expansion of the heart led to the left ventricle's becoming filled with pneuma, so the right ventricle becomes filled with blood which comes from the liver via the vena cava. With every contraction of the heart the blood is pumped into the lungs through the so-called φλὲψ ἀρτηριώδης, "the vein resembling an artery," i.e., the pulmonary artery, and thence, presumably, distributed by other "veins" to the rest of the organs. The return of the blood from the heart to the vena cava is prevented by the tricuspid valve and from the pulmonary artery to the heart by the semilunar valves. As is the case with the pneuma, no blood is returned to the heart; nor is there any communication between veins and arteries in the heart, the left and right ventricles being quite separate.

Erasistratus, then, wrongly attributed functions of the auricles to the two pulmonary vessels and incorrectly believed, like Galen after him, that the blood was manufactured in the liver and distributed throughout the body by the veins.

The structure of the brain is described by Erasistratus with a greater accuracy than Herophilus had achieved. Galen tells us that in his old age Erasistratus had leisure for research and made his dissections more accurate. The implication of Galen's evidence here is that Erasistratus had not systematically examined the structure of the brain until he was an old man. As a result of these dissections he distinguished the cerebrum (ἐγκέφαλος) from the cerebellum (which

he called the ἐπεγκρανίς, not the παρεγκεφαλίς, as Herophilus had done). He also gave a detailed description of the cerebral ventricles or cavities within the brain and of the meninges or membranes that cover the brain.

Erasistratus rejected the view of those thinkers who, like Empedocles, Aristotle, Diocles, Praxagoras, and the Stoics, had maintained that the heart was the seat of the central intelligence. In agreement with Alcmaeon, the author of the Hippocratic treatise "On the Sacred Disease," and Plato, he placed this central organ in the brain. Herophilus had clearly shown by dissection that the nerves originated in the brain and had specified the "fourth ventricle" of the cerebellum as the seat of *hegemonikon,* or organ of thought. Erasistratus was most probably in agreement with this viewpoint. For, as was seen above, his observations that the cerebellum of the brain of man had more convolutions than that of other animals had led him to the conclusion that the number of convolutions varied with the degree of intellectual development.

He also agreed with Herophilus that the brain was the starting point of all the nerves. Originally he held that they sprang from dura mater (παχεία μήνιγξ). He had discovered through vivisection that incisions into this membrane adversely affected the motor ability of living creatures. But later—probably as a result of his more accurate dissections—he succeeded in tracing the nerves into the interior of the brain and discovered the origin of each type of sensory nerve in the cerebrum.

Again in agreement with Herophilus, Erasistratus recognized the difference between sensory and motor nerves, the νεῦρα αισθητικά and κινητικά (Herophilus, however, called the latter προαιρητικά). Erasistratus was of the opinion that the nerve fibers were formed "like a sail" out of three different, imperceptible strands of artery, vein, and nerve. He also believed that the nerves were filled with psychic pneuma drawn, most probably, from the ventricles of the cerebrum. The widespread belief that Erasistratus renounced this theory in his old age and maintained that the nerves contained not pneuma but marrow or brain substance[3] is inaccurate and based upon a misunderstanding of the text.[4] Motion, he held, was effected through the agency of the muscles, which were formed from a texture of vein, artery, and nerve. The pneuma, led to the muscles via the arteries, invested them with the capacity of synchronous expansion of length and contraction of breadth and vice versa, which resulted in the movement of the bodily parts containing them.

In pathology Erasistratus rejected the influential humoral theory or, at any rate, the abuses of this theory practiced in the school of Praxagoras. He considered research into the formation of the humors to be superfluous and, according to Galen, did not mention black bile at all in his writings. However, he could not dispense entirely with the assumption of morbid changes in the bodily humors ($\kappa\alpha\kappa o\chi v\mu i\alpha$); he considered apoplexy, for example, to be a disease of the brain caused by an excessive secretion of cold, viscous, and glutinous humors which prevented the psychic pneuma from passing into the nerves. But the main cause of disease he held to be plethora ($\pi\lambda\eta\vartheta\acute{\omega}\rho\alpha$ or $\pi\lambda\tilde{\eta}\vartheta os\ \tau\rho o\varphi\tilde{\eta}s$), i.e., the flooding of the veins with a superfluity of blood engendered by an excessive intake of nourishment. As the plethora increases, the limbs begin to swell, then become sore, more sluggish, and harder to move. If the plethora increases still more, the superfluous blood is then discharged through the synanastomoses into the arteries, where it is compressed by the pneuma which is constantly pumped from the heart. This compressed blood collects in the extremities of the arteries and causes local inflammation ($\varphi\lambda\epsilon\gamma\mu o\nu\acute{\eta}$) accompanied by fever. Moreover, since the flow of the pneuma is impeded by the presence of this blood in the arteries, it cannot perform its natural functions. As examples of diseases brought about in this way by plethora he mentions, among others, ailments of the liver, spleen, and stomach, coughing of blood, phrenitis, pleuritis, and peripneumonia.

Erasistratus' treatment for plethora consisted primarily of starvation ($\dot{\alpha}\sigma\iota\tau\acute{\iota}\alpha$), on the ground that the veins, when emptied, would more easily receive back the blood which had been discharged into the arteries. Unlike many of his contemporaries he did not resort freely to phlebotomy but employed it only upon rare occasions. Erasistratus preferred prevention to therapy and in a separate treatise stressed the importance of hygiene. In general he was opposed to violent remedies, especially purgatives, preferring in their stead carefully regulated exercise and diet and the vapor bath.

Although Erasistratus founded a school of medicine, none of his successors seems to have made any significant mark in the history of medicine. His true importance lies in the fact that he, together with Herophilus, laid the foundations for the scientific study of anatomy and physiology, and their careful dissections provided a basis and stimulus for the anatomical investigations undertaken by Galen over four centuries later.

NOTES

1. There is also a strong tradition that he had heard Theophrastus himself; see Diogenes Laërtius, 5.57, and Galen, 4.729K. For the tradition linking Erasistratus with the Lyceum generally, see Galen, 2.88K; for Strato's influence, see below.
2. See F. Wehrli, *Straton von Lampsakos, die Schule des Aristoteles,* V (Basel, 1950), commentary on fr. 108, p. 71; and F. Solmsen, "Greek Philosophy and the Discovery of the Nerves," in *Museum Helveticum,* **18** (1961), 183.
3. See M. Wellmann, "Erasistratus," in Pauly-Wissowa, VI, 1 (Stuttgart, 1907), 343; and G. Verbeke, *L'évolution de la doctrine du pneuma* (Paris–Louvain, 1945), p. 185.
4. With Solmsen, *op. cit.,* p. 188.

BIBLIOGRAPHY

For a collection of the evidence of Erasistratus' views, see R. Fuchs's dissertation *Erasistratea* (Leipzig, 1892). References in Galen are cited according to C. G. Kuhn, ed., *Claudii Galeni Opera omnia,* 20 vols. (Leipzig, 1821–1833).

See also C. Allbutt, *Greek Medicine in Rome* (London, 1921); H. Diels, "Über das physikalische System des Straton," in *Sitzungsberichte der Preussischen Akademie der Wissenschaften zu Berlin,* **1** (1893), 101 ff.; J. F. Dobson, "Erasistratus," in *Proceedings of the Royal Society of Medicine,* **20** (1926–1927), 825 ff.; R. Fuchs, "De Erasistrato capita selecta," in *Hermes,* **29** (1894), 171–203; W. W. Jaeger, "Das Pneuma im Lykeion," *ibid.,* **48** (1913), 29–72, and *Scripta minora,* I (Rome, 1960), 57–102; *Diokles von Karystos, die griechische Medizin und die Schule des Aristoteles* (Berlin, 1938); "Vergessene Fragmente des Peripatetikers Diokles von Karystos nebst zwei Anhaengen zur Chronologie der dogmatischen Aerzteschule," in *Abhandlungen der Preussischen Akademie der Wissenschaften,* Phil.-hist. Kl. (1938), no. 3, 1–46, and in *Scripta minora,* II (Rome, 1960), 185–241; and "Diocles of Carystus: A New Pupil of Aristotle," in *Philosophical Review,* **49** (1940), 393–414, and in *Scripta minora,* II, 243–265; W. H. S. Jones, *The Medical Writings of Anonymus Londinensis* (Cambridge, 1947); I. M. Lonie, "Erasistratus, the Erasistrateans and Aristotle," in *Bulletin of the History of Medicine,* **38** (1964), 426–443; F. Solmsen, "Greek Philosophy and the Discovery of the Nerves," in *Museum Helveticum,* **18** (1961), 150 ff.; F. Steckerl, *The Fragments of Praxagoras of Cos and His School* (Leiden, 1958); G. Verbeke, *L'évolution de la doctrine du pneuma* (Paris–Louvain, 1945); F. Wehrli, *Straton von Lampsakos, die Schule des Aristoteles,* V (Basel, 1950); M. Wellmann, "Erasistratus," in Pauly-Wissowa, VI, 1 (Stuttgart, 1907), 333–350; and L. G. Wilson, "Erasistratus, Galen, and the Pneuma," in *Bulletin of the History of Medicine,* **33** (1959), 293–314.

I am grateful to the University of Newcastle upon Tyne Research Fund for a grant which enabled me to consult works in London libraries.

JAMES LONGRIGG

ERASTUS (LIEBER), THOMAS (*b.* Baden, Switzerland, 1523; *d.* Basel, Switzerland, 1 January 1583), *medicine, natural philosophy, theology.*

Erastus studied theology and philosophy at Basel

from 1540 to 1544 and medicine at Bologna and Padua from 1544 to 1555, receiving the M.D. in 1552. While in Bologna he married Isotta a Canonici; they had no children. At Meiningen he was physician to Count William of Henneberg in 1557, and the following year he became professor of medicine at Heidelberg. His anti-Calvinist attitude—he believed in granting the state supremacy in all ecclesiastical affairs, a doctrine that came to be called Erastianism—led to his fall from the favor of Frederick III, elector palatine. In 1580, therefore, he left Heidelberg for Basel, where he became professor of theology and moral philosophy.

Erastus is remembered chiefly as an inexorable and abusive critic of astrology, natural magic, and particularly of Paracelsus and iatrochemistry. He condemned such superstitious practices as the curative use of human blood and parts of corpses and of amulets in the cure of epilepsy, but he firmly believed in Satan, demons, and witches—accusing the witches, against the claims of their defenders (Wierus and others) that they were victims of drug-induced hallucinations, of true cohabitation with the devil. A successful physician with great experience in the use of watering places (Bad Kissingen), he strictly adhered to traditional humoralism and ancient medical practice but nevertheless criticized Galen.

Erastus laced his rational arguments heavily with theological dogma and polemics, yet on occasion he made sound observations, such as tracing vitriol and alum in mineral water with oak gall water. Opposing traditional ideas about the brain, he insisted that what mattered was not its substance but its function and its production of impulses ("spirits"). Erastus also demolished the Galenic theory that epilepsy was caused by obstruction of pathways by viscid mucus, since such a theory could not explain why sensation was disturbed but motility was not. He recognized forms of petit mal manifested by dizziness, hiccups, or excessive sneezing.

No two events in nature, Erastus contended, are equal in cause and effect: hence the futility of forecasting. According to him, heaven acts in conformity with a general plan and does not interfere with the course of events that are specific for the individual object. For example, it supplies heat and moisture in spontaneous generation, but the process is due entirely to the specific disposition of certain parts of matter. Attempts at astrological divination work through the invocation of demons and therefore are damnable heresy. Furthermore, Erastus said, comets do not foretell evil events, such as wars, pestilence, and the deaths of kings; they are merely terrestrial exhalations and, as such, produce drought and

heat—factors not conducive to the outbreak of epidemics.

Erastus' criticism of Paracelsus was fivefold. First, he criticized Paracelsus' denial of the existence and universal significance of the elements and humors established in ancient science and medicine and their replacement with the three principles, salt, sulfur, and mercury; these had never been isolated from any object by heat or chemical manipulation. Solidity, inflammability, and volatility were not caused by the presence of any of the three Paracelsian principles but by the proportions in which the four elements of the ancients (air, water, fire, earth) were "mixed" in an object. It was the special mixture of water and subtle earth particles that made mercury prone to go up in smoke. Sulfur was inflammable because of the fire and warm air it contained. Without this air, sulfur would become inert and lose its "sulfurousness." The solidity of salt showed its kinship with the earth.

Second, Erastus angrily repudiated the significance attributed by Paracelsus to the power of imagination and the conversion of something spiritual into matter; this he regarded as the main part of the natural magic—the "Neoplatonic fallacy"—practiced by Paracelsus. The concepts of microcosm and quintessence were sheer nonsense: how could the human body contain the virtues and materials of all parts of the outside world? Who could show that bread already and actually contains human blood instead of being converted into it when consumed?

Third, disease, according to Paracelsus, was not the disturbance of humoral balance in an individual man (as the ancients rightly taught); man was merely the passive recipient of an outside agent that, like a parasite, takes possession and inflicts its own schedule of life on the organism, thereby consuming it. (This was the parasitistic or ontological concept.) In this, Erastus said, Paracelsus confused disease with its cause and disregarded the functions of organs, which alone decide the character of a disease.

Fourth, the chemical and notably the metallic (mainly mercury) remedies recommended by Paracelsus were nonassimilable poisons. No therapy could work except through the humors. Therefore, potable gold was a magician's swindle.

Fifth, said Erastus, Paracelsus was an ignorant man, a "grunting swine" who, driven by ambition and vanity, replaced sane teaching with insane delusion, the comprehensible with the incomprehensible, truth with falsehood, and salubrious medicine with pestilential poison. His "cures" were at best temporary and, as a rule, injurious. Although not devoid of some knowledge of chemistry, he was largely a magus informed by the devil and evil spirits.

Many of Erastus' arguments must have seemed unanswerable to his contemporaries, but he fought a losing battle against iatrochemistry, which through Paracelsians (Turquet de Mayerne, inspired by Oswald Croll) appeared in the first British pharmacopoeia (1618). The Paracelsian concept of disease was further developed by J. B. van Helmont and Harvey and finally became the historical root of the modern concept established in the nineteenth century.

BIBLIOGRAPHY

I. Original Works. Erastus' main writings are *Disputationes de medicine nova Paracelsi,* 4 pts. (Basel, 1572–1573), pt. 2 with supp. against alchemy; *De occultis pharmacorum potestatibus* (Basel, 1574); *Repetitio disputationis de lamiis seu strigibus* (Basel, 1578; repr. Hamburg, 1606); *De astrologia divinatrice epistolae D. Thomae Erasti ad diversos scriptae* (Basel, 1580); and *De auro potabili,* followed by *De cometarum significationibus sententia* (Basel, 1584—*De auro* text at end dated 1576 and preface dated 1578).

II. Secondary Literature. Accounts of Erastus' life are Melchior Adam, *Vitae Germanorum medicorum* (Heidelberg, 1620), pp. 242–246; and J. Karcher, "Thomas Erastus (1524–1583), der unversöhnliche Gegner des Theophrastus Paracelsus," in *Gesnerus,* **14** (1957), 1–13. Erastus' criticism of astrology is discussed in Lynn Thorndike, *A History of Magic and Experimental Science,* V (New York, 1941), 652–667; that of Paracelsus and iatrochemistry in Walter Pagel, *Paracelsus. An Introduction to Philosophical Medicine in the Era of the Renaissance* (Basel–New York, 1958), pp. 311–333. On Erastianism, see R. Wiesel-Roth, *Thomas Erastus. Beitrag zur Geschichte der reformierten Kirche und zur Lehre von der Staatssouveränität* (Baden, Switzerland, 1953).

The following, although of no necessity, are a helpful amenity for the student. For Erastiana (correspondence and biographical notes), see Henricus a Leda Smetius, *Miscellanea medica* (Frankfurt, 1611), and H. Pfister, *Bad Kissingen vor Vierhundert Jahren* (Würzburg, 1954), pp. 17–26.

Walter Pagel

ERATOSTHENES (*b.* Cyrene [now Shahhat, Libya], *ca.* 276 B.C.; *d.* Alexandria, *ca.* 195 B.C.), *geography, mathematics.*

Eratosthenes, son of Aglaos, was born in Cyrene but spent most of his working life in Alexandria, where he was head of the library attached to the famous Museum from *ca.* 235 until his death. At some period during his early manhood he went to Athens for the ancient equivalent of a university education, and there he associated with the Peripatetic Ariston of Chios, Arcesilaus and Apelles of the Academy, and Bion the Cynic (Strabo, *Geography,* 15). When he was

about thirty, he was invited to Alexandria by King Ptolemy III (Euergetes I), possibly at the instigation of Eratosthenes' fellow countryman Callimachus, who had already been given a post in the library by Ptolemy II (Philadelphus). On the death of the first chief librarian, Zenodotus, *ca.* 235, Eratosthenes was appointed to the post, Callimachus having died *ca.* 240 (*Suda Lexicon, s.v.,* calls Eratosthenes a pupil of Callimachus). At some time during his stay in Alexandria he became tutor to Euergetes' son and remained in favor with the royal court until his death. (See the anecdote related in Athenaeus, *Deipnosophistai,* VII, 276a, concerning Eratosthenes and Queen Arsinoe III.)

The above represents the most probable account of Eratosthenes' life according to the consensus of scholarly opinion, but the exact dates of the stages of his career are disputed and certainty is unattainable. In particular, Knaack puts the date of his birth back to *ca.* 284, and Jacoby (*Fragmente der griechischen Historiker,* IIB [1930], 704) even as far as 296 (suggesting that in the *Suda Lexicon, s.v.,* ρκς′ is a copyist's error for ρκα′, which then refers to the 121st olympiad, i.e., 296–293, not the 126th, i.e., 276–273), while the date of his death becomes either about 203 (Knaack) or 214 (Jacoby), both scholars accepting the testimony of our sources that Eratosthenes died at eighty (the *Suda Lexicon*) or eighty-one (Censorinus, *De die natali,* p. 15) or eighty-two (Pseudo-Lucian, Μακρόβιοι, p. 27). The reason for supposing that he must have been born earlier than 276 is that Strabo calls him γνώριμος of Zeno of Citium (the founder of Stoicism), a word that often means "pupil" in such a context; but Zeno died in 262, and Eratosthenes could hardly have studied under him at the tender age of fourteen. To this it may be answered that γνώριμος can also mean simply "acquainted with," and that the date of Zeno's death may be as late as 256 (see Diogenes Laertius, VII, 6:28). There is also considerable doubt about the order of succession of the early librarians at Alexandria. A papyrus fragment (*Oxyrhynchus papyri,* X, 1241, col. 2) lists them as Zenodotus (whose name is presumed to have occurred at the damaged end of the previous column), Apollonius Rhodius, Eratosthenes, Aristophanes of Byzantium, Aristarchus of Samothrace, and another Apollonius; but there are several mistakes and chronological difficulties in this list (*cf.* Grenfell and Hunt, *ad loc.*), and it is by no means certain that Apollonius Rhodius succeeded Zenodotus directly—the *Suda Lexicon* (*s.v.* "Apollonius") has him succeeding Eratosthenes, although this may arise from confusion with the later Apollonius (if he is correctly placed).

Eratosthenes was one of the foremost scholars of his time and produced works (of which only fragments remain) on geography, mathematics, philosophy, chronology, literary criticism, and grammar as well as writing poetry. According to the *Suda Lexicon,* he was described as Πένταθλος ("All-Rounder"), "another Plato," and "Beta"—the last possibly because, working in so many fields (and polymathy was greatly admired by the Alexandrians), he just failed to achieve the highest rank in each (see Strabo's remark that Eratosthenes was a mathematician among geographers and a geographer among mathematicians: *Geography,* 94; *cf.* 15), or perhaps simply because he was the second chief librarian. His most enduring work was in geography (particularly notable is his measurement of the circumference of the earth), but he himself seems to have taken most pride, as regards his scientific work, in his solution to the famous problem of doubling the cube, to celebrate which he composed an epigram disparaging previous solutions and dedicated to Euergetes and his son; the authenticity of this poem has been questioned (by Hiller and by Powell), but on inadequate grounds. As a mathematician, Eratosthenes ranked high enough in the estimation of the great Archimedes to have one of the latter's treatises, the *Method,* dedicated to him and to be the recipient of a difficult problem in indeterminate analysis, known as the "Cattle Problem," for communication to the mathematicians of Alexandria. In philosophy, Eratosthenes was an eclectic and, according to Strabo (*Geography,* 15), somewhat of a dilettante. He was the first Greek writer to make a serious study of chronological questions and established the system of dating by olympiads, while as an authority on Old Comedy he is constantly cited in the scholia to Aristophanes' plays.

Eratosthenes' *Geography* (Γεωγραφικά) was in three books, as we learn from Strabo, who quotes from it frequently and is, in fact, the chief source of our knowledge of it. It long remained a prime authority on geographical matters; Julius Caesar evidently consulted it, since in his description of the Germans he mentions that Eratosthenes knew of the Hercynian Forest (*De bello Gallico,* VI, 24), and Strabo (writing around the turn of the Christian era) admits that for the southeastern quarter of the inhabited world (*oikoumene*) he has no better authority than Eratosthenes (*Geography,* 723). The work was the first scientific attempt to put geographical studies on a sound mathematical basis, and its author may be said to have been the founder of mathematical geography. It was concerned with the terrestrial globe as a whole, its division into zones, changes in its surface, the position of the *oikoumene* as then known, and the actual mapping of it, with numerous estimates of distances along a few roughly defined parallels and meridians; but it also contained a certain amount of material descriptive of peoples and places.

Strabo, who disliked the mathematical side of the subject and much preferred purely descriptive geography (see *Geographical Fragments of Hipparchus,* pp. 36, 162, 164, 171, 191), several times complains that Eratosthenes put too much emphasis on mathematical topics such as the above (*Geography,* 48–49, 62, 65). Hipparchus (second century B.C.), on the other hand, criticizes his predecessor for not making sufficient use of astronomical data in fixing the reference lines of his map and not treating the subject in a mathematical enough manner. (Hipparchus wrote a work in three books, *Against the Geography of Eratosthenes,* of which we have substantial fragments quoted by Strabo, often inextricably mingled with citations from Eratosthenes himself—see *Geographical Fragments of Hipparchus.*) One of Eratosthenes' main purposes was to correct the traditional Ionian map, which had a round *oikoumene* with Delphi at the center, wholly surrounded by a circular ocean (as envisaged, e.g., by Anaximander and Hecataeus and already ridiculed by Herodotus, *History,* IV, 36, 2), and to sketch a better one (Strabo, *Geography,* 68), making use of all the data at his command—which, as head of the largest library in antiquity, must have been considerable (*ibid.,* 69).

Eratosthenes used as his base line a parallel running from Gibraltar through the middle of the Mediterranean and Rhodes, to the Taurus Mountains (Toros Dağlari, in Turkey), which were extended due east to include the Elburz range (south of the Caspian), the Hindu Kush, and the Himalayas, which formed the northern boundary of India (such a line, approximately bisecting the known world, had already been suggested in the previous century by Dicaearchus, a pupil of Aristotle—see *Geographical Fragments of Hipparchus,* p. 30). Intersecting this main parallel at right angles was a meridian line taken as passing through Meroë, Syene (modern Aswan, on the Tropic of Cancer), Alexandria, Rhodes, and the mouth of the Borysthenes (modern Dnieper—*ibid.,* pp. 146–147). Wherever Eratosthenes found in his sources data (such as distances in stades, similarities in fauna, flora, climate, or astronomical phenomena, lengths of the longest days, etc., recorded at different places) that he could correlate with one or both of the above base lines, he was enabled to sketch in other parallels. In addition, he divided at least the southeastern quarter of the *oikoumene* (we have no information about his treatment of the remainder) into rough geometrical figures shaped like parallelograms, which he called "seals" (σφραγῖδες), forming the first "seal" out of India and working westward (*ibid.,* pp. 128–129).

Naturally, the data at his disposal, mainly travelers' estimates of days' voyages and marches, which are notoriously unreliable—the only scientific data available were the gnomon measurements of Philo, prefect of Ptolemy, at Meroë (Strabo, *Geography,* 77), of Eratosthenes himself at Alexandria, and of Pytheas at Marseilles (*ibid.,* 63), together with some sun heights recorded by the latter (*Geographical Fragments of Hipparchus,* p. 180)—were of dubious accuracy, and any mapping done on the basis of them was bound to be largely guesswork. Hipparchus has no difficulty in showing that the figures and distances given by Eratosthenes are mathematically inconsistent with each other, and he therefore rejects them, together with some of the sensible alterations proposed by Eratosthenes for the traditional map, thus demonstrating that inspired guesswork sometimes gives better results than scientific caution (*ibid.,* pp. 34–35).

It is uncertain whether the measurement of the earth's circumference was first published in the *Geography* or in a separate treatise; if the latter, it would at any rate have been mentioned in the larger work. The method is described in detail by Cleomedes (*De motu circulari,* I, 10), the only ancient source to give it. Assuming that Syene was on the Tropic of Cancer (because there, at midday on the summer solstice, the gnomon—i.e., a vertical pointer set upright on a horizontal base—cast no shadow and a well, especially dug for this purpose [according to Pliny, *Natural History,* II, 73] was illuminated to its bottom by the sun's rays), and that this town and Alexandria were on the same meridian, Eratosthenes made a measurement of the shadow cast at Alexandria at midday on the solstice by a pointer fixed in the center of a hemispherical bowl, known as a "scaphe" (σκάφη—presumably he used this form of gnomon because the shadow of a thin stylus would be better defined than that of a large pillar or post) and estimated that the shadow amounted to 1/25 of the hemisphere, and thus 1/50 of the whole circle. Since the rays of the sun can be regarded as striking any point on the earth's surface in parallel lines, and the lines produced through the vertical gnomons at each place meet at the center of the earth, the angle of the shadow at Alexandria (*ABC* in Figure 1) is equal to the alternate angle (*BCD*) subtended by the arc *BD,* which is the distance along the meridian between Alexandria and Syene, estimated by Eratosthenes at 5,000 stades; and since it is 1/50 of the whole circle, the total circumference must be 250,000 stades. This

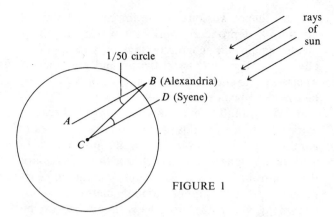

FIGURE 1

is the figure reported by Cleomedes. Hipparchus accepts a figure of 252,000 stades as Eratosthenes' measurement (Strabo, *Geography,* 132, corroborated by Pliny, *Natural History,* II, 247, whose further statement that Hipparchus added 26,000 stades to Eratosthenes' figure is incorrect—see *Geographical Fragments of Hipparchus,* p. 153), and it seems fairly certain that Eratosthenes himself added the extra 2,000 in order to obtain a number readily divisible by 60; he divided the circle into sixtieths only (Strabo, *Geography,* 113–114), the familiar division into 360° being unknown to him and first introduced into Greek science by Hipparchus (*Geographical Fragments of Hipparchus,* pp. 148–149; D. R. Dicks, "Solstices, Equinoxes, and the Pre-Socratics," in *Journal of Hellenic Studies,* **86** [1966], 27–28).

The method is sound in theory, as Hipparchus recognized, but its accuracy depends on the precision with which the basic data could be determined. The figure of 1/50 of the circle (equivalent to 7° 12′) for the difference in latitude is very near the truth, but Syene (lat. 24° 4′ N.) is not directly on the tropic (which in Eratosthenes' time was at 23° 44′ N.), Alexandria is not on the same meridian (lying some 3° to the west), and the direct distance between the two places is about 4,530 stades, not 5,000. Probably Eratosthenes himself was aware that this last figure was doubtful (without trigonometrical methods, which he certainly did not know, it would have been impossible to measure the distance accurately), and so felt at liberty to increase his final result by 2,000. Nonetheless, the whole measurement was a very creditable achievement and one that was not bettered until modern times. On the most probable value of the stade Eratosthenes used (on this vexed question, see *Geographical Fragments of Hipparchus,* pp. 42–46), 252,000 stades are equivalent to about 29,000 English miles, which may be compared with the modern figure for the earth's circumference of a little less than 25,000 miles.

He obtained a value for the obliquity of the eclip-

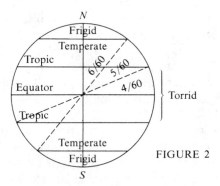

FIGURE 2

tic equivalent to 23; 51, 20°; a figure accepted as accurate by both Hipparchus and Ptolemy. Apparently he estimated the arc between the greatest and least meridian altitudes of the sun (at summer and winter solstices) to be 11/83 of a great circle. This value, which is twice that for the obliquity of the ecliptic, is 47; 42, 39°+. How he discovered this curious ratio (if he did) is not clear (*ibid.*, fr. 41 and comment, pp. 167–168), and whether this measurement was fully described in the *Geography* or elsewhere cannot be determined—Strabo does not mention it, and he was undoubtedly writing with a copy of the *Geography* before him. What certainly would have found a place in this work was Eratosthenes' division of the terrestrial globe into zones. Of these he envisaged five (see Figure 2): a frigid zone around each pole, with a radius of 6/60 each, or 25,200 stades on the meridian circle (in his division of the circle into sixtieths, each sixtieth = 252,000 ÷ 60 = 4,200 stades), a temperate zone between each frigid zone and the tropics, with a radius of 5/60 or 21,000 stades, and a torrid zone comprising the two areas from the equator to each tropic, with a radius of 4/60 or 16,800 stades each (4/60 is equivalent to 24°, an approximate figure for the obliquity of the ecliptic known probably from the time of Eudoxus and used occasionally even by Hipparchus, e.g., *Commentarii in Arati et Eudoxi Phaenomena*, I, 10, 2)—making a total of 126,000 stades from pole to pole, i.e., half the whole circumference (see Geminus, *Isagoge*, XVI, 6 f.; V, 45 f.; Strabo, *Geography*, 113–114; *cf.* 112). The frigid zones were arbitrarily defined by the "arctic" and "antarctic" circles of an observer on the main parallel of latitude (roughly 36° N.), i.e., the circles marking the limits of the circumpolar stars that never rise or set and the stars that are never visible at that latitude (see *Geographical Fragments of Hipparchus*, pp. 165–166). Within this framework the *oikoumene*, according to Eratosthenes, has a "breadth" (north–south, as always in Greek geography) of 38,000 stades from the Cinnamon country (south of Meroë) to Thule, and a "length" (east–west) of 77,800 stades from the further

side of India to beyond the Straits of Gibraltar (Strabo, *Geography*, 62–63, 64).

Although it is clear from the *Geography* that Eratosthenes was familiar with the concept of the celestial sphere, he does not seem to have done any original work in astronomy apart from the above measurements made in a geographical context; his name is not connected with any purely astronomical observation (figures for the distance and size of the sun attributed to him by Eusebius of Caesarea, *Praeparatio evangelica*, XV, 53, and Macrobius, *In somnium Scipionis*, I, 20, 9, are worthless, coming from these sources), he does not appear among the authorities cited by Ptolemy in the *Phaseis* for data relating to the parapegmata or astronomical calendars (see *Geographical Fragments of Hipparchus*, pp. 111–112), and only one astronomical title is attributed to him (and that wrongly): the fragmentary *Catasterismoi* (Robert, ed. [Berlin, 1878]; see Maass, "Analecta Eratosthenica," in *Philologische Untersuchungen*, **6** [1883], 3–55), which tells how various mythical personages were placed among the stars and gave their names to the different constellations, descriptions of which are given. It is possible that an inferior second-century compilation of the same nature, called *Poetica astronomica* (Bunte, ed. [Leipzig, 1875]) and going under the name of the Augustan scholar Hyginus, is based partly on a work of Eratosthenes, who is cited some twenty times (as against, e.g., ten times for Aratus), but this is hardly serious astronomy (see Rose, *Handbook of Latin Literature*, 3rd ed. [1954], p. 447).

In mathematics, Eratosthenes' chief work seems to have been the *Platonicus*, of which we have a few extracts given by Theon of Smyrna, who wrote in the second century (*Expositio rerum mathematicarum ad legendum Platonem utilium*, Hiller, ed. [Leipzig, 1878], pp. 2, 127, 129, 168). In this work, Eratosthenes apparently discussed from a mathematical and philosophical point of view such topics as proportion and progression (essential tools in Greek mathematics) and, arising from this, the theory of musical scales (Ptolemy, *Harmonica*, II, 14, Düring, ed. [Göteborg, 1930], pp. 70 f.; see Düring's ed. of Porphyry's commentary on this [1932], p. 91). Also in this work he gave his solution of the famous Delian problem of doubling the cube and described a piece of apparatus by which a solution could be obtained by mechanical means; the description is preserved for us by Eutocius, a sixth-century commentator on the works of Archimedes, and includes Eratosthenes' epigram (mentioned above) commemorating his achievement (*Eutocii commentarii in libros de sphaera et cylindro*, II, 1, in *Archimedes opera omnia*, Heiberg ed., III,

88 f.; epigram, p. 96); Pappus also describes the apparatus and the method (*Collectio*, III, F. Hultsch, ed. [Berlin, 1876], 22–23, 56–58). Eutocius gives his information in the form of a letter from Eratosthenes to King Ptolemy Euergetes; the "letter" is almost certainly not genuine, but there is no reason to doubt that the contents represent the matter of Eratosthenes' solution (perhaps at least partly in his own words) or that the epigram is his.

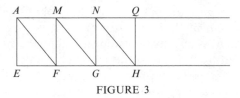

FIGURE 3

The history of the problem of doubling the cube and the various solutions proposed are fully discussed by Heath (*History of Greek Mathematics,* I, 244–270). Briefly, the problem resolves itself into finding two mean proportionals in continued proportion between two given straight lines: if a and b are the two given straight lines and we find x and y such that $a:x = x:y = y:b$, then $y = x^2/a = ab/x$; eliminating y, we have $x^3 = a^2b$, and in the case where b is twice a, $x^3 = 2a^3$, and thus the cube is doubled. Eratosthenes' mechanical solution envisaged a framework of two parallel rulers with longitudinal grooves along which could be slid three rectangular (or, according to Pappus, *loc. cit.*, triangular) plates (marked with their diagonals parallel—see Figure 3) moving independently of each other and able to overlap; if one of the plates remains fixed and the other two are moved so that they overlap as in Figure 4, it can easily be shown that points A, B, C, D lie on a straight line in such a way that AE, BF, CG, DH are in continued proportion, and BF and CG are the required mean proportionals between the given straight lines AE and DH.

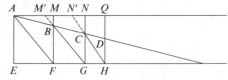

FIGURE 4. Only diagonals and righthand edges of movable plates marked.

In arithmetic Eratosthenes invented a method called the "sieve" (κόσκινον) for finding prime numbers (Nicomachus, *Introductio arithmetica,* I, 13, 2–4). According to this, one writes down consecutively the odd numbers, starting with 3 and continuing as long as desired; then, counting from 3, one passes over

two numbers and strikes out the third (a multiple of 3 and hence not prime) and continues to do this until the end—thus 3 5 7 9̸ 11 13 1̸5̸ 17 19 2̸1̸ 23 25 2̸7̸ 29 31 3̸3̸ 35, etc. The same process is gone through with 5, but this time passing over four numbers and striking out the fifth (a multiple of 5)—3 5 7 9̸ 11 13 1̸5̸ 17 19 2̸1̸ 23 2̸5̸ 2̸7̸ 29 31 3̸3̸ 3̸5̸, etc. The process may be repeated with consecutive odd numbers as many times as one likes, on each occasion, if n is the odd number, $n-1$ numbers being passed over and the next struck out; the remaining numbers will all be prime. Pappus (late third century) also attributes to Eratosthenes a work *On Means* (Περὶ μεσοτήτων), the contents of which are a matter of conjecture but which was important enough to form part of what Heath calls the *Treasury of Analysis* (ἀναλυόμενος τόπος), comprising works by Euclid, Apollonius, Aristaeus, and Eratosthenes (Pappus, *Collectio*, Hultsch, ed., VII, 3, p. 636, 24; see Heath, *History of Greek Mathematics,* II, pp. 105, 399 ff.).

In chronology, Eratosthenes apparently wrote two works, *Chronography* (Χρονογραφίαι) and *Olympic Victors* (Ὀλυμπιονῖκαι); both must have entailed considerable original research (he was the first Greek writer we know to have made a scientific study of the dating of events), and the former seems to have been a popularizing work containing a number of anecdotes, several of which are repeated by Plutarch (e.g., "Demosthenes," Loeb ed., IX, 4; Teubner ed., XXX, 3; "Alexander," Loeb ed., III, 2; Teubner ed., XXXI, 2). Eratosthenes' datings remained authoritative throughout antiquity and in many cases cannot be improved upon today—e.g., the fall of Troy, 1184/1183 B.C.; the Dorian migration, 1104/1103; the first olympiad, beginning 777/776; the invasion of Xerxes, 480/479; the outbreak of the Peloponnesian War, 432/431.

In literary criticism Eratosthenes wrote a work in not less than twelve books entitled *On the Old Comedy,* the contents of which ranged over textual criticism, discussion of the authorship of plays from the dates of performances, and the meanings and usages of words; it was highly thought of by ancient scholars, being frequently cited, and its loss is greatly to be regretted. He also seems to have written a separate work on grammar. Finally, as befitted an Alexandrian polymath, he had a not inconsiderable reputation as a poet; his three main poetical works were *Hermes, Erigone,* and *Anterinys* or *Hesiod* (apparently alternative titles). The first had the same theme at the beginning as the well-known Homeric hymn but went on to draw a picture of the ascent of Hermes to the heavens and to give a vividly imaginative description of the zones of the earth as seen from there (Achilles

Tatius, *Isagoge,* p. 153c in Petavius' *Uranologion* [1630]—the lines are reprinted by Hiller and by Powell); this passage was copied by Vergil (*Georgics,* I, 233–239). The *Erigone* was a star legend dealing with the story of Icarius, his daughter Erigone, and her dog, all of whom in this version were translated to the heavens as Boötes, Virgo, and Sirius, the Dog Star. The subject matter of the third poem is unknown. Only a few fragments of Eratosthenes' poetry are extant (the longest, some sixteen lines, being the passage from the *Hermes* mentioned above), and it is impossible to judge its intrinsic merit from these.

BIBLIOGRAPHY

The only published collection of all Eratosthenes' fragments is G. Bernhardy, *Eratosthenica* (Berlin, 1822), which is now greatly out of date. I have been much indebted in the preparation of this article to R. M. Bentham's unpublished Ph.D. thesis (London) entitled "The Fragments of Eratosthenes of Cyrene." It was made available to me through the kindness of his supervisor, Prof. E. H. Warmington (formerly of Birkbeck College, University of London), following the unfortunate death of the author before he submitted his thesis.

See also G. Knaack, "Eratosthenes," in Pauly-Wissowa, VI (1907), cols. 358–388; E. H. Bunbury, *History of Ancient Geography,* I (London, 1879), ch. 16; E. H. Berger, *Die geographische Fragmente des Eratosthenes* (Leipzig, 1880); A. Thalamas, *La géographie d'Ératosthène* (Versailles, 1921); D. R. Dicks, *The Geographical Fragments of Hipparchus* (London, 1960); E. Hiller, *Eratosthenis carminum reliquiae* (Leipzig, 1872); and J. U. Powell, *Collectanea Alexandrina* (Oxford, 1925).

D. R. DICKS

ERCKER (also **ERCKNER** or **ERCKEL**), **LAZARUS** (*b.* Annaberg, Saxony, *ca.* 1530; *d.* Prague, Bohemia, 1594), *chemistry, metallurgy.*

Ercker was the son of Asmus Erckel. After finishing school at Annaberg, he studied at the University of Wittenberg in 1547–1548. He married Anna Canitz on 7 October 1554, and through the help of one of his wife's relatives, a doctor named Johann Neef, he was appointed assayer at Dresden by Elector Augustus, an enthusiastic admirer of alchemy and metallurgy. A year later he became chief consultant and supervisor in all matters relating to the mineral arts and mint affairs for Freiberg, Annaberg, and Schneeberg but soon was demoted, for unknown reasons, to warden of the mint at Annaberg. In the spring of 1558 he made a trip to the Tyrol to become acquainted with its mines and foundries, and in the autumn of the same year Prince Henry of Brunswick made

him first warden and then master of the mint at Goslar.

After the death of his wife in 1567 Ercker returned to Dresden, where he sought a position with Elector Augustus of Saxony but failed because of intrigue and an unsuccessful attempt to obtain silver from poor ores. He then went to Prague, where his brother-in-law, Casper Richter, was a minter and, through the latter's support, was appointed control tester at Kutna Hora.

In 1574 Ercker published (at Prague) his famous book *Beschreibung allerfürnemisten mineralischen Ertzt.* This brought him to the attention of Emperor Maximilian II, who named him his courier for mining affairs and clerk in the Supreme Office of the Bohemian crown.

During the reign of the next emperor, Rudolf II, a well-known patron of alchemists, Ercker became chief inspector of mines and was knighted on 10 March 1586, receiving the title von Schreckenfels. The motto on his coat of arms was "Erst Prob's dann Lob's." Ercker's second wife, Susanna, for many years managed the mint at Kutna Hora and had the title manager-mistress of the mint. Both of his sons, Joachim and Hans, were assayers.

Through his various posts Ercker acquired extensive experience in chemistry and metallurgy. His first work was *Probierbuchlein* (1556), dedicated to Augustus. In 1563 he wrote *Münzbuch* and in 1569 a book on the testing of ores, *Zkouseni rud.* In 1574 Ercker published his magnum opus, *Beschreibung allerfürnemisten mineralischen Ertzt.* The only one of Ercker's works to contain many drawings, it presents a systematic review of the methods of testing alloys and minerals of silver, gold, copper, antimony, mercury, bismuth, and lead; of obtaining and refining these metals, as well as of obtaining acids, salts, and other compounds. The last chapter is devoted to saltpeter. Ercker described laboratory procedures and equipment, gave an account of preparing the cupel, of constructing furnaces, and of the assaying balance and the method of operating it. He used as his model Agricola's *De re metallica,* yet he was quite original and included only the procedures he himself had tested. Ercker was so hostile to alchemy that he did not use alchemical symbols, although his *Probierbuchlein* (1556) included a full list of them.

Ercker's *Beschreibung* may be regarded as the first manual of analytical and metallurgical chemistry. Of particular interest to the historian of science is his observation that a cupel containing copper and lead weighs more after roasting in a furnace than before, which, says Ercker, although it is of no importance to the assayer, is surprising (bk. I).

Ercker maintained that precipitating copper from a solution by means of iron does not mean that iron becomes copper and that transmutation takes place; that copper sets silver free from a solution, and if one wants to precipitate copper and silver from solutions of nitrates, one should use iron plates and copper plates; and that iron reduces copper from its solution. This had already been written by Alexander Suchten in *Tractatus secundum de antimonio* (*ca.* 1570, published posthumously in 1604): "Venus so das Eisen aus dem Vitriol reducirt hat." Ercker's account of the fact that zinc precipitates other metals from solutions, cited by J. R. Partington, is to be found only in the 1684 and later editions and therefore was added by an unknown commentator.

Ercker also discovered a new method of refining gold; an exact description was sent to Augustus but is not extant. Ercker's book inspired Löhneyss, Glauber, and others in their own writings on assaying.

BIBLIOGRAPHY

I. ORIGINAL WORKS. Ercker's writings include *Münzbuch, wie es mit den Münzen gehalten sind* (1563); *Beschreibung allerfürnemisten mineralischen Ertzt und Berckwercksarten . . .* (Prague, 1574; Frankfurt, 1580, 1598, 1623, 1629); and *Aula subterranea alias Probierbuch Herrn Lasari Erckers* (Frankfurt, 1672, 1684, 1703, 1736). The *Beschreibung* was translated into English by John Pettus as *Fleta minor, the Laws of Art and Nature* (London, 1683, 1686, 1689) and appeared in a more modern version as *Treatise on Ores and Assaying* (Chicago, 1951); into modern German as *Beschreibung der allevornehmsten mineralischen Erze und Bergwerksarten vom Jahre 1580* (Berlin, 1960), Freiberger Forschungshefte D34; and into Dutch as *Uytvoerige Operinge der onderaarolsche Wereld* (The Hague, 1745).

The MS of the *Probierbuchlein* is in the Sachsisches Landesbibliothek, Dresden, MS J343; that of the *Münzbuch,* in the Herzog Augustus Bibliothek, Wolfenbüttel, MS 2728; and that of *Zkouśeni rud* in the National Archives, Prague, MS 3053.

II. SECONDARY LITERATURE. On Ercker's life and work, see *Allgemeine deutsche Biographie,* VI (1885), 214; E. V. Armstrong and H. S. Lukens, "Lazarus Ercker and His 'Probierbuch.' Sir John Pettus and His 'Fleta Minor,'" in *Journal of Chemical Education,* **16** (1939), 553–562; P. R. Beierlein, *Lazarus Ercker, Bergmann, Hüttenmann und Münzmeister in 16. Jahrhundert* (Berlin, 1955); and in *Beschreibung der allervornehmsten mineralischen Erze und Bergwerksarten vom Jahre 1580* (Berlin, 1960), *passim.* Beierlein's is the best biography (although he consulted only the Landesarchiv in Dresden), but the chemical commentary on Ercker's treatise is incomplete and often in error. See also J. Ferguson, in *Bibliotheca chemica,* I (Glasgow, 1906), 242–245; J. R. Partington, *A History of Chemistry,* II (London, 1961), 104–107; and A. Wrany, in *Geschichte der Chemie* (Prague, 1902), p. 91.

Much material concerning Ercker's activity in Bohemia in 1583–1593, none of which has yet been used, is in the National Archives, Prague, Prazska Mincownia collection, boxes 17–25 (1583–1593). Several letters from Ercker to Wilhelm Rosenberg, burgrave of Prague, are in the Archives of Třeboň, Czechoslovakia. There should also be documents concerning him in the archives of Goslar, Brunswick, and Wolfenbüttel. All the material in the Landesarchiv, Dresden, was used by Beierlein in his study of Ercker.

WŁODZIMIERZ HUBICKI

ERDMANN, OTTO LINNÉ (*b.* Dresden, Germany, 11 April 1804; *d.* Leipzig, Germany, 9 October 1869), *chemistry.*

Erdmann was the son of the physician and botanist Carl Gottfried Erdmann. In 1820, after apprenticeship to a pharmacist, he studied medicine at the Medical-Surgical Academy in Dresden; in 1822 he entered the University of Leipzig, where his interest in chemistry was stimulated by L. W. Gilbert, professor of physics. After graduating in medicine in 1824 and qualifying as a university lecturer in 1825, Erdmann devoted the rest of his life to chemistry. In 1827, after a year directing a nickel mine and foundry at Hasserode, he was appointed extraordinary professor, and in 1830 professor, of technical chemistry at Leipzig, where he established his reputation as a teacher and researcher. Erdmann was Rektor Magnificus of Leipzig from 1848 to 1849, and from 1835 he was a director, and eventually chairman, of the Leipzig-Dresden Railway Company. A prominent Freemason, he devoted much time to the improvement of the cultural facilities and technological prosperity of the city of Leipzig. He married Clara Jungnickel, by whom he had three sons and a daughter.

The Saxon government was persuaded by Erdmann to build chemical laboratories at the university; and after they were opened in 1842[1] Erdmann was able to compete with Liebig at Giessen and attract large numbers of students, many of whom achieved eminence, e.g., C. F. Gerhardt.[2] He toured Germany and France in 1836 in order to meet other chemists, including his future collaborator, R. F. Marchand. Erdmann visited England in 1842, and he was a voluble spokesman for noninterference with the individual chemist's right to freedom of choice between atomic and equivalent weights at the important Karlsruhe Conference in 1860.[3] He greatly enriched chemical communications by the creation in 1834 of the *Journal für praktische Chemie.* His textbooks, and especially his encyclopedia of industrial chemistry, helped to educate the revolutionary generation of Kolbe and Kekulé. To this younger German generation, however, he came to typify the stereotyped, unimaginative chemistry against which they rebelled so passionately and fruitfully.

Erdmann's researches, which spanned mineralogical, industrial, inorganic, and organic chemistry, were primarily descriptive and analytical. In organic chemistry, between 1840 and 1841 (simultaneously with Laurent, who corrected him), he investigated the nature of indigotin and prepared a number of derivatives that were important later, including isatin and tetrachloro-*p*-benzoquinone.[4] He subsequently investigated and isolated hematoxylin from logwood[5] and euxanthic acid from Indian yellow.[6]

Erdmann's confusion over the empirical formula of isatin led him skeptically to redetermine the atomic weight of carbon in 1841. In collaboration with Marchand he supported Dumas and Stas in lowering its atomic weight from Berzelius' value of 76.43 ($O = 100$) to 75.08.[7] Subsequently, until the death of Marchand in 1850, they made a number of accurate redeterminations.[8] In most cases they obtained values significantly different from those established by Berzelius and sufficiently close to whole numbers to persuade them that there might be some truth in Prout's hypothesis that atomic weights were multiples of a common unit. There followed a dispute with Berzelius, who abhorred *Multiplenfieber,* in which Erdmann maintained an empirical position that chemists should be guided only by accurate experiments.[9]

NOTES

1. O. L. Erdmann, "Das chemische Laboratorium der Universität Leipzig," in *Journal für praktische Chemie,* **31** (1844), 65–75, with plans.
2. E. Grimaux and C. Gerhardt, *Charles Gerhardt, sa vie, son oeuvre, sa correspondance 1816–1856* (Paris, 1900), pp. 19–21, 85, 218, 264–265, 449, 452.
3. R. Anschütz, *August Kekulé* (Berlin, 1929), I, 671–688.
4. O. L. Erdmann, "Untersuchungen über den Indigo," in *Journal für praktische Chemie,* **19** (1840), 321–362; **22** (1841), 257–299.
5. O. L. Erdmann, "Ueber das Hämatoxylin," ibid., **26** (1842), 193–216, also in *Reports of the British Association for the Advancement of Science,* **11** (1842), 33–34.
6. O. L. Erdmann, "Ueber das Jaune indien und die darin enthaltene organische Säure (Euxanthinsäure)," in *Journal für praktische Chemie,* **33** (1844), 190–209.
7. O. L. Erdmann and R. F. Marchand, "Ueber das Atomgewicht des Kohlenstoffes," ibid., **23** (1841), 159–189.
8. O. L. Erdmann, "Ueber das Atomgewicht des Wasserstoffes und Calciums," ibid., **26** (1842), 461–478; ". . . Calciums, Chlors, Kaliums und Silbers," ibid., **31** (1844), 257–279; ". . . Kupfers, Quecksilbers, und Schwefels," ibid., 385–402; ". . . Eisens," ibid., **33** (1844), 1–6.

9. O. L. Erdmann, "Rechfertigung einiger Atomgewichtsbestimmungen," ibid., **37** (1846), 65–80; "Einige Bemerkungen über die Atomgewichte der einfachen Körper," ibid., **55** (1852), 193–203.

BIBLIOGRAPHY

I. ORIGINAL WORKS. The *Royal Society Catalogue of Scientific Papers* assigns 96 papers to Erdmann, of which 17 were joint researches with Marchand. Unrecorded are several papers on artistic subjects and his many editorial notes to his journals: *Die neuesten Forschungen im Gebiete der technischen und ökonomischen Chemie,* 18 vols. (1828–1833), more familiarly known, through its second title page, as *Journal für technische und ökonomische Chemie.* In 1834 this amalgamated with the well-established *Journal für Chemie und Physik* and was edited jointly by Erdmann and F. W. Schweigger-Seidel as *Journal für praktische Chemie,* **1–9** (1834–1836). Volumes **10–15** (1837–1838) were edited by Erdmann alone, vols. **16–50** (1839–1850) jointly with Marchand, vols. **51–57** (1850–1852) alone, and vols. **58–108** (1853–1869) jointly with G. Werther. The journal was continued after Erdmann's death by Kolbe. Other papers by Erdmann, together with extensive analyses of their contents, may be traced in Berzelius' *Jahres-Bericht über die Fortschritte der physischen Wissenschaften,* **7** (1828)**–25** (1846) and its continuation, *Jahresbericht über die Fortschritte der reinen, pharmaceutischen und technischen Chemie* (1849–1870); see annual indexes.

Erdmann published the following books: *Ueber das Nickel* (Leipzig, 1827); *Lehrbuch der Chemie* (Leipzig, 1828; 3rd ed., 1840; 4th ed., 1851), trans. into Dutch (Amsterdam, 1836); *Grundriss der allgemeinen Waarenkunde* (Leipzig, 1833; 2nd ed., 1852; 3rd ed., 1857), with many posthumously revised eds. (after the 7th ed., 1871, edited by C. R. Krönig, it became known as "Erdmann-Krönig" and, with a succession of editors, reached the 17th and final ed. in 1925); and *Ueber das Studium der Chemie* (Leipzig, 1861). He was an editor of *Universel-Lexicon der Handelswissenschaften,* 3 vols. (Leipzig, 1837–1839).

The Karl Marx University, Leipzig, holds 19 of Erdmann's letters, written between 1828 and 1869. There are several letters to Berzelius in the Royal Swedish Academy of Sciences, Stockholm.

II. SECONDARY LITERATURE. The basic sources of information concerning Erdmann's career are the unctuous obituaries by H. Kolbe, in *Journal für praktische Chemie,* **108** (1869), 449–458—an adapted and unacknowledged English version by A. W. Williamson, in *Journal of the Chemical Society,* **23** (1870), 306–310—and in *Berichte der Deutschen chemischen Gesellschaft,* **3** (1870), 374–381. See also *Sitzungsberichte der K. Bayerischen Akademie der Wissenschaften zu München* (1870), pt. 1, 415–417; and J. R. Partington, *A History of Chemistry,* IV, 397. An assessment of Erdmann's atomic weights may be made from G. F. Becker, *Atomic Weight Determinations: A Digest of the Investigations Published Since 1814,* Smithsonian Miscellaneous Collections: Constants of Nature, pt. IV (Washington, D.C., 1880); and F. W. Clarke, *A Recalculation of the Atomic Weights,* Constants of Nature, pt. V (Washington, D.C., 1882; 2nd ed., 1897). For a very important critique of *Grundriss der allgemeinen Waarenkunde,* see R. A. C. E. Erlenmeyer, in *Zeitschrift für Chemie und Pharmacie,* **4** (1861), 217–220, 251–256, 284–287, 320–323, 385–386.

Contemporary opinions of Erdmann may be found in R. Anschütz, *August Kekulé* (Berlin, 1929), I, *passim;* E. Grimaux and C. Gerhardt, *Charles Gerhardt, sa vie, son oeuvre, sa correspondance 1816–1856* (Paris, 1900), pp. 19–21, 85, 218, 264–265, 449, 452; and O. Wallach, ed., *Briefwechsel zwischen J. Berzelius und F. Wöhler,* 2 vols. (Leipzig, 1901), *passim.*

W. H. BROCK

ERIUGENA, JOHANNES SCOTTUS (*b.* Ireland, first quarter ninth century; *d.* England [?], last quarter ninth century), *philosophy.*

Nothing is known of Eriugena's life before 847, by which time he had already left Ireland and had been living for some years in France. By 851 the reputation for learning he had acquired was sufficient for his being asked to give his views on the dispute that had arisen over the interpretation of Augustine's teaching on predestined grace (Ebo of Grenoble, *Liber de tribus epistolis,* XXXIX; Migne, *Patrologia,* CXXII, 1052A). In his reply, *De praedestinatione,* he revealed a critical understanding of the relevant texts of Augustine and adopted his precept that the seven liberal arts should be applied to the solution of theological problems. He also gave early evidence of a knowledge of Greek that was to become exceptional, if not unique, in ninth-century Europe.

Eriugena specifically attributed to an inadequate understanding of Greek and the liberal arts the failure of his contemporaries to understand Augustine's teaching (*De praedestinatione,* XVIII; *Patrologia,* CXXII, 403C10–D1) and made these two disciplines the principal subjects in the curriculum of the palace school at Laon, over which he presided with the assistance of his fellow countryman Martin. Here he restored to the arts their ancient classical function of propaedeutic to philosophy and theology. He taught them through the medium of a book that had been forgotten since the end of the ancient world, the *De nuptiis* of Martianus Capella, and used another forgotten work, Boethius' *Consolation of Philosophy,* for more advanced studies. To these texts he and his colleagues appended commentaries that, although not certainly extant in complete form today, established the matter and method of teaching in schools throughout Europe from the ninth to the twelfth centuries. Eriugena and his colleagues at Laon thus founded the educational system of the later Middle Ages and perpetuated the Carolingian renaissance.

The fame of the Greek scholarship at Laon was such that Charles the Bald commissioned Eriugena to translate into Latin the treatises of the pseudo-Dionysius and the *First Ambigua* of Maximus the Confessor. These labors, to which he added for his own purposes translations of the *Quaestiones ad Thalassium* of Maximus and the *De hominis opificio* of Gregory of Nyssa, occupied the years between 860 and 864. They brought Eriugena, already inclined toward Platonism by his reading of Augustine, into direct contact with the fully developed post-Plotinian Neoplatonism which had been absorbed by the Greek Fathers but until then had been a closed book for the Latin West. The immediate consequence of this contact was the composition, between 864 and 866, of Eriugena's greatest work, the *Periphyseon* or *De divisione naturae,* in which the Western and Eastern forms of Neoplatonism are synthesized within a Christian context. In his subsequent writings—the *Expositiones super Ierarchiam caelestem,* his commentary on St. John's Gospel (of which only three fragments survive), and his homily on St. John's Prologue—he enunciates the theories of the *Periphyseon* with greater conviction and expresses them in more precise language, but nowhere does he change or abandon them.

These last works were written between 866 and 870, after which nothing further is known of Eriugena; his end is as obscure as his beginning.

BIBLIOGRAPHY

I. ORIGINAL WORKS. Eriugena's principal works, edited by H. J. Floss, are collected in volume CXXII of J. P. Migne's *Patrologiae cursus completus; Series latina,* 2nd ed. (Paris, 1865); but, as will be seen from the following list, the collection is far from complete. It also includes two spurious works, *Expositiones super ecclesiasticam s. Dionysii* and *Expositiones seu Glossae in mysticam Theologiam s. Dionysii.*

De praedestinatione is in *Patrologia . . .,* 355A–440A.

"Iohannis Scotti annotationes in Marcianum," in *Annotationes in Marcianum,* Cora E. Lutz, ed. (Cambridge, Mass., 1939), is an edition of an anonymous commentary on the *De nuptiis* preserved in Paris, Bibliothèque Nationale, MS lat. 12960, one of a group of commentaries that contain Eriugena material and are probably derived from Eriugena's own lost commentary.

H. Silvestre, ed., "Le commentaire inédit de Jean Scot Érigène au mètre ix du livre iii du De consolatione philosophiae de Boèce," in *Revue d'histoire ecclésiastique,* **47** (1952), 44–122, is an edition of an anonymous commentary preserved in Brussels, Bibliothèque Royale, MS 10066–

10067. Silvestre today is less certain that Eriugena is the author, although it is certainly based on his teaching.

Translations of the pseudo-Dionysian works are in *Patrologia . . .,* 1023–1194.

The translation of the *First Ambigua* of Maximus the Confessor is in *Patrologia . . .,* 1193–1222 and 1023A–1024B (introduction and chs. 1–4, sec. 3 only, together with a fragment from a later part printed under the title *Liber de egressu et regressu animae*). The full text is preserved in Paris, MS Mazarine 561, from which one folio has become detached and is now fol. 9 of MS Vat. Reg. lat. 596; Paris, MS Arsenal 237, a contemporary copy of Mazarine MS that lacks the last three folios; and Cambridge, MS Trinity College 0.9, 5, a transcription of the Mazarine MS by Mabillon. Preparatory work for an edition by the late Raymond Flambard based on the two Paris MSS is preserved in Paris, Archives Nationales, AB xxviii[100].

"Scolia Maximi," an unpublished translation of the *Quaestiones ad Thalassium* of Maximus, is preserved in MSS Monte Cassino 333 and Troyes, Bibliothèque Municipale 1234.

De imagine, edited by M. Cappuyns as "Le De imagine de Grégoire de Nysse traduit par Jean Scot Érigène," in *Recherches de théologie ancienne et médiévale,* **32** (1965), 205–262, is an edition of Eriugena's translation of the *De hominis opificio* from the unique MS Bamberg Staatsbibliothek Patr. 78.

Periphyseon (*De divisione naturae*) is in *Patrologia . . .,* 439–1022. Of the new edition in preparation under the auspices of the Dublin Institute for Advanced Studies, the first volume has been published as vol. VII of the series Scriptores latini Hiberniae: I. P. Sheldon-Williams and Ludwig Bieler, eds., *Iohannis Scotti Eriugenae Periphyseon* (*De divisione naturae*) *liber primus* (Dublin, 1968).

Expositiones super Ierarchiam caelestem can be found in *Patrologia . . .,* 125–265 and in H. F. Dondaine, "Les Expositiones super Ierarchiam caelestem de Jean Scot Érigène," in *Archives d'histoire doctrinale et littéraire du moyen âge,* **18** (1951), 245–302—Eriugena's commentary on the *Celestial Hierarchy* of the pseudo-Dionysius, chs. 1, 2, 7–14 in *Patrologia . . .,* the rest in Dondaine.

The *Commentary on St. John's Gospel* is in *Patrologia . . .,* 297A–348B; three fragments are preserved in MS Laon 81.

The homily on the Prologue of St. John's Gospel, was edited by Edouard Jeauneau as *Jean Scot, Homélie sur le Prologue de Jean,* Sources chrétiennes no. 151 (Paris, 1969) and is also in *Patrologia . . .,* 283B–296D.

His poems were edited by L. Traube, in *Monumenta Germaniae historica, Poetae latini aeui Caroli,* III (Berlin 1896), 518–556; an incomplete collection is in *Patrologia . . .,* 1221C–1240C.

II. Secondary Literature. The most important and recent monographs on Eriugena are H. Bett, *Johannes Scotus Erigena: A Study in Mediaeval Philosophy* (Cambridge, 1925; repr. New York, 1964); M. Cappuyns, *Jean Scot Érigène, sa vie, son oeuvre, sa pensée* (Louvain–Paris, 1933; repr. Brussels, 1965), still the best work on Eriugena;

and M. Dal Pra, *Scoto Eriugena,* 2nd ed. (Milan, 1951). John J. O'Meara, *Eriugena* (Cork, 1969), is of exceptional importance and resumes the findings of more recent research, to which it makes its own valuable contribution. Extensive bibliographies are given in Cappuyns, *op. cit.,* pp. xi–xvii; I. P. Sheldon-Williams, "A Bibliography of the Works of Johannes Scottus Eriugena," in *Journal of Ecclesiastical History,* **10,** no. 2 (1960), 223–224; and in Jeauneau's translation of the "Homily," pp. 171–198.

I. P. Sheldon-Williams

ERLANGER, JOSEPH (*b.* San Francisco, California, 5 January 1874; *d.* St. Louis, Missouri, 5 December 1965), *physiology.*

In his late years Erlanger wrote a short but delightful autobiography in which he minimizes his scientific achievements with characteristic modesty but gives interesting details on his early family life. Erlanger's father was born in Württemberg, Germany, and in 1842, at the age of sixteen, landed alone in New York, went to New Orleans, and then became an itinerant peddler along the Mississippi Valley. The gold rush drew him to California, where he became a businessman in San Francisco after having tried his luck at mining. He married the daughter of his business partner. A large family was born to them, Erlanger being their sixth child.

After two years at the San Francisco Boys' High School—during which he acquired a sound knowledge of German and Latin—Erlanger was admitted to the University of California. He enrolled in the college of chemistry to prepare for the medical career that he already had in mind. His native abilities for observation and experimentation were demonstrated in a thesis, written in his senior year, on the development of the eggs of the newt *Amblystoma.*

Erlanger then attended the newly founded Johns Hopkins Medical School in Baltimore. During his medical studies he found time for research, especially during vacations, which he could not spend at home because of the cost of travel. In Lewellys Barker's laboratory Erlanger attacked a fundamental problem of neurophysiology. In 1900 he succeeded in localizing the exact position in the spinal cord of the motor nerve cells that innervate a given muscle, by means of a delicate histological study based on the alterations undergone by the motor nerve cells of the rabbit after the excision of the corresponding muscle. Erlanger's results were the first decisive experimental confirmation of F. Sano's views, according to which each muscle is activated by definite motor nerve cells. Barker's treatise "The Nervous System and Constituent Neurons" describes these findings in detail.

A year later Erlanger published his first paper,

which came to the attention of William H. Howell, professor of physiology at Johns Hopkins, who, shortly after Erlanger took his medical degree, offered him an assistant professorship. In this first paper, "A Study of the Metabolism in Dogs With Shortened Small Intestines," Erlanger sought to ascertain the extent of intestine that could safely be excised in surgical operations. This early research is marked by the dual concern for physiology and medicine with which Erlanger was always to be occupied. This is perhaps the reason why he devoted the major part of his career to the study of circulation and of cardiac physiology.

In 1904 Erlanger imagined and built with his own hands a sphygmomanometer; a form of this instrument bears his name, although others later devised similar apparatuses without mentioning Erlanger's priority. This instrument allowed him to demonstrate that the pulse pressure can also give the precise volume of the pulse wave, a result that was to have an immediate application in the separation of the effects of the pulse pressure from those of the arterial pressure. Erlanger was thus able to demonstrate that in patients affected by albuminuria the discharge of albumin depends much more on the volume of the pulse wave than on the arterial pressure.

From 1904 on Erlanger concerned himself with the conduction of excitation in the heart. He proved that the Stokes-Adams syndrome resulted from an impaired conduction between the auricles and ventricles, similar to the effect obtained through the experimental exercise of pressure on the auriculoventricular junction of the turtle's heart. The fainting spells that characterize the syndrome occur when the partial block of auriculoventricular conduction temporarily becomes complete. The German anatomist Wilhelm His had previously described the only conducting muscular connection between the auricles and ventricles, the narrow auriculoventricular bundle that bears his name. Erlanger devised a clamp with which controlled pressure could be reversibly applied to the His bundle of the beating heart in a dog. He thus produced all degrees of auriculoventricular block, from the normal 1:1 sequence to complete block, through the partial blocks characterized by two or more auricular beats for a simple ventricular contraction. These pioneering experiments are the basis of current knowledge of intracardiac conduction; the finer features of this conduction were to be analyzed many years later by Frank Schmitt.

In 1906 Erlanger was offered the chair of physiology at the University of Wisconsin, an assignment worthy of his abilities. Here he was asked to equip a modern laboratory, and he became responsible for the teaching of the entire field of physiology.

In 1910, when Washington University in St. Louis completely reorganized its medical school (soon to be a research center of worldwide reputation), Erlanger became its professor of physiology. New laboratories devoted to the major fundamental sciences were built close to each other in the vicinity of large hospitals, reinforced by an excellent library.

World War I diverted Erlanger's activity to quite different problems. Among them was the treatment of wound shock, for which he proposed the administration of a solution of glucose and gum acacia, a procedure that was used successfully by the U.S. Army during its campaign in France; this was the first example of treatment by an artificial serum containing a component of large molecular weight—that is, a high polymer.

He also became interested in the problem of blind landing of airplanes. After numerous flights he proposed a new design of the instrument panel so that the major instruments would always remain in the pilot's visual field.

As soon as the war ended, Erlanger resumed his work on circulation. He investigated the mechanism that produces the sounds of Korotkoff (the sounds that are detected by a stethoscope placed on the skin over an arterial region above which a controlled pressure is applied through a pneumatic cuff—the regular procedure for the measurement of arterial pressure). Erlanger showed that these sounds pose a difficult problem of fluid mechanics, which he solved. Working with J. C. Bramwell, he demonstrated, with an elegant and precise technique, that the crest of the pulse wave is unstable. The pulse wave breaks, as does a sea wave on a beach, because its crest dilates the artery and thus proceeds with a higher velocity than the foot of the wave. These two components of the wave can be separated by the observer as corresponding to sharp and dull sounds, respectively. When the dull sound occurs, the pressure applied in the cuff indicates exactly the diastolic pressure.

In 1921 Erlanger and his colleague Herbert Gasser, professor of pharmacology at Washington University, became associated in a new field of research. In about ten years they created, with George Bishop, modern neurophysiology with the use of the cathode-ray oscillograph (then called the Braun tube, after its inventor). Under their able hands this dim and fragile ancestor of the brilliant oscillograph of today immediately proved itself a remarkable instrument. By coupling it with amplifying vacuum tubes they obtained, for the first time, an exact picture of the action potentials that are the electric signs of the nervous impulses. Because of the smallness and brief duration of these action potentials no other instrument could record them; the cathode-ray oscillograph revealed that the nerve action potential is formed by several component waves traveling with unequal velocities. When Gasser showed these records to Louis Lapicque, the professor of physiology at the Sorbonne, Lapicque perceived their significance immediately. Ten years earlier, Lapicque and René Legendre had observed that nerves of slow excitability (that is, of long chronaxies) were constituted of smaller fibers than the nerves of fast excitability (or brief chronaxies). Lapicque had then assumed that the impulse travels more rapidly in large fibers than in small ones. A histological investigation by Lapicque, Gasser, and Henri Desoille immediately showed that multifunctional (motor and sensory) nerves that display a multiwave action potential contain two or three groups of fibers, each of which is characteristically of a different diameter. On the other hand, a unifunctional nerve—the phrenic nerve, for example—that innervates only one muscle and contains no sensory fibers is made up of fibers of uniform diameter. These results led Erlanger and Gasser to formulate their law by which nervous impulse velocity is directly proportional to fiber diameter.

Many further discoveries in neurophysiology arose from Erlanger and Gasser's joint work. They were awarded the Nobel Prize in 1944. The disclosure of the time-course of the excitability cycle of nerve that has had a decisive impact upon all further theoretical attempts toward the formulation of excitatory processes is derived, however, from the work of Erlanger and E. A. Blair.

Although he reached retirement age Erlanger did not cease working. He resumed teaching in the medical school of Washington University during World War II, while his younger colleagues were called to military duties. He remained active after the war and was in close contact with the members of the laboratory, who benefited from his profound knowledge of all the domains of physiology. He also devoted much time to the history of this science, to the profit of the Medical School library. That he was an able and elegant historical writer is testified to by, among other things, his account of William Beaumont's experiments, in which he interpreted Beaumont's observations of the digestive process in the human stomach in the light of modern knowledge and showed how they constitute a most excellent experimental work a century ahead of its time.

Erlanger was a family man. In his last years he sustained with courage the losses of his devoted wife Aimée Hirstel, his only son, Herman, and his son-in-law. His reserve at first approach quickly gave way to his natural kindness and to his generous and smiling inclinations. He was an invaluable source of inspiration for both American and foreign physiologists and especially for those who had the privilege of working under his guidance in his laboratory.

BIBLIOGRAPHY

Erlanger's works include "A Study of the Metabolism in Dogs With Shortened Small Intestines," in *American Journal of Physiology,* **6** (1901), 1–30, written with W. Hewlett; "An Experimental Study of Blood Pressure and of Pulse Pressure in Man," in *Johns Hopkins Hospital Reports,* **12** (1904), 147–378, written with D. R. Hooker; "On the Physiology of Heart-block in Mammals, With Especial Reference to the Causation of Stokes-Adams Disease," in *Journal of Experimental Medicine,* **7** (1905), 675–724, and **8** (1906), 8–58; "Studies in Blood Pressure Estimations by Indirect Methods. I. The Mechanism of Oscillatory Criteria," in *American Journal of Physiology,* **39** (1916), 401–446; "Studies in Blood Pressure Estimations by Indirect Methods. II. The Mechanism of the Compression Sound of Korotkoff," *ibid.,* **40** (1916), 82–125; "The Compound Nature of the Action Current of Nerve as Disclosed by the Cathode-ray Oscillograph," *ibid.,* **70** (1924), 624–666, written with H. S. Gasser; "The Action Potential Waves Transmitted Between the Sciatic Nerve and Its Spinal Roots," *ibid.,* **78** (1926), 574–591, written with G. H. Bishop and H. S. Gasser; "The Effects of Polarization Upon the Activity of Vertebrate Nerve," *ibid.,* 630–657, written with G. H. Bishop; "The Role Played by the Sizes of the Constituent Fibres of a Nerve Trunk in Determining the Form of Its Action Potential," *ibid.,* **80** (1927), 1522–1547, written with H. S. Gasser; "Directional Differences in the Conduction of the Impulse Through Heart Muscle and Their Possible Relation to Extra Systolic and Fibrillary Contractions," *ibid.,* **87** (1928), 326–347, written with F. O. Schmitt; "The Irritability Changes in Nerve in Response to Subthreshold Induction Shocks and Constant Currents," *ibid.,* **99** (1931), 108–155, written with E. A. Blair; "William Beaumont's Experiments and Their Present Day Value," in *Bulletin of the St. Louis Medical Society* (8 Dec. 1933); and *Electrical Signs of Nervous Activity* (Philadelphia, 1937).

For further details of Erlanger's life, see his autobiographical "A Physiologist Reminisces," the prefatory chapter to *Annual Review of Physiology,* **26** (1964), 1–14.

A. M. MONNIER

ERLENMEYER, RICHARD AUGUST CARL EMIL

(*b.* Wehen, Germany, 28 June 1825; *d.* Aschaffenburg, Germany, 22 January 1909), *chemistry.*

Erlenmeyer was one of the earliest disciples of Kekulé and advocated Kekulé's views on the constitution of organic compounds at a time when many of the leading chemists still adhered to dualistic or to type theories. Erlenmeyer himself was converted from the old chemical types to the newer views on valence and structure. He entered the University of

Giessen in 1845 as a medical student, but on hearing Liebig lecture he decided to study chemistry, first at Giessen and then at Heidelberg, where he became one of Kekulé's first private students. He was professor of chemistry at the Munich Polytechnic School from 1868 until his retirement in 1883. In addition to teaching and publishing many research papers, Erlenmeyer was an editor of the *Zeitschrift für Chemie und Pharmazie* and of Liebig's *Annalen der Chemie*. He was coauthor of the three-volume *Lehrbuch der organischen Chemie* (1867–1894).

Erlenmeyer published important work in both experimental and theoretical organic chemistry. His researches were mostly in the synthesis and constitution of aliphatic compounds. In 1865 he discovered and synthesized isobutyric acid. He synthesized guanidine in 1868 and gave the first correct structural formulas of guanidine, creatine, and creatinine. He prepared several hydroxy acids and explained the formation and structure of the lactones derived from them in 1880. He synthesized tyrosine in 1883. Erlenmeyer invented the conical flask that bears his name (1861).

Erlenmeyer also dealt with many theoretical problems, and his remarks on valence and structure were fundamental to the development of these new ideas. He introduced the term "Strukturchemie" as well as the designations "monovalent," "divalent," and so on, which he employed in place of "monoatomic" and "diatomic."

Alexander Crum Brown in 1864 depicted the structures of organic compounds by drawing chemical bonds with dotted lines and enclosing the atomic symbols in circles. Chemists were hesitant to accept and use these graphic representations until Erlenmeyer in 1866 abandoned the old type formulas and adopted the new structural ones. By modifying Crum Brown's graphic formulas, he introduced the modern structural notation.

Another central problem in the new structural theory concerned the constitution of ethylene and other unsaturated compounds. Crum Brown suggested that their unique feature was the sharing of two valence units by each of two carbon atoms. Erlenmeyer not only adopted the double bond for ethylene but also introduced the triple bond to represent acetylene. His formulas, using lines to represent chemical bonds, proved convincing, and chemists adopted his notation.

Erlenmeyer investigated constitutional problems and proposed structural formulas for many organic substances. He immediately adopted Kekulé's ring structure for benzene and proposed the modern naphthalene formula of two benzene rings with two carbon atoms in common.

In 1880 he formulated what is known as the Erlenmeyer rule: All alcohols in which the hydroxyl group is attached directly to a double-bonded carbon atom become aldehydes or ketones. He had attempted to prepare such alcohols but obtained the isomeric carbonyl compounds in every case. Erlenmeyer concluded that such alcohols were incapable of existence, being converted at the instant of their formation into aldehydes or ketones by an intramolecular rearrangement.

BIBLIOGRAPHY

Erlenmeyer, with others, wrote the *Lehrbuch der organischen Chemie*, 3 vols. (Leipzig–Heidelberg, 1867–1894). He also wrote a small treatise, *Über den Einfluss des Freiherrn J. von Liebig auf die Entwicklung der reinen Chemie* (Munich, 1874), as a tribute to Liebig. His new graphic formulas, the triple bond, and the naphthalene structure are found in his "Studien über die s.g. aromatischen Säuren," in *Annalen der Chemie*, **137** (1866), 327–359; his rule on vinyl alcohols is in "Über Phenylbrommilchsäure," in *Berichte der Deutschen chemischen Gesellschaft*, **13** (1880), 305–310. There is a bibliography of his papers with a detailed account of his life and work by M. Conrad, *ibid.*, **43** (1910), 3645–3664.

William Henry Perkin wrote an interesting brief account of Erlenmeyer's work in *Journal of the Chemical Society*, **99** (1911), 1649–1651.

ALBERT B. COSTA

ERMAN, GEORG ADOLPH (*b.* Berlin, Germany, 12 May 1806; *d.* Berlin, 12 July 1877), *physics, meteorology, geophysics, geography, geology, paleontology.*

Erman was the son of Paul Erman, professor of physics at the University of Berlin. He himself earned the doctorate at that institution in 1826 with a dissertation in physics. In 1832 he became *Privatdozent* there and, in 1834, assistant professor of physics. He married Marie Bessel, daughter of the astronomer; one of their ten children, J. P. A. Erman, became a renowned Egyptologist.

In 1828 Erman accompanied a Norwegian expedition to Russia and Siberia. Leaving the expedition, he began a journey around the world that lasted until 1830 and took him from Kamchatka to San Francisco, Cape Horn, Rio de Janeiro, Portsmouth, and St. Petersburg. The primary purpose of the voyage, which he undertook on a Russian corvette, was geographic and geodesic surveying; Erman made altitude determinations, measurements of terrestrial magnetism, and meteorological observations and correlated these with the corresponding data that he had gathered in Russia and northern Asia. He also made numerous notes on natural history, general geography, ethnology, sociology, and economics; these he combined with an account of his travels in the first five volumes of his *Reise um die Erde* (1833–1848), the second section of which, in two volumes, contains his more purely scientific data.

From 1841 until 1867 Erman edited the *Archiv für wissenschaftliche Kunde von Russland,* a periodical designed to propagate Russian belles lettres, which in addition contained articles on science and the arts, as well as reports on economic and social conditions and events in Russia. Erman himself made many contributions to the journal, some original and some reportorial, on many topics. In particular, he wrote on the earth sciences, and his articles include surveys of the geology of European Russia (1841) and of northern Asia (1842), each illustrated with a geological map. He also wrote on the Tertiary of East Prussia, the Cretaceous of northern Spain, and on the mammalian remains from the Baumann cave in the Harz mountains.

BIBLIOGRAPHY

I. Original Works. Erman's major work is *Reise um die Erde durch Nord-Asien und die beiden Oceane in den Jahren 1828, 1829 und 1830,* 7 vols. (Berlin, 1833–1848). His articles include the series "Beiträge zur Klimatologie des Russischen Reiches," in *Archiv für wissenschaftliche Kunde von Russland,* **1** (1841), 562–579; **3** (1843), 365–438; **4** (1845), 617–640; **6** (1848), 441–488; **9** (1851), 33–130; **12** (1853), 645–665; "Über den dermaligen Zustand und die allmälige Entwickelung der geognostischen Kenntnisse vom Europäischen Russland," *ibid.,* **1** (1841), 59–108, 254–313, with a geological map; "Ueber die geognostischen Verhältnisse von Nord-Asien in Beziehung auf das Gold-Vorkommen in diesem Erdtheile," *ibid.,* **2** (1842), 522–556, 712–789, 808–809, with a geological map, and **3** (1843), 121–177, 185–186; "Bemerkungen über einem am Ural gebräuchlichen Seilbohrapparat," *ibid.,* **12** (1853), 333–357; and "Bemerkungen über ein bei den Jakuten und in Andalusien gebräuchliches Feuerzeug," *ibid.,* **19** (1860), 298–326.

II. Secondary Literature. On Erman and his work, see Poggendorff and *Neue deutsche Biographie,* IV (1959), 598–600; Wilhelm Erman, "Paul Erman. Ein Berliner Gelehrtenleben 1764–1851," in *Schriften des Vereins für die Geschichte Berlins,* **53** (1927), 104–105, 169–172, 187–192, 199–209, 216–222, and 227; a portrait is on p. 184.

Heinz Tobien

ERRERA, LÉO-ABRAM (*b.* Laeken, Belgium, 4 September 1858; *d.* Uccle, Belgium, 1 August 1905), *botany, biology, philosophy.*

The son of a distinguished Venetian banker, Giacomo Errera, and of Marie Oppenheim, who was of German origin, Errera must have spent his early years in an exceptional environment. His father, who was the Italian consul general in Belgium, was an ardent patriot who had fiercely defended Venice against the Austrians in 1849. His maternal grandfather was a revolutionary who fought for new political ideas in 1830 in Frankfurt. Errera was a brilliant student at the Faculty of Letters of the University of Brussels, where he received his baccalaureate. He continued his studies at the Faculty of Sciences of the same university, and was awarded the doctorate in 1879. After spending some time abroad, in 1884 Errera was named university lecturer in anatomy and plant physiology at the University of Brussels. In 1885, he married his first cousin, Rose-Eugénie May. He was elected in 1887 to the Royal Academy of Sciences, Letters, and Fine Arts of Belgium.

In 1894 he succeeded J. E. Bommer in the chair of general botany, a position that he held until his death in 1905 and that he enhanced by such achievements as the creation of the first botany laboratories for students. Among his many students those who were to become outstanding in Belgian biological science were Émile Laurent, Jean Massart, Émile Marchal, and Émile de Wildeman. Errera was, first and foremost, a remarkable teacher and was responsible for countless academic and pedagogical reforms. The first to publish (in 1897) wall charts for the teaching of plant physiology, Errera throughout his life displayed an interest in pedagogy, and his lecture entitled "The Utility of Superfluous Studies" shows clearly the direction of his didactic ideas.

In spite of his early death, Errera left a body of astonishingly varied scientific work. Although not a taxonomist, he undertook, while still very young, studies on the genus *Epilobium* and the phylogeny of the *Salix.* This work led to the publication of a paper entitled "Routines et progrès de la botanique systématique," in which he defined with clarity and foresight what was to become the taxonomy of today. In ethology Errera appears as a precursor because of his research work on the heterostylism of the *Primula*—revived by Jules MacLeod—on the fertilization of *Pentastemon* and *Geranium.* But it is in the domain of plant physiology that Errera's contribution is most striking. He first studied the alkaloids, precisely describing their microchemical characteristics and their localization. He was the first to point out the presence of glycogen, forming protoplasm, in the Ascomycetes. Subsequently he discovered this polysaccharide—which had previously been thought to exist only in animals—in a series of microorganisms. Again, in physiology, Errera was a pioneer in the area of physicochemical analysis. In fact, he was the first, making use of the remarkable work of Henri Devaux, to explain the arrangement of cellular walls by utilizing surface tension. Errera again touched upon biophysics when he investigated the mechanisms of the rising of sap and of the growth of the sporangiferous filaments of the *Phycomycetes*. Errera also took up the study of the transmission of acquired characteristics by the *Aspergillus* in its adaptation to concentrated solutions. Errera seems to have been one of the first biologists to undertake the study of life from a strictly physicochemical perspective, and the title of one of his posthumous works, *Cours de physiologie moléculaire,* is a good illustration.

Errera was a controversialist of the first order and became internationally known for his condemnation of anti-Semitism and for two courageous articles, "L'acte de tolérance" and "Six sermons sur les juifs." A warmhearted man and a poet in his leisure time (he left several collections of verse), Errera may be considered one of the most authentic humanists of the late nineteenth century.

BIBLIOGRAPHY

I. ORIGINAL WORKS. The edition *Recueil des oeuvres de Léo Errera* includes *Botanique générale I* (Brussels, 1908); *Mélanges, vers et prose* (Brussels, 1908); *Botanique générale II* (Brussels, 1909); *Physiologie générale, Philosophie* (Brussels, 1910); and *Pédagogie, Biographies* (Brussels, 1922).

II. SECONDARY LITERATURE. On Errera and his work see L. Frédericq and J. Massart, "Notice sur L. Errera," in *Annuaire. Académie royale de Belgique* (1908), 131–279; J. Massart, *Léo Errera,* Hayez, ed. (Brussels, 1905); and University of Brussels, Gutenberg, ed., *Commémoration Léo Errera* (Brussels, 1960).

P. E. PILET

ESCHER VON DER LINTH, HANS CONRAD (*b.* Zurich, Switzerland, 24 August 1767; *d.* Zurich, 9 March 1823), *geology, hydraulics.*

Escher came from an old Zurich family. His father was an administrator of the canton of Zurich and ran a prosperous textile factory. Escher, who had eleven brothers and sisters, took over his father's business in 1788, after traveling in France, England, Austria, and Italy, and studying for a year in Göttingen, Germany. In the following years he became involved in politics. His judgment, strength of character, and patriotism gained him responsible administrative positions in his native canton; and in 1798, during the French occupation, he was the head of the Great Council of Switzerland.

Escher began his most important work in hydraulics in 1803 on measures to control the devastation caused by the flooding of the Linth River. The Linth was at that time a rapid mountain river that flowed into the Lake of Zurich and caused heavy high-water damage all year long. According to Escher's plan the Linth River would be conducted into the neighboring Lake of Walen to the east and thereby be rendered harmless. The connection to the Lake of Zurich would be provided by an artificial canal. For this work, largely completed in 1811 and entirely finished in 1823—half a year after his death—Escher and his male descendants obtained the surname "von der Linth." In the last decade of his life Escher was again active in the politics and administration of the canton of Zurich. His only son, Arnold (1807–1872), was an important Alpine geologist.

Soon after he returned from his travels, Escher began the geological investigations of the Alps which occupied him for many years. As early as 1796 he published a geological survey of the Swiss Alps, which was later followed by a series of geological profiles from Zurich to the St. Gotthard Pass. In 1809, in the course of his wanderings in the upper Linth Valley, Escher made an observation that became of great importance for later conceptions of the geological structure of the Alps. He found an older "graywacke formation" (later known as the Permian Verrucano) that lies above the younger "Alpine Limestone Formation" (later known as the Jurassic *Lochseiten* limestone). Escher did not pursue the consequences of this inverse stratification—one reason was undoubtedly the sharp criticism of Leopold von Buch, who rejected his interpretation. Escher's view, that here occurs a tectonic phenomenon connected with the tectonic nappe structure widespread in the Swiss Alps, has long been confirmed, however.

A controversy between Escher and Buch developed on another point. Buch and many other geologists of the time found it difficult to accept water erosion as the fundamental cause of the formation of the great Alpine valleys. They attributed that process instead to ancient tectonic rifts and subsidence. Escher rejected this view, taking for his example the valley of Valais, the widest in the Alps. The direction of this valley does not follow the course and strike of the rocks in the valley walls, but rather intersects them at an angle of thirty to forty degrees. Escher attributed a major influence in the formation of valleys to erosion by rivers (1818).

He further recognized that the distribution of erratic boulders not of local origin in the northern foreland of the Alps corresponds to the watersheds of the great Alpine rivers—the Rhine, the Aare, the Reuss, and the Rhone (1822). Escher thought these boulders were transposed from their watershed in catastrophic floods—today we know that Pleistocene glaciers in the same valleys transported rocks characteristic of the substratum in their moraines into the northern foreland. The occasionally diverse rocks present in the watersheds of the glaciers were thereby distributed separately. Escher's observation, however, was fundamentally sound.

In his geological works Escher showed himself to be a precise, thorough, and critical observer who shied away from hypotheses and the propounding of theories. His modesty led him to publish only a few of his geological studies and investigations.

BIBLIOGRAPHY

I. ORIGINAL WORKS. Escher's works include *Geognostische Übersicht der Alpen in Helvetien* (Zurich, 1796); *Alpina,* 2 vols. (Zurich, 1806–1807); "Geognostische Beschreibung des Linthtales," in *Leonard's Taschenbuch für die gesamte Mineralogie,* **3** (1809), 339–354; "Über die geognostischen Verhältnisse der Gebirge der Linthtäler," *ibid.,* **6** (1812), 369–394; "Die Bildungsart der Täler betreffend," *ibid.,* **12** (1818), 199–221; and "Beiträge zur Naturgeschichte der freiliegenden Felsblöcke in der Nähe des Alpen-Gebirges," *ibid.,* **16** (1822), 631–676.

II. SECONDARY LITERATURE. On Escher and his work see H. Hölder, *Geologie und Paläontologie in Texten und ihrer Geschichte* (Freiburg im Breisgau, 1960), pp. 65–68, 73; J. J. Hottinger, *Hans Konrad Escher von der Linth, Charakterbild eines Republikaners* (Zurich, 1852); R. Lauterborn, "Der Rhein. Naturgeschichte eines deutschen Stromes," in *Berichte der Naturforschenden Gesellschaft zu Freiburg im Breisgau,* **33** (1934), 105–107; G. Meyer von Knonau, in *Allgemeine deutsche Biographie,* VI (Munich, 1877), 365–372; and R. Wolf, "H. C. Escher von der Linth," in *Biographien zur Kulturgeschichte der Schweiz,* IV (Zurich, 1862), 317–348.

HEINZ TOBIEN

ESCHERICH, THEODOR (*b.* Ansbach, Germany, 29 November 1857; *d.* Vienna, Austria, 15 February 1911), *pediatrics.*

Escherich was a pioneer pediatrician whose clinical insights and organizational abilities—linked to profound interests in bacteriology, immunology, and biochemistry—were devoted to improving child care, particularly infant hygiene and nutrition. He was born in a manufacturing town in Franconia, the younger son of *Kreismedizinalrat* Ferdinand Escherich, a medical statistician. His mother was Maria Sophie Frieder, daughter of Baron Carl Stromer von Reichenbach, a Bavarian army colonel. Because of his prankish tendencies in early schooldays, Escherich was sent to the great Jesuit seminary Stella Matutina, in Feldkirch, Austria. It apparently had no repressive effect upon him.

Escherich began his academic and medical education in 1876 at Strasbourg, continued at Kiel, Berlin, and Würzburg, and qualified at Munich in 1881. His doctoral dissertation was entitled "Die marantische Sinusthrombose bei Cholera infantum." In 1882 he joined the medical clinic of the Julius Hospital, Würzburg, becoming first assistant to its director, Karl Gerhardt, a well-known internist with an outstanding knowledge of pediatrics. His interest was thus aroused in this specialty, but since Germany lacked the necessary training facilities, Escherich had to seek them elsewhere, first in Paris and then in Vienna, where he worked for some months under Hermann Widerhofer at the St. Anna Children's Hospital. In 1885 he obtained clinical assistantships in Munich at the Children's Polyclinic of the Reisingerianum and at the Hauner Children's Hospital under Heinrich von Ranke. He habilitated himself at the University of Munich and became *Privatdozent* in pediatrics in 1886.

The increasing impact of Robert Koch's discoveries and his own experiences as scientific assistant in the 1884 cholera epidemic at Naples (to which Gerhardt had sent him) persuaded Escherich that bacteriology could solve or illuminate many pediatric problems. Circumstances at Munich fostered this belief. Koch's pupil, Wilhelm Frobenius, taught him pure culture techniques and methods of bacterial characterization; and he had access to Max von Pettenkofer's hygienic institute, Otto von Bollinger's bacteriological laboratory, Carl von Voit's physiological institute, and Franz von Soxhlet's dairy industry facilities. Escherich's work on cholera had drawn his attention to the bacterial flora of the intestine in infants, and after a further year of intensive laboratory investigations he published a monograph on the relationship of intestinal bacteria to the physiology of digestion in the infant. This work, *Die Darmbakterien des Säuglings und ihre Beziehungen zur Physiologie der Verdauung* (1886), established its author as the leading bacteriologist in the field of pediatrics. During the ensuing years he began studies of artificial nutrition, which led him to formulate a new system of prescribing cow's milk and to become a resolute advocate of breast-feeding for infants.

In 1890, when he was only thirty-three, Escherich was called to Graz to succeed Rudolf von Jaksch as extraordinary professor of pediatrics and director of the provincial children's clinic. Four years later he was promoted to ordinary professor; at that time he also refused a call to Leipzig as Otto Heubner's successor. His happiest years were spent in Graz, where he married Margaretha Pfaundler, daughter of the physicist Leopold Pfaundler. They had two children, a son and a daughter.

In Graz, Escherich instituted a broad program of clinical and laboratory researches and found scope for his organizational talents. He extended the diphtheria investigations that he had already launched in Munich and summarized the findings in two monographs, *Ätiologie und Pathogenese der epidemischen Diphtherie* (1892) and *Diphtherie, Croup, Serumtherapie* (1895). In 1890 he developed an interest in tetany of infants. He became the leading authority on this disease and in his final monograph, *Die Tetanie der Kinder* (1909), correctly ascribed it to parathyroid insufficiency. In 1891, one year after Koch discovered tuberculin, Escherich reported disappointing results in extensive trials of this product on tuberculous children. Thereafter problems of childhood tuberculosis remained among his chief concerns.

Escherich persuaded the Styrian government to build and maintain an infants' division as a branch of the provincial orphanage, attached to the children's clinic. He personally chose the furnishings and laboratory equipment for the expanded institution, designed the auditorium, founded a library, and established a diphtheria division for conducting bacteriological studies on suspected cases. The patient load trebled and a small provincial hospital was transformed into an important scientific and teaching institute.

When Widerhofer died in 1902, Escherich was appointed to his chair at Vienna. Although promised a new clinic, he himself had to draw plans, raise money, and negotiate with the government for this project, which was not completed during his lifetime. Meanwhile he renovated the venerable St. Anna Children's Hospital, making changes whenever he could extract sufficient funds from governmental and charitable sources. In 1903, determined to reduce the capital's infant mortality, Escherich appealed for support to the women of Vienna. The response was such that in the following year he established, with imperial patronage and civic approval, the Infants' Care Association (Verein Säuglingsschutz).

In the St. Anna Hospital, Escherich set up an infants' division and started an exemplary school for infant nursing. Medical students later received clinical instruction in this previously neglected field. The infant care headquarters on the hospital grounds became an educational center for mothers and a distributing point for cow's milk preparations and breast-feeding propaganda. In 1908, the year of the Emperor Franz Joseph's sixtieth jubilee, Escherich again drew attention to the inexcusably high national rate of infant mortality. His efforts resulted in eventual construction of the Imperial Institute for Maternal and Infant Care.

Escherich formed a pediatric section of the Viennese Society for Internal Medicine—serving indefatigably as chairman—and founded the Austrian Society for Child Research. In 1908 he was president of the German Pediatric Association. He coedited several well-known journals, held honorary membership in many foreign medical associations—including the American Pediatric Society—and was the only European pediatrician to address the International Congress of Arts and Sciences at the St. Louis World's Fair in 1904. He received the title *Hofrat* in 1906.

Throughout this decade in Vienna, Escherich worked relentlessly. Consulted professionally by royal families, he attended congresses abroad, published reports on various topics, and encouraged the researches of pupils such as Clemens von Pirquet and Béla Schick, who had accompanied him from Graz. Escherich's health deteriorated when his young son died from appendicitis. About five years later, in February 1911, a succession of cerebral attacks culminated in fatal apoplexy. The Kinderklinik, built to his plans, was officially dedicated soon afterward.

Idealism and progressiveness animated and gave purpose to Escherich's zealous, unbounded industry. Distinguished looking and always meticulously dressed, yet genial and approachable, he disliked intrigue or spiteful gossip. To young patients he showed affection, to older ones respectful candor. His stimulating lectures stressed pathology and reflected a preventive outlook.

More than one quarter of his publications relate to bacteriology. His earliest monograph (1886) included classic descriptions of *Bacterium coli commune* (later eponymously designated *Escherichia coli*) and *Bacterium coli aërogenes*. Although his claims that *B. coli* could cause cystitis and other localized infections were undisputed, his contention that some virulent strains provoked infantile diarrhea and gastroenteritis was verified only after sixty years. He narrowly missed discovering Sonne dysentery bacilli, of which he isolated several cultures, only to discard them because they failed to produce gas in carbohydrate-containing media.

In 1889, Escherich confirmed the causal role of the Klebs-Löffler bacillus in a diphtheria epidemic. He instituted antitoxin therapy in his clinic patients in 1894 and recorded exceptionally favorable results in a subsequent monograph. Through experiments on healthy children he showed the futility of attempting to prevent the disease by oral or rectal administration of antitoxin. In Vienna, he vigorously sponsored Paul Moser's antistreptococcus serum in scarlet fever treatment.

Escherich was intensely interested in the diagnosis,

pathogenesis, and control of tuberculosis. He pioneered in X-ray detection of the disease in children. In his last years he advocated construction of sanatoria, emphasized the tuberculous nature of scrofula, and reinvestigated the *Stichreaktion*—the swelling and redness at the site of subcutaneous injection—observed by him during early therapeutic trials of tuberculin. The diagnostic importance of this test was effectively realized only after von Pirquet modified the technique.

The scope of Escherich's clinical reports ranged from chorea to status lymphaticus, but his biochemical investigations focused on the physiology and pathology of infant nutrition. His breast-feeding edicts stemmed partly from recognition that improper milk formulas induced nutritional disorders, and partly from his observations that whereas healthy mother's milk was bacteriologically sterile, cow's milk might convey scarlet fever as well as intestinal infections. After disproving the prevailing dogma that cow's milk casein was indigestible and showing that bowel fermentation could be influenced by withholding dietary carbohydrates, Escherich evolved new dietary formulas for infants of various ages and weights, based on volumetric intakes by breast-fed counterparts. Characteristically, he devised a sterilizing apparatus to render the mixtures safe and arranged free distribution of the treated material to needy mothers.

Escherich's inspired common sense and technical ingenuity were directed into many channels. Thus his first building in Vienna, which housed separate laboratories for bacteriological, chemical, and X-ray activities, had a flat roof on which infants could lie or children play. The infants' wards were equipped with many novel facilities of his design, such as fully air-conditioned *couveuses* which were used either to protect especially vulnerable infants or to isolate infected ones. His acute social awareness and flair for innovation and coordination, combined with a bent for bacteriological and biochemical research, made Escherich the acknowledged leader of pediatrics in his day. Less versatility and longer life might have won him greater celebrity and more durable renown.

BIBLIOGRAPHY

I. ORIGINAL WORKS. The only bibliography of Escherich's works, listing about 170 items, is the inaccurate and incomplete one provided in an obituary by his former pupil and successor, Clemens von Pirquet (see below). Among his most important published works were four monographs, *Die Darmbakterien des Säuglings und ihre Beziehungen zur Physiologie der Verdauung* (Stuttgart, 1886); *Ätiologie und Pathogenese der epidemischen Diphtherie. I. Der Diphtherie-bacillus* (Vienna, 1892); *Diphtherie, Croup, Serumtherapie nach Beobachtungen an der Universitäts-Kinderklinik in Graz* (Vienna, 1895); and *Die Tetanie der Kinder* (Vienna, 1909). Escherich was also coauthor of a posthumous monograph, *Scharlach* (Vienna, 1912), written with Béla Schick; and of a long chapter on *Bacterium coli commune* in W. Kolle and A. Wassermann's *Handbuch der pathogenen Mikroorganismen,* II (Jena, 1903), 334–474, written with M. Pfaundler.

Many of his shorter contributions appeared in two or more journals. Among the more original and characteristic of these were "Die marantische Sinusthrombose bei Cholera infantum," in *Jahrbuch für Kinderheilkunde,* 19 (1883), 261–274; "Klinisch-therapeutische Beobachtungen aus der Cholera-Epidemie in Neapel," in *Münchener medizinische Wochenschrift,* 31 (1884), 561–564; "Bakteriologische Untersuchungen über Frauenmilch," in *Fortschritte der Medizin,* 3 (1885), 231–236; "Ueber Darmbakterien im allgemeinen und diejenigen der Säuglinge im Besonderen, sowie die Beziehungen der letzteren zur Aetiologie der Darmerkrangungen," in *Centralblatt für Bacteriologie,* 1 (1887), 705–713; "Die normale Milchverdauung des Säuglings," in *Jahrbuch für Kinderheilkunde,* 27 (1888), 100–112; "Zur Reform der künstlichen Säuglingsernährung," in *Wiener klinische Wochenschrift,* 2 (1889), 761–763; "Zur Aetiologie der Diphtherie," in *Centralblatt für Bakteriologie,* 7 (1890), 8–13; "Ueber Milchsterilisirung zum Zwecke der Säuglingsernährung mit Demonstration eines neuen Apparates," in *Berliner klinische Wochenschrift,* 27 (1890), 1029–1033; "Idiopathische Tetanie im Kindesalter," in *Wiener klinische Wochenschrift,* 3 (1890), 769–774; "Die Resultate der Koch'schen Injektionen bei Skrofulose und Tuberculose," in *Jahrbuch für Kinderheilkunde,* 33 (1891–1892), 369–426; "Ueber einen Schutzkörper im Blute der von Diphtherie geheilten Menschen," in *Centralblatt für Bakteriologie,* 13 (1893), 153–161, written with R. Klemensiewicz; "Bemerkungen über den Status lymphaticus der Kinder," in *Berliner klinische Wochenschrift,* 33 (1896), 645–650; "Begriff und Vorkommen der Tetanie im Kindesalter," *ibid.,* 34 (1897), 861–866; "Versuche zur Immunisirung gegen Diphtherie auf dem Wege des Verdauungstractes," in *Wiener klinische Wochenschrift,* 10 (1897), 799–801; "Kritische Stimmen zum gegenwärtigen Stande der Heilserumtherapie," in *Heilkunde* (Vienna), 2 (1897–1898), 593–606; "La valeur diagnostique de la radiographie chez les enfants," in *Revue mensuelle des maladies de l'enfance,* 16 (1898), 233–242; "Pyocyaneusinfectionen bei Säuglingen," in *Centralblatt für Bakteriologie,* 25 (1899), 117–120; "Zur Aetiologie der Dysenterie," *ibid.,* 26 (1899), 385–389; "Zur Kenntniss der Unterschiede zwischen der naturlichen und künstlichen Ernährung des Säuglings," in *Wiener klinische Wochenschrift,* 13 (1900), 1183–1186; "Die Erfolge der Serumbehandlung des Scharlach an der Universitäts-Kinderklinik in Wien," *ibid.,* 16 (1903), 663–668; *Bitte an die Wiener Frauen* [a pamphlet] (Vienna, 1903); "Die Grundlage und Ziele der modernen Kinderheilkunde," in *Wiener klinische Wochenschrift,* 17 (1904), 1025–1027; "Die neue Säuglingsabteilung im St.

Anna-Kinderspital in Wien," *ibid.,* **18** (1905), 977–982; "Antrag auf Einsetzung eines Komitees behufs Ausarbeitung von Vorschlägen zur Förderung der Brusternährung," *ibid.,* **18** (1905), 572–575; "Der Verein 'Säuglingsschutz' auf der hygienischen Ausstellung in der Rotunde 1906," *ibid.,* **19** (1906), 871–875; "Zur Kenntnis der tetanoiden Zustände des Kindesalters," in *Münchener medizinische Wochenschrift,* **54** (1907), 2073–2074; "Hermann Freiherr von Widerhofer 1832–1901," in *Wiener klinische Wochenschrift,* **20** (1907), 1510–1513; "Die Bedeutung des Schularztes in der Prophylaxe der Infectionskrankheiten," in *Monatsschrift für Gesundheitspflege,* **26** (1908), 117–130; "Was nennen wir Skrofulose?," in *Wiener klinische Wochenschrift,* **22** (1909), 224–228; "Die Infektionswege der Tuberkulose, insbesondere im Säuglingsalter," *ibid.,* 515–522; and "Ueber Indikationen und Erfolge der Tuberkulintherapie bei der kindlichen Tuberculose," *ibid.,* **23** (1910), 723–730.

II. Secondary Literature. Obituaries include "Theodor Escherich, M.D.," in *Boston Medical and Surgical Journal,* **164** (1911), 474–475; "Death of Professor Escherich," in *Lancet* (1911), **1,** 626; H. Finkelstein, "Theodor Escherich†," in *Deutsche medizinische Wochenschrift,* **37** (1911), 604–605; I. Fischer, "Escherich, Theodor," in *Biographisches Lexikon der hervorragenden Aerzte der letzten 50 Jarhe,* I (Berlin-Vienna, 1932), 375; F. Hamburger, "Theodor Escherich†" in *Wiener klinische Wochenschrift,* **24** (1911), 263–266; W. Katner, "Theodor Escherich," in *Neue deutsche Biographie,* IV (Berlin, 1959), 649–650; M. von Pfaundler, "Theodor Escherich†," in *Münchner medizinische Wochenschrift,* **58** (1911), 521–523; and C. von Pirquet, "Theodor Escherich," in *Zeitschrift für Kinderkrankheiten,* **1** (1910–1911), 423–441, which includes a bibliography—the same text, without bibliography, may be found in *Mittheilungen der Gesellschaft für innere Medizin und Kinderheilkunde, Beilage VIII,* **9** (1911), 82–93.

Other references to Escherich's life and work are A. Gronowicz, *Béla Schick and the World of Children* (New York, 1954); K. Kundratitz, "Professor Dr. Theodor Escherichs Leben und Wirken," in *Wiener klinische Wochenschrift,* **73** (1961), 722–725; Erna Lesky, *Die Wiener medizinische Schule im 19. Jahrhundert* (Graz, 1965); M. Neuberger, "Zur Geschichte der Wiener Kinderheilkunde," in *Wiener medizinische Wochenschrift,* **85** (1935), 197–203, trans. by R. Rosenthal as "The History of Pediatrics in Vienna," in *Medical Record,* **156** (1943), 746–751; B. Schick, "Pediatrics in Vienna at the Beginning of the Century," in *Journal of Pediatrics,* **50** (1957), 114–124; L. Schönbauer, *Das medizinisches Wien,* 2nd ed. (Vienna, 1947); and R. Wagner, *Clemens von Pirquet. His Life and Work* (Baltimore, 1968).

Claude E. Dolman

ESCHOLT, MIKKEL PEDERSÖN (*b. ca.* 1610; *d.* Christiania [now Oslo], Norway, 1669), *geology.*

The date and place of Escholt's birth are unknown; he first appears in the register of Copenhagen Uni-

versity for 1628 as coming from Malmö. He studied theology in Copenhagen and became chaplain of the castle at Akershus Castle in Christiania in 1646. He seems to have acted as an intelligence officer during the campaigns to reconquer the provinces lost by Denmark-Norway to Sweden by the treaty of Brömsebro (1645). In 1660 he was rewarded with the parish of Vestby in Östfold. When he died his oldest son inherited the parish.

His present scientific reputation stems from his book *Geologia norvegica* (Oslo, 1657). It is the first scientific treatise printed in Norway and also one of the first books printed in Norwegian. The book was written to calm the populace who felt doom approaching because of the slight but distinctly felt earthquake of 24 April 1657. The book gives a clear and surprisingly modern view of geological phenomena, with numerous apt references from both classic and recent literature. Escholt demonstrated the rather unusual regularity of the earthquakes (two each century) in the Oslo region and was aware of the relationship of earthquakes to volcanism. He was the first to use the word "geology" in the modern sense—as the science of the earth. Through an English translation of his book (1662), the word came into use in the scientific literature in the following decade.

His only other known works are brilliantly written but highly polemic theological papers. Escholt does not seem to have influenced or been in contact with contemporary scientists in Copenhagen, and he is barely mentioned in Garboe's exhaustive history of geology in Denmark.

BIBLIOGRAPHY

I. Original Works. Escholt's only scientific work is his *Geologia norvegica* (Oslo, 1657; English trans., 1662). A facsimile ed. was published in 1957.

II. Secondary Literature. See A. Garboe, *Geologiens historie i Danmark,* I (Copenhagen, 1959), 11, 47. A number of short notes, especially concerning Escholt's early use of the word "geology," have appeared in Scandinavian journals and newspapers.

Nils Spjeldnaes

ESCHSCHOLTZ, JOHANN FRIEDRICH (*b.* Dorpat, Russia [now Tartu, Estonian S.S.R.], 1 November 1793; *d.* Dorpat, 7 May 1831), *medicine, zoology.*

Eschscholtz received a medical education at Dorpat University. He took part, as a physician and naturalist, in voyages around the world on the brig *Rurik* from 1815 to 1818, under the command of Captain

O. Kotzebue. His collections, which he made together with A. Chamisso, were given to Dorpat University and the Moscow Society of Naturalists. From 1819 he was extraordinary professor of medicine and dissector at Dorpat University, where from 1822 he was director of the zoological cabinet and, from 1828, ordinary professor of anatomy. He was a member of the Moscow Society of Naturalists, the Deutsche Akademie der Naturforscher Leopoldina and the Swiss Society of Natural Science. A bay in Alaska, an atoll in the Marshall Islands, a genus of plants of the family Papaveraceae, and a genus of ctenophora are named in honor of Eschscholtz.

BIBLIOGRAPHY

I. ORIGINAL WORKS. Eschscholtz' writings include *Ideen zur Aneinanderreihung der rückgratigen Thiere auf vergleichenden Anatomie begründet* (Dorpat, 1819); *Beschreibung des inneren Skelets einiger Insekten (Gryllotalpa)* (Dorpat, 1820); *Species insectorum novae descriptae (Carabicini)* (Moscow, 1823); *Entomographien* (Berlin, 1824); "Dissertatio de coleopterorum genere Passalus," in *Nouveaux Mémoires de la Société Impériale des Naturalistes de Moscou,* **1** (1829), 13–28; *Zoologisches Atlas enthaltend Abbildungen und Beschreibungen neuer Thierarten, während des Flottenkapitains von Kotzebue zweiter Reise um die Welt in den Jahren 1823–1826 beobachtet* (Berlin, 1829–1833); "Übersicht der zoologischen Ausbeute," in *Reise um die Welt in den Jahren 1823–1826* (Weimar, 1830); and *Beschreibung der Anchinia, einer neuen Gattung der Mollusken* (St. Petersburg, 1835).

II. SECONDARY LITERATURE. On Eschscholtz and his work, see J. F. Recke and C. E. Napiersky, *Allgemeine Schriftsteller- und Gelehrten-Lexicon der Provinzen Livland, Esthland und Kurland* (Mitau, 1827), p. 523; and T. Beise and C. E. Napiersky, *Nachträge und Fortsetzungen* (1859), p. 173.

L. J. BLACHER

ESCLANGON, ERNEST BENJAMIN (*b.* Mison, France, 17 March 1876; *d.* Eyrenville, France, 28 January 1954), *astronomy, mathematics, physics.*

Esclangon came from a family of landed proprietors. The practical attitudes of his class are apparent in his realistic approach to problems in pure mathematics, applied celestial mechanics, relativity, observational astronomy, instrumental astronomy, astronomical chronometry, aerodynamics, interior and exterior ballistics, and aerial and underwater acoustic detection. He contributed to all of these fields to a greater or lesser degree, but always effectively.

Esclangon's first training was as a mathematician. As a student at the École Normale Supérieure (1895–1898) and an *agrégé* in mathematics (1898), he took up the problem of quasi-periodic functions. (Quasi-periodic functions, newly introduced, constitute a remarkable class among the almost periodic functions; their Fourier expansion is formed by a limited number of terms.) Esclangon elaborated a theory for these functions, studied their differentiation and integration, and examined the differential equations which allow them as coefficients. His doctoral thesis established a basis for their employment at a time when their role in mathematical physics was only beginning to be developed.

Esclangon's subsequent career as an astronomer and teacher was the result of chance—in the form of a vacant position—and of his own curiosity that led him to accept it. He was an astronomer at Bordeaux, beginning in 1899, then director of the observatories of Strasbourg (1918) and Paris (1929–1944). In addition, he taught mathematics at the Bordeaux Faculty of Science (from 1902), then became professor of astronomy at Strasbourg (in 1919) and then at Paris (1930–1946).

For fifty years Esclangon explored all the branches of fundamental astronomy. He devoted special attention to perfecting instruments, with a view to increasing the precision of observations. Of particular interest is his solution to a critical problem in positional astronomy, the rigorous definition of the axis of rotation of a transit instrument. Esclangon demonstrated that by fitting an objective in one of the extremities of this axis, which is hollow, and fitting a reticle to the other, the observer is permitted to measure the displacement of the instantaneous axis of rotation continuously throughout the course of the observation.

Esclangon's work in ballistics began in 1914 when, at the beginning of World War I, he proposed to French military authorities that they employ sound-ranging techniques to localize enemy artillery. He was charged with organizing the experimental study of this method; he was thus able to analyze the two components of the wave emitted by the projectile, the conical shock wave and the spherical wave centered on the point of emission. Esclangon then succeeded in 1916 in eliminating the registration of the shock wave and thereby assured a great precision in pinpointing enemy gun locations.

As director of the Bureau International de l'Heure (1929–1944), Esclangon was led to devote himself to problems of time. In addition to making studies on the astronomical determination of time and on its conservation and diffusion, he devised the "talking clock" (employing time signals from an observatory clock) that has made telephonic announcements of the exact time available to the Paris public since 1933.

Esclangon's practical bias and his inclination toward solid demonstrations (whether mathematical or experimental) caused him to be critical of the general theory of relativity. In a memoir of 1937, "La notion de temps. Temps physique et relativité . . .," he discusses the restrictions necessary to certain conclusions that have been stated too absolutely and states how, for example, it is possible to conceive of phenomena faster than light and why the ordinary formulas are not strictly applicable to the motion of masses at great speeds.

Esclangon was a member of the Académie des Sciences (1939) and the Bureau des Longitudes (1932). He served as president of the Union Astronomique Internationale from 1935 to 1938. He assumed his official functions with simplicity and amiability; he was affable and loved to joke, and did not deny himself leisure time. It would almost seem that he accomplished his body of important work without effort.

BIBLIOGRAPHY

I. Original Works. Esclangon published 247 memoirs, monographs, and articles, of which some of the most important are, in mathematical analysis, "Les fonctions quasi-périodiques," his doctoral thesis, in *Annales de l'Observatoire de Bordeaux,* **11** (1904), 1–276; and "Nouvelles recherches sur les fonctions quasi-périodiques," *ibid.,* **16** (1917), 51–176.

His astronomical works include "Sur les transformations de la comète Daniel . . .," in *Bulletin astronomique,* **25** (1908), 81–91; "Mémoire sur la réfraction astronomique," in *Bulletin de Comité international permanent pour l'exécution photographique de la carte du ciel,* **6** (1913), 319–389; "Sur la précision des observations méridiennes et des mesures de longitudes," in *Annales de l'Observatoire de Strasbourg,* **1** (1926), 373–405; "Mémoire sur l'amélioration des observations méridiennes," in *Bulletin astronomique,* **6** (1930), 229–260; "L'horloge parlante de l'Observatoire de Paris," in *L'astronomie,* **47** (1933), 145–155; "Horloges indiquant simultanément le temps moyen et le temps sidéral," in *Bulletin astronomique,* **11** (1938), 181–189; and "Sur la transformation en satellites permanents de la terre de projectiles auto-propulsés," in *Comptes rendus hebdomadaires des séances de l'Académie des sciences,* **225** (1947), 513–515.

In theoretical physics, Esclangon wrote "Mémoire sur les preuves astronomiques de la relativité," in *Bulletin astronomique,* **1** (1920), 303–329; and "La notion de temps. Temps physique et relativité . . .," *ibid.,* **10** (1937), 1–72.

His work in applied physics includes "Le vol plané sans force motrice," in *Comptes rendus hebdomadaires des séances de l'Académie des sciences,* **147** (1908), 496–498; "Sur un régulateur rotatif de vitesse," *ibid.,* **152** (1911), 32–35; "Sur un régulateur thermique de précision," *ibid.,* **154** (1912), 178–181, 495–497; "Sur un nouveau régulateur

de température . . .," *ibid.,* **156** (1913), 1667–1670; "Mémoire sur l'intensité de la pesanteur," in *Annales de l'Observatoire de Bordeaux,* **15** (1915), 99–314; and "Le vol plané sans force motrice," in *Comptes rendus hebdomadaires des séances de l'Académie des sciences,* **177** (1923), 1102–1104.

His publications in military science comprise *Mémoire sur la détection sous-marine . . .* (Paris, 1918), in the Archives de la Marine de Guerre; and *L'acoustique des canons et des projectiles* (Paris, 1925).

II. Secondary Literature. On Esclangon and his work, see J. Chazy, "Notice nécrologique sur Ernest Esclangon," in *Comptes rendus hebdomadaires des séances de l'Académie des sciences,* **238** (1954), 629–632; and "Ernest Esclangon (1876–1954)," in *Annuaire du Bureau des longitudes* (1955), C1–C6; A. Danjon, "Obituary Notice: Ernest Esclangon," in *Monthly Notices of the Royal Astronomical Society,* **115** (1955), 124; J. Jackson, "Obituaries: Prof. E. Esclangon," in *Nature,* **173** (1954), 567; and A. Pérard, "Quelques mots de l'oeuvre scientifique d'Ernest Esclangon," in *L'astronomie,* **68** (1954), 201–204.

Jacques R. Lévy

ESKOLA, PENTTI ELIAS (*b.* Lellainen, Honkilahti, Finland, 8 January 1883; *d.* Helsinki, Finland, 6 December 1964), *petrology, mineralogy, geology.*

The son of a farmer, Eskola enrolled at the University of Helsinki in 1901 and in 1906 obtained his candidate degree (about equivalent to an M.S.) in chemistry. He obtained his Ph.D. at the same university in 1914 with a dissertation entitled "On the Petrology of the Orijärvi Region in Southwestern Finland." By 1915 he had embarked on his lifework, the study of the mineral facies of rocks. During stays in Norway and the United States in 1920–1921 he worked specifically on eclogites; he spent 1922–1924 as a geologist of the Finnish Survey; and in 1924 he was named extraordinary professor and in 1928 ordinary professor of geology at the University of Helsinki, a position he held until 1953. Eskola was one of the generation of petrologists confronted by the complexities of the Fennoscandian crystalline complex who were inspired by a famous paper of J. J. Sederholm on granites and gneiss to develop a structural-metasomatic school of petrology, influenced also by the application to petrology of physical chemistry by J. H. L. and T. Vogt.

Eskola was a leading petrologist in these and other subjects (mentioned below), within his own field. He held honorary degrees from the universities of Oslo (1938), Padua (1942), Bonn (1943), and Prague (1948). He was honorary president of the Geological Society of Finland and an honorary or corresponding member of many learned societies and academies of science. In 1964 he received the Vetlesen Prize.

Eskola married Mandi Wiiro in 1914. They had two children; a son, Matti, born in 1916 and killed in World War II in 1941; and a daughter, Päivätär, born in 1920, who became a teacher of chemistry. Eskola and his wife were known for their warmth and cooperation in the scientific and personal care of their students, many of whom held important positions in Finland and abroad.

The success of Eskola's scientific work was probably based on the coincidence of three major factors: his thoroughness and steadiness, evident in all his writings and his notable care in rewriting manuscripts; the breadth of his topical experience, including his degree in chemistry, displayed not only in his specific papers but also, especially, in his various textbooks in geology and mineralogy, some of which were written in German; and his constant striving to combine laboratory results with field data, evident in his work in chemistry, his stay at the Geophysical Laboratory in Washington from March 1921 to November 1922, his own fieldwork, and his published work.

Eskola's major contribution to the earth sciences and an idea that he developed throughout his life—indeed, from his first work on solid-state reactions in 1904 to his last years—was the concept of mineral facies, essentially a continuation of Ulrich Grubenmann's assignment of metamorphic rocks to epi-, meso-, and katazones. In 1914 he wrote on p. 114 of his Ph.D. thesis a definition that is still applicable today: "In any rock of a metamorphic formation which has arrived at a chemical equilibrium through metamorphism at constant temperature and pressure conditions, the mineral composition is controlled only by the chemical composition." Originally he differentiated between five separate facies, stressing their independence of mode of formation. He named them, according to a mineral typical of each, the sanidine, hornfels, greenschist, amphibolite, and eclogite facies. By 1939 in *Die Entstehung der Gesteine* this nomenclature had evolved into a two-dimensional temperature-pressure classification further differentiated into metamorphic and magmatic facies. An early summary of this basic idea was published in 1920 under the title "On Mineral Facies of Rocks" (*Norsk geologisk tidsskrift,* **6,** 143–194).

Through his own work and that of many other colleagues—specifically Paul Niggli, V. M. Goldschmidt, T. F. W. Barth, F. J. Turner, N. L. Bowen, and H. Yoder—the equilibrium boundaries and stability fields were increasingly specified and modified, until after 1955 an intensive search for new metamorphic facies standards began on a grand scale. As a corollary, Eskola centered repeated efforts on the highest-grade facies, the eclogite problem. An introductory account is his paper "On the Eclogites of Norway" (*Skrifter utg. af Videnskabsselskabet i Kristiania,* **1,** no. 8 [1921]). The modern concept that deeper layers of the earth do not necessarily differ in composition but rather in the density of their minerals largely originated in Eskola's high-pressure facies idea.

As in his early work on the petrology of Orijärvi, he often interpreted the composition of his mineral facies as the result of metasomatism, specifically of a "replacement of lime, soda, and potash by iron oxides and magnesia." Eskola originally considered the intrusion of granites to be principally responsible for such ionic migrations but was open to a later interpretation by some of his students who derived the same elements from a process of metamorphic differentiations and tectonic energies. A syngenetic interpretation (*in situ* formation) favored more recently by some ore geneticists seems not to have been considered during Eskola's lifetime.

The problem of the origin of metamorphic rocks is indigenous to the geology of Finland. Likewise, the granite problem is a typical Fennoscandian study, and it consequently received almost as much attention from Eskola as the metamorphic rock enigmas. As early as 1932 he summarized his ideas on granites in a paper entitled "On the Origin of Granitic Magmas." He recognized various possible origins, including anatectic processes by which a "pore magma" may form and migrate to produce migmatites. In his later years he agreed that palingenesis or anatexis may have played a more important role than was recognized at first; but he did not share the extreme interpretations of some transformist schools.

Because the problems mentioned so far could be investigated largely through work in Scandinavia, Eskola's contributions to the advancement of Fennoscandian geology were numerous. He was particularly active in the interpretation of pre-Cambrian stratigraphy; for example, the term "Karelian" goes back to his work. In connection with this work, and as a man with a thorough philosophical mind, in 1954 he also presented a book on the possible cosmogenic origin of the earth and of life (*In Quest of a Picture of the World*).

BIBLIOGRAPHY

I. ORIGINAL WORKS. From a list of about 170 original publications (articles, books, and monographs), the following perhaps best represent Eskola's lifework: "The Silicates of Strontium and Barium," in *American Journal of Science,* **4** (1922), 331–375; "On the Origin of Granitic Magmas," in *Mineralogische und petrographische Mit-*

teilungen, n.s. **42** (1932), 455–481; "Wie ist die Anordnung der äusseren Erdsphären nach der Dichte zustande gekommen?," in *Geologische Rundschau,* **27** (1936), 61–73; "Die metamorphen Gesteine," in T. F. W. Barth, Carl W. Correns, and Pentti Eskola, *Die Entstehung der Gesteine; Ein Lehrbuch der Petrogenese* (Berlin, 1939; repr., 1960), pp. 263–407; "Einführung, 'Finnlandheft der geologischen Rundschau,'" in *Geologische Rundschau,* **32** (1941), 401–414; "Kern und Schichten der Erde," in *Sitzungsberichte der Finnischen Akademie der Wissenschaften* (1945), 218–228 (Finnish ed. 1946); *Kristalle und Gesteine. Ein Lehrbuch der Kristallkunde und allgemeinen Mineralogie* (Finnish ed. 1939; repr. Vienna, 1946); "About the Granite Problem and Some Masters of the Study of Granite," in *Comptes rendus de la Sociéte géologique de Finlande,* **28** (1955), 117–130, also in *Bulletin de la Commission géologique de la Finlande,* **168** (1955), 117–130; "On the Mineral Facies of Charnockites," in *Journal of the Madras University,* **27** (1957), 101–119; and "Granitentstehung bei Orogenese und Epirogenese," in *Geologische Rundschau,* **50** (1960), 105–113.

II. SECONDARY LITERATURE. On Eskola and his work, see T. F. W. Barth, "Memorial to Pentti Eskola (1883–1964)," in *Bulletin of the Geological Society of America,* **76,** no. 9, 117–120; Vladi Marmo, "Pentti Eskola," in *Bulletin de la Commission géologique de la Finlande,* **218** (1965), 20–53; Toini Mikkola, "Memorial of Pentti Eskola," in *American Mineralogist,* **53,** 544–548; and T. G. Sahama, "Pentti Eskola," memorial address given in Helsinki, 10 November 1965.

G. C. AMSTUTZ

ESPY, JAMES POLLARD (*b.* Washington County, Pennsylvania, 9 May 1785; *d.* Cincinnati, Ohio, 24 January 1860), *meteorology.*

Espy was educated at Transylvania University in Lexington, Kentucky, and taught school before embarking upon a full-time career as a meteorologist in the mid-1830's. He did his earliest known work in the field in 1825 while teaching at the Franklin Institute in Philadelphia, his interest stemming from the writings of Dalton and Daniell.

The most common kind of meteorological activity in the antebellum United States was the gathering of observations. From these observations, physical explanations were sometimes deduced (e.g., William Redfield) or the data were analyzed mathematically in some fashion (e.g., Elias Loomis). From such roots arose movements to develop networks of observers and regular systems for processing the resulting data. Espy participated in this tradition in the founding of a system of meteorological observations in Pennsylvania in 1836 and in his labors (*ca.* 1840–1852) to erect a national system of volunteer weather observers which was supplanted by Joseph Henry's telegraph-linked corps of observers.

Espy's principal significance in the history of meteorology arises from a less typical kind of research for his time and place. By direct experimentation, he tried to derive physical concepts supported by quantitative data. Others might talk of the role of atmospheric electricity; Espy flew giant kites.

Espy's most notable experimental work centered on heat effects. He devised an instrument, the "nephelescope," to simulate, as it were, the behavior of clouds and, particularly, to measure the dry and moist adiabatic cooling rates. While the resulting data varied from the correct values, Espy displayed great physical insight in deducing the role of latent heat in cloud formation and rainfall. He was, apparently, the first to point out that the latent heat released by condensation of the vapor in clouds resulted in a considerable expansion of the air, the latent heat, therefore, providing the energy for continued rain and upward movement of the cloud.

As the concept of the saturated adiabatic expansion of rising air currents is basic to meteorology, Espy clearly merits recognition as an important pioneer. Lacking any sophisticated mathematical apparatus or a knowledge of modern thermodynamics and other factors involved in cloud dynamics, Espy's work did not lead directly to the work of Kelvin and others from which the modern theory stems.

It is possible, however, that his enthusiastic proselytizing for his views helped pave the way for the acceptance of the later work. In 1840 Espy addressed the Glasgow meeting of the British Association for the Advancement of Science. An account of his theories sent to the French Academy was favorably reviewed in the *Comptes rendus* in 1841 by a committee whose members were D. F. J. Arago, C. G. M. Pouillet, and J. Babinet. Espy lectured widely in the United States, undoubtedly deserving credit for stirring up popular interest and support for meteorology.

Working against recognition of his theories, however, especially their very real contributions, was this same quality of enthusiastic commitment. Espy was most contentious and not always receptive to criticism. Time would prove W. C. Redfield, with whom Espy was involved in a controversy, correct on the motion of storms; from his results Espy deduced spectacular conclusions, some unconvincing or apparently refutable. (His suggested burning of forests to produce rainfall was disregarded by narrow minds immune to the need for controlled experiments.)

In short, because of his aprofessional behavior the emerging community of professional scientists was inclined to overlook his real contributions, which have been rediscovered periodically by historically inclined meteorologists.

BIBLIOGRAPHY

I. Original Works. A satisfactory list of Espy's articles is in the *Royal Society Catalogue of Scientific Papers,* II, 522–523. For the full flavor of the man and his ideas, it is necessary to consult his monographic works. The best known is *The Philosophy of Storms* (Boston, 1841). Less known but also quite valuable is *Report on Meteorology,* 4 vols. (Washington, D.C., 1843–1857), submitted by Espy in his anomalous role as the national meteorologist.

Friends of his succeeded in attaching riders to bills authorizing funds for Espy's work. None of the executive establishments requested this work, and as a result these reports were issued separately under rather odd circumstances. The first (1843) is a report to the surgeon general of the army and is fairly rare. The second (1850) and third (1851) are reports to the secretary of the navy. Both are nos. 559 and 560 of the congressional series (Senate executive document no. 30, 31st Congress, 1st session); their being bound together produces problems in determining the dates of the two reports. The fourth and last report was simply submitted by the president to Congress in 1857, but most of the work is of an earlier date. It too is in the congressional series (no. 889) as Senate executive document no. 65, 34th Congress, 3rd session. (For a modern comment on the four *Reports,* see *Meteorological Abstracts* [February 1955], p. 143.)

MS sources for Espy may be found in the archives of the American Philosophical Society and the Franklin Institute, both in Philadelphia. There is much on Espy's activities in the papers of his two leading American contemporaries in meteorology, Elias Loomis and W. C. Redfield, in the Beinecke Library, Yale University, New Haven. The personal papers of Joseph Henry and the archives of the Smithsonian Institution in Washington, D.C., during Henry's secretariat have documents relating to Espy and on the development of American work in meteorology.

II. Secondary Literature. The earliest authoritative statement on Espy is the report in the *Comptes rendus hebdomadaires de l'Académie des sciences,* **12** (1841), 454–462, referred to above. After Espy's death, Alexander Dallas Bache, an old friend, wrote a necrology which can stand as a good summary of how Espy was regarded by the professional scientists. It appears in the Smithsonian Institution *Annual Report* for 1859 (Washington, 1860), pp. 108–111. In 1894 William Morris Davis, a fine geographer and geologist, presented a rather superficial view of Espy's period in American meteorological work in "The Redfield and Espy Period," a paper presented at the International Meteorological Congress, Chicago, 21–24 August 1894 (U.S. Weather Bureau, *Bulletin* no. 11, pp. 305–316).

Espy is briefly referred to in W. J. Humphreys, "A Review of Papers on Meteorology and Climatology Published by the American Philosophical Society Prior to the Twentieth Century," in *Proceedings of the American Philosophical Society,* **86** (Sept. 1942), 29–33.

Part of the modest revival of interest in Espy in recent years is undoubtedly the result of his appearance in K. Schneider-Carius, *Wetterkunde Wetterforschung, Geschichte ihrere Probleme und Erkenntnisse in Dokumenten aus drei Jahrtausenden* (Munich, 1956), pp. 192–196. Espy is discussed and some of his unpublished documents are printed in Nathan Reingold, *Science in Nineteenth Century America: A Documentary History* (New York, 1964), pp. 92–107, 128–134.

J. E. McDonald, "James Espy and the Beginnings of Cloud Thermodynamics," in *Bulletin of the American Meteorological Society,* **44** (1963), 633–641, is an important recent appraisal. W. E. Knowles Middleton, *A History of the Theories of Rain and Other Forms of Precipitation* (London, 1965), pp. 155–160, is the most recent retrospective appraisal of Espy.

Nathan Reingold

ESSON, WILLIAM (*b.* Carnoustie, Scotland, 17 May 1839 [perhaps 1838]; *d.* Oxford, England, 25 August 1916), *chemistry.*

Esson's father, also William Esson, was a bridge-building engineer, and consequently the family often moved. Esson was an only child. He attended the Royal Academy in Inverness and the grammar school in Cheltenham, then studied mathematics at St. John's College in Oxford. In 1860 he was elected a fellow of Merton College, later becoming a tutor of Merton, where he was a lecturer in mathematics. His teaching was so successful that other colleges sent their students to hear him. In 1872 he married Elisabeth Meek, a pastor's daughter; they had two sons and one daughter. Esson became a deputy professor in 1894 and in 1897 obtained a professorship of geometry. Between 1898 and 1913 he was chairman of the board of the faculty of natural science. He had been a fellow of the Royal Society since 1869. He was a passionate mountain climber and a member of an alpine club.

Although Esson was a mathematician and a professor of geometry, he published almost nothing in his own field; he contributed greatly, however, to the employment of higher mathematics in chemistry (this science had so far been satisfied with the application of arithmetic). Esson worked in his youth as a chemistry demonstrator under Vernon Harcourt, a professor of chemistry at Oxford, and the collaboration between the two researchers continued after Esson himself rose to the rank of professor. Together they considered many problems of chemical kinetics.

In 1864 they investigated the reaction of potassium permanganate with oxalic acid and, in the course of their study, nearly succeeded in formulating the law of mass action ("On the Laws of Connexion Between the Conditions of a Chemical Change and Its Amount," in *Proceedings of the Royal Society,* **14** [1865], 470–475). (The law was formulated at the same time, but in a simpler and more general form, by the

Norwegians C. M. Guldberg and Peter Waage, who are generally credited with it.)

By examining the reaction process, Esson and Harcourt reached the conclusion that "in unit volume of a dilute solution at constant temperature the rate of chemical change varies directly with the mass of each of the interacting substances," or "the velocity of chemical change is directly proportional to the quantity of substance undergoing change." They presented the law of reaction velocity in exponential form: $y = ae^{-\alpha x}$, where x is the time, a the initial concentration and α a constant. Or, in the case of a binary reaction where $u = ae^{-(\alpha+\beta)x}$, it follows that it takes an infinite time for the chemical reaction to go to completion.

In a later article ("On the Observation of the Course of Chemical Change," in *Philosophical Transactions of the Royal Society,* **157** [1867], 117–137), Esson and Harcourt discussed the reaction between hydrogen peroxide and hydroiodic acid. They established that "in the presence of a large excess of iodide, the reaction is of the first order in respect of the hydrogen peroxide." In studying the effect of temperature on the reaction they found that K and K_0 (the velocity coefficient at the two absolute temperatures T and T_0) are related to the temperatures

$$\frac{k}{k_0} = \left(\frac{T}{T_0}\right)^m.$$

To be sure, Esson's formulations were rather too complicated for his contemporaries. Their merits were not recognized until the laws of chemical kinetics were fully stated on the basis of other considerations and the work of many other researchers.

BIBLIOGRAPHY

I. ORIGINAL WORKS. Besides the articles in the text, Esson's writings, all with Harcourt, include "Characters of Plane Curves, a Law of Connexion Between Two Phenomena Which Influence Each Other," in *International Congress of Mathematicians* (1912); and "On the Variation With Temperature of the Rate of a Chemical Change," in *Philosophical Transactions of the Royal Society,* **212** (1913), 187–204, a summary of their work in this field. A more detailed version of the 1865 paper is in *Philosophical Transactions of the Royal Society,* **156** (1866), 193–221. It appears under Harcourt's name only, but Esson's participation is mentioned in the text.

II. SECONDARY LITERATURE. See the article "William Esson," in *Proceedings of the Royal Society,* **93A** (1917), 54; and J. R. Partington, *A History of Chemistry,* IV (London, 1964), 585–587.

F. SZABADVÁRY

ESTIENNE (STEPHANUS), CHARLES (*b.* Paris, France, *ca.* 1505; *d.* Paris, 1564), *anatomy, natural history, scientific publication.*

The son of Henri I Estienne and younger brother of François and Robert I, Charles belonged to the famous dynasty of Parisian printers and publishers. His father died in 1520, and his mother married Simon de Colines, another printer, who was later to publish Estienne's anatomical atlas. After learning Greek under Jean Lascaris, he extended his knowledge of classical philology at the University of Padua. While in Italy (1530–1534) he became interested in botany, horticulture, and medicine. After returning to Paris he followed the extracurricular courses (*cours libres*) in anatomy and medicine given by Jacques Dubois (Sylvius) at the Collège de Tréguier. His literary activity started in 1535 with three abstracts based on the works of the diplomat Lazare de Baïf. Estienne then published several treatises on gardening and the names of plants and birds. *De re hortensi libellus* (1535) and *Seminarium* (1536) were favorably received and republished. In these books Estienne showed himself to be a scholar of importance but a mediocre naturalist in sacrificing observation to history and philology.

In 1536 Charles Estienne had printed by his brother a short treatise entitled *Anatomia.* (This volume is listed in the Estienne firm's catalogues, but no copy seems to have survived.) Around 1538 Estienne married Geneviève de Verley, daughter of Gilles de Verley, surgeon to the king. Estienne was at the time studying medicine but without being formally registered at the Faculté de Médecine. However, the faculty recognized his diligence, and in 1540 he received the degree of bachelor; on 20 June 1542 he was promoted to *docteur régent.* From then on he seems to have devoted himself to the practice of medicine. He taught anatomy from 1544 to 1547 at the Faculté de Médecine in Paris, where he was given the title of *lector ordinarius.* During this period the Latin and French editions of his manual of anatomy were published, although much of this work had been drafted before 1539. In November 1550 Estienne brought out a treatise on diet and a classification of foods, dedicated to the inquisitor Guillaume de Bailly. (This dedication is more easily understood in light of the dangers that threatened the Estienne publishing house; earlier that year Robert, accused of Protestantism, had been obliged to seek refuge in Geneva.) Charles Estienne then gave up his medical practice in order to manage the family business. He printed Belon's zoological treatise, and edited several dictionaries, geographical guides, and works in classical philology, grammar, and history. He translated into

French the treatise of Vegetius on the diseases of horses and also wrote a rural encyclopedia (*Praedium rusticum,* 1554), which when revised by his son-in-law, Jean Liébault, had tremendous success in the book trade. Nevertheless, Estienne was a poor businessman. Accused of having squandered the inheritance of Robert's children and heavily in debt, he was imprisoned in 1561 at the Châtelet, where he spent the last four years of his life.

Estienne's main scientific work, *De dissectione partium corporis humani* (1545), poses a particular problem: although it was published two years after the *Fabrica* of Vesalius, it antedates it in actual composition. Estienne worked it out in collaboration with the surgeon Estienne de la Rivière (Riverius), who had probably done some dissection and also helped in preparing the plates. Estienne first used woodcuts executed by Jollat and stored in his father-in-law's office (four plates are dated from 1530 to 1532); he then used various erotic engravings—among others—and inserted a special section that included internal anatomical details; finally (from 1534 to 1539), he had some original anatomical engravings made. In 1539 Estienne de la Rivière lodged a complaint at the Parlement against Estienne's claiming his rights as author. The lawsuit delayed the publication of the work, two-thirds of which had already been printed and was subsequently submitted in 1541 to the Faculté de Médecine for approval.

In the *De dissectione,* Estienne stated at the outset the principle of the new anatomical method: "One should not believe in books on anatomy but far more in one's own eyes." The book's many original observations include the morphology and physiological significance of the "feeding holes" of bones, the cartilaginous meniscus of the temporomandibular joint, the orbicular ligament of the radius, the three-part composition of the sternum, the path of the trigeminal and phrenic nerves, a sharp distinction between the sympathetic chain (considered as a nerve) and the vagus, the canal of the ependyma and the enlargements of the spinal cord, the cerebrospinal fluid, the valvulae in the hepatic veins, and the scrotal septum. Estienne also described the ideal anatomical theater and expounded the technique of dissecting cadavers and wiring skeletons.

BIBLIOGRAPHY

I. ORIGINAL WORKS. *De dissectione partium corporis humani libri tres, a Carolo Stephano doctore medico editi, una cum figuris et incisionum declarationibus a Stephano Riuerio chirurgo compositis* (Paris, 1545), trans. into French as *La dissection des parties du corps humain* (Paris, 1546;

facsimile ed., Paris, 1965); *De nutrimentis* (Paris, 1550); *Praedium rusticum* (Paris, 1554), trans. into French and revised by Liébault as *L'agriculture et maison rustique* (Paris, 1564).

II. SECONDARY LITERATURE. On Estienne and his work see R. Herrlinger, "Carolus Stephanus and Stephanus Riverius," in *Clio medica,* **2** (1967), 275–287; E. Lau, *Charles Estienne* (Wertheim, 1930); G. Rath, "Charles Estienne, Anatom in Schatten Vesals," in *Sudhoffs Archiv für Geschichte der Naturwissenschaften,* **39** (1955), 35–43; and E. Wickersheimer, *La médecine et les médecins en France à l'époque de la Renaissance* (Paris, 1906). A facsimile edition of *La dissection* with a historical introduction by P. Huard and M. D. Grmek, *L'oeuvre de Charles Estienne et l'école anatomique parisienne* (Paris, 1965), includes a bibliography. A bibliography may also be found in A. A. Renouard, *Annales de l'imprimerie des Estienne* (Paris, 1843). For the anatomical iconography of Estienne, see the manual of L. Choulant and several articles of C. E. Kellett, particularly his brochure *Mannerism and Medical Illustration* (Newcastle, 1961).

M. D. GRMEK

EUCKEN, ARNOLD THOMAS (*b.* Jena, Germany, 3 July 1884; *d.* Chiemsee, Germany, 16 June 1950), *physical chemistry.*

Eucken did experimental and administrative work in physical chemistry. Much of his research was associated with projects of Walther Nernst, and he shared some of Nernst's attitudes toward physical chemistry. His father, Rudolf C. Eucken, was a philosophy professor who received the Nobel Prize for literature in 1908. After graduating from the Gymnasium of Jena in 1902, Eucken entered the University of Kiel, where the inorganic chemistry course, taught by Heinrich Blitz, was oriented to new ideas in physical chemistry. Eucken also studied at the University of Jena. He entered the University of Berlin in 1905 to work in Nernst's laboratory, and he completed his doctorate the following year. From 1908 Eucken was an assistant in Nernst's laboratory. He was *Privatdozent* from 1911 and was in charge of some of the physical chemistry laboratories from 1913. In 1919 he became director of physical chemistry at the Technische Hochschule of Breslau and from 1930 was director of the Institute for Physical Chemistry at the University of Göttingen. Eucken received many academic honors.

After some early work in electrochemistry Eucken became very much involved with heat theory, and conducted experimental research associated with the determination of specific heats. In 1909 he measured specific heats by a method of Nernst's; in the following decades he tested many heat laws experimentally. His work included experiments on the specific heats of hydrogen at low temperatures, Debye's law for specific heats of solids at low temperatures, the range of applicability of Nernst's heat theorem, the difference (predicted by quantum theory) between the specific heats of ortho- and para-hydrogen, and Einstein's theory of specific heats. In the context of his later research on reaction kinetics, Eucken studied contact catalysis and the exchange of vibrational energy between gas molecules.

Eucken was eager to discourage overspecialization in science. For example, he felt that physical chemistry should be studied in terms of physics and mathematics. (He believed, however, that for satisfactory teaching, physical chemistry should remain a separate subject because it is not at all obvious how to apply physics to chemical phenomena.) Similarly, Eucken hoped to increase the interaction between science and technology; some of the research in his laboratory was chosen for the sake of its practical application, and Eucken was chairman of a committee to promote theoretical study of industrial processes.

BIBLIOGRAPHY

I. ORIGINAL WORKS. Eucken wrote many physical chemistry texts. An early English version is *Fundamentals of Physical Chemistry,* trans. and adapted by E. R. Jette and V. K. LaMer (New York, 1925). For a listing of Eucken's papers, consult *Chemical Abstracts* for the period 1900–1950.

II. SECONDARY LITERATURE. See E. Bartholomé, in *Die Naturwissenschaften,* **37** (1950), 481–483; and R. Oesper, "Arnold Eucken," in *Journal of Chemical Education,* **27** (1950), 540–541.

SIGALIA DOSTROVSKY

EUCLID (*fl.* Alexandria [and Athens?], *ca.* 295 B.C.), *mathematics.* The following article is in two parts: Life and Works; Transmission of the Elements.

Life and Works.

Although Euclid (Latinized as Euclides) is the most celebrated mathematician of all time, whose name became a synonym for geometry until the twentieth century,[1] only two facts of his life are known, and even these are not beyond dispute. One is that he was intermediate in date between the pupils of Plato (*d.* 347 B.C.) and Archimedes (*b. ca.* 287 B.C.); the other is that he taught in Alexandria.

Until recently most scholars would have been content to say that Euclid was older than Archimedes on the ground that Euclid, *Elements* I.2, is cited in Archimedes, *On the Sphere and the Cylinder* I.2; but in 1950 Johannes Hjelmslev asserted that this reference was a naïve interpolation. The reasons that he gave are not wholly convincing, but the reference is certainly contrary to ancient practice and is not unfairly characterized as naïve; and although it was already in the text in the time of Proclus, it looks like a marginal gloss which has crept in.[2] Although it is no longer possible to rely on this reference,[3] a general consideration of Euclid's works such as that presented here still shows that he must have written after such pupils of Plato as Eudoxus and before Archimedes.

Euclid's residence in Alexandria is known from Pappus, who records that Apollonius spent a long time with the disciples of Euclid in that city.[4] This passage is also attributed to an interpolator by Pappus' editor, Friedrich Hultsch, but only for stylistic reasons (and these not very convincing); and even if the Alexandrian residence rested only on the authority of an interpolator, it would still be credible in the light of general probabilities. Since Alexander ordered the foundation of the town in 332 B.C. and another ten years elapsed before it began to take shape, we get as a first approximation that Euclid's Alexandrian activities lay somewhere between 320 and 260 B.C. Apollonius was active at Alexandria under Ptolemy III Euergetes (acceded 246) and Ptolemy IV Philopator (acceded 221) and must have received his education about the middle of the century. It is likely, therefore, that Euclid's life overlapped that of Archimedes.

This agrees with what Proclus says about Euclid in his commentary on the first book of the *Elements.* The passage, which is contained in Proclus' summary of the history of geometry,[5] opens:

> Not much younger than these [Hermotimus of Colophon and Philippus of Medma, two disciples of Plato] is Euclid, who put together the elements, arranging in order many of Eudoxus' theorems, perfecting many of Theaetetus', and also bringing to irrefutable demonstration the things which had been only loosely proved by his predecessors. This man lived[6] in the time of the first Ptolemy;[7] for Archimedes, who followed closely upon the first [Ptolemy], makes mention of Euclid,[8] and further they say that Ptolemy once asked him if there were a shorter way to the study of geometry than the *Elements,* to which he replied that there was no royal road to geometry. He is therefore younger than Plato's circle, but older than Eratosthenes and Archimedes; for these were contemporaries, as Eratosthenes somewhere says.[9] In his aim he was a Platonist, being in sympathy with this philosophy, whence he made the end of the whole *Elements* the construction of the so-called Platonic figures.

Since Plato died in 347, Ptolemy I ruled from 323 and reigned from 304 to 285, and Archimedes was born in 287, the chronology of this passage is self-consistent but allows a wide margin according to whether Euclid flourished in Ptolemy's rule or reign. It is clear, however, that Proclus, writing over six centuries later, had no independent knowledge, obviously relying upon Archimedes for his lower date. The story about the royal road is similar to a tale that Stobaeus tells about Menaechmus and Alexander.[10] Euclid may very well have been a Platonist, for mathematics received an immense impetus from Plato's encouragement; what Proclus says about his relationship to Plato's associates, Eudoxus and The-aetetus, is borne out by his own works; and if he were a Platonist, he would have derived pleasure from making the *Elements* end with the construction of the five regular solids. The testimony of so zealous a Neoplatonist as Proclus is not, however, necessarily conclusive on this point.

Confirmation of Proclus' upper date comes from the relationship of Euclid to Aristotle, who died in 322 B.C. Euclid's postulates and axioms or "common notions" undoubtedly show the influence of Aristotle's elaborate discussion of these topics.[11] Aristotle, on the other hand, shows no awareness of Euclid, and he gives a proof of the proposition that the angles at the base of an isosceles triangle are equal which is pre-Euclidean and would hardly have been cited if *Elements* I.5 had been at hand.[12]

If exact dates could be assigned to Autolycus of Pitane, greater precision would be possible, for in his *Phaenomena* Euclid quotes (but without naming his source) propositions from Autolycus' *On the Moving Sphere*. Autolycus was the teacher of Arcesilaus, who was born about 315 B.C. It would be reasonable to suppose that Autolycus was at the height of his activities about 300 B.C., and the date that would best fit the middle point of Euclid's active career is about 295 B.C.; but the uncertainties are so great that no quarrel can be taken with the conventional round date of 300 B.C.[13]

He is therefore a totally different person from Euclid of Megara, the disciple of Plato, who lived about a hundred years earlier.[14] His birthplace is unknown,[15] and the date of his birth can only be guessed. It is highly probable, however, quite apart from what Proclus says about his Platonism, that he attended the Academy, for Athens was the great center of mathematical studies at the time; and there he would have become acquainted with the highly original work of Eudoxus and Theaetetus. He was probably invited to Alexandria when Demetrius of Phalerum, at the direction of Ptolemy Soter, was setting up the great library and museum. This was shortly after 300 B.C., and Demetrius, then an exile

from Athens, where he had been the governor, would have known Euclid's reputation. It is possible that this had already been established by one or more books, but the only piece of internal or external evidence about the order in which Euclid wrote his works is that the *Optics* preceded the *Phaenomena* because it is cited in the preface of the latter. Euclid must be regarded as the founder of the great school of mathematics at Alexandria, which was unrivaled in antiquity. Pappus or an interpolator[16] pays tribute to him as "most fair and well disposed toward all who were able in any measure to advance mathematics, careful in no way to give offense, and although an exact scholar not vaunting himself," as Apollonius was alleged to do; and although the object of the passage is to denigrate Apollonius, there is no reason to reject the assessment of Euclid's character. It was presumably at Alexandria, according to a story by Stobaeus,[17] that someone who had begun to learn geometry with Euclid asked him, after the first theorem, what he got out of such things. Summoning a slave, Euclid said, "Give him three obols, since he must needs make gain out of what he learns." The place of his death is not recorded—although the natural assumption is that it was Alexandria—and the date of his death can only be conjectured. A date about 270 B.C. would accord with the fact that about the middle of the century Apollonius studied with his pupils.

Arabic authors profess to know a great deal more about Euclid's parentage and life, but what they write is either free invention or based on the assumption that the so-called book XIV of the *Elements,* written by Hypsicles, is a genuine work of Euclid.

Geometry: Elements (Στοιχεῖα). Euclid's fame rests preeminently upon the *Elements,* which he wrote in thirteen books[18] and which has exercised an influence upon the human mind greater than that of any other work except the Bible. For this reason he became known in antiquity as Ὁ Στοιχειωτής, "the Writer of the Elements," and sometimes simply as Ὁ Γεωμέτρης, "the Geometer." Proclus explains that the "elements" are leading theorems having to those which follow the character of an all-pervading principle; he likens them to the letters of the alphabet in relation to language, and in Greek they have the same name.[19] There had been *Elements* written before Euclid— notably by Hippocrates, Leo, and Theudius of Magnesia—but Euclid's work superseded them so completely that they are now known only from Eudemus' references as preserved by Proclus. Euclid's *Elements* was the subject of commentaries in antiquity by Hero, Pappus, Porphyry, Proclus, and Simplicius; and Geminus had many observations about it in a work now lost. In the fourth century Theon of Alexandria reedited it, altering the language in some places with a view to greater clarity, interpolating intermediate steps, and supplying alternative proofs, separate cases, and corollaries. All the manuscripts of the *Elements* known until the nineteenth century were derived from Theon's recension. Then Peyrard discovered in the Vatican a manuscript, known as *P,* which obviously gives an earlier text and is the basis of Heiberg's definitive edition.

Each book of the *Elements* is divided into propositions, which may be theorems, in which it is sought to prove something, or problems, in which it is sought to do something. A proposition which is complete in all its parts has a general enunciation (πρότασις); a setting-out or particular enunciation (ἔκθεσις), in which the general enunciation is related to a figure designated by the letters of the alphabet; a definition (διορισμός),[20] which is either a closer statement of the object sought, with the purpose of riveting attention, or a statement of the conditions of possibility; a construction (κατασκευή), including any necessary additions to the original figure; a proof or demonstration (ἀπόδειξις); and a conclusion (συμπέρασμα), which reverts to the language of the general enunciation and states that it has been accomplished. In many cases some of these divisions may be missing (particularly the definition or the construction) because they are not needed, but the general enunciation, proof, and conclusion are always found. The conclusion is rounded off by the formulas ὅπερ ἔδει δεῖξαι ("which was to be proved") for a theorem and ὅπερ ἔδει ποιῆσαι ("which was to be done") for a problem, which every schoolboy knows in their abbreviated Latin forms as Q.E.D. and Q.E.F. These formal divisions of a proposition in such detail are special to Euclid, for Autolycus before him—the only pre-Euclidean author to have any work survive entire—had normally given only a general enunciation and proof, although occasionally a conclusion is found; and Archimedes after him frequently omitted the general or particular enunciation.

The Greek mathematicians carefully distinguished between the analytic and the synthetic methods of proving a proposition.[21] Euclid was not unskilled in analysis, and according to Pappus he was one of the three writers—the others being Apollonius and Aristaeus the Elder—who created the special body of doctrine enshrined in the *Treasury of Analysis.* This collection of treatises included three by Euclid: his *Data, Porisms,* and *Surface Loci.* But in the *Elements* the demonstrations proceed entirely by synthesis, that is, from the known to the unknown, and nowhere is appeal made to analysis, that is, the assumption of the thing to be proved (or done) and the deduction of the consequences until we reach something already accepted or proved true. (Euclid does, however, make frequent use of *reductio ad absurdum* or *demonstratio per impossibile,* showing that if the conclusion is not accepted, absurd or impossible results follow; and this may be regarded as a form of analysis. There are also many pairs of converse propositions, and either one in a pair could be regarded as a piece of analysis for the solution of the other.) No hint is given by Euclid about the way in which he first realized the truth of the propositions that he proves. Majestically he proceeds by rigorous logical steps from one proved proposition to another, using them like stepping-stones, until the final goal is reached.

Each book (or, in the case of XI–XIII, group of books) of the *Elements* is preceded by definitions of the subjects treated, and to book I there are also prefixed five postulates (αἰτήματα) and five common notions (κοίναι ἔννοιαι) or axioms which are the foundation of the entire work. Aristotle had taught that to define an object is not to assert its existence; this must be either proved or assumed.[22] In conformity with this doctrine Euclid defines a point, a straight line, and a circle, then postulates that it is possible

1. To draw a straight line from any point to any point
2. To produce a finite straight line continuously in a straight line
3. To describe a circle with any center and radius.

In other words, he assumes the existence of points, straight lines, and circles as the basic elements of his geometry, and with these assumptions he is able to prove the existence of every other figure that he defines. For example, the existence of a square, defined in I, definition 22, is proved in I.46.

These three postulates do rather more, however, than assume the existence of the things defined. The first postulate implies that between any two points only one straight line can be drawn; and this is equivalent to saying that if two straight lines have the same extremities, they coincide throughout their length, or that two straight lines cannot enclose a space. (The latter statement is interpolated in some of the manuscripts.) The second postulate implies that a straight line can be produced in only one direction at either end, that is, the produced part in either direction is unique, and two straight lines cannot have a common segment. It follows also, since the straight line can be produced indefinitely, or an indefinite number of times, that the space of Euclid's geometry is infinite in all directions. The third postulate also implies the infinitude of space because no limit is placed upon the radius; it further implies that space is continuous, not discrete, because the radius may be indefinitely small.

The fourth and fifth postulates are of a different order because they do not state that something can be done. In the fourth the following is postulated:

4. All right angles are equal to one another.

This implies that a right angle is a determinate magnitude, so that it serves as a norm by which other angles can be measured, but it is also equivalent to an assumption of the homogeneity of space. For if the assertion could be proved, it could be proved only by moving one right angle to another so as to make them coincide, which is an assumption of the invariability of figures or the homogeneity of space. Euclid prefers to assume that all right angles are equal.

The fifth postulate concerns parallel straight lines. These are defined in I, definition 23, as "straight lines which, being in the same plane and being produced indefinitely in both directions, do not meet one another in either direction." The essential characteristic of parallel lines for Euclid is, therefore, that they do not meet. Other Greek writers toyed with the idea, as many moderns have done, that parallel straight lines are equidistant from each other throughout their lengths or have the same direction,[23] and Euclid shows his genius in opting for nonsecancy as the test of parallelism. The fifth postulate runs:

5. If a straight line falling on two straight lines makes the interior angles on the same side less than two right angles, the two straight lines, if produced indefinitely, will meet on that side on which are the angles less than two right angles.

In Figure 1 the postulate asserts that if a straight line (*PQ*) cuts two other straight lines (*AB, CD*) in *P, Q* so that the sum of the angles *BPQ, DQP* is less than two right angles, *AB, CD* will meet on the same side of *PQ* as those two angles, that is, they will meet if produced beyond *B* and *D*.

There was a strong feeling in antiquity that this postulate should be capable of proof, and attempts to prove it were made by Ptolemy and Proclus, among others.[24] Many more attempts have been made in modern times. All depend for their apparent success on making, consciously or unconsciously, an assumption which is equivalent to Euclid's postulate. It was Saccheri in his book *Euclides ab omni naevo vindicatus* (1733) who first asked himself what would be the consequences of hypotheses other than that of Euclid, and in so doing he stumbled upon the possibility of non-Euclidean geometries. Being convinced, as all mathematicians and philosophers were until the

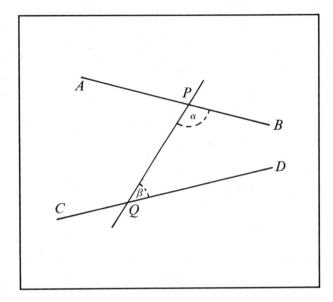

FIGURE 1. Book I, Postulate 5, $\alpha + \beta < 2$ Right Angles

nineteenth century, that there could be no geometry besides that delineated by Euclid, he did not realize what he had done; and although Gauss had the first understanding of modern ideas, it was left to Lobachevski (1826, 1829) and Bolyai (1832), on the one hand, and Riemann (1854), on the other, to develop non-Euclidean geometries. Euclid's fifth postulate has thus been revealed for what it really is—an unprovable assumption defining the character of one type of space.

The five common notions are axioms, which, unlike the postulates, are not confined to geometry but are common to all the demonstrative sciences. The first is "Things which are equal to the same thing are also equal to one another," and the others are similar.

The subject matter of the first six books of the *Elements* is plane geometry. Book I deals with the geometry of points, lines, triangles, squares, and parallelograms. Proposition 5, that in isosceles triangles the angles at the base are equal to one another and that, if the equal straight lines are produced, the angles under the base will be equal to one another, is interesting historically as having been known (except in France) as the *pons asinorum;* this is usually taken to mean that those who are not going to be good at geometry fail to get past it, although others have seen in the figure of the proposition a resemblance to a trestle bridge with a ramp at each end which a donkey can cross but a horse cannot.

Proposition 44 requires the student "to a given straight to apply in a given rectilineal angle a parallelogram equal to a given triangle," that is, on a given straight line to construct a parallelogram equal to a given area and having one of its angles equal to a

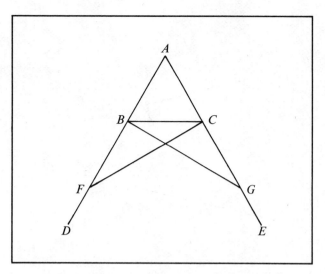

FIGURE 2.   Book I, Proposition 5, Pons Asinorum

given angle. In Figure 3, *AB* is the given straight line and the parallelogram *BEFG* is constructed equal to the triangle *C* so that ∠ *GBE* = ∠ *D*. The figure is completed, and it is proved that the parallelogram *ABML* satisfies the requirements. This is Euclid's first example of the application of areas,[25] one of the most powerful tools of the Greek mathematicians. It is a geometrical equivalent of certain algebraic operations. In this simple case, if $AL = x$, then $x \cdot AB \cos D = C$, and the theorem is equivalent to the solution of a first-degree equation. The method is developed later, as it will be shown, so as to be equivalent to the solution of second-degree equations.

Book I leads up to the celebrated proposition 47,

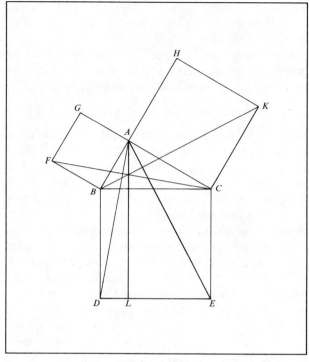

FIGURE 4.   Book I, Proposition 47, "Pythagoras' Theorem," $BC^2 = CA^2 + AB^2$

"Pythagoras' theorem," which asserts: "In right-angled triangles the square on the side subtending the right angle is equal to the [sum of the] squares on the sides containing the right angle." In Figure 4 it is shown solely by the use of preceding propositions that the parallelogram *BL* is equal to the square *BG* and the parallelogram *CL* is equal to the square *AK*, so that the whole square *BE* is equal to the sum of the squares *BG, AK*. It is important to notice, for

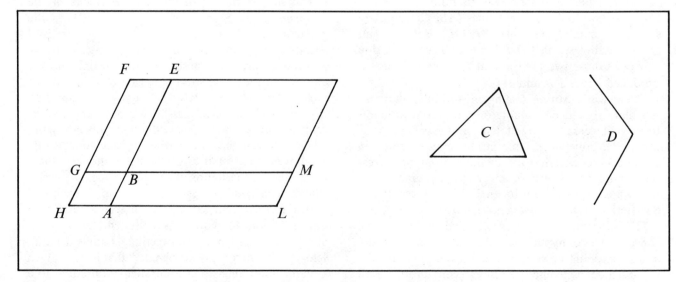

FIGURE 3.   Book I, Proposition 44, Application of Areas

a reason to be given later, that no appeal is made to similarity of figures. This fundamental proposition gives Euclidean space a metric, which would be expressed in modern notation as $ds^2 = dx^2 + dy^2$. It is impossible not to admire the ingenuity with which the result is obtained, and not surprising that when Thomas Hobbes first read it he exclaimed, "By God, this is impossible."

Book II develops the transformation of areas adumbrated in I.44, 45 and is a further exercise in geometrical algebra. Propositions 5, 6, 11, and 14 are the equivalents of solving the quadratic equations $ax - x^2 = b^2$, $ax + x^2 = b^2$, $x^2 + ax = a^2$, $x^2 = ab$. Propositions 9 and 10 are equivalent to finding successive pairs of integers satisfying the equations $2x^2 - y^2 = \pm 1$. Such pairs were called by the Greeks side numbers and diameter numbers. Propositions 12 and 13 are equivalent to a proof that in any triangle with sides a, b, c, and angle A opposite a,

$$a^2 = b^2 + c^2 - 2\,bc \cos A.$$

It is probably not without significance that this penultimate proposition of book II is a generalization of "Pythagoras' theorem," which was the penultimate proposition of book I.

Book III treats circles, including their intersections and touchings. Book IV consists entirely of problems about circles, particularly the inscribing or circumscribing of rectilineal figures. It ends with proposition 16: "In a given circle to inscribe a fifteen-angled figure which shall be both equilateral and equiangular." Proclus asserts that this is one of the propositions that Euclid solved with a view to their use in astronomy, "for when we have inscribed the fifteen-angled figure in the circle through the poles, we have the distance from the poles both of the equator and the zodiac, since they are distant from one another by the side of the fifteen-angled figure"—that is to say, the obliquity of the ecliptic was taken to be $24°$, as is known independently to have been the case up to Eratosthenes.[26]

Book V develops the general theory of proportion. The theory of proportion as discovered by the Pythagoreans applied only to commensurable magnitudes because it depends upon the taking of aliquot parts, and this is all that was needed by Euclid for the earlier books of the *Elements*. There are instances, notably I.47, where he clearly avoids a proof that would depend on similitude or the finding of a proportional, because at that stage of his work it would not have applied to incommensurable magnitudes. In book V he addresses himself at length to the general theory. There is no book in the *Elements* that has so won

the admiration of mathematicians. Barrow observes: "There is nothing in the whole body of the *Elements* of a more subtle invention, nothing more solidly established and more accurately handled, than the doctrine of proportionals." In like spirit Cayley says, "There is hardly anything in mathematics more beautiful than this wondrous fifth book."[27]

The heart of the book is contained in the definitions with which it opens. The definition of a ratio as "a sort of relation in respect of size between two magnitudes of the same kind" shows that a ratio, like the elephant, is easy to recognize but hard to define. Definition 4 is more to the point: "Magnitudes are said to have a ratio one to the other if capable, when multiplied, of exceeding one another." The definition excludes the infinitely great and the infinitely small and is virtually equivalent to what is now known as the axiom of Archimedes. (See below, section on book X.) But it is definition 5 which has chiefly excited the admiration of subsequent mathematicians: "Magnitudes are said to be in the same ratio, the first to the second and the second to the fourth, when, if any equimultiples whatever be taken of the first and third, and any equimultiples whatever of the second and fourth, the former equimultiples alike exceed, are alike equal to, or alike fall short of, the latter equimultiples respectively taken in corresponding order." It will be noted that the definition avoids mention of parts of a magnitude and is therefore applicable to the incommensurable as well as to the commensurable. De Morgan put its meaning very clearly: "Four magnitudes, A and B of one kind, and C and D of the same or another kind, are proportional when all the multiples of A can be distributed among the multiples of B in the same intervals as the corresponding multiples of C among those of D"; or, in notation, if m, n are two integers, and if mA lies between nB and $(n + 1)B$, mC lies between nD and $(n + 1)D$.[28] It can be shown that the test proposed by Euclid is both a necessary and a sufficient test of proportionality, and in the whole history of mathematics no equally satisfactory test has ever been proposed. The best testimony to its adequacy is that Weierstrass used it in his definition of equal numbers;[29] and Heath has shown how Euclid's definition divides all rational numbers into two coextensive classes and thus defines equal ratios in a manner exactly corresponding to a Dedekind section.[30]

The remaining definitions state the various kinds of transformations of ratios—generally known by their Latin names: *alternando, invertendo, componendo, separando, convertendo, ex aequali,* and *ex aequali in proportione perturbata*—and with remorseless logic the twenty-five propositions apply these various operations to the objects of Euclid's definitions.

It is a sign of the abiding fascination of book V for mathematicians that in 1967 Friedhelm Beckmann applied his own system of axioms, set up in close accordance with Euclid, in such a way as to deduce all definitions and propositions of Euclid's theory of magnitudes, especially those of books V and VI. In his view magnitudes, rather than their relation of "having a ratio," form the base of the theory of proportions. These magnitudes represent a well-defined structure, a so-called Eudoxic semigroup, with the numbers as operators. Proportion is interpreted as a mapping of totally ordered semigroups. This mapping proves to be an isomorphism, thus suggesting the application of the modern theory of homomorphism.

Book VI uses the general theory of proportion established in the previous book to treat similar figures. The first and last propositions of the book illustrate the importance of V, definition 5, for by the method of equimultiples it is proved in proposition 1 that triangles and parallelograms having the same height are to one another as their bases, and in proposition 33 it is proved that in equal circles the angles at the center or circumference are as the arcs on which

they stand. There are many like propositions of equal importance. Proposition 25 sets the problem "To construct a rectilineal figure similar to one, and equal to another, given rectilineal figure."[31] In propositions 27–29 Euclid takes up again the application of areas. It has been explained above that to apply ($\pi\alpha\rho\alpha\beta\acute{\alpha}\lambda\lambda\epsilon\iota\nu$) a parallelogram to a given straight line means to construct on that line a parallelogram equal to a given area and having a given angle. If the straight line is applied to only part of the given line, the resulting figure is said to be deficient ($\dot{\epsilon}\lambda\lambda\epsilon\acute{\iota}\pi\epsilon\iota\nu$); if to the straight line produced, it is said to exceed ($\dot{\upsilon}\pi\epsilon\rho\beta\acute{\alpha}\lambda\lambda\epsilon\iota\nu$). Proposition 28 is the following problem: "To a given straight line to apply a parallelogram equal to a given rectilineal figure and deficient by a parallelogrammatic figure similar to a given one." (It has already been shown in proposition 27 that the given rectilineal figure must not be greater than the parallelogram described on half the straight line and similar to the defect.) In Figure 5 let the parallelogram TR be applied to the straight line AB so as to be equal to a given rectilineal figure having the area S and deficient by the parallelogram PB, which is similar to the given parallelogram D. Let $AB = a$, $RP = x$; let the angle of D be α and the ratio of its sides $b:c$. Let E be the midpoint of AB

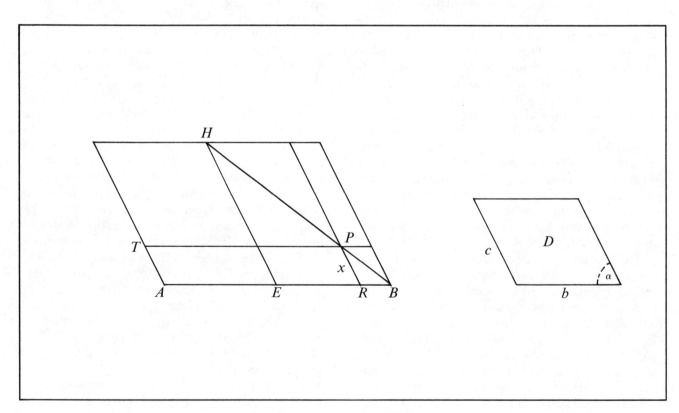

FIGURE 5.   Book VI, Proposition 28

420

and let EH be drawn parallel to the sides. Then

$$\text{(the parallelogram } TR) = \text{(the parallelogram } TB)$$
$$- \text{(the parallelogram } PB)$$
$$= ax \sin \alpha - \frac{b}{c} x \cdot x \sin \alpha.$$

If the area of the given rectilineal figure is S, this may be written

$$S = ax \sin \alpha - \frac{b}{c} x^2 \sin \alpha.$$

Constructing the parallelogram TR is therefore equivalent to solving geometrically the equation

$$ax - \frac{b}{c} x^2 = \frac{S}{\sin \alpha}.$$

It can easily be shown that Euclid's solution is equivalent to completing the square on the left-hand side. For a real solution it is necessary that

$$\frac{S}{\sin \alpha} \geq \frac{c}{b} \cdot \frac{a^2}{4}$$

i.e., $S \geq \left(\frac{c}{b} \cdot \frac{a}{2} \right) (\sin \alpha) \left(\frac{a}{2} \right)$

i.e., $S \geq HE \sin \alpha \cdot EB$

i.e., $S \geq$ parallelogram HB,

which is exactly what was proved in VI.27. Proposition 29 sets the corresponding problem for the excess: "To a given straight line to apply a parallelogram equal to a given rectilineal figure and exceeding by a parallelogrammatic figure similar to a given one." This can be shown in the same way to be equivalent to solving geometrically the equation

$$ax + \frac{b}{c} x^2 = \frac{S}{\sin \alpha}.$$

In this case there is always a real solution. No διορισμός or examination of the conditions of possibility is needed, and Euclid's solution corresponds to the root with the positive sign.

This group of propositions is needed by Euclid for his treatment of irrationals in book X, but their chief importance in the history of mathematics is that they are the basis of the theory of conic sections as developed by Apollonius. Indeed, the very words "parabola," "ellipse," and "hyperbola" come from the Greek words for "to apply," "to be deficient," and "to exceed."

It is significant that the antepenultimate proposition of the book, proposition 31, is a generalization of "Pythagoras' theorem": "In right-angled triangles any [literally "the"] figure [described] on the side subtending the right angle is equal to the [sum of the] similar and similarly described figures on the sides containing the right angle."

Books VII, VIII, and IX are arithmetical; and although the transition from book VI appears sharp, there is a logical structure in that the theory of proportion, developed in all its generality in book V, is applied in book VI to geometrical figures and in book VII to numbers. The theory of numbers is continued in the next two books. The theory of proportion in book VII is not, however, the general theory of book V but the old Pythagorean theory applicable only to commensurable magnitudes.[32] This return to an outmoded theory led both De Morgan and W. W. Rouse Ball to suppose that Euclid died before putting the finishing touches to the *Elements*,[33] but, although the three arithmetical books seem trite in comparison with those that precede and follow, there is nothing unfinished about them. It is more likely that Euclid, displaying the deference toward others that Pappus observed, thought that he ought to include the traditional teaching. This respect for traditional doctrines can be seen in some of the definitions which Euclid repeats even though he improves upon them or never uses them.[34] Although books VII–IX appear at first sight to be a reversion to Pythagoreanism, it is Pythagoreanism with a difference. In particular, the rational straight line takes the place of the Pythagorean monad;[35] but the products of numbers are also treated as straight lines, not as squares or rectangles.

After the numerical theory of proportion is established in VII.4–19, there is an interesting group of propositions on prime numbers (22–32) and a final group (33–39) on least common multiples. Book VIII deals in the main with series of numbers "in continued proportion," that is, in geometrical progression, and with geometric means. Book IX is a miscellany and includes the fundamental theorem in the theory of numbers, proposition 14: "If a number be the least that is measured by prime numbers, it will not be measured by any other prime number except those originally measuring it," that is to say, a number can be resolved into prime factors in only one way.

After the muted notes of the arithmetical books Euclid again takes up his lofty theme in book X, which treats irrational magnitudes. It opens with the following proposition (X.1): "If two unequal magnitudes be set out, and if there be subtracted from the greater a magnitude greater than its half, and from that which is left a magnitude greater than its half, and so on continually, there will be left some magnitude less than the lesser magnitude set out." This is the basis of the "method of exhaustion," as later used by Euclid in book XII. Because of the use made of it by Archimedes, either directly or in an equivalent form, for the purpose of calculating areas and volumes, it has become known, perhaps a little unreasonably, as the axiom of Archimedes. Euclid needs the axiom at this point as a test of incommensurability, and his next proposition (X.2) asserts: "If the lesser of two unequal magnitudes is continually subtracted from the greater, and the remainder never measures that which precedes it, the magnitudes will be incommensurable."

The main achievement of the book is a classification of irrational straight lines, no doubt for the purpose of easy reference. Starting from any assigned straight line which it is agreed to regard as rational—a kind of datum line—Euclid asserts that any straight line which is commensurable with it in length is rational, but he also regards as rational a straight line commensurable with it only in square. That is to say, if m, n are two integers in their lowest terms with respect to each other, and l is a rational straight line, he regards $\sqrt{m/n} \cdot l$ as rational because $(m/n)l^2$ is commensurable with l^2. All straight lines not commensurable either in length or in square with the assigned straight line he calls irrational. His fundamental proposition (X.9) is that the sides of squares are commensurable or incommensurable in length according to whether the squares have or do not have the ratio of a square number to each other, that is to say, if a, b are straight lines and m, n are two numbers, and if $a:b = m:n$, then $a^2:b^2 = m^2:n^2$ and conversely. This is easily seen in modern notation, but was far from an easy step for Euclid. The first irrational line which he isolates is the side of a square equal in area to a rectangle whose sides are commensurable in square only. He calls it a medial. If the sides of the rectangle are l, $\sqrt{k}\, l$, the medial is $k^{1/4}l$. Euclid next proceeds to define six pairs of compound irrationals (the members of each pair differing in sign only) which can be represented in modern notation as the positive roots of six biquadratic equations (reducible to quadratics) of the form

$$x^4 \pm 2alx^2 \pm bl^4 = 0.$$

The first pair are given the names "binomial" (or "biterminal") and "apotome," and Euclid proceeds to define six pairs of their derivatives which are equivalent to the roots of six quadratic equations of the form

$$x^2 + 2alx + bl^2 = 0.$$

In all, Euclid investigates in the 115 propositions of the book (of which the last four may be interpola-

tions) every possible form of the lines which can be represented by the expression $\sqrt{(\sqrt{a} \pm \sqrt{b})}$, some twenty-five in all.[36]

The final three books of the *Elements,* XI–XIII, are devoted to solid geometry. Book XI deals largely with parallelepipeds. Book XII applies the method of exhaustion, that is, the inscription of successive figures in the body to be evaluated, in order to prove that circles are to one another as the squares on their diameters, that pyramids of the same height with triangular bases are in the ratio of their bases, that the volume of a cone is one-third of the cylinder which has the same base and equal height, that cones and cylinders having the same height are in the ratio of their bases, that similar cones and cylinders are to one another in the triplicate ratio of the diameters of their bases, and that spheres are in the triplicate ratio of their diameters. The method can be shown for the circle. Euclid inscribes a square in the circle and shows that it is more than half the circle. He bisects each arc and shows that each triangle so obtained is greater than half the segment of the circle about it. (In Figure 6, for example, triangle EAB is greater than half the segment of the circle EAB standing on AB.) If the process is continued indefinitely, according to X.1, we shall be left with segments of the circle smaller than some assigned magnitude, that is, the circle has been exhausted. (A little later Archimedes was to refine the method by also circum-

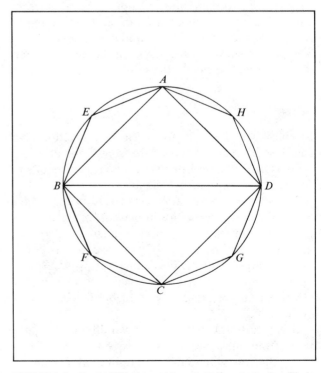

FIGURE 6. Book XII, Proposition 2, Exhaustion of a Circle

scribing a polygon, and so compressing the figure, as it were, between inscribed and circumscribed polygons.) Euclid refrains from saying that as the process is continued indefinitely, the area of the polygon will in the limit approach the area of the circle, and rigorously proves that if his proposition is not granted, impossible conclusions would follow.

After some preliminary propositions book XIII is devoted to the construction in a sphere of the five regular solids: the pyramid (proposition 13), the octahedron (14), the cube (15), the icosahedron (16), and the dodecahedron (17). These five regular solids had been a prime subject of investigation by the Greek mathematicians, and because of the use made of them by Plato in the *Timaeus* were known as the Platonic figures.[37] The mathematical problem is to determine the edge of the figure in relation to the radius of the circumscribing sphere. In the case of the pyramid, octahedron, and cube, Euclid actually evaluates the edge in terms of the radius, and in the case of the icosahedron and dodecahedron he shows that it is one of the irrational lines classified in book X—a minor in the case of the icosahedron and an apotome in the case of the dodecahedron. In a final splendid flourish (proposition 18), Euclid sets out the sides of the five figures in the same sphere and compares them with each other, and in an addendum he shows that there can be no other regular solids. In Figure 7, $AC = CB$, $AD = 2DB$, $AG = AB$, $CL = CM$, and BF is divided in extreme and mean ratio at $N(BF:BN = BN:NF)$. He proves that AF is the side of the pyramid, BF is the side of the cube, BE is the

side of the octahedron, BK is the side of the icosahedron, and BN is the side of the dodecahedron; their values, in terms of the radius r, are respectively $2/3 \sqrt{6} \cdot r$, $\sqrt{2} \cdot r$, $2/3 \sqrt{3} \cdot r$, $r/5 \sqrt{10(5 - \sqrt{5})}$, $r/3(\sqrt{15} - \sqrt{3})$.

Proclus, as already noted, regarded the construction of the five Platonic figures as the end of the *Elements,* in both senses of that ambiguous word. This is usually discounted on the ground that the stereometrical books had to come last, but Euclid need not have ended with the construction of the five regular solids; and since he shows the influence of Plato in other ways, this splendid ending could easily be a grain of incense at the Platonic altar.

Proclus sums up Euclid's achievement in the *Elements* in the following words:[38]

> He deserves admiration preeminently in the compilation of his *Elements of Geometry* on account of the order and selection both of the theorems and of the problems made with a view to the elements. For he included not everything which he could have said, but only such things as were suitable for the building up of the elements. He used all the various forms of deductive arguments,[39] some getting their plausibility from the first principles, some starting from demonstrations, but all irrefutable and accurate and in harmony with science. In addition he used all the dialectical methods, the *divisional* in the discovery of figures, the *definitive* in the existential arguments, the *demonstrative* in the passages from first principles to the things sought, and the *analytic* in the converse process from the things sought to the first principles. And the various species of conversions,[40] both of the simpler (propositions) and of the more complex, are in this treatise accurately set forth and skillfully investigated, what wholes can be converted with wholes, what wholes with parts and conversely, and what as parts with parts. Further, we must make mention of the continuity of the proofs, the disposition and arrangement of the things which precede and those which follow, and the power with which he treats each detail.

This is a fair assessment. The *Elements* is on the whole a compilation of things already known, and its most remarkable feature is the arrangement of the matter so that one proposition follows on another in a strictly logical order, with the minimum of assumption and very little that is superfluous.

If we seek to know how much of it is Euclid's own work, Proclus is again our best guide. He says, as we have seen, that Euclid "put together the elements, arranging in order many of Eudoxus' theorems, perfecting many of Theaetetus', and also bringing to irrefutable demonstration the things which had been only loosely proved by his predecessors."[41] According to a scholiast of book V, "Some say that this book

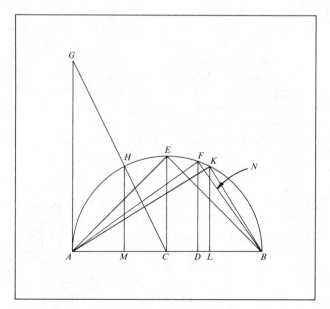

FIGURE 7. Book XIII, Proposition 18, Inscription of Regular Solids in a Sphere

is the discovery of Eudoxus, the disciple of Plato."[42] Another scholiast confirms this, saying, "This book is said to be the work of Eudoxus of Cnidus, the mathematician, who lived about the times of Plato."[43] He adds, however, that the ascription to Euclid is not false, for although there is nothing to prevent the discovery from being the work of another man, "The arrangement of the book with a view to the elements and the orderly sequence of theorems is recognized by all as the work of Euclid." This is a fair division of the credit. Eudoxus also, as we can infer from Archimedes,[44] is responsible for the method of exhaustion used in book XII to evaluate areas and volumes, based upon X.1, and Archimedes attributes to Eudoxus by name the theorems about the volume of the pyramid and the volume of the cone which stand as propositions 7 and 10 of book XII of the *Elements.* Although Greek tradition credited Hippocrates with discovering that circles are to one another as the squares on their diameters,[45] we can be confident that the proof as we have it in XII.2 is also due to Eudoxus.

The interest of Theaetetus in the irrational is known from Plato's dialogue,[46] and a commentary on book X which has survived in Arabic[47] and is attributed by Heiberg[48] to Pappus credits him with discovering the different species of irrational lines known as the medial, binomial, and apotome. A scholium to X.9[49] (that squares which do not have the ratio of a square number to a square number have their sides incommensurable) attributes this theorem to Theaetetus. It would appear in this case also that the fundamental discoveries were made before Euclid but that the orderly arrangement of propositions is his work. This, indeed, is asserted in the commentary attributed to Pappus, which says:

> As for Euclid he set himself to give rigorous rules, which he established, relative to commensurability and incommensurability in general; he made precise the definitions and the distinctions between rational and irrational magnitudes, he set out a great number of orders of irrational magnitudes, and finally he clearly showed their whole extent.[50]

Theaetetus was also the first to "construct" or "write upon" the five regular solids,[51] and according to a scholiast[52] the propositions concerning the octahedron and the icosahedron are due to him. His work therefore underlies book XIII, although the credit for the arrangement must again be given to Euclid.

According to Proclus,[53] the application of areas, which, as we have seen, is employed in I.44 and 45, II.5, 6, and 11, and VI.27, 28, and 29, is "ancient, being discoveries of the muse of the Pythagoreans."

A scholiast to book IV[54] attributes all sixteen theorems (problems) of that book to the Pythagoreans. It would appear, however, that the famous proof of what is universally known as "Pythagoras' theorem," I.47, is due to Euclid himself. It is beyond doubt that this property of right-angled triangles was discovered by Pythagoras, or at least in his school, but the proof was almost certainly based on proportions and therefore not applicable to all magnitudes. Proclus says:

> If we give hearing to those who relate things of old, we shall find some of them referring this discovery to Pythagoras and saying that he sacrificed an ox upon the discovery. But I, while marveling at those who first came to know the truth of this theorem, hold in still greater admiration the writer of the *Elements,* not only because he made it secure by a most clear proof, but because he compelled assent by the irrefutable reasonings of science to the still more general proposition in the sixth book. For in that book he proves generally that in right-angled triangles the figure on the side subtending the right angle is equal to the similar and similarly situated figures described on the sides about the right angle.[55]

On the surface this suggests that Euclid devised a new proof, and this is borne out by what Proclus says about the generalization. It would be an easy matter to prove VI.31 by using I.47 along with VI.22, but Euclid chooses to prove it independently of I.47 by using the general theory of proportions. This suggests that he proved I.47 by means of book I alone, without invoking proportions in order to get it into his first book instead of his sixth. The proof certainly bears the marks of genius.

To Euclid also belongs beyond a shadow of doubt the credit for the parallel postulate which is fundamental to the whole system. Aristotle had censured those "who think they describe parallels" because of a *petitio principii* latent in their theory.[56] There is certainly no *petitio principii* in Euclid's theory of parallels, and we may deduce that it was post-Aristotelian and due to Euclid himself. In nothing did he show his genius more than in deciding to treat postulate 5 as an indemonstrable assumption.

The significance of Euclid's *Elements* in the history of thought is twofold. In the first place, it introduced into mathematical reasoning new standards of rigor which remained throughout the subsequent history of Greek mathematics and, after a period of logical slackness following the revival of mathematics, have been equaled again only in the past two centuries. In the second place, it marked a decisive step in the geometrization of mathematics.[57] The Pythagoreans and Democritus before Euclid, Archimedes in some of his works, and Diophantus afterward showed that

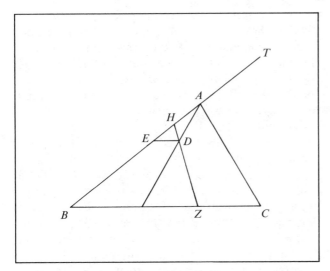

FIGURE 8. *On the Division of Figures,* Proposition 19

Greek mathematics might have developed in other directions. It was Euclid in his *Elements,* possibly under the influence of that philosopher who inscribed over the doors of the Academy "God is for ever doing geometry," who ensured that the geometrical form of proof should dominate mathematics. This decisive influence of Euclid's geometrical conception of mathematics is reflected in two of the supreme works in the history of thought, Newton's *Principia* and Kant's *Kritik der reinen Vernunft.* Newton's work is cast in the form of geometrical proofs that Euclid had made the rule even though Newton had discovered the calculus, which would have served him better and made him more easily understood by subsequent generations; and Kant's belief in the universal validity of Euclidean geometry led him to a transcendental aesthetic which governs all his speculations on knowledge and perception.

It was only toward the end of the nineteenth century that the spell of Euclidean geometry began to weaken and that a desire for the "arithmetization of mathematics" began to manifest itself; and only in the second quarter of the twentieth century, with the development of quantum mechanics, have we seen a return in the physical sciences to a neo-Pythagorean view of number as the secret of all things. Euclid's reign has been a long one; and although he may have been deposed from sole authority, he is still a power in the land.

The Data (Δεδομένα). The *Data,* the only other work by Euclid in pure geometry to have survived in Greek, is closely connected with books I–VI of the *Elements.* It is concerned with the different senses in which things are said to be given. Thus areas, straight lines, angles, and ratios are said to be "given in magnitude" when we can make others equal to them.

Rectilineal figures are "given in species" or "given in form" when their angles and the ratio of their sides are given. Points, lines, and angles are "given in position" when they always occupy the same place, and so on. After the definitions there follow ninety-four propositions, in which the object is to prove that if certain elements of a figure are given, other elements are also given in one of the defined senses.

The most interesting propositions are a group of four which are exercises in geometrical algebra and correspond to *Elements* II.28, 29. Proposition 58 reads: "If a given area be applied to a given straight line so as to be deficient by a figure given in form, the breadths of the deficiency are given." Proposition 84, which depends upon it, runs: "If two straight lines contain a given area in a given angle, and if one of them is greater than the other by a given quantity, then each of them is given." This is equivalent to solving the simultaneous equations

$$y - x = a$$
$$xy = b^2,$$

and these in turn are equivalent to finding the two roots of

$$ax + x^2 = b^2.$$

Propositions 59 and 85 give the corresponding theorems for the excess and are equivalent to the simultaneous equations

$$y + x = a$$
$$xy = b^2$$

and the quadratic equation

$$ax - x^2 = b^2.$$

A clue to the purpose of the *Data* is given by its inclusion in what Pappus calls the *Treasury of Analysis.*[58] The concept behind the *Data* is that if certain things are given, other things are necessarily implied, until we are brought to something that is agreed. The *Data* is a collection of hints on analysis. Pappus describes the contents of the book as known to him;[59] the number and order of the propositions differ in some respects from the text which has come down to us.

Marinus of Naples, the pupil and biographer of Proclus, wrote a commentary on, or rather an introduction to, the *Data.* It is concerned mainly with the different senses in which the term "given" was understood by Greek geometers.

On Divisions of Figures (Περὶ διαιρεσεων βιβλίον). Proclus preserved this title along with the titles of other works of Euclid,[60] and gives an indication of its contents: "For the circle is divisible into parts unlike by definition, and so is each of the rectilineal figures; and this is indeed what the writer of the *Elements* himself discusses in his *Divisions,* dividing given figures now into like figures, now into unlike."[61] The book has not survived in Greek, but all the thirty-six enunciations and four of the propositions (19, 20, 28, 29) have been preserved in an Arabic translation discovered by Woepcke and published in 1851; the remaining proofs can be supplied from the *Practica geometriae* written by Leonardo Fibonacci in 1220, one section of which, it is now evident, was based upon a manuscript or translation of Euclid's work no longer in existence. The work was reconstructed by R. C. Archibald in 1915.

The character of the book can be seen from the first of the four propositions which has survived in Arabic (19). This is "To divide a given triangle into two equal parts by a line which passes through a point situated in the interior of the triangle." Let D be a point inside the triangle ABC and let DE be drawn parallel to CB so as to meet AB in E. Let T be taken on BA produced so that $TB \cdot DE = 1/2 \, AB \cdot BC$ (that is, let TB be such that when a rectangle having TB for one of its sides is applied to DE, it is equal to half the rectangle $AB \cdot BC$). Next, let a parallelogram be applied to the line TB equal to the rectangle $TB \cdot BE$ and deficient by a square, that is, let H be taken on TB so that

$$(TB - HT) \cdot HT = TB \cdot BE.$$

HD is drawn and meets BC in Z. It can easily be shown that HZ divides the triangle into two equal parts and is the line required.

The figures which are divided in Euclid's tract are the triangle, the parallelogram, the trapezium, the quadrilateral, a figure bounded by an arc of a circle and two lines, and a circle. It is proposed in the various cases to divide the given figure into two equal parts, into several equal parts, into two parts in a given ratio, or into several parts in a given ratio. The propositions may be further classified according to whether the dividing line (transversal) is required to be drawn from a vertex, from a point within or without the figure, and so on.[62]

In only one proposition (29) is a circle divided, and it is clearly the one to which Proclus refers. The enunciation is "To draw in a given circle two parallel lines cutting off a certain fraction from the circle." In fact, Euclid gives the construction for a fraction of one-third and notes a similar construction for a quarter, one-fifth, "or any other definite fraction."[63]

Porisms (Πορίσματα). It is known both from Pappus[64] and from Proclus[65] that Euclid wrote a three-book work called *Porisms.* Pappus, who includes the work in the *Treasury of Analysis,* adds the information that it contained 171 theorems and thirty-eight lemmas. It has not survived—most unfortunately, for it appears to have been an exercise in advanced mathematics;[66] but the account given by Pappus encouraged such great mathematicians as Robert Simson and Michel Chasles to attempt reconstructions, and Chasles was led thereby to the discovery of anharmonic ratios.

The term "porism" commonly means in Greek mathematics a corollary, but that is not the sense in which it is used in Euclid's title. It is clearly derived from πορίζω, "I procure," and Pappus explains that according to the older writers a porism is something intermediate between a theorem and a problem: "A theorem is something proposed with a view to the proof of what is proposed, and a problem is something thrown out with a view to the construction of what is proposed, [and] a porism is something proposed with a view to the finding [πορισμόν] of the very thing proposed."[67] Proclus reinforces the explanation. The term, he says, is used both for "such theorems as are established in the proofs of other theorems, being windfalls and bonuses of the things sought, and also for such things as are sought, but need discovery, and are neither pure bringing into being nor pure investigation."[68] As examples of a porism in this sense, Proclus gives two: first, the finding of the center of a circle, and, second, the finding of the greatest common measure of two given commensurable magnitudes.

Pappus says that it had become characteristic of porisms for the enunciation to be put in shortened form and for a number of propositions to be comprehended in one enunciation. He sets out twenty-nine different types in Euclid's work (fifteen in book I, six in book II, and eight in book III). His versions suggest that the normal form of Euclid's porisms was to find a point or a line satisfying certain conditions. Pappus says that Euclid did not normally give many examples of each case, but at the beginning of the first book he gave ten propositions belonging to one class; and Pappus found that these could be comprehended in one enunciation, in this manner:

> If in a quadrilateral, whether convex or concave, the sides cut each other two by two, and the three points in which the other three sides intersect the fourth are given, and if the remaining points of intersection save one lie on straight lines given in position, the remaining point will also lie on a straight line given in position.[69]

Pappus proceeds to generalize this theorem for any system of straight lines cutting each other two by two.

In modern notation, let there be n straight lines, of which not more than two pass through one point and no two are parallel. They will intersect in $1/2n(n-1)$ points. Let the $(n-1)$ points in which one of the lines is intersected be fixed. This will leave $1/2n(n-1) - (n-1) = 1/2(n-1)(n-2)$ other points of intersection. If $(n-2)$ of these points lie on straight lines given in position, the other $1/2(n-1)(n-2) - (n-2) = 1/2(n-2)(n-3)$ points of intersection will also lie on straight lines given in position, provided that it is impossible to form with these points of intersection any triangle having for sides the sides of the polygon.[70]

Pappus adds: "It is unlikely that the writer of the *Elements* was unaware of this result, but he would have desired only to set out the first principle. For he appears in all the porisms to have laid down only the first principles and seminal ideas of the many important matters investigated."[71]

Pappus' remarks about the definition of porisms by the "older writers" have been given above. He—or an interpolator—censures more recent writers who defined a porism by an incidental characteristic: "a porism is that which falls short of a locus theorem in respect of its hypotheses."[72] What this means is far from clear, but it led Zeuthen[73] to conjecture that Euclid's porisms were a by-product of his researches into conic sections—which, if true, would be a happy combination of the two meanings of porism. Zeuthen takes the first proposition of Euclid's first book as quoted by Pappus: "If from two points given in position straight lines be drawn so as to meet on a straight line given in position, and if one of them cuts off from a straight line given in position a segment measured toward a given point on it, the other will also cut off from another straight line a segment having to the first a given ratio."[74] He notes that this proposition is true if a conic section, regarded as a "locus with respect to four lines" (see below), is substituted for the first given straight line, with the two given points as points on it.[75] It will be convenient to turn immediately to Euclid's investigations into conic sections and the "three- and four-line locus," noting that, from one point of view, his *Porisms* would appear to have been the earliest known treatise on projective geometry and transversals.

Conics. We know from Pappus that Euclid wrote a four-book work on conic sections, but it has not survived even in quotation. The relevant passage in the *Collection* reads: "Apollonius, having completed Euclid's four books of conics and added four others, handed down eight volumes of conics."[76] The work was probably lost by Pappus' time, for in the next sentence he mentions as still extant the five books

of Aristaeus on "solid loci." Aristaeus preceded Euclid, for Euclid, according to Pappus or an interpolator, thought that Aristaeus deserved the credit for the discoveries in conics he had already made, and neither claimed originality nor wished to overthrow what he had already done. (It is at this point that Pappus contrasts Euclid's character with that of Apollonius, noted above.) In particular, Euclid wrote as much about the three- and four-line locus as was possible on the basis of Aristaeus' conics without claiming completeness for his proofs.[77]

Euclid doubtless shared the early Greek view that conic sections were generated by the section of a cone by a plane at right angles to a generator, and he would have used the names "section of a right-angled cone," "section of an acute-angled cone," and "section of an obtuse-angled cone," which were in use until Apollonius established the terms "parabola," "ellipse," and "hyperbola"; but he was aware that an ellipse can be obtained by any section of a cone or cylinder not parallel to the base, for in his *Phaenomena* he says: "If a cone or cylinder be cut by a plane not parallel to the base, the section is a section of an acute-angled cone which is like a shield [θυρεός]."[78]

Furthermore, Euclid was aware of the focus-directrix property (that a conic section is the locus of a point whose distance from a fixed point bears a constant relation to its distance from a fixed straight line), even though it is nowhere mentioned by Apollonius: Pappus cites the property as a lemma to Euclid's *Surface Loci*,[79] from which it is clear that it was assumed in that book without proof. It is likely, therefore, that it was proved either in Euclid's *Conics* or by Aristaeus.

Euclid was also aware that a conic may be regarded as the locus of a point having a certain relationship to three or four straight lines. He discussed this locus in his *Conics,* and he may be the original author to whom Pappus thinks Apollonius should have deferred.[80] The locus is thus defined by Pappus:

If three straight lines be given in position, and from one and the same point straight lines be drawn to meet the three straight lines at given angles, and if the ratio of the rectangle contained by two of the straight lines toward the square on the remaining straight line be given, then the point will lie on a solid locus given in position, that is, on one of the three conic sections. And if straight lines be drawn to meet at given angles four straight lines given in position, and the ratio of the rectangle contained by two of the straight lines so drawn toward the rectangle contained by the remaining two be given, then likewise the point will lie on a conic section given in position.[81]

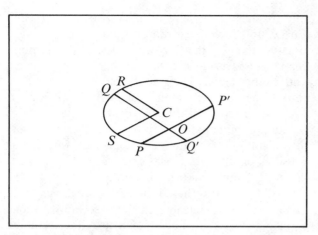

FIGURE 9. $QO \cdot OQ' : PO \cdot OP' = CR^2 : CS^2$

The three-line locus is clearly a special case of the four-line locus in which two of the straight lines coincide. The general case is the locus of a point whose distances x, y, z, u from four straight lines (which may be regarded as the sides of a quadrilateral) have the relationship $xy : zu = k$, where k is a constant.

From the property that the ratio of the rectangles under the segments of any intersecting chords drawn in fixed directions in a conic is constant (being equal to the ratio of the squares on the parallel semidiameters), it is not difficult to show that the distances of a point on a conic from an inscribed trapezium bear the above relationship $xy : zu = k$, and it is only a further step to prove that this is true of any inscribed quadrilateral. It is rather more difficult to prove the converse theorem—that the locus of a point having this relationship to the sides, first of a trapezium, then of any quadrilateral, is a conic section—but it would have been within Euclid's capacity to do so. Apollonius says of his own *Conics:*

> The third book includes many remarkable theorems useful for the synthesis of solid loci and for determining limits of possibility. Most of these theorems, and the most elegant, are new, and it was their discovery which made me realize that Euclid had not worked out the synthesis of the locus with respect to three and four lines, but only a chance portion of it, and that not successfully; for the synthesis could not be completed without the theorems discovered by me.[82]

In the light of this passage Zeuthen conjectured that Euclid (and the other predecessors of Apollonius) saw that a point on a conic section would have the four-line property with respect first to an inscribed trapezium and then to any inscribed quadrilateral, but failed to prove the converse, even for a trapezium; they failed because they did not realize that the hyperbola is a curve with two branches.[83] It is an attractive suggestion.

Pappus exonerates Euclid from blame on the ground that "he wrote so much about the locus as was possible by means of the *Conics* of Aristaeus but did not claim finality for his proofs" and that "neither Apollonius himself nor anyone else could have added anything to what Euclid wrote, using only those properties of conics which had been proved up to Euclid's time."[84] Since Apollonius implies that he had worked out a complete theory, it is curious that he does not set it out in his treatise; but book III, propositions 53–56 of his *Conics,* when taken together, give what is in effect the converse of the three-line locus: "If from any point of a conic there be drawn three straight lines in fixed directions to meet respectively two fixed tangents to the conic and their chord of contact, the ratio of the rectangles contained by the first two lines so drawn to the square on the third is constant."

Apollonius in his first preface claims originality for his book IV and for parts of book III. He regarded the first four books as an introduction concerned with the elements of the subject. Since Pappus says that Apollonius completed the first four books of Euclid's *Conics,* we may infer from the two passages taken together that Euclid's work covered the same ground as Apollonius' first three books, but not so completely. It would appear that Euclid's work was no advance on that of Aristaeus, which would account for the fact that the latter's *Conics* was still extant, although that of Euclid had been lost, by the time of Pappus.

What Pappus calls "those properties of conics which had been proved up to Euclid's time" can be conjectured from references by Archimedes to propositions not requiring demonstration "because they are proved in the elements of conics" or simply "in the *Conics.*"[85] This would imply that the proofs were given by Aristaeus or Euclid or both. In addition, Archimedes assumes without proof the fundamental properties of the parabola, ellipse, and hyperbola in the form, for the ellipse, of

$$PN^2 : AN \cdot A'N = CB^2 : CA^2,$$

where AA' is the major axis, BB' the minor axis, C the center, P any point on the curve, and N the foot of the perpendicular from P to AA'. More generally he assumes that if QV is an ordinate of the diameter PCP of an ellipse (with corresponding formulas for the parabola and hyperbola), the ratio $QV^2 : PV \cdot P'V$ is constant. It would appear that Euclid must have treated the fundamental characteristics of the curves as proportions, and it was left to Apollonius to develop, by means of the application of areas, the fundamental properties of curves as equations between areas.[86]

Surface Loci (Τόποι πρὸς ἐπιφανείᾳ). *Surface Loci,* a work in two books, is attributed to Euclid by Pappus and included in the *Treasury of Analysis.*[87] It has not survived, and its contents can be conjectured only from remarks made by Proclus and Pappus about loci in general and two lemmas given by Pappus to Euclid's work.

Proclus defines a locus as "the position of a line or surface having one and the same property,"[88] and he says of locus theorems (τοπικά) that "some are constructed on lines and some on surfaces." It would appear that loci on lines are loci which are lines and loci on surfaces are loci which are surfaces. But Pappus says that the equivalent of a quadratrix may be obtained geometrically "by means of loci on surfaces as follows," and he proceeds to use a spiral described on a cylinder (the cylindrical helix).[89] The possibility that loci on surfaces may be curves, of a higher order than conic sections, described on surfaces gets some support from an obscure passage in which Pappus divides loci into fixed, progressive, and reversionary, and adds that linear loci are "demonstrated" (δείκνυνται) from loci on surfaces.[90]

Of the two lemmas to the *Surface Loci* which Pappus gives, the former[91] and the attached figure are unsatisfactory as they stand; but if Tannery's restoration[92] is correct, one of the loci sought by Pappus contained all the points on the elliptical parallel sections of a cylinder and thus was an oblique circular cylinder; other loci may have been cones.

It is in the second lemma that Pappus states and proves the focus-directrix property of a conic, which implies, as already stated, that Euclid must have been familiar with it. Zeuthen, following an insight by Chasles,[93] conjectures that Euclid may have used the property in one of two ways in the *Surface Loci:* (1) to prove that the locus of a point whose distance from a given straight line is in a given ratio to its distance from a given plane is a cone; or (2) to prove that the locus of a point whose distance from a given point is in a given ratio to its distance from a given plane is the surface formed by the revolution of a conic about an axis. It seems probable that Euclid's *Surface Loci* was concerned not merely with cones and cylinders (and perhaps spheres), but to some extent with three other second-degree surfaces of revolution: the paraboloid, the hyperboloid, and the prolate (but not the oblate) spheroid. If so, he anticipated to some extent the work that Archimedes developed fully in his *On Conoids and Spheroids.*

Book of Fallacies (Ψευδάρια). Proclus mentions a book by Euclid with this title which has not survived but is clearly identical with the work referred to as *Pseudographemata* by Michael Ephesius in his commentary on the *Sophistici elenchi* of Aristotle.[94] It obviously belonged to elementary geometry and is sufficiently described in Proclus' words:

> Do you, adding or subtracting accidentally, fall away unawares from science and get carried into the opposite error and into ignorance? Since many things seem to conform with the truth and to follow from scientific principles, but lead astray from the principles and deceive the more superficial, he has handed down methods for the clear-sighted understanding of these matters also, and with these methods in our possession we can train beginners in the discovery of paralogisms and avoid being misled. The treatise in which he gave this machinery to us he entitled [the *Book of*] *Fallacies,* enumerating in order their various kinds, exercising our intelligence in each case by theorems of all sorts, setting the true side by side with the false, and combining the refutation of the error with practical illustration. This book is therefore purgative and disciplinary, while the *Elements* contains an irrefutable and complete guide to the actual scientific investigation of geometrical matters.[95]

Astronomy: Phaenomena (Φαινόμενα). This textbook of what the Greeks called *sphaeric,* intended for use by students of astronomy, survives in two recensions, of which the older must be the nearer to Euclid's own words.[96] It was included in the collection of astronomical works which Pappus calls Ὁ Ἀστρονομούμενος τόπος, *The Treasury of Astronomy,* alternatively known as *The Little Astronomy,* in contrast with Ptolemy's *Syntaxis,* or *Great Astronomy.* In the older, more authentic recension it consists of a preface and sixteen propositions.

The preface gives reasons for believing that the universe is a sphere and includes some definitions of technical terms. Euclid in this work is the first writer to use "horizon" absolutely—Autolycus had written of the "horizon (i.e., bounding) circle"—and he introduces the term "meridian circle." The propositions set out the geometry of the rotation of the celestial sphere and prove that stars situated in certain positions will rise or set at certain times. Pappus comments in detail on certain of the propositions.[97]

It is manifest that Euclid drew on Autolycus, but both of them cite without proof a number of propositions, which suggests that they had in their hands a still earlier textbook of sphaeric, which Tannery conjectured to have been the composition of Eudoxus. Many of the propositions are proved in the *Sphaerica* of Theodosius, written several centuries later. He naturally uses the theorems of Euclid's *Elements* in his proofs. By examining the propositions assumed by Autolycus, and by further considering what other propositions are needed to establish them, it is thus possible to get some idea of how much of Euclid's

Elements was already known in the fourth century before Christ.[98]

Optics: Optica ('Οπτικά). The *Optica*, which is attributed to Euclid by Proclus, is also attested by Pappus, who includes it, somewhat curiously, in the *Little Astronomy*.[99] It survives in two recensions; there is no reason to doubt that the earlier one is Euclid's own work, but the later appears to be a recension done by Theon of Alexandria in the fourth century, with a preface which seems to be a pupil's reproduction of explanations given by Theon at his lectures.[100]

The *Optica,* an elementary treatise in perspective, was the first Greek work on the subject and remained the only one until Ptolemy wrote in the middle of the second century.[101] It starts with definitions, some of them really postulates, the first of which assumes, in the Platonic tradition, that vision is caused by rays proceeding from the eye to the object. It is implied that the rays are straight. The second states that the figure contained by the rays is a cone which has its vertex in the eye and its base at the extremities of the object seen. Definition 4 makes the fundamental assumption that "Things seen under a greater angle appear greater, and those under a lesser angle less, while things seen under equal angles appear equal." When he comes to the text, Euclid makes a false start in proposition 1—"Of things that are seen, none is seen as a whole"—because of an erroneous assumption that the rays of light are discrete; but this does not vitiate his later work, which is sound enough. From proposition 6 it is easy to deduce that parallel lines appear to meet. In the course of proposition 8 he proves the equivalent of the theorem

$$\frac{\tan \alpha}{\tan \beta} < \frac{\alpha}{\beta},$$

where α and β are two angles and $\alpha < \beta < 1/2\pi$. There are groups of propositions relating to the appearances of spheres, cones, and cylinders. Propositions 37 and 38 prove that if a straight line moves so that it always appears to be the same size, the locus of its extremities is a circle with the eye at the center or on the circumference. The book contains fifty-eight propositions of similar character. It was written before the *Phaenomena,* for it is cited in the preface of that work. Pappus adds twelve propositions of his own based on those of Euclid.

Catoptrica (Κατοπτρικά). Proclus also attributes to Euclid a book entitled *Catoptrica,* that is, on mirrors. The work which bears that name in the editions of Euclid is certainly not by him but is a later compilation, and Proclus is generally regarded as having made a mistake. If the later compilation is the work

of Theon, as may well be the case, it would have been quite easy for Proclus to have assigned it to Euclid inadvertently.

Music: Elements of Music (Αἱ κατὰ μουσικὴν στοιχειώσεις). Proclus[102] attributes to Euclid a work with this title and Marinus,[103] in his preface to Euclid's *Data,* refers to it as Μουσικῆς στοιχεῖα. Two musical treatises are included in the editions of Euclid's works, but they can hardly both be by the same author, since the *Sectio canonis,* or *Division of the Scale* (Κατατομὴ κανόνος), expounds the Pythagorean doctrine that the musical intervals are to be distinguished by the mathematical ratio of the notes terminating the interval, while the *Introduction to Harmony* (Εἰσαγωγὴ ἁρμονική) is based on the contrary theory of Aristoxenus, according to which the scale is formed of notes separated by a tone identified by the ear. It is now universally accepted that the *Introduction to Harmony* is the work of Cleonides, the pupil of Aristoxenus, to whom it is attributed in some manuscripts; but there is no agreement about the *Sectio canonis,* except that such a trite exposition of the Pythagorean theory of musical intervals is hardly worthy to be dignified with the name *Elements of Music.* The strongest argument for its authenticity is that Porphyry in his commentary on Ptolemy's *Harmonica* quotes almost the whole of it except the preface and twice, or perhaps three times, refers to Euclid's *Sectio canonis* as though it is the work he is quoting;[104] but the passages cited by Porphyry differ greatly from the text in dispute. Gregory, who was the first to question the attribution to Euclid, would have assigned it to Ptolemy,[105] along with the *Introduction to Harmony;* but his main reason, that it is not mentioned before Ptolemy, is not sufficiently strong to outweigh the primitive character of the work. Tannery thinks that the two last propositions, 19 and 20, which specially justify the title borne by the treatise, may have been added by a later editor who borrowed from Eratosthenes, but that the rest of the work must have been composed before 300 B.C. and would attribute it to the school of Plato.[106] This is not convincing, however, for we have seen how Euclid perpetuated arithmetical theories that had become outmoded, and he could have done likewise for the Pythagorean musical theory. (It is of no significance that there are three arithmetical propositions in the *Sectio canonis* not found in the *Elements.*) As for Platonism, Euclid was himself a Platonist. Jan, who included the book under Euclid's name in his *Musici scriptores Graeci,* takes the view that it was a summary of a longer work by Euclid himself.[107] Menge, who edited it for *Euclidis opera omnia,* considers that it contains some things unworthy of Euclid and is of the opinion that it was extracted by some other writer from the authentic *Elements of Music,* now lost.[108] All that it seems possible to say with certainty is that Euclid wrote a book entitled *Elements of Music* and that the *Sectio canonis* has some connection with it.

Mechanics. No work by Euclid on mechanics is extant in Greek, nor is he credited with any mechanical works by ancient writers. According to Arabic sources, however, he wrote a *Book on the Heavy and the Light,* and when Hervagius was about to publish his 1537 edition there was brought to him a mutilated fragment, *De levi et ponderoso,* which he included as one of Euclid's works. In 1900 Curtze published this side by side with a *Liber Euclidis de gravi et levi et de comparatione corporum ad invicem* which he had found in a Dresden manuscript. It is clearly the same work expressed in rather different language and, as Duhem observed, it is the most precise exposition that we possess of the Aristotelian dynamics of freely moving bodies. Duhem himself found in Paris a manuscript fragment of the same work, and in 1952 Moody and Clagett published a text, with English translation, based chiefly on a manuscript in the Bodleian Library at Oxford. A little earlier Sarton had expressed the view that "It contains the notion of specific gravity in a form too clear to be pre-Archimedean"; but it is not all that clear, and there is no reason to think that Archimedes was the first Greek writer to formulate the notion of specific gravity. It is no objection that the dynamics is Aristotelian, for, as Clagett points out, "The only dynamics that had been formulated at all, in the time in which Euclid lived, was the dynamics of Aristotle." In Clagett's judgment, "No solid evidence has been presented sufficient to determine the question of authenticity one way or the other."[109]

In 1851 Woepcke published under the title *Le livre d'Euclide sur la balance* an Arabic fragment that he had discovered in Paris. The fact that the letters used in the figures follow each other in the Greek order suggests a Greek origin. It contains a definition, two axioms, and four propositions and is an attempt to outline a theory of the lever, not on the basis of general dynamical considerations, as in Aristotle, but on the basis of axioms which may be regarded as self-evident and are confirmed by experience. It is therefore Euclidean in character; and since it falls short of the finished treatment of the subject by Archimedes, it could very well be an authentic work of Euclid, although it owes its present form to some commentator or editor. Woepcke found confirmation of its authenticity in a note in another Paris manuscript, *Liber de canonio.* After citing the proposition that the lengths of the arms of a lever parallel to the

horizon are reciprocally proportional to the weights at their extremities, the author adds: "Sicut demonstratum est ab Euclide et Archimede et aliis." Heiberg and Curtze were unwilling to ascribe to the *Book on the Balance* an earlier origin than the Arabs, but Duhem accepted its authenticity. Clagett, after first allowing as "quite likely that the text was translated from the Greek and that in all probability there existed a Greek text bearing the name of Euclid," has more recently expressed the opinion that it "may be genuine and is of interest because, unlike the statement of the law of the lever in the Aristotelian *Mechanics,* its statement on the subject is proved on entirely geometrical grounds."[110]

Duhem's researches among the Paris manuscripts led him to discover a third mechanical fragment attributed to Euclid under the title *Liber Euclidis de ponderibus secundum terminorum circumferentiam.* It contains four propositions about the circles described by the ends of the lever as it rises and falls. As it stands, it is unlikely to be a direct translation of a Euclidean original, but it could derive from a work by Euclid. Duhem noticed how these three fragmentary works fill gaps in each other and conjectured that they might be the debris of a single treatise. This indeed seems probable, and although Duhem was inclined to identify the treatise with Ptolemy's lost work *On Turnings of the Scale* (Περὶ ῥοπῶν), the ultimate author from whom all three fragments spring could have been Euclid. In view of the Arabic traditions, the high probability that the work on the balance is derived from Euclid, the way in which the fragments supplement each other, and the fact that Euclid wrote on all other branches of mathematics known to him and would hardly have omitted mechanics, this is at least a hypothesis that can be countenanced.

An amusing epigram concerning a mule and an ass carrying burdens that the ass found too heavy is attributed to Euclid. It was first printed by Aldus in 1502 and is now included in the appendix to the *Palantine Anthology,* which Melancthon rendered into Latin verse.[111]

NOTES

1. The identification began in ancient times, for Aelian (second/third century), *On the Characteristics of Animals* VI.57, Scholfield ed., II (London–Cambridge, Mass., 1959), 76.26–78.10, notes that spiders can draw a circle and "lack nothing of Euclid" (Εὐκλείδου δέονται οὐδέν).
2. The reference is Archimedes, *On the Sphere and the Cylinder* I.2, Heiberg ed., I, 12.3: διὰ τὸ β τον α τῶν Εὐκλείδου—"by the second [proposition] of the first of the [books] of Euclid." This is the proposition "To place at a given point a straight

line equal to a given straight line." Johannes Hjelmslev, "Über Archimedes' Grössenlehre," in *Kongelige Danske Videnskabernes Selskabs Skrifter,* Matematisk-fysiske Meddelelser, **25,** 15 (1950), 7, considers that the reference should have been to Euclid I.3—"Given two unequal straight lines, to cut off from the greater a straight line equal to the less"—but the reference to Euclid I.2 is what Archimedes needed at that point. Hjelmslev also argues that the reference is inappropriate because Archimedes is dealing with magnitudes, but for Archimedes magnitudes (in this instance, at any rate) can be represented by straight lines to which Euclid's propositions apply. He is on stronger ground, however, in arguing that "Der Hinweis ist aber jedenfalls vollkommen naiv und muss von einem nicht sachkundigen Abschreiber eingesetzt worden sein." Hjelmslev receives some encouragement from E. J. Dijksterhuis, *Archimedes,* p. 150, note, who justly observes: "It might be argued against this that, all the same, Euclidean constructions can be applied to these line segments functioning as symbols. For the rest, the above doubt as to the genuineness of the reference is in itself not unjustified. Archimedes never quotes Euclid anywhere else; why should he do it all at once for this extremely elementary question?" Jean Itard, *Les livres arithmétiques d'Euclide,* pp. 9–10, accepts Hjelmslev's contentions wholeheartedly, and concludes, "Il y a certainement interpolation par quelque scoliaste ou copiste obscur."

The reference was certainly in Proclus' text, for Proclus says that Archimedes mentions Euclid, and nowhere else does he do so. If the reference were authentic, it would be relevant that *On the Sphere and the Cylinder* was probably the fourth of Archimedes' works (T. L. Heath, *The Works of Archimedes,* p. xxxii).

3. It is possible that when Archimedes says in *On the Sphere and the Cylinder* I.6, Heiberg ed., I, 20.15–16, ταῦτα γὰρ ἐν τῇ Στοιχειώσει παραδέδοται, "for these things have been handed down in the *Elements,*" he may be referring to Euclid's *Elements,* particularly XII.2 and perhaps also X.1; but since there were other *Elements,* and the term was also applied to a general body of doctrine not attributable to a particular author, the reference cannot be regarded as certain.

4. Pappus, *Collection* VII.35, Hultsch ed., II, 678.10–12: σχολάσας (Hultsch συσχολάσας) τοῖς ὑπὸ Εὐκλείδου μαθηταῖς ἐν Ἀλεξανδρείᾳ πλεῖστον χρόνον.

5. Proclus, *In primum Euclidis,* Friedlein ed., p. 68.6–23.

6. The word is γέγονε. It literally means "was born"; but E. Rohde in the article "Γέγονε in den Biographica des Suidas," in *Rheinisches Museum für Philologie,* n.s. **33** (1878), 161–220, shows that out of 129 instances in the *Suda* it is certainly equivalent to "flourished" in eighty-eight cases, and probably in another seventeen. This must be the meaning in Proclus, for his anecdote implies that Euclid was not younger than Ptolemy.

7. This was Ptolemy I, commonly called Ptolemy Soter, who was born in 367 or 366 B.C., became ruler of Egypt in 323, declared himself king in 304, effectively abdicated in 285, and died in 283 or 282.

8. The Greek text as printed by Friedlein, p. 68.11–13 from the surviving manuscript M (Monacensis) is γέγονε δὲ οὗτος ὁ ἀνὴρ ἐπὶ τοῦ πρώτου Πτολεμαίου· καὶ γὰρ ὁ Ἀρχιμήδης ἐπιβαλὼν καὶ τῶ πρώτω μνημονεύει τοῦ Εὐκλείδου. The second καί is clearly superfluous, or else a miscopying of some other word (to substitute ἐν would ease the problem of interpretation); and ἐπιβαλὼν is not easy to understand. Grynaeus and August printed the words as Ἀρχιμήδης καὶ ἐν τῷ πρώτῳ in their editions (1533; 1826), and the manuscript Z, which is the basis of Zamberti's Latin translation (1539) did not have ἐπιβαλὼν. Since Heiberg's discussion in *Litterärgeschichtliche Studien über Euklid,* pp. 18–22, the words ὁ Ἀρχιμήδης ἐπιβαλὼν καὶ τῷ πρώτῳ have generally been understood to mean "Archimedes, following closely on the first [Ptolemy]," but Peter Fraser in *Alexandria,* I, 386–388 and II, note 82, offers a new interpretation. He interprets ἐπιβαλὼν as meaning "over-

lapping" and thinks it refers not to Ptolemy but to Euclid, with αὐτῷ understood; he sees τῷ πρώτῳ, understood as ἐν τῷ πρώτῳ, as a reference to the first work in the Archimedean corpus, that is, *On the Sphere and the Cylinder.* His translation is therefore "This man flourished under the first Ptolemy; for Archimedes, who overlapped with him, refers to him in his first book [?]." The theory is attractive, but I do not agree with Fraser that there is any awkwardness in τῷ πρώτῳ referring to Ptolemy so soon after ἐπὶ τοῦ πρώτου, nor do I see any difficulty in saying that Archimedes (*b.* 287) followed closely on Ptolemy I (abdicated 285, *d.* 283/282). On the whole, therefore, I prefer Heiberg's intepretation, but Fraser's full discussion merits careful study.

9. Archimedes died in the siege of Syracuse in 212 B.C., according to Tzetzes, at the age of seventy-five; if so, he was born in 287. Eratosthenes, to whom Archimedes dedicated *The Method,* was certainly a contemporary, but the work in which he said so has not survived. The *Suda* records that he was born in the 126th olympiad (276–273 B.C.).

10. Stobaeus, *Eclogues* II, 31.115, *Anthologium,* Wachsmuth and Hense, eds., II (Berlin, 1884); 228.30–33.

11. See T. L. Heath, *The Thirteen Books of Euclid's Elements,* 2nd ed., I, 117–124, 146–151. "On the whole I think it is from Aristotle that we get the best idea of what Euclid understood by a postulate and an axiom or common notion" (*ibid.,* p. 124). See also T. L. Heath, *Mathematics in Aristotle* (Oxford, 1949), pp. 53–57.

12. Aristotle, *Prior Analytics* I, 24, 41b13–22.

13. Hultsch, in Pauly-Wissowa, VI, col. 1004, also gives 295 B.C. as the date of Euclid's ἀκμή. The latest and most thorough discussion is by Peter Fraser, in *Alexandria,* I, 386–388, with notes in II, especially note 82. He concludes that Euclid may have been born about 330–320 B.C. and did not live much, if at all, after about 270 B.C. This would give him a middle date of 300–295. The round figure of 300 B.C. is given by Hankel, Gow, Zeuthen, Cantor, Loria, Hoppe, Heath, and van der Waerden. Michel gives the same date "au plus tard." Thaer puts Euclid's productive period (*Wirksamkeit*) in the last decade of the fourth century B.C. Heinrich Vogt, "Die Entdeckungsgeschichte des Irrationalen nach Plato usw.," in *Biblioteca mathematica,* 3rd ser., **10** (1910), 155, and—in greater detail—in "Die Lebenzeit Euclids," *ibid.,* **13** (1913), 193–202, puts Euclid's birth at about 365 and the composition of the *Elements* at 330–320. Similar dates are given by Max Steck in his edition of P. L. Schönberger, *Proklus Diadochus* (Halle, 1945), and in *Forschungen und Fortschritte,* **31** (1957), 113; but these authors, as Fraser rightly notes, do not pay sufficient attention to the links of Euclid with Ptolemy Soter and of his pupils with Apollonius. The latest date suggested for Euclid's *floruit* is 280 B.C., given in the brief life prefixed by R. N. Adams to the twenty-first and subsequent editions of Robert Simson's *Elements* (London, 1825), but it rests on no reasoned argument.

14. It would hardly be necessary to mention this confusion, of which the first hint is found in Valerius Maximus VIII.12, Externa 1, in the reign of Tiberius (14–37), were it not common in the Middle Ages and repeated in all the printed editions of Euclid from 1482 to 1566. Karl R. Popper, *Conjectures and Refutations,* 3rd ed. (London, 1969), p. 306, has revived it ("Euclid the Geometrician . . . you don't mean the man from Megara, I presume"), but only, it must be assumed, in jest.

Jean Itard has recently advanced the theory that Euclid may not have been an individual but a school. In *Les livres arithmétiques d'Euclide,* p. 11, he advances three hypotheses: (1) that Euclid was a single individual who composed the various works attributed to him; (2) that Euclid was an individual, the head of a school which worked under him and perhaps continued after his death to produce books to which they gave his name; (3) that a group of Alexandrian mathematicians issued their works under the name of Euclid of Megara, just as (he alleges) the chemists of the same period attributed their works to Democritus. Of these speculations he thinks the second "paraît être la plus raisonnable." Itard exaggerates "les difficultés qui surgissent à chaque instant dans la chronologie lorsque l'on admet l'existence d'un seul Euclide"; there is a lack of precise information but no difficulty about Euclid's chronology. No one has hitherto seen any reason for thinking that the author of the *Elements* could not also have been the author of the other books attributed to him. There are differences within the books of the *Elements* themselves, notably the difference between books VII–IX, with which Itard is particularly concerned, and books V and X; but these are explicable by less drastic suggestions, as will be shown later. The reason why the name of Euclid the Geometer ever came to be confused with Euclid of Megara, who lived a century earlier, is clear from the passage of Valerius Maximus cited above. Valerius says that Plato, on being asked for a solution to the problem of making an altar double a cubical altar, sent his inquiries to "Euclid the geometer." One early commentator wished to alter this to "Eudoxus," which is probably right. The first specific identification of Euclid the Geometer with Euclid of Megara does not occur until Theodorus Metochita (*d.* 1332), who writes of "Euclid of Megara, the Socratic philosopher, contemporary with Plato" as the author of works on plane and solid geometry, data, optics, and so on. Euclid was a common Greek name; Pauly-Wissowa lists no fewer than eight Eukleides and twenty Eukleidas.

15. The idea that he was born at Gela in Sicily springs from the same confusion. Diogenes Laertius II.106 says that he was "of Megara, or according to some, of Gela, as Alexander says in the *Diadochai.*"

16. The passage is bracketed by Hultsch, but he brackets with frequency and not always with convincing reason.

17. Stobaeus, *Eclogues* II, 31.114, Wachsmuth and Hense eds., II, 228.25–29.

18. The so-called books XIV and XV are not by Euclid. Book XIV is by Hypsicles, probably in the second century B.C.; book XV, by a pupil of Isidore of Miletus in the sixth century.

19. Proclus, *In primum Euclidis,* Friedlein ed., p. 72.6–13.

20. This is one of two mathematical uses of the term διορισμός, the other being a determination of the conditions of possibility. In the present sense it is almost part of the particular enunciation. Proclus, Friedlein ed., pp. 203.1–205.12, explains these formal divisions of a proposition.

21. The fullest discussion is in Pappus, *Collection* VII, Pref. 1–3, Hultsch ed., II, 634.3–636.30. It was James Gow, *A Short History of Greek Mathematics,* p. 211, note 1, who first recognized that the correct translation of τόπος ἀναλυόμενος was "storehouse (or treasury) of analysis."

22. Aristotle, *Posterior Analytics* I, 10, 76a31–77a4.

23. The equidistance theory was represented in antiquity by Posidonius, as quoted by Proclus, *In primum Euclidis,* Friedlein ed., p. 176.7–10; Geminus, also as quoted by Proclus, *ibid.,* p. 177.13–16; and Simplicius as quoted by al-Nayrīzī, Curtze ed., pp. 25.8–27.14. (The "philosopher Aganis" also quoted in this passage must be Geminus.) The direction theory is represented by Philoponus in his comment on Aristotle, *Posterior Analytics* II, 16, 65a4 and was probably held by Aristotle himself.

24. For Ptolemy's attempt, see Proclus, *In primum Euclidis,* Friedlein ed., pp. 365.5–367.27; for Proclus' own attempt, *ibid.,* pp. 368.24–373.2.

25. Proclus, *In primum Euclidis,* Friedlein ed., pp. 419.15–420.23, explains at some length what is meant by the application of areas, their exceeding, and their falling short.

26. *Ibid.,* p. 269.8–21.

27. Isaac Barrow, in lecture VIII of 1666, *Lectiones habitae in scholis publicis academiae cantabrigiensis* (Cambridge, 1684), p. 336, states, "Cum hoc elegio praefixum hanc disputationem claudo, nihil extare (me judice) in toto Elementorum opere proportionalitatum doctrinam subtilius inventum, solidius stabilitum, accuratius pertractatum." The English translation

is that of Robert Simson, at the end of his notes to book V of *The Elements of Euclid,* 21st ed. (London, 1825), p. 294. Simson "most readily" agrees with Barrow's judgment.

28. Augustus De Morgan, "Proportion," in *The Penny Cyclopaedia,* XIX, 51. Oskar Becker's theory (see Bibl.) that there was an earlier general theory of proportion hinted at by Aristotle is discussed in the article on Theaetetus.

29. H. G. Zeuthen, *Lehre von den Kegelschnitten im Altertum,* p. 2.

30. Heath, *The Thirteen Books of Euclid's Elements,* I, 124–126.

31. Plutarch, *Quaestiones conviviales* VIII, 2, 4.720a—compare *Non posse suaviter vivi secundum Epicurum,* 11, 1094b—says that the discovery of this proposition, rather than the one about the square on the hypotenuse of a right-angled triangle, was the occasion of a celebrated sacrifice by Pythagoras.

32. According to VII, definition 21 (20), "Numbers are proportional when the first is the same multiple, or the same part, or the same parts, of the second as the third is of the fourth." H. G. Zeuthen, *Histoire des mathématiques,* p. 128, comments: "Sans doute, en ce qui concerne l'égalité des rapports, cette définition ne renferme rien d'autre que ce qu'impliquait déjà la cinquième définition du cinquième Livre." The same author, in his article "Sur la constitution des livres arithmétiques des Éléments d'Euclide," in *Oversigt over det K. Danske Videnskabernes Selskabs Forhandlinger* (1910), 412–413, sees a point of contact between the special and general theories in VII.19, since it shows that the definition of proportion in V, definition 5, has, when applied to numbers, the same significance as in VII, definition 21 (20); and we can henceforth borrow any of the propositions proved in book V.,

33. "This book has a completeness which none of the others (not even the fifth) can boast of: and we could almost suspect that Euclid, having arranged his materials in his own mind, and having completely elaborated the tenth Book, wrote the preceding books after it, and did not live to revise them thoroughly" (Augustus De Morgan, "Eucleides," in *Dictionary of Greek and Roman Biography and Mythology,* William Smith, ed., II [London, 1846], 67). See also W. W. Rouse Ball, *A Short Account of the History of Mathematics,* 3rd ed. (London, 1901), pp. 58–59.

34. After defining a point as "that which has no part" (I, definition 1), Euclid says, "The extremities of a line are points" (I, definition 3), which serves to link a line with a point but is also a concession to an older definition censured by Aristotle as unscientific. So for lines and surfaces (I, definition 6). Among unused definitions are "oblong," "rhombus," and "rhomboid," presumably taken over from earlier books.

35. Paul-Henri Michel, *De Pythagore à Euclide,* p. 92.

36. See Augustus De Morgan, "Eucleides," in *Smith's Dictionary of Greek and Roman Biography,* p. 67; and "Irrational Quantity," in *The Penny Cyclopaedia,* XIII, 35–38.

37. See Eva Sachs, *Die fünf Platonischen Körper* (Berlin, 1917).

38. Proclus, *In primum Euclidis,* Friedlein ed., p. 69.4–27.

39. In Greek, συλλογισμοί, but the word can hardly be used here in its technical sense. Two attempts have been made to turn the *Elements* into syllogisms!

40. Geometrical conversion is discussed by Proclus, *In primum Euclidis,* Friedlein ed., pp. 252.5–254.20.

41. *Ibid.,* p. 68.7–10.

42. Scholium 1, book V, *Euclidis opera omnia,* Heiberg and Menge, eds., V, 280.7–9.

43. Scholium 3, book V, *ibid.,* p. 282.13–20.

44. Archimedes, *On the Sphere and the Cylinder* I, preface, *Archimedis opera omnia,* Heiberg ed., I, 2–13; compare *Quadrature of the Parabola,* preface, *ibid.,* 262–266.

45. Simplicius, *Commentary on Aristotle's Physics* A2, 185a14, Diels ed. (Berlin, 1882), 61.8–9.

46. Plato, *Theaetetus,* 147D ff.

47. Franz Woepcke, in *Mémoires présentés à l'Académie des sciences,* **14** (1856), 658–720; W. Thomson, *Commentary of Pappus on Book X of Euclid's Elements,* Arabic text and trans., remarks, notes, glossary by G. Junge and Thomson (Cambridge, Mass., 1930; repr., 1968).

48. J. L. Heiberg, *Litterärgeschichtliche Studien über Euklid,* pp. 169–171.

49. Scholium 62, book X, *Euclidis opera omnia,* Heiberg and Menge, eds., V, 450.16.

50. T. L. Heath, *The Thirteen Books of Euclid's Elements,* III, 3–4; for Thomson's trans., see *op. cit.,* pp. 63–64.

51. *Suda Lexicon, s.v.,* Adler ed., I.2 (Leipzig, 1931), Θ 93, p. 689.6–8.

52. Scholium 1, book XIII, *Euclidis opera omnia,* Heiberg and Menge, eds., V, 654.5–6.

53. Proclus, *In primum Euclidis,* Friedlein ed., p. 419.15–18.

54. Scholium 4, book IV, *Euclidis opera omnia,* Heiberg and Menge, eds., V, 273.13–15.

55. Proclus, *In primum Euclidis,* Friedlein ed., p. 426.6–18.

56. Aristotle, *Prior Analytics* II, 16, 65a4.

57. Compare Karl Popper, *Conjectures and Refutations,* 3rd ed., pp. 88–89: "Ever since Plato and Euclid, but not before, geometry (rather than arithmetic) appears as the fundamental instrument of all physical explanations and descriptions, in the theory of matter as well as in cosmology." Popper has no doubt that Euclid was a Platonist and that the closing of the *Elements* with the construction of the Platonic figures is significant.

58. Pappus, *Collection* VII.3, Hultsch ed., II, 636.19.

59. *Ibid.,* pp. 638.1–640.3.

60. Proclus, *In primum Euclidis,* Friedlein ed., p. 69.4.

61. *Ibid.,* p. 644.22–26.

62. A detailed classification is made in R. C. Archibald, *Euclid's Book on Divisions of Figures,* pp. 15–16.

63. Hero, *Metrica* III.8, Schöne ed., III, *Heronis Alexandrini opera quae supersunt omnia* (Leipzig, 1903), 172.12–174.2, considers the related problem "To divide the area of a circle into three equal parts by two straight lines." "That this problem is not rational," he notes, "is clear"; but because of its utility he proceeds to give an approximate solution.

64. Pappus, *Collection* VII.13, Hultsch ed., II, 648.18–19.

65. Proclus, *In primum Euclidis,* Friedlein ed., p. 302.12–13.

66. If it had survived, it might have led B. L. van der Waerden to modify his judgments in *Science Awakening,* 2nd ed. (Groningen, undated [1956?]), p. 197: "Euclid is by no means a great mathematician . . . Euclid is first of all a pedagogue, not a creative genius. It is very difficult to say which original discoveries Euclid added to the work of his predecessors."

67. Pappus, *Collection* VII.14, Hultsch ed., II, 650.16–20.

68. Proclus, *In primum Euclidis,* Friedlein ed., p. 301.22–26.

69. Pappus, *Collection* VII.16, Hultsch ed., II, 652.18–654. Some words have been added in the translation for the sake of clarity.

70. Robert Simson, *De porismatibus tractatus* in *Opera quaedam reliqua* (Glasgow, 1776), pp. 392–393, elucidated this passage in elegant Latin, which Gino Loria, in *Le scienze esatte nell'antica Grecia,* 2nd ed. (Milan, 1914), pp. 256–257, first put into modern notation.

71. Pappus, *Collection* VII.17, Hultsch ed., II, 654.16–19.

72. *Ibid.,* VII.14, Hultsch ed., II, 652.2.

73. H. G. Zeuthen, *Die Lehre von den Kegelschnitten im Altertum,* pp. 165–184.

74. Pappus, *Collection* VII.18, Hultsch ed., II, 656.1–4.

75. H. G. Zeuthen, *op. cit.,* p. 152.

76. Pappus, *Collection* VII.30, Hultsch ed., II, 672.18–20.

77. *Ibid.,* pp. 676.25–678.15.

78. Euclid, *Phaenomena,* preface, in *Euclidis opera omnia,* Heiberg and Menge, eds., VIII, 6.5–7.

79. Pappus, *Collection* VII.312, Hultsch ed., II, 1004.23–1006.2.

80. τῷ πρώτῳ γράψαντι, *ibid.,* p. 678.14.

81. *Ibid.,* p. 678.15–24.

82. Apollonius, *Conics* I, preface, *Apollonii Pergaei quae Graece exstant,* Heiberg ed., I (Leipzig, 1891), 4.10–17.

83. H. G. Zeuthen, *Die Lehre von den Kegelschnitten im Altertum,* pp. 136–139.

84. Pappus, *Collection* VII.35, Hultsch ed., II, 678.4–6; *ibid.,* VII.33, p. 676.21–24.

85. Archimedes, *Quadrature of a Parabola,* proposition 3, in *Archimedes opera omnia,* II, 2nd Heiberg ed. (Leipzig, 1910–1915), 268.3; *On Conoids and Spheroids,* proposition 3, Heiberg ed., I, 270.23–24; *ibid.,* p. 274.3. But when the Latin text of *On Floating Bodies,* II.6, Heiberg ed., II, 362.10–11, says of a certain proposition, "Demonstratum est enim hoc per sumpta," it probably refers to a book of lemmas rather than to Euclid's *Conics.*

86. For a full discussion of the propositions assumed by Archimedes, the following works may be consulted: J. L. Heiberg, "Die Kenntnisse des Archimedes über die Kegelschnitte," in *Zeitschrift für Mathematik und Physik,* Jahrgang 25, Hist.-lit. Abt. (1880), 41–67; and T. L. Heath, *Apollonius of Perga,* pp. l–lxvi; *The Works of Archimedes,* pp. lii–liv; *A History of Greek Mathematics,* II (Oxford, 1921), 121–125.

87. Pappus, *Collection* VII.3, Hultsch ed., II, 636.23–24.

88. Proclus, *In primum Euclidis,* Friedlein ed., p. 394.17–19.

89. Pappus, *Collection* IV.51, Hultsch ed., I, 258.20–262.2.

90. *Ibid.,* VII.21, Hultsch ed., II, 660.18–662.22. A large part of the passage is attributed to an interpolator by Hultsch, but without reasons.

91. *Ibid.,* VII.312, p. 1004.17–22.

92. Paul Tannery, review of J. L. Heiberg's *Litterärgeschichtliche Studien über Euklid,* in *Bulletin des sciences mathématiques,* 2nd ser., **6** (1882), 149–150; reprinted in *Mémoires scientifiques,* XI (Toulouse–Paris, 1931), 144–145.

93. Michel Chasles, "Aperçu historique," pp. 273–274; H. G. Zeuthen, *Die Lehre von den Kegelschnitten im Altertum,* pp. 423–431. J. L. Heiberg takes a different view in his *Litterärgeschichtliche Studien über Euklid,* p. 79.

94. Alexander (?), *Commentary on Aristotle's Sophistici elenchi,* Wallies ed., (Berlin, 1898), p. 76.23.

95. Proclus, *In primum Euclidis,* Friedlein ed., pp. 69.27–70.18.

96. The older recension is, however, best illustrated in a Vienna manuscript of the twelfth century; the later recension is found in a Vatican manuscript of the tenth century.

97. Pappus, *Collection* VI.104–130, Hultsch ed., II, 594–632.

98. The task was attempted by Hultsch, *Berichte über die Verhandlungen der Kgl. Sächsischen Gesellschaft der Wissenschaften zu Leipzig,* Phil.-hist. Classe, **38** (1886), 128–155. The method definitely establishes as known before Euclid the following propositions: I.4, 8, 17, 19, 26, 29, 47; III.1–3, 7, 10, 16 (corollary), 26, 28, 29; IV.6; XI.3, 4, 10, 11, 12, 14, 16, 19, and 38 (interpolated). But Hultsch went too far in adding the whole chain of theorems and postulates leading up to these propositions, for in some cases (e.g., I.47) Euclid worked out a novel proof.

99. Proclus, *In primum Euclidis,* Friedlein ed., p. 69.2; Pappus, *Collection* VI.80–103, Hultsch ed., II, 568.12–594.26.

100. Only the later recension was known until the end of the nineteenth century, but Heiberg then discovered the earlier one in Viennese and Florentine manuscripts. Both recensions are included in the Heiberg-Menge *Opera omnia.*

101. See A. Lejeune, *Euclide et Ptolémée: Deux stades de l'optique géométrique grecque.*

102. Proclus, *In primum Euclidis,* Friedlein ed., p. 69.3.

103. Marinus, *Commentary on Euclid's Data,* preface, in *Euclidis opera omnia,* Heiberg and Menge, eds., VIII, 254.19.

104. Καὶ αὐτὸς ὁ Στοιχειωτὴς Εὐκλείδης ἐν τῇ τοῦ Κανόνος κατατομῇ, Porphyry, *Commentary on Ptolemy's Harmonies,* Wallis ed., *Opera mathematica,* III (Oxford, 1699), 267.31–32; ἐν τῇ τοῦ Κανόνος Κατατομῇ Εὐκλείδου, *ibid.,* 272.26–27; Καὶ αὐτῷ τῷ Στοιχειωτῇ καὶ ἄλλοις πολλοῖς κανονικοῖς, *ibid.,* 269.5–6.

105. David Gregory, *Euclidis quae supersunt omnia,* preface.

106. Paul Tannery, "Inauthenticité de la *Division du canon* attribuée à Euclide," in *Comptes rendus des séances de l'Académie des inscriptions et belles-lettres,* **4** (1904), 439–445; also in his *Mémoires,* III, 213–219.

107. *Excerpta potius dicas quam ipsa verba hominis sagacissimi,* in C. Jan, ed., *Musici scriptores Graeci,* p. 118.

108. *Euclidis opera omnia,* Heiberg and Menge, eds., VIII, xxxvii–xlii.

109. J. Hervagius, ed., "Euclidis de levi et ponderoso fragmentum," in *Euclidis Megarensis mathematici clarissimi Elementorum geometricorum libri xv* (Basel, 1537), pp. 585–586, and foreword; M. Curtze, "Zwei Beiträge zur Geschichte der Physik im Mittelalter," in *Biblioteca mathematica*, 3rd ser., **1** (1900), 51–54; P. Duhem, *Les origines de la statique*, I, 61–97; and George Sarton, *Introduction to the History of Science*, I, 156.

110. F. Woepcke, "Notice sur des traductions arabes de deux ouvrages perdus d'Euclide," in *Journal asiatique*, 4th ser., **18** (1851), 217–232; M. Curtze, "Das angebliche Werk des Eukleides über die Waage," in *Zeitschrift für Mathematik und Physik*, **19** (1874), 262–263; P. Duhem, *op. cit.*, pp. 61–97; Marshall Clagett, *The Science of Mechanics in the Middle Ages*, p. 28; and *Greek Science in Antiquity* (London, 1957), p. 74.

111. *Anthologia palatina, Appendix nova epigrammatum*, Cougny ed. (Paris, 1890), 7.2; *Euclidis opera omnia*, Heiberg and Menge, eds., VIII.285, with Melancthon's rendering on p. 286.

BIBLIOGRAPHY

I. ORIGINAL WORKS. The definitive edition of Euclid's extant works is *Euclidis opera omnia*, J. L. Heiberg and H. Menge, eds., 8 vols. plus suppl., in the Teubner Classical Library (Leipzig, 1883–1916). It gives a Latin translation of Euclid's works opposite the Greek text and includes the spurious and doubtful works, the scholia, Marinus' commentary on the *Data*, and the commentary on books I-X of the *Elements* by al-Nayrīzī in Gerard of Cremona's Latin translation. The details are: I, *Elementa I–IV*, J. L. Heiberg, ed. (1883); II, *Elementa V–IX*, J. L. Heiberg, ed. (1884); III, *Elementa X*, J. L. Heiberg, ed. (1886); IV, *Elementa XI–XIII*, J. L. Heiberg, ed. (1885); V, *Elementa qui feruntur XIV–XV. Scholia in Elementa*, J. L. Heiberg, ed. (1888); VI, *Data cum commentario Marini et scholiis antiquis*, H. Menge, ed. (1896); VII, *Optica, Opticorum recensio Theonis, Catoptrica cum scholiis antiquis*, J. L. Heiberg, ed. (1895); VIII, *Phaenomena et scripta musica*, H. Menge, ed., *Fragmenta*, collected and arranged by J. L. Heiberg (1916); suppl., *Anaritii in decem libros priores Elementorum Euclidis commentarii*, M. Curtze, ed. (1899).

The *Sectio canonis* and the (Cleonidean) *Introductio harmonica* are also included in *Musici scriptores Graeci*, C. Jan, ed. (Leipzig, 1895; repr. Hildesheim, 1962), pp. 113–166 and 167–208, respectively.

The text of Heiberg's edition of the *Elements* has been reproduced by E. S. Stamatis in four volumes (Athens, 1952–1957), with a trans. into modern Greek, introductions, and epexegeses. Stamatis is also bringing out a new edition of the *Elements* in the Teubner series reproducing Heiberg's text and variant readings. Heiberg's Latin translation is omitted but the notes to it are reproduced and assigned to the corresponding place in the Greek text. The variant readings and notes take account of critical editions later than those available to Heiberg and there are additional notes on the mathematics, ancient testimonies, a bibliography, and relevant papyrus fragments. The first vol. is *Euclidis Elementa I, Libri I–IV, cum appendicibus, post I. L. Heiberg*, E. S. Stamatis, ed. (Leipzig, 1969).

The first printed Latin translation of the *Elements* appeared at Venice in 1482; the first edition of the Greek text, edited by Simon Grynaeus, at Basel in 1533. The first complete edition of the works of Euclid in Greek, edited by David Gregory, was published at Oxford in 1703, and it remained the only complete edition until that of Heiberg and Menge. Most of the early editions and translations are listed in the following works: Thomas L. Heath, *The Thirteen Books of Euclid's Elements*, 3 vols. (Cambridge, 1905, 1925; New York, 1956), I, 91–113—97 titles from 1482 to 1820; Charles Thomas Stanford, *Early Editions of Euclid's Elements* (London, 1926)—84 titles from 1482 to 1600, with bibliographical illustrations; Max Steck, "Die geistige Tradition der frühen Euklid-Ausgaben," in *Forschungen und Fortschritte*, **31** (1957), 113–117—60 titles to 1600; F. J. Duarte, *Bibliografia in Eucleides, Arquimèdes, Newton* (Caracas, 1967)—123 titles of editions of the *Elements*, with bibliographical illustrations.

II. SECONDARY LITERATURE. Two ancient works are of prime importance: the commentary of Proclus on the first book of the *Elements* and the *Collection* of Pappus. Both are available in good editions. Proclus: *Procli Diadochi in primum Euclidis Elementorum librum commentarii*, G. Friedlein, ed. (Leipzig, 1883, repr. 1967); Thomas Taylor, *The Philosophical and Mathematical Commentaries of Proclus on the First Book of Euclid's Elements* (London, 1788–1789, 1791), is superseded by G. R. Morrow, *Proclus. A Commentary on the First Book of Euclid's Elements* (Princeton, N.J., 1970); more useful trans. of the most relevant passages are scattered T. L. Heath, *op cit.*; a German trans. and commentary, *Proklus Diadochus 410–485 Kommentar zum ersten Buch von Euklids Elementen*, P. Leander Schönberger, trans., intro. by Max Steck (Halle, 1945); and a French trans., Paul ver Eecke, *Proclus de Lycie. Les commentaires sur le premier livre des Éléments d'Euclide* (Paris-Bruges, 1948). Pappus: *Pappi Alexandrini Collectionis quae supersunt*, F. Hultsch, ed., 3 vols. (Berlin 1876–1878), with a French trans. by Paul ver Eecke, *Pappus d'Alexandrie: La Collection mathématique*, 2 vols. (Paris-Bruges, 1933).

An extensive modern literature has grown around Euclid. The older works which have not been superseded and the chief recent literature may be classified as follows:

General: J. L. Heiberg, *Litterärgeschichtliche Studien über Euklid* (Leipzig, 1882); F. Hultsch, "Autolykos und Euklid," in *Berichte der Verhandlung der Kgl. Sächsischen Gesellschaft der Wissenschaften zu Leipzig*, Phil-hist. Classe, **38** (1886), 128–155; Max Simon, "Euclid und die sechs planimetrischen Bücher," in *Abhandlungen zur Geschichte der mathematischen Wissenschaften*, **11** (1901); Thomas L. Heath, *The Thirteen Books of Euclid's Elements* (see above), I, 1–151; F. Hultsch, "Eukleides 8," in Pauly-Wissowa, II (Leipzig, 1907), cols. 1003–1052; Estelle A. DeLacy, *Euclid and Geometry* (London, 1965); E. J. Dijksterhuis, *De Elementen van Euclides*, 2 vols. (Groningen, 1929–1930); Jürgen Mau, "Eukleides 3," in *Die kleine Pauly*, II (Stuttgart, 1967), cols. 416–419.

Elements: General—Max Simon, "Euclid und die sechs planimetrischen Bücher"; T. L. Heath, *The Thirteen Books of Euclid's Elements*; E. J. Dijksterhuis, *De Elementen van Euclides*; Clemens Thaer, *Die Elemente von Euklid*, 5 pts. (Leipzig, 1933–1937); and A. Frajese and L. Maccioni, *Gli Elementi di Euclide* (Turin, 1970).

Postulates and axioms—Girolamo Saccheri, *Euclides ab omni naevo vindicatus* (Milan, 1733), and an English trans. of the part relating to postulate 5, George Bruce Halstead, *Girolamo Saccheri's Euclides Vindicatus Edited and Translated* (Chicago-London, 1920); B. L. van der Waerden, *De logische grondslagen der Euklidische meetkunde* (Groningen, 1937); A. Frenkian, *Le postulat chez Euclide et chez les modernes* (Paris, 1940); Cydwel A. Owen, *The Validity of Euclid's Parallel Postulate* (Caernarvon, 1942); A. Szabó, "Die Grundlagen in der frühgriechischen Mathematik," in *Studi italiani di filologia classica,* n.s. **30** (1958), 1–51; "Anfänge des euklidischen Axiomensystems," in *Archive for History of Exact Sciences,* **1** (1960), 37–106; "Was heisst der mathematische Terminus ἀξίωωα," in *Maia,* **12** (1960), 89–105; and "Der älteste Versuch einer definitorisch-axiomatischen Grundlegung der Mathematik," in *Osiris,* **14** (1962), 308–309; Herbert Meschkowski, *Grundlagen der Euklidischen Geometrie* (Mannheim, 1966); G. J. Pineau, *The Demonstration of Euclid's Fifth Axiom, The Treatment of Parallel Lines Without Euclid's Fifth Axiom, The Self-contradiction of Non-Euclidean Geometry, The Fault of Euclid's Geometry* (Morgan Hill, Calif.); Imre Tóth, "Das Parallelproblem in Corpus Aristotelicum," in *Archive for History of Exact Sciences,* **3** (1967), 1–422; and N. C. Zouris, *Les demonstrations du postulat d'Euclide* (Grenoble, 1968). See also Frankland (below).

Book I—William Barrett Frankland, *The First Book of Euclid's Elements With a Documentary Based Principally Upon That of Proclus Diadochus* (Cambridge, 1905).

Book V—Augustus De Morgan, "Proportion," in *The Penny Cyclopaedia,* XIX (London, 1841), 49–53; O. Becker, "Eudoxus-Studien I. Eine voreudoxische Proportionenlehre und ihre Spuren bei Aristoteles und Euklid," in *Quellen und Studien zur Geschichte der Mathematik,* **2** (1933), 311–333; Friedhelm Beckmann, "Neue Gesichtspunkte zum 5. Buch Euklids," in *Archive for History of Exact Sciences,* **4** (1967), 1–144.

Books VII–IX—Jean Itard, *Les livres arithmétiques d'Euclide* (Paris, 1962).

Book X—[Augustus De Morgan], "Irrational Quantity," in *The Penny Cyclopaedia,* XIII (London, 1839), 35–38; H. G. Zeuthen, "Sur la constitution des livres arithmétiques des Eléments d'Euclide et leur rapport à la question de l'irrationalité," in *Oversigt over det K. Danske Videnskabernes Selskabs Forhandlinger* (1915), pp. 422 ff.; William Tomson (and Gustav Junge), *The Commentary of Pappus on Book X of Euclid's Elements: Arabic Text and Translation,* Harvard Semitic Series, VIII (Cambridge, Mass., 1930; repr. 1968).

Data: M. Michaux, *Le commentaire de Marinus* (Paris, 1947).

On Divisions of Figures: Franz Woepcke, "Notice sur des traductions arabes de deux ouvrages perdus d'Euclide," in *Journal asiatique,* 4th ser., **18** (1851), 233–247; R. C. Archibald, *Euclid's Book on Divisions of Figures With a Restoration Based on Woepcke's Text and the Practica Geometriae of Leonardo Pisano* (Cambridge, 1915).

Porisms: Robert Simson, *De Porismatibus tractatus,* in *Opera quaedam reliqua* (Glasgow, 1776), pp. 315–594; Michel Chasles, *Les trois livres de Porismes d'Euclide* (Paris,

1860); H. G. Zeuthen, *Die Lehre von den Kegelschnitten im Altertum* (Copenhagen, 1886; repr. Hildesheim, 1966), pp. 160–184.

Conics: H. G. Zeuthen, *Die Lehre von den Kegelschnitten im Altertum* (see above), pp. 129–130; T. L. Heath, *Apollonius of Perga* (Cambridge, 1896), pp. xxxi-xl.

Surface Loci: Michel Chasles, "Aperçu historique sur l'origine et le développement des méthodes en géométrie," in *Mémoires couronnés par l'Académie royale des sciences et des belles-lettres de Bruxelles,* II (Brussels, 1837), note 2, "Sur les lieux à la surface d'Euclide," 273–274; H. G. Zeuthen, *op. cit.,* pp. 423–431; T. L. Heath, *The Works of Archimedes* (Cambridge, 1897), pp. lxi-lxvi.

Optics: Giuseppe Ovio, *L'Ottica di Euclide* (Milan, 1918); Paul ver Eecke, *Euclide: L'Optique et la Catoptrique* (Paris-Bruges, 1938); Albert Lejeune, *Euclide et Ptolemée: Deux stades de l'optique géométrique grecque* (Louvain, 1948).

Catoptrica: Paul ver Eecke, *Euclide: L'Optique et la Catoptrique* (see above); Albert Lejeune, "Les 'Postulats' de la Catoptrique dite d'Euclide," in *Archives internationales d'histoires des sciences,* no. 7 (1949), 598–613.

Mechanics: Franz Woepcke, "Notice sur des traductions arabes de deux ouvrages perdus d'Euclide," in *Journal asiatique,* 4th ser., **18** (1851), 217–232; M. Curtze, "Zwei Beiträge zur Geschichte der Physik im Mittelalter," in *Biblioteca mathematica,* 3rd ser. **1** (1900), 51–54; P. Duhem, *Les origines de la statique* (Paris, 1905), pp. 61–97; E. A. Moody and Marshall Clagett, *The Medieval Science of Weights* (Madison, Wis., 1952), which includes *Liber Euclidis de ponderoso et levi et comparatione corporum ad invicem,* Marshall Clagett, ed., with intro., English trans., and notes by Ernest A. Moody; Marshall Clagett, *The Science of Mechanics in the Middle Ages* (Madison, Wis., 1959), which contains "*The Book on the Balance* Attributed to Euclid," trans. from the Arabic by Clagett.

Euclid and his works occupy a prominent place in many histories of mathematics, or Greek mathematics, including Jean Étienne Montucla, *Histoire des mathématiques,* 2nd ed., I (Paris, 1798), 204–217; Hermann Hankel, *Zur Geschichte der Mathematik in Alterthum und Mittelalter* (Leipzig, 1874), pp. 381–404; James Gow, *A Short History of Greek Mathematics* (Cambridge, 1884), pp. 195–221; Paul Tannery, *La géométrie grecque* (Paris, 1887), pp. 142–153, 165–176; H. G. Zeuthen, *Histoire des mathématiques dans l'Antiquité et le Moyen Age* (Paris, 1902)—a translation of a Danish original (Copenhagen, 1893) with additions and corrections—pp. 86–145; Moritz Cantor, *Vorlesungen über Geschichte der Mathematik,* 3rd ed., I (Leipzig, 1907), 258–294; Edmund Hoppe, *Mathematik und Astronomie im klassischen Altertum* (Heidelberg, 1911), pp. 211–239; Gino Loria, *Le scienze esatte nell' antica Grecia,* 2nd ed. (Milan, 1914), pp. 188–268; T. L. Heath, *A History of Greek Mathematics,* I (Oxford, 1921), 354–446; Paul-Henri Michel, *De Pythagore à Euclide* (Paris, 1950), pp. 85–94; and B. L. van der Waerden, *Science Awakening* (Groningen, undated [1956?])—a translation with revisions and additions of a Dutch original, *Ontwakende Wetenschap*—pp. 195–200. Shorter perceptive assessments of his work are in J. L. Heiberg, *Naturwissenschaften*

Mathematik und Medizin im klassischen Altertum, 2nd ed. (Leipzig, 1920), translated by D. C. Macgregor as *Mathematics and Physical Science in Classical Antiquity* (London, 1922), pp. 53–57; J. L. Heiberg, *Geschichte der Mathematik und Naturwissenschaften im Altertum* (Munich, 1925), pp. 13–22, 75–76, 81; George Sarton, *Ancient Science and Modern Civilization* (Lincoln, Nebr., 1954); and Marshall Clagett, *Greek Science in Antiquity* (London, 1957), pp. 58–59. The text of the most important passages, with an English translation opposite and notes, is given in vol. I of the 2-vol. Loeb Classical Library *Selections Illustrating the History of Greek Mathematics* (London–Cambridge, Mass., 1939), especially pp. 436–505.

IVOR BULMER-THOMAS

EUCLID: Transmission of the Elements.

Any attempt to plot the course of Euclid's *Elements* from the third century B.C. through the subsequent history of mathematics and science is an extraordinarily difficult task. No other work—scientific, philosophical, or literary—has, in making its way from antiquity to the present, fallen under an editor's pen with anything like an equal frequency. And with good reason: it served, for almost 2,000 years, as the standard text of the core of basic mathematics. As such, the editorial attention it constantly received was to be expected as a matter of course. The complexity of the history of this attention is, moreover, not simply one of a multiplicity of translations; it includes an amazing variety of redactions, emendations, abbreviations, commentaries, scholia, and special versions for special purposes.

The Elements in Greek Antiquity. The history of the *Elements* properly begins within later Greek mathematics itself. Comments on Euclid's major work were evidently far from uncommon. Indeed, Proclus (410–485), the author of the major extant Greek commentary on the *Elements,* several times refers to similar efforts by his predecessors in a way that makes it clear that the production of works or glosses on or about Euclid was a frequent—even all too frequent and not particularly valuable—activity. It would seem that Proclus had in mind an already considerable body of scholia and remarks (largely, perhaps, in various separate philosophical and scientific works) on the *Elements,* as well as other commentaries specifically devoted to it. We know that at least four such commentaries, or at least partial commentaries, existed. The earliest was written by Hero of Alexandria, but we know of its contents only through the few references in Proclus himself and through fragments preserved in the Arabic commentary of al-Nayrīzī (*d. ca.* 922). Far more important and extensive was the commentary of Pappus of Alexandria, a work whose Greek text is also lost but of which we possess an

Arabic translation of the comments on book X. Proclus also mentions the Neoplatonist Porphyry (*ca.* 232–304), although it is doubtful that his work on Euclid would have been as systematic and penetrating as those of Hero and Pappus. Finally, although he did not compose a commentary specifically on the *Elements* itself, mention should be made of Geminus of Rhodes, whose lost work on the order or doctrine of mathematics (its exact title is uncertain) so often served Proclus with valuable source material. In the period following Proclus, it should be noted that Simplicius, in addition to his well-known commentaries on a number of works of Aristotle, also wrote a *Commentary on the Premises* [or *the Proemium*] *of the Book of Euclid.* Again we are indebted to al-Nayrīzī, who preserved fragments of the work.

To these more formal works on the *Elements,* one should add the substantial number of Greek scholia. Many derive from the commentaries of Proclus and Pappus, the latter being especially significant when they derive from the lost books of his work. Others are of a much later date, to say nothing of an inferior quality, and reach all the way to the fourteenth century (where the arithmetical comments of the monk Barlaam to book II of the *Elements* stand as the most extensive so-called scholium of all).

The event, however, that had the most enduring effect within the Greek phase of the transmission of the *Elements* was the edition and slight emendation it underwent at the hands of Theon of Alexandria (fourth century; not to be confused with the second-century Neoplatonist, Theon of Smyrna). The result of Theon's efforts furnished the text for every Greek edition of Euclid until the nineteenth century. Fortunately, in his commentary to Ptolemy's *Almagest,* Theon indicates that he was responsible for an addendum to the final proposition of book VI in his "edition (ἔκδοσις) of the *Elements*"; for it was this confession that furnished scholars with their first clue in unraveling the problem of the pre-Theonine, "pristine" Euclid. In 1808 François Peyrard noted that a Vatican manuscript (Vat. graec. 190) which Napoleon had appropriated for Paris did not contain the addition Theon had referred to. This, coupled with other notable differences from the usual Theonine editions of the *Elements,* led Peyrard to conclude that he had before him a more ancient version of Euclid's text. Accordingly, he employed the Vatican codex, as well as several others, in correcting the text presented by the *editio princeps* of Simon Grynaeus (Basel, 1533). Others, utilizing occasional additional (but always Theonine) manuscripts or earlier editions, continued to improve Peyrard's text, but it was not until J. L. Heiberg began the reconstruction of the text anew on the basis of the Vatican and almost all other known manuscripts that a critical edition of the *Elements* was finally (1883–1888) established. Heiberg not only in great measure succeeded in getting behind the numerous Theonine alterations and additions, but also was able to sift out a considerable number of pre-Theonine interpolations. In addition to the authority of the non-Theonine Vatican manuscript, he culled papyri fragments, scholia, and every known ancient quotation of, or reference to, the *Elements* for evidence in his construction of the "original" Euclid. The result still stands.

The Medieval Arabic Euclid. A most appropriate introduction to the dissemination of the *Elements* throughout the Islamic world can be had by quoting the entry on Euclid in the *Fihrist* ("Index") of the tenth-century biobibliographer Muḥammad ibn Isḥāq ibn Abī Yaʿqūb al-Nadīm:

> A geometer, he was the son of Naucrates, who was in turn the son of B[a]r[a]nīq[e]s. He taught geometry and is found as an author in this field earlier than Archimedes and others; he belonged among those called mathematical philosophers. On his book *Of the Elements of Geometry:* Its title is στοιχεῖα, which means "elements of geometry." It was twice translated by al-Ḥajjāj ibn Yūsuf ibn Maṭar: one translation, the first, is known under the name of Hārūnian, while the other carries the label Maʾmūnian and is the one to be relied and depended upon. Furthermore, Isḥāq ibn Ḥunayn also translated the work, a translation in turn revised by Thābit ibn Qurra al-Ḥarrānī. Moreover, Abū ʿUthman al-Dimashqī translated several books of this same work; I have seen the tenth in Mosul, in the library of ʿAlī ibn Aḥmad al-ʿImrānī (one of whose pupils was Abuʾl-Ṣaqr al-Qabīṣī who in turn in our time lectures on the *Almagest*). Hero commented upon this book [i.e. the *Elements*] and resolved its difficulties. Al-Nayrīzī also commented upon it, as did al-Karābīsī, of whom further mention will be made later. Further, al-Jawharī (who will also be treated below) wrote a commentary on the whole work from beginning to end. Another commentary on book V was done by al-Māhānī. I am also informed by the physician Naẓīf that he saw the Greek of book X of Euclid and that it contained forty more propositions than that which we have (109 propositions) and that he had decided to translate it into Arabic. It is also reported by Yūḥannā al-Qass [i.e., the priest] that he saw the proposition which Thābit claimed to belong to book I, maintaining that it was in the Greek version; and Naẓīf said that he had shown it to him [Yūḥannā?]. Furthermore, Abū Jaʿfar al-Khāzin al-Khurāsānī (who will be mentioned again below) composed a commentary on Euclid's book, as did Abuʾl-Wafāʾ, although the latter did not finish his. Then a man by the name of Ibn Rāhiwayh al-Arrajānī commented on book X, while Abuʾl-Qāsim al-Anṭāqī commented on the whole work and this has come out [been published?]. Further, a commentary was made by Sanad ibn ʿAlī (nine books of which, and a part of the tenth, were seen by Abū ʿAlī) and book X was commented upon by Abū Yūsuf al-Rāzī at the instance of Ibn al-ʿAmīd. In his treatise *On the Aims of Euclid's Book* al-Kindī mentioned that this book had been composed by a man by the name of Apollonius the Carpenter and that he drafted it in fifteen parts. Now, at the time when this composition had already become obsolete and in need of revision, one of the kings of Alexandria became interested in the study of geometry. Euclid was alive at this time and the king commissioned him to rework the book and comment upon it; this Euclid did and thus it came about that it was ascribed to him [as author]. Later, Hypsicles, a pupil of Euclid, discovered two further books, the fourteenth and the fifteenth; he brought them to the king and they were added to the others. And all this took place in Alexandria. Among Euclid's other writings belong: The book *On Appearances* [i.e., the *Phaenomena*]. The book *On the Difference of Images* [i.e., the *Optica*]. The book *On Given Magnitudes* [i.e., the *Data*]. The book *On Tones*, known under the title *On Music* (spurious). The book *On Division,* revised by Thābit. The book *On Practical Applications* [i.e., the *Porisma*] (spurious). The book *On the Canon.* The book *On the Heavy and the Light.* The book *On Composition* (spurious). The book *On Resolution* (spurious).

Al-Nadīm's report immediately reveals the extensive attention Euclid had already received by the end of the tenth century: two complete translations, each in turn revised, perhaps two partial translations, and an amazing variety of commentaries. What is more, this flurry of activity over the *Elements* was to continue for at least 300 years more. But before recounting the more salient aspects of this later history, it will be necessary to expand certain facets of al-Nadīm's account of the earlier efforts to work Euclid into the mainstream of Islamic mathematics. By way of introduction it may be worth indicating that the totally fanciful account reported from al-Kindī of how Euclid came to compose the *Elements* may well have derived, as Thomas Heath has maintained, from a confusing misinterpretation of the Greek preface to book XIV by Hypsicles. More important than Islamic beliefs as to the origin of the *Elements,* however, is the history of how and when this work was introduced to the Arabic-speaking world. Here al-Nadīm is more reliably informed. The first translation by al-Ḥajjāj (*fl. ca.* 786–833) to which he refers was made, as the label he assigns it indicates, under the ʿAbbāsid caliphate of Hārūn al-Rashīd (786–809), at the instance of his vizier Yaḥyā ibn Khālid ibn Barmak. We also know that a manuscript of the *Elements* was obtained from the Byzantine emperor by an earlier caliph, al-Manṣūr (754–775), although apparently without then occasioning its translation into Arabic. And this patronage of science by the 'Abbāsid caliphs is even more in evidence in Ḥajjāj's realization that a second, shorter recension of his translation would be likely to gain the favor of Ma'mūn (813–833). It is this version alone which appears to be extant (books I–VI, XI–XIII only). The first six books exist in a Leiden manuscript conjoined with al-Nayrīzī's commentary, and from the prefatory remarks of this work we learn that, in preparing his second version of the *Elements*, Ḥajjāj "left out the superfluities, filled up the gaps, corrected or removed the errors, until he had perfected the book and made it more certain, and had summarized it, as it is found in the present version. This was done for specialists, without changing any of its substance, while he left the first version as it was for the vulgar." Although what we have of Ḥajjāj's second version has not yet undergone a thorough analysis, that it was composed with something of the notion of a school text in mind seems evident. For, to cite several instances, the tendency to distinguish separate cases of a proposition and the use of numerical examples to illustrate various proofs point toward a preoccupation with pedagogical concerns that was to become fairly characteristic of the Arabic Euclid and of the medieval Latin versions that derived from it.

The second, largely new translation of Euclid was accomplished, as al-Nadīm tells us, by Isḥāq ibn Ḥunayn, son of Ḥunayn ibn Isḥāq, the most illustrious of all translators of Greek works into Arabic. Again a second recension was prepared, in this instance by a scholar who in his own right holds a major position within the history of Islamic mathematics, Thābit ibn Qurra. Although no copies of Isḥāq's initial version appear to have survived, we do possess a number of manuscripts of the Isḥāq-Thābit recension. Further study of these manuscripts is needed to say much in detail of the character of this translation, but we do know that Thābit utilized Greek manuscripts in whatever reworking he did of the text (as stated in a marginal note to a Hebrew translation of the *Elements* and confirmed by Thābit's own reference to a Greek text). Whether Isḥāq (or even Thābit) relied to any great extent on one of the Ḥajjāj versions for any sort of guidance is problematic. For, in a comparison of a single manuscript of what are presumably books XI–XIII of Ḥajjāj with their corresponding parts in the Isḥāq-Thābit redaction, Martin Klamroth, the first scholar to examine the two translations in depth, confessed that the difference was slight. But perhaps, assuming the ascription of Klamroth's manuscript of XI–XIII to Ḥajjāj correct, this lack of variation occurs only in the later books.

It is at this point perhaps noteworthy that Klamroth was of the opinion that the Arabic tradition as a whole is closer, as we have it, to the original Euclid than the text presented by extant Greek manuscripts. Heiberg, however, marshaled a considerable amount of evidence against Klamroth's contention and clearly confirmed the superior reliability of the Greek tradition. At the same time, he established the filiation of the Isḥāq-Thābit version and a particular divergent Greek manuscript.

To complete al-Nadīm's account of translations, mention should be made that Abū 'Uthman al-Dimashqī (*fl. ca.* 908–932) not only translated parts of the *Elements* but also the commentary of Pappus to book X (the latter alone being extant). Furthermore, al-Nadīm's report of the intention of Naẓīf ibn Yumn (*d. ca.* 990) to translate book X appears to be reflected in various additions and modifications deriving from the Greek that are extant in Arabic under Naẓīf's name. Finally, although it escaped al-Nadīm's notice, the spurious books XIV and XV of the *Elements* were translated by the Baghdad mathematician and astronomer Qusṭā ibn Lūqā.

The full roster of Arabic translations of Euclid's major work only begins to sketch the program of activity concerning the *Elements* within Islamic mathematics and science. The numerous commentaries mentioned by al-Nadīm are adequate testimony to that. But even before one turns to these, attention should be drawn to yet other forms that found expression among Arabic treatments of Euclid. Quite distinct from translations proper (*naql*) there are a number of epitomes or summaries (*ikhtiṣār* or *mukhtaṣar*), recensions (*taḥrīr*), and emendations (*iṣlaḥ*) of the *Elements*.

Undoubtedly the most famous of the epitomes is that included by the Persian philosopher Avicenna (Ibn Sīnā) in the section on geometry in his voluminous philosophical encyclopedia, the *Kitāb al-Shifā'*. All fifteen books of the *Elements* are present, but with abbreviated proofs. Nor was Avicenna alone in his attempt to distill Euclid into a more compact dosage; we have already seen that Ḥajjāj considered one of the primary virtues of his second version of the *Elements* to be its shorter length, and other summaries were composed by Muẓaffar al-Asfuzārī (*d.* before 1122), a colleague of Omar Khayyām (al-Khayyāmī), and also, if we can believe a report by the fourteenth-century historian Ibn Khaldūn (*Muqaddima*, VI, 20), by one Ibn al-Ṣalt (presumably the Hispano-Muslim physician, astronomer, and logician Abū'l-Ṣalt [1067/1068–1134]).

More significant within the history of Islamic mathematics are the various recensions or *taḥrīr* of the *Elements*. The best known is that of the Persian philosopher and scientist Nāṣir al-Dīn al-Ṭūsī, who composed similar editions of many other Greek mathematical, astronomical, and optical works. We know that at least one *Taḥrīr Uṣūl Uqlīdis* ("Recension of Euclid's Elements") was completed by al-Ṭūsī in 1248. It covered all fifteen books and made use of both the Ḥajjāj and Isḥāq-Thābit translations. There is, however, yet another *Taḥrir* of the *Elements* that is traditionally ascribed to al-Ṭūsī. Although it covers only books I–XIII, it is considerably more detailed than the more frequently appearing 1248 version. Printed in Rome in 1594, we know of only two extant manuscripts (both at the Biblioteca Medicea-Laurenziana in Florence) of this thirteen-book *Taḥrir*. However, one of these codices explicitly asserts that the work was completed on 10 Muḥarram 1298. Since al-Ṭūsī died in 1274, this gives grounds (and there appear to be other reasons as well) for seriously doubting the ascription to him. Yet whatever conclusion may finally be reached concerning its authorship, the preface to this *Taḥrir* is particularly instructive with respect to the reason for composing such redactions of the *Elements* and with regard to the kind of added material they would be likely to contain. Beginning with a few remarks specifying the place of geometry within the classification of the sciences and several fanciful statements about Euclid's biography, this preface makes special note of the two previously executed translations by Ḥajjāj and (revised by) Thābit and then launches into a more elaborate description of all else Islamic scholars had done with, and to, the *Elements*. This interim "history" of Euclides Arabus tells us that much effort had been spent in removing all difficulties from the text and in clarifying its numerous obscurities. Examples were inserted to make complex things more obvious and, moving in the opposite direction, some things that were too obvious were left out. Some related propositions were combined and treated as one, implicit assumptions were made explicit, and care was taken to specify (at least by number) just which previous theorems were being utilized in a particular proof. And all of this was done, our preface continues, not just in the body of the text of these versions of the *Elements,* but everywhere in the margins and even between lines. The varieties of information produced in such a fashion are now, the author of the present *Taḥrir* submits, sorely in need of proper arrangement and clarification, and he goes on to reveal his intention of satisfying this need through the presentation as a unified whole of the original text, together with relevant commentary. His resulting *Taḥrir* needs much closer scrutiny in order to set forth

the complete spectrum of all of the types of added material it contains, but it is clear from the preface we have been summarizing that it presumably includes, in addition to its own original contributions, many features similar to those its author has just recounted among the works of his predecessors.

One other *Taḥrir* of the *Elements* bears specific mention: that of Muḥyi 'l-Dīn al-Maghribī (*fl.* thirteenth century), a mathematician and astronomer who worked in both Syria and Marāgha and to whom we owe editions (literally "purifications," *tahdhīb*) of Greek works on spherical trigonometry (Theodosius and Menelaus) and of Apollonius' *Conics,* and a similar work entitled *The Essence* (*Khulāṣa*) *of the Almagest.* His *Taḥrir* may have been written shortly after the genuine fifteen-book *Taḥrir* of al-Ṭūsī, since it is found in a manuscript dated 1260/1261. It contains, on the other hand, a preface that is similar in many ways to that found in the later (1298) *Taḥrir,* wrongly, it appears, ascribed to al-Ṭūsī. It also complains of the faults in previous attempts to treat Euclid, but it is more specific in assigning at least some of the blame to Avicenna, a certain al-Nīsābūrī, and Abū Ja'far al-Khāzin (cited in al-Nadīm's chronicle). Al-Maghribī's work sets out to remedy these faults and especially to explain all of the puzzles (*shukūk*) occasioned by Euclid and to supply the added lemmas (*muqaddamāt*) necessary for various proofs. In sum, one can say that al-Maghribī's *Taḥrir,* as well as the others we have mentioned above, began from the existing translations of the *Elements* and, through the incorporation (albeit in revised form) of presumably a good many of the notions contained in earlier commentaries as well as through the creation of much original material, proceeded to the preparation of an improved Euclid that may well have been ultimately intended to serve more adequately than Euclid himself as a school text. Exactly what this improved *Elements* contains as its most salient characteristics will be revealed only after a great deal more analysis of the relevant texts. And the same must be said for the translations of Ḥajjāj and Isḥāq-Thābit.

The third type of redaction of the *Elements* mentioned above, those labeled "Emendations" (*Iṣlāḥ*), is difficult to characterize beyond what is revealed by the title, since no known copies have survived of those to which reference is made by Islamic scholars. We are told, for example, that al-Kindī composed an *Iṣlāḥ* of the *Elements* in addition to his work *On the Aims* (*Aghrāḍ*) *of Euclid's Book.* Similarly, *Iṣlāḥ*'s were written by the astronomer al-Jawharī and the Persian philosopher and scientist Athīr al-Dīn al-Abharī (d. 1265), but we know of them only through fragmentary quotations in other works. Further, in a way related to the emending of Euclid, it should be mentioned that the contribution of at least a few Islamic mathematicians to the transmission of the *Elements* appeared in the form of specific additions (*ziyādāt*), often merely to particular propositions within the text.

There remain the substantial number of Arabic commentaries, alternatively entitled *tafsīr* or *shurūḥ*, on Euclid. One can, indeed, extend their sequence considerably beyond that revealed in the *Fihrist*. In another passage of that work al-Nadīm notes what would be, were the reference correct, the very first such commentary on the *Elements:* one ascribed to the central figure of Arabic alchemy, Jābir ibn Ḥayyān. But this clearly seems to be an error, introduced by a later scribal addition, for the thirteenth-century astronomer Jābir ibn Aflaḥ. When we turn, however, to the list of genuine commentaries in al-Nadīm and supplement it with information drawn from later sources, the number becomes so considerable (nearly fifty, of which more than half are extant in some form or another) that only the most notable can be mentioned here. Among the most significant recorded by al-Nadīm is that by the Persian mathematician and astronomer al-Māhānī, who commented on book X and on book V, and that by the somewhat later al-Nayrīzī. The latter, which is often a source for comments from lost Greek works on Euclid, was translated into Latin in the twelfth century by Gerard of Cremona. When one pushes beyond the Euclid entry of the *Fihrist,* note should be made of the particularly astute commentary on book V written by the Andalusian mathematician Ibn Muʿādh al-Jayyānī. It contains, apart from Greek mathematics itself, the first known comprehension of the brilliant definition of the equality of ratios formulated by Eudoxus. In fact, apart from several brief glosses in the medieval Latin Euclid, this definition was seldom properly understood in the West before Isaac Barrow in the seventeenth century. Finally, some note should be made of the fact that figures in Islam who derived appreciable eminence from other pursuits also saw fit to expend time in commenting on the *Elements.* Thus, one might cite the philosophers al-Kindī and al-Fārābī, who commented on books I and V. And similar attention should be drawn to the treatises on Euclid written by Alhazen (Ibn al-Haytham), author of the extremely significant textbook on optics, *Kitāb al-Manāẓir,* and to the commentary dealing with the problems of parallels, ratios, and proportion by the even more famous Persian mathematician and poet Omar Khayyām.

A somewhat more informative outline of the commentaries can be had if one turns from their authors to the questions and subjects they treat. Although so few have been edited, to say nothing of studied, that only the most tentative attempt can be made to assay the contents of these commentaries, it is nevertheless possible to see at least some of the areas of major concern. To begin with, it should be made clear that the commentaries were more often than not on parts, and not the whole, of the *Elements.* Thus, as one expects within almost any body of Euclidean commentarial literature, considerable effort was spent in mulling over premises, i.e., definitions, postulates, and axioms (for example, in the treatises of al-Karābīsī [see al-Nadīm's report], al-Fārābī, Ibn al-Haytham, and Omar Khayyām referred to above). As a subclass of this genre of concern, emphasis should be placed upon special tracts, or passages in more general commentaries, that carried on the series of attempts already made in Greek mathematics to prove the parallels postulate (thus, to cite but a portion of the literature, we have two separate treatises on this topic written by Thābit ibn Qurra, a separate work dealing with it by al-Nayrīzī, and treatments of it in the *Taḥrīr* of both al-Maghrībi and al-Ṭūsī [both the genuine and the spurious *Taḥrīr* of the latter]).

Moving beyond the concern expressed over premises, one is immediately struck by the unusually high proportion of commentaries on books V and X. Although further investigation is needed to establish all of the motives behind the larger share of attention received by these books, a preliminary conjecture can easily be made. On the one hand, the extreme complexity of the treatment of irrational magnitudes in book X undoubtedly required more exposition and explanation to assure comprehension. On the other hand, the central role played by the theory of proportion contained in book V throughout all geometry probably caused Islamic mathematicians, rightly, to view this book as more fundamental than others. This, coupled with the consideration that some trouble was had in appreciating the Eudoxean definition of equal ratios that is included in book V, most likely gave it a position of some priority in the eyes of potential commentators.

One feature of the series of Arabic commentaries on the *Elements* should be recorded: Although the greater number of such commentators were mathematicians, astronomers, or physicians (or some combination thereof), a minority were not that, but rather philosophers. Of course, a philosopher of the mark of al-Kindī was as much concerned with things scientific as he was with things philosophical. But others, such as al-Fārābī and Avicenna, did not have his scientific interests or acumen. Yet they too commented on, or epitomized, the *Elements.* We are also informed that the philosopher and Shāfiʿite theologian Fakhr al-Dīn al-Rāzī (1149–1210) wrote on Euclid's premises and that the Cordovan philosopher, physician, and Aristotelian commentator par excellence Averroës (Ibn Rushd) wrote a treatise on what was needed from Euclid for the study of Ptolemy's *Almagest*. It is likely, to be sure, that such works on the *Elements* written by philosophers (most, unfortunately, are lost) were less penetrating and exacting than the more mathematical product of other commentators; they are, nonetheless, still significant as a measure of the extent to which the importance of Euclid had penetrated Islamic thought. In sum, the Arabic phase of the *Elements'* history may well prove to be not merely the most manifold but, even mathematically, the most creative of all.

Other medieval Near Eastern translations of the *Elements* all seem to have been based on one or another of the Arabic versions already mentioned. This is certainly the case with the Persian translation (completed in 1282–1283) of al-Ṭūsī's fifteen-book *Taḥrīr,* ostensibly made by his pupil Quṭb al-Dīn al-Shīrāzī (1236–1311). Similarly, although there are a fair number of medieval Hebrew compendia and special recensions of the *Elements,* the basic thirteenth-century Hebrew translation (or translations) appears to derive from the revised Isḥāq-Thābit version but contains marginal reference to the Ḥajjāj translation as well. It is still problematic whether we have here two distinct Hebrew translations or the collaborative effort over a number of years (*ca.* 1255–1270) of the two scholars involved: Moses ibn Tibbon and Jacob ben Maḥir ibn Tibbon.

Even more debatable is the issue of the Syriac version of Euclid. It was frequently the case that Arabic translations of Greek works were executed via a Syriac intermediary. It is, however, rather doubtful that this was true with the *Elements.* We do possess fragments of a Syriac redaction in a fifteenth- or sixteenth-century manuscript, and comparison of these fragments with the Arabic tradition clearly indicates a filiation, although without any absolute evidence of the direction in which the parentage must have run. If one asks how early the Syriac edition must be dated, present evidence necessitates moving it back to the eleventh or twelfth century. For instance, we know that the Syriac polymath Abu'l-Faraj (Bar Hebraeus, 1226–1286) lectured on Euclid at Marāgha in 1268. Furthermore, reference to a Syriac version of the *Elements* is made in the 1298 pseudo-al-Ṭūsī *Taḥrīr* and in mathematical opuscula of Ibn al-Sarī (*d.* 1153). Finally, note should be made of fragments of an Armenian version of the *Elements,* for it too appears to be related to the Arabic (Isḥāq-

Thābit) tradition. It seems most probable that this Armenian Euclid was the work of Gregory Magistros (*d.* 1058), in one of whose letters we find the announcement that he had begun a translation of the *Elements.* If to this we add the fact, as one scholar has urged, that Gregory knew only Greek and Syriac, but no Arabic, it would appear that he based his translation in some way or another on the Syriac version under discussion. This gives us a terminus ante quem of the first half of the eleventh century for this version, but there is no other evidence on the basis of which we can, with any certainty, assign it an earlier date. One can merely indicate that the editor of these Syriac fragments, G. Furlani, judged them to have a very close relation to the Arabic text of Ḥajjāj and that they were, in his view, in some way derived from this text. He dismissed the apparent contrary evidence one might derive from the Syriac transcription of Greek terms, since this often occurs in Syriac works that we know were based on Arabic originals. However, the second scholar to examine the fragments, Mlle. Claire Baudoux, claimed a definite link with the Isḥāq-Thābit translation (not investigated by Furlani) and concluded that the Syriac redaction preceded Isḥāq and served as an intermediary between it and the Greek original. Nevertheless, it would seem that the issue must stand unresolved until a fresh comparison is made with both Arabic translations and all relevant evidence is presented in detail. Until then, it would seem more plausible to hold the tentative conclusion that the Syriac version had an Arabic source, and not vice versa.

As the article we have quoted above from the *Fihrist* already indicates, Euclid's other works also existed in Arabic, although al-Nadīm has omitted the names of their translators. Indeed, we are still not able to identify translators in all instances. Thus, although the original translator of *On the Division of Figures* remains unknown, we do have information that Thābit ibn Qurra revised the translation, and it is, as a matter of fact, on the basis of this revision that, together with other Latin material drawn from the work of the thirteenth-century mathematician Leonardo Fibonacci, we have been able to reconstruct the contents of this Euclidean treatise. Similarly, we know that Thābit also corrected the translation of the work *On the Heavy and the Light.* There is also a treatise extant in Arabic called *The Book of Euclid on the Balance,* but there is no further information concerning its provenance.

Three other minor works, the *Data,* the *Phaenomena,* and the *Optica* (the Arabs were not aware of the pseudo-Euclidean *Catoptrica*), have a similar Islamic history. All three were part of that collection of shorter works known as the "middle books" (*muta-wassiṭāt*), which functioned as appropriate texts for the segment of mathematics falling between the *Elements* and Ptolemy's *Almagest*. Both the *Data* and the *Optica* underwent Isḥāq–Thābit translation-revisions and later *Taḥrīr* at the hands of al-Ṭūsī. Of the *Phaenomena* we are reliably informed only of the recension done by al-Ṭūsī.

The Medieval Latin Euclid: The Greek–Latin Phase. The first known Latin reference to Euclid is found in Cicero (*De oratore*, III, 132)—surely a good number of years before any attempt was made to translate the *Elements*. This latter aspect of the Latin history of Euclid begins, as far as extant sources tell us, with a fragment attributed to the third-century astrologer Censorinus. What we have in this fragment that gives excerpts from the *Elements* might also be reflected in the Euclid passages in the *De nuptiis* of Martianus Capella, although some historians feel that Martianus may have been utilizing a Greek source as well as some Latin adaptation of (or at least of parts of) the *Elements.*

The second piece of evidence in the history of Latin renditions of Euclid is found in a fifth-century palimpsest in the Biblioteca Capitolare at Verona. Treated with chemicals in the nineteenth century, it is now all but impossible to decipher. We can establish, however, that it contains fragments of a translation from books XII–XIII of the *Elements*. Very little else can be said with any surety of the translation, although its most recent editor, M. Geymonat, has urged that the palimpsest be dated slightly later and has suggested that Boethius was the author of the translation of the fragments that it contains.

Whether or not this suggestion is correct, it is to the problem of the Boethian Euclid that we must now turn. We do know that Boethius made such a translation because Cassiodorus refers to it in his *Institutiones* (II, 6, 3: "ex quibus Euclidem translatum Romanae linguae idem vir magnificus Boethius edidit") and also preserved a letter from Theodoric to Boethius himself (*Variae*, I, 45, 4) in which the existence of the translation is again attested. However, we are far less well informed of the extent and nature of this translation, for the "Boethian" geometries—or better, geometrical materials—that have come down to us are in a late fragmentary form. Basically, the excerpts we possess of Boethius' translation derive from four sources, each considerably later than the date of his actual translating efforts: (1) excerpts in the third recension of book II of Cassiodorus' *Institutiones* (eighth or ninth century); (2) excerpts in a number of manuscripts of a later redaction of the *Agrimensores,* a collection (made *ca.* 450) of materials concerned with surveying, land division, mapmaking, the rules of land tenure, etc. (*ca.* ninth century); (3) excerpts within the so-called five-book "Boethian" geometry (eighth century); (4) excerpts within the so-called two-book "Boethian" geometry (eleventh century). Special note might be taken of the full content of the last two sources, inasmuch as they appear in the literature under Boethius' name. The earlier of the two compilations, in five books, consists of gromatic material in book I and in part of book V, of excerpts from Boethius' *Arithmetica* in book II, and of excerpts from his translation of Euclid in books III–IV and in the initial section of book V. The two-book version of the "Boethian" geometry seems to have been compiled by a Lotharingian scholar without especially acute mathematical ability and contains its excerpts from Boethius' translation in book I, as well as a brief preface and a concluding section on the abacus, while book II consists largely of *Agrimensores* material. If one combines the extracts of Boethius' *Elements* from these two works with the extracts found in the Cassiodorus and *Agrimensores* sources listed above, the total schedule, as it were, of translated Euclid amounts to (a) almost all the definitions, postulates, and axioms of books I–V of the *Elements;* (b) the enunciations of almost all the propositions of books I–IV; and (c) the proofs for book I, propositions 1–3. The above four sources containing these extracts often overlap in the items they include, but it is notable that a sequence of the enunciations of propositions from book III (i.e., 7–22) is found only in the five-book "Boethian" geometry, while the definitions of book V appear only in the recension of Cassiodorus. (The relation of the four sources can be seen in the chart below.)

The ninth- through fifteenth-century manuscripts in which these sources (especially the last three) of Euclidean excerpts appear are, for the most part, collections containing other material pertinent to the quadrivium. But even when, with new and more complete translations of the *Elements* in the twelfth century, this kind of collection began to lose the dominant position it once held in medieval Latin mathematics, traces of the Boethian Euclid linger on through occasional conjunction with the newly translated material. Thus, we know of at least two different mélanges of parts of the Boethian excerpts with one of the translations of the *Elements* from the Arabic by Adelard of Bath (that labeled Adelard II below). One of these mélanges dates from about 1200 and seems to have been compiled by a North German scholar with appreciably more mathematical wit than, for example, the author of the two-book "Boethius" discussed above. It is preserved in a single thirteenth-century manuscript: Lüneburg, Ratsbücherei MS miscell. D 4°48. The second mélange occurs in four manuscripts, three of them of the twelfth century, but little has been done to attempt to determine the provenance of its author. Further, cognizance should be taken of the fact that the Boethian "source" of both mélanges seems to have been the two-book *Geometry*. Finally, although we do have these attempts to combine the Greek–Latin Boethian extracts of the *Elements* with the Arabic–Latin tradition deriving from Adelard, it should be made clear that they constituted but a minor part of the medieval Euclid in the West; the Adelardian-based tradition was soon to hold all in sway.

However, before we move to this tradition and to the Arabic–Latin Euclid in general, two other Greek–Latin medieval versions must be mentioned. Of the first we have but a fragment (I, 37–38 and II, 8–9). It exists in a single tenth-century manuscript in Munich. Although extremely literal, its translator, an Italian, knew little of what he was doing, since he translated as numbers the letters designating geometrical figures.

The second Greek–Latin Euclid we must discuss constitutes the most exact translation ever made of the *Elements*, being a *de verbo ad verbum* rendering in which the order of words and occasionally the syntax itself are often more Greek than Latin. Based solely on a Theonine text, the translation is known from two extant manuscripts and covers books I–XIII and XV. Neither manuscript names the translator, but a stylistic analysis of the text has established that he is identical with the anonymous twelfth-century translator of Ptolemy's *Almagest* from the Greek. A preface fortunately attached to the latter translation informs us that our nameless author was a one-time medical student at Salerno who, learning of the existence in Palermo (*ca.* 1160) of a Greek codex of Ptolemy, journeyed to Sicily in order to see this treasure and, after a period of further scientific preparation, set himself to putting it in Latin. Presumably our translator did the same for the *Elements* shortly thereafter (since no mention of such an effort is made in his description of other of his activities in his preface to his version of Ptolemy). When one turns to the translation itself, it is immediately evident that its author was extremely acute, both as an editor and as a mathematician. Not only does he give an extraordinarily exact rendering of the Greek, but on occasion he also employs brackets to indicate several passages in an alternate Greek manuscript he was using. What is more, several times he employs these same brackets to improve the logic of a proof. Unfortunately, the superb Latin Euclid he produced

exerted very little, if any, influence upon his medieval successors. (It might also be indicated that one manuscript of this translation contains a pastiche of books XIV–XV in place of the missing book XIV; it too derives from Greek sources and even castigates translators from the Arabic for being insufficiently careful.)

The Greek–Latin phase of the medieval Euclid is also, perhaps, the most appropriate point at which mention should be made of the minor Euclidean works during this period. For, contrary to what proved to be true for the *Elements,* these shorter works have a medieval Latin history that derives predominantly—in all instances through anonymous translators—from the Greek. Thus, in place of the apparently lost version of the *Data* from the Arabic by Gerard of Cremona (who also translated the *Elements*), we possess several codices of an accurate rendering made in the twelfth century directly from the Greek. Similarly, although there do exist copies of Gerard's Arabic–Latin *Optica,* they are overwhelmingly outnumbered by manuscripts containing Greek–Latin translations. Indeed, there appear to be two distinct versions of the *Optica* from the Greek, some manuscripts of which are so variant as to lead one to expect an even more complicated history. There also seem to be several versions of the pseudo-Euclid *Catoptrica* made from the Greek. What is more, there is a totally separate *De speculis* translated from the Arabic (by Gerard?) and ascribed to Euclid. We know of no Greek or Arabic original from which it may have derived, although it does exist in Hebrew in several manuscripts. The *Sectio canonis* had several propositions from it transmitted through the medium of Boethius' *De institutione musica.* The *Phaenomena,* on the other hand, was not put into Latin before the Renaissance.

The Medieval Latin Euclid: The Arabic–Latin Phase. Once integral translations of the *Elements* from the Arabic were available to the medieval scholar, all Greek–Latin fragments and versions receded into the background. The new, dominant tradition was, however, twofold; one wing derived from the Ḥajjāj Euclid, the other from that of Isḥāq–Thābit, the recensions of al-Ṭūsī and al-Maghribī coming too late, of course, to enter into the competition of translating activity in the twelfth century.

The Latin *Elements* based upon the Isḥāq–Thābit text was the accomplishment of Gerard of Cremona, the most industrious of all translators of scientific, philosophical, and medical works from the Arabic. We know that he translated the *Elements* from its citation in his *Vita,* written by one of his pupils and appended to one of his translations of Galen. Identi-

444

fied among extant manuscripts in 1901, Gerard's Euclid was soon realized to be the closest to the Greek tradition of all Arabic-Latin versions. It alone contains Greek material—for example, the preface to the spurious book XIV—absent from the other versions. Ironically, however, it clearly seems to have been less used, and less influential, than the somewhat more inaccurate (Adelardian-based) editions. It derives its more faithful reflection of the Greek original from the fact that the Arabic of Isḥāq-Thābit, upon which it was based, is itself a more exact reproduction of the Greek. We have no explicit ascription stating that Gerard worked from this particular Arabic translation, but even the most preliminary examination of Gerard's text reveals that this was in all probability the case. For instance, the phrase "Thebit dixit" occurs frequently throughout the body of the translation. At least some of these occurrences—perhaps almost all of them—are not due to Gerard's reflecting on the text he was rendering, but to a direct translation of that text itself, since several citations that have been published by Klamroth from the Arabic and are reproduced in Gerard indicate that Thābit is named therein as well. (Third-person references by an author to himself are, of course, quite common.) While awaiting evidence that will issue from a direct comparison of the Isḥāq-Thābit and Gerard texts, note should be made of the fact that Gerard has the two propositions added after VIII, 25, and the corollaries to VIII, 14–15, which are characteristic of Isḥāq-Thābit. Yet this is not the only Arabic version Gerard had before him, at least not in its pure form. For he includes VIII, 16, which is not, according to Klamroth, in the Isḥāq-Thābit manuscripts examined. Furthermore, after having followed these manuscripts by reproducing VIII, 11–12, as two separate propositions, at the conclusion of book VIII Gerard has an addendum claiming that these two propositions were found as one *in alio libro;* the addendum continues by reproducing this combined version, proof and all. Exactly who found this combined version of VIII, 11–12, in another book is problematic; use of the first person in this passage in Gerard is not conclusive, since it could derive directly from his Arabic text. We do know, however, that the Adelardian tradition ostensibly based on the Ḥajjāj Euclid (of which book VIII is not extant) does conflate the two propositions in question. Therefore, either Gerard utilized texts of both Isḥāq-Thābit and Ḥajjāj in making his translation or, which seems more likely, he based his labors on an Isḥāq-Thābit text that contained material drawn from one or another of the Ḥajjāj versions.

Gerard also contributed to the literature of Euclides Latinus by translating the commentary of al-Nayrīzī on the *Elements,* the commentary of Muḥammad ibn ʿAbd al-Bāqī (*fl. ca.* 1100) on book X, and at least part of Dimashqī's translation of Pappus' commentary on book X.

By far the most important share of the medieval Arabic-Latin Euclid belongs to the English translator, mathematician, and philosopher Adelard of Bath. Not only was Adelard himself the author of at least three versions of the *Elements,* but he served as the point of departure for numerous offspring redactions and revisions as well. Taken together as the Adelardian tradition, they soon gained a virtual monopoly when it came to using and quoting Euclid in the Latin Middle Ages.

The first version due to Adelard himself (hereafter specified by the Roman numeral I) is the only one within the whole tradition that is, properly speaking, a translation. As such, there is clear indication that it was based on the Arabic of Ḥajjāj. Thus, the proofs in Adelard I appear to correspond quite well with what we have of Ḥajjāj. Further, Adelard I carries the same three added definitions (at least two, however, going back in some way to the Greek) to book III found in Ḥajjāj and agrees with him in reproducing the maximum of six separate cases in the proof of III, 35. But it also seems clear that Adelard I did not utilize Ḥajjāj as we have him today. The fact that he does not reproduce the arithmetical examples in books II and VI, or the added propositions in book V, that are present in the extant Ḥajjāj is not the point at issue, since these features are most likely from the commentary of al-Nayrīzī to which our Ḥajjāj text is attached. What is significant is the fact that Adelard I does contain the corollaries to II, 4, and VI, 8, which are not present in our Ḥajjāj and, even more suggestive, does not include VI, 12, which is contained in our Ḥajjāj. If one couples the latter fact with the statement in the pseudo-Ṭūsī *Taḥrīr* that Ḥajjāj did not include VI, 12, it then becomes most probable that there existed another Ḥajjāj, slightly variant from the text we possess, and that it was this Arabic version which Adelard I employed. One could argue that it would have been the first version Ḥajjāj prepared under Hārūn al-Rashīd, but we know that this earlier redaction was somewhat longer than the second version—which, presumably, is the one we possess. And Adelard I in other respects (particularly in the length and detail of the proofs) appears to correspond well enough with the second extant version to make derivation from a longer Arabic original unlikely. A variant of Ḥajjāj's second redaction seems a more plausible source.

Adelard II, on the other hand, does not give rise to similar problems, since it is not a translation but an abbreviated edition or, as Adelard himself calls it (in his version III), a *commentum*. Although the briefest of Adelard's efforts in putting Euclid into Latin, it was unquestionably the most popular. This is clear not merely on grounds of the far greater number of extant copies of Adelard II, but also because the translations given here of the "enunciations" (of definitions, postulates, axioms, and propositions) were subsequently appropriated for a good many other versions by editors other than Adelard. Indeed, the diversity thus growing out of Adelard II appears to be present within it as well, since there is considerable variation among the almost fifty manuscript copies thus far identified. The earliest extant codex (Oxford, Trinity College 47) presents, for example, a text that is more concise than any consistently presented by other manuscripts.

The characteristic feature of Adelard II lies in its proofs, which are not truly proofs at all, but *commenta* furnishing relevant directions in the event one should wish to carry out a proof. One is constantly reminded, for instance, of just which proposition or definition or axiom one is building upon, or whether the argument—should it be carried out—is direct or indirect. The *commenta* talk about the proposition and its potential proof; the language is, in our terms, metamathematical. Moreover, this talk about the proof often puts greater emphasis upon the constructions to be utilized than on the proof proper.

Adelard III, referred to by Roger Bacon as an *editio specialis,* continues this fondness for the metamathematical remark, but now embeds them within and throughout full proofs as such. In the bargain, one often finds that such reflections veer from the proposition and the proof at hand to external mathematical matters.

Adelard II and III have much in common besides their author. Both contain Arabisms; both contain Grecisms; and III quotes II. More important, however, both make use of original Latin material: they employ notions drawn from Boethian arithmetic, use classical expressions, and even (Adelard III) allude to Ovid.

A fourth major constituent of the Adelard tradition is the version of the *Elements* prepared by Campanus of Novara. It too takes over the Adelard II enunciations and, through the formulation of proofs that seem largely independent of Adelard, fashions what is, from the mathematical standpoint, the most adequate Arabic–Latin Euclid of all (its earliest dated extant copy being that of 1259). The *additiones* Campanus made to his basic Euclidean text are particularly notable. With an eye to making the *Elements* as self-contained as possible, he devoted considerable care to the elucidation and discussion of what he felt to be obscure and debatable points. He also attempted to work Euclid more into the current of thirteenth-century mathematics by relating the *Elements* to, and even supplementing it with, material drawn from the *Arithmetica* of Jordanus de Nemore.

Furthermore, Campanus and the three versions of Adelard (especially II and III) served as sources for an amazing multiplication of other versions of the *Elements.* Although the extent to which they diverge from one or another of these sources may not be great or marked with much originality, and although they frequently seem to concentrate on selected books of Euclid, one can still discern among extant manuscripts some fifteen or more additional "editions" belonging to the Adelardian tradition. They range from the thirteenth through the fifteenth centuries and none, as far as a cursory investigation has shown, bears the name of an author or a compiler.

As a whole the Adelardian tradition formed the dominant medieval Euclid. Further, although by far the greatest share of this tradition was not a strict translation from its Arabic source, the divergence from the original Greek thus occasioned caused little difficulty. The few missing propositions were easily remedied, and changes in the order of theorems gave rise to no mathematical qualms at all. Misunderstanding of what Euclid intended also seems to have been quite infrequent, save for the always problematic criterion of Eudoxus for the equality of ratios, which was ensconced in book V, definition 5. Here the most influential medieval interpretation—that of Campanus—curiously seems to have been conditioned by a strange quirk in transmission. For in place of the genuine V, definition 4, Campanus (and all other constituents of the Adelardian tradition) has another, mathematically useless, definition; and in his attempt to make sense of this, Campanus formulated a mechanics of explanation that he in turn extended to his discussion, and consequent misunderstanding, of the "Eudoxean" V, definition 5. Thus, we are witness to a unique instance in which the existence of a spurious fragment within the textual tradition seriously affected interpretation of something genuine.

The most impressive characteristic of the Adelardian–Campanus *Elements* is not, however, to be found in missing or misunderstood fragments of the text, but rather in the frequent additions made to it, additions which often take the form of supplementary propositions or premises but also occur as reflective remarks within the proofs to standard Euclidean theorems. It is not possible to tabulate even a small fraction of these additions, but it is important to realize something of the basis of their concern. To begin with, the motive behind many, indeed most, of them was to render the whole of their Euclid more didactic in tone. The trend toward a "textbook" *Elements,* noted in its Arabic history (and even, in a way, in Theon of Alexandria's new Greek redaction), was being extended. The reflections mentioned above about the structure of proofs are surely part of this increased didacticism. The labeling of the divisions within a proof, express directions as to how to carry out required constructions and how to draw three-dimensional figures, indications of what "sister" propositions can be found elsewhere in the *Elements,* and even clarifying references to notions from astronomy and music are all evidence of the same. We are also witness to the erosion (again pedagogically helpful) of the strict barrier fixed by the Greeks between number and magnitude. For the care not to employ the general propositions of book V in the arithmetical books VII–IX has been pushed out of sight, and one can find admittedly insufficient numerical proofs in propositions (especially in books V and X) dealing with general magnitude.

Of even greater interest is an ever-present preoccupation with premises and with what is fundamental. Axioms are everywhere added (even in the middle of proofs) to cover all possible gaps in the chain of reasoning, and considerable attention is paid to the logic of what is going on. Once again one sees a fit with didactic aims. But this emphasis on basic notions and assumptions was directed not only toward making the geometry of the *Elements* more accessible to those toiling in the medieval faculty of arts; it was also keyed to the bearing of issues within this geometry upon external, largely philosophical problems. Most notable in this regard is the time spent in worrying over the conceptions of incommensurability, of the so-called horn angle (between the circumference of a circle and a tangent to it), and of the divisibility of magnitudes. For these conceptions all relate, at bottom, to the problems of infinity and continuity that so often exercised the wits of medieval philosophers. The Adelardian tradition furnished, as it were, a schoolbook Euclid that admirably fit Scholastic interests both within and beyond the bounds of medieval mathematics.

There is one other medieval Latin version of the *Elements* that is connected, more tenuously to be sure, with the Adelard versions. Its connection derives from its appropriation of most of the Adelard II enunciations, although frequently with substantial change. It exists in anonymous form in a single manuscript (Paris, BN lat. 16646). We know, however, that this codex was willed to the Sorbonne in 1271 by Gérard d'Abbeville and is in all probability identical with a manuscript described in the *Biblionomia* (*ca.* 1246) of Richard de Fournival. Richard, however, identifies the manuscript as "Euclidis geometria, arismetrica et stereometria ex commentario Hermanni secundi." This version is, therefore, presumably the work of Hermann of Carinthia (*fl. ca.* 1140–1150), well-known translator of astronomical and astrological texts from the Arabic. Its proofs differ from those of the Adelardian tradition, and the occurrence of Arabisms not in this tradition has been viewed as evidence for Hermann's use of another Arabic text in compiling his redaction. The recent suggestion that this text was the Isḥāq–Thābit translation seems doubtful, however, for unlike Gerard of Cremona's version, ostensibly based on that Arabic translation, Hermann not only lacks the references to "Thebit" but also does not have the additions in book VIII (see above) characteristic of Isḥāq–Thābit. On the other hand, it is true that Hermann does show variations from the text of Ḥajjāj as we have it; but all of these, it appears, are also in Adelard I, which clearly derives, as we have seen, from some Ḥajjāj text or another. Hermann repeats the differences noted above in discussing Adelard I and with him alone, among all Latin versions, carries the full six separate cases for III, 35. One other piece of evidence might be noted: In a series of propositions (V, 20–23) dealing with proportion, the Greek Euclid specifies only one (V, 22) for "any number of magnitudes whatever," the others being stated merely for three magnitudes. Now all versions save Adelard I and Hermann, including our Ḥajjāj and, to judge from Gerard's translation, Isḥāq–Thābit, adopt the policy of stating all four propositions in general form. Hermann and Adelard I, on the other hand, retain the "three magnitudes" version of the Greek for V, 20–21, 23 but also substitute *tres* for *quotlibet* in V, 22. The filiation of these two is, therefore, quite close. One can conclude, then, that Hermann used at least both Adelard II (from which he derives many enunciations) and presumably the same version of Ḥajjāj used by Adelard I (and possibly also Adelard I itself). Finally, it has been noticed that Hermann contains the Arabicism *aelman geme* (corresponding to ʿilm jāmiʿ) to refer to axioms or common notions, while both Ḥajjāj and, to judge from Klamroth, Isḥāq–Thābit employ *al-ʿulūm al-mutaʿārafa.* Yet here Hermann could still be following some Ḥajjāj text, since ʿilm jāmiʿ (also used, incidentally, in Avicenna's epitome of Euclid) occurs as a marginal alternative in our manuscript of Ḥajjāj.

Occasional claims have also been made for two other identifiable medieval Latin versions of the *Elements.* One, purported to be by Alfred the Great, has been shown to derive from erroneous marginal ascriptions (in a single manuscript) of the Adelard II version to "Alfredus." The second, a seventeenth-century manuscript catalogue, refers to a version of the *Elements* as "ex Arab. in Lat. versa per Joan. Ocreatum." We do know of passages at the beginning of book V and in book X (props. 9, 23, 24) in certain manuscripts of Adelard II that do appeal to "Ocrea Johannis," but in such a way that this may be but a reference to separate (marginal?) comments on Euclid or to some other mathematical treatise, rather than to a distinct translation or version of the *Elements*.

If, in conclusion, one compares the very substantial amount of material constituting the Arabic–Latin Euclid with the equally extensive history the *Elements* had in Islam, several rather striking differences are apparent, even at the present, extremely preliminary stage of investigation of these two traditions. In both one finds an overwhelming number of versions, editions, and variants of Euclid proper. Although here one has a common trait, there is, on the other hand, no doubt that the number of commentaries composed in Arabic far exceeds those in Latin. Indeed, there seems to be but one "original" commentary proper in Latin, questionably ascribed to the thirteenth-century Dominican philosopher Albertus Magnus and in any event greatly dependent upon earlier translated material (notably the commentary of al-Nayrīzī). When we do find *Questiones super Euclidem,* they treat more of general problems within geometry, mathematics, or natural philosophy than they do of issues specifically tied to particular Euclidean premises or theorems. In point of fact, this bears upon a second element of contrast between the Arabic and Latin traditions—that the latter seems to have moved more rapidly toward serving the interests of philosophy, while the former remained more strictly mathematical in its concerns. One does not find, for example, anything like the Arabic debate about the parallels postulate in the Latin texts. But then Islamic mathematics itself was more lively and creative than that of the medieval Latin West.

The Renaissance and Modern Euclid. Four events seem to have been the most outstanding in determining the course of the *Elements* in the sixteenth and succeeding centuries: (1) the publication of the medieval version of Campanus of Novara, initially as the first printed Euclid at Venice (1482) by Erhard Ratdolt, and at many other places and dates in the ensuing 100 years; (2) a new Latin translation from the Greek by Bartolomeo Zamberti in 1505; (3) the *editio princeps* of the Greek text by Simon Grynaeus at Basel in 1533; (4) another Greek-Latin translation made in 1572 by Federico Commandino. The publications result-

ing from these four versions show their effect in almost all later translations and versions, be they Latin or vernacular.

Of Campanus we have spoken above. The printed Euclid following his was, ignoring the publication in various forms of the "Boethian" fragments, not the influential one of Zamberti but only portions of the *Elements* included in the gigantic encyclopedia *De expetendis et fugiendis rebus* (Venice, 1501) of Georgius Valla (*d.* 1499). This was not, to be sure, an easily accessible Euclid; for in addition to being an extraordinarily cumbersome book to use, the selections from the *Elements* are scattered among materials translated from other Greek mathematical and scientific texts. The first publication of a Greek-based Latin *Elements* as an integral whole was that at Venice in 1505 prepared by Bartolomeo Zamberti (*b. ca.* 1473). His translation derived from a strictly Theonine Greek text, a factor which has Zamberti attributing the proofs to this Alexandrian redactor (*cum expositione Theonis insignis mathematici*). The work also contains translations of the minor Euclidean works (which were also, in part, in Valla's encyclopedia).

Zamberti was most conscious of the advantages he believed to accrue from his working from a Greek text. This enabled him, he claimed, to add things hitherto missing and properly to arrange and prove again much found in the version of Campanus. Indeed, his animus against his medieval predecessor is far from gentle: his Euclid was, Zamberti complains, replete with "wondrous ghosts, dreams and fantasies" (*miris larvis, somniis et phantasmatibus*). Campanus himself he labels *interpres barbarissimus.*

The attack thus launched by Zamberti was almost immediately answered by new editions of Campanus, the most notable of them being that prepared at Venice in 1509 by the Franciscan Luca Pacioli. Pacioli regarded himself as a corrector (*castigator*) who freed Campanus from the errors of copyists, especially in the matter of incorrectly drawn figures. In direct reply to Zamberti, Campanus was now presented as *interpres fidissimus.*

A kind of détente was subsequently reached between the Campanus and Zamberti camps, for there was soon a series of published *Elements* reproducing the editions of both *in toto,* the first appearing at Paris in 1516. Each theorem and proof first occurs *ex Campano* and is immediately followed by its mate and proof *Theon ex Zamberto.* The *additiones* due to Campanus appear in place but are appropriately set off and indicated as such.

The end of the first third of the sixteenth century brought with it the first publication of the Greek text of the *Elements.* The German theologian Simon

Grynaeus (d. 1541) accomplished this, working from two manuscripts, with an occasional reference to Zamberti's Latin. His edition, which included the text of Proclus' *Commentary* as well, was the only complete one of the Greek before the eighteenth century. The other Greek Euclids of the Renaissance were all partial, most frequently offering only the enunciations of the propositions in Greek (usually with accompanying translation). The most significant of such "piecemeal" *Elements* is unquestionably that of the Swiss mathematician and clockmaker Conrad Dasypodius, or Rauchfuss (1532–1600). Dasypodius makes it abundantly clear that his edition, issued in three parts at Strasbourg in 1564, was intended as a school-text Euclid. For this reason he believed it more convenient to give merely the enunciations of books III–XIII to accompany the full text of books I–II, all in Greek with Latin translation. The pedagogical design of his publication is also seen from the fact that in spite of his exclusion of a great deal of genuine Euclid, he nevertheless saw fit to include the text of the more readily comprehensible arithmetical version of book II that was composed by the Basilian monk Barlaam (d. ca. 1350).

In any event, the printing of the complete Greek text in 1533, plus the earlier appearance of both Campanus and Zamberti, provided the raw material, as it were, for the first, pre-Commandino, phase of the Renaissance Euclid. The irony is that Campanus and Zamberti, and not the Greek *editio princeps,* played the dominant role. Some note of the most significant of the many early Renaissance printed editions of the *Elements* will make this clear. (The substantial, but totally uninvestigated, manuscript material of this period is here excluded from consideration.)

If one focuses, to begin with, upon Latin Euclids, the first important "new" version (Paris, 1536) is that of the French mathematician Oronce Fine (1494–1555). Yet his contribution seems to have been to insert the Greek text of the enunciations for the *libri sex priores* in the appropriate places in Zamberti's Latin translation of the whole of these books. Similarly, his compatriot Jacques Peletier du Mans brought out another six-book Latin version (Lyons, 1557), this time based, as Commandino noted, more on Campanus' Arabic–Latin edition than upon anything Greek. (It should be recorded, however, that Peletier supplemented what he took from Campanus' *additiones* with some interesting ones of his own.) At Paris in 1566 yet a third French scholar, Franciscus Flussatus Candalla (François de Foix, Comte de Candale, 1502–1594) produced a Latin *Elements.* Covering all fifteen books, and appending three more

on the inscription and circumscription of solids, the appeal is once again not to the Greek text as such, but to Zamberti and Campanus. And when there is something not derived from these two, it seems as often as not to have been Candalla's own invention.

A contemporary summary view of the status of Euclid scholarship was revealed when, a few years before Candalla's expanded *Elements,* Johannes Buteo published his *De quadratura circuli* at Lyons in 1559. This work contained as an appendix Buteo's *Annotationum opuscula in errores Campani, Zamberti, Orontii, Peletarii . . . interpretum Euclidis.* Campanus was, he felt, the best of these editors, for his errors derive from his Arabic source and not from an ineptitude in mathematics. Zamberti, on the other hand, although he worked directly from the Greek, showed less acumen in geometry. Even less adequate, in Buteo's judgment, were the works of Fine and Peletier, the latter taking the greatest liberties with the text and ineptly adding or omitting as he saw fit.

We have thus far spoken merely of sixteenth-century Latin translations, but the same pattern reflecting the central impact of Campanus and Zamberti can also be discerned in the most notable vernacular renderings. The earliest of these to be printed was the Italian translation by the mathematician, mechanician, and natural philosopher Niccolò Tartaglia. Its first edition appeared at Venice in 1543. When Tartaglia submits that his redaction was made *secondo le due tradittioni,* there is no question that Campanus—who appears to be heavily favored—and Zamberti are meant. When Campanus has added propositions or premises, Tartaglia has appropriately translated them and noted their absence *nella seconda tradittione,* while things omitted by Campanus but included by Zamberti receive the reverse treatment.

The next languages to receive the privilege of displaying Euclid among their goods were French (by Pierre Forcadel at Paris in 1564) and German (at the hands of Johann Scheubel and Wilhelm Holtzmann in 1558 and 1562). We are better informed, however, of the circumstances surrounding the production of the more elaborate, first complete English edition. Yet before we describe this, it will be well to note an even earlier intrusion of Euclidean materials into English. This is found in *The Pathway to Knowledg* (London, 1551) of the Tudor mathematical practitioner Robert Recorde. Recorde fully recognized the ground he was breaking, for in anticipation of the dismay even Euclid's opening definitions would likely cause in the "simple ignorant" who were to be his readers, he cautioned: "For nother is there anie matter more straunge in the englishe tungue, then this whereof never booke was written before now, in that tungue."

Recorde's purpose was distinctly practical, and he expressly mentioned the significance of geometry for surveying, land measure, and building. The *Pathway* contains the enunciations of books I–IV of Euclid, reworked and reordered to serve his practical aims.

The first proper English translation was the work of Sir Henry Billingsley, later lord mayor of London (*d.* 1606), and appeared at London in 1570 with a preface by John Dee, patron and sometime practitioner of the mathematical arts. A truly monumental folio volume, Billingsley's translation contains "manifolde additions, Scholies, Annotations and Inventions . . . gathered out of the most famous and chiefe Mathematiciens, both of old time and in our age" and even includes pasted flaps of paper that can be folded up to produce three-dimensional models for the propositions of book XI. Each book begins with a summary statement that includes considerable commentary and often an assessment of the views of Billingsley's predecessors, most notably those of Campanus and Zamberti. The role these two scholars played in Billingsley's labors is confirmed in yet another way. There exists in the Princeton University Library a copy of the 1533 *editio princeps* of the Greek text of the *Elements* bound together with a 1558 Basel "combined" edition of Campanus and Zamberti. It is not known how these volumes came into Princeton's possession, but both contain manuscript notes in Billingsley's hand. The fact that these notes are found on only five pages of the Greek text, but on well over 200 of both Campanus and Zamberti, is clearly suggestive of Billingsley's major source. Once again, the two basic Latin versions, one medieval and one Renaissance, have exhibited the considerable extent of their influence.

However, the better part of this influence was interrupted suddenly and decisively by the fourth major version listed above: the publication at Pesaro in 1572 of the Latin translation by Federico Commandino of Urbino. Commandino—who, in addition to the place he holds in the history of physics deriving from his *Liber de centro gravitatis* (Bologna, 1565), prepared exacting Latin versions of many other Greek mathematical works—was clearly the most competent mathematician of all Renaissance editors of Euclid. He was also most astute in his scholarship, for we know that in addition to the 1533 *editio princeps,* he employed at least one other Greek manuscript in establishing the text for his translation. For the first time, save for the anonymous translation in the twelfth century, we now have a version (no matter what language) of the *Elements* that is solidly based on a tolerably critical Greek original. It even includes, also for the first time, a rendering of numerous Greek scholia.

Aware, but critical, of the efforts of his predecessors, Commandino leaves no doubt of the advantage of staying closer to the Greek sources so many of them had minimized, if not ignored. The result of his labors may prove to be of less fascination than other versions, since it so closely follows the Greek we already know, but the importance it held for the subsequent modern history of the *Elements* is immeasurable. It came to serve, in sum, as the base of almost all other proper translations before Peyrard's discovery of the "pristine" Euclid in the early nineteenth century. Thus, to cite only the most notable cases in point, Greek texts of the *Elements* with accompanying Latin translation frequently based the latter on Commandino: for example, Henry Briggs's *Elementorum Euclidis libri VI priores* (London, 1620) and even David Gregory's 1703 Oxford edition of Euclid's *Opera omnia* (which was the standard, pre-nineteenth-century source for the Greek text). Commandino was also followed in later strictly Latin versions: that of Robert Simson, simultaneously issued in English at Glasgow in 1756; and even that of Samuel Horsley, appearing at London in 1802. Vernacular translations often followed a similar course, beginning with the Italian translation, revised by Commandino himself, appearing at Urbino in 1575 and extending to and beyond the English version by John Keill, Savilian professor of astronomy at Oxford, in 1708.

In all translations based heavily on Commandino, one naturally remained close to the (Theonine) Greek tradition; but there were also other efforts after, as well as before, Commandino that did not stay so nearly on course. These were the numerous commentaries on Euclid, the various schoolbook *Elements,* and, in a class by itself, the edition of Christopher Calvius.

The commentaries of the sixteenth through eighteenth centuries were almost always limited to specific books or parts of the *Elements.* We have already noted the 1559 *Annotationum opuscula* of Buteo, but a considerable amount of related commentarial literature began to flourish around the same time. Giovanni Battista Benedetti (1530–1590) brought out his *Resolutio omnium Euclidis problematum . . . una tantummodo circini data apertura* at Venice in 1553 in response to a controversy that had recently arisen out of some reflection by several Italian scholars on Euclid. Petrus Ramus, who had previously produced a Latin version of the *Elements* in 1545, published at Frankfurt in 1559 his *Scholae mathematicae,* in which he scrutinized the structure of Euclid from the standpoint of logic. Along related lines, mention might be made of the curious *Euclideae demonstrationes in syllogismos resolutae* (Strasbourg, 1564) of Conrad Dasypodius and Christianus Herlinus. Such works as these were in a way extensions, perhaps fanciful ones, of the medieval Scholastic concern with the logic of the *Elements*. Yet another development can be seen in the various attempts to reduce the *Elements* to practice. We have already noticed this standpoint in Robert Recorde, and to this one could add the first German translation of books I–VI —published by Wilhelm Holtzmann (Xylander) in 1562—which was written with the likes of painters, goldsmiths, and builders in mind; and the Italian version (1613–1625) of Antonio Cataldi, which expressly declared itself to be an *Elementi ridotti alla practica.*

On a more specific plane, commentaries on book V, and particularly upon the Eudoxean definition of equal ratios that we have already seen to be problematic, continued in the sixteenth and seventeenth centuries. Beginning with the almost totally unknown works of Giambattista Politi, *Super definitiones et propositiones quae supponuntur ab Euclide in quinto Elementorum eius* (Siena, 1529), and of Elia Vineto Santone, *Definitiones elementi quincti et sexti Euclidis* (Bordeaux, 1575), the issue was also broached by Galileo in the added "Fifth Day" of his *Discorsi . . . a due nuove scienze* (an addendum first published at Florence in 1674 in Vincenzo Viviani's *Quinto libro degli Elementi d'Euclide*). Finally, note should be made of two of the most impressive early modern commentaries on selected aspects of Euclid. The first is Henry Savile's lectures *Praelectiones tresdecim in principium Elementorum Euclidis Oxoniae habitae MDCXX* (Oxford, 1621), which cover only the premises and first eight propositions of book I but do so in an extraordinarily penetrating, and still valuable, way. The last work to be mentioned is so famous that one often forgets that it formed part of the commentarial literature on the *Elements*—the *Euclides ab omni naevo vindicatus* (Milan, 1733) of Girolamo Saccheri, in which this Jesuit mathematician and logician fashioned the attempt to prove Euclid's parallels postulate that has won him so prominent a place in the histories of non-Euclidean geometry.

Closely connected with the commentarial literature we have sampled is the magisterial Latin version of the *Elements* composed by another, much earlier Jesuit scholar, Christopher Clavius (1537–1612). The first edition of his *Euclidis Elementorum libri XV* appeared at Rome in 1574. Not, properly speaking, a translation, as Clavius himself admitted, but a personal redaction compiled from such earlier authors as Campanus, Zamberti, and Commandino, the work is chiefly notable, to say nothing of immensely valuable, for the great amount of auxiliary material it

contains. Separate *praxeis* are specified for the constructions involved in the problems, long *excursus* appear on such debatable issues as the horn angle, and virtually self-contained treatises on such topics as composite ratios, mean proportionals, the species of proportionality not treated in Euclid, and the quadratrix are inserted at appropriate places. Indeed, by Clavius' own count, to the 486 propositions he calculated in his Greek-based Euclid, he admits to adding 671 others of his own; "in universum ergo 1234 propositiones in nostro Euclide demonstrantur," he concludes. And the value of what he has compiled matches, especially for the historian, its mass.

The final segment of the modern history of Euclid that requires description is what might most appropriately be called the handbook tradition, both Latin and vernacular, of the *Elements*. Many of the briefer Renaissance versions already mentioned are properly part of this tradition, and if one sets no limit on size, editions like that of Clavius would also qualify. In point of fact, the undercurrent of didacticism we have seen to be present in the medieval Arabic and Latin versions of the *Elements* can justly be regarded as the beginning of this handbook, or school text, tradition.

In the seventeenth century, however, the tradition takes on a more definite form. Numerous examples could be cited from this period, but all of them show the tendency to shorten proofs, to leave out propositions—and even whole books—of little use, and to introduce symbols wherever feasible to facilitate comprehension. This did not mean, to be sure, the disappearance of the sorts of supplementary material characteristic of so many translations and redactions of the *Elements*. On the contrary, such material was often rearranged and retained, and even created anew, when it seemed to be fruitful from the instructional point of view. For instance, one of the most popular (some twenty editions through the first few years of the nineteenth century) handbooks, the *Elementa geometriae planae et solidae* (Antwerp, 1654) of André Tacquet (1612–1660), covers books I–VI and XI–XII, with added material from Archimedes. Its proofs are compendious, but it makes up for its gain in this regard through the addition of a substantial number of pedagogically useful corollaries and scholia. On the other hand, the *Euclidis Elementorum libri XV breviter demonstrati* (Cambridge, 1655) of Isaac Barrow (1630–1677) stubbornly holds to its status as an epitome. Producer of perhaps the shortest handbook of all of books I–XV, Barrow achieved this maximum of condensation by appropriating the symbolism of William Oughtred (1574–1660) that the latter employed in a *declaratio* of book X of Euclid in his *Clavis mathematicae* (Oxford, 1648 ff.). In his preface, Barrow claimed that his goal was "to conjoin the greatest Compendiousness of Demonstration with as much perspicuity as the quality of the subject would admit." Although his success struck some (for example, John Keill) as producing a somewhat obscure compendium, this did not prevent the appearance of numerous (some ten) editions, several of them in English. Vernacular handbooks appeared in other languages as well, perhaps the most notable being *Les Elémens d'Euclide* (Lyons, 1672) of the French Jesuit Claude-François Milliet de Chales (1621–1678). Appearing earlier in Latin (Lyons, 1660), this handbook, covering, like Tacquet's, books I–VI and XI–XII, went through some twenty-four subsequent editions, including translations into English and Italian.

The next stage in the handbook tradition belongs to the nineteenth century, where there occurred a veritable avalanche of Euclid primers, frequently radically divergent from any imaginable text of the *Elements*. Quite separate from these attempts to make Euclid proper for the grammar schools, lycées, and Gymnasia of the 1800's, the rise of classical philology carried with it the efforts to establish a sound and critical text of the *Elements*. These efforts, in turn, gave rise to the annotated translations of the present century, with an audience primarily the historian and the classicist, rather than the mathematician.

Only a paltry few of the almost innumerable versions of the *Elements* dating from the Renaissance to the present have even been mentioned above—most are merely listed in bibliographies and remain totally unexamined. Even the few titles of this period that have been cited have received little more than fleeting attention—often limited to their prefaces—from historians. Further study will, one feels certain, reveal much more of the significance this mass of Euclidean material holds for the history of mathematics and science as a whole.

BIBLIOGRAPHY

Abbreviations of frequently cited works:

Clagett, *Medieval Euclid* = Marshall Clagett, "The Medieval Latin Translations From the Arabic of the *Elements* of Euclid, With Special Emphasis on the Versions of Adelard of Bath," in *Isis,* **44** (1953), 16–42.

Curtze, *Supplementum* = *Anaritii in decem libros priores Elementorum Euclidis ex interpretatione Gherardi Cremonensis,* Maximilian Curtze, ed. (Leipzig, 1899), supplement to *Euclidis Opera omnia,* Heiberg and Menge, eds.

Heath, *Euclid* = Thomas L. Heath, *The Thirteen Books of Euclid's Elements Translated From the Text of Heiberg With Introduction and Commentary,* 3 vols. (2nd ed., Cambridge, 1925; repr. New York, 1956).

Heiberg, *Euclides* = *Euclidis Opera omnia,* J. L. Heiberg and H. Menge, eds., 8 vols. (Leipzig, 1883–1916).

Heiberg, *Litt. Stud.* = J. L. Heiberg, *Litterärgeschichtliche Studien über Euklid* (Leipzig, 1882).

Heiberg, *Paralipomena* = J. L. Heiberg, "Paralipomena zu Euklid," in *Hermes,* **38** (1903), 46–74, 161–201, 321–356.

Klamroth, *Arab. Euklid* = Martin Klamroth, "Ueber den arabischen Euklid," in *Zeitschrift der Deutschen morgenländischen Gesellschaft,* **35** (1881), 270–326, 788.

Sabra, *Simplicius* = A. I. Sabra, "Simplicius's Proof of Euclid's Parallels Postulate," in *Journal of the Warburg and Courtauld Institutes,* **32** (1969), 1–24.

Sabra, *Thābit* = A. I. Sabra, "Thābit ibn Qurra on Euclid's Parallels Postulate," in *Journal of the Warburg and Courtauld Institutes,* **31** (1968), 12–32.

General Euclidean Bibliographies

The most complete bibliography of Euclid is still that of Pietro Riccardi, *Saggio di una bibliografia Euclidea,* in 5 pts., (Bologna, 1887–1893); this work also appeared in the *Memorie della Reale Accademia delle Scienze dell' Istituto di Bologna,* 4th ser., **8** (1887), 401–523; **9** (1888), 321–343; 5th ser., **1** (1890), 27–84; **3** (1892), 639–694. More complete bibliographic information on pre-1600 eds. of the *Elements,* and works dealing with Euclid, can be found in Charles Thomas-Sanford, *Early Editions of Euclid's Elements* (London, 1926). Other bibliographies are listed in the bibliography to pt. I of the present article.

The Elements in Greek Antiquity

1. *Establishment of the "pristine" Greek text.* A history of the text in capsule form was first given in Heiberg, *Litt. Stud.,* pp. 176–186. Heiberg, *Euclides,* V (Leipzig, 1888), xxiii–lxxvi gives a more complete analysis of the Theonine and pre-Theonine texts, together with an outline of the criteria and methods used in establishing the latter. Further material relevant to the textual problem is found in Heiberg, *Paralipomena,* pp. 47–53, 59–74, 161–201. Heath, *Euclid,* I, 46–63, gives a summary of all of the Heiberg material above.

2. *Greek commentaries.* The ed. of the Greek text of Proclus by G. Friedlein is noted in pt. I of the present article, together with several trans. To this one should now add the English trans. of Glenn Morrow (Princeton, 1970). Of the literature on Proclus, the most useful to cite is J. G. van Pesch, *De Procli fontibus* (Leiden, 1900). For the commentary of Pappus, extant only in Arabic, see the following section. Heath, *Euclid,* I, 19–45, gives a convenient summary of Greek commentarial literature. To this one should add Sabra, *Simplicius,* for material on this commentary, extant only in Arabic fragments (and Latin trans. thereof). Finally, Heiberg has treated the commentaries as well as the citations of Euclid in all other later

Greek authors (notably commentators on Aristotle); this material is assembled in Heiberg, *Litt. Stud.,* pp. 154–175, 186–224; and *Paralipomena,* pp. 352–354.

3. *Greek scholia.* Most of these are published in Heiberg, *Euclides,* V (Leipzig, 1888), 71–738, supplemented by Heiberg, *Paralipomena,* pp. 321–352. Scholia to the minor Euclidean works are published in vols. VI–VIII of Heiberg, *Euclides.* The most complete discussion of the scholia is J. L. Heiberg, "Om Scholierne til Euklids Elementer" (with a French résumé), in *Kongelige Danske Videnskabernes Selskabs Skrifter,* Hist.-philosofisk afdeling II, 3 (1888), 227–304. Once again there is a summary of this Danish article in Heath, *Euclid,* I, 64–74.

The Medieval Arabic Euclid

1. *General works.* Serious study of the Arabic Euclid began with J. C. Gartz, *De interpretibus et explanatoribus Euclidis arabicis schediasma historicum* (Halle, 1823) and was continued in J. G. Wenrich, *De auctorum graecorum versionibus et commentariis syriacis, arabicis, armeniacis persicisque commentatio* (Leipzig, 1842), pp. 176–189. The problem of the reports of Euclid in Arabic literature was broached in Heiberg, *Litt. Stud.,* pp. 1–21; but the major step was taken in Klamroth, *Arab. Euklid.* Klamroth's contentions concerning the superiority of the Arabic tradition were answered by Heiberg in "Die arabische Tradition der Elemente Euklid's," in *Zeitschrift für Mathematik und Physik,* Hist.-lit. Abt., 29 (1884), 1–22. This was followed by the summary article, which included material on Arabic commentators, of Moritz Steinschneider, "Euklid bei den Arabern: Eine bibliographische Studie," in *Zeitschrift für Mathematik und Physik,* Hist.-lit. Abt., 31 (1886), 81–110. Cf. Steinschneider's *Die arabischen Uebersetzungen aus dem Griechischen* (Graz, 1960), pp. 156–164 (originally published in *Centralblatt für Bibliothekswesen,* supp. 5, 1889). See also the article by A. G. Kapp in section 6 below. A summary view of our knowledge (as of the beginning of the present century) of the *Elements* in Islam can be found in Heath, *Euclid,* I, 75–90. M. Klamroth has also published a translation of some of the summaries of Greek works by the ninth-century historian al-Yaʿqūbī which includes a résumé of the *Elements:* "Ueber die Auszüge aus griechischen Schriftstellern bei al-Jaʿqūbī," in *Zeitschrift der Deutschen morgenländischen Gesellschaft,* 42 (1888), 3–9. The standard bibliography of Arabic mathematics and mathematicians is Heinrich Suter, *Die Mathematiker und Astronomen der Araber und ihre Werke, Abhandlungen zur Geschichte der mathematischen Wissenschaften,* X (Leipzig, 1900), with *Nachträge und Berichtigungen, op. cit.,* XIV (Leipzig, 1902), 155–185. The most recent history of Islamic mathematics, with appended bibliography, is contained in A. P. Juschkewitsch [Youschkevitch], *Geschichte der Mathematik im Mittelalter* (Basel, 1964; original Russian ed., Moscow, 1961).

2. *The translation of al-Ḥajjāj.* We know of but a single MS containing (presumably) the second Ḥajjāj version together with Nayrīzī's commentary for books I–VI (and a few lines of VII) alone: Leiden, 399, 1. This has been ed. with a modern Latin trans. by J. L. Heiberg, R. O.

Besthorn, *et al., Codex Leidensis 399, 1: Euclidis Elementa ex interpretatione al-Hadschdschadschii cum commentariis al-Narizii,* in 3 pts. (Copenhagen, 1893–1932). Confirmation is needed of the report of two further MSS containing a Ḥajjāj version of books XI–XII: MSS Copenhagen LXXXI and Istanbul, Fātiḥ 3439. There is no secondary literature specifically devoted to the Ḥajjāj *Elements,* but information is contained in Klamroth, *Arab. Euklid.*

3. *The translation of Isḥāq–Thābit.* The most frequently cited MS of this version is Oxford, Bodleian Libr., MS Thurston 11 (279 in Nicoll's catalogue), dated 1238. This was one of the two basic codices employed in Klamroth, *Arab. Euklid.* The literature also makes continual reference to MS Bodl. Or. 448 (280 in Nicoll) as an Isḥāq–Thābit text; it is not this, but rather a copy of al-Maghribī's *Taḥrīr* (the error derives from a marginal misascription to Thābit that was reported by Nicoll in his catalogue). There are, however, a number of other extant copies of Isḥāq–Thābit. Intention to edit the Isḥāq–Thābit trans. was announced (but apparently abandoned) by Claire Baudoux, "Une édition polyglotte orientale des Eléments d'Euclide: La version arabe d'Ishāq et ses derivées," in *Archeion,* 19 (1937), 70–71. Of the literature on the reviser of this trans., Thābit ibn Qurra, see Eilhard Wiedemann, "Ueber Thābit, sein Leben und Wirken," in *Sitzungsberichte der physikalisch-medizinischen Sozietät zu Erlangen,* 52 (1922), 189–219; A. Sayili, "Thābit ibn Qurra's Generalization of the Pythagorean Theorem," in *Isis,* 51 (1960), 35–37; and section 7 below. An integral ed. and Russian trans. of Thābit's mathematical works is in preparation.

4. *The epitomes of Avicenna and others.* A. I. Sabra has edited Avicenna's compendium of the *Elements,* and it will appear in the Cairo ed. of Avicenna's *Kitāb al-Shifaʾ.* A brief description of this compendium was published by Karl Lokotsch, *Avicenna als Mathematiker, besonders die planimetrischen Bücher seiner Euklidübersetzung* (Erfurt, 1912). A copy of a poem praising Euclid, ascribed to Avicenna in a MS found in the Topkapi Museum at Istanbul, is the subject of A. S. Unver, "Avicenna's Praise of Euclid," in *Journal of the History of Medicine,* 2 (1947), 198–200 (other occurrences of the poem, however, disagree with this ascription). The only other Euclid compendium treated in the literature is that of al-Asfuzārī, in L. A. Sédillot, "Notice de plusieurs opuscules mathématiques: V. Quatorzième livre de l'épitome de l'Imam Muzhaffar-al-Isferledi sur les Elements d'Euclide," in *Notices et extraits des manuscrits de la Bibliothèque du Roi,* 13 (1838), 146–148.

5. *The Taḥrīr of al-Ṭūsī, pseudo-Ṭūsī, and al-Maghribī.* The genuine, fifteen-book *Taḥrīr* of al-Ṭūsī exists in an overwhelming number of MSS and has also been frequently printed (Istanbul, 1801; Calcutta, 1824; Lucknow, 1873–1874; Delhi, 1873–1874; Tehran, 1881). Indication of the spurious nature of the thirteen-book *Taḥrīr* usually ascribed to al-Ṭūsī is established by Sabra, *Thābit,* n. 11, and *Simplicius,* postscript, p. 18; doubt is also raised in B. A. Rozenfeld, A. K. Kubesov, and G. S. Sobirov, "Kto by avtorom rimskogo izdania 'Izlozhenia Evklida Nasir ad-Dina at-Tusi'" ("Who Was the Author of the Rome

Edition 'Recension of Euclid by Naṣīr al-Dīn al-Ṭūsī'?"), in *Voprosy istorii estestvoznaniya i tekhniki,* **20** (1966), 51–53. The spurious *Taḥrīr* was printed at Rome in 1594, and we know of only two extant MSS: Bibl. Laur. Or. 2. and Or. 51; the latter carries the 1298 date, causing, among other factors, the problems with al-Ṭūsī's authorship. Almost all of the literature dealing with al-Ṭūsī and Euclid treats of the spurious *Taḥrīr:* H. Suter, "Einiges von Naṣīr el-Dīn's Euklid-Ausgabe," in *Bibliotheca mathematica,* 2nd ser., **6** (1892), 3–6; E. Wiedemann, "Zu der Redaktion von Euklids Elementen durch Naṣīr al Din al Ṭūsī," in *Sitzungsberichte der physikalisch-medizinischen Sozietät zu Erlangen,* **58/59** (1926/1927), 228–236; C. Thaer, "Die Euklid-Überlieferung durch al-Ṭūsī," in *Quellen und Studien zur Geschichte der Mathematik, Astronomie und Physik,* Abt. B, Studien, **3** (1936), 116–121. More general works on al-Ṭūsī as a mathematician include an ed. of the Arabic text of the *Rasā ʿil al-Ṭūsī,* 2 vols. (Hyderabad, 1939–1940); E. Wiedemann, "Naṣīr al Din al Ṭūsī," in *Sitzungsberichte der physikalisch-medizinischen Sozietät zu Erlangen,* **60** (1928), 289–316; and B. A. Rozenfeld, "O matematicheskikh rabotakh Nasireddina Ṭūsī" ("On the Mathematical Works of Naṣīr al-Dīn al-Ṭūsī"), in *Istoriko-matematicheskie issledovaniya,* **4** (1951), 489–512. See also section 7 below. The *Taḥrīr* of al-Maghribī is found in the thirteenth-century MS Bodl. Or. 448, as well as two later codices in Istanbul. It is identified and discussed, together with the ed. and trans. of a fragment from it in Sabra, *Simplicius,* pp. 13–18, 21–24.

6. *Commentaries. The Arabic translation of the commentary of Pappus on book X.* Extracts of the Arabic text, together with a French trans., were first published by Franz Woepcke in "Essai d'une restitution de travaux perdus d'Apollonius sur les quantités irrationnelles," in *Mémoires présentés par divers savants à l'Académie des sciences,* **14** (1856), 658–720 (also published separately). Woepcke also published the full text of the commentary without date or place of publication (Paris, 1855[?]). This was in turn trans. into German with comments by H. Suter, "Der Kommentar des Pappus zum X Buche des Euklides aus der arabischen Übersetzung des Abū ʿOthmān al-Dimashḳī ins Deutsche übertragen," in *Abhandlungen zur Geschichte der Naturwissenschaften und der Medizin,* **4** (1922), 9–78. A new ed. of the Arabic text with notes and English trans. was published by William Thomson and Gustav Junge, *The Commentary of Pappus on Book X of Euclid's Elements* (Cambridge, Mass., 1930). Critical remarks on this text were published by G. Bergstrasser, "Pappos Kommentar zum Zehnten Buch von Euklid's Elementen," in *Der Islam,* **21** (1933), 195–222. A fragment of Gerard of Cremona's trans. of this Pappus text is printed in G. Junge, "Das Fragment der lateinischen Übersetzung des Pappus-Kommentars zum 10. Buche Euklids," in *Quellen und Studien zur Geschichte der Mathematik, Astronomie und Physik,* Abt. B, **3,** Studien (1936), 1–17.

Arabic commentaries. Very few of the great number of these have been published or studied. A list, quite complete in terms of present knowledge, giving brief indications of author, subject, and relevant bibliography, can be found in E. B. Plooij, *Euclid's Conception of Ratio and his Definition of Proportional Magnitudes as Criticized by Arabian Commentators* (Rotterdam, n.d.), pp. 3–13. More elaborate is A. G. Kapp, "Arabische Übersetzer und Kommentatoren Euklids, sowie deren math.-naturwiss. Werke auf Grund des Ta'rīkh al-Ḥukamā' des Ibn al-Qifṭī," in *Isis,* **22** (1934), 150–172; **23** (1935), 54–99; **24** (1935), 37–79; as the title indicates, this extensive article contains much material trans. from the biobibliographical work of Qifṭī (*ca.* 1172–1248). Arabic commentaries on book X are treated in G. P. Matvievskaya, *Uchenie o chisle na srednevekovom blizhnem i srednem vostoke* ("Studies on Number in the Medieval Near and Middle East"; Tashkent, 1967), pp. 191–229; The following commentaries (listed in approximate chronological order) have been ed. or trans. (if only partially) and analyzed: (1) al-Nayrīzī: books I–VI in Arabic in the Heiberg-Besthorn ed. of Ḥajjāj cited in section 2 above; books I–X (incomplete[?]) in the Latin trans. of Gerard of Cremona in Curtze, *Supplementum,* pp. 1–252 only (the remainder of this volume containing the commentary not of al-Nayrīzī, but of ʿAbd al-Baqi; see below). Determination through examination of al-Nayrīzī of various interpolations in the text of the *Elements* was done in Heiberg, *Paralipomena,* pp. 54–59. (2) Al-Fārābī on books I and V has been trans. into Russian on the basis of its two Hebrew copies (MSS Munich 36 and 290, not edited): M. F. Bokshteyn and B. A. Rozenfeld, "Kommentarii Abu Nasra al-Farabi k trudnostyam vo vvedeniakh k pervoy i pyatoy knigam Evklida" ("The Commentary of Abū Naṣr al-Fārābī on the Difficulties in the Introduction to Books I and V of Euclid"), in *Akademiya nauk SSR, Problemy vostokovedeniya,* no. 4 (1959), 93–104. The Arabic text of this brief work of al-Fārābī has now also apparently been discovered: Escorial MS Arab. 612, 109r–111v. A fragment of this, or of another Euclidean opusculum by al-Fārābī, is Tehran, Faculty of Theology, MS 123-D, 80v–82r. See also A. Kubesov and B. A. Rozenfeld, "On the Geometrical Treatise of al-Fārābī," in *Archives internationales d'histoire des sciences,* **22** (1969), 50. (3) Ibn al-Haytham, *On the Premises of Euclid,* has also received a (partial) Russian trans. by B. A. Rozenfeld as "Kniga kommentariev k vvedeniam knigi Evklida 'Nachala'" ("Book of Commentaries to Introductions to Euclid's *Elements*"), in *Istoriko-matematicheskie issledovaniya,* **11** (1958), 743–762. (4) Ibn Muʿādh al-Jayyānī on book V has been reproduced in facsimile with accompanying English trans. in the book of E. B. Plooij cited above. This book also contains a trans. of passages relevant to book V from the commentaries of al-Māhānī, al-Nayrīzī, Ibn al-Haytham, and Omar Khayyām. (5) Omar Khayyām's work on Euclid has received the most attention of all. A. I. Sabra has published a critical Arabic text (without trans.) as *Explanation of the Difficulties in Euclid's Postulates* (Alexandria, 1961). There is an earlier ed., on the basis of a single MS, by T. Erani (Tehran, 1936). A Russian trans., with commentary, has been published by B. A. Rozenfeld and A. P. Youschkevitch in *Istoriko-matematicheskie issledovaniya,* **6** (1953), 67–107, 143–168; repr. with a MS facsimile in Omar Khayyām, *Traktaty* (Moscow, 1961). The

English trans. by Amir-Móez in *Scripta mathematica,* **24** (1959), 272–303, must be used with great care. (6) ʿAbd al-Bāqī's commentary on book X in Gerard of Cremona's Latin trans. is printed in Curtze, *Supplementum,* pp. 252–386. H. Suter has given corrections to Curtze's text in "Ueber den Kommentar des Muḥ b. ʿAbdelbāqi zum 10 Buche des Euklides," in *Bibliotheca mathematica,* 3rd ser., **7** (1907), 234–251. See the following section for literature on yet other commentarial material.

7. *On the parallels postulate.* The importance of this postulate in the history of mathematics is reflected not merely in the frequency of its discussion by Islamic authors, but also by the attention it has received from modern historians. As an introduction, see B. A. Rozenfeld and A. P. Youschkevitch, *The Prehistory of Non-Euclidean Geometry in the Middle East,* XXV International Congress of Orientalists, *Papers Presented by the USSR Delegation* (Moscow, 1960). Compare B. A. Rozenfeld, "The Theory of Parallel Lines in the Medieval East," in *Actes du XIᵉ Congrès International d'Histoire des Sciences, Varsovie-Cracovie 1965,* **3** (Warsaw, etc., 1968), 175–178. More specifically, two treatments of Thābit ibn Qurra are trans. and analyzed in Sabra, *Thābit.* The problem is also the subject of Sabra, *Simplicius.* In fact, these two articles contain a mine of information pertinent to the issue throughout Islamic mathematics. The two Thābit treatises have also been analyzed and trans. into Russian by B. A. Rozenfeld and A. P. Youschkevitch in *Istoriko-matematicheskie issledovaniya,* **14** (1961), 587–597; and **15** (1963), 363–380. Extracts from the treatments of the postulate by al-Jawharī, Qayṣar ibn Abi 'l-Qāsim, and al-Maghribī are found in the two Sabra articles. The greatest amount of attention has been paid to al-Ṭūsī's struggles with the problem, beginning with a trans. into Latin by Edward Pocock of the proof of the postulate in the pseudo-Ṭūsī *Taḥrīr;* this was printed in John Wallis, *Opera mathematica,* II (Oxford, 1693), 669–673. Both this proof and that in the genuine fifteen-book al-Ṭūsī *Taḥrīr* were published and analyzed in Arabic by A. I. Sabra, "Burhān Naṣīr al-Dīn al-Ṭūsī ʿalā muṣādarat Uqlīdis al-khāmisa," in *Bulletin of the Faculty of Arts of the University of Alexandria,* **13** (1959), 133–170. Russian treatment again occurs in G. D. Mamedbeili, *Mukhammad Nasureddin Tusi o teorii parallelnykh liny i teorii otnosheny* ("Muḥammad Naṣīr al-Dīn al-Ṭūsī on the Theory of Parallel Lines and the Theory of Proportion"; Baku, 1959), and in *Istoriko-matematicheskie issledovaniya,* **13** (1960), 475–532. Finally, the article of H. Dilgan, "Demonstration du Vᵉ postulat d'Euclide par Schams-ed-Din Samarkandi, Traduction de l'ouvrage Aschkal-ut-tessis de Samarkandi," in *Revue d'histoire des sciences,* **13** (1960), 191–196, does not contain a proof by Samarqandī, but rather one by Athīr al-Dīn al-Abharī that was reproduced in a commentary to Samarqandī's work.

8. *Translations into other Near Eastern languages.* The most adequate account of Hebrew versions and commentaries is in Moritz Steinschneider, *Die hebräischen Übersetzungen des Mittelalters und die Juden als Dolmetscher* (Berlin, 1893; repr. Graz, 1956), 503–513. The fragments of the Syriac version were published and trans. by G. Furlani,

"Bruchstücke eine syrischen Paraphrase der 'Elemente' des Eukleides," in *Zeitschrift für Semitistik und verwandte Gebiete,* **3** (1924), 27–52, 212–235. Furlani held that the paraphrase was derived from the Arabic of Ḥajjāj. This was questioned, and the opposing view placing it before, and as a source of, the Isḥāq trans., by C. Baudoux, "La version syriaque des 'Eléments' d'Euclide," in *IIᵉ Congrès national des sciences* (Brussels, 1935), pp. 73–75. The fragments of the early Armenian version were published and trans. (into Latin) by Maurice Leroy, "La traduction arménienne d'Euclide," in *Annuaire de l'Institut de philologie et d'histoire orientales et slaves* (*Mélanges Franz Cumont*), **4** (1936), 785–816. The letter of Gregory Magistros announcing his translating activity with respect to Euclid was published and analyzed by Leroy in the same *Annuaire,* **3** (1935), 263–294. Additional material on Armenian Euclids can be found in the article (not presently examined) by T. G. Tumanyai, "'Nachala' Evklida po drevnearmyanskim istochnikam" ("Euclid's *Elements* in Ancient Armenian Sources"), in *Istoriko-matematicheskie issledovaniya,* **6** (1953), 659–671, and in G. B. Petrosian and A. G. Abramyan, "A Newly Discovered Armenian Text of Euclid's Geometry," in *Proceedings of the Tenth International Congress of the History of Science, Ithaca, 1962,* II (Paris, 1964), 651–654.

9. *Euclid's minor works.* The Arabic trans. of the Euclidean opuscula, together with a discussion of their role as "middle books" in Islamic mathematics and astronomy, was first examined by M. Steinschneider, "Die 'mittleren' Bücher der Araber und ihre Bearbeiter," in *Zeitschrift für Mathematik und Physik,* **10** (1865), 456–498. There is little material dealing specifically with these shorter works, but in addition to the general literature in section 1 above, see Clemens Thaer, "Euklids Data in arabischer Fassung," in *Hermes,* **77** (1942), 197–205. The prolegomena in vols. VI–VIII of Heiberg, *Euclides,* also contains information on the Arabic phase of these opuscula. For literature dealing with the Islamic role in the work *On the Division of Figures* and the works on mechanics, see pt. I of the present article.

The Medieval Latin Euclid: The Greek–Latin Phase

1. *Euclidean material in Roman authors.* The fragments in Censorinus are appended in F. Hultsch's ed. (Leipzig, 1867) of the *De die natali,* pp. 60–63. For Euclid in Martianus Capella, see the *De nuptiis philologiae et mercurii,* VI, 708 ff.

2. *The Verona palimpsest.* The Euclid fragments have recently been edited, with facsimile and notes, by Mario Geymonat, *Euclidis latine facti fragmenta Veronensia* (Milan, 1964). This work contains references to all other previous literature on the palimpsest, both of paleographers and historians of mathematics.

3. *The Boethian Euclid excerpts.* The best account of all of the variables involved is the absolutely fundamental work of Menso Folkerts, *"Boethius" Geometrie II: Ein mathematisches Lehrbuch des Mittelalters* (Wiesbaden, 1970). This contains a critical ed. of (a) the two-book "Boethian" *Geometry;* (b) the Euclid excerpts preserved in all four earlier medieval sources. The Boethian–Adelard

mélanges in Ratsbücherei Lüneburg MS miscell. D 4° 48 have been treated and ed. by Folkerts in *Ein neuer Text des Euclides Latinus: Faksimiledruck der Handschrift Lüneburg D 4° 48, f. 13r-17v* (Hildesheim, 1970), and, together with a consideration of the mélanges in the Paris and Munich MSS (see following section), in "Anonyme lateinische Euklidbearbeitungen aus dem 12. Jahrhundert," in *Denkschriften der Österreichischen Akademie der Wissenschaften,* Math.-naturwiss. Klasse (1970), 5–42. See also Folkerts' earlier article, "Das Problem der pseudo-boethischen Geometrie," in *Sudhoff's Archiv für Geschichte der Medizin und der Naturwissenschaften,* **52** (1968), 152–161. An earlier work that also attempted, as a tangential problem, to sort out the threads of the "Boethian" Euclid is Nicolaus Bubnov, *Gerberti Opera mathematica* (Berlin, 1899; repr. Hildesheim, 1963). Before Folkerts the standard ed. of the two-book geometry was that of Gottfried Friedlein in his text of Boethius' *De institutione arithmetica . . . de institutione musica . . . accedit geometria quae fertur Boetii* (Leipzig, 1867), pp. 372–428. The five-book "Boethian" geometry still does not exist in a critical ed. The first two books have appeared, however, among Boethius' works in J. P. Migne, *Patrologia Latina,* vol. LXIII, cols. 1352–1364 (cols. 1307–1352 contain the two-book geometry now in Folkerts). Books I, III, IV, and part of V are in F. Blume, K. Lachmann, and A. Rudorff, *Die Schriften der römischen Feldmesser,* I (Berlin, 1848), 377–412. The remaining section of book V is unedited. Although the five-book geometry has therefore not received adequate editing, a most exacting analysis of its MS sources and history has been made by C. Thulin, *Zur Überlieferungsgeschichte des Corpus Agrimensorum. Exzerptenhandschriften und Kompendien* (Göteborg, 1911). The Euclid excerpts found in Cassiodorus have been edited by R. A. B. Mynors in his text of the *Institutiones* (Oxford, 1937), pp. 169–172. A recent important article that treats of the role of "Boethian" geometry in the earlier Middle Ages is B. L. Ullman, "Geometry in the Medieval Quadrivium," in *Studi di bibliografia e di storia in onore di Tammaro de Marinis,* IV (Verona, 1964), 263–285. Among the earlier literature on the problems of Boethius and the *Elements* are H. Weissenborn, "Die Boetius-Frage," in *Abhandlungen zur Geschichte der Mathematik,* **2** (1879), 185–240; J. L. Heiberg, "Beiträge zur Geschichte der Mathematik im Mittelalter, II. Euklid's Elemente im Mittelalter," in *Zeitschrift für Mathematik und Physik,* Hist.-lit. Abt., **35** (1890), 48–58, 81–100; Georg Ernst, *De geometricis illis quae sub Boëthii nomine nobis tradita sunt, quaestiones* (Bayreuth, 1903); M. Manitius, "Collationen aus einem geometrischen Tractat," in *Hermes,* **39** (1904), 291–300, and "Collationen aus der *Ars geometrica," ibid.,* **41** (1906), 278–292; and several pieces by Paul Tannery, now included in his *Mémoires scientifiques,* V (Paris, 1922), 79–102, 211–228, 246–250.

4. *Boethian–Adelardian mélanges.* The Lüneburg MS mélange has been ed. by Folkerts (see above). A second mélange exists in the four MSS Paris, BN lat. 10257; Oxford, Bodl. Digby 98; Munich, CLM 13021, and CLM 23511. The Paris MS has been ed. in the unpublished dissertation of George D. Goldat, "The Early Medieval Traditions of Euclid's Elements" (Madison, Wisc., 1956).

5. *Munich manuscript fragment.* This has been ed., from MS Univ. Munich 2° 757, by Curtze, *Supplementum,* pp. xvi–xxvi. Corrections to Curtze's text can be found in Heiberg, *Paralipomena,* pp. 354–356, and *Bibliotheca mathematica,* 3rd ser., **2** (1901), 365–366. A new edition of the text has been prepared by Mario Geymonat in "Nuovi frammenti della geometria 'Boeziana' in un codice del IX secolo?" in *Scriptorium,* **21** (1967), 3–16. Geymonat dates the fragment as of the ninth century rather than the tenth; whether this be correct or not, the question that he poses of Boethius' authorship for this fragment should in all probability be answered negatively.

6. *Twelfth-century Greek–Latin translation.* This is found in only two extant MSS: Paris, BN lat. 7373, and Florence, Bib. Naz. Centr. Fondo Conventi Soppressi C I 448. The trans. has been analyzed in full in John Murdoch, "Euclides Graeco-Latinus. A Hitherto Unknown Medieval Latin Translation of the *Elements* Made Directly from the Greek," in *Harvard Studies in Classical Philology,* **71** (1966), 249–302. The Greek-Latin version of Ptolemy's *Almagest* made by the same translator is discussed, and its preface edited, in C. H. Haskins, *Studies in the History of Mediaeval Science* (Cambridge, Mass., 1924), ch. 9.

7. *Medieval Latin versions of the Euclidean opuscula.* General information on the trans. of these minor works can be found in Heiberg, *Euclides,* prolegomena to vols. VI–VIII. In fact, the text of one of the Greek-Latin renderings of the *Optica* has been ed. in Heiberg, *Euclides,* VII (1895), 3–121. The Greek-Latin version of the *Data* was first noted by A. A. Björnbo, "Die mittelalterlichen lateinischen Übersetzungen aus dem Griechischen auf dem Gebiete der mathematischen Wissenschaften," in *Festschrift Moritz Cantor* (Leipzig, 1909), p. 98. It has since been edited in the unpublished dissertation of Shuntaro Ito, "The Medieval Latin Translation of the Data of Euclid" (Madison, Wisc., 1964). The pseudo-Euclidean *De speculis*—to be distinguished from the equally spurious *Catoptrica*—has been edited in A. A. Björnbo and S. Vogl, *Alkindi, Tideus und Pseudo-Euklid. Drei optische Werke, Abhandlungen zur Geschichte der mathematischen Wissenschaften,* vol. XXVI, pt. 3 (Leipzig, 1911); cf. S. Vogl in *Festschrift Moritz Cantor* (Leipzig, 1909), pp. 127–143.

The Medieval Latin Euclid: The Arabic–Latin Phase

1. *General.* A brief resumé of our earlier knowledge of this wing of the medieval Latin Euclid can be found in Heath, *Euclid,* I, 93–96. The fundamental comprehensive description is now Clagett, *Medieval Euclid,* which includes appendices that present sample texts from all of the basic twelfth-century versions constituting the Arabic-Latin *Elements*.

2. *The translation of Gerard of Cremona.* The most complete discussion, including a listing of MSS, is Clagett, *Medieval Euclid,* pp. 27–28, 38–41. See also A. A. Björnbo, "Gerhard von Cremonas Uebersetzung von Alkwarizmis Algebra und von Euklids Elementen," in *Bibliotheca mathematica,* 3rd ser., **6** (1905), 239–248. Still useful for Gerard's life and career is B. Boncompagni, "Della vita

e delle opere di Gherardo cremonense," in *Atti dell' Accademia pontificia de' Nuovi Lincei,* 1st ser., **4** (1851), 387–493. A more critical text of Gerard's *vita et libri translati* appended in a number of MSS to his trans. of Galen's *Ars parva* has been given, with annotations to the list of works trans., by F. Wüstenfeld, *Die Übersetzungen arabischer Werke in das Lateinische seit dem XI Jahrhundert. Abhandlungen der Königlichen Gesellschaft der Wissenschaften zu Göttingen* (Göttingen, 1877), pp. 57–81. For Gerard's trans. from the Arabic of commentaries on the *Elements,* see section 6 of the Arabic Euclid bibliography above.

3. *The Adelardian tradition.* The first extensive article treating of Adelard's role in the transmission of Euclid was that of Hermann Weissenborn, "Die Übersetzung des Euklid aus dem Arabischen in das Lateinische durch Adelhard von Bath . . .," in *Zeitschrift für Mathematik und Physik,* Hist.-lit. Abt., **25** (1880), 143–166. It was Clagett, *Medieval Euclid,* pp. 18–25, who first distinguished the three separate versions to be ascribed to Adelard. This article also lists a good portion of extant MSS of the three recensions. A more detailed analysis of the nature of these three versions, together with that of Campanus of Novara, is given in J. Murdoch, "The Medieval Euclid: Salient Aspects of the Translations of the *Elements* by Adelard of Bath and Campanus of Novara," in XIIe Congrès International d'Histoire des Sciences, Colloques, in *Revue de synthèse,* **89** (1968), 67–94. For the misinterpretation within the Adelard tradition and within medieval mathematics in general of the Eudoxean definition of equal ratios, see J. Murdoch, "The Medieval Language of Proportions: Elements of the Interaction With Greek Foundations and the Development of New Mathematical Techniques," in A. C. Crombie, ed., *Scientific Change* (London, 1963), pp. 237–271, 334–343. The erroneous ascription of an Adelard version to Alfred the Great was set forth in Edgar Jorg, *Des Boetius und des Alfredus Magnus Kommentar zu den Elementen des Euklid (Nach dem Codex [Z. L. CCCXXXII] B. der Bibliotheca Nazionale di S. Marco zu Venedig), Zweites Buch* (Bottrop, 1935); that this particular MS contains merely both an Adelard II (Boethius) and an Adelard III (Alfred) version was established by M. Clagett, "King Alfred and the *Elements* of Euclid," in *Isis,* **45** (1954), 269–277. Works on Adelard himself are C. H. Haskins, *Studies in the History of Mediaeval Science* (Cambridge, Mass., 1924), ch. 2, and the frequently over-enthusiastic book of Franz Bliemetzrieder, *Adelard von Bath* (Munich, 1935). On the version of Campanus, in addition to the article of Murdoch cited above, see Hermann Weissenborn, *Die Uebersetzungen des Euklid durch Campano und Zamberti* (Halle, 1882). Further biobibliographical information on Campanus is contained in the text and trans. of his *Theorica planetarum,* as edited by Francis S. Benjamin, Jr., and G. J. Toomer (in press).

4. *The translation of Hermann of Carinthia.* See Clagett, *Medieval Euclid,* pp. 26–27, 38–42, and the ed. of books I–VI by H. L. L. Busard, "The Translation of the *Elements* of Euclid From the Arabic Into Latin by Hermann of Carinthia (?)," in *Janus,* **54** (1967), 1–142.

5. *Other translations and commentaries.* The supposed reference to a pre-Adelardian *Elements* in England "Yn tyme of good kyng Adelstones day" (as stated by a fourteenth-century verse) has been shown to apply to masonry, and not geometry, by F. A. Yeldham, "The Alleged Early English Version of Euclid," in *Isis,* **9** (1927), 234–238. For the problem of references to a trans. by Johannes Ocreat, see Clagett, *Medieval Euclid,* pp. 21–22. A commentary on books I–IV of the *Elements* that exists in a single MS (Vienna, Dominik. 80/45) and is there ascribed to Albertus Magnus is discussed by J. E. Hoffmann, "Ueber eine Euklid-Bearbeitung, die dem Albertus Magnus zuschrieben wird," in *Proceedings of the International Congress of Mathematicians, Cambridge, 1958,* pp. 554–566; and by B. Geyer, "Die mathematischen Schriften des Albertus Magnus," in *Angelicum,* **35** (1958), 159–175. An example of later medieval *questiones super Euclidem* are those of Nicole Oresme, recently edited (Leiden, 1961) by H. L. L. Busard; cf. J. E. Murdoch, in *Scripta mathematica,* **27** (1964), 67–91.

The Renaissance and Modern Euclid

1. *General.* There exists very little literature dealing with the transmission of Euclid from 1500 to the present; even bibliographies have not received much attention since the nineteenth century. And there is absolutely no work covering the fairly extensive body of MS materials from the sixteenth and seventeenth centuries. For the printed materials, the most adequate general works are the bibliographies cited above of Riccardi and Sanford, together with the brief survey of the principal eds. of the *Elements* in Heath, *Euclid,* I, 97–113. Dates and places of the versions of the *Elements* that are mentioned above have been given in the body of the text and will not be repeated here.

2. *Latin and Greek editions in the Renaissance and early modern period.* An outline of the major eds. is given in Heiberg, *Euclides,* V (1888), ci–cxiii. Heiberg has also treated of the significance for Euclid and Greek mathematics of Giorgio Valla and his encyclopedic *De expetendis et fugiendis rebus* in "Philologischen Studien zu griechischen Mathematikern: III. Die Handschriften George Vallas von griechischen Mathematikern," in *Jahrbuch für classische Philologie,* **12,** supp. (1881), 337–402; and *Beiträge zur Geschichte Georg Valla's und seiner Bibliothek, Centralblatt für Bibliothekswesen,* Beiheft 16 (Leipzig, 1896). On Zamberti see the monograph of Weissenborn on Campanus and this author that is cited in the section on the Adelardian tradition, above. There is no adequate work on Commandino or Clavius, especially concerning their role in the trans. and dissemination of Greek mathematics. Note has been taken, however, that in the seventeenth century the Jesuit Ricci, a student of Clavius, was instrumental in effecting a Chinese version of the latter's *Elements:* see L. Vanhee, "Euclide en chinois et mandchou," in *Isis,* **30** (1939), 84–88. The possibility of an earlier, thirteenth-century translation of Euclid into Chinese has been briefly discussed by Joseph Needham and Wang Ling, *Science and Civilisation in China,* III (Cambridge, 1959), 105.

Girolamo Saccheri's *Euclidis ab omni naevo vindicatus*

has received a modern ed. and English trans. by G. B. Halsted (Chicago, 1920). But this contains only book I of Saccheri's treatise. For book II (dealing with the theory of proportion) see Linda Allegri, "Book II of Girolamo Saccheri's *Euclides ab omni naevo vindicatus,*" in *Proceedings of the Tenth International Congress of the History of Science, Ithaca, 1962,* II (Paris, 1964), 663–665; an English trans. is to be found in the same author's unpublished dissertation, "The Mathematical Works of Girolamo Saccheri, S. J. (1667–1733)" (Columbia University, 1960).

3. *Vernacular translations.* A recent detailed treatment of the appearance of the *Elements* in England up to *ca.* 1700 (in both English and Latin) is Diana M. Simpkins, "Early Editions of Euclid in England," in *Annals of Science,* **22** (1966), 225–249. Robert Recorde and his inclusion of Euclidean material in *The Pathway to Knowledg* is the subject of Joy B. Easton, "A Tudor Euclid," in *Scripta mathematica,* **27** (1964), 339–355. Most extensive attention has been paid to the 1570 English trans. by Sir Henry Billingsley. In addition to the work of Simpkins, above, see G. B. Halsted, "Note on the First English Euclid," in *American Journal of Mathematics,* **2** (1879), 46–48 (which contains the first notice of the volumes at Princeton with marginalia in Billingsley's hand); W. F. Shenton, "The First

English Euclid," in *American Mathematical Monthly,* **35** (1928), 505–512; R. C. Archibald, "The First Translation of Euclid's *Elements* Into English and Its Source," in *American Mathematical Monthly,* **57** (1950), 443–452. See also Edward Rosen, "John Dee and Commandino," in *Scripta mathematica,* **28** (1970), 325. An annotated bibliography of French trans. of Euclid is Marie Lacoarret, "Les traductions françaises des oeuvres d'Euclide," in *Revue d'histoire des sciences,* **10** (1957), 38–58. I. J. Depman has written of unnoticed Russian eds. of the *Elements* in *Istoriko-matematicheskie issledovaniya,* **3** (1950), 467–473.

4. *The nineteenth and twentieth centuries.* The most complete survey of the great number of nineteenth-century school-text Euclids is still to be found in the Riccardi bibliography (introductory section above). The most notable twentieth-century trans. that contain considerable historical and analytic annotation are, in English, Heath, *Euclid;* in Italian, Federigo Enriques, *et al.,* eds., *Gli Elementi d'Euclide e la critica antica e moderna,* 4 vols. (Rome-Bologna, 1925–1935); in Dutch, E. J. Dijksterhuis, *De Elementen van Euclides,* 2 vols. (Groningen, 1929–1930); in Russian, D. D. Morduchai-Boltovskogo, *Nachala Evklida,* 3 vols. (Moscow–Leningrad, 1948–1950); in French, of books VII–IX only, Jean Itard, *Les livres arithmétiques d'Euclide* (Paris, 1961).

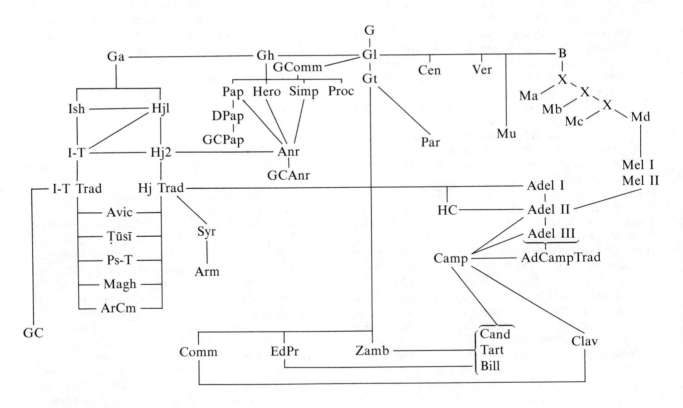

Filiation of the Major Versions of Euclid's *Elements* in the Middle Ages and the Renaissance. Note: Some of the lines of filiation indicated are conjectural. No attempt has been made to differentiate them from those definitely known to be true.

AdCampTrad = Variant versions deriving from those of Adelard and Campanus

Adel I = First translation of Adelard of Bath

Adel II = *Commentum* of Adelard of Bath

Adel III = *Editio specialis* of Adelard of Bath

Anr = Commentary of al-Nayrīzī

ArCm = Arabic commentaries

Arm = Armenian version presumably made by Gregory Magistros

Avic = Epitome of Avicenna

B = Translation of Boethius

Bill = English translation of Henry Billingsley

Camp = Version of Campanus of Novara

Cand = Version of Franciscus Flussatus Candalla

Cen = Fragments in Censorinus

Clav = Edition of Christopher Clavius

Comm = Translation of Federico Commandino

DPap = Translation by al-Dimashqī of book X of Pappus' commentary

EdPr = *Editio princeps* of Greek text by Simon Grynaeus

G = Original Greek text

Gl = Pre-Theonine Greek text

Ga = Greek text employed for Arabic translations

GC = Translation by Gerard of Cremona of Isḥāq–Thābit text

GCAnr = Gerard of Cremona's translation of al-Nayrīzī's commentary

GComm = Greek commentaries

GCPap = Gerard of Cremona's translation of (part of) Dimashqī's translation of Pappus on book X

Gh = Greek text utilized by Hero of Alexandria

Gt = Redaction of Greek text by Theon of Alexandria

HC = Version of Hermann of Carinthia

Hero = Hero of Alexandria's commentary

Hj1 = First version of Ḥajjāj

Hj2 = Second version of Ḥajjāj

Hj Trad = Ḥajjāj Arabic tradition

Ish = Translation of Isḥāq ibn Ḥunayn

I-T = Translation of Isḥāq as revised by Thābit ibn Qurra

I-T Trad = Isḥāq–Thābit tradition

Ma = Boethian excerpts preserved in Cassiodorus

Magh = *Taḥrīr* of al-Maghribī

Mb = Boethian excerpts preserved in *Agrimensores* material

Mc = Boethian excerpts in five-book geometry of "Boethius"

Md = Boethian excerpts in two-book geometry of "Boethius"

Mel I = Boethian–Adelardian mélanges in MS Lüneburg D 4° 48

Mel II = Boethian–Adelardian mélanges in Paris and Munich MSS

Mu = Fragments of translation in MS Univ. Munich 2° 757

Pap = Greek commentary of Pappus of Alexandria

Par = Anonymous Greek–Latin translation of twelfth century

Proc = Greek commentary of Proclus on book I

Ps-T = Thirteen-book *Taḥrīr* erroneously ascribed to al-Ṭūsī

Simp = Commentary of Simplicius on the premises

Syr = Syriac redaction

Tart = Italian translation by Niccolò Tartaglia

Ṭūsī = Genuine fifteen-book *Taḥrīr* by al-Ṭūsī

Ver = Verona palimpsest of fifth century

X = Ancestors not further specified here

Zamb = Translation of Bartolomeo Zamberti of Theonine text

JOHN MURDOCH

EUCTEMON (*fl.* Athens, fifth century B.C.), *astronomy.*

Euctemon is cited (with Meton) by Ptolemy for observations of the summer solstice, including that of 27 June 432 B.C., the reliability of which was doubted by both Ptolemy (*Almagest,* III, 1) and Hipparchus (modern calculations show that the solstice actually occurred about a day and a half later) but which was still used, as the earliest observation available, to confirm the final Hipparchian-Ptolemaic figure of 365.25 less 1/300 of a day for the length of the solar year. Euctemon collaborated with Meton in suggesting a regular intercalation cycle of nineteen years (the Metonic cycle) to correlate the lunar month with the solar year (see B. L. Van der Waerden, in *Journal of Hellenic Studies,* **80** [1960],

170); this cycle contained 235 lunar months (seven of which were intercalary) and 6,940 days, giving a mean lunar month about two minutes too long and a solar year of 365 5/19 days (some thirty minutes too long). According to Geminus (*Isagoge,* VIII, 50), who connects Euctemon but not Meton with this cycle (Meton's name may have dropped out of the text; see Manitius, *ad loc.*), 110 of the months were "hollow" (i.e., twenty-nine days each) and 125 "full" (thirty days).

Euctemon is frequently cited—some forty-five times in the calendar attached to Geminus' *Isagoge* (Manitius, ed., pp. 210 ff.), over fifty times in Ptolemy's *Phaseis,* and often in the other Greek calendars (published in *Sitzungsberichte der Heidelberger Akademie der Wissenschaften,* phil.-hist. Klasse, **1** [1910]; **2** [1911]; **3A** [1913], which contains a conjectural restoration by Rehm of Euctemon's "parapegma"; **4** [1914]; and **5** [1920])—for "weather prognostications" (ἐπισημασίαι) such as formed part of a parapegma, which was a type of almanac, originally engraved on stone or wood, and later transmitted in manuscript form, giving astronomical and meteorological phenomena for the days of each month. His and Meton's parapegma may well have been the first influential text of this kind in Greece (see A. Rehm, "Parapegmastudien," in *Abhandlungen der Bayerischen Akademie der Wissenschaften,* phil.-hist. Abt., n.s. **19** [1941]).

Euctemon is the earliest name mentioned in connection with equinoxes in the extant parapegmata and, according to a second-century B.C. papyrus known as the *Ars Eudoxi* (F. Blass, ed. [Kiel, 1887]—perhaps a student's exercise, containing many errors, partly based on Eudoxus' work with later material added), he gave the lengths of the astronomical seasons, starting from the summer solstice, as ninety, ninety, ninety-two, and ninety-three days, respectively. This shows that he was aware of the nonuniformity of the sun's course round the earth, but it is unlikely that his parapegma was arranged according to zodiacal months, as later ones were. Ptolemy says (*Phaseis,* in *Claudii Ptolemaei opera quae extant omnia,* J. L. Heiberg, ed., II [Leipzig, 1907], 67) that Euctemon made observations at Athens, in the Cyclades, and in Macedonia and Thrace. Rehm thinks that the pseudo-Theophrastian treatise *De signis* (*Theophrasti Eresii opera quae supersunt omnia,* F. Wimmer, ed., III [Leipzig, 1862], fr. 6, pp. 115–130) conceals an original meteorological work by Euctemon, but this is pure conjecture.

Euctemon also did some work in geography and is cited by Avienus (fourth century A.D., but using much older sources) for information concerning the straits of Gibraltar (*Ora maritima,* 337, 350 ff.). Avienus calls him both an Athenian (47–48, 350) and an inhabitant of Amphipolis (337), and he may have been among the Athenian colonists who established a new foundation there in 437 B.C.

BIBLIOGRAPHY

In addition to the works cited in the text, see A. Rehm, "Euktemon 10," in Pauly-Wissowa, XI (1907), cols. 1060–1061.

D. R. Dicks

EUDEMUS OF RHODES (*b.* Rhodes; *fl.* second half of fourth century B.C.), *philosophy, history of science.*

From the title so often given in antiquity to Eudemus the Peripatetic philosopher, it is a fair deduction that he was born at Rhodes; and this is specifically attested by Strabo.[1] The dates of his birth and death are unknown, but his links with Aristotle and Theophrastus show when he flourished.

Nothing is known of his background save that he had a brother Boethus, who had a son, Pasicles.[2] He became a pupil of Aristotle, although whether first at Assos, Mitylene, or Athens must remain uncertain.[3] He won the master's good opinion to such an extent that he and Theophrastus of Lesbos were known as Aristotle's "companions."[4] It is disputed whether it was to him or to Eudemus of Cyprus that Aristotle addressed the moving verses, generally known as the "altar elegy," in which he expressed his veneration for Plato at a time when he had felt compelled to diverge from the Platonic philosophy, but on balance Eudemus of Rhodes would seem to be thus favored. (It is, however, the Cypriot and not the Rhodian Eudemus in whose honor Aristotle's early philosophical dialogue "Eudemus" is named.)

Aulus Gellius recounts that as Aristotle approached death, his disciples gathered round him and asked him to choose his successor; they agreed that Theophrastus and Eudemus were preeminent among them. A little later, when they were again assembled, Aristotle asked for some Rhodian and some Lesbian wine to be brought to him. He pronounced both to be civilized wines—the Rhodian strong and joyful, but the Lesbian sweeter, thus indicating Theophrastus as his successor.[5] Eudemus took the choice of Theophrastus in good part, for Andronicus of Rhodes, in a lost work quoted by Simplicius, records a letter that Eudemus wrote to Theophrastus asking that an accurate copy of passages in the fifth book of Aristotle's *Physics* be sent to him on account of errors in his own manuscript.[6] It is usually deduced from this passage that after Aristotle's death Eudemus set up his own school elsewhere (perhaps in his native Rhodes).

The main importance of Eudemus in the history of thought is that he, Theophrastus, Strato, Phanias, and others brought Aristotle's lectures, lecture notes, their own records, and the recollections of themselves and others to a state fit for publication, thus making the works of Aristotle available to the world. One of the three ethical works in the Aristotelian corpus, the *Eudemian Ethics,* actually bears Eudemus' name,[7] but the significance of the title, which is first attested by Atticus Platonicus[8] in the age of the Antonines, is still an open question, complicated by the fact that books IV–VI are identical with books V–VII of the *Nicomachean Ethics.* At various times it was thought that the treatise was a genuine work of Aristotle dedicated to Eudemus, or that Eudemus was himself its author, or that it was a work of Aristotle edited by Eudemus (opening to discussion whether Eudemus cited the master's words exactly or used them as the basis for what is substantially a work of his own, as he did with the *Physics,* see below).

But in 1841 L. Spengel pronounced the *Eudemian Ethics* to be a restatement of Aristotle's teaching with extensive additions by Eudemus. This view so prevailed that the Greek texts of Fritzsche (1851) and Susemihl (1884) were both entitled *Eudemi Rhodii Ethica,* and the English commentaries on the *Nicomachean Ethics* by Sir Alexander Grant (1857), J. A. Stewart (1892), and J. Burnet (1900) all took it for granted that Eudemus was the author.

Later P. von der Mühl (1909), E. Kapp (1912), and, most notably, W. Jaeger (1923) sought to restore the authenticity of the *Eudemian Ethics.*[9] Jaeger considered it to be intermediate to the ethics *more geometrico demonstrata* in the *Protrepticus,* as recovered from Iamblichus, and the final version of Aristotle's moral teaching in the *Nicomachean Ethics.* This notion of three stages in the development of Aristotle's ethics no longer convinces, and it is generally held nowadays that the differences between the *Eudemian Ethics,* the *Nicomachean Ethics,* and the *Magna moralia* are to be explained by the audiences to which they were addressed; but the belief that the *Eudemian Ethics* is a genuine work of Aristotle has been reinforced by a detailed examination of its language. Its style, in the nature of lecture notes with no literary graces, supports the hypothesis of Aristotelian authorship. It is cited or referred to in other works of Aristotle, notably the *Politics.*[10] If Eudemus were the author of the *Eudemian Ethics,* it is hardly conceivable that he would have allowed such an expression as καθάπερ διαιρούμεθα καὶ ἐν τοῖς ἐξωτερικοῖς

λόγοις, "as we have distinguished in the published writings."[11]

No recent commentators have produced a convincing solution; and it may now be regarded as certain, on grounds of style apart from other considerations, that Eudemus was not the author. In all probability it should be regarded as an authentic work of Aristotle, possibly edited after his death as Gigon believes;[12] why Eudemus' name came to be attached to it remains a puzzle.

Although the moral teaching of the *Eudemian Ethics* is fundamentally the same as that of the other ethical treatises, the final book differs in that it holds up as the ideal τὴν τοῦ θεοῦ θεωρίαν, "the contemplation of God." This has led to a picture of the Rhodian philosopher as the "pious Eudemus" which would lose its force if von Arnim is right in detecting the substitution of θεός ("God") for νοῦς ("mind") by a Christian interpolator.[13]

The sixth-century commentator Asclepius, noting the lack of orderliness and continuity in Aristotle's *Metaphysics,* relates that Aristotle, being himself conscious of these faults, sent the work to Eudemus for his opinion. Eudemus judged it unsuitable to publish such a work to all and sundry—thus implying a belief in an esoteric Aristotelian doctrine—and Asclepius adds that after Aristotle's death, when parts of the work were found to be missing, the school's survivors filled the gaps with extracts from his other works.[14] This is improbable, since the esoteric doctrine did not arise until later and, moreover, such a story would imply that Eudemus left Athens while Aristotle was still head of the school—a contradiction of evidence already given; nor does the *Metaphysics* draw on other works. Another commentator, Alexander of Aphrodisias, implies that Eudemus did some editorial work on the treatise; this is more likely, and accords with a scholium to one of the oldest manuscripts stating that most scholars attributed the second book, α minor, to his nephew Pasicles.[15]

Eudemus wrote a *Physics,* in four books, that covered the same ground as Aristotle's treatise, the first book corresponding to Aristotle's I and II, the second to Aristotle's III and V, the third to Aristotle's IV, and the fourth to Aristotle's VI and VIII, which confirms the belief that VII is not genuine. Simplicius used Eudemus' work extensively in his elucidation of Aristotle; some ninety fragments are gathered together by Wehrli. It is thus possible substantially to reconstruct Eudemus' treatise, but as it so largely overlaps that of Aristotle it is not necessary to discuss the contents here.

Eudemus made contributions of his own to the Aristotelian logic. He wrote a book—or possibly two separate books[16]—on analytics and the categories and another entitled *On Discourse,* which seems to have dealt with the same topics as Aristotle's *De interpretatione.* That Galen wrote a commentary on it is evidence that it had some vogue in antiquity.[17]

According to Boethius, Theophrastus and Eudemus (in one place "or Eudemus") added five moods to the four in the first syllogistic figure,[18] and a Greek fragment of unknown authorship adds that they were later made into a fourth figure.[19] The four moods of the first figure are those known since Peter of Spain as Barbara, Celarent, Darii, and Ferio. Boethius explains that the five new moods are obtained by conversion of the terms of the four original moods. Thus, if *A* is in all *B* and *B* is in all *C,* it follows that *A* is in all *C* (Barbara); and by conversion, if *A* is in all *B* and *B* is in all *C,* we may conclude that *C* is in some *A.* This is the fifth mood, Bramantip, and in the same way the sixth, seventh, eighth, and ninth moods (Carmenes, Dimaris, Fesapo, and Fresison) may be obtained. (It has sometimes been queried why Aristotle himself did not group these last five moods in a fourth figure, for they are implicit in his work; Fesapo and Fresison are specifically mentioned by him, and he explicitly states that a syllogism always results from conversion of the premises.)[20]

The work of Theophrastus and Eudemus on the new moods is bound up with the distinction that they developed between necessary and merely factual premises and conclusions. Aristotle believed that there were combinations of an apodeictic and an assertoric premise which led to an apodeictic conclusion. For the first figure he laid down the rule that an apodeictic major and an assertoric minor may lead to an apodeictic conclusion, while the combination of an assertoric major and an apodeictic minor cannot. According to Alexander of Aphrodisias, the followers of Eudemus and Theophrastus took the opposite view, holding that if either the major or the minor premise is assertoric the conclusion must also be assertoric. Similarly they held that if either premise is negative the conclusion must also be negative, and if either premise is particular the conclusion must be particular. They summarized their doctrine in the saying that the conclusion must be like the "inferior premise," [21] or as it was later put into Latin, *peiorem semper sequitur conclusio partem.*

Another divergence between Aristotle and his two leading pupils arose over problematic syllogisms. For Aristotle, the proposition "That all *B* should be *A* is contingent" entails "That no *B* should be *A* is contingent"; and "That some *B* should not be *A* is contingent," with related propositions; and the proposition "That no *B* should be *A* is contingent" does not imply "That no *A* should be *B* is contingent." According to

Alexander, Theophrastus and Eudemus rejected this departure from the general principle that universal negative propositions are simply convertible and particular negative propositions not convertible. They have found a supporter in modern times in H. Maier, but W. D. Ross regards Aristotle as completely justified.[22] It depends upon what Aristotle is understood to mean by contingency, and it is unfortunate that Alexander's book, *On the Disagreement Concerning Mixed Moods . . .* has not survived.

It is remarkable that in their development of Aristotle's logic the names of Theophrastus and Eudemus are so often conjoined. Although there are many references to Theophrastus alone, only one to Eudemus alone is recorded; it may thus rightly be inferred that Theophrastus had the major share in the work. Bochenski supposes that the *Organon* represents Aristotle's earlier logical thinking and that in his later lectures he advanced beyond it; and that Theophrastus and Eudemus, who were present at these lectures, separately represent the mature development of Aristotle's logical thought. The coincidence of their views, he thinks, cannot be explained by chance or close and prolonged collaboration.[23]

From a long passage in Damascius[24] it may be inferred that Eudemus wrote a history of theology that appears to have dealt with the origins of the universe and to have ranged over the views of the Babylonians, Egyptians, and Greeks. A single reference in Proclus[25] establishes that Eudemus wrote one purely mathematical work—*On the Angle*—in which he took the view that angularity is a quality rather than a quantity (since angularity arises from an inclination of lines, and since both straightness and inclination are qualities, so also must angularity be).

Eudemus is also important for his studies in the history of science. He wrote three works—a history of arithmetic, a history of geometry, and a history of astronomy—which are of capital value for the transmission of the facts about early Greek science. Although, like all of Eudemus' works, they have been lost, it is mainly through the use made of them by later writers that we possess any knowledge of the rise of Greek geometry and astronomy. Eudemus is not known to have had any predecessors in this field,[26] and he may justly be regarded as the father of the history of science, or at the least as sharing the paternity with his fellow Peripatetics Theophrastus, author of *Views of the Physicists,* and Menon, author of a history of medicine.

The *History of Arithmetic* is known from only one reference, made by Porphyry in his commentary on Ptolemy's *Harmonics,*[27] stating that in the first book Eudemus dealt with the Pythagorean correlation of numbers with musical intervals.

The *History of Geometry*, in at least two books, is known from many ancient references and citations. According to Simplicius, it was written in a summary style like a memorandum.[28] A passage in Proclus' commentary on the first book of Euclid's *Elements*[29] was formerly known as "the Eudemian summary" in the belief that it was an extract from this work. This cannot be so, since it leads up to the work of Euclid, who was later than Eudemus, and there is no stylistic break in the narrative. The earlier part—up to the sentence where Proclus writes, "Those who have compiled histories carry the development of this science up to this point" (*sc.* Philippus of Opus, who lived just before Euclid)—would appear to be a condensation of Eudemus' narrative, written soon after his death, for it is unlikely that a later writer would have stopped at that precise date. The summary tells how Thales introduced the study of geometry from Egypt into Greece and recounts the work of his successors, without, however, ever referring to Democritus. This omission offers further proof that the passage cannot be taken directly from Eudemus, since he would certainly have mentioned this mathematical pioneer who was held in high esteem by Aristotle (although Proclus might not).

One of the most important chapters in the history of Greek mathematics—the work of Hippocrates of Chios on the quadrature of lunes—is known through Eudemus. It is known from the use made of it by Simplicius in his commentary on Aristotle's *Physics*.[30] Simplicius reproduces passages from Eudemus, who may himself be giving the words of Hippocrates along with comments of his own; and many scholars have addressed themselves to the task of separating what Eudemus wrote from what Simplicius added.

From surviving references, Eudemus is also known to have recorded in his *History of Geometry* the theorem that if two triangles have two angles and one side equal, the remaining angle and sides will also be equal (Euclid I.26), discovered by Thales and used by him to find the distances of ships from the shore;[31] the theorem that if two straight lines intersect the vertical and opposite angles are equal (Euclid I.15), discovered but not proved by Thales;[32] the theorem that the interior angles of a triangle are equal to two right angles (Euclid I.32), first proved by the Pythagoreans by means of a line drawn parallel to the base;[33] the "application of areas" (i.e., the erection on a straight line, or on a segment thereof, or on the straight line produced, of a parallelogram with a given angle equal to a given area, which is a species of geometrical algebra), also the discovery of the Pythagoreans;[34] the problem of drawing a straight line perpendicular to a given straight line from a point outside it (Euclid I.12), first investigated by Oeno-

pides,[35] who also first discovered the Euclidean method of constructing a rectilinear angle equal to a given rectilinear angle (Euclid I.23).[36]

Tannery thought that the *History of Geometry* was already lost by the time of Pappus, and that for his knowledge of such matters as the quadrature of the circle and the duplication of the cube Pappus relied on a compilation entitled Ἀριστοτελικὰ κήρια ("Aristotelian apiary"), drawn up, perhaps toward the end of the third century B.C., by his older contemporary Sporus of Nicaea, who in turn would have drawn on Eudemus. This failed to convince Heiberg, who made out a strong case for believing that both Pappus and Eutocius had the text of Eudemus before them.[37]

Eudemus' *History of Astronomy*, also in at least two books, is of further value, through its use by later writers, as a source book. It is, for example, through this work that Oenopides is known to have discovered the obliquity of the ecliptic; and Eudemus recorded its value as being that of the side of a fifteen-sided polygon, that is, twenty-four degrees.[38] Eudemus' history is also the ultimate source, through its use by the Peripatetic philosopher Sosigenes (second century A.D.), of Simplicius' account of Eudoxus' system of concentric spheres on which the poles of the heavenly bodies rotate—the first attempt to account mathematically for the solar, lunar, and planetary motions.[39] Among other topics known to have been dealt with by Eudemus are solar eclipses, particularly Thales' prediction of the eclipse of 28 May 585 B.C.; the cycle of the great year after which all the heavenly bodies are found in the same relative positions; the realization by Anaximander that the earth is a heavenly body moving about the middle of the universe; the discovery by Anaximenes that the moon reflects the light of the sun and his explanation of lunar eclipses; and the inequality of the times between the solstices and the equinoxes.[40]

Aelian, writing in the second or third century A.D., has seven references to a work on animals written by Eudemus,[41] but it has been questioned whether he is to be identified with Eudemus of Rhodes. Apuleius mentions "Aristotle and Theophrastus and Eudemus and Lyco and other lesser Platonists" as having written on the birth and nourishment of animals,[42] and as, in the context, Eudemus of Rhodes must be understood, this would support the identification; but the citations given by Aelian are of fabulous stories about animals which do not fit in well with the serious scientific character of Eudemus of Rhodes.

A history of Lindos was written by a certain Eudemus. Wilamowitz was prepared to believe that this was Eudemus of Rhodes, but Wehrli thinks it highly improbable. There is no evidence on which the question can be settled, as Felix Jacoby sees it;

but since Lindos was a port, with a famous temple, in Eudemus' native Rhodes, there is nothing improbable in the suggestion that so prolific an author as Eudemus should have recorded its history, perhaps after his return from Athens.[43]

NOTES

1. Strabo, XIV 2, 13.
2. Asclepius, *In Aristotelis Metaphysica,* I.1, 980ª 22, Hayduck ed., 4.18–22; Scholium to Aristotle's *Metaphysics α,* 993ª 30, *Scholia in Aristotelem (Aristotelis opera IV),* Brandis ed., 589 a 41–43.
3. W. Jaeger, *Aristoteles,* 109 n. 2, takes the view that he (and Theophrastus) became students of Aristotle at Assos; but this seems to be bound up with his view of the date of the *Eudemian Ethics,* and there is no real evidence one way or the other.
4. Ammonius, *In Aristotelis Analytica priora,* I.9, 30ª 15, Wallies ed., 38.38–39, οἱ δ'ἑταῖροι αὐτοῦ, Θεόφραστος καὶ Εὔδημος, is one of many passages in which the expression is used.
5. Aulus Gellius, *Noctes Atticae,* XIII.5, Marshall ed., II. 387, 1–29; Boethius, *De syllogismo hypothetico,* I, in Migne ed., *Patrologia latina,* LXIV, 831D, noted that Theophrastus was a man capable of all learning but tackling only the peaks, whereas Eudemus followed a broader road of learning; but it was as though he scattered seminal ideas without gathering any great harvest. This passage of Boethius, referring to the *Analytica,* is the only evidence of any difference between Theophrastus and Eudemus in the field of logic.
6. Simplicius, *In Aristotelis Physica,* VI, proemium, Diels ed., 923.7–16.
7. It has almost universally been assumed that "Eudemian" refers to Eudemus of Rhodes, but, as Dirlmeier points out, there is no precise evidence whether Eudemus of Rhodes or Eudemus of Cyprus is indicated by the title. He himself makes the suggestion—which has not found favor—that as *Eudemian Ethics* I.5 is a pessimistic reflection on the theme, "It is best not to be born," which plays an impressive part in the dialogue "Eudemus" but not in the other two ethical works, the *Eudemian Ethics* was intended to be a posthumous tribute to his friend the Cypriot (F. Dirlmeier, *Aristoteles magna moralia,* 2nd ed., 1966, p. 97).
8. As preserved by Eusebius, *Praeparatio evangelica XV,* in *Patrologia graeca,* Migne ed., XXI, 1305 A, Dindorf ed., I.344. 24–26: αἱ γοῦν Ἀριστοτέλους περὶ ταῦτα πραγματείαι, Εὐδήμειοί τε καὶ Νικομάχειοι καὶ Μεγάλων Ἠθικῶν ἐπιγραφόμεναι.
9. L. Spengel, "Über die unter dem Namen des Aristoteles erhaltenen ethischen Schriften," in *Abhandlungen der Bayrischen Akademie der Wissenschaften,* **3** (Munich, 1841), 439–551; P. von der Mühl, *De Aristotelis Ethicorum Eudemiorum auctoritate* (Göttingen, 1909); E. Kapp, *Das Verhaltnis der eudemischen zur nikomachischen Ethik* (Freiburg, 1912); W. Jaeger, *Aristoteles* (Berlin, 1923; 2nd ed., 1955), pp. 237–270, trans. by R. Robinson (Oxford, 1934; 2nd ed., 1948), pp. 228–258.
10. F. Dirlmeier, *Aristoteles Eudemische Ethik,* pp. 112–115; for quotations in the treatise see the same author, "Merkwurdige Zitate in der Eudemischen Ethik des Aristoteles" in *Sitzungsberichte der Heidelberger Akademie der Wissenschaften,* phil.-hist. Klasse (1962), Abh. 2.
11. *Ethica Eudemia,* 1218ᵇ 33–34, Susemihl ed., 16, B. 3–4.
12. O. Gigon, *Aristoteles Die Nikomachische Ethik,* p. 39.
13. H. von Arnim, *Die drei Aristotelischen Ethiken,* p. 68.
14. Asclepius, *In Aristotelis Metaphysica,* I.1, 980ª 22, Hayduck ed., 4.4–15.
15. See note 2. Asclepius in the passage referred to in the same note says that it was book A which Pasicles was alleged to have written, but this, he adds, is untrue.
16. Alexander, *In Aristotelis Analytica priora,* Wallies ed., 31.4–10,

124.8–15,   126.29–127.2,   141.1–5,   173.32–174.3,   220.9–16, 389.31–390.3; Alexander, *In Aristotelis Topica,* Wallies ed., 131.14–19; Philoponus, *In Aristotelis Analytica priora,* Wallies ed., 48.12–18, 123.12–20, 129.15–19; Ammonius, *In Aristotelis Analytica priora,* Wallies ed., 38.38–39.2, 45.42–45, 49.6–12; Olympiodorus, *Prolegomena,* Busse ed., 13.24–25. Eudemus is named by Philoponus, *In Aristotelis Categorias,* proemium, Busse ed., 7.16, along with Phanias and Theophrastus as the author of Κατηγορίαι, Περὶ ἑρμηνείας, and Ἀναλυτικά. He is named by David, *In Porphyrii Isagogen,* Busse ed., 102.4 with Theophrastus alone as the author of Κατηγορίαι; but the individual citations catalogued above make it likely that the Ἀναλυτικά and the Κατηγορίαι are the same work.

17. Alexander, *In Aristotelis Topica,* Wallies ed., 69.13–16, *In Aristotelis Analytica priora,* Wallies ed., 16.12–17, *In Aristotelis Metaphysica,* Hayduck ed., 85.9–11; scholium *In Aristotelis Analytica priora* 1, codex 1917, Brandis ed., 146a 24. As seen in the previous note, Philoponus refers to this work by the same title as that of Aristotle, Περὶ ἑρσηνείας, but this would appear to refer to its subject matter and the title Περὶ λεξέως is better attested.
18. Boethius, *De syllogismo categorico,* II, Migne ed., *Patrologia latina,* LXIV, 813 C ("Theophrastus vel Eudemus"), 814 C ("Theophrastus et Eudemus"), 815 B ("Theophrastus et Eudemus"). Alexander, *In Analytica priora,* I.4, 26ᵇ 30, Wallies ed., 69. 26, attributes the five additional moods to Theophrastus without mention of Eudemus.
19. For references to later publications and discussions of this anonymous Greek fragment see N. Rescher, *Galen and the Syllogism* (Pittsburgh, 1966), p. 2, n. 9.
20. Aristotle, *Analytica priora,* I, 29ª 23.
21. Alexander, *In Aristotelis Analytica priora,* Wallies ed., 124 8–127.16. W. D. Ross, *Aristotle's Prior and Posterior Analytics,* pp. 41–42, suggests that the distinction between Aristotle and his followers is not so sharp as might at first appear.
22. Alexander, *In Aristotelis Analytica Priora,* Wallies ed., 159.8–13, 220.9–221.5; H. Maier, *Die Syllogistik des Aristoteles,* IIa, 37–47; W. D. Ross, *Aristotle's Prior and Posterior Analytics,* p. 45.
23. I. M. Bochenski, *La logique de Théophraste,* p. 125.
24. Damascius, *Dubitationes et solutiones de primis principiis* 124–125, Ruelle ed., I, 319.8–323.17.
25. Proclus, *In primum Euclidis,* definition 8, Friedlein ed., 125.6.
26. It is reading too much into Proclus, *In primum Euclidis,* Friedlein ed., 65.14, to suppose that Hippias wrote a history of mathematics.
27. Porphyry, *In Ptolemaei Harmonica,* Düring ed., in *Göteborgs högskolas Årsskrift,* **38** (1932), 114.23–115.9.
28. Simplicius, *In Aristotelis Physica,* Diels ed., 60.42–44.
29. Proclus, *In primum Euclidis,* Friedlein ed., 64.16–70.18, trans. by Glenn R. Morrow as *Proclus' A Commentary on the First Book of Euclid's Elements* (Princeton, 1970), pp. 51–70.
30. Simplicius, *In Aristotelis Physica,* Diels ed., 60.22–68.32.
31. Proclus, *In primum Euclidis,* Friedlein ed., 352.14–16, the application of the theorem to the distances of ships raises problems, for which see Thomas Heath, *A History of Greek Mathematics,* I, 131–133.
32. Proclus, *In primum Euclidis,* Friedlein ed., 299.1–5.
33. *Ibid.,* 379.2–16.
34. *Ibid.,* 419.15–18.
35. *Ibid.,* 283.7–8. In this case Eudemus is not specifically mentioned as the source, although he must be.
36. *Ibid.,* 333.5–9.
37. Paul Tannery, "Sur les fragments d'Eudème de Rhodes relatifs à l'histoire des mathématiques," in *Annales de la Faculté des Lettres de Bordeaux,* **4** (1882), 70–76, reprinted in *Mémoires scientifiques,* I (Toulouse–Paris, 1912), 156–177; "Sur Sporos de Nicée," in *Annales de la Faculté des Lettres de Bordeaux,* **4** (1882), 257–261, repr. in *Mémoires scientifiques,* I (Toulouse–Paris, 1912), 178–184; *Bulletin des sciences,* **7** (1883), 283–284, repr. in *Mémoires scientifiques,* II (Toulouse–Paris, 1912), 4–5; J. L. Heiberg, in *Philologus,* **43** (1884), 345–346.
38. Theon of Smyrna, *Expositio rerum mathematicarum,* Hiller ed.,

198.14–15, 199.6–8, with Diels's conjecture of Λόξωσιν for διάζωσιν.

39. Simplicius, *In Aristotelis De caelo,* Heiberg ed., 488.18–24, 493.4–506.18. A brief account is given by Aristotle, *Metaphysics,* Λ8, 1073ᵇ17–1074ᵃ14. For translations and explanations see G. Schiaparelli, *Le sfere omocentriche di Eudosso, di Callippo e di Aristotele,* and T. L. Heath, *Aristarchus of Samos,* 193–211; *A History of Greek Mathematics,* I, 329–334; and *Greek Astronomy,* 65–70.

40. Clement of Alexandria, *Stromata,* I, 14, 65.1, Stählin ed., II (Berlin, 1960), 41.8–15; Diogenes Laertius, I, 23, Long ed., 9.18–21; Theon of Smyrna, Hiller ed., 198.16–199.2; Simplicius, *In Aristotelis De caelo,* Heiberg ed., 471.1–6.

41. Aelian, *On the Characteristics of Animals,* III, 20, 21; IV, 8, 45, 53, 56; V, 7.

42. Apuleius, *Apologia,* 36.

43. *Die Fragmente der griechischen Historiker,* Jacoby, ed., IIIB (Leiden, 1950), nos. 524, 532B-C 10, C32, D1, D2, Εὔδημος ἐν τῷ Λινδιακῷ (*sc.* λόγῳ) pp. 503, 508, 510, 512, 513, *Kommentar,* IIIb (Leiden, 1955), 441–442, and *Noten,* IIIb (Leiden, 1955), 259. Jacoby's conclusion is, "Bei Eudemos denkt man natürlich zuerst an der Schüler des Aristoteles . . . Aber der Name ist gewöhnlich, und es fehlt an entscheidenden Gründen für die Identifikation." For Wilamowitz see "Nachrichten über Versammlungen," reporting a paper by U. von Wilamowitz-Moellendorff to the Archäologische Gesellschaft zu Berlin, 4 March 1913, in *Berliner philologische Wochenschrift* (1913), col. 1372.

BIBLIOGRAPHY

I. Original Works. From references in ancient writers Eudemus is believed to have written the following works. None has survived except in quotations or paraphrases. There are variants for the Greek titles, but those given are the most probable. A few of the titles are uncertain and are preceded by a question mark: (1) Ἀναλυτικά (?Κατηγορίαι)—possibly a separate work; (2) Περὶ λέξεως; (3) Φυσικά; (4) History of Theology; (5) Περὶ γωνίας; (6) Ἀριθμητικὴ ἱστορία; (7) Γεωμετρικὴ ἱστορία; (8) Ἀστρολογικὴ ἱστορία; (9) (?) Stories of Animals.

In addition, for a reason which cannot now be ascertained, the name of Eudemus is attached to the Ἠθικὰ Εὐδήμεια (or Εὐδημία) of Aristotle. The best text of the Eudemian ethics is still that of F. Susemihl, [*Aristotelis Ethica Eudemia*] *Eudemi Rhodii Ethica* (Leipzig, 1884), but a new critical ed. by R. Walzer for the Oxford Classical Texts series is now in press.

Quotations from Eudemus by ancient writers have been collated in L. Spengel, *Eudemi Rhodii Peripatetici Fragmenta quae supersunt* (Berlin, 1866; 2nd ed., 1870); F. W. A. Mullach, in *Fragmenta philosophorum Graecorum,* III (Paris, 1881), 222–292; and, most recently and most satisfactorily, in Fritz Wehrli, *Eudemos von Rhodos,* in *Die Schule des Aristoteles, Texte und Kommentar,* VIII (Basel, 1955; 2nd ed., 1969).

II. Secondary Literature. According to Simplicius, a life of Eudemus was written in antiquity by an otherwise unknown Damas. It has not survived. Modern studies of Eudemus by A. T. H. Fritzsche, *De Eudemi Rhodii philosophi Peripatetici vita et scriptis* (Regensburg, 1851); C. A. Brandis, *Handbuch der Geschichte der Griechisch-Römischen Philosophie,* III, 1 (Berlin, 1860), 215–250; E. Zeller, *Die Philosophie der Griechen in ihrer geschichtlichen*

Entwicklung, II, 2, 3rd ed. (Leipzig, 1879; Obraldruck-Leipzig, 1921), 869–881; and E. Martini in Pauly-Wissowa, VI (Stuttgart, 1907), cols. 895–901, are now superseded by Wehrli (see above), who has also written a new article, "Eudemus von Rhodos," in Pauly-Wissowa, supp. XI (Stuttgart, 1968), cols. 652–658.

For the *Eudemian Ethics* see P. von der Mühl, *De Aristotelis Ethicorum Eudemiorum auctoritate* (dissertation, Göttingen, 1909); E. Kapp, *Der Verhältnis der eudemischen zur nikomachischen Ethik* (dissertation, Freiburg, 1912); W. Jaeger, *Aristoteles, Grundlegung einer Geschichte seiner Entwicklung* (Berlin, 1923; 2nd ed., 1955), 237–270, trans. by R. Robinson as *Aristotle, Fundamentals of the History of His Development* (Oxford, 1934, 2nd ed., 1948), 228–258; H. von Arnim, "Die drei Aristotelischen Ethiken," in *Sitzungsberichte der Akademie der Wissenschaften in Wien,* Phil.-hist. Klasse, **202** (Vienna, 1921); Franz Dirlmeier, *Aristoteles Eudemische Ethik,* in *Aristoteles Werke in Deutscher Übersetzung,* Grumach ed., VII (Berlin, 1962; 2nd ed., 1969) with elaborate intro. and notes (see especially introduction, 110–143), and Franz Dirlmeier, *Aristoteles Magna moralia* in *Aristoteles Werke,* Grumach ed., VIII (Berlin 1958; 2nd ed., 1966), 97–99.

The contribution of Eudemus to the development of Aristotelian logic cannot easily be separated from that of Theophrastus, and may be studied in I. M. Bochenski, *La logique de Théophraste* (Fribourg, 1947) and *Ancient Formal Logic* (Amsterdam, 1951), pp. 72–76; W. D. Ross, *Aristotle's Prior and Posterior Analytics* (Oxford, 1949), pp. 41–42, 45–47; Jan Łukasiewicz, *Aristotle's Syllogistic* (Oxford, 1953; 2nd ed., 1957), pp. 25–28, 38–42; Storrs McCall, *Aristotle's Modal Syllogisms* (Amsterdam, 1963), pp. 2, 15–16.

Eudemus' studies in the history of mathematics are the subject of the following papers by Paul Tannery: "Sur les fragments d'Eudème de Rhodes relatifs à l'histoire des mathématiques," in *Annales de la Faculté des Lettres de Bordeaux,* **4** (1882), 70–76, repr. in *Mémoires scientifiques,* I (Toulouse–Paris, 1912), 168–177; "Le fragment d'Eudème sur la quadrature des lunules," in *Mémoires de la Société des sciences physiques et naturelles de Bordeaux,* 2nd ser., **5** (1883), 217–237, repr. in *Mémoires scientifiques,* I (Toulouse–Paris, 1912), 339–370. His papers on Hippocrates of Chios are also relevant. For the question whether Eudemus' history was directly available to Simplicius and Eutocius, see J. L. Heiberg, *Philologus,* **43** (1884), 345–346. Eudemus' contributions to the history of science are summarized in Thomas Heath, *A History of Greek Mathematics* (Oxford, 1921), I, 118–120; II, 244.

U. Schoebe has written a Latin dissertation on the first book of Eudemus' *Physics* under the title *Quaestiones Eudemeae de primo Physicorum libro* (Halle, 1931).

Ivor Bulmer-Thomas

EUDOXUS OF CNIDUS (*b.* Cnidus, *ca.* 400 B.C.; *d.* Cnidus, *ca.* 347 B.C.), *astronomy, mathematics.*

A scholar and scientist of great eminence, Eudoxus, son of a certain Aischines, contributed to the development of astronomy, mathematics, geography, and philosophy, as well as providing his native city with laws. As a young man he studied geometry with Archytas of Tarentum, from whom he may well have taken his interest in number theory and music; in medicine he was instructed by the physician Philiston; and his philosophical inquiries were stimulated by Plato, whose lectures he attended as an impecunious student during his first visit to Athens. Later his friends in Cnidus paid for a visit to Egypt, where he seems to have had diplomatic dealings with King Nekhtanibef II on behalf of Agesilaus II of Sparta.

Eudoxus spent more than a year in Egypt, some of the time in the company of the priests at Heliopolis. He was said to have composed his *Oktaeteris,* or eight-year calendric cycle, during his sojourn with them. Next he settled at Cyzicus in northwestern Asia Minor and founded a school. He also visited the dynast Mausolus in Caria. A second visit to Athens, to which he was followed by some of his pupils, brought a closer association with Plato, but it is not easy to determine mutual influences in their thinking on ethical and scientific matters. It is unlikely that Plato had any influence upon the development of Eudoxian planetary theory or much upon the Cnidian's philosophical doctrine of forms, which recalls Anaxagoras; but it is possible that Plato's *Philebos* was written with the Eudoxian view of *hedone* (that pleasure, correctly understood, is the highest good) in mind.

Back in Cnidus, Eudoxus lectured on theology, cosmology, and meteorology, wrote textbooks, and enjoyed the respect of his fellow citizens. In mathematics his thinking lies behind much of Euclid's *Elements,* especially books V, VI, and XII. Eudoxus investigated mathematical proportion, the method of exhaustion, and the axiomatic method—the "Euclidean" presentation of axioms and propositions may well have been first systematized by him. The importance of his doctrine of proportion lay in its power to embrace incommensurable quantities.

It is difficult to exaggerate the significance of the theory, for it amounts to a rigorous definition of real number. Number theory was allowed to advance again, after the paralysis imposed on it by the Pythagorean discovery of irrationals, to the inestimable benefit of all subsequent mathematics. Indeed, as T. L. Heath declares (*A History of Greek Mathematics,* I [Oxford, 1921], 326–327), "The greatness of the new theory itself needs no further argument when it is remembered that the definition of equal ratios in Eucl. V, Def. 5 corresponds exactly to the modern theory of irrationals due to Dedekind, and that it is word for word the same as Weierstrass's definition of equal numbers."

Eudoxus also attacked the so-called "Delian problem," the traditional one of duplicating the cube; that is, he tried to find two mean proportions in continued proportion between two given quantities. His strictly geometrical solution is lost, and he may also have constructed an apparatus with which to describe an approximate mechanical solution; an epigram ascribed to Eratosthenes (who studied the works of Eudoxus closely) refers to his use of "lines of a bent form" in his solution to the Delian problem: the "organic" demonstration may be meant here. Plato is said to have objected to the use by Eudoxus (and by Archytas) of such devices, believing that they debased pure or ideal geometry. Proclus mentions "general theorems" of Eudoxus; they are lost but may have embraced all concepts of magnitude, the doctrine of proportion included. Related to the treatment of proportion (as found in *Elements* V) was his method of exhaustion, which was used in the calculation of the volume of solids. The method was an important step toward the development of integral calculus.

Archimedes states that Eudoxus proved that the volume of a pyramid is one-third the volume of the prism having the same base and equal height and that the volume of a cone is one-third the volume of the cylinder having the same base and height (these propositions may already have been known by Democritus, but Eudoxus was, it seems, the first to prove them). Archimedes also implies that Eudoxus showed that the areas of circles are to each other as the squares on their respective diameters and that the volumes of spheres to each other are as the cubes of their diameters. All four propositions are found in *Elements* XII, which closely reflects his work. Eudoxus is also said to have added to the first three classes of mathematical mean (arithmetic, geometric, and harmonic) two more, the subcontraries to harmonic and to geometric, but the attribution to him is not quite certain.

Perhaps the most important, and certainly the most influential, part of Eudoxus' lifework was his application of spherical geometry to astronomy. In his book *On Speeds* he expounded a system of geocentric, homocentric rotating spheres designed to explain the irregularities in the motion of planets as seen from the earth. Eudoxus may have regarded his system simply as an abstract geometrical model, but Aristotle took it to be a description of the physical world and complicated it by the addition of more spheres; still more were added by Callippus later in the fourth century B.C. By suitable combination of spheres the periodic motions of planets could be represented approximately, but the system is also, as geometry, of intrinsic merit because of the hippopede, or "horse fetter," an eight-shaped curve, by which Eudoxus represented a planet's apparent motion in latitude as well as its retrogradation.

Eudoxus' model assumes that the planet remains at a constant distance from the center, but in fact, as critics were quick to point out, the planets vary in brightness and hence, it would seem, in distance from the earth. Another objection is that according to the model, each retrogradation of a planet is identical with the previous retrogradation in the shape of its curve, which also is not in accord with the facts. So, while the Eudoxian system testified to the geometrical skill of its author, it could not be accepted by serious astronomers as definitive, and in time the theory of epicycles was developed. But, partly through the blessing of Aristotle, the influence of Eudoxus on popular astronomical thought lasted through antiquity and the Middle Ages. In explaining the system, Eudoxus gave close estimates of the synodic periods of Saturn, Jupiter, Mars, Mercury, and Venus (hence the title of the book, *On Speeds*). Only the estimate for Mars is seriously faulty, and here the text of Simplicius, who gives the values, is almost certainly in error (Eudoxus, frag. 124 in Lasserre).

Eudoxus was a careful observer of the fixed stars, both during his visit to Egypt and at home in Cnidus, where he had an observatory. His results were published in two books, the *Enoptron* ("Mirror") and the *Phaenomena*. The works were criticized, in the light of superior knowledge, by the great astronomer Hipparchus two centuries later, but they were pioneering compendia and long proved useful. Several verbatim quotations are given by Hipparchus in his commentary on the astronomical poem of Aratus, which drew on Eudoxus and was also entitled *Phaenomena*. A book by Eudoxus called *Disappearances of the Sun* may have been concerned with eclipses, and perhaps with risings and settings as well. The statement in the *Suda Lexicon* that he composed an astronomical poem may result from a confusion with Aratus, but a genuine *Astronomia* in hexameters, in the Hesiodic tradition, is a possibility. A calendar of the seasonal risings and settings of constellations, together with weather signs, may have been included in the *Oktaeteris*. His observational instruments included sundials (Vitruvius, *De architectura* 9.8.1).

Eudoxus' knowledge of spherical astronomy must have been helpful to him in the geographical treatise *Ges periodos* ("Tour [Circuit] of the Earth"). About 100 fragments survive; they give some idea of the plan of the original work. Beginning with remote Asia, Eudoxus dealt systematically with each part of the known world in turn, adding political, historical,

and ethnographic detail and making use of Greek mythology. His method is comparable with that of such early Ionian logographers as Hecataeus of Miletus. Egypt was treated in the second book, and Egyptian religion, about which Eudoxus could write with authority, was discussed in detail. The fourth book dealt with regions to the north of the Aegean, including Thrace. In the sixth book he wrote about mainland Hellas and, it seems, North Africa. The discussion of Italy in the seventh book included an excursus on the customs of the Pythagoreans, about whom Eudoxus may have learned much from his master Archytas of Tarentum (Eudoxus himself is sometimes called a Pythagorean).

It is greatly to be deplored that not a single work of Eudoxus is extant, for he was obviously a dominant figure in the intellectual life of Greece in the age of Plato and Aristotle (the latter also remarked on the upright and controlled character of the Cnidian, which made people believe him when he said that pleasure was the highest good).

BIBLIOGRAPHY

The biography of Eudoxus in Diogenes Laertius 8.86–8.90 is anecdotal but not worthless. The fragments have been collected, with commentary, in F. Lasserre's book *Die Fragmente des Eudoxos von Knidos* (Berlin, 1966). Eudoxian parts of Euclid's *Elements* are discussed by T. L. Heath in his edition of that work, 2nd ed., 3 vols. (Cambridge, 1926). The mathematical properties of the hippopede have been much studied; see especially O. Neugebauer, *The Exact Sciences in Antiquity*, 2nd ed. (Providence, R. I., 1957), 182–183. On the *Ges periodos*, see F. Gisinger, *Die Erdbeschreibung des Eudoxos von Knidos* (Leipzig–Berlin, 1921). A chronology of Eudoxus' life and travels is G. Huxley, "Eudoxian Topics," in *Greek, Roman and Byzantine Studies*, 4 (1963), 83–96.

See also Oskar Becker, "Eudoxos-Studien," in *Quellen und Studien zur Geschichte der Mathematik, Astronomie und Physik*, Abt. B, Studien, 2 (1933), 311–333, 369–387, and 3 (1936), 236–244, 370–410; Hans Künsberg, *Der Astronom, Mathematiker und Geograph Eudoxos von Knidos*, 2 pts. (Dinkelsbühl, 1888–1890); and G. Schiaparelli, *Scritti sulla storia della astronomia antica*, II (Bologna, 1926), 2–112.

G. L. HUXLEY

EULER, LEONHARD (*b.* Basel, Switzerland, 15 April 1707; *d.* St. Petersburg, Russia, 18 September 1783), *mathematics, mechanics, astronomy, physics.*

Life. Euler's forebears settled in Basel at the end of the sixteenth century. His great-great-grandfather, Hans Georg Euler, had moved from Lindau, on the Bodensee (Lake Constance). They were, for the most part, artisans; but the mathematician's father, Paul

Euler, graduated from the theological department of the University of Basel. He became a Protestant minister, and in 1706 he married Margarete Brucker, daughter of another minister. In 1708 the family moved to the village of Riehen, near Basel, where Leonhard Euler spent his childhood.

Euler's father was fond of mathematics and had attended Jakob Bernoulli's lectures at the university; he gave his son his elementary education, including mathematics. In the brief autobiography dictated to his eldest son in 1767, Euler recollected that for several years he diligently and thoroughly studied Christoff Rudolf's *Algebra,* a difficult work (dating, in Stifel's edition, from 1553) which only a very gifted boy could have used. Euler later spent several years with his maternal grandmother in Basel, studying at a rather poor local Gymnasium; mathematics was not taught at all, so Euler studied privately with Johann Burckhardt, an amateur mathematician. In the autumn of 1720, being not yet fourteen, Euler entered the University of Basel in the department of arts to get a general education before specializing. The university was small; it comprised only a few more than a hundred students and nineteen professors. But among the latter was Johann I Bernoulli, who had followed his brother Jakob, late in 1705, in the chair of mathematics. During the academic year, Bernoulli delivered daily public lectures on elementary mathematics; besides that, for additional pay he conducted studies in higher mathematics and physics for those who were interested. Euler laboriously studied all the required subjects, but this did not satisfy him. According to the autobiography:

> . . . I soon found an opportunity to be introduced to a famous professor Johann Bernoulli. . . . True, he was very busy and so refused flatly to give me private lessons; but he gave me much more valuable advice to start reading more difficult mathematical books on my own and to study them as diligently as I could; if I came across some obstacle or difficulty, I was given permission to visit him freely every Saturday afternoon and he kindly explained to me everything I could not understand . . . and this, undoubtedly, is the best method to succeed in mathematical subjects.[1]

In the summer of 1722, Euler delivered a speech in praise of temperance, "De temperantia," and received his *prima laurea,* a degree corresponding to the bachelor of arts. The same year he acted as opponent (*respondens*) at the defense of two theses—one on logic, the other on the history of law. In 1723 Euler received his master's degree in philosophy. This was officially announced at a session on 8 June 1724; Euler made a speech comparing the philosophical ideas of Descartes and Newton. Some time earlier,

in the autumn of 1723, he had joined the department of theology, fulfilling his father's wish. His studies in theology, Greek, and Hebrew were not very successful, however; Euler devoted most of his time to mathematics. He finally gave up the idea of becoming a minister but remained a wholehearted believer throughout his life. He also retained the knowledge of the humanities that he acquired in the university; he had an outstanding memory and knew by heart the entirety of Vergil's *Aeneid.* At seventy he could recall precisely the lines printed at the top and bottom of each page of the edition he had read when he was young.

At the age of eighteen, Euler began his independent investigations. His first work, a small note on the construction of isochronous curves in a resistant medium,[2] appeared in *Acta eruditorum* (1726); this was followed by an article in the same periodical on algebraic reciprocal trajectories (1727).[3] The problem of reciprocal trajectories was studied by Johann I Bernoulli, by his son Nikolaus II, and by other mathematicians of the time. Simultaneously Euler participated in a competition announced by the Paris Académie des Sciences which proposed for 1727 the problem of the most efficient arrangement of masts on a ship. The prize went to Pierre Bouguer, but Euler's work[4] received the *accessit.* Later, from 1738 to 1772, Euler was to receive twelve prizes from the Academy.

For mathematicians beginning their careers in Switzerland, conditions were hard. There were few chairs of mathematics in the country and thus little chance of finding a suitable job. The income and public recognition accorded to a university professor of mathematics were not cause for envy. There were no scientific magazines, and publishers were reluctant to publish books on mathematics, which were considered financially risky. At this time the newly organized St. Petersburg Academy of Sciences (1725) was looking for personnel. In the autumn of that year Johann I Bernoulli's sons, Nikolaus II and Daniel, went to Russia. On behalf of Euler, they persuaded the authorities of the new Academy to send an invitation to their young friend also.

Euler received the invitation to serve as adjunct of physiology in St. Petersburg in the autumn of 1726, and he began to study this discipline, with an effort toward applying the methods of mathematics and mechanics. He also attempted to find a job at the University of Basel. A vacancy occurred in Basel after the death of a professor of physics, and Euler presented as a qualification a small composition on acoustics, *Dissertatio physica de sono* (1727).[5] Vacancies were then filled in the university by drawing lots among the several chosen candidates. In spite of a recommendation from Johann Bernoulli, Euler was not chosen as a candidate, probably because he was too young—he was not yet twenty. But, as O. Spiess has pointed out, this was in Euler's favor;[6] a much broader field of action lay ahead of him.

On 5 April 1727 Euler left Basel for St. Petersburg, arriving there on 24 May. From this time his life and scientific work were closely connected with the St. Petersburg Academy and with Russia. He never returned to Switzerland, although he maintained his Swiss citizenship.

In spite of having been invited to St. Petersburg to study physiology, Euler was at once given the chance to work in his real field and was appointed an adjunct member of the Academy in the mathematics section. He became professor of physics in 1731 and succeeded Daniel Bernoulli, who returned to Basel in 1733 as a professor of mathematics. The young Academy was beset with numerous difficulties, but on the whole the atmosphere was exceptionally beneficial for the flowering of Euler's genius. Nowhere else could he have been surrounded by such a group of eminent scientists, including the analyst, geometer, and specialist in theoretical mechanics Jakob Hermann, a relative; Daniel Bernoulli, with whom Euler was connected not only by personal friendship but also by common interests in the field of applied mathematics; the versatile scholar Christian Goldbach, with whom Euler discussed numerous problems of analysis and the theory of numbers; F. Maier, working in trigonometry; and the astronomer and geographer J.-N. Delisle.

In St. Petersburg, Euler began his scientific activity at once. No later than August 1727 he started making reports on his investigations at sessions of the Academy; he began publishing them in the second volume of the academic proceedings, *Commentarii Academiae scientiarum imperialis Petropolitanae* (*1727*) (St. Petersburg, 1729). The generous publication program of the Academy was especially important for Euler, who was unusually prolific. In a letter written in 1749 Euler cited the importance that the work at the Academy had for many of its members:

> . . . I and all others who had the good fortune to be for some time with the Russian Imperial Academy cannot but acknowledge that we owe everything which we are and possess to the favorable conditions which we had there.[7]

In addition to conducting purely scientific work, the St. Petersburg Academy from the very beginning was also obliged to educate and train Russian scientists, and with this aim a university and a Gymnasium were organized. The former existed for nearly fifty years and the latter until 1805. The Academy was also charged to carry out for the government a study of Russian territory and to find solutions for various technological problems. Euler was active in these projects. From 1733 on, he successfully worked with Delisle on maps in the department of geography. From the middle of the 1730's he studied problems of shipbuilding and navigation, which were especially important to the rise of Russia as a great sea power. He joined various technological committees and engaged in testing scales, fire pumps, saws, and so forth. He wrote articles for the popular periodical of the Academy and reviewed works submitted to it (including those on the quadrature of the circle), compiled the *Einleitung zur Rechen-Kunst*[8] for Gymnasiums, and also served on the examination board.

Euler's main efforts, however, were in the mathematical sciences. During his fourteen years in St. Petersburg he made brilliant discoveries in such areas as analysis, the theory of numbers, and mechanics. By 1741 he had prepared between eighty and ninety works for publication. He published fifty-five, including the two-volume *Mechanica*.[9]

As is usual with scientists, Euler formulated many of his principal ideas and creative concepts when he was young. Neither the dates of preparation of his works nor those of their actual publication adequately indicate Euler's intellectual progress, since a number of the plans formulated in the early years in St. Petersburg (and even as early as the Basel period) were not realized until much later. For example, the first drafts of the theory of motion of solid bodies, finished in the 1760's, were made during this time. Likewise Euler began studying hydromechanics while still in Basel, but the most important memoirs on the subject did not appear until the middle of the 1750's; he imagined a systematic exposition of differential calculus on the basis of calculus of finite differences in the 1730's but did not realize the intention until two decades later; and his first articles on optics appeared fifteen years after he began studying the subject in St. Petersburg. Only by a complete study of the unpublished Euler manuscripts would it be possible to establish the progression of his ideas more precisely.

Because of his large correspondence with scientists from many countries, Euler's discoveries often became known long before publication and rapidly brought him increasing fame. An index of this is Johann I Bernoulli's letters to his former disciple—in 1728 Bernoulli addressed the "most learned and gifted man of science Leonhard Euler"; in 1737 he wrote, the "most famous and wisest mathematician";

and in 1745 he called him the "incomparable Leonhard Euler" and *"mathematicorum princeps."* Euler was then a member of both the St. Petersburg and Berlin academies. (That certain frictions between Euler and Schumacher, the rude and despotic councillor of the St. Petersburg Academy, did Euler's career no lasting harm was due to his tact and diplomacy.) He was later elected a member of the Royal Society of London (1749) and the Académie des Sciences of Paris (1755). He was elected a member of the Society of Physics and Mathematics in Basel in 1753.

At the end of 1733 Euler married Katharina Gsell, a daughter of Georg Gsell, a Swiss who taught painting at the Gymnasium attached to the St. Petersburg Academy. Johann Albrecht, Euler's first son, was born in 1734, and Karl was born in 1740. It seemed that Euler had settled in St. Petersburg for good; his younger brother, Johann Heinrich, a painter, also worked there. His quiet life was interrupted only by a disease that caused the loss of sight in his right eye in 1738.

In November 1740 Anna Leopoldovna, mother of the infant Emperor Ivan VI, became regent, and the atmosphere in the Russian capital grew troubled. According to Euler's autobiography, "things looked rather dubious."[10] At that time Frederick the Great, who had succeeded to the Prussian throne in June 1740, decided to reorganize the Berlin Society of Sciences, which had been founded by Leibniz but allowed to degenerate during Frederick's father's reign. Euler was invited to work in Berlin. He accepted, and after fourteen years in Russia he sailed with his family on 19 June 1741 from St. Petersburg. He arrived in Berlin on 25 July.

Euler lived in Berlin for the next twenty-five years. In 1744 he moved into a house, still preserved, on the Behrenstrasse. The family increased with the birth of a third son, Christoph, and two daughters; eight other children died in infancy. In 1753 Euler bought an estate in Charlottenburg, which was then just outside the city. The estate was managed by his mother, who lived with Euler after 1750. He sold the property in 1763.

Euler's energy in middle age was inexhaustible. He was working simultaneously in two academies—Berlin and St. Petersburg. He was very active in transforming the old Society of Sciences into a large academy—officially founded in 1744 as the Académie Royale des Sciences et des Belles Lettres de Berlin. (The monarch preferred his favorite language, French, to both Latin and German.) Euler was appointed director of the mathematical class of the Academy and member of the board and of the committee directing the library and the publication of scientific works. He also substituted for the president, Maupertuis, when the latter was absent. When Maupertuis died in 1759, Euler continued to run the Academy, although without the title of president. Euler's friendship with Maupertuis enabled him to exercise great influence on all the activities of the Academy, particularly on the selection of members.

Euler's administrative duties were numerous: he supervised the observatory and the botanical gardens; selected the personnel; oversaw various financial matters; and, in particular, managed the publication of various calendars and geographical maps, the sale of which was a source of income for the Academy. The king also charged Euler with practical problems, such as the project in 1749 of correcting the level of the Finow Canal, which was built in 1744 to join the Havel and the Oder. At that time he also supervised the work on pumps and pipes of the hydraulic system at Sans Souci, the royal summer residence.

In 1749 and again in 1763 he advised on the organization of state lotteries and was a consultant to the government on problems of insurance, annuities, and widows' pensions. Some of Euler's studies on demography grew out of these problems. An inquiry from the king about the best work on artillery moved Euler to translate into German Benjamin Robins' *New Principles of Gunnery.* Euler added his own supplements on ballistics, which were five times longer than the original text (1745).[11] These supplements occupy an important place in the history of ballistics; Euler himself had written a short work on the subject as early as 1727 or 1728 in connection with the testing of guns.[12]

Euler's influence upon scientific life in Germany was not restricted to the Berlin Academy. He maintained a large correspondence with professors at numerous German universities and promoted the teaching of mathematical sciences and the preparation of university texts.

From his very first years in Berlin, Euler kept in regular working contact with the St. Petersburg Academy. This contact was interrupted only during military actions between Prussia and Russia in the course of the Seven Years' War—although even then not completely. Before his departure from the Russian capital, Euler was appointed an honorary member of the Academy and given an annual pension; on his part he pledged to carry out various assignments of the Academy and to correspond with it. During the twenty-five years in Berlin, Euler maintained membership in the St. Petersburg Academy *à tous les titres,* to quote N. Fuss. On its commission he finished the books on differential calculus and navigation begun before his departure for Berlin; edited the mathematical section of the Academy journal; kept the Academy apprised, through his letters, of scientific and technological thought in Western Europe; bought books and scientific apparatus for the Academy; recommended subjects for scientific competitions and candidates to vacancies; and served as a mediator in conflicts between academicians.

Euler's participation in the training of Russian scientific personnel was of great importance, and he was frequently sent for review the works of Russian students and even members of the Academy. For example, in 1747 he praised most highly two articles of M. V. Lomonosov on physics and chemistry; and S. K. Kotelnikov, S. Y. Rumovski, and M. Sofronov studied in Berlin under his supervision for several years. Finally, Euler regularly sent memoirs to St. Petersburg. About half his articles were published there in Latin, and the other half appeared in French in Berlin.

During this period, Euler greatly increased the variety of his investigations. Competing with d'Alembert and Daniel Bernoulli, he laid the foundations of mathematical physics; and he was a rival of both A. Clairaut and d'Alembert in advancing the theory of lunar and planetary motion. At the same time, Euler elaborated the theory of motion of solids, created the mathematical apparatus of hydrodynamics, successfully developed the differential geometry of surfaces, and intensively studied optics, electricity, and magnetism. He also pondered such problems of technology as the construction of achromatic refractors, the perfection of J. A. Segner's hydraulic turbine, and the theory of toothed gearings.

During the Berlin period Euler prepared no fewer than 380 works, of which about 275 were published, including several lengthy books: a monograph on the calculus of variations (1744);[13] a fundamental work on calculation of orbits (1745);[14] the previously mentioned work on artillery and ballistics (1745); *Introductio in analysin infinitorum* (1748);[15] a treatise on shipbuilding and navigation, prepared in an early version in St. Petersburg (1749);[16] his first theory of lunar motion (1753);[17] and *Institutiones calculi differentialis* (1755).[18] The last three books were published at the expense of the St. Petersburg Academy. Finally, there was the treatise on the mechanics of solids, *Theoria motus corporum solidorum seu rigidorum* (1765).[19] The famous *Lettres à une princesse d'Allemagne sur divers sujets de physique et de philosophie,* which originated in lessons given by Euler to a relative of the Prussian king, was not published until Euler's return to St. Petersburg.[20] Written in an absorbing and popular manner, the book was an unusual success and ran to twelve editions in the original

French, nine in English, six in German, four in Russian, and two in both Dutch and Swedish. There were also Italian, Spanish, and Danish editions.

In the 1740's and 1750's Euler took part in several philosophical and scientific arguments. In 1745 and after, there were passionate discussions about the monadology of Leibniz and of Christian Wolff. German intellectuals were divided according to their opinions on monadology. As Euler later wrote, every conversation ended in a discussion of monads. The Berlin Academy announced as the subject of a 1747 prize competition an exposé and critique of the system. Euler, who was close to Cartesian mechanical materialism in natural philosophy, was an ardent enemy of monadology, as was Maupertuis. It should be added that Euler, whose religious views were based on a belief in revelation, could not share the religion of reason which characterized Leibniz and Wolff. Euler stated his objections, which were grounded on arguments of both a physical and theological nature, in the pamphlet *Gedancken von den Elementen der Cörper...* (1746).[21] His composition caused violent debates, but the decision of the Academy gave the prize to Justi, author of a rather mediocre work against the theory of monads.

In 1751 a sensational new argument began when S. König published some critical remarks on Maupertuis's principle of least action (1744) and cited a letter of Leibniz in which the principle was, in König's opinion, formulated more precisely. Submitting to Maupertuis, the Berlin Academy rose to defend him and demanded that the original of Leibniz' letter (a copy had been sent to König from Switzerland) be presented. When it became clear that the original could not be found, Euler published, with the approval of the Academy, "Exposé concernant l'examen de la lettre de M. de Leibnitz" (1752),[22] where, among other things, he declared the letter a fake. The conflict grew critical when later in the same year Voltaire published his *Diatribe du docteur Akakia, médecin du pape,* defending König and making laughingstocks of both Maupertuis and Euler. Frederick rushed to the defense of Maupertuis, quarreling with his friend Voltaire and ordering the burning of the offensive pamphlet. His actions, however, did not prevent its dissemination throughout Europe. The argument touched not only on the pride of the principal participants but also on their general views: Maupertuis and, to a lesser degree, Euler interpreted the principle of least action theologically and teleologically; König was a follower of Wolff and Voltaire— the greatest ideologist of free thought.

Three other disputes in which Euler took part (all discussed below) were much more important for the development of mathematical sciences: his argument with d'Alembert on the problem of logarithms of negative numbers, the argument with d'Alembert and Daniel Bernoulli on the solution of the equation of a vibrating string, and Euler's polemics with Dollond on optical problems.

As mentioned earlier, after Maupertuis died in 1759, Euler managed the Berlin Academy, but under the direct supervision of the king. But relations between Frederick and Euler had long since spoiled. They differed sharply, not only in their views but in their tastes, treatment of men, and personal conduct. Euler's bourgeois manners and religious zeal were as unattractive to the king as the king's passion for bons mots and freethinking was to Euler. Euler cared little for poetry, which the king adored; Frederick was quite contemptuous of the higher realms of mathematics, which did not seem to him immediately practical. In spite of having no one to replace Euler as manager of the Academy, the king, nonetheless, did not intend to give him the post of president. In 1763 it became known that Frederick wanted to appoint d'Alembert, and Euler thus began to think of leaving Berlin. He wrote to G. F. Müller, secretary of the St. Petersburg Academy, which had tried earlier to bring him back to Russia. Catherine the Great then ordered the academicians to send Euler another offer.

D'Alembert's refusal to move permanently to Berlin postponed for a time the final decision on the matter. But during 1765 and 1766 grave conflicts over financial matters arose between Euler and Frederick, who interfered actively with Euler's management of the Academy after the Seven Years' War. The king thought Euler inexperienced in such matters and relied too much on the treasurer of the Academy. For half a year Euler pleaded for royal permission to leave, but the king, well-aware that the Academy would thus lose its best worker and principal force, declined to grant his request. Finally he had to consent and vented his annoyance in crude jokes about Euler. On 9 June 1766, Euler left Berlin, spent ten days in Warsaw at the invitation of Stanislas II, and arrived in St. Petersburg on 28 July. Euler's three sons returned to Russia also. Johann Albrecht became academician in the chair of physics in 1766 and permanent secretary of the Academy in 1769. Christoph, who had become an officer in Prussia, successfully resumed his military career, reaching the rank of major-general in artillery. Both his daughters also accompanied him.

Euler settled in a house on the embankment of the Neva, not far from the Academy. Soon after his return he suffered a brief illness, which left him almost completely blind in the left eye; he could not now read and could make out only outlines of large objects. He could write only in large letters with chalk and slate. An operation in 1771 temporarily restored his sight, but Euler seems not to have taken adequate care of himself and in a few days he was completely blind. Shortly before the operation, he had lost his house and almost all of his personal property in a fire, barely managing to rescue himself and his manuscripts. In November 1773 Euler's wife died, and three years later he married her half sister, Salome Abigail Gsell.

Euler's blindness did not lessen his scientific activity. Only in the last years of his life did he cease attending academic meetings, and his literary output even increased—almost half of his works were produced after 1765. His memory remained flawless, he carried on with his unrealized ideas, and he devised new plans. He naturally could not execute this immense work alone and was helped by active collaborators: his sons Johann Albrecht and Christoph; the academicians W. L. Krafft and A. J. Lexell; and two new young disciples, adjuncts N. Fuss, who was invited in 1772 from Switzerland, and M. E. Golovin, a nephew of Lomonosov. Sometimes Euler simply dictated his works; thus, he dictated to a young valet, a tailor by profession, the two-volume *Vollständige Anleitung zur Algebra* (1770),[23] first published in Russian translation.

But the scientists assisting Euler were not mere secretaries; he discussed the general scheme of the works with them, and they developed his ideas, calculated tables, and sometimes compiled examples. The enormous, 775-page *Theoria motuum lunae* . . . (1772)[24] was thus completed with the help of Johann Albrecht, Krafft, and Lexell—all of whom are credited on the title page. Krafft also helped Euler with the three-volume *Dioptrica* (1769–1771).[25] Fuss, by his own account, during a seven-year period prepared 250 memoirs, and Golovin prepared seventy. Articles written by Euler in his later years were generally concise and particular. For example, the fifty-six works prepared during 1776 contain about the same number of pages (1,000) as the nineteen works prepared in 1751.

Besides the works mentioned, during the second St. Petersburg period Euler published three volumes of *Institutiones calculi integralis* (1768–1770),[26] the principal parts of which he had finished in Berlin, and an abridged edition of *Scientia navalis—Théorie complette de la construction et de la manoeuvre des vaisseaux* (1773).[27] The last, a manual for naval cadets, was soon translated into English, Italian, and Russian, and Euler received for it large sums from the Russian and French governments.

The mathematical apparatus of the *Dioptrica* remained beyond the practical opticist's understanding; so Fuss devised, on the basis of this work, the *Instruction détaillée pour porter les lunettes de toutes les différentes espèces au plus haut degré de perfection dont elles sont susceptibles* . . . (1774).[28] Fuss also aided Euler in preparing the *Éclaircissemens sur les établissemens publics* . . . (1776),[29] which was very important in the development of insurance; many companies used its methods of solution and its tables.

Euler continued his participation in other functions of the St. Petersburg Academy. Together with Johann Albrecht he was a member of the commission charged in 1766 with the management of the Academy. Both resigned their posts on the commission in 1774 because of a difference of opinion between them and the director of the Academy, Count V. G. Orlov, who actually managed it.

On 18 September 1783 Euler spent the first half of the day as usual. He gave a mathematics lesson to one of his grandchildren, did some calculations with chalk on two boards on the motion of balloons; then discussed with Lexell and Fuss the recently discovered planet Uranus. About five o'clock in the afternoon he suffered a brain hemorrhage and uttered only "I am dying," before he lost consciousness. He died about eleven o'clock in the evening.

Soon after Euler's death eulogies were delivered by Fuss at a meeting of the St. Petersburg Academy[30] and by Condorcet at the Paris Academy of Sciences.[31] Euler was buried at the Lutheran Smolenskoye cemetery in St. Petersburg, where in 1837 a massive monument was erected at his grave, with the inscription, "Leonhardo Eulero Academia Petropolitana." In the autumn of 1956 Euler's remains and the monument were transferred to the necropolis of Leningrad.

Euler was a simple man, well disposed and not given to envy. One can also say of him what Fontenelle said of Leibniz: "He was glad to observe the flowering in other people's gardens of plants whose seeds he provided."

Mathematics. Euler was a geometer in the wide sense in which the word was used during the eighteenth century. He was one of the most important creators of mathematical science after Newton. In his work, mathematics was closely connected with applications to other sciences, to problems of technology, and to public life. In numerous cases he elaborated mathematical methods for the direct solution of problems of mechanics and astronomy, physics and navigation, geography and geodesy, hydraulics and ballistics, insurance and demography. This practical orientation of his work explains his tendency to prolong his investigations until he had derived a convenient formula for calculation or an immediate solution in numbers or a table. He constantly sought algorithms that would be simple to use in calculation and that would also assure sufficient accuracy in the results.

But just as his friend Daniel Bernoulli was first of all a physicist, Euler was first of all a mathematician. Bernoulli's thinking was preeminently physical; he tried to avoid mathematics whenever possible, and once having developed a mathematical device for the solution of some physical problem, he usually left it without further development. Euler, on the other hand, attempted first of all to express a physical problem in mathematical terms; and having found a mathematical idea for solution, he systematically developed and generalized it. Thus, Euler's brilliant achievements in the field are explained by his regular elaboration of mathematics as a single whole. Bernoulli was not especially attracted by more abstract problems of mathematics; Euler, on the contrary, was very much carried away with the theory of numbers. All this is manifest in the distribution of Euler's works on various sciences: twenty-nine volumes of the *Opera omnia* (see Bibliography [1]) pertain to pure mathematics.

In Euler's mathematical work, first place belongs to analysis, which at the time was the most pressing need in mathematical science; seventeen volumes of the *Opera omnia* are in this area. Thus, in principle, Euler was an analyst. He contributed numerous particular discoveries to analysis, systematized its exposition in his classical manuals, and, along with all this, contributed immeasurably to the founding of several large mathematical disciplines: the calculus of variations, the theory of differential equations, the elementary theory of functions of complex variables, and the theory of special functions.

Euler is often characterized as a calculator of genius, and he was, in fact, unsurpassed in formal calculations and transformations and was even an outstanding calculator in the elementary sense of the word. But he also was a creator of new and important notions and methods, the principal value of which was in some cases properly understood only a century or more after his death. Even in areas where he, along with his contemporaries, did not feel at home, his judgment came, as a rule, from profound intuition into the subject under study. His findings were intrinsically capable of being grounded in the rigorous mode of demonstration that became obligatory in the nineteenth and twentieth centuries. Such standards were not, and could not be, demanded in the mathematics of the eighteenth century.

It is frequently said that Euler saw no intrinsic

impossibility in the deduction of mathematical laws from a very limited basis in observation; and naturally he employed methods of induction to make empirical use of the results he had arrived at through analysis of concrete numerical material. But he himself warned many times that an incomplete induction serves only as a heuristic device, and he never passed off as finally proved truths the suppositions arrived at by such methods.

Euler introduced many of the present conventions of mathematical notation: the symbol e to represent the base of the natural system of logarithms (1727, published 1736); the use of letter f and of parentheses for a function $f([x/a] + c)$ (1734, published 1740); the modern signs for trigonometric functions (1748); the notation $\int n$ for the sum of divisors of the number n (1750); notations for finite differences, Δy, $\Delta^2 y$, etc., and for the sum Σ (1755); and the letter i for $\sqrt{-1}$ (1777, published 1794).

Euler had only a few immediate disciples, and none of them was a first-class scientist. On the other hand, according to Laplace, he was a tutor of all the mathematicians of his time. In mathematics the eighteenth century can fairly be labeled the Age of Euler, but his influence upon the development of mathematical sciences was not restricted to that period. The work of many outstanding nineteenth-century mathematicians branched out directly from the works of Euler.

Euler was especially important for the development of science in Russia. His disciples formed the first scientific mathematical school in the country and contributed to the rise of mathematical education. One can trace back to Euler numerous paths from Chebyshev's St. Petersburg mathematical school.

[In the following, titles of articles are not, as a rule, cited; dates in parentheses signify the year of publication.]

Theory of Numbers. Problems of the theory of numbers had attracted mathematicians before Euler. Fermat, for example, established several remarkable arithmetic theorems but left almost no proofs. Euler laid the foundations of number theory as a true science.

A large series of Euler's works is connected with the theory of divisibility. He proved by three methods Fermat's lesser theorem, the principal one in the field (1741, 1761, 1763); he suggested with the third proof an important generalization of the theorem by introducing Euler's function $\varphi(n)$, denoting the number of positive integers less than n which are relatively prime to n: the difference $a^{\varphi(n)} - 1$ is divisible by n if a is relatively prime to n. Elaborating related ideas,

Euler came to the theory of n-ic residues (1760). Here his greatest discovery was the law of quadratic reciprocity (1783), which, however, he could not prove. Euler's discovery went unnoticed by his contemporaries, and the law was rediscovered, but incompletely proved, by A. M. Legendre (1788). Legendre was credited with it until Chebyshev pointed out Euler's priority in 1849. The complete proof of the law was finally achieved by Gauss (1801). Gauss, Kummer, D. Hilbert, E. Artin, and others extended the law of reciprocity to various algebraic number fields; the most general law of reciprocity was established by I. R. Shafarevich (1950).

Another group of Euler's works, in which he extended Fermat's studies on representation of prime numbers by sums of the form $mx^2 + ny^2$, where m, n, x, and y are positive integers, led him to the discovery of a new efficient method of determining whether a given large number N is prime or composite (1751, et seq.). These works formed the basis for the general arithmetic theory of binary quadratic forms developed by Lagrange and especially by Gauss.

Euler also contributed to so-called Diophantine analysis, that is, to the solution, in integers or in rational numbers, of indeterminate equations with integer coefficients. Thus, by means of continued fractions, which he had studied earlier (1744, et seq.), he gave (1767) a method of calculation of the smallest integer solution of the equation $x^2 - dy^2 = 1$ (d being a positive nonsquare integer). This had been studied by Fermat and Wallis and even earlier by scientists of India and Greece. A complete investigation of the problem was soon undertaken by Lagrange. In 1753 Euler proved the impossibility of solving $x^3 + y^3 = z^3$ in which x, y, and z are integers, $xyz \neq 0$ (a particular case of Fermat's last theorem); his demonstration, based on the method of infinite descent and using complex numbers of the form $a + b\sqrt{-3}$, is thoroughly described in his *Vollständige Anleitung zur Algebra,* the second volume of which (1769) has a large section devoted to Diophantine analysis.

In all these cases Euler used methods of arithmetic and algebra, but he was also the first to use analytical methods in number theory. To solve the partition problem posed in 1740 by P. Naudé, concerning the total number of ways the positive integer n is obtainable as a sum of positive integers $m < n$, Euler used the expansions of certain infinite products into a power series whose coefficients give the solution (1748). In particular, in the expansion

$$\prod_{r=1}^{\infty} (1 - x^r) = \sum_{k=-\infty}^{\infty} (-1)^k x^{(3k^2 - k)/2}$$

474

the right-hand series is one of theta functions, introduced much later by C. Jacobi in his theory of elliptic functions. Earlier, in 1737, Euler had deduced the famous identity

$$\sum_{n=1}^{\infty} \frac{1}{n^s} = \prod_p \left[1 \Big/ \left(1 - \frac{1}{p^s} \right) \right],$$

where the sum extends over all positive integers n and the product over all primes p (1744), the left-hand side is what Riemann later called the zeta-function $\zeta(s)$.

Using summation of divergent series and induction, Euler discovered in 1749 (1768) a functional equation involving $\zeta(s)$, $\zeta(1 - s)$, and $\Gamma(s)$, which was rediscovered and established by Riemann, the first scientist to define the zeta-function also for complex values of the argument. In the nineteenth and twentieth centuries, the zeta-function became one of the principal means of analytic number theory, particularly in the studies of the laws of distribution of prime numbers by Dirichlet, Chebyshev, Riemann, Hadamard, de la Vallée-Poussin, and others.

Finally, Euler studied mathematical constants and formulated important problems relevant to the theory of transcendental numbers. His expression of the number e in the form of a continued fraction (1744) was used by J. H. Lambert (1768) in his demonstration of irrationality of the numbers e and π. F. Lindemann employed Euler's formula $\ln(-1) = \pi i$ (discovered as early as 1728) to prove that π is transcendental (1882). The hypothesis of the transcendence of a^b, where a is any algebraic number $\neq 0,1$ and b is any irrational algebraic number—formulated by D. Hilbert in 1900 and proved by A. Gelfond in 1934—presents a generalization of Euler's corresponding supposition about rational-base logarithms of rational numbers (1748).

Algebra. When mathematicians of the seventeenth century formulated the fundamental theorem that an algebraic equation of degree n with real coefficients has n roots, which could be imaginary, it was yet unknown whether the domain of imaginary roots was restricted to numbers of the form $a + bi$, which, following Gauss, are now called complex numbers. Many mathematicians thought that there existed imaginary quantities of another kind. In his letters to Nikolaus I Bernoulli and to Goldbach (dated 1742), Euler stated for the first time the theorem that every algebraic polynomial of degree n with real coefficients may be resolved into real linear or quadratic factors, that is, possesses n roots of the form $a + bi$ (1743). The theorem was proved by d'Alembert (1748) and

by Euler himself (1751). Both proofs, quite different in ideas, had omissions and were rendered more precise during the nineteenth century.

Euler also aspired—certainly in vain—to find the general form of solution by radicals for equations of degree higher than the fourth (1738, 1764). He elaborated approximating methods of solutions for numerical equations (1748) and studied the elimination problem. Thus, he gave the first proof of the theorem, which was known to Newton, that two algebraic curves of degrees m and n, respectively, intersect in mn points (1748, 1750). It should be added that Euler's *Vollständige Anleitung zur Algebra,* published in many editions in English, Dutch, Italian, French, and Russian, greatly influenced nineteenth- and twentieth-century texts on the subject.

Infinite Series. In Euler's works, infinite series, which previously served mainly as an auxiliary means for solving problems, became a subject of study. One example, his investigation of the zeta-function, has already been mentioned. The point of departure was the problem of summation of the reciprocals of the squares of the integers

$$\sum_{n=1}^{\infty} \frac{1}{n^2} = \zeta(2),$$

which had been vainly approached by the Bernoulli brothers, Stirling, and other outstanding mathematicians. Euler solved in 1735 a much more general problem and demonstrated that for any even integer number $2k > 0$,

$$\zeta(2k) = a_{2k} \pi^{2k},$$

where a_{2k} are rational numbers (1740), expressed through coefficients of the Euler-Maclaurin summation formula (1750) and, consequently, through Bernoulli numbers (1755). The problem of the arithmetic nature of $\zeta(2k + 1)$ remains unsolved.

The summation formula was discovered by Euler no later than 1732 (1738) and demonstrated in 1735 (1741); it was independently discovered by Maclaurin no later than 1738 (1742). The formula, one of the most important in the calculus of finite differences, represents the partial sum of a series, $\sum_{n=1}^{m} u(n)$, by another infinite series involving the integral and the derivatives of the general term $u(n)$. Later Euler expressed the coefficients of the latter series through Bernoulli numbers (1755). Euler knew that although this infinite series generally diverges, its partial sums under certain conditions might serve as a brilliant means of approximating the calculations shown by James Stirling (1730) in a particular case of

$$\sum_{n=2}^{m} \log (n!).$$

By means of the summation formula, Euler in 1735 calculated (1741) to sixteen decimal places the value of Euler's constant,

$$C = 0.57721566\cdots,$$

belonging to an asymptotic formula,

$$\sum_{n=1}^{m} \frac{1}{n} \simeq \ln m + C,$$

which he discovered in 1731 (1738).

The functions studied in the eighteenth century were, with rare exceptions, analytic, and therefore Euler made great use of power series. His special merit was the introduction of a new and extremely important class of trigonometric Fourier series. In a letter to Goldbach (1744), he expressed for the first time an algebraic function by such a series (1755),

$$\frac{\pi}{2} - \frac{x}{2} = \sin x + \frac{\sin 2x}{2} + \frac{\sin 3x}{3} + \cdots.$$

He later found other expansions (1760), deducing in 1777 a formula of Fourier coefficients for expansion of a given function into a series of cosines on the interval $(0,\pi)$, pointing out that coefficients of expansion into a series of sines could be deduced analogously (1798). Fourier, having no knowledge of Euler's work, deduced in 1807 the same formulas. For his part, Euler did not know that coefficients of expansion into a series of cosines had been given by Clairaut in 1759.

Euler also introduced expansion of functions into infinite products and into the sums of elementary fractions, which later acquired great importance in the general theory of analytic functions. Numerous methods of transformation of infinite series, products, and continued fractions into one another are also his.

Eighteenth-century mathematicians distinguished convergent series from divergent series, but the general theory of convergence was still missing. Algebraic and analytic operations on infinite series were similar to those on finite polynomials, without any restrictions. It was supposed that identical laws operate in both cases. Several tests of convergence already known found almost no application. Opinions, however, differed on the problem of admissibility of divergent series. Many mathematicians were radically against their employment. Euler, sure that important correct results might be arrived at by means of divergent series, set about the task of establishing the legitimacy of their application. With this aim, he

suggested a new, wider definition of the concept of the sum of a series, which coincides with the traditional definition if the series converges; he also suggested two methods of summation (1755). Precise grounding and further development of these fruitful ideas were possible only toward the end of the nineteenth century and the beginning of the twentieth century.[32]

The Concept of Function. Discoveries in the field of analysis made in the middle of the eighteenth century (many of them his own) were systematically summarized by Euler in the trilogy *Introductio in analysin infinitorum* (1748),[15] *Institutiones calculi differentialis* (1755),[18] and *Institutiones calculi integralis* (1768–1770). The books are still of interest, especially the first volume of the *Introductio*. Many of the problems considered there, however, are now so far developed that knowledge of them is limited to a few specialists, who can trace in the book the development of many fruitful methods of analysis.

In the *Introductio* Euler presented the first clear statement of the idea that mathematical analysis is a science of functions; and he also presented a more thorough investigation of the very concept of function. Defining function as an analytic expression somehow composed of variables and constants—following in this respect Johann I Bernoulli (1718)—Euler defined precisely the term "analytic expression": functions are produced by means of algebraic operations, and also of elementary and other transcendental operations, carried out by integration. Here the classification of functions generally used today is also given; Euler speaks of functions defined implicitly and by parametric representation. Further on he states his belief, shared by other mathematicians, that all analytic expressions might be given in the form of infinite power series or generalized power series with fractional or negative exponents. Thus, functions studied in mathematical analysis generally are analytic functions with some isolated singular points. Euler's remark that functions are considered not only for real but also for imaginary values of independent variables was very important.

Even at that time, however, the class of analytic functions was insufficient for the requirements of analysis and its applications, particularly for the solution of the problem of the vibrating string. Here Euler encountered "arbitrary" functions, geometrically represented in piecewise smooth plane curves of arbitrary form—functions which are, generally speaking, nonanalytic (1749). The problem of the magnitude of the class of functions applied in mathematical physics and generally in analysis and the closely related problem of the possibility of analytic expression of nonanalytic functions led to a lengthy polemic involving many mathematicians, including Euler, d'Alembert, and Daniel Bernoulli. One of the results of this controversy over the problem of the vibrating string was the general arithmetical definition of a function as a quantity whose values somehow change with the changes of independent variables; the definition was given by Euler in *Institutiones calculi differentialis*.[18] He had, however, already dealt with the interpretation of a function as a correspondence of values in his *Introductio*.

Elementary Functions. The major portion of the first volume of the *Introductio* is devoted to the theory of elementary functions, which is developed by means of algebra and of infinite series and products. Concepts of infinitesimal and infinite quantity are used, but those of differential and integral calculus are lacking. Among other things, Euler here for the first time described the analytic theory of trigonometric functions and gave a remarkably simple, although nonrigorous, deduction of Moivre's formula and also of his own (1743),

$$e^{\pm xi} = \cos x \pm i \sin x.$$

This was given earlier by R. Cotes (1716) in a somewhat different formulation, but it was widely used only by Euler. The logarithmic function was considered by Euler in the *Introductio* only for the positive independent variable. However, he soon published his complete theory of logarithms of complex numbers (1751)—which some time before had ended the arguments over logarithms of negative numbers between Leibniz and Johann Bernoulli and between d'Alembert and Euler himself in their correspondence (1747–1748). Euler had come across the problem (1727–1728) when he discussed in his correspondence with Johann I Bernoulli the problem of the graphics of the function $y = (-1)^x$ and arrived at the equality $\ln(-1) = \pi i$.

Functions of a Complex Variable. The study of elementary functions brought d'Alembert (1747–1748) and Euler (1751) to the conclusion that the domain of complex numbers is closed (in modern terms) with regard to all algebraic and transcendental operations. They both also made early advances in the general theory of analytic functions. In 1752 d'Alembert, investigating problems of hydrodynamics, discovered equations connecting the real and imaginary parts of an analytic function $u(x,y) + iv(x,y)$. In 1777 Euler deduced the same equations,

$$\frac{\partial u}{\partial x} = \frac{\partial v}{\partial y}, \qquad \frac{\partial u}{\partial y} = -\frac{\partial v}{\partial x},$$

from general analytical considerations, developing a new method of calculation of definite integrals $\int f(z)\, dz$ by means of an imaginary substitution

$$z = x + iy$$

(1793, 1797). He thus discovered (1794) that

$$\int_0^\infty \frac{\sin x}{x}\, dx = \frac{\pi}{2}.$$

Euler also used analytic functions of a complex variable, both in the study of orthogonal trajectories by means of their conformal mapping (1770) and in his works on cartography (1778). (The term *projectio conformis* was introduced by a St. Petersburg academician, F. T. Schubert [1789].) All of these ideas were developed in depth in the elaboration of the general theory of analytic functions by Cauchy (1825) and Riemann (1854), after whom the above-cited equations of d'Alembert and Euler are named.

Although Euler went from numbers of the form $x + iy$ to the point $u(x,y)$ and back, and used a trigonometric form $r(\cos \varphi + i \sin \varphi)$, he saw in imaginary numbers only convenient notations void of real meaning. A somewhat less than successful attempt at geometric interpretation undertaken by H. Kühn (1753) met with sharp critical remarks from Euler.

Differential and Integral Calculus. Both branches of infinitesimal analysis were enriched by Euler's numerous discoveries. Among other things in the *Institutiones calculi differentialis,* he thoroughly elaborated formulas of differentiation under substitution of variables; revealed his theorem on homogeneous functions, stated for $f(x,y)$ as early as 1736; proved the theorem of Nikolaus I Bernoulli (1721) that for $z = f(x,y)$

$$\frac{\partial^2 z}{\partial x \partial y} = \frac{\partial^2 z}{\partial y \partial x};$$

deduced the necessary condition for the exact differential of $f(x,y)$; applied Taylor's series to finding extrema of $f(x)$; and investigated extrema of $f(x,y)$, inaccurately formulating, however, sufficient conditions.

The first two chapters of the *Institutiones* are devoted to the elements of the calculus of finite differences. Euler approached differential calculus as a particular case, we would say a limiting case, of the method of finite differences used when differences of the function and of the independent variable approach zero. During the eighteenth century it was often said against differential calculus that all its formulas were incorrect because the deductions were based on the principle of neglecting infinitely small summands, e.g., on equalities of the kind $a + \alpha = a$, where α is infinitesimal with respect to a. Euler thought that such criticism could be obviated only by supposing all infinitesimals and differentials equal to zero, and therefore he elaborated an original calculus of zeroes. This concept, although not contradictory in itself, did not endure because it proved insufficient in many problems; a strict grounding of analysis was possible if the infinitesimals were interpreted as variables tending to the limit zero.

The methods of indefinite integration in the *Institutiones calculi integralis* (I, 1768) are described by Euler in quite modern fashion and in a detail that practically exhausts all the cases in which the result of integration is expressible in elementary functions. He invented many of the methods himself; the expression "Euler substitution" (for rationalization of certain irrational differentials) serves as a reminder of the fact. Euler calculated many difficult definite integrals, thus laying the foundations of the theory of special functions. In 1729, already studying interpolation of the sequence $1!, 2!, \cdots, n!, \cdots$, he introduced Eulerian integrals of the first and second kind (Legendre's term), today called the beta- and gamma-functions (1738). He later discovered a number of their properties.

Particular cases of the beta-function were first considered by Wallis in 1656. The functions B and Γ, together with the zeta-function and the so-called Bessel functions (see below), are among the most important transcendental functions. Euler's main contribution to the theory of elliptic integrals was his discovery of the general addition theorem (1768). Finally, the theory of multiple integrals also goes back to Euler; he introduced double integrals and established the rule of substitution (1770).

Differential Equations. The *Institutiones calculi integralis* exhibits Euler's numerous discoveries in the theory of both ordinary and partial differential equations, which were especially useful in mechanics.

Euler elaborated many problems in the theory of ordinary linear equations: a classical method for solving reduced linear equations with constant coefficients, in which he strictly distinguished between the general and the particular integral (1743); works on linear systems, conducted simultaneously with d'Alembert (1750); solution of the general linear equation of order n with constant coefficients by reduction to the equation of the same form of order $n - 1$ (1753). After 1738 he successfully applied to second-order linear equations with variable coefficients a method that was highly developed in the nineteenth century; this consisted of the presentation of particular solutions in the form of generalized power series. Another Eulerian device, that of expressing solutions by definite integrals that depend

on a parameter (1763), was extended by Laplace to partial differential equations (1777).

One can trace back to Euler (1741) and Daniel Bernoulli the method of variation of constants later elaborated by Lagrange (1777). The method of an integrating factor was also greatly developed by Euler, who applied it to numerous classes of first-order differential equations (1768) and extended it to higher-order equations (1770). He devoted a number of articles to the Riccati equation, demonstrating its involvement with continued fractions (1744). In connection with his works on the theory of lunar motion, Euler created the widely used device of approximating the solution of the equation $dy/dx = f(x,y)$, with initial condition $x = x_0, y = y_0$ (1768), extending it to second-order equations (1769). This Euler method of open polygons was used by Cauchy in the 1820's to demonstrate the existence theorem for the solution of the above-mentioned equation (1835, 1844). Finally, Euler discovered tests for singular solutions of first-order equations (1768).

Among the large cycle of Euler's works on partial differential equations begun in the middle of the 1730's with the study of separate kinds of first-order equations, which he had encountered in certain problems of geometry (1740), the most important are the studies on second-order linear equations—to which many problems of mathematical physics may be reduced. First was the problem of small plane vibrations of a string, the wave equation originally solved by d'Alembert with the so-called method of characteristics. Given a general solution expressible as a sum of two arbitrary functions, the initial conditions and the boundary conditions of the problem admitted of arriving at solutions in concrete cases (1749). Euler immediately tested this method of d'Alembert's and further elaborated it, eliminating unnecessary restrictions imposed by d'Alembert upon the initial shape and velocity of the string (1749). As previously mentioned, the two mathematicians engaged in an argument which grew more involved when Daniel Bernoulli asserted that any solution of the wave equation might be expressed by a trigonometric series (1755). D'Alembert and Euler agreed that such a solution could not be sufficiently general. The discussion was joined by Lagrange, Laplace, and other mathematicians of great reputation and lasted for over half a century; not until Fourier (1807, 1822) was the way found to the correct formulation and solution of the problem. Euler later developed the method of characteristics more thoroughly (1766, 1767).

Euler encountered equations in other areas of what became mathematical physics: in hydrodynamics; in the problem of vibrations of membranes, which he reduced to the so-called Bessel equation and solved (1766) by means of the Bessel functions $J_n(x)$; and in the problem of the motion of air in pipes (1772). Some classes of equations studied by Euler for velocities close to or surpassing the velocity of sound continue to figure in modern aerodynamics.

Calculus of Variations. Starting with several problems solved by Johann and Jakob Bernoulli, Euler was the first to formulate the principal problems of the calculus of variations and to create general methods for their solution. In *Methodus inveniendi lineas curvas . . .*[13] he systematically developed his discoveries of the 1730's (1739, 1741). The very title of the work shows that Euler widely employed geometric representations of functions as flat curves. Here he introduced, using different terminology, the concepts of function and variation and distinguished between problems of absolute extrema and relative extrema, showing how the latter are reduced to the former. The problem of the absolute extremum of the function of several independent variables,

$$\int_a^b F(x, y, y') \, dx,$$

where F is the given and $y(x)$ the desired minimizing or maximizing function, is treated as the limiting problem for the ordinary extremum of the function

$$W_n(y_0, y_1, \cdots, y_n) = \sum_{k=0}^{n-1} F\left(x_k, y_k, \frac{y_{k+1} - y_k}{\Delta x}\right) \Delta x,$$

where $x_k = a + k \, \Delta x$, $\Delta x = (b - a)/n$, $k = 0$, 1, $\cdots$, n (and $n \to \infty$). Thus Euler deduced the differential equation named after him to which the function $y(x)$ should correspond; this necessary condition was generalized for the case where F involves the derivatives y', y'', $\cdots$, $y^{(n)}$. In this way the solution of a problem in the calculus of variations might always be reduced to integration of a differential equation. A century and a half later the situation had changed. The direct method imagined by Euler, which he had employed only to obtain his differential equation, had (together with similar methods) acquired independent value for rigorous or approximate solution of variational problems and the corresponding differential equations.

In the mid-1750's, after Lagrange had created new algorithms and notations for the calculus of variations, Euler abandoned the former exposition and gave instead a detailed and lucid exposition of Lagrange's method, considering it a new calculus—which he called variational (1766). He applied the calculus of variations to problems of extreme values

of double integrals with constant limits in volume III of the *Institutiones calculi integralis* (1770); soon thereafter he suggested still another method of exposition of the calculus, one which became widely used.

Geometry. Most of Euler's geometrical discoveries were made by application of the methods of algebra and analysis. He gave two different methods for an analytical exposition of the system of spherical trigonometry (1755, 1782). He showed how the trigonometry of spheroidal surfaces might be applied to higher geodesy (1755). In volume II of the *Introductio* he surpassed his contemporaries in giving a consistent algebraic development of the theory of second-order curves, proceeding from their general equation (1748). He constituted the theory of third-order curves by analogy. But Euler's main achievement was that for the first time he studied thoroughly the general equation of second-order surfaces, applying Euler angles in corresponding transformations.

Euler's studies of the geodesic lines on a surface are prominent in differential geometry; the problem was pointed out to him by Johann Bernoulli (1732, 1736, and later). But still more important were his pioneer investigations in the theory of surfaces, from which Monge and other geometers later proceeded. In 1763 Euler made the first substantial advance in the study of the curvature of surfaces; in particular, he expressed the curvature of an arbitrary normal section by principal curvatures (1767). He went on to study developable surfaces, introducing Gaussian coordinates (1772), which became widely used in the nineteenth century. In a note written about 1770 but not published until 1862 Euler discovered the necessary condition for applicability of surfaces that was independently established by Gauss (1828). In 1775 Euler successfully renewed elaboration of the general theory of space curves (1786), beginning where Clairaut had left off in 1731.

Euler was also the author of the first studies on topology. In 1735 he gave a solution to the problem of the seven bridges of Königsberg: the bridges, spanning several arms of a river, must all be crossed without recrossing any (1741). In a letter to Goldbach (1750), he cited (1758) a number of properties of polyhedra, among them the following: the number of vertices, S, edges, A, and sides, H, of a polyhedron are connected by an equality $S - A + H = 2$. A hundred years later it was discovered that the theorem had been known to Descartes. The Euler characteristic $S - A + H$ and its generalization for multidimensional complexes as given by H. Poincaré is one of the principal invariants of modern topology.

Mechanics. In an introduction to the *Mechanica* (1736) Euler outlined a large program of studies embracing every branch of the science. The distinguishing feature of Euler's investigations in mechanics as compared to those of his predecessors is the systematic and successful application of analysis. Previously the methods of mechanics had been mostly synthetic and geometrical; they demanded too individual an approach to separate problems. Euler was the first to appreciate the importance of introducing uniform analytic methods into mechanics, thus enabling its problems to be solved in a clear and direct way. Euler's concept is manifest in both the introduction and the very title of the book, *Mechanica sive motus scientia analytice exposita.*

This first large work on mechanics was devoted to the kinematics and dynamics of a point-mass. The first volume deals with the free motion of a point-mass in a vacuum and in a resisting medium; the section on the motion of a point-mass under a force directed to a fixed center is a brilliant analytical reformulation of the corresponding section of Newton's *Principia;* it was sort of an introduction to Euler's further works on celestial mechanics. In the second volume, Euler studied the constrained motion of a point-mass; he obtained three equations of motion in space by projecting forces on the axes of a moving trihedral of a trajectory described by a moving point, i.e., on the tangent, principal normal, and binormal. Motion in the plane is considered analogously. In the chapter on the motion of a point on a given surface, Euler solved a number of problems of the differential geometry of surfaces and of the theory of geodesics.

The *Theoria motus corporum solidorum,*[19] published almost thirty years later (1765), is related to the *Mechanica.* In the introduction to this work, Euler gave a new exposition of punctual mechanics and followed Maclaurin's example (1742) in projecting the forces onto the axes of a fixed orthogonal rectilinear system. Establishing that the instantaneous motion of a solid body might be regarded as composed of rectilinear translation and instant rotation, Euler devoted special attention to the study of rotatory motion. Thus, he gave formulas for projections of instantaneous angular velocity on the axes of coordinates (with application of Euler angles), and framed dynamical differential equations referred to the principal axes of inertia, which determine this motion. Special mention should be made of the problem of motion of a heavy solid body about a fixed point, which Euler solved for a case in which the center of gravity coincides with the fixed point. The law of motion in such a case is, generally speaking, expressed by means of elliptic integrals. Euler was led to this problem by the study of precession of the equinoxes

and of the nutation of the terrestrial axis (1751).[33] Other cases in which the differential equations of this problem can be integrated were discovered by Lagrange (1788) and S. V. Kovalevskaya (1888). Euler considered problems of the mechanics of solid bodies as early as the first St. Petersburg period.

In one of the two appendixes to the *Methodus...*[13] Euler suggested a formulation of the principle of least action for the case of the motion of a point under a central force: the trajectory described by the point minimizes the integral ∫ *mv ds*. Maupertuis had stated at nearly the same time the principle of least action in a much more particular form. Euler thus laid the mathematical foundation of the numerous studies on variational principles of mechanics and physics which are still being carried out.

In the other appendix to the *Methodus,* Euler, at the insistence of Daniel Bernoulli, applied the calculus of variations to some problems of the theory of elasticity, which he had been intensively elaborating since 1727. In this appendix, which was in fact the first general work on the mathematical theory of elasticity, Euler studied bending and vibrations of elastic bands (either homogeneous or nonhomogeneous) and of a plate under different conditions; considered nine types of elastic curves; and deduced the famous Euler buckling formula, or Euler critical load, used to determine the strength of columns.

Hydromechanics. Euler's first large work on fluid mechanics was *Scientia navalis.* Volume I contains a general theory of equilibrium of floating bodies including an original elaboration of problems of stability and of small oscillations in the neighborhood of an equilibrium position. The second volume applies general theorems to the case of a ship.

From 1753 to 1755 Euler elaborated in detail an analytical theory of fluid mechanics in three classic memoirs—"Principes généraux de l'état d'équilibre des fluides"; "Principes généraux du mouvement des fluides"; and "Continuation des recherches sur la théorie du mouvement des fluides"—all published simultaneously (1757).[34] Somewhat earlier (1752) the "Principia motus fluidorum" was written; it was not published, however, until 1761.[35] Here a system of principal formulas of hydrostatics and hydrodynamics was for the first time created; it comprised the continuity equation for liquids with constant density; the velocity-potential equation (usually called after Laplace); and the general Euler equations for the motion of an incompressible liquid, gas, etc. As has generally been the case in mathematical physics, the main innovations were in the application of partial differential equations to the problems. At the beginning of the "Continuation des recherches" Euler emphasized that he had reduced the whole of the theory of liquids to two analytic equations and added:

> However sublime are the researches on fluids which we owe to the Messrs. Bernoulli, Clairaut and d'Alembert, they flow so naturally from my two general formulae that one cannot sufficiently admire this accord of their profound meditations with the simplicity of the principles from which I have drawn my two equations, and to which I was led immediately by the first axioms of mechanics.[36]

Euler also investigated a number of concrete problems on the motion of liquids and gases in pipes, on vibration of air in pipes, and on propagation of sound. Along with this, he worked on problems of hydrotechnology, discussed, in part, above. Especially remarkable were the improvements he introduced into the design of a hydraulic machine imagined by Segner in 1749 and the theory of hydraulic turbines, which he created in accordance with the principle of action and reaction (1752–1761).[37]

Astronomy. Euler's studies in astronomy embraced a great variety of problems: determination of the orbits of comets and planets by a few observations, methods of calculation of the parallax of the sun, the theory of refraction, considerations on the physical nature of comets, and the problem of retardation of planetary motions under the action of cosmic ether. His most outstanding works, for which he won many prizes from the Paris Académie des Sciences, are concerned with celestial mechanics, which especially attracted scientists at that time.

The observed motions of the planets, particularly of Jupiter and Saturn, as well as the moon, were evidently different from the calculated motions based on Newton's theory of gravitation. Thus, the calculations of Clairaut and d'Alembert (1745) gave the value of eighteen years for the period of revolution of the lunar perigee, whereas observations showed this value to be nine years. This caused doubts about the validity of Newton's system as a whole. For a long time Euler joined these scientists in thinking that the law of gravitation needed some corrections. In 1749 Clairaut established that the difference between theory and observation was due to the fact that he and others solving the corresponding differential equation had restricted themselves to the first approximation. When he calculated the second approximation, it was satisfactorily in accordance with the observed data. Euler did not at once agree. To put his doubts at rest, he advised the St. Petersburg Academy to announce a competition on the subject. Euler soon determined that Clairaut was right, and on Euler's recommendation his composition received

the prize of the Academy (1752). Euler was still not completely satisfied, however. In 1751 he had written his own *Theoria motus lunae exhibens omnes ejus inaequalitates* (published in 1753), in which he elaborated an original method of approximate solution to the three-body problem, the so-called first Euler lunar theory. In the appendix he described another method which was the earliest form of the general method of variation of elements. Euler's numerical results also conformed to Newton's theory of gravitation.

The first Euler lunar theory had an important practical consequence: T. Mayer, an astronomer from Göttingen, compiled, according to its formulas, lunar tables (1755) that enabled the calculation of the position of the moon and thus the longitude of a ship with an exactness previously unknown in navigation. The British Parliament had announced as early as 1714 a large cash prize for the method of determination of longitude at sea with error not to exceed half a degree, and smaller prizes for less exact methods. The prize was not awarded until 1765; £3,000 went to Mayer's widow and £300 to Euler for his preliminary theoretical work. Simultaneously a large prize was awarded to J. Harrison for his construction of a more nearly perfect chronometer. Lunar tables were included in all nautical almanacs after 1767, and the method was used for about a century.

From 1770 to 1772 Euler elaborated his second theory of lunar motion, which he published in the *Theoria motuum lunae, nova methodo pertractata* (1772).[24] For various reasons, the merits of the new method could be correctly appreciated only after G. W. Hill brilliantly developed the ideas of the composition in 1877–1888.

Euler devoted numerous works to the calculation of perturbations of planetary orbits caused by the mutual gravitation of Jupiter and Saturn (1749, 1769) as well as of the earth and the other planets (1771). He continued these studies almost to his death.

Physics. Euler's principal contribution to physics consisted in mathematical elaboration of the problems discussed above. He touched upon various physical problems which would not yield to mathematical analysis at that time. He aspired to create a uniform picture of the physical world. He had been, as pointed out earlier, closer to Cartesian natural philosophy than to Newtonian, although he was not a direct representative of Cartesianism. Rejecting the notion of empty space and the possibility of action at a distance, he thought that the universe is filled up with ether—a thin elastic matter with extremely low density, like super-rarified air. This ether contains material particles whose main property is impenetrability. Euler thought it possible to explain the diversity of the observed phenomena (including electricity, light, gravitation, and even the principle of least action) by the hypothetical mechanical properties of ether. He also had to introduce magnetic whirls into the doctrine of magnetism; these are even thinner and move more quickly than ether.

In physics Euler built up many artificial models and hypotheses which were short-lived. But his main concept of the unity of the forces of nature acting deterministically in some medium proved to be important for the development of physics, owing especially to *Lettres à une princesse d'Allemagne.* Thus, his views on the nature of electricity were the prototype of the theory of electric and magnetic fields of Faraday and Maxwell. His theory of ether influenced Riemann.

Euler's works on optics were widely known and important in the physics of the eighteenth century. Rejecting the dominant corpuscular theory of light, he constructed his own theory in which he attributed the cause of light to peculiar oscillations of ether. His *Nova theoria lucis et colorum* (1746)[38] explained some, but not all, phenomena. Proceeding from certain analogies that later proved incorrect, Euler concluded that the elimination of chromatic aberration of optic lenses was possible (1747); he conducted experiments with lenses filled with water to confirm the conclusion. This provoked objections by the English optician Dollond, who, following Newton, held that dispersion was inevitable. The result of this polemic, in which both parties were partly right and partly wrong, was the creation by Dollond of achromatic telescopes (1757), a turning point in optical technology. For his part, Euler, in his *Dioptrica,* laid the foundations of the calculation of optical systems.

NOTES

All works cited are listed in the Bibliography. References to Euler's *Opera omnia* (see [1] in Bibliography) include series and volume number.

1. 20, p. 75.
2. "Constructio linearum isochronarum in medio quocunque resistente," in 1, 2nd ser., VI, p. 1.
3. "Methodus inveniendi traiectorias reciprocas algebraicas," in 1, 1st ser., XXVII, p. 1.
4. To be published in 1, 2nd ser., XX.
5. 1, 3rd ser., I, p. 181.
6. 26, p. 51.
7. 13, II, p. 182.
8. *Einleitung zur Rechen-Kunst zum Gebrauch des Gymnasii bey der Kayserlichen Academie der Wissenschafften in St. Petersburg* (St. Petersburg, 1738–1740). See 1, 3rd ser., II, 1–303.
9. *Mechanica sive motus scientia analytice exposita,* 2 vols. (St. Petersburg, 1736). See 1, 2nd ser., I and II.
10. 20, p. 77.

11. *Neue Grundsätze der Artillerie aus dem Englischen des Herrn Benjamin Robins übersetzt und mit vielen Anmerkungen versehen* (Berlin, 1745). See 1, 2nd ser., XIV.
12. See 1, 2nd ser., XIV, 468–477.
13. *Methodus inveniendi lineas curvas maximi minimive proprietate gaudentes* (Lausanne–Geneva, 1744). See 1, 1st ser., XXIV.
14. *Theoria motuum planetarum et cometarum* (Berlin, 1744). See 1, 2nd ser., XXVIII, 105–251.
15. *Introductio in analysin infinitorum,* 2 vols. (Lausanne, 1748). See 1, 1st ser., VIII and IX.
16. *Scientia navalis,* 2 vols. (St. Petersburg, 1749). See 1, 2nd ser., XVIII and XIX.
17. *Theoria motus lunae* (Berlin, 1753). See 1, 2nd ser., XXIII, 64–336.
18. *Institutiones calculi differentialis cum eius usu in analysi finitorum ac doctrina serierum* (Berlin, 1755). See 1, 1st ser., X.
19. *Theoria motus corporum solidorum seu rigidorum ex primis nostrae cognitionis principiis stabilita . . .* (Rostock–Greifswald, 1765). See 1, 2nd ser., III and IV.
20. The work, which comprises 234 letters, was published at St. Petersburg in 3 vols. The first two vols. (letters 1–154) appeared in 1768; vol. III appeared in 1772. See 1, 3rd ser., XI and XII.
21. *Gedancken von den Elementen der Cörper, in welchen das Lehr-Gebäude von den einfachen Dingen und Monaden geprüfet und das wahre Wesen der Cörper entdecket wird* (Berlin, 1746). See 1, 3rd ser., II, 347–366.
22. "Exposé concernant l'examen de la lettre de M. de Leibnitz, alléguée par M. le Professeur Koenig, dans le mois de mars 1751 des Actes de Leipzig, à l'occasion du principe de la moindre action." See 1, 2nd ser., V, 64–73.
23. The work was first published at St. Petersburg in Russian (vol. I, 1768; vol. II, 1769). It then appeared in a two-volume German edition (St. Petersburg, 1770). See 1, 1st ser., I.
24. *Theoria motuum lunae, nova methodo pertractata* (St. Petersburg, 1772). See 1, 2nd ser., XXII.
25. The work was published sequentially, in 3 vols., at St. Petersburg. Vol. I deals with principles of optics (1769); vol. II with construction of telescopes (1770); and vol. III with construction of microscopes (1771). See 1, 3rd ser., III and IV.
26. The work's 3 vols. were published sequentially in St. Petersburg in 1768, 1769, and 1770. See 1, 1st ser., XI, XII, and XIII.
27. To be published in 1, 2nd ser., XXI.
28. See 1, 3rd ser., VII, 200–247.
29. *Éclaircissemens sur les établissemens publics en faveur tant des veuves que des morts, avec la déscription d'une nouvelle espèce de tontine aussi favorable au public qu'utile à l'état* (St. Petersburg, 1776). See 1, 1st ser., VII, 181–245.
30. See 17.
31. Condorcet's *éloge* was first published in *Histoire de l'Académie royale des sciences pour l'année 1783* (Paris, 1786), pp. 37–68. It is reprinted in 1, 3rd ser., XII, 287–310.
32. See 50, chs. 1–2.
33. "Recherches sur la précession des équinoxes et sur la nutation de l'axe de la terre." See 1, 2nd ser., XXIX, 92–123.
34. See 1, 2nd ser., XII, 2–132.
35. See 1, 2nd ser., XII, 133–168.
36. See 1, 2nd ser., XII, 92, for the original French.
37. See 1, 2nd ser., XV, pt. 1, 1–39, 80–104, 157–218.
38. See 1, 3rd ser., V, 1–45.

BIBLIOGRAPHY

I. ORIGINAL WORKS. Euler wrote and published more than any other mathematician. During his lifetime about 560 books and articles appeared, and he once remarked to Count Orlov that he would leave enough memoirs to fill the pages of publications of the St. Petersburg Academy for twenty years after his death. Actually the publication of his literary legacy lasted until 1862. N. Fuss published about 220 works, and then the work was carried on by V. Y. Buniakovsky, P. L. Chebyshev, and P.-H. Fuss. Other works were found still later. The list compiled by Eneström (25) includes 856 titles and 31 works by J.-A. Euler, all written under the supervision of his father.

Euler's enormous correspondence (approximately 300 addressees), which he conducted from 1726 until his death, has been only partly published. For an almost complete description, with summaries and indexes, see (37) below. For his correspondence with Johann I Bernoulli, see (2) and (3); with Nikolaus I Bernoulli (2) and (4); with Daniel Bernoulli (2) and (3); with C. Goldbach (2) and (5); with J.-N. Delisle (6); with Clairaut (7); with d'Alembert (3) and (8); with T. Mayer (9); with Lagrange (10); with J. H. Lambert (11); with M. V. Lomonosov (12); with G. F. Müller (13); with J. D. Schumacher (13); with King Stanislas II (14); and with various others (15).

1. Euler's complete works are in the course of publication in a collection that has been destined from the outset to become one of the monuments of modern scholarship in the historiography of science: *Leonhardi Euleri Opera omnia* (Berlin-Göttingen-Leipzig-Heidelberg, 1911–   ). The *Opera omnia* is limited for the most part to republishing works that Euler himself prepared for the press. All texts appear in the original language of publication. Each volume is edited by a modern expert in the science it concerns, and many of the introductions constitute full histories of the relevant branch of science in the seventeenth and eighteenth centuries. Several volumes are in course of preparation. The work is organized in three series. The first series (*Opera mathematica*) comprises 29 vols. and is complete. The second series (*Opera mechanica et astronomica*) is to comprise 31 vols. and still lacks vols. XVI, XVII, XIX, XX, XXI, XXIV, XXVI, XXVII, and XXXI. The third series (*Opera physica, Miscellanea, Epistolae*) is to comprise 12 vols. and still lacks vols. IX and X. Euler's correspondence is not included in this edition.

2. P.-H. Fuss, ed., *Correspondance mathématique et physique de quelques célèbres géomètres du XVIIIe siècle*, 2 vols. (St. Petersburg, 1843). See vol. I for correspondence with Goldbach. For correspondence with Johann I Bernoulli, see II, 1–93; with Nikolaus I Bernoulli, II, 679–713; and with Daniel Bernoulli, II, 407–665.

3. G. Eneström, ed., *Bibliotheca mathematica*, 3rd ser., **4** (1903), 344–388; **5** (1904), 248–291; and **6** (1905), 16–87; for correspondence with Johann I Bernoulli. For Euler's correspondence with Daniel Bernoulli, see **7** (1906–1907); 126–156. See **11** (1911), 223–226, for correspondence with d'Alembert.

4. *Opera postuma*, I (St. Petersburg, 1862), 519–549.

5. A. P. Youschkevitch and E. Winter, eds., *Leonhard Euler und Christian Goldbach. Briefwechsel 1729–1764* (Berlin, 1965).

6. A. T. Grigorian, A. P. Youschkevitch, *et. al.*, eds., *Russko-frantsuskie nauchnye svyazi* (Leningrad, 1968), pp. 119–279.

7. G. Bigourdan, ed., "Lettres inédites d'Euler à

Clairaut," in *Comptes rendus du Congrès des sociétés savantes, 1928* (Paris, 1930), pp. 26–40.

8. *Bullettino di bibliografia e di storia delle scienze matematiche e fisiche,* **19** (1886), 136–148.

9. Y. K. Kopelevich and E. Forbs, eds., *Istoriko-astronomicheskie issledovania*, V (1959), 271–444; X (1969), 285–308.

10. J. L. Lagrange, *Oeuvres*, J. A. Serret and G. Darboux, eds., XIV (Paris, 1892), 135–245.

11. K. Bopp, "Eulers und J.-H. Lamberts Briefwechsel," in *Abhandlungen der Preussischen Akademie der Wissenschaften* (1924), 7–37.

12. M. V. Lomonosov, *Sochinenia*, VIII (Moscow-Leningrad, 1948); and *Polnoe sobranie sochineny*, X (Moscow-Leningrad, 1957).

13. A. P. Youschkevitch, E. Winter, *et. al.*, eds., *Die Berliner und die Petersburger Akademie der Wissenschaften im Briefwechsel Leonhard Eulers*, 2 vols. See vol. I (Berlin, 1959) for letters to G. F. Müller; vol. II (Berlin, 1961) for letters to Nartov, Schumacher, Teplov, and others.

14. T. Kłado and R. W. Wołoszyński, eds., "Korrespondencja Stanisława Augusta z Leonardem Eulerem . . ." in *Studia i materiały z dziejów nauki polskiej*, ser. C, no. 10 (Warsaw, 1965), pp. 3–41.

15. V. I. Smirnov *et. al.*, eds., *Leonard Euler. Pisma k uchenym* (Moscow-Leningrad, 1963). Contains letters to Bailly, Bülfinger, Bonnet, C. L. Ehler, C. Wolff, and others.

II. Secondary Literature.

16. J. W. Herzog, *Adumbratio eruditorum basilensium meritis apud exteros olim hodieque celebrium* (Basel, 1778), pp. 32–60.

17. N. Fuss, *Éloge de Monsieur Léonard Euler* (St. Petersburg, 1783). A German trans. of this is in (1), 1st ser., I, xliii–xcv.

18. Marquis de Condorcet, *Éloge de M. Euler*, in *Histoire de l'Académie royale des sciences pour l'année 1783* (Paris, 1786), pp. 37–68.

19. R. Wolf, *Biographien zur Kulturgeschichte der Schweiz*, IV (Zurich, 1862), 87–134.

20. P. Pekarski, "Ekaterina II i Eyler," in *Zapiski imperatorskoi akademii nauk*, **6** (1865), 59–92.

21. P. Pekarski, *Istoria imperatorskoi akademii nauk v Peterburge*, **1** (1870), 247–308. See also index.

22. M. I. Sukhomlinov, ed., *Materialy dlya istorii imperatorskoi akademii nauk, 1716–1760*, 10 vols. (St. Petersburg, 1885–1900). See indexes.

23. *Protokoly zasedany konferentsii imperatorskoi akademii nauk s 1725 po 1803 god*, 4 vols. (St. Petersburg, 1897–1911). See indexes.

24. A. Harnack, *Geschichte der königlichen preussischen Akademie der Wissenschaften*, I–III (Berlin, 1900).

25. G. Eneström, "Verzeichnis der Schriften Leonhard Eulers," in *Jahresbericht der Deutschen Mathematiker-Vereinigung*, Ergänzungsband **4** (Leipzig, 1910–1913). An important bibliography of Euler's works in three parts, listed in order of date of publication, in order of date of composition, and by subject. The first part is reprinted in (35), I, 352–386.

26. O. Spiess, *Leonhard Euler. Ein Beitrag zur Geistesgeschichte des XVIII. Jahrhunderts* (Frauenfeld-Leipzig, 1929).

27. G. Du Pasquier, *Léonard Euler et ses amis* (Paris, 1927).

28. W. Stieda, *Die Übersiedlung Leonhard Eulers von Berlin nach Petersburg* (Leipzig, 1931).

29. W. Stieda, *J. A. Euler in seinen Briefen, 1766-1790* (Leipzig, 1932).

30. A. Speiser, *Die Basler Mathematiker* (Basel, 1939).

31. E. Fueter, *Geschichte der exakten Wissenschaften in der Schweizerischen Aufklärung, 1680-1780* (Aarau, 1941).

32. Karl Euler, *Das Geschlecht Euler-Schölpi. Geschichte einer alten Familie* (Giessen, 1955).

33. E. and M. Winter, eds., *Die Registres der Berliner Akademie der Wissenschaften, 1746-1766. Dokumente für das Wirken Leonhard Eulers in Berlin* (Berlin, 1957). With an intro. by E. Winter.

34. *Istoria akademii nauk SSSR,* I (Moscow-Leningrad, 1958). See index.

35. Y. K. Kopelevich, M. V. Krutikova, G. M. Mikhailov, and N. M. Raskin, eds., *Rukopisnye materialy Leonarda Eylera v arkhive akademii nauk SSR,* 2 vols. (Moscow-Leningrad, 1962-1965). Vol. I contains an index of Euler's scientific papers, an index of official and personal documents, summaries of proceedings of conferences of the Academy of Sciences of St. Petersburg with respect to Euler's activities, an index of Euler's correspondence, a reedited version of the first part of (24), and many valuable indexes. Vol. II contains 12 of Euler's papers on mechanics published for the first time. See especially I, 120-228.

36. G. K. Mikhailov, "K pereezdu Leonarda Eylera v Peterburg" ("On Leonhard Euler's Removal to St. Petersburg," in *Izvestiya Akademii nauk SSSR. Otdelenie tekhnicheskikh nauk,* no. 3 (1957), 10-38.

37. V. I. Smirnov and A. P. Youschkevitch, eds., *Leonard Eyler. Perepiska. Annotirovannye ukazateli* (Leningrad, 1967).

38. F. Dannemann, *Die Naturwissenschaften in ihrer Entwicklung und in ihrem Zusammenhänge,* II-III (Leipzig, 1921). See indexes.

39. R. Taton, ed., *Histoire générale des sciences,* II (Paris, 1958). See index.

40. I. Y. Timchenko, *Osnovania teorii analiticheskikh funktsy. Chast I. Istoricheskie svedenia* (Odessa, 1899).

41. M. Cantor, *Vorlesungen über Geschichte der Mathematik,* III-IV (Leipzig, 1898-1908). See indexes.

42. H. Wieleitner, *Geschichte der Mathematik,* II (Berlin-Leipzig, 1911-1921). See indexes.

43. D. J. Struik, *A Concise History of Mathematics,* 2 vols. (New York, 1948; 2nd ed., London, 1956).

44. J. E. Hofmann, *Geschichte der Mathematik,* pt. 3 (Berlin, 1957). See index.

45. A. P. Youschkevitch, *Istoria matematika v Rossii do 1917 goda* (Moscow, 1968). See index.

46. Carl B. Boyer, *A History of Mathematics* (New York, 1968).

47. L. E. Dickson, *History of the Theory of Numbers,* 3 vols. (Washington, 1919-1927; 2nd ed., 1934). See indexes.

48. D. J. Struik, "Outline of a History of Differential Geometry," in *Isis,* **19** (1933), 92-120; **20** (1933), 161-191.

49. J. L. Coolidge, *A History of Geometrical Methods* (Oxford, 1940).

50. G. H. Hardy, *Divergent Series* (Oxford, 1949).

51. A. I. Markuschevitsch, *Skizzen zur Geschichte der analytischen Funktionen* (Berlin, 1955).

52. Carl B. Boyer, *History of Analytic Geometry* (New York, 1956). See index.

53. N. I. Simonov, *Prikladnye metody analiza u Eylera* (Moscow, 1957).

54. A. T. Grigorian, *Ocherki istorii mekhaniki v Rossii* (Moscow, 1961).

55. C. Truesdell, "The Rational Mechanics of Flexible or Elastic Bodies," in (1), 2nd ser., XI, pt. 2.

56. S. Timoschenko, *History of the Strength of Materials* (New York-Toronto-London, 1953).

57. A. P. Mandryka, *Istoria ballistiki* (Moscow-Leningrad, 1964).

58. N. N. Bogolyubov, *Istoria mekhaniki mashin* (Kiev, 1964).

59. F. Rosenberger, *Die Geschichte der Physik in Grundzügen,* II (Brunswick, 1884). See index.

60. V. F. Gnucheva, *Geografichesky departament akademii nauk XVIII veka* (Moscow-Leningrad, 1946).

61. E. Hoppe, *Die Philosophie Leonhard Eulers* (Gotha, 1904).

62. A. Speiser, *Leonhard Euler und die deutsche Philosophie* (Zurich, 1934).

63. G. Kröber, *L. Euler. Briefe an eine deutsche Prinzessin. Philosophische Auswahl* (Leipzig, 1965), pp. 5-26. See also intro.

Many important essays on Euler's life, activity, and work are in the following five memorial volumes.

64. *Festschrift zur Feier 200. Geburtstages Leonhard Eulers* (Leipzig-Berlin, 1907), a publication of the Berliner Mathematische Gesellschaft.

65. A. M. Deborin, ed., *Leonard Eyler, 1707-1783* (Moscow-Leningrad, 1935).

66. E. Winter, et. al., eds., *Die deutsch-russische Begegnung und Leonhard Euler . . .* (Berlin, 1958).

67. M. A. Lavrentiev, A. P. Youschkevitch, and A. T. Grigorian, eds., *Leonard Eyler. Sbornik statey* (Moscow, 1958). See especially pp. 268-375 and 377-413 for articles on Euler's work in astronomy and his physical concepts.

68. K. Schröder, ed., *Sammelband der zu Ehren des 250. Geburtstages Leonhard Eulers . . . vorgelegten Abhandlungen* (Berlin, 1959).

69. *Istoriko-matematicheskie issledovania* (Moscow, 1949-1969). For articles on Euler, see II, V-VII, X, XII, XIII, XVI, and XVII.

70. G. K. Mikhailov, "Leonard Eyler," in *Izvestiya akademii nauk SSSR. Otdelenie tekhnicheskikh nauk,* no. 1 (1955), 3-26, with extensive bibliography.

A. P. YOUSCHKEVITCH

EULER-CHELPIN, HANS KARL AUGUST SIMON VON (*b.* Augsburg, Germany, 15 February 1873; *d.* Stockholm, Sweden, 6 November 1964), *biochemistry.*

Hans von Euler-Chelpin was the son of Rigas von Euler-Chelpin, a captain in the Royal Bavarian Regiment, and Gabrielle Furtner and was of the same family lineage as the Swiss mathematician Leonhard Euler. He attended schools in Munich, Würzburg, and Ulm, then from 1891 to 1893 studied art at the Munich Academy of Painting. His concern with the theory of colors caused him to become interested in the spectrum, and he turned his attention to science.

Euler-Chelpin enrolled at the then University of Berlin, where he studied physics under Emil Warburg and Max Planck and organic chemistry under Emil Fischer and A. Rosenheim. During the next two years he worked with W. Nernst in Göttingen. In the summer of 1897 he became an assistant to Svante Arrhenius in Stockholm, where he qualified as *Privatdozent* in physical chemistry at the University of Stockholm in 1898; he spent the summers of 1899 and 1900 with J. H. van't Hoff in Berlin.

Until this time Euler-Chelpin had concentrated on physical chemistry, a subject being developed with much enthusiasm in Germany and Sweden. He now turned toward organic chemistry, visiting the laboratories of Arthur Hantzsch at Würzburg and Leipzig and Johannes Thiele at Strasbourg. He began research in the field at this time, partly in collaboration with his wife, Astrid Cleve. His visits to the laboratories of E. Buchner in Berlin and G. Bertrand in Paris reflected a developing interest in fermentation.

He became professor of general and organic chemistry at the University of Stockholm in 1906. All of his remaining professional work was carried out in Sweden, of which country he became a citizen in 1902. Nevertheless, in World War I he reported for service in the German army, serving in the artillery and, after 1915, in the air force. In the winter of 1916–1917 he was assigned to a military mission in Turkey to stimulate production of munitions and alcohol. He then returned to the air force, where he became commander of a bomber squadron. During this period he had an arrangement with the University of Stockholm that permitted him to compress his teaching activities into a half year. During World War II Euler-Chelpin again made himself available to Germany, but in a diplomatic capacity.

In 1929, the year in which he shared the Nobel Prize in chemistry with Arthur Harden for studies on fermentation, Euler-Chelpin became director of the Vitamin Institute and Institute of Biochemistry founded at the University of Stockholm through the joint support of the Kurt and Alice Wallenburg Foundation and the International Education Board of the Rockefeller Foundation. In 1941 he retired from teaching but continued his research activities almost to the end of his life. He was twice married: to Astrid Cleve (1902–1912), daughter of P. T. Cleve, professor of chemistry at the University of Uppsala, by whom he had five children; and Elisabeth, Baroness Ugglas (1913–1964), by whom he had four children. Both women were associated with him in some of his investigations. His son Ulf Svante von Euler shared the Nobel Prize in medicine or physiology in 1970.

Euler-Chelpin's early interest in inorganic catalysis was soon transferred to biochemical studies and particularly to the enzymes associated with fermentation. His studies on the chemistry of plants led him to concentrate his interest on those fungi that lend themselves to the study of metabolic problems. His studies on vitamins were not really a diversion; most of this work contributed to the understanding of enzyme cofactors. His late work on cancer was also an extension of his work on enzymes.

The work for which Euler-Chelpin received the Nobel Prize in 1929 was closely associated with Buchner's discovery that cell-free yeast juice was still able to ferment sugar, and the observation by Harden and Young that such juice, when passed through an ultrafilter, was separated into two fractions, neither of which alone had the power to ferment sugar but which on mixing again showed normal fermenting activity. Euler-Chelpin studied the low molecular-weight fraction—named cozymase—for more than a decade, starting in 1923. By 1929 he and his associates, particularly K. Myrbäck and R. Nilsson, had clarified the role of cozymase in fermentation.

Harden had shown that phosphoric acid played a role in fermentation by giving rise to certain sugar phosphates. Euler-Chelpin and Nilsson developed the use of inhibitors whereby certain stages in enzyme-catalyzed reactions can be blocked by use of a toxic substance, using fluoride to block that phase of fermentation in which cozymase functions. With Myrbäck, Euler-Chelpin showed that when glucose reacts with phosphoric acid it splits into two three-carbon fragments, one of which remains combined with phosphate. The two other fragments then combine to form glucose diphosphate, while the non-phosphorylated fragment undergoes further degradation. The reaction thereby shows that the sugar molecule undergoing fermentation splits into an energy-rich and an energy-poor fragment.

Euler-Chelpin also investigated the chemical nature of cozymase. Although cozymase is widely distributed in the plant and animal world, Euler-Chelpin and his associates found yeast to be the most practical source for its preparation. Starting with a crude extract having 200 units of activity, they concentrated this into a product having a specific activity of 85,000 units. This product corresponded to a nucleotide, containing sugar, a purine base, and a phosphate; it was clearly related to adenylic acid, which had been isolated by others from muscle. When Warburg showed nicotinamide to be a cofactor in erythrocytes, Euler-Chelpin tested for nicotinamide in cozymase with positive results. Soon thereafter Euler-Chelpin, Fritz Schlenk, and their co-workers showed the chemical structure of cozymase to be that of diphosphopyridine nucleotide (DPN).

In his work on vitamins, Euler-Chelpin assisted in clarifying the role of nicotinamide and thiamine (B$_1$) in metabolically active compounds. Somewhat earlier, in association with the Swiss chemist Paul Karrer, he had helped clarify the vitamin A activity of the carotenoid pigments. His work on tumors dealt particularly with the role of nucleic acids.

BIBLIOGRAPHY

I. ORIGINAL WORKS. There is no collected bibliography of Euler-Chelpin's more than 1,100 research papers, but see the listings in Poggendorff and in the author indexes of *Chemical Abstracts*. His Nobel Prize lecture, "Fermentation of Sugars and Fermentative Enzymes," is available in *Nobel Lectures, Including Presentation Speeches and Laureates' Biographies, Chemistry, 1922–1941* (Amsterdam, 1966), pp. 144–155. His work is dealt with in detail in his books *Grundlagen und Ergebnisse der Pflanzenchemie*, 3 vols. (Brunswick, 1908–1909); *Chemie der Hefe und der alkoholischen Gärung* (Leipzig, 1915); *Chemie der Enzym*, 2 vols. (Munich-Wiesbaden, 1920–1927); *Biokatalysatoren* (Stuttgart, 1930); *Homogene Katalyse* (Stuttgart, 1931); *Biochemie der Tumoren* (Stuttgart, 1942), written with B. Skarzynski; *Reductone, ihre chemischen Eigenschaften und biochemischen Wirkungen* (Stuttgart, 1950); and *Chemotherapie und Prophylaxe des Krebses* (Stuttgart, 1962).

II. SECONDARY LITERATURE. The best biographical sketch is Feodor Lynen's obituary "Hans von Euler-Chelpin," in *Bayerische Akademie der Wissenschaften, Jahrbuch* (1965), pp. 206–212. Also see R. Lepsius, "Hans Karl August von Euler-Chelpin zum Gedächtnis," in *Chemikerzeitung*, **88** (1964), 933–936. Memoirs on his eightieth birthday are B. Eistert, "Hans von Euler-Chelpin zum 80 Geburtnis," *ibid.*, **77** (1953), 65; and W. Franke,

"Zu Hans von Euler's 80 Geburtstag," in *Naturwissenschaften*, **40** (1953), 177–180. Also see the sketch accompanying his lecture in *Nobel Lectures,* pp. 156–158.

AARON J. IHDE

EUSTACHI, BARTOLOMEO (*b.* San Severino, Ancona, Italy, *ca.* 1500–1510; *d.* on the Via Flaminia en route to Fossombrone, Italy, 27 August 1574), *medicine.*

Bartolomeo was the son of Mariano, a physician, and Francesca (Benvenuti) Eustachi. He had a good humanistic education, in the course of which he acquired such an excellent knowledge of Greek, Hebrew, and Arabic that he was able to edit an edition of the Hippocratic glossary of Erotian (1566) and is said to have made his own translations of Avicenna (Ibn Sīnā) from the Arabic. He appears to have studied medicine at the Archiginnasio della Sapienza in Rome, but it is not known precisely when. He began to practice medicine in his native land about 1540. He was thence invited to be physician first to the duke of Urbino, and then, in 1547, to the duke's brother, Cardinal Giulio della Rovere, whom Eustachi followed to Rome in 1549. There he was invited to join the medical faculty of the Sapienza as the equivalent of professor of anatomy, and to this end he was permitted to obtain cadavers for dissection from the hospitals of Santo Spirito and Consolazione. With advancing years Eustachi was so severely afflicted by gout that he was compelled to resign his chair. He continued, however, to serve Cardinal della Rovere, and it was in response to the cardinal's summons to Fossombrone in 1574 that he set forth, only to die on the way.

Eustachi's first works were *Ossium examen* and *De motu capitis,* both written in 1561 and directed against the anti-Galenism of Vesalius, for whom he had developed a unilateral hostility. Otherwise his researches had a more unbiased scientific purpose and displayed his notable ability as an anatomist.

In 1562 and 1563 Eustachi produced a remarkable series of treatises on the kidney, *De renum structura;* the auditory organ, *De auditus organis;* the venous system, *De vena quae azygos graecis dicitur;* and the teeth, *De dentibus.* These were published, together with the two earlier defenses of Galen, in *Opuscula anatomica* (1564), although the *De dentibus* has a separate title page bearing the date 1563. The treatise on the kidney was the first work specifically dedicated to that organ—it displays a detailed knowledge of the kidney superior to that of any earlier work and contains the first account of the suprarenal gland and a correct determination of the relative levels of the kidneys. It was also in this treatise that Eustachi for the first time emphasized the problem of anatomical variation, which had been previously touched upon briefly by Vesalius.

The second treatise on the auditory organ provides a correct account of the tube (*tuba auditiva*) that is still referred to eponymously by Eustachi's name, and contains a description of the tensor tympani and stapedius muscles. Eustachi's claim to discovery of the stapes is inadmissible, however, since it was mentioned orally by Giovanni Filippo Ingrassia in 1546 and in print by Pedro Jimeno (1549), Luis Collado (1555), and Falloppio (1561).

Eustachi, basing his work on the dissection of fetuses and newborn children, was also the first to make a study of the teeth in any considerable detail. He provided an important description of the first and second dentitions and, in some respects preceded by the account of Falloppio, described the hard outer tissue and soft inner structure of the teeth. He further attempted an explanation of the problem, not yet completely solved, of the sensitivity of the tooth's hard structure. In his work on the azygos vein and its ramifications Eustachi described the thoracic duct and indicated a careful and relatively advanced knowledge of the heart's structure.

In 1552 Eustachi, with the help of Pier Matteo Pini, a relative and an artist, prepared a series of forty-seven anatomical illustrations; these were engraved, two on the obverse and reverse of a single copper plate, by Giulio de' Musi of Rome. The illustrations were prepared for a book entitled *De dissensionibus ac controversiis anatomicis* but were never published. The first eight large octavo plates, labeled Tabula Prima–Octava, were used in the *Opuscula anatomica* to portray aspects of the kidneys, the azygos vein and its ramifications, the veins of the arm, the heart, and the Eustachian valve (*valvula venae cavae* in the right auricle) which is illustrated in Tabula Octava. Somewhat curiously the stapes is illustrated on Tabula Septima with the kidney, perhaps a last-minute addition since this ossicle is also portrayed and more correctly located on one of the plates (XXXXI) discussed below.

Since Eustachi mentioned forty-seven plates (that is, forty-seven copperplate engravings) in the *Opuscula anatomica* but actually made use of only eight of them in that work, the remainder seemed to have been lost after his death and were sought for long and unsuccessfully—by Marcello Malpighi, among others. Ultimately the missing thirty-nine engravings (in folio size and differently labeled Tabula IX–XXXXVII) were discovered in the early eighteenth century in the possession of a descendant of Pier Matteo Pini, to whom Eustachi had, as it was learned, bequeathed them. They were purchased by Pope Clement XI for 600 scudi and presented to Giovanni Maria Lancisi, his physician and a successor to Eustachi in the chair of anatomy at the Sapienza. Lancisi published the plates, together with the eight smaller ones that had already appeared in 1564, under the title *Tabulae anatomicae Bartholomaei Eustachi quas a tenebris tandem vindicatas* (1714). Although devoid of Eustachi's planned text, the plates alone assure him a distinguished position in the history of anatomy. They are not the first copper-engraved anatomical illustrations to be produced, as has sometimes been declared, however, but rather the third, following those of Giambattista Canano (1541?) and those of Thomas Geminus, *Compendiosa totius anatomie delineatio aere exarata* (1545). Nevertheless, they are strikingly modern in appearance, clearly produced without decorative accompaniment. Sometimes, as in the instance of the "musclemen," they display both sides of the body in juxtaposition, with a numbered rule on three sides of the figures to which numbered references are made in the text for identification of detail.

Despite such modern effects the plates are, oddly enough, arranged in a way that suggests the pattern of dissection that had been followed from medieval times up to that of Vesalius, that is, beginning with the most corruptible parts and continuing thence to the least corruptible. Thus the Eustachian plates begin with the abdominal structures, then those of the thorax, followed by the nervous system, vascular system, muscles, and finally the bones. Despite the apparent detail and precision of representation within the illustrations, their arrangement suggests some sparsity of dissection material—unlike the relative wealth of it available to Vesalius which permitted him to discard the traditional organization of anatomical treatises.

A possible paucity of cadavers is also suggested by a kind of economy of detail in some of the Eustachian figures of the whole body, such as the "musclemen," except in those areas meant expressly for representation of a specific structure. Lack of information on Eustachi's activities prevents more than such surmise of limited dissection material. Whatever the case, examination of the individual plates reveals him to have had remarkable powers of observation. As an example, Tabula XVIII, displaying the base of the brain and in particular the sympathetic nervous system, surpasses in accuracy any similar delineation produced during the sixteenth century. In fact, the illustration of the sympathetic system is generally considered to be one of the best ever produced. The other illustrations of the nervous system are, however, of lesser quality, perhaps inferior to those of Vesalius. Similarly Tabula XXVI, illustrating the vascular system and the relationships of vessels to muscles, is also of notably superior quality, and this may likewise be said of Tabula XXXXII, which represents the dissection of the laryngeal structures. Had the Eustachian anatomical illustrations not been lost to the medical world for over a century, it seems likely that anatomical studies would have reached maturity in the seventeenth rather than the eighteenth century.

BIBLIOGRAPHY

The *Opuscula anatomica* (Venice, 1564) is an exceedingly rare book; it was reprinted in Leiden, 1707, and Delft, 1726. The *Tabulae anatomicae* (Rome, 1714) was republished in Amsterdam, 1722, but with copies of the original plates; in Rome, 1728, with the original plates again used; in Leiden, 1744, with newly engraved copies of the plates accompanied by separate outline plates of equal size on which explanatory letters were engraved. This edition, edited by B. S. Albinus, is the most desirable one for purposes of study. Further editions of the *Tabulae* were published in Venice, 1769; Amsterdam, 1798, in German translation; and Amsterdam, 1800. Finally, there is a commentary as well as an edition of the plates by Gaetano Petrioli, to whom Lancisi bequeathed them, *Riflessioni anatomiche sulle note di Lancisi fatte sopra le tavole del cel. B. Eustachio* (Rome, 1740). It is chiefly of significance for the attached biography of Eustachi by Barnardo Gentili.

There is a biographical study of Eustachi by G. Bilancioni, *Bartolomeo Eustachi* (Florence, 1913), and a collection of documents, *Memorie e documenti riguardanti Bartolomeo Eustachio pubblicati nel quarto centenario dalla nascita* (Fabriano, 1913). The plates as anatomical illustrations are discussed by Ludwig Choulant, Mortimer Frank, trans., *History and Bibliography of Anatomic Illustration* (Chicago, 1920), pp. 200–204, and by Robert Herrlinger, *Geschichte der medizinischen Abbildung* (Munich, 1967), pp. 133–137.

C. D. O'MALLEY

EUTOCIUS OF ASCALON

EUTOCIUS OF ASCALON (*b.* Palestine, *ca.* A.D. 480), *mathematics.*

Eutocius was the author of commentaries on three works by Archimedes. He also edited and commented on the first four books of the *Conics* of Apollonius.

His commentary on the first book of Archimedes' *On the Sphere and Cylinder* was dedicated to Ammonius, who was a pupil of Proclus and the teacher of Simplicius and many other sixth-century philosophers, and who could not have lived long after 510. Eutocius' four commentaries on the *Conics* are dedicated to Anthemius of Tralles, the architect of Hagia Sophia in Constantinople, who died about 534. For these reasons the central point of Eutocius' activities may be put about 510, and it has become conventional to date his birth about 480.

The old belief that Eutocius flourished about fifty years later arose from a note at the end of three of his Archimedean commentaries—on *On the Sphere and Cylinder*, Books I and II, and on the *Measurement of a Circle*—to the effect that each of them was "edited by Isidorus, the mechanical engineer, our teacher." These words, bracketed by Heiberg, cannot refer to Eutocius because they are not compatible with his relationship to Ammonius, for Isidorus of Miletus continued the construction of Hagia Sophia after the death of Anthemius about 534 and could not have been Eutocius' teacher. The words are best understood as an interpolation by a pupil of Isidorus and contain the interesting information that Isidorus revised the commentaries in question. Similarly, a reference in the commentary on *On the Sphere and Cylinder*, Book II (Archimedes, Heiberg ed., III, 84.8–11) to an instrument for drawing parabolas invented by "Isidorus, the mechanical engineer, our teacher" is also best understood as an interpolation. Tannery mentions the possibility that the Isidorus in question may have been a nephew of the successor of Anthemius, who supervised the reconstruction of Hagia Sophia after an earthquake in 557. Ascalon (now Ashkelon), where Eutocius was born, lay between Azotus (now Ashdod) and Gaza on the coast of Palestine; it is the city made famous in the lament of David over Saul and Jonathan: "Tell it not in Gath, publish it not in the streets of Askelon" (II Samuel 1:20). The *Suda Lexicon* relates an unedifying story of a Thracian mercenary named Eutocius who made a lot of money and tried to buy himself into society, first at Eleutheropolis (now Beyt Guvrin, Israel), then at Ascalon, but few have followed Tannery in seeing an ancestor; it seems more probable that Ascalon has been introduced into this story by reason of the mathematician's name and fame.

In his preface to his interpretation of the *Measurement of a Circle* Eutocius refers to his earlier commentaries on *On the Sphere and Cylinder*, and in the commentary on Book I he asks Ammonius to bear with him if he should have erred through youth (Archimedes, Heiberg ed., III, 2.13). He explains that he has found no satisfactory commentaries on Archimedes before his own time and promises further elucidation of the master if his work should meet with the approval of Ammonius. Apart from the *Measurement of a Circle*, he later wrote commentaries on both books of Archimedes' treatise *On Plane Equilibria*. The commentary on Book I was dedicated to an otherwise unknown Peter, whose name reveals him to have been a Christian. It is a fair inference that Eutocius did not know the works of Archimedes entitled *Quadrature of a Parabola* and *On Spirals,* for if he had, he would have referred to them at certain points of his commentary (Archimedes, Heiberg ed., III, 228.25; 278.10; 280.4; 286.13) instead of making less suitable references. Presumably the commentaries on Apollonius' *Conics* were written later than those on Archimedes' works, but there is no direct evidence. All these commentaries have survived. It has been debated whether he also wrote a commentary on Ptolemy's *Syntaxis,* but there is no suggestion of one in a passage of his commentary on the *Measurement of a Circle* (Archimedes, Heiberg ed., III, 232.15–17), where he mentions "Pappus and Theon and many others" as having interpreted that work.

Eutocius is not known to have done any original mathematical work, and his elucidations of Archimedes and Apollonius do not add anything of mathematical significance. Nevertheless, the examples of long multiplication in his commentary on the *Measurement of a Circle* are the best available evidence of the way in which the Greeks handled such operations, and he preserves solutions of mathematical problems by the earlier Greek geometers that are sometimes the sole evidence for their existence and are therefore of major importance for the historian of mathematics.

It is through Eutocius that we have a valuable collection of solutions by Greek geometers of the problem of finding two mean proportionals to two given straight lines, that is, if a and b are two given straight lines, to find two other straight lines x and y such that $a:x = x:y = y:b$. It was to this that a problem which attracted the best Greek mathematicians for several centuries—how to find a cube double another cube—had been reduced by Hippocrates, for if $a:x = x:y = y:b$, then $a^3:x^3 = a:b$, and if $b = 2a$, then x is the side of a cube double a cube of side a. From that time the problem appears to have been attacked exclusively in this form.

The first proposition of Archimedes' *On the Sphere and Cylinder,* Book II, is "Given a cone or cylinder, to find a sphere equal to the cone or cylinder." He shows as analysis that this can be reduced to the problem of finding two mean proportionals and then, in the synthesis, says: "Between the two straight lines, let two mean proportionals be found." It is at this point that Eutocius begins an extended comment (Archimedes, Heiberg ed., III, 54.26–106.24). After noting that the method of finding two mean proportionals is in no way explained by Archimedes, he observes that he had found the subject treated by many famous men, of whom he omits Eudoxus be-

cause in his preface he said he had solved the problem by curved lines but had not used them in the proof and had, moreover, treated a certain discrete proportion as though it were continuous, which a mathematician of his caliber would not have done. In order that the thinking of those men whose solutions have been handed down might be manifest, Eutocius sets out the manner of each discovery. He gives a solution attributed to Plato (but almost certainly wrongly attributed), followed by solutions given by Hero in his *Mechanics* and *Belopoeïca,* by Philo of Byzantium, by Apollonius, by Diocles in his work *On Burning Mirrors,* by Pappus in his *Introduction to Mechanics,* by Sporus of Nicaea, by Menaechmus (two solutions), by Archytas as related by Eudemus, by Eratosthenes, and by Nicomedes in his book *On Conchoidal Lines.* (This is not a chronological order; chronologically the order would probably be Archytas, Eudoxus, Menaechmus, the pseudo-Plato, Eratosthenes, Nicomedes, Apollonius, Philo, Diocles, Hero, Sporus, and Pappus. There is, indeed, no discernible order in Eutocius' list.)

Hero's solution is given in his *Mechanics* I, 11, which has survived only in an Arabic translation, and in his *Belopoeïca,* and is reproduced by Pappus, *Collection* III, 25–26. Pappus' solution is given in *Collection* III, 27 and VIII, 26; it is the latter passage that Eutocius has in mind. The conchoid is described by Pappus, *Collection* IV, 39–40, and he mentions that it was used by Nicomedes for finding two mean proportionals but does not give a proof. The other solutions would not be known but for their preservation by Eutocius. It is a pity that he did not include what purported to be Eudoxus' solution despite the obvious errors in transmission, but for what he has preserved he deserves the gratitude of posterity. The solution ascribed to Eratosthenes is prefaced by a letter, allegedly from Eratosthenes to Ptolemy Euergetes, giving the history of the problem of doubling the cube and its reduction to the problem of finding two mean proportionals; the letter is not authentic, but it closes with a genuine condensed proof and an epigram that Eratosthenes put on a votive monument. The solution attributed to Plato is probably not authentic because, among other reasons, it is mechanical, but the solutions of Eudoxus and Menaechmus show that the problem was studied in the Academy and may be Platonic. According to Eutocius, Nicomedes was exceedingly vain about his solution and derided that of Eratosthenes as impractical and lacking in geometrical sense. The solutions of Diocles, Sporus, and Pappus are substantially identical and so are those of Apollonius, Hero, and Philo.

It is only a little later, in commenting on the fourth

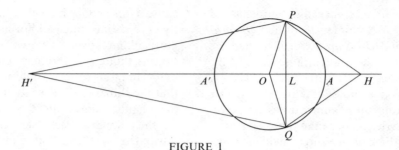

FIGURE 1

proposition of *On the Sphere and Cylinder,* Book II, that Eutocius gives a further precious collection of solutions that would not otherwise be known. Proposition 4 is the problem "To cut a given sphere by a plane so that the volumes of the segments are to one another in a given ratio." In Proposition 2 Archimedes had shown that a segment of a sphere is equal to a cone with the same base as the segment and height $h(3r - h)/(2r - h)$, where r is the radius of the sphere and h is the height of the segment (*LA* in the figure). In Proposition 4 he proves geometrically that if h, h' are the heights of the two segments, so that $h + h' = 2r$, and they stand in the ratio $m:n$, then

$$h\frac{(3r - h)}{(2r - h)} = \frac{m}{n} h'\frac{(3r - h')}{(2r - h')}.$$

By the elimination of h' this becomes the cubic equation

$$h^3 - 3h^2r + \frac{4m}{m + n} r^3 = 0.$$

The problem is thus reduced (in modern notation) to finding the solution of a cubic equation that can be written

$$h^2 (3r - h) = \frac{4m}{m + n} r^3.$$

Archimedes preferred to treat this as a particular case of a general equation

$$x^2(a - x) = bc^2,$$

where b is a given length and c^2 a given area. For a real solution it is necessary that

$$bc^2 > \frac{4}{27} a^3.$$

In the particular case of II, 4, there are always two real solutions.

  Before proceeding to the synthesis of the main problem, Archimedes promised to give the analysis and synthesis of this subsidiary problem at the end, but Eutocius could not find this promise kept in any

of the texts of Archimedes. He records that after an extensive search he found in an old book a discussion of some theorems that seemed relevant. They were far from clear because of errors and the figures were faulty, but they seemed to give the substance of what he wanted. The language, moreover, was in the Doric dialect and kept the names for the conic sections that had been used by Archimedes. Eutocius was therefore led to the conclusion, as we also must be, that what he had before him was in substance the missing text of Archimedes, and he proceeded to set it out in the language of his own day. The problem is solved, in modern notation, by the intersection of the parabola and rectangular hyperbola

$$x^2 = \frac{c^2}{a} y, (a - x)y = ab.$$

  Others before Eutocius had noticed the apparent failure of Archimedes to carry out his promise, and Eutocius also reproduced solutions by Dionysodorus and Diocles. Dionysodorus solved the particular case of the cubic equation to which II, 4 reduces, that is,

$$(3r - x):\frac{m}{m + n}r = 4r^2:x^2.$$

His solution is the intersection of the parabola and rectangular hyperbola

$$y^2 = \frac{m}{m + n}r (3r - x) \text{ and}$$

$$xy = \frac{m}{m + n}2r^2.$$

Diocles solved not the subsidiary equation but the original problem, II, 4, by means of the intersection of an ellipse

$$(y + a - x)^2 = \frac{n}{m}\{ (a + b)^2 - x^2\}$$

and the rectangular hyperbola

$$(x + a)(y + b) = 2ab.$$

  It is clear that the *Measurement of a Circle* was

490

already reduced to three propositions, with the second and third in the wrong order, when Eutocius had it before him. The chief value of his commentary is that he works out in detail the arithmetical steps where Archimedes merely gives the results. Archimedes requires a number of square roots. Eutocius excuses himself from working them out, on the ground that the method is explained by Hero and by Pappus, Theon, and other commentators on Ptolemy's *Syntaxis,* but he multiplies the square root by itself to show how close the approximation is. At the end Eutocius reveals that Apollonius in a work called Ὠκυτόκιον (*Formula for Quick Delivery*) found a closer approximation to the ratio of a circumference of a circle to its diameter than did Archimedes; and he exculpates Archimedes from the censure of Sporus of Nicaea, whose own teacher, Philo of Gadara, also found a more exact value, on the ground that Archimedes was looking for a figure useful in daily life.

Apart from what has been noted above, Eutocius' comments on Archimedes do not add much of value to the text, and occasionally he errs, as in saying that two parabolic segments in Proposition 8 of *On Plane Equilibria* are similar (Archimedes, Heiberg ed., III, 290.23–24). In a commentary on the difficult lemma that is Proposition 9 of the same book and leads to the location of the center of gravity of a portion of a parabola cut off by parallel chords, he admits himself forced to paraphrase.

The commentaries on the *Conics* display more mathematical acumen. In his preface to Book I, Apollonius explains how uncorrected copies came to be in circulation before he had completed his revision. It is therefore probable that there were variant readings and alternative proofs in the manuscripts from earliest days. It is clear that when Eutocius came to comment on Apollonius, he had before him differing versions, and he found it necessary to prepare a recension for his own purposes; in two manuscripts the four books of his comment have the heading "A Commentary of Eutocius of Ascalon on the First (Second, Third, Fourth) of the *Conics* of Apollonius as Edited by Himself." Eutocius' edition suffered at the hands of interpolators, probably in the ninth century, when mathematics had a renaissance at Constantinople under Leo the Mathematician. The best manuscript of the commentary (W, Cod. Vat. gr. 204) was copied in the tenth century, and at a number of points Eutocius' citations from Apollonius are clearly nearer to the original than is the text of the *Conics* as we have it today. In commenting on Apollonius, Eutocius had been preceded by Serenus and Hypatia. The most interesting features of the commentary are in the early pages, where Eutocius emphasizes the generality of Apollonius' method of producing conic sections from any cone.

All the books by Archimedes on which Eutocius commented have survived, and his elucidations may have contributed to their survival. There must also be some significance in the fact that the four books of the *Conics* on which he commented have survived in Greek, whereas Books V–VII have survived only in Arabic and Book VIII is entirely lost. His commentaries on Archimedes were translated into Latin along with the parent works by William of Moerbeke in 1269. The commentaries have usually been printed with the editions of Archimedes and Apollonius and have never been printed separately. The definitive text is to be found in Heiberg's editions of Archimedes and Apollonius with a Latin translation and valuable prolegomena and notes.

BIBLIOGRAPHY

I. ORIGINAL WORKS. Eutocius' commentaries can be found in *Commentarii in libros Archimedis De sphaera et cylindro, in Dimensionem circuli et in libros De planorum aequilibris,* in *Archimedis opera omnia,* J. L. Heiberg, ed., 2nd ed., III (Leipzig, 1915), 1–448; and *Commentaria in Conica,* in *Apollonii Pergaei quae graece exstant,* J. L. Heiberg, ed., II (Leipzig, 1893), 168–361.

II. SECONDARY LITERATURE. See Paul Tannery, "Sur l'histoire des lignes et surfaces courbes dans l'antiquité," in *Bulletin des sciences mathématiques,* 2nd ser., **7** (1883), 278–291; and "Eutocius et ses contemporains," *ibid.,* **8** (1884), 315–329, repr. in *Mémoires scientifiques,* II (Toulouse–Paris, 1912), 1–47, 118–136; and Sir Thomas Heath, *A History of Greek Mathematics* (Oxford, 1921), I, 52, 57–58; II, 25, 45, 126, 518, 540–541.

IVOR BULMER-THOMAS

EVANS, FREDERICK JOHN OWEN (*b.* London [?], England, 9 March 1815; *d.* London, 20 December 1885), *hydrography, geomagnetism.*

Evans came of a naval family, his father, John Evans, being master, R.N. He volunteered for the navy himself at the age of thirteen. Having served on H.M.S. *Rose* and *Winchester* on the American station for five years, he was transferred to the survey ship *Thunder* under Captain Richard Owen. Here he began a long lifetime of work devoted to exact surveying at sea and to geomagnetism.

After three years around the coasts of Central America and the Bahamas, Evans served in a succession of ships in the Mediterranean; in the master's line, he had responsibility for navigation. In 1841 he became master and senior surveying officer of H.M.S. *Fly,* assigned to exploration in the Coral Sea, around the Great Barrier Reef of Australia, and in the Torres Strait; his hydrographic work revealed a safe and easy passage through the strait and was an important contribution to the development of New South Wales. On 12 November 1846 he married Elizabeth Mary Hall of Plymouth, daughter of a naval captain. In 1847 Evans joined the *Acheron* and returned to the antipodes, where for four years he did hydrographic work on the coasts of New Zealand.

After distinguished service in the Baltic during the Crimean War, in 1855 Evans was appointed superintendent of the Compass Department of the Navy; he was promoted to staff commander in 1863, staff captain in 1867, and full captain in 1872. He became chief naval assistant to the hydrographer to the navy in 1865 and himself occupied that important post from 1874 to 1884. He was appointed a companion of the Order of the Bath and knight commander in 1881.

Evans' recognition as an outstanding scientist comes from his solution of the problems associated with compass navigation in iron and armor-plated ships and from his observations leading to the publication of a chart of curves of equal magnetic declination for the navigable world. In his work on compass errors, he had the collaboration of the eminent mathematician Archibald Smith; Evans as experimenter and Smith as theoretician made a formidable team. Together they solved a problem of great importance to the British Navy and to navigation in general at the time when iron ships were coming into wide use.

Some of Evans' experiments were carried out on board the pioneer Atlantic steamship *Great Eastern*. The results led to proposals for the proper placing of the needles in the compass in relation to the soft iron magnets and in relation to the ship itself. Both induction effects and the magnetic field created by the metal were fully considered. An important indirect contribution to oceanography was the compilation of the magnetic instructions made by Evans and Smith for the great voyage of the *Challenger* in 1872–1876.

Evans' contribution to science was recognized by his election as fellow of the Royal Society in 1862. He was also a fellow of the Royal Astronomical Society and a fellow and council member of the Royal Geographical Society. In 1884 he represented Britain at the Congress of Washington for the establishment of a prime meridian.

BIBLIOGRAPHY

I. ORIGINAL WORKS. Evans' works include "Reduction and Discussion of the Deviations of the Compass Observed on Board of All the Iron-built Ships, and a Selection of the Wood-built Steamships in Her Majesty's Navy," in *Philosophical Transactions of the Royal Society,* **150** (1860), 337–378; "On the Effect Produced on the Deviations of the Compass by the Length and Arrangement of the Compass-needles; and a New Mode of Correcting the Quadrantal Deviation," *ibid.,* **151** (1861), 161–182, written with A. Smith; "On the Magnetic Character of the Armour-plated Ships of the Royal Navy, and on the Effect on the Compass of Particular Arrangements of Iron in a Ship," *ibid.,* **155** (1865), 263–324, written with A. Smith; "On the Amount and Changes of the Polar Magnetism in Her Majesty's Iron-built and Armour-plated Ship 'Northumberland,'" *ibid.,* **158** (1868), 487–504; *Admiralty Manual for Deviation of the Compass* (London, 1862; 2nd ed., 1863; 3rd ed., 1869); *Elementary Manual for Deviations of the Compass* (London, 1870); and "On the Present Amount of Westerly Magnetic Declination on the Coast of Great Britain, and Its Annual Changes," in *Philosophical Transactions of the Royal Society,* **162** (1872), 319–330.

II. SECONDARY LITERATURE. On Evans and his work see "Sir F. J. O. Evans," in *Nature,* **33** (1886), 246–248; "Captain Sir Frederick J. O. Evans," in *Proceedings of the Royal Geographical Society,* **8** (1886), 112–113; J. B. Jukes, *Narrative of the Surveying Voyage of H.M.S. "Fly." Commanded by Capt. F. P. Backwood, in Torres Strait, New Guinea etc. During the Years 1842–1846,* 2 vols. (London, 1847).

K. C. DUNHAM

EVANS, LEWIS (*b.* Llangwnadl, Carnarvonshire, Wales, 1700; *d.* New York, N.Y., 11 June 1756), *cartography, geography, geomorphology, geology.*

Evans came to Philadelphia sometime before 1736 and became known as surveyor, draftsman, and mapmaker. He also gave lectures on electricity and wrote on climatology. He was a friend and associate of Benjamin Franklin, John Bartram, Governor Thomas Pownall, and Cadwallader Colden, and was very helpful to the visiting Swedish scientist Peter Kalm, who referred to him as "an ingenious engineer."

Based on his own surveys and explorations he made maps of land tracts and boundaries. His two great published maps are "A Map of Pennsylvania, New-Jersey, New-York, And the Three Delaware Counties" (1749; a revision was published in 1752) and "A General Map of the Middle British Colonies in America" (1755). In the booklet that accompanies the latter map Evans not only describes the geography, geomorphology, and some geology and other natural features of the region, but also makes a vigorous attack on the contemporary permissive policy of the British administrators toward the French encroachment in the Ohio Valley, which he vehemently insists must be preserved for English settlement. After a second fiery pamphlet, which even hinted at treasonous collusion with France, Governor

Robert Hunter Morris of Pennsylvania secured Evans' imprisonment in New York, to which he had moved. He was released from jail only three days before he died, leaving his motherless eleven-year-old daughter to the care of friends.

In 1743 Evans traveled with John Bartram and Conrad Weiser to Lake Ontario. In his journal (published in 1776 by Pownall) he recorded observations on raised beaches of the once higher Great Lakes and speculated penetratingly on the drainage of the earlier lakes and the consequent rise of the land because "This part of America was disburthened of such a Load of Waters."

Evans filled in the blank spaces of his maps of 1749 and 1752 with notes on weather, roads, streams, and geology. His notes on the Endless (Appalachian) Mountains were not only descriptive but were also the first analysis of their origin, based on the fossils preserved in their strata and on the erosion of valleys from a former plain (peneplain in modern terms) to form the ridges.

Evans' 1755 "Map of the Middle British Colonies" is partly a geologic map, showing not only the location of economic minerals—"coals," "freestone," pottery clay, and petroleum—but also the trends of the mountains and some indication of rock types. He was keenly aware of the three-dimensional nature of geology and constructed map profiles and sections of strata with their "particular fossils" to accompany the map. Evans said, however, that for "want of room in the plate," these sections would be published on later maps when he had more space. Evans died before these later maps could be published.

The thirty-six-page *Analysis of a General Map of the Middle British Colonies . . . and a Description of the Face of the Country . . .* that accompanied the 1755 map contains (in addition to the discussion of the administration of the Ohio Valley) a long and clear statement of the geomorphic and geologic provinces of the eastern United States. In this prototypical work of American geomorphology, divisions now known as the New England Upland, the Coastal Plain, the Fall Line, the Piedmont, the Blue Ridge, the Folded Appalachians, the Allegheny Front, and the Allegheny Plateau are first delineated. Evans realized that these regions were different because of differing rocks and structures.

Evans' great map was reprinted and copied at least twenty-seven times in the next fifty years. His regional geologic and physiographic classification provided in greater or lesser part the geologic framework for later writers in the sixty years following his death. He was a great cartographer and an early student not only of landscape but also of fossils and the relation of bedrock to surface morphology. He was the first in America to recognize the principles of isostasy.

BIBLIOGRAPHY

I. ORIGINAL WORKS. Evans' maps and publications, now very rare, are reproduced in facsimile, with his extensive unpublished notes, in Laurence H. Gipson, *Lewis Evans* (Philadelphia, 1939). Evans' 1743 journal is published in Thomas Pownall, *A Topographical Description of Such Parts of North America . . .* (London, 1776; Pittsburgh, 1949).

II. SECONDARY LITERATURE. For Evans' life and the history of his maps see Gipson, above; H. N. Stevens, *Lewis Evans, His Map of the Middle British Colonies* (London, 1905, 1924, 1929); and L. C. Wroth, *An American Bookshelf* (Philadelphia, 1934).

For discussion of Evans' geological observations see G. W. White, "Lewis Evans' Early American Notice of Isostasy," in *Science,* **114** (1951), 302–303; "Lewis Evans' Contributions to Early American Geology—1743–1755," in *Transactions of the Illinois Academy of Science,* **44** (1951), 152–158; and "Lewis Evans (1700–56): A Scientist in Colonial America," in *Nature,* **177** (1956), 1055–1056.

GEORGE W. WHITE

EVANS, WILLIAM HARRY (*b.* Shillong, Assam, India, 22 July 1876; *d.* Church Whitfield, Dover, England, 13 November 1956), *entomology.*

Evans' parents were both of military families; his father was General Sir Horace M. Evans (related by marriage to Charles Dickens) and his mother—"the best woman I have known and the greatest influence in my life"—was a keen naturalist, Elizabeth Annie, daughter of Surgeon General T. Tresidder. Conventionally educated at King's School, Canterbury, and the Royal Military Academy, Woolwich, Evans was commissioned an officer in the Royal Engineers at the age of twenty and retired at fifty-five. Apart from the Somaliland Campaign (1903–1904) and World War I (1914–1918), in each of which he was both wounded and decorated, his entire army career was spent in India. He rose to be a distinguished staff officer and was coauthor of several textbooks on administration and military engineering. It is as a naturalist, however, that he will be remembered.

His first tour of army duty (1900–1901) was in Chitral, on the North-West Frontier of India. His hobby there, with a fellow sapper, Major (later Major General) G. A. Leslie, was collecting butterflies. A joint paper, Evans' first publication, listed 139 species, of which, despite the attentions of Lionel de Nicéville, the foremost authority in India, ten could be named only doubtfully and nineteen could not be named at all. Although the relevant literature was voluminous and described a great number of individual genera and species, it contained little analytical or comparative data and did not help in naming fresh material. The concept of linking isolated species as geographically separated races or subspecies of one extensive species had not then been adopted.

Evans resolved to document the butterflies of the Indian region in a way that he himself, with no biological training, could understand. He devoted his spare time to study and collecting; his vacations to touring and visiting museums. He searched the literature for all original descriptions and examined practically every type specimen. From 1910 on he published subregional lists for areas not previously covered, and then in 1923 he began to publish a series of papers on classification in the *Journal of the Bombay Natural History Society.* They were such a success that the Society published them as a volume, *Identification of Indian Butterflies* (Madras, 1927; 2nd ed., rev., 1932).

Here were gathered 695 generic and nearly 4,400 specific names under 320 valid genera and 1,442 species, all readily identifiable through remarkably concise and practical keys. The various subspecies and synonyms were noted, and the whole was well indexed. An informative introduction dealt comprehensively with structure, classification, collecting, and study. This handy volume, in which not a word was wasted, remains the only work dealing fully with the subject of identification, and many popular treatments have stemmed from it. Thus, by the time he retired, Evans had achieved his ambition and published a standard work of the greatest value to oriental entomologists. But now he planned a greater one.

The Hesperiidae (skippers), one of the largest cosmopolitan families of butterflies, had long been neglected owing to their smallness and drab appearance. Evans undertook to reclassify them. He settled near the British Museum (Natural History) in London and worked there as regularly as the staff. In 1937 the Museum published his *Catalogue of the African Hesperiidae . . . in the British Museum,* which, in little more than 200 pages, classified naturally through concise keys 421 species, illustrated 116 species for the first time, and for every species gave diagrams of the male genitalia—features of essential value in distinguishing among this family.

During World War II Evans continued his work in the museum; his deafness was aggravated when a bomb detonated on the road outside the room where he was working on a drawer of specimens, the blast shattering the window and clearing his table, leaving him holding a bare pin. Undeterred, he pub-

lished further catalogues of the Hesperiidae, of which one (1949) covered those of Europe, Asia, and Australia, and a final four volumes (1951–1955) were devoted to those of the Americas. Thus, a year before he died, Evans had established a complete classification of the Hesperiidae of the world, in which he had marshaled 747 published generic names (over a hundred his own) in 525 recognized genera with 3,000 species and nearly 2,000 subspecies, placing a further 3,300 names as synonyms. Not only had he provided comparative keys for the essential features of each natural subfamily, group, and genus, down to subspecies, but he had also given diagrams of the male genitalia of every recognized species and illustrated the majority of the least-known ones in color, for the first time.

BIBLIOGRAPHY

Besides the main works discussed, Evans' many shorter papers are well listed in *The Lepidopterist's News,* **10** (1957), 197–199. Two additions to that list are "Revisional Notes on African Hesperiidae," in *Annals and Magazine of Natural History,* ser. 12, **8,** pt. 4 (1956), 881–885; and "A Revision of the *Arhopala* group of Oriental Lycaenidae," in *Bulletin of the British Museum* (*Natural History*), **5** (1957), 85–141.

C. F. Cowan

EVELYN, JOHN (*b.* Wotton, Surrey, England, 31 October 1620; *d.* London, England, 27 February 1706), *arboriculture, horticulture.*

Evelyn was the grandson of George Evelyn, principal manufacturer of gunpowder under Queen Elizabeth, and the second son of Richard Evelyn, high sheriff of Surrey and Sussex in 1633–1634. He was at Balliol College, Oxford, from 1637 to 1640. On account of the political situation in England he left the country in November 1643 and traveled through France and Italy for the next three years. From June 1645 to April 1646 he was mostly in Padua, studying anatomy and physiology. He brought back anatomical tables which he presented to the Royal Society; these are now in the Royal College of Surgeons.

In July 1646 Evelyn returned to Paris, where he attended courses in chemistry by Nicasius Le Fèvre. In 1649 he went through another course in chemistry at Sayes Court in England. In June 1647 he married Mary, the daughter of Sir Richard Browne, Charles I's diplomatic agent in France. The marriage was a happy one. Of their five sons and three daughters, only one daughter survived her father.

Evelyn spent the last year of the Civil War in England, and his first work, *Of Liberty and Servitude,* a translation of the French treatise against tyranny of F. de la Mothe le Vayer, appeared in January 1649. It was during his last stay in Paris, from 1649 to 1652, that Nanteuil engraved his portrait (1650). Before leaving, he wrote a short treatise on *The State of France, As It Stood in the IXth Yeer of This Present Monarch, Lewis XIIII* (London, 1652). In February 1652 he finally returned to England and settled at Sayes Court, his father-in-law's estate at Deptford in Kent. This was to be his home for the next forty years. After the death of his brother George in 1699, he succeeded to the family estate of Wotton, where he took up residence in 1700.

During his travels Evelyn visited hospitals and was interested in their organization. He showed he had a notion of the importance of isolation during the plague by suggesting the construction of an infirmary. His *Diary* contains a description of touching for the king's evil in 1660 and notices of treatments, medicinal springs, and surgical operations (particularly an amputation of the leg and cutting for the stone). He was present at several dissections and in 1683 attended Walter Charleton's lecture on the heart. He was concerned with hygiene and in *Fumifugium* (1661), a work on the pollution of the air in London, he proposed removing certain trades and planting a green belt of fragrant trees and shrubs around the city. He also possessed some knowledge of zoology.

Horticulture was an enduring interest throughout Evelyn's life and at the beginning of 1653 he started laying out the gardens at Sayes Court, which were to become famous. He began making notes for a vast projected work on horticulture, *Elysium britannicum.* The work, to which Sir Thomas Browne contributed, was never completed and only a synopsis was printed in 1659. But Evelyn continued adding to his notes throughout his life. He also offered valuable practical information to gardeners by publishing translations of important French works, particularly *The Compleat Gard'ner* from Jean de La Quintinie (1693).

Evelyn's principal work, *Sylva,* was the outcome of his association with the Royal Society. Following inquiries made in September 1662 by the commissioners of the navy to the Royal Society concerning timber trees, he drew up a report which he enlarged and presented to the Royal Society on 16 February 1664. *Sylva* was the first book published by order of the Society. It was an immediate success, and more than a thousand copies were sold in less than two years. Evelyn received special thanks from the king and the work appears to have had considerable influence on the propagation of timber trees throughout the kingdom. *Sylva* is not a scientific work but the exhortation of a lover of trees to his countrymen to

repair the damage caused by the Civil War. It contains practical information interspersed haphazardly with classical references. To *Sylva* was annexed *Pomona,* a discourse on the cultivation of fruit trees for the production of cider, and *Kalendarium hortense,* a gardener's almanac, being a chapter of the unfinished *Elysium britannicum. A Philosophical Discourse of Earth* appeared in 1676; it was added to *Sylva* in 1679, as *Terra.* His *Acetaria. A Discourse of Sallets,* also part of *Elysium britannicum,* was published separately in 1699, then added to the 1706 edition of *Sylva. Sylva* was advertised in 1670 and 1671 in the autumn catalogue of books at the Frankfurt fair. Alexander Hunter popularized *Sylva* with an extensively annotated edition, collated from the five original editions, in 1776.

To familiarize his countrymen with the philosophy of Epicurus, Evelyn published his translation of the first book of Lucretius' *De rerum natura* in 1656, followed by a commentary on the works of Gassendi and atomism. Evelyn had taken no part in the affairs of state during the Interregnum, but at the end of 1659 he published an anonymous pamphlet, *An Apologie for the Royal Party,* to induce Colonel Morley, later lieutenant of the Tower, to declare for the king. This proved unsuccessful but may have eased the way for the return of Charles II, to whom Evelyn presented a *Panegyric* on his coronation. In this he suggested that Charles should become the founder of a body for the furthering of experimental knowledge. In 1654, at Oxford, Evelyn had met John Wilkins, the leader of an active group of men interested in science; he thus met Christopher Wren, with whom he collaborated several times during his lifetime. In 1659 he sent Robert Boyle a suggestion for the foundation of a "Mathematical College," or community for scientific study. In December 1660 Evelyn was proposed a member of the society for "the promoting of experimental philosophy," then meeting at Gresham College.

Evelyn was instrumental in obtaining royal patronage and the name of "Royal Society" for the group in 1662. He attended the meetings regularly, served on the council frequently, and was offered the presidency. In January 1661 he drew up a "History of Arts Illiberal and Mechanick" (Royal Society Archives). He was appointed a member of several committees of inquiry, including that for agriculture, and contributed papers on various subjects. In 1665 he sat on the committee for the improvement of the English language.

The fourteen years following the king's return were those of Evelyn's greatest public activity, although the offices he held were only temporary appointments.

He served on several commissions from 1660 to 1674—for the improvement of London streets in 1662, for the Royal Mint in 1663, and for the repair of St. Paul's Cathedral in 1666, during which he worked with Christopher Wren. On 13 September 1666 Evelyn presented his plan for the rebuilding of the city, together with a discourse on the problems involved. But the entire replanning soon appeared impracticable. During the two Dutch wars (1664–1667 and 1672–1674) he was commissioner for the sick and wounded mariners and prisoners of war, his most responsible appointment. From 1671 to 1674 he was a member of the Council for Foreign Plantations, later the Council of Trade and Foreign Plantations. In 1674 *Navigation and Commerce* appeared, being the introduction to a history of the Dutch war that Charles had asked Evelyn to write, which was never finished.

In January 1667 Evelyn obtained for the library of the Royal Society the famous collection of books and manuscripts of the earl of Arundel. The collection of stones bearing Greek and Latin inscriptions was also secured through his good offices for the university of Oxford. *Sculptura: or the History, and Art of Chalcography and Engraving in Copper* (1662) was the outcome of a paper read before the Royal Society. His artistic interests also led him to translate two books from the French of Roland Fréart de Chambray, *A Parallel of the Antient Architecture with the Modern* (1664) and *An Idea of the Perfection of Painting* (1668). To the *Parallel* Evelyn added an *Account of Architects and Architecture,* which he dedicated in the second edition to Christopher Wren. The book appears to have been an indispensable work for later architects. His *Numismata. A Discourse of Medals, Antient and Modern* (1697) closed with a discussion of character as derived from effigies.

Evelyn's translations also include *Instructions Concerning Erecting of a Library* from the French of Gabriel Naudé and Jansenist writings against the Jesuits, for in spite of his tolerance he was hostile to Catholicism. He was a staunch and devout Anglican and found a spiritual advisor in Jeremy Taylor. In 1672 he formed a pious friendship with Margaret Blagge, later Mrs. Godolphin, a maid of honor to the queen, and wrote her *Life* to commemorate her virtues. Among his closest friends was Samuel Pepys, the diarist.

Evelyn lacked detachment and a methodical training to make his contributions scientifically valid. His activity was guided by religious and patriotic motives. His various publications were intended to "give ferment to the curious." His *Diary,* which he kept throughout his life, is his greatest contribution, albeit to letters rather than to science.

BIBLIOGRAPHY

I. ORIGINAL WORKS. Evelyn's works—some of which were published anonymously or pseudonymously—include his trans. of F. de la Mothe le Vayer's *Of Liberty and Servitude* (London, 1649); *The State of France, As It Stood in the IXᵗʰ Yeer of This Present Monarch, Lewis XIIII* (London, 1652); *An Essay on the First Book of T. Lucretius Carus De rerum natura* (London, 1656); as "Philocepos," *The French Gardiner: Instructing How to Cultivate all Sorts of Fruit-Trees, and Herbs for the Garden,* trans. from N. de Bonnefons (London, 1658); *The Golden Book of St. John Chrysostom, Concerning the Education of Children,* trans. from the Greek (London, 1659); *A Character of England, As It Was Lately Presented in a Letter, to a Noble Man of France* (London, 1659); *An Apologie for the Royal Party: Written in a Letter to a Person of the Late Councel of State* (London, 1659); *Elysium britannicum* (London, ca. 1659), a synopsis of proposed work on gardening, presumably a table of contents, British Museum Add. MS. 15950, f. 143; *The Late News or Message From Bruxels Unmasked, and His Majesty Vindicated, From the Base Calumny and Scandal Therein Fixed on Him* (London, 1660); *The Manner of Ordering Fruit-Trees,* trans. from "Le Sieur Le Gendre" (London, 1660) [attributed to Evelyn by F. E. Budd, in *Review of English Studies,* **14** (1938), 285–297]; *A Panegyric to Charles the Second, Presented to His Majestie the XXIII. of April, Being the Day of His Coronation. MDCLXI* (London, 1661); *Fumifugium: or the Inconveniencie of the Aer and Smoak of London Dissipated* (London, 1661); *Instructions Concerning Erecting of a Library . . .* trans. from Gabriel Naudé (London, 1661); *Tyrannus or the Mode: in a Discourse of Sumptuary Lawes* (London, 1661); *Sculptura: or the History, and Art of Chalcography and Engraving in Copper* (London, 1662); *Sylva, or A Discourse of Forest-Trees, and the Propagation of Timber in His Majesties Dominions . . . To Which Is Annexed Pomona; or, An Appendix Concerning Fruit-Trees in Relation to Cider; The Making and Several Ways of Ordering It . . . Also Kalendarium Hortense; or, Gard'ners Almanac; Directing What He Is to Do Monethly Throughout the Year* (London, 1664); *A Parallel of the Antient Architecture With the Modern,* trans. from the French of R. Fréart de Chambray (London, 1664); Μυστήριον τῆς 'Ανομίας. *That Is, Another Part of the Mystery of Jesuitism,* trans. from the French of A. Arnauld and P. Nicole (London, 1664); *The Pernicious Consequences of the New Heresie of the Jesuites Against the King and the State,* trans. from the French of P. Nicole (London, 1666); *The English Vineyard Vindicated by John Rose Gard'ner to His Majesty* (London, 1666) [Evelyn's authorship identified by G. Keynes]; *Publick Employment and an Active Life Prefer'd to Solitude* (London, 1667); *An Idea of the Perfection of Painting,* trans. from R. Fréart de Chambray (London, 1668); *The History of the Three Late Famous Impostors* (London, 1669); *Navigation and Commerce, Their Original and Progress* (London, 1674); *A Philosophical Discourse of Earth, Relating to the Culture and Improvement of It for Vegetation, and the Propagation of Plants, & c. as It Was Presented to the Royal Society, April 29.1675* (London,

1676) [called *Terra* in later editions]; *The Compleat Gard'ner; or, Directions for Cultivating and Right Ordering of Fruit-Gardens and Kitchen-Gardens; With Divers Reflections on Several Parts of Husbandry,* trans. from J. de La Quintinie (London, 1693); *Numismata. A Discourse of Medals, Antient and Modern* (London, 1697); and *Acetaria. A Discourse of Sallets* (London, 1699).

His occasional contributions include "An Account of Snow-Pits in Italy," in R. Boyle, *New Experiments and Observations Touching Cold* (London, 1665), pp. 407–409; "An Advertisement of a Way of Making More Lively Counterfeits of Nature in Wax, Than Are Extant in Painting: And of a New Kind of Maps in a Low Relievo. Both Practised in France," in *Philosophical Transactions of the Royal Society,* **1** (1665), 99–100; "A Letter . . . Concerning the Spanish Sembrador or New Engin for Ploughing . . . Sowing . . . and Harrowing, at Once," *ibid.,* **5** (1670), 1055–1057; "Panificium, or the Several Manners of Making Bread in France. Where, by Universal Consent, the Best Bread in the World Is Eaten," in J. Houghton, *A Collection of Letters for the Improvement of Husbandry & Trade,* no. 12 (16 Jan. 1683), 127–136; "An Abstract of a Letter From the Worshipful John Evelyn Esq; Sent to One of the Secretaries of the R. Society Concerning the Dammage Done to his Gardens by the Preceding Winter," in *Philosophical Transactions of the Royal Society,* **14** (1684), 559–563; and "Letter to William Cowper Relating to the Anatomical Tables [acquired by Evelyn in Padua]," *ibid.,* **23** (1702), 1177–1179.

Evelyn's shorter works are collected in William Upcott, ed., *The Miscellaneous Writings of John Evelyn, Esq., F. R. S.,* (London, 1825). Posthumous publications are R. M. Evanson, ed., *The History of Religion. A Rational Account of the True Religion* (London, 1850); Geoffrey Keynes, ed., *Memoires for My Grand-son* (Oxford, 1926) and *Directions for the Gardiner at Says-Court But Which May Be of Use for Other Gardens* (Oxford, 1932); Walter Frere, ed., *A Devotionarie Book of John Evelyn* (London, 1936); E. S. de Beer, ed., *London Revived* (Oxford, 1938), the discourse on the replanning of the City of London after the Great Fire; and Harriet Sampson, ed., *The Life of Mrs. Godolphin* (London, 1939).

Evelyn's diary was published first in shortened form by William Bray as *Memoirs* (London, 1818); of many later editions E. S. de Beer, ed., *The Diary of John Evelyn,* 6 vols. (Oxford, 1955), is definitive. Selections from the *Diary* containing important notes may be found in *Voyage de Lister à Paris en MDCXCVIII. . . . On y a joint des Extraits des ouvrages d'Evelyn relatifs à ses voyages en France de 1648 à 1661* (Paris, 1873); H. Maynard Smith, ed., *John Evelyn in Naples 1645* (Oxford, 1914); and *The Early Life and Education of John Evelyn* (Oxford, 1920); and Howard C. Levis, ed., *Extracts from the Diaries and Correspondence of John Evelyn and Samuel Pepys Relating to Engraving* (London, 1915).

Evelyn's correspondence may be found in William Bray, ed., *Diary and Correspondence of John Evelyn . . . To Which Is Subjoined, The Private Correspondence Between King Charles I and Sir Edward Nicholas,* in Bohn's Historical Library (London, 1859 and later issues); F. E. Rowley Heygate, ed., *Seven Letters of John Evelyn, 1665–1703* (London, 1914); and Clara Marburg, *Mr. Pepys and Mr. Evelyn* (Philadelphia, 1935).

II. SECONDARY LITERATURE. Books on Evelyn include Helen Evelyn, *The History of the Evelyn Family* (London, 1915); Florence Higham, *John Evelyn, Esquire. An Anglican Layman of the Seventeenth Century* (London, 1968); W. G. Hiscock, *John Evelyn and Mrs. Godolphin* (London, 1951) and *John Evelyn and His Family Circle* (London, 1955), which present an adverse view of Evelyn; Geoffrey Keynes, *John Evelyn. A Study in Bibliophily With a Bibliography of His Writings* (Oxford, 1968); and Arthur Ponsonby, *John Evelyn. Fellow of the Royal Society; Author of "Sylva"* (London, 1933).

Lectures and articles that deal with Evelyn and his work are Jackson I. Cope, "Evelyn, Boyle and Dr. Wilkinson's 'Mathematico-Chymico-Mechanical School,'" in *Isis,* **50** (1959), 30–32; Edward Gordon Craig, "John Evelyn and the Theatre in England, France and Italy," in *The Mask,* **10** (1924), repr. in *Books and Theatres* (1925), pp. 3–68; E. S. de Beer, "John Evelyn, F.R.S. (1620–1706)," in *Notes and Records of the Royal Society of London,* Tercentenary Number, **15** (1960), 231–238; Margaret Denny, "The Early Program of the Royal Society and John Evelyn," in *Modern Language Quarterly,* **1** (1940), 481–497; Leonard Guthrie, *The Medical History of John Evelyn, D.C.L., F.R.S., and of His Time 1620–1706* (London, 1905), two lectures delivered before the Harveian Society of London and the King's College Medical Society in October 1902 and October 1903, respectively; George B. Parks, "John Evelyn and the Art of Travel," in *The Huntington Library Quarterly,* **10** (1947), 251–276; W. Barclay Squire, "Evelyn and Music," in *The Times Literary Supplement* (17 Apr. 1924; 16 Oct. 1924; 14 May 1925; 10 Dec. 1925; and 14 Oct. 1926); and F. Sherwood Taylor, "The Chemical Studies of John Evelyn," in *Annals of Science,* **8** (1952), 285–292.

COLETTE AVIGNON

EVERSHED, JOHN (*b.* Gomshall, Surrey, England, 26 February 1864; *d.* Ewhurst, Surrey, England, 17 November 1956), *solar physics.*

The seventh son of John Evershed and Sophia, daughter of David Brent Price of Portsmouth, Evershed came of a family long established in Surrey and was educated at a private Unitarian school in Brighton and later at Croydon. In 1906 he married Mary Acworth Orr, who assisted him in his work until her death in 1949. A year later Evershed married Margaret Randall, who survived him; there were no children from either marriage.

Evershed was introduced into scientific circles by his elder brother Sydney, an inventor of electrical apparatus for the Royal Navy and a researcher in permanent magnetism. As a young man he studied solar spectroscopy and carried out experiments in the production of continuous and absorption spectra of heated gases at his private observatory at Kenley,

Surrey. He was fortunate in inheriting in 1894 the instruments of A. Cowper Ranyard, the distinguished amateur astronomer; these included an eighteen-inch refractor and a spectrograph. A liberal-minded employer (he was engaged in the analysis of oils and other products for a London firm) granted him leave to go on several total solar eclipse expeditions. Professor H. H. Turner provided introductions which led him in 1898 to travel for this purpose to India via the United States (where Evershed spent a month with George E. Hale) and Japan. It was on this expedition that he photographed for the first time the continuous spectrum to the ultraviolet of the Balmer series limit at λ 3646.

Correspondence and meetings with Sir William Huggins, then president of the Royal Society, resulted in Evershed's appointment as assistant to C. Michie Smith, director of the Kodaikanal and Madras observatories in India. Making full use of the high altitude of Kodaikanal (2,343 meters) and of his skill in designing and building instruments, Evershed began a long series of spectroheliograms of the sun's disk. He further began the research which led in 1909 to the discovery in sunspots of the small radial motions of gases parallel to the sun's surface, now known as the Evershed effect. In 1911 he succeeded to the directorship of the Kodaikanal and Madras observatories. During his period of office he carried out a great deal of chromospheric research, made early use of hydrogen α spectroheliograms, and, in 1915, led an expedition to Kashmir, where exceptionally good viewing conditions made it possible for him to measure the small shift to the red of spectrum lines in connection with Einstein's predictions.

Evershed retired in 1923 and returned to England. He settled at Ewhurst, Surrey, where with undiminished enthusiasm he again equipped a private observatory. There he carried out high-dispersion work (employing large liquid prisms) to determine the exact wavelengths of the solar spectrum, sunspots, prominences, and minute line-shifts, and to study the Zeeman effect in assessing the strength of magnetic fields of sunspots. The death of his wife deprived him of her practical assistance, but he continued to make observations until 1953. He then presented some of his instruments to the Royal Greenwich Observatory.

An ingenious designer of optical instruments and an indefatigable and meticulous observer, Evershed contributed much to the knowledge of solar physics during the early decades of this century.

BIBLIOGRAPHY

F. J. M. Stratton, "John Evershed," in *Biographical Memoirs of Fellows of the Royal Society* (1957) contains a full bibliography of Evershed's work. Some representative examples are "Experiments on the Radiation of Heated Gases," in *Philosophical Magazine,* **39** (1895), 460; "Wave-length Determinations and General Results Obtained From a Detailed Examination of Spectra Photographed at the Solar Eclipse of January 22, 1898," in *Philosophical Transactions of the Royal Society,* **197A** (1901), 381; "Solar Eclipse of 1900 May 28—General Discussion of Spectroscopic Results," *ibid.,* **201** (1903), 457; "Radial Motions in Sunspots," in *Monthly Notices of the Royal Astronomical Society,* **70** (1910), 217; "The Spectrum of Nova Aquilae," *ibid.,* **79** (1919), 468; "The Solar Rotation and Shift Towards the Red Derived From the H and K Lines in Prominences," *ibid.,* **95** (1935), 503; "Note on the Zeeman Effect in Sunspot Spectra," *ibid.,* **99** (1939), 217; and "Measures on the Relative Shift of the Line 5250.218 and Neighboring Lines in Mt. Wilson Solar Magnetic Field Spectra," *ibid.,* **99** (1939), 438.

An autobiographical notice is "Recollections of Seventy Years of Scientific Work," in *Vistas in Astronomy,* I (London–New York, 1955), 33.

P. S. LAURIE

EWING, JAMES (*b.* Pittsburgh, Pennsylvania, 25 December 1866; *d.* New York, N.Y., 16 May 1943), *pathology.*

Ewing was the son of Thomas and Julia Hufnagel Ewing, members of a prominent western Pennsylvania family. He completed a classical education at Amherst College, from which he received the A.B. degree in 1888 and the M.A. in 1891. In 1891 he obtained a medical doctorate from the College of Physicians and Surgeons of Columbia University; he subsequently returned there as a tutor in histology (1893–1897), a Clark fellow (1896–1899), and an instructor in clinical pathology (1897–1898). Ewing's mentors at the College of Physicians and Surgeons were Francis Delafield and T. Mitchell Prudden. He also served a brief apprenticeship with another eminent pathologist of the era, Alexander Kolisko, at the Vienna Clinic.

In 1899, following a period of voluntary service as a contract surgeon in the Spanish-American War, Ewing accepted a professorship in clinical pathology at the Medical College of Cornell University in New York. In 1932 he assumed the newly created chair in oncology there, which position he occupied until his retirement in 1939.

A review of Ewing's earlier works reveals the underlying influences of his preceptors at the College of Physicians and Surgeons, and especially the inspiration of Prudden, to whom Ewing dedicated the first and second editions of his *Clinical Pathology of the Blood* (1901; 1903). His contributions of this period include significant reports on the pathogenesis of infectious diseases (see, for example, his Wesley

M. Carpenter Lecture of 1900), immunity and blood serum reactions, and medicolegal questions. Ewing's connection with Cornell University allowed him to do research at the Loomis Laboratory for Research in Experimental Pathology, where an experimental cancer program was begun in 1902 under the auspices of the New York Memorial Hospital (Collis P. Huntington Fund). In 1906 Ewing and his associates published a significant finding on lymphosarcoma in dogs. This investigation showed that the disease was transmitted from one animal to another during coitus by the transfer of viable tumor cells. By virtue of this and other important laboratory discoveries, Ewing soon became one of the foremost American spokesmen in experimental oncology.

By 1910 Ewing recognized the need for a comprehensive organization of anticancer activities. Ewing's scheme for such a center was characterized about 1950 by Leonard Scheele, then surgeon-general of the U.S. Public Health Service, as a plan for "a cancer institute in the modern sense—an institution where scientists of many disciplines combine their efforts and resources in a common mission, cancer research." Ewing was able to implement his idea in 1913, when he was elected president of the Medical Board of the General Memorial Hospital for the Treatment of Cancer and Allied Diseases. In this position and later as first director of research and director of Memorial Hospital (from 1931 to 1939), Ewing supervised the creation of a primary cancer facility—the present Memorial Sloan-Kettering Cancer Center in New York City.

Under Ewing's direction Memorial Hospital entered a new era, one especially fruitful for the clinical management of neoplastic disorders through radiation therapy. As in his studies on the fundamental aspects of cancer research, he brought to the problems of radiology and radiotherapy his own creative and systematic intellect (see his Mutter Lecture of 1922 and the Caldwell Lecture of 1925), and he imparted the wealth of his practical experiences to a younger generation of clinicians. Ewing's stature as the medical administrator of Memorial Hospital is measured precisely in Emerson's dictum that "Every institution is but the lengthened shadow of some man."

The early death of his wife, Catherine Halsted, in 1902 evoked reclusive and eccentric tendencies in Ewing's personality. In later years he suffered the agonizing discomforts of tic doloreux (trigeminal neuralgia), which curtailed his professional activities. He remained an avid sports enthusiast nevertheless, with a marked preference for tennis and baseball, and he possessed a keen artistic temperament.

Ewing's works include several monographs and textbooks. *Clinical Pathology of the Blood* is a rich source on hematologic disorders, while *Neoplastic Diseases* is the cornerstone of modern oncology. In the latter work Ewing recorded a number of significant discoveries in tumor morphology and distinguished that form of malignant osteoma now called "Ewing's sarcoma."

Ewing was a founder and charter member of the American Association for Cancer Research (1907) and the American Cancer Society (1913), and an appointee to the first National Advisory Cancer Council (1937). His services to pathology were acknowledged by numerous international tributes and by his election to the National Academy of Sciences.

BIBLIOGRAPHY

A complete list of Ewing's publications through 1930 (with some biographical detail) appears in a special cancer edition of *The Annals of Surgery*, **93** (1931), xi–xv. See also Frank E. Adair, ed., *Cancer in Four Parts . . . Comprising International Contributions to the Study of Cancer* (Philadelphia, 1931), pp. xi–xv.

Ewing's books and monographs include *Clinical Pathology of the Blood: A Treatise on the General Principles and Special Applications of Hematology* (Philadelphia–New York, 1901; 2nd ed., 1903); *Neoplastic Diseases: A Textbook on Tumors* (Philadelphia–London, 1919; 2nd ed., 1922; 3rd ed., 1928; 4th ed., 1940); and *Causation, Diagnosis and Treatment of Cancer* (Baltimore, 1931).

Ewing's contributions to periodicals include "Conjugation in the Asexual Cycle of the Tertian Malarial Parasite (The Wesley M. Carpenter Lecture)," in *New York Medical Journal*, **74** (1901), 145–151; "A Study of the So-Called Infectious Lymphosarcoma of Dogs," in *Journal of Medical Research*, **15** [n.s. **10**] (1906), 209–228, written with Silas Beebe; "Cancer Problems (The Harvey Society Lecture)," in *Archives of Internal Medicine*, **1** (1908), 175–217; "An Analysis of Radiation Therapy in Cancer (The Mutter Lecture)," in *Transactions of the College of Physicians of Philadelphia*, 3rd ser., **44** (1922), 190–235; and "Tissue Reactions to Radiation (The Caldwell Lecture)," in *American Journal of Roentgenology*, **15** (1926), 93–115.

See also Ewing's articles on "Identity" (pp. 62–103), "The Signs of Death" (pp. 104–137), and "Sudden Death" (pp. 138–160) in Frederick Peterson and Walter S. Haines, eds., *A Textbook of Legal Medicine and Toxicology*, vol. I (Philadelphia–London, 1903); and "Identity" (pp. 132–174), "The Signs of Death" (pp. 175–208), and "Sudden Death" (pp. 209–233) in Frederick Peterson, Walter S. Haines, and Ralph Webster, eds., *Legal Medicine and Toxicology, by Many Specialists*, vol. I (Philadelphia–London, 1923). In addition see Hans Schmaus, trans. by A. E. Thayer, ed. with additions by Ewing, *A Textbook of Pathology and Pathological Anatomy* (Philadelphia–New York, 1902).

Although no formal biography of Ewing exists, a number of Ewing documents are held in the Hayes Martin Collection at the Memorial Sloan-Kettering Cancer Center Library, New York, N. Y. Three peripheral works which in part discuss Ewing's contributions are Victor A. Triolo and Ilse L. Riegel, "The American Association for Cancer Research, 1907–1940: Historical Review," in *Cancer Research,* **21** (1961), 137–167; Victor A. Triolo, "Nineteenth Century Foundations of Cancer Research: Origins of Experimental Research," in *Cancer Research,* **24** (1964), 4–27; and Victor A. Triolo and Michael B. Shimkin, "The American Cancer Society and Cancer Research: Origins and Organization, 1913–1943," in *Cancer Research,* **29** (1969), 1615–1641.

VICTOR A. TRIOLO

EWING, JAMES ALFRED (*b.* Dundee, Scotland, 27 March 1855; *d.* Cambridge, England, 7 January 1935), *physics.*

Ewing's most important research dealt with magnetism. He was one of the first to observe the phenomenon of hysteresis, which he named and studied both experimentally and theoretically. He also did research in seismography and thermodynamics. Although his research was important, Ewing was probably more influential through his continuing efforts to establish engineering education. In three quite different positions—as professor of mechanical engineering in Tokyo, as director of naval education, and as professor at Cambridge—he was involved with the teaching of engineering. In addition, he published many papers and books and participated in numerous committees dealing with engineering problems and the application of science.

Ewing's father was a minister of the Free Church of Scotland; both his brothers became clergymen. Ewing studied at the Dundee high school and then went to the University of Edinburgh on an engineering scholarship. A good student, he came under the influence of Peter Tait and Fleeming Jenkin. Ewing did some early research with Jenkin on the harmonic analysis of vowel sounds (using the traces produced by Edison's phonograph). Through Jenkin, Ewing came in contact with William Thomson and participated in three expeditions for laying transatlantic telegraph cables.

Following Jenkin's recommendation, Ewing went to Japan in 1878 as professor of mechanical engineering at the University of Tokyo. The university provided Ewing with the means to establish a seismological observatory and, beginning in the winter of 1879–1880, he erected instruments and recorded earthquakes. Ewing was especially eager to obtain a continuous record of motion during an earthquake, and he devised a new type of seismograph for this purpose. In the latter part of his stay in Tokyo he was involved in teaching physics and began his experimental study of magnetism. He later received the Japanese Order of the Precious Treasure.

Ewing returned to England in 1883, after five years in Japan. At first he held the chair of engineering at the University of Dundee and continued his research on magnetism. In 1890 Ewing was made professor of mechanism and applied mechanics at Cambridge University. There was at the time disagreement within the university over its involvement with engineering education, and there were some who believed that the subject had no place at Cambridge. (The engineering professorship held by James Stuart, Ewing's predecessor, was not supposed to be renewed.) During Ewing's tenure engineering became accepted, and in 1892 the mechanical sciences tripos was established.

Ewing was director of naval education from 1903 to 1916. Lord Selborne and Admiral John Fisher appointed him to this position as part of their program to reform education in the British navy and to provide training in science and engineering. During World War I he was in charge of "Room 40," a group that intercepted and deciphered German messages.

From 1916 to 1929 Ewing was principal and vice-chancellor of the University of Edinburgh and was active in its expansion—constructing new buildings, founding new chairs, and enlarging the staff. He received honorary degrees from the universities of Oxford, Cambridge, Durham, and St. Andrews. He was made a fellow of the Royal Society in 1887 and received the Royal Medal for his research on magnetism in 1895. He was knighted in 1911 and made honorary member of the Institution of Civil Engineers in 1929 and of the Institution of Mechanical Engineers in 1932.

Ewing began his research on magnetic hysteresis through a project to study the effect of stress on the thermoelectric properties of metals. In 1881 he discovered that the thermoelectric effect lags behind the applied stress. He next suggested that other pairs of variables might also be related in such a cyclic manner, and he studied the transient currents produced by twisting a magnetized wire. He found a lag here also and introduced the term *hysteresis,* from the Greek word meaning "to be late," to describe it. Ewing then turned to the study of hysteresis in magnetization. He observed in 1882 that the area enclosed by the hysteresis loop is proportional to the work done during a complete cycle of magnetization and demagnetization. In 1885 he presented an important paper on this topic to the Royal Society.

The lag, in some processes, between a force and its effect was known in Germany before Ewing's experiments, and Kohlrausch had invented the term *elastische Nachwirkung* for it in 1866. In his 1885 paper Ewing noted that Emil Warburg had independently discovered magnetic hysteresis and had emphasized the physical importance of the area of the hysteresis loop ("Magnetische Untersuchungen," in *Annalen der Physik und Chemie,* **13** [1881], 141–164).

BIBLIOGRAPHY

I. ORIGINAL WORKS. Ewing's writings include "On Friction Between Surfaces Moving at Low Speeds," in *Proceedings of the Royal Society,* **26** (1877), 93–94, written with Fleeming Jenkin; "On the Harmonic Analysis of Certain Vowel Sounds," in *Transactions of the Royal Society of Edinburgh,* **28** (1878), 745–777, written with Fleeming Jenkin; "On a New Seismograph," in *Proceedings of the Royal Society,* **31** (1881), 440–446; "Effects of Stress on the Thermoelectric Quality of Metals," *ibid.,* **32** (1881), 399–402; "On the Production of Transient Electric Currents in Iron and Steel Conductors by Twisting Them When Magnetised or by Magnetising Them When Twisted," *ibid.,* **33** (1881–1882), 21–23; "On Effects of Retentiveness in the Magnetisation of Iron and Steel," *ibid.,* **34** (1882–1883), 39–45; "Earthquake Measurement," in *Memoirs of the Science Department of the University of Tokyo,* **9** (1883), 1–92; "Experimental Researches in Magnetism," in *Philosophical Transactions of the Royal Society,* **176** (1885), 523–640; "Contributions to the Molecular Theory of Induced Magnetism," in *Proceedings of the Royal Society,* **48** (1890), 342–358; *Magnetic Induction in Iron and Other Metals* (London, 1892); *Steam Engine and Other Heat Engines* (Cambridge, 1894); *The Strength of Materials* (Cambridge, 1899); *The Mechanical Production of Cold* (Cambridge, 1908); *Thermodynamics for Engineers* (Cambridge, 1920); and "An Engineer's Outlook," in *Nature,* **130** (1932), 341–350.

For a more extensive listing of Ewing's papers, see the *Catalogue of Scientific Papers of the Royal Society,* IX (London, 1891), and XIV (London, 1915).

II. SECONDARY LITERATURE. On Ewing's life and work, see R. T. Glazebrook, "James Alfred Ewing," in *Obituary Notices of the Royal Society,* I (London, 1932–1935), 475–492; and E. Griffiths, "Sir Alfred Ewing," in *Proceedings of the Physical Society,* **47** (1935), 1135.

SIGALIA DOSTROVSKY

EYTELWEIN, JOHANN ALBERT C. (*b.* Frankfurt am Main, Germany, 31 December 1764; *d.* Berlin, Germany, 18 August 1848), *hydraulic engineering, mechanics.*

Eytelwein, the son of an impoverished tradesman, joined the Prussian artillery at the age of fifteen. Realizing that an army career held little promise, he studied civil engineering privately, passing the state examination for surveyor in 1786. In 1790 he qualified as civil engineer and left the army with the rank of lieutenant to enter the Prussian civil service. His first assignment was to Küstrin [now Kostrzyn] as regional superintendent of dikes of the Oderbruch, the low fertile land on the west bank of the Oder between Lebus and Schwedt.

A concern about the lack of a school or training program for engineers to staff government bureaus led Eytelwein to publish a collection of problems in applied mathematics for surveyors and engineers (*Sammlung . . .,* 1793). Eytelwein, like his French contemporary M. R. de Prony, was one of the first to write on the application of mechanics and mathematics to the design of structures and machines in order to bring rational methods to both the practicing engineer and the student. Called to Berlin in 1794 as director of the Board of Public Works, he became responsible for the regulation of many rivers of eastern Germany, including the Oder, Warthe, and connecting waterways; he also shared in planning the harbors of Memel, Pillau [Baltiysk], and Swinemunde [now Swinoujście]. In 1797 he was a cofounder of a civil engineering journal, the first in Germany, that was later carried on by Crelle as *Journal für die Baukunst.*

Eytelwein's efforts on behalf of an engineering institution were realized with the founding of the Berlin Bauakademie in 1799; this was the first German engineering school of university stature and one of the two nuclei of the later Technische Hochschule-Berlin. He was the first director of the Bauakademie and held that post for seven years; he also lectured on mechanics, hydraulics, hydrostatics, machine design, dike embankments, and stream regulation. In addition to writing books and articles on numerous technical topics, he served on commissions such as that which established a definitive set of weights and measures for Prussia. His *Handbuch der Mechanik . . .* (1801) was the most important book of this era, for it was the first to combine practice and theory. He was elected member of the Academy of Sciences in 1803 and lectured (1810–1815) at the recently founded University of Berlin. A man whose energy matched his ability, he was appointed director of the Prussian Public Works Deputation in 1809, to become in 1816 chief commissioner in charge of all hydraulic works of the kingdom.

Eytelwein's health began to fail in 1825; he retired in 1830 on the fortieth anniversary of his entry into the civil service. Nevertheless, he remained active and published a major work on analytical methods in his seventy-third year. Although he became blind at eighty and deaf before his death, he continued to be concerned in the mathematical instruction of his grandchildren.

Throughout his life Eytelwein was a strong influence in the elevation of the standards of engineering education, bringing to it the analytical methods of the time. His writings were distinguished for their clarity and sweep, practice being viewed and upgraded by developing analysis.

BIBLIOGRAPHY

I. ORIGINAL WORKS. Eytelwein's major works are *Sammlung von Aufgaben aus der angewandten Mathematik für Feldmesser, Ingenieure and Baumeister* (Berlin, 1793); *Grundlehren der Hydraulik* (Berlin, 1796); *Vergleichung der in den Preussischen Staaten eingeführten Maasse und Gewichte* (Berlin, 1798; 2nd ed., 1817); *Anweisung zum Zeichnen* (Berlin, 1799); *Anweisung zur Construction von Faschinenwerken* (Berlin, 1799); *Handbuch der Mechanik fester Körper und Hydraulik* (Berlin, 1801; 3rd ed., Leipzig, 1842); *Handbuch der Statik fester Körper,* 3 vols. (Berlin, 1808; 2nd ed., 1832); *Handbuch der Perspektive,* 2 vols. (Berlin, 1810); *Grundlehren der höheren Analysis,* 2 vols. (Berlin, 1824); *Handbuch der Hydrostatik* (Berlin, 1826); and *Anweisung zur Lösung höherer numerischer Gleichungen* (Berlin, 1837).

II. SECONDARY LITERATURE. Biographical sketches may be found in C. von Hoyer, *Allgemeine deutsche Biographie,* XLVIII (1904), 462; Löbe, *Allgemeine deutsche Biographie,* VI (1877), 464; C. Matschoss, *Männer der Technik* (Düsseldorf, 1925), p. 69; M. Rühlmann, *Vorträge über Geschichte der technischen Mechanik* (Berlin, 1885), pp. 284–286; R. Schröder, *Neue deutsche Biographie,* IV (1959), 714; and S. Timoshenko, *History of Strength of Materials* (New York, 1953), p. 101. For a list of his work see Poggendorff, I, 708.

R. S. HARTENBERG

IBN EZRA, ABRAHAM BEN MEIR, also known as **Abū Isḥāq Ibrāhim al-Mājid ibn Ezra,** or **Avenare** (*b.* Toledo, Spain, *ca.* 1090; *d.* Calahorra, Spain, *ca.* 1164–1167 [?]), *mathematics, astronomy.*

A versatile genius with a charming Hebrew style, Ibn Ezra disseminated rationalistic and scientific Arabic learning in France, England, and Italy. From about 1140 to 1160 he traveled continually, and it was in this last period of his life that his works were written. Ibn Ezra was a Hebrew grammarian, exegete, astrologer, translator from Arabic into Hebrew, and poet, as well as a scientist. His work as a Jewish biblical commentator was much admired by Spinoza. Ibn Ezra considered the physical sciences and astrology fundamental for every branch of Jewish learning.

Three of his treatises were devoted to numbers. *Sefer ha-eḥad* ("Book of the Unit") describes the theory of numbers from one to nine; *Sefer ha-mispar* ("Book of the Number") is on the fundamental operations of arithmetic. The latter describes the decimal system for integers with place value of the numerals from left to right, and the zero is given as *galgal* ("wheel" or "circle") in the preface. In the body of the treatise, however, Ibn Ezra returns to use of the letters of the Hebrew alphabet as numerals. The Indian influence is, nevertheless, unmistakable. The third book, *Yesod mispar* ("The Foundation of Numerals"), is concerned with grammatical peculiarities.

In Ibn Ezra's translation of al-Bīrūnī's *Ta'amē lūḥōt al-Chowārezmī* ("Commentary on the Tables of al-Khwārizmī"; the Arabic original is lost) there is interesting information on the introduction of Indian mathematics and astronomy into Arabic science during the eighth century.

Ibn Ezra was concerned with permutations and combinations, as is shown in his *Sefer ha-'olam* ("Book of the World"). In addition to treatises on the calendar, *Shalosh she'elot* ("Three Chronological Questions") and *Sefer ha-'ibbur* ("Book on Intercalation"), and the astrolabe, *Keli ha-neḥoshet* ("The Astrolabe"), Ibn Ezra wrote a number of astrological works (Steinschneider lists more than fifty) that were very popular and were translated into many languages. Two were printed in Latin in 1482 and 1485, respectively; and all of them appeared in Latin in 1507. Only two of the Hebrew originals have been printed, both in modern times. They are rich in original ideas and in the history of scientific subjects. The astrological works were translated into French in 1213 by Hagin, a Jew in the employ of Henry Bate at Malines (Mechelen), who in turn translated the French into Latin. Both the French and the Catalan translations are of great philological interest.

BIBLIOGRAPHY

Works dealing with Ibn Ezra and his writings are Henry Bate *et al., De luminaribus et diebus criticis* (Padua, 1482–1483); H. Edelmann, *Keli neḥoshet* (Königsberg, 1845); J. L. Fleischer, *Sefer ha-mōrōt* (Bucharest, 1932); Yekuthiel Ginsburg, "Rabbi Ben Ezra on Permutations and Combinations," in *The Mathematics Teacher,* **15** (1922), 347–356, text from *Sefer ha-'olam;* S. J. Halberstam, *Sefer ha-'ibbur* (Lyck [Elk], Poland, 1874); D. Kahana, *Rabi Abraham ibn Ezra,* II (Warsaw, 1894), 107–111; Martin Levey, *Principles of Hindu Reckoning* (Madison, Wis., 1965), pp. 8, 35; Raphael Levy, *The Astrological Works of Abraham ibn Ezra. A Literary and Linguistic Study With Special Reference to the Old French Translation of Hagin* (Baltimore, 1927); Alexander Marx, "The Scientific Work of Some Outstanding Mediaeval Jewish Scholars," in *Essays and Studies in Memory of Linda R. Miller* (New York, 1938), pp. 138–140; Ernst Müller, *Abraham ibn Esra Buch der Einheit aus dem Hebräischen übersetzt nebst Parallelstellen und Erläuterun-*

gen zur Mathematik Ibn Esras (Berlin, 1921); Samuel Ochs, "Ibn Esras Leben und Werke," in *Monatsschrift für Geschichte und Wissenschaft des Judentums,* **60** (1916), 41–58, 118–134, 193–212; M. Olitzki, "Die Zahlensymbolik des Abraham ibn Esra," in *Jubelschrift Hildesheimer* (Berlin, 1890), pp. 99–120; S. Pinsker, *Yesod mispar* (Vienna, 1863), and *Abrahami Ibn Esra, Sepher ha-echad, liber de novem numeris cardinalibus cum Simchae Pinsker interpretatione primorum quatuor numerorum. Reliquorum numerorum interpretationem et proemium addidit M. A. Goldhardt* (Odessa, 1867); George Sarton, *Introduction to the History of Science,* II, pt. 1 (Baltimore, 1931), 187–189; M. Silberberg, ed., *Sefer ha-mispar. Das Buch der Zahl, ein hebräisch-arithmetisches Werk des R. Abraham ibn Esra . . .* (Frankfurt, 1895); D. E. Smith and Yekuthiel Ginsburg, "Rabbi Ben Ezra and the Hindu-Arabic Problem," in *American Mathematical Monthly,* **25** (1918), 99–108; and the following by M. Steinschneider: "Abraham Judaeus-Savasorda und Ibn Esra," in *Zeitschrift für Mathematik,* **12** (1867), 1–44, and **25** (1880), supp. 57–128; *Verzeichniss der hebräische Handschriften der K. Bibliothek zu Berlin* (Berlin, 1897; 1901); *Die hebräischen Übersetzungen . . .* (repr. Graz, 1956), p. 869; *Die arabische Literatur der Juden* (repr. Hildesheim, 1964), p. 156; and *Mathematik bei den Juden* (repr. Hildesheim, 1964), pp. 87–91.

MARTIN LEVEY

FABBRONI (or erroneously **FABRONI**), **GIOVANNI VALENTINO MATTIA** (*b.* Florence, Italy, 13 February 1752; *d.* Florence, 17 December 1822), *economics, physics.*

The son of Orazio Fabbroni, who came from a noble family, and of Rosalind Werner, from Heidelberg, Fabbroni showed his ability so early that in 1768 he was made assistant to Felice Fontana in the Museum of Physics and Natural Sciences in Florence, where he became assistant director in 1780 and director (for a year) in 1805. From 1776 to 1778 he lived in Paris, and then in London. He frequented the enlightened and radical circles of the two capitals, meeting Benjamin Franklin and corresponding with Thomas Jefferson.

Upon his return to Florence in 1782, Fabbroni married the patrician Teresa Ciamignani, from Grosseto, continued his economic studies defending free trade, entered politics, and published many works on agriculture, botany, chemistry, physics, archaeology, and philology. In 1798 he participated in the work of the Commission on Weights and Measures in Paris; in 1802 he was named an honorary professor at the University of Pisa and in 1803, director of the mint. Fabbroni was a member of the Accademia dei Georgofili, of the Società Italiana delle Scienze dell'Accademia dei XL, and of thirty or more other academies, both Italian and foreign.

In a memoir read in 1792 to the Accademia dei Georgofili of Florence (published in 1801) and reworked in a new memoir published in 1799 in the *Journal de physique,* Fabbroni maintained that the phenomena discovered by Galvani (1791) were not due to the action of an electric fluid, but to the reciprocal action of dissimilar metals upon contact, in the presence of moisture. Because of these writings he is often considered the originator of the chemical theory of the battery, even though at the time the two memoirs were written, the battery had not been invented and Fabbroni maintained that the galvanic phenomena were independent of the production of electricity. Nonetheless, his ideas certainly influenced the emergence in the first years of the nineteenth century of the chemical theory of the battery.

BIBLIOGRAPHY

I. ORIGINAL WORKS. A collection of Fabbroni's writings is *Scritti di pubblica economia,* 2 vols. (Florence, 1847–1848). See also "Sur l'action chimique des différens métaux," in *Journal de physique . . .,* **49** (1799), 348–357; and "Dell'azione dei metalli nuovamente avvertita," in *Atti della R. Società economica di Firenze ossia de' Georgofili,* **4** (1801), 349–370. Letters and documents concerning Fabbroni are in the Italian State Archives, the Biblioteca Nazionale in Florence, and the Bibliothèque Nationale in Paris.

II. SECONDARY LITERATURE. On Fabbroni and his work, see Mario Gliozzi, "Giovanni Fabbroni e la teoria chimica della pila," in *Archeion,* **18** (1936), 160–165; Andrea Mustoxidi, "Giovanni Fabbroni," in Emilio de Tipaldo, ed., *Biografia degli italiani illustri,* I (Venice, 1834), 337–345, with list of published and unpublished works; Poggendorff, I, cols. 709–710; Ugo Schiff, "Il museo di storia naturale," in *Archeion,* **9** (1928), 296–297, 318–320; and Franco Venturi, "Giovanni Fabbroni," in *Illuministi italiani,* III (Milan, 1958), 1081–1134, with selections from his works.

MARIO GLIOZZI

FABRE, JEAN HENRI (*b.* Saint-Léons, Aveyron, France, 22 December 1823; *d.* Sérignan, Vaucluse, France, 11 October 1915), *entomology, natural history.*

Fabre was the son of Antoine Fabre, an *homme de chicane* (a sort of law officer), and of Victoire Salgues. He began his studies at the parochial school of his native village, then continued them, beginning in 1833, at the *collège* of Rodez. A scholarship student at the École Normale Primaire in Avignon, he obtained his *brevet supérieur* in 1842 and was appointed a teacher at the lycée of Carpentras in the same year. At Montpellier he prepared for his *baccalauréat,* which he passed, and then earned a double *licence ès sciences* in mathematics and physics. He next went to the lycée of Ajaccio, Corsica, as a physics teacher, remaining there until December 1851. Following this he taught in the lycée of Avignon (1853), then received the *licence* in natural history at Toulouse, and finally defended his thesis for the *doctorat ès sciences naturelles* at Paris in 1854. Henceforth, Fabre devoted himself almost exclusively to the research on the biology and behavior of insects that was to make him one of the great figures of entomology.

In 1855 Fabre published his first work on a hymenopterous vespid (*Cerceris*) that paralyzes its prey (beetles). His second memoir (1857) was concerned with the hypermetamorphosis of the *Meloidae* (coleoptera). In 1856 Fabre was awarded the Prix Montyon (for experimental physiology) of the Institut de France, and in 1859 Charles Darwin cited him in his *Origin of Species,* a valuable encouragement for a poorly paid young teacher.

In an attempt to improve his financial situation, Fabre undertook a research on the coloring principle of madder (alizarin), which he succeeded in isolating in 1866. This discovery resulted in his being awarded the Legion of Honor, and he was received in Paris by Napoleon III. But on his return to Avignon, Fabre learned that alizarin had just been obtained from coal tars and that his process had been superseded. He turned to writing textbooks and gave a free course on the sciences, at the same time forming a friendship with the philosopher John Stuart Mill, who was then living in Avignon. A victim of various jealousies and vexations, Fabre left that city in November 1870 and moved to Orange, and then in 1879 to Sérignan, where he devoted all his time to observations on the life and habits of insects. On 11 July 1887 he was elected a corresponding member of the Académie des Sciences, and his jubilee was celebrated on 3 April 1910.

Fabre's first marriage was to Marie Villard (30 October 1844); they had many children, including three sons and one daughter. Having become a widower shortly after moving to Sérignan, he remarried and had a son and two daughters by his second wife. One of his daughters married the physician G. V. Legros, who was his first biographer.

Fabre's scientific work includes the ten-volume *Souvenirs entomologiques* (1879–1907), which presents a considerable number of original observations on the behavior of insects (and also of arachnids); these had been preceded by various memoirs published as books or periodical articles (1855–1879).

It is the latter group of publications that contains Fabre's principal discoveries: hypermetamorphosis of the *Meloidae;* the relationship between the sex of the egg and the dimensions of the cell among the solitary bees; the habits of the dung beetles; and the paralyzing instinct of the solitary wasps *Cerceris, Sphex, Tachytes, Ammophila,* and *Scolia.*

These last researches, which posed the problem of instinct and its acquisition by insects, were much discussed and were the object of lively criticism by E. Rabaud. Recent works, such as those of A. Steiner (1962) on the wasp *Liris nigra,* which preys on crickets, confirm Fabre's observations and show that the prey is a checkerboard of stimulating zones, each of which provokes a precise and practically unalterable response by the predator.

Although his works were admired by Darwin, Fabre was all his life opposed to evolution, remaining convinced of the fixity of species. For him, each animal species was created as we see it today, with the same instinctual equipment (whereas the modern explanation of instinct draws on the notion of natural selection).

Fabre had the great merit of demonstrating the importance of instinct among the insects, while certain of his predecessors (J. C. W. Illiger, Jean Th. Lacordaire) supposed that insects are endowed "with reasoning or inventing faculties comparable to those of the higher animals, and of man" (J. Rostand, "Jean-Henri Fabre," p. 157).

Responsible for significant discoveries concerning the lives and habits of insects, Fabre remains especially important in the history of science because of the popularity of his *Souvenirs entomologiques;* reading them led more than one person to become a naturalist.

In addition, in order to earn a living, Fabre, between 1862 and 1901, wrote some forty works of scientific popularization, designed chiefly for the young and ranging from mathematics and physics to natural history.

He also composed poems in French and in Provençal; the latter resulted in his being called *felibre di Tavan.*

Fabre remains the very model of the self-taught scientist—solitary, poor, proud, and independent. He was also an attentive and minute observer and a writer of unquestionable talent.

BIBLIOGRAPHY

I. ORIGINAL WORKS. Fabre's major work is *Souvenirs entomologiques,* 10 vols. (Paris, 1879–1907), trans. into English by A. Teixiera de Mattos as *The Works of J. Henri Fabre* (London, 1912 *et seq.*). Among his other writings are two theses presented to the Faculty of Sciences in Paris in 1855, "Recherches sur les tubercules de l'*Himantoglossom hircinum,*" in *Annales des sciences naturelles,* Botanique, ser. 4, **3** (1855), 253–291, with two plates, and "Recherches sur l'anatomie des organes reproducteurs et sur le développment des Myriapodes," *ibid.,* Zoologie, ser. 4, **3** (1855), 257–316, with four plates. See also "Observations sur les moeurs des *Cerceris* et sur la cause de la longue conservation des Coléoptères dont ils approvisionnent leurs larves," *ibid.,* **4** (1855), 129–150; "Recherches sur la cause de la phosphorescence de l'Agaric de l'olivier," *ibid.,* Botanique, ser. 4, **4** (1855), 179–197; "De la germination des Ophrydées et de la nature de leurs tubercules," *ibid.,* **5** (1856), 163–186, with one plate; "Étude sur l'instinct et les métamorphoses des Sphégiens," *ibid.,* Zoologie, **6** (1857), 137–183; "Mémoire sur l'hypermétamorphose et les moeurs des Méloïdes," *ibid.,* **7** (1857), 299–365; "Nouvelles observations sur l'hypermétamorphose et les moeurs des Méloïdes," *ibid.,* **9** (1858), 265–276; *Mémoire sur la recherche des corps étrangers introduits frauduleusement dans la garance en poudre* (Avignon, 1859); *Note sur le mode de reproduction des truffes* (Avignon, 1859); "Rapport sur l'alizarine artificielle de M. Roussin," in *Bulletin de la Société d'agriculture et d'horticulture de Vaucluse,* **10** (Aug. 1861), 235–248; "Étude sur le rôle du tissu adipeux dans la sécrétion urinaire chez les insectes," in *Annales des sciences naturelles,* Zoologie, ser. 4, **19** (1863), 351–389; *Insectes Coléoptères observés aux environs d'Avignon* (Avignon, 1870); and "Étude sur les moeurs et la parthénogenèse des Halictes," in *Annales des sciences naturelles,* Zoologie, ser. 6, **9** (1879), art. 4. His poetry is collected in *Oubreto provençalo dôu Felibre di Tavan* (Avignon, 1909). A list of forty of his textbooks, published between 1862 and 1901, may be found in the biography by Cuny (below), pp. 184–185.

II. SECONDARY LITERATURE. On Fabre and his work see M. Coulon, *Le génie de J. H. Fabre* (Paris, 1924); H. Cuny, *Jean-Henri Fabre et les problèmes de l'instinct* (Paris, 1967); Augustin Fabre, *The Life of Jean Henri Fabre, the Entomologist* (London, 1921), B. Miall, trans.; G. V. Legros, *La vie de J. H. Fabre naturaliste par un disciple* (Paris, 1913); E. Rabaud, *J. H. Fabre et la science* (Paris, 1924); and J. Rostand, "Jean-Henri Fabre," in *Hommes de vérité,* ser. 2 (Paris, 1948), pp. 109–168.

JEAN THÉODORIDÈS

FABRI, HONORÉ, or **HONORATUS FABRIUS** (*b.* Virieu-le-Grand, Dauphiné, France, 5 April 1607; *d.* Rome, Italy, 8 March 1688), *mathematics, natural philosophy.*

Fabri came from a family of judges in Valromey that was probably related to the Vaugelas family.[1] Following his studies at the *institut* in Belley, he entered the Jesuit novitiate in Avignon on 18 October 1626, remaining until 1628. In the fall of that year he went to the Collège de la Trinité[2] in Lyons, where he completed his course in Scholastic philosophy under Claude Boniel. After teaching for two years at the *collège* in Roanne,[3] he returned to Lyons in 1632 in order to begin his course in theology, which

he finished—following his ordination as a priest in 1635—in 1636. In the latter year he was named professor of logic at the *collège* in Arles, where for two years he gave lectures on philosophy that included natural philosophy as well. It was at this time that he discovered—independently of Harvey—the circulation of the blood, which he taught publicly.[4]

Besides being prefect at the *collège* in Aix-en-Provence (1638–1639), Fabri was leader of a sort of circle that, among other things, brought him the acquaintance of—and a long-lasting correspondence with—Gassendi. He was then recalled to Lyons to finish his third year of probation under P. Barnaud and in 1640 was promoted to professor of logic and mathematics, and also to dean, at the Collège de la Trinité.

During the following six years Fabri taught metaphysics, astronomy, mathematics, and natural philosophy. This period was the most brilliant and fruitful of his life; several books that he published later were developed from lectures delivered during this time. Fabri was the first of many famous professors produced by the Collège de la Trinité: his students included Pierre Mousnier, who later edited many of his teacher's lectures; the mathematician François de Raynaud,[5] who became famous through his friendship with Newton; Jean-Dominique Cassini; and Philippe de La Hire. Claude Dechales[6] and the astronomer and mathematician Berthet were also members of this circle. Among these scholars and the two Huygenses (father and son), Leibniz, Descartes, Mersenne, and others an active correspondence developed.

The foci of Fabri's tremendous activity were almost all urgent questions of the science of his day: heliocentrism, Saturn's rings, the theory of the tides, magnetism, optics, and kinematics. In mathematics, infinitesimal methods and the continuum problem were most prominent.

Fabri's favorable reception of certain Cartesian conceptions[7] embroiled him in an intense controversy with his superiors, which finally led to his expulsion from Lyons and his transfer to Rome, where he arrived on 12 September 1646. Although his stay was supposed to be only provisional, he was made a member in the same year of the Penitentiary College (the Inquisition). He served on that body, finally as Grand Inquisitor, for thirty-four years. Despite his important work in Church politics and theology—Fabri was considered the first expert on Jansenism—there was still time for his wide-ranging scientific research.

In mathematics, Fabri showed that despite the influence of Cavalieri and Torricelli, he was an independent and original thinker. This is clear from his principal mathematical work, *Opusculum geometricum*. Through the functional reinterpretation of Cavalieri's concept of indivisibles by means of a dynamically formulated concept of *fluxus*, Fabri approached similar ideas put forth by Newton. Fabri, however, was not able to free himself of a rather cumbersome, purely geometrical representation. In his *Synopsis geometrica* he developed a method of teaching based on his concept of *fluxus* and was not unsuccessful in using his somewhat inadequately formulated principle of homogeneity in his investigations on infinitesimals.

The *Opusculum geometricum* contains, besides an ingenious quadrature of a cycloid which Leibniz found inspiring, various quadratures and cubatures that amount to special cases of $\int x^n \sin x \, dx$, $\int \sin^p x \, dx$, and $\int\int \arcsin x \, dx \, dy$, as well as centroid determinations of sinusoidal and cycloidal segments together with their elements of rotation about both axes. The book doubtless originated in connection with the controversy over cycloids and Pascal's challenge.

In Rome, Fabri became acquainted with Michel Angelo Ricci, who recommended him to the Medici Grand Duke Leopold II. The latter made Fabri a corresponding member of the Accademia del Cimento. In 1660, with an anonymous work,[8] Fabri opened the controversy with Huygens over Saturn's rings which, after five years and a great expenditure of energy, was decided in Huygens' favor. Fabri was a fair opponent: he apologized and openly adopted Huygens' opinion. In the *Brevis annotatio* is a note that reads, more or less, "As long as no strict proof for the motion of the earth has been found, the Church is competent to decide [the issue]. If the proof, however, is found, then there should be no difficulty in explaining that the relevant passages in the Bible must be interpreted in a more symbolic sense." This statement would perhaps have been tolerated later, under Pope Clement IX, on whom Fabri had a strong influence; under Alexander VII, however, it brought Fabri (as a member of the Holy Office) fifty days in prison, and his release was effected only through the intervention of Leopold II. Yet this did not prevent the combative Jesuit from inserting into his *Dialogi physici* (1665) a chapter entitled "De motu terrae." It was also in 1665 that Fabri discovered the Andromeda nebula, which he at first thought was a new comet.

In natural philosophy Fabri was less fortunate than in mathematics. Nevertheless, the following achievements are noteworthy: the constant use of the concept of the static moment; an attempted explanation of tidal phenomena based on the action of the moon,

even though it involved air pressure as the medium; an explanation of the blue color of the sky based on the principle of dispersion; and investigations on capillarity. His attempted explanation of cohesion, however, was completely unsuccessful.

In 1668 Fabri began a year's sick leave in Virieu-le-Grand, where he supervised the publication of various of his works. He continued to work in the Holy Office in Rome until 1680. The last eight years of his life he spent a short distance outside the city, devoting himself to historical studies.[9]

Aside from the individual scientific achievements mentioned, Fabri's efforts to introduce a priori methods in natural philosophy as well as in philosophy are important historically,[10] as is his lasting influence on Leibniz, who richly recompensed Fabri with his friendship. Newton, for his part, mentioned in his second paper on light and colors that he first learned of Grimaldi's experiments through the medium of "some Italian author," whom he identified as Fabri in his dialogue *De lumine*.[11]

NOTES

1. See Gassendi's letter to Mousnier of 1 October 1665, published in Fabri's *Cours de philosophie* (Lyons, 1646).
2. The Collège de la Trinité was transferred by the aldermen of Lyons to the Jesuits in 1565 but was closed in 1594, following the attempt by the Jesuits' pupil Jean Châtel on the life of Henry IV. The Jesuits were expelled from France but returned in 1604 and reestablished the college; it was not completed, however, until 1660.
3. In 1634, at the age of ten, François de La Chaise (later the Jesuit Père Lachaise) entered this *collège;* he was later bound to Fabri by close friendship and an active correspondence.
4. See *Journal des sçavans* (1666), pp. 395–400.
5. Also known as Regnauld and Reynaud.
6. Also known as de Chales.
7. See D.-G. Morhof, *Polyhistori litterarum,* II (Lübeck, 1690), 115; and Adrien Baillet, *La vie de Descartes,* II (Paris, 1691), 299.
8. *Eustachii de divinis septempedani brevis annotatio in systema Saturnium Christiani Hugenii* (Rome, 1660).
9. The manuscripts of the last creative period, most of them unpublished, are in the library of the city of Lyons.
10. See Leibniz's letter to Johann Bernoulli, dated 15 October 1710, in *Leibniz's mathematische Schriften,* C. I. G. Gerhardt, ed., III (Halle, 1856), 856.
11. Newton to Oldenburg, 7 December 1675, in H. W. Turnbull, ed., *The Correspondence of Isaac Newton,* I (Cambridge, 1959), 384.

BIBLIOGRAPHY

I. ORIGINAL WORKS. Fabri's mathematical writings include *Opusculum geometricum de linea sinuum et cycloide* (Rome, 1659), written under the pseudonym "Antimus Farbius"; and *Synopsis geometrica* (Lyons, 1669), to which is appended *De maximis et minimis in infinitum proposi-*

tionum centuria, written in 1658/1659. These works, as well as the minor *Brevis synopsis trigonometriae planae* (1658/1659) are discussed in Fellmann, below.

Fabri's works in natural philosophy are *Tractatus physicus* . . . (Lyons, 1646); *Dialogi physici* . . . (Lyons, 1665); *Synopsis optica* (Lyons, 1667); *Dialogi physici* . . . (Lyons, 1669); and *Physica* (Lyons, 1669).

II. SECONDARY LITERATURE. On Fabri's work, see Carlos Sommervogel, *Bibliothèque de la Compagnie de Jésus,* III (Paris-Brussels, 1892), 512–522, which contains an extensive bibliography; and E. A. Fellmann, "Die mathematischen Werke von Honoratus Fabry," in *Physis* (Florence), **1-2** (1959), 6–25, 69–102.

E. A. FELLMANN

FABRI, NICOLAS DE PEIRESC. *See* **Peiresc, Nicolas Fabri de.**

FABRICI, GIROLAMO (or **FABRICIUS AB AQUAPENDENTE, GERONIMO FABRIZIO**) (*b.* Aquapendente, near Orvieto, Italy, *ca.* 1533; *d.* Padua, Italy, 21 May 1619), *anatomy, physiology, embryology, surgery.*

Fabrici was born of a noble and once-wealthy family; that he was the eldest son is indicated by his having been named for his paternal grandfather. Around 1550 his family sent him to Padua where, under the patronage of a patrician Venetian family named Lippomano or Lipamano, he studied Greek and Latin, then logic and philosophy. He went on to medicine and took his degree in medicine and philosophy at Padua in about 1559.

Fabrici studied with Gabriele Falloppio, whom he succeeded as teacher of anatomy upon the latter's death in 1562; from 1563 to 1565 he devoted himself to giving private anatomy lessons. In April 1565 he was nominated by the university to lecture on both anatomy and surgery; the position brought him an annual salary of 100 florins and entailed additional responsibilities in anatomical work. He presented his first lecture on 18 December 1566; he was repeatedly reconfirmed in his academic position (with appropriate raises in pay) and in 1600 was given life tenure, with the title *sopraordinario.* From 1609 on anatomy and surgery were given separately, and Fabrici became *sopraordinario* lecturer in anatomy only, retaining his full salary, however, which by that time amounted to 100 scudi a year. He retired from teaching in 1613, having served the University of Padua for nearly fifty years.

Fabrici's long academic career was not without strife. In 1588 he was publicly accused by his students of neglecting his teaching—a charge that would seem to have some ground in truth, but which may be explained in part by Fabrici's repeated illnesses. Certainly he was of difficult character, as may be seen by his clash with his German students, whom he ridiculed in the course of a public lecture in February 1589 because of their slow and harsh speech—the quarrel was reconciled only in October of that year. He further became embroiled in a protest in 1597 about having been placed after the professors of philosophy on the *Rotula* of the university; had an argument in 1608 with Eustachio Rudio; became involved in a dispute about the schedule of courses with his colleague Annibale Bimbiolo in 1611; and in 1613 attempted to prevent the nomination of a German councillor of the university because he was annoyed with the German students for attending the private anatomy classes given by Giulio Casseri.

It is likely, too, that Fabrici slighted his teaching duties in the interest of scientific research. He did, however, make substantial contributions to the university; among other things, the construction of a permanent anatomical theater, built in 1594 and inaugurated by him in 1595 (still preserved, and now bearing Fabrici's name), was in large part due to his efforts. His merit as a teacher was publicly acknowledged; if some of these acknowledgments are of a formal nature (as for example, those given on the occasions of his academic reconfirmations), others are undoubtedly sincere (the gratitude expressed by the fractious German students for the course in surgery that he conducted in 1606).

Fabrici further took active part in other matters concerning the university: in 1574 he was instrumental in securing the acquittal of a German student from a charge of homicide; in 1591 he intervened on behalf of some German students who had been arrested for carrying arms; in 1592–1593 he concerned himself with the reconstruction of the temporary anatomical theater and in 1595 with free admission to the permanent theater; in 1606 he again acted on behalf of an arrested German student; and in the winter of 1608–1609 he gave a cadaver to the German students (among whom were Olaus Worm and Caspar Bartholin) so that they could prepare the skeleton. It is thus clear that his relations with his students improved with the passage of time.

As a surgeon and physician Fabrici enjoyed high professional acclaim and the patronage of many eminent people. In 1581 he attended a brother of the duke of Mantua; in 1591 he was consulted by the duke of Urbino about the cure for certain fevers that were rampant in Pesaro; and in 1594 he corresponded with Mercuriale and Tagliacozzi about a case of rectogenital fistula. He went to Florence in 1604 to treat Carlo de' Medici, the son of Ferdinand I and Christina di Lorena, while in 1606 he visited Galileo, who subsequently became his patient. He visited Venice with Spigelio on 9 October 1607, and while he was there took care of Paolo Sarpi, who had been wounded a few days before; for these services he was made a knight of St. Mark by the Republic of Venice.

At some unknown time Fabrici married Violante Vidal; they had no children and she died in 1618. He did have an illegitimate son, Francesco, probably born before his marriage. Francesco also took his degree in medicine but was a source of little pleasure or pride to his father—in fact, a quarrel over money brought father and son into legal confrontation; Fabrici had serious disagreements with other close relatives as well. The person to whom he was closest was his great-grandniece, Semidea, whom he adopted on the death of her father and raised as his daughter in Padua. He married her to Daniele Dolfin on 9 May 1619; on 13 May he fell ill and died a few days later, almost certainly at his house in Padua. His funeral took place on 23 May, in the Franciscan church; the oration was given by Giovanni Tuilio, and he was buried, *sine titulo,* in the west cloister.

As a scientist, Fabrici was an indefatigable and scrupulous observer, describing his results with exactitude. His interpretation of observed phenomena was often shaped by tradition, however, and he may not be considered a comparative anatomist in the modern sense because he made no studies of homologous structures and did not attempt to analyze relationships and affinities of the organs that he studied. His primary purpose in his studies of fetal anatomy, for example, was to prepare a tool for the interpretation of the purpose and end of the organs under consideration; he was more concerned with finding philosophically based principles than with morphological detail and tended to modify observations that did not verify such principles. Thus he often failed to pursue his own discoveries to their logical conclusions. His interpretation of nature was, then, a teleological one, and his methods of observation derived largely from Galen.

Fabrici published his results in several volumes, including *De visione, voce, auditu* (Venice, 1600); *De locutione et ejus instrumentis liber* (Venice, 1601); *De brutorum loquela* (1603); *De venarum ostiolis* (1603); *De musculi artificio, ossium de articulationibus* (1614); *De respiratione et eius instrumentis, libri duo . . .* (1615); *De gula, ventriculo, intestinis tractatus* (1618); *De motu locali animalium secundum totum* (1618); and *Hieronymi Senis De totius animalis integumentis opusculum* (1618)—all of which may be considered as parts of the uncompleted but monumental *Totius animalis fabricae theatrum* which he meant to publish

and to which he devoted many years. In addition, there are in the St. Mark's library in Venice 167 *Tabulae anatomicae,* collected in eight volumes, part of the 300 color plates that Fabrici finished in 1600 as his major purely anatomical work.

One of the most famous (and most thoroughly studied) of Fabrici's works is *De venarum ostiolis.* The treatise, published in Padua, consists of twenty-three folio pages, supplemented by eight beautiful plates. In it Fabrici reports that he had first observed the valves of the veins in 1574 (the first demonstration to his students was in 1578 or 1579), as was recognized by his student Salomon Alberti, who published, with Fabrici's permission, a preliminary illustration of the venous valves (Nuremberg, 1585). Although the valves of the veins had been studied previously by G. B. Canano and by Amato Lusitano (indeed, a dispute arising therefrom had involved Vesalius, Eustachi, and Falloppio), Fabrici made no use of their contributions, perhaps intentionally; he describes the venous valves *ex novo,* systematically and accurately.

His interest in reconciling his observations with the traditional Galenic concepts of function misled Fabrici into missing the real significance of the venous valves, however. He accepts the notion of the blood flowing centrifugally, drawn by the viscera, and interprets the function of the venous valves to be the slowing down of the influx of the blood to provide for its even distribution to various parts of the body. He thus gives a teleological account of the number, alternate positioning, and conformation of the valves, pointing out that they are not present in the large veins of the trunk, such as the vena cava, in which the blood flows directly to the viscera and vital organs; they are found instead in the veins of the limbs, where they prevent an excessive inflow of blood, which would both cause swelling and deprive the vital organs of nourishment. He describes the valves as corresponding to the openings of collateral branches of the veins and calls them *ostiola.* In addition to thus regulating the flow of the blood mechanically, the valves also serve to prevent excessive stretching of the blood vessels and to reinforce the walls of the veins.

This is demonstrated by the formation of varicose veins in those who do heavy work; the blood of such persons is more dense and held longer by the valves, which become dilated, then subside as the veins dilate. The valves are further demonstrable by the application of a tourniquet to the upper arm; they then appear as a series of regularly spaced knots on its surface. Fabrici observed that if, after ligating the vein, one pressed upon it with a finger, one could observe the valves in action—acting, he thought, to retard the progress of the blood; his misinterpretation may have been due in part to his confusion of laboratory observation with the clinical symptoms of valvular insufficiency that he had noted, particularly in cases of varicosity.

Perhaps the most notable contribution of *De venarum ostiolis* is that William Harvey drew upon it in beginning his studies of the circulation of blood. Harvey was the pupil of Fabrici—indeed, he even lived for a while in his house—and from Fabrici's work he obtained the illustrations that would, with substantial modifications, serve him for his *De motu cordis.*

Fabrici's embryological studies were written concurrently with his later anatomical works. They include *De formato foetu* (1604) and *De formatione ovi et pulli* (published posthumously in 1621); and these two treatises in themselves would assure Fabrici's place among the most important biologists of his time.

In his introduction to *De formato foetu* (his last embryological treatise, despite its earlier publication), Fabrici divides his studies on generation into three parts. The first of these, dealing with the propagation of the seed and the organs that produce it, is presumably *De instrumentis seminis,* which was never published and is probably lost; the second, his work concerning the nature and properties of the seed and the generation and formation of the fetus, is *De formatione ovi et pulli;* while the third, his treatment of the fetus itself, is the *De formato foetu.* Both of the extant works are of a rather narrative character and are written in a somewhat inelegant Latin; many reputable scholars suggest that these works grew out of Fabrici's classroom lectures.

De formatione ovi et pulli is divided into two parts. The first, in three chapters, deals with the formation of the egg. The first chapter discusses the three bases of animal generation given by Aristotle (the egg, the seed, and spontaneously from decomposing materials); Fabrici differs from Aristotle, however, in asserting that most insects are born from eggs in which there is no differentiation between the formative and the nutritive elements; and in specific opposition to Aristotle he classifies Testacea as oviparous. In some respects, Fabrici's classification approaches the *ex ovo omnia* of Harvey; he excludes from his list of oviparous creatures only mammals and those insects that he believes to have been the products of spontaneous generation. His discussion of the generation of birds involves two aspects, that of the egg (whose *uterus* is the ovaries and oviducts) and that of the chicken (whose *uterus* is the egg itself).

Although Fabrici's embryological studies often surpassed those of Aldrovandi and Coiter, he here makes two mistakes. He interprets the germinal disc of the hen's egg as the scar left on the yolk by the detachment of the peduncle that had attached it to the ovary during its development, and he states that the function of the cloacal bursa in the hen (the bursa of Fabricius, which he discovered) is to store the semen of the rooster.

In the second chapter of *De formatione ovi et pulli* Fabrici states two functions of the "uterus": the formation of the egg and, immediately thereafter, its nutrition. The yolk of the egg is formed in the "upper uterus" (the *ovarium* of the yolk), while the remaining part is formed in the "lower uterus" (the *uterus,* or *ovarium* of the whole egg). The egg thus leaves the ovary as a naked yolk; and the chalazae, the albumen, the two membranes of the shell, and the shell itself develop subsequently in the oviduct. The yolk grows as it is nourished by material brought to it through the blood vessels that run through the ovisac while the egg is still attached to the ovary; after detachment the egg ceases to grow and the albumen, adhering to the yolk, grows by apposition. All parts of the egg are therefore derived from the blood, although from different portions of it. The chapter closes with a discussion of the formation of the shell; the third chapter concerns the usefulness of the uterus.

The second part of the treatise, also in three chapters, is concerned with the generation of the chick within the egg and begins with a description of the eggs of various species. Many of the notions and arguments set forth in the first part of the book are then summarized.

The second chapter of the second part deals with the three basic functions of the egg: the formation, growth, and nutrition of the chick. These considerations draw Fabrici into complex and difficult problems that he is unable to resolve, not because of any inadequacy as an observer but rather because of the science that he has inherited. He concludes his discussion with the trophic functions of both yolk and albumen and goes on to demonstrate that the chalaza attached to the thicker part of the egg is the only possible source of formative material. He considers semen to act as only the effective cause of generation; it never enters the egg, being prevented from doing so by the depth and plication of the uterus, which combine to keep the semen from reaching the upper oviduct—and by the time the egg has reached the lower oviduct it is encased in its protective shell. Fabrici postulates that semen is collected in the cloacal bursa of the hen and thence, through its radiant or spiritual powers, fertilizes the entire egg and uterus.

Thus, in oviparous animals the material and the agent of generation are not only distinct but separated by a notable physical distance; this view in no way contradicts Aristotle's doctrine that all the material for generation is contained in the female. (In dealing with viviparous animals, however, Fabrici adopts the Galenic interpretation by which the male seed is both material and efficient cause of generation.)

Fabrici then speculates further on the various possible causes and conditions of generation, including a discussion of the order in which various parts of the embryo are formed during its development. This question had been debated since Aristotle's time, and Fabrici affirmed that certain structures constituting the "carina" (or "keel") can be seen prior to the development of the heart and viscera; it is probable that his "carina" is in fact the whole one- or two-day-old embryo, in which the head, vertebral column, and ribs are visible to the naked eye. H. B. Adelmann (in *The Embryological Treatises of Hieronymus Fabricius of Aquapendente* [Ithaca, 1942]) maintains, on the basis of Fabrici's rather complicated discussion, that he did not, like Aristotle, consider the heart to be the first organ formed, nor yet, like Galen, the liver; rather, he may be thought to have preceded Harvey in giving priority to the blood.

The last chapter of the treatise returns to teleology to consider the utility of both the egg and the semen of the rooster.

De formatione ovi et pulli is illustrated with seven plates, of which only the first three are labeled. The last five plates are the most significant since they represent the first printed figures of the development of the chick, beginning with the third or fourth day of incubation. Some of these figures—especially those illustrating embryonic appendages—are difficult to interpret, although all are admirable for their subtlety of detail (obtained without magnification). The representation of vascularization at the third and sixth days of incubation is perhaps typical of the series. The time of hatching is, however, somewhat oddly given as twenty-four days, perhaps as the result of retardation induced by the experimental incubation.

Fabrici's other major embryological work, *De formato foetu,* illustrates the way in which nature provides for the necessities of the fetus during its intrauterine life. It treats specifically of the umbilical vessels, the urachus, the fetal membranes, fetal waste products, the "carnea substantia" (placenta), and the uterus. The treatise includes comparative studies of morphological details in dogs, cats, rabbits, mice, guinea pigs, sheep, cattle, goats, roebuck, horses, pigs, birds, sharks, and man. Fabrici's description of the umbilical cord and its vessels is accurate, as is his differentiation of the action of the umbilical vessels in various animals; he also provides an adequate description of the right and left atria of the heart, the foramen ovale and the ductus arteriosus, the vena cava, and the pulmonary vein in the fetus.

The value of Fabrici's observations is, however, lessened by his need to impose a Galenic interpretation upon them. He posits that no fetal organ exercises any "public" action—that is, any function for the benefit of the whole organism—but only "private" ones; each attracts, utilizes, and voids nutriment for itself alone. The whole fetus needs only nutriment and vital spirit (which helps to digest the nutriment) in order to grow; the nutriment and vital spirit reach the fetus through the umbilical veins and arteries, with the maternal uterus thus doing the work of the fetal heart and liver. (If the umbilical vessels are ligated just above the umbilicus the fetal heart and arteries cease to pulsate, thereby demonstrating that the flow of vital spirit has been interrupted.) Although Fabrici differs from Galen on occasion (as when, for instance, he maintains that the blood which reaches the fetal liver does not need purification), his embryological theories are most often in agreement with him.

Fabrici champions Galen against Aranzio in the question of the relationship between the maternal and fetal blood vessels; he maintains that during pregnancy the uterine vessels terminate in apertures to which the fetal vessels are in some way united (although he does not specify the nature of this union). These connections are always of vein to vein and artery to artery to prevent confusion in the distribution of the vital spirit. Fabrici presents observations of the chorionic villi and the crypts of the placenta and interprets them respectively as the terminals of umbilical and uterine vessels. He considered the chorionic villi as patent, however, whereas Harvey thought them to be blind. Aranzio and Harvey, in their belief in the separateness of maternal and fetal circulation, were closer to the truth than Fabrici. Fabrici's explanation of the atrophy of the umbilical cord following birth is Aristotelian; having served its purpose, it is destined to decay.

Fabrici describes the relationship between the chorion and the allantois in some species in a fuller and more accurate manner than Vesalius, Colombo, or even Harvey. He follows tradition in considering the amnion to be a receptacle for fetal sweat and, in agreement with Falloppio, says that fetal urine is (except in species provided with allantois) stored in the chorion. He next examines the embryonic appendages of some herbivorous animals in light of this theory and states that Aranzio is in error when he says that the human fetus lacks a urachus and discharges urine into the amniotic cavity.

Fabrici then discusses fetal waste products and the Galenic principle whereby there are only six of them: sweat (in the amnion); urine (in the chorion or allantois); bile; phlegm; feces; and the white, caseous residue adhering to the skin, cast off by the fetus in the course of assimilating nutrition.

Although Fabrici's work on the umbilical vessels, the fetal membranes, and fetal waste products are of only limited (if any) originality, he does draw some original, if faulty, conclusions about the significance of the placenta, which he studied more fully than any of his predecessors, including Vesalius and Falloppio. Fabrici is the first to give a reasoned classification of the various forms of placentas and to attempt to correlate these forms with the various types of animals; he limits the term "placenta" (introduced by Colombo) to refer to the discoidal type of placenta found in humans and in some animals (including rabbits, mice, rats, and guinea pigs). He is also the first to study human decidua and the subplacenta of the guinea pig and to print illustrations of the chorionic villi and uterine crypts of horses, pigs, and ruminant animals. In passing he deals with the complex problem of the cotyledons, which had been controversial since the work of Praxagorus and Aristotle.

Fabrici contests Aranzio's view of the function of the placenta—that it acts as a uterine liver to purify the blood of the fetus—although he admits that a small amount of blood is purified by the fleshy placental substance (but only to provide for its own nutrition and hence as a "private" and not a "public" action). The work ends with a chapter containing a highly traditional account of how much nature does to ensure the safe birth of the fetus.

De formato foetu is, like its predecessor, illustrated. It contains thirty-four plates of great interest which illustrate, in some instances for the first time, various aspects of the anatomy of the uterus and of the fetus in humans and in sheep, cows, horses, pigs, dogs, rats, mice, guinea pigs, and sharks. As Fabrici was the first to study he was also the first to illustrate the decidua of the human uterus, the uterine crypts in animals (interpreted as the open ends of uterine vessels), and the subplacenta in guinea pigs. In addition, the work contains interesting plates of the venous and arterial ducts and of the omphalomesenteric vein and artery in the dog; the last plate in the book illustrates the development of serpents, a topic on which Fabrici does not touch in the text. The plates, the work of an unknown artist, are well executed although sometimes lacking clarity (although many fine details are shown), and they are better integrated into the text

than those of *De formatione ovi et pulli*. Adelmann discusses the embryological color illustrations.

Fabrici's surgical works are gathered in the *Pentateuchos cheirurgicum* (printed in Frankfurt am Main in 1592, and edited by a pupil of Fabrici, Johann Hartman Beyer, apparently without his consent) and in the *Operationes chirurgicae,* published in Venice in 1619 as an addendum to the *Pentateuchos.*

The five books of the *Pentateuchos* are primarily devoted to the description of tumors, wounds, ulcers and fistulas, fractures, and dislocations; to these the *Operationes* adds a description of surgical instruments (some of which are illustrated) and classic surgical techniques, including a discussion of particular technical expedients devised by Fabrici himself and emphasizing some differences between Fabrici's technique and that of others. Of particular interest are two plates illustrating an orthopedic device, in the shape of a man, designed to combine in one apparatus the principles for all existing devices for the correction of orthopedic injuries and deformities. A passage by Antonio Vallisneri indicates that this device was actually built and used.

Although Fabrici's surgical works have not yet been studied in any detail it is clear that they rely on both Hippocrates and Galen in diagnostics and therapy. (The medications that Fabrici prescribes are, for example, traditional ones.) Yet the books had great success and went through many editions in many languages; the versification of the first book of the *Pentateuchos* by Antonio Filippo Ciucci (Rome, 1653) can be taken as an exemplar of Fabrici's fame as a surgeon.

BIBLIOGRAPHY

I. ORIGINAL WORKS. Fabrici's works were collected into *Opera omnia anatomica et physiologica* (Leipzig, 1687); a later ed. (Leiden, 1738) is more complete. Individual works are cited in the text; unless otherwise specified, all were published in Padua.

Modern eds. are *Delle valvole delle vene,* with trans. and intro. by Felice Grondona (Milan, 1966); and *Dell'orecchio, organo dell'udito* and *Della laringe, organo della voce,* trans. and commentary by Luigi Stroppiana (Rome, 1967).

II. SECONDARY LITERATURE. On Fabrici's life and work see H. B. Adelmann, *The Embryological Treatises of Hieronymus Fabricius of Aquapendente* (Ithaca, N.Y., 1942); L. Belloni, "Di una avvenuta chiamata di Gaspare Tagliacozzi allo studio di Padova (1594) e di un consulto epistolare tra G. Mercuriali, G. Tagliacozzi e G. Fabrici d'Acquapendente sovra un caso di fistola retto-genitale," in *Rivista di storia delle scienze mediche e naturali,* **43** (1952); "Die deutsche Aussprache in einer kurzen Abhandlung von Conrad Hofmann an Hieronymus Fabrici ab Aquapendente," in *Sudhoffs Archiv für Geschichte der Medizin und der Naturwissenschaften,* **37** (1953); and "Valvole venose e flusso centrifugo del sangue. Cenni storici," in *Simposi Clinici Ciba,* **5** (1968); A. F. Ciucci, *L'Ospidale di Parnaso,* Bruno Zanobio, ed. (Milan, 1962); G. Favaro, "Contributi alla biografia di Girolamo Fabrici di Acquapendente," in *Memorie e documenti per la storia della Università di Padova* (Padua, 1922); "L'insegnamento anatomico di Girolamo Fabrici d'Acquapendente," in *Monografie storiche sullo studio di Padova. Contributo del R. Istituto Veneto di scienze, lettere ed arti alla celebrazzione del VII centenario della università* (Venice, 1922); K. J. Franklin, *De venarum ostiolis of Hieronymus Fabricius of Aquapendente* (Baltimore, 1933); E. Gurlt, *Geschichte der Chirurgie,* II (Berlin, 1898), 445–481; G. Sterzi, "Le 'Tabulae Anatomicae' ed i Codici marciani con note autografe di Hieronymus Fabricius ab Aquapendente," in *Anatomischer Anzeiger,* **35** (1910); and L. Stroppiana, "Realtà scomparse. Divagando tra G. Fabrizi d'Acquapendente e Antonio Vallisneri," in *Humana studia,* **4** (1952).

BRUNO ZANOBIO

FABRICIUS AB AQUAPENDENTE. *See* **Fabrici, Girolamo.**

FABRICIUS, JOHANN CHRISTIAN (*b.* Tønder, South Jutland, Denmark, 7 January 1745; *d.* Kiel, Germany, 3 March 1808), *entomology.*

Fabricius was undoubtedly one of the most distinguished entomologists and ranks with Carl de Geer, P. A. Latreille, A. G. Oliver, and other prominent specialists of earlier times. In many respects he surpassed them, especially as a theoretical natural scientist. Linnaeus, whose most important contribution hardly lay in the field of entomology, was full of admiration for him—a rather unusual attitude for the Nestor of Swedish science—and his colleagues throughout the world expressed great respect for his work. Quantitatively speaking, the most important part of Fabricius' work was concentrated in the field of descriptive systematics (taxonomy); qualitatively speaking, a very important section of his work fell into the advanced theoretical area of natural history. This is shown clearly in his *Philosophia entomologica* (1778), *Betrachtungen über die allgemeinen Einrichtungen in der Natur* (1781), and *Resultate natur-historischer Vorlesungen* (1804).

Two basic principles guided Fabricius' approach to entomological systematics: he distinguished, on the one hand, between the artificial and the natural characters; and, on the other hand, he stressed the importance of the various structures of the mouth. The terminology he applied to categories of higher systems differed somewhat from modern terminology: he used the words "classis" for "order" and "ordo" for what

we call "family"; furthermore, he founded his system on the genus and the species, which in his opinion constituted the main bases. It seemed especially important to Fabricius that genera were the natural combinations of related species. He believed that "classes" and "ordines" were artificial concepts. He seems to have understood that even genera can be classified into the natural system—the nearest equivalent to our present "families"—but he probably understood that the time was not yet ripe, that scientific knowledge and general outlook were too narrow for such classification. He thought (not without hesitation) of one large system based on the structure of the mouth organs as being the natural system (see *Philosophia entomologica,* p. 85; *cf.* p. 97).

It was Fabricius' greatest ambition to build a system based on the naturally defined genera, without doubt a definite and new contribution to insect systematics. He considered this more important than a dry description of the various species. In the latter area, however, his contributions are imposing: he named and described some 10,000 insects.

Less known than Fabricius' contributions in the field of insect systematics are his evolutionistic ideas and speculations. He considered systematics to be a means to understanding important scientific functions and phenomena in general. In a frequently quoted sentence he said: "As we would not call a man learned because he can read, so we would not call a man a scientist who knows nothing but the system" (*Resultate natur-historischer Vorlesungen,* p. 138). Many of his ideas concerning evolution sound amazingly modern. For instance, he considered it possible that a species could be formed through mixing existing species (i.e., some form of hybridization) and through morphological adaptation and modification. In his opinion, such phenomena caused an unbelievable wealth of forms and species of living organisms. Fabricius could not believe in haphazard creation and definitely thought that man originated from the great apes (*Resultate,* p. 208). He also discussed the influence of environment on the development of the species, as well as some selective phenomena (females prefer the strongest males, etc.). Henriksen even called Fabricius the "Father of Lamarckism" (1932, p. 80).

Fabricius did not lead the life of a sedentary scientist. He did much traveling, both on the Continent and in Great Britain. He studied for two years under Linnaeus in Uppsala (1762–1764), traveled in Germany on several occasions, in Holland (1766–1767), in Scotland, France, and Italy (1768–1769), and visited London during the summers of 1772–1775 and even later. He also went to Norway, Austria, Switzerland, and Russia. On these trips he came in close contact with the best-known scientists of his time and visited the greatest museums. Fabricius was an extrovert who was liked and appreciated everywhere, and his personality helped to create a fruitful mutual exchange of information and ideas.

Fabricius was professor of natural science and economics, first at the University of Copenhagen, then at Kiel. His extensive collections, as well as the material he described and named, are in the Fabricius collections of the Zoological Museum of Copenhagen, a great part on loan from Kiel. Other collections are in Paris at the Muséum National d'Histoire Naturelle (Bosc Collection), the British Museum (Natural History [Banks Collection]), and in Glasgow (Hunter Collection).

BIBLIOGRAPHY

I. ORIGINAL WORKS. Fabricius' major writings are *Systema entomologiae* (Flensburg–Leipzig, 1775); *Genera insectorum* (Kiel, n.d. [preface dated 26 Dec. 1776]); *Philosophia entomologica* (Hamburg–Kiel, 1778); *Betrachtungen über die allgemeinen Einrichtungen in der Natur* (Hamburg, 1781); *Species insectorum,* 2 vols. (Hamburg–Kiel, 1781); *Mantissa insectorum,* 2 vols. (Copenhagen, 1787); *Entomologia systematica,* 4 vols. and supp. (Copenhagen, 1792–1798); *Systema eleutheratorum,* 2 vols. (Kiel, 1801); *Systema rhyngotorum* (Brunswick, 1803); *Resultate naturhistorischer Vorlesungen* (Kiel, 1804; repr. page for page, Kiel, 1818); *Systema piezatorum* (Brunswick, 1804); *Systema antliatorum* (Brunswick, 1805); *Systema glossatorum* (Brunswick, 1807), only 112 pp. printed, three known copies; facs. ed. by F. Bryk (Neubrandenburg, 1938); and "Autobiographie des Naturforschers Fabricius," in *Kieler Blätter,* 1 (Kiel, 1819), 88–117, trans. from the Danish, with notes and commentary by F. W. Hope, in *Transactions of the Entomological Society of London,* 4 (1845), 1–16.

II. SECONDARY LITERATURE. On Fabricius and his work, see K. L. Henriksen, "Oversigt over Dansk Entomologis Historie," in *Entomologiske Meddelelser* (Copenhagen), 15 (1922–1937); J. Schuster, "Linné und Fabricius. Zu ihrem Leben und Werk," in *Münchener Beiträge zur Geschichte und Literatur der Naturwissenschaften und Medizin,* 4 (1928); R. A. Staig, *The Fabrician Types of Insects in the Hunterian Collection at Glasgow University,* 2 vols. (Cambridge, 1931–1940); S. L. Tuxen, "The Entomologist, J. C. Fabricius," in *Annual Review of Entomology,* 12 (1967), 1–14; and E. Zimsen, *The Type Material of J. C. Fabricius* (Copenhagen, 1964).

BENGT-OLOF LANDIN

FABRY, CHARLES (*b.* Marseilles, France, 11 June 1867; *d.* Paris, France, 11 December 1945), *physics.*

Charles Fabry and his brothers, Eugène, a mathe-

matician, and Louis, an astronomer, all graduated from the École Polytechnique in Paris. Fabry became *agrégé de physique* in 1889 and *docteur ès sciences* in physics in 1892 (University of Paris). After the customary assignment teaching at various lycées in France, Fabry returned to Marseilles to teach and do research at the university; he remained there from 1894 to 1920.

Fabry worked primarily on the precise measurement of optical interference effects, an interest already apparent in his thesis, "Théorie de la visibilité et de l'orientation des franges d'interférence" (Marseilles, 1892). He joined the laboratory of Macé de Lepinay, where this branch of optics was of primary concern. The majority of Fabry's research projects involved an interferometer that he invented with Alfred Pérot.

First devised in 1896, the Fabry-Pérot interferometer is based upon multiple reflection of light between two plane parallel half-silvered mirrors. The distribution of light produced by interference of rays that have undergone different numbers of reflections is characterized by extremely well defined maxima and minima, and monochromatic light produces a set of sharp concentric rings. Different wavelengths in the incident light can be distinguished by the sets of rings produced. This instrument produced sharper fringes than that devised by the American, Albert Michelson. For spectroscopy, their apparatus cheaply duplicates the advantages of the diffraction grating. Fabry and Pérot continued to work together; for about a decade they applied their interferometer to spectroscopy and metrology; an important project, for example, involved determining a series of standard wavelengths.

From 1906 Fabry worked with Henri Buisson on similar experiments and applications of the interference technique. In 1912 they verified for helium, neon, and krypton the Doppler-broadening of emission lines predicted by the kinetic theory of gases—an effect that Michelson had verified for metallic vapors at low pressure. A simple method, devised in 1914, enabled Fabry and Buisson to confirm experimentally in the laboratory the Doppler effect for light (this measurement had previously been made using stellar sources). By their technique a horizontal rotating white disk is illuminated so that points at opposite ends of a diameter constitute equal sources of light moving in opposite directions; the disk is viewed at an oblique angle, and the interferometer then detects the difference in position of the sets of rings produced by light from the two ends of the diameter.

Fabry's interest in astronomy—developed while observing with his brothers when they were students—led him to use the interferometer to study the spectra of the sun and stars, as well as to improve photometric techniques to measure the brightness of the nocturnal sky. As part of this work he showed that the ultraviolet absorption in the upper atmosphere is due to ozone.

As first director of the Institute of Optics and professor of physics both at the Sorbonne and at the École Polytechnique, Fabry spent the latter part of his life mainly in Paris, where he was elected to the Academy of Sciences in 1927. He was also a member of the International Committee on Weights and Measures and the Bureau of Longitudes; he received medals from the Royal Society, the Franklin Institute, and the National Academy of Sciences. Fabry was interested in the popularization of science; he taught a large public course in electrotechnology and wrote some popular works.

BIBLIOGRAPHY

I. ORIGINAL WORKS. Fabry's works include *Les applications des interférences lumineuses* (Paris, 1923); *Optique* (Paris, 1926; 4th ed., 1934), lectures given at the Sorbonne, Jean Mallassez and Maurice Virlogeux, eds.; "Histoire de la physique," in G. Hanotaux, ed., *Histoire de la nation française,* I (Paris, 1924), 165–418; "Mesure de petites épaisseurs en valeur absolue," in *Comptes rendus hebdomadaires des séances de l'Académie des sciences,* **123** (1896), 802–805, written with A. Pérot; "Sur une nouvelle méthode de spectroscopie intérferentielle," *ibid.,* **126** (1898), 34–36, written with A. Pérot; "Sur la largeur des raies spectrales et la production d'interférences à grande différence de marche," *ibid.,* **154** (1912), 1224–1227, written with H. Buisson; and "Vérification expérimentale du principe de Doppler-Fizeau," in *Journal de physique,* 5th ser., **9** (1919), 234–239, written with H. Buisson. Some of Fabry's work is repr. in his *Oeuvres choisies* (Paris, 1938), which includes a list of his publications (pp. 669–689).

II. SECONDARY LITERATURE. On Fabry and his work see Louis de Broglie, "Charles Fabry," in *Obituary Notices of the Royal Society,* V (1945–1948), 445–450; Maurice Caullery, "Notice nécrologique sur Charles Fabry," in *Comptes rendus hebdomadaires des séances de l'Académie des sciences,* **221** (1945), 721–724; and F. A. Jenkins and H. E. White, *Fundamentals of Optics* (New York, 1957), ch. 14.

SIGALIA DOSTROVSKY

FABRY, LOUIS (*b.* Marseilles, France, 20 April 1862; *d.* Les Lecques, near Toulon, France, 26 January 1939), *astronomy, applied celestial mechanics.*

Fabry, the older brother of the physicist Charles Fabry, was admitted to the École Polytechnique in 1880. He obtained his *licence ès sciences* at the Faculté de Marseille in 1883 and the following year was accepted in the school of practical astronomy recently created at the Observatoire de Paris. During his course of study, while practicing on the *équatorial coudé* that had just been placed in use, he had the extraordinary luck of discovering a comet.

The origin of these heavenly bodies was then much discussed—the existence of comets with hyperbolic orbits made it possible to hold the opinion that they originated outside of the solar system. Fabry, in his doctoral thesis in 1893, proved by statistical methods that this hypothesis was not compatible with the distribution presented by the elements of the orbits. He established in particular that the distribution of motions does not yield the dissymmetry that the motion of the sun would introduce if the comets came from infinity.

The subject, however, was not exhausted. Why, in fact, are certain orbits hyperbolic? To solve this problem, the Académie des Sciences posed it as a competition. The prize was shared by Fabry and one of his colleagues, Gaston Fayet; both showed that the orbits known as hyperbolic had become so by the action of planetary perturbations—all of them were originally elliptical.

The observation of the minor planets and the elaboration of rapid methods for identifying them and for calculating and improving their ephemerides constitute, along with his researches on the comets, the essential portion of Fabry's work. He was of that generation of astronomers who strove to cultivate observations for the use of their successors; he lived in a period in which new technology served observation, while the means of calculation were practically nonexistent.

Fabry was named in 1886 to the Observatoire de Nice and then in 1890 to that of Marseilles, where he remained until his retirement in 1925. He was elected a corresponding member of the Académie des Sciences in 1919.

BIBLIOGRAPHY

I. ORIGINAL WORKS. Fabry's principal works on comets are "Découverte d'une comète à l'Observatoire de Paris," in *Comptes rendus hebdomadaires des séances de l'Académie des sciences,* **101** (1885), 1121–1125; "Sur le calcul du grand axe des orbites cométaires," in *Bulletin astronomique,* **11** (1894), 485–488; "Études sur la probabilité des comètes hyperboliques et l'origine des comètes," his doctoral thesis of 1893, in *Annales de la Faculté des Sciences de Marseille,* **4** (1895), 1–214; and "Sur la véritable valeur du grand axe d'une orbite cométaire," in *Comptes rendus hebdomadaires des séances de l'Académie des sciences,* **138** (1904), 335–337.

His writings on the minor planets are "Tables numériques destinées à faciliter le calcul des éphémérides," in

Bulletin astronomique, **2** (1885), 453–463; "Procédé abrégé pour rectifier les éphémérides," *ibid.,* **20** (1903), 243–250; "L'identification des petites planètes," *ibid.,* **30** (1913), 49–64; "Sur la rectification des orbites des planètes," *ibid.,* **31** (1914), 68–79; "Sur l'emploi des latitudes géocentriques pour faciliter l'identification des petites planètes," in *Comptes rendus hebdomadaires des séances de l'Académie des sciences,* **172** (1921), 27–31; and "Nouvelles formules pour le calcul de la ligne de recherche," *ibid.,* **173** (1921), 892–894.

His observations, as well as the numerous orbits of minor planets that he determined, were generally published in *Bulletin astronomique,* beginning in 1885. His seismological writings appeared in *Comptes rendus hebdomadaires des séances de l'Académie des sciences,* **149-152** (1909–1911).

II. SECONDARY LITERATURE. On Fabry's life and work see G. Fayet, "Notice nécrologique sur L. Fabry," in *Comptes rendus hebdomadaires des séances de l'Académie des sciences,* **208** (1939), 545–547.

JACQUES R. LÉVY

AL-FADL IBN HĀTIM AL-NAYRĪZĪ. *See* al-Nayrīzī.

FAGNANO DEI TOSCHI, GIOVANNI FRANCESCO (*b.* Sinigaglia, Italy, 31 January 1715; *d.* Sinigaglia, 14 May 1797), *mathematics.*

Giovanni Francesco was the only child of Giulio Carlo Fagnano to show an interest in mathematics. He was ordained a priest and in 1752 was appointed canon of the cathedral of Sinigaglia. Three years later he was made archpriest. He wrote an unpublished treatise on the geometry of the triangle that was inspired by a similar work of his father's. Fagnano made several important contributions to the subject, among them the theorem that the triangle which has as its vertices the bases of the altitudes of any triangle has these altitudes as its bisectors. He also contributed several analytical communications to the *Acta eruditorum* and to other Italian and foreign reviews.

Among his most important results is that, given

$$S_n = \int x^n \sin x \, dx, \quad C_n = \int x^n \cos x \, dx,$$

we obtain

$$S_n = -x^n \cos x + nC_{n-1},$$
$$C_n = x^n \sin x - nS_{n-1}.$$

He also calculated the integrals:

$$\int \tan x \, dx = -\log \cos x, \quad \int \cot x \, dx = \log \sin x.$$

BIBLIOGRAPHY

Among Fagnano's writings is *Nova acta eruditorum* (1774), pp. 385–420. Two further sources of information on

his mathematical work are *Enciclopedia delle matematiche elementari,* I, pt. 2 (1932), 491–492; and Gino Loria, *Storia delle matematiche,* 2nd ed. (Milan, 1950), p. 664.

A. NATUCCI

FAGNANO DEI TOSCHI, GIULIO CARLO (*b.* Sinigaglia, Italy, 6 December 1682; *d.* Sinigaglia, 26 September 1766), *mathematics.*

Fagnano, the son of Francesco Fagnano and Camilla Bartolini, was born into a noble family that had included Pope Honorius II and had been established in his native town for nearly 350 years. In 1723 he was appointed *gonfaloniere* of Sinigaglia; while he held this office he was subjected to calumny by envious fellow citizens. He was the father of many children, among them Giovanni Francesco, a distinguished mathematician.

Fagnano began to study mathematics after reading the first volume of Malebranche's *Recherche de la verité*; and although he was self-educated, he soon made such progress that he became famous both in Italy and abroad. In 1721 Louis XV conferred upon him the title of count; in 1745 Pope Benedict XIV made him a marquis of Sant' Onofrio. He belonged to the Royal Society of London and the Berlin Academy of Sciences, and at his death he had been nominated for membership in the Paris Academy of Sciences. Fagnano maintained correspondence with many contemporary mathematicians, especially Grandi, Riccati, Leseur, and Jacquier; he was praised by Euler and Fontenelle, the permanent secretary of the French Academy. Lagrange, at the age of twenty, turned to him for help in publishing his first work.

Fagnano's works were published at intervals in the *Giornale dei letterati* and in the Raccolta Calogera. These were later collected and with other, unpublished works included in *Produzioni matematiche.*

In algebra Fagnano suggested new methods for the solution of equations of the second, third, and fourth degrees. He also organized in a rational manner the knowledge that scientists had of imaginary numbers, establishing for them a special algorithm that was far better than Bombelli's primitive one. In this field he established the well-known formula

$$\frac{\pi}{4} = \log\left(\frac{1-i}{1+i}\right)^{1/2}.$$

This is reminiscent of Euler's celebrated formula

$$e^{\pi i} = -1,$$

which unites the four most important numbers in mathematics.

In geometry Fagnano formulated a general theory

of geometric proportions that is more noteworthy than the countless writings, published previously, that were intended to illustrate book V of Euclid's *Elements*. Much more important, however, is his work on the triangle, for which he may well be considered the founder of the geometry of the triangle. Some of the problems solved are as follows:

To find in the plane of a triangle, *ABC*, a point, *P*, that will reduce to the minimum the sum *PA* + *PB* + *PC* or the sum $PA^2 + PB^2 + PC^2$.

To find in the plane of a quadrangle, *ABCD*, a point that will render minimum the sum *PA* + *PB* + *PC* + *PD*.

Two of Fagnano's major findings are (1) that the sum of the squares of the distances of the center of gravity of a triangle from the vertices equals one-third the sum of the squares of the sides and (2) that given a triangle, *ABC*, for every point *P* of *BC* we may construct an inscribed triangle, with its vertex at *P*, of minimum perimeter. He also solved the problem proposed by Simon Lhuilier: Draw through a given point the straight line on which two given straight lines cut off the minimum segment. This leads analytically to a third-degree equation.

The most important results achieved by Fagnano, however, were in analytical geometry and in integral calculus. He rectified the ellipse $x^2 + 2y^2 = a^2$, which has as its major axis the mean proportion between the minor axis and its double. The equation

$$(x^2 + y^2)^2 - 2a^2(x^2 - y^2) = 0$$

represents a fourth-degree curve that, owing to its shape, is called "lemniscate," a term derived from the Greek *lemniscata*. This was first studied by Jakob I Bernoulli (1694), but it was made famous by Fagnano's research. He established its rectification and demonstrated that each of its arcs may be divided with ruler and compass into *n* equal parts when *n* is of the form $2 \cdot 2^m$, $3 \cdot 2^m$, $5 \cdot 2^m$.

He gave the name "elliptical integrals" to integrals of the form

$$\int f\left(x_1 \sqrt{P[x]}\right) dx$$

in which $P(x)$ is a polynomial of the third or fourth degree. Euler found a basic result in their theory, known as the theorem of addition, that includes the results first found by Fagnano in the lemniscate arcs. For this reason some have considered Fagnano's work the forerunner of the theory of elliptic functions—a claim undoubtedly put forward by Legendre.

Disputes, encouraged by Fagnano's uncle Giovanni, arose over who deserved the credit of priority in these studies—Fagnano or Nikolaus I Bernoulli. On the advice of his friend Riccati, Fagnano soon put an end to these arguments.

Fagnano also found the area of the lemniscate, thus demonstrating that Tschirnhausen's opinion was erroneous and that it was impossible to square the area composed of several leaves.

In 1714 Fagnano proposed the following problem: Given a biquadratic parabola of the form $y = x^4$, and given a portion of it, determine another portion of the same curve in such a manner that the difference between the portions is rectifiable. Since no one replied, Fagnano himself published the solution, thus extending the method to infinite species of rectifiable parabolas. Fagnano also studied the problem of squaring hyperbolic spaces.

BIBLIOGRAPHY

I. ORIGINAL WORKS. Among Fagnano's works are *Giornale dei letterati,* **19** (1714), 438; *Produzioni matematiche,* 2 vols. (Pesaro, 1750); and *Opere matematiche del marchese G. C. de' Toschi di Fagnano,* V. Volterra, G. Loria, and D. Gambioli, eds. (Rome, 1911).

II. SECONDARY LITERATURE. On Fagnano or his work, see Luigi Bianchi, *Lezioni sulla teoria delle funzioni di variabile complessa e delle funzioni ellittiche* (Pisa, 1901), p. 250; Gino Loria, *Curve piane speciali algebriche e trascendenti. Teoria e storia* (Milan, 1930), I, 257 ff., and *Storia delle matematiche,* 2nd ed. (Milan, 1950), pp. 664–666; and A. Natucci, "Anton Maria Legendre inventore della teoria delle funzioni ellittiche," in *Archimede,* **4,** no. 6 (1952), 261.

There is an article on Fagnano in *Enciclopedia italiana* (1932), XIV, in which he is called Giulio Cesare. He is mentioned twice by Eugenio G. Togliatti in *Enciclopedia delle matematiche elementari:* II, pt. 1, 184, and in "Massimi e minimi," II, pt. 2. See also *Elogi e biografie d'illustri italiani del conte Giuseppe Mamiani della Rovere* (Florence, 1845), pp. 63–104.

A. NATUCCI

FAHRENHEIT, DANIEL GABRIEL (*b.* Danzig [Gdansk], Poland, 24 May 1686; *d.* The Hague, Netherlands, 16 September 1736), *experimental physics.*

Fahrenheit was the scion of a wealthy merchant family that had come to Danzig from Königsberg in the middle of the seventeenth century. His father, Daniel, married Concordia Schumann, the daughter of a Danzig wholesaler. From this union there were five children, three girls and two boys, of whom Daniel was the eldest.

In 1701 Fahrenheit's parents died suddenly, and his guardian sent him to Amsterdam to learn business. It was there, apparently, that Fahrenheit first became acquainted with, and then fascinated by, the rather specialized and small but rapidly growing business of making scientific instruments. About 1707 he began his years of wandering, during which he acquired the techniques of his trade by observing the practices of other scientists and instrument makers. He traveled throughout Germany, visiting his native city of Danzig as well as Berlin, Halle, Leipzig, and Dresden. He met Olaus Roemer in Copenhagen in 1708, and in 1715 he entered into correspondence with Leibniz about a clock for determining longitude at sea. In 1714 Christian von Wolff published a description of one of Fahrenheit's early thermometers in the *Acta eruditorum.* Fahrenheit returned to Amsterdam in 1717 and established himself as a maker of scientific instruments. There he became acquainted with three of the greatest Dutch scientists of his era: W. J. 'sGravesande, Hermann Boerhaave, and Pieter van Musschenbroek. In 1724 he was admitted to the Royal Society, and in the same year he published in the *Philosophical Transactions* his only scientific writings, five brief articles in Latin. Just before his death in 1736, Fahrenheit took out a patent on a pumping device that he hoped would be useful in draining Dutch polders.

Fahrenheit's most significant achievement was his development of the standard thermometric scale that bears his name. Nearly a century had passed since the construction of the first primitive thermometers, and although many of the basic problems of thermometry had been solved, no standard thermometric scale had been developed that would allow scientists in different locations to compare temperatures. About 1701 Olaus Roemer had constructed a spirit thermometer based upon two universal fiducial points. The upper fixed point, determined by the temperature of boiling water, was labeled 60°; the lower fixed point, determined by the temperature of melting ice, was set at 7-1/2°. This latter, seemingly arbitrary, number was chosen to allow exactly 1/8 of the entire scale to stand below the freezing point. Since 0° on the Roemer scale approximated the temperature of an ice and salt mixture (which was widely considered to be at the coldest possible temperature), all readings on Roemer's thermometer were assumed to be positive.

Roemer did not publish anything about his thermometer, and its existence was unknown to most of his contemporaries except Fahrenheit, who thought mistakenly that his own thermometric scale was patterned after Roemer's. In 1708, while visiting Roemer, Fahrenheit watched the Danish astronomer as he graduated several thermometers. These particular instruments were being graduated to a scale of 22-1/2°, or 3/8 of Roemer's standard scale of 60°.

Since most of the scale would then be in the temperate range, it is probable that Roemer was designing them for meteorological purposes. In a letter addressed to Boerhaave, Fahrenheit gave the following description of Roemer's procedure.

> I found that he had stood several thermometers in water and ice, and later he dipped these in warm water, which was at blood-heat [*welches blutwarm war*], and after he had marked these two limits on all the thermometers, half the distance between them was added below the point in the vessel with ice, and the whole distance divided into 22-1/2 parts, beginning with 0 at the bottom then 7-1/2 for the point in the vessel with ice and 22-1/2 degrees for that at blood-heat.[1]

The problem with Fahrenheit's account is that he took Roemer's "blood-warm" (22-1/2°) to be a primary fiducial point, fixed quite literally at the temperature of the human blood. In fact, 22-1/2° on the Roemer scale is considerably below body temperature (by about 15° on the modern Fahrenheit thermometer). Furthermore, Roemer used boiling water (set at 60°), not blood temperature, as his upper fixed point. The simplest explanation for Fahrenheit's misunderstanding of the Roemer scale seems to lie in the ambiguity of the term "blood-warm." It can mean either a tepid heat or the exact temperature of the human blood.[2] Roemer probably intended to convey the former meaning, and Fahrenheit obviously understood the latter one.

When Fahrenheit began producing thermometers of his own, he graduated them after what he believed were Roemer's methods. The upper fixed point (labeled 22-1/2°) was determined by placing the bulb of the thermometer in the mouth or armpit of a healthy male.[3] The lower fixed point (labeled 7-1/2°) was determined by an ice and water mixture. In addition, Fahrenheit divided each degree into four parts, so that the upper point became 90° and the lower one 30°. Later (in 1717) he moved the upper point to 96° and the lower one to 32° in order to eliminate "inconvenient and awkward fractions."[4]

In an article on the boiling points of various liquids, Fahrenheit reported that the boiling temperature of water was 212° on his thermometric scale. This figure was actually several degrees higher than it should have been. After Fahrenheit's death it became standard practice to graduate Fahrenheit thermometers with the boiling point of water (set at 212°) as the upper fixed point. As a result, normal body temperature became 98.6° instead of Fahrenheit's 96°. This variant of the Fahrenheit scale became standard throughout Holland and Britain. Today it is used for meteorological purposes in most English-speaking countries.

Fahrenheit knew that the boiling temperature of water varied with the atmospheric pressure, and on this principle he constructed a hypsometric thermometer that enabled one to determine the atmospheric pressure directly from a reading of the boiling point of water. He also invented a hydrometer that became a model for subsequent developments.

In the early eighteenth century, it was not at all unusual for a person without formal scientific training to be admitted to the Royal Society. Makers of scientific instruments could be particularly valuable members because they often operated on the farthest frontiers of scientific knowledge, defining universal constants on which to scale their instruments and isolating the variables that affected their operation. In order to make reliable instruments that would be useful to the scientific community as a whole, Fahrenheit was obliged to concern himself with a wide variety of scientific problems: measuring the expansion of glass, assessing the thermometric behavior of mercury and alcohol, describing the effects of atmospheric pressure on the boiling points of liquids and establishing the densities of various substances. His direct contributions, it is true, were small, but in raising appreciably the level of precision that was obtainable in many scientific observations, Fahrenheit affected profoundly the course of experimental physics in the eighteenth century.

NOTES

1. Quoted from W. E. Knowles Middleton, *A History of the Thermometer and Its Use in Meteorology* (Baltimore, 1966), p. 71.
2. J. U. W. Grimm's *Deutsches Wörterbuch* gives "tepid" as the definition of *blutwarm*. *Blutwärme* ("blood-heat") is defined as the temperature of the blood. Francis Hauksbee, in his *Physico-Mechanical Experiments,* wrote: "I caus'd some water to be heated about Blood-warm" (quoted in the *Oxford English Dictionary,* I, 933). The French translation (Paris, 1754, I, 385) reads ". . . je fis chauffer de l'eau jusqu'à ce qu'elle fût un peu plus que tiède." ("I caused some water to be heated until it was a little more than tepid.")
3. It seems curious that Fahrenheit should have stipulated that the subject be a man. Perhaps this is owing to some remnant of the once widely held notion that women naturally have a lower body temperature than men. (See Jacques Roger, *Les sciences de la vie dans la pensée française du xviiie siècle* [Paris, 1963], pp. 84 ff.)
4. Middleton, *loc. cit.*

BIBLIOGRAPHY

I. Original Works. All of Fahrenheit's published works appeared in the *Philosophical Transactions of the Royal Society,* **33** (1724): "Experimenta circa gradum caloris liquorum nonnullorum ebullientium instituta," pp. 1–3; "Experimenta & observationes de congelatione aquae in vacuo factae," pp. 78–84; "Materium quarundum gravitates specificae diversis temporibus ad varios scopos exploratae," pp. 114–115; "Araeometri novi descriptio & usus," pp. 140–141; "Barometri novi descriptio," pp. 179–180. These are translated into German in Ostwalds Klassiker der Exacten Wissenschaften, no. 57 (Leipzig, 1954), pp. 3–18.

II. Secondary Literature. There is an enormous quantity of literature on Fahrenheit. The following list is limited to recent articles of importance: Florian Cajori, "Note on the Fahrenheit Scale," in *Isis,* **4** (1921), 17–22; Ernst Cohen and W. A. T. Cohen-De Meester, "Daniel Gabriel Fahrenheit," in *Verhandelingen der Konenklijke akademie van wetenschappen,* sec. 1, **16,** no. 2 (1937), 1–37; N. Ernest Dorsey, "Fahrenheit and Roemer," in *Journal of the Washington Academy of Sciences,* **36,** no. 11 (1946), 361–372; and W. E. Knowles Middleton, *A History of the Thermometer and Its Use in Meteorology* (Baltimore, 1966), pp. 66–79.

J. B. Gough

FALCONER, HUGH (*b.* Forres, Scotland, 29 February 1808; *d.* London, England, 31 January 1865), *paleontology, botany.*

After studying successively at the universities of Aberdeen (where he took the M.A.) and Edinburgh (where he obtained the M.D.) Falconer went to India in 1830 as surgeon with the East India Company. In 1832 he was appointed superintendent of the botanic garden at Saharanpur, at the foot of the Siwalik Hills, part of the sub-Himalayan range. He returned to England in 1842 and was appointed in 1844 to superintend the arrangement of Indian fossils for the British Museum. In 1848 he went to Calcutta as superintendent of the botanic garden there and professor of botany at the Calcutta Medical College. He returned again to England in 1855, his health impaired.

Falconer's official posts were thus botanical, and in the course of his duties he explored mountainous country and made immense collections of plants, including new species, the genus *Falconeria* (Scrophulariaceae) being named for him in 1839. He was largely responsible for starting the cultivation of Indian tea (while at Saharanpur) and for the introduction into India of the quinine-bearing plant (while at Calcutta). But his scientific fame rests chiefly on his researches among the vertebrate fossils, particularly the mammals, which he and Captain (later Sir) Proby Cautley brought to light from among the late Tertiary rocks of the Siwalik Hills. He investigated these with extraordinary energy and skill, hunting the living animals around him and preparing their skeletons for comparison with the fossils. This vertebrate fossil fauna was unexampled for extent and richness in any region then known. It included species of mastodon, elephant, rhinoceros, hippopotamus, and giraffe, as well as some reptiles (crocodiles and tortoises) and fishes. This great work accomplished by Falconer and Cautley was recognized in England in 1837 by the bestowal on them jointly of the Geological Society's highest honor, the Wollaston Medal. Unfortunately, their discoveries were never fully described and illustrated.

During the last ten years of his life Falconer made researches into Pleistocene mammals and the evidences of prehistoric man, both in Britain and in various parts of Europe. At the time of his death he was foreign secretary of the Geological Society and a vice-president of the Royal Society.

BIBLIOGRAPHY

I. ORIGINAL WORKS. Falconer's paleontological writings, published and in manuscript, were gathered together and edited, with a biographical sketch, by Charles Murchison in *Palaeontological Memoirs and Notes of the Late Hugh Falconer, A.M., M.D.,* 2 vols. (London, 1868). Of these works the chief was the unfinished *Fauna antiqua Sivalensis,* written with P. T. Cautley, which began publication in 1846 (London). His botanical papers are listed by Murchison, I, lv–lvi.

II. SECONDARY LITERATURE. In addition to Murchison, above, see also W. J. Hamilton, obituary notice in *Proceedings of the Geological Society of London,* **21** (1865), xlv–xlix; Charles Lyell, remarks in the course of the president's anniversary address in *Proceedings of the Geological Society of London,* **2** (1837), 508–510; obituary notice in *Proceedings of the Royal Society,* **15** (1866–1867), 14–20; and H. B. Woodward, *The History of the Geological Society of London* (London, 1908), pp. 128–129.

JOHN CHALLINOR

FALLOPPIO, GABRIELE (*b.* Modena, Italy, 1523 [?]; *d.* Padua, Italy, 3 October 1562), *medicine.*

Gabriele Falloppio, son of Geronimo and Caterina Falloppio, was first educated in the classics, but after the death of his father and ensuing financial difficulties, he was directed toward a career in the church. With improvement in the family's finances he turned to medicine, studying in Modena under Niccolo Machella and, according to the records, dissecting a body for his teacher in December 1544. Although still a student, but perhaps in need of funds, Falloppio began the practice of surgery but displayed so little aptitude for that subject—as demonstrated by the fatal outcome of a number of his cases—that he soon thereafter abandoned it and returned wholly to the study of medicine. There is a possibility that he spent some time at Padua under Giambattista da Monte and Matteo Realdo Colombo, the successor of Vesalius; and it can be stated with certainty that he studied for a period, about 1548, in Ferrara under the direction of "my teacher" Antonio Musa Brasavola and Giambattista Canano.

Falloppio was appointed to the chair of pharmacy in Ferrara and in 1549 accepted the chair of anatomy at the University of Pisa, where he was wrongfully accused of practicing human vivisection. During this period he spent some time in Florence dissecting the bodies of lions in the Medici zoo and thereby disproving Aristotle's statement that the bones of lions are wholly solid and without marrow. Despite the charges against him, he was offered and accepted the famous chair of anatomy at Padua as a successor to Colombo. He took up his duties toward the end of 1551 and lectured and demonstrated with such success as to attract a number of later to be distinguished students, including the comparative anatomist Volcher Coiter. Falloppio was fully appreciated by the university's authorities; he was regularly reappointed to the chair of anatomy until advancing pulmonary tuberculosis first limited his activities and finally killed him.

Of the various works by and attributed to Falloppio only the *Observationes anatomicae* (1561) was published during his lifetime and can be said with certainty to be fully authentic. It is not, however, a general and systematic textbook of anatomy but an unillustrated commentary or series of observations on the *De humani corporis fabrica* of Vesalius, in which Falloppio sought to correct errors committed by his illustrious predecessor and to present new material hitherto overlooked. His criticism, contrary to a characteristic of that age, is temperate and friendly, so that it is not uncommon to find the object of the criticism referred to as the "divine Vesalius," upon whose scientific foundations Falloppio, as a worthy successor, was willing to admit that he had based his own work. Since the *Observationes anatomicae* is not an all-inclusive study of anatomy, it never received the popular acclaim given, for example, to the *De re anatomica* (1559) of Colombo. It is, nevertheless, a work of greater originality.

Falloppio's investigations were the consequence of dissection not only of adult human bodies but also of fetuses, newborn infants, and children "up to the first seven months, and in several beyond" (fol. 17*v*). He was thus able to make a number of observations and contributions to knowledge of primary and secondary centers of ossification. His most notable contributions of this nature were his descriptions of the ossification of the occiput (fols. 21*r* ff.), of the sternum (fols. 51r–52v), and of the primary centers of the innominate bone (fols. 59r–60r). In his studies of the teeth Falloppio provided for the first time a clear description of primary dentition, the follicle of the tooth bud, and the manner of growth and replacement of the primary by the secondary tooth, as well as the first denial of the belief that teeth and bones are derived from the same tissues (fols. 39r–42v). Falloppio's description of the auditory apparatus was superior to that of Vesalius and includes the first clear account of the round and oval windows, the cochlea, the semicircular canals, and the scala vestibuli and tympani (fols. 27r–30v). He also referred to the third ossicle of the ear, the stapes, actually already mentioned in print (Pedro Jimeno, 1549), but he declared that it had been first described orally in lectures by Giovanni Filippo Ingrassia during a visit to Rome in 1546 (fols. 25r–27r).

Not the least important of Falloppio's contributions were those dealing with the muscles, among which were his relatively detailed account of the subcutaneous muscles of the scalp and face (fols. 62r–v, 63v–64r, 66v–68r) and his first description of the arrangement of the extrinsic muscles of the ear (fols. 62v–63r). In his investigation of the muscles of the head and neck he discovered and described the external pterygoid muscle (fol. 72v), analyzed the functions of the muscles of mastication (fols. 71v–73r), described the *tensor* and *levator veli palati* (fols. 76v–77r), and redescribed with greater clarity some of the intrinsic muscles of the larynx (fols. 77v–79r). His greatest contribution to the study of the muscles of the head, however, was his account of the arrangement and functions of the muscles of the orbit (fols. 64r–66v, 68r–71v). For the first time, he described the *levator palpebrae,* even though this honor was later to be claimed by Giulio Cesare Aranzi (1587). He observed the nictitating membrane of mammals, first described by Aristotle and thereafter seemingly disregarded. He recognized the compound action of the oblique muscles, and he was the first to describe and provide the name for the trochlea of the superior oblique muscle. In addition to further, lesser contributions to the study of the muscles of the trunk, he added notably to knowledge of the intrinsic muscles of the hand and of their action (fols. 101v–108v), separated the adductor mass of the thigh into its three elements, and noted the *quadratus femoris* (fol. 101v), which had been previously overlooked.

In considering the vascular system, Falloppio denied the long-held belief that the walls of the vessels were composed of fibers which by their direction controlled the flow of blood (fol. 114v). Curiously enough, however, he denied the existence of the venous valves (fol. 118v), which were actually known as early as 1546 and described by Vesalius (1555), and he failed to refer to the description of the pulmonary transit of the blood provided in detail by Colombo (1559). He did, on the other hand, give the first relatively adequate account of the distribution of the carotid arteries and of the cerebral circulation (fols. 121v–126r). He made a major contribution to knowledge of the nervous system through his clear distinction and description of the trochlear nerve (hitherto known only through the briefest mention by Alessandro Achillini [1520]); he traced it to its origin in the brain stem, demonstrated its exclusive termination in the superior oblique muscle of the eye, and satisfied himself beyond any doubt that this nerve was an entity deserving separate classification, and that it was "reflected on a cartilaginous pulley, and it turns the eye inwards" (fols. 155r–156r). Unwilling to upset the classic number (seven) and arrangement of the cranial nerves, he increased that number only to eight, although in fact he recognized eleven of the twelve cranial nerves.

Falloppio's most important contribution to urology is his account of the kidneys, although it is always difficult to determine whether the priority is properly that of Falloppio or of his contemporary Bartolomeo Eustachi. With this understood, attention may be called to what seems to have been the earliest account of a case of bilateral duplication of the ureter and renal vessels: "Here at Padua I have observed and pointed out to my spectators double urinary passages and double sinuses in the middle of each human kidney, as well as many other things departing from the normal" (fols. 179v–180r). Falloppio seems, moreover, to have been the first to observe the straight tubules (fol. 180r–v) that are, however, eponymously named from Bellini's more detailed description of 1662, and he noted the multiple calyxes of the human kidney (fols. 180v–181r). It was in the course of these remarks on the kidney that Falloppio criticized Vesalius for describing and illustrating in the *Fabrica* (1543) the unipapillary kidney of the dog instead of that of man—although he readily recognized Vesalius' need to use the less fatty kidney of the dog in order to permit a better illustration of that organ's structure. Falloppio further proposed the comparison of the renal papillae to small stills distilling off the urine from the blood (fols. 181v–182r). He also first described the three muscle coats of the urinary bladder: "It possesses three tunics, as do the stomach and intestines"; and the bladder's internal sphincter "formed by nature to contain the urine and prevent its being strained out" (fol. 182r).

Falloppio's name is perhaps most closely associated

with his description of the uterine or fallopian tubes, which in fact he described correctly as resembling small trumpets: "[The extremity] resembles the bell of a brass trumpet, wherefore the seminal passage, with or without its windings, resembles a kind of trumpet" (fol. 197*r*). Owing, however, to incorrect interpretation of Falloppio's word *tuba,* some of the descriptive meaning has been lost in English. His description of the uterine tubes is sufficiently accurate in detail to justify their bearing his name; he furthermore described the clitoris (fol. 193*r–v*), asserted the existence of the hymen in virgins, a matter long under dispute (fol. 194*r*), coined the word "vagina" (192*r*) for what had previously been called the cervix or neck of the uterus, and disproved the popular notion that the penis entered the uterus during coition (fol. 192*v*). He described certain vesicle-like structures filled with an aqueous fluid and others with a yellow humor (fol. 195*v*)—these may represent Graafian follicles or possibly a corpus luteum, and Falloppio's would therefore be the second mention of these structures after a somewhat similar account by Vesalius (1555).

In the preface to the *Observationes anatomicae* there is promise of a much larger, more detailed, illustrated work, mentioned as if it were well on the way toward completion. No trace of it remains today.

Falloppio may be called a student of Vesalius through the latter's books, and both his life and his book indicate a spiritual relationship to the earlier anatomist. In the *Observationes anatomicae* the Vesalian obligation to dissect, observe, and weigh one's findings by independent judgment is everywhere apparent—as well as being consciously expressed by the author, to whose criticisms Vesalius replied in *Anatomicarum Gabrielis Falloppii observationum examen* (1564).

The remainder of Falloppio's writings, originally lecture notes, were edited for publication at various times after his death and may therefore represent more or less than the original content. Of these, *Expositio in librum Galeni de ossibus* (1570), *Observationes de venis* (1570), *De humani corporis anatome compendium* (1571), and *De partibus similaribus humani corporis* (1575) deal with anatomy. Further works are concerned with syphilis, balneology, surgery, and the composition of drugs. The popular *Secreti diversi et miracolosi* (1563), often attributed to Falloppio, is spurious.

BIBLIOGRAPHY

I. ORIGINAL WORKS. Falloppio's works are most accessible in the collected editions *Opera omnia* (Venice, 1584; 2 vols., Frankfurt, 1600; 3 vols., Venice, 1606). The *Observationes anatomicae* has been reproduced in facsimile with Italian translation and notes, *Observationes anatomicae a cura di Gabriella Righi e Pericle Di Pietro,* 2 vols. (Modena, 1964).

II. SECONDARY LITERATURE. The best biographical study remains that of Giuseppe Favaro, *Gabrielle Falloppia Modenese* (*MDXXIII–MDLXII*) (Modena, 1928). Several studies of particular aspects of Falloppio's work include P. Di Pietro and G. Cavazzuti, "La descrizione falloppiana delle tube uterine," in *Acta medicae historiae patavina,* **11** (1964–1965), 51–60; Pietro Franceschini, "Luci e ombre nella storia delle trombe di Falloppia," in *Physis,* **7** (1965), 215–250; C. D. O'Malley, "Gabriele Falloppia's Account of the Cranial Nerves," in *Medizingeschichte im Spektrum. Festschrift zum fünfundsechzigsten Geburtstag von Johannes Steudel* (Wiesbaden, 1966), pp. 132–137; and "Gabriele Falloppia's Account of the Orbital Muscles," in *Medicine, Science and Culture. Historical Essays in Honor of Owsei Temkin* (Baltimore, 1968), pp. 77–85. On the literary relations of Falloppio and Vesalius, see C. D. O'Malley, *Andreas Vesalius of Brussels* (Berkeley-Los Angeles, 1964), pp. 289 ff.

C. D. O'MALLEY

FANKUCHEN, ISIDOR (*b.* Brooklyn, New York, 19 July 1905; *d.* Brooklyn, 28 June 1964), *crystallography.*

Fankuchen's parents were of modest means; he had two brothers and certainly was not spoiled in his youth. Intelligent and hard-working, he put himself through school by running a radio repair shop. Having obtained the B.S. from Cooper Union in 1926, he entered Cornell University as Hecksher fellow in 1929, married Dina Dardik in 1931, and received the Ph.D. under C. C. Murdock in 1933. In England as a fellow of the Schweinburg Foundation, he worked under Sir Lawrence Bragg in Manchester (1934–1936), then under J. D. Bernal at the Crystallographic Laboratory in Cambridge (1936–1938) and Birkbeck College in London (1938–1939).

On his return to the United States, Fankuchen held a national research fellowship in protein chemistry at the Massachusetts Institute of Technology (1939–1941) and served briefly as associate director of the Anderson Institute for Biological Research in Red Wing, Minnesota (1941–1942). In 1942, the year Cambridge University awarded him his second Ph.D., he joined the faculty of the Polytechnic Institute of Brooklyn, where he soon became head of the division of applied physics (1946) and where he remained until his death.

Fankuchen exerted a great influence on the teaching of crystallography and X-ray diffraction. By 1950 the Polytechnic Institute of Brooklyn had been turned into a teeming center for crystallographic research, as Fankuchen and H. F. Mark were joined by P. P. Ewald, Rudolf Brill, and David Harker. The monthly meetings of the "Point Group," Fankuchen's seminar, regularly attracted a number of crystallographers from outside the New York area. Fankuchen also organized intensive summer courses in X-ray diffraction, intended primarily for scientists in related disciplines, and thereby furthered the dissemination of crystallographic concepts in the scientific community. The Polycrystal Book Service, from which any crystallographic book can be purchased directly, is one of his creations.

From 1948 until his death, Fankuchen was the first American editor of *Acta crystallographica;* he fulfilled his editorial duties with unusual distinction, thanks to his keen critical sense and absolute scientific integrity. He belonged to many scientific societies; he was a charter member of both the American Society for X-Ray and Electron Diffraction and the Crystallographic Society of America; in 1950, when these two organizations merged to form the American Crystallographic Association, he became its first president. At the time of his death, he was chairman of the National Committee for Crystallography.

The mark of Fankuchen's scientific production is its diversity. His interests ranged widely through physics, chemistry, biology, and even mineralogy and metallurgy. He applied X-ray diffraction to new problems and refined or developed the necessary techniques and apparatus, as, for example, his very ingenious condensing monochromator to provide the intense X-ray beam required by his work on tomato mosaic virus. Of the crystal structures that he and his co-workers determined, the most memorable (of which he published a description in 1938, with Bernal and D. P. Riley) is that of the tomato bushy-stunt virus, a living crystalline substance.

For three years Fankuchen was an active member of the Bernal group and contributed to their results in the field of macromolecular compounds. He shared in the determination of the molecular weight of a tobacco-seed globulin (with Dorothy Crowfoot [Hodgkin], 1938) and the taking of the striking 5-degree-oscillation X-ray pattern of a single crystal of wet chymotrypsin (set forth in a paper that also dealt with hemoglobin, with Bernal and M. Perutz, 1938). He collaborated in the 1940 monograph on steroids (with Bernal and Crowfoot), in which crystal data are listed for more than eighty sterol derivatives, and in a famous paper on plant virus preparations (with Bernal, 1941).

In a series of publications, with Mark and others (1943–1949), Fankuchen studied fibers—chrysotile, chain polymers, fibrous proteins, and so forth. For this delicate work he devised a microcamera, in which the bore of a thermometer provided the collimator for the desired microbeam. With M. Bergmann he adapted this microcamera to the study of long spacings (1949), while he simultaneously investigated small-angle scattering from metal films (with B. Carroll, 1948). From 1947 to 1953, with a large number of collaborators, he conducted many studies of bones and teeth, in which investigations the microcamera proved its worth. First with H. S. Kaufman (1949), then with B. Post and R. S. Schwartz (1951), he successfully used diffraction at low temperatures utilizing a clever technique to prevent ice formation.

Fankuchen had a warm and buoyant personality. His kindness and helpfulness were legendary. His boisterous friendliness and deeply human qualities endeared him to all. During his last three years— while he was ill with cancer and knew it—he took great suffering in stride, never permitting it to interfere with his work. The Fankuchen Memorial Lectures perpetuate his name.

BIBLIOGRAPHY

I. ORIGINAL WORKS. A complete bibliography of Fankuchen's works may be found in J. D. H. Donnay, "Memorial of Isidor Fankuchen," in *American Mineralogist,* **50** (1965), 539–547.

II. SECONDARY LITERATURE. On Fankuchen's life and works see also J. D. Bernal, "Prof. Isadore Fankuchen," in *Nature,* **203** (1964), 916–917; J. D. H. Donnay, "Isidor Fankuchen, 1905–1964," in *Bulletin de la Société française de minéralogie (et de cristallographie)*, **87** (1964), 299; P. P. Ewald, "I. Fankuchen," in *Acta crystallographica,* **17** (1964), 1091–1093; and E. Ubell, "A Crystal Grows in Brooklyn," in *Norelco Reporter,* **10** (1963), 3, 39.

J. D. H. DONNAY

FANO, GINO (*b.* Mantua, Italy, 5 January 1871; *d.* Verona, Italy, 8 November 1952), *geometry.*

Fano was the son of Ugo and Angelica Fano; his father, a Garibaldian, had independent means. Gino studied from 1888 to 1892 at the University of Turin under Corrado Segre; while there, he met Guido Castelnuovo and specialized in geometry. In 1893–1894, while at Göttingen, he met Felix Klein, whose Erlangen program he had translated into Italian (*Annali di matematica,* 2nd ser., **17** [1889–1890], 307–343). From 1894 to 1899 Fano was assistant to Castelnuovo in Rome, was at Messina from 1899 to 1901, and in 1901 became professor at the University of Turin, where he taught until the Fascist laws of 1938 deprived him of his position. During World War

II he taught Italian students at an international camp near Lausanne. After 1946 Fano lectured in the United States and Italy. In 1911 he married Rosetta Cassin; two sons became professors in the United States.

Fano worked mainly in projective and algebraic geometry of *n*-space S_n. Early studies deal with line geometry and linear differential equations with algebraic coefficients; he also pioneered in finite geometry. Later work is on algebraic and especially cubic surfaces, as well as on manifolds with a continuous group of Cremona transformations. He showed the existence of irrational involutions in three-space S_3, i.e., of "unirational" manifolds not birationally representable on S_3. He also studied birational contact transformations and non-Euclidean and non-Archimedean geometries.

BIBLIOGRAPHY

I. ORIGINAL WORKS. Among Fano's many textbooks are *Lezioni di geometria descrittiva* (Turin, 1914; 3rd ed., 1925) and *Lezioni di geometria analitica e proiettiva* (Turin, 1930; 3rd ed., 1958), written with A. Terracini. Two of his articles appeared in *Encyclopädie der mathematischen Wissenschaften* (Leipzig, 1898–1935): "Gegensatz von synthetischer und analytischer Geometrie in seiner historischen Entwicklung im XIX. Jahrhundert," in III (Leipzig, 1907), 221–288; and "Kontinuierliche geometrische Gruppen," *ibid.,* 289–388.

II. SECONDARY LITERATURE. See A. Terracini, "Gino Fano," in *Bollettino dell'Unione matematica italiana,* 3rd ser., **7** (1952), 485–490; and "Gino Fano, 1871–1952, cenni commemorative," in *Atti dell'Accademia delle scienze* (Turin), classe di scienze fisiche, **87** (1953), 350–360. A bibliography, compiled by the editorial board, is in *Rendiconti del Seminario matematico, Università e politecnico di Torino,* **9** (1950), on pp. 33–45 of this issue dedicated to Fano (with portrait).

DIRK J. STRUIK

AL-FĀRĀBĪ, ABŪ NAṢR MUḤAMMAD IBN MUḤAMMAD IBN ṬARKHĀN IBN AWZALAGH

(Latin **Alf[h]arabius, Abunazar,** among other forms) (*b.* Wasīj, district of Fārāb, *ca.* 870; *d.* Damascus, 950), *philosophy, music.*

The district of Fārāb—on both sides of the middle Jaxartes (now the Syr Darya) at the mouth of its tributary, the Aris—was conquered and Islamized by the Samanids in 839–840, and al-Fārābī's grandfather may have been a pagan convert. His father is said to have been an army officer of noble Persian descent, apparently in the service of the Samanid emirs, who claimed descent from the old Sassanid emperors of Persia and patronized the emerging New Persian literature; but the family almost certainly spoke Sogdian or a Turkic dialect and exhibited Turkish manners and habits of dress. Al-Fārābī probably commenced his study of Islamic sciences (mainly law—the residents of Fārāb followed the legal school of al-Shāfiʿī) and music at Bukhara before going to Marv, where he seems to have begun his study of logic with the Syriac-speaking Nestorian Christian Yūḥannā ibn Ḥaylān, who was to continue teaching him in Baghdad and perhaps in Haran and whom he later acknowledged as his main teacher.

In the caliphate of al-Muʿtaḍid (892–902) both teacher and disciple went to Baghdad. Ibn Ḥaylān devoted himself to his religious duties, either monastic administration or theological instruction in Nestorian monasteries. Al-Fārābī was his only prominent student in logic and philosophy, and his only Muslim student. The complete silence of Arabic sources about Ibn Ḥaylān in any connection except as the teacher of al-Fārābī; Ibn Ḥaylān's isolation from the intellectual life of Baghdad, where Arabic was the main language of instruction in philosophy; and the report that al-Fārābī arrived at Baghdad knowing Turkish and a number of other languages but not Arabic (that is, he did not know Arabic well enough to study philosophy in that language) all indicate that he must have studied with Ibn Ḥaylān in Syriac or Greek or both. It is unlikely that the language of instruction (which included elaborate commentaries on Aristotle's *Organon*) could have been in any of the Turkic dialects, in Sogdian, or even in New Persian. In Baghdad, al-Fārābī set about perfecting his knowledge of Arabic (including the study of advanced Arabic grammar) in about 900 with the well-known philologist Ibn al-Sarrāj, in exchange for lessons in logic and music; he mastered it so well that his writings became a model of simple and clear Arabic philosophic prose. His newly acquired knowledge of Arabic enabled him to participate more fully in the philosophic circles in Baghdad (he is said to have attended the lecture courses of his older contemporary, the Nestorian Christian Mattā ibn Yūnus) and to make fuller use of the extensive body of scientific literature that existed in that language.

In the caliphate of al-Muktafī (902–908) or early in the caliphate of al-Muqtadir (908–932), al-Fārābī left Baghdad to continue his studies in Constantinople. Apparently he traveled first to Haran in the company of Ibn Ḥaylān. "After this [that is, after completing the study of Aristotle's *Posterior Analytics* with Ibn Ḥaylān] he traveled to the land of the Greeks and stayed in their land for eight years until he completed [the study of the] science[s] and learned the entire philosophic syllabus." This report is quoted by al-Khaṭṭābī (931–998) from al-Fārābī's own account of his studies. Al-Fārābī's linguistic interests, his contacts with the Syriac- and Greek-speaking teachers in Baghdad who could have provided him with the incentive and necessary information for the trip, and the relative ease with which a determined Muslim scholar (for example, the historian al-Masʿūdī) could visit Constantinople during this period make it difficult to doubt the authenticity of the report, which helps to explain a number of facets of al-Fārābī's works and thought, such as his access to certain traditions and texts and the character of his Platonism. Al-Fārābī's works, in turn, can now provide us with a better understanding of the course of philosophic studies at the University of Constantinople in the period between Photius and Michael Psellus.

Sometime between 910 and 920 al-Fārābī returned to Baghdad to spend more than two decades teaching and writing, which established his reputation as the foremost Muslim philosopher and the greatest philosophic authority after Aristotle. His teacher Ibn Ḥaylān died in Baghdad sometime before 932. Although al-Fārābī must have had a number of students who later spread his works and teachings in Persia and Syria, his only students who are known by name are the prominent Jacobite Christian theologian and philosopher Yaḥyā ibn ʿAdī, who headed an active but hardly brilliant philosophic school in Baghdad until his death in 975, and his brother Ibrāhīm, who was still with al-Fārābī in Aleppo shortly before the latter's death. Al-Fārābī's true disciples, however, were men like Ibn Sīnā, Ibn Rushd, and Maimonides, and his influence persisted in the learned tradition of the study of and commentary on Aristotle and Plato in Arabic, Hebrew, and Latin.

While defending the claims of philosophy and the philosophic way of life, al-Fārābī carefully avoided the theological, sectarian, and political controversies that raged in Baghdad during this period. He was not a member of the religious or scribal class. He must have had a number of friends among the many officers from his native land who formed the elite corps of the army and occupied high positions as bodyguards of the caliphs. Through them he probably came in contact with the prominent scribes and viziers who patronized the philosophic sciences, such as Ibn al-Furāt, ʿAlī ibn ʿĪsā, and Ibn Muqlah. He wrote his major work on music at the request of Abū Jaʿfar al-Karkhī, who became vizier in 936. This work was of great importance in the history of music theory and science. It is treated in the next section of the article.

It is unlikely that, at the age of seventy, al-Fārābī would have chosen to leave Baghdad merely in search of additional fame. By 942 the internal political confusion and the threat to the safety and well-being of the city's inhabitants had become extremely grave. The caliph, his viziers, and his bodyguard were so menaced by the rebellion of a former tax collector from the south that the caliph fled and took refuge with the Ḥamdānid prince of Mosul. Al-Fārābī departed to an area which, in 942, seemed more peaceful and was governed by a dynasty more congenial to him than the Hamdānids of Mosul. The Ikhshīdids, who ruled Egypt and Syria, were originally army officers from Farghānah, not far from al-Fārābī's birthplace in central Asia; and the Nubian slave Kāfūr, who held the power as regent, was a liberal patron of the arts. Al-Fārābī stayed in Damascus for about two years (during which he perhaps visited Aleppo) and then went to Egypt, no doubt driven there by the conflict in Syria between the Ikhshīdids and the Ḥamdānids, which was to last until 947. In the meantime, the Ḥamdānid prince Sayf al-Dawlah occupied Aleppo and Damascus and began to surround himself with a circle of learned men, whom he supported liberally. About a year before his death, al-Fārābī left Egypt to join Sayf al-Dawlah's circle. When he died in Damascus in 950, the prince and his courtiers performed the funeral prayers for him. He was buried outside the southern, or minor, gate of the city.

Al-Fārābī believed that science (that is, philosophy) had reached its highest development in the Socratic tradition, as embodied in the writings of Plato and Aristotle, their Greek commentators, and others who developed or made independent contributions to the natural and mathematical sciences. This tradition, which had declined in its original home and the spirit or purpose of which had become extinct or confused, must now find a new home in the civilization of Islam, wherein a new tradition of learning had been developing for more than two centuries; must reassert its claim as the supreme wisdom available to man; must infuse the new learning with critical understanding of its foundations and a sense of harmony, order, and purpose; and must clarify the principles and presuppositions of man's view of himself and the natural whole of which he is a part. Al-Fārābī's effort to recover, explain, defend, and reestablish this view of science as the highest stage of human wisdom took into account the gulf that separated the cultural environment of Greek science from the new Islamic environment in such matters as language, political and legal traditions, and characteristic habits of thought, and especially the pervasive impact of the revealed religions on the character and direction of political life and scientific thought. With persistence and skill, he set about teaching others what must have been the core of his own experience: the reconversion of man and his thought to the natural understanding, as distinguished from the multiplicity of customs, legal and political opinions, and religious beliefs.

Al-Fārābī's teaching activity followed an elaborate philosophic syllabus developed on a number of levels and based on the writings of Aristotle, a number of Platonic dialogues, and the works of Hippocrates and Galen, Euclid and Ptolemy, Plotinus and Porphyry, and the Greek commentators of the schools of Athens and Alexandria.

It began with introductory accounts of the opinions and writings of these authors, comprehensive accounts of the organization of the sciences, and epitomes of individual works. These were followed by a group of paraphrases of individual works, glosses on special difficulties in them, and expositions of particular themes. These led, finally, to a smaller group of lengthy commentaries in which the basic works of Aristotle were explored in great detail, taking into account the contributions, criticisms, and comments of earlier commentators.

Although al-Fārābī wrote commentaries on Euclid's *Elements* and Ptolemy's *Almagest*, the mathematical art to which he devoted particular attention was music. He wrote extensively on its history, theory, and instruments; and it is significant that his chosen art was the practice of music rather than medicine. Unlike the expository and didactic style of his mathematical writings, his specialized writings on natural science are for the most part polemical: against Galen's interpretations of Aristotle's views on the parts of animals; against John Philoponus' criticism of Aristotle's views on the eternity of the world and movement; against the physician al-Rāzī's views on matter, time, place, and atoms; against the theologian Ibn al-Rāwandī's account of dialectic, which was the method used by the theologians in natural science; against the doctrines of the theologians in general concerning atoms and vacuum; and finally against the scientific claims of astrology and alchemy. Judged from two of these writings which have been edited and studied (*On Vacuum* and *Against John Philoponus*), al-Fārābī's intention was not primarily to defend the doctrines of Aristotle against his critics, but rather to clarify the questions at issue, to ascertain the assumptions, coherence, and relevance of the arguments against Aristotle's natural science, and to determine whether they are based on genuine differences between Aristotle and his opponents or merely on a misunderstanding or misinterpretation of Aristotle, overconfidence in the theoretical implications

of certain experiments, or eagerness to support a religious doctrine. Al-Fārābī's openness regarding the foundations of Aristotle's natural science was restrained, however, because of his awareness of the decline of scientific learning since Aristotle's time and the overwhelming odds against free scientific inquiry in the new religious environment.

Al-Fārābī's departure from Aristotle is explicit in his writings on political science, which are inspired by a comprehensive view of Plato's philosophy and modeled after the *Republic* and the *Laws*. The intention of these works is both theoretical and practical. The theoretical intention emerges as al-Fārābī brings together the views of Plato and Aristotle and attempts to harmonize them without removing the underlying polarity between their two philosophies, leaving the reader with the conviction that the residual disagreement between the two leading philosophers may constitute the fundamental unresolved questions of science. The practical intention is expressed through the construction of constitutions proposed for cities whose institutions, doctrines, and practices are meant to promote, support, or at least not inhibit the development of scientific inquiry.

[A full bibliography follows the section below.]

MUHSIN MAHDI

AL-FĀRĀBĪ: Music.

Apart from a brief section in the *Iḥṣāʾ al-ʿulūm* (*Enumeration of the Sciences*), which provided medieval European theorists with one or two definitions but is otherwise of little interest, only one of al-Fārābī's musicological works has been edited, *Kitāb al-mūsīqā al-kabīr*. This is, however, probably the greatest Arabic treatise on music, and in it al-Fārābī not only demonstrates his mastery over the corpus of theory inherited from the Greeks but also justifies his reputation as an executant musician by giving a comprehensive account of some of the main features of contemporary practice.

For his methodology and definitions (set forth in the introduction) al-Fārābī draws upon the techniques of Greek philosophical inquiry. As for subject matter, however, some aspects with which the Greeks were concerned, such as ethical theories, receive scant attention, and the cosmological implications that had been of considerable interest to al-Kindī are ignored. The main theoretical section of the work begins with the physics of sound. Here al-Fārābī follows Aristotle—not uncritically, but nevertheless without any great originality: it was left to the late tenth-century Ikhwān al-Ṣafāʾ (Brethren of Purity) to introduce the concept of the spherical propagation of sound (which had been put forward in Aetius' *De placitis philosophorum*). There follow definitions of the basic elements of note, pitch, and interval, and then a detailed exposition of various tetrachord species (diatonic, chromatic, and enharmonic), not all of which are taken from Greek theorists; and of the structure of the Greek two-octave Greater Perfect System. This theoretical section concludes with a highly abstract analysis of rhythm based upon the concept of the *chronos protos*.

Musical practice is examined separately. Particular attention is paid to presenting the differing scales obtainable on the main melody instruments, especially the lutes (*ʿud, ṭunbūr khurāsānī, ṭunbūr baghdādī*), certain intervals being defined empirically (by halving distances between frets) rather than by ratios, as they are in the theoretical section. A second and less recondite passage on rhythm discusses the various forms taken by the most commonly used rhythmic cycles, and another important aspect of contemporary practice (generally ignored by other theorists) is illuminated by an account of various types of voice production and ornamentation and the way these should be utilized in the course of a composition.

The *Kitāb al-mūsīqā al-kabīr* appears to owe little to the musical treatises of al-Kindī, even when covering the same ground; and although later theorists recognize its importance, they in turn show considerable independence of mind. What influence al-Fārābī had is most apparent on the theoretical side: certain texts reproduce some of the more abstruse passages (especially on scale), but in the main it is his definitions that are cited—and then discussed, enlarged upon, and sometimes modified. Subsequent accounts of musical practice generally employ different descriptive methods and concentrate on other aspects of a complex and constantly evolving musical system.

Al-Fārābī's *Kitāb al-mūsīqā al-kabīr* is translated in R. d'Erlanger, *La musique arabe,* I and II (Paris, 1930–1935).

O. WRIGHT

BIBLIOGRAPHY

I. ORIGINAL WORKS. The following are the main text editions and translations that have appeared since the compilation of Rescher's bibliographies (see below), and many of them give further bibliographical information (especially about MSS) not contained in the bibliographies listed below: *La statistique des sciences,* Osman Amine, ed., 3rd ed. (Cairo, 1968); *Kitāb al-mūsīqā al-kabīr* ("The Great Book on Music"), Ghaṭṭās Khashaba, ed. (Cairo, 1967); "Le Kitāb al-ḥaṭāba," ("Commentary on the Rhetoric"), Jacques Langhade, ed. and trans., in *Mélanges de l'Université Saint-Joseph* (Beirut), **43** (1968), 61–177; Ralph Lerner and Muhsin Mahdi, eds., *Medieval Political Philosophy: A Sourcebook* (New York, 1963), pp. 22–94, which contains selections in English from "Enumeration of the Sciences," "Political Regime," "Attainment of Happiness," and "Plato's Laws"; *Book of Letters: Commentary on Aristotle's "Metaphysics,"* Muhsin Mahdi, ed. (Beirut, 1969); *Book of Religion and Related Texts,* Muhsin Mahdi, ed. (Beirut, 1968); *Utterances Employed in Logic,* Muhsin Mahdi, ed. (Beirut, 1968); *The Political Regime,* Fauzi Najjar, ed. (Beirut, 1964); "Fārābī'nin Peri Hermeneias Muhtasarı," Mübahat Türker-Küyel, ed. and trans., in *Ardṣtirma* (Ankara), **4** (1966), 1–85; and "Un petit traité attribué à al-Fārābī," Mübahat Türker-Küyel, ed. and trans., *ibid.,* **3** (1965), 25–63.

II. SECONDARY LITERATURE. Al-Fārābī's life and works are discussed in Moritz Steinschneider, *Al-Fārābī (Alfarabius)* (St. Petersburg, 1869; repr. Amsterdam, 1966), which gives a comprehensive account of the sources, both printed and MS, available in Europe at that date and which is still indispensable for Hebrew and Latin translations and fragments; and Nicholas Rescher, *Al-Fārābī: An Annotated Bibliography* (Pittsburgh, 1962) and *The Development of Arabic Logic* (Pittsburgh, 1964), pp. 122–128, which list most of the texts and studies that have appeared since Steinschneider, but not the important uncataloged and recently cataloged MSS in the libraries of Turkey, Iran, etc., most of which will be listed in vol. III of Fuat Sezgin, *Geschichte des arabischen Schrifttums* (Leiden, 1967–   ). See also Richard Walzer, "Al-Fārābī," in *Encyclopaedia of Islam,* 2nd ed. (Leiden–London, 1960–   ), II, 778–781; and *Index Islamicus,* J. D. Pearson, ed. (Cambridge, 1958–   ), sec. 4b. Max Meyerhof—"Von Alexandrien nach Baghdad," in *Sitzungsberichte der Preussischen Akademie der Wissenschaft,* Phil.-hist. Kl., **23** (1930), 389–429—studied the few autobiographical fragments that have survived from al-Fārābī's history of philosophy. These should now be supplemented by the fragments reported by al-Khaṭṭābī, in Kabul (Afghanistan) Library of the Ministry of Information, Arabic MS 217, fol. 154r.

The following are the main studies that have appeared since Rescher, many of which provide bibliographical information not found in the works listed above: Muhsin Mahdi, "Alfarabi," in Leo Strauss and Joseph Cropsey, eds., *History of Political Philosophy* (Chicago, 1963), pp. 160–180; "Alfarabi Against Philoponus," in *Journal of Near Eastern Studies,* **26** (1967), 223–260 (the appendix, 253–260, contains a trans. and notes); "The *editio princeps* of Fārābī's *Compendium legum Platonis,*" *ibid.,* **20** (1961), 1–24; Fauzi Najjar, "Fārābī's Political Philosophy and Schīʿism," in *Studia Islamica,* **14** (1961), 57–72; and Richard Walzer, *Greek Into Arabic* (Cambridge, Mass., 1962), *passim;* and "Early Islamic Philosophy," in A. H. Armstrong, ed., *Cambridge History of Later Greek and Early Medieval Philosophy* (Cambridge, 1967), pp. 641–669, 689–691. See also A. Kubesov and B. A. Rosenfeld, "On the Geometrical Treatise of al-Fārābī," in *Archives internationales d'histoire des sciences,* **22** (1969), 50.

MUHSIN MAHDI

FARADAY, MICHAEL (*b.* Newington, Surrey [now part of Southwark, London], England, 22 September 1791; *d.* Hampton Court, Middlesex, England, 25 August 1867), *chemistry, physics.*

Early Life and Education. Michael Faraday was born into a poor family, of which he was the third of four children. His father, James, was a blacksmith who had left his own smithy in Outhgill, near Kirkby Stephen, early in 1791 to seek work in London. His increasing ill health prevented him from providing more than the bare necessities for his family. Faraday later recalled that he was once given a loaf of bread, which was to feed him for a week. James Faraday died in 1809. Michael's mother, the former Margaret Hastwell, was the mainstay of the family. She made do with what she had for material needs, but clearly offered her younger son that emotional security which gave him the strength in later life to reject all social and political distinctions as irrelevant to his own sense of dignity. She died in 1838.

There are no sources for Faraday's early years, so it is impossible to say what effects they had on him; we can only infer them from his adult life. His contemporaries uniformly described him as kind, gentle, proud, and simple in both manner and attitude. He loved children, although he had none of his own, and never lost his enthusiasm for natural beauty, especially such grandiose spectacles as a thunderstorm or an alpine waterfall. Such traits, when taken in conjunction with his solicitude for the success of his older brother, Robert, as a gas fitter and for the education of his younger sister, Margaret, bespeak a close-knit family, enjoying simple pleasures but raised in all propriety by stern but loving parents.

The one early influence of which there can be no doubt was that of religion. Faraday's parents were members of the Sandemanian Church, and Faraday was brought up within its discipline. The Sandemanian religion is a peculiar offshoot of Protestant Christianity. It is fundamentalist in the sense that Sandemanians believe in the literal truth of Scripture, but its emphasis is not on the fire and brimstone of most fundamentalist sects. Rather, it stresses that love and sense of community which marked the primitive Christian Church. It was this love and sense of community which were to sustain Faraday throughout his life. His friend and close associate at the Royal Institution, John Tyndall, a self-styled agnostic, wrote in some puzzlement in his journal for Sunday, 24 October 1852: "I think that a good deal of Faraday's week-day strength and persistency might be referred to his Sunday Exercises. He drinks from a fount on Sunday which refreshes his soul for a week" (Royal Institution, "Tyndall's Journals," MS, V, 163).

Faraday drew more than strength from his religion.

It gave him both a sense of the necessary unity of the universe derived from the unity and benevolence of its Creator and a profound sense of the fallibility of man. Both are worth stressing. The origins of field theory are to be found in Faraday's detailed experimental researches on electricity, but the speculations and imaginings which led him to the experiments and the courage which permitted him to publish physical heresies owe something to his unquestioning belief in the unity and interconnections of all phenomena. This belief, in turn, derived from his faith in God as both creator and sustainer of the universe. The fallibility of man was clearly described in the Book of Job, and this was the book in Faraday's Bible which he had marked the most with marginal emphases. Faraday never engaged in scientific polemics. He presented his results to his peers, having done his best to assure that his experiments were accurate and his reasoning sound. If he were wrong, better experiments and sounder reasoning would prevail and his beloved science would progress. He considered himself to be merely an instrument by which truth was revealed. To insist upon his infallibility would border on blasphemy. He was content to publish his results and let posterity judge how close he had come to being right.

These qualities of Faraday became apparent only when he was in the full tide of fame. In his youth, it seemed unlikely that he would ever have the opportunity to exercise them, at least in the pursuit and presentation of scientific truths. His formal education was almost nil, consisting of the rudiments of reading, writing, and ciphering. When he was thirteen years old, Faraday helped contribute to the family earnings by delivering newspapers for a Mr. G. Riebau of 2 Blandford Street. Riebau was an émigré from France who had fled the political maelstrom of the French Revolution. He not only let out newspapers but also sold and bound books. When Faraday turned fourteen, he was apprenticed to Riebau to learn the art of bookbinding. It was in the seven years of his apprenticeship that Faraday developed the extraordinary manual dexterity that was to distinguish his later experimental researches. The proximity of books stocked for sale and brought in to be rebound stimulated his mind. He became an omnivorous reader, absorbing fact and fancy in equal amounts. The result was a severe case of intellectual indigestion as Faraday became the repository of hosts of unconnected statistics and ideas.

His condition was relieved by the discovery of an elementary treatise, *The Improvement of the Mind,* written by an eighteenth-century clergyman, Isaac Watts. This treatise, as Riebau reported, was "frequent took in his Pocket," and Faraday followed each of Watts's suggestions for self-improvement. He began to keep a commonplace book, in which he could record ideas and interesting observations; he attended lectures and took notes; he began a correspondence with a young man, Benjamin Abbott, with the express hope of improving himself; he later helped found a discussion group devoted to the exchange of ideas. All these things Watts strongly recommended. But Watts went further than providing mechanical aids to learning; he also presented a philosophy which appeared to protect its adherents from false theories and intellectual delusions. Accurate observation of facts and precision in language would prevent a philosopher from premature generalization, which had led many an unwary student astray. This advice, together with his deep sense of human fallibility, reinforced the caution with which Faraday later approached natural philosophy. He never embarked upon an explanation without first testing for himself the facts that needed explanation. Mere reading of the results of others never satisfied him. Some of his most brilliant investigations owed their origin to a casual observation of an anomaly when Faraday checked his facts. And it was not until every fact had been checked and rechecked that Faraday would generalize. Watts had a very attentive pupil.

Faraday's passion for science was first aroused by a chance reading of the article "Electricity" in a copy of the *Encyclopaedia Britannica* which he was rebinding. His curiosity was piqued, and he set out to check what facts he could by means of a small electrostatic generator which he constructed out of some old bottles and waste lumber. The article had been written by one James Tytler, who espoused a somewhat idiosyncratic view of the nature of electricity. For Tytler, all electrical effects could be explained by assuming the existence of a peculiar fluid whose various modes of motion would account for optical and thermal, as well as electrical, phenomena. Electricity was likened by Tytler to a vibration, rather than a flow of material particles through space, and this view permitted him to suggest answers to such thorny questions as why conductors conduct electricity and insulators do not. In the course of his article, Tytler also took the opportunity to demolish the two-fluid theory of electricity then current on the Continent and the Franklinian one-fluid theory popular in Great Britain.

One should be wary of drawing too far-reaching consequences from any single influence, but there are a number of aspects of Tytler's article and its possible effects on Faraday that deserve mention. Tytler was, first of all, a scientific heretic presenting a view that had few supporters. As such, he was both belligerent and challenging, and undoubtedly stirred Faraday to

exercise his own judgment more than if the subject had been presented calmly, without controversy, as Truth. Tytler's defensive posture also forced him to attack the more orthodox theories and underline their very real weaknesses. If nothing else, these criticisms must have made Faraday skeptical of the accepted theories and forced him to keep an open mind on controversial points. When he later challenged both the one-fluid and the two-fluid theories, he was able to do so in precisely those areas where Tytler had first sown doubts. His challenge to orthodox electrical theories in 1838 was couched in Tytlerian terms. Current electricity, for Faraday as for Tytler, was a vibration, not a material flow. How much Faraday owed to Tytler is impossible to assess, but the influence was strong enough to lead Faraday to refer to Tytler's article a number of times in the laboratory diary that he kept from 1820 on.

With the reading of Tytler's article, Faraday began the pursuit of science in earnest. In the London of the early nineteenth century, however, it was difficult for an apprentice bookbinder to find many sources of scientific enlightenment. He could and did attend public lectures, but there were no night schools, no correspondence courses, no public libraries from which he could gain scientific enlightenment. He was, therefore, doubly fortunate to fall in with a group of young men with a common passion for science. They had come together as the City Philosophical Society in 1808 and were led by John Tatum, at whose house they met every Wednesday night. Tatum would deliver a lecture on a scientific subject and then throw open his library to the members of the society.

Faraday was introduced into this company in February 1810, when he attended his first lecture. At the City Philosophical Society, he received a basic education in the sciences, attending and taking careful notes on lectures on electricity, galvanism, hydrostatics, optics, geology, theoretical mechanics, experimental mechanics, chemistry, aerology (pneumatic chemistry), astronomy, and meteorology. These lectures should not be overestimated, for they were often mere catalogs of facts (mineralogy is an example) but, when possible, Tatum illustrated them with experiments and introduced his avid listeners to interesting pieces of scientific apparatus. It was at the City Philosophical Society, for example, that Faraday first saw a voltaic pile in operation.

Faraday's interest was increasingly focused on science. It was this interest that led him to the discovery of a work which he lauded throughout his life, Jane Marcet's *Conversations on Chemistry*. Mrs. Marcet's treatment of chemistry differed considerably from other contemporary, more technical accounts. She had written it for those people, like herself, who had been entranced by the lectures of Humphry Davy at the Royal Institution of Great Britain. Davy approached chemistry as if it were the key to the ultimate mysteries of nature, and Mrs. Marcet did likewise. Here was no dry catalog of chemical facts, or recipes, but a grand scheme which tied together chemical reactions, electrical relations, and thermal and optical phenomena. The impact on Faraday was considerable. The simplicity of his views on electricity was forever destroyed, his thoughts were directed specifically to chemistry, and, most important of all, he was introduced to the thoughts of Humphry Davy, who became, for him, the example of what he would like to be.

Faraday and Humphry Davy. There was, seemingly, little chance that Michael Faraday, bookbinder's apprentice, would ever become anything other than Michael Faraday, bookbinder. But one of Riebau's customers offered him tickets to Davy's lectures at the Royal Institution. He went, took careful notes, and copied them out in a clear hand. Each scientific point made by Davy was grasped eagerly and recounted to his friends at the City Philosophical Society. But this was mere playing at science, and it was with a heavy heart that Faraday, in October 1812, accepted the end of his apprenticeship and prepared to devote himself to bookbinding instead of his beloved chemistry. An accident changed his life and the life of science. In late October, while examining chloride of nitrogen, a very unstable substance, Davy was temporarily blinded by an explosion. Faraday was recommended to Davy as an amanuensis, and Davy was pleased to have him. In December, Faraday sent Davy the carefully bound notes he had taken at Davy's lectures. Davy was flattered but could do nothing at that time to help his young admirer. In February 1813, however, an assistant in the laboratory of the Royal Institution was fired for brawling. Davy immediately sent for Faraday and on 1 March 1813, Faraday took up his new position at the Royal Institution.

Davy exerted the most important influence on Faraday's intellectual development. Davy's mind was both penetrating and wide-ranging. He seems to have been an omnivorous reader interested in metaphysics as well as chemistry, poetry as well as physics. His science was characterized by brilliant flashes of insight soundly supported by experimental evidence. Although he never committed himself to a specific theoretical or metaphysical viewpoint, he was aware of and often used those that offered clues to the nature of matter and its forces. In the early nineteenth century, there were a number of points of view which could provide the chemist with guidelines worth following. There was, first of all, an English tradition in chemistry which could be traced back to the patron saint of English science, Sir Isaac Newton. It took Newton's work on force and raised it to the level of a universal science. The classic example in the eighteenth century is Gowin Knight's *An Attempt to Demonstrate That All the Phenomena in Nature May Be Explained by Two Simple Active Principles, Attraction and Repulsion* (1748), whose message can be seen from the title. There seems little reason to doubt that both Faraday and Davy had read Knight's work. It was a part of the English scientific tradition and, more important, was easily available in the library of the Royal Institution.

But Knight's work was relatively crude. Although he reduced matter to the forces of attraction and repulsion, he separated the two, hypothesizing one kind of matter with attractive force only and another consisting solely of repulsive force. The association of these two forms of matter gave rise to the phenomena of the sensible world. A more subtle solution to the problem of complexity and of the nature of matter was provided by the Jesuit Rudjer Bošković in his *Philosophiae naturalis theoria redacta ad unicam legem virium in natura existentium,* which first appeared in 1758. Like Knight, Bošković dismissed the reality of matter and substituted forces but, unlike Knight, was able to combine the forces of attraction and repulsion in one "atom." In Figure 1, the pattern of forces of a Boscovichean atom is represented graphically. The point at O is a mathematical point which serves merely as the center of the forces of which the atom consists. Beyond H, the atomic force is attractive, decreasing inversely as the square of the distance, thus satisfying the Newtonian principle of universal attraction. From H to A, the force varies, according to the distance from O, in a continuous fashion from attractive to repulsive and back to attractive. The number of such variations can be multiplied at will to account for phenomena. From A to O, the force becomes increasingly repulsive, reaching infinite repulsion at O and thereby preserving impenetrability as a characteristic of "matter." These point-atoms were likened by Bošković to the points that make up the lines of the letters of the alphabet. A combination of the point-atoms gave the chemical elements, just as a combination of points made up the letters. Combinations of elements yielded the chemical compounds, and so on. Ultimately, then, all "matter" is one; observable complexities were the result of successive levels of complexity of particulate arrangements.

This system was particularly appealing to chemists

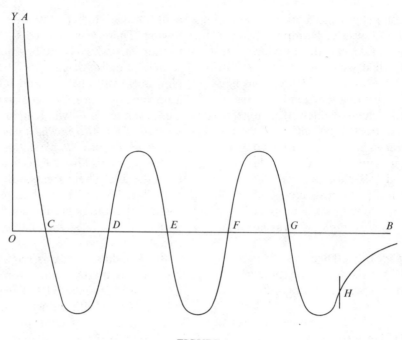

FIGURE 1

of philosophical disposition. It reconciled simplicity and complexity, providing a fundamental order in place of the taxonomic confusion engendered by the myriad chemical compounds and the ever-growing number of chemical elements. It also offered some insight into such specifically chemical problems as elective affinity by referring to the patterns of forces which could interact to create stable compounds. Davy remarked in his last years that he had found Bošković's theory of some importance in the development of his ideas and, in 1844, Faraday publicly declared his preference for Boscovichean atoms over those suggested by Dalton. Just how important Bošković's theory was for Faraday is currently a matter of scholarly dispute. I shall here insist that it was fundamental, but the reader is warned that my discussion of Faraday's use of it is controversial.

Knight and Bošković provided the justification for the unity of matter in which both Faraday and Davy believed. They did not, however, provide the concept of the unity of force. Both Knight and Bošković, after having composed their primary matter(s) from attractive and repulsive forces, gave material explanations for such phenomena as light and electricity. It may, of course, be argued that since all phenomena are ultimately the results of attractive and repulsive forces, then all phenomena are convertible into one another. But no such argument was ever made, to my knowledge, by those within the English tradition or by those who were partial to Bošković. Yet Faraday wrote in 1845: "I have long believed in the unity of

force," and his work from 1831 on was devoted to the conversion of one force into another. Whence came this conviction of the unity and convertibility of forces? Again, what follows is controversial, but it seems to make sense of what otherwise appears inexplicable and is, therefore, worth putting forward, albeit with considerable caution.

The doctrine of the unity and convertibility of forces is the product of German philosophy at the end of the eighteenth century. It was first suggested by Immanuel Kant in his *Metaphysische Anfangsgründe der Naturwissenschaft* (1786) and then developed by F. J. Schelling in his *Naturphilosophie.* The disciples of Schelling—the *Naturphilosophen,* among whom may be counted Hans Christian Oersted, the discoverer of the first force conversion (electromagnetism)—used this idea as the basic guiding thread in their researches. What is extremely difficult is to connect this movement in Germany and Scandinavia to Davy and Faraday. The one possible link, since neither Davy nor Faraday read German and translations were unavailable, is the poet and metaphysician Samuel Taylor Coleridge. He visited Germany in 1798–1799 and came back to England with his head overflowing with *Naturphilosophie.* He and Davy were close friends in the early 1800's and it seems likely, indeed inevitable, given Coleridge's enthusiasm and love of conversation, that he and Davy discussed *Naturphilosophie* in some detail. The unity and convertibility of forces was the kind of idea Davy always liked to keep in the back of his mind without committing himself to it. It might suggest some experiments and was, therefore, a potentially useful way of looking at the world. If he passed it on to Faraday, it fell on eager ears, for there can be no doubt of Faraday's commitment to it. The difference in their reaction may have lain in their different attitudes toward religion rather than toward science. Both men were religious, but Faraday was by far the more intense in his religious feelings and more determined in his devotion. One of the reasons Coleridge had been so enthusiastic in his acceptance of *Naturphilosophie* was that it appeared to contradict materialism and offer a place to spirit in the world. Faraday may have felt the same. Certainly the unity and convertibility of force offered a more appealing view of God's creation than the chaotic interplay of material atoms in which most scientists believed.

If we accept this admittedly hypothetical reconstruction, we can suggest the course of Faraday's development under Davy's tutelage. The unity and convertibility of force would first appeal to the apprentice chemist, for it bound the universe together in an aesthetically and theologically satisfying way. From here, it was a short step to the acceptance of, or at least the willingness to think in terms of, Boscovichean atoms. The English tradition of dynamic atomism made such a step intellectually respectable, for this tradition had based its insistence on the ultimate reality of force on sound empirical arguments which Faraday was to repeat in 1844 and 1846. It is within this intellectual and philosophic framework that his work should be viewed.

There was ample opportunity for Davy and Faraday to discuss these fundamental aspects of the nature and structure of physical reality. Soon after Faraday's employment at the Royal Institution, Davy decided to visit the Continent, and he asked Faraday to accompany him and Lady Davy. The party set out on 13 October 1813 and during 1813 and 1814 visited France and Italy, where Faraday met many of the leading scientists of the day. During this tour, Davy discoursed on every scientific subject under the sun, and Faraday eagerly drank it all in. Faraday became fully conscious, during this period, of the philosophical complexities underlying the apparent simplicity of chemical science and even more aware of the dangers of both methodological and metaphysical complacency.

Journeyman Chemist. Upon his return to London in April 1815, Faraday threw himself into chemistry with the enthusiasm that marked his whole scientific life. He assiduously searched all the scientific journals available to him, keeping a detailed bibliographical record of everything he read and carefully noting everything of interest to him. In 1816 he published his first paper, "Analysis of Caustic Lime of Tuscany," which was followed by a series of short (and inconsequential) articles on subjects suggested to him by the researches of Davy and William T. Brande, Davy's successor as professor of chemistry at the Royal Institution. By 1820 Faraday had established a modest but solid reputation as an analytical chemist, so much so that he was involved in court cases requiring expert testimony. One such case involved the question of the ignition point of heated oil vapor. This, together with his brother's involvement in the new gas lighting of London, led Faraday to investigate the general properties of the various oils used for heating and lighting. From these researches came the discovery of benzene in 1825.

But Faraday's mind ranged far wider than the composition of lime or illuminating gas. One of his earliest chemical enthusiasms had been Davy's work on chlorine, in which Davy had exploded Lavoisier's theory of acids by showing that not all acids contain oxygen, as the name of that element implied. Hydrochloric acid consisted solely of hydrogen and chlorine. Davy's demonstration that chlorine supported combustion also destroyed the unique place among the elements allotted to oxygen as the sole supporter of combustion, and thereby tended to weaken Lavoisier's magistral chemical synthesis. But this demonstration raised an important problem, and it was to this problem that Faraday turned. If chlorine were a supporter of combustion, as the almost explosive nature of the combination of iron with chlorine clearly demonstrated, then why did chlorine not combine with carbon, the combustible substance par excellence? All attempts at "burning" carbon in chlorine were unsuccessful, but Faraday was convinced that compounds of chlorine and carbon must exist. In 1820 he produced the first known compounds of chlorine and carbon, C_2Cl_6 and C_2Cl_4. These compounds had been produced by the substitution of chlorine for hydrogen in "olefiant gas," our modern ethylene. This was the first substitution reaction; such reactions, in the hands of Charles Gerhardt and Augustin Laurent in the 1840's, were to be used as a serious challenge to the dualistic electrochemical theories of J. J. Berzelius.

Beyond his work in analytical and pure chemistry, Faraday showed himself to be a pioneer in the application of chemistry to problems of technology. In 1818, together with James Stodart, a cutler, he began a series of experiments on the alloys of steel. Although he was able to produce alloys of superior quality, they were not capable of commercial production because they required the use of such rare metals as platinum, rhodium, and silver. Nevertheless, Faraday demonstrated that the increasingly urgent problem of producing higher-grade steels could be attacked by science. The later work on steel of Henry Sorby, Henry Bessemer, and Robert A. Hadfield was based directly on Faraday's work in the early part of the century.

In 1824 Faraday was asked by the Royal Society to conduct experiments on optical glass. Again his researches were inconclusive, but he paved the way for later improvements in glass manufacture. It was in these experiments that he produced a glass containing borosilicate of lead and with a very high refractive index. This was the glass he used in 1845 when he discovered the rotation of the plane of polarization of a light ray in an intense magnetic field.

The 1820's were busy years for Faraday beyond his chemical researches. In 1821 he married Sarah Barnard, the sister of one of the friends he had made at the City Philosophical Society. Their marriage was an extremely happy one and sustained Faraday throughout the extraordinary mental exertions of the next forty years. Sarah Barnard was not an intellectual. Once, when she was asked why she did not study chemistry, she replied, "Already it is so absorbing, and exciting to him that it often deprives him of his sleep and I am quite content to be the pillow of his mind." Faraday needed no one to whom to talk about his research. His dialogue was with nature. "I do not think I could work in company, or think aloud, or explain my thoughts . . .," he remarked near the end of his life. He founded no school and left no disciple who had been formed in his laboratory.

In the 1820's the precarious financial position of the Royal Institution also demanded Faraday's attention. He helped support the institution by performing chemical analyses, and he contributed to its fiscal stability by instituting the Friday evening discourses in 1825. He gave more than a hundred of these lectures before his retirement in 1862. They served to educate the English upper class in science, and part of the growing support for science in Victorian England may legitimately be attributed to the efforts of Faraday and his fellow lecturers to popularize science among those with influence in government and the educational establishment.

Early Researches on Electricity. During these busy years, Faraday was forced to push his love for electricity into the background. Yet it was never far from his thoughts, as the events of 1821 and 1831 reveal. In 1821 a series of brilliant researches culminated in the discovery of electromagnetic rotation; in 1831, seemingly out of nowhere, came the discovery of electromagnetic induction and the beginning of the experimental researches in electricity which were to

lead Faraday to the discovery of the laws of electrochemistry, specific inductive capacity, the Faraday effect, and the foundations of classical field theory. These researches, in both 1821 and 1831, are all of a piece; and it would be well here to make explicit what I think is the theoretical thread that holds them together.

By 1821 we know that Faraday was aware of and even toying with the concept of Boscovichean point-atoms. This is not to say that Faraday was a disciple of Bošković, for he certainly did not follow the Boscovichean system in his work. But the notion of atoms as centers of force had a strong appeal for him, and it is this notion that, I should like to suggest, provided the conceptual framework for his work on electricity. In particular, there were two specific consequences of the theory of point-atoms which were to be fundamental to Faraday's discoveries. The first was the emphasis upon complex patterns of force that followed logically from the consideration of the interaction of numbers of Boscovichean particles. It should be remembered that Boscovichean atoms were not the atoms of early nineteenth-century chemistry. Rather, the chemical elements were agglomerates of Boscovichean atoms whose specific chemical properties were the direct result of the patterns of force produced by the intimate association of point-atoms. When Faraday thought of material particles, therefore, he did not envision them as submicroscopic billiard balls with which certain rather simple forces were associated, but rather as centers of a complex web of forces.

Put another way, the orthodox scientist of the 1820's tended to think in terms of central forces emanating from particles, and central forces always act in straight lines between particles. Faraday's vision was far more intricate, and it permitted him to contemplate forces in manifold ways. An example may be useful here. Oersted's discovery of a circular magnetic "force" around a current-carrying wire was explained by André-Marie Ampère as the resultant of central forces emanating from current elements in the wire. To Faraday, the circular "force" was simple and could be used to explain the apparent polarity (or central forces) of magnetic poles. No one but Faraday could or did take a circular "force" seriously, but this was the germ of the idea of the line of force which was to be central to the development of Faraday's theories.

The second consequence involved a subtle shift in point of view away from the orthodox physics of the day. Boscovichean atoms, as force, were infinite in extent because the forces associated with atoms extended to infinity. Thus all material associations on the molecular level were really interpenetrations of fields of force, if I may be excused the anachronism. All particles, then, were associated with one another; the only differences were in the degrees of approximation of molecular or atomic centers of force. A displacement of a particle, anywhere in the universe, ought to affect every other particle. The same conclusion, of course, may be drawn from orthodox action-at-a-distance physics, but it is neither obvious nor obtrusive. In the theory of point-atoms it is both.

Furthermore, it suggests a mode of action which Faraday seized upon in his researches on the nature of electricity. Orthodox theory assumed the existence of one or two "imponderable" electrical fluids with whose particles specific electrical forces were associated. Electrical energy was transferred from place to place by the translation of these particles. The interlocking of force particles permitted energy to be transferred without the permanent displacement of the particles by means of the vibration of the particles. If a line of particles could be put under a strain, this line could then transmit energy either by the rapid breakdown and buildup of the strain or by vibrating transversally to the direction along which the strain was exerted. The first kind of "vibration" was that suggested by Faraday in 1838 to explain spark discharge, electrochemical decomposition, and ordinary electrical conduction. The second sort of vibration was tentatively put forward by Faraday in 1846 in his "Thoughts on Ray Vibrations" to account for the transmission of light through a vacuum without having recourse to a vibrating medium such as the luminiferous ether.

All this was far in the future in 1821, when Faraday was persuaded to take up the subject of electromagnetism by his friend Richard Phillips, an editor of *Philosophical Magazine*. Ever since Hans Christian Oersted's announcement of the discovery of electromagnetism in the summer of 1820, editors of scientific journals had been inundated with articles on the phenomenon. Theories to explain it had multiplied, and the net effect was confusion. Were all the effects reported real? Did the theories fit the facts? It was to answer these questions that Phillips turned to Faraday and asked him to review the experiments and theories of the past months and separate truth from fiction, groundless speculation from legitimate hypothesis. Faraday agreed to undertake a short historical survey but he did so reluctantly, since his attention was focused on problems of chemistry rather remote from electromagnetism.

His enthusiasm was aroused in September 1821, when he turned to the investigation of the peculiar nature of the magnetic force created by an electrical current. Oersted had spoken of the "electrical conflict" surrounding the wire and had noted that "this conflict performs circles," but this imprecise description had had little impact upon Faraday. Yet as he experimented he saw precisely what was happening. Using a small magnetic needle to map the pattern of magnetic force, he noted that one of the poles of the needle turned in a circle as it was carried around the wire. He immediately realized that a single magnetic pole would rotate unceasingly around a current-carrying wire so long as the current flowed. He then set about devising an instrument to illustrate this effect. (See Figure 2.) His paper "On Some New Electro-Magnetical Motions, and on the Theory of Magnetism" appeared in the 21 October 1821 issue of the *Quarterly Journal of Science*. It records the first conversion of electrical into mechanical energy. It also contained the first notion of the line of force. Faraday, it must be remembered, was a mathematical illiterate. His experiments revealed a circular "line" of magnetic force around a current-carrying wire and he found no difficulty in accepting this as a simple fact. To his mathematically trained contemporaries, such a force could not be simple but must be resolved into central forces. This is what Ampère had done so ingeniously in his early papers on electromagnetism. Ampère's sophisticated mathematical reasoning could have no

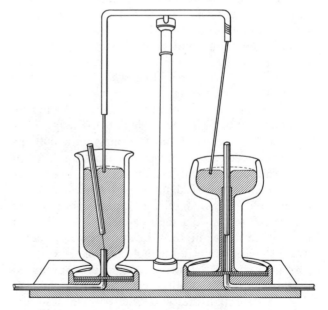

FIGURE 2. Faraday's apparatus for illustrating electromagnetic rotation. At left, a cylindrical bar magnet, plunged into a beaker of mercury (which was part of the electrical circuit), rotated around the end of a current-carrying wire that made contact with the mercury. At right, the magnet was fixed and the wire was so mounted that it could turn about the point of suspension, and thus rotate around the magnetic pole.

effect on Faraday and he refused to move from what he considered the bedrock of experiment. Stubbornness, however, is no virtue in science and he was forced to face the problem of deriving central forces from his circular force in order to explain the "simple" attractions and repulsions of magnetic poles. His solution was both simple and elegant. If a straight current-carrying wire were bent into a loop, then the circular lines of magnetic force would be arranged in such a way as to concentrate them within the loop and the magnetic "polarity" of the loop would reflect this concentration. That there were no "poles" or termini of the magnetic force, Faraday showed by another simple demonstration. He wound a glass tube with insulated wire and then placed it half-submerged in water. A long magnetic needle was affixed to a cork so that it could float freely on the water. When a current was passed through the wire surrounding the glass tube, "poles" were formed at opposite ends of the tube and the north pole of the needle was attracted to the south pole of the helix. If an ordinary magnet had been used, the needle would have approached the electromagnet and clung to it, giving the illusion that the south pole attracted the north pole.

But in Faraday's setup the result was surprising. The needle moved toward the helix, entered the glass tube and continued through it until the north pole of the needle was situated at the north pole of the helix. The result was as Faraday had expected; it was merely another example of his electromagnetic rotations. A single magnetic "pole" would continue to move around and through the helix and never come to rest. The line of force along which it moved was the resultant of the circular magnetic forces surrounding the wires of the helix and did not emanate in straight lines from the "poles" of the magnet.

Thus Faraday's work on electromagnetic rotations led him to take a view of electromagnetism different from that of most of his contemporaries. Where they focused on the electrical fluids and the peculiar forces engendered by their motion (Ampère's position), he was forced to consider the line of force. He did not know what it was in 1821, but he suspected that it was a state of strain in the molecules of the current-carrying wire and the surrounding medium produced by the passage of an electrical "current" (whatever that was) through the wire. Such a state of strain, he knew, was transmitted some distance from the source of the strain, the current-carrying wire. Might it not be legitimate to speculate that if the strain could be intensified and concentrated, it might induce a similar state in a neighboring wire?

In the years between 1821 and 1831, Faraday re-

turned sporadically to this question. He attempted to detect the strain by passing plane polarized light through an electrolyte through which a current was passing; he queried the best form of magnet to produce the maximum strain, concluding that it might be "a very thick ring"; he attempted to induce an electrical current by means of static electricity; but all to no avail. Yet in these years his ideas gradually developed and clarified. The wave theory of light revealed how strains could transmit energy; the work of his friend Charles Wheatstone on sound, particularly on Chladni figures, which Faraday illustrated to audiences at the Royal Institution, showed how vibrations could produce symmetrical arrangements of particles; the discovery of "magnetism by rotation" by François Arago in 1825 (Arago's wheel) revealed to him the insufficiency of Ampère's electrodynamic theory.

In 1831 Faraday learned of Joseph Henry's experiments in Albany, New York, with powerful electromagnets in which the polarity could be reversed almost instantaneously by a simple reversal of "current" direction. The stage was set. An electromagnet, in the shape of a thick iron ring, wound on one side with insulated wire should set up the powerful strain; the strain should be conducted through the particles of the ring, which would then be thrown into a peculiar arrangement, much as the particles of dust were affected on a neighboring iron plate when one close by was set into vibration by a violin bow; the resultant arrangement would distort the intermolecular forces by which the iron molecules of the ring cohered, and this strain should then be detected by a secondary winding on the other side of the iron ring. On 29 August 1831, Faraday tried the experiment. When the primary circuit was closed, the galvanometer in the secondary circuit moved. An electrical "current" had been induced by another "current" through the medium of an iron ring. This discovery is always called the discovery of electromagnetic induction, but it should be noted that it is no such thing.

It was not until some weeks later that Faraday discovered the conditions under which a permanent magnet could generate a current of electricity (17 October 1831). It was at this date that he could declare that he had demonstrated the reverse of Oersted's effect, namely, the conversion of magnetic force into electrical force. Further investigation led him to the invention of the first dynamo, whereby the reverse of his 1821 discovery of electromagnetic rotations could be accomplished. Mechanical force could be converted into electrical force by a simple machine. A copper disk, rotated between the poles of a magnet, produced a steady electrical current in a circuit run-

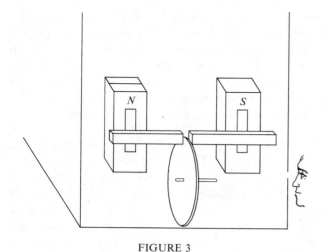

FIGURE 3

ning from the center of the disk through a wire to the edge of the disk. (See Figure 3.)

The concept of the line of force now moved to the very center of Faraday's thought, where it was to remain for the rest of his life. It was the line of force which tied all his researches on electricity and magnetism together. It was therefore with some embarrassment that he confessed, in his second series of *Experimental Researches in Electricity,* that he did not know what the line of force was. In the first series, he had described it as a strain imposed upon the molecules of the conducting wire and the surrounding medium by the passage of the electric "current." This state of strain he christened "the electrotonic state," but it defied every attempt at detection. His abandonment of the electrotonic state in the second series was only temporary, for without it Faraday was deprived of his basic concept. The electrotonic state had to exist, for there could be no doubt of the reality of the line of force. It was his attempts to relate the two which led him onward through the brilliant series of researches which culminated in his general theory of electricity in 1838.

Electrochemistry. The first and second series of *Experimental Researches* had been concerned with the relations between electricity and magnetism. In the summer of 1832 Faraday appeared to go off on a tangent, with an investigation into the identity of the electricities produced by the various means then known. His commitment to the unity of force led him to believe that the electricities produced by electrostatic generators, voltaic cells, thermocouples, dynamos, and electric fishes were identical, but belief was no substitute for proof. Furthermore, this identity had been challenged by Sir Humphry Davy's brother John, who insisted that electrical effects were not produced by a single agent but were the complex

results of a combination of powers. There was little sense in continuing to work on electricity until this question was settled. It seemed like a routine investigation involving mere manipulative skill to demonstrate that electricity, no matter how produced, had the same effects. But in the course of this investigation Faraday was to be led to the laws of electrolysis and, more important, to challenge the concept of action at a distance.

Faraday's attack was straightforward. Searching the literature, he found reports of the similar effects of the various forms of electricity. The only real difficulties arose with static electricity. There were published reports that electrostatic discharges had both magnetic and electrochemical effects but, upon repeating the experiments, Faraday found them equivocal. Electrostatic discharge, for example, could magnetize needles, but Faraday found it impossible to produce a magnet in an electrostatic field. Similarly, William Hyde Wollaston had reported the decomposition of water by an underwater spark in 1801, but it was possible to view this decomposition as the result of the mechanical shock, the heat generated by the discharge, or both. To remove all possible doubts, Faraday turned his experimental skill to the examination of these two effects.

The magnetic effect of an electrostatic discharge was examined by means of a simple galvanometer. The discharge was "slowed down" by passing it through wet string. The galvanometer was deflected, thus settling the question once and for all. Here Faraday might have stopped, but he saw the opportunity to compare static and voltaic electricity quantitatively and was quick to seize it. An electrometer gave him a relative reading of static charge; the deflection of the galvanometer permitted him to correlate the charge with the discharge. Fortunately (and unbeknown to Faraday), his galvanometer here acted like a ballistic galvanometer, and he was able to conclude that "if the same absolute quantity of electricity pass through the galvanometer, whatever may be its intensity, the deflecting force upon the magnetic needle is the same." Faraday immediately proceeded to devise an apparatus by which he could compare, quantitatively, the currents produced by electrostatic and voltaic discharge. Insofar as electricity affected a galvanometer needle, then, Faraday had given conclusive proof of the identity of static and voltaic electricity. He had, furthermore, devised an instrument for the measurement of relative quantities of electricity.

The remaining problem was that of electrochemical decomposition by electrostatic discharge. Once again, the desired effect was produced without the ambiguity of Wollaston's experiment. Faraday might have rested here, since the identity of electricities was what he had set out to prove, but the opportunity for further discoveries was clear and he set out to exploit it. The course was obvious. Both electrostatic and current electricity decomposed water; the "throw" of the galvanometer permitted the accurate measure of electrical quantity. Could not the quantity of electricity be correlated with the products of electrochemical decomposition? In answering this question, Faraday enunciated his two laws of electrochemistry: (1) Chemical action or decomposing power is exactly proportional to the quantity of electricity which passes in solution; (2) The amounts of different substances deposited or dissolved by the same quantity of electricity are proportional to their chemical equivalent weights. Thus, Faraday had not only proved the identity of electricities; he had added another link in the chain of the convertibility of forces. Electricity was not only involved in chemical affinity, as the invention of the voltaic cell had shown, but it was the force of chemical affinity. In 1881 Hermann von Helmholtz was to use Faraday's 1834 papers on electrochemistry as the experimental basis for his suggestion that electricity must be particulate, or Faraday's laws of electrochemistry would make no sense.

In the course of his electrochemical researches, Faraday made a discovery with revolutionary implications. As he varied the conditions under which electrostatic discharge produced electrochemical decomposition, he found to his surprise that no "poles" were required. Ever since the invention of the voltaic cell, electrochemists had assumed that the + and − terminals of the circuit acted as centers of force, which force, acting at a distance upon the molecules in solution, literally tore them apart. Hence the term "poles." But when Faraday passed an electrostatic discharge through some blotting paper soaked with a solution of potassium iodide into the air, the potassium iodide decomposed. Where, now, were the centers of force of orthodox theory? More important, what was "acting at a distance" upon the potassium iodide molecules? The mere passage of the "current" was sufficient to decompose the potassium iodide. The experiment with electrostatic discharge suggested to Faraday that decomposition was not effected by action at a distance.

In a series of ingenious experiments, Faraday went on to show that the molecules were not "torn apart" at all. Instead, the two components of a binary salt seemed to migrate in opposite directions through the solution, without ever becoming free chemical agents, until they reached the terminals upon which they were deposited. Faraday accounted for this strange behavior by claiming that the electric "current" exalted the affinities of the components of a salt on opposite sides of the compound molecules, thus permitting each component to leave its original partner and join with another close by. The electrical force determined the direction of this recombination, one component moving toward the "positive" terminal, the other toward the "negative." The "exaltation" was passed along from one molecule to the next, beginning from the terminals and moving out into the solution. There was no action at a distance, but only intermolecular forces created by the strain imposed by the electrical force. It is difficult to visualize this process without having recourse to point-atoms and the patterns of force which produced their chemical identity.

Such patterns of force could be distorted by the imposition of other forces, and this appears to be what Faraday meant by his use of the term "exaltation." Under the strain of electrical force, the affinities of the molecular components were both "exalted" and aligned, permitting their transfer through the solution. As in an American square dance "grand right and left," where each partner passes around the square by taking the hands of people passing in the opposite direction, the chemical elements in solution passed through the solution, ever bound to a partner, until they were freed at the termini. The process involved three steps: the creation of the initial strain by the imposition of the electrical force; the exaltation of the affinities along the direction of the electrical force, which exaltation caused the component atoms of the decomposing substances to shift in opposite directions and to be bound by partners moving the other way; and the "shift" in which the strain was momentarily relieved, only to be reimposed immediately by the constant application of the electric force at the termini. The electric force was transmitted by this rapid series of buildups and breakdowns of strain, and electrical energy could be transmitted in this fashion without the transfer of a material agent. It was even possible to deduce the second law of electrochemistry from this scheme. Each shift required the breaking of a chemical bond of specific strength, so it was to be expected that the total force employed (quantity of electricity) should bear some specific and simple relation to the total quantity of matter decomposed by this force.

In his published accounts, Faraday only hinted at what has been presented here as his theory of electrochemical decomposition. The conceptual framework was too conjectural for Faraday to present it to his co-workers in electrochemistry. But what he could and did do was to publish his factual results and also prepare the way for a successful challenge to the prevalent theory by introducing a new nomenclature which was theoretically neutral. Instead of poles, which implied centers of force, Faraday used the term "electrode," which had no such implication. Similarly "cathode," "anode," "electrolysis," "electrolyte," "anion," and "cation" were merely descriptive terms. William Whewell of Trinity College, Cambridge, was the source of most of these neologisms.

Faraday's Theory of Electricity. Faraday's electrochemical researches suggested to him a new perspective on electrostatics. If electrochemical forces did not act at a distance, was it preposterous to think that electrostatic forces also were intermolecular? The researches of Charles Coulomb in the 1780's had appeared to settle that question once and for all in favor of action at a distance, but Faraday drew courage from his electrochemical work and sought to find experimental confirmation for his new point of view. Two consequences flowed logically from the substitution of intermolecular forces for action at a distance. First, the electrostatic force ought to vary if it depended upon the ability of the molecules of a medium to transmit it and, second, this force ought to be transmitted in curved lines, since the transmitting molecules occupied a volume of space, rather than in the straight lines assumed by action-at-a-distance physics. Experimental confirmation of both these conclusions was quick in coming. The inductive force did vary when different substances were used to transmit it. The discovery of specific inductive capacity was an important structural element in the construction of Faraday's novel electrical theory. The inductive force was also shown to be transmitted in curved, not straight, lines, thus confirming once again Faraday's belief in intermolecular forces.

By 1838 Faraday was in a position to put all the pieces together into a coherent theory of electricity. The particles of matter were composed of forces arranged in complex patterns which gave them their individuality. These patterns could be distorted by placing the particles under strain. Electrical force set up such a strain. In electrostatics, the strain was imposed on molecules capable of sustaining large forces; when the line of particulate strain gave way, it did so with the snap of the electric spark. Lightning was the result of the same process on a larger scale. In electrochemistry, the force of the "breaking" strain was that of the chemical affinities of the elements of the chemical compound undergoing electrochemical decomposition. The shift of the particles of the elements toward the two electrodes momentarily relaxed the strain, but it was immediately re-created by the constant application of electric force at the electrodes upon the nearest particles of the electrolyte. This buildup and breakdown of interparticulate strain, passing through the electrolyte, constituted the electrical "current." It was a transfer of energy which did not entail a transfer of matter; Faraday's caution in adopting the term "current" appeared justified. The same situation obtained in ordinary conduction through a wire. The molecules of good conductors could not sustain much of a strain at all, so here the buildup and breakdown of the strain was exceptionally rapid and the "conduction" was therefore correspondingly good.

The theory was elegant, firmly based on experiment, and complete. It was also heretical, challenging almost all the fundamental concepts of orthodox electrical science. Faraday knew this and put it forward with appropriate caution. The experimental results were clearly and firmly reported; the theoretical aspect was hedged with fuzzy and tentative language, hesitantly and sometimes confusedly presented. It is, I think, fair to say that no one in the 1830's took the theory seriously. Even Faraday was unable to advocate it with the necessary vigor. The strain of eight years of unremitting intellectual effort at the farthest frontier of electrical theory ultimately broke his powerful mind. In 1839 he suffered a nervous breakdown, from which he never really recovered. For five years he was unable to concentrate his mental faculties on the problems of electricity and magnetism. He passed this time by devoting himself to the affairs of the Royal Institution and such other researches as did not require his total intellectual commitment. It was in these years, for example, that he extended his earlier work on the condensation of gases. He was able here to use his experimental talents to the full without being forced to focus his mind on the consequences.

Even during these years, Faraday kept returning to his electrical theory. In 1844 he published a small paper entitled "Speculation Touching Electrical Conduction and the Nature of Matter," in which he "proved" to his own satisfaction that only Boscovichean atoms were compatible with the observed conduction and nonconduction of electricity through material bodies. Again, it seems unlikely that Faraday convinced anyone, but this exercise did serve to stimulate him to return to his old preoccupation with the nature of electricity and magnetism. In 1846 he was led, in his "Thoughts on Ray Vibrations," to an embryonic form of the electromagnetic theory of light, later developed by James Clerk Maxwell. In both these essays Faraday was, as it were, conducting a dialogue with himself, attempting to clarify his own ideas and to grasp the full implications of his own speculative hypotheses. These works therefore are of importance more because they reveal Faraday's mind to us than because they are important steps in the progress of electrical and magnetic science.

Last Researches: The Origins of Field Theory. The last, and in many ways the most brilliant, of Faraday's series of researches was stimulated by the quite specific comments of one of the few people who thought his theory of electricity worthy of serious attention. On 6 August 1845, William Thomson, the future Lord Kelvin, addressed a lengthy letter to Faraday, describing his success with the mathematical treatment of the concept of the line of force. At the end of the letter Thomson listed some experiments to test the results of his reasonings on Faraday's theory, and it was this that pushed Faraday once more into active scientific research. One of Thomson's suggestions was that Faraday test the effect of electrical action through a dielectric on plane-polarized light. As Thomson wrote:

> It is known that a very well defined action, analogous to that of a transparent crystal, is produced upon polarized light when transmitted through glass in any ordinary state of violent constraint. If the constraint, which may be elevated to be on the point of breaking the glass, be produced by electricity, it seems probable that a similar action might be observed.

The effect predicted by Thomson was one which Faraday had been seeking to detect since the 1820's, but with no success. Thomson's belief that it should exist reinforced Faraday's, and he returned to the laboratory to find it. As in the 1820's, his search was fruitless, but this time, instead of abandoning his search, he altered the question he put to nature. His own work in the 1830's had illustrated the convertibility of electrical and magnetic force. The failure to detect an effect of electrical force on polarized light might only reflect the fact that electrical force produced a very small effect which he could not detect. The force of an electromagnet was far stronger and might, therefore, be substituted in order to make the expected effect manifest.

On 13 September 1845 his efforts finally bore fruit. The plane of polarization of a ray of plane-polarized light was rotated when the ray was passed through a glass rhomboid of high refractive index in a strong magnetic field. The angle of rotation was directly proportional to the strength of the magnetic force and, for Faraday, this indicated the direct effect of magnetism upon light. "That which is magnetic in the forces of matter," he wrote, "has been affected, and in turn has affected that which is truly magnetic

in the force of light." The fact that the magnetic force acted through the mediation of the glass suggested to Faraday that magnetic force could not be confined to iron, nickel, and cobalt but must be present in all matter. No body should be indifferent to a magnet, and this was confirmed by experiment. Not all bodies reacted in the same way to the magnetic force. Some, like iron, aligned themselves along the lines of magnetic force and were drawn into the more intense parts of the magnetic field. Others, like bismuth, set themselves across the lines of force and moved toward the less intense areas of magnetic force. The first group Faraday christened "paramagnetics"; the second, "diamagnetics."

The discovery of diamagnetism stimulated the production of theories to account for this new phenomenon. Ever since the work of Coulomb in the 1780's, most physicists had assumed (with Coulomb) the existence of polar molecules to account for magnetism. The simple thing to do, when faced with the apparent repulsion of diamagnetic substances by magnetic poles, was simply to assume some kind of "reverse" polarity leading to repulsion rather than attraction. Since such explanations necessarily involved the existence of magnetic or electrical "fluids," Faraday was skeptical. Furthermore, Faraday's attention was increasingly focused on the line of force, rather than on the particles of matter affected by the line of force. In his experiment on the rotation of the plane of polarization of a light ray, Faraday had noted that the "polarity" involved was in the line of magnetic force, not in the interposed glass. Experiments with diamagnetics further convinced him that there were no poles in diamagnetics but only reactions to the line of magnetic force.

He therefore rejected the polar theories of his contemporaries and substituted one of his own. Paramagnetics were substances that conducted the magnetic force well, thereby concentrating lines of force through them; diamagnetic substances were poor conductors of magnetism, thus diverging the lines of magnetic force passing through them. (See Figure 4.) A glance at the patterns of the lines of force was sufficient to disprove the polar theory: the lower figure is *not* the opposite of the top one. There are, in fact, no poles in diamagnetics. The top figure also indicates that there are no "poles" in paramagnetics either, if poles be defined as the termini of the magnetic force. As Faraday went on to show, the lines of magnetic force, unlike their electrostatic cousins, are continuous curves having no termini. They cannot be accounted for in terms of force-atoms under strain, and Faraday ignored his earlier model of interparticulate strain for the transmission of magnetic force.

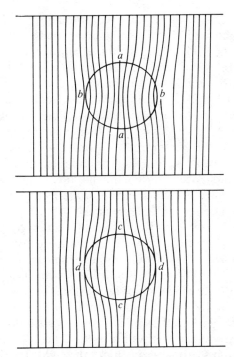

FIGURE 4. Diagrammatic representations of a paramagnetic substance (top) and a diamagnetic substance (bottom) in a uniform magnetic field. The "polarity" of the paramagnetic substance is represented by the compression of the lines of force at *aa*. There is no such compression in the diamagnetic substance; *cc* does not represent polarity opposite to that at *aa*.

Instead, he spoke of a "flood of power" marked out by the lines of force or compared a magnet to a galvanic circuit in which the magnet was the source of power; but the surrounding medium played the part of the connecting wires to transmit the magnetic "current." A magnet was described as the "habitation of lines of force."

Such explanations were manifestly unsatisfactory, for they provided no mechanism whatsoever for magnetic phenomena. They expressed metaphorically what Faraday felt the phenomena to be, but they gave little insight into their underlying causes. Only one point emerged clearly, and this point was of fundamental importance. Whatever the cause of magnetism, the manifestation of magnetic force took place in the medium surrounding the magnet. This manifestation was the magnetic field and the energy of the magnetic system was in the field, not in the magnet. By extension, the same could be said (and was so said by Faraday) of electrical and gravitational systems. This is the fundamental axiom of classic field theory.

By the mid-1850's Faraday had gone as far as he could go. He had provided a new perspective for those who would look on all manifestations of force in the phenomenal world. His description of this perspective

was fuzzy and imprecise but capable of clarification and precision if taken up by someone who could share Faraday's vision. Such a man was James Clerk Maxwell, who, in the 1850's and 1860's, built field theory on the foundations Faraday had laid.

Faraday was unable to appreciate what his young disciple was doing. His mind deteriorated rapidly after the mid-1850's, and even if he had been able to understand Maxwell's mathematics, it is doubtful that he would have been able to follow Maxwell's chain of reasoning. As his mental faculties declined, Faraday gracefully retreated from the world. He resigned from all social clubs in the 1850's, concentrating what remained of his energies on his teaching functions at the Royal Institution. His Christmas lectures for a juvenile audience for 1859–1860, on the various forces of matter, and for 1860–1861, on the chemical history of a candle, were edited by William Crookes and have become classics. But even his lecturing abilities began to fade, and he was forced to abandon the lectern in 1861. In 1862 he resigned his position at the Royal Institution, retiring to a house provided for him by Queen Victoria at Hampton Court. On 25 August 1867 he died.

BIBLIOGRAPHY

I. ORIGINAL WORKS. Faraday collected his papers in four vols.: *Experimental Researches in Electricity*, 3 vols. (London, 1839–1855), and *Experimental Researches in Chemistry and Physics* (London, 1859). The course of his thought may be followed in Thomas Martin, ed., *Faraday's Diary, Being the Various Philosophical Notes of Experimental Investigation Made by Michael Faraday*, 7 vols. and index (London, 1932-1936). For a complete list of Faraday's lectures and writings, see Alan Jeffreys, *Michael Faraday, A List of His Lectures and Published Writings* (London, 1960). Henry Bence Jones was Faraday's close friend and, after his death, collected a large number of letters to and from Faraday, which, together with excerpts from diaries, etc., were published as *Life and Letters of Faraday*, 2 vols. (London, 1870). This volume must be used with great caution, since the editor was not averse to correcting and amending Faraday's language. See also *The Selected Correspondence of Michael Faraday*, L. Pearce Williams, ed., 2 vols. (Cambridge, 1971). There are important MS collections at the Royal Institution of Great Britain, the Royal Society of London, the Institution of Electrical Engineers, London, and the Wellcome Medical Historical Library.

II. SECONDARY LITERATURE. The most recent and detailed biography is L. Pearce Williams, *Michael Faraday, A Biography* (London–New York, 1965). A complete list of books on Faraday is to be found in the bibliography compiled by M. Lukomskaya as an appendix to the Russian trans. of Faraday's *Experimental Researches in Electricity*

(Moscow, 1951). John Tyndall, *Faraday as a Discoverer* (London, 1869, and many subsequent eds.) reveals what Faraday's orthodox friends thought of his physical heresies. Silvanus P. Thompson, *Michael Faraday. His Life and Work* (London, 1898) depicts Faraday as seen by a famous electrical engineer to whom Faraday's electrical discoveries far outweighed his work on field theory.

There are portraits of Faraday in the Royal Institution of Great Britain and the National Portrait Gallery.

L. Pearce Williams

FAREY, JOHN (*b.* Woburn, Bedfordshire, England, 1766; *d.* London, England, 6 January 1826), *geology.*

Farey was educated in a local school and then sent at the age of sixteen to Halifax, Yorkshire, where he studied mathematics, drawing, and surveying. He married about 1790 and had a large family. In 1792 he was appointed land steward of the Woburn estates belonging to Francis, fifth duke of Bedford; but when the duke died suddenly in 1802, his brother John dismissed Farey, who then went to London, where he resided until his death. There he wrote articles for the new *Cyclopaedia* of Abraham Rees and contributed many papers, mostly on geology but also on the theory of music and decimal coinage, to scientific and other journals. In 1807 he began a survey of Derbyshire for the Board of Agriculture, which occupied him for several years; this was followed by employment as a "mineral surveyor," as he entitled himself. His eldest son, John, Jr., born in 1791, from the age of fourteen assisted his father by making drawings and plans and later became well known as an engineer.

Farey contributed to geology in two ways, one of them indirect. He was taught the principles of geology by William Smith, who was employed by the duke of Bedford in October 1801 to construct water meadows in a boggy part of his estate. Farey was already interested in the distribution of local soils and rocks, and during the next few months he eagerly absorbed Smith's ideas on stratification, which the latter seems to have imparted to him readily. From this time on, Farey took every opportunity to promulgate Smith's claim to be a pioneer in English geology, writing to Sir Joseph Banks on the subject in 1802 and, from 1806, referring to Smith's ideas in the many articles he wrote for the *Philosophical Magazine* and Rees's *Cyclopaedia.* By constantly urging the importance of Smith's discoveries and stressing his priority, Farey undoubtedly helped to make Smith's name better known.

As a geologist, however, Farey is entitled to respect for the work he carried out himself, although it has scarcely been noticed in the standard histories of geology. In 1806 and 1807, while going from London to Brighton to visit his brother, he made a study of the strata visible along the route; and the application of Smith's principles enabled him to construct a geological section from London to Brighton on a scale of an inch to the mile (thus over five feet long), which he presented to Sir Joseph Banks. Farey recognized the anticlinal structure of the district and realized that denudation had removed the overlying Chalk formation between the North Downs and the South Downs, a very advanced concept for his time. He also had a good idea of the succession of the rocks under the Chalk. The section unfortunately was not published but must have been well known, for a few years later he stated that copies were "in the hands of many." Some of the strata shown in the section are described in his article "Clay," in Rees's *Cyclopaedia* (vol. VIII).

Farey's investigation of Derbyshire soils and strata began in 1807, and about this time he drew up a detailed geological section of the succession of strata from Ashover in Derbyshire to the Lincolnshire coast (i.e., from Carboniferous rocks through Triassic and Jurassic to the Cretaceous), quite a remarkable achievement. This section, like the London–Brighton one, was not published, although several manuscript copies exist.

The first volume of Farey's report on Derbyshire appeared in 1811. It contains his detailed account of the soils and rocks, together with a colored geological map of Derbyshire and the adjacent counties, on a scale of six miles to the inch. (The area covered is approximately sixty miles from north to south and forty from east to west.) On this map he depicted with considerable accuracy the geological series now known as Carboniferous Limestone, Millstone Grit, Coal Measures, Magnesian Limestone, and Keuper Marl. His only major error lay in regarding the Bunter Sandstone (Triassic) as alluvial gravel. This, the first geological map of an English county actually published, deserves to be better known. In the same volume are two colored plates illustrating what are now known as block diagrams, perspective drawings showing in two dimensions the effects of different kinds of faulting on bedded strata (a subject that Farey had studied in considerable detail), together with the effects of denudation on the faulted rocks. These alone indicate Farey's capabilities as a geologist. The second and third volumes of the report are devoted mostly to agriculture and transport and have little geological interest.

While the first volume was in press, Farey sent to Sir Joseph Banks "An Account of the Great Derbyshire Denudation," which was published in *Philosophical Transactions of the Royal Society* (1811). A more detailed account of the area, "On the Ashover Denudation in the County of Derby," was read to the Geological Society of London (of which Farey was never a member) in April 1813. It was accompanied by a detailed section and a large map. Unfortunately it was never published, possibly because of its length, and this gave rise to much ill feeling on Farey's part toward the officers of the society.

Farey was an inveterate compiler of lists and indexes, some of which, listing localities of fossils named in Sowerby's *Mineral Conchology* and William Smith's works on fossils, were published in the *Philosophical Magazine.* As a mineral surveyor he visited many parts of the British Isles, and his knowledge was drawn on by G. B. Greenough in the compilation of his *Geological Map of England and Wales* (1819).

BIBLIOGRAPHY

I. ORIGINAL WORKS. *A General View of the Agriculture and Minerals of Derbyshire,* 3 vols. (London, 1811–1817), is Farey's main published work. His signed scientific papers are listed in the Royal Society's *Catalogue of Scientific Papers (1800–1863),* II (London, 1868), 561–563, but this list is probably far from complete, since he wrote for the *Monthly Magazine* and agricultural magazines and did not always sign his work. MS geological sections by Farey are in the British Museum (Natural History), the Institute of Geological Sciences (South Kensington), and the Sheffield Central Reference Library. Farey's personal papers were probably destroyed in a fire at his son's house in 1850.

II. SECONDARY LITERATURE. An obituary notice in the *Monthly Magazine,* n.s. **1** (1826), 430, was drawn on by W. S. Mitchell in his "Biographical Notice of John Farey," in *Geological Magazine,* **10** (1873), 25–27. L. R. Cox, in "New Light on William Smith and His Work," in *Proceedings of the Yorkshire Geological Society,* **25** (1942), 1–99, describes Farey's relations with Smith and lists a number of the articles in which he refers to Smith; John Challinor, "From Whitehurst's 'Inquiry' to Farey's Derbyshire," in *Transactions and Annual Report. North Staffordshire Field Club,* **81** (1947), 52–88, provides a valuable commentary on Farey's work in Derbyshire; Trevor D. Ford describes and discusses Farey's MS sections in "The First Detailed Geological Sections Across England, by John Farey (1806–8)," in *Mercian Geologist,* **2** (1967), 41–49, with reproductions (redrawn) of three of Farey's sections.

JOAN M. EYLES

AL-FARGHĀNĪ, ABU'L-ʿABBĀS AḤMAD IBN MUḤAMMAD IBN KATHĪR (*b.* Farghāna, Transoxania; *d.* Egypt, after 861), *astronomy.*

Al-Farghānī was one of the astronomer-astrologers employed by the Abbasid caliph al-Ma'mūn, who reigned in Baghdad from 813 to 833. His name some-times occurs in the Arabic sources as Muḥammad ibn Kathīr, sometimes as Aḥmad ibn Muḥammad ibn Kathīr, and it was probably this variation (in addition to variations of the title of his best-known book—see below) that led Ibn al-Qifṭī to assume the existence of two Farghānīs, a father and a son. But this assumption has now been generally dismissed as very likely no more than a misunderstanding.[1]

Al-Farghānī's activities extended to engineering, and it is in connection with his efforts as an engineer that we have some biographical information about him. According to Ibn Taghrībirdī, he supervised the construction of the Great Nilometer (*al-miqyās al-kabīr*), also known as the New Nilometer (*al-miqyās al-jadīd*), at al-Fusṭāṭ (Old Cairo). It was completed in 861, the year in which the caliph al-Mutawakkil, who ordered the construction, died. (The *Wafayāt al-aʿyān* of Ibn Khallikān reports the event but, in the Cairo edition, gives the name of the engineer as Aḥmad ibn Muḥammad al-Qarṣānī, the last word being no doubt a corruption of "al-Farghānī"—see bibliography.) But engineering was not al-Farghānī's forte, as appears from the following story, which Ibn Abī Usaybiʿa transcribed from the *Kitāb al-Mukāfaʾa* of Aḥmad ibn Yūsuf,[2] who heard it from Abū Kāmil.

Al-Mutawakkil had charged the two sons of Mūsā ibn Shākir, Muḥammad and Aḥmad, with supervising the digging of a canal named al-Jaʿfarī. They delegated the work to "Aḥmad ibn Kathīr al-Farghānī who constructed the New Nilometer," thus deliberately ignoring a better engineer, Sanad ibn ʿAlī, whom, out of professional jealousy, they had caused to be sent to Baghdad, away from al-Mutawakkil's court in Sāmarrā. (The caliphal capital had been transferred from Baghdad to Sāmarrā by al-Muʿtaṣim in 836.) The canal was to run through the new city, al-Jaʿfariyya, which al-Mutawakkil had built near Sāmarrā on the Tigris and named after himself. Al-Farghānī committed a grave error, making the beginning of the canal deeper than the rest, so that not enough water would run through the length of the canal except when the Tigris was high. News of this angered the caliph, and the two brothers were saved from severe punishment only by the gracious willingness of Sanad ibn ʿAlī to vouch for the correctness of al-Farghānī's calculations, thus risking his own welfare and possibly his life. As had been correctly predicted by astrologers, however, al-Mutawakkil was murdered, shortly before the error became apparent.[3] The explanation given for al-Farghānī's mistake is that being a theoretician rather than a practical engineer, he never successfully completed a construction (*wa-kānat maʿrifatuhu awfā min tawfīqihi li-annahu mā tamma lahu ʿamalun qaṭṭu*).

Al-Yaʿqūbī (d. 897) gives a more charitable reason for al-Farghānī's failure: the stony ground chosen for al-Jaʿfariyya, a place called al-Māḥūza, was simply too hard to dig. He does not mention al-Farghānī by name, but says that work on the canal was entrusted to "Muḥammad ibn Mūsā al-Munajjim and those geometers who associated themselves with him" (Kitāb al-buldān, p. 267).

The Fihrist of Ibn al-Nadīm, written in 987, ascribes only two works to al-Farghānī: (1) "The Book of [the thirty?] Chapters, a summary of the Almagest" (Kitāb al-Fuṣul, ikhtiyār[4] al-Majisṭī), and (2) a "Book on the Construction of Sundials" (Kitāb ʿAmal al-rukhāmāt). Ibn al-Qifṭī (d. 1248) reproduces the same list under Muḥammad ibn Kathīr (the name that occurs in the Fihrist) but splits the first title into two: Kitāb al-Fuṣul and Kitāb Ikhtiṣār [sic] al-Majisṭī. To Aḥmad ibn Muḥammad ibn Kathīr he attributes one work, entitled Al-Madkhal ilā ʿilm hayʾat al-aflāk wa-ḥarakāt al-nujūm ("Introduction to the Science of the Structure of the Spheres and of the Movements of the Stars"), which he describes as consisting of thirty chapters (singular, bāb) presenting a summary (jawāmiʿ) of the book by Ptolemy. This is the only title assigned to al-Farghānī by Ibn Ṣāʿid (d. 1244) and Bar-Hebraeus (d. 1286). As has been noted, the two Farghānīs are in fact one; and the same work that Ibn al-Qifṭī mistakenly believed to be two has in fact been known by a variety of titles: Jawāmiʿ ʿilm al-nujūm wa 'l-ḥarakāt al-samāwiyya, Uṣūl ʿilm al-nujūm, Kitāb al-Fuṣul al-thalāthīn, ʿIlal al-aflāk, and so on. This takes us back to the list in Ibn al-Nadīm; but other works must be added to it, notably two(?) treatises on the astrolabe that have come down to us and a commentary on the astronomical tables of al-Khwārizmī.

The Jawāmiʿ, or the Elements, as we shall call it here, was al-Farghānī's best-known and most influential work. He wrote it after the death of al-Maʾmūn in 833 but before 857. Abu'l-Ṣaqr al-Qabīsī (d. 967) wrote a commentary on it which is preserved in the Istanbul manuscript, Aya Sofya 4832, fols. 97v–114v. Two Latin translations of the Elements were made in the twelfth century, one by John of Spain (John of Seville) in 1135[5] and the other by Gerard of Cremona before 1175. Printed editions of the first translation appeared in 1493, 1537, and 1546. (Gerard's translation was not published until 1910.) Jacob Anatoli made a Hebrew translation of the book that served as a basis for a third Latin version, which appeared in 1590, and Jacob Golius published a new Latin text together with the Arabic original in 1669. (For particulars of these editions, see bibliography.) The influence of the Elements on medieval Europe

is clearly attested by the existence of numerous Latin manuscripts in European libraries. References to it in medieval writers are many, and there is no doubt that it was greatly responsible for spreading knowledge of Ptolemaic astronomy, at least until this role was taken over by Sacrobosco's Sphere. But even then, the Elements of al-Farghānī continued to be used, and Sacrobosco's Sphere was clearly indebted to it. It was from the Elements (in Gerard's translation) that Dante derived the astronomical knowledge displayed in the Vita nuova and in the Convivio. The following is a summary of the contents of the thirty chapters constituting the Elements.

Chapter 1, to which nothing corresponds in the Almagest, describes the years of the Arabs, the Syrians, the Romans, the Persians, and the Egyptians, giving the names of their months and days and the differences between their calendars. Chapters 2–5 expound the basic concepts of Almagest I.2–8: sphericity of the heaven and of the earth, the central position of the earth, and the two primary movements of the heavens. In chapter 5 al-Farghānī gives the Ptolemaic value for the inclination of the ecliptic as 23°51′, and reports the value determined at the time of al-Maʾmūn as 23°35′.[6] (In one of his treatises on the astrolabe he states a different value observed at a later date.) Chapters 6–9 give a description of the inhabited quarter and list the seven climes and the names of well-known lands and cities. In chapter 8 al-Farghānī gives the Maʾmūnic measurements of the circumference and the diameter of the earth: 20,400 miles and approximately 6,500 miles, respectively. Chapters 10–11 discuss ascensions of the signs of the zodiac in the direct spheres, al-aflāk al-mustaqīma (i.e., horizons of the equator), and oblique spheres, al-aflāk al-māʾila (i.e., horizons of the climes), and equal and unequal (zamāniyya, temporal) hours.

There follow descriptions of the spheres of each of the planets and their distances from the earth (chapter 12); movements of the sun, moon, and fixed stars in longitude (chapter 13); movements of the five planets in longitude (chapter 14); retrograde motions of the wandering planets (chapter 15); magnitudes of eccentricities and of the epicycles (chapter 16); and revolutions of the planets in their orbs (chapter 17). The assertion of chapters 13 and 14 is that the slow eastward motion of the sphere of the fixed stars about the poles of the ecliptic through one degree every 100 years (the Ptolemaic value) is shared by the spheres (the apogees) of the sun, as well as of those of the moon and the five planets.

Chapter 18 concerns movements of the moon and of the planets in latitude; chapter 19, the order of the fixed stars in respect of magnitude and the positions of the most remarkable among them (al-Farghānī counts fifteen); chapter 20, lunar mansions; chapter 21, the distances of the planets from the earth (Ptolemy had stated only the distances of the sun and the moon); chapter 22, the magnitudes of the planets compared with the magnitude of the earth ("Ptolemy only showed the magnitude of the sun and of the moon, but not that of the other planets; it is, however, easy to know the latter by analogy with what he did for the sun and the moon"); chapter 23, rising and setting; chapter 24, ascension, descension, and occultation; chapter 25, phases of the moon; chapter 26, emergence of the five planets; chapter 27, parallax; chapters 28–30, solar and lunar eclipses and their intervals.

Al-Farghānī's *Jawāmic* thus gives a comprehensive account of the elements of Ptolemaic astronomy that is entirely descriptive and nonmathematical. These features, together with the admirably clear and well-organized manner of presentation, must have been responsible for the popularity this book enjoyed. It must be noted that, as far as numerical values are concerned, the early printed editions show significant divergences. For example, Mercury's diameter is given no fewer than four different values: 1/28, 1/20, 1/10, and 1/18 the diameter of the earth. Only one edition (Frankfurt, 1590) has the first correct value.[7] And in Golius' 1669 Arabic-Latin edition, which is generally superior to the earlier ones, the value of the same diameter differs in the Latin translation (where it is given as 1/18 the diameter of the earth) from that in the Arabic text (1/28 the diameter of the earth).

Al-Farghānī's writings on the astrolabe survive in a number of manuscripts bearing different titles: *Fī Ṣancat al-asṭurlāb, al-Kāmil fi 'l-asṭurlāb, Kitāb cAmal al-asṭurlāb*. The thirteenth-century manuscript at the British Museum (Or. 5479)[8] is a substantial work of forty-eight folios (37v–85r) that ought to be counted among the more respectable treatises devoted to this subject in Arabic. Addressed to the scholar who has reached an "intermediate stage in the knowledge of geometry and the computation of the stars" (fol. 38r), it deals at length with the mathematical theory of the astrolabe and purports to correct faulty constructions which were current at the time of its writing. It is no mere rule-of-thumb manual and was in fact intended to resolve doubts and difficulties created by such manuals. In this work al-Farghānī states the inclination of the ecliptic to be 23°33', "as we found by observation in our time" (fol. 46v). On page 49v "our time" is given as the year 225 of Yazdegerd, i.e., A.D. 857–858.

Al-Bīrūnī in his treatise *On the Calculation of Chords in Circles* assigns to al-Farghānī a work entitled *cIlal Zīj al-Khwārizmī*, in which, apparently, al-Farghānī gave explanations (*cilal*, reasons) for al-Khwārizmī's computational procedures.[9] This work has been lost. But in addition to its having been available to and made use of by al-Bīrūnī in the eleventh century, it had been carefully studied by Aḥmad ibn al-Muthannā ibn cAbd al-Karīm in the tenth. Ibn al-Muthannā, whose commentary on al-Khwārizmī's tables survives in Hebrew and Latin translations, tells us that he found al-Farghānī's book lacking in proofs and altogether suffering from omissions and redundancies. But his remarks would suggest that his own book was either based on al-Farghānī's treatise or at least took its starting point from it. The Latin translation, made by Hugo of Santalla in the second quarter of the twelfth century, was reported by C. H. Haskins but, following Suter, was wrongly identified as a commentary on al-Farghānī by al-Bīrūnī.[10] Two Hebrew versions of Ibn al-Muthannā have recently been published with English translation.[11]

NOTES

1. See H. Suter's art. on al-Farghānī in the 1st ed. of the *Encyclopaedia of Islam* and the rev. art. by J. Vernet in the 2nd ed. See also C. Nallino, "Astrologia e astronomia presso i musulmani," in *Raccolta di scritti editi e inediti*, V (Rome, 1944), 135.

2. Ibn Abī Uṣaybica, *Ṭabaqāt al-aṭibbāʾ*, p. 207, refers to *Kitāb Ḥusn al-cuqbā*, the title of a ch. in *Kitāb al-Mukāfaʾa*.

3. According to the same story, going back to Abū Kāmil, another victim of the intrigues of the two sons of Mūsā was the philosopher al-Kindī, whom they had caused to be estranged from al-Mutawakkil and whose library they had confiscated. Sanad's condition for getting them out of their difficulty was that the library be restored to al-Kindī.

4. *Ikhtiyār* (selection) is found in G. Flügel's ed. of the *Fihrist* and in the (undated) Cairo ed. But the word should no doubt be read *ikhtiṣār* (summary).

5. See F. Woepcke, "Notice sur quelques manuscrits arabes relatifs aux mathématiques...," pp. 116–117.

6. Ibn Yūnus reports that the mission ordered by al-Maʾmūn to prepare the so-called *Mumtaḥan* or *Maʾmūnic zīj* recorded two values of the obliquity at two different places and times: 23°33' at Baghdad in A.H. 214 (A.D. 829–830), and 23°33'52" at Damascus in A.H. 217 (A.D. 832–833). According to the Princeton University Library MS Yahuda 666 (fol. 37v), al-Farghānī reported two values from the *Mumtaḥan*: one equal to the Baghdadian determination of 23°33' and the other the same as that stated in the *Elements*: 23°35'. For Ibn Yūnus, see *Notices et extraits des manuscrits de la Bibliothèque Nationale...*, VII (Paris, 1803), 56–57.

7. See P. J. Toynbee, "Dante's Obligations to Alfraganus...," p. 424, n. 1.

8. Copies of the same work are in the Berlin MSS nos. 5790, 5791, and 5792. A fourth MS at Berlin, no. 5793, fols. 1r–97v, not seen by the present writer, seems to be a different work. See W. Ahlwardt, *Verzeichnis der arabischen Handschriften der Königlichen Bibliothek zu Berlin*, V (Berlin, 1893), 226–227.

9. See "Risāla fī istikhrāj al-awtār fi 'l-dāʾira," in *Rasāʾil al-Bīrūnī*, I (Hyderabad, 1949), pp. 128, 168.

10. See C. H. Haskins in *Romanic Review*, **2** (1911), esp. 7–9, and

his *Studies in the History of Mediaeval Science,* 2nd ed. (Cambridge, Mass., 1927), p. 74, where the same mistaken identification is repeated. But see Millás Vallicrosa, *Estudios sobre Azarquiel* (Madrid–Granada, 1943–1950), pp. 25–26.

11. See Bernard R. Goldstein, *Ibn al-Muthannā's Commentary on the Astronomical Tables of al-Khwārizmī* (New Haven–London, 1967). Hugo's Latin text is edited by Eduardo Millás Vendrell, S. I., in *El comentario de Ibn al-Muṭannāʾ a las Tablas Astronómicas de al-Jwārizmī* (Madrid–Barcelona, 1963).

BIBLIOGRAPHY

I. ORIGINAL WORKS. The Latin trans. of the *Elements* by John of Spain was first printed at Ferrara in 1493: *Breuis ac perutilis compilatio Alfragani astronomorum pertissimi totum id continens quod ad rudimenta astronomica est opportunum.* This was reprinted at Nuremberg in 1537 as part of *Continentur in hoc libro Rudimenta astronomica Alfragani. Item Albategnius. . . . De motu stellarum, ex observationibus tum proprijs, tum Ptolemaei, omnia cum demonstrationibus geometricis & additionibus Ioannis de Regiomonte. Item Oratio introductoria in omnes scientias mathematicas Ioannis de Regiomonte. . . . Eiusdem introductio in Elementa Euclidis. Item epistola Philippi Melanthonis nuncupatoria, etc.* A second reprint, giving the name of the translator for the first time in print, appeared at Paris in 1546: *Alfragani astronomorum pertissimi compendium, id omne quod ad Astronomica rudimenta spectat complectens, Ioanne Hispalensi interprete, nunc primum peruetusto exemplari consulto, multis locis castigatus redditum.* Francis J. Carmody's ed., *Alfragani Differentie in quibusdam collectis scientie astrorum* (Berkeley, Calif., 1943), gives a critical representation of John's version based on some of the extant MSS.

The Latin trans. by the Heidelberg professor Jacob Christmann, published at Frankfurt in 1590, made use of John's version as well as of a Hebrew trans. by Jacob Anatoli: *Muhamedis Alfragani Arabis Chronologica et astronomica elementa, e Palatinae Bibliothecae verteribus libris versa, expleta, et scholiis expolita. Additus est Commentarius, qui rationem calendarii Romani, Aegyptiaci, Arabici, Persici, Syriaci & Hebraei explicat. . . .* According to Woepcke (see below), p. 120, this version was reprinted in 1618.

Gerard of Cremona's trans., made before 1175, was not printed until 1910: *Alfragano (Al-Farġānī) Il 'Libro dell' aggregazione delle stelle' (Dante, Convivio, II, vi–134) secondo il Codice Mediceo-Laurenziano, Pl. 29, Cod. 9 contemporaneo a Dante,* introduction and notes by Romeo Campani (Città de Castello, 1910).

An ed. of the Arabic text was prepared by Jacob Golius on the basis of a Leiden MS. It was published (Amsterdam, 1669) after Golius' death with a Latin trans. and copious notes covering only the first nine chs. of al-Farghānī's book: *Muhammedis Fil. Ketiri Ferganensis, qui vulgo Alfraganus dicitur, Elementa Astronomica, Arabice & Latine. Cum notis ad res exoticas siue Orientales, quae in iis occurrunt.*

Ch. 24 of the *Elements, De ortu et occasu Planetarum, et de occultationibus eorum sub radiis solis,* was twice printed together with Sacrobosco's *Sphere: Sphaera*

Ioannis de Sacro Bosco emendata, etc. (Paris, 1556), fols. 53r–54v; (Paris, 1564), fols. 58v–60r.

For the Arabic MSS of al-Farghānī's works, see C. Brockelmann, *Geschichte der arabischen Literatur,* I, 2nd ed. (Leiden, 1943), 249–250; supp. vol. I (Leiden, 1936), 392–393. See also H. Suter, *Die Mathematiker und Astronomen der Araber und ihre Werke* (Leipzig, 1900), pp. 18–19.

MSS of Jacob Anatoli's Hebrew trans. of the *Elements* are listed in M. Steinschneider, *Die hebraeischen übersetzungen des Mittelalters* (repr. Graz, 1956), pp. 554–559 (sec. 343).

For Latin MSS of the *Elements,* see F. Woepcke, "Notice sur quelques manuscrits arabes relatifs aux mathématiques, et récemment acquis par la Bibliothèque impériale," in *Journal asiatique,* 5th ser., **19** (1862), 101–127, esp. 114–120; F. J. Carmody, *Arabic Astronomical and Astrological Sciences in Latin Translation, A Critical Bibliography* (Berkeley–Los Angeles, 1959), pp. 113–116.

A brief but useful description of the early European eds. of the *Elements* is in P. J. Toynbee, "Dante's Obligations to Alfraganus in the *Vita Nuova* and *Convivio,*" in *Romania,* **24** (1895), 413–432, esp. 413–417.

II. SECONDARY LITERATURE. Biographical and bibliographical information is in Ibn al-Nadīm, *al-Fihrist,* G. Flügel, ed., I (Leipzig, 1871), 279; Ibn al-Qifṭī, *Taʾrīkh al-ḥukamāʾ,* J. Lippert, ed. (Leipzig, 1930), pp. 78, 286; Ibn Abī Uṣaybiʿa, *Ṭabaqāt al-aṭibbāʾ,* A. Müller, ed., I (Cairo, 1882), 207–208; Abu 'l-Faraj ibn al-ʿIbrī (Bar-Hebraeus), *Taʾrīkh mukhtaṣar al-duwal,* A. Ṣālḥānī, ed. (Beirut, 1890), pp. 236–237; Ibn Ṣāʿid al-Andalusī, *Ṭabaqāt al-umam,* L. Cheikho, ed. (Beirut, 1912), pp. 54–55; Ibn Khallikān, *Wafayāt al-aʿyān,* I (Cairo, 1882), 483–485—the relevant passage in the ch. on Abu 'l-Raddād is missing from F. Wüstenfeld's ed. of the *Wafayāt,* fasc. 4 (Göttingen, 1837), no. 362, p. 53, and from de Slane's trans., *Ibn Khallikan's Biographical Dictionary,* II (Paris, 1843), 75; Ibn Taghrībirdī, *al-Nujūm al-zāhira,* T. G. J. Juynboll and B. E. Mathes, eds., I (Leiden, 1851), 742–743; Aḥmad ibn Yūsuf, *Kitāb al-Mukāfaʾa,* Aḥmad Amīn and ʿAlī al-Jārim, eds. (Cairo, 1941), pp. 195–198. Gaston Wiet discusses al-Farghānī's construction of the "New Nilometer" in "Une restauration du Nilomètre de l'île de Rawda sous Mutawakkil (247/861)," in *Comptes rendus de l'Académie des inscriptions et belles-lettres* (1924), pp. 202–206. Here Wiet cites a reference in Ibn al-Zayyāt's *al-Kawākib al-sayyāra* to the tomb of al-Farghānī in the *qarāfa* of Cairo, thus giving evidence that al-Farghānī died in Egypt. Material relevant to the episode concerned with the al-Jaʿfarī project is to be found in al-Yaʿqūbī, *Kitāb al-buldān,* in *Bibliotheca geographorum arabicorum,* M. J. De Goeje, ed., 7 (Leiden, 1892), 266–267. See also Yāqūt, *Muʿjam al-buldān,* F. Wüstenfeld, ed., II (Leipzig, 1867), 86–87; III (Leipzig, 1868), 17; and IV (Leipzig, 1869), 413.

A trans. of al-Farghānī's intro. to his treatise on the astrolabe (Berlin MS no. 5790) is included in Eilhard Wiedemann, "Einleitungen zu arabischen astronomischen Werken," in *Weltall,* **20** (1919–1920), 21–26, 131–134; see also Wiedemann's "Zirkel zur Bestimmung der Gebetszeiten," in *Beiträge zur Geschichte der Naturwissenschaften*

62, in *Sitzungsberichte der Physikalish-medizinischen Sozietät in Erlangen,* **52** (1922), 122–125. J. B. J. Delambre, in *Histoire de l'astronomie du moyen-âge* (Paris, 1819), pp. 63–73, gives a detailed account of al-Farghānī's *Elements,* chapter by chapter. See also J. L. E. Dreyer, *History of the Planetary Systems from Thales to Kepler* (Cambridge, 1906), *passim;* P. Duhem, *Le système du monde,* II (Paris, 1914), 206–214; and, concerning the relation of Sacrobosco to al-Farghānī, Lynn Thorndike, *The Sphere of Sacrobosco and Its Commentators* (Chicago, 1949), pp. 15–19. Brief accounts of al-Farghānī are to be found in the *Encyclopaedia of Islam* and in Sarton's *Introduction to the History of Science,* I (Baltimore, 1927), 567.

A. I. SABRA

FARKAS, LASZLO (LADISLAUS) (*b.* Dunaszerdahely, Hungary [now Dunajska Streda, Czechoslovakia], 10 May 1904; *d.* near Rome, Italy, 31 December 1948), *physical chemistry.*

Farkas was the son of a pharmacist and the eldest of several children. After finishing secondary school he studied chemistry from 1922 to 1924 at the Technische Hochschule in Vienna, and then at the Technische Hochschule (now Technische Universität) in Berlin. In 1927 he received his doctorate, and in the same year he entered the Kaiser Wilhelm Institut in Berlin, where he worked under Haber's guidance. When the Nazis came to power, Farkas left Germany and went to England, where he taught colloid chemistry at Cambridge. In 1936 he accepted an offer to serve as professor of physical chemistry at the newly organized Faculty of Sciences of Hebrew University, Jerusalem, and he remained there until his death. He died in a plane crash on his way to the United States to seek support for acquisition of scientific instruments.

Farkas began his scientific activity with work in the field of photochemical sensitizing in the region of the ultraviolet. He then turned to a study of the equilibrium distribution of the two forms of molecular hydrogen—ortho-hydrogen and para-hydrogen.

During this period deuterium was discovered, and Farkas saw in the substance a valuable aid for his investigations into the homogeneous catalysis of ortho- and para-hydrogen conversion. With his brother he developed an electrolytic method which resulted in the simplest known procedure for producing heavy water.

In the following years various investigations involving deuterium constituted the central portion of Farkas' scientific activity. He studied thoroughly the equilibrium of the reaction

$$H_2O + HD \rightleftharpoons HOD + H_2$$

and its role in the separation of the hydrogen isotope.

He also investigated the various exchange reactions of heavy hydrogen. Moreover, he established the ratio of heavy to light water in the liquid and vapor phases over an extensive temperature range. Farkas also determined the catalytic activity of heavy hydrogen in various processes. His work was, in many respects, of pioneering importance in the field of deuterium and heavy-water research.

Farkas also left many publications on reaction kinetics and a few on analytical chemistry. During World War II he was secretary of the Scientific Advisory Committee of the Middle East Supply Center. He also contributed to the organization of scientific research in the new state of Israel.

BIBLIOGRAPHY

A list of Farkas' 100 or so publications in German, English, and Israeli journals is in *L. Farkas Memorial Volume* (Jerusalem, 1952), pp. 305–309.

On Farkas, see two articles by E. K. Rideal: "Prof. L. Farkas," in *Nature,* **162** (1949), 313; and "Ladislaus Farkas," in *L. Farkas Memorial Volume* (Jerusalem, 1952), pp. 1–2.

F. SZABADVÁRY

FARMER, JOHN BRETLAND (*b.* Atherstone, Warwickshire, England, 5 April 1865; *d.* Exmouth, Devon, England, 26 January 1944), *botany.*

Farmer was the only son of John Henry and Elizabeth Farmer. After education at grammar school and private tutoring, he went to Oxford in 1883 and took a first-class degree in natural sciences in 1887. He was then appointed demonstrator in botany and was elected to a fellowship in 1889. In 1892 Farmer became assistant professor of botany at the Royal College of Science (which later became the Imperial College of Science and Technology); from 1895 until his retirement in 1929 he was a full professor, playing an active part in the development of the college as one of its governors.

His visit to India and Ceylon in 1892–1893 so impressed Farmer with the need for applied biologists to work in underdeveloped countries that he set about encouraging both instruction and research that would be useful in the tropics: his students worked all over the world, and he was active in advisory work on colonial administration and in setting up research institutes.

Farmer's own research was wide-ranging in pure botany and cytology. His early papers were on morphology and physiology, and it was not until 1893 that he published his first cytological work on nuclear division in the spore mother cells of *Lilium martagon.*

It was followed by several other papers, mostly on spore formation, including a demonstration in 1894 that chromosome reduction was an essential preliminary to fertilization in *Hepaticae,* which provided material for many later studies. With the zoologist J. E. S. Moore, Farmer showed many features in common between reduction division in plant and animal cells, and in 1903 he went on to demonstrate similarities in cell division of malignant and normal growths. Their paper of 1904 introduced the term "maiotic phase" (later changed to "meiotic phase") and illustrated reduction division in species as diverse as a lily, a cockroach, and a fish, discussing its occurrence at different points in the life histories of organisms. Later Farmer worked on the centrosphere and kinoplasm, the dimensions of chromosomes, and water utilization in plants.

He also found time to write a textbook on practical botany and a popular introduction to botany, and to translate, with A. D. Darbishire, de Vries's *Mutationstheorie.* He edited a six-volume work on nature study and, for shorter or longer periods, the journals *Annals of Botany, Science Progress,* and *Gardeners' Chronicle.*

In 1892 he married Edith Mary Pritchard, and they had one daughter. He was elected a fellow of the Royal Society in 1900 and knighted in 1926.

BIBLIOGRAPHY

I. ORIGINAL WORKS. Papers referred to in the text are "On Nuclear Division in the Pollen-Mother-Cells of *Lilium martagon,*" in *Annals of Botany,* **7** (1893), 392–396; "Studies in Hepaticae: On *Pallavicinia decipiens,*" *ibid.,* **8** (1894), 35–52; "On the Resemblance Between the Cells of Malignant Growths in Man and Those of Normal Reproductive Tissues," in *Proceedings of the Royal Society,* **72** (1903), 499–504, written with J. E. S. Moore and C. E. Walker; and "On the Maiotic Phase (Reduction Division) in Animals and Plants," in *Quarterly Journal of Microscopical Science,* **48** (1904), 489–557, written with J. E. S. Moore.

II. SECONDARY LITERATURE. The most important evaluation of Farmer's scientific work is V. H. Blackman's article in *Obituary Notices of Fellows of the Royal Society of London,* **5** (1945–1948), 17–31. The bibliography is comprehensive for papers but omits the 2nd ed. of *A Practical Introduction to the Study of Botany: Flowering Plants* (London, 1902); "The Structure of Animal and Vegetable Cells," in E. Ray Lankester, ed., *A Treatise on Zoology,* I (London, 1903), 1–46; and his popular introduction to botany, *Plant Life* (London, 1913). Blackman also wrote the entry on Farmer in *Dictionary of National Biography, Supplement, 1941–1950,* pp. 245–246.

Other obituaries are an unsigned one in *Gardeners' Chronicle,* 3rd ser., **115** (1944), 64; and R. J. Tabor, in *North Western Naturalist,* **19** (1944), 310–311.

DIANA M. SIMPKINS

FARRAR, JOHN (*b.* Lincoln, Massachusetts, 1 July 1779; *d.* Cambridge, Massachusetts, 8 May 1853), *mathematics, physics, education.*

Farrar was responsible for conceiving and carrying through a sweeping modernization of the science and mathematics curriculum at Harvard College, his alma mater. He brought in the best French and other European writings on introductory mathematics, most of them unknown and unused in the United States. Much of the responsibility for shifting from the Newtonian fluxional notations to Leibniz's algorithm for the calculus was his. In natural philosophy Farrar also relied heavily upon French authors. He introduced current concepts in mechanics, electricity and magnetism, optics, and astronomy.

As the foundation of his curricular reform, Farrar carried through the translation of many French works between 1818 and 1829. Published in separate, topical volumes, they became elements of two series: Cambridge Mathematics and Cambridge Natural Philosophy. He selected and combined the writings most suitable to the needs of his students. The burden of Farrar's presentation in mathematics was carried by Lacroix, Euler, Legendre, and Bézout, but he also drew from John Bonnycastle and Bowditch. In natural philosophy he relied most heavily upon Biot, but used Bézout, Poisson, Louis-Benjamin Francoeur, Gay-Lussac, Ernst Gottfried Fischer, Whewell, and Hare as well.

Farrar's translations provided an excellent introductory program that was used not only at Harvard but also at West Point and other colleges; they went through several editions. Farrar was a fine teacher and, as Hollis professor of mathematics and natural philosophy, played an important role throughout Harvard College. One of his major aspirations, the establishment of an astronomical observatory at Harvard, was not attained until after his death.

In the larger community Farrar made similar contributions. He was active in the American Academy of Arts and Sciences and occasionally translated such topical works as Arago's 1832 *Tract on Comets,* written in preparation for the comet of that year. He published essay reviews in the *North American Review* and occasional observations and a few scientific papers on astronomy, meteorology, and instruments in the *Memoirs of the American Academy of Arts and Sciences* and the *Boston Journal of Philosophy and the Arts.*

BIBLIOGRAPHY

I. ORIGINAL WORKS. Farrar wrote few scientific papers. In the *Memoirs of the American Academy of Arts and Sciences,* he published "An Account of the Violent and Destructive Storm of the 23d of September 1815," **4** (1821), 92–97, and "An Account of a Singular Electrical Phenomenon," *ibid.,* 98–102. His "Account of an Apparatus for Determining the Mean Temperature and the Mean Atmospherical Pressure for Any Period" appeared in *Boston Journal of Philosophy and the Arts,* **1** (1823–1824), 491–494. In the *North American Review* he published several review essays, all of them unsigned: **6** (1817–1818), 205–224; **8** (1818–1819), 157–168; **12** (1821), 150–174; **14** (1822), 190–230; and a few observations: **3** (1816), 36–40, 285–287; **6** (1817–1818), 149, 292.

His primary publishing activity lay in translating and combining French writings with a few others in a manner that effectively produced good college textbooks which were abreast of recent advances. Some appeared without any indication of Farrar's role; others did not name the authors on the title page but always scrupulously noted them at some point. The first editions (often of many) are a translation of S. F. Lacroix, *An Elementary Treatise on Arithmetic* (Boston, 1818); a translation of S. F. Lacroix, *Elements of Algebra* (Cambridge, Mass., 1818); a translation of L. Euler, *An Introduction to the Elements of Algebra* (Cambridge, Mass., 1818); a translation of A. M. Legendre, *Elements of Geometry* (Boston, 1819); translations of S. F. Lacroix and E. Bézout, *An Elementary Treatise on Plane and Spherical Trigonometry* (Cambridge, Mass., 1820); *An Elementary Application of Trigonometry* (Cambridge, Mass., 1822); translations of E. Bézout, *First Principles of the Differential and Integral Calculus* (Cambridge, Mass., 1824); *An Elementary Treatise on Mechanics* (Cambridge, Mass., 1825); *An Experimental Treatise on Optics* (Cambridge, Mass., 1826); *Elements of Electricity, Magnetism, and Electro-Magnetism* (Cambridge, Mass., 1826); *An Elementary Treatise on Astronomy* (Cambridge, Mass., 1827); a translation of E. G. Fischer, *Elements of Natural Philosophy* (Boston, 1827); and a translation of F. Arago, *Tract on Comets* (Boston, 1832).

Letters and other MS records are held by the Harvard University Archives and the Massachusetts Historical Society, and a few by the Boston Public Library.

II. SECONDARY LITERATURE. On Farrar or his work, see Mrs. John Farrar, *Recollections of Seventy Years* (Boston, 1866); Dirk J. Struik, *Yankee Science in the Making* (New York, 1962), pp. 227–229, *passim;* and [John Gorham Palfrey], *Notice of Professor Farrar* (Boston, 1853).

BROOKE HINDLE

IBN AL-FARRUKHĀN. *See* 'Umar ibn al-Farrukhān.

FATOU, PIERRE JOSEPH LOUIS (*b.* Lorient, France, 28 February 1878; *d.* Pornichet, France, 10 August 1929), *mathematics.*

Fatou attended the École Normale Supérieure from 1898 to 1901. The scarcity of mathematical posts in Paris led him to accept a post at the Paris observatory, where he worked until his death. He received his doctorate in 1907 and was appointed titular astronomer in 1928. Fatou worked in practical astronomy: on determining the absolute positions of stars and planets, on instrumental constants, and on measurements of twin stars.

In order to calculate the secular perturbations produced on a planet P' through the movement of another planet, P, Gauss had had the idea of spreading the mass of P' over its orbit, so that the mass of each arc is proportional to the time it takes for the planet to trace it. This proposition is valid only when the distinction between periodic and secular perturbations does not apply, i.e., when n' is very small (n and n' are the mean motions of the planets P and P', respectively). By means of general existence theorems of solutions of differential equations, Fatou studied these motions of material systems subjected to forces whose periods tend to zero. Gauss's intuitive result had often been used in practice but had never been rigorously justified.

Fatou also studied the movement of a planet in a resistant medium. This work was based on the probability that stellar atmospheres had previously been far more extensive than they are now and would thus have given rise to capture phenomena that can be used to explain the origins of twin stars and of certain satellites.

Along with this work Fatou did both related and general mathematical research. He contributed important results on the Taylor series, the theory of the Lebesgue integral, and the iteration of rational functions of a complex variable. When studying the circle of convergence of the Taylor series, several points of view are possible: (1) one can look for criteria of convergence or divergence of the series itself on the circumference; (2) one can consider the limit values of the circle of the analytic function represented by the series and try to determine where these limit values are finite or infinite, as well as the properties of the functions of the argument represented by the real and imaginary parts of the series when these functions are well defined; (3) one can consider what points on the circumference, singular in the Weierstrass sense, also determine the analytic extension of the series. The link between these problems led Fatou to formulate a fundamental theorem in the theory of the Lebesgue integral. He found that the theory of the Lebesgue integral allowed the first two of the above problems to be treated with more precision and more generality, with the following general result: If

$f_n(x) \geqslant 0$ for all values of n, $x \in E$ and $f_n(x) \to f(x)$ as $n \to \infty$, then

$$\int_E f(x)dx \leqslant \lim_{n \to \infty} \int_E f(x)dx.$$

The theorem implies that if the right-hand side is finite, then $f(x)$ is finite almost everywhere and integrable; if $f(x)$ is not integrable or is infinite in a set of positive measure, then

$$\lim_{n \to \infty} \int_E f_n(x)dx = \infty.$$

This work was advanced by Carathéodory, Friedrich and Marcel Riesz, Griffith Evans, Leon Lichtenstein, Gabor Szegö and Nicolas Lusin.

Fatou also showed ways in which the algebraic signs of a_n affect the number and character of singularities in the Taylor series. Given the series $\Sigma a_n x^n$ $0 < R < \infty$, a sequence $\{\lambda_n\}$ exists such that the series obtained by changing the signs of a_{λ_n} has the circle of convergence as a cut. This theorem was proved in general by Hurwitz and George Pólya.

BIBLIOGRAPHY

I. ORIGINAL WORKS. Fatou's writings include "Séries trigonométriques et séries de Taylor," in *Acta mathematica,* **30** (1906), 335–400; "Sur la convergence absolue des séries trigonométriques," in *Bulletin de la Société mathématique de France,* **41** (1913), 47–53; "Sur les lignes singulières des fonctions analytiques," *ibid.,* 113–119; "Sur les fonctions holomorphes et bornées à l'intérieur d'un cercle," *ibid.,* **51** (1923), 191–202; "Sur l'itération analytique et les substitutions permutables," in *Journal de mathématiques pures et appliquées,* 9th ser., **2** (1923), 343–384, and **3** (1924), 1–49; "Substitutions analytiques et équations fonctionnelles à deux variables," in *Annales scientifiques de l'École normale supérieure,* 3rd ser., **41** (1924), 67–142; "Sur l'itération des fonctions transcendantes entières," in *Acta mathematica,* **47** (1926), 337–370; and "Sur le mouvement d'un système soumis à des forces à courte période," in *Bulletin de la Société mathématique de France,* **56** (1928), 98–139. The Société Mathématique de France expects to publish Fatou's papers.

II. SECONDARY LITERATURE. On Fatou or his work, see Jean Chazy, "Pierre Fatou," in *Bulletin astronomique,* **8,** fasc. 7 (1934), 379–384; Griffith Evans, *The Logarithmic Potential* (New York, 1927); P. Fatou, *Notice sur les travaux scientifiques de M. P. Fatou* (Paris, 1929); A. Hurwitz and G. Pólya, "Zwei Beweise eines von Herrn Fatou vermuteten Satzes," in *Acta mathematica,* **40** (1916), 179–183; S. Mandelbrojt, *Modern Researches on the Singularities of Functions Defined by the Taylor Series* (Houston, Tex., 1929), chs. 9, 12; and M. Riesz, "Neuer Beweis des Fatouschen Satzes," in *Göttingensche Nachrichten* (1916);

and "Ein Konvergenz satz für Dirichletsche Reihen," in *Acta mathematica,* **40** (1916), 349–361.

HENRY NATHAN

FAUJAS DE SAINT-FOND, BARTHÉLEMY (*b.* Montélimar, Dauphiné, France, 17 May 1741; *d.* Saint-Fond, Dauphiné, 18 July 1819), *geology.*

Faujas (who took his full name from the family estate at Saint-Fond in Dauphiné) was for some years a successful lawyer but, possessed by an overwhelming passion for natural history and coming under the influence of Buffon, he abandoned his legal career and was appointed assistant naturalist at the Muséum d'Histoire Naturelle in Paris in 1778. In 1785 he became royal commissioner of mines and in 1793 was made professor of geology at the museum, a post he held until his death.

Faujas was a wide-ranging naturalist, as was common in his day. His most continuous and concentrated attention was given to rocks, minerals, and fossils (i.e., to geology, a name adopted in the late 1770's), but he was also a physicist and a chemist. He applied his discoveries and investigations to practical affairs: for instance, in 1775 he found, analyzed, and opened up a deposit of a volcanic tuff similar to the Italian pozzolana, which was used industrially in France in the making of cement; he also wrote treatises on the construction and navigation of balloons, a practical scientific activity fashionable at that time.

The existence of a group of old volcanoes in central France had been ascertained by Guettard in 1752 through observations begun at Montélimar itself, but Guettard wrote later that basalt, with its prevalent columnar structure, was a crystallization from water. In the 1760's Desmarest found the true explanation (not published until 1774): this basalt was of volcanic origin. Meanwhile, Faujas had been exploring the hilly districts of Vivarais and Velay in east-central France and found that the basalt there was also volcanic. (It resulted from regional volcanic activity in the Tertiary period which produced as its latest manifestation, particularly in central France, the very new-looking cones of ashes and associated lava flows.) It is not clear to what extent Faujas was familiar with Desmarest's work, but his discoveries were at any rate independent, and he embodied them in 1778 in a great folio work on the ancient volcanoes of Vivarais and Velay (accounts of other researches were included). This work established once and for all that basalt, a rock important scientifically because of its distinctive characteristics, its widespread occurrence, and the manner of its association with other kinds of rock, was the product of volcanic action. The controversy over the origin of basalt was, however, by no means settled; the Wernerian (neptunist) view that it was an aqueous precipitate was vigorously advocated until well into the nineteenth century. In fact, Faujas himself was, except on this question, generally a neptunist.

In 1784 Faujas journeyed through England to Scotland; a full account was published in 1797. He narrated entertaining details of his travels and described arts, industries, and customs, but his most important observations were geological. He realized the volcanic nature of the basalt of the inner Hebrides and paid special attention to the spectacular columnar occurrence on the isle of Staffa, about which his curiosity had been aroused by Banks's account in Thomas Pennant's *Second Tour of Scotland . . .* (1774). (The Staffa basalt was recognized as of volcanic origin by Bishop Uno von Troil in his *Letters From Iceland,* 1780.) He also recognized the volcanic nature of the terraced hills in central Scotland, but he had no idea that these were vastly older (Paleozoic) than those of the western islands (Tertiary). His discrimination between the various kinds of dark, fine-grained rocks was faulty. In particular, Faujas contradicted Whitehurst's correct hypothesis regarding the basaltic nature of the Derbyshire toadstones, and he identified as old lavas some rocks that were unquestionably sedimentary in origin. Unfortunately, his specimens had been lost in a shipwreck on the way back to France; probably a more careful scrutiny of them would have prevented some of his mistakes.

In his monograph on the chalk of Maastricht, Faujas described a huge reptilian skull which he thought was that of a crocodile. Cuvier discussed this at length in his *Ossemens fossiles* (V, pt. 2, [1824]), calling it a "marine serpent-like reptile"; the name mosasaur was proposed by the English geologist W. D. Conybeare in Cuvier's volume. It is now placed as the representative of an extinct group among the lizards. This was perhaps the most notable discovery in the field of vertebrate paleontology up to that time.

BIBLIOGRAPHY

I. ORIGINAL WORKS. Faujas's main geological works are *Recherches sur les volcans éteints du Vivarais et du Velay* (Grenoble, 1778); *Minéralogie des volcans* (Paris, 1784); *Voyage en Angleterre, en Écosse et aux Îles Hébrides,* 2 vols. (Paris, 1797), also trans. into English, 2 vols. (London, 1799) and, later, with notes and a memoir by Archibald Geikie, 2 vols. (Glasgow, 1907); *Histoire naturelle de la montagne de Saint-Pierre de Maestricht* (Paris, 1799); and *Essai de géologie, ou Mémoires pour servir à l'histoire naturelle du globe,* 2 vols. (Paris, 1803–1809).

There is a very full bibliography of Faujas's works,

including his papers in the *Annales du Muséum d'histoire naturelle* (Paris), in *Nouvelle biographie générale,* XVII (Paris, 1856), 167–171; the work was reprinted at Copenhagen in 1963–1969. The list in the *British Museum General Catalogue of Printed Books,* photolith. ed., LXXI (London, 1960), cols. 274–276, is also very full as regards books. The reader should note, though, that in each source there are one or two items that are not in the other.

II. SECONDARY LITERATURE. The chief source in English is A. Geikie, "Memoir of the Author," in his ed. of the *Voyage* (see above). See also *British Museum General Catalogue of Printed Books;* J. Challinor, "The Early Progress of British Geology—III," in *Annals of Science,* 10 (1954), 107–148, esp. 126–129; and T. D. Ford, "Barthélemy Faujas de St. Fond," in *Bulletin of the Peak District Mines Historical Society,* 2 (1965), 236–240.

JOHN CHALLINOR

FAULHABER, JOHANN (*b.* Ulm, Germany, 5 May 1580; *d.* Ulm, 1635), *mathematics.*

The Faulhaber family lived in Ulm from the middle of the fifteenth century and had been vassals of the abbot of Fulda from 1354 to 1461. Like his father, who died in 1593, Faulhaber first learned weaving. Ursula Esslinger of Ravensburg, whom Faulhaber married in 1600, bore him nine children, several of whom distinguished themselves as mathematicians. His son Johann Matthäus, born in 1604, learned weaving from his father before he turned to mathematics. In 1622 he accompanied his father to Basel to survey the fortifications and became director of the assignment following his father's death. A second son, also named Johann, born in 1609, was captain in the Corps of Engineers at Ulm.

His natural abilities led Faulhaber from weaving to mathematics. His first teacher in Ulm was the writing and arithmetic teacher David Saelzlin. The mathematicians of the sixteenth century concerned with algebra called themselves Cossists (from the Italian *cosa,* or "thing," which was used to designate the quantity being sought). Max Jaehns calls Faulhaber one of the most significant of the Cossists and the first to take algebra into equations higher than the third degree.[1]

His education did not make Faulhaber proficient in Latin, but with laborious effort he translated the Latin texts that he needed, lent by Michael Maestlin in Tübingen, as we learn from his letter of 16 April 1617 to Matthäus Beger, in Reutlingen:

. . . since then I have taken the trouble to get the most distinguished books in German . . . as a person who never studied Latin and only now have attained some understanding of the language . . . to translate from Latin into simple German so that I now have at hand in German the books of Euclid, Archimedes,

Apollonius, Serenus, Theodosius, Regiomontanus, Cardano. . . .

After Faulhaber had helped Johann Kraft, arithmetic master in Ulm, to publish an arithmetic text, he founded his own school in Ulm in 1600.[2]

From 1604 on, Faulhaber received a salary of 30 guldens for running this school, but it was withdrawn in 1610 for a few months because he was concerning himself more and more with physical and technical inventions and developing an extensive literary activity that took him away from his pedagogic duties. Above and beyond this, he incurred the displeasure of the municipal council because he published *Neu erfundener Gebrauch eines Niederländischen Instruments zum Abmessen und Grundlegen, mit sehr geschwindem Vortheil zu practiciren* without the permission of the office responsible for supervision of the schools. About this time Faulhaber set up the formulas for the sum of the powers for natural numbers up to the thirteenth power, a problem with which Leonhard Euler was later concerned in a general way. He knew of the expression for the final difference of the arithmetic series obtained by raising the terms of an arithmetic series of the first order to a higher power.[3] More and more his school became an educational institute for higher mathematical sciences, and an artillery and engineering school was later added. In the pedagogic field Faulhaber's particular merit was in the dissemination of mathematical knowledge for general use. His arithmetic text, *Arithmetischer Wegweiser zu der hochnutzlichen freyen Rechenkunst* (1614), is a very clear textbook for the period. In the early editions he got as far as the "rule of three"; in later editions he treated all of the computations for ordinary use and even the fundamentals of equations. His writings on algebra are difficult to interpret because he used symbols that are no longer common. His particular concerns in these works are the theory of progressions, theory of magic squares, and the nature of numbers.

Like Michael Stifel, the Augustinian monk and promotor of calculation with logarithms, Faulhaber was noted for his mystical consideration of pseudomathematical problems. He attempted to interpret future events from numbers in the Bible: from Genesis, Jeremiah, Daniel, and Revelation. Together with the master baker from Ulm, Noah Kolb, he predicted the end of the world by 1605 and was put in jail for this in 1606. On the basis of his confession that he had not acted with evil intent, but from an irresistible impulse of conscience, he was released. As early as seven years later he again believed that he could see "numeri figurati"—figured numbers—in certain numbers from the Bible, and his view that God had used pyramidal numbers in the prophecies of the Bible was expressed in *Neuer mathematischer Kunstspiegel* (1612). Faulhaber meditated on the numbers 2,300 (Daniel VIII: 14), 1,335 (Daniel XII: 12), 1,290 (Daniel XII: 11), 1,260 (Revelation XI: 3), and 666 (Revelation XX: 2). These are the same numbers with which Stifel concerned himself.[4]

The extent to which this mystic arithmetic had affected Faulhaber can be seen in his books *Andeutung einer unerhörten neuen Wunderkunst . . .* (1613) and *Himmlische geheime Magia . . .* (1613). In the latter book he attempted to solve the hidden riddles of his sealed numbers by a peculiar transposition of the German, Latin, Greek, and Hebrew alphabets, a puzzle in which he refers to the tribes Gog and Magog mentioned in Revelation XX: 8. This biblical interpretation being contrary to Christian teachings, he drew the enmity of the clergy upon himself, and at their instance he was warned by the magistrate in Ulm that he should no longer print such interpretations without the knowledge and permission of the censor, upon pain of losing his civil rights. Since other theologians, such as Hasenreffer, the chancellor of Tübingen University, also sent warning letters to the Ulm city council, the prohibition of *Himmlische geheime Magia* was intensified.

Faulhaber also devoted himself to alchemy, which he practiced as a believer in Johann Valentin Andreae's *Chymische Hochzeit des Christiani Rosencreutz,* first published anonymously about 1604. On 21 January 1618 he wrote to Rudolph von Bünau: ". . . I am not sparing any efforts in inquiring about the commendable Rosicrucian Society . . ."; and on 21 March 1621, to Bünau: ". . . with the help of God, I have come to the point where I can make 2 grains of gold out of 1 grain of gold in a few days, which is why I give praise and thanks to the Almighty, and although one-tenth is supposed to become 10, up to now, I have not been able to get it any further and have worked it with my own hands."

These mysterious arts brought Faulhaber into contact with Duke Johann Friedrich von Württemberg. In 1619 he obtained permission to teach his arts and sciences freely in the duchy, and he continued to have that permission until after he again distributed his forbidden writings about Gog and Magog.

The reputation of Faulhaber's mathematics school extended so far that Descartes studied with him in 1620.[5] According to Veesenmayer, Descartes had already corresponded with Faulhaber concerning questions of plane analytic geometry[6] and had been stimulated to write *Discours sur la méthode . . .* (1637).[7] Descartes called him a "mathematicum insignem et imprimis in numerorum doctrina versatum et praeceptorem."[8]

Faulhaber's lasting accomplishment was the dissemination and explanation of the logarithmic method of calculation. The dissemination of the logarithms associated with Stifel, Bürgi, and Napier occurred through his chief work, *Ingenieurs-Schul*, the *Appendix oder Anhang . . . Ingenieurs-Schul*, and the *Zehntausend Logarithmi . . . .* He gives the logarithms of the numbers 1–10,000 to seven places and the values of the six natural goniometric functions to ten places. Along with the solution of plane and spherical triangles, with the applications to fortification and astronomical geography, we find the reason in the *Appendix:* ". . . that the entire foundation and correct basis of the logarithms from which they originate and are made, are briefly indicated and explained. . . ." In addition, it was the first publication of the Briggs logarithms in German.[9] Faulhaber devoted himself to the stereometric analogue to the Pythagorean theorem, which he found and to which he was led by an apocalyptic number, 666. He first published this theorem as a numerical example in his "Miracola arithmetica," which is part of the *Continuatio des neuen mathematischen Kunstspiegel* (1620). Descartes, who probably learned it from Faulhaber, reproduced it in 1620: "In tedraedro rectangula basis potentia aequalis est potentistrium facierum simul." If one imagines a rectangular system of coordinates intersected by an inclined plane; if A, B, and C are the areas of the right triangles that occur on the planes of the coordinates; and if D is the area of the triangle determined by the intersections of the axes on this intersecting plane, then $D^2 = A^2 + B^2 + C^2$.

Faulhaber usually gives the solutions of his mathematical problems only in hints. Among the problems treated in the first part of his *Ingenieurs-Schul*, the question concerning an irregular circle-heptagon (p. 168) attained some measure of fame because the well-known nineteenth-century mathematicians Moebius and Siebeck concerned themselves with it. Faulhaber inscribed within a circle a heptagon with sides of lengths 2,300, 1,600, 1,290, 1,000, 666, 1,260, and 1,335, then asked how the radius of this circle could be found and how many degrees and minutes each angle contained. The numbers again are those of Michael Stifel. Faulhaber does not say how he solved this problem, but in the *Ingenieurs-Schul* (ch. 13, p. 157), he gives the result according to which the radius of the circle is "found to be $1582\frac{6223}{10,000}$."[10]

Faulhaber's prestige as a fortifications engineer is based on his many assignments in this field. Besides Duke August von Brunswick-Lüneburg, his services were sought by Duke Johann Friedrich von Württemberg, Cardinal Dietrichstein of Nicholsburg (near Vienna), King Gustavus Adolphus of Sweden, and the cities of Randegg, Schaffhausen, and Fürstenberg. Unfortunately, in this area too his full development was hindered by his religious fanaticism. On 5 December 1618 he entered the service of Landgrave Philipp von Butzbach, who sought him as an adviser, but he continued to live in Ulm. He wanted to make "all his inventions" known to the landgrave except his work in "munitions," i.e., in fortifications, which he was forbidden to divulge by the municipal council of Ulm.[11] Soon thereafter Faulhaber concerned himself anew with his interpretations of biblical numbers, and in April 1619 his mathematical and astronomical writings to the landgrave suddenly ended. Perhaps it was because the prince, who was firmly grounded in Christian teachings, believed that he had deceived himself about Faulhaber, or perhaps it was, as Faulhaber claims, that he was supposed to give the prince secrets which he had to consider in total confidence. In spite of this, the landgrave remained interested in his former adviser. In 1622 he received news about Faulhaber through Konrad Dietrich, superintendent in Ulm; and we learn from his letter of 23 March 1625 that Faulhaber "has been reconciled with an honorable councilman in Ulm and has promised in the name of the Almighty to let his whims fall."[12]

The full picture of Faulhaber's character is revealed from the controversies into which he was drawn. To be sure, we do not find his scientific importance reduced, but he nevertheless appears shady in view of the interplay of serious perception and speculative fantasy. Having been ordered by the municipal council of Ulm to publish an almanac for the year 1618, he used the ephemerides of Johannes Kepler. In it Kepler listed two rare constellations that were supposed to appear before and after 1 September, which means that the appearance of a comet cannot be regarded as excluded. In his almanac Faulhaber predicted a comet for 1 September 1618, on the basis of a consideration of the longitude and latitude of Mars and the moon. When one of the greatest comets of that era appeared in November 1618, Faulhaber no longer doubted the efficacy of his secret numbers and had this opinion published through his friend J. G. Goldtberg.[13] Attacked as vehemently by Hebenstreit and Zimpertus Wehe in Ulm as he was defended by Matthäus Beger in Reutlingen, Faulhaber appealed to the municipal council in Ulm and the ecclesiastical authorities, who decided in his favor.

He had a lively contact with Johannes Kepler. Upon the order of the magistrate, in 1622 he and Kepler designed a gauging kettle for the measurement of length, volume, and weight, which was cast by Hans Braun in 1627.[14]

NOTES

1. *Geschichte der Kriegswissenschaften,* vol. XXI of *Geschichte der Wissenschaften in Deutschland* (Munich–Leipzig, 1890), sec. 2, ch. 4, p. 1115, par. 118.
2. Although he was the author, it did not appear under his name. Georg Veesenmayer, *De Ulmensium in arithmeticam meritis* (Ulm, 1794), p. 6.
3. L. E. Ofterdinger, "Beiträge zur Geschichte der Mathematik in Ulm," in *Programmschrift des Königlichen Gymnasiums zu Ulm* (Ulm, 1866–1867).
4. Joseph E. Hofmann, "Michael Stifel," in *Sudhoffs Archiv für Geschichte der Medizin und Naturwissenschaften,* supp. 9 (1968), 2–5.
5. J. G. Doppelmayer, *Historische Nachricht von den Nürnbergischen Mathematicis und Künstlern* (Nuremberg, 1730), p. 209.
6. Veesenmayer, *op. cit.,* p. 7.
7. Christian Thomasius, *Historia sapientiae et stultitiae* (Halle, 1693), II, 113.
8. Ofterdinger, *op. cit.,* p. 5.
9. MS notes for and drafts of this work are in the Ulm municipal archives.
10. A. Germann, "Das irreguläre Siebeneck des Ulmer Mathematikers Joh. Faulhaber," in *Programmschrift des Königlichen Gymnasiums zu Ulm* (Ulm, 1875–1876), pp. 3–13.
11. *Ulmer Ratsprotokolle,* no. 61 (1611), fol. 674b.
12. Wilhelm Diehl, *Landgraf Philipp von Butzbach,* no. 4 of the series *Aus Butzbachs Vergangheit* (Giessen, 1922), pp. 37 ff.
13. Conrad Holzhalbius, *Herrn Faulhabers . . . Continuatio seiner neuen Wunderkunsten oder arithmetischen Wunderwerken* (Zurich, 1617).
14. K. E. Haeberle, *10,000 Jahre Waage* (Balingen, 1966), pp. 109–111.

BIBLIOGRAPHY

I. ORIGINAL WORKS. Faulhaber's writings are *Arithmetischer cubicossischer Lustgarten mit neuen Inventionibus gepflanzet* (Tübingen, 1604, 1708); *Neu erfundener Gebrauch eines Niederländischen Instruments zum Abmessen und Grundlegen . . .* (Augsburg, 1610); *Neue geometrische und perspectivische Inventiones etlicher sonderbarer Instrument . . .* (Frankfurt, 1610); *Neue geometrische und perspectivische Inventiones zu Grundrissen der Pasteyen und Vestungen* (Frankfurt, 1610); *Neuer mathematischer Kunstspiegel . . .* (Ulm, 1612); *Andeutung einer unerhörten neuen Wunderkunst . . .* (Nuremberg, 1613), Latin ed., *Ansa inauditae et novae artis . . .* (Ulm, 1613); *Himmlische geheime Magia oder neue cabalistische Kunst und Wunderrechnung vom Gog und Magog* (Nuremberg, 1613), trans. by Johannes Remmelin as *Magia arcana coelestis, sive Cabalisticus novus, artificiosus et admirandus computus de Gog et Magog . . .* (Nuremberg, 1613); *Arithmetischer Wegweiser zu der hochnutzlichen freyen Rechenkunst . . .* (Ulm, 1614, 1615, 1675, 1691, 1708, 1736, 1762, 1765), the last two eds. entitled *Arithmetischer Tausendkünstler . . . ; Neue Invention einer Haus- und Handmühle* (Ulm, 1617); *Solution, wie man die Fristen, welche ohne Interesse auf gewisse Ziel zu bezahlen verfallen . . .* (Ulm, 1618); *Continuatio des neuen mathematischen Kunstspiegel . . .* (Tübingen, 1620), which contains thirty-two inventions; *Zweiundvierzig Secreta . . .* (Augsburg, 1621), which contains the inventions in the *Continuatio* plus ten new ones; *Appendix oder Anhang der Continuation des neuen mathematischen Kunstspiegel . . .* (Augsburg, 1621); *Erste deutsche Lection . . . das Prognosticon vom Gog und Magog . . .* (Augsburg, 1621); *Tarif über das kurze und lange Brennholz . . .* (Ulm, 1625); *Ingenieurs-Schul . . .,* 4pts. (Frankfurt, 1630–1633), which is partly extracted from the works of Adrian Vlacq, John Napier, and Matthias Bernegg; *Appendix oder Anhang des ersten Theils der Ingenieurs-Schul . . .* (Augsburg, 1631); and *Zehntausend Logarithmi der absoluten oder ledigen Zahlen . . .* (Augsburg, 1631). A holograph MS of Faulhaber's, entitled "Beobachtungen von Mund- und Sonnenringen" (1619), is in the Darmstadt Landesbibliothek, 4° 3044. Ten letters from Faulhaber to Philipp von Butzbach are in the Darmstadt Staatsarchiv, 55, 1618–1619.

II. SECONDARY LITERATURE. On Faulhaber's work, see Matthias Bernegger, *Sinum, tangentium et secantium canon . . .* (Strasbourg, 1619[?]), which refers to *Neue geometrische und perspectivische Inventiones etlicher sonderbarer Instrument . . . ;* and *Phantasma qua Joh. Faulhaber de ansa inauditae et admirabilis artis . . .* (Strasbourg, 1614), a refutation of *Andeutung einer unerhörten neuen Wunderkunst;* Benjamin Bramer, *Beschreibung eines sehr leichten Perspectiv und grundreissenden Instruments auf einem Stande . . .* (Kassel, 1630), which refers to *Appendix oder Anhang der Continuation des neuen mathematischen Kunstspiegel . . . ;* Georg Galgemair, *Centiloquium circini proportionum. Ein neuer Proportionalzirkel von 4, 5, 6 oder mehr Spitzen* (Nuremberg, 1626), which refers to *Neue geometrische und perspectivische Inventiones etlicher sonderbarer Instrument . . . ;* J. G. Goldtberg, *Fama syderea nova. Gemein offentliches Ausschreiben . . .* (Nuremberg, 1618, 1619); *Expolitio famae sidereae novae Faulhabereanae . . .* (Prague, 1619); *Postulatum aequitatis plenissimum, Das ist: Ein billiges und rechtmässiges Begehren, die Expolitionem famae Faulhaberianae betreffend . . .* (Prague, 1619); *Fama syderea nova, das ist weitere Continuatio der Göttlichen neuen Wunderzeichen und grossen Miraculn . . .* (Nuremberg, 1620); and *Vindiciarium Faulhaberianum continuatio . . .* (Ulm, 1620); Conrad Holzhalbius, *Herrn Faulhabers . . . Continuatio seiner neuen Wunderkünsten oder arithmetischen Wunderwerken* (Zurich, 1617); Petrus Roth, *Arithmetica philosophica . . .,* II (Nuremberg, 1608), which refers to *Arithmetischer cubicossischer Lustgarten mit neuen Inventionibus gepflanzet;* and David Verbez, *Miracula arithmetica zu der Continuation des arithmetischen Wegweisers* (Augsburg, 1622).

Biographical literature includes C. G. Jöcher, in *Allgemeinen Gelehrten-Lexikon,* II (Leipzig, 1750), col. 527; Hermann Keefer, "Johannes Faulhaber, der bedeutendste Ulmer Mathematiker und Festungsbaumeister," in *Württembergische Schulwarte,* 4 (1928), 1–12; Emil von Loeffler, "Ein Ingenieur und Artillerie-Offizier der Festung Ulm in 30-jährigen Kriege," in *Ulmer Tagblatt* (1886), Sonntagsbeilage no. 52 and (1887), Sonntagsbeilage nos. 1–6, also in *Allgemeine Militarzeitung,* **60** (1885), which refers to Faulhaber and Joseph Furtenbach; Max Schefold, "Ein Zyklus von Faulhaberbildnissen," in *Ulmer Tagblatt* (30 Apr. and 7 May 1926); Albrecht Weyermann, *Nachrichten von Gelehrten, Künstlern und andere merkwürdigen Per-*

sonen aus Ulm (Ulm, 1798), pp. 206–215; and J. H. Zedler, in *Universal-Lexikon,* IX (Halle–Leipzig, 1735), col. 317.

Documents concerning Faulhaber are in the Darmstadt Staatsarchiv, 55 XVII. The Faulhaber family's coat of arms is reproduced in J. F. Schannat, *Fuldischer Lehnhof, sive de clientela Fuldensi beneficiaria nobili et equestri tractatus historico-iuridicus . . .* (Frankfurt, 1736), pp. 83, 91.

PAUL A. KIRCHVOGEL

FAVORSKY, ALEXEI YEVGRAFOVICH (*b.* Pavlovo, Russia, 6 March 1860; *d.* Leningrad, U.S.S.R., 8 August 1945), *chemistry.*

One of A. M. Butlerov's outstanding students, Favorsky graduated from St. Petersburg University in 1882. A professor from 1896, in 1921 he became an associate member, and in 1929 a full member, of the Soviet Academy of Sciences. His entire career was devoted to the study of the reactions of organic unsaturates, primarily the acetylenic hydrocarbons. The results of his work form the basis of many general methods of synthesis, including a number that are of industrial significance.

In 1884 Favorsky discovered the isomerization phenomena of acetylenic hydrocarbons (e.g., $C-C-C\equiv C \rightarrow C-C\equiv C-C$) and explained their mechanism, advancing a hypothesis concerning the intermediary formation of derivatives of allene ($CH_2=C=CH_2$) and vinyl ethers ($C=C-OR$). In 1891 he confirmed the latter experimentally with the reaction

$$CH_3-C\equiv CH + ROH \xrightarrow{KOH} CH_3-\underset{\underset{OR}{|}}{C}=CH_2.$$

Subsequently, Favorsky and his students broadly applied the "vinylization of alcohols" as a quantitative method for obtaining vinyl ethers. From these ethers they prepared aldehydes, acids, and polymers related to the balsams. Verification of the hypothesis concerning the formation of allenes led to the development of methods for the synthesis of dienes:

$$\underset{\underset{C}{|}}{C}-\underset{\underset{C}{|}}{C}-C\equiv C \longrightarrow \underset{\underset{C}{|}}{C}-C=C=C \longrightarrow C=\underset{\underset{C}{|}}{C}-C=C.$$

As a result, Favorsky's school was the first to synthesize isoprene ($CH_2=\underset{\underset{CH_3}{|}}{C}-CH=CH_2$; V. N. Ipatiev, 1897) and butadiene 1,3 ($CH_2=CHCH=CH_2$; S. V. Lebedev, 1928). These were recognized as intermediates in the synthesis of rubber.

Between 1905 and 1907 Favorsky studied the condensation of ketones with acetylenic hydrocarbons:

$$CH_3-\underset{\underset{CH_3}{|}}{C}O + HC\equiv C-R \longrightarrow CH_3-\underset{\underset{CH_3}{|}}{C}OH-C\equiv C-R.$$

He used this reaction in the development of a simple method for synthesizing isoprene (1932):

$$CH_3-\underset{\underset{CH_3}{|}}{C}OH-C\equiv CH \longrightarrow$$

$$CH_3-\underset{\underset{CH_3}{|}}{C}OH-CH=CH_2 \longrightarrow CH_2=\underset{\underset{CH_3}{|}}{C}-CH=CH_2.$$

More recently, Walter Reppe has used an analogous technique to synthesize acrylic acid ($CH_2=CHCOOH$) and its derivatives by condensing ketones with hydrogen cyanide.

In 1900–1910 Favorsky established the reversibility of the isomeric conversions (or transformations) for a series of acetylene, allene, and diene compounds and explained the phenomena of tautomerism and reversible isomerization in one set of reactions. He discovered the simultaneous isomerization of bromine derivatives into six isomeric forms, given an equilibrium system. Developing one of the most effective theories of affinity capacity, Favorsky pointed out the stability of free radicals among metal ketyls of the fatty series and substantiated this experimentally. Between 1933 and 1936 he studied the limits of dehydrogenation of carbocyclic hydrocarbons and stated the maximum possible nonsaturation of $C-C$ bonds for each isomer from C_3- to C_8-. This work led Favorsky's school to the synthesis of a great many thermodynamically unstable compounds, such as cyclopropene (by I. A. Dyakonov). In 1891 Favorsky predicted the existence of polyene compounds (cumulenes), the stability of which, according to his theory, must increase with an increase in methylation of the end group; these substances were discovered in the middle of the twentieth century by F. Bohlmen:

$$(CH_3)_3C-C\equiv C-C\equiv C-C(CH_3)_3.$$

As a result of the systematic investigation of compounds found in an unstable state, Favorsky concluded that isomerization, polymerization, and cracking could all be reduced to a common cause. This conclusion allowed him to determine the path taken by the original reagent activated by means of heat or a catalyst and to explain the action of catalysts applied to acetylene and diene compounds. Favorsky concluded that prototropic transfer, or the "migration of hydrogen," is elicited by alkaline catalysts in the initial act of isomerization, polymerization, and cracking. In addition, he developed general methods for the synthesis of various unsaturated alcohols, α-keto alcohols, dichloro ketones, displaced derivatives of the acrylic acids, and dioxane—a solvent for many organic compounds and completely miscible with water.

Favorsky was responsible for a large school of chemists, including V. N. Ipatiev, S. V. Lebedev, I. N. Nazarov, A. E. Poray-Koshits, Z. Jotsich, Y. S. Zalkind, and M. F. Shostakovsky. From 1900 to 1930 he was editor-in-chief of the *Zhurnal Russkago fiziko-khimicheskago obshchestva.*

BIBLIOGRAPHY

I. ORIGINAL WORKS. Favorsky's selected writings are in *Izbrannye Trudy* ("Selected Works," Moscow–Leningrad, 1940; 2nd ed., 1960).

II. SECONDARY LITERATURE. On Favorsky or his work, see V. I. Kuznetsov, *Razvitie issledovany polimerizatsii nepredelnykh soedineny v SSSR* (Moscow, 1959), issued on the centennial of Favorsky's birth; *Voprosy teorii stroenia organicheskikh soedineny* (Leningrad, 1960), a collection honoring the centennial of Favorsky's birth; and M. F. Shostakovsky, *Akademik Alexei Yevgrafovich Favorsky* (Moscow–Leningrad, 1953).

V. I. KUZNETSOV

FAVRE, PIERRE ANTOINE (*b.* Lyons, France, 20 February 1813; *d.* Marseilles, France, 17 February 1880), *chemistry.*

Favre received a medical degree from the Faculty of Medicine in Paris in 1835. Inspired by Jean Dumas's lectures in chemistry at the School of Medicine in 1840, he turned to chemistry. He was admitted to Eugène Peligot's private laboratory and helped in the latter's classic work on uranium compounds; he became Peligot's *préparateur* at the Conservatory of Arts and Manufactures. He was named fellow of the Faculty of Medicine of Paris in 1843 and worked in G. Andral's laboratory on physiological problems. In 1851 he became head of the analytical chemistry laboratory of the Central School of Arts and Manufactures, while continuing as a fellow of the Faculty of Medicine. In 1853 he received the degree of Doctor of Physical Science. He was named professor of chemistry of the Faculty of Science of Besançon in 1854 and was called to the newly created Faculty of Science at Marseilles in 1856.

Twice laureate of the French Academy of Sciences, Favre was elected a correspondent of the chemistry section in 1864 and a correspondent of the physics section in 1868. Favre became dean of the Faculty of Science in Marseilles in 1872, retiring in 1878 because of ill health.

Favre's earliest independent researches were determinations of the equivalent weight of zinc and studies of cupric carbonate and the ammonium carbonates of zinc and magnesium. He wrote several papers on such physiologically important compounds as lactic acid, mannitol, and the constituents of human perspiration.

His interests turned to thermochemistry about 1848—indeed, Favre is perhaps best-known for using the term "calorie" (1853) to denote the unit of heat. Between 1845 and 1853, he and Johann T. Silbermann, a French physicist, collaborated in a series of important thermochemical researches. They demonstrated the falsity of Dulong's rule, which states that the heat of combustion of a compound composed of carbon and hydrogen is the sum of the heats of combustion of the elements it contains. Particularly valuable was their study of the heats of combustion and formation of a large number of substances using a newly devised "mercury calorimeter." The instrument was somewhat inaccurate, and their results were superseded after some years but were nevertheless widely used. They also showed that the heat of combustion of carbon in oxygen is less than that of carbon in nitrous oxide, evidence that helped to strengthen the case for the diatomicity of the oxygen molecule. The collaboration between Favre and Silbermann was influential in replacing the vague notion of chemical affinity with more precise thermodynamic expressions.

In 1857 Favre elegantly substantiated Joule's ideas about the conservation of energy by means of a voltaic battery operating an electric motor which raised a weight. He showed that the total heat evolved in the battery and the circuit, when added to the equivalent in heat required to raise the weight, was equal to that evolved by the battery alone when it was short-circuited.

Shortly before retirement Favre and the mathematician Claude Valson determined both the heats and the volume changes of solution of many salts. This study was cited by Arrhenius in defense of the theory of electrolytic dissociation. Favre was often called upon by the state and by commercial interests; he served as a consultant in the preparation of canned foods, salt, and petroleum distillates for house and street illumination.

Favre was a careful and skillful experimenter. Highly regarded by his contemporaries for his diligence and chemical ability, his mission was to gather data, not to devise the bold new hypotheses which alter the scientific paradigm.

BIBLIOGRAPHY

I. ORIGINAL WORKS. Favre's works include more than seventy papers (alone or with collaborators), which are listed by Poggendorff and in the first series of the *Catalogue of Scientific Papers* of the Royal Society of London.

II. SECONDARY LITERATURE. J. S. Partington gives a brief analysis of Favre's work in *A History of Chemistry,* IV (London, 1964), 691 and *passim.* A short biography by F. LeBlanc, "Notice nécrologique sur P. A. Favre . . .," is in *Bulletin de la Société chimique de Paris,* **33** (1880), 390–400; see also *Comptes rendus hebdomadaires des séances de l'Académie des sciences,* **90** (1880), 329. There is no detailed study of his life and achievement.

LOUIS KUSLAN

FAYE, HERVÉ (*b.* St. Benoît-du-Sault, France, 1 October 1814; *d.* Paris, France, 4 July 1902), *astronomy, geodesy.*

Son of a civil engineer, Faye entered the École Polytechnique in 1832. His vocation for astronomy emerged after a few years of work in France under the supervision of his father. He entered the Paris Observatory in 1836. While working there, under Arago's direction, Faye discovered the periodic comet of 1843 (since known by his name) and computed its orbit. His career thereafter was manifold. He taught geodesy at the École Polytechnique as lecturer from 1848 to 1854 and became full professor in 1873; meanwhile he was professor of astronomy at Nancy. As academic administrator he was rector of the Academy at Nancy and general inspector of secondary schools. He was honored for his achievements throughout his life, beginning with his election to the Académie des Sciences at the age of thirty-three. A member of the Bureau des Longitudes, he served as president for more than twenty years.

Most of Faye's research is contained in the more than 200 notes that he published in the *Comptes rendus hebdomadaires des séances de l'Académie des sciences.* These researches were essentially theoretical in character, including, among other things, an explanation of the tails of comets as well as a discussion of the discrepancies produced in normal Newtonian orbits by radiation pressure from the sun (*Comptes rendus,* **47** [1858], 836). He understood that meteorites follow cometary orbits and are therefore related objects (*ibid.,* **64** [1867], 549). He improved observational techniques in astronomy, advocating the use of photography, designing a zenith telescope (*ibid.,* **23** [1846], 872), and studying carefully refraction as the major cause of errors (*ibid.,* **39** [1854], 381).

Faye's theory of the sun was widely adopted. He considered the sun to be a gaseous sphere with large convective motions (*ibid.,* **60** [1865], 89 and 138), the sunspots being holes (*ibid.,* **61** [1865], 1082) with internal cyclonic motions (*ibid.,* **76** [1873], 509). He also studied earth cyclones. In a book on the origin of worlds (1884) he developed and improved Laplace's cosmological theory.

Faye also spent much effort in developing geodetic projects in France and all over the world. He first introduced an idea close to isostasy, that the figure of the earth is almost an equilibrium figure, continents being lighter than the crust under the oceans (*ibid.,* **90** [1880], 1185).

Faye's ideas were widely publicized during his lifetime. Some of them remain valid to this day, while others contributed significantly to the development of science in his time.

BIBLIOGRAPHY

I. ORIGINAL WORKS. Faye's most important works are listed in the text. His theory of the sun is described in *Annuaire du Bureau des longitudes* (1873), p. 443, and (1874) p. 407; his theory of comets is in the same publication, (1883), p. 717. See also *Cours d'astronomie,* 2 vols. (Paris, 1882), which includes his most important ideas, and *Sur l'origine des mondes* (Paris, 1884) for his cosmological theory.

II. SECONDARY LITERATURE. Several notices on Faye may be found in *Annuaire du Bureau des longitudes* for 1903.

J. KOVALEVSKY

AL-FAZĀRĪ, MUḤAMMAD IBN IBRĀHĪM (*fl.* second half of the eighth century), *astronomy.*

Al-Fazārī came from an old Arab family (his genealogy is traced back twenty-seven generations by Yāqūt) which had settled in Kūfa. He is first heard of in connection with the building of Baghdad in the latter half of 762, when he was associated with the other astrologers—Nawbakht, Māshāʾallāh and ʿUmar ibn al-Farrukhān al-Ṭabarī—who were involved in that work. He apparently remained at the Abbasid court; for, when an embassy arrived from Sind which included an Indian astronomer (whose identity is unknown, although it was certainly not Kanaka), the Caliph al-Manṣūr asked al-Fazārī to work with this Indian on an Arabic translation of a Sanskrit astronomical text. The date of this embassy is variously given as 771 or 773. Another Arab astronomer who worked with this Indian was Yaʿqūb ibn Ṭāriq.

The Sanskrit astronomical text that was translated with the assistance of al-Fazārī was apparently entitled *Mahāsiddhānta* and belonged to what later became known as the *Brahmapakṣa* (see essay IV on Indian astronomy in supplement); its most immediate cognates were the *Paitāmahasiddhānta* of the *Viṣṇudharmottarapurāṇa* and the *Brāhmasphuṭasiddh-*

ānta of Brahmagupta; but the Indian astronomer evidently also conveyed to his Arab collaborators information about the Āryabhaṭīya of Āryabhata I. The Arabic translation of this Sanskrit text was entitled Zīj al-Sindhind; from it descends a long tradition within Islamic astronomy, which survived in the East until the early tenth century and in Spain until the twelfth. The first derivative work was evidently the Zīj al-Sindhind al-kabīr of al-Fazārī himself.

Already in this work the elements of the Brahmapaksa begin to be contaminated with those of other schools. Although the system of the kalpa and the mean motions of the planets, their apogees, and their nodes remain within the tradition of the Zīj al-Sindhind, the maximum equations are derived primarily from the Zīj al-Shāh, which represents the ārdharātrika school in Indian astronomy (see essay VI), and also from the Āryabhaṭīya; the geographical section of the work also reveals the influence of the Āryabhaṭīya and of a Sassanian tradition ascribed to Hermes. Moreover, al-Fazārī allows great inconsistencies in this zīj, as he extracted convenient rules from one source or another without trying to make them coincide. Thus, he displays three values of R—3,438 (from the Āryabhaṭīya), 3,270 (from the Zīj al-Sindhind), and 150 (from the Zīj al-Shāh)—and two values of the maximum equation of the sun—2;11,15° and 2;14° (from the Zīj al-Shāh).

After writing this zīj al-Fazārī composed another, probably about 790, called the Zīj ʿalā sinī al-ʿArab ("Astronomical Tables According to the Years of the Arabs"). In this zīj he apparently tabulated the mean motions of the planets for one to sixty saura days, 1,0 to 6,0 saura days (6,0 saura days being equal to one sidereal year), one to sixty sidereal years, and an unknown number of sixty-year periods; and he evidently added tables for converting kalpa aharganas into Hegira dates. Of this latter set of tables we still have copies of the Mujarrad table for finding the day of the week with which each Arab year and month begins. Moreover, we have al-Fazārī's list of the countries of the world and their dimensions from this zīj; the dimensions presuppose a much larger earth than that allowed by the circumference of the earth which he introduced into his Zīj al-Sindhind al-kabīr from the Āryabhaṭīya.

Very little else is known of al-Fazārī's works. A few lines of his Qaṣīda fī ʿilm al-nujūm ("Poem on the Science of the Stars") are preserved by Yāqūt and al-Ṣafadī, and the bibliographers record books on the use of the plane astrolabe (al-Fazārī is said to have been the first in Islam to construct one) and the armillary sphere, and on the measurement of noon. But we do have enough of his zījes to know that his work was almost entirely derivative and that he could not even combine his disparate sources into a unified system. His significance lies entirely in that he helped to introduce a large body of Indian astronomical parameters and computational techniques to Islamic scientists.

BIBLIOGRAPHY

The numerous references to al-Fazārī are collected and discussed in D. Pingree, "The Fragments of the Works of al-Fazārī," in Journal of Near Eastern Studies, 29 (1970), 103–123.

DAVID PINGREE

FECHNER, GUSTAV THEODOR (b. Gross-Särchen, near Halle, Germany, 19 April 1801; d. Leipzig, Germany, 18 November 1887), psychology.

Fechner was the second of five children of Samuel Traugott Fechner, a rural, innovative Lutheran preacher, and Johanna Dorothea Fischer Fechner. The precocious child had learned Latin from his father by the time of the latter's death, when Fechner was five. After attending the Gymnasium at Soran (near Dresden, where the family moved in 1815), in 1817 Fechner matriculated at the University of Leipzig, where he spent the rest of his life. He took the M.D. there in 1822 but never practiced medicine. In 1833 he married Clara Volkmann, the sister of his colleague and friend A. W. Volkmann, a physiologist in vision.

Fechner's first writings were satirical pieces that he published under the pseudonym "Dr. Mises." The first of these was written in 1821; they appeared sporadically over the next twenty-five years. Fantastical and by turns strained or brilliant, these pieces usually attack the materialism popular in Germany early in the nineteenth century—or Nachtansicht, as Fechner called it—in contrast with his own Tagesansicht, in which life and consciousness are coequal with matter.

Fechner's first scientific work was in physics, lecturing on it in 1824 (as ordinarius, without pay), translating physics and chemistry texts from the French (by which he earned his living), and conducting investigations in electricity, particularly on Ohm's law. In 1831 he published Massbestimmungen über die galvanische Kette, a paper of great importance on quantitative measurements of the galvanic battery. This made his reputation as a physicist, and he was appointed professor of physics in 1834. During this period, the only indications of his future interest in psychological problems were his satires, two papers on complementary colors and subjective colors (1838), and his famous paper on subjective afterimages, published in 1840.

Fechner then plunged into a long, serious neurotic illness which necessitated his resignation from his chair of physics in 1839. This began somatically with a partial blindness brought on by gazing at the sun through colored glasses in the experiments on colors and afterimages; it then deepened psychologically into an inability to take food, various psychotic symptoms, and a year of severe autistic thinking. Then on 5 October 1843, having lived for three years in the dark and despairing of ever seeing again, Fechner ventured into his garden, unwound the bandages he wore around his eyes, and found his vision not only regained but abnormally powerful, since he had semihallucinatory experiences of seeing the souls of flowers. His recovery was then slow and progressive.

This peak experience in the garden is reflected in his next work, *Nanna oder über das Seelenleben der Pflanzen* ("Nanna, or the Soul Life of Plants," 1848). In this philosophically diffuse book as well as in his 1851 book, *Zend-Avesta oder über die Dinge des Himmels und des Jenseits* ("Zend-Avesta, or Concerning Matters of Heaven and the World to Come"), Fechner developed what has been called his panpsychism, a development of his *Tagesansicht:* since mind and matter were two aspects of the same thing, the entire universe could be looked at from the point of view of its mind.

But how could this be made scientific? On the morning of 22 October 1850 (called commemoratively by psychophysicists Fechner Day), while Fechner was awaking in bed, the solution came. It was to make the relative increase of stimulation the measure of the increase of the corresponding sensation; and this suggested that the arithmetical series of perceived intensities might correspond to a geometrical series of external energies.

In part, this solution to Fechner's problem was based upon Helmholtz's famous "On the Conservation of Force," published three years earlier. Since energy could neither be created nor be destroyed, all energy impinging on sense organs traversed the nervous system and ended in effectors. Sensation was the mental aspect of this, which had to be just as orderly and related to these physical events in an orderly manner. The relation between sensation and these neurological events he called "inner psychophysics." This was impossible to study. It was, therefore, the relationship of sensation to the external stimulus energy, or outer psychophysics, that could alone be studied.

These ideas, after a decade of thought and experiment, resulted in 1860 in Fechner's classic work, the *Elemente der Psychophysik,* a text of the "exact science of the functional relations or relations of dependency between body and mind." Through its sometimes redundant details are developed three of the basic methods of a new science to be called psychophysics:

1. The method of just noticeable differences, later called the method of limits. The difference between two discriminable stimuli is gradually decreased until discrimination is just lost; and conversely, the difference between two stimuli that are not discriminable is gradually increased until discrimination is barely possible: the average of these two determinations is a measure of the just noticeable difference.

2. The method of right and wrong cases, later called the method of constant stimuli, or simply the constant method, which has become, since the work of G. E. Müller and F. M. Urban, the most important. A range of stimuli are used, none of which is adjustable; a standard stimulus is compared in some given respect with each of a series of similar stimuli, presented in chance order: from the percentage of correct judgments with each comparison stimulus, a threshold is mathematically determined.

3. The method of average error. Whereas the two foregoing methods are partly systematizations of work by others, this method was original with Fechner in collaboration with his brother-in-law, Volkmann. A variable stimulus is adjusted sundry times to apparent equality with a standard, the average adjustment being the "constant error" and its standard deviation the sensitivity. These methods were used by Fechner in classical experiments in lifted weights, visual brightnesses, tactual and visual distances, temperature sensitivity, and even a classification of stars by magnitude following Steinheil.

But this methodology is mere apparatus to carry the central and pervading conception of the *Elemente,* Fechner's fascinating development of the Weber fraction into what has come to be known as the Weber-Fechner law. E. H. Weber, a senior colleague of Fechner at Leipzig, concluded after a series of elegant studies on lifted weights, judged line lengths, and various tactual sensations, that the just noticeable difference in stimulus intensity is a constant fraction of the total intensity at which it is measured. These experiments were first described in Latin in 1834 but achieved attention only when brought together with other facts in Weber's famous chapter on touch in Rudolph Wagner's *Handwörterbuch der Physiologie* (Brunswick, 1846). This may be expressed as

$$\frac{\Delta R}{R} = \text{constant},$$

where R is *Reiz*, or stimulus, and ΔR is the amount of increase in R necessary for a subject to see any difference. This is approximately true for the middle range of sensory stimulation in any modality in men, or in animals where a behavioral response takes the place of an introspected difference in sensation.

Fechner assumed that on the mental side there is a corresponding increase in sensation, ΔS, and that all such ΔS's are equal and can be treated as units, whence

$$\Delta S = C \frac{\Delta R}{R},$$

where C is the constant of proportionality. Integrating, and solving for the constant of integration at threshold where $S = 0$,

$$S = C \log R,$$

where R is measured in units of its threshold value. This is the fundamental relation between mind on the left-hand side of the equation, and matter on the right. It is now known as the Weber-Fechner law, although Fechner with confusing generosity called it Weber's law.

While the methodology of the *Elemente* is sound and permanent, its theoretical purpose and its working out into Fechner's law kindled immediate controversy, which is still far from being resolved. Even before 1860, Fechner's ideas resulted in papers by Helmholtz and Mach on the new psychophysics. And Wilhelm Wundt, in his first psychological publications from 1862 on, made Fechner's work centrally important. His detractors, on the other hand, claimed that Fechner had not measured sensation at all. Their fundamental objections were (1) that it is meaningless to say that one ΔS equals another unless S is independently measurable, and (2) what has been called the quantity objection: in experience, pink is not part of scarlet, nor a thunderclap a summation of murmurs.

As Fechner founded psychophysics with this decade of work, so in the next decade (1865–1876) he founded experimental aesthetics, publishing his *Vorschule der Ästhetik* in 1876. This work treats of its methods, principles, and problems, particularly that of the "golden section," or most aesthetically pleasing relation of length to breadth of an object, a kind of Weber's fraction for aesthetics. He endlessly measured the dimensions of pictures, cards, books, snuffboxes, writing paper, and windows, among other things, in an attempt to develop experimental aesthetics "from below," in rebellion against the Romantic attempt "from above down" first to formulate abstract principles of beauty.

In the final decade of his life, the turbulent wake of his *Elemente* drew Fechner back into psychophysics. In 1882 he answered his critics with his last important book, *Revision der Hauptpunkte der Psychophysik*. This helped to place Fechner's psychophysics even more securely as a cornerstone of the new so-called experimental psychology as it was to be developed in the latter part of the century by Wundt and others.

BIBLIOGRAPHY

I. ORIGINAL WORKS. Fechner's chief works are *Massbestimmungen über die galvanische Kette* (Leipzig, 1831), which is available on microfilm in the "Landmarks of Science" series, I. Bernard Cohen, Charles C. Gillispie, *et al.,* eds. (New York, 1967); *Das Büchlein vom Leben nach dem Tode* (Dresden, 1836; Leipzig, 1841, 1915, 1922; Hamburg, 1887, 1906), English trans. by H. Wernekke as *On Life After Death* (London, 1882; Chicago, 1906, 1914), another English trans. by M. C. Wadsworth, intro. by W. James, as *The Little Book of Life After Death* (Boston, 1904; New York, 1943); *Nanna oder über das Seelenleben der Pflanzen* (Leipzig, 1848, 1920; Hamburg, 1903); *Zend-Avesta oder über die Dinge des Himmels und des Jenseits,* 3 vols. (Leipzig, 1851; Hamburg, 1906); *Elemente der Psychophysik,* 2 vols. (Leipzig, 1860, 1889)—of which vol. I of the 1889 ed. includes a bibliography, originally compiled by R. Müller, of 175 of Fechner's publications—English trans. of vol. I only by H. E. Adler, as *Elements of Psychophysics,* E. G. Boring and D. Howes, eds. (New York, 1966); *Vorschule der Ästhetik* (Leipzig, 1876); *Die Tagesansicht gegenüber der Nachtansicht* (Leipzig, 1879, 1904); and *Revision der Hauptpunkte der Psychophysik* (Leipzig, 1882).

II. SECONDARY LITERATURE. On Fechner's life, see the sympathetic biography by his nephew, J. E. Kuntze, *Gustav Theodor Fechner (Dr. Mises), Ein deutsches Gelehrtenleben* (Leipzig, 1892), which reprints the bibliography of Fechner's works cited above. See also K. Lasswitz, *Gustav Theodor Fechner* (Stuttgart, 1896, 1910); and G. S. Hall, *Founders of Modern Psychology,* (New York, 1912), pp. 123–177.

On Fechner's scientific contributions, see, in addition to the foregoing, T. Ribot, *German Psychology of Today* (New York, 1886), pp. 134–187; W. Wundt, *Gustav Theodor Fechner* (Leipzig, 1901); and M. Wentscher, *Fechner und Lotze* (Munich, 1925). Other bibliographical items on Fechner's scientific work, as well as a good short introduction to him, are given in E. G. Boring, *A History of Experimental Psychology* (New York, 1929), pp. 265–287. William James's scornful evaluation of Fechner and his work may be found in his *Principles of Psychology,* I (New York, 1890), 533–549. See also R. I. Watson, *The Great Psychologists* (Philadelphia, 1968), pp. 229–241.

Of recent revisions of the Weber-Fechner law, the most important are H. Helson and W. C. Michels, "A Reformulation of the Fechner Law in Terms of Adaptation Level Applied to Rating Scale Data," in *American Journal of Psychology,* **62** (1949), 355–368; S. S. Stevens, "On the Psychophysical Law," in *Psychological Review,* **64** (1957), 153–181; R. D. Luce and W. Edwards, "The Derivation of Subjective Scales From Just Noticeable Differences," *ibid.,* **65** (1958), 222–237; and R. D. Luce and E. Galanter, "Discrimination," in R. D. Luce, R. R. Bush, and E. Galanter, eds., *Handbook of Mathematical Psychology,* I (New York, 1963), 191–243, esp. 206–213.

See also H. Eisler, "A General Differential Equation in Psychophysics: Derivation and Empirical Test," in *Scandinavian Journal of Psychology,* **4** (1963), 1–8; M. Mashhour, "On Eisler's General Psychophysical Differential Equation and His Fechnerian Integration," *ibid.,* **5** (1964), 225–233, a highly critical paper on Eisler; and J. C. Falmagne, "The Generalized Fechner Problem and Discrimination," in *Journal of Mathematical Psychology,* in press.

On the quantity objection specifically, see E. B. Titchener, *Experimental Psychology,* II, pt. 2 (London, 1905), xlvii–lxviii. E. G. Boring, "The Stimulus Error," in *American Journal of Psychology,* **32** (1921), 449–471, particularly pp. 451–460, contains other references on the problem.

On Fechner's more philosophical thought, see W. James, *A Pluralistic Universe* (New York, 1909), pp. 133–177; R. B. Perry, *Philosophy of the Recent Past* (New York, 1926), pp. 81–86; and G. S. Brett, *History of Psychology,* ed. and abridged by R. S. Peters (Cambridge, Mass., 1965), pp. 580–590.

JULIAN JAYNES

FEDDERSEN, BEREND WILHELM (*b.* Schleswig, Germany, 26 March 1832; *d.* Leipzig, Germany, 1 July 1918), *physics.*

Not much is known of Feddersen's early life or parentage, except that he was an only child. He studied in Göttingen, Berlin, Leipzig, and Kiel, where he received the doctorate in 1858 with a dissertation on the nature of electric-spark discharges, which he studied by improving a rotating-mirror technique of Wheatstone's. In the same year he moved to Leipzig, where he spent the rest of his life.

By his early and subsequent investigations, Feddersen showed that the discharge of a Leiden flask produces a train of damped oscillations, which he contrived to record in a series of splendid photographs. The finding that a circuit made up of a capacitance, a resistance, and an inductance produces oscillations whose frequency and amplitude depend on these components also proved to be of considerable technological importance. Feddersen's photographs served to confirm the 1853 theory of William Thomson (the future Lord Kelvin), who had been occupied—as had Faraday—with the analysis of long-distance signaling in connection with the first attempt to lay a transatlantic cable and had developed the formula for the frequency of a damped resonant circuit; Thomson's public acknowledgment of his debt to Feddersen brought the latter worldwide renown. The beginnings of radiotelegraphy likewise depended on spark-discharge techniques, which dominated radio transmission well on into the twentieth century.

Feddersen is also remembered for his contributions to scientific bibliography, because of his personal participation in (and financial support of) Poggendorff's *Biographisch-literarisches Handwörterbuch* (now in its seventh edition). He undertook the editorship of the third volume after the death of the first editor, Johann Christian Poggendorff, in 1877; but its appearance was delayed because of a quarrel with the publisher, and the work was finally taken over by a third physicist, Arthur Joachim von Oettingen. (All three died octogenarians, a useful trait for bibliographers.) Feddersen supported the publication of volumes III and IV by donating the substantial sum of 30,000 marks and a few weeks before his death sought to assure the appearance of further volumes by setting up, jointly with his wife, a 100,000-mark endowment that was unfortunately wiped out by the runaway German inflation of 1919–1923.

In 1866 Feddersen married a distant cousin, Dora Feddersen, who was sickly and on whose account he practically withdrew from scientific activity. She died in 1889, and in 1890 he married Helga Kjär, who survived him until 1936. There were no children from either marriage.

BIBLIOGRAPHY

I. ORIGINAL WORKS. A list of Feddersen's publications appears in Poggendorff's *Biographisch-literarisches Handwörterbuch,* vols. I–V. His diss., *Beiträge zur Kenntniss des elektrischen Funkens,* was published by the University of Kiel (a copy is in the British Museum). The article based on the diss. and his subsequent papers in *Annalen der Physik,* 2nd ser., **103, 108, 112, 113, 115, 116,** form vol. CLXVI of Ostwald's Klassiker der exakten Wissenschaften, T. Des Coudres, ed. (Leipzig, 1908), which also contains a portrait.

II. SECONDARY LITERATURE. Obituaries are A. von Oettingen, in *Berichte der königlichen sächsischen Gesellschaft der Wissenschaften,* **70** (1918), 353; T. Des Coudres, in *Physikalische Zeitschrift,* **19** (1918), 393; and G. von Eichhorn, in *Jahrbuch der drahtlosen Telegraphie,* **13** (1918/ 1919), 345. An appreciation on the centenary of his birth by W. Dudensing is in *Hochfrequenztechnik und Elektroakustik,* **39** (1932), 77. For the subsequent history of Poggendorff's bibliography, see *Isis,* **57** (1966), 389.

CHARLES SÜSSKIND

FÉE, ANTOINE-LAURENT-APOLLINAIRE (*b.* Ardentes, France, 7 November 1789; *d.* Paris, France, 21 May 1874), *botany.*

From 1809 to 1815 Fée was a pharmacist with the French army, mostly in Spain. For about nine years he operated his own pharmacy in Paris, but in 1825 he resumed his military career and in succeeding years received appointments to the teaching staffs of military hospitals in Lille, Paris, and Strasbourg. While in Strasbourg, he earned an M.D. degree from the Faculty of Medicine in 1833 and shortly thereafter obtained a professorship of botany at that institution. Fée had many interests—literature and the humanities, natural history, pharmacy—but botany remained his chief preoccupation throughout his life. In 1824 he was elected to membership in the Paris Academy of Medicine, and in 1874 he became president of the Société Botanique de France.

Fée's major contribution was to cryptogamic botany and, to a lesser extent, plant physiology. His most ambitious work was an extensive descriptive study of ferns published in eleven memoirs between 1844 and 1866. This investigation won praise from Adolphe Brongniart (1868), who characterized it as "un travail considérable," although he felt that Fée had relied too heavily on veins in differentiating genera. In plant physiology Fée conducted research on movement in plants, especially as such movements were affected by light. Also noteworthy was his work on lichens and cryptogams occurring on medicinal barks imported into France, such as cinchona, angostura, and cascarilla. His early enthusiasm for classical literature prompted him to write about the plants mentioned by Vergil, Theocritus, Pliny, and other literary figures of antiquity.

Among Fée's publications were a number of biographies of prominent botanists. During the last few years of his life, Fée's botanical investigations were concerned mainly with the cryptogams of Brazil.

BIBLIOGRAPHY

I. ORIGINAL WORKS. Fée's writings include *Flore de Virgile . . .* (Paris, 1822); *Essai sur les cryptogames des écorces exotiques officinales* (Paris, 1824; supp. and rev., 1837); *Méthode lichénographique et genera . . .* (Paris, 1824); *Cours d'histoire naturelle pharmaceutique . . .,* 2 vols. (Paris, 1828); *Mémoires sur la famille des fougères,* 11 pts. (Strasbourg–Paris, 1844–1866), repr. as no. 52 of Historia Naturalis Classica (Codicote, England–New York, 1966); *Cryptogames vasculaires du Brésil* (Strasbourg, 1869; supp. and revs., Paris–Nancy, 1872–1873). For a comprehensive listing of Fée's publications in botany and other fields, see *Bulletin. Société botanique de France,* **21** (1874), 173–178.

II. SECONDARY LITERATURE. For an evaluation of Fée's work on ferns, see A. T. Brongniart, *Rapport sur les progrès de la botanique phytographique . . .* (Paris, 1868), pp. 39–41; Fée's research in plant physiology is described by P. E. S. Duchartre, *Rapport sur les progrès de la botanique physiologique . . .* (Paris, 1868), pp. 343 ff. Biographical material about Fée is in J. F. C. Hoefer, ed., *Nouvelle biographie générale . . .,* XVII (Paris, 1858), 255–259. See also A. Balland, *Les pharmaciens militaires français* (Paris, 1913), pp. 169–171, 302–304, and "Quelques pensées et quelques opinions littéraires de Fée," in *Bulletin de la Société d'histoire de la pharmacie,* **4** (1926), 314–315; *Bulletin. Société botanique de France,* **21** (1874), 168–178; and P. Durrieu, "Les goûts archéologiques d'un pharmacien militaire de l'armée française en Espagne sous le Premier Empire," in *Journal des savants,* n.s. **13** (1915), 364–373.

ALEX BERMAN

FEIGL, GEORG (*b.* Hamburg, Germany, 13 October 1890; *d.* Wechselburg, Germany, 25 April 1945), *mathematics.*

Feigl was the son of Georg Feigl, an importer, and Maria Pinl, from Bohemia. He attended the Johanneum in Hamburg and began to study mathematics and physics at the University of Jena in 1909. A severe chronic stomach disorder forced him to interrupt his studies several times, and he did not finish them until 1918, when he received the doctorate with a dissertation on conformal mapping that was supervised by Paul Koebe. In 1919 Feigl became a teaching assistant to Erhard Schmidt, a well-known mathematician at the University of Berlin. Schmidt was the scientist who most influenced Feigl and also developed his gift for teaching. Generations of students in mathematics at the University of Berlin took the introductory course "Einführung in die höhere Mathematik," which Feigl created and which after his death was published, in enlarged form, as a textbook (1953) by Hans Rohrbach.

In 1925 Feigl married Maria Fleischer, daughter of Paul Fleischer, an economist and member of the Reichstag. In 1927 he became assistant professor and in 1933 associate professor at the University of Berlin. From 1928 to 1935 he was, by appointment of the Prussian Academy of Sciences in Berlin, the managing editor of the *Jahrbuch über die Fortschritte der Mathematik,* at that time the only periodical that reviewed papers on mathematics.

Feigl's field of research was geometry, especially the foundations of geometry and topology. But his scientific activity was rather limited because of his illness, and he soon had to choose between research and teaching. His talents led Feigl to devote himself to a reform of the teaching of mathematics. He became a leading member of the National Council of

German Mathematical Societies, and it was essentially through him that the new fundamental concepts of Felix Klein and David Hilbert and the modern mode of mathematical thinking based on axioms and structures were introduced into universities and even high schools.

In 1935 Feigl was called as full professor to the University of Breslau. There during World War II he formed a computing team that worked for the German Aeronautic Research Institute. In January 1945, when the Russians marched into Breslau, he moved with his team to the castle of Graf Schönburg at Wechselburg, Saxony, near Chemnitz (now Karl-Marx-Stadt). There it proved impossible to maintain his necessary medical supervision, a circumstance that led to Feigl's death a few months later.

BIBLIOGRAPHY

Feigl's writings include "Elementare Anordnungssätze der Geometrie," in *Jahresbericht der Deutschen Mathematikervereinigung,* **33** (1924), 2–24; "Zum Archimedesschen Axiom," in *Mathematische Zeitschrift,* **25** (1926), 590–601; "Eigenschaften der einfachen stetigen Kurven," *ibid.,* **27** (1927), 162–168; "Fixpunktsätze für spezielle n-dimensionale Mannigfaltigkeiten," in *Mathematische Annalen,* **98** (1927–1928), 355–398; "Erfahrungen über die mathematische Vorbildung der Mathematik-Studierenden des 1. Semesters," in *Jahresbericht der Deutschen Mathematikervereinigung,* **37** (1928), 187–199; "Geschichtliche Entwicklung der Topologie," *ibid.,* 273–286, repr. in the series Wege der Forschung, vol. CLXXVII (Darmstadt, in press); "Das Unendliche im Schulunterricht," in *Zeitschrift für mathematischen und naturwissenschaftlichen Unterricht,* **60** (1929), 385–393; "Der Übergang von der Schule zur Hochschule," in *Jahresbericht der Deutschen Mathematikervereinigung,* **47** (1937), 80–88; "Ausbildungsplan für Lehramtsanwärter in der Fächern reine Mathematik, angewandte Mathematik und Physik," in *Deutsche Mathematik,* **4** (1939), 98–108, 135–136, written with Georg Hamel; "Erfahrungen über das Mathematikstudium der Lehramtsanwärter nach der neuen Ausbildungsordnung," *ibid.,* **6** (1942), 467–471. His textbook is *Einführung in die höhere Mathematik,* Hans Rohrbach, ed. (Berlin-Göttingen-Heidelberg, 1953).

Some biographical information may be found in *Neue deutsche Biographie,* V (1961), 57.

Hans Rohrbach

FEJÉR, LIPÓT (*b.* Pécs, Hungary, 9 February 1880; *d.* Budapest, Hungary, 15 October 1959), *mathematics.*

Fejér became interested in mathematics while in the higher grades of the Gymnasium, and in 1897 he won a prize in one of the first mathematical competitions held in Hungary. From 1897 to 1902 he studied mathematics and physics at the universities of Budapest and Berlin. During the academic year 1899–1900 H. A. Schwarz directed his attention, through a suggestion made by C. Neumann concerning Dirichlet's problem, to the theory of Fourier series. Later in 1900 Fejér published, in the *Comptes rendus* of the Paris Academy, the fundamental summation theorem that bears his name and was also the basis of his doctoral dissertation at Budapest (1902). After participating in mathematical seminars in Göttingen and Paris, he taught at the University of Budapest from 1902 to 1905 and at that of Kolozsvár (now Cluj, Rumania) from 1905 to 1911. He was professor of higher analysis at the University of Budapest from 1911 until his death. Collaborating with F. and M. Riesz, A. Haar, G. Pólya, G. Szegö, O. Szász, and other mathematicians of international rank, Fejér became the head of the most successful Hungarian school of analysis.

Fejér was a vice-chairman of the International Congress of Mathematicians held at Cambridge in 1912. In 1933 he and Niels Bohr, two of the four European scientists invited to the Chicago World's Fair, were awarded honorary doctorates by Brown University. Fejér was elected to the Hungarian Academy of Sciences in 1908 and was also a member of several foreign academies and scientific societies. Besides receiving a number of state and academic prizes for his work, he was honorary chairman of the Bolyai Mathematical Society from its founding and the holder of an honorary doctorate from Eötvös University, Budapest (1950).

Fejér's main works deal with harmonic analysis. His classic theorem on (C, 1) summability of trigonometric Fourier series (1900) not only gave a new direction to the theory of orthogonal expansions but also, through significant applications, became a starting point for the modern general theory of divergent series and singular integrals. Through a Tauberian theorem of G. H. Hardy's the convergence theory of Fourier series was considerably affected by Fejér's theorem as well; it is closely connected with Weierstrass' approximation theorems and with the more advanced theory of power series and harmonics (potential theory), and makes possible a number of analogues for related series, such as Laplace series. In 1910 Fejér found a new method of investigating the singularities of Fourier series that was suitable for a unified discussion of various types of divergence phenomena. These results were continued and generalized in several directions by Fejér himself, by Lebesgue (1905), by M. Riesz and S. Chapman (1909–1911), by Hardy and Littlewood (1913), by T. Carleman (1921), and others.

Fejér's contributions to approximation theory and the constructive theory of functions are of great importance. In 1918 he solved Runge's problem on complex Lagrange interpolation relating to an arbitrary Jordan curve, and in the following decades he enriched the field of real Lagrange and Hermite interpolation and mechanical quadrature by introducing new procedures. His work in mechanical quadrature produced wide response in the literature (Akhiezer, Erdös, Grünwald, Natanson, Pólya, J. A. Shohat, Szegö, and Turán, among others). As for Fejér's results in complex analysis, particular stress may still be laid on a joint paper with Carathéodory (1907), of which the basic ideas influenced considerably the literature on entire functions, and a new standard proof of the fundamental theorem of conformal mappings, found in 1922 with F. Riesz.

BIBLIOGRAPHY

I. ORIGINAL WORKS. Fejér's collected works, with bilingual comments in Hungarian and German, are *Fejér Lipót Összegyüjtött Munkái / Leopold Fejér, Gesammelte Arbeiten,* 2 vols. (Budapest, 1970). The summation theorem was first printed as "Sur les fonctions bornées et intégrables," in *Comptes rendus hebdomadaires des séances de l'Académie des sciences,* **131** (1900), 984–987; the author's name is here printed incorrectly as "Tejér."

II. SECONDARY LITERATURE. The proof found with F. Riesz was published (with the permission of the authors) in T. Rado, "Über die Fundamentalabbildung schlichter Gebiete," in *Acta litterarum ac scientiarum R. Univertatis hungarica Francisco-Josephina,* **1** (1922), 240–251. See also C. Carathéodory, "Bemerkungen zu dem Existenztheorem der konformen Abbildung," in *Bulletin of the Calcutta Mathematical Society,* **20** (1930), 125–134.

Other works on Fejér are Émile Borel, *Leçons sur les séries divergentes* (Paris, 1901), p. 88, n.; the article in *Encyclopaedia Brittanica,* 12th ed. (1922), XXXI, 877; and P. Turán, "Fejér Lipót matematikai munkássága" ("The Mathematical Work of Lipót Fejér"), in *Matematikai lapok,* **1** (1950), 160–170.

An obituary by P. Szász, G. Szegö, and P. Turán is in *Magyar tudományos akadémia III osztályának kőzleményei,* **10** (1960), 103–148; and in *Matematikai lapok,* **11** (1960), 8–18, 225–228.

MIKLÓS MIKOLÁS

FENNEMAN, NEVIN MELANCTHON (*b.* Lima, Ohio, 26 December 1865; *d.* Cincinnati, Ohio, 4 July 1945), *geology.*

Fenneman was the son of William Henry Fenneman, a Reformed Church clergyman, and Rebecca Oldfather. He attended Heidelberg College in Tiffin, Ohio, and then taught high school. He joined the faculty of the Colorado State Normal School in 1892, and in 1893 he married Sarah Alice Glisan, also a faculty member. In 1900 Fenneman undertook graduate work under T. C. Chamberlin, whom he had met with C. R. Van Hise while doing fieldwork in Wisconsin ten years before. Meanwhile, a Harvard summer course in 1895 had brought him under the influence of William Morris Davis. At the University of Chicago he was also a protégé of R. D. Salisbury, receiving the Ph.D. in the near-record time of three semesters.

Returning to Colorado in 1902, Fenneman served as professor of geology at the University of Colorado, where he studied the Boulder oil fields, which were opening at this time. After three semesters at Colorado, Fenneman was called to the geology faculty of the University of Wisconsin, where he taught for four years. Here he began pioneering work in applying regionally the scientific principles of landform study.

In 1907 Fenneman went to the University of Cincinnati, where he started a department of geology and geography. His tenure at Cincinnati may be said to have ended only with his death. Even during his eight years as professor emeritus (he retired as professor of geology and head of the department in 1937, at the age of seventy-three) he continued his studies and publications and taught advanced courses. He spent every summer with geological parties in the field.

From 1900 to 1902 Fenneman was geologist with the Wisconsin Geological and Natural History Survey; from 1906 to 1908, with the Illinois State Geological Survey; and from 1914 to 1916, with the Ohio Geological Survey. He served with the U.S. Geological Survey as assistant geologist (1901–1919), associate geologist (1919–1924), and geologist (for a period from 1924 on), publishing a steady flow of papers and reports. These included the most famous work of his career, the map "Physiographic Divisions of the United States" (1916), which was the original for all such maps and was adopted by the U.S. Geological Survey and all other government agencies, as well as serving as the basis of regional work in the United States and university courses on the subject.

Primarily a geographer who approached his subject from the point of view of the college teacher of introductory courses, Fenneman had long been concerned with the search for classificatory principles in what was largely a descriptive field. By 1915 the proposal for a physiographic and topographic classification of the United States based on natural subdivisions led to the formation of a committee consisting of Eliot Blackwelder, Marius R. Campbell, Douglas Johnson, and F. E. Matthes. It was headed by Fenneman, who went on leave to work in Washington for the year. Fenneman published the *Physiography of Western United States* (New York, 1931) and *Physiography of Eastern United States* (New York, 1938). The two works were the definitive genetic description and analysis of the physiography of the United States, subdivision by subdivision.

His fields of research also included the action of waves and currents on shores and surveys of the Wisconsin lakes and the oil fields of Colorado, Missouri, and Ohio.

Fenneman's scientific work brought him many honors, among them the presidency of the American Association of Geographers in 1918 and of the Geological Society of America in 1935. In 1938 he was awarded the gold medal of the Geographical Society of Chicago "for eminent achievements in physiography."

His colleagues regarded him as "the last of the great trio of American physiographers—Davis, Fenneman, and Johnson—. . . who . . . developed a rigid application of logic to the study of land forms and their evolution" (Raymond Walters, "Memorial," p. 142).

BIBLIOGRAPHY

See Walter Bucher, "Memorial to Nevin M. Fenneman," in *Proceedings. Geological Society of America*. Annual Report for 1945, pp. 215–228, which has a complete bibliography; and Raymond Walters, "Obituary. Nevin M. Fenneman," in *Science,* **102** (10 Aug. 1945), 142–143.

GEORGE B. BARBOUR

FENNER, CLARENCE NORMAN (*b.* near Clifton, New Jersey, 19 July 1870; *d.* near Clifton, 24 December 1949), *petrology, volcanology.*

Fenner was the son of William Griff Fenner and Elmina Jane Carpenter Fenner. He received the degree of Engineer of Mines from the School of Mines of Columbia University in 1892. After fifteen years of experience in the field he returned to Columbia to earn the M.A. in 1909 and the Ph.D. in 1910. On graduation he joined the staff of the Geophysical Laboratory, Carnegie Institution of Washington, remaining there until his retirement in 1938. Fenner then returned to his childhood home, where he lived with his brother Herbert and continued his petrological studies until his death.

Fenner's principal contributions to petrology are his experimental determination of the thermal stability of the various polymorphs of silica; field description, chemical analysis, and structural and theoretical study of the great eruption of Mt. Katmai, in Alaska

(1912); recognition of a type of basalt crystallization leading to iron enrichment; and a physicochemical theory of rock solution and of ore deposition by gaseous emanations. In addition he investigated uranium and thorium minerals bearing on the age of the earth and devised chemical methods for their separation and analysis. During World War I, Fenner was in charge of the optical glass plant of the Spencer Lens Company at Hamburg, New York. He succeeded in putting optical glass on a production basis, thereby helping to establish in the United States that industry which had formerly been reliant on Germany.

In the laboratory Fenner investigated silica, the principal constituent of silicates and of the earth's crust. He showed that the lowest-temperature form of SiO_2—quartz—inverts to tridymite at $870° ± 10°$ C. and that tridymite inverts to cristobalite at $1470° ± 10°$ C. He found that the velocity of transformation of one form of silica into another was very sluggish and that one stable form did not always pass directly into the next most stable form, but progressed through successive steps. Fenner attributed the appearance of tridymite and cristobalite in some natural occurrences to this process. The transformations in the laboratory were speeded up by the use of a sodium tungstate flux. Although it has been suggested that this technique may have led to contamination of the material and hence to incorrect values for pure SiO_2, the values of transformation ascertained by Fenner have been confirmed and accepted. His concern with equilibrium, high precision, and detailed definition of products—as well as the application of laboratory data to natural occurrences—set a sound foundation for subsequent study of all silicate systems.

Fenner was a member of the 1919 National Geographic Society expedition to Mt. Katmai, Alaska, to study the violent eruption of 1912, believed to be the second largest energy release by a volcano in historic time. In 1923 he returned to the Katmai region as the leader of an expedition sponsored by the Geophysical Laboratory, Carnegie Institution of Washington.

The principal deduction that Fenner drew from these field studies and subsequent laboratory work was that assimilation of andesitic wall rock by rhyolitic magma had proceeded by means of escaping volatile substances whose exothermic reactions underwent continually increasing acceleration which finally led to explosion (1950, p. 604). The large volume of ejecta was laid down mainly in the Valley of Ten Thousand Smokes as dust and gas mixtures—incandescent tuff flows—issuing through fissures in the valley from a sill or very similar body of magma in the underlying sedimentary strata. The many fumaroles, after which the valley was named, were presumed by Fenner to be due to the continued evolution of gases from that body of magma.

Chemical analyses of the various rocks carried out by Fenner demonstrated that the quantitative variation of each chemical oxide is essentially linear for the series of intermediate rocks. He regarded the nearly linear relationship as evidence of the primary role of assimilation in the formation of rocks intermediate in composition between the end members andesite and rhyolite. On a much later occasion he agreed in part with Norman L. Bowen that the deviations from linearity for some of the rocks suggest that crystal differentiation had played some part in the production of the Katmai rocks.

Fenner's emphasis on assimilative processes was also apparent in his study (1938) of a rhyolite flow that he described as following a valley in an eroded basalt surface on Gardiner River, Yellowstone Park, Wyoming. The older basalts were impregnated and mobilized, according to Fenner, by the rhyolite magma, and the compositions of the resulting soaked rocks were represented by straight lines between the basalt and rhyolite. It was his belief that in their origin in the depths of the earth, the two magmas would have formed a conjugate pair of immiscible liquids (1948, p. 500). (It is interesting to note that at an early stage of his field studies, in 1914, he had been impressed by the apparent chemical attack of magma injected into country rock in New Jersey.)

Fenner assigned volatiles a major role in the assimilation process, a view no doubt acquired from his experimental work with George W. Morey in water-containing systems (1917). He also assigned a primary role to gaseous emanations in transporting metals from a magma into surrounding rocks, thereby producing some types of ore deposits (1933). Further concern with volatiles led to studies (1934, 1936) in the geyser basins of Yellowstone Park, Wyoming, where he deduced the manner in which emanations modified the composition of the rocks through which they had passed and were in turn modified. He made a major contribution to the understanding of ash flow mechanics through his emphasis of their high mobility because of entrapped gases.

Fenner was a leading antagonist of Norman L. Bowen's prevailing theory that magmatic fractionation and differentiation lead to silica and alkali enrichment. His own arguments led Fenner to believe that differentiation of magma proceeded in the direction of iron enrichment, producing ferrogabbro, not granite. Later work showed that both Bowen and Fenner were correct in their views, if the partial pressure of oxygen under which the magma crystallizes is taken into consideration. The "Fenner trend" obtains where the partial pressure of oxygen is relatively low in a magma, whereas the "Bowen trend" obtains if the same magma crystallizes under a relatively high partial pressure of oxygen. The "Fenner trend" was particularly well illustrated in the layered intrusion of Skaergaard, Greenland. Fortunately, both men remained friends; their discussion of the subject in debate was carried out only in print after thoughtful review.

Although Fenner was most experienced as a field petrologist, he had a deep understanding of the limitations of experimental studies and theory that purported to elucidate natural phenomena. He applied the data on simplified systems bearing on rocks and related theoretical deductions with great caution, relying mainly on geological field relations and petrographic observations. Even his own strong support of the roles played by gaseous emanations and by assimilation was tempered by the belief that other processes may be of quantitative importance.

Fenner was elected, with Norman L. Bowen and Joseph P. Iddings, to the Petrologists' Club of Washington in the year of its founding, 1910. Although he was a quiet and unassuming man, the records of the club indicate that he was not reluctant to debate the issues with other petrological leaders such as Whitman Cross, Esper S. Larsen, Jr., Frederick E. Wright, Adolf Knopf, and Henry S. Washington.

BIBLIOGRAPHY

Fenner published widely, on a variety of subjects. On silica polymorphs, see "The Stability Relations of the Silica Minerals," in *American Journal of Science*, **4** (1913), 331–384; and "The Relations Between Tridymite and Cristobalite," in *Transactions of the Society of Glass Technology*, **3** (1919), 116–125.

The Katmai area of Alaska is discussed in "The Katmai Region, Alaska, and the Great Eruption of 1912," in *Journal of Geology*, **28** (1920), 569–606; "The Origin and Mode of Emplacement of the Great Tuff Deposit in the Valley of Ten Thousand Smokes," in *National Geographic Society, Contributed Technical Papers*, **1** (1923), 1–74; "Earth Movements Accompanying the Katmai Eruption," in *Journal of Geology*, **34** (1926), 673–772; and "The Chemical Kinetics of the Katmai Explosion," in *American Journal of Science*, **248** (1950), 593–627, 697–725.

Two articles on the Gardiner River area of Wyoming are "Contact Relations Between Rhyolite and Basalt on Gardiner River, Yellowstone Park," in *Bulletin of the Geological Society of America*, **49** (1938), 1441–1484; and "Rhyolite Basalt Complex on Gardiner River, Yellowstone

Park, Wyoming: A Discussion," *ibid.,* **55** (1944), 1081–1096.

Additional field studies are "The Watchung Basalt and the Paragenesis of Its Zeolites and Other Secondary Minerals," in *Annals of the New York Academy of Sciences,* **20** (1910), 93–187; "The Mode of Formation of Certain Gneisses in the Highlands of New Jersey," in *Journal of Geology,* **22** (1914), 594–612, 694–702; and "Pleistocene Climate and Topography of the Arequipa Region, Peru," in *Bulletin of the Geological Society of America,* **59** (1948), 895–917.

The basalt fractionation trend is discussed in "The Crystallization of Basalts," in *American Journal of Science,* **18** (1929), 225–253; "The Residual Liquids of Crystallizing Magmas," in *Mining Magazine,* **22** (1931), 539–560; and "A View of Magmatic Differentiation," in *Journal of Geology,* **45** (1937), 158–168.

On immiscibility in magmas, see "Immiscibility in Igneous Magmas," in *American Journal of Science,* **246** (1948), 465–502.

Articles on the role of volatiles include "The Ternary System $H_2O-K_2SiO_3-SiO_2$," in *Journal of the American Chemical Society,* **39** (1917), 1173–1229, chemical study by George W. Morey, microscopic study by Fenner; "Hydrothermal Metamorphism in Geyser-Basins of Yellowstone Park, as Shown by Deep Drilling," in *Transactions of the American Geophysical Union,* **15** (1934), 240–243; and "Bore-Hole Investigations in Yellowstone Park," in *Journal of Geology,* **44** (1936), 225–315.

Ore deposits are discussed in "Study of a Contact Metamorphic Ore-Deposit. The Dolores Mine at Matehuala, S. L. P., Mexico," in *Economic Geology,* **7** (1912), 444–484, written with J. E. Spurr and G. H. Garrey; "Pneumatolytic Processes in the Formation of Minerals and Ores," in American Institute of Mining and Metallurgical Engineers, *Ore Deposits of the Western States* (New York, 1933), pp. 58–106; and "The Nature of the Ore-Forming Fluid: A Discussion," in *Economic Geology,* **35** (1940), 883–904.

On the optical glass industry, see "The Technique of Optical Glass Smelting," in *Journal of the American Ceramic Society,* **2** (1919), 102–145; and "The Use of Optical Pyrometers for Control of Optical Glass Furnaces," in *Bulletin of the American Institute of Mining and Metallurgical Engineers,* no. 151 (1919), 1001–1011, and in *Pyrometry* (New York, 1920), pp. 495–505.

HATTEN S. YODER, JR.

FERCHAULT, RENÉ ANTOINE. *See* **Réaumur, René Antoine Ferchault de.**

FERGUSON, JAMES (*b.* near Rothiemay, Banffshire, Scotland, 25 April 1710; *d.* London, England, 16 November 1776), *astronomy, instrument making.*

Son of tenant farmer John Ferguson and his wife, Elspet Lobban, James was the second of six children. His formal education consisted of three months at Keith Grammar School in 1717. While working at a variety of domestic jobs from 1720 until 1735, he

mastered the elements of surveying, horology, astronomy, and portraiture. In 1739 he married Isabella Wilson, and they lived in Edinburgh until sailing for London in 1743.

Colin Maclaurin discovered Ferguson's mechanical abilities and introduced him to Martin Folkes, who encouraged Ferguson to lecture to the Royal Society about his astronomical contrivances. A skilled designer of clocks and planispheres (as well as a "solar eclipsareon"), he became an accomplished public lecturer and expounder of Newtonian ideas, especially after the publication of his *Astronomy Explained Upon Sir Isaac Newton's Principles* (1756), which went through seventeen editions. He lectured extensively in London and the provinces (including Bath, Bristol, Derby, Leeds, Liverpool, and Newcastle) and was unofficial "popularizer in residence" to the court of George III. Elected fellow of the Royal Society in 1763, Ferguson spent his last years in London, pained by an unhappy marriage and the disgrace of the prostitution of his only daughter. He wrote a short, partial autobiography, which served as the preface to his *Select Mechanical Exercises* (1773).

Ferguson's scientific work, while both careful and extensive, was neither original nor distinguished. His forte was popularization, and his confessedly weak mathematical background stood him in good stead in writing books for the lay public, particularly his classic *Young Gentleman's and Lady's Astronomy* (1768). He published several technical papers in the *Philosophical Transactions* on eclipses, celestial globes, hygrometers, and horological instruments. His models of the planetary system were classics of engineering design whose accuracy far surpassed anything previously available. Several of his books were used in British grammar schools as late as the 1840's.

BIBLIOGRAPHY

I. ORIGINAL WORKS. Ferguson's major works include *The Use of a New Orrery* (London, 1746); *A Brief Description of the Solar System* (Norwich, 1753); *An Idea of the Material Universe* (London, 1754); *Astronomy Explained Upon Sir Isaac Newton's Principles* (London, 1756); *Lectures on Select Subjects in Mechanics, Hydrostatics, Pneumatics, and Optics* (London, 1760); *Analysis of a Course of Lectures* (London, 1761); *Syllabus of a Course of Lectures* (Edinburgh, 1768); *The Young Gentleman's and Lady's Astronomy* (London, 1768); *An Introduction to Electricity* (London, 1770); and *Select Mechanical Exercises* (London, 1773).

II. SECONDARY LITERATURE. F. Henderson, *Life of James Ferguson* (Edinburgh, 1867) contains Ferguson's short autobiographical notice as well as a very useful discussion of Ferguson's many tracts and shorter works. Some useful information is also found in H. Mayhew, *The Story of the Peasant-Boy Philosopher* (London, 1857).

LAURENS LAUDAN

FERMAT, PIERRE DE (*b.* Beaumont-de-Lomagne, France, 20 August 1601; *d.* Castres, France, 12 January 1665), *mathematics.*

Factual details concerning Fermat's private life are quite sparse.[1] He apparently spent his childhood and early school years in his birthplace, where his father, Dominique Fermat, had a prosperous leather business and served as second consul of the town. His uncle and godfather, Pierre Fermat, was also a merchant. To the family's firm financial position Fermat's mother, Claire de Long, brought the social status of the parliamentary *noblesse de robe.* Hence, his choice of law as his profession followed naturally from the social milieu into which he was born. Having received a solid classical secondary education locally, Fermat may have attended the University of Toulouse, although one can say with certainty only that he spent some time in Bordeaux toward the end of the 1620's before finally receiving the degree of Bachelor of Civil Laws from the University of Orleans on 1 May 1631.

Returning to Toulouse, where some months earlier he had purchased the offices of *conseiller* and *commissaire aux requêtes* in the local *parlement,* Fermat married his mother's cousin, Louise de Long, on 1 June 1631. Like his in-laws, Fermat enjoyed as *parlementaire* the rank and privileges of the *noblesse de robe;* in particular he was entitled to add the "de" to his name, which he occasionally did. Fermat's marriage contract, the price he paid for his offices, and several other documents attest to the financial security he enjoyed throughout his life.

Five children issued from Fermat's marriage. The oldest, Clément-Samuel, apparently was closest to his father. As a lawyer he inherited his father's offices in 1665 and later undertook the publication of his father's mathematical papers.[2] Fermat's other son, Jean, served as archdeacon of Fimarens. The oldest daughter, Claire, married; her two younger sisters, Catherine and Louise, took holy orders. These outward details of Fermat's family life suggest that it followed the standard pattern for men of his social status. The direct male line ended with the death of Clément-Samuel's son, Jean-François, from whom Claire's grandson inherited the offices originally bought by Fermat.

As a lawyer and *parlementaire* in Toulouse, Fermat seems to have benefited more from the high rate of mortality among his colleagues than from any outstanding talents of his own. On 16 January 1638 he rose to the position of *conseiller aux enquêtes* and in 1642 entered the highest councils of the *parlement:* the criminal court and then the Grand Chamber. In 1648 he acted as chief spokesman for the *parlement* in negotiations with the chancellor of France, Pierre Séguier. However, Fermat's letters to Séguier and to his physician and confidant, Marin Cureau de La Chambre,[3] suggest that Fermat's performance in office was often less than satisfactory; and a confidential report by the *intendant* of Languedoc to Colbert in 1664 refers to Fermat in quite deprecatory terms. A staunch Catholic, Fermat served also—again probably by reason of seniority—as member and then president of the Chambre de l'Édit, which had jurisdiction over suits between Huguenots and Catholics and which convened in the Huguenot stronghold of Castres.

In addition to his fame as a mathematician, Fermat enjoyed a modest reputation as a classical scholar. Apparently equally fluent in French, Italian, Spanish, Latin, and Greek, he dabbled in philological problems and the composition of Latin poetry (see appendixes to his *Oeuvres,* I).

Except for an almost fatal attack of the plague in 1652, Fermat seems to have enjoyed good health until the years immediately preceding his death. He died in Castres, two days after having signed his last *arrêt* for the Chambre de l'Édit there. At first buried in Castres, his remains were brought back to the family vault in the Church of the Augustines in Toulouse in 1675.

The Development of Fermat's Mathematics. Fermat's letters and papers, most of them written after 1636 for friends in Paris, provide the few available hints regarding his development as a mathematician. From them one can infer that his stay in Bordeaux in the late 1620's most decisively shaped his approach to mathematics; almost all of his later achievements derived from research begun there. It was apparently in Bordeaux that Fermat studied in depth the works of François Viète. From Viète he took the new symbolic algebra and theory of equations that served as his basic research tools. More important, however, Viète's concept of algebra as the "analytic art" and the program of research implicit in that concept largely guided Fermat's choice of problems and the manner in which he treated them. Fermat himself viewed his work as a continuation of the Viètan tradition.

From Viète, Fermat inherited the idea of symbolic algebra as a formal language or tool uniting the realms of geometry and arithmetic (number theory). An algebraic equation had meaning in both realms, depending only on whether the unknowns denoted line segments or numbers. Moreover, Viète's theory of equations had shifted attention away from solutions of specific equations to questions of the relationships between solutions and the structures of their parent equations or between the solutions of one equation and those of another. In his own study of the application of determinate equations to geometric constructions, Viète laid the groundwork for the algebraic study of solvability and constructibility. Fermat sought to build further on this foundation. An overall characteristic of his mathematics is the use of algebraic analysis to explore the relationships between problems and their solutions. Most of Fermat's research strove toward a "reduction analysis" by which a given problem could be reduced to another or identified with a class of problems for which the general solution was known. This "reduction analysis," constituted from the theory of equations, could be reversed in most cases to operate as a generator of families of solutions to problems.

At first Fermat, like Viète, looked to the Greek mathematicians for hints concerning the nature of mathematical analysis. Believing that the so-called "analytical" works cited by Pappus in book VII of the *Mathematical Collection,* most of which were no longer extant,[4] contained the desired clues, Fermat followed Viète and others in seeking to restore those lost texts, such as Apollonius' *Plane Loci* (*Oeuvres,* I, 3–51) and Euclid's *Porisms* (*Oeuvres,* I, 76–84). Another supposed source of insight was Diophantus' *Arithmetica,* to which Fermat devoted a lifetime of study. These ancient sources, together with the works of Archimedes, formed the initial elements in a clear pattern of development that Fermat's research followed. Taking his original problem from the classical sources, Fermat attacked it with the new algebraic techniques at his disposal. His solution, however, usually proved more general than the problem that had inspired it. By skillful application of the theory of equations in the form of a "reduction analysis," Fermat would reformulate the problem in its most general terms, often defining thereby a class of problems; in many cases the new problem structure lost all contact with its Greek forebear.

In Fermat's papers algebra as the "analytic art" achieved equal status with the traditional geometrical mode of ancient mathematics. With few exceptions he presented only the algebraic derivation of his results, dispensing with their classical synthetic proofs. Convinced that the latter could always be provided, Fermat seldom attempted to carry them out, with the result in several cases that he failed to see how the use of algebra had led to the introduction of concepts quite foreign to the classical tradition.

In large part Fermat's style of exposition charac-

terized the unfinished nature of his papers, most of them brief essays or letters to friends. He never wrote for publication. Indeed, adamantly refusing to edit his work or to publish it under his own name, Fermat thwarted several efforts by others to make his results available in print. Showing little interest in completed work, he freely sent papers to friends without keeping copies for himself. Many results he merely entered in the margins of his books; e.g., his "Observations on Diophantus," a major part of his work on number theory, was published by his son on the basis of the marginalia in Fermat's copy of the Bachet edition of the *Arithmetica.* Some other work slipped into print during Fermat's lifetime, although only by virtue of honoring his demand for anonymity. This demand allows no clear or obvious explanation. Fermat knew of his reputation and he valued it. He seemed to enjoy the intellectual combat of the several controversies to which he was a party. Whatever the reason, anonymity and refusal to publish robbed him of recognition for many striking achievements and toward the end of his life led to a growing isolation from the main currents of research.

Fermat's name slipped into relative obscurity during the eighteenth century. In the mid-nineteenth century, however, renewed interest in number theory recalled him and his work to the attention of mathematicians and historians of mathematics. Various projects to publish his extant papers culminated in the four-volume edition by Charles Henry and Paul Tannery, from which the extent and importance of Fermat's achievements in fields other than number theory became clear.

Analytic Geometry. By the time Fermat began corresponding with Mersenne and Roberval in the spring of 1636, he had already composed his "Ad locos planos et solidos isagoge" (*Ouevres,* I, 91–103), in which he set forth a system of analytic geometry almost identical with that developed by Descartes in the *Géométrie* of 1637. Despite their simultaneous appearance (Descartes's in print, Fermat's in circulated manuscript), the two systems stemmed from entirely independent research and the question of priority is both complex and unenlightening. Fermat received the first impetus toward his system from an attempt to reconstruct Apollonius' lost treatise *Plane Loci* (loci that are either straight lines or circles). His completed restoration, although composed in the traditional style of Greek geometry, nevertheless gives clear evidence that Fermat employed algebraic analysis in seeking demonstrations of the theorems listed by Pappus. This application of algebra, combined with the peculiar nature of a geometrical locus and the slightly different proof procedures required by locus demonstrations, appears to have revealed to Fermat that all of the loci discussed by Apollonius could be expressed in the form of indeterminate algebraic equations in two unknowns, and that the analysis of these equations by means of Viète's theory of equations led to crucial insights into the nature and construction of the loci. With this inspiration from the *Plane Loci*, Fermat then found in Apollonius' *Conics* that the *symptomata*, or defining properties, of the conic sections likewise could be expressed as indeterminate equations in two unknowns. Moreover, the standard form in which Apollonius referred the *symptomata* to the cone on which the conic sections were generated suggested to Fermat a standard geometrical framework in which to establish the correspondence between an equation and a curve. Taking a fixed line as axis and a fixed point on that line as origin, he measured the variable length of the first unknown, *A*, from the origin along the axis. The corresponding value of the second unknown, *E*, he constructed as a line length measured from the end point of the first unknown and erected at a fixed angle to the axis. The end points of the various lengths of the second unknown then generated a curve in the *A,E* plane.

Like Descartes, then, Fermat did not employ a coordinate system but, rather, a single axis with a moving ordinate; curves were not plotted, they were generated. Within the standard framework

> Whenever two unknown quantities are found in final equality, there results a locus [fixed] in place, and the end point of one of these unknown quantities describes a straight line or a curve ["Isagoge," *Oeuvres*, I, 91].

The crucial phrase in this keystone of analytic geometry is "fixed in place";[5] it sets the task of the remainder of Fermat's treatise. Dividing the general second-degree equation $Ax^2 + By^2 + Cxy + Dx + Ey + F = 0$ into seven canonical (irreducible) forms according to the possible values of the coefficients, Fermat shows how each canonical equation defines a curve: $Dx = Ey$ (straight line), $Cxy = F$ (equilat-

eral hyperbola), $Ax^2 \pm Cxy = By^2$ (straight lines), $Ax^2 = Ey$ (parabola), $F - Ax^2 = Ay^2$ (circle), $F - Ax^2 = By^2$ (ellipse), and $F + Ax^2 = By^2$ (axial hyperbola). In each case he demonstrates that the constants of the equation uniquely fix the curve defined by it, i.e., that they contain all the data necessary to construct the curve. The proof relies on the construction theorems set forth in Euclid's *Data* (for the straight line and circle, or "plane loci") or Apollonius' *Conics* (for the conic sections, or "solid loci"). In a corollary to each case Fermat employs Viète's theory of equations to establish the family of equations reducible to the canonical form and then shows how the reduction itself corresponds to a translation (or expansion) of the axis or the origin or to a change of angle between axis and ordinate. In the last theorem of the "Isagoge," for example, he reduces the equation $b^2 - 2x^2 = 2xy + y^2$ to the canonical form $2b^2 - u^2 = 2v^2$, where $u = \sqrt{2}x$ and $v = x + y$. Geometrically, the reduction shifts the orthogonal x,y system to a skew u,v system in which the u-axis forms a 45° angle with the x-axis and the v-ordinate is erected at a 45° angle on the u-axis. The curve, as Fermat shows, is a uniquely defined ellipse.

Although the analytic geometries of Descartes and Fermat are essentially the same, their presentations differed significantly. Fermat concentrated on the geometrical construction of the curves on the basis of their equations, relying heavily on the reader's knowledge of Viète's algebra to supply the necessary theory of equations. By contrast, Descartes slighted the matter of construction and devoted a major portion of his *Géométrie* to a new and more advanced theory of equations.

In the years following 1636, Fermat made some effort to pursue the implications of his system. In an appendix to the "Isagoge," he applied the system to the graphic solution of determinate algebraic equations, showing, for example, that any cubic or quartic equation could be solved graphically by means of a parabola and a circle. In his "De solutione problematum geometricorum per curvas simplicissimas et unicuique problematum generi proprie convenientes dissertatio tripartita" (*Oeuvres,* I, 118–131), he took issue with Descartes's classification of curves in the *Géométrie* and undertook to show that any determinate algebraic equation of degree $2n$ or $2n - 1$ could be solved graphically by means of curves determined by indeterminate equations of degree n.

In 1643, in a memoir entitled "Isagoge ad locos ad superficiem" (*Oeuvres,* I, 111–117), Fermat attempted to extend his plane analytic geometry to solids of revolution in space and perhaps thereby to restore the content of Euclid's *Surface Loci*, another

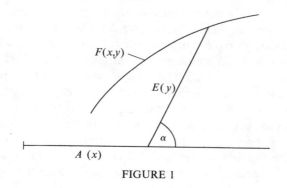

FIGURE I

text cited by Pappus. The effort did not meet with success because he tried to reduce the three-dimensional problem to two dimensions by determining all possible traces resulting from the intersection of a given solid by an arbitrary plane. The system required, first of all, an elaborate catalog of the possible traces for various solids. Second, the manipulation of the equation of any trace for the purpose of deriving the parameters that uniquely determine the solid requires methods that lay beyond Fermat's reach; his technique could at best define the solid qualitatively. Third, the basic system of the 1636 "Isagoge," lacking the concept of coordinates referred to two fixed orthogonal axes, presented substantial hurdles to visualizing a three-dimensional correlate.

Although Fermat never found the geometrical framework for a solid analytic geometry, he nonetheless correctly established the algebraic foundation of such a system. In 1650, in his "Novus secundarum et ulterioris ordinis radicum in analyticis usus" (*Oeuvres,* I, 181–188), he noted that equations in one unknown determine point constructions; equations in two unknowns, locus constructions of plane curves; and equations in three unknowns, locus constructions of surfaces in space. The change in the criterion of the dimension of an equation—from its degree, where the Greeks had placed it, to the number of unknowns in it—was one of the most important conceptual developments of seventeenth-century mathematics.

The Method of Maxima and Minima. The method of maxima and minima, in which Fermat first established what later became the algorithm for obtaining the first derivative of an algebraic polynomial, also stemmed from the application of Viète's algebra to a problem in Pappus' *Mathematical Collection.* In a lemma to Apollonius' *Determinate Section,* Pappus sought to divide a given line in such a way that certain rectangles constructed on the segments bore a minimum ratio to one another,[6] noting that the ratio would be "singular." In carrying out the algebraic analysis of the problem, Fermat recognized that the division of the line for rectangles in a ratio greater than the minimum corresponded to a quadratic equation that would normally yield two equally satisfactory section points. A "singular" section point for the minimum ratio, he argued, must mean that the particular values of the constant quantities of the equation allow only a single repeated root as a solution.

Turning to a simpler example, Fermat considered the problem of dividing a given line in such a way that the product of the segments was maximized. The algebraic form of the problem is $bx - x^2 = c$, where b is the length of the given line and c is the product of the segments. If c is the maximum value of all possible products, then the equation can have only one (repeated) root. Fermat then sought the value for c in terms of b for which the equation yielded that (repeated) root. To this end he applied a method of Viète's theory of equations called "syncrisis," a method originally devised to determine the relationships between the roots of equations and their constant parameters. On the assumption that his equation had two distinct roots, x and y, Fermat set $bx - x^2 = c$ and $by - y^2 = c$, whence he obtained $b = x + y$ and $c = xy$. Taking these relationships to hold generally for any quadratic equation of the above form, he next considered what happened in the case of a repeated root, i.e., when $x = y$. Then, he found, $x = b/2$ and $c = b^2/4$. Hence, the maximum rectangle results from dividing the given line in half, and that maximum rectangle has an area equal to one-quarter of the square erected on the given line b.

Amending his method in the famous "Methodus ad disquirendam maximam et minimam" (*Oeuvres,* I, 133–136), written sometime before 1636, Fermat expressed the supposedly distinct roots as A and $A + E$ (that is, x and $x + y$), where E now represented the difference between the roots. In seeking, for example, the maximum value of the expression $bx^2 - x^3$, he proceeded as follows:

$$bx^2 - x^3 = M^3$$
$$b(x + y)^2 - (x + y)^3 = M^3,$$
whence $\quad 2bxy + by^2 - 3x^2y - 3xy^2 - y^3 = 0.$

Division by y yields the equation

$$2bx + by - 3x^2 - 3xy - y^2 = 0,$$

which relates the parameter b to two roots of the equation via one of the roots and their difference. The relation holds for any equation of the form $bx^2 - x^3 = M^3$, but when M^3 is a maximum the equation has a repeated root, i.e., $x = x + y$, or $y = 0$. Hence, for that maximum, $2bx - 3x^2 = 0$, or $x = 2b/3$ and $M^3 = 4b^3/27$.

Fermat's method of maxima and minima, which is clearly applicable to any polynomial $P(x)$, originally rested on purely finitistic algebraic foundations.[7] It assumed, counterfactually, the inequality of two equal roots in order to determine, by Viète's theory of equations, a relation between those roots and one of the coefficients of the polynomial, a relation that was fully general. This relation then led to an extreme-value solution when Fermat removed his counterfactual assumption and set the roots equal. Borrowing a term from Diophantus, Fermat called this counterfactual equality "adequality."

Although Pappus' remark concerning the "singularity" of extreme values provided the original inspiration for Fermat's method, it may also have prevented him from seeing all its implications. Oriented toward unique extreme values and dealing with specific problems that, taken from geometrical sources and never exceeding cubic expressions, failed to yield more than one geometrically meaningful solution, Fermat never recognized the distinction between global and local extreme values or the possibility of more than one such value. This block to an overall view of the problem of maxima and minima vitiates an otherwise brilliant demonstration of Fermat's method, which he wrote for Pierre Brûlard de St.-Martin in 1643 (*Oeuvres,* supp., 120–125) and which employs the sophisticated theory of equations of Descartes's *Géométrie.* There Fermat established what today is termed the "second derivative criterion" for the nature of an extreme value ($f''(x) < 0$ for a maximum, $f''(x) > 0$ for a minimum), although his lack of a general overview forestalled investigation of points of inflection ($f''(x) = 0$).

The original method of maxima and minima had two important corollaries. The first was the method of tangents[8] by which, given the equation of a curve, Fermat could construct the tangent at any given point on that curve by determining the length of the subtangent. Given some curve $y = f(x)$ and a point (a,b) on it, Fermat assumed the tangent to be drawn and

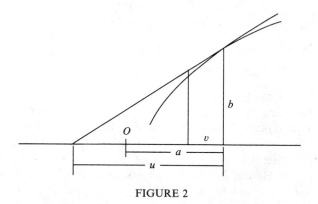

FIGURE 2

to cut off a subtangent of length u on the x-axis. Taking an arbitrary point on the tangent and denoting the difference between the abscissa of that point and the abscissa a by v, he counterfactually assumed that the ordinate to the point on the tangent was equal to the ordinate $f(a - v)$ to the curve, i.e., that the two ordinates were "adequal." It followed, then, from similar triangles that

$$\frac{b}{u} \approx \frac{f(a - v)}{u - v}.$$

Fermat removed the adequality, here denoted by $\approx$, by treating the difference v in the same manner as in the method of maxima and minima, i.e., by considering it as ultimately equal to zero. His method yields, in modern symbols, the correct result, $u = f(a)/f'(a)$, and, like the parent method of maxima and minima, it can be applied generally.

From the method of maxima and minima Fermat drew as a second corollary a method for determining centers of gravity of geometrical figures (*Oeuvres,* I, 136–139). His single example—although again the method itself is fully general—concerns the center of gravity of a paraboloidal segment. Let CAV be the generating parabola with axis AI and base CV. By symmetry the center of gravity O of the paraboloidal segment lies on axis $AI = b$ at some distance $AO = x$

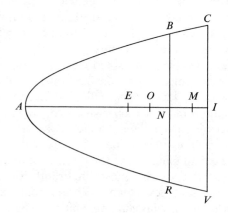

FIGURE 3

from the vertex A. Let the segment be cut by a plane parallel to the base and intersecting the axis at an arbitrary distance y from point I. Let E and M denote the centers of gravity of the two resulting subsegments. Since similar figures have similarly placed centers of gravity (Archimedes), $b/x = (b - y)/AE$, whence $EO = x - AE = xy/b$. By the definition of the center of gravity and by the law of the lever, segment $CBRV$ is to segment BAR as EO is to OM. But, by Archimedes' *Conoids and Spheroids,* proposition 26, paraboloid CAV is to paraboloid BAR as AI^2 is to AN^2, or as b^2 is to $(b - y)^2$, whence

$$\frac{EO}{OM} = \frac{CBRV}{BAR} = \frac{b^2 - (b - y)^2}{(b - y)^2}.$$

Here Fermat again employed the notion of adequality to set OM counterfactually equal to OI, whence

$$OI = b - x \approx OM = \left(\frac{xy}{b}\right)\left(\frac{b^2 - 2by + y^2}{2by - y^2}\right).$$

He removed the adequality by an application of the method of maxima and minima, i.e., by dividing through by y and then setting $y\,(=OI-OM)$ equal to zero, and obtained the result $x=2b/3$. In applying his method to figures generated by curves of the forms $y^q=kx^p$ and $x^py^q=k$ (p,q positive integers), Fermat employed the additional lemma that the similar segments of the figures "have the same proportion to corresponding triangles of the same base and height, even if we do not know what that proportion is,"[9] and argued from that lemma that his method of centers of gravity eliminated the problem of quadrature as a prerequisite to the determination of centers of gravity. Such an elimination was, of course, illusory, but the method did not depend on the lemma. It can be applied to any figure for which the general quadrature is known.

Fermat's method of maxima and minima and its corollary method of tangents formed the central issue in an acrid debate between Fermat and Descartes in the spring of 1638. Viewing Fermat's methods as rivals to his own in the *Géométrie*, Descartes tried to show that the former were at once paralogistic in their reasoning and limited in their application. It quickly became clear, however, that, as in the case of their analytic geometries, Fermat's and Descartes's methods rested on the same foundations. The only substantial issue was Descartes's disapproval of mathematical reasoning based on counterfactual assumptions, i.e., the notion of adequality. Although the two men made formal peace in the summer of 1638, when Descartes admitted his error in criticizing Fermat's methods, the bitterness of the dispute, exacerbated by the deep personal hatred Descartes felt for Fermat's friend and spokesman, Roberval, poisoned any chance for cooperation between the two greatest mathematicians of the time. Descartes's sharp tongue cast a pall over Fermat's reputation as a mathematician, a situation which Fermat's refusal to publish only made worse.[10] Through the efforts of Mersenne and Pierre Hérigone, Fermat's methods did appear in print in 1642, but only as bare algorithms that, by setting the difference y of the roots equal to zero from the start, belied the careful thinking that originally underlay them. Moreover, other mathematicians soon were publishing their own, more general algorithms; by 1659, Huygens felt it necessary to defend Fermat's priority against the claims of Johann Hudde. In time, Fermat's work on maxima and minima was all but forgotten, having been replaced by the differential calculus of Newton and Leibniz.

Methods of Quadrature. Fermat's research into the quadrature of curves and the cubature of solids also had its beginnings in the research that preceded his introduction to the outside mathematical world in 1636. By that time, he had taken the model of Archi-

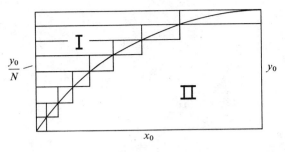

FIGURE 4

medes' quadrature of the spiral[11] and successfully extended its application to all spirals of the forms $\rho=(a\vartheta)^m$ and $R/R-\rho=(\alpha/\vartheta)^m$. Moreover, he had translated Archimedes' method of circumscription and inscription of sectors around and within the spiral into a rectangular framework. Dividing a given ordinate y_0 (or the corresponding abscissa x_0) of a curve $y=f(x)$ into N equal intervals and drawing lines parallel to the axis, Fermat determined that Area I in Figure 4 lay between limits $\dfrac{y_0}{N}\displaystyle\sum_{i=1}^{N}x_i$ and $\dfrac{y_0}{N}\displaystyle\sum_{i=1}^{N-1}x_i$, where x_i is the abscissa that corresponds to ordinate $(i/N)y_0$. Since he possessed a recursive formula for determining $\displaystyle\sum_{=1}^{N}i^m$ for any positive integer m, Fermat could prove that

$$\frac{1}{N^{m+1}}\sum_{i=1}^{N}i^m>\frac{1}{m+1}>\frac{1}{N^{m+1}}\sum_{i=1}^{N-1}i^m$$

for all values of N. In each case the difference between the bounds is $1/N$, which can be made as small as one wishes. Hence, for any curve of the form $y^m=kx$, Fermat could show that the curvilinear Area I $=[1/(m+1)]x_0y_0$ and the curvilinear Area II $=[m/(m+1)]x_0y_0$. As an immediate corollary, he found that he could apply the same technique to determine the volume of the solid generated by the rotation of the curve about the ordinate or axis, with the restriction in this case that m be an even integer.

Sometime before 1646 Fermat devised a substantially new method of quadrature, which permitted the treatment of all curves of the forms $y^q=kx^p$ and $x^py^q=k$ (p,q positive integers; in the second equation $p+q>2$). The most striking departure from the earlier method is the introduction of the concept of adequality, now used in the sense of "approximate equality" or "equality in the limiting case." In the first example given in his major treatise on quadrature[12] Fermat derives the shaded area under the curve $x^2y=k$ in Figure 5 as follows (we use modern

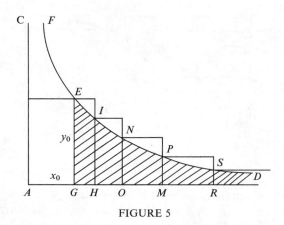

FIGURE 5

notation to abbreviate Fermat's lengthy verbal description while preserving its sense): the infinite x-axis is divided into intervals by the end points of a divergent geometric sequence of lengths AG, AH, AO, $\cdots$, or x_0, $(m/n)x_0$, $(m/n)^2x_0$, $\cdots$, where $m > n$ are arbitrary integers. Since $(m/n)^i - (m/n)^{i-1} = (m/n)^{i-1}(m/n - 1)$, each interval can, by suitable choice of m and n, be made as closely equal to another as desired and at the same time can be made as small as desired. Fermat has, then, $GH \approx HO \approx OM \approx \cdots$ and $GH \to 0$. From the curve and the construction of the intervals, it follows directly that the approximating rectangles erected on the intervals form the convergent geometric series

$$(m/n - 1)x_0y_0, \ (n/m)(m/n - 1)x_0y_0,$$
$$(n/m)^2(m/n - 1)x_0y_0, \ \cdots.$$

Its sum is $(m/n - 1)x_0y_0 + x_0y_0$, which is "adequal" to the shaded area. It approaches the curved area ever more closely as the size of the intervals approaches zero, i.e., as $m/n \to 1$. In the limiting case, the sum will be x_0y_0, which in turn will be the exact area of the shaded segment. Generalizing the procedure for any curve $x^py^q = k$ and a given ordinate y_0, Fermat determined that the area under the curve from y_0 on out is $[q/(p - q)]x_0y_0$. Adapting the procedure to curves $y^q = kx^p$ (by dividing the finite axis from 0 to x_0 by a convergent geometric sequence of intervals), he was also able to show that the area under the curve is $[q/(p + q)]x_0y_0$.

In the remainder of his treatise on quadrature, Fermat shifted from the geometrical style of exposition to the algebraic and, on the model of Viète's theory of equations, set up a "reduction analysis" by which a given quadrature either generates an infinite class of quadratures or can be shown to be dependent on the quadrature of the circle. To carry out the project he introduced a new concept of "application of all y^n to a given segment," by which he meant the limit-sum of the products $y^n \triangle x$ over a given

segment b of the x-axis as $\triangle x \to 0$ (in the absence of any notation by Fermat, we shall borrow from Leibniz and write $Omn_b\,y^n$ to symbolize Fermat's concept). Fermat then showed by several concrete examples that for any curve of the form $y^n = \Sigma a_ix^i/b_jx^j$ the determination of $Omn_b\,y^n$ follows directly from setting $y^n = \Sigma u_i$, where $u_i = a_ix^i/b_jx^j$. For each i the resulting expression $u_i = f(x)$ will denote a curve of the form $u_i^q = kx^p$ or the form $x^pu_i^q = k$. For each curve, the determination of $Omn_b u_i$ corresponds to the direct quadrature set forth in the first part of the treatise, and hence it is determinable. Therefore $Omn_b y^n = \Sigma Omn_b u_i$ is directly determinable.

Fermat next introduced the main lemma of his treatise, an entirely novel result for which he characteristically offered no proof. For any curve $y = f(x)$ decreasing monotonically over the interval 0,b, where $f(0) = d$ and $f(b) = 0$, $Omn_b y^n = Omn_d nxy^{n-1}$. This result is equivalent to the modern statement

$$\int_0^b y^n \, dx = n \int_0^d xy^{n-1} \, dy.$$

One example from Fermat's treatise on quadrature suffices to display the subtlety and power of his reduction analysis. Can the area beneath the curve $b^3 = x^2y + b^2y$ (i.e., the "witch of Agnesi") be squared algebraically? Two transformations of variable and an application of the main lemma supply the answer. From $by = u^2$ and $bv = xu$, it follows first that $Omn_x y = 1/b \ Omn_x u^2 = 2/b \ Omn_u xu = 2 \ Omn_u v$. Hence, the quadrature of the original area depends on that of the transformed curve $F(u,v)$. But substitution of variables yields $b^2 = u^2 + v^2$, the equation of a circle. Therefore, the quadrature of the original area depends on the quadrature of the circle and cannot be carried out algebraically.

Fermat's treatise first circulated when it was printed in his *Varia opera* of 1679. By then much of its contents had become obsolescent in terms of the work of Newton and Leibniz. Even so, it is doubtful what effect the treatise could have had earlier. As sympathetic a reader as Huygens could make little sense of it.[13] In addition, Fermat's method of quadrature, like his method of tangents, lacks even the germ of several concepts crucial to the development of the calculus. Not only did Fermat not recognize the inverse relationship between the two methods, but both methods, conceptually and to some extent operationally, steered away from rather than toward the notion of the tangent or the area as a function of the curve.

Fermat's one work published in his lifetime, a treatise on rectification appended to a work on the cycloid by Antoine de La Loubère,[14] was a direct corollary of the method of quadrature. Cast, however, in the strictly geometrical style of classical Greek mathematics, it hid all traces of the underlying algebraic analysis. In the treatise Fermat treated the length of a curve as the limit-sum of tangential segments $\triangle S$ cut off by abscissas drawn through the end points of intervals $\triangle y$ on a given y ordinate. In essence, he showed that for any curve $y = y(x)$,

$$\frac{\triangle S^2}{\triangle y^2} = [x'(y)]^2 + 1.$$

Taking $u^2 = [x'(y)]^2 + 1$ as an auxiliary curve, Fermat used the relation $S = Omn_y u$ to reduce the problem of rectification to one of quadrature. He used the same basic procedure to determine the area of the surface generated by the rotation of the curve about an axis or an ordinate, as the results in a 1660 letter to Huygens indicate.

Number Theory. As a result of limited circulation in unpublished manuscripts, Fermat's work on analytic geometry, maxima and minima and tangents, and quadrature had only moderate influence on contemporary developments in mathematics. His work in the realm of number theory had almost none at all. It was neither understood nor appreciated until Euler revived it and initiated the line of continuous research that culminated in the work of Gauss and Kummer in the early nineteenth century. Indeed, many of Fermat's results are basic elements of number theory today. Although the results retain fundamental importance, his methods remain largely a secret known only to him. Theorems, conjectures, and specific examples abound in his letters and marginalia. But, except for a vague outline of a method he called "infinite descent," Fermat left no obvious trace of the means he had employed to find them. He repeatedly claimed to work from a method, and the systematic nature of much of his work would seem to support his claim.

In an important sense Fermat invented number theory as an independent branch of mathematics. He was the first to restrict his study in principle to the domain of integers. His refusal to accept fractional solutions to problems he set in 1657 as challenges to the European mathematics community (*Oeuvres*, II, 332–335) initiated his dispute with Wallis, Frénicle, and others,[15] for it represented a break with the classical tradition of Diophantus' *Arithmetica*, which served as his opponents' model. The restriction to integers explains one dominant theme of Fermat's work in number theory, his concern with prime numbers and divisibility. A second guiding theme of his

research, the determination of patterns for generating families of solutions from a single basic solution, carried over from his work in analysis.

Fermat's earliest research, begun in Bordeaux, displays both characteristics. Investigating the sums of the aliquot parts (proper divisors) of numbers, Fermat worked from Euclid's solution to the problem of "perfect numbers"—$\sigma(a) = 2a$, where $\sigma(a)$ denotes the sum of all divisors of the integer a, including 1 and a—to derive a complete solution to the problem of "friendly" numbers—$\sigma(a) = \sigma(b) = a + b$—and to the problem $\sigma(a) = 3a$. Later research in this area aimed at the general problem $\sigma(a) = (p/q)a$, as well as $\sigma(x^3) = y^2$ and $\sigma(x^2) = y^3$ (the "First Challenge" of 1657). Although Fermat offered specific solutions to the problem $\sigma(a) = na$ for $n = 3, 4, 5, 6$, he recorded the algorithm only for $n = 3$. The central role of primeness and divisibility in such research led to several corollaries, among them the theorem (announced in 1640) that $2^k - 1$ is always a composite number if k is composite and may be composite for prime k; in the latter case, all divisors are of the form $2mk + 1$.

Fermat's interest in primeness and divisibility culminated in a theorem now basic to the theory of congruences; as set down by Fermat it read: If p is prime and a^t is the smallest number such that $a^t = kp + 1$ for some k, then t divides $p - 1$. In the modern version, if p is prime and p does not divide a, then $a^{p-1} \equiv 1 \pmod{p}$. As a corollary to this theorem, Fermat investigated in depth the divisibility of $a^k \pm 1$ and made his famous conjecture that all numbers of the form $2^{2^n} + 1$ are prime (disproved for $n = 5$ by Euler). In carrying out his research, Fermat apparently relied on an extensive factual command of the powers of prime numbers and on the traditional "sieve of Eratosthenes" as a test of primeness. He several times expressed his dissatisfaction with the latter but seems to have been unable to find a more efficient test, even though in retrospect his work contained all the necessary elements for one.

A large group of results of fundamental importance to later number theory (quadratic residues, quadratic forms) apparently stemmed from Fermat's study of the indeterminate equation $x^2 - q = my^2$ for nonsquare m. In his "Second Challenge" of 1657, Fermat claimed to have the complete solution for the case $q = 1$. Operating on the principle that any divisor of a number of the form $a^2 + mb^2$ (m not a square) must itself be of that form, Fermat established that all primes of the form $4k + 1$ (but not those of the form $4k + 3$) can be expressed as the sum of two squares, all primes of the form $8k + 1$ or $8k + 3$ as the sum of a square and the double of a square, all primes of the form $3k + 1$ as $a^2 + 3b^2$, and that the product of any two primes of the form $20k + 3$ or $20k + 7$ is expressible in the form $a^2 + 5b^2$.

Another by-product of this research was Fermat's claim to be able to prove Diophantus' conjecture that any number can be expressed as the sum of at most four squares. Extending his research on the decomposition of numbers to higher powers, Fermat further claimed proofs of the theorems that no cube could be expressed as the sum of two cubes, no quartic as the sum of two quartics, and indeed no number a^n as the sum of two powers b^n and c^n (the famous "last theorem," mentioned only once in the margin of his copy of Diophantus' *Arithmetica*). In addition, he claimed the complete solution of the so-called "four-cube problem" (to express the sum of two given cubes as the sum of two other cubes), allowing here, of course, fractional solutions of the problem.

To prove his decomposition theorems and to solve the equation $x^2 - 1 = my^2$, Fermat employed a method he had devised and called "infinite descent." The method, an inverse form of the modern method of induction, rests on the principle (peculiar to the domain of integers) that there cannot exist an infinitely decreasing sequence of integers. Fermat set down two rather vague outlines of his method, one in his "Observations sur Diophante" (*Oeuvres*, I, 340–341) and one in a letter to Carcavi (*Oeuvres*, II, 431–433). In the latter Fermat argued that no right triangle of numbers (triple of numbers a, b, c such that $a^2 + b^2 = c^2$) can have an area equal to a square ($ab/2 = m^2$ for some m), since

> If there were some right triangle of integers that had an area equal to a square, there would be another triangle less than it which had the same property. If there were a second, less than the first, which had the same property, there would be by similar reasoning a third less than the second which had the same property, and then a fourth, a fifth, etc., ad infinitum in decreasing order. But, given a number, there cannot be infinitely many others in decreasing order less than it (I mean to speak always of integers). From which one concludes that it is therefore impossible that any right triangle of numbers have an area that is a square [letter to Carcavi, *Oeuvres*, II, 431–432].

Fermat's method of infinite descent did not apply only to negative propositions. He discovered that he could also show that every prime of the form $4k + 1$ could be expressed as the sum of two squares by denying the proposition for some such prime, deriving another such prime less than the first, for which the proposition would again not hold, and so on. Ultimately, he argued, this decreasing sequence of primes would arrive at the least prime of the form $4k + 1$—

namely, 5—for which, by assumption, the proposition would not hold. But $5 = 2^2 + 1^2$, which contradicts the initial assumption. Hence, the proposition must hold. Although infinite descent is unassailable in its overall reasoning, its use requires the genius of a Fermat, since nothing in that reasoning dictates how one derives the next member of the decreasing sequence for a given problem.

Fermat's letters to Jacques de Billy, published by the latter as *Doctrinae analyticae inventum novum*,[16] form the only other source of direct information about Fermat's methods in number theory. In these letters Fermat undertook a complete treatment of the so-called double equations first studied by Diophantus. In their simplest form they required the complete solution of the system $ax + b = \square$, $cx + d = \square$. By skillful use of factorization to determine the base solution and the theorem that, if a is a solution, then successive substitution of $x + a$ for x generates an infinite family of solutions, Fermat not only solved all the problems posed by Diophantus but also extended them as far as polynomials of the fourth degree.

The importance of Fermat's work in the theory of numbers lay less in any contribution to contemporary developments in mathematics than in their stimulative influence on later generations. Much of the number theory of the nineteenth century took its impetus from Fermat's results and, forced to devise its own methods, contributed to the formulation of concepts basic to modern algebra.

Other Work. *Probability*. Fermat shares credit with Blaise Pascal for laying the first foundations of the theory of probability. In a brief exchange of correspondence during the summer of 1654, the two men discussed their different approaches to the same solution of a problem originally posed to Pascal by a gambler: How should the stakes in a game of chance be divided among the players if the game is prematurely ended? In arriving at specific, detailed solutions for several simple games, Fermat and Pascal operated from the basic principle of evaluating the expectation of each player as the ratio of outcomes favorable to him to the total number of possible outcomes. Fermat relied on direct computations rather than general mathematical formulas in his solutions, and his results and methods quickly became obsolete with the appearance in 1657 of Christiaan Huygens' mathematically more sophisticated *De ludo aleae*.

Optics (Fermat's Principle). In 1637, when Fermat was engaged with traditional and rather pedestrian problems in geostatics, he read Descartes's *Dioptrique*. In a letter to Mersenne, which opened the controversy between Descartes and Fermat mentioned above,

Fermat severely criticized the work. Methodologically, he could not accept Descartes's use of mathematics to make a priori deductions about the physical world. Philosophically, he could not agree with Descartes that "tendency to motion" (Descartes's basic definition of light) could be understood and analyzed in terms of actual motion. Physically, he doubted both the assertion that light traveled more quickly in a denser medium (he especially questioned the meaning of such a statement together with the assertion of the instantaneous transmission of light) and Descartes's law of refraction itself. Mathematically, he tried to show that Descartes's demonstrations of the laws of reflection and refraction proved nothing that Descartes had not already assumed in his analysis, i.e., that Descartes had begged the question. The ensuing debate in the fall of 1637 soon moved to mathematics as Descartes launched a counterattack aimed at Fermat's method of tangents, and Fermat returned to the original subject of optics only in the late 1650's, when Claude Clerselier reopened the old argument while preparing his edition of Descartes's *Lettres*.

Fermat, who in his earlier years had fervently insisted that experiment alone held the key to knowledge of the physical world, nonetheless in 1662 undertook a mathematical derivation of the law of refraction on the basis of two postulates: first, that the finite speed of light varied as the rarity of the medium through which it passed and, second, that "nature operates by the simplest and most expeditious ways and means." In his "Analysis ad refractiones" (*Oeuvres*, I, 170–172), Fermat applied the second postulate (Fermat's principle) in the following manner: In Figure 6 let the upper half of the circle represent the rarer of two media and let the lower half represent the denser; further, let *CD* represent a given incident ray. If the "ratio of the resistance of the denser medium to the resistance of the rarer medium" is expressed as the ratio of the given line *DF* to some line *M*, then "the motions which occur along lines *CD* and *DI* [the refracted ray to be determined] can be measured with the aid of the lines *DF* and *M;* that is, the motion that occurs along the two lines is represented comparatively by the sum of two rectangles, of which one is the product of *CD* and *M* and the other the product of *DI* and *DF*" ("Analysis ad refractiones," pp. 170–171). Fermat thus reduces the problem to one of determining point *H* such that that sum is minimized. Taking length *DH* as the unknown *x*, he applies his method of maxima and minima and, somewhat to his surprise (expressed in a letter to Clerselier), arrives at Descartes's law of refraction.

Although Fermat took the trouble to confirm his derived result by a formal, synthetic proof, his interest in the problem itself ended with his derivation. Physical problems had never really engaged him, and he had returned to the matter only to settle an issue that gave rise to continued ill feeling between him and the followers of Descartes.

In fact, by 1662 Fermat had effectively ended his career as a mathematician. His almost exclusive interest in number theory during the last fifteen years of his life found no echo among his junior contemporaries, among them Huygens, who were engaged in the application of analysis to physics. As a result Fermat increasingly returned to the isolation from which he had so suddenly emerged in 1636, and his death in 1665 was viewed more as the passing of a grand old man than as a loss to the active scientific community.

NOTES

1. All published modern accounts of Fermat's life ultimately derive from Paul Tannery's article in the *Grande encyclopédie*, repr. in *Oeuvres*, IV, 237–240. Some important new details emerged from the research of H. Blanquière and M. Caillet in connection with an exhibition at the Lycée Pierre de Fermat in Toulouse in 1957: *Un mathématicien de génie, Pierre de Fermat 1601–1665* (Toulouse, 1957).

2. *Diophanti Alexandrini Arithmeticorum libri sex et de numeris multangulis liber unus. Cum commentariis C. G. Bacheti V. C. et observationibus D. P. de Fermat Senatoris Tolosani* (Toulouse, 1670); *Varia opera mathematica D. Petri de Fermat Senatoris Tolosani* (Toulouse, 1679; repr. Berlin, 1861; Brussels, 1969).

3. Cureau shared Fermat's scientific interests and hence provided a special link to the chancellor. There is much to suggest that the *parlement* of Toulouse took advantage of Fermat's ties to Cureau.

4. Regarding book VII and its importance for Greek geometrical analysis, see M. S. Mahoney, "Another Look at Greek Geometrical Analysis," in *Archive for History of Exact Sciences*, **5** (1968), 318–348. On its influence in the early seventeenth century, see Mahoney, "The Royal Road" (diss., Princeton, 1967), ch. 3.

5. Fermat's original Latin reads: *fit locus loco.* The last word is not redundant, as several authors have thought; rather, the phrase is elliptic, lacking the word *datus.* Fermat's terminology here comes directly from Euclid's *Data* (*linea positione data:* a line given, or fixed, in position).

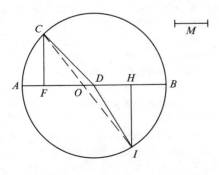

FIGURE 6

Regarding the algebraic symbolism that follows here and throughout the article, note that throughout his life Fermat employed the notation of Viète, which used the capital vowels for unknowns and the capital consonants for knowns or parameters. To avoid the confusion of an unfamiliar notation, this article employs Cartesian notation, translating Fermat's *A* uniformly as *x*, *E* as *y*, etc.

6. Pappus, *Mathematical Collection* VII, prop. 61. The geometrical formulation is too complex to state here without a figure and in addition requires some interpretation. In Fermat's algebraic formulation, the problem calls for the determination of the minimum value of the expression

$$\frac{bc - bx + cx - x^2}{ax - x^2},$$

where *a*, *b*, *c* are given line segments.

7. The modern foundation of Fermat's method is the theorem that if $P(x)$ has a local extreme value at $x = a$, then $P(x) = (x - a)^2 R(x)$, where $R(a) \neq 0$.

8. Fermat's original version of the method is contained in the "Methodus ad disquirendam maximam et minimam" (*Oeuvres*, I, 133–136); in its most finished form it is described in a memoir sent to Descartes in June 1638 (*Oeuvres*, II, 154–162).

9. Fermat to Mersenne, 15 June 1638 (*Oeuvres*, supp., pp. 84–86).

10. Descartes's most famous remark, made to Frans van Schooten, who related it to Huygens (*Oeuvres*, IV, 122), was the following: "Monsieur Fermat est Gascon, moi non. Il est vrai, qu'il a inventé plusieurs belles choses particulières, et qu'il est homme de grand esprit. Mais quant à moi j'ai toujours étudié à considérer les choses fort généralement, afin d'en pouvoir conclure des règles, qui aient aussi ailleurs de l'usage." The connotation of "troublemaker" implicit in the term "Gascon" is secondary to Descartes's charge, believed by some of his followers, that Fermat owed his reputation to a few unsystematic lucky guesses.

11. In his treatise *On Spirals*.

12. "De aequationum localium transmutatione et emendatione ad multimodam curvilineorum inter se vel cum rectilineis comparationem, cui annectitur proportionis geometricae in quadrandis infinitis parabolis et hyperbolis usus" (*Oeuvres*, I, 255–285). The treatise was written sometime between 1657 and 1659, but at least part of it dates back to the early 1640's.

13. Huygens to Leibniz, 1 September 1691 (*Oeuvres*, IV, 137).

14. "De linearum curvarum cum lineis rectis comparatione dissertatio geometrica. Autore M.P.E.A.S." The treatise was published with La Loubère's *Veterum geometria promota in septem de cycloide libris, et in duabus adjectis appendicibus* (Toulouse, 1660).

15. The dispute is recorded in Wallis' *Commercium epistolicum de quaestionibus quibusdam mathematicis nuper habitum* (Oxford, 1658). The participants were William Brouncker, Kenelm Digby, Fermat, Bernard Frénicle, Wallis, and Frans van Schooten.

16. Published as part of Samuel Fermat's edition of Diophantus in 1670 (see note 2).

BIBLIOGRAPHY

I. ORIGINAL WORKS. The modern edition of the *Oeuvres de Fermat*, Charles Henry and Paul Tannery, eds., 4 vols. (Paris, 1891–1912), with supp. by Cornelis de Waard (Paris, 1922), contains all of Fermat's extant papers and letters in addition to correspondence between other men concerning Fermat. The edition includes in vol. III French translations of those papers and letters that Fermat wrote in Latin and also a French translation of Billy's *Inventum novum*. English translations of Fermat's "Isagoge" and "Methodus ad disquirendam maximam et minimam" have been published in D. J. Struik's *A Source Book in Mathematics, 1200–1800* (Cambridge, Mass., 1969).

II. SECONDARY LITERATURE. The two most important summaries of Fermat's career are Jean Itard, *Pierre Fermat, Kurze Mathematiker Biographien*, no. 10 (Basel, 1950); and J. E. Hofmann, "Pierre Fermat—ein Pionier der neuen Mathematik," in *Praxis der Mathematik*, **7** (1965), 113–119, 171–180, 197–203. Fermat's contributions to analytic geometry form part of Carl Boyer, *History of Analytic Geometry* (New York, 1956), ch. 5; and the place of Fermat in the history of the calculus is discussed in Boyer's *Concepts of the Calculus* (New York, 1949), pp. 154–165. The most detailed and enlightening study of Fermat's work in number theory has been carried out by J. E. Hofmann; see, in particular, "Über zahlentheoretische Methoden Fermats und Eulers, ihre Zusammenhänge und ihre Bedeutung," in *Archive for History of Exact Sciences*, **1** (1961), 122–159; and "Studien zur Zahlentheorie Fermats," in *Abhandlungen der Preussischen Akademie der Wissenschaften*, Mathematisch–Naturwissenschaftliche Klasse, no. 7 (1944). Fermat's dispute with Descartes on the law of refraction and his own derivation of the law are treated in detail in A. I. Sabra, *Theories of Light From Descartes to Newton* (London, 1967), chs. 3–5.

MICHAEL S. MAHONEY

FERMI, ENRICO (*b.* Rome, Italy, 29 September 1901; *d.* Chicago, Illinois, 28 November 1954), *physics*.

His father, Alberto Fermi, was an administrative employee of the Italian railroads; his mother, Ida de Gattis, was a schoolteacher. Fermi received a traditional education in the public schools of Rome, but his scientific formation was due more to the books he read than to personal contacts. It is possible to gather exact information on his readings from an extant notebook, and later in life he mentioned having studied such works as Poisson's *Traité de mécanique*, Richardson's *Electron Theory of Matter*, Planck's *Vorlesungen über Thermodynamik*, and several by Poincaré.

Fermi was fundamentally an agnostic, although he had been baptized a Catholic. In 1928 he married Laura Capon, the daughter of an admiral in the Italian navy. His wife's family was Jewish and was severely persecuted during the Nazi-Fascist period. Fermi, who enjoyed excellent health until his fatal illness, led a very simple, frugal life with outdoor activities as his main recreations. His unusual physical strength and endurance enabled him to hike, play tennis, ski, and swim; although in none of these sports was he outstanding.

Fermi was a member of a great many academies and scientific societies, including the Accademia dei Lincei, the U.S. National Academy of Sciences, and the Royal Society of London. He received the Nobel Prize in 1938 and the Fermi Prize, named for him, a few days before his death. His prominent part in the development of atomic energy involved him in numerous extrascientific activities that were not particularly attractive to him. He undertook administrative duties conscientiously and ably but without great enthusiasm.

Fermi's scientific accomplishments were in both theoretical and experimental physics, an unusual feat in the twentieth century, when increasing specialization tends to narrow the field of study. Fermi's statistics (independently found also by Paul Dirac) and his theory of beta decay were his greatest theoretical contributions. Artificial radioactivity produced by neutron bombardment, slow neutrons, and the realization of a nuclear chain reaction were his greatest experimental achievements. These highlights and his many other results have left their imprint on the most diverse parts of physics.

When Fermi attended school, humanistic literary studies were emphasized: Italian, Latin, and Greek. He was a model student and obtained consistently top marks. But he was primarily self-taught, from a very early age, and he said later that by age ten he had succeeded by concentrated effort in understanding practically unaided why the equation $x^2 + y^2 = r^2$ represents a circle. His older sister, Maria, and his older brother, Giulio, contributed to this early schooling. In grade school he also exhibited a prodigious memory for poems. In 1915 Giulio died unexpectedly, and the sad event left a deep mark on Enrico. Fortunately he then struck up a friendship with a schoolmate, Enrico Persico, and the two boys' common scientific interests led to a lifelong friendship. They became the first two professors of theoretical physics in Italy.

A colleague of Fermi's father, Adolfo Amidei, was perhaps the first adult to recognize Fermi's unusual talent. He lent him books on mathematics and physics in a pedagogically graduated progression. Fermi himself acquired some secondhand books on mechanics and mathematical physics, mainly Poisson's *Mécanique* and A. Caraffa's *Elementa physicae mathematicae* (Rome, 1840). By his seventeenth year, while still in high school, he had acquired a thorough knowledge of classical physics, comparable to that of an advanced graduate student in a university. Furthermore, Fermi and Persico had performed many experiments with apparatus they had built themselves, and thus had acquired an excellent grasp of contemporary experimental physics. For example, they determined precisely the value of the acceleration of gravity at Rome, the density of Rome tap water, and the earth's magnetic field. Fermi was also proficient at building small electric motors.

At Amidei's suggestion, Fermi competed for a fellowship at the Scuola Normale Superiore in Pisa, where he could acquire an education at no expense to his family. He had to write an assigned essay for the competition, "Distinctive Characters of Sound." (Fortunately the essay was preserved and after his death was found in the archives of the school.) On the first page is the partial differential equation for vibrating reeds, followed by about twenty pages for its solution through eigenfunctions, the determination of the characteristic frequencies, etc. One can easily imagine the surprise of the examiner who received this essay from a seventeen-year-old boy just out of high school. He was convinced of the candidate's genius and assured Fermi of that fact when he met him.

From the Scuola Normale Superiore, Fermi wrote regularly to Persico, and the letters give a vivid insight into his life there. He was by no means bashful, and after about a year in Pisa he said that he was the authority on quantum theory and that everybody, including the professors, depended on him to teach them the new physics. At Pisa he became a close friend of Franco Rasetti, another physics student of great ability.

Fermi received his doctorate from the University of Pisa in 1922 and then returned to Rome. At that time he met Orso Mario Corbino, director of the physics laboratory at the University of Rome. Corbino immediately recognized Fermi's talent and became his lifelong friend and patron. Above all, he saw Fermi as the instrument for one of his fondest aspirations—the rebirth of Italian physics.

To acquire a direct knowledge of modern physics beyond the provincial state of Italian physics, Fermi had to see the world, and to be in touch with foreign scientists. He therefore competed for a foreign fellowship and spent some time at Max Born's institute at Göttingen and then with Paul Ehrenfest at Leiden. With the latter he struck up a warm friendship, and Ehrenfest greatly encouraged him. He and Arnold Sommerfeld helped to introduce Fermi to other physicists. When the fellowship expired in 1924, Fermi returned to Florence with the post of lecturer. He then competed for a chair of mathematical physics; he did not win although he was supported by Tullio Levi-Civita and Vito Volterra.

Up to this time Fermi's work had been primarily in general relativity (tensor analysis), where he had developed a theorem of permanent value: that in the vicinity of a world line, space can always be approximated by a pseudo-Euclidean metric (FP no. 3 in *Collected Papers*). In statistical mechanics he had written subtle papers on the ergodic hypothesis (FP no. 11) and on quantum theory. Here he had developed an original form of analyzing collisions of charged particles. He developed the field produced by the charged particle by the Fourier integral and used the information from optical processes to determine the result of the collision (FP no. 23). This method was later refined and better justified on the basis of quantum mechanics and is generally known as the Weizsäcker-Williams method. Other studies on the entropy constant of a perfect gas are historically important as preparation for things to come. An experiment done with Rasetti, who was also in Florence, on the depolarization of resonance light in an alternating magnetic field was the subject of Fermi's first important experimental paper (FP no. 28). This experiment was the first of a series that, in subsequent years, was to become extremely important in the hands of other physicists.

In 1925 Wolfgang Pauli discovered the exclusion principle, which in the language of the old quantum theory prevents more than one electron from occupying an orbit completely defined by its quantum numbers. This principle had far-reaching consequences in statistical mechanics for a particulate gas when its temperature and density are such that the cube of the de Broglie wavelength is large compared with the total volume divided by the number of particles in the formula $N/V \gg (2\pi kmT)^{3/2} h^{-3}$. Peculiar phenomena, comprised under the technical name of "degeneracy," then appear: for instance, the specific heat of the gas vanishes. The problems of gas degeneracy had been known for many years. Bose and Einstein had shown in 1924 that they could be solved by a modification of classical statistical mechanics (Bose-Einstein statistics). Bose-Einstein statistics are applicable to light quanta and account for Planck's radiation formula. But Bose-Einstein statistics are not applicable to particles obeying Pauli's principle, for which one needs the new type of statistics discovered by Fermi early in 1926 (FP no. 30). Dirac independently found the same result a few months later and connected it to the new quantum mechanics. Fermi statistics, which are applicable to electrons, protons, neutrons, and all particles of half integral spin, have a pervading importance in atomic and nuclear physics and in solid-state theory. The importance of Fermi statistics was immediately appreciated by physicists and established Fermi as a leader in the international community of theoreticians, as was obvious at the International Conference in Physics held at Como in 1927.

In 1927, mainly through the efforts of Corbino, a chair in theoretical physics, the first such in Italy, was established at the University of Rome. In the competition for the position Fermi placed first and Persico second. Fermi then came to Rome, to the Physics Institute in Via Panisperna, to join Corbino. He had friends there among the mathematicians also, and he was soon joined by Rasetti.

They strove to establish a modern school of physics in Rome. The first task was to recruit students suitable for advanced training and capable of later becoming independent scientists. The first was Emilio Segrè, who was then an engineering student but who had always had a strong interest in physics. When he became acquainted with Rasetti and Fermi through mutual friends, he enthusiastically joined the group as an advanced student. Segrè informed his schoolmate and friend Ettore Majorana of the new opportunity and introduced him to Fermi and Rasetti. Majorana soon (1928) transferred from engineering to physics. Edoardo Amaldi was recruited directly from undergraduate work in physics. Later they were joined by many others. Some came as temporary visitors: Giulio Racah, later rector of the University of Jerusalem and known mainly for his studies on the Racah coefficients and atomic spectra; Giovanni Gentile, Jr., later professor of theoretical physics at Milan; Gilberto Bernardini, later an experimental physicist and director of the Scuola Normale Superiore; and Bruno Rossi, who was a pioneer in the study of cosmic rays. Others joined them as students or fellows: Bruno Pontecorvo; Ugo Fano, later professor of theoretical physics at the University of Chicago; Eugenio Fubini; Renato Einaudi, later professor of mechanics at the University of Turin; and Leo Pincherle, later professor of physics at London University. Gian Carlo Wick came at a later date as an assistant professor. The activity in Rome helped to reanimate several other centers, including Florence (Rossi, Bernardini, Giuseppe Occhialini, Racah) and Turin (Gleb, Wataghin, Wick), and brought about a notable rebirth of physics in Italy.

Fermi's next important study was the application of his statistics to an atomic model. This had been anticipated, however, by L. H. Thomas, who was working independently. The Thomas-Fermi atom gives very good approximations in a great number of problems. The fundamental idea was to compute the density of the electronic cloud around the nucleus as an atmosphere of a totally degenerate gas of electrons attracted by the nucleus. Fermi made numerous applications of his method to X-ray spectroscopy, to the periodic system of the elements, to optical spectroscopy, and later to ions. This work required a considerable amount of numerical calculation, which he performed with a primitive desk calculator. Many of these results were summarized in a paper he read at the University of Leipzig in 1928 (FP no. 49). Other studies in atomic and molecular physics followed.

Another important group of papers which were

devoted to the reformulation of quantum electro-dynamics made this important subject accessible to many physicists (FP no. 67). Dirac had written a fundamental but difficult paper on the subject. Fermi, after reading the paper, decided to obtain the same results by more familiar methods. He developed by Fourier analysis the electromagnetic field which obeys Maxwell's equations, and he quantized the single harmonic components as oscillators. He thus wrote the Hamiltonian of the free field, giving a Hamiltonian form to Maxwell's theory. To this Hamiltonian, he added the Hamiltonian of an atom and a term representing the interaction between atom and radiation. The complete system was then treated by perturbation theory.

Corbino did not miss any occasion to extol Fermi's work, and early official recognition followed. Mussolini named Fermi to the newly created Accademia d'Italia. He was the only physicist so honored, and this singular recognition was also accompanied by a substantial stipend. He was subsequently elected to the Accademia dei Lincei, at an unusually young age. In spite of some grumbling by older professors, it became clear to the academic world and to the culti-vated public that Fermi was indeed the leading Ital-ian physicist. His economic position became com-fortable although not affluent.

At this time quantum mechanics had reached its full development; nonrelativistic problems, at least in principle, were soluble except for mathematical difficulties. In this sense atomic physics was showing signs of exhaustion, and one could expect the next really important advances to be in the study of the nucleus. Realizing this, Fermi decided to switch to nuclear physics. He initially investigated the theory of the hyperfine structure of the spectral lines and the nuclear magnetic momenta (FP no. 57), a suitable subject for making the transition from atomic to nuclear physics.

The development of experimental physics at Rome presented greater problems than had theoretical physics. In the latter Fermi was a leader of worldwide reputation at the peak of his powers; no substantial amount of money was needed; and it was relatively easy to attract young people from Italy and from abroad. Indeed, very early promising physicists des-tined to leave their mark in science came to Rome. Besides the Italians mentioned earlier, several foreign physicists studied there, among them Hans Bethe, Edward Teller, Rudolf Peierls, Fritz London, Felix Bloch, George Placzek, and Homi Bhabha. The state of experimental physics was different. Rasetti was the senior man, and although he had outstanding ability he was no Fermi. The only techniques known locally were spectroscopic, and therefore the equipment available was predominantly spectroscopic. Shops were poor and money was scarce. In this situation, with the object of widening the techniques available and ultimately turning to nuclear physics, Rasetti, Amaldi, and Segrè spent periods of about a year in the laboratories of Lise Meitner, Peter Debye, and Otto Stern, respectively, learning various experi-mental techniques.

In Rome, about 1929, Rasetti and Fermi began experiments on nuclear subjects. In the meantime the great discoveries of the early 1930's, the portents of the impending revolution in nuclear physics, were being made: positron, neutron, deuterium, and arti-ficial acceleration. The Solvay Conference of 1933 was devoted to the nucleus, and shortly thereafter Fermi developed the theory of beta decay, based on the hypothesis of the neutrino formulated for the first time in 1930 by Wolfgang Pauli. Beta decay—the spontaneous emission of electrons by nuclei—presented major theoretical difficulties. Apparently, energy and momentum were not conserved. There were also other difficulties with angular momentum and the statistics of the nuclei. Pauli sought a way out of the apparent paradoxes by postulating the simultaneous emission of the electron and of a practi-cally undetectable particle, later named "neutrino" by Fermi. There remained the task of giving substance to this hypothesis and of showing that it could account quantitatively for the observed facts.

An entirely new type of force had to be postulated, the so-called weak interaction. This new force, to-gether with gravity, electromagnetism, and the strong interaction which binds the particles of the nucleus, constitutes the family of forces presently known in physics. They should account for the whole universe. Weak interactions occur between all parti-cles and are thus unlike electromagnetic or strong interactions, which are restricted to certain particles. The first manifestation of the weak interaction to be treated in detail was the beta decay. The treatment was accomplished by applying second quantization and destruction and creation operators for fermions and by adopting (or better, guessing) a Hamiltonian for weak interactions on the basis of formal criteria, such as relativistic invariance, linearity, and absence of derivatives. Of the five possible choices which satisfied the formal requirements, Fermi treated the vector interaction in detail, mainly because of its analogy with electromagnetism.

In his paper on beta decay, Fermi also introduced a new fundamental constant of nature, the Fermi constant, G, which plays a role analogous to that of the charge of the electron in electromagnetism. This constant has been experimentally determined from the energy available in beta decay and the mean life of the decaying substance. Its value is 1.415×10^{-49} erg cm^3. To clarify its significance we point out that the electromagnetic interaction is of the order of 10^{12} times stronger than the weak interaction. More precisely, the dimensionless number $e^2/\hbar c = 1/137$ is to be compared with $G^2(\hbar c)^{-2}(\hbar/mc)^{-4} \sim 5.10^{-14}$, where m is the mass of the pion. The famous paper in which Fermi developed this theory had far-reaching consequences for the future development of nuclear and particle physics (FP no. 80). For instance, it served as an inspiration to Hideki Yukawa in his theory of the nuclear forces. It is probably the most important theoretical paper written by Fermi.

Soon thereafter Frédéric Joliot and Irène Joliot-Curie discovered artificial radioactivity—the creation of radioactive isotopes of stable nuclei by alpha particle bombardment. The Joliots' discovery provided the occasion for experimental activity which Fermi continued for the rest of his life. Fermi reasoned that neutrons should be more effective than alpha particles in producing radioactive elements because they are not repelled by the nuclear charge and thus have a much greater probability of entering the target nuclei.

Acting on this idea, Fermi bombarded several elements of increasing atomic numbers with neutrons. He hoped to find an artificial radioactivity produced by the neutrons. His first success was with fluorine. The neutron source was a small ampul containing beryllium metal and radon gas. The detecting apparatus consisted of rather primitive Geiger-Müller counters. Immediately thereafter Fermi, with the help of Amaldi, D'Agostino, Rasetti, and Segrè, carried out a systematic investigation of the behavior of elements throughout the periodic table. In most cases they performed chemical analysis to identify the chemical element that was the carrier of the activity. In the first survey, out of sixty-three elements investigated, thirty-seven showed an easily detectable activity. The nuclear reactions of (n, α), (n, p), and (n, γ) were then identified, and all available elements, including uranium and thorium, were irradiated. In uranium and thorium the investigators found several forms of activity after bombardment but did not recognize fission. Fermi and his collaborators, having proved that no radioactive isotopes were formed between lead and uranium, put forward the natural hypothesis that the activity was due to transuranic elements. These studies, which were continued by Otto Hahn, Lise Meitner, Irène Joliot-Curie, Frédéric Joliot, and Savitch, culminated in 1938 in the discovery of fission by Hahn and Fritz Strassmann.

In October 1935 Fermi and his collaborators, now including Pontecorvo, observed that neutrons passed through substances containing hydrogen have in-creased efficiency for producing artificial radioactivity. Fermi interpreted this effect as due to the slowing down of the neutrons by elastic collisions with hydrogen atoms. Thus slow neutrons were discovered. The study of slow neutrons was to form the main object of Fermi's work for several years thereafter. Among other things, Fermi and his collaborators showed that the neutrons reached thermal energy and that neutrons of a few electron volts of energy could show sharp peaks (resonances) in the curve of the collision and absorption cross section, versus neutron energy. Fermi then developed a mathematical theory of the slowing down of neutrons, and he tested it experimentally in considerable detail. This work lasted until about 1936. All the neutron work, which cost approximately a thousand American dollars, was supported by the Consiglio Nazionale delle Ricerche of Italy. The tremendous experimental activity of the years 1934–1938 brought a considerable change in the working habits at the Rome Institute. Because of the lack of time, it became impossible for Fermi to follow all the developments in physics as he had done before. He was forced to curtail the extracurricular teaching of promising young men, nor could he spare time for foreign visitors, who practically stopped coming.

The Ethiopian War marked the beginning of the decline of the work at the Institute, and the death of Corbino on 23 January 1937 brought further serious complications. The deteriorating political situation also materially hampered the work, and finally the Fascist racial laws of 1938 directly affected Fermi's wife. The foregoing problems and his deep, although mute, resentment against injustice were the final arguments that convinced Fermi to leave Fascist Italy. He passed the word to Columbia University, where he had been previously, that he was willing to accept a position there. In December 1938 he received the Nobel Prize in Stockholm. He then proceeded directly from there to New York. He was not to return to Italy until 1949.

Fermi had barely settled himself at Columbia when Bohr brought to the United States the news of the discovery of fission. This discovery made a tremendous impression on all physicists. Fermi and others immediately saw the possibility of the emission of secondary neutrons and perhaps of a chain reaction; he started at once to experiment in this direction.

In the early period at Columbia, Fermi was helped by H. L. Anderson, a graduate student who later took his Ph.D. under Fermi and remained a close collaborator and friend to the end of Fermi's life. The young physicist Walter Zinn was also associated with Fermi for an extended period. Leo Szilard was independently pursuing similar studies, and there were active interchanges of ideas and even some collaboration with other Columbia and Princeton University groups during early research on the chain reaction.

The first problem was to investigate whether on the fission of uranium secondary neutrons were in fact emitted—as was expected because the fragments have excess neutrons for their stability. If such did occur, it might be possible to use these neutrons to produce further fission, and under favorable circumstances one could obtain a chain reaction. To make this possible it is necessary to use the fission neutrons economically, i.e., to employ the neutrons to produce other fissions and not to lose them in parasitic captures by uranium, by other materials used as a moderator to slow down the neutrons, or by escape from the body of the reactor. If one uses natural uranium and a graphite moderator with unseparated isotopes, the margin by which one can obtain a chain reaction is very small and utmost care is needed in husbanding the neutrons. It soon became apparent that more than two neutrons were emitted per fission. This is a necessary but not sufficient condition for a chain reaction. But the number, now known to be about 2.5, is small enough to create an extremely difficult technical problem.

Of the two isotopes contained in natural uranium, only the isotope of mass 235, present in one part out of 140, is fissionable by slow neutrons and the cross section is large at low energy. On the other hand, most of the fission neutrons are unable to produce fission in the abundant isotope uranium 238, but if they are slowed down they are easily captured by U^{235} and produce fission. Neutrons must therefore be slowed down, but in the collisions that reduce the energy of the neutrons there is always a fraction of neutrons which are captured without producing fission. The moderator must thus be carefully chosen. Hydrogen, the first obvious choice, captures too many neutrons, and deuterium was unavailable in sufficient quantities. So for practical purposes the only suitable, and available, substance in 1939 was graphite, and several physicists independently suggested its use. In a long series of measurements of great ingenuity, Fermi and his collaborators studied the purity of the materials (impurities were often important neutron capturers) and the best configuration in which to assemble them. In order to analyze the problems facing him, Fermi needed a great amount of quantitative information on cross sections, delayed neutrons, branching ratios of the fission reactions, and nuclear properties of several nuclei to be used in a future reactor. This information was not available. He then proceeded to collect it with the help of many collabo-

rators. Other independent groups were of course working on the same problems, but exchange of information was limited by self-imposed secrecy.

The potential overwhelming practical importance of this work was clear to physicists, and Fermi, together with George B. Pegram, chairman of the physics department at Columbia and a close personal friend, tried to alert the U.S. government to the implications of the recent discoveries. A small subsidy for further research was obtained from the U.S. Navy, and the studies that were to culminate in the atomic bomb were initiated. At the beginning the staff and equipment were completely inadequate. Perhaps Fermi thought he might be able to repeat, on a somewhat larger scale, work similar to the neutron research in Rome. He certainly did not realize, as very few scientists did, the project's colossal requirements of manpower and means for its successful completion. Fermi was always reluctant to take administrative responsibilities and he concentrated his efforts on the scientific side, leaving to others the staggering problems of organization and procurement. As an expert of exceptional ability and great authority, he naturally helped; but his activity was directed primarily to the scientific aspects of the problems. It must also be remembered that his position—first as an alien, and later, after the United States entered World War II, as an enemy alien—rendered his situation difficult.

Fermi concentrated his efforts on obtaining a chain reaction using ordinary uranium of normal isotopic composition. As soon as it was established that of the two isotopes present in natural uranium, only U^{235} is fissionable by slow neutrons, it became apparent that if one could obtain pure U^{235} or even enrich the mixture in U^{235}, the making of a reactor or possibly even of an atom bomb would be comparatively easy. Still, the isotope separation was such a staggering task that it discouraged most physicists. By the end of the war even this task had been mastered, and isotope separation was soon a normal industrial operation.

In 1939 and 1940, however, isotope separation was very uncertain and other avenues had to be explored. In December 1940, Fermi and Segrè discussed another possibility: the use of the still undiscovered element 94 (plutonium) of mass 239 (Pu^{239}). This substance promised to undergo slow neutron fission and thus to be a replacement for U^{235}. If it could be produced by neutron capture of U^{238} in a natural-uranium reactor, followed by two beta emissions, one could separate it chemically and obtain a pure isotope with, it was hoped, a large slow-neutron cross section. Similar ideas had independently occurred in England and Germany. J. W. Kennedy, Glenn Seaborg, Segrè, and Arthur Wahl undertook the preparation and measurement of the nuclear properties of Pu^{239}, using the Berkeley cyclotron. The favorable results of these experiments (January–April 1941) added impetus to the chain-reaction project because it opened another avenue for the realization of a nuclear bomb. By December 1941, the whole world was engulfed in war, and military applications were paramount. The United States developed, under government supervision, an immense organization, which evolved according to the technical necessities and led to the establishment of the Manhattan Engineer District (MED). The purpose of the MED was to make an atomic bomb in time to influence the course of the war. The history of this development is admirably recounted in the Smyth report. Fermi had a technically prominent part in the whole project. His work at Columbia was still on a small scale, but in 1942 he transferred to Chicago, where it was expanded. It culminated on 2 December 1942 with the first controlled nuclear chain reaction at Stagg Field at the University of Chicago.

The industrial and military developments of the release of nuclear energy, which are of immense importance, will not be treated here. The nuclear reactor, however, is also a scientific instrument of great capabilities, and these were immediately manifest to Fermi. Even during the war, under the extreme pressure of the times, he took advantage of these capabilities to begin research on neutron diffraction, neutron reflection and polarization, measurements of scattering lengths, etc. These investigations, developed later by other physicists, opened up whole new areas of a science sometimes called neutronology, i.e., application of neutronic methods to solid state and various other branches of physics.

When his work at Chicago was finished, Fermi went to Los Alamos, New Mexico, where the Los Alamos Laboratory of the Manhattan Engineer District, under the direction of J. R. Oppenheimer, had the assignment of assembling an atomic bomb. Fermi spent most of the period from September 1944 to early 1946 at Los Alamos, where he served as a general consultant. He also collaborated in the building of a small chain reactor using enriched uranium in U^{235} and heavy water. Fermi actively participated in the first test of the atomic bomb in the desert near Alamogordo, New Mexico, on 16 July 1945.

Following the successful test of the bomb, he was appointed by President Truman to the interim committee charged with advising the president on the use of the bomb and on many fundamental policies concerning atomic energy.

In 1946 the University of Chicago created the Institute for Nuclear Studies and offered a professorship to Fermi. The Institute was promising in its financing and organization; and Fermi, although very influential in its direction, would be spared administrative duties. The offer proved attractive to Fermi, and early in 1946 he and his family left Los Alamos for Chicago. He remained at the University of Chicago for the rest of his life.

The new institute had Samuel K. Allison as director and a faculty in which the new generation of physicists who had been active in the Manhattan Project was strongly represented: Herbert Anderson, Maria Goeppert Mayer, Edward Teller, and Harold C. Urey were among the first members. At Chicago, Fermi rapidly formed a school of graduate students whom he instructed personally, in a fashion reminiscent of his earlier days in Rome. Among those who later distinguished themselves as physicists were Richard L. Garwin, Murray Gell-Mann, Geoffrey Chew, Owen Chamberlain, Marvin L. Goldberger, Leona Marshall, Darragh E. Nagle, T. D. Lee, and C. N. Yang. Chicago thus became an extremely active center in many different areas of physics.

Fermi himself had concluded, at the end of the war, that nuclear physics was reaching a stage of maturity and that the future fundamental developments would be in the study of elementary particles. He thus prepared himself for this new field by learning as much as possible of the theory and by fostering the building of suitable accelerators with which to perform experiments. We have a hint of his effort to assimilate the theory in his Silliman lectures at Yale in the spring of 1950, which were published as *Elementary Particles* (New Haven, 1951). He systematically organized a great number of calculations on all subjects, numerical data, important reprints, etc., which he called the "artificial memory." This material was a daily working tool for him and substituted for books, which he scarcely used any more. It also helped his memory, which, although still excellent, was not as amazing as in his early youth and could not cope with the avalanche of new results.

During the postwar Chicago years, Fermi traveled a good deal, particularly to research centers, where he could meet young, active physicists. He repeatedly visited the Brookhaven National Laboratory, the Radiation Laboratory in Berkeley, the Los Alamos Laboratory, and many universities. He was welcomed everywhere, especially by the younger men who profited from these contacts with him and, in turn, helped Fermi to preserve his youthful spirit. He attended all the Rochester conferences on high-energy physics and taught in several summer schools. In 1949 he revisited Italy, where he was very well received by his former colleagues and by the new generation of physicists who had heard of him as an almost legendary figure.

As soon as the Chicago cyclotron was ready for operation, Fermi again started experimental work on pion-nucleon scattering. (He had coined the word "pion" to indicate pi-mesons.) He found experimentally the resonance in the isotopic spin 3/2, ordinary spin 3/2 state, which had been predicted by Keith Bruckner. The investigation became a major one. With H. L. Anderson and others, Fermi worked out the details up to an energy of about 400 MeV lab of the nucleon-pion interaction. The methods and techniques employed, including the extensive use of computers, were for many years models for the subsequent host of investigators of particle resonances.

In addition to this experimental activity, Fermi did theoretical work on the origin of cosmic rays, devising a mechanism of acceleration by which each proton tends to equipartition of energy with a whole galaxy. These ideas had an important influence on the subsequent studies on cosmic rays. He also developed a statistical method for treating high-energy collision phenomena and multiple production of particles. This method has also received wide and useful applications.

In 1954 Fermi's health began to deteriorate, but with great will power he carried on almost as usual. He spent the summer in Europe, where he taught at summer schools in Italy and France, but on his return to Chicago in September he was hospitalized. An exploratory operation revealed an incurable stomach cancer. Fully aware of the seriousness of his illness and his impending death, he nevertheless maintained his remarkable equanimity and self-control. He died in November and was buried in Chicago.

It is too early to give a historically valid assessment of Fermi's place in the history of physics. He was the only physicist in the twentieth century who excelled in both theory and experiment, and he was one of the most versatile. His greatest accomplishments are (chronologically) the statistics of particles obeying the exclusion principle, the application of these statistics to the Thomas-Fermi atom, the recasting of quantum electrodynamics, the theory of beta decay, the experimental study of artificial radioactivity produced by neutron bombardment and the connected discovery of slow neutrons and their phenomenology, the experimental realization of a nuclear chain reaction, and the experimental study of pion-nucleon collision. In addition there are Fermi's innumerable, apparently isolated contributions to atomic, molecular, nuclear, and particle physics, cosmic rays, relativity, etc., many of which initiated whole new chapters of physics.

BIBLIOGRAPHY

Fermi's *Collected Papers,* E. Segrè, E. Amaldi, H. L. Anderson, E. Persico, F. Rasetti, C. S. Smith, and A. Wattenberg, eds., 2 vols. (Chicago, 1962–1965), contains most of Fermi's papers and a complete bibliography, a biographical introduction by E. Segrè, introductions to the various papers by members of the editorial committee, a chronology of Fermi's life, and subsidiary material.

Secondary literature includes Laura Fermi, *Atoms in the Family* (Chicago, 1954), a biography by Fermi's wife emphasizing the human aspects of their life; Emilio Segrè, *Enrico Fermi, Physicist* (Chicago, 1970), a scientific biography; and H. D. Smyth, *Atomic Energy for Military Purposes* (Princeton, 1945), which gives an excellent account of the history of the development of atomic energy up to 1945.

See also R. G. Hewlett and O. E. Anderson, Jr., *The New World* (University Park, Pa., 1962); R. G. Hewlett and F. Duncan, *Atomic Shield* (University Park, Pa., 1969); and *Review of Modern Physics,* 27 (1955), 249–275, which contains the memorial symposium in honor of Fermi held at Washington, D.C. (Apr. 1955).

EMILIO SEGRÈ

FERNALD, MERRITT LYNDON (*b.* Orono, Maine, 5 October 1873; *d.* Cambridge, Massachusetts, 22 September 1950), *botany.*

Fernald achieved a complete revision in 1950 of Asa Gray's *Manual of the Botany of the Northern United States* (1908), the most critical comprehensive floristic work ever published for any part of North America, and propounded the theory of persistence of plants on nunataks—"the largest single contribution to the science of phytogeography since the time of Darwin" (Merrill, p. 53). The son of Merritt Caldwell Fernald, president of Maine State College of Agriculture and Mechanic Arts (later the University of Maine), and Mary Lovejoy Heyward Fernald, he published his first botanical paper at seventeen. About 830 titles were to follow, chiefly concerning the identities, accurate definitions, and verified distributions of plants of the northeastern United States. Fernald's taxonomic papers were carefully prepared and provocative, although sometimes they were more commentary than conclusion. His approach was to trace types back, often to pre-Linnaean botanists, then to search for clarifying evidence in the field. His masterly acquaintance with botanical literature led him along old paths to fresh decisions.

A short, stout man, Fernald was tireless in the field, boyishly joyous, given to punning, and optimistic throughout his life. His tremendous industry and total absorption with systematic botany were the mainsprings of his success. A "mere grind" was his own appraisal, but his friend Ludlow Griscom called him a "one-pointed, one-sided botanical machine." On the invitation of Sereno Watson, Fernald had become an assistant in the Gray Herbarium early in 1891 and enrolled that fall in Harvard's Lawrence Scientific School. He graduated in 1897 with the B.S. degree, *magna cum laude,* his only earned degree.

Fernald wrote monographs on such genera as *Potamogeton* and *Draba,* which in turn led to his classic paper "Persistence of Plants in Unglaciated Areas of Boreal America" (1925), a documented rebuttal to the generally held view that a moving ice sheet had annihilated all the plants and animals before it. His "nunatak theory" excited debate among geologists and biologists. It stands, somewhat sculptured, like Botanist's Dome of the Gaspé Peninsula, a landmark of plant geography.

"His trenchant criticism . . . [assisted] in maintaining the standards of American botanical scholarship" (Merrill, p. 54). His humor enhances the descriptions and recipes of *Edible Wild Plants of Eastern North America* (1943), which he wrote with A. C. Kinsey.

Fernald married Margaret Howard Grant of Providence, Rhode Island, on 5 April 1907, and a son and two daughters were born to them. His association with Harvard spanned nearly sixty years. He was unforgettable to his students, and his work was avidly followed by readers of *Rhodora,* which he edited for thirty-two years. Fernald was acknowledged doyen in the study of the flora of the eastern United States. "When he was formed," wrote Merrill (p. 61), "the mold was destroyed; there never can be another Fernald."

BIBLIOGRAPHY

I. Original Works. A bibliography of Fernald's publications, by Katherine Fernald Lohnes and Lazella Schwarten, forms an appendix to Merrill's sketch (see below). Among his writings are "Persistence of Plants in Unglaciated Areas of Boreal America," in *Memoirs of the American Academy of Arts and Sciences,* **15** (1925), 239–342; and *Edible Wild Plants of Eastern North America* (Cornwall, N.Y., 1943), written with A. C. Kinsey, revised ed. by Reed C. Rollins (New York, 1958).

II. Secondary Literature. The fullest, and an eminently fair, appraisal of Fernald is by Elmer D. Merrill, in *Biographical Memoirs. National Academy of Sciences,* **28** (1954), 45–98. Other sketches are Arthur Stanley Pease, in *Rhodora,* **53** (1951), 33–39; John M. Fogg, Jr., *ibid.,* 39–43; Harley Harris Bartlett, *ibid.,* 44–55; Reed C. Rollins, *ibid.,* 55–61; and Ludlow Griscom, *ibid.,* 61–65. Rollins published a shortened version of his appraisal in *Bulletin of the Torrey Botanical Club,* **78** (1951), 270–272. A few salient comments appear in Una F. Weatherby, *Charles Alfred Weatherby* (Cambridge, Mass., 1951), pp. 128, 144, 178.

JOSEPH EWAN

FERNEL, JEAN FRANÇOIS (*b.* Montdidier, France, 1497 [?]; *d.* Fontainebleau, France, 26 April 1558), *medicine.*

Fernel's year of birth was probably 1497, according to Sherrington's scrutiny of the various reports available. The son of a well-to-do innkeeper at Montdidier, he was twelve years old when the family moved to Clermont, twenty miles from Paris. In his writings Fernel calls himself "Ambianus," apparently because Montdidier was within the diocese of Amiens.

After schooling at Clermont, Fernel went to the Collège de Ste. Barbe in Paris (1519) and, at the age of twenty-two, took his M.A. degree. For the next five years he was virtually a recluse, feeling that it was necessary to improve his mind and extend his knowledge, particularly in philosophy, astronomy, and mathematics. These studies were interrupted in 1524 by a serious illness ("quartan fever") that forced him to go to the country for a period of convalescence. After that time Fernel's father ceased to support his studies because of his duties to his other children. Obliged to support himself in Paris, Fernel lectured on philosophy and began studying medicine, apparently halfheartedly at first. In 1527 he published his first book, *Monalosphaerium,* which was followed in 1528 by *Cosmotheoria,* both of them mathematical and astronomical. At the time, astrology occupied an important position in mathematics and astronomy; the *Cosmotheoria,* however, contained measurements made by Fernel—his estimate of a degree of meridian was good enough to be in close agreement with that of Jean Picard 140 years later and thus was an important contribution to geophysics.

In the meantime Fernel had married, and he was now severely criticized by his father-in-law, a senator of Paris, for neglecting his medical studies and his duties as head of a family in favor of these unprofitable interests. The young man had done well as a teacher of philosophy at his college and in astronomy. He had also acquired a collection of instruments, among them an astrolabe of his own design for finding the hour and for measuring time, but he was now compelled to sell these instruments and to take his medical studies seriously. These were completed in 1530, when he obtained his *venia practicandi.*

Within six years Fernel became one of the most famous physicians in France. Students flocked to his lectures, and "from his School there went forth skilled physicians more numerous than soldiers from the Trojan horse, and spread over all regions and quarters of Europe" (Plancy, in C. S. Sherrington, *Endeavour of Jean Fernel*). His reputation at the court of the dauphin (later Henry II) became firmly established when he saved the life of Henry's mistress, Diane de

Poitiers. The prince wanted to keep him at Fontaine-bleau as court physician, but Fernel begged "in all charity" to be allowed to return to Paris, to his books, his students, and his patients. Fernel was less successful with Francis I, Henry's father, who died in 1547. He had treated the king for syphilis with a decoction of his own, although the established cure at that time was treatment with mercury. Fernel had criticized this method and later wrote a book on his cure of syphilis. Popular though he was at the court and in the city, he had many enemies at the university; he was, however, too powerful to be suppressed. In 1534 he was appointed professor of medicine.

In 1536, while teaching medicine at the Collège de Cornouailles, Fernel began writing his *De naturali parte medicinae* (1542), addressed to medical subjects that he later named "physiology," thus introducing this term for the science of the functions of the body. The new title was destined to remain, and the book was read for a century, until Harvey's discovery of the circulation of the blood (1628) gave physiology its present experimental direction. Fernel's physiology was still the humoral medicine of his time. It did not discuss respiration, circulation, digestion, and such; the six chapters following that on anatomy concern the elements, the temperaments, the spirits, the innate heat, the faculties, the humors, and the procreation of man. The spirit is said to enter the fetus on the fortieth day of pregnancy; the substance of the soul and its faculties are hidden from us, and therefore we must treat its instruments as "immediate causes" in studying the body.

In spite of his orthodox Galenic physiology Fernel had something new to offer his contemporaries. The medicine of that time acknowledged the influence of magic and sorcery on the origin and development of disease, and people of sufficient means employed private astrologers. Fernel, who had believed in astrology, of which there was still a trace in his *De abditis rerum causis* (1548, but begun before his "Physiologiae"), gradually came to the view that the "whole book of healing was nothing other than a copy of inviolable laws observable in Nature," as formulated in his unfinished last work. His first biographer, Guillaume Plancy (1514–1568), explained Fernel's change of attitude toward contemporary medicine by his respect for facts. Fernel was an observer who emphasized the value of practice and experience; and the astrological predictions did not agree with the lessons of these masters. In the end he utterly condemned astrology. To the young he must have seemed a reformer; to his Scholastically trained colleagues at the University of Paris, a nonsensical if not a dangerous heretic.

With his observant mind, breadth of knowledge, and new attitude toward his profession, Fernel was a man of the Renaissance, which was well under way both at the court of Francis I and among educated citizens, scholars, architects, and painters of Paris. The university, which had remained the stronghold of the old type of scholarship, conservative and against innovation, did not honor its great son until long after his death.

Fernel's *De abditis rerum causis* is written as a dialogue among three characters: Brutus, a cultured man of the sixteenth century; Philiatros, whose name denotes a senior candidate for the doctor's degree; and Eudoxus, a physician older than his two friends and speaking with the voice of Fernel himself. It is an exposition of the beliefs of the educated citizen of that period, what he thinks about God, nature, the soul, matter, medicine, the preternatural, etc., as well as a plea for observation and common sense in the experienced world of nature, but it also admits the existence of a world of incorporeal beings between earth and heaven. "God" may have meant the Supreme Being, but the other words had different meanings. Matter, for instance, was substance composed of the traditional four elements; the soul was the principle of life and mind and had come from the stars. There were three kinds of soul, as Aristotle had taught: the soul of plants, which was nutritive and reproductive; the soul of animals, which was sentient and vegetative; and the soul of man, which partaking of these qualities incorporates reason also with them in a unified way. This book had great appeal for the educated citizens in European cultural centers and went through at least thirty editions. Yet today it seems of less importance than Fernel's contributions to medicine and astronomy (geophysics).

Fernel worked tirelessly to complete his textbook *Medicina,* first published in 1554. His future biographer, Plancy, who lived in his household from 1548 until Fernel's death, tells of the struggle of his last years, torn as he was between a great practice, the writing of his books, and, from 1556, the service of Henry II as physician to the court after the death of his substitute, Maître de Bourges. Fernel was then about sixty and counted on a measure of peace at Fontainebleau. Wars with Spain and England interfered with this expectation. He was compelled to follow the king to the battlefield, all the while trying to write his *Febrium curandarum methodus generalis* ("Treatment of Fevers"). He witnessed the capture of Calais, which the English had held for some two hundred years, then finally settled at the court in Fontainebleau, bringing his wife with him. She died a few months later. This was a severe shock, and he was soon taken ill and died, in spite of the ministrations of all the other physicians at the court.

On his deathbed Fernel was greatly worried that he had found no time to put the finishing touches to his *Medicina*. It fell upon Plancy to edit the full text of the *Universa medicina* (1567), which contained chapters on physiology, pathology, therapeutics, and such. Fernel's latest biographer, Sir Charles Sherrington, has raised the question of whether there were any original observations of value in Fernel's *Medicina*. The important contribution was undoubtedly the "Physiologiae," in which he had noted peristalsis and the systole and diastole of the heart; he did not, however, realize that the veins and arteries were connected by capillaries. Also of interest is his notion that the veins hinder clotting. Fernel's anatomical observations, among them the earliest description of the spinal canal, were good and clearly presented, before or simultaneous with Vesalius' *De humani corporis fabrica* (1543), the shadow of which may well have lain too heavily over significant contributions from contemporaries and predecessors. In medicine Fernel gave early descriptions of appendicitis and endocarditis. His ranking in the history of medicine, however, rests mainly upon his role as a reformer fighting to replace magic, sorcery, and astrology with observations at the sickbed.

BIBLIOGRAPHY

I. ORIGINAL WORKS. Some of Fernel's books are *Monalosphaerium* (Paris, 1527); *Cosmotheoria* (Paris, dated 1527 but apparently not issued until March 1528); *De abditis rerum causis* (Paris, 1548); *Medicina* (Paris, 1554), of which the first seven chapters, called collectively "Physiologiae," represent a reedited version of *De naturali parte medicinae,* also trans. into French as *Les VII livres de la physiologie* (Paris, 1655); *Universa medicina,* Guillaume Plancy, ed. (Paris, 1567), which also includes Plancy's *Vita Fernelii; Febrium curandarum methodus generalis* (Frankfurt am Main, 1577); and *De luis venerae perfectissima cura liber* (Antwerp, 1579).

II. SECONDARY LITERATURE. A scholarly appraisal of Fernel's work, together with an English trans. of Plancy's *Vita Fernelii,* is in C. S. Sherrington, *Endeavour of Jean Fernel* (Cambridge, 1946).

RAGNAR GRANIT

FERRARI, LUDOVICO (*b.* Bologna, Italy, 2 February 1522; *d.* Bologna, October 1565), *algebra.*

Little is known of Ferrari's life. His father, Alessandro, was the son of a Milanese refugee who had settled in Bologna. Following his father's death Ferrari went to live with his uncle Vincenzo. In November 1536 he was sent to Milan by his uncle to join the household of Girolamo Cardano, replacing his uncle's son Luca, who was already in Cardano's service. Although he had not received a formal education, Ferrari was exceptionally intelligent. Cardano therefore instructed him in Latin, Greek, and mathematics and employed him as amanuensis. In Cardano's autobiography, written many years later, Ferrari is described as having "excelled as a youth all my pupils by the high degree of his learning" (*De vita propria liber* [1643], p. 156).

In 1540 Ferrari was appointed public lecturer in mathematics in Milan, and shortly afterward he defeated Zuanne da Coi, a mathematician of Brescia, at a public disputation. He also collaborated with Cardano in researches on the cubic and quartic equations, the results of which were published in the *Ars magna* (1545). The publication of this book was the cause of the celebrated feud between Ferrari and Niccolò Tartaglia of Brescia, author of *Quesiti et inventioni diverse* (1546). In the wake of the resulting public disputation, Ferrari received offers of employment from many persons of importance, including Emperor Charles V, who wanted a tutor for his son, and Ercole Gonzaga, cardinal of Mantua. He accepted Gonzaga's offer and, at the request of the cardinal's brother, Ferrante, then governor of Milan, he carried out a survey of that province. After this he was in the cardinal's service for some eight years. On his retirement because of ill health Ferrari went to Bologna to live with his sister. From September 1564 until his death in October 1565, he held the post of *lector ad mathematicam* at the University of Bologna.

When Ferrari went to live with Cardano, the latter was earning his livelihood by teaching mathematics. Although Cardano was a qualified physician, he had not yet been accepted by the College of Physicians and was then preparing his first works on medicine and mathematics for publication. It is likely that Ferrari was introduced to mathematics through Cardano's *Practica arithmetice* (1539). While this work was in preparation, news reached Cardano that a method of solving the cubic equation of the form $x^3 + ax = b$, where a and b are positive, was known to Niccolò Tartaglia of Brescia. Until then Cardano had accepted Luca Pacioli's statement in the *Summa de arithmetica, geometria, proportioni et proportionalita* (1494) that the cubic equation could not be solved algebraically. On learning that Tartaglia had solved the equation in the course of a disputation with Antonio Maria Fiore in 1535, Cardano probably tried to find the solution himself, but without success. In 1539, before his book was published, he asked Tartaglia for the solution, offering to include it in his forthcoming book under Tartaglia's name. Tartaglia refused, on the ground that he wished to publish his discovery himself. But when he visited Cardano in Milan in March 1539, he gave him the solution on the solemn promise that it would be kept secret. In 1542, however, Cardano learned that the cubic equation had been solved several years before Tartaglia by Scipione Ferro, *lector ad mathematicam* at the University of Bologna from 1496 to 1526. During a visit to Bologna, Cardano and Ferrari were shown Ferro's work, in manuscript, by his pupil and successor Annibale dalla Nave. After this Cardano did not feel obliged to keep his promise.

Having learned the method of solving one type of cubic equation, Cardano and Ferrari were encouraged to extend their researches to other types of cubics and to the quartic. Ferrari found geometrical demonstrations for Cardano's formulas for solving $x^3 + ax = bx^2 + c$ and $x^3 + ax^2 = b$; he also solved the quartic of the form $x^4 + ax^2 + b = cx$ where a, b, c, are positive. The results were embodied in Cardano's *Ars magna* (1545). In it he attributed the discovery of the method of solving the equation $x^3 + ax = b$ to Scipione Ferro and its rediscovery to Tartaglia. That this apparent breach of secrecy angered Tartaglia is evident from book IX of his *Quesiti et inventioni diverse* (1546), where he recounted the circumstances in which he had made his discovery and Cardano's attempts to obtain the solution from him. He also gave a verbatim account of the conversation at their meeting in Milan, along with his comments.

Ferrari, loyal to his master and impetuous by nature, reacted quickly. In February 1547 he wrote to Tartaglia, protesting that the latter had unjustly and falsely made statements prejudicial to Cardano. Having criticized the mathematical content of Tartaglia's work and accused him of repetition and plagiarism, Ferrari challenged him to a public disputation in geometry, arithmetic, and related disciplines. Scholarly disputations, common in those days, were often the means of testing the professional ability of the participants. Since both Ferrari and Tartaglia were engaged in the public teaching of mathematics, a disputation was a serious matter. In his reply Tartaglia, while insisting that Cardano had not kept his promise, said that he had used injurious words in order to provoke Cardano to write to him. He asked Ferrari to leave Cardano to fight his own battles; otherwise, Ferrari should admit that he was writing at Cardano's instigation. Saying that he would accept the challenge if Cardano at least countersigned Ferrari's letter, Tartaglia went on to raise objections to the conditions of the proposed disputation—the subjects, the location, the amount of caution money to be deposited, and the judges.

Twelve letters were exchanged, full of charges and insults, each party trying to justify his position. Tartaglia maintained that Cardano had broken his promise and that Ferrari was writing at Cardano's instance. Ferrari asserted that the solution of the cubic equation was known to both Scipione Ferro and Antonio Maria Fiore long before Tartaglia had discovered it and that it was magnanimous of Cardano to mention Tartaglia in the *Ars magna*. He also denied that he was writing on Cardano's behalf. In the course of this correspondence each party issued a series of thirty-one problems for the other to solve. Tartaglia sent his problems in a letter dated 21 April 1547. The problems were no more difficult than those found in Pacioli's *Summa*. On 24 May 1547 Ferrari replied with thirty-one problems of his own but did not send the solutions to those set by Tartaglia. In his reply (July 1547) Tartaglia sent the solutions to twenty-six of Ferrari's problems, leaving out those which led to cubic equations; a month later he gave his reasons for not solving these five problems. In a letter dated October 1547 Ferrari replied, criticizing Tartaglia's solutions and giving his solutions to the problems set by the latter. Tartaglia, replying in June 1548, said he had not received Ferrari's letter until January and that he was willing to go to Milan to take part in the disputation. In July 1548 both parties confirmed their acceptance.

There is no record of what happened at the meeting except for scattered references in Tartaglia's *General trattato di numeri, et misure* (1556–1560). The parties met on 10 August 1548 in the church of Santa Maria del Giardino dei Minori Osservanti in the presence of a distinguished gathering that included Ferrante Gonzaga, governor of Milan, who had been named judge. Tartaglia says that he was not given a chance to state his case properly. Arguments over a problem of Ferrari's that Tartaglia had been unable to resolve lasted until suppertime, and everyone was obliged to leave. Tartaglia departed the next day for Brescia, and Ferrari was probably declared the winner.

Ferrari's method of solving the quartic equation $x^4 + ax^2 + b = cx$ was set out by Cardano in the *Ars magna*. It consists of reducing the equation to a cubic. The discovery was made in the course of solving a problem given to Cardano by Zuanne da Coi: "Divide 10 into three proportional parts so that the product of the first and second is 6." If the mean is x, it follows that $x^4 + 6x^2 + 36 = 60x$, or $(x^2 + 6)^2 = 60x + 6x^2$. This last equation can be put in the form

$$(x^2 + 6 + y)^2 = 6x^2 + 60x + y^2 + 12y + 2yx^2$$
or
$$(x^2 + 6 + y)^2 = (2y + 6)x^2 + 60x + (y^2 + 12y),$$

where y is a new unknown. If y is chosen so that the

right-hand side of the equation is a perfect square, then y satisfies the condition

$$60^2 = 4(2y + 6)(y^2 + 12y),$$

which can be reduced to the cubic equation

$$y^3 + 15y^2 + 36y = 450.$$

That Ferrari's method of solution is applicable to all cases of the quartic equation was shown by Rafael Bombelli in his *Algebra* (1572).

BIBLIOGRAPHY

I. ORIGINAL WORKS. The letters exchanged by Ferrari and Tartaglia were printed, and copies were sent to several persons of influence in Italy. (A complete set of these letters is in the Department of Printed Books of the British Museum.) They have been published by Enrico Giordani in *I sei cartelli di matematica disfida, primamente intorno alla generale risoluzione delle equazioni cubiche, di Lodovico Ferrari, coi sei contro-cartelli in risposta di Nicolò Tartaglia, comprendenti le soluzioni de' quesiti dall'una e dall'altra parte proposti* (Milan, 1876). Ferrari's work on the cubic and quartic equations is described in Cardano's *Artis magnae, sive de regulis algebraicis* (Nuremberg, 1545).

II. SECONDARY LITERATURE. Cardano wrote a short biography of Ferrari, "Vita Ludovici Ferrarii Bononiensis," in his *Opera omnia* (Lyons, 1663), IX, 568–569. References to Ferrari in Cardano's other works are cited in J. H. Morley, *Life of Girolamo Cardano of Milan, Physician* (London, 1854), I, 148–149, 187. The history of mathematics in sixteenth-century Italy is outlined in Ettore Bortolotti, *Storia della matematica nella Università di Bologna* (Bologna, 1947), pp. 35–80. Arnaldo Masotti, "Sui cartelli di matematica disfida scambiati fra Lodovico Ferrari e Niccolò Tartaglia," in *Rendiconti dell'Istituto lombardo di scienze e lettere,* **94** (1960), 31–41, cites the important secondary literature on Ferrari.

S. A. JAYAWARDENE

FERRARIS, GALILEO (*b.* Livorno Vercellese, Italy, 31 October 1847; *d.* Turin, Italy, 7 February 1897), *electrical engineering, physics.*

One of four children of a pharmacist, Ferraris became one of the prime electrical innovators of the 1880's. At age ten he went to live in Turin with a physician uncle, who guided the boy's education in the sciences and classics. He subsequently spent three years at the University of Turin and two years at the Scuola d'Applicazione di Torino, graduating in 1869 with the title engineer. His doctoral thesis at the university in 1872 was *Teoria matematica della propagazione dell'elettricità nei solidi omogenei.*

Ferraris then taught technical physics at the Regio Museo Industriale in Turin and also investigated light waves and the optical characteristics of telescopes, especially the phase difference of two waves in sinusoidal motion. This led to the concept of phase-displaced electrical waves and a rotating electromagnetic field. Further studies in polyphonic acoustics and interference in telephone circuits sharpened Ferraris' grasp of coacting forces in and out of phase. His continued interest in optics resulted in the publication, in 1876, of the *Proprietà degli strumenti diottrici.* . . .

Ferraris represented the Italian government on the awards jury at the 1881 International Electricity Exposition at Paris, where he learned of the Deprez system of high-voltage alternating current transmission and low-voltage distribution. He was also a delegate to the Paris conference of 1882 to determine standard electrical units, and in 1883 he was his country's delegate to the electrical exposition in Vienna. These duties prepared Ferraris for his service in 1883 as president of the international section of the Electricity Exposition at Turin, where he saw the Gaulard-Gibbs transformer. A paper on the transformer, presented before the Academy of Science at Turin in 1885, led to an intensive study of the interlocking relationships of electrical and magnetic forces in the primary and secondary circuits of the transformer system, and he drew heavily on the optical analogy of light polarized elliptically and circularly.

Carrying the notion further, Ferraris visualized the placing of two electromagnets, each fed by a current displaced 90° out of phase, at right angles to each other, thereby producing the equivalent of a revolving magnetic field. This could induce currents in an included copper drum (or rotor), and the resulting torque would be equivalent to the power of an alternating current electric motor—then still the missing unit in the production of an alternating current system. Ferraris constructed such a device and tested it in August–September 1885 by feeding one coil with current from a small Siemens alternator and the second coil with current from a Gaulard transformer. Switching the currents reversed the direction of rotation. Ferraris freely discussed his principle and openly showed his models in classroom and laboratory. He did not apply for patents because he felt a professional pride in discovery and the extension of all knowledge. This was indicated when he wrote: "Above industrial importance I perceive scientific importance, above material use, intellectual use."

The Ferraris principle led to the design and construction of an alternating-current motor without commutator or brushes, which had a squirrel-cage copper rotor revolving by induction from its surrounding "rotating" stator field; it was asynchronous and self-starting. This type of motor today is responsible for the bulk of conversion of electrical power to mechanical power. Ferraris announced his discovery before the Royal Academy of Sciences at Turin on 18 March 1888. Others later claimed priority for the concept of the rotating field—especially Deprez, Walter Baily, and Nikola Tesla. In litigation in German and U.S. courts between 1895 and 1900, it was established that Ferraris had anticipated the principle but that Tesla had applied it, independently, to motor design. The original Ferraris devices are still preserved at the Istituto Elettrotecnico Nazionale Galileo Ferraris in Turin, an institute inaugurated in 1935 as a center for all forms of electrical research and study.

Ferraris participated in the AEG-Oerlikon effort to extend alternating current systems, as demonstrated in the 175-kilometer Lauffen–Frankfurt transmission line that inaugurated the Frankfurt Electrical Exposition of 1891 (at which he was awarded highest honors). He represented his government and was elected vice-president of the electrical exposition in Chicago in 1893, where the standards for the henry, the joule, and the watt were adopted.

BIBLIOGRAPHY

Ferraris' *Opere* were published in 3 vols. by the Associazione Elettrotecnica Italiana (Milan, 1902–1904); see also *Sulla illuminazione elettrica* (Turin, 1879); "Rotazioni elettrodinamiche prodotte per mezzo di correnti alternate," in *Atti dell'Accademia della scienze*, **23** (1888), 360–363; and *Lezioni di elettrotecnica dettate nel R. Museo industriale italiano in Torino* (Turin-Rome, 1897; 2nd ed., 1904).

BERN DIBNER

FERREIN, ANTOINE (*b.* Frespech, near Agen, Lot-et-Garonne, France, 25 October 1693; *d.* Paris, France, 28 February 1769), *anatomy.*

Ferrein was the son of Antoine Ferrein and Françoise d'Elprat, both members of old Agenois families. At his father's wish he began legal studies and did so at Cahors, although he was much more interested in mathematics and the natural sciences. After reading a work by Borelli, in which physiological propositions were purportedly derived from anatomical information by means of mathematical procedures, Ferrein decided to devote himself entirely to medical and anatomical research. Certain ideas of iatromechanics deeply influenced his thinking throughout his life. He followed the idea of an *anatomie subtile* which would seek out in the *petites machines* of the body the explanation of most physiological and pathological phenomena.

In 1714 Ferrein left Cahors to go to Montpellier, where he studied medicine under Raymond Vieussens and Antoine Deidier. In 1716 he received his bachelor's degree, but family obligations forced him to interrupt his studies and move to Marseilles, where he gave private classes in anatomy, physiology, and surgery. He later returned to Montpellier, and on 27 September 1728 he received the title of Doctor of Medicine. He then taught in the Montpellier Faculty of Medicine as *suppléant* to Astruc. After his applications for the chairs of medicine and chemistry were refused (1731–1732), however, he left Montpellier for Paris.

Since he had no right to practice medicine in Paris, Ferrein gave public instruction in anatomy there. Later he became the chief medical officer of the French army in Italy (1733–1735). During this period he sought to combat several epidemics of miliary fever. He finally met the requirements of the Paris Faculty of Medicine, and although he was fully accredited by Montpellier, he requested and obtained another bachelor's degree in 1736 and that of Doctor of Medicine in 1738. From then on, Ferrein, an ambitious, tireless worker and brilliant speaker, made an extraordinary career for himself. On 22 February 1741 he was elected to the Academy of Sciences as assistant anatomist; in 1742 he became associate; and on 21 May 1750, pensioner.

The decade 1740–1750 was the most fruitful of Ferrein's life. He published a series of memoirs on the structure and function of several organs. In 1742 he was named professor of medicine at the Collège Royal and also became professor of surgery at the Faculty of Medicine. He was awarded the chair of pharmacy in 1745. Ferrein's courses became famous, but more for the clarity and order of his exposition than for the originality of his ideas. In 1751, in addition to all his teaching duties and an exhausting medical practice, Ferrein replaced Winslow as professor of anatomy at the Jardin du Roi. He died following a stroke.

In 1731, while he was competing for the chair of medicine at Montpellier, Ferrein propounded a theory on the shape of the heart during systole that was the origin of a long dispute within several learned societies. Against his rival Antoine Fizès and an opinion then generally prevalent, Ferrein maintained that the heart shrank during systole and that its tip curled over and forward. This was a new and accurate explanation of the heart's beating against the thoracic wall. In 1733 Ferrein published the results of his microscopic research on the parenchymatous and vascular structure of the liver. He was the first to glimpse certain anatomical peculiarities of the hepatobiliary system, but unfortunately he drew erroneous physiological conclusions. Ferrein's researches on lymph ducts, hepatic inflammation, and the movements of the jaw were little valued by subsequent generations.

In 1741 Ferrein reviewed and modified Dodart's theory of phonation. According to Ferrein, the lips of the glottis form two true "vocal cords"; sounds arise solely from the vibration of these cords, which is produced by the stream of exhaled air. Thus the air performs the same function as a violin bow. In this hypothesis the larynx is considered to be a combination of wind and string instrument. Apart from Leonardo da Vinci's experiments, Ferrein was the first to study phonation experimentally by forcing air through the detached larynxes of various animals.

According to his histological researches (1749), the kidney is not composed of glomerules, as Malpighi believed, nor are the blood vessels coiled, as was taught by Ruysch; rather, it is made up of a collection of "white tubes." Ferrein described the "pyramids" and the tubular structure of the kidneys, but he misconstrued their function.

It was also Ferrein who formulated the rules for examination of the abdominal organs by palpation. He also denied the existence of true hermaphroditism.

BIBLIOGRAPHY

I. ORIGINAL WORKS. Almost all Ferrein's scientific studies were published in the *Mémoires de l'Académie royale des sciences;* of particular interest are "De la formation de la voix de l'homme" (1741), p. 50; and "Sur la structure des viscères nommés glanduleux, et particulièrement sur celle des reins et du foie" (1749), pp. 489–530. The most famous of his competition theses is *Quaestiones medicae duodecim* (Montpellier, 1732). The great success of his courses led some of his students to publish them directly from the original MSS or their class notes—these publications include *Cours de médecine pratique rédigé d'après les principes de M. Ferrein par M. Arnault de Nobleville* (Paris, 1769); *Matière médicale,* published by Andry (Paris, 1770); and *Éléments de chirurgie pratique,* published from Ferrein's MSS by H. Gauthier (Paris, 1771). Some of the original MSS are in the library of the Paris Faculty of Medicine.

II. SECONDARY LITERATURE. The biography by Grandjean de Fouchy, "Éloge de M. Ferrein," in *Histoire de l'Académie royale des sciences pour l'année 1768* (1772), pp. 151–162, is the basic secondary source. Biographical information is also in N. F. J. Eloy, "Ferrein," in *Dictionnaire historique de la médecine,* II (Mons, 1778), 223–224; J. R. Marboutin, "Antoine Ferrein," in *Revue de l'Agenais,* **61** (1934), 309–311; and A. Portal, *Histoire de l'anatomie et de la chirurgie,* V (Paris, 1770). A concise appraisal of Ferrein's publications is in J.-E. Dezeimeris, *Dictionnaire historique de la médecine,* II (Paris, 1834), 297–300. His researches on the kidney are analyzed in F. Grondona, "La struttura dei reni da F. Ruysch à W. Bowman," in *Physis,* **7** (1965), 281–316.

M. D. GRMEK

FERREL, WILLIAM (*b.* Bedford [now Fulton] County, Pennsylvania, 29 January 1817; *d.* Maywood, Kansas, 18 September 1891), *mathematical geophysics.*

After Laplace, Ferrel was the chief founder of the subject now known as geophysical fluid dynamics. He gave the first general formulation of the equations of motion for a body moving with respect to the rotating earth and drew from them the consequences for atmospheric and oceanic circulation. He contributed to meteorological and tidal theory and to the problem of "earth wobble" (changes in the axis and speed of the earth's rotation).

Born in remote south-central Pennsylvania, Ferrel was the eldest of six boys and two girls born to Benjamin Ferrel and his wife, whose maiden name was Miller. In 1829 the family moved across Maryland into what is now West Virginia, where Ferrel received the usual rudimentary education during a couple of winters in a one-room schoolhouse. A shy and solitary boy, he avidly devoured the few scientific books he acquired by arduous trips to Martinsburg, West Virginia, or Hagerstown, Maryland. Stimulated in 1832 by a partial solar eclipse, by 1835 he had taught himself, with only a crude almanac and a geography book as guides, to predict eclipses. Not until 1837, when he was twenty, did he learn "the law of gravitation, and that the moon and planets move in elliptic orbits."[1] With money saved from schoolteaching, in 1839 Ferrel entered Marshall College, Mercersburg, Pennsylvania (later merged with Franklin College in Lancaster), where he "saw [for] the first time a treatise on algebra."[2] Lack of money forced him to leave after two years of study, and he returned home to teach school for two years. He completed his degree in 1844 at the newly founded Bethany College in Bethany, West Virginia.

Ferrel then went west to teach school, first in Missouri and then in Kentucky. In Liberty, Missouri, he found a copy of Newton's *Principia* (presumably the Glasgow edition, with the 1740 tidal papers added), and later he sent to Philadelphia for a copy of Laplace's *Mécanique céleste* (in Bowditch's translation). In 1853, aged thirty-six, Ferrel wrote his first scientific paper. He moved to Nashville, Tennessee, the first city in which he had ever lived, in 1854, and there, while teaching school, he became an important contributor to the *Nashville Journal of Medicine and Surgery.* Through Benjamin Apthorp Gould, in whose *Astronomical Journal* he had published his first and some subsequent papers, Ferrel was offered his first scientific post, on the *American Nautical Almanac* staff. He remained with the *Almanac* in Cambridge, Massachusetts, from 1858 to 1867, when Benjamin

Peirce of Harvard persuaded Ferrel to go to Washington to join the U.S. Coast Survey, of which Peirce was the new superintendent. In 1882 Ferrel joined the U.S. Army's Signal Service (predecessor of the Weather Bureau), where he remained until 1886. On his retirement at age seventy he moved to Kansas City, Kansas, to be with his brother Jacob and other relatives, but the lack of "scientific associations and access to scientific libraries"[3] in the West led him to return to Martinsburg, West Virginia, in 1889 and 1890. He died in Maywood, Kansas, at the age of seventy-four.

A painfully shy man, Ferrel never married, nor did he found a school in his subject. He did not apply for any of his scientific positions, yet he became a member of the National Academy of Sciences (1868), an associate fellow of the American Academy of Arts and Sciences, an honorary member of the meteorological societies of Austria, Britain, and Germany, and a recipient of the honorary degrees of A.M. and Ph.D.

Ferrel's career as a scientist began about 1850 with his study of Newton's *Principia*. Concentrating on tidal theory—in which Newton's work had been extended in papers presented to the French Academy in 1740 by Daniel Bernoulli, Euler, and Maclaurin and published in editions of the *Principia* after Newton's death—Ferrel conjectured "that the action of the moon and sun upon the tides must have a tendency to retard the earth's rotation on its axis."[4] In his *Mécanique céleste* Laplace had discounted any effect of the tides on the earth's rate of axial rotation. In his first published paper (1853) Ferrel showed that Laplace had neglected the second-order terms that should cause tidal retardation. Since Laplace had claimed to account for all the observed acceleration in the moon's orbit without tidal friction, Ferrel suggested that the latter might be counteracted by the earth's shrinking as it cooled. When about 1860 it became clear that Laplace's theory could not account for the observed value of the moon's acceleration, Ferrel returned to the problem in a paper read to the American Academy of Arts and Sciences in 1864. Although others reached the same general conclusion independently, Ferrel's was the first quantitative treatment of tidal friction, a problem that continues to be of scientific interest.

After three more papers published locally, Ferrel returned to tidal theory in 1856 with his second paper in Gould's *Astronomical Journal*. In it he suggested that Laplace was in error when he claimed that the diurnal tide would vanish in an ocean of uniform depth. Ferrel's criticisms were parallel to Airy's, and both were strongly opposed by Kelvin. The problem

of "oscillations of the second kind" to which they relate remains of current scientific interest.

In both these early papers Ferrel established the basis of his contributions to the theory of tides. Laplace had ignored fluid friction, which was not successfully treated mathematically until Navier and Poisson in the 1820's and Saint-Venant and Stokes in the 1840's inaugurated the modern theory. In tidal studies Airy (1845) assumed friction to be proportional to the first power of the velocity, in which case (as in Laplace's) the equations are linear. Thomas Young (1823), although he assumed friction to be proportional to the square of the velocity, failed to introduce the required equation of continuity. Ferrel's major contribution to tidal theory was thus to begin the full nonlinear treatment necessitated by realistic assumptions concerning friction.

After joining the Coast Survey, Ferrel made important contributions to the techniques of tidal prediction. He extended the nonharmonic developments of the tide-producing potential beyond the points reached by Laplace and Lubbock, and he gave the first reasonably complete harmonic development. Here his endeavors were parallel to those of Kelvin, who was responsible for the first tide-predicting machine (probably the earliest piece of large-scale computing machinery). In 1880 Ferrel, too, designed a tide predictor, which went into service in 1883. Although it was an analogue machine like Kelvin's, Ferrel's gave maxima and minima rather than a continuous curve as its output. Ferrel also made considerable progress in dealing with the shallow-water tidal components and in using tidal data to calculate the mass of the moon.

His studies of astronomical and geophysical tides established Ferrel's claim to a modest place in the history of science. His claim to a major place in this history lies elsewhere: He was the first to understand in mathematical detail the significance of the earth's rotation for the motion of bodies at its surface, and his application of this understanding to the motions of ocean and atmosphere opened a new epoch in meteorology. From Maury's *Physical Geography of the Sea* (1855) Ferrel learned of the belts of high pressure at 30° latitude and of low pressure at the equator and the poles. Looking for the cause of this distribution of pressure, Ferrel realized that since Laplace's tidal equations were of general application, both winds and currents must be deflected by the earth's rotation.

Pressed to write a critical review of Maury's book, Ferrel instead put his own ideas into "An Essay on the Winds and Currents of the Ocean" (*Nashville Journal*, October 1856), a precise but nonmathematical account of the general circulation, and on joining the *Nautical Almanac* staff he began to develop his ideas in mathematical form. In Gould's *Astronomical Journal* early in 1858 Ferrel made explicit the notion of an inertial circle of motion on the earth and used it to explain the gyratory nature of storms (although purely inertial motions are now known to be common in the ocean but almost absent from the atmosphere). In a series of papers published in his colleague J. D. Runkle's *Mathematical Monthly* in 1858 and 1859, then collected to form a separate pamphlet published in New York and London in 1860, Ferrel developed a general quantitative theory of relative motion on the earth's surface and applied it to winds and currents. His result, now known as Ferrel's law, was "that if a body is moving in any direction, there is a force, arising from the earth's rotation, which always deflects it to the right in the northern hemisphere, and to the left in the southern" (1858).[5] This theory and its derivation were carried to a wider audience by a summary article in the *American Journal of Science* for 1861.

Like others who treated relative motion and its geophysical consequences at about the same time, Ferrel appears to have been indebted to Foucault's pendulum (1851) and gyroscope (1852). Ferrel's treatment was remarkable for its clarity and generality, and by continuing to develop his ideas in a series of publications extending over thirty years, he pioneered in the development of meteorology from a descriptive science to a branch of mathematical physics.

When he began, meteorological thought was dominated by the unphysical ideas of Dove, who, drawing on Hadley's explanation of the trade winds (1735), insisted that the earth's rotation acted only on meridional atmospheric motions to deflect them only zonally. Although Ferrel agreed that temperature differences between equatorial and polar regions drove both atmosphere and ocean, he supported by mathematical deduction his insistence that all atmospheric motions, whatever their direction, were deflected by the earth's rotation. His application of this principle to explain both the general circulation and the rotary action of cyclonic storms began to be generally accepted in the 1870's, as weather forecasting services spread over Europe and North America. In his theory of the general circulation Ferrel developed the basic principle that on the rotating earth, convection between equator and pole must be chiefly by westerly winds. He gave the traditional three-cell diagram of the circulation, abandoned only since about 1950, as it has become clear that this scheme of an average circulation pattern along any meridian, although

heuristically useful, is not supported by the data. Ferrel modified Espy's convection-condensation theory of cyclonic storms, and he gave a plausible account of the great force of tornadoes.

By 1880 meteorology seems to have caught up with Ferrel's ideas, and he was not always able to accept the advances of the following decade that built upon his innovations. The Ferrel-Espy convection-condensation theory explains tropical hurricanes but not midlatitude storms, yet about 1890 Ferrel argued vigorously against Hann's ideas on the latter type of storm. He was also unwilling to admit the role of wind stress in the generation of ocean currents. Yet Ferrel had led in bringing sound physical principles, expressed with the tools of mathematical analysis, to bear on the largest problems of oceanic and atmospheric motion: thus at his death he was called "the most eminent meteorologist and one of the most eminent scientific men that America has produced."[6] This eminence came to Ferrel for "having given to the science of meteorology a foundation in mechanics as solid as that which Newton laid for astronomy."[7]

NOTES

1. MS autobiography, printed with minor changes in *Biographical Memoirs. National Academy of Sciences,* **3** (1895), 291.
2. *Ibid.,* 292.
3. Quoted from a letter of Ferrel by Frank Waldo in *American Meteorological Journal,* **8** (1891), 360.
4. Autobiography, 294.
5. "The Influence of the Earth's Rotation Upon the Relative Motion of Bodies Near Its Surface," in *Astronomical Journal,* **5** (1858), 99.
6. W. M. Davis, *American Meteorological Journal,* **8** (1891), 359.
7. Cleveland Abbe, *Biographical Memoirs. National Academy of Sciences,* **3** (1895), 281.

BIBLIOGRAPHY

I. ORIGINAL WORKS. Ferrel's most significant paper, "The Motions of Fluids and Solids Relative to the Earth's Surface," appeared originally in *Mathematical Monthly,* **1** and **2** (Jan. 1859–Aug. 1860) and was republished with notes by Waldo as Professional Papers of the U.S. Signal Service, no. 8 (Washington, D.C., 1882). Other papers on meteorology, including his 1856 "Essay on the Winds and Currents of the Ocean," were reprinted as no. 12 of the same series (Washington, D.C., 1882). Ferrel also wrote three treatises: "Meteorological Researches for the Use of the Coast Pilot," published in three pts. as appendixes to *Report of the Superintendent of the U.S. Coast Survey for 1875* (Washington, D.C., 1878), *1878* (Washington, D.C., 1881), and *1881* (Washington, D.C., 1883); "Recent Advances in Meteorology," published as app. 71 to *Report of the Chief Signal Officer to the Secretary of War for 1885* (Washington, D.C., 1886), as Professional Papers of the

Signal Service, no. 17, and as House Executive Document no. 1, pt. 2, 49th Congress, 1st session; *A Popular Treatise on the Winds* (New York, 1889). Ferrel's major work on tides is his *Tidal Researches,* appended to the *Coast Survey Report for 1874* (Washington, D.C., 1874); and he described his tide predictor in app. 10 to the *Coast Survey Report for 1883* (Washington, D.C., 1884). Ferrel's bibliography, in *Biographical Memoirs. National Academy of Sciences,* **3** (1895), 300–309, lists more than 100 items; it is preceded (287–299) by an edited version of his autobiography, the holograph MS of which is in the Harvard College Library.

II. SECONDARY LITERATURE. Cleveland Abbe's memoir, in *Biographical Memoirs. National Academy of Sciences,* **3** (1895), 267–286, is the fullest; more concise is William M. Davis' in *Proceedings of the American Academy of Arts and Sciences,* **28** (1893), 388–393. Alexander McAdie wrote a biographical article, accompanied by a portrait, in *American Meteorological Journal,* **4** (1888), 441–449; memorial articles by Simon Newcomb, Abbe, Davis, Waldo, and others are *ibid.,* **8** (1891), 337–369. K. Schneider-Carius, *Wetterkunde. Wetterforschung* (Freiburg, 1955), a sourcebook in the Orbis Academicus series, is useful on the history of meteorology. Among the older works, Frank Waldo, *Modern Meteorology* (London, 1893), in the Contemporary Science Series, gives—if anything—too much attention to Ferrel. Ferrel's place in the history of tidal theory is easier to assess, thanks to Rollin A. Harris, *Manual of Tides—Part I,* app. 8 to *Coast Survey Report for 1897* (Washington, D.C., 1898); pp. 455–462 of this excellent history are devoted to Ferrel.

HAROLD L. BURSTYN

FERRIER, DAVID (*b.* Aberdeen, Scotland, 13 January 1843; *d.* London, England, 19 March 1928), *neurophysiology, neurology.*

Ferrier was the second son of David Ferrier and Hannah Bell. His early education was at the Aberdeen Grammar School and later at Aberdeen University, where he was graduated M.A. in 1863. He won first-class honors in classics and philosophy and thus the Ferguson scholarship, which was open to all Scottish students of philosophy and was considered their premier award. It was at this time that he came under the influence of Alexander Bain, the famous logician and psychologist, and at his suggestion went to Heidelberg in 1864 to study psychology for a year.

In the following year Ferrier entered the medical school of the University of Edinburgh, and after receiving the M.B. degree in 1868 with all possible distinction, he served for a brief period as assistant to Thomas Laycock, professor of practical medicine at the university and the man who had influenced the young Hughlings Jackson. Ferrier supplemented his income by teaching but, finding this distasteful, he spent two years as assistant to a general practitioner, a Dr. Image of Bury St. Edmunds, and used

his spare time to study comparative anatomy. The latter provided his M.D. thesis, "The Comparative Anatomy and Intimate Structure of the Corpora Quadrigemina," for which he was awarded a gold medal in 1870.

Since he disliked general practice, in 1870 Ferrier obtained an appointment as lecturer on physiology at the medical school of the Middlesex Hospital; one year later he moved to King's College Hospital and Medical School, where he remained for the rest of his professional life. At first he was demonstrator in physiology, but in 1872 he succeeded to the chair of forensic medicine vacated by William Augustus Guy and held it until 1889; Ferrier helped Guy to compile a popular textbook of medical jurisprudence that bore both their names. In 1874 he was elected assistant physician to the hospital and became physician in charge of outpatients and full physician in 1890. His last academic appointment was as professor of neuropathology, the chair having been specially instituted for him in 1889. Ferrier also held appointments at the West London Hospital and, from 1880 to 1907, at the National Hospital, Queen Square, where it is said that he was one of the last physicians to conduct his ward rounds wearing the traditional top hat and black tailcoat. He retired from King's College in 1908, when he was elected emeritus professor of neuropathology and consulting physician to the hospital.

Ferrier was one of the original members of the Physiological Society, founded in 1876, and he was made an honorary member in 1927. He was also a founding editor of *Brain,* along with J. C. Bucknill, J. Crichton-Browne, and Hughlings Jackson; the first number appeared in April 1878. In 1876 he was elected a fellow of the Royal Society and in the following year a fellow of the Royal College of Physicians of London. Ferrier received a number of medals from these societies, and he gave several of their important lectures. Many other honors were bestowed upon him, including *lauréat* of the Institut de France, honorary degrees from the universities of Cambridge and Birmingham, and in 1911 a knighthood.

Ferrier was quiet and reserved and disliked controversy. He possessed an outstandingly active and agile mind, and his philosophical training stood him in good stead in his scientific work. He had an unquenchable thirst for knowledge but lacked the patience and powers of observation of some of his contemporaries in clinical neurology. It was ironic that although Ferrier was exceedingly fond of animals, he was accused, along with Gerald Yeo, of cruelty to experimental subjects. At the trial in 1882 he successfully upheld animal experimentation and won his case by proving that his colleague Yeo, who had carried out the operations on living animals, possessed a license to do so. Ferrier was a lover of classical literature, art, and the sea, and it is recorded that he "remained alert and dapper to the end." In 1874 he married Constance Waterlow; they had a son, Claude, who became a well-known architect, and a daughter.

Ferrier's work as a clinical neurologist was not outstanding, although he always had a large private practice. In this field he was dwarfed by such famous contemporaries as Jackson, W. R. Gowers, H. C. Bastian, and E. F. Buzzard, who were busy creating the British school of neurology at the National Hospital, Queen Square. On the other hand, Ferrier excelled all of these in experimental physiology, which he pursued together with medical practice. In this field he will always be remembered for his contributions to the problem of the localization of function in the cerebral cortex.

In the 1860's Jackson had suggested that the cerebral cortex must represent bodily function in an orderly fashion, but he based his contentions on clinical observation and hypothesis alone. The French clinicians J. B. Bouillaud, Aubertin, and P. P. Broca had already put forward this idea, but again without experimental proof. The first experimental support came from G. T. Fritsch and E. Hitzig in 1870, and early in 1873 Ferrier discussed it with his friend and fellow student at Edinburgh, Sir James Crichton-Browne, director of the West Riding Lunatic Asylum at Wakefield. As a result, during the spring and summer of that year Ferrier carried out investigations in the laboratory recently installed at the asylum. So began his detailed, systematic exploration of the cerebral cortex in different vertebrates, ranging from the lowest to the highest and including the ape, which was conducted over the next decade and more. He set about this work with the express purpose of confirming or refuting the theoretical suggestions made by Jackson with respect to the localized cortical areas of function. But whereas Fritsch and Hitzig had used only the dog and galvanic electric current, Ferrier employed faradic stimulation to the cortex, an important technical advance that remained universally popular until the introduction of improved methods in the 1920's. Ferrier, moreover, studied mainly primates and his researches were more fully and methodically planned.

He mapped much of the cerebral cortex and carefully delineated the "motor-region," as he termed it; the scheme of localized function that he put forward was based on the concept of "motor" and "sensory" regions. Like Fritsch and Hitzig, Ferrier carried out ablations of local areas of cerebral cortex as well as

stimulation and observed the resulting functional deficit. Jackson's concept of "discharging" and "destroying" lesions was therefore reproduced experimentally and his theories put to the test. As far as primates were concerned, they were shown to be correct.

Ferrier's fame as an experimental neurologist was made by these studies, which he first published in the *West Riding Lunatic Asylum Reports* of 1873 and later in book form as *The Function of the Brain* (1876). The latter, which contains the substance of his Croonian lecture of the Royal Society given in February 1874, is one of the most significant publications in the field of cortical localization. It was supplemented by a later and detailed review of his results and those of others, which he delivered as *The Croonian Lectures* [of the Royal College of Physicians of London] *on Cerebral Localisation* in the same year. Wide publicity was given to Ferrier's findings at the International Medical Congress of 1880, in London.

Ferrier was one of the contributors to the spectacular advances made in the neurological sciences toward the end of the nineteenth century, although he was occasionally in error. For instance, he considered that the cortical visual area was in the angular gyrus rather than in the calcarine cortex; no doubt his cortical lesions had involved the nearby optic radiation. He was also guilty of unwarranted extrapolation from his findings in animals to the human brain, although he was by no means alone in the procedure. Thus, he transferred the results from his monkey experiments to a diagram of the human brain, and this was widely accepted. In the first case the interchange of data among the different species is now known to be impossible, and in the second, the belief in well-defined "centers," to which Ferrier's work contributed, is no longer acceptable.

The influence of Ferrier's work was widespread, and he and Fritsch and Hitzig inspired many attempts to chart the cortex. In addition he had an important influence on the embryonic field of brain surgery, for he urged his surgical colleagues to attack cerebral lesions operatively. Throughout his life he was passionately fond of laboratory work, and in addition to his classic studies of the cerebral cortex, he also carried out investigations of the cerebellum, the limb plexuses, and further studies on the quadrigeminal bodies, thereby extending his earlier work.

BIBLIOGRAPHY

I. ORIGINAL WORKS. Ferrier published many articles on the physiology of the nervous system and on clinical neurological topics. The most important are "Experimental

Researches in Cerebral Physiology and Pathology," in *West Riding Lunatic Asylum Medical Reports,* **3** (1873), 30–96; *The Function of the Brain* (London, 1876; 1886); *The Localisation of Cerebral Disease* (London, 1878), the Gulstonian lectures for 1878; *The Croonian Lectures on Cerebral Localisation* (London, 1890); and "The Regional Diagnosis of Cerebral Disease," in *A System of Medicine,* C. Allbutt and H. D. Rolleston, eds., VIII (London, 1911), 37–162.

II. SECONDARY LITERATURE. Obituaries are C. S. S[herrington], "Sir David Ferrier, 1843–1928," in *Proceedings of the Royal Society,* **103B** (1928), viii–xvi; *British Medical Journal* (1928), **1**, 525–526, 574–575; and *Lancet* (1928), **1**, 627–629. Each includes a portrait. See also H. W. Lyle, *King's and Some King's Men* (London, 1935), pp. 279–281; H. R. Viets, "West Riding, 1871–1876," in *Bulletin of the History of Medicine,* **6** (1938), 477–487; and David McK. Rioch, in *The Founders of Neurology,* W. Haymaker, ed. (Springfield, Ill., 1953), pp. 122–125, with portrait. For Ferrier's contribution to knowledge of the visual pathway, see S. Polyak, *The Vertebrate Visual System* (Chicago, 1957), pp. 147–149.

EDWIN CLARKE

FERRO (or FERREO, DAL FERRO, DEL FERRO), SCIPIONE (*b.* Bologna, Italy, 6 February 1465; *d.* Bologna, between 29 October and 16 November 1526), *mathematics.*

Scipione was the son of Floriano Ferro (a papermaker by trade) and his wife, Filippa. He was a lecturer in arithmetic and geometry at the University of Bologna from 1496 to 1526, except for a brief stay in Venice during the last year. In 1513 he is recorded as an "arithmetician" by Giovanni Filoteo Achillini, in the poem *Viridario.* After Scipione's death, the same subjects were taught by his disciple, Annibale dalla Nave (or della Nave). Nave married Scipione's daughter, Filippa, and inherited his father-in-law's surname, thereby calling himself dalla Nave, alias dal Ferro. Scipione's activity as a businessman is demonstrated by various notarial documents from the years 1517–1523.

No work of Scipione's, either printed or in manuscript, is known. It is known from several sources, however, that he was a great algebraist. We are indebted to him for the solution of third-degree, or cubic, equations, which had been sought since antiquity. As late as the end of the fifteenth century, Luca Pacioli judged it "impossible" by the methods known at that time (*Summa,* I, dist. VIII, tractate 5). Scipione achieved his solution in the first or second decade of the sixteenth century, as is known from the texts of Tartaglia and Cardano. He did not print any account of his discovery, but divulged it to various people and expounded it in a manuscript that came into the possession of Nave, but which today is unknown.

In 1535 a disciple of Scipione, named Antonio Maria Fiore, in a mathematical dispute with Tartaglia proposed some problems leading to cubic equations lacking the second-degree term, of which he claimed to know the method of solution, having learned "such a secret," thirty years before, from a certain "great mathematician" (Tartaglia, *Quesiti,* bk. ix, question 25). Tartaglia (who had concerned himself with cubic equations as early as 1530) was now (1535) induced to seek and find the solution to them. Some years later, yielding to entreaties, he communicated his solution to Cardano (1539).

Still later (1542), in Bologna, Nave made known to Cardano the existence of the aforementioned manuscript by Scipione: this is attested by Ludovico Ferrari, Cardano's famous disciple, who was with the master (Ferrari to Tartaglia, *Secondo cartello,* p. 3; see also Tartaglia to Ferrari, *Seconda risposta,* p. 6). Cardano, who published the *Ars magna* (1545) somewhat later, represented the solution of cubic equations as the distinguished discovery of Scipione and Tartaglia. The following is taken from the *Ars magna* (chs. 1, 11):

> Scipione Ferro of Bologna, almost thirty years ago, discovered the solution of the cube and of things equal in number [that is, the equation $x^3 + px = q$, where p and q are positive numbers], a really beautiful and admirable accomplishment. In distinction this discovery surpasses all mortal ingenuity and all human subtlety. It is truly a gift from heaven, although at the same time a proof of the power of reason, and so illustrious that whoever attains it may believe himself capable of solving any problem. In emulation of him Niccolò Tartaglia of Brescia, a friend of ours, in order not to be conquered when he entered into competition with a disciple of Ferro, Antonio Maria Fiore, came upon the same solution, and revealed it to me because of his many entreaties to him.

The coincidence of Scipione's and Tartaglia's rules was confirmed by Ettore Bortolotti by means of an ancient manuscript (MS 595N of the Library of the University of Bologna), which reproduces Scipione's rule, which had been obtained from him by Pompeo Bolognetti, lecturer *ad praxim mathematicae* at Bologna in the years 1554–1568. Basing his conclusion on another text by Cardano and on a manuscript of the *Algebra* of Rafael Bombelli (MS B.1569 of the Library of the Archiginnasio of Bologna, assignable to about 1550), Bortolotti concludes, with plausibility, that Scipione did indeed solve both the equations $x^3 + px = q$, and $x^3 = px + q$.

How Scipione arrived at the solution of cubic equations is not known. But there is no lack of attempts at reconstructing his method. For example, in his examination of the *Liber abbaci* of Leonardo Fibonacci and of the *Algebra* of Bombelli, Giovanni Vacca seemed to be able to reproduce, in the following simple procedure, the one used by Scipione: If the sum of square roots is expressed as

$$x = \sqrt{a + \sqrt{b}} + \sqrt{a - \sqrt{b}}, \tag{1}$$

it can be seen, by raising to the square, that this satisfies the second-degree equation (lacking the term in x)

$$x^2 = (2\sqrt{a^2 - b}) + 2a. \tag{2}$$

Analogously, if the sum of cube roots is expressed as

$$X = \sqrt[3]{a + \sqrt{b}} + \sqrt[3]{a - \sqrt{b}}, \tag{3}$$

then it can be seen, raising to the cube, that this satisfies the third-degree equation (lacking the term in X^2)

$$X^3 = (3\sqrt[3]{a^2 - b})X + 2a. \tag{4}$$

Therefore, equation (4) is solved by means of (3). If one writes

$$p = 3\sqrt[3]{a^2 - b}, \quad q = 2a \tag{a}$$

or

$$a = \frac{q}{2}, \quad b = \frac{q^2}{4} - \frac{p^3}{27}, \tag{b}$$

the cubic equation (4) and the formula which solves it (3) assume the accustomed form:

$$X^3 = pX + q, \tag{5}$$

$$X = \sqrt[3]{\frac{p}{2} + \sqrt{\frac{q^2}{4} - \frac{p^3}{27}}} + \sqrt[3]{\frac{p}{2} - \sqrt{\frac{q^2}{4} - \frac{p^3}{27}}}. \tag{6}$$

This latter is called Cardano's formula, but incorrectly because Cardano does not deserve credit for having discovered it but only for having published it for the first time.

Another of Scipione's contributions to algebra concerns fractions having irrational denominators. The problem of rationalizing the denominator of such a fraction, when there are square roots intervening, goes back to Euclid. In the sixteenth century, the same problem appears with roots of greater index. And here one can single out the case of fractions of the type

$$\frac{1}{\sqrt[3]{a} + \sqrt[3]{b} + \sqrt[3]{c}},$$

which Scipione was the first to deal with, as can be seen by the manuscript of Bombelli, cited above. In that manuscript, Bombelli calls Scipione "a man uniquely gifted in this art."

Finally, it should be noted that Scipione also applied himself to the geometry of the compass with a fixed opening, although this theory was ancient, the first examples going back to Abu'l-Wafa' (tenth century). In the first half of the sixteenth century, this question arose again, particularly because of Tartaglia, Cardano, and Ferrari. And it is because of the testimony of the last-named (Ferrari to Tartaglia, *Quinto cartello,* p. 25) that we can state that Scipione also took up this problem, but we know nothing about his researches and his contributions.

BIBLIOGRAPHY

I. ORIGINAL WORKS. We possess no original works by Scipione Ferro, and the sources from which knowledge of his activity is derived are indicated in the text. Of the printed sources, the following later eds. are more accessible than the originals: Cardano, *Ars magna,* in *Opera omnia,* C. Spon, ed., vol. IV (Lyons, 1663); facs. repr. of the *Opera omnia,* with intro. by A. Buck (Stuttgart–Bad Cannstatt, 1966); Tartaglia, *Quesiti et inventioni diverse,* facsimile of the edition of 1554, sponsored by the Atheneum of Brescia, A. Masotti, ed. (Brescia, 1959); and Ferrari, *Cartelli,* and Tartaglia, *Risposte,* in the autograph ed. by E. Giordani (Milano, 1876), or in the facs. ed. of the original, sponsored by the Atheneum of Brescia, A. Masotti, ed. (in press). Concerning the two Bolognese manuscripts which have been cited, see E. Bortolotti, "L'algebra nella scuola matematica bolognese del secolo XVI," in *Periodico di matematiche,* 4th ser., **5** (1925), 147–184, as well as the collection of extracts entitled *Studi e ricerche sulla storia della matematica in Italia nei secoli XVI e XVII* (Bologna, 1928).

II. SECONDARY LITERATURE. See C. Malagola, *Della vita e delle opere di Antonio Urceo detto Codro* (Bologna, 1878), pp. 352–355, and app. XXVII, pp. 574–577, which contains "Documenti intorno a Scipione dal Ferro"; L. Frati, "Scipione dal Ferro," in *Bollettino di bibliografia e storia delle scienze matematiche,* **12** (1910), 1–5; and in *Studi e memorie per la storia dell'Università di Bologna,* **2** (1911), 193–205; E. Bortolotti, "I contributi del Tartaglia, del Cardano, del Ferrari, e della scuola matematica bolognese alla teoria algebrica delle equazioni cubiche," in *Studi e memorie per la storia dell'Università di Bologna,* **10** (1926), 55–108, thence in the aforementioned volume of *Studi e ricerche;* and G. Vacca, "Sul commento di Leonardo Pisano al Libro X degli Elementi di Euclide e sulla risoluzione delle equazioni cubiche," in *Bollettino dell'Unione matematica italiana,* **9** (1930), 59–63.

On Annibale dalla Nave, see A. Favaro, note in Eneström, *Bibliotheca mathematica,* 3rd ser., **2** (1901), 354.

On A. M. Fiore, see A. Masotti, note in *Atti* of the

meeting in honor of Tartaglia, held at the Atheneum of Brescia in 1959, p. 42.

Ferro's role in the solution of cubic equations, is discussed in all histories of mathematics. See, for example, M. Cantor, *Vorlesungen über Geschichte der Mathematik;* J. Tropfke, *Geschichte der Elementar-Mathematik;* and D. E. Smith, *History of Mathematics, passim.*

Also of interest are D. E. Smith, *A Source Book in Mathematics* (New York, 1929; repr. New York, 1959), pp. 203–206, where one can read the solution of the cubic equation given in Cardano, *Ars magna,* ch. 11, English trans. by R. B. McClenon; O. Ore, *Cardano, the Gambling Scholar* (Princeton, 1953), pp. 62–107, where can be found the history of the solution of cubic equations, with English translations of various texts by Tartaglia, Cardano, and Ferrari, and various mentions of Ferro; and, finally, G. Sarton, *Six Wings: Men of Science in the Renaissance* (Bloomington, Ind., 1957), pp. 28–36, 246–249.

ARNALDO MASOTTI

FERSMAN, ALEKSANDR EVGENIEVICH (*b.* St. Petersburg, Russia, 8 November 1883; *d.* Sochi, U.S.S.R., 20 May 1945), *mineralogy, geochemistry.*

Fersman's father, Evgeny Aleksandrovich Fersman, was an architect and later a soldier. The atmosphere of his home, which encouraged both art and thought, was unusual in the military environment of that day. Fersman's mother, Maria Eduardovna Kessler, was a talented pianist and painter; her brother, A. E. Kessler, who studied under the well-known chemist A. M. Butlerov, was also an important influence on the boy's education.

The Fersman family usually spent the summer holidays on Kessler's estate near Simferopol, and there, in the Crimean mountains, young Fersman was attracted to mineralogy and began his first mineral collection. The development of his interests was furthered by a trip to Czechoslovakia, to which the family was obliged to go because of the mother's illness. There, in Karlovy Vary (Carlsbad), an old mining area that was no longer prosperous, he could purchase crystals and druses to fill out his mineral collection. Thus, by the time he graduated from the Odessa Classical Gymnasium in 1901, with a gold medal, Fersman's interests had already been formed; he was very much interested in mineralogy, he had a good mineralogical collection, and he had accumulated a substantial store of personal observations.

In Novorossisk University, which Fersman entered, the lecture course in descriptive mineralogy was extremely boring, and Fersman at first wished to give up mineralogy and study the history of art instead. Friends of his family, Professor P. G. Melikashvili and the chemist A. I. Gorbov, advised him to give up this idea and to study the structure of matter and questions of molecular chemistry. To B. P. Veynberg, a student of D. I. Mendeleev and a specialist in physical chemistry, Fersman owed his acquaintance with ideas on the nature of crystalline substances, such as ice and frost patterns.

In 1903 Fersman's father was given command of the First Moscow Cadet Corps, and the son transferred to Moscow University. Here he approached the head of the department of mineralogy, V. I. Vernadsky, who found him a place in his laboratory. There Fersman mastered the goniometric method of measuring crystals. He worked persistently, and while still a student (1904–1907) he published his first seven scientific works, devoted to crystallography and the mineralogy of stolpenite, gmelinite, and other substances.

When Fersman graduated from the university in 1907, Vernadsky retained him in his department to prepare to become a professor. In 1908 Fersman worked in Victor Goldschmidt's laboratory at Heidelberg University, where he perfected his crystallographic and optical methods. He was commissioned by Goldschmidt to make a tour of the most important jewelers of western Europe and select the most interesting crystals of natural diamonds for study. In Frankfurt, Hanau, and Berlin tens of thousands of carats of diamonds were displayed before him on special tables. As a result of these observations Goldschmidt and Fersman wrote a joint monograph on the crystallography of the diamond (1911) that is still significant.

At Heidelberg, Fersman attended Rosenbusch's lectures on petrography. In France he visited Lacroix's laboratory in Paris and made a trip to study the pegmatites of the islands in the Elbe. This trip played a large role in determining his scientific interests, for Fersman later dedicated many years to research on pegmatites.

In 1912 Fersman became senior curator in the mineralogical section of the Geological Museum of the Russian Academy of Sciences. In the same year, for the first time he gave a course in geochemistry at Shanyavsky University in Moscow. He also took part in the organization of a popular scientific journal, *Priroda,* to which he gave considerable attention throughout his life. In it he published major articles and notes on geochemistry and mineralogy, diamonds, alloys of radium, emeralds, zeolites, platinum, gases, and other useful minerals found in Russia and other countries.

In 1914 the first period of Fersman's scientific activity came to a close, when his great gift for and inclination toward scientific synthesis and theoretical generalization became apparent.

At the beginning of 1915 a commission was organized in the Academy of Sciences for the study of the natural resources of Russia, and Fersman was elected scientific secretary. In connection with the work of the commission he studied the deposits of various useful minerals in the Crimea, Mongolia, Trans-Baykal, the Urals, the Altai, and various regions of European Russia.

During World War I, Fersman traveled to the front and compiled geological maps showing the location of construction materials and water-bearing and waterproof horizons, knowledge of which was important for successful military operations. During this period Fersman faced, in the broadest form, the problems of use of mineral raw materials. He was interested in economics and technology, as well as in mineralogy and geology.

Soon after the Russian Revolution, Lenin turned to the Academy of Sciences for definitions of the new problems facing science. He talked with Fersman, who was impressed with his concern for efficient placement of industry nearer to mineral raw materials and for guaranteeing the Soviet Republic a domestic supply of raw materials. All this awakened Fersman's interest and influenced the direction and planning of his scientific research.

In 1919 Fersman was elected to the Academy of Sciences and was chosen director of its mineralogical museum. Besides imparting his own enthusiasm for science to his students and colleagues, he was modest and encouraged the progress of other researchers. In the winter of 1919–1920 Fersman gave a course of lectures at Petrograd (now Leningrad) University on the geochemistry of Russia, and in the following year repeated it at the Geographical Institute of Petrograd.

Fersman made sizable contributions to the solution of an important theoretical problem of geochemistry: the frequency of distribution of the chemical elements in the rocks of the earth's crust (clarkes). The term "clarke" (the concentration of an element in the earth's crust) was proposed by Fersman in honor of the American scientist F. W. Clarke, one of the first to consider this problem in his fundamental work, *The Relative Abundance of the Chemical Elements* (1889). Fersman calculated the clarkes for most of the elements. Before Fersman, clarkes were expressed in weight percentages. He showed that for geochemical purposes the atomic percentages were more important, thus introducing into science the concept of "atomic clarkes." As a result he discovered the independence of geochemical abundances from the positions of the elements in the periodic system and the concentration and depletion of the various elements. He showed that abundances within the earth's crust

were determined by the effects of the migration of the elements, while abundances in space were related to the stability of the atomic nucleus. He was the first to consider the problem of regional geochemistry and the division of European Russia into geochemical districts, and he provided a classification of hypogene processes. An expanded course of these lectures was published in 1922.

Fersman published the monograph *Dragotsennye i tsvetnye kamni Rossii* ("Precious and Colored Stones of Russia," 1920), as well as works on feldspar, fuller's earth, and saline mud.

Noteworthy among his numerous investigations and expeditions in the 1920's and 1930's is his work in the Khibiny Mountains (Kola Peninsula). He first traveled to the Khibiny Mountains in May 1920, as a member of the commission of the Murmansk railroad, which was headed by the president of the Academy of Sciences, A. P. Karpinsky, and the geologist A. P. Gerasimov. Fersman later expended much of his creative energy in the study of the Khibiny Mountains. He led many expeditions, and his research enabled him to combine separate facts and observations into a coherent system providing an integrated view of the formation processes of the geological structures of the Fenno-Scandinavian shield.

The mineralogical and geochemical research in the Khibiny was crowned by the discovery of great deposits of apatite. Fersman was not only a scientist but also the developer of this inhospitable region. At his initiative the Khibiny mining station was opened in 1937, and Fersman was its first director. This station later grew into an important scientific institution: the Kola branch of the Soviet Academy of Sciences.

Fersman's study of central Asia began in 1924 and continued until the 1940's. At the beginning of his work it was believed that central Asia had few deposits of useful minerals. He carried out considerable scientific research to show the mineral riches there and thus refuted that erroneous belief. Fersman discovered in the Karakum Desert deposits of virgin sulfur; with his help a sulfur refinery was built on the site of the discovery and has supplied the Soviet Union with sulfur ever since.

In the Urals, Fersman investigated pegmatite, rare elements, and deposits of copper, chromium, and other useful minerals. In Siberia he began research showing the value of further study and the great richness of the deposits there. Fersman wrote the important works *Geokhimicheskie problemy Sibiri* ("Geochemical Problems of Siberia") and *Geokhimicheskie problemy Soyuza* ("Geochemical Problems of the [Soviet] Union," 1931).

During World War II, Fersman was concerned with military geology and problems of securing strategic materials. He headed the Commission for the Geological-Geographical Services of the Red Army, to which he attracted many important specialists. He traveled to the front many times with reports and lectures on strategic materials and military geology. In this period he wrote several books and articles on strategic materials of the Soviet Union and Germany, and by comparing them showed that the Soviet Union's military potential guaranteed its victory.

Fersman gave much attention to the history of science. He tried to show the origins of scientific ideas and the achievements of researchers, in particular his predecessors D. I. Mendeleev and V. I. Vernadsky. With particular warmth he wrote of those who, with him, created the new science of geochemistry and of his teachers Vernadsky, Goldschmidt, and G. Hevesy, among others. Fersman had the ability to write sketches that give clear pictures of scientists. Reading his *Zanimatelnaya geokhimia* ("Entertaining Geochemistry," 1948), one can learn of the remarkable work of Marie Curie, A. P. Karpinsky, N. S. Kurnakov, P. I. Preobrazhensky, V. G. Khlopin, and many other distinguished scientists. Through his popular articles, sketches, and books, such as *Zanimatelnaya mineralogia* ("Entertaining Mineralogy," 1928), *Puteshestvia za kamnem* ("Traveling for Rocks," 1956), *Vospominania o kamne* ("Recollections About Rocks," 1940), *Rasskazy o samotsvetakh* ("Tales of Semiprecious Stones," 1957), Fersman helped explain the practical significance of theoretical research in geology.

Fersman gave much attention to his students, many of whom became outstanding scientists: D. I. Shcherbakov, A. A. Saukov, V. V. Shcherbina, and O. A. Vorobeva, among others.

For his achievements and services in geochemistry, mineralogy, geology, and geography Fersman was elected member or corresponding member of sixteen scientific organizations and societies in his native country and abroad. He was awarded the Lenin Prize (1929), the State Prize of the U.S.S.R. of the First Degree (1941), medals from the University of Belgium (1936) and the Wollaston Medal (1942), and the order of the Red Banner of Labor.

Fersman was an active leader of the Academy of Sciences of the U.S.S.R., occupying at various times the posts of vice-president, member of the Presidium, academician-secretary of the Section of Mathematics and Natural Sciences, president of the Council for the Study of Natural Resources, and director of publications.

His scientific creativity was characterized by an exceptionally broad scope and an integral view of nature. With a good understanding of the underlying relationships between various phenomena, he was a master of theoretical generalization and scientific synthesis.

Of major significance were Fersman's works in geochemistry, which he, like his teacher Vernadsky, understood more deeply and more broadly than his contemporaries. According to his definition, geochemistry should concern the history of atoms of chemical elements in the earth's crust and their behavior under various thermodynamic, physical, and chemical conditions of nature. Fersman showed with great clarity the significance of Mendeleev's periodic law for geochemistry.

All his life Fersman did research in mineralogy and geochemistry; he showed graphically, vividly, and in a fascinating way that these sciences do not consist of dry ideas, of inanimate, dead objects of nature; rather, they are sciences of the origins and history of natural phenomena, the complex chemical processes that form the face of the earth and that slowly but inexorably transform what appears to be lifeless stone into new chemical compounds. The idea of geochemical character lay at the basis of all his further work, which was closely connected with the study of the useful minerals of the U.S.S.R. It appeared to him that at the foundation of all surrounding life, all surrounding transformations, and even of the very life processes themselves lay the laws of the dispersion and combination of ninety chemical elements, from which the earth and all of space are constructed, and that one cannot, by studying it only in the laboratory, tear lifeless stone away from the great laboratory of nature in which its transformations take place.

In *Khimicheskie elementy zemli i kosmosa* ("The Chemical Elements of the Earth and Space," 1923), Fersman extended the problem of the history of the elements to the universe; in *Geokhimia Rossii* ("Geochemistry of Russia," 1922) he had tried to apply these ideas to the understanding of those different phenomena which take place in widely distant regions of the U.S.S.R. Most important in Fersman's work is his constant recurrence to the basic problems posed in the past, introducing the study of chemical processes into the chemistry of space while still taking each element back from space to earth and giving attention to its use by man.

In 1932 Vernadsky, following the publication of the monograph *Pegmatity* ("Pegmatites"), expressed pleasure with this new and important work, through which scientists have come to a deeper understanding of the world's structure and of the role of atoms in that structure, about which C. F. Shönbein and Faraday had theorized in the late 1830's. The periodicity of properties in space pointed to a spiral pattern of phenomena, the more so since for the periodic system the spiral was very important.

Fersman asserted that the whole course of chemical processes in space is simply a great Mendeleevian system, in which the laws of energetics and the level of energy govern separate cells, moving elements and combinations of elements about in time and space. The places of the elements in the periodic system reflected a definite step in the chemical history of earth and the universe, between which there is an inner connection.

Fersman's ideas frequently were ahead of his time, and in many of his works he foresaw the future, describing future science and technology. In a special sketch in *Zanimatelnaya geokhimia* attention is drawn to future achievements: the use of atmospheric gases, the ozone screen, the warmth of the earth's depths, atomic energy, the energy of ocean waves and winds, and of new synthetic carbon compounds, and of man's penetration into space.

In calling the geochemical activity of mankind "technogenesis," Fersman meant the economic and industrial activity of man, according to his own scale and significance, compared with the processes of nature herself. Technogenesis basically leads to the extraction of chemical elements from the earth, the redistribution of elements from the depths on the earth's surface, and the agricultural and engineering regrouping of elements. Analyzing the pattern of use of separate elements, its connection with clarkes, and the role of clarkes of concentration, Fersman showed that man concentrates certain elements (gold, platinum, silver, and so forth) and disperses others (carbon, tin, magnesium, silicon, and so forth). As a result he defined the basic geochemical relations between man and nature and noted that the laws of geochemistry force man to seek technical solutions in the use of poor lodes with scattered and rare elements. Technogenesis represents a distinguished theoretical and practical achievement of science, especially in the light of contemporary achievements (atomic experiments and space research). Fersman's work in this area will long light the way for new research, inventions, and the conservation of natural resources.

BIBLIOGRAPHY

I. ORIGINAL WORKS. Fersman's writings include *Der Diamant* (Heidelberg, 1911), written with V. Goldschmidt; *Dragotsennye i tsvetnye kamni Rossii* ("Precious and Colored Stones of Russia," Petrograd, 1920); *Geokhimia Rossii* ("Geochemistry of Russia," Petrograd, 1922); *Puti k nauke budushchego* ("Paths Toward the Science of the Future," Petrograd, 1922); *Khimicheskie elementy zemli i kosmosa* ("The Chemical Elements of the Earth and Space," Petrograd, 1923); *Khimia mirozdania* ("The Chemistry of the Universe," Petrograd, 1923); *Istoria almaznogo fonda* ("History of Diamond Stocks," Moscow, 1924); *Zanimatelnaya mineralogia* ("Entertaining Mineralogy," Leningrad, 1928); *Pegmatity* ("Pegmatites," Leningrad, 1931); *Geokhimicheskie problemy Sibiri* ("Geochemical Problems of Siberia," Moscow–Leningrad, 1931); *Geokhimicheskie problemy Soyuza* ("Geochemical Problems of the [Soviet] Union," Moscow–Leningrad, 1931); *Geokhimia* ("Geochemistry"), 4 vols. (Leningrad, 1933–1939); *Vospominania o kamne* ("Recollections about Rocks," Moscow, 1940); *Voyna i strategicheskoe syre* ("The War and Strategic Raw Material," Krasnoufimsk, 1940); *Geologia i voyna* ("Geology and War," Moscow–Leningrad, 1943); *Khimia zemli na novykh putyakh* ("The Chemistry of the Earth on New Paths," Moscow, 1944); *Mineralnoe syre zarubezhnykh stran* ("The Mineral Raw Materials of Foreign Countries," Moscow–Leningrad, 1947); *Zanimatelnaya geokhimia* ("Entertaining Geochemistry," Moscow–Leningrad, 1948); *Ocherki po istorii kamnya* ("Essays on the History of Rocks"), 2 vols. (Moscow, 1954–1961). Many of his works were brought together as *Izbrannye trudy* ("Selected Works"), 7 vols. (Moscow, 1952–1962).

II. SECONDARY LITERATURE. On Fersman or his works, see G. P. Barsanov, "Kharakternye cherty tvorchestva akademika A. E. Fersmana i ego raboty po mineralogii" ("Characteristic Features of the Creative Work of Academician A. E. Fersman and His Work in Mineralogy"), in *Trudy mineralogicheskogo muzeya Akademii nauk, USSR* no. 5 (1953), 7–18; R. F. Gekker, "Akademik A. E. Fersman i ego rabota vo Vserossyskom obshchestve okhrany prirody" ("Academician A. E. Fersman and His Work in the All-Russian Society for the Conservation of Nature"), in *Okhrana prirody,* no. 3 (1948), 113–119; D. P. Grigoriev and I. I. Shafranovsky, *Vydayushchiesya russkie mineralogi* ("Distinguished Russian Mineralogists," Moscow–Leningrad, 1949), pp. 196–233; O. V. Isakova, ed., *Aleksandr Evgenievich Fersman* (Moscow, 1940), a bibliographical collection; O. Pisarzhevsky, *Fersman* (Moscow, 1959); and A. A. Saukov, "Raboty A. E. Fersmana po geokhimii" ("The Work of A. E. Fersman in Geochemistry"), in *Yubileyny sbornik, posvyashchenny tridtsatiletiyu Velikoy Oktyabrskoy sotsialisticheskoy revolyutsii* ("Jubilee Collection, Dedicated to the Thirtieth Anniversary of the Great October Socialist Revolution"), I (Moscow–Leningrad, 1947), 57–60.

See also I. I. Shafranovsky, "Trudy A. E. Fersman po kristallografii" ("The Work of A. E. Fersman in Crystallography"), in *A. E. Fersman. Kristallografia almaza* ("A. E. Fersman. Crystallography of the Diamond," Moscow, 1955), pp. 532–546; D. I. Shcherbakov, "Aleksandr Evgenievich Fersman i ego tvorchestvo" ("Aleksandr Evgenievich Fersman and His Work"), *ibid.,* pp. 490–531; "A. E. Fersman i ego puteshestvia" ("A. E. Fersman and His Travels," Moscow, 1953); "Osnovnye cherty tvorchestva A. E. Fersmana i drugikh" ("Basic Features of the Creative Work of A. E. Fersman and Others," in *Voprosy geokhimii i mineralogii* ("Questions of Geochemistry and Mineralogy," Moscow, 1956), pp. 1–175; and *Aleksandr Evgenievich Fersman. Zhizn i deyatelnost* ("Aleksandr Evgenievich Fersman. Life and Work"), of which Shcherbakov was editor (Moscow, 1965).

Other sources include V. V. Shcherbina *et al.,* in *Byulleten Moskovskogo obshchestva ispytateley prirody,* n.s. **51**, no. 1 (1946), 90–97; O. M. Shubnikova, "Ocherk zhizni i deyatelnosti A. E. Fersman i drugikh" ("Essay on the Life of A. E. Fersman and Others"), in *Zapiski Vserossyskogo mineralogicheskogo obshchestva,* 2nd ser., **75**, no. 1 (1946), 55–64; A. V. Sidorenko, "Issledovania A. E. Fersmana v Turkmenii i ikh znachenie" ("The Research of A. E. Fersman in Turkmen and Its Significance"), in *Izvestiya Turkmenistanskogo filiala Akademii nauk USSR,* no. 1 (1950), 28–39; and N. D. Zelinsky, "Pamyati akademika A. E. Fersmana" ("Memories of Academician A. E. Fersman"), in *Uspekhi khimii,* **14**, no. 6 (1945), 463–467.

A. MENIAILOV

FESSENDEN, REGINALD AUBREY (*b.* Milton, Quebec, 6 October 1866; *d.* Hamilton, Bermuda, 22 July 1932), *radio engineering.*

Fessenden was the son of Rev. E. J. Fessenden and Clementina Trenholme Fessenden; his father had charge of a small parish in East Bolton, Quebec. When the boy was nine, the family moved to Niagara Falls, Ontario, where he entered De Veaux Military College; he later attended Trinity College School at Port Hope, Ontario, and Bishop's College at Lennoxville, Quebec. Fessenden's first position was as principal of Whitney Institute in Bermuda, but he gave it up after two years and took up a relatively lowly job as tester in the New York factory of Thomas Edison, who was his idol. He soon graduated to Edison's New Jersey laboratory, where he was encouraged to specialize in solving chemical problems. In 1890 Fessenden went to work for the Westinghouse Electric and Manufacturing Co.

In 1892 Fessenden was named professor of electrical engineering at Purdue University; after a year he moved to a similar position at the Western University of Pennsylvania (now the University of Pittsburgh), where he remained for seven years. He next served as a special agent for the U.S. Weather Bureau from 1900 to 1902; his assignment was to adapt radiotelegraphy to weather forecasting and storm warning.

Fessenden's first contribution was the development of the electrolytic detector in 1900 (patent granted in 1903), a device sufficiently more sensitive than the primitive radiotelegraphy detectors of the day to make radiotelephony feasible for the first time. This invention led to other ideas, such as the use of a specially designed alternator as the source of high-frequency oscillations (one machine produced 50,000 cycles per second) and the invention of the heterodyne receiver, forerunner of the superhet. Many of his ideas were in advance of the times and were not elaborated until many years later, by others.

In 1902 two Pittsburgh financiers, Thomas H. Given and Hay Walker, Jr., formed the National Electric Signalling Company to exploit Fessenden's ideas and made him general manager. The firm made many contributions during the eight years of its existence; its station at Brant Rock, Massachusetts, transmitted the first voice signals over long distances in 1906, and the company manufactured radio equipment. But its dreams of competing with the American Marconi Company in establishing an international communications network came to naught when Fessenden demanded that a Canadian subsidiary controlled by himself should run a link with Britain, an arrangement opposed by his backers. Fessenden sued and won a judgment of $406,000, sending the company into bankruptcy.

During his career Fessenden obtained some 300 patents. Not a few became the subject of litigation; in one case, he sued the Radio Corporation of America for $60 million, asserting that he was being prevented from selling devices based on his own patents. The suit was settled out of court. He remained a controversial figure. Among his admirers was Elihu Thomson, himself a prominent inventor and engineer, who is said to have described Fessenden as "the greatest wireless inventor of the age—greater than Marconi." It is difficult to escape the conclusion that many of the fights in which Fessenden became involved were traceable to a choleric temperament and a persistent fear that men of business were getting the best of him—a fear not entirely without justification in the early days of radio.

BIBLIOGRAPHY

There is a biography by his widow, Helen M. Fessenden, *Fessenden, Builder of Tomorrows* (New York, 1940). For an account of Fessenden's role in the development of radio, see G. L. Archer, *History of Radio to 1926* (New York, 1938), pp. 67 ff. An obituary is in the *New York Times* (24 July 1932), p. 22.

CHARLES SÜSSKIND

FEUERBACH, KARL WILHELM (*b.* Jena, Germany, 30 May 1800; *d.* Erlangen, Germany, 12 March 1834), *mathematics.*

Karl Wilhelm was the third of the eleven children of Eva Wilhelmine Maria Troster and the famed

German jurist Paul Johann Anselm Feuerbach. By the age of twenty-two, the gifted young mathematician had been awarded the Ph.D., had made a significant contribution to a pleasant and active branch of mathematical research, and had been named professor of mathematics at the Gymnasium at Erlangen.

Feuerbach's scientific output was small, and his fame as a mathematician rests entirely upon three publications, which constitute the total output of his scientific career. His most important contribution was a theorem in Euclidean geometry, the theorem of Feuerbach:

> The circle which passes through the feet of the altitudes of a triangle touches all four of the circles which are tangent to the three sides of the triangle; it is internally tangent to the inscribed circle and externally tangent to each of the circles which touch the sides of the triangle externally [*Eigenschaften*].

In this statement one recognizes the nine-point circle of a triangle, which had been fully described though not named by Brianchon and Poncelet in 1821. The proof of this theorem was presented with a number of other conclusions on the geometry of the triangle in his small book *Eigenschaften einiger merkwürdigen Punkte* . . ., published in 1822. In this work Feuerbach developed a number of algebraic identities involving the lengths of the sides and other parts of a triangle and then proved that the two circles in question were tangent by showing that the distance between their centers was equal to the sum of their radii. He used as a model for this investigation Euler's "Solutio facilis problematum," a paper that had been published in 1765. Recognition came slowly, but many years after his death a number of papers appeared devoted to a discussion of the nine-point circle of a triangle and the theorem of Feuerbach.

In 1827 Feuerbach brought out the results of his second investigation. After an exhaustive analysis of this work, Moritz Cantor concluded that Feuerbach had proved to be an independent co-discoverer with Moebius of the theory of the homogeneous coordinates of a point in space. In the meantime, however, Feuerbach's teaching career was beset by difficulties and his health had become seriously impaired. At the age of twenty-eight he retired permanently and spent the rest of his life in Erlangen as a recluse.

BIBLIOGRAPHY

I. ORIGINAL WORKS. Feuerbach's works are *Eigenschaften einiger merkwürdigen Punkte des geradlinigen Dreiecks und mehrerer durch sie bestimmten Linien und Figuren. Eine analytisch-trigonometrische Abhandlung* (Nuremberg, 1822), a portion of which was ed. and trans. by R. A. Johnson as "Feuerbach on the Theorem Which Bears His Name," in D. E. Smith, ed., *A Source Book in Mathematics* (New York–London, 1929), paperback repr. (New York, 1959); "Einleitung zu dem Werke Analysis der dreyeckigen Pyramide durch die Methode der Coordinaten und Projectionen. Ein Beytrag zu der analytischen Geometrie," in L. Oken, *Isis*, VI (Jena, 1826), 565; and *Grundriss zu analytischen Untersuchungen der dreyeckigen Pyramide* (Nuremberg, 1827). An unpublished MS on the theory of the triangular pyramid, dated 7 July 1826, is in the Feuerbach family archives.

II. SECONDARY LITERATURE. On the man and his work, see C. J. Brianchon and J. V. Poncelet, "Géométrie des courbes: Recherches sur la détermination d'une hyperbole équilatère, au moyen de quatre conditions données," in *Annales de mathematiques,* **11** (1 Jan. 1821); M. Cantor, "Karl Wilhelm Feuerbach," in *Sitzungsberichte der Heidelberger Akademie der Wissenschaften,* Math.-naturwissen. Klasse, Abh. 25 (1910); L. Euler, "Solutio facilis problematum quorumdam geometricorum difficillimorum," in *Novi commentarii academiae scientiarum imperialis Petropolitanae,* **11** (1765), 103; H. Eulenberg, *Die Familie Feuerbach* (Stuttgart, 1924); L. Feuerbach, *Anselm Ritter von Feuerbachs biographischer Nachlass* (Leipzig, 1853); L. Guggenbuhl, "Karl Wilhelm Feuerbach, Mathematician," in *Scientific Monthly,* **81** (1955), 71; R. A. Johnson, *Modern Geometry: An Elementary Treatise on the Geometry of the Triangle and the Circle* (Boston–New York, 1929), repr. in paperback as *Advanced Euclidean Geometry* (New York, 1960); J. Lange, "Geschichte des Feuerbachschen Kreises," in *Wissenschaftliche Beilage zum Jahresbericht der Friedrichs Wederschen Ober-Realschule zu Berlin Programme No. 114* (1894); J. S. Mackay, "History of the Nine Point Circle," in *Proceedings of the Edinburgh Mathematical Society,* **11** (1892), 19; G. Radbruch, *Paul Johann Anselm Feuerbach* (Vienna, 1934) and *Gestalten und Gedanken. Die Feuerbachs, Eine geistige Dynastie* (Leipzig, 1948); T. Spoerri, *Genie und Krankheit* (Basel–New York, 1952); J. Steiner, *Die geometrischen Constructionen ausgefuhrt mittelst der geraden Linie und eines festen Kreises* (Berlin, 1833); and Olry Terquem, "Considération sur le triangle rectiligne," in *Nouvelles annales de mathématiques,* **1** (1842), 196.

LAURA GUGGENBUHL

FEUILLÉE, LOUIS (*b.* Mane, Basses-Alpes, France, 1660; *d.* Marseilles, France, 18 April 1732), *astronomy, botany.*

Feuillée made astronomical observations when he was only ten years old. He was then in the service of the Minims, having entered that order in Avignon on 2 March 1680. In 1699 he accompanied Jacques Cassini on his exploration of the Greek coast and in 1703 went to the Antilles and the South American coast, returning to Brest in 1706. As a result of this mission he was appointed royal mathematician and corresponding member of the Académie des Sciences in Paris.

In 1707 Feuillée sailed for the west coast of South America, via Cape Horn, landing in Chile, where he remained until 1709. He explored the Chilean and Peruvian coasts, returning to Brest by 27 August 1711. Louis XIV placed in his care an observatory built at Marseilles. In 1724 he was again commissioned by the Academy to establish the longitude of the meridian of Hierro Island in the Canaries.

The work of Feuillée is interesting and reliable, although it was marred by his controversy with A. F. Frézier, who explored the west coast of South America between 1711 and 1714. His description of the flora of the coasts of Peru and Chile is still much appreciated.

BIBLIOGRAPHY

Feuillée's astronomical observations were published in the *Mémoires de l'Académie des sciences* (1699–1710). His best-known work is the *Journal des observations physiques, mathématiques, et botaniques, faites sur les côtes orientales de l'Amérique Méridionale et dans les Indes Occidentales de 1707 à 1712* (Paris, 1714), which was followed by the *Suite du journal . . .* (Paris, 1725). In the *Histoire des plantes médicinales qui son les plus d'usage aux royaumes du Pérou et du Chili* (Paris, 1714–1725) he incorporated the description, illustration, and uses of 100 medicinal plants.

Francisco Guerra

FEULGEN, ROBERT JOACHIM (*b.* Werden, Germany, 2 September 1884; *d.* Giessen, Germany, 24 October 1955), *biochemistry, histochemistry.*

The son of a textile worker, Feulgen was educated in Werden, Essen, and Soest; in 1905 he entered the medical faculty of the University of Freiburg im Breisgau. For the completion of his training he went to Kiel, and while working in the city hospital he prepared his dissertation on the purine metabolism of patients afflicted with chronic gout. The years 1912–1918 were spent in Berlin at the Physiological Institute, the chemistry section of which was headed by the nucleic acid chemist Hermann Steudel. Feulgen's extension and criticism of Steudel's work formed the subject of his *Habilitationsschrift* in 1919. The following year he was appointed to the Physiological Institute at Giessen, where he spent the rest of his life, rising from assistant professor in 1923 to associate professor in 1927 and director of the Physiological-Chemical Institute in 1931.

In Berlin, Feulgen improved on Steudel's extraction technique for thymonucleic acid (DNA), so that the product gave no biuret reaction and dissolved readily in water to give a colorless solution. By combining nucleic acid with Congo red and malachite green, extracting the salts formed, and subjecting these to an elementary analysis, he believed he had obtained more reliable percentage compositions. His matching of the resulting nitrogen : phosphorus ratios with the ratios predicted from various molecular structures strengthened rather than shook his confidence in the tetranucleotide hypothesis.

Feulgen's major discovery came in 1914, when he took up Steudel's observation (1908) of the reducing action of apurinic acid (then called thymic acid) on Fehling's solution. Feulgen found that phenylhydrazine reacted with this acid, indicating the presence of aldehyde groups, and that Schiff's reagent gave the magenta color indicative of furan. By the use of aldehyde blocking controls Feulgen was led to the correct conclusion that in the mild hydrolysis of thymonucleic acid to apurinic acid, loss of purines exposes aldehyde groups, which easily give rise to the furan structure. By treating thymonucleic acid with NHCl for ten minutes (optimum is pH 7) before application of Schiff's reagent, he obtained the magenta color. Untreated thymonucleic acid failed to produce it, as did the RNA of yeast nucleic acid. It was already known that the carbohydrates in thymonucleic and yeast nucleic acid differ, the latter being a pentose sugar and the former probably a hexose sugar, in Feulgen's opinion glucal—a compound discovered that year by Fischer—because it contains the furan structure, and he called his discovery the nucleal reaction.

In 1923, nine years after his discovery of the nucleal reaction, Feulgen applied it as a histochemical stain. In this way he was able to show that thymonucleic acid is found only in the nucleus and that both plant and animal cells give a positive nucleal reaction. Although he failed to detect the nuclei in yeast cells, he rightly concluded that the pentose nucleic acid of yeast is localized in the cytoplasm. At Feulgen's first demonstration of these results, at the Congress of Physiology held at Tübingen in 1923, Albrecht Kossel was impressed, but otherwise there was skepticism. By a thorough examination of the technique and by the use of aldehyde blocking, Feulgen established his test. In 1937 he succeeded in isolating rye germ nuclei that gave the nucleal reaction. This work effectively banished the old division of nucleic acids into the thymonucleic acids of animals and the yeast (pentose) nucleic acids of plants and established in its place the occurrence of both in the same cell. Although modern research has shown the presence of RNA in the nucleus and of DNA in the cytoplasm, the major part of these acids is still distributed in the way shown by Feulgen's test.

In 1924 Feulgen and Voit discovered a positive nucleal reaction in the cytoplasm without previous mild hydrolysis. This "plasmal" reaction they showed to be due to aldehyde groups; and finding that lipide solvents negative the test, they concluded that a lipide precursor is responsible. In 1928 they isolated "plasmalogen," and eleven years later they identified it as an acetal phosphatide.

Feulgen was orthodox in his acceptance of the tetranucleotide hypothesis, according to which thymonucleic acid is an oligonucleotide formed from a nucleotide of each of the bases thymine, cytosine, adenine, and guanine. Later he expressed reservations, and in 1936 it was clear to him that the undegraded material is a polymer and that the usual extractive procedures yield a mixture of depolymerized fragments. Studies of viscosity and optical activity gave him evidence of this change from what he termed the *a* form to the *b* form. He also discovered that conversion of *a* to *b* can be achieved by the action of a commercial preparation of the pancreatic juice. The depolymerizing enzyme that he believed to be present he called nucleogelase. From the same material M. McCarthy later obtained DNA depolymerase. These findings received little attention, and today Feulgen is remembered not for them but for his introduction of the nucleal reaction, which transformed nucleic acid cytochemistry. He was a skillful experimentalist whose certainty of the specific character of his nucleal reaction has been justified despite nearly thirty years of disputation over its nature and specificity.

BIBLIOGRAPHY

I. ORIGINAL WORKS. Feulgen and his co-workers published seventy original papers, four contributions to biochemical textbooks, and a scholarly review of nucleic acid chemistry: "Chemie und Physiologie der Nucleinstoffe," in A. Kanitz, ed., *Die Biochemie in Einzeldarstellungen* (Berlin, 1923). A complete list of his publications will be found in F. H. Kasten and in K. Felix (see below). The majority of his papers appeared in *Hoppe-Seyler's Zeitschrift für physiologische Chemie;* the most important are "Über die 'Kohlenhydratgruppe' in der echten Nucleinsäure. Vorläufige Mitteilungen," **92** (1914), 154–158, announcing the nucleal reaction; "Mikroskopisch-chemischer Nachweis einer Nucleinsäure vom Typus der Thymonucleinsäure und die darauf beruhende elektive Färbung von Zellkernen in mikroscopischen Präparaten," **135** (1924), 203–248, on the Feulgen stain, written with H. Rossenbeck; and "Die Darstellung der b-Thymonucleinsäure mittels der Nucleogelase," **238** (1936), 105–110. The plas-

mal reaction was described in "Über einen weitverbreiteten Stoff (Plasmal, Plasmalogen), seinen histologischen Nachweis und seiner Beziehungen zum Geruch des gekochten Fleisches," in *Klinische Wochenschrift,* **4** (1925), 1330, written with K. Voit.

II. SECONDARY LITERATURE. The best scientific biography of Feulgen is K. Felix, "Robert Feulgen zum Gedächtnis," in *Hoppe-Seyler's Zeitschrift für physiologische Chemie,* **307** (1957), 1–13, with portrait and full bibliography. An English summary of this essay plus additional biographical data is in F. H. Kasten, "Robert Feulgen," in W. Sandritter, ed., *Hundred Years of Histochemistry in Germany* (Stuttgart, 1964), pp. 97–101, also with portrait and bibliography. For brief details of Feulgen's life see *Leopoldina,* **1** (1955), 52–53; and Poggendorff, VIIa, pt. 2 (Berlin, 1958), 31. The most authoritative account of the debate over the mechanism and specificity of the Feulgen reaction is A. G. E. Pearse, *Histochemistry Theoretical and Applied,* 2nd ed. (London, 1960), pp. 193–201.

ROBERT OLBY

FIBONACCI, LEONARDO, or **LEONARDO OF PISA** (*b.* Pisa, Italy, *ca.* 1170; *d.* Pisa, after 1240), *mathematics.*

Leonardo Fibonacci, the first great mathematician of the Christian West, was a member of a family named Bonacci, whose presence in Pisa since the eleventh century is documented. His father's name is known to have been Guilielmo. It is thus that Fibonacci is to be understood as a member of the Bonacci family and not as "son of a father of the name of Bonacci," as one might suppose from the words "filio Bonacij" or "de filiis Bonacij," which appear in the titles of many manuscripts of his works. The sobriquet "Bigollo" (from *bighellone,* loafer or ne'er-do-well), used by Leonardo himself, remains unexplained. Did his countrymen wish to express by this epithet their disdain for a man who concerned himself with questions of no practical value, or does the word in the Tuscan dialect mean a much-traveled man, which he was?

Life. Leonardo himself provides exact details on the course of his life in the preface to the most extensive and famous of his works, the book on calculations entitled *Liber abbaci* (1202). His father, as a secretary of the Republic of Pisa,[1] was entrusted around 1192 with the direction of the Pisan trading colony in Bugia (now Bougie), Algeria. He soon brought his son there to have him learn the art of calculating, since he expected Leonardo to become a merchant. It was there that he learned methods "with the new Indian numerals," and he received excellent instruction (*ex mirabili magisterio*). On the business trips on which his father evidently soon sent him and which took him to Egypt, Syria, Greece (Byzantium), Sicily, and

604

Provence, he acquainted himself with the methods in use there through zealous study and in disputations with native scholars. All these methods, however—so he reports—as well as "algorismus" and the "arcs of Pythagoras" (apparently the abacus of Gerbert) appeared to him as in "error" in comparison with the Indian methods.[2] It is quite unclear what Leonardo means here by the "algorismus" he rejects; for those writings through which the Indian methods became known, especially after Sacrobosco, a younger contemporary of Leonardo, bear that very name. Could he mean the later *algorismus linealis,* reckoning with lines, the origin of which is, to be sure, likewise obscure?

Around the turn of the century Leonardo returned to Pisa. Here for the next twenty-five years he composed works in which he presented not only calculations with Indian numerals and methods and their application in all areas of commercial activity, but also much of what he had learned of algebraic and geometrical problems. His inclusion of the latter in his own writings shows that while the instruction of his countrymen in the solution of the problems posed by everyday life was indeed his chief concern, he nevertheless also wished to provide material on theoretical arithmetic and geometry for those who were interested in more advanced questions. He even speaks once of wanting to add the "subtleties of Euclid's geometry";[3] these are the propositions from books II and X of the *Elements,* which he offers to the reader not only in proofs, in Euclid's manner, but in numerical form as well. His most important original accomplishments were in indeterminate analysis and number theory, in which he went far beyond his predecessors.

Leonardo's importance was recognized at the court of the Hohenstaufen emperor Frederick II. Leonardo's writings mention the names of many of the scholars of the circle around the emperor, including Michael Scotus, a court astrologer whom Dante (*Inferno,* XX, 115 ff.) banished to hell; the imperial philosopher, Master Theodorus; and Master Johannes of Palermo. Through a Master Dominicus, probably the Dominicus Hispanus mentioned by Guido Bonatti (see Boncompagni, *Intorno ad alcune opere di Leonardo Pisano,* p. 98, n.), Leonardo was presented to the emperor, who evidently desired to meet him, when Frederick held court in Pisa about 1225.[4] After 1228 we know almost nothing more concerning Leonardo's activity in Pisa. Only one document has survived, from 1240, in which the Republic of Pisa awards the "serious and learned Master Leonardo Bigollo" (*discretus et sapiens*) a yearly *salarium* of "libre XX denariorem" in addition to the usual allowances, in recognition of his usefulness to the city and its citizens through his teaching and devoted services. He evidently had advised the city and its officials, without payment, on matters of accounting, a service the city expected him to continue. This decree of the city, which was inscribed on a marble tablet in the Pisa city archives in the nineteenth century,[5] is the last information we have on Leonardo's life.

Writings. Five works by Leonardo are preserved:

1. The *Liber abbaci* (1202, 1228);
2. The *Practica geometriae* (1220/1221);
3. A writing entitled *Flos* (1225);
4. An undated letter to Theodorus, the imperial philosopher;
5. The *Liber quadratorum* (1225). We know of further works, such as a book on commercial arithmetic, *Di minor guisa;*[6] especially unfortunate is the loss of a tract on book X of the *Elements,* for which Leonardo promised a numerical treatment of irrationals instead of Euclid's geometrical presentation.[7]

Leonardo's works have been collected in the edition by Boncompagni; in 1838 Libri edited only one chapter of the *Liber abbaci.* Boncompagni, however, provides only the Latin text without any commentary. Hence, despite much specialized research on the *Flos* and on the *Liber quadratorum,* which Ver Eecke has translated into French, there is still no exhaustive presentation of Leonardo's problems and methods. The most detailed studies of the substance of the works are those by Cantor, Loria, and Youschkevitch.

Liber abbaci. The word *abacus* in the title does not refer to the old abacus, the sand board; rather, it means computation in general, as was true later with the Italian masters of computation, the *maestri d'abbaco.* Of the second treatment of 1228, to which "new material has been added and from which superfluous removed," there exist twelve manuscript copies from the thirteenth through the fifteenth centuries; but only three of these from the thirteenth and the beginning of the fourteenth centuries are complete. Leonardo divided this extensive work, which is dedicated to Michael Scotus, into fifteen chapters; it will be analyzed here in four sections.

Section 1 (chapters 1–7; *Scritti,* I, 1–82). Leonardo refers to Roman numerals and finger computation, which the student still needs for marking intermediate results.[8] Then the Indian numerals are introduced; following the Arabic manner, the units stand "in front" (on the right), and the fractions are on the left of the whole numbers. In addition, he introduces the fraction bar. All the computational operations are taught methodically through numerous examples and the results are checked, mostly by the method of casting out nines (seven and eleven are also used in this way). Rules are developed for the factoring of fractions into sums of unit factors. Various symbols are introduced for the representation of fractions. Thus, for example, $\dfrac{6\ \ 2}{7\ \ 5}$ is to be read as $\dfrac{2}{5} + \dfrac{6}{7 \cdot 5}$; $0\dfrac{6\ \ 2}{7\ \ 5}$ means $\dfrac{2}{5} \cdot \dfrac{6}{7}$; and $\dfrac{6\ \ 2}{7\ \ 5}0$ is to be understood as $\dfrac{2}{5} + \dfrac{6}{7}$. Finally, $\dfrac{1\ \ 1\ \ 1\ \ 5}{5\ \ 4\ \ 3\ \ 9}$ signifies $\dfrac{5}{9} + \left(\dfrac{1}{3} + \dfrac{1}{4} + \dfrac{1}{5}\right) \cdot \dfrac{5}{9}$. The first—and the most frequently employed—of these representational methods corresponds to the ascending continued fraction $\dfrac{2 + \dfrac{6}{7}}{5}$. Numerous tables (for multiplication, prime numbers, factoring numbers, etc.) complete the text.

Section 2 (chapters 8–11; *Scritti,* I, 83–165). This section contains problems of concern to merchants, such as the price of goods, calculation of profits, barter, computation of interest, wages, calculations for associations and partnerships, metal alloys, and mixture calculations; the computations of measurements and of currency conversions in particular reflect the widespread trade of the medieval city with the lands bordering the Mediterranean. One of the mixture problems included is known from Chinese mathematics, the "problem of the 100 birds"; a problem in indeterminate analysis, it requires that one purchase for 100 units of money 100 birds of different sorts, the price of each sort being different.

Section 3 (chapters 12 and 13; *Scritti,* I, 176–351). This is the most extensive section and contains problems of many types, which are called *erraticae questiones.* They are mostly puzzles, such as are found in the mathematical recreations of all times. Among them are the "cistern problems" (A spider climbing the wall of a cistern advances so many feet each day and slips back so many feet each night. How long will it take it to climb out?) and, from Egyptian mathematics, the famous so-called "hau calculations," which can be expressed in the form $ax \pm b/c \cdot x = s$. Leonardo calls them *questiones arborum* after the first example, in which a tree is supposed to stand twenty-one ells above the ground with 7/12 of its length in the earth; therefore, $x - 7/12\ x = 21$. Another group are "motion problems," involving either pursuit (as in the famous "hare and hound" problem, in which one must determine how long it will take a hound chasing a hare at a proportional speed to catch the hare) or opposite movements. In both cases the motions can be delayed through backward movements. Since in many problems the speed is not con-

stant, but increases arithmetically, rules for the summation of series are given at the beginning of chapter 12. A group of problems that had already appeared in the epigrams of *The Greek Anthology* (a recent edition is W. R. Paton, ed. [Cambridge, Mass., 1953], V, 25 ff.) and can be designated as "giving and taking," is called by Leonardo *de personis habentibus denarios;* in these there are two or more people, each of whom demands a certain sum from one or several of the others and then states what the proportion now is between his money and that of the others. A simple example is (1) $x + 7 = 5 \cdot (y - 7)$; (2) $y + 5 = 7 \cdot (x - 5)$. In the problem of "the found purse" (*de inventione bursarum*) two or more people find a purse, and we are told for each individual what ratio the sum of his money and the total money in the purse has to the sum of the remaining individuals' monies; for example, with three people the modern arrangement would be (1) $x + b = 2 \cdot (y + z)$; (2) $y + b = 3 \cdot (x + z)$; (3) $z + b = 4(x + y)$. They are, therefore, problems in indeterminate analysis.

Another very extensive group, "one alone cannot buy," takes the form of "horse buying" (*de hominibus equum emere volentibus*). In this case it is given that one of those concerned can buy an object only if he receives from the other (or others) a portion of his (or their) cash.[9] Variations are also given that involve up to seven people and five horses; in these cases, if the price of the horse is not known, the problem is indeterminate. A problem of this type involving three people, where the equation would be

$$x + \frac{y + z}{3} = y + \frac{z + x}{4} = z + \frac{x + y}{5} = s,$$

corresponds to Diophantus II, 24. A further group treats the business trips of a merchant, which are introduced as *de viagiis.* These are the famous problems of the "gate-keeper in the apple garden." It is here that the problems involving mathematical nesting of the form $\langle [(a_1 x - b_1) \cdot a_2 - b_2] \cdot a_3 - \cdots \rangle \cdot a_n - b_n = s$ are to be solved.[10] Of the multitude of other problems treated in the *Liber abbaci,* the following should be mentioned: numerous remainder problems, in which, for example, a number n is sought with the property $n \equiv 1$ (mod. 2, 3, 4, 5, and 6) $\equiv 0$ (mod. 7); the Chinese remainder problem *Ta yen,*[11] the finding of perfect numbers; the summation of a geometric series; the ancient Egyptian problem of the "seven old women" (to find $\sum\limits_{i=1}^{n} 7^i$, the seven wives of St. Ives); the Bachet weight problem; the chess problem (to find $\sum\limits_{i=0}^{64} 2^i$); and the rabbit problem. This last problem assumes that a pair of rabbits requires one month to mature and thereafter reproduces itself once each month. If one starts with a single pair, how many pairs will one have after n months? The answer leads to the famous Fibonacci series, the first recurrent series. Its general form for any term k_n is $k_n = k_{n-1} + k_{n-2}$; and, in this case, it can be expressed as

$$k_n = \frac{1}{\sqrt{5}} \cdot \left[\left(\frac{1 + \sqrt{5}}{2} \right)^n - \left(\frac{1 - \sqrt{5}}{2} \right)^n \right].$$

Leonardo demonstrates an astonishing versatility in the choice of methods of solution to be used in particular instances; he frequently employs a special procedure, for which he usually has no specific name and which has been tailored with great skill to fit the individual problem. He also shows great dexterity in the introduction of an auxiliary unknown; in this he is like Iamblichus, who demonstrated the same talent in his explanation of the *Epanthema* of Thymaridas of Paros. At other times Leonardo makes use of definitely general methods. These include the simple false position, as in the "hau calculations" and the *regula versa,* in which the calculation is made in reverse order in the nesting problems in *de viagiis;* there is also the double false position, to which the whole of chapter 13 is devoted and which is called—as in Leonardo's Arabic models—*regula elchatayn.* With this rule, linear and pure quadratic problems can be solved with the aid of two arbitrarily chosen quantities, a_1 and a_2, of unknown magnitude and the resulting errors, f_1 and f_2. Leonardo knew this procedure, but he generally used a variation. The latter consists in ascertaining from the two errors how much closer one has come to the true answer (*veritati appropinquinare*)[12] in the second attempt and then determining the number that one must now choose in order to obtain the correct solution. A special solution for an indeterminate problem is provided by the *regula proportionis.* If, for example, in the final equation of a problem $63/600\, x = 21/200\, b$, then, according to this rule, $x = 21/200$ and $b = 63/600$ or, in whole numbers, $x = 63$ and $b = 63$.

Leonardo also employed, as easy mechanical solutions, formulas (especially in the "horse buying," "found purse," and "journey" problems) that can have been obtained only by means of algebra. He knew the algebraic methods very well; he called them *regulae rectae* and stated that they were used by the Arabs and could be useful in many ways. He called the unknown term *res* (Arabic *shaiᵓ,* "thing"); and since he used no operational symbols and no notations for further unknowns here (see, however, under *Practica geometriae*), he had to designate them as

denarii secundi or, as the case might be, *denarii tercii hominis* and take the trouble to carry them through the entire problem. For most of the problems Leonardo provided two or more methods of solution.

An example of the "giving and taking" type is the following, in which the system named above—(1) $x + 7 = 5 \cdot (y - 7)$; (2) $7 \cdot (x - 5) = y + 5$—is involved. First, $x + y$ is presented as a line segment; at the point of contact $y = 7$ and $x = 5$ is marked off along both sides. Then the segment $y - 7$ (or $x - 5$) is equal to 1/6 (or 1/8) of the whole segment $x + y$, and together, therefore, the two segments equal $7/24 \cdot (x + y)$. The further solution is achieved by means of the simple false position $x + y = 24$. There follows still another algebraic solution, this one using the *regulae rectae.* First, $y - 7$ is designated as *res;* then (1) $x = 5\ res - 7$ and (2) $res + 12 = 7 \cdot (5\ res - 12)$.[13]

In some cases the problem is not solvable because of mutually contradictory initial conditions. In other cases the problem is called *insolubilis, incongruum,* or *inconveniens,*[14] unless one accepts a "debit" as a solution. Leonardo is here thinking of a negative number, with which he also makes further calculations. Our operations $22 + (-9) = 22 - 9$ and $-1 + 11 = +10$ he represents with *adde denarios,* as 22 *cum debito secundi* $(=9)$ *scilicet extrahe* 9 *de* 22 and *debitum primi* $(= -1)$ *cum bursa* $(=11)$ *erunt* 10.[15]

Section 4 (chapters 14 and 15; *Scritti,* I, 352–387). Leonardo here shows himself to be a master in the application of algebraic methods and an outstanding student of Euclid. Chapter 14, which is devoted to calculations with radicals, begins with a few formulas of general arithmetic. Called "keys" (*claves*), they are taken from book II of Euclid's *Elements.* Leonardo explicitly says that he is forgoing any demonstrations of his own since they are all proved there. The fifth and sixth propositions of book II are especially important; from them, he said, one could derive all the problems of the *Aliebra* and the *Almuchabala.* Square and cube roots are taught numerically according to the Indian-Arabic algorithm, which in fact corresponds to the modern one.

Leonardo also knew the procedure of adding zeros to the radicands in order to obtain greater exactness; actually, this had already been done by Johannes Hispalensis (*fl.* 1135–1153) and al-Nasawī (*fl. ca.* 1025). Next, examples are given that are illustrative of the ancient methods of approximation. For $\sqrt{A} = \sqrt{a^2 + r}$ the first approximation is $a_1 = a + r/2a$. With $r_1 = a_1^2 - A$, the second approximation is then $a_2 = a_1 - r_1/2a_1$. With the cube root $\sqrt[3]{A} = \sqrt[3]{a^3 + r}$, the first approximation is

$$a_1 = a + \frac{r}{(a+1)^3 - a^3} = a + \frac{r}{3a^2 + 3a + 1}.$$

For a second approximation Leonardo now set $r_1 = A - a_1^3$ and

$$a_2 = a_1 + \frac{r_1}{3a_1 \cdot (a+1)}.$$

He was no doubt thinking of this further approximation when he spoke of his own achievement,[16] for the first approximation was already known to al-Nasawī. The chapter then goes on systematically to carry out complete operations with Euclidean irrationals. There are expressions such as

$$a \pm \sqrt{b}, \quad \sqrt{a} \pm \sqrt{b}, \quad \sqrt{a} \cdot \sqrt{b}, \quad \sqrt{a} \cdot \sqrt[4]{b},$$

$$\sqrt[4]{a} \cdot \sqrt[4]{b}, \quad \sqrt[4]{a} \pm \sqrt[4]{b}, \quad (a + \sqrt{b}) \cdot (c \pm \sqrt{d}),$$

$$(a + \sqrt[4]{b}) \cdot (\sqrt{c} \pm \sqrt[4]{d}), \quad \frac{a - \sqrt{b}}{\sqrt{c} + \sqrt{d}}, \quad \sqrt{a + \sqrt{b}},$$

$$\frac{a}{b + \sqrt{c} + \sqrt[4]{d}}, \quad \sqrt[3]{a} + \sqrt[3]{b}.$$

The proof, which is never lacking, of the correctness of the calculations is presented geometrically. On one occasion the numbers are represented as line segments, for example, in the computation of

$$4 + \sqrt[4]{10} = \sqrt{16 + \sqrt{10} + 8 \cdot \sqrt[4]{10}},$$

where proposition 4 of book II of the *Elements* is used as a "key." On the other hand, the proof is made by means of rectangular surfaces. An example is $\sqrt{4 + \sqrt{7}} + \sqrt{4 - \sqrt{7}} = \sqrt{14}$. Here $(4 + \sqrt{7})$ is conceived as the area of a square, to which at one corner, through the elongation of the two intersecting sides, a square of area $(4 - \sqrt{7})$ is joined. Thus $\sqrt{4 + \sqrt{7}} + \sqrt{4 - \sqrt{7}}$ is the side of a larger square, which consists of the squares $4 + \sqrt{7}$ and $4 - \sqrt{7}$ and the two rectangles each equal to $\sqrt{4 + \sqrt{7}} \cdot \sqrt{4 - \sqrt{7}}$.

With respect to mathematical content Leonardo does not surpass his Arab predecessors. Nevertheless, the richness of the examples and of their methodical arrangement, as well as the exact proofs, are to be emphasized. At the end of chapter 15, which is divided into three sections, one sees particularly clearly what complete control Leonardo had over the geometrical as well as the algebraic methods for solving quadratic equations and with what skill he could use them in applied problems. The first section is concerned with proportions and their multifarious transformations. In one problem, for example, it is given

that (1) $6 : x = y : 9$ and (2) $x + y = 21$. From (1) it is determined that $xy = 54$; then, using Euclid II, 5,

$$\left(\frac{x - y}{2}\right)^2 = \left(\frac{21}{2}\right)^2 - 54$$

and

$$x - y = 15.$$

From this follow the solutions 3 and 18.[17] The end points of the segments are denoted by letters of the alphabet (*abgd*··· or *abcd*···); for example, .*a.b* signifies a segment. Leonardo, however, also speaks about the numbers *a.b.c.d.*, by which he means $(ab) \cdot (cd)$. Sometimes, though, only a single letter is given for the entire segment.

The second section first presents applications of the Pythagorean theorem, such as the ancient Babylonian problem of a pole leaning against a wall and the Indian problem of two towers of different heights. On the given line joining them (i.e., their bases) there is a spring which shall be equally distant from the tops of the towers. The same problem was solved in chapter 13 by the method of false position. Many different types of problems follow, such as the solution of an indeterminate equation $x^2 + y^2 = 25$, given that $3^2 + 4^2 = 25$; or problems of the type *de viagiis*, in which the merchant makes the same profit on each of his journeys. Geometric and stereometric problems are also presented; thus, for example, the determination of the amount of water running out of a receptacle when various bodies, including a sphere (with $\pi = 3\frac{1}{7}$), are sunk in.

The third section contains algebraic quadratic problems (*questiones secundum modum algebre*). First, with reference made to "Maumeht," i.e., to al-Khwārizmī, the six normal forms $ax^2 = bx$, $ax^2 = c$, $bx = c$, $ax^2 + bx = c$, $ax^2 + c = bx$ (here Leonardo is acquainted with both solutions), and $ax^2 = bx + c$ are introduced; they are then exactly computed in numerous, sometimes complicated, examples. Frequently what is sought is the factorization of a number, usually 10, for example,

$$\frac{x}{10 - x} + \frac{10 - x}{x} = \sqrt{5}.$$

Another problem is $\sqrt{8x} \cdot \sqrt{3x} + 20 = x^2$, and still another is $x^2 - 2x - 4 = \sqrt{8x^2}$. Here Leonardo represents x^2 as a square divided into three rectangular parts: $\sqrt{8x^2}$, $2x$, and 4. With the aid of Euclid II, 6 he then obtains $x = \sqrt{7 + \sqrt{8}} + (1 + \sqrt{2})$. Since in the problem a "fortune" ($x^2 = $ avere) was sought, the final solution is $x^2 = 10 + 2 \cdot \sqrt{8} + \sqrt{116 + 40 \cdot \sqrt{8}}$. Leonardo also includes equations of higher degrees that can be reduced to quadratics. For example, it is given that (1) $y = 10/x$; (2) $z = y^2/x$; and (3) $z^2 = x^2 + y^2$. This leads to $x^8 + 100x^4 = 10,000$. The numerical examples are taken largely from the algebra of al-Khwārizmī and al-Karajī,[18] frequently even with the same numerical values. In this fourth section of the *Liber abbaci* there also appear further names for the powers of the unknowns.

When several unknowns are involved, then (along with *radix* and *res* for x) a third unknown is introduced as *pars* ("part," Arabic, *qasm*); and sometimes the sum of two unknowns is designated as *res*. For x^2, the names *quadratus, census,* and *avere* ("wealth," Arabic, *māl*)[19] are employed; for x^3, *cubus;* for x^4, *census de censu* and *censuum census;* and for x^6, *cubus cubi.* The constant term is called *numerus, denarius,* or *dragma.*

Practica geometriae (*Scritti*, II, 1–224). This second work by Leonardo, which he composed in 1220 or 1221, between the two editions of the *Liber abbaci,* is dedicated to the Magister Dominicus mentioned above. Of the nine extant manuscripts one is in Rome, which Boncompagni used, and two are in Paris.[20] In this work Leonardo does not wish to present only measurement problems for the layman; in addition, for those with scientific interests, he considers geometry according to the method of proof. Therefore, the models are, on the one hand, Hero and the *Agrimensores,* and Euclid and Archimedes on the other. Leonardo had studied the *Liber embadorum* of Plato of Tivoli (1145) especially closely and took from it large sections and individual problems with the same numerical values. This work by Plato was a translation of the geometry of Savasorda (Abraham bar Hiyya), written in Hebrew, which in turn reproduced Arabic knowledge of the subject.

The *Practica* is divided into eight chapters (*distinctiones*), which are preceded by an introduction. In the latter the basic concepts are explained, as are the postulates and axioms of Euclid (including the spurious axioms 4, 5, 6, and 9) and the linear and surface measures current in Pisa.[21] The first chapter presents, in connection with the surfaces of rectangles, examples of the multiplication of segments, each of which is given in a sum of various units (rod, foot, ounce, etc.). The propositions of book II of the *Elements* are also recalled. The second chapter and the fifth chapter treat, as a preparation for the following problems, square and cube roots and calculation with them in a manner similar to that of the *Liber abbaci.* Next, the duplications of the cube by Archytas, Philo of Byzantium, and Plato, which are reported by Eutocius, are demonstrated, without reference to their source. The solutions of Plato and Archytas, Leonardo

took from the *Verba filiorum* of the Banū Mūsā, a work translated by Gerard of Cremona. That of Philo appears also in Jordanus de Nemore's *De triangulis,* and probably both Leonardo and Jordanus took it from a common source. (See M. Clagett, *Archimedes in the Middle Ages,* I, 224, 658–660.) The third chapter provides a treatment with exact demonstrations of the calculation of segments and surfaces of plane figures: the triangle, the square, the rectangle, rhomboids (*rumboides*), trapezoids (*figurae quae habent capita abscisa*), polygons, and the circle; for the circle, applying the Archimedean polygon of ninety-six sides, π is determined as $864:275 \sim 3.141818 \cdots$. In addition, Leonardo was acquainted with quadrilaterals possessing a reentrant angle (*figura barbata*) in which a diagonal falls outside the figure.

Many of the problems lead to quadratic equations, for which the formulas of the normal forms are used. They are given verbally. Hence, for example, in the problem $4x - x^2 = 3$, we are told: If from the sum of the four sides the square surface is subtracted, then three rods remain. Attention is also drawn here to the double solution. Along with this, Leonardo gives practical directions for the surveyor and describes instrumental methods, such as can be used in finding the foot of the altitude of a triangular field or in the computation of the projection of a field lying on a hillside. Among the geodetic instruments was an archipendulum. With the help of it and a surveyor's rod, the horizontal projections of straight lines lying inclined on a hillside could be measured. For the surveyor who does not understand the Ptolemaic procedure of determining half-chords from given arcs, appropriate instructions and a table of chords are provided. This is the only place where the term *sinus versus arcus,* certainly borrowed from Arabic trigonometry, appears. The fourth chapter is devoted to the division of surfaces; it is a reworking of the *Liber embadorum,* which ultimately derives from Euclid's lost *Book on Divisions of Figures;* the latter can be reconstructed (see Archibald) from the texts of Plato of Tivoli and of Leonardo and from that of an Arabic version. In the sixth chapter Leonardo discusses volumes, including those of the regular polyhedrons, in connection with which he refers to the propositions of book XIV of Euclid. The seventh chapter contains the calculation of the heights of tall objects, for example, of a tree, and gives the rules of surveying based on the similarity of triangles; in these cases the angles are obtained by means of a quadrant.

The eighth chapter presents what Leonardo had termed "geometrical subtleties" (*subtilitates*) in the preface to the *Liber abbaci.* Among those included is the calculation of the sides of the pentagon and the decagon from the diameter of circumscribed and inscribed circles; the inverse calculation is also given, as well as that of the sides from the surfaces. There follow two indeterminate problems: $a^2 + 5 = b^2$ and $c^2 - 10 = d^2$. The *Liber quadratorum* treats a similar problem: $a^2 + 5 = b^2$, together with $a^2 - 5 = c^2$. Finally, to complete the section on equilateral triangles, a rectangle and a square are inscribed in such a triangle and their sides are algebraically calculated, with the solution given in the sexagesimal system.

Flos (*Scritti*, II, 227–247). The title of this work, which—like two following ones—is preserved in a Milanese manuscript of 1225, is *INCIPIT flos Leonardi bigoli pisani super solutionibus quarumdam questionum ad numerum et ad geometriam vel ad utrumque pertinentium.* Sent to Frederick II, it contains the elaboration of questions that Master Johannes of Palermo posed in the emperor's presence in Pisa. The work had been requested by Cardinal Raniero Capocci da Viterbo; Leonardo, moreover, provided him with additional problems of the same type. For the first problem (involving the equations $x^2 + 5 = y^2$ and $x^2 - 5 = z^2$) only the solution is presented; it is treated in the *Liber quadratorum*. The second question that Master Johannes had posed concerns the solution of the cubic equation $x^3 + 2x^2 + 10x = 20$. Leonardo, who knew book X of the *Elements* thoroughly, demonstrates that the solution can be neither a whole number, nor a fraction, nor one of the Euclidean irrational magnitudes. Consequently, he seeks an approximate solution. He gives it in sexagesimal form as $1° 22' 7'' 42''' 33^{IV} 4^V 40^{VI}$, the 40 being too great by about 1 1/2. We are not told how the result was found. We know only that the same problem appears in the algebra of al-Khayyāmī,[22] where it is solved by means of the intersection of a circle and a hyperbola. One may suppose that the solution follows from the Horner method, which was known to the Chinese and the Arabs.

Next Leonardo presents a series of indeterminate linear problems. If the first of these (*tres homines pecuniam communem habentes*),[23] which had already been solved by various methods in the *Liber abbaci*, was really posed by Master Johannes, then he must have taken it from the algebra of al-Karajī.[24] The following examples are well-known from the *Liber abbaci* as "the found purse" and "one alone cannot buy" problems. Here, too, negative solutions are given. In one problem with six unknowns, one of them is chosen arbitrarily, while *causa* and *res* are taken for two of the others.

Letter to Master Theodorus (undated; *Scritti*, II, 247–252). The principal subject of the letter is the "problem of the 100 birds," which Leonardo had

already discussed in the *Liber abbaci*. This time, however, Leonardo develops a general method for the solution of indeterminate problems. A geometrical problem follows that is reminiscent of the conclusion of the *Practica geometriae*. A regular pentagon is to be inscribed in an equilateral triangle. Leonardo's treatment is a model for the early application of algebra in geometry. The solution is carried through to the point where a quadratic equation is reached, and then an approximate value is determined—again sexagesimally. The letter concludes with a linear problem with five unknowns; instead of a logically constructed calculation, however, only a mechanical formula is given.

Liber quadratorum (*Scritti*, II, 253–279). This work, composed in 1225, is a first-rate scientific achievement and shows Leonardo as a major number theorist. Its subject, which had already appeared among the Arabs[25] and was touched upon at the end of the *Practica geometriae* and in the introduction to the *Flos,* is the question, proposed by Master Johannes, of finding the solution of two simultaneous equations $x^2 + 5 = y^2$ and $x^2 - 5 = z^2$, or $y^2 - x^2 = x^2 - z^2 = 5$. The problem itself does not appear until late in the text; before that Leonardo develops propositions for the determination of Pythagorean triples. He knows that the sum of the odd numbers yields a square. He first considers the odd numbers from 1 to $(a^2 - 2)$ for odd a; the sum is $\left(\dfrac{a^2 - 1}{2}\right)^2$. If a^2 is added to this expression, then another square results, $\left(\dfrac{a^2 + 1}{2}\right)^2$. For even a the corresponding relation is

$$\left[1 + 3 + \cdots \left(\frac{a^2}{2}\right) - 3\right] + \left[\frac{a^2}{2} - 1 + \frac{a^2}{2} + 1\right]$$
$$= \left[\left(\frac{a}{2}\right)^2 - 1\right]^2.$$

Leonardo was acquainted with still further number triples, such as the Euclidean: $2pq$, $p^2 - q^2$, $p^2 + q^2$; and he had already given another one in the *Liber abbaci*.[26]

He obtains still more triples in the following manner: if $(a^2 + b^2)$ and $(x^2 + y^2)$ are squares and if, further, $a:b \neq x:y$ and $a:b \neq y:x$, then it is true that $(a^2 + b^2) \cdot (x^2 + y^2) = (ax + by)^2 + (bx - ay)^2 = (ay + bx)^2 + (by - ax)^2$. The problem was known to Diophantus, and a special case exists in a cuneiform text from Susa. Next Leonardo introduces a special class of numbers: $n = ab \cdot (a + b) \cdot (a - b)$ for even $(a + b)$, and $n = 4ab \cdot (a + b) \cdot (a - b)$ for odd $(a + b)$. He names such a number *congruum* and demonstrates that it must be divisible by 24. He finds that $x^2 + h$ and $x^2 - h$ can be squares simultaneously only if h is a *congruum*. For $a = 5$ and $b = 4$, $h = 720 = 5 \cdot 12^2$. The problem now, therefore, is to obtain two differences of squares $y^2 - x^2 = x^2 - z^2 = 720$. He determines that $2401 - 1681 = 1681 - 961$, or $49^2 - 41^2 = 41^2 - 31^2$. Following division by 12^2 he gets

$$\left(3\frac{5}{12}\right)^2 + 5 = \left(4\frac{4}{12}\right)^2$$

and

$$\left(3\frac{5}{12}\right)^2 - 5 = \left(2\frac{7}{12}\right)^2.$$

One does not learn how Leonardo obtains the squares 961, 1681, and 2401; however, one can ascertain it from a procedure in Diophantus.[27] Leonardo then proves a further series of propositions in number theory, such as that a square cannot be a *congruum*, that $x^2 + y^2$ and $x^2 - y^2$ cannot simultaneously be squares, that $x^4 - y^4$ cannot be a square, etc. Next Leonardo considers expressions such as the following: $x + y + z + x^2$, $x + y + z + x^2 + y^2$, and $x + y + z + x^2 + y^2 + z^2$. They are all to be squares and they are to hold simultaneously. This was another of Master Theodorus' questions. In the questions treated in the *Liber quadratorum,* Leonardo was long without a successor.

In surveying Leonardo's activity, one sees him decisively take the role of a pioneer in the revival of mathematics in the Christian West. Like no one before him he gave fresh consideration to the ancient knowledge and independently furthered it. In arithmetic he showed superior ability in computations. Moreover, he offered material to his readers in a systematic way and ordered his examples from the easier to the more difficult. His use of the chain rule in the "Rule of Three" is a new development; and in the casting out of nines he no longer finds the remainder solely by division, but also employs the sum of digits. His rules for factoring numbers and the formation of perfect numbers are especially noteworthy, as is the recurrent series in the "rabbit problem." He treated indeterminate equations of the first and second degrees in a manner unlike that of anyone before him; ordinarily he confines himself to wholenumber solutions—in contrast with Diophantus—where such are required. In geometry he demonstrates, unlike the *Agrimensores,* a thorough mastery of Euclid, whose mathematical rigor he is able to recapture, and he understands how to apply the new methods of algebra to the solution of geometric problems. Moreover, in his work a new concept of number

seems to be emerging, one that recognizes negative quantities and even zero as numbers. Thus, on one occasion[28] he computes $360 - 360 = 0$ and $0:2 = 0$. Especially to be emphasized is his arithmetization of the Euclidean propositions and the employment of letters as representatives for the general number.

Leonardo's Sources. Early in his youth Leonardo already possessed the usual knowledge of a merchant of his time, as well as that preserved from the Roman tradition (abacus, surveying, formulas, etc.). Then came his journeys. What he absorbed on them cannot in most cases be determined in detail. The knowledge of the Greeks could have reached him either from the already existing Latin translations of the Arabic treatments or in Constantinople, where he had been. One can, to be sure, establish where individual problems and methods first appear, but one cannot decide whether what is involved is the recounting of another's work or an original creation of Leonardo's. The only clear cases are those in which a problem is presented with the same numerical values or when the source itself is named.

Leonardo is fully versed in the mathematics of the Arabs; for example, he writes mixed numbers with the whole numbers on the right. Algebra was available to him in the translations of the works of al-Khwārizmī by Adelard of Bath, Robert of Chester, and Gerard of Cremona or in the treatment by Johannes Hispalensis. The numerical examples are frequently taken directly from the algebra of al-Khwārizmī or from the *Liber embadorum* of Plato of Tivoli, e.g., the paradigm $x^2 + 10x = 39$. The calculation with irrationals and the relevant examples correspond to those in the commentary on Euclid by al-Nayrīzī (Anaritius), which Gerard of Cremona had translated. Countless problems are taken, in part verbatim, from the writings of Abu-Kāmil and of al-Karajī. The cubic equation in the *Flos* stems from al-Khayyāmī. Leonardo readily refers to the Arabs and to their technical words, such as *regula elchatayn* (double false position), *numerus asam* (the prime number), and *figura cata* (which he uses in connection with the chain rule); this is the "figure of transversals" in the theorem of Menelaus of Alexandria.

The geometry of the Greeks had become known through the translations from the Arabic of Euclid's *Elements* by Adelard of Bath, Hermann of Carinthia, and Gerard of Cremona, through al-Nayrīzī's commentary, and perhaps to some extent through the anonymous twelfth-century translation of the *Elements* from the Greek (see *Harvard Studies in Classical Philology,* **71** [1966], 249–302); for the measurement of circles there existed the translations of Archimedes' work by Plato of Tivoli and Gerard of

Cremona.[29] For the geometric treatment of the cone and sphere, of the measurement of the circle and triangle (with Hero's formula), and of the insertion of two proportional means, the *Verba filiorum* of the Banū Mūsā was available in Gerard of Cremona's translation and was used extensively by Leonardo. On the other hand, problems from the arithmetic of Diophantus could have come only from Arabic mathematics or from Byzantium. On this subject Leonardo had obtained from the "most learned Master Muscus" a complicated problem of the type "one alone cannot buy," which is also represented in Diophantus. (That Leonardo actually had access to the Greek is shown by his rendering of ῥητοί as "riti.") Other problems that point to Byzantium are those of the type "giving and taking" and the "well problems," which had already appeared in the arithmetical epigrams of *The Greek Anthology*.

Leonardo also includes problems whose origin lies in China and India, such as the *Ta yen* rules, remainder problems, the problem of the "100 birds," and others. Concerning the course of their transmission, nothing definitive can be said. Nevertheless, they were most likely (like the "100 birds" problem found in Abū Kāmil) transmitted through the Arabs. Problems that appeared in ancient Babylonia (quadratic equations, Pythagorean number triples) or in Egypt (unit fraction calculations, "the seven old women") had been borrowed from the Greeks.

Influence. With Leonardo a new epoch in Western mathematics began; however, not all of his ideas were immediately taken up. Direct influence was exerted only by those portions of the *Liber abbaci* and of the *Practica* that served to introduce Indian-Arabic numerals and methods and contributed to the mastering of the problems of daily life. Here Leonardo became the teacher of the masters of computation (the *maestri d'abbaco*) and of the surveyors, as one learns from the *Summa* of Luca Pacioli, who often refers to Leonardo. These two chief works were copied from the fourteenth to the sixteenth centuries. There are also extracts of the *Practica,* but they are confined to the chapters on plane figures and surveying problems; they dispense with exact proofs and with the *subtilitates* of the eighth chapter.

Leonardo was also the teacher of the "Cossists," who took their name from the word *causa,* which was used for the first time in the West by Leonardo in place of *res* or *radix.* His alphabetical designation for the general number or coefficient was first improved by Viète (1591), who used consonants for the known quantities and vowels for the unknowns.

Many of the problems treated in the *Liber abbaci,* especially some of the puzzle problems of recreational

arithmetic, reappeared in manuscripts and then in printed arithmetics of later times: e.g., the problem types known as "giving and taking," "hare and hound," "horse buying," "the found purse," "number guessing," "the twins' inheritance," and the indeterminate problem of the "100 birds," which reappeared as the "rule of the drinkers" (*regula coecis, regula potatorum*) and whose solution Euler established in detail in his algebra (1767). Cardano, in his *Artis arithmeticae tractatus de integris,* mentions appreciatively Leonardo's achievements when he speaks of Pacioli's *Summa.* One may suppose, he states, that all our knowledge of non-Greek mathematics owes its existence to Leonardo, who, long before Pacioli, took it from the Indians and Arabs.[30]

In his more advanced problems of number theory, especially in the *Liber quadratorum,* Leonardo at first had no successor. This situation lasted until the work of Diophantus became available in the original text and was studied and edited by Bachet de Méziriac (1621); he, and then Fermat, laid the foundation for modern number theory. Leonardo, however, remained forgotten. Commandino's plan to edit the *Practica* was not carried out. While the historians Heilbronner (1742) and Montucla (1758) showed their ignorance of Leonardo's accomplishments, Cossali (1797) placed him once more in the proper light; however, since the texts themselves could not be found, Cossali had to rely on what was available in Pacioli. It is thanks to Libri and Boncompagni that all five of Leonardo's works are again available.

NOTES

1. In Italian his title is *deputato della patria pubblico* (Biblioteca Magliabechiana, Florence, Palchetto III, no. 25) and *pubblico cancelliere* (Biblioteca Comunale, Siena, L.IV.21).
2. *Scritti,* I, 1: "quasi errorem computavi respectu modi indorum."
3. *Ibid.:* "quedam etiam ex subtilitatibus euclidis geometrice artis apponens."
4. On the dating, see Cantor, *Vorlesungen über Geschichte der Mathematik,* II, 41.
5. Illustration in Arrighi, *Leonardo Fibonacci,* p. 15.
6. Boncompagni, *Intorno ad alcune opere di Leonardo Pisano,* p. 248.
7. *Ibid.,* p. 246: "ideo ipsum X^m librum glosare incepi, reducens intellectum ipsius ad numerum qui in eo per lineas et superficies demonstratur."
8. He should "hold the numbers in his hand" ("retinere in manu," *Scritti,* I, 7). Thus the student "should bring memory and understanding into harmony with the hands and the numerals" (*Scritti,* I, 1).
9. Youschkevitch, *Geschichte der Mathematik im Mittelalter,* p. 377; Vogel, "Zur Geschichte der linearen Gleichungen mit mehreren Unbekannten."
10. See Vogel, *Ein byzantinisches Rechenbuch des frühen 14. Jahrhunderts,* p. 157.
11. See Cantor, *op. cit.,* p. 26.

12. *Scritti,* I, 318: "adpropinquacio veritati."
13. *Ibid.,* 191: "una res et denarii 12 sunt septuplum quinque rerum et de denariis 12."
14. *Ibid.,* 228, 351.
15. *Ibid.,* 228, 352.
16. *Ibid.,* 378: "inveni hunc modum reperiendi radices."
17. *Ibid.,* 396.
18. A list of the common examples is in Woepke, *Extrait du Fakhrī, Traité d'algèbre par Aboū Bekr Mohammed ben Alhaçan Alkarkhī,* p. 29.
19. *Scritti,* I, 442 ff.
20. See Libri, *Histoire des sciences mathématiques en Italie,* II, 305.
21. See Arrighi, *op. cit.,* p. 18.
22. See Woepke, *L'algèbre d'Omar Alkayyāmī,* p. 78.
23. See Cantor, *op. cit.,* pp. 48 f.
24. See Woepke, *Extrait du Fakhrī . . .,* pp. 141 ff.
25. See Youschkevitch, *op. cit.,* p. 235.
26. *Scritti,* I, 402. The example there involves $x^2 + y^2 = 41$.
27. See Ver Eecke, *Léonard de Pise,* p. 44.
28. *Scritti,* I, 296.
29. See Clagett, *Archimedes in the Middle Ages,* I, *The Arabo-Latin Tradition,* chs. 1, 2.
30. G. Cardano, *Opera,* X (Lyons, 1663), 118, col. 2.
31. See Loria, *Storia delle matematiche,* I, 383.

BIBLIOGRAPHY

I. ORIGINAL WORKS. The only complete edition of Leonardo's works is *Scritti di Leonardo Pisano,* B. Boncompagni, ed., 2 vols. (Rome, 1857–1862). Earlier, G. Libri, in his *Histoire des sciences mathématiques en Italie,* 4 vols. (Paris, 1838–1841), published the introduction and ch. 15 of *Liber abbaci* (II, note 1, 287 ff.; note 3, 307 ff.) and the introduction of *Practica geometriae* (II, note 2, 305 f.). Also, B. Boncompagni published three short works in *Opuscoli di Leonardo Pisano* (Florence, 1852). The *Liber quadratorum* was translated into French by P. Ver Eecke as *Léonard de Pise. Le livre des nombres carrés* (Bruges, 1952). An Italian adaptation of the *Practica geometriae* of 1442 is G. Arrighi, *Leonardo Fibonacci. La pratica di geometria, volgarizzata da Cristofano di Gherardo di Dino cittadino pisano. Dal codice 2186 della Biblioteca Riccardiana di Firenze* (Pisa, 1966). There are also two Italian translations of the introduction to the *Liber abbaci* in the MSS cited in note 1.

II. SECONDARY LITERATURE. General criticism includes B. Boncompagni, "Della vita e delle opere di Leonardo Pisano matematico del secolo decimoterzo," in *Atti dell'Accademia pontificia dei Nuovi Lincei,* **5** (1851–1852), 5–91, 208–246; and *Intorno ad alcune opere di Leonardo Pisano matematico del secolo decimoterzo, notizie raccolte* (Rome, 1854); M. Cantor, *Vorlesungen über Geschichte der Mathematik,* II (Leipzig, 1913), 3–53; G. Loria, "Leonardo Fibonacci," in *Gli scienziati italiani,* Aldo Mieli, ed. (Rome, 1923), pp. 4–12; and *Storia delle matematiche,* I (Turin, 1929), 379–410; G. Sarton, *Introduction to the History of Science,* II (Baltimore, 1931), 611–613; D. E. Smith, *History of Mathematics,* 2 vols. (New York, 1958), *passim.;* J. Tropfke, *Geschichte der Elementarmathematik* (Berlin–Leipzig: I, 3rd ed., 1930; II, 3rd ed., 1933; III, 3rd ed., 1937; IV, 3rd ed., 1939; V, 2nd ed., 1923; VI, 2nd ed., 1924; VII, 2nd ed., 1924), *passim ;* and A. P. Youschkevitch, *Ge-schichte der Mathematik im Mittelalter* (Leipzig, 1964), 371–387, trans. from the Russian (1961).

Special criticism includes the following on the *Liber abbaci:* A. Agostini, "L'uso delle lettere nel *Liber abaci* di Leonardo Fibonacci," in *Bollettino dell'Unione matematica italiana,* 3rd ser., **4** (1949), 282–287; and K. Vogel, "Zur Geschichte der linearen Gleichungen mit mehreren Unbekannten," in *Deutsche Mathematik,* **5** (1940), 217–240.

On *Practica geometriae:* R. C. Archibald, *Euclid's Book on Divisions of Figures* (Cambridge, 1915); M. Curtze, "Der *Liber embadorum* des Abraham bar Chijja Savasorda in der Übersetzung des Plato von Tivoli," in *Abhandlungen zur Geschichte der mathematischen Wissenschaften mit Einschluss ihrer Anwendungen,* **12** (1902), 3–183; and J. Millás-Vallicrosa, *Abraam bar Hiia. Llibre de Geometriá* (Barcelona, 1931).

On *Flos:* F. Woepke, "Sur un essai de déterminer la nature de la racine d'une équation du troisième degré, contenue dans un ouvrage de Léonard de Pise," in *Journal de mathématiques pures et appliquées,* **19** (1854), 401–406.

On the *Liber quadratorum* and number theory: L. E. Dickson, *History of the Theory of Numbers,* vols. I and II (1919–1920), *passim.;* F. Lucas, "Recherches sur plusieurs ouvrages de Léonard de Pise et sur diverses questions d'arithmétique supérieure," in *Bullettino di bibliografia e storia delle scienze matematiche e fisiche,* **10** (1877), 129–193, 239–293; and R. B. McClenon, "Leonardo of Pisa and His 'Liber Quadratorum,'" in *American Mathematical Monthly,* **26** (1919), 1–8.

On the history of the problems: E. Bortolotti, "Le fonti arabe di Leonardo Pisano," in *Memorie. R. Accademia delle scienze dell'Istituto di Bologna,* fis.-mat. cl., 7th ser., **8** (1929–1930), 1–30; Marshall Clagett, *Archimedes in the Middle Ages,* I, *The Arabo-Latin Tradition* (Madison, Wis., 1964), see index under "Leonardo Fibonacci"; P. Cossali, *Origine, trasporto in Italia, primi progressi in essa dell'algebra,* I (Parma, 1797), ch. 5, 96–172; II (Parma, 1799), 41, l. 16; M. Dunton and R. E. Grimm, "Fibonacci on Egyptian Fractions," in *The Fibonacci Quarterly,* **4** (1966), 339–354; V. Sanford, *The History and Significance of Certain Standard Problems in Algebra* (New York, 1927); K. Vogel, *Die Practica des Algorismus Ratisbonensis* (Munich, 1954), index, p. 267; and *Ein byzantinisches Rechenbuch des frühen 14. Jahrhunderts* (Vienna, 1968), pp. 153 ff.; J. Weinberg, *Die Algebra des Abū Kāmil Sogā'ben Aslam* (Munich, 1935); and F. Woepke, *L'algèbre d'Omar Alkayyāmī* (Paris, 1851); *Extrait du Fakhrī, Traité d'algèbre par Aboū Bekr Mohammed ben Alhaçan Alkarkhī* (Paris, 1853); and "Recherches sur plusieurs ouvrages de Léonard de Pise et sur les rapports qui existent entre ces ouvrages et les travaux mathématiques des Arabes," in *Atti dell'Accademia pontificia dei Nuovi Lincei,* **10** (1856–1857), 236–248; **14** (1860–1861), 211–356.

See also Archibald; Boncompagni, *Opuscoli;* Ver Eecke; and G. Loria, "Leonardo Fibonacci." Sarton, II, 613, cites B. Boncompagni, *Glossarium ex libro abbaci* (Rome, 1855), not known to be in German or Italian libraries.

KURT VOGEL

FICHOT, LAZARE-EUGÈNE (*b.* Le Creusot, Saône-et-Loire, France, 18 January 1867; *d.* Tabanac, Gironde, France, 17 July 1939), *marine hydrography.*

Fichot entered the École Polytechnique in 1884, and upon graduation in 1886 he joined the Marine Corps of Hydrographic Engineers. In November 1926 he was made chief of the Marine Hydrographic Service, and at the time of his death he was its director. In 1912 Fichot received the Binoux Award of the Académie des Sciences for his contributions to geography and navigation, and in 1925 he was elected a member of the academy. He was made an officer of the Legion of Honor in 1911 and a member of the Bureau des Longitudes in 1923. He was elected president of the Geodesy Section of the National Committee on Geodesy and Geophysics in 1923.

Fichot's main interests were in hydrography, geodesy, geography, terrestrial physics, and the theory of the tides. His major contributions, however, were in hydrography and tidal theories. During his work in the hydrographic service he made numerous hydrographic, geologic, and meteorologic observations along the coasts of France and the French colonies in Asia and Africa and prepared maps of the coastal regions. In 1908 he discovered a new navigational route along the coast of Indochina by accurately mapping obstructions between Cam Ranh Bay and Nha Hang. In 1912 he prepared hydrographic maps of the Gironde estuary, which became an important navigational route during World War I.

One of Fichot's major works, *Les marées et leur utilisation industrielle* (1923), contained a comprehensive synthesis of existing knowledge on tides as well as results of some of his own research on the subject. Later, in collaboration with Henri Poincaré, he tried to expand his work on tides. Parts of their research were published as *Exposé critique de la théorie des marées* (1938–1941). Nevertheless, the study remained unfinished at the time of Fichot's death.

BIBLIOGRAPHY

Fichot's writings include "Rapport sur la reconnaissance hydrographique de l'embouchure de la Gironde en 1912," in *Recherches hydrographiques sur le régime des côtes,* XIX (Paris, 1911–1914), 12–82; "Rapport sur les travaux de la Mission hydrographique de l'Indo-Chine en 1909–1910," in *Annales hydrographiques,* **33**, no. 2 (1913), 357–393; "Marées océaniques et marées internes," in *Encyclopédie des sciences mathématiques pures et appliquées,* VI, pt. 8 (Leipzig-Paris, 1916), 1–96; *Les marées et leur utilisation industrielle* (Paris, 1923); and "L'influence de la rotation terrestre sur la physionomie des marées," in *Annuaire publié par le Bureau des longitudes* (Paris, 1925), pp. A.1–A.71.

ASIT K. BISWAS
MARGARET R. PEITSCH

FICK, ADOLF EUGEN (*b.* Kassel, Germany, 3 September 1829; *d.* Blankenberge, Belgium, 21 August 1901), *physiology, physical medicine.*

Fick's father, Friedrich Fick, a senior municipal architect in Kassel, was instrumental in reorganizing street construction in that city. His mother, Nanni Sponsel, had nine children, of whom Adolf was the youngest. Two of his brothers were professors (one of anatomy and the other of law) at Marburg when Fick began to study there. In the winter semester of 1847–1848, he began his work in mathematics and physics, fields for which he possessed great aptitude, although he soon switched to medicine.

The descendant of Protestant émigrés from Salzburg, Fick was raised in that faith. He was a man of high moral sense, pious but without any formal church affiliation. He married Emilie von Coelln on 24 October 1862. They had five children, two of whom died in early childhood. One son became a jurist; another became an anatomist in Berlin.

As a medical student at Marburg, Fick was guided in anatomy not only by his brother Ludwig but also by his early friendship with Carl Ludwig, then lecturing at Marburg on anatomy and physiology. Fick became one of the main proponents (with Carl Ludwig) of the new orientation of physiology toward physics, to which Hermann von Helmholtz, Ernst Brücke, and Emil du Bois-Reymond also subscribed. It was their objective to determine quantitatively, whenever possible, the fundamental capabilities of the organism's components and to explain them on the basis of general physicochemical laws of nature. Fick wrote his first scientific paper, an investigation on the torque exerted by the motor muscles of the femur in the hip joint (1849), when he was still a student. In this he demonstrated his gift for planning, executing, and mathematically evaluating mathematical-physical research of physiological processes. In the fall of 1849 he continued his studies at Berlin, where he became friends with Helmholtz and du Bois-Reymond. In Berlin the anatomist Johannes Müller, because of his entirely different approach in theory and practice, was unable to exert any great influence on Fick, who was said, moreover, to have been then a genius in the art of enjoying a carefree student life.

Upon his return to Marburg, Fick obtained his doctorate on 27 August 1851, with a thesis on visual errors due to astigmatism. He then accepted the position of a prosector in anatomy with his brother but followed Carl Ludwig to Zurich six months later. This was the beginning of a scientific career characterized by an unusually diversified scientific output. Fick remained in Zurich from 1852 until 1868, first as prosector in anatomy with Ludwig until 1855, subsequently as associate professor of the anatomical and physiological auxiliary sciences until 1861, and later as full professor of physiology. In the winter semester of 1868 he assumed the same position in the Faculty of Medicine at Würzburg, which he held for more than thirty years. In 1883 he moved into a new physiological institute.

During 1878–1879 Fick was rector of the University of Würzburg. He retired at the age of seventy, in 1899. On 21 August 1901 he died from the aftereffects of a cerebral hemorrhage. In 1929 his sons founded the Adolf Fick Fund, which every five years awards a prize for an outstanding contribution to physiology.

Fick's lifework is concerned primarily with problems on the borderline between medicine, physiology, and physics. Accordingly, his monograph *Medizinische Physik,* published in 1856, when he was twenty-six years old, is most characteristic of the problems he preferred, particularly (1) molecular physics: diffusion of gases and water, filtration, endosmosis, and porous diffusion; (2) mechanics of solids, including the geometry of articulations, and the statics and dynamics of muscles; (3) hydrodynamics, as applied to the motion of fluids in rigid and/or elastic vessels (blood vessels), and pulse variations and their accurate recording; (4) sound; (5) the theory of heat in physics and physiology, the origin of heat, and the law of the conservation of energy in the body; (6) optics: the path of rays in the eye and ophthalmoscopy, the microscope, the horopter, and color perception and the theory of color sense; and (7) the theory of electricity: the origin, derivation, and measurement of bioelectric phenomena.

All of these fields were treated with great success by Fick. He showed himself to be an analyst who reflected in strict conformity with physical laws and arrived at the precise mathematical expression of physiological processes. In similar form he analyzed the mechanics of the saddle joint of the thumb (1854) and, in the same year, the torsional movement of the several muscles that move the bulbus oculi. In addition, he studied molecular biophysics, e.g., the expansion of bodies by heat (1854). He then examined, on the basis of the investigations of Brücke, Jolly, Ludwig, and Max Cloëtta, the process of diffusion and developed a differential equation for the flow between a saturated sodium chloride solution and distilled water (1855). In experimental analysis, he introduced the collodion membrane into the study of porous diffusion in endosmosis (1857).

The physics of the vision process had occupied Fick since his doctoral thesis, "De errore optico," which was concerned with perceptual illusions due to astigmatism (1851). He also analyzed the blind spot in the eye and the phenomenon of monocular polyopia through unequal ratios of curvature in the refractive media (1854). Beginning in 1853, Fick became interested in the sequence of excitation in the retina, in adaptation, in the phenomenon of the latent stage, and then in the slow initiation of retinal stimulation (1864). He reflected on the phenomena of color vision and arrived at a confirmation of Helmholtz' three-component theory (1880) but raised essential reservations in regard to Ewald Hering's theory. Later, he was successful in building the first practical instrument for measurement of intraocular pressure, the ophthalmotonometer (1888). Fick had a particular gift for and ingenuity in constructing physiological measuring devices and developed, among other apparatus, an improved aneroid manometer for measurement of the pressure gradient in the vascular system and a pneumograph for recording peripheral variations of the thorax in breathing (1869–1897).

Fick also made outstanding contributions to hemodynamics. He developed a principle, which came to be called Fick's law (1870), that permits calculation of the cardiac output from the measurement of the minute volume of oxygen consumption and arteriovenous oxygen difference in the living organism. The arteriovenous differences of oxygen level indicate the amount of oxygen per 100 cc. carried off by the blood flowing through the lung, and the oxygen consumption per minute indicates how many times per minute 100 cc. of blood have circulated through heart and lung; this result is called the minute volume, and its division by the heart rate yields the stroke volume.

Fick considered mathematical expression to be the most exact and, indeed, the only adequate language of science. In this spirit he theorized on the speed of the flow of blood in the vascular system, on its conditions, and on its measurability. He also introduced the principle of plethysmography. In this procedure the arm, for example, is placed in a rigid, water-filled vessel to which a narrow upright tube is attached. If the supply of blood to the arm exceeds the outflow of blood within the same interval, water is displaced and the meniscus in the tube will rise (1869). Fick also constructed a model of the blood-vessel ramifications from rigid and elastic tubes in order to simulate and analyze the pressure drop in the circulation, especially in the capillaries (1888).

Fick's gift for mathematical-physical thinking is most clearly exhibited in his numerous studies on the nature of heat, on the causes of thermal expansion of bodies (1854–1855), and on the nature, magnitude, and origin of body heat. These problems occupied him again and again, and he maintained that questions of thermal production and consumption should be kept separate from temperature topography. He ascribed all processes of heat generation to the expenditure of chemical energy, which takes place, for instance, to provide the mechanical energy for muscular effort. This raises the question of the substances supplying the energy through their decomposition. Liebig declared them to be the proteins. In 1865 Fick decided the question during a climb of the Faulhorn, in Switzerland, by determining the energy consumed (in m. kg. of climbing effort) and through calculation of protein catabolism from the nitrogen passed in the urine. He found that protein is insufficient to provide the energy requirements.

Fick further investigated the generation of heat in the muscle with sensitive thermometers and related its magnitude to the effort of contraction produced. In order to increase accuracy in such investigations, he introduced the concept and the methodology of isotonic and isometric determination of the process of muscular contraction (1867, 1882, 1889). He constructed an apparatus for measuring the output of m. kg. with the aid of a rotating wheel, the dynamometer (1891). He was always concerned with testing the validity of the law of the conservation of energy in the body. In this connection, Fick also encountered the need for adequate and graduated stimulation of muscles and nerves.

His physiology of stimulation attempted in this connection to find new quantifiable parameters. In contrast with generally held views, Fick found that contraction is a function not only of the intensity of nerve stimulation but also the duration of the stimulus. He demonstrated this specifically for the smooth muscle, in this case the shell adductor of mussels (1863). Fick further investigated electrotonus under optimum exact conditions (1866–1869) and the effect of transversal nerve stimulation. Notable was his successful attempt to demonstrate, against the opinion held by French scientists, that electrical stimulation of the anterior marginal bundles of the spinal cord was possible. He made this an occasion (1869) for an interesting discussion of his research principle of clarifying the fundamental properties of the elemental components of the animal organism, as opposed to a method of research preferring observations and tests of the entire animal.

The similarity of the physiological objectives of Helmholtz and Fick is astonishing. Both were particularly interested in the clarification of the physico-chemical and quantitatively determinable processes. Both performed outstandingly in this field. The prerequisites here were always to work under the simplest possible arrangements, under constant or known conditions, or both, and to test the consequences produced by a measurable variation of one of the conditions. Causal analysis within the context of the exact natural sciences, quantification, construction of measuring devices and apparatus—these were the objectives and paths of the generation of German physiologists in the second half of the nineteenth century which included, besides Adolf Fick and Helmholtz, Carl Ludwig and many others.

Fick's acumen was not concentrated in scientific analysis alone. He was also greatly interested in the methodology of scientific investigation, as his philosophical articles demonstrate. He dismissed the romantic physiology of Schelling and Hegel as "superior nonsense" (1870). Like Helmholtz, he followed Kant and believed the empiricism of John Stuart Mill to be inadequate. His epistemological attempt at refutation of Kirchhoff (1882) and his *Philosophischer Versuch über die Wahrscheinlichkeiten* (Würzburg, 1883) reflect these views.

Fick possessed a crystal-clear manner of thinking and writing. He was modest, without intellectual arrogance, and stood by his convictions fearlessly. When he realized the consequences of alcohol abuse, he ceaselessly called public attention to this problem. Among Fick's disciples and collaborators were Friedrich Schenck, Magnus Blix, Johannes Gad, A. Gürber, and Jacques Loeb.

BIBLIOGRAPHY

I. ORIGINAL WORKS. All of Fick's publications were collected and published posthumously in *Adolf Fick. Gesammelte Abhandlungen*, 4 vols. (Würzburg, 1903–1905), with portraits, bibliographic data, and a biography by F. Schenck. Among his writings are *Medizinische Physik* (Brunswick, 1856); *Compendium der Physiologie des Menschen mit Einschluss der Entwicklungsgeschichte* (Vienna, 1860); *Beiträge zur vergleichenden Physiologie der irritablen Substanzen* (Brunswick, 1863); *Lehrbuch der Anatomie und Physiologie der Sinnesorgane* (Lahr, 1864); *Untersuchungen über elektrische Nervenreizung* (Brunswick, 1864); "Über die Entstehung der Muskelkraft," in *Vierteljahrsschrift der Naturforschenden Gesellschaft in Zürich*, **10** (1865), 317 ff.; "Spezielle Bewegungslehre," in L. Hermann, *Handbuch der Physiologie*, I, pt. 2 (Leipzig, 1879); "Dioptrik und Lichtempfindung," *ibid.*, III, pt. 1 (Leipzig, 1879); *Mechanische Arbeit und Wärmeentwicklung bei der Muskeltätigkeit* (Leipzig, 1882); and *Myothermische Untersuchungen aus* *dem physiologischen Laboratorium zu Zürich und Würzburg* (Wiesbaden, 1889).

The journal publications are easily accessible in *Gesammelte Abhandlungen* and need not be listed here in detail.

II. SECONDARY LITERATURE. Notices and obituaries are René du Bois-Reymond, "Nachruf," in *Naturwissenschaftliche Rundschau* (Stuttgart), **16** (1901), 576–577; Friedrich Fick, "Adolf Fick, Professor der Physiologie (1829–1901)," in *Lebensläufe aus Franken,* I (1919), 94 ff., a biography; Max von Frey, "Gedächtnisrede," in *Sitzungsberichte der Physikalische-medizinischen Gesellschaft zu Würzburg* (1901), 65 ff., with references to the literature; Adam Kunkel, "Nachruf," in *Münchener medizinische Wochenschrift,* **48** (1901), 1705–1708; Kurt Quecke, "Adolf Fick (1829–1901). Physiologe," in *Lebensbilder aus Kurhessen und Waldeck,* **4** (1950), 82–90; K. E. Rothschuh. *Geschichte der Physiologie* (Berlin-Göttingen-Heidelberg, 1953), pp. 150–153; F. Schenck, "Zum Andenken an A. Fick," in *Pflügers Archiv für die gesamte Physiologie des Menschen und der Tiere,* **90** (1902), 313–361, with portraits and biography, a good analysis of his work; Dietrich Trincker, "Adolf Fick," in *Neue deutsche Biographie,* V (Berlin, 1961), 127–128; and Edgar Wöhlisch, "A. Fick und die heutige Physiologie," in *Die Naturwissenschaften,* **29** (1938), 585–591; and *Biographisches Lexikon der hervorragenden Ärzte . . .,* II (Berlin-Vienna, 1930), 515–516.

K. E. ROTHSCHUH

FIELDS, JOHN CHARLES (*b.* Hamilton, Ontario, Canada, 14 May 1863; *d.* Toronto, Ontario, Canada, 9 August 1932), *mathematics, education.*

Fields was the son of John Charles Fields and Harriet Bowes. His father died when the boy was eleven, and his mother, when he was eighteen. Fields matriculated at the University of Toronto in 1880 and received the B.A. in 1884, with a gold medal in mathematics. Johns Hopkins University awarded him a Ph.D. in 1887. He was appointed professor of mathematics at Allegheny College in 1889 and resigned in 1892 in order to continue his studies in Europe. The next decade found Fields primarily in Paris and Berlin, where associations with Fuchs, Frobenius, Hensel, Schwarz, and Max Planck contributed to his intellectual growth. In 1902 he was appointed special lecturer at the University of Toronto, where he remained until his death. He was appointed research professor in 1923.

Fields's lifelong interest in algebraic functions is first evident in his papers of 1901–1904. His treatment is completely algebraic, without recourse to geometric intuition. The structure has both elegance and generality; its machinery is simple, its parts coordinated.

His involvement in mathematical societies was of an international nature. Fields was elected a fellow of the Royal Society of Canada (1907) and of London

(1913). He held various offices in the British and American Associations for the Advancement of Science and the Royal Canadian Institute (of which he was president from 1919 to 1925). He was also a corresponding member of the Russian Academy of Sciences and the Instituto de Coimbra (Portugal). The success of the International Congress of Mathematicians at Toronto in 1924 was due to his untiring efforts as president.

Fields conceived the idea of establishing an international medal for mathematical distinction and provided funds for this purpose in his will. The International Congress of Mathematicians at Zurich in 1932 adopted his proposal, and the Fields Medal was first awarded at the next congress, held at Oslo in 1936.

BIBLIOGRAPHY

I. ORIGINAL WORKS. Fields's writings include "Symbolic Finite Solutions by Definite Integrals of the Equation $d^n y/dx^n = x^m y$," in *American Journal of Mathematics,* **8** (1886), 178–179, his Ph.D. thesis; and *Theory of the Algebraic Functions of a Complex Variable* (Berlin, 1906), which establishes a general plan for proving the Riemann-Roch theorem. With the assistance of J. Chapelon, Fields edited the *Proceedings* of the 1924 International Congress of Mathematicians (Toronto, 1928).

Fields's papers are held by the Rare Books and Special Collections Department of the University of Toronto. They include reprints of some of his published speeches and papers, as well as notebooks of lectures and seminars that he attended in Berlin. In addition, the collection contains two bound volumes of notes made by students of the lectures of Weierstrass, *Theorie der elliptischen Functionen* (recorded by A. Darendorff) and *Theorie der hyperelliptischen Functionen* (taken down by an anonymous auditor in the summer semester of 1887).

II. SECONDARY LITERATURE. J. L. Synge, "Obituary Notice of John Charles Fields," in *Obituary Notices of Fellows of the Royal Society of London,* **2** (1933), 129–135 (with portrait), is quite extensive. It contains a full bibliography of Fields's publications (39 titles) and an analysis of the works by his former pupil and colleague, S. Beatty. It also includes the final form of his theorems leading up to and including the proof of the Riemann-Roch theorem. See also Synge, "John Charles Fields," in *Journal of the London Mathematical Society,* **8,** pt. 2 (1933), 153–160. A short statement in *The Royal Canadian Institute Centennial Volume 1849–1949* (Toronto, 1949), William Stewart Wallace, ed., p. 163, gives evidence of Fields's personal dedication to the Royal Canadian Institute.

HENRY S. TROPP

FIESSINGER, NOËL (*b.* Thaon-les-Vosges, France, 24 December 1881; *d.* Paris, France, 15 January 1946), *medicine, biology.*

Fiessinger was descended from a family of Alsatian physicians. His great-grandfather served as field surgeon at Waterloo, and his father, Charles Fiessinger, was a celebrated cardiologist and author. Fiessinger was recognized early as a brilliant physician and a renowned biologist and biochemist. The highest responsibilities in the Paris hospitals were entrusted to him, particularly the great laboratory of the Hôpital Beaujon and the chair of experimental medicine and the renowned chair of clinical medicine at the Hôtel-Dieu. He was prominent among those who brought about a major revolution in clinical thinking—that is, the idea that clinical medicine must be closely associated with biological research.

He elucidated the histogenesis of cirrhosis. This degenerative process of the liver cells is the same whatever the conditions, pathological or otherwise, which determine it (1908). Fiessinger demonstrated the existence of enzymes in the white cells of the blood. He showed that these cells, according to their type, contain either protease or lipase. The presence of protease accounts for the dissolution of internal blood clots or purulent collections, while lipase weakens the lipidic membrane of the Koch bacilli, thus permitting their attack by the protease-carrying white cells.

World War I turned Fiessinger's efforts away from this pioneer work in biochemistry. He made major observations in the biology of war wounds, observations gathered under the precarious conditions in field hospitals, often under severe artillery fire.

After the war Fiessinger revealed himself to be an eminent physiologist. He was among the first to define the principles of functional exploration of an organ, which he applied most successfully to the liver, through such new tests as galactose and Bengal pink dye. Fiessinger's achievements as a biologist are matched by his many contributions to clinical medicine, especially by his discovery of the Fiessinger-Leroy-Retter disease, which up to that time was undefined. His influence as a renowned teacher was considerable, and numerous prominent physicians in many countries are his former students.

BIBLIOGRAPHY

Fiessinger's papers appeared in various medical journals, but all his essential contributions are included in the last editions of his books, which were standard texts for the medical profession.

See *Nouveaux procédés d'exploration fonctionnelle du foie* (Paris, 1934), written with H. Walter; *Syndromes et maladies* (Paris, 1942); *Diagnostics difficiles* (Paris, 1943); and *Diag-*

nostics biologiques (Paris, 1944), written with H. R. Oliver and M. Herbain.

A. M. MONNIER

FINE, HENRY BURCHARD (*b.* Chambersburg, Pennsylvania, 14 September 1858; *d.* Princeton, New Jersey, 22 December 1928), *mathematics.*

Fine was the son of Lambert Suydam Fine, a Presbyterian minister. After the death of his father, his mother settled in Princeton, where Fine attended the university. During Fine's undergraduate years, his interest in mathematics was awakened by the young instructor George Halstead, who promoted the study of non-Euclidean geometry in the United States. After a year as an assistant in physics and three years as a mathematics tutor at Princeton (1880–1884), Fine, like many of his colleagues, went to Germany to study. At Leipzig he attended Felix Klein's lectures and in 1885 wrote a dissertation on an algebraic geometric problem suggested by Eduard Study, with whom he became friendly. After a summer in Berlin attending Leopold Kronecker's lectures, Fine returned to Princeton, where he taught mathematics until his death. In 1888 Fine married Philena Forbes.

In 1903, Woodrow Wilson's first year in the presidency of Princeton, Fine was appointed dean of the faculty; and when Wilson resigned to run for governor of New Jersey, Fine acted as president of the university until a successor was named in 1912. He then became dean of the departments of science, a post he held until his death. Fine was a founding member of the American Mathematical Society in 1891 and its president in 1911–1912.

Fine's impact on science lies mainly in his support of science and mathematics at Princeton. As dean of the faculty he promoted the mathematician Luther Eisenhart and brought in Oswald Veblen, G. A. Bliss, George Birkhoff, and J. H. M. Wedderburn. A professorship of mathematics and a mathematics building at Princeton were named for Fine.

Among his few contributions to mathematics were an expansion of his dissertation; several papers on differential equations; and, most important, a paper on Newton's method of approximation (1916) and an exposition of a theorem of Kronecker's on numerical equations (1914). Fine was the author of several undergraduate textbooks and an exposition of the number system of algebra.

BIBLIOGRAPHY

For a complete bibliography of Fine's publications and related secondary sources, see "Henry Burchard Fine," in

American Mathematical Society Semicentennial Publication, I (New York, 1938), 167–169. Also consult Oswald Veblen, "Henry Burchard Fine," in *Bulletin of the American Mathematical Society,* **35** (1929), 726–730.

C. S. FISHER

FINK (FINCKE), THOMAS (*b.* Flensburg, Denmark [now Germany], 6 January 1561; *d.* Copenhagen, Denmark, 24 April 1656), *mathematics, astronomy, medicine.*

The son of Raadmand Jacob Fincke and Anna Thorsmede (who died six days after his birth), Fink studied from 1577 to 1582 in Strasbourg. Afterward he attended many universities: Jena, Wittenberg, Heidelberg (matriculated 6 February 1582), Leipzig (matriculated summer of 1582), Basel (studied medicine in 1583), and Padua (from 6 November 1583 to 1587). This varied education led to his receipt of the M.D. at Basel on 24 August 1587. After three years of traveling through Germany and Austria, Fink became physician-in-ordinary to Duke Philip of Holstein-Gottorp. When Philip died in 1591, Fink was appointed professor of mathematics at Copenhagen, his field of instruction being changed to rhetoric in 1602 and to medicine in 1603. He held high university posts and carried out his duties until only a few years before his death at the age of ninety-five.

Fink's most famous book is the *Geometriae rotundi* (1583), published when he was twenty-two. This important work is divided into fourteen books. The elementary theses on the circle are collected in the four opening books and the remaining books treat trigonometry, the last three being devoted to spherical trigonometry. A central place is occupied by Rheticus' goniometric tables, but here Fink took a step backward, giving the tables for each function separately and always from 0° to 90°, rather than using the complementary character of the functions, as Rheticus had done. In Strasbourg, Fink had been a pupil of the mathematician Dasypodius but seems to have learned mainly astrology from him. He makes it clear that he was an autodidact in mathematics. His inspiration and guide was not Euclid's *Elements*—this work disturbed him—but Ramus' *Geometria* (1569). Therefore the *Geometriae rotundi* is based mainly on Ramus, many proofs being comprehensible only after consulting the *Geometria*. Even the word "rotundum" in Fink's title, meaning both circle and sphere, was introduced by Ramus. Fink also adopted the term "radius" from him and himself introduced such terms as "tangent" and "secans." He devised new formulas, such as the law of tangents, and proved in this work that he was abreast of the mathematics of his time.

The *Geometriae rotundi* was meant as a textbook, since it treats basic formulas and refers the reader to Regiomontanus for more detail. As a textbook it was very influential. Such mathematicians as Lansbergen, Clavius, Napier, and Pitiscus recommended the work and adopted much from it. Fink's other works show his interest in astrology and astronomy. He was in contact with Tycho Brahe and Magini. But never again in a long series of publications did he reach the level of the *Geometriae rotundi*.

BIBLIOGRAPHY

I. ORIGINAL WORKS. For a bibliography see H. Ehrencron-Müller, *Forfatterlexikon omfattende Danmark . . .,* III (Copenhagen, 1926), 46–49; this work does not mention *Methodica tractatio doctrinae sphaericae* (Coburg, 1626); *Theses logicae* (Copenhagen, 1594); C. Ostenfeld, *Oratio in orbitum T. Finckii* (Copenhagen, 1656). Fink's most important works are *Geometriae rotundi libri XIIII* (Basel, 1583); and *Horoscopographia sive de inveniendo stellarum situ astrologia* (Schleswig, 1591), which includes a horoscope of Heinrich Graf von Rantzau.

II. SECONDARY LITERATURE. On Fink or his work, see Niels Nielsen, *Matematikken i Danmark 1528–1800* (Copenhagen, 1912), pp. 69–70; and H. F. Rördam, *Kjöbenhavns Universitets historie fra 1537 til 1621* (Copenhagen, 1873–1877), III, 550–562. On *Geometriae rotundi,* see A. von Braunmühl, *Vorlesungen über Geschichte der Trigonometrie,* I (Leipzig, 1900), 186–193; and J. Tropfke, *Geschichte der Elementar-Mathematik,* vols. IV (Berlin, 1922; new ed., 1940) and V (Berlin, 1923), see index in vol. VII (Berlin, 1927).

J. J. VERDONK

FINLAY, CARLOS JUAN (*b.* Puerto Príncipe [now Camagüey], Cuba, 3 December 1833; *d.* Havana, Cuba, 20 August 1915), *medicine.*

Finlay was the son of a Scotch father and a French mother whose family lived in Trinidad. An aunt who had had a school in Edinburgh taught him at home until he was eleven; he then went to France for further, more formal schooling. There he developed severe chorea which left him with a speech impediment—a lisp—that he never lost. In 1851, having returned home to Cuba, he nearly died of typhoid fever. Undaunted, Finlay became and remained all his life an avid sportsman, swimmer, and horseback rider. Besides Spanish, he became fluent in English, French, and German.

Finlay attended Jefferson Medical College in Philadelphia, where he studied under Robley Dunglison and John K. Mitchell and his son, Weir. He graduated in 1855, rejecting lucrative offers to practice in the Spanish colony of New York City. After

619

a brief trip to Peru, he settled in Havana, where he practiced general medicine and ophthalmology.

In Philadelphia, John Mitchell taught that malaria and other epidemic fevers were caused by living organisms. In 1879 the U.S. Yellow Fever Commission in Havana concluded that yellow fever was transmissible and that its vector was probably airborne, would attack a person once only, and produced a specific, self-limiting disease. Finlay had written much about yellow fever as arising from telluric influences, miasmata, and meteorological conditions. He had theorized that filth was converted into some hypothetical vegetable-animal germ and had suggested that alkalinity of air caused yellow fever. Working closely with the Commission, he shortly suggested that the disease was transmitted by the household mosquito, *Culex fasciatus,* now called *Aedes aegypti.*

Finlay thought that the mosquito's bill acted in transferring virus in the same way as a dirty needle acts in transferring hepatitis. He considered that the morbific cause of the disease was carried from the blood of an infected patient to a healthy person, but did not mention any change in the material thus transferred. From 1881 until 1898 he conducted 103 experiments wherein he induced mosquitoes to bite yellow fever patients and then bite healthy recent immigrants (who volunteered for the experiment, knowing that they would eventually get yellow fever anyway, since everyone did). The experiments lacked control, because none of Finlay's subjects was kept within screens or away from patients who had yellow fever. From the protocols we know that yellow fever probably was not transmitted; nor were the experiments accepted by physicians and students of the disease in Cuba or elsewhere. Finlay became the laughingstock of the orthodox physicians of Havana.

Finlay thought that a mosquito which drew only a little blood and was only slightly infected would produce mild disease which would confer immunity. Although a shrewd observer and a splendid and kindly physician, he was not trained as an investigator and that he experimented at all was remarkable. When the Yellow Fever Board—Reed, Lazear, Agramonte, and Carroll—came to Havana in 1900 Finlay provided them with mosquitoes, eggs, and instructions for raising mosquitoes. He was also one of a team of physicians who verified the diagnosis of epidemic and experimental yellow fever, an essential function, since there was no laboratory test.

In 1900 Walter Reed and the Board excluded filth as the route for infection, found that Sanarelli's yellow fever bacillus was the familiar hog cholera organism, and showed that the virus was transmittable to the female mosquito from an affected patient only during the first two to three days of the course of the illness. The mosquito then must incubate the virus for about two weeks before her bite could infect a susceptible person.

Finlay was exactly right in naming the mosquito as the vector of the disease and in identifying the variety of mosquito. The precision of his hypothesis is admirable, but his ideas were neglected—as were the similar proposals made by Josiah Nott in Alabama in 1854. Perhaps in atonement for their rejection of his ideas, Cubans have made Finlay a national hero, an honor well deserved for the brilliant hypothesis that he staunchly stuck to against universal disbelief. Happily he lived to see it proved correct.

BIBLIOGRAPHY

I. ORIGINAL WORKS. Finlay's works have been collected as *Obras completas,* 4 vols. (Havana, 1965–1970).

II. SECONDARY LITERATURE. The best biographical source is his son, Carlos E. Finlay, *Carlos Finlay and Yellow Fever* (New York, 1940). See also William B. Bean, "Carlos Finlay," in *Current Medical Digest,* 37 (1970), 366–367; S. Bloom, *Dr. Carlos J. Finlay* (Havana, 1959); J. A. Del Regato, "Carlos Finlay and the Carrier of Death," in *Américas* (May 1968); "Editorial. Carlos J. Finlay (1833–1915). Student of Yellow Fever," in *Journal of the American Medical Association,* **198** (1966), 188–189. A further source of information on Finlay's life and work is Cesar Rodriguez Exposito, *Centenary of the Graduation of Dr. Carlos J. Finlay in Jefferson Medical College* (Havana, 1956).

WILLIAM B. BEAN

FINSEN, NIELS RYBERG (*b.* Thorshavn, Faeroe Islands, 15 December 1860; *d.* Copenhagen, Denmark, 24 September 1904), *therapeutic medicine.*

Finsen, a descendant of the Viking Icelanders and son of a governor of the Faeroe Islands, was educated at Reykjavik and obtained a medical degree at the University of Copenhagen in 1891. He was subsequently appointed a demonstrator in anatomy at the surgical academy of the same university; but his interest in the therapeutic uses of light became his exclusive professional preoccupation by 1892 and led him shortly thereafter to abandon a career in academic medicine. This decision appears to have been prompted by his affinity for a sunlit environment, which he valued especially because of his attachment to outdoor life.

Yet Finsen's earliest studies were not centered on the salubrious effects of sunlight. His first investigations were devoted to an antithetical question: the nature of light-induced inflammations. Such inflammations were exhibited in patients with smallpox after prolonged exposures to solar radiations, which gave rise to severe blistering of the irritated skin. As a consequence, the patients became more susceptible to infection, secondary fever, and excessive scar formation. Finsen found that the injurious influences of light were produced by the so-called chemical rays placed in the blue, violet, and especially the ultraviolet (actinic) parts of the spectrum. The other extremity of the spectrum presented the opposite phenomenon: the red and ultrared rays, which gave a minimum chemical effect, were found to promote rapid healing of the smallpox lesions and to obviate the unfortunate complications of the disease under the conditions of ordinary light. These initial results in photobiology, including the "red room" (red light) treatment of variola, were issued by Finsen in 1893 and 1894.

Finsen's investigations were sufficiently advanced by 1894 to permit the conclusion that light harbored a direct therapeutic quality. Evidence for this argument was gathered from the contemporary researches of Émile Duclaux and others on the lethal effects of light upon bacteria. In 1895 Finsen applied this finding to the treatment of lupus vulgaris, an intractable and highly disfiguring form of tuberculosis of the skin. He employed for this purpose a powerful source of light from a carbon arc, filtered through a quartz prism, whereby the diseased tissues were exposed to high concentrations of the "incitant," or ultraviolet, rays. A series of publications reporting successful cures of lupus vulgaris by means of the concentrated-light treatment brought Finsen an international reputation and inaugurated the modern era of phototherapy.

This discovery was especially acclaimed in Copenhagen, where an institute for the study of phototherapy was founded in 1896 through philanthropic efforts and was placed under Finsen's direction. He and his associates in subsequent years reported from this Lysinstitut numerous clinical experiences with the "light-bath" method.

Finsen's health had begun to fail in his twenty-third year. By the age of thirty he was almost totally incapacitated from a constrictive pericarditis, which he attributed to a hydatid infection contracted during his student days in Iceland. This led him to perform a self-study on the problem of water and salt metabolism, the results of which laid the scientific foundations for the low-fluid and low-salt-intake therapy. Despite his rapid decline, Finsen labored with formidable vigor. He lived to witness a final tribute in the form of the Nobel Prize for Medicine or Physiology, awarded to him in 1903.

BIBLIOGRAPHY

I. ORIGINAL WORKS. Finsen's writings on photobiology include "Om lysets indvirkninger paa huden," in *Hospitalstidende*, **1** (1893), 721–728; "Om de kemiske straalers skadelige virkning paa den dryiske organisme," *ibid.*, 1069–1083; "Endnu et par ord om koppebehandling," *ibid.*, 1269–1273; "Les rayons chimiques et la variole," in *La semaine médicale*, **14** (1894), 483–488; and "The Red Light Treatment of Smallpox," in *British Medical Journal* (1895), **2**, 1412–1414.

His reports on the treatment of lupus vulgaris are *Om anvendelse i medicinen af koncentrerede kemiske lystraaler* (Copenhagen, 1896); *Über die Anwendung von concentrirten chemischen Lichtstrahlen in der Medicin* (Leipzig, 1899); *La photothérapie* (in three parts: "Les rayons chimiques et la variole;" "La lumière comme agent d'excitabilité;" and "Traitement du *Lupus vulgaire* par les rayons chimiques concentrés") (Paris, 1899), trans. from the German ed. with an app. on the light treatment of *Lupus* by James H. Sequeira, as *Phototherapy* (London, 1901); *Om bekaempelse af Lupus vulgaris med en rede gørelse for de i Denmark opennaaede resultaten* (Copenhagen, 1902); *Die Bekampfung des Lupus vulgaris* (Jena, 1903); and *La lutte contre le Lupus vulgaire* (Paris, 1903).

II. SECONDARY LITERATURE. A good popular account of Finsen's work appears in Paul De Kruif, *Men Against Death* (New York, 1932), ch. 10, pp. 283–299. For Finsen's treatment of his own pericarditis, see Hugo Roesler, "Niels Ryberg Finsen's Disease and His Self-Instituted Treatment," in *Annals of Medical History*, n.s. **8** (1936), 353–356, esp. 356 for a list of Finsen's publications on this subject.

VICTOR A. TRIOLO

FIRMICUS MATERNUS (*b.* Sicily, *fl.* A.D. 330–354), *astrology.*

Our only information about Firmicus' life comes from his two extant works, the *Mathesis,* a popular handbook on astrology, and the *De errore profanarum religionum,* an attack upon pagan cults. Nearly all scholars accept his authorship of both works, but doubts still remain about the date of composition of the *Mathesis.* The author of the *De errore* was a Christian, and the seeming pagan character of the *Mathesis* suggests that Firmicus was converted to Christianity before he composed the *De errore* (*ca.* 346). It is, however, quite possible to reconcile the two works from a religious standpoint, particularly since they were written at a time when pagan and Christian doctrines were being freely intermingled in philosophical and religious literature. Firmicus dedicated the *Mathesis* to Lollianus Mavortius as ordinary consul elect, an office that we know Lollianus held in 355. Book I was composed in Constantine's lifetime (*d.* 337); and since Firmicus informs us that he was engaged for a long time in writing the work, it is reasonable to suppose that it was composed intermittently over a period of nearly twenty years before 354.

The *Mathesis* has been called "the most comprehensive handbook of astrology to come down to us from antiquity" (Franz Boll). Compiled as a handy guide for practitioners of the art, it best represents popular traditions of the previous four centuries and bears little resemblance to Ptolemy's quasi-scientific manual of astrology, the *Quadripartitum*. Sources for such compilations cannot be assigned with any assurance; citations are traditional and wholly unreliable. Firmicus' citations include the legendary Hermes, Orpheus, Abraham, Petosiris, Nechepso, and Aesculapius.

Book I presents a defense of astrology and book II a preliminary conspectus of the elements. Book III deals with the *thema mundi* (the aspect of the heavens at the beginning of the present cosmos) and with the effects of each of the seven planets in the twelve *loci;* book IV, with the relations of the moon with the other planets; book V, with the effects of the planets in the signs, together with houses and decans; and book VI, with planets in trine and quartile aspect and in opposition and conjunction, with the horoscopes of such notables as Paris, Oedipus, Homer, and Archimedes, and with more precise definitions of *loci.* Book VII takes up the horoscopes of individual types and occupations and is marked by undue attention to sexual and moral deviates. Book VIII presents a composite of the traditional Mesopotamian and Egyptian "barbaric" spheres. Prepared by an admitted amateur, the *Mathesis* contains many gross errors in astronomical knowledge, such as a

nocturnal culmination of Mercury and an elongation of 90° for Venus.

Firmicus' injunctions to astrologers to pronounce their responses in public in a loud voice indicate the effectiveness of the measures of Christian emperors to curb divinatory activities. Firmicus is mentioned only once, by Sidonius Apollinaris, before the eleventh century, at which time his book appears to have begun to enjoy a vogue.

BIBLIOGRAPHY

I. ORIGINAL WORKS. The standard editions of Firmicus are *Matheseos libri VIII,* W. Kroll and F. Skutsch, eds., 2 vols. (Leipzig, 1897–1913); and *De errore profanarum religionum: Traduction nouvelle avec texte et commentaire,* G. Heuten, ed. (Brussels, 1938).

II. SECONDARY LITERATURE. For the best general introduction, see Franz Boll, "Firmicus Maternus," in Pauly-Wissowa, *Real-Encyclopädie,* VI (Stuttgart, 1909), cols. 2365–2379. For a detailed account of Latin astrological literature, including the "barbaric" spheres, and for comparisons between the *Mathesis* and Manilius' *Astronomica,* see Boll's *Sphaera* (Leipzig, 1903). The most complete account in English of Latin astrology and its Hellenistic backgrounds is in F. H. Cramer, *Astrology in Roman Law and Politics* (Philadelphia, 1954). The article on Firmicus in Schanz-Hosius, *Geschichte der römischen Literatur,* IV, pt. 1 (Munich, 1914), 129–137, is valuable for bibliography and documentation. L. Thorndike, *A History of Magic and Experimental Science,* I (New York, 1923), 525–538, argues cogently for a single authorship of both works and a long period of composition for the *Mathesis.*

WILLIAM H. STAHL